General Chair's Welcome

It is with my greatest pleasure that I welcome you to ISCAS 2004. This is the third ISCAS for Canada after ISCAS 1973, Toronto, and ISCAS 1984, Montreal. As in previous ISCAS's, the Organizing Committee has orchestrated a broad selection of functions and conference activities which, I am fully convinced, will meet with your approval.

We have chosen the broad theme "World Prosperity Through Information Processing, Enhancement, and Distribution" for ISCAS 2004 because what drives today's way of life, generation of wealth, education, and almost anything else imaginable has to do with information one way or another. And what we do best as circuits and systems practitioners is to facilitate the processing, enhancement, archiving, and transmission of information be it through communications or image processing systems. Keeping this theme in mind, we have invited world-class circuits and systems visionaries to look into the future and make prognostications. An incisive keynote lecture will start each of the three days of the conference. In addition to the regular papers, in lecture or poster format, invited papers and tutorials, there will also be two thought provoking forum debates. Ljiljana Trajkovic will supply more details in her welcome note as the Technical Program Chair.

Last but not least, we are putting together a social program that is second to none. There will be the traditional Welcome Reception, on Sunday, May 23, and the Banquet on Tuesday, May 25. We are also planning a Farewell Reception for the evening of Wednesday, May 26, in the hope of retaining more delegates for the last session of the last conference day. Rabab Ward, Social Program Chair, has some other surprises for us that will have to remain under wraps at the present time.

Cool Vancouver is ready for you. Come and get it! And to the members of the Organizing Committee and the staff of Conference Management Services, thank you, thank you, thank you ... for making ISCAS 2004 possible.

Andreas Antoniou
General Chair, ISCAS 2004

Technical Program Chair's Message

On behalf of the Technical Program Committee, it is my pleasure to welcome you to ISCAS 2004.

ISCAS 2004's technical program covers topics in all areas of interest to the Circuits and Systems Society. A record number of 2,134 papers were submitted. The Technical Program Committee selected 1,262 papers (59%) to be presented at ISCAS 2004. Papers are organized into 16 technical tracks consisting of 17 parallel sessions (14 lecture and 3 poster). In total, 168 lecture and 90 poster sessions, including 19 Invited Sessions, are scheduled to be held during the three days of the symposium.

ISCAS 2004 also features three keynote and two forum plenary sessions. The topics of the keynote plenary sessions are consistent with the theme of the conference: "Genomic Signal Processing" by Dr. P. P. Vaidyanathan, "CNN Technology for Brain-like Spatial-Temporal Sensory Computing - Present and Future" by Dr. T. Roska, and "Practical Applications of 3-D and 4-D Filters" by Dr. L. T. Bruton. The two plenary forum sessions address challenges in the areas of new technologies of interest to the CAS Society: the forum "New Era of Technology" will address new technologies and the opportunities they bring to the research community, while the forum "The Future of Circuits and Systems" will host past and current presidents of the CAS Society.

Eight half-day and two full-day tutorials/short courses will be offered by leading experts from industry and academia.

The technical program has been prepared by the ISCAS 2004 Technical Program Committee, consisting of 25 Committee Members, 267 Review Committee Members, and almost 1,400 reviewers. 7,448 reviews were received and each paper submitted to ISCAS this year received at least two reviews.

I would like to offer special thanks to the reviewers, the Review Committee Members, and to the Technical Program Committee Members for their dedicated service. I also would like to thank the Plenary Session Chair and the plenary speakers, the Forum Sessions Chair and the forum participants, the Invited Sessions Co-Chairs and the organizers of invited sessions, and the Tutorials and Short Courses Co-Chairs and their presenters, for their fine contributions to the technical program. Finally, my sincere thanks also go to the authors for their excellent contributions and participation in ISCAS 2004. Without them, this conference could not be the success it is.

Ljiljana Trajković
Technical Program Chair, ISCAS 2004

2004 IEEE International Symposium on Circuits and Systems

Proceedings

Volume V of V

Blind Signal Processing
Computer-Aided Network Design
Invited Sessions
Neural Systems and Applications
Power Systems and Power Electronic Circuits

May 23 - 26, 2004
Sheraton Vancouver Wall Centre Hotel
Vancouver, British Columbia, Canada

Sponsored by

The Institute of Electrical and Electronics Engineers
Circuits and Systems Society

IEEE

CAS

IEEE Catalog Number 04CH37512
ISBN 0-7803-8251-X
Library of Congress: 80-646530

Editorial production by Billene Mercer and Conference Management Services

Cover Photo Credits: Tourism Vancouver / Colin Jewall
Cover production by Lance Cotton

IEEE

ORGANIZING COMMITTEE

General Chair
Andreas Antoniou
University of Victoria

Vice General Chair & Technical Program Chair
Ljiljana Trajković
Simon Fraser University

Vice General Chair & Social Program Chair
Rabab K. Ward
University of British Columbia

Finance Chair
Pan Agathoklis
University of Victoria

Invited Sessions Co-Chairs
Majid Ahmadi
University of Windsor
Paulo Diniz
Federal University of Rio de Janeiro

Tutorials and Short Courses Co-Chairs
Andreas Andreou
Johns Hopkins University
Mohamad Sawan
École Polytechnique de Montréal

Local Arrangements Co-Chairs
Panos Nasiopoulos
University of British Columbia
Resve Saleh
University of British Columbia

Forum Sessions Chair
Mona Zaghloul
George Washington University

Exhibits Chair
Graham Jullien
University of Calgary

Publications Chair
Wu-Sheng Lu
University of Victoria

Plenary Sessions Chair
Len T. Bruton
University of Calgary

Publicity Chair
Andreas Spanias
Arizona State University

Conference Advisors
Vijay K. Bhargava
University of British Columbia
William A. Gruver
Simon Fraser University
Tor Sverre Lande
University of Oslo
Michael H. Smith
University of British Columbia

International Coordinators
Marcello de Campos
South America
Anthony C. Davies
Europe
Takao Hinamoto
Asia
M. N. Srikanta Swamy
Canada
Alan N. Willson, Jr.
U.S.A.

ISCAS Steering Committee Chair
Peter Wu
National Chiao Tung University

IEEE CAS Society Executive Director
Heidi Zazza

Conference Management
Billene Mercer
Conference Management Services, Inc.

Technical Program Committee

Technical Program Chair
Ljiljana Trajković
Simon Fraser University

TPC Members

Analog Signal Processing
Wouter A. Serdijn
Delft University of Technology
Igor Filanovsky
University of Alberta
Tuna Tarim
Texas Instruments

Biomedical Circuits and Systems
Tor Sverre Lande
University of Oslo

Blind Signal Processing
Ruey-Wen Liu
University of Notre Dame

Cellular Neural Networks and Array Computing
Bertram Shi
Hong Kong University of Science and Technology

Circuits and Systems for Communications
Magdy Bayoumi
University of Louisiana
Tony Ng
University of Hong Kong
Nam Ling
Santa Clara University

Computer-Aided Network Design
Sung Kyu Lim
Georgia Institute of Technology

Digital Signal Processing
Tapio Saramäki
Tampere University of Technology

Graph Theory and Computing
K. Thulasiraman
University of Oklahoma

Multimedia Systems and Applications
(Winston) Wai-Chi Fang
NASA's Jet Propulsion Laboratory, California Institute of Technology
Oscar C. Au
Hong Kong University of Science and Technology

Nanoelectronics and Gigascale Systems
Chin-Teng Lin
National Chiao Tung University

Neural Systems and Applications
H. K. Kwan
University of Windsor

Nonlinear Circuits and Systems
Gianluca Setti
University of Ferrara

Power Systems and Power Electronic Circuits
Ian A. Hiskens
University of Wisconsin-Madison
Marian Kazimierczuk
Wright State University

Sensory Systems
Orly Yadid-Pecht
Ben-Gurion University

Visual Signal Processing and Communications
Tihao Chiang
National Chiao Tung University

VLSI Systems and Applications
Chein-Wei Jen
National Chiao Tung University
Mircea Stan
University of Virginia

Invited Sessions
Majid Ahmadi
University of Windsor
Paulo Diniz
Federal University of Rio de Janeiro

Review Committee

Esam Abdel-Raheem
Eduard Alarcón
Phil Allen
Andreas Andreou
Andreas Antoniou
Alyssa Apsel
Kaoru Arakawa
Paolo Arena
Peter Aronhime
Oscar Au
Wael Badawy
Salvatore Baglio
Diego Barrettino
Geoffrey Barrows
Mark Barry
Magdy Bayoumi
Murat Becer
Maurice Bellanger
Jinian Bian
Elaheh Bozorgzadeh
Alison Burdett
Sergio Callegari
Francky Catthoor
Gert Cauwenberghs
Gustavo Liñán Cembrano
Enis Cetin
Shing-Chow Chan
Robert C. Chang
Lap-Pui Chau
Chang Wen Chen
Charlie Chen
Guanrong Chen
Jie Chen
Liang-Gee Chen
Oscal T.-C. Chen
Tsu-Han Chen
Hui Cheng
Tihao Chiang
Jun Dong Cho
Ki Young Choi
Malgorzata Chrzanowska-Jeske
Chris Chu
Henry Chung
Andrzej Cichocki
Pedro Correia
Dariusz Czarkowski
Jeff Davis
Tobi Delbruck
Manuel Delgado-Restituto
Uday Desai
Mario Di Bernardo
Zhi Ding
Ian Dobson
Karen Egiazarian
William Eisenstadt
Mourad El-Gamal
Ezz I. El-Masry
Ralph Etienne-Cummings
Wai-Chi (Winston) Fang
Orla Feely
Igor Filanovsky
Daniel Foty
Rob Fox
Moncef Gabbouj
Roman Genov
Julio Georgiou
Marco Gilli
Leonid Goldgeisser
Michael Green
Patrick Groeneveld
Qianping Gu
Francisco Guinjoan
Dong S. Ha
Philipp Häfliger
Ramesh Harjani
John Harris
Paul Hasler
Yun He
Yuwen He
Zhihai He
Markus Helfenstein
Luis Hernandez
Dwight HIll
Takao Hinamoto
Timmer Hiourichi
Ian A. Hiskens
K. C. Ho
Xianlong Hong
Timothy Horiuchi
Jiang Hu
Hsiang-Cheh Huang
Giacomo Indiveri
Yujiro Inouye
Mohammed Ismail
Vadim Ivanov
Mohsin Jamali
Hakan Johansson
Pedro Julian
Christian Jutten
Yoji Kajitani
Lina Karam
Karim Karim
Ryan Kastner
Raj Katti
Masayuki Kawamata
JongWon Kim
Taewhan Kim
Yong Chang Kim
Ljupco Kocarev
Geza Kolumban
Alexander Korotkov
Alex Kot
Ivan S. Kourtev
Anil Kulkarni
C.-C. Jay Kuo
H. K. Kwan
John Lach
Gwo Giun Lee
Shipeng Li
Yong Lian
Vinicius Licks
Wen-Nung Lie
Sung Kyu Lim
Yong-Ching Lim
Chia-Wen Lin
Ching-Yung Lin
Chin-Teng Lin
David Lin
Yuan-Pei Lin
Bernabe Linares-Barranco
Nam Ling
Bin-Da Liu
Derong Liu
Ruey-Wen Liu
Shih-Chii Liu
Chun-Shien Lu
Wu-Sheng Lu
Rastislav Lukac

Felix Lustenberger
Patrick Madden
Gian Mario Maggio
Enrico Magli
Diana Marculescu
Martin Margala
Wolfgang Mathis
Pinaki Mazumder
Dinesh P. Mehta
Janet Meiling
Bradley Minch
Karen Miu
Yoshikazu Miyanaga
Un-Ku Moon
P. R. Mukund
Tertulien Ndjountche
Sergio Netto
Robert Newcomb
Tony Ng
Jitkasame Ngarmnil
Truong Nguyen
Mohammed Niamat
Yoshifumi Nishio
Josef Nossek
Maciej Ogorzalek
Kohshi Okumura
Soontorn Oraintara
Ari Paasio
Gaetano Palumbo
David Pan
Peter Jeng-Shyang Pan
Rajendran Panda
Keshab Parhi
Belén Pérez-Verdú
See-May Phoong
Ioannis Pitas
Alessandro Piva
K. N. Plataniotis
Ting Chuen Pong
Miodrag Potkonjak
P. K. Rajan
Ravi P. Ramachandran
Jaime Ramirez-Angulo
Csaba Rekeczky
Markku Renfors
Gabriel Rincon-Mora
Alessandro Rizzo
Gordon Roberts
Angel Rodriguez-Vazquez
Fabrizio Rovati
Riccardo Rovatti
Jaijeet Roychowdhury
Toshimichi Saito
Fathi M. Salem
Edgar Sanchez-Sinencio
Tapio Saramäki
Mohamad Sawan
Thomas Schimming
Hanspeter Schmid
Wolfgang Schwarz
Radu M. Secareanu
Wouter A. Serdijn
Gianluca Setti
Naresh Shanbhag
Goubin Shen
Bertram Shi
Yun Q. Shi
Tadashi Shibata
Jose Silva-Martinez
Man-Hung Siu
Gerald Sobelman
Samer Sonkusale
Dimitrios Soudris
Andreas Spanias
Ramalingam Sridhar
Mircea Stan
Alan Stocker
Thanos Stouraitis
Tadashi Suetsugu
Yichuang Sun
Wonyong Sung
Johan Suykens
Harold Szu
Ioan Tabus
Sheldon Tan
Tuna Tarim
Ronald Tetzlaff
Sergios Theodoridis
K. Thulasiraman
Murat Torlak
Ljiljana Trajković
Shuichi Ueno
S. Ulukus
Fabrizio Vacca
P. P. Vaidyanathan
Olli Vainio
Andre van Schaik
Miroslav Velev
Anastasios N. Venetsanopoulos
Khurram Waheed
Albert Wang
Chung-Neng Wang
Jun Wang
Michael Weeks
Foo Say Wei
Gu-Yeon Wei
Denise Wilson
Martin Wong
Apisak Worapishet
An-Yeu (Andy) Wu
Dapeng Oliver Wu
Peter (Chung-Yu) Wu
Orly Yadid-Pecht
Jar-Ferr Yang
Xiaokang Yang
Yuh-Shyong Yang
Mona Zaghloul
Wei Xing Zheng
Lin Zhiping
Wenwu Zhu
Takis Zourntos

Reviewers

John Abcarius
Hisham Abdalla
Khalid Abed
Andreas Abel
Saman Abeysekera
Antonio J. Acosta Jimenez
George Adam
Michael Adams
Cyrus (Morteza) Afghahi
Arshan Aga
Ameya Agnihotri
Ruth Aguilar
M. Omair Ahmad
Sabbir Ahmad
Majid Ahmadi
Arshad Ahmed
Gil-Cho Ahn
Hee Tae Ahn
Otmane Ait Mohamed
Venkataramana Ajjarapu
Asim J. Al Khalili
Eduard Alarcón
Ken Albin
Felix Albu
Ahmed Ali
Massimo Alioto
Ahmed Allam
David Allstot
Elad Alon
Corinne Alonso
Oguz Altun
Mezyad Amourah
Bruno Andò
Boris Andreev
Andreas Andreou
Panagiotis Androutsos
Fabiola Angulo
Mohab Anis
Alyssa Apsel
Kaoru Arakawa
Shoko Araki
Paolo Arena
Sabri Arik
Mark Arnold
Peter Aronhime
David Arrowsmith
Tughrul Arslan
Ali Assi
Oscar Au
Arlo Aude
Yves Audet
Jose L. Ausin
Julio Ayala
Karsilayan Aydin
Nadine Azmard
Wael Badawy
Yong-Cheol Bae
Sumit Bagga
Faramarz Bahmani
Donghyun Baik
Ovidiu Bajdechi
Rola Baki
Ganesh Balachandran
Ganesh Balamurugan
David Balya
Abhishek Bandyopadhyay
Alexander Baranovski
Francisco Barat
James Barby
Mauro Barni
Paola Barrera
Diego Barrettino
Geoffrey Barrows
Mark Barry
Franco Bartolini
Adriano Basile
Anuj Batra
Andreas Bauer
Buyurman Baykal
Murat Becer
Fernando Beltran Blazquez
Igor Belykh
Zina Ben Miled
Philippe Benabes
Krzysztof Berezowski
Stuart Bergen
Subhra Bhattacharya
Mitun Bhattacharyya
Jinian Bian
Armin Biere
Mario Biey
Neil Bird
Luiz W. P. Biscainho
Mainak Biswas
Ben Blalock
Shannon Blunt
Itsda Boonyaroonate
Shekhar Borkar
Edoardo Botti
Robert Bowman
Elaheh Bozorgzadeh
Marco Branciforte
Victor Manuel Brea
Robert Bregovic
Paul Brennan
J. W Bruce
Erik Bruun
Brad Bryant
Linkai Bu
Antonio Buonomo
Alison Burdett
Rod Burt
Phil Burton
Karen Butler-Purry
Jianfei Cai
D. E. Calbaza
Luiz Caloba
Bruce Calvert
Francisco Javier Cañete
Claudio Canizares
Antonio Cantoni
Jun Cao
Alfonso Carlosena
Ricardo Carmona-Galán
Tony Chan Carusone
Pedro Carvalho
Marco Cassia
Salvatore Castorina
Gert Cauwenberghs
Daniele Caviglia
Abdullah Celik
Mehmet Celik
Gustavo Liñán Cembrano
Ediz Cetin
Shantanu Chakrabartty
Sudipto Chakraborty
Arthur Chan
Din-Yuen Chan
Shing-Chow Chan
Thomas Chan
Yui-Lam Chan
Tony Chan Carusone
Mukul Chandorkar
R. Chandramouli
Hemender R Chandupatla
Chip Hong Chang
Dong-Young Chang
Feng-Cheng Chang
Lung-Hao Chang
Pao-Chi Chang
Soon-Jyh Chang
Tae-Gyu Chang
Yun-Nan Chang
Chantana Chantrapornchai
Kishore Chatterjee
Shouri Chatterjee
Lap-Pui Chau
Ravi Chawla
Paolo Checco
Yuen-Hui Chee
Matthew Cheely
Bor-sen Chen
Charlie Chen
Chuen-Yau Chen
Chunhong Chen

H. H. Chen
Hung-Ming Chen
Jerry Chen
Jiann-Jone Chen
Jie Chen
Kevin Chen
Liang Chen
Liang-Gee Chen
Lulin Chen
Mei-Juan Chen
Nancy F. Chen
Qiang Chen
Sau-Gee Chen
Shiun Chen
Tao Chen
Trista Chen
Wei-Zen Chen
Xinyu Chen
Yingda Chen
Ying-Jui Chen
Yueh-Hong Chen
Yung-Chang Chen
Zhibo Chen
Kuo-Hsing Cheng
Samuel Cheng
Chong-Yung Chi
Jen-Shiun Chiang
Tihao Chiang
Elisabetta Chicca
Archana Chidanandan
Lih-Yih Chiou
Changhyuk Cho
Jun Dong Cho
Yu-Ju Cho
Bo-Kyung Choi
Ki Young Choi
Kyu-Won Choi
Yun-Ho Choi
John Chong
Mitchai Chongcheawchumnan
Bhaskar Choubey
Hwang-Cherng Chow
Ernst Christen
Chris Chu
Hsiao-Chiang Chuang
Sean Chuang
Henry Chung
Pau-Choo Chung
Andrzej Cichocki
Maciej Ciesielski
Mathew Clapp
Martin Clara
Marc Cohen
Robert Colbert
Jeffrey Coleman
Steve Collins
Francisco Colodro
Vittorio Colonna
Timothy Constandinou
Nicolas Constantin
Massimo Conti
Geoffrey Coram
Fernando Corinto
Pasquale Corsonello
Juan Cousseau
Sergio A. Cruces-Alvarez
György Cserey
Arpad I. Csurgay
Eugenio Culurciello
Paolo Cusinato
Dariusz Czarkowski
Frank Dachselt
S. Dasgupta
Minas Dasygenis
Jeff Davis
Oscar De Feo
Jose M. de la Rosa
Enno de Lange
Jeroen De Maeyer
Wim De Wilde
Debatosh Debnath
Jamal Deen
Levent Degertekin
Armin Deiss
Paolo Del Giudice
Rocio Del Rio
Tobi Delbruck
Jose Delgado-Frias
Manuel Delgado-Restituto
Andreas Demosthenous
Yunbin Deng
Tracy Denk
Evangelos Dermatas
Uday Desai
Mohamed Dessouky
Mario Di Bernardo
Antonio Di Giandomenico
Marco Di Mauro
Alejandro Diaz-Sanchez
Huseyin Dinc
Yongwang Ding
Zhi Ding
Hormoz Djahanshahi
Ian Dobson
Alen Docef
Akimitsu Doi
Carlos Dominguez Matas
Vaibhav Donde
Zhao Dong
Zhiwei Dong
Konstantinos Doris
Diann Dow
Piotr Dudek
Chris Duffy
Benoit Dufort
Jeff Dugger
Fikret Dulger
Adriana Dumitras
Günhan Dündar
Antoine Dupret
Francisco Duque-Carrillo
Sumantra Dutta Roy
Carl Ebeling
Udaykiran Eduri
Karen Egiazarian
William Eisenstadt
Muhieddine El Kaissi
Imtinan Elahi
Praveen Elakkumanan
Mourad El-Gamal
Mohamed Elgamel
Walid Elgharbawy
Duncan Elliott
Ahmed Elwakil
Tetsuro Endo
Guillermo Espinosa
Bruno Estibals
Brian L. Evans
Robert Ewing
Rubem Dutra R. Fagundes
Nikos Fakotakis
Jing-Jing Fang
Tao Fang
Yanmei Fang
Yao-Chun Fang
Shahin Farahani
Ethan Farquhar
Kamran Farzan
M. Faulkner
Christian Jesus B. Fayomi
Orla Feely
Yan Feng
F. V. Fernandez
Vittorio Ferrari
Jose Ferreira de Jesus
Giuseppe Ferri
Claudio Fiegna
Joan Figueras
Miguel Figueroa
Hani Fikry
Igor Filanovsky
Fabio Filicori
Franco Fiori
Timothy Fischer
Michael Flynn
Peter Foldesy
David J. Foley
Say Wei Foo
Gary Ford
Mauro Forti
Luigi Fortuna
Daniele Fournier-Prunaret
Rob Fox
Guinjoan Francesc

Kenneth Francken
Mattia Frasca
Douglas Frey
Matthias Frey
Jessica Fridrich
Dan Friedman
Eby Friedman
Dongdong Fu
Zhongtao Fu
Koji Fujii
Minoru Fujishima
Marco Fumagalli
Paul Furth
Roberto Gaddi
Gilles Gagnon
Viktor Gal
Zbigniew Galias
Krzysztof Galkowski
Jason Gallas
Carlos Galup-Montoro
Kun Gao
Xiqi Gao
Yunlong Gao
Gabriel Garcera
David Garrett
Wesley Gee
Roman Genov
Pando Gr. Georgiev
Julio Georgiou
Andrea Gerosa
Soheil Ghiasi
Reza Ghodssi
Arindam Ghosh
Soumik Ghosh
Georges Gielen
Marco Gilli
Gianluca Giustolisi
Antonio Gnudi
Joao Goes
Chee Kiang Goh
Eugene Goldberg
Leonid Goldgeisser
Danian Gong
Wenrui Gong
Jose Luis Gonzalez
Anand Gopalan
Atanas Gotchev
Costas Goutis
David Graham
Elena Gramatova
Marco Grangetto
Giuseppe Grassi
Michael Green
Stefano Gregori
Tom Griffin
Warren Gross
Viktor Gruev
Limin Gu
Qianping Gu
Yong Liang Guan
Drew Guckenberger
Davide Guermandi
Thomas Guionnet
Xiaochuan Guo
Oscar Gustafsson
Bah Hwee Gwee
Dong S. Ha
Soonhoi Ha
Sandro Haddad
Christoforos Hadjicostis
Predrag Hadzibabic
Martin Haenggi
Christoph Hagleitner
David Haigh
Kari Halonen
Nozomu Hamada
Shy Hamami
Hsueh-Ming Hang
Pavan Hanumolu
Ramesh Harjani
Ian Harris
Jay H. Harris
John Harris
Reid Harrison
Iiro Hartimo
Jackson Harvey
Martin Hasler
Paul Hasler
James Haslett
Jun He
Ming He
Xin He
Yuwen He
Zhihai He
Markus Helfenstein
Arman Hematy
Mark Hempstead
Rahmi Hezar
Asai Hideki
Charles M. Higgins
Takashi Hisakado
Ian A. Hiskens
Daniel Hissel
Anthony Ho
K. C. Ho
James Hoe
Jeremy Holleman
Johnny Holmberg
David Hong
Xianlong Hong
Yoshihiko Horio
Timothy Horiuchi
Anup Hosangadi
Michael Hsiao
Hong-Yean Hsieh
Ming-ta Hsieh
Chiou-Ting Hsu
David Hsu
Huai-Yi Hsu
June-Ming Hsu
Ming-Feng Hsu
Spencer Hsu
Jiang Hu
Yongjian Hu
Chung-Lin Huang
Chun-Yueh Huang
Han Huang
Hong-Yi Huang
Hsiang-Cheh Huang
Hsiang-Chun Huang
Ing-Jer Huang
Jiwu Huang
Ming-Yen Huang
Zhijun (Jerry) Huang
Kevin Huckins
Jenq-Neng Hwang
Luigi Iannelli
Tohru Ikeguchi
Giacomo Indiveri
Kris Iniewski
Yujiro Inouye
Adrian Ioinovici
Abraham Islas
Mohammed Ismail
Yehea Ismail
Misha Ivanov
Vadim Ivanov
Roozbeh Jafari
Lakhmi Jain
Uday Jain
Chein-Wei Jen
Okjune Jeon
Mieczyslaw Jessa
Kai-Yuan Jheng
Zhu Ji
Minqiang Jiang
Ruoxin Jiang
Xiaohong Jiang
Craig Jin
Hua Jin
Xuan Jing
Kenya Jinno
Hakan Johansson
David Johns
Mark Johnson
Alex Jones
Youngjoong Joo
Ivan Jorgensen
Jer-Min Jou
Shyh-Jye Jou
Jorge Juan-Chico
Pedro Julian
Graham Jullien
Byughoo Jung

Sungyong Jung
Christian Jutten
Yoshinobu Kajikawa
Yoji Kajitani
Gregory Kalivas
Priyank Kalla
Asko Kananen
M. Kandemir
Sung-Mo Kang
Rafal Karakiewicz
Juha Karhunen
Karim Karim
Tanja Karp
Soundarapandian Karthikeyan
Raj Katti
Shoji Kawahito
Masayuki Kawamata
Mitsuru Kawamoto
Ian Kearney
Laszlo Kek
Kristina Kelber
Michael Peter Kennedy
Mustafa Keskin
S. A. Khaparde
Sunil P. Khatri
Roni Khazaka
Abdelhakim Khouas
Phanumas Khumsat
Wing-Hung Ki
Jose Kilchoer
Bruce Kim
Chang-Su Kim
Chang-sung Kim
Hyung Kim
Jaeha Kim
JongWon Kim
Mingyu Kim
Soo-Won Kim
Taewhan Kim
Yong Chang Kim
Peter Kinget
Gabor Kis
Peter Kiss
Jacques-Olivier Klein
Bendik Kleveland
Hung-Yang Ko
Hyungjong Ko
Ljupco Kocarev
Fatih Kocer
J. Kodate
Wern Koe
Eleftherios Kofidis
Teruhiko Kohama
Tohru Kohda
Geza Kolumban
Keiji Konishi
Sunil Kopparapu
Ivica Kopriva
Alexander Korotkov
Alex Korshak
Alex Kot
Yajun Kou
Odysseas Koufopavlou
Christos Koukourlis
Georgios Kouroupetroglou
Ivan S. Kourtev
George Koutroumpezis
Michiel Kouwenhoven
Piotr Kowalczyk
Slobodan Kozic
Shoba Krishnan
Nikolaos Kroupis
Chung-Wei Ku
Shiann-Rong Kuang
Matt Kucic
Satish Kulkarni
Ajay Kumar
Ashok Kumar
Sunil Kumar
Ravi Kummaraguntla
Deepa Kundur
Weiying Kung
Pavani Kuntala
C.-C. Jay Kuo
Chih-Hung Kuo
Lun-Chia Kuo
WenKuang Kuo
Christoph Kuratli
Azadeh Kushki
Do-Kyoung Kwon
Lee-Kyung (Winnie) Kwon
John Lach
Shuvendu Lahiri
Shang-Hong Lai
Yeong-Kang Lai
Yung-Yuan Lai
Mika Laiho
Kabada R. Lakshmikumar
Hylas Y. H. Lam
Andy Lambrechts
Hui Lan
Tor Sverre Lande
Andreas Larsson
Laurent Latorre
Charlotte Lau
Francis C. M. Lau
Anthony J. Lawrance
Changjin Lee
Chen-Yi Lee
Ching-ren Lee
Chunyi Lee
Gwo Giun Lee
Haechang Lee
Heung-Kyu Lee
Jang Lee
Seok-jun Lee
Seung-Hoon Lee
Tai-cheng Lee
Tsung-Sum Lee
Wei-Rong Lee
Domine Leenaerts
Michael Lehmann
Sheau-Fang Lei
Zhijun Lei
Jin-Jang Leou
Adrian Leuciuc
S. H. Leung
Peter Levine
Anthony Lewis
Steve Lewis
Chih-Hung Li
Fan-Min Li
Gang Li
Gary Li
Jiang Li
Jin Li
Jipeng Li
Ping Li
Qing Li
Shengyuan Li
Shipeng Li
Sing-Rong Li
Xiang Li
Yijun Li
Yunchu Li
Zhen Li
Zhengguo Li
Chung-Jr Lian
Yong Lian
Jing Liang
Sheng-Fu Liang
Xiaofeng Liao
Valentino Liberali
Patrick Lichtsteiner
John Lidgey
Wen-Nung Lie
Jenn-Jier James Lien
Sung Kyu Lim
Sotirios Limotyrakis
Bor-Ren Lin
Chia-Wen Lin
Chih-Hsiu Lin
Ching-Yung Lin
Chin-Teng Lin
David Lin
Fang Lin
Guo-Shiang Lin
Hongchin Lin
Hsin-Lei Lin
Weisi Lin
Xiao Lin
Yao-Chung Lin
Yuan-Pei Lin
Zhiping Lin

Alejandro Linares-Barranco
Bernabé Linares-Barranco
Erik Lindberg
Saska Lindfors
Leif Lindgren
Nam Ling
Bin-Da Liu
Der-Jenq Liu
Derong Liu
Mingjian Liu
Ruey-Wen Liu
Shen Iuan Liu
Shih-Chii Liu
Zheng Liu
Paula Lopez Martinez
David Lopez Vilariño
Jose M. Lopez Villegas
Rafael Lopez-Ahumada
Antonio Lopez-Martin
Jesus Lopez-Soto
Flavio Lorenzelli
Seo-How Low
Per Lowenborg
Ruibing Lu
Shih-lien Lu
Wu-Sheng Lu
Yan Lu
Zhe-Ming Lu
Zhongkang Lu
Patanè Luca
Rastislav Lukac
Heather Luman
Hui Luo
Qiang Luo
Zuying Luo
Felix Lustenberger
Hsi-Pin Ma
Shao Dan Ma
Shyh-Yi Ma
Benoit Macq
Patrick Madden
Gian Mario Maggio
Sebastian Magierowski
J. Mahattanakul
Brian Mak
Pui-In Mak
Yuri Makarov
Shoji Makino
Marek Makowski
Piero Malcovati
Arto Malinen
Franco Maloberti
Hong Man
Nicolò Manaresi
Sabato Manfredi
Yiannos Manoli
Makram Mansour
Mohammad Mansour
Irena Maravic
Mario Marchesoni
Diana Marculescu
Theo Marescaux
Martin Margala
Janos Markus
Joao Marques Silva
Maurizio Martina
Herminio Martinez-Garcia
Luis Martinez-Salamero
Pina Marziliano
Saverio Mascolo
Guido Masera
James Mason
Mark Massie
Sanu Mathew
Hervé Mathias
Wolfgang Mathis
Toshimasa Matsuoko
Brent Maundy
Kartikeya Mayaram
Joceli Mayer
Pinaki Mazumder
Gianluca Mazzini
Jim McCalley
Ryan McGinnis
Michael McGuire
Cyril Measson
Dinesh P. Mehta
Swati Mehta
Bingfeng Mei
Yan Meng
Scott Meninger
Ricardo Merched
Patrick Merkli
Christian Merkwirth
Alfred Mertins
Dimitris Metafas
Tetsuhisa Mido
Jean-Yves Mignolet
Jovica Milanovic
Miro Milanovic
Bradley Minch
Brian Minnis
Shahriar Mirabbasi
Chinmaya Mishra
Rosario Mita
Joydeep Mitra
Subhasish Mitra
Teruyuki Miyajima
Yoshikazu Miyanaga
Akio Miyazaki
Daisuke Miyazaki
Prabhu Mohan
Ahmed Nader Mohieldin
Philip K. T. Mok
Oystein Moldsvor
Andrea Molino
Antonio Mondragon
Bert Monna
Un-Ku Moon
Paul Moore
Eric Moreau
Carl Moreland
Kazuyoshi Mori
Takashi Morie
Khurran Muhammad
Jayanta Mukherjee
P. R. Mukund
Jan Mulder
Mitsuji Muneyasu
Fernando Muñoz
Tadao Murata
Alan Murray
Shinichiro Mutoh
Mike Myers
Azad Naeemi
Oivind Naess
Sam Naffziger
Angelo Nagari
Kavita Nair
Hiroyuki Nakajima
Bharat R Nallamilli
KiYoung Nam
Alberto Nannarelli
Ashok Narasimhan
Amit Narayan
G. Narayanan
Rafael de Jesus Navas Gonzalez
Ghanshyam Nayak
Tertulien Ndjountche
Ashok Nedungadi
Joe Neff
Louis Nervegna
Sergio Netto
Shawn Neugebauer
Simon Neukom
Andrea Neviani
Robert Newcomb
Chiu Wa Ng
Hiok-Tiaq Ng
Tony Ng
King Ngi Ngan
Jitkasame Ngarmnil
Thao Nguyen
Truong. T. Nguyen
Chrys Ngwa
Zhicheng Ni
Mario Nicola
Ali Niknejad
Spyros Nikolaidis
Borivoje Nikolic
Dragan Nikolik
Peter Nilsson
Tetsuo Nishi
Akinori Nishihara

Shotaro Nishimura
Yoshifumi Nishio
Rajeev Nongpiur
Ladislav Novak
Peter Novak
Tony Nygard
Douglas O'Shaughnessy
Yasunari Obuchi
Hiroshi Ochi
Maciej Ogorzalek
Seda Ogrenci Memik
Chanhee Oh
Takahiro Oie
Hiroyuki Okada
Garret Okamoto
Kohshi Okumura
Omid Oliaei
Ashkan Olyaei
Takao Onoye
Job Oostveen
Maurits Ortmanns
Laszlo Orzo
Elaine Ou
Mourad Oulmane
Fabricio Ourique
Sule Ozev
Serdar Ozoguz
Ari Paasio
K. R. Padiyar
Fabrice Paillet
Antonio Palacios
Vassilis Paliouras
Gaetano Palumbo
Feng Pan
Peter Jeng-Shyang Pan
Zhen Gang Pan
Marios C. Papaefthymiou
Anand Pappu
S. Young Park
Nelson Passos
Stefano Pastore
Susana Paton
Bipul Paul
Eduarda Pedro
Soo-Chang Pei
Wen-Hsiao Peng
Salvatore Pennisi
Raquel Perez-Aloe
Fernando Perez-Gonzalez
Belén Pérez-Verdú
David Perreault
Michael Perrott
Pasi Pertila
Gene Petilli
Istvan Petras
Marong Phadoongsidhi
Dinh Tuan Pham
Khoman Phang
Ralf Philipp
Maria Cristina Piccirilli
Petri Piiroinen
Aggelos Pikrakis
Alessandro Piovaccari
Tuomo Pirinen
Peter Pirsch
Bardia Pishdad
Ioannis Pitas
Jim Plusquellic
Jonne Poikonen
Massimo Poli
Robi Polikar
Gabriel Popescu
Dragana Popovic
Bernd Porr
Veikko Porra
Sonia Porta
Philippe Pouliquen
Salil Prabhakar
Mohit Prasad
Enrique Prefasi
Benoit Provost
Xin Qi
Tongfeng Qian
Qinru Qiu
Gang Qu
Davide Quaglia
Federico Quaglio
Michele Quarantelli
Ricardo Queiroz
Maria Paula Queluz
Shakeel Qureshi
Sridhar Radhakrishnan
Luigi Raffo
Susanto Rahardja
Manoj Rajagopalan
P. K. Rajan
Jaime Ramirez-Angulo
Hanming Rao
Hamid R. Rategh
Ruwan Ratnayake
Cassiano Rech
Gunther Reiszig
Csaba Rekeczky
Branimir Reljin
Markku Renfors
Mehrdad Reshadi
Woogeun Rhee
Cassio Ribeiro
David Rich
Brad Riel
Tapani Ristaniemi
Alessandro Rizzo
Gordon Roberts
Elisenda Roca
Angel Rodriguez-Vazquez
Pieter Rombouts
Riccardo Rovatti
Jaijeet Roychowdhury
Ken Ruan
Adoracion Rueda
Lawrence Ryan
Justo Sabadell
Ebrahim Saberinia
Enrico Sacchi
Vincenzo Sacco
Manoj Sachdev
Mona Safi-Harb
Sadiq Sait
Toshimichi Saito
Philippe Salib
Roghoyeh Salmeh
Marcello Salmeri
Stefano Santi
Sachin Sapatnekar
Tapio Saramäki
Fausto Sargeni
Daisuke Satoh
Farook Sattar
Nicolò Savalli
Yvon Savaria
Mohamad Sawan
Prashant Saxena
Rolf Schaumann
Thomas Schimming
Ken Schindler
Hanspeter Schmid
Stefan Schneider
Wolfgang Schwarz
Radu M. Secareanu
Ankit Seedher
Mark Seidel
Jose Manoel Seixas
Hiroo Sekiya
Luca Selmi
Inchang Seo
Wee Ser
Wouter A. Serdijn
Francisco Serra-Graells
Guillermo J. Serrano
Teresa Serrano-Gotarredona
Gianluca Setti
Jack Sewell
Mohsen Shaaban
Arvin Shahani
Behnam Shahrrava
Kaveh Shakeri
Ying Shan
Gaurav Sharma
Day-Fann Shen
Hung-Da Shen
Jacky Shen
Jizhong Shen
Meiyin Shen
Pei-Ling Shen

Wein Tung Shen
Zion Shen
Bing Sheu
Bertram Shi
Rock Shi
Weiping Shi
Yun Q. Shi
Tadashi Shibata
Chin-Shiuh Shieh
John Shields
Hyunchul Shin
Jitae Shin
Isao Shirakawa
Doron Shmilovitz
Omid Shoaei
Haiyan Shu
Sandeep Shukla
Maher Sid-Ahmed
Svante Signell
Christopher Silva
L. Miguel Silveira
Calvin Sim
Montek Singh
Sameep Singh
Konstantinos Siozios
Stelios Siskos
Athanassios Skodras
Michael Small
Paul Smith
Bogdan Smolka
H. C. So
Gerald Sobelman
Ahmed Soliman
S. A. Soman
Han-Woong Son
Hongjiang Song
Hwangjun Song
Xiaodan Song
Samer Sonkusale
Theerachet Soorapanth
Dimitrios Soudris
Andreas Spanias
Ramalingam Sridhar
Srinivasa Sridhara
Srivathsan Sridharan
Venkatesh Srinivasan
S. C. Srivastava
Ian St. John
Mircea Stan
Milutin Stanacevic
Alexander Stankovic
Robert (Bogdan) Staszewski
Jesper Steensgaard
D'Amico Stefano
Alan Stocker
Viera Stopjakova
Marco Storace
Haihua Su
Hsuan-Jung Su
Mu-Chun Su
Simon Su
Tzu-Liang Su
Tadashi Suetsugu
Goel Sumeer
Ming-Ting Sun
Qibin Sun
Shih-Wei Sun
Xiaoling Sun
Xiaoyan Sun
Yi-Ran Sun
Prasanna Sundararajan
Lars Sundstrom
Wonyong Sung
Steve Sunter
Myung Sunwoo
Johan Suykens
Masakiyo Suzuki
Francesco Svelto
Istvan Szatmari
Zoltan Szlavik
Michael Tadeusiewicz
Akira Taguchi
Chris Taillefer
Atsushi Takahashi
Hiroshi Takahashi
Norikazu Takahashi
Mike Talley
Jean-Pierre Talpin
Vincent H. S. Tam
Arunas Tamasevicius
Eric Tan
Sheldon Tan
Yap-Peng Tan
Mamoru Tanaka
Tetsuro Tanaka
Jun Tang
Wallace K. S. Tang
Xiaoou Tang
Yonghui Tang
Tanes Tanitteerapan
Liang Tao
Satoshi Taoka
Aleksandar Tasic
Baris Taskin
Konstantinos Tatas
Ayman Tawfik
David Tay
J. Luis Tecpanecatl-Xihuitl
Attila Telegdy
Gabor Temes
Francesco Tenore
Kemal Tepe
Takahide Terada
A. Thanachayanont
Bharath Kumar Thandri
George Theodoridis
Sergios Theodoridis
Evaggelia Theohari
Patrick Thiran
D. Thukaram
K. Thulasiraman
Parimala Thulasiraman
Gergely Timar
Saibun Tjuatja
Andrew Tohmc
Pasquale Tommasino
C. S. Tong
Guido Torelli
Hiroyuki Torikai
Murat Torlak
Levente Torok
Paola Tortori
Ramón Tortosa-Navas
Laszlo Toth
Nur Touba
Michel Toulouse
Chris Toumazou
Alexandros Tourapis
Ljiljana Trajković
Alessandro Trifiletti
Joe Tritschler
Chia-Yang Tsai
Sam S. Tsai
Tsung-Han Tsai
Yao-Hong Tsai
Tommy Tsang
Chi K. Tse
Sofia Tsekeridou
Vladi Tsipenyuk
Yannis Tsividis
Tadashi Tsubone
Kai-man Tsui
Yi-Shin Tung
John Tuthill
Shuichi Ueno
Takeshi Ueno
Tetsushi Ueta
Akio Ushida
Toshimitsu Ushio
P. P. Vaidyanathan
Olli Vainio
Alberto Valdes-Garcia
Maurizio Valle
Jan van der Spiegel
Johan van der Tang
Koen van Hartingsveldt
Arthur van Roermund
Andre van Schaik
Jan Vandenbussche
Joos Vandewalle
Bart Vanhoof
Vijay Varadarajan
Prab Varma
Gabor Vattay

Benoit Veillette
Miroslav Velev
Srinivasa Vemuru
Andreas Veneris
Giri Venkataramanan
Sathya Venkatesh
Ingrid Verbauwhede
Haridimos T. Vergos
Anthony Vetro
Eva Vidal
Fernando Vidal-Verdú
Valerio Vignoli
Ari Visa
Natarajan Viswanathan
Stefano Vitali
Konstantinos Vitoroulis
Jacob Vogelstein
Vladimir Vukadinovic
Khurram Waheed
Baohua Wang
Chris Wang
Chua-Chin Wang
Chung-Neng Wang
Feng-Hsing Wang
Gang Wang
Jinn-Shyan Wang
Lei Wang
Li-C. Wang
Maogang Wang
Nanyan Wang
Shih-Hao Wang
Wei Wang
Wen-Hao Wang
Xianmin Wang
Xu Fang Wang
Yanjie Wang
Yi-Chuan Wang
Yuke Wang
Lars Wanhammar
Toshimasa Watanabe
Michael Weeks
Gu-Yeon Wei
Hsiao Wei Su
Laurens Weiss
Changyun Wen
Ro-Min Weng
Holger Wenske
Neil Weste
Jurjen Westra
I-Chyn Wey
Jacob White
Benjamin Widdup
Andreas Wiesbauer
Sinbad Wilmot
Denise Wilson
Dag T. Wisland
Florentin Woergoetter
Christopher Wolff
Dicky Chi Wah Wong
Kahou Wong
Kainam Thomas Wong
Man-Chung Wong
Ngai Wong
Roger Woods
Apisak Worapishet
An-Yeu (Andy) Wu
Bing-Fei Wu
Chai Wah Wu
Chung-Yu Wu
H. Wu
H. R. Wu
Ja-Ling Wu
Min Wu
Qing Wu
Shaoquan Wu
Tsung-Han Wu
Yang Wu
Bin Xia
Bo Xia
X.-G Xia
Ying Xiao
Fan Xiaohua
Zhijie Xiong
Zixiang Xiong
JiZheng Xu
Li Xu
Qinwei Xu
Shengyuan Xu
Orly Yadid-Pecht
Isao Yamada
Kiyotaka Yamamura
Toshihiko Yamasaki
Chunzhu Yang
Huazhong Yang
Jar-Ferr Yang
Xiaojian Yang
Yaohua Yang
Yuh-Shyong Yang
Ji Yao
Libin Yao
Susu Yao
Kim Hui Yap
Terry Tao Ye
Chia-Hung Yeh
Jih-Chiang Yeo
Mehmet Can Yildiz
Peng Yin
Dominic Wing Kin Yip
Kun Wah Yip
Jianghong Yu
Lu Yu
Sung-Nien Yu
Yajun Yu
Mehmet Yuce
Ilya Yusim
Mona Zaghloul
Regan Zane
Akos Zarandy
Mauricio Zavaleta
Bing Zeng
Wenjun Zeng
Zhang Zengyan
Cishen Zhang
Gang Zhang
Haoran Zhang
Hui Zhang
Liping Zhang
Ming Zhang
Tao Zhang
Tong Zhang
Xian-Da Zhang
Yi Zhang
Zhishou Zhang
Peiyi Zhao
Wen-yi Zhao
Jinghong Zheng
Sanbao Zheng
Wei Xing Zheng
Weiguo Zheng
Yuanjin Zheng
Guoan Zhong
Dian Zhou
Sean Zhou
Yi Qing Zhou
Bin Zhu
Ce Zhu
Tong Zhu
Zhanybai Zhusubaliyev
Zeljko Zilic
Yann Zinzius
Ljiljana Zivanov
Ahmed Zobaa
Vladimir Zolotov
Dekun Zou
Takis Zourntos

TABLE OF CONTENTS

Volume I

ASP-L1: PIPELINED ADC

ASP-L1.1: DIGITAL ERROR CORRECTION AND CALIBRATION OF GAIN NON-LINEARITIES IN A PIPELINED ADC .... I - 1
Arun Ravindran, University of North Carolina at Charlotte, United States; Anup Savla, Jennifer Leonard, The Ohio State University, United States

ASP-L1.2: A NEW DIGITAL BACKGROUND CALIBRATION TECHNIQUE FOR PIPELINED ADC .... I - 5
Kamal El-Sankary, Mohamad Sawan, École Polytechnique de Montréal, Canada

ASP-L1.3: A 1.8V 8-BIT 250MSAMPLE/S NYQUIST-RATE CMOS PIPELINED ADC .... I - 9
Tae-Hwan Oh, Ho-Young Lee, Ho-Jin Park, Jae-Whui Kim, Samsung Electronics, Republic of Korea

ASP-L1.4: DIGITAL BACKGROUND AUTO-CALIBRATION OF DAC NON-LINEARITY IN PIPELINED ADCs .... I - 13
Martin Kinyua, Texas Instruments, United States; Franco Maloberti, University of Texas, Dallas, United States; William Gosney, Southern Methodist University, United States

ASP-L1.5: A 28mW 10b 80MS/s PIPELINED ADC IN 0.13µm CMOS .... I - 17
Peter Bogner, Infineon Technologies AG, Austria

ASP-L2: ANALOG FILTERING TECHNIQUES I

ASP-L2.1: ON DESIGNING LINEARLY TUNABLE HIGH-Q OTA-C FILTERS WITH LOW SENSITIVITY .... I - 21
Jader A. De Lima, Fernando M. Alcaide, University of Sao Paulo State, Brazil

ASP-L2.2: EXACT DESIGN OF ALL-MOS LOG FILTERS .... I - 25
Francisco Serra-Graells, Xavier Redondo, Institut de Microelectrònica de Barcelona - CNM, Spain

ASP-L2.3: ADAPTIVE LOG DOMAIN FILTERS USING FLOATING GATE TRANSISTORS .... I - 29
Pamela Abshire, Eric Liu Wong, Yiming Zhai, Marc Cohen, University of Maryland, United States

ASP-L2.4: A FULLY PROGRAMMABLE LOG-DOMAIN BANDPASS FILTER USING MULTIPLE-INPUT TRANSLINEAR ELEMENTS .... I - 33
Ravi Chawla, Haw-Jing Lo, Arindam Basu, Paul Hasler, Brad Minch, Georgia Institute of Technology, United States

ASP-L2.5: 6.61 M TO 317M Hz NTH-ORDER CURRENT-MODE LOW-PASS AND HIGH-PASS OTA-ONLY-WITHOUT-C FILTER .... I - 37
Chun-Ming Chang, Chung Yuan Christian University, Taiwan

ASP-L3: VOLTAGE REFERENCES

ASP-L3.1: A CMOS BANDGAP REFERENCE CIRCUIT FOR SUB-1-V OPERATION WITHOUT USING EXTRA LOW THRESHOLD-VOLTAGE DEVICE .... I - 41
Ming-Dou Ker, Jung-Sheng Chen, Ching-Yun Chu, National Chiao Tung University, Taiwan

ASP-L3.2: DESIGN AND OPTIMIZATION OF A HIGH PSRR CMOS BANDGAP VOLTAGE REFERENCE .... I - 45
Armin Tajalli, Sharif University of Technology / Microelectronics Research Center, Iran; Seyed Mojtaba Atarodi, MixeCore Design, United States; Abbas Khodaverdi, Microelectronics Research Center, Iran; Farzad Sahandi Esfanjani, Tripath, United States

ASP-L3.3: DESIGN OF A 1.5-V HIGH-ORDER CURVATURE-COMPENSATED CMOS BANDGAP REFERENCE .... I - 49
Chi Yat Leung, Ka Nang Leung, Philip K. T. Mok, The Hong Kong University of Science and Technology, Hong Kong SAR of China

ASP-L3.4: TOWARDS A SUB-1 V CMOS VOLTAGE REFERENCE .... I - 53
Laleh Najafizadeh, Igor Filanovsky, University of Alberta, Canada

ASP-L3.5: A LOW-POWER LOW-VOLTAGE MOSFET-ONLY VOLTAGE REFERENCE .... I - 57
Ferdinando Bedeschi, STMicroelectronics, Italy; Edoardo Bonizzoni, Andrea Fantini, University of Pavia, Italy; Claudio Resta, STMicroelectronics, Italy; Guido Torelli, University of Pavia, Italy

ASP-L4: PIPELINED AND FOLDED ADC

ASP-L4.1: A BACKGROUND DIGITAL SELF-CALIBRATION SCHEME FOR PIPELINED ADCS BASED ON TRANSFER CURVE ESTIMATION .... I - 61
Hanjun Jiang, Haibo Fei, Iowa State University, United States; Degang Chen, Iowa State Univerisity, United States; Randall Geiger, Iowa State University, United States

ASP-L4.2: A NOVEL QUEUEING ARCHITECTURE FOR BACKGROUND CALIBRATION OF PIPELINE ADCS .... I - 65
Anup Savla, Jennifer Leonard, The Ohio State University, United States; Arun Ravindran, University of North Carolina at Charlotte, United States

ASP-L4.3: A 12-BIT 40MSPS 3.3-V 56-mW PIPELINED A/D CONVERTER IN 0.25-um CMOS .... I - 69
Reza Lotfi, Mohammad Taherzadeh-Sani, Omid Shoaei, University of Tehran, Iran

ASP-L4.4: A 10-BIT, 3.3-V, 60MSAMPLE/S, COMBINED RADIX<2 AND 1.5-BIT/STAGE PIPELINED ANALOG-TO-DIGITAL CONVERTER .... I - 73
Babak Nejati, Uuniversity of California at San Diego, United States; Omid Shoaei, University of Tehran, Iran

ASP-L4.5: A LOW VOLTAGE LOW POWER 8-BIT FOLDING/INTERPOLATING ADC WITH RAIL-TO-RAIL INPUT RANGE .... I - 77
Hamid Movahedian, Meysam Azin, Mehrdad Sharif Bakhtiar, Sharif University of Technology, Iran

ASP-L5: ANALOG FILTERING TECHNIQUES II

ASP-L5.1: BEHAVIORAL MODELING OF RF FILTERS IN VHDL-AMS FOR AUTOMATED ARCHITECTURAL AND PARAMETRIC OPTIMIZATION .... I - 81
Tom J. Kazmierski, Fazrena Hamid, University of Southampton, United Kingdom

ASP-L5.2: A DESIGN STRATEGY FOR AREA EFFICIENT HIGH-ORDER HIGH-Q SC FILTERS .... I - 85
Jose L. Ausín, J. Francisco Duque-Carrillo, University of Extremadura, Spain; Guido Torelli, University of Pavia, Italy; Miguel Angel Domínguez, University of Extremadura, Spain

ASP-L5.3: LOW ACTIVE-SENSITIVITY ALLPOLE ACTIVE-RC FILTERS USING IMPEDANCE TAPERING .... I - 89
Dražen Jurišic, University of Zagreb, Croatia; George S. Moschytz, Swiss Federal Institute of Technology (ETH), Zürich, Switzerland; Neven Mijat, University of Zagreb, Croatia

ASP-L5.4: DIGITAL TUNING OF ANALOG BANDPASS FILTERS BASED ON ENVELOPE DETECTION .... I - 93
Taner Sümesaglam, Aydin I. Karsilayan, Texas A&M University, United States

ASP-L5.5: A PROGRAMMABLE BANDPASS ARRAY USING FLOATING-GATE ELEMENTS .... I - 97
David Graham, Paul Smith, Richard Ellis, Ravi Chawli, Paul Hasler, Georgia Institute of Technology, United States

ASP-L6: BROADBAND AND UWB CIRCUITS

ASP-L6.1: DESIGN OF A SiGe LOW-NOISE AMPLIFIER FOR 3.1-10.6 GHz ULTRA-WIDEBAND RADIO .... I - 101
Bo Shi, Michael, Yan Wah Chia, Institute for Infocomm Research, Singapore

ASP-L6.2: A FEEDFORWARD COMPENSATED HIGH-LINEARITY DIFFERENTIAL TRANSCONDUCTOR FOR RF APPLICATIONS .... I - 105
Su-Tarn Lim, John R. Long, Delft University of Technology, Netherlands

ASP-L6.3: A PPM GAUSSIAN PULSE GENERATOR FOR ULTRA-WIDEBAND COMMUNICATIONS .... I - 109
Sumit Bagga, Delft University of Technology, Netherlands; Giuseppe de Vita, Università di Pisa, Italy; Sandro A. P. Haddad, Wouter A. Serdijn, John R. Long, Delft University of Technology, Netherlands

ASP-L6.4: A HIGH IIP3 X-BAND BICMOS MIXER FOR RADAR APPLICATIONS .... I - 113
Xuejin Wang, Arizona State University, United States; Aykut Dengi, Neolinear, Inc., United States; Sayfe Kiaei, Arizona State University, United States

ASP-L6.5: A 12.5GHZ RF MATRIX AMPLIFIER IN 180NM SOI CMOS .... I - 117
Jinho Park, David Allstot, University of Washington, United States

ASP-L7: FLASH ADC

ASP-L7.1: TERMINATION OF AVERAGING NETWORKS IN FLASH ADCS .... I - 121
Pedro Figueiredo, João Vital, Chipidea Microelectrónica, Portugal

ASP-L7.2: A STATISTICAL BACKGROUND CALIBRATION TECHNIQUE FOR FLASH ANALOG-TO-DIGITAL CONVERTERS .... I - 125
Chun-Cheng Huang, Jieh-Tsorng Wu, National Chiao Tung University, Taiwan

ASP-L7.3: A STUDY OF DIGITAL DECODERS IN FLASH ANALOG-TO-DIGITAL CONVERTERS .... I - 129
Erik Säll, Mark Vesterbacka, K. Ola Andersson, Linköping University, Sweden

ASP-L7.4: OPTIMAL OFFSET AVERAGING FOR FLASH AND FOLDING A/D CONVERTERS .... I - 133
Ovidiu Carnu, Adrian Leuciuc, State University of New York at Stony Brook, United States

ASP-L7.5: 6-BIT LOW POWER LOW AREA FREQUENCY MODULATION BASED FLASH ADC .... I - 137
Quentin Diduck, Martin Margala, University of Rochester, United States

ASP-L8: ANALOG SIGNAL PROCESSING I

ASP-L8.1: SYNTHESIS OF AMPLIFIER TRANSFER FUNCTION USING TIME-DOMAIN RESPONSE .... I - 141
Igor Filanovsky, University of Alberta, Canada; Arie van Staveren, Chris Verhoeven, TU Delft, Netherlands

ASP-L8.2: OPTIMIZED DYNAMIC TRANSLINEAR IMPLEMENTATION OF THE GAUSSIAN WAVELET TRANSFORM .... I - 145
Sandro A. P. Haddad, Nanko Verwaal, Delft University of Technology, Netherlands; Richard Houben, Medtronic, Netherlands; Wouter A. Serdijn, Delft University of Technology, Netherlands

ASP-L8.3: METHOD FOR EQUALIZER DESIGN BASED ON TIME-DOMAIN SYMMETRY .... I - 149
Mladen Vucic, Goran Molnar, Hrvoje Babic, Faculty of Electrical Engineering and Computing, Croatia

ASP-L8.4: EXACT SYNTHESIS OF BANDPASS RESPONSES USING AN ISOLATED CASCADE CONNECTION OF SAME ORDER LOWPASS AND HIGHPASS FILTERS .... I - 153
Roberto Gómez-García, José I. Alonso, Universidad Politécnica de Madrid, Spain

ASP-L8.5: MIXING DOMAINS IN SIGNAL PROCESSING .... I - 157
Yannis Tsividis, Columbia University, United States

ASP-L9: OSCILLATORS

ASP-L9.1: PHASE NOISE AND ACCURACY IN QUADRATURE OSCILLATORS .... I - 161
Luca Romanò, Salvatore Levantino, Andrea Bonfanti, Carlo Samori, Politecnico di Milano, Italy; Andrea Lacaita, Politecnico di Milano / IFN-CNR Sez. Milano, Italy

ASP-L9.2: A 2.4 GHZ CMOS QUADRATURE LC-OSCILLATOR/MIXER .... I - 165
Luís Oliveira, Jorge Fernandes, IST/INESC-ID Lisboa, Portugal; Igor Filanovsky, University of Alberta, Canada; Chris Verhoeven, Delft University of Technology, Netherlands

ASP-L9.3: QUADRATURE VCO BASED ON DIRECT SECOND HARMONIC LOCKING .... I - 169
Paola Tortori, Davide Guermandi, Eleonora Franchi, Antonio Gnudi, Università di Bologna, Italy

ASP-L9.4: A WIDEBAND LC-VCO WITH ENHANCED PSRR FOR SOC APPLICATIONS .... I - 173
Sebastian Magierowski, University of Calgary, Canada; Krzysztof Iniewski, Simon Fraser University, Canada; Stefan Zukotynski, University of Toronto, Canada

ASP-L9.5: DCS1800/WCDMA ADAPTIVE VOLTAGE-CONTROLLED OSCILLATOR .... I - 177
Aleksandar Tasic, Wouter A. Serdijn, John R. Long, TU Delft, Netherlands

ASP-L10: CIRCUIT THEORY IN ELECTRONICS

ASP-L10.1: A DYNAMIC ANALYSIS OF A LATCHED CMOS COMPARATOR .... I - 181
Lourans Samid, Patrick Volz, Yiannos Manoli, University of Freiburg, Germany

ASP-L10.2: GENERALIZATIONS OF ADJOINT NETWORKS TECHNIQUES FOR RLC INTERCONNECTS MODEL-ORDER REDUCTIONS .... I - 185
Herng-Jer Lee, Chia-Chi Chu, Wu-Shiung Feng, Chang Gung University, Taiwan

ASP-L10.3: SYNTHESIS OF STATIC MULTIPLE INPUT MULTIPLE OUTPUT MITE NETWORKS .... I - 189
Shyam Subramanian, David Anderson, Paul Hasler, Georgia Institute of Technology, United States

ASP-L10.4: INFORMATION CAPACITY AND POWER EFFICIENCY IN OPERATIONAL TRANSCONDUCTANCE AMPLIFIERS .... I - 193
Makeswaran Loganathan, Suvarcha Malhotra, Pamela Abshire, University of Maryland, United States

ASP-L10.5: A POWER-PERFORMANCE ADAPTIVE LOW VOLTAGE ANALOG CIRCUIT DESIGN USING INDEPENDENTLY CONTROLLED DOUBLE GATE CMOS TECHNOLOGY .... I - 197
Arvind Kumar, Sandip Tiwari, Cornell University, United States

ASP-L11: CMOS ADC

ASP-L11.1: A 89dB LOW-POWER CMOS SIGMA-DELTA MODULATOR FOR BLUETOOTH APPLICATION .... I - 201
Haibin Huang, Ezz I. El-Masry, Dalhousie University, Canada

ASP-L11.2: A 0.35μm CMOS 17-BIT@40kS/s SENSOR A/D INTERFACE BASED ON A PROGRAMMABLE-GAIN CASCADE 2-1 SIGMA-DELTA MODULATOR ..... I - 205
José M. García-González, Sara Escalera, José M. de la Rosa, Oscar Guerra, Fernando Medeiro, Rocío del Río, Belén Pérez-Verdú, Angel Rodríguez-Vázquez, Instituto de Microelectrónica de Sevilla (IMSE), Spain

ASP-L11.3: A CMOS LOW-POWER ADC FOR DVB-T AND DVB-H SYSTEMS ..... I - 209
Olujide Adeniran, Sony Semiconductor and Electronic Solutions / University College London, United Kingdom; Andreas Demosthenous, University College London, United Kingdom; Chris Clifton, Sam Atungsiri, Randeep Soin, Sony Semiconductor and Electronic Solutions, United Kingdom

ASP-L11.4: A 0.5μm CMOS PROGRAMMABLE DISCRETE-TIME DELTA-SIGMA MODULATOR WITH FLOATING GATE ELEMENTS ..... I - 213
Angelo Pereira, Daniel Allen, Paul Hasler, Georgia Institute of Technology, United States

ASP-L11.5: A 50-MHZ CMOS QUADRATURE CHARGE SAMPLING CIRCUIT WITH 66 DB SFDR ..... I - 217
Sami Karvonen, Tom Riley, Juha Kostamovaara, University of Oulu, Finland

ASP-L12: ANALOG SIGNAL PROCESSING II

ASP-L12.1: SPECTRAL ANALYSIS OF SIGNALS EXPERIENCING NONSTATIONARY STOCHASTIC TIME-SHIFTS ..... I - 221
Omid Oliaei, Motorola Labs, United States

ASP-L12.2: POWER-CONSCIOUS DESIGN METHODOLOGY FOR CLASS-A SWITCHED-CURRENT WAVE FILTERS ..... I - 225
Reuben Wilcock, Bashir Al-Hashimi, University of Southampton, United Kingdom

ASP-L12.3: LOW POWER CURRENT RECTIFIERS FOR LARGE-SCALE CURRENT-MODE SIGNAL PROCESSING ..... I - 229
Ralf Philipp, Ralph Etienne-Cummings, Johns Hopkins University, United States

ASP-L12.4: AN ULTRA-LOW POWER ANALOGUE DIRECTIONALITY SYSTEM FOR DIGITAL HEARING AIDS ..... I - 233
Phil Corbishley, Esther Rodriguez-Villegas, Chris Toumazou, Imperial College, United Kingdom

ASP-L12.5: NTH ORDER CIRCULAR SYMMETRY PATTERN AND HEXOGONAL TESSELATION: TWO NEW LAYOUT TECHNIQUES CANCELLING NONLINEAR GRADIENT ..... I - 237
Chengming He, Kuangming Yap, Degang Chen, Randall Geiger, Iowa State University, United States

ASP-L13: MIXED SIGNAL TESTING

ASP-L13.1: LUMPED PASSIVE CIRCUITS FOR 5GHZ EMBEDDED TEST OF RF SOCS ..... I - 241
Jangsup Yoon, William Eisenstadt, University of Florida, United States

ASP-L13.2: A TRANSLINEAR-BASED RF RMS DETECTOR FOR EMBEDDED TEST ..... I - 245
Qizhang Yin, William Eisenstadt, Robert M. Fox, University of Florida, United States

ASP-L13.3: THE SRE/SRM APPROACH FOR SPECTRAL TESTING OF AMS CIRCUITS ..... I - 249
Zhongjun Yu, Degang Chen, Randall Geiger, Iowa State University, United States

ASP-L13.4: A CALIBRATION TECHNIQUE FOR A HIGH-RESOLUTION FLASH TIME-TO-DIGITAL CONVERTER ..... I - 253
Peter Levine, Gordon Roberts, McGill University, Canada

ASP-L13.5: AN ALTERNATIVE DFT METHODOLOGY TO TEST HIGH-RESOLUTION SIGMA DELTA MODULATORS ..... I - 257
Sara Escalera, José M. García-González, Oscar Guerra, José M. de la Rosa, Fernando Medeiro, Belén Pérez-Verdú, Angel Rodríguez-Vázquez, IMSE - CNM - CSIC, Spain

ASP-L14: GENERAL CIRCUIT THEORY

ASP-L14.1: REACTANCE NETWORK SHAPING A SINUSOIDAL PULSE WITH SINUSOIDAL ENVELOPE OF FINITE DURATION .... I - 261
Igor Filanovsky, University of Alberta, Canada

ASP-L14.2: ACTIVE RC NETWORKS: WHAT BENEFIT IS THERE IN HAVING CONJUGATE REAL ZEROS IN THE SECOND-ORDER TRANSFER FUNCTION? .... I - 265
Ernst Goepel, Siemens Building Technologies, Germany

ASP-L14.3: MIXED-MODE PARAMETER ANALYSIS OF FULLY DIFFERENTIAL CIRCUITS .... I - 269
Timo Rahkonen, Jyri Kortekangas, University of Oulu, Finland

ASP-L14.4: DYNAMIC MODELING OF ANALOG INTEGRATED FILTERS FOR THE STABILITY STUDY OF ON-CHIP AUTOMATIC TUNING LOOPS .... I - 273
Herminio Martínez-Garcia, Eva Vidal, Eduard Alarcón, Alberto Poveda, Technical University of Catalonia (UPC), Spain

ASP-L14.5: SIMULTANEOUS REALIZATION OF Z21/Z11 AND Y21 OF AN RC TWO-PORT .... I - 277
Tetsuo Nishi, Masato Ogata, Kyushu University, Japan

ASP-L15: CURRENT-STEERING DAC

ASP-L15.1: AN ON-CHIP DNL ESTIMATION AND RECONFIGURATION FOR IMPROVED LINEARITY IN CURRENT STEERING DAC .... I - 281
Sunil Rafeeque, Vinita Vasudevan, Indian Institute of Technology, India

ASP-L15.2: NONLINEAR DISTORTION IN CURRENT-STEERING D/A-CONVERTERS DUE TO ASYMMETRICAL SWITCHING ERRORS .... I - 285
Martin Clara, Andreas Wiesbauer, Wolfgang Klatzer, Infineon Technologies AG, Austria

ASP-L15.3: A CALIBRATION METHOD FOR CURRENT STEERING DIGITAL TO ANALOG CONVERTERS IN CONTINUOUS TIME MULTI-BIT SIGMA DELTA MODULATORS .... I - 289
Matthias Keller, Yiannos Manoli, Albert-Ludwigs-University, Germany; Friedel Gerfers, Philips Semiconductors GmbH, Germany

ASP-L15.4: MODELING OF THE IMPACT OF THE CURRENT SOURCE OUTPUT IMPEDANCE ON THE SFDR OF CURRENT-STEERING CMOS D/A CONVERTERS .... I - 293
Tao Chen, Georges Gielen, Katholieke Universiteit Leuven, Belgium

ASP-L15.5: A DIGITAL CALIBRATION FOR A 16-BIT, 400-MHZ CURRENT-STEERING DAC .... I - 297
Jussi Pirkkalaniemi, Marko Kosunen, Helsinki University of Technology, Finland; Mikko Waltari, Conexant Systems, Inc., United States; Kari Halonen, Helsinki University of Technology, Finland

ASP-L16: BIOINSPIRED CIRCUITS

ASP-L16.1: HARDWARE IMPLEMENTATION OF COMPLEX REACTION-DIFFUSION NEURAL NETWORKS USING LOG-DOMAIN TECHNIQUES .... I - 301
Teresa Serrano-Gotarredona, Bernabé Linares-Barranco, Instituto de Microelectrónica de Sevilla (IMSE), Spain

ASP-L16.2: TIME-BASED ARITHMETIC USING STEP FUNCTIONS .... I - 305
Vishnu Ravinuthula, John Harris, University of Florida, United States

ASP-L16.3: A BIO-PHYSICALLY INSPIRED SILICON NEURON .... I - 309
Ethan Farquhar, Paul Hasler, Georgia Institute of Technology, United States

ASP-L16.4: A RECONFIGURABLE BIDIRECTIONAL ACTIVE 2 DIMENSIONAL DENDRITE MODEL .... I - 313
Ethan Farquhar, David Abramson, Paul Hasler, Georgia Institute of Technology, United States

ASP-L16.5: A FAMILY OF FLOATING-GATE ADAPTING SYNAPSES BASED UPON TRANSISTOR CHANNEL MODELS .... I - 317
Christal Gordon, Ethan Farquhar, Paul Hasler, Georgia Institute of Technology, United States

ASP-L17: VOLTAGE AND CURRENT SOURCES I

ASP-L17.1: A NEW CHARGE PUMP CIRCUIT DEALING WITH GATE-OXIDE RELIABILITY ISSUE IN LOW-VOLTAGE PROCESSES .... I - 321
Ming-Dou Ker, Shih-Lun Chen, Chia-Sheng Tsai, National Chiao Tung University, Taiwan

ASP-L17.2: A PROCESS, VOLTAGE, AND TEMPERATURE COMPENSATED CMOS CONSTANT CURRENT REFERENCE .... I - 325
Susanta Sengupta, Georgia Institute of Technology, United States; Kanan Saurabh, Texas Instruments, India; Phillip Allen, Georgia Institute of Technology, United States

ASP-L17.3: HIGH-PSR BIAS CIRCUITRY FOR NTSC SYNC SEPARATION .... I - 329
Chua-Chin Wang, National Sun Yat-Sen University, Taiwan; Yih-Long Tseng, National Yun Yat-Sen University, Taiwan; Tzung-Je Lee, National Sun Yat-Sen University, Taiwan; Ron Hu, Asuka Semiconductor, Inc., Taiwan

ASP-L17.4: DESIGN OF AN ULTRA-LOW-POWER CURRENT SOURCE .... I - 333
Edgar Mauricio Camacho-Galeano, Carlos Galup-Montoro, Márcio Cherem Schneider, Universidade Federal de Santa Catarina, Brazil

ASP-L17.5: BIAS CURRENT GENERATORS WITH WIDE DYNAMIC RANGE .... I - 337
Tobi Delbrück, Univ/ETH Zürich, Switzerland; André van Schaik, University of Sydney, Australia

ASP-L18: DIGITAL-TO-ANALOG CONVERTERS

ASP-L18.1: HIGH-PRECISION DAC BASED ON A SELF-CALIBRATED SUB-BINARY RADIX CONVERTER .... I - 341
Marc Pastre, Maher Kayal, Swiss Federal Institute of Technology (EPFL), Switzerland

ASP-L18.2: ADDRESSING STATIC AND DYNAMIC ERRORS IN BANDPASS UNIT ELEMENT MULTIBIT DAC'S .... I - 345
Jeroen De Maeyer, Pieter Rombouts, Ludo Weyten, Ghent University, Belgium

ASP-L18.3: A 1-V 400MS/S 14BIT SELF-CALIBRATED CMOS DAC WITH ENHANCED DYNAMIC LINEARITY .... I - 349
Saeed Saeedi, Saeid Mehrmanesh, Sharif University of Technology / Microelectronics Research Center, Iran; Seyed Mojtaba Atarodi, MixeCore Design, United States; Hesam Aslanzadeh, Sharif University of Technology, Iran

ASP-L18.4: A 14-BIT, 200 MS/S DIGITAL-TO-ANALOG CONVERTER WITHOUT TRIMMING .... I - 353
Kuo-Hsing Cheng, Tsung-Shen Chen, Chia-Ming Tu, Tamkang University, Taiwan

ASP-L18.5: A FLOATING-GATE DAC ARRAY .... I - 357
Guillermo Serrano, Paul Hasler, Georgia Institute of Technology, United States

ASP-L19: DESIGN TECHNIQUES FOR ASP

ASP-L19.1: ON LEAKAGE CURRENT TEMPERATURE CHARACTERIZATION USING SUB-PICO-AMPERE CIRCUIT TECHNIQUES .... I - 361
Bernabé Linares-Barranco, Teresa Serrano-Gotarredona, Rafael Serrano-Gotarredona, Luis A. Camuñas, Instituto de Microelectrónica de Sevilla (IMSE), Spain

ASP-L19.2: COMPARISON OF QUASI-/PSEUDO-FLOATING GATE TECHNIQUES .... I - 365
Inchang Seo, Robert M. Fox, University of Florida, United States

ASP-L19.3: A GENERALIZED TIMING-SKEW-FREE, MULTI-PHASE CLOCK GENERATION PLATFORM FOR PARALLEL SAMPLED-DATA SYSTEMS .... I - 369
Sai-Weng Sin, Seng-Pan U., Rui Paulo Martins, University of Macau, Macao SAR of China

ASP-L19.4: AUTOMATIC RAPID PROGRAMMING OF LARGE ARRAYS OF FLOATING-GATE ELEMENTS .... I - 373
Guillermo Serrano, Paul Smith, Haw-Jing Lo, Ravi Chawla, Tyson Hall, Christopher Twigg, Paul Hasler, Georgia Institute of Technology, United States

ASP-L19.5: ON MISMATCH PROPERTIES OF MOS AND RESISTORS CALIBRATED LADDER STRUCTURES .... I - 377
Bernabé Linares-Barranco, Teresa Serrano-Gotarredona, Rafael Serrano-Gotarredona, Gustavo Vicente-Sánchez, Instituto de Microelectrónica de Sevilla (IMSE), Spain

ASP-L20: VOLTAGE AND CURRENT SOURCES II

ASP-L20.1: A LOW VOLTAGE CMOS CONSTANT CURRENT-VOLTAGE REFERENCE CIRCUIT .... I - 381
Ilkka Nissinen, Juha Kostamovaara, University of Oulu, Finland

ASP-L20.2: A LOW-VOLTAGE CMOS LOW-DROPOUT REGULATOR WITH ENHANCED LOOP RESPONSE .... I - 385
Ka Nang Leung, Philip K. T. Mok, Sai Kit Lau, The Hong Kong University of Science and Technology, Hong Kong SAR of China

ASP-L20.3: LOW-VOLTAGE LINEAR VOLTAGE REGULATOR SUITABLE FOR MEMORIES .... I - 389
Walter Aloisi, Stello Matteo Billé, Gaetano Palumbo, Università di Catania, Italy

ASP-L20.4: HIGH-VOLTAGE-TOLERANT POWER SUPPLY IN A LOW-VOLTAGE CMOS TECHNOLOGY .... I - 393
Vladislav Potanin, Elena Potanina, National Semiconductor, United States

ASP-L20.5: A CMOS BANDGAP REFERENCE WITH CORRECTION FOR DEVICE-TO-DEVICE VARIATION .... I - 397
Preetam Tadeparthy, Texas Instruments, India

ASP-L21: SIGMA-DELTA CONVERTERS I

ASP-L21.1: SYSTEMATIC DESIGN OF DOUBLE-SAMPLING SIGMA DELTA ADC'S WITH MODIFIED NTF .... I - 401
Pieter Rombouts, Jeroen De Maeyer, Johan Raman, Ludo Weyten, Ghent University, Belgium

ASP-L21.2: A CASCADED CONTINUOUS-TIME SIGMA-DELTA MODULATOR WITH 80 dB DYNAMIC RANGE .... I - 405
Maurits Ortmanns, Markus Kuderer, Yiannos Manoli, University of Freiburg, Germany; Friedel Gerfers, Philips Semiconductors GmbH, Germany

ASP-L21.3: DIRECT RF SAMPLING CONTINUOUS-TIME BANDPASS DELTA-SIGMA A/D CONVERTER DESIGN FOR 3G WIRELESS APPLICATIONS .... I - 409
Ut-Va Koc, Jaesik Lee, Bell Labs, Lucent Technologies, United States

ASP-L21.4: FEED-FORWARD PATH AND GAIN SCALING - A SWING AND DISTORTION REDUCTION SCHEME FOR SECOND ORDER SIGMA-DELTA MODULATOR .... I - 413
Wern Ming Koe, Texas Instruments, United States; Franco Maloberti, University of Texas, Dallas, United States

ASP-L21.5: A LOW-POWER 10-BIT CONTINUOUS-TIME CMOS SIGMA-DELTA A/D CONVERTER .... I - 417
Jannik Hammel Nielsen, Erik Bruun, Technical University of Denmark, Denmark

ASP-L22: FILTER APPLICATIONS

ASP-L22.1: CMOS DIGITALLY PROGRAMMABLE CELL FOR HIGH FREQUENCY AMPLIFICATION AND FILTERING .... I - 421
Belén Calvo, Santiago Celma, Maria Teresa Sanz, University of Zaragoza, Spain

ASP-L22.2: QUASI-PARALLEL MULTI-PATH DETECTION ARCHITECTURE USING FLOATING-GATE-MOS-BASED CDMA MATCHED FILTERS .... I - 425
Tomoyuki Nakayama, Toshihiko Yamasaki, Tadashi Shibata, University of Tokyo, Japan

ASP-L22.3: A 0.18μm CMOS SC LOWPASS FILTER FOR BLUETOOTH CHANNEL SELECTION .... I - 429
Jacqueline Pereira, Antonio Petraglia, Federal University of Rio de Janeiro, Brazil; Franco Maloberti, University of Texas, Dallas, United States

ASP-L22.4: A TUNABLE DUTY-CYCLE-CONTROLLED SWITCHED-R-MOSFET-C CMOS FILTER FOR LOW-VOLTAGE AND HIGH-LINEARITY APPLICATIONS .... I - 433
Shelly Xiao, José Silva, Un-Ku Moon, Gábor Temes, Oregon State University, United States

ASP-L22.5: A 50-MHZ BICMOS QUADRATURE CHARGE SAMPLER AND COMPLEX BANDPASS SC FILTER FOR NARROWBAND APPLICATIONS .... I - 437
Sami Karvonen, Tom Riley, Sami Kurtti, Juha Kostamovaara, University of Oulu, Finland

ASP-L23: SENSOR INTERFACE CIRCUITS

ASP-L23.1: NEW CMOS CURRENT-MODE AMPLITUDE SHIFT KEYING DEMODULATOR (ASKD) DEDICATED FOR IMPLANTABLE ELECTRONIC DEVICES .... I - 441
Abdelouahab Djemouai, Mohamad Sawan, École Polytechnique de Montréal, Canada

ASP-L23.2: A FULLY INTEGRATED 0.5 -7 Hz CMOS BANDPASS AMPLIFIER. .... I - 445
Alfredo Arnaud, Universidad de la República, Uruguay; Carlos Galup-Montoro, Univ. Fed. de Santa Catarina, Brazil

ASP-L23.3: A NANO-POWER TUNEABLE EDGE-DETECTION CIRCUIT .... I - 449
Timothy G. Constandinou, Imperial College of Science, Technology & Medicine, United Kingdom; Julius Georgiou, Imperial College of Science, Technology & Medicine / Geosilicon Limited, United Kingdom; Chris Toumazou, Imperial College of Science, Technology & Medicine / Toumaz Technology Limited, United Kingdom

ASP-L23.4: A HIGH SPEED AND LOW POWER CMOS CURRENT COMPARATOR FOR PHOTON COUNTING SYSTEMS .... I - 453
Fausto Borghetti, University of Pavia, Italy; Lorenzo Farina, STMicroelectronics, Italy; Piero Malcovati, University of Pavia, Italy; Franco Maloberti, University of Pavia, Italy / University of Texas at Dallas, United States

ASP-L23.5: VLSI POTENTIOSTAT FOR AMPEROMETRIC MEASUREMENTS FOR ELECTROLYTIC REACTIONS .... I - 457
Harpreet Narula, John Harris, University of Florida, United States

ASP-L24: SIGMA-DELTA CONVERTERS II

ASP-L24.1: LOW POWER IMPLEMENTATION OF AN N-TONE $\Sigma\Delta$ CONVERTER .... I - 461
Shubha Bommalingaiahnapallya, Ramesh Harjani, University of Minnesota, United States

ASP-L24.2: LOW-VOLTAGE SIGMA-DELTA MODULATOR TOPOLOGIES FOR BROADBAND APPLICATIONS .... I - 465
Mohammad Yavari, Omid Shoaei, University of Tehran, Iran

ASP-L24.3: FAST AND EFFICIENT ALGORITHM TO DESIGN NOISE-SHAPING FIR FILTERS FOR HIGH-ORDER OVERLOAD-FREE STABLE SIGMA-DELTA MODULATORS .... I - 469
Mitsuhiko Yagyu, Tokyo University of Agriculture and Technology, Japan; Akinori Nishihara, Tokyo Institute of Technology, Japan

ASP-L24.4: DOMINO FREE 4-PATH TIME-INTERLEAVED SECOND ORDER SIGMA-DELTA MODULATOR .... I - 473
Kye-Shin Lee, Yunyoung Choi, Franco Maloberti, University of Texas, Dallas, United States

ASP-L24.5: AN RF SUB-SAMPLING MIXER, PGA AND SD ADC FOR CONVERSION AT 900 MHZ .... I - 477
Ralph Mason, Chris DeVries, Eugene Ivanov, Carleton University, Canada

ASP-L25: HIGH GAIN AMPLIFIERS

ASP-L25.1: ROBUST DESIGN OF HIGH GAIN AMPLIFIERS USING DYNAMICAL SYSTEMS AND BIFURCATION THEORY .... I - 481
Chengming He, Le Jin, Degang Chen, Randall Geiger, Iowa State University, United States

ASP-L25.2: HARMONIC DISTORTION IN THREE-STAGE NESTED-MILLER-COMPENSATED AMPLIFIERS .... I - 485
Gaetano Palumbo, Salvatore Pennisi, University of Catania, Italy

ASP-L25.3: A HIGH-GAIN OFFSET-COMPENSATED DIFFERENTIAL AMPLIFIER .... I - 489
Craig Petrie, Tianxue Sun, Brigham Young University, United States; Matt Miller, Motorola, Inc., United States

ASP-L25.4: SINGLE MILLER CAPACITOR COMPENSATED MULTISTAGE AMPLIFIERS FOR LARGE CAPACITIVE LOAD APPLICATIONS .... I - 493
Xiaohua Fan, Chinmaya Mishra, Edgar Sánchez-Sinencio, Texas A&M University, United States

ASP-L25.5: AN IMPROVED FREQUENCY COMPENSATION TECHNIQUE FOR LOW POWER, LOW VOLTAGE CMOS AMPLIFIERS .... I - 497
Preetam Tadeparthy, Texas Instruments, India

ASP-L26: MIXED SIGNAL CIRCUITS

ASP-L26.1: A TEMPERATURE COMPENSATED DIGITALLY TRIMMABLE ON-CHIP IC OSCILLATOR WITH LOW VOLTAGE INHIBIT CAPABILITY .... I - 501
Andre Vilas Boas, Alfredo Olmos, Motorola - Brazil Semiconductor Technology Center, Brazil

ASP-L26.2: HIGH SPEED ARRAY OF OSCILLATOR-BASED TRULY BINARY RANDOM NUMBER GENERATORS .... I - 505
Nikolaos Stefanou, Sameer Sonkusale, Texas A&M University, United States

ASP-L26.3: A CMOS TIME AMPLIFIER FOR FEMTO-SECOND RESOLUTION TIMING MEASUREMENT .... I - 509
Mourad Oulmane, Gordon Roberts, McGill University, Canada

ASP-L26.4: FAULT MODELING OF RF BLOCKS BASED ON NOISE ANALYSIS .... I - 513
Jerzy Dabrowski, Linköping University, Sweden

ASP-L26.5: ACCURATE FAULT DETECTION IN SWITCHED-CAPACITOR FILTERS USING STRUCTURALLY ALLPASS BUILDING BLOCKS .... I - 517
Antonio Petraglia, Jorge Cañive, Mariane Petraglia, Federal University of Rio de Janeiro, Brazil

ASP-L27: ADC CIRCUITS

ASP-L27.1: A LOW THERMAL ERROR SAMPLING COMPARATOR FOR ACCURATE SETTLING MEASUREMENTS .... I - 521
David I. Bergman, Bryan Waltrip, National Institute of Standards and Technology, United States

ASP-L27.2: OP-AMP SWING REDUCTION IN SIGMA-DELTA MODULATORS .... I - 525
Sunwoo Kwon, University of Texas, Dallas, United States; Franco Maloberti, University of Texas, Dallas, United States / University of Pavia, Italy

ASP-L27.3: FLOATING GATE COMPARATOR WITH AUTOMATIC OFFSET MANIPULATION FUNCTIONALITY .... I - 529
Eric Liu Wong, Pamela Abshire, Marc Cohen, University of Maryland, College Park, United States

ASP-L27.4: A NOVEL DELTA-SIGMA MODULATOR USING RESONANT-TUNNELING QUANTIZERS .... I - 533
Masaru Chibashi, Keisuke Eguchi, Takao Waho, Sophia University, Japan

ASP-L27.5: LOW KICKBACK NOISE TECHNIQUES FOR CMOS LATCHED COMPARATORS .... I - 537
Pedro Figueiredo, João Vital, Chipidea Microelectrónica, Portugal

ASP-L28: HIGH SPEED AMPLIFIERS

ASP-L28.1: A LOW DISTORTION AND FAST SETTLING AUTOMATIC GAIN CONTROL AMPLIFIER IN CMOS TECHNOLOGY .... I - 541
Chung-Wei Lin, Yen-Zen Liu, Klaus Y.J. Hsu, National Tsing Hua University, Taiwan

ASP-L28.2: A CMOS HIGH-SPEED MULTISTAGE PREAMPLIFIER FOR COMPARATOR DESIGN .... I - 545
Xian Ping Fan, Pak Kwong Chan, Nanyang Technological University, Singapore

ASP-L28.3: LINEARIZED CMOS OTA USING ACTIVE-ERROR FEEDFORWARD TECHNIQUE .... I - 549
Stanislaw Szczepanski, Slawomir Koziel, Gdansk University of Technology, Poland; Edgar Sánchez-Sinencio, Texas A&M University, United States

ASP-L28.4: AN ANALYTICAL MODEL FOR THE SLEWING BEHAVIOR OF CMOS TWO-STAGE OPERATIONAL TRANSCONDUCTANCE AMPLIFIERS .... I - 553
Nima Maghari, Mohammad Yavari, Omid Shoaei, University of Tehran, Iran

ASP-L28.5: A PROGRAMMABLE GAIN AMPLIFIER BUFFER DESIGN FOR VIDEO APPLICATIONS .... I - 557
Gonggui Xu, Haydar Bilhan, Texas Instruments, United States

ASP-L29: RF FRONTEND CIRCUITS

ASP-L29.1: A 1.5 V HIGH-LINEARITY CMOS MIXER FOR 2.4 GHZ APPLICATIONS .... I - 561
Hung-Che Wei, Ro-Min Weng, Kun-Yi Lin, National Dong Hwa University, Taiwan

ASP-L29.2: FULLY-INTEGRATED DECT/BLUETOOTH MULTI-BAND LNA IN 0.18 um CMOS .... I - 565
Vojkan Vidojkovic, Johan van der Tang, Eric Hanssen, Eindhoven University of Technology, Netherlands; Arjan Leeuwenburgh, National Semiconductor, Netherlands; Arthur van Roermund, Eindhoven University of Technology, Netherlands

ASP-L29.3: LOW VOLTAGE, LOW POWER FOLDED-SWITCHING MIXER WITH CURRENT-REUSE IN 0.18 μm CMOS .... I - 569
Vojkan Vidojkovic, Johan van der Tang, Eindhoven University of Technology, Netherlands; Arjan Leeuwenburgh, National Semiconductor, Netherlands; Arthur van Roermund, Eindhoven University of Technology, Netherlands

ASP-L29.4: A DECT/BLUETOOTH MULTI-STANDARD FRONT-END WITH ADAPTIVE IMAGE REJECTION IN 0.18 μm CMOS .... I - 573
Vojkan Vidojkovic, Johan van der Tang, Eindhoven University of Technology, Netherlands; Arjan Leeuwenburgh, National Semiconductor, Netherlands; Arthur van Roermund, Eindhoven University of Technology, Netherlands

ASP-L29.5: DIRECT RF SAMPLING MIXER WITH RECURSIVE FILTERING IN CHARGE DOMAIN .... I - 577
Khurram Muhammad, Robert Staszewski, Texas Instruments, Inc., United States

ASP-L30: ANALOG-TO-DIGITAL CONVERTERS

ASP-L30.1: A NOVEL SINGLE AMPLIFIER ARCHITECTURE FOR SECOND ORDER NOISE SHAPING .... I - 581
Rahmi Hezar, Oguz Altun, Texas Instruments, United States

ASP-L30.2: A PROGRAMMABLE ANALOG-TO-DIGITAL CONVERTER .... I - 585
Gang Xu, Jiren Yuan, Lund University, Sweden

ASP-L30.3: AN IMPROVED ALGORITHMIC ADC CLOCKING SCHEME .... I - 589
Min Gyu Kim, Gil-Cho Ahn, Un-Ku Moon, Oregon State University, United States

ASP-L30.4: ANALYSIS AND COMPENSATION OF NONLINEARITY MISMATCHES IN TIME-INTERLEAVED ADC ARRAYS .... I - 593
Christian Vogel, Gernot Kubin, Graz University of Technology, Austria

ASP-L30.5: MULTI-STEP ANALOG-TO-DIGITAL CONVERTERS WITH TRAPPING WINDOW .... I - 597
Masaaki Naka, Toshimichi Saito, HOSEI University, Japan

ASP-L31: HIGH-FREQUENCY AMPLIFIERS I

ASP-L31.1: DESIGN OF A 1.8 V 4. 9 ~ 5.9 GHZ CMOS BROADBAND LOW NOISE AMPLIFIER WITH 0.28 DB GAIN FLATNESS USING AMER INDUCTOR LOADS .... I - 601
Yin-Lung Lu, Yi-Cheng Wu, Kyung-Wan Yu, Wei-Li Chen, M. Frank Chang, University of California, Los Angeles, United States

ASP-L31.2: A LOW-POWER DC-7-GHZ SOI CMOS DISTRIBUTED AMPLIFIER .... I - 605
Ertan Zencir, Ahmet Tekin, Numan S. Dogan, North Carolina A&T State University, United States; Ercument Arvas, Syracuse University, United States

ASP-L31.3: BANDWIDTH ENHANCEMENT OF MULTI-STAGE AMPLIFIERS USING ACTIVE FEEDBACK .... I - 609
Reza Samadi, Aydin I. Karsilayan, José Silva-Martínez, Texas A&M University, United States

ASP-L31.4: A NOVEL NON-UNIFORM DISTRIBUTED AMPLIFIER .... I - 613
Ahmad Yazdi, Payam Heydari, University of California, Irvine, United States

ASP-L31.5: A 0.1-12-GHZ FULLY DIFFERENTIAL CMOS DISTRIBUTED AMPLIFIER EMPLOYING A FEEDFORWARD DISTORTION CANCELLATION TECHNIQUE .... I - 617
Ziad El-Khatib, Leonard MacEachern, Samy A. Mahmoud, Carleton University, Canada

ASP-L32: RF POWER AMPLIFIERS

ASP-L32.1: GAIN/BANDWIDTH PROGRAMMABLE PA CONTROL LOOP FOR GSM/GPRS QUAD-BAND APPLICATIONS .... I - 621
Paolo Cusinato, Stefano Cipriani, Guglielmo Sirna, Gianni Puccio, Eric Duvivier, Texas Instruments, Inc., France

ASP-L32.2: AN INTEGRATED LINEAR RF POWER DETECTOR .... I - 625
Suhas Kulhalli, Sumantra Seth, Shih-Tsang Fu, Texas Instruments, India

ASP-L32.3: KNOWLEDGE- AND OPTIMIZATION-BASED DESIGN OF RF POWER AMPLIFIERS .... I - 629
João Ramos, Kenneth Francken, Georges Gielen, Michiel Steyaert, Katholieke Universiteit Leuven, Belgium

ASP-L32.4: TRANSMITTER UTILISING BANDPASS DELTA-SIGMA MODULATOR AND SWITCHING MODE POWER AMPLIFIER .... I - 633
Jaakko Ketola, Johan Sommarek, Jouko Vankka, Kari Halonen, Helsinki University of Technology, Finland

ASP-L32.5: INTERMODULATION PRODUCTS IN THE EER TECHNIQUE APPLIED TO CLASS-E AMPLIFIERS .... I - 637
Dusan Milosevic, Johan van der Tang, Arthur van Roermund, Eindhoven University of Technology, Netherlands

ASP-L33: DATA CONVERTERS I

ASP-L33.1: A COMPARISON OF TOTALLY DIGITAL ADCS FOR SOCS .... I - 641
Josias Mainardi, Adão Souza Jr., Luigi Carro, Altamiro Susin, Universidade Federal do Rio Grande do Sul, Brazil

ASP-L33.2: ACCURATE TESTING OF ADC'S SPECTRAL PERFORMANCE USING IMPRECISE SINUSOIDAL EXCITATIONS .... I - 645
Zhongjun Yu, Degang Chen, Randall Geiger, Iowa State University, United States

ASP-L33.3: USE OF DYNAMIC ELEMENT MATCHING IN A MULTI-PATH SIGMA-DELTA MODULATOR .... I - 649
Vincenzo Ferragina, University of Pavia / STMicroelectronics, Italy; Andrea Fornasari, University of Pavia, Italy; Umberto Gatti, Siemens Mobile Communications, Italy; Piero Malcovati, University of Pavia, Italy; Franco Maloberti, University of Pavia, Italy / University of Texas at Dallas, United States; Luigi Marco Athos Monfasani, University of Pavia, Italy

ASP-L33.4: DYNAMIC ELEMENT MATCHING TECHNIQUES FOR DELTA-SIGMA ADCS WITH LARGE INTERNAL QUANTIZERS .... I - 653
Brent Nordick, Craig Petrie, Yongjie Cheng, Brigham Young University, United States

ASP-L33.5: DECIMATING SC FILTER FOR HIGH SPEED SIGMA-DELTA D/A CONVERTERS .... I - 657
Mikael Gustavsson, Nianxiong Nick Tan, GlobeSpanVirata Inc., United States

ASP-L34: OPERATIONAL AMPLIFIERS

ASP-L34.1: A FIXED TRANSCONDUCTANCE BIAS TECHNIQUE FOR CMOS ANALOG INTEGRATED CIRCUITS .... I - 661
Shanthi Pavan, Indian Institute of Technology, India

ASP-L34.2: IMPROVEMENTS IN BIASING AND COMPENSATION OF CMOS OPAMPS .... I - 665
Sean Nicolson, Khoman Phang, University of Toronto, Canada

ASP-L34.3: A COMPENSATION-BASED OPTIMIZATION METHODOLOGY FOR GAIN-BOOSTED OPAMP .... I - 669
Jie Yuan, Nabil Farhat, University of Pennsylvania, United States

ASP-L34.4: A LOW-VOLTAGE CMOS RAIL-TO-RAIL OPERATIONAL AMPLIFIER USING DOUBLE P-CHANNEL DIFFERENTIAL INPUT PAIRS .... I - 673
Chun-Jen Huang, Hong-Yi Huang, Fu-Jen Catholic University, Taiwan

ASP-L34.5: A 87DB, 2.3 GHZ, SIGE BICMOS OPERATIONAL TRANSCONDUCTANCE AMPLIFIER .... I - 677
Siddharth Devarajan, Ronald Gutmann, Kenneth Rose, Rensselaer Polytechnic Institute, United States

ASP-L35: SIGNAL PROCESSING BUILDING BLOCKS I

ASP-L35.1: LOW-VOLTAGE CMOS ANALOG FOUR QUADRANT MULTIPLIER BASED ON FLIPPED VOLTAGE FOLLOWERS .... I - 681
Jaime Ramírez Angulo, Shanta Thoutam, New Mexico State University, United States; Antonio J. López-Martín, Public University of Navarra, Spain; Ramon Gonzalez Carvajal, Universidad de Sevilla, Spain

ASP-L35.2: COMPACT CMOS LINEAR TRANSCONDUCTOR AND FOUR-QUADRANT ANALOGUE MULTIPLIER .... I - 685
Mladen Panovic, Andreas Demosthenous, University College London, United Kingdom

ASP-L35.3: HIGHLY-LINEAR, CURRENT-FEEDBACK RESISTIVE SOURCE-DEGENERATED MOS TRANSCONDUCTOR .... I - 689
Phanumas Khumsat, Prince of Songkla University, Thailand; Apisak Worapishet, Mahanakorn University of Technology, Thailand

ASP-L35.4: COMPACT TUNABLE CMOS OTA WITH HIGH LINEARITY .... I - 693
Meghraj Kachare, Jaime Ramírez Angulo, New Mexico State University, United States; Antonio J. López-Martín, Public University of Navarra, Spain; Ramón G. Carvajal, University of Seville, Spain

ASP-L35.5: THREE-DECADE PROGRAMMABLE FULLY DIFFERENTIAL LINEAR OTA .... I - 697
Yunbin Deng, Shantanu Chakrabartty, Gert Cauwenberghs, Johns Hopkins University, United States

ASP-L36: COMPUTER ANALYSIS AND SYNTHESIS

ASP-L36.1: SYSTEMATIC SYNTHESIS METHOD FOR ANALOGUE CIRCUITS - PART I NOTATION AND SYNTHESIS TOOLBOX .... I - 701
David Haigh, Imperial College, United Kingdom; Paul Radmore, University College London, United Kingdom

ASP-L36.2: SYSTEMATIC SYNTHESIS METHOD FOR ANALOGUE CIRCUITS - PART II ACTIVE-RC CIRCUIT SYNTHESIS .... I - 705
David Haigh, Fang Tan, Christos Papavassiliou, Imperial College, United Kingdom

ASP-L36.3: SYSTEMATIC SYNTHESIS METHOD FOR ANALOGUE CIRCUITS - PART III ALL-TRANSISTOR CIRCUIT SYNTHESIS .... I - 709
David Haigh, Fang Tan, Christos Papavassiliou, Imperial College, United Kingdom

ASP-L36.4: APPLICATIONS OF TREE/LINK PARTITIONING FOR MOMENT COMPUTATIONS OF GENERAL LUMPED RLC NETWORKS WITH RESISTOR LOOPS .... I - 713
Herng-Jer Lee, Ming-Hong Lai, Chia-Chi Chu, Wu-Shiung Feng, Chang Gung University, Taiwan

ASP-L36.5: NEW RELAXATION-BASED CIRCUIT SIMULATOR BASED ON FAST AUTOMATIC DIFFERENTIATION .... I - 717
Ganesh Kumar Basnet, Masayuki Yamauchi, Tsuyoshi Otake, Mamoru Tanaka, Sophia University, Japan

ASP-L37: CLASS AB AMPLIFIERS

ASP-L37.1: DESIGN OF CLASS AB OUTPUT STAGES USING THE STRUCTURAL METHODOLOGY .... I - 721
Vadim Ivanov, Texas Instruments, Inc., United States; Igor Filanovsky, University of Alberta, Canada

ASP-L37.2: A NOVEL LOW-VOLTAGE LOW-POWER CLASS-AB LINEAR TRANSCONDUCTOR .... I - 725
Marta Laguna Garcia, Universidad de Sevilla, Spain; Carlos Aristoteles De la Cruz-Blas, Public University of Navarra, Spain; Antonio Torralba, Ramon Gonzalez Carvajal, Universidad de Sevilla, Spain; Antonio J. López-Martín, Alfonso Carlosena, Public University of Navarra, Spain

ASP-L37.3: A NOVEL FAMILY OF LOW-VOLTAGE VERY LOW POWER SUPER CLASS AB OTAS WITH SIGNIFICANTLY ENHANCED SLEW RATE AND BANDWIDTH .... I - 729
Sushmita Baswa, Jaime Ramírez Angulo, New Mexico State University, United States; Antonio J. López-Martín, Public University of Navarra, Spain; Ramón G. Carvajal, University of Seville, Spain

ASP-L37.4: POWER EFFICIENT FULLY DIFFERENTIAL LOW-VOLTAGE TWO STAGE CLASS AB/AB OP-AMP ARCHITECTURES .... I - 733
Shanta Thoutam, Jaime Ramírez Angulo, New Mexico State University, United States; Antonio J. López-Martín, Universidad Pública de Navarra, Spain; Ramon Gonzalez Carvajal, Universidad de Sevilla, Spain

ASP-L37.5: A PSEUDO-CLASS-AB TELESCOPIC-CASCODE OPERATIONAL AMPLIFIER .... I - 737
Mohammad Taherzadeh-Sani, Reza Lotfi, Omid Shoaei, University of Tehran, Iran

ASP-L38: SIGNAL PROCESSING BUILDING BLOCKS II

ASP-L38.1: CONFIGURABLE DIRECT-CONVERSION / SUPERHETERODYNE BASEBAND DOWN-LINK CHANNEL FOR W-CDMA APPLICATIONS .... I - 741
Paolo Cusinato, Texas Instruments, Inc., France

ASP-L38.2: REALIZATION OF AN ANALOG CURRENT-MODE 2-DIMENSIONAL DCT .... I - 745
Mikko Pänkäälä, Jonne Poikonen, Laura Vesalainen, Ari Paasio, University of Turku, Finland

ASP-L38.3: A BUMP-CIRCUIT-BASED MOTION DETECTOR USING PROJECTED-ACTIVITY HISTOGRAMS .... I - 749
Masayuki Umejima, Toshihiko Yamasaki, Tadashi Shibata, University of Tokyo, Japan

ASP-L38.4: A PRECISE CMOS MISMATCH MODEL FOR ANALOG DESIGN FROM WEAK TO STRONG INVERSION .... I - 753
Teresa Serrano-Gotarredona, Bernabé Linares-Barranco, Jesús Velarde-Ramírez, Instituto de Microelectrónica de Sevilla (IMSE), Spain

ASP-L38.5: PROGRAMMABLE MULTIPLE INPUT TRANSLINEAR ELEMENTS .... I - 757
Haw-Jing Lo, Guillermo Serrano, Paul Hasler, David Anderson, Georgia Institute of Technology, United States; Bradley Minch, Cornell University, United States

ASP-P1: ANALOG CIRCUITS IN SYSTEMS I

ASP-P1.1: EFFECTS OF NOISE AND JITTER ON ALGORITHMS FOR BANDPASS SAMPLING IN RADIO RECEIVERS .... I - 761
Yi-Ran Sun, Svante Signell, Royal Institute of Technology (KTH), Sweden

ASP-P1.2: PERFORMANCE ANALYSIS OF SAMPLING SWITCHES IN VOLTAGE AND FREQUENCY DOMAINS USING VOLTERRA SERIES .... I - 765
Andreas Gothenberg, Tokyo Institute of Technology, Japan / KTH-Royal Institute of Technology, Sweden; Hannu Tenhunen, Royal Institute of Technology (KTH), Sweden

ASP-P1.3: A LOW POWER THYRISTOR-BASED CMOS PROGRAMMABLE DELAY ELEMENT .... I - 769
Junmou Zhang, University of Rochester, United States; Simon R. Cooper, Andrew R. LaPietra, Michael W. Mattern, Robert M. Guidash, Eastman Kodak Company, United States; Eby G. Friedman, University of Rochester, United States

ASP-P1.4: JITTER IN OSCILLATORS WITH 1/F NOISE SOURCES .... I - 773
Chengxin Liu, John McNeill, Worcester Polytechnic Institute, United States

ASP-P1.5: A DUAL-SLOPE PHASE FREQUENCY DETECTOR AND CHARGE PUMP ARCHITECTURE TO ACHIEVE FAST LOCKING OF PHASE-LOCKED LOOP .... I - 777
Kuo-Hsing Cheng, National Central University, Taiwan; Wei-Bin Yang, Industrial Technology Research institute, Taiwan; Shu-Chang Kuo, Tamkang University, Taiwan

ASP-P2: ANALOG CIRCUITS IN SYSTEMS II

ASP-P2.1: A DUTY CYCLE CONTROL CIRCUIT FOR HIGH SPEED APPLICATIONS .... I - 781
Armin Tajalli, Sharif University of Technology / Microelectronics Research Center, Iran; Saeid Mehrmanesh, Microelectronics Research Center, Iran; Seyed Mojtaba Atarodi, MixeCore Design, United States

ASP-P2.2: A WIDE-RANGE AND FAST-LOCKING CLOCK SYNTHESIZER IP BASED ON DELAY-LOCKED LOOP .... I - 785
Chorng-Sii Hwang, National Taiwan University, Taiwan; Poki Chen, National Taiwan University of Science and Technology, Taiwan; Hen-Wai Tsao, National Taiwan University, Taiwan

ASP-P2.3: A LOW-VOLTAGE LOW-SENSITIVITY SINUSOIDAL VCO FOR DPLL REALIZATIONS .... I - 789
Jader A. De Lima, Peterson R. Agostinho, University of Sao Paulo State (UNESP), Brazil

ASP-P2.4: HIGH-SPEED HIGH-PRECISION ANALOG RANK ORDER FILTER IN CMOS TECHNOLOGY .... I - 793
Ramon Gonzalez Carvajal, Universidad de Sevilla, Spain; Jaime Ramírez Angulo, Gladys Omayra Ducoudray, New Mexico State University, United States; Antonio J. López-Martín, Public University of Navarra, Spain

ASP-P2.5: A Gm-C BUMP EQUALIZER FOR LOW-VOLTAGE LOW-POWER APPLICATIONS .... I - 797
Renato Galembeck, UFSC, Brazil; Jader Lima, UNESP, Brazil; Marcio Schneider, UFSC, Brazil

ASP-P3: CONTROLLED AMPLIFIERS

ASP-P3.1: PROGRAMMABLE-GAIN AMPLIFIERS BASED ON AC COUPLINGS FOR CONTINUOUS RECEPTION .... I - 801
Jarkko Jussila, Kari Halonen, Helsinki University of Technology, Finland

ASP-P3.2: HIGHLY LINEAR WIDE TUNING RANGE CMOS TRANSCONDUCTOR OPERATING IN MODERATE INVERSION .... I - 805
Jaime Ramírez Angulo, Chandrika Durbha, New Mexico State University, United States; Antonio J. López-Martín, Public University of Navarra, Spain; Ramón G. Carvajal, University of Seville, Spain

ASP-P3.3: A LOW VOLTAGE CMOS TRANSRESISTANCE-BASED VARIABLE GAIN AMPLIFIER .... I - 809
Seoung-Jae Yoo, Arun Ravindran, Mohammed Ismail, The Ohio State University, United States

ASP-P3.4: A CMOS DB-LINEAR VGA WITH PRE-DISTORTION COMPENSATION FOR WIRELESS COMMUNICATION APPLICATIONS .... I - 813
Yuanjin Zheng, Institute of Microelectronics, Singapore; Jiangnan Yan, Yong Ping Xu, National University of Singapore, China

ASP-P3.5: FULLY DIFFERENTIAL FLOATING-GATE PROGRAMMABLE OTAS WITH NOVEL COMMON-MODE FEEDBACK .... I - 817
Ravi Chawla, Guillermo Serrano, Daniel Allen, Angelo Pereira, Paul Hasler, Georgia Institute of Technology, United States

ASP-P4: CURRENT AMPLIFIERS

ASP-P4.1: A HIERARCHY OF INPUT STAGES FOR CURRENT FEEDBACK OPERATIONAL AMPLIFIERS .... I - 821
Amr Tammam, Khaled Hayatleh, Bryan Hart, John Lidgey, Oxford Brookes University, United Kingdom

ASP-P4.2: HIGH PERFORMANCE CURRENT-FEEDBACK OP-AMPS .... I - 825
Amr Tammam, Khaled Hayatleh, Bryan Hart, John Lidgey, Oxford Brookes University, United Kingdom

ASP-P4.3: SENSITIVITY-ENHANCED OEIC WITH CAPACITANCE MULTIPLIER FOR REDUCED LOWER-CUTOFF FREQUENCY .... I - 829
Jürgen Leeb, Johannes Knorr, Horst Zimmermann, Vienna University of Technology, Austria

ASP-P4.4: ON THE SECOND HARMONIC CONTROL REQUIREMENTS IN BALANCED COMMON-EMITTER BJT LOW NOISE AMPLIFIER .... I - 833
Antti Heiskanen, Timo Rahkonen, University of Oulu, Finland

ASP-P4.5: A CURRENT-MODE INTERFACE CASCADE ON COG(CICC) FOR TFT-LCD SYSTEMS .... I - 837
Il Kwon Chang, Yong Weon Jeon, Jang Sub Lee, Kyung Wol Kim, Jin Tae Kim, Samsung Electronics, Republic of Korea

ASP-P5: ANALOG FILTERING I

ASP-P5.1: SWITCHED CAPACITORS: A BRIDGE BETWEEN ANALOG AND DIGITAL SP .... I - 841
Roxana Saint-Nom, Daniel Jacoby, Instituto Tecnologico de Buenos Aires, Argentina

ASP-P5.2: A VARIABLE GAIN HIGH LINEARITY LOW POWER BASEBAND FILTER FOR WLAN .... I - 845
Sachin Ranganathan, Terri Fiez, Oregon State University, United States

ASP-P5.3: ANALOGUE ADAPTIVE FILTERS USING WAVE SYNTHESIS TECHNIQUE .... I - 849
Yan Xie, Bashir M. Al-Hashimi, University of Southampton, United Kingdom

ASP-P5.4: A 1.5 V CLASS A 5TH ORDER LOG DOMAIN FILTER IN SIGE TECHNOLOGY .... I - 853
Franklin Baez, Jon Duster, Kevin Kornegay, Cornell University, United States

ASP-P5.5: A FULLY-DIFFERENTIAL SWITCHED CAPACITOR CHOPPER STABILIZED HIGH-PASS FILTER (MIRRORED INTEGRATOR) .... I - 857
Zeljko Ignjatovic, Mark Bocko, University of Rochester, United States

ASP-P6: ANALOG FILTERING II

ASP-P6.1: A FIVE-TRANSISTOR BANDPASS FILTER ELEMENT .... I - 861
Paul Smith, David Graham, Ravi Chawla, Paul Hasler, Georgia Institute of Technology, United States

ASP-P6.2: OPTIMUM AREA ALLOCATION FOR RESISTORS AND CAPACITORS IN CONTINUOUS-TIME MONOLITHIC FILTERS .... I - 865
Haibo Fei, Randall Geiger, Degang Chen, Iowa State University, United States

ASP-P6.3: A 0.8 UM PROGRAMMABLE IIR SC FILTER .... I - 869
Joarez Monteiro, Antonio Petraglia, Federal University of Rio de Janeiro, Brazil

ASP-P6.4: LOW-POWER COMPACT Gm-C FILTERS STRUCTURES .... I - 873
Stefano D'Amico, Andrea Baschirotto, University of Lecce, Italy

ASP-P6.5: LOW-IF COMPLEX FILTER WITH TRANSCONDUCTANCE NETWORKS .... I - 877
Chao Yang, Theng Tee Yeo, Masaaki Itoh, OKI Techno Center (Singapore) Pte Ltd, Singapore

ASP-P7: ANALOG CIRCUITS AND TECHNOLOGY I

ASP-P7.1: SERIES-PARALLEL ASSOCIATION OF TRANSISTORS FOR THE REDUCTION OF RANDOM OFFSET IN NON-UNITY GAIN CURRENT MIRRORS .... I - 881
Rafaella Fiorelli, Alfredo Arnaud, Universidad de la República, Uruguay; Carlos Galup-Montoro, Universidade Federal de Santa Catarina, Brazil

ASP-P7.2: AN ADAPTIVE CIRCUITS CONCEPT TO ADDRESS MISMATCH IN ANALOG CIRCUITS .... I - 885
Radu Secareanu, Bill Peterson, Motorola, Inc., United States

ASP-P7.3: OPTIMAL DISTRIBUTION OF THE RF FRONT-END SYSTEM SPECIFICATIONS TO THE RF FRONT-END CIRCUIT BLOCKS .... I - 889
Aleksandar Tasic, Wouter A. Serdijn, John R. Long, TU Delft, Netherlands

ASP-P7.4: A POOR MAN'S BiCMOS USING STANDARD CMOS .... I - 893
Farshid Rezaei, Ken Martin, University of Toronto, Canada

ASP-P7.5: A NOVEL 5GHZ RF POWER DETECTOR .... I - 897
Tao Zhang, William Eisenstadt, Robert M. Fox, University of Florida, United States

ASP-P8: ANALOG CIRCUITS AND TECHNOLOGY II

ASP-P8.1: MARGIN NORMALIZATION AND PROPAGATION IN ANALOG VLSI .... I - 901
Shantanu Chakrabartty, Gert Cauwenberghs, Johns Hopkins University, United States

ASP-P8.2: A MINIMIZATION OF THE CHARGE INJECTION IN SWITCHED-CURRENT CIRCUITS .... I - 905
Chunyan Wang, Concordia University, Canada

ASP-P8.3: A SIMPLE VOLTAGE REFERENCE USING TRANSISTOR WITH ZTC POINT AND PTAT CURRENT SOURCE .... I - 909
Laleh Najafizadeh, Igor Filanovsky, University of Alberta, Canada

ASP-P8.4: AN INPUT-FREE NMOS VT EXTRACTOR CIRCUIT IN PRESENCE OF BODY EFFECTS .... I - 912
Susanta Sengupta, Georgia Institute of Technology, United States

ASP-P8.5: A LOW VOLTAGE RAIL TO RAIL V-I CONVERSION SCHEME FOR APPLICATIONS IN CURRENT MODE A/D CONVERTERS .... I - 916
Rahul Shukla, Jaime Ramírez Angulo, New Mexico State University, United States; Antonio J. López-Martín, Universidad Pública de Navarra, Spain; Ramon Gonzalez Carvajal, Universidad de Sevilla, Spain

ASP-P9: TESTING OF MIXED SIGNAL CIRCUITS

ASP-P9.1: TESTING HIGH RESOLUTION ADCS USING DETERMINISTIC DYNAMIC ELEMENT MATCHING .... I - 920
Beatriz Olleta, Hanjun Jiang, Degang Chen, Iowa State University, United States; Randall Geiger, Iowa State Univeristy, United States

ASP-P9.2: PARAMETER OPTIMIZATION OF DETERMINISTIC DYNAMIC ELEMENT MATCHING DACS FOR ACCURATE AND COST-EFFECTIVE ADC TESTING .... I - 924
Hanjun Jiang, Beatriz Olleta, Degang Chen, Randall Geiger, Iowa State University, United States

ASP-P9.3: AN SOC COMPATIBLE LINEARITY TEST APPROACH FOR PRECISION ADCS USING EASY-TO-GENERATE SINUSOIDAL STIMULI .... I - 928
Le Jin, Chengming He, Degang Chen, Randall Geiger, Iowa State University, United States

ASP-P9.4: FAST IMPLEMENTATION OF A LINEARITY TEST APPROACH FOR HIGH-RESOLUTION ADCS USING NON-LINEAR RAMP SIGNALS .... I - 932
Le Jin, Chengming He, Degang Chen, Randall Geiger, Iowa State University, United States

ASP-P9.5: A DIGITAL TUNING ALGORITHM FOR ON-CHIP RESISTORS .... I - 936
Ayman Fayed, Texas Instruments, Inc., United States; Mohammed Ismail, The Ohio State University, United States

ASP-P10: MIXED SIGNAL AND SENSOR INTERFACE CIRCUITS

ASP-P10.1: AN INTEGRATED A-SI TFT DEMULTIPLEXER FOR DRIVING GATE LINES IN ACTIVE-MATRIX ARRAYS .... I - 940
Kambiz Moez, University of Waterloo, Canada

ASP-P10.2: ON-CHIP CALIBRATION TECHNIQUE FOR DELAY LINE BASED BIST JITTER MEASUREMENT .... I - 944
Bryan Nelson, Mani Soma, University of Washington, United States

ASP-P10.3: A SINGLE-ENDED TO DIFFERENTIAL CAPACITIVE SENSOR INTERFACE CIRCUIT DESIGNED IN CMOS TECHNOLOGY .... I - 948
Tajeshwar Singh, Trond Ytterdal, Norwegian University of Science and Technology, Norway

ASP-P10.4: 4/2 PAM SERIAL LINK TRANSMITTER WITH TUNABLE PRE-EMPHASIS .... I - 952
Chih-Hsien Lin, Chang-Hsiao Tsai, Chih-Ning Chen, Shyh-Jye Jou, National Central University, Taiwan

ASP-P11: RF CIRCUITS I

ASP-P11.1: POST-OPTIMIZATION DESIGN CENTERING FOR RF INTEGRATED CIRCUITS .... I - 956
Kiyong Choi, David Allstot, University of Washington, United States

ASP-P11.2: RF CMOS FULLY-INTEGRATED HETERODYNE FRONT-END RECEIVERS DESIGN TECHNIQUE FOR 5 GHZ APPLICATIONS .... I - 960
Rola A. Baki, Mourad El-Gamal, McGill University, Canada

ASP-P11.3: A LOW VOLTAGE DUAL GATE INTEGRATED CMOS MIXER FOR 2.4GHZ BAND APPLICATIONS .... I - 964
Jiqing Cui, Yong Lian, Ming Fu Li, National University of Singapore, Singapore

ASP-P11.4: COMPARISON OF THE IMAGE REJECTION BETWEEN THE PASSIVE AND THE GILBERT MIXER .... I - 968
Xiaoyan Wang, Pietro Andreani, Technical University of Denmark, Denmark

ASP-P11.5: A TUNABLE TRANSMISSION LINE PHASE SHIFTER (TTPS) .... I - 972
Taeik Kim, David Allstot, University of Washington, United States

ASP-P12: RF CIRCUITS II

ASP-P12.1: A LATERAL-BJT-BIASED CMOS VOLTAGE-CONTROLLED OSCILLATOR .... I - 976
Sankaran Aniruddhan, Min Chu, David Allstot, University of Washington, United States

ASP-P12.2: A NEW SOFTWARE TOOL TO MODEL MEASURED RF-DATA WITH OPTIMUM CIRCUIT TOPOLOGY .... I - 980
B. Siddik Yarman, Ali Kilinc, Ahmet Aksen, Isik University, Turkey

ASP-P12.3: NOVEL INTERDIGITAL MICROSTRIP BANDPASS FILTER WITH IMPROVED SPURIOUS RESPONSE .... I - 984
Sio-Weng Ting, Kam-Weng Tam, Rui Paulo Martins, University of Macau, China

ASP-P12.4: RF ROBUSTNESS ENHANCEMENT THROUGH STATISTICAL ANALYSIS OF CHIP-PACKAGE CO-DESIGN .... I - 988
Xinzhong Duo, Li-Rong Zheng, Hannu Tenhunen, Royal Institute of Technology (KTH), Sweden

ASP-P12.5: DESIGN AND ANALYSIS OF A DISTRIBUTED REGENERATIVE FREQUENCY DIVIDER USING DISTRIBUTED MIXER .... I - 992
Amin Q. Safarian, Payam Heydari, University of California, Irvine, United States

ASP-P13: ANALOG CIRCUITS I

ASP-P13.1: CURRENT MODE SQUARE ROOT-DOMAIN PALMO CELL FOR PROGRAMMABLE ANALOG VLSI .... I - 996
Yaxiong Zhang, Alister Hamilton, University of Edinburgh, United Kingdom

ASP-P13.2: SYNTHESIS OF LOG-DOMAIN FILTER WITH WELL-DEFINED OPERATING POINT .... I - 1000
Zhan Xu, Ezz I. El-Masry, Dalhousie University, Canada

ASP-P13.3: I/Q IMBALANCE OF TWO-PATH LADDER FILTERS .... I - 1004
Peter Kiss, Vladimir Prodanov, Agere Systems, United States

ASP-P13.4: ANALOG ENVELOPE CONSTRAINED FILTER WITH INPUT UNCERTAINTY .... **I - 1008**
Ba-Ngu Vo, Alex Leong, Pok Iu, University of Melbourne, Australia

ASP-P13.5: A 1.8 V CMOS LINEAR TRANSCONDUCTOR AND ITS APPLICATION TO CONTINUOUS-TIME FILTERS .... **I - 1012**
Xuguang Zhang, Ezz I. El-Masry, Dalhousie University, Canada

ASP-P14: ANALOG CIRCUITS II

ASP-P14.1: 1.2V LOW-POWER FOUR-QUADRANT CMOS TRANSCONDUCTANCE MULTIPLIER OPERATING IN SATURATION REGION .... **I - 1016**
Stanislaw Szczepanski, Slawomir Koziel, Gdansk University of Technology, Poland

ASP-P14.2: A NOVEL APPROACH TO THE LINEARIZATION OF THE DIFFERENTIAL TRANSCONDUCTORS .... **I - 1020**
Hamid Reza Sadr M. N., Islamic Azad University, Iran

ASP-P14.3: DESIGN OF CONSTANT-DELAY SYSTEMS BASED ON SYMMETRY OF TIME-DOMAIN RESPONSE .... **I - 1024**
Goran Molnar, Mladen Vucic, Faculty of Electrical Engineering and Computing, Croatia

ASP-P14.4: WINNER-TAKE-ALL CLASS AB INPUT STAGE: A NOVEL CONCEPT FOR LOW-VOLTAGE POWER-EFFICIENT CLASS AB AMPLIFIERS .... **I - 1028**
Jaime Ramírez Angulo, Sushmita Baswa, New Mexico State University, United States; Antonio J. López-Martín, Public University of Navarra, Spain; Ramón G. Carvajal, University of Seville, Spain

ASP-P14.5: DESIGN OF ARBITRARY-ORDER MINIMAL OPERATIONAL-AMPLIFIER BIBO STABLE BODY-TYPE VARIABLE-AMPLITUDE ACTIVE-RC EQUALIZERS .... **I - 1032**
Behrouz Nowrouzian, iHear Unicare, Inc., Canada; Arthur T.G. Fuller, Nortel Networks, Canada; M. N. S. Swamy, Concordia University, Canada

ASP-P15: HIGH-FREQUENCY AMPLIFIERS II

ASP-P15.1: A FILTER FREE CLASS D AUDIO AMPLIFIER WITH 86% POWER EFFICIENCY .... **I - 1036**
Patrick Muggler, Wayne Chen, Clif Jones, Paras Dagli, Texas Instruments, United States; Navid Yazdi, Corning IntelliSene, United States

ASP-P15.2: A RATIONAL FUNCTION BASED PREDISTORTER FOR HIGH POWER AMPLIFIER .... **I - 1040**
Dongliang Huang, Henry Leung, University of Calgary, Canada; Xinping Huang, Communications Research Centre Canada, Canada

ASP-P15.3: A 1V 4.2MW FULLY INTEGRATED 2.5GB/S CMOS LIMITING AMPLIFIER USING FOLDED ACTIVE INDUCTORS .... **I - 1044**
Chia-Hsin Wu, Jieh-Wei Liao, Shen-Iuan Liu, National Taiwan University, Taiwan

ASP-P15.4: A NOVEL MATRIX-BASED LUMPED-ELEMENT ANALYSIS METHOD FOR CMOS DISTRIBUTED AMPLIFIERS .... **I - 1048**
Kambiz Moez, Mohammad Ibrahim Elmasry, University of Waterloo, Canada

ASP-P16: DATA CONVERTERS II

ASP-P16.1: AN IF-SAMPLING TIMING SKEW-INSENSITIVE PARALLEL S/H CIRCUIT .... **I - 1052**
Mikko Aho, Väinö Hakkarainen, Helsinki University of Technology, Finland; Lauri Sumanen, Swiss Center for Electronics and Microtechnology, Switzerland; Mikko Waltari, Conexant Systems, Inc., United States; Kari Halonen, Helsinki University of Technology, Finland

ASP-P16.2: A VERY LOW-POWER CMOS PARALLEL A/D CONVERTER FOR EMBEDDED APPLICATIONS .... I - 1056
Jorge Fernandes, Manuel Silva, Instituto Superior Técnico/INESC-ID Lisboa, Portugal

ASP-P16.3: A FREQUENCY DIGITIZER BASED ON THE CONTINUOUS TIME PHASE DOMAIN NOISE SHAPING .... I - 1060
Mohammad Sharifkhani, University of Waterloo, Canada

ASP-P16.4: PROGRAMMABLE VOLTAGE-OUTPUT, FLOATING-GATE DIGITAL-ANALOG CONVERTER .... I - 1064
Erhan Ozalevli, Paul Hasler, Farhan Adil, Georgia Institute of Technology, United States

ASP-P16.5: AN I/Q-MULTIPLEXED AND OTA-SHARED CMOS PIPELINED ADC WITH AN A-DQS S/H FRONT-END FOR TWO-STEP-CHANNEL-SELECT LOW-IF RECEIVER .... I - 1068
Pui-In Mak, Kin-Kwan Ma, Weng-Ieng Mok, Chi-Sam Sou, Kit-Man Ho, Cheng-Man Ng, University of Macau, Macao SAR of China; Seng-Pan U., University of Macau / Chipidea Microelectronics Ltd., Macao SAR of China; Rui Paulo Martins, University of Macau, Macao SAR of China

ASP-P17: DATA CONVERTERS III

ASP-P17.1: MODELLING AND OPTIMIZATION OF LOW PASS CONTINUOUS-TIME SIGMA-DELTA MODULATORS FOR CLOCK JITTER NOISE REDUCTION .... I - 1072
Luis Hernández, Universidad Carlos III de Madrid, Spain; Andreas Wiesbauer, Infineon Technologies AG, Austria; Susana Patón, Universidad Carlos III de Madrid, Spain; Antonio Di Giandomenico, Infineon Technologies AG, Austria

ASP-P17.2: DESIGN ISSUES AND PERFORMANCE LIMITATIONS OF A CLOCK JITTER INSENSITIVE MULTIBIT DAC ARCHITECTURE FOR HIGH-PERFORMANCE LOW-POWER CT ?? MODULATORS .... I - 1076
Friedel Gerfers, Philips Semiconductors GmbH, Germany; Maurits Ortmanns, Yiannos Manoli, Albert-Ludwigs-University Freiburg, Germany

ASP-P17.3: SIGMA-DELTA MODULATORS OPERATED IN OPTIMIZATION MODE .... I - 1080
Shiang-Hwua Yu, National Sun Yat-Sen University, Taiwan; Jwu-Sheng Hu, National Chiao Tung University, Taiwan

ASP-P17.4: REDUCING QUANTIZATION NOISE WITH RECURSIVE SIGMA-DELTA MODULATORS .... I - 1084
Daniël Schinkel, Ed van Tuijl, Anne-Johan Annema, University of Twente, Netherlands

ASP-P17.5: TAKING ADVANTAGE OF LVDS INPUT BUFFERS TO IMPLEMENT SIGMA-DELTA A/D CONVERTERS IN FPGAS .... I - 1088
Fabio Sousa, Volker Mauer, Altera Europe, United Kingdom; Neimar Duarte, PI Componentes / Altera, Brazil; Ricardo Jasinski, Volnei Pedroni, Federal Center of Technological Education of Paran, Brazil

ASP-P18: ANALOG CIRCUITS III

ASP-P18.1: A CONTINUOUS-TIME FIELD PROGRAMMABLE ANALOG ARRAY (FPAA) CONSISTING OF DIGITALLY RECONFIGURABLE GM-CELLS. .... I - 1092
Joachim Becker, Yiannos Manoli, Albert-Ludwigs-University, Germany

ASP-P18.2: DISTORTION CANCELLATION VIA POLYPHASE MULTIPATH CIRCUITS .... I - 1096
Eisse Mensink, Eric Klumperink, Bram Nauta, University of Twente, Netherlands

ASP-P18.3: LOG-DOMAIN WAVELET BASES .... I - 1100
Sandro A. P. Haddad, Sumit Bagga, Wouter A. Serdijn, Delft University of Technology, Netherlands

ASP-P19: ANALOG CIRCUITS IV

ASP-P19.1: CHAOS IN A DIFFERENTIAL FOURTH-ORDER LOG-DOMAIN BAND-PASS FILTER .... I - 1104
Alon Ascoli, Orla Feely, University College Dublin, Ireland

ASP-P19.2: A NOVEL FOUR QUADRANT CMOS ANALOG MULTIPLIER/DIVIDER .... I - 1108
Gang Li, Brent Maundy, University of Calgary, Canada

ASP-P19.3: A 0.35um CMOS CURRENT-MODE T/H WITH -81dB THD .... I - 1112
Madhu Chennam, Silicon Laboratories, United States; Terri Fiez, Oregon State University, United States

ASP-P19.4: A 0.8V CMOS ANALOG DECODER FOR AN (8,4,4) EXTENDED HAMMING CODE .... I - 1116
Nhan Nguyen, Chris Winstead, Vincent Gaudet, Christian Schlegel, University of Alberta, Canada

ASP-P19.5: LOW POWER LVDS TRANSMITTER WITH LOW COMMON MODE VARIATION .... I - 1120 FOR 1/GB/S-PER PIN OPERATION
Gunjan Mandal, Pradip Mandal, Alliance Semiconductor, India

ASP-P20: SIGMA-DELTA CONVERTERS III

ASP-P20.1: LOOP DELAY COMPENSATION IN BANDPASS CONTINUOUS-TIME .... I - 1124 SIGMA-DELTA MODULATORS WITHOUT ADDITIONAL FEEDBACK COEFFICIENTS
Hassan Aboushady, Marie-Minerve Louërat, University of Paris VI, France

ASP-P20.2: A MULTIRATE BASED BAND-PASS SIGMA-DELTA MODULATOR .... I - 1128
Francisco Colodro, Antonio Torralba, Alfredo Perez VegaLeal, Francisco Perez Ridao, University of Seville, Spain

ASP-P20.3: SIGMA-DELTA ANALOG TO DIGITAL CONVERTER ARCHITECTURE BASED UPON .... I - 1132 A MODULATOR DESIGN EMPLOYING A MIRRORED INTEGRATOR
Zeljko Ignjatovic, Mark Bocko, University of Rochester, United States

ASP-P20.4: MULTIBIT QUADRATURE SIGMA-DELTA MODULATOR WITH DEM SCHEME .... I - 1136
Roberto Maurino, Analog Devices, United Kingdom; Christos Papavassiliou, Imperial College, United Kingdom

ASP-P20.5: AN IF-SAMPLING SC COMPLEX LOWPASS ∑Δ MODULATOR WITH HIGH IMAGE .... I - 1140 REJECTION BY CAPACITOR SHARING
Wang Tung Cheng, Kong-Pang Pun, Cheong-Fat Chan, Chiu-Sing Choy, The Chinese University of Hong Kong, Hong Kong SAR of China

ASP-P21: SIGMA-DELTA CONVERTERS IV

ASP-P21.1: LOW-DISTORTION DELTA-SIGMA TOPOLOGIES FOR MASH ARCHITECTURES .... I - 1144
José Silva, Un-Ku Moon, Gábor Temes, Oregon State University, United States

ASP-P21.2: A PROGRAMMABLE COEFFICIENT CONTINUOUS-TIME A/D ∑-Δ MODULATOR .... I - 1148
Daniel Allen, Angelo Pereira, Paul Hasler, Georgia Institute of Technology, United States

ASP-P21.3: A 0.8-μW SWITCHED-CAPACITOR SIGMA-DELTA MODULATOR USING A CLASS-C .... I - 1152 INVERTER
Youngcheol Chae, Minho Kwon, Gunhee Han, Yonsei University, Republic of Korea

ASP-P21.4: A 4TH ORDER SINGLE-LOOP DELTA-SIGMA ADC WITH 8-BIT TWO-STEP FLASH .... I - 1156 QUANTIZATION
Yongjie Cheng, Craig Petrie, Brent Nordick, Brigham Young University, United States

Volume II

MMSA-L1: ADVANCED MULTIMEDIA SYSTEMS: END-TO-END FRAMEWORKS

MMSA-L1.1: A STEGANOGRAPHIC FRAMEWORK FOR DUAL AUTHENTICATION AND COMPRESSION OF HIGH RESOLUTION IMAGERY ..... II - 1
Deepa Kundur, Texas A&M University, United States; Yang Zhao, University of Toronto, Canada; Patrizio Campisi, Universita degli Studi di Roma ``Roma Tre', Italy

MMSA-L1.2: CONTENT-BASED IMAGE RETRIEVAL WITH AUTOMATED RELEVANCE FEEDBACK OVER DISTRIBUTED PEER-TO-PEER NETWORK ..... II - 5
Ivan Lee, University of Sydney, Australia; Ling Guan, Ryerson University, Canada

MMSA-L1.3: A UNIFIED FRAMEWORK FOR SIMILARITY CALCULATION BETWEEN IMAGES ..... II - 9
Azadeh Kushki, Panagiotis Androutsos, Konstantinos N. Plataniotis, Anastasios N. Venetsanopoulos, University of Toronto, Canada

MMSA-L1.4: AN EFFICIENT TRANSMISSION FRAMEWORK OF DIGITAL MULTIMEDIA BROADCASTING (DMB) SYSTEMS ..... II - 13
Bong-Ho Lee, So Ra Park, Young Kwon Hahm, Soo In Lee, ETRI, Republic of Korea

MMSA-L1.5: A SCALABLE VIDEO TRANSMISSION SYSTEM USING BANDWIDTH INFERENCE IN CONGESTION CONTROL ..... II - 17
Qiang Liu, Jenq-Neng Hwang, University of Washington, United States

MMSA-L2: MULTIMEDIA WATERMARKING AND DATA HIDING

MMSA-L2.1: CORRELATION-BASED WATERMARKING FOR HALFTONE IMAGES ..... II - 21
Ming Sun Fu, Oscar C Au, Hong Kong University of Science and Technology, Hong Kong SAR of China

MMSA-L2.2: CDMA-BASED WATERMARKING RESISTING TO CROPPING ..... II - 25
Yanmei Fang, Jiwu Huang, Shaoquan Wu, Sun Yat-Sen University, China

MMSA-L2.3: HIGH CAPACITY LOSSLESS DATA HIDING BASED ON INTEGER WAVELET TRANSFORM ..... II - 29
Guorong Xuan, Tongji University, China; Yun Q. Shi, Zhicheng Ni, New Jersey Institute of Technology, United States; Jidong Chen, Chengyun Yang, Yizhan Zhen, Junxiang Zheng, Tongji University, China

MMSA-L2.4: LOSSLESS DATA HIDING: FUNDAMENTALS, ALGORITHMS AND APPLICATIONS ..... II - 33
Yun Q. Shi, Zhicheng Ni, Dekun Zou, New Jersey Institute of Technology, United States; Changyin Liang, Shenzhen Polytechnic, China; Guorong Xuan, Tongji University, China

MMSA-L2.5: PERCEPTION BASED BINARY IMAGE WATERMARKING ..... II - 37
Anthony T.S. Ho, Niladri B. Puhan, Pina Marziliano, Anamitra Makur, Yong Liang Guan, Nanyang Technological University, Singapore

MMSA-L3: MULTIMEDIA DATABASE AND RETRIVAL SYSTEMS

MMSA-L3.1: RELEVANCE FEEDBACK USING RANDOM SUBSPACE METHOD ..... II - 41
Wei Jiang, Tsinghua University, China; Mingjing Li, Hong-Jiang Zhang, Microsoft Research Asia, China; Jie Zhou, Tsinghua University, China

MMSA-L3.2: INTERACTIVE CONTENT-BASED IMAGE RETRIEVAL USING LAPLACIAN MIXTURE MODEL IN THE WAVELET DOMAIN ..... II - 45
Tahir Amin, Ling Guan, Ryerson University, Canada

MMSA-L3.3: DESIGN ISSUES ON REQUEST MIGRATION FOR VIDEO-ON-DEMAND SERVICES .... II - 49
Yinqing Zhao, C.-C. Jay Kuo, University of Southern California, United States

MMSA-L3.4: KEYWORD PROPAGATION FOR IMAGE RETRIEVAL .... II - 53
Feng Jing, Tsinghua University, China; Mingjing Li, Hong-Jiang Zhang, Microsoft Research Asia, China; Bo Zhang, Tsinghua University, China

MMSA-L3.5: A NOVEL APPROACH TO VIDEO SEQUENCE MATCHING USING COLOR AND EDGE FEATURES WITH THE MODIFIED HAUSDORFF DISTANCE .... II - 57
Sang Hyun Kim, LG Electronics, Inc., Republic of Korea; Rae-Hong Park, Sogang University, Republic of Korea

MMSA-L4: MULTIMEDIA COMMUNICATION AND TRANSMISSION

MMSA-L4.1: A NOVEL APPROACH TO REAL-TIME MULTIMEDIA FORWARDING OVER HETEROGENEOUS NETWORKS .... II - 61
Keman Yu, Jiang Li, Shipeng Li, Microsoft Research Asia, China

MMSA-L4.2: PRIORITIZED RETRANSMISSION FOR ERROR PROTECTION OF VIDEO STREAMING OVER WLANS .... II - 65
Hsiao-Cheng Wei, Yuh-Chou Tsai, Chia-Wen Lin, National Chung Cheng University, Taiwan

MMSA-L4.3: VIDEO ERROR CONCEALMENT BY USING KALMAN-FILTERING TECHNIQUE .... II - 69
Zhi-Wei Gao, Wen Nung Lie, National Chung Cheng University, Taiwan

MMSA-L4.4: CHANNEL DECODER ARCHITECTURE OF OFDM BASED DMB SYSTEM .... II - 73
Bontae Koo, Jinkyu Kim, Juhyun Lee, Nakwoong Eum, Jongdae Kim, ETRI, Republic of Korea; Hyunmook Cho, Kongju National University, Republic of Korea

MMSA-L4.5: MODELLING POWER CONSUMPTION OF A H.263 VIDEO ENCODER .... II - 77
Xiaoan Lu, Thierry Fernaine, Yao Wang, Polytechnic University, United States

MMSA-L5: MULTIMEDIA CODING AND SEGMENTATION

MMSA-L5.1: A NEW MARKER-BASED WATERSHED ALGORITHM .... II - 81
Hai Gao, Ping Xue, Nanyang Technological University, Singapore; Weisi Lin, Institute for Infocomm Research, Singapore

MMSA-L5.2: OUT-OF-LOOP RATE CONTROL FOR VIDEO CODEC HARDWARE/SOFTWARE CO-DESIGN .... II - 85
Ching-Ho Chen, Chun-Jen Tsai, National Chiao Tung University, Taiwan

MMSA-L5.3: AN AUDIO-SCENE CUT DETECTION METHOD USING FUZZY C-MEANS ALGORITHM FOR AUDIO-VISUAL INDEXING .... II - 89
Naoki Nitanda, Miki Haseyama, Hideo Kitajima, Hokkaido University, Japan

MMSA-L5.4: REGION-OF-INTEREST VIDEO CODING BY FUZZY CONTROL FOR H.263+ STANDARD .... II - 93
Ming-Chieh Chi, Mei-Juan Chen, Ching-Ting Hsu, Nation Dong-Hwa University, Taiwan

MMSA-L5.5: NEW ADAPTIVE PARTIAL DISTORTION SEARCH USING CLUSTERED PIXEL MATCHING ERROR CHARACTERISTIC .... II - 97
Ko-Cheung Hui, Wan-Chi Siu, Yui-Lam Chan, The Hong Kong Polytechnic University, Hong Kong SAR of China

MMSA-L6: MULTIMEDIA UNDERSTANDING AND RECOGNITION

MMSA-L6.1: A REAL-TIME AUTOMATIC LIPREADING SYSTEM .... II - 101
Shilin Wang, Wing Hong Lau, S. H. Leung, H. Yan, City University of Hong Kong, China

MMSA-L6.2: MODEL-BASED VIDEO SCENE CLUSTERING WITH NOISE ANALYSIS ..... II - 105
Hong Lu, Zhenyan Li, Yap-Peng Tan, Nanyang Technological University, Singapore

MMSA-L6.3: FACE RECOGNITION WITH THE ROBUST FEATURE EXTRACTED BY THE GENERALIZED FOLEY-SAMMON TRANSFORM ..... II - 109
Guang Dai, Yuntao Qian, Zhejiang University, China

MMSA-L6.4: TWO-SENSOR NOISE ROBUST ASR WITH MISSING FRAMES FOR AURORA2 TASK ..... II - 113
Cenk Demiroglu, David Anderson, Georgia Institute of Technology, United States

MMSA-L6.5: INTRINSIC MODE FUNCTIONS FOR GAIT RECOGNITION ..... II - 117
Prem Kuchi, Sethuraman Panchanathan, Arizona State University, United States

MMSA-L7: MULTIMEDIA SYSTEMS AND APPLICATIONS: ADVANCED TECHNIQUES

MMSA-L7.1: A NEW CFA INTERPOLATION TECHNIQUE FOR SINGLE-SENSOR DIGITAL CAMERAS ..... II - 121
Rastislav Lukac, Konstantinos N. Plataniotis, University of Toronto, Canada

MMSA-L7.2: QOS-DRIVEN SCHEDULING FOR MULTIMEDIA APPLICATIONS ..... II - 125
Shaoxiong Hua, Gang Qu, University of Maryland, United States

MMSA-L7.3: TRACE-PATH ANALYSIS AND PERFORMANCE ESTIMATION FOR MULTIMEDIA APPLICATION IN EMBEDDED SYSTEM ..... II - 129
Nelson Chang, Kun-Bin Lee, Chien-Wei Jen, National Chiao Tung University, Taiwan

MMSA-L7.4: A TEMPORAL ERROR CONCEALMENT ALGORITHM FOR H.264 USING LAGRANGE INTERPOLATION ..... II - 133
Jinghong Zheng, Lap-Pui Chau, Nanyang Technological University, Singapore

MMSA-L7.5: AN ADAPTIVE SPATIAL FILTER FOR EARLY DEPTH TEST ..... II - 137
Chang-Hyo Yu, Lee-Sup Kim, KAIST, Republic of Korea

MMSA-L8: VLSI/SOC IMPLEMENTATION FOR MULTIMEDIA SYSTEMS I

MMSA-L8.1: A POWER-AWARE IP CORE DESIGN FOR THE VARIABLE-LENGTH DCT/IDCT TARGETING AT MPEG4 SHAPE-ADAPTIVE TRANSFORMS ..... II - 141
Kuan-Hung Chen, Jiun-In Guo, Jinn-Shyan Wang, Ching-Wei Yeh, Tien-Fu Chen, National Chung Cheng University, Taiwan

MMSA-L8.2: MPEG4 AVC/H.264 DECODER WITH SCALABLE BUS ARCHITECTURE AND DUAL MEMORY CONTROLLER ..... II - 145
Hae-Yong Kang, Kyung-Ah Jeong, Jung-Yang Bae, Young-Su Lee, Seung-Ho Lee, C&S Technology, Republic of Korea

MMSA-L8.3: VLSI ARCHITECTURE DESIGN OF MOTION ESTIMATOR AND IN-LOOP FILTER FOR MPEG-4 AVC/H.264 ENCODERS ..... II - 149
Yueh-yi Wang, Yan-Tsung Peng, Chun-Jen Tsai, National Chiao Tung University, Taiwan

MMSA-L8.4: DIVISION-FREE RASTERIZER FOR PERSPECTIVE-CORRECT TEXTURE FILTERING ..... II - 153
Donghyun Kim, Lee-Sup Kim, KAIST, Republic of Korea

MMSA-L8.5: AN ADAPTIVE DSP PROCESSOR FOR HIGH-EFFICIENCY COMPUTING MPEG-4 VIDEO ENCODER ..... II - 157
Li-Hsun Chen, Oscal T.-C. Chen, National Cheng Chung University, Taiwan; Ting-Yi Wang, Chi-Lung Wang, Industrial Technology Research Institute, Taiwan

MMSA-P1: DATA HIDING AND WATERMARKING TECHIQUES FOR MULTIMEDIA SYSTEMS

MMSA-P1.1: ON IMPROVING THE ITERATIVE WATERMARK EMBEDDING TECHNIQUE FOR JPEG-TO-JPEG WATERMARKING .... II - 161
Peter H. W. Wong, Andy Chang, Oscar C. Au, Hong Kong University of Science and Technology, Hong Kong SAR of China

MMSA-P1.2: DATA HIDING FOR MULTIMODAL BIOMETRIC RECOGNITION .... II - 165
Alexia Giannoula, Dimitrios Hatzinakos, University of Toronto, Canada

MMSA-P1.3: ON THE SECURITY OF STRUCTURAL INFORMATION EXTRACTION/EMBEDDING FOR IMAGES .... II - 169
Chun-Shien Lu, Institute of Information Science, Academia Sinica, Taiwan

MMSA-P1.4: A ROBUST WATERMARKING METHOD BASED ON WAVELET AND ZERNIKE TRANSFORM .... II - 173
Jie Chen, Hongxun Yao, Wen Gao, Shaohui Liu, Harbin Institute of Technology, China

MMSA-P1.5: A TEXTURE-BASED TAMPER DETECTION SCHEME BY FRAGILE WATERMARK .... II - 177
Yazhou Liu, Wen Gao, Hongxun Yao, Shaohui Liu, Harbin Institute of Technology, China

MMSA-P2: INTELLIGENT PROCESSING TECHNIQUES FOR MULTIMEDIA SYSTEMS AND APPLICATIONS

MMSA-P2.1: A NEURAL NETWORK APPROACH FOR HUMAN EMOTION RECOGNITION IN SPEECH .... II - 181
Muhammad Waqas Bhatti, University of Sydney, Australia; Yongjin Wang, Ling Guan, Ryerson University, Canada

MMSA-P2.2: A PROBABILISTIC REASONING APPROACH TO CLOSED-ROOM PEOPLE MONITORING .... II - 185
Ji Tao, Yap-Peng Tan, Nanyang Technological University, Singapore

MMSA-P2.3: ADAPTIVE MULTIMEDIA CONTENT PERSONALIZATION .... II - 189
Nikolaos Doulamis, National Technical University of Athens, Greece; Pavlos Georgilakis, Technical University of Crete, Greece

MMSA-P2.4: PROBABILISTIC APPROACH TO K-NEAREST NEIGHBOR VIDEO RETRIEVAL .... II - 193
Nai-xiang Lian, Yap-Peng Tan, Nanyang Technological University, Singapore

MMSA-P2.5: VIDEO SUMMARIZATION BY SPATIAL-TEMPORAL GRAPH OPTIMIZATION .... II - 197
Shi Lu, Michael Lyu, Irwin King, The Chinese University of Hong Kong, Hong Kong SAR of China

MMSA-P3: COMMUNICATION AND CODING TECHNIQUES FOR MULTIMEDIA SYSTEMS

MMSA-P3.1: MULTI-SERVER OPTIMAL BANDWIDTH MONITORING FOR COLLABORATIVE DISTRIBUTED RETRIEVAL .... II - 201
Lihang Ying, Anup Basu, University of Alberta, Canada; Satish Tripathi, University of California, Riverside, United States

MMSA-P3.2: MULTIPLE DESCRIPTION VIDEO CODING BASED ON ZERO PADDING .... II - 205
Dong Wang, Nishan Canagarajah, David Redmill, David Bull, University of Bristol, United Kingdom

MMSA-P3.3: LOW-VARIANCE TCP-FRIENDLY THROUGHPUT ESTIMATION FOR CONGESTION CONTROL OF LAYERED VIDEO MULTICAST .... II - 209
Kitae Nahm, C.-C. Jay Kuo, University of Southern California, United States

MMSA-P3.4: HIGH PERFORMANCE SPATIAL-TEMPORAL DE-INTERLACING TECHNIQUE USING INTERFIELD INFORMATION .... II - 213
Ching-Ting Hsu, Mei-Juan Chen, Chin-Hui Huang, National Dong Hwa University, Taiwan

MMSA-P3.5: CROSS-LAYER DESIGN FOR QOS WIRELESS COMMUNICATIONS ..... II - 217
Jie Chen, Tiejun Lv, Brown University, United States; Haitao Zheng, Lucent Technologies, United States

MMSA-P4: VLSI/SOC IMPLEMENTATION FOR MULTIMIDIA SYTEMS II

MMSA-P4.1: A NEW CRYPTOGRAPHIC SYSTEM AND ITS VLSI IMPLEMENTATION ..... II - 221
Jui-Cheng Yen, Hun-Chen Chen, Shin-Shian Jou, National United University, Taiwan

MMSA-P4.2: ANALYSIS AND ARCHITECTURE DESIGN FOR HIGH PERFORMANCE JPEG2000 COPROCESSOR ..... II - 225
Bing-Fei Wu, Chung-Fu Lin, National Chiao Tung University, Taiwan

MMSA-P4.3: A COMPACT BLOCK-MATCHING CELL FOR ANALOGUE MOTION ESTIMATION PROCESSORS ..... II - 229
Mladen Panovic, Andreas Demosthenous, University College London, United Kingdom

MMSA-P4.4: A PERFORMANCE-DRIVEN CONFIGURABLE MOTION ESTIMATOR FOR FULL-SEARCH BLOCK-MATCHING ALGORITHM ..... II - 233
Yeong-Kang Lai, Lien-Fei Chen, National Chung Hsing University, Taiwan

MMSA-P4.5: A NOVEL EMBEDDED MEMORY ARCHITECTURE FOR REAL-TIME MESH-BASED MOTION ESTIMATION ..... II - 237
Mohammed Sayed, Wael Badawy, University of Calgary, Canada

VLSI-L1: LOW POWER CIRCUITS AND ARCHITECTURE

VLSI-L1.1: TRANSFERRING PERFORMANCE GAIN FROM SOFTWARE PREFETCHING TO ENERGY REDUCTION ..... II - 241
Deepak Agarwal, Sumitkumar Pamnani, AMD, United States; Gang Qu, Donald Yeung, University of Maryland, College Park, United States

VLSI-L1.2: A LOW-POWER DECIMATION FILTER FOR A SIGMA-DELTA CONVERTER BASED ON A POWER-OPTMIZED SINC FILTER ..... II - 245
Andrea Gerosa, Andrea Neviani, University of Padova, Italy

VLSI-L1.3: EMPIRICAL EVALUATION OF TIMING AND POWER IN RESONANT CLOCK DISTRIBUTION ..... II - 249
Juang-ying Chueh, Conrad Ziesler, Marios Papaefthymiou, University of Michigan, United States

VLSI-L1.4: A POWER AND AREA EFFICIENT MULTI-MODE FEC PROCESSOR ..... II - 253
Yi-Chen Tseng, Chien-Ching Lin, Hsie-Chia Chang, Chen-Yi Lee, National Chiao Tung University, Taiwan

VLSI-L1.5: ASYNCHRONOUS, QUASI-ADIABATIC (ASYNCHROBATIC) LOGIC FOR LOW-POWER VERY WIDE DATA WIDTH APPLICATIONS ..... II - 257
David J. Willingham, Izzet Kale, University of Westminster, United Kingdom

VLSI-L2: VIDEO

VLSI-L2.1: A DISTRIBUTED TS-MUX ARCHITECTURE FOR MULTI-CHIP EXTENSION BEYOND THE HDTV LEVEL ..... II - 261
Takayuki Onishi, Mitsuo Ikeda, Jiro Naganuma, Makoto Endo, Yoshiyuki Yashima, NTT Corporation, Japan

VLSI-L2.2: AN EFFICIENT ARCHITECTURE FOR COLOR SPACE CONVERSION USING DISTRIBUTED ARITHMETIC ..... II - 265
Faycal Bensaali, Abbes Amira, Ahmed Bouridane, Queen's University Belfast, United Kingdom

VLSI-L2.3: HARDWARE ARCHITECTURE DESIGN FOR H.264/AVC INTRA FRAME CODER ..... II - 269
Yu-Wen Huang, Bing-Yu Hsieh, Tung-Chien Chen, Liang-Gee Chen, National Taiwan University, Taiwan

VLSI-L2.4: ANALYSIS AND DESIGN OF MACROBLOCK PIPELINING FOR H.264/AVC VLSI ARCHITECTURE .... II - 273
Tung-Chien Chen, Yu-Wen Huang, Liang-Gee Chen, National Taiwan University, Taiwan

VLSI-L2.5: LOW-POWER DA-CONVERTERS FOR DISPLAY APPLICATIONS USING STEPWISE CHARGING AND CHARGE RECOVERY .... II - 277
Christoph Saas, Artur Wróblewski, Josef A. Nossek, Munich University of Technology, Germany

VLSI-L3: LOW POWER ARITHMETIC

VLSI-L3.1: GLITCH-CONSCIOUS LOW-POWER DESIGN OF ARITHMETIC CIRCUITS .... II - 281
Henrik Eriksson, Per Larsson-Edefors, Chalmers University of Technology, Sweden

VLSI-L3.2: REDUCING MULTIPLIER ENERGY BY DATA-DRIVEN VOLTAGE VARIATION .... II - 285
Tomoyuki Yamanaka, Vasily Moshnyaga, Fukuoka University, Japan

VLSI-L3.3: A NOVEL FAST LOW VOLTAGE DYNAMIC THRESHOLD TRUE SINGLE PHASE CLOCKING ADIABATIC CIRCUIT .... II - 289
Michael Yang, James Barby, University of Waterloo, Canada

VLSI-L3.4: A LOW POWER AND FAST WAKE UP CIRCUIT .... II - 293
Hung-wei Chen, National United University, Taiwan; Wen-Cheng Yen, Faraday Technology Corporation, Taiwan

VLSI-L3.5: A LEAKAGE ESTIMATION AND REDUCTION TECHNIQUE FOR SCALED CMOS LOGIC CIRCUITS CONSIDERING GATE-LEAKAGE .... II - 297
Hafijur Rahman, Chaitali Chakrabarti, Arizona State University, United States

VLSI-L4: MPEG

VLSI-L4.1: HARDWARE ARCHITECTURE FOR GLOBAL MOTION ESTIMATION FOR MPEG-4 ADVANCED SIMPLE PROFILE .... II - 301
Ching-Yeh Chen, Shao-Yi Chien, Wei-Min Chao, Yu-Wen Huang, Liang-Gee Chen, National Taiwan University, Taiwan

VLSI-L4.2: QME: AN EFFICIENT SUBSAMPLING-BASED BLOCK MATCHING ALGORITHM FOR MOTION ESTIMATION .... II - 305
Kun-Bin Lee, Hao-Yun Chin, Hui-Cheng Hsu, Chein-Wei Jen, National Chiao Tung University, Taiwan

VLSI-L4.3: DIGITAL CONTROLLED ANALOG ARCHITECTURE FOR DCT AND DST USING CAPACITOR SWITCHING .... II - 309
Arindam Basu, Ashis Kumar Mal, Anindya Sundar Dhar, Indian Institute of Technology, Kharagpur, India

VLSI-L4.4: LOW-POWER PARALLEL TREE ARCHITECTURE FOR FULL SEARCH BLOCK-MATCHING MOTION ESTIMATION .... II - 313
Siou-Shen Lin, Po-Chih Tseng, Liang-Gee Chen, National Taiwan University, Taiwan

VLSI-L4.5: A FAST DUAL SYMBOL CONTEXT-BASED ARITHMETIC CODING FOR MPEG-4 SHAPE CODING .... II - 317
Kun-Bin Lee, National Chiao Tung University, Taiwan; Jih-Yiing Lin, Sunplus Technology, Taiwan; Chein-Wei Jen, National Chiao Tung University, Taiwan

VLSI-L5: LOW POWER BUSES AND CIRCUITS

VLSI-L5.1: DYNAMIC RECONFIGURABLE BUS ENCODING SCHEME FOR REDUCING THE ENERGY CONSUMPTION OF DEEP SUB-MICRON INSTRUCTION BUS .... II - 321
Siu-Kei Wong, Chi-Ying Tsui, Hong Kong University of Science and Technology, Hong Kong SAR of China

VLSI-L5.2: LOW POWER COUPLING-BASED ENCODING FOR ON-CHIP BUSES .... II - 325
Maged Ghoneima, Yehea Ismail, Northwestern University, United States

VLSI-L5.3: UNIFIED BUS ENCODING BY STREAM RECONSTRUCTION WITH VARIABLE STRIDES ..... II - 329
Tien-Fu Chen, Tsung-Ming Hsieh, Chun-Li Wei, National Chung Cheng University, Taiwan

VLSI-L5.4: A NEW PARALLEL ARCHITECTURE FOR LOW POWER LINEAR FEEDBACK SHIFT REGISTERS ..... II - 333
Abdullah Mamun, Rajendra Katti, North Dakota State University, United States

VLSI-L5.5: A LOW-POWER GROUP-BASED VLD DESIGN ..... II - 337
Cheng-Hung Liu, National Chiao Tung University, Taiwan; Bai-Jue Shieh, Sunplus Technology Co., Ltd., Taiwan; Chen-Yi Lee, National Chiao Tung University, Taiwan

VLSI-L6: IMAGE PROCESSING

VLSI-L6.1: EXTENDED INTELLIGENT EDGE-BASED LINE AVERAGE WITH ITS IMPLEMENTATION AND TEST METHOD ..... II - 341
Yu-Lin Chang, Shyh-Feng Lin, Liang-Gee Chen, National Taiwan University, Taiwan

VLSI-L6.2: A LOW POWER CURRENT-MODE PIXEL WITH ON-CHIP FPN CANCELLATION AND DIGITAL SHUTTER ..... II - 345
Amine Bermak, Hong Kong University of Science and Technology, China; Farid Boussaïd, Abdesselam Bouzerdoum, Edith Cowan University, Australia

VLSI-L6.3: A REAL-TIME VLSI MEDIAN FILTER EMPLOYING TWO-DIMENSIONAL BIT-PROPAGATING ARCHITECTURE ..... II - 349
Hideo Yamasaki, Tadashi Shibata, University of Tokyo, Japan

VLSI-L6.4: RECONFIGURABLE DISCRETE COSINE TRANSFORM PROCESSOR FOR OBJECT-BASED VIDEO SIGNAL PROCESSING ..... II - 353
Po-Chih Tseng, Chao-Tsung Huang, Liang-Gee Chen, National Taiwan University, Taiwan

VLSI-L6.5: AN INSTRUCTION SET FOR THE EFFICIENT IMPLEMENTATION OF THE CORDIC ALGORITHM ..... II - 357
Sven Simon, Matthias Müller, Hochschule Bremen, Germany; Holger Gryska, Infineon Technologies AG, Germany; Andreas Wortmann, Hochschule Bremen, Germany; Steffen Buch, Infineon Technologies AG, Germany

VLSI-L7: LOW POWER CODES AND CRYPTOGRAPHY

VLSI-L7.1: ENERGY-PERFORMANCE TRADEOFFS FOR THE SHARED MEMORY IN MULTI-PROCESSOR SYSTEMS-ON-CHIP ..... II - 361
Kimish Patel, Enrico Macii, Politecnico di Torino, Italy; Massimo Poncino, Università di Verona, Italy

VLSI-L7.2: LEFT-TO-RIGHT BINARY SIGNED-DIGIT RECODING FOR ELLIPTIC CURVE CRYPTOGRAPHY ..... II - 365
Rajendra Katti, Xiaoyu Ruan, North Dakota State University, United States

VLSI-L7.3: LOW POWER DECODING OF BCH CODES ..... II - 369
Yuejian Wu, Nortel Networks, Canada

VLSI-L7.4: A HEURISTIC APPROACH TO SYNTHESIZE BOOLEAN FUNCTIONS USING TANT NETWORK ..... II - 373
Hafiz Md. Hasan Babu, Md. Rafiqul Islam, Rumana Nazmul, Md. Anwarul Haque, Ahsan Raja Chowdhury, Dhaka University, Bangladesh

VLSI-L7.5: POWER CONSUMPTION OPTIMIZATION FOR LOW LATENCY VITERBI DECODER ..... II - 377
Mario Steinert, Stefano Marsili, Infineon Technologies AG, Germany

VLSI-L8: CODING

VLSI-L8.1: A FAST REED-SOLOMON PRODUCT-CODE DECODER WITHOUT REDUNDANT COMPUTATIONS .... II - 381
Hyun-Yong Lee, In-Cheol Park, KAIST, Republic of Korea

VLSI-L8.2: NOVEL BIT MANIPULATION UNIT FOR COMMUNICATION DIGITAL SIGNAL PROCESSORS .... II - 385
Sung Dae Kim, Sug Hyun Jeong, Myung Hoon Sunwoo, Ajou university, Republic of Korea; Kyung Ho Kim, Samsung Electronics, Republic of Korea

VLSI-L8.3: JOINT CODE-ENCODER-DECODER DESIGN FOR LDPC CODING SYSTEM VLSI IMPLEMENTATION .... II - 389
Hao Zhong, Tong Zhang, Rensselaer Polytechnic Institute, United States

VLSI-L8.4: MULTI-LEVEL MEMORY SYSTEMS USING ERROR CONTROL CODES .... II - 393
Hsie-Chia Chang, Chien-Ching Lin, Tien-Yuan Hsiao, Jieh-Tsorng Wu, Ta-Hui Wang, National Chiao Tung University, Taiwan

VLSI-L8.5: MEMORY-BASED LOW DENSITY PARITY CHECK CODE DECODER ARCHITECTURE USING LOOSELY COUPLED TWO DATA-FLOWS .... II - 397
Se-Hyeon Kang, In-Cheol Park, KAIST, Republic of Korea

VLSI-L9: HIGH PERFORMANCE LOW POWER

VLSI-L9.1: PSEUDO-FOOTLESS CMOS DOMINO LOGIC CIRCUITS FOR HIGH-PERFORMANCE VLSI DESIGNS .... II - 401
Jinn-Shyan Wang, Shang-Jyh Shieh, Ching-Wei Yeh, Yuan-Hsun Yeh, National Chung Cheng University, Taiwan

VLSI-L9.2: A TECHNIQUE FOR HIGH-SPEED CIRCUITS ON SOI USING LOOK-AHEAD TYPE ACTIVE BODY BIAS CONTROL .... II - 405
Masaaki Iijima, Katsuya Fujita, Kazuki Fukuoka, Masahiro Numa, Keisuke Yamamoto, Kobe University, Japan; Kengo Takata, Mitsubishi Electric Corp., Japan

VLSI-L9.3: FAST THERMAL ANALYSIS FOR VLSI CIRCUITS VIA SEMI-ANALYTICAL GREEN'S FUNCTION IN MULTI-LAYER MATERIALS .... II - 409
Baohua Wang, Pinaki Mazumder, University of Michigan, United States

VLSI-L9.4: B-DTNMOS: A NOVEL BULK DYNAMIC THRESHOLD NMOS SCHEME .... II - 413
Walid Elgharbawy, Magdy Bayoumi, University of Louisiana at Lafayette, United States

VLSI-L9.5: ENERGY EFFICIENT DUAL THRESHOLD VOLTAGE DYNAMIC CIRCUITS EMPLOYING SLEEP SWITCHES TO MINIMIZE SUBTHRESHOLD LEAKAGE .... II - 417
Volkan Kursun, Eby G. Friedman, University of Rochester, United States

VLSI-L10: ARITHMETIC

VLSI-L10.1: A SCALABLE COMPACT ARCHITECTURE FOR THE COMPUTATION OF INTEGER BINARY LOGARITHMS THROUGH LINEAR APPROXIMATION .... II - 421
Christophe Layer, Hans-Jörg Pfleiderer, University of Ulm, Germany; Christoph Heer, Infineon Technologies AG, Germany

VLSI-L10.2: A METHOD FOR INCREASING THE THROUGHPUT OF FIXED COEFFICIENT DIGIT-SERIAL/PARALLEL MULTIPLIERS .... II - 425
Magnus Karlsson, University of Kalmar, Sweden; Mark Vesterbacka, Linköping University, Sweden; Wlodek Kulesza, University of Kalmar, Sweden

VLSI-L10.3: MODULO DEFLATION IN (2n +1, 2n, 2n -1) CONVERTERS .... II - 429
Shaoqiang Bi, Concordia University, Canada; Wei Wang, University of Western Ontario, Canada; Asim Al-Khalili, Concordia University, Canada

VLSI-L10.4: GEOMETRIC-MEAN INTERPOLATION FOR LOGARITHMIC NUMBER SYSTEMS .... II - 433
Mark Arnold, Lehigh University, United States

VLSI-L10.5: A LOW POWER 16-BIT BOOTH LEAPFROG ARRAY MULTIPLIER USING DYNAMIC ADDERS .... II - 437
Kwen-Siong Chong, Bah-Hwee Gwee, Joseph Sylvester Chang, Nanyang Technological University, Singapore

VLSI-L11: SOC DESIGN TECHNOLOGY

VLSI-L11.1: ARBITRATE-AND-MOVE PRIMITIVES FOR HIGH THROUGHPUT ON-CHIP INTERCONNECTION NETWORKS .... II - 441
Aydin Balkan, Gang Qu, Uzi Vishkin, University of Maryland, United States

VLSI-L11.2: SUBSTRATE NOISE-AWARE FLOORPLANNING FOR MIXED-SIGNAL SOCS .... II - 445
Marcin Jeske, Grzegorz Blakiewicz, Malgorzata Chrzanowska-Jeske, Benyi Wang, Portland State University, United States

VLSI-L11.3: PARAMETERIZED SOC DESIGN FOR PORTABLE SYSTEMS .... II - 449
Sumant Bhutoria, Chaitali Chakrabarti, Arizona State University, United States

VLSI-L11.4: A 0.18µm IMPLEMENTATION OF A FLOATING-POINT UNIT FOR A PROCESSING-IN-MEMORY SYSTEM .... II - 453
Taek-Jun Kwon, USC Information Sciences Institute, United States; Joong-Seok Moon, Apple Computer, Inc., United States; Jeff Sondeen, Jeff Draper, USC Information Sciences Institute, United States

VLSI-L11.5: POWER-AWARE IMPLEMENTATION OF ASIC/SOC IN 0.13 MICRON CMOS TECHNOLOGY .... II - 457
Aleksandar Pance, Madan Mohan, Paul Master, QuickSilver Technology, Inc., United States

VLSI-L12: ADDERS AND MULTIPLIERS

VLSI-L12.1: DYNAMIC PASS-TRANSISTOR DOT OPERATORS FOR EFFICIENT PARALLEL-PREFIX ADDERS .... II - 461
Henrik Eriksson, Per Larsson-Edefors, Chalmers University of Technology, Sweden

VLSI-L12.2: A GATE-LEVEL STRATEGY TO DESIGN CARRY SELECT ADDERS .... II - 465
Massimo Alioto, Università di Siena, Italy; Gaetano Palumbo, Massimo Poli, Università di Catania, Italy

VLSI-L12.3: SWITCHING ACTIVITY IN BIT-SERIAL CONSTANT-COEFFICIENT MULTIPLIERS .... II - 469
Kenny Johansson, Oscar Gustafsson, Lars Wanhammar, Linköping University, Sweden

VLSI-L12.4: MULTIPLIER BLOCKS USING CARRY-SAVE ADDERS .... II - 473
Oscar Gustafsson, Linköping University, Sweden; Andrew G. Dempster, University of Westminster, United Kingdom; Lars Wanhammar, Linköping University, Sweden

VLSI-L12.5: A LOW LATENCY AND LOW POWER DYNAMIC CARRY SAVE ADDER .... II - 477
Ramyanshu Datta, Jacob A. Abraham, University of Texas, Austin, United States; Robert Montoye, Wendy Belluomini, Hung Ngo, Chandler McDowell, Jente B. Kuang, Kevin Nowka, IBM Austin Research Laboratory, United States

VLSI-L13: NOISE IN DIGITAL CIRCUITS

VLSI-L13.1: AN EXPERIMENTAL COMPARISON OF SUBSTRATE NOISE GENERATED BY CMOS AND BY LOW-NOISE DIGITAL CIRCUITS .... II - 481
Edgar Albuquerque, INESC-ID, Portugal; Manuel Silva, IST/INESC-ID, Portugal

VLSI-L13.2: AN ACCURATE AND EFFICIENT ESTIMATION OF SWITCHING NOISE IN SYNCHRONOUS DIGITAL CIRCUITS ..... II - 485
Husni Habal, Terri Fiez, Kartikeya Mayaram, Oregon State University, United States

VLSI-L13.3: AN IMPROVED TECHNIQUE TO INCREASE NOISE-TOLERANCE IN DYNAMIC DIGITAL CIRCUITS ..... II - 489
Fernando Mendoza-Hernandez, Universidad de Sonora, Mexico; Mónico Linares-Aranda, Victor H. Champac-Vilela, Instituto Nacional de Astrofísica, Óptica y Electrónica, Mexico

VLSI-L13.4: THE NOISE IMMUNITY OF DYNAMIC DIGITAL CIRCUITS WITH TECHNOLOGY SCALING ..... II - 493
Fernando Mendoza-Hernandez, Universidad de Sonora, Mexico; Mónico Linares-Aranda, Victor H. Champac-Vilela, Instituto Nacional de Astrofísica, Óptica y Electrónica, Mexico

VLSI-L13.5: A NEW NON-ITERATIVE MODEL FOR SWITCHING WINDOW COMPUTATION WITH CROSSTALK NOISE ..... II - 497
Omar Hafiz, University of Arizona, United States; Pinhong Chen, Cadence Design Systems, United States; Janet Meiling Wang, University of Arizona, United States

VLSI-L14: TURBO AND VITERBI ALGORITHMS

VLSI-L14.1: NOVEL PIPELINING OF MSB-FIRST ADD-COMPARE SELECT UNIT STRUCTURE FOR VITERBI DECODERS ..... II - 501
Keshab K. Parhi, University of Minnesota, United States

VLSI-L14.2: PIPELINED PARALLEL ARCHITECTURE FOR HIGH THROUGHPUT MAP DETECTORS ..... II - 505
Ruwan Ratnayake, Gu-Yeon Wei, Aleksandar Kavcic, Harvard University, United States

VLSI-L14.3: PARALLEL TURBO DECODING ..... II - 509
Yuping Zhang, Keshab K. Parhi, University of Minnesota, United States

VLSI-L14.4: VLSI ARCHITECTURE EXPLORATION FOR SLIDING-WINDOW LOG-MAP DECODERS ..... II - 513
Chien-Ming Wu, Chip Implementation Center (CIC), Taiwan; Ming-Der Shieh, National Cheng Kung University, Taiwan; Chien-Hsing Wu, National Chung Cheng University, Taiwan; Ying-Tsung Hwang, National Yunlin University of Science and Technology, Taiwan; Jun-Hong Chen, National Cheng King University, Taiwan; Hsin-Fu Lo, Chip Implementation Center (CIC), Taiwan

VLSI-L14.5: AN ASYNCHRONOUS SOVA DECODER FOR WIRELESS COMMUNICATION APPLICATION ..... II - 517
Wing-Kin Chan, Chiu-Sing Choy, Cheong-Fat Chan, Kong-Pang Pun, The Chinese University of Hong Kong, Hong Kong SAR of China

VLSI-L15: INTERCONNECT

VLSI-L15.1: DECOUPLING TECHNIQUE AND CROSSTALK ANALYSIS FOR COUPLED RLC INTERCONNECTS ..... II - 521
Junmou Zhang, Eby G. Friedman, University of Rochester, United States

VLSI-L15.2: EFFECT OF RELATIVE DELAY ON THE DISSIPATED ENERGY IN COUPLED INTERCONNECTS ..... II - 525
Maged Ghoneima, Yehea Ismail, Northwestern University, United States

VLSI-L15.3: EFFECT OF SHIELD INSERTION ON REDUCING CROSSTALK NOISE BETWEEN COUPLED INTERCONNECTS ..... II - 529
Junmou Zhang, Eby G. Friedman, University of Rochester, United States

VLSI-L15.4: ELECTRICAL ISOLATION AND FANOUT IN INTRA-CHIP OPTICAL INTERCONNECTS .... II - 533
Anand Pappu, Alyssa Apsel, Cornell University, United States

VLSI-L15.5: MANAGING INDUCTIVE COUPLING IN WIDE SIGNAL BUSSES .... II - 537
Bassel Soudan, University of Sharjah, United Arab Emirates

VLSI-L16: CRYPTOGRAPHY

VLSI-L16.1: A HIGH SPEED ASIC IMPLEMENTATION OF THE RIJNDAEL ALGORITHM .... II - 541
Refik Sever, TUBITAK-BILTEN, Turkey; A. Neslin Ismailoglu, TÜBITAK-BILTEN, Turkey; Yusuf Cagatay Tekmen, Murat Askar, Middle East Technical University, Turkey

VLSI-L16.2: DESIGN OF A RECONFIGURABLE AES ENCRYPTION/DECRYPTION ENGINE FOR MOBILE TERMINALS .... II - 545
Thilo Pionteck, Thorsten Staake, Thomas Stiefmeier, Lukusa D. Kabulepa, Manfred Glesner, Darmstadt University of Technology, Germany

VLSI-L16.3: HIGH-SPEED HARDWARE IMPLEMENTATIONS OF THE KASUMI BLOCK CIPHER .... II - 549
Paris Kitsos, Michalis Galanis, Odysseas Koufopavlou, University of Patras, Greece

VLSI-L16.4: NEW CLASS OF THE FPGA EFFICIENT CRYPTOGRAPHIC PRIMITIVES .... II - 553
Nikolay Moldovyan, Ma Eremeev, Specialized Center of Program Systems, Russian Federation; Nicolas Sklavos, Odysseas Koufopavlou, University of Patras, Greece

VLSI-L16.5: TWO FAST RSA IMPLEMENTATIONS USING HIGH-RADIX MONTGOMERY ALGORITHM .... II - 557
Soner Yesil, A. Neslin Ismailoglu, TÜBITAK-BILTEN, Turkey; Yusuf Cagatay Tekmen, Middle East technical University, Turkey; Murat Askar, Middle East Technical University, Turkey

VLSI-L17: I/O CIRCUITS

VLSI-L17.1: LOW COMPLEXITY DIGITAL PLL FOR INSTANT ACQUISITION CDR .... II - 561
Gordon Allan, John Knight, Carleton University, Canada

VLSI-L17.2: A TECHNIQUE TO DESKEW DIFFERENTIAL PCB TRACES .... II - 565
Amer Atrash, Georgia Institute of Technology, United States; Brian Butka, Integrated Device Technologies, United States

VLSI-L17.3: A NEW INTERPOLATED SYMBOL TIMING RECOVERY METHOD .... II - 569
Xiong Liu, Alan N. Willson, Jr., University of California, Los Angeles, United States

VLSI-L17.4: A NEW SCHMITT TRIGGER CIRCUIT IN A 0.13 um 1/2.5 V CMOS PROCESS TO RECEIVE 3.3 V INPUT SIGNALS .... II - 573
Shih-Lun Chen, Ming-Dou Ker, National Chiao Tung University, Taiwan

VLSI-L17.5: DESIGN ON MIXED-VOLTAGE-TOLERANT I/O INTERFACE WITH NOVEL TRACKING CIRCUITS IN A 0.13-μM CMOS TECHNOLOGY .... II - 577
Che-Hao Chuang, Industrial Technology Research Institute, Taiwan; Ming-Dou Ker, National Chiao Tung University, Taiwan

VLSI-L18: FIELD PROGRAMMABLE AND RECONFIGURABLE

VLSI-L18.1: AUTONOMOUS MEMORY BLOCK FOR RECONFIGURABLE COMPUTING .... II - 581
Wim J.C. Melis, Peter Y.K. Cheung, Wayne Luk, Imperial College London, United Kingdom

VLSI-L18.2: COMPACT HAMMING-COMPARATOR-BASED RANK ORDER FILTER FOR DIGITAL VLSI AND FPGA IMPLEMENTATIONS .... II - 585
Volnei Pedroni, CEFET-PR, Brazil

VLSI-L18.3: APPLICATION PERFORMANCE OF ELEMENTS IN A FLOATING-GATE FPAA ....II - 589
Tyson Hall, Christopher Twigg, Paul Hasler, David Anderson, Georgia Institute of Technology, United States

VLSI-L18.4: A NOVEL FPGA ARCHITECTURAL IMPLEMENTATION OF PIPELINED THINNING ALGORITHM ....II - 593
Pei-Yung Hsiao, Chun-Ho Hua, Chang Gung University, Taiwan; Chien-Chen Lin, National Taiwan Normal University, Taiwan

VLSI-L18.5: AN EFFICIENT FPGA IMPLEMENTATION OF ADVANCED ENCRYPTION STANDARD ALGORITHM ....II - 597
Shuenn-Shyang Wang, Wan-Sheng Ni, Tatung University, Taiwan

VLSI-L19: CLOCKING

VLSI-L19.1: EXPONENTIALLY TAPERED H-TREE CLOCK DISTRIBUTION NETWORKS ....II - 601
Magdy El-Moursy, Eby G. Friedman, University of Rochester, United States

VLSI-L19.2: A NEW MESOCHRONOUS CLOCKING SCHEME FOR SYNCHRONIZATION IN SOC ....II - 605
Behzad Mesgarzadeh, Christer Svensson, Atila Alvandpour, Linköping University, Sweden

VLSI-L19.3: THE IMPACT OF CLOCK GATING SCHEMES ON THE POWER DISSIPATION OF SYNTHESIZABLE REGISTER FILES ....II - 609
Matthias Mueller, Andreas Wortmann, Sven Simon, Hochschule Bremen, Germany; Michael Kugel, Tim Schoenauer, Infineon Technologies AG, Germany

VLSI-L19.4: LEAKAGE POWER REDUCTION FOR CLOCK GATING SCHEME ON PD-SOI ....II - 613
Kazuki Fukuoka, Masaaki Iijima, Kenji Hamada, Masahiro Numa, Kobe University, Japan; Akira Tada, Renesas Technology Corp., Japan

VLSI-L19.5: TIME BORROWING AND CLOCK SKEW SCHEDULING EFFECTS ON MULTI-PHASE LEVEL-SENSITIVE CIRCUITS ....II - 617
Baris Taskin, Ivan Kourtev, University of Pittsburgh, United States

VLSI-L20: MEMORY

VLSI-L20.1: A NEW DUAL PUMPING CIRCUIT WITHOUT BODY EFFECTS FOR LOW SUPPLY VOLTAGE ....II - 621
Ming-chih Hsieh, Zheng-hong Wang, Hongchin Lin, National Chung Hsing University, Taiwan; Yen-Tai Lin, eMemory Technology Inc., Taiwan

VLSI-L20.2: A FULLY SYMMETRICAL SENSE AMPLIFIER FOR NON-VOLATILE MEMORIES ....II - 625
Ferdinando Bedeschi, STMicroelectronics, Italy; Edoardo Bonizzoni, University of Pavia, Italy; Osama Khouri, Claudio Resta, STMicroelectronics, Italy; Guido Torelli, University of Pavia, Italy

VLSI-L20.3: STATIC DIVIDED WORD MATCHING LINE FOR LOW-POWER CONTENT ADDRESSABLE MEMORY DESIGN ....II - 629
Kuo-Hsing Cheng, National Central University, Taiwan; Chia-Hung Wei, Tamkang University, Taiwan; Shu-Yu Jiang, National Central University, Taiwan

VLSI-L20.4: LOW POWER DUAL MATCHLINE TERNARY CONTENT ADDRESSABLE MEMORY ....II - 633
Nitin Mohan, Manoj Sachdev, University of Waterloo, Canada

VLSI-L20.5: EFFICIENT HARDWARE IMPLEMENTATION OF A CRYTO-MEMORY BASED ON AES ALGORITHM AND SRAM ARCHITECTURE ....II - 637
Anna Labbé, Annie Pérez, Jean-Michel Portal, L2MP-Polytech, France

VLSI-L21: CURRENT-MODE AND SENSING

VLSI-L21.1: POSITIVE-FEEDBACK SOURCE-COUPLED LOGIC: A DELAY MODEL .... II - 641
Massimo Alioto, Ada Fort, Luca Pancioni, Santina Rocchi, Valerio Vignoli, University of Siena, Italy

VLSI-L21.2: MULTI-GHZ ENERGY-EFFICIENT ASYNCHRONOUS PIPELINED CIRCUITS IN MOS CURRENT MODE LOGIC .... II - 645
Tin Wai Kwan, Maitham Shams, Carleton University, Canada

VLSI-L21.3: A NEW LEVEL RESTORATION CIRCUIT FOR MULTI-VALUED LOGIC .... II - 649
Avni Morgul, Turgay Temel, Bogaziçi University, Turkey

VLSI-L21.4: IMPLEMENTATION OF MCML UNIVERSAL LOGIC GATE FOR 10 GHz-RANGE IN 0.13 μm CMOS TECHNOLOGY .... II - 653
Shahnam Khabiri, Maitham Shams, Carleton University, Canada

VLSI-L21.5: HIGH INPUT RANGE SENSE COMPARATOR FOR MULTILEVEL FLASH MEMORIES .... II - 657
Alessandro Cabrini, University of Pavia, Italy; Rino Micheloni, Osama Khouri, STMicroelectronics, Italy; Stefano Gregori, University of Texas, Dallas, United States; Guido Torelli, University of Pavia, Italy

VLSI-L22: FLIP-FLOPS

VLSI-L22.1: DUAL-EDGE TRIGGERED LEVEL CONVERTING FLIP-FLOPS .... II - 661
Hamid Mahmoodi-Meimand, Kaushik Roy, Purdue University, United States

VLSI-L22.2: A NOVEL CMOS DOUBLE-EDGE TRIGGERED FLIP-FLOP FOR LOW-POWER APPLICATIONS .... II - 665
Yu-Yin Sung, Robert C. Chang, National Chung Hsing University, Taiwan

VLSI-L22.3: CONTENTION REDUCED/CONDITIONAL DISCHARGE FLIP-FLOPS FOR LEVEL CONVERSION IN CVS SYSTEMS .... II - 669
Peiyi Zhao, Golconda Pradeep Kumar, Magdy Bayoumi, University of Louisiana at Lafayette, United States

VLSI-L22.4: AN EFFICIENT IMPLEMENTATION OF D-FLIP-FLOP USING THE GDI TECHNIQUE .... II - 673
Arkadiy Morgenshtein, Technion, Israel; Alexander Fish, Ben-Gurion University, Israel; Israel A. Wagner, IBM Research Laboratory, Israel

VLSI-L22.5: DATA-RETENTION FLIP-FLOPS FOR POWER-DOWN APPLICATIONS .... II - 677
Hamid Mahmoodi-Meimand, Kaushik Roy, Purdue University, United States

VLSI-L23: TEST, VERIFICATION AND SIGNAL PROCESSING

VLSI-L23.1: MIXED RL-HUFFMAN ENCODING FOR POWER REDUCTION AND DATA COMPRESSION IN SCAN TEST .... II - 681
Mohammad Tehranipour, Mehrdad Nourani, University of Texas, Dallas, United States; Karim Arabi, PMC Sierra, Canada; Ali Afzali-Kusha, University of Tehran, Iran

VLSI-L23.2: ASSERTION-BASED ON-LINE VERIFICATION AND DEBUG ENVIRONMENT FOR COMPLEX HARDWARE SYSTEMS .... II - 685
Kevin Peterson, Yvon Savaria, École Polytechnique de Montréal, Canada

VLSI-L23.3: LOW POWER PATTERN GENERATION FOR BIST ARCHITECTURE .... II - 689
Nisar Ahmed, Mohammad Tehranipour, Mehrdad Nourani, University of Texas, Dallas, United States

VLSI-L23.4: A DESIGN COMPLEXITY COMPARISON METHOD FOR LOOP-BASED SIGNAL PROCESSING ALGORITHMS: PARTICLE FILTERS .... II - 693
Sangjin Hong, Miodrag Bolic, Petar Djuric, State University of New York at Stony Brook, United States

VLSI-L23.5: AN EFFICIENT ARCHITECTURE FOR 1-D DISCRETE BIORTHOGONAL WAVELET TRANSFORM .... II - 697
Isa Servan Uzun, Abbes Amira, Ahmed Bouridane, Queen's University Belfast, United Kingdom

VLSI-L24: FREQUENCY SYNTHESIS, ESD

VLSI-L24.1: LEAST SQUARES APPROXIMATION-BASED ROM-FREE DIRECT DIGITAL FREQUENCY SYNTHESIZER .... II - 701
Ching-Hua Wen, Huai-Yi Hsu, Hung Yang Ko, An-Yeu (Andy) Wu, National Taiwan University, Taiwan

VLSI-L24.2: A NOVEL DDS ARCHITECTURE USING NONLINEAR ROM ADDRESSING WITH IMPROVED COMPRESSION RATIO AND QUANTISATION NOISE .... II - 705
Malinky Ghosh, Lakshmi Chimakurthy, Foster Dai, Richard Jaeger, Auburn University, United States

VLSI-L24.3: QUADRATURE DIRECT DIGITAL FREQUENCY SYNTHESIS USING FINE-GRAIN ANGLE ROTATION .... II - 709
Sung-Won Lee, In-Cheol Park, KAIST, Republic of Korea

VLSI-L24.4: A CORDIC LIKE PROCESSOR FOR COMPUTATION OF ARCTANGENT AND ABSOLUTE MAGNITUDE OF A VECTOR .... II - 713
Koushik Maharatna, University of Bristol, United Kingdom; Alfonso Troya, Miloš Krstic, Eckhard Grass, Ulrich Jagdhold, IHP, Germany

VLSI-L24.5: ESD PROTECTION DESIGN FOR IC WITH POWER-DOWN-MODE OPERATION .... II - 717
Ming-Dou Ker, National Chiao Tung University, Taiwan; Kun-Hsien Lin, Industrial Technology Research Institute, Taiwan

VLSI-P1: LOW POWER DESIGN AND IMPLEMENTATION I

VLSI-P1.1: DYNAMIC POWER OPTIMIZATION OF THE TRACE-BACK PROCESS FOR THE VITERBI ALGORITHM .... II - 721
Mihail Petrov, Tudor Murgan, Abdulfattah Obeid, Cristian Chitu, Peter Zipf, Darmstadt University of Technology, Germany; Jörg Brakensiek, Nokia Research Center, Germany; Manfred Glesner, Darmstadt University of Technology, Germany

VLSI-P1.2: LOW POWER IMPLEMENTATION OF POLYPHASE FILTERS IN QUADRATIC RESIDUE NUMBER SYSTEM .... II - 725
GianCarlo Cardarilli, Andrea Del Re, University of Rome Tor Vergata, Italy; Alberto Nannarelli, Technical University of Denmark, Denmark; Marco Re, University of Rome Tor Vergata, Italy

VLSI-P1.3: ULTRA LOW-VOLTAGE LOW-POWER EXPONENTIAL VOLTAGE-MODE CIRCUIT WITH TUNABLE OUTPUT RANGE .... II - 729
Quoc-Hoang Duong, Trung-Kien Nguyen, Sang-Gug Lee, Information and Communications University, Republic of Korea

VLSI-P1.4: GLITCHING POWER REDUCTION THROUGH SUPPLY VOLTAGE ADAPTATION MECHANISM FOR LOW POWER ARRAY STRUCTURE DESIGN .... II - 733
Sangjin Hong, Shu-Shin Chin, Magesh Sadasivam, State University of New York at Stony Brook, United States

VLSI-P1.5: IMPLEMENTATION OF A LOW POWER MOTION DETECTION CAMERA PROCESSOR USING A CMOS IMAGE SENSOR .... II - 737
Suh-Ho Lee, Seon-Wook Kim, Suki Kim, Korea University, Republic of Korea

VLSI-P2: LOW POWER DESIGN AND IMPLEMENTATION II

VLSI-P2.1: HIGH PERFORMANCE SENSE AMPLIFIER CIRCUIT FOR LOW POWER SRAM APPLICATIONS .... II - 741
Hwang-Cherng Chow, Shu-Hsien Chang, Chang Gung University, Taiwan

VLSI-P2.2: EVALUATION OF POWER CUT-OFF TECHNIQUES IN THE PRESENCE OF GATE LEAKAGE .... II - 745
Mindaugas Draždžiulis, Per Larsson-Edefors, Chalmers University of Technology, Sweden

VLSI-P2.3: CROSSTALK ENERGY REDUCTION BY TEMPORAL SHIELDING .... II - 749
Sabino Salerno, Enrico Macii, Politecnico di Torino, Italy; Massimo Poncino, Università di Verona, Italy

VLSI-P2.4: A POWER-OPTIMIZED 64-BIT PRIORITY ENCODER UTILIZING PARALLEL PRIORITY LOOK-AHEAD .... II - 753
Cheong Kun, Shaolei Quan, Andrew Mason, Michigan State University, United States

VLSI-P2.5: A MAXIMUM TOTAL LEAKAGE CURRENT ESTIMATION METHOD .... II - 757
Yongjun Xu, Chinese Academy of Sciences, China; Zuying Luo, Tsinghua University, China; Xiaowei Li, Chinese Academy of Sciences, China

VLSI-P3: VIDEO IP CORES

VLSI-P3.1: ARCHITECTURE DESIGN OF MDCT-BASED PSYCHOACOUSTIC MODEL CO-PROCESSOR IN MPEG ADVANCED AUDIO CODING .... II - 761
Tsung-Han Tsai, National Central University, Taiwan; Shih-Way Huang, National Taiwan University, Taiwan; Yi-Wen Wang, National Central University, Taiwan

VLSI-P3.2: A LOW-POWER DCT IP CORE BASED ON 2D ALGEBRAIC INTEGER ENCODING .... II - 765
Minyi Fu, University of Windsor, Canada; Graham Jullien, Vassil Dimitrov, University of Calgary, Canada; Majid Ahmadi, University of Windsor, Canada

VLSI-P3.3: A PARAMETERIZED POWER-AWARE IP CORE GENERATOR FOR THE 2-D 8x8 DCT/IDCT .... II - 769
Rei-Chin Ju, Jia-Wei Chen, Jiun-In Guo, Tien-Fu Chen, National Chung Cheng University, Taiwan

VLSI-P3.4: VLSI DESIGN OF DUAL-MODE VITERBI/TURBO DECODER FOR 3GPP .... II - 773
Kai Huang, Fan-Min Li, Pei-Ling Shen, An-Yeu (Andy) Wu, National Taiwan University, Taiwan

VLSI-P3.5: A COST-EFFECTIVE MPEG-4 SHAPE-ADAPTIVE DCT WITH AUTO-ALIGNED TRANSPOSE MEMORY ORGANIZATION .... II - 777
Kun-Bin Lee, Hui-Cheng Hsu, Chein-Wei Jen, National Chiao Tung University, Taiwan

VLSI-P4: ARITHMETIC MODULE IMPLEMENTATION

VLSI-P4.1: A LOW-POWER 1.85 GHz 32-BIT CARRY LOOKAHEAD ADDER USING DUAL PATH ALL-N-LOGIC .... II - 781
Ge Yang, University of California, Santa Cruz, United States; Seong-Ook Jung, T-RAM, United States; Kwang-Hyun Baek, Rockwell Scientific, United States; Soo Hwan Kim, Suki Kim, Korea University, Republic of Korea; Sung-Mo Kang, University of California, Santa Cruz, United States

VLSI-P4.2: GIGAHERTZ-RANGE MCML MULTIPLIER ARCHITECTURES .... II - 785
Venkat Srinivasan, Dong Ha, Jos Sulistyo, Virginia Tech, United States

VLSI-P4.3: A PROGRAMMABLE BASE 2D-LNS MAC WITH SELF-GENERATED LOOK-UP TABLES .... II - 789
Wenjing Zhang, Graham Jullien, Vassil Dimitrov, University of Calgary, Canada

VLSI-P4.4: A DESIGN OF 4-OPERAND REDUNDANT BINARY PARALLEL ADDER USING NEURON MOS .... II - 793
Masahiro Sakamoto, Shuusaku Mizukami, Daisuke Hamano, Hisato Fujisaka, Hiroshima City University, Japan

VLSI-P4.5: A LOW POWER HIGH SPEED ACCUMULATOR FOR DDFS APPLICATIONS .... II - 797
Michael Chappell, Alistair McEwan, Oxford University, United Kingdom

VLSI-P5: VLSI ARCHITECTURES

VLSI-P5.1: REDUCED BINARY TREE FIR FILTERS .... II - 801
Artur Wróblewski, Marek Wróblewski, Christoph Saas, Josef A. Nossek, Munich University of Technology, Germany

VLSI-P5.2: PHASED TAG CACHE: AN EFFICIENT LOW POWER CACHE SYSTEM .... II - 805
Rui Min, Wenben Jone, Yiming Hu, University of Cincinnati, United States

VLSI-P5.3: MICROSYSTEM CONTROLLER FOR SENSOR NETWORK CONTROL AND DATA CORRECTION .... II - 809
Prasanna Balasundaram, Kartik Vaidyanathan, Andrew Mason, Michigan State University, United States

VLSI-P5.4: A DYNAMIC TASK SCHEDULING ALGORITHM FOR BATTERY POWERED DVS SYSTEMS .... II - 813
Jameel Ahmed, Chaitali Chakrabarti, Arizona State University, United States

VLSI-P6: ARITHMETIC AND DSP IMPLEMENTATION

VLSI-P6.1: VLSI-EFFICIENT IMPLEMENTATION OF FULL ADDER-BASED MEDIAN FILTER .... II - 817
Adrian Burian, Jarmo Takala, Tampere University of Technology, Finland

VLSI-P6.2: STATIC FLOATING-POINT UNIT WITH IMPLICIT EXPONENT TRACKING FOR EMBEDDED DSP .... II - 821
Hung-Yueh Lin, Tay-Jyi Lin, Chie-Min Chao, Yen-Chin Liao, Chih-Wei Liu, Chein-Wei Jen, National Chiao Tung University, Taiwan

VLSI-P6.3: SAMPLED ANALOG ARCHITECTURE FOR DCT AND DST .... II - 825
Ashis Kumar Mal, Arindam Basu, Anindya Sundar Dhar, Indian Institute of Technology, Kharagpur, India

VLSI-P6.4: B-SPLINE FACTORIZATION-BASED ARCHITECTURE FOR INVERSE DISCRETE WAVELET TRANSFORM .... II - 829
Chao-Tsung Huang, Po-Chih Tseng, Liang-Gee Chen, National Taiwan University, Taiwan

VLSI-P6.5: DESIGN OF AN EFFICIENT VARIABLE-LENGTH FFT PROCESSOR .... II - 833
Chung-Ping Hung, Sau-Gee Chen, Kun-Lung Chen, National Chiao Tung University, Taiwan

VLSI-P7: ARITHMETIC AND CRYPTOGRAPHY

VLSI-P7.1: A NOVEL RADIX-4 BIT-LEVEL MODULAR MULTIPLIER FOR FAST RSA CRYPTOSYSTEM .... II - 837
Jin-Hua Hong, Bin-Yan Tsai, Liang-Te Lu, National University of Kaohsiung, Taiwan; Shao-Hui Shieh, National Tsing-Hua University, Taiwan

VLSI-P7.2: DESIGN OF RESIDUE-TO-BINARY CONVERTER FOR A NEW 5-MODULI SUPERSET RESIDUE NUMBER SYSTEM .... II - 841
Bin Cao, Thambipillai Srikanthan, Chip-Hong Chang, Nanyang Technological University, Singapore

VLSI-P7.3: AN ERROR PATTERN ROM COMPRESSION METHOD FOR CONTINUOUS DATA .... II - 845
Byung-Do Yang, Lee-Sup Kim, KAIST, Republic of Korea

VLSI-P7.4: GF(2K) MULTIPLIERS BASED ON MONTGOMERY MULTIPLICATION ALGORITHM ..... II - 849
Apostolos Fournaris, Odysseas Koufopavlou, University of Patras, Greece

VLSI-P7.5: LOGICALLY REVERSIBLE ARITHMETIC CIRCUIT USING PASS-TRANSISTOR ..... II - 853
Takashi Hisakado, Hiroyoshi Iketo, Kohshi Okumura, Kyoto University, Japan

VLSI-P8: CIRCUIT DESIGN I

VLSI-P8.1: MAX AND MIN FUNCTIONS USING MULTIPLE-VALUED RECHARGED SEMI-FLOATING GATE CIRCUITS ..... II - 857
Henning Gundersen, Yngvar Berg, University of Oslo, Norway

VLSI-P8.2: A NOVEL SERIAL MULTIPLIER USING FLOATING-GATE TRANSISTORS ..... II - 861
Luis Fortino Cisneros Sinencio, INAOE, Mexico; Alejandro Díaz Sánchez, Instituto Tecnológico de Puebla, Mexico; Jaime Ramírez Angulo, New Mexico State University, United States

VLSI-P8.3: CMOS EXPONENTIAL CURRENT-TO-VOLTAGE CIRCUIT BASED ON NEWLY PROPOSED APPROXIMATION METHOD ..... II - 865
Quoc-Hoang Duong, Trung-Kien Nguyen, Sang-Gug Lee, Information and Communications University, Republic of Korea

VLSI-P8.4: DELAY ANALYSIS AND OPTIMAL BIASING FOR HIGH SPEED LOW POWER CURRENT MODE LOGIC CIRCUITS ..... II - 869
Vasanth Kakani, Foster Dai, Richard Jaeger, Auburn University, United States

VLSI-P8.5: FAST AND PRECISE INTERCONNECT CAPACITIVE COUPLING NOISE MODEL ..... II - 873
Young Jun Lee, Yong-Bin Kim, Northeastern University, United States

VLSI-P9: FPGA AND PLA

VLSI-P9.1: A NOVEL SCHEME OF IMPLEMENTING HIGH SPEED AWGN COMMUNICATION CHANNEL EMULATORS IN FPGAS ..... II - 877
Yongquan Fan, Zeljko Zilic, McGill University, Canada

VLSI-P9.2: A HIGH PERFORMANCE LOW POWER DYNAMIC PLA WITH CONDITIONAL EVALUATION SCHEME ..... II - 881
Kwang-Il Oh, Lee-Sup Kim, KAIST, Republic of Korea

VLSI-P9.3: AN EMBEDDED FLEXIBLE CONTENT-ADDRESSABLE MEMORY CORE FOR INCLUSION IN A FIELD-PROGRAMMABLE GATE ARRAY ..... II - 885
Steven J.E. Wilton, University of British Columbia, Canada; Christopher W. Jones, Greenville, SC, United States; Julien Lamoureux, University of British Columbia, Canada

VLSI-P9.4: A UNIFIED ARCHITECTURE OF MD5 AND RIPEMD-160 HASH ALGORITHMS ..... II - 889
Chiu-Wah Ng, Tung-Sang Ng, Kun-Wah Yip, University of Hong Kong, Hong Kong SAR of China

VLSI-P9.5: WHIRLPOOL HASH FUNCTION: ARCHITECTURE AND VLSI IMPLEMENTATION ..... II - 893
Paris Kitsos, Odysseas Koufopavlou, University of Patras, Greece

VLSI-P10: IMAGE PROCESSING AND IMPLEMENTATION

VLSI-P10.1: AN ARCHITECTURE FOR FRACTAL IMAGE COMPRESSION USING QUAD-TREE MULTIRESOLUTION ..... II - 897
Alejandro Martínez-Ramírez, Alejandro Díaz Sánchez, INAOE, Mexico; Mónico Linares-Aranda, Instituto Nacional de Astrofísica, Óptica y Electrónica, Mexico; Javier Vega-Pineda, ITCh, Mexico

VLSI-P10.2: ON BOARD PROCESSOR DEVELOPMENT FOR NASA'S SPACEBONE IMAGING RADAR WITH VLSI SYSTEM-ON-CHIP TECHNOLOGY .... II - 901
Wai-Chi Fang, NASA's Jet Propulsion Laboratory, United States; Michael Jin, NASA's Jet Propulsion Lab, United States

VLSI-P10.3: LOW COST AND LATENCY EMBEDDED 3D GRAPHICS RECIPROCATION .... II - 905
Dan Crisu, Stamatis Vassiliadis, Sorin Cotofana, Delft University of Technology, Netherlands; Petri Liuha, Nokia Research Center, Finland

VLSI-P10.4: SYSTOLIC COUNTERS WITH UNIQUE ZERO STATE .... II - 909
Mircea Stan, University of Virginia, United States

VLSI-P10.5: SOBEL EDGE DETECTION PROCESSOR FOR A REAL-TIME VOLUME RENDERING SYSTEM .... II - 913
Natalia Kazakova, University of Alberta, Canada; Martin Margala, University of Rochester, United States; Nelson Durdle, University of Alberta, Canada

VLSI-P11: CIRCUIT DESIGN II

VLSI-P11.1: FORWARD BODY BIASED KEEPER FOR ENHANCED NOISE IMMUNITY IN DOMINO LOGIC CIRCUITS .... II - 917
Volkan Kursun, Eby G. Friedman, University of Rochester, United States

VLSI-P11.2: MODELING AND DESIGNING ENERGY-DELAY OPTIMIZED WIDE DOMINO CIRCUITS .... II - 921
Christine Kwong, Bhaskar Chatterjee, Manoj Sachdev, University of Waterloo, Canada

VLSI-P11.3: AN ALL-DIGITAL 50% DUTY-CYCLE CORRECTOR .... II - 925
Yi-Ming Wang, Jinn-Shyan Wang, Chung-Cheng University, Taiwan

VLSI-P11.4: DESIGN TECHNIQUES FOR PULSED STATIC CMOS .... II - 929
Kavitha Seshadri, Adrianne Pontarelli, Gauri Joglekar, Gerald Sobelman, University of Minnesota, United States

VLSI-P11.5: A PHASE-ADJUSTABLE NEGATIVE PHASE SHIFTER USING A SINGLE-SHOT LOCKING METHOD .... II - 933
Chua-Chin Wang, Ya-Hsin Hsueh, Sen-Fu Hong, National Sun Yat-Sen University, Taiwan; Rong-Sui Kao, VIA Technologies, Inc., Taiwan

VLSI-P12: ARRAY ARCHITECTURE AND SOC

VLSI-P12.1: VLSI ARCHITECTURE OF THE RECONFIGURABLE COMPUTING ENGINE FOR DIGITAL SIGNAL PROCESSING APPLICATIONS .... II - 937
Lien-Fei Chen, Yeong-Kang Lai, National Chung Hsing University, Taiwan

VLSI-P12.2: APPLICATION-SPECIFIC CONFIGURATION OF MULTITHREADED PROCESSOR ARCHITECTURE FOR EMBEDDED APPLICATIONS .... II - 941
Mary Kiemb, Kiyoung Choi, Seoul National University, Republic of Korea

VLSI-P12.3: RLC EFFECTS ON WORST-CASE SWITCHING PATTERN FOR ON-CHIP BUSES .... II - 945
Shang-Wei Tu, Jing-Yang Jou, National Chiao Tung University, Taiwan; Yao-Wen Chang, National Taiwan University, Taiwan

VLSI-P12.4: FAST RECONFIGURING MESH-CONNECTED VLSI ARRAYS .... II - 949
Jigang Wu, Thambipillai Srikanthan, Nanyang Technological University, Singapore

Volume III

CNN-L1: BIO-INSPIRED AND NEUROMORPHIC ARRAY COMPUTERS

CNN-L1.1: AN ANALOG CMOS CHIP IMPLEMENTING A CNN-BASED LOCOMOTION CONTROLLER FOR QUADRUPED WALKING ROBOTS ..... III - 1
Kazuki Nakada, Hokkaido University, Japan; Tetsuya Asai, Hokkaido Univeristy, Japan; Yoshihito Amemiya, Hokkaido University, Japan

CNN-L1.2: A BIOMIMETIC VLSI ARCHITECTURE FOR SMALL TARGET TRACKING ..... III - 5
Vivek Pant, Charles Higgins, University of Arizona, United States

CNN-L1.3: ANALOG INTEGRATED 2-D OPTICAL FLOW SENSOR WITH PROGRAMMABLE PIXELS ..... III - 9
Alan Stocker, New York University, United States; Rodney J. Douglas, University of Zürich / ETH Zürich, Switzerland

CNN-L1.4: A MULTI-CHIP IMPLEMENTATION OF CORTICAL ORIENTATION HYPERCOLUMNS ..... III - 13
Thomas Choi, Bertram Shi, Hong Kong University of Science and Technology, Hong Kong SAR of China; Kwabena Boahen, UPenn, United States

CNN-L1.5: AN ARTIFICIAL IMMUNE SYSTEM FOR VISUAL APPLICATIONS WITH CNN-UM ..... III - 17
Gyürgy Cserey, University of Notre Dame, United States; András Falus, Semmelweis University of Medicine, Hungary; Wolfgang Porod, University of Notre Dame, United States; Tamás Roska, Pázmány University, Hungary

CNN-L2: IMPLEMENTATION OF CNNS AND ARRAY COMPUTERS

CNN-L2.1: HIGH DENSITY VLSI IMPLEMENTATION OF A BIPOLAR CNN WITH REDUCED PROGRAMMABILITY ..... III - 21
Ari Paasio, University of Turku, Finland; Jacek Flak, Mika Laiho, Kari Halonen, Helsinki University of Technology, Finland

CNN-L2.2: ON THE IMPLEMENTATION OF RTD BASED CNNS ..... III - 25
Sing-Rong Li, Pinaki Mazumder, Leon O. Chua, University of Michigan, United States

CNN-L2.3: A SIMPLICIAL CNN ARCHITECTURE FOR ON-CHIP IMAGE PROCESSING ..... III - 29
Pablo Sergio Mandolesi, Pedro Julian, Universidad Nacional del Sur, Argentina; Andreas G. Andreou, Johns Hopkins University, United States

CNN-L2.4: EFFICIENT HARDWARE-ORIENTED CELLULAR ACTIVE CONTOURS ..... III - 33
David L. Vilariño, University of Santiago de Compostela, Spain; Csaba Rekeczky, Hungarian Academy of Sciences, Hungary

CNN-L2.5: HIGH-LEVEL DESIGN ENVIRONMENT FOR MASSIVE PARALLEL VLSI-IMPLEMENTATIONS OF STATISTICAL SIGNAL- AND IMAGE PROCESSING MODELS ..... III - 37
Stephan Stilkerich, Imperial College, Germany; Joachim K. Anlauf, University of Bonn, Germany

CNN-L3: COMPLEX SPATIO-TEMPORAL DYNAMICS IN MULTI-LAYER CNNS

CNN-L3.1: RICH DYNAMICS IN WEAKLY-COUPLED FULL-RANGE CELLULAR NEURAL NETWORKS ..... III - 41
Mauro Di Marco, Mauro Forti, Università di Siena, Italy; Alberto Tesi, Università di Firenze, Italy

CNN-L3.2: ARRAYS OF SWITCHED CHUA'S CIRCUITS ..... III - 45
Riccardo Caponetto, Luigi Fortuna, Mattia Frasca, Sebastiano Guzzardi, University of Catania, Italy; Alessandro Rizzo, Politecnico di Bari, Italy

CNN-L3.3: INVESTIGATION OF PHASE-WAVE PROPAGATION PHENOMENA IN SECOND ORDER CNN ARRAYS ..... III - 49
Zonghuang Yang, Kazuya Tsuruta, Yoshifumi Nishio, Tokushima University, Japan; Akio Ushida, Tokushima Bunri University, Japan

CNN-L3.4: PATTERN FORMATION ON THE PROTOTYPE COMPLEX-CELL CNN-UM CHIP (CACE1K) ..... III - 53
Dávid Bálya, Hungarian Academy of Sciences, Hungary; István Petrás, Csaba Rekeczky, AnaLogic and Neural Computing Systems Laboratory, Hungary

CNN-L3.5: A 32 x 32 FOUR LAYER REACTION-DIFFUSION CNN CHIP ..... III - 57
Bertram Shi, Tao Luo, Hong Kong University of Science and Technology, Hong Kong SAR of China

CNN-L4: ANALYSIS AND APPLICATIONS OF CNNS

CNN-L4.1: ANALYSIS AND DESIGN OF CELLULAR NEURAL NETWORKS ..... III - 61
Fernando Corinto, Marco Gilli, Pier Paolo Civalleri, Politecnico di Torino, Italy

CNN-L4.2: A UNIFYING PROOF OF GLOBAL ASYMPTOTICAL STABILITY OF NEURAL NETWORKS WITH DELAY ..... III - 65
Chai Wah Wu, IBM T. J. Watson Research Center, United States; Ying Sue Huang, Pace University, United States

CNN-L4.3: REGULAR SMALL-WORLD CELLULAR NEURAL NETWORKS: KEY PROPERTIES AND EXPERIMENTS ..... III - 69
Gergely Tímár, Dávid Bálya, Hungarian Academy of Sciences, Hungary

CNN-L4.4: FINGERPRINT FEATURE MATCHING USING CNNS ..... III - 73
Qun Gao, EMPA, Switzerland; George S. Moschytz, ETH Zurich, Switzerland, Switzerland

CNN-L4.5: CNN-BASED REAL-TIME VIDEO DETECTION OF PLASMA INSTABILITY IN NUCLEAR FUSION APPLICATIONS ..... III - 77
Paolo Arena, Adriano Basile, Luigi Fortuna, University of Catania, Italy; Giuseppe Mazzitelli, Associazione Euratom-ENEA sulla Fusione, Italy; Alessandro Rizzo, Politecnico di Bari, Italy; Maria Zammataro, University of Catania, Italy

CNN-P1: CIRCUIT DESIGN FOR ARRAY COMPUTERS

CNN-P1.1: A GRAY-CODE CURRENT-MODE ADC FOR MIXED-MODE CELLULAR COMPUTER ..... III - 81
Laura Vesalainen, Jonne Poikonen, Mikko Pänkäälä, Ari Paasio, University of Turku, Finland

CNN-P1.3: AN ON-OFF TEMPORAL FILTER CIRCUIT FOR VISUAL MOTION ANALYSIS ..... III - 85
Bertram Shi, Eric Tsang, Philip Au, Hong Kong University of Science and Technology, Hong Kong SAR of China

CNN-P1.4: DESIGN OF A PIXEL ARRAY CIRCUIT FOR THINNING PROCESS ..... III - 89
Chunyan Wang, Kuo-Ting Wu, Concordia University, Canada

CNN-P1.5: IMPROVED CELL CORE FOR A MIXED-MODE POLYNOMIAL CNN ..... III - 93
Mika Laiho, Helsinki University of Technology / University of Turku, Finland; Ari Paasio, University of Turku, Finland; Kari Halonen, Helsinki University of Technology, Finland

CNN-P2: APPLICATIONS OF CELLULAR NEURAL NETWORKS AND ARRAY COMPUTERS

CNN-P2.1: VERY HIGH SPEED VITERBI DECODER WITH CIRCULARLY CONNECTED ANALOG CNN CELL ARRAY ..... III - 97
Hyongsuk Kim, Hongrak Son, Chonbuk National University, Republic of Korea; Tamás Roska, Hungarian Academy of Sciences, Hungary; Leon Chua, University of California, Berkeley, United States

CNN-P2.2: LOSSLESS IMAGE CODING BASED ON LIFTING WAVELET USING DISCRETE-TIME CELLULAR NEURAL NETWORK WITH MULTI-TEMPLATES ..... III - 101
Hisashi Aomori, Tsuyoshi Otake, Sophia University, Japan; Nobuaki Takahashi, IBM Japan, Ltd., Japan; Mamoru Tanaka, Sophia University, Japan

CNN-P2.3: DESIGN METHOD FOR ORIENTATION-SELECTIVE CNN FILTERS ..... III - 105
Radu Matei, Technical University of Iasi, Romania

CNN-P2.4: A NEW MAXNET ..... III - 109
Yi C. Chang, National Chung Cheng University / Wu-Feng Institute of Technology, Taiwan; Sung-Nien Yu, National Chung Cheng University, Taiwan; Chung J. Kuo, Delta Electronics, Inc., Taiwan

CNN-P2.5: JOINT ESTIMATION OF DOA AND ANGULAR SPREAD USING POLARIMETRIC COMPLEX MUSIC ..... III - 113
Thanat Sooknuan, Raungrong Suleesathira, King Mongkut's University of Technology Thonburi, Thailand

DSP-L1: FIR DIGITAL FILTERS

DSP-L1.1: MINIMAX DESIGN OF FIR FILTERS WITH LOW GROUP DELAY USING ENHANCED SEQUENTIAL QUADRATIC PROGRAMMING ..... III - 117
Wu-Sheng Lu, University of Victoria, Canada

DSP-L1.2: CONSTRAINED EIGENFILTER DESIGN ..... III - 121
Ying-Man Law, Chi-Wah Kok, Hong Kong University of Science and Technology, Hong Kong SAR of China

DSP-L1.3: ON THE DESIGN OF REAL AND COMPLEX FIR FILTERS WITH FLATNESS AND PEAK ERROR CONSTRAINTS USING SEMIDEFINITE PROGRAMMING ..... III - 125
S. C. Chan, K. M. Tsui, University of Hong Kong, Hong Kong SAR of China

DSP-L1.4: MTH-BAND LINEAR-PHASE FIR FILTER INTERPOLATORS AND DECIMATORS UTILIZING THE FARROW STRUCTURE ..... III - 129
Håkan Johansson, Oscar Gustafsson, Linköping University, Sweden

DSP-L1.5: ON OPTIMAL IFIR FILTER DESIGN ..... III - 133
Alireza Mehrnia, Alan N. Willson, Jr., University of California, Los Angeles, United States

DSP-L2: IIR DIGITAL FILTERS

DSP-L2.1: MINIMIZATION OF L2-SENSITIVITY FOR STATE-SPACE DIGITAL FILTERS SUBJECT TO L2-SCALING CONSTRAINTS ..... III - 137
Takao Hinamoto, Hiroaki Ohnishi, Hiroshima University, Japan; Wu-Sheng Lu, University of Victoria, Canada

DSP-L2.2: JOINT OPTIMIZATION OF ERROR FEEDBACK AND COORDINATE TRANSFORMATION FOR ROUNDOFF NOISE MINIMIZATION IN 2-D STATE-SPACE DIGITAL FILTERS ..... III - 141
Takao Hinamoto, Hiroaki Ohnishi, Hiroshima University, Japan; Wu-Sheng Lu, University of Victoria, Canada

DSP-L2.3: DESIGN OF MAXIMALLY FLAT GROUP DELAY FILTERS ..... III - 145
Alfonso Fernandez-Vazquez, Gordana Jovanovic-Dolecek, INAOE, Mexico

DSP-L2.4: A SEMI-DEFINITE PROGRAMMING (SDP) METHOD FOR DESIGNING IIR SHARP CUT-OFF DIGITAL FILTERS USING FREQUENCY-RESPONSE MASKING ..... III - 149
H. H. Chen, S. C. Chan, K. L. Ho, University of Hong Kong, China

DSP-L2.5: AN ALGORITHM FOR THE OPTIMIZATION OF ADJUSTABLE FRACTIONAL-DELAY ALL-PASS FILTERS ..... III - 153
Juha Yli-Kaakinen, Tapio Saramäki, Tampere University of Technology, Finland

DSP-L3: DIGITAL FILTERS

DSP-L3.1: FPGA-BASED 3D MEDIAN FILTERING USING WORD-PARALLEL SYSTOLIC ARRAYS .... III - 157
Carlos Castro-Pareja, Jogikal Jagadeesh, Sharmila Venugopal, The Ohio State University, United States; Raj Shekhar, The Cleveland Clinic Foundation, United States

DSP-L3.2: A SHIFTED PERMUTED DIFFERENCE COEFFICIENT METHOD .... III - 161
Henrik Ohlsson, Oscar Gustafsson, Lars Wanhammar, Linköping University, Sweden

DSP-L3.3: USING ALL SIGNED-DIGIT REPRESENTATIONS TO DESIGN SINGLE INTEGER MULTIPLIERS USING SUBEXPRESSION ELIMINATION .... III - 165
Andrew G. Dempster, University of Westminster, United Kingdom; Malcolm Macleod, QinetiQ Ltd, United Kingdom

DSP-L3.4: DIGITAL FILTER DESIGN USING SUBEXPRESSION ELIMINATION AND ALL SIGNED-DIGIT REPRESENTATIONS .... III - 169
Andrew G. Dempster, University of Westminster, United Kingdom; Malcolm Macleod, QinetiQ Ltd, United Kingdom

DSP-L3.5: ON THE DESIGN OF IIR BANDPASS FILTERS WITH AN ADJUSTABLE BANDWIDTH AND CENTRE FREQUENCY .... III - 173
Håkan Johansson, Linköping University, Sweden

DSP-L4: DIGITAL FILTER BANKS

DSP-L4.1: ON COMPLETION OF M-CHANNEL PERFECT RECONSTRUCTION FILTER BANKS WITH PRESCRIBED ADMISSIBLE FIR SCALING FILTER .... III - 177
Ying-Jui Chen, Kevin Amaratunga, Massachusetts Institute of Technology, United States

DSP-L4.2: A 3D POLYPHASE-DFT CONE FILTER BANK FOR BROAD BAND PLANE WAVE FILTERING .... III - 181
Leonard Bruton, University of Calgary, Canada

DSP-L4.3: CHARACTERIZATION OF REGULAR LINEAR-PHASE PARAUNITARY FILTER BANKS USING DYADIC-BASED STRUCTURES .... III - 185
Kevin Amaratunga, Ying-Jui Chen, Massachusetts Institute of Technology, United States; Soontorn Oraintara, University of Texas, Arlington, United States

DSP-L4.4: TIME-DOMAIN CONSTRAINTS FOR THE DESIGN OF FRM-BASED COSINE-MODULATED AND MODIFIED DFT FILTER BANKS WITH LARGE NUMBER OF BANDS AND ZERO INTERSYMBOL INTERFERENCE .... III - 189
Miguel Furtado, Paulo Diniz, Sergio Netto, Federal University of Rio de Janeiro, Brazil; Tapio Saramäki, Tampere University of Technology, Finland

DSP-L4.5: M-CHANNEL LIFTING STRUCTURE FOR UNIMODULAR FILTER BANK .... III - 193
Rohit Kumar, University of Texas, Arlington, United States; Ying-Jui Chen, Massachusetts Institute of Technology, United States; Soontorn Oraintara, University of Texas, Arlington, United States; Kevin Amaratunga, Massachusetts Institute of Technology, United States

DSP-L5: DIGITAL SIGNAL PROCESSING APPLICATIONS I

DSP-L5.1: ACTIVE ARRAY BEAMFORMING USING THE FREQUENCY-RESPONSE MASKING TECHNIQUE .... III - 197
Yongzhi Liu, Zhiping Lin, Nanyang Technological University, Singapore

DSP-L5.2: AUDIO WATERMARKING USING TIME-FREQUENCY COMPRESSION EXPANSION .... III - 201
Say Wei Foo, Shuet Mun Ho, Ling Mei Ng, Nanyang Technological University, Singapore

DSP-L5.3: AN EQUATION ERROR APPROACH FOR THE DESIGN OF DIGITAL IIR COMPENSATION FILTERS IN IQ MODULATOR ..... III - 205
Anthony Lim, Victor Sreeram, Guo-Qing Wang, University of Western Australia, Australia

DSP-L5.4: FULL PARALLEL PROCESS FOR MULTIDIMENSIONAL WAVE DIGITAL FILTERING VIA MULTIDIMENSIONAL RETIMING TECHNIQUE ..... III - 209
Chien-Hsun Tseng, Stuart Lawson, University of Warwick, United Kingdom

DSP-L5.5: MULTIDIMENSIONAL WAVE DIGITAL FILTERING APPROACH FOR NUMERICAL INTEGRATION OF NON-LINEAR SHALLOW WATER EQUATIONS ..... III - 213
Chien-Hsun Tseng, Stuart Lawson, University of Warwick, United Kingdom

DSP-L6: DIGITAL SIGNAL PROCESSING APPLICATIONS II

DSP-L6.1: A HIGH-SPEED LOW-LATENCY DIGIT-SERIAL HYBRID ADDER ..... III - 217
Krister Landernäs, Johnny Holmberg, Mälardalen University, Sweden; Mark Vesterbacka, Linköping University, Sweden

DSP-L6.2: RECONSTRUCTION OF NON-UNIFORMLY SAMPLED SIGNAL USING TRANSPOSED FARROW STRUCTURE ..... III - 221
Djordje Babic, Markku Renfors, Tampere University of Technology, Finland

DSP-L6.3: DIRECT VERSUS ITERATIVE METHODS FOR FIXED-POINT IMPLEMENTATION OF MATRIX INVERSION ..... III - 225
Mikko Ylinen, Adrian Burian, Jarmo Takala, Tampere University of Technology, Finland

DSP-L6.4: A NEW SEGMENTATION TECHNIQUE FOR NOISY MULTI-COMPONENT SIGNALS USING WAVELET TRANSFORM ..... III - 229
Farook Sattar, Nanyang Technological University, Singapore; Rajamani Doraiswami, University of New Brunswick, Canada; Moe Pwint, Nanyang Technological University, Singapore

DSP-L6.5: A NOVEL COMBINED FIRST AND SECOND ORDER LAGRANGE INTERPOLATION SAMPLING PROCESS FOR A DIGITAL CLASS D AMPLIFIER ..... III - 233
Victor Adrian, Bah-Hwee Gwee, Joseph Sylvester Chang, Nanyang Technological University, Singapore

DSP-L7: MULTIDIMENSIONAL SIGNAL PROCESSING I

DSP-L7.1: COMPUTING THE TRANSFER FUNCTION FOR SECOND-ORDER 2D SYSTEMS ..... III - 237
George Antoniou, Marinos Michael, Montclair State University, United States

DSP-L7.2: 2D QUATERNION FOURIER SPECTRAL ANALYSIS AND ITS APPLICATIONS ..... III - 241
Ja-Han Chang, Soo-Chang Pei, Jian-Jiun Ding, National Taiwan University, Taiwan

DSP-L7.3: AN APPROACH TO AUTOMATIC GENERATION OF WAVE DIGITAL STRUCTURES FROM PDES ..... III - 245
Michael Vollmer, Ruhr-Universität Bochum, Germany

DSP-L7.4: MINIMIZATION OF L2-SENSITIVITY FOR 2-D STATE-SPACE DIGITAL FILTERS SUBJECT TO L2-SCALING CONSTRAINTS ..... III - 249
Takao Hinamoto, Hiroaki Ohnishi, Hiroshima University, Japan; Wu-Sheng Lu, University of Victoria, Canada

DSP-L7.5: BAYER PATTERN BASED DIGITAL ZOOMING APPROACH ..... III - 253
Rastislav Lukac, Konstantinos N. Plataniotis, University of Toronto, Canada

DSP-L8: ADAPTIVE SIGNAL PROCESSING I

DSP-L8.1: SUBBAND ADAPTIVE FILTERING WITH CRITICAL SAMPLING USING THE DATA SELECTIVE AFFINE PROJECTION ALGORITHM .... III - 257
Rogerio Guedes Alves, Clarity Technologies Inc., United States; Jose Antonio Apolinário Jr., Instituto Militar de Engenharia, Brazil; Mariane Petraglia, Federal University of Rio de Janeiro, Brazil

DSP-L8.2: TRACKING PERFORMANCE OF AN FDLMS NEAR-END CROSSTALK CANCELLER FOR xDSL SYSTEMS .... III - 261
Rajeev Nongpiur, Dale Shpak, Andreas Antoniou, University of Victoria, Canada

DSP-L8.3: AN ADAPTIVE BUTLER-CANTONI BASED TIME DELAY ESTIMATION (ABCTDE) METHOD - IIR WHITENING FILTERING APPROACH .... III - 265
Jonah Gamba, Yusuke Tsuda, Tetsuya Shimamura, Saitama University, Japan

DSP-L8.4: MSE ANALYSIS OF AN ALLPASS FILTER-BASED ADAPTIVE IIR NOTCH FILTER WITH A NORMALIZED ALGORITHM .... III - 269
Aloys Mvuma, Hiroshima University, Japan; Shotaro Nishimura, Shimane University, Japan; Takao Hinamoto, Hiroshima University, Japan

DSP-L8.5: ROBUST ADAPTIVE BEAMFORMING FOR LARGE STEERING ANGLE ERROR .... III - 273
Changzheng Ma, Zhongyuan Institute of Technology, China; Boon Poh Ng, Haoji Bao, Xuebin Yang, Nanyang Technological University, Singapore

DSP-L9: MULTIDIMENSIONAL SIGNAL PROCESSING II

DSP-L9.1: DESIGN OF MULTIDIMENSIONAL FILTER BANKS USING GRÃ–BNER BASES: A SURVEY .... III - 277
Zhiping Lin, Nanyang Technological University, Singapore; Li Xu, Akita Prefectural University, Japan; Qinghe Wu, Beijing Institute of Technology, China

DSP-L9.2: A DIRECTIONAL DECOMPOSITION: THEORY, DESIGN AND IMPLEMENTATION .... III - 281
Truong T. Nguyen, Soontorn Oraintara, University of Texas, Arlington, United States

DSP-L9.3: EFFICIENT OUTPUT-PRUNING OF THE 2-D FFT ALGORITHM .... III - 285
Saad Bouguezel, M. Omair Ahmad, M. N. S. Swamy, Concordia University, Canada

DSP-L9.4: REDUCED-ORDER REALIZATION OF FORNASINI-MARCHESINI MODEL FOR 2D SYSTEMS .... III - 289
Li Xu, Akita Prefectural University, Japan; Liankui Wu, Qinghe Wu, Beijing Institute of Technology, China; Zhiping Lin, Nanyang Technological University, Singapore; Yegui Xiao, Hiroshima Prefectural Women's University, Japan

DSP-L9.5: IMMITTANCE AND TELEPOLATION-BASED PROCEDURES TO TEST STABILITY OF CONTINUOUS-DISCRETE BIVARIATE POLYNOMIALS .... III - 293
Yuval Bistritz, Tel Aviv University, Israel

DSP-L10: ADAPTIVE SIGNAL PROCESSING II

DSP-L10.1: EFFICIENT ADAPTIVE VOLTERRA FILTERS FOR ACTIVE NONLINEAR NOISE CONTROL WITH A LINEAR SECONDARY-PATH .... III - 297
Dayong Zhou, Victor DeBrunner, University of Oklahoma, United States

DSP-L10.2: A CONVERGENCE MODEL FOR A CORDIC-BASED ARMA LATTICE FILTER .... III - 301
Shin'ichi Shiraishi, Miki Haseyama, Hideo Kitajima, Hokkaido University, Japan

DSP-L10.3: TRACKING OF LINEAR TIME VARYING SYSTEMS BY STATE-SPACE RECURSIVE LEAST-SQUARES .... III - 305
Mohammad Malik, Hafsa Qureshi, Rashid Bhatti, National University of Sciences and Technology, Pakistan

DSP-L10.4: A NEW APPROACH FOR COMPUTING CANONICAL CORRELATIONS AND COORDINATES ..... III - 309
Mohammed Hasan, University of Minnesota, Duluth, United States

DSP-L10.5: FAST ADAPTIVE IDENTIFICATION OF AUTOREGRESSIVE SIGNALS SUBJECT TO NOISE ..... III - 313
Wei Xing Zheng, University of Western Sydney, Australia

DSP-L11: DIGITAL SIGNAL PROCESSING

DSP-L11.1: A HARDWARE EFFICIENT IMPLEMENTATION OF AN ADAPTIVE SUBSAMPLE DELAY ESTIMATOR ..... III - 317
Douglas Maskell, National Taiwan University, Singapore; Graham Woods, Andrew Kerans, JCU, Australia

DSP-L11.2: DISCRETE PDF ESTIMATION IN THE PRESENCE OF NOISE ..... III - 321
Byung-Jun Yoon, Palghat P. Vaidyanathan, California Institute of Technology, United States

DSP-L11.3: ITERATIVE ALGORITHM FOR THE DESIGN OF OPTIMAL FIR ANALYSIS/SYNTHESIS FILTERS FOR OVERDECIMATED FILTER BANKS ..... III - 325
Andre Tkacenko, Palghat P. Vaidyanathan, California Institute of Technology, United States

DSP-L11.4: BEAMPATTERN SYNTHESIS FOR CONCENTRIC CIRCULAR RING ARRAY USING MMSE DESIGN ..... III - 329
Yunhong Li, Dominic K. C. Ho, University of Missouri-Columbia, United States; Chiman Kwan, Intelligent Automation Inc., United States

DSP-L11.5: ROBUST LOCAL POLYNOMIAL REGRESSION USING M-ESTIMATOR WITH ADAPTIVE BANDWIDTH ..... III - 333
S. C. Chan, Zhiguo Zhang, University of Hong Kong, Hong Kong SAR of China

DSP-L12: DIGITAL SIGNAL PROCESSING FOR COMMUNICATIONS I

DSP-L12.1: AN INTER-CARRIER INTERFERENCE SUPPPRESSION SCHEME FOR OFDM SYSTEMS IN TIME-VARYING FADING CHANNELS ..... III - 337
Shaoping Chen, South-Central University for Nationalities / Huazhong University of Science and Technology, China; Tianren Yao, Huazhong University of Science and Technology, China

DSP-L12.2: A TWO-DIMENSIONAL DOA ESTIMATION ALGORITHM FOR CDMA SYSTEM WITH PLANE ANTENNA ARRAY ..... III - 341
Wei Yang, Shiming Li, Zhenhui Tan, Northern Jiaotong University, China

DSP-L12.3: A CROSSCORRELATION PREDISTORTER USING MEMORY POLYNOMIALS ..... III - 345
André Kokkeler, University of Twente, Netherlands

DSP-L12.4: A FREQUENCY DOMAIN APPROACH FOR BLIND IDENTIFICATION WITH FILTER BANK PRECODERS ..... III - 349
Palghat P. Vaidyanathan, Bojan Vrcelj, California Institute of Technology, United States

DSP-L12.5: IMPULSIVE NOISE REDUCTION IN DSL USING A NONLINEAR TEQ ..... III - 353
Fernando Gregorio, Jose Figueroa, Juan Cousseau, Universidad Nacional del Sur, Argentina

DSP-L13: IMPLEMENTATION OF DSP ALGORITHMS

DSP-L13.1: HIGH-SPEED AREA-EFFICIENT RECURSIVE DFT/IDFT ARCHITECTURES ..... III - 357
Lan-Da Van, Chih-Chyau Yang, National Chip Implementation Center (CIC), Taiwan

DSP-L13.2: HARDWARE EFFICIENT FAST PARALLEL FIR FILTER STRUCTURES BASED ON ITERATED SHORT CONVOLUTION ..... III - 361
Chao Cheng, VIA Technologies (China), Ltd., China; Keshab K. Parhi, University of Minnesota, United States

DSP-L13.3: 2D-DCT ON FPGA BY POLYNOMIAL TRANSFORMATION IN TWO-DIMENSIONS ..... III - 365
Arturo Méndez Patiño, Instituto Tecnológico de Morelia, Mexico; Marcos Martínez Peiró, Francisco Ballester, Guillermo Payá, Universidad Politécnica de Valencia, Spain

DSP-L13.4: ACCELERATING MUSIC METHOD ON RECONFIGURABLE HARDWARE FOR SOURCE LOCALISATION ..... III - 369
Aziz Ahmedsaid, Abbes Amira, Ahmed Bouridane, Queen's University Belfast, United Kingdom

DSP-L13.5: A PERTURBATION THEORY ON STATISTICAL QUANTIZATION EFFECTS IN FIXED-POINT DSP WITH NON-STATIONARY INPUTS ..... III - 373
Changchun Shi, Robert W. Brodersen, University of California, Berkeley, United States

DSP-L14: DIGITAL SIGNAL PROCESSING FOR COMMUNICATIONS II

DSP-L14.1: THE PERFORMANCE AND ROBUST IMPLEMENTATION OF A BLIND CMOE RECEIVER FOR MC-CDMA SYSTEMS ..... III - 377
Hui Cheng, S. C. Chan, University of Hong Kong, China

DSP-L14.2: A SUBSPACE MULTIUSER BEAMFORMING ALGORITHM FOR DOWNLINK IN MOBILE COMMUNICATION SYSTEMS ..... III - 381
Nanyan Wang, Panajotis Agathoklis, Andreas Antoniou, University of Victoria, Canada

DSP-L14.3: IMPLEMENTATION OF A ZERO-FORCING RESIDUE EQUALIZER USING A LAGUERRE FILTER ARCHITECTURE ..... III - 385
Saman Abeysekera, Nanyang Technological University, Singapore

DSP-L14.4: SPURS MODELING IN DIRECT DIGITAL PERIOD SYNTHESIZERS RELATED TO PHASE ACCUMULATOR TRUNCATION ..... III - 389
Badre Izouggaghen, Abdelhakim Khouas, Yvon Savaria, École Polytechnique de Montréal, Canada

DSP-L14.5: JOINTLY MINIMUM SER TRANSMITTER AND RECEIVER FIR MIMO FILTERS FOR QAM SIGNALLING ..... III - 393
Are Hjørungnes, University of Oslo, Norway

DSP-L15: DISCRETE-TIME TRANSFORMS AND WAVELETS

DSP-L15.1: QUANTUM CIRCUIT DESIGN OF 8x8 DISCRETE HARTLEY TRANSFORM ..... III - 397
Chien-Cheng Tseng, Tsung-Ming Hwang, National Kaohsiung First University of Science and Technology, Taiwan

DSP-L15.2: ERROR ANALYSIS AND COMPLEXITY OPTIMIZATION FOR THE MULTIPLIER-LESS FFT-LIKE TRANSFORMATION (ML-FFT) ..... III - 401
K. M. Tsui, S. C. Chan, K. W. Tse, University of Hong Kong, Hong Kong SAR of China

DSP-L15.3: CONDENSED RECURSIVE STRUCTURES FOR COMPUTING MULTI-DIMENSIONAL DCT WITH ARBITRARY LENGTH ..... III - 405
Che-Hong Chen, Bin-Da Liu, Jar-Ferr Yang, National Cheng Kung University, Taiwan

DSP-L15.4: A RATIONAL SUBDIVISION SCHEME USING COSINE-MODULATED WAVELETS ..... III - 409
S. C. Chan, Xuemei Xie, University of Hong Kong, Hong Kong SAR of China

DSP-L15.5: A NOVEL PREFILTER DESIGN FOR HIGHER MULTIPLICITY DISCRETE MULTIWAVELET TRANSFORMS ..... III - 413
Tai-Chiu Hsung, Daniel Pak-Kong Lun, The Hong Kong Polytechnic University, Hong Kong SAR of China

DSP-L16: AUDIO AND SPEECH PROCESSING I

DSP-L16.1: SCALABLE ARCHITECTURE FOR WORD HMM-BASED SPEECH RECOGNITION .... III - 417
Shingo Yoshizawa, Naoya Wada, Noboru Hayasaka, Yoshikazu Miyanaga, Hokkaido University, Japan

DSP-L16.2: AN AUDIO SIGNAL SCALING TECHNIQUE USING HARMONIC GROUPING AND SHIFTING .... III - 421
Saman Abeysekera, Nanyang Technological University, Singapore; Kabi Padhi, Javed Absar, Sapna George, ST Microelectronics Pte. Ltd., Singapore

DSP-L16.3: NOISE-ROBUST AUTOMATIC SPEECH RECOGNITION USING MAINLOBE-RESILIENT TIME-FREQUENCY QUANTILE-BASED NOISE ESTIMATION .... III - 425
Siu Wa Lee, Pak-Chung Ching, Tan Lee, The Chinese University of Hong Kong, Hong Kong SAR of China

DSP-L16.4: SPEECH PITCH DETECTION IN NOISY ENVIRONMENT USING MULTI-RATE ADAPTIVE LOSSLESS FIR FILTERS .... III - 429
Dong-Yan Huang, Weisi Lin, Susanto Rahardja, Institute for Infocomm Research, Singapore

DSP-L16.5: ADAPTIVE MICROPHONE ARRAY WITH NOISE STATISTICS UPDATES .... III - 433
Hai Quang Dam, Siow Yong Low, Sven Nordholm, Hai Huyen Dam, WATRI, Australia

DSP-L17: DETECTION AND ESTIMATION

DSP-L17.1: SIGNAL DETECTION BASED ON PATTERN CLASSIFICATION FOR USE IN WIRELESS CPFSK RECEIVERS .... III - 437
Dieter Brückmann, University of Wuppertal, Germany; André Neubauer, Infineon Technologies AG, Germany

DSP-L17.2: ITERATIVE MAP MULTI-USER DETECTION OF SYNCHRONOUS CDMA WITH CHANNEL DISTORTION .... III - 441
Shannon Blunt, Naval Research Laboratory, United States; Dominic K. C. Ho, University of Missouri-Columbia, United States

DSP-L17.3: A NOVEL WIDEBAND DOA ESTIMATION TECHNIQUE BASED ON HARMONIC SOURCE MODEL FOR A UNIFORM LINEAR ARRAY .... III - 445
Yegui Xiao, Hiroshima Prefectural Women's University, Japan; Liying Ma, K. Khorasani, Concordia University, Canada

DSP-L17.4: ON NOISY FIR FILTERING VIA TOTAL LEAST SQUARES ESTIMATION .... III - 449
Wei Xing Zheng, University of Western Sydney, Australia

DSP-L17.5: SOURCE LOCALIZATION USING TDOA WITH ERRONEOUS RECEIVER POSITIONS .... III - 453
Dominic K. C. Ho, L. Kovavisaruch, H. Parikh, University of Missouri-Columbia, United States

DSP-L18: AUDIO AND SPEECH PROCESSING II

DSP-L18.1: A NOVEL METHOD TO REPRESENT THE SPEECH SIGNALS BY USING LANGUAGE AND SPEAKER INDEPENDENT PREDEFINED FUNCTIONS SETS .... III - 457
Ümit Güz, Hakan Gürkan, B. Siddik Yarman, Isik University, Turkey

DSP-L18.2: AN IMPROVED CRITICAL-BAND TRANSFORM PROCESSOR FOR SPEECH APPLICATIONS .... III - 461
Chao Wang, Yit-Chow Tong, Nanyang Technological University, Singapore

DSP-L18.3: BOOSTING AS A DIMENSIONALITY REDUCTION TOOL FOR AUDIO CLASSIFICATION .... III - 465
Sourabh Ravindran, David Anderson, Georgia Institute of Technology, United States

DSP-L18.4: MULTIPLE PITCH ESTIMATION OF POLY-PHONIC AUDIO SIGNALS IN A FREQUENCY-LAG DOMAIN USING THE BISPECTRUM ..... III - 469
Saman Abeysekera, Nanyang Technological University, Singapore

DSP-L18.5: COMPATIBLE PROBABILITY MEASURES OF THE OUTPUTS OF TEMPLATE-BASED SPEAKER IDENTIFICATION CLASSIFIERS FOR DATA FUSION ..... III - 473
Guangyu Zhou, Wasfy Mikhael, University of Central Florida, United States

DSP-P1: DIGITAL FILTERS

DSP-P1.1: DESIGN OF VARIABLE FRACTIONAL DELAY FIR FILTER USING SYMMETRY ..... III - 477
Chien-Cheng Tseng, National Kaohsiung First University of Science and Technology, Taiwan

DSP-P1.2: A NEW CONTENTION RESOLUTION ALGORITHM FOR THE DESIGN OF MINIMAL LOGIC DEPTH MULTIPLIERLESS FILTERS ..... III - 481
Fei Xu, Chip-Hong Chang, Ching-Chuen Jong, Nanyang Technological University, Singapore

DSP-P1.3: AN DFII BASED STRUCTURE FOR 2-D SEPARABLE-DENOMINATOR DIGITAL FILTERS WITH VERY LOW L2-SENSITIVITY MEASURE ..... III - 485
Zixue Zhao, Gang Li, Tao Fang, Nanyang Technological University, Singapore

DSP-P1.4: HARDWARE REDUCTION BY COMBINING PIPELINED A/D CONVERSION AND FIR FILTERING FOR CHANNEL EQUALIZATION ..... III - 489
Shahriar Shahramian, Tony Chan Carusone, University of Toronto, Canada

DSP-P1.5: DESIGN OF MULTIPLIERLESS PROGRAMMABLE LINEAR PHASE NARROWBAND-BANDPASS FIR FILTERS ..... III - 493
Kamakshi Sivaramakrishnan, Ivan Linscott, Leonard Tyler, Stanford University, United States

DSP-P2: DIGITAL SIGNAL PROCESSING I

DSP-P2.1: FAST RLS FOURIER ANALYZERS IN THE PRESENCE OF FREQUENCY MISMATCH ..... III - 497
Yegui Xiao, Hiroshima Prefectural Women's University, Japan; Liying Ma, Concordia University, Canada; Rabab K. Ward, University of British Columbia, Canada; Li Xu, Akita Prefectural University, Japan

DSP-P2.2: CHANNEL EQUALIZATION USING THE G-PROBE ..... III - 501
Constantinos Panayiotou, Andreas Spanias, Kostas Tsakalis, Arizona State University, United States

DSP-P2.3: COMBINED ECHO AND NOISE CANCELLATION BASED ON GAUSS-SEIDEL PSEUDO AFFINE PROJECTION ALGORITHM ..... III - 505
Felix Albu, University of Bucharest, Romania; H. K. Kwan, University of Windsor, Canada

DSP-P2.4: FPGA MONTGOMERY MODULAR MULTIPLICATION ARCHITECTURES SUITABLE FOR ECCS OVER GF(p) ..... III - 509
Ciaran McIvor, Máire McLoone, John McCanny, Queen's University Belfast, United Kingdom

DSP-P2.5: LOW-ORDER MODELING OF HEAD-RELATED TRANSFER FUNCTIONS USING WAVELET TRANSFORMS ..... III - 513
Julio Torres, Mariane Petraglia, Roberto Tenenbaum, Federal University of Rio de Janeiro, Brazil

DSP-P3: DIGITAL FILTERS AND FILTER BANKS

DSP-P3.1: AN EFFICIENT VLSI/FPGA ARCHITECTURE FOR COMBINING AN ANALYSIS FILTERBANK FOLLOWING A SYNTHESIS FILTERBANK ..... III - 517
Ravindra Sande, Samsung India Software Operations, India; Anantharaman Balasubramanian, Texas A&M University, United States

DSP-P3.2: ON THE PERFORMANCE OF DISCRETE-TIME HYBRID FILTER BANKS WITH COEFFICIENT ERRORS AND ARMA STOCHASTIC PROCESS INPUT .... III - 521
Marcos Aurélio de Andrade Pinheiro, CNEN/IEN, Brazil; Antonio Petraglia, UFRJ/COPPE/EE, Brazil

DSP-P3.3: ALTERNATIVE SUBBAND SIGNAL STRUCTURES FOR COMPLEX MODULATED FILTER BANKS WITH PERFECT RECONSTRUCTION .... III - 525
Ari Viholainen, Markku Renfors, Tampere University of Technology, Finland

DSP-P3.4: A NEW WINDOW FOR THE DESIGN OF COSINE-MODULATED MULTIRATE SYSTEMS .... III - 529
Pilar Martin, Fernando Cruz-Roldán, Universidad de Alcalá, Spain; Tapio Saramäki, Tampere University of Technology, Finland

DSP-P3.5: DESIGN OF COMPLEX POLYPHASE IIR MULTI-FLATTOP FILTERS .... III - 533
Artur Krukowski, Izzet Kale, University of Westminster, United Kingdom

DSP-P4: DIGITAL SIGNAL PROCESSING II

DSP-P4.1: SELECTING BETTER EEG CHANNELS FOR CLASSIFICATION OF MENTAL TASKS .... III - 537
Kouhyar Tavakolian, University of Northern British Columbia, Canada; Ali Motie Nasrabadi, Amirkabir University of Technology, Canada; Siamak Rezaei, University of Northern British Columbia, Canada

DSP-P4.2: A SIMPLE APPROACH TO THE DESIGN OF ONE-DIMENSIONAL SPARSE ARRAYS .... III - 541
Sanjit K. Mitra, University of California, Santa Barbara, United States; Mikhail Tchobanou, Moscow Power Engineering Institute, Russian Federation; Gordana Jovanovic-Dolecek, INAOE, Mexico

DSP-P4.3: MULTIDIMENSIONAL STABILITY TEST USING SUM-OF-SQUARES DECOMPOSITION .... III - 545
Bogdan Dumitrescu, Tampere University of Technology, Finland

DSP-P4.4: GRADIENT-BASED DEPTH ESTIMATION FROM 4D LIGHT FIELDS .... III - 549
Donald Dansereau, Leonard Bruton, University of Calgary, Canada

DSP-P4.5: HIGH PERFORMANCE VITERBI DECODER USING MODIFIED REGISTER EXCHANGE METHODS .... III - 553
Jae-Sun Han, Taejin Kim, Chanho Lee, Soongsil University, Republic of Korea

DSP-P5: DISCRETE-TIME TRANSFORMS AND WAVELETS

DSP-P5.1: UNIFIED SELECTABLE FIXED-COEFFICIENT RECURSIVE STRUCTURES FOR COMPUTING DCT, IMDCT, AND SUBBAND SYNTHESIS FILTERING .... III - 557
Zhan-Yuan Cheng, Che-Hong Chen, Bin-Da Liu, Jar-Ferr Yang, National Cheng Kung University, Taiwan

DSP-P5.2: IMPROVED RADIX-4 AND RADIX-8 FFT ALGORITHMS .... III - 561
Saad Bouguezel, M. Omair Ahmad, M. N. S. Swamy, Concordia University, Canada

DSP-P5.3: AN EFFICIENT SPLIT-RADIX FHT ALGORITHM .... III - 565
Saad Bouguezel, M. Omair Ahmad, M. N. S. Swamy, Concordia University, Canada

DSP-P5.4: MATHEMATICAL PROPERTIES OF THE TWO-PARAMETER FAMILY OF 9/7 BIORTHOGONAL FILTERS .... III - 569
David Tay, LaTrobe University, Australia; Slaven Marusic, Marimuthu Palaniswami, University of Melbourne, Australia; Guang Deng, LaTrobe University, Australia

DSP-P6: DIGITAL SIGNAL PROCESSING III

DSP-P6.1: IMPROVED CHANNEL ESTIMATION OVER FREQUENCY-FLAT AND RAPID FADING CHANNELS ..... III - 573
Tae Jin Hwang, Heung-Ki Baik, Chonbuk National University, Republic of Korea

DSP-P6.2: AN IMPROVED WIGNER DISTRIBUTION SYNTHESIS METHOD FOR SEPARATION OF MULTIPLE NONSTATIONARY SIGNALS ..... III - 577
Yichang Tsai, National Chung Cheng University, Taiwan; Chung J. Kuo, Delta Electronics, Inc., Taiwan

DSP-P6.3: IMPLEMENTATION OF FARROW STRUCTURE BASED INTERPOLATORS WITH SUBFILTERS OF ODD LENGTH ..... III - 581
Ali Shahed hagh ghadam, Djordje Babic, Vesa Lehtinen, Markku Renfors, Tampere University of Technology, Finland

DSP-P6.4: AN ENSEMBLE AVERAGE APPROACH TO REMOVE ADVERSE EFFECTS ON POWER SPECTRAL ESTIMATION DUE TO SAMPLING JITTERS ..... III - 585
Taikang Ning, Trinity College, United States

DSP-P6.5: A LOW-POWER FRACTIONAL DECIMATOR ARCHITECTURE FOR AN IF-SAMPLING DUAL-MODE RECEIVER ..... III - 589
Riku Uusikartano, Jarmo Takala, Tampere University of Technology, Finland

DSP-P7: DIGITAL SIGNAL PROCESSING FOR COMMUNICATIONS

DSP-P7.1: NEW COMPLEX-ARITHMETIC HETERODYNE FILTER ..... III - 593
Grace Cho, Verizon Wireless, United States; Louis Johnson, Oklahoma State University, United States; Michael Soderstrand, University of California, United States

DSP-P7.2: A GROUP-BLIND MULTIUSER RECEIVER FOR MC-CDMA SYSTEMS ..... III - 597
Hui Cheng, S. C. Chan, University of Hong Kong, China

DSP-P7.3: SWITCHING METHODS FOR LINEAR TURBO EQUALIZATION ..... III - 601
Seok-Jun Lee, Naresh Shanbhag, Andrew Singer, University of Illinois at Urbana-Champaign, United States

DSP-P7.4: BLIND MLSE BASED ON THE MATCHED FILTER ESTIMATION USING THE CMA ..... III - 605
Izzet Ozcelik, Izzet Kale, University of Westminster, United Kingdom; Buyurman Baykal, Middle East Technical University, Tunisia

DSP-P7.5: EQUALIZATION OF OFDM SYSTEMS IN TIME-VARYING CHANNELS USING FREQUENCY DOMAIN REDUNDANCY ..... III - 609
Shaoping Chen, South-Central University for Nationalities, China; Tianren Yao, Huazhong University of Science and Technology, China

DSP-P8: DIGITAL SIGNAL PROCESSING IV

DSP-P8.1: FPGA ARCHITECTURES FOR REAL-TIME 2D/3D FIR/IIR PLANE WAVE FILTERS ..... III - 613
Arjuna Madanayake, Leonard Bruton, Chris Comis, University of Calgary, Canada

DSP-P8.2: ADAPTIVE LMS PROCESSING ARCHITECTURES EMPLOYING FREQUENCY DOMAIN SUB-CONVOLUTION ..... III - 617
Andrew Gray, JPL / California Institute of Technology, United States; Scott Hoy, Parminder Ghuman, NASA / Goddard Space Flight Center, United States

DSP-P8.3: A REUSABLE IP FFT CORE FOR DSP APPLICATIONS ..... III - 621
Evaggelia Theochari, Konstantinos Tatas, Dimitrios Soudris, Democritus University of Thrace, Greece; Konstantinos Masselos, Konstantinos Potamianos, Spyros Blionas, Intracom SA, Hellenic Telecommunications Industry, Greece; Antonios Thanailakis, Democritus University of Thrace, Greece

DSP-P8.4: VLSI PROCESSOR ARCHITECTURE FOR REAL-TIME GA PROCESSING AND PE-VLSI DESIGN .... III - 625
Tetsuya Imai, Masaya Yoshikawa, Hidekazu Terai, Hironori Yamauchi, Ritsumeikan University, Japan

DSP-P8.5: AN EXTENDED GENERALIZED SIDELOBE CANCELLER IN TIME AND FREQUENCY DOMAIN .... III - 629
Zhu Liang Yu, Meng Hwa Er, Nanyang Technological University, Singapore

DSP-P9: DIGITAL SIGNAL PROCESSING V

DSP-P9.1: LINEAR DECODING ALGORITHM OBTAINED FROM NONLINEAR ANALYSIS OF SIGMA-DELTA MODULATORS .... III - 633
Ingo Wiemer, Wolfgang Schwarz, Dresden University of Technology, Germany

DSP-P9.2: A POWER-AWARE IP CORE GENERATOR FOR THE ONE-DIMENSIONAL DISCRETE FOURIER TRANSFORM .... III - 637
Chih-Da Chien, Chien-Chang Lin, Jiun-In Guo, Tien-Fu Chen, National Chung Cheng University, Taiwan

DSP-P9.3: DYNAMIC NONLINEAR THRESHOLD DECOMPOSITION ALGORITHM FOR IMPLEMENTING STACK FILTERS .... III - 641
Guangming Shi, Liya Sun, Honghua Liu, Daojun Huang, Xidian University, China

DSP-P9.4: COMPATIBLE DESIGN OF CCMP AND OCB AES CIPHER USING SEPARATED ENCRYPTOR AND DECRYPTOR FOR IEEE 802.11I .... III - 645
Ho Yung Jang, Joon Hyoung Shim, Jung Hee Suk, In Cheol Hwang, Jun Rim Choi, Kyungpook National University, Republic of Korea

DSP-P9.5: LOW-COMPLEXITY BIT-SERIAL CONSTANT-COEFFICIENT MULTIPLIERS .... III - 649
Kenny Johansson, Oscar Gustafsson, Lars Wanhammar, Linköping University, Sweden

DSP-P10: DIGITAL SIGNAL PROCESSING VI

DSP-P10.1: RECONSTRUCTION OF BANDLIMITED SIGNALS FROM NOISY DATA BY ADAPTIVE THRESHOLDING .... III - 653
Luu Nguyen, Miroslaw Pawlak, University of Manitoba, Canada

DSP-P10.2: MINIMAX DESIGN OF LINEAR-PHASE FIR FILTERS WITH ADJUSTABLE BANDWIDTHS .... III - 657
Per Löwenborg, Håkan Johansson, Linköping University, Sweden

DSP-P10.3: AN ENERGY-EFFICIENT RECONFIGURABLE ANGLE-ROTATOR ARCHITECTURE .... III - 661
Guichang Zhong, Fan Xu, Dengwei Fu, Alan N. Willson, Jr., University of California, Los Angeles, United States

DSP-P10.4: IMPLEMENTATION OF APPLICATION-SPECIFIC DSP FOR OFDM SYSTEMS .... III - 665
Myung Hoon Sunwoo, Kyung Lan Heo, Seung Keun Oh, Jeong Hoo Lee, Jong Ha Moon, Ajou university, Republic of Korea; In Ho Kim, Korea Institute of S&T Evaluation and Planning, Republic of Korea

DSP-P10.5: WIDEBAND AUDIO COMPRESSION USING WRAPPED LINEAR PREDICTION AND THE DISCRETE WAVELET TRANSFORM .... III - 669
Mohamed Deriche, King Fahd University of Petroleum & Minerals, Saudi Arabia; Daryl Ning, Queensland University of Technology, Australia

NEGS-L1: NANOELECTRONICS AND GIGASCALE SYSTEMS

NEGS-L1.1: INTEGRATED MEMS STRUCTURES AND CMOS CIRCUITS FOR BIOELECTRONIC INTERFACE WITH SINGLE CELLS .... III - 673
Natasha Reeves, Yingkai Liu, Nicole Nelson, Suvarcha Malhotra, Makeswaran Loganathan, Jean-Marie Lauenstein, Jack Chaiyupatumpa, Elisabeth Smela, Pamela Abshire, University of Maryland, United States

NEGS-L1.2: ACCURATE SIMULATION OF PHASE NOISE IN RF MEMS VCOs .... III - 677
Manas Behera, Volodymyr Kratyuk, Yutao Hu, Kartikeya Mayaram, Oregon State University, United States

NEGS-L1.3: CHARACTERIZATION OF A 16-BIT THRESHOLD LOGIC SINGLE-ELECTRON TECHNOLOGY ADDER .... III - 681
Mawahib Sulieman, Valeriu Beiu, Washington State University, United States

NEGS-L1.4: ROBUST AND FAULT-TOLERANT CIRCUIT DESIGN FOR NANOMETER-SCALE DEVICES AND SINGLE-ELECTRON TRANSISTORS .... III - 685
Alexandre Schmid, Yusuf Leblebici, Swiss Federal Institute of Technology (EPFL), Switzerland

NEGS-L1.5: SELF-ORGANIZATION AND EMERGENT MODELS IN BACTERIAL ADHESION ON ENGINEERED POLYMER SURFACES .... III - 689
Maide Bucolo, University of Catania, Italy; Santina Carnazza, University of Messina, Italy; Luigi Fortuna, Mattia Frasca, University of Catania, Italy; Salvatore Guglielmino, University of Messina, Italy; Giovanni Marletta, Cristina Satriano, University of Catania, Italy

NEGS-P1: NANOELECTRONICS AND NANOARCHITECTURE

NEGS-P1.1: ANALYSIS OF ANALOG TO DIGITAL CONVERTER BASED ON SINGLE-ELECTRON TUNNELING TRANSISTORS .... III - 693
Chaohong Hu, Delft University of Technology, Netherlands / Shanghai Jiao Tong University, China; Sorin Cotofana, Delft University of Technology, Netherlands; Jianfei Jiang, Shanghai Jiao Tong University, China

NEGS-P1.2: PROGRAMMABLE LOGIC GATE BASED ON RESONANT TUNNELLING DEVICES .... III - 697
José María Quintana, María José Avedillo, Héctor Pettenghi, Centro Nacional de Microelectrónica (CNM), Spain

NEGS-P1.3: POWDER-BASED FABRICATION TECHNIQUES FOR SINGLE-WALL CARBON NANOTUBE CIRCUITS .... III - 701
Daniel Dai, Yehea Ismail, Northwestern University, United States; Wei Wang, University of Western Ontario, Canada; Hanif Ladak, University of Western Ontario, Robarts Institute, Canada

NEGS-P1.4: RESONANT TUNNELING DIODE BASED QMOS EDGE TRIGGERED FLIP-FLOP DESIGN .... III - 705
Hui Zhang, Pinaki Mazumder, University of Michigan, United States; Kyounghoon Yang, KAIST, Republic of Korea

NEGS-P1.5: EFFECTIVENESS OF ENERGY RECOVERY TECHNIQUES IN REDUCING ON-CHIP POWER DENSITY IN MOLECULAR NANO-TECHNOLOGIES .... III - 709
Myeong-Eun Hwang, Arijit Raychowdhury, Kaushik Roy, Purdue University, United States

NEGS-P2: MODELING AND SIMULATION

NEGS-P2.1: THE TUNNELING FIELD EFFECT TRANSISTOR (TFET): THE TEMPERATURE DEPENDENCE, THE SIMULATION MODEL, AND ITS APPLICATION .... III - 713
Thomas Nirschl, Peng-Fei Wang, Walter Hansch, Doris Schmitt-Landsiedel, Munich University of Technology, Germany

NEGS-P2.2: ELECTRO-THERMAL STUDY OF NANO-INDUCTORS FOR INTEGRATED LOW POWER CONVERTERS .... III - 717
Alain Salles, Bruno Estibals, Corinne Alonso, LAAS-CNRS, France

NEGS-P2.3: AN ON-CHIP DELAY MEASUREMENT MODULE FOR NANOSTRUCTURES CHARACTERIZATION ..... III - 721
Olivier Duval, Yvon Savaria, École Polytechnique de Montréal, Canada

VSPC-L1: MOTION ESTIMATION

VSPC-L1.1: FAST VARIABLE BLOCK-SIZE MOTION ESTIMATION ALGORITHMS BASED ON MERGE AND SPLIT PROCEDURES FOR H.264/MPEG-4 AVC ..... III - 725
Zhi Zhou, Ming-Ting Sun, University of Washington, United States; Yuh-Feng Hsu, CCL/ITRI, Taiwan

VSPC-L1.2: A NOVEL KITE-CROSS-DIAMOND SEARCH ALGORITHM FOR FAST BLOCK MATCHING MOTION ESTIMATION ..... III - 729
Chi-Wai Lam, Lai-Man Po, City University of Hong Kong, Hong Kong SAR of China; Chun-Ho Cheung, Hong Kong Institute of Technology, Hong Kong SAR of China

VSPC-L1.3: A NEW SUCCESSIVE ELIMINATION ALGORITHM FOR FAST BLOCK MATCHING IN MOTION ESTIMATION ..... III - 733
Ce Zhu, Wei-Song Qi, Wee Ser, Nanyang Technological University, Singapore

VSPC-L1.4: EFFICIENT MULTI-FRAME MOTION ESTIMATION ALGORITHMS FOR MPEG-4 AVC/JVT/H.264 ..... III - 737
Mei-Juan Chen, Nationl Dong Hwa University, Taiwan; Yi-Yen Chiang, Hung-Ju Li, Ming-Chieh Chi, National Dong Hwa University, Taiwan

VSPC-L1.5: VARIABLE BLOCK SIZE MOTION ESTIMATION ALGORITHM AND ITS HARDWARE ARCHITECTURE FOR H.264/AVC ..... III - 741
Jae Hun Lee, Samsung Electronics Co., Ltd., Republic of Korea; Nam Suk Lee, Samsung Electronics Co., ltd., Republic of Korea

VSPC-L2: VIDEO OVER NETWORKS

VSPC-L2.1: ERROR RESILIENT METHODS FOR REAL-TIME MPEG-4 VIDEO STREAMING ..... III - 745
Wei-Ying Kung, Hao-Song Kong, Anthony Vetro, Huifang Sun, Mitsubishi Electric Research Laboratories, United States

VSPC-L2.2: CONTENT-BASED ADAPTIVE MEDIA PLAYER FOR NETWORK VIDEO ..... III - 749
Chih-Hao Liang, Chung-Lin Huang, National Tsing-Hua University, Taiwan

VSPC-L2.3: MACROBLOCK-BASED REVERSE PLAY ALGORITHM FOR MPEG VIDEO STREAMING ..... III - 753
Chang Hong Fu, Hong Kong Polytechnic University, Hong Kong SAR of China; Yui-Lam Chan, Wan-Chi Siu, The Hong Kong Polytechnic University, Hong Kong SAR of China

VSPC-L2.4: ERROR-RESILIENT ROI CODING USING PRE- AND POST-PROCESSING FOR VIDEO SEQUENCES ..... III - 757
Ali Jerbi, Jian Wang, UB Video Inc., Canada; Shahram Shirani, McMaster University, Canada

VSPC-L2.5: ANALYSIS OF MULTI-HYPOTHESIS MOTION COMPENSATED PREDICTION FOR ROBUST VIDEO TRANSMISSION ..... III - 761
Wei-Ying Kung, University of Southern California, United States; Chang-Su Kim, The Chinese University of Hong Kong, China; C.-C. Jay Kuo, University of Southern California, United States

VSPC-L3: TRANSCODING

VSPC-L3.1: TRANSCODING FOR PROGRESSIVE FINE GRANULARITY SCALABLE VIDEO CODING ..... III - 765
Jizheng Xu, Feng Wu, Shipeng Li, Microsoft Research Asia, China

VSPC-L3.2: THE FAST CLOSE-LOOP VIDEO TRANSCODER WITH LIMITED DRIFTING ERROR .... III - 769
Lujun Yuan, Chinese Academy of Sciences, China; Feng Wu, Qi Chen, Shipeng Li, Microsoft Research Asia, China; Wen Gao, Chinese Academy of Sciences, China

VSPC-L3.3: FRAME-SKIPPING TRANSCODING WITH MOTION CHANGE CONSIDERATION .... III - 773
Haiyan Shu, Lap-Pui Chau, Nanyang Technological University, Singapore

VSPC-L3.4: ERROR-RESILIENT TRANSCODING USING ADAPTIVE INTRA REFRESH FOR VIDEO STREAMING .... III - 777
Hong-Jyh Chiou, Yuh-Ruey Lee, Chia-Wen Lin, National Chung Cheng University, Taiwan

VSPC-L3.5: ADAPTIVE INTRA-FRAME QUANTIZATION FOR VERY LOW BIT RATE VIDEO CODING .... III - 781
Feng Pan, Z. G. Li, K. P. Lim, Xiao Lin, Susanto Rahardja, D. J. Wu, S. Wu, Institute for Infocomm Research, Singapore

VSPC-L4: ADVANCED VIDEO CODING

VSPC-L4.1: ENHANCED DIRECT MODE CODING FOR BI-PREDICTIVE PICTURES .... III - 785
Xiangyang Ji, Chinese Academy of Sciences, China; Yan Lu, Debin Zhao, Harbin Institute of Technology, China; Wen Gao, Siwei Ma, Chinese Academy of Sciences, China

VSPC-L4.2: WEIGHTED PREDICTION IN THE H.264/MPEG AVC VIDEO CODING STANDARD .... III - 789
Jill Boyce, Thomson, United States

VSPC-L4.3: SUPER HIGH RESOLUTION VIDEO CODEC SYSTEM WITH MULTIPLE MPEG-2 HDTV CODEC LSI's .... III - 793
Ken Nakamura, Takeshi Yoshitome, Yoshiyuki Yashima, NTT Corporation, Japan

VSPC-L4.4: A DATA REUSING ARCHITECTURE FOR MPEG VIDEO CODING .... III - 797
Vasily Moshnyaga, Koichi Masunaga, Naoki Kajiwara, Fukuoka University, Japan

VSPC-L4.5: A NOVEL RESYNCHRONIZATION MARKER POSITIONING APPROACH FOR ROBUST VIDEO TRANSMISSION .... III - 801
Tao Fang, Lap-Pui Chau, Nanyang Technological University, Singapore

VSPC-L5: ENCODER OPTIMIZATION

VSPC-L5.1: LOCAL VISUAL PERCEPTUAL CLUES AND ITS USE IN VIDEOPHONE RATE CONTROL .... III - 805
Xiaokang Yang, Weisi Lin, Zhongkang Lu, Xiao Lin, Susanto Rahardja, Ee Ping Ong, Susu Yao, Institute for Infocomm Research, Singapore

VSPC-L5.2: A NOVEL RATE CONTROL FOR H.264 .... III - 809
Jianfeng Xu, Yun He, Tsinghua University, China

VSPC-L5.3: IMPROVED FRAME-LAYER RATE CONTROL FOR H.264 USING MAD RATIO .... III - 813
Minqiang Jiang, Xiaoquan Yi, Nam Ling, Santa Clara University, United States

VSPC-L5.4: FAST MULTI-BLOCK SELECTION FOR H.264 VIDEO CODING .... III - 817
Andy Chang, Peter H. W. Wong, Y. M. Yeung, Oscar C. Au, Hong Kong University of Science and Technology, Hong Kong SAR of China

VSPC-L5.5: A POWER-AWARE ME ARCHITECTURE USING SUBSAMPLE ALGORITHM .... III - 821
Hsien-Wen Cheng, Lan-Rong Dung, National Chiao Tung University, Taiwan

VSPC-L6: SCALABLE VIDEO CODING

VSPC-L6.1: MULTIPLE DESCRIPTION MOTION COMPENSATION VIDEO CODING WITH LEAKY PREDICTION .... III - 825
Chien-Min Chen, Chih-Ming Chen, Yung-Chang Chen, National Tsing Hua University, Taiwan

VSPC-L6.2: STACK ROBUST FINE GRANULARITY SCALABILITY .... III - 829
Hsiang-Chun Huang, Tihao Chiang, National Chiao Tung University, Taiwan

VSPC-L6.3: AN IMPROVEMENT TO FINE GRANULARITY SCALABILITY BASED ON H.26L .... III - 833
Bin Wang, Xiaodong Gu, Hong-Jiang Zhang, Microsoft Research Asia, China

VSPC-L6.4: ADAPTIVE FINE GRANULARITY SCALABLE CODING FOR VIDEO STREAMING .... III - 837
Yunlong Gao, Lap-Pui Chau, Nanyang Technological University, Singapore

VSPC-L6.5: AN EMBEDDED WAVELET IMAGE CODER WITH PARALLEL ENCODING AND SEQUENTIAL DECODING OF BIT-PLANES .... III - 841
Yufei Yuan, Mrinal Mandal, University of Alberta, Canada

VSPC-L7: VIDEO PROCESSING

VSPC-L7.1: A FRAMEWORK FOR FULLY AUTOMATIC MOVING VIDEO-OBJECT SEGMENTATION BASED ON GRAPH PARTITIONING .... III - 845
Ibrahim Karliga, Jenq-Neng Hwang, University of Washington, United States

VSPC-L7.2: A MPEG-7-AIDED SEGMENTATION TOOL FOR CONTENT-BASED VIDEO CODING .... III - 849
Patrick Ndjiki-Nya, Oleg Novychny, Heinrich-Hertz-Institut, Germany

VSPC-L7.3: A LOOK-AHEAD METHOD FOR PAN AND ZOOM DETECTION IN VIDEO SEQUENCES USING BLOCK-BASED MOTION VECTORS IN POLAR COORDINATES .... III - 853
Adriana Dumitras, Barry G. Haskell, Apple Computer, Inc., United States

VSPC-L7.4: A VLC/FLC DATA PARTITIONING SCHEME FOR MPEG-4 .... III - 857
Jie Li, King Ngi Ngan, City University of Hong Kong, Hong Kong SAR of China; Chengji Zhao, HK Applied Science&Technology Research Institute, Hong Kong SAR of China

VSPC-L7.5: INERTIAL AND VISION HEAD TRACKER SENSOR FUSION USING A PARTICLE FILTER FOR AUGMENTED REALITY SYSTEMS .... III - 861
Fakhr-eddine Ababsa, Malik Mallem, Laboratoire Systemes Complexes, France

VSPC-L8: IMAGE COMPRESSION

VSPC-L8.1: HIGH-SPEED EBCOT WITH DUAL CONTEXT-MODELING CODING ARCHITECTURE FOR JPEG2000 .... III - 865
Jen-Shiun Chiang, Chun-Hau Chang, Yu-Sen Lin, Chang-You Hsieh, Chih-Hsien Hsia, Tamkang University, Taiwan

VSPC-L8.2: REGION OF INTEREST DETERMINED BY PERCEPTUAL-QUALITY AND RATE-DISTORTION OPTIMIZATION IN JPEG 2000 .... III - 869
Chih-Chang Chen, Oscal T.-C. Chen, National Chung Cheng University, Taiwan

VSPC-L8.3: JPEG2000 HIGH-SPEED PROGRESSIVE DECODING SCHEME .... III - 873
Hiroaki Sugita, Minh Vu, Takahiko Masuzaki, Hiroshi Tsutsui, Tomonori Izumi, Takao Onoye, Yukihiro Nakamura, Kyoto University, Japan

VSPC-L8.4: FAST SEARCH ALGORITHMS FOR ECVQ USING PROJECTION PYRAMIDS AND VARIANCE OF CODEWORDS .... III - 877
Ahmed Swilem, Kousuke Imamura, Hideo Hashimoto, Kanazawa University, Japan

VSPC-L8.5: LAPLACIAN-MODEL BASED INVERSE QUANTIZATION FOR DCT-BASED IMAGE CODEC SYSTEM .... III - 881
Kwang-deok Seo, LG Electronics, Inc., Republic of Korea; Jae-kyoon Kim, KAIST, Republic of Korea; Kook-yeol Yoo, Yeungnam University, Republic of Korea

VSPC-P1: 3-D AND IMAGE PROCESSING

VSPC-P1.1: 2D PHYSICS-BASED DEFORMABLE OBJECTS USING FAST FREE VIBRATION MODAL ANALYSIS .... III - 885
Stelios Krinidis, Ioannis Pitas, Aristotle University of Thessaloniki, Greece

VSPC-P1.2: FAIR QOS RESOURCE MANAGEMENT AND NON-LINEAR PREDICTION OF 3D RENDERING APPLICATIONS .... III - 889
Anastasios Doulamis, National Technical University of Athens, Greece

VSPC-P1.3: MULTI-REFERENCE OBJECT POSE INDEXING AND 3-D MODELING FROM VIDEO USING VOLUME FEEDBACK .... III - 893
Alireza Nasiri Avanaki, Babak Hamidzadeh, Faouzi Kossentini, Rabab K. Ward, University of British Columbia, Canada

VSPC-P1.4: PERCEPTUAL VIDEO QUALITY EVALUATION USING FUZZY INFERENCE SYSTEM .... III - 897
Susu Yao, Weisi Lin, Zhongkang Lu, Ee Ping Ong, Xiaokang Yang, Institute for Infocomm Research, Singapore

VSPC-P1.5: ITERATED CONDITIONAL MODES FOR INVERSE HALFTONING .... III - 901
Ken-Chung Ho, National United University, Taiwan

VSPC-P2: GENERAL IMAGE AND VIDEO PROCESSING I

VSPC-P2.1: THE PLENOPTIC VIDEOS: CAPTURING, RENDERING AND COMPRESSION .... III - 905
S. C. Chan, University of Hong Kong, Hong Kong SAR of China; King-To Ng, Microsoft Research Asia, China; Zhi-Feng Gan, Kin-Lok Chan, University of Hong Kong, Hong Kong SAR of China; Heung-Yeung Shum, Microsoft Research Asia, China

VSPC-P2.2: A NEW, EFFICIENT, AND FLEXIBLE ERROR DETECTION APPROACH FOR COMPRESSED VISUAL CONTENTS .... III - 909
Hua Cai, Microsoft Research Asia, China; Bing Zeng, The Hong Kong University of Science and Technology, Hong Kong SAR of China

VSPC-P2.3: COLOR IMAGE FILTERING AND ENHANCEMENT BASED ON GENETIC ALGORITHMS .... III - 913
Rastislav Lukac, Konstantinos N. Plataniotis, University of Toronto, Canada; Bogdan Smolka, Silesian University of Technology, Poland; Anastasios N. Venetsanopoulos, University of Toronto, Canada

VSPC-P2.4: TRACKING MULTIPLE POINT TARGETS USING GENETIC INTERACTING MULTIPLE MODEL BASED ALGORITHM .... III - 917
Mukesh Zaveri, Shabbir Merchant, Uday Desai, Indian Institute of Technology, Bombay, India

VSPC-P2.5: AUTOMATIC WHITE BALANCING USING STANDARD DEVIATION OF RGB COMPONENTS .... III - 921
Hong-Kwai Lam, Oscar C. Au, Chi-Wah Wong, The Hong Kong University of Science and Technology, Hong Kong SAR of China

VSPC-P3: GENERAL IMAGE AND VIDEO PROCESSING II

VSPC-P3.1: A LOCALLY-ADAPTIVE ALGORITHM FOR MEASURING BLOCKING ARTIFACTS IN IMAGES AND VIDEOS .... III - 925
Feng Pan, Xiao Lin, Susanto Rahardja, Weisi Lin, Ee Ping Ong, Susu Yao, Zhongkang Lu, Xiaokang Yang, Institute for Infocomm Research, Singapore

VSPC-P3.2: EDGE MAP GUIDED ADAPTIVE POST-FILTER FOR BLOCKING AND RINGING ARTIFACTS REMOVAL .... **III - 929**
Hao-Song Kong, Anthony Vetro, Huifang Sun, Mitsubishi Electric Research Laboratories, United States

VSPC-P3.3: IN-CAMERA DETECTION OF FABRIC DEFECTS .... **III - 933**
Ibrahim Baykal, Graham Jullien, University of Calgary, Canada

VSPC-P3.4: ADAPTIVE COLOR FILTER ARRAY DEMOSAICKING WITH ARTIFACT SUPPRESSION .... **III - 937**
Lanlan Chang, Yap-Peng Tan, Nanyang Technological University, Singapore

VSPC-P3.5: REDUCTION OF GRAY LEVEL DISTURBANCES IN PLASMA DISPLAY PANELS .... **III - 941**
Chang-Su Kim, The Chinese University of Hong Kong, Hong Kong SAR of China; Sang-Uk Lee, Seoul National University, Republic of Korea

VSPC-P4: IMAGE AND VIDEO COMPRESSION

VSPC-P4.1: RATE-DISTORTION OPTIMIZED BIT ALLOCATION FOR ERROR RESILIENT VIDEO TRANSCODING .... **III - 945**
Minghui Xia, Princeton University, United States; Anthony Vetro, Huifang Sun, Mitsubishi Electric Research Laboratories, United States; Bede Liu, Princeton University, United States

VSPC-P4.2: A NEW QUANTIZED DCT AND ITS IMPLEMENTATION ON MULTIMEDIA PROCESSOR .... **III - 949**
Ming-Yan Chan, Chi-Wah Kok, Hong Kong University of Science and Technology, Hong Kong SAR of China

VSPC-P4.3: CONTEXT-BASED BINARY ARITHMETIC CODING WITH STOCHASTIC BIT RESHUFFLING FOR ADVANCED FINE GRANUALITY SCALABILITY .... **III - 953**
Wen-Hsiao Peng, Chung-Neng Wang, Tihao Chiang, Hsueh-Ming Hang, National Chiao Tung University, Taiwan

VSPC-P4.4: A NEW SEARCHLESS FRACTAL IMAGE ENCODING METHOD FOR A REAL-TIME IMAGE COMPRESSION DEVICE .... **III - 957**
Songpol Ongwattanakul, Xianwei Wu, David Jackson, University of Alabama, United States

VSPC-P4.5: ADAPTIVE ARITHMETIC CODING FOR IMAGE PREDICTION ERRORS .... **III - 961**
Nobutaka Kuroki, Takahiro Manabe, Masahiro Numa, Kobe University, Japan

VSPC-P5: VIDEO CODING

VSPC-P5.0: A NEW DIGITAL IMAGE SCRAMBLING METHOD BASED ON FIBONACCI NUMBERS .... **III - 965**
Jiancheng Zou, Rabab K. Ward, University of British Columbia, Canada; Dongxu Qi, North China University of Technology, China

VSPC-P5.1: AN ADAPTIVE MACROBLOCK-GROUP CODING ALGORITHM FOR PROGRESSIVE AND INTERLACED VIDEO .... **III - 969**
Guoping Li, Yun He, Tsinghua University, China

VSPC-P5.2: OBJECT-BASED RATE CONTROL FOR MPEG-4 VIDEO OBJECT CODING .... **III - 973**
Zhenzhong Chen, King Ngi Ngan, The Chinese University of Hong Kong, Hong Kong SAR of China

VSPC-P5.4: A NOVEL ADAPTIVE MULTI-MODE SEARCH ALGORITHM FOR FAST BLOCK-MATCHING MOTION ESTIMATION .... **III - 977**
Yilong Liu, Soontorn Oraintara, University of Texas, Arlington, United States

Volume IV

BIO-L1: IMPLANTABLE ELECTRONICS

BIO-L1.1: A MATCHED BIPHASIC MICROSTIMULATOR FOR AN IMPLANTABLE RETINAL PROSTHETIC DEVICE ....IV - 1
Praveen Singh, North Carolina State University, United States; Wentai Liu, University of California, Santa Cruz, United States; Mohanasankar Sivaprakasam, North Carolina State University, United States; Mark Humayun, James Weiland, University of Southern California, United States

BIO-L1.2: LOW-POWER IMPLANTABLE MICROSYSTEM INTENDED TO MULTICHANNEL CORTICAL RECORDING ....IV - 5
Benoit Gosselin, Virginie Simard, Mohamad Sawan, École Polytechnique de Montréal, Canada

BIO-L1.3: DESIGN OF A LOW-POWER, IMPLANTABLE ELECTROMYOGRAM AMPLIFIER ....IV - 9
Ravi Ananth, Alfred Mann Foundation for Scientific Research, United States; Edward Lee, University of Southern California, United States

BIO-L1.4: A WAVELET BASED R-WAVE DETECTOR FOR CARDIAC PACEMAKERS IN 0.35 CMOS TECHNOLOGY ....IV - 13
Joachim Neves Rodrigues, Viktor Öwall, Leif Sörnmo, Lund University, Sweden

BIO-L1.5: A CLOSED LOOP TRANSCUTANEOUS POWER TRANSFER SYSTEM FOR IMPLANTABLE DEVICES WITH ENHANCED STABILITY ....IV - 17
Guoxing Wang, Wentai Liu, University of California, Santa Cruz, United States; Rizwan Bashirullah, Mohanasankar Sivaprakasam, Gurhan Alper Kendir, Ying Ji, North Carolina State University, United States; Mark Humayun, James Weiland, University of Southern California, United States

BIO-L2: MEDICAL SENSORS AND AMPLIFIERS

BIO-L2.1: A LOW NOISE CMOS AMPLIFIER FOR ENG SIGNALS ....IV - 21
Arantxa Uranga, Natalia Lago, Xavier Navarro, Nuria Barniol, Universitat Autònoma de Barcelona, Spain

BIO-L2.2: INTEGRATED MULTI-ELECTRODE FLUIDIC NITRIC-OXIDE SENSOR AND VLSI POTENTIOSTAT ARRAY ....IV - 25
Mihir Naware, Abhishek Rege, Johns Hopkins University, United States; Roman Genov, University of Toronto, Canada; Milutin Stanacevic, Gert Cauwenberghs, Nitish Thakor, Johns Hopkins University, United States

BIO-L2.3: A LOW-POWER CMOS NEURAL AMPLIFIER WITH AMPLITUDE MEASUREMENTS FOR SPIKE SORTING ....IV - 29
Timothy K. Horiuchi, Thomas Swindell, University of Maryland, United States; David Sander, Pennsylvania State University, United States; Pamela Abshire, University of Maryland, United States

BIO-L2.4: LOW NOISE AMPLIFIER FOR RECORDING ENG SIGNALS IN IMPLANTABLE SYSTEMS ....IV - 33
Jordi Sacristán, M Teresa Osés, CNM-IMB, Spain

BIO-L2.5: A LOW-POWER IMPLANTABLE PSEUDO-BJT-BASED SILICON RETINA WITH SOLAR CELLS FOR ARTIFICIAL RETINAL PROSTHESES ....IV - 37
Chung-Yu Wu, Felice Cheng, Cheng-Ta Chiang, National Chiao Tung University, Taiwan; Po-Kang Lin, Taipei Veterans General Hospital, Taiwan

BIO-P1: BIOMEDICAL CIRCUITS AND SYSTEMS I

BIO-P1.1: AN EFFICIENT INDUCTIVE POWER LINK DESIGN FOR RETINAL PROSTHESIS ....IV - 41
Gurhan Alper Kendir, North Carolina State University, United States; Wentai Liu, University of California, Santa Cruz, United States; Rizwan Bashirullah, North Carolina State University, United States; Guoxing Wang, University of California, Santa Cruz, United States; Mark Humayun, James Weiland, University of Southern California, United States

BIO-P1.2: CRANIOFACIAL LANDMARKS EXTRACTION BY PARTIAL LEAST SQUARES REGRESSION ....IV - 45
Idris El-Feghi, Yasser Alginahi, Maher Sid-Ahmed, Majid Ahmadi, University of Windsor, Canada

BIO-P1.3: FUZZY VECTOR FILTERS FOR MICROARRAY IMAGE ENHANCEMENT ....IV - 49
Rastislav Lukac, Konstantinos N. Plataniotis, University of Toronto, Canada; Bogdan Smolka, Silesian University of Technology, Poland; Anastasios N. Venetsanopoulos, University of Toronto, Canada

BIO-P1.4: CMOS MICROELECTRODE ARRAY FOR EXTRACELLULAR STIMULATION AND RECORDING OF ELECTROGENIC CELLS ....IV - 53
Flavio Heer, Wendy Franks, Ian McKay, Stefano Taschini, Andreas Hierlemann, Henry Baltes, ETH Zuerich, Switzerland

BIO-P1.5: A C-LESS ASK DEMODULATOR FOR IMPLANTABLE NEURAL INTERFACING CHIPS ....IV - 57
Chua-Chin Wang, Ya-Hsin Hsueh, U. Fat Chio, Yu-Tzu Hsiao, National Sun Yat-Sen University, Taiwan

BIO-P2: BIOMEDICAL CIRCUITS AND SYSTEMS II

BIO-P2.1: A COMPARISON OF TWO SIMILARITY MEASURES IN INTENSITY-BASED ULTRASOUND IMAGE REGISTRATION ....IV - 61
Shuang Gao, Yang Xiao, Shao-hai Hu, Beijing Jiaotong University, China

BIO-P2.2: NONLINEAR TECHNIQUES AND NEURAL ACTIVITY: EMERGENT TRENDS IN MEG DATA ....IV - 65
Manuela La Rosa, Maide Bucolo, University of Catania, Italy; Gea Bucolo, Luigi Fortuna, Università di Catania, Italy; Mattia Frasca, University of Catania, Italy; David Shannahoff-Khalsa, University of California, San Diego, United States; Massimiliano Sorbello, University of Catania, Italy

BIO-P2.3: A NOVEL REPRESENTATION METHOD FOR ELECTROMYOGRAM (EMG) SIGNAL WITH PREDEFINED SIGNATURE AND ENVELOPE FUNCTIONAL BANK ....IV - 69
Hakan Gürkan, Ümit Güz, B. Siddik Yarman, Isik University, Turkey

COMM-L1: COMMUNICATION ARCHITECTURES

COMM-L1.1: MEMORY OPTIMIZATION TECHNIQUES FOR UMTS CODE GENERATION ....IV - 73
Daniele Lo Iacono, Ettore Messina, Giuseppe Avellone, Agostino Galluzzo, STMicroelectronics, Italy

COMM-L1.2: LOW COMPLEXITY SYSTEM-ON-CHIP ARCHITECTURES OF PARALLEL-RESIDUE-COMPENSATION IN CDMA SYSTEMS ....IV - 77
Yuanbin Guo, Dennis McCain, Nokia, Inc., United States; Joseph Cavallaro, Rice University, United States

COMM-L1.3: POWER EFFICIENT ARCHITECTURE FOR (3,6)-REGULAR LOW-DENSITY PARITY-CHECK CODE DECODER ....IV - 81
Yijun Li, Mahmoud Elassal, Magdy Bayoumi, University of Louisiana at Lafayette, United States

COMM-L1.4: AN EFFICIENT ARCHITECTURE FOR PEAK-TO-AVERAGE POWER RATIO REDUCTION IN OFDM SYSTEMS IN THE PRESENCE OF PULSE-SHAPING FILTERING ....IV - 85
Theodoros Giannopoulos, Vassilis Paliouras, University of Patras, Greece

COMM-L1.5: AN EFFICIENT MEMORY COMPRESSION SCHEME FOR 8k FFT IN A DVB-T RECEIVER AND THE CORRESPONDING ERROR MODEL .... IV - 89
Stamatis Krommydas, Vassilis Paliouras, University of Patras, Greece

COMM-L2: CDMA SYSTEMS

COMM-L2.1: A NOVEL MULTIPATH SEARCHER IMPLEMENTATION FOR WCDMA RECEIVERS .... IV - 93
Eugene Grayver, Eugene Grayver, United States; Ahmed ElTawil, Jean-François Frigon, University of California, Los Angeles, United States; Kambiz Shoarinejad, Ali-Azam Abbasfar, Danijela Cabric, Innovics Wireless, United States

COMM-L2.2: LOW POWER FLEXIBLE RAKE RECEIVERS FOR WCDMA .... IV - 97
Boris Andreev, Edward Titlebaum, Eby G. Friedman, University of Rochester, United States

COMM-L2.3: A SIMPLE STF-OFDM TRANSMISSION SCHEME WITH MAXIMUM FREQUENCY DIVERSITY GAIN .... IV - 101
Sang-Soon Park, Han-Kyoung Kim, Heung-Ki Baik, Chonbuk National University, Republic of Korea

COMM-L2.4: LOW-POWER DESIGN FOR CELL SEARCH IN W-CDMA .... IV - 105
Chi-Fang Li, Yuan-Sun Chu, National Chung Cheng University, Taiwan; Wern-Ho Sheen, National Chiao Tung University, Taiwan

COMM-L2.5: IMPROVEMENT OF FUZZY POWER CONTROL FOR DS-CDMA CELLULAR MOBILE SYSTEM HAVING VARIABLE NUMBER OF USERS .... IV - 109
Wasimon Panichpattanakul, Watit Bejapolakul, Chulalongkorn University, Thailand

COMM-L3: ULTRA WIDE BAND SYSTEMS

COMM-L3.1: ANALYSIS OF HIGHER-ORDER N-TONE SIGMA-DELTA MODULATORS FOR ULTRA WIDEBAND COMMUNICATIONS .... IV - 113
Kai-Chuan Chang, Gerald Sobelman, Ebrahim Saberinia, Ahmed Tewfik, University of Minnesota, United States

COMM-L3.2: 3-STATE PSEUDORANDOM NOISE (PN) SEQUENCE 3-PULSE REFERENCE SHARING ULTRA WIDEBAND SYSTEM .... IV - 117
Chun Yi Lee, Chris Toumazou, Imperial College London, United Kingdom

COMM-L3.3: REFERENCE SHARING ULTRA WIDEBAND COMMUNICATION SYSTEM .... IV - 121
Chun Yi Lee, Chris Toumazou, Imperial College London, United Kingdom

COMM-L3.4: DESIGN OF A LOW-COMPLEXITY RECEIVER FOR IMPULSE-RADIO ULTRA-WIDEBAND COMMUNICATION SYSTEMS .... IV - 125
Chia-Hsiang Yang, Yu-Hsuan Lin, Shih-Chun Lin, Tzi-Dar Chiueh, National Taiwan University, Taiwan

COMM-L3.5: A CMOS IMPULSE GENERATOR FOR UWB WIRELESS COMMUNICATION SYSTEMS .... IV - 129
Youngkyun Jeong, University of Texas, Arlington, United States; Sungyong Jung, Universtiy of Texas at Arlington, United States; Jin Liu, University of Texas, Dallas, United States

COMM-L4: OSCILLATOR DESIGN

COMM-L4.1: THE DESIGN OF A 14GHZ I/Q RING OSCILLATOR IN 0.18μm CMOS .... IV - 133
Yalcin Eken, John Uyemura, Georgia Institute of Technology, United States

COMM-L4.2: A MUTUAL-NEGATIVE-RESISTANCE QUADRATURE CMOS LC OSCILLATOR .... IV - 137
Apisak Worapishet, S. Virunphun, M. Chongcheawchamnan, S. Srisathit, Mahanakorn University of Technology, Thailand

COMM-L4.3: ANALYSIS OF OSCILLATOR AMPLITUDE CONTROL, AND ITS APPLICATION TO AUTOMATIC TUNING OF QUALITY FACTOR FOR ACTIVE LC FILTERS ....IV - 141
Shaorui Li, Yannis Tsividis, Columbia University, United States

COMM-L4.4: A WIDE TUNING RANGE VCO USING CAPACITIVE SOURCE DEGENERATION....................IV - 145
Byunghoo Jung, Ramesh Harjani, University of Minnesota, United States

COMM-L4.5: BEHAVIOR MODELLING FOR LC TANK BASED OSCILLATORS..IV - 149
Tamer Riad, Raafat Mansour, University of Waterloo, Canada

COMM-L5: PHASE LOCKED LOOPS (PLL) CIRCUITS AND ARCHITECTURES

COMM-L5.1: ON THE DESIGN OF AN OFFSET-PLL MODULATION LOOP FOR THE EGSM BAND ....IV - 153
Amr Hafez, Cairo University, Egypt; Waleed Aboueldahab, Ahmed Helmy, MEMSCAP Egypt, Egypt

COMM-L5.2: PHASE LOCKED LOOP GAIN SHAPING FOR GIGAHERTZ OPERATION..................................IV - 157
Krzysztof Iniewski, Simon Fraser University, Canada; Sebastian Magierowski, University of Toronto, Canada; Marek Syrzycki, Simon Fraser University, Canada

COMM-L5.3: PHASE-LOCKED LOOP ARCHITECTURE FOR ADAPTIVE JITTER OPTIMIZATION ....IV - 161
Socrates Vamvakos, University of California, Berkeley, United States; Carl Werner, Rambus, Inc., United States; Borivoje Nikolic, University of California, Berkeley, United States

COMM-L5.4: FAST-SWITCHING ANALOG PLL WITH FINITE-IMPULSE RESPONSE.....................................IV - 165
Salvatore Levantino, Luca Romanò, Carlo Samori, Politecnico di Milano, Italy; Andrea Lacaita, Politecnico di Milano / IFN-CNR Sez. Milano, Italy

COMM-L5.5: A NOVEL ULTRA HIGH-SPEED FLIP-FLOP-BASED FREQUENCY DIVIDER............................IV - 169
Ravindran Mohanavelu, Payam Heydari, University of California, Irvine, United States

COMM-L6: DECODING FOR COMMUNICATION SYSTEMS

COMM-L6.1: REED-SOLOMON BEHAVIORAL VIRTUAL COMPONENT FOR COMMUNICATION SYSTEMS ....IV - 173
Emmanuel Casseau, Bertrand Le Gal, UBS University, France; Christophe Jego, ENST Bretagne, France; Nathalie Le Heno, Turbo Concept, France; Eric Martin, UBS University, France

COMM-L6.2: A HIGH THROUGHPUT LIMITED SEARCH TRELLIS DECODER FOR CONVOLUTIONAL CODE DECODING ....IV - 177
Tong Zhang, Rensselaer Polytechnic Institute, United States

COMM-L6.3: A LOW-POWER, HARD-DECISION ANALOGUE CONVOLUTIONAL DECODER USING THE MODIFIED FEEDBACK DECODING ALGORITHM ....IV - 181
Billy Tomatsopoulos, Andreas Demosthenous, University College London, United Kingdom

COMM-L6.4: A NEW DIGITAL SIGNATURE SCHEME..IV - 185
Alaa Eldin Fahmy, Wael Badawy, University of Calgary, Canada

COMM-L6.5: SCALEABLE CHECK NODE CENTRIC ARCHITECTURE FOR LDPC DECODER......................IV - 189
Rohit Singhal, Gwan Choi, Nathan Mickler, Prabhavati Koteeswaran, Texas A&M University, United States

COMM-L7: CIRCUITS FOR COMMUNICATIONS

COMM-L7.1: DIRECT SIGMA DELTA GMSK MODULATOR MODELING AND DESIGN FOR 2.5G TX APPLICATIONS ....IV - 193
Nicola Lofù, Gianfranco Avitabile, Politecnico di Bari, Italy; Biagio Bisanti, Stefano Cipriani, Texas Instruments France, France

COMM-L7.2: A RADIO-FREQUENCY CMOS ACTIVE INDUCTOR AND ITS APPLICATION IN DESIGNING HIGH-Q FILTERS ....IV - 197
Haiqiao Xiao, Rolf Schaumann, W. Robert Daasch, Phillip Wong, Branimir Pejcinovic, Portland State University, United States

COMM-L7.3: IMPLEMENTATION OF NOVEL SAMPLING AND RECONSTRUCTION CIRCUITS IN DIGITAL RADIOS ....IV - 201
Yefim Poberezhskiy, Rockwell Scientific, United States; Gennady Poberezhskiy, Raytheon Company, United States

COMM-L7.4: A HIGH SPEED TRANS-IMPEDANCE AMPLIFIER USING 0.13 μM TRIPLE-WELL CMOS TECHNOLOGY ....IV - 205
Noushin Riahi, Alzahra University, Iran; Ali Fotowat Ahmady, Unistar Micro Technology Inc., Canada; Lawrence Loh, Silicon Bridge, Inc., United States

COMM-L7.5: 6.8MW 2.5GB/S AND 42.5MW 5GB/S 1:8 CMOS DEMULTIPLEXERS ....IV - 209
Shanfeng Cheng, José Silva-Martínez, Texas A&M University, United States

COMM-L8: OPTICAL COMMUNICATION

COMM-L8.1: FOUR-CHANNEL SiGe TRANSIMPEDANCE AMPLIFIER ARRAY FOR PARALLEL OPTICAL INTERCONNECTS ....IV - 213
Sung Min Park, Ewha Womans University, Republic of Korea

COMM-L8.2: CMOS LIMITING OPTICAL PREAMPLIFIERS USING DYNAMIC BIASING FOR WIDE DYNAMIC RANGE ....IV - 217
Sharon Goldberg, Stephen Liu, Sean Nicolson, Khoman Phang, University of Toronto, Canada

COMM-L8.3: IMPROVED MULTICAST SWITCH ARCHITECTURE FOR OPTICAL CABLE TELEVISION AND VIDEO SURVEILLANCE NETWORKS ....IV - 221
Heikki Kariniemi, Jari Nurmi, Tampere University of Technology, Finland

COMM-L8.4: RAPID BIT-ERROR-RATE MEASUREMENTS OF INFRARED COMMUNICATION SYSTEMS ....IV - 225
Meng-Lin Hsia, Oscal T.-C. Chen, National Chung Cheng University, Taiwan; Huang-Tsung Chan, Sun-Chen Wang, Yaw-Tyng Wu, Chung-Cheng University, Taiwan

COMM-L8.5: EFFICIENT ANALYSIS OF MULTILAYERED BROADSIDE EDGE-COUPLED ANISOTROPIC STRUCTURES FOR MICROWAVE APPLICATIONS ....IV - 229
Mohamed Lamine Tounsi, Abdfelhamid Khodja, USTHB University, Algeria; Mustapha Chérif-Eddine Yagoub, University of Ottawa, Canada

COMM-L9: FREQUENCY SYNTHESIZERS

COMM-L9.1: DIRECT DIGITAL FREQUENCY SYNTHESIZER WITH MULTI-STAGE LINEAR INTERPOLATION ....IV - 233
Hiroomi Hikawa, Oita University, Japan

COMM-L9.2: A 5.1-GHz CMOS PLL BASED INTEGER-N FREQUENCY SYNTHESIZER WITH RIPPLE-FREE CONTROL VOLTAGE AND IMPROVED ACQUISITION TIME ....IV - 237
Sadeka Ali, Martin Margala, University of Rochester, United States

COMM-L9.3: HIGH-SPEED RF MULTI-MODULUS PRESCALER ARCHITECTURE FOR SIGMA-DELTA FRACTIONAL-N PLL FREQUENCY SYNTHESIZERS ....IV - 241
Ahmed Wafa, Ayman Ahmed, MEMSCAP Egypt, Egypt

COMM-L9.4: A NEW PRESCALER FOR FULLY INTEGRATED 5-GHZ CMOS FREQUENCY SYNTHESIZER ....IV - 245
Chin-Sheng Chen, Robert C. Chang, National Chung Hsing University, Taiwan

COMM-L9.5: A 5-GHZ DELTA-SIGMA PLL FREQUENCY SYNTHESIZER FOR WLAN APPLICATIONS .... IV - 249
Sau-Mou Wu, Wei-Liang Chen, Yuan Ze University, Taiwan

COMM-L10: RECEIVERS ARCHITECTURE AND DESIGN

COMM-L10.1: CMOS INTEGRATED TRANSFORMER-FEEDBACK Q-ENHANCED LC BANDPASS FILTER FOR WIRELESS RECEIVERS .... IV - 253
Wesley Gee, Phillip Allen, Georgia Institute of Technology, United States

COMM-L10.2: AN IIP2 CALIBRATION TECHNIQUE FOR DIRECT CONVERSION RECEIVERS .... IV - 257
Mikko Hotti, Jussi Ryynänen, Helsinki University of Technology, Finland; Kalle Kivekäs, Nokia Research Center, Finland; Kari Halonen, Helsinki University of Technology, Finland

COMM-L10.3: UMTS/GSM MULTI MODE RECEIVER DESIGN .... IV - 261
Horst Fischer, Frank Henkel, Michael Engels, Peter Waldow, IMST GmbH, Germany

COMM-L10.4: A CMOS MULTI-STANDARD RECEIVER ARCHITECTURE FOR ISM AND UNII BAND APPLICATIONS .... IV - 265
Ho-Kwon Yoon, Mohammed Ismail, The Ohio State University, United States

COMM-L10.5: A 5GHZ DIRECT-CONVERSION RECEIVER WITH DC OFFSET CORRECTION .... IV - 269
Paul Laferriere, Carleton University, Canada; Dave Rahn, Cognio Canada, Inc., Canada; Calvin Plett, John Rogers, Carleton University, Canada

COMM-L11: RF AMPLIFIERS FOR COMMUNICATIONS

COMM-L11.1: GAIN BANDWIDTH CONSIDERATIONS IN FULLY INTEGRATED DISTRIBUTED AMPLIFIERS IMPLEMENTED IN SILICON .... IV - 273
Rony Amaya, Jorge Aguirre, Calvin Plett, Carleton University, Canada

COMM-L11.2: AN EXACT ANALYSIS OF CLASS-E POWER AMPLIFIERS FOR RF COMMUNICATIONS .... IV - 277
Brett Klehn, Syed Islam, Rochester Institute of Technology, United States

COMM-L11.3: A POWER CONSTRAINED SIMULTANEOUS NOISE AND INPUT MATCHING LOW NOISE AMPLIFIER DESIGN TECHNIQUE .... IV - 281
Trung-Kien Nguyen, Yang-Moon Su, Sang-Gug Lee, Information and Communications University, Republic of Korea

COMM-L11.4: INDUCTIVE PEAKING IN WIDEBAND CMOS CURRENT AMPLIFIERS .... IV - 285
Bendong Sun, University of Waterloo, Canada; Fei Yuan, Ryerson University, Canada; Ajoy Opal, University of Waterloo, Canada

COMM-L11.5: A DC-6 GHZ, 50 DB DYNAMIC RANGE, SIGE HBT TRUE LOGARITHMIC AMPLIFIER .... IV - 289
Chris Holdenried, TRLabs, Canada; Jim Haslett, University of Calgary, Canada

COMM-L12: CLOCK/DATA RECOVERY

COMM-L12.1: A 5-Gb/s 1/8-RATE CMOS CLOCK AND DATA RECOVERY CIRCUIT .... IV - 293
Jin Kyu Kwon, Tae Kwan Heo, Sang-Bock Cho, University of Ulsan, Republic of Korea; Sung Min Park, Ewha Womans University, Republic of Korea

COMM-L12.2: A HIGH-SPEED LOW-VOLTAGE PHASE DETECTOR FOR CLOCK RECOVERY FROM NRZ DATA .... IV - 297
Francesco Centurelli, Università di Roma, Italy; Massimo Pozzoni, STMicroelectronics, Italy; Giuseppe Scotti, Alessandro Trifiletti, Università di Roma, Italy

COMM-L12.3: A CMOS 10GB/S CLOCK AND DATA RECOVERY CIRCUIT WITH A NOVEL ADJUSTABLE KPD PHASE DETECTOR .... IV - 301
Xinyu Chen, Michael Green, University of California, Irvine, United States

COMM-L12.4: A 1/8-RATE CLOCK AND DATA RECOVERY ARCHITECTURE FOR HIGH-SPEED COMMUNICATION SYSTEMS .... IV - 305
Pedram Sameni, University of British Columbia, Canada; Shahriar Mirabbasi, Univeristy of British Columbia, Canada

COMM-L12.5: AN INTERPOLATION FILTER BASED ON SPLINE FUNCTIONS FOR NON-SYNCHRONIZED TIMING RECOVERY .... IV - 309
Afshin Haftbaradaran, Ken Martin, Universtiy of Toronto, Canada

COMM-L13: RF CIRCUITS AND SYSTEMS

COMM-L13.1: A LINEARIZATION TECHNIQUE FOR RF LOW NOISE AMPLIFIER .... IV - 313
Chunyu Xin, Edgar Sánchez-Sinencio, Texas A&M University, United States

COMM-L13.2: A SELF-SYNCHRONIZED RF-INTERCONNECT FOR 3-DIMENSIONAL INTEGRATED CIRCUITS .... IV - 317
Qun Gu, Zhiwei Xu, Jenwei Ko, Szukang Hsien, M. Frank Chang, University of California, Los Angeles, United States

COMM-L13.3: 2.4 GHz RF DOWN-CONVERSION MIXERS IN STANDARD CMOS TECHNOLOGY .... IV - 321
Arif Siddiqi, Tadeusz Kwasniewski, Carleton University, Canada

COMM-L13.4: A 0.18um CMOS 900 MHZ RECEIVER FRONT-END USING RF Q-ENHANCED FILTERS .... IV - 325
Chris DeVries, Ralph Mason, Carleton University, Canada

COMM-L13.5: ADAPTABLE MOS CURRENT MODE LOGIC FOR USE IN A MULTI-BAND RF PRESCALER .... IV - 329
Mark Houlgate, Daniel Olszewski, Karim Abdelhalim, Leonard MacEachern, Carleton University, Canada

COMM-P1: COMMUNICATION ARCHITECTURES AND SYSTEMS I

COMM-P1.1: A NOVEL MEMORYLESS AES CIPHER ARCHITECTURE FOR NETWORKING APPLICATIONS .... IV - 333
Yeong-Kang Lai, Li-Chung Chang, National Chung Hsing University, Taiwan

COMM-P1.2: COUPLING EFFECTS IN AN INTEGRATED BEAM-FORMING TRANSMITTER .... IV - 337
Svetoslav Gueorguiev, Saska Lindfors, Aalborg University, Denmark

COMM-P1.3: ENHANCED PARALLEL INTERFERENCE CANCELLATION USING DECORRELATOR FOR THE BASE-STATION RECEIVER .... IV - 341
Archana Chidanandan, Rose-Hulman Institute of Technology, United States; Magdy Bayoumi, University of Louisiana at Lafayette, United States

COMM-P1.4: DIGITAL RECEIVER ARCHITECTURES FOR THE 802.15.4 STANDARD .... IV - 345
Nicola Scolari, Swiss Federal Institute of Technology (EPFL), Switzerland; Christian Enz, Swiss Federal Institute of Technology (EPFL) / Centre Suisse d'Electronique et de Microtechnique, Switzerland

COMM-P2: COMMUNICATION CIRCUITS DESIGN I

COMM-P2.1: TRANSCEIVER CIRCUITS FOR PULSE-BASED ULTRA-WIDEBAND .... IV - 349
Takahide Terada, Shingo Yoshizumi, Yukitoshi Sanada, Tadahiro Kuroda, Keio University, Japan

COMM-P2.2: DESIGN AND COMPARISON OF CMOS CURRENT MODE LOGIC LATCHES .... IV - 353
Muhammad Usama, Tadeusz Kwasniewski, Carleton University, Canada

COMM-P2.3: A FAST-LOCK DLL WITH POWER-ON RESET CIRCUIT .... IV - 357
Kuo-Hsing Chen, Yu-Lung Lo, National Central University, Taiwan

COMM-P2.4: HIGH-SPEED AND WIDE-TUNING-RANGE LC FREQUENCY DIVIDERS .... IV - 361
Ken Yamamoto, Takayasu Norimatsu, Minoru Fujishima, University of Tokyo, Japan

COMM-P2.5: IMPROVING THE ACQUISITION TIME OF A PLL-BASED, INTEGER-N FREQUENCY SYNTHESIZER .... IV - 365
Syed Irfan Ahmed, Ralph Mason, Carleton University, Canada

COMM-P3: COMPUTATION KERNELS AND IP FOR COMMUNICATION SYSTEMS I

COMM-P3.1: SHUNT-PEAKING IN MCML GATES AND ITS APPLICATION IN THE DESIGN OF A 20 GB/S HALF-RATE PHASE DETECTOR .... IV - 369
Hung Tien Bui, Yvon Savaria, École Polytechnique de Montréal, Canada

COMM-P3.2: A NOVEL 1.5V CMFB CMOS DOWN-CONVERSION MIXER DESIGN FOR IEEE 802.11A WLAN SYSTEMS .... IV - 373
Xuezhen Wang, Robert Weber, Degang Chen, Iowa State University, United States

COMM-P3.3: A CMOS LOW POWER BUFFER BASED BANDPASS IF FILTER FOR BLUETOOTH .... IV - 377
Hussain Al-Zaher, Mohammad K. Al-Ghamdi, King Fahd University of Petroleum & Minerals, Saudi Arabia

COMM-P3.4: BANDPASS SIGMA-DELTA (S-D) ARCHITECTURE BASED EFFICIENT FM DEMODULATOR FOR SOFTWARE RADIO .... IV - 381
Saman Abeysekera, Nanyang Technological University, Singapore

COMM-P3.5: A LOW-POWER LOW-VOLTAGE NMOS BULK-MIXER WITH 20 GHZ BANDWIDTH IN 90 NM CMOS .... IV - 385
Christoph Kienmayer, Technical University of Vienna, Austria / Infineon Technologies AG, Germany; Marc Tiebout, Werner Simbürger, Infineon Technologies AG, Germany; Arpad L. Scholtz, Technical University of Vienna, Austria

COMM-P4: COMMUNICATION CIRCUITS DESIGN II

COMM-P4.1: A SIMPLE AND ROBUST SUPER-REGENERATIVE OSCILLATOR FOR THE 2.4GHZ ISM BAND .... IV - 389
Pere Palà-Schönwälder, F. Xavier Moncunill-Geniz, Francisco del Águila López, Jordi Bonet-Dalmau, M. Rosa Giralt-Mas, Universitat Politècnica de Catalunya, Spain

COMM-P4.2: A 10GBASE-LX4 RECEIVER FRONT END TRANSIMPEDANCE AMPLIFIER AND LIMITING AMPLIFIER .... IV - 393
Hung-Chieh Tsai, Jyh-Yih Yeh, Wei-Hsuan Tu, Tai-Cheng Lee, Chorng-Kuang Wang, National Taiwan University, Taiwan

COMM-P4.3: SIMULTANEOUS BI-DIRECTIONAL SIGNALING WITH ADAPTIVE PRE-EMPHASIS .... IV - 397
Ming-ta Hsieh, Gerald Sobelman, University of Minnesota, United States

COMM-P4.4: 1000BASE-T GIGABIT ETHERNET BASEBAND DSP IC DESIGN .... IV - 401
Hsiu-ping Lin, Nancy Fang-Yih Chen, Jyh-Ting Lai, An-Yeu (Andy) Wu, National Taiwan University, Taiwan

COMM-P4.5: A NEW FULL CMOS 2.5-V TWO-STAGE LINE DRIVER WITH VARIABLE GAIN FOR ADSL APPLICATIONS ....IV - 405
Saeid Mehrmanesh, Sharif University of Technology, Iran; Seyed Mojtaba Atarodi, MixeCore Design, United States; Hesam Aslanzadeh, Saeed Saeedi, Amin Quasem Safarian, Sharif University of Technology, Iran

COMM-P5: COMPUTATION KERNELS AND IP FOR COMMUNICATION SYSTEMS II

COMM-P5.1: A CLASS A/B LOW POWER AMPLIFIER FOR WIRELESS SENSOR NETWORKS ....IV - 409
Yuen-Hui Chee, Jan Rabaey, Ali M. Niknejad, University of California, Berkeley, United States

COMM-P5.2: CHARACTERISTICS AND MODELING OF A BROADBAND TRANSMISSION-LINE TRANSFORMER ....IV - 413
Arto Malinen, Kari Stadius, Kari Halonen, Helsinki University of Technology, Finland

COMM-P5.3: A LOW-IF/ZERO-IF RECONFIGURABLE RECEIVER WITH TWO-STEP CHANNEL SELECTION TECHNIQUE FOR MULTISTANDARD APPLICATIONS ....IV - 417
Pui-In Mak, University of Macau, Macao SAR of China; Seng-Pan U., University of Macau / Chipidea Microelectronics Ltd., Macao SAR of China; Rui Paulo Martins, University of Macau, Macao SAR of China

COMM-P5.4: A LOW-POWER, 10 GHZ BACK-GATED TUNED VOLTAGE CONTROLLED OSCILLATOR WITH AUTOMATIC AMPLITUDE AND TEMPERATURE COMPENSATION ....IV - 421
Rizwan Murji, Jamal Deen, McMaster University, Canada

COMM-P5.5: JITTER IN HIGH-SPEED SERIAL AND PARALLEL LINKS ....IV - 425
Pavan Kumar Hanumolu, Oregon State University, United States; Bryan Casper, Randy Mooney, Intel Corporation, United States; Gu-Yeon Wei, Harvard University, United States; Un-Ku Moon, Oregon State University, United States

COMM-P6: WIRELESS SYSTEMS AND HIGH SPEED SYSTEMS

COMM-P6.1: EXPERIMENTAL RESULTS OF A TYPE-BASED PREDISTORTER FOR SSPA LINEARIZATION ....IV - 429
Xinping Huang, Pierre Tardif, Mario Caron, Communications Research Centre Canada, Canada

COMM-P6.2: R&D PLATFORM FOR MAN SCALE WIRELESS AD HOC NETWORKS - TOWARDS A HARDWARE DEMONSTRATOR FOR 4G+ SYSTEMS - ....IV - 433
Peter Rauschert, Arasch Honarbacht, Anton Kummert, University of Wuppertal, Germany

COMM-P6.3: OPTIMIZATION METHOD FOR DESIGNING FILTER BANK CHANNELIZER OF A SOFTWARE DEFINED RADIO USING VERTICAL COMMON SUBEXPRESSION ELIMINATION ....IV - 437
A.p. Vinod, E. M.-K. Lai, A. B. Premkumar, C. T. Lau, Nanyang Technological University, Singapore

COMM-P6.4: A LOW-POWER CROSSTALK-INSENSITIVE SIGNALING SCHEME FOR CHIP-TO-CHIP COMMUNICATION ....IV - 441
Kamran Farzan, David Johns, University of Toronto, Canada

COMM-P6.5: POWER SPECTRAL DENSITY OF TRANSMIT REFERENCE DOUBLET TRAINS AND REFERENCE SHARING DOUBLET TRAINS IN ULTRA WIDEBAND SYSTEMS ....IV - 445
Chun Yi Lee, Chris Toumazou, Imperial College London, United Kingdom

COMM-P7: COMMUNICATION ARCHITECTURES AND SYSTEMS II

COMM-P7.1: BURST RECEIVER FOR UPSTREAM COMMUNICATIONS OVER TWISTED PAIR LINES ....IV - 449
Alex Paek, Hichem Besbes, Yu Zhang, Ted Burk, Saf Asghar, Celite Milbrandt, Bruce Webb, Celite Systems, United States

COMM-P7.2: A NOVEL I/Q MISMATCH COMPENSATION SCHEME FOR A LOW-IF RECEIVER FRONT-END ..... IV - 453
Jérémie Chabloz, Christian Enz, CSEM, Switzerland

COMM-P7.3: TUNABLE MATCHING NETWORKS FOR FUTURE MEMS-BASED TRANSCEIVERS ..... IV - 457
Amro Elshurafa, Ezz I. El-Masry, Dalhousie University, Canada

COMM-P7.4: ON VARIOUS LOW-HARDWARE-COMPLEXITY LMS ALGORITHMS FOR ADAPTIVE I/Q CORRECTION IN QUADRATURE RECEIVERS ..... IV - 461
Ediz Çetin, Izzet Kale, Richard Morling, University of Westminster, United Kingdom

COMM-P7.5: A NEW NDA TIMING ERROR DETECTOR FOR BPSK AND QPSK WITH AN EFFICIENT HARDWARE IMPLEMENTATION FOR ASIC-BASED AND FPGA-BASED WIRELESS RECEIVERS ..... IV - 465
Yair Linn, University of British Columbia, Canada

COMM-P8: CIRCUITS AND NETWORKS FOR COMMUNICATIONS

COMM-P8.1: A CMOS-MEMS MAGNETIC THIN-FILM INDUCTOR FOR RADIO FREQUENCY AND INTERMEDIATE FREQUENCY FILTER CIRCUITS ..... IV - 469
Gang Zhang, Richard Carley, Carnegie Mellon University, United States

COMM-P8.2: DIFFERENTIATING NETWORK CONVERSATION FLOW FOR INTRUSION DETECTION AND DIAGNOSTICS ..... IV - 473
John McEachen, Naval Postgraduate School, United States; John Zachary, University of South Carolina, United States; Daniel Ettlich, Naval Postgraduate School, United States

COMM-P8.3: VLSI IMPLEMENTATION ISSUES OF LATTICE DECODERS FOR MIMO SYSTEMS ..... IV - 477
Zhan Guo, Peter Nilsson, Lund University, Sweden

COMM-P8.4: A CALIBRATION METHOD FOR PLLS BASED ON TRANSIENT RESPONSE ..... IV - 481
Marco Cassia, Technical University of Denmark, Denmark; Peter Shah, RF Magic, United States; Erik Bruun, Technical University of Denmark, Denmark

GTC-L1: GRAPH ALGORITHMS AND APPLICATIONS I

GTC-L1.1: MODELING AND DIAGNOSIS OF ANALOG CIRCUITS WITH PROBABILISTIC GRAPHICAL MODELS ..... IV - 485
Christian Borgelt, University of Magdeburg, Germany; Daniela Girimonte, Giuseppe Acciani, Politechnic of Bari, Italy

GTC-L1.2: A FAST ALGORITHM FOR CROSSPOINT ASSIGNMENT UNDER CROSSTALK CONSTRAINTS WITH SHIELDING EFFECTS ..... IV - 489
Keiji Kida, Xiaoke Zhu, Changwen Zhuang, SII EDA Technologies, Inc., Japan; Yasuhiro Takashima, Shigetoshi Nakatake, University of Kitakyushu, Japan

GTC-L1.3: ON RELATION BETWEEN NON-DISJOINT DECOMPOSITION AND MULTIPLE-VERTEX DOMINATORS ..... IV - 493
Elena Dubrova, Maxim Teslenko, Andrés Martinelli, Royal Institute of Technology (KTH), Sweden

GTC-L1.4: STUDIES ON DISTRIBUTED ALGORITHM FOR NETWORK FLOW OPTIMIZATION PROBLEM BASED ON TIE-SET FLOW VECTOR SPACE ..... IV - 497
Yuki Shibata, Haruki Kubo, Hitoshi Watanabe, Soka University, Japan

GTC-L1.5: UNFOLDING OF PETRI NETS WITH SEMILINEAR REACHABILITY SET ..... IV - 501
Atsushi Ohta, Kohkichi Tsuji, Aichi Prefectural University, Japan

GTC-P1: GRAPH ALGORITHMS AND APPLICATIONS II

GTC-P1.1: S-SEQUENCE : A NEW FLOORPLAN REPRESENTATION METHOD PRESERVING ROOM ABUTMENT RELATIONSHIPS .... IV - 505
Yohei Ishimaru, Keishi Sakanushi, Shinsuke Kobayashi, Yoshinori Takeuchi, Masaharu Imai, Osaka University, Japan

GTC-P1.2: CALCULATION OF NON-MIXED SECOND DERIVATIVES IN MULTIRATE SYSTEMS THROUGH SIGNAL FLOW GRAPH TECHNIQUES .... IV - 509
Andrea Arcangeli, Stefano Squartini, Francesco Piazza, Università Politecnica delle Marche, Italy

GTC-P1.3: CHAOTIC SEQUENCES IN ACO ALGORITHMS .... IV - 513
Flavio Cannavó, Luigi Fortuna, Mattia Frasca, Luca Patané, University of Catania, Italy

GTC-P1.4: CONSTRUCTION OF LINEARLY TRANSFORMED PLANAR BDD BY WALSH COEFFICIENTS .... IV - 517
Mark Karpovsky, Boston University, United States; Radomir Stankovic, Faculty of Electronics, Yugoslavia; Jaakko Astola, Tampere University of Technology, Finland

GTC-P1.5: NEW GRAPH-BASED ALGORITHMS FOR PARTITIONING VLSI CIRCUITS .... IV - 521
Christopher Augeri, United States Air Force Academy, United States; Hesham Ali, University of Nebraska-Omaha, United States

GTC-P2: GRAPH ALGORITHMS AND APPLICATIONS III

GTC-P2.1: MULTI-LAYER CONSTRAINED VIA MINIMIZATION WITH CONJUGATE CONFLICT CONTINUATION GRAPHS .... IV - 525
Rung-Bin Lin, Shu-Yu Chen, Yuan Ze University, Taiwan

GTC-P2.2: A COMPARISON OF GRAPH COLORING HEURISTICS FOR REGISTER ALLOCATION BASED ON COALESCING IN INTERVAL GRAPHS .... IV - 529
Thomas Zeitlhofer, Bernhard Wess, Vienna University of Technology, Austria

NLCS-L1: OSCILLATORS DESIGN AND IMPLEMENTATION

NLCS-L1.1: A LOW PHASE NOISE 2.0V 900MHZ CMOS VOLTAGE CONTROLLED RING OSCILLATOR .... IV - 533
Dean Badillo, Motorola, Inc., United States; Sayfe Kiaei, Arizona State University, United States

NLCS-L1.2: SINGLE ENDED RAIL-TO-RAIL CMOS OTA BASED VARIABLE-FREQUENCY RING-OSCILLATOR .... IV - 537
Hervé Barthélemy, L2MP-Polytech, France; Stéphane Meillére, L2MP-Toulon, France; Sylvain Bourdel, L2MP-Polytech, France

NLCS-L1.3: A PRECISE 90° QUADRATURE OTA-C VCO BETWEEN 50-130 MHz .... IV - 541
Bernabé Linares-Barranco, Teresa Serrano-Gotarredona, J. Ramos-Martos, J. Ceballos-Cáceres, J. M. Mora, Alejandro Linares-Barranco, Instituto de Microelectrónica de Sevilla (IMSE), Spain

NLCS-L1.4: A 2.4GHZ LOW-POWER LOW-PHASE-NOISE CMOS VCO USING SPIRAL INDUCTORS AND JUNCTION VARACTORS .... IV - 545
Jie Long, Robert Weber, Iowa State University, United States

NLCS-L1.5: FULLY CURRENT CONTROLLABLE AM/FM MODULATOR AND QUADRATURE SINUSOIDAL OSCILLATOR BASED ON CCCIIs .... IV - 549
Montree Siripruchyanun, Poolsak Koseeyaporn, King Mongkut's Institute of Technology North Bangkok, Thailand; Jeerasuda Koseeyaporn, Paramote Wardkein, King Mongkut's Institute of Technology Ladkrabang, Thailand

NLCS-L2: PLLS DESIGN, IMPLEMENTATION AND APPLICATION

NLCS-L2.1: MODELING, DESIGN AND CHARACTERIZATION OF A NEW LOW JITTER ANALOG DUAL TUNING LC-VCO PLL ARCHITECTURE ....IV - 553
Roberto Nonis, University of Udine, Italy; Nicola Da Dalt, Infineon Technologies AG, Austria; Pierpaolo Palestri, Luca Selmi, University of Udine, Italy

NLCS-L2.2: SPECTRAL SHAPING BY GENERALIZED TRANSFER FUNCTION DESIGN IN FREQUENCY MODULATION SIGMA DELTA SYNTHESIZERS ....IV - 557
Soeren Sappok, Andre Kruth, Guerkan Ordu, Ralf Wunderlich, Stefan Heinen, RWTH Aachen, Germany

NLCS-L2.3: THE DESIGN OF A DIFFERENTIAL CMOS CHARGE PUMP FOR HIGH PERFORMANCE PHASE-LOCKED LOOPS ....IV - 561
Bortecene Terlemez, John Uyemura, Georgia Institute of Technology, United States

NLCS-L2.4: FREQUENCY SYNTHESIZER FOR ON-CHIP TESTING AND AUTOMATED TUNING ....IV - 565
Ari Y. Valero-López, Alberto Valdes-Garcia, Edgar Sánchez-Sinencio, Texas A&M University, United States

NLCS-L2.5: AN ADPLL CIRCUIT USING A DDPS FOR GENLOCK APPLICATIONS ....IV - 569
Dorin Emil Calbaza, Ioan Cordos, Nigel Seth-Smith, Gennum Corporation, Canada; Yvon Savaria, École Polytechnique de Montréal, Canada

NLCS-L3: CHAOS-BASED METHODOLOGIES FOR SECURITY

NLCS-L3.1: RESONANCE PROPERTIES OF CHEBYSHEV CHAOTIC SEQUENCES ....IV - 573
Tomohiro Yoshimura, Tohru Kohda, Kyushu University, Japan

NLCS-L3.2: RSA ENCRYPTION ALGORITHM BASED ON TORUS AUTOMORPHISMS ....IV - 577
Ljupco Kocarev, Marjan Sterjev, University of California, San Diego, United States; Paolo Amato, STMicroelectronics, Italy

NLCS-L3.3: A DOUBLE SCROLL BASED TRUE RANDOM BIT GENERATOR ....IV - 581
Müstak E. Yalçin, Johan Suykens, Joos Vandewalle, Katholieke Universiteit Leuven, Belgium

NLCS-L3.4: POST-PROCESSING OF DATA GENERATED BY A CHAOTIC PIPELINED ADC FOR THE ROBUST GENERATION OF PERFECTLY RANDOM BITSTREAMS ....IV - 585
Stefano Poli, Sergio Callegari, Riccardo Rovatti, University of Bologna, Italy; Gianluca Setti, University of Ferrara, Italy

NLCS-L3.5: A NOVEL CFAR INTRUSION DETECTION METHOD USING CHAOTIC STOCHASTIC RESONANCE ....IV - 589
Di He, Henry Leung, University of Calgary, Canada

NLCS-L4: NONLINEAR DYNAMICS AND CHAOS IN COMMUNICATIONS I: MODULATION AND CODING

NLCS-L4.1: OPTIMAL CHAOS SHIFT KEYING COMMUNICATIONS WITH CORRELATION DECODING ....IV - 593
Ji Yao, University of Birmingham, United Kingdom

NLCS-L4.2: OPTIMUM NONCOHERENT FM-DCSK DETECTOR: APPLICATION OF CHAOTIC GML DECISION RULE ....IV - 597
Géza Kolumbán, Gábor Kis, Budapest University of Technology and Economics, Hungary; Francis C. M. Lau, Chi K. Michael Tse, The Hong Kong Polytechnic University, Hong Kong SAR of China

NLCS-L4.3: GENERALIZED CORRELATION-DELAY-SHIFT-KEYING SCHEME FOR NONCOHERENT CHAOS-BASED COMMUNICATION SYSTEMS ....IV - 601
Wai M. Tam, Francis C. M. Lau, Chi K. Michael Tse, The Hong Kong Polytechnic University, Hong Kong SAR of China

NLCS-L4.4: AN IMPROVED MULTIPLE ACCESS SCHEME FOR CHAOS-BASED DIGITAL COMMUNICATIONS USING ADAPTIVE RECEIVERS ....IV - 605
Wai M. Tam, Francis C. M. Lau, Chi K. Michael Tse, The Hong Kong Polytechnic University, Hong Kong SAR of China

NLCS-L4.5: ON DISTRIBUTIONS OF MULTIPLE ACCESS INTERFERENCE FOR SPREAD SPECTRUM COMMUNICATION SYSTEMS USING M-PHASE SPREADING SEQUENCES OF MARKOV CHAINS ....IV - 609
Hiroshi Fujisaki, Kanazawa University, Japan

NLCS-L5: NONLINEAR DYNAMICS AND CHAOS IN COMMUNICATIONS II: CODING AND TRAFFIC

NLCS-L5.1: PULSE SHAPING AND SIR-ENERGY TRADE-OFF IN CHAOS-BASED ASYNCHRONOUS DS-CDMA ....IV - 613
Riccardo Rovatti, University of Bologna, Italy; Gianluca Setti, Gianluca Mazzini, University of Ferrara, Italy

NLCS-L5.2: MARKOVIAN SS CODES IMPLY INVERSION-FREE CODE ACQUISITION IN ASYNCHRONOUS DS/CDMA SYSTEMS ....IV - 617
Nobuoki Eshima, Oita Medical University, Japan; Yutaka Jitsumatsu, Tohru Kohda, Kyushu University, Japan

NLCS-L5.3: A PRACTICAL ALGORITHM FOR TURBO-DECODING ENHANCEMENT ....IV - 621
Dejan Spasov, University of California, San Diego, United States; Gian Mario Maggio, STMicroelectronics, United States; Ljupco Kocarev, University of California, San Diego, United States

NLCS-L5.4: ANALYSIS AND EFFECTS OF RETRANSMISSION MECHANISMS ON DATA NETWORK PERFORMANCE ....IV - 625
Sabato Manfredi, Franco Garofalo, University of Naples Federico II, Italy; Mario di Bernardo, University of Sannio, Italy

NLCS-L5.5: ANALYSIS OF INTERNET TOPOLOGY DATA ....IV - 629
Johnson Chen, Ljiljana Trajkovic, Simon Fraser University, Canada

NLCS-L6: NONLINEAR CIRCUITS MODELLING

NLCS-L6.1: DISTORTION ANALYSIS OF NONLINEAR NETWORKS BASED ON SPICE-ORIENTED HARMONIC BALANCE METHOD ....IV - 633
Yoshihiro Yamagami, Hiroo Yabe, Yoshifumi Nishio, Tokushima University, Japan; Akio Ushida, Tokushima Bunri University, Japan

NLCS-L6.2: A TOOL FOR THE INTEGRATION OF NEW VHDL-AMS MODELS IN SPICE ....IV - 637
Marco Zorzi, Francesco Franzè, Nicolo Speciale, Guido Masetti, University of Bologna, Italy

NLCS-L6.3: EVENT-DRIVEN SIMULATION AND MODELING OF AN RF OSCILLATOR ....IV - 641
Robert Staszewski, Chan Fernando, Texas Instruments, Inc., United States; Poras Balsara, University of Texas, Dallas, United States

NLCS-L6.4: FITTING OF 2-DIMENSIONAL POLYNOMIAL DEVICE MODEL BASED ON SIMULATED VOLTAGE AND CURRENT SPECTRA ....IV - 645
Janne Aikio, Timo Rahkonen, University of Oulu, Finland

NLCS-L6.5: SIMPLE RC MODELS OF DISTRIBUTED RC LINES IN CONSIDERATION WITH THE DELAY TIME ....IV - 649
Masato Ogata, Yoshiaki Okabe, Tetsuo Nishi, Kyushu University, Japan

NLCS-L7: MODELLING AND ANALYSIS OF NONLINEAR SYSTEMS

NLCS-L7.1: POLYPHASE REPRESENTATION OF MULTIRATE VOLTERRA FILTERS ....IV - 653
David Schwingshackl, Infineon Technologies AG, Austria; Gernot Kubin, Graz University of Technology, Austria

NLCS-L7.2: NONLINEAR DYNAMICS OF GEAR-SHIFTING DIGITAL PHASE-LOCKED LOOPS ....IV - 657
Raymond Flynn, Orla Feely, UCD, Ireland

NLCS-L7.3: A NOVEL METHOD FOR IDENTIFYING CUBICALLY NONLINEAR SYSTEMS USING MINIMALLY BANDPASS SAMPLED DATA ....IV - 661
Ching-Hsiang Tseng, National Taiwan Ocean University, Taiwan

NLCS-L7.4: PWL IDENTIFICATION OF DYNAMICAL SYSTEMS: SOME EXAMPLES ....IV - 665
Oscar De Feo, Swiss Federal Institute of Technology, Lausanne, Switzerland; Marco Storace, University of Genova, Italy

NLCS-L7.5: PATTERN DETECTION IN NOISY SIGNALS..IV - 669
Markus Christen, Albert Kern, Jan-Jan van der Vyver, Ruedi Stoop, University of Zürich / ETH Zürich, Switzerland

NLCS-L8: NONLINEAR CIRCUITS ANALYSIS AND DESIGN

NLCS-L8.1: SPIKE POSITION MAP WITH QUANTIZED STATE AND ITS APPLICATION TO ALGORITHMIC A/D CONVERTER ....IV - 673
Hiroshi Hamanaka, Hiroyuki Torikai, Toshimichi Saito, HOSEI University, Japan

NLCS-L8.2: SIGMA-DELTA A/D FUZZY CONVERTER..IV - 677
Gianluca Giustolisi, Gaetano Palumbo, Università degli Studi di Catania, Italy

NLCS-L8.3: NOISE ROBUSTNESS CONDITION FOR CHAOTIC MAPS WITH PIECEWISE CONSTANT INVARIANT DENSITY ....IV - 681
Fabio Pareschi, Gianluca Setti, University of Ferrara, Italy; Riccardo Rovatti, University of Bologna, Italy

NLCS-L8.4: STRUCTURE AND INFORMATION CONTENT IN SEQUENCES FROM THE SINGLE-LOOP SIGMA-DELTA MODULATOR WITH DC INPUT ....IV - 685
Frank Dachselt, Stefan Quitzk, Dresden University of Technology, Germany

NLCS-L8.5: SPECTRAL ALIASING EFFECTS OF PWM SIGNALS WITH TIME-QUANTIZED SWITCHING INSTANTS ....IV - 689
Stefano Santi, Riccardo Rovatti, University of Bologna, Italy; Gianluca Setti, University of Ferrara, Italy

NLCS-L9: SWITCHING CIRCUITS AND SYSTEMS: BIFURCATION ANALYSIS AND CONTROL

NLCS-L9.1: HOPF-LIKE TRANSITIONS IN NONSMOOTH DYNAMICAL SYSTEMS ....IV - 693
Gerard Olivar, Universita Politecnica de Catalunya, Spain; Fabiola Angulo, Universidad Nacional de Colombia, Colombia; Mario di Bernardo, University of Sannio, Italy

NLCS-L9.2: GRAZING BIFURCATIONS IN PERIODIC HYBRID SYSTEMS...IV - 697
Vaibhav Donde, University of Illinois at Urbana-Champaign, United States; Ian Hiskens, University of Wisconsin-Madison, United States

NLCS-L9.3: ANALYSIS OF BIFURCATION IN SWITCHED DYNAMICAL SYSTEMS WITH PERIODICALLY MOVING BORDERS: APPLICATION TO POWER CONVERTERS ....IV - 701
Yue Ma, Hiroshi Kawakami, University of Tokushima, Japan; Chi K. Michael Tse, Hong Kong Polytechnic University, Hong Kong SAR of China

NLCS-L9.4: CONTROLLING LIMIT CYCLES IN PLANAR DYNAMICAL SYSTEMS: A NONSMOOTH BIFURCATION APPROACH ....IV - 705
Fabiola Angulo, Universidad Nacional de Colombia, Colombia; Enric Fossas, Institut d'Organitzacio i Control, Spain; Gerard Olivar, Universitat Politècnica de Catalunya, Spain; Mario di Bernardo, University of Bristol, United Kingdom / University of Sannio, Italy

NLCS-L9.5: DITHER FOR CHATTERING REDUCTION IN SLIDING MODE CONTROL SYSTEMS ....IV - 709
Luigi Iannelli, University of Napoli Federico II, Italy; Francesco Vasca, University of Sannio, Italy

NLCS-L10: NONLINEAR CIRCUITS AND ARRAYS

NLCS-L10.1: PULSE-EXCITED RC NONAUTONOMOUS CHAOTIC OSCILLATOR STRUCTURES ....IV - 713
Ahmed Elwakil, Sharjah University, United Arab Emirates; Serdar Özoguz, Istanbul Technical University, Turkey

NLCS-L10.2: TOWARDS FULL CHARACTERIZATION OF CONTINUOUS SYSTEMS IN TERMS OF PERIODIC ORBITS ....IV - 716
Zbigniew Galias, AGH University of Science and Technology, Poland

NLCS-L10.3: A BINARY-QUANTIZED PSEUDO-DIFFUSION SYSTEM ....IV - 720
Hisato Fujisaka, Daisuke Hamano, Masahiro Sakamoto, Takeshi Kamio, Hiroshima City University, Japan

NLCS-L10.4: SYNCHRONIZATION IN SYSTEMS COUPLED VIA COMPLEX NETWORKS ....IV - 724
Chai Wah Wu, IBM T. J. Watson Research Center, United States

NLCS-L10.5: GLOBAL ADAPTIVE SYNCHRONIZATION BASED UPON POLYTOPIC OBSERVERS ....IV - 728
Floriane Anstett, Gilles Millerioux, Gerard Bloch, Centre de Recherche en Automatique de Nancy, France

NLCS-L11: APPLICATIONS OF NONLINEAR CIRCUITS

NLCS-L11.1: CHAOS CONTROL IN PERMANENT MAGNET SYNCHRONOUS MOTOR ....IV - 732
Ding Liu, Haipeng Ren, Xi'an University of Technology, China; Xiaoyan Liu, University of Waterloo, Canada

NLCS-L11.2: SELF-INDUCED OSCILLATIONS IN COUPLED FLUXGATE MAGNETOMETER: A NOVEL APPROACH TO OPERATING THE MAGNETIC SENSORS ....IV - 736
Visarath In, Andy Kho, Adi Bulsara, SPAWAR Systems Center San Diego, United States; Antonio Palacios, San Diego State University, United States; Salvatore Baglio, B. Ando, University of Catania, Italy; Patrick Longhini, San Diego State University, United States; Joseph D. Neff, Brian Meadows, SPAWAR Systems Center San Diego, United States

NLCS-L11.3: DESIGN OF BURSTING IN TWO-DIMENSIONAL DISCRETE-TIME NEURON MODELS ....IV - 740
Hiroto Tanaka, Toshimitsu Ushio, Osaka University, Japan

NLCS-L11.4: LOW POWER REAL TIME ELECTRONIC NEURON VLSI DESIGN USING SUBTHRESHOLD TECHNIQUE ....IV - 744
Young Jun Lee, Jihyun Lee, Yong-Bin Kim, Joseph Ayers, Northeastern University, United States; A. Volkovskii, A. Selverston, H. Abarbanel, M. Rabinovich, University of California, San Diego, United States

NLCS-L11.5: LINEARIZATION OF CMOS LNA'S VIA OPTIMUM GATE BIASING ....IV - 748
Vladimir Aparin, Gary Brown, Qualcomm, Inc., United States; Lawrence E. Larson, University of California, San Diego, United States

NLCS-P1: NONLINEAR CIRCUITS ANALYSIS AND DESIGN I

NLCS-P1.1: A LOW POWER AND HIGH SPEED CMOS VOLTAGE-CONTROLLED RING OSCILLATOR ....IV - 752
Mónico Linares-Aranda, Daniel Pacheco, Instituto Nacional de Astrofísica, Óptica y Electrónica, Mexico

NLCS-P1.2: DESIGN AND IMPLEMENTATION OF CMOS DISTRIBUTED MIXERS AND OSCILLATORS FOR WIDE-BAND RF FRONT-END ....IV - 756
Yogesh Ramadass, Nirmal Chakrabarti, Indian Institute of Technology, Kharagpur, India

NLCS-P1.3: NOVEL ARCHITECTURES OF CLASS AB CMOS MIRRORS WITH PROGRAMMABLE GAIN .... IV - 760
Chandrika Durbha, Jaime Ramírez Angulo, New Mexico State University, United States; Antonio J. López-Martín, Public University of Navarra, Spain; Ramon Gonzalez Carvajal, Escuela Superior de Ingenieros, Spain

NLCS-P1.4: TWO LOW-VOLTAGE HIGH-SPEED CMOS FREQUENCY-INSENSITIVE PWM SIGNAL GENERATORS BASED ON RELAXATION OSCILLATOR .... IV - 764
Montree Siripruchyanun, Poolsak Koseeyaporn, King Mongkut's Institute of Technology North Bangk, Thailand; Jeerasuda Koseeyaporn, Paramote Wardkein, King Mongkut's Institute of Technology Ladkrabang, Thailand

NLCS-P1.5: A CLOSED-FORM PHASE NOISE SOLUTION FOR AN IDEAL LC OSCILLATOR .... IV - 768
Hua Zhang, University of Texas, Dallas, United States; Jianzhong Zhang, Nokia Research Center, United States; Dian Zhou, Jin Liu, University of Texas, Dallas, United States; Liangjun Jiang, University of Central Florida, United States; Yan Pan, Texas Instruments, Inc., United States

NLCS-P2: NONLINEAR CIRCUITS ANALYSIS AND DESIGN II

NLCS-P2.1: STABILITY ANALYSIS OF POWER CIRCUIT COMPRISING VIRTUAL INDUCTANCE .... IV - 772
Octavian Dranga, Hirohito Funato, Satoshi Ogasawara, Utsunomiya University, Japan; Chi K. Michael Tse, Hong Kong Polytechnic University, Hong Kong SAR of China; Herbert Iu, University of Western Australia, Australia

NLCS-P2.2: AN UNLIMITED LOCK RANGE DLL FOR CLOCK GENERATOR .... IV - 776
Kwangoh Kim, Nohman Park, Taekyu Kim, Dawintech, Inc., Republic of Korea

NLCS-P2.3: DESIGN AND SIMULATION OF FRACTIONAL-N PLL FREQUENCY SYNTHESIZERS .... IV - 780
Mücahit Kozak, Eby G. Friedman, University of Rochester, United States

NLCS-P2.4: MAINTAINING CHAOS IN AN ASSOCIATIVE CHAOTIC NEURAL NETWORK EXHIBITING INTERMITTENCY .... IV - 784
Masaharu Adachi, Tokyo Denki University, Japan

NLCS-P2.5: A MIXED PLL/DLL ARCHITECTURE FOR LOW JITTER CLOCK GENERATION .... IV - 788
Yong-Cheol Bae, Gu-Yeon Wei, Harvard University, United States

NLCS-P3: NONLINEAR CIRCUITS AND SYSTEMS ANALYSIS AND APPLICATION I

NLCS-P3.1: EXPERIMENTAL EVIDENCE FOR AMPLITUDE DEATH INDUCED BY DYNAMIC COUPLING: VAN DER POL OSCILLATORS .... IV - 792
Keiji Konishi, Future University, Hakodate, Japan

NLCS-P3.2: BIFURCATION ANALYSIS IN A PIECEWISE-SMOOTH SYSTEM WITH PERIODIC THRESHOLD .... IV - 796
Takuji Kousaka, Masahiko Mori, Fukuyama University, Japan

NLCS-P3.3: ON THE SYNCHRONIZATION REGION IN NETWORKS OF COUPLED OSCILLATORS .... IV - 800
Paolo Checco, Politecnico di Torino, Italy; Ljupco Kocarev, University of California, San Diego, United States; Gian Mario Maggio, STMicroelectronics / University of California, United States; Mario Biey, Politecnico di Torino, Italy

NLCS-P3.4: NONLINEAR DYNAMICS OF A NONIDEAL SIGMA-DELTA MODULATOR WITH PERIODIC INPUT .... IV - 804
Emer Condon, Orla Feely, University College Dublin, Ireland

NLCS-P4: NONLINEAR CIRCUITS AND SYSTEMS ANALYSIS AND APPLICATION II

NLCS-P4.1: HYPERREAL OPERATING POINTS FOR NONLINEAR TRANSFINITE NETWORKS .... IV - 808
Armen Zemanian, State University of New York at Stony Brook, United States

NLCS-P4.2: ROBUST THREE-STATE PFD ARCHITECTURE WITH ENHANCED FREQUENCY ACQUISITION CAPABILITIES .... IV - 812
Francesco Centurelli, Stefano Costi, Mauro Olivieri, Università di Roma, Italy; Salvatore Pennisi, Università di Catania, Italy; Alessandro Trifiletti, Università di Roma, Italy

NLCS-P4.3: TWO-SIDED PROJECTION METHOD IN VARIATIONAL EQUATION MODEL ORDER REDUCTION OF NONLINEAR CIRCUITS .... IV - 816
Lihong Feng, Xuan Zeng, Jiarong Tong, Fudan University, China; Charles Chiang, Synopsys, Inc., United States; Dian Zhou, University of Texas, Dallas, United States

NLCS-P4.4: SMALL WORLD EFFECTS IN NETWORKS: AN ENGINEERING INTERPRETATION .... IV - 820
Sabato Manfredi, University of Naples Federico II, Italy; Mario di Bernardo, University of Sannio, Italy; Franco Garofalo, University of Naples Federico II, Italy

SEN-L1: IMAGE SENSORS I

SEN-L1.1: A TIME-TO-FIRST SPIKE CMOS IMAGER .... IV - 824
Xin Qi, Xiaochuan Guo, John Harris, Univerisity of Florida, United States

SEN-L1.2: SECOND GENERATION OF HIGH DYNAMIC RANGE, ARBITRATED DIGITAL IMAGER .... IV - 828
Eugenio Culurciello, Ralph Etienne-Cummings, Johns Hopkins University, United States

SEN-L1.3: ANALOG INTEGRATED CIRCUIT FOR DETECTION OF APPROACHING OBJECT AGAINST MOVING BACKGROUND BASED ON LOWER ANIMAL VISION .... IV - 832
Kimihiro Nishio, Hiroo Yonezu, Shinya Sawa, Yoichi Yoshikawa, Yuzo Furukawa, Toyohashi University of Technology, Japan

SEN-L1.4: ADAPTIVE MULTIPLE RESOLUTION CMOS ACTIVE PIXEL SENSOR .... IV - 836
Evgeny Artyomov, Orly Yadid-Pecht, Ben-Gurion University, Israel

SEN-L1.5: A TIME-BASED CMOS IMAGE SENSOR .... IV - 840
Qiang Luo, National Semiconductor, United States; John Harris, University of Florida, United States

SEN-L2: VISION SENSORS

SEN-L2.1: SELF-BIASING LOW POWER ADAPTIVE PHOTORECEPTOR .... IV - 844
Tobi Delbrück, Daniel Oberhoff, ETH/Univ. Zürich, Switzerland

SEN-L2.2: NORMAL OPTICAL FLOW MEASUREMENT ON A CMOS APS IMAGER .... IV - 848
Swati Mehta, Ralph Etienne-Cummings, Johns Hopkins University, United States

SEN-L2.3: HIGH SPEED AND HIGH RESOLUTION CURRENT WINNER-TAKE-ALL CIRCUIT IN CONJUNCTION WITH ADAPTIVE THRESHOLDING .... IV - 852
Alexander Fish, Vadim Milrud, Orly Yadid-Pecht, Ben-Gurion University, Israel

SEN-L2.4: AN ELECTRONIC CALIBRATION SCHEME FOR LOGARITHMIC CMOS PIXELS .... IV - 856
Bhaskar Choubey, University of Oxford, United Kingdom; Satoshi Aoyama, Renesas Technology Corp., Japan; Dileepan Joseph, Stephen Otim, Steve Collins, University of Oxford, United Kingdom

SEN-L2.5: LINEAR CURRENT MODE IMAGER WITH LOW FIX PATTERN NOISE .... **IV - 860**
Viktor Gruev, Ralph Etienne-Cummings, Johns Hopkins University, United States; Timothy K. Horiuchi, University of Maryland, United States

SEN-L3: MEMS AND SENSORY SYSTEMS

SEN-L3.1: A CIRCUIT TO MODEL AN IONIC POLYMER-METAL COMPOSITE AS ACTUATOR .... **IV - 864**
Claudia Bonomo, Luigi Fortuna, Salvatore Graziani, Dino Mazza, University of Catania, Italy

SEN-L3.2: HIGHLY SENSITIVE SILICON MICRO-G ACCELEROMETERS WITH OPTICAL OUTPUT .... **IV - 868**
Salvatore Baglio, Salvatore Castorina, University of Catania, Italy; Jaume Esteve, Centro Nacional de Microelectrónica (CNM), Spain; Nicolò Savalli, University of Catania, Italy

SEN-L3.3: DETECTION AND COMPENSATION OF SENSOR MALFUNCTION IN TIME DELAY BASED DIRECTION OF ARRIVAL ESTIMATION .... **IV - 872**
Tuomo Pirinen, Jari Yli-Hietanen, Pasi Pertilä, Ari Visa, Tampere University of Technology, United States

SEN-L3.4: AN RF POWERED, WIRELESS TEMPERATURE SENSOR IN QUARTER MICRON CMOS .... **IV - 876**
Fatih Kocer, Paul M. Walsh, Michael P. Flynn, University of Michigan, United States

SEN-L3.5: SILICON ON SAPPHIRE CMOS ARCHITECTURES FOR INTERFEROMETRIC ARRAY READOUT .... **IV - 880**
Francisco Tejada, Andreas G. Andreou, Johns Hopkins University, United States; Joseph Miragliotta, Robert Osiander, Danielle Wesolek, Johns Hopkins University Applied Physics Lab, United States

SEN-L4: CHEMICAL, ACOUSTIC, OLFACTORY AND NEUROMORPHIC SENSORS

SEN-L4.1: BEARING ANGLE ESTIMATION FOR SONAR MICRO-ARRAY USING ANALOG VLSI SPATIOTEMPORAL PROCESSING .... **IV - 884**
Matthew Clapp, Ralph Etienne-Cummings, Johns Hopkins University, United States

SEN-L4.2: A MONOLITHIC FULLY-DIFFERENTIAL CMOS GAS SENSOR MICROSYSTEM FOR MICROHOTPLATE TEMPERATURES UP TO 450°C .... **IV - 888**
Diego Barrettino, Markus Graf, Kay-Uwe Kirstein, Andreas Hierlemann, Henry Baltes, Swiss Federal Institute of Technology (ETH), Zürich, Switzerland

SEN-L4.3: A LOW-VOLTAGE, CHEMICAL SENSOR INTERFACE FOR SYSTEMS-ON-CHIP: THE FULLY-DIFFERENTIAL POTENTIOSTAT .... **IV - 892**
Steven Martin, Fadi Gebara, Timothy Strong, Richard Brown, University of Michigan, United States

SEN-L4.4: A MIXED-MODE TEMPERATURE CONTROL CIRCUIT FOR GAS SENSORS .... **IV - 896**
Raimon Casanova, José Luis Merino, Ángel Dieguez, University of Barcelona, Spain; Sebastià Bota, Universitat de les Illes Balears, Spain; Josep Samitier, University of Barcelona, Spain

SEN-L4.5: A VLSI MODEL OF THE BAT LATERAL SUPERIOR OLIVE FOR AZIMUTHAL ECHOLOCATION .... **IV - 900**
R. Z. Shi, Timothy K. Horiuchi, University of Maryland, United States

SEN-P1: NETWORK SENSORS AND MEMS

SEN-P1.1: CLUSTERING ROUTING ALGORITHM USING GAME-THEORETIC TECHNIQUES FOR WSNS .... **IV - 904**
Zeng-Wei Zheng, Zhao-Hui Wu, Huai-Zhong Lin, Zhejiang University, China

SEN-P1.2: TOWARD A CIRCUIT THEORY FOR SENSOR NETWORKS WITH FADING CHANNELS ..... IV - 908
Martin Haenggi, University of Notre Dame, United States

SEN-P1.3: A 2-MB/S, UHF-BAND, WIRELESS SENSOR NODE FOR REAL-TIME MULTI-POINT SENSING ..... IV - 912
Koji Fujii, Jun Terada, Junichi Kodate, Tsuneo Tsukahara, Shin'ichiro Mutoh, Yuichi Kado, NTT Corporation, Japan

SEN-P1.4: ELECTRICAL CONNECTIVITY ANALYSIS OF A MEMS MICROPACKAGE ..... IV - 916
Sazzadur Chowdhury, Majid Ahmadi, University of Windsor, Canada; William Miller, University of WIndsor, Canada

SEN-P1.5: SURFACE MICROMACHINING IN SILICON ON SAPPHIRE CMOS TECHNOLOGY ..... IV - 920
Francisco Tejada, Andreas G. Andreou, Johns Hopkins University, United States; Dennis Wickenden, Arthur Francomacaro, Johns Hopkins University Applied Physics Lab, United States

SEN-P2: NEUROMORHIC AND SENSORY SYSTEMS

SEN-P2.1: A 2-DIMENSIONAL ACTIVE COCHLEAR MODEL FOR ANALOG VLSI IMPLEMENTATION ..... IV - 924
Hisako Shiraishi, André van Schaik, University of Sydney, Australia

SEN-P2.2: AN ALTERNATIVE ANALOG VLSI IMPLEMENTATION OF THE MEDDIS INNER HAIR CELL MODEL ..... IV - 928
Alistair McEwan, Oxford University, United Kingdom; André van Schaik, University of Sydney, Australia

SEN-P2.3: ELIMINATION OF QUANTIZATION EFFECTS IN MEASURED TEMPORAL NOISE ..... IV - 932
Leif Lindgren, Linköping University, Sweden

SEN-P2.4: CHARGE-BASED PREDICTION CIRCUITS FOR FOCAL PLANE IMAGE COMPRESSION ..... IV - 936
Walter Leon, Sina Balkir, Khalid Sayood, Michael Hoffman, University of Nebraska-Lincoln, United States

SEN-P2.5: MINIMIZATION OF POWER DISSIPATION OF ANALOG CHANNEL-SELECT FILTER AND NYQUIST-RATE A/D CONVERTER IN UTRA/FDD ..... IV - 940
Jarkko Jussila, Kari Halonen, Helsinki University of Technology, Finland

SEN-P3: IMAGE SENSORS II

SEN-P3.1: A SINGLE-CHIP CMOS VISUAL ORIENTATION SENSOR ..... IV - 944
Reid Harrison, University of Utah, United States

SEN-P3.2: A HIGH SPEED CENTROID COMPUTATION CIRCUIT IN ANALOG VLSI ..... IV - 948
Ananth Bashyam, Paul Furth, Michael Giles, New Mexico State University, United States

SEN-P3.3: MINIPIX: FOCAL PLANE TEMPORAL FREQUENCY IMAGE PROCESSOR ..... IV - 952
Alasdair Sutherland, Alister Hamilton, David Renshaw, Yaxiong Zhang, University of Edinburgh, United Kingdom; Mark Glover, QinetiQ Ltd, United Kingdom

SEN-P3.4: ALOHA CMOS IMAGER ..... IV - 956
Eugenio Culurciello, Andreas G. Andreou, Johns Hopkins University, United States

SEN-P3.5: CMOS APS IMAGER EMPLOYING 3.3V 12 BIT 6.3 MS/S PIPELINED ADC ..... IV - 960
Shy Hamami, Leonid Fleshel, Orly Yadid-Pecht, Ben-Gurion University, Israel

Volume V

BSP-L1: BLIND SIGNAL PROCESSING I

BSP-L1.1: FREQUENCY DOMAIN BLIND SOURCE SEPARATION USING SMALL AND LARGE SPACING SENSOR PAIRS .... V - 1
Ryo Mukai, Hiroshi Sawada, Shoko Araki, Shoji Makino, NTT Corporation, Japan

BSP-L1.2: GRADIENT FLOW BEARING ESTIMATION WITH BLIND IDENTIFICATION OF MULTIPLE SIGNALS AND INTERFERENCE .... V - 5
Milutin Stanacevic, Gert Cauwenberghs, Johns Hopkins University, United States; Larry Riddle, Signal Systems Corporation, United States

BSP-L1.3: NONLINEAR ICA SOLUTIONS FOR CONVOLUTIVE MIXING OF PNL MIXTURES .... V - 9
Daniele Vigliano, Aurelio Uncini, Raffaele Parisi, Università degli Studi di Roma, Italy

BSP-L1.4: BLIND SEPARATION WITH GAUSSIAN MIXTURE MODEL FOR CONVOLUTIVELY MIXED SOURCES .... V - 13
Masashi Ohata, Toshiharu Mukai, RIKEN BMC Research Center, Japan; Kiyotoshi Matsuoka, Kyushu Institute of Technology, Japan

BSP-L1.5: BLIND ESTIMATION AND EQUALIZATION OF TIME-VARYING CHANNELS USING THE INTERACTING MULTIPLE MODEL ESTIMATOR .... V - 17
Ziauddin Kamran, Thiagalingam Kirubarajan, Alex Gershman, McMaster University, Canada

BSP-P1: BLIND SIGNAL PROCESSING II

BSP-P1.1: NEW ALGORITHM FOR BLIND ADAPTIVE EQUALIZATION BASED ON CONSTANT MODULUS CRITERION .... V - 21
Yajun Kou, Wu-Sheng Lu, Andreas Antoniou, University of Victoria, Canada

BSP-P1.2: SEQUENTIAL BLIND EXTRACTION OF MIXED SOURCE SIGNALS WITH GUARANTEED CONVERGENCE .... V - 25
Derong Liu, Sanqing Hu, University of Illinois, United States

BSP-P1.3: A VERTICAL LAYERED SPACE-TIME CODE AND ITS BLIND SYMBOL DETECTION .... V - 29
Ke Deng, Qinye Yin, Ming Luo, Zheng Zhao, Xi'an Jiaotong University, China

BSP-P1.4: BLIND SOURCE RECOVERY FOR NON-MINIMUM PHASE SURROUNDINGS .... V - 33
Khurram Waheed, San Diego State University, United States; Fathi M. Salem, Michigan State University, United States

BSP-P1.5: SPARSE COMPONENT ANALYSIS OF OVERCOMPLETE MIXTURES BY IMPROVED BASIS PURSUIT METHOD .... V - 37
Pando Georgiev, Andrzej Cichocki, Brain Science Institute, RIKEN, Japan

BSP-P2: BLIND SIGNAL PROCESSING III

BSP-P2.1: A DUAL MODE DECISION FEEDBACK EQUALIZER EMPLOYING THE CONJUGATE GRADIENT ALGORITHM .... V - 41
Yihai Zhang, Aaron Gulliver, University of Victoria, Canada

BSP-P2.2: BLIND UPLINK SPACE-TIME CHANNEL ESTIMATION FOR SPACE-TIME CODED MULTICARRIER CODE DIVISION MULTIPLE ACCESS SYSTEMS .... V - 45
Yanxing Zeng, Qinye Yin, Le Ding, Xi'an Jiaotong University, China; Ke Deng, Xi' an Jiaotong University, China

BSP-P2.3: A POLYNOMIAL METHOD FOR BLIND IDENTIFICATION OF MIMO CHANNELS .... V - 49
Weizhou Su, Wei Xing Zheng, University of Western Sydney, Australia

CAD-L1: PLACEMENT AND ROUTING I

CAD-L1.1: FAST FORCE-DIRECTED /SIMULATED EVOLUTION HYBRID FOR MULTIOBJECTIVE VLSI CELL PLACEMENT .... V - 53
Sadiq Sait Mohammed, King Fahd University of Petroleum & Minerals, Saudi Arabia; Junaid Asim Khan, University of British Columbia, Canada

CAD-L1.2: SIMULTANEOUS DELAY AND POWER OPTIMIZATION IN GLOBAL PLACEMENT .... V - 57
Mongkol Ekpanyapong, Karthik Balakrishnan, Vidit Nanda, Sung Kyu Lim, Georgia Institute of Technology, United States

CAD-L1.3: MODULE PLACEMENT BASED ON QUADRATIC PROGRAMMING AND RECTANGLE PACKING USING LESS FLEXIBILITY FIRST PRINCIPLE .... V - 61
Sheqin Dong, Zhong Yang, Xianlong Hong, Tsinghua University, China; Yu-liang Wu, The Chinese University of Hong Kong, Hong Kong SAR of China

CAD-L1.4: PERFORMANCE AND RLC CROSSTALK DRIVEN GLOBAL ROUTING .... V - 65
Ling Zhang, Tong Jing, Xianlong Hong, Jingyu Xu, Tsinghua University, China; Jinjun Xiong, Lei He, University of California, Los Angeles, United States

CAD-L1.5: MULTI-LAYER FLOORPLANNING FOR RELIABLE SYSTEM-ON-PACKAGE .... V - 69
Pun Hang Shiu, Ramprasad Ravichandran, Siddharth Easwar, Sung Kyu Lim, Georgia Institute of Technology, United States

CAD-L2: PLACEMENT AND ROUTING II

CAD-L2.1: ROUTING RESOURCES CONSUMPTION ON M-ARCH AND X-ARCH .... V - 73
Bo-Kyung Choi, University of California, Los Angeles, United States; Charles Chiang, Jamil Kawa, Synopsys, Inc., United States; Majid Sarrafzadeh, University of California, Los Angeles, United States

CAD-L2.2: A PLACEMENT ALGORITHM FOR IMPLEMENTATION OF ANALOG LSI/VLSI SYSTEMS .... V - 77
Lihong Zhang, Rabindra Raut, Concordia University, Canada; Yingtao Jiang, University of Nevada, United States

CAD-L2.3: RECURSIVELY COMBINE FLOORPLAN AND Q-PLACE IN MIXED MODE PLACEMENT BASED ON CIRCUIT'S VARIETY OF BLOCK CONFIGURATION .... V - 81
Changqi Yang, Xianlong Hong, Tsinghua University, China; Hannah Honghua Yang, Intel Corporation, United States; Qiang Zhou, Yici Cai, Yongqiang Lu, Tsinghua University, China

CAD-L2.4: LAYER ASSIGNMENT ALGORITHM FOR RLC CROSSTALK MINIMIZATION .... V - 85
Bin Liu, Yici Cai, Qiang Zhou, Xianlong Hong, Tsinghua University, China

CAD-L2.5: CROSSTALK DRIVEN ROUTING RESOURCE ASSIGNMENT .... V - 89
Hailong Yao, Qiang Zhou, Xianlong Hong, Yici Cai, Tsinghua University, China

CAD-L3: ANALOG MODELING, SYNTHESIS & OPTIMIZATION I

CAD-L3.1: RF CIRCUIT SYNTHESIS USING PARTICLE SWARM OPTIMIZATION .... V - 93
Jinho Park, David Allstot, University of Washington, United States

CAD-L3.2: AN OPTIMIZATION-BASED TOOL FOR THE HIGH-LEVEL SYNTHESIS OF DISCRETE-TIME AND CONTINUOUS-TIME SIGMA-DELTA MODULATORS IN THE MATLAB/SIMULINK ENVIRONMENT .... V - 97
Jesús Ruiz-Amaya, José M. de la Rosa, Fernando Medeiro, Francisco V. Fernández, Rocío del Río, Belén Pérez-Verdú, Angel Rodríguez-Vázquez, Instituto de Microelectrónica de Sevilla (IMSE), Spain

CAD-L3.3: A NOVEL ANALOG LAYOUT SYNTHESYS TOOL .... V - 101
Lihong Zhang, Ulrich Kleine, Otto-von-Guericke University of Magdeburg, Germany

CAD-L3.4: BEHAVIORAL MODELING OF ANALOG CIRCUITS BY DYNAMIC SEMI-SYMBOLIC ANALYSIS .... V - 105
Junjie Yang, Sheldon Tan, University of California, Riverside, United States

CAD-L3.5: COMPACT SEMICONDUCTOR DEVICE MODELING USING HIGHER LEVEL METHODS .... V - 109
Matt Francis, Vivek Chaudhary, Alan Mantooth, University of Arkansas, United States

CAD-L4: ANALOG MODELING, SYNTHESIS & OPTIMIZATION II

CAD-L4.1: CONSISTENT MODEL FOR DRAIN CURRENT MISMATCH IN MOSFETS USING THE CARRIER NUMBER FLUCTUATION THEORY .... V - 113
Hamilton Klimach, Universidade Federal de Santa Catarina / Universidade Federal do Rio Grande do Sul, Brazil; Alfredo Arnaud, Universidad de la República, Uruguay; Marcio Schneider, Carlos Galup-Montoro, Universidade Federal de Santa Catarina, Brazil

CAD-L4.2: SCALING RULES AND PARAMETER TUNING PROCEDURE FOR ANALOG DESIGN REUSE IN TECHNOLOGY MIGRATION .... V - 117
Alessandro Savio, Luigi Colalongo, Zsolt Miklos Kovács-Vajna, University of Brescia, Italy; Michele Quarantelli, PDF Solutions, Italy

CAD-L4.3: FAST SENSITIVITY ANALYSIS OF TRANSMISSION LINE NETWORKS .... V - 121
Natalie Nakhla, Anestis Dounavis, Ram Achar, Michel Nakhla, Carleton University, Canada

CAD-L4.4: FAST TIME-DOMAIN SYMBOLIC SIMULATION FOR SYNTHESIS OF SIGMA-DELTA ANALOG-DIGITAL CONVERTERS .... V - 125
Hui Zhang, Alex Doboli, State University of New York at Stony Brook, United States

CAD-L4.5: AN EFFICIENT ALGORITHM FOR TRANSIENT AND DISTORTION ANALYSIS OF MILDLY NONLINEAR ANALOG CIRCUITS .... V - 129
Junjie Yang, Sheldon Tan, University of California, Riverside, United States

CAD-L5: DIGITAL CIRCUITS SYNTHESIS & OPTIMIZATION

CAD-L5.1: GENERATION OF DISJOINT CUBES FOR MULTIPLE-VALUED FUNCTIONS .... V - 133
Bogdan Falkowski, Cicilia Lozano, Nanyang Technological University, Singapore; Susanto Rahardja, Institute for Infocomm Research, Singapore

CAD-L5.2: OUTPUT-PATTERN DIRECTED DECOMPOSITION FOR LOW POWER DESIGN .... V - 137
Chi-Wei Hu, TingTing Hwang, National Tsing Hua University, Taiwan

CAD-L5.3: FPGA IMPLEMENTATION OF CONTROLLER-DATAPATH PAIR IN CUSTOM IMAGE PROCESSOR DESIGN .... V - 141
Hongtu Jiang, Viktor Öwall, Lund University, Sweden

CAD-L5.4: LEAST LEAKAGE VECTOR ASSISTED TECHNOLOGY MAPPING FOR TOTAL POWER OPTIMIZATION .... V - 145
Yi-Ching Au, Chi-Ying Tsui, Hong Kong University of Science and Technology, Hong Kong SAR of China

CAD-L5.5: A PREDICTIVE METHODOLOGY FOR ACCURATE SUBSTRATE PARASITIC EXTRACTION .... V - 149
Ajit Sharma, Chenggang Xu, Oregon State University, United States; Wen Kung Chu, Nishath Verghese, Cadence Design Systems, United States; Terri Fiez, Kartikeya Mayaram, Oregon State University, United States

CAD-L6: MIXED SIGNAL CIRCUIT AND DEVICE MODELING

CAD-L6.1: MODELING AND OPTIMIZATION OF ON-CHIP SPIRAL INDUCTOR IN S-PARAMETER DOMAIN .... V - 153
Kenichi Okada, Tokyo Institute of Technology, Japan; Hiroaki Hoshino, Hidetoshi Onodera, Kyoto University, Japan

CAD-L6.2: A PHYSICAL AND ANALYTICAL MODEL FOR SUBSTRATE NOISE COUPLING ANALYSIS .... V - 157
Robert Shreeve, Terri Fiez, Kartikeya Mayaram, Oregon State University, United States

CAD-L6.3: AN IMPROVED Z-PARAMETER MACRO MODEL FOR SUBSTRATE NOISE COUPLING .... V - 161
Chenggang Xu, Terri Fiez, Kartikeya Mayaram, Oregon State University, United States

CAD-L6.4: A COUPLED ITERATIVE/DIRECT METHOD FOR EFFICIENT TIME-DOMAIN SIMULATION OF NONLINEAR CIRCUITS WITH POWER/GROUND NETWORKS .... V - 165
Zhao Li, C.-J. Richard Shi, University of Washington, United States

CAD-L6.5: MULTIGRID-BASED SUBSTRATE COUPLING MODEL EXTRACTION .... V - 169
João Silva, Luís Silveira, IST/INESC-ID/Cadence European Labs, Portugal

CAD-L7: INTERCONNECT AND CLOCK DISTRIBUTION

CAD-L7.1: PARTIAL RANDOM WALK FOR LARGE LINEAR NETWORK ANALYSIS .... V - 173
Weikun Guo, Sheldon Tan, University of California, Riverside, United States; Zuying Luo, Xianlong Hong, Tsinghua University, China

CAD-L7.2: MATRIX PENCIL BASED REALIZABLE REDUCTION FOR DISTRIBUTED INTERCONNECTS .... V - 177
Janet Meiling Wang, Omar Hafiz, University of Arizona, United States

CAD-L7.3: FREQUENCY DRIVEN REPEATER INSERTION FOR DEEP SUBMICRON .... V - 181
Nisar Ahmed, Mohammad Tehranipour, Dian Zhou, Mehrdad Nourani, University of Texas, Dallas, United States

CAD-L7.4: MODELING AND IMPLEMENTATION OF TWISTED DIFFERENTIAL ON-CHIP INTERCONNECTS FOR CROSSTALK NOISE REDUCTION .... V - 185
Ilhan Hatirnaz, Yusuf Leblebici, Swiss Federal Institute of Technology (EPFL), Switzerland

CAD-L7.5: DELAY BOUND BASED CMOS GATE SIZING TECHNIQUE .... V - 189
Alexandre Verle, Xavier Michel, Philippe Maurine, Nadine Azémard, Daniel Auvergne, Université de Montpellier II, France

CAD-L8: FUNDAMENTALS OF CAD ALGORITHMS

CAD-L8.1: IMPROVING SYMBOLIC ANALYSIS IN CMOS ANALOG INTEGRATED CIRCUITS .... V - 193
Jorge Aguila-Meza, Leticia Torres-Papaqui, Esteban Tlelo-Cuautle, INAOE, Mexico

CAD-L8.2: A SYSTEMATIC APPROACH FOR ANALYZING FAST ADDITION ALGORITHMS USING COUNTER TREE DIAGRAMS .... V - 197
Naofumi Homma, Jun Sakiyama, Taihei Wakamatsu, Takafumi Aoki, Tohoku University, Japan; Tatsuo Higuchi, Tohoku Institute of Technology, Japan

CAD-L8.3: HWP: A NEW INSIGHT INTO CANONICAL SIGNED DIGIT .... V - 201
Fei Xu, Chip-Hong Chang, Ching-Chuen Jong, Nanyang Technological University, Singapore

CAD-L8.4: FAST MULTILEVEL FLOORPLANNING FOR LARGE SCALE MODULES .... V - 205
Ching-Chung Hu, De-Sheng Chen, Yi-Wen Wang, Feng Chia University, Taiwan

CAD-L8.5: HOW MANY SOLUTIONS DOES A SAT INSTANCE HAVE? .... V - 209
Pushkin Pari, Lin Yuan, Gang Qu, University of Maryland, United States

CAD-L9: VERIFICATION, TESTING, AND VALIDATION

CAD-L9.1: A NEW GENERATION OF ISCAS BENCHMARKS FROM FORMAL VERIFICATION OF HIGH-LEVEL MICROPROCESSORS .... V - 213
Miroslav Velev, Carnegie Mellon University, United States

CAD-L9.2: ESDINSPECTOR: A NEW LAYOUT-LEVEL ESD PROTECTION CIRCUITRY DESIGN VERIFICATION TOOL USING A SMART-PARAMETRIC CHECKING MECHANISM .... V - 217
Rouying Zhan, Haigang Feng, Haolu Xie, Albert Wang, Illinois Institute of Technology, United States

CAD-L9.3: FAULT EQUIVALENCE AND DIAGNOSTIC TEST GENERATION USING ATPG .... V - 221
Andreas Veneris, Robert Chang, University of Toronto, Canada; Magdy S. Abadir, Motorola, Inc., United States; Mandana Amiri, University of British Columbia, Canada

CAD-L9.4: A HYBRID-TYPE TEST PATTERN GENERATING MECHANISM .... V - 225
Chuen-Yau Chen, An-Chi Hsu, National Yunlin University of Science and Technology, Taiwan

CAD-L9.5: PLACEMENT AND ROUTING OPTIMIZATION FOR CIRCUITS DERIVED FROM BDDS .... V - 229
Thomas Eschbach, Albert-Ludwigs-University, Germany; Rolf Drechsler, University of Bremen, Germany; Bernd Becker, Albert-Ludwigs-University, Germany

CAD-L10: NEW AREAS IN CAD I

CAD-L10.1: L-SIMULATOR: A MAGPEEC-BASED NEW CAD TOOL FOR SIMULATING MAGNETIC-ENHANCED IC INDUCTORS OF 3D ARBITRARY GEOMETRY .... V - 233
Haibo Long, Tsinghua Unversity, China; Zhenghe Feng, Tsinghua University, China; Haigang Feng, Albert Wang, Illinois Institute of Technology, United States; Tianling Ren, Tsinghua University, China

CAD-L10.2: AN EFFICIENT APPROACH FOR HIERARCHICAL SUBMODULE EXTRACTION .... V - 237
Yi-Wei Lin, Jing-Yang Jou, National Chao Tung University, Taiwan

CAD-L10.3: ALGORITHM FOR YIELD DRIVEN CORRECTION OF LAYOUT .... V - 241
Yang Wang, Yici Cai, Xianlong Hong, Qiang Zhou, Tsinghua University, China

CAD-L10.4: SYMBOLIC NOISE ANALYSIS IN ANALOG INTEGRATED CIRCUITS .... V - 245
Carlos Sánchez-López, Esteban Tlelo-Cuautle, INAOE, Mexico

CAD-L10.5: AN EFFICIENT LOGIC EXTRACTION ALGORITHM USING PARTITIONING AND CIRCUIT ENCODING .... V - 249
Lily Huang, Tai-Ying Jiang, Jing-Yang Jou, National Chiao Tung University, Taiwan; Heng-Liang Huang, Legend Design Technology, Taiwan

CAD-L11: NEW AREAS IN CAD II

CAD-L11.1: FREQUENCY-DOMAIN ERROR ANALYSIS OF LINEAR MULTISTEP METHODS .... V - 253
Giorgio Casinovi, Georgia Institute of Technology, United States; Giuseppe Veca, Università di Roma "La Sapienza", Italy

CAD-L11.2: ENHANCING SCHEDULING SOLUTIONS THROUGH ANT COLONY OPTIMIZATION .... V - 257
Shekhar Kopuri, Nazanin Mansouri, Syracuse University, United States

CAD-L11.3: SYNTHESIS SCHEME FOR LOW POWER DESIGNS WITH MULTIPLE SUPPLY VOLTAGES BY TABU SEARCH .... V - 261
Ling Wang, Yingtao Jiang, Henry Selvaraj, University of Nevada Las Vegas, United States

CAD-L11.4: GRAPHICS PROCESSOR UNIT (GPU) ACCELERATION OF FINITE-DIFFERENCE TIME-DOMAIN (FDTD) ALGORITHM .... V - 265
Sean Krakiwsky, Laurence Turner, Michal Okoniewski, University of Calgary, Canada

CAD-L11.5: A NEW MULTI-RAMP DRIVER MODEL WITH RLC INTERCONNECT LOAD .... V - 269
Lakshmi Kalpana Vakati, Janet Meiling Wang, University of Arizona, United States

CAD-P1: CAD FOR ANALOG AND MIXED SIGNAL CIRCUITS

CAD-P1.1: A COMPACT OPTIMIZATION METHODOLOGY FOR SINGLE-ENDED LNA .... V - 273
Gülin Tulunay, Sina Balkir, University of Nebraska-Lincoln, United States

CAD-P1.2: X RAY AND BLUE PRINT: TOOLS FOR MOSFET ANALOG CIRCUIT DESIGN ADDRESSING SHORT-CHANNEL EFFECTS .... V - 277
Rodrigo Oliveira Pinto, Texas A&M University, United States; Franco Maloberti, University of Texas, Dallas, United States

CAD-P1.3: SEAMS - A SYSTEMC ENVIRONMENT WITH ANALOG AND MIXED-SIGNAL EXTENSIONS .... V - 281
Hessa Aljunaid, Tom J. Kazmierski, University of Southampton, United Kingdom

CAD-P1.4: PROPERTIES OF FASTEST LINEARLY INDEPENDENT TRANSFORMS OVER GF(3) .... V - 285
Bogdan Falkowski, Cheng Fu, Nanyang Technological University, Singapore

CAD-P1.5: MULTI-POLARITY HELIX TRANSFORM OVER GF(3) .... V - 289
Cheng Fu, Bogdan Falkowski, Nanyang Technological University, Singapore

CAD-P2: CAD FOR DIGITAL CIRCUITS

CAD-P2.1: A DESIGN FLOW FOR MULTIPLIERLESS LINEAR-PHASE FIR FILTERS: FROM SYSTEM SPECIFICATION TO VERILOG CODE .... V - 293
Kai-Yuan Jheng, National Taiwan University, Taiwan; Shyh-Jye Jou, National Central University, Taiwan; An-Yeu (Andy) Wu, National Taiwan University, Taiwan

CAD-P2.2: SHIELDING AREA OPTIMIZATION UNDER THE SOLUTION OF INTERCONNECT CROSSTALK .... V - 297
Xin Zhao, Yici Cai, Qiang Zhou, Xianlong Hong, Tsinghua University, China; Lei He, Jinjun Xiong, University of California, Los Angeles, United States

CAD-P2.3: PERFORMANCE METRICS FOR ASYNCHRONOUS DIGITAL CIRCUITS APPLICABLE TO COMPUTER-AIDED DESIGN .... V - 301
Rajani Parthasarthy, Ivan Kourtev, University of Pittsburgh, United States

CAD-P2.4: RTL/ISS CO-MODELING METHODOLOGY FOR EMBEDDED PROCESSOR USING SYSTEMC .... V - 305
Yoichi Yuyama, Masao Aramoto, Kyoto University, Japan; Kazutoshi Kobayashi, University of Tokyo, Japan; Hidetoshi Onodera, Kyoto University, Japan

CAD-P3: TESTING, VERIFICATION, AND SIMULATION

CAD-P3.1: TEST VECTOR GENERATION AND CLASSIFICATION USING FSM TRAVERSALS .... V - 309
Ralph Marczynski, Mitchell Thornton, Stephen Szygenda, Southern Methodist University, United States

CAD-P3.2: FORMAL VERIFICATION OF AN SOC PLATFORM PROTOCOL CONVERTER .... V - 313
Jounaïdi Ben Hassen, Sofiène Tahar, Concordia University, Canada

CAD-P3.3: FPGA BASED ACCELERATOR FOR FUNCTIONAL SIMULATION ..... V - 317
Mohamed Wageeh, Ayman Wahba, Ashraf Salem, Mentor Graphics Egypt, Egypt; Mohamed Sheirah, Ain Shams University, Egypt

CAD-P3.4: FREQUENCY DOMAIN WAVELET METHOD WITH GMRES FOR LARGE-SCALE LINEAR CIRCUIT SIMULATION ..... V - 321
Jian Wang, Xuan Zeng, Fudan University, China; Wei Cai, University of North Carolina at Charlotte, United States; Charles Chiang, Synopsys, Inc., United States; Jiarong Tong, Fudan University, China; Dian Zhou, University of Texas, Dallas, United States

CAD-P3.5: EVENT-DRIVEN DYNAMIC POWER MANAGEMENT BASED ON WAVELET FORECASTING THEORY ..... V - 325
Ali Abbasian, Safar Hatami, Ali Afzali-Kusha, University of Tehran, Iran; Mehrdad Nourani, University of Texas, Dallas, United States; Caro Lucas, University of Tehran, Iran

CAD-P4: NEW IDEAS IN PHYSICAL DESIGN

CAD-P4.1: A NOVEL ENCODING METHOD INTO SEQUENCE-PAIR ..... V - 329
Chikaaki Kodama, Kunihiro Fujiyoshi, Tokyo University of Agriculture and Technology, Japan; Teppei Koga, Japan Advanced Institute of Science and Technology, Japan

CAD-P4.2: COMPUTING LARGE-CHANGE SENSITIVITY OF PERIODIC RESPONSES OF NONLINEAR CIRCUITS USING REDUCTION TECHNIQUES ..... V - 333
Praveen Pai, Carleton University, Canada; Emad Gad, University of Ottawa, Canada; Ram Achar, Carleton University, Canada; Roni Khazaka, McGill University, Canada; Michel Nakhla, Carleton University, Canada

CAD-P4.3: QUICK AND EFFECTIVE BUFFERED LEGITIMATE SKEW CLOCK ROUTING ..... V - 337
Meng Zhao, Xinjie Wei, Yici Cai, Xianlong Hong, Tsinghua University, China

CAD-P4.4: THEORY OF T-JUNCTION FLOORPLANS IN TERMS OF SINGLE-SEQUENCE ..... V - 341
Xuliang Zhang, SII EDA Technologies, Inc., Japan; Yoji Kajitani, University of Kitakyushu, Japan

CAD-P4.5: GENERATING RANDOM BENCHMARK CIRCUITS FOR FLOORPLANNING ..... V - 345
Tao Wan, Malgorzata Chrzanowska-Jeske, Portland State University, United States

INV-1: SPIKING NEURAL NETWORKS I

INV-1.1: A PROGRAMMABLE ARRAY OF SILICON NEURONS FOR THE CONTROL OF LEGGED LOCOMOTION ..... V - 349
Francesco Tenore, Johns Hopkins University, United States; Ralph Etienne-Cummings, Johns Hopkins University / University of Maryland, College Park, United States; M. Anthony Lewis, Iguana-Robotics, Inc., United States

INV-1.2: SIGNAL RECONSTRUCTION FROM SPIKING NEURON MODELS ..... V - 353
Dazhi Wei, John Harris, University of Florida, United States

INV-1.3: AN EVENT-BASED VLSI NETWORK OF INTEGRATE-AND-FIRE NEURONS ..... V - 357
Elisabetta Chicca, Giacomo Indiveri, Rodney J. Douglas, University of Zürich / ETH Zürich, Switzerland

INV-1.4: A TIME DOMAIN WINNER-TAKE-ALL NETWORK OF INTEGRATE-AND-FIRE NEURONS ..... V - 361
Jens Petter Abrahamsen, Philipp Häfliger, Tor Sverre Lande, University of Oslo, Norway

INV-2: ULTRA WIDEBAND SYSTEMS

INV-2.1: MULTI-BAND OFDM: A NEW APPROACH FOR UWB ..... V - 365
Anuj Batra, Jaiganesh Balakrishnan, Anand Dabak, Texas Instruments, United States

INV-2.2: PULSED OFDM MODULATION FOR ULTRA WIDEBAND COMMUNICATIONS .... V - 369
Ebrahim Saberinia, Jun Tang, Ahmed Tewfik, Keshab K. Parhi, University of Minnesota, United States

INV-2.3: INTERFERENCE RESILIENT TRANSMISSION SCHEME FOR MULTIBAND OFDM SYSTEM IN UWB CHANNELS .... V - 373
Seung Young Park, Samsung Advanced Institute of Technology, Republic of Korea; Gadi Shor, Wisair, Israel; Yong Suk Kim, Samsung Advanced Institute of Technology, Republic of Korea

INV-2.4: HIGH PERFORMANCE SOLUTION FOR INTERFERING UWB PICONETS WITH REDUCED COMPLEXITY SPHERE DECODING .... V - 377
Jun Tang, Ahmed Tewfik, Keshab K. Parhi, University of Minnesota, United States

INV-2.5: CHANNEL ESTIMATION AND SYNCHRONIZATION WITH SUB-NYQUIST SAMPLING AND APPLICATION TO ULTRA-WIDEBAND SYSTEMS .... V - 381
Irena Maravic, Martin Vetterli, Swiss Federal Institute of Technology, Lausanne, Switzerland; Kannan Ramchandran, University of California, Berkeley, United States

INV-3: SPIKING NEURAL NETWORKS II

INV-3.1: SILICON SPIKE-BASED SYNAPTIC ARRAY AND ADDRESS-EVENT TRANSCEIVER .... V - 385
R. Jacob Vogelstein, Udayan Mallik, Gert Cauwenberghs, Johns Hopkins University, United States

INV-3.2: BIOLOGICALLY INSPIRED ARTIFICIAL NEURAL NETWORK ALGORITHM WHICH IMPLEMENTS LOCAL LEARNING RULES .... V - 389
Ausra Saudargiene, Bernd Porr, Florentin Wörgötter, University of Stirling, United Kingdom

INV-3.3: SPIKE BASED LEARNING WITH WEAK MULTI-LEVEL STATIC MEMORY .... V - 393
Håvard Kolle Riis, Philipp Häfliger, University of Oslo, Norway

INV-3.4: SPIKE SYNCHRONIZATION IN A NETWORK OF SILICON INTEGRATE-AND-FIRE NEURONS .... V - 397
Shih-Chii Liu, Rodney J. Douglas, University of Zürich / ETH Zürich, Switzerland

INV-3.5: A SPIKE-BASED ANALOGUE CIRCUIT THAT EMPHASISES TRANSIENTS IN AUDITORY STIMULI. .... V - 401
Natasha Chia, Steve Collins, University of Oxford, United Kingdom

INV-4: MULTIRATE SYSTEMS FOR COMMUNICATIONS

INV-4.1: TRANSMULTIPLEXERS AS PRECODERS IN MODERN DIGITAL COMMUNICATION: A TUTORIAL REVIEW .... V - 405
Palghat P. Vaidyanathan, Bojan Vrcelj, California Institute of Technology, United States

INV-4.2: ZERO-FORCING EQUALIZATION FOR TIME-VARYING SYSTEMS WITH MEMORY .... V - 413
Cássio Ribeiro, Marcello de Campos, Paulo Diniz, COPPE/Poli/Federal University of Rio de Janeiro, Brazil

INV-4.3: ON PILOT PATTERN DESIGN FOR PSAM-OFDM SYSTEM .... V - 417
Wei Zhang, The Chinese University of Hong Kong, Hong Kong SAR of China; Xiang-Gen Xia, University of Delaware, United States; Pak-Chung Ching, The Chinese University of Hong Kong, Hong Kong SAR of China

INV-4.4: COMPLETE CHARACTERIZATION OF CHANNEL INDEPENDENT GENERAL DMT SYSTEMS WITH CYCLIC PREFIX .... V - 421
Soura Dasgupta, University of Iowa, United States; Ashish Pandharipande, University of Florida, United States

INV-4.5: ANTIPODAL PARAUNITARY PRECODING FOR OFDM APPLICATION .... V - 425
See-May Phoong, Kai-Yen Chang, National Taiwan University, Taiwan; Yuan-Pei Lin, National Chiao Tung University, Taiwan

INV-5: DIGITAL SIGNAL PROCESSING FOR SMART MULTI-MEDIA SYSTEMS

INV-5.1: NONLINEAR DIGITAL FILTERS FOR BEAUTIFYING FACIAL IMAGES IN MULTI-MEDIA SYSTEMS .... V - 429
Kaoru Arakawa, Meiji University, Japan

INV-5.2: VIDEO ERROR CONCEALMENT TECHNIQUES USING PROGRESSIVE INTERPOLATION AND BOUNDARY MATCHING ALGORITHM .... V - 433
Tsung-Han Tsai, Yu-Xuan Lee, Yu-Fong Lin, National Central University, Taiwan

INV-5.3: JPEG BASED IMAGE COMPRESSION WITH ADAPTIVE RESOLUTION CONVERSION SYSTEM .... V - 437
Kazuhiro Shimauchi, Masahiro Ogawa, Akira Taguchi, Musashi Institute of Technology, Japan

INV-5.4: PERCEPTUAL CODING OF DIGITAL COLOUR IMAGES BASED ON A VISION MODEL .... V - 441
C. S. Tan, D. M. Tan, H. R. Wu, Monash University, Australia

INV-5.5: SMART NOISE REDUCTION SYSTEM BASED ON ALE AND NOISE RECONSTRUCTION SYSTEM .... V - 445
Naoto Sasaoka, Yoshio Itoh, Tottori University, Japan; Kensaku Fujii, Himeji Institute of Technology, Japan; Yutaka Fukui, Tottori University, Japan

INV-6: SILICON IMPLEMENTATIONS OF CNN AND PROGRAMMABLE MIXED-SIGNAL VISION I

INV-6.1: A 39x48 GENERAL-PURPOSE FOCAL-PLANE PROCESSOR ARRAY INTEGRATED CIRCUIT .... V - 449
Piotr Dudek, UMIST, United Kingdom

INV-6.2: AN IMPROVED ANALOG COMPUTATION CELL FOR PARIS II, A PROGRAMMABLE VISION CHIP .... V - 453
Sébastien Moutault, Hervé Mathias, Jacques-Olivier Klein, Antoine Dupret, Université de Paris 11-Orsay, France

INV-6.3: A CNN-DRIVEN LOCALLY ADAPTIVE CMOS IMAGE SENSOR .... V - 457
Ricardo Carmona, Carlos M. Domínguez-Matas, Jorge Cuadri, Francisco Jiménez-Garrido, Angel Rodríguez-Vázquez, Instituto de Microelectrónica de Sevilla (IMSE), Spain

INV-6.4: Nx16 CELLULAR TEST CHIPS FOR LOW-PASS FILTERING LARGE IMAGES .... V - 461
Asko Kananen, Mika Laiho, Kari Halonen, Helsinki University of Technology, Finland; Ari Paasio, University of Turku, Finland

INV-6.5: A MIXED-SIGNAL CMOS DTCNN CHIP FOR PIXEL-LEVEL SNAKES .... V - 465
V. M. Brea, David L. Vilariño, D. Cabello, University of Santiago de Compostela, Spain

INV-7: NONLINEAR DYNAMICS AND COMPLEXITY IN NETWORK TRAFFIC MODELING AND CONTROL

INV-7.1: AN OVERVIEW AND COMPARISON OF ANALYTICAL TCP MODELS .... V - 469
Inas Khalifa, Ljiljana Trajkovic, Simon Fraser University, Canada

INV-7.2: SELF-SIMILARITY IN MAX/AVERAGE AGGREGATED PROCESSES .... V - 473
Gianluca Mazzini, University of Ferrara, Italy; Riccardo Rovatti, University of Bologna, Italy; Gianluca Setti, University of Ferrara, Italy

INV-7.3: MODELLING OF TCP PACKET TRAFFIC IN A LARGE INTERACTIVE GROWTH NETWORK .... V - 477
David K. Arrowsmith, Matthew Woolf, Queen Mary, University of London, United Kingdom

INV-7.4: ON THE CORRELATION OF TCP TRAFFIC IN BACKBONE NETWORKS .... V - 481
Hung Xuan Nguyen, Patrick Thiran, Swiss Federal Institute of Technology (EPFL), Switzerland; Chadi Barakat, INRIA, France

INV-7.5: A ROBUST APPROACH TO ACTIVE QUEUE MANAGEMENT CONTROL IN NETWORKS .... V - 485
Sabato Manfredi, Franco Garofalo, University of Naples Federico II, Italy; Mario di Bernardo, University of Sannio, Italy

INV-8: BIONICS AND THEORY OF CELLULAR NEURAL NETWORKS

INV-8.1: ON DYNAMIC BEHAVIOR OF WEAKLY CONNECTED CELLULAR NEURAL NETWORKS .... V - 489
Marco Gilli, Fernando Corinto, Politecnico di Torino, Italy

INV-8.2: TOWARDS A BIO-INSPIRED MIXED-SIGNAL RETINAL PROCESSOR .... V - 493
Timothy G. Constandinou, Julius Georgiou, Chris Toumazou, Imperial College of Science, Technology & Medicine, United Kingdom

INV-8.3: BRAIN-MACHINE INTERFACES USING THIN-FILM SILICON MICROELECTRODE ARRAYS .... V - 497
Daryl R. Kipke, University of Michigan, United States

INV-8.4: CNN WAVE BASED COMPUTATION FOR ROBOT NAVIGATION PLANNING .... V - 500
Paolo Arena, Adriano Basile, Università degli Studi di Catania, Italy; Luigi Fortuna, Mattia Frasca, Universitá degli Studi di Catania, Italy

INV-8.5: FINITE ITERATION DT-CNN - NEW DESIGN AND OPERATING PRINCIPLES .... V - 504
Christian Merkwirth, Max-Planck-Institut für Informatik, Germany; Jochen Bröcker, Maciej Ogorzalek, Jorg D. Wichard, AGH University of Science and Technology, Poland

INV-9: CURRENT CHALLENGES IN MIXED-SIGNAL/RF DESIGN AND CAD

INV-9.1: A NUMERICAL DESIGN APPROACH FOR HIGH SPEED, DIFFERENTIAL, RESISTOR-LOADED, CMOS AMPLIFIERS .... V - 508
Ethan Crain, Michael Perrott, Massachusetts Institute of Technology, United States

INV-9.2: ANALYSIS TECHNIQUES FOR OBTAINING THE STEADY-STATE SOLUTION OF MOS LC OSCILLATORS .... V - 512
Makram Mansour, Amit Mehrotra, University of Illinois at Urbana-Champaign, United States; William Walker, Fujitsu Laboratories of America, Inc., United States; Amit Narayan, Berkeley Design Automation, Inc., United States

INV-9.3: MACROMODELING OF DIGITAL LIBRARIES FOR SUBSTRATE NOISE ANALYSIS .... V - 516
Zhe Wang, University of Minnesota, United States; Rajeev Murgai, Fujitsu Laboratories of America, Inc, United States; Jaijeet Roychowdhury, University of Minnesota, United States

INV-9.4: CONTINUATION METHOD IN MULTITONE HARMONIC BALANCE .... V - 520
Suihua Lu, Amit Narayan, Amit Mehrotra, Berkeley Design Automation, Inc., United States

INV-9.5: MILLIMETER-WAVE CMOS DEVICE MODELING AND SIMULATION .... V - 524
Chinh H. Doan, Sohrab Emami, Ali M. Niknejad, Robert W. Brodersen, University of California, Berkeley, United States

INV-10: FREQUENCY-RESPONSE MASKING TECHNIQUES

INV-10.1: IMPROVED DESIGN OF FREQUENCY-RESPONSE-MASKING FILTERS USING ENHANCED SEQUENTIAL QUADRATIC PROGRAMMING .... V - 528
Wu-Sheng Lu, University of Victoria, Canada; Takao Hinamoto, Hiroshima University, Japan

INV-10.2: FREQUENCY-RESPONSE MASKING TECHNIQUE INCORPORATING EXTRAPOLATED IMPULSE RESPONSE BAND-EDGE SHAPING FILTER .... V - 532
Ya Jun Yu, Nanyang Technological University, Singapore; Yong Ching Lim, Nayang Technological University, Singapore; Kok Lay Teo, Guohui Zhao, The Hong Kong Polytechnic University, Hong Kong SAR of China

INV-10.3: FREQUENCY-RESPONSE MASKING BASED FILTERS WITH THE EVEN-LENGTH BANDEDGE SHAPING FILTER .... V - 536
Jianghong Yu, Yong Lian, National University of Singapore, Singapore

INV-10.4: AN EFFICIENT ALGORITHM FOR THE OPTIMIZATION OF FIR FILTERS SYNTHESIZED USING THE MULTISTAGE FREQUENCY-RESPONSE MASKING APPROACH .... V - 540
Juha Yli-Kaakinen, Tapio Saramäki, Tampere University of Technology, Finland; Ya Jun Yu, National University of Singapore, Singapore

INV-10.5: A CONSTANT-Q SPECTRAL TRANSFORMATION WITH IMPROVED FREQUENCY RESPONSE .... V - 544
Danillo Graziosi, Cristiano Santos, Sergio Netto, Luiz Biscainho, Federal University of Rio de Janeiro, Brazil

INV-11: RECENT ADVANCES IN THE CONTROL OF POWER ELECTRONICS

INV-11.1: A ZERO-VOLTAGE AND ZERO-CURRENT SWITCHING THREE-LEVEL DC-DC CONVERTER WITH SECONDARY-ASSISTED REGENERATIVE PASSIVE SNUBBER .... V - 548
Tingting Song, Nianci Huang, Sichuan University, China; Adrian Ioinovici, Holon Academic Institute of Technology, Israel

INV-11.2: HYSTERETIC CONTROLLER FOR CMOS ON-CHIP SWITCHING POWER CONVERTERS .... V - 552
Gerard Villar, Eduard Alarcon, Herminio Martinez, Eva Vidal, Francesc Guinjoan, Technical University of Catalunya (UPC), Spain; Sonia Porta, Universidad Pública de Navarra, Spain; Alberto Poveda, Technical University of Catalunya (UPC), Spain

INV-11.3: STATE TRAJECTORY PREDICTION CONTROL FOR BOOST CONVERTERS .... V - 556
Kelvin Leung, Shu Hung Henry Chung, City University of Hong Kong, Hong Kong SAR of China

INV-11.4: DESIGN EQUATIONS FOR SUB-OPTIMUM OPERATION OF CLASS E AMPLIFIER WITH NONLINEAR SHUNT CAPACITANCE .... V - 560
Tadashi Suetsugu, Fukuoka University, Japan; Marian Kazimierczuk, Wright State University, United States

INV-11.5: TOWARD THE INTEGRATION OF MICROSYSTEMS SUPPLY .... V - 564
Bruno Estibals, Corinne Alonso, Alain Salles, Angel Cid-Pastor, Henri Camon, LAAS-CNRS, France; Luis Martínez Salamero, ETSE-URV, Spain

INV-12: INFORMATION ASSURANCE AND DATA HIDING I

INV-12.1: ENHANCED MULTIPLE HUFFMAN TABLE (MHT) ENCRYPTION SCHEME USING KEY HOPPING .... V - 568
Dahua Xie, C.-C. Jay Kuo, University of Southern California, United States

INV-12.2: VISUAL CRYPTOGRAPHY FOR PRINT AND SCAN APPLICATIONS .... V - 572
Wei-Qi Yan, Duo Jin, Mohan Kankanhalli, National University of Singapore, Singapore

INV-12.3: WEB SEARCH STEGANALYSIS: SOME CHALLENGES AND APPROACHES .... V - 576
Chandramouli Rajarathnam, Stevens Institute of Technology, United States

INV-12.4: ROBUST VQ-BASED DIGITAL WATERMARKING FOR MEMORYLESS BINARY SYMMETRIC CHANNEL .... V - 580
Jeng-Shyang Pan, Min-Tsang Sung, National Kaohsiung University of Applied Sciences, Taiwan; Hsiang-Cheh Huang, National Chiao Tung University, Taiwan; Bin-Yih Liao, National Kaohsiung University of Applied Sciences, Taiwan

INV-12.5: USING INVISIBLE WATERMARKS TO PROTECT VISIBLY WATERMARKED IMAGES .... V - 584
Yongjian Hu, South China University of Technology, China; Sam Kwong, City University of Hong Kong, Hong Kong SAR of China; Jiwu Huang, Sun Yat-Sen University, China

INV-13: HETEROGENEOUS SYSTEMS

INV-13.1: A 2.5 MILLIWATT SOS CMOS RECEIVER FOR OPTICAL INTERCONNECT .... V - 588
Alyssa Apsel, Zhongtao Fu, Cornell University, United States

INV-13.2: ACTIVE SUBSTRATES FOR OPTOELECTRONIC INTERCONNECT .... V - 592
Donald Chiarulli, Steven Levitan, Jason Bakos, University of Pittsburgh, United States; Charlie Kuznia, Peregrine Semiconductor, United States

INV-13.3: INTEGRATED RADIOACTIVE THIN FILMS FOR SENSING SYSTEMS .... V - 596
Amit Lal, Hui Li, Hang Guo, Cornell University, United States

INV-13.4: WIDENING THE DYNAMIC RANGE OF THE READOUT INTEGRATION CIRCUIT FOR UNCOOLED MICROBOLOMETER INFRARED SENSORS .... V - 600
Alexander Belenky, Alexander Fish, Shy Hamami, Vadim Milrud, Orly Yadid-Pecht, Ben-Gurion University, Israel

INV-13.5: A 16 X 16 PIXEL SILICON ON SAPPHIRE CMOS PHOTOSENSOR ARRAY WITH A DIGITAL INTERFACE FOR ADAPTIVE WAVEFRONT CORRECTION .... V - 604
Eugenio Culurciello, Andreas G. Andreou, Johns Hopkins University, United States

INV-14: ADVANCES IN SPEECH PROCESSING WITH APPLICATIONS

INV-14.1: COMPENSATION FOR CLOCK SKEW IN VOICE OVER PACKET NETWORKS BY SPEECH INTERPOLATION .... V - 608
Tõnu Trump, Ericsson AB, Sweden

INV-14.2: FAST ADAPTIVE COMPONENT WEIGHTED CEPSTRUM POLE FILTERING FOR SPEAKER IDENTIFICATION .... V - 612
Arthur Swanson, L-3 Communications, United States; Ravi Ramachandran, Steven Chin, Rowan University, United States

INV-14.3: A LOUDNESS ENHANCEMENT TECHNIQUE FOR SPEECH .... V - 616
Marc Boillot, John Harris, University of Florida, United States

INV-14.4: USABLE SPEECH DETECTION USING A CONTEXT DEPENDENT GAUSSIAN MIXTURE MODEL CLASSIFIER .... V - 620
Robert Yantorno, Brett Smolenski, Ananth Iyer, Jashmin Shah, Temple University, United States

INV-14.5: DECOMPOSITION AND RECOGNITION OF A MULTI-CHANNEL AUDIO SOURCE USING MATCHING PURSUIT ALGORITHM .... V - 624
David Bjonrberg, Sedig Agili, Aldo Morales, Pennsylvania State University, Harrisburg, United States

INV-15: BEHAVIORAL MODELING AND ANALOG AND MIXED SIGNAL SIMULATION

INV-15.1: MIXED SIGNAL ASPECTS OF BEHAVIORAL MODELING AND SIMULATION .... V - 628
Gabriel Popescu, Leonid Goldgeisser, Synopsys, Inc., United States

INV-15.2: BEHAVIOURAL MODELLING OF ANALOGUE FAULTS IN VHDL-AMS — A CASE STUDY .... V - 632
Mark Zwolinski, Andrew Brown, University of Southampton, United Kingdom

INV-15.3: A NEW MODEL ARCHITECTURE FOR CUSTOMER SOFTWARE INTEGRATION .... V - 636
Ken Ruan, Synopsys, Inc., United States

INV-15.4: PROGRAMMING INTERFACE REQUIREMENTS FOR AN AMS SIMULATOR .... V - 640
Martin Vlach, Lynguent, Inc., United States

INV-15.5: MULTIPLE DOMAIN BEHAVIORAL MODELING USING VHDL-AMS .... V - 644
Peter Wilson, Neil Ross, Andrew Brown, Andrew Rushton, University of Southampton, United Kingdom

INV-16: NONLINEARITY: COMPLEXITY AND NOISE

INV-16.1: STATISTICAL PROPERTIES OF CHAOTIC SEQUENCES GENERATED BY JACOBIAN ELLIPTIC CHEBYSHEV RATIONAL MAPS .... V - 648
Tohru Kohda, Kyushu University, Japan

INV-16.2: OUTAGE IN CHAOS COMMUNICATION .... V - 652
Anthony Lawrance, University of Birmingham, United Kingdom; Gan Ohama, Shiga University, Japan

INV-16.3: ANALYSIS OF HYBRID SYSTEMS BY MEANS OF EMBEDDED RETURN MAPS .... V - 656
Joerg Krupar, Andreas Mögel, Wolfgang Schwarz, Dresden University of Technology, Germany

INV-16.4: ANDRONOV-HOPF BIFURCATION OF SINUSOIDAL OSCILLATORS UNDER THE INFLUENCE OF NOISE .... V - 660
Wolfgang Mathis, University of Hannover, Germany

INV-16.5: COMPUTATION BY NATURAL SYSTEMS DEFINED .... V - 664
Ruedi Stoop, Norbert Stoop, ETH and University of Zürich, Switzerland

INV-17: BLIND SIGNAL PROCESSING: BSS AND ICA

INV-17.1: AUDIO SOURCE SEPARATION BASED ON INDEPENDENT COMPONENT ANALYSIS .... V - 668
Shoji Makino, Shoko Araki, Ryo Mukai, Hiroshi Sawada, NTT Corporation, Japan

INV-17.2: ICAR: INDEPENDENT COMPONENT ANALYSIS USING REDUNDANCIES .... V - 672
Laurent Albera, I3S, France; Anne Ferréol, Pascal Chevalier, Thales Communications, France; Pierre Comon, I3S, France

INV-17.3: BLIND SIMO CHANNEL ESTIMATION FOR CPM USING THE LAURENT APPROXIMATION .... V - 676
Shawn Neugebauer, Zhi Ding, University of California, Davis, United States

INV-17.4: ADAPTIVE SUPER-EXPONENTIAL ALGORITHMS FOR BLIND DECONVOLUTION OF MIMO SYSTEMS .... V - 680
Kiyotaka Kohno, Yujiro Inouye, Mitsuru Kawamoto, Tetsuya Okamoto, Shimane University, Japan

INV-17.5: BEYOND ICA: ROBUST SPARSE SIGNAL REPRESENTATIONS .... V - 684
Andrzej Cichocki, Yuanqing Li, Pando Georgiev, Shun-ichi Amari, RIKEN, Brain Science Institute, Japan

INV-18: INFORMATION ASSURANCE AND DATA HIDING II

INV-18.1: BLIND DETECTION OF PHOTOMONTAGE USING HIGHER ORDER STATISTICS .... V - 688
Tian-Tsong Ng, Shih-Fu Chang, Columbia University, United States; Qibin Sun, Institute for Infocomm Research, Singapore

INV-18.2: DATA HIDING FOR BI-LEVEL DOCUMENTS USING SMOOTHING TECHNIQUE .... V - 692
Huijuan Yang, Alex C. Kot, Nanyang Technological University, Singapore

INV-18.3: A ROBUST AUDIO WATERMARKING SCHEME .... V - 696
Hyoung Joong Kim, Taehoon Kim, Kangwon National University, Republic of Korea; In-Kwon Yeo, Chonbuk National University, Republic of Korea

INV-18.4: SLANT TRANSFORM WATERMARKING FOR TEXTURED IMAGES .... V - 700
Anthony T.S. Ho, Xunzhan Zhu, Yong Liang Guan, Pina Marziliano, Nanyang Technological University, Singapore

INV-18.5: A REVIEW OF VIDEO REGISTRATION METHODS FOR WATERMARK DETECTION IN DIGITAL CINEMA APPLICATIONS .... V - 704
Hui Cheng, Sarnoff Corporation, United States

INV-19: SILICON IMPLEMENTATIONS OF CNN AND PROGRAMMABLE MIXED-SIGNAL VISION II

INV-19.1: 3-NEIGHBORHOOD MOTION ESTIMATION IN CNN SILICON ARCHITECTURES .... V - 708
Lauri Koskinen, Ari Paasio, Kari Halonen, Helsinki University of Technology, Finland

INV-19.2: SELF-SYNCHRONIZED AUDIO WATERMARK IN DWT DOMAIN .... V - 712
Shaoquan Wu, Jiwu Huang, Daren Huang, Sun Yat-Sen University, China; Yun Q. Shi, New Jersey Institute of Technology, United States

INV-19.3: FEATURE DIFFERENCE ANALYSIS IN VIDEO AUTHENTICATION SYSTEM .... V - 716
Dajun He, Institute for Infocomm Research, Singapore; Zhiyong Huang, National University of Singapore, Singapore; Ruihua Ma, Qibin Sun, Institute for Infocomm Research, Singapore

INV-19.4: PREDICTION OF BRAIN ELECTRICAL ACTIVITY IN EPILEPSY USING HIGHER-DIMENSIONAL PREDICTION ALGORITHM FOR DISCRETE TIME CELLULAR NEURAL NETWORKS (DTCNN) .... V - 720
Frank Gollas, Christian Niederhöfer, Ronald Tetzlaff, Johann Wolfgang Goethe University, Germany

NSA-L1: NEURAL NETWORK ALGORITHMS AND ARCHITECTURES

NSA-L1.1: FUZZY NEURAL NETWORK CLASSIFICATION DESIGN USING SUPPORT VECTOR MACHINE .... V - 724
Chin-Teng Lin, Chang-Moun Yeh, Chun-Fei Hsu, National Chiao Tung University, Taiwan

NSA-L1.2: GLOBAL CONVERGENCE ANALYSIS OF DECOMPOSITION METHODS FOR SUPPORT VECTOR MACHINES .... V - 728
Norikazu Takahashi, Tetsuo Nishi, Kyushu University, Japan

NSA-L1.3: SELF-ORGANIZING TOPOLOGICAL TREE .... V - 732
Pengfei Xu, Chip-Hong Chang, Nanyang Technological University, Singapore

NSA-L1.4: SENSITIVITY ANALYSIS OF LOW-COMPLEXITY VECTOR QUANTIZERS FOR FOCAL-PLANE IMAGE COMPRESSION .... V - 736
José Gabriel R. C. Gomes, Sanjit K. Mitra, University of California, Santa Barbara, United States

NSA-L1.5: CHAOTIC COMMUNICATIONS USING NONLINEAR TRANSFORM-PAIRS .... V - 740
W. P. Tang, H. K. Kwan, University of Windsor, Canada

NSA-L2: NEURAL NETWORK CIRCUITS AND SYSTEMS I

NSA-L2.1: A NEW CHARGE-PACKET DRIVEN MISMATCH-CALIBRATED INTEGRATE-AND-FIRE NEURON FOR PROCESSING POSITIVE AND NEGATIVE SIGNALS IN AER BASED SYSTEMS .... V - 744
Bernabé Linares-Barranco, Teresa Serrano-Gotarredona, Rafael Serrano-Gotarredona, Jesús Costas-Santos, Instituto de Microelectrónica de Sevilla (IMSE), Spain

NSA-L2.2: CAN SPIKE TIMING DEPENDENT PLASTICITY COMPENSATE FOR PROCESS MISMATCH IN NEUROMORPHIC ANALOGUE VLSI? .... V - 748
Katherine Cameron, Alan Murray, University of Edinburgh, United Kingdom

NSA-L2.3: AN MDAC SYNAPSE FOR ANALOG NEURAL NETWORKS ..... V - 752
Ryan Kier, Reid Harrison, University of Utah, United States; Randall Beer, Case Western Reserve University, United States

NSA-L2.4: SUPERVISED LEARNING IN A TWO-INPUT ANALOG FLOATING-GATE NODE ..... V - 756
Jeff Dugger, Paul Hasler, Georgia Institute of Technology, United States

NSA-L2.5: MIXED-SIGNAL REAL-TIME ADAPTIVE BLIND SOURCE SEPARATION ..... V - 760
Abdullah Celik, Milutin Stanacevic, Gert Cauwenberghs, Johns Hopkins University, United States

NSA-P1: NEURAL SYSTEMS AND APPLICATIONS II

NSA-P1.1: FAST NEURAL NETWORKS FOR SUB-MATRIX (OBJECT/FACE) DETECTION ..... V - 764
Hazem El-Bakry, Herbert Stoyan, Friedrich Alexander University of Erlangen-Nürnberg, Germany

NSA-P1.2: NEURAL NETWORK SYSTEM FOR FACE RECOGNITION ..... V - 768
Ernst Kussul, Tatiana Baidyk, CCADET, UNAM, Mexico; Maksym Kussul, Beenet Computer Systems, Inc., Canada

NSA-P1.3: NO-REFERENCE QUALITY ASSESSMENT OF JPEG IMAGES BY USING CBP NEURAL NETWORKS ..... V - 772
Paolo Gastaldo, Rodolfo Zunino, Genoa University, Italy

NSA-P1.4: A NEURAL SYSTEM FOR RADIATION DISCRIMINATION IN NUCLEAR FUSION APPLICATIONS ..... V - 776
Basilio Esposito, ENEA - Centro Ricerche Frascati, Italy; Luigi Fortuna, Università degli Studi di Catania, Italy; Alessandro Rizzo, Politecnico di Bari, Italy

NSA-P1.5: MIXED ANALOG-DIGITAL IMAGE PROCESSING CIRCUIT BASED ON HAMMING ARTIFICIAL NEURAL NETWORK ARCHITECTURE ..... V - 780
Stéphane Badel, Alexandre Schmid, Yusuf Leblebici, Swiss Federal Institute of Technology (EPFL), Switzerland

NSA-P2: NEURAL NETWORK CIRCUITS AND SYSTEMS II

NSA-P2.1: ON SYNTHETIC AER GENERATION ..... V - 784
Alejandro Linares-Barranco, Gabriel Jimenez-Moreno, Antón Civit-Ballcels, Arquitectura y Tecnología de Computadores. ETSI In, Spain; Bernabé Linares-Barranco, Instituto de Microelectrónica de Sevilla (IMSE), Spain

NSA-P2.2: LOGIC COMPUTATION USING COUPLED NEURAL OSCILLATORS ..... V - 788
Dongming Xu, Jose Principe, John Harris, University of Florida, United States

NSA-P2.3: C4 BAND-PASS DELAY FILTER FOR CONTINUOUS-TIME SUBBAND ADAPTIVE TAPPED-DELAY FILTER ..... V - 792
Heejong Yoo, David Graham, David Anderson, Paul Hasler, Georgia Institute of Technology, United States

NSA-P2.4: NEURAL MODELING OF THE LARGE-SIGNAL DRAIN CURRENT OF THE DUAL-GATE MESFET WITH DC AND PULSED I-V MEASUREMENTS ..... V - 796
Mohammad Abdeen, Mustapha Chérif-Eddine Yagoub, University of Ottawa, Canada

NSA-P2.5: HOPFIELD ASSOCIATIVE MEMORY ON MESH ..... V - 800
Rafic Ayoubi, University of Balamand, Lebanon; Haissam Ziade, Lebanese University, Lebanon; Magdy Bayoumi, University of Louisiana, United States

NSA-P3: NEURAL SYSTEMS AND APPLICATIONS I

NSA-P3.1: FREQUENCY SENSITIVE SELF-ORGANIZING MAPS AND ITS APPLICATION IN COLOR QUANTIZATION ..... V - 804
Chip-Hong Chang, Pengfei Xu, Nanyang Technological University, Singapore

NSA-P3.2: LEARNING VECTOR QUANTIZATION: CLUSTER SIZE AND CLUSTER NUMBER .... V - 808
Christian Borgelt, University of Magdeburg, Germany; Daniela Girimonte, Giuseppe Acciani, Politechnic di Bari, Italy

NSA-P3.3: FAST LEARNING ALGORITHMS FOR NEW L2 SVM BASED ON ACTIVE SET ITERATION METHOD .... V - 812
Juan-juan Gu, Hefei Association University, China; Liang Tao, Anhui University, China; H. K. Kwan, University of Windsor, Canada

NSA-P3.4: DESIGN AND SENSITIVITY ANALYSIS OF FEED-FORWARD NEURAL ADC'S .... V - 816
Hamid Movahedian, Mehrdad Sharif Bakhtiar, Sharif University of Technology, Iran

NSA-P3.5: GLOBAL ASYMPTOTIC STABILITY OF A CLASS OF NEURAL NETWORKS WITH TIME VARYING DELAYS .... V - 820
Tolga Ensari, Sabri Arik, Istanbul University, Turkey; Vedat Tavsanoglu, Yildiz Technical University, Turkey

PSEC-L1: POWER INTEGRATED CIRCUITS

PSEC-L1.1: A 1.2-V BUCK CONVERTER WITH A NOVEL ON-CHIP LOW-VOLTAGE CURRENT-SENSING SCHEME .... V - 824
Chi Yat Leung, Philip K. T. Mok, Ka Nang Leung, The Hong Kong University of Science and Technology, Hong Kong SAR of China

PSEC-L1.2: LOOP GAIN ANALYSIS AND DEVELOPMENT OF HIGH-SPEED HIGH-ACCURACY CURRENT SENSORS FOR SWITCHING CONVERTERS .... V - 828
Hylas Y. H. Lam, Wing-Hung Ki, Hong Kong University of Science and Technology, Hong Kong SAR of China; Dongsheng Ma, Louisiana State University, United States

PSEC-L1.3: 5V-ONLY, STANDARD .5UM CMOS PROGRAMMABLE AND ADAPTIVE FLOATING-GATE CIRCUITS AND ARRAYS USING CMOS CHARGE PUMPS .... V - 832
Mark Hooper, Matt Kucic, Paul Hasler, Georgia Institute of Technology, United States

PSEC-L1.4: ANALYSIS OF SWITCHED CAPACITOR DC-DC STEP DOWN CONVERTER .... V - 836
Andrabadu Viraj, Gehan Amarathunga, University of Cambridge, United Kingdom

PSEC-L1.5: OPTIMIZATION DESIGN OF THE DICKSON CHARGE PUMP CIRCUIT WITH A RESISTIVE LOAD .... V - 840
Ming Zhang, Nicolas Llaser, University of South Paris, France

PSEC-L2: POWER CONVERTER CONTROL

PSEC-L2.1: NOVEL D^2T CONTROL FOR SINGLE-SWITCH DUAL-OUTPUT SWITCHING POWER CONVERTERS .... V - 844
Chi K. Michael Tse, Siu Chung Wong, K. C. Tam, Hong Kong Polytechnic University, Hong Kong SAR of China

PSEC-L2.2: TSK-FUZZY CONTROLLER DESIGN FOR A PWM BOOST DC-DC SWITCHING REGULATOR OPERATING AT DIFFERENT STEADY STATE OUTPUT VOLTAGES .... V - 848
Spartacus Gomáriz, Eduard Alarcón, Francisco Guinjoan, Universitat Politècnica de Catalunya, Spain; Enric Vidal-Idiarte, Luis Martínez Salamero, Universitat Rovira i Virgili, Spain; Domingo Biel, Universitat Politècnica de Catalunya, Spain

PSEC-L2.3: SPURIOUS MODULATION ON CURRENT-MODE CONTROLLED DC/DC CONVERTERS: AN EXPLANATION FOR INTERMITTENT CHAOTIC OPERATION .... V - 852
Siu Chung Wong, Chi K. Michael Tse, K. C. Tam, Hong Kong Polytechnic University, Hong Kong SAR of China

PSEC-L2.4: SMALL-SIGNAL DUTY CYCLE TO INDUCTOR CURRENT TRANSFER FUNCTION FOR BOOST PWM DC-DC CONVERTER IN CONTINUOUS CONDUCTION MODE .... V - 856
Brad Bryant, Marian Kazimierczuk, Wright State University, United States

PSEC-L2.5: SAMPLE AND HOLD EFFECT IN PWM DC-DC CONVERTERS WITH PEAK CURRENT-MODE CONTROL .... V - 860
Brad Bryant, Marian Kazimierczuk, Wright State University, United States

PSEC-L3: POWER AMPLIFIERS

PSEC-L3.1: THE IMPLEMENTATION OF A TRANSIENT DC-LINK BOOST BASED DIGITAL AMPLIFIER FOR ELIMINATING PULSE-DROPPING DISTORTION .... V - 864
Y. C. Julian Chiu, Bin Zhou, Shu Hung Henry Chung, Ricky Lau, City University of Hong Kong, Hong Kong SAR of China

PSEC-L3.2: ANALYSIS OF CLASS D INVERTER WITH IRREGULAR DRIVING PATTERNS .... V - 868
Hirotaka Koizumi, Kosuke Kurokawa, Tokyo University of Agriculture and Technology, Japan; Shinsaku Mori, Nippon Institute of Technology, Japan

PSEC-L3.3: A NEW TOPOLOGY FOR A SIGMA-DELTA AUDIO POWER AMPLIFIER .... V - 872
Antonio Zorzano Martínez, University of La Rioja, Spain; Fernando Beltrán Blázquez, University of Zaragoza, Spain; José Ramón Beltrán Blázquez, University of Zarzagoza, Spain

PSEC-L3.4: SPECTRAL ANALYSIS OF A NOVEL TRANSIENT DYNAMIC BOOST PWM INVERTER CONTROL FOR POWER AMPLIFIERS .... V - 876
Bin Zhou, Y. C. Julian Chiu, Wing Hong Lau, Shu Hung Henry Chung, City University of Hong Kong, Hong Kong SAR of China

PSEC-L3.5: CHARACTERISTICS AND MODELING OF PEM FUEL CELLS .... V - 880
Subbaraya Yuvarajan, Dachuan Yu, North Dakota State university, United States

PSEC-L4: MULTILEVEL POWER CONVERTERS

PSEC-L4.1: A BOOST - SWITCHED CAPACITOR - INVERTER WITH A MULTILEVEL WAVEFORM .... V - 884
Boris Axelrod, Yefim Berkovich, Adrian Ioinovici, Holon Academic Institute of Technology, Israel

PSEC-L4.2: NOVEL MULTILEVEL CONVERTER FOR POWER FACTOR CORRECTION .... V - 888
Bor-Ren Lin, Tsung-Yu Yang, Yung-Chuan Lee, National Yunlin University of Science and Technology, Taiwan

PSEC-L4.3: A COMPACT GENERALIZED SOLUTION TO THE DETERMINATION OF SPECTRAL COMPONENTS FOR MULTILEVEL UNIFORMLY SAMPLED PWM .... V - 892
Bin Zhou, Wing Hong Lau, Shu Hung Henry Chung, City University of Hong Kong, Hong Kong SAR of China

PSEC-L4.4: CURRENT-MODE CONTROL TO ENHANCE CLOSED-LOOP PERFORMANCE OF ASYMMETRICAL HALF-BRIDGE DC-TO-DC CONVERTERS .... V - 896
Wonseok Lim, Byungcho Choi, Jiemyung Ko, Kyungpook National University, Republic of Korea

PSEC-L4.5: STEP-UP VERSUS STEP-DOWN DC/DC CONVERTERS FOR RF-POWERED SYSTEMS .... V - 900
Sean Nicolson, Khoman Phang, University of Toronto, Canada

PSEC-L5: SYSTEMS THEORY FOR POWER

PSEC-L5.1: HILBERT SPACE TECHNIQUES FOR REACTIVE POWER COMPENSATION WITH LIMITED CURRENT BANDWIDTH .... V - 904
Hanoch Lev-Ari, Alex Stankovic, Northeastern University, United States

PSEC-L5.2: A GEOMETRICAL STUDY ON VOLTAGE COLLAPSE MECHANISMS OF POWER SYSTEMS .... V - 908
Yongqiang Liu, South China University of Technology, China; Zheng Yan, Yixin Ni, Felix Wu, University of Hong Kong, China

PSEC-L5.3: PROBABILISTIC LOAD-DEPENDENT CASCADING FAILURE WITH LIMITED COMPONENT INTERACTIONS .... V - 912
Ian Dobson, University of Wisconsin, United States; Benjamin Carreras, Oak Ridge National Laboratory, United States; David Newman, University of Alaska, United States

PSEC-L5.4: METER PLACEMENT FOR LOAD ESTIMATION IN RADIAL POWER DISTRIBUTION SYTEMS .... V - 916
Jie Wan, Areva T&D, United States; Karen Miu, Drexel University, United States

PSEC-L5.5: A TOPOLOGICAL MEASUREMENTS AND RTUS DESIGN AGAINST A CONTINGENCY .... V - 920
Garng Huang, Jiansheng Lei, Texas A&M University, United States

PSEC-P1: POWER ELECTRONICS AND SYSTEMS

PSEC-P1.1: SINGLE-PHASE CAPACITOR CLAMPED INVERTER WITH SIMPLE STRUCTURE .... V - 924
Bor-Ren Lin, Chun-Hao Huang, National Yunlin University of Science and Technology, Taiwan

PSEC-P1.2: A GENERALIZED AVERAGING MODEL FOR FIXED FREQUENCY CONVERTERS .... V - 928
Shek-Wai Ng, Yim-Shu Lee, The Hong Kong Polytechnic University, Hong Kong SAR of China

PSEC-P1.3: CHARACTERIZING NONLINEAR LOAD HARMONICS USING FRACTAL ANALYSIS .... V - 932
Kingsley Umeh, Azah Mohamed, Ramizi Mohamed, Aini Hussain, National University of Malaysia (UKM), Malaysia

PSEC-P1.4: ANALYSIS OF NONLINEAR DYNAMICS IN DELTA MODULATORS FOR PWM CONTROL .... V - 936
Hiroshi Shimazu, Toshimichi Saito, HOSEI University, Japan

PSEC-P2: CONTROL OF POWER CONVERTERS

PSEC-P2.1: A CAD SIMULATOR BASED ON LOOP GAIN MEASUREMENT FOR SWITCHING CONVERTERS .... V - 940
Dongsheng Ma, Louisiana State University, United States; Vincent Tam, Artesyn Technologies Asia-Pacific Ltd., Hong Kong SAR of China; Wing-Hung Ki, Hylas Y. H. Lam, Hong Kong University of Science and Technology, Hong Kong SAR of China

PSEC-P2.2: DIGITALLY CONTROLLED BUCK CONVERTER .... V - 944
Dušan Gleich, Miro Milanovic, Suzana Uran, Franc Mihalic, University of Maribor, Slovenia

PSEC-P2.3: A 3-D PWM CONTROL, H-BRIDGE TRI-LEVEL INVERTER FOR POWER QUALITY COMPENSATION IN THREE-PHASE FOUR-WIRED SYSTEMS .... V - 948
Pui-In Mak, Man-Chung Wong, Seng-Pan U., University of Macau, Macao SAR of China

PSEC-P2.4: AUTOMATED STATE-VARIABLE FORMULATION FOR POWER ELECTRONIC CIRCUITS AND SYSTEMS .... V - 952
Juri Jatskevich, Tarek Aboul-Seoud, University of British Columbia, Canada

PSEC-P2.5: GLOBAL BEHAVIOR ANALYSIS OF A DC-DC BOOST POWER CONVERTER OPERATING WITH CONSTANT POWER LOAD .... V - 956
Claudio Rivetta, Geoffrey Williamson, Illinois Institute of Technology, United States

PSEC-P3: POWER ELECTRONICS CIRCUITS

PSEC-P3.1: SINGLE-PHASE THREE-LEVEL CONVERTER FOR POWER FACTOR CORRECTION .... V - 960
Bor-Ren Lin, Tsung-Yu Yang, National Yunlin University of Science and Technology, Taiwan

PSEC-P3.2: CHARACTERIZATION OF CHARGE-PUMP RECTIFIERS FOR STANDARD SUBMICRON CMOS PROCESSES .... V - 964
Mark Hooper, Matt Kucic, Paul Hasler, Georgia Institute of Technology, United States

PSEC-P3.3: SUPPRESSION OF HARMONIC SPIKES IN ASYNCHRONOUS SIGMA DELTA MODULATION BY RANDOMIZING HYSTERESIS WINDOW .... V - 968
Apinan Aurasopon, Pinit Kumhom, Kosin Chamnongthai, King Mongkut's University of Technology Thonburi, Thailand

PSEC-P3.4: A NEW RELIABLE SELF SUPPLIED GATE DRIVE CIRCUIT FOR SCRS WITH BREAKOVER DIODES FOR PROTECTION .... V - 972
Jun Zhang, Renjie Ding, Haitao Song, Tsinghua University, China

PSEC-P4: ANALYSIS OF POWER SYSTEMS

PSEC-P4.1: HILBERT SPACE TECHNIQUES FOR EVALUATING TRADE-OFFS IN REACTIVE POWER COMPENSATION .... V - 976
Hanoch Lev-Ari, Alex Stankovic, Kevin Xu, Milun Perišic, Northeastern University, United States

PSEC-P4.2: PETRI NET BASED TRANSFORMER FAULT DIAGNOSIS .... V - 980
Pavlos Georgilakis, John Katsigiannis, Kimon Valavanis, Technical University of Crete, Greece

PSEC-P4.3: EQUILIBRIUM ANALYSIS OF VOLTAGE-FED FIELD ORIENTED CONTROLLED INDUCTION MOTORS .... V - 984
Rubén Salas-Cabrera, Instituto Tecnologico de Cd. Madero, Mexico; Claudio A. Cañizares, University of Waterloo, Canada

PSEC-P4.4: NEURAL NETWORK APPROACH FOR ESTIMATION OF LOAD COMPOSITION .... V - 988
Jiwu Duan, Dariusz Czarkowski, Zivan Zabar, Polytechnic University, United States

PSEC-P4.5: MODIFYING EIGENVALUE INTERACTIONS NEAR WEAK RESONANCE .... V - 992
Vincent Auvray, Ian Dobson, University of Wisconsin, United States; Louis Wehenkel, Université de Liège, Belgium

ISCAS 2004

Technical Program

FREQUENCY DOMAIN BLIND SOURCE SEPARATION USING SMALL AND LARGE SPACING SENSOR PAIRS

Ryo Mukai   Hiroshi Sawada   Shoko Araki   Shoji Makino

NTT Communication Science Laboratories, NTT Corporation
2–4 Hikaridai, Seika-cho, Soraku-gun, Kyoto 619–0237, Japan
{ryo,sawada,shoko,maki}@cslab.kecl.ntt.co.jp

ABSTRACT

This paper presents a method for solving the permutation problem of frequency domain blind source separation (BSS) when the number of source signals is large, and the potential source locations are omnidirectional. We propose a combination of small and large spacing sensor pairs with various axis directions in order to obtain proper geometrical information for solving the permutation problem. Experimental results show that the proposed method can separate a mixture of six speech signals that come from various directions, even when two of them come from the same direction.

1. INTRODUCTION

Blind source separation (BSS) is a technique for estimating original source signals using only observed mixtures. When the source signals are $s_i(t)(i = 1, ..., N)$, the signals observed by sensor j are $x_j(t)(j = 1, ..., M)$, and the separated signals are $y_k(t)(k = 1, ..., N)$, the BSS model can be described as: $x_j(t) = \sum_{i=1}^{N}(h_{ji} * s_i)(t)$, $y_k(t) = \sum_{j=1}^{M}(w_{kj} * x_j)(t)$, where h_{ji} is the impulse response from source i to sensor j, w_{kj} are the separating filters, and $*$ denotes the convolution operator.

Figure 1 shows a flow of the BSS in frequency domain. A convolutive mixture in the time domain is converted into multiple instantaneous mixtures in the frequency domain. Therefore, we can apply an ordinary independent component analysis (ICA) algorithm [1] in the frequency domain to solve a BSS problem in a reverberant environment. Using a short-time discrete Fourier transform, the model is approximated as: $\mathbf{X}(\omega, n) = \mathbf{H}(\omega)\mathbf{S}(\omega, n)$, where, ω is the angular frequency, and n represents the frame index. The separating process can be formulated in each frequency bin as: $\mathbf{Y}(\omega, n) = \mathbf{W}(\omega)\mathbf{X}(\omega, n)$, where $\mathbf{S}(\omega, n) = [S_1(\omega, n), ..., S_N(\omega, n)]^T$ is the source signal in frequency bin ω, $\mathbf{X}(\omega, n) = [X_1(\omega, n), ..., X_M(\omega, n)]^T$ denotes the observed signals, $\mathbf{Y}(\omega, n) = [Y_1(\omega, n), ..., Y_N(\omega, n)]^T$ is the estimated source signal, and $\mathbf{W}(\omega)$ represents the separating matrix. $\mathbf{W}(\omega)$ is determined so that $Y_i(\omega, n)$ and $Y_j(\omega, n)$ become mutually independent.

The ICA solution suffers permutation and scaling ambiguities. This is due to the fact that if $\mathbf{W}(\omega)$ is a solution, then $\mathbf{D}(\omega)\mathbf{P}(\omega)\mathbf{W}(\omega)$ is also a solution, where $\mathbf{D}(\omega)$ is a diagonal complex valued scaling matrix, and $\mathbf{P}(\omega)$ is an arbitrary permutation matrix. We thus have to solve the permutation and scaling problems to reconstruct separated signals in the time domain.

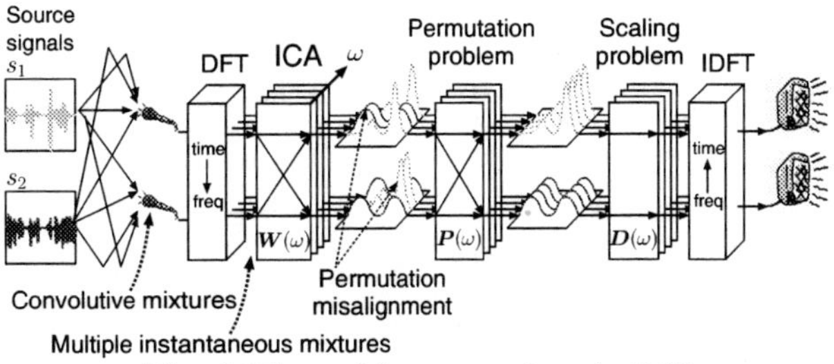

Fig. 1. Flow of frequency domain BSS

There is a simple and reasonable solution for the scaling problem: $\mathbf{D}(\omega) = \text{diag}\{[\mathbf{P}(\omega)\mathbf{W}(\omega)]^{-1}\}$, which is obtained by the minimal distortion principle (MDP) [2], and we can use it. On the other hand, the permutation problem is complicated, especially when the number of source signals is large. Time domain BSS does not suffer the permutation problem, however it takes much computational time as compared with frequency domain BSS [3]. Therefore we adopt the frequency domain approach.

Many methods have been proposed for solving the permutation problem, and the use of geometric information, such as beam patterns [4, 5, 6], direction of arrival (DOA) and source locations[7], is effective approach. We have proposed a robust method by combining the correlation based method [8] and the DOA based method [4, 5], which almost completely solves the problem for 2-source cases [9]. However it is insufficient when the number of signals is large or when the signals come from a similar direction. In this paper, we propose a method for obtaining proper geometric information for solving the permutation problem in such cases.

2. GEOMETRIC INFORMATION FOR SOLVING PERMUTATION PROBLEM

2.1. Invariant in ICA solution

If a separating matrix $\mathbf{W}$ is calculated successfully and it extracts source signals with scaling ambiguity, $\mathbf{D}(\omega)\mathbf{W}(\omega)\mathbf{H}(\omega) = \mathbf{I}$ holds (except for singular frequency bins). Because of the scaling ambiguity, we cannot obtain $\mathbf{H}$ simply from the ICA solution. However, the ratio of elements in the same column $H_{ji}/H_{j'i}$ is invariable in relation to $\mathbf{D}$, and given by

$$\frac{H_{ji}}{H_{j'i}} = \frac{[\mathbf{W}^{-1}\mathbf{D}^{-1}]_{ji}}{[\mathbf{W}^{-1}\mathbf{D}^{-1}]_{j'i}} = \frac{[\mathbf{W}^{-1}]_{ji}}{[\mathbf{W}^{-1}]_{j'i}}, \qquad (1)$$

where $[\cdot]_{ji}$ denotes ji-th element of the matrix. We can estimate several types of geometric information related to source signals by using this invariant. The estimated information is utilized for solving the permutation problem.

2.2. DOA estimation with ICA solution

We can estimate the DOA of source signals by using the above invariant [10]. With a farfield model, a frequency response is formulated as:

$$H_{ji}(\omega) = e^{\jmath\omega c^{-1}\mathbf{a}_i^T\mathbf{p}_j}, \tag{2}$$

where c is the speed of wave propagation, $\mathbf{a}_i$ is a unit vector that points to the direction of source i, and $\mathbf{p}_j$ represents a location of sensor j. According to this model, we have

$$\begin{aligned} H_{ji}/H_{j'i} &= e^{\jmath\omega c^{-1}\mathbf{a}_i^T(\mathbf{p}_j-\mathbf{p}_{j'})} \\ &= e^{\jmath\omega c^{-1}\|\mathbf{p}_j-\mathbf{p}_{j'}\|\cos\theta_{i,jj'}}, \end{aligned} \tag{3}$$

where $\theta_{i,jj'}$ is the direction of source i relative to the sensor pair j and j'. By using the argument of (3) and (1), we can estimate:

$$\begin{aligned} \hat{\theta}_{i,jj'} &= \cos^{-1}\frac{\arg(H_{ji}/H_{j'i})}{\omega c^{-1}\|(\mathbf{p}_j-\mathbf{p}_{j'})\|} \\ &= \cos^{-1}\frac{\arg([\mathbf{W}^{-1}]_{ji}/[\mathbf{W}^{-1}]_{j'i})}{\omega c^{-1}\|(\mathbf{p}_j-\mathbf{p}_{j'})\|}. \end{aligned} \tag{4}$$

This procedure is valid for sensor pairs with a small spacing.

2.3. Estimation of sphere with ICA solution

Interpretation of the ICA solution by a nearfield model yields other geometric information. When we adopt the nearfield model, including the attenuation of the wave, $H_{ji}(\omega)$ is formulated as:

$$H_{ji}(\omega) = \frac{1}{\|\mathbf{q}_i-\mathbf{p}_j\|}e^{\jmath\omega c^{-1}(\|\mathbf{q}_i-\mathbf{p}_j\|)} \tag{5}$$

where $\mathbf{q}_i$ represents the location of source i. By taking the ratio of (5) for a pair of sensors j and j' we obtain:

$$H_{ji}/H_{j'i} = \frac{\|\mathbf{q}_i-\mathbf{p}_{j'}\|}{\|\mathbf{q}_i-\mathbf{p}_j\|}e^{\jmath\omega c^{-1}(\|\mathbf{q}_i-\mathbf{p}_j\|-\|\mathbf{q}_i-\mathbf{p}_{j'}\|)}. \tag{6}$$

By using the modulus of (6) and (1), we have:

$$\frac{\|\mathbf{q}_i-\mathbf{p}_{j'}\|}{\|\mathbf{q}_i-\mathbf{p}_j\|} = \left|\frac{[\mathbf{W}^{-1}]_{ji}}{[\mathbf{W}^{-1}]_{j'i}}\right|. \tag{7}$$

By solving (6) for $\mathbf{q}_i$, we have a sphere whose center $O_{i,jj'}$ and radius $R_{i,jj'}$ are given by:

$$O_{i,jj'} = \mathbf{p}_j - \frac{1}{r_{i,jj'}^2-1}(\mathbf{p}_{j'}-\mathbf{p}_j), \tag{8}$$

$$R_{i,jj'} = \|\frac{r_{i,jj'}}{r_{i,jj'}^2-1}(\mathbf{p}_{j'}-\mathbf{p}_j)\|, \tag{9}$$

where $r_{i,jj'} = |[\mathbf{W}^{-1}]_{ji}/[\mathbf{W}^{-1}]_{j'i}|$. Thus, we can estimate a sphere $(\hat{O}_{i,jj'}, \hat{R}_{i,jj'})$ on which $\mathbf{q}_i$ exists by using the result of ICA $\mathbf{W}$ and the locations of the sensors $\mathbf{p}_j$ and $\mathbf{p}_{j'}$. Figure 2 shows an example of the spheres determined by (7) for various ratios $r_{i,jj'}$. This procedure is valid for sensor pairs with a large spacing.

The models (2) and (5) are simple approximation without the multi-path propagation and reverberation, however we can obtain information for classifying signals by using them.

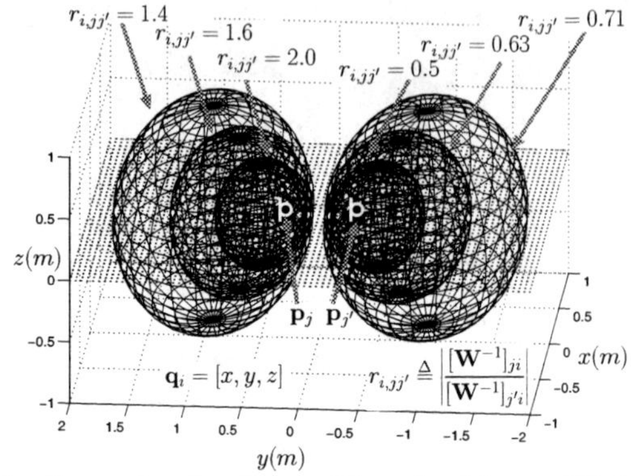

Fig. 2. Example of spheres determined by eq.(7) ($\mathbf{p}_j = [0, 0.3, 0]$, $\mathbf{p}_{j'} = [0, -0.3, 0]$)

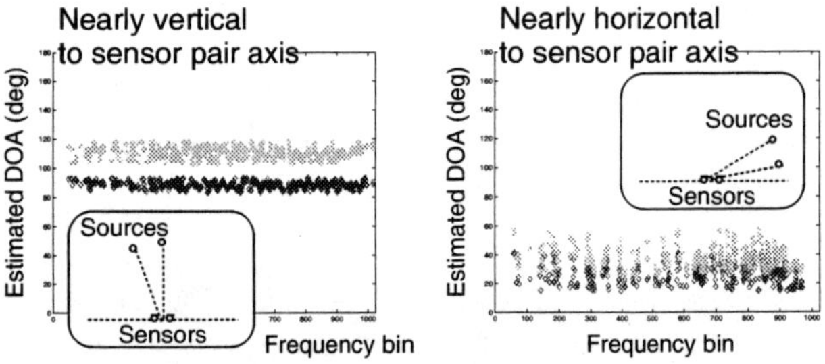

Fig. 3. Source locations and estimated DOAs

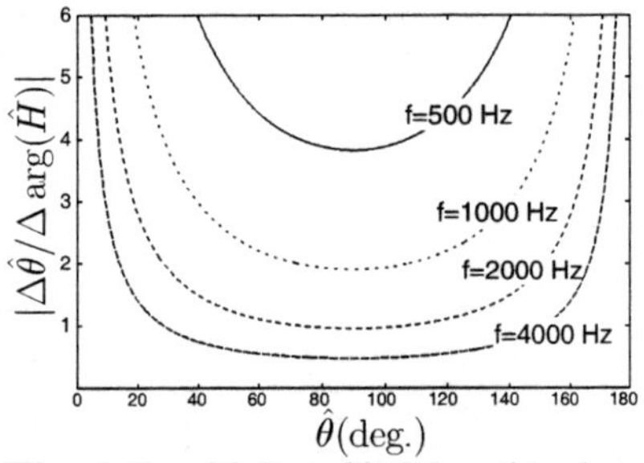

Fig. 4. Sensitivity of DOA estimation

3. SENSITIVITY AND AMBIGUITY IN SOURCE LOCATION ESTIMATION

3.1. Sensitivity of DOA estimation

DOA estimation is sensitive to source locations. Figure 3 shows examples of DOA estimation with two different source locations. When the source signals are almost in front of a sensor pair, their directions can be estimated robustly. However, when the signals are nearly horizontal to the axis of the pair, the estimated directions tend to have large errors. This can be explained as follows.

When we denote an error in calculated $\arg(H_{ji}/H_{j'i})$ as $\Delta\arg(\hat{H})$, and an error in $\hat{\theta}_{i,jj'}$ as $\Delta\hat{\theta}$, the ratio $|\Delta\hat{\theta}/\Delta\arg(\hat{H})|$ can be approximated by the partial derivative of (4):

$$\left|\frac{\Delta\hat{\theta}}{\Delta\arg(\hat{H})}\right| \approx \left|\frac{1}{\omega c^{-1}\|\mathbf{p}_j-\mathbf{p}_{j'}\|\sin(\hat{\theta}_{i,jj'})}\right|. \tag{10}$$

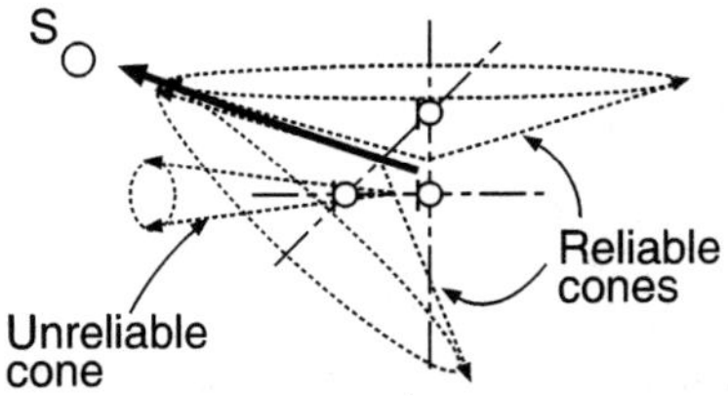

Fig. 5. Combination of small spacing sensor pairs with different axes

Figure 4 shows examples of this value for several frequency bins. We can see that $\Delta \arg(\hat{H})$ causes a large error in the estimated DOA when the direction is near the axis of the sensor pair. Therefore, we consider the estimated DOA to be unreliable in such cases.

3.2. Ambiguity of estimated DOA

Another problem regarding DOA estimation is ambiguity. When we use only one pair of sensors or a linear array, we estimate a cone rather than a direction. If we assume a plane on which sources exist, the cone is reduced to two half-lines. However, the ambiguity of two directions that are symmetrical with respect to the axis of the sensor pair still remains.This is a fatal problem when the source locations are omnidirectional.

When the spacing between sensors is larger than half a wavelength, spatial aliasing causes another ambiguity, but we do not consider this here.

3.3. Resolving sensitivity and ambiguity

A nonlinear arrangement of sensors is suitable for resolving both sensitivity and ambiguity. We propose a combination of small and large spacing sensor pairs that have various axis directions.

By using the DOA estimation described in Sec.2.2 with the small spacing sensor pairs that have different axis directions, we can estimate cones which have various vertex angles for one source direction. Because of the sensitivity explained in Sec.3.1, we assume that obtuse cones are reliable, and acute cones are unreliable. Then, we can determine a bearing line pointing to a source direction by using the reliable cones (Fig. 5).

Even when some signals come from the same or a similar direction, we can distinguish between them by using the information provided by the large spacing sensor pair described in Sec.2.3. The source locations can be estimated by combining the estimated direction and spheres (Fig. 6). Then, we can classify separated signals in the frequency domain according to the estimated source locations.

4. EXPERIMENTS

We carried out experiments for 6 sources and 8 microphones using speech signals convolved with impulse responses measured in a room. The room layout is shown in Fig. 7. Other conditions are summarized in Table 1. The experimental procedure is as follows.

First, we apply ICA to $x_j(t) (j = 1, ..., 8)$, and calculate separating matrix $\mathbf{W}(\omega)$ for each frequency bin. Then we

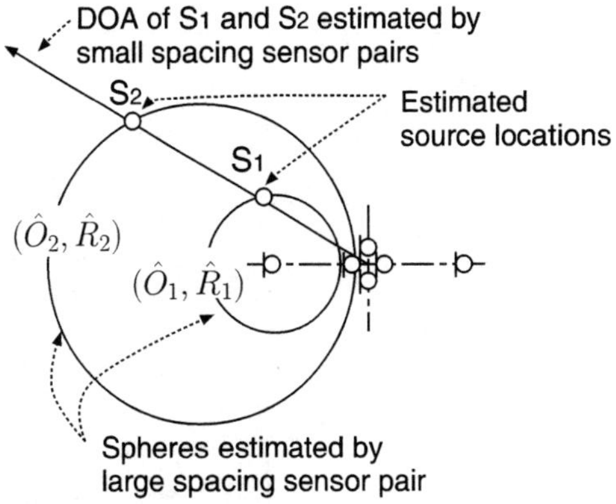

Fig. 6. Combination of small spacing sensor pairs and a large spacing sensor pair

estimate DOAs by using the rows of $\mathbf{W}^{-1}(\omega)$ corresponding to the small spacing microphone pairs (1-3, 2-4, 1-2 and 2-3). Figure 8 shows a histogram of estimated DOAs. We can find five clusters in this histogram, and one cluster is twice the size of the others. This implies that these are six source signals, and two of them come from the same direction (about 150°). We can solve permutation for four sources by using this information (Fig. 9).

Then, we apply the estimation of spheres to the signals that belong to the large cluster by using the rows of $\mathbf{W}^{-1}(\omega)$ corresponding to the large spacing microphone pairs (7-5, 7-8, 6-5 and 6-8). Figure 10 shows estimated radiuses for S_4 and S_5 regarding the microphone pair 7-5. Finally, we can classify the signals into six clusters.

Unfortunately, the classification by the estimated location tends to be inconsistent especially in a reverberant environment. In many frequency bins, several signals are assigned to the same cluster, and such classification is inconsistent. We solve the permutation only for frequency bins with a consistent classification, and we employ a correlation based method [9] for the rest. The correlation based method solves the permutation so that the inter-frequency correlation for neighboring or harmonic frequency bins becomes maximized. In addition, we use the spectral smoothing method proposed in [11] to make separating filters in the time domain from the result of ICA $\mathbf{W}(\omega)$.

The performance is measured from the signal-to-inference ratio (SIR). The portion of $y_k(t)$ that comes from $s_i(t)$ is calculated by $y_{ki}(t) = \sum_{j=1}^{M} (w_{kj} * h_{ji} * s_i)(t)$. If we solve the permutation problem so that $s_i(t)$ is output to $y_i(t)$, the SIR for $y_k(t)$ is defined as:

$$\text{SIR}_k = 10 \log[\textstyle\sum_t y_{kk}(t)^2 / \sum_t (\sum_{i \neq k} y_{ki}(t))^2] \text{ (dB)}.$$

We measured SIRs for three permutation solving strategies: only the correlation based method ("C"), the estimated DOAs and correlation ("D+C"), and the combination of estimated DOAs, spheres and correlation ("D+S+C", proposed method). We also measured input SIRs by using the mixture observed by microphone 1 for the reference ("Input SIR"). The results are summarized in Table 2.

Our proposed method succeeded in separating six

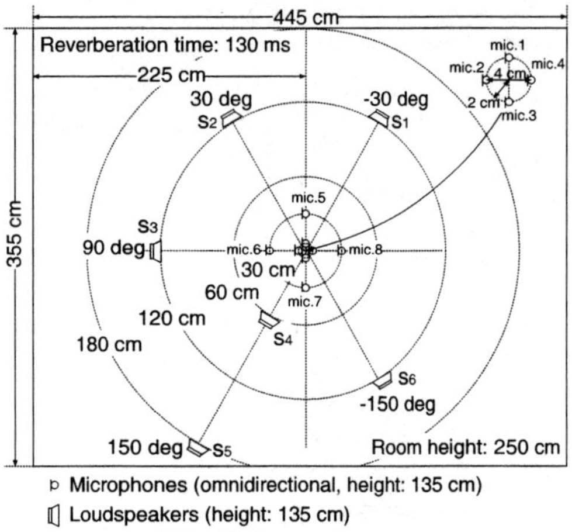

Fig. 7. Room layout

Table 1. Experimental conditions

Sampling rate	8 kHz
Data length	6 s
Frame length	2048 point (256 ms)
Frame shift	512 point (64 ms)
ICA algorithm	Infomax (complex valued)

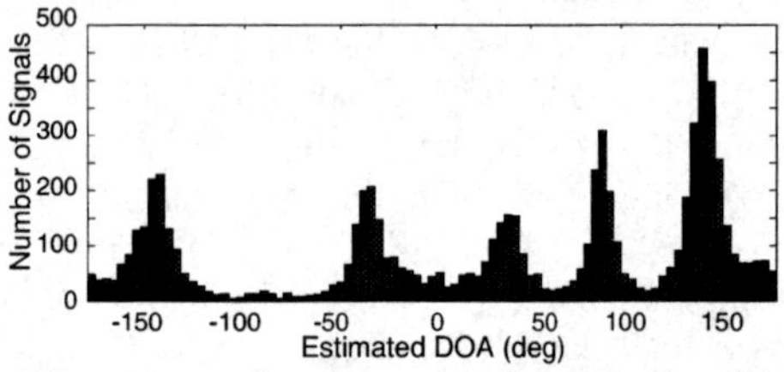

Fig. 8. Histogram of estimated DOAs obtained by using small spacing microphone pairs

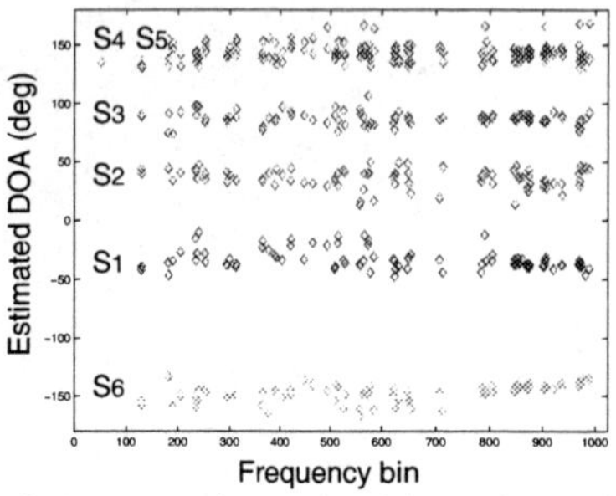

Fig. 9. Permutation solved by using DOAs

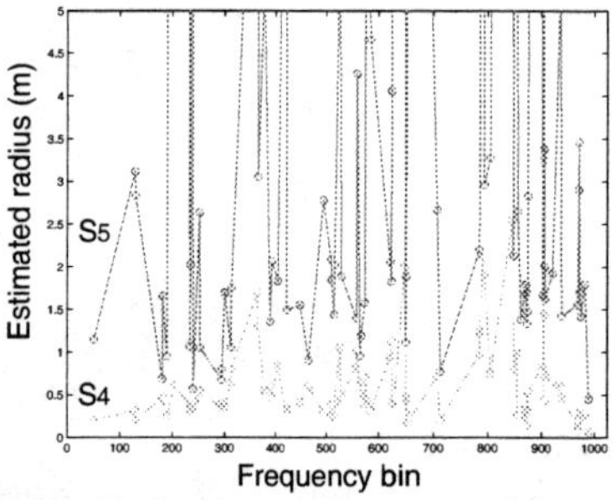

Fig. 10. Estimated radiuses for S_4 and S_5

Table 2. Experimental results (dB)

	SIR_1	SIR_2	SIR_3	SIR_4	SIR_5	SIR_6	ave.
Input SIR	-8.3	-6.8	-7.8	-7.7	-6.7	-5.2	-7.1
C	4.4	2.6	4.0	9.2	3.6	-2.0	3.7
D+C	4.5	10.8	14.4	4.5	5.4	8.8	8.1
D+S+C	12.3	5.6	14.5	7.6	8.9	10.8	10.0

speech signals. We can see that the discrimination obtained by using estimated spheres is effective in improving the separation performance for signals coming from the same direction.

5. CONCLUSION

We proposed the combination of small and large spacing microphone pairs with various axis directions in order to obtain proper geometrical information for solving the permutation problem in frequency domain BSS. In experiments, our method succeeded in the separation of six speech signals, even when two come from the same direction. The computation time was about 1 min. for 6 s. data.

6. REFERENCES

[1] A. Hyvärinen, J. Karhunen, and E. Oja, *Independent Component Analysis*, John Wiley & Sons, 2001.

[2] K. Matsuoka and S. Nakashima, "Minimal distortion principle for blind source separation," in *Proc. of Intl. Workshop on Independent Component Analysis and Blind Signal Separation (ICA'01)*, 2001, pp. 722–727.

[3] K. Matsuoka, Y. Ohba, Y. Toyota, and S. Nakashima, "Blind separation for convolutive mixture of many voices," in *Proc. IWAENC 2003*, 2003, pp. 279–282.

[4] S. Kurita, H. Saruwatari, S. Kajita, K. Takeda, and F. Itakura, "Evaluation of blind signal separation method using directivity pattern under reverberant conditions," in *Proc. of ICASSP'00*, 2000, pp. 3140–3143.

[5] M. Z. Ikram and D. R. Morgan, "A beamforming approach to permutation alignment for multichannel frequency-domain blind speech separation," in *Proc. of ICASSP'02*, 2002, vol. 1, pp. 881–884.

[6] L. C. Parra and C. V. Alvino, "Geometric source separation: Merging convolutive source separation with geometric beamforming," *IEEE Trans. on Speech and Audio Processing*, vol. 10, no. 6, pp. 352–362, Sep. 2002.

[7] V. C. Soon, L. Tong, Y. F. Huang, and R. Liu, "A robust method for wideband signal separation," in *Proc. of ISCAS '93*, 1993, pp. 703–706.

[8] F. Asano, S. Ikeda, M. Ogawa, H. Asoh, and N. Kitawaki, "A combined approach of array processing and independent component analysis for blind separation of acoustic signals," in *Proc. ICASSP 2001*, May 2001, pp. 2729–2732.

[9] H. Sawada, R. Mukai, S. Araki, and S. Makino, "A robust and precise method for solving the permutation problem of frequency-domain blind source separation," in *Proc. Intl. Symp. on Independent Component Analysis and Blind Signal Separation (ICA2003)*, 2003, pp. 505–510.

[10] H. Sawada, R. Mukai, and S. Makino, "Direction of arrival estimation for multiple source signals using independent component analysis," in *Proc. of ISSPA'03*, 2003, vol. 2, pp. 411–414.

[11] H. Sawada, R. Mukai, S. de la Kethulle de Ryhove, S. Araki, and S. Makino, "Spectral smoothing for frequency-domain blind source separation," in *Proc. IWAENC 2003*, 2003, pp. 311–314.

GRADIENT FLOW BEARING ESTIMATION WITH BLIND IDENTIFICATION OF MULTIPLE SIGNALS AND INTERFERENCE

Milutin Stanacevic, Gert Cauwenberghs

ECE Department, Johns Hopkins University
Baltimore, MD 21218, USA
E-mail: {*miki,gert*}@*jhu.edu*

Larry Riddle

Signal Systems Corporation
Severna Park, MD 21146, USA
E-mail: *lriddle@signalsystemscorp.com*

ABSTRACT

We present a technique for reliably estimating the 3-D direction cosines of a broadband traveling wave impinging on an array of four sensors of dimensions smaller than the shortest wavelength in the source. Gradient flow converts observed mixtures of delayed source signals into instantaneous linear mixtures of their temporal derivatives through observation of spatial gradients of the field. This formulation is equivalent to independent component analysis (ICA), where the mixing matrix directly yields the direction cosines. Experiments with acoustic data from a microphone array show improved bearing accuracy through second order blind identification (SOBI) of band-limited Gaussian signal from interfering sources.

1. INTRODUCTION

Formulating source localization algorithms that perform robustly with sub-wavelength dimensions of the sensor array is a challenging problem introduced by miniaturization of integrated sensors. It is well known that the precision of delay-based bearing estimation degrades with shrinking dimensions (aperture) of the sensor array [1]. Classical time-difference of arrival estimation techniques [2] based on cross-correlation of the signals require high oversampling ratios for estimating small time delays.

Gradient flow [3] avoids the problem of estimating small time delays between sensor observations by relating amplitudes of spatial and temporal gradients in the signal across the sensor array, or equivalently resolve terms in a Taylor expansion of the field [4]. The idea of wavefront sensing in space for localizing sound was first introduced by Blumlein in the 1930s [5].

Section 2 presents the gradient flow approach for blind separation and localization of multiple sources. In Section 3, we show that the problem of estimating time-delays simplifies to least-square problem in the case of one source. Section 4 describes second-order blind identification (SOBI) algorithm [6, 7, 8] used for localization in the case of multiple source and how the assumptions of the algorithm are met in the obtained model. In Section 5 we compare the performance of least-mean-square (LMS) and SOBI algorithm for the case of one bandlimited Gaussian source signal.

This work was partly supported by ONR N00014-99-1-0612, ONR/DARPA N00014-00-C-0315 and N00014-00-1-0838.

2. GRADIENT FLOW LOCALIZATION

A traveling wave emitted by a source is observed over a distribution of sensors in space, which here we consider to be discrete but which could be continuous. We define $\tau(\mathbf{r})$ as the time lag between the wavefront at point $\mathbf{r}$ and the wavefront at the center of the array, *i.e.,* the propagation time $\tau(\mathbf{r})$ is referenced to the center of the array.

For an integrated MEMS or VLSI array with dimensions typically smaller than 1 cm, the distance from the source is much larger than the dimensions of the sensor array, and the *far-field* approximation is a sensible approximation. In the far field, the wavefront delay $\tau(\mathbf{r})$ is approximately linear in the projection of $\mathbf{r}$ on the unit vector $\mathbf{u}$ pointing towards the source,

$$\tau(\mathbf{r}) \approx \frac{1}{c}\mathbf{r}\cdot\mathbf{u} \tag{1}$$

where c is the speed of (acoustic or electromagnetic) wave propagation.

Let $x(\mathbf{r},t)$ be the signal picked up by a sensor at position $\mathbf{r}$. As one special case we will consider a two-dimensional array of sensors, with position coordinates p and q so that $\mathbf{r}_{pq} = p\mathbf{r}_1 + q\mathbf{r}_2$ with orthogonal vectors $\mathbf{r}_1$ and $\mathbf{r}_2$ in the sensor plane. In the *far-field* approximation (1), the sensor observations of the source are advanced in time by $\tau_{pq} = p\tau_1 + q\tau_2$, where

$$\begin{aligned}\tau_1 &= \frac{1}{c}\mathbf{r}_1\cdot\mathbf{u}\\ \tau_2 &= \frac{1}{c}\mathbf{r}_2\cdot\mathbf{u}\end{aligned} \tag{2}$$

are the inter-time differences (ITD) of source between adjacent sensors on the grid along the p and q place coordinates, respectively. Knowledge of the *angle coordinates* τ_1 and τ_2 uniquely determines, through (2), the direction vector $\mathbf{u}$ along which source impinges the array, in reference to the $\{p,q\}$ plane.

The signal observed at sensor with position coordinates p and q can be expressed as

$$x_{pq}(t) = \sum_{\ell=1}^{\mathcal{L}} s^\ell(t) + \tau_{pq}^\ell \dot{s}^\ell(t) + \tfrac{1}{2}(\tau_{pq}^\ell)^2 \ddot{s}^\ell(t) + \ldots + n_{pq}(t) \quad (3)$$

where $n_{pq}(t)$ represent additive noise in the sensor observations. A gradient flow formulation is obtained by evaluating spatial gradients of x_{pq} along the p and q position coordinates, around the origin $p = q = 0$:

$$\begin{aligned} \xi_{ij}(t) &\equiv \left. \frac{\partial^{i+j}}{\partial^i p \partial^j q} x_{pq}(t) \right|_{p=q=0} \\ &= \sum_\ell (\tau_1^\ell)^i (\tau_2^\ell)^j \frac{d^{i+j}}{d^{i+j}t} s^\ell(t) \; + \; \nu_{ij}(t), \end{aligned} \quad (4)$$

where ν_{ij} are the corresponding spatial derivatives of the sensor noise n_{pq} around the center. Taking spatial derivatives ξ_{ij} of order $i + j \leq k$, and differentiating ξ_{ij} to order $k - (i + j)$ in time yields a number of different linear observations in the kth-order time derivatives of the signals s.

As an example, consider the first-order case $k = 1$, corresponding to (3):

$$\begin{aligned} \xi_{00}(t) &= \textstyle\sum_\ell s^\ell(t) + \nu_{00}(t), \\ \xi_{10}(t) &= \textstyle\sum_\ell \tau_1^\ell \dot{s}^\ell(t) + \nu_{10}(t), \\ \xi_{01}(t) &= \textstyle\sum_\ell \tau_2^\ell \dot{s}^\ell(t) + \nu_{01}(t). \end{aligned} \quad (5)$$

Taking the time derivative of ξ_{00}, we thus obtain from the sensors a linear instantaneous mixture of the time-differentiated source signals,

$$\begin{bmatrix} \dot{\xi}_{00} \\ \xi_{10} \\ \xi_{01} \end{bmatrix} \approx \begin{bmatrix} 1 & \cdots & 1 \\ \tau_1^1 & \cdots & \tau_1^{\mathcal{L}} \\ \tau_2^1 & \cdots & \tau_2^{\mathcal{L}} \end{bmatrix} \begin{bmatrix} \dot{s}^1 \\ \vdots \\ \dot{s}^{\mathcal{L}} \end{bmatrix} + \begin{bmatrix} \dot{\nu}_{00} \\ \nu_{10} \\ \nu_{01} \end{bmatrix}, \quad (6)$$

an equation in the standard form $\mathbf{x} = \mathbf{A}\mathbf{s} + \mathbf{n}$, where $\mathbf{x}$ is given and the mixing matrix $\mathbf{A}$ and sources $\mathbf{s}$ are unknown. Under the assumptions that the source signals are independent, this formulation is equivalent to standard independent component analysis (ICA), and a number of approaches exist for solving this problem [9]. ICA produces, at best, an estimate $\hat{\mathbf{s}}$ that recovers the original sources $\mathbf{s}$ up to arbitrary scaling and permutation. The direction cosines τ_i^ℓ are found from the ICA estimate of $\mathbf{A}$, after first normalizing each column (*i.e.,* , each source estimate) so that the first row of the estimate $\hat{\mathbf{A}}$, like the real $\mathbf{A}$ according to (6), contains all ones. This simple procedure together with (2) yields estimates of the direction vectors $\hat{\mathbf{u}}^\ell$ along with the source estimates $\hat{s}^\ell(t)$, which are obtained by integrating the components of $\hat{\mathbf{s}}$ over time and removing the DC components.

The proposed gradient flow technique requires computation of temporal derivative and first-order spatial gradients along p and q directions of the signal impinging on the sensor array. Estimates of ξ_{00}, ξ_{10} and ξ_{01} are obtained by finite-difference gradient approximation on a grid (precise up to terms at least of third order), using a planar array of four sensors, illustrated in Figure 1:

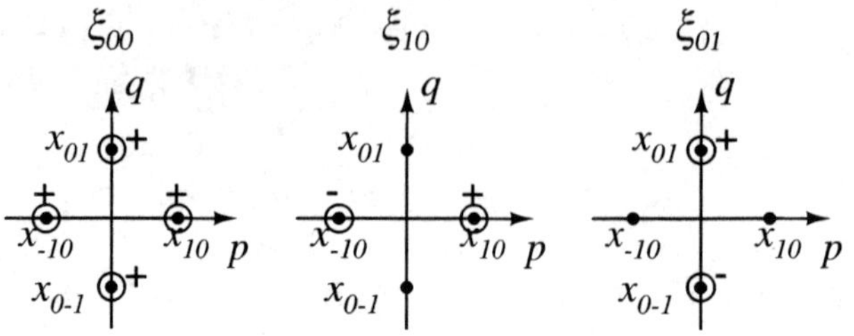

Fig. 1. *Geometry of a planar four sensor array*

$$\begin{aligned} \xi_{00} &\approx \tfrac{1}{4}\big(x_{-1,0} + x_{1,0} + x_{0,-1} + x_{0,1}\big) \\ \xi_{10} &\approx \tfrac{1}{2}\big(x_{1,0} - x_{-1,0}\big) \\ \xi_{01} &\approx \tfrac{1}{2}\big(x_{0,1} - x_{0,-1}\big) \end{aligned} \quad (7)$$

3. SINGLE SOURCE LOCALIZATION

In the case of one directional source, the equation (6) simplifies to

$$\begin{aligned} \dot{\xi}_{00}(t) &= \dot{s}(t) + \dot{\nu}_{00}(t), \\ \xi_{10}(t) &= \tau_1 \dot{s}(t) + \nu_{10}(t), \\ \xi_{01}(t) &= \tau_2 \dot{s}(t) + \nu_{01}(t), \end{aligned} \quad (8)$$

and the problem of estimating time delays converts to standard least-square problem in the unknown delays τ_1 and τ_2, with estimates

$$\begin{aligned} \hat{\tau}_1 &= \frac{\mathrm{E}[\dot{\xi}_{00}\xi_{10}]}{\mathrm{E}[\dot{\xi}_{00}^2]} \\ &= \frac{\mathrm{r}_{\dot{\xi}_{00}\xi_{10}}(0)}{\mathrm{r}_{\dot{\xi}_{00}\dot{\xi}_{00}}(0)} \\ \hat{\tau}_2 &= \frac{\mathrm{E}[\dot{\xi}_{00}\xi_{01}]}{\mathrm{E}[\dot{\xi}_{00}^2]} \\ &= \frac{\mathrm{r}_{\dot{\xi}_{00}\xi_{01}}(0)}{\mathrm{r}_{\dot{\xi}_{00}\dot{\xi}_{00}}(0)}. \end{aligned} \quad (9)$$

From least-square estimates of the time delays, we can directly obtain estimates of azimuth angle θ and elevation angle ϕ angle according to (2):

$$\begin{aligned} \tau_1 &= \frac{1}{c}|\mathbf{r}_1| \, \sin\theta \sin\phi \\ \tau_2 &= \frac{1}{c}|\mathbf{r}_2| \, \cos\theta \sin\phi \end{aligned} \quad (10)$$

An interesting observation is that the estimate of azimuth angle is independent of the speed of sound as it involves spatial gradients only; estimation of the elevation angle on the other hand requires knowledge of the speed of sound in relating spatial and temporal derivatives. The estimate of azimuth angle can be obtained simply from the ratio of delay estimates $\hat{\tau}_1$ and $\hat{\tau}_2$:

$$\hat{\theta} = \arctan\frac{\hat{\tau}_1}{\hat{\tau}_2} \quad (11)$$

or by finding the null of the expression

$$\hat{\tau}_1\cos(\theta) - \hat{\tau}_2\sin(\theta) \ . \quad (12)$$

4. MULTIPLE SOURCE LOCALIZATION

The ICA algorithm we have chosen to implement for separation and localization of multiple sources is second-order blind identification (SOBI) algorithm [6, 7, 8]. SOBI deals effectively with non-white stationary and non-stationary statistics for sources and noise in the ICA model (6). The unknown sources are temporal derivatives of impinging signals, leading to temporal structural information of sources that have to be separated. The noise signals in the ICA model can be expanded to sensor noise term and dispersive ambient noise term. The sensor noise contributions are

$$\begin{aligned} \dot{\nu}_{e00} &= \tfrac{1}{4}(\dot{e}_{-10} + \dot{e}_{10} + \dot{e}_{0-1} + \dot{e}_{01}) \\ \nu_{e10} &= \tfrac{1}{2}(e_{10} - e_{-10}) \\ \nu_{e01} &= \tfrac{1}{2}(e_{01} - e_{0-1}) \end{aligned} \quad (13)$$

where e_{10}, e_{-10}, e_{01} and e_{0-1} represent sensor noise at corresponding sensors. Since the cross-correlation of the signal and its derivative is zero and under the assumption that sensors noise is uncorrelated across the sensor array, the sensor noise contribution in ICA model becomes spatially white. The disperse noise covariance matrix, under the assumption that correlation between signals coming from different directions is zero, also becomes diagonal, leading to diagonal covariance matrix of complete noise in observations.

The SOBI algorithm is based on a joint diagonalization of a set of covariance matrices obtained at different time lags. The covariance matrix of observation signals at time lag τ is

$$\mathrm{R}_x(\tau) = \mathrm{E}[\mathbf{x}(\mathrm{t}+\tau)\mathbf{x}^{\mathrm{T}}(\mathrm{t})] = \mathbf{A}\mathrm{R}_s(\tau)\mathbf{A}^{\mathrm{T}} + \sigma(\tau)^2\mathrm{I}, \quad (14)$$

where we used the assumption that noise term is spatially white. After estimating the covariance matrices at different time lags and subtracting the estimated noise contributions, by jointly diagonalizing the obtained set of matrices the mixing matrix $\mathbf{A}$ is estimated. The time delays have to be chosen in such a way that covariance matrices carry maximally different information.

The use of only second-order statistics makes the algorithm more robust than higher-order statistics ICA algorithms. It also allows separation of Gaussian sources. By observing the equation (9), we can notice that LMS solution represents a special case of SOBI, as only a covariance matrix of zero time lag is used for bearing estimation.

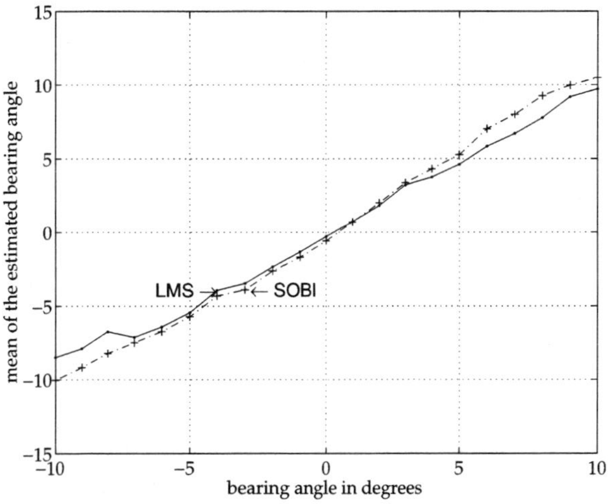

Fig. 2. *Mean value of estimated angle using LMS and SOBI as bearing angle is swept from from* $-10°$ *to* $10°$.

5. EXPERIMENTAL SETUP AND RESULTS

To quantify the performance of gradient flow bearing estimation, the experimental setup with one directional source in open-field environment was used. The effective distance between microphones in the planar array of four sensors was 15.87 cm. The sound source was bandlimited (20-300Hz) Gaussian signal presented through a loudspeaker. Data was sampled at 2048 samples per seconds. The distance between loudspeaker and microphone array was approximately 18 m. Signal-to-noise (SNR) ratio was around 25-30 dB. The experiments were performed for bearing angles from $-10°$ to $10°$ in increments of $1°$. The data was played for 30 seconds and the bearing estimates were obtained for 1 second data.

For a localization of a single source, simple expressions can be obtained for the Cramer-Rao lower bound on the variance of bearing angle, assuming Gaussian univariate distributions for the source and noise components [10]. In this experimental setup, the Cramer-Rao bound was around 1 degree. The assumption of uncorrelated noise is violated for subwavelength sensor geometries, and gradient flow exploits correlated noise and temporal dependencies to obtain superior bearing accuracies.

Before bearing estimation of direction cosines using temporal and spatial gradients, common mode offset correction is performed on the estimated spatial gradients. Common mode offsets arise from gain mismatch errors in the sensors. Since the correlation between any stationary signal and its time-derivative is zero, the correlation between common-mode and gradient variable is also zero. Therefore, using only second-order statistics, we can estimate the leakage of common-mode component in gradient estimates and compensate for it.

The estimates of bearing angle were obtained with both LMS and SOBI. The mean of estimators for bearing angles

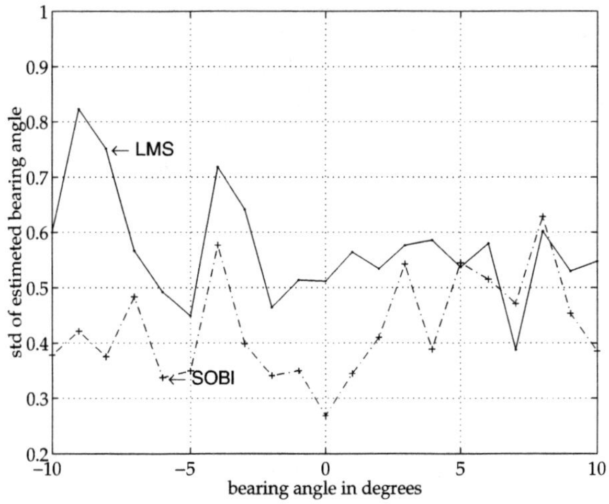

Fig. 3. *Standard deviation of estimated angle using LMS and SOBI as bearing angle is swept from from* $-10°$ *to* $10°$.

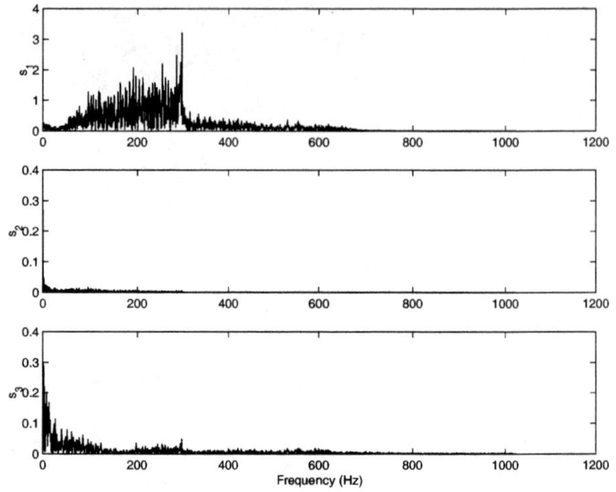

Fig. 4. *Spectrum of the estimated sources obtained by the SOBI algorithm for one second of data and one bearing angle.*

from $-10°$ to $10°$ is shown in Figure 2, and the standard deviation is shown in Figure 3. As expected, the estimators obtained using ICA algorithm have smaller bias error and variance achieving sub-degree accuracy.

In Figure 4 we show the frequency characteristics of the estimated source signals obtained with SOBI for one second of data and one bearing angle. Since we have three observations in our ICA model, we can estimate up to three sources. The first estimated source is the temporal derivative of the bandlimited (20-300Hz) Gaussian signal presented through the loudspeaker, while the second and third, much smaller in amplitude, represent some interfering background directional sources, or wind noise.

6. CONCLUSION

Gradient flow offers a framework in which ICA can be applied directly to bearing estimation. We obtained improvements in accuracy by modeling signal, noise and interference using second-order temporal decorrelations in the SOBI ICA framework, exploiting temporal structure or non-stationarity in the signal and interfering noise sources. Experimental results demonstrate angular resolutions better than predicted by the Cramer-Rao lower bound for maximum-likelihood estimation assuming stationary uncorrelated Gaussian noise components [10].

7. REFERENCES

[1] S. Haykin, *Adaptive Filter Theory*, Prentice-Hall, 2nd edition, 1991.

[2] C.H. Knapp and G.C. Carter, "The generalized correlation method for estimation of time delay," *IEEE Trans. on Acoustics, Speech, and Signal Processing*, vol. 24, no. 4, pp. 320–327, 1976.

[3] G. Cauwenberghs, M. Stanacevic, and G.Zweig, "Blind broadband source localization and separation in miniature sensor arrays," in *Proc. IEEE Int. Symp. Circuits and Systems (ISCAS'2001)*, Sydney, Australia, 2001, vol. 3, pp. 193–196.

[4] J. Barrère and G. Chabriel, "A compact sensor array for blind separation of sources," *IEEE Trans. Circuits and Systems, Part I*, vol. 49, no. 5, pp. 565–574, 2001.

[5] A.D. Blumlein, "Improvements in and relating to sound-transmission, sound-recording and sound-reproducing systems," *British Patent*, , no. 394325, 1933.

[6] A. Belouchrani, K. Abed-Meraim, J. Cardoso, and E. Mouline, "A blind source separation technique using second-order statistics," *IEEE Transactions on Signal Processing*, vol. 45, no. 2, pp. 434–444, 1997.

[7] L. Molgedey and G. Schuster, "Separation of a mixture of independent signals using time delayed correlations," *Physical Review Letters*, vol. 72, no. 23, pp. 3634–3637, 1994.

[8] A. Ziehe and K. Muller, "Tdsep - an efficient algorithm for blind separation using time structure," in *Proc. Int. Conf. on Artificial Neural Networks(ICANN'98)*, Skovde, Sweden, 1998, pp. 675–680, Springer Verlag.

[9] A. Cichocki and S. Amari, *Adaptive Blind Signal and Image Processing: Learning Algorithms and Applications*, Wiley, 2002.

[10] B. Friedlander, "On the cramer-rao bound for time delay and doppler estimation," *IEEE Transactions on Information Theory*, vol. 30, no. 3, pp. 575–580, 1984.

NONLINEAR ICA SOLUTIONS FOR CONVOLUTIVE MIXING OF PNL MIXTURES

Daniele Vigliano[*]; *Aurelio Uncini*[*] ; *Raffaele Parisi* [*]

[*]Dipartimento INFOCOM, Università di Roma "La sapienza" – Italy
Via Eudossiana, 18, 00184 Roma – Italy
d.vigliano@inwind.it; aurel@ieee.org; parisi@infocom.uniroma1.it;

ABSTRACT

This paper introduces an ICA approach to a novel nonlinear convolutive BSS problem. The mixing model considered here is an evolution of the Post Nonlinear one: it is the convolutive mixing of a PNL mixture. The main aim of this paper is to enlarge the set of blind sources separation problems that can be approached by Nonlinear ICA with some stricter mixing environments than the one just widely described in literature. The Flexibility of the algorithm is given by the on line estimation of the score function performed by Spline Neurons.

1. INTRODUCTION

The Scientific Community's interest for blind signal processing and in particular for Blind Sources separation (BSS), performed through Independent Component Analysis (ICA) is growing. A large number of problems in biomedical or communication field now can be approached by ICA. The first studies about Independent Component Analysis aim only at resolving the famous Cocktail party problem first in static then in reverberant environments. Recently the so called flexible ICA improves the pdf matching into the neural network processing and it provides a faster learning by the estimation of parameters related with the pdf of the signals. A critical issue about the BSS algorithms is that linear mixing model is too unrealistic and "poor" in a lot of real situations.

In [4] Hyvarinen and Pajunen give an important result in nonlinear ICA theory, exploring the existence of a solution of a nonlinear problem, underlining the hard non uniqueness of the provided solution and also proposing some constraints about the mixing model granting the uniqueness of the solution. Now a large number of papers explore solutions to Post Nonlinear Mixing problem (PNL) [6][3][2]. It is possible to find only few results of the convolutive post nonlinear problem ([9][10]) and of some static nonlinear problems [5] more complex than PNL. In [1] Jutten and Karhunen review the recent advances in BSS of nonlinear mixing models. This paper explores the solution of the BSS problem in a novel convolutive nonlinear mixing environment, stricter than the one just widely diffused in literature. The new mixing environment is composed by several mixing block: a convolutive mixing channel follows a PNL block.

2. THE NONLINEAR ISSUE

The aim of this section is at introducing BSS problem in nonlinear environment underlining the approaches and exploring the existing algorithms in term of uniqueness and existence of the solution.

Considering an N vector of independent sources $\mathbf{s}[n]=\{s_1[n],..,s_N[n]\}$; considering a vector of signals received by a N-sensor array after a generally nonlinear convolutive mixing $\mathbf{x}[n]=\{x_1[n],..,x_N[n]\}$ of the original sources. The nonlinear convolutive environment was introduced in order to achieve a mixing model closer to the real one than the others just explored in literature. The general formulation of the hidden mixing model is:

$$\mathbf{x}(n)=\mathcal{F}\{\mathbf{s}(n),...,\mathbf{s}(n-L)\} \qquad (1)$$

in which $\mathcal{F}\{.\}$ is a dynamic nonlinear distorting function.

The solution of the BSS problem is expressed as: $\mathbf{y}[n]=\mathcal{G}\circ\mathcal{F}\{\mathbf{s}(n)\}$.

Resolving the blind sources separation problem in this context means to recover, making no particular a-priori assumptions, the original sources from the observation only of $\mathbf{x}[t]$. Into performing the separation, ICA recovers the original sources up to some trivial non-uniqueness[1]. The desired solution can be expressed in a closed form as:

$$\mathbf{y}[n]=\mathbf{P\Lambda D s}[n] \qquad (2)$$

in which $\mathbf{P}$ is a permutation matrix, $\mathbf{\Lambda}$ is a diagonal scaling matrix and $\mathbf{D}$ is a diagonal delay matrix.

The issue of separating mixture from models (1) with the only constraint of independent output signals and no other a priori assumption is affected by a strong non uniqueness, it can be shown with the following well known example. Considering two random variables s_1, s_2,

[1] Non-uniqueness consists into permutation, scaling factor and time delay of the original sources.

with the joint pdf $p_{s_1,s_2}(s_1,s_2)=\begin{cases} s_1 e^{-s_1^2} & s_1\in\mathbb{R}^+ \\ 1/2\pi & s_2\in[0,2\pi[\end{cases}$ processed by the nonlinear transform: $[y_1,y_2]=\mathcal{H}(\mathbf{s})=[s_1\cos(s_2),s_1\sin(s_2)]$. As is well known in the literature the resulting random variables y_1, y_2, are independent too and have gaussian distribution. This shows that independence conservation constraint alone is not strong enough to recover original sources from a generic nonlinear mixing environment [2]. If the transform $\mathcal{H}(.)$ has no particular structure or there are no more assumptions about the mixing (and demixing) model, the provided results could not be the desired solution. The main issue for generic nonlinear problems is to ensure the presence of conditions (in term of sources, mixing environment, recovering structure) granting at least theoretically the possibility to achieve the desired solution; that is the same as investigating the existence and uniqueness of the solution to the given problem. In [4] authors proposed a constructive way (a Gram-Schmidt like method) to obtain solutions of the separation problem in a static nonlinear mixing environment. In order to grant the uniqueness of the solutions it was applied some constraints to the mixing environment. The idea introduced, is general: adding some "soft" constraint to the problem produces the uniqueness of the solution.

In [2] a theoretical demonstration of the existence and uniqueness of the solution is given when the problem is PNL with convolutive mixing. It is possible to find in [1] a short explanation of many other ICA approach to nonlinear BSS. In [5] was performed the solution in the case of PNL static model followed by static mixing, stricter static nonlinear model that adds static mixing after the PNL mixing model. Thus the output independence is a weak approach to the problem of sources separation in a general nonlinear environment. In static and convolutive nonlinear mixing it is impossible to perform the recovering of the desired signals making no other assumption [1].

3. THE MIXING-DEMIXING STRUCTURE

This section explores the recovery of original sources from nonlinear convolutive mixing assuming a-priori information about the mixing model; the information assumed for the mixing model are used to design the recovering network. The mixing environment modelled in this paper is represented in Fig. 1.

In which **A** is a NxN static matrix, $\mathbf{F}[\mathbf{r}(n)]=\left[f_1\left[r_1(n)\right],\ f_N\left[r_N(n)\right]\right]^T$ is the Nx1 vector of nonlinear distorting functions, one for each channel, and **Z**[k] is a FIR matrix with L taps filters.

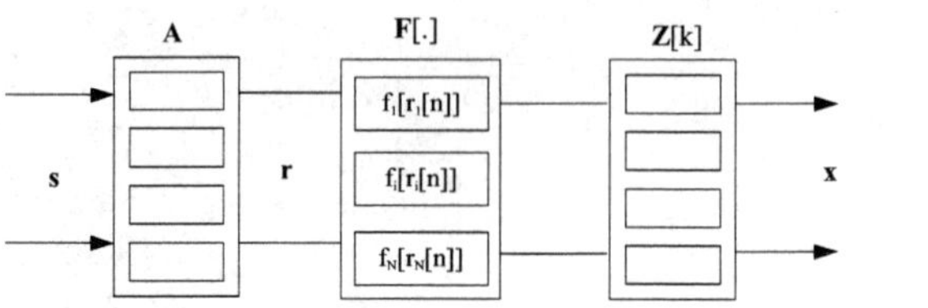

Fig. 1 The Block diagram of the convolutive nonlinear mixing model.

The mixing model written in a closed form is:

$$\mathbf{x}[n]=\mathcal{F}[\mathbf{s}]=\sum_{k=0}^{L-1}\mathbf{Z}[k]\mathbf{F}[\mathbf{A}\mathbf{s}[n-k]] \tag{3}$$

This mixing model enlarges the set of possible mixing environments from which is possible to recover separated signals. The recovering structure mirrors the mixing model.

The close form for the recovered outputs is:

$$\mathbf{y}[n]=\mathcal{G}[\mathbf{x}]=\mathbf{BG}\left[\sum_{k=0}^{L-1}\mathbf{W}[k]\mathbf{x}[n-k]\right] \tag{4}$$

In which **B** is a NxN static matrix, $\mathbf{G}\left[\mathbf{W}[n]*\mathbf{x}[n]\right]$ is the Nx1 vector of nonlinear compensating functions, one for each channel, and **W**[k] is a FIR matrix with K taps filters. Introducing the knowledge about the particular kind of mixing model is the key to avoid the strict nonuniqueness of the solution; such assumption limits the weakness of the output independence condition reducing the cardinality of all possible independent output solutions; with this constraint the problem of recovery the original sources isn't ill posed any more.

4. THE DEMIXING ALGORITHM AND NETWORK MODEL

This section explores the blind demixing algorithm, the adaptive network and the network used to compensate the nonlinear distortion. The blind algorithm performs an on-line adaptive learning of the network parameters on the base of the output independence estimation. Considering N observations $[x_1(k),\dots,x_N(k)]$ over a (M+1)-point time block and the corresponding N outputs over the same time block, defining the vectors $\mathcal{X}=\left[\mathbf{x}(0),\dots,\mathbf{x}(M)\right]^T$, $\mathcal{Y}=\left[\mathbf{y}(0),\dots,\mathbf{y}(M)\right]^T$. In this work the Kullback-Leibler divergence between $p_{\mathcal{Y}}[y]$ and $\tilde{p}_{\mathcal{Y}}[y]=\prod_{i=1;j=1}^{N,M}p_{y_i}\left[y_i[j]\right]$ is used as a way to quantify the output independence. Considering a specific demixing model with parameters $\mathbf{\Phi}$, the cost function of the algorithm is:

$$\Im\{y[n],\mathbf{\Phi}\} = KL\left(p_y, \prod_{i=1;j}^{N} p_{y_i}\left[y_i[j]\right]\right) = \qquad (5)$$

$$= \int_{\Im} p_y(y)\,\log\left(\frac{p_y(y)}{\prod_{i,j} p_{y_i}\left[y_i[j]\right]}\right) dy$$

The (5) is function of both output signals and model parameters $\mathbf{\Phi}$. KL divergence is minimized with respect of the model's parameters using both steepest descendent (stochastic gradient) and Natural gradient. Minimizing the cost function (5) with respect of the algorithm's parameters shows some terms like:

$$\frac{\partial}{\partial\Phi}\log\left[p_{y_i}(y_i)\right] = \frac{\dot{p}_{y_i}(y_i)}{p_{y_i}(y_i)}\frac{\partial y_i}{\partial\Phi} = \psi_i(y_i)\frac{\partial y_i}{\partial\Phi} \qquad (6)$$

in which $\psi_i(y_i) = \dot{p}_i(y_i)/p_i(y_i)$ are the so called Score Functions[2]. In this paper, the Spline Neurons are used to perform the on-line estimation of both Score function and nonlinear compensating functions (for a detail about the Spline Neurons see [5][6][7][8]). The most suitable property of the Spline Neurons, as function estimator, is theirs local learning: for each learning step only the four control points nearest to the training input are considered; no matter how many control points the Spline curve has. In this paper the learning of the score function is performed in a direct way (see [2]) minimizing, with respect to the spline control points $\mathbf{Q}$, the:

$$\varepsilon_j = \frac{1}{2}E\left\{\left[\tilde{\psi}_j(y_j,\mathbf{\Phi}) - \frac{\dot{p}_{y_j}(y_j)}{p_{y_j}(y_j)}\right]^2\right\} \quad j = 1..N \qquad (7)$$

$\tilde{\psi}_j$ is the Spline model of the score function. The resulting rule for the Spline Neuron is:

$$\frac{\partial\varepsilon}{\partial \mathbf{Q}^{\psi}_{\mathbf{i},(\mathbf{i+m})}} = \left[\frac{1}{4}\mathbf{T_u M}_m \mathbf{T_u M Q_i^{g_j}} + \frac{1}{\Delta y}\dot{\mathbf{T}}_{\mathbf{u}}\mathbf{M}_m\right] \quad m = 0..3 \qquad (8)$$

in which $\mathbf{M}_m$ is the m-th column of the matrix $\mathbf{M}$ (matrix of coefficients), $\mathbf{T}$ is the local abscissa vector.

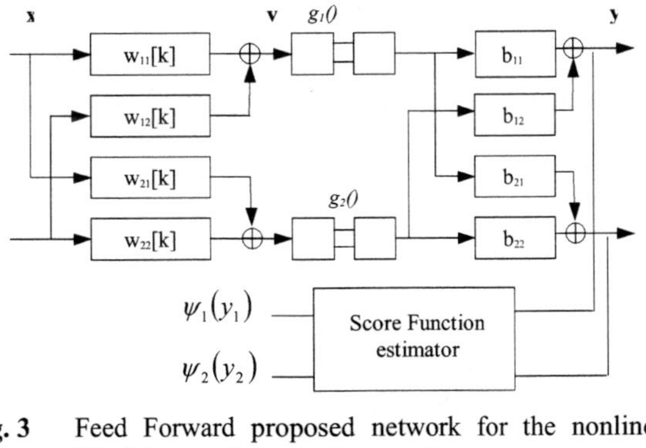

Fig. 3 Feed Forward proposed network for the nonlinear blind deconvolution and separation.

Fig 3 shows the network used to perform the separation; this network is a cascade of blocks, each of one is well described in literature and previously used to resolve more simple problems. The output of the channel i-th is expressed in close form as:

$$y_i[n] = \sum_{j=1}^{N} b_{ij} g_j\left[\sum_{h=1}^{N}\sum_{k=0}^{L-1} w_{jh}[k]\,x_h[n-k]\right] \quad i = 1..N \qquad (9)$$

Considering the set of the learning parameters $\mathbf{\Phi} = \left\{b_{ij}, w_{pq}[k], \mathbf{Q}^g, \mathbf{Q}^{\Psi}\right\}$ in which $\mathbf{Q}^g$ and $\mathbf{Q}^{\Psi}$ are the parameters of the Spline networks. Deriving the cost function (5) with respect the learning parameter results:

$$\frac{\partial\Im\{y[n],\mathbf{\Phi}\}}{\partial\mathbf{\Phi}} = \frac{\partial}{\partial\mathbf{\Phi}}\sum_{n=0}^{M}\Big[-\log\left|\det\mathbf{B}\right| + \qquad (10)$$

$$-\log\prod_{i=1}^{N} g_i\left[v_i[n]\right] - \log\left|\mathbf{W}(0)\right| - \sum_{i=1}^{N}\log p_{y_i}(y_i)\Big]$$

In (10) the expected value of the signals has been replaced by the instantaneous value. In order to minimize $\Im\{\mathbf{\Phi}, y\}$ the general rule provided by the steepest descendent method is:

$$\mathbf{\Phi}(k+1) = \mathbf{\Phi}(k) - \eta_{\mathbf{\Phi}}(k)\frac{\partial}{\partial\mathbf{\Phi}}\Im\{\mathbf{\Phi}, y\} \qquad (11)$$

The learning rule for the elements of the matrix $\mathbf{B}$ is:

$$\frac{\partial}{\partial\mathbf{\Phi}}\Im\{\mathbf{\Phi}(k), y\} = -\left(I - \mathbf{\Psi}_{\mathbf{y}}\mathbf{y}^T\right)\mathbf{B} \qquad (12)$$

The learning rule for the elements of the FIR matrix $\mathbf{W}$ is:

$$\frac{\partial\Im}{\partial\mathbf{W}[k]} = -\mathbf{W}[0]^{-T}\delta_k - \left\{\left[\ddot{g}_1(v_1)/\dot{g}_1(v_1) \;\; \cdots \;\; \ddot{g}_N(v_N)/\dot{g}_N(v_N)\right]^T + \qquad (13)$$

$$+\begin{bmatrix}\mathbf{\Psi}_{\mathbf{y}}^T(\mathbf{B})_1 & 0\\ 0 & \mathbf{\Psi}_{\mathbf{y}}^T(\mathbf{B})_N\end{bmatrix}\begin{bmatrix}\dot{g}_1(v_1)\\ \dot{g}_N(v_N)\end{bmatrix}\right\}\mathbf{x}^T(n-k)$$

in the last equation $(\mathbf{B})_m$ is used as the m-th column of the matrix $\mathbf{B}$. The learning rules for the control points $\mathbf{Q}^g$ of the Spline neurons that compensate the nonlinear distorting functions are:

$$\frac{\partial\Im\{\mathbf{\Phi}(k), y\}}{\partial\mathbf{Q}^{\mathbf{g}}_{\mathbf{i},(\mathbf{i+m})}} = -\left[\frac{\dot{\mathbf{T}}_{\mathbf{u}}\mathbf{M}_m}{\dot{\mathbf{T}}_{\mathbf{u}}\mathbf{M}\mathbf{Q}_{\mathbf{i}}^{\mathbf{g_j}}} + \mathbf{\Psi}_{\mathbf{y}}^T(\mathbf{B})_j\,\mathbf{T_u M}_m\right] \quad m = 0,..,3 \qquad (14)$$

in which $\mathbf{M}_m$ and $\mathbf{T}$ have the same meaning as in (8). The learning rule for the Spline SG neurons dedicated to Score Functions is (8). The use of FIR allows the inverting also quite non-minimum phase systems (with zeros near to the unit circle) that could have convergence problem in other situations. One of the main problem using FIR is the length of filters: to resolve real problems or simply some non trivial ones a large number of filter taps is required; must be noted that learning time grow in a nonlinear way with the FIR length.

[2] Sometimes in literature Score Function is defined considering the opposite of the function (6).

4. EXPERIMENTAL RESULTS

This section collects the experimental result of the proposed architectures. Although the algorithm is able to perform the separation of n-channel mixture in this test has been considered the mixture of a male voice (speaking: "Le donne i cavalier l'arme") and white noise.

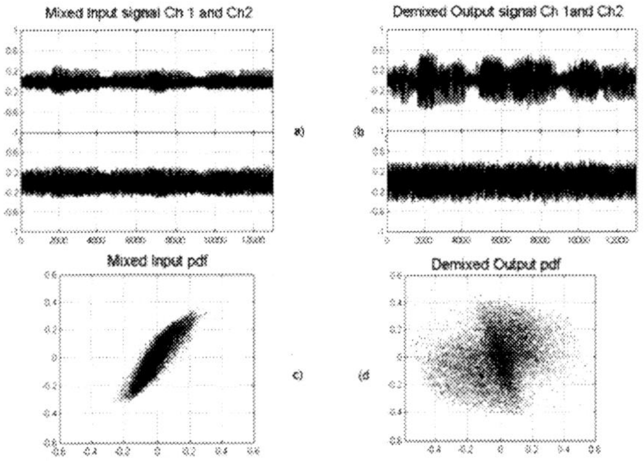

Fig.4 a) Input mixture; b) Output demixed signals; c) Input mixd Pdf; d) Output demixed Pdf.

Fig. 4 shows the resulting signals after a training of 900 epochs with the learning rat e set $\mu=7*10^{-7}$; the recovering network has 103 control points Spline neurons and a 31 taps Fir matrix. The mixing environment applied, with respect to fig.1, is: **A**=[0.7, -0.3; 0.5, 0.7];

$$\mathbf{Z}\left[z^{-1}\right]=\begin{bmatrix} 0.7-0.3z^{-1}+0.6z^{-2} & 0.3-0.2z^{-1}-0.06z^{-2} \\ -0.3+0.3z^{-1}+0.11z^{-2} & 0.7+0.3z^{-1}-0.06z^{-2} \end{bmatrix}.$$

The nonlinear distortions applied in this test are:

$$F\left[f_1(r_1), f_2(r_2)\right]=\left[r_1+2r_1^3, 0.5r_2+\tanh(5r_2)\right].$$

The Signal Interface Ratio (SIR dB) introduced in [11] measures the performance of the proposed algorithm.

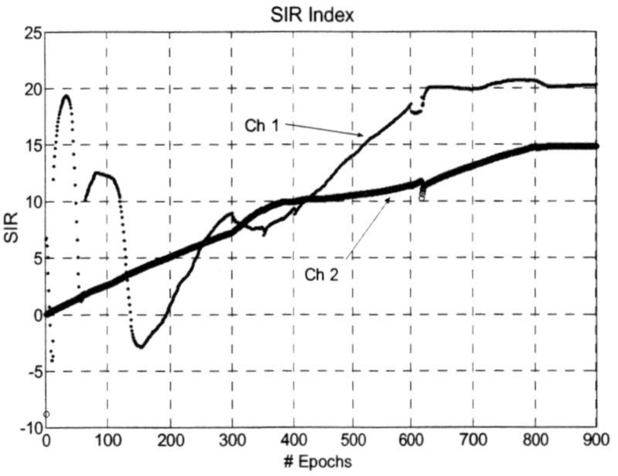

Fig. 5 Signal interference ratio index during the training

The SIR index trend during the learning shows the performance of the algorithm in a quantitative way.

Fig 5 shows how, after a first period, the algorithm perform the separation of the output signals. The reason of the starting transient has been the bad starting condition and the number of block each of one separately have to converge to the optimum values.

5. CONCLUSION

This paper introduces a novel mixing model for which the BSS performed by ICA is granted. The algorithm widely explored in this paper is able to separate the convolutive mixing of a PNL mixture. The FIR recovering network performs the on line estimation of the score function by the Spline Neurons. Spline Neurons perform also the nonlinear compensating function estimation.

REFERENCES

[1] C. Jutten, J. Karhunen, "Advances in Nonlinear Blind Sources Separation", 4th International Symposium on Independent Component Analysis and B lind Signal Separation (ICA2003), April 2003, Nara, Japan.

[2] A. Taleb, "A Generic Framework for Blind Sources Separation in Structured Nonlinear Models", in IEEE transaction on signal processing, vol. 50. no 8 August 2002.

[3] A. Taleb, C. Jutten, "Sources Separation in post nonlinear mixtures", in transaction on signal processing, vol. 47. no 10 August 1999.

[4] A. Hyvarinen, P. Pajunen, "Nonlinear Independent Component Analysis: Existence and Uniqueness Results", Neural Networks 12(2): 429-439, 1999.

[5] M. Solazzi, F. Piazza, A. Uncini, "Nonlinear Blind Source Separation by Spline neural Network", ICASSP 2001, Salt Lake city, USA May 8-11, 2001.

[6] M. Solazzi, R. Parisi, A. Uncini, "Blind Source separation in nonliner mixtures by adaptive spline neural network", ICA 2001 in Proc. of the 3rd Workshop on Independent Component Analysis and Signal Separation (ICA2001), San Diego (California, USA), 2001.

[7] A. Uncini, L. Vecci, F. Piazza, "Learning and approximation capabilities of adaptive Spline activation function neural network", NN, Vol. 11, no. 2, pag. 259-270 March 1998.

[8] M.Solazzi, F. Piazza, A. Uncini, "An adaptive Spline Nonlinear Function for Blind Signal Proessing", proc. of IEEE Whorkshop on neural networks for signal Processing X, pp396-404, December, 2000.

[9] F. Milani, M. Solazzi, A. Uncini, "Blind Source Separation of convolutive nonlinear mixtures by flexible spline nonlinear functions", proc. of IEEE Int. Conference on Acoustic Speech and Signal Processing, ICASSP'02, Orlando, USA, May, 2002.

[10] M. B. Zade, C. Jutten, K. Najeby, "Blind Separating, Convolutive Post nonlinear Mixture", ICA 2001 in Proc. of the 3rd Workshop on Independent Component Analysis and Signal Separation (ICA2001), San Diego (California, USA), 2001, pp. 138–143.

[11] D. Shobben, K. Torkkola, and P. Smaragdis, "Evaluation of blind signal separation methods", in proceeding of ICA and BSS, Aussois, France, January 11-15, 1999.

Blind Separation with Gaussian Mixture Model for Convolutively Mixed Sources

Masashi OHATA[*1] Toshiharu MUKAI[*1] Kiyotoshi MATSUOKA[*2]

[*1]Biologically Integrative Sensors Lab., RIKEN BMC Research Center, Japan

[*2] Dep. of Brain Science and Engineering, Kyushu Institute of Technology, Japan

E-mail address: ohatama@bmc.riken.jp

Abstract

This paper proposes an online blind separation algorithm with Gaussian mixture model for convolutively mixed sources. Although similar algorithms were proposed, they were derived for independent and identically distributed (iid) sources. They may not work for sources which are not made iid by any linear filte. From the theoretical viewpoint, our algorithm also works well for the sources. Furthermore, since it can estimate the distributions of sources and search for an optimal separator simultaneously, it can be applied to the case where their statistical properties are quite unknown, except that sources are non-Gaussian.

1. Introduction

Blind source separation (BSS) is a technique to estimate N sources $s_i(t)$ (i = 1, ..., N) from the observations $x_j(t)$ (j = 1, ..., M) of their convolutive mixture measured by M sensors without any information of the mixture ($N \leq M$). We assume that sources are mutually independent, non-Gaussian signals with zero mean and each source is stationary.

The relationship between $x_j(t)$ and $s_i(t)$ is written as

$$x_j(t) = \sum_{n=1}^{N}\sum_{\tau=0}^{\infty} a_{jn,\tau} s_n(t-\tau) = \sum_{n=1}^{N} a_{jn}(z) s_n(t), \quad (1)$$

where $a_{jn}(z) \triangleq \sum_{\tau=0}^{\infty} a_{jn,\tau} z^{-\tau}$ is a transfer function from $s_n(t)$ to $x_j(t)$ and z denotes the time-shift operator: $z^{-k}s_n(t) = s_n(t-k)$. Defining $w_{km}(z) = \sum_{\tau=-\infty}^{\infty} w_{km,\tau} z^{-\tau}$, we can obtain the set of source signals from the observations by

$$y_k(t) = \sum_{m=1}^{M} w_{km}(z) x_m(t) \quad (k = 1, ..., N). \quad (2)$$

Equations (1) and (2) can be expressed in the form of vectors and matrices as follows:

$$\mathbf{x}(t) = \mathbf{A}(z)\, \mathbf{s}(t), \quad (3)$$

$$\mathbf{y}(t) = [\mathbf{w}_1^T(z), ..., \mathbf{w}_N^T(z)]^T\, \mathbf{x}(t) = \mathbf{W}(z)\mathbf{x}(t). \quad (4)$$

Here $\mathbf{A}(z) = [a_{ij}(z)]$ is an $M\times N$ matrix and $\mathbf{w}_k(z) = [w_{k1}(z), ..., w_{kM}(z)]$ is a row vector ($\mathbf{W}(z)$ can be also expressed as $\mathbf{W}(z) = \sum_{l=-\infty}^{\infty} \mathbf{W}_l z^{-l}$: $\mathbf{W}_l = [w_{ij,l}]$). Entries of $\mathbf{y}(t)$ are given by $y_k(t) = \mathbf{w}_k(z)\mathbf{x}(t)$. In this paper, for mathematical simplicity, we deal with the case of $M = N$. We assume that $\mathbf{A}(z)$ is nonsingular for $|z| = 1$ and $\sum_k \|\mathbf{A}_k\| < \infty$, and its inverse also is an element of the same class as $\mathbf{A}(z)$. Then $\mathbf{x}(t)$ also stationary.

Our goal is to estimate the separator such as $\mathbf{W}(z) = \mathbf{A}^{-1}(z)$, from the observation and statistical independence of sources. Note that $\mathbf{A}(z)$ is not assumed to be nonminimum phase system and thus $\mathbf{W}(z)$ is given by a noncausal filter. We usually refer to $\mathbf{W}(z)$ as separator.

Due to the schema of BSS, The separator has two indeterminacies: permutation matrix $\mathbf{P}$ and nonsingular diagonal transfer matrix $\mathbf{D}(z) = \mathrm{diag}\{d_1(z), ..., d_N(z)\}$. Accordingly, the obtained separator forms $\mathbf{W}(z) = \mathbf{P}\mathbf{D}(z)\mathbf{A}^{-1}(z)$. Such a separator is valid in the context of BSS. Since we do not deal with the ambiguity of $\mathbf{P}$ in this paper, bellow we discuss the BSS problem as setting $\mathbf{P} = \mathbf{I}$. Let $\mathcal{S}$ be the set of all valid separators: its elements take a form of $\mathbf{D}(z)\mathbf{A}^{-1}(z)$.

The indeterminacy on $\mathbf{D}(z)$ is crucial for the problem of BSS. In Amari et al. [2], $\mathbf{D}(z)$ is determined such that each entry of $\mathbf{D}(z)\mathbf{s}(t)$ become independent and identically distributed (iid). We refer to, a signal made iid by an appropriate linear filter $d_i(z)$, as a *liner signal* and all other sources as *nonlinear signals* in this paper [a linear signal is equivalent to an iid signal in the context of BSS for convolutive mixture].

For linear signals, the traditional determination is adequate. However, for nonlinear signals, e.g. speech signal, the determination is unreasonable. Ohata and Matsuoka [8] showed that BSS algorithms based on the information-theoretic approach are unstable at valid separators for some nonlinear source. The stability conditions shown in the paper are related with power spectra of independent signals. The conditions imply that the algorithms are unstable at desired separators in the case where sources are signals with finite power in limited frequency range. That implies that the stability of the algorithms depends on how to eliminate the indeterminacy on $\mathbf{D}(z)$.

In order to overcome the instability of the algorithms, Matsuoka and Nakashima [6] proposed a new way for determining $\mathbf{D}(z)$. The criterion is referred to as 'minimal distortion principle (MDP).' The strategy has a favorite property for application of BSS, especially, biomedical signal processing [9]. As mentioned in Section 3, the output of the obtained separator is a set of independent factor signals, which constructs the observations.

2. Evaluation function

As a evaluation value of the statistical independency among $\mathbf{y}(t)$, the mutual information is known. For convolutive mixture, it is defined by

$$I_m(\mathbf{W}(z)) = \sum_{i=1}^{N} h[y_i(t)] - h[\mathbf{y}(t)], \quad (5)$$

where $h[y_i(t)]$ represents the entropy rate of the time series $\{..., y_i(-1), y_i(0), y_i(1), ...\}$. We can obtain an valid separator by minimizing the information with $\mathbf{W}(z)$. Since $\mathbf{x}(t)$ is stationary, $h[\mathbf{y}(t)]$ is reexpressed as

$$h[\mathbf{y}(t)] = h[\mathbf{x}(t)] + (2\pi j)^{-1}\int_{|z|=1} \log|\det \mathbf{W}(z)| z^{-1} dz. \quad (6)$$

On the other hand, $h[y_i(t)]$ is not explicitly expressed as a

0-7803-8251-X/04/$17.00 ©2004 IEEE

function of $\mathbf{W}(z)$. Accordingly, the minimization is not executed in actual. If the rough statistical property of each source (concretely speaking, which each source is sub-Gaussian or super-Gaussian) and the sources are all linear signals, we often replace $h[y_i(t)]$ in $I_m(\mathbf{W}(z))$ with $-E[\log r_i(y_i(t))]$ where $r_i(s_i)$ denotes a model of the probability density function (pdf) of source $s_i(t)$ and $E[\cdot]$ represents the expectation operator. Then the renovated function can be minimized with respect to $\mathbf{W}(z)$. This method is known as the maximum likelihood approach. Note that a fixed model is chosen, according as a source signal is sub-Gaussian or super-Gaussian (see [2]).

However, a prior knowledge about the statistics of sources is not given or forecasted generally in actual situations. It is impossible to determine an optimal combination of source pdf models. From our experience, the choice of the fixed models seems to have an effect on the convergence of BSS algorithms

In order to overcome these difficulties, we introduce Gaussian mixture model (GMM) as a parametric pdf model of source. The model is given by

$$q_i(s_i \mid \theta_i) = \textstyle\sum_{k=1}^{K} c_{ik}\phi(s_i \mid \mu_{ik}, \sigma_{ik}), \qquad (7)$$

where $\theta_i = \{ c_{ik}, \mu_{ik}, \sigma_{ik} ; k = 1, ..., K\}$ and $\phi(u \mid \mu, \sigma)$ denotes a Gaussian distribution with mean μ and standard deviation σ. Here coefficients c_{ik} must satisfy

$$0 \le c_{ik} \le 1 \text{ and } \textstyle\sum_{k=1}^{K} c_{ik} = 1 \quad (i = 1, ..., N), \qquad (8)$$

and σ_{ik} must be more than zero.

If all the source signals are linear signals and the true pdfs of $y_i(t)$ are expressed by model (7), the mutual information is evaluated by the function

$$L(\mathbf{W}(z), \Theta) = -\sum_{i=1}^{N} E[\log q_i(y_i(t) \mid \theta_i)] - h[\mathbf{y}(t)], \qquad (9)$$

where $\Theta = \{\theta_1, ..., \theta_N\}$. This function can be reexpressed as

$$L(\mathbf{W}(z), \Theta) = I'_m(\mathbf{W}(z)) + \sum_{i=1}^{N} D[p_i(\cdot) \,||\, q_i(\cdot \mid \theta_i)]. \qquad (10)$$

$I'_m(\mathbf{W}(z))$ and $D[p_i(\cdot) \mid q_i(\cdot \mid \theta_i)]$ are defined as follows:

$$I'_m(\mathbf{W}(z)) \triangleq \textstyle\sum_{i=1}^{N} H[y_i(t)] - h[\mathbf{y}(t)],$$

$$D[p_i(\cdot) \,||\, q_i(\cdot \mid \theta_i)] \triangleq E\left[\log \frac{p_i(y_i(t))}{q_i(y_i(t) \mid \theta_i)}\right].$$

Here $p_i(y_i)$ and $H[y_i(t)]$ denote the true pdf of $y_i(t)$ and its entropy, respectively. $I'_m(\mathbf{W}(z))$ is the mutual information for all iid signals. $D[p_i(\cdot) \| \; q_i(\cdot \mid \theta_i)]$ represents the Kullbak-Leibler divergence between $p_i(\cdot)$ and $q_i(\cdot \mid \theta_i)$. The divergence is always nonnegative. It is zero if and only if $p_i(y_i)$ is equal to $q_i(y_i \mid \theta_i)$.

If all the true pdfs of the output signals belong to the set of all Gaussian mixture distributions, there exist the set $\{\theta_i^*\}$ such as $p_i(y_i) = q_i(y_i \mid \theta_i^*)$. Setting θ_i to θ_i^*, $L(\mathbf{W}(z), \Theta^*)$ is equal to $I'_m(\mathbf{W}(z))$.

3. An online algorithm

Let $\Delta\theta_i$ and $\Delta\mathbf{W}(z)$ be the update values for θ_i and $\mathbf{W}(z)$, respectively. By the natural gradient method, which is proposed by Amari [1], the learning rules for θ_i and $\mathbf{W}(z)$ are given by

$$\Delta\theta_i \propto -I(\theta_i)^{-1} \frac{\partial}{\partial \theta_i} L(\mathbf{W}(z), \Theta), \qquad (11)$$

$$\Delta\mathbf{W}(z) \propto -\frac{\partial}{\partial \mathbf{W}(z)} L(\mathbf{W}(z), \Theta)\mathbf{W}^H(z)\mathbf{W}(z), \qquad (12)$$

where $I(\theta_i)$ denotes the Fisher information matrix of GMM: $I(\theta_i) \triangleq -E[\partial^2 \log q_i(s_i \mid \theta_i) / \partial\theta_i^2]$. Here $\partial L(\mathbf{W}(z),\Theta)/\partial\mathbf{W}(z)$ is defined by $\Sigma_k (\partial L(\mathbf{W}(z),\Theta)/\partial\mathbf{W}_k) z^{-k}$.

3.1 online algorithm for Θ

Since matrix $I(\theta_i)$ becomes very complicated, it is not suitable for an online algorithm. Thus we use the method showed in [7]. We describe the short explain of the method below.

Introduce random variables $\mathbf{z}_i(t) = [z_{i1}(t), \cdots, z_{iK}(t)]^T$ $(i=1, \cdots, N)$, indicating which is chosen from K Gaussian distributions. Let $\mathbf{e}_k$ be a K-dimensional vector whose k-th entry is unity and others are all zero. Variable $\mathbf{z}_i$ takes an element $\mathbf{e}_k$ of the set $\{\mathbf{e}_1, \cdots, \mathbf{e}_K\}$ with probability c_{ik}. Then, the joint pdf of $s_i(t)$ and $z_i(t)$ becomes

$$q_i(s_i, \mathbf{z}_i \mid \theta_i) = \textstyle\sum_{l=1}^{K} c_{il} z_{il} \phi(s_i \mid \mu_{il}, e^{\xi_{il}}). \qquad (13)$$

For the Fisher information matrix, we use the joint distribution $q_i(s_i, \mathbf{z}_i \mid \theta_i)$ instead of the marginal distribution $q_i(s_i \mid \theta_i)$. This modification makes the calculation of the Fisher information be very simple. We omit their derivation in this paper. Our algorithm is given by the update rules (14)–(16).

$$\Delta c_{ik} = \alpha_\theta \{ h_{ik}(t) - c_{ik} \}, \qquad (14)$$

$$\Delta\mu_{ik} = \alpha_\theta \frac{h_{ik}(t)}{c_{ik}} \{ y_i(t) - \mu_{ik} \}, \qquad (15)$$

$$\Delta\xi_{ik} = \alpha_\theta \frac{h_{ik}(t)}{2c_{ik}} \left\{ \frac{(y_i(t) - \mu_{ik})^2}{e^{2\xi_{ik}}} - 1 \right\}. \qquad (16)$$

Here α_θ is a learning rate and $h_{ik}(t)$ is defined as

$$h_{ik}(t) \triangleq \frac{c_{ik}\phi(y_i(t) \mid \mu_{ik}, e^{\xi_{ik}})}{\sum_{l=1}^{N} c_{il}\phi(y_i(t) \mid \mu_{il}, e^{\xi_{il}})}. \qquad (17)$$

Note that we set $\sigma_{ik} = \exp(\xi_{ik})$ and treat ξ_{ik} instead of σ_{ik} as a parameter because σ_{ik} has to be always positive.

Although the parameters of $q_i(s_i \mid \theta_i)$ can be estimated by the update rule, the algorithm sometimes causes the estimated parameters be in disorder. The cause of the phenomenon is that one of $\{c_{ik}\}$ is close to one and the others are close to all zero. If once they are zero, it is difficult that they are adaptive to the change of statistical property of the sources. In order to avoid the phenomenon on $\{c_{ik}\}$, we modify the rule (14) partly as

$$\Delta c_{ik} = \alpha_\theta \left\{ h_{ik}(t) - c_{ik} + \gamma\left(\frac{1}{K} - c_{ik}\right) \right\}, \qquad (18)$$

where γ is a parameter taking a value in the interval (0, 1). The term $\gamma(1/K - c_{ik})$ in the bracket plays the role of suppressing that some of $\{c_{ik}\}$ become close to zero. If the initial values of c_{ik} satisfy (8), the learning rule satisfies the condition for $0 < \gamma < 1$.

3.2 online algorithm for W(z)

At the beginning of this subsection, we describe some mathematical notations. Let $d\mathbf{W}(z)$ be a tangent vector $\mathbf{W}(z)$ on the manifold formed by all $N \times N$ nonsingular transfer matrices. Given a square matrix $\mathbf{X}$, diag $\mathbf{X}$ (off-diag $\mathbf{X}$) denotes the operator seting its off-diagonal

(diagonal) entries to zeros. We introduce the following two nonholonomic constraints:

$$\text{diag } d\mathbf{W}(z)\mathbf{W}^{-1}(z) = \mathbf{O}, \quad (19)$$

$$\text{off-diag } d\mathbf{W}(z)\mathbf{W}^{-1}(z) = \mathbf{O}. \quad (20)$$

These constrains plays the important role in the MDP.

Applying constraint (19) to (12), we have

$$\Delta\mathbf{W}(z) \propto -\text{off-diag } E[\varphi(\mathbf{y}(t)|\Theta)\mathbf{y}^H(t, z)]\cdot\mathbf{W}(z), \quad (21)$$

where $\varphi(\mathbf{y}(t)|\Theta) = [\varphi_1(y_1(t)|\theta_1), ..., \varphi_N(y_N(t)|\theta_N)]^T$ and $\varphi_i(y_i(t)|\theta_i) = -d \log q_i(y_i(t)|\theta_i) / dy_i = \sum_{k=1}^{K} h_{ik}(t) r_{ik}(t)$. $r_{ik}(t)$ is defined by $r_{ik}(t) = \exp(-2\xi_{ik}(t))\{y_i(t) - \mu_{ik}\}$. $\mathbf{y}(t, z)$ is defined by $\mathbf{y}(t, z) = \sum_{\tau=-\infty}^{\infty} \mathbf{y}(t+\tau)z^{-\tau}$ and $\mathbf{y}^H(t, z) = \sum_{\tau=-\infty}^{\infty} \mathbf{y}^T(t+\tau)z^{\tau}$ is its Hermitian transpose.

The equilibrium point of this rule is given by off-diag $E[\varphi(\mathbf{y}(t)|\Theta)\mathbf{y}^H(t, z)] = \mathbf{O}$. If $\mathbf{W}(z)$ is an element of $\mathcal{S}$, in other words, elements of $\mathbf{y}(t)$ are mutually independent, then this condition holds. Since $d_i(z)s_i(t)$ ($i = 1, ..., N$) also are mutually independent, $\mathbf{D}(z)$ is not uniquely determined by the rule (15)-(18) and (21).

In order to eliminate the indeterminacy in $\mathbf{D}(z)$, the MDP defines the optimal separator such as an element minimizing

$$f(\mathbf{W}(z)) = E[\, \| \mathbf{x}(t) - \mathbf{y}(t) \|^2] \quad (22)$$

in the set $\mathcal{S}$. The optimal separator is given by

$$\mathbf{W}^*(z) = \text{diag}\mathbf{A}(z)\cdot\mathbf{A}^{-1}(z). \quad (23)$$

Let $\mathbf{E}(z)$ be a diagonal nonsingular matrix. If the mixing process is given by $\mathbf{A}(z)\mathbf{E}(z)$, the optimal separator is calculated as $\text{diag}\mathbf{A}(z)\mathbf{E}(z)\cdot(\mathbf{A}(z)\mathbf{E}(z))^{-1} = \text{diag } \mathbf{A}(z)\cdot\mathbf{E}(z)\mathbf{E}^{-1}(z)\mathbf{A}^{-1}(z) = \mathbf{W}^*(z)$. Thus, the separator is uniquely determined in the principle and makes the output signals not be iid.

The separator gives us independent signals $\mathbf{y}^*(t) = \mathbf{W}^*(z)\mathbf{x}(t) = [a_{11}(z)s_1(t), ..., a_{nn}(z)s_n(t)]^T$.

By applying constraint (20) to the steepest descent gradient algorithm for (22), we have

$$\Delta\mathbf{W}(z) \propto -\text{diag } E[(\mathbf{y}(t) - \mathbf{x}(t))\mathbf{y}^H(t, z)]\cdot\mathbf{W}(z). \quad (24)$$

$\mathbf{W}^*(z)$ is obtained by executing this algorithm in the set $\mathcal{S}$. Tangent vectors satisfying the nonholonomic constraints (19) and (20) are orthogonal to each other. Thus, the optimal separator can be obtained by combing both algorithms (21) and (24):

$$\Delta\mathbf{W}(z) \propto -\text{off-diag } E[\varphi(\mathbf{y}(t) \mid \Theta)\mathbf{y}^H(t, z)]\cdot\mathbf{W}(z)$$
$$-\eta \text{ diag } E[(\mathbf{y}(t) - \mathbf{x}(t))\mathbf{y}^H(t, z)]\cdot\mathbf{W}(z). \quad (25)$$

Here η is a weighting factor.

According to the stochastic gradient method, the expectation operation is removed from this update rule. Setting $\mathbf{W}(z)$ to $\sum_{l=-L}^{L} \mathbf{W}_l z^{-l}$ and defining matrices $\Phi_\eta(\mathbf{y}(t), \mathbf{x}(t)| \Theta)$ and $\mathbf{U}(t)$ as (28) and (29), respectively, we have the update rule (30). α_W is a learning rate for updating $\mathbf{W}(z)$.

At time t, the output $\mathbf{y}(t)$ and $u_{ij}(t)$ are calculated by only

$$\mathbf{y}(t-L) = \sum_{k=-L}^{L} \mathbf{W}_k \mathbf{x}(t-L-k) \quad (26)$$

and

$$u_{ij}(t-2L) = \sum_{r=-L}^{L} y_i(t-2L+r) w_{ij,r} \,. \quad (27)$$

Although our algorithm is derived on the assumption that source signals are all stationary, it also is available for nonstationary sources.

4. Results of Experiments

In order to show the validity of our algorithm, we applied it to two observations of mixture of two speech signals ($N = 2$) in a reverberant environment: the sampling frequency was 16[kHz] and the number of the observation points was 1.6×10^5 (Its reverberation time was about 0.5[s]). Figures 1 and 2 show waveforms of the source signals and the observations, respectively. We set $L = 1000$ in the separator $\mathbf{W}(z)$, $\alpha_\theta = 5.0\times10^{-4}$, $\alpha_W = 1.0\times10^{-7}$ and $\eta = 1.0$. The initial value of $\mathbf{W}(z)$ was set to the identity matrix $\mathbf{I}$ and The initial values of parameter θ_i was set such that $q_i(s_i \mid \theta_i)(i = 1, 2)$ were close to the uniform distributions in the interval $(-1.5, 1.5)$. The obtained output signals are shown in Figure 3.

These figures shows that $y_1(t)$ and $y_2(t)$ correspond to $s_1(t)$ and $s_2(t)$, respectively. This result implies that our algorithm is available.

5. Discussion

BSS algorithms with GMM was proposed ([5], [11]) and $L(\mathbf{W}(z), \Theta)$ is also used as a cost function. However, they have two problems. One is that the magnitudes of the outputs are not uniquely determined. Define $\mathbf{G} = \text{diag}\{g_1, ..., g_N\}$ ($g_i > 0$) and $\theta'_i = \{c_{ik}, g_i \mu_{ik}, g_i \sigma_{ik}; k = 1, ..., K\}(\Theta' = \{\theta'_1, ..., \theta'_N\})$. It is impossible to distinguish $(\mathbf{W}^*(z), \Theta^*)$ from $(\mathbf{GW}^*(z), \Theta^{*\prime})$ by value of $L(\mathbf{W}(z), \Theta)$. This implies that the separator obtained by minimizing (9) has ambiguity of a diagonal scaling matrix. If magnitudes of the separator's outputs are determined, Θ is fixed at a value. Therefore, in [7](for instantaneous mixture), the additional term which makes the variances of the output $y_i(t)$ be unity was introduced for resolving this problem. However, this approach is unfavorable for convolutive mixture of nonstationary sources.

The other problem is to assume that sources are all linear signals. If those algorithms are applied in the case where one of sources is a periodical signal, which is classified into nonlinear signal, those algorithms are unstable at a valid separator. The same phenomenon may also cause in the case of convolutively mixed speech signals.

We resolve these problems by introducing the MDP approach. The second term of (25) plays the role to eliminate the indeterminacy on $\mathbf{D}(z)$ in a valid separator and to automatically determine the magnitude of $\mathbf{y}(t)$ (because $\mathbf{G}$ is considered as a kind of the indeterminacy). Therefore, the advantages of our algorithm are listed as follows:

- The scaling of the output is uniquely determined.
- Our algorithm is applicable to convolutive mixture of nonlinear sources, which are either of stationary and nonstationary signals.
- Furthermore, it does not require the rough statistical properties of the sources.
- Both the updates of θ_i and $\mathbf{W}(z)$ are online executed simultaneously.

$$\Phi_\eta(\mathbf{y}(t),\mathbf{x}(t)\mid\Theta)=\begin{bmatrix}\eta(y_1(t)-x_1(t)) & \varphi_1(y_1(t)\mid\theta_1) & \cdots & \varphi_1(y_1(t)\mid\theta_1)\\ \varphi_2(y_2(t)\mid\theta_2) & \eta(y_1(t)-x_1(t)) & & \varphi_2(y_2(t)\mid\theta_2)\\ \vdots & & \ddots & \vdots\\ \varphi_N(y_N(t)\mid\theta_N) & \varphi_N(y_N(t)\mid\theta_N) & \cdots & \eta(y_1(t)-x_1(t))\end{bmatrix}. \quad (28)$$

$$\mathbf{U}(t)=[u_{ij}(t)] \text{ and } u_{ij}(t)\triangleq\textstyle\sum_{r=-L}^{L} y_i(t+r)w_{ij,r}\ . \quad (29)$$

$$\Delta\mathbf{W}_\tau=-\alpha_W\Phi_\eta(\mathbf{y}(t-3L),\ \mathbf{x}(t-3L)\mid\Theta)\mathbf{U}(t-3L-\tau). \quad (30)$$

References

[1] S. Amari, "Natural gradient learning works efficiently in learning," Neural Computation, vol.10, pp. 51-276, 1998.

[2] S. Amari, S.C. Douglas, A. Cichocki and H.H. Yang, "Multichannel blind deconvolution and equalization using the natural gradient," in Proc. IEEE International Workshop on Wireless Communication, pp. 101-104, 1997.

[3] S. Amari, T.P. Chen, and A. Cichocki, "Nonholonomic orthogonal learning algorithms for blind source separation," Neural Computation, vol.12, no.6, pp.1463-1484, 2000.

[4] S. Choi, S. Amari, A. Cichocki, and R. Liu, "Natural gradient learning with a nonholonomic constraint for blind deconvolution of multiple channels," in Proc. International Workshop on Independent Component Analysis and Blind Signal Separation, pp.371-376, 1999.

[5] S. Deligne and R. Gropinath "An EM algorithm for convolutive independent component analysis," Neurocomputing, Vol. 49, pp.187-211, 2002.

[6] K. Matsuoka and S. Nakashima, "Minimal distortion principle for blind source separation," in Proc. International Workshop on Independent Component Analysis and Blind Signal Separation, pp.722-727, 2001.

[7] M. Ohata, T. Tokunari and K. Matsuoka, "An online algorithm for blind source separation with Gaussian mixture model," in Proc. of IEEE 2000 Adaptive System for Signal Processing, Communication and Control,pp.367-376, 2000.

[8] M. Ohata and K. Matsuoka, "Stability analyses of information-theoretic blind separation algorithms in the case where the sources are nonlinear processes," IEEE Trans. on Signal Processing, Vol.50, No.1, 2002.

[9] M. Ohata, T. Matsumoto, A. Shigematsu and K. Matsuoka, "Independent Component Analysis of Electrogastrogram Data," in Proc. of ICA2003, pp.53-58, 2003.

[10] D.-T. Pham, "Mutual information approach to blind separation of stationary sources," IEEE Trans. on Information Theory, Vol. 48, No.7, pp.1935-1946, 2002.

[11] Yu Xiao and Hu Guangrui, "Speech Separation based on the GMM pdf Estimation and the Feedback Architecture", Proc. Of ICA International Workshop on Theory II, pp. 353-357, 1999.

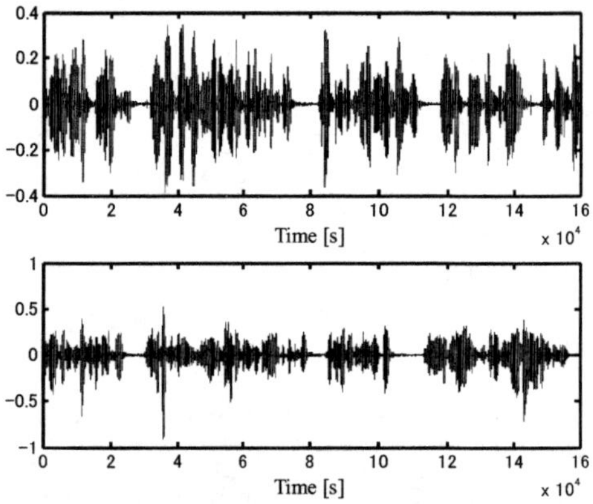

Figure 1: Source signals

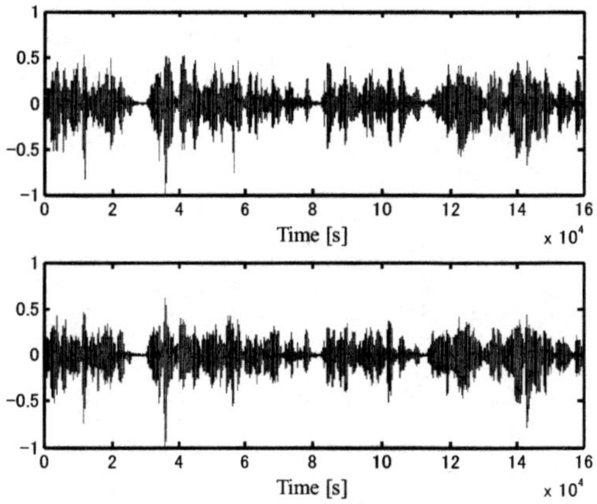

Figure 2: Observations

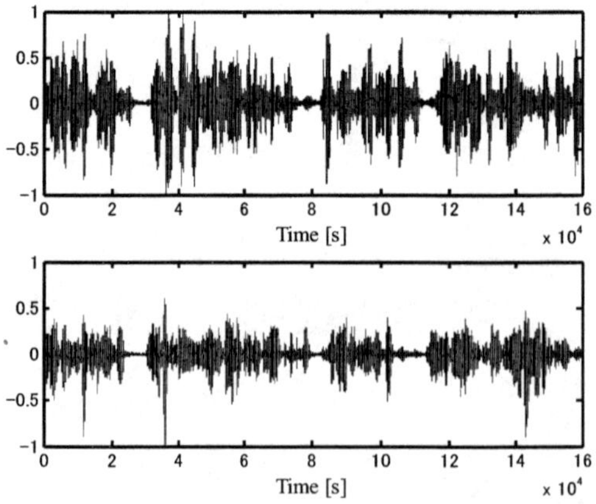

Figure 3: Seperator's output signals

BLIND ESTIMATION AND EQUALIZATION OF TIME-VARYING CHANNELS USING THE INTERACTING MULTIPLE MODEL ESTIMATOR

Ziauddin M. Kamran, T. Kirubarajan, and Alex B. Gershman

Department of Electrical & Computer Engineering, McMaster University
1280 Main Street West, Hamilton, Ontario, Canada L8S 4K1
kamranzm@mcmaster.ca, kiruba@mcmaster.ca, gershman@mail.ece.mcmaster.ca

ABSTRACT

This paper presents an adaptive multiple model blind equalization algorithm based on the Interacting Multiple Model (IMM) estimator to estimate the channel and the transmitted sequence corrupted by intersymbol interference (ISI) and noise. A computationally feasible implementation based on a weighted sum of Gaussian approximation of the density functions of the data signals is introduced. The proposed method avoids the exponential growth of the number of terms used in the weighted Gaussian sum approximation of the plant noise making it practical for real-time processing. Simulations demonstrate that the proposed IMM equalizer yields substantially improved performance compared with the recently proposed equalizer based on a (non-interacting) network of extended Kalman filters.

1. INTRODUCTION

Blind equalization of time-varying channels has attracted much research interest in wireless communications and related fields over the past few decades. Various approaches to data detection can be broadly divided into symbol-by-symbol and sequence estimation [10]. Since a linear equalizer may perform poorly in a severe ISI channel, some recent work has addressed nonlinear blind equalization techniques. For instance, the algorithms in [6], [11] employ a sequence estimator and a bank of channel estimators and alternatively optimize with respect to data and channel. Sequence estimation is performed by a blind search of a modified trellis and channel estimation is accomplished by conditioning on survivor sequences in the trellis and constructing the corresponding maximum likelihood or minimum mean-square error channel estimate. Symbol-by-symbol maximum *a posteriori* probability (MAP)-based blind equalization schemes have also been reported. For example, in [7], Bayesian symbol-by-symbol blind equalization techniques combine recursive channel estimation with the MAP and the Bayesian decision equalization methods for known channels discussed in [1]. Another Bayesian blind equalization method using Kalman filter for joint symbol and channel estimation is introduced in [8].

Recently, a blind Network of Extended Kalman Filters (NEKF) has been proposed for linear channel equalization in [3]. The network relies on the assumption that the density functions of the data signals can be represented by a weighted Gaussian sum (WGS) [2], [9]. A serious drawback of this approach is that the number of Gaussian terms in the sum increases exponentially through iterations. In order to make it computationally feasible, the Gaussian sum in the equalizer is truncated after each iteration, which results in significant performance degradation. This approach computes the state estimate that accounts for each possible current model without considering any possible switching between the models. In this paper, it is demonstrated that the NEKF-based solution can be further improved by using a *switching* or *interacting* multiple model estimation approach that computes the state estimate accounting for possible transitions in the models from one time to another. In addition, the IMM algorithm is *decision free* [4] in the sense that at each time only the probabilities (conditioned on the available data) of each model being the prevailing one are evaluated. Simulation examples are presented to demonstrate the effectiveness of the proposed approach.

The rest of the paper is organized as follows. In Section 2, we formulate the equalization problem as an augmented state estimation under model uncertainty problem. In Section 3, we develop the proposed IMM-based equalizer. Simulation results are presented in Section 4. Section 5 concludes this paper.

2. PROBLEM FORMULATION

2.1. Channel model

In this paper, the transmission of digital data over a baseband channel is considered. Thus, the baseband channel output $z(k)$ at time k can be represented by the following state space model

$$\mathbf{D}(k+1) = \widetilde{\mathbf{F}}\mathbf{D}(k) + \widetilde{\mathbf{G}}d(k+1) \quad (1)$$

$$z(k) = \mathbf{H}^T\mathbf{D}(k) + n(k) \quad (2)$$

where $\widetilde{\mathbf{F}}$ is the $N \times N$ one-step transition matrix with all elements being zero except those in positions $(i+1, i)$, $i = 1, \ldots, N-1$, which are 1, $\widetilde{\mathbf{G}} = [1, 0, \ldots, 0]^T$ is the $N \times 1$ vector, $\mathbf{H}$ is the channel parameter vector of length N, and $\mathbf{D}(k) = [d(k), d(k-1), \ldots, d(k-N+1)]^T$ consists of the N last transmitted symbols. The transmitted sequence at the channel input $\{d(k)\}$ is composed of independent and identically distributed (i.i.d.) symbols from a finite alphabet $\gamma = \{d_i, i = 1, \ldots, q\}$ that is specific to the type of modulation, and $n(k)$ represents an additive Gaussian noise sequence $\mathcal{N}[0, \sigma_n^2]$.

Under the assumption that the channel coefficients are perfectly known, a network of Kalman filters was introduced for stationary channel equalization in [9]. Unfortunately, in many practical communication systems, the channel coefficients are unknown and/or time-varying. To account for unknown coefficients, we augment the state vector to include the channel parameters as states. Denoting the unknown time-varying channel parameters as a vector $\mathbf{H}(k) = [h_0(k), \ldots, h_{N-1}(k)]^T$, we define the augmented state vector $\mathbf{Y}(k) \triangleq \left[\mathbf{D}^T(k) \quad \mathbf{H}^T(k)\right]^T$. We adopt a (discrete time) Wiener process to model the time-varying channel given by

$$\mathbf{H}(k+1) = \mathbf{H}(k) + \mathbf{v}(k) \tag{3}$$

where $\mathbf{v}(k)$ is an i.i.d. zero-mean Gaussian vector with covariance matrix $\sigma_v^2 \mathbf{I}_{N\times N}$. The augmented state equation is then

$$\mathbf{Y}(k+1) = \mathbf{F}\mathbf{Y}(k) + \mathbf{w}(k) \tag{4}$$

where $\mathbf{F} = \begin{bmatrix} \widetilde{\mathbf{F}} & \mathbf{0}_{N\times N} \\ \mathbf{0}_{N\times N} & \mathbf{I}_{N\times N} \end{bmatrix}$ and the augmented process noise $\mathbf{w}(k) = \begin{bmatrix} \widetilde{\mathbf{G}}d(k+1) \\ \mathbf{v}(k) \end{bmatrix}$. The nonlinearity of this formulation arises in the observation equation

$$z(k) = f\left[\mathbf{Y}(k)\right] + n(k) \tag{5}$$

where $f\left[\mathbf{Y}(k)\right] = \sum_{j=1}^{N}\left[\mathrm{Y}_j(k)\mathrm{Y}_{j+N}(k)\right] = \mathbf{D}^T(k)\mathbf{H}(k)$ and Y_{j+N} is the channel coefficient corresponding to the jth input Y_j.

2.2. Weighted Gaussian sum approximation

It is well known that if the plant noise $\mathbf{w}(k)$, the observation noise $n(k)$, and the initial estimate of the state are Gaussian and mutually independent, a standard extended Kalman filter (EKF) yields the suboptimal estimate of the state [4]. However, if the plant noise is non-Gaussian, which is the case in the context of data channel equalization, the EKF performs poorly. This problem can be resolved by approximating the density function of $\mathbf{w}(k)$ by a WGS [2] as given below

$$p(\mathbf{w}(k)) \triangleq \sum_{i=1}^{q} p_i \mathcal{N}\left[\mathbf{w}(k); \mathbf{w}_i, \mathbf{Q}_i\right] \tag{6}$$

where $\sum_{i=1}^{q} p_i = 1$, $p_i \geq 0$ for all i, $\mathbf{w}_i = \mathbf{G}d_i$, $\{d_i, i = 1,\ldots,q\}$ are the q values that $d(k)$ can take associated with the probabilities $\{p_i, i = 1,\ldots,q\}$, $\mathbf{G} = \left[\widetilde{\mathbf{G}}^T \ \mathbf{0}_N^T\right]^T$, $\mathbf{Q}_i = \xi\mathbf{I}_{2N}$, $\mathbf{I}_{2N}$ being the identity matrix, and ξ chosen small enough so that each Gaussian density function is located on a neighborhood of $\mathbf{w}_i$ with a probability mass equal to p_i.

It can be shown [2] that $p(\mathbf{w}(k))$ converges uniformly to any density function of practical concern as the number of terms q increases and the covariance $\mathbf{Q}_i$ approaches the zero matrix.

3. IMM ESTIMATOR FOR BLIND EQUALIZATION

In this section, a hybrid system for blind equalization with Markovian switching coefficients is presented by multiple models with a given probability of switching between the models (or modes). The model is one of r hypothesized models, $M_1,\ldots,M_r$ for the system and the event that model j is in effect during the sampling period ending at time k (i.e., the sampling period $(k-1,k]$) will be denoted by $M_j(k)$. Such systems are also called hybrid systems [4], because they have both continuous noise uncertainties (i.e., plant and measurement noises) and discrete uncertainties (i.e., model uncertainties). As the plant noise in (6) is approximated by a weighted q Gaussian terms, it can be assumed that one is dealing with a hybrid system with q modes of operation

$$\mathbf{Y}(k+1) = \mathbf{F}\mathbf{Y}(k) + \mathbf{w}_j(k) \quad \forall j \in \mathcal{S} \tag{7}$$

$$z(k) = f[\mathbf{Y}(k)] + n(k) \tag{8}$$

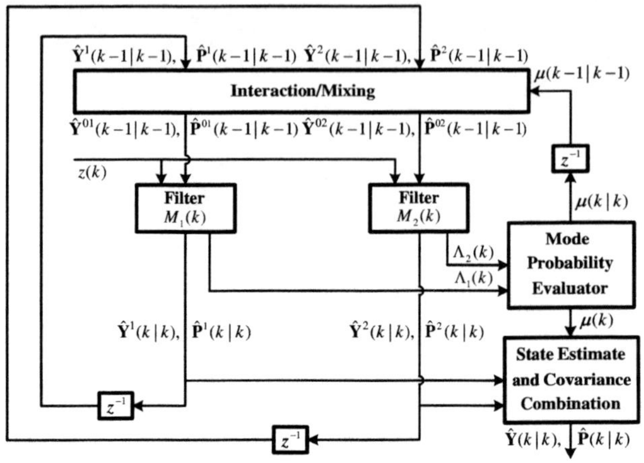

Fig. 1. One cycle of the IMM estimator.

where $\mathcal{S} = \{1,\ldots,q\}$ is the set of possible modes. Hence such a hybrid system state is dependent on a Markovian switching process generated by the symbol sequence and can be estimated effectively using an IMM algorithm [4]. For q models, the IMM algorithm is implemented with q EKFs operating in parallel at each cycle. The structure of the IMM algorithm is

$$(N_e; N_f) = (q; q) \tag{9}$$

where N_e is the *number of estimates* at the start of the cycle of the algorithm and N_f is the *number of filters* in the algorithm. The switching between the models is assumed to be governed by a finite-state Markov chain according to

$$\pi_{ij} \triangleq P\{M_j(k)|M_i(k-1)\} \tag{10}$$

where π_{ij} are the transition probabilities of switching from model $M_i(k-1)$ to model $M_j(k)$. The transition probabilities π_{ij} are assumed known, time invariant and independent of the base state. One cycle of the IMM-based blind equalization as depicted in Fig. 1 is summarized below. At every cycle, each EKF will produce a model conditioned state estimate $\hat{\mathbf{Y}}^j(k|k)$ and the associated covariance $\hat{\mathbf{P}}^j(k|k)$ based on its input state $\hat{\mathbf{Y}}^{0j}(k-1|k-1)$, covariance $\hat{\mathbf{P}}^{0j}(k-1|k-1)$ and the current observation $z(k)$.

1. *Calculation of the mixing probabilities*: This is the probability that the symbol corresponding to mode M_i was in effect at $k-1$ given that M_j is in effect at k conditioned on the measurements up to time $k-1$, Z^{k-1}, for all $i,j = 1,\ldots,q$. Here symbols corresponding to i and j takes on values in γ and the calculation is done for all q values in γ.

$$\begin{aligned} \mu_{i|j}(k-1|k-1) &\triangleq P\{M_i(k-1)|M_j(k), Z^{k-1}\} \quad (11) \\ &= \frac{1}{\bar{c}_j} P\{M_j(k)|M_i(k-1), Z^{k-1}\} \\ &\quad .P\{M_i(k-1)|Z^{k-1}\} \\ &= \frac{1}{\bar{c}_j}\pi_{ij}\mu_i(k-1) \quad i,j = 1,\ldots,q \end{aligned}$$

where $\bar{c}_j = \sum_{i=1}^{q}\pi_{ij}\mu_i(k-1)$ is the normalizing constant and $\mu_i(k-1)$ is the posterior symbol probabilty at time $(k-1)$.

2. *Mixing*: Starting with $\hat{\mathbf{Y}}^i(k-1|k-1)$, state estimate of the EKF matched to mode i at time $k-1$, the mixed initial condition for the filter matched to each symbol at time k is computed as follows

$$\begin{aligned}&\hat{\mathbf{Y}}^{0j}(k-1|k-1) \qquad (12)\\ &= \sum_{i=1}^{q} \hat{\mathbf{Y}}^i(k-1|k-1)\mu_{i|j}(k-1|k-1) \quad j=1,\ldots,q\end{aligned}$$

and the associated covariance is

$$\begin{aligned}&\hat{\mathbf{P}}^{0j}(k-1|k-1) \qquad (13)\\ &= \sum_{i=1}^{q} \mu_{i|j}(k-1|k-1)\Big\{\hat{\mathbf{P}}^i(k-1|k-1)\\ &+\big[\hat{\mathbf{Y}}^i(k-1|k-1)-\hat{\mathbf{Y}}^{0j}(k-1|k-1)\big]\\ &.\big[\hat{\mathbf{Y}}^i(k-1|k-1)-\hat{\mathbf{Y}}^{0j}(k-1|k-1)\big]^T\Big\}\\ &\qquad\qquad j=1,\ldots,q\end{aligned}$$

3. *Mode-matched filtering*: The estimate (12) and covariance (13) are used as input to the filter matched to $M_j(k)$, which uses observation at time k, $z(k)$ to yield $\hat{\mathbf{Y}}^j(k|k)$ and $\hat{\mathbf{P}}^j(k|k)$. These are the outputs of the EKF and the EKF equations can be found in [4], [5]. The likelihood functions corresponding to the q filters are computed as

$$\begin{aligned}\Lambda_j(k) &\triangleq p\big[z(k)|M_j(k), Z^{k-1}\big] \qquad (14)\\ &= p\big[z(k)|M_j(k), \hat{\mathbf{Y}}^{0j}(k-1|k-1),\\ &\qquad \hat{\mathbf{P}}^{0j}(k-1|k-1)\big]\\ &= p\big[\nu_j(k)\big]\\ &= \mathcal{N}\big[\nu_j(k); 0, S_j(k)\big] \quad j=1,\ldots,q\end{aligned}$$

where $\nu_j(k)$ and $S_j(k)$ are the innovation and its covariance from the mode-matched filter corresponding to mode j.

4. *Mode probability update*: The probability that the mode j ($j = 1,\ldots,q$) is in effect is updated as follows

$$\begin{aligned}\mu_j(k) &\triangleq P\{M_j(k)|Z^k\}\\ &= \frac{1}{c}p\big[z(k)|M_j(k), Z^{k-1}\big]P\{M_j(k)|Z^{k-1}\}\\ &= \frac{1}{c}\Lambda_j(k)\sum_{i=1}^{q} P\{M_j(k)|M_i(k-1), Z^{k-1}\}\\ &\quad .P\{M_j(k-1)|Z^{k-1}\}\\ &= \frac{1}{c}\Lambda_j(k)\sum_{i=1}^{q}\pi_{ij}\mu_i(k-1)\\ &= \frac{1}{c}\Lambda_j(k)\bar{c}_j \; j=1,\ldots,q \qquad (15)\end{aligned}$$

where the normalizing constant $c = \sum_{j=1}^{q}\Lambda_j(k)\bar{c}_j$.

5. *Estimate and covariance combination*: Finally, combination of the mode-conditioned estimates and covariances is done using the following mixture equations

$$\hat{\mathbf{Y}}(k|k) = \sum_{j=1}^{q}\hat{\mathbf{Y}}^j(k|k)\mu_j(k) \qquad (16)$$

$$\begin{aligned}\hat{\mathbf{P}}(k|k) &= \sum_{j=1}^{q}\mu_j(k)\Big\{\hat{\mathbf{P}}^j(k|k) \qquad (17)\\ &+\big[\hat{\mathbf{Y}}^j(k|k)-\hat{\mathbf{Y}}(k|k)\big]\big[\hat{\mathbf{Y}}^j(k|k)-\hat{\mathbf{Y}}(k|k)\big]^T\Big\}\end{aligned}$$

This combination is not part of the algorithm recursions whereas it is used only for output purposes. The estimated state $\hat{\mathbf{Y}}(k|k)$ is the concatenation of the so-resulted estimations of both the symbol vector $\hat{\mathbf{D}}(k|k)$ and the channel coefficents $\hat{\mathbf{H}}(k|k)$. It corresponds to a blind estimation of the channel and the transmitted data sequence corrupted by ISI and noise.

4. SIMULATION RESULTS

Computer simulations are conducted to illustrate the effectiveness of the proposed approach. In the simulations, the transmitted data sequence is an i.i.d. binary sequence. The initial model probabilities are $\mu(0) = [0.5 \;\; 0.5]^T$ and the mode-switching probability matrix is given by

$$\begin{bmatrix}\pi_{11} & \pi_{12}\\ \pi_{21} & \pi_{22}\end{bmatrix} = \begin{bmatrix}0.5 & 0.5\\ 0.5 & 0.5\end{bmatrix}$$

A time-varying channel is selected for simulation with channel coefficients $\mathbf{H}(0) = [1 \;\; 0.2 \;\; 0.5]^T$ having $\sigma_v^2 = 5\times10^{-5}$. A realization of the nonstationary channel modeled by the Wiener process is shown in [3]. In the WGS approximation, the covariance matrix $\mathbf{Q}_i$ is taken the same for all i and equal to $\sigma_v^2\,\mathbf{I}_{2N\times2N}$. Decisions on the estimated transmitted symbols are done with a delay $l = 2$. In Fig. 2, the performance of the proposed approach is compared with that of the NEKF-based equalizer by computing the corresponding bit error rates (BER). For each signal to noise ratio (SNR), the BER is computed over 100 Monte Carlo runs of length 10000 symbols. This thus reflects the effects of estimator convergence. From Fig. 2, it can be noticed that the equalizer based on IMM algorithm improves the performance compared to the NEKF-based equalizer.

The convergence of channel coefficients is shown in Fig. 3 where we compare the mean square error (MSE) of the channel estimates as a function of the number of iterations for SNR = 20 dB. Fig. 4 depicts the convergence of channel coefficients with $\mathbf{H}(0) = [0.62 \;\; 0.56 \;\; 0.48 \;\; 0.46 \;\; 0.22]^T$ having $\sigma_v^2 = 6\times10^{-6}$ at different SNRs. For both cases, the MSE is evaluated over 100 Monte Carlo runs. As seen in these figures, the IMM-based equalizer performs better than the NEKF-based approach. During the simulations, we have observed that the IMM equalizer is more stable than the NEKF equalizer where instability arise when the real symbol predicted states tend more towards N-binary vectors. The superior performance of IMM equalizer is due to its "adaptive bandwidth" capability. The IMM-based approach is also robust against inaccurate prior information about the plant noise. Suppose the exact probability of the data sequence is not known for the binary data sequence. In this case, the IMM equalizer continues to give acceptable results because it will adjust the mode probabilities accordingly based on the received data.

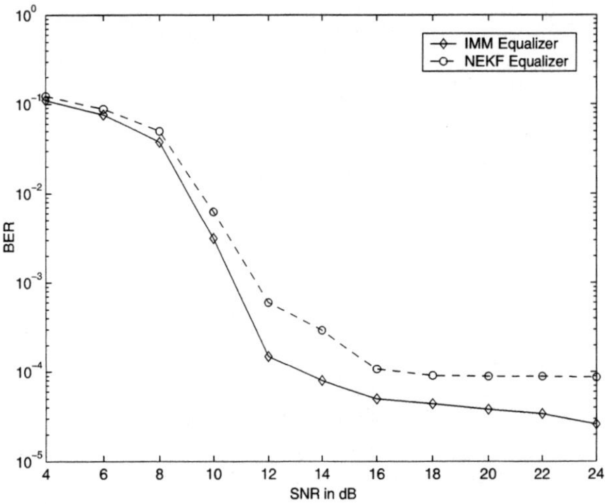

Fig. 2. BER as a function of SNR.

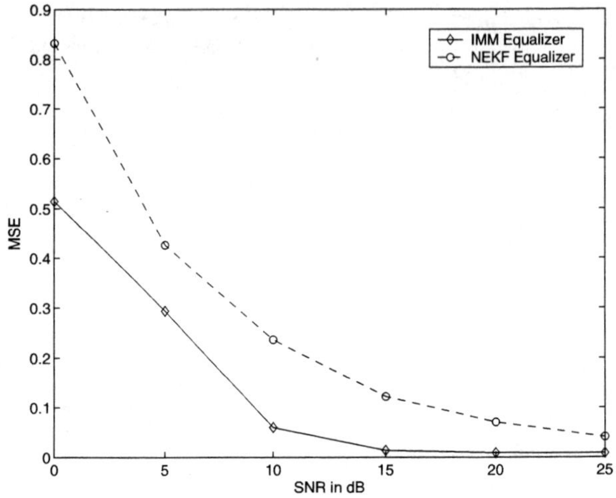

Fig. 4. MSE as a function of SNR with $l = 4$.

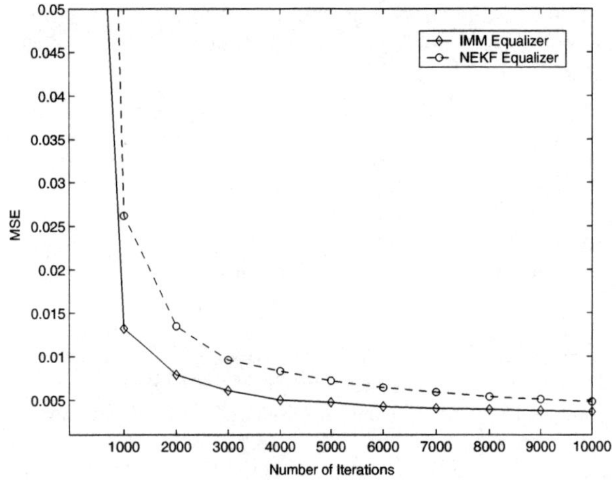

Fig. 3. MSE with varying number of iterations.

5. CONCLUSIONS

In this paper, a novel interacting multiple model based nonstationary channel estimator and equalizer has been presented for systems with non-Gaussian plant noise. The proposed equalizer can handle changes in the system structure as well as in the noise statistics. It was shown that IMM equalizer results in superior performance compared with the previous equalizer consisting of a (static or non-interacting) network of extended Kalman filters. The major advantage of the the IMM equalizer is that, unlike the NEKF equalizer, it avoids the number of terms in the WGS approximation of the plant noise which increases dramatically through iterations. It also avoids the exponential growth of the state complexity caused by increasing channel memory length in [7]. The IMM equalizer is more stable in high SNR and robust against inaccurate *a priori* information about the plant noise.

6. REFERENCES

[1] K. Abend and B. D. Fritchman, "Statistical detection for communication channels with intersymbol interferance," *Proc. IEEE*, vol. 58, pp. 779-785, May 1970.

[2] D. L. Alspach and H. W. Sorenson, "Nonlinear Bayesian estimation using Gaussian sum approximations," *IEEE Trans. Automat. Contr.*, vol. AC-17, pp. 439-448, Aug. 1972.

[3] R. Amara and S. Marcos, "A blind network of extended Kalman filters for nonstationary channel equalization," in *Proc. IEEE ICASSP*, Salt Lake City, Utah, vol. 4, pp. 2117-2120, May 2001.

[4] Y. Bar-Shalom, X. R. Li, and T. Kirubarajan, *Estimation with Applications to Tracking and Navigation: Theory, Algorithms, and Software.* New York: John Wiley & Sons, Inc., 2001.

[5] S. Haykin, *Adaptive Filter Theory.* New Jersey: Prentice Hall, 2002.

[6] R. A. Iltis, "A Bayesian maximum-likelihood sequence estimation algorithm for a priori unknown channels and symbol timing," *IEEE J. Select. Areas in Commun.*, vol. 10, pp. 579-588, Apr. 1992.

[7] R. A. Iltis, J.J. Shynk, and K. Giridhar, "Bayesian algorithms for blind equalization using parallel adaptive filtering ," *IEEE Trans. Commun.*, vol. 42, pp. 1017-1032, Feb./Mar./Apr. 1994.

[8] G.-K. Lee, S. B. Gelfand, and M. P. Fitz, "Bayesian techniques for blind deconvolution," *IEEE Trans. Commun.*, vol. 44, pp. 826-835, July 1996.

[9] S. Marcos, "A network of adaptive Kalman filters for data channel equalization," *IEEE Trans. Signal Processing*, vol. 48, no. 9, pp. 2620-2627, Sept. 2000.

[10] J. G. Proakis, *Digital Communications.* New York: McGraw-Hill, 2001.

[11] N. Seshadri, "Joint channel and data estimation using blind trellis search techniques," *IEEE Trans. Commun.*, vol. 42, pp. 1000-1016, Feb./Mar./Apr. 1994.

NEW ALGORITHM FOR BLIND ADAPTIVE EQUALIZATION BASED ON CONSTANT MODULUS CRITERION

Y. J. Kou, W.-S. Lu, and A. Antoniou

Department of Electrical and Computer Engineering
University of Victoria, Victoria, B.C., Canada V8W 3P6
{ykou, wslu,aantoniou}@ece.uvic.ca

ABSTRACT

Constant modulus (CM) based algorithms for blind channel equalization are well known for their effectiveness and simplicity. Recently, new CM-based equalization algorithms with improved performance have been proposed. In this paper, a new blind adaptive CM equalization algorithm using a quasi-Newton optimization method is proposed. Simulation results are presented which demonstrate that the proposed algorithm leads to an improved convergence rate as well as reduced computational complexity relative to those of some existing algorithms.

1. INTRODUCTION

Adaptive channel equalization techniques have been widely used in communication systems to deal with intersymbol interference (ISI) caused by channel distortion or multipath transmission. Conventional equalization algorithms require the transmission of a training signal to update the parameters of the equalizer. This inevitably reduces channel capacity. In addition, the inclusion of a training signal increases the complexity of the transceiver significantly. Therefore, blind adaptive equalization algorithms that do not require a training phase are often preferred. Among various blind equalization algorithms, constant-modulus (CM) based algorithms are well known for their effectiveness and simplicity [1]. However, these are usually implemented in terms of gradient based algorithms which are usually quite slow [2][3]. Recently, several improved CM-based blind adaptive equalization algorithms have been proposed. In [4], a blind equalization algorithm based on stochastic gradient decent minimization of order-α Renyi's entropy was proposed. A fast recursive constant modulus algorithm (RCMA) based on the recursive least square (RLS) algorithm was proposed in [5]. These algorithms reduce the time required for convergence at the cost of increased computational complexity.

In this paper, a new blind adaptive CM equalization algorithm using a quasi-Newton optimization method [6] is derived. Simulation results are presented to demonstrate that the proposed algorithm outperforms the algorithms in [1][5] in terms of convergence rate and achieves reduced compuatational complexity relative to that of RCMA algorithm.

The authors are grateful to Micronet, NCE Program, and the Natural Sciences and Engineering Research Council of Canada for supporting this work.

2. PROBLEM FORMULATION

Consider the digital communication system depicted in Fig. 1, where a_i, d_i, x_i and n_i represent the CM input signal, channel output signal, received signal, and additive white Gaussian noise (AWGN), respectively. In communication systems, the channel characteristics are far from ideal and a channel equalizer is ofter needed to combat ISI especially in the case of wireless communications.

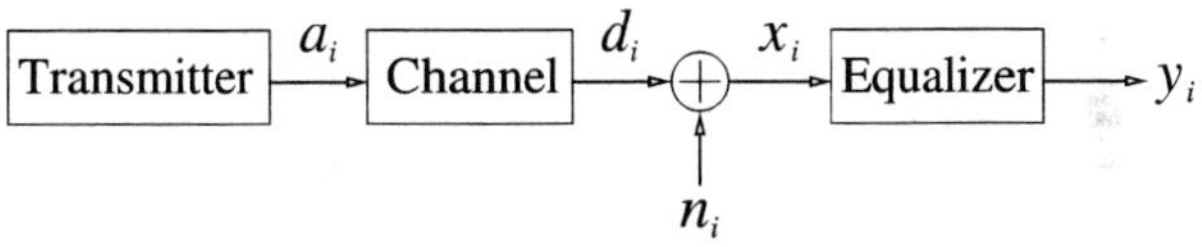

Fig. 1. Block diagram of a digital communication system.

The output signal of the equalizer in Fig. 1 can be expressed as

$$y_i = \mathbf{w}^H \mathbf{x}_i \tag{1}$$

where $\mathbf{x}_i = [x_{i-n+1}\ x_{i-n+2}\ \cdots\ x_i]^T$ is a block of input samples available at time instant i, $\mathbf{w}$ is an n-dimensional weight vector, and n is the length of the equalizer. If perfect equalization is achieved, then y_i has a constant instantaneous modulus. The aim of a constant-modulus based equalizer, therefore, is to minimize modulus variations of sequence y_i. Mathematically, the optimization problem for the equalizer can be formulated as

$$\underset{\mathbf{w}}{\text{minimize}} \quad f(\mathbf{w}) = \sum_{i=1}^{N} (|y_i|^2 - 1)^2 \tag{2}$$

where N is the length of sequence y_i.

3. QUASI-NEWTON ALGORITHM

The objective function in (2) is a fourth-order polynomial of variable $\mathbf{w}$ and is not in general convex. A commonly used optimization method to solve the problem in (2) is the steepest descent method (SDM) [6]. In each iteration, the SDM uses the gradient $\nabla f(\mathbf{w})$ to compute a search direction which in conjunction with a line search step determines the next iterate. Various least-mean-squares (LMS) algorithms for the channel equalization are essentianlly different implementations of the SDM proposed in the past [7]. A serious drawback of the SDM is its slow convergence, especially when the condition number of the Hessian matrix of

$f(\mathbf{w})$ is large. The Newton method along with Hessian matrix manipulation to ensure its positive definiteness solves the problem in (2) significantly faster at the cost of a considerable increase in computational complexity [6]. The main computational burden in the Newton method is the evaluation of the inverse of a possibly modified Hessian matrix of $f(\mathbf{w})$. Recursive least-squares algorithms are essentially adaptive implementations of the Newton method [7]. The class of quasi-Newton methods, which does not require the evaluation of the Hessian matrix and its inverse, offers a quadratic convergence rate with much reduced computational effort relative to that of the Newton method. Moreover, because the approximate inverse of the Hessian matrix is always positive definite, quasi-Newton algorithms are descent algorithms in that the objective function decreases monotonically as iteration continues. One of the most frequently used quasi-Newton algorithms is the Broyden-Fletcher-Goldfrab-Shanno (BFGS) algorithm [6] which is summarized below.

Table 1. BFGS algorithm

Step 1
Input initial $\mathbf{w}_0$ and stopping tolerance ϵ.
Set $k = 0$ and $\mathbf{S}_0 = \mathbf{I}_n$.
Compute $\mathbf{g}_k = \nabla f(\mathbf{w}_k)$.

Step 2
Set $\mathbf{d}_k = -\mathbf{S}_k\mathbf{g}_k$ and find α_k that minimizes $f(\mathbf{w}_k + \alpha\mathbf{d}_k)$.
Set $\mathbf{w}_{k+1} = \mathbf{w}_k + \alpha\mathbf{d}_k$.

Step 3
If $||\alpha_k\mathbf{d}_k|| < \epsilon$, output $\mathbf{w}^* = \mathbf{w}_{k+1}$ and stop.
Otherwise go to Step 4.

Step 4
Compute $\mathbf{g}_{k+1}$ and set $\gamma_k = \mathbf{g}_{k+1} - \mathbf{g}_k$.
Update matrix $\mathbf{S}_k$ using

$$\mathbf{S}_{k+1} = \mathbf{S}_k + \frac{\delta_k\delta_k^T}{\delta_k^T\gamma_k} - \frac{\mathbf{S}_k\gamma_k\gamma_k^T\mathbf{S}_k}{\gamma_k^T\mathbf{S}_k\gamma_k} \quad (3)$$

Set $k = k+1$ and repeat from Step 2.

From (2), it is clear that the optimized weight vector $\mathbf{w}$ depends on the data set $\{\mathbf{x}_k,\ k = 1, \ldots, N\}$. If we refer to this data block as *block l*, then the minimizer of the problem in (2) can be denoted as $\mathbf{w}_l^*$. In the next section, we derive an explicit expression for $f(\mathbf{w})$ for a complex-valued input signal and weights, an efficient line search method, and an adaptive implementation of the equalizer that generates a good approximation of $\mathbf{w}_l^*$ in real time.

4. NEW ADAPTIVE CM EQUALIZER

4.1. Data structure

For real-time channel equalization, the input data is processed block by block. A new data block for the next round of processing is generated by including a certain number of new input samples, say N_v while discarding N_v old samples. The data structure is illustrated in Fig. 2, where N_B is the size of one data block.

4.2. Objective function

When the lth block of data is processed, vector $\mathbf{x}_1$ in (2), which represents the first n samples of the data, assumes the form $[x_{lN_v+1} \cdots x_{lN_v+n}]^T$ and vector $\mathbf{x}_N$ in (2), which represents the last n

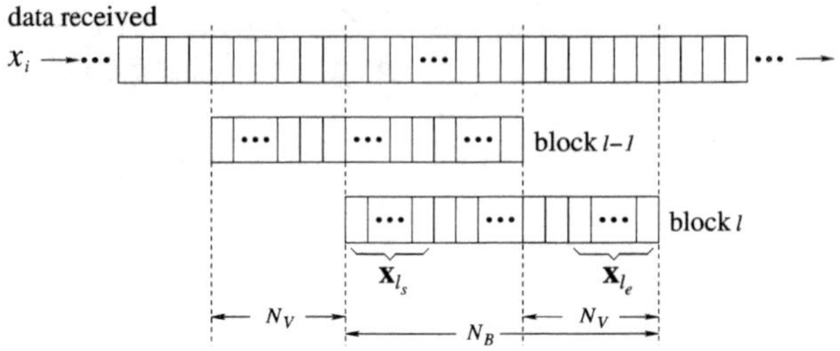

Fig. 2. Data structure of the adaptive CM equalization algorithm.

samples of the data, is given by $[x_{lN_v+N_B-n+1} \cdots x_{lN_v+N_B}]^T$. Therefore, the objective function for the lth data block in (2) becomes

$$f_l(\mathbf{w}) = \sum_{i=l_s}^{l_e} (\mathbf{w}^H\mathbf{x}_i\mathbf{x}_i^H\mathbf{w} - 1)^2 \quad (4)$$

where $l_s = lN_v + n$ and $l_e = lN_v + N_B$. Since both $\mathbf{w}$ and $\mathbf{x}_i$ are complex-valued, in general, we need to write $\mathbf{w} = \mathbf{w}_r + j\mathbf{w}_j$ and $\mathbf{x}_i = \mathbf{x}_{r,i} + j\mathbf{x}_{j,i}$, and express (4) as

$$\begin{aligned} f_l(\hat{\mathbf{w}}) &= \sum_{i=l_s}^{l_e} \left[\left(\hat{\mathbf{w}}^T\hat{\mathbf{x}}_{1,i}\right)^2 + \left(\hat{\mathbf{w}}^T\hat{\mathbf{x}}_{2,i}\right)^2 - 1\right]^2 \\ &= \sum_{i=l_s}^{l_e} (A_i^2 + B_i^2 - 1)^2 \end{aligned} \quad (5)$$

where $A_i = \hat{\mathbf{w}}^T\hat{\mathbf{x}}_{1,i}$, $B_i = \hat{\mathbf{w}}^T\hat{\mathbf{x}}_{2,i}$, and

$$\hat{\mathbf{w}} = \begin{bmatrix}\mathbf{w}_r \\ \mathbf{w}_j\end{bmatrix},\ \hat{\mathbf{x}}_{1,i} = \begin{bmatrix}\mathbf{x}_{r,i} \\ \mathbf{x}_{j,i}\end{bmatrix},\ \hat{\mathbf{x}}_{2,i} = \begin{bmatrix}-\mathbf{x}_{j,i} \\ \mathbf{x}_{r,i}\end{bmatrix}.$$

The gradient of $f_l(\hat{\mathbf{w}})$ is found to be

$$\mathbf{g}_l(\hat{\mathbf{w}}) = 4\sum_{k=l_s}^{l_e} \left(A_i^2 + B_i^2 - 1\right)\left(A_i\hat{\mathbf{x}}_{1,i} + B_i\hat{\mathbf{x}}_{2,i}\right). \quad (6)$$

Using the BFGS algorithm, the weight vector is updated to

$$\hat{\mathbf{w}}_{k+1} = \hat{\mathbf{w}}_k + \alpha_k\hat{\mathbf{d}}_k \quad (7a)$$

with

$$\hat{\mathbf{d}}_k = -\mathbf{S}_k\mathbf{g}_l(\hat{\mathbf{w}}_k). \quad (7b)$$

where $\hat{\mathbf{w}}_k$ denotes the current weight vector, matrix $\mathbf{S}_{k+1}$ is obtained using (3), and α_k is a positive scalar that minimizes $f_l(\hat{\mathbf{w}}_k + \alpha\hat{\mathbf{d}}_k)$. The process of finding α_k is known as *line search* and can in the present case be carried out accurately and efficiently as described next.

4.3. Line search

We can write $f_l(\hat{\mathbf{w}}_k + \alpha\hat{\mathbf{d}}_k)$ as a fourth-order polynomial of scalar variable α as

$$f_l(\hat{\mathbf{w}}_k + \alpha\hat{\mathbf{d}}_k) = c_4\alpha^4 + c_3\alpha^3 + c_2\alpha^2 + c_1\alpha + c_0 \quad (8)$$

where

$$c_4 = \sum_{i=l_s}^{l_e} (C_i^2 + D_i^2)^2$$

$$c_3 = \sum_{i=l_s}^{l_e} 4(A_iC_i + B_iD_i)(C_i^2 + D_i^2)$$

$$c_2 = \sum_{i=l_s}^{l_e} \left[2(C_i^2 + D_i^2)(A_i^2 + B_i^2 - 1) + 4(A_iC_i + B_iD_i)^2\right]$$

$$c_1 = \sum_{i=l_s}^{l_e} 4(A_iC_i + B_iD_i)(A_i^2 + B_i^2 - 1)$$

$$c_0 = \sum_{i=l_s}^{l_e} (A_i^2 + B_i^2 - 1)^2$$

and $C_i = \hat{\mathbf{d}}_l^T \hat{\mathbf{x}}_{1,i}$, $D_i = \hat{\mathbf{d}}_l^T \hat{\mathbf{x}}_{2,i}$. The minimizer α_k must satisfy the condition $df_l(\hat{\mathbf{w}}_k + \alpha\hat{\mathbf{d}}_k)/d\alpha = 0$, i.e.,

$$4c_4\alpha^3 + 3c_3\alpha^2 + 2c_2\alpha + c_1 = 0. \tag{9}$$

If there is only one real root for (9), then it is the minimizer α_k. Otherwise, there are three real roots for (9), two of which satisfy the second-order sufficiency condition

$$6c_4\alpha^2 + 3c_3\alpha + c_2 > 0. \tag{10}$$

and we choose the one that achieves the absolute minimum value of the function $f_l(\hat{\mathbf{w}}_k + \alpha\hat{\mathbf{d}}_k)$ as α_k.

4.4. Adaptive implementation

A real-time implementation of the algorithm can be carried out as follows. At any given instant, a block of input samples of size N_B is used to construct vectors $\{\mathbf{x}_i,\ i = l_s, \ldots, l_e\}$ in (4) and a certain number of iterations of the BFGS algorithm are applied to minimize $f_l(\hat{\mathbf{w}})$ in (4) to obtain an improved weight vector $\hat{\mathbf{w}}_l^*$. This approximate solution is utilized for channel equalization for the next N_v samples periods. The data set is then updated by including N_v new samples and excluding N_v old samples, and the BFGS algorithm is applied again. The initial weight vector $\hat{\mathbf{w}}_0$ and matrix $\mathbf{S}_0$ for a given data block l are the vector $\hat{\mathbf{w}}_{l-1}^*$ and matrix $\mathbf{S}_k$ obtained in the previous iteration for the proceeding data block, respectively, except the case of $l = 0$ in that a reasonable $\hat{\mathbf{w}}_0$ and $\mathbf{S}_0 = \mathbf{I}_{2n}$ are utilized. The algorithm is summarized in Table 2.

5. SIMULATIONS

The proposed equalization algorithm was applied to the baseband communication system shown in Fig. 1 and its performance was evaluated and compared with that of the LMS algorithm proposed in [1] and the RCMA algorithm proposed in [5]. A commonly used performance measure for equalization algorithms is the residual ISI which is defined as

$$ISI = 10\log_{10}\frac{||\mathbf{f}||_2^2 - ||\mathbf{f}||_\infty^2}{||\mathbf{f}||_\infty^2}$$

where $\mathbf{f} = \mathbf{h} * \mathbf{w}$ denotes the convolution of the channel impulse response and the weight vector of the equalizer, and $||\mathbf{f}||_2$ and $||\mathbf{f}||_\infty$ represent the 2-norm and infinity-norm of vector $\mathbf{f}$, respectively. In all simulations, $k_{max} = 1$ was used.

Table 2. Blind CM equalization adaptation algorithm

Outer Loop:
Step $\mathcal{O}1$
Set $l = 0$ and $\mathbf{S}_0 = \mathbf{I}_{2n}$.
Input an initial $\hat{\mathbf{w}}_0$ and parameters N_B and N_v.
Step $\mathcal{O}2$
Form data block l.
Inner Loop:
Step $\mathcal{I}1$
Set the maximum number of iterations to k_{max}.
Set $k = 0$.
Compute $\mathbf{g}_l(\hat{\mathbf{w}}_k)$ using (6).
Step $\mathcal{I}2$
Compute $\hat{\mathbf{d}}_k$ using (7b).
Find α_k using the method in Sec. 4.3.
Update $\hat{\mathbf{w}}_k$ to $\hat{\mathbf{w}}_{k+1}$ using (7a).
Step $\mathcal{I}3$
Compute $\mathbf{g}_l(\hat{\mathbf{w}}_{k+1})$ and update $\mathbf{S}_k$ to $\mathbf{S}_{k+1}$ using (3).
Step $\mathcal{I}4$
If $k + 1 = k_{max}$,
then set $\hat{\mathbf{w}}_l^* = \hat{\mathbf{w}}_{k+1}$ and go to Step $\mathcal{O}3$.
Otherwise, go to Step $\mathcal{I}5$.
Step $\mathcal{I}5$
Set $k = k + 1$ and go to Step $\mathcal{I}2$.
Step $\mathcal{O}3$
Set $\hat{\mathbf{w}}_0 = \hat{\mathbf{w}}_l^*$, $\mathbf{S}_0 = \mathbf{S}_{k+1}$.
Set $l = l + 1$ and repeat from Step $\mathcal{O}2$.

Example 1: Each two consecutive input bits were mapped into a four-quadrature amplitude-modulation (4-QAM) symbol and then the modulation symbols were transmitted through a channel with the channel impulse response given by $\mathbf{h} = [0.815\ 0.419\ 0.419]^T$ [5]. Additive white Guassian noise (AWGN) was added at the output of the channel and the signal-to-noise ratio (SNR) of the received signal was set to 20 dB. At the receiver, a baud-spaced equalizer was implemented. For all algorithms, the length of the equalizer n was chosen to be 9 and the equalizer was initialized with $\mathbf{w}_0 = [0\ 0\ 0\ 0\ 1\ 0\ 0\ 0\ 0]^T$. For the proposed algorithm, N_B and N_v were chosen to be 100 and 20, respectively. The step size was set to 5×10^{-3} for the LMS algorithm, and the forgetting factor was set to 0.99 for the RCMA algorithm. The performance of the equalization algorithms was evaluated and averaged over 50 trials. The residual ISI of the proposed algorithm versus the sample index is plotted in Fig. 3 as the solid curve while those of the algorithms [1] and [5] are plotted as the dashed and dot-dashed curves, respectively. It is observed that the proposed algorithm reaches convergence at -16 dB residual ISI within 200 samples, whereas the algorithms in [1] and [5] require 1300 and 600 samples, respectively, to achieve the same level of residual ISI. The amount of computation per sample required by the proposed algorithm is approximately 25% less than that required by the RCMA algorithm in [5].

Example 2: This example is concerned with a telephone channel with channel impulse response vector $\mathbf{h} = [0.04\ -0.05\ 0.07\ -0.21\ -0.5\ 0.72\ 0.36\ 0.21\ 0.03\ 0.07]^T$ [8]. Using the same settings as in Example 1, the proposed algorithm and the algorithms

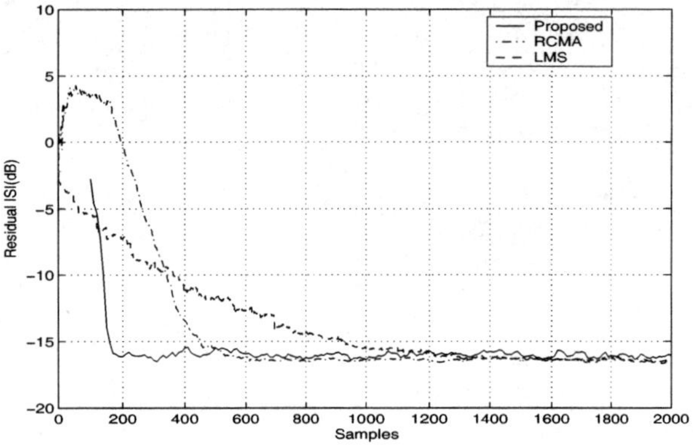

Fig. 3. Performance comparison of various equalization algorithms.

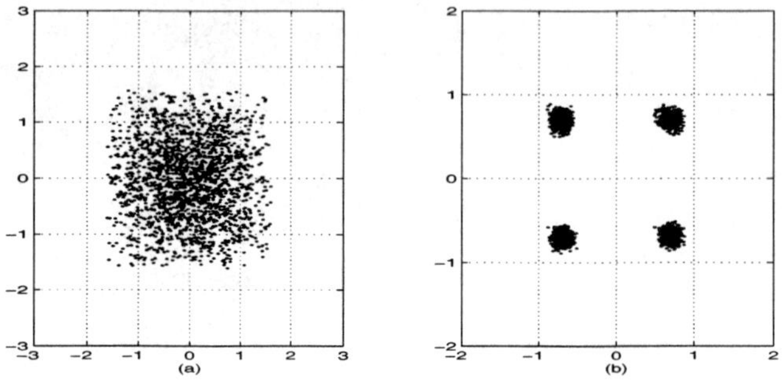

Fig. 5. (a) Constellation of the received signal before equalization; (b) Constellation of the signal after equalization.

in [1][5] were applied to equalize the channel. The performance of the equalization algorithms was evaluated and averaged over 50 trials. The residual ISI of the proposed algorithm versus the sample index is plotted in Fig. 4 as a solid curve while those of the algorithms [1] and [5] are plotted as dashed and dot-dashed curves, respectively. It is observed that the proposed algorithm reaches convergence at -20 dB residual ISI within 200 samples, whereas the algorithms in [1] and [5] require 600 and 500 samples, respectively, to achieve the same level of performance. The constellations of the input and output signals of the equalizer are plotted in Fig. 5(a) and 5(b), respectively. As can be observed, the constellation of the input signal is fairly disturbed which would lead to a poor bit-error-rate. The application of the proposed algorithm led to a discernible constellation for the output signal.

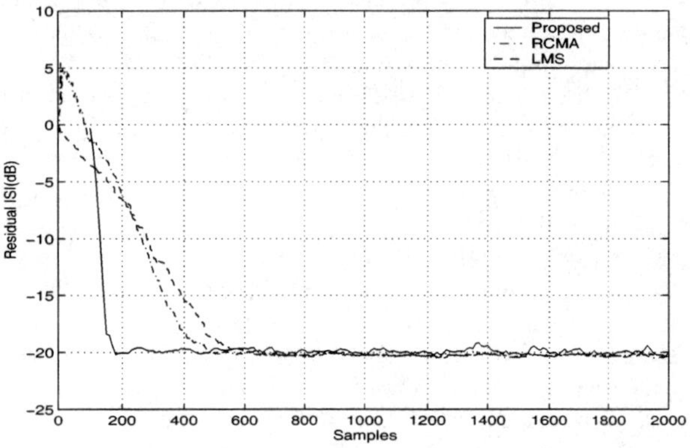

Fig. 4. Performance comparison of various equalization algorithms.

6. CONCLUSIONS

An efficient CM adaptation algorithm for blind channel equalization has been proposed. The algorithm is based on a quasi-Newton optimization method where the computation intensive update of the inverse of the Hessian matrix is carried out using the rank-two BFGS formula. As a result, the proposed adaptation algorithm offers fast convergence rate and reduced computational complexity relative to those of some existing algorithms.

7. REFERENCES

[1] J. R. Treichler and M. G. Agee, "A new approach to multipath correction of constant modulus signals," *IEEE Trans. on Acoustics, Speech, and Signal Processing*, vol. 31, pp. 459-472, Apr. 1983.

[2] J. R. Treichler, V. Wolff, and C. R. Johnson, Jr., "Observed misconvergence in the constant modulus adaptive algorithm," *Proc. of 25rd Asilomar Conference on Signals, Systems and Computers*, pp. 663-667, Nov. 1991.

[3] C. R. Johnson, Jr., P. Schniter, T. J. Endres, J. D. Behm, and R. A. Casas, "Blind equalization using the constant modulus criterion: A review," *Proc. IEEE*, vol. 86, no. 10, pp. 1927-1950, Oct. 1998.

[4] I. Santamaria, C. Pantaleon, L. Vielva, and J. C. Principe, "A fast algorithm for adaptive blind equalization using order-α Renyi's entropy," *IEEE International Conference on Acoustics, Speech and Signal Processing*, vol.3, p.p. 2657-2660, May 2002.

[5] R. Pickholtz and K. Elbarbary, "The recursive constant modulus algorithm; a new approach for real-time array processing," *Proc. of 27rd Asilomar Conference on Signals, Systems and Computers*, pp. 627-632, Nov. 1993.

[6] R. Fletcher, *Practical Methods of Optimization*, 2nd ed., Wiley, New York, 1987.

[7] S. Haykin, *Adaptive Filter Theory*, 4th ed., Prentice-Hall, 2000.

[8] J. Proakis, *Digital Communications*, 4th ed., McGraw-Hill, 2000.

SEQUENTIAL BLIND EXTRACTION OF MIXED SOURCE SIGNALS WITH GUARANTEED CONVERGENCE

Derong Liu and Sanqing Hu
Department of Electrical and Computer Engineering
University of Illinois at Chicago, Chicago, IL 60607
E-mail: dliu@ece.uic.edu, shu@cil.ece.uic.edu

Abstract— **This paper presents a gradient-based method for sequential blind extraction of mixed source signals. In our approach the number of unknowns to be determined is only $2(m-1)$ and there is no need for computing any matrix inversion. Our algorithm is guaranteed to converge. In addition, the convergence speed of our algorithm can be improved by adjusting a parameter. Simulation results show the operation characteristic, the effectiveness of our method, and the advantages over the existing algorithm.**

I. INTRODUCTION

Blind separation of independent source signals from their mixtures has received considerable attention in recent years. This class of signal processing techniques has applications in many technical areas such as communications [4], signal processing [6], speech processing [1], and image restoration [9], to name a few.

The objective of blind source separation is to recover sources from their mixtures without the prior knowledge of sources and the mixing channel. Recently, independent component analysis has been used to deal with the case of instantaneous mixtures [4], [8]. In general, there are two classes of approaches for recovering original sources from instantaneous mixtures: the simultaneous separation approach [5], [7], [8], [10], [14] and the extraction approach [2], [3], [11]–[13]. In the separation approach, all separable sources are separated simultaneously, whereas in the extraction approach sources are extracted one by one. Simultaneous separation, if possible, is of course desirable. In some ill-conditioned cases, simultaneous blind separation is not obtainable; but sequential blind extraction is since sequential blind extraction requires weaker solvability conditions than simultaneous blind separation [11].

Consider a general linear case of instantaneously mixing n sources with m observable mixtures

$$y(t) = Ax(t) \tag{1}$$

where $x(t) = [x_1(t), \cdots, x_n(t)]^T$ is a vector of mutually independent unknown sources with zero means, $y(t) = [y_1(t), \cdots, y_m(t)]^T$ is a vector of mixed signals (as a result, $y(t)$ has zero means since $x(t)$ has zero means) and $A = \{a_{ij}\}$ is an $m \times n$ unknown constant matrix referred to as the mixing matrix. The task of blind extraction is to recover sources one by one from the available mixtures $y_1, \cdots, y_m$.

Most of the existing studies are based on the assumptions that $n = m$ and A is nonsingular or A has full column rank. In practice, however, the number of sources may not be known *a priori,* and the mixing matrix A may be rectangular, or singular even though $n = m$. In general, there are four ill-conditioned cases as enumerated in [11].

We use the following general blind extraction model:

$$z = By = Cx \tag{2}$$

where z is an m-dimensional output vector, $B = \{b_{ij}\}$ is an $m \times m$ blind extraction matrix, and $C = BA$. The task of blind source extraction is to determine B such that one component in z corresponds to a source signal up to a scale. The commonly used simultaneous blind separation model is the same as (2), where B is called a separation matrix. Under the condition that at most one source is Gaussian, the choice of B is generally based on the principle that all outputs are mutually independent such that $C = DP$, where D is a diagonal matrix and P is a permutation matrix.

The fourth-order cumulant (e.g., kurtosis) is used in the present paper to form a cost function. Only one output (the extracted signal) is required to be pairwise independent with other outputs in the extraction step. This is in line with the spirit from a recent work [11]. In [11], the Gauss-Newton algorithm is employed for computing the blind extraction matrix. In that algorithm, the number of unknowns to be determined equals m^2, i.e., equals the dimension of B. In addition, one has to compute a matrix inversion in each step which increases the computational complexity. Moreover, the convergence of the algorithm of [11] can only be guaranteed by carefully adjusting two parameters (β and μ in (15) of [11]). The present paper will develop a gradient-based method for computing the blind extraction matrix. In our approach the number of unknowns to be determined is only $2(m-1)$ and there is no need for computing any matrix inversion.

0-7803-8251-X/04/$17.00 ©2004 IEEE

II. Problem Formulation

Following similar arguments as in [12], we can show that if the conditions in Theorem 1 of [11] hold for $p = 1$, the nonsingular matrix B will have the following form:

$$B = \begin{bmatrix} 1 & 0 & \cdots & 0 \\ -f_{21} & 1 & \cdots & 0 \\ \vdots & \vdots & \ddots & \vdots \\ -f_{m1} & 0 & \cdots & 1 \end{bmatrix} \begin{bmatrix} 1 & -f_{12} & \cdots & -f_{1m} \\ 0 & 1 & \cdots & 0 \\ \vdots & \vdots & \ddots & \vdots \\ 0 & 0 & \cdots & 1 \end{bmatrix}$$

$$= \begin{bmatrix} 1 & -f_{12} & \cdots & -f_{1m} \\ -f_{21} & 1+f_{21}f_{12} & \cdots & f_{21}f_{1m} \\ \vdots & \vdots & \ddots & \vdots \\ -f_{m1} & f_{m1}f_{12} & \cdots & 1+f_{m1}f_{1m} \end{bmatrix} \quad (3)$$

and B satisfies

$$BA = \begin{bmatrix} c_1 & 0 & \cdots & 0 \\ 0 & a_{22} & \cdots & a_{2n} \\ \vdots & \vdots & \cdots & \vdots \\ 0 & a_{m2} & \cdots & a_{mn} \end{bmatrix}$$

where $c_1 \neq 0$.

Without loss of generality, let z_1 be the signal to be extracted. Based on the idea in [14], a cost function is defined using the fourth-order cumulants as

$$J = \sum_{j=2}^{m} \mathrm{Cum}_{2,2}^2(z_1, z_j). \quad (4)$$

The properties of cumulants can be found in many references (see, e.g., [7]). Based on (2), we have

$$\mathrm{Cum}_{2,2}(z_1, z_j) = \mathrm{Cum}_{2,2}\left(\sum_{l=1}^{m} b_{1l}y_l, \sum_{l=1}^{m} b_{jl}y_l\right)$$

$$= \sum_{l_1,l_2,l_3,l_4=1}^{m} b_{1l_1}b_{1l_2}b_{jl_3}b_{jl_4}\mathrm{Cum}(y_{l_1}, y_{l_2}, y_{l_3}, y_{l_4}). \quad (5)$$

The next lemma provides conditions under which the signal z_1 is extracted.

Lemma 1: (Theorem 4 [11]). Suppose that $x_1(t), \cdots, x_n(t)$ are sup-Gaussian (or sub-Gaussian) and mutually independent stationary sources with zero means. If there exists a nonsingular $m \times m$ extraction matrix B such that $J = 0$ and $z_1 \not\equiv 0$, then z_1 is an extracted signal.

In light of Lemma 1, blind extraction using (2) is converted to solving the following constrained minimization problem:

$$\min_{\det(B)\neq 0} J. \quad (6)$$

In [11], the Gauss-Newton algorithm is employed for solving the above optimization problem. The algorithm has three properties as follows: 1) The number of unknowns to be determined in the algorithm is m^2, which is the dimension of B; 2) The algorithm involves the computation of the inversion of an $m^2 \times m^2$ matrix which results in a high computational load; 3) The convergence of the algorithm of [11] can only be guaranteed by carefully adjusting two parameters (i.e., β and μ in (15) of [11]).

III. The Gradient-Based Method

Note that B is nonsingular if B has the form of (3). The constrained minimization problem (6) becomes

$$\min J \text{ subject to } B \text{ having the form of (3).} \quad (7)$$

In this case, $b_{11} = 1, b_{1i} = -f_{1i}, b_{i1} = -f_{i1}, b_{ii} = 1 + f_{i1}f_{1i}$ for $i = 2, 3, \cdots, m$ and $b_{ij} = f_{i1}f_{1j}$ for $i, j = 2, 3, \cdots, m, i \neq j$. Then, from (5), it is easy to see that J can be rewritten as $J(f_{12}, f_{13}, \cdots, f_{1m}, f_{21}, f_{31}, \cdots, f_{m1})$. The constrained minimization problem (6) can be converted to the following unconstrained minimization problem:

$$\min J(f_{12}, f_{13}, \cdots, f_{1m}, f_{21}, f_{31}, \cdots, f_{m1}) \triangleq \min J(f) \quad (8)$$

where $f \triangleq [f_{12}, f_{13}, \cdots, f_{1m}, f_{21}, f_{31}, \cdots, f_{m1}]^T$. Letting the time derivative of the variable be directly proportional to the negative gradient of $J(f)$ with respect to the vector variable f, we have the following dynamic equation

$$\frac{df}{dt} = -\mu\frac{\partial J(f)}{\partial f}, \quad f_0 = f(0) \quad (9)$$

where μ is a positive scaling constant.

Equation (9) is our gradient-based method for the minimization of cost function J. It can easily be shown that the function J defined in (8) or (4) is an energy function. To see this, we compute the time derivative of $J(f)$ along the positive half trajectory of (9). We obtain $\forall t \geq 0$

$$\frac{dJ(f)}{dt} = \left[\frac{\partial J(f)}{\partial f}\right]^T \frac{df}{dt} = -\mu\left[\frac{\partial J(f)}{\partial f}\right]^T \frac{\partial J(f)}{\partial f} \leq 0.$$

The above shows that for any initial condition f_0, the gradient-based algorithm in (9) for minimization of J will be convergent. By Lemma 2, if the point of convergence f^* is such that $J(f^*) = 0$, then the obtained matrix B can lead to an extracted signal z_1. In all examples used in our simulation studies in the next section, we were able to obtain $J(f^*) \approx 0$ every time.

Remark 1: The number of unknowns to be determined in equation (9) is only $2(m-1)$. This number is less than m^2 which is the number of unknowns to be determined in the algorithm of [11]. Moreover, in equation (9) we do not need to compute any matrix inversion, unlike in the Gauss-Newton algorithm in [11].

IV. Simulation Results

To verify extraction results and show convergence behavior of the approach, the following performance index is introduced. Let B be given as in (3). Denote

$$\mathrm{ISI} = \frac{\sum_{i=1}^{m} |f_{1i}| + |f_{i1}| - \max_{1\le i\le m}\{|f_{1i}|, |f_{i1}|\}}{\max_{1\le i\le m}\{|f_{1i}|, |f_{i1}|\}}. \quad (10)$$

Example 1: (Example 1 in [11]). Consider three sources $x_1(t) = 5\sin(n_1(t) - 0.5), x_2(t) = \cos(n_2(t)\pi), x_3(t) = \sin(3(n_3(t) - 0.5))$, where n_1, n_2, n_3 are independent uniform white noise with values in $[0, 1]$. The kurtoses of x_1, x_2 and x_3 are $-4.3867, -0.3848$, and -0.3371, respectively. Obviously, these three sources are sub-Gaussian. The mixing matrix is assumed to be

$$A = \begin{bmatrix} 1.0 & 0.3 & 0.1 \\ 0.3 & 0.6 & 0.2 \\ 0.1 & 0.3 & 0.1 \end{bmatrix}.$$

As shown in Example 1 of [11], only the source x_1 can be extracted. To apply our approach, the extraction matrix B has the following form:

$$B = \begin{bmatrix} 1 & -f_{12} & -f_{13} \\ -f_{21} & 1 + f_{21}f_{12} & f_{21}f_{13} \\ -f_{31} & f_{31}f_{12} & 1 + f_{31}f_{13} \end{bmatrix}. \quad (11)$$

The initial value of f_0 is chosen randomly as

$$(f_{12}, f_{13}, f_{21}, f_{31})^T = (-0.7794, 0.5245, -0.8542, 0.7845)^T.$$

Letting $\mu = 2 \times 10^9$ and using the gradient-based method, we obtain the blind extraction matrix B and, in turn, the matrix $C = BA$ as follows:

$$B = \begin{bmatrix} 1.0000 & -0.0611 & -0.8773 \\ -0.3356 & 1.0205 & 0.2944 \\ -0.1119 & 0.0068 & 1.0981 \end{bmatrix}$$

$$C = \begin{bmatrix} 0.8939 & 0.0002 & 0.0000 \\ -0.0000 & 0.5999 & 0.2000 \\ -0.0000 & 0.2999 & 0.1000 \end{bmatrix}.$$

In Figure 1, three sources are given in the three subplots of the first row and three observable mixtures in the second row. The extracted signal z_1 is shown in the first subplot of the third row. The mixtures of the remaining sources z_2 and z_3 are shown in the second and third subplots, respectively, of the third row. The first subplot in the fourth row shows the calibrated deviation $z_1 - 0.8939x_1$, which shows that the source x_1 is extracted. To show advantage of our approach over the algorithm in [11], we also depict the calibrated deviation $z_1 + 0.1167x_1$ obtained in [11] by using the Gauss-Newton algorithm in the second subplot of the fourth row. Comparing these two subplots, one can see that the computed calibrated deviation of our approach is much smaller than that of [11]. To show convergence

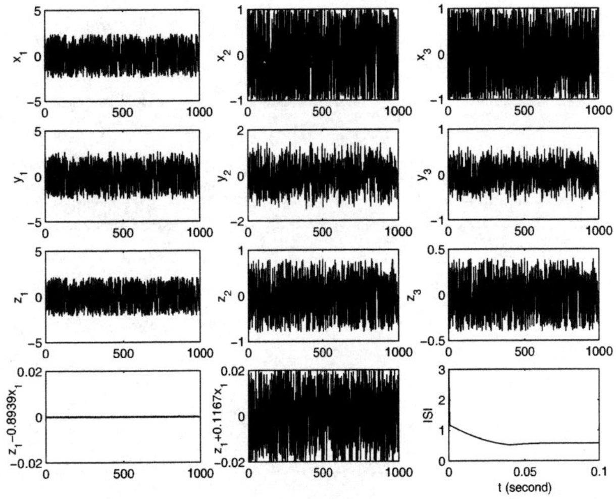

Fig. 1. Blind extraction from mixtures of three sources in Example 1.

of our approach, the performance index ISI is depicted in the last subplot of Figure 1. One can see that equation (9) converges to an equilibrium in about 0.1 seconds. In fact, to speed up the convergence speed of equation (9), we only need to increase the parameter μ. This property is specifically useful for problems with a slowly time-varying mixing matrix.

Example 2: (Example 3 in [11]). The three sources are the same as those in Example 1. The mixing matrix is assumed to be

$$A = \begin{bmatrix} 1 & 0.3 & 0.5 \\ 0.3 & 0.4 & 0.2 \\ 0.4 & 0.3 & 0.6 \end{bmatrix}.$$

Since A is nonsingular, all sources can be extracted one by one via sequential blind extraction shown in Example 3 of [11]. In the first round of blind extraction, we let the extraction matrix B_1 have the same form as in (11). The initial value of vector $(f_{12}, f_{13}, f_{21}, f_{31})^T$ is chosen randomly as in Example 1. Letting $\mu = 2 \times 10^9$ and using the gradient-based method, we can obtain the blind extraction matrix B_1 and the resulting $C_1 = B_1A$ as follows:

$$B_1 = \begin{bmatrix} 1.0000 & -0.1546 & -0.7841 \\ -0.4687 & 1.0724 & 0.3676 \\ -0.6250 & 0.0966 & 1.4901 \end{bmatrix}$$

$$C_1 = \begin{bmatrix} 0.6400 & 0.0029 & -0.0014 \\ 0.0001 & 0.3986 & 0.2007 \\ 0.0000 & 0.2982 & 0.6009 \end{bmatrix}.$$

The three sources and three mixtures are shown in the subplots of the first and second rows of Figure 2, respectively. The extracted signal z_1 is shown in the first subplot of the third row. The mixtures of the remaining sources

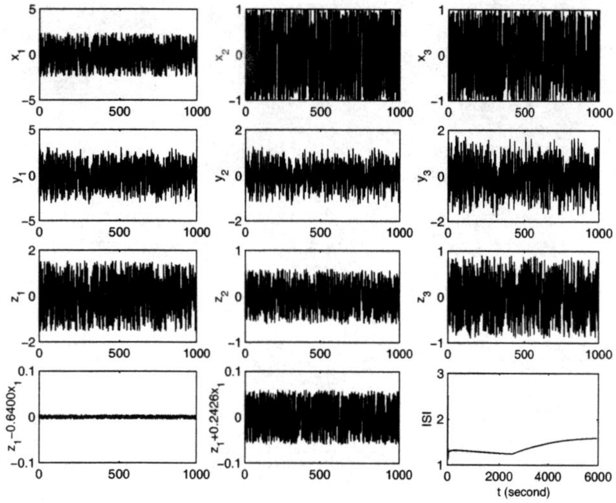

Fig. 2. First round of blind extraction from mixtures of three sources in Example 3.

z_2 and z_3 are shown in the second and third subplots, respectively, of the third row. The first subplot in the fourth row shows the calibrated deviation $z_1 - 0.6400x_1$, which shows that the source x_1 is extracted. We also depict the calibrated deviation $z_1 + 0.2426x_1$ obtained in [11] by using the Gauss-Newton algorithm in the second subplot of the fourth row. One can see that the computed calibrated deviation of our approach is much smaller than that of [11]. Convergence of the performance index ISI is shown in the last subplot of Figure 2. One can see that equation (9) converges to an equilibrium in 0.2 seconds.

In the second round of blind extraction, the mixtures are the remainders z_2 and z_3 of the first round. In this step, the extraction matrix B_2 has the following form:

$$B_2 = \left[\begin{array}{cc} 1 & -f_{12} \\ -f_{21} & 1 + f_{12}f_{21} \end{array} \right].$$

The initial value of vector $(f_{12}, f_{21})^T$ is chosen randomly as $(-0.0810, 0.3015)^T$. Letting $\mu = 2 \times 10^9$ and using the gradient-based approach, we can obtain for second round of blind extraction matrix B_2 and the resulting $C_2 = B_2\bar{C}_1$ as follows:

$$B_2 = \left[\begin{array}{cc} 1.0000 & -0.3335 \\ -0.9899 & 1.3301 \end{array} \right]$$

$$C_2 = \left[\begin{array}{ccc} 0.0001 & 0.2992 & 0.0003 \\ -0.0000 & 0.0020 & 0.6006 \end{array} \right]$$

where $\bar{C}_1$ is a 2×3 matrix composed of the second and third rows of C_1.

Figure 3 presents the results of the second round of blind extraction. In Figure 3, zz_2 is the extracted signal, and zz_3 is the remainder of the second round of blind extraction. The first two subplots of the second row show

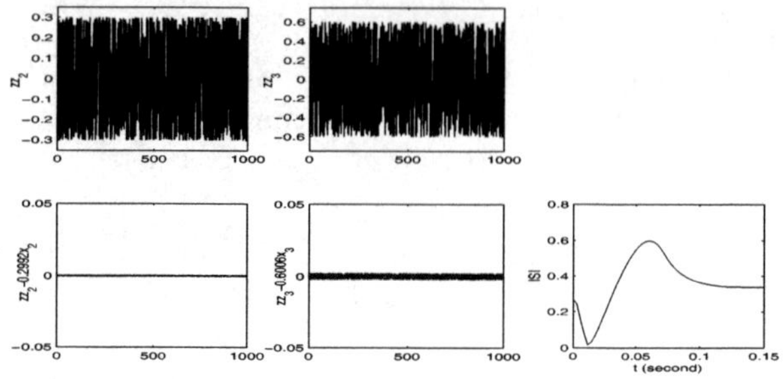

Fig. 3. Second round of blind extraction from mixtures of two remaining sources in Example 3.

the deviations $zz_2 - 0.2992x_2$ and $zz_3 - 0.6006x_3$, respectively. The third subplot of the second row shows the performance index ISI of the second round of extraction. Compare the present results with that obtained in [11] by using the Gauss-Newton algorithm, one can see that the deviations by our approach are much smaller than that of [11].

ACKNOWLEDGMENT

This work was supported by the National Science Foundation under Grant ANI-0203063.

REFERENCES

[1] A. J. Bell, T. J. Sejnowski, An information-maximization approach to blind separation and blind devolution, Neural Comput. 7 (6) (1995) 1004–1034.

[2] A. Cichocki, S. I. Amari, R. Thawonmas, Blind signal extraction using self-adaptive nonlinear nonlinear Hebbian learning rule, in Proc. Int. Symp. Nonlinear Theory Appl., Kochi, Japan, (1996) 377–380.

[3] A. Cichocki, R. Thawonmas, S. I. Amari, Sequential blind signal extraction in order specifi ed by stochastic properties, Electron. Lett. 33 (1) (1997) 64–65.

[4] P. Comon, Independent component analysis, a new concept? Signal Process. 36 (3) (1994) 287–314.

[5] P. Comon, C. Jutten, J. Herault, Blind separation of sources, Part II: Problems statement, Signal Process. 24 (1) (1991) 11–20.

[6] A. Hyvarinen, E. Oja, Independent component analysis: algorithms and applications, Neural Network. 13 (4) (2000) 411-430.

[7] B. C. Ihm, D. J. Park, Blind separation of sources using higher-order cumulants, Signal Process. 73 (3) (1999) 267–276.

[8] C. Jutten, J. Herault, Blind separation of sources, Part I: An adaptive algorithm based on neuromimetic architecture, Signal Process. 24 (1) (1991) 1–10.

[9] D. Kundur, D. Hatzinakos, A novel blind devolution scheme for image restoration using recursive fi ltering, IEEE Trans. Signal Process. SP-46 (2) (1998) 375–390.

[10] U. A. Lindgren, H. Broman, Source separation using a criterion based on second-order statistics, IEEE Trans. Signal Process. SP-46 (7) (1998) 1837-1850.

[11] Y. Li, J. Wang, Sequential blind extraction of instantaneously mixed sources, IEEE Trans. Signal Process. 50 (5) (2002) 997-1006.

[12] Y. Li, J. Wang, J. M. Zurada, Blind extraction of singularly mixed source signals, IEEE Trans. Neural Network. 11 (6) (2000) 1413–1422.

[13] Z. Malouche, O. Macchi, Adaptive unsupervised extraction of one component of a linear mixture with a single neuron, IEEE Trans. Neural Network. 9 (1) (1998) 123–135.

[14] A. Mansour, C. Jutten, Forth-order criteria for blind sources separation, IEEE Trans. Signal Process. SP-43 (8) (1995) 2022-2025.

A VERTICAL LAYERED SPACE-TIME CODE AND ITS BLIND SYMBOL DETECTION

Ke Deng, Qinye Yin, Ming Luo, Zheng Zhao

Institute of Information Engineering, School of Electronics and Information Engineering
Xi'an Jiaotong University, Xi'an 710049, China
E-mail: denke@mail.china.com, qyyin@xjtu.edu.cn, {cnmingluo,zhaozhengwan}@hotmail.com

ABSTRACT

Among various space-time codec schemes, the capacity provided by V-BLAST is very close to the channel capacity and its structure is relatively simple, therefore it is included by some standards. However, there are little research on the blind symbol detection without knowing the CSI and blind channel identification. By introducing some redundancy, a new V-BLAST scheme is proposed in this paper so that both the original transmitted symbol can be detected and the channel can be identificated via an ESPRIT-like method. Furthermore, this scheme can also eliminate the constraint that the number of receive antennas is more than transmit ones. Computer simulations evaluate the robustness and effectiveness of the propose algorithms.

1. INTRODUCTION

Because of the rapid growth demand for high data rate transmission in the wireless communication systems, one of the goals of third- and fourth- generation cellular is to provide higher data transmission rate than before. Deploying multiple antennas at the transmitter and/or receiver can dramatically improve the system capacity without increasing the power and bandwidth consumption[1]. Space-time coding, exploiting the multiple antennas, is thought as an effective way of improving the capacity, and has been accepted in many communication standards.

Among various space-time codec schemes, V-BLAST is relatively simple and able to implement high data rate in wideband wireless communication systems. And it was advised to some protocols used in WLAN (wireless local area network). [2] pointed out the capacity provided by V-BLAST can be very close to the channel capacity when the transmitting symbols from every transmit antenna is independent. The proposed symbol detection algorithms for V-BLAST include the V-BLAST OPT detection algorithm[3], efficient square-root method[4], sphere decoding algorithm[5]. All the above algorithms require the knowledge of the accurate channel state information (CSI), and to a multi-transmit multi-receive antennas system like V-BLAST, the normal training sequences will cost much more channel resource. Hence the research of the blind symbol detection and the channel identification is very urgent. [6] can decode the symbols without the knowledge of CSI, but its computational complexity is very high.

In this paper, we propose a Rotational Invariant V-BLAST (RI-VBLAST) codec scheme, in addition, the blind symbol detection and the blind channel identification algorithms are also given in flat-fading channel. By introducing the reasonable redundancy into V-BLAST system, the CSI and the original transmit signal can be estimated blindly via an ESPRIT-like method.

2.SYSTEM MODEL

2.1. RI-VBLAST Scheme

The scheme of RI-VBLAST is depicted in fig.1. The whole system includes N transmit antennas and M receive antennas. The original information symbols c_i, passed a serial/parallel converter and coder, are transmitted from the antennas.

Just like the [2], we focus on the channel with the scatter richness and Rayleigh fading, so the subchannel of every transmit-receive antenna pair is independent.

As described in fig.1, h_{mn} is the channel fading between the m-th receive antenna and the n-th transmit antenna. In a symbol periodic k, the received signal of the m-th receiver is:

$$r_m(k) = \sum_{n=1}^{N} h_{mn} s_n(k) \tag{1}$$

where $s_n(k)$ denotes the transmit symbol from antenna n in periodic k. The vector form of whole received signals in period k is

$$\mathbf{r}(k) = \begin{bmatrix} r_1(k) \\ \vdots \\ r_M(k) \end{bmatrix} = \mathbf{H} \begin{bmatrix} s_1(k) \\ \vdots \\ s_N(k) \end{bmatrix} + \mathbf{n}(k) \tag{2}$$

where the dimension of $\mathbf{r}(k)$ is $M \times 1$, $\mathbf{H}$ is the channel fading matrix with dimensions $M \times N$ and defined as:

$$\mathbf{H} = [h_{mn}] \tag{3}$$

* Partially supported by the National Natural Science Foundation (No. 60272071) and the Research Fund for Doctoral Program of Higher Education (No. 20020698024) of China.

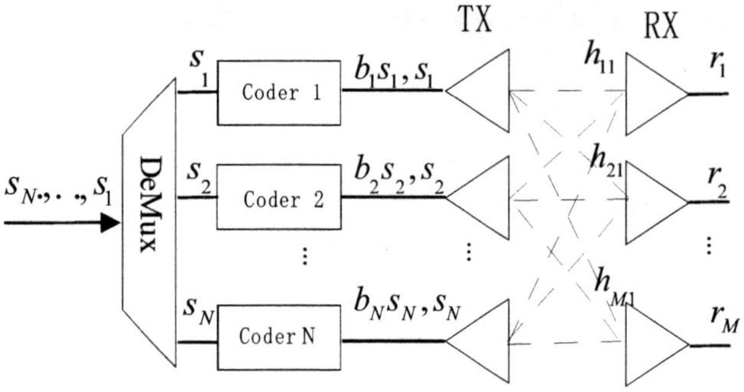

Fig. 1 The Discrete base band model of RI-VBLAST scheme

$\mathbf{n}(t)$ is a vector with dimensions $M \times 1$ which represents the additive white Gaussian noise.

2.2. The Signal Model

Most conventional decoding algorithms of V-BLAST [2][3][4][5] assume that the received antennas is more than transmit ones. Nevertheless, this constraint can be eliminated by introducing some redundancy and generating the different code structure in the different relations between M and N.

Now we describe a new codec scheme which is able to get rid of this constraint for different cases:

Case 1: $N \leq M$

The code scheme of RI-VBLAST in this case is also depicted in fig.1. The original input symbols after a serial/parallel converter are input into the N coders. We denote the input of the i-th coder is s_i, then the output symbols after coding are s_i, $b_i s_i$. In this case, one symbol input will cause two symbols output.

The effect of b_i here is a transmit antenna identifier, that is, the different antenna has the different b_i. In order to keep the dynamic range, the modulus of b_i is 1. ($|b_i| = 1$) In addition, the more distance among the b_i, the better performance can be achieved, hence the final result of selecting b_i is uniformly distributing on the unit circle.

Suppose the length of a output symbol block is L ($L = 2K$, K is the number of the original input symbols). Notice the equation $s_i(2k) = b_i s_i(2k-1)$ and then the two matrices $\mathbf{X}$ and $\mathbf{Y}$ can be constructed:

$$\mathbf{X} = [\,\mathbf{r}(1) \;\; \mathbf{r}(3) \;\; \ldots \;\; \mathbf{r}(L-1)\,] \tag{4}$$

$$\mathbf{Y} = [\,\mathbf{r}(2) \;\; \mathbf{r}(4) \;\; \ldots \;\; \mathbf{r}(L)\,] \tag{5}$$

$\mathbf{X}$ and $\mathbf{Y}$ can be expressed as:

$$\mathbf{X} = \mathbf{H} \begin{bmatrix} s_1(1) & s_1(2) & \ldots & s_1(K) \\ \vdots & \vdots & \ddots & \vdots \\ s_N(1) & s_N(2) & \ldots & s_N(K) \end{bmatrix} + \mathbf{N}_1 \tag{6}$$

$$\mathbf{Y} = \mathbf{H} \begin{bmatrix} b_1 s_1(1) & \ldots & b_1 s_1(K) \\ \vdots & \ddots & \vdots \\ b_N s_N(1) & \ldots & b_N s_N(K) \end{bmatrix} + \mathbf{N}_2 \tag{7}$$

If we let

$$\mathbf{S} = \begin{bmatrix} s_1(1) & s_1(2) & \ldots & s_1(K) \\ \vdots & \vdots & \ddots & \vdots \\ s_N(1) & s_N(2) & \ldots & s_N(K) \end{bmatrix} \tag{8}$$

and

$$\mathbf{\Phi} = \mathrm{diag}\{\, b_1 \;\; b_2 \;\; \ldots \;\; b_N \,\} \tag{9}$$

then the two following equations exist:

$$\mathbf{X} = \mathbf{HS} + \mathbf{N}_1 \tag{10}$$

$$\mathbf{Y} = \mathbf{H\Phi S} + \mathbf{N}_2 \tag{11}$$

Case 2: $2M \geq N > M$

In this case, the main differences are: one symbol input causes 3 symbols output, the corresponding outputs of s_i is s_i,$b_i s_i$, $b_i^2 s_i$. Here the b_i and b_i^2 also increase the numbers of equivalent receive antennas.

Suppose the block length is L ($L = 3K$). Similarly, two equations $s_i(3k+2) = b_i s_i(3k+1)$, $s_i(3k+3) = b_i^2 s_i(3k+1)$ exist. The X and Y are constructed as (12) and (13):

$$\mathbf{X} = \begin{bmatrix} \mathbf{r}(1) & \mathbf{r}(4) & \ldots & \mathbf{r}(L-2) \\ \mathbf{r}(2) & \mathbf{r}(5) & \ldots & \mathbf{r}(L-1) \end{bmatrix} = \begin{bmatrix} h_{11} & h_{12} & \ldots & h_{1N} \\ \vdots & \vdots & \ddots & \vdots \\ h_{M1} & h_{M2} & \ldots & h_{MN} \\ b_1 h_{11} & b_2 h_{12} & \ldots & b_N h_{1N} \\ \vdots & \vdots & \ddots & \vdots \\ b_1 h_{M1} & b_2 h_{M2} & \ldots & b_N h_{MN} \end{bmatrix} \times \begin{bmatrix} s_1(1) & s_1(2) & \ldots & s_1(K) \\ \vdots & \vdots & \ddots & \vdots \\ s_N(1) & s_N(2) & \ldots & s_N(K) \end{bmatrix} + \mathbf{N}_1 \tag{12}$$

$$\mathbf{Y} = \begin{bmatrix} \mathbf{r}(2) & \mathbf{r}(5) & \ldots & \mathbf{r}(L-1) \\ \mathbf{r}(3) & \mathbf{r}(6) & \ldots & \mathbf{r}(L) \end{bmatrix} = \begin{bmatrix} h_{11} & h_{12} & \ldots & h_{1N} \\ \vdots & \vdots & \ddots & \vdots \\ h_{M1} & h_{M2} & \ldots & h_{MN} \\ b_1 h_{11} & b_2 h_{12} & \ldots & b_N h_{1N} \\ \vdots & \vdots & \ddots & \vdots \\ b_1 h_{M1} & b_2 h_{M2} & \ldots & b_N h_{MN} \end{bmatrix} \times \begin{bmatrix} b_1 s_1(1) & \ldots & b_1 s_1(K) \\ \vdots & \ddots & \vdots \\ b_N s_N(1) & \ldots & b_N s_N(K) \end{bmatrix} + \mathbf{N}_2 \tag{13}$$

Consequently (10) and (11) also exist and the dimension of $\mathbf{H}$ is $2M \times N$ in this case, which means the equivalent receive antennas doubled.

This method of increasing equivalent antennas can be further generalized: In the case of $kM \geq N > (k-1)M$, the (10) and (11) can be constructed if the one symbol input cause $(k+1)$ symbols output. The constraints of VBLAST is totally eliminated.

In fact, this scheme exploits the equivalence of the space diversity and the time diversity. By introducing the redundancy in the time domain, we obtain some time diversity in the receive side. We can exploit the equivalence and look on the time diversity as the space diversity, that means the equivalent number of receive antennas increases.

3. BLIND ALOGRITHM

(10) and (11) are the basic equations of ESPRIT[7], which can estimate Φ directly. But the estimation of $\mathbf{H}$ and $\mathbf{S}$ is more important in this case, hence we use an ESPRIT-like method — DOA-Matrix method[8] to estimate the $\mathbf{H}$ and $\mathbf{S}$ directly, which also exploits the rotation invariant of the subspace.

3.1. Blind symbol detection

In order to estimate the transmit symbols, two auxiliary matrices are constructed:

$$\begin{aligned}\mathbf{U} &= \mathbf{X}^H = \mathbf{S}^H\mathbf{H}^H + \mathbf{N}_1 \\ \mathbf{V} &= \mathbf{Y}^H = \mathbf{S}^H\mathbf{\Phi}^H\mathbf{H}^H + \mathbf{N}_2\end{aligned} \tag{14}$$

Now compute the autocorrelation matrix of $\mathbf{U}$ and the correlation matrix of $\mathbf{V}$ and $\mathbf{U}$:

$$\mathbf{R}_{uu} = \mathbf{U}\mathbf{U}^H = \mathbf{S}^H\mathbf{R}_{hh}\mathbf{S} + \sigma_n^2\mathbf{I} = \mathbf{R}_{uuo} + \sigma_n^2\mathbf{I} \tag{15}$$

$$\mathbf{R}_{vu} = \mathbf{V}\mathbf{U}^H = \mathbf{S}^H\mathbf{\Phi}^H\mathbf{R}_{hh}\mathbf{S} \tag{16}$$

where $\mathbf{R}_{hh} = \mathbf{H}^H\mathbf{H}$ is a Hermitian matrix with dimensions $N \times N$. The rank of $\mathbf{R}_{uuo}$ is N. Calculate the eigen value decomposition (EVD) of it, we get:

$$\mathbf{R}_{uuo} = \sum_{l=1}^{N} \mu_l \nu_l \nu_l^H \tag{17}$$

where μ_l and ν_l are the nonzero eigen values and its corresponding eigen vectors respectively. Then the Penrose-Moore inverse of $\mathbf{R}_{uuo}$ is calculated:

$$\mathbf{R}_{uuo}^{+} = \sum_{l=1}^{N} \mu_l^{-1} \nu_l \nu_l^H \tag{18}$$

Finally we construct $\mathbf{R}$:

$$\mathbf{R} = \mathbf{R}_{vu}\mathbf{R}_{uuo}^{+} \tag{19}$$

By calculating the eigen values and corresponding eigen vectors of $\mathbf{R}$, the matrix $\mathbf{S}$ and $\mathbf{\Phi}$ can be estimated, we have the following theorem:

Theorem 1.: Given $\mathbf{S}$ *is full row rank,* $\mathbf{R}_{hh}$ *is nonsingular, and* $\mathbf{\Phi}$ *has no equal elements in the main diagonal line, the* N *nonzero eigen values of matrix* $\mathbf{R}$ *equal the* N *diagonal elements of matrix* $\mathbf{\Phi}^H$*, and the corresponding eigen vector equals the columns of* $\mathbf{S}^H$ *respectively, namely:*

$$\mathbf{R}\mathbf{S}^H = \mathbf{S}^H\mathbf{\Phi}^H \tag{20}$$

The proof is in the appendix.

After $\mathbf{S}$ and $\mathbf{\Phi}$ are obtained, the ordinal of the transmit antenna can be decided from b_i and $\mathbf{\Phi}$, then the transmitted symbols $\mathbf{c}_i$ can be detected.

3.2. Blind channel identification

The channel fading matrix $\mathbf{H}$ is also able to be estimated via the DOA-Matrix method. The autocorrelation matrix of $\mathbf{X}$ and the correlation matrix between $\mathbf{X}$ and $\mathbf{Y}$ are also needed:

$$\mathbf{R}_{xx} = \mathbf{X}\mathbf{X}^H = \mathbf{H}\mathbf{R}_{ss}\mathbf{H}^H + \sigma_n^2\mathbf{I} = \mathbf{R}_{xxo} + \sigma_n^2\mathbf{I} \tag{21}$$

$$\mathbf{R}_{yx} = \mathbf{Y}\mathbf{X}^H = \mathbf{H}\mathbf{\Phi}\mathbf{R}_{ss}\mathbf{H}^H \tag{22}$$

where $\mathbf{R}_{ss} = \mathbf{S}\mathbf{S}^H$ is the autocorrelation matrix of $\mathbf{S}$. The rest steps is like section 3.1, finally an important matrix $\mathbf{R}$ is constructed:

$$\mathbf{R} = \mathbf{R}_{yx}\mathbf{R}_{xxo}^{+} \tag{23}$$

A similar theorem also holds:

Theorem 2.: Given $\mathbf{H}$ *is full column rank,* $\mathbf{R}_{ss}$ *is nonsingular, and* $\mathbf{\Phi}$ *has no equal elements in the diagonal line, the* N *nonzero eigen values of matrix* $\mathbf{R}$ *equal the* N *diagonal elements of matrix* $\mathbf{\Phi}^H$*, and the corresponding eigen vector equals the columns of* $\mathbf{H}$ *respectively, namely:*

$$\mathbf{R}\mathbf{H} = \mathbf{H}\mathbf{\Phi} \tag{24}$$

After the similar process, the channel fading matrix is estimated finally.

4. SIMULATION RESULTS

Extensive computer simulations have been conducted to demonstrate the effectiveness of our symbol detection and channel identification algorithms for the RI-VBLAST scheme. We use the differential binary phase shift keying (DBPSK) modulation mode in our simulations.

The performance of the symbol detection algorithm is evaluated by bit error rate (BER) and that of the channel identification algorithm is evaluated by root mean square error (RMSE) which is defined as:

$$RMSE = \frac{1}{\|\hat{\mathbf{H}}\|}\sqrt{\frac{1}{N_t}\sum_{i=1}^{N_t}\|\hat{\mathbf{H}}(i) - \mathbf{H}\|^2} \tag{25}$$

where N_t is the number of Monte-Carlo trials, and $\| \bullet \|$ represents the Frobenius norm.

Test Case1: Fig.2 shows the performance of DOA-Matrix algorithm and the V-BLAST OPT algorithm[3] with (N,M)=(2,2) and (N,M)=(3,2) antennas. In order to compare the performance of the cases which has the different N, here we define $SNR = NE_b/n_0$, that is to say, the total power of all transmit antennas is identical with the cases which has the different N and the same SNR. In addition, the conventional V-BLAST OPT algorithm is unable to handle the case of (N,M)=(3,2), we expand it by exploiting the equivalent receive antenna thought proposed in this paper. The length of one block once handling is

32. With the simulation, we observe that the performance of DOA-Matrix method is close to the V-BLAST OPT algorithm which needs the accurate CSI.

Test Case2: Fig.3 shows the performance of our channel identification algorithm with $(N,M)=(2,2)$ and $(N,M)=(3,2)$ antennas. L denotes the length of one block handling once. We can observe that the performance turns better with longer L. Of course, the performance of the case of $(N,M)=(3,2)$ is better than that of $(N,M)=(2,2)$.

5. CONCLUSION

We expand the V-BLAST scheme to the case that the number of transmit antennas is more than the receive ones by introducing some redundancy. Furthermore, we proposed the blind symbol detection and channel identification algorithms for the new V-BLAST scheme — RI-VBLAST by exploiting an ESPRIT-like method. The simulation results show that the performance of proposed symbol decoding algorithm is close to the V-BLAST OPT one which needs the accurate CSI.

REFERNCES

[1] G. J. Foschini, Jr. And M. J. Gans, "On limits of wireless communication in a fading environment when using multiple antennas," *Wireless Personal Commun.*, vol. 6, no. 3, pp. 311-335, Mar. 1998.

[2] G. J. Foschini, G. D. Golden, R. A. Valenzuela and P. W. Wolniansky, "Simplified processing for high spectral efficiency wireless communication employing multi-element arrays," *IEEE J. Select. Areas Commun.*, vol. 17, pp. 1841-1852, Nov. 1999.

[3] G. D. Golden, C. J. Foschini, R. A. Velenzuela and P. W. Wolniasky, "Detection Algorithm and Initial Labora- tory Results Using V-BLAST Space-Time Communications Architecture, *IEE Elec.Letters*, Vol. 35, No. 1, 1999.

[4] B. Hassibi, "An efficient square-root algorithm for BLAST," [Online] Available:http://mars.bell-labs.com.

[5] O.Damen, A. Chkeif, and J.-C. Belfiore, "Lattice code decoder for space-time codes," *IEEE Commun. Letters*, vol. 4, pp. 161-163, May 2000.

[6] A. Shokrollahi, B. Hassibi, B. Hochwald, and W. Sweldens, "Representation theory for high-rate multiple-antenna code desigh," *IEEE Trans. Inform. Theory*, vol.47, pp. 2335-2367, Sept. 2001.

[7] R. Roy and T. Kailath, "ESPRIT — Estimation of signal parameters via rotational invariance techniques," *Optical engineering*, 29(4), pp. 984-995, 1990

[8] Q. Yin, R. Newcomb, L. Zou, "Estimating 2-D angle of arrival via two parallel linear array," *Proc. IEEE ICASSP'89*, Vol. 3, pp. 2803-2806, 1989

[9] Z. Zhao, Q. Yin, A. Feng, K. Deng, "An vertical layered space-time coding and the symbol detection," *Proc. MWSCAS-2002*, Vol. 2, pp.597-600, August 2002.

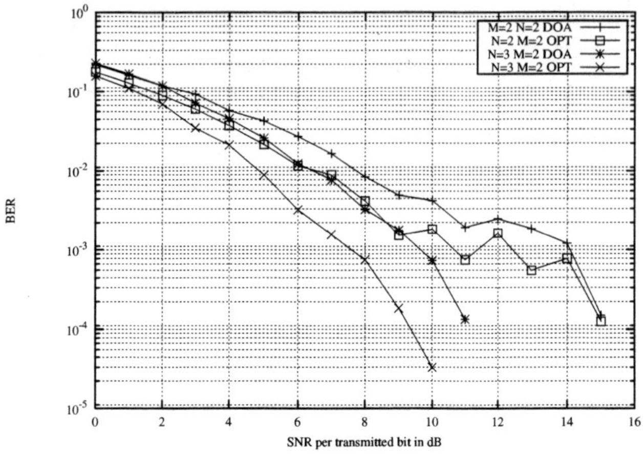

Fig.2 The performance of symbol detection algorithm

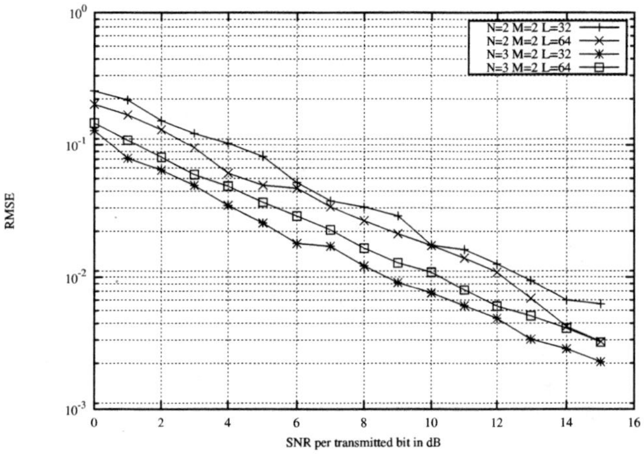

Fig.3 The performance of channel identification algorithm

APPENDIX

The proof of (20):
From (15) we have:

$$\mathbf{R}_{hh}\mathbf{S} = (\mathbf{S}\mathbf{S}^H)^{-1}\mathbf{S}\mathbf{R}_{uuo}$$

Substitute it into (16), $\mathbf{R}_{vu}$ can be written as:

$$\mathbf{R}_{vu} = \mathbf{S}^H\mathbf{\Phi}^H(\mathbf{S}\mathbf{S}^H)^{-1}\mathbf{S}\mathbf{R}_{uuo}$$

Finally, we have:

$$\begin{aligned}\mathbf{R}\mathbf{S}^H &= \mathbf{R}_{vu}\mathbf{R}_{uuo}^{+}\mathbf{S}^H \\ &= \mathbf{S}^H\mathbf{\Phi}^H(\mathbf{S}\mathbf{S}^H)^{-1}\mathbf{S}\mathbf{R}_{uuo}\mathbf{R}_{uuo}^{+}\mathbf{S}^H \\ &= \mathbf{S}^H\mathbf{\Phi}^H\end{aligned}$$

Blind Source Recovery for Non-minimum Phase Surroundings

Khurram Waheed[1] *and Fathi M. Salem*[2]

[1]San Diego State University, San Diego CA 92182-1309
[2]Michigan State University, East Lansing, MI 48824-1226

ABSTRACT

This paper revisits the issue of Blind Source Recovery (BSR) of non-minimum phase linear MIMO systems and provides a new rigorous derivation of the proposed state space algorithms [4,6,7,9]. Using state space representation for both the mixing and convolving (or reverberative) surrounding environment and the recovery (or demixing and deconvolving) networks, we systematically derive update laws based on the minimization of Kullback-Liebler divergence using the theory of optimization and the calculus of variations along the Riemannian contra-variant (or the natural) gradient. A simulation example for recovery of signals from a challenging non-minimum environment is also included in this paper.

1. INTRODUCTION

Most mixing surroundings in real-life are non-minimum phase, i.e., the cumulative transfer function of the corrupting environment contains transmission zeros that are outside the unit circle. In this case, the theoretical environment inverse (or the intended recovery network) is unstable due to the presence of poles outside the unit circle. These unstable poles are required to cancel out the non-minimum phase transmission zeros of the environment. In order to avoid instability due to the existence of these poles outside the unit circle, the natural gradient algorithm may be derived with the constraint that the demixing system is a double sided FIR filter, i.e., instead of trying to determine the IIR inverse of the environment, we will approximate the inverse using an all zero extended non-causal filter [6, 8]. The use of the state space warrants that the derived framework is rich in structure while at the same time compact in representation. This paper extends our earlier results for such systems, see [3, 4].

The recovery of original sources from observations made from a non-minimum phase system is a challenging task. Firstly the update laws for the non-minimum phase networks tend to be more computationally intensive as they require computation of both forward proceeding states and backward propagating co-states. Note that this method of computation also introduces additional delay in the recovered sources. Secondly, it has been observed that the algorithm convergence is more sensitive to the selection of an appropriate size of the demixing network, the choice of the learning rate and its adaptation over time as well as the adopted score function for adaptation [7, 6].

The choice of the state space representation has several advantages, which include a compact representation capable of handling both time delayed and filtered versions of signals [2, 7, 6], possibility of multiple efficient and generalized internal structural representations which allows for proficient *identifiability* [2]. Further, The inverse for a state space representation is well defined and ensures the existence of a solution for the BSR structure or *recoverability* [6].

2. STATE SPACE NON-MINIMUM PHASE BLIND SOURCE RECOVERY FRAMEWORK

In the linear dynamic case, the environment models is assumed to be of the state space form

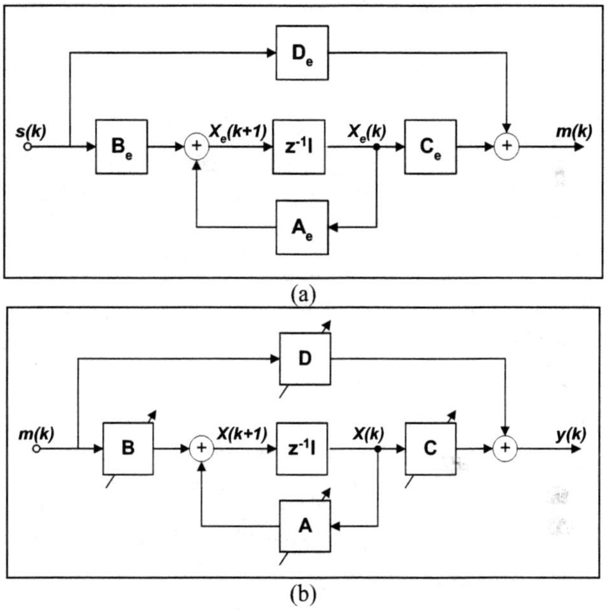

Figure 1. Linear Dynamic State Space (a) Environment Network Model, (b) Recovery Network

$$X_e(k+1) = A_e X_e(k) + B_e s(k) \quad (1)$$

$$m(k) = C_e X_e(k) + D_e s(k) \quad (2)$$

where s(k) denotes the vector of actual source signals at time instant k, $m(k)$ denotes the vector of measurements at time instant k, $X_e(k)$ specify the internal states of the unknown mixing system and $C_e\ (zI-A_e)^{-1}B_e+D_e$ represent the discrete-time transfer function of the demixing network.

The corresponding state space feedforward recovery network has the form

$$X(k+1) = A\,X(k) + B\,m(k) \quad (3)$$

$$y(k) = C\,X(k) + D\,m(k) \quad (4)$$

where $y(k)$ is the output of the feedforward network at the same instant k, $X(k)$ specify the internal states of the demixing network and $C(zI-A)^{-1}B+D$ represents the discrete-time transfer function of the demixing network. For convenience, we will assume that the demixing network is represented in Canonical Form I (or the controller canonical form) [2, 4].

However, for the case of non-minimum phase environment the demixing network is constrained (due to stability issues) to be a non-causal FIR filter. Therefore the parametric matrices A and B are kept fixed in canonical form. The update laws for the remaining parametric state space matrices C and D is according to the theorem (and the subsequent proof) given below.

2.1. Theorem

Assume the MIMO (mixing) environment modeled by a non-minimum phase transfer function. Then, the update law for the zeros of the state space FIR demixing network using the natural gradient is given by

$$\Delta C_i = \eta(k)\left[C_i - \varphi(y(k))u(k-i)^T\right], i = 1,2,\cdots,L-1 \quad (5)$$

$$\Delta D = \eta(k)\left[D - \varphi(y(k))u(k)^T\right] \quad (6)$$

where the state space matrix C is defined as

$$C = \begin{bmatrix} C_1 & C_2 & \cdots & C_{L-1} \end{bmatrix} \quad (7)$$

C_i being the MIMO FIR filter coefficients corresponding to delay z^{-i}, L being the total number of taps in the filter and u(k) represents an information back-propagation filter, given as

$$u(k) = \sum_{i=1}^{L-1} C_i^T y(k+i) + D^T y(k)$$

or equivalently in the adjoint state space as

$$\lambda(k-1) = A^T \lambda(k) + C^T y(k)$$

$$u(k) = B^T \lambda(k) + D^T y(k)$$

where T represents the matrix Hermitian transpose operator, $\varphi(y)$ is the appropriate score function given by

$$\varphi(y_i) = -\frac{\partial \log p(y_i)}{\partial y_i} = -\frac{\partial p(y_i) / \partial y_i}{p(y_i)},$$

and A and B are given by (43), with $A_{1j} = 0$.

2.2. Proof

On the outset, we state that the algorithm requires a frame of $3L$-2 samples of the input signal to initiate the computations of the update of the parameter matrices in (5)-(6). Note also that the time index of the parameter update (5)-(6) is not necessarily the same time index of the signals. In practice, it may be chosen so that the time index of the parameters is a delayed version of the signal time index so that no issues of non-causality would arise. Also in practice, the initialization of the update law (5)-(6) uses an identity (or a nonsingular matrix) for the center tap matrix (specifically, $j = \lfloor (L-1)/2 \rfloor$) and small random or zero values for the remaining tap matrices.

As mentioned before, for the non-minimum phase mixing case, we constrain the demixing network to be a double-sided FIR filter so as to approximate the intended unstable IIR inverse. Using the MIMO controller canonical form, the matrices A and B of dimensions $(L\text{-}1)m \times (L\text{-}1)m$ and $(L\text{-}1)m \times m$, respectively take the form

$$A = \begin{bmatrix} O_m & O_m & \cdots & O_m & O_m \\ I_m & O_m & \cdots & O_m & O_m \\ O_m & I_m & \cdots & O_m & O_m \\ \vdots & \vdots & \vdots & \vdots & \vdots \\ O_m & O_m & \cdots & I_m & O_m \end{bmatrix}, \quad B = \begin{bmatrix} I_m \\ O_m \\ O_m \\ \vdots \\ O_m \end{bmatrix}$$

where

I_m is an Identity matrix of dimension $m \times m$ and O_m is a Zero matrix of dimension $m \times m$

For compactness, we can represent (3)-(4) using the compact (mixed time-frequency) transfer function notation of the MIMO FIR filter as

$$y_k = \left[\overline{W}(z)\right] m_k \quad (8)$$

where, $\overline{W}(z)$ is the transfer function of the demixing MIMO FIR filter of the form,

$$\overline{W}(z) = W_0 + W_1 z^{-1} + \ldots + W_{L-1} z^{-L+1} \quad (9)$$

Further we can define a matrix $\overline{W}$, comprised of the coefficients of the filter $\overline{W}(z)$. Note that each constituent sub-matrix W_i contains co-efficients pertaining to the ith filter lag acting on the time instant k-i. This matrix $\overline{W}$ is related to the state space matrices C and D as given below.

$$\overline{W} \triangleq \begin{bmatrix} W_0 & W_1 & \cdots & W_{L-1} \end{bmatrix} = \begin{bmatrix} D & C \end{bmatrix}$$

where the state space matrices C and D in terms of the sub-matrices W_i are represented as

$$D = W_0$$
$$C = [W_1 \quad W_2 \quad \cdots \quad W_{L-1}] \quad (10)$$

For the MIMO transfer function $\overline{W}(z)$, the conjugate (or Hermitian) transpose of the transfer function is given by

$$\begin{aligned} \left[\overline{W}(z)\right]^T &= \left[W_0 + W_1 z^{-1} + \ldots + W_{L-1} z^{-L+1}\right]^T \\ &= W_0^T + W_1^T z^1 + \ldots + W_{L-1}^T z^{L-1} \\ &= \overline{W}^T(z^{-1}) \end{aligned} \quad (11)$$

where, T is the Hermitian transpose operator. Notice that again we can define a matrix of coefficients $\overline{W}^T$ comprised of the transpose of the constituent sub-matrices acting in a time reversed fashion, i.e., the ith lag of this time reversed filter acts on the future (non-causal) time instant k+i, i.e.,

$$\overline{W}^T \triangleq \left[W_0^T \quad W_1^T \quad \cdots \quad W_{L-1}^T\right] \quad (12)$$

Using the definition of $\overline{W}$, we can express the network output equation (8) in the time domain as

$$y_k = \overline{W} \underline{m}_k \equiv D m_k + \sum_{i=1}^{L-1} C_i m_{k-i} \quad (13)$$

where $\underline{m}_k$ represents an $Lm \times 1$ matrix of observations at the kth iteration, which includes the kth and previous L-1 m-d measurement (or observation) vectors, i.e.,

$$\underline{m}_k \triangleq \left[m_k^T \quad m_{k-1}^T \quad \cdots \quad m_{k-L+1}^T\right]^T$$

and y_k represents the output vector at the kth iteration

For the derivation of the update laws, we minimize the Kullback-Liebler divergence (or relative entropy) of the output with the constraints of an FIR representation. Using the maximization of entropy approach, the Hamiltonian can be defined [2, 4] as

$$H^k = -\log \det\left|W_j\right| - \sum_i \log\left(q_i\left(y_i, W(z)\right)\right) \tag{14}$$

where $q_i\left(y_i, W(z)\right)$ is the online estimated marginal distribution of $y_i(k)$ and $j \in [0, L-1]$.

Using the ordinary stochastic gradient update laws see [1, 2, 4], the update law for the sub-matrices W_i is given by

$$\Delta W_i = -\eta_k \frac{\partial H^k}{\partial W_i} = \eta_k \left[\left(W_j^{-1}\right)^T \delta_{ji} - \varphi(y_k) m_{k-i}^T \right] \tag{15}$$

where δ_{ji} equals 1 only when $i = j$, and $j \in [0, L-1]$. It should be noted that W_j is required here to be nonsingular. For the non-minimum phase case, j is appropriately chosen to be an intermediate tap, e.g. $j = \lfloor (L-1)/2 \rfloor$, in order to render an appropriate double-sided FIR filter. In any event, it will be seen below that using the natural gradient, the resulting parameter update laws (24) or (25) do not depend on the explicit nonsingularity of the j-th tap matrix and will include all tap matrices. The Stochastic Gradient (SG) law for the complete MIMO filter matrix is given by

$$\begin{aligned}\left[\Delta \overline{W}(z)\right]_{SG} &= \sum_{i=0}^{L-1} \Delta W_i z^{-i} \\ &= \eta_k \left[\left(W_j^{-1}\right)^T \delta_{ji} - \sum_{i=0}^{L-1} \varphi(y_k) m_{k-i}^T z^{-i} \right]\end{aligned} \tag{16}$$

Using the definition of the natural gradient in the systems space [1], we can modify (16) to have the update of the demixing filter parameters according to the Riemannian contra-variant gradient as

$$\begin{aligned}\Delta \overline{W}(z) &= \left[\Delta \overline{W}(z)\right]_{SG} \overline{W}^T(z^{-1}) \overline{W}(z) \\ &= \eta_k \left[\left(W_j^{-1}\right)^T \delta_{ji} - \sum_{i=0}^{L-1} \varphi(y_k) m_{k-i}^T z^{-i} \right] \overline{W}^T(z^{-1}) \overline{W}(z)\end{aligned} \tag{17}$$

where, the first term on the right hand side can be simplified as

$$\begin{aligned}\left(W_j^{-1}\right)^T \delta_{ji} \overline{W}^T(z^{-1}) \overline{W}(z) &= \left(W_j^{-1}\right)^T \left(\delta_{ji} \overline{W}^T(z)\right) \overline{W}(z) \\ &= \left(W_j^{-1}\right)^T W_j^T \overline{W}(z) = I^T \overline{W}(z) = \overline{W}(z)\end{aligned} \tag{18}$$

while the second term can be simplified as

$$\begin{aligned}&\sum_{i=0}^{L-1} \varphi(y_k) m_{k-i}^T z^{-i} \overline{W}^T(z^{-1}) \overline{W}(z) \\ &= \varphi(y_k) \sum_{i=0}^{L-1} \left(\left[\overline{W}(z)\right] m_{k-i}\right)^T \overline{W}(z) z^{-i} = \varphi(y_k) \sum_{i=0}^{L-1} y_{k-i}^T \overline{W}(z) z^{-i} \\ &= \varphi(y_k) \sum_{i=0}^{L-1} \left(\left[\overline{W}^T(z^{-1})\right] y_{k-i}\right)^T z^{-i} = \varphi(y_k) \sum_{i=0}^{L-1} u_{k-i}^T z^{-i}\end{aligned} \tag{19}$$

where, we used equality (8) and u_k^T is the transpose of the non-causal back-propagation filter output, defined as

$$u_k \triangleq \left[\overline{W}^T(z^{-1})\right] y_k \equiv D^T y_k + \sum_{i=1}^{L-1} C_i^T y_{k+i} \tag{20}$$

Let $\underline{y}_k$ represent an $NL \times 1$ matrix of outputs at the kth iteration, which includes the kth and the future L-1 N-d output vectors, i.e.,

$$\underline{y}_k \triangleq \left[y_k^T \quad y_{k+1}^T \quad \cdots \quad y_{k+L-1}^T \right]^T$$

then the back-propagation filter can also be expressed in the compact form

$$u(k) = \overline{W}^T \underline{y}_k \tag{21}$$

Using (18) and (19) in (17), we get the following update laws for the MIMO FIR transfer Function

$$\Delta \overline{W}(z) = \eta_k \left[\overline{W}(z) - \sum_{i=0}^{L-1} \varphi(y_k) u_{k-i}^T z^{-i} \right] \tag{22}$$

The update law (22) is non-causal as computation of $u_{k-i}^T; i = 0,1,2,\ldots,L-1$ requires up to $L-1$ future outputs to be available. Practically the update law can be implemented (or made causal) by introducing a delay of $L-1$ iterations between the update of the parameters and the computation of the output at the kth iteration (this in fact amount of using a frame-buffering in order to initialize the update law as is used in practice). Thus, a causal version of (85) becomes

$$\left[\Delta \overline{W}(z)\right]_C = \eta_k \left[\overline{W}(z) - \sum_{i=0}^{L-1} \varphi(y_{k-L+1}) u_{k-L-i+1}^T z^{-i} \right] \tag{23}$$

Writing down the update laws in the time domain for the ith component coefficient matrix W_i, we have

$$\left[\Delta W_i\right]_C = \eta_k \left[W_i - \varphi(y_{k-L+1}) u_{k-L-i+1}^T \right] \tag{24}$$

Defining the new time index $\overline{k} = k - L + 1$, we have

$$\left[\Delta W_i\right]_C = \eta_k \left[W_i - \varphi\left(y\left(\overline{k}\right)\right) u^T\left(\overline{k} - i\right) \right] \tag{25}$$

where the computation of the network output $y\left(\overline{k}\right)$ and the back-propogation filter output $u\left(\overline{k}\right)$ is computed as in the following expressions

$$\begin{aligned}y\left(\overline{k}\right) &\triangleq y(k-L+1) = \overline{W} \underline{m}_{k-L+1} \\ &= \sum_{i=1}^{L-1} C_i m(k-L+1-i) + Dm(k-L+1)\end{aligned} \tag{26}$$

$$\begin{aligned}u\left(\overline{k}\right) &\triangleq u(k-L+1) = \overline{W}^T \underline{y}_{k-L+1} \\ &= \sum_{i=1}^{L-1} C_i^T y(k-L+1+i) + D^T y(k-L+1)\end{aligned} \tag{27}$$

Therefore the explicit causal update laws for the state space matrices C and D, using the time index $\overline{k}$ are

$$\Delta C_i = \eta_k \left[C_i - \varphi(y(\overline{k})) u(\overline{k} - i)^T \right], i = 1,2,\cdots,L-1 \tag{28}$$

$$\Delta D = \eta_k \left[D - \varphi(y(\overline{k})) u(\overline{k})^T \right] \tag{29}$$

In the state space regime, the information back-propagation filter assumes the form

$$\lambda(\overline{k} - 1) = A^T \lambda(\overline{k}) + C^T y(\overline{k}) \tag{30}$$

$$u(\overline{k}) = B^T \lambda(\overline{k}) + D^T y(\overline{k}) \tag{31}$$

The non-minimum natural gradient algorithms [4-6, 8, 9] require both forward and backward in time propagation by the adaptive MIMO FIR filter. One would require a buffer of $3L$-2 input samples to initiate the adaptation according to the update laws.

The non-minimum phase state space BSR update laws, being more general, also encompass the domain of the minimum phase mixing network case. This can be verified conveniently by using definition (20) in the update law for the matrix D, see [6].

3. SIMULATION

For the environment, a 3 × 3 non-minimum phase IIR filtering environment model with 3 zeros outside the unit circle was employed. Two of the non-minum phase zeros are complex while one is real. The environment parameters are

$$\sum_{j=0}^{m-1} \mathcal{A}_i m(k-i) = \sum_{i=0}^{n-1} \mathcal{B}_i s(k-i) + v(k) \qquad (32)$$

where

$$\mathcal{A}_0 = \begin{bmatrix} 1 & -1 & -1 \\ 1 & -1 & 1 \\ 1 & -1 & 1 \end{bmatrix}, \mathcal{A}_1 = \begin{bmatrix} 0.5 & -0.8 & -0.7 \\ 0.8 & 0.3 & -0.2 \\ -0.2 & -0.5 & 0.7 \end{bmatrix}, \mathcal{A}_2 = \begin{bmatrix} 0.06 & 0.3 & -0.5 \\ 0.2 & -0.1 & -0.4 \\ -0.5 & -0.6 & 0.3 \end{bmatrix}$$

$$\mathcal{B}_0 = \begin{bmatrix} 1 & 0.6 & -0.6 \\ 0.5 & -0.9 & 0.7 \\ 0.6 & 0.8 & 1 \end{bmatrix}, \mathcal{B}_1 = \begin{bmatrix} 0.5 & 0.9 & 0.5 \\ -0.2 & 0.7 & -0.6 \\ -0.7 & -0.7 & 0.6 \end{bmatrix}, \mathcal{B}_2 = \begin{bmatrix} .8 & 0.3 & 0.7 \\ -0.1 & 0.6 & 0.4 \\ 0.5 & 0.6 & 0.4 \end{bmatrix}$$

and $v(k)$ is additive gaussian noise of variance σ^2.

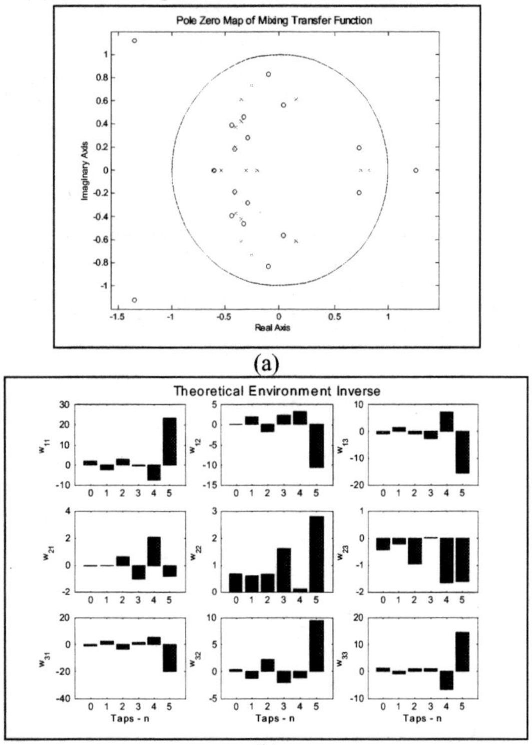

Figure 3. (a) Pole zero map of the environment transfer function, (b) Theoretical inverse of the environment

For the purpose of simulation the sources were assumed to be laplacian in nature, noise σ^2=0.1. The score function used in the simulation is the hyperbolic tangent [5] and demixing network was assumed to comprise of 41 taps per filter. The algorithm was able to converge in approximately 12000 online iterations. The learning rate was initialized at 0.0008 and then decayed exponentially. The global transfer function achieved is scaled and the second separated component also has a sign inversion. The results are summarized in Figures 3 and 4.

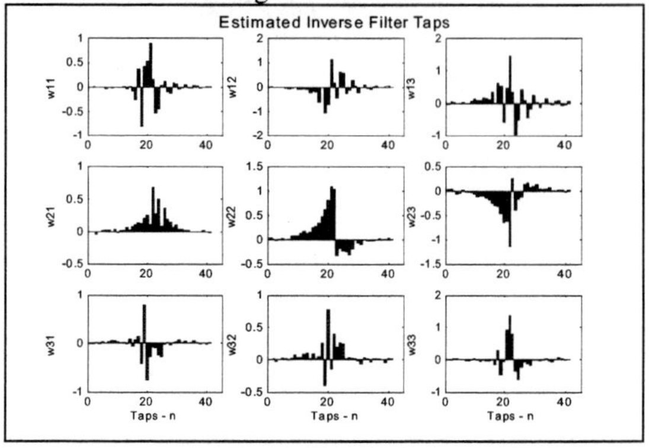

Figure 4. Estimated all zero demixing network

4. CONCLUSIONS

The paper presents a meticulous proof for blind recovery from non-minimum phase environments utilizing the theory of optimization and the contra-variant Riemannian (or the natural) gradient. The demixing network, in this case, has been restricted to be an all zero non-causal filter. The derivation uses the compact, hybrid FIR notation and the results are cast into the state space framework. An illustrative simulation example exhibits the online adaptation capabilities of the algorithms.

5. REFERENCES

1. Amari, S., *Natural Gradient Works Efficiently in Learning.* Neural Computation, 1998. 10: p. 251–276.
2. Salam, F.M. and G. Erten, *The State Space Framework for Blind Dynamic Signal Extraction and Recovery*, in *Proceedings of the IEEE International Symposium on Circuits and Systems.* 1999. p. 66–69.
3. Waheed, K. and F.M. Salam, *Blind Source Recovery: Some Implementation and Performance Issues*, in *Proceedings of the 44th IEEE Midwest Symposium on Circuits and Systems.* 2001: Dayton, OH. p. 694–697.
4. Waheed, K. and F.M. Salam, *State Space Blind Source Recovery of Non-minimum Phase Environments*, in *45th IEEE International Midwest Symposium on Circuits and Systems.* 2002: Tulsa, OK. p. 422-425.
5. Waheed, K. and F.M. Salam, *State-Space Blind Source Recovery for Mixtures of Multiple Source Distributions*, in *Proc. of IEEE Int'l Symposium on Circuits and Systems.* 2002: Scottsdale, Arizona. p. 197–200.
6. Waheed, K. and F.M. Salem, *Blind Source Recovery: A Framework in the State Space,* Journal of Machine Learning Research, Special Issue on Independent Component Analysis, the MIT Press, Cambridge, MA, USA, 4(Dec):1411-1446, 2003
7. Waheed, K. and F.M. Salem, *Blind Source Recovery: A State Space Formulation*, in *Contemporary Issues in System Stability and Control with Applications*, D. Liu and P.J. Antsaklis, Editors. 2003, Birkhauser: Boston, MA. p. 171-196.
8. Zhang, L.Q. and A. Cichocki, *Blind Deconvolution of Dynamical Systems: A State Space Approach.* Journal of Signal Processing, 2000. 4(2): p. 111–130.
9. Zhang, L.Q., A. Cichocki, and S. Amari, *Multichannel Blind Deconvolution of Nonminimum Phase Systems using Information Backpropagation*, in *Proc. of 5th Int'l Conf. on Neural Information Processing.* 1999: Perth, Australia. p. 210–216.

SPARSE COMPONENT ANALYSIS OF OVERCOMPLETE MIXTURES BY IMPROVED BASIS PURSUIT METHOD

Pando Georgiev and Andrzej Cichocki

Brain Science Institute, RIKEN
Lab. for Advanced Brain
Signal Processing
2-1, Hirosawa, Wako-shi
Saitama, 351-0198, Japan
georgiev,cia@bsp.brain.riken.go.jp

ABSTRACT

We formulate conditions under which we can solve precisely the Blind Source Separation problem (BSS) in the under-determined case (less sensors than sources), up to permutation and scaling of sources. Under these conditions, which include information about *sparseness* of the sources (and hence we call the problem *sparse component analysis* (**SCA**)), we can 1) identify the mixing matrix uniquely (up to scaling and permutation) and 2) recover uniquely the original sources. We present a new algorithm for estimation of the mixing matrix, as well as an algorithm for **SCA** (estimation of sparse sources), which improves the standard basis pursuit method of S. Chen, D. Donoho and M. Sounders (when the mixing matrix is known or correctly estimated). Our methods are illustrated with examples.

1. INTRODUCTION

Consider the following linear representation problem (matrix factorization): represent given data in matrix form $\mathbf{X} \in \mathbb{R}^{m\times N}$ as

$$\mathbf{X} = \mathbf{AS}, \quad \mathbf{A} \in \mathbb{R}^{m\times n}, \mathbf{S} \in \mathbb{R}^{n\times N}, \tag{1}$$

where the unknown matrices $\mathbf{A} \in \mathbb{R}^{m\times n}$ (dictionary, n means usually the number of sources and m is the number of observations, $m \leq n$) and $\mathbf{S} \in \mathbb{R}^{n\times N}$ (signals, N is the number of samples) have some specific properties, for instance:

- the rows of $\mathbf{S}$ are statistically independent as much as possible - this is *Independent Component Analysis* (ICA) problem;

- $\mathbf{S}$ contains as many zeros as possible - this is sparse representation problem or *Sparse Component Analysis* (SCA) problem;

- the elements of $\mathbf{X}, \mathbf{A}$ and $\mathbf{S}$ are nonnegative - this is *nonnegative matrix factorization* (NMF), with several potential applications including decomposition of objects into "natural" components, learning the parts of the objects (e.g. learns from set of faces the parts a face consists of, i.e. eyes, nose, mouth, etc.), redundancy and dimensionality reduction, micro-array data mining, enhancement of images in nuclear medicine, etc. (see [8]).

There is a large amount of papers devoted to ICA problems (see for instance [4], [7] and references therein) but mostly in the complete case (when the number of sources is equal to the number of sensors). We refer to [9], [1], [10], [3] and reference therein for some recent papers on SCA and overcomplete ICA.

A more general related problem is called Blind Source Separation (BSS) problem, in which we know a priory that such a representation like (1) exists and the task is to recover the sources (and the mixing matrix) as correctly as possible. A fundamental property of BSS problem (which makes it so attractive) is that such recovering is possible up to permutation and scaling of the sources.

In this paper we consider the BSS problem in the overcomplete case (more sources than sensors), as additional information compensating the lack of sources is *sparseness*. The task is to estimate the unknown sources $\mathbf{S}$ (and the mixing matrix $\mathbf{A}$) using the available data matrix $\mathbf{X}$ only. We describe conditions under which this is possible uniquely (up to permutation and scaling of the sources, which is an usual condition in the complete BSS problems).

We present a new algorithm for identification of the mixing matrix, which works correctly under some conditions (see conditions 1) and 2) of Theorem 1.)

We develop also an improvement of the basis pursuit method of Chen, Donoho and Saunders [5], (which in fact is l_1 norm minimization problem), when the mixing matrix is known or estimated. This improvement is also reduced to a linear programming problem, but we are able to find the sparsest solution of a linear overcomplete system. We present examples which illustrate our methods.

2. SPARSE COMPONENT ANALYSIS

In this section we develop a method for a solution of the BSS problem in case when the following assumptions are satisfied:
A1) the mixing matrix $\mathbf{A} \in \mathbb{R}^{m\times n}, m \leq n$ has the property that any square $m \times m$ submatrix of it is nonsingular, and
A2) the sources are sufficiently sparse, more precisely, when each column of the source matrix $\mathbf{S}$ has no more than $m/2$ nonzero components.

These conditions are crucial for uniqueness of the solution of the linear system overcomplete system $\mathbf{As}(k) = \mathbf{x}(k), k = 1, ..., N$. It is based on the following simple

Observation: *If there exists a solution $\mathbf{s}_*$ of the system $\mathbf{As} = \mathbf{x}$ with no more than $m/2$ nonzero elements, then this solution $\mathbf{s}_*$ is the unique sparsest solution of $\mathbf{As} = \mathbf{x}$.* [1]

[1]The number of the nonzero elements can be relaxed to $m - 1$. In this case the uniqueness is guaranteed for almost all $\mathbf{x}$ (in measure sense).

0-7803-8251-X/04/$17.00 ©2004 IEEE

Proof. If there are two solutions $\mathbf{s}_1$ and $\mathbf{s}_2$ with this property, then $\mathbf{A}(\mathbf{s}_1 - \mathbf{s}_2) = 0$ and the vector $\mathbf{s}_1 - \mathbf{s}_2$ contains no more than m nonzero elements, say $q \leq m$, which means that q columns of $\mathbf{A}$ are linearly dependent, a contradiction with A1). ∎

We will say that the source signals $\mathbf{s}(k), k = 1, ..., N$ have *level of sparseness* L, if for any k they have at most L nonzero elements. It is clear from what we said above, that the sources are identifiable precisely, if we know the mixing matrix $\mathbf{A} \in \mathbb{R}^{m \times n}$ and the level of sparseness of the sources is less than or equal to $m/2$.

2.1. Matrix identification

In this section we describe conditions under which we can identify the mixing matrix in the sparse BSS problem.

Theorem 1 *Assume that the number of sources is unknown and*

1) for each source $s_i(k), k = 1, ..., N$ there are at least two time instances when all the signals are zero except s_i (so each source is uniquely present at least twice), and

2) $\mathbf{A}(\mathbf{s}(k) - M\mathbf{s}(q)) \neq 0$ for any $M \in \mathbb{R}$, any $k = 1, ..., N$ and any $q = 1, ..., N, k \neq q$ for which $\mathbf{s}(k)$ has more that one nonzero element.

Then the number of sources is recoverable and the matrix $\mathbf{A}$ *is identifiable up to permutation and scaling.*

Proof. We cluster in groups all nonzero normalized column vectors of $\mathbf{X}$ such that each group consists of vectors which differ only by sign. From conditions 1) and 2) it follows that the number of the groups containing more that one element is precisely the number of sources n, and that each such group will represent a normalized column of $\mathbf{A}$ (up to sign). ∎

Below, we include an algorithm for identification of the mixing matrix.

Algorithm for identification of the mixing matrix

1) Remove all zero columns of $\mathbf{X}$ (if any) and obtain a matrix $\mathbf{X}_1 \in \mathbb{R}^{m \times N_1}$.

2) Normalize the columns $\mathbf{x}_i, i = 1, \ldots, N_1$ of $\mathbf{X}_1 : \mathbf{y}_i = \mathbf{x}_i / \|\mathbf{x}_i\|$ and put $i = 1, j = 2, k = 1$.

3) if either $\mathbf{y}_i = \mathbf{y}_j$ or $\mathbf{y}_i = -\mathbf{y}_j$, then put $\mathbf{a}_k = \mathbf{y}_i$, increase i, k with 1, put $j = i + 1$ and if $i < N_1$, repeat 3) (otherwise stop). Otherwise: if $j < N_1$, increase j by 1 and repeat 3). If $j = N_1$, increase i by 1, put $j = i + 1$ and repeat 3). Stop when $i = N_1$.

2.2. Identification of sources

Improved basis pursuit (BP) method

The famous basis (**BP**) pursuit method of S. Chen, D. Donoho and M. Sounders [5] is a principle for decomposing a signal into an "optimal" superposition of dictionary elements, where optimal means having the smallest l_1 norm of coefficients among all such decompositions. So, it consists of finding a minimum l_1 solution of a linear overcomplete system. Such minimality of the l_1 norm ensures sparseness of the coefficients of the solution. Namely, if $\mathbf{A} \in \mathbb{R}^{m \times n}$, $m < n$, then the minimum l_1 solution of the system $\mathbf{As} = \mathbf{x}$ has at most m nonzero elements for almost all $\mathbf{x} \in \mathbb{R}^m$ - a well know fact (see [6] for instance). This problem can be reduced to a standard linear programming problem as follows:

minimize $\quad \sum_{i=1}^{n} u_i$

under constraints:

$$u_i \geq s_i, \quad u_i \geq -s_i, \quad \mathbf{As} = \mathbf{x}. \tag{2}$$

A disadvantage of this method is that it not always finds the sparsest solution. For a comprehensive discussion of this topic see [6], where the authors give a sufficient condition for this.

Simple Example. Let $\mathbf{s}_* = (0, 2, -3, 0, 0, 0, 0, 0, 0, 0, 0, 0)^T$ be a solution of the system $\mathbf{As} = \mathbf{x}$, where $\mathbf{A} \in \mathbb{R}^{5 \times 12}$ is randomly generated. In large number of cases, when we generate random matrix $\mathbf{A}$, the **BP** method doesn't find the sparsest solution $\mathbf{s}_*$.

A=

Columns 1 through 7

0.2974	0.5527	0.3759	0.9200	0.1939	0.5488	0.6273
0.0492	0.4001	0.0099	0.8447	0.9048	0.9316	0.6991
0.6932	0.1988	0.4199	0.3678	0.5692	0.3352	0.3972
0.6501	0.6252	0.7537	0.6208	0.6318	0.6555	0.4136
0.9830	0.7334	0.7939	0.7313	0.2344	0.3919	0.6552

Columns 8 through 12

0.8376	0.7165	0.7006	0.1146	0.8230
0.3716	0.5113	0.9827	0.6649	0.6739
0.4253	0.7764	0.8066	0.3654	0.9994
0.5947	0.4893	0.7036	0.1400	0.9616
0.5657	0.1859	0.4850	0.5668	0.0589

Solution by **BP**:

[−0.9977 0.0000 −0.9640 0.8411 0.0000 −0.0000
0.0000 0.0000 −0.0000 0.0000 0.4042 −0.2228]

Improved BP method with several Degrees of Freedom (DOF)

We assume that the matrix $\mathbf{A}$ is known (or estimated correctly) and it satisfies assumption A1). Assume that there is a solution $\mathbf{s}_*$ of the system $\mathbf{As} = \mathbf{x}$ with no more than $m/2$ nonzero elements. Recall that in this case (see **Observation**) the sparsest solution is equal to $\mathbf{s}_*$ and is unique. So, the criterion for finding the sparsest solution is to find one with no more than $m/2$ nonzero elements. We propose the following modification of the **BP** method:

Solve the following set of minimization problems (where $\mathbf{e} = (1, 1, \ldots, 1) \in \mathbb{R}^n$):

$$\text{minimize} \quad \sum_{i=1, i \neq j}^{n} e_i u_i \quad , j = 1, ..., n \tag{3}$$

under constraints:

$$u_i \geq s_i, \quad u_i \geq -s_i, \quad \mathbf{As} = \mathbf{x}. \tag{4}$$

It the sparsest solution is not found (which has no more than $m/2$ nonzero elements), we replace pairs of the coefficients e_i in (3) with zeros and solve consecutively these problems until obtaining the sparsest solution. If it is not found again, proceed analogically with the triples of zeros and so on. For small m this procedure is effective up to level 2, i.e. taking pairs of zeros in the coefficients of (3). This of course is a combinatorial problem which increases the computational time, but not so dramatically when the solution is very sparse.

The reason why our improvement works, is clear: suppose for instance that the sparsest solution $\mathbf{s}_*$ has 2 nonzero elements s_i and s_j. Putting $e_i = e_j = 0$ in some step of the algorithm, it will find this solution, since the minimum of the cost function is zero and it is obtained exactly at $\mathbf{s}_*$. In most cases the sparsest solution is obtained putting only one coefficient e_i zero.

In a similar way, as Theorem 1, we can prove the following its generalization.

Theorem 2 *Assume that*

1) for each source $s_i(k), k = 1, ..., N, i = 1, ..., n$ there are $k_i \geq 2$ time instances when all of the source signals are zero except s_i (so each source is uniquely present k_i times), and

2) the set $S_p = \{j \in \{1, ..., N\} : \mathbf{A}(\mathbf{s}(p) - M\mathbf{s}(j)) = 0$ for some $M \in \mathbb{R}\}$, contains less than $\min_{1 \le i \le m} k_i$ elements for any $p \in \{1, ..., N\}$ for which $\mathbf{s}(p)$ has more that one nonzero element.

Then the matrix $\mathbf{A}$ is identifiable up to permutation and scaling.

3. COMPUTER SIMULATION EXAMPLES

First example. We generated artificially sources, shown in Fig.1. They have level of sparseness 2 and each source is uniquely active (nonzero) at only 10 time instants. For instance, $s_4(k) = 0$ for $k = 211, ..., 220$, as unique nonzero source in this period is s_3, but with very small amplitude. Nevertheless, our algorithm is capable to estimate precisely any randomly chosen matrix after the linear mixture of the sources. For instance, we generated randomly a matrix $\mathbf{A}46 \in \mathbb{R}^{4\times 6}$, and mixed the sources by it. The mixed sources are shown in Fig. 2. We run our algorithm for estimating the mixing matrix (shown below as $\mathbf{A}$) and run the standard **BP** method - the results of separation are shown in Fig. 4. Our method **basis pursuit with DOF** gives excellent results (shown in Fig. 3), much better than those obtained by the standard **BP** method.

Normalized initial matrix: A46N

$\mathbf{A}46N =$

0.5545	0.2548	0.4418	−0.6681	0.1204	0.2668
0.6600	−0.3980	−0.1768	−0.5911	0.1578	0.9094
0.2303	−0.7329	−0.3451	−0.4261	−0.8536	−0.3189
−0.4516	0.4893	−0.8090	−0.1508	0.4815	−0.0042

Estimated matrix (normalized)

$\mathbf{A} =$

−0.2668	−0.1204	0.6681	−0.4418	0.2548	0.5545
−0.9094	−0.1578	0.5911	0.1768	−0.3980	0.6600
0.3189	0.8536	0.4261	0.3451	−0.7329	0.2303
0.0042	−0.4815	0.1508	0.8090	0.4893	−0.4516

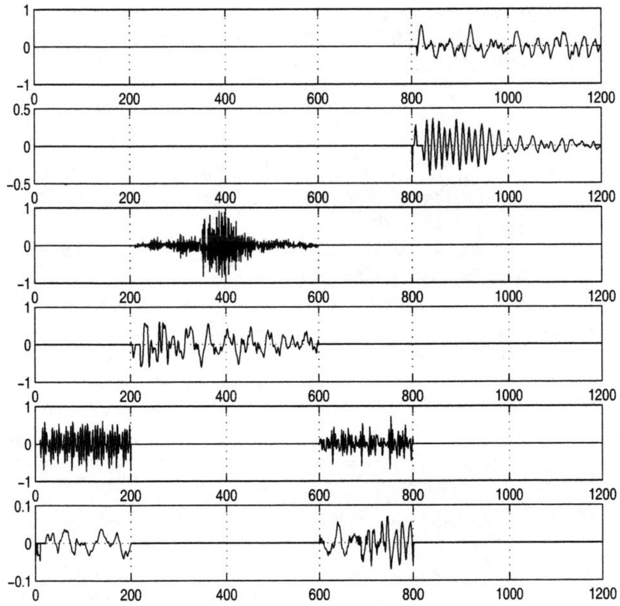

Fig. 1. Original sources.

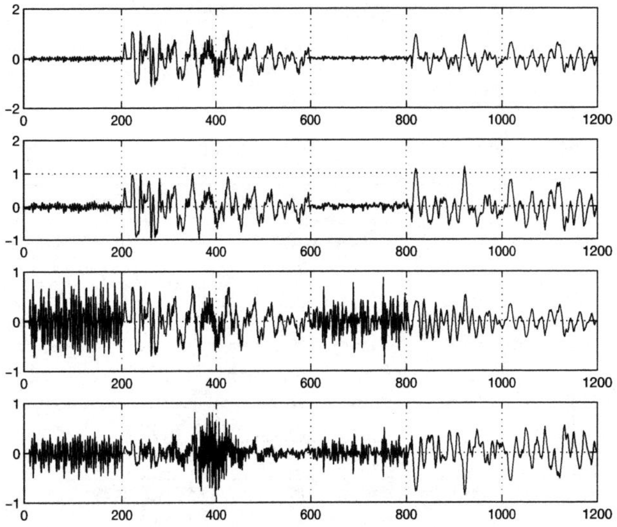

Fig. 2. Mixed (observed) signals.

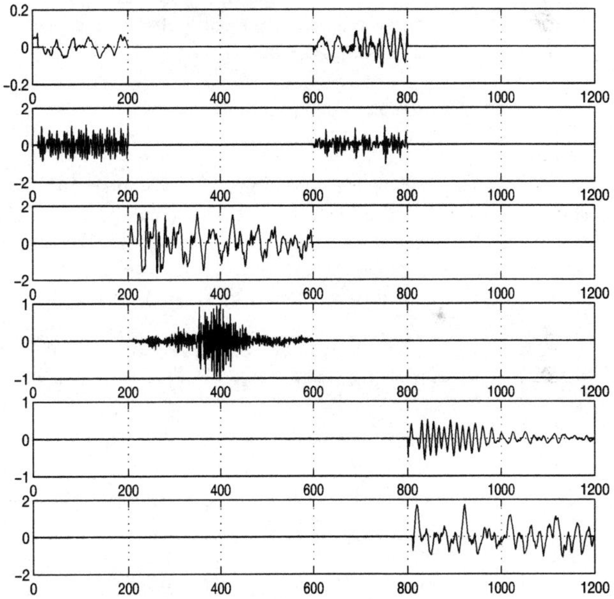

Fig. 3. Estimated sources by the new method: **basis pursuit with DOF** using the estimated matrix.

Second example.

In this example for the complete case ($m = n$) of instantaneous mixtures, we demonstrate the effectiveness of our algorithm for identification of the mixing matrix in the case considered in Theorem 1. We mixed 3 images of landscapes (shown in the first row of Fig. 6) with a 3-dimensional randomly generated matrix $\mathbf{A}$ ($\det\mathbf{A} = 0.0016$). We transformed the three mixtures by two dimensional discrete Haar wavelet transform. As a result, since this transform is linear, the high frequency components of the Haar wavelet transform of the source images become very sparse (see

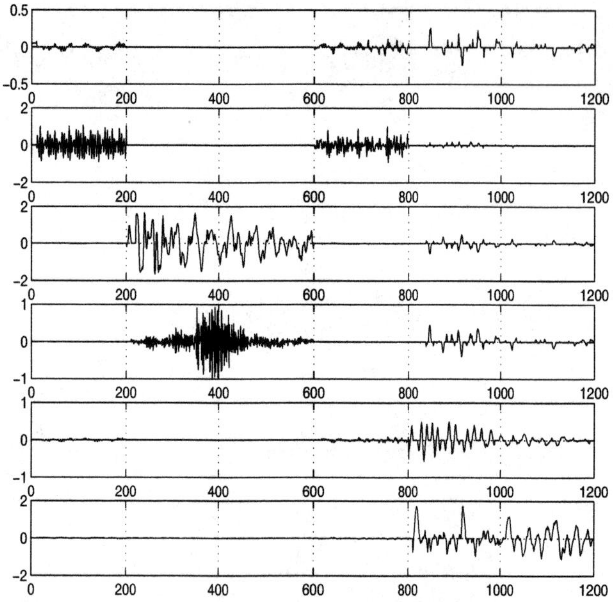

Fig. 4. Estimated sources by the standard basis pursuit using the estimated matrix. Some artifacts are clearly observed, especially for the first and fourth estimated sources.

Fig. 5) and they satisfy the conditions of Theorem 1. We used only the 10-th row (160 points) of the diagonal coefficients of this Haar wavelet transform and estimated very precisely the mixing matrix. The mixed images and the recovered images are shown in Fig. 6 (the second and the third row respectively).

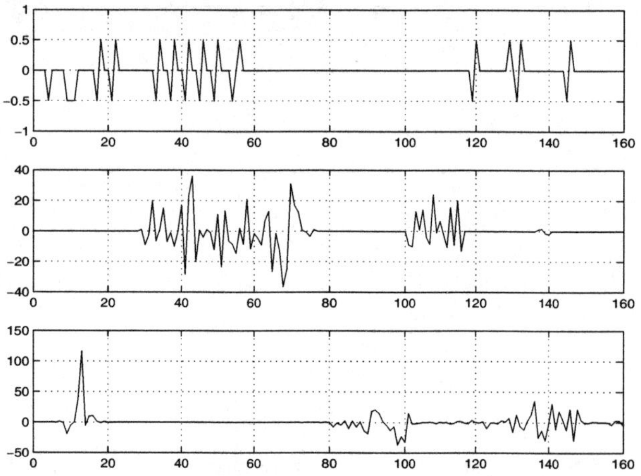

Fig. 5. Diagonal wavelet coefficients of the original images (here shown only the 10-th row of each of the three (120×160) matrixes). They satisfy the conditions of Theorem 1 and this is the reason for the perfect reconstruction of the original images, since our algorithm uses only the tenth row of each of the mixed images.

4. CONCLUSION

We presented sufficient conditions for BSS of sparse signals and developed two new algorithms: for identification of the mixing matrix, and for improvement of the basis pursuit method of Chen, Donoho and Sounders. We presented two experiments. The first one is for separation of artificially created signals with sufficient level of sparseness, which gives excellent results, much better than those obtained by the standard basis pursuit method. The second one demonstrates the effectiveness of our matrix identification algorithm in practical situations. We separated perfectly randomly mixed landscape images using our algorithm, after applying 2-dimensional discrete Haar wavelet transform.

Fig. 6. Original images (first row), the normalized mixed images (second row) and separated normalized images (third row).

5. REFERENCES

[1] A.M. Bronstein, M.M. Bronstein, M. Zibulevsky, Y. Y. Zeevi, " Blind Separation of Reflections using Sparse ICA", in Proc. Int. Conf. ICA2003, Nara, Japan, pp.227-232.

[2] S. Amari and A. Cichocki "Adaptive blind signal processing - neural network approaches". *Proceedings IEEE* (invited paper), 86(10): 2016-2048, 1998.

[3] P. Bofill and M. Zibulevsky, " Underdetermined Blind Source Separation using Sparse Representation", Signal Processing, vol. 81, no. 11, pp. 2353-2362, 2001.

[4] A. Cichocki and S. Amari. *Adaptive Blind Signal and Image Processing*. John Wiley, Chichester, 2003.

[5] Chen, Scott Shaobing; Donoho, David L.; Saunders, Michael A., " Atomic decomposition by basis pursuit", SIAM J. Sci. Comput. 20 (1998), no. 1, 33–61.

[6] D. Donoho and M. Elad, "Optimally sparse representation in general (nonorthogonal) dictionaries via l^1 minimization", Proc. Nat. Acad. Sci., March 4, 2003, vol.100, no.5, 2197-2202.

[7] A. Hyvarinen, J. Karhunen and E. Oja, " Independent Component Analysis", John Wiley & Sons, 2001.

[8] D. D. Lee and H. S. Seung *Learning the parts of objects by non-negative Matrix Factorization*, Nature, Vol. 40, pp. 788-791, 1999.

[9] K. Waheed, F. Salem, "Algebraic Overcomplete Independent Component Analysis", in Proc. Int. Conf. ICA2003, Nara, Japan, pp.1077-1082.

[10] M. Zibulevsky, and B. A. Pearlmutter, Blind source separation by sparse decomposition, Neural Comp. 13(4), 2001.

A DUAL MODE DECISION FEEDBACK EQUALIZER EMPLOYING THE CONJUGATE GRADIENT ALGORITHM

Yihai Zhang and T. Aaron Gulliver
Department of Electrical and Computer Engineering
University of Victoria
P.O. Box 3055 STN CSC
Victoria, B.C. V8W 3P6 CANADA

ABSTRACT

Decision feedback equalizers are commonly employed to reduce the error caused by intersymbol interference. In this paper, we propose a decision feedback equalizer (DFE) that automatically switches between the constant modulus algorithm (CMA) and the decis ion-directed algorithm (DD). Performance is compared with a CMA-DFE. To improve the slow convergence characteristics of the LMS algorithm, the Conjugate Gradient (CG) algorithm is employed. Simulation results show that the proposed decision feedback equalizer provides better performance and faster convergence with reasonable computational complexity.

1. INTRODUCTION

In modern digital communication systems, digital signals are transmitted at a high speed through band-limited time dispersive channels, which cause multi-path fading and signal distortion, and result in intersymbol interference (ISI). Channel equalization is an effective technique to remove ISI from the received signal.

Generally, equalizers can be classified into two types: linear equalizers and nonlinear equalizers [1]. In practice, a linear equalizer does not perform very well on channels with spectral nulls in their frequency response. In this case, a nonlinear equalizer should be used. The most popular nonlinear equalizer is the decision feedback equalizer (DFE).

Decision feedback equalization uses the fact that the output signal of the forward filter contains ISI caused by previously detected symbols and undetected symbols. If the previous decisions are correct, their ISI contribution can be subtracted from the current received signal [2]. A DFE will often provide better ISI cancellation than a linear equalizer, especially if the channel has spectral nulls.

Equalizing a channel without a training sequence is known as blind equalization. The most popular blind equalization algorithms are the Constant Modulus Algorithm (CMA) and the Decision-Directed algorithm (DD). In this paper, we propose an equalizer that automatically switches between the CMA and DD algorithms. Similar work has been done in recent years [3] [4]. To speed up the convergence rate of the LMS algorithm, we employ a modified Conjugate Gradient (CG) algorithm to update the coefficients of the equalizer. This paper is organized as follows. In Section 2 we describe the system mo del. Section 3 gives a brief description of the CMA and DD algorithms, and presents our proposed equalizer structure. Performance results are shown in Section 4, and finally some conclusions are presented in Section 5.

2. SYSTEM MODEL

We assume a sequence of input symbols $\{s_k\}$, which are independent, identically distributed (i.i.d.) BPSK signals {+1, -1} with zero mean, transmitted through a linear transversal time dispersive channel. The channel output can be expressed as

$$x_k = \sum_{i=0}^{M} h_i s_{k-i} + n_k , \qquad (1)$$

where $H(z) = [h_0 \quad h_1 \dots \quad h_M]^T$ is the equivalent impulse response of the transmitter filter, unknown channel and receiver filter, $\{n_k\}$ is assumed to be additive white Gaussian noise (AWGN) with zero mean which is independent of the input signal. We also assume that the coefficients of the channel and equalizer are real, and the previously detected symbols are assumed to be correct. Since the transmitted signal is BPSK, the phase recovery algorithm is not considered in this system.

The structure of the proposed nonlinear decision feedback equalizer is shown in Fig. 1. The operation of the decision feedback equalizer is characterized by

$$z_k = \sum_{i=1}^{K_1} a_i x_{k-i} - \sum_{j=1}^{K_2} b_j \hat{s}_{k-j} , \qquad (2)$$

where z_k is the input signal of the decision device, $A(z) = [a_1 \quad a_2 \dots \quad a_{K_1}]^T$ is the weight vector of the

0-7803-8251-X/04/$17.00 ©2004 IEEE

feedforward equalizer, and $B(z)=[b_1 \ \ b_2 \dots \ \ b_{K_2}]^T$ is the weight vector of the feedback filter. $\{\hat{s}_k\}$ is the previous decision at the output of the decision device, and with BPSK modulation, $\hat{s}_k = \mathrm{sgn}(z_k)$. We also define the weight vector and input signal vector of the equalizer as

$W(z)=[a_1 \ \ a_2 \dots \ \ a_{K_1} \ \ b_1 \ \ b_2 \dots \ \ b_{K_2}]^T$, and

$U(z)=[x_{k-1} \ \ x_{k-2} \dots \ \ x_{k-K_1} \ \ -\hat{s}_{k-1} \ \ -\hat{s}_{k-2} \dots \ \ -\hat{s}_{k-K_2}]^T$,

respectively. The total length of the weight vector is *m*. Then we have that $z_k = W^T U$.

3. A DUAL MODE DECISION FEEDBACK EQUALIZER

3.1 Decision-Directed Algorithm

Generally, an estimate of the transmitted signal can be obtained from the DFE output. If this estimate is good enough (identical to the transmitted signal), then it can be used to replace the training sequence and adjust the parameters of the equalizer. This is known as the decision directed (DD) algorithm because the equalizer learns by using its own decisions [5]. The general form of the cost function is

$$J_{DD} = E\left[\left(|z_k - \hat{s}_k|\right)^2\right], \tag{3}$$

and the gradient of this cost function with respect to the weight vector of the equalizer can be expressed as

$$g_{DD} = 2(z_k - \hat{s}_k)u_k. \tag{4}$$

in the DD algorithm, the error signal is calculated by

$$e_{DD} = z_k - \hat{s}_k. \tag{5}$$

The DD algorithm is not suitable for a cold start initialization since the resulting high decision error rate can prevent convergence. Therefore another blind algorithm should be used to open the eye for the DD algorithm.

3.2 Constant Modulus Algorithm

The Constant Modulus Algorithm (CMA) was first proposed by Godard [6] for QAM signals. The cost function of the CMA criterion is based on the fact that the magnitude of the transmitted signal constellation is constant. The general form of the CMA cost function is

$$J_{CM} = E\left[\left(|z_k|^2 - R_2\right)^2\right]. \tag{6}$$

In (6), R_2 is the dispersion constant of the source sequence $\{s_k\}$, so that $R_2 = E\left[|s_k|^4\right] / E\left[|s_k|^2\right]$. Since we use BPSK modulation, $R_2 = 1$. The gradients of the CMA cost function with respect to the weight vector of the equalizer can be expressed as

$$g_{CMA} = 4\left(|z_k|^2 - 1\right)z_k u_k, \tag{7}$$

and the error signal of the CM criterion is

$$e_{CM} = \left(|z_k|^2 - 1\right)z_k. \tag{8}$$

Since the CMA algorithm cost function is not a convex function, it may converge to different local minimums depending on the initial parameters of the equalizer. However, this ill-convergence can be avoided by proper initialization with the center tap one and all others zero, and with small step sizes in the adaptation loops [6] [7].

The CMA algorithm is robust and easy to implement, and can converge to a local minima even when the eye is not initially open. However, it has a large steady state error once the equalizer has converged. Therefore it is necessary to switch to the DD algorithm to improve convergence speed and the output error rate once the eye pattern of the equalizer has been opened by the CMA algorithm.

3.3 A Dual Mode DFE Employing the Conjugate Gradient Algorithm

Usually, the least mean squares (LMS) algorithm is used to update the equalizer weights because it is simple and easy to implement. The major disadvantage of LMS is its slow convergence. The conjugate gradient (CG) algorithm is an established optimization technique that has been applied to numerous signal processing problems [8]. The computational complexity of the CG algorithm lies between that of the LMS and Recursive Least Squares (RLS) algorithms, and provides comparable convergence to the RLS algorithm.

The conjugate gradient algorithm updates the filter coefficients with a set of directions which are conjugate to each other. The first step in each iteration initializes the LMS direction; each subsequent step moves in a direction that is a linear combination of the current gradient vector and the previous direction vector. The direction is reinitialized every *m* steps using just the gradient. The step size is determined using a line search or calculated using the Hessian. However, an alternate technique was derived in [9] which does not require knowledge of the Hessian or a line search. Boray and Srinath [8] proposed a modified CG algorithm which averages the instantaneous gradient estimates over a specified number "n_w" of past values, instead of using the instantaneous value of the gradient as in the LMS algorithm. The gradient estimate is given by

$$g_k = \left(\frac{1}{n_w}\right) \bullet \sum_{j=i-n_w+1}^{i} g_{inst}\left(w_k(i), u(j)\right) \tag{9}$$

where g_{inst} is the instantaneous gradient, i.e. the gradient of the CMA or DD algorithms, calculated with the most recent weight vector $w_k(i)$ and the past input value $u(j)$. Usually $n_w \le m$. A constant or a normalized step size can be used to simplify the algorithm.

The proposed equalizer initially adjusts the weight vector using the CMA algorithm. Once reliable decisions are being made (the eye is open), the equalizer switches to the DD algorithm to improve steady state performance. However, if future decisions are not reliable, the equalizer can switch back to the CMA algorithm. In the proposed dual mode blind equalization algorithm, the DFE automatically switches between the CMA and DD algorithms based on the MSE of the output signal. When the equalizer output MSE is below a certain threshold, the DD algorithm is used; otherwise, the CMA algorithm is used.

The proposed algorithm is summarized as follow:

For each input $u(i)$:

Step 1: Set the initial weight w_0, compute the gradient estimate g_0, if MSE greater than the threshold, $g_0 = g_{CMA}$; otherwise, $g_0 = g_{DD}$. Set $d_0 = -g_0$.

Step 2: Repeat for $k = 0,1,\ldots n_w - 1$

a) Set $w_{k+1} = w_k + \alpha d_k$ (10)

b) Compute an estimate of the gradient g_{k+1} at w_{k+1}. If MSE greater than a certain threshold (i.e. -15dB)

$$g_{k+1} = \left(\frac{1}{n_w}\right) \bullet \sum_{j=i-n_w+1}^{i} g_{CMA}(w_{k+1}(i), u(j)), \quad (11)$$

else

$$g_{k+1} = \left(\frac{1}{n_w}\right) \bullet \sum_{j=i-n_w+1}^{i} g_{DD}(w_{k+1}(i), u(j)). \quad (12)$$

c) Unless $k = n_w - 1$, set $d_{k+1} = -g_{k+1} + \beta_k d_k$, where

$$\beta_k = \frac{g_{k+1}^T g_{k+1}}{g_k^T g_k}. \quad (13)$$

Step 3: Replace w_0 by w_k, shift in a new input vector, and return to Step 1.

Each iteration, the system poles can be checked to ensure they lie within the unit cycle. If some poles are outside the unit cycle, the system may become unstable. In this situation, the weight update is ignored i.e., $w_{k+1} = w_k$.

4. SIMULATION RESULTS

In this section, we investigate the performance of the blind decision feedback equalizer proposed above. The channel used in this simulation was obtained from [1, Chap. 10], *H(z)* = [0.04 -0.05 0.07 -0.21 -0.5 0.72 0.36 0.21 0.03 0.07]. A 21-tap forward transversal filter and a 5-tap feedback filter were employed. The forward filter was initialized so that the center tap of the filter is one and the other taps are zero, and the feedback filter was initialized to zero. BPSK modulation $\{\pm 1\}$ and additive white Gaussian noise were used with the SNR set to 30dB. All previous decisions were assumed to be correct. The step size used to update the weight vector was 0.005, and the simulations were repeated for 200 trials.

We consider three issues in this section. First, we compare the proposed decision feedback equalizer with the CMA-DFE equalizer, with coefficients updated using the LMS algorithm. We see from Fig. 2 that the proposed algorithm has better MSE and convergence than the CMA-DFE algorithm. This is because the output error level of the CMA algorithm is larger than that of the DD algorithm after the equalizer converges, and the DD algorithm provides faster convergence after the eye is open. Clearly the performance of the DFE has been enhanced by automatically switching between the CMA and DD algorithms.

Another issue is the computational complexity and convergence behavior of the proposed DFE as n_w is varied. Note that the conjugate gradient algorithm is employed, and $n_w = 1$ means there is no averaging in the gradient estimate, so the CG algorithm reverts to the LMS algorithm. Fig. 3 shows that the equalizer converges faster as the value of n_w increases, and the computational complexity of the CG algorithm is $O(mn_w^2)$ [8]. Thus there is a tradeoff between convergence rate and computational complexity. It was found that $n_w = 3$ provides the best balance between computational complexity and convergence.

Fig. 4 shows how the convergence of the proposed equalizer is affected by changes in the step size. We considered three step sizes (0.001, 0.005, and 0.01). It was found that the algorithm provides better convergence with a larger step size. However, the equalizer may become unstable and not converge when the step size is close to 1.

The decision feedback equalizer converged poorly when the signal to noise ratio (SNR) was too low. This is because at low SNRs, there is significant error propagation and large numbers of decision errors. Conversely, when the SNR is high, the decision errors caused by noise are infrequent and can be ignored.

5. CONCLUSIONS

The convergence rate and MSE performance are two important issues in equalization. In this paper, a new dual mode DFE was introduced which is based on the constant modulus algorithm and the decision-directed algorithm. It was shown that this CMA-DD-DFE has better intersymbol interference cancellation performance than a CMA-DFE. To improve the convergence rate, a modified conjugate gradient algorithm was employed. It was shown that the CG algorithm provides faster convergence. The penalty is a slight increase in the computational complexity.

REFERENCES

[1] J.G. Proakis, Digital Communications, 4th Ed., McGraw-Hill, New York, NY, 2001.

[2] D.A. George, R.R. Bowen and J.R. Storey, “An adaptive decision feedback equalizer,” *IEEE Trans. Commun. Technol.,* vol. 19, pp. 281-293, June 1971.

[3] R.A. Casas, C.R. Johnson, J. Harp and S. Caffee, “ On initialization strategies for blind adaptive DFEs,” Proc. *Wireless Commun. and Networking Conference*, vol. 2, pp. 792-796, Sep. 1999.

[4] N. McGinty, “ Strategy to transition from the constant modulus algorithm to a decision feedback blind equalizer,” Proc. *IEEE Int. Conf. Acoustics, Speech & Signal Proc.*, vol. 3, pp. 2661-2664, May 2002.

[5] S. Haykin, Adaptive Filter Theory, 3rd Ed., Prentice Hall, 1996.

[6] D.N. Godard, “ Self-recovering equalization and carrier tracking in two dimensional data communication systems,” *IEEE Trans. Commun.,* vol. 28, pp. 1867-1875, Nov. 1980.

[7] C.R. Johnson, P. Schniter, T.J. Enders, J.D. Behm, D.R. Brown and R.A. Casas, “ Blind equalization using the Constant modulus criterion: a review,” *Proc. IEEE*, vol. 86, pp. 1927-1950, Oct. 1998.

[8] G.K. Boray and M.D. Srinath, “ Conjugate gradient techniques for adaptive filtering,” *IEEE Trans. Circuits and Systems-I: Fundamental Theory and Applications,* vol. 39, pp. 1-10, Jan. 1992.

[9] D.G. Leuenberger, Introduction to Linear and Nonlinear Programming, Addison-Wesley, Reading, MA, 1984.

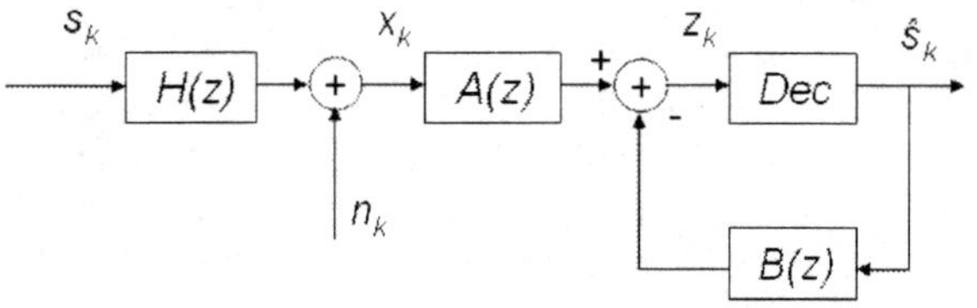

Figure 1: The proposed decision feedback equalizer structure.

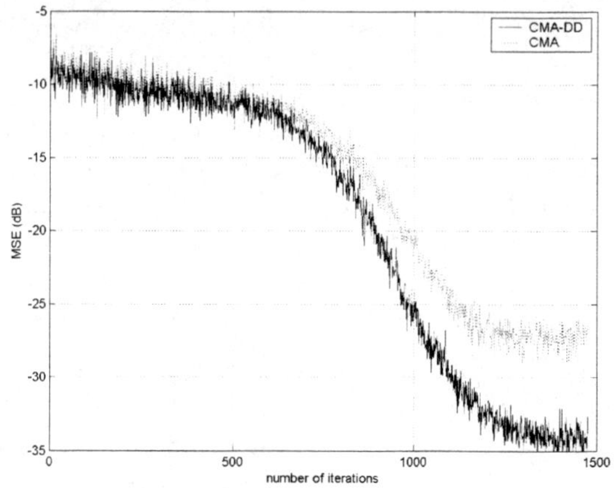

Figure 2: Performance of CMA-DD-DFE and CMA-DFE with the LMS algorithm, SNR=30dB, step size = 0.005.

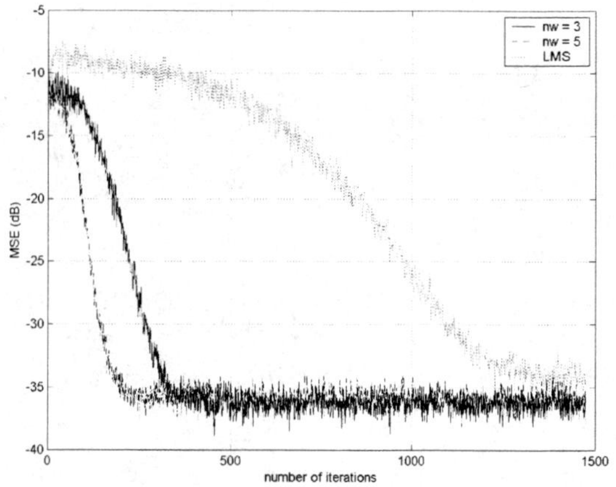

Figure 3: Equalizer performance with different window sizes, SNR = 30dB, step size = 0.005.

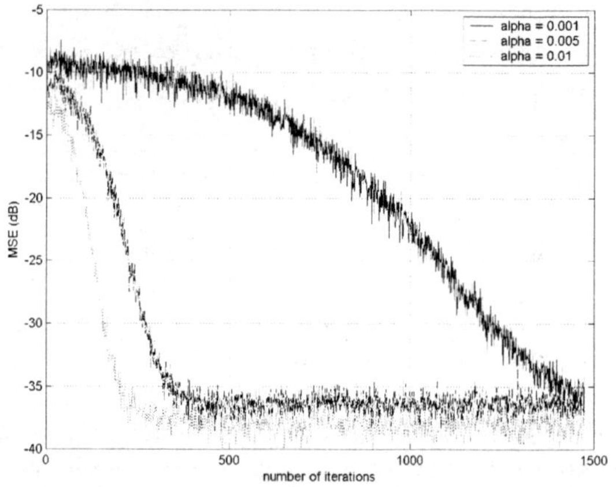

Figure 4: Equalizer performance with different step sizes, SNR = 30dB, $n_w = 3$.

BLIND UPLINK SPACE-TIME CHANNEL ESTIMATION FOR SPACE-TIME CODED MULTICARRIER CODE DIVISION MULTIPLE ACCESS SYSTEMS*

Yanxing Zeng, Qinye Yin, Le Ding, Ke Deng

School of Electronics and Information Engineering, Xi'an Jiaotong University, P. R. China, 710049

ABSTRACT

The uplink of a space-time coded multicarrier code division multiple access (MC-CDMA) system equipped with a uniform linear array (ULA) at the base station is investigated. Transmission takes place over frequency-selective fading channels and the goal is to blindly estimate the space-time channels and the directions of arrival (DOA) for all active users in one macrocell. An auxiliary matrix including both uplink space-time channels and DOAs is constructed, on which the eigen decomposition is performed to acquire the DOAs for multiple users. Then, the singular value decomposition (SVD) is applied on the eigenvectors corresponding to the DOA of individual user to obtain a closed-form solution of the uplink space-time vector channel. Computer simulations demonstrate the effectiveness of the proposed scheme.

1. INTRODUCTION

Space-time coding (STC) is an effective coding technique that uses transmit diversity to combat the detrimental effects in wireless fading channels by combining signal processing at the receiver with coding techniques appropriate to multiple transmit antennas to achieve higher date rates. The matured MC-CDMA is another technique that has properties desirable for high-data-rate wireless communications, such as insensitivity to frequency-selective channel and frequency diversity. Thus, the combination of STC and MC-CDMA is one of the most promising schemes for next-generation wireless communications[1,2,3]. However, for MC-CDMA systems with STC transmitter diversity, each co-channel signal source will produce extra interference signals that appear independent to antennas at the receiver[1], which results in severe multiuser interference (MUI). So, it is meaningful to exploit spatial processing techniques with antenna array at the base station to resist MUI and enhance system capacity.

On the other hand, in various STC schemes, the channel state information (CSI) is crucial in the decoding process. The system performance will be seriously degraded if CSI is not available or is not accurate. Hence, the channel estimation is needed in practical space-time coded MC-CDMA systems.

In this paper, we apply a ULA to the base station of space-time coded MC-CDMA systems over frequency-selective fading channels. We develop a closed-form blind space-time vector channel and DOA estimator for all active users in one macrocell. Our estimator employs an ESPRIT-like method, namely, direction of arrival Matrix (DOA-Matrix) method[5], to separate multiple co-channel users with different impinging DOAs. Then, by exploiting the subspace decomposition technique in combination with the finite alphabet property of transmitted symbols, this estimator provides closed-form solutions of the uplink space-time vector channel and DOA for multiple users. Performance of our estimator is evaluated by extensive computer simulations.

Notation: $[\bullet]^*$, $[\bullet]^T$ and $[\bullet]^H$ denote the complex conjugate, the transpose and the conjugate transpose of matrix.

2. SYSTEM MODEL

2.1 Baseband Model

It is assumed that K users uniformly distributed around a macrocell site. All K active users share the same set of subcarriers. The number of subcarriers equals to the length of spreading code G. Fig. 1 displays the baseband model of the space-time coded MC-CDMA systems equipped with a ULA at the base station.

As shown in Fig. 1, the input data symbol of each user is coded by Alamouti's space-time block encoder [2,3] with two transmit antennas. However, it is not too difficult to extend our algorithm to arbitrary STC schemes. When MC-CDMA systems over wireless finite impulse response (FIR) channels, a usual approach for combating the resultant inter-block interference (IBI) is via adding cyclic prefix (ACP) to each transmitted data block. Meanwhile, by removing CP (RCP) at the beginning of each received data block, the IBI can be eliminated.

The MC-CDMA scheme does the spreading spectrum operation in the frequency domain [4]. We define the m-th assigned frequency domain spreading code for the k-th user as a vector $\mathbf{c}_k^{(m)}$ ($k=1,\cdots,K$; m=1,2), which can be written as

$$\mathbf{c}_k^{(m)} = \left[c_k^{(m)}(1) \quad c_k^{(m)}(2) \quad \cdots \quad c_k^{(m)}(G) \right]^T \tag{1}$$

Given the length G of frequency domain spreading codes, the maximum number of active users is therefore determined to be

$$K_{\max} = \lfloor G/2 \rfloor \tag{2}$$

* Partially supported by the National Natural Sciences Foundation (No.60272071) and the Research Fund for Doctoral Program of Higher Education (No.20020698024) of China.

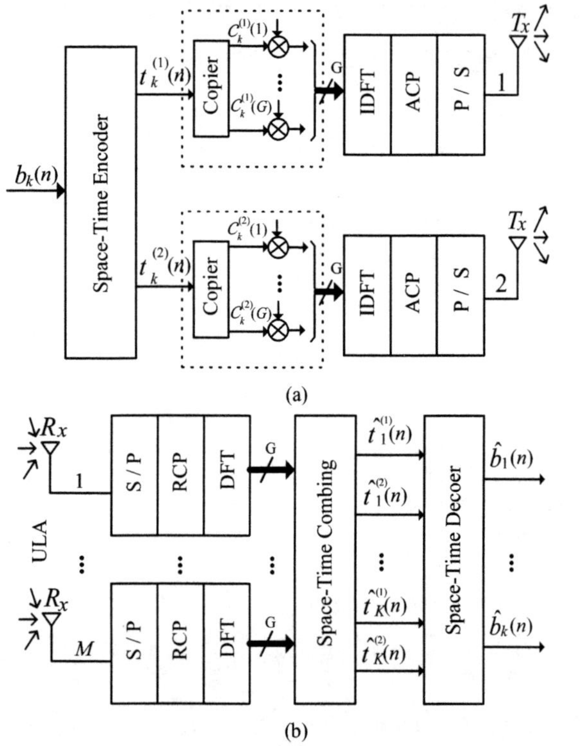

Fig. 1 The baseband model of the space-time block coded MC-CDMA systems with a ULA at the base station (a) user k's transmitter (b) base station receiver

2.2 Numerical Model

We assume that symbols from different users are synchronized using timing synchronization techniques for the uplink to simplify the analysis.

For clarity, we describe the uplink wireless finite impulse response (FIR) channel between the reference element 1 of the receive antenna array and the *m*-th (*m*=1, 2) transmit antenna of the *k*-th user as

$$\mathbf{h}_k^{(m)} = \left[h_k^{(m)}(0) \quad h_k^{(m)}(1) \quad \ldots \quad h_k^{(m)}(L) \right]^{\mathrm{T}} \tag{3}$$

where L expresses the maximum common FIR channel length of all uplink channels between receive antennas and transmit antennas. Without loss of generality, we assume $L < G$ and the channel is constant during several tens of MC-CDMA symbols.

The frequency domain attenuations in each subcarrier caused by above dispersive channels can be obtained by performing the Discrete Fourier Transform (DFT) on the aforementioned time domain FIR vector $\mathbf{h}_k^{(m)}$, which is depicted as a vector $\boldsymbol{\eta}_k^{(m)}$

$$\begin{aligned}\boldsymbol{\eta}_k^{(m)} &= \left[\eta_k^{(m)}(1) \quad \eta_k^{(m)}(2) \quad \cdots \quad \eta_k^{(m)}(G) \right]^{\mathrm{T}} \\ &= \mathbf{F}_{\mathrm{DFT}}(:,1:(L+1))\,\mathbf{h}_k^{(m)} = \mathbf{F}_{\mathrm{FRO}}\,\mathbf{h}_k^{(m)}\end{aligned} \tag{4}$$

where the matrix $\mathbf{F}_{\mathrm{DFT}}$ with dimensions $G \times G$ is a DFT matrix, and its every entry is $(\mathbf{F}_{\mathrm{DFT}})_{l,l'} = \frac{1}{\sqrt{G}}\exp(-j\frac{2\pi}{G}(l-1)(l'-1))$ (l,l' =1,⋯,G); the matrix $\mathbf{F}_{\mathrm{FRO}}$ with dimensions $G \times (L+1)$ is named as the frequency response operator (FRO) and is defined as $\mathbf{F}_{\mathrm{FRO}} = \mathbf{F}_{\mathrm{DFT}}(:,\ 1:L+1)$,which means that matrix $\mathbf{F}_{\mathrm{FRO}}$ consists of the first (L+1) column vectors of the DFT matrix $\mathbf{F}_{\mathrm{DFT}}$.

We define the frequency domain IBI-free uplink received data vector on the reference element 1 of the base station antenna array from the *m*-th transmit antenna of the *k*-th user as

$$\widetilde{\mathbf{x}}_{1,k}^{(m)} = diag(\mathbf{c}_k^{(m)})\,\boldsymbol{\eta}_k^{(m)}\, t_k^{(m)} = \mathbf{f}_k^{(m)}\mathbf{h}_k^{(m)} t_k^{(m)} \tag{5}$$

where $\mathbf{f}_k^{(m)} = diag(\mathbf{c}_k^{(m)})\mathbf{F}_{\mathrm{FRO}}$, $t_k^{(m)}$ is the transmitted symbol from the *m*-th transmit antenna of the *k*-th user.

Successively, on the reference element 1, the received data vector consisting of received signals from all K active users over each subcarrier is

$$\widetilde{\mathbf{x}}_1 = \sum_{k=1}^{K}\sum_{m=1}^{2}\widetilde{\mathbf{x}}_{1,k}^{(m)} + \widetilde{\mathbf{n}}_1 = F_1\mathbf{t} + \widetilde{\mathbf{n}}_1 \tag{6}$$

where $\mathbf{t}$ consists of transmitted symbols from all transmit antennas of K users, namely, $\mathbf{t} = \left[t_1^{(1)} \quad t_1^{(2)} \quad \cdots \quad t_K^{(1)} \quad t_K^{(2)} \right]^{\mathrm{T}}$; $\widetilde{\mathbf{n}}_1$ is a vector of independent identically distributed (i.i.d) complex zero-mean Gaussian noises with variance σ_n^2 ; F_1 is a $G \times 2K$ matrix, and is defined as $\mathbf{F}_1 = \left[\mathbf{f}_1^{(1)}\mathbf{h}_1^{(1)} \quad \mathbf{f}_1^{(2)}\mathbf{h}_1^{(2)} \quad \cdots \quad \mathbf{f}_K^{(1)}\mathbf{h}_K^{(1)} \quad \mathbf{f}_K^{(2)}\mathbf{h}_K^{(2)} \right]$.

Likewise, on the *r*-th element of receive antenna array, the received data vector is written as

$$\widetilde{\mathbf{x}}_r = \sum_{k=1}^{K}\sum_{m=1}^{2}\widetilde{\mathbf{x}}_{r,k}^{(m)} + \widetilde{\mathbf{n}}_r = F_r\mathbf{t} + \widetilde{\mathbf{n}}_r \tag{7}$$

where the matrix F_r is defined as $\mathbf{F}_r = \left[\mathbf{f}_1^{(1)}\mathbf{h}_1^{(1)} e^{j\chi(r-1)\sin\theta_1} \quad \mathbf{f}_1^{(2)}\mathbf{h}_1^{(2)} e^{j\chi(r-1)\sin\theta_1} \quad \cdots \quad \mathbf{f}_K^{(1)}\mathbf{h}_K^{(1)} e^{j\chi(r-1)\sin\theta K} \quad \mathbf{f}_K^{(2)}\mathbf{h}_K^{(2)} e^{j\chi(r-1)\sin\theta K} \right]$, and $\chi = 2\pi\, d/\lambda$. Notations λ , d and θ_k represent the wavelength of Radio Frequency (RF) carrier, the inter-element spacing and the incident angle with respect to the array normal, respectively. Note that, like the assumption in [1], since the angle spread in macrocell case is very small, the DOAs of different multipaths for all subcarriers associate with user k is approximately identical. Therefore, θ_k practically denotes the DOA of the cluster consisting of different multipaths for all sub-carriers associated with user *k'* two transmit antennas.

3. DOA-MATRIX METHOD FOR SPACE-TIME CHANNEL ESTIMATION

3.1Dual Subarrays and Extended Received Data Vectors

Concatenating uplink received data vectors from element 1 to $M-1$, an extended data vector can be obtained, that is

$$\mathbf{u} = \left[\widetilde{\mathbf{x}}_1^{\mathrm{T}} \quad \widetilde{\mathbf{x}}_2^{\mathrm{T}} \quad \cdots \quad \widetilde{\mathbf{x}}_{M-1}^{\mathrm{T}} \right]^{\mathrm{T}} = \mathbf{A}\mathbf{t} + \widetilde{\mathbf{n}}_{\mathrm{head}} \tag{8}$$

where $\mathbf{u}$ is a $(M-1)G \times 1$ vector; $\widetilde{\mathbf{n}}_{\mathrm{head}}$ is given by $\widetilde{\mathbf{n}}_{\mathrm{head}} = \left[\widetilde{\mathbf{n}}_1^{\mathrm{T}} \quad \widetilde{\mathbf{n}}_2^{\mathrm{T}} \quad \cdots \quad \widetilde{\mathbf{n}}_{M-1}^{\mathrm{T}} \right]^{\mathrm{T}}$; $\mathbf{A}$ is a $(M-1)G \times 2K$ matrix, and is given by

$$\begin{aligned}\mathbf{A} &= \left[\mathbf{F}_1^{\mathrm{T}} \quad \mathbf{F}_2^{\mathrm{T}} \quad \cdots \quad \mathbf{F}_{M-1}^{\mathrm{T}} \right]^{\mathrm{T}} \\ &= \left[(\mathbf{I}_{M-1} \otimes \mathbf{f}_1^{(1)})(\mathbf{a}_1 \otimes \mathbf{h}_1^{(1)}) \quad (\mathbf{I}_{M-1} \otimes \mathbf{f}_1^{(2)})(\mathbf{a}_1 \otimes \mathbf{h}_1^{(2)}) \quad \cdots \right.\end{aligned}$$

$$(\mathbf{I}_{M-1}\otimes\mathbf{f}_K^{(1)})(\mathbf{a}_K\otimes\mathbf{h}_K^{(1)}) \quad (\mathbf{I}_{M-1}\otimes\mathbf{f}_K^{(2)})(\mathbf{a}_K\otimes\mathbf{h}_K^{(2)})] \\ =[\mathbf{A}_1^{(1)}\mathbf{g}_1^{(1)} \quad \mathbf{A}_1^{(2)}\mathbf{g}_1^{(2)} \quad \cdots \quad \mathbf{A}_K^{(1)}\mathbf{g}_K^{(1)} \quad \mathbf{A}_K^{(2)}\mathbf{g}_K^{(2)}] \tag{9}$$

where $\mathbf{I}_{M-1}$ is an $(M-1)\times(M-1)$ identity matrix; $\otimes$ denotes the Kronecker product; $\mathbf{A}_k^{(m)}=\mathbf{I}_{M-1}\otimes\mathbf{f}_k^{(m)}$, $\mathbf{g}_k^{(m)}=\mathbf{a}_k\otimes\mathbf{h}_k^{(m)}$ ($k=1,\cdots,K$; $m=1,2$); $\mathbf{a}_k$ is an $(M-1)\times1$ steering vector for the k-th user, and is defined as $\mathbf{a}_k=[1 \quad e^{j\chi\sin\theta k} \quad \cdots \quad e^{j\chi(M-2)\sin\theta k}]^{\mathrm{T}}$. As shown in equation (9), $\mathbf{g}_k^{(m)}$ is the Kronecker product between the user's steering vector and corresponding FIR channel, so it is named as the space-time channel vector.

When concatenating uplink-received data vectors from element 2 to M, another extended data vector can be obtained, i.e.

$$\mathbf{y}=[\widetilde{\mathbf{x}}_2^{\mathrm{T}} \quad \widetilde{\mathbf{x}}_3^{\mathrm{T}} \quad \cdots \quad \widetilde{\mathbf{x}}_M^{\mathrm{T}}]^{\mathrm{T}}=\mathbf{A}\,\mathbf{\Phi}\mathbf{t}+\widetilde{\mathbf{n}}_{\text{tail}} \tag{10}$$

where $\mathbf{\Phi}$ is a $2K\times2K$ diagonal matrix, and is given by

$$\Phi=diag(e^{j\chi\sin\theta_1},e^{j\chi\sin\theta_1},\cdots,e^{j\chi\sin\theta K},e^{j\chi\sin\theta K}) \tag{11}$$

Terms on the main diagonal of the matrix $\mathbf{\Phi}$ are associated with users' DOAs. Hence, we call them DOA items. $\widetilde{\mathbf{n}}_{\text{tail}}$ is given by $\widetilde{\mathbf{n}}_{\text{tail}}=[\widetilde{\mathbf{n}}_2^{\mathrm{T}} \quad \widetilde{\mathbf{n}}_3^{\mathrm{T}} \quad \cdots \quad \widetilde{\mathbf{n}}_M^{\mathrm{T}}]^{\mathrm{T}}$.

3.2 Auxiliary Matrix Model for Joint DOA and Space-Time Vector Channel Estimation

Now, the auto-correlation matrix of $\mathbf{u}$ and the cross-correlation matrix between $\mathbf{y}$ and $\mathbf{u}$ are defined as

$$\mathbf{R}_{\mathbf{uu}}=\mathrm{E}[\mathbf{uu}^{\mathrm{H}}]=\mathbf{A}\mathrm{E}[\mathbf{tt}^{\mathrm{H}}]\mathbf{A}^{\mathrm{H}}+\sigma_n^2\mathbf{I}_{(M-1)G} \\ =\mathbf{A}\mathbf{R}_{tt}\mathbf{A}^{\mathrm{H}}+\sigma_n^2\mathbf{I}_{(M-1)G}=\mathbf{R}_{\mathbf{uuo}}+\sigma_n^2\mathbf{I}_{(M-1)G} \tag{12}$$

$$\mathbf{R}_{\mathbf{yu}}=\mathrm{E}[\mathbf{yu}^{\mathrm{H}}]=\mathbf{A\Phi}\mathrm{E}[\mathbf{tt}^{\mathrm{H}}]\mathbf{A}^{\mathrm{H}}+\mathrm{E}[\widetilde{\mathbf{n}}_{\text{tail}}\widetilde{\mathbf{n}}_{\text{head}}^{\mathrm{H}}] \\ =\mathbf{A\Phi}\mathbf{R}_{tt}\mathbf{A}^{\mathrm{H}}+\sigma_n^2\mathbf{J}_{(M-1)G}=\mathbf{R}_{\mathbf{yuo}}+\sigma_n^2\mathbf{J}_{(M-1)G} \tag{13}$$

where $\mathbf{R}_{tt}$ denotes the auto-correlation matrix of the symbol vector $\mathbf{t}$; $\mathbf{I}$ is an $(M-1)G\times(M-1)G$ identity matrix; $\mathbf{J}$ is an $(M-1)G\times(M-1)G$ matrix whose terms on the G-th diagonal above the main diagonal are all ones.

When original symbols from different users are uncorrelated, $\mathbf{R}_{tt}$ is a nonsingular matrix, and the rank of $\mathbf{R}_{\mathbf{uuo}}$ equals to $2K$. Performing eigen decomposition on $\mathbf{R}_{\mathbf{uuo}}$ can obtain

$$\mathbf{R}_{\mathbf{uuo}}=\sum_{l=1}^{(M-1)G}\mu_l\mathbf{v}_l\mathbf{v}_l^{\mathrm{H}} \tag{14}$$

where μ_l and $\mathbf{v}_l$ are eigenvalues and corresponding eigenvectors of $\mathbf{R}_{\mathbf{uuo}}$, respectively.

When $\mathbf{A}$ is full column-rank, $\mathbf{R}_{tt}$ is nonsingular, and $(M-1)G>2K$, two following properties hold:

1) $\mu_1\geq\cdots\geq\mu_{2K}>\mu_{2K+1}=\cdots=\mu_{(M-1)G}=0$, and we have $\mathbf{R}_{\mathbf{uuo}}=\sum_{l=1}^{2K}\mu_l\mathbf{v}_l\mathbf{v}_l^{\mathrm{H}}$;

2) $Span\{\mathbf{v}_{2K+1},\mathbf{v}_{2K+2},\cdots,\mathbf{v}_{(M-1)G}\}\perp Range\{\mathbf{A}\}$.

Based on the above two properties, we can obtain following equations:

$$\mathbf{A}^{\mathrm{H}}[\sum_{l=2K+1}^{(M-1)G}\mathbf{v}_l\mathbf{v}_l^{\mathrm{H}}]\mathbf{A}=\mathbf{O}_{2K} \tag{15}$$

$$\mathbf{A}^{\mathrm{H}}[\sum_{l=1}^{2K}\mathbf{v}_l\mathbf{v}_l^{\mathrm{H}}]\mathbf{A}=\mathbf{A}^{\mathrm{H}}[\sum_{l=1}^{2K}\mathbf{v}_l\mathbf{v}_l^{\mathrm{H}}+\sum_{l=2K+1}^{(M-1)G}\mathbf{v}_l\mathbf{v}_l^{\mathrm{H}}]\mathbf{A}=\mathbf{A}^{\mathrm{H}}\mathbf{A} \tag{16}$$

We define an auxiliary matrix by $\mathbf{R}_{\mathbf{uuo}}$ and $\mathbf{R}_{\mathbf{yuo}}$ as in [5]

$$\mathbf{R}=\mathbf{R}_{\mathbf{yuo}}\mathbf{R}_{\mathbf{uuo}}^{+} \tag{17}$$

where $\mathbf{R}_{\mathbf{uuo}}^{+}$ is the Penrose-Moore pseudo-inverse of $\mathbf{R}_{\mathbf{uuo}}$, and is defined by $\mathbf{R}_{\mathbf{uuo}}^{+}=\sum_{l=1}^{2K}\frac{1}{\mu_l}\mathbf{v}_l\mathbf{v}_l^{\mathrm{H}}$.

Theorem 1 Given $\mathbf{A}$ is column full-rank, $\mathbf{R}_{tt}$ is nonsingular, and there are not identical terms in the main diagonal of $\mathbf{\Phi}$, the eigenvalues and corresponding eigenvectors of the matrix $\mathbf{R}$ are DOA items and the space-time channel vectors, respectively, namely, $\mathbf{RA}=\mathbf{A\Phi}$.

The detailed proof of this theorem is similar to that presented in [5]. For briefness, it is omitted here.

Base on Theorem 1, $2K$ eignevalues can be obtained via eigen decomposition on matrix $\mathbf{R}$. These eigenvalues are DOA items associated with K different users. As we have assumed that DOAs associated with two transmit antennas of the k-th user are identical, $2K$ eigenvalues in deed include K different values, which are $e^{j\chi\sin\theta_1},\cdots,e^{j\chi\sin\theta K}$, and each of them with multiplicity of two. Meanwhile, the eigenvectors corresponding to the same eigenvalue are no longer the space-time channel vectors associated with two transmit antennas of a user respectively, but the linear combination of them. We define a matrix consisting of the two eigenvectors corresponding to the same eigenvalue as $\widetilde{\mathbf{A}}_k$ ($k=1,\cdots,K$), and the following relationship exists

$$\widetilde{\mathbf{A}}_k=[\mathbf{A}_k^{(1)}\mathbf{g}_k^{(1)} \quad \mathbf{A}_k^{(2)}\mathbf{g}_k^{(2)}]\,\mathbf{F} \tag{18}$$

where $\widetilde{\mathbf{A}}_k$ is a column full-rank $(M-1)G\times2$ matrix; F is an unknown full-rank 2×2 matrix.

Thus, by performing SVD on $\widetilde{\mathbf{A}}_k$, we can obtain

$$\widetilde{\mathbf{A}}_k=[\mathbf{U}_{k,s} \quad \mathbf{U}_{k,o}]\begin{bmatrix}\mathbf{\Sigma}_k\\ \mathbf{0}\end{bmatrix}\mathbf{V}_{k,s}{}^{\mathrm{H}} \tag{19}$$

where $\mathbf{U}_{k,s}$ is an $(M-1)G\times2$ matrix; $\mathbf{U}_{k,o}$ is an $(M-1)G\times((M-1)G-2)$ matrix; $\mathbf{\Sigma}_k$ is a 2×2 matrix; $\mathbf{0}$ is an $((M-1)G-2)\times2$ zero matrix; $\mathbf{V}_{k,s}{}^{\mathrm{H}}$ is a 2×2 matrix.

Because $\mathbf{U}_{k,o}{}^{\mathrm{H}}\perp Range\{\widetilde{\mathbf{A}}_k\}$, we have

$$\mathbf{U}_{k,o}{}^{\mathrm{H}}\mathbf{A}_k{}^{(m)}\mathbf{g}_k{}^{(m)}=\mathbf{0} \tag{20}$$

where $m=1,2$; 0 is an $((M-1)G-2)\times1$ zero vector.

Clearly, in this matrix equation, there are $(M-1)(L+1)$ unknowns while there are $((M-1)G-2)$ linear equations. When $((M-1)G-2)\geq(M-1)(L+1)$, the above linear equation set is overdeterminated. For different matrix $\mathbf{A}_k{}^{(m)}$, by solving (20), we can estimate the corresponding uplink space-time channel vector $\hat{\mathbf{g}}_k{}^{(m)}$ for $k=1,\cdots,K$, and $m=1,2$, respectively. It should be noted that, an arbitrary complex coefficient exits between the estimated space-time channel vector $\hat{\mathbf{g}}_k^{(m)}$ with the original one, which can be removed by use of the finite alphabet property of transmitted symbols. The detailed algorithm can be found in [6]. For briefness, it is omitted here.

4. SIMULATION RESULTS

Extensive computer simulations have been conducted to demonstrate the performance of our algorithm. The performance is evaluated by the mean square error (MSE) of channel identification, which is defined as

$$MSE = \frac{1}{N_t}\sum_{i=1}^{N_t}\left\|\hat{\mathbf{H}}(i)-\mathbf{H}\right\|_F^2 \Big/ \left\|\mathbf{H}\right\|_F^2$$

where N_t is the number of Monte-Carlo trials in the simulation; the channel matrix H with dimensions $(M-1)(L+1)\times 2K$ is defined as $\mathbf{H} = \left[\mathbf{g}_1^{(1)} \quad \mathbf{g}_1^{(2)} \quad \cdots \quad \mathbf{g}_K^{(1)} \quad \mathbf{g}_K^{(2)}\right]$; the matrix $\hat{\mathbf{H}}(i)$ is the estimate of H from the i-th trial; the operator $\|\bullet\|_F$ depicts the Frobenius norm.

We use the MC-CDMA system with the differential four-phase shift keying (D4PSK) modulation mode in our simulations, and 50 Monte-Carlo trials are performed for each simulation. Hadamard codes with length G = 32 are assigned to different users. The channel impulse responses are generated from independent complex Gaussian random variables. We use samples within 50 MC-CDMA symbols to estimate the auto- and the cross-correlation matrices of the uplink-received data sequence, which are the approximate estimation of auto- and cross-correlation matrices in ensemble-average sense.

The SNR per receive antenna is fixed to 18 dB, and the number of active users changes from 1 to 15. Fig. 2 (a) shows the MSE versus the number of users with antenna array of 2 and 4 elements, respectively. Fig. 2 (b) shows the DOA estimation error versus the number of active users. Clearly, as the number of array elements increases, the system performance increases. Nevertheless, the system performance is degraded as the number of user increases.

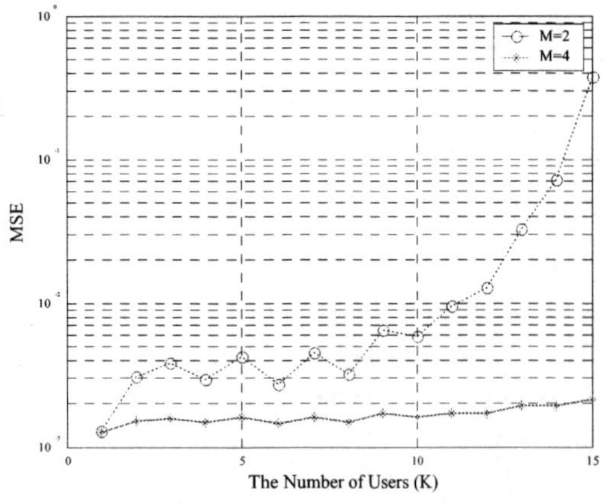

Fig.2 (a) the MSE versus the number of users

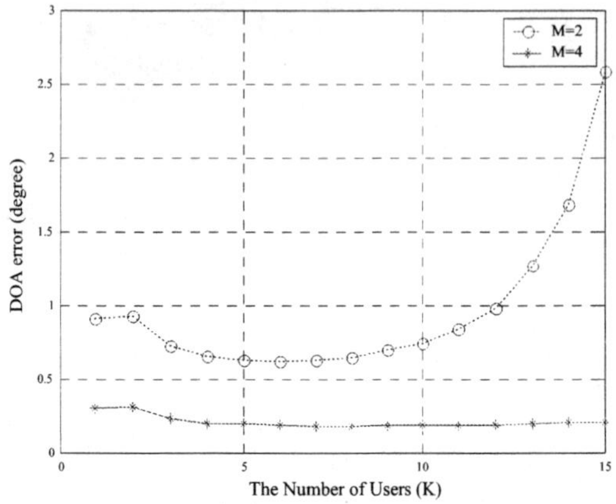

Fig.2 (b) the estimation error of DOA versus the number of users

5. CONCLUTIONS

We have presented a blind joint estimation algorithm of space-time vector channel and DOA for space-time coded MC-CDMA system equipped with a ULA at the base station. Via this algorithm, the closed-form solutions of the space-time channels and DOAs can be obtained without training sequences. Moreover, comparing with the MC-CDMA system with a single antenna at the base station, our algorithm is more robust against noise and can well suppress MUI in multiuser scenarios. In summary, with the proposed channel estimator, combing space-time coded MC-CDMA with bade station antenna arrays is a promising technique for efficiently high-data-rate transmission over mobile wireless channels.

6. REFERENCES

[1] J. Li, K. B. Letaief and Z. Cao, "Co-channel interference cancellation for space-time coded OFDM systems," IEEE Trans. Wireless Commun, vol.2, pp41-49, Jan. 2003.

[2] W. Sun and H.B. Li, "A subspace-based channel identification algorithm for forward link in space-time coded MC-CDMA systems," Proc.IEEE WCNC 2002, Orlando, USA, vol.1, pp. 445-448, Mar. 2002.

[3] X.Y. Hu and Y.H. Chew, "Performance of space-time block coded MC-CDMA system over frequency selective fading channel using semi-blind channel estimation technique," Proc.IEEE WCNC 2003, New Orleans,USA, vol. 1, pp. 414-419, Mar. 2003.

[4] S. Hara and R. Prasad, "Design and performance of multicarrier CDMA systems in frequency-selective Reyleigh fading channels," IEEE Trans. Veh. Technol., vol. 48, pp 1584-1594, Sep. 1999.

[5] Q.Y. Yin, R.W. New comb and L.H. Zou, "Estimating 2-D angles of arrival via two parallel linear array," Proc. IEEE ICASSP 1989, Glasgow, Scotland, pp. 2803-2806, May 1989.

[6] X.J. Wu, Q.Y. Yin and A.G. Feng, "Equivalently blind time-domain channel estimation for MC-CDMA system over frequency-selective fading channels in multiuser scenario," Proc. IEEE VTC 2001 Fall, Atlantic city, USA, vol. 4, pp. 2687-2691, Nov. 2001.

A POLYNOMIAL METHOD FOR BLIND IDENTIFICATION OF MIMO CHANNELS

Weizhou Su and Wei Xing Zheng

School of QMMS, University of Western Sydney
Penrith South DC NSW 1797, Australia
Email: w.su@uws.edu.au w.zheng@uws.edu.au

ABSTRACT

In this paper, the blind identification problem for a multiple-input multiple-output (MIMO) channel is investigated by applying the spectral factorization method. The existing conditions on blind identification from signal's second order statistics (SOS) are relaxed by introducing an MIMO pre-filter in the transmitting side of the channel. A design criterion of the pre-filter in the frequency domain is given. Based on the properties of this pre-filter, a blind identification algorithm from SOS is developed. Simulation results show that this algorithm is effective in modelling MIMO FIR channels.

1. INTRODUCTION

Recently, much research work has been done on blind identification in signal process and telecommunication areas (see, [10] and [14]). This problem is to model data transmission channels from received signals. Usually, the data transmission channels in telecommunication systems are MIMO channels (i.e., there are multiple users and multiple sensors or multiple received signals by oversampling in the systems) and have a finite impulse response (FIR). These features may cause two kinds of interferences - multiple access interference (MAI) and intersymbol interference (ISI). With rapid increase of the data transmission rate and the number of users in communication systems, these interferences seriously affect the quality of signal transmission. The conventional method of suppressing these interferences is to model the channel by using a "training sequence" and then to design an equalizer for the channel. However, in real applications, such as mobile communication systems, transmission channels are time-varying channels so that considerable training data have to be transmitted for estimating channel model, which is practically undesirable. To improve the quality and efficiency of signal transmission, several blind identification and equalization methods (see, [10] and [14]) are developed in the channel modelling by only using received signals.

This work was supported by a Research Grant from the Australian Research Council.

In [1], [3], [13] and [15], blind identification from second order statistics (SOS) of received signals is discussed. It is shown that if the transfer function of an MIMO FIR channel is irreducible and column-reduced, it can be determined up to a unitary matrix. On the other hand, Liu and his colleagues propose a new method, so-called *Auto-Correlation Matching Principle* in the time domain, for blind equalization in [7]-[9]. The key idea of this method is to introduce a pre-filter in the transmitting side of a channel. If the pre-filter is designed properly, then the channel can be equalized from the information of the pre-filter and SOS of the received signals. It should be noted that lots of researches on blind identification and blind equalization of MIMO FIR systems with colored inputs using second-order statistics are available (see, e.g., [4], [6]). Moreover, the recent work on optimum prefiltering can be found in [11]-[12].

In this paper, the conditions on blind identification from the received signal's SOS are reviewed by using the spectral factorization method. Then a pre-filter is introduced in the transmitting side of a channel. To provide sufficient information for channel identification, a design criterion of the pre-filter is presented in the frequency domain. It is shown that, with this pre-filter, the identifiable conditions can be significantly relaxed. Finally, a blind identification algorithm is developed. Simulation results are presented to illustrates the effectiveness of this algorithm.

2. PROBLEM AND PRELIMINARY

A multi-user communication system can be modelled as follows:

$$y(n) = \sum_{l=0}^{L} H(l)v(n-l) + w(n), \quad t \in \mathbb{Z} \qquad (1)$$

where $y(n) \in \mathbb{R}^M$ is a received signal vector from M sensors, $v(n) \in \mathbb{R}^m$ is a transmitted signal vector from m users, $w(n) \in \mathbb{R}^M$ is an additive white noise, $\{H(0), H(1), \cdots, H(L)\}$, $H(l) \in \mathbb{R}^{M\times m}$ is the impulse response of the channel. Denote the i-th user's signal by $v_i(n)$. The transmission channel from $v_i(n)$ to $y(n)$ is called the i-th subchannel of the channel (1). Let the impulse response of the i-th subchannel be $\{h_i(0), h_i(1), \cdots, h_i(L_i)\}$. Then the transfer functions of the channel (1) and the i-th subchannel are

$$\mathbf{H}(z) = \sum_{l=0}^{L} H(l) z^{-l} \quad (2)$$

and

$$\mathbf{h_i}(z) = \sum_{l=0}^{L_i} h_i(l) z^{-l}, \quad i = 1, \cdots, m$$

respectively. Here $L_i \leq L$, $i = 1, \cdots, m$. So $\mathbf{H}(z) = [\mathbf{h}_1(z), \mathbf{h}_2(z), \cdots, \mathbf{h}_m(z)]$ and $H(l) = [h_1(l), h_2(l), \cdots, h_m(l)]$, $l = 0, 1, \cdots, L$.

There exist various methods in identifying the parameter matrices $H(0), \cdots, H(L)$. One of the efficient ways is the subspace method based on second order statistics of the received signals. In this case, it is assumed that

A1 *The transmitted signal $v(n)$, $n \in \mathbb{Z}$ is an* i.i.d. *(independent and identically distributed) signal and $w(n) \in \mathbb{R}^M$ is a white noise.*

A2 *$\mathbf{H}(z)$ is irreducible, i.e., $\mathbf{H}(z)$ has full column rank for all $z \neq 0$.*

A3 *$\mathbf{H}(z)$ is column-reduced, i.e., $[h_1(L_1), \cdots, h_m(L_m)]$ has full column rank.*

Suppose that $\mathcal{R}_M$ is the set of all $M \times 1$ rational functions. Then $\mathcal{R}_M$ is an M-dimensional space. For any m-dimensional subspace $\mathcal{H}$ of $\mathcal{R}_M$, there always exist some polynomial bases $\{\mathbf{b}_1(z), \cdots, \mathbf{b}_m(z)\}$. If the sum of degree $\{\mathbf{b}_1(z), \cdots, \mathbf{b}_m(z)\}$, i.e., $\sum_{i=1}^{m} \deg(\mathbf{b}_i(z))$, is minimal, it is called a *minimal polynomial basis* (MPB) of $\mathcal{H}$. In fact, if the polynomial matrix $\mathbf{H}(z)$ satisfies A2 and A3, then the columns of $\mathbf{H}(z)$ is an MPB of the subspace *span* $\{\mathbf{h}_1(z), \cdots, \mathbf{h}_m(z)\}$.

The following lemmas show some useful properties of the MPB.

Lemma 2.1. *Suppose $\mathbf{B}_1(z)$ and $\mathbf{B}_2(z)$ are $M \times m$ polynomial matrices and columns of $\mathbf{B}_1(z)$ and $\mathbf{B}_2(z)$ are two MPB of $\mathcal{H}$. $L_1 \leq \cdots \leq L_m$ are the degrees of the columns of $\mathbf{B}_1(z)$ and $\mathbf{B}_2(z)$. $L_1, \cdots, L_m$ take different values $\bar{L}_1, \cdots, \bar{L}_d$. ν_i, $i = 1, \cdots, d$ are the numbers of $L_1, \cdots, L_m$ equal to $\bar{L}_i$, $i = 1, \cdots, d$. Then there exists $\mathbf{T}(z)$ such that*

$$\mathbf{B}_1(z) = \mathbf{B}_2(z)\mathbf{T}(z) \quad (3)$$

where

$$\mathbf{T}(z) = \begin{bmatrix} \mathbf{T}_{11}(z) & \mathbf{T}_{12}(z) & \cdots & \mathbf{T}_{1d}(z) \\ 0 & \mathbf{T}_{22}(z) & \cdots & \mathbf{T}_{2d}(z) \\ & \cdots & \cdots & \\ 0 & 0 & \cdots & \mathbf{T}_{dd}(z) \end{bmatrix} \quad (4)$$

and $\mathbf{T}_{ij}(z)$, $i, j = 1, \cdots, d$ are $\nu_i \times \nu_j$ matrices, the degree of $\mathbf{T}_{ij}(z)$ is equal to $\bar{L}_j - \bar{L}_i$.

Proof: See [3]. □

This lemma leads to a well known result.

Lemma 2.2. *Assume that the noise in the communication system (1) is null; the channel transfer function $\mathbf{H}(z)$ satisfies assumptions A1-A3; and the degrees of all subchannels are known and different. Then, for an $M \times m$ irreducible and column-reduced spectral factor $\hat{\mathbf{H}}(z)$ of $\mathbf{S}_{yy}(z)$, there exists a unique factorization such that*

$$\mathbf{S}_{yy}(z) = \hat{\mathbf{H}}(z)\hat{\mathbf{H}}^{\sim}(z) \quad (5)$$

where $\hat{\mathbf{H}}^{\sim}(z) = \hat{\mathbf{H}}^{*}(1/\bar{z})$.

If the channel is not column-reduced, we have the following result:

Lemma 2.3. *Suppose that $\mathbf{H}_1(z)$ and $\mathbf{H}_2(z)$ are bases of an m-dimensional subspace $\mathcal{H}$ but not MPBs of $\mathcal{H}$. Let the degrees of $\mathbf{H}_1(z)$ and $\mathbf{H}_2(z)$ be L. Then there exists a unimodular matrix $\mathbf{T}(z)$ such that*

$$\mathbf{H}_1(z) = \mathbf{H}_2(z)\mathbf{T}(z). \quad (6)$$

Moreover, the degree of $\mathbf{T}(z)$ satisfies

$$\deg(\mathbf{T}(z)) \leq m(L - L_1) \quad (7)$$

where L_1 is the lowest degree in an MPB of $\mathcal{H}$.

Proof: See [5]. □

Lemmas 2.2 and 2.3 show that assumption A3 is necessary in identifying an MIMO channel. However, in practice, this assumption would not be satisfied. To avoid this problem, some interesting methods are proposed. Intuitively, if a proper pre-filter in the transmitting side of the channel is introduced, then the knowledge from the pre-filter can provide more information for channel identification. In this paper, we introduce a pre-filter $\mathbf{P}(z)$ in the transmitting side of the channel. More precisely, the data transmission channel is modelled as

$$\mathbf{y}(z) = \mathbf{H}(z)\mathbf{s}(z) \quad (8)$$

and

$$\mathbf{s}(z) = \mathbf{P}(z)\mathbf{v}(z) \quad (9)$$

where $\mathbf{y}(z)$, $\mathbf{s}(z)$ and $\mathbf{v}(z)$ are z-transformation of the received signal $y(n)$, the pre-filter output $s(n)$ and the transmitted signal $v(n)$, respectively. The problem under consideration in this paper is how to design this pre-filter and apply the information from this filter to channel identification.

3. BLIND IDENTIFICATION VIA PRE-FILTERS

In this section, we propose a design criterion of the pre-filter $\mathbf{P}(z)$. Then some channel identifiable conditions are discussed and a blind identification algorithm is developed.

Now, select the polynomials $\mathbf{r}_1(z), \cdots, \mathbf{r}_m(z)$ and $\mathbf{p}_1(z), \cdots, \mathbf{p}_m(z)$ such that

C1 *The degrees of polynomial $\mathbf{r}_i(z)$, $i = 1, \cdots, m$ are K_i, $i = 1, \cdots, m$, and $\mathbf{r}_i(z)$, $i = 1, \cdots, m$ are mutually coprime.*

C2 *$\mathbf{r}_i^{\sim}(z)$ and $\mathbf{r}_j(z)$, $j \neq i$ are coprime.*

C3 *The polynomials $\mathbf{p}_1(z), \cdots, \mathbf{p}_m(z)$ are given by*

$$\mathbf{p}_i(z) = \prod_{j \neq i} \mathbf{r}_j(z), \quad i = 1, \cdots, m.$$

Then the pre-filter is chosen as follow:

$$\mathbf{P}(z) = \operatorname{diag}\{\mathbf{p}_1(z), \cdots, \mathbf{p}_m(z)\}.$$

Lemma 3.1. *Suppose that the polynomials $\mathbf{r}_1(z), \cdots, \mathbf{r}_m(z)$ and $\mathbf{p}_1(z), \cdots, \mathbf{p}_m(z)$ satisfy C1 and C3, respectively. If polynomials $\mathbf{q}_1(z), \cdots, \mathbf{q}_m(z)$ satisfy*

$$\mathbf{q}_1(z)\mathbf{p}_1(z) + \cdots + \mathbf{q}_m(z)\mathbf{p}_m(z) = 0, \quad (10)$$

and the degrees of the polynomials $\mathbf{q}_1(z), \cdots, \mathbf{q}_m(z)$ are smaller than $K_1, \cdots, K_m$, respectively, then $\mathbf{q}_1(z) = \cdots = \mathbf{q}_m(z) \equiv 0$, $\forall z \in \mathbb{C}$.

Proof: See Theorem 2.6.1 in [2]. □

Lemma 3.2. *Suppose that the polynomials* $\mathbf{r}_1(z), \cdots, \mathbf{r}_m(z)$ *and* $\mathbf{p}_1(z), \cdots, \mathbf{p}_m(z)$ *satisfy C1, C2 and C3, respectively;* $\mathbf{T}(z)$ *is an* $m \times m$ *polynomial matrix; the degrees of the columns in* $\mathbf{T}(z)$ *are* $K_i - 1$, $i = 1, \cdots, m$*; and the matrix* $\mathbf{S}_{vv}(z) = \text{diag}\{\mathbf{p}_1(z)\mathbf{p}_1^{\sim}(z) \cdots \mathbf{p}_m(z)\mathbf{p}_m^{\sim}(z)\}$*. Then the equation*

$$\mathbf{T}(z)\mathbf{S}_{vv}(z)\mathbf{T}^{\sim}(z) = \mathbf{S}_{vv}(z) \tag{11}$$

has a unique solution $\mathbf{T}(z) = I$.

Proof: Since the diagonal terms of both sides in (11) are equal, there holds

$$\sum_{j=1}^{m} \mathbf{t}_{ij}(z)\mathbf{p}_j(z)\mathbf{p}_j^{\sim}(z)\mathbf{t}_{ij}^{\sim}(z) = \mathbf{p}_i(z)\mathbf{p}_i^{\sim}(z), \quad i=1,\cdots,m \tag{12}$$

where $\mathbf{t}_{ij}(z)$ is the ij-th element of the matrix $\mathbf{T}(z)$.

Applying Lemma 3.1 to (12), it can be easily verified that $\mathbf{t}_{ij}(z) = 0$, $j \neq i$ and $\mathbf{t}_{ii}(z) = 1$, i.e., $\mathbf{T}(z) = I$. □

Note that Lemma 3.2 is a polynomial form of the *Linear Shift-Independence Condition* proposed in [8] and [9]. This lemma leads to one of our main results. It shows that the pre-filter reduces the sensitivity of an identification algorithm to the knowledge of the degrees of the channel.

Theorem 3.1. *Suppose that the FIR transfer function* $\mathbf{H}(z)$ *of the channel (1) satisfies assumptions A1-A3; the lowest and highest degrees of* $\mathbf{H}(z)$, $\bar{L}_1$ *and* $\bar{L}_m$*, are unknown and only an upper bound* $\bar{L}$ *of* $\bar{L}_m - \bar{L}_1$ *is known; and the noise in the channel is null. If the pre-filter* $\mathbf{P}(z) = \text{diag}\{\mathbf{p}_i(z)\}$*; the polynomials* $\mathbf{p}_i(z)$, $i = 1, \cdots, m$ *and* $\mathbf{r}_i(z)$, $i = 1, \cdots, m$ *satisfy conditions C1-C3, respectively; and the degrees of* $\mathbf{r}_i(z)$, $i = 1, \cdots, m$ *are chosen to be* $(\bar{L}+1)$*, then the transfer function of the channel (1) can be completely determined from SOS of the received signals and* $\mathbf{P}(z)$.

Proof: This theorem is proven by straightforwardly applying Lemma 2.1 and 3.2. Details are omitted due to limited space. □

In Theorem 3.1 and Lemma 2.2, the channel transfer function $\mathbf{H}(z)$ is assumed to be irreducible and column-reduced. The following theorem shows that these conditions can be relaxed via selecting a high order pre-filter $\mathbf{P}(z)$.

Theorem 3.2. *Suppose that the transmitted signal satisfies assumption A1; the upper bound of an MIMO FIR channel* $\mathbf{H}(z)$ *is* $L \leq M$*; and the noise in the channel is null. If the pre-filter* $\mathbf{P}(z) = \text{diag}\{\mathbf{p}_i(z)\}$*; the polynomials* $\mathbf{p}_i(z)$, $i = 1, \cdots, m$ *and* $\mathbf{r}_i(z)$, $i = 1, \cdots, m$ *satisfy conditions C1-C3, respectively; the degree of* $\mathbf{r}_i(z)$, $i = 1, \cdots, m$ *are chosen to be* $L+2$ *and* $\mathbf{r}_i(z)$ *has no multiple roots. Then the transfer function* $\mathbf{H}(z)$ *of the channel (1) can be completely determined up to a complex scalar from SOS of the received signals and* $\mathbf{P}(z)$.

Proof: Since by (9) the signal $s(n)$ is generated by the pre-filter P(z) in which the input signal $v(n)$ is a transmitted signal and is *i.i.d.*, we have

$$\mathbf{S}_{vv}(z) = \text{diag}\{\mathbf{p}_1(z)\mathbf{p}_1^{\sim}(z), \cdots, \mathbf{p}_m(z)\mathbf{p}_m^{\sim}(z)\}.$$

Denote the roots of $\mathbf{r}_i(z)$ by $z_{i1}, \cdots, z_{i(L+2)}$. For simplicity, all these roots are selected from on the unit circle. According to conditions C1-C3, it holds

$$\mathbf{S}_{vv}(z_{ik}) = \text{diag}\{0, \cdots, 0, \mathbf{p}_i(z_{ik})\mathbf{p}_i^*(z_{ik}), 0, \cdots, 0\}.$$

Consider the channel (1) in the case where the noise is null. Then the power spectral density of the signal y is given by

$$\mathbf{S}_{yy}(z_{ik}) = \mathbf{h}_i(z_{ik})\mathbf{p}_i(z_{ik})\mathbf{p}_i^*(z_{ik})\mathbf{h}_i^*(z_{ik}), \quad k = 1, \cdots, L+2. \tag{13}$$

Since the rank of $\mathbf{S}_{yy}(z_{ik})$ is one, there exists a vector a_{ik} such that

$$\mathbf{S}_{yy}(z_{ik}) = a_{ik}a_{ik}^*. \tag{14}$$

It follows from (13) and (14) that

$$\mathbf{h}_i(z_{ik})\mathbf{p}_i(z_{ik}) = a_{ik}c_{ik} \tag{15}$$

where c_{ik} is a scalar to be determined with $c_{ik}^*c_{ik} = 1$. Moreover, the parameter vectors of the vector polynomial $\mathbf{h}_i(z)$ is given by

$$[h_i(0), \cdots, h_i(L+1)]U_i = \left[\frac{a_{i1}c_{i1}}{\mathbf{p}_i(z_{i1})}, \frac{a_{i2}c_{i2}}{\mathbf{p}_i(z_{i2})}, \cdots, \frac{a_{i(L+1)}c_{i(L+1)}}{\mathbf{p}_i(z_{i(L+1)})}\right] \tag{16}$$

where

$$U_i = \begin{bmatrix} 1 & 1 & \cdots & 1 \\ z_{i1}^{-1} & z_{i2}^{-1} & \cdots & z_{i(L+1)}^{-1} \\ \vdots & \vdots & \cdots & \vdots \\ z_{i1}^{-L} & z_{i2}^{-L} & \cdots & z_{i(L+1)}^{-L} \end{bmatrix}.$$

To determine the parameters c_{ik}, $k = 1, \cdots, L+1$, let us consider the singular value decomposition of $\mathbf{S}_{yy}(z_{i(L+2)})$. By the same reason as that for $\mathbf{S}_{yy}(z_{ik})$, $k = 1, \cdots, L+1$, there exists a vector $a_{i(L+2)}$ such that $\mathbf{S}_{yy}(z_{i(L+2)}) = a_{i(L+2)}a_{i(L+2)}^*$ and

$$\mathbf{h}_i(z_{i(L+2)})\mathbf{p}_i(z_{i(L+2)}) = a_{i(L+2)}c_{i(L+2)} \tag{17}$$

where $c_{i(L+2)}$ is an undetermined parameter with $c_{i(L+2)}^*c_{i(L+2)} = 1$. Substituting (16) into (17) leads to

$$\left[\frac{a_{i1}c_{i1}}{\mathbf{p}_i(z_{i1})}, \frac{a_{i2}c_{i2}}{\mathbf{p}_i(z_{i2})}, \cdots, \frac{a_{iL_i}c_{i(L+1)}}{\mathbf{p}_i(z_{i(L+1)})}\right] u_i = \frac{a_{i(L+2)}c_{i(L+2)}}{\mathbf{p}_i(z_{i(L+2)})} \tag{18}$$

where $u_i = [u_{i1}, \cdots, u_{i(L+1)}]^T = U_i^{-1}[1 \;\; z_{i(L+2)}^{-1} \; \cdots \; z_{i(L+2)}^{-L}]^T$. Consequently, (18) can be written as

$$\left[\frac{a_{i1}u_{i1}}{\mathbf{p}_i(z_{i1})}, \frac{a_{i2}u_{i2}}{\mathbf{p}_i(z_{i2})}, \cdots, \frac{a_{iL_i}u_{i(L+1)}}{\mathbf{p}_i(z_{i(L+1)})}\right] \begin{bmatrix} c_{i1} \\ \vdots \\ c_{i(L+1)} \end{bmatrix} = \frac{a_{i(L+2))}c_{i(L+2)}}{\mathbf{p}_i(z_{i(L+2)})}.$$

Then, the parameters c_{ik}, $k = 1, \cdots, L+1$ are determined up to a scalar $c_{i(L+2)}$.

Noting that the polynomial $\mathbf{r}_i(z)$ has no multiple roots, the matrix U_i is nonsingular. From (16), the parameters of $\mathbf{h}_i(z)$ are determined up to a scalar $c_{i(L+2)}$. □

Algorithm of blind identification via spectral factorization

Step 1: Calculate $\mathbf{S}_{yy}(z)$ at the roots z_{ik}, $k = 1, \cdots, L+2$ of the polynomial $\mathbf{r}_i(z)$.

Step 2: Get the eigenvectors a_{ik} of $\mathbf{S}_{yy}(z_{ik})$, $k = 1, \cdots, L+2$ associated with the maximum eigenvalues.

Step 3: Determine the parameters $c_{i1}/c_{i(L+2)}$, $\cdots$, $c_{i(L+1)}/c_{i(L+2)}$ by solving (18).

Step 4: An estimate $[\bar{h}_i(0), \cdots, \bar{h}_i(L)]$ of the coefficient matrix of $\mathbf{h}_i(z)$ is obtained by solving (16).

Step 5: For a real coefficient channel, select any element a_0 of the estimation result $[\bar{h}_i(0), \cdots, \bar{h}_i(L)]$. Then an estimate of the coefficient matrix is given by

$$\mathrm{Re}\left\{[\bar{h}_i(0), \cdots, \bar{h}_i(L)]|a_0|/a_0\right\}.$$

There exists an important feature which leads to asymptotical convergence of the developed algorithm.

Theorem 3.3. *The estimate of a_{ik}, $i = 1, \cdots, m$, $k = 1, \cdots, L+2$ asymptotically converges to $\mathbf{h}_i(z_{ik})$ up to a complex scalar.*

Next the case where the additive noise $w(n)$ in (1) is not free is considered. Suppose that $w(n)$ is white noise with zero mean and $\mathbf{E}(ww^*) = \delta^2 I$. The power spectral density of the received signal satisfies

$$\mathbf{S}_{yy}(z) = \mathbf{S}_{xx}(z) + \delta^2 I \tag{19}$$

where $\mathbf{S}_{xx}(z)$ is the power spectral density of the desired received signal $x(n) = \sum_{l=0}^{L} H(l)v(n-l)$. Following the above discussion, for any roots z_{ik}, $k = 1, \cdots, L+2$ of $\mathbf{r}_i(z)$, $\mathbf{S}_{yy}(z_{ik})$ has eigenvalues $\lambda_{ik} + \delta^2, \delta^2, \cdots, \delta^2$, where λ_{ik} is the eigenvalue of $\mathbf{S}_{xx}(z_{ik})$. Hence, the eigenvector of $\mathbf{S}_{yy}(z_{ik})$ associated with $\lambda_{ik} + \delta^2$ is the same as that of $\mathbf{S}_{xx}(z_{ik})$ associated with λ_{ik}. Therefore, the algorithm for the noise-free case is still available for the case where the received signals are corrupted by an additive white noise.

4. SIMULATIONS

Consider an MIMO FIR transmission channel as follows:

$$\begin{bmatrix} x_1(n) \\ x_2(n) \\ x_3(n) \end{bmatrix} = \begin{bmatrix} v(n) - 1.6v(n-1) + v(n-2) \\ v(n) - v(n-1) \\ v(n) + 2.2v(n-1) + v(n-2) \end{bmatrix.$$
$$\left. \begin{matrix} v(n) + 2.2v(n-1) + v(n-2) \\ v(n) - v(n-1) \\ v(n) - 1.6v(n-1) + v(n-2) \end{matrix} \right] . \tag{20}$$

The pre-filters are selected as

$$\mathbf{p}_1(z) = (1 - z^{-2})(1 + z^{-1} + z^{-2})$$

and

$$\mathbf{p}_2(z) = (1 + z^{-2})(1 - z^{-1} + z^{-2}).$$

Then applying the developed blind identification algorithm, we obtain

$$[h_1(0)\; h_1(1)\; h_1(2)] = \begin{bmatrix} 0.9999 & -1.6075 & 1.0002 \\ 1.0228 & -1.0187 & 0.0235 \\ 1.0402 & 2.1763 & 1.0396 \end{bmatrix}$$

and

$$[h_2(0)\; h_2(1)\; h_2(2)] = \begin{bmatrix} 1.0507 & 2.2334 & 0.9846 \\ 0.9737 & -1.0111 & 0.0353 \\ 0.9490 & -1.6265 & 1.0578 \end{bmatrix}.$$

5. CONCLUSIONS

We have studied the blind identification problem for an FIR multiuser channel in this paper. By introducing a proper MIMO pre-filter in the transmitting side of the channel, a large class of MIMO-FIR channels can be identified from SOS of the received signals and the information of the pre-filter. A blind identification algorithm has been presented. And its asymptotical convergence has been proven. Finally, the preliminary simulation results have indicated that the developed blind identification algorithm is efficient.

6. REFERENCES

[1] K. Abed-Meriam, P. Loubaton and E. Moulines, "A subspace algorithm for certain blind identification problems," *IEEE Trans. Information Theory*, vol. 43, pp. 499-511, 1997.

[2] P. A. Fuhrmann, *A Polynomial Approach to Linear Algebra*. New York: Springer, 1996.

[3] A. Gorokhov and P. Loubaton, "Subspace-based techniques for blind separation of convolutive mixtures with temporally correlated source," *IEEE Trans. Circuit Syst. I*, vol. 44, pp. 813-820, 1997.

[4] Y. Hua and J. Tugnait, "Blind identifiability of FIR-MIMO systems with colored inputs using second order statistics," *IEEE Signal Processing Letter*, vol. 7, pp. 348-350, 2000.

[5] T. Kailath, *Linear Systems*. Eaglewood Cliffs, NJ: Prentice-Hall, 1980.

[6] M. Kawamoto and Y. Inouye, "Blind deconvolution of MIMO-FIR systems with colored inputs using second-order statistics," *IEICE Trans. on Fundamentals of Electronics, Communications and Computer Sciences*, vol. E86-A, pp. 597-604, 2003.

[7] R. Liu, Y. Inouye and H. Luo, "A system-theoretic foundation for blind signal separation of MIMO-FIR convolutuve mixtures - A review," *Second International Workshop on Independent Component Analysis & Blind Signal Separation*, Helsinki, Finland, June 2000.

[8] H. Luo and R. Liu, "Blind equalization for MIMO-FIR channels based only on second order statistics by use of pre-filters," in *Proc. 2nd IEEE Workshop on Signal Processing Advances in Wireless Communications*, Annapolis, MD, May 1999, pp. 106-109.

[9] H. Luo, R. Liu, X. T. Ling and X. Li, "The autocorrelation matching method for distributed MIMO communications over unknown FIR channels," in *Proc. 2001 IEEE Int. Conf. Acoust., Speech, Signal Process.*, Salt Lake City, UT, May 2001, vol. 4, pp. 2161-2164.

[10] H. Liu, G. Xu, L. Tong and T. Kailath, "Recent developments in blind channel equalization: From cyclostationarity to subspace," *Signal Processing*, vol. 50, pp. 83-99, 1996.

[11] A. Scaglione, G. B. Giannakis and S. Barbarossa, "Redundant filterbank precoders and equalizers, I: Unification and optimal designs," *IEEE Trans. Signal Processing*, vol. 47, pp. 1988-2006, 1999.

[12] A. Scaglione, G. B. Giannakis and S. Barbarossa, "Redundant filterbank precoders and equalizers, II: Blind channel estimation, synchronization, and direct equalization," *IEEE Trans. Signal Processing*, vol. 47, pp. 2007-2022, 1999

[13] D. T. M. Slock, "Blind fractionally-spaced equalization, perfect-reconstruction filter-banks and multichannel linear prediction," in *Proc. 1994 IEEE Int. Conf. Acoust., Speech, Signal Process.*, Adeliade, Australia, Apr. 1994, vol. 4, pp. 585-588.

[14] L. Tong and S. Perreau, "Multichannel blind identification: From subspace to maximum likelihood methods," *Proceedings of the IEEE*, vol. 86, pp 1951-1968, 1998.

[15] L. Tong, G. Xu and T. Kailath, "A new approach to blind identification and equalization of multipath channels,", in *Proc. 25th Asilomar Conf. Signals, Syst., Comput.*, Pacific Grove, CA, Nov. 1991, vol. 2, pp. 856-860.

FAST FORCE-DIRECTED/SIMULATED EVOLUTION HYBRID FOR MULTIOBJECTIVE VLSI CELL PLACEMENT

Sadiq M. Sait

Dept. of Computer Engineering,
King Fahd University of Petroleum & Minerals,
Dhahran-31261, Saudi Arabia.
email: sadiq@ccse.kfupm.edu.sa

Junaid Asim Khan

Dept. of Electrical & Computer Engineering,
The University of British Columbia,
Vancouver, Canada.
email: junaidk@ece.ubc.ca

ABSTRACT

VLSI Standard Cell Placement is a hard optimization problem, which is further complicated with new issues such as power dissipation and performance. In this work, a fast hybrid algorithm is designed to address this problem. The algorithm employs Simulated Evolution (SE), an iterative search heuristic that comprises three steps: evaluation, selection and allocation. Solution quality is a strong function of the allocation procedure which is both time consuming and difficult. In this work a force directed approach in the allocation step of SE is used to both accelerate and improve the solution quality. Due to the imprecise nature of design information at the placement stage, objectives to be optimized are expressed in the fuzzy domain. The search evolves towards a vector of fuzzy goals. The proposed heuristic is compared with a previously presented SE approach. It exhibits significant improvement in terms of runtime for the same quality of solution.

1. INTRODUCTION

In VLSI physical design, the standard cell placement step consists of assigning modules (typically several thousands) to locations on the silicon surface under numerous design constraints while trading-off several objectives. In general, placement in VLSI physical design is a multiobjective optimization problem [1]. The most important objectives are power dissipation, delay, wirelength and area (width) of the chip [2, 3]. Several attempts using SE (Simulated Evolution) as a search heuristic and fuzzy logic to cope with the multiobjective nature of the problem have been attempted [4, 5, 6]. However, these exhibited unreasonably large runtime requirements.

In this paper, we revisit the algorithm from [6] and show how it can be accelerated in order to run on large circuits in a reasonable amount of time. The basic SE algorithm comprises three steps: *evolution*, *selection*, and *allocation*. Of these, the allocation step is the slowest, with a complexity of $O(n^2)$. In this paper we integrate a new allocation scheme into SE based on a force-directed algorithm that has a complexity of $O(n)$. We show that this hybrid approach gives results that are comparable (or of better quality especially for large test cases) but with much smaller runtimes [6].

The paper is organized as follows. Section 2 covers the problem formulation. In Section 3 the proposed algorithm is discussed. Experimental results are presented in Section 4 and conclusions in Section 5.

2. PROBLEM FORMULATION

SE has proved to be an excellent heuristic for standard cell placement problem in terms of solution quality. However, in terms of runtime, it is not as efficient as other well known deterministic placement algorithms. In this paper we have targeted the problem of decreasing the runtime of SE without degrading the overall solution quality. After inspecting the previous SE-based techniques it is observed that the main time consuming step is the allocation step [7, 4, 6]. The asymptotic time complexity of this step in these algorithms is $O(n^2)$. The runtime of SE can be considerably decreased if the allocation step is modified to consume only $O(n)$ time. In this work, we present a fuzzy, forced-directed approach to solve allocation in $O(n)$ time. For comparison we target the same standard cell placement problem and cost function presented in [1, 4, 6].

3. PROPOSED ALGORITHM

In this section we describe our Simulated Evolution based hybrid search algorithm. We begin with a brief discussion of the basic SE heuristic.

3.1. Basic Simulated Evolution (SE)

The general SE algorithm is illustrated in Figure 1 and comprises three main steps: **evaluation**, **selection**, and **alloca-**

ALGORITHM $Simulated\_Evolution(B, \Phi_{initial}, StoppingCondition)$
NOTATION
B= Bias Value. Φ= Complete solution.
m_i= Module i. g_i= Goodness of m_i.
$ALLOCATE(m_i, \Phi_i)$=Function to allocate m_i in partial solution Φ_i
Begin
Repeat
EVALUATION:
ForEach $m_i \in \Phi$ evaluate g_i;
/* Only elements that were affected by moves of previous */
/* iteration get their goodnesses recalculated*/
SELECTION:
ForEach $m_i \in \Phi$ **DO**
begin
IF $Random > min(g_i, 1)$
THEN
begin
$S = S \cup m_i$; Remove m_i from Φ
end
end
Sort the elements of S
ALLOCATION:
ForEach $m_i \in S$ **DO**
begin
$ALLOCATE(m_i, \Phi_i)$
end
Until *Stopping Condition is satisfied*
Return Best solution.
End $(Simulated\_Evolution)$

Fig. 1. Structure of the Simulated Evolution algorithm [7].

tion. In the **evaluation** step the **goodness** of each cell in its current location, in the range $[0, 1]$, is computed using some measure.

In the **selection** step, the algorithm probabilistically selects unfit elements. Elements with low goodness values have higher probabilities of getting selected for relocation. These selected elements are identified as the selection set and are removed from the solution. These selected elements are one by one reassigned to new locations in a constructive **allocation** step. The objective of this step is to improve their goodness values, thereby reducing the overall cost of the solution.

Different constructive allocation schemes are proposed in literature [8, 7]. One such scheme is **sorted individual best fit**, where all the selected elements are sorted in descending order with respect to their connectivity with the partial solution and placed in a queue. The sorted elements are removed one at a time and *trial* moves are carried out for all the available empty positions. The element is *finally* placed in a position where maximum reduction in cost for the partial solution is achieved. This process is continued until the selected queue is empty. The overall complexity of this algorithm is $O(n^2)$ where n is the number of selected elements. Other more elaborate schemes are **weighted bipartite matching allocation** and **branch-and-bound search allocation** [8]. However, these allocation strategies are more complex than "sorted individual best fit", while the quality of solution remains comparable [8]. In summary, selection and allocation steps determine and dictate the search strategy, while evaluation provides feedback to the search scheme.

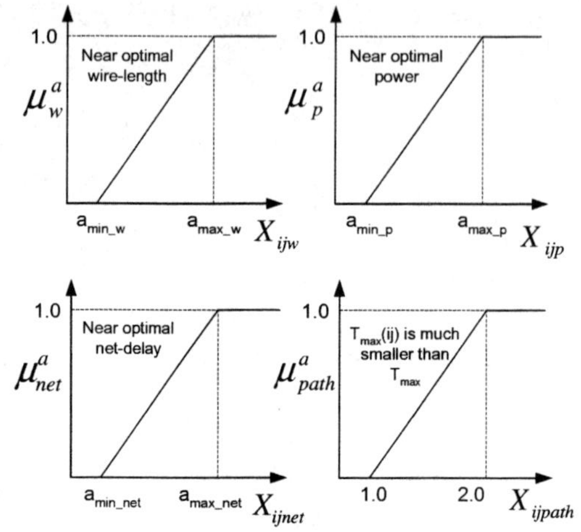

Fig. 2. Membership functions used in fuzzy allocation.

The main contribution in this paper is a new *allocation* scheme; this will be discussed in Section 3.2. However, the evaluation and selection schemes are same as in [6], except that OWA-operators for fuzzy aggregation are replaced by the new fuzzy aggregating functions proposed in [9].

3.2. Fuzzy Force Directed Allocation

In the allocation stage, the selected cells are to be reassigned to best available locations. We consider selected cells as movable modules and remaining cells as fixed modules. In previous works [4, 5, 6], **sorted individual best fit** scheme was employed. For large circuits the run time was unreasonably high. To address this problem, a force directed allocation is proposed in this work. According to this approach optimal x-position and y-position of the cell under consideration are found. The y-position indicates the row to which the cell should be relocated. If the y-position is in between two rows then the row nearest to y-position is selected. In order to satisfy the width constraint, if the width of selected row after adding the cell is more than the maximum allowable width then the next nearest row that satisfies the width constraint is chosen. The x-position indicates the exact location of the cell in the selected row.

The basic idea behind the force directed method is that cells connected by a net exert forces on each other. Suppose a cell a is connected to another cell b by a net of weight w_{ab}. Let d_{ab} represents the distance between a and b. Then the force of attraction between the cells is proportional to the product $w_{ab} \times d_{ab}$. A cell i connected to several cells j at distance d_{ij} by wires of weights w_{ij}, experiences a total force F_i given by

$$F_i = \sum_j w_{ij} \cdot d_{ij} \qquad (1)$$

The best location for a cell i is where the x-component

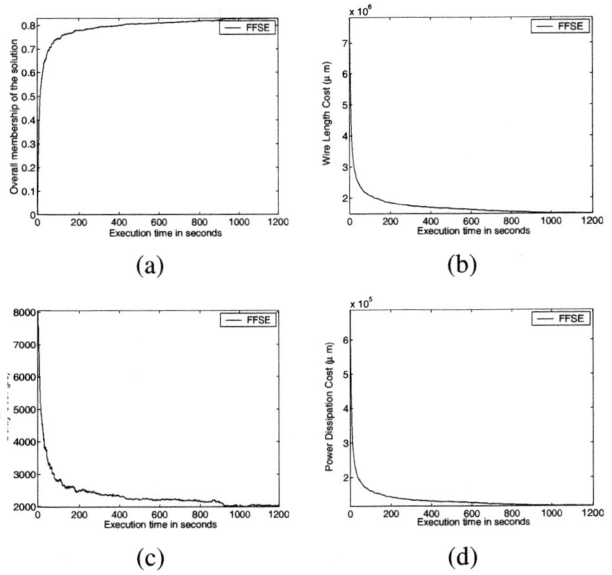

Fig. 3. (a) Overall membership; (b) Wirelength; (c) Delay; and (d) Power; versus execution time in Seconds (S15850).

and y-component of F_i are both zero. We can write these conditions as follows,

$$\sum_j w_{ij} \cdot (x_j - x_i) = 0; \quad \& \quad \sum_j w_{ij} \cdot (y_j - y_i) = 0 \qquad (2)$$

Solving the above equations for x_i and y_i we have

$$x_i = \frac{\sum_j w_{ij} \cdot x_j}{\sum_j w_{ij}} \qquad y_i = \frac{\sum_j w_{ij} \cdot y_j}{\sum_j w_{ij}} \qquad (3)$$

Values x_i and y_i are the optimal x-position and y-position for a cell i with respect to current x and y positions of all the cells j connected to it [1]. They point to the new location that is better in terms of all objectives. For this purpose, proper weights to each of the nets connecting cell i and cell j are to be chosen. A good way to choose these weights is to use fuzzy logic. The following fuzzy rule is used to find these weights:

Rule R1: IF a net is *good in wire-length* AND *good in power* AND *good in delay* **THEN** it has a low weight.

According to this rule, a net will have a smaller weight only if it is good in terms of all the objectives. In fact weight signifies a badness factor (opposite of goodness in evaluation). The cell will try to move in the directions of those nets that have higher weight (higher badness). We use the membership functions in Figure 2, with the following base values.

$$X_{ijw}(x) = \frac{l^*_{ij}}{l_{ij}} \quad X_{ip}(x) = \frac{l^*_{ij}}{(1+S_{ij})\,l_{ij}}$$
$$X_{inet}(x) = \frac{ID^*_{ij}}{ID_{ij}} \quad X^e_{ipath}(x) = \frac{T_{max}}{T_{max}(ij)} \qquad (4)$$

where l_{ij} represents wire-length of a net ij and l^*_{ij} is its estimated lower bound. S_{ij} is its switching probability (required to estimate power in CMOS circuits). ID_{ij} is interconnect delay of net ij and ID^*_{ij} is its estimated lower bound. T_{max} is the delay of longest path and $T_{max}(ij)$ is the delay of longest path traversing net ij.

Using these base values and corresponding μ^a_i where superscript a denotes allocation, we find a goodness factor g_{ij}, using AFA and OFA operators proposed in [9], for the net connecting cells i and j, as follows,

$$g_{ij} = \mu^a_{ij} = 1 - \frac{\sum_{k=w,p,d} \bar{\mu^a}^2_{k,ij}}{\sum_{j=w,p,d} \mu^a_{k,ij}} \qquad (5)$$

where

$$\mu^a_{d,ij} = \frac{\mu^{a\,2}_{net,ij} + \mu^{a\,2}_{path,ij}}{\mu^a_{net,ij} + \mu^e_{path,ij}} \qquad (6)$$

Now the weight of the net w_{ij} is calculated as follows,

$$w_{ij} = 1 - g_{ij} \qquad (7)$$

In this proposed allocation schemes it is clear that for each cell we have to find the best location only once, therefore the complexity of the proposed allocation scheme is $O(n)$ where n is the number of cells selected in selection stage of the algorithm. All other issues such as already occupied zero-force locations, cells already in their zero-force locations, etc., are resolved using the previous ad-hoc approaches available in the literature [1].

4. EXPERIMENTS AND RESULTS

Fast Fuzzy Force Directed Simulated Evolution (FFSE) and Biasless Fuzzy Simulated Evolution (BLFSE) [6], were applied on 12 ISCAS benchmark circuits. In BLFSE, execution is aborted when no improvement is observed in the last 500 iterations (maximum of 5000 iterations), whereas for FFSE the algorithm is run for a fixed 5000 iterations. The 0.25 micron CMOS digital low power standard cell library for MOSIS is used.

Table 1 compares the quality of the final solution generated by BLFSE and FFSE. The circuits are listed in order of their size (136-10383 modules). From the results, it is clear that FFSE outperformed BLFSE for all circuits in terms of execution time. Also, in most cases, FFSE shows minimal degradation in terms of solution quality but with a significant improvement in runtime. For larger circuits (S3330 & S5378), FFSE performed better than BLFSE in terms of both the final solution quality and the runtime.

Observe that the algorithm converges very fast. This behavior can be observed in Figure 3, where convergence is achieved after approximately 400 seconds (6.6 minutes), and the remaining time is spent in fine tuning the solution

Circuit		BLFSE				FFSE			
	# of Cells	L (μm)	P (μm)	D (ps)	T (s)	L (μm)	P (μm)	D (ps)	T(s)
S298	136	4548	915	139	46	4975	999	135	4.8
S386	172	8357	2036	203	117	9422	2169	213	6.8
S832	310	23140	5251	416	192	26112	5863	400	11
S641	433	12811	3072	687	175	12485	2897	674	24
S953	440	29576	5025	223	351	29988	4683	244	17
S1238	540	41318	12303	363	699	41362	12934	377	20
S1196	561	35810	11276	360	613	38282	12363	350	22
S3330	1961	183288	24797	459	5351	163756	24112	483	87
S5378	2993	326840	48360	435	11823	243721	41560	376	149
S9234	5844	UH	UH	UH	UH	655370	114231	908	440
S13207	8651	UH	UH	UH	UH	1339837	144189	1604	885
S15850	10383	UH	UH	UH	UH	1477662	115049	2006	1202

Table 1. Layout found by BLFSE, and FFSE. 'L', 'P' and 'D' represent the wire-length, power, and delay costs respectively. 'T' is the execution time in Seconds. The last 3 circuits were not tested for BLFSE because of large runtime requirement. 'UH' indicates Unreasonably High runtime requirement.

quality. It is also obvious from these results that FFSE totally avoids early random walk, which is a problem in other non-deterministic heuristics such as Simulated Annealing. Memory requirement of SE is also considerably low when compared to other iterative heuristics such as Genetic algorithm as SE keeps only one solution in its memory at a time.

5. CONCLUSION

In this paper, we have proposed Fuzzy Force Directed Simulated Evolution Algorithm for multiobjective VLSI standard cell placement. The allocation stage is improved by using a force directed strategy to reduce the execution time from $O(n^2)$ in previous SE based approaches to $O(n)$.

Fuzzy logic is used to overcome the multiobjective nature of the problem. It is employed at evaluation and allocation stages and in the choice of the best solution from the set of generated solutions.

The proposed scheme is compared with BLFSE. It is observed that FFSE performs much better than BLFSE in terms of runtime, with no significant degradation in solution quality. FFSE can be used for large circuits whereas BLSFE performs poorly for circuits with more than $2000 - 3000$ cells.

Acknowledgment: The authors and the research team acknowledge King Fahd University of Petroleum & Minerals for its support under research project COE/ITERATE/221.

6. REFERENCES

[1] Sadiq M. Sait and Habib Youssef. VLSI Physical Design Automation: Theory and Practice. *World Scientific Publications, Singapore*, 2001.

[2] Glenn Holt and Akhilesh Tyagi. EPNR: An Energy-Efficient Automated Layout Synthesis Package. *IEEE International Conference on VLSI in Computers and Processors*, pages 224–229, October 1995.

[3] Glenn Holt and Akhilesh Tyagi. GEEP: A Low Power Genetic Algorithm Layout System. *IEEE 39th MWSCAS*, 3:1337–1340, August 1996.

[4] Sadiq M. Sait, Habib Youssef, and Junaid A. Khan. Fuzzy Evolutionary Algorithm for VLSI Placement. *GECCO-2001*, July 2001.

[5] Sadiq M. Sait, Habib Youssef, Junaid A. Khan, and Aiman Al-Maleh. Fuzzy Simulated Evolution for Power and Performance Optimization of VLSI Placement. *INNS-IEEE, IJCNN2001*, July 2001.

[6] Junaid A. Khan, Sadiq M. Sait, and M. R. Minhas. Fuzzy Biasless Simulated Evolution for Multiobjective VLSI Placement. *IEEE Congress on Evolutionary Computation, CEC2002, Honolulu*, May 2002.

[7] Sadiq M. Sait and Habib Youssef. *Iterative Computer Algorithms with Applications in Engineering: Solving Combinatorial Optimization Problems.* IEEE Computer Society Press, California, December 1999.

[8] Ralph M. Kling and Prithviraj Banerjee. ESP: Placement by Simulated Evolution. *IEEE Transactions on Computer-Aided Design*, 8(3):245–255, March 1989.

[9] Junaid A. Khan and Sadiq M. Sait. Fuzzy Aggregating Functions for Multiobjective VLSI Placement. *FUZZ-IEEE 2002, Honolulu*, May 2002.

SIMULTANEOUS DELAY AND POWER OPTIMIZATION IN GLOBAL PLACEMENT

Mongkol Ekpanyapong, Karthik Balakrishnan, Vidit Nanda, and Sung Kyu Lim
School of Electrical and Computer Engineering, Georgia Institute of Technology
{pop,gte245v,gte272u,limsk}@ece.gatech.edu

ABSTRACT

Delay and power minimization are two important objectives in the current circuit designs. Retiming is a very effective way for delay optimization for sequential circuits. In this paper we propose a framework for multi-level global placement with retiming, targeting simultaneous delay and power optimization. We propose GEO-P for power optimization and GEO-PD algorithm for simultaneous delay and power optimization and provide smooth wirelength, power and delay tradeoff. In GEO-PD, we use retiming based timing analysis and visible power analysis to identify timing and power critical nets and assign proper weights to them to guide the multi-level optimization process. We show an effective way to translate the timing and power analysis results from the original netlist to a coarsened sub-netlist for effective multi-level delay and power optimization. Our GEO-P achieves 27% average power improvement and our GEO-PD provides gains in both delay and power improvement. To the best of our knowledge, this is the first paper addressing simultaneous delay and power optimization in multi-level global placement.

1. INTRODUCTION

Delay minimization and power minimization are two important objectives in the design of the high-performance, portable, and wireless computing and communication systems. Thus, a considerable research effort has been made in trying to find power and delay-efficient solutions to circuit design problems. One such procedure that is applied at the logic level is circuit placement.

The placement problem for a given sequential netlist involves global placement and detailed placement. Global placement identifies the partition block-level location for cells, whereas detailed placement provides complete location information for each cell while preserving the global placement. Recently, global placement has attracted significant attention due to tighter circuit constraints and increasing complexities. There are three major approaches to global placement: min-cut based algorithms, analytical approaches, and simulated annealing techniques. The min-cut based approach uses top-down methods to recursively partition a circuit into smaller sub-netlists. Due to the high flexibility and small runtime of this approach, it has been adopted in many modern state-of-the-art placement algorithms.

In this paper we propose a framework for mincut-based global placement with retiming, simultaneously optimizing delay and power. We first discuss the importance of retiming delay and visible power as opposed to the conventional static delay and total power for sequential circuits. Then we propose GEO-P, the modified version of GEO targeting power optimization. We use visible power analysis to guide the partitioner to group gates such that long wires are not driven by the gates with high switching activity. We also propose GEO-PD algorithm for simultaneous delay and power optimization. In GEO-PD, we use retiming based timing analysis and visible power analysis to identify timing and power critical nets and assign proper weights to them to guide the multi-level optimization process. In general, timing and power analysis are done at the original netlist while a recursive multi-level approach performs partitioning and placement on the sub-netlist as well as its coarsened representations. We show an effective way to translate the timing and power analysis results from the original netlist to a coarsened sub-netlist for effective multi-level delay and power optimization.

The organization of this paper is as follows. Section 2 describes problem formulation. Section 3 is devoted to our algorithm. Section 4 presents our experimental result and analysis. Finally, the last section presents our conclusions.

2. PROBLEM FORMULATION

Given a sequential gate-level netlist $NL(C, N)$, where C is the set of cells representing gates and flip-flops, and N is the set of nets connecting the cells, the purpose of the Performance driven Global Placement with Retiming (PGPR) problem is to assign cells in NL to $m \times n$ (= K) blocks while area constraint for each block is satisfied. In other words, the placement region is divided into $m \times n$ tiles, and we perform cell placement at the center of these tiles. Given a PGPR solution B, let $\omega(B)$ and $\phi(B)$ respectively denote the wirelength and retiming delay. The formal definition of PGPR is as follows:

PGPR Problem: the Performance driven Global Placement with Retiming (PGPR) problem under the given area constraints $A = (L_i, U_i)$ has a solution P: $C \rightarrow B$, wherein each cell in C is assigned to a unique block, where $B = \{B_1(x_1,y_1), B_2(x_2,y_2),..., B_K(x_K,y_K)\}$ denotes the set of blocks and (x_i,y_i) represents the geometric location of B_i. B is feasible if it satisfies the following conditions: i) $B_i \subset C$, $1 \le i \le K$, ii) $L_i \le |B_i| \le U_i$, $1 \le i \le K$, iii) $B_1 \cup B_2 \cup ... \cup B_K = C$, iv) $B_i \cap B_j = \emptyset$ for all $i \ne j$. The objective is to minimize $\phi(B)$ while maintaining an acceptable $\omega(B)$.

2.1. Delay Objective

By employing the concept of retiming graph, we model NL using a directed graph $R = (V, E)$. Each vertex v has delay $d(v)$ and each edge $e=(u,v)$ has delay $d(e)$. We assume $d(e)$ is proportional to the Manhattan distance between u and v. The edge weight $w(e)$ of $e=(u,v)$ denotes the number of flip-flops between gate u and v. The path weight can be calculated by $w(p)=\sum_{e \in p} w(e)$. Let $w^r(e)$ denote edge weight after retiming r, i.e. number of flip-flops on the edge after retiming. Then, $w^r(p)=\sum_{e \in p} w^r(e)$. A circuit is retimed to a delay ϕ by a retiming r if the following conditions are satisfies; (i) $w^r(e) \ge 0$ for each e, (ii) $w^r(p) \ge 1$ for each path p such that $d(p) > \phi$. We define the edge length of $e=(u,v)$ as $l(e)=-\phi w(e)+d(v)+d(e)$, and the path length of p as $l(p)=\sum_{e\in p} l(e)$. The *sequential arrival time* [3] of vertex v, denote $l(v)$, is maximum path length from PIs or FFs to v. If the sequential arrival time of all POs or FFs are less than or equal to ϕ, the target delay ϕ is called *feasible*. Let $q(e)=\phi w(e)-d(u)-d(e)$ be the required edge length of e. The required path length $q(p)=\sum_{e\in p} q(e)$. The *sequential required time* of vertex v, denote $q(v)$ is the minimum required path length from v to POs or FFs, when $q(PO)$ or $q(FF) = \phi$. Then slack of v is given by $q(v)-l(v)$. Let D_g be the maximum $d(v)$ among all v in V. Then, the *retiming delay* $\phi(B)$ of a PGPR solution B is the minimum feasible $\phi + D_g$.

2.2. Wirelength Objective

We model netlist NL using a hypergraph $H=(V, E_H)$, where the vertex set V represents cells, and the hyperedge set E_H represents nets in NL. Each hyperedge is a non-empty subset of V. The x-span of hyperedge h, denoted h_x, is defined as $h_x = \max_{c\in h}\{x_i|c\in B_i\} - \min_{c\in h}\{x_i|c\in B_i\}$. The y-span, denoted h_y, is calculated using the y-coordinates. The sum of x-span and y-span of each hyperedge h is the half-parameter of the bounding block (HPBB) of h and denoted $HPBB(h)$. The *wirelength* $\omega(B)$ of global placement solution B is the sum of HPBB of all hyperedges in H.

2.3. Power Objective

For power objective, we model NL as hypergraph $H=(V, E_H)$ as discussed in Section 2.2. Let V_{dd} denote the supply voltage, f is the global clock frequency, $C_g(v)$ and $C_w(v)$ represent the gate capacitance and wire capacitance seen by gate v, and $SA(v)$ is switching activity of v. $C_g(v)$ is the sum of the input capacitance of all sink gates driven by v. Let n_v denote the net whose driving gate is v. Let VG be the set of *visible gates* that is defined as $VG=\{v|s(n_v)=1\}$, if n_v is cut. Then, the *visible power consumption* $\pi(B)$ of global placement solution B is calculated as follows: $P_v=(V^2_{dd}f\cdot\sum_{v\in VG}(C_g(v)+C_w(v))\cdot SA(v))/2$. The rationale is that the power consumption by the gate driving a long wire is much larger than that of short wire. We note that $C_w(v)=HPBB(n_v)\cdot C_g(v)$, the wire capacitance $C_w(v)$ is the only factor that changes based on a placement solution. Thus, we attempt to minimize the visible power in our algorithms.

3. METHODOLOGY

3.1. Overview of GEO-PD Algorithm

An overview of the GEO-PD algorithm is shown in Figure 1. GEO-PD is a multi-level global placement for simultaneous delay and power optimization. GEO-PD places the given netlist NL into $K=n\times m$ dimension using a top-down recursive bipartitioning approach. `GEO-PD` consists of two subroutines: `GEO-PD-2way` recursively bipartitions NL, whereas `GEO-PD-Kway` refines these partitioning results occasionally. `GEO-PD-2way` is performed on the sub-netlist, whereas `GEO-PD-Kway` is performed on the entire netlist. Initially, the partitioning tree T has only root node R, and all cells in NL are inserted into R. The FIFO (First In First Out) queue Q is used to support the recursive breadth-first cut sequence.

`GEO-PD-2way` first generates the sub-netlist from the given partition tree node and performs multi-level clustering on it. We use ESC clustering algorithm [1] for this purpose. Then we obtain a random initial partitioning B among the clusters at the top level of the hierarchy. The subsequent top-down multi-level refinement is used to improve B in terms of delay and power. We perform retiming based timing analysis RTA [2] to identify timing critical nets. We also perform power analysis [4] to identify power critical nets. Then we compute the delay and power weights for the nets in the sub-netlist. The subsequent iterative improvement through cluster move tries to minimize the weighted cutsize. Finally we project the current solution to the next level coarser netlist for multi-level optimization. At the end of `GEO-PD-2way`, two new children nodes are inserted into T based on B.

`GEO-PD-Kway` refinement is performed when we obtain 2^j partitions ($j > 1$) from `GEO-PD-2way` (4, 8, 16 partitions, etc). We first perform a restricted multi-level clustering, where grouping among cells in different partition is prohibited. This allows the partitioner to preserve the initial partitioning results. Then we again perform multi-level partitioning in the same way as in `GEO-PD-2way` for additional delay and power improvement. `GEO-PD-Kway` is applied onto the global netlist for more global level optimization.

```
===========================================
GEO-PD(NL,K)
insert all cells in NL to root node R in T
insert R into Q (= FIFO queue)
while (leaf nodes in T < K)
     N = remove front element in Q
     GEO-PD-2way(N) (= bipartitioning on N)
     split cells in N into N1 and N2
     insert N1 and N2 into Q and T
     if (2^j leaf nodes exists in T, j>1)
          GEO-PD-Kway(T)
return T
-------------------------------------------
GEO-PD-2way(N)
NL' = sub-netlist containing cells in N
ESC(NL') (= multi-level clustering on NL')
h = height of the cluster hierarchy
B = random partitioning for clusters at h
for (i = h downto 0)
     NL'(i) = coarsened NL' at level i
     while (gain)
          DELAY-WEIGHT(NL'(i))
          POWER-WEIGHT(NL'(i))
          net weight = power + delay weight
          while (gain)
               move cells in NL'(i)
               update B
     project B to level i-1
return B
-------------------------------------------
GEO-PD-Kway(T)
B = initial partitioning for NL from T
ESC'(NL) (= restricted clustering)
perform multi-level partitioning
update T
===========================================
```

Figure 1. Overview of the GEO-PD algorithm

3.2. Weight Computation

For simultaneous delay and power optimization, we first identify timing and power critical nets and assign proper weights to them to guide the optimization process. A net is *timing critical* if it lies along a critical path and *power critical* if it has high fanout with large wirelength and is driven by a gate with high switching activity. In GEO-PD, retiming delay and visible power are minimized through retiming based timing analysis [2] and visible power analysis [4]. We use *sequential slack* to compute how much time slack exists before timing violation occurs after retiming. These values are then used to compute the delay weights of the nets for retiming delay minimization. In case of power optimization, we use switching activity and gate/wire capacitance to compute power weights of the nets for visible power minimization. Both delay and power weights are added together, and GEO-PD performs multi-level partitioning to minimize the total weighted wirelength.

We note that the multi-level approach [1] is very effective in minimizing the weighted cutsize and wirelength. However, timing and power analysis is typically done at the original netlist while a recursive multi-level approach performs partitioning and placement on the sub-netlist as well as its coarsened representations. Thus, it is crucial that we have an effective way to translate the timing and power analysis results from the original netlist to a coarsened sub-netlist.

```
==========================================
DELAY-WEIGHT(NL')
set delay of edges in R (= retiming G)
perform RTA(R) (= timing analysis)
compute sequential slack for nodes in R
for each cluster C in NL'
   C(R) = all cells in R grouped into C
   slack(C) = min among cells in C(R)
X = top x% clusters with small slack
for each net N in NL'
   if (all clusters in N are in X)
      compute delay-weight(N) using Eqn1
------------------------------------------
POWER-WEIGHT(NL')
for each net Nv in NL'
   Nv' = corresponding net in NL
   compute HPBB(Nv')
   compute power-weight(Nv) using Eqn2
==========================================
```

Figure 2. Delay and power weight computation in GEO-PD

3.2.1. Delay Weight Computation

Figure 2 shows `DELAY-WEIGHT(NL')` algorithm. Before we perform retiming based timing analysis (RTA), we initialize the edge delay in *R* (= retiming graph) based on the current placement results. We set the delay of edges to their Manhattan distances. Then, a Bellman-Ford variant RTA is performed from a given feasible delay to compute sequential slack. For each cluster *C* from the given coarsened sub-netlist `NL'`, we compute *C(R)*, the set of all the nodes in *R* that are grouped into *C*. We use the minimum slack among all cells in *C(R)* as the slack for *C*. The reason we use the minimum slack value is since the critical path information is preserved regardless of multi-level clustering results (we have also performed experiments using average slack value instead of minimum. But the minimum slack method generated better delay results).

After the cluster slack computation is finished, we sort the clusters in a non-decreasing order of their slack values. We store the top *x*% (we use 3% in our experiment) into a set *X*. For each net that contains *only* the clusters in *X*, we use the following equation to compute the delay weight:

$$dwgt(n) = \alpha\left(1 - \frac{\min\{slack(v) \mid v \in n\}}{\max\{slack(w) \mid w \in NL'\}}\right)^{p1} \quad (1)$$

This equation gives higher weights to the nets that contain smaller minimum cluster slack, thus giving higher priority to the nets containing more timing critical clusters. For those clusters that is not in top x%, we give dwgt(n) = 0 and performing partitioning using only cutsize weight as in ESC [1]. Instead of requiring *all* clusters in a net to be timing critical, we tried another scheme where we give delay weights to the nets with 2 or more timing critical clusters. Our related experiment indicates that this approach produced worse results. Our extensive experiments indicate that α=25, $p1$=1, and x=3% are an excellent empirical choice.

3.2.2. Power Weight Computation

Figure 2 shows `POWER-WEIGHT(NL')`, our power weight calculator. As discussed earlier in Section 2.3, our goal is to minimize visible power consumption i.e. power consumption by the gate driving a long wire (among blocks). Then our goal is to minimize the weighted wirelength. For a net driven by a gate *v*, we use the following equation to assign power weight:

$$pwgt(n_v) = \beta\left(\frac{SA(v)[C_g(v) + C_w(v)]}{\max\{SA(u)[C_g(u) + C_w(u)] \mid u \in V\}}\right)^{p2} \quad (2)$$

where $SA(v)$, $C_g(v)$ and $C_w(v)$ respectively represent the switching activity, gate capacitance and wire capacitance seen by gate *v*. We use $C_w(v)= HPBB(n_v)\cdot Cg(v)$. This equation gives higher weights to the nets that have high fanout, larger wirelength, and source gate with high switching activity. In a multi-level approach, each net in the original netlist *NL* is transformed depending on the given sub-netlist *NL'* and its multi-level clustering information. For example, $n_a=\{a,b,c,d\}$ in *NL* becomes $n_{C1}=\{C1,C2\}$, if *NL'* contains *a* and *b* only and *a* is clustered into *C*1 and *b* into *C*2. In this case, we compute $HPBB(n_a)$ based on the location of *C*1, *C*2, *c*, and *d*, and use *SA*(*a*) in our power weight equation. Our extensive experiments indicate that β=25 and $p2$=0.3 are an excellent empirical choice. Since GEO-PD algorithm aims for simultaneously delay and power optimization, by disable retiming analysis and setting delay weight to zero, our algorithm GEO-P can target only for power optimization.

4. EXPERIMENTAL RESULTS

Our algorithms are implemented in C++/STL, compiled with gcc v2.96, and run on Pentium III 746 MHz machine. The benchmark set consists of six big circuits from ISCAS89 [5] and four big circuits from ITC99 [6] suites. We generate random switching activity values for these circuits since such information is not available. The sis package from the university of California at Bekeley can compute the switching activity for sequential circuits, but it takes a prohibited amount of runtime even for a circuit with a few thousand gates. We assume unit delay for all gates in the circuits. Table 1 shows the statistical information of benchmark circuits. We provide the number of gates, PI, PO, and FF for each circuit. Dr and Ds represent the lower bound on retiming delay and static delay, which are calculated by assigning zero delay to all edges and performing retiming and static timing analysis. Gr and Gs represent retiming delay and static delay from our GEO-PD. We note that retiming can improve the delay results significantly. For example, delay can be reduced by 32% from ESC for s38417 with retiming, which makes retiming a very attractive choice for delay optimization. This explains why our

GEO-PD algorithm focuses on retiming delay as opposed to static delay.

Table 1 Benchmark circuit characteristics. Dr and Ds show the lower bound on retiming delay and static delay, and Gr and Gs show the retiming delay and static delay from our GEO-PD.

ckt	gate	FF	Dr	Ds	Gr	Gs
b17o	22854	1414	38	44	61	99
b20o	11979	490	44	74	72	110
b21o	12156	490	43	74	70	113
b22o	17351	703	46	79	76	124
s5378	2828	163	32	33	57	69
s9234	5597	211	39	58	48	95
s13207	8027	669	50	59	91	102
s15850	9786	597	62	82	100	140
s38417	22397	1636	32	47	41	67
s38584	19407	1452	47	56	69	84

We conduct experiments using ESC [1], GEO [2], and our GEO-P and GEO-PD algorithms. ESC is a state-of-the-art cutsize driven multi-level algorithm, and GEO is a state-of-the-art simultaneous cutsize and delay driven multi-level algorithm. GEO-P is obtained by setting delay weights of GEO-PD to zero for power optimization only. Lastly, GEO-PD is a simultaneous power and delay driven multi-level algorithm. We report wirelength, retiming delay, and visible power. Note that the delay and power results are based on block location. We report 8×8 global placement results. We report average improvement ratio normalized comparing with ESC (lower than unity means improvement). We also report the average runtime of each algorithm measured in second.

Table 2 shows the results among ESC, GEO, GEO-P, and GEO-PD. GEO has 10% better retiming delay than ESC at the cost of 16% increase in wirelength. Our GEO-P has 27% better visible power than ESC at the cost of 10% increase in wirelength. Finally, GEO-PD has 5% better retiming delay and 14% better visible power than ESC at the cost of 25% increase in wirelength. GEO-PD improves the retiming delay of s38584 by 21%. The visible power improvement is as much as 31% for s9234. Moreover, the retiming delay and visible power improvement is consistent among all 10 circuits. In overall, GEO-PD reveals a smooth wirelength, delay, and power tradeoff curve and improves both delay and power results of ESC at the cost of increase in wirelength.

5. CONCLUSIONS

To the best of our knowledge, this is the first paper addressing both delay and power optimization in multi-level placement. In addition, we demonstrated the importance of optimizing the retiming delay and visible power as opposed to the conventional static delay and total power. We demonstrated how wirelength has conflicting objectives against power and delay and proposed an effective algorithm GEO-PD for smooth delay, power, and wirelength tradeoff. We also propose GEO-P, which achieve 27% improvement in terms of power.

6. REFERENCE

[1] J. Cong and S. K. Lim, Edge separability based circuit clustering with application to circuit partitioning, *To appear in TCAD.*

[2] J. Cong and S. K. Lim, Physical Planning with Retiming, *In IEEE International Conference in Computer Aided Design*, page 2-7, 2000.

[3] P. Pan, A. K. Karandikar, and C. L. Liu, Optimal clock period clustering for sequential circuits with retiming, *IEEE Trans on Computer-Aided Design, pages 489-498,1998.*

[4] H. Vishnu and M. Pedram, Delay-Optimal Clustering Targeting Low-Power VLSI Circuits., *IEEE Trans on Computer-Aided Design*, page 639-643, 1995.

[5] http://www.cbl.ncsu.edu

[6] http://www.cad.polito.it/tools/9.html

Table 2 Comparison among ESC, GEO, GEO-P, and GEO-PD on 8×8 global placement. Each algorithm reports wirelength, retiming delay, visible power, and runtime.

	ESC			GEO			GEO-P			GEO-PD		
ckt	wire	r-dly	v-pow	wire	r-dly	v-pow	wire	r-dly	v-pow	wire	r-dly	v-pow
b17o	9629	70	5232	10451	63	5697	9982	63	4604	10468	61	4938
b20o	5772	72	3335	6730	79	3660	6450	71	3101	7277	72	3145
b21o	6357	79	3458	6618	65	3468	6703	75	2863	7491	70	3235
b22o	7243	77	4076	7724	69	4473	8570	83	3879	8685	76	4211
s5378	1502	60	384	1462	45	389	1539	57	234	1597	57	269
s9234	1425	50	427	1685	48	476	1510	52	292	1683	48	296
s13207	1525	91	747	1925	77	900	1803	91	536	2367	91	634
s15850	1587	99	584	2085	90	814	1720	96	395	2236	100	517
s38417	2032	41	1158	2695	41	1483	2524	43	963	2819	41	1088
s38584	2973	87	1950	3663	68	2091	3061	79	1619	3546	69	1766
Ratio	1.00	1.00	1.00	1.16	0.90	1.14	1.10	0.98	**0.79**	1.25	**0.95**	**0.88**
Time	104			2231			121			2257		

MODULE PLACEMENT BASED ON QUADRATIC PROGRAMMING AND RECTANGLE PACKING USING LESS FLEXIBILITY FIRST PRINCIPLE *

Sheqin Dong[1] *Zhong Yang*[1] *Xianlong Hong*[1] *Yuliang Wu*[2]

[1]Department of Computer Science and Technology, Tsinghua University, Beijing, P.R.China, 100084
Tel:+86-010-62785564 Fax:+86-010-62781489 Email : dongsq@mail.tsinghua.edu.cn
[2]Department of Computer Science and Engineering, The Chinese University of Hong Kong

ABSTRACT

We formulate the floorplanning or placement problem as fixed die and connectivity-oriented problem. To solve such a problem, we first use quadratic programming to optimize the total wire-length of the placement and then using a deterministic rectangle packing algorithm based on less flexibility first principle to fulfill the placement in an estimated fixed die area. Experimental results demonstrated that our method is promising for practical use.

1. INTRODUCTION

Floorplan representations have evolved from dual graphs to slicing tree [1], to sequence pair [2], bounded-slicing grids[3], O-tree [4], CBL [5], TCG[6] etc. This evolution has been driven by the quest for a complete and irredundant representation that allows efficient heuristic search for good packs. Complex objectives such as path-delay or wire-length seem difficult to be dealt with using annealing of sequence-pairs[7].

The classical floorplaning literature treats the die area as having constrained aspect ratio, but unbounded size. The objective is to "find the packing with smallest containing die". In reality, the use of floorplanning or placement during the chip synthesis process nearly often comes after the die size and package have been chosen. Thus floorplanning or placement should be cast as a fixed-die problem. To handle objectives such as wire-length and path-delay, one must focus on connectivity. The correct approach must be connectivity-oriented, rather than packing-oriented. Current floorplaning formulations do not enforce this constraint[7].

In this paper, we first formulate the VLSI module placement problem as an interconnect centric fixed die placement problem. To solve such a problem, we propose a algorithm imitates the manual placement with two steps. In the first step, quadratic programming[8][9] is used to optimize the total wire length of the floorplan and estimate the aspect ratio range of the target chip. In the second step, given an area usage by user, we use a quasi-human rectangle packing algorithm to fulfill the placement while with the local interconnect and global interconnect optimization.

The rest of the paper is composed as follows: Sec. 2 gives the problem formulation. Sec. 3 is the details of the proposed algorithm. Sec. 4 is experimental results and the comparison with other method. We draw our conclusions in Sec. 5.

*This work is supported by the NSFC and HK RGC joint Grant No.60218004, NSFC 90307005 and NSF of USA CCR-0096383, and also is partly supported by the 863 Hi-Tech Research & Development Program of China 2002AA1Z1460

2. PROBLEM FORMULATION

Given a set of circuit building blocks, the interconnect central fixed die placement problem can be defined as follows:

A set of rectangular blocks $M=\{M_1, M_2, \ldots, M_n\}$, Each M_i is defined by a tuple (h_i, w_i), where h_i and w_i are the height and the width of block M_i, respectively .. The area of block M_i is denoted as $AREA(M_i)$.

A set of nets specifying the interconnections between pins of blocks and a set of pads (external pins) are also given. Target chip area: H×W. The target chip area usage, i.e.$\Sigma AREA(M_i)/$ H×W, and the aspect ratio (H/W) of the chip are given.

A placement $P=\{Mi\ (x_i, y_i) \mid 1 <= i <= n\}$ is an assignment of coordinates to the lower left corners of n rectangular blocks in the given chip area such that there is no two rectangular blocks overlapping. The objective of the placement is to find an assignment so that the interconnection wire-length between blocks is minimized while satisfying the given constraints, if any.

3. PROPOSED ALGORITHM

Our algorithm uses two steps to optimize the total wire-length and the chip area. In the first step, a quadratic programming approach is used to minimize a 2-order objective function of the total wire-length, the result of this iterative process is an initial floorplan that has minimum total wire length. Based on the initial floorplan, the aspect ratio of the chip can be estimated, after given the area usage of the placement, the chip area can be calculated easily. In the second step, we use the efficient rectangle-packing algorithm based on the less flexibility first principle to fulfill the placement, while in the block packing (or placement) process, the local interconnect and the global interconnect optimization are both taken into consideration.

3.1. Quadratic Programming for Wire-length Minimization

The objective of quadratic programming is the minimization of the following formula:

$$W(x,y) = \sum_{i=1}^{n}\sum_{j=1}^{n} \omega_{ij} \cdot \frac{1}{2}[((x_i - x_j)^2 + (y_i - y_j)^2]$$

subject to

$$g(x,y) = (r_i + r_j)^2 - [(x_i - x_j)^2 + (y_i - y_j)^2] \le 0$$

where $W(x,y)$ is the object function of the total wire length; n is the number of the blocks of the placement instance; (x_i, y_i) is the center coordinates of the circle that represent block i ; w_{ij} is the number of the nets between block i and block j; r_i is the radius of block i . $g(x, y)$ is the constrained condition that any two circles can not be overlap.

Our goal to use quadratic programming is to get an initial floorplanning with its total wire length is minimized. So we simply use the minimum circle that contain block i to represent block i. To optimize $W(x,y)$ subject to $g(x, y)$, we use penalty function method to convert this problem to an objective function without constrained conditions as follow[8]:

$$P(x,y,c_k) = W(x,y) + c_k \sum_{i=1}^{n}\sum_{j=1}^{n} Max(0, g(x,y))$$

where $c_0=1$, $c_{k+1} = 10*c_k$.

To solve $P(x, y, c_k)$ and get its global optimization solution, we need fix some blocks position in advance. Suppose there is at least one block is fixed (such as boundary constraints), for every c_k , the coefficient matrix of the first order derivative of $P(x, y, c_k)$ is a diagonally dominant matrix, it is also a positively definite matrix. So, for a c_k , there is an optimization solution for $P(x, y, c_k)$. In such a way, we have a convergence sequence $\{(x, y)^k\}$. The limit of the sequence $\{(x, y)^k\}$ is the optimization solution of the problem, it is an initial floorplan with minimum total wire length. Many methods have been proposed to solve such a problem, such as Newton method, conjugated grads and sequential quadratic programming etc.

We use Sequential Quadratic Programming algorithm based on Matlab to solve the problem of quadratic programming for total wire length minimization.

3.2. Placement Based on The Less Flexibility First Principle

For thousands of years, Chinese masons have been using the following rule-of-thumb in their daily stone packing work: "Golden are the corners; silvery are the sides; and straw are the hollows". The principle is honored that when a new packing step is taken, the deformation of the packing area should be kept as minimal as possible. There are three kinds of empty space a rectangle can choose to pack in: corners, space sides, and the rest, which are the "hollow" areas. Intuitively speaking, the deformation of the working space (i.e. placement area) caused by corner packing is minimal, then the deformation caused by side packing, then that of space packing. Basically, to obtain final better results, we should try to use up the more restricted resource and fulfill the requests with more restricted requirements earlier so as to leave the more "flexible" resources or requests for the getting harder later stages of the processing. This schedule sequence of packing tasks can be named as the *Less Flexibility First* (LFF) principle.

Similarly, the rectangles to be packed also have different degrees of flexibility depending on their shapes and sizes. It is generally agreed that a rectangle with large size or with larger longer-side/short-side ratio tend to possess a less packing flexibility. i.e. a rectangle of larger size or "slimmer" shape will have a higher priority of being packed earlier.

Interconnect wiring length optimization among VLSI blocks is also very important. In a placement, the more number of nets between two blocks also reflect the flexibility of their being bundled together. Actually, a packing step always involves more than two blocks that will be the neighbors of the block to be packed. In this case, it is reasonable to use interconnect flexibility among these blocks rather than interconnect flexibility between two blocks to measure the interconnect flexibility of the block to be packed in a corner position.

3.3. Implementation of Placement Based on the Less Flexibility First Principle

In a given rectangle chip area, it is reasonable that packed rectangle (VLSI module) must occupy one empty corner of the current chip packing configuration. When an unpacked block is selected for a corner of the current chip packing configuration and is packed in the selected corner, a new current chip packing configuration is formed by packing the block in the selected corner. There are many candidate corners for a packing step and one corner has four packing orientations for a rectangle. For example, in Figure.1, there are total 7 corners in the current chip packing configuration, and a block B can be packed at any one of them with four possible orientations. As a result, there are 28 packing choices for block B. In Figure 1, the real line block form the current chip packing configuration.

In terms of the flexibility definitions discussed above, we define one choice of this kind as a Candidate Corner Packing Step (CCPS). A CCPS can be represented by a six-tuple as

< block_id, length, width, orientation, x, y>

The block_id is block name, the length and width are values of the longer and shorter sides of the packed block, orientation indicate if the rectangle is placed with its longer side laid horizontally or vertically. (x, y) is the left-lower corner coordinate of the suggested location.

For a CCPS, a fitness value is calculated as follows:

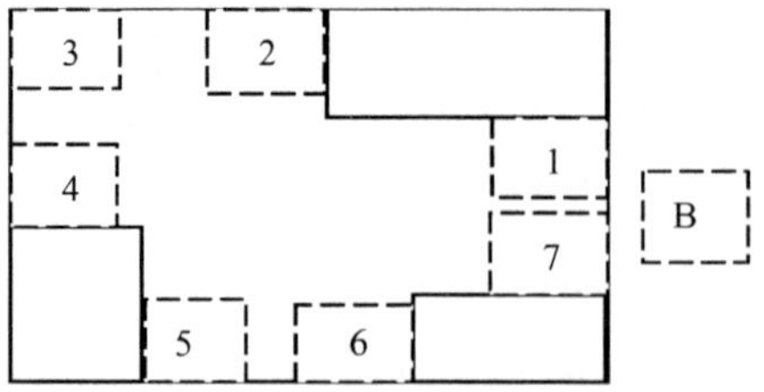

Figure. 1 Candidate packing corners for a block B

$$FV_i = \mu_1 * A_p + \mu_2 * (1/C_{net}^{\ i}) + \mu_3 *(1/C_{net}^{\ P})$$
$$- \mu_4 *(|x_{i\text{-}cur}\text{-}x_{i\text{-}ori}|^2 + |y_{i\text{-}cur}\text{-}y_{i\text{-}ori}|^2)$$

where μ_1, μ_2 , μ_3 , μ_4 are the weights.

Under the current chip packing configuration, for an unpacked block *i* we have a list of candidate CCPSs. For example, in the current chip packing configuration in Figure 1, unpacked block B has 28 CCPSs. We sort all CCPSs of all unpacked blocks for the current chip packing configuration in a list according to the definition of the flexibility of a rectangle in the last section. In the list, an earlier CCPS possesses a higher priority. Then the unpacked blocks are pseudo-packed with a greedy strategy following the list order until no rectangle can be packed in the present chip packing configuration. A_p is the total area of all rectangles had been packed, $C_{net}^{\ i}$ is the interconnect flexibility among the block i that to be packed and the already packed blocks in one corner. $C_{net}^{\ P}$ is the interconnect cost among all blocks had been packed. To ensure that $C_{net}^{\ i}$ and $C_{net}^{\ P}$ have high precision, we use half perimeter model to calculate the net length in the program. $C_{net}^{\ i}$ is the local connectivity of block i, while $C_{net}^{\ P}$ is the total connect cost for a CCPS. We combine the local connect optimization and global connect optimization in this implementation manner.

In the initial floorplan, block i has an "original" position $(x_{i\text{-}ori}, y_{i\text{-}ori})$. While in a pseudo-packing step, block i will have a current packing position $(x_{i\text{-}cur}, y_{i\text{-}cur})$. The optimization strategy is to minimize the distance between the two points such that the total wire length minimized in the quadratic programming step could be maintained. So we add a penalty term ($- \mu_4 *(|x_{i\text{-}cur}\text{-}x_{i\text{-}ori}|^2 + |y_{i\text{-}cur}\text{-}y_{i\text{-}ori}|^2)$) to FV_i . The negative of this term is to ensure that a CCPS of block i that have larger distance away from its original position in the initial floorplan will have more opportunity to be discarded. In Figure 2, estimated chip area is divided into grid and the position of block i is represented by its center. The location of block center is measured by the grid coordinates.

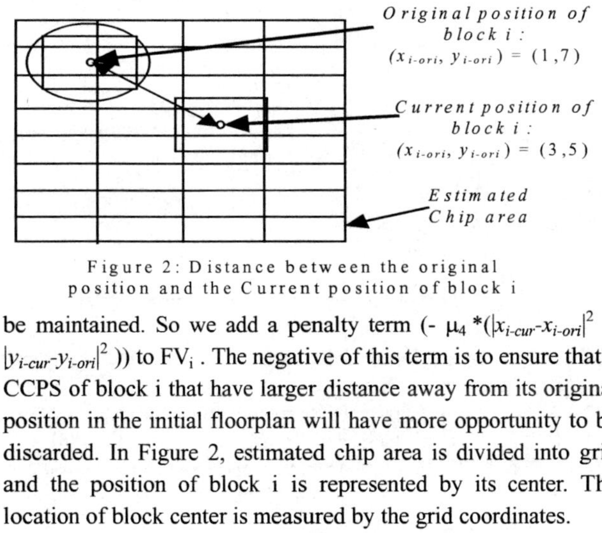

Figure 2: Distance between the original position and the Current position of block i

Under the current chip packing configuration, block that its CCPS has the highest FV_i will be selected as the next packing step. After this step, a new current chip packing configuration is formed and the rest unpacked blocks again have a new set of CCPSs. This placement process iterative to have all modules placed in the chip area.

3.4. Summary of The Proposed Algorithm

Our algorithm combine quadratic programming and rectangle packing based on the Less flexibility first principle. We name it as QP-LFF(). The algorithm can be summarized as follows.

Procedure QP-LFF()

1. solve $P(x, y, c_k)$ iteratively and get its global optimization solution using Matlab SQP;
2. estimate the aspect ratio range of the initial floorplan;
3. set the area usage range;
4. calculate a set of chip area according to 2 and 3;
5. for a chip area
6. set the current packing configuration as blank chip;
7. for all unpacked blocks under the current packing configuration
8. produce the set of CCPSs;
9. sort the set of CCPSs according to the definition of the flexibility of a rectangle;
10. Psudo-packing in a greedy manner according to the list order and calculate FV for every CCPS;
11. Select the block that its CCPS has the highest FV value and packed the block according to the CCPS, form the new current packing configuration;
12. End for
13. End for

4. EXPERIMENTAL RESULTS

We use Matlab Sequential Quadratic Programming to get the initial floorplan which has the minimum connect cost. The initial floorplan will be the input of the rectangle packing algorithm. We have implemented the rectangle packing in the given chip area algorithm in C programming language in a PC with Pentium 4 1.8G CPU, the operating system is Redhat Linux 7.2.

Most floorplan algorithm published in recent years are based on a specific floorplan representation and use Simulated Annealing to search an optimal floorplan. These algorithms search a floorplan with constrained aspect ratio, but unbounded size. With these conditions, they have more opportunity to get placement with higher area usage and shorter total wire length. So it is unfair to compare our experimental results directly with these results, we just list our experimental results together with the results given in the recent published papers to demonstrate the solution quality of our method.

In Table 1, we list the experimental results on MCNC benchmarks and Playout with simultaneously optimize chip area and total wire length. Note that the results of Enhanced O-tree(on Sun Ultra60), TCG(on Sun Ultra60), SA-LP(on Pentium III 500MHz processor)[10], QP-LFF(on Pentium IV 1.8G Redhat Linux 7.2) are obtained on different machines, the experiment results of others may be the best of them, while ours is the average of the top five results . We get our experimental results with area usage varied in the range of 90%--99%, aspect ratio of the chip is set to 1:1.5, 1:1.4, 1:1.3, 1:1.2, 1:1.1, 1:1, 1.1:1, 1.2:1, 1.3:1, 1.4:1, 1.5:1. In table 1, we only list one run time of rectangle packing of QP-LFF. For run time of quadratic programming using Matlab, the time for Apte is about several seconds while ami49 need about half an hour. We run quadratic programming for a benchmark only once.

In table 2, we list the best area value with only optimize the chip area for enhanced O-tree, TCG, and FAST-SP[11], while our results are selected from the results with optimize the chip area and total wire length simultaneously. We only list the results for ami33 and ami49. Figure 3 is the initial floorplan of Playout, and Figure 4 is the placement of Playout based on the initial floorplan in Figure 3.

5. CONCLUSIONS

In practical use, users often give the range of aspect ratio and the range of the chip size in advance according to the placement instance. For such a placement problem, simulated annealing based on a specific floorplan representation may need very long time to get a feasible solution or may fail. However, our proposed algorithm can get feasible solutions for the given range or prove that there are no feasible solutions for the given range in a reasonable shorter time. The proposed algorithm imitates the manual placement with two steps. Because we emphasize interconnect in the two steps, our method is obviously an interconnect centric placement algorithm while with the chip size and aspect ratio are given in advance. Such a placement manner is more close to the practical placement process. Experimental results demonstrated the effective of the proposed method.

Table 1 Comparison of QP-LFF with other method for simultaneously Optimize area and total wire length

		Ami33	Ami49	Playout
Enhanced O-tree	Area(mm^2)	1.299	39.92	-
	Wire(mm)	52.13	702.8	-
	Time(s)	205	700	-
TCG	Area(mm^2)	1.237	38.20	
	Wire(mm)	50.29	663.1	-
	Time(s)	939	3613	-
SA-LP	Area(mm^2)	1.227	37.54	-
	Wire(mm)	46.14	752.4	-
	Time(s)	2902.1	16198.4	-
QP-LFF	Area(mm^2)	1.232	39.18	91.200
	Wire(mm)	45.3	879.9	6523.65
	Time(s)	6.42	53.07	63.43

Table 2 Comparison of minimum: other methods only optimize for area while QP-LFF optimize area and total wire length simultaneously

		Ami33	Ami49	Playout
Enhanced O-tree	Area(mm^2)	1.24	37.73	-
	Time(s)	118	406	-
FAST-SP	Area(mm^2)	1.205	36.50	-
	Time(s)	20	31	-
TCG	Area(mm^2)	1.20	36.77	-
	Time(s)	306	434	-
QP-LFF	Area(mm^2)	1.177	36.60	89.30
	Time(s)	6.90	69.17	32.00

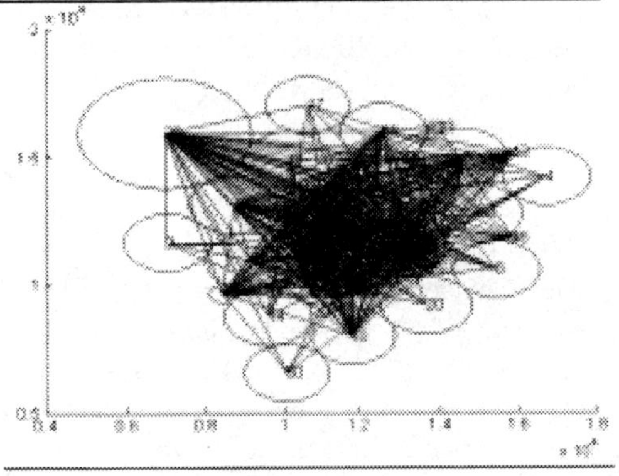

Figure3: initial floorplan of Playout

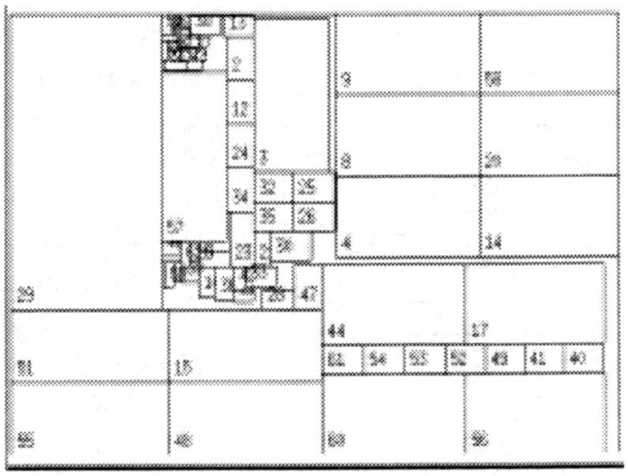

Figure4: placement of Playout

REFERENCES

[1] D.F.Wong, C.L. Liu, A new algorithm for floorplan design, Proc. of 23rd ACM/IEEE DAC'86, 101-107, 1986

[2] H.Murata, K.Fujyoshi, S.Nakatake, and Y.Kajitani, Rectangular-packing-based module placement, ICCAD'95, 472-479, 1995

[3] S.Nakatake, K.Fujiyoshi, H.Murata, and Y.Kajitani, Module placement on BSG-structure and IC layout applications, ICCAD'96, 484-491, 1996

[4] P. -N.Guo, C.-K.Cheng, T.Yoshimura, An O-tree representation of non-slicing floorplan and its applications , DAC'99, 1999

[5] Xianlong Hong, Gang Huang, Yici Cai, Jiangchun Gu, Sheqin Dong, C, -K Cheng, Jun Gu, Corner Block List: An Efficient and Effective Topological Representation of Non-Slicing Floorplan, ICCAD'00, 8-12.

[6] Jai-Ming Lin, Yao-Wen Chang: TCG: A Transitive Closure Graph-Based Representation for Non-Slicing Floorplans. DAC 2001: 764-769

[7] A. B. Kahng. Classical Floorplanning Harmful?: Proc. ACM Intl. Symp. On Physical Design, 207-213, April, 2000.

[8] L Sha, R W Dutton. An analytical algorithm for placement of arbitrarily sized rectangular blocks. In: Proceedings of the 22th Design Automation Conference, Las Vegas, NV, 1985, 602-608

[9] Takumi Okamoto , Takeshi Yoshimura "A New Approach to VLSI Floorplanning based on Quadratic Programming and Rectangle Packing" SASIMI 2001

[10]Jae-Gon Kim, Yeong-Dae Kim, "A linear programming-based algorithm for floorplanning in vlsi design" IEEE Trans. Computer-Aided Design. vol. 22, pp.584–592, May.2003.

[11] X. Tang and D.F. Wong. FAST-SP, ASP-DAC'2001, pp 521-526, Yokohama, Japan, Jan. 2001

PERFORMANCE AND RLC CROSSTALK DRIVEN GLOBAL ROUTING*

Ling Zhang[1], *Tong Jing*[1], *Xianlong Hong*[1], *Jingyu Xu*[1], *Jinjun Xiong*[2], *Lei He*[2]

[1] Dept. of CST, Tsinghua Univ., Beijing 100084, P. R. China
[2] Dept. of EE, UCLA, Los Angeles, CA 90095-1594, USA
E-mail: {zhangling, jingtong}@tsinghua.org.cn; jinjun@ucla.edu; lhe@ee.ucla.edu

ABSTRACT

This paper presents a global routing algorithm that minimizes total wire length and satisfies RLC crosstalk constraints specified at sinks. Our algorithm is based on critical network concept and search space traversing technology (SSTT) for global routing synthesis and Tabu search for shield insertion and net ordering (SINO) to eliminate noise. The algorithm achieves about 20x speedup compared with a recent work using iterative deletion based global routing and simulated annealing based SINO. Furthermore, our algorithm increases the wire length by 4% compared with global routing without crosstalk constraints, achieving a 2.5x reduction compared with the aforementioned recent work.

1. INTRODUCTION

With the progress of VDSM technology and giga-hertz clock frequencies, performance optimization becomes an increasingly dominant factor in global routing [1]. One of the major concerns is coupling noise elimination. There are some works focusing on noise reduction, which mainly fall into two categories, noise modeling [2-3] and noise minimization [4-7]. Among noise minimization algorithms, some post optimizations are performed after global routing. Ref. [4] described a two-part algorithm of region-based crosstalk risk estimation and crosstalk reduction. In [5], the iSINO algorithm is proposed, which eliminates crosstalk by inserting shields.

Researchers find that it is more flexible if they reduce noise in the global routing phase. In [6], it constructs Steiner tree with a cost function including crosstalk consideration. If the crosstalk of initial routing solution still exceeds the given bound, then do rip up. Ref. [7] proposed the GSINO Algorithm. Since the simulated annealing (SA) method is used, it takes long running time. Meanwhile, the objective of these two algorithms is to minimize crosstalk. They do not take timing performance and routability into consideration.

This paper studies RLC coupling noise elimination problem in the process of global routing. The main contribution of this paper is that an efficient crosstalk elimination algorithm based on Tabu search is proposed. Moreover, timing performance and routability are simultaneously considered at global routing level. That is, it regards wire length as the objective and considers timing, RLC coupling noise, and routability as the constraints. Then, the performance optimization is performed throughout the global routing phase under multi-constraints.

The remainder of this paper is organized as follows. Section 2 gives necessary preliminaries. In Section 3, the coupling noise elimination algorithm based on Tabu search is described in detail. In Section 4, we discuss global routing with performance optimization. Section 5 shows experimental results. Section 6 concludes and gives some possibilities for future work.

* This work was partially supported by the SRFDP 20020003008, the NSFC (China) 60176016, the NSF CAREER Award (USA) CCR-0093273, Hi-Tech Research and Development (863) Program of China 2002AA1Z1460, and the Key Faculty Support Program of Tsinghua Univ. [2002] 4.

2. PRELIMINARIES

2.1. Global Routing Problem

With the progress in multi-layer routing technology, routing area is a whole chip plane. Thus, a net can be specified as a set of nodes in global routing graph (GRG). Then, the problem of routing a net can be described as a rectilinear Steiner tree (RST) problem of specified nodes in GRG [9].

2.2. RLC Noise Model

The *LSK* model for RLC crosstalk [3, 7] is used in this paper. Different from earlier noise model [2], the *LSK* model considers coupling inductance between adjacent and non-adjacent sensitive nets. For any two segments N_{it} and N_{jt} in region R_t, the inductive coupling coefficient between them is

$$k_{it,jt} = \frac{L_{it,jt}}{\sqrt{L_{it} \cdot L_{jt}}} \quad (1)$$

where $L_{it,jt}$ is the mutual inductance between N_{it} and N_{jt}, and L_{it} and L_{jt} are the self inductance for N_{it} and N_{jt}, respectively. A formula-based K_{eff} model has been developed in [3] to calculate the coupling coefficients $k_{it,jt}$. Furthermore, the total amount of inductive coupling induced on N_{it} can be represented by the sum of the inductive coupling coefficients $K_{it} = \sum_{j \neq i} k_{it,jt}$ for all net segments N_{jt} that are sensitive to N_{it}.

To consider the effect of interconnect length and the general case where the total coupling is not uniform in all routing regions, a length-scaled K_{eff} (*LSK*) model was proposed in [7], where the *LSK* value is defined as

$$LSK = \sum_t l_t \cdot K_{it} \quad (2)$$

where l_t is length of R_t and K_{it} is total coupling for N_{it} in region t.

2.3. Tabu Search

Tabu search has been widely used to cope with the overwhelming computational intractability of NP-hard combinatorial optimization problems since firstly proposed by

Glover in 1986 [8]. The basic idea of this technology is simple, which records and taboos the local minimum points that has been reached so as to avoid getting stuck at these points and finds out new search ways that could lead to the global minimum point eventually. The outline of Tabu search algorithm can be described in Fig.1.

Step1. Select an initial solution x^{now}, and set Tabu list H=empty;
Step2. While not meet the stop conditions do
Generate a candidate list Can_N(x^{now}) from the neighborhood N(x^{now},H) of x^{now} that doesn't conflict with H;
Select the best solution from Can_N(x^{now}):x^{next};
x^{now}=x^{next};
Update Tabu list H;
End While

Fig.1. Outline of Tabu search algorithm.

Key factors of Tabu search are neighborhood, Tabu object, Tabu length and aspiration rule. The following are some concerns in applying Tabu search method. (1) How to choose proper Tabu object and Tabu length. (2) How to search efficiently in neighborhood. (3) How to set the reasonable aspiration rule.

3. NOISE ESTIMATION AND ELIMINATION

3.1. The Three-Step Method

The flow chart of crosstalk estimation and elimination is shown in Fig.2. There are three main steps described as follows.

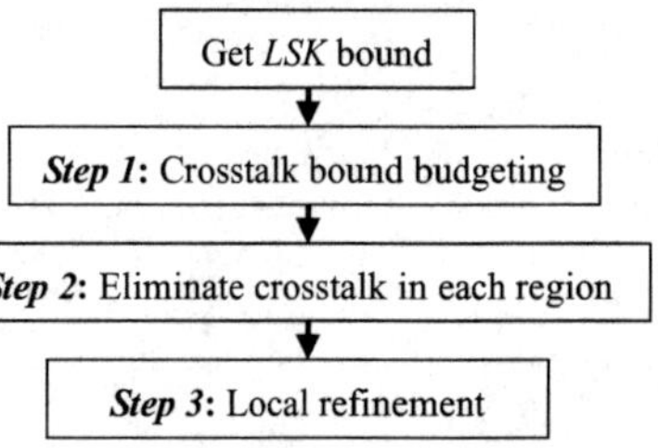

Fig.2. The three-step method.

Step 1: *LSK* bound budgeting

We partition the *LSK* bound at each sink of a net into the GRG edges belonging to the source-sink paths by using CBUD (uniform distributed crosstalk budgeting) strategy. This strategy uniformly partitions the *LSK* bound into edges according to their length. Let $\overline{LSK_{ij}}$ be the crosstalk bound at sink p_{ij} for net N_i, *len* be the total length from the source p_{io} to sink p_{ij}, each routing region (one GRG edge) R_t on the path is then assigned a uniform crosstalk sub-bound $\overline{K_{it}}$:

$$\overline{K_{it}} = \frac{\overline{LSK_{ij}}}{len} \tag{3}$$

If the segment N_{it} is shared with multiple paths starting from the same source to different sinks, we use the minimum value computed for these paths according to equation (3). If we focus on the same region and compare the K_{eff} with the sub-bound of one net $\overline{K_{it}}$, $\overline{K_{it}}$ will be expressed as K_{th}.

Step 2: Crosstalk elimination in each region

According to each $\overline{K_{it}}$ computed in ***Step 1***, this step applies optimization method in each region to insert shields, so that the crosstalk of all regions is within the given bound.

Ref. [3] and [7] introduced the strategy of using SA method to insert shields in each region. SA method could obtain good results. But it takes comparative long runtime. In the following, we will introduce the strategy of using Tabu search method for this step, which obtains similar performance as SA while greatly shortening runtime (see Section 5).

Step 3: Local refinement

Check each net to eliminate possible remnant crosstalk and delete unnecessary shields so that the final area is minimized.

First, to eliminate remnant crosstalk, the net N_{it} with most critical crosstalk violation is chosen, and shields will be inserted in the least congested region R_t on N_i's path.

Second, to reduce total area, the most congested region R_t is chosen, and the slack K_{it} - K_{th} of all nets in R_t is computed. If possible, shields could be deleted when K_{th} increases properly.

3.2. Crosstalk Elimination Based on Tabu Search

We use Tabu search in ***Step 2*** to reduce runtime. The method, described in Fig.3, is much faster than SA but can obtain similar performance.

Set the global solution in one GRG edge as initial solution x^{cur};
Set Tabu list H=empty; a=0; c=0;
While(a < N_a)
tmpcost = ∞ ;
b = 0;
While (b < N_b)
x^{new} = x^{cur};
randommove (x^{new});
If cost (x^{new}) is in H
c++;
If c < N_c, then continue;
Else c = 0;
If cost (x^{new}) < tmpcost, then
x^{tmp} = x^{new};
tmpcost = cost (x^{new});
b++;
Insert x^{cur} into H;
x^{cur}= x^{tmp};
If cost (x^{cur}) < cost (x^{min}), then x^{min} = x^{cur}; a = 0;
Else a++;
Update H;

Fig.3. Tabu search in ***Step 2***.

In the crosstalk elimination problem, a solution is a sequence of net ordering in a certain GRG edge. There are often hundreds of GRG edges in middle-scale circuits. Meanwhile, Tabu search algorithm is used for each GRG edge. So, the runtime of Tabu search greatly affects the efficiency of the whole routing algorithm.

In Fig.3, we can see that Tabu search finds a best legal candidate in a candidate set of x^{cur}'s neighborhood, taboos the current solution x^{cur}, and accepts that best legal candidate as new x^{cur}. It records the best solution x^{min} throughout the whole search process. Following such a flow, this method has the ability to traverse from the local minimal solution in the search space and can record the best solution it has ever reached. For convenience, we use x^{new} as a copy of x^{cur} at the beginning of iteration one time and do random movement on x^{new}. We record the best neighbor solution of x^{cur} in x^{tmp}.

We use the cost of each solution as the Tabu object because it is convenient to taboo a set of solutions having the same cost. The cost function includes the following four factors: (1) Total number of nets that are adjacent to their sensitive nets; (2) Total number of shields in a GRG edge; (3) Summation of $(K_{eff} - K_{th})$ for all nets with $K_{eff} > K_{th}$ in a GRG edge; (4) Total number of nets with $K_{eff} > K_{th}$ in a GRG edge.

We use four kinds of random movements to find a neighbor solution: swap two net randomly, move one net randomly, insert one shield randomly and remove one shield randomly. Each of these random movements has restrictions so that a neighbor solution is still a feasible solution (that is, to exclude the cases such as two shields are next to each other) and each movements has some certain possibility to be conducted with the control of different weights on them.

Four parameters, N_a, N_b, N_c, and Tabu length, could affect the running time and performance of Tabu search method. N_a is the total iteration times if Tabu search couldn't find a new best solution. N_b is the total number of neighbor solutions that Tabu search method regards them as legal candidates. N_c is the times that this method will try to find one legal candidates. N_c is also a kind of aspiration criterion for that if there are no more legal neighbor solution after N_c times search, this method will accept the last solution as a legal candidates, even if it has been tabooed actually. Tabu length is the times that one cost value is labeled as illegal.

Some of the parameters should fit for the scale of search space, which means for a larger search space, we need larger N_a , N_b , and Tabu length to obtain better results, and for a smaller search space, we need smaller N_a , N_b , and Tabu length to shorten the running time. In our problem formulation, the scale of search space directly depends on the number of nets in one GRG edge. When the number of nets is from 20 to 50, Tabu length and N_c do not have great effects on the final results. But all these parameters have effects on the runtime.

Based on large number of experiments, we find that the proper value of these parameters are as follows: N_a=350, N_b=20, N_c=10, Tabu length=3.

4. GLOBAL ROUTING WITH PERFORMANCE OPTIMIZATION

Besides the above RLC crosstalk elimination, we also perform performance optimization in global routing phase, which includes timing performance, routability, and coupling noise. It has been implemented as the performance and RLC crosstalk driven global router, called PO-GR, which consists of the following two parts.

(1) ***Part 1***: timing performance and routability

(2) ***Part 2***: crosstalk estimation and elimination

In ***Part 1***, it firstly generates an initial routing solution considering congestion and timing optimization. The timing analysis and optimization method follows the critical network concept introduced in [9] and the congestion reduction uses the search space traversing technology (SSTT) introduced in [10]. Then, ***Part 2*** eliminates the crosstalk from the solution by inserting shields and gets a mid-result. Finally, regard the mid-result as input and send it to ***Part 1*** for iterations. The flow chart and pseudo code of PO-GR are shown in Fig.4 and Fig.5, respectively.

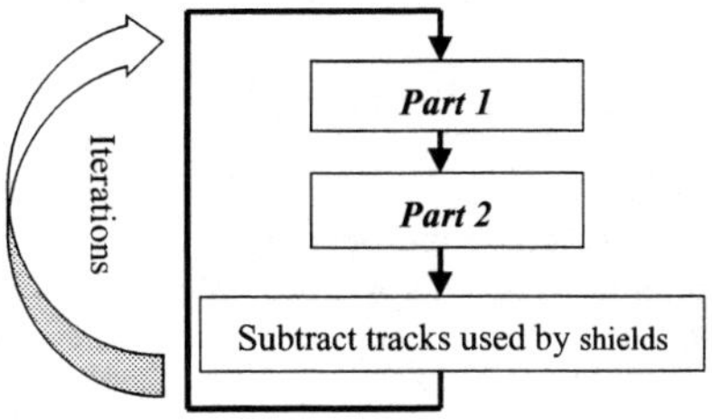

Fig.4. The flow chart of PO-GR.

*1. Call **Part 1** to generate a minimum wire length initial solution X^0 without congestion and timing violation;*
*2. Call **Part 2** to obtain $X^1 = CEE(X^0)$;*
3. If(no edge overflow in X^1) do go to step 4.;
Else do go back to step 1. to generate a new solution;
*4. Call **Part 1** again to obtain congestion and timing optimized solution X^2 from X^1;*

Fig.5. The pseudo code of PO-GR.

5. EXPERIMENTAL RESULTS AND DISCUSSIONS

The PO-GR is implemented in the C language. It runs on a SUN Enterprise 450 workstation with Unix OS. There are two running modes. *T* mode is timing-driven mode. *W* mode is non-timing-driven mode. We only reduce congestion in *W* mode, and consider both congestion and timing performance in *T* mode. We compared our crosstalk elimination algorithm with the recent work [3, 7] based on SA algorithm.

5.1. Benchmark Data

We tested three MCNC benchmarks under 0.2*um* technology. Table 1 summarizes the benchmark data sets. Sensitivity rate of 0.5 is given to all nets and a random sensitivity matrix is created. *LSK* bound at each sink is set to be 1000. N_a=350, N_b=20, N_c=10, and Tabu length=3.

Table 1 Benchmark data

Circuits	Number of nets	Grids
C2	745	9×11
C5	1764	16×18
C7	2356	16×18

5.2. Results

The experimental results are shown in Table 3, Table 4, and Table 5, respectively. Table 2 shows the needed notations.

5.3. Discussions

(1) From row 4 in Table 3, we can see that Tabu search sharply decreases runtime in XSINO step (about 20x speedup) compared with simulated annealing method [3, 7]. From row 10 in Table 3, we also can see that the total runtime is greatly shortened. The local refinement step has not been changed, but its runtime has been decreased slightly (row 7 in Table 3), which means Tabu search doesn't make any bad effects on succeeding optimization.

(2) From Table 4, we can find that Tabu search obtains similar results in routing area compared with simulated annealing method [3, 7]. The shielding number only increases a little.
(3) Row 4 in Table 5 shows that the wire length increment of PO-GR is no more than 4.65%. So, crosstalk estimation and elimination does slight bad effects on wire length. Furthermore, the wire length increment is about 10% in [7]. Therefore, we achieve 2.5x wire length reduction.
(4) The minimum delay slack (i.e., required delay – current delay), denoted as Min-R and shown in row 9 and row 10 in Table 5, is almost unaffected. So, PO-GR keeps the effectiveness in timing optimization in ***Part 1*** (i.e., P1 in Table 5).

Table 2 Needed notations

SA	The existing crosstalk elimination algorithm based on simulated annealing in [3, 7]
Tabu search	Our crosstalk elimination algorithm based on Tabu search
XSINO	***Step 2***: Crosstalk elimination in each region
LR	***Step 3***: Local refinement
TT	Total runtime (XSINO + LR)
Sn	Shield number
PO-GR	Our two-step router with performance optimization
P1	***Part 1***: timing and congestion optimization
Min-R	Minimum delay slack (required delay–current delay)

Table 3 Comparison of runtime (*s*) between Tabu search and SA

Circuits		C2	C5	C7
XSINO	SA	901.97	2140.36	3748.78
	Tabu search	45.75	112.87	237.80
	Runtime reduction	856.22	2027.49	3510.98
LR	SA	153.53	56.36	453.70
	Tabu search	91.44	34.08	227.50
	Runtime reduction	62.09	22.28	226.20
TT (XSINO + LR)	SA	1055.50	2196.72	4202.48
	Tabu search	137.19	146.95	465.30
	Total runtime reduction	918.31	2049.77	3737.18

Table 4 Comparison of results between Tabu search and SA

Circuits		C2	C5	C7
Area	SA	149×196	271×301	342×395
	Tabu search	149×202	273×307	346×393
Sn	SA	158	460	589
	Tabu search	165	501	621
	Sn increment	7	41	32

6. CONCLUSIONS AND FUTURE WORK

A performance and RLC crosstalk driven global routing algorithm is presented. The experimental results show that this algorithm is able to: (1) Preserve the good routing result and greatly decrease the running time. (2) Tackle coupling noise, timing performance and routability simultaneously. It efficiently eliminates crosstalk throughout the process of global routing by inserting shields, which has little influence on wire length and timing performance. (3) Take coupling inductance into consideration.

For future work, we plan to improve the timing efficiency of this algorithm. We will try to reduce the running time of local refinement step and find better strategies for crosstalk partitioning.

Table 5 Comparison of results between P1 and PO-GR

Circuits		C2	C5	C7
W mode	(P1) Wire length (*um*)	480350	1307456	1552916
	(PO-GR) Wire length (*um*)	477326	1368198	1575922
	Wire length increment	-0.63%	4.65%	1.48%
T mode	(P1) Wire length (*um*)	476424	1346876	1569366
	(PO-GR) Wire length (*um*)	479100	1280352	1567818
	Wire length increment	0.56%	-4.94%	-0.10%
	(P1) Min-R	-0.009243	0.012124	0.000034
	(PO-GR) Min-R	-0.007195	0.003439	0.001243

7. REFERENCES

[1] R. Kastner, E. Bozorgzadeh, and M. Sarrafzadeh, "An exact algorithm for coupling-free routing", in: *Proc. ACM ISPD*, Sonoma, CA, USA, 2001, pp.10-15.
[2] T. Sakurai, S. Kobayashi, and M. Node. "Simple expressions for interconnecting delay, coupling and crosstalk in VLSI's", in: *Proc. IEEE ISCAS*, Singapore, 1991, pp.2375-2378.
[3] L. He and K. M. Lepak. "Simultaneous shield insertion and net ordering for capacitive and inductive coupling minimization", in: *Proc. ACM ISPD*, San Diego, CA, USA, 2000, pp.56-61.
[4] T. X. Xue, E. S. Kuh, and D. S. Wang. "Post global routing crosstalk synthesis", *IEEE Trans on CAD*, 1997, 16(12): pp.1418-1430.
[5] J. J. Xiong, J. Chen, J. Ma, and L. He. "Post global routing RLC crosstalk budgeting", in: *Proc IEEE/ACM ICCAD*, San Jose, CA, USA, 2002, pp.504-509.
[6] H. Zhou and D. F. Wong. "Global routing with crosstalk constraints", *IEEE Trans on CAD*, 1999, 18(11): pp.1683-1688.
[7] J. Ma and L. He. "Towards Global routing with RLC crosstalk constraints", in: *Proc ACM/IEEE DAC*, New Orleans, Louisiana, USA, 2002, pp.669-672.
[8] F. Glover, "Future paths for integer programming and links to artificial intelligence", *Computers and Operations research*, 1986, 13(5): pp.533-549.
[9] T. Jing, X. L. Hong, H. Y. Bao, *et al*, "A novel and efficient timing-driven global router for standard cell layout design based on critical network concept", in: *Proc. IEEE ISCAS*, Scottsdale, Arizona, USA, 2002, pp.I165-I168.
[10] T. Jing, X. L. Hong, H. Y. Bao, *et al*, "SSTT: Efficient Local Search for GSI Global Routing", *J. Comput. Sci. & Technol.*, 2003, 18(5): pp.632-639.

Multi-layer Floorplanning for Reliable System-on-Package

Pun Hang Shiu, Ramprasad Ravichandran, Siddharth Easwar, and Sung Kyu Lim
School of Electrical and Computer Engineering
Georgia Institute of Technology, Atlanta, GA 30332-0250
{pshiu,raam,seaswar,limsk}@ece.gatech.edu

Abstract

Physical design automation for the new emerging mixed-signal System-on-Package (SOP) technology requires a new kind of floorplanner—it must place both active components such as digital IC, analog ICs, memory modules, MEMS, and opto-electronic modules, and embedded passive components such as capacitors, resistors, and inductors in a multi-layer packaging substrate while considering various signal integrity issues. We propose a new interconnect-centric multi-layer floorplanner named MF-SOP, which is based on a multiple objective stochastic Simulated Annealing method. The contribution of this work is first to formulate this new kind of floorplanning problem and then to develop an effective algorithm that handles various design constraints unique to SOP. The related experiments show that the area reduction of MF-SOP compared to its 2-D counterpart is on the order of O(k) and wirelength reduction is 39% average for k-layer SOP, while satisfying design constraints.

I. Introduction

The next generation electronic packaging technology called System-on-Package (SOP) [1] integrates both active components (such as digital IC, analog ICs, memory modules, MEMS, and opto-electronic modules) and embedded passive components (such as capacitors, resistors, and inductors) all into a single high speed/density multi-layer packaging substrate. SOP is more advanced than PCB, MCM, or SIP (System-in-Package) since MCM handles the integration of digital ICs only and SIP handles digital components and passive elements only. Moreover, the SOP design paradigm facilitates rapid reengineering via reuse libraries. Therefore, SOP promises a high return on investment at a very low risk within shorter time-to-market cycle compared to the System-On-Chip (SOC) paradigm.

A high performance mixed signal system employs a lot of passive components—up to 30 passive components per an IC. For example, Sony Handy Cam DCR-PC7 has 43 ICs and 1329 passive elements. Such passive components continue to take up much circuit board real estate. Therefore, rigorous attempts have been made to replaces them with so-called embedded passive components (EPC), which are small and flat enough to be inserted between package layers. EPCs allow devices to get smaller or designers to fit more functionality in the same space; eliminate the costs currently needed to purchase and solder on discrete devices; allow for more design flexibility; and derive electrical benefits from the different current path that would be traveled. EPCs can also be used for simultaneous switching noise reduction, cross talk reduction, network matching, and signal integrity. The complexity of a radio frequency front-end IC is considerably simpler with high quality passive components. However, EPC placement needs to be done carefully while considering design constraints.

The physical layout resource environment of SOP is multi-layer in nature—the top layer is mainly used to accommodate active components, the middle layers are mainly for passive components, and the I/O pins are located at the bottom of the SOP package. Therefore, all layers are used for both placement and routing unlike PCB or MCM. Therefore, the existing design tools for PCB or MCM can not be used directly for the design of SOP. The existing work on multi-layer floorplanning is very few. Authors in [2] solved multi-layer floorplanning for vertically stacked digital systems. However, this work does not address the mixed-signal integration issues existing in SOP technology. Therefore, SOP technology requires a new kind of multi-layer floorplanner. We propose a new interconnect-centric multi-layer floorplanner named MF-SOP, which is based on a multiple objective stochastic Simulated Annealing method. The contribution of this work is first to formulate this new kind of floorplanning problem and then to develop an effective algorithm that handles various design constraints unique to SOP. The related experiments show that the area reduction of MF-SOP compared to its 2-D counterpart is on the order of O(k) and wirelength reduction is 48% average for k-layer SOP, while satisfying design constraints.

This paper organization is as follows. The problem formulation is given in Section II. SOP constraints are described in Section III. Experimental results and conclusions are given in Section IV and V, respectively.

II. Problem Formulation

A multi-layer SOP floorplan consists of a set $B=\{b_i | 0 \leq i < n\}$ of n blocks and a set $L=\{l_i | 0 \leq i < k\}$ of k layers. A block is either an active component or embedded passive component (EPC). We assume rectangular shape for all these blocks. Each floorplan f_i has a set of blocks B_i, which is a non-empty proper subset of B. A SOP floorplan F is represented by a set $F=\{f_0, f_1, \ldots, f_{k-1}\}$, where a floorplan f_i is a 2-dimensional placement of blocks in B_i. A SOP floorplan F is *feasible* if (i) F is free of overlap among block location, (ii) F satisfies the layer and geometric constraints specified by the user. The goal is to minimize the following cost function based on an SOP floorplan solution F: $C(F) = c_1 \cdot area(F) + c_2 \cdot wire(F) + c_3 \cdot via(F) + c_4 \cdot penalty(F)$. The first term $area(F)$ is the final footprint area of SOP package. The minimization of this objective results in a minimal overall SOP package area. The second term $wire(F)$ is the half-perimeter bounding box (HPBB) based estimation of wirelength. We ignore the height (z-dimension) of the *bounding cube* and use only the x and y-dimension for the computation of the wirelength of a net. Instead, the z-dimension has a direct impact on $via(F)$. If a net n spans from layer i to layer j, then $via(n) = |i - j|$. The sum of via(n) for all nets is via(F). The penalty term $penalty(F) = 0$ when there is no constraint violation in F.

We observe from related experiments that adding the following components to $C(F)$ results in a more compact multi-layer floorplan: *total flatten area flat*(F) and *dimension deviation*

0-7803-8251-X/04/$17.00 ©2004 IEEE

dev(F). *flat(F)* is the sum of all floorplans, $flat(F) = \sum a(f_i)$. The minimization of this objective results in a highly compact floorplan for each layer. *dev(F)* measures how much the upper right corner (URC) of a floorplan deviates from the average URC. We compute the average URC (u_x, u_y) by $u_x = \Sigma u_x(f_i)/k$, where $u_x(f_i)$ denotes the x-coordinate of the URC of a floorplan f_i. We compute $u_y(f_i)$ using y-coordinates instead. Let $d(f_i) = |u_x - u_x(f_i)| + |u_y - u_y(f_i)|$ be the dimension deviation of a floorplan of f_i. Then *dev(F)*, the dimension deviation of SOP floorplan *F* is simply the sum of all $d(f_i)$. The minimization of this objective results in a more dimension-balanced floorplan among all layers. It may seem redundant to have all three area-related objectives *area(F)*, *flat(F)*, and *dev(F)* in *C(F)*. However, our related experiments indicate that each of these three objectives contribute to the minimization of not only the final footprint area *area(F)* but also the wirelength estimation *wire(F)*.

Among many proposed methods to represent 2-dimensional floorplanning, we extend the sequence pair (SP) [4] to represent the multi-layer SOP floorplan solution. Our multi-layer sequence pair is represented by ($SP_0|SP_1|...|SP_{k-1}$), where SP_i contains the positive and negative sequence for the blocks contained in layer *i*. For a faster area evaluation for a given multi-layer SP, we use longest common subsequences (LCS) [3] method. A recent effort [5,6,7] uses various floorplanning representations to impose design constraints for 2-dimensional constraints.

Authors in [4] propose three types of moves for solution perturbation during Simulated Annealing: M1 (swap two modules in positive sequence), M2 (swap two modules from both positive and negative sequence), and M3 (rotate). We add two moves M4 and M5 to search the solution of multi-layer floorplanning effectively: M4 is similar to M1, except that the two blocks are from positive sequences in different layers. M5 selects a block from layer *i* and moves it to another layer *j*. The location in positive and negative sequence from SP_j is again randomly chosen.

III. SOP Geometric Constraints

A. SOP Geometric Constraints

We categorize the geometric constraints among active and passive components introduced in Section II into the following 6 types: (i) *noise*: decoupling capacitors are placed nearby I/Os or active components, (ii) *thermal*: some active/passive components are placed in certain layers, (iii) *power*: digital and analog ICs are placed in different voltage islands, (iv) *timing*: blocks from a critical path are placed closer, (v) *interface*: I/O blocks are placed near the bottom layer, (vi) *cluster*: functionally dependant blocks are placed close together.

Most of the active components are required to be placed on the top layer due to heat dissipation requirement. However, some active components that do not generate too much heat can be placed in the middle layers. EPCs can be placed at any layers, but using middle layers is the most beneficial in reducing the overall footprint area of SOP. However, some EPCs are required to be placed on the top layer due to thermal and/or noise issues. Third, some active components need to be placed nearby together or apart from each other due to several reasons including signal/power integrity, performance optimization, etc. Lastly, most EPCs need to be placed closer to the related active components. Handling the layer constraints is straightforward, but the geometric constraints are harder to satisfy.

Table 1. Geometric Constraints for SOP Floorplanning

type	*method*	*syntax*	*meaning*
noise	point	$[b_i\|(x,y,z)]$	b_i touches (x,y,z)
thermal	layer	$[B_i\|l]$	B_i in layer l
power	region	$[B_i\|(x,y,w,h)]$	B_i intersects with region (x,y) and $(x+w,y+h)$
timing	abutment	$[B_i]$	B_i abutted
interface	boundary	$[B_i\|TBLR/l]$	B_i near boundary of layer l
cluster	group	$[B_i\|(x,y,z)]$	B_i within a distance of (x,y,z)

Table 1 describes these 6 geometric constraint types we consider in SOP floorplanning. A prior timing analysis or signal integrity analysis is performed by the user to identify (i) the source of timing, noise, thermal, and power supply problem, and (ii) ways to fix these problems in a form of constraint. Each constraint is then translated into a geometric form so that our multi-level floorplanner attempts to satisfy this geometric constraint. Our strategy is to quantify the amount of violation of the constraints specified, and guide Simulated Annealing-based optimization so that the amount of violation is minimized or completely removed if possible.

Our strategy for effective solution space search during Simulated Annealing is as follows:

1. construction of initial solution: we first assign all blocks under layer constraints to the target layers and fix them during the annealing. For the remaining blocks, we randomly and evenly distribute them into all layers.
2. solution perturbation: we perform more inter-layer moves (M4 and M5 discussed in Section II.A) during high temperature annealing and more intra-layer moves (M1, M2, and M3) during low temperature annealing.
3. weighting constants in *C(F)*: we focus more on penalty(*F*) and via(*F*) during high temperature annealing and more on area(*F*) and wire(*F*) during low temperature annealing.

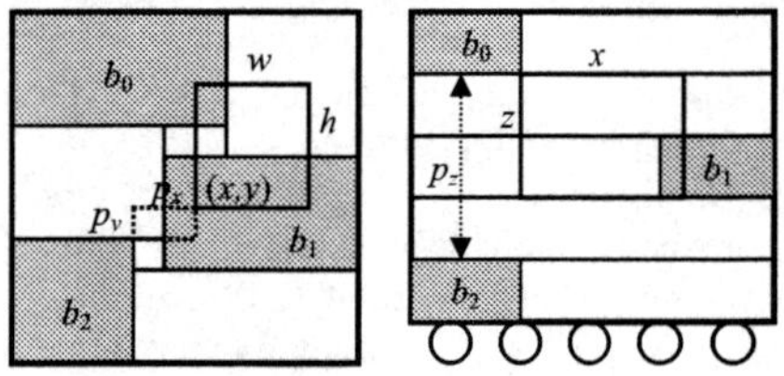

(a) (b)

Figure 1. Constraint Examples. (a) region constraint $r_1=[\{b_0,b_1\}|(x,y,w,h)]$ and $r_2=[\{b_1,b_2\}|(x,y,w,h)]$. r_1 is satisfied and r_2 has penalty of p_x+p_y. (b) group constraint $g_1=[\{b_0,b_1\}|(x,y,z)]$ and $g_2=[\{b_0,b_2\}|(x,y,z)]$. g_1 is satisfied and g_2 has penalty of p_z. *y*-dimension is not shown.

An example of region constraint is given in Figure 1(a). First, consider $r_1=[\{b_0,b_1\}|(x,y,w,h)]$. Since both b_0 and b_1 are intersecting with the region defined by (x,y,w,h), we see that r_1 is satisfied and the penalty is zero. Now consider $r_2=[\{b_1,b_2\}|(x,y,w,h)]$. Since b_2 is completely outside the region, r_2 is not satisfied and its penalty is computed by the sum of p_x and p_y. An example of group constraint is given in Figure 1(b). First, consider $g_1=[\{b_0,b_1\}|(x,y,z)]$. Since the distance between b_0

and b_1 is within the 3-dimensional distance (x,y,z), we see that g_1 is satisfied and the penalty is zero. Now consider $g_2=[\{b_0,b_2\}|(x,y,z)]$. Since the z-distance between b_0 and b_2 is bigger than z, g_2 is not satisfied and its penalty is p_z.

B. Penalty Computation

The penalty computation for constraint violation is summarized in Table 2. Penalty computation for x-dimension (p_x) is shown only, but other dimensions (p_y) and (p_z) can be computed similarly using y/z-coordinates and height/layer information. The overall penalty $p=p_x+p_y+p_z$. Note that p_z contributes to our via cost and usually carries more weights than p_x or p_y. The point, layer, and region constraints are *intersection-based*—these constraints are violated if there is no intersection between the blocks and the region given. The abutment, boundary, and group constraints are *distance-based*—these constraints are violated if the distance among the blocks is bigger than the given threshold. We specify absolute coordinates for the intersection-based constraints, whereas relative distance information is given in distance-based constraints. Finally, the overall penalty function *penalty(F)* for a given SOP floorplanning solution F is the sum of the penalty among all constraints given.

Table 2. Penalty Computation for x-dimension (p_x).

method	*syntax*	*penalty* (p_x)
point	$p=[b_i\|(x,y,z)]$	$\min\{\|x-x_i\|, \|x-(x_i+w_i)\|\}$
layer	$l=[B_i\|l]$	$\sum\|l(b_i)-l\|$
region	$r=[B_i\|(x,y,w,h)]$	$\sum\min\{\|x-(x_i+w_i)\|, \|(x+w)-x_i)\|\}$
abutment	$a=[B_i]$	$\sum[(x_i+w_i)-x_j]$, b_i and b_i separated
boundary	$b=[B_i\|TBLR/l]$	$\sum[w(f_i)-(x_i+w_i)]$ for R boundary
group	$g=[B_i\|(x,y,z)]$	$\sum[x-\|(x_i+w_i)-x_j\|]$, if $\|(x_i+w_i)-x_j\|>x$

In an example shown in Figure 2, we use the following 6 constraints for 4-layer SOP floorplanning with 10 blocks: $p=[b_0|(10,10,3)]$, $l=[\{b_1\}|0]$, $r=[\{b_2\}|(3,3,5,5)]$, $a=[\{b_3,b_4\}]$, $b=[\{b_6\}|L]$, $g=[\{b_7,b_8\}|(5,5,5)]$. This example considers all six types of SOP constraints given in Table 1. Figure 2 shows a solution F that includes several constraint violations. In the top layer (layer 0) we have two active components b_0 and b_5 while other layers contain embedded passive components. First, the point constraint $p=[b_0|(10,10,3)]$ is not satisfied in F since b_0 is in layer 0 instead of layer 3 although b_0 contains the point (10,10) in x/y dimension. This increases the via cost by 3. Second, the layer constraint $l=[\{b_1\}|0]$ is not satisfied since b_1 is in layer 2 instead of layer 0. This also increases the via cost by 2. Third, the region constrain $r=[\{b_2\}|(3,3,5,5)]$ is satisfied in F since b_2 intersects with the given region (= rectangle labeled r). Thus the penalty is zero. Fourth, the abutment constraint $a=[\{b_3,b_4\}]$ is satisfied in F since b_3 and b_4 in layer 3 are abutted. Thus the penalty is zero. Fifth, the boundary constraint $b=[\{b_6\}|L]$ is satisfied in F since b_6 is in contact with the left boundary of layer 2. Thus the penalty is zero. Lastly, the group constraint $g=[\{b_7,b_8\}|(5,5,5)]$ is satisfied in F since the distance between b_7 and b_8 in all three dimension is smaller than the size of the given cube (= rectangle labeled g). Thus the penalty is zero.

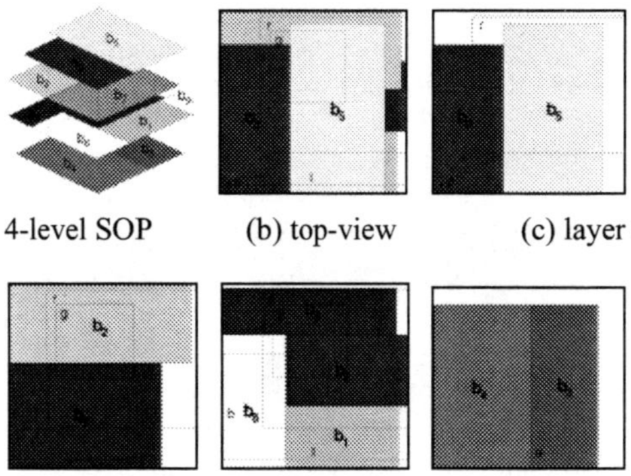

(a) 4-level SOP (b) top-view (c) layer 0

(d) layer 1 (e) layer 2 (f) layer 3

Figure 2. A 4-layer SOP floorplanning with 10 blocks with the following 6 geometric constraints: $p=[b_0|(10,10,3)]$, $l=[\{b_1\}|0]$, $r=[\{b_2\}|(3,3,5,5)]$, $a=[\{b_3,b_4\}]$, $b=[\{b_6\}|L]$, $g=[\{b_7,b_8\}|(5,5,5)]$.

IV. Experimental Results

We implemented our algorithm MF-SOP in C++/STL and ran on a Dell Dimension 8800 Linux box. We used GSRC floorplanning benchmark circuits. We report the area, wirelength, inter-layer via, and runtime for 4-layer SOP in all of our experiments. Table 3 shows the comparison among (i) single-layer floorplanning, (ii) 4-layer SOP floorplanning without geometric constraints, and (iii) 4-layer SOP floorplanning with geometric constraints. We randomly select constraints from 6 types for each circuit, and we impose more constraints for bigger circuits. We summarize our observations here:

1. compared to the single layer floorplanning, the final package area for 4-layer floorplanning is reduced by 73% on the average (order of $O(k)$ reduction). This indicates that the floorplan for all 4 layers is highly compact and their shapes are similar. The impact of geometric constraint on final area was not significant—only 2% increase on the average. This shows the effectiveness our MF-SOP in obtaining high quality multi-layer SOP floorplanning solutions in the presence of complex design constraints in SOP.
2. the wirelength reduction for 4-layer floorplanning is 35% on the average compared to the single-layer case. Since the wirelength in z-direction is not considered (this is actually our via cost), the 35% saving mainly comes from the final package area reduction. The impact of geometric constraint on final wirelength was not significant—only 4% average increase.
3. The impact of geometric constraint on via results was not significant—only 4% average increase. In some cases MF-SOP was able to find a solution with smaller wirelength and via. This again shows the effectiveness our MF-SOP in handling complex design constraints in SOP.
4. The runtime has been increased by 9x with 4-layer floorplanning. The runtime slightly increased when MF-SOP considers geometric constraints. There are several factors that contribute to the runtime increase: (i) we need highly compact floorplan for all 4 layers and their shapes need to be similar, (ii) we need to minimize 2-dimensional wirelength and via cost simultaneously.
5. We observe that abutment (a), boundary (b), and group (g) constraints are easier to satisfy than point (p) and region (r)

constraints. We note that the distance-based constraints are easier to handle than the intersection-based constraints. This indicates that specifying the absolute location is a stronger constraint than the relative distance. Layer constraint is always satisfied since our initial solution satisfy the layer constraint before Simulated Annealing, and we lock all blocks under layer constraints and do not move.

Figure 3. A snapshot of 4-layer n300 floorplanning

V. Conclusions

In this paper, we proposed a new multi-layer floorplanner MF-SOP for the new emerging mixed-signal System-on-Package (SOP) technology. MF-SOP places both active components and embedded passive components in a multi-layer SOP substrate. MF-SOP considers various types of geometric constraints in order to address various signal, thermal, and power integrity issues existing in the design of reliable SOP. We assume in this paper that the geometric constraints are specified by the user as an input to our multi-layer SOP floorplanner. Our ongoing research effort attempts to integrate STA (Static Timing Analysis), SIA (Signal Integrity Analysis), and TPA (Thermal and Power Analysis) engines into our floorplanner so that the geometric constraints are also automatically generated. The goal is to develop built-in STA/SIA/TPA that runs fast but with high fidelity so that they will not slow down the optimization process while guiding the optimization.

Reference

[1] Rao Tummala and Vijay Madisetti, “System on Chip or System on Package?,” IEEE Design & Test of Computers, pp 48-56, 1999.

[2] Y. Deng and W. Maly, ``Interconnect characteristics of 2.5-D system integration scheme," Int. Symposium on Physical Design, 2001, Apr. 2001.

[3] Xiaoping Tang, Ruiqi Tian, D.F. Wong, ``Fast Evaluation of Sequence Pair in Block Placement by Longest Common Subsequence Computation," Design, Automation and Test in Europe Conference, 2000. pp 106-111.

[4] H. Murata, K. Fujiyoshi, et al., ``VLSI module placement based on rectangle-packing by the sequence pair," IEEE Transaction on CAD, vol. 15:12, pp. 1518-1524, 1996.

[5] E. F. Y. Young, C. C. N. Chu, M. L. Ho, ``A unified method to handle different kinds of placement constraints in floorplan design," ACM Design Automation Conf, 2002. pp. 661-667

[6] F. Y. Young, D. F. Wong, H. H. Yang, ``Slicing floorplans with boundary constraints," IEEE Trans on Computer-Aided Design of Integrated Circuits and Systems, pp. 1385-1389

[7] Xiaoping Tang, D. F. Wong, ``Floorplanning width Alignment and Performance Constraints," ACM Design Automation Conf., 2002.

Table 3. Comparison among single-layer floorplanning and 4-layer SOP floorplanning with/without geometric constraints. We report the final package area, wirelength, total number of vias used, and total runtime. The total number of initial and final constraints, and the number of failed constraints for each constraint type are also reported.

	k=1		*k*=4, no constraints			*k*=4, with constraints			# constr		failed constr type					
	area	wire	area	wire	via	area	wire	via	init	fin	*p*	*l*	*r*	*a*	*b*	*g*
n10	258152	18164	66740	8395	43	72850	12827	41	6	2	1	0	0	0	0	1
n10b	251778	15128	74469	10198	45	80496	10252	39	6	1	1	0	0	0	0	0
n10c	268865	19880	71760	11640	37	83200	12422	28	6	2	1	0	0	0	0	1
n30	245115	54586	68420	38106	114	68796	43456	137	10	4	1	0	1	0	0	2
n30b	234574	45931	60984	34850	138	61102	17043	141	9	3	2	0	0	0	0	1
n30c	233867	55979	69153	33353	144	77407	45156	154	9	3	1	0	0	0	0	2
n50	231431	104395	60973	44749	292	64428	55664	312	12	4	2	0	2	0	0	0
n50b	237266	94790	61650	34019	291	63928	104567	317	13	4	2	0	0	0	0	2
n50c	234567	106562	60960	46619	284	67554	40091	314	12	6	4	0	1	0	0	1
n100	210378	180413	56166	54347	455	56852	76069	519	14	7	3	2	2	0	0	0
n100b	185868	169767	50400	58270	500	53074	61863	496	14	7	2	1	2	0	0	2
n100c	208616	185215	53760	56278	495	58322	82858	557	14	6	3	2	0	0	0	1
n200	214349	393644	56977	123765	1098	58548	127741	1074	14	5	1	0	1	0	2	1
n200b	208960	336236	56635	119849	931	56723	119919	915	14	6	2	1	2	0	0	1
n200c	206954	394358	53345	148634	1039	56296	129221	1120	14	6	3	1	1	0	1	0
n300	329589	658162	83232	165433	1354	85200	174274	1413	14	6	2	1	2	0	0	1
ave	235021	177076	62852	61782	454	67799	69589	474								
ratio	1.00	1.00	0.27	0.35	1.00	0.29	0.39	1.04								
runtime	150		1201			1409										

ROUTING RESOURCES CONSUMPTION ON M-ARCH AND X-ARCH

B. Choi†, C. Chiang‡, J. Kawa‡, M. Sarrafzadeh†

†Computer Science Dept., UCLA, Los Angeles, CA 90095
‡Synopsys, Inc., 700 E. Middlefield Road, Mountain View, CA 94043

ABSTRACT

In this paper we study the difference in the usage of routing resources during detailed routing between a router using a Manhattan (M-arch) scheme and another using the X-architecture (X-arch) one. We compare the wirelength, number of vias, and routability between the M-arch and X-arch for two terminal nets (best case for X-arch). The experiments show that the X-arch produced on average net wirelength that were 13.7% less than the corresponding ones produced by the M-arch. But the X-arch also produced 18% less completed routes and needed 1.62 times as many vias as the M-arch when experiments were done under the condition of preferred direction routing. When a semi-preferred routing condition was chosen the X-arch used 13.5% less wirelength but with an additional cost of 11.5% in unrouted nets and a 1.61 times in the number of vias needed compared to the M-arch router under the same condition.

1. INTRODUCTION

In this paper we study the difference in the utilization of routing resources between two detailed-routing architectures: the Manhattan-architecture (*M-arch*) and the X-architecture (*X-arch*). The M-arch is the traditional routing architecture which uses orthogonal horizontal and vertical wires only. On the other hand the X-arch routing architecture uses horizontal, vertical, 45^0, and 135^0 wires.

In an attempt to improve the routability of the M-arch, researchers expanded their effort from the M-arch (*rectilinear*) to the X-arch (*octilinear*) [1, 2, 3]. Initially, empirical results indicated that the X-arch was on an average 17% more efficient than the M-arch [4] for two terminal nets. Then, multi-terminal net routing in the X-arch was addressed as a Steiner tree problem [5]. Recently, [6] gave an exact algorithm to construct X-arch Steiner tree. It concluded that the total wirelength reduction was about 10% only for multi-terminal nets routing.

It appears that most research conducted on the X-arch addressed wirelength reduction by using a scheme based on the planar formula for a Steiner tree. Construction of the Steiner tree for each net is done in the global routing phase. For any comparison of the two architectures to be relevant both M-arch and X-arch implementations must have the same chip size, routing resources(routing pitches), and they must follow the same design rules at all stages of the detailed routing. There are several papers [7, 8] that studied the X-arch channel routing in the late 80s and early 90s. Here we study the routing resources consumption of both the X-arch and the M-arch in detailed routing with an area routing setting.

We discuss the motivation of this research in Section 2. Then we describe the experiments we preformed in Section 3. In Section 4 we tabulate the results and explain them.

2. MOTIVATION

Using Steiner tree the reduction in wirelength achieved by moving from the M-arch to the X-arch was 10% on an average for multiple-terminal nets and 17% for 2-terminal nets.

However, both results cited in the last paragraph were obtained using a planar formulation. This approach and the resulting savings in wirelength cannot be directly transferred to the equivalent detailed routing procedure. There are at least two more steps needed: layer assignment and track assignment, from the planar topology connection to the final three dimensional detailed routing.

Here we study the difference in end results stemming from moving from a planar connection to a final detailed routing one for both M-arch and X-arch architectures. In this study we focus on three specific issues: reduction in wirelength, routability, and quality of routing. First, we investigate the reduction in wirelength achieved by the X-arch in detailed routing. Second, we study the routability of the X-arch to see the relation between planar wirelength reduction and routability. Third, we compare the quality of routing results obtained from M-arch to those of the X-arch.

The quality of routability represents the correlation quality between global routing and detailed routing. A high correlation means that the detailed router is able to closely follow through from the results of the global router. Hence, the wirelength reduction from global routing can be realized in detailed routing. On the other hand, if the correlation is poor the detailed router won't be able to properly utilize the global router's results and wirelength reduction from global routing will not be fully realized.

In global routing, the chip is first divided into global cells. Then the capacity on the boundaries of each global cell is calculated. The accuracy of the capacity calculation is crucial for the realization of good global routing results. For example, assume each of the global cell in Figure 1 can have four nets route through based on the capacities calculation before global routing. Due to poor correlation, the router can actually handle three nets only in detailed routing. In the global routing in Figure 1, three global nets and one local net route through each of the top two rows. But the detailed routing of the two local nets have to have the routing topology changed from global routing due to insufficient routing tracks. The wirelength of the two local nets becomes three times, and two times of its global routing wirelength respectively.

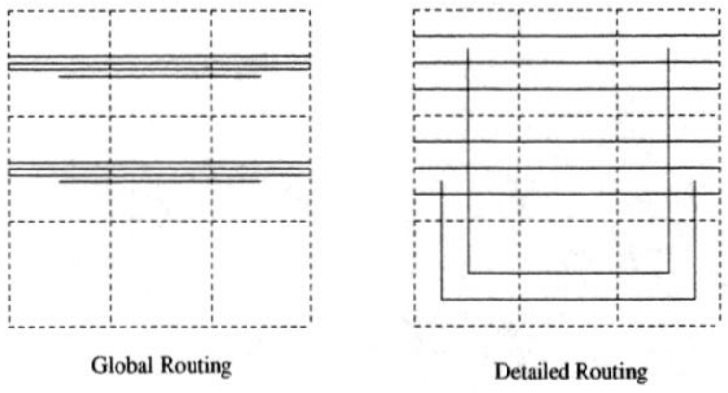

Fig. 1. Correlation

The M-arch has worked very well for calculating the capacity of each of the global cells boundaries. This is the case because the calculation is straight forward. For each global cell, its horizontal and vertical capacities can be calculated from Equation 1 and Equation 2, where κ is the blockage inside the global cell, ι is the connections end inside the global cell, and ρ is the pitch (wire width + spacing) of the layer. Both α and β are constants. Note that the capacity determines the number of nets that can be routed through the boundary.

$$C_h = (height - \alpha \times \kappa - \beta \times \iota)/\rho \qquad (1)$$

$$C_v = (width - \alpha \times \kappa - \beta \times \iota)/\rho \qquad (2)$$

For the X-arch different shapes has been proposed for the global cell implementation [1, 9]. We have no knowledge of any previously published work on the capacity calculation of the X-arch and the correlation between global routing results and detailed routing results under X-arch. In our test cases we model the correlation of the X-arch as its routability.

To compare the quality of detailed routing results between the M-arch and the X-arch, we count the number of vias such routing created in each of them. The via count is a manufacturiability, thus yield, issue. Due to the deep sub-micro technology lithography issues, high via count is more likely to reduce yield. Reducing the number of vias is one of the key challenges for today's router. Here we show how both M-arch and X-arch connect a two pins net in the preferred direction for each layer(Figure 2). For the total wire length, the X-arch uses $2 + 3\sqrt{2} \approx 6.243$ unit lengths and M-arch uses 8 unit lengths. But the X-arch needs 3 vias while the M-arch needs only one via (Figure 2). Therefore the M-arch routing is better than X-arch routing if manufacturiability is given its due importance.

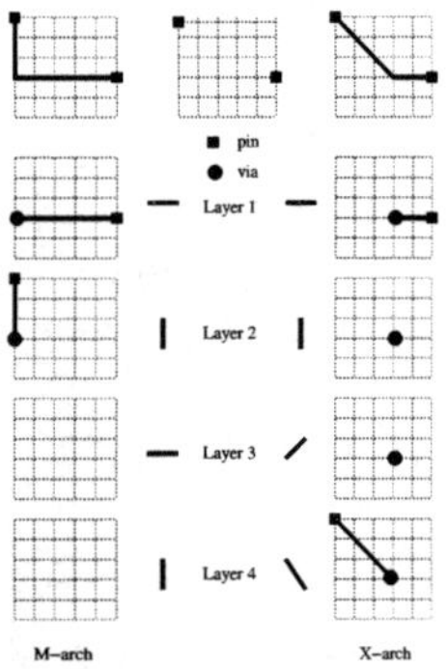

Fig. 2. Examples

3. MODELS OF TESTING

We use a routing area with four layers, 100 by 100 grid each to do all our experiments. For M-arch, the first and third layers use "horizontal" as the preferred direction of routing and the second and fourth layers use "vertical" as the preferred direction of routing. For the X-arch, the preferred direction(s) for each layer is horizontal, vertical, 45^0, and 135^0.

To cover more routing cases in our experiments, we included the two widely used routing environments : grid-based and gridless. In the grided environment each grid point is λ apart from its immediate neighbors, where λ is the spacing requirement dictated by the physical design rules. For M-arch, this constitutes a prefect environment because there is at least λ distance between every gridpoint. But for X-arch, this commonly used grid-based model has a major drawback: if a gridpoint has a $\pm 45^0$ wire pass through, topological design rules of minimum spacing requirement dictates that the adjacent gridpoints not be used for routing anymore (Figure 3). However, if the X-arch routing is under a gridless based model then the other wire can be routed with λ distance from the existing 45^0 wire and not the next gridpoint. Please see Figure 3 for the difference between grided and gridless routing. To compensate the gridless model in our grid-based router, we routed our X-arch using both schemes. In the X-arch we block the adjacent gridpoints of an $\pm 45^0$ wire, but we allow the adjacent gridpoints to be routed under X-arch non-block (XN) model. Note that the XN model is less restrict than gridless based X-arch router. Therefore, the results from XN model should be better than a gridless X-arch router.

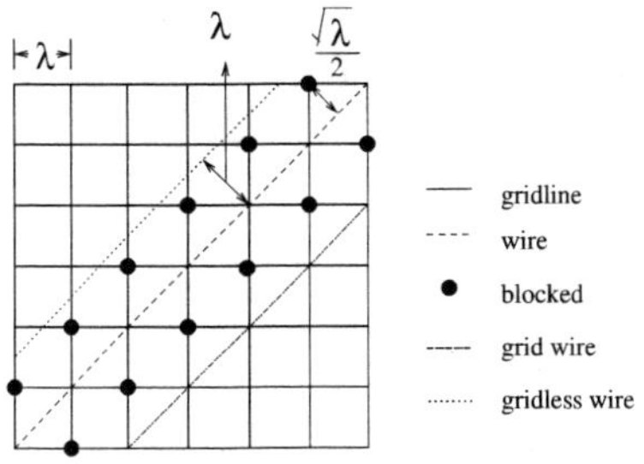

Fig. 3. Grid in One Layer

4. EXPERIMENTAL RESULTS

To compare the routability of different routing schemes, we built a simple maze router and ran it on 100 × 100 grids with randomly generated sequences of two terminal nets. The 100 different sequence of 100 randomly generated 2-terminal nets were then fed as an input to the maze router. Then we counted the number of the nets each model was able to route. We also compared the average wirelength and number of vias per net across all nets.

In Figure 4, we show the difference in routing results between M-arch and X-arch for preferred direction using the maze router. That is, the router has to obey preferred layer direction on each layer and route accordingly. The preferred direction for each layer is the same as described in Section 3. We show the results using different via cost on the Y axis. When via cost is n, it means it is more expensive to use a via to change layer than route $n - \epsilon$ grid on the same layer. For X axis, it is the ratio to the M-arch.

Routing using X-arch completed the routing of about 82% of the nets M-arch routed. To our surprise, the XN model resulted in only 1% more routed nets than the M-arch. This means that routability of M-arch is better than X-arch. From a wirelength per net reference point the X-arch and XN model resulted in about 84% to 94% of the wirelngth of the M-arch (depending on the via cost). This ratio is smaller than the 17% reported in previous research when via cost is higher. Therefore, the reduction in wirelength degraded when we moved from planar to three dimensional detailed routing.

Next, we show the number of vias per net in Figure 4. It clearly shows the X-arch or the XN model consumes more vias than the M-arch. Even when the via cost is 100, the number of vias per net from X-arch is about 21% more than that of the M-arch results. Therefore, the quality of routing of the X-arch is worse than the quality of routing of the M-arch.

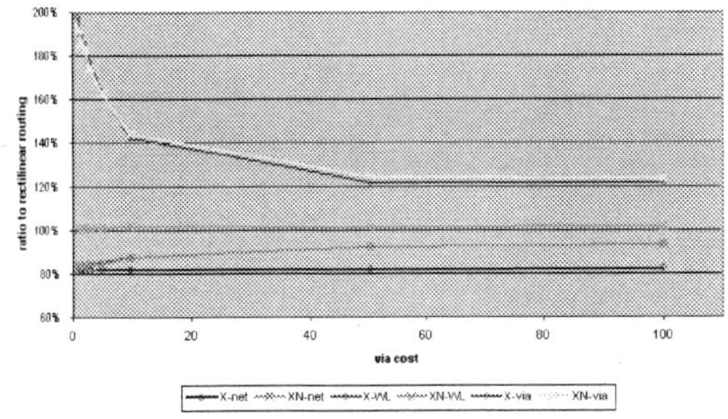

Fig. 4. Preferred Direction Routing

The routers in the industry are all semi-preferred direction routers. The semi-preferred feature gives routers more flexibility. We also routed our experiments in a semi-preferred routing environment. In our experiments, the cost of every non-preferred direction routing is 3 times more than preferred direction routing during the routing path search. We compared the results in Figure 5. The wirelength reduction, routability, and routing quality are similar to that of preferred direction routing for M-arch and X-arch.

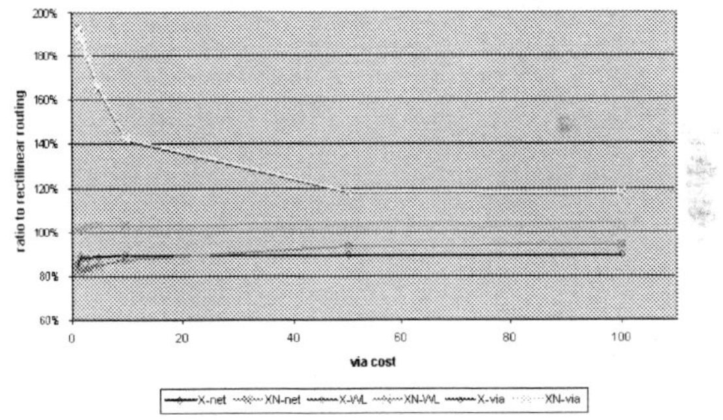

Fig. 5. Semi-preferred Direction Routing

We also compare the routability, wirelength reduction, and routing quality for all six models – M-arch, X-arch, and XN model under both preferred direction routing and semi-preferred direction routing. From Figure 6, we find that the semi-preferred enlivenment provides better routability for all the M-arch, X-arch, and XN models. Under semi-preferred direction the M-arch achieved about 96 routed nets out of 100 nets, X-arch finished 85 nets and XN model finished 98 nets. For preferred direction routing, the router achieved 95, 77, and 96 nets for M-arch, X-arch and XN models respectively.

Next we check the wirelength reduction in Figure 7.

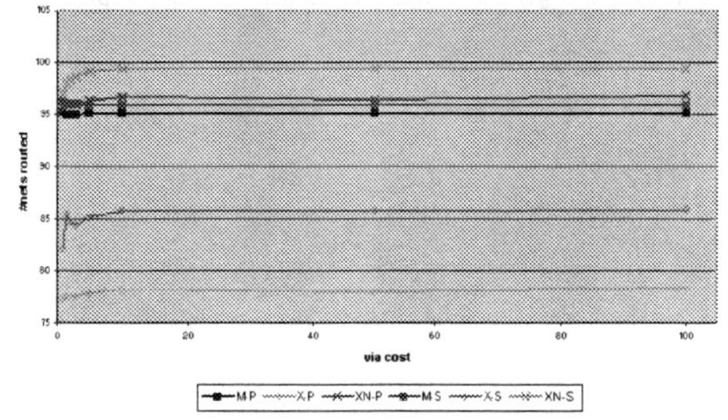

Fig. 6. Routability

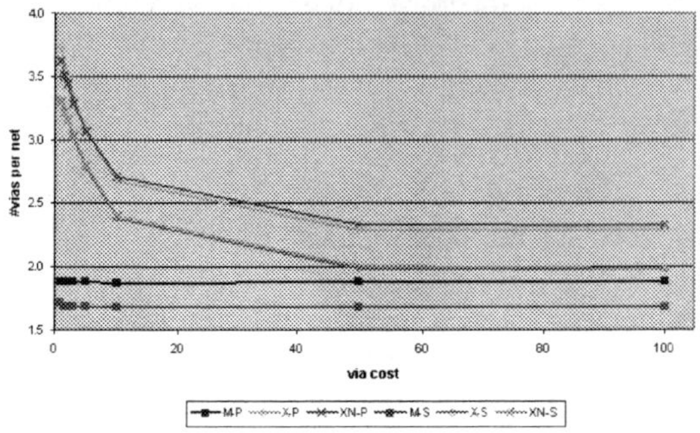

Fig. 8. Via Per Net

The preferred and semi-preferred show similar wirelength reductions. When the via cost is low, the wirelength reduction is close to 17% as previous report. When the via cost goes up, the reduction is reduced to about 6%. Note that, vias are one of the factors that causes yeild problem. Therefore, vias are always more expensive than wires. Hence, it is highly unlikely it can achieve 17% reduction in real designs for two terminal nets.

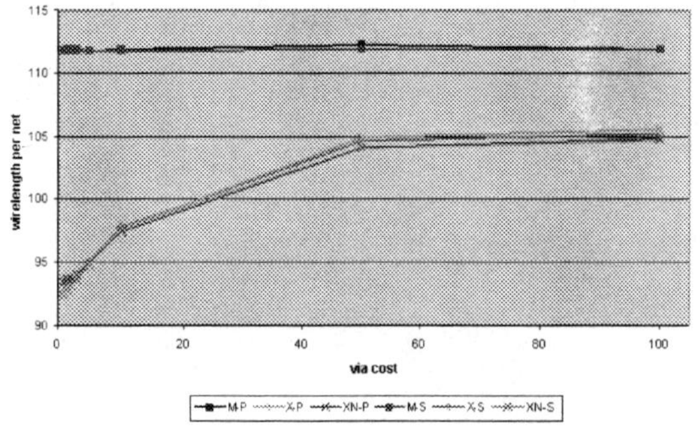

Fig. 7. Wirelength

In Figure 8 we show vias per net for all six routing schemes. It shows that the M-arch for both preferred and semi-preferred direction routing have the lowest number of vias per net. For preferred direction routing, the X-arch has 1.21 times to 1.96 times as many vias per net as the M-arch's vias per net. For the XN model, the range is from 1.24 to 1.91 that of the M-arch's results. For the semi-preferred routing, X-arch is between 1.17 to 1.93 times, and XN model is between 1.17 to 1.91 times that of the M-arch's results.

5. CONCLUSIONS

In this paper, we studied the routability of the X-arch. We modeled the routability as the number of nets that can complete through a particular area and we analyzed the routability of each different routing scheme. Experimental results showed that X-arch routing can save wirelength by 13% on averge for two terminal nets. However, to get the wirelength saving by using $\pm 45^0$ wires, we needed more vias than in M-arch routing. We observed from the experiment that X-arch routing required at least 18% more vias than M-arch routing and had about 17% worse routability.

6. REFERENCES

[1] C. Chiang and M. Sarrafzadeh, "Wirability of Knock-Knee Layouts with 45^0 Wires," *IEEE TCAS*, vol. 38, no. 6, pp. 613–624, June 1991.

[2] M. Sarrafzadeh and C. K. Wong, "Hierarchical Steiner Tree Construction in Uniform Orientations," *IEEE TCAD*, vol. 11, no. 8, pp. 1095–1103, 1992.

[3] S. Teig, "The X Architecture: Not Your Father's Diagonal Wiring," in *Proc. SLIP*, 2002, pp. 33–37.

[4] L. P. Seiler, *A Hardware Assisted Methodology for VLSI Design Rule Checking*, Ph.D. thesis, EECS, MIT, 1985.

[5] A. Kahng et al., "Highly Scalable Algorithms for Rectilinear and Octilinear Steiner Tree," in *Proc. ASP-DAC*, 2003, pp. 827–833.

[6] C. Coulston, "Constructing Exact Octagonal Steiner Minimal Trees," in *Proc. GLSVLSI*, 2003, pp. 1–6.

[7] K. K. Lee and H. W. Leong, "SOAR: A Channel Router for Octilinear Routing Model," in *Proc. ASP-DAC*, 1992, pp. 346–351.

[8] E. Lodi et al., "A 2d Channel Router for a Diagonal Model," *INTEGRATION*, pp. 111–125, 1991.

[9] H. Chen et al., "Estimation of Wirelength Reduction for *lamda*-Geometry vs. Manhattan Placement and Routing," in *Proc. SLIP*, 2003, pp. 71–76.

A PLACEMENT ALGORITHM FOR IMPLEMENTATION OF ANALOG LSI/VLSI SYSTEMS

Lihong Zhang(1), Rabin Raut(1), Yingtao Jiang(2)

(1) ECE, Concordia University, 1455 de Maisonneuve Blvd. West, Montreal, Canada
(2) ECE, University of Nevada, 4505 Maryland Parkway, Las Vegas, NV, USA

ABSTRACT

Analog macro-cell placement by nature is an NP-complete (Non-deterministic Polynomial-time) problem. In this paper, we present a novel approach following the optimization flow of normal genetic algorithm (GA) controlled by the methodology of simulated annealing. The bit-matrix representation is employed to improve the search efficiency. Moreover, a cell-slide based flat placement style satisfying the symmetry constraints is developed to drastically reduce the configuration space without degrading search opportunities. Furthermore, the dedicated cost function covers the special requirements of analog integrated circuits, including area, net length, aspect ratio, proximity, parasitic effect, etc. The algorithm parameters are studied using fractional factorial experiments and a meta-GA approach. The proposed algorithm has been tested using several analog circuits, and appears superior to the simulated-annealing approaches mostly used for analog macro-cell placement nowadays.

1. INTRODUCTION

In the recently growing mixed-signal designs, the digital portion can be tackled using modern cell-based tools for synthesis, mapping and physical design. The design of analog layouts is intrinsically more difficult than that of digital layouts due to design complexity, fallibility and low productivity. Moreover, it has been long recognized that for a successful analog/mixed-signal design, various layout constraints have to be seriously addressed [1]. Therefore, the analog portion has to be still routinely designed by hand, which costs extraordinarily disproportional amount of effort and time compared with only a small fraction occupied in the total chip area. Furthermore, because the device properties get less reliable for analog integrated circuits (ICs) as device size decreases, the electrical effects caused by layouts are indispensable even during the circuit design. Therefore, the automation of analog layout synthesis is strongly required.

The analog macro-cell placement, which has attracted great interests in industry and academia, is one of the most significant stages in the analog layout synthesis [2]. Due to the device-matching constraints in the design of analog layouts, it is usually preferable to cluster devices to form macro-cells or modules, which are parameterized for sub-circuits. The objective of the analog macro-cell placement is to position macro-cells appropriately so that the chip area and the total interconnections can be minimized under the given constraints.

Among the placement strategies, simulated annealing (SA) and genetic algorithm (GA) are two promising approaches. The SA has been widely used for the layout synthesis of both digital and analog circuits [2]-[5]. Although a SA-based approach can typically yield good placement solutions, to guarantee the convergence to the global minimum, an all-out effort is required to tune the suitable definitions. In contrast, the GA has been mainly applied for the design of digital circuits [6][7][9]. This paper presents a novel algorithm with the combination of GA and SA. Some statistical and optimization techniques are deployed to study the sensitivity, correlation and the optimal values of the GA parameters.

The rest of the paper is organized as follows. Section 2 briefly reviews prior work. In Section 3 our novel adaptive algorithm is described. In Section 4 we give the parameter optimization using fractional factorial experiments and meta-GA. In Section 5 the experimental results are shown and in Section 6 the conclusion is drawn.

2. REVIEW OF PRIOR WORK

The placement problem in the VLSI layout is NP-complete [2]. Although the constructive, force-directed, quadratic, branch-and-bound and min-cut techniques have some successful applications in the different types of digital layout designs, they are not well suited for analog layout designs [3]. On the other hand, SA and GA belonging to iterative placement techniques have shown their promising potentials for the macro-cell placement problems.

Rijmenants employed a SA placement approach in their analog layout tool ILAC [4]. Cohn used a SA Jepsen-Gellatt flat model in their placement tool KOAN for analog circuits [2]. Similarly, Lampaert based their performance driven placement on SA in the analog layout tool LAYLA [3]. Both KOAN and LAYLA used simple module generators and their placements were based on single devices. As a matter of fact, however complex a placement algorithm is, some analog circuit structures can never be constructed using geometry sharing techniques (e.g., interdigital cascode structures [4]) based on single devices. On the other hand, Shahookar employed a GA for standard cell placement [6]. The first work on the digital macro-cell placement using a GA was reported in [7]. However, thus far GAs have not widely been applied in the layout generation of analog ICs.

Some works have been done to combine two or more stochastic techniques for placement problems. Handa developed a polycell placement algorithm for analog LSI chips and combined the GA and Tabu Search (TS) in an entire placement process [8]. Esbesene developed a unification algorithm of GA and SA for the digital macro-cell placement [9]. Both Handa and Esbensen based their algorithm designs on the empirical argument that GA could converge very fast in the initial phase but poorer in the late phase. Nonetheless, these combination methods have changed the nature of GA, i.e., the population evolution. Thus, the final result tends to be a local minimum.

3. ALGORITHM DESCRIPTION

To solve the analog macro-cell placement problem, in this section, a genetic placement approach enhanced with simulated annealing (called GASA) is described.

3.1. Genetic Representation and Operators

The conventional chromosomal representation of GAs is based on bit-string [6]. Chan developed a flexible bit-matrix chromosomal representation for digital circuit placement [7], where a chromosome is obtained by rearranging genes from the bit-string to a two-dimensional array. This divide-and-conquer technique allows the GA to generate new configurations faster without degrading its search efficiency. The similar idea with some modifications [10] is adopted in this study.

The crossover operation is performed on two chosen parents. Based on horizontal and vertical random cut lines, the combination of (i) the left-top and right-bottom parts in one parent and (ii) the right-top and left-bottom parts in the other parent produces an offspring. The mutation operator inverses one bit at random in the bit-matrix, which provides a way to widen the range of genes in a population. The inversion operator shuffles the arrangement of genes in a chromosome by exchanging its rows and columns randomly, which weakens the linkage among genes in a chromosome.

3.2. Algorithm Description

The general GA normally employs a fixed mutation rate that is quite small, typically ranging from 0.001 to 0.03. However, according to our experiment, in the population the lack of diversity inevitably occurs, especially during the initial search phase. Therefore, we design a variant mutation rate strategy. A simple linear decrement of the mutation rate is insufficient for this purpose because the acceptance of any mutated result should also be regulated across different evolution stages. Actually SA is a general methodology rather than a completely specified algorithm [5]. The inherent features of SA make it particularly suited to fill the gap because random chromosomal changes in GAs could be considered as random moves in terms of SA. Thus, the mutation rate can be controlled by a temperature following a certain cooling schedule. In this way, the random chromosomal change of the GA is greater in the initial phase and becomes less and less until the search finally converges. Since only one parameter is involved, a Cauchy cooling schedule is applied [5].

Inspired by natural phenomena, the similarity between two mating parents is checked so that the mating of two similar parents is avoided. The similarity check is performed by calculating Haming Distance between two solutions. Moreover, the similarity between an offspring and each of its parents is also checked to foster diversity in the new generation. The algorithm outline of GASA is listed in Fig. 1.

3.3. Cost Function and Cell Slide

The cost function is a weighted function of three components as given in Eq. (1) used to evaluate a placement state,

```
Algorithm GASA
Begin
  initialize the first population randomly and sample several
  random placement states to setup the temperature T;
  evaluate the fitness of each placement state, i.e., the
  reciprocal of the corresponding cost; // Eq. (1).
  while not (stopCriterion()) // setting when to terminate an
  evolution process.
    while (innerLoopCriterion()) //seting a local equilibrium.
      foreach (N * crossover-rate) // N: population size.
        while (similarity-checking1(two chosen parents))
          choose the first parent using the rank selection;
          choose the second parent randomly;
        endwhile
        do crossover and inversion to generate one offspring;
        if (similarity-checking2(offspring and each parent))
          replace the similar parent with the offspring;
        endif
      endfor
      choose the best N members as the new generation;
      foreach (N/2)
        do mutation on the clone (β) of one member (α) based
        on the mutation-rate; // α, β: member variables.
        ΔC = Cost(β) - Cost(α);
        if ( U(0,1) ≤ e^(-ΔC/T) ) //U(0,1): random uniform generator.
          α = β;
        endif
      endfor
    endwhile
    update the mutation-rate and T;
  endwhile
  output the best member;
End
```

Fig. 1: Outline of GASA.

$$C = \alpha_{area} C_{area} + \alpha_{nets} C_{nets} + \alpha_{size} C_{size}, \quad (1)$$

where α_* are the weight factors for the corresponding costs C_*. C_{area} is the area cost that is made up of the whole area, NWELL and PWELL region areas, and the analog and digital region areas. It could make NWELL/PWELL regions relatively concentrated and make analog/digital regions isolated from each other as required in mixed-signal circuits. C_{nets} is the net-length cost, in which priority coefficient can be specified for each net. C_{size} is the size cost, which is used to control the shape of the final layout.

A placement algorithm is actually an optimization flow following a certain geometric construction strategy. As for the geometric construction manner, based on the normal Gellat-Jepsen flat placement style [2], we developed a cell-slide based flat placement style [11]. After all the macro-cells have been absolutely placed, a process of cell slide is executed to determine the relative positions of the macro-cells. Thus, overlaps among macro-cells are avoided deliberately so that the relative position instead of the absolute coordinates becomes the ultimate focus of the search. An algorithm with polynomial time complexity is designed. In addition to the complexity reduction, this approach can also fulfill the symmetry constraints simultaneously.

Table 1: Design of the fractional factorial experiment.

Exp. No.	1	2	3	4	5	6	7	8	9	10	11	12	13
	cr	mr	cr*mr		ir	cr*ir		mr*ir	M	cr*M		mr*M	ir*M
1	1	1	1	1	1	1	1	1	1	1	1	1	1
...	...	...	...	...	...	...	...	...	...	...	...	...	...
27	3	3	2	1	3	2	1	2	1	3	1	3	2

cr: crossover rate; mr: attenuation factor of mutation rate; ir: inversion rate; M: population size; *: grouping of two factors.

4. PARAMETER OPTIMIZATION

Since the genetic operators are of paramount importance to the overall performance of the algorithm, their parameters have to be carefully investigated. The flow of the parameter optimization is depicted in Fig. 2, consisting of two major steps: the parameter analysis and the parameter determination.

First the fractional factorial experiment is employed for the parameter analysis. In our study, a Taguchi orthogonal array $L_{27}(3^{13})$ [12] was employed to construct the fractional factorial experiment. The population size, crossover rate, inversion rate and attenuation factor of mutation rate were taken as the factors. The experiment design is depicted in Table. 1. Three columns (4, 7 and 11) were left for error estimation. After conducting a group of experiments, the data from all experiments are analyzed to determine the effects of the various parameters based on the statistical methods including analysis of mean (ANOM) and analysis of variance (ANOVA) [12].

Although the optimal parameter ranges are derived using the fractional factorial experiment and thus the explored configuration space is reduced, due to the limited number of levels (i.e., only 3) in the orthogonal array, this array would not be sufficient for the parameter determination. Therefore, a meta-GA [6] was employed to optimize GASA to obtain the exact values based on the result of the fractional factorial experiment. Although this process is time-consuming, it runs only once during the algorithm tuning. The meta-GA is itself a genetic optimization process, which runs the GASA to manipulate its parameters to optimize the fitness. The individuals in the meta-GA population consisted of three integers in the range of [0, 10], representing the GASA crossover rate, inversion rate and attenuation factor of mutation rate.

5. EXPERIMENTAL RESULTS

To testify the performance of the proposed GASA algorithm, it has been coded in C++ and integrated into an automated layout tool ALADIN (Automatic Layout Design Aid for Analog Integrated Circuits) [11]. Different from KOAN and LAYLA that use non-optimal single devices, ALADIN provides enough freedom to analog circuit designers who can construct the layouts of complex sub-circuits (i.e., macro-cells) in a technology independent way [11].

Since analog benchmark circuits are still unavailable for synthesis purposes in both industry and academia, our test circuits are collected from a variety of sources, including five different kinds of operational amplifiers and one comparator. The first three circuits are used to determine the algorithm parameters and evaluate the GASA, while the latter three circuits are then used to examine the general applicability of the optimal parameters. Each algorithm was executed ten times in a Sun-Ultra60 workstation. To demonstrate the efficiency of GASA, three other approaches coded in C++ in the same platform have been included for comparison: i) CA, one optimization using simulated annealing where the Cauchy cooling schedule and the cell slide operation were used; ii) BMGA1, a simpler GA implementation that imitates [7] (i.e., without the cell slide operation and similarity checks); iii) BMGA2: includes the cell slide operation and similarity checks based on BMGA1.

The comparison results are given in Table 2, where C_{mean} and T_{mean} are the normalized percentages of mean cost and execution time respectively. As a whole, GASA outperforms the other three algorithms. To examine the general applicability of the obtained optimal parameters on other circuits, we also

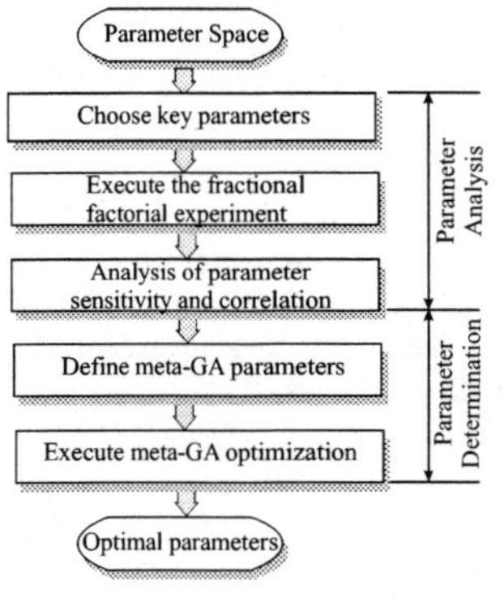

Fig. 2: Flow of the parameter optimization.

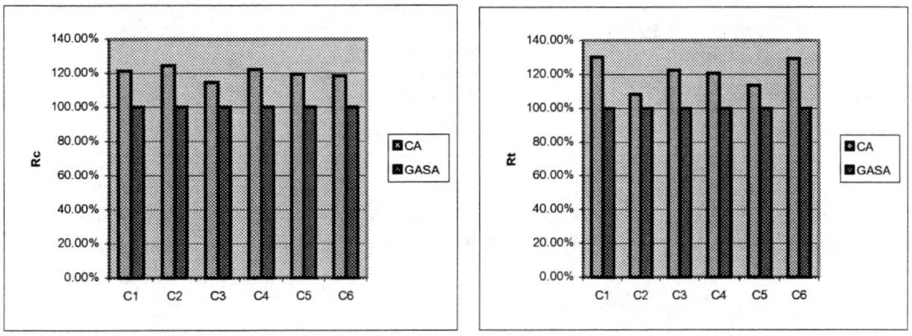

(a) cost ratio

(b) execution time ratio

Fig. 3: Performance comparison of CA and GASA on six circuits (C1-C6).

run CA and GASA to compare their performance on the other three circuits. The bar charts of the experimental results are depicted in Fig. 3 on the normalized cost ratio and normalized execution time ratio between CA and GASA on different circuits. On average, compared with CA, which is widely used in the placement problems of digital and analog circuits, GASA reduced cost by 20.1% and execution time by 20.8%.

Table 2: Comparison among different algorithms.

		CA	BMGA1	BMGA2	GASA
Cir-	C_{mean}	121.4%	154.4%	119.4%	100%
cuit1	T_{mean}	130.2%	111.5%	111.8%	100%
Cir-	C_{mean}	124.5%	143.8%	102.9%	100%
cuit2	T_{mean}	108.3%	168.1%	156.9%	100%
Cir-	C_{mean}	114.7%	161.1%	114.7%	100%
cuit3	T_{mean}	122.5%	129.9%	134.1%	100%

Fig. 4a shows the schematic of a CMOS high speed comparator abounding in symmetry and matching constraints, with the partitioning indicated by rectangles. As a comparison, the placement result obtained using a performance-driven placement algorithm in the state-of-the-art layout tool LAYLA [3] is depicted in Fig. 4b. A denser placement layout result, which is obtained using the GASA, is depicted in Fig. 4c. Because in ALADIN the placement phase is followed by a compaction-based constructive detailed routing phase that automatically minimizes the channel space, the placement output is an abutting structure of macro-cells. In Fig. 4c, the transistor H has been separated from other analog transistors. The PMOS and NMOS regions are respectively concentrated and separated. The post-layout simulation results show the conformance to the original specifications.

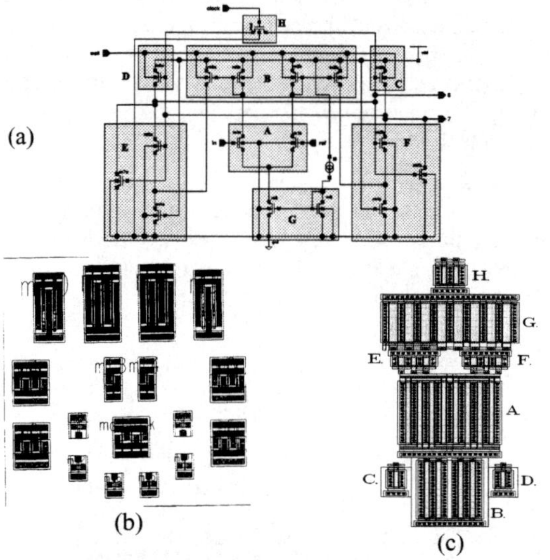

Fig. 4: Schematic (a), a placement solution from LAYLA (b) and our placement solution from GASA (c) of the CMOS comparator.

6. CONCLUSIONS

In this paper the analog macro-cell placement problem has been solved following the optimization flow of a hybrid genetic algorithm controlled by the methodology of simulated annealing. The proposed scheme employs a two-dimensional bit-matrix representation and flexible operators. A cell-slide based flat placement style, which helps to satisfy the symmetry constraints of analog circuits, is developed. To study the algorithm parameters, the fractional factorial experiment using an orthogonal array has been employed, followed by a meta-GA to determine the exact parameter values. Experimental results have shown that this proposed algorithm along with the optimized parameters, with very high computation efficiency, can generate placement of macro-cells, comparable to the ones manually generated by experts.

ACKNOWLEDGEMENT

The research was in part supported by a grant awarded to Dr. R. Raut by the Natural Science and Research Council (NSERC) of Canada, and a grant from SRC and UNLV to Dr. Y. Jiang in USA.

REFERENCES

[1] R. A. Rutenbar and J. Cohn, "Layout Tools for Analog ICs and Mixed-signal SoCs: a Survey," *Proc. International Symposium on Physical Design*, pp. 76-83, 2000.

[2] John M. Cohn, David J.Garrod, Rob A.Rutenbar, and L.Richard Carley, Analog Device-level Layout Automation, Boston: Kluwer Academic Publishers, 1994.

[3] K. Lampaert, G. Gielen, and W. Sansen, Analog Layout Generation for Performance and Manufacturability, Boston: Kluwer Academic Publishers, 1999.

[4] J. Rijmenants, J. B. Litsios, T. R. Schwarz, and M. G. R. Degrauwe, "ILAC: An Automated Layout Tool for Analog CMOS Circuits," *IEEE J. Solid-State Circuits*, vol. 24, no. 2, pp. 417-425, Apr. 1989.

[5] D. F. Wong, H. W. Leong, and C. L. Liu, Simulated Annealing for VLSI Design, Boston: Kluwer Academic Publisher, 1988.

[6] P. Mazumder and E. M. Rudnick, Genetic Algorithms (for VLSI Design, Layout & Test Automation), Now York: Prentice Hall PTR, 1999.

[7] H. Chan, P. Mazumder, and K. Shahookar. "Macro-cell and Module Placement by Genetic Adaptive Search with Bitmap-represented Chromosome," *Integration, the VLSI Journal*, pp. 49-77, Dec. 1991.

[8] K. Handa and S. Kuga, "Polycell Placement for Analog LSI Chip Designs by Genetic Algorithms and Tabu Search," *Proc. IEEE Conference on Evolutionary Computation*, vol. 2, pp. 716-721, 1995.

[9] H. Esbensen and P. Mazumder, "SAGA: A Unification of the Genetic Algorithm with Simulated Annealing and Its Application to Macro-Cell Placement," *Proc. 7th International Conference on VLSI Design*, pp. 211-214, 1994.

[10] L. Zhang and U. Kleine, "A Genetic Approach to Analog Macro-cell Placement with Bit-matrix Representation," *Proc. 2nd Joint Symposium on Opto- & Microelectronic Device and Circuits*, pp. 173-178, March 2002.

[11] L. Zhang and U. Kleine, "An Automatic Layout Design Aid for Analog Integrated Circuits," *Proc. 7. GMM/ITG Diskussionssitzung "Analog '03*, pp. 109-114, Sept. 2003.

[12] G. Taguchi, Taguchi Methods: Research and Development, Quality Engineering Series, vol. 1, Spiral-bound, 1993.

RECURSIVELY COMBINE FLOORPLAN AND Q-PLACE IN MIXED MODE PLACEMENT BASED ON CIRCUIT'S VARIETY OF BLOCK CONFIGURATION

Changqi Yang[1], Xianlong Hong[1], Hannah Honghua Yang[2], Qiang Zhou[1], Yici Cai[1], Yongqiang Lu[1]

[1]Department of Computer Science and Technology, Tsinghua University
[2]Strategic CAD Labs, Intel Corporation
Email: yangchangqi99@mails.tsinghua.edu.cn; hxl-dcs@mail.tsinghua.edu.cn

ABSTRACT

Mixed Mode Placement (MMP) is characterized by a number of same-height standard cells mixed with scattering big blocks in a fixed die. The variety of size and number of blocks introduces challenges to existing algorithms in achieving reasonable solution quality and running time. A new design flow named Recursive Mixed Mode Placement (RMMP) is presented in this paper to provide a solution of MMP with this circuit's variety of block configuration taken into account. It starts from recursively partitioning circuits to form a tree of virtual blocks in the different condition of the size and number of blocks as well as the logical or physical hierarchy. Then it combines floorplan on block level and Quadratic-Place (Q-Place) on cell level to complete the global placement. Our approach takes advantage of combining floorplan and Q-Place algorithms to fit the variety of circuit's components. The combined approach improves the algorithm efficiency and obtains satisfactory results of MMP in terms of wire length and running time on various industry and academia test cases.

1. INTRODUCTION

As VLSI/ULSI technology has profoundly advanced during recent years, more and more circuit cells are integrated into one single chip (such as in SOC or in complex microprocessor designs). To improve design process and time-to-market, some critical parts of circuit, such as *Datapath, Memory* and *Intellectual Property (IP),* are designed separately as sub circuits and then merged into one chip. Therefore, placement tool should handle not only hundreds of thousands of small same-height standard cells but also blocks inside which little details are concerned. Blocks can be classified into two types: One is hard block that has fixed geometrical shape but can be placed at random location of the chip area with various orients such as IP, etc. The other is soft block (flexible block) as custom-designed circuits. It can change its aspect ratio of shape in a given range with area fixed. Thus, placement is performed on three different types of circuit objects: 1) Hard Blocks, 2) Soft Blocks and 3) Standard cells. As we have known, the floorplan tools can perform placement of macro blocks very well, but do not scale to standard cells of large amount because of its high time complexity. On the other hand, traditional placement tools, such as Quadratic-Place (Q-Place) based method, cannot obtain satisfactory result because they have no knowledge about features of macro blocks to eliminate overlap. In some cases, blocks are not even allowed in placement flow. A new design methodology is needed to address the problem of *Mix Mode Placement (MMP)* which can handle both small standard cells and large macro blocks and meet requirement of minimum wire length, timing performance and congestion constraints.

The circuit's variety is another important issue about MMP. Here, the variety is limited to the fact that the size and the number of blocks vary greatly from one circuit to another. Table 1 in the section of experimental results shows the number of blocks in different circuits varies from 2 to 424, and the size of block normalized by cell varies from 57 to 2045. The circuit's variety impacts the practicability of existing MMP algorithms because they are refined to fit circuits with a special block configuration and have difficulties in extending to other styles of mixed circuit. The ideal solution should be the design methodology with higher feasibility, fitness and robust which concerns the variety of circuit's block configuration.

The previous work about MMP can be found in [1][2][3][4]. They use some heuristic method to handle the blocks in circuits. The main disadvantage of them is that they place cells and blocks separately which can hardly obtain an overlap-free placement with optimum performance. Recently, researches on top down algorithm about mix mode placement become popular which involves a hierarchical design flow. A Capo+Parquet+Capo design flow presented by S.N.Adya in [5] and the algorithm HMMP presented by W.M.Wu in [6] generated the two-level hierarchy of circuit according to physical and logical relationship respectively among cells and blocks. Then floorplan is adopted on block level and placement on cell level inside each of blocks. They both fit the block configuration of few but huge blocks. However, for the circuit with lots of medium-sized blocks, they are believed to fall into the trap of running out of time in floorplan stage because of the high time complexity of simulated annealing (SA) method. Moreover, there is room for improvement in the algorithms in each design stage of them. The algorithm MPG-MS presented by C.C Chang and J.Cong in [7] generated multi-level hierarchy followed by

^This work is supported by the National Natural Science Foundation of China 60121120706 and National Natural Science Foundation of USA CCR-0096383, the National Foundation Research (973) Program of China G1998030403, and 863 Hi-Tech Research & Development Program of China 2002AA1Z1460
*This work is also supported in part by a grant from Intel Corporation.

the SA based floorplan on each level. Though it can result in near-global optimum especially with the configuration of numerous blocks, it consumes much more time due to its SA process. For the circuit mixed with little blocks, it would delay the design period more than that of HMMP.

This paper addresses a new design methodology, named *Recursive Mixed Mode Placement (RMMP)*, for MMP which also has a top down design flow of: *Partition, Floorplan, Global Placement and Detailed Placement*. It starts from recursively partitioning circuits to form a tree of virtual blocks in the different condition of the size and number of blocks as well as the logical hierarchy. Then it combines floorplan on block level and Quadratic-Place (Q-Place) on cell level to complete the global placement. Final detailed placement is done to remove all the tiny overlaps and dispose components exactly into the rows. Our approach takes advantage of floorplan and Q-Place to fit the variety of circuit's components. The combined approach improves the algorithm efficiency and can handle circuits with various block configurations and obtain satisfactory results. We compare RMMP with HMMP and MPG-MS on wire length and running time in the section of experimental result, and the results prove the fitness and feasibility of our approach. The remaining part of the paper is organized as follows: Section 2 will introduce our design flow and some details about algorithms in the flow will discussed in Section 3, 4 and 5. Section 6 presents empirical validation of our work, and we come to the final conclusion in Section 7.

2. RECURSIVE DESIGN FLOW

Given a set of placement components, including standard cells, PADs and macro blocks, the mixed mode placement problem is defined as follows: Place the standard cells and macro blocks on the fix-sized chip with no overlap, the standard cells are guaranteed to be on the legal rows and positions, while the total length of all nets are minimized. Moreover, blocks should be aligned properly within rows because many of them (such as Datapath, Memory, etc.) are composed of standard cells which will be mapped into the chip according to the relative locality to the origin of blocks.

The presented design flow RMMP completes placement through 3 routines: partition, global placement and detailed placement. In partition phase, the components and netlist of circuit are analyzed and partitioned recursively to construct a tree of virtual blocks with the root pointing to the whole circuit on the chip. Each node in the tree is a block of "*virtual*" because it does not exist physically but is the combination of standard cells, physical blocks or its child virtual blocks according to the variety of block configuration of the circuit. In global placement phase, an operation-selector will travel along through all the nodes of the tree in the order of Deep First Search (DFS). For the virtual block composed by physical blocks or by child virtual blocks, the selector performs a fixed-outline floorplanner to determine the position and the shape of these components. Otherwise, it performs a Q-Place based placer for that composed by standard cells. In the phase of detailed placement, a min-cut based algorithm is adopted over the whole chip to eliminate the overlap between random standard cells and blocks and to locate the standard cells on the legal positions (in row and aligned with site). Figure 1 shows the outline of our design flow.

RMMP takes the variety of circuit into consideration by generating different virtual block tree for different block

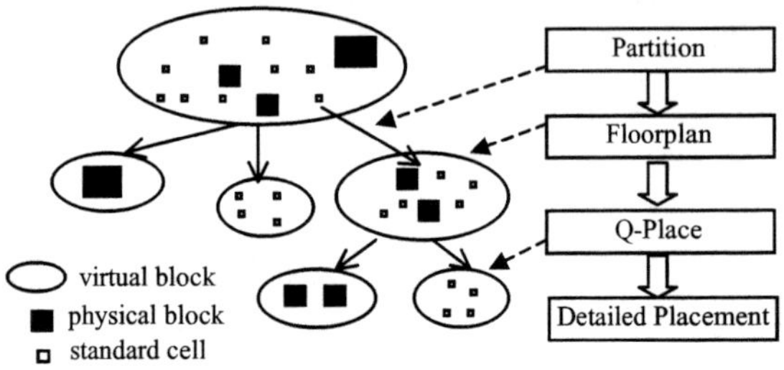

Figure 1: The virtual block tree and the design flow

configuration. For instance, for the circuit with little but huge blocks, the physical blocks are isolated to construct a virtual block one by one; and for circuits consisting of numerous blocks, some of blocks are gathered together to construct a virtual block followed by the floorplanning within it. RMMP is feasible because most of overlaps among physical blocks are eliminated by using floorplan which can also obtain a near-global optimum of performance; and it is efficient because the Q-Place adopted in standard cell placement is a rapid but efficient determinate algorithm which can obtain a satisfactory result in very short time.

3. PARTITION

Assuming the whole cells and blocks as the root of the virtual block tree, partition is done recursively in each node of the tree to create its children with two issues taken into account: 1) minimize the number of interconnection between the final blocks, and 2) determine the area size and shape aspect ratio range of virtual soft blocks. To control the deepness of the recursion, 3 parameters are presented here: P is the number of parts the circuit would be partitioned, C is the number of standard cells who would stay within one leaf virtual block, and B is the number of physical or virtual blocks who are in one virtual block. The partition algorithm can be outlined briefly as figure 2 shown:

Algorithm Partition ()
Begin
Use a bottom up cluster algorithm to create clusters each of which consists of no more than C standard cells;
If the number of physical blocks (plus clusters) N is less than B then
Skip;
Else
Use a top down partition algorithm to create partitions with the number of Min (P, N/P);

Convert all its partitions into the virtual blocks by modifying the netlist;
For each of its virtual blocks do
Partition ();
End

Figure 2: Outline of Partition

The partition algorithm consists of two sub-processes: a bottom up clustering and a top down multiple level partitioning. An improved algorithm based on MFFC named IMFFC is adopted to create small clusters composed of tightly connected cells [8]. And then it constructs a hypergraph which considers clusters as vertices and nets outside clusters as hyper-edges. After that, with geometrical constraints applied, the multiple level partition package hMETIS is applied on the hypergraph to complete the partitioning stage [9][10].

The experiment shows that: if cells or clusters which connect to the same original block are partitioned into the same virtual block, we can obtain better floorplan result with shorter wire length because such a virtual block trends to be placed abutting to its neighborhood. Therefore, the floorplan can benefit from partition that satisfies such the connectivity constraint. Two heuristic methods are suitable to apply the connectivity constraint: 1) pull the cells or clusters connected to a block to form virtual blocks directly and 2) add extra-nets to these cells and attach them with heavier weight. The former has the simplest time complexity, but maybe impact the partition quality as it ignores the objective of minimizing the cut number between blocks. However, the latter is the time consuming method but can result in better performance. We adopt the latter approach.

The post-process of the partition is the determination of geometrical size of virtual blocks which is the necessary condition for successive floorplan and placement. Even in placement, blocks size is more important because cell level placement must be done in the die area determined by blocks size. If the Too small block area is too small (such as the sum of the area of cells within it), it will make the placement trapped into the congested condition. Therefore, the net congestion in virtual block should be estimated to guide expanding the block area to the appropriate value. Though it is hard to estimate congestion before placement is done to determine the exact location of cells, a simple approach can be applied as: counting the number of the inner nets, when it exceeds a custom threshold the area of virtual block should be scaled larger in a proper ratio.

4. GLOBAL PLACEMENT

The global placement is the combination of a floorplanner and a Q-Placer which are performed in the charge of an operation selector.

4.1 Fixed-Outline Floorplan

The objective of floorplan is minimizing the wire length of nets connected between blocks. However, the area packed by blocks does not have to be minimized as an objective because of congestion in the successive cell level placement. We introduce the boundary constraint that makes blocks placed within the chip to obtain a legal floorplan. Therefore, the objective function is defined as:

$$\Phi = W_L * Len + W_{R1} * Max(ChipX - PX, 0) + W_{R2} * Max(ChipY - PY, 0) \quad (1)$$

where Len is the total net length, PX and PY are width and height in packing, ChipX and ChipY are width and height of the chip, w_L and w_{R1} as well as w_{R2} are weighting factors.

Two issues are the most important to floorplan: one is topological representation and the other is algorithm for optimizing. Corner Block List (CBL) is presented as a more efficient topological representation which has a smaller upper bound on the number of possible configurations, needs only linear computation effort to generate a corresponding placement and decreases the redundancies in the previous representations such as SP, BSG, and TCG. The optimization process is based on SA method and is carried on towards the objective of minimizing the wire length of block nets with new solution generated through the move operations on CBL. The details can be found in [11][12].

In the end of floorplanning, pin assignment is adopted to create virtual primary input/output for each virtual block. The fixed PADs are needed by the Q-Place method. Simply placing virtual PADs on the boundaries evenly helps cell level placement but it might increase wire length and congestion between blocks. We place the PADs on the boundary as follows: we draw a line between the center points of the virtual block and the bounding box of the net connected to this virtual pin. The intersecting point between the line and boundaries of the virtual block is the location where the virtual pin should be assigned.

4.2 Q-Place

Q-Place is one of the deterministic methods. It setups a quadratic programming problem, whose objective is to minimize the total wire length and whose constraints make the cells distribute on the chip evenly. As we known, the objective function of quadratic optimization is the weighted sum of the squared length of nets as:

$$\Phi = \sum_{n\in N} L_n \bullet W_n = \sum_{n\in N} w_n \sum_{i,j\in n} ((x_i - x_j)^2 + (y_i - y_j)^2) \quad (2)$$

The placement algorithm tries to minimize the objective subjected to the distribution constraints which make cells scattering on the chip evenly and do this by recursively partitioning the cells and chip into sub regions. In each partition level, the distribution constraints can be fixed as:

$$\frac{\sum_{i\in Tr} x_i}{|T_r|} = \mu_r, \frac{\sum_{i\in Tr} y_i}{|T_r|} = \upsilon_r \quad (3)$$

Here, T_r is the set of cells in region r, and (μ_r, υ_r) is the center of it.

Lagrangian successive over-relaxation iterative method can be used to solve such a problem. But the biggest of its disadvantages is iterative adjustment of the Lagrangian relaxation factors by result of previous solution, which may introduce more run-time complexity and lack of stableness. We adopt a faster and more stable method FaSa based on Lagrange multipliers to solve the linear constraint quadratic programming problem [13]. Instead of iterative adjustment with Lagrangian relaxation factors, we regard the Lagrangian factors as variables and solve theirs values at the same time as other x_i and y_i. FaSa can surely guarantee such equation-like linear constraints like Equ. (3) and accelerates the process of placement by avoiding adjustment of all the factors.

5. DETAILED PLACEMENT

After global placement in all virtual blocks, all standard cells and original blocks are placed near their optimum locations. Detailed placement is adopted throughout the whole chip with original blocks aligned to rows previously.

Detailed placement is done through three steps: (1) An initial detailed placement is obtained by inheriting cell positions determined by global placement procedure. (2) Row assignment refining and row evening are interlaced to reduce the wire length in y and x directions respectively, while rows evening and overlap removing are also done within interlaces. (3) The wire length is reduced further by cell permutation within rows.

Before detailed placement, areas occupied by hard blocks should be determined and filled with dummy cells to prevent other real cells from entering.

6. EXPERIMENTAL RESULTS

Table 1. Characteristics of test circuits

Circuits	#cells	#macro blocks	#nets	Block area (average) /cell area(average)
block2	7094	2	10049	2045
block6	5996	6	10049	872
block8	5662	8	10049	695
block9	5895	9	10049	751
block10	5151	10	10049	676
ibm01	12260	246	14111	57
ibm02	19071	271	19584	157
ibm11	69779	373	81454	166
ibm13	83285	424	99666	162

The proposed algorithm is implemented in C/C++ and works on a SUN v880 workstation (750MHz UltraSparc, 4GB RAM). The test circuits are provided by ARCADIA design systems Inc. and downloaded from [14], whose characteristics are shown in table 1. The columns of "macro blocks" and " Block area (average)/cell area (average)" show the circuit's variety of block configuration. We compare RMMP with HMMP and MPG-MS on wire length and running time which have published in [6] and [7], and the results are shown in table 2 and table 3.

Table 2. RMMP versus HMMP

Circuits	HMMP RunTime (s)	RMMP RunTime (s)	HMMP WireLen (um)	RMMP WireLen (um)	Impr (wl %)
block2	358	320	1. 692e6	1.662e6	1. 74
block6	382	335	2.029e6	1.556e6	23. 30
block8	405	314	2.022e6	1.522e6	24. 74
block9	388	295	2.345e6	2.216e6	5. 5
block10	388	356	2.131e6	1.376e6	35.4

Table 3. RMMP versus MPG-MS

Circuits	MPG WireLen (um)	RMMP WireLen (um)	MPG RunTime (min)	RMMP RunTime (min)	Impr (time %)
ibm01	3.01e6	3.16e6	18	11	39
ibm02	7.42e6	6.59e6	32	18	44
ibm11	2.65e7	2.96e7	112	82	27
ibm13	3.77e7	3.94e7	151	105	30

In table 2, RMMP can obtain results with shorter wire length than HMMP while the running time does not increase. The improvement ratio ranges between 1.74% and 35.4%. The reason is that we have adopted a very effective CBL model in floorplan as well as some heuristic means in partition which can result in a more reasonable partition than the simply FM algorithm. Furthermore, RMMP can be extended to the cases with numerous blocks such as ibm01, ibm02, ibm11 and ibm13 which cannot be solved by HMMP. In table 3, RMMP cannot obtain much better results than MPG-MS because MPG-MS has more opportunities in finding near-global optimum due to its SA based floorplan. However, the results of wire length are still comparable while the running time consumed by RMMP is reduced much more than that of MPG-MS. It is because of the involvement of rapid efficient Q-Place algorithm. RMMP can improve the design period in the range of 27%-44% which has the great sense of feasibility and practice.

7. CONCLUSIONS AND FUTURE WORK

RMMP is an efficient and feasible algorithm and is competent for MMP. It takes advantage of floorplan and Q-Place to fit the variety of circuit components and achieve very good placement quality. However, we expect to extend it to solve timing-driven mixed mode placement problem in the future.

8. REFERENCES

[1] A Feller and R. Noto, "A speed-oriented, fully-automatic layout program for random logic VLSI devices", in National Computer Conf. Proc., pp.303-311, 1978.

[2] Y. S. Baek, B. Y. Cheon, K. S. Kim, H. C. Lee and C. D. Lee, "Cell Designer: An automatic placement and routing tool for the mixed design of macro and standard cells", IEICE Trans. Fundamentals, 1992, 75(2): 224-231.

[3] Hong Yu, Xianlong Hong, Yici Cai, "MMP: A novel placement algorithm for combined macro blocks and standard cell layout design", IEEE ASP-DAC2000, Japan.

[4] M. Upton, K. Samii, S. Sugiyama, "Integrated placement for mixed macro cell and standard cell", in 27th ACM/IEEE Design Automation Conference, 1990, pp32-35

[5] S.N. Adya and I.L. Markov, "Consistent Placement of Macro-Blocks Using Floorplanning and Standard-Cell Placement", Proceeding of ISPD, 2002.

[6] Wu Weimin, Hong Xianlong, Cai Yici, Yang Changqi, Gu Jun, "A Mixed Mode Placement Algorithm for Combined Design of Macro Blocks and Standard Cells", Proceeding of ASICON 2001, Shanghai, China

[7] Chin-Chih Chang, Jason Cong, "Multi-level Placement for Large-Scale Mixed-Size IC Designs", Proceeding of ASP-DAC2002, Japan.

[8] Yongqiang Lu, Xianlong Hong, "Combining clustering and partitioning in Quadratic Placement", ISCAS'2003.

[9] George Karypis, Rajat Aggarwal, Vipin Kumar and Shashi Shekhar, "Multilevel Hypergraph Partitioning: Application in VLSI Domain", Proceeding of DAC 97.

[10] George Karypis and Vipin Kumar, "Multilevel k-way Hypergraph Partitioning", Proceeding of DAC 99.

[11] Hong Xianlong, Huang Gang et al. "Corner Block List: An Effective and Efficient Topological Representation of Non-slicing Floorplan" *ICCAD*'2000.

[12] Yuchun Ma, Xianlong Hong, Sheqin Dong, Yici Cai, Chung-Kuan Cheng, Jun Gu, "Stairway Compaction using Corner Block List and Its Applications with rectilinear blocks", Proceeding of 7th IEEE/ACM Asia & South Pacific Design Automation Conference (ASP-DAC2002), Bangalore, India, 2002.1

[13] Wenting Hou, Xianlong Hong, et al., "FaSa: A Fast and Stable Quadratic Placement Algorithm", International Conference on Communications Circuits and Systems and West Sino Expositions Proceedings, Chengdu, China, Vol 2, pages 1391-1395.

[14] http://vlsicad.eecs.umich.edu/BK/ISPD02bench

LAYER ASSIGNMENT ALGORITHM FOR RLC CROSSTALK MINIMIZATION[1]

Bin Liu, Yici Cai, Qiang Zhou, Xianlong Hong
Department of Computer Science and Technology
Tsinghua University, China, 100084

ABSTRACT

With advances in VLSI technology, crosstalk is becoming a vital factor in high performance designs, making noise mitigation in early design stages necessary. In this paper, we propose a layer assignment algorithm optimizing congestion and crosstalk. A new model for crosstalk noise measuring is developed where wire length is used as a scale for the potential noise immunity, and both capacitive and inductive coupling between sensitive nets are considered. We also take shield insertion into account for further crosstalk mitigation. Experimental results show that our approach could efficiently reduce maximum noise voltage and the number of shields needed compared to the algorithm proposed in [12].

1. INTRODUCTION

With the advent of deep-submicron era and the unceasing advances in VLSI technology, feature size keeps shrinking, leading to significant decrease in wire width and inter-wire space. Crosstalk has become a vital factor and bottleneck in high performance VLSI designs. Capacitive coupling has gained attention for some years, and most recent research shows that with clock frequency increasing, inductive coupling has become rather important in the timing and noise analysis for a growing number of signal lines [1].

Many crosstalk mitigation techniques have been developed. Early works mostly deal with the crosstalk problem in detailed routing stage [2] [3] [4] [5] [6]. There have been a few researches on the coupling problem in global routing recently [7], but the accuracy of such method is usually unsatisfactory. Zhou and Wong use traditional RC model and solve a global routing-layer/track assignment problem under crosstalk constrains [8]. A recent work by James and He solve the problem with budget-modulate strategy in view of both capacitive and inductive crosstalk [9]. These works combine layer/track assignment in global routing, and achieve fairly good performances.

In fact, both global routing and detailed routing are not the most effective stage to deal with crosstalk. It is too early to handle crosstalk during global routing since the relative positions of nets are undetermined, causing much trouble to accurate noise estimation; conversely, it is too late in detailed routing since flexibility in this stage is much limited. Layer assignment is a stage between global routing and detailed routing. This is just the moment to handle crosstalk effectively because while we can get the wire adjacency information, flexibility to separate sensitive nets to different layers is preserved. Furthermore, the problem size in layer assignment is usually smaller than in global routing so that it is easier to get a better solution. An optimal layer assignment algorithm is proposed in [10]. This algorithm considers only capacitive noise in VHV layers and assumes the tracks in the H layer have been determined. Kay and Rutenbar suggested an integer linear programming based layer/track assignment method [11]. However, this work ignores the difference between long and short wires, and the time-consuming nature of ILP makes it unsuitable for large and complex designs. In [12], an efficient layer assignment algorithm optimizing congestion is proposed; however, this work ignores other factors like crosstalk.

In this paper, we introduce an efficient method to measure crosstalk in the early stage of the design cycle, and an effective algorithm is proposed. Our algorithm aims at minimizing the multiplex cost of global noise severity and congestion of each layer and is able to handle both capacitive and inductive coupling for arbitrary number of layers. We take special notice of the difference in noise immunity between nets of different lengths and try to prevent a long wire from suffering too much interference instead of simply minimizing the total noise voltage of all nets. This algorithm does not only optimize the most severely coupled nets, rather, global optimization is performed to reduce crosstalk as greatly as possible. We also preserve empty tracks around severely coupled areas for shield insertion in succeeding stages for further crosstalk optimization.

2. PRELIMINARIES AND MODEL

[1] This paper is supported by NSFC(60176016) and Specialized Research Fund for the Doctoral Program of Higher Education:SRFDP-20020003008

In our discussion, layer assignment is a self-existent stage between global routing and detailed routing. After global routing, the net is routed on the GRG grid, which passed on some GRG edges called segment. The capacity of each GRG-Edge in each layer is estimated previously, and crosstalk tolerance is given as the maximum noise voltage allowed at every sink of a net. For simple presentation, one net is assigned to only one layer pair in this paper, but the model and algorithm here also allows long nets to be divided into several parts and assigned to different layer pairs.

Two signal nets, N_1 and N_2, are sensitive to each other if the noise on N_2 imposed by N_1 at a switching moment may cause logic malfunction. What we are mostly concerned about is the crosstalk noise caused by sensitive nets. The sensitive rate of a net is defined as the ratio of number of nets sensitive with it to the total number of signal nets. Sensitivity of all nets is predetermined according to logical and timing constraints before routing.

Consider all the segments $\{S_i/i=1...n\}$ in a wire along the path from source to sink. The peak noise voltage at the sink can be figured as the total of maximum noise voltages added on each segment given the following preliminaries: the additive noise voltage generated on each segment, NV_i, is a random variable within the range $[-M_i, M_i]$ obeying some distribution, and is independent on each other. The p.d.f for noise on S_i is

$$f(NV_i)=\begin{cases} f_i(x_i), x_i\in[-M_i,M_i] \\ 0, else \end{cases} \quad (2.1)$$

Therefore the additive noise at the sink t is

$$N=\sum_i NV_i \quad (2.2)$$

According to Statistic theory, the p.d.f for N is the convolution of all the p.d.f's for the noise on one segment

$$f(N)=f(NV_1)*f(NV_2)*\cdots=f_1*f_2*\cdots(N) \quad (2.3)$$

Note that $f_i(x_i)$ is zero in the range $[-\infty,-M_i]\cup[M_i,\infty]$, the property of convolution tells us that $f(n)$ is nonzero in the range $[-\infty,-\Sigma M_i]\cup[\Sigma M_i,\infty]$. That is, the peak noise voltage at the sink is $\sum M_i$.

Based on the preliminaries above, we can formulate a cost function to measure the severity of crosstalk in each coplanar structure. Consider all nets routed in a coplanar structure $\{N_{jk1}, N_{jk2}...N_{jkx}\}$. For net N_i, all pins on which are $\{p_{i0}, p_{i1}...p_{is}\}$, with p_{i0} being the source and $p_{ij}(j=1...s)$ being sinks, the Manhattan distance along the path between p_{i0} and sink p_{ij} is L_{ij}. The maximum bearable noise voltage at p_{ij}, t_{ij}, is distributed uniformly as a budget to each segment on the path from p_{i0} to p_{ij}. If a segment has more than one noise budgets toward different sinks, the budget used is the minimum.

The crosstalk cost for the segment of net N_i on GRG-Edge E_m, $CRC(N_i, E_m)$, is defined as the ratio of actual noise voltage to the noise budget.

$$CRC(N_i,E_m)=\frac{\sum_{k\neq l}N(k,l)}{t_{ij}/L_{ij}}=\frac{L_{ij}}{t_{ij}}\cdot\sum_{k\neq l}N(k,l) \quad (2.4)$$

Here l is the track number N_i occupies, and $N(k,l)$ represents the noise voltage on N_i imposed by the net on Track k. p_{ij} is the sink that bottlenecks the noise budget on the current segment. The crosstalk cost for GRG-Edge e on Layer la, $Cross(e,la)$, is the sum of the crosstalk cost for every segment in this coplanar structure.

Definition 2.4 can also be explained from another perspective: since peak noise voltage cumulates with wire length, the same noise voltage on a longer wire is more dangerous than on a shorter one, then we can use the ratio of Manhattan distance between source and sink to the noise tolerance as a weight factor for noise severity in the cost function. If the segment is on the more than one such paths, the weight is the maximum of all possible weights so that the noise constraint on the least crosstalk-immune sink is satisfied.

The noise voltage $\Sigma N(k,l)$ in Definition 2.4 is calculated under consideration of both capacitive and inductive coupling. Various noise models can be used in this calculation as long as they can work out the noise on a segment. Since this calculation is usually performed in the inner loop of optimization, and detailed information about net order in a coplanar structure is undetermined, a fast probabilistic method is used in our algorithm. Work in [7] has proved that capacitive noise voltage generated on a segment is directly proportional to the number of nets sensitive with it and inversely proportional to the total resource in the coplanar structure. For inductive noise, an efficient Keff model [9] is used. Statistical experiment under randomly generated net order shows that inductive noise a segment suffers under Keff model follows similar rule to that of capacitive noise when congestion rate is relatively high. On the whole, the total noise on a segment can be estimated using the ratio of sensitive nets to total resource of a coplanar structure.

3. ALGORITHM

In [12], the layer assignment algorithm optimizes wire congestion. Of course in an area with higher congestion rate, crosstalk is more severe, but it's rather inaccurate to use congestion to measure the severity of crosstalk because it ignores the difference in sensitivity and crosstalk tolerances between nets. Here we propose a new algorithm for both congestion and crosstalk optimization. The crosstalk problem may not be completely solved here because it is neither necessary nor efficient. Instead, a tradeoff between congestion and crosstalk is performed to preserve empty tracks for shields in subsequent stages. Our algorithm can be described as follows,

Input:

Number of routing layers n;

Routing resource i.e. capacity on Layer la of GRG-Edge e, $R(e,la)$;

Result of global routing, i.e. all GRG-Edges that Net N_i passes;

For every net N_i, with source and sinks $\{p_{i0}, p_{i1} \ldots p_{is}\}$, the crosstalk tolerance t_{ik} at each sink;

A sensitive matrix SM, $SM[x][y]=1$ if N_x and N_y are sensitive to each other, else $SM[x][y]=0$.

Output:

The layer pair $L[i]$ for each net N_i.

The goal of this algorithm is to minimize the global cost composite of congestion and crosstalk. Crosstalk cost has been formulated in Section 2. As to congestion, not only the foregone signal wires, but also the shielding wires to be inserted in subsequent stages are considered. The number of shields needed is estimated according to the total number of sensitive net pairs in the coplanar structure with a conservatively simple formula.

$$Shield(e,la) = \delta \cdot TotalSensPairs(e,la) \quad (3.1)$$

Experiments show that δ is about 0.5-0.7. Congestion is calculated with the total of foregone signal wires and expected shields as required resource.

$$Cong(e,la) = \left(Shield(e,la) + Signal(e,la)\right) / R(e,la) \quad (3.2)$$

The global cost is then synthesized from global congestion and global crosstalk cost as follows,

$$GlobalCost = \sum_{e}\sum_{la} (\alpha \cdot Cong(e.la) + \beta \cdot Ramp\left(Cong(e,la) - MaxCon\right) + \gamma \cdot Cross(e,la)) \quad (3.3)$$

$MaxCon$ is a predefined congestion ratio that should not be exceeded, typically 0.8-0.9. α and γ are the weight of congestion and crosstalk, respectively, and β is a big punitive weight preventing overuse of resources. $Ramp(x)$ is 0 when x is no greater than 0 and x otherwise. Typical value for $MaxCon$ is 0.8-0.9, and α 1, β 20, γ 3 here.

In order to minimize the global cost, we get an initial solution obeying the following orders:

- ✓ Areas with larger congestion should take priority over less congested areas.
- ✓ Areas with more sensitive net pairs should take priority over less sensitivity-congested areas.
- ✓ Nets with higher sensitive rate should take priority over less sensitive nets.
- ✓ Nets with lower noise tolerance should take priority over more tolerant nets.
- ✓ Longer nets should take priority over shorter ones.

Succinctly, all the above is trying to determine the layer for more critical and influential nets earlier. This strategy can effectively prevent the algorithm entering an impasse, and reduce the drawbacks of the order sensitive nature of heuristics.

Following these strategies, priority weight is defined for every GRG-Edge and also every net. For a GRG-Edge,

$$W(e) = \sum_{la} Cong(e,la) + \lambda \cdot SensPairs(e) \quad (3.4)$$

λ is the relative weight of the number of sensitive net pairs in the coplanar structure, 0.3 in our algorithm. $Cong(e.\ la)$ indicates the congestion of e on layer la, and $SensPairs(e)$ means the number of sensitive net pairs in the GRG-Edge. For a net,

$$W(N_i) = Length(N_i) \cdot SensRate(N_i) \Big/ \frac{1}{m}\sum_{j=1}^{m} t_{ij} \quad (3.5)$$

Here $Length(N_i)$ is the longest Manhattan Distance from source to sinks of net N_i. $SensRate(Ni)$ is the sensitive rate of Ni defined in Section 2.

There are two loops when selecting a net to process. The outer loop process every GRG-Edge following descending order of $W(e)$, and the inner loop process every unassigned net on current GRG-Edge following descending order of $W(N_i)$. When deciding the layer pair for a specific net, all GRG-Edges that the net passes are taking into account and a tentative Pre-assign strategy is developed. We calculate the costs for all possible assignment, and the assignment with least cost is adopted.

Simulated annealing is performed to optimize the initial solution. A new solution can be generated from current solution with 2 methods: moving the lower layer of a randomly selected net up by 2 layers, or moving the upper layer down by 2 layers. Obviously every possible assignment can be reached from any initial solution with these methods. The initial temperature for simulated annealing is not very high because the initial solution is usually quite close to satisfactory.

```
Algorithm for Layer Assignment:
Repeat
  E_i is the unprocessed GRG-Edge with max W(e)
  Repeat
  N_j is the unassigned net on E_i with max W(N_j)
    For all layer pairs
      Assuming N_j is assigned to current layer pair
      Calculate global cost
    End for
    Assign N_j to the layer pair with min global cost
    Update the congestion of the layer pair
  Until all nets that pass E_i has been assigned
Unitl all GRG-Edges has been processed
Optimize the solution with simulated annealing
Output final solution
```

4. EXPERIMENTAL RESULTS

The algorithm has been implemented in C on SUN OS 5.8. We have tested the performance with several benchmarks modified from industrial instance and the

IBM benchmarks. In each benchmark, nets are distributed into 6 layers. The global crosstalk is measured with the total number of sensitive net pairs in the same coplanar structure, and also the maximum crosstalk of all nets measured by coupling length. Results of our algorithm are compared with those of the algorithm in [12]. Global maximum congestions in a coplanar structure are also compared.

Table 1:information of test benchmarks

	Gdc	mibm01	mibm02	mibm07	mibm08	mibm10
Net	1754	8753	5551	3780	8712	52917
GRG-Edge	11026	5624	3748	662	4302	13200

Table 2:total crosstalk (coupling length) , ours/[12]

SensRate	20%	30%	40%	50%
gdc	2798/6430	6006/10670	9972/14830	14676/19702
mibm01	938/2448	1646/3442	2482/4800	3910/6318
mibm02	278/1062	498/1090	782/1978	1432/2626
mibm07	76/126	360/462	416/656	816/1004
mibm08	262/792	959/1850	1780/2548	2554/3054
mibm10	3676/11738	8164/17736	13336/25886	19480/31290

Table 3:max noise (coupling length), ours/[12]

SensRate	20%	30%	40%	50%
Gdc	52/99	73/116	96/166	145/182
mibm01	74/123	74/137	83/224	125/240
mibm02	26/48	29/67	50/95	74/118
mibm07	32/54	62/67	78/91	129/213
mibm08	55/79	73/151	146/251	176/233
mibm10	91/135	130/171	155/233	216/340

Table 4:max congestion ratio, ours/[12]

SensRate	20%	30%	40%	50%
Gdc	0.875/0.8	0.875/0.8	0.8/0.8	0.8/0.8
mibm01	0.55/0.45	0.5/0.45	0.45/0.45	0.5/0.45
mibm02	0.78/0.67	0.78/0.67	0.89/0.67	0.78/0.67
mibm07	0.64/0.56	0.62/0.56	0.60/0.56	0.62/0.56
mibm08	0.71/0.57	0.71/0.57	0.71/0.57	0.71/0.57
mibm10	0.57/0.57	0.63/0.57	0.6/0.57	0.6/0.57

From the results above, we can see that in a comprehensive range of sensitive rates, our algorithm can reduce crosstalk effectively and the congestion is generally even distributed to each layer. It is expected that with number of layer pairs increased, the performance of our algorithm will be still better.

5. CONCLUSION

We present a new scale to measure global crosstalk severity and propose an effective algorithm to optimize a multiplex goal of congestion and crosstalk concerning both capacitive and inductive coupling. Experimental Results has demonstrated that our algorithm works quite well in a comprehensive range of sensitive rates. Future works following this research can add track assignment, wire spacing and other such techniques into consideration and use more accurate noise estimation to achieve better results.

6. REFERENCES

[1] X. Huang, Y. Cao, D. Sylvester, S. Lin, T. King, and C. Hu, "RLC Signal Integrity Analysis of High-Speed Global Interconnects", IEDM'00, pp. 731-734, Dec. 2000

[2] S.S. Sapatnekar, "RC Interconnect Optimization under the Elmore Delay Model", DAC'94, pp. 387-391, Jun. 1994

[3] Yehea I. Ismail and Eby G. Friedman, "Effects of inductance on the propagation delay and repeater insertion in VLSI circuits", DAC'99, pp. 721-724, Jan. 1999

[4] M. Becer, V. Zolotov, D. Blaauw, R. Panda and I. Hajj, "Analysis of Noise Avoidance Techniques in DSM Inter-connects using a Crosstalk Noise Model", DATE'02, pp. 456-463, Mar. 2002

[5] J. Cong , D.Z. Pan and P.V. Srinivas, "Improved Crosstalk Modeling for Noise Constrained Interconnect Optimization", ASPDAC'01, pp. 373-378, Jan. 2001

[6] L. He and K.M. Lepak, "Simultaneous shield insertion and net ordering for capacitive and inductive coupling minimization", ISPD'00, pp. 55-60, Apr. 2000

[7] M.R. Becer, D. Blaauw, I.N. Hajj and R.Panda, "Early probabilistic noise estimation for capacitively coupled inter-connects", SLIP'02, pp. 77-83, Apr. 2002

[8] H. Zhou and D.F. Wong, "Global Routing with Crosstalk Constraints", DAC'98, pp. 374-377, June.1998

[9] James D.Z. Ma and Lei He, "Towards Global Routing with RLC Crosstalk Constraints", DAC'02, pp. 669-672, Jan. 2002

[10] S. Thakur, K.Y. Chao and D.F. Wong, "An optimal layer assignment algorithm for minimizing crosstalk for three layer VHV channel routing", ISCAS'95, pp. 207-210, May 1995

[11] R. Kay and R.A. Rutenbar, "Wire packing - a strong formulation of crosstalk-aware chip-level track/layer assignment with an efficient integer programming solution", ISPD'00, pp. 61-68, Apr. 2000

[12] Y. Zhou, Q. Zhou, Y.C. Cai and X.L. Hong, "Congestion Based Layer Assignment of Global Routing", ASICON'03, pp. 216-220, Oct. 2003,

CROSSTALK DRIVEN ROUTING RESOURCE ASSIGNMENT

Hailong Yao, Qiang Zhou, Xianlong Hong, Yici Cai

EDA Lab., Dept. of Computer Science & Technology
Tsinghua University, Beijing, 100084, China

ABSTRACT

Crosstalk noise is one of the emerging issues in deep submicrometer technology which causes many undesired effects on the circuit performance. In this paper, a CDRRA algorithm, which integrates the routing layers and tracks to address the crosstalk noise issue during the track/layer assignment stage, is proposed. The CDRRA problem is formulated as a weighted bipartite matching problem and solved using the linear assignment algorithm. The crosstalk risks between nets are represented by an undirected graph and the maximum number of the concurrent crosstalk risking nets is computed as the max-clique of the graph. Then the nets in each max-clique are assigned to disadjacent tracks. Thus the crosstalk noise can be avoided based on the clique concept. The algorithm is tested by a set of bench mark examples and the experimental results show that it can improve the final routing layout a lot with little loss of the completion rate.

1. INTRODUCTION

As VLSI technology advances, crosstalk noise has emerged to be critical in determining the performance of the overall chip. Crosstalk noise profoundly affects the performance of a circuit because it causes signal delays, logic hazards and even circuit malfunctioning [1].

Traditionally, the routing problem is divided into two sequential stages: global routing and detailed routing. The global router determines wirings for each net in a rough scale across the Global Routing Cells (GRC), while the detailed router decides the exact connections for nets inside the individual GRCs each at a time. Most previous works on crosstalk control are performed during the detailed routing stage [2-5] where the estimation of crosstalk can be accurate but the flexibility to control it is restricted. However, the crosstalk control during the global routing stage can not be accurate enough though it has more freedom. With enough flexibility and fairly accurate net routing information, an intermediate stage proves to be an ideal place to solve the problem [6]. In literature, Track Assignment (TA) and Cross Point Assignment (CPA) stages have been taken as the intermediate ones. In [6], the track assignment problem is studied to improve the routing results and the running time of the whole routing stage, but it does not consider the crosstalk issue. Some works on crosstalk avoidance during the CPA stage have been done in [7-9]. Then the crosstalk-aware track/layer assignment which can estimate the crosstalk more accurately is study in [10-12]. In [10-11], the track/layer assignment is integrated into global routing and in [12] it is formulated as an ILP problem and solved using a constructive randomized rounding technique.

In this paper, a new crosstalk-aware track/layer assignment heuristic algorithm called Crosstalk Driven Routing Resource Assignment (CDRRA) is proposed. It fully utilizes the routing resources including the routing layers and tracks to resolve the crosstalk noise issue. The rest of the paper is organized as follows. In Section 2, the crosstalk noise model is given. And in Section 3, details about the CDRRA algorithm are discussed. Then in Section 4, the experimental results are presented. Finally, a conclusion to our work is drawn in Section 5.

2. CROSSTALK NOISE MODEL

2.1 Preliminaries

After global routing, the routing area is divided into m×n global routing cells. Then during the CDRRA stage, the GRCs are merged into horizontal and vertical slices, which can be represented as $HS_i = \{grc_{i1}, grc_{i2}, ...grc_{in}\}$ and $VS_j = \{grc_{1j}, grc_{2j}, ...grc_{mj}\}$ respectively. Then the slices are dissected into routing tracks which serve as the routing resources onto which net segments can be assigned. Thus a slice S_k with r routing tracks can be represented as $S_k = \{t_{k1}, t_{k2}, ..., t_{kr}\}$. Fig.1 gives an example of a horizontal slice with 4 GRCs and 6 routing tracks.

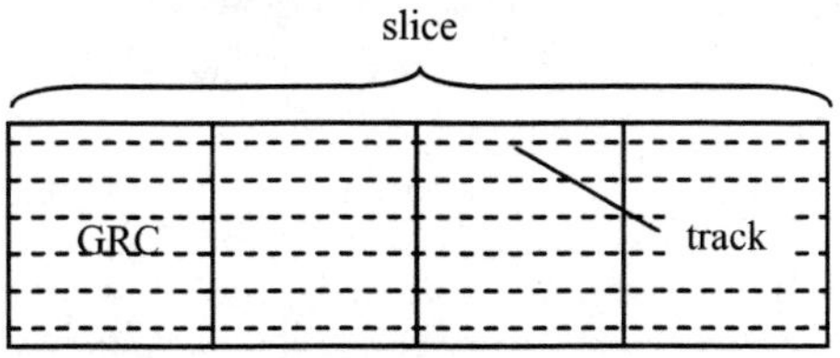

Fig. 1 A horizontal slice: A row of GRCs

For each slice, the net segments are constructed from the global nets across different GRCs within the slice. Since the detailed routing results are not decided, the end points of the net segments are set to the centers of the GRCs. In Fig. 2, a Steiner-tree based global route for net ABCDE spans through three slices, two vertical (4, 8) and one horizontal (7). Then the global net is decomposed into three net segments: AC, BD and DE. Note that A, B, C, D and E are the center points of the corresponding GRCs.

This work is supported by the National Natural Science Foundation of China (NSFC) 60176016, High-Tech Research & Development (863) Program Of China 2002AAIZ1460 and Specialized Research Fund for the Doctoral Program of Higher Education (SRFDP) 20020003008

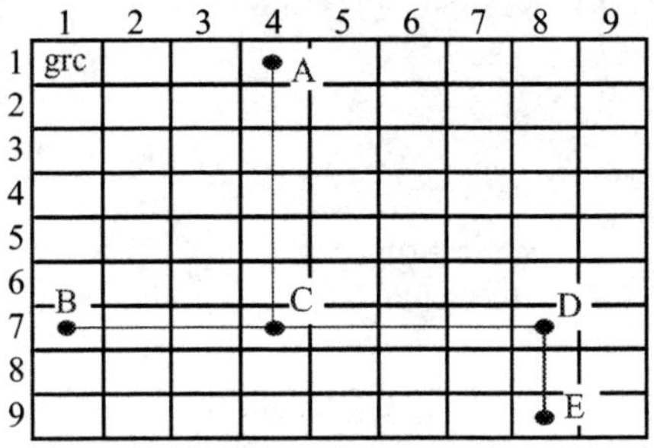

Fig. 2 Decomposition of a global net in the slices

2.2 Crosstalk Model

According to [13], if a switch event on signal net N_1 causes signal net N_2 to malfunction, then N_1 and N_2 are regarded to be sensitive to each other, where N_1 is called the aggressor and N_2 the victim. While the sensitivity rate is defined as the ratio of the number of aggressors for N_i to the total number of signal nets. Then we can define a sensitivity matrix $S = [s_{i,j}]_{N\times N}$ where N is the total number of the signal nets and $s_{i,j} = 1$ if N_i and N_j are sensitive to each other, otherwise $s_{i,j} = 0$. Since when N_i is sensitive to N_j, N_j is also sensitive to N_i, the sensitivity matrix is symmetric. In the implementation, we randomly generate the sensitive nets for each signal net for a given sensitivity rate and then construct the sensitivity matrix.

Crosstalk noise depends on the coupling capacitances, the driver resistances, the load capacitances and the input waveforms. In the CDRRA algorithm, only the capacitive crosstalk noise is considered. Here we adopt the coupling model same as [5] to calculate the coupling capacitance between sensitive nets.

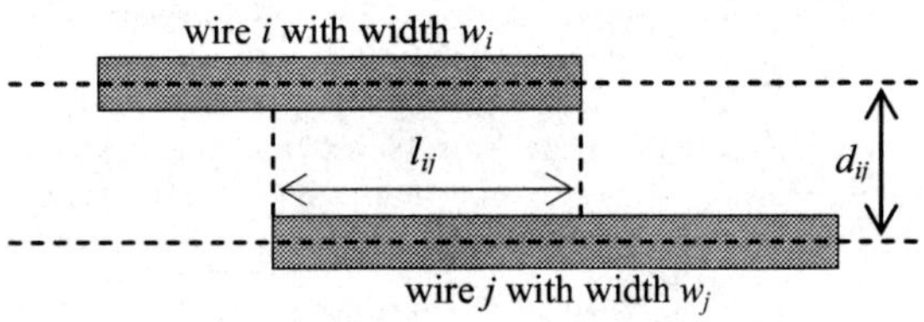

Fig. 3 The coupling capacitance between two nets

In Fig. 3, the coupling capacitance between the two nets i and j can be calculated as formula (1) where w_i and w_j are the widths of wires i and j ($w_i, w_j > 0$), f_{ij} is the unit length fringing capacitance between wires i and j, l_{ij} is the overlap length of wires i and j and d_{ij} is the distance from the center line of wire i to the center line of wire j.

$$C_c(i,j) = \frac{f_{ij} \times l_{ij}}{d_{ij}} \times \frac{1}{1 - \frac{w_i + w_j}{2d_{ij}}} \quad (1)$$

In formula (1), we assume that w_i, w_j and f_{ij} are constant parameters. So in order to achieve capacitive crosstalk-free layouts, we should control l_{ij} and d_{ij}, which means to avoid adjacent sensitive nets from running in parallel for a long distance. In the implementation, we assume that only adjacent sensitive nets will violate the crosstalk constraint when their overlap length exceeds a predefined constant MAXOL and disadjacent nets will never run into trouble with crosstalk violation.

3. THE CDRRA ALGORITHM

3.1 Problem Formulation

For the CDRRA algorithm, only the global nets which span more than one GRC are considered. The algorithm is performed in slice-by-slice manner. For each slice S_k, the assignment of net segments to the routing tracks can be formulated as follows:

$$\Phi: N \times T \rightarrow C \quad (2)$$

where N is the set of net segments inside the slice S_k and T the set of the routing tracks of S_k. C is the set of costs which indicates the consumption of the assignment pairs $<n_i,t_j>$ ($1 \le i \le |N|$, $1 \le j \le |T|$). The track assignment problem is to find a feasible set $\Phi' = \{c_{i,j} \mid c_{i,j} = \Phi(<n_i,t_j>), 1 \le i \le |N|, 1 \le j \le |T|\}$ for all the elements in N, where the objective

$$\sum_{c_{i,j} \in \Phi'} c_{i,j} \text{ is minimized.}$$

The cost for assigning net segment n_i to the routing track t_j is mainly composed of the following items:

- **Layer Cost:** During the CDRRA procedure, the tracks on all the routing layers with the same routing direction (horizontal or vertical) are projected onto one plane. A net segment can be assigned to any of the tracks with the same routing direction. Since longer net segments on lower routing layers will block the connections between other pins and net segments, higher costs are assigned to longer net segments for lower routing layers.
- **Obstacle Cost:** In order to make the CDRRA result feasible for the detailed router, we should consider the via draw-up issue during the assignment procedure. When there is an obstacle between the pins of a net and a routing track, then the assignment of the net's segments onto the routing track should be avoided because it will trouble via draw-up procedure in the following detailed routing.
- **Net Length Cost:** When the segments of a net are not assigned to the routing tracks simultaneously, the pre-routed net components make the routing tracks in the neighborhood more preferable for the minimization of the total net length. So we assign lower costs for these routing tracks.

Besides the items mentioned above, the cost matrix also plays a role in preventing the already assigned nets from coupling with the latter ones, which will be covered in section 3.3. When the cost matrix is constructed with the above components, a linear assignment algorithm [14] is adopted to find a minimum cost matching for the assignment problem.

3.2 The Underlying Graph Model

- **The Crosstalk Graph (*XG*)**

According to the sensitivity rate and the sensitivity matrix defined in section 2.2, we can get a Crosstalk Graph $XG(V, E_{xg})$ where V represents the set of all the nets and E_{xg} represents the sensitive relationship between the corresponding nets. Where there are two nets sensitive to each other, there is an edge in E_{xg} between their corresponding vertexes in V.

- **The Interval Graph (*IG*)**

After global routing, each global net is represented by a set of global routing cells which indicates its routing path. To store the net segments' overlap information, we introduce the Interval Graph (*IG*). Different from *XG* which stores the information for all the nets, *IG* is slice-based and only has information for the net segments in the current slice. Each time a new slice is being processed, *IG* should be reconstructed. *IG* can be denoted as $IG(V_{ig}, E_{ig})$ where V_{ig} is the set of the net segments inside the current slice and E_{ig} stores the overlap information between each two net segments. Where there is overlap between two net segments, there is an edge in E_{ig} between the corresponding vertexes in V_{ig}. Note that *IG* is a weighted graph and the weights on the edges in E_{ig} are the values of the overlap lengths.

- **The Real Crosstalk Graph (*RXG*)**

According to the crosstalk model introduced in section 2.2, only adjacent sensitive nets with their overlap length exceeding the constant MAXOL will violate the crosstalk constraint. So we introduce another graph called the Real Crosstalk Graph (*RXG*) to represent the real crosstalk risks. In fact, *RXG* is the subgraph of *XG* and *IG*. In $RXG(V_{rxg}, E_{rxg})$, each vertex in V_{rxg} represents a net segment and each edge in E_{rxg} represents a crosstalk risk between the corresponding net segments. When two net segments are sensitive to each other according to *XG* and their overlap length exceeds MAXOL according to *IG*, then there is an edge in E_{rxg} between the corresponding vertexes in V_{rxg}. When such pair of net segments is assigned to adjacent tracks, a crosstalk violation is assumed to occur. By calculating the max-clique of *RXG*, the maximum set of concurrent crosstalk risking net segments can be acquired. These net segments should not be assigned adjacent to each other to observe the crosstalk constraint.

- **The Tracks' Adjacency Graph (*TAG*)**

When *RXG* has been constructed in a slice, the net segments in the max-clique of *RXG* should first be assigned to disadjacent tracks to avoid the crosstalk violation. To store the adjacency information of the routing tracks in a slice, we introduce the Tracks' Adjacency Graph (*TAG*). Like *IG* and *RXG*, *TAG* is also slice-based and should be reconstructed for each new slice. In $TAG(V_{tag}, E_{tag})$, V_{tag} is the set of routing tracks in the current slice and E_{tag} stores the adjacency information between the routing tracks. There is an edge in E_{tag} if and only if the two routing tracks of the corresponding vertexes are adjacent to each other. Then it is obvious that the maximum independent set in *TAG* stores the maximum number of the disadjacent routing tracks, to which real crosstalk risking net segments in *RXG* should be assigned.

Due to the characteristics of the routing tracks, the task of calculating the maximum independent set in *TAG* can be saved by a simple numbering technique. The routing tracks in each slice are numbered in the sequential order according to their relative positions and each time the real crosstalk risking net segments are assigned to the even numbered tracks or the odd ones iteratively. Thus, the crosstalk noise can be eliminated.

Fig. 4 illustrates the graphs defined above. The *XG* for all the nets named *a*, *b*, *c*, *d* and *e* is shown in a). Since net *a* does not pass through the current slice, it does not appear in the *IG* of b). In the *IG*, the overlap lengths are marked as the weights of the corresponding edges. We assume the constant MAXOL is set to 50, and then the *RXG* of c) is constructed from the *XG* and the *IG* of the current slice. f) gives the *TAG* of the tracks in the slice of d). From the *TAG*, we can compute the maximum independent set *MIS*={1, 3, 5}. Then we calculate the max-clique of the *RXG* as *MC*={*b*, *c*, *d*}. Now we can assign *MC* to *MIS* using the linear assignment algorithm. Finally we assign the net segment *e* to a routing track under the guidance of the *IG* and the cost matrix. Note that net segments that do not overlap with each other can share the same track, while those which overlap can not. Now we get the crosstalk-free assignment result shown in d).

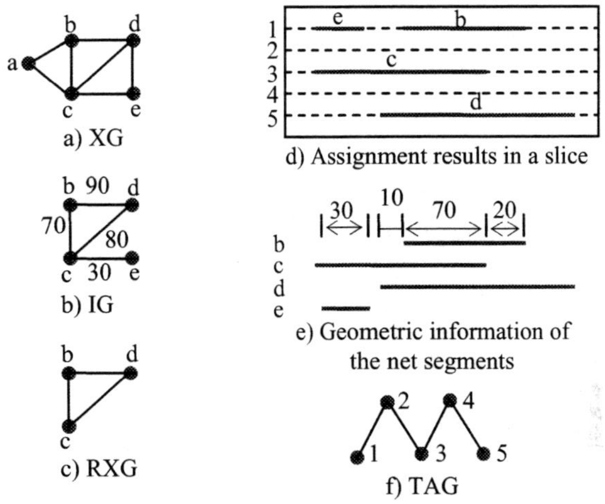

Fig. 4 Illustration of the Graph Model

3.3 Overview of the CDRRA Algorithm

The main steps of our algorithm can be summarized as in Fig. 5. For each slice, the assignment is an iterative procedure where the updated cost matrix eliminates the crosstalk risks between the already assigned net segments and the latter sensitive ones. After processing the net segments from the *RXG*, the algorithm tries to assign net segments as many as possible. Since the max-clique

(1) Read in the sensitivity rate and construct the crosstalk graph (XG).
(2) For all the horizontal and vertical slices, DO
(3) Construct the Interval Graph (IG).
(4) Construct the Real Crosstalk Graph (RXG).
(5) Construct the tracks' adjacency graph (TAG).
(6) Construct the cost matrix for the assignments of net segments onto the routing tracks.
(7) Compute the maximum clique in RXG and the maximum independent set in TAG. Calculate the minimum cost assignment solution using the linear assignment algorithm.
(8) Update IG, RXG, TAG and the cost matrix according to the assignment results. If RXG is NULL, then go to (9), else go to (7).
(9) Compute the maximum clique from IG and assign the net segments onto the remaining routing tracks using the same algorithm until all the net segments are assigned or the routing tracks are not available.
(10)END For

Fig. 5 Main Steps of the CDRRA Algorithm

problem and its equivalent one maximum independent set problem are both NP-complete [15], we adopt some heuristics to find the near optimal solution to save the running time. After the CDRRA procedure, the assignment results are fed to the detailed router as pre-routes. Note that the assignment results serve as a guide but not a constraint for the follow-up detailed router. During detailed routing stage, any pre-routed net segment can be ripped up and re-routed with a higher cost than ordinary rip-ups.

4. EXPERIMENTAL RESULTS

We have implemented the CDRRA algorithm in C programming language on SUN Enterprise E450 and have tested it by a set of IBM benchmarks. Parts of the experimental results are shown in Table 1, where the numbers of the routing layers and tracks are denoted in the column under "L/T". The total numbers of the net segments and nets are shown below "Segs" and "Nets" respectively. In the implementation, the sensitivity rate is set to be 0.5, and then the numbers of the sensitive net segment pairs and the sensitive net pairs are calculated as in column "S-SP's" and "S-NP's". The CDRRA algorithm runs in two modes named "XC" and "NXC". "XC" indicates that the crosstalk control is considered and "NXC" denotes no crosstalk control. Fig. 5 gives the main steps for the "XC" running mode. In "NXC" mode, only the Interval Graph is considered in the algorithm and step (9) in Fig. 5 is repeated until all the net segments have been assigned or no routing resources are available. From the experimental results, we can see that the overflow, i.e. the number of net segments or the number of nets failed in the assignment, is small compared to the total number to be assigned. However, at the expense of the little overflow, the CDRRA algorithm can greatly improve the routing layout in the crosstalk avoidance aspect. From the columns below "X-SP's" and "X-NP's", which means the number of the net segment pairs and the number of the net pairs violating the crosstalk constraints, we can see that the crosstalk violation pairs are greatly reduced. The improvement ratios are 37.5-69.8% for the sensitive net segment pairs and 36.8-68.8% for the sensitive net pairs respectively.

5. CONCLUSION

In this paper, a CDRRA algorithm is proposed to address the crosstalk problem between the global routing and detailed routing stage. The basic idea of the algorithm is to calculate the crosstalk risking net segments by the clique heuristic and then assign them to disadjacent tracks using the minimum weighted bipartite matching problem formulation. The experimental results indicate that the CDRRA algorithm can greatly improve the final routing layout and eliminate most of the crosstalk violations.

6. REFERENCES

[1] H.B. Bakoglu, "Circuits, Interconnections, and Packaging for VLSI," Addison Wesley, 1990.

[2] T. Gao and C. L. Liu, "Minimum crosstalk channel routing," In Proc. Intl. Conf. on Computer-Aided Design, pages 692-696, Santa Clara, CA, November 1993.

[3] T. Gao and C. L. Liu, "Minimum crosstalk switchbox routing," In Proc. Intl. Conf. on Computer-Aided Design, pages 610-615, San Jose, CA, November 1994.

[4] H. Zhou and D. F. Wong, "An optimal algorithm for river routing with crosstalk constraints," In Proc. Intl. Conf. on Computer-Aided Design, San Jose, CA, 1996.

[5] R. Kastner, E. Bozorgzadeh and M. Sarrafzadeh, "An Exact Algorithm for Coupling-Free Routing," in Proc. International Symposium on Physical Design, April 2001.

[6] S. H. Batterywala, N. Shenoy, W. Nicholls, and Hai Zhou, "Track Assignment: A Desirable Intermediate Step Between Global Routing and Detailed Routing," IEEE International Conference on Computer Aided Design, San Jose, CA, 2002.

[7] H.-P. Tseng, L. Scheffer, and C. Sechen, "Timing and crosstalk driven area routing," in Proc. 35th ACM/IEEE Design Automation Conf., June 1998, pp. 378–381.

[8] C.-C. Chang and J. Cong, "Pseudo pin assignment with crosstalk noise control," in Proc. Int. Symp. on Physical Design, pp. 41-47, 2000.

[9] H.L. Yao, Q. Zhou, X.L. Hong and Y.C. Cai, "Cross Point Assignment Algorithm with Crosstalk Constraint," Proceedings of the 5th International Conference on ASIC, October, 2003, pp. 352-355.

[10] T. Xue and E.S. Kuh, "Post global routing crosstalk synthesis," TCAD, pp. 1418-1430, Dec. 1997.

[11] H. Zhou and D.F. Wong, "Global Routing with Crosstalk Constraints," in Proc. ACM/IEEE Design Automation Conference, June 1998.

[12] R. Kay, and R.A. Rutenbar, "Wire Packing: A Strong Formulation of Crosstalk-aware Chip-level Track/Layer Assignment with an Efficient Integer Programming Solution," ISPD '00, 61-68.

[13] J.D. Ma and L. He., "Towards Global Routing with RLC Crosstalk Constraints," Proceedings. IEEE/ACM Design Automation Conference, New Orleans, Louisiana, USA, June 10-14 2002, pp. 669-62.

[14] R.E. Burkard and U. Derigs, Assignment and Matching Problem: Solution Methods with Fortran-Programs, New York: Springer-Verlag, 1980.

[15] M.R. Garey, D.S. Johnson, Computers and Intractability: A Guide to the Theory of NP-completeness (Freeman, New York, 1979), p.53-56.

Table 1 Experimental Results of the CDRRA Algorithm

circuit	L/T	Segs	overflow		S-SP's	X-SP's		Imp. (%)	Nets	overflow		S-NP's	X-NP's		Imp. (%)
			XC	NXC		XC	NXC			XC	NXC		XC	NXC	
ibm1	4/5055	61310	0	0	113072	5233	17323	69.8	11753	0	0	48340678	5211	16681	68.8
ibm2	5/7318	109839	3	0	445515	28538	45667	37.5	18688	3	0	122232602	28091	44425	36.8
ibm3	5/10652	218784	23	0	776259	49827	85059	41.4	44681	23	0	698710308	49330	83483	40.9
ibm4	5/ 11231	248462	3	0	717704	36244	83724	56.7	50678	3	0	898885822	35833	81637	56.1
ibm5	5/14390	372000	4	0	1234360	80716	136888	41.0	64971	4	0	1477391812	79668	133512	40.3

L/T: No. of Layers/Tracks, Segs: No. of the net segments, XC: with crosstalk control, NXC: with no crosstalk control
S-SP's: No. of sensitive net segment pairs, X-SP's: No. of net segment pairs with crosstalk violations
S-NP's: No. of sensitive net pairs, X-NP's: No. of net pairs with crosstalk violations, Imp.: improvement ratio

RF Circuit Synthesis using Particle Swarm Optimization

Jinho Park and David J. Allstot[1]

Dept. of Electrical Engineering, Box 352500, University of Washington, Seattle, WA 98195-2500

Abstract — **A parasitic-aware design methodology based on particle swarm optimization is presented. It is shown that parasitic-aware design using the particle swarm rather than using the simulated annealing heuristic gives better performance. The proposed parasitic-aware design methodology is advantageous due to its inherent efficiency to synthesize RF circuits with parasitic-laden passive components, and it regains the performance of RF circuits with ideal components.**

Index Terms — **Design centering, integrated circuit reliability, Monte Carlo methods, optimization methods, power amplifiers, RF integrated circuits, simulated annealing.**

I. INTRODUCTION

The main principle of the parasitic-aware optimization methodology is that device and package parasitics are considered as a natural part of the design process from the beginning of the overall design phase. When all parasitic effects are considered, the complete circuit becomes much too complicated for hand analysis, even with help of circuit simulators, so finding the optimum solution is problematic as demonstrated in [1].

A few CMOS RF and analog circuits have been synthesized using the parasitic-aware optimization methodology implemented with the simulated annealing heuristic [2]. However, it is found that simulated annealing for these applications requires too much computing resources in terms of executing the total number of iterations of the optimization loop; this drawback impedes the widespread use of this technique. On the other hand, the particle swarm optimization algorithm is known to be fast and robust owing to its unique *competition* and *cooperation* properties between search agent *particles*.

In this paper, a parasitic-aware design and optimization technique for RF integrated circuits based on particle swarm optimization (PSO) is discussed and comparative results with simulated annealing are presented.

II. STRUCTURE OF PARASITIC-AWARE SYNTHESIS

The general parasitic-aware synthesis flow for RF integrated circuits includes three major functional blocks as illustrated in Fig. 1: Optimization core, RF circuit simulator, and parasitic model generator.

The optimization core first generates a netlist by modifying the design parameters to be optimized in accordance with the optimization algorithm; conventionally, simulated annealing has been used, but particle swarm is used in this work. Then, the netlist invokes the parasitic modeling to generate the specific parasitic values. The complete *netlist* with all parasitics considered is simulated by an RF circuit simulator. After the completion of circuit simulation, the netlist is evaluated according to a predetermined cost function and the cost of the *netlist* is computed. Iterations around the optimization loop continue in a similar manner until a predetermined cost function objective is met or a maximum number of iterations are executed.

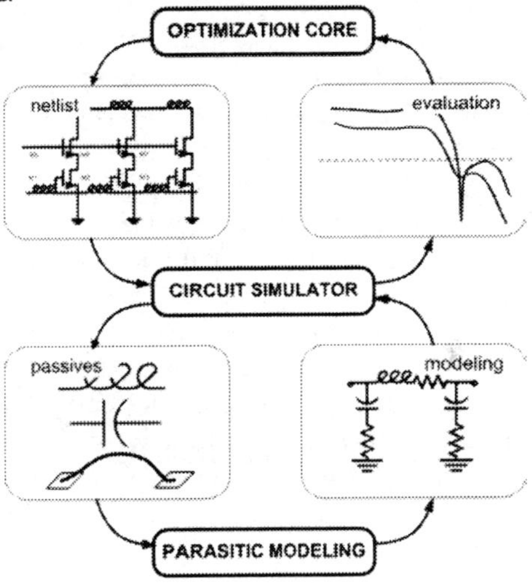

Fig. 1. The parasitic-aware design and optimization flow.

III. PARTICLE SWARM OPTIMIZATION (PSO)

A number of social scientists found that *swarming* differs from other group behaviors of animals in that individual members of the group can profit from the discoveries and previous experiences of all other members of the group during the search for food [3]. The swarming behaviors are usually found in a flock of birds, butterflies, schools of fish, etc. The particle swarm algorithm basically mimics this swarming behavior and has been implemented successfully in several problems requiring optimization.

[1] Research supported by National Science Foundation grants CCR-0086032 and CCR-0120255, and Semiconductor Research Corp. grants 2001-HJ-926 and 2003-TJ-1093.

The particle swarm algorithm works on a population of potential solution candidates called *particles*. The movements of the particles are governed by updating the position and velocity vectors of each individual particle in an effort to find the optimum solution. Therefore, the key to success of the algorithm is how effectively each individual particle is moved through the multi-dimensional design hyperspace.

The obvious advantage of the algorithm is its simplicity. It is easy to understand and simple to implement since it is explained using only the two straightforward equations shown in Fig. 2.

$$\vec{x}(t+1) = \vec{x}(t) + \vec{v}(t)$$

$$\vec{v}(t+1) = \underbrace{w.\vec{v}(t)}_{\text{Inertia}} + \underbrace{c_1.rand(0,1).(\vec{x}_{Pbest}(t) - \vec{x}(t))}_{\text{Competition}} + \underbrace{c_2.Rand(0,1).(\vec{x}_{Gbest}(t) - \vec{x}(t))}_{\text{Cooperation}}$$

Fig. 2. The PSO describing equations.

Generally, any motion of an object is expressed by two vector components: the present position and present velocity vectors. The position vector of an object at time t is simply a vector sum of these two vectors. Similarly, PSO uses present position and velocity vector components to determine the next position vector of each particle as described in the first line of the equations. The second line illustrates the determination of the next velocity vector. The velocity update process uses three different vectors called *inertia*, *competition*, and *cooperation*. *Cooperation* is the vector between the globally best position that one member of the group has found and the current position of the particle weighted by a uniformly distributed random function. Since each particle has knowledge of the globally best position (x_{Gbest}) by cooperating and communicating with each other, this is called the *cooperation* factor. *Competition* is the vector between the personally best position (x_{Pbest}) that each individual member has experienced and its current position. It is called *competition* factor because it describes the tendency of each particle to explore its personally experienced best position more often; it does not require any communication with the others, unlike the *cooperation* factor. Finally, *inertia* is simply the previous velocity weighted by a constant, w. Since it is the tendency to maintain the previously direction, it is called *inertia*. The particle swarm algorithm combines these factors, namely, *inertia*, *competition*, and *cooperation* in a manner that enables the particles to swarm towards the optimum solution.

At a first glance, it may be unclear the *inertia* and *competition* factors are needed while *cooperation* is indispensable in finding the optimum solution. The main idea of adding inertia and competition is to avoid trapping in local minima. Let us examine the simple example in Fig. 3, which demonstrates the algorithm using only the *cooperation* factor. The probability of finding the global optimum without being trapped in local minima is strongly dependent on the initial location of each particle because the particles always try to explore only the vicinity of the position known to be the best so far. This is very similar to gradient decent optimization since it always tries to find the best solution near the starting point of the simulation. This problem is solved using the complete particle swarm algorithm as shown in Fig. 4. It increases the probability of finding the global optimum without becoming trapped in local minima. In addition to increased robustness, it also improves the speed because of the *competition* factor.

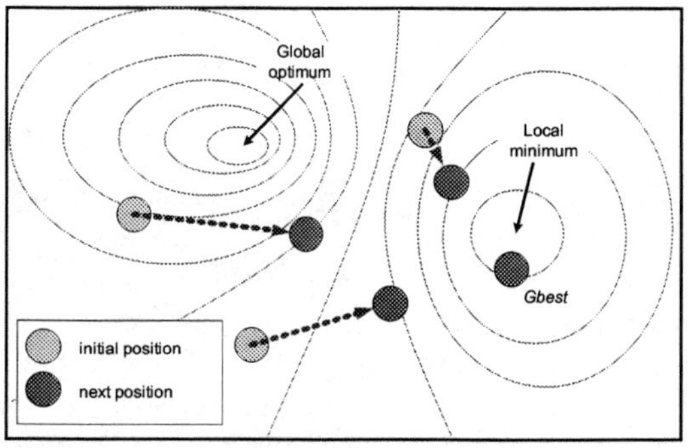

Fig. 3. A four-particle PSO example with the cooperation factor enabled and the inertia and competition factors disabled.

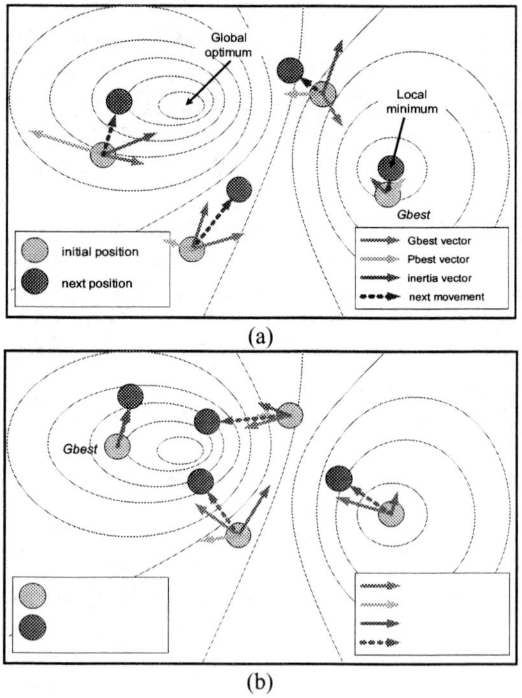

Fig. 4. PSO example.

The manner in which the three factors are combined is critical in determining the efficiency and robustness of the PSO algorithm. It is reasonable to assign the same weight for each factor in general since the problem is initially assumed to be unknown; in other words, the possibility of getting close to the global solution contributed by each factor is the same. Therefore, the weights of the factors, w, $c_1 rand(0,1)$, and $c_2 Rand(0,1)$ are chosen to be 1 on average, and hence c_1 and c_2 are 2. However, the inertia weight, w, is sometimes adjusted to be less than 1 to make the particles converge in time. This issue is discussed in more detail in the following sections.

Because of the simplicity of the PSO algorithm, it requires only primitive mathematical operators, and is computationally inexpensive in terms of both computer memory requirements and computing cycles.

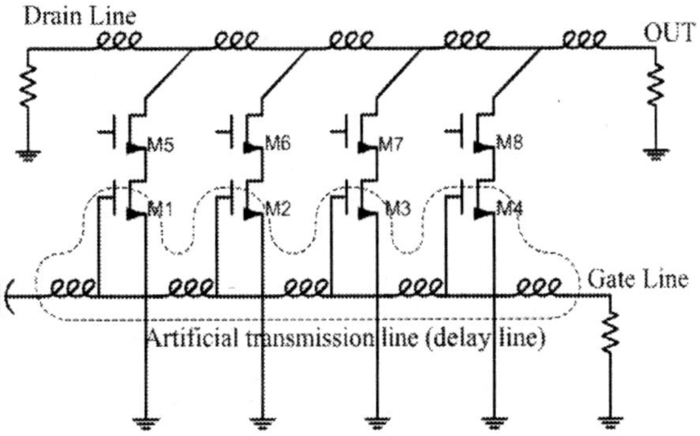

Fig. 5. Four-stage CMOS distributed amplifier with lumped LC gate and drain delay lines.

IV. COMPARATIVE RESULTS

The simulated annealing technique is widely used for CMOS RF and analog circuit synthesis [1]. Other specific areas include the design of very-large-scale-integrated (VLSI) mixed-signal circuits, analog circuit design, model parameter fitting, image processing, and molecular physics. Unlike gradient decent algorithms, simulated annealing avoids local minima by using a hill-climbing process; that is, it conditionally accepts poorer solutions in order to escape local minima while searching for the global minimum. However, the main drawback of the hill-climbing process is the long run times required to obtain a solution. To demonstrate the viability of the synthesis technique based on particle swarm optimization, we now compare performance of the conventional simulated annealing and particle swarm synthesis tools.

In order to investigate the impact that the parameters of particle swarm optimization have on the performance of parasitic-aware RF design and optimization, practical RF IC design problems, a four-stage distributed amplifier and an RF power, are chosen as test vehicles. The detailed design and specifications of these circuits are shown in Figs. 5 and 6. Both designs use a 350nm CMOS process and a single power supply voltage of 3.3V.

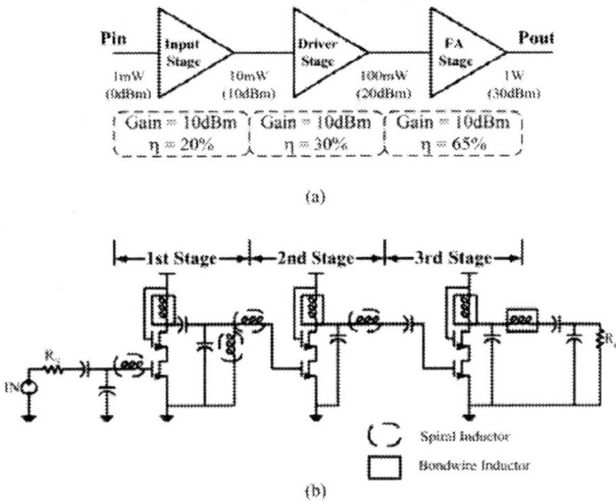

Fig. 6. (a) Three-stage (class-AB, class-E, class-E) power amplifier showing the desired gain and drain efficiency distributions. (b) CMOS implementation.

For this purpose, the best parameter settings are used for all optimization runs. Since the value of *Temp* or *T* is critical for the conventional simulated annealing, the adaptive *Temp* coefficient algorithm is used, which guarantees the optimal *Temp* coefficient condition by estimating the slope of the cost function. While *PSO* has no specific starting point since the system is initialized with a random population in nature, a randomly generated starting point was also used for the simulated annealing.

The simulated annealing technique is widely used for CMOS RF and analog circuit synthesis [1]. Other specific areas include the design of very-large-scale-integrated (VLSI) mixed-signal circuits, analog circuit design, model parameter fitting, image processing, and molecular physics. Unlike gradient decent algorithms, simulated annealing avoids local minima by using a hill-climbing process; that is, it conditionally accepts poorer solutions in order to escape local minima while searching for the global minimum. However, the main drawback of the hill-climbing process is the long run times required to obtain a solution. To demonstrate the viability of the synthesis technique based on particle swarm optimization, we now compare performance of the conventional simulated annealing and particle swarm synthesis tools.

In order to investigate the impact that the parameters of particle swarm optimization have on the performance of parasitic-aware RF design and optimization, practical RF IC design problems, a four-stage distributed amplifier and an RF power, are chosen as test vehicles. The detailed design and specifications of these circuits are shown in Figs. 5 and 6. Both designs use a 350nm CMOS process and a single power supply voltage of 3.3V.

The best parameter settings are used for all optimization runs. Since the value of *Temp* or *T* is critical for the conventional simulated annealing, the adaptive *Temp* coefficient algorithm is used, which determines the optimum *Temp* coefficient by estimating the cost function slope. Since *PSO* is initialized with a random population, a randomly generated starting point was also used for the simulated annealing.

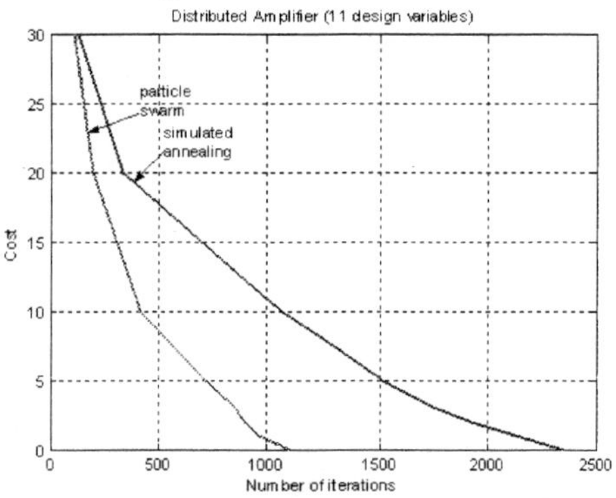

(a)

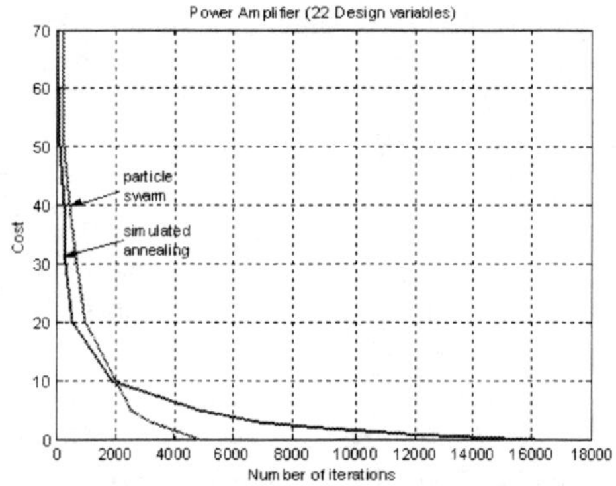

(b)

Fig. 7. Costs. (a) Four-stage DA, and (b) a three-stage PA.

Figure 8 illustrates key final results for the parasitic-aware design and optimization of the CMOS distributed amplifier and power amplifier. Substantial performance improvements are achieved using the parasitic-aware synthesis methodology. For the distributed amplifier, the optimized circuit has a much better frequency response than the un-optimized one. Note that the optimizer has cleverly regained phase linearity by eliminating internal parasitic mismatch effects. Loss compensation and improved phase linearity also drastically improve the gain flatness as shown in Fig. 8(a). The gain roughness in the passband before optimization was greater than ±2.4dB, but after optimization it was reduced to less than ±1.0dB.

The desirability of parasitic-aware synthesis is also vividly illustrated in Fig. 8(b). With ideal parasitic-free inductors on all three stages, the output power is 30dBm and the drain efficiency is 58% for an input power level of 0dBm. Using parasitic-laden spiral and bond wire inductors, but before optimization, the output power drops to 27dBm and the drain efficiency decreases to only 30% with 0dBm of input power. Clearly, the parasitics of the passive components significantly degrade the performance of the power amplifier. However, both the output power and the drain efficiency are restored to the ideal specifications after parasitic-aware optimization.

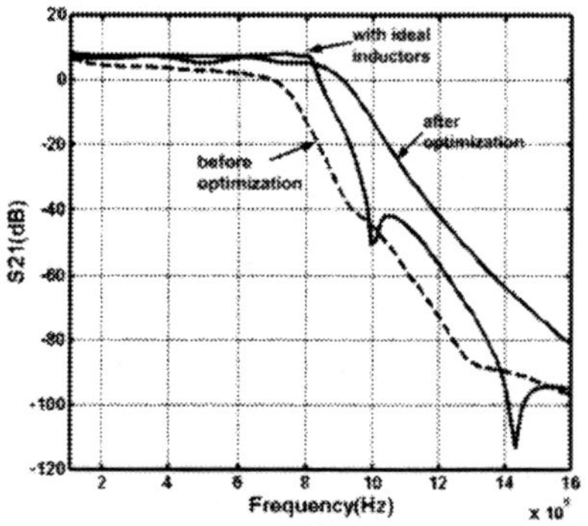

(a)

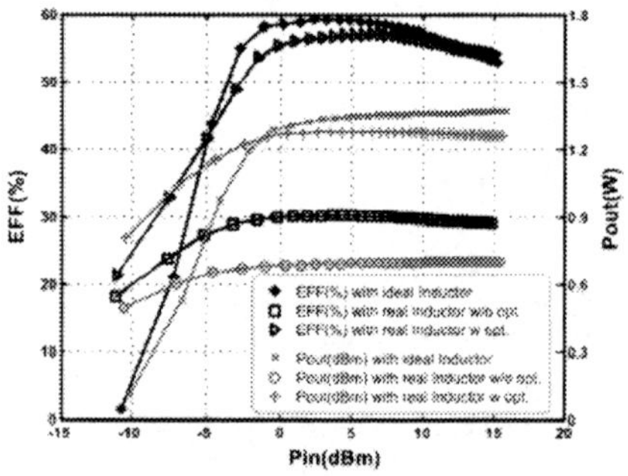

(b)

Fig. 8. Optimization results. (a) DA, and (b) PA.

V. CONCLUSIONS

An RF IC synthesis methodology based on particle swarm optimization is introduced. It is more suitable for complex circuits such as power amplifiers and ultra-wideband amplifiers than optimization using the simulated annealing heuristic, and it has advantages in efficiency of synthesis.

REFERENCES

[1] D.J. Allstot, et al., *Parasitic-aware RF IC Design and Optimization*, Boston: Kluwer Academic Pub., 2003.

[2] B.M. Ballweber, et al., "A fully integrated 0.5-5.5GHz CMOS distributed amplifier," *IEEE J. Solid-State Circuits*, vol. 35, pp. 231-239, Feb. 2000.

[3] J. Kennedy, et al., "Particle swarm optimization," *IEEE Intl. Conf. Neural Networks*, vol. 4, pp. 1942-1948, 1995.

AN OPTIMIZATION-BASED TOOL FOR THE HIGH-LEVEL SYNTHESIS OF DISCRETE-TIME AND CONTINUOUS-TIME ΣΔ MODULATORS IN THE MATLAB/SIMULINK ENVIRONMENT

Jesús Ruiz-Amaya, José M. de la Rosa, Fernando Medeiro, Francisco V. Fernández, Rocío del Río, Belén Pérez-Verdú and Angel Rodríguez-Vázquez

Instituto de Microelectrónica de Sevilla – IMSE-CNM (CSIC)
Edificio CICA-CNM, Avda. Reina Mercedes s/n, 41012- Sevilla, SPAIN
Phone: +34 95 5056666, Fax: +34 95 5056686, E-mail: ruiz@imse.cnm.es

ABSTRACT

This paper presents a MATLAB toolbox for the automated high-level sizing of ΣΔ Modulators (ΣΔMs) based on the combination of an accurate time-domain behavioural simulator and a statistical optimizer. The implementation on the well-known MATLAB/SIMULINK platform brings numerous advantages in terms of data manipulation, flexibility and simulation with other electronic subsystems. Moreover, this is the first tool dealing with the synthesis of ΣΔMs using both Discrete-Time (DT) and Continuous-Time (CT) circuit techniques.(*)

1. INTRODUCTION

ΣΔ Analog-to-Digital Converters (ADCs) have demonstrated to be an attractive solution for the implementation of analog-digital interfaces in *systems-on-chip* [1][2][3]. However, the need to design high-performance ΣΔ ADCs in adverse digital technologies together with the vertiginous rate imposed by the technology evolution has motivated the interest for CAD tools which can optimize the design procedure in terms of efficiency and short time-to-market. For this purpose, several tools for oversampling converter synthesis have been reported in the last years [1]-[8]. Among them, most succesful approaches belong to the so-called optimization-based synthesis tools [6][7].

The conceptual block diagram of this class of tools is shown in Fig.1. The design process of a ΣΔM starts from the high-level modulator specifications (resolution, signal bandwidth, etc). The objective is to get the building block specifications (design parameters) that optimize the performance of the modulator; that is, those specifications which satisfy the modulator-level specifications with the minimum power consumption and silicon area. At each iteration of the optimization procedure, circuit performances are evaluated at a given point of the design parameter space. According to such an evaluation, a movement in the design parameter space is generated and the process is repeated again.

The iterative nature of the optimization procedure requires a very

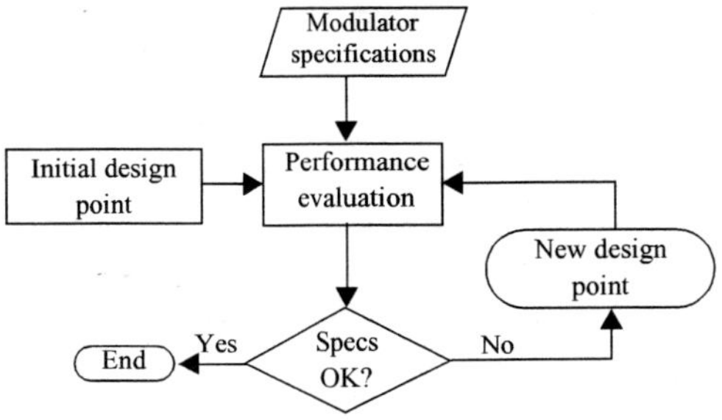

Figure 1. Block diagram of an optimization-based ΣΔM synthesis tool.

(*)This work has been supported by the EU ESPRIT IST Project 2001-34283/TAMES-2 and the Spanish CICYT Project TIC2001-0929/ADAVERE.

efficient mechanism for performance evaluation. Behavioural simulation is used in the synthesis approaches in [6][7]. This technique enables very efficient analysis while providing high accuracy levels.

In the tools described in [6][7], both simulation engine and models, are implemented using a programming language like C. Modulator libraries are usually available, containing a limited number of architectures implemented by Switched-Capacitor (SC) circuits. Although a text or graphical interface is usually provided to create new architectures, block models cannot be easily changed without the qualified contribution of a specialist programmer. On the other hand, the possible circuit techniques used to implement the modulators are constrained by the capabilities of the simulation engine and the available block models.

To overcome these problems the proposed ΣΔ synthesis tool has been implemented using the MATLAB/SIMULINK platform [9][10]. The embedded behavioural simulator is able to efficiently evaluate the performances of LowPass (LP) or BandPass (BP) ΣΔMs implemented using either SC, SwItched current (SI) or CT circuit techniques. This enables the synthesis tool to deal with all those types of ΣΔMs.

The implementation on the MATLAB/SIMULINK platform provides a number of advantages: (a) it is a widely used platform, familiar to a large number of engineers, whereas special-purpose tools [6][7] require to learn a proprietary text-based or graphical interface; (b) it has direct access to very powerful tools for signal processing and data manipulation; (c) it has complete flexibility to create new ΣΔM architectures, and even to include different blocks, either CT or DT; and (d) it enables a high flexibility for the extension of the block library.

The synthesis toolbox in this paper includes a complete library of building blocks for all circuit techniques (SC, CT and SI), considering the most important error mechanisms [1][2][3]. The behavioural models of such building blocks are incorporated as SIMULINK S-functions [11], which – compared to the use of SIMULINK library blocks as in [12][13][14] – decreases the computational cost to acceptable levels for synthesis purposes. The optimization core contains adaptive statistical optimization techniques for wide space exploration and deterministic techniques for fine tuning [1]. Besides, the addition of knowledge about specific architectures has been enabled. Such knowledge, which can be coded using a standard programming language (C or C++), is compiled at run-time and incorporated into the optimization process. This makes the presented MATLAB toolbox an optimization-based synthesis tool but with the appealing features of knowledge-based systems.

2. DESCRIPTION OF THE TOOL

The presented synthesis tool uses a statistical optimizer for design parameter selection and a time-domain behavioural simulator for performance evaluation. Both tools are integrated in the MATLAB/SIMULINK environment as described below.

2.1 Optimization core

Deterministic optimization methods, like those available in the MATLAB standard distribution [9], are not suitable because initially we may have little or no idea of an appropriate design point. Therefore, the optimization procedure is quickly trapped in a local minimum. For this reason, we developed an optimizer which combines an adaptive statistical optimization algorithm inspired in simulated annealing (local minima of the cost function can then be avoided) with a design-oriented formulation of the cost function (which accounts for the modulator performances). This optimizer has been integrated in the MATLAB/SIMULINK platform by using the MATLAB engine library [9], so that the optimization core runs in background while MATLAB acts as a computation engine.

Fig.2 shows the flow diagram of the optimizer where starting from a modulator topology, e.g., a modulator whose design parameters (building block specifications) are not known and arbitrary initial conditions, a set of design parameter perturbations is generated. With the new design parameters, a set of simulations are done to evaluate the modulator performance. From the simulation results, it automatically builds a cost function (that has to be minimized). The type and value of the perturbations as well as the iteration acceptance or rejection criteria depend on the selected optimization method. The optimization process is divided into two steps:

- The first step explores the design space by dividing it into a multi-dimensional coarse grid, resulting in a mesh of hypercubes (*main optimization*). A statistical method is usually applied in this step.
- Once the optimum hypercube has been obtained, a final optimization is performed inside this hypercube (*local optimization*). A deterministic method is usually used in this step.

In addition, the optimization core is very flexible, in so far as the cost function formulation is very versatile: multiple targets with several weights, constraints, dependent variables, and logarithmic grids are permitted. This optimization procedure has been extensively tested with design problems involving behavioural simulators as well as electrical simulators [1].

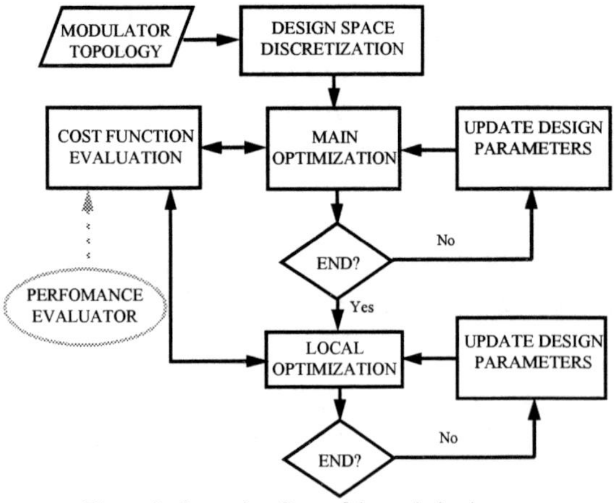

Figure 2. Operation flow of the optimization core.

2.2 Time-domain behavioural simulator

The simulation of ΣΔMs using transistor-level SPICE-like simulators lead to excessive CPU times (typically from days to weeks) [15]. For that reason, different alternatives have been developed, which at the price of losing some accuracy in their results, reduce the simulation time [1]. One of the best accuracy-speed trade-offs is achieved by using the so-called *behavioural simulation* technique using functional models [6][7][16][17][18]. This has been the technique used in our simulator, which has been implemented as a toolbox in the MATLAB/SIMULINK platform.

Modelling and simulation of ΣΔMs in SIMULINK was first reported in [12], although limited to SC architectures. Although very intuitive, the implementation of the behavioral models of each basic building block requires several sets of elementary SIMULINK blocks. This means a penalty in computation time which may become critical in an optimization-based synthesis process in which hundreds or thousands of simulations must be executed.

To overcome this problem, in the simulator in this paper, behavioural models have been incorporated in SIMULINK by using C-coded S-functions [11]. As a consequence, the CPU time for the time-domain simulation of a DT/CT ΣΔM involving 65536 samples is typically a few seconds[†1]. Besides, the proposed simulator is able to deal with any circuit technique: SC, SI or CT. Table 1 summarizes the basic blocks modelled as well as its non-idealities. A detailed description of these non-idealities and their behavioural models – beyond the scope of this paper – can be found in [1], [2] and [3] for SC, CT and SI circuits, respectively.

2.3 The MATLAB ΣΔM synthesis toolbox

The proposed tool has been conceived as a MATLAB toolbox for the simulation and synthesis of ΣΔMs. The Graphical User Interface (GUI) included in the toolbox allows to navigate easily through all steps of the simulation, synthesis and post-processing of results. For illustration purposes, Fig.3 shows part of the toolbox GUI for architecture description. By using this GUI, the user

Table 1: Building blocks and non-idealities modelled in the simulator.

Circuit technique	Block		Non-idealities
SC	Integrators	Opamps	Finite and non-linear gain, dynamic limitations (incomplete settling error, harmonic distortion), output range, thermal noise.
		Switches	Thermal noise, finite and non-linear resistance.
		Capacitors	Non-linearity, mismatching.
	Resonators		Non-idealities associated to the integrators.
SI	Integrators		Finite and non-linear gain, finite output and input conductance, dynamic limitations (incomplete settling, harmonic distortion, charge injection), thermal noise.
	Resonators		Feedback gain error, non-idealities associated to the integrators.
CT	Integrators		Finite and non-linear gain, dynamic limitations (parasitic capacitors, high and low frequency poles), thermal noise, output range and lineal input range, offset.
	Resonators		Non-idealities associated to the integrators.
ALL	Clock		Jitter.
	Comparators		Offset, hysteresis.
	Quantizers /DACs		Integral non-linearity, gain error, offset, jitter, delay time.

†1. All simulations shown in this paper were done using a PC with an AMD XP2400 CPU@2GHz @512MB-RAM.

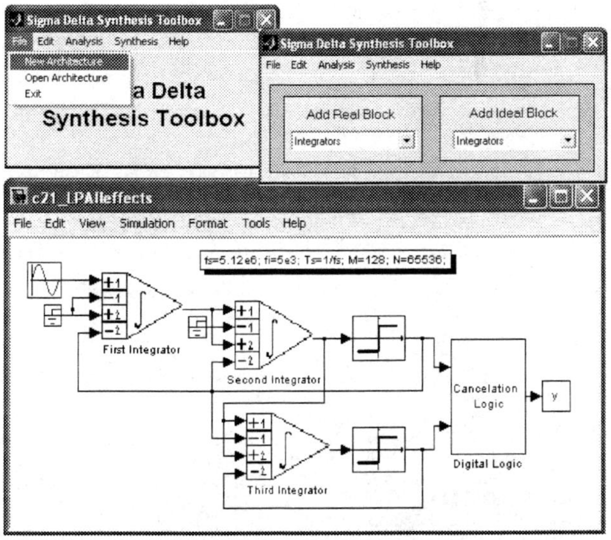

Figure 3. Illustrating the GUI for editing ΣΔMs of the Synthesis Toolbox.

can either open an existing ΣΔM architecture or create a new one in the SIMULINK platform by connecting the building-block available in the simulator. When a simulation is finished, different performance figures such as output spectrum, in-band noise power, harmonic distortion, etc. can be computed from the output data through the analysis/data processing menu. In addition, parametric analysis and MonteCarlo simulations can be performed. High-level synthesis is started from the synthesis menu, where constraints, performance specifications, design parameters, optimization algorithms, etc., can be specified. Then, the optimization core starts the exploration of the design space to find out the optimum solution by using the simulation results for performance evaluation.

3. ΣΔM SYNTHESIS EXAMPLES

To illustrate the simulation and synthesis capabilities of this toolbox two ΣΔM architectures have been selected:

- An SC 2-1 cascade single bit ΣΔM (SC 2-1 sb) (Fig.4(a)).
- A CT 5th-order LP ΣΔM (CT 5th-order LP) (Fig.4(b)).

One of the most important degrading factors in SC cascade ΣΔMs is the mismatch error. This is illustrated in the MonteCarlo simulation of Fig.5(a), where a random Gaussian mismatch error with zero mean and 0.5% standard deviation has been assumed. Each plot corresponds to a parametric analysis of the SNR versus input amplitude. About 450 simulations with 32768 samples were run to obtain this figure and it only took 9.1 minutes of CPU time.

On the other hand, one of the major advantages of CT ΣΔMs lies in that they can achieve higher sampling frequencies. However, CT ΣΔMs degrade drastically their performance as a result of two important errors: clock jitter noise and delay time between the quantizer clock edge and DAC response. The effect of this later is illustrated in Fig.5(b). Two different cases have been considered: a fixed delay, which is independent on quantizer input voltage magnitude; and a signal-dependent delay, which is practically constant for large quantizer input voltages, but rises for decreasing inputs [2].

To show the capabilities of the Synthesis Toolbox, the high-level sizing of the modulators in Fig.4 is performed. The modulator specifications are: 15bits@20kHz for the SC 2-1 sb ΣΔM and

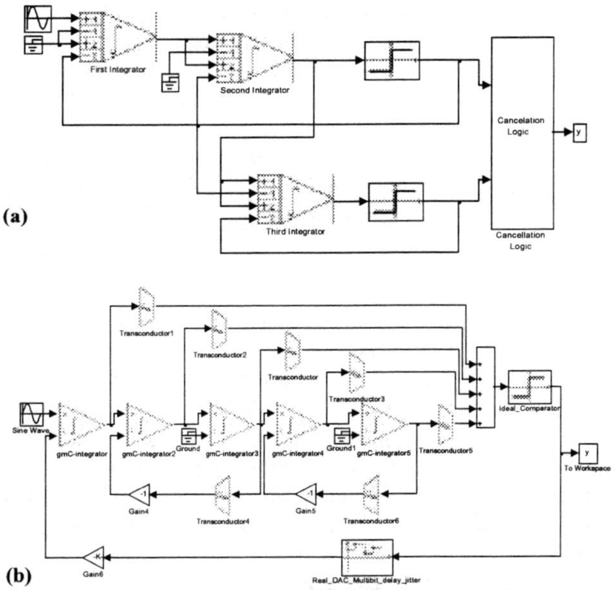

Figure 4. Block diagrams of the (a) SC 2-1 sb ΣΔM and (b) CT 5th-order LP ΣΔM in the SIMULINK environment.

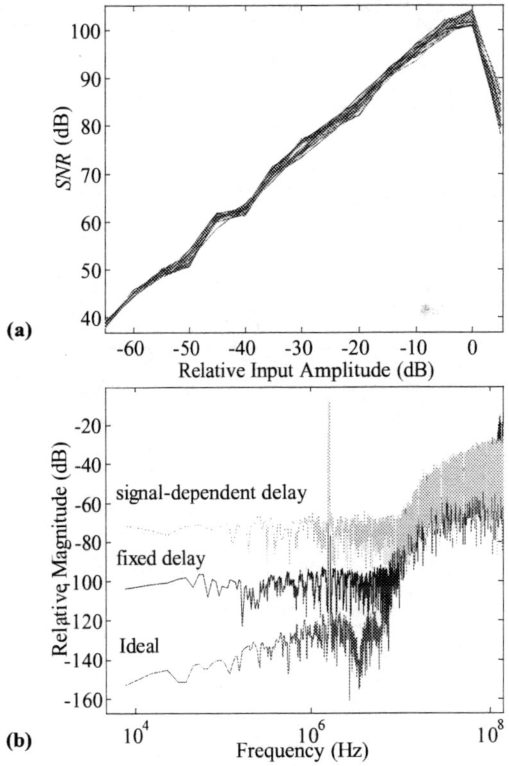

Figure 5. Simulation examples. Effect of (a) mismatch error on SC 2-1 sb ΣΔM and (b) loop delay time on CT 5th-LP ΣΔM.

12bits@6.25MHz for the CT 5th-order LP ΣΔM. The objective is to meet those specifications with the minimum power consumption and silicon area. Once design parameters, design specifications, and constraints have been specified through the toolbox GUI, a wide exploration of the design space is performed by the optimizer. At each point of the design space a simulation is done to evaluate the modulator performances.

Table 2 and Table 3 show the results of the high-level synthesis for both modulators. The optimization procedures required 817 iterations for the SC 2-1 sb modulator and 674 iterations for the CT 5th-LPΣΔ taking 16.4 minutes and 52.1 minutes of CPU time, respectively. Finally, Fig.6 illustrates both output spectra (indicating the Signal-to-Noise plus Distortion Ratio *(SNDR)*) corresponding to the high-level sizing provided by the synthesis toolbox.

CONCLUSIONS

A MATLAB toolbox for the synthesis of CT and DT ΣΔMs has been described. Based on the combination of an accurate and efficient SIMULINK-based time-domain behavioural simulator and an advanced statistical optimizer, the proposed toolbox allows to efficiently map the modulator specifications into building-block specifications in reasonable computation times. To the best of our knowledge, this is the first tool that is able to synthesize an arbitrary ΣΔM architecture using any circuit technique (SC, SI or CT). In addition, the implementation in the MATLAB/SIMULINK platform brings also numerous advantages with a relatively low penalty in computation time.

REFERENCES

[1] F. Medeiro, B. Pérez-Verdú, and A. Rodríguez-Vázquez: *Top-Down Design of High-Performance ΣΔ Modulators*, Kluwer, 1999.

[2] J. A. Cherry and W. M. Snelgrove: *Continuous-Time Delta-Sigma Modulators for High-Speed A/D Conversion: Theory, Practice and Fundamental Performance Limits*, Kluwer, 2000.

[3] J. M. de la Rosa, B. Pérez-Verdú and A. Rodríguez-Vázquez: *Systematic Design of CMOS Switched-Current Bandpass Sigma-Delta Modulators for Digital Communications Chips*, Kluwer, 2002.

[4] G. F.M. Beenker, J.D. Conway, G. G. Schrooten and A. G.J. Slenter: "Analog CAD for Consumer ICs", *in Proc. Workshop on Advances in Analog Circuit Design*, pp. 343-355, 1992.

[5] M.F. Mar and R.W. Brodersen: "A Design System for On-Chip Oversampling A/D Interfaces", *IEEE Transactions on VLSI Systems*, Vol. 3, pp. 345-354, September 1995.

[6] K. Francken, P. Vancorenland and G. Gielen: "DAISY: A Simulation-Based High-Level Synthesis Tool for ΔΣ Modulators", *Proc. IEEE Int. Conf. Computer-Aided Design*, pp. 188-192, 2000.

[7] F. Medeiro, B. Pérez-Verdú, A. Rodríguez-Vázquez and J.L. Huertas: "A vertically Integrated tool for Automated Design of ΣΔ Modulators", *IEEE Journal of Solid-State Circuits*, Vol. 30, No. 7, July 1995.

[8] G.G. E. Gielen and R.A. Rutenbar: "Computer-Aided Design of Analog and Mixed-Signal Integrated Circuits", *Proceedings of the IEEE*, Vol. 88, pp. 1825-1852, December 2000.

[9] The MathWorks Inc.:Using MATLAB Version 6", July 2002.

[10] The MathWorks Inc.: "Using Simulink Version 5", July 2002.

[11] J. Moreno-Reina, J.M. de la Rosa, F. Medeiro, R. Romay, R. del Río, B. Pérez-Verdú and A. Rodríguez-Vázquez: "A Simulink-based Approach for Fast and Precise Simulation of Switched-Capacitor, Switched-Current and Continuous-Time ΣΔ Modulators", *Proc. IEEE Int. Symp. Circuits and Systems*, pp. IV.620-623, 2003.

[12] S. Brigati, F. Francesconi, P. Malcovati, D. Tonieto, A. Baschirotto and F. Maloberti: "Modeling Sigma-Delta Modulator Non-idealities in SIMULINK", *Proc. IEEE Int. Symp. Circuits and Systems*, pp. II.384-387, 1999.

[13] P. Malcovati, S. Brigati, F. Francesconi, F. Maloberti, P. Cusitano and A. Baschirotto: "Behavioural Modeling of Switched-Capacitor Sigma-Delta Modulators", *IEEE Trans. on Circuits and Systems-I*, pp. 352-364, March 2003.

[14] N. Chandra and G. Roberts: "Top-Down Analog Design Methodology using MATLAB and SIMULINK", *Proc. IEEE Int. Symp. Circuits and Systems*, pp. IV.319-322, 2001.

[15] V. F. Dias, V. Liberali and F. Maloberti: "Design Tools for Oversampling Data Converters: Needs and Solutions", *Microelectronics Journal*, Vol. 23, pp. 641-650, 1992.

[16] C. H. Wolff and L. Carley: "Simulation of Δ−Σ Modulators Using Behavioral Models", *Proc. IEEE Int. Symp. Circuits and Systems*, pp. 376-379, 1990.

[17] V. Liberali, V.F. Dias, M. Ciapponi and F. Maloberti, "TOSCA: a Simulator for Switched-Capacitor Noise-Shaping A/D Converters," *IEEE Trans. Computer-Aided Design*, Vol. 12, pp. 1376-1386, Sept. 1993.

[18] K. Francken, M. Vogels, E. Martens and G. Gielen: "A Behavioral Simulation Tool for Continuous-Time ΔΣ Modulators," *Proc. IEEE Int. Conf. Computer-Aided Design*, pp. 234-239, 2002.

Table 2: High-level synthesis results for SC 2-1 sb ΣΔM.

OPTIMIZED SPECS FOR: 15bits@20kHz		Integ. I	Integ. II-III
Modulator	Sampling frequency (MHz)	5.12	
	Oversampling ratio	128	
Integrators	Sampling capacitor C_i(pF)	6	1.5
	Feed-back capacitor C_o(pF)	24	3
	MOS switch-ON resistance (kΩ)	≤ 0.84	≤ 1.7
Opamps	DC-gain (dB)	≥ 58.5	≥ 56
	DC-gain non-linearity (V^{-2})	≤ 22%	≤ 22%
	Output swing (V)	2.7	
	Input noise PSD (nV/sqrt(Hz))	≤ 8.1	≤ 278
	Output current (mA)	≥ 0.5	≥ 0.23
	Input transconductance (mA/V)	≥ 0.5	≥ 0.14
Comparators	Hysteresis (V)	≤ 0.2	
Technology	Cap. non-linearity (ppm/V^2)	≤ 89	

Table 3: High-level synthesis results for CT 5th-order LP ΣΔM.

OPTIMIZED SPECS FOR: 12bits@6.25MHz		Integ. I	Other Integ.
Modulator	Sampling frequency (MHz)	300	
	Oversampling ratio	40	
Transconductors	Transconductance (mA/V)	0.6	0.15
	DC-gain (dB)	≥ 34	≥ 42
	Parasitic output capacitor (pF)	≤ 0.66	≤ 0.04
	Input linear swing (V)	≥ 0.5	≥ 0.32
	HD3 (dB)	≤ −50	≤ −30
DAC	Clock jitter (ps)	≤ 0.5	
	Excess loop delay time (ns)	≤ 0.77	

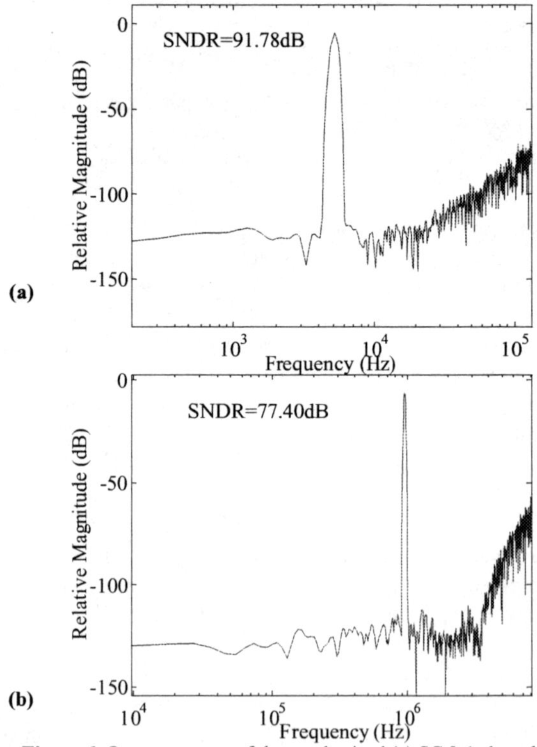

Figure 6. Output spectra of the synthesized (a) SC 2-1 sb and (b) CT 5th-order LP ΣΔMs.

A NOVEL ANALOG LAYOUT SYNTHESYS TOOL

Lihong Zhang, Ulrich Kleine

IESK, Otto-von-Guericke University of Magdeburg
PO Box 4120, D-39016 Magdeburg, Germany

ABSTRACT

This paper presents a layout synthesis tool for analog integrated circuits called ALADIN. The layout generation is based on relatively complex sub-circuits rather than non-optimal single devices. A flexible module generator environment is developed for designers to write and maintain technology and application independent module generators for sub-circuits. A two-stage placement technique is proposed, which dramatically improves the placement accuracy without compromising the efficiency compared with the one-stage placement. The analog module routing consists of two phases including global routing and detailed routing. The minimum-Steiner-tree based global routing can be integrated into the placement procedure to improve the routability of placement solutions. The compaction based constructive detailed routing finally realizes the layout of the whole circuit. The benefit of ALADIN providing layouts comparable to expert manual ones is demonstrated with several circuits showing its competition compared with several existing tools.

1. INTRODUCTION

Unlike the digital portions, most of the analog circuits in the increasingly important mixed-signal chips are still routinely designed by hand, which always accounts for design errors and expensive design iterations. Because CAD tools for analog circuits are still in their infancy, analog circuits have become the design bottleneck of the integrated mixed-signal systems. Furthermore, because of technology scaling, the device electrical properties get worse in analog circuits and the design window decreases. Due to some special and necessary constraints of analog layouts (e.g., large variation of MOS transistor sizes, sensitivity to parasitic capacitance, crosstalk, device matching, symmetrical requirements, etc.), the analog layout design is intrinsically more difficult than the digital counterpart [1]. For a reliable optimization, electrical parameters caused by different layout topologies have already become necessary for high performance circuits even during circuit design.

Thus far some tools have been developed to automate the generation of analog layouts using procedure generators, template-driven, rule-based, constructive or algorithm-based techniques [2]. ILAC is a process-independent tool that automatically generates layout for analog CMOS circuits [3]. Its major drawback is that the applied certain features of the digital layout styles limit the ability to achieve high-quality dense analog circuit-level layouts. Cohn introduced KOAN-ANAGRAM II for analog device-level layout automation [4]. Lampaert developed a performance driven analog layout tool LAYLA [2]. Their distinguished feature is the simplicity of their device generation library. Nonetheless, since the construction of the entire analog layout is normally based on non-optimal single devices rather than decomposed sub-circuits (also called modules in this paper), the complexity dramatically increases especially for large circuits. On the other hand, irrespective of the complexity of an algorithm, some structures can never be constructed using geometry sharing techniques (e.g., interdigital cascode structures [3]). Therefore, to obtain the best results, a number of sub-circuit generators will have to be constructed.

Unlike these automated layout tools that are mainly designed for system designers and therefore more or less bypass the knowledge of circuit designers, a novel Automatic Layout Design Aid for Analog Integrated Circuits (ALADIN) is currently being developed for analog circuit experts who can bring their specific knowledge into the synthesis process to create high quality layouts. Designers can construct layouts of parameterizable modules in a technology and application independent way. The placement and routing of modules are performed automatically under the constraints defined by designers.

The rest of the paper describes the entire ALADIN system in detail. Section 2 introduces the structure and layout synthesis flow of ALADIN. The technique of module generation is explained in Section 3. The placement strategies and routing approaches are described in Sections 4 and 5 respectively. Section 6 shows the experimental results and the conclusion is drawn in Section 7.

2. STRUCTURE AND DESIGN FLOW

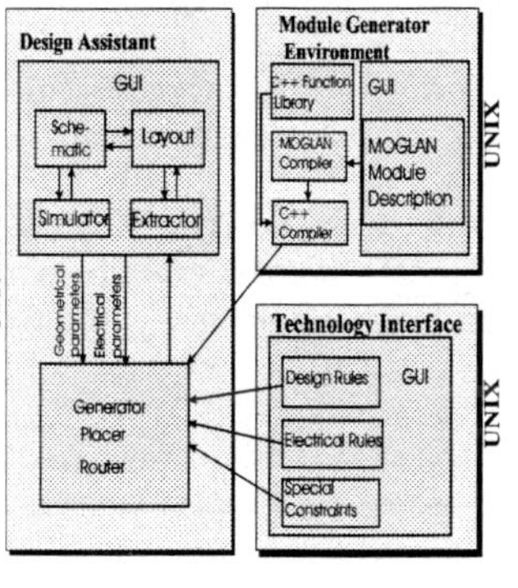

Fig. 1: Structure overview of ALADIN.

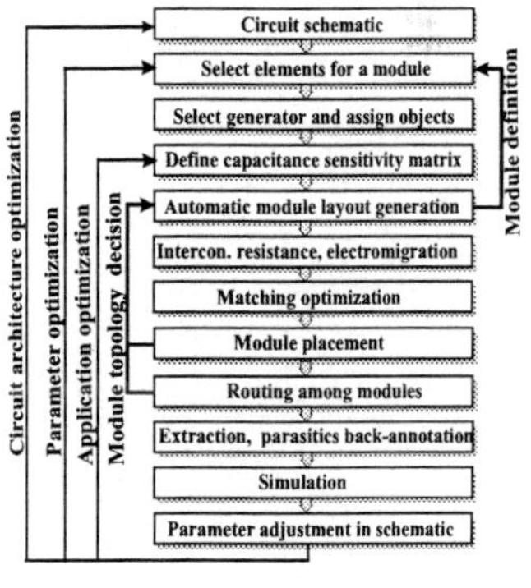

Fig. 2: Circuit design flow of ALADIN.

In Fig. 1 a structure overview of ALADIN is given. The entire system mainly consists of three components: Design Assistant, Module Generator Environment and Technology Interface. The Design Assistant is integrated into Cadence Design Framework II and can be easily adapted to other commercial software. It provides a graphical user interface to optimize an analog circuit from schematic to layout. The Module Generator Environment allows designers to write technology and application independent parameterizable modules as complex as needed [5]. The technology interface eases the input of description rules [6].

The circuit design flow of ALADIN is illustrated in Fig. 2. In a schematic editor, a simulated circuit is partitioned into several modules. The steps from selecting elements of a module to layout generation are repeated until all the modules are defined. During the placement of all the modules, the appropriate topology is chosen from possible alternatives for each module. The global routing is performed simultaneously along with the placement procedure followed by the detailed routing. After the layout generation, an extraction is performed and the parasitic elements are automatically annotated back into the schematic. The entire circuit is then simulated with a good parasitic estimation and the optimization loop starts by changing parameters of the circuit (i.e., parameter optimization) and/or by redefining constraints (i.e., application optimization).

3. ANALOG MODULE GENERATION

The novel concept of module generation in ALADIN is that the layout synthesis is not bound to a fixed generator library. Analog circuit designers can easily modify the existing topologies or create new generators. ALADIN presents a natural description language (MOGLAN) [5], which enables designers to describe parameterizable modules hierarchically and design-rules independently. The MOGLAN description is translated into C++ code, which is then compiled into an executable technology and application independent generator with a C++ function library. This language features loops, conditional statements and a set of functions to create and to wire primitive geometries without considering absolute coordinates. Moreover, the design rules are automatically evaluated and fulfilled.

Basic geometric objects are generated by calling primitive functions. Each geometric object contains special properties that define whether its edges are fixed or variable for moving inwards or outwards. If an edge is variable, it can be moved automatically in the following stages to improve the layout. Complex modules are constructed by compacting either geometric primitives or hierarchically built objects to an existing structure. In our compaction method, only outer edges of the main structure are stored in the database and no general edge graph need creating, which dramatically speeds up the operation. Furthermore, it is much easier for designers to predict the result.

Furthermore, electrical properties, e.g., parasitic capacitance, RC time constants of gates, electromigration, etc. are automatically regarded during the module generation. The matching of two circuit elements can numerically be calculated in the Design Assistant using the standard deviation of electrical parameters and simplified device models.

4. ANALOG MODULE PLACEMENT

Based on the study of prior work, we have developed a novel two-stage placement technique in ALADIN.

4.1. Cost Function and Slide-Based Flat Placement

The cost function is used to evaluate the quality of searched states. It consists of three parts given in Eq. (1),

$$C = \alpha_{area} C_{area} + \alpha_{nets} C_{nets} + \alpha_{size} C_{size}, \quad (1)$$

where α_* are the weight factors for the corresponding cost C_*. C_{area} is the area cost that is used to decrease the entire area, make NWELL and PWELL regions relatively concentrated and isolate analog and digital regions as required in mixed-signal circuits. C_{nets} is the net-length cost, in which priority coefficient can be specified for each net. C_{size} is the size cost used to control the shape of the final layout. Graphic input windows of the cost function are provided in the Design Assistant.

A placement algorithm is actually an optimization flow following a certain geometric construction strategy. As for the geometric construction manner, based on the normal Gellat-Jepsen flat placement style [4], we developed a module-slide based flat placement style. After all the modules have been absolutely placed, a process of module slide is executed to determine the relative positions of the modules. Thus, overlaps among modules are avoided so that relative position instead of absolute coordinates becomes the ultimate focus of the search. An algorithm with polynomial time complexity is designed. In addition to the complexity reduction, this approach can also fulfill the symmetry constraints simultaneously.

4.2. Two-Stage Placement Technique

Conventionally the placement and the global routing are always performed sequentially. Thus, the result of the global routing can only be used for the detailed routing and useless for the placement procedure. Although the simultaneous performance of the placement and the global routing can lead to a more accurate search, it is normally very time-consuming. Furthermore, the placement using an inaccurate estimator, e.g., half-perimeter, obtains the final solution that can only roughly assure the short net length. To address this problem, we propose a two-stage placement technique depicted in Fig. 3 based on our two currently developed innovative placement algorithms GASA and VFSRA_P. GASA manages the unification of genetic algorithm (GA) and simulated annealing (SA) [7]. VFSRA_P is a very fast simulated re-annealing placement algorithm following the cooling schedule Eq. (2) [8], which is exponentially faster than Cauchy Annealing (CA) widely used in [2][3][4],

$$T_i(k) = T_{0i} e^{-c_i k^{1/D}}, \quad (2)$$

where T_{0i} is the initial temperature, C_i is the control factor, D is the dimension of configuration space and k is the current annealing step. According to our experiment, GASA can perform much better than simple GA or SA approaches (e.g., 14.1% better than VFSRA_P in terms of cost values), while VFSRA_P works 35.5% faster than CA and 16.2% faster than GASA.

We divide the entire placement procedure into two stages: global placement and detailed placement. Based on the evolution of populations, GASA, as the search strategy during the global placement, can find near-optimal solutions efficiently if using the half-perimeter estimator. The search in this stage is relatively rough and quick. Then the optimization transfers to the detailed placement stage controlled by VFSRA_P. VFSRA_P is different from GASA and based on an initial state provided by the global placement stage. Moreover, since the VFSRA_P works with relatively low initial temperature, the modification of states during the detailed placement stage is slight and careful. During

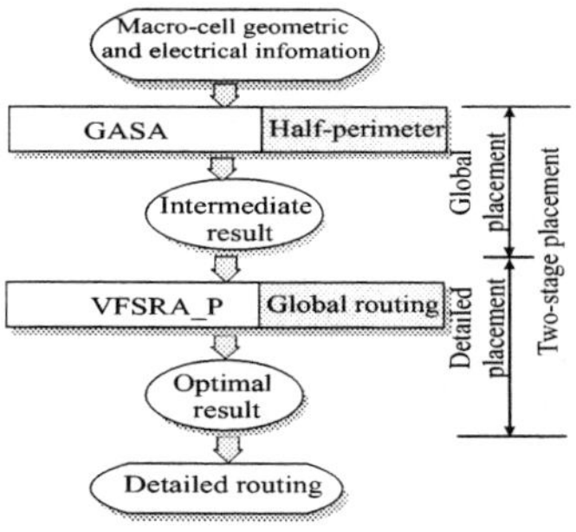

Fig. 3: Two-stage placement flow.

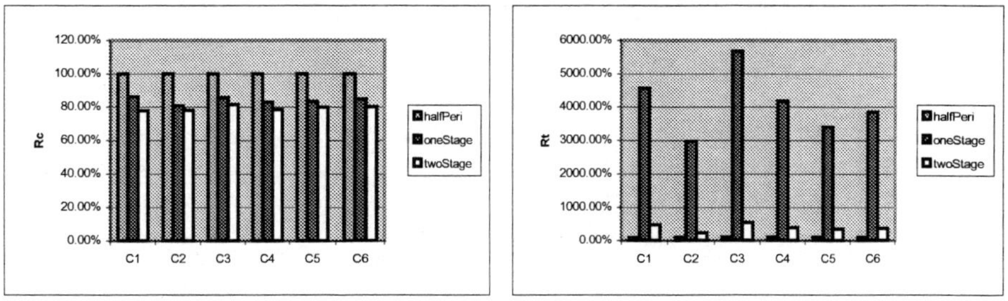

(a) cost ratio (b) execution time ratio

Fig. 4: Comparison among halfPeri, oneStage and twoStage on six circuits (C1-C6).

the detailed placement stage, the minimum-Steiner-tree based global routing is performed.

Because the global placement has provided a near-optimal solution, the optimization difficulty during the detailed placement is considerably alleviated. Therefore, compared with the simultaneous performance of the placement and the global routing in one single stage, the execution time of this two-stage placement technique becomes acceptable. Fig. 4 shows the comparison among GASA placement strategy using a half-perimeter estimator (named halfPeri), one-stage placement strategy (i.e., GASA placement simultaneously running the minimum-Steiner-tree based global routing, named oneStage) and two-stage placement strategy (named twoStage) on six circuits. The experimental results show the twoStage approach on average improves the search quality by 4.6% and speeds up nearly 10 times compared with the oneStage approach.

5. ANALOG MODULE ROUTING

The area routing is widely applied in the analog circuit layout automation tools [2][4], where the routability is largely dependent on the routing sequence of nets. Furthermore, the placement algorithm is always responsible for allocating the routing area. Although LAYLA proposed one dynamic routing area estimation technique [2], 10% to 15% area redundancy still exists when comparing their results with manual layouts. In ALADIN two routing phases including the global routing and detailed routing are employed. The global routing is integrated into the placement procedure to improve the accuracy of routing estimation. The compaction based constructive detailed routing generates the final layout based on the output of the placement procedure.

Because nets play a critical role in analog circuits due to parasitic effects, crosstalk, etc., the minimum-Steiner-tree based global routing is developed. Not only is the minimum-Steiner-tree method used for the net-length estimation, it also elaborates the routing plan taking the net sensitivity and channel congestion into account. A weighted graph is used to model the routing regions. A rectilinear channel graph is formed by passing channels (or edges) through critical regions and forming vertices at their intersections. Finding a global route becomes equivalent to finding an optimal subtree (the minimum-Steiner-tree) in the routing pin graph that spans the terminal vertices. A Dijksta shortest path algorithm is applied to solve this minimum-Steiner-tree problem [9].

A technique of simultaneous execution between the placement and global routing has been developed, where the global routing is executed for each intermediate placement solutions. It makes better search results without losing the solvability of the problem. This global-routing-driven placement strategy is especially effective for the analog layout designs where the number of nets is relatively small but with complex constraints [10].

A potential problem in the traditional placement and global routing procedures is whether the estimated channel width is accurate. So a post-processing procedure, such as compaction, is required. However, a different strategy is used in ALADIN, where the placement phase is followed by a compaction-based constructive detailed routing phase that automatically minimizes the channel space. The width of channels need not be regarded during the placement except for the channel congestion to avoid overburdened channels. The congestion degree of a channel is represented by the number of the passed nets in this channel. It is taken as a weight in the weighted graph of the global routing model, apart from the channel length. The Dijksta shortest path algorithm optimizes the routing paths and finds the one that is balanced and the shortest. In this way, the conventional problems of routability and post-processing are avoided.

In the detailed routing, for each module, the interconnections within the module are first wired densely around the module boundary using a ring router [11]. Then this dense module is compacted towards others according to the position relationship extracted from the placement output solution. The interconnections between the compacting module and the reference module are routed within a relatively small scope using a modified maze router [11]. In this way, the whole layout area remains as small as possible because the compaction can assure high density. Moreover, no estimation of routing area is needed and the problem of routability can be solved during each compaction step.

6. IMPLEMENTATION AND EXPERIMENTAL RESULTS

ALADIN package comprises 77,724 lines of C++ code, 5,354 lines of Cadence SKILL code and 5,327 lines of Tcl/Tk code. To demonstrate its effectiveness, we present some placed and routed layout examples compared with expert manual layouts or the state-of-the-art automated layout tool LAYLA.

In Fig. 5a the schematic of the rail-to-rail operational amplifier is depicted. The partitioning of the modules is indicated by the shaded rectangles. The die photograph of the result from ALADIN fabricated using 0.8μm CMOS technology is depicted in Fig. 5b. Its core layout area is 210μm * 167μm. As a comparison, the die photograph of a manual layout using the same technology is depicted in Fig. 5c. The core layout area is

208μm * 159μm. Measurement results show that the performance of the layout generated by ALADIN is comparable to that of the manual layout.

(a)

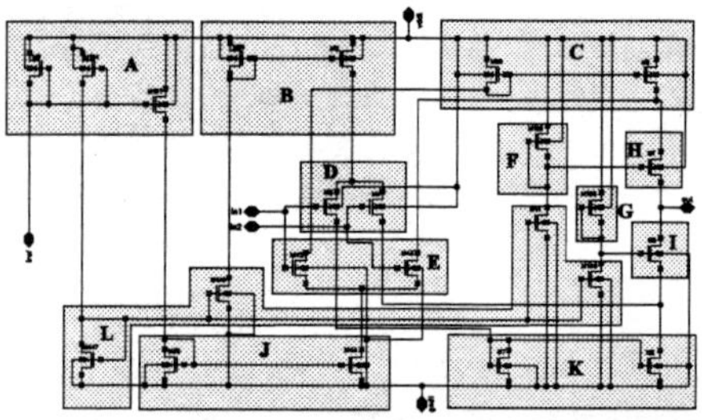

(b) (c)

Fig. 5: Schematic (a), die photographs of ALADIN layout (b) and manual layout (c) for the rail-to-rail operational amplifier.

In Fig. 6a the schematic of an one-stage operational amplifier is depicted. Its layout is generated by ALADIN and the corresponding die photograph is shown in Fig. 6b. As a comparison, the routed result generated by LAYLA is depicted in Fig. 6c [2]. The measurement results reveal very good compliance with the simulation results.

(a)

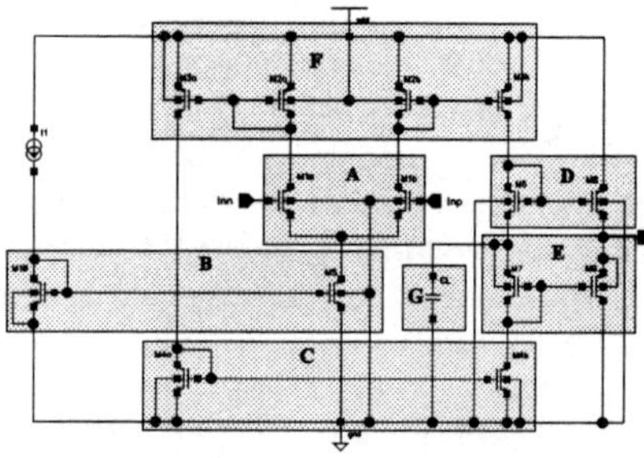

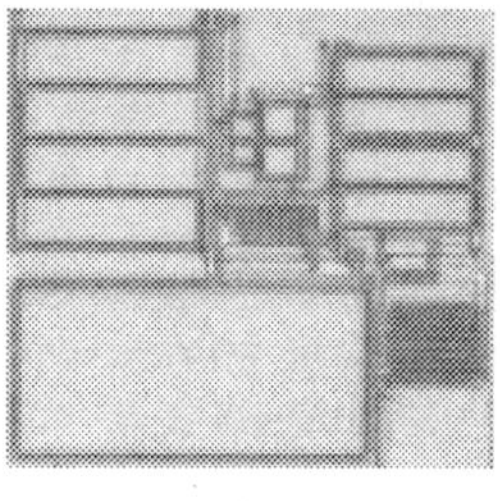

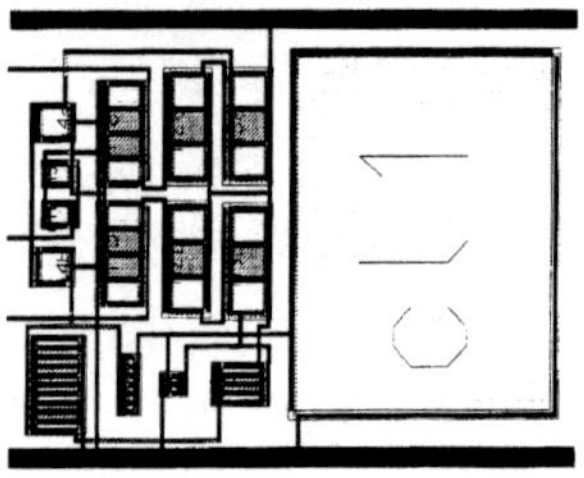

(b) (c)

Fig. 6: Schematic of an one-stage operational amplifier (a), the die photograph generated by ALADIN (b) and the routed layout generated by LAYAL (c).

7. CONCLUSIONS

In this paper a layout synthesis tool, ALADIN, for analog integrated circuits has been introduced. Designers can write and maintain their own technology and application independent module generators for sub-circuits. Different from other systems, layout generation is based on optimal complex modules rather than single devices. A two-stage placement technique, which is more suited for the analog module placement, is proposed. The minimum-Steiner-tree based global routing is integrated into the placement procedure to improve the routability of placement solutions. The compaction based constructive detailed routing finally realizes the layout of the entire circuit. The experimental results show that this tool is competitive and can provide results comparable to expert manual layouts.

8. ACKNOWLEDGEMENT

The authors would like to thank Prof. Georges Gielen at Katholieke Universiteit Leuven in Begium and Prof. Rob A. Rutenbar at Carnegie Mellon University in USA for their help during this research.

9. REFERENCES

[1] G. Gielen and R. A. Rutenbar, "Computer-Aided Design of Analog and Mixed-Signal Integrated Circuits," *Proc. IEEE*, vol. 88, pp. 1825-1852, Dec. 2000.

[2] K. Lampaert, G. Gielen, and W. Sansen, Analog Layout Generation for Performance and Manufacturability, Boston: Kluwer Academic Publishers, 1999.

[3] J. Rijmenants, J. B. Litsios, T. R. Schwarz, and M. G. R. Degrauwe, "ILAC: An Automated Layout Tool for Analog CMOS Circuits," *IEEE J. Solid-State Circuits*, vol. 24, no. 2, pp. 417-425, April 1989.

[4] J. M. Cohn, D. J. Garrod, R. A. Rutenbar, and L. R. Carley, "KOAN/ANAGRAM II: New Tools for Device-Level Analog Placement and Routing," *IEEE J. Solid-State Circuits*, vol. 26, pp. 330-342, March 1991.

[5] M. Wolf, U. Kleine, and J. Schulze, "New Description Language and Graphical User Interface for Module Generation in Analog Layouts," *Proc. International Symposium of Circuit and System*, vol. VI, pp. 290-293, June 1998.

[6] L. Zhang, U. Kleine, T. Rudolph, and M. Wolf, "A New Design Rule Description for Automated Layout Tools," *Proc. 7th IEEE International Conference on Electronics, Circuits and System*, pp. 988-992, Dec. 2000.

[7] L. Zhang, and U. Kleine, "A Genetic Approach to Analog Module Placement with Simulated Annealing," *Proc. IEEE International Symposium on Circuits and Systems*, pp. 345-348, May 2002.

[8] L. Ingber, "Simulated Annealing: Practice versus Theory," *Mathematical and Computer Modeling,* vol. 18, no. 11, pp. 29-57, 1993.

[9] S. M. Sait and H. Youssef, VLSI Physical Design Automation (Theory and Practice), McGraw-Hill, 1995.

[10]J. A. Prieto, A. Rueda, J. M. Quintana, and J. L. Huertas, "A Performance-Driven Placement Algorithm with Simultaneous Place&Route Optimization for Analog IC's," *Proc. European Design and Test Conference*, pp. 389-394, 1997.

[11]L. Zhang, U. Kleine, and M. Wolf, "Automatic Inner Wiring for Integrated Analog Modules," *Proc. 8th International Conference on Mixed Design of Integrated Circuits and Systems*, pp. 109-114, June 2001.

Behavioral Modeling of Analog Circuits by Dynamic Semi-Symbolic Analysis

Junjie Yang
Department of Electrical Engineering
University of California, Riverside, CA 92521, USA
jyang@ee.ucr.edu

Sheldon X.-D. Tan
Department of Electrical Engineering
University of California, Riverside, CA 92521, USA
stan@ee.ucr.edu

Abstract— *The paper presents a novel approach to behavioral modeling of analog circuits by dynamic semi-symbolic analysis, where some circuit parameters are kept as symbols and the others are given as numeric values. Our new method is based on the determinant decision diagram (DDD) representation of small-signal characteristics of linear analog circuits. The basic idea is to dynamically reorder DDD vertices such that all the DDD vertices corresponding to symbolic parameters are separated from DDD vertices for numerical parameters. In this way, DDD sizes of symbolic portion of DDD can be significantly reduced by suppressing numerical DDD nodes. Our new approach is different from the existing MTDDD based semi-symbolic analysis method where reordering is done before DDD is constructed and DDD-based graph operations are still valid in the new method. The proposed dynamic ordering algorithm, which is based on swap of adjacent variables, also improves the existing DDD-based vertex sifting algorithm as no special sign rule is required after DDD vertices are swapped. Experimental results have demonstrated that the proposed dynamic semi-symbolic method leads to up to 30% symbolic DDD node reduction compared MTDDD method on real analog circuits and can be performed very efficiently.*

I. Introduction

Symbolic analysis is devoted to the analysis of integrated circuits in which part or all the circuit parameters and the complex frequency variable are represented by symbols. As illustrated in [1], simple yet accurate symbolic expressions can also be interpretable by analog designers to gain insight into circuit behavior, performance and stability, and are important for many applications in circuit design such as transistor sizing and optimization, topology selection, sensitivity analysis [2]. But its applications have traditionally suffered from the exponential growth of the complexity of the symbolic results with the circuit size. This has motivated many approximate techniques [3]. But the approximated expressions will lose some information which are crucial in some applications.

From analog circuit design standpoints, most of the time, only some circuit parameters are of interest for a specific circuit design task. This leads to the semi-symbolic analysis where only a small portion of circuit parameters are kept as symbols and the rest of parameters are given as numerical values. This semi-symbolic analysis brings the benefits of reduced symbolic expressions at no cost of accuracy loss.

Semi-symbolic analysis was studied in the past by numerical interpolation method [4], by parameter extraction [5], and by using multi-terminal DDD concepts [6]. But existing approaches either have very restrictive usage or suffer circuit-size limitation problems. Recently, a multi-terminal DDD (MTDDD) graph was proposed to represent the semi-symbolic expressions [6]. The idea is to order the complex DDD nodes before DDD constructions such that the DDD nodes, whose matrix elements contain symbolic circuit parameters, are positioned above all DDD nodes whose matrix elements contain only numerical circuit parameters when the complex DDD graph is constructed. The s-expanded MTDDDs are then constructed in a bottom up fashion. But the resulting MTDDDs do not hold many DDD properties and they are difficult to perform many DDD graph operations, which are crucial to many symbolic analysis tasks, on the MTDDD graphs. Worse, some numerical MTDDD nodes may appear above symbolic MTDDD nodes in the resulting MTDDDs. As a result, size of MTDDDs is not adequately reduced and benefits of MTDDDs are not fully explored. For example, in Fig. 1, numerical node **2** will be the parent node of symbolic node sC_1 in the resulting MTDDD graph. As a result, the MTDDD may still be large as those numerical DDD nodes can be suppressed.

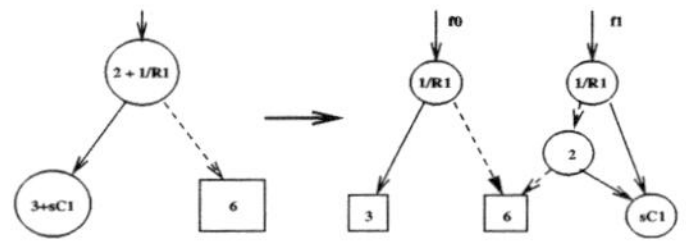

Fig. 1. A MTDDD graph which has numerical DDD nodes.

In this paper, we propose a new approach to dynamic semi-symbolic analysis of analog circuits. The new method also starts with exact symbolic representations of small-signal circuit characteristics of linear analog circuit via determinant decision diagrams [7–9]. The new method does not explicitly construct the MTDDDs, instead it first constructs normal s-expanded DDDs from complex DDDs. Then a dynamic variable ordering procedure is performed to move all the symbolic s-expanded DDD nodes to the top of the DDD graph. After this, all the bottom numerical DDD nodes can be suppressed implicitly to reduce overall DDD sizes. The resulting s-expanded DDDs are constructed such that the MTDDD graphs are implicitly embedded in the top portion of the s-expanded DDD graphs. The new approach offers several advantages over MTDDD-based method. First, all the DDD-based graph operations can still be performed on the resulting s-expanded DDD graphs as not MTDDD graphs are constructed. Second, it will result in smaller size of DDD graphs. Third, with the freedom of moving nodes to different positions of a DDD graph, the new method is able to change the set of symbolic parameters dynamically without re-performing the symbolic analysis each time. We will show how dynamic variable ordering is performed on the DDD graphs via adjacent variable order swap, which does not require any new sign rule for the moved DDD vertices. This is in contrast to the DDD size optimization method based on a vertex swap method in [10], where a special sign rule was developed for the swapped DDD vertices and the proposed sign rule has significant impacts on the effects of DDD minimization.

II. DDDs and DDD Implementations

In this section, we first review the notion of determinant decision diagrams. Then we present some implementation details that are important for the new dynamic variable ordering algorithm.

Determinant Decision Diagrams [7] are compact and canonical graph-based representation of determinants. It represents a determinant that has an associated variable x_i and points to two other nodes (cofactors of the determinant) in the graph. The node D is written as a tuple $(I(x_i), D_{x_i}, D_{\overline{x}_i})$ where x_i is called the *top* variable of the determinant D, D_{x_i}, $D_{\overline{x}_i}$ are the cofactor and remainder of D with respect to x_i respectively. $I(x_i)$ is the index of x_i. For convenience, D_{x_i}, $D_{\overline{x}_i}$ are also written as X_1 and X_0 in the sequel. DDD is a ordered graph where an index is assigned to each variable and the variable must appear in descending order in terms of their indices along each path in the DDD. The index of variable is also called *level* of the variable.

Following a similar implementation of BDDs, a global hash table, called *unique* table, allows a node of DDD node $(I(x_i), X_1, X_0)$ to be found in a constant time. The hash table typically is implemented by an array of hash-trees – *unique[]* where all level-j DDD nodes are stored in *unique[j]* in a tree structure based on their hash values of tuple $(I(x_j), X_1, X_0)$. The advantage of such two-level hashing implementation is that we can access all the nodes at level j without walking the entire unique table. This is important for the dynamic variable ordering as it will become clear later.

III. Dynamic Variable Ordering For DDDs

It has been shown that the order swap of two adjacent variables in an OBDD affects only the BDD nodes at the two levels and all other nodes remain unchanged [11–13]. It turns out that this is also true for DDD graph as each DDD node represent a determinant.

As we mentioned before that we can access all the nodes at level i directly. This makes the variable swap very memory efficient as we don't need to touch nodes at other levels. Suppose we want to swap the variable x at level i and y at level j with $i > j$. i.e. $I(x) = i$ and $I(y) = j$. After the swap, the indices of the two variables will become: $I_a(x) = j$ and $I_a(y) = i$. Suffix a means the index after the order swap.

The basic idea behind variable swap is that we maintain an identical determinant represented by each node even though indices of the nodes are changed. Meanwhile, we also make sure that descending variable order is enforced after swap. The major difference between BDD and DDD is that we need to determine the signs of swapped DDD nodes, It turns out that the sign rule is still valid. We have the following lemma without proof.

Lemma 1 *The sign rule proposed in [7] is still applicable to the swaped DDD nodes after the swap of their variable orders.*

To perform the swap operation, we first consider all the DDD nodes at level i. Those nodes can be further classified into two types: (1) node x in which the top variables of both its two child nodes X_1 and X_0 are not y and (2) nodes otherwise. For the first type nodes, the reordering procedure is shown in Fig. 2. The new node DDD node y with its 0-edge pointing to x and 1-edge pointing to 0 terminal is created first in step (b). Since for a DDD graph, node pointing to 1 terminal has to suppressed (zero suppression rule), we end up with a DDD graph shown in step (c), the index of x becomes j after the swap, so the whole operation is equivalent to overwriting node x with a new index $I_a(x) = j$.

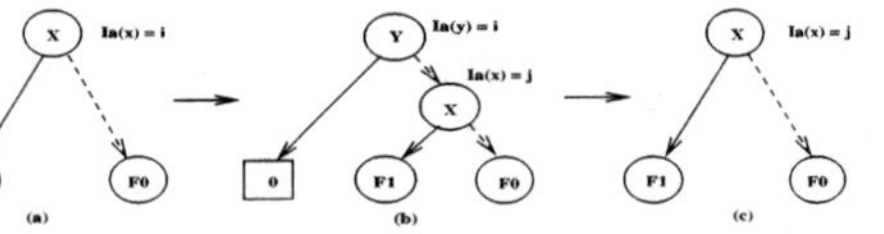

Fig. 2. Variable swap for the first type of the nodes at level i.

For the second type nodes, we follow the same method in [13] as follows: For each node $x = (I(x), X_1, X_0)$ at level i, we create a new node, $(I_a(y), (I_a(x), F_{11}, F_{01}), (I_a(x), F_{01}, F_{00}))$, to overwrite it, where X_{11} and X_{10} which are the cofactor and remainder of X_1 with respect to y respectively, and X_{01} and X_{00} are the cofactor and remainder of X_0 with respect to y respectively. X_{11} and X_{10} and X_{01} and X_{00} can be computed very easily by using COFACTOR(D,P) and REMAINDER(D,P) operations [7] and they take a constant time in this particular case as x and y are adjacent variables.

For each node y at level j, we can simply overwrite its index with $I_a(y) = i$ after all the level i nodes are processed. Note that after we change the index of a node, we need to put it into a new position in the unique table. Let i and j be the adjacent indices with $i > j$. The pseudo algorithm of the adjacent variable swap method is shown in Fig. 3.

```
SWAPADJACENTVARIABLEORDER(j,i)
  for each node x = (i,X1,X0) in unique[i] do // process type 1 nodes
    if(I(X1) is not j and I(X0) is not j)
      Overwrite the node with (j,X1,X0) and put it into a hash tree Hj
  for each node x in unique[i] do // process type 1 nodes
    if(I(X1) = j or I(X0) = j)
      X11 = COFACTOR(X1, j)
      X10 = REMAINDER(X1, j)
      X01 = COFACTOR(X0, j)
      X00 = REMAINDER(X0, j)
      Create nodes F1 = (j,X11,X01) and F0 = (j,X10,X00)
      Check and put new node (j,F1,F0) into the Hj
  for each node y = (j,Y1,Y0) in unique[j] do
    Overwrite the node with (i,Y1,Y0) and put it into the hash tree Hi
  unique[j] = Hj
  unique[i] = Hi
```

Fig. 3. Dynamic adjacent variable swap algorithm for DDD graphs

It is easily to see that time complexity of the algorithm is linearly proportional to the number of DDD nodes in level i and level j. As a result, garbage collection is also performed to reduce the unreferenced nodes at level i and j before the swap operations.

IV. Semi-Symbolic Analysis via Dynamic Variable Ordering

With the efficient adjacent variable swap algorithm, semi-symbolic analysis essentially boils down to moving all the symbolic DDD nodes to the top of the DDD graph and suppress all unused numerical DDD nodes if necessary. The concept is best illustrated using a simple RC filter circuit shown in Fig. 4. Its system equations can be written as

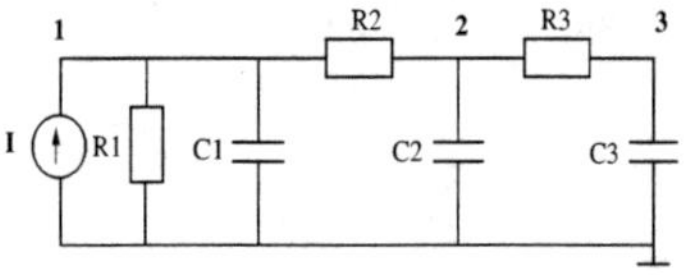

Fig. 4. A simple RC circuit.

$$\begin{bmatrix} \frac{1}{R_1} + s\mathbf{C_1} + \frac{1}{R_2} & -\frac{1}{R_2} & 0 \\ -\frac{1}{R_2} & \frac{1}{R_2} + sC_2 + \frac{1}{R_3} & -\frac{1}{R_3} \\ 0 & -\frac{1}{R_3} & \frac{1}{R_3} + s\mathbf{C_3} \end{bmatrix} \begin{bmatrix} v_1 \\ v_2 \\ v_3 \end{bmatrix} = \begin{bmatrix} I \\ 0 \\ 0 \end{bmatrix}$$

Let C_1 and C_3 be two symbolic parameters in the circuit. Matrix entries $\frac{1}{R_1} + sC_1 + \frac{1}{R_2}$ and $\frac{1}{R_3} + sC_3$ will be assigned indices larger than the indices of all other entries to make them appear on the top of the corresponding DDD graph. Let T be the 3×3 system matrix and we are interested in the following transfer function

$$H(s) = \frac{V_3(s)}{I(s)} = \frac{(-1)^{1+3} det(T_{13})}{det(T)}, \qquad (1)$$

where T_{13} is the matrix obtained by removing row 1 and column 3 from T. The resulting determinant of the system matrix, its cofactor T_{13}, and their DDD representations are shown in Fig. 5, where each non-zero element is designated by a symbol and is assigned a unique index in parentheses. The index of each symbol is also marked along each DDD node in the resulting DDD graph. It can be seen that symbolic nodes A and G appear above all the other numerical DDD nodes.

Once complex DDDs are obtained, s-expanded DDDs are can be computed very efficiently [8]. Consider again the circuit in Fig. 4 and its system determinant. Let us introduce a unique symbol for each circuit parameter in its admittance form. Specifically, we introduce $a = \frac{1}{R_1}, b = f = \frac{1}{R_2}, d = e = -\frac{1}{R_2}, g = k = \frac{1}{R_3}, i = j = -\frac{1}{R_3}$,

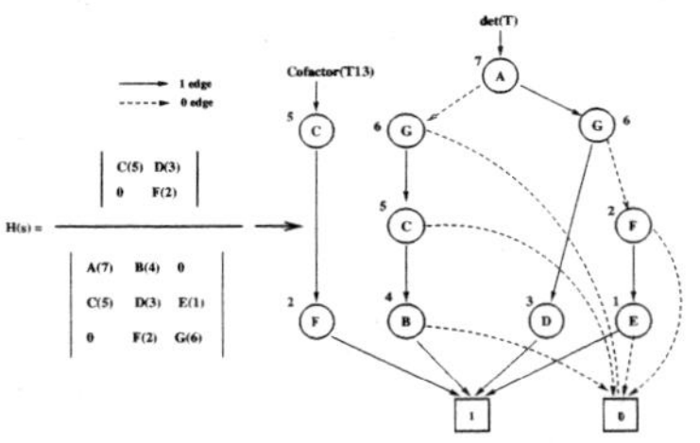

Fig. 5. A complex pre-ordered DDD for the transfer function

$C_1 = c, h = C_2, l = C_3$. Then the circuit matrix can be rewritten as

$$\begin{bmatrix} a+b+\mathbf{c}s & d & 0 \\ e & f+g+hs & i \\ 0 & j & k+\mathbf{l}s \end{bmatrix}$$

The corresponding s-expanded DDDs are shown in Fig. 6. If a circuit parameter is symbolic, its corresponding DDD nodes are symbolic, otherwise they are numerical DDD nodes. For example, c represents a symbolic term C_1, so its DDD nodes are symbolic. All the symbolic DDD nodes are represented by a shaded circle in Fig. 6. We notice that numerical DDD nodes a, b still appear above symbolic DDD nodes c and l as complex DDD node of index A is above complex node of index D and s-expanded DDD construction process does not change the relative order of the resulting s-expanded DDD nodes.

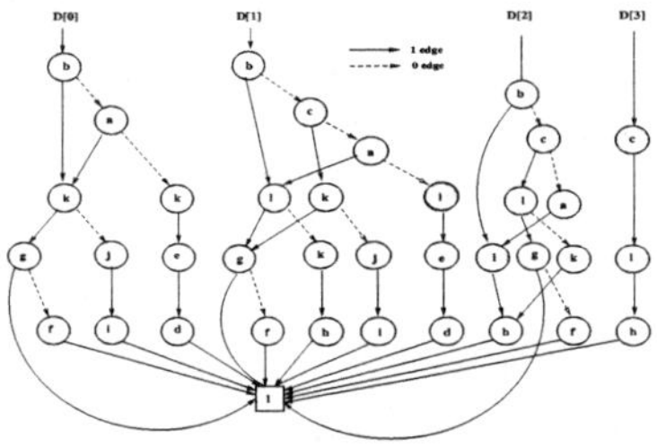

Fig. 6. Semi-symbolic s-expanded DDDs for $det(T)$

We then perform the dynamic variable swap to move all the symbolic DDD nodes to the top of DDD graph such that all symbolic nodes are above numerical DDD nodes. The process is done for each index whose nodes are symbolic. Each time, we swap two adjacent variables at a time until the symbolic index reach to a position where no other numerical indices are larger than it. After this, we process next symbolic index until all the symbolic indices are above numerical indices. The resulting re-ordered s-expanded DDD graphs are shown in Fig. 7.

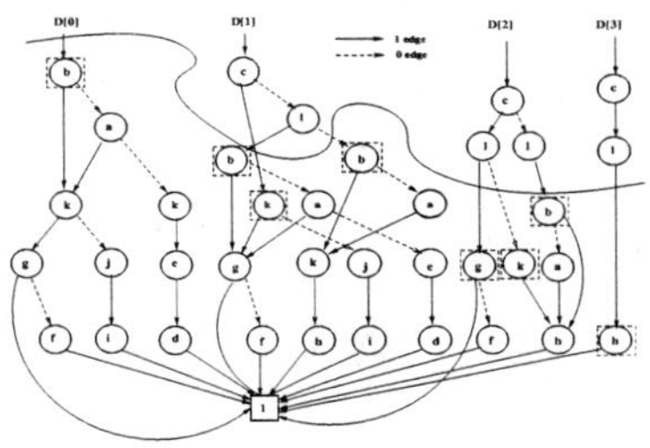

Fig. 7. Semi-symbolic s-expanded DDDs for $det(T)$ after dynamic variable ordering

After the dynamic variable ordering, numerical and symbolic DDD nodes are completely separated. A line is draw between the two portions of the DDD nodes in Fig. 7. If we only keep all symbolic nodes and those nodes whose parent is a symbolic node, designated by a dotted box, we end up with a MTDDD graph where the left numeral nodes are the numerical terminals in the MTDDD. The numerical values at those embedded numerical terminals can be easily obtained by a DFS traversal of the DDD tree rooted at those nodes and all the other numerical DDD nodes can be implicitly removed by garbage collection. Moreover many DDD graph operations can still be applied if we still keep the numerical nodes. For example, we can find the dominant terms in the semi-symbolic DDD by using the fastest incremental shortest path algorithm [14]. The difference is that once we find one shortest path, we need to subtract all the paths which go through the numerical terminal on the shortest path.

Another advantage of the new algorithm over MTDDD method is that we can dynamically change the symbolic circuit parameter sets without reconstructing the DDDs and sDDDs. As a result, the new algorithm is very amenable to analog circuit design and optimization applications where such changes are encountered very frequently.

V. Experimental Results

The proposed approach has been implemented and tested on a number of practical analog circuits ranging from more regularly structured ladder circuits to less regularly structured such as *Cascode* and $\mu A741$ amplifiers. The algorithms described in [7, 8] are used to construct complex DDDs and s-expanded DDDs for a transfer function for each circuit. A Linux system with dual 500Mhz Intel Celeron CPUs and 300M memory is used for all the following experiments.

Table II summarizes statistics of semi-symbolic s-expanded symbolic expressions at different stages of operations. In Table II, column 1, 2 and 3 list for each circuit, respectively, its name (*Circuit*), the number of nodes (*#nodes*), and the number of symbolic branches selected and total number of branches of the circuit (*Symb #bch/#bch*). Columns 4 to 5 show, respectively, the number of s-expanded DDD nodes ($|sDDD|$ *before DVO*) before dynamic variable ordering (DVO) for a transfer function, and the number of s-expanded DDD nodes after DVO ($|sDDD|$ *after DVO*). Columns 6 to 7 present, respectively, the number of symbolic DDD nodes (*Symb* $|sDDD|$), and number of embedded numerical terminals (*Num #terminal*). Column 8, ($|MTDDD|$), shows the number of MTDDD nodes, which consists of both numerical and symbolic DDD nodes. Those numerical nodes are parent nodes of some symbolic nodes. They are obtained by only pre-ordering complex DDD variables without dynamic variable ordering. The last column in the table presents the DDD size reduction percentage of symbolic $|sDDD|$ over $|MTDDD|$.

From the table we can observe sizes of symbolic sDDD are smaller than sizes of MTDDDs for all circuits except for *rclad21*. The reduction percentage ranges from 10% to 30%. This is significantly improvement especially for real analog circuits like *Cascode* and $\mu A741$ Opamps. For *rclad21* circuit, the reason that the size of MTDDDs is smaller than that of symbolic sDDDs is that it is a ladder circuit and the resulting sDDD has a variable ordering close to optimal ordering, dynamic variable ordering will increase the DDD size which offsets the DDD size reduction provided by bringing symbolic DDD nodes together. This also reflects the fact that after dynamic variable ordering, the total sDDD sizes get increased for all the circuits. But the increase will not be significant if the number of symbolic branches are not large. Also we notice that if the number of symbolic branches become larger (more than 10), the benefit of semi-symbolic analysis will begin to lose as the symbolic DDD portion itself will become significant large.

Table I shows the CPU time for constructing and evaluating the semi-symbolic DDD for the same transfer function of each circuit. For comparison reason, we also list the CPU time for constructing complex DDDs when the best variable ordering is used [7] in Column 2 (*cDDD (w/o preorder)*). Column 3 lists the CPU time of complex DDD construction when pre-ordering is used to put all the symbolic complex DDDs to the top of complex DDD tree (*cDDD (w preorder)*). It appears that the CPU time does increase with preordering, but it become less significantly for large less-structured circuits like $\mu A741$. Therefore, the CPU time for complex DDD construction is still very fast in the presence of a few symbolic branches.

TABLE I
CPU TIME STATISTICS ON SEMI-SYMBOLIC DDD OPERATIONS.

Circuit	cDDD (w/o preorder)	cDDD (w preorder)	sDDD	DVO	Eval of sDDD	Eval of Symb DDD
rlc6	0.02	0.03	0.06	0.02	0.10	0.02
rlc14	0.16	1.48	6.65	15.09	34.8	1.4
Cascode	0.32	0.95	2.49	4.88	3.50	0.04
$\mu A741$	1.44	1.59	5.76	0.21	15.22	0.06
rctreeA	0.10	0.14	0.24	0.57	1.85	0.24
rctreeB	0.17	0.31	0.74	0.24	6.49	0.41
bigtst	0.34	0.38	0.43	0.19	2.27	0.14
rclad21	0.02	0.03	0.03	< 0.01	0.05	0.01

Column 4 to 5 gives the CPU time for constructing the s-expanded DDD and for performing the DVO. Both CPU times are essentially proportional to the size of s-DDDs and can be performed very efficiently. The last two columns present the CPU time for taking the numerical evaluation of all the coefficients for the whole s-expanded DDD graph (*Eval of sDDD*) and for only the symbolic portion of the s-expanded DDD graph (*Eval of Symb DDD*) after DVO. So the benefits of evaluating only symbolic sDDD are obvious.

VI. CONCLUSIONS

This paper proposes a novel approach to semi-symbolic analysis of large analog circuits based on DDD graphs. We show how DDD nodes in adjacent variables in DDD graphs can be swapped in order, which in turn can be used to reduce the DDD presentation of a linearized analog circuits via dynamic variable ordering. The new dynamic variable ordering method does not require special sign rule for order-swapped DDD vertices compared with the existing method and is able to dynamically change symbolic circuit parameter sets. The new method also preserve all the DDD graph operation capabilities compared to MTDDD method. Experimental results have demonstrated that the proposed semi-symbolic method leads to up to 30% symbolic DDD node reduction compared with MTDDD method on real analog circuits and can be performed very efficiently.

TABLE II
STATISTICS ON SEMI-SYMBOLIC s-EXPANDED DDDS

Circuit	#nodes	Symb #bch/#bch	$\|sDDD\|$ before DVO	$\|sDDD\|$ after DVO	Symb $\|sDDD\|$	Num #terminal	$\|MTDDD\|$	Imprv(%)
rlc6	13	8/19	2466	2697	639	260	700	8.7%
rlc14	29	8/45	2.84×10^5	3.51×10^5	13198	11720	18174	27.4%
Cascode	14	5/78	74778	81577	942	813	1311	28.2%
$\mu A741$	23	5/93	2.14×10^5	2.15×10^5	693	441	1098	36.9%
rctreeA	40	8/80	13027	13211	1629	170	1785	8.7%
rctreeB	53	8/106	36407	36579	2246	238	2450	8.3%
bigtst	32	5/51	20082	20514	1230	474	1396	11.9%
rclad21	22	8/41	630	660	184	21	173	-6.4%

REFERENCES

[1] G. Gielen and W. Sansen, *Symbolic Analysis for Automated Design of Analog Integrated Circuits*. Kluwer Academic Publishers, 1991.

[2] G. Gielen, P. Wambacq, and W. Sansen, "Symbolic analysis methods and applications for analog circuits: A tutorial overview," *Proc. of IEEE*, vol. 82, pp. 287–304, Feb. 1994.

[3] F. V. Fernández, A. Rodríguez-Vázquez, J. L. Hertas, and G. Gielen, *Symbolic Analysis Techniques: Application to Analog Design Automation*. IEEE Press, 1998.

[4] K. Singhal and J. Vlach, "Generation of immittance functions in symbolic form for lumped distributed active networks," *IEEE Trans. on Circuits and Systems I: Fundamental Theory and Applications*, vol. CAS-21, pp. 39–45, 1974.

[5] P. Sannuti and N. N. Puri, "Symbolic network analysis—an algebraic formulation," *IEEE Trans. on Circuits and Systems I: Fundamental Theory and Applications*, vol. 27, pp. 679–687, Aug. 1980.

[6] T. Pi and C.-J. Shi, "Multi-terminal determinant decision diagrams: a new approach to semi-symbolic analysis of analog integrated circuits," in *Proc. Design Automation Conf. (DAC)*, pp. 19–22, June 2000.

[7] C.-J. Shi and X.-D. Tan, "Canonical symbolic analysis of large analog circuits with determinant decision diagrams," *IEEE Trans. on Computer-Aided Design of Integrated Circuits and Systems*, vol. 19, pp. 1–18, Jan. 2000.

[8] C.-J. Shi and X.-D. Tan, "Compact representation and efficient generation of s-expanded symbolic network functions for computer-aided analog circuit design," *IEEE Trans. on Computer-Aided Design of Integrated Circuits and Systems*, vol. 20, pp. 813–827, April 2001.

[9] W. Verhaegen and G. Gielen, "Efficient ddd-based symbolic analysis of large linear analog circuits," in *Proc. Design Automation Conf. (DAC)*, pp. 139–144, June 2001.

[10] A. Manthe and C.-J. R. Shi, "Lower bound based ddd minimization for efficient symbolic circuit analysis," in *Proc. IEEE Int. Conf. on Computer Design (ICCD)*, pp. 374–379, Nov. 2001.

[11] M. Fujita, Y. Matsunaga, and T. Kauda, "On variable ordering of binary decision diagrams for the application of multi-level logic synthesis," in *Proc. European Design Automation Conf.*, pp. 40–54, March 1991.

[12] N. Ishiura, H. Sawada, and S. Yajima, "Minimization of binary decision diagrams based on exchanges of variables," in *Proc. Int. Conf. on Computer Aided Design (ICCAD)*, pp. 472–475, Nov. 1991.

[13] R. Rudell, "Dynamic variable ordering for ordered binary decision diagrams," in *Proc. Design Automation Conf. (DAC)*, pp. 42–47, June 1993.

[14] S. X.-D. Tan and C.-J. Shi, "Efficient ddd-based term generation algorithm for analog circuit behavioral modeling," in *Proc. Asia South Pacific Design Automation Conf. (ASPDAC)*, pp. 789–794, Jan. 2003.

COMPACT SEMICONDUCTOR DEVICE MODELING USING HIGHER LEVEL METHODS

Matt Francis, Vivek Chaudhary, H. Alan Mantooth

University of Arkansas, 3217 Bell Engineering Center, Fayetteville, AR 72701

ABSTRACT

This paper describes an approach for generating compact semiconductor device models in SPICE (C/C++) from a higher-level representation in the form of an XML-based schema for device modeling‡. The methodology undertaken includes generation of low level C/C++ code for SPICE and SPICE-like simulators from higher level languages such as HDLs (VHDL-AMS, MAST and Verilog-A) or directly from XML utilizing existing graphical tools (Paragon†). As such, the aim of the paper is to introduce the proposed XML-based standard and its associated methods‡, as well as highlight the modeling process through an instructive example.

1. INTRODUCTION

Semiconductor device modeling takes a huge amount of time and resource at present. This is primarily due to the lack of modeling tools to facilitate the research, implementation and characterization of these complex models, and the fact that device model implementation directly in SPICE-like simulators is cumbersome and error-prone at best. Without advanced modeling tools the ability to compile other types of models, such as behavioral models of circuits, into advanced simulators is effectively disabled.

The objective of this paper is to define a scalable, comprehensive, extensible metafile for representing compact models, describe a reference implementation of an API that acts as a neutral interface between hardware description languages (HDL) compilers or advanced modeling tools and simulators, and illustrate a concrete example for validation and demonstration.

The concept and benefit of a compact model compiler based on standard high-level behavioral languages has been described by a number of investigators [2-4]. The major obstacle for model compilers is the generated code efficiency, which has been reported to be 100 to 1000 times slower than hand-developed code. A development from the University of Washington indicates this is not a problem [5].

† This work is sponsored by DARPA/MTO NeoCAD Program under Grant No. N66001-01-1-8919, by Office of Naval Research under Subaward No. USC 01-636, and by an NSF CRCD award (EEC-0088011).

‡This work is sponsored by the Semiconductor Research Corporation (SRC)

2. ABSTRACT MODEL REPRESENTATION

In order to facilitate easily created, modified, and maintained compact device models, a compact yet extensible format for model representation is necessary. The format designed strives to capture all of the information necessary to create a semiconductor device model, with the emphasis being on the data specific to the model, removing simulator dependence, facilitating rapid adoption, and supporting enhancements through use of the eXtensible Markup Language [6]. The choice of XML can be justified as such:

1. The use of XML enables the description of model information in a simple, flexible and extensible structured text format.
2. XML lends itself to standardization and open sourcing for expressing model descriptions, which will lead model developers to share models among different modeling environments in a format that is independent of any simulator.
3. The technologies and applications built on XML provide a powerful and efficient way of expressing and manipulating almost all types of data, including complex mathematics [7].
4. XML's html-like syntax leads to self-documenting and easily readable models.

The model expressions and equations are expressed in MathML [8], which is an XML format for describing mathematical notation. It is a popular standard for encoding the structure of mathematical expressions and has been adopted by several mathematical software vendors. Thus, integrating complex mathematical expressions from these packages into a device model (and vice versa) is enabled.

2.1. XML Schema

The XML model format can best be explained using a Document Type Definition (DTD), which is a formal description in XML declaration syntax. It defines the different legal building blocks (elements) of the database, describing where they may occur and how they all fit together. The DTD is given below as Fig. 1.

```
<!-- model declararations -->
<!ELEMENT model (comment?, interface, body+)>
<!ATTLIST model name CDATA #REQUIRED version CDATA
#REQUIRED>
<!ELEMENT comment (#PCDATA)>
<!-- model interface -->
<!ELEMENT interface (comment?, parameter*, port*)>
<!-- model parameters -->
<!ELEMENT parameter (comment?, validity*)>
<!ATTLIST parameter default CDATA #IMPLIED name CDATA
#REQUIRED nature (real | integer | time) #REQUIRED unit CDATA
#REQUIRED type (instance | process) #REQUIRED>
<!-- parameter validity -->
<!ELEMENT validity (range+)>
<!ATTLIST validity message CDATA #REQUIRED type (note | error |
warning) #REQUIRED>
<!ELEMENT range EMPTY>
<!-- range of validity -->
<!ATTLIST range min CDATA #REQUIRED max CDATA
#REQUIRED exclude_max (yes | no) #IMPLIED exclude_min (yes | no)
#IMPLIED>
<!-- model interface ports -->
<!ELEMENT port (comment?)>
<!ATTLIST port name CDATA #REQUIRED mode (in | out | inout)
#REQUIRED nature (electrical | mechanical_angular_speed |
mechanical_angular_displacement | mechanical_translational_speed |
mechanical_translational_displacement | thermal | optical | magnetic)
#REQUIRED type (terminal | quantity | signal | logic) #REQUIRED>
<!-- end of model interface declaration -->
<!-- model body -->
<!ELEMENT body (branch*, equation*, piecewise*, eqblock*,
macromodel*)>
<!ATTLIST body name CDATA #REQUIRED>
<!-- model body symbol -->
<!ELEMENT symbol (vector_graphics)>
<!-- model body branch -->
<!ELEMENT branch (connection+, quantity+, equation*, piecewise*,
eqblock*, comment?)>
<!ATTLIST branch name CDATA #IMPLIED>
<!-- model body branch quantity -->
<!ELEMENT quantity EMPTY>
<!ATTLIST quantity name CDATA #REQUIRED nature (through |
across) #REQUIRED type CDATA #IMPLIED unit CDATA
#IMPLIED>
<!-- model body macromodel -->
<!ELEMENT macromodel (connection*, comment?,
macromodel.parameter, macromodel.architecture)>
<!ATTLIST macromodel name CDATA #REQUIRED filename CDATA
#REQUIRED>
<!ELEMENT macromodel.parameter EMPTY>
<!ATTLIST macromodel.parameter name CDATA #REQUIRED value
CDATA #REQUIRED>
<!ELEMENT macromodel.architecture EMPTY>
<!ATTLIST macromodel.architecture name CDATA #REQUIRED>
<!-- model body branch/macromodel connection -->
<!ELEMENT connection EMPTY>
<!ATTLIST connection name (pos | neg | CDATA) #REQUIRED
mappedto CDATA #REQUIRED type (port | param) #REQUIRED>
<!-- model/branch equation blocks -->
<!ELEMENT eqblock (equation*, piecewise*, comment*)>
<!ATTLIST eqblock nature (simultaneous | sequential) #REQUIRED>
<!ELEMENT piecewise (if, elseif*, else, comment*)>
<!ELEMENT if (condition, equation+, piecewise*, comment*)>
<!ELEMENT elseif (condition, equation+, piecewise*, comment*)>
<!ELEMENT else (condition, equation+, piecewise*, comment*)>
<!ELEMENT condition (math)>
<!-- model/branch equations -->
<!ELEMENT equation (math)>
<!ATTLIST equation type (sequential | simulataneous | conditional)
#REQUIRED>
```

Fig. 1. DTD showing the syntax of XML model database.

Each model document has an interface and a body. The model interface consists of the model name, connection points and parameters. The body contains the model topology and equations. The topology consists of branches and instances of other models. The branches are in turn defined by their 'through' and 'across' variables and mathematical expressions involving these variables. The topology and mathematical expressions collectively define the model behavior. Note that while the focus of the schema is compact semiconductor modeling, it is generic enough to support multi-domain modeling, such as thermal, optical, fluidic, and mechanical systems.

2.1. Analysis and Utility Methods

To aid in the dissemination and adoption of the XML standard, several analysis and utility methods were created. Many of these methods operate upon the XML model to give CAD tool developers a rich tool set for model analysis, creation, and new code generator development. Other methods provide ways of developing new models by facilitating the creation of an XML model file from HDLs, including Verilog-A [9,10], VHDL-AMS [11-14] and MAST [15], or from proprietary languages and interfaces (Fig. 2). For example, the developed C++ code generation method can generate code from Verilog-A by analyzing the XML model file utilizing analysis routines coupled with the Verilog-A import method. Another example of a use of these methods is found in the Paragon modeling tool [1], which can create the XML model file directly from a graphical interface or derive one from a circuit netlist [16]. Some of the important analysis and utility methods that accompany the XML schema are:

1. Creation of an Abstract Syntax Tree (AST), an internal data representation obtained by parsing the MathML expression trees. The AST represents the inter-relationships among the variables and constants in the model equations and expressions.
2. Analysis of the AST for determining the functional dependency and time dependency characteristics. This enables the generation of efficient code by

distinguishing constants, time-varying variables, and "post-iterative" calculations (e.g., power).

3. Methods for verifying model robustness such as checking for discontinuities in the model.
4. Methods to maximize a model's performance such as reducing the number of FLOPS (floating point operations) in an expression.
5. Mathematical methods to ease C/C++ generation, including derivation and integration.
6. Methods to generate C/C++ SPICE code.
7. Methods to generate HDLs, including Verilog-A, VHDL-AMS and MAST.
8. Methods to import HDLs (Verilog-A, VHDL-AMS, MAST) to the XML format.

The methods and algorithms described highlight some of the most relevant for model development. Many more specific methods exist to address simulator and language-specific issues.

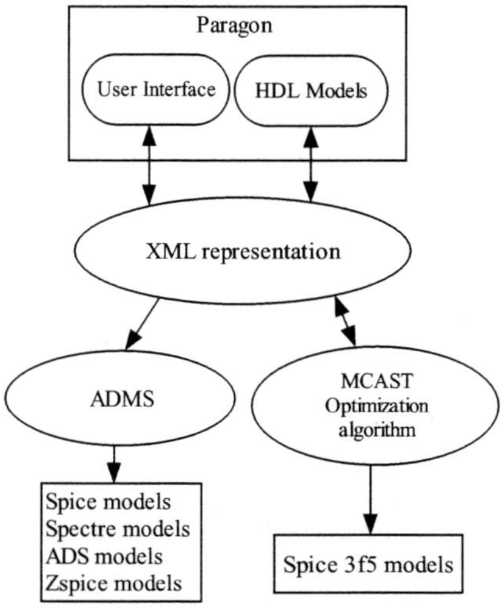

Fig. 2. Model creation process utilizing the XML schema and the associated methods.

3. MODEL COMPILATION

Model compilation, in the context of this paper, is the process of automatically generating compact semiconductor device models in C/C++ for SPICE and SPICE-like simulators from the abstract XML representation of the model. The advantages of using a model compiler to develop these models over hand-written models are the following:

1. The model development time is dramatically reduced as the model developer does not need to manually write low level C/C++ code.
2. The generated model is easier to maintain and reuse as the modeler does not have to read and modify the low level C/C++ code.
3. The same abstract representation of the model can be used by a model compiler to generate low level C/C++ code for different simulators.

Some of the model compilers developed in the past include ADMIT [20], iSMILE [21], MAST/Saber [22], ADMS [4] and MCAST [5]. The biggest challenge has been developing model compilers capable of generating compact low level C/C++ code comparable in speeds to the hand-generated models. MCAST has developed industry-grade device models comparable in simulation speeds to the hand-written models. Paragon is capable of generating low-level C/C++ code for the fREEDA [23] simulator. ADMS is a freely available model compiler available through the open-source process [24] and it supports popular simulators such as Spectre, ADS, McSPICE, and NanoSim. The overall model creation process has been illustrated in Fig 2.

4. SEMICONDUCTOR DEVICE MODEL EXAMPLE

To illustrate the usefulness of compact semiconductor device modeling using higher-level tools, the compact C/C++ model for the EKV MOSFET was generated in Paragon. The details of the Paragon environment and its usage can be found at [1]. In order to create the EKV MOSFET model, a higher-level description of the model is first entered through the Paragon environment (Fig 3).

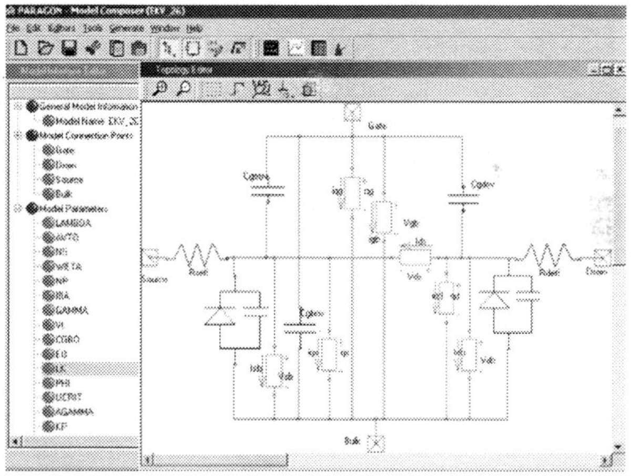

Fig. 3. Screenshot of Paragon showing the creation of the EKV MOSFET model.

Next, all model information is saved in XML. As the EKV model is quite large, for illustrative purposes only the XML code corresponding to the capacitive branch *Cgsov* between the gate and source is included in Fig 4. The remaining details of the model can be viewed in the Technical Report on the EKV MOSFET model [18].

The final step in the model creation process is to generate the model code. Using Paragon the compact EKV MOSFET model was generated for the fREEDA simulator. The generated model was simulated utilizing fREEDA and the simulation results of the generated model were compared with the results of the hand-written model for speed and accuracy. The automatically generated C code performed accurately and simulated in the same time as the hand-written model. Of note is that implementation and validation using Paragon took less than one week, while implementation and validation of the hand-written model in fREEDA took nearly eight weeks.

```
<branch name="Cgsov">
 <connection mappedto="gate" name="pos" type="port"/>
 <connection mappedto="source" name="neg" type="port"/>
 <quantity default="" name="Igsov" nature="through" type="current"
unit="ampere"/>
 <quantity default="" name="Vgsov" nature="across" type="voltage"
unit="volt"/>
 <equation><mrow> <mi> Igsov </mi> <mo> = </mo> <mi> Cgsov
</mi> <mo> * </mo> <mo> d_by_dt </mo> <mo> ( </mo> <mi> Vgs
</mi> <mo>) </mo> </mrow> </equation>
</branch>
```

Fig. 4. XML content for branch *Cgsov* in the EKV MOS model.

The compact model creation methodology presented here can be easily extended to support multi-domain modeling being limited by such support in existing commercial simulators.

5. CONCLUSION

The modeling methodology described in this paper enables the user to quickly and correctly create new compact semiconductor device models for various SPICE and SPICE-like simulators. The generated C code can be at least as fast or faster than hand written codes, while the XML model file from which it is generated remains maintainable and reusable. The project partners University of Arkansas, University of Washington and Motorola plan to place the modeling methodology, XML schema and the associated methods in the public domain [24] as a part of the open-source initiative sponsored by the Semiconductor Research Corporation. It is hoped that this will lead to rapid and pervasive adoption of these standard in the academia and the industry.

6. REFERENCES

[1] V. Chaudhary, M. Francis, X. Huang, H. A. Mantooth, "Paragon - A mixed-signal behavioral modeling environment," *IEEE Int. Conf. on Communications, Circuits, & Syst. (ICCCAS)*, pp. 1315-1321, Chengdu, China, June 30, 2002.

[2] R. V. H. Booth, "An extensible compact model description language and compiler," *Proc. IEEE BMAS*, pp. 39-44, Oct. 2001.

[3] M. Zorzi, N. Speciale, G. Masetti, "Automatic embedding of a ferroelectric capacitor model in Eldo," *Proc. IEEE BMAS*, pp. 97-101, Oct. 2001.

[4] L. Lemaitre, C. McAndrew, S. Hamm, "ADMS – Automatic Device Model Sythesizer," *Proc. IEEE Custom Int. Circ. Conf.*, pp. 27-30, 2002.

[5] B. Wan, B. P. Hu, L. Zhou, C.-J. Shi, "MCAST – An abstract-syntax-tree based model compiler for circuit simulation," *IEEE Custom Integrated Circuits Conf. (CICC)*, pp. 249-252, Sept. 2003.

[6] Extensible Markup Language (XML), http://www.w3.org/XML

[7] Extensible Stylesheet Language, http://www.w3.org/TR/xslt

[8] Mathematical Markup Language (MathML), http://www.w3.org/Math

[9] D. Fitzpatrick, I. Miller, *Analog Behavioral Modeling with Verilog-A*, Kluwer Academic Publishers, Norwell, MA, 1997.

[10] P. Frey, D. O'Riordan," Verilog-AMS: Mixed-signal simulation and cross domain connect modules", *Proc. IEEE BMAS*, pp. 103 –108, Oct. 2000.

[11] *1076.1-1999 IEEE Standard VHDL Analog and Mixed-Signal Extensions Language Reference Manual*, IEEE Press, ISBN 0-7381-1640-8.

[12] P. Ashenden, G. D. Peterson, D. A. Teegarden, *The Systems Designer's Guide to VHDL-AMS*, Morgan-Kaufmann, San Francisco, CA, 2003.

[13] R. S. Cooper, *The Designer's Guide to Analog and Mixed-Signal Modeling*, Avant! Press, 2001.

[14] E. Christen, K. Bakalar, "VHDL-AMS a hardware description language for analog and mixed-signal applications," *IEEE Trans. on Circuits and Systems, part II*, Vol. 46 Issue: 10, pp. 1263–1272, Oct. 1999.

[15] H. A. Mantooth, M. Fiegenbaum, *Modeling with an Analog Hardware Description Language*, Kluwer Academic Publishers, Norwell, MA, 1995.

[16] X. Huang, C. Gathercole and H. A. Mantooth, "Modeling nonlinear dynamics in analog circuits via root localization," *IEEE Trans. Computer-Aided Design*, vol. 22, no. 7, pp. 895-907, July 2003.

[17] C. C. Enz, F. Krummenacher, E. A. Vittoz, "An analytical MOS transistor model valid in all regions of operation and dedicated to low-voltage and low-current applications," *J. Analog Integrated Circuits and Signal Processing*, vol. 8, no. 1, pp. 83-114, July 1995.

[18] M. Bucher, C. Lallement, C. Enz, F. Théodoloz, F. Krummenacher, "The EPFL-EKV MOSFET Model Equations for Simulation", Technical Report, Model Version 2.6, June 1997. Revision I, September, 1997, Revision II, July, 1998.

[19] H. A. Mantooth, M. Vlach, "Beyond SPICE with Saber and MAST," *IEEE Proc. of Int. Symposium on Circuits Syst.*, vol. 1, pp. 77-80, May 1992.

[20] S. Liu, K.C. Hsu, P. Subramaniam, "ADMIT-ADVICE Modeling Interface Tool", *IEEE Custom Integrated Circuits Conference*, 1988.

[21] A.T. Yang, and S.M. Kang, "iSMILE: A Novel Circuit Simulation Program with emphasis on New Device Model Development", *26th Design Automation Conference, 1989.*

[22] MAST/Saber User Manual, Synopsys, Inc.

[23] fREEDA Circuit Simulator, http://guppie.egrc.ncsu.edu/freeda

[24] ADMS Model Compiler, http://sourceforge.net/projects/mot-adms

CONSISTENT MODEL FOR DRAIN CURRENT MISMATCH IN MOSFETS USING THE CARRIER NUMBER FLUCTUATION THEORY

H. Klimach[1,2], *A. Arnaud*[3], *M. C. Schneider*[1], *and C. Galup-Montoro*[1]

[1] Universidade Federal de Santa Catarina, Brazil, [2] Universidade Federal do Rio Grande do Sul, Brazil,
[3] Universidad de la Republica, Uruguay

ABSTRACT

This paper presents a new approach for accurate MOS transistor matching calculation. Our model, which is based on an accurate physics-based MOSFET model, allows the assessment of mismatch from process parameters and is valid for any operating region. Experimental results taken on a test set of transistors implemented in a 1.2 μm CMOS technology corroborate the theoretical development of this work.

Index Terms – MOSFET, analog design, matching, compact models.

1. INTRODUCTION

It is widely recognized that the performance of most analog or even digital circuits is limited by MOS transistor matching [1-4]. The shrinkage of the dimensions of MOSFETs and the reduction in the supply voltage make matching limitations even more important to such an extent that several new studies have been published in recent years [5-8]. Existing mismatch models use either simple drain current models limited to a specific operating region [1, 2, 4, 7, 8] or complex expressions [6] like that of BSIM. In general, however, the applicability of dc current models to characterize mismatch is not questioned. It is widely accepted that matching can be modeled by the random variations in geometric, process and/or device parameters. The effect of the random parameters on the drain current is quantified using the dc model of the transistor. As pointed out in [7, 8] there is a fundamental flaw in this approach that results in inconsistent modeling of matching. In effect, mismatch models implicitly assume that the actual values of the lumped model parameters can be obtained integrating the position-dependent distributed models over the areas of the channel region of the device, e.g., for the threshold voltage V_T

$$V_T = \frac{1}{WL} \iint_{channel-area} V_T(x,y)\,dx\,dy \tag{1}$$

where W and L are the width and the length of the transistor.

As analyzed in [7, 8], the application of (1) to series or parallel association of transistors leads to an inconsistent model of matching owing to the nonlinear nature of MOSFETs. Consequently, the simple consideration of random fluctuations in the lumped parameters of the dc current model is not appropriate to develop matching models and new formulas must be derived from basic principles. Fortunately, the formalism needed to model matching is already available in low frequency (LF) noise modeling. In this paper, we will show that the carrier number fluctuation theory [9], employed to derive LF transistor noise, can be adapted to model current matching in MOSFETs. To obtain general results for all bias regions of the transistor we have used the Advanced Compact MOSFET (ACM) model, a physics-based one-equation all-region model [10].

2. THE ACM MODEL

According to ACM [10], the drain current in a long-channel transistor is given by,

$$I_D = \frac{\mu W}{nC'_{ox}}\left(-Q'_I + nC'_{ox}\phi_t\right)\frac{dQ'_I}{dx} \tag{2}$$

where Q'_I is the inversion charge density, n is the slope factor, C'_{ox} is the oxide capacitance per unit area, μ is the effective mobility and ϕ_t is the thermal voltage.

The other specificity of the ACM model is the use of the unified charge control model (UCCM) [12] to link the carrier charge density with the applied voltages

$$V_P - V_X = \phi_t\left[\frac{Q'_I}{Q'_{IP}} - 1 + \ln\left(\frac{Q'_I}{Q'_{IP}}\right)\right] \tag{3}$$

where $Q'_{IP} = -nC'_{ox}\phi_t$, $V_P = (V_{GB}-V_T)/n$ is the pinch-off voltage, and V_X is the channel potential. As shown in [13] the use of (2) in conjunction with (UCCM) (3) gives

$$I_D = -\mu\frac{W}{dx}Q'_I dV_X \; . \tag{4}$$

Consequently, the ACM model is fully consistent with the quasi-Fermi potential formulation for the drain current [11].

3. CONSISTENT SMALL-SIGNAL MODEL OF THE MOSFET CHANNEL

To calculate the effect of the fluctuations on the drain current along the channel, we split the transistor into 3 series elements: the upper transistor, the lower transistor, and a small channel element of length Δx and area $\Delta A = W\Delta x$ (Fig. 1(a)).

Small-signal analysis allows one to calculate the effect of the local current fluctuation ($i_{\Delta A}$) on the drain current (ΔI_d), as shown in Fig. 1(b). The current division between the channel element and the equivalent small-signal resistance of the rest of the channel gives $\Delta I_d = (\Delta x/L)i_{\Delta A}$. This very simple result for the current division, proportional to a geometric ratio, is a consequence of the quasi-Fermi potential formulation for the drain current, i.e., the conductance of the channel element and the transconductances of the upper and lower transistors are proportional to the local charge density. Thus, the square of the total drain current fluctuation is

$$(\Delta I_D)^2 = \sum_L (\Delta I_d)^2 = \lim_{\Delta x \to 0} \sum \left[(\Delta x / L) i_{\Delta A} \right]^2 = \frac{1}{L^2} \int_0^L \left[\Delta x (i_{\Delta A})^2 \right] dx . \quad (5)$$

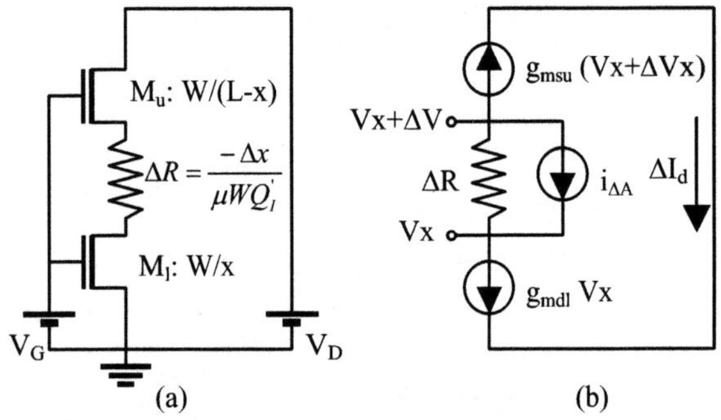

Fig. 1. Splitting of a transistor into three series elements (a) Transistor equivalent circuit (b) Small-signal equivalent circuit

4. NUMBER FLUCTUATION MISMATCH MODEL

The relationship between local fluctuation of the inversion charge density and the local current fluctuation that follows from expression (4) is

$$i_{\Delta A} = I_D \frac{\Delta Q_I'}{Q_I'} \quad (6)$$

where $\Delta Q'_I$ is the fluctuation of the inversion charge density in the channel element of area ΔA. For the sake of simplicity we will consider fluctuation in the number of carriers only, but the analysis can also be extended to include mobility fluctuation for the computation of the local current fluctuation.

From UCCM (3), one can readily derive the relation between local charge density and threshold voltage fluctuations

$$\Delta Q_I' = -C_{ox}' \frac{Q_I'}{Q_I' - nC_{ox}'\phi_t} \Delta V_T . \quad (7)$$

The local fluctuation of the threshold voltage ΔV_T is calculated from the conventional expression for the standard deviation [2]

$$\overline{\Delta V_T^2} = \sigma_{VT}^2 = \frac{A_{VT}^2}{W \Delta x} . \quad (8)$$

Using (6), (7), and (8) we calculate $(i_{\Delta A})^2$ and obtain the expression of $\overline{\Delta I_D^2}$ inserting this resultant into (5). With the aid of (2) the integration over the channel length in (5) is changed to the integration over the channel charge density given by

$$\sigma_{I_D}^2 = \overline{\Delta I_D^2} = \frac{\mu C_{ox}' I_D A_{VT}^2}{nL^2} \int_{Q_{IS}'}^{Q_{ID}'} \frac{1}{nC_{OX}'\phi_t - Q_I'} dQ_I' . \quad (9)$$

Assuming, as in [14], Poisson statistics for the depletion charge fluctuations, then

$$A_{VT}^2 = \frac{q^2}{C_{ox}'^2}(N x_D) = \frac{q^2}{C_{ox}'^2} N_{oi} \quad (10)$$

where N is the average number of impurities per unit volume in the depletion region, x_D is the depletion deep, and $N_{oi}=Nx_D$ is the effective number of impurities per unit area.

Finally, using (10) and integrating (9) from source to drain results

$$\frac{\sigma_{I_D}^2}{I_D^2} = \frac{q^2 N_{oi} \mu}{L^2 n C_{ox}' I_D} \ln \left(\frac{nC_{ox}'\phi_t - Q_{IS}'}{nC_{ox}'\phi_t - Q_{ID}'} \right) . \quad (11)$$

The result in (11) is essentially the same as that derived for flicker noise in MOS transistors in [13]. This is because mismatch is a "dc noise" and the physical origin of both mismatch and flicker noise is fluctuation of either fixed charges or localized states along the channel.

5. MISMATCH MODEL IN TERMS OF INVERSION LEVELS

A useful alternative expression for (11) is obtained if the charge densities at source and drain are expressed in terms of the normalized forward and reverse currents i_f and i_r. In the ACM model [10], the drain current is expressed as the difference between forward (I_F) and reverse (I_R) components

$$I_D = I_F - I_R = I(V_G, V_S) - I(V_G, V_D) = I_S(i_f - i_r) \quad (12)$$

where $I_S = ½\mu C'_{ox} n\phi_t^2 (W/L)$ is the specific current, which is proportional to the geometric ratio W/L of the transistor. V_G, V_S, and V_D are the gate, source, and drain voltages, respectively. i_f and i_r are the normalized forward and reverse currents or inversion levels at source and drain, respectively. Note that in the saturation region, the drain current is almost independent of V_D; therefore, $i_f >> i_r$ and $I_D \cong I_F$. On the other hand, if V_{DS} is low, then $i_f \cong i_r$. Using the relationship between inversion charge densities and currents [10], expression (11) can be rewritten as

$$\frac{\sigma_{I_D}^2}{I_D^2} = \frac{N_{oi}}{WLN^{*2}} \frac{1}{i_f - i_r} \ln \left(\frac{1 + i_f}{1 + i_r} \right) \quad (13)$$

where we define N^* as in [13, 15]

$$N^* = \frac{-Q_{IP}'}{q} = \frac{nC_{ox}'\phi_t}{q} . \quad (14)$$

From weak to strong inversion in the linear region, $i_f \cong i_r$ and (13) reduces to

$$\frac{\sigma_{I_D}^2}{I_D^2} = \frac{N_{oi}}{WLN^{*2}} \frac{1}{1 + i_f} . \quad (15)$$

In weak inversion, $i_f << 1$; thus, the first order series expansion of (13) leads to

$$\frac{\sigma_{I_D}^2}{I_D^2} = \frac{N_{oi}}{WLN^{*2}} \quad (16)$$

for either saturation or nonsaturation.

In saturation ($i_r \to 0$), expression (13) can be written as

$$\frac{\sigma_{I_D}^2}{I_D^2} = \frac{N_{oi}}{WLN^{*2}} \frac{\ln(1 + i_f)}{i_f} . \quad (17)$$

In weak inversion, $i_f << 1$ and (17) is almost insensitive to the current. In strong inversion, the normalized mismatch given by (17) reduces to the conventional expression if the logarithmic term is assumed to be constant.

Finally, the random errors due to edge effects can be modeled as in [1] resulting in a complete mismatch model

$$\frac{\sigma_{I_D}^2}{I_D^2} = \frac{1}{WL}\left[\frac{N_{oi}}{N^{*2}}\frac{1}{i_f - i_r}\ln\left(\frac{1+i_f}{1+i_r}\right) + \frac{B_L}{L} + \frac{B_W}{W}\right] \quad (18)$$

where B_L and B_W are the channel length and width mismatch factors due to edge effects.

6. MEASUREMENTS

Intradie current mismatch of 24 NMOS 30μm x 1.2μm transistors was measured on a test circuit in the ES2 1.2μm CMOS DLM process, using the circuit shown in figure 2. M_{REF} was kept the same for all measurements while the remaining 23 transistors were used as M_i for data acquisition. Five dies were measured, presenting the same matching behaviour.

Figure 3 presents the mismatch power normalized to the dc power $(SD^2(I_D)/I_D^2)$ for drain-to-source voltage ranging from 20mV (linear region) to 1V (saturation), for different inversion levels. Simulated curves (dotted lines) have been determined from expression (13), with i_r calculated through ACM long channel model expressions [10]. The average number of impurities per unit area (N_{oi}) is estimated as $6.1\text{x}10^{12}$ cm^{-2}; the resulting A_{VT} calculated from (10) is about 29mV-μm. Specific current (I_S) for the devices under test is 1.2μA. It should be emphasized that drain current mismatch results from geometrical and technological fluctuations; however, for most of the cases, the dominant factor that affects current mismatch can be associated with V_T mismatch.

In weak inversion (i_f = 0.01 and 0.1), mismatch is constant from linear to saturation region and independent of the inversion level, as predicted by (16). For i_f = 1, mismatch is approximately one half of the value measured for very weak inversion, as predicted in (15). For higher inversion levels (i_f = 10 and 100) and operation in the linear region, mismatch reduces by a factor of approximately $1/(1+i_f)$, compared to weak inversion. This value increases up to a plateau when saturation is reached, presenting a good agreement with expression (17) of our model.

At this point one should compare our mismatch model with Pelgrom's model [2]. If we were to use Pelgrom's mismatch model together with the ACM model, the expression for the normalized mismatch power would be

$$\frac{\sigma_{I_D}^2}{I_D^2} = \frac{N_{oi}}{WLN^{*2}}\left(\frac{2}{1+\sqrt{1+i_f}}\right)^2 \quad (19)$$

where the dependence on i_f is the same as that of $(g_m/I_D)^2$. Expressions (17) and (19) agree in weak inversion ($i_f << 1$), but for i_f=1000 (17) predicts a value 80% greater than (19). Explanation for this difference, arises from the distributed nature of the MOSFET. While Pelgrom's model assumes a lumped V_T for the MOSFET, our model assumes a distributed V_T along the channel. As a consequence, for strong inversion and saturation, the part of the channel closer to the drain plays a less important role in the charge fluctuation along the channel than the part of the channel closer to the source.

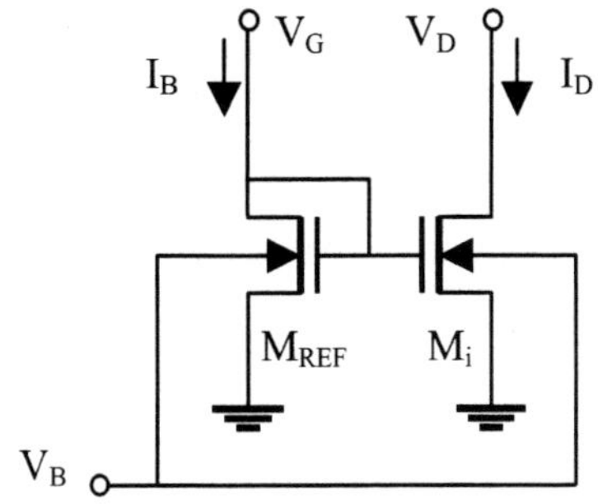

Fig. 2. Test circuit: M_{REF} is the reference transistor while M_i is the transistor under test. I_B (V_B, V_D) is a current (voltage) source.

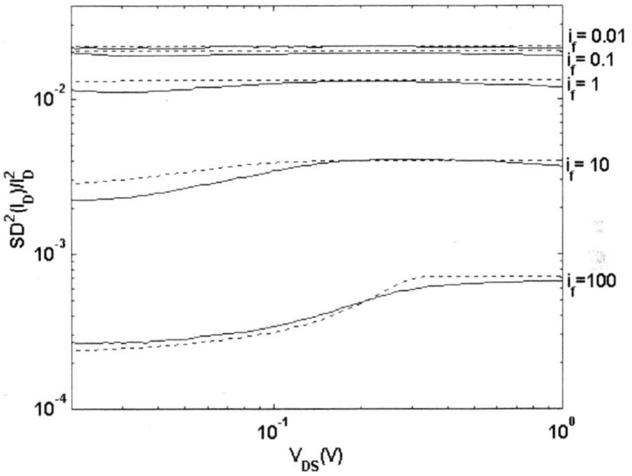

Fig. 3. Normalized current mismatch power. Bulk was kept on zero voltage. (Measurements: —; simulation: - - -.)

Figure 4 shows the measured and simulated dependence of current matching on inversion level for linear (V_{DS}=50mV) and saturation (V_{DS}=1V) regions. For this measurement, transistors were paired in series for reduction of short channel effects, resulting in 12 transistor pairs per die, with 30μm x 2.4μm equivalent dimensions. N_{oi} was estimated as $7.6\text{x}10^{12}$ cm^{-2}, resulting in A_{VT} of 32mV-μm. In agreement with our model, measurements show that matching is identical for either linear or saturated conditions under weak inversion. For inversion levels greater than one, mismatch decreases with inversion level more intensively for linear than for saturated condition. As can be seen, expression (13) describes current mismatch accurately for any inversion condition. In the bias range from 10μA to 100μA mismatch seems to increase for the linear region. This behavior is attributed to an effective reduction of the drain-to-source voltage due to voltage drops in contact and diffusion resistances.

It is well know that bulk voltage also affects matching of MOS transistors [4]. Figure 5 shows relative mismatch for different values of bulk voltage for saturation (V_{DS}=1V). Transistors were paired in series for reduction of short channel effects. In this case, one can suppose that N_{oi} is modulated by V_{GB}. So, when bulk is more reverse biased, effective number of impurities per unit area increases under the gate, thus making N_{oi} higher. It is clear in this figure that the greater the reverse bulk voltage, the greater the mismatch. As can be observed in Fig. 5, a forward-biased bulk improves matching. When the bulk is forward biased at 0.3V, an apparent reduction of mismatch occurs for bias current lower than 10nA. Such behavior is attributed to the action of the parasitic lateral bipolar transistor. The solid lines in figure 5 represent expression (13). Different values of N_{oi} were chosen to fit the mismatch characteristics in weak inversion.

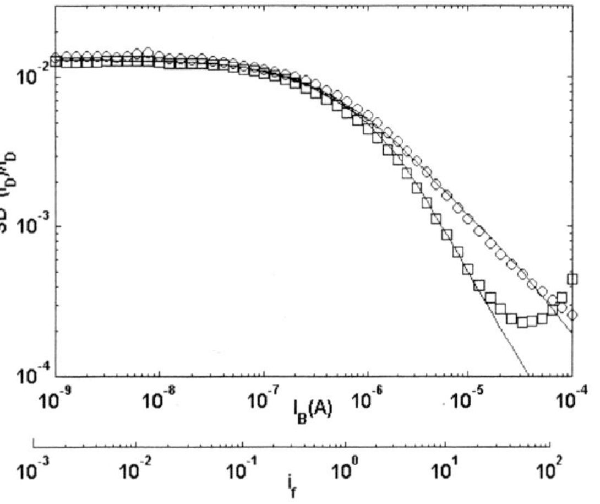

Fig. 4. Dependence of measured current matching on inversion level for linear (□ - V_{DS}=50mV) and saturation (○ - V_{DS}=1V) regions. Bulk was kept at zero volt. Solid lines show theoretical expression (13).

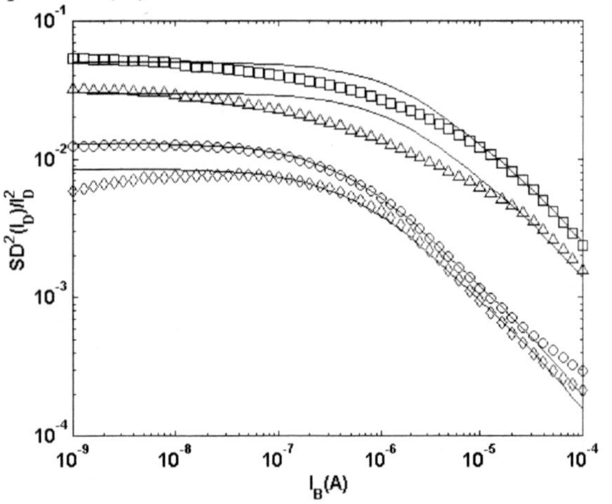

Fig. 5. Dependence of mismatch on bulk-to-source voltage in saturation ($V_{DS}=1V$). □ = -1.8V; △ = -1.2V; ○ = 0V; ◇ =+0.3V. Solid lines represent expression (13).

7. CONCLUSIONS

A mismatch model for the MOS transistor, continuous in all operation regions, has been developed. A physics based approach, based on fluctuation of carrier number, was used to integrate all the contributions of small mismatch elements along the transistor channel. This approach along with the description of the dc characteristics of MOSFET's from the ACM model resulted in a compact easy-to-use formula for mismatch that covers any operating region. Results obtained are quite similar to those derived in [13] for *1/f* noise, since the physical mechanisms at the origin of both phenomena are similar. Experimental results confirmed the accuracy of our model under various bias conditions. It is expected that this work will help designers understand and predict transistor mismatch in an accurate and easy way.

ACKNOWLEDGMENT

The authors are grateful to CNPq and CAPES, Brazilian agencies for scientific development, for their financial support.

REFERENCES

[1] J-B Shyu, G. C. Temes, and F. Krummenacher, "Random error effects in matched MOS capacitors and current sources", *IEEE J. Solid-State Circuits*, vol. 19, no. 6, pp. 948-955, Dec. 1984.

[2] M. J. M. Pelgrom, A. C. J. Duinmaijer, and A. P. G. Welbers, "Matching properties of MOS transistors", *IEEE J. Solid-State Circuits*, vol. 24, no. 5, pp. 1433-1440, Oct. 1989.

[3] F. Forti and M. E. Wright, "Measurements of MOS current mismatch in the weak inversion region", *IEEE J. Solid-State Circuits*, vol. 29, no. 2, pp. 138-142, Feb. 1994.

[4] M. J. Chen, J. S. Ho, and T. H. Huang, "Dependence of current match on back-gate bias in weakly inverted MOS transistor and its modeling", *IEEE J. Solid-State Circuits*, vol. 31, no. 2, pp. 259-262, Feb. 1996.

[5] J. A. Croon et al., "A comparison of extraction techniques for threshold voltage mismatch", *Proc. IEEE 2002 Int. Conference on Microelectronic Test Structures*, pp. 235-240, 2002.

[6] P. G. Drennan, and C. C. McAndrew, "Understanding MOSFET mismatch for analog design", *IEEE J. Solid-State Circuits*, vol. 38, no. 3, pp. 450-456, March 2003.

[7] M-F. Lan and R. Geiger, "Impact of model errors on predicting performance of matching-critical circuits, *43rd. IEEE Midwest Symp. on Circuits and Systems*, pp.1324-1328, 2000.

[8] M-F. Lan and R. Geiger, "Modeling of random channel parameter variations in MOS transistors", *IEEE International Symposium on Circuits and Systems (ISCAS)*, vol. I, pp.85-88, 2001.

[9] S.Cristensson, I.Lundstrom, C.Svensson, "Low frequency noise in MOS transistors", Solid-State Electron, vol.11, pp.797-812, 1968.

[10] A. I. A. Cunha, M. C. Schneider, C. Galup-Montoro, "An MOS transistor model for analog circuit design", *IEEE J Solid-State Circuits*, vol.33, no.10, pp.1510-1519, Oct.1998.

[11] Tsividis, Y. P., *"Operation and Modeling of the MOS Transistor"*. McGraw Hill, 1999

[12] Y. Byun, K. Lee and M. Shur, "Unified charge control model and subthreshold current in heterostructure field effect transistors," *IEEE Electron Device Letters*, vol. 11, no. 1, pp. 50-53, Jan. 1990.

[13] A. Arnaud and C. Galup-Montoro, "A compact model for flicker noise in MOS transistors for analog circuit design", *IEEE Trans. Electron Devices*, vol. 50, no.8, pp. 1815-1818, August 2003.

[14] M. J. M. Pelgrom, "Low-power CMOS data conversion, Chap. 14 in *Low-Voltage/Low-Power Integrated Circuits and Systems*, E. Sánchez-Sinencio, A. G. Andreou, Eds. IEEE Press, York, 1999.

[15] K. K. Hung, P. K. Ko, C. Hu, and Y. C. Cheng, "A physics-based MOSFET noise model for circuit simulators," IEEE Trans. Electron Devices, vol. 37, pp. 1323-1333, May 1990.

SCALING RULES AND PARAMETER TUNING PROCEDURE FOR ANALOG DESIGN REUSE IN TECHNOLOGY MIGRATION

A. Savio, L. Colalongo, Zs. M. Kovács-Vajna

Dept. of Electronics-DEA
University of Brescia
via Branze 38, 25123 Brescia, Italy

M. Quarantelli

PDF Solutions
Desenzano del Garda, Brescia, Italy

ABSTRACT

In this paper a methodology for analog design reuse during technology scaling is proposed. First of all, analytical resizing rules are derived for MOS transistors working in saturation and triode regions. These rules, however, do not account for parasitic effects that, especially in submicron technologies, lead to inaccurate scaling. For this reason, for example, DC gain and unity gain frequency may substantially differ from the original. In order to compensate these inaccuracies, a tuning procedure based on SPICE simulations is proposed. Finally, the migration and tuning procedures are validated and simulation results are compared scaling down a Miller OTA from 0.25μm to 0.15μm technology.

1. INTRODUCTION

Nowadays, there is an increasing interest in design porting due to the growing gap between the circuit complexity and designers productivity. Moreover, the time to market pressure and the evolution of system on chip (SoC) design, enabled by the advent of deep submicron processes, have enlarged this gap. Today, the synthesis of digital circuits is a highly automated phase whereas the design of analog blocks is still a hand based procedure. So it is clear that the analog section of a Soc represents the bottleneck of the whole project.

Analog automatic synthesis is much less developed than digital counterpart due to the larger number of variables, specifications and interactions that lead to a much harder problem. In a standard digital design, the only degree of freedom is the transistor width and the major goal is the tradeoff between speed and power consumption. When an analog circuit is designed, an additional number of variables, such as transistor lengths and bias currents, and different specifications, such as unity gain frequency (UGF), voltage gain, signal to noise ratio (SNR) and so on, have to be considered. Furthermore, each specific application can lead to specific requirements that can increase the problem complexity.

Technology scaling has reduced the power delay product of digital circuits improving global digital performances but, on the other hand, the dimension scaling in analog circuits doesn't produce evident benefits and often circuit behaviors are modified and performances are reduced. The design reuse involves two different aspects that lead to poorer results. First of all, the technology migration modifies technology parameters such as mobility, gate oxide capacitance, voltage supply, ..., and, in the second place, the scaling can modify environmental conditions (e.g. output capacitance). Previous considerations show that the analog design's migration is not a straightforward procedure and it is a challenging issue which often involves a complete redesign.

Different solutions have been proposed recently in order to solve these problems both at research level [1], [2], [3] and at commercial level [4]. The last two approaches ([3] and [4]) are based on optimization loops and massive SPICE simulations in order to suitably modify the circuit parameters, such as transistor dimensions and bias currents, and check the obtained performances against the goal specifications. Such an approach has the drawback of a complete redesign of the analog block. This means that all the considerations done by a skilled designer in the original project and all the strategies that an expert designer might have followed are completely lost.

The first solution proposed in [1] and [2], instead, is an analytical approach dealing with the issue of technology scaling making use of the original circuit characteristics and the technology parameters keeping the original design strategies. This analytical procedure is supported by simulation [1], [2] and experimental [5] results in the case of a complete Miller OTA reuse. These results are obtained only for a circuit with all transistors working in the saturation region and they don't consider the possibility that the goal specifications (original circuit performances) may not sufficiently accurate. For example it is possible that the voltage gain decreases during the scaling.

In our work we present the analytical approach which includes the case of an analog circuit with MOSFETs in triode and saturation regions. Moreover, in order to compensate possible inaccurate solutions, a tuning procedure based on an optimization loop is also presented.

This paper is organized as follows. In Section 2, we summarize the resizing procedure as presented in [1] and we extend the rules including MOS transistors in triode region. Section 3 describes the tuning procedure we have adopted to accurately meet the specifications. Section 4 shows the simulation results of a Miller OTA resizing from 0.25μm technology to 0.15μm technology and the results of the tuning procedure. Finally, in Section 5 conclusions are drawn.

2. RESIZING RULES

Resizing rules [1] are based on the application of the ACM (Advanced Compact Model) MOSFET model described in [6]. In [6] very accurate expressions for large and small signal characteristics are derived. These equations are valid for a MOSFET working in linear and saturation regions and in weak, moderate and strong inversion.

0-7803-8251-X/04/$17.00 ©2004 IEEE

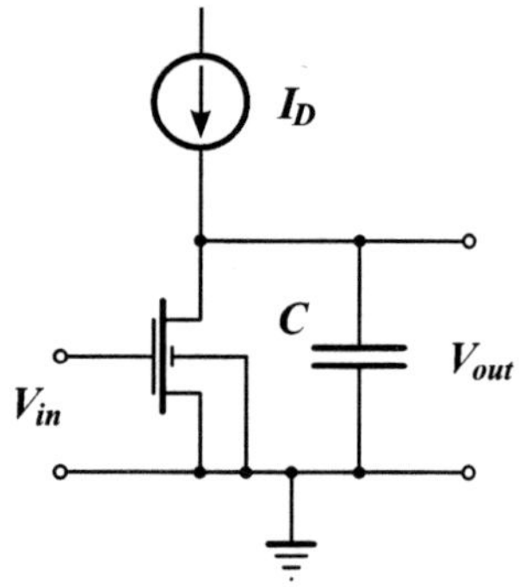

Fig. 1. Common source amplifier.

In the ACM model, the drain current I_D of the MOSFET is given by:

$$I_D = I_S \cdot (i_f - i_r) \tag{1}$$

where i_f and i_r are the forward and reverse normalized currents (or inversion levels) respectively, and I_S is the normalization current defined by:

$$I_S = \mu n C'_{ox} \frac{\phi_t^2}{2} \frac{W}{L} \tag{2}$$

where μ is the carrier mobility, n is the slope factor, C'_{ox} is the gate oxide capacitance per unit area, ϕ_t the thermal voltage and W/L the aspect ratio of the transistor.

The gate transconductance is given by:

$$g_m = \frac{g_{ms} - g_{md}}{n} \tag{3}$$

where g_{ms} and g_{md} are the source and drain transconductances (i.e. derivatives of the drain current with respect to the source or drain voltage) tied to previous parameters as follows:

$$\frac{i_f I_S}{\phi_t g_{ms}} = \frac{\sqrt{1+i_f}+1}{2} \tag{4a}$$

$$\frac{i_r I_S}{\phi_t g_{md}} = \frac{\sqrt{1+i_r}+1}{2} \tag{4b}$$

When we scale down a circuit from a technology 1 to a technology 2, we can define the following scaling factors:

$$\begin{aligned} V_{DD2} &= V_{DD1}/K_V \\ C'_{ox2} &= C'_{ox1}K_{ox} \\ L_{min2} &= L_{min1}/K_L \\ \mu_2 &= \mu_1/K_\mu \\ V_{E2} &= V_{E1}K_E \end{aligned} \tag{5}$$

with V_{DD} the supply voltage, L_{min} the minimum transistor channel length and V_E the Early voltage per unit length.

From these scaling factors and from previous equations (1-4), as reported in [1], it is possible to derive the resizing rules in the case of a MOS transistor in a common source amplifier configuration (Figure 1). These rules preserve the original gain bandwidth (GBW) product and the original dynamic range (DR) defined as the maximum signal to noise ratio at the output node.

In [1], two different strategies are obtained: constant inversion level scaling and channel length scaling that allows taking advantages from the smaller dimensions of a new generation technology.

The drawback of these procedures is that they regard only the case of a saturated MOSFET when the reverse inversion level (i_r) is zero.

In this paper a MOS transistor in triode region has been also considered: in this case $i_r \neq 0$ therefore the complete set of equations 1-4 has to be taken into account. As in [1], in order to keep the original DR and GBW product, the output capacitance and transconductance scaling factors must be K_V^2 ($C_2 = C_1 K_V^2$ and $g_{m2} = g_{m1}K_V^2$).

If we consider the transistor's intrinsic cutoff frequency f_T:

$$f_T = \frac{g_m}{2\pi(C_{gs} + C_{gb} + C_{gd})} \tag{6}$$

when the MOS transistor is in triode region ($C_{gs} = C_{gd} \simeq C'_{ox}WL/2$ and $C_{gb} \simeq 0$), we can find a good approximation for f_T:

$$f_T \simeq \frac{g_m}{2\pi C'_{ox} WL} \tag{7}$$

Now using equations 2, 3 and 4 into 7 we are able to write the relation between f_T and inversion levels i_f and i_r:

$$f_T \simeq \frac{\mu\phi_t}{2\pi L^2} \cdot \left[\frac{i_f}{1+\sqrt{1+i_f}} - \frac{i_r}{1+\sqrt{1+i_r}}\right] \tag{8}$$

As in the original work, two different strategies can be derived, namely constant inversion level scaling and channel length scaling.

2.1. Constant inversion level scaling

If we want to preserve the original inversion levels ($i_{f2} = i_{f1}$ and $i_{r2} = i_{r1}$), from equation 8 it follows that the ratio μ_2/L_2^2 must remain the same, hence:

$$L_2 = \frac{L_1}{\sqrt{K_\mu}} \tag{9}$$

The difference:

$$g_{ms} - g_{md} = \mu n C'_{ox}\phi_t \frac{W}{L} \cdot \left[\frac{i_f}{1+\sqrt{1+i_f}} - \frac{i_r}{1+\sqrt{1+i_r}}\right]$$

together with the relation $g_{m2} = g_{m1}K_V^2$ let us to write the new transistor width (assuming an invariant slope factor n):

$$W_2 = W_1 \frac{K_V^2}{K_{ox}} \sqrt{K_\mu} \tag{10}$$

Finally, substituting equation 2 into 1 the drain current scaling factor can be found:

$$I_{D2} = I_{D1} K_V^2 \tag{11}$$

2.2. Channel length scaling

As in [1], we first scale down the transistor channel length by the factor K_L defined by the technology scaling. Then the new length becomes:

$$L_2 = \frac{L_1}{K_L} \tag{12}$$

From equation 7, in order to preserve the cutoff frequency f_T, it follows immediately that the new transistor width has to be:

$$W_2 = W_1 K_L \frac{K_V^2}{K_{ox}} \tag{13}$$

Starting from equations 1-4 and assuming that g_{ms} and g_{md} scaled up like g_m ($g_{ms2} = g_{ms1}K_V^2$ and $g_{md2} = g_{md1}K_V^2$), in some steps we can find the target inversion levels:

$$i_{f2} = \left[\frac{K_\mu}{K_L^2}(\sqrt{1+i_{f1}}-1)+1\right]^2 - 1 \tag{14a}$$

$$i_{r2} = \left[\frac{K_\mu}{K_L^2}(\sqrt{1+i_{r1}}-1)+1\right]^2 - 1 \tag{14b}$$

and hence the bias current of the transistor:

$$I_{D2} = I_{D1}K_V^2 \frac{K_\mu K_L^{-2}\left(\sqrt{1+i_{f1}}+\sqrt{1+i_{r1}}-2\right)+2}{\sqrt{1+i_{f1}}+\sqrt{1+i_{r1}}} \tag{15}$$

or the equivalent expression:

$$I_{D2} = I_{D1}\frac{K_L^2 K_V^2}{K_\mu}\frac{i_{f2}-i_{r2}}{i_{f1}-i_{r1}} \tag{16}$$

The above equations are valid not only in the case of a MOSFET working in a triode region but also in the case of a saturated MOSFET. In the case of constant inversion level scaling, instead, new scaling factors are like those reported in [1].

Without loss of generality thanks to simple algebraical manipulations, the extended resizing rules, valid for any region of polarization and for any inversion level, can substitute for the original scaling factors reported in [1]. In fact, we can impose in equations 14b, 15 and 16 the condition $i_{r1} = 0$ in order to consider a MOS transistor that operates in the saturation region.

3. TUNING PROCEDURE

Simulation and measurement results [2], [5] show a good compliance between original and scaled circuits but, unfortunately, this is true just for a few technologies.

In many applications as, for instance, operational amplifiers the DC voltage gain depends linearly on Early voltage scaling factor K_E ([2]) and analytical calculations can't directly control it through other technology parameters. For example, if we consider a saturated MOSFET that operates in weak inversion, DC gain scaling factor ([2]) is K_E/K_L. If this ratio is less than one, the scaled amplifier will have a lower voltage gain. In other words, DC gain is technology dependent and it can be substantially different from the original one.

Furthermore, the scaling procedure does not account for parasitic effects (e.g. parasitic capacitance, ...) therefore, especially in recent technologies where these effects are stronger, the frequency behavior (e.g. UGF) may be altered. Moreover, short channel effects modify the theoretical mobility, that depends on bias and MOSFET feature size, so these variations lead to an inaccurate mobility scaling factor (K_μ). The method becomes inaccurate as well.

Finally, in order to preserve the DR, the load capacitor scaling factor must be $C_2 = C_1K_V^2$. This is a good approximation only if the following stage scales down with the same rules. In this case the load capacitor C accounts for the input gate capacitance (WLC'_{ox}) which, as equations 9 and 10 or 12 and 13 show, scales up by K_V^2. However, C could be imposed by other constraints and, hence, it should scale in a different manner. For example, if C is the capacitive load of an output stage (pad + bonding + package capacitance), it substantially remains unchanged.

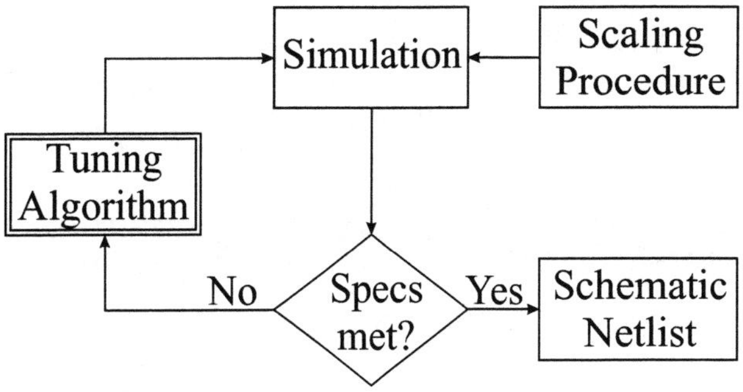

Fig. 2. Flow of the tuning procedure.

These and other constraints may lead to a scaled circuit with an altered frequency response. A tuning procedure based on SPICE simulation can adjust circuit parameters such as bias currents and MOSFET aspect ratios in order to obtain a frequency response close to the original. The basic idea is to modify the bias current to compensate the UGF, usually increasing I_D implies increasing UGF also, and to modify the channel length L to compensate the DC gain. Obviously, the channel width W must be suitably scaled in order to preserve the inversion levels i_f and i_r.

As Figure 2 shows, the procedure starts with an initial point provided by constant inversion level scaling or channel length scaling methods. If this starting point is not a good solution (i.e. it does not meet the specifications) the tuning algorithm provides a modified dimensioning. The tuning loop ends when specifications are met or when the maximum number of iterations is reached.

The method tries to fit the original transfer function using the Levenberg-Marquardt algorithm ([7]) that has proved to be an effective way to solve nonlinear least square problems. In order to apply this algorithm, the original (goal) and scaled frequency responses (gain curves) have to be suitably sampled (e.g. 10 points/decade). Jacobian matrixes are calculated by using a finite difference method and all circuit parameters (transistor widths, lengths and bias currents) are modified at each single step preserving inversion levels provided by the analytical resizing rules.

When the gain curve is sufficiently fitted, also the phase response is correct. When the phase margin shows a good agreement with the original one, the procedure leads also to an accurate compensation network.

4. SIMULATION RESULTS

The resizing rules and the tuning procedure have been applied to a Miller OTA. The operational amplifier of Figure 3 has been originally designed in 0.25μm technology with a supply voltage of 2.5V and then scaled in 0.15μm technology with a 1.2V supply voltage using channel length scaling resizing rules. Technology scaling factors are: K_V=2.08, K_L=1.67, K_{ox}=1.83, $K_{\mu n}$=0.77 and $K_{\mu p}$=1.04.

Simulation results are reported in Figure 4, whereas a performance comparison is also shown in Table 1.

The DC gain is reduced by about 9dB when the channel length scaling is used. This requires the adoption of a tuning procedure in order to compensate this gain reduction. The UGF (and phase margin) mismatch between original and scaled circuits is mainly due to the capacitive load assumed to be invariant (C=5pF) as discussed in Section 3.

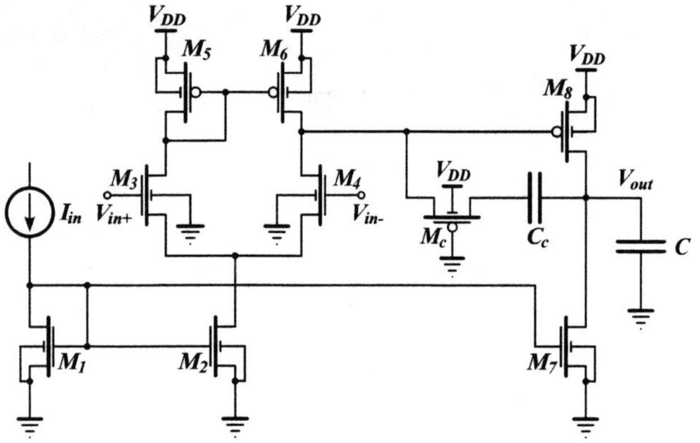

Fig. 3. Miller OTA schematic.

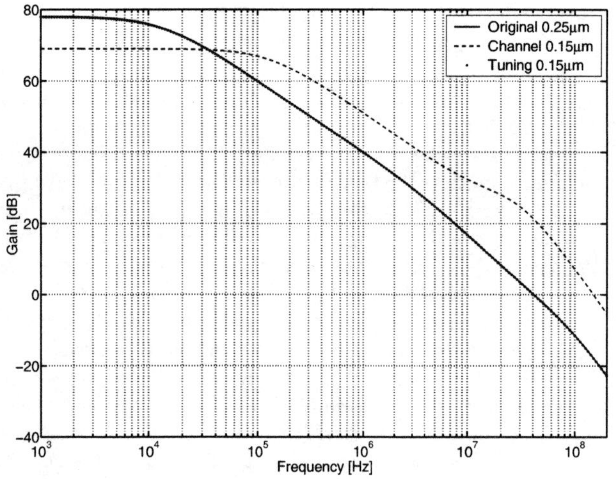

Fig. 4. Simulation results (gain) of a Miller OTA with invariant (C=5pF) load capacitor: original circuit (0.25μm technology, solid line), channel length scaling (0.15μm technology, dashed line) and tuning procedure (0.15μm technology, dots).

The tuning procedure allows us to suitably modify the scaling results to obtain a solution close to the original. Figure 4 shows two overlapping gain curves while Table 1 highlights this good agreement: similar results have been obtained for the phase curves which are not shown. An advantage of this procedure is the automatic sizing of the compensation network (M_c and C_c).

In this particular case only AC analysis is required. The overall number of iterations is 50 and the overall simulation time is about 200 seconds. Finally, the power consumption of the scaled amplifier is much lower than the channel length scaling result. This is mainly due to the reduction of the bias currents to obtain a lower UGF. Multiplying equation 9 by eq. 10 or 12 by 13 it is clear that the MOSFET gate area scales up by K_V^2/K_{ox}. When $K_V^2/K_{ox} > 1$ the scaled circuit area is larger than the original one. The further area growth, after the tuning procedure, is mainly due to the channel length and width increase necessary to compensate the gain reduction. Moreover, there are more valid solutions with different area in the complex solution space. It is also possible to adopt a constrained fitting procedure to find a solution with smaller area.

	Original Circuit 0.25μm	Ch. Length Scaling 0.15μm	Tuning Procedure 0.15μm
DC Gain [dB]	78.0	68.9	77.9
UGF [MHz]	40.8	150.8	40.6
Phase Margin [°]	46.5	1.8	46.1
Power [μW]	750	652	136
Area [μm^2]	112	276	386

Table 1. Simulation result comparison of a Miller OTA with invariant capacitive load (C=5pF) scaled down from 0.25μm technology to 0.15μm technology.

5. CONCLUSIONS

In this paper the resizing rules recently proposed by Galup-Montoro and Schneider in [1] and [2] have been extended to the case of a MOSFET working in triode region.

In order to compensate the parasitic effects and DC gain drops we have proposed a tuning procedure based on SPICE simulations that acts on bias currents and on transistor sizes. This procedure tries to fit, using the Levenberg-Marquardt algorithm, the original transfer function therefore, theoretically, it can be applied to meet other circuit specifications as well.

As an example, a Miller OTA scaled down from a 0.25μm technology to a 0.15μm technology has been used to validate our procedure. Results show that with a small number of iterations we can obtain a good compliance between original and scaled performances largely exceeding the analytical method inaccuracies.

6. REFERENCES

[1] C. Galup-Montoro and M. C. Schneider, "Resizing Rules for the Reuse of MOS Analog Designs," in *Proceedings SBCCI 2000: XIII Symposium on Integrated Circuits and Systems Design, Manaos, Brazil*, September 2000, pp. 89–93.

[2] C. Galup-Montoro, M.C. Schneider, R. M. Coitinho, "Resizing Rules for MOS Analog-Design Reuse," *IEEE Design & Test of Computers*, pp. 50–58, March-April 2002.

[3] S. Funaba, A. Kitagawa, T. Tsukada, G. Yokomizo, "A Fast and Accurate Method of Redesigning Analog Subcircuits for Technology Scaling," *Analog Integrated Circuits and Signal Processing*, vol. 25, pp. 299–307, 2000.

[4] *http://www.neolinear.com.*

[5] R. Acosta, F. Silveira, P. Aguirre, "Experiences on Analog Circuit Technology Migration and Reuse," in *Proceedings XV Symposium on Integrated Circuits and Systems Design, Porto Alegre, Brazil*, September 2002, pp. 169–174.

[6] A. I. A. Cunha, M. C. Schneider, C. Galup-Montoro, "An MOS Transistor Model for Analog Circuit Design," *IEEE Journal of Solid-State Circuits*, vol. 33, no. 10, pp. 1510–1519, October 1998.

[7] W. H. Press, S. A. Teukolsky, W. T. Vetterling, B. P. Flannery, *Numerical Recipes in C*, Cambridge University Press, second edition, 2002.

Fast Sensitivity Analysis of Transmission Line Networks

Natalie Nakhla, Anestis Dounavis, Ram Achar and Michel Nakhla
Department of Electronics, Carleton University, Ottawa, Ontario, K1S 5B6
Tel (613) 520-5780; Fax: (613) 520-5708; Email: msn@doe.carleton.ca

Abstract --This paper describes an efficient approach for sensitivity analysis of lossy transmission lines in the presence of nonlinear terminations. Sensitivity information is derived from the macromodels based on delay extraction and matrix-rational approximation of Telegrapher's equations. The method enables sensitivity analysis of interconnect structures with respect to both electrical and physical parameters. An important advantage of the proposed approach is that the derivatives of the stamp of the transmission line with respect to per-unit-length parameters are obtained analytically.

I. Introduction

The ever increasing quest for higher-operating speeds, miniature devices and denser layouts has made the signal integrity analysis a challenging task. As signal frequencies are approaching the GHz range, the interconnect effects such as delay, crosstalk, ringing and distortion become the dominant factors limiting the overall performance of microelectronic/microwave systems. At higher frequencies, the length of the interconnect becomes a significant fraction of the operating wavelength, and conventional lumped impedance models become inadequate in describing the interconnect performance and transmission line (TL) models become necessary [1]-[9]. The major difficulty usually encountered while linking distributed transmission lines and nonlinear simulators is the problem of mixed frequency/time [1]-[7]. This is because, distributed elements are usually characterized in the frequency-domain where as nonlinear components such as drivers and receivers are represented only in time-domain.

To address this issue, several algorithms were proposed for time-domain macromodeling of distributed transmission line networks [2]-[7]. For such macromodels, apart from accuracy, the issues of interest are passivity and causality of the macromodel. This is because, a stable but non-passive macromodel when connected to the rest of the circuitry can lead to unstable networks [3]. Also, a non-causal model can cause signals to appear even before the application of the signal [2]. Recently a new macromodel was suggested based on delay extraction and matrix-rational approximation (DeMRA) to preserve both the causality and passivity properties of transmission line macromodels [7]. Application of this new MTL model to the sensitivity analysis for optimization of high-speed interconnects in the presence of nonlinear terminations is presented in this paper.

II. Review of DeMRA Based TL Macromodels

DeMRA based algorithm provides closed-form (analytical) computation of time-domain macromodels for transmission line networks, while preserving its causality and passivity properties. It employs a mechanism for delay reduction prior to performing matrix rational approximation (MRA) [6] of the exponential matrix describing Telegrapher's equations. It also leads to significantly lower order macromodels for long lossy coupled delay lines. A brief review of the algorithm is given below.

Consider the matrix exponential form of Telegrapher's equations [2]

$$\begin{bmatrix} \boldsymbol{V}(d,s) \\ -\boldsymbol{I}(d,s) \end{bmatrix} = e^{(\boldsymbol{A}+s\boldsymbol{B})} \begin{bmatrix} \boldsymbol{V}(0,s) \\ \boldsymbol{I}(0,s) \end{bmatrix};$$

$$\boldsymbol{A}(s) = \begin{bmatrix} \boldsymbol{0} & -\boldsymbol{R} \\ -\boldsymbol{G} & \boldsymbol{0} \end{bmatrix} d; \qquad \boldsymbol{B}(s) = \begin{bmatrix} \boldsymbol{0} & -\boldsymbol{L} \\ -\boldsymbol{C} & \boldsymbol{0} \end{bmatrix} d \tag{1}$$

where $\boldsymbol{R}, \boldsymbol{L}, \boldsymbol{G}, \boldsymbol{C}$ are the per-unit-length (PUL) parameter matrices of the transmission-line, $\boldsymbol{V}(s)$, $\boldsymbol{I}(s)$ are the terminal voltage and current vectors of the transmission line and d is the length of the line. Using perturbation and assuming that $\|\boldsymbol{A}\| \ll \|s_{max}\boldsymbol{B}\|$ (where s_{max} corresponds to the maximum frequency of interest), (1) can be approximated as [10]

$$e^{(\boldsymbol{A}+s\boldsymbol{B})} \approx e^{\frac{s\boldsymbol{B}}{m}} \prod_{k=1}^{m} e^{\boldsymbol{C}_k}; \qquad (\boldsymbol{C}_k \equiv f(\boldsymbol{A}, \boldsymbol{B})) \tag{2}$$

where $\|\boldsymbol{C}_1\| \gg \|\boldsymbol{C}_2\| \gg \ldots \|\boldsymbol{C}_m\|$. It was shown that the product [7]

$$\prod_{k=1}^{m} Q_k + \varepsilon_m; \qquad Q_k \equiv e^{\frac{A}{2m}} e^{\frac{sB}{m}} e^{\frac{A}{2m}} \tag{3}$$

converges asymptotically to $e^{(A+sB)}$ as $m \to \infty$. The associated error (ε_m) in this case is given by

$$\|\varepsilon_m\| = \max_{0 \le s \le s_{max}} \left\| e^{(A+sB)} - \prod_{k=1}^{m} Q_k \right\| \cong O\left(\frac{1}{m^2}\right) \tag{4}$$

If $\|A\| \ll \|s_{max}B\|$ (which is the case for long low lossy lines), then an alternative form for (3) can be used with a better error-bound, and is given by:

$$\prod_{k=1}^{m} Q_k + \varepsilon_m; \qquad Q_k \equiv e^{\frac{sB}{2m}} e^{\frac{A}{m}} e^{\frac{sB}{2m}} \tag{5}$$

Also it can be proved that, the average of the approximations in (3) and (5) as given by

$$\prod_{k=1}^{m} Q_k + \varepsilon_m; \quad Q_k \equiv \frac{1}{2}\left(e^{\frac{A}{2m}} e^{\frac{sB}{m}} e^{\frac{A}{2m}} + e^{\frac{sB}{2m}} e^{\frac{A}{m}} e^{\frac{sB}{2m}} \right) \tag{6}$$

further reduces the error. The products represented by (3) or (5) can be viewed as a cascade of m transmission lines. In addition, each of the kth product term can be viewed as a cascade of lossy and lossless transmission lines. The lossy terms are macromodeled using the passive matrix rational approximation [6]. The resulting macromodels are of significant lower orders (since a significant delay portion is already extracted from these terms). They are later combined with the lossless terms using the method of characteristics approach [4]-[5]. For example, each Q_k in (5) can be realized as shown in Fig. 1. It is to be noted that passivity of the entire macromodel is now guaranteed as the passivity of each sub-line in (5) (see Fig. 1) is preserved. Also since the delay of the line is represented by an explicit delay element, the causality property of the line is also automatically preserved. The next section describes the sensitivity analysis of transmission lines using DeMRA based macromodels.

III. Development of The Sensitivity Analysis Algorithm

From (3) and (5) and Fig. 1, it is obvious that the sensitivity analysis of DeMRA based macromodels require derivatives of the MRA based transmission line stamp and also that of the delay element, with respect to a particular varying interconnect parameter of interest (represented by λ). Computation of derivatives for MRA based macromodels has been well described in the literature [9]. Hence this section focusses on the computation of derivatives of the stamp of the delay element. For simplicity, the main concept is illustrated using a single lossless line. However, it should be emphasized that the presented algorithm can be easily extended to multiconductor lines.

Using the Method of Characteristics [4], the solution of equations in (1), for a lossless line can be written in the form

$$V_1 = Z_0 I_1 + e^{-\gamma d}[2V_2 - e^{-\gamma d}(Z_0 I_1 + V_1)] \tag{7}$$

$$V_2 = Z_0 I_2 + e^{-\gamma d}[2V_1 - e^{-\gamma d}(Z_0 I_2 + V_2)] \tag{8}$$

where γ is the propagation constant, Z_0 is the characteristic impedance, V_1 and I_1 are the terminal voltage and current at the near end of the line, and V_2 and I_2 are the terminal voltage and current at the far end of the line. Next, (7) and (8) can re-written as

$$V_1 - Z_0 I_1 = W_1 \tag{9}$$

$$V_2 - Z_0 I_2 = W_2 \tag{10}$$

where

$$W_1 = e^{-\gamma d}[2V_2 - e^{-\gamma d}(Z_0 I_1 + V_1)] \tag{11}$$

$$W_2 = e^{-\gamma d}[2V_1 - e^{-\gamma d}(Z_0 I_2 + V_2)] \tag{12}$$

Using (7)-(12), a relation between W_1 and W_2 can be obtained as

$$W_1 = e^{-\gamma d}[2V_2 - W_2] \tag{13}$$

$$W_2 = e^{-\gamma d}[2V_1 - W_1] \tag{14}$$

Next, (7)-(10) can be expressed in the time domain as

$$v_1(t) - Z_0 i_1(t) = w_1(t) \tag{15}$$

$$v_2(t) - Z_0 i_2(t) = w_2(t) \tag{16}$$

$$w_1(t) = 2v_2(t-\tau) - w_2(t-\tau) \tag{17}$$

$$w_2(t) = 2v_1(t-\tau) - w_1(t-\tau) \tag{18}$$

As seen, since (15)-(18) are in the time-domain, they can be solved simultaneously with the time-domain modified nodal analysis (MNA) equations [11] describing the rest of the circuit.

Next, for sensitivity analysis, derivatives of (15)-(18) are needed with respect to a particular interconnect parameter of interest. Based on (15)-(18), it can be proven that time-domain sensitivity with respect to a parameter λ, can be presented by the following delay equations

$$\tilde{v}_1(t) - Z_0\tilde{i}_1(t) - \frac{\partial Z_0}{\partial \lambda} i_1(t) = \tilde{w}_1(t) \quad (19)$$

$$\tilde{v}_2(t) - Z_0\tilde{i}_2(t) - \frac{\partial Z_0}{\partial \lambda} i_2(t) = \tilde{w}_2(t) \quad (20)$$

$$\tilde{w}_1(t) = 2\tilde{v}_2(t-\tau) - \tilde{w}_2(t-\tau) - \frac{d\tau}{d\lambda}\left[2\frac{\partial}{\partial t}v_2(t-\tau) - \frac{\partial}{\partial t}w_2(t-\tau)\right] \quad (21)$$

$$\tilde{w}_2(t) = 2\tilde{v}_1(t-\tau) - \tilde{w}_1(t-\tau) - \frac{d\tau}{d\lambda}\left[2\frac{\partial}{\partial t}v_1(t-\tau) - \frac{\partial}{\partial t}w_1(t-\tau)\right] \quad (22)$$

where

$$\tilde{v}_1(t) = \frac{\partial}{\partial \lambda}v_1(t); \qquad \tilde{v}_2(t) = \frac{\partial}{\partial \lambda}v_2(t) \quad (23)$$

$$\tilde{i}_1(t) = \frac{\partial}{\partial \lambda}i_1(t); \qquad \tilde{i}_2(t) = \frac{\partial}{\partial \lambda}i_2(t) \quad (24)$$

$$\tilde{w}_1(t) = \frac{\partial}{\partial \lambda}w_1(t); \qquad \tilde{w}_2(t) = \frac{\partial}{\partial \lambda}w_2(t) \quad (25)$$

It is important to note that (19)-(22) form a similar set of time-domain equations to that of original delay equations represented by (15)-(18), with the exception that they are associated with new set of variables defined by derivatives in (23)-(25) and also with an additional derivative term in each of them. It is to be noted here that:
$\frac{\partial}{\partial t}v_1(t-\tau)$, $\frac{\partial}{\partial t}v_2(t-\tau)$, $\frac{\partial}{\partial t}w_1(t-\tau)$ and $\frac{\partial}{\partial t}w_2(t-\tau)$ are known from the solution of (15)-(18). These delay sensitivity equations can be solved simultaneously with the MNA sensitivity equations representing the rest of the circuit. Details of the derivation and implementation aspects are not given here due to the lack of space.

IV. Numerical Examples

A coupled interconnect system with a nonlinear diode is shown in Fig. 2. Fig. 3 shows transient responses of the far-end voltages corresponding to a 5 Volt input pulse with rise/fall times 0.1ns and a pulse width of 1ns. Fig. 4 shows the sensitivities with respect to L_{11} for the active line. The results of the proposed method are compared with the perturbation of the lumped segment model [2] (referred to as SPICE Perturbation). Both the proposed method and the perturbation results are in good agreement.

It is to be noted that using the proposed method provides the following advantages. (i) Using the closed-form matrix-rational approximation based macromodel provides significant CPU advantages compared to lumped segmentation model [2]. (ii) Perturbation based techniques can lead to inaccurate results (depending on the magnitude of the perturbation). (iii) In addition, the nonlinear differential equations representing the perturbed network must be solved separately for every parameter of interest. However, in the proposed approach, the sensitivity information with respect to all the parameters can be essentially obtained from the solution of the original network.

V. Conclusions

A new approach for sensitivity analysis of lossy multiconductor transmission lines in the presence of nonlinear terminations is described. Sensitivity information is derived from the recently developed closed-form delay extraction and matrix-rational approximation based transmission-line model. The method enables sensitivity analysis of interconnect structures with respect to both electrical and physical parameters, while providing significant computational cost advantages. The proposed algorithm can easily be extended to include multiconductor transmission lines and also frequency-dependent parameters.

References

[1] A. Deustsch, "Electrical characteristics of interconnections for high-performance systems," *Proc. of IEEE*, vol. 86, pp. 315-355, Feb. 1998.

[2] C. R. Paul, Analysis of Multiconductor Transmission Line. New York: Wiley, 1994.

[3] A. Odabasioglu, M. Celik and L. T. Pilleggi, "PRIMA: Passive Reduced-Order Interconnect Macromodeling Algorithm, "*IEEE Trans. on CAD*, pp. 645-653, August 1998.

[4] F. Y. Chang, "The generalized method of characteristics for waveform relaxation analysis of lossy coupled transmission lines," *IEEE Trans. MTT*, pp. 2028-2038, Dec. 1989.

[5] S. Grivet-Talocia, F. Canavero, "Topline: a delay pole-residue method for simulation of disperive interconnetcs, *Proceedings EPEP-2002*, pp. 359-362. Monterery, CA, Nov. 2003.

[6] A. Dounavis, R. Achar and M. Nakhla "A General Class of Passive Macromodels for Lossy Multiconductor Transmission Lines," *IEEE Transactions on Microwave Theory and Techniques*, pp. 1686 -1696, October. 2001.

[7] A. Dounavis, N. Nakhla, R. Achar and M. Nakhla, "Delay Extraction and Passive Macromodeling of Lossy Coupled Transmission Lines", accepted for publication in *IEEE 12th Topical Meeting on EPEP*, New Jersy, Oct. 2003.

[8] C. Jiao, A. Cangellaris, A. Yaghmour, J. Prince, "Sensitivity Analysis of MTLs and optimization for high-speed Circuit Design", *IEEE Trans. on Advanced Packaging*, pp. 132-142, May. 2000.

[9] A. Dounavis, R. Achar and M. Nakhla "Efficient Sensitivity Analysis of Lossy Multiconductor Transmission Lines with Nonlinear Terminations," *IEEE Trans. on MTT*, pp. 2292 - 2299, Dec. 2001.

[10] F. Fer, "Resolution de l'equation matricielle dU/dt = pU par produit infini d'exponentielles matricielles", *Acad. Roy. Belg. Cl. Sci.*, vol. 44, no. 5, pp. 818-829, 1958.

[11] J. Vlach, K. Singhal, *Computer Methods for Circuit Analysis and Design*, NY, Van Nostrand Reinhold.

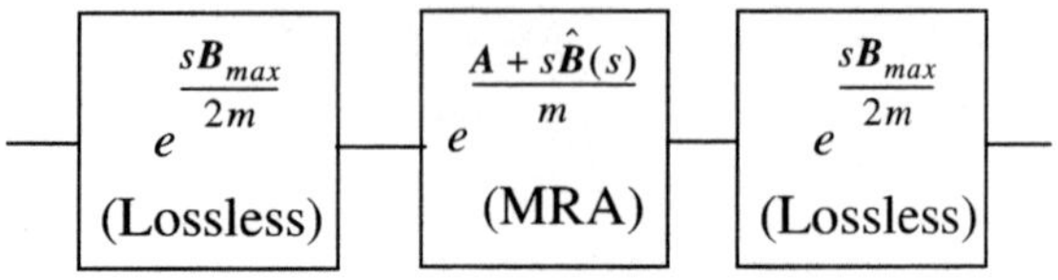

Fig. 1: Macromodel Realization of the Product Terms in (5).

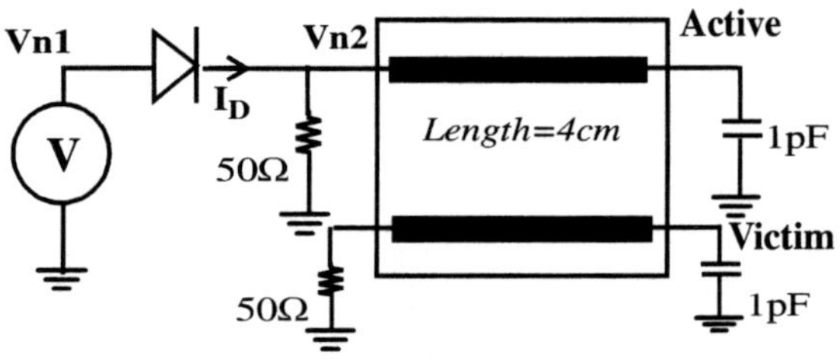

$I_D = 6.9e\text{-}16(e^{40(Vn1-Vn2)} - 1)A$ $\quad R = \begin{bmatrix} 0.14 & 0 \\ 0 & 0.14 \end{bmatrix} \Omega/cm$

$C = \begin{bmatrix} 1.1 & -0.45 \\ -0.45 & 1.1 \end{bmatrix} pF/cm$ $\quad L = \begin{bmatrix} 6.3 & 2.9 \\ 2.9 & 6.3 \end{bmatrix} nH/cm$

Fig. 2: Coupled interconnect system

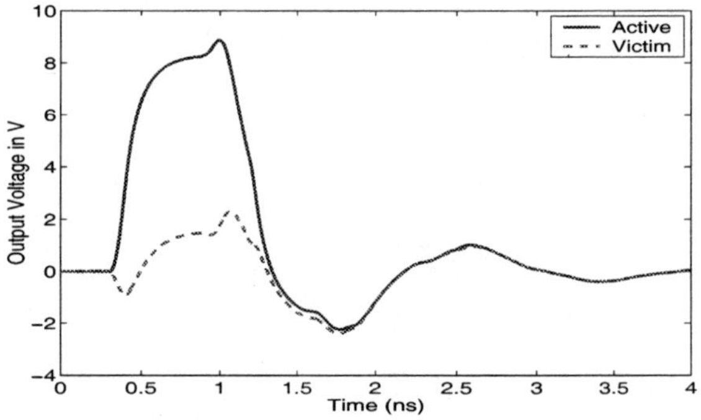

Fig. 3: Output transient response of circuit

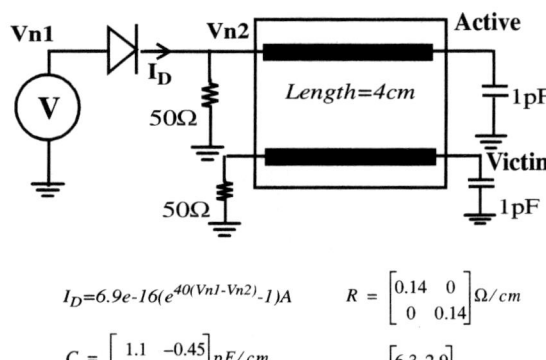

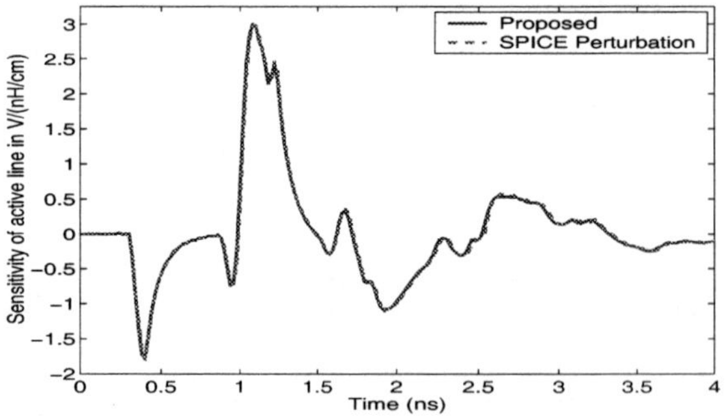

Fig. 4: Sensitivity of active line response w.r.t. L_{11}

FAST TIME-DOMAIN SYMBOLIC SIMULATION FOR SYNTHESIS OF SIGMA-DELTA ANALOG-DIGITAL CONVERTERS

Hui Zhang and Alex Doboli

Electrical and Computer Engineering Department
State University of New York at Stony Brook, Stony Brook, NY, 11794-2350
Email: {huizhang, adoboli}@ece.sunysb.edu

ABSTRACT

This paper presents a fast time-domain simulation method for automated synthesis of Σ–Δ ADC. At the system-level, the method uses a symbolic description of the ADC netlist. Building blocks in the netlist are macromodels, which include circuit non-idealities and nonlinear behavior. As compared to existing simulation methods for Σ–Δ ADC, this technique is fully automated and uses more complex circuit macromodels.

1. INTRODUCTION

Systems-on-chip (SOC) include a large variety of RF, analog and digital IP blocks fabricated on the same silicon die, so that demanding requirements on area, timing, power consumption, reliability, and accuracy are met [13]. IP blocks include RF front-ends, high-frequency filters, analog to digital (ADC) and digital to analog (DAC) converters, digital processors, DSP, co-processors, and memory circuits. SOC are employed for a large variety of applications in areas like telecommunication, mobile and wireless computing, military systems, smart sensors, and so on.

Design of mixed-signal SOC is a complex challenge demanding expertise in many areas. Initially, SOC functionality is described at an abstract level using specification languages and notations like MATLAB, VHDL-AMS, or SystemC. Then, analog and digital IP cores are selected, and system functionality is mapped to the cores. Next, system performance trade-off analysis will contemplate different IP core performance constraints for meeting the overall system performance. Finally, IP cores are synthesized. Digital IP cores are synthesized using well-known digital high-level and logic-level synthesis environments [17]. Bus architecture synthesis, IP core placement, and bus routing complete the tasks of IP integration and design at the physical level. Currently, the SOC design flow has a bottleneck in the step of analog IP core synthesis [13]. This paper proposes a fast simulation method for synthesis of analog IP cores.

Analog synthesis methods are still in their infancy. Research in academia [4,8,12,14,15] (to mention a few), and industry (Neolinear and Analog Design Automation Inc) offers interesting CAD approaches for design of analog circuits, like opamps, operational transconductor circuits (OTA), comparators, and so on. More recently, synthesis tools are proposed for RF circuits, like low noise amplifiers, oscillators, and so on [16]. The next challenge is to address automated synthesis of complex analog IP blocks, like ADC, DAC, PLL, transceivers, and high-frequency filters. This would complete the automated SOC synthesis flow from abstract specifications, and give the benefits of quick time-to-market and superior performance due top the capability of analyzing more design points. Design will be cheaper, as it would demand lesser designer expertise.

This paper presents a fast time-domain simulation method for automated synthesis of Σ–Δ ADC. Fast simulation is critical for any automated synthesis technique, as a large number of design points needs to be analyzed in reasonably long time. For example, about 30,000 points were contemplated during the design of a fourth-order Σ–Δ ADC [9,10]. Even for the most simplistic models, SPICE simulation was still too long.

The proposed time-domain simulation method uses a symbolic description of the ADC netlist, so that Kirchhoff's laws are satisfied at the system level. Building blocks are circuit macromodels (for OTA, opamp, and comparator), which include many non-idealities, like finite gain, poles and zeros, CMRR, PM, fall and rise time, and so on. Nonlinear behavior is also modeled in the approach. Macromodel parameters are automatically linked to the design points obtained with a commercial analog circuit synthesis tool. As compared to other simulation methods for Σ–Δ ADC [12,18], our technique is fully automated, and can uses more complex macromodels. This offers the benefit of having a more realistic simulation, thus the advent of faster design closure. The method is much faster than SPICE simulation, without necessarily trading-off accuracy. Also, our approach does not require extensive expertise in ADC, and does not compromise accuracy, like behavioral models. This work extends to time domain a method we previously suggested for AC simulation of large linear systems [5].

This paper is organized as six sections. Section 2 presents the simulation methodology. Section 3 presents macromodeling of the ADC building blocks. Next, system modeling is detailed, and simulation results are given in Section 5. Section 6 discusses our conclusions.

2. SIMULATION METHODOLOGY

Figure 1 shows the high-level synthesis flow for Σ–Δ ADC [9,10]. Figure 2 presents the generic structure of certain Σ–Δ ADC [11]. These structures are inputs for the tool. First,

0-7803-8251-X/04/$17.00 ©2004 IEEE

in a bottom-up step, macromodels are created for all ADC blocks, like opamps, OTA, and comparators. The right part of the figure exemplifies this step. Using a commercial analog circuit synthesis tool, circuit transistors are sized, and circuit performances are characterized using SPICE and Spectre simulation. This is a one step process, and simulation results are stored in a database. For system simulation, each of the circuits is replaced in the system model by a macromodel. Section 3 details the used macromodels. Then, performance parameters are linked to the simulation data of each circuit.

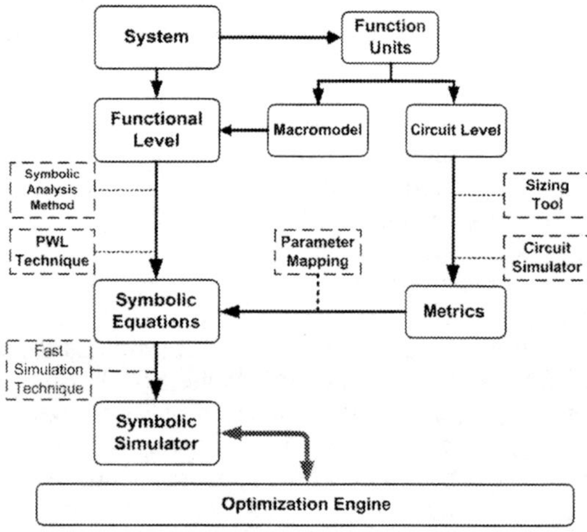

Figure 1: Synthesis environment.

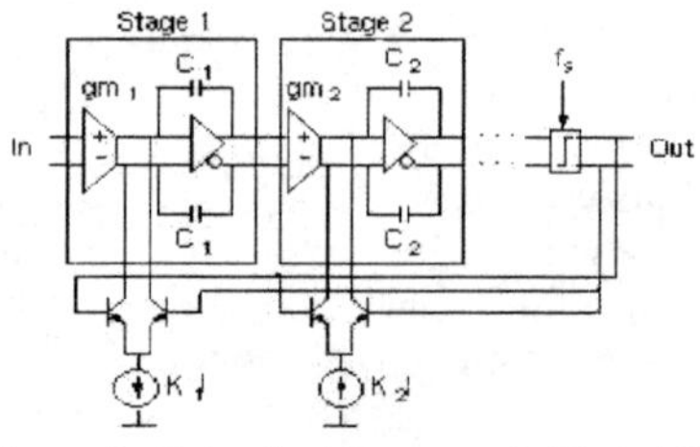

Figure 2: Generic structure of a Σ–Δ ADC

System-level simulation symbolically composes the macromodel parameters into a mathematical model, which describes the steps for calculating the Σ–Δ ADC output voltage over time. The output voltage is used to find typical ADC parameters, like SNR and DR. Figure 3 shows the structure of the simulation model for a third-order Σ–Δ ADC.

3. CIRCUIT MACROMODELING

A. OTA modeling: For OTA modeling, we started from the macromodels proposed by Gomez *et al* [1]. We extended these models to fully differential mode (DM) by duplicating the single end stages, the common mode stages, the intermediate and output stages, and the dominant pole stage. The resulting macromodel is shown in Figure 4.

Next, we related the data collected during circuit synthesis and SPICE/Spectre simulation to the values of the devices in the macromodel. We used the relationships proposed by Gomez *et al* [1]:

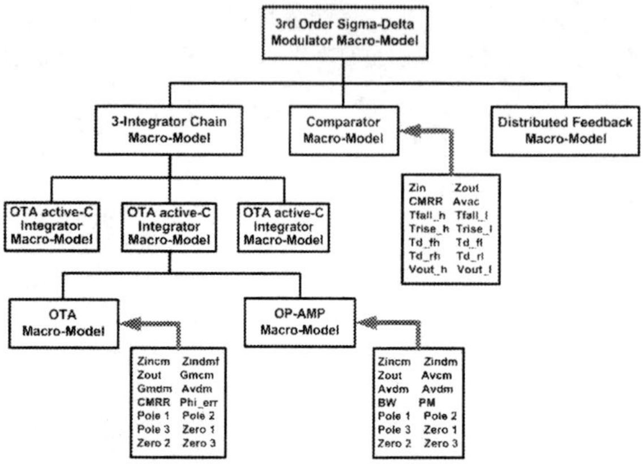

Figure 3: Simulation model for 3rd order Σ–Δ ADC

- V_{os} was obtained through SPICE/Spectre simulation.
- $C_{cm} = 1 / (4\,\pi|Z_{icm}(f_1)|)$.
- $C_d = 1/(2\pi f_1 Z_{idm}[I_m](f_1)) - C_{cm}$.
- $R_d = Z_{idm}[Re](C_d + C_{cm})^2 / C_d^2$.
- C_3 depends on the position of the first pole, as given by SPICE simulation.
- L_4 relates to the dominant zero, as offered by SPICE simulation.
- $1/(R_o C_o)$ is the frequency of the dominant pole.
- R_o results from SPICE/Spectre simulation directly
- R_1, R_2, L_1 and L_2 are determined by common mode zeros.
- $R_3 = R_4 = R_5 = 1$k Ohm are constant.
- $g_{m4} = gm_o = 1$mmhos are fixed.
- $L_1 = L_2 = 0$ are constant.
- g_m was obtained from SPICE/Spectre simulation.

So far, we used a linear model: in the gain stage, currents I_{dm} and I_{cm} are proportional to the voltages V_{icm} and V_{idm}. We will use a nonlinear model in the future. Nonlinearities will be tackled by using piecewise linear (PWL) approximations that are obtained through our technique for PWL model extraction from trained neural networks [6,7].

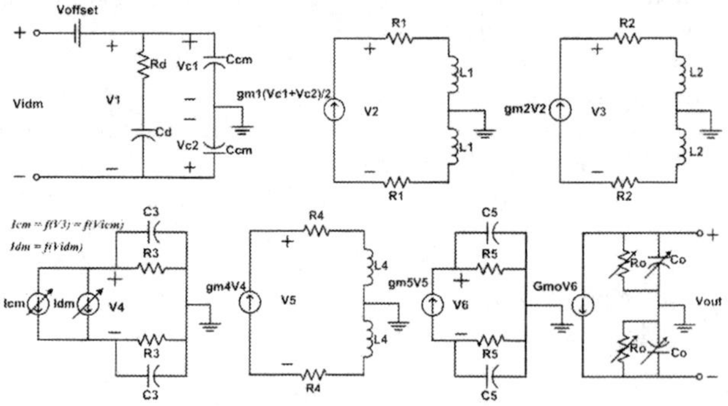

Figure 4: OTA macromodel

B. Opamp modeling: Figure 5 presents the macromodel used for opamps. The model includes three stages: (1) The input stage of the fully differential opamp macromodel is the same as the input stage of OTA marcomodel. (2) The intermediate stage describes the two dominant poles in differential mode. For accurate modeling, like for the OTA circuit, we could add more RC circuits to express additional poles, and more RL circuits to describe zeros. (3) In the output stage, we added a dc bias voltage to the differential output voltages. The bias voltage is needed for transient analysis.

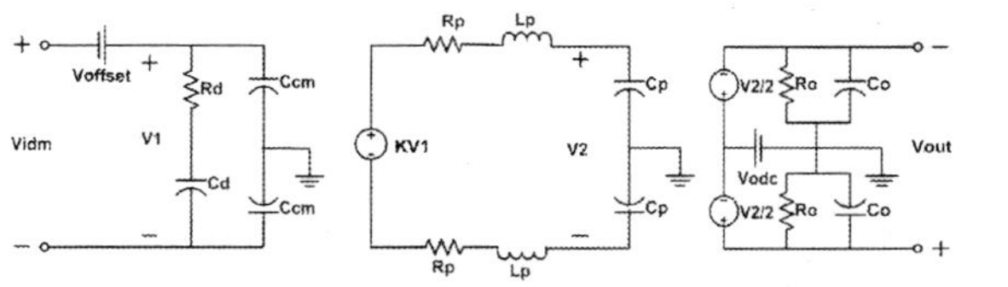

Figure 5: Opamp macromodel

Following relationships were used to relate the macromodel parameters to data collected from circuit synthesis and SPICE/Spectre simulation:

- R_d, C_d, C_{cm} are the same as for OTA parameter mapping.
- R_p, L_p are related to the first and second poles found through SPICE simulation.
- V_{os}, V_{odc}, R_o, and C_o are offered by SPICE/Spectre simulation, directly.

At present, the OTA and opamp macromodels don't model the self-limiting behavior of the differential pair, which is a limitation of the current models. However, the limitation can be easily removed by expressing this behavior through a hyperbolic tangent function. A similar modeling was already used for the comparator (please see the next subsection).

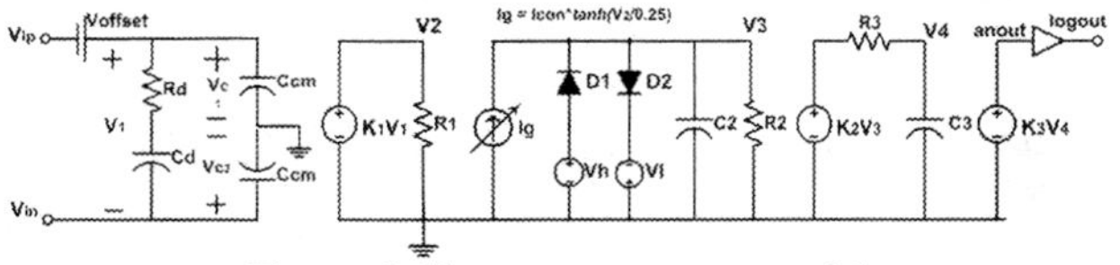

Figure 6: Comparator macromodel

C. Comparator modeling: Figure 6 presents the comparator circuit macromodel. This model is based on the model presented by Moscovici [2] and Connelly [3]. The comparator model has the same input stage as the OTA and opamp models. The nonlinear *Gm* stage expresses the self-limiting behavior of the differential pair by using a hyperbolic tangent function. *Gm* is bounded to the range $-I_{con}$ and I_{con}. As motivated by Moscovici [2], the two diodes specify a certain time for the slew limited mode of the circuit. The I-V characteristic of the diodes is modeled as PWL functions using our model extraction techniques [6,7]. Finally, R_3C_3 and R_2C_2 are the two poles of the comparator, and R_3C_3 is the delay time for large input overdrive voltages.

Following relationships were used to relate the macromodel parameters to data collected from circuit synthesis and SPICE/Spectre simulation:

- R_d, C_d, and C_{cm} are obtained using the same formulas as for the OTA and opamp input stages.
- K_1, R_1 was set to 1.
- R_2 C_2 corresponds to the 2nd pole.
- R_3 C_3 are the 1st pole/ delay time.
- V_h, V_l are the minimum and maximum output voltages.

4. SYSTEM MODELING

Figure 3 presents the overall system model for simulating a third order Σ–Δ ADC. This model can be easily extended for simulating higher order Σ–Δ ADC or ADC of different topologies. The system-level model includes interconnected macrocells, where a macrocell might be the macromodel of a basic circuit (for example, OTA or comparator), or a functional block, like an opamp with integration capacitors. Symbolic equations were set up for the system network by formulating Kirchhoff's laws for the interconnected macrocells. After symbolically solving these equations, a set of mathematical expressions was found for relating over time the unknown voltages and currents to the known signals and circuit parameters. Expressions were encoded as a C++ program, which was then optimized for fast execution. Well-known compiler optimizations were applied, such as common expression naming, common expression elimination, and constant folding.

Assuming a linear OTA model, the 4-port macromodel is composed of two sub-matrixes: (a) *function sub-matrix* and (b) *state sub-matrix*. The entries of the function sub-matrix are determined by the values of the components. The state sub-matrix is related to state variables and the previous state of the circuit. For an OTA circuit, we assumed V_i and V_o as *known*, and then I_i and I_o are the *unknowns*. The opposite reasoning would have been also correct. Following symbolic equations describe the OTA behavior at its four ports:

$$Iip = TAp1Vip + TAp2Vin + TAp3Vop + TAp4Von + TAp5,$$
$$Iin = TAn1Vip + TAn2Vin + TAn3Vop + TAn4Von + TAn5,$$
$$Iop = TBp1Vip + TBp2Vin + TBp3Vop + TBp4Von + TBp5,$$
$$Ion = TBn1Vip + TBn2Vin + TBn3Vop + TBn4Von + TBn5,$$

Symbols TA_{x1}~TA_{x4} and TB_{x1}~TB_{x4} are symbolic functions of the OTA macromodel parameters (shown in Figure 4). These functions were obtained using Mathematica software for symbolically solving the nodal equations of the OTA macromodel. Symbols TA_{p5}, TA_{n5}, TB_{p5}, TB_{n5} form the state matrix. They are functions of voltages V_{cd0}, V_{cm10}, V_{cm20}, V_{c30}, V_{c50}, V_{op0}, V_{on0}, and current I_{40}. A similar symbolic matrix was set-up for opamp circuits.

The integrator consists of two linear models, an OTA macrocell and an opamp-C macrocell. Figure 7 shows how the models for the composing cells are related to form the model for the integrator. Then, the same reasoning is used to calculate the model for an integrator chain. Assuming that V_i and I_f (the DAC output currents) are known, and that I_i and V_o are unknown, then following symbolic matrix will describe the 4-port behavior of an integrator:

$$I_{ip} = A_{p1}V_{ip} + A_{p2}V_{in} + A_{p3}I_{f1} + A_{p4}I_{f2} + A_{p5}(V_{ip0}, V_{in0}, V_{xp0}, V_{xn0})$$
$$I_{in} = A_{n1}V_{ip} + A_{n2}V_{in} + A_{n3}I_{f1} + A_{n4}I_{f2} + A_{n5}(V_{ip0}, V_{in0}, V_{xp0}, V_{xn0})$$
$$V_{op} = B_{p1}V_{ip} + B_{p2}V_{in} + B_{p3}I_{f1} + B_{p4}I_{f2} + B_{p5}(V_{ip0}, V_{in0}, V_{xp0}, V_{xn0})$$
$$V_{on} = B_{n1}V_{ip} + B_{n2}V_{in} + B_{n3}I_{f1} + B_{n4}I_{f2} + B_{n5}(V_{ip0}, V_{in0}, V_{xp0}, V_{xn0})$$

The values of the OTA and opamp state variables are updated over time using following equations:

$$V_{xp} = D_{p1}V_{ip} + D_{p2}V_{in} + D_{p3}I_{f1} + D_{p4}I_{f2} + D_{p5}(V_{ip0}, V_{in0}, Vxp0, V_{xn0})$$
$$V_{xn} = D_{n1}V_{ip} + D_{n2}V_{in} + D_{n3}I_{f1} + D_{n4}I_{f2} + D_{n5}(V_{ip0}, V_{in0}, V_{xp0}, V_{xn0})$$

where A_{p1}~A_{p4}, A_{n1}~A_{n4}, B_{p1}~B_{p4}, B_{n1}~B_{n4} are functions of $TA_{p(1\text{-}4)}$, $TA_{n(1\text{-}4)}$, $TB_{p(1\text{-}4)}$, $TB_{n(1\text{-}4)}$ (these are OTA macromodel parameters), and $SA_{p(1\text{-}2)}$, $SA_{n(1\text{-}2)}$, $SB_{p(1\text{-}2)}$, $SB_{n(1\text{-}2)}$ (these are opamp macromodel parameters). Similarly, the state functions A_{p5}, A_{n5}, B_{p5}, B_{n5} are related to the state variables TA_{p5}, TA_{n5}, TB_{p5}, TB_{n5} of the OTA, and the state variables SA_{p3}, SA_{n3}, SB_{p3}, SB_{n3} of the opamp.

During the Σ–Δ synthesis step [9,10] optimization process, the function sub-matrix is kept constant, once the macrocell parameters are instantiated. This avoids repeating a large set of computations during the lengthy time domain simulations needed to calculate SNR and DR of a solution point. Only the state sub-matrix must be updated for each new time step. The function sub-matrix is modified only when parameters are changed during the synthesis cycle. In our experience, this greatly reduced the total simulation time of a converter.

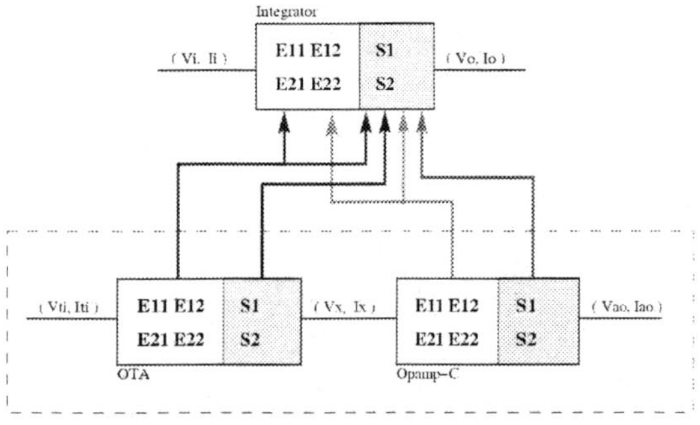

Figure 7: Integrator model

5. EXPERIMENTAL RESULTS

The suggested simulation method was successfully used for automated synthesis of 3rd and 4th order Σ–Δ ADC [9,10]. The goal was to maximize the ADC signal to noise ratio (SNR) and dynamic range (DR). The presented time-domain simulator took around 1 sec on a SUN Blade100 workstation, while HSPICE simulation (using much simpler circuit models) would still take about 10 time longer. The synthesis flow explored about 30,000 solution points in three days. For a sin wave signal of 625kHz at the input of the ΣΔ modulator, the maximum DR and the output spectrum are plotted in Figure 8. The maximum SNR is 69db, and DR is 77db. The design quality is similar to that of the modulator reported in [12]. However, the proposed method is more flexible than that in [12], because it does not assume a working design beforehand. Also, we were successful in synthesizing a higher-order converter, which is more challenging to design.

6.CONCLUSIONS

This paper presents a fast time-domain simulation method for automated synthesis of Σ–Δ ADC. At the system-level, the method uses a symbolic description of the ADC netlist. Building blocks in the netlist are macromodels, which include circuit non-idealities and nonlinear behavior. As compared to existing simulation methods for Σ–Δ ADC, this technique is fully automated and uses more complex circuit macromodels.

ACKNOWLEDGEMENTS

Hui Zhang acknowledges Yi Zhang for kindly helping him understanding the OTA circuit functioning, and setting-up the testbench for OTA performance measurement.

This work has been supported by Defense Advanced Research Projects Agency (DARPA) and managed by the Sensors Directorate of the Air Force Research Laboratory, USAF, Wright-Patterson AFB, OH 45433-6543.

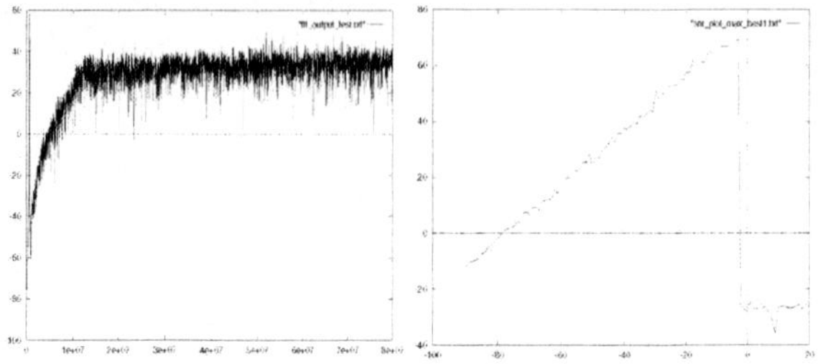

Figure 8: SNR and DR plots for ΣΔ ADC

REFERENCES

[1] G. J. Gomez, S. H. K. Embabi, E. Sanchez-Sinencio, M. C. Lefevre, ``*A Generic Parameterizable CMOS OTA Macromodel*'', IEEE Trans. C & S, 1995.

[2] A. Moscovici, ``*High Speed A/D converters - understanding data converters through SPICE*'', Kluwer, 1999.

[3] L. A. Connelly, P. Choi, ``Macromodeling with SPICE'', Prentice Hall.

[4] A.Doboli et al, ``Behavioral Synthesis of Analog Systems using Two-Layered Design Space Exploration'', Proc. of DAC, 1999.

[5] A. Doboli et al, ``A Regularity-based Hierarchical Symbolic Analysis Method for Large-scale Analog Networks'', IEEE Trans. C & S - II, No 11, 2001.

[6] S. Doboli, A. Doboli, ``Piecewise-Linear Modeling of Analog Circuits using Trained Feed-Forward Neural Networks and Adaptive Clustering of Hidden Neurons", Intern'l Joint Conf. on Neural Networks, 2003.

[7] S. Doboli, G. Gothoskar, A. Doboli, "Extraction of Piecewise-Linear Analog Circuit Models from Trained Neural Networks using Hidden Neuron Clustering", Proc. of DATE Conf., 2003.

[8] A. Doboli et al, ``Exploration-Based High-Level Synthesis of Linear Analog Systems Operating at Low/Medium Frequencies", IEEE Trans. CAD, No. 11, 2003.

[9] H. Tang, H. Zhang, A. Doboli, ``Towards High-Level Synthesis of Analog and Mixed-Signal Systems from VHDL-AMS Specifications - A Case Study for a Sigma-Delta Analog-Digital Converters'', Forum on Specification and Design Languages, 2003.

[10] H. Tang, H. Zhang, A. Doboli, ``Synthesis of Continuous-Time Filters and Analog to Digital Converters by Integrated Constraint Transformation, Floorplanning and Routing'', Proc. of GLSVLSI, 2003.

[11] J. Cherry, W. M. Snelgrove, ``Continuous-Time Delta-Sigma Modulators for High-Speed A/D Conversion'', Kluwer, 2000.

[12] K. Franken et al, ``DAISY: A Simulation-Based High-level Synthesis Tool for Sigma-delta Modulators'', Proc. ICCAD, 2002.

[13] G. Gielen, R. Rutenbar, ``Computer Aided Design of Analog and Mixed-signal Integrated Circuits'', Proc. of IEEE, No 12, 2000.

[14] M. Hershenson et al, "Optimal design of a CMOS op-amp via Geometric Programming", IEEE Trans. CAD, No.1, 2001.

[15] M. Krasnicki et al, ``MAELSTROM: Efficient Simulation-Based Synthesis for Custom Analog Cells'', Proc. DAC, 1999.

[16] P. Vancorenland et al, ``A Layout-Aware Synthesis Methodology for RF Circuits'', Proc. of ICCAD, 2001.

[17] G. DeMicheli, ``Synthesis and Optimization of Digital Circuits'', McGraw Hill, 1994.

[18] F. Medeiro, ``Top-Down Design of High Performance Sigma-Delta Modulators'', Kluwer, 1998.

An Efficient Algorithm For Transient and Distortion Analysis of Mildly Nonlinear Analog Circuits

Junjie Yang

Department of Electrical Engineering
University of California, Riverside, CA 92521, USA
jyang@ee.ucr.edu

Sheldon X.-D. Tan

Department of Electrical Engineering
University of California, Riverside, CA 92521, USA
stan@ee.ucr.edu

Abstract— *This paper presents an effective approach to transient and distortion analyses for mildly nonlinear analog circuits. Our method is based on Volterra functional series representation of nonlinear circuits. It computes the nonlinear responses using nonlinear current method which recursively solves a series of linear Volterra circuits to obtain linear and higher-order responses of a nonlinear circuit. The linear Volterra circuits are solved in frequency domain using an efficient graph-based technique, which can derive transfer functions of any large linear network very efficiently. The harmonic distortion can be directly obtained in the frequency domain while the transient responses are obtained via inverse Laplace transformation. The new algorithm takes advantage of the identical Volterra circuits for second and higher order response, which results in significant saving in deriving the transfer functions. Experimental results for a number of nonlinear circuits are obtained and compared with SPICE3 to validate the effectiveness of this method.*

I. Introduction

Transient analysis of nonlinear circuits is the most computationally intensive analysis. Linear multistep formulas (LMS) based on backward difference formulas [1] are widely used methods for transient simulation of nonlinear circuits due to their robustness. The so-called predictor-corrector algorithms where explicit LMS formulas for the predictor and an implicit LMS formula for the corrector can be used to further speed up the transient simulation. These methods are general enough for both wildly and hard nonlinear circuits. Since Newton-Raphson iterations are carried out at every time step of integration, these algorithms are, however, very time consuming. If only steady state response is required, some special analysis methods for nonlinear circuits were developed such as harmonic balance methods in frequency domain and shooting methods in time domain [2].

For wireless/communication applications, a number of circuits which operate at radio frequencies (RF) typically exhibits so-called mildly or weakly nonlinear properties where devices typically have a fixed dc operating point and the inputs are ac signals. When the amplitude of these input signals is small (such that their operation points do not change too much), the nonlinearities in these circuit can be approximated adequately using the truncated Taylor series expansion of the nonlinear devices at their dc operating points [3].

The mildly nonlinearities can be exploited to speed up the transient simulation for such nonlinear circuits [4]. Examples are linear centric method for nonlinear distortion analysis [5] and sampled-data simulation method using Volterra functional series [6]. Volterra functional series can represent a weakly nonlinear function in terms of a number of linear functions called Volterra kernels. From circuit theory's perspective, it leads to a set of linear circuits, called Volterra circuits, whose responses can adequately approximate the response of the original nonlinear circuit. In [6], a sampled-data simulation method is used where the simulation errors are dependent on sampling intervals and sampling window sizes. As a result, the runtime is dependent on the accuracy requirements. It is also difficult to obtain frequency domain information such as harmonic distortions as the algorithm operates in the time domain.

In this paper, we propose a new approach to transient and distortion analysis of mildly nonlinear analog circuits. Our method is also based on the Volterra functional series. But instead of solving the Volterra circuits in time domain as done by traditional methods like SPICE3 or by the sampled-data method [6], we solve the Volterra circuits in frequency domain by using a graph-based symbolic analysis method [7, 8]. Once frequency domain responses are obtained, transient responses can be obtained by fast numerical inverse Laplace transformation [1]. One important benefit for doing frequency domain analysis is that we can easily obtain the frequency-domain characteristics like harmonic distortions as they can be easily computed from the frequency responses of various order Volterra circuits. Experimental results for some real nonlinear circuits are studied and compared with SPICE3 to validate the new method. Both transient and second harmonic distortion (HD2) results are computed for each nonlinear circuit to show the effectiveness of the new method.

This paper is organized as follows: Section II reviews the concept of the Volterra circuit models for nonlinear circuits and DDD graphs. Section III presents the new simulation algorithm and illustrates the algorithm using a simple nonlinear circuit. Section IV gives the experimental results for a low noise RF amplifier and compares them with that of SPICE3. Section V concludes the paper.

II. Volterra Circuits and DDD Graphs

A. Volterra Circuits

A nonlinear circuit in time domain can be expressed by the following differential equations

$$Gv(t)+C\frac{dv(t)}{dt}=Dw(t)+i_{non}(v(t)) \quad (1)$$

Here G and C represent, respectively, the conductance and capacitance matrices whose elements are made of the linear devices and first-order terms of the Taylor series expansion of the nonlinear devices. D is the position vector for input $w(t)$ and $i_{non}(v(t))$ represents the second and higher-order currents generated by the nonlinear devices. By substituting Volterra functional series of $v(t)$ and $i(t)$ into the equation, we will obtain a set of linear differential equations [4, 6]:

$$\begin{aligned} Gv_1(t)+C\frac{dv_1(t)}{dt} &= Dw(t) \\ Gv_2(t)+C\frac{dv_2(t)}{dt} &= i_2(v_1(t)) \\ Gv_3(t)+C\frac{dv_3(t)}{dt} &= i_3(v_1(t),v_2(t)) \\ &\cdots \\ Gv_m(t)+C\frac{dv_m(t)}{dt} &= i_m(v_1(t),v_2(t),\ldots,v_{m-1}(t)) \end{aligned} \quad (2)$$

where $v_m(t)$ is the m-th order term of Volterra series expansion of $v(t)$ and $i_m(t)$ is the input of the m-th order Volterra circuit and can be obtained from lower-order responses: $v_{m-1}(t)$, $v_{m-2}(t)$,...,$v_1(t)$.

For a given nonlinear circuit, we assume that currents are nonlinear functions of voltages for nonlinear devices, the i-v characteristic of the nonlinear device can be expanded at the DC operating point as a Taylor series

$$\begin{aligned} I(V) &= I_0+i \\ &= f(V_0)+\sum_{n=1}^{\infty}\frac{f^n(V_0)}{n!}v^n = I_0+\sum_{n=1}^{\infty}a_nv^n \end{aligned} \quad (3)$$

Here I_0 and V_0 represent the DC current and voltage values over the nonlinear device, i and v represent the corresponding small signal voltage and current values respectively. Hence i can be expressed as a polynomial function of v with coefficients $a_i, i=1...n$. For Volterra functional series, if the input is changed from $v(t)$ to $\lambda v(t)$, we have [4]:

$$v(t)=\sum_{m=1}^{\infty}v_m(t)\lambda^m, i(t)=\sum_{m=1}^{\infty}i_m(t)\lambda^m \quad (4)$$

Substituting Eq.(3) into Eq.(4), we have

$$\sum_{m=1}^{\infty}i_m(t)\lambda^m=\sum_{n=1}^{\infty}a_n(\sum_{m=1}^{\infty}v_m(t)\lambda^m)^n. \quad (5)$$

Equating terms of the same order in λ, we have

$$i_m(t)=a_1v_m(t)+J_m(t), \quad (6)$$

where $J_m(t)$ is the contribution from the responses of low order (less than m) Volterra circuits. But for $m=1, J_m(t)=0$. Eq.(6) essentially reflects the fact that each Volterra circuit is a linear circuit (a_1 is used for all the Volterra circuits for the nonlinear device) and higher order responses can be computed in an order-increasing way starting from the first order. The response of the whole circuit will be the sum of the responses from all the Volterra circuits.

B. The DDD Graph based Method for Deriving Transfer Functions

In this subsection, we briefly review a graph-based method, called determinant decision diagrams (DDDs), to derive the exact transfer functions of a linear circuit [9].

Determinant Decision Diagrams [9] are compact and canonical graph-based representation of determinants. A DDD graph is similar to binary decision diagrams (BDDs) except that a sign is associated wit each node to represent the sign of product terms from expansion of determinants. Also like BDDs, DDDs are very capable of representing huge number of symbolic terms from a determinant. Most importantly, it can derive the s-expanded polynomial of a determinant symbolically via *s*-expanded DDDs [7]. The recent hierarchical approach using DDD graphs can essentially derive transfer functions for almost arbitrary large networks [8], which makes the solving of linear networks in frequency domain much easy and efficient.

III. New Approach to Transient Analysis of Nonlinear Circuits

From previous analysis on Volterra functional series, we know that all Volterra circuits are linear circuits similar to the original circuit. All second and higher-order Volterra circuits are the same except the input current sources are different.

As a result, the new method consists of following steps to obtain the transient response and harmonic distortions of a nonlinear circuit: (1) Based on the nonlinear analytical expressions of i-v curve for each nonlinear device in the nonlinear circuit, derive the corresponding relationship between i_m and v_m for each of them and generate the corresponding Volterra circuits; (2) Compute the transfer functions for each Volterra circuit using the DDD-based method. Note that only two linear circuits are required, i.e. the circuit for the first order response and the circuit for the higher order responses. (3) Add all the frequency/transient responses of different-order Volterra circuits to obtain the frequency/transient responses of the original nonlinear circuit. Specific harmonic distortions can be obtained by computing the transfer functions in the Volterra circuits of different orders.

In the following we use a simple nonlinear circuit to demonstrate the new method. The circuit is shown on the top part of Fig. 1 which consists of a nonlinear diode. The devices take the following values: $R=20k\Omega, Is=1\times 10^{-6}A, C_0=1.0\times 10^{-12}F$, where Is is the saturation current of the diode. The input is $V_{in}=0.7+0.1sin(2\pi ft)$ at $f=1Mhz$. For a diode, we have the following equation characterizing its i-v relationship:

$$i_D=I_s(e^{\frac{v_D}{V_T}}-1) \quad (7)$$

where i_D and v_D are the current and voltage of the diode respectively, and V_T is the thermal voltage. Let i_d and v_d be the AC components of i_D and v_D, then we have the following Taylor expansion

$$i_d=I_D[\frac{v_d}{V_T}+\frac{v_d^2}{2V_T^2}+\frac{v_d^3}{6V_T^3}+\frac{v_d^4}{24V_T^4}+\frac{v_d^5}{120V_T^5}+\ldots] \quad (8)$$

where $I_D=I_S(e^{\frac{V_D}{V_T}}-1)$. The simulated DC values are $V_D=0.08926V, I_D=30.385uA$. Replace i_d and v_d with its Volterra series, we obtain

$$i_{d,1} = \frac{I_D}{V_T}v_{d,1} \quad (9)$$

$$i_{d,2} = \frac{I_D}{V_T}v_{d,2}+\frac{1}{2V_T^2}v_{d,1}^2I_D \quad (10)$$

$$i_{d,3} = \frac{I_D}{V_T} v_{d,3} + \left(\frac{1}{2V_T^2}(2v_{d,1}v_{d,2} + \frac{1}{6V_T^3}v_{d,1}^3\right) I_D \tag{11}$$

$$i_{d,4} = \frac{I_D}{V_T} v_{d,4} + \left(\frac{1}{2V_T^2}(2v_{d,1}v_{d,3} + v_{d,2}^2)\right.$$
$$\left. + \frac{1}{6V_T^3}(3v_{d,1}^2 v_{d,2}) + \frac{1}{24V_T^4}v_{d,1}^4\right) I_D \tag{12}$$

$$\dots$$

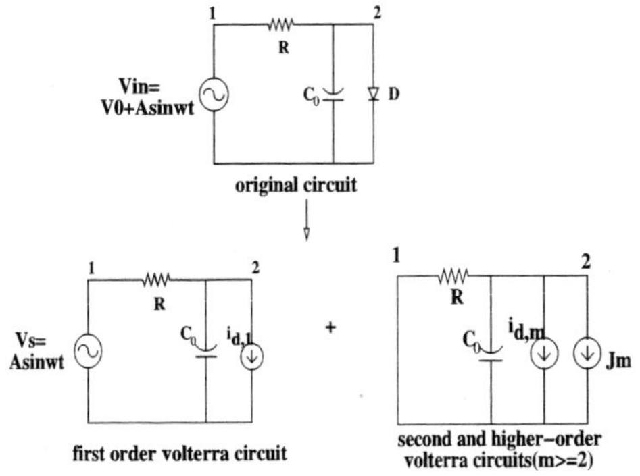

Fig. 1. A simple diode circuit and its Volterra circuits

The Volterra circuits of different orders are shown in Fig. 1. For the first-order Volterra circuit, the voltage transfer function from the input node 1 to the output node 2 is computed by DDD-based method as

$$H_{12}(s) = \frac{V_2(s)}{V_1(s)} = \frac{-5.0 \times 10^{-5}}{0.00123 + 1.0 \times 10^{-12}s} \tag{13}$$

For the second order or higher order Volterra circuits, the impedance driving-point function is computed from node 2 to node 2 as $J_m(t)$ becomes the new stimulus. As a result, we have

$$H_{22}(s) = \frac{V_2(s)}{J_2(s)} = \frac{1.0}{0.00123 + 1.0 \times 10^{-12}s} \tag{14}$$

After this, we use inverse Laplace transformation to obtain the transient response for each of the Volterra circuits.

$$v_{d,1}(t) = \mathcal{L}^{-1}(V_{in}(s)H_{12}(s))$$
$$J_2(s) = \mathcal{L}\left(\frac{I_D}{2V_T^2}(\mathcal{L}^{-1}(V_{in}(s)H_{12}(s)))^2\right)$$
$$v_{d,2}(t) = \mathcal{L}^{-1}(J_2(s)H_{22}(s))$$
$$\dots$$
$$V_{out}(s) = V_{in}(s)H_{12}(s) + J_2(s)H_{22}(s) + J_3(s)H_{22}(s) + \dots$$

The transient response at node 2 is shown in Fig. 2.

For distortion analysis, let's consider the second harmonic distortion (HD2) since it is the most dominant harmonic distortion in most cases. It is defined as follows:

$$HD2 = 20log\left(\frac{|A_{2w_0}|}{|A_{w_0}|}\right), \tag{15}$$

where A_{w_0} is the amplitude of the input signal and A_{2w_0} is the second harmonic amplitude at the output node which can be easily computed from the second order Volterra circuit. The computed HD2 is shown in Fig. 3. Similarly, we can obtain HD3, HD4 ... etc.

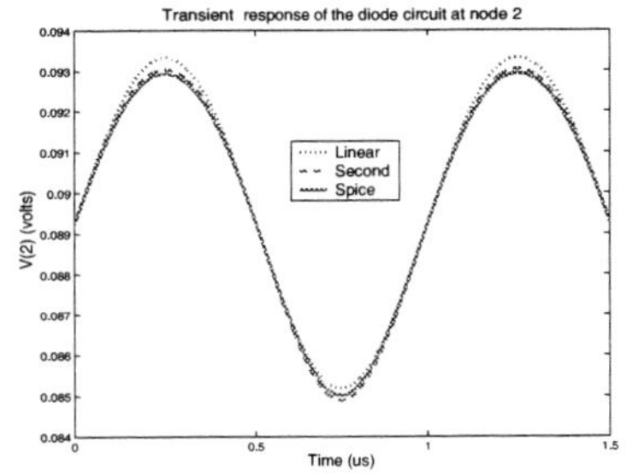

Fig. 2. Transient response for the diode circuit

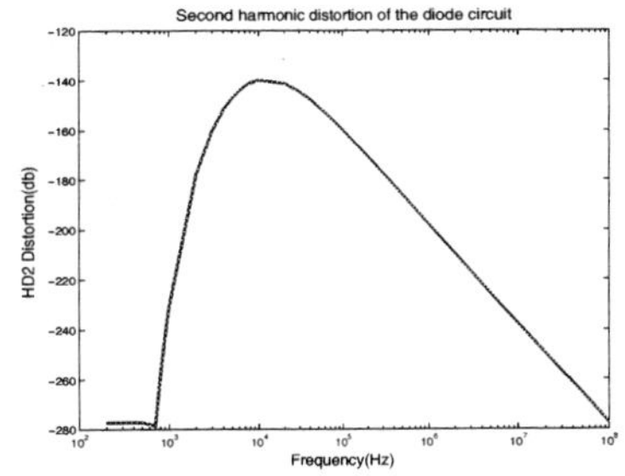

Fig. 3. Second harmonic distortion (HD2) for the diode circuit

IV. Experimental Results

The proposed algorithm has been implemented using C++. We simulate a number of nonlinear analog circuits using the new algorithm. The experimental results are collected on a PC with 2.4*Ghz* P-IV CPU and 484MB memory. But due to the space limitation, we only report the detailed simulation results for one bipolar low-noise amplifier (LNA) shown in Fig. 4 in the following.

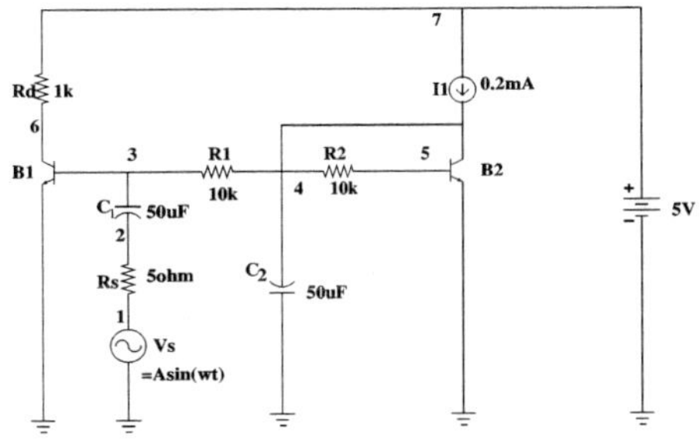

Fig. 4. A simple low-noise amplifier circuit

Firstly, we obtain the DC conditions from SPICE3 for the nonlinear device. We get $V_B = V(3) = 0.73V, V_C = V(6) = 4.8V, I_{C1} = 0.19mA$, and then the AC parameters are computed as follows: $r_b = 1.0\Omega, r_\pi = 13.16k\Omega, C_u = 133.8fF, C_\pi = 20.66fF, g_m = \frac{I_C}{V_t} = 0.0076A/V, r_o = 510k\Omega$. The AC equivalent circuit is shown in Fig. 5 along with the second and higher order Volterra circuits.

By using DDD-based method, we obtain all the required transfer functions

$$H_{14}(s) = \frac{V_4(s)}{V_1(s)} = \frac{0.0015 - 4.16 \times 10^{-15}s}{0.0002 + 4.60 \times 10^{-15}s + 4.58 \times 10^{-27}s^2},$$

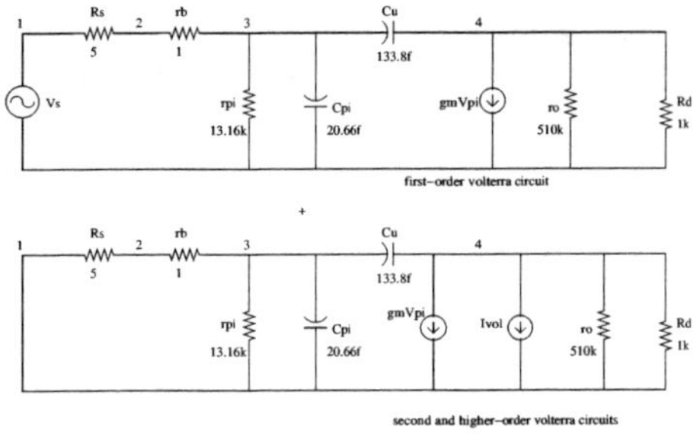

Fig. 5. The Volterra circuits for the low-noise amplifier circuit

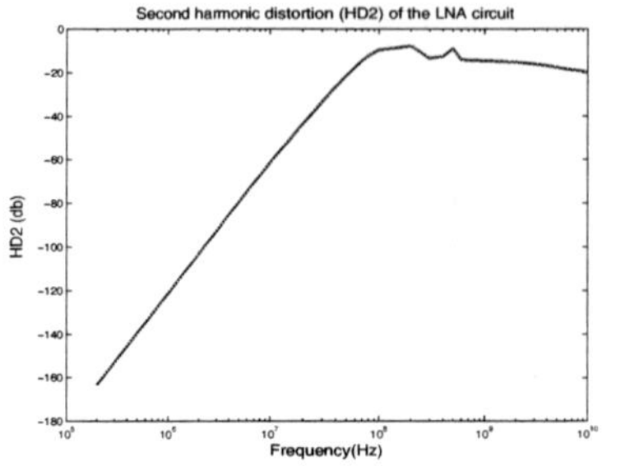

Fig. 7. Second harmonic distortion (HD2) for the LNA circuit

$$H_{13}(s) = \frac{V_3(s)}{V_1(s)} = \frac{0.0002 + 4.16 \times 10^{-15} s}{0.0002 + 4.60 \times 10^{-15} s + 4.58 \times 10^{-27} s^2},$$

where $H_{14}(s)$ and $H_{13}(s)$ are the transfer functions from input node 1 to output node 4 and node 3 respectively. For the second and higher-order Volterra circuit, we have

$$H_{44}(s) = \frac{V_4(s)}{I_{vol}(s)} = \frac{0.20 + 2.45 \times 10^{-13} s}{0.0002 + 4.60 \times 10^{-15} s + 4.58 \times 10^{-27} s^2},$$

$$H_{43}(s) = \frac{V_3(s)}{I_{vol}(s)} = \frac{-2.496e-14s}{0.0002 + 4.60 \times 10^{-15} s + 4.58 \times 10^{-27} s^2},$$

where $H_{44}(s)$ and $H_{43}(s)$ are the transfer functions from Volterra current source of different orders to output node 4 and node 3 respectively, I_{vol} is the current source for different order Volterra circuits.

Following the same calculations as shown for the diode circuit, we obtain the transient response shown in Fig. 6. The SPICE3 simulated results are also shown and compared with our results. It can be clearly seen that the results are in good agreement with each other when the second and the third order Volterra circuits are considered. HD2 distortion is also calculated and shown in Fig. 7.

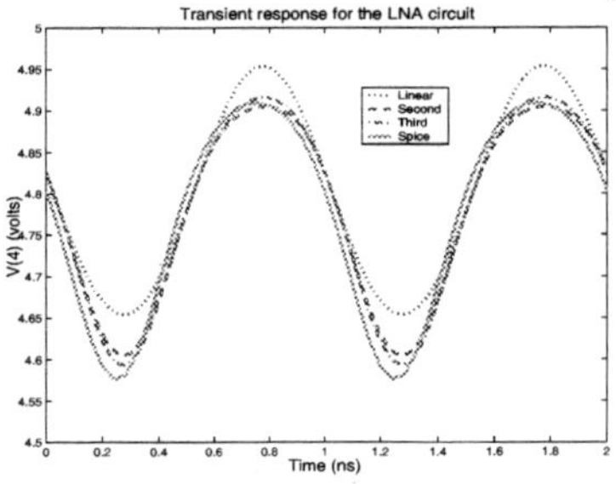

Fig. 6. Transient response for the LNA circuit

The whole computation for deriving transient response takes 1.1 seconds for the new algorithm. While SPICE3 take 2.0 seconds to finish the same task. We notice that if we increase the time interval and number of time steps, SPICE3 simulation time will go up accordingly, while the new method will still take the same time (1.1 seconds) as the new approach computes responses in the frequency domain and is independent of time steps and time intervals. Note that if truncation is carried out for very high order transfer functions, Hurwitz polynomial [10] can be applied to enforce the stability of the transfer functions if stability is required.

V. Conclusion

In this paper, we have proposed a new approach to transient and distortion analyses of mildly nonlinear analog circuits. The new method is based on Volterra functional series. But instead of solving the Volterra circuits numerically in time domain as traditional methods do, we use a graph-based symbolic method to obtain the responses of Volterra circuits of various orders in frequency domain directly. The new method exploits the identical Volterra circuit structures for higher order nonlinear responses and the efficiency of determinant decision diagrams based method for deriving transfer functions. A number of nonlinear analog circuits have been simulated using the new method and the results have been compared with that of SPICE3 to demonstrate the effectiveness of the proposed method.

References

[1] J. Vlach and K. Singhal, *Computer Methods for Circuit Analysis and Design*. New York, NY: Van Nostrand Reinhold, 1995.

[2] K. S. Kundert, J. K. White, and A. Sangiovanni-Vincentelli, *Steady-State Methods for Simulating Analog and Microwave Circuits*. Kluwer Academic Publishers, 1990.

[3] P. Wambacq and W. Sansen, *Distortion Analysis of Analog Integrated Circuits*. Kluwer Academic Publishers, 1998.

[4] M. Schetzen, *The Volterra and Wiener Theory of Nonlinear Systems*. New York: Wiley, 1981.

[5] P. Li and L. Pileggi, "Nonlinear distortion analysis via linear-centric models," in *Proc. Asia South Pacific Design Automation Conf. (ASPDAC)*, pp. 897–903, 2003.

[6] F. Yuan and A. Opal, "An efficient transient analysis algorithm for mildly nonlinear circuits," *IEEE Trans. on Computer-Aided Design of Integrated Circuits and Systems*, vol. 21, no. 6, pp. 662–673, 2002.

[7] C.-J. Shi and X.-D. Tan, "Compact representation and efficient generation of s-expanded symbolic network functions for computer-aided analog circuit design," *IEEE Trans. on Computer-Aided Design of Integrated Circuits and Systems*, vol. 20, pp. 813–827, April 2001.

[8] S. X.-D. Tan, "A general s-domain hierarchical network reduction algorithm," in *Proc. Int. Conf. on Computer Aided Design (ICCAD)*, pp. 650–657, 2003.

[9] C.-J. Shi and X.-D. Tan, "Canonical symbolic analysis of large analog circuits with determinant decision diagrams," *IEEE Trans. on Computer-Aided Design of Integrated Circuits and Systems*, vol. 19, pp. 1–18, Jan. 2000.

[10] S. X.-D. Tan and J. Yang, "Hurwitz stable model reduction for non-tree structured RLCK circuits," in *IEEE Int. System-on-Chip Conf. (SOC)*, pp. 239–242, 2003.

GENERATION OF DISJOINT CUBES FOR MULTIPLE-VALUED FUNCTIONS

Bogdan J. Falkowski   Cicilia C. Lozano

School of Electrical and Electronic Engineering
Nanyang Technological University
Block S1, 50 Nanyang Avenue, Singapore 639798

Susanto Rahardja

Institute for Infocomm Research
21 Heng Mui Keng Terrace
Singapore 119613

ABSTRACT

In this paper, an algorithm to generate disjoint cubes representation of a multiple-valued logic function is presented. The algorithm converts an array of non-disjoint multiple-valued cubes to an array of disjoint multiple-valued cubes by applying different multiple-valued cube calculus operations. Efforts to reduce the calculation time and the number of resulting disjoint cubes have been integrated into the presented algorithm. Experimental results of the algorithm for several test files are also shown.

1. INTRODUCTION

Different reduced representations of logic functions have been developed in order to simplify and improve logic minimization algorithms. One of such important representations is an array of cubes that in worst case when all the cubes are minterms is equal to the truth vector of a given logic function. Many logic minimization algorithms require operations on cubes that are disjoint i.e. they do not have any overlapping minterms [1-4]. Such representations are also important for efficient calculations of various spectra of logic functions [5]. Preprocessing of an input data to generate disjoint cubes representation of a Boolean function is very useful since number of disjoint cubes is usually much less than number of minterms and hence the logic function can be stored in much reduced size [1, 4]. Some algorithms have been developed to convert an array of non-disjoint cubes of a binary function to its equivalent array of disjoint cubes [5, 6]. Recent developments in efficient calculation of spectra for multiple-valued functions [7] also require efficient algorithms to calculate disjoint cubes for multiple-valued logic functions. It should be noticed that there is an increasing research interest in multiple-valued logic design due to the fact that multiple-valued circuits require less number of interconnections that is an important problem in current Ultra Large Scale Integration (ULSI) circuits.

In this paper, an algorithm that generates an array of disjoint cubes from an array of non-disjoint cubes representing a multiple-valued input multiple-valued output logic function is given. The algorithm uses several basic cube calculus operations described in [1-4] and modified here for multiple-valued case. The presented approach applies to any multiple-valued logic function and experimental results for most important multiple-valued cases: ternary and quaternary for several test files are also shown.

2. BASIC DEFINITIONS

Definition 1: Let $\boldsymbol{X} = (X_1, X_2, \ldots, X_n)$ be a set of n multiple-valued variables, then an n-variable multiple-valued input p-multiple-valued output function can be defined as a mapping $f(\boldsymbol{X}) : R_1 \times R_2 \times \ldots \times R_n \rightarrow P$, where X_i, $1 \leq i \leq n$ takes the values from the set $R_i = \{ 0,1, \ldots, r_i\text{-}1 \}$ and $P = \{ 0,1,\ldots,p-1,- \}$ (where '–' denotes a don't care value).
If $R_1 = R_2 = \ldots = R_n = \{0, 1, \ldots, r\text{-}1 \}$ then $f(\boldsymbol{X})$ is said to be an n-variable r-valued input p-multiple-valued output logic function.

Definition 2: A literal of multiple-valued input variable X_i, denoted by X_i^{Si}, is defined as follows:

$$X_i^{S_i} = \begin{cases} 1 \text{ if } X_i \in S_i \\ 0 \text{ if } X_i \notin S_i \end{cases}$$

where $S_i \subseteq R_i$. S_i is called true set of literal X_i.

Definition 3: A cube is a product of literals, $X_1^{S_1}, X_2^{S_2}, \ldots, X_j^{S_j}$, $j \leq n$. A minterm is a cube that includes literals of all function variables $X_1, X_2, \ldots, X_n$ in which every set S_i contains only one element.

Definition 4: A cube intersection operation is defined as

$$A \cap B = \begin{cases} X_1^{S_1A \cap S_1B} X_2^{S_2A \cap S_2B} \ldots X_n^{S_nA \cap S_nB} \text{ if } S_iA \cap S_iB \neq \varnothing \text{ for all } i \\ \varnothing \text{ otherwise} \end{cases}$$

where A and B are cubes and S_iA and S_iB are true sets of literal X_i in A and B, respectively.

Definition 5: Two cubes A and B are disjoint if $A \cap B = \varnothing$. Otherwise, A and B are non-disjoint (overlapped).

Definition 6: Disjoint sharp operation on cubes A and B, $A \#_d B$ is defined as

$$A \#_d B = \begin{cases} A \text{ if } A \cap B = \varnothing \\ \varnothing \text{ if } A \subseteq B \\ K \text{ otherwise} \end{cases},$$

where K is a set of disjoint cubes.
Let $M = \{i \mid S_iA \not\subset S_iB\}$ where the members of M are arranged in ascending order and $1 \leq i \leq n$. Then the number of disjoint cubes in K is equal to the number of elements in M.
If m_j is the j-th element of M then the j-th disjoint cube of K is:

$$K_j = X_1^{S_1A} \ldots X_{m_j-1}^{S_{m_j-1}A} X_{m_j}^{S_{m_j}A \cap (\bar{S}_{m_j}B)} X_{m_j+1}^{S_{m_j+1}A \cap S_{m_j+1}B} \ldots X_n^{S_nA \cap S_nB}$$

Example 1: Let cubes A and B be $X_1^{\{0,2,3\}} X_2^{\{1,2\}} X_3^{\{1\}}$ and $X_1^{\{1,2\}} X_2^{\{2,3\}} X_3^{\{1\}}$, respectively. Then according to Definition 6, $M = \{1, 2\}$ and $K = \{ X_1^{\{0,3\}} X_2^{\{2\}} X_3^{\{1\}}, X_1^{\{0,2,3\}} X_2^{\{1\}} X_3^{\{1\}} \}$.

Definition 7: Cube A is said to absorb cube B if $A \cap B = B$.

3. ALGORITHM TO GENERATE DISJOINT CUBES

The algorithm presented here takes an array of non-disjoint cubes representing a multiple-valued function as its input. The cubes can be coded using integer numbers (e.g. cube 12-0, which represents three minterms $X_1^1X_2^2X_3^0X_4^0$, $X_1^1X_2^2X_3^1X_4^0$ and $X_1^1X_2^2X_3^2X_4^0$ for ternary functions) or by recoding the integer numbers into their binary positional notation [2] (e.g. 010 001 111 100 to represent the same ternary cube).

The disjoint cubes generation algorithm for any *p*-multiple-valued output logic function is given in Figure 1.

The variables and procedures used in the algorithm are:

- *InArray* = the array used to store the input to the algorithm that are in the form of non-disjoint cubes.
- *Disjoint*= the array used to store the output of the algorithm that are in the form of disjoint cubes.
- *Cube1* and *Cube2* = *Disjoint* index numbers of the pair of cubes being compared.
- *n* = contains current number of cubes in *Disjoint.*
- *n*1, *n*2 ,..., *n*<*p*-1> = contain number of disjoint cubes for output value 1, 2 ... , *p*-1 respectively.
- *d* = number of disjoint cubes resulting from *Cube2* $\#_d$ *Cube1.*
- *D* = a cube array that stores the results of *Cube2* $\#_d$ *Cube1.*
- *i, j, k* = looping variables.
- *Sort(a)* = the procedure that sorts the cubes having output value *a* in *Disjoint* such that cubes having more missing literals are located in front of (having smaller index numbers than) cubes with fewer number of missing literals.
- *Disjoints(b)* = the procedure that processes the non-disjoint cubes with output value *b* in *Disjoint* array and generates their equivalent disjoint cubes representation using the cube calculus operations described in Section 2. This procedure updates the value of the variable *n* accordingly.
- *Merge(c)* = the procedure that scans the disjoint cubes produced by *Disjoints(c)* and merges some of the cubes, if possible. As such, this procedure also changes the values of *n* and *n*<*c*> variables.

It can be seen that the algorithm processes non-disjoint cubes having different output values separately. Because of this sequential way of processing, at the end of the program the disjoint cubes in *Disjoint* array are ordered in such a way that the first disjoint cube in *Disjoint* is located at *Disjoint*[0] and disjoint cubes having a particular output value *i* are located between *Disjoint* [*n*1 + *n2* + ... + *n*<*i* -1> -1] and *Disjoint* [*n*1+*n*2+...+ *n*< *i* >-1]. Thus, it is not necessary to store the output values of the disjoint cubes. The output value of a particular disjoint cube in *Disjoint* array can be calculated from the index value of the disjoint cube and values of *n*1, *n*2, ... and *n*<*p*-1>. Such an approach reduces memory space required for storage.

The execution time of the algorithm increases with the number of non-disjoint cubes which is partly due to the increasing number of cube calculus operations done within function *Disjoint( )*. In order to speed up the execution time of function *Disjoint( )* partitioning can be implemented as has been done for binary case [8]. With partitioning, when the number of non-disjoint cubes having a particular output value is greater than a certain value called *part_threshold* the non-disjoint array is partitioned into *p* sub-arrays when an appropriate partitioning variable is found. As long as the size of the largest sub-array is greater than a number called *size_limit* and appropriate partitioning variable exists, each sub-array is further divided into *p* sub-arrays. The total number of non-disjoint cubes with don't care as literal for the partitioning variable must be less than the value of constant *dcmin.* At the end of partitioning, the non-disjoint cubes are divided into sub-arrays whereby cubes located at different sub-arrays are guaranteed to be disjoint of each other and so the *Disjoint( )* function are applied separately for each sub-array. Since the number of non-disjoint cubes in a sub-array is smaller than the total number of non-disjoint cubes, the total number of cube pairs to be considered within *Disjoint( )* function is also fewer. This leads to smaller number of cube calculus operations involved, which in turn leads to shorter execution time.

Figure 1. Disjoint cubes generation algorithm for *p*-valued output logic functions.

```
For each output i
{ Initialize all variables to zero;
  For each nonzero output value j
  { For each  cube in InArray
    { if the cube has output value j for output i)
     { Copy the cube to Disjoint[n];
       n++; }}
  if(no. of cubes with output value j for output i is greater than 1)
   { Disjoints(j);
     n<j> = n – (n1 + n2 + ... + n<j-1>);
     Merge(j); }
  if(j is not the last nonzero output value)
     Cube1 = n;
  else
     exit the program;}}
--------------------------------------------------------------------------------
Disjoints(b)
{ For Cube1 =Cube1 to n-2
  { Sort(b);
    Cube2=n-1;
    while(Cube2 != Cube1)
    { calculate D and d of Cube2#d Cube1;
       if(d = 0) //Cube 1 absorbs Cube 2
      { for(k = Cube2+1 to n-1)
          Disjoint[k-1]=Disjoint[k];
        n=n-1;
       Cube2 = Cube2-1;}
      Else if(d = 1 and D[0]=Disjoint[Cube2])
        Cube2=Cube2-1;
      Else
      { for(k = Cube2+1 to n-1)
          Disjoint[k+d-1]=Disjoint[k];
        For(k = 0 to d-1)
          Disjoint[Cube2+k]=D[k];
        Cube2=Cube2-1;
        n=n+d-1;}}} }
```

4. IMPLEMENTATION OF THE ALGORITHM

The algorithm described in Section 3 has been implemented using Microsoft Visual C++. In the program, each cube is coded using integer numbers and implemented as a Cube object. The *Disjoint* array is implemented simply by giving each Cube pointers that point to the next and previous Cube in the array. The properties of the array itself are maintained by a Cubelist object that stores the variables *n*1, *n*2, *n*3, ...*n*<*p*-1> and the pointers to the first and last Cube in the array.

One advantage of implementing the *Disjoint* array as a list of pointers to Cube object is that it simplifies the process of removing and inserting a cube to *Disjoint* (which is often required in function *Disjoints*) as well as changing the order of the cubes in *Disjoint* (which is necessary for *Sort( )* function). All of them can be done by simply updating the next and previous properties of affected Cube objects and deleting unused Cube objects so no copying of cube literals are needed. Another advantage is that it is easier to integrate various variables associated with a cube by adding the variables as properties of the Cube object. One example of such a variable is *merge_indicator*. It is a variable that is used to help speed up the execution of *Merge( )* function.

In case of binary functions, the *Merge( )* function needs to consider only one pair of disjoint cubes at a time to see whether they are adjacent and hence can be merged. In multiple-valued functions, the *Merge( )* function needs to compare a set of *p* disjoint cubes at one time. Only when a set of *p* disjoint cubes that are adjacent to each other at a particular position is found then a merging can happen. This means that the number of comparisons involved in *Merge( )* is increasing with *p*, which translates to longer execution time.

The steps of the function *Merge(c)* are given below. In the function description, the variable *cubeptr[ ]* is a cube pointer array of size *p* that stores the addresses of the cubes that may possibly be merged, hence '*cubeptr[k]*' can be read as 'the cube whose address is stored in *cubeptr[k]*'. Also, '*cubeptr[k]* and *cubeptr[l]* are adjacent at *position*' means that the literals of *cubeptr[k]* and *cubeptr[l]* are same for all variables except for the *position*-th variable.

Step 1: *Set cubeptr[0] to the first disjoint cube with output value c in the Cubelist; i = 0;*
Step 2: *If cubeptr[i] is not the last cube in Cubelist: go to step 3;*
else: go to step 11;
Step 3: *If i = 0:go to step 4;*
else if i = 1:go to step 6;
else if i = p – 1: go to step 7;
else: go to step 8;
Step 4: *cubeptr[0]->merge_indicator =2;*
cubeptr[1]=cubeptr[0]->next;
if cubeptr[1] is not the last cube in Cubelist:go to step 5;
else: {i= 1;go to step 2};
Step 5: *If cubeptr[0]->merge_indicator=2 and cubeptr[0] and cubeptr[1] are adjacent: cubeptr[0]->merge_indicator=1;*
If cubeptr[1]->merge_indicator=2: cubeptr[1]=cubeptr[1]->next;
i= 1;go to step 2;
Step 6: *If cubeptr[0] & cubeptr[1] are adjacent:*
{ position = the different literal position;go to step 10;}
else: go to step 9;
Step 7: *If cubeptr[i=1] & cubeptr[i] are adjacent at position:*
{ change cubeptr[0]->literal[position] to '-';
move cubeptr[0] to the front of all disjoint cubes with output value c;
remove & delete cubeptr[1] to cubeptr[p-1];
adjust n<c>;i=0;go to step 2;}
else: go to step 9;
Step 8: *if cubeptr[i=1] & cubeptr[i] are adjacent at position:*
{if cubeptr[i]->merge_indicator=2:
{cubeptr[I]=cubeptr[I]->next; i=1; go to step 2;}
else:go to step 10;}
else:go to step 9;
Step 9 : *cubeptr[i]->cubeptr[i]->next; go to step 2;*
Step 10: *cubeptr[i+1]=cubeptr[i]->next;i=i+1;go to step 2;*
Step 11: *if i > 1:{ i =1; cubeptr[1]=cubeptr[1]->next;*
go to step 2;}
else if i = 1: { i=0; cubeptr[0]=cubeptr[0]->next;
go to step 2;}
else: exit Merge(c);

Basically what the *Merge( )* function does is that it first compares *cubeptr[0]* and *cubeptr[1]* to see if they are adjacent. If they are, then a set of *p*-2 disjoint cubes that can be merged with *cubeptr[0]* and *cubeptr[1]* can be generated. The *Merge( )* function then tries to find all the members of the set one by one. If all *p*-2 components of the set exist then the disjoint cubes in the set together with *cubeptr[0]* and *cubeptr[1]* are merged and the merged cube is put in front of the Cubelist. Otherwise the cubes *cubeptr[0]* and *cubeptr[1]* cannot be merged and next pair of *cubeptr[0]* and *cubeptr[1]* is considered.

The value of *merge_indicator* ranges from 0 to 2. At the instance when a Cube object is created, the variable value is set to 0. As soon as Cube object is pointed by *cubeptr[0]* its *merge_indicator* value is changed to 2. Finally when a Cube object that is adjacent to the Cube object is found the value of its *merge_indicator* is set to 1. Let the *cubeptr[0]*, *cubeptr[1]* and the *p*-2 disjoint cubes that can be merged with them belong to the set *P*. Since a Cube object with *merge_indicator* value 2 is not adjacent to any of the Cube objects located behind it, if i ($2 \le i \le p$-1) members of the set *P* have been found and the *merge_indicator* of the latest member of the set *P* that has been found is 2, then it can be concluded straight away that no other components of the set *P* can be found in the Cubelist and hence the search for the remaining disjoint cubes of set *P* can be skipped. By skipping the search for the rest of the *P* members, the number of unnecessary comparisons is reduced which improves execution time.

5. EXPERIMENTAL RESULTS

The Microsoft Visual C++ implementation of the algorithm has been run on several binary benchmark functions that have been modified to represent multiple-valued (ternary and quaternary) functions instead of original benchmark binary functions. The translation from binary to ternary or quaternary cases has been done by changing every 2 input (output) bits in binary files to an input (output) symbol in multiple-valued files. If the number of input and /or output variables is odd, then a zero bit is added behind the binary cubes to make it even. For input (output), -- is taken as -, 00 is taken as 0, 01 is taken as 1 and 10 is taken as 2, whereas 11 is taken as 3 for quaternary case (p = 4) and not used (taken as 0) for ternary case (p = 3). With these conversions, the binary benchmark files become an array of ternary or quaternary cubes and/or minterms.

Notice that for the binary cubes that contain unpaired missing literal, the missing literal is replaced by the corresponding 0 and 1 literals first before converting them to the ternary or quaternary minterms or cubes.

Table 1 shows the number of non-disjoint cubes for each multiple-valued output for several binary benchmarks when they are translated to ternary files. Table 2 lists the final number of '1' and '2' disjoint cubes (*n*1 and *n*2) of each output of the same ternary test files. The same results for quaternary case are given in Tables 3 and 4. Similar to the binary results discussed in [1], it can be seen from the tables that the total number of disjoint cubes are generally much less than the number of non-disjoint cubes for both ternary and quaternary cases.

Table 1: Numbers of non-disjoint '1' and '2' cubes of several benchmarks when they are used as ternary test files.

Input files	Output 1		Output 2		Output 3		Output 4		Output 5	
	*n*1	*n*2	*n*1	*n*2	*n*1	*n*2	*n*1	*n*2	*n*1	*n*2
9sym	0	153	-	-	-	-	-	-	-	-
clip	88	60	120	90	0	76	-	-	-	-
con1	20	8	-	-	-	-	-	-	-	-
ex1010	29	34	40	23	26	27	26	32	24	31
inc	14	14	15	18	10	24	5	8	0	6
misex1	6	1	11	8	7	7	0	11	-	-
squar5	6	0	6	4	5	4	3	6	-	-
z5xp1	21	0	18	13	15	14	12	18	18	18
z9sym	0	120	-	-	-	-	-	-	-	-
alu4	111	15	862	66	570	985	418	260	-	-
b12	36	20	24	1384	12	52	626	716	0	82
t481	0	3856	-	-	-	-	-	-	-	-

Table 2: Numbers of resulting disjoint '1' and '2' cubes for several ternary test files.

Input files	Output 1		Output 2		Output 3		Output 4		Output 5	
	*n*1	*n*2	*n*1	*n*2	*n*1	*n*2	*n*1	*n*2	*n*1	*n*2
9sym	0	62	-	-	-	-	-	-	-	-
clip	42	23	37	40	0	14	-	-	-	-
con1	13	8	-	-	-	-	-	-	-	-
ex1010	25	34	38	23	26	25	26	30	22	31
inc	8	6	13	14	6	12	5	6	0	2
misex1	5	1	9	7	6	6	0	7	-	-
squar5	2	0	6	4	3	4	1	2	-	-
z5xp1	5	0	18	5	15	14	4	6	2	2
z9sym	0	72	-	-	-	-	-	-	-	-
alu4	50	29	117	100	119	139	66	83	-	-
b12	12	8	24	50	18	8	38	76	0	94
t481	0	1009	-	-	-	-	-	-	-	-

Table 3: Numbers of non-disjoint '1', '2' and '3' cubes of several benchmarks when they are used as quaternary test files.

Input files	Output 1 *n*1/*n*2/*n*3	Output 2 *n*1/*n*2/*n*3	Output 3 *n*1/*n*2/*n*3	Output 4 *n*1/*n*2/*n*3
9sym	0/570/0	-	-	-
clip	326/148/0	292/251/0	0/148/0	-
con1	26/20/0	-	-	-
misex1	12/2/0	11/11/0	14/9/0	0/12/0
squar5	7/5/4	9/6/5	8/8/4	8/8/0
z9sym	0/420/0	-	-	-
alu4	301/106/0	1673/1245/0	3112/5674/0	2593/1074/0

Table 4: Numbers of resulting disjoint '1', '2' and '3' cubes for several quaternary test files.

Input files	Output 1 *n*1/*n*2/*n*3	Output 2 *n*1/*n*2/*n*3	Output 3 *n*1/*n*2/*n*3	Output 4 *n*1/*n*2/*n*3
9sym	0/207/0	-	-	-
clip	157/64/0	112/124/0	0/73/0	-
con1	22/20/0	-	-	-
misex1	8/2/0	11/9/0	10/8/0	0/8/0
squar5	7/5/4	9/6/5	5/8/4	2/2/0
z9sym	0/210/0	-	-	-
alu4	125/236/0	824/837/0	857/1031/0	597/419/0

6. CONCLUSIONS

A new efficient algorithm and its implementation for the conversion of non-disjoint cubes representation of a multiple-valued input multiple-valued output function into its equivalent disjoint cubes representation have been presented and some experimental results have been shown. The algorithm integrates efforts to minimize the number of resulting disjoint cubes through function *Sort()* and *Merge()*. It also tries to obtain smaller execution time by cutting down the number of cube calculus operations through reducing the number of unnecessary comparisons through *merge*_indicator variable. Further reduction of the execution time may be obtained if partitioning is used. The research summarized here will have an impact on the application of multiple-valued logic functions not only in the synthesis, analysis, and testing of multiple-valued logic circuits but also in efficient calculation of spectra for multiple-valued functions.

7. REFERENCES

[1] M. J. Ciesielski, S. Yang and M. A. Perkowski, "Multiple-valued minimization based on graph coloring", *Proc. IEEE International Conference on Computer Design*, Cambridge, Massachusetts, pp. 262-265, Oct. 1989.

[2] S. J. Hong, R. G. Cain and D. L. Ostapko, "MINI: A heuristic approach for logic minimization", *IBM Journal of Research and Development*, Vol. 18, No. 5, pp. 443-458, Sept. 1974.

[3] K. Kinoshita, K. Asada and O. Karatsu, *Logic Design for VLSI*. Tokyo: Iwanami Shoten Publishers, 1985.

[4] A. R. Newton, *Selected Papers on Logic Synthesis for Integrated Circuits Design*. New York: IEEE Press, 1987.

[5] B. J. Falkowski, I. Schafer and M. A. Perkowski, "Effective computer methods for the calculation of Rademacher-Walsh spectrum for completely and incompletely specified Boolean functions", *IEEE Trans. on Computer-Aided Design*, Vol. 11, No. 10, pp. 1207-1226, Oct. 1992.

[6] L. Shivakumaraiah and M. A. Thornton, "Computation of disjoint cube representations using a maximal binate variable heuristic", *Proc. 34th South-Eastern Symposium on System Theory*, Huntsville, Alabama, pp. 417-421, March 2002.

[7] D. Jankovic, R. S. Stankovic and R. Drechsler, "Efficient calculation of fixed-polarity polynomial expressions for multiple-valued logic functions", *Proc. 32nd IEEE International Symposium on Multiple-Valued Logic*, Boston, Massachusetts, pp. 76-82, May 2002.

[8] S. Aborhey, "Functional complexity estimation for large combinational circuits", *IEE Proc. Computers and Digital Techniques,* Vol. 149, No. 2, pp. 39-45, March 2002.

Output-Pattern Directed Decomposition for Low Power Design

Chi-Wei Hu and TingTing Hwang
Department of Computer Science, National Tsing Hua University, HsinChu, Taiwan 30043

Abstract

It is observed that in some circuits, highly active output-patterns fall into a small set. Based on this observation, we propose an output directed decomposition architecture and an algorithm to synthesize the decomposed logic. The results on several MCNC benchmark circuits shows that the proposed method can achieve 34.1% reduction in power as compared to circuits without decomposition.

1 Introduction

Many power minimization techniques have been proposed [1]. Among them, one effective approach is to decompose a circuit so that most of time only a small circuit is active [2, 3, 4, 5]. In this paper, we also take the decomposition approach to reduce power consumption. Our decomposition is based on the observation that highly active output-patterns fall into a small set.

Table 1 shows some experimental results on several MCNC benchmarking circuits. The columns labeled **Total OP** and **Frequent OP** are total number of possible output-patterns, and the number of most frequently occurred output-patterns. The column labeled **Pr.(%)** shows the probability of high occurrence output-patterns. It can be seen from the table that a few output-patterns have very high probabilities. Based on the above observations, we propose a decomposition architecture and algorithm to synthesize the decomposed logic. In the decomposed architecture, one small part is used to compute a few highly occurred output-patterns, and the other large part is used for all the other infrequently occurred output-patterns.

Our approach is similar to the decomposed architecture proposed in [6]. The major differences between these two methods are as follows. First, in Ruan's architecture, only pipelined structures can be handled while we can handle all types of circuits. Second, Ruan's algorithm have to enumerate all the input-patterns to know the correlation between the input-patterns and the output-patterns. It is infeasible to handle large size circuits. We propose to use an OBDD-based heuristic algorithm to compute the output-pattern frequency. Third, we propose a fine-tune logic partitioning phase to achieve more power reduction in our algorithm.

Table 1: The Distribution of Output-Patterns

Circuits	Total OP	Frequent OP	Pr.(%)
sao2	16	3	96.6
misex3	16384	2	80.8
table5	32768	4	90.6
table3	16384	4	92.5
tbk	256	4	91.6

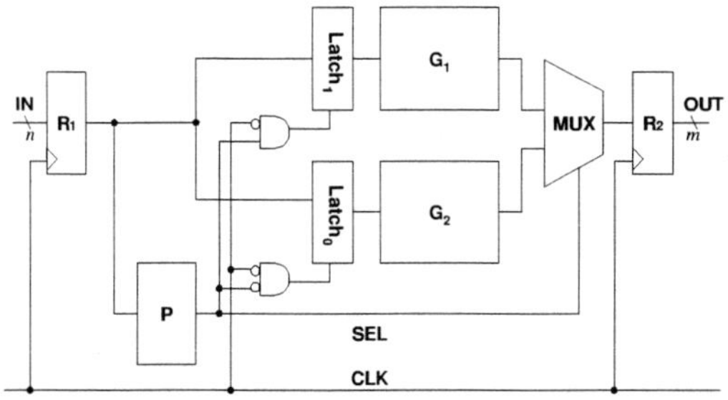

Figure 1: A Decomposition Architecture

2 Decomposition Architecture

Consider a combinational circuit separated by registers R_1 and R_2. Our decomposition architecture shown in Figure 1 is to decompose the combinational circuit into two subcircuits, $group_1$ circuit, G_1 and $group_2$ circuit, G_2. The precomputation circuit P decides which part is activated and generates the signal SEL. When $SEL = 1$, G_1 is selected and G_2 is blocked and when $SEL = 0$, G_2 is selected and G_1 is blocked. SEL is also used to select correct output by controlling the MUX.

Notice that there are two latchs in front of G_1 and G_2. $Latch_1$ is transparent when $SEL = 1$ and $CLK = 0$, and $Latch_0$ is transparent when $SEL = 0$ and $CLK = 0$. They are needed not only to block the inputs to the unselected group, but also to block the glitches propagating to G_1 and G_2 when P is evaluating the SEL.

The behavior is described as follows. When CLK is high, P receives new inputs from register R_1. During the time of P evaluating SEL, the two low-enable latches are not transparent because $CLK = 1$. After CLK is low, the stable SEL will turn on one of the two subcircuits. Note that from the above operation, one important rule for our architecture is that SEL must be stable before CLK is falling. To further improve the architecture for

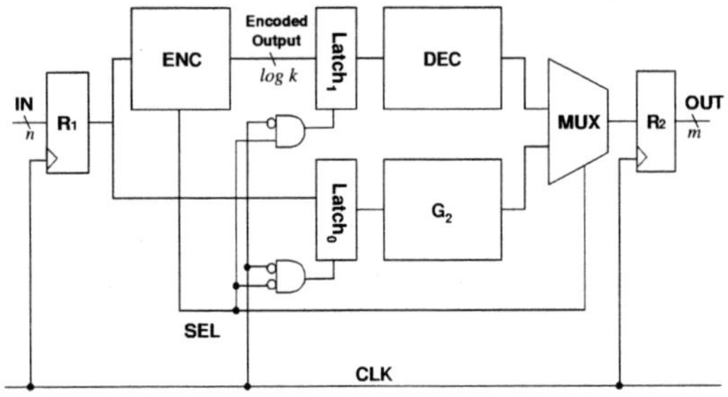

Figure 2: A Decomposition with Codec Architecture

low power, we apply the codec architecture from [7] to the highly active subcircuit G_1 as shown in Figure 2. In this architecture, k is the number of output-patterns in G_1. The reason to adopt codec architecture is because the precomputation circuit P is the logic to determine if an input is evaluated in G_1 (SEL signal). Since there are just a few output patterns in G_1, only a small amount of logic gates is needed for the precomputation circuit P to generate the encoded output patterns. By adding this logic in P, we save the logic in G_1 to determine the individual input patterns. Now the new precomputation circuit, ENC, generates not only signal SEL, but also the encoding bits for different output-patterns. The decoding circuit, DEC, decodes the encoded output-pattern and produces the correct outputs.

3 Decomposition Algorithm

Based on the architecture proposed in Section 2, we propose a two-phase synthesis algorithm. In the first phase we will cluster a small set of output-patterns with high probability into a small circuit G_1. In the second phase, to reduce the size of select logic ENC, we move some logic from G_1 to G_2. The detailed descriptions of the two phases are described in the following sections. We assume in the following input-patterns are evenly distributed.

3.1 OBDD-Based Output-Pattern Directed Partitioning of Input Pattern

The objective of the partition algorithm is to partition a combinational circuit into G_1 and G_2. G_1 is small but contains only a few highly active output-patterns, and G_2 is large but contains a lot of low active output-patterns. Before we perform the partition, we must first compute the occurrence probability of output-patterns. If a combinational circuit has n inputs and m outputs, there are 2^n input-patterns and at most 2^m different output-patterns. A straightforward enumeration of all 2^n input-patterns as proposed in [6] to get output-patterns is obviously not feasible for large size of circuit. Therefore, we porpose an OBDD-based heuristic to solve the problem.

Given a combinational circuit with n inputs and m outputs. OBDD[8] is used to represent the circuit. For m outputs, there are $BDD_1, BDD_2, ..., BDD_m$. These BDDs are used to count the number of output-pattern. Before we present the algorithm, we define $count\_onset(M)$ as the number of different input-patterns mapping to a particular output-pattern M.

The algorithm proceeds as follows. First, for each $OBDD$, we compute the numbers of input-patterns that evaluate to 1 and to 0, respectively. Next, for each output, we select the output phase that has more input-patterns mapping to it. Then, we compute the *and* operation of OBDDs with the selected output phases. This output-pattern is put into our candidate set and will be used as our seed output-pattern to search for other output-patterns. An output-pattern is denoted as $(o_1, o_2, o_3, ..., o_m)$, where o_x is the phase of BDD_x, and o_x=1 and o_x=0 denote the positive and negative phase, respectively.

Based on the seed pattern, we search other patterns with high input probability. The search begins with computing the *gain* of inverting output phase for each output bit. The *gain* is defined as follows. Given an output-pattern $(o_1, o_2, ..., o_m)$,

$$gain(o_x) = count\_onset((o_1, o_2, ..., o_x{}', ..., o_m)).$$

Of all output bits, we select the bit o_x with the maximum *gain*, invert the output o_x, produce a new output-pattern, and lock the bit. If the output-pattern has $gain > 0$, it is inserted to the candidate set. If there is a tie, all output-patterns generated by inverting the bit with the maximum *gain* and *gain* > 0 are inserted to the candidate set. However, only the output-pattern where $o_x{}'$ has the highest probability will be selected as the seed pattern for the next iteration. Based on the new seed output-pattern, we repeat the inverse-output computation for all unlock outputs to obtain other output-patterns. The procedure continues until all outputs are locked.

Once we have the probabilities of output-patterns, we select output-patterns to G_1 based on a threshold probability. The threshold is set by users, for example, 3% or 5%. The output-pattern in candidate set that has input probability higher than the threshold is selected in G_1 and the rest will be left in G_2. Figure 3 shows the algorithm.

After performing phase 1 of our algorithm, we have the initial ENC circuit and G_2 circuit which will be fine-tuned to further reduce power.

3.2 ENC Simplification

In the second phase, logic is moved from G_1 to G_2 to simplify the selection logic of ENC, which has the major power consumption of the whole decomposition architecture. By doing so, the area of G_2 may also be increased. However, G_2 contains only the output patterns of low occurrence probability. Hence, the increased power of G_2 will be much smaller than the reduced power of ENC.

The simplification algorithm is to find input-pattern in ENC that does not have many adjacent inputs evaluating to the same output-pattern. This input-pattern may result in a cube with many literals and thus consume more

PROCEDURE OBDD-Based Partitioning
Input: A circuit with n inputs, m outputs
Output: S, which is the output-pattern set of G_1
BEGIN
  Build BDDs for each output, $o_1, o_2, ..., o_m$;
  For each OBDD,
    Compute the phase that has higher input probability;
  Initialize output pattern solution t;
  Count the probability of this output-pattern t;
  Add t to C;
  WHILE there are any unlocked output
    Change the phase of o_x which has max *gain*;
    Lock o_x;
    Count the probability of this output-pattern t;
    Add t to C;
  END WHILE
  For $\forall t \in C$, if $prob(t)$ >threshold then add t to S;
END PROCEDURE

Figure 3: Output-Pattern Directed Partition.

area. The algorithm begins with calling ESPRESSO to simplify the SEL circuit in two-level logic. Then each cube in two-level form is ordered to cube-list by its cost in decreasing order. The cost of each cube is defined to be the number of literals of the cube. If the numbers of literals are equal, then the cost is determined by the number of *essential minterm*. The *essential minterm* of a cube is defined as the minterm that is covered only by this cube. The less number of *essential minterm* of this cube, the higher the cost.

The next step is to move the cubes in cube-list to G_2 one after the other taking estimated *power* into consideration. The *power* will be defined later. The moving cube step will repeat until the cube-list is empty. Then, the best subsequence that results in the best *power* will be selected.

In each cube moving step, for the new decomposed logic, *power gain* has to be evaluated. *Power* is defined to reflect the reduced logic of ENC and the increased logic of G_2 after a cube is moved from ENC to G_2.

We first show how to estimate the decreased logic of ENC and then the increased logic of G_2. Now ENC is estimated as

$$new\_enc = org\_enc \times \frac{new\_sel}{org\_sel}$$

where org_enc is the number of literals in minimized factored form of the ENC circuit after phase 1, new_sel and org_sel are the number of literals in a sum-of-product form of the new SEL circuit, and of the SEL circuit generated by phase 1, respectively. The estimation is defined based on the assumption that the area of ENC circuit is proportional to its number of literals in a sum-of-product form of SEL circuit.

Next, area of G_2 is estimated as

$$new\_g_2 = org\_g_2 + (org - org\_g_2) \times \frac{dc(org\_g_2) - dc(new\_g_2)}{dc(org\_g_2)}$$

PROCEDURE ENC Simplification
Input: circuit ENC and G_2
Output: new circuit of ENC and G_2
BEGIN
  Simplify SEL by ESPRESSO;
  Order cube-list of SEL by cost in decreasing order;
  WHILE not empty cube-list
    Move cube with max cost to G_2;
    Compute *power*;
    Update the *essential minterms* of cube in SEL;
  END WHILE
  Select min power subsequence of cubes $c_1, c_2, ..., c_k$;
  Move $c_1, c_2, ..., c_k$ to G_2;
END PROCEDURE

Figure 4: Fine-Tune Algorithm.

Table 2: Output-patterns Directed Partitioning

Cir.	In	Out	G_1 Pat.	G_1 Pr(%)	CPU(s)
sao2	10	4	3	96.6	0.03
misex3	14	14	2	80.8	0.53
table5	17	15	4	90.6	0.33
table3	14	14	4	92.5	0.35
frg1	28	3	5	91.7	0.06
tbk	11	8	4	91.6	0.16
toolarge	38	3	2	96.6	0.81
apex2	38	3	2	96.6	0.62

where org_g_2 and org are the number of literals in a factored form of the G_2 circuit generated by phase 1 and orignial circuit, respectively. $dc(org\_g_2)$ and $dc(new\_g_2)$ are the size of *don't-care* set of the original G_2 circuit after phase 1, and the new G_2 circuit after one cube is added, respectively. The estimation is defined based on the assumption that the area of G_2 circuit is disproportional to its size of *don't-care* set of G_2 circuit.

Finally, let $prob(G_2)$ denote the probabilities that G_2 is active and note that ENC is active all the time. The power cost is modeled as follows:

$$\begin{aligned} & power(ENC + G_2) \\ & = area(ENC) + area(G_2) \times prob(G_2) \\ & = new\_enc + new\_g_2 \times prob(G_2) \end{aligned}$$

Figure 4 shows the algorithm.

4 Experimental Results

Our algorithm, presented in Section 3, is implemented in the C language and executed on a SUN Blade Workstation. Several MCNC benchmark circuits were tested to show the performance. Each circuit was first synthesized by the script of SIS, *script.rugged*. In order to have more accurate estimation of area and timing, Synopsys

Table 3: Comparisons of Area and Power After Phase 1 and Phase 2

Cir.	Orig.		Decomp							
			Phase 1				Phase 2			
	A	P	A	AI(%)	P	PR(%)	A	AI(%)	P	PR(%)
sao2	1415	35.7	2412	66.2	20.7	41.9	2295	58.1	18.6	47.9
misex3	7154	136.5	7821	9.1	106.3	22.1	8184	14.2	99.2	27.3
table5	10322	104.8	14365	38.2	127.7	-21.8	12199	17.3	88.3	15.8
table3	9291	105.4	12169	30.1	98.1	6.9	12342	32.0	62.3	40.9
frg1	1693	32.6	4273	152.3	31.5	3.3	4267	152.0	30.2	7.3
tbk	5610	100.4	7460	32.9	110.0	-9.6	7552	34.6	67.8	32.4
toolarge	4976	87.5	7960	59.9	32.4	64.1	9298	86.8	30.7	64.9
apex2	4210	77.4	6743	60.1	51.6	33.3	7060	67.1	47.8	36.9
Avg.	-	-	-	56.1	-	17.5	-	57.7	-	34.1

Table 4: Area Overhead After Phase 1 and Phase 2

Cir.	Orig.	Decomp	
	A	OA	OAI(%)
sao2	1415	990	68.2
misex3	7154	1704	23.8
table5	10322	1941	18.8
table3	9291	1704	18.3
frg1	1693	2143	126.5
tbk	5610	1238	22.1
toolarge	4976	2660	53.4
apex2	4210	2660	63.1

Design_Analyzer was used to map circuits to TSMC .25 library. At last, power estimation was performed by Synopsys Prime_Power. In these experiments, 2.5V supply voltage and a clock frequency of 20 MHz was assumed. The first set of experiments was performed to show the effectiveness of our OBDD-based output-pattern directed partitioning algorithm. Table 2 shows the results. The column labeled **Pat.** shows the number output-patterns. The column labeled **Pr(%)** shows the sum of occurrence probabilities of these output-patterns in G_1. The results show that our OBDD-based algorithm can find highly occurred output-patterns in very short period.

The second experiment is to understand how effective of our proposed algorithm. The circuits are selected by our OBDD-based algorithm performed in phase 1. Area and power of circuit are recorded after phase 1 and phase 2 are performed. The results are shown in Table 3. The columns labeled **Orig** and **Decomp** are the area and power of original circuit without decomposition and after decomposition, respectively. The columns labeled **A** and **P** show the area and the power consumption (in μW) of the circuit, respectively. The columns labeled **AI(%)** show the percentage of area increased and is computed by: $AI = \frac{(Decomp\_Area - Orig\_Area)}{Orig\_Area}$ Similarly, the columns labeled **PR(%)** show the percentage of power reduction computed by: $PR = \frac{(Orig\_Power - Decomp\_Power)}{Orig\_Power}$

From Table 3, we can see that after performing phase 1, most circuits have power reduction but at the expense of area increase. On the average, area is increased about 56.1% and power is reduced about 17.5%. After performing phase 2, power consumption is further reduced for all the circuits with very little area overhead. On the average, area increase is 57.7% and power reduction is 34.1%.

Table 4 shows the area overhead (latches and one MUXs) after decomposition. The columns labeled **OA** show the area of the overhead. The columns labeled **OAI(%)** show the percentage of overhead area increased and is computed by: $OAI = \frac{(Overhead\_Area - Orig\_Area)}{Orig\_Area}$
For the two small circuits *sao2* and *frg1*, area overhead for decomposed architecture become significant because the original circuit is too small.

References

[1] Chandrakasan A.P.,et.al., "Low-power CMOS Digital Design," *IEEE Journal of Solid-State Circuits*, Vol. 27, pp. 473-484, April 1992.

[2] Alidina M., et.al., "Precomputation-based Sequential Logic Optimization For Low Power," *IEEE Trans on VLSI Systems*, pp. 426-436, Vol. 2, Issue 4, Dec. 1994.

[3] G. Lakshminarayana,et.al., "Common-case Computation: a High-level Technique for Power and Performance Optimization," *Proc. of DAC*, pp. 56-61, June 1999.

[4] Sue-Hong Chow, et.al., "Low Power Realization of Finite State Machines—A Decomposition Approach," *ACM TODAES*, Vol. 1, Issue 3, pp. 315-340, July 1996.

[5] Shanq-Jang Ruan,et.al., "A Bipartition-Codec Architecture to Reduce Power in Pipelined Circuits," *Proc. of ICCD*, pp. 84-90, Nov. 1999.

[6] Shanq-Jang Ruan, et.al., "ENPCO: An Entropy-Based Partition-Codec Algorithm to Reduce Power for Bipartition-Codec Architecture in Pipelined Circuits," *IEEE Trans on VLSI Systems*, Vol. 10, No.1, pp. 942-949, Dec. 2002.

[7] Shanq-Jang Ruan, et.al., "A Bipartition-Codec Architecture to Reduce Power in Pipelined Circuits," *IEEE Transactions on CAD*, Vol. 20, pp. 343-348, Feb. 2001.

[8] Karl S. Brace, et.al., "Efficient Implementation of a BDD Package," *Proc. of DAC*, pp. 40-45, June, 1990.

[9] E. Sentovich et al., "SIS: A System for Sequential Circuit synthesis," *Technology Report UCB/ERL M92/41, ERL, Department of EECS, University of California, Berkeley*, May 1992.

FPGA IMPLEMENTATION OF CONTROLLER-DATAPATH PAIR IN CUSTOM IMAGE PROCESSOR DESIGN

Hongtu Jiang and Viktor Öwall

CCCD, Department of Electroscience
Lund University, Lund 22100, Sweden

ABSTRACT

In order to reduce the effort of the controller design in the customized image convolution processor, a controller synthesis tool is developed based on [9] to support the design flow from a system or algorithm specification to RTL level VHDL. Architecture extensions to basic FSMs structures are implemented with the purpose of optimizing controller design for area and power consumption. Together with controller implementation, a custom datapath architecture with three level memory hierarchies is developed aiming at a real-time power efficient image processing solution with low I/O bandwidth requirements. The complete design is prototyped on Xilinx Virtex 2 platform with comparable performance with that of TI C64x processor at only 2/15 of its clock frequency.

1. INTRODUCTION

Efficient CAD tools are desired to reduce the increasing design efforts when algorithms implemented on ASICs are getting more complicated. Compared to hardware accelerators by direct algorithm mapping, microprogrammed structures prevail in non performance critical applications since they take up less design resources while providing user flexibility to a certain extent. Usually controller design in such structures, identified as controller-datapath pairs, comprises manual design of basic FSMs in the form of HDLs or state tables which is followed by commercial synthesis tools. Such methods could work well for simple control tasks until more complicated controller design is required. In the implementation of large control dominated algorithms, state coding itself could increase design effort substantially. On the other hand, although basic FSM structures may provide functionality for any applications, it fails to utilize the intrinsic structural advantages existing in specific algorithms to optimize the design in certain dimensions in the design space, e.g. power consumption, speed. All these requirements accounts for the necessity of high level synthesis tools that can take an algorithm from a high level simulation to synthesizable HDLs at RTL level. This is an important area in the research community since current design tools at physical and logic level are widely available and extensively used in industry [1]. With the presented synthesis tool, algorithm in the form of microprograms could be easily changed and the new controller is synthesized automatically.

The target application is a custom image convolution processor. Two dimensional image convolution is one of the fundamental processing elements among many image applications and has the following mathematical form:

$$y(K_1, K_2) = \sum\sum x(K_1 - m_1, K_2 - m_2)h(m_1, m_2),$$

where x and h denotes the image to be filtered and kernel functions respectively. Due to its nature of computational intensity and a consequent high I/O data transfer, several dedicated solutions are adopted to fulfill the demanding processing task. For example, as part of the image processing library for TI's high performance DSP C64x, such fundamental function is implemented as high performance library routines written in optimized assembly language to take advantage of the specific DSP architectures[2]. Alternatively, many FPGA implementations of image convolutions has also been reported[3, 4, 5]. Focusing mainly on architecture optimizations for arithmetic operations to enhance calculation performance, few of these implementations address the combination of data flow architecture which is crucial to sustained calculation capability with substantial I/O datarates reductions. In this paper, a streamlined dataflow architecture composed of three level of memory hierarchy are implemented on Xilinx Virtex 2 FPGAs. To demonstrate the functionality, an edge detection operator named Mexican hat is used. Its distribution in two dimensions may be expressed in terms of the radial distance r from the origin,

$$z(r) = \nabla^2 G(r) = K(1 - \frac{r^2}{2\sigma^2})e^{-\frac{r^2}{2\sigma^2}}.$$

Combined with a dedicated controller synthesized by the tool, a sustained calculation capacity of 4.8G MAC/s is achieved at the clock frequency of 80Mhz. Furthermore, potential power savings are also identified by such memory hierarchy for the future ASIC solutions. It is estimated that

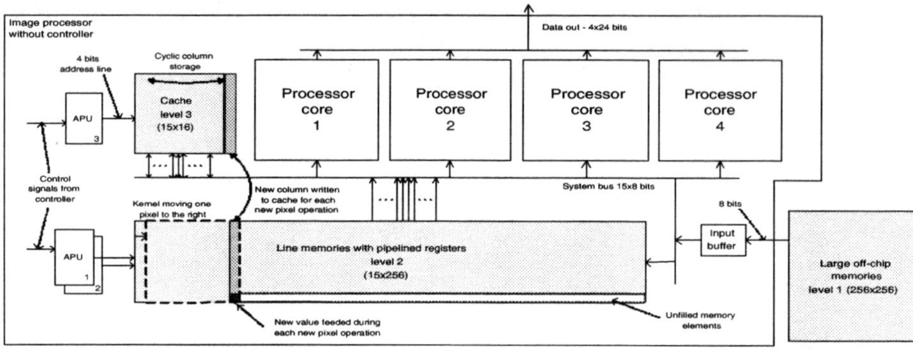

Fig. 1. Block diagram of image processor datapath with 3 level memory hierarchies

power consumed by memory operations in three level hierarchies can be over 35 times lower than that of the one level direct memory scheme.

2. DATAPATH ARCHITECTURE

As a consequence of the image size, the complete image has to be stored on off-chip memory. During the convolving operations, each kernel position requires 15×15 pixel values from off-chip memory in a direct way. When this is performed in real-time, a very high data throughput is needed. Furthermore, accessing large off-chip memory is expensive in terms of power consumption and so is the signaling due to the large capacitance of package pins and PCB wires.

In [6], a tailored architecture with a streamlined dataflow was proposed to achieve the desired filtering while keeping a low I/O bandwidth. By the observation that each pixel value, except the one in the extreme corners, is used in several calculations, a two level memory hierarchy were introduced. instead of accessing off-chip memories 225 times for each pixel operation, 14 full rows of pixel values and 15 values on the 15th row are filled into 15 on-chip line memories before any convolution starts. After the first pixel operation, for each successive pixel position, the value in the upper left is discarded and a new value is read from the off-chip memory. As a result, pixel values from the off-chip memories are read only once for the whole convolving process from upper left to the lower right. Thus the input datarates are reduced from 15×15 accesses/pixel to only 1 read.

In light of the fact that using a two level memory structure can reduce both power consumptions and I/O bandwidth requirements, one extra cache level is introduced in this paper to further optimize for power consumptions during memory operations, Fig. 1. Since accessing large memory consumes more power, 15 small cache memories are added to the two level memory hierarchy. The small cache memories are composed of 15 separate memories to provide one column pixel values during each clock cycle. Instead of reading pixel values directly from line memories 15 times for each pixel operation, one column of pixel values are read to cache memories first from the line memories for each new pixel operation except for the first one. During each pixel operation, direct line memory access is replaced by small caches read. As a result, reading pixel values from line memories 15 times could be replaced by only once plus 15 times small cache accesses. Under the assumption that the embedded memories are implemented in a 0.35μm process CMOS, by the calculation method in [7], it is shown in Table. I that the power consumption for the total memory operations could be reduced to over 2.5 times compared to that of a two level hierarchy. In addition to the new cache memory, one extra address calculator is synthesized by proposed synthesis tool to handle the cache address calculation, the architecture of which is shown in the following section. In order to simplify address calculation, the depth for each cache is set to 16. In doing so, incrementing the largest address represented in 4 bits binaries will reset it back to the beginning. This will allow circular operations on the cache memories. During new pixel operations each new column of data are filled into the cache in a cyclic manner. However, the achieved low power solution has to be traded for extra clock cycles introduced during the initial filling of the third level cache memories. In the case of an image size of 256×256, this will contribute 61696 clock cycles in total. Compared with the total clock cycles in the magnitude of 10^7, such a side effect is negligible.

In fact, alternative schemes exists by different combinations of the three memories in the hierarchy. All these possible solutions differ substantially in memory accesses. In Table. I, different memory scheme is shown where M0, C0, C1 denotes off-chip memory, line memories and smaller caches respectively. Although the number of data provided to the datapath remains the same for all four schemes, the access counts to the large off-chip memories varies. For the two and three level hierarchy structures, the counts to the large memory M0 are reduced by nearly 225 times compared to

Table 1. memory hierarchy schemes and access counts for an image of 256×256 (M0=off-chip 0.18μm, C0 and C1=0.35μm)

Scheme	M0	C0	C1	energy cost
A: M0	13176900			790 mJ
B: M0→ C0	65536	13176900		56.6 mJ
C: M0→ C1	929280		13176900	68.9 mJ
D: M0→C0→C1	65536	929280	13176900	20.8 mJ

that of the one level solution. Between scheme B and D, the least access counts to both external memories and line memories are identified in three level memory structure, but this is achieved at the cost of extra clock cycles introduced during each pixel operation. Thus, trade off should be made when different implementation techniques are used. For the FPGAs where power consumption is considered less significant, two level hierarchies prevails due to its lower clock cycle counts. While for ASIC solutions, three level hierarchy is more preferable for it results in reasonable power reduction.

In addition to memory schemes, a pipelined adder tree is implemented on four processor cores to fulfil the sustained calculation capacity. Four cores have been chosen to be able to perform several convolutions in parallel. Each core contains 15 multipliers with adjoining RAMs filled with the kernel function. During each clock cycle, one column of data comprising 15 pixel values are read to the corresponding multipliers. The multiplication are performed in parallel and the products are summed in the subsequent adder tree. Combined with memory hierarchy, a very high sustained calculation capacity is obtained.

3. CONTROLLER SYNTHESIS

The processor cores require a very simple controller with only 2 control signals. However, the line memories and caches, plus RAMs accommodating kernel function values require extensive address calculations and loop control. In order to reduce the effort of the controller design, a synthesis tool is developed based on [9] to support synthesis process from high level c-like algorithm specification to RTL controller module in VHDL format. The modified tool takes in a behavioral description of the datapath architecture defining the available set of micro-operations, and a microprogram written in C-like input syntax that contains the algorithm with additional declarations such as memories. As an output a complete controller with module descriptions and interconnection specifications is generated. Since architecture extensions to basic FSM could result in optimized controllers in specific application, a range of controller architectures are supported and can be accessed through architecture option specification before the synthesis starts. For the current image processor, a controller architecture with incremental circuitry and a DFSM is implemented, Fig. 3. The DFSM is a decision finite state machine used as an assistance to the controller for conditional statements. In this architecture, the branch address calculation within the same block of code, composed of only sequential statements, is performed by the hardware incrementer. At the end of a block a non-incremental branch address is calculated by the control logic and a select branch signal is set. This architecture is particularly suitable for algorithms that have long stretches of sequential statements, i.e. the next state generation logic in the basic FSM architecture can be replaced by an incrementer and MUX. With the tool, more complicated architectures can be implemented freely based on user requirements and algorithm structures since it is a fully automated synthesis process.

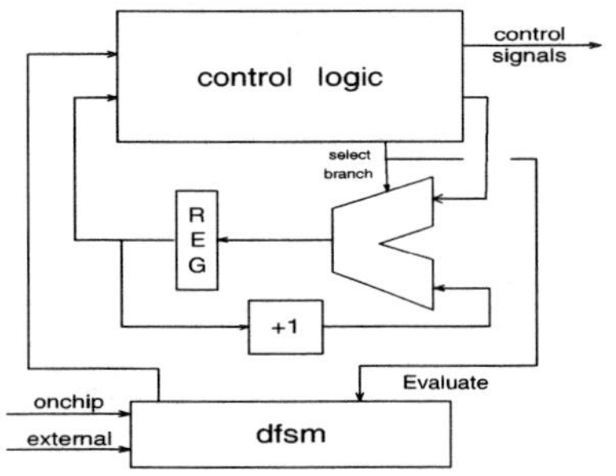

Fig. 2. Controller Architecture with incremental circuitry

In addition to the control structure, address processing units (APU) are also synthesized by the tool to perform address calculation. Due to the considerable data flow in calculating the memory addresses, APUs are usually implemented separately from controllers to reduce control logic. Declared in the microprogram, like variables in C, APUs are implemented in a common structure as shown in Fig. 5. To map the address calculations in the microprogram into hardware, constant value assignments to the APUs are implemented as one of the MUX constants. Additions and subtractions are performed in the adder and the results are stored in one of the register banks on either sides. Parameters such as image size is provided through the external

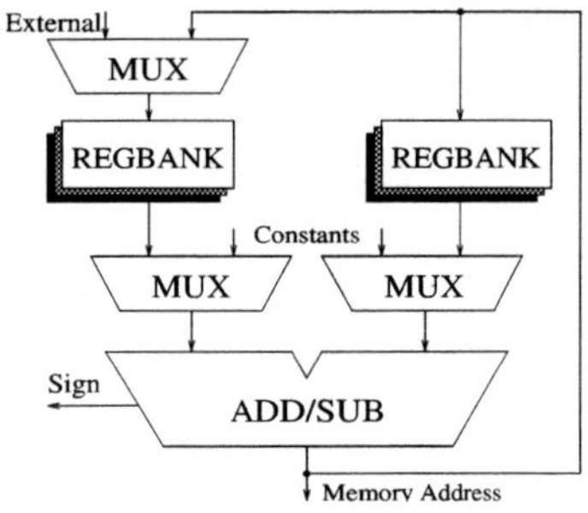

Fig. 3. Address processing unit

input. To facilitate simple calculation descriptions in a microprogram, several sets of micro operations are predefined by the synthesis tool so that the user of the microprogram could use calculation functions like "++" and "--" with no knowledge of the implementation details. This will simplify the microprogram substantially in an address calculation intensive tasks.

Besides calculating addresses, APU architectures can also facilitate loop control in the microprogram. Similar to address pointers, loop variables are declared and used in the microprograms like ordinary C variables. After synthesis, each variable is implemented in one APU, and the value is incremented or decremented until the sign bit changes, which indicates the completion of the loop. Depending on whether the loop iterations are constant, the value are stored in the one of the registers in the APU architecture or read from external input to initiate the loop. The sign bit is connected to either the scheduler of the controller architecture or to a DFSM and used to affect the control flow of the microprogram.

In the future, system level modelling language like SystemC is within special interest for its ability to specify both algorithm and hardware. With this support, future designs could go up further to the system level to enable more complicated system design and early verification.

4. RESULTS

The control logic is generated in the format of espresso [8]– a PLA design format supported by Synopsys design compiler. Other formats are also likely in the future since the information stored in the database are truth tables and could be easily modified for other formats. The associated controller architecture as well as the datapath are implemented in VHDL. The whole system has been prototyped on Xilinx Virtex 2 XC2V1000-4FG456 FPGAs with 4454 CLBs reported. The whole system performs four sperate convolutions in parallel with calculation capacity of 4.8 GMAC/s at a clock frequency of 80 MHz. This is comparable with the 4.8 GMAC/s calculation capacity of C64x with 8 multipliers of 8 bits running at 600 Mhz. With the simulation result for higher speed greed of -6, a working frequency of over 100Mhz are obtained. This will outperform the most up to date TI C64x running at 720Mhz today. With the image size of 256×256, a frame rate of 81.4 frames/s is achieved using kernel size of 15×15.

5. CONCLUSIONS

Controller synthesis tools are powerful for the design of controller-datapath systems. By supporting higher level C-like control algorithm specification, the effort of controller design is reduced substantially. Memory hierarchy schemes are important factors for optimizing datapath designs. In this image application, a sustained calculation capacity comparable to TI high performance C64x series as well as substantial I/O reductions are obtained. Furthermore, the potential power consumption reductions are envisioned to be over 35 times compared to that of direct implementation.

6. REFERENCES

[1] P. Eles, K. Kuchcinski, and Z. Peng, *System Synthesis with VHDL*, Kluwer Academic Publishers, 1998.

[2] F. K.karadayi et al., *Imaging with Texas Instruments' TMS320C64x*, Texas Instruments White paper, Dec. 2001

[3] E.Jamro et al., *Convolution operation implemented in FPGA structures for real-time image processing*, Proceedings of the 2001 ISPA.

[4] E.Jamro et al., *FPGA implementation of addition as a part of the convolution* , Proceedings of Euromicro Symposium on Digital Systems, Design, 2001

[5] D.C.M.Bilsby et al., *Comparison of a programmable DSP and a FPGA for real-time multiscale convolution* , IEE Colloquium on High Performance Architectures for Real-Time Image Processing, 1998

[6] V. Öwall, et al., "A Custom Image Convolution DSP with a Sustained Calculation Capacity of $>$ 1 GMAC/s and Low I/O Bandwidth", *The J. of VLSI Signal Processing*, 335-350; Nov 1999.

[7] F. Catthoor et al., *Custom memory management methodology*, Kluwer acdemic Publishers, 1998

[8] R. Rudell, "ESPRESSO - Boolean minimization program", *Technical report*, University of California at Berkeley, Dec 1985.

[9] V. Öwall, *Synthesis of Controllers from a Range of Controller Architectures*, Ph.D. Thesis, Lund University, Dec 1994

LEAST LEAKAGE VECTOR ASSISTED TECHNOLOGY MAPPING FOR TOTAL POWER OPTIMIZATION*

Yi-Ching Au and Chi-Ying Tsui

Department of Electrical and Electronic Engineering
Hong Kong University of Science and Technology
Clear Water Bay, Hong Kong SAR, China
E-mail: eerobert@ust.hk

ABSTRACT

As the feature size continues to shrink and the threshold voltage keeps on reducing, leakage power becomes a significant component of the total power consumption. The cost function used in the conventional logic synthesis tools was over simplified in modeling the total power consumption of the logic network. On the other hand, recent works on reducing leakage power mainly focused on an already synthesized network. In this paper, we propose a complete model of the total power consumption of the logic network which includes the leakage power and also takes into account the operating duty cycle of the applications. In addition, we propose a least leakage vector (LLV) assisted technology mapping to optimize the total power of the final mapped network. The LLV used during technology mapping phase is obtained from the technology decomposed network. Experimental results show that an average of 20% reduction in total power consumption is obtained comparing with the conventional low power technology mapping algorithm.

1. INTRODUCTION

The total power consumption P_{total} of a logic network is given by,

$$P_{total} = P_{dynamic} + P_{leakage} + P_{static}$$

where $P_{dynamic}$ is the dynamic power consumed due to the switching of logic state, $P_{leakage}$ is leakage power and P_{static} is instantaneous short circuit power consumed during the gate switching. Previous works on low power synthesis were mainly on optimization of dynamic power, as it is assumed the other component ($P_{leakage}$, P_{static}) are usually order of magnitude smaller than $P_{dynamic}$. This assumption, however, is no longer valid as the technology advances to the very deep submicron era. From the BSIM3v3 model [1], the sub-threshold drain current model is given by,

$$I_{ds0} = \mu_0 C_{ox} \frac{W}{L} \cdot v_t^2 \cdot \left(1 - e^{-V_{ds}/v_t}\right) \cdot e^{(V_{gs}-V_{th}-V_{off})/n \cdot v_t}$$

where μ_0 is the zero bias mobility, C_{ox} is the gate oxide capacitance per unit area, v_t is the thermal voltage, W and L are the width and length of transistor respectively, V_{ds}, V_{gs} and V_{th} are the drain-source voltage, gate voltage and the threshold voltage of the transistor respectively. V_{off} is the offset voltage at sub-threshold region, and n is sub-threshold slope. It can be seen that, the drain current increases exponentially with the decrease in V_{th}. From the Berkeley Predictive Technology Model [2], there is a 0.4V drop in V_{th} when the technology scales from 0.18um to 0.07um and this results in a 40 times increase in sub-threshold leakage current.

On the other hand, the conventional logic synthesis tools only focus in reducing the active dynamic power consumption. The power consumed during the standby time is not considered. With the increase in leakage current, the leakage power consumed in standby time is no longer negligible and has to be considered during the design stage in order to achieve an overall optimal power solution for the applications. This is true especially for those portable applications, such as PDAs and mobile phones, of which the battery serving time stays in standby mode for most of the time.

1.1. Previous works in standby power reduction

Chen *et al.* [3] proposed to use the transistors stacking effect to reduce the leakage power. This is based on the fact that the leakage current is significantly reduced when two or more transistors are "off" in a series path. This finding implies that the leakage power can be minimized by just applying an appropriate input vector to the primary inputs of the network so that the total leakage power is minimized.

Finding the least leakage vector for a given logic network is proved to be a NP-hard problem. In order to solve the problem, heuristic algorithms have been proposed. Abdollahi *et al.* [4] formulated the problem as a Boolean satisfiability problem and solved it by using commercial tools, while Halter *et al.* [5] used statistical approach to find the vector under some pre-defined confidence levels. Johnson *et al.* [6] defined a new measure called leakage observability to estimate the leakage power of a given network under one bit change in the input. An algorithm was then proposed to find the least leakage vector by iteratively flipping the input bit.

1.2. Low power logic synthesis

Tsui *et al.* [7] proposed a low power logic synthesis which takes the dynamic power into account as the cost function during technology mapping. A power-delay trade-off curve is used to store the trade-off design points during local matching and a pre-order traversal is performed after the whole network is matched to read out the best mapping for a given timing constraint.

* This work was supported in part by Hong Kong RGC CERG under Grant HKUST6214/03E and HKUST HIA02/03.EG03.

Kang *et al.* [8] extend the dynamic power logic synthesis to consider the leakage power. They include the active leakage power in the cost function of the technology mapping by weighting the leakage power fro different input combinations with their corresponding occurrence probabilities. However they did not consider the standby leakage power during optimization

1.3. Contributions of this work

Previous works on leakage power reduction, e.g. least leakage vector, focus on minimizing the leakage power for a given synthesized/mapped network. However it does not have the flexibility to find the optimal total power solution which includes the leakage power. Given a mapped network MN_i, we can find the corresponding least leakage vector LV_i. However there may exist another implementation combination (MN_j, LV_j) which has less total power than (MN_i, LV_i). Therefore we should work on the least leakage problem in an early phase, i.e. the logic synthesis phase. Given a Boolean network, we synthesize the network into a circuit and at the same time find the least-leakage vector for the circuit such that the leakage power is minimized subject to some performance constraints. At the same time, if we only minimize the dynamic power or the leakage power alone, for applications that have bursty operating characteristic, it may not give the optimal result. Instead, we need a combined optimization effort, i.e. reducing the dynamic and active and standby leakage power at the same time.

In this work, we propose a leakage-aware technology mapping algorithm. In addition to the dynamic power consumption, we also include leakage power during the active mode which is modeled by the weighted average leakage power of the gates, and the standby leakage power which is determined by a LLV, in the cost function. We propose an input-vector assisted technology mapping method to calculate the cost of the standby leakage power. The input vector is found in the technology decomposition stage. Experimental results showed that up to 45% reduction can be achieved and on average 20% reduction is achieved over the conventional dynamic power optimization approach.

2. TECHNOLOGY MAPPING TARGETING TOTAL POWER OPTIMIZATION

Our leakage-aware technology mapper follows a procedure similar to that proposed in [7], with the difference that leakage power of the gates is also considered. In each node a total power-delay curve is stored. During the mapping procedure, the power cost of each node matching is calculated and the non-infereior solution is stored in the power-delay tradeoff curve. In [7], the power function of the node n that has a gate g matched is given by,

$$P(n,g) = 0.5C_{diff}(n,g)\frac{V_{dd}^2}{T_{cycle}}E_n + \sum_{n_i \in inputs(n,g)} \left(0.5C_{n_i}\frac{V_{dd}^2}{T_{cycle}}E_{n_i} + \frac{P(n_i,g_i)}{fanout(n_i)} \right)$$

where $C_{diff}(n, g)$ is the internal capacitance for node n and matching gate g, V_{dd} is the supply voltage, T_{cycle} is cycle time, E_n is the average switching activity at node n, E_{ni} is the average switching activity at the output of node n_i, C_{ni} is the output load capacitance seen at n_i, and $P(n_i, g_i)$ is the accumulated dynamic power cost function at node n_i with matching gate g_i. fanout(n_i) is the number of fanout count at node n_i.

if leakage power is to be considered in the tree mapping procedure, we need to add te leakage power term $P_{leakage}(n, g)$ in the total power cost function P(n, g). $P_{leakage}$ consists of two parts, the active leakage and the standby leakage. The active $P_{leakage}$ not only accounts for the leakage power in active mode when there is no switching in the current cycle but also the leakage power consumed during the period after the output of the gate stablized and before the next output switching. In [8], the authors used the input pattern occurrence probabilities weighted with the corresponding leakage current at the matching gate g to model the average active $P_{leakage}$. The active $P_{leakage}$ is given by,

$$P_{leakage}^{active}(n,g) = V_{dd} \cdot \sum_{U \in Input} I_{leakage}^{active}(U,g) \cdot p(U)$$

where active $I_{leakage}(U, g)$ is the leakage current of gate g when the input vector is U and p(U) is the occurrence probability of U.

The standby $P_{leakage}$ is the leakage power consmption when the network is in the standby mode. Here the average leakage power overall possible inputs of the gate cannot be used as the power model since a single least-leakage vector is used to minimize the leakage power during standby time. The standby $P_{leakage}$ is then given by,

$$P_{leakage}^{standby}(V,n,g) = V_{dd} \cdot I_{leakage}^{standby}(V,g)$$

where V is the input vector to the gate g when the LLV is used as the primary input vector.

The logic network either works in the active mode or the standby mode. To cope with the characteristics of the operation duty cycle of the applications, a duty cycle parameter, α, is introduced which is defined as the ratio of the time that the network is in the active mode. The final total power cost function is then given by,

$$P(V,n,g) = \alpha \cdot 0.5C_{diff}(n,g)\frac{V_{dd}^2}{T_{cycle}}E_n + \sum_{n_i \in inputs(n,g)} \left(\alpha \cdot 0.5C_{n_i}\frac{V_{dd}^2}{T_{cycle}}E_{n_i} + \frac{P(n_i,g_i)}{fanout(n_i)} \right) + \alpha \cdot P_{leakage}^{active}(n,g) + (1-\alpha) \cdot P_{leakage}^{standby}(V,n,g)$$

The standby $P_{leakage}$ is highly depends on proper choose of input vecotr, V. Carefully choosing the V will make large difference in the leakage power of the final mapped result. Consider an example of a 4-input NAND gate, the leakage power with the corresponding input vectors are shown in Table 1. The leakage power of the input vectors 0000 and 1111 have a 25 times difference.

In order to have a better power cost for the mapping solution, input pattern dependent technology mapping can be used. The principle is that during the gate matching at node n, a mapping is found together with the corresponding vector at the input of the gate such that the total power (including the leakage power and dynamic power) is minimized. The standby $P_{leakage}$ cost is obtained form the vector found in this process. Using this pattern dependent technology mappping, however, makes the mapping problem more complicated especially when there are a lot of reconvergent nodes. This is because the input vector found for matching a local node may have conflict with that obtained during matching for other nodes which share the same reconvergent fanout points. This is illustrated in Figure 1. The best mapping solution of node 1 for least leakage is NAND2 with a LLV at (A, B) equal to (1, 1). The best mapping for node 4 is NOR2 and INV1X with LLV (A, B) equal to (0, 0). Conflicts occur at the input vector and in order to find an optimal solution, four combinations (0, 0), (0, 1), (1, 0), (1, 1) have to be tried for (A, B). The complexity to do this conflict resolution during mapping is then exponential to the number of conflict inputs.

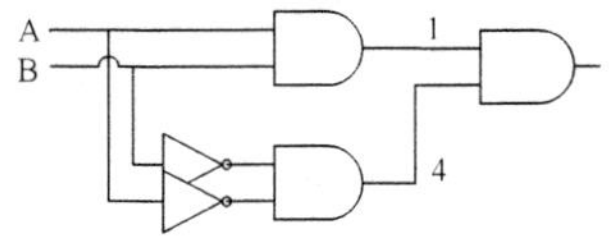

Fig. 1: Conflict Example

INPUT	NAND4	TREE	PARA
0000	0.43	7.34	7.34
0001	0.65	12.31	12.31
0010	0.65	10.53	12.31
0011	1.26	17.20	17.28
0100	0.65	12.31	12.31
0101	1.26	17.28	17.28
0110	1.26	15.50	17.28
0111	11.18	22.18	22.25
1000	0.64	10.53	10.53
1001	1.21	15.50	15.50
1010	1.21	13.71	15.50
1011	7.62	20.39	20.47
1100	1.20	15.42	15.42
1101	7.26	20.39	20.39
1110	7.07	18.61	20.32
1111	10.89	21.95	21.95

Tab. 1: Leakage Power of NAND4 and Its Decomposition (Units in nW)

Instead of using pattern dependent mapping, here we propose a input-vector assisted technology mapping. Assume that the least-leakage vector of the final mapped circuit is known, the logic value of internal nodes can be propagated from the primary input of the technology-decomposed network. Using these internal logic values, an accurate leakage power cost function can be used during the mapping process. The first assumption, however, is not valid since the least-leakage vector can only be found after the network is mapped. The issue is then, can an input vector be found without the knowledge of the final mapped circuit such that it is close to the LLV of the final mapped circuit? Here a heuristic is proposed which use the least leakage input vector obtained form the technology-decomposed network. This is based on an observation that for most of the gate primitives in the cell library, the LLV for the primitive is the same as that of its decomposed network. For example a NAND4 gate has two different pattern tree decompositions using NAND2 and INV1X called PARA and TREE. The PARA is balance tree decomposition while the TREE is a three level decomposition. The corresponding leakage power of all input vectors for the two decomposition is shown in Table 1. Notice that the least two leakage vectors of NAND4 are 0000 and 1000, while both of its decompositions have the same vectors. On the other hand, the most two leakage vectors are 0111 followed by 1111 and it is also the same in TREE and PARA decompositions. This trend is valid for most of the gate primitives in a typical cell library.

By using the above heuristic in technology mapping, the existing low power technology mapping algorithm is modified. Figure 2 shows the design flow of the proposed leakage-aware technology mapping algorithm with the modification is shown in shaded color. The LLV obtained from the decomposed network is only an approximation to the final mapped network. It is not necessary equivalent to the LLV of the final resultant network. Hence, after obtaining the final mapped network, the LLV is found again.

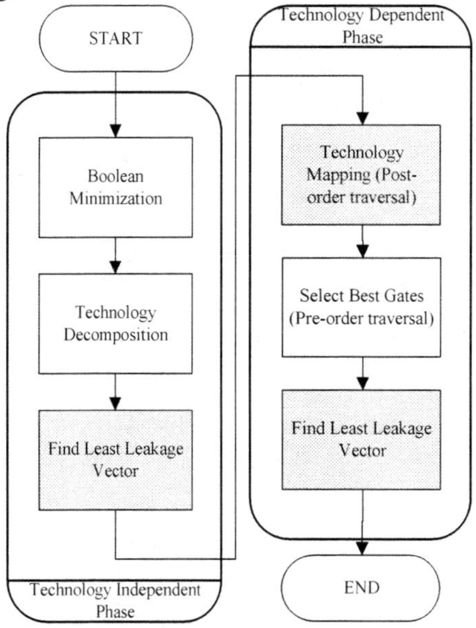

Fig. 2: Leakage Aware Logic Synthesis Flow

3. EXPERIMENTAL RESULTS

In order to verify the effectiveness of the proposed scheme, experiments using the MCNC91 benchmarks were carried out. The proposed algorithm was implemented in the Berkeley SIS platform [9] with USC POSE extension [10] and custom modifications. A 0.07um technology library based on Berkeley's predictive technology model [1] was used. A 1V power supply was used. The leakage current was simulated with the temperature 110°C and 25°C for active and standby mode respectively. The internal capacitance and timing information was also built by HSPICE. The algorithm in [5] was used to find the LLV for both the decomposed network and the final mapped network under a confidence level of 98%. The duty cycle parameter, α, was set to 1% and all inputs signal probabilities were set to 0.5. The operating frequency of the network is assumed to be 500MHz.

The experimental results and comparisons between proposed method and conventional power-delay technology mapping [7] are summarized in Table 2. Columns 2-7 show the area, delay and different power consumption components obtained by [7] while columns 8-13 are the corresponding normalized values of proposed method over the results using the algorithm in [7]. It should be noted that for both cases the final standby leakage power is obtained by finding the LLV of the final mapped circuits using the algorithm proposed in [5]. It can be seen that the total power is reduced by 20% on average when the leakage power is taken into account. For circuits such as, *majority*, the total power reduction can be as high as 45%. It can also be seen that the standby $P_{leakage}$ is reduced by 44% and the active $P_{leakage}$ is reduced by 20% while the $P_{dynamic}$ is increased by 16% on average. In general there is also a reduction in area. It is because the reduction of the leakage power is achieved by mapping some of the gates using large fan-in primitives. This will also result in fewer gates and less area. An average of 13% reduction is achieved. The average delay of the two schemes is about the same.

Besides, we also compared the LLV obtained from the decomposed network and that obtained from the final mapped circuit. It is observed that the two vectors are almost the same. This justifies the use of the technology-decomposed network to obtain the LLV for the mapping process.

4. CONCLUSION

In this paper, a total power model which includes active and standby leakage power was described. ALLV assisted technology mapping algorithm was presented to minimize the total power during logic synthesis. It uses the LLV obtained from the technology-decomposed network to act as an approximation of the LLV of the resultant network to calculate the standby power cost function during the technology mapping process. Experimental results show that significant improvement on total power optimization over traditional dynamic power technology mapping technique is achieved.

5. REFERENCES

[1] W. Liu, *et al.*, "BSIM3v3.2.2 MOSFET Model Users' Manual," Department of EECS, UC Berkeley, 1999.

[2] "Predictive Technology Model." http://www-device.eecs.berkeley.edu/~ptm.

[3] Z. Chen, M.C. Johnson, L. Wei, and K. Roy, "Estimation of standby leakage power in CMOS circuits considering accurate modeling of transistors stacks," in *Proc. Symp. Low Power Design Electron*, 1998, pp. 239-244.

[4] A. Abdollahi, et al., "Runtime mechanisms for leakage current reduction in CMOS VLSI circuits," in *Proc. Int. Symp. Low Power Electronics and Design*, 2002, pp. 213-218.

[5] J.P. Halter, F.M. Najm, "A gate-level leakage power reduction method for ultra-low-power CMOS circuits," in *Proc of the IEEE Custom Integrated Circuits Conference*, 1997, pp. 475-478.

[6] M.C. Johnson, et al., "Leakage control with efficient use of transistor stacks in single threshold CMOS," *in IEEE Trans. on Very Large Scale Integration Systems*, vol. 10, pp 1-5, Feb, 2002.

[7] C.Y. Tsui, *et al.*, "Technology decomposition and mapping targeting low power dissipation," in *Proc. of the Design Automation Conference*, pp. 68-73, 1993.

[8] C.W. Kang and M. Pedram, "Technology mapping for low leakage power and high speed with hot-carrier effect consideration," in *Proc. of the Design Automation Conference*, pp. 203-208, 2003

[9] "SIS: A system for sequential circuit synthesis," Report M92/41, UC Berkeley, 1992

[10] "POSE: Power optimization and synthesis environment," USC, 1995.

Benchmarks	Conventional Dynamic Power Mapping [7]						Proposed Method / Dynamic Power Mapping					
	Area	Delay	Dynamic	Active	Standby	Total	Area	Delay	Dynamic	Active	Standby	Total
C499	789276	556.85	89757.36	23986.84	1693.08	2685.62	0.91	1.01	1.08	0.88	0.71	0.84
C5315	3341452	873.68	484535.88	98864.09	7005.09	12468.21	0.89	1.19	1.13	0.83	0.65	0.85
con1	34408	117.77	4415.64	1048.73	62.32	116.34	0.86	0.93	1.13	0.86	0.71	0.88
cordic	238272	265.72	29239.00	6794.96	430.08	786.11	0.87	0.94	1.17	0.82	0.67	0.87
count	234872	609.32	25893.50	6780.88	428.13	617.72	1.09	0.95	1.21	1.00	0.58	0.83
e64	3348694	530.65	331937.21	70581.90	3578.81	7568.21	0.74	1.03	1.20	0.50	0.24	0.68
ex4	765816	290.04	101897.92	21607.06	1477.47	2697.75	0.89	0.97	1.21	0.85	0.63	0.87
example2	533324	365.48	54132.13	16205.03	1088.26	1640.37	0.85	1.00	1.21	0.75	0.61	0.81
i9	1490050	2128.37	155996.13	47885.39	2409.88	5058.92	0.97	1.11	1.05	0.93	0.73	0.87
lal	259828	233.68	32778.82	7204.73	398.64	755.37	0.84	0.92	1.24	0.72	0.42	0.79
majority	19380	84.90	2434.52	605.58	37.41	37.69	0.61	1.02	1.14	0.60	0.15	0.55
my_adder	294236	839.30	45113.50	9078.93	644.90	1163.42	0.94	1.00	1.12	0.95	0.66	0.86
rd73	275672	331.11	36667.07	8104.16	552.00	994.19	0.90	1.04	1.17	0.85	0.47	0.76
rd84	320484	352.44	42769.06	9159.12	605.33	1118.56	0.82	1.05	1.12	0.75	0.62	0.82
vg2	366724	307.08	34341.80	10404.15	621.36	1062.60	0.82	1.08	1.24	0.71	0.47	0.74
x1	652528	312.93	84287.67	18109.27	1120.39	2012.61	0.84	1.10	1.18	0.75	0.49	0.78
x4	825860	445.94	112073.34	22657.64	1401.79	2698.31	0.90	1.04	1.13	0.85	0.58	0.83
Z9sym	208760	372.77	27227.70	5947.98	397.21	724.99	0.85	1.08	1.11	0.77	0.65	0.83
						Average	0.87	1.03	1.16	0.80	0.56	0.80

Tab. 2: Simulation Results and Comparisons

A Predictive Methodology for Accurate Substrate Parasitic Extraction

Ajit Sharma, Chenggang Xu
School of EECS
Oregon State University
Corvallis, OR 97331

Wen Kung Chu, Nishath K. Verghese
Cadence Design Systems
San Jose, CA 95134

Terri S. Fiez, Kartikeya Mayaram
School of EECS
Oregon State University
Corvallis, OR 97331

***Abstract*—A methodology for determining the substrate profile for accurate prediction of parasitics using Green's function based substrate extractors is presented. The technique requires fabrication of only a few test structures and results in an accurate three layered approximation. The substrate resistances are accurate to within 10% of measurements. This methodology can be used along with a scalable macromodel for a qualitative pre-design and pre-layout estimation of the digital switching noise that couples though the substrate to sensitive analog/RF circuits.**

I. INTRODUCTION

With the continued scaling of CMOS processes, it is possible to integrate digital, analog and RF circuitry on a single chip. The digital circuits inject noise that is sensed by the sensitive analog and RF blocks through the common substrate. This noise coupling from the substrate can severely degrade the performance of noise sensitive circuits. For this reason, it is essential that substrate noise coupling analysis is included in the design flow of mixed-signal integrated circuits.

Considerable work has been done in modeling the substrate. Green's function based substrate parasitic extractors [1 - 6] and scalable macro-models for the silicon substrate [7, 8] are two approaches that have received considerable attention over the last few years.

In this paper, an optimization-based method is used to calibrate the substrate profile for use with substrate parasitic extractors based on the Green's function approach. The proposed technique requires fabrication of only a few test structures for z-parameter measurements. This approach leads to a simple description of the layered substrate which is derived from the doping profiles. Substrate resistances obtained from Green's function based extractors using this calibration approach are found to be in close agreement with measurements.

The paper is organized as follows. The relationships between the substrate z-parameters and resistances are explained in Section II. The motivation for optimization of substrate resistivities and thicknesses is given in Section III. The calibration methodology is described in Section IV. Section V provides measurement results that validate this optimization procedure. Finally conclusions are provided in Section VI.

Ajit Sharma is now with the School of ECE, Georgia Institute of Technology, Atlanta, GA.

This work is supported in part by NSF grant CCR-0096176, by SRC under contract 2001-TJ-911 and by the DARPA TEAM project under contract F33615-02-1-1179.

II. SUBSTRATE RESISTANCES AND Z-PARAMETERS

For a generic two-port as shown in Figure 1, the z-parameters are obtained by forcing a current though one port and measuring the voltage at the other port which is an open circuit.

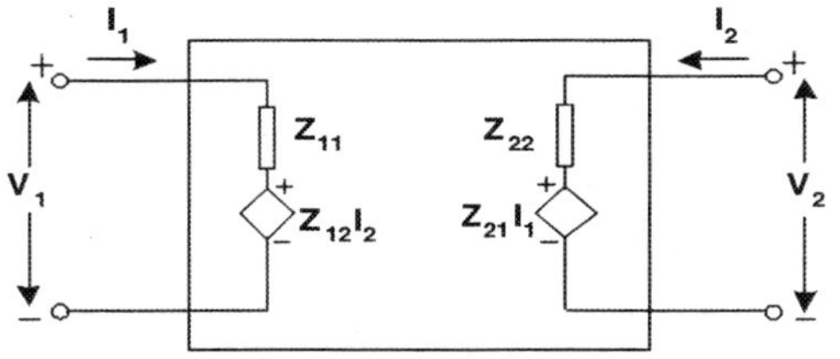

Fig. 1. The z-parameter representation of a two-port system.

The substrate can be modeled as in [2], [3], [8] by a simple two-port π-network of resistances for frequencies less than a few GHz. Figure 2 shows the π-resistive network. Between two ports[1] of interest there is a cross coupling resistance, R_{12}. R_{11} and R_{22} account for the coupling from each of the ports to the backplane.

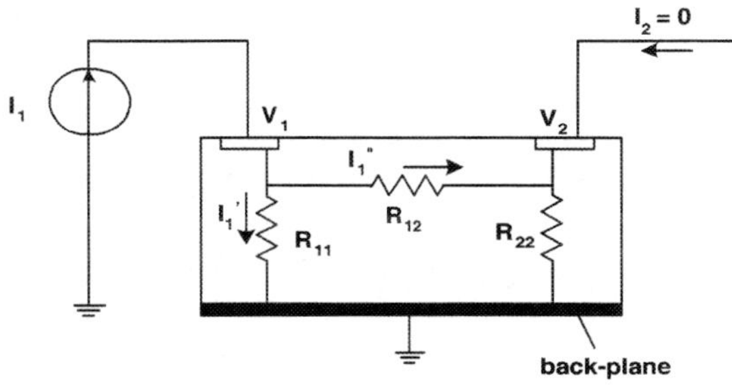

Fig. 2. The π resistance network for the substrate parasitics between two ports.

Eqs. (1) and (2) give the relationship between the resistance values and the equivalent z-parameters. Here a current I_1 is injected into the substrate at port 1, and port 2 is left open. From circuit analysis,

$$Z_{11} = R_{11} \mid\mid (R_{12} + R_{22}) \quad (1)$$

[1] A P^+ diffusion region on the substrate is referred to as a P^+ contact, or *port*. Examples of ports in real circuits are substrate taps and active regions of transistors.

$$Z_{12} = \frac{R_{11}R_{22}}{R_{11} + R_{12} + R_{22}} \quad (2)$$

From a circuit design point of view, the noise transfer function (NTF) given by Eq. (3) is of interest.

$$\frac{v_{out}}{v_{in}} = \frac{v_2}{v_1} = \frac{R_{22}}{R_{12} + R_{22}} \quad (3)$$

Eqs. (1), (2) and (3), show that there exists a close relationship between the substrate resistances, the equivalent z-parameter representation and the amount of noise transferred from the injector to the sensor. Hence, an error in the prediction of the substrate resistances will result in a corresponding error in the calculated noise voltage at the node of interest.

III. Motivation for Optimization of Substrate Resistivities and Thicknesses

A generic flow for substrate noise coupling analysis is shown in Figure 3. In this flow the substrate parasitic extractor is an important block. 3-D Green's function based substrate parasitic

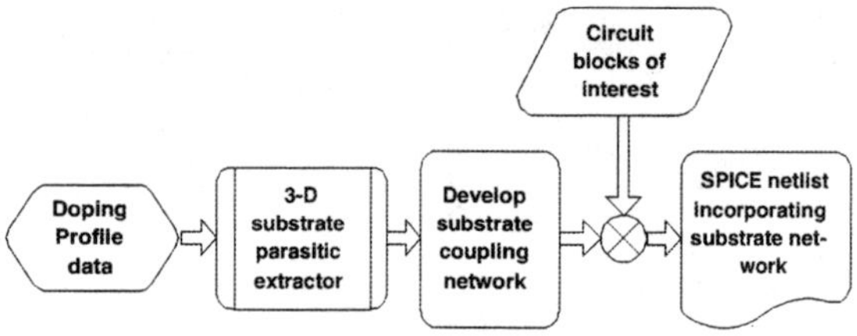

Fig. 3. Generic flow for substrate noise analysis.

extractors and most commercially available substrate analysis tools (e.g., Cadence SeismIC, Agilent/EEsof-Momentum) require the substrate to be described as layers of uniform resistivities. This layered description of the silicon substrate must be derived from the actual spreading resistance profile (SRP) data of the particular process run, as shown in Figure 4. Typically

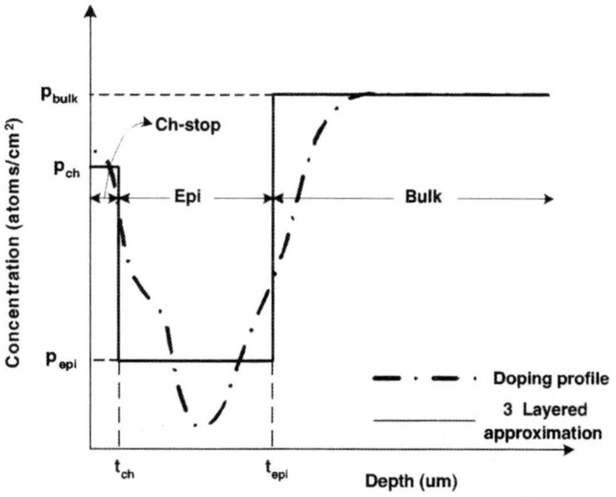

Fig. 4. Actual doping profile and a 3 layered representation of the silicon substrate.

a heavily doped substrate is approximated by a 3 layered structure and a lightly doped substrate by a 2 layered structure [2]. Therefore, the accurate determination of the equivalent resistivities and the thicknesses of the layers is important.

The resistivities and thicknesses used for the layers have been found to impact the values of the extracted substrate parasitics significantly. Despite accurate information of the doping profiles, it is difficult to heuristically determine a good approximation to the substrate using only a few layers of uniform resistivities.

A very fine discretization of the SRP data may yield accurate results, but the associated computational costs are enormous. For instance, if the number of layers is increased from 3 to 6 for a single injector-sensor pair, the computation time for extraction increases by about 4 times. The computation time also increases with the number of contacts which indicates that such an approach will be impractical for large circuits. For this reason, an optimization based approach to obtain a simple layered description of the substrate has been developed.

The empirical macro-model proposed in [8], is extracted using a least squares fit to data. The accuracy of the model, the efficiency of its extraction and its range of application can be improved by using an optimum set of test structures. In [8] z-parameter data was obtained from measurements from test structures fabricated on silicon. This can, however, be expensive and time consuming.

If a fast and accurate alternative to measurement data is available, then there is a minimal dependence on fabricated test structures.

The proposed procedure, uses only a few test structures to obtain optimized substrate layer resistivities and thicknesses. The Green's function based extractors can be used with these "calibrated" layer resistivities and thicknesses to obtain the z-parameters. This provides a cost effective, reliable and fast means of obtaining data for model generation and validation.

IV. Calibration Methodology

After the optimization, the resulting 3 layered approximation is accurate (as shown later) and speeds up computation of the substrate parasitic network significantly. Figure 5 shows the flowchart for the calibration procedure which was implemented as a stand-alone application. The calibration procedure requires the following:

1) A fast 3-D substrate parasitic extractor.
2) An appropriate calibration metric, which depends on the substrate being calibrated.
3) An appropriate initial guess for the variables being optimized.
4) An appropriate termination criterion (e.g. tolerances on the variable values or gradients).

EPIC [6], a fast 3-D Green's function based substrate parasitic extractor has been used in this work. The Z_{11} values of eight test structures obtained from measurement are used as a reference for this calibration. These structures were fabricated on a test chip in the TSMC $0.35\mu m$ CMOS heavily doped process. The set of structures chosen for calibration must span the range of equivalent port sizes that can occur in the layout. Also, since both square and rectangular ports can occur in a layout, the calibration set must include square and rectangular shaped structures as well.

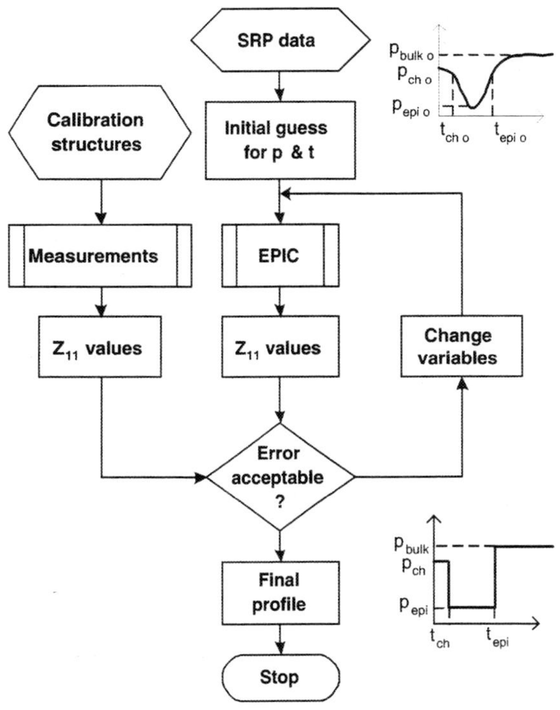

Fig. 5. Flow-chart for substrate calibration.

As explained earlier, a simple description of the vertical stratification of the substrate is to be determined. Z_{11}, the open-circuit parameter that represents the impedance from the port (at the surface) to the back-plane, is the *only* parameter that is used in this calibration procedure. There are two reasons for using Z_{11}. First, Z_{11} can be extracted from measurement independent of other contacts and second, Z_{11} from Eq. (1), contains contributions of R_{11} and R_{12} - both of which affect the degree of signal coupling.

The optimization loop is based on the Levenberg-Marquardt algorithm [9]. An arbitrary initial guess for the layer resistivities and thicknesses is made based on inspection of the actual SRP data. The variables are the resistivities and thicknesses of the channel stop (ρ_{ch}, t_{ch}) and the epitaxial (ρ_{epi}, t_{epi}) regions. The resistivity of the heavily doped bulk was assumed constant (since the doping level is constant) and taken from the SRP data. The thickness of the fabricated die was $200\mu m$. The thickness of the bulk region is obtained by subtracting the thicknesses of the channel stop and epi layers from the wafer thickness.

The optimization problem that has to be solved is given in Eq. (4)

$$min \quad \frac{1}{2}\sum_{i}\left\{\frac{Z_{11,i\ EPIC}(\rho_{ch}, t_{ch}, \rho_{epi}, t_{epi})}{Z_{11,i\ measured}} - 1\right\}^2 \qquad (4)$$

where *i* refers to the ith test structure used for the calibration.

A simple weighting of the values is inherently achieved by the normalization used in the above objective function. The Z_{11} values from EPIC are divided by the corresponding measured Z_{11} values. This normalization is needed because some of the ports are very small ($0.7\mu m \times 0.7\mu m$) and hence they have much larger Z_{11}'s as compared to other large ports ($60\mu m \times 60\mu m$) which have small Z_{11}'s. This can cause the resistivities and thicknesses to be biased and the normalization ensures that all the test cases are weighted equally.

Table I shows the values of Z_{11} obtained after the optimization based on measurements from the eight test structures used in the calibration phase. Table II shows the values of the resistivities and thicknesses obtained before and after the calibration procedure.

TABLE I

Z_{11} VALUES FROM EPIC AFTER OPTIMIZATION ARE IN GOOD AGREEMENT WITH MEASUREMENTS FOR THE STRUCTURES USED IN THE CALIBRATION STEP.

Contact Size	Z_{11} (Ω) *Measurement*	Z_{11}(Ω) *After Optimization*
$0.7\mu m \times 0.7\mu m$	1830.9341	1837.411
$0.85\mu m \times 3.1\mu m$	1016.689	977.430
$1.5\mu m \times 1.5\mu m$	1138.192	1157.296
$2.3\mu m \times 2.3\mu m$	887.186	904.536
$3.1\mu m \times 3.1\mu m$	738.027	742.528
$6\mu m \times 6\mu m$	488.034	477.487
$20\mu m \times 40\mu m$	96.961	100.727
$60\mu m \times 60\mu m$	37.535	36.150

TABLE II

RESISTIVITIES AND THICKNESSES OF LAYERS BEFORE AND AFTER OPTIMIZATION.

Parameter	*Initial guess*	*Value After Optimization*
ρ_{ch}	$0.6\Omega cm$	$0.205\Omega cm$
t_{ch}	$1.8\mu m$	$0.9525\mu m$
ρ_{epi}	$6\Omega cm$	$6.587\Omega cm$
t_{epi}	$3.2\mu m$	$3.235\mu m$

V. EXPERIMENTAL VERIFICATION

Measurements for the validation of the optimization procedure have been made on the $0.35\mu m$ test-chip described above. Figure 6 compares the pre- and post-optimization Z_{11} values for 7 *different* structures that were not used in the calibration step. The contact sizes are shown in the figure as well. As can be seen from the figure, considering all cases, the maximum error after optimization is approximately 10 %.

Figure 7 shows the Z_{12} values, for a pair of $0.85\mu m \times 1.5\mu m$ contacts as a function of mutual separation. The Z_{12} values obtained from the calibrated layer thicknesses and resistivities are more accurate than those obtained by using the initial guess. For noise estimation purposes, the values of R_{11} and R_{12} are important and these can be easily calculated from the values of Z_{11} and Z_{12}. Figures 8 and 9 show the R_{11} and R_{12} values respectively, of a pair of $0.85\mu m \times 1.5\mu m$ P^+ contacts as the separation between them is increased from $0.6\mu m$ to $100\mu m$. Using the calibrated layer resistivities and thicknesses, EPIC is able to accurately match measurements. The maximum error is

7 % in R_{11} and the cross coupling resistance R_{12} is predicted accurately to within 10 %.

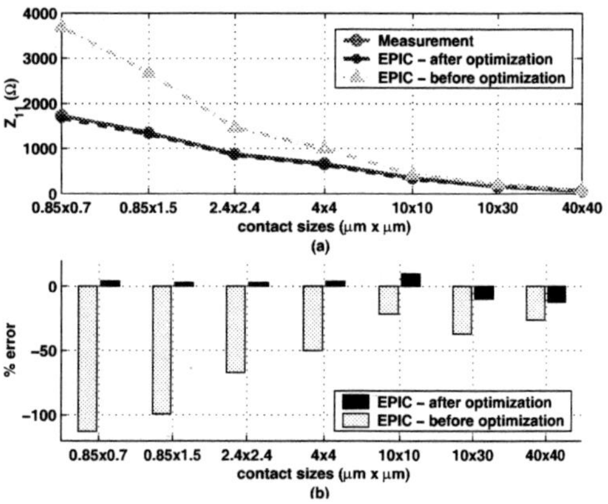

Fig. 6. Validation step with Z_{11} for different contact sizes: (a) Values of Z_{11} before and after optimization, (b) Percentage error relative to measurements for the test structures used for validation.

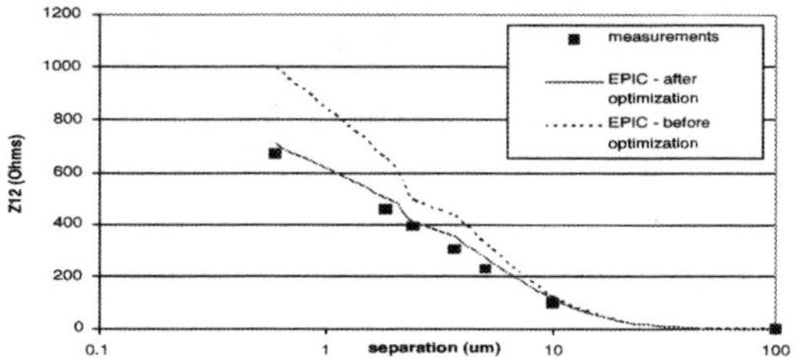

Fig. 7. Z_{12} values (before and after optimization) of two $0.85\mu m \times 1.5\mu m$ contacts as a function of separation.

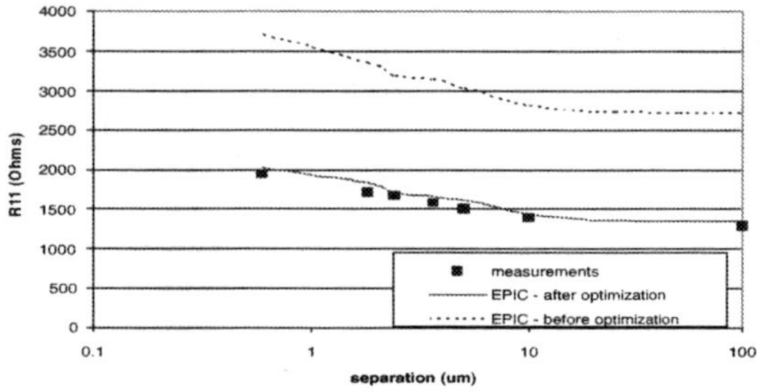

Fig. 8. R_{11} values (before and after optimization) of two $0.85\mu m \times 1.5\mu m$ contacts as a function of separation.

The above results lead to an accurate prediction of the noise transfer function (NTF) as shown in Figure 10. Using the values of R_{11} and R_{12} obtained from EPIC and the initial guess (i.e. layer resistivities and thicknesses before optimization), the NTF predicted underestimates the noise by up to 40 %. As can be seen, as the separation between the ports increases, the value of NTF decreases. A higher value of NTF implies greater coupling. The graph in Figure 10 implies coupling between the ports decreases monotonically with increase in separation - which is what is expected.

VI. Conclusions

The results presented in this paper show that (i) it is possible to accurately model a heavily doped epitaxial substrate as a 3 layered structure, with uniform layer resistivities and (ii) the Z_{11} of a single contact is sufficient for use in the proposed calibration procedure. In summary, a systematic calibration of the substrate has been shown to be an effective and seamless means of obtaining accurate substrate parasitic extraction, thus enabling substrate noise coupling analysis in large mixed-signal SoCs.

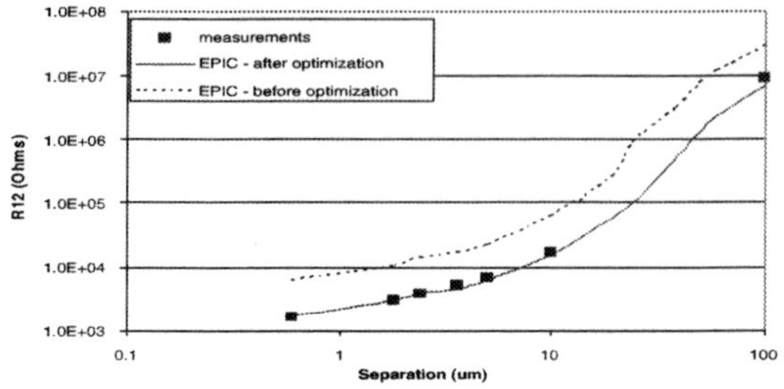

Fig. 9. R_{12} values (before and after optimization) of two $0.85\mu m \times 1.5\mu m$ contacts as a function of separation.

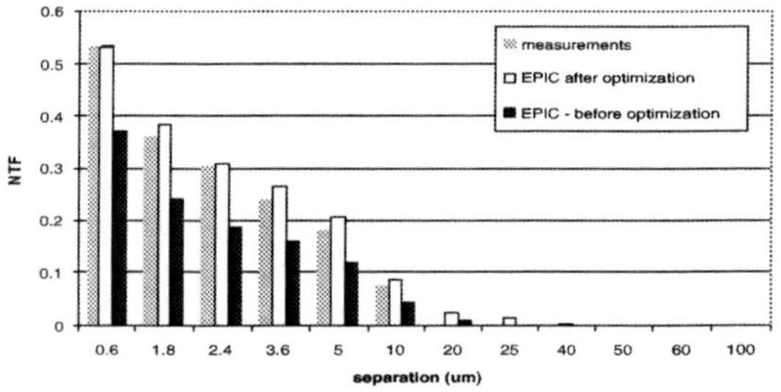

Fig. 10. Noise transfer function (NTF) is predicted accurately after the calibration procedure for two $0.85\mu m \times 1.5\mu m$ contacts as a function of separation.

References

[1] N. K. Verghese, T. J. Schmerbeck and D. J. Allstot, "Simulation techniques and solutions for mixed-signal coupling in integrated circuits," Kluwer Academic Publishers, 1995.

[2] R. Gharpurey and R. G. Meyer, "Modeling and analysis of substrate coupling in integrated circuits,"*IEEE Journal of Solid-State Circuits*, vol. 31, no. 3, pp. 344-352, March 1996.

[3] N. K. Verghese, D. J. Allstot and M. A. Wolfe, "Fast parasitic extraction for substrate coupling in mixed-signal ICs,"*IEEE Custom Integrated Circuits Conference*, pp. 121-124, May 1995.

[4] J. P. Costa, M. Chou and L. M. Silveira, "Efficient techniques for accurate modeling and simulation of substrate coupling in mixed-signal IC's,"*IEEE Trans. Computer-Aided Design*, pp. 597-607, May 1999.

[5] T. Smedes, N. P. van der Meijs and A. J. van Genderen, "Boundary element methods for 3D capacitance and substrate resistance calculations in inhomogeneous media in a VLSI layout verification package", *Advances in Engineering Software,* vol. 20, no. 1, pp. 19-27, 1994.

[6] C. G. Xu, K. Mayaram, T. S. Fiez, "EPIC: A program for extraction of the resistance and capacitance of substrate with the Green's Function method," Dept. of ECE, Oregon State University, 2001.

[7] A. J. van Genderen, N. P. van der Meijs and T. Smedes, "Fast computation of substrate resistances in large circuits," *Proc. European Design and Test Conf.*, pp. 560-565, March 1996.

[8] D. Ozis, T. Fiez, and K. Mayaram, "A comprehensive geometry-dependent macromodel for substrate noise coupling in heavily doped CMOS processes," *IEEE Custom Integrated Circuits Conference*, May 2002.

[9] D. Marquardt, "An algorithm for least-squares estimation of nonlinear parameters," *SIAM J. Appl. Math.*, vol. 11, pp 431-441, 1963.

Modeling and Optimization of On-Chip Spiral Inductor in S-parameter Domain

Kenichi Okada
Dept. Communications &
Computer Engineering
Kyoto University
kokada@vlsi.kuee.kyoto-u.ac.jp

Hiroaki Hoshino
Dept. Communications &
Computer Engineering
Kyoto University
hoshino@vlsi.kuee.kyoto-u.ac.jp

Hidetoshi Onodera
Dept. Communications &
Computer Engineering
Kyoto University
onodera@i.kyoto-u.ac.jp

ABSTRACT

This paper presents a methodology for optimizing the layout of on-chip spiral inductors using structural parameters and design frequency in a response surface method. The proposed method uses scattering parameters (S-parameter) to express inductor characteristics, and hence is independent of spiral geometries and equivalent circuit models. The procedure of inductor optimization is described, and a design example is presented.

1. INTRODUCTION

Radio-frequency integrated circuits (RF ICs) require high-quality on-chip inductors. Spiral inductors are key components in voltage-controlled oscillators (VCO) and low-noise amplifiers (LNA)[1, 2]. At high frequency, several effects degrade the quality factor (Q) of the inductor, such as skin effects, eddy current, substrate loss, and self resonance[1]. The magnitude of these effects depends heavily on the layout structure of the spiral inductor. The design of spiral inductors represents an area of major interest in RF circuit design, and several structures for high-Q inductors have been proposed[3, 4].

A practical problem in the design of spiral inductors is developing an appropriate design that achieves high-Q at a specific inductance. This process is commonly conducted by trial and error through measurement or simulation, and typically takes considerable time to arrive at an optimal structure. Another approach is to use an analytical equation[5, 6, 7], which eliminates the need for fabrication or simulation, but requires a new equation model to be developed for each technology, spiral structure and equivalent circuit model. Moreover, it is difficult to construct a model that can satisfactorily express the characteristics for a wide frequency range, because the models use component values(L, R, etc.) of equivalent circuits. In our method, spiral inductors are hence characterized by S-parameters instead of equivalent circuits.

In this paper, a methodology for designing an optimal layout structure is presented. The methodology is based on a response surface method (RSM), and the inductor characteristics are described by the response surface functions of scattering parameters (S-parameters). The response surface function is an equation involving physical layout parameters such as line width, line spacing and number of turns. Exploiting the well-established theory of experimental design[9], the proposed method significantly reduces the number for trial fabrications and simulations required to achieve the desired result. The model by the proposed method is constructed entirely in the S-parameter domain, and hence is independent of equivalent circuit models and spiral geometries. The proposed method therefore can be applied to any type of inductor, and various equivalent circuit models can be readily obtained from the same S-parameter-based equations with little numerical calculation.

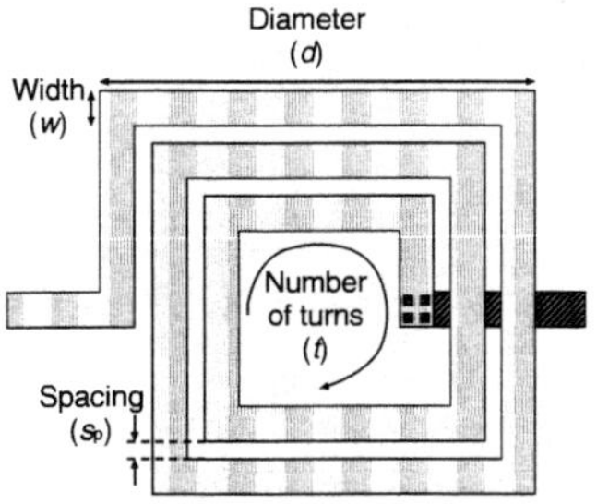

Figure 1: A rectangular layout structure of spiral inductor.

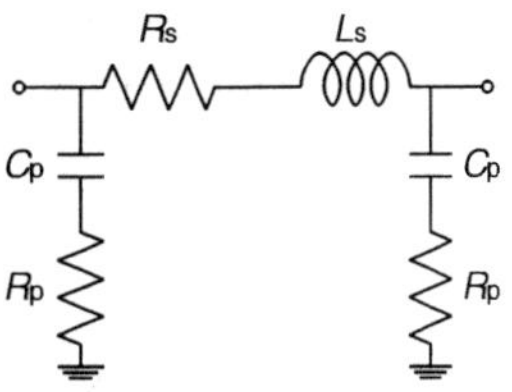

Figure 2: An example of equivalent-circuit model.

Section 2 explains the required characteristics of spiral inductors and discusses conventional design methodologies for optimizing the structure. The proposed design optimization methodology is then described and applied to a design example in Sect. 3 and Sect. 4. Section 5 evaluates its performance in terms of modeling accuracy of the designed spiral inductor. Finally, Sect. 6 concludes this paper.

2. CHARACTERISTICS OF INDUCTORS AND DESIGN PROBLEMS

In this section, the required performance of spiral inductors in the design of RF circuits is described and the difficulties in obtaining optimal structures are explained.

Several spiral geometries with high-Q values have been proposed, including rectangular (Fig. 1), octagonal, circular, and symmetric designs[1, 4]. Corresponding equivalent-circuit models have also been proposed, and example of which is shown in Fig. 2. The equivalent-circuit model consists of several passive components; series metal resistance R_s, capacitance between metal and substrate C_p, and resistance of the silicon substrate R_p, in addition to inductance L_s.

The layout geometry shown in Fig. 1 and the equivalent circuit model described in Fig. 2 are adopted in this study as an example. The proposed method can deal with other layout geometries and equivalent-circuit models with no theoretical limitations. The layout of spiral inductors can be determined by the diameter (d), metal width (w), number of turns (t), and spacing between lines (s_p), collectively referred to as "structural parameters". The value of each component in the equivalent circuit can then be calculated from the S-parameters with a straightforward translation.

The inductor structure is usually designed so as to maximize the Q value of the inductor, under constraints of the required inductance. The trial-and-error approach is time-consuming, and does not guarantee that an optimal structure will be found in the limited design term. Without experienced guidance, such attempts require an impractical number of iterations and tend to arrive at a local optimum.

When an analytical model for a spiral inductor is known, the structure can be easily and quickly optimized analytically or numerically. The analytical model basically expresses the Q value in terms of component values (L_s, R_s, etc.) of the equivalent circuit[5, 6, 7], and the equations are usually polynomials of the structural parameters. There are two types of analytical models; physical[5, 6] and numerical[7]. The former is based on physical analysis, while the latter is derived by curve fitting from measured values. For high frequencies, skin effects and eddy current must also be considered. Moreover, a new set of equations needs to be derived for every layout strategy and every technology. For the numerical model, a refined equation model needs to be derived for each equivalent circuit model, and the equations need to be reconstructed for different equivalent circuit models.

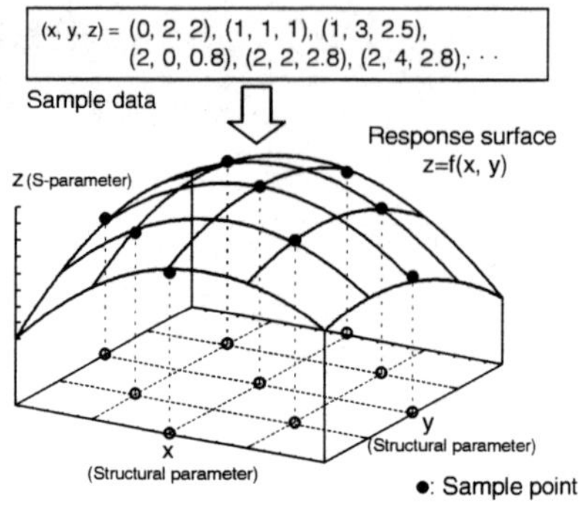

Figure 3: Example of response surface function with two parameters

3. OPTIMIZATION METHODOLOGY

In this section, a design methodology for optimizing the layout of spiral inductors is proposed. Spiral inductors are modeled in the S-parameter domain using a response surface method. Inductance and Q values are derived from S-parameters. By modeling in the S-parameter domain, circuit designers are able to readily change the equivalent circuit model for specific applications, and any spiral geometry can be handled. Optimization by the proposed method is also of low computational cost because the inductor characteristics are expressed as polynomial equations. The models are characterized based on the theory of experimental design[9].

The proposed methodology adopts the response surface method to construct a model from measured and/or simulated inductor characteristics. The spiral inductor layout is then optimized using the developed model.

The response surface method constructs a simple model representing the relationship between structural parameters and S-parameters. The model is derived from measured and/or simulated sample data. Figure 3 shows an example of a response surface function (RSF). The functions are equations with variables of structural parameters and frequency. The design constraints and objective functions are determined uniquely from the functions of S-parameters given the design equivalent circuit model. Constraint conditions are typically applied in terms of inductance, maximum parasitic capacitance, or maximum layout area. The objective function is usually Q or parasitic capacitance, but other objectives are possible. These values are then calculated from S-parameters, and the appropriate component values for the equivalent circuit can also be calculated from the S-parameters. Although the transformation is dependent on the equivalent circuit model, the derivation is straightforward. This approach therefore allows the inductor structure to be optimized purely through functions of S-parameters.

The values of objective functions can be obtained from the response surface functions for any set of structural parameters, whereas the response surface functions themselves are constructed from discrete sample data. With smart experimental design, the response surface method can provide a continuous equation from limited sample data through statistical theory. Using such a continuous objective function, the proposed method can be used to systematically search for an optimal structure analytically and/or numerically, significantly reducing the cost of optimization. Furthermore, the proposed method can incorporate frequency as one of the variables in the response surface functions, allowing the frequency-dependence of spiral inductors to be taken into consideration.

The use of S-parameters to characterize the inductor allows the proposed method to be applied to all spiral geometries and equivalent circuits. S-parameters are independent of the equivalent circuit model, and can be measured for all spiral geometries.

4. DESIGN EXAMPLE

In this section, the proposed method is applied to the optimization of a spiral inductor in a 0.18μm CMOS process. A commercial three-dimensional field-solver[10] is employed to evaluate S-parameters. In this optimization, the Q value of an inductor of 3 nH at 2.5 GHz is maximized. Diameter (d), metal width (w), and number of turns (t), as shown in Fig. 1, are adopted as structural parameters with the following ranges.

$$150\mu m \leq d \leq 250\mu m \quad (1)$$

$$10\mu m \leq w \leq 20\mu m \quad (2)$$

$$t = 2.5, 3.5 \text{ and } 4.5 \quad (3)$$

Table 1 lists the 26 points taken as sample data. The parameter sets for each sample point are determined by the experimental design. Top metal is used. It is well known that the metal spacing (s_p) should be minimal to reduce fringing capacitance between metal lines[1].

To evaluate the S-parameters, the frequency f is swept from 2 to 3 GHz in 0.1-GHz increments, and the inductor characteristics are simulated for each sample point listed in Table 1. The response surface model of S-parameters is built from the sample data in Table 1. Four response surface functions are necessary in this example, for the real and imaginary parts of s_{11} and s_{21}. The four variables are diameter (d), metal width (w), number of turns (t) and frequency (f).

In this example, the response surface functions are assumed to be cubic, however, higher polynomials may also be used if necessary. The functions are calculated in this case from the S-parameters evaluated by the electromagnetism simulator. The response surfaces are shown in Fig. 4. A spiral structure cannot be realized in the region of $d < 2(t \cdot w + (t-1) \cdot s_p)$, as indicated by the broken line in Fig. 4(a). Figure 5 shows the response surface functions in the frequency domain, demonstrating that the functions effectively express the frequency dependence of the S-parameters. The accuracy of the response surface functions is discussed in Sect. 5.

In this design example, the constraint condition is L and the ob-

Table 1: Sample spiral inductors structures for creating the response surface

Diameter [μm]	Width [μm]	# turns
150	10	2.5, 3.5, 4.5
150	15	2.5, 3.5
200	10	2.5, 3.5, 4.5
200	15	2.5, 3.5, 4.5
200	20	2.5, 3.5
250	10	2.5, 3.5, 4.5
250	15	2.5, 3.5, 4.5
250	20	2.5, 3.5, 4.5
100	10	3.5
300	15	3.5
200	5	3.5
250	25	3.5

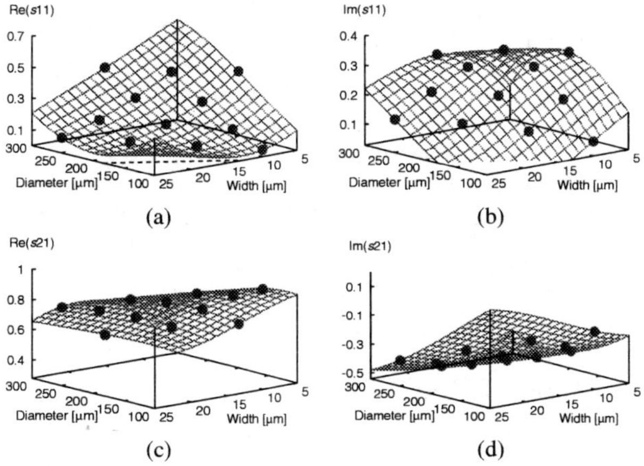

Figure 4: Plot of sample data and response surface functions of S-parameters at t = 3.5 and f = 2.5GHz for (a) Re(s_{11}), (b) Im(s_{11}), (c) Re(s_{21}), and (d) Im(s_{21}).

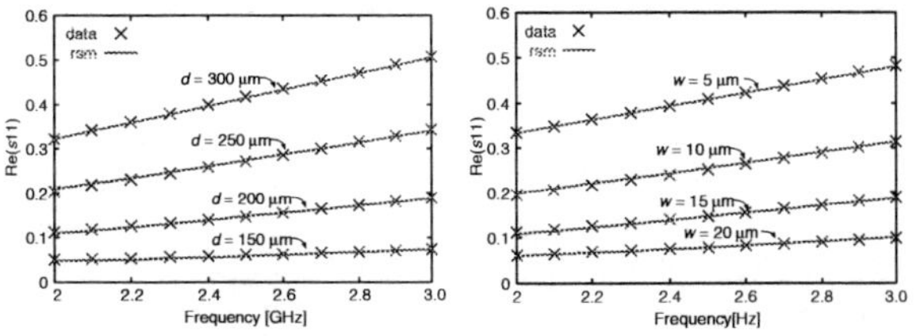

(a) $w = 10\mu m$ (b) $d = 200\mu m$

Figure 5: Sample data and response surface functions for Re(s_{11}) vs. frequency at t = 3.5. (a) w = 10μm, and (b) d = 200μm.

jective function is Q, as calculated using the following equations.

$$Q = -\frac{\mathrm{Im}(y_{11})}{\mathrm{Re}(y_{11})} \quad (4)$$

$$L = -\frac{1}{2\pi f} \cdot \mathrm{Im}\left(\frac{1}{y_{12}}\right) \quad (5)$$

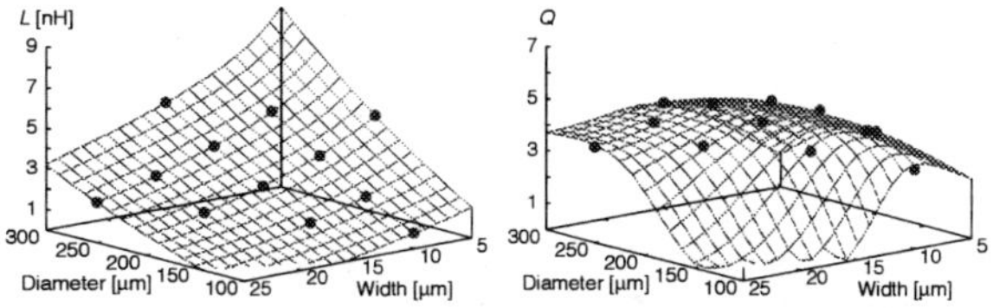

Figure 6: Plot of L and Q at f = 2.5 GHz and t = 3.5

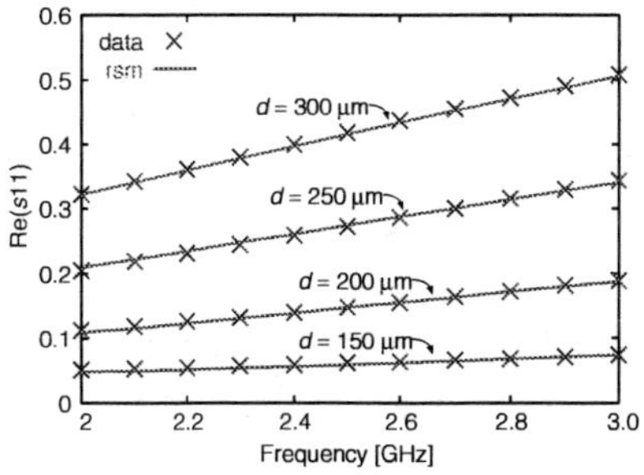

Figure 7: The range of structures that satisfy the required inductance of 3 ± 0.05 nH

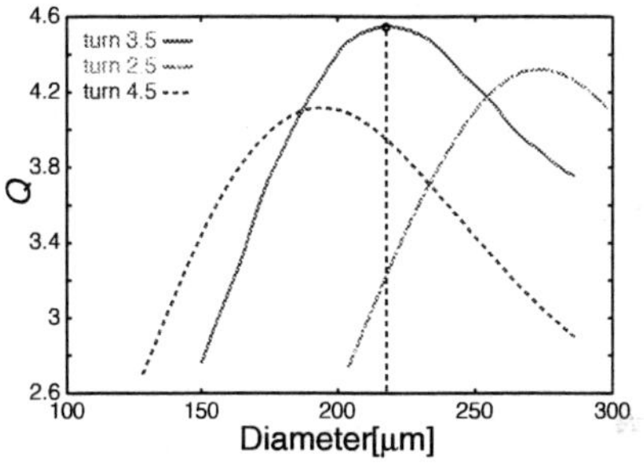

Figure 8: Q values of the structures shown in Fig. 7

Y-parameters can then be calculated from the S-parameters. Plots of L and Q are shown in Figs. 6.

The range of structures satisfying the required inductance of 3 ± 0.05 nH at 2.5 GHz in the three-parameter space of diameter d, width w and number of turns t is shown in Fig. 7. The Q values of the structures are shown in Fig. 8. The curve for t = 3.5 realizes the maximum Q value, and this t value is selected as optimal. Thus, the optimal structure derived from Figs. 7 and 8 provides a Q value of 4.5 with structural parameters of d = 220 μm, w = 14 μm, and t = 3.5.

5. ACCURACY OF RESPONSE SURFACE FUNCTIONS

In this section, the modeling error of the response surface functions is investigated. Figure 9 (solid circle) and Table 2 shows the error between simulated and modeled values at the optimal point. The simulated values are derived from electromagnetic simulations, and the modeled values are derived from response surface model. All errors are less than 1% in the case of S-parameters and L, and

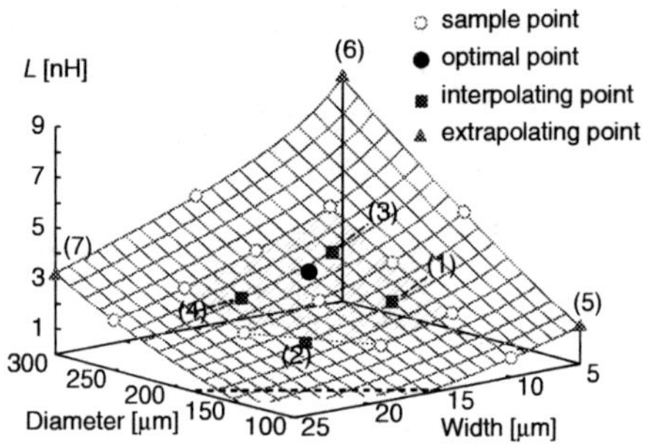

Figure 9: Sample points and data point for evaluation of model accuracy. Shaded area represents the interpolation region.)

Table 2: Accuracy of the proposed model at the optimal point

	Error at 2.5 GHz [%]	Average error 2–3 GHz [%]	Maximum error 2–3 GHz [%]
Re(s_{11})	0.5	0.3	0.6
Im(s_{11})	0.1	0.3	0.6
Re(s_{21})	0.1	0.1	0.2
Im(s_{21})	0.2	0.4	0.7
L	0.6	0.5	0.6
Q	2.1	2.2	3.3

Table 3: Accuracy of the proposed model at sample points

	Average error [%]	Maximum error [%]
Re(s_{11})	2.1	27
Im(s_{11})	1.2	10
Re(s_{21})	0.4	6.2
Im(s_{21})	0.7	5.9
L	1.0	8.3
Q	2.9	14

around 3% for Q.

The average and maximum error of the response surface functions at the sample points listed in Table 1 are shown in Fig. 9 (open circles) and Table 3 for the frequency range of 2–3 GHz. The maximum errors of Re(s_{11}) and Im(s_{11}) are 10% or more at the edge of the frequency range. The average errors are less than 3%, and the response surface functions are satisfactorily accurate at the sample points.

The error at a number of interpolated and extrapolated points is shown in Fig. 9 (squares and triangles) and Table 4. The errors in S-parameters, L, and Q are less than 4% in the interpolation region, which is considered sufficiently accurate for practical inductor optimization.

The error at the extrapolated data points is quite large, as expected. This is very natural because the response surface model does not aim to extrapolate sample data. Therefore, when using the RSM, care should be taken to choose parameter ranges such that all the points of interest will be included. If the optimal structure falls in the extrapolation region, new response surface functions should be derived from new sample points that include all the points of interest.

Table 4: Accuracy of the proposed model at other points

	Structure (d, w, t)	Max. ave. error [%]	Average L error [%]	Average Q error [%]
(1)	(175,12,3.5)	2.2 in Re(s_{11})	1.1	2.7
(2)	(175,18,3.5)	3.3 in Im(s_{11})	2.5	3.2
(3)	(225,12,3.5)	1.8 in Re(s_{11})	1.2	0.8
(4)	(225,18,3.5)	3.3 in Im(s_{11})	1.8	2.2
(5)	(100, 5,3.5)	6.7 in Im(s_{11})	2.3	16.8
(6)	(300, 5,3.5)	45 in Im(s_{11})	9.5	118
(7)	(300,25,3.5)	4.3 in Im(s_{11})	1.4	3.3

Max. ave. error is the maximum average error for four RSFs (Re(s_{11}), Im(s_{11}), Re(s_{21}) and Im(s_{21})). (1)–(4) are interpolation points, (5)-(7) are extrapolation points.

6. CONCLUSION

We propose a methodology to optimize a spiral inductor systematically. Our model employs S-parameters instead of inductance and Q value, so the model is applicable to many spiral geometries and equivalent-circuit models. A response surface model of S-parameters is built using structural parameters and frequency, the optimized layout is determined from the optimized structural parameters, and the equivalent-circuit model is calculated from the S-parameters. Inductor characteristics such as Q are optimized using the derived polynomial functions of S-parameters. The method was applied to an example spiral inductor problem. The response surface functions were calculated from electromagnetic simulations. The average error of the model was less than 4% in the interpolation region.

7. REFERENCES

[1] J. Craninckx and M. Steyaert, "A 1.8-GHz Low-Phase-Noise CMOS VCO using Optimized Hollow Inductors," IEEE J. Solid-State Circuits, Vol. 32, pp. 736–744, May 1997.

[2] Y. Ho, K. Kim, B. A. Floyd, C. Wann, Y. Taur, I. Lagnado and K. K. O, "4- and 13-GHz Tuned Amplifiers Implemented in a 0.1-μm CMOS Technology on SOI, SOS, and Bulk Substrates," IEEE J. Solid-State Circuits, Vol. 33, pp. 2066–2073, 1998.

[3] Kenneth O, "Estimation Methods for Quality Factors of Inductors Fabricated in Silicon Integrated Circuit Process Technologies," IEEE J. Solid-State Circuits, Vol. 33, pp. 1249–1252, Aug. 1998.

[4] A. M. Niknejad and R. G. Meyer, "Analysis, Design, and Optimization of Spiral Inductors and Transformers for Si RF IC's," IEEE J. Solid-State Circuits, Vol. 33, pp. 1470–1481, Oct. 1998.

[5] H. R. Rategh and T. H. Lee, "Multi-GHz Frequency Synthesis & Division," Kluwer Academic Publishers, 2001.

[6] J. Crols and M. Steyaert, "CMOS Wireless Transceiver Design," Kluwer Academic Publishers, 2000.

[7] M. Park, S. Lee, C.S. Kim, H.K. Yu, J.G. Koo and K.S. Nam, "The Detailed Analysis of High Q CMOS-Compatible Microwave Spiral Inductors in Silicon Technology," IEEE Trans. on Electron Devices, Vol. 45, pp. 1953–1959, 1998.

[8] S. S. Mohan, M. M. Hershenson, S. P. Boyd and T. H. Lee, "Simple Accurate Expressions for Planar Spiral Inductances" *IEEE J. Solid-State Circuits*, Vol. 34, pp.1419–1424, Oct. 1999.

[9] G. E. P. Box and N. R. Draper, "Empirical Model-Building and Response Surfaces" John Wiley & Sons, 1987.

[10] Ansoft Corp., "Ansoft HFSS Manual," 2001.

A PHYSICAL AND ANALYTICAL MODEL FOR SUBSTRATE NOISE COUPLING ANALYSIS[1]

R. Shreeve, T. S. Fiez, and K. Mayaram

School of EECS, Oregon State University, Corvallis, OR 97331-3211

ABSTRACT

This paper presents a new approach for computing the equivalent circuit models for substrate noise coupling in mixed-signal integrated circuits. An analytical model is derived from a physical understanding of the various coupling paths in a heavily doped silicon substrate. The approach has been validated with measured data from a 0.35μm CMOS process and demonstrates that the coupling mechanisms are complex whereby simple resistance expressions cannot be used. Since the model is physical, it can be used to predict substrate noise coupling without the need for extensive computer simulations using three-dimensional finite difference or Green's function solvers.

1. INTRODUCTION

Accurate prediction of substrate coupling enables the integration of multiple digital and analog cells onto a single integrated circuit. This reduces the cost of the overall system by reducing printed circuit board costs and eliminating IC packaging costs.

Existing models for predicting substrate coupling require three dimensional finite difference or Green's function solvers to define the substrate coupling network as in [1 ,2, 3]. Highly trained experts and simulation times of several days can be required to apply these tools. Macro models on the other hand use curve fitting to match empirical results but provide a limited physical understanding of the conduction mechanisms as in [4, 5].

A direct analytical approach provides two significant advantages. First, the analytical approach gives the designer insight into the key factors affecting substrate coupling. Second, the analytical equation enables an immediate prediction of the coupling between circuits using a programmable calculator. This capability significantly enhances decision making during the IC floor planning stage. Analytical methods have been introduced previously [6] but are not sufficiently accurate to capture the complex conduction mechanisms. The analytical model presented in this paper, overcomes prior limitations since it is derived from a physical understanding of the coupling paths.

2. COUPLING BETWEEN SIMILAR ISLANDS

A typical MOS transistor layout in a heavily doped silicon process is represented in Fig. 1. To simplify the measurement of substrate coupling resistance, the island regions are implemented as P+ island regions. The coupling between two island regions is shown in Fig. 2. Two conduction paths are identified in this figure. These correspond to surface and bulk conduction paths. The channel stop layer under the field oxide is important in determining the surface conduction. Measurements and device simulations have confirmed that bulk conduction dominates for injector-sensor separations greater than 100μm. For separations less than 2μm surface conduction dominates.

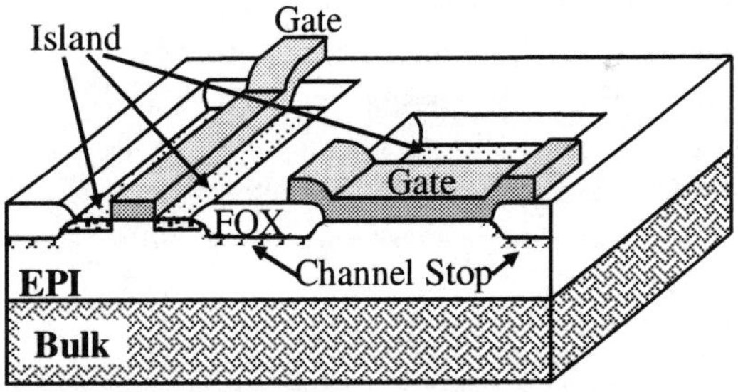

Fig. 1: Heavily doped silicon cross section.

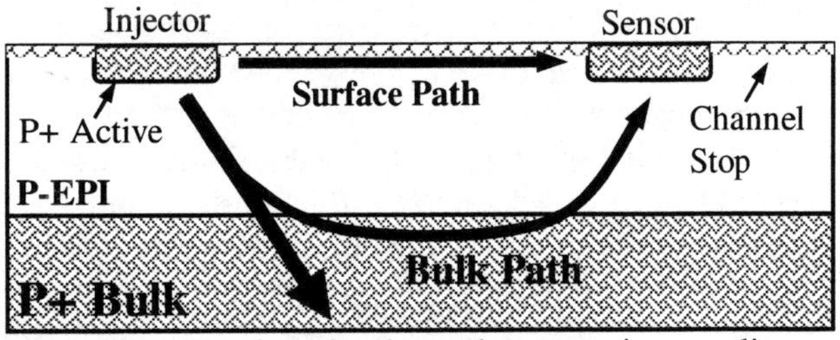

Fig. 2: Cross section showing substrate noise coupling mechanisms between an injector and sensor.

---

[1] This work was sponsored in part through NSF, SRC and DARPA TEAM under contracts CCR-0096176, 2001-TJ-911, and F33615-02-1-1179, respectively.

An equivalent circuit representation for these coupling paths is shown in Fig. 3, where the surface conduction is shown as a block that is described in detail later. In this figure, R_{Gnd} is the resistance from the p+ bulk to ground and is defined by the die perimeter ring, substrate taps, and possibly a backside contact. Hence, R_{Gnd} is determined by the chip layout and is typically a few ohms. The small R_{Gnd} value is difficult to measure but is critical for determining the other resistances. For our measurements a backside Kelvin contact was used resulting in a predictable R_{Gnd} of 1.8Ω.

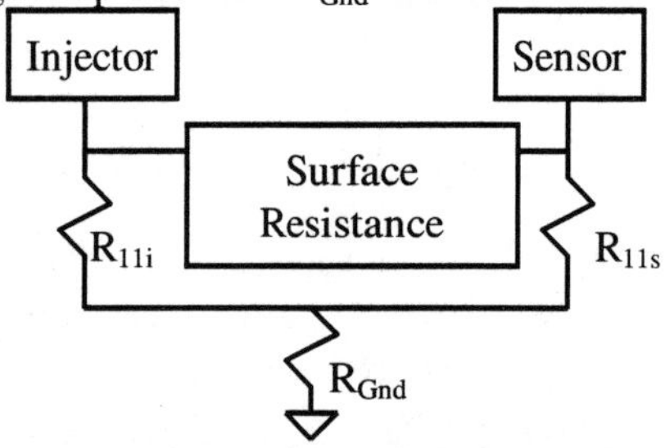

Fig. 3: Equivalent circuit model for injector to sensor substrate noise coupling.

2.1. Bulk Conduction: R11

The bulk conduction resistances (R_{11i} and R_{11s} in Fig. 3) are calculated from the depth and resistivity of the channel stop and epi layers. The current path starts at the surface and spreads out as it propagates across the surface layer. This resistance is analytically calculated by assuming multiple parallel paths to the bulk. Each path consists of a surface resistor and a bulk resistor. These paths are combined in parallel by starting furthest from the contact and working toward the contact. The bulk resistance can be calculated using the traditional resistance formula,

$$\Delta R_{bulk} = \frac{pd}{A} \tag{1}$$

where p is the resistivity, d is the bulk depth, and A is the cross sectional area. The cross sectional area, A, of ΔR_{Bulk} is defined by a wedge created by making W2 larger than W1 as shown in Fig. 4. The surface resistance, ΔR_{surf}, is a wedge on the surface created on each of the four sides of the original island area. This surface resistance is calculated using the wedge equation:

$$\Delta R_{surf} = \frac{pL \bullet \ln(W_2/W_1)}{(W_2 - W_1)} \tag{2}$$

where w_1 is the narrow side of the wedge and w_2 is the wide side of the wedge. Starting a long distance from the contact, the accumulated resistance is calculated recursively by Equation (3).

$$R_{Accum} = (R_{Accum} + \Delta R_{surf}) \,||\, \Delta R_{bulk} \tag{3}$$

where R_{accum} is the accumulated resistance, ΔR_{surf} is the incremental surface resistance, and ΔR_{bulk} is the incremental bulk resistance.

A test chip was fabricated in a 0.35μm heavily doped process to verify these results. A comparison of the measured values from the test chip and calculated values of R_{11} are shown in Fig. 4. The error between the computed and measured results is less than 5%, which shows that our analytical model is accurate.

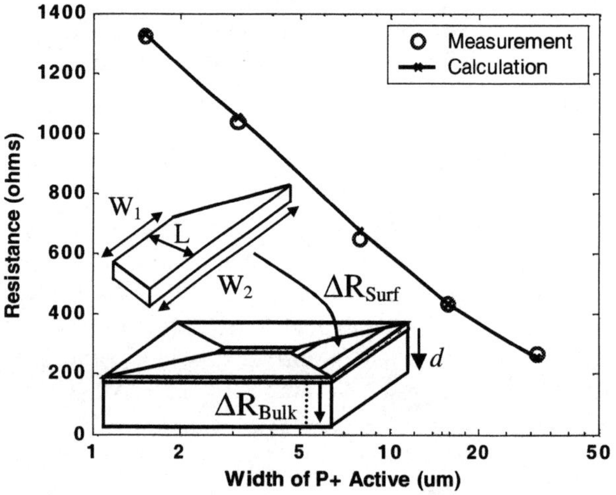

Fig. 4: Resistance to bulk as a function of active region width.

2.2. Surface Conduction

Surface conduction is much more complex than bulk conduction. For small injector-sensor separations, the resistance calculation between two parallel conductors [7] can be used as seen in Fig. 5. However, large errors start to appear at separations above 5μm.

Clearly, a simple conduction model cannot predict surface resistance and an alternate model is needed to predict the surface conduction component for these separations. As the carriers traverse from the injector to sensor some of them flow through the epi layer into the bulk. For this reason, a distributed ladder network consisting of channel stop and epi resistances as shown in Fig. 6 is required for accurately modeling the surface conduction.

The results of an eight-section ladder network calculation are compared with measurements in Fig. 7 and a good agreement is observed for separations up to 2.4μm (the error is less than 5%). A discrepancy remains for separations between 2.4μm and 100μm.

To address this discrepancy, insight is derived from three-dimensional device simulations using ATLAS. Fig. 8 shows the constant surface current density contours for an injector-sensor pair at a spacing of 10μm. The current density distribution at a cutline placed midway between the injector and sensor is shown in Fig. 9. The variation

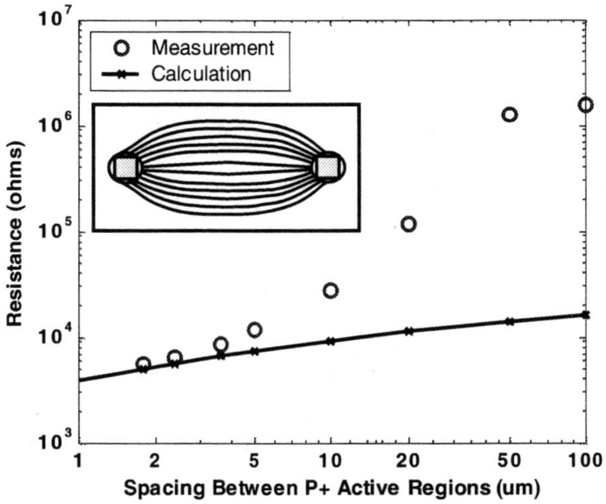

Fig. 5: Resistance from injector to sensor using equations for two parallel conductors.

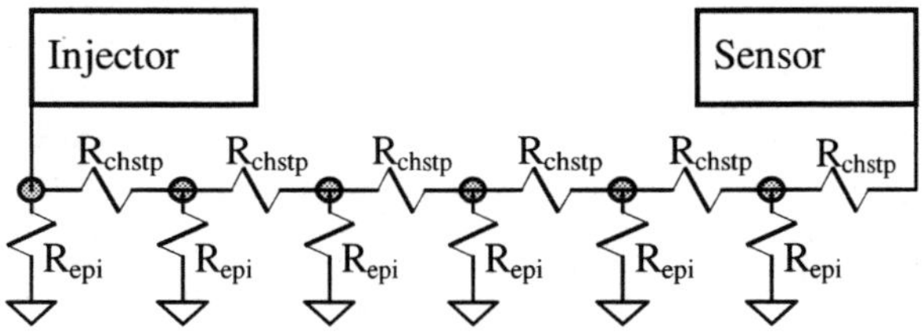

Fig. 6: A distributed ladder network models the coupling between the injector and sensor through the surface conduction path.

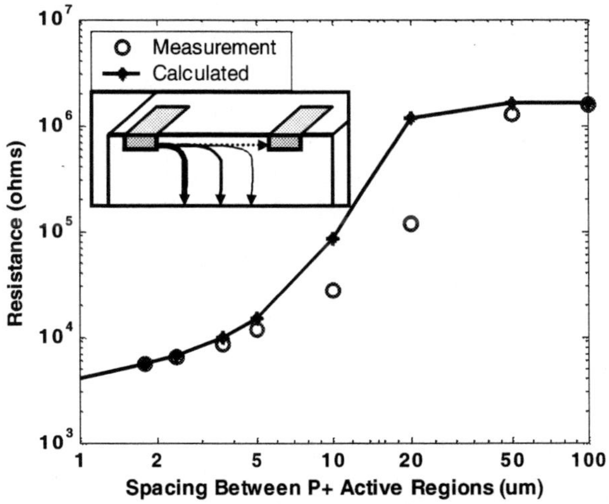

Fig. 7: Resistance from injector to sensor using the ladder network and parallel conductor approximation.

in current density indicates that current flow between the injector and sensor is not uniform. This non-uniform current distribution can be accounted for by segmenting the current flow through paths of different resistances between the injector and the sensor. In our calculations eleven segments were found to provide sufficient accuracy. Fig. 10 demonstrates that the error is less than 9% for all separations using this approach.

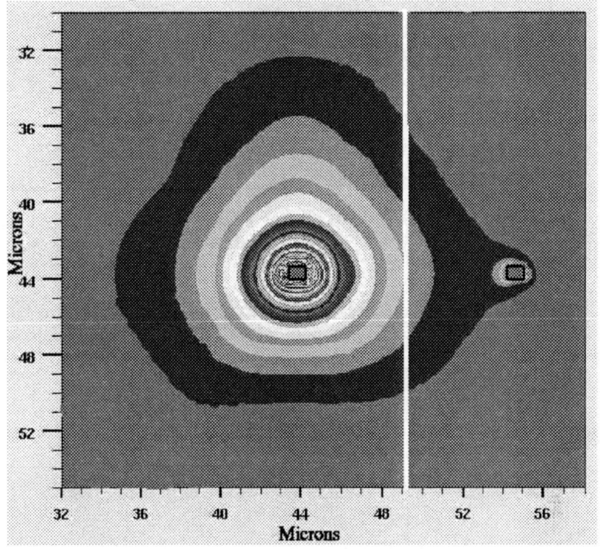

Fig. 8: Surface constant current density contours and a cutline at the midpoint between injector and sensor.

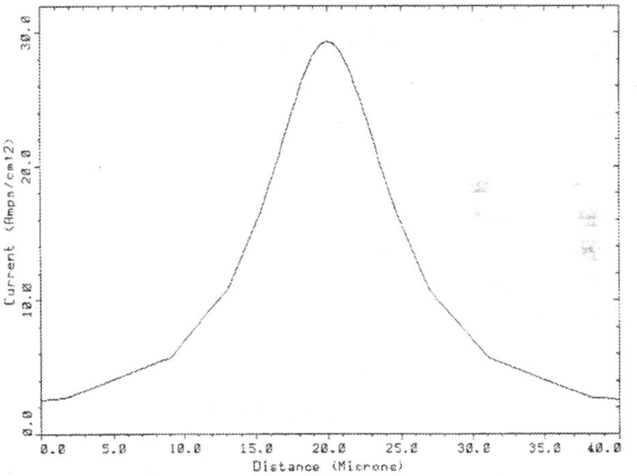

Fig. 9: Surface current density as a function of distance along the cutline between the injector and sensor.

3. COUPLING BETWEEN DIFFERENT ISLANDS

So far we have demonstrated results for identical injector and sensor contacts. Next, we show that our approach is general and can be used for contacts of different sizes. We consider one, two, and four contacts. A single contact is 0.7μm × 0.85μm, two contacts are 0.7μm × 1.5μm, and four contacts are 0.7μm × 3.1μm. Fig. 11 shows that the error is less than 10% for all separations. Additional details are provided in Table 1, where the errors for different separations are tabulated.

4. CONCLUSION

An analytical approach has been described that improves the efficiency of substrate parasitic model extraction over traditional finite difference or Green's function methods. Three key observations lead to a model that is accurate in predicting the substrate network elements. The resulting model has an error of less than 10% when compared with measurements on test structures fabricated in a 0.35μm heavily doped CMOS process.

This model provides insight into the effect of silicon process improvements. For example, reductions in channel stop concentration would increase R_{11} by reducing surface spreading. R_{12} increases for moderate separations because of increased current draw from the ladder network into the bulk.

Projections can also be made for design of isolation structures. For example, an isolation structure placed midway between the active areas would significantly increase R_{surf} for separations of less than 10μm. This is a result of interrupting the channel stop region. Current in this region is shunted to the bulk.

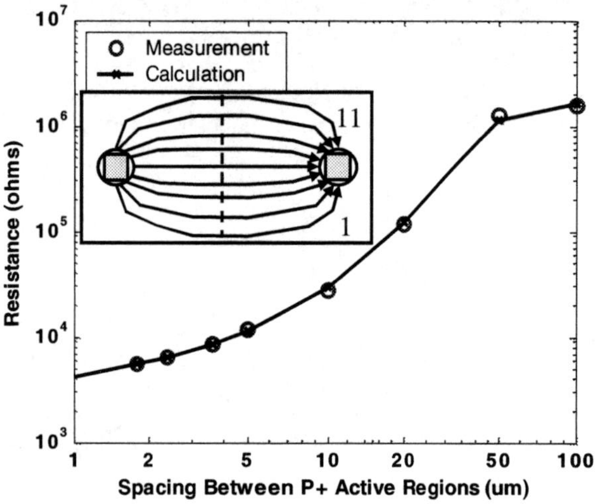

Fig. 10: Surface resistance from injector to sensor using eleven segments to model the surface non-uniform current distribution.

Table 1. Relative error for the model with respect to measured data for surface resistance between the injector and sensor.

Spacing (μm)	One Contact	Two Contacts	Four Contacts
1.8	-0.14%	0.07%	0.14%
2.4	0.51%	2.99%	-3.77%
3.6	0.62%	0.34%	-3.07%
5	-2.26%	3.03%	1.40%
10	7.13%	9.46%	10.23%
20	2.77%	-0.85%	0.00%
50	-8.59%	-7.12%	9.25%
100	2.11%	0.25%	-0.80%

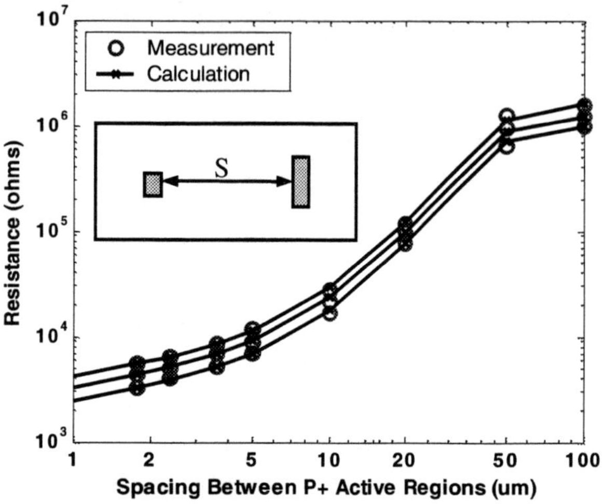

Fig. 11: Resistance from the model compared with the measured data for surface resistance between the injector and sensor. The top curve is for 1-contact the middle curve is for 2-contacts, and the bottom curve is for 4-contacts.

5. REFERENCES

[1] D. K. Su, M. J. Loinaz, S. Masui and B. A. Wooley, "Experimental results and modeling techniques for substrate noise in mixed-signal integrated circuits," *IEEE J. Solid-State Circuits*, vol. 28, pp. 420-430, April 1993.

[2] N. K. Verghese, T. J. Schmerbeck and D. J. Allstot, *Simulation Techniques and Solutions for Mixed-signal Coupling in Integrated Circuits*, Kluwer Academic Publishers, 1995.

[3] A. M. Niknejad, R. Gharpurey, R. G. Meyer, "Numerically stable green function for modeling and analysis of substrate coupling in integrated circuits", IEEE Trans. on CAD of IC and Systems, Vol. 17, pp. 305-315, April 1998.

[4] D. Ozis, T. Fiez, and K. Mayaram, "A comprehensive geometry-dependent macromodel for substrate noise coupling in heavily doped CMOS processes," *Proceedings of IEEE Custom Integrated Circuits Conference*, May 2002, pp. 497-500.

[5] A. J. Van Genderen and N. P. Van der Meijs, "Modeling substrate coupling effects using a layout-to-circuit extraction program," *Proceedings of IEEE Benelux Workshop on Circuits*, Systems and Signal Processing, November 1997, pp 193-200.

[6] S. *Kristiansson, S. P. Kagganti, T. Ewert, F. Ingvarson, J. Olsson, K.O. Jeppson,* "Substrate resistance modeling for noise coupling analysis", *International Conference on Microelectonic Test Structures*, pp. 124-129, March 2003

[7] M. Plonus, *Applied Electromagnetics*, McGraw-Hill p.168, 1978.

An Improved Z-parameter Macro Model for Substrate Noise Coupling

Chenggang Xu, Terri Fiez and Karti Mayaram
School of EECS, Oregon State University, Corvallis, OR 97331
E-mail: karti@ece.orst.edu

***Abstract* — A new formulation of Z-parameter based macro models for substrate noise coupling is presented. With this new formulation, the variation of the self Z-parameter at small contact separations can be simulated. Instead of calculating the Z-parameters directly by two-contact models, the Z-parameters are calculated as a least-square solution to an over-determined set of equations. In this way, the interaction between neighboring contacts can be easily accounted for in the Z-parameter calculation. The method can be applied to both the lightly and the heavily doped substrates. Examples for a lightly doped substrate are provided.**

I. INTRODUCTION

With the increasing integration of analog and digital circuits on a chip, circuit isolation has become a serious problem. Significant effort has been made to model the substrate parasitics by numerical simulation [1]-[4]. An alterative to the numerical simulation approach is to develop scalable or parametric macro models for substrate coupling [5]-[7]. These models can be established through detailed numerical simulations, careful curve fitting, and experimental verifications. Once these models have been established, they can be easily used to extract parasitics with several orders of magnitude of speedup. In addition, these empirical macro models can provide designers with more insight into the placement of various components before the final layout is done.

Fig. 1 shows the resistance network for a two contact problem. In [5] and [6], models for the resistance network were established from numerical simulation and curve fitting. These models, however, were found to have difficulties when applied to multi-contact cases, and a Z-parameter-based approach was proposed in [7] for a heavily doped substrate. This approach is based on the assumption that the self Z-parameter, i.e., Z_{ii}, depends only on the geometry of contact-*i*, and the mutual Z-parameter, i.e., Z_{ij}, depends only on the geometry and separation of contact-*i* and contact-*j*. This assumption, however, is not always accurate. As shown in Section 2, the self Z-parameter Z_{ii}, in general, depends not only on the geometry of contact-*i*, but also on other neighboring contacts. Experimental results in [8] show that the self Z-parameter decreases when the separation between two contacts is small. The mutual Z-parameter Z_{ij}, on the other hand, also has a dependence on other neighboring contacts in addition to contact-*i* and contact-*j*.

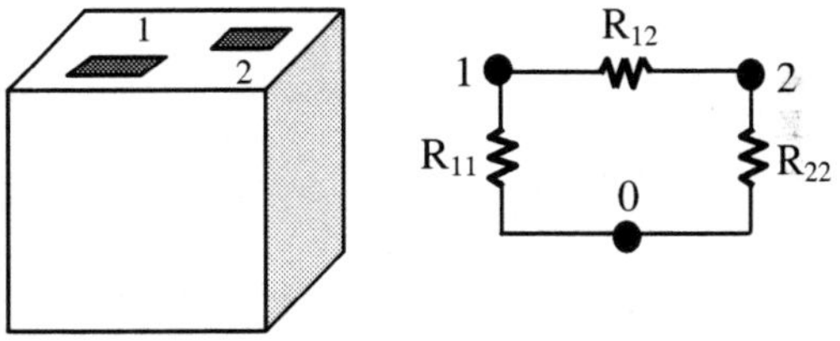

Fig. 1: Circuit models for substrate parasitics between two contacts.

In this paper, we present a new formulation for the Z-parameter-based macro models that accounts for the effects of neighboring contacts. Section 2 is devoted to numerical simulation and the formulation of Z-parameter models for two contacts. An application example of the new formulation is provided in Section 3. Conclusions are presented in Section 4.

II. MACRO MODEL FORMULATION

For a two-contact problem, there are three Z-parameters, i.e. Z_{11}, Z_{22} and Z_{12}. Since the self Z-parameter was assumed to depend only on contact geometry in [7], the self and mutual Z-parameter models can be determined independently. The self and mutual Z-parameter models for a heavily doped substrate were expressed in [7] as

$$Z_{11} = \frac{1}{c_0 + c_1 P_c + c_2 A_c} \tag{1}$$

$$Z_{12} = \alpha e^{-\beta \cdot d} \tag{2}$$

where c_1, c_2, c_3 , α and β are fitting constants, d is the separation between the two contacts, and P_c and A_c are the contact perimeter and area, respectively.

Numerical simulation shows that for a two contact problem, the two self Z-parameters are correlated by contact geometry and separation. The self and mutual Z-parameters are also related. As such, instead of separately building self and mutual Z-parameter models, we establish models for two dimensionless compound Z-parameters. The two parameters are:

$$Z_{c1} = \frac{Z_{12}}{Z_{11} + Z_{22}} \tag{3}$$

$$Z_{c2} = \frac{Z_{11} + Z_{22}}{\hat{Z}_{11} + \hat{Z}_{22}} \tag{4}$$

where $\hat{Z}_{11}$ and $\hat{Z}_{22}$ are self Z-parameters of contact *1* and *2* when they were assumed to be the only contact in the substrate, respectively. Based on our simulation results, this Z-parameter can be calculated by the following equation

$$\hat{Z}_{ii} = \frac{1}{(c_0 + c_1 P_c + c_2 A_c)} \times \frac{1}{\sqrt{c_3 + c_4\left[\left(x_c - \frac{L_x}{2}\right)^2 + \left(y_c - \frac{L_y}{2}\right)^2\right]}} \times \frac{1}{1 + c_5\left[\left(\frac{l_x}{l_y} - \frac{L_x}{L_y}\right)^2 + \left(\frac{l_y}{l_x} - \frac{L_y}{L_x}\right)^2\right]} \tag{5}$$

where c_i, i = 0, 1, ..., 5, are fitting parameters. The definitions for the other terms in this equation are shown in Fig. 2, where L_x and L_y defines the chip boundary in the x and y directions, respectively. l_x and l_y are the contact dimensions in the x and y direction, respectively. $(x_c,\ y_c)$ are the coordinates of the contact center relative to the lower left corner of the substrate.

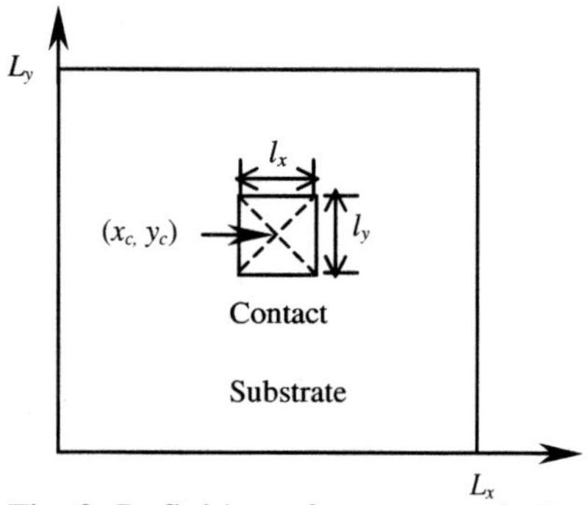

Fig. 2: Definition of parameters in Eq. (5).

The first term in the above equation addresses the dependence on contact area and perimeter, which is the same as Eq. (1). The second term addresses the effect of contact position in the substrate, and the third term addresses the contact orientation.

Fig. 3 shows a comparison between the simulation and model for a 10μmx10μm contact at different positions in a lightly doped substrate. We see when the contact is close to the edge or corner areas of the substrate, the self *Z*-parameter is strongly position dependent. The model and the simulation are in good agreement except at the edges and corners.

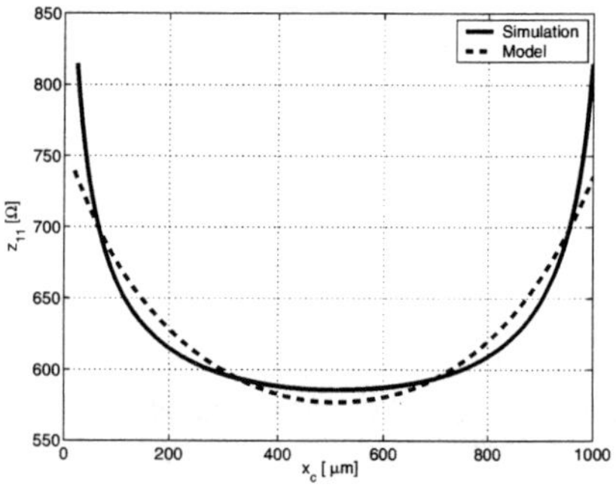

(a)

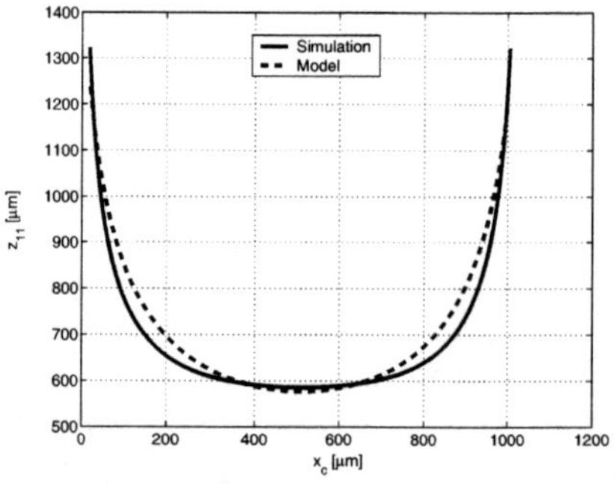

(b)

Fig. 3: Comparison on self impedance between the model (Eq. (5)) and simulation results for a 10μmx10μm contact at different positions in a 1024μmx1024μm substrate. (a) Different positions along the center line of the substrate, and (b) different positions along the diagonal of the substrate.

From a comprehensive set of simulation results, we find Z_{c1} and Z_{c2} can be expressed as

$$Z_{c1} = \frac{Z_{12}}{Z_{11} + Z_{22}} = \frac{1}{\alpha_0 + \alpha_1 \cdot \exp\left(\beta \cdot \bar{d}^{\alpha_2}\right)} \tag{6}$$

$$Z_{c2} = \frac{Z_{11} + Z_{22}}{\hat{Z}_{11} + \hat{Z}_{22}} = \left[1 - \exp\left(\lambda_1 \bar{d}\right)\right]^{\lambda_2} \tag{7}$$

where α 's are fitting constants, and λ's and β are functions of contact size, separation and orientation, and $\bar{d}$ is a dimensionless effective separation between two contacts. Fig. 4 shows the definition of contact separation for two generic contacts. The virtual contacts (shown by dashed lines) have the same area as the corresponding real contacts (shown in solid line) with a height of the real contact seen along the direction connecting the two contact centroids. When the two contacts are aligned to each other, the effective separation d_{eff} becomes equal to the inner edge separation between the two contacts. Eqs. (8) - (12) show the calculation of $\bar{d}$.

$$\bar{d} = \sqrt{\left(\frac{d_x}{L_x}\right)^2 + \left(\frac{d_y}{L_y}\right)^2} \tag{8}$$

$$d_x = d_{eff} \cdot \cos\theta \qquad d_y = d_{eff} \cdot \sin\theta \tag{9-10}$$

$$d_{eff} = d_c - \left(l_{\xi,1} + l_{\xi,2}\right)/2 \tag{11}$$

$$l_{\xi,i} = \frac{A_i}{l_{x,i}\sin\theta + l_{y,i}\cos\theta}, \quad i = 1, 2. \tag{12}$$

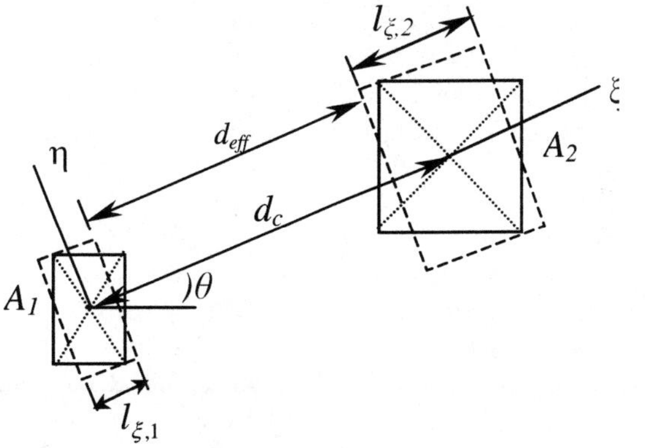

Fig. 4: Two generic contacts. A_1 and A_2 are the areas of contact 1 and contact 2, respectively.

As mentioned in Section 1, the approach in [7] is based on the assumption that the self Z-parameter Z_{ii} depends only on the geometry of contact-*i*, and the mutual Z-parameter Z_{ij} depends only on the geometry and separation of contact-*i* and contact-*j*. As such, the macro models established in two-contact cases can be directly applied to multi-contact cases. This assumption, however, is accurate only for contacts at relatively large separations. To show this, we do two calculations. The first is for 5 contacts at different *d* values, as shown in Fig. 5. In the second case, we do the same calculation with contacts 3, 4 and 5 removed. We then compare Z_{11} and Z_{22} for the two cases, respectively. If the approach in [7] were applied to the two cases, Z_{11} and Z_{12} in the two cases would be the same. In addition, Z_{11} would be a constant. However, as shown in Fig. 6 for a lightly doped substrate, when the separation is small, Z_{11} and Z_{12} in the two cases are quite different. Numerical simulations also show that this difference increases as the number of contacts increases. Similar results were also obtained for the heavily doped substrate. Therefore, there is a need to include the effects from neighboring contacts into the Z-parameter calculations.

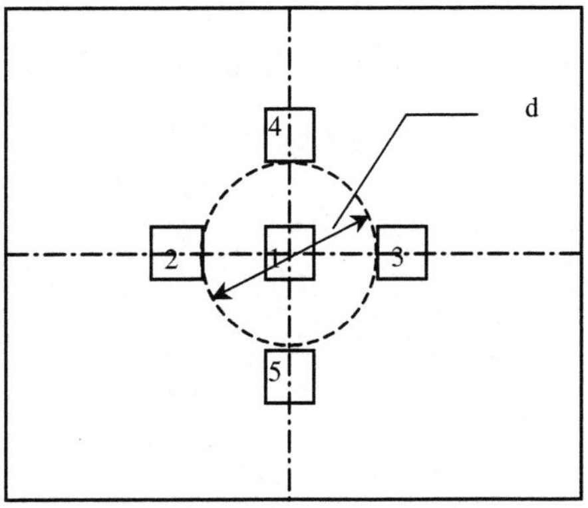

Fig. 5: 5 contacts in a lightly doped substrate. All contacts are of size 10μm×10μm.

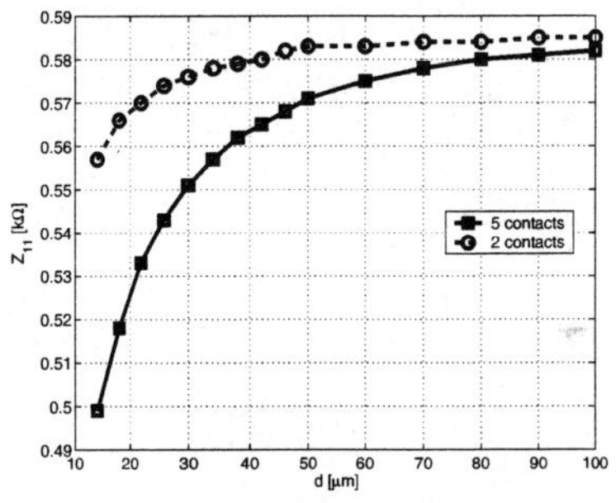

(a)

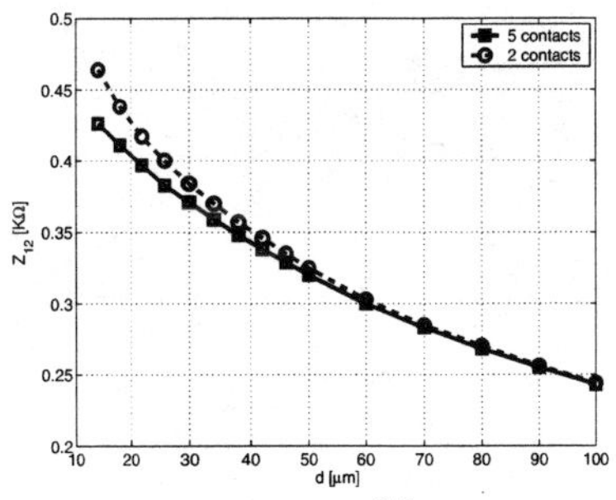

(b)

Fig. 6: Effect of neighboring contacts on Z_{11} and Z_{12}. (a) Z_{11} for a lightly doped substrate, (b) Z_{12} for a lightly doped substrate.

When the self Z-parameter Z_{ii} is no longer considered as a parameter depending only on the geometry of contact-*i*, different two-contact combinations of contact-*i* may result in different Z_{ii}. We assume that, the Z-parameters in multi-contact cases are such that Eqs. (6) and (7) are satisfied with a minimum total error for all two-contact combinations.

For a *n*-contact problem, there are $n(n-1)/2$ mutual coupling Z-parameters and *n* self coupling Z-parameters in the Z-matrix. In other words, there are $n(n+1)/2$ unknown Z-parameters to be calculated. Applying Eqs. (6) and (7) to all the two-contact combinations gives a set of $n(n-1)$ equations with the self and the mutual Z-parameters as unknowns (parameter $\hat{Z}_{ii}$ can be directly calculated from Eq. (5), and, are not unknowns.) When $n > 3$, this equation set becomes over-determined, and its solution is obtained by the least-square method.

III. APPLICATION EXAMPLE

We take the 5-contact problem shown in Fig. 5 again as an example to demonstrate the application of the new Z-parameter-based formulation for the lightly doped substrate.

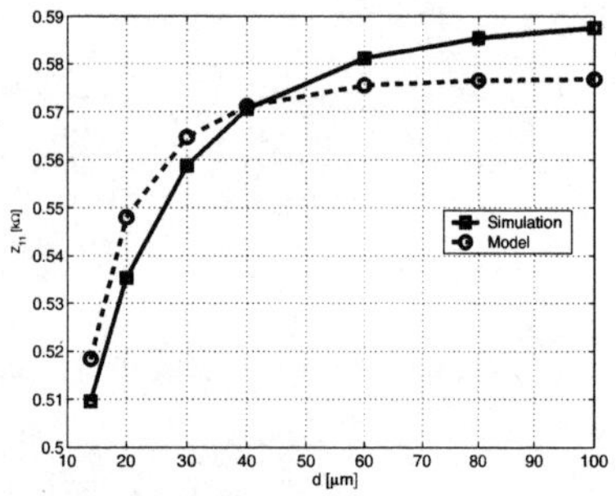

(a)

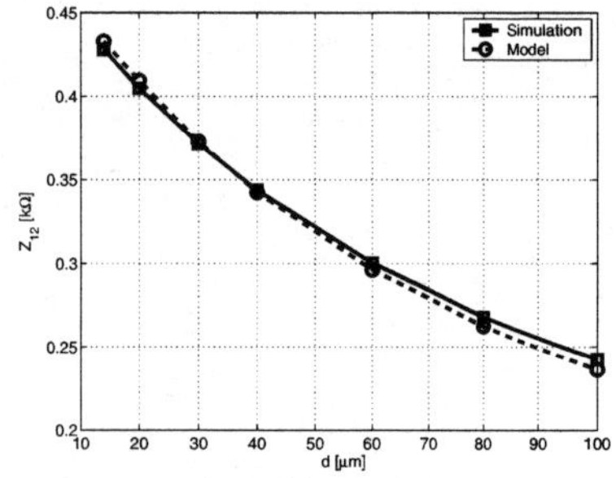

(b)

Fig. 7: Comparison between simulation and model for 10μm×10μm contacts. (a) Z_{11}, and (b) Z_{12}.

Fig. 7 shows comparisons of Z_{11} and Z_{12} between simulation and the model for 10μm×10μm contacts. We see that the variation of the self Z-parameter is well modeled, and the mutual Z-parameter is also in good agreement with simulation results. The variation of the self Z-parameter cannot be modeled by the method of [7].

IV. CONCLUSIONS

A new Z-parameter-based macro model formulation is presented in this paper. Together with a least-square approximation, Z-parameters accounting for the interaction among neighboring contacts can be calculated more accurately. This results in an accurate substrate noise coupling model for small and large separations. The same method can also be applied to the heavily doped substrates.

ACKNOWLEDGMENTS

This work is supported in part by NSF grant CCR-0096176, by SRC under contract 2001-TJ-911 and by the DARPA TEAM project under contract F33615-02-1-1179.

REFERENCES

[1] B. R. Stanisic, N. K. Verghess, R. A. Rutenbar, R. L. Carley, and D. J. Allstot, "Addressing substrate coupling in mixed-mode IC's and power distribution synthesis," *IEEE J. Solid-State Circuits*, vol. 29, pp. 226-237, March 1994.

[2] L. M. Silveira and N. Vargas, "Characterizing substrate coupling in deep-submicron design," *IEEE Design & Test of Computers*, vol. 19, pp. 4-14, Jan. 2002.

[3] R. Gharpurey and R. G. Meyer, "Modeling and analysis of substrate coupling in integrated circuits," *IEEE J. Solid-State Circuits*, vol. 31, pp. 344 - 352, March. 1996.

[4] A. M. Niknejad, R. Gharpurey, and R. G. Meyer, "Numerical stable Green function for modeling and analysis of substrate coupling in integrated circuits," *IEEE Trans. on Computer-Aided Design,* vol. 17, pp. 305-315, April 1998.

[5] A. Samavedam, A. Sadate, K. Mayaram and T. Fiez, "A scalable substrate noise coupling model for design of mixed-signal IC's," *IEEE J. Solid-State Circuits*, vol. 35, pp. 895-905, June 2000.

[6] A. J. van Genderen and N. P. van der Meijs and T. Smedes, "Fast computation of substrate resistances in large circuits," in *Proc. European Design and Test Conf.* pp. 560-565, March 1996.

[7] D. Ozis, T. Fiez, and K. Mayaram, "A comprehensive geometry-dependent macro model for substrate coupling in heavily doped CMOS processes," *Proc. IEEE Custom Integrated Circuits Conf.*, pp. 497-500, May 2002.

[8] S. Kristiansson, S. P. Kagganti, T. Ewert, F. Ingvarson, J. Olsson and K. O. Jeppson, "Substrate resistance modeling for noise coupling analysis," *Proc. IEEE Int. Conf. on Microelectronic Test Structures,* pp. 124-129, 2003.

A COUPLED ITERATIVE/DIRECT METHOD FOR EFFICIENT TIME-DOMAIN SIMULATION OF NONLINEAR CIRCUITS WITH POWER/GROUND NETWORKS*

Zhao Li and C.-J. Richard Shi
Department of Electrical Engineering, University of Washington
Seattle, WA 98195
{lz2000, cjshi}@ee.washington.edu

Abstract: A coupled iterative/direct circuit analysis method is proposed for efficient SPICE-accurate time-domain simulation of nonlinear circuits with large-scale power/ground networks. The system under study is partitioned into a linear part including power/ground networks, a nonlinear part, and an interface between them. The part of power/ground networks is formulated by nodal analysis based on *RCLK* elements, and solved by an efficient conjugate gradient iterative method with an incomplete Cholesky decomposition preconditioner. The nonlinear circuit part is formulated by modified nodal analysis, and solved by the direct method as in SPICE. The iterative method and the direct method are coupled by a Gauss-Seidel like relaxation scheme with SPICE built-in varying time step-size numerical integration. How the condition number of a circuit matrix changes with time step-sizes is further studied. Experimental results on digital circuits with power/ground networks demonstrate that the proposed coupled iterative/direct method yields SPICE-like accuracy with orders of magnitude speedup for circuits with tens of thousands elements.

1. INTRODUCTION

With the increasing operation frequency, lower supply voltage and smaller device feature size, the effects of power/ground networks, such as *Ldi/dt* drop, *IR* drop, resonance, are becoming more and more pronounced [5]. An improper circuit design neglecting power/ground networks and packaging will result in excessive voltage drops and fluctuations in circuit supply nodes. The noise margin for digital circuits is therefore reduced, which may unfortunately disturb gate delays or even produce logic errors. The increasing demand to integrate digital, analog and radio frequency (RF) circuits into one single chip requires accurate analysis of VLSI circuits together with power/ground networks [1][3][5][9]. For such purposes as well as high fidelity coupled circuit and electromagnetic modeling [7], SPICE-like simulators are desirable for accurate transistor-level time-domain simulation.

However, efficient simulation of such systems presents a complexity challenge to SPICE [4]. To accomplish transient simulation, SPICE uses numerical integration formulae at each time point and applies the Newton-Raphson (NR) method to linearize nonlinear devices. Then the circuit system is simulated at each time point by solving a system of linear equations $\boldsymbol{Ax} = \boldsymbol{b}$, where $\boldsymbol{A}$ is typically in the form of a so-called modified nodal analysis (MNA) circuit matrix. It is well known that device evaluation dominates the simulation of small to medium scale circuits and can be speeded up using table-lookup nonlinear device models or parallel computation techniques. However, for large scale nonlinear circuits coupled with power/ground networks, the per-iteration cost of transient simulation with SPICE is dominated by LU factorization of the circuit matrix $\boldsymbol{A}$.

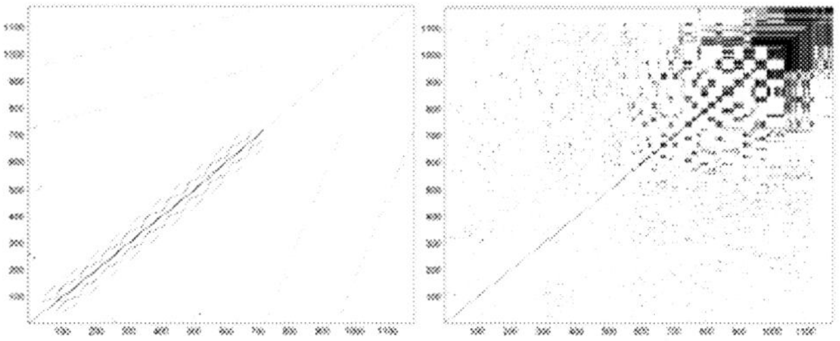

(a) (b)

Figure 1. The circuit matrix structure of a power/ground example (a) before LU factorization and (b) after LU factorization.

Figure 1 (a) and Figure 1 (b) show the circuit matrix structure before and after LU factorization for a power/ground analysis example in Section 4. It can be seen that the original circuit matrix before factorization is very regular and sparse (9618 elements in a 1177x1177 matrix, which means the sparsity is 0.70%), while the matrix after factorization becomes irregular and much denser (89733 elements, the element number is increased 9.33X due to fill-ins and the sparsity becomes 6.68%). Therefore, a key idea to achieve speedup and save memory is to apply efficient krylov-subspace based iterative methods [1][5][9] on power/ground networks analysis since those methods only require matrix-vector multiplications on the original sparse circuit matrix.

Although iterative methods have been shown to be efficient for transient simulation of large-scale power/ground networks [1], their application to general nonlinear circuits is limited. The reason is that the circuit matrix for a nonlinear circuit is typically not symmetric positive definite, which prohibits the usage of efficient preconditioners for iterative methods. Iterative methods without good preconditioners are well known to have the convergence problem. The direct method based on the Newton-Raphson iteration as in SPICE is still the most efficient way for general nonlinear circuit simulation.

Noticing different application areas of iterative and direct methods, we present a new coupled iterative/direct method capable of analyzing nonlinear circuits with power/ground networks *in SPICE-like accuracy yet orders of magnitude speedup*. The system under study is partitioned into three parts – a linear part including power/ground networks, a nonlinear part and an interface between them. Two key ideas are:

1) For power/ground networks, nodal analysis (NA) formulation of *RCLK* elements is applied so that an efficient iterative conjugate gradient method with an incomplete Cholesky decomposition preconditioner [1][6] can be used. For different circuit formulation methods, how the condition number of a circuit matrix changes with time step-sizes is further studied.
2) For nonlinear circuits, the modified nodal analysis (MNA) formulation is applied and the direct method as in SPICE is

---

* This research was supported by DARPA NeoCAD Program under Grant No. N66001-01-8920, NSF-SRC Joint Initiative on Mixed Signal Electronic Technologies under Grant No. CCR-0120371, and NSF CAREER Award under Grant No. 9985507.

used. The iterative method and the direct method are coupled together by a Gauss-Seidel style relaxation scheme [8].

This paper is organized as follows. Section 2 proposes the new coupled iterative/direct method. The NA formulation for iterative methods and the condition number variation with time step-sizes are described in Section 3. Experimental results on digital circuits with power/ground networks are shown in Section 4. Section 5 concludes this paper.

2. THE COUPLED ITERATIVE/DIRECT METHOD

The system under study is shown in Fig. 2, in which power/ground networks are coupled with nonlinear circuits through a linear interface. It can be seen that parasitic coupling effects between the power network and the ground network are also incorporated. The linear interface is constructed so that only a few linear elements (such as resistors connecting grid nodes of power/ground networks and supply nodes of nonlinear circuits) are introduced.

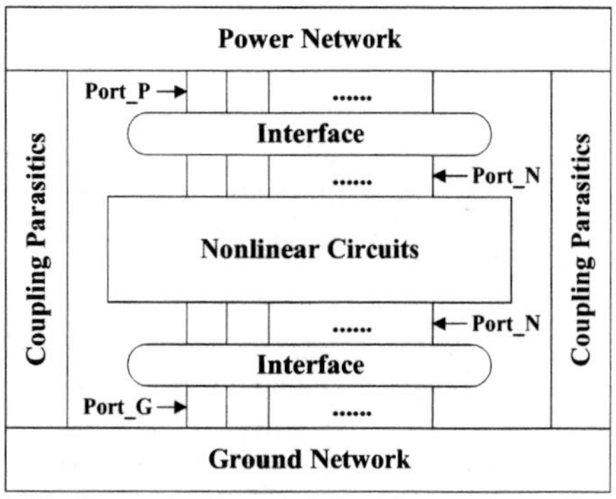

Figure 2. Nonlinear circuits coupled with power/ground networks.

Figure 3 shows the related circuit matrix structure for the system in Fig. 2. Y_{PG}, Y_N and Y_I represent circuit matrices of power/ground networks, nonlinear circuits, and the interface, respectively. C_{PG} and C_{PG}^T are coupling matrices between power/ground networks and the interface. C_N and C_N^T are coupling matrices between nonlinear circuits and the interface. All other parts in the circuit matrix are zero. The unknown variables $\boldsymbol{v}$ and the right hand side (RHS) vectors $\boldsymbol{b}$ are also labeled accordingly.

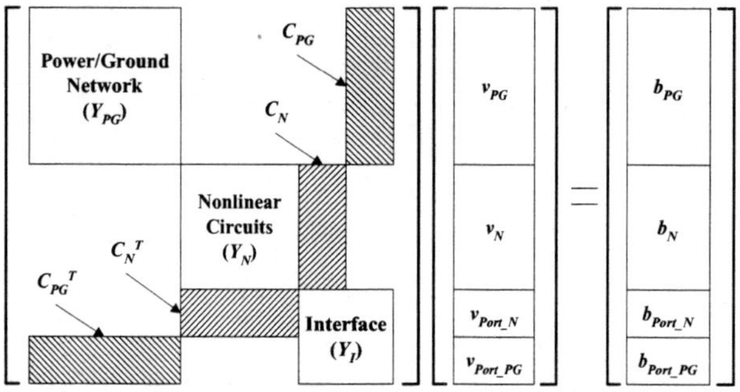

Figure 3. The circuit matrix structure for a system in Fig. 2.

The circuit matrix Y_I for the interface can be further partitioned as below,

$$Y_I = \begin{bmatrix} Y_{INN} & Y_{INP} \\ Y_{IPN} & Y_{IPP} \end{bmatrix}$$

where Y_{INN} and Y_{IPP} are self-admittance matrices for the ports of nonlinear circuits and those of power/ground networks, respectively, Y_{IPN} and Y_{INP} are coupling matrices between the ports of nonlinear circuits and those of power/ground networks. In general, $Y_{IPN} = Y_{INP}^T$.

To introduce the Gauss-Seidel style relaxation scheme [8], we regroup circuit sub-matrices and sub-vectors as below,

$$Y_{PG}^* = \begin{bmatrix} Y_{PG} & C_{PG} \\ C_{PG}^T & Y_{IPP} \end{bmatrix} \quad Y_N^* = \begin{bmatrix} Y_N & C_N \\ C_N^T & Y_{INN} \end{bmatrix}$$

$$v_{PG}^* = \begin{bmatrix} v_{PG} \\ v_{Port-PG} \end{bmatrix} \quad v_N^* = \begin{bmatrix} v_N \\ v_{Port-N} \end{bmatrix}$$

$$b_{PG}^* = \begin{bmatrix} b_{PG} \\ b_{Port-PG} \end{bmatrix} \quad b_N^* = \begin{bmatrix} b_N \\ b_{Port-N} \end{bmatrix}$$

Therefore, the circuit matrix in Fig. 3 can be further written in the following format,

$$Y_{PG}^* v_{PG}^* = b_{PG}^* - Y_{IPN} v_{Port-N} \quad (1)$$

$$Y_N^* v_N^* = b_N^* - Y_{INP} v_{Port-PG} \quad (2)$$

According to Eq. (1), once $v_{Port\text{-}N}$ is fixed, v_{PG}^* can be solely solved. After v_{PG}^* (and therefore $v_{Port\text{-}PG}$) is given, v_N^* (and therefore $v_{Port\text{-}N}$) can be determined by Eq. (2). Then, the new $v_{Port\text{-}N}$ is compared to the old $v_{Port\text{-}N}$ used during solving Eq. (1) to check if this *Gauss-Seidel relaxation* is converged. The coupled iterative/direct method is summarized in Table I.

Table I. The coupled iterative/direct method.

```
INITIALIZATION:
    Construct Y_INP and Y_IPN
    t=0
WHILE (t<T_final){
  OUTER LOOP: do{
      Construct matrix Y_PG* and vector b_PG*
      Apply ICD-CG to compute v_PG* based on v_Port-N using Eq. (1)
      INNER LOOP: do{
          Construct matrix Y_N* and vector b_N*
          Apply NR linearization and solve Eq. (2)
      } while (v_N* not converge)
  } while (v_Port-N not converge)
  Determine the next time step-size h_n
  t = t + h_n
}
```

It can be seen that the costly simulation of power/ground networks is in the outer loop, while the cheap simulation of nonlinear circuits is in the inner loop. The number of inner nonlinear iterations under the Gauss-Seidel relaxation scheme is generally higher than that with SPICE, since several outer iterations may be required to achieve the final convergence. Even so, a great simulation speedup is still achievable since the cost of each inner nonlinear iteration is much lower than that of one SPICE nonlinear iteration. The reason is that the size of nonlinear circuits is reduced greatly with power/ground networks decoupled in our scheme. Further, the simulation of nonlinear circuits can be speeded up using table-lookup nonlinear device models or parallel computation techniques.

3. ITERATIVE METHODS WITH NA FORMULATION

The MNA formulation for circuit elements is widely used in modern circuit simulators based on direct methods. However, as shown in the Section 1, the simulation of power/ground networks presents a challenge for direct methods. Therefore, the NA formulation of *RCL* elements has been applied for power/ground network simulation based on iterative methods [1]. The NA formulation of *R* and *C* is the same as their MNA formulation. The

NA formulation of L with the trapezoid numerical integration formula is shown in Fig. 4. It can be seen that the equivalent conductance is $h_n/(2L)$ rather than $(2L)/h_n$ in the MNA formulation. The mutual inductance can be incorporated easily by so called K-elements [2]. It has been proved that the circuit matrix with *RCLK* elements based on the NA formulation is symmetric positive definite [1][2]. Therefore, we have implemented the conjugate gradient (CG) method with an incomplete Cholesky decomposition preconditioner [1][6] (named by ICD-CG).

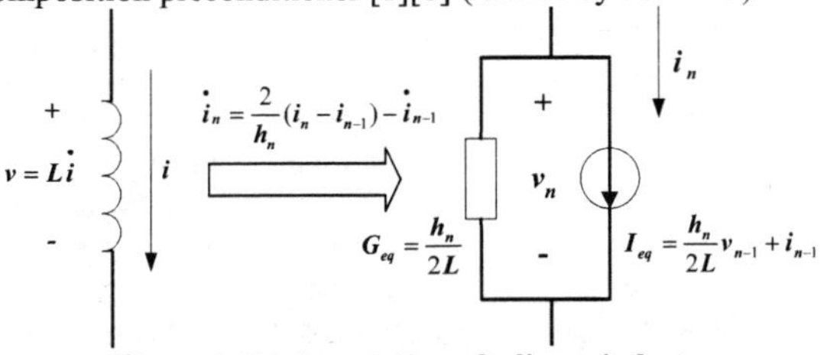

Figure 4. NA formulation of a linear inductor.

The *RCL* circuit example in Fig. 5, the structure of which is typical in power/ground networks, is used to study how the condition number of a circuit matrix changes with time step-sizes. As shown in Fig. 6, the condition number for the MNA formulation is becoming worse as the time step-size decreases (h_n is less than 1). The reasons are: *1)* The MNA formulation of voltage sources introduces zero diagonal elements; *2)* The self-admittance matrix element at node 1 is only contributed by a fixed resistor. Therefore, a tighter tolerance is required when time step-size becomes smaller with iterative methods [5]. If the voltage source E and the serial resistor R_1 in Fig. 5 are replaced by an equivalent Norton current source and a parallel resistor, the condition number is kept relatively small with time step-sizes cahnged, as shown in Fig. 6. Unfortunately, it is not suitable for iterative methods since the MNA formulation of a linear inductor either introduces a negative diagonal element or causes the circuit matrix asymmetric.

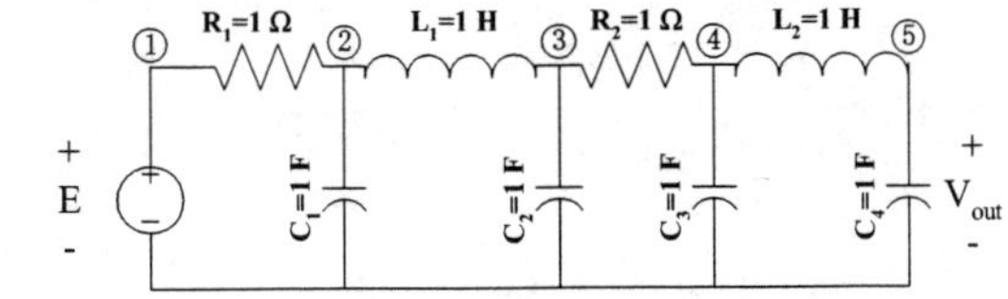

Figure 5. A *RCL* circuit example.

As mentioned previously, the circuit matrix with the NA formulation is symmetric positive definite, which should be suitable for iterative methods. However, in Fig. 6 the condition number for the NA formulation is becoming worse with the time step-size increased (h_n is larger than 1). The reason is that the effects of linear inductors become ignorable with an enlarged time step-size – an enlarged equivalent conductance of $h_n/(2L)$ means a reduced equivalent resistance. In this case, linear inductors are close to short branches, which will cause excessive numerical errors with the NA formulation. Therefore, proper window-based truncation techniques on inductance matrices [2] should be applied before using the NA formulation so that ignorable (mutual) inductors are not present.

Once ignorable (mutual) inductances are truncated, it will be safe enough to use the NA formulation since the time step-size h_n is determined by time constants with relatively small values in a circuit. Figure 7 shows the histogram of SPICE time step-sizes for the *RCL* circuit in Fig. 5. It can be seen that most time step-sizes are less than 1 and some are even less than 0.1. Therefore, the condition number of the circuit matrix is required to be relatively small for h_n less than 1. According to Fig. 6, the NA formulation does ensure the condition number of the circuit matrix relatively small when the time step-size is decreased for h_n less than 1.

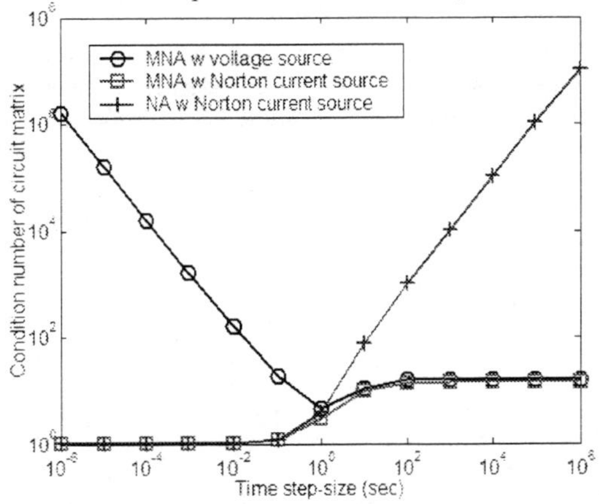

Figure 6. The condition number variation with time step-sizes.

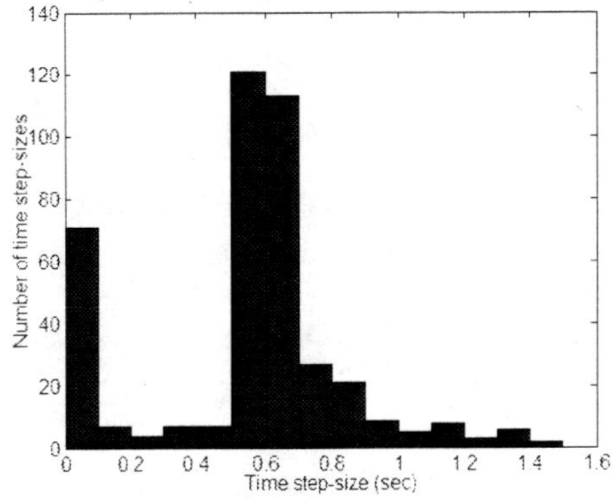

Figure 7. The histogram of time step-sizes.

4. EXPERIMENTAL RESULTS

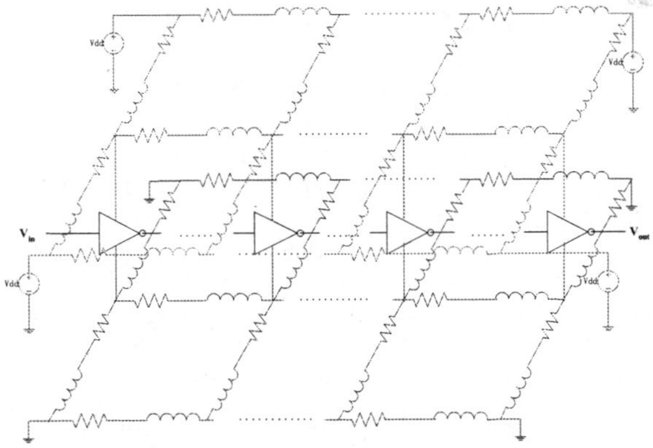

Figure 8. The power/ground analysis example.

In Fig. 8. the power and ground supply networks are modeled as two *RCL* mesh layers (parasitic coupling capacitors are not shown in Fig. 8). Between these two layers is a 20-stage inverter chain, different inverters of which are connected to different power/ground grid nodes. Furthermore, *RCL* loads are added for each inverter to model interconnect lines between adjacent stages.

Figure 9 shows the transient output waveform of the inverter chain when the output signal is digital "1" (the high voltage level). The "1" signal has been disturbed due to the *IR*-drop (the input

Vdd is 3.3v) and $L*dI/dt$ effects of the power/ground network. Table II shows the simulation results with varied numbers of elements modeling the power/ground network. In our experiments, the size of two *RCL* meshes is changed to vary the number of elements. The run time comparison between SPICE3 and the proposed method with the tolerance of the iterative method set to 1e-6 and 1e-8 is shown in Fig. 10. We can see that the coupled iterative/direct method achieves more speedup for larger circuits. The maximum overall speed-up reach *85.39X* and *16.74X* (with about 60 thousand elements) with the tolerance set to 1e-6 and 1e-8, respectively. The speedup is comparable to a recent explored direct method [3].

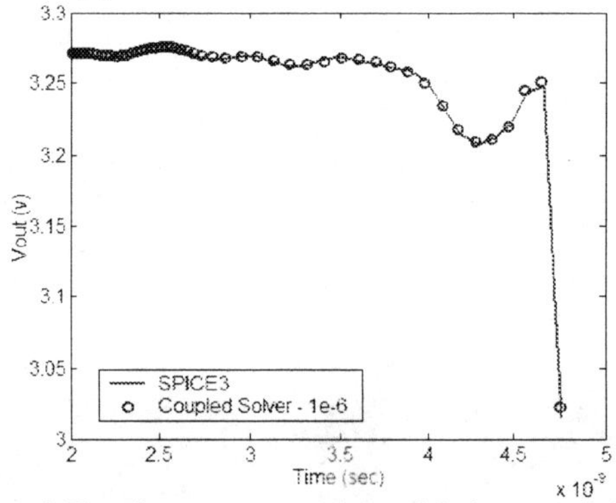

Figure 9. Transient output waveform of the inverter chain for power/ground analysis example.

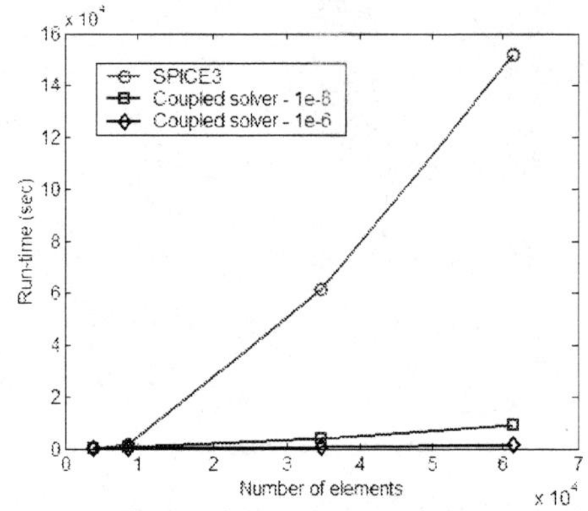

Figure 10. Run time comparison.

It can be seen from Table II that the number of outer Gauss-Seidel iterations is typically increased to *4X* to *5X* of that of SPICE nonlinear iterations. When the tolerance is set to 1e-6, the average number of CG iterations for each Gauss-Seidel step is 6.5 to 9, and it becomes 22 to 45 if the tolerance is set to 1e-8. The number of CG iterations increases dramatically to achieve high accuracy, i.e., when the tolerance is set to 1e-8. One way to improve the performance of iterative methods on large-scale power/ground networks for high accuracy is to apply multigrid-like methods [9].

5. CONCLUSION

A coupled iterative/direct time-domain circuit analysis method has been proposed for nonlinear circuits coupled with large-scale power/ground networks. Nodal analysis formulation of *RCLK* elements is applied on power/ground networks and an efficient iterative conjugate gradient method with an incomplete Cholesky decomposition preconditioner is used. Modified nodal analysis formulation is applied on nonlinear circuits and the direct method based on the Newton-Raphson iteration is used. The iterative method and the direct method are coupled together by a Gauss-Seidel style relaxation scheme. We further studied how the condition number of a circuit matrix changes with time step-sizes. Experimental results on digital circuits with power/ground networks show that the proposed method yields SPICE-like accuracy with orders of magnitude speedup over SPICE3.

REFERENCES

[1] T. Chen and C. C.-P. Chen, "Efficient Large-Scale Power Grid Analysis based on Preconditioned Krylov-subspace Iterative Methods", *Proc. IEEE/ACM Design Automation Conference*, pp. 559-562, June 2001.

[2] T. Chen, C. Luk, and C. C.-P. Chen, "INDUCTWISE: Inductance-Wise Interconnect Simulator and Extractor", *IEEE Trans. on CAD*, vol. 22, no. 7, pp.884-894, July 2003.

[3] Z. Li and C.-J. R. Shi, "SILCA: Fast-Yet-Accurate Time-Domain Simulation of VLSI Circuits with Strong Parasitic Coupling Effects", *Proc. IEEE/ACM Int. Conf. on Computer-Aided Design*, pp. 793-799, Nov. 2003.

[4] L. W. Nagel, *SPICE: A Computer Program to Simulate Semiconductor Circuits*, University of California, Berkeley, Tech. Rep., UCB/ERL M520, May 1975.

[5] J. R. Phillips and L. M. Silveira, "Simulation Approaches for Strongly Coupled Interconnect Systems", *Proc. IEEE/ACM Int. Conf. on Computer-Aided Design*, pp. 430-437, November 2001.

[6] Y. Saad, "*Iterative Methods for Sparse Linear Systems*", 2nd Edition, SIAM, 2003.

[7] Y. Wang, V. Jandhyala, and C.-J. R. Shi, "Coupled Electromagnetic-Circuit Simulation of Arbitrarily-Shaped Conducting Structures", *Proc. IEEE Conf. on Electrical Performance of Electronic Packaging*, pp. 233-236, October 2001.

[8] J. K. White and A. Sangiovanni-Vincentelli, *Relaxation Techniques for the Simulation of VLSI Circuits*, Kluwer Academic Publishers, 1987.

[9] J. N. Kozhaya, S. R. Nassif, and F. N. Najm, "A Multigrid-like Technique for Power Grid Analysis", *IEEE Trans. on CAD*, vol. 21, no. 10, pp. 1148-1160, Oct. 2002.

Table II. Simulation results for the power/ground analysis example.

#Elems	SPICE3		Coupled Solver						Speedup	
			ε=1e-6			ε=1e-8				
	#Iter	Overall (sec)	#GS Iter	#CG Iter	Overall (sec)	#GS Iter	#CG Iter	Overall (sec)	ε= 1e-6	ε= 1e-8
4002	4016	456.51	17363	113832	79.30	17350	384444	242.43	5.76	1.88
8851	4171	1.75e3	16432	128561	183.34	16205	475429	647.25	9.55	2.70
34802	3986	6.17e4	17679	156038	869.80	17868	772263	3.88e3	70.94	15.90
61602	4377	1.52e5	20208	162416	1.78e3	22335	1002260	9.08e3	85.25	16.74

MULTIGRID-BASED SUBSTRATE COUPLING MODEL EXTRACTION

João M. S. Silva and L. Miguel Silveira

Instituto Superior Técnico
INESC ID Lisboa / Cadence European Laboratories
R. Alves Redol, 9, 1000-029 Lisboa, Portugal
{jmss,lms}@algos.inesc-id.pt

ABSTRACT

Substrate noise is one of the most eminent problems in high-frequency mixed-signal designs, such as communication, biomedical and analog signal processing circuits and systems. Fast-switching digital blocks inject noise into the common substrate hindering the performance of high-precision sensible analog circuitry. Miniaturization effects on IC's complexity, inevitably make the accuracy requirements for substrate coupling simulation increase. However, model extraction and evaluation should not increase, leading to the need for fast and accurate substrate model extraction tools.

In this paper, we propose an extraction methodology based on a 3D Finite Difference formulation. The resulting 3D mesh is efficiently reduced to a circuit-level contact-based model by means of a fast Multilevel algorithm. Extraction results show the proposed method to be very efficient, achieving linear complexity independently of accuracy and discretization and outperforming competing algorithms that show quadratic complexity.

1. INTRODUCTION

Substrate bulk behavior in IC's is not perfectly insulator so that undesired coupling between different devices occurs [1, 2]. Coupling manifests itself by current injection / reception into / from the substrate through active and channel areas, as well as through substrate and well contact ties. As deep sub-micron MOS processes reach further in miniaturization and with increasing frequencies of operation, faster-switching digital blocks inject high-frequency noise into the substrate. In purely digital circuits or blocks, this can be troublesome for millions of logic gates switching noise can cause power supply voltage level fluctuations, affecting logic gates delay and circuit overall time performance. It is however in the context of mixed-signal design that the issue of substrate coupling is most prominent. Industry trends aimed at integrating higher levels of circuit functionality have triggered a proliferation of mixed analog-digital systems. The design of such systems is an increasingly difficult task owing to the various coupling problems that result from the combined requirements for high-speed digital and high-precision analog components. Analog circuitry relies on accurate levels of currents and voltages, so that analog transistors are correctly biased and projected performance is met. When substrate injected currents migrate through the substrate, substrate voltages fluctuate, causing havoc in sensitive analog transistors and possibly leading to malfunctioning circuitry [1, 2, 3, 4].

Analyzing the effects of substrate coupling requires that a model of such couplings is obtained and used in a verification framework. Typically such a verification is done at the electrical level by means of a circuit simulator which is fed the substrate model together with the models of the devices. In this paper, an efficient methodology is proposed for generating arbitrarily accurate substrate coupling models. The methodology proposed for model extraction, based on a Finite Difference formulation and Multilevel solution of the resulting mesh networks, is detailed and several methods for model extraction are compared. In Section 2 we review background work in the area of substrate model extraction. In Section 3 the proposed model extraction algorithm is presented and performance results are presented in Section 4. Finally in Section 5 some conclusions are drawn.

2. PREVIOUS WORK

Several extraction methodologies were previously studied and, based on them, several extracting tools were developed. The simplest modeling methodologies consist on finding coupling elements based on heuristic rules. Such methods are very attractive since the extraction overhead is minimal and they lead to simple first order models which also have low simulation costs [1, 4, 5, 6]. These models are, however, generally very imprecise. Furthermore, heuristic models are only really useful to the designer, for they are unable to account for higher order effects and, in fact, rely on designer's experience to prune out the expected relevant couplings [7]. On the other hand, methodologies that avoid a-priori heuristic pruning and work at the electrical level directly are typically based on a full description of the media and all the possible couplings. A problem that arises from model extraction in those cases is the extraction time and the size of the final model. There are two major classes of methods which have been proposed to generate such a model: Boundary Element Methods (BEM) and Finite Difference (FD) or Finite Element Methods (FEM).

In BEM methods, only the surface of the substrate contacts is discretized which leads to a system of equations that

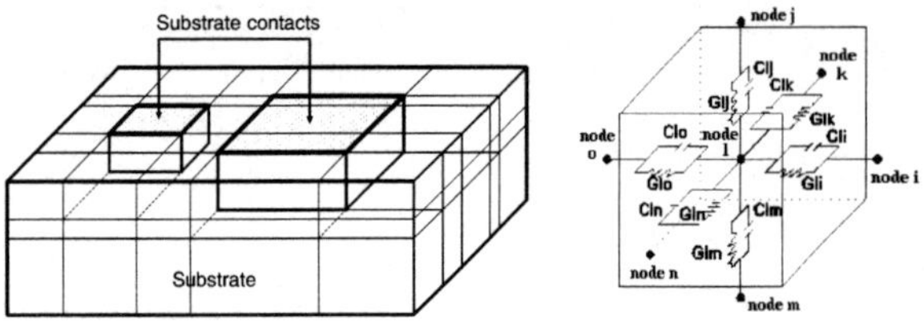

(a) (b)

Fig. 1. (a) FD method discretization and (b) resulting substrate resistive-capacitive (RC) model.

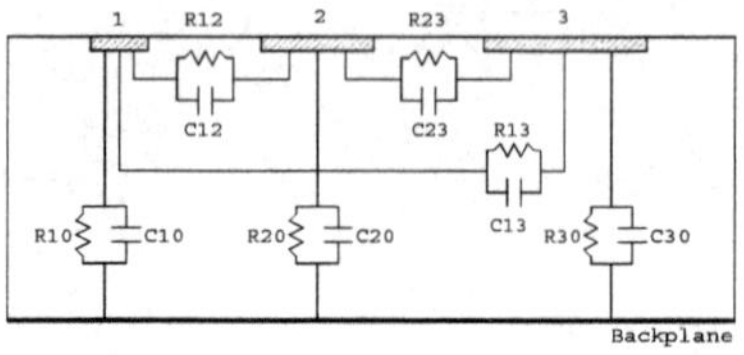

Fig. 2. RC model for a three contact configuration.

corresponds to small but full matrices. Extraction of such models requires intensive computations which restrains the range of applicability of this method to small and medium sized circuits [8, 2, 3]. Fortunately, significant progress in BEM methods performance has been lately achieved [9, 10]. Particularly, in [10] a multigrid-based algorithm was developed for solving the integral equations associated with the substrate problem. In FD or FEM methods, the whole 3D volume of the substrate is discretized leading to large but sparse matrices. Contrarily to BEM methods, FD/FEM methods produce large yet sparse matrices which nonetheless require extensive memory resources. This type of methods has also been recently enhanced with fast solution techniques [11, 12, 13].

3. SUBSTRATE MODEL EXTRACTION

In this work, extraction of a model of the couplings between substrate contacts is performed using a FD-based method. The large 3D mesh resulting from the discretization is solved using a fast Multilevel algorithm.

3.1. Finite Difference Tridimensional Model

Finite Difference methods imply a discretization of the substrate volume into a large number of small cuboid elements. Obviously the finer the discretization, a more accurate model is obtained. An example of such a discretization is shown in Figure 1-(a).

Applying Gauss' Law to a volume V containing an element (node) of the resulting discretization and then using the divergence theorem, results in:

$$\int_V \nabla E \, dV = \int_S E \, dS = \int_V \frac{\rho}{\varepsilon} \, dV, \qquad (1)$$

where E is the electric field, ρ and ε are, respectively, the resistivity and permitivity of the medium, and S the surface of the cuboid. Assuming an homogeneous medium in each substrate layer and if E_{ij} denotes the electrical field normal to the cuboid side surface between nodes i and j, results in:

$$\nabla E = \frac{\sum_j E_{ij} S_{ij}}{V_{cuboid}}, \qquad \sum_j E_{ij} S_{ij} = \frac{\rho}{\varepsilon} V_{cuboid}. \qquad (2)$$

Approximating the electrical field by [1] $E_{ij} = (V_i - V_j)/l_{ij}$ where l_{ij} is the distance between adjacent nodes i and j, and V_i, V_j the scalar potential at those nodes, we obtain:

$$\sum_j [G_{ij}(V_i - V_j) + C_{ij}(\frac{\partial V_i}{\partial t} - \frac{\partial V_j}{\partial t})] = 0 \qquad (3)$$

where $G_{ij} = \sigma S_{ij}/l_{ij}$, being σ the conductivity of the medium, and $C_{ij} = \varepsilon S_{ij}/l_{ij}$. Eqn. (3), derived from Maxwell's Laws, is in fact Kirchoff's Current Law (KCL) applied to node i. If node i had a current source connected to it, the right-hand side of Eqn. (3) would equal the value of the current source.

For typical values of σ and ϵ, the dielectric relaxation time of the substrate is on the order of tens of picoseconds, much smaller than the typical time scales of the circuit. Thus, it is reasonable to neglect intrinsic substrate capacitances for frequencies of operation up to a few GHz. Experimental comparisons conducted with detailed device simulators have shown that such an approximation does not affect the precision of the results for frequencies in the GHz range [11]. Mixed-mode systems with aggressive fast digital components may require more accurate modeling. In [14] a method for substrate dynamic models extraction is proposed, and it is shown that dynamic RC model extraction increases extraction complexity only by a constant factor. For the sake of simplicity, in this paper we will focus on resistive model extraction only. The methodology proposed here can however be applied to RC model extraction.

3.2. Circuit-level Model Extraction

Using the 3D mesh model from Eqn. (3) directly into any electrical simulator is prohibitive. Instead we propose to use the typical substrate contacts-based model which is depicted in Figure 2 for a simple configuration. Considering a system with m contacts and using Nodal Analysis (NA), the corresponding system of equations can be written as $G_c U = J$, where $G_c \in \mathbb{R}^{m \times m}$ is the matrix of resistive coupling elements between the m *contacts*, and $U, J \in \mathbb{R}^m$ are, respectively, the vectors of *contact* voltages and *contact* injected currents. Similarly applying NA to the 3D mesh model, assuming a discretization leading to n mesh nodes, leads to a similar system of equations, $GV = I$, where now $G \in \mathbb{R}^{n \times n}$. $V, I \in \mathbb{R}^n$ are, respectively the voltage vector

[1] This approximation leads to infinite precision when mesh spacing tends to zero.

for all nodes in the discretization mesh and the corresponding injected currents. This system is naturally analog to the previous one, but much larger.

The substrate contact model, defined by G_c, can readily be obtained from the 3D model by means of simple computations. Suppose we set the voltage in contact k to $1V$. In that case the k-th component of U is at $1V$ and thus J equals the k-th column of G_c. However, setting a contact's voltage to some value is equivalent to setting the voltages of all mesh nodes that fall within the contact to that value. As nodal analysis (NA) is used, the inputs to the 3D mesh should be currents, applied to nodes adjacent to the contact nodes. The values of such currents can easily be obtained from the corresponding Norton equivalent circuits seen by those nodes, leading to $I = Y_{adj}U$ where $Y_{adj} \in \mathbb{R}^{n \times m}$ is a matrix, describing the Norton equivalent admittances seen by the mesh nodes that are adjacent to nodes on the contacts. Clearly most of the entries in Y_{adj} are zero, with the exception of lines related to the nodes adjacent to the contacts. On the other hand, the output of the system is given by the currents on the contacts. Therefore, it is easy to see that these can be obtained as

$$J = Y_{adj}^T V = \underbrace{Y_{adj}^T G^{-1} Y_{adj}}_{G_c(s)} U \tag{4}$$

which exposes the conductance model of the system of contacts. This process can be repeated as many times as the number of contacts so that the full conductance matrix G_c is formed, one column at a time. The cost of computing the contact model, G_c, for a system of m contacts is thus equal to m times the cost of solving the 3D mesh to determine the node voltages. This can be performed very efficiently by means of a fast Multigrid algorithm with a cost of $\mathcal{O}(N)$ per solve as we shall see in Section 4.

3.3. Finite Difference Mesh Solving Methods

Several methods were considered for solving the large 3D system which is required to obtain the reduced circuit-level model. In this work we propose to use Multigrid (MG) methods [15] and to determine their efficiency we compare them to Krylov-subspace methods [16], such as Conjugate Gradient (CG). Since it is well known that the performance of Krylov-subspace methods can be further improved if a good preconditioner can be obtained [16]. For our work, we have determined that incomplete Cholesky factorization demonstrated to yield the most efficient behavior and was thus used as a preconditioner (PCG). Actually the Multigrid method itself can also be used a preconditioner (and we term that MGPCG).

MG methods are based on the existence of different accuracy grids. Starting from a fine 3D mesh, the initial problem recursively projected to coarser grids until direct resolution (by Gaussian elimination, for instance) is efficient. Solutions obtained at any coarser level are then interpolated

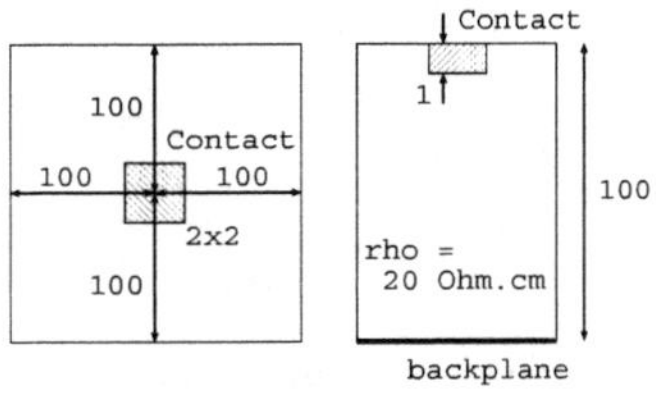

Fig. 3. Experimental layout and corresponding substrate profile (units in microns).

Method	33×33×17	65×65×33	129×129×65
CG	132	189	290
PCG	29	46	88
MG	7	4	3
MGPCG	4	3	3

Table 1. Iteration count for increasing $x \times y \times z$ discretizations.

to finer levels where local solutions are adjusted. This is called a MG V-Cycle. An explicit algorithm and details can be obtained from [15].

4. RESULTS

The proposed methodology was implemented in an extraction tool named SMX (Substrate Model eXtractor) [17]. SMX reads contact layout in CIF format and outputs the matrix of conductance couplings, G_c, in a form that can be easily included into a Spice description. The model obtained by SMX has been compared to those obtained using Xtract [9] and is shown to be of similar accuracy, depending on discretization parameters of both tools. No prototype circuit has been manufactured which could lead us to assert about extractors absolute accuracy.

In the following sections results are presented concerning model extraction complexity using CG, PCG, MG and MGPCG as FD mesh solvers. Results shown here refer to extraction on a simple one contact configuration with backplane shown in Figure 3, although the tool was tested with many more configurations, including portions of industrial designs. Experiments were conducted on the test configuration using finer and finer discretizations, leading to larger matrices.

4.1. Iteration Complexity

The first issue we look into is iteration count, which is directly related to computational complexity. As can been seen from Table 1 and Figure 4, which depict the number of iterations and the residue norm plots, while CG and PCG show rapidly degrading linear iteration complexity, MG-based methods converge in a constant number of iterations. This is the main advantage of using Multilevel methods to solve the substrate coupling problem.

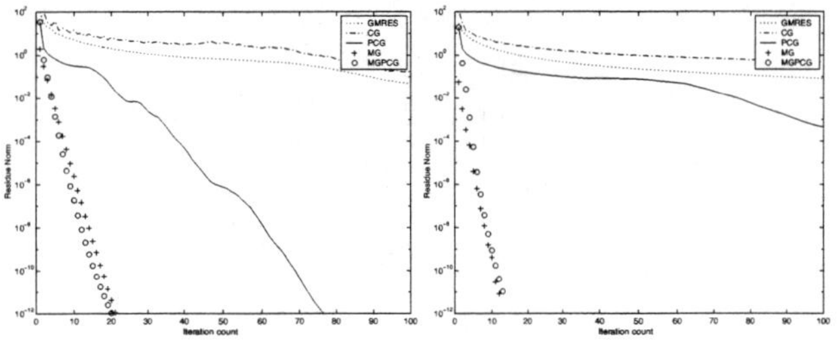

(a) (c)

Fig. 4. Residue norm evolution for mesh discretizations of (a) $33 \times 33 \times 17$, and (b) $129 \times 129 \times 65$.

Method	33×33 ×17	65×65 ×33	129×129×65
CG	2.55	31.20	369.14
PCG	1.09	13.52	189.30
MG	2.65	18.55	140.20
MGPCG	2.26	17.09	136.08

Table 2. Total extraction time (includes setup time) for increasing $x \times y \times z$ discretizations (in seconds).

4.2. Time Complexity

In spite of their superior convergence rate, MG-based have to pay some penalty for setting up the projection and interpolation operators. As such, when setup times are taken into account the break-even points are likely to occur only for high levels of discretization (i.e. for high accuracy models). As can be seen from Table 2, MGPCG shows the best results followed closely by MG and by large surpassing unpreconditioned Krylov-subspace methods. SMX with MGPCG runs, for the presented example, in about 70% the time of SMX with PCG. As Krylov-subspace methods have $\mathcal{O}(N)$ time complexity, this time saving ratio will quickly increase for finer and finer discretizations.

4.3. Memory Complexity

The drawback of using MG-based methods is their memory requirements. These methods need to keep not only projection and interpolation operators in memory but all levels matrices and vectors also. It can be shown that these structures memory requirements are limited to the double amount of the finner level structures memory [15]. This leads to MG-based methods showing approximately double memory requirements of Krylov-subspace methods.

5. CONCLUSIONS

A methodology for the extraction of substrate coupling models has been presented. Contact-based models, obtained from a formulation based on a Finite Difference discretization, were computed using a fast Multigrid algorithm. Comparison tests reveal that using Multigrid based methods offers constant iteration complexity which in the substrate problem leads to faster model extraction times and enables extraction of very accurate models resulting from fine discretizations.

6. REFERENCES

[1] David K. Su, Marc J. Loinaz, Shoichi Masui, and Bruce A. Wooley. Experimental results and modeling techniques for substrate noise in mixed-signal integrated circuits. *IEEE Journal of Solid-State Circuits*, 28(4):420–430, April 1993.

[2] Ranjit Gharpurey and Robert G. Meyer. Modeling and analysis of substrate coupling in integrated circuits. *IEEE Journal of Solid-State Circuits*, 31(3):344–353, March 1996.

[3] Nishath K. Verghese, David J. Allstot, and Mark A. Wolfe. Verification techniques for substrate coupling and their application to mixed-signal IC design. *IEEE Journal of Solid-State Circuits*, 31(3):354–365, March 1996.

[4] Bram Nauta and Gian Hoogzaad. How to deal with substrate noise in analog CMOS circuits. In *European Conference on Circuit Theory and Design*, pages Late 12:1–6, Budapest, Hungary, September 1997.

[5] A. J. van Genderen, N. P. van der Meijs, and T. Smedes. Fast computation of substrate resistances in large circuit. In *European Design and Test Conference*, Paris, February 1996.

[6] Sujoy Mitra, R. A. Rutenbar, L. R. Carley, and D. J. Allstot. A methodology for rapid estimation of substrate-coupled switching noise. In *IEEE 1995 Custom Integrated Circuits Conference*, pages 129–132, 1995.

[7] Joel R Phillips and L. Miguel Silveira. Simulation approaches for strongly coupled interconnect systems. In *International Conference on Computer Aided-Design*, November 2001.

[8] T. Smedes, N. P. van der Meijs, and A. J. van Genderen. Extraction of circuit models for substrate cross-talk. In *International Conference on Computer Aided-Design*, San Jose, CA, November 1995.

[9] João P. Costa, Mike Chou, and L. Miguel Silveira. Efficient techniques for accurate modeling and simulation of substrate coupling in mixed-signal IC's. In *DATE'98 - Design, Automation and Test in Europe, Exhibition and Conference*, pages 892–898, Paris, France, February 1998.

[10] Mike Chou and Jacob White. Multilevel integral equation methods for the extraction of substrate coupling parameters in mixed-signal ic's. In 35th *ACM/IEEE Design Automation Conference*, pages 20–25, June 1998.

[11] F. J. R. Clement, E. Zysman M. Kayal, and M. Declercq. Layin: Toward a global solution for parasitic coupling modeling and visualization. In *Proc. IEEE Custom Integrated Circuit Conference*, pages 537–540, May 1994.

[12] Balsha Stanisic, Nishath K. Verghese, Rob A. Rutenbar, L. Richard Carley, and David J. Allstot. Addressing substrate coupling in mixed-mode IC's: Simulation and power distribution systems. *IEEE Journal of Solid-State Circuits*, 29(3):226–237, March 1994.

[13] J. Kanapka, J. Phillips, and J. White. Fast methods for extraction and sparsification of substrate coupling. In *Proc. 37th Design Automation Conference*, June 2000.

[14] João M. S. Silva and L. Miguel Silveira. Dynamic models for substrate coupling in mixed-mode systems. In *VLSI-SOC'2003 XII IFIP International Conference on VLSI*, Darmstadt, Germany, December 2003.

[15] William L. Briggs. *A Multigrid Tutorial*. Society for Industrial and Applied Mathematics, Philadelphia, Pennsylvania, 1987.

[16] Yousef Saad. *Iterative Methods for Sparse Linear Systems*. Pws Publishing Co., 1996.

[17] João M. S. Silva. Modeling substrate coupling in mixed analog-digital circuits (in portuguese). Master's thesis, Instituto Superior Técnico, Technical University of Lisbon, Lisboa, Portugal, May 2003.

Partial Random Walk For Large Linear Network Analysis

Weikun Guo, Sheldon X.-D. Tan, †Zuying Luo, †Xianlong Hong
Department of Electrical Engineering, University of California, Riverside, CA 92521
†Department of Computer Science and Technology, Tsinghua University, Beijing 100084,China

ABSTRACT

This paper proposes a new simulation algorithm for analyzing large power distribution networks, modeled as linear RLC circuits, based on a novel partial random walk concept. The random walk simulation method has been shown to be an efficient way to solve for a small number of nodes in a lager power distribution network [6], but the algorithm becomes expensive to solve for nodes that are more than a few. In this paper, we combine direct methods like LU factorization with the random walk concept to solve power distribution networks when a significant number of node waveforms is required. We also apply an equivalent circuit modeling method to speed up the direct simulation of subcircuits. Experimental results show that the resulting algorithm, called partial random walk (PRW), has significant advantages over the pure random walk method especially when the VDD/GND nodes are sparse and accuracy requirement is high.

1. INTRODUCTION

Signal integrity on the on-chip power distribution networks has become a limiting factor for designing high performance VLSI systems in today deep submicron technology. The challenges for designing and verifying a reliable on-chip power deliver network lie in the increasing sizes of the network circuits that typically contain millions RLC components. Conventional circuit analyzers (such as SPICE) cannot meet such demanding simulation tasks and efficient simulations are highly required to reduce the increasing design productivity gap in deep submicron design regime.

Different methods have been proposed in the past to address this problem[1][2][3][4][5][7][8]. These existing approaches include frequency domain analysis method[1], the hierarchical method [7], the multi-grid method [5]. Recently Qian *et al* proposes a new statistical method to solve the power/ground (P/G) networks. The new algorithm exploits the fact that there exist some VDD/GND nodes evenly distributed for some mesh-structured P/G grid networks (due to advanced packaging technologies like IBM C4 package). A random walk process is applied to solve the node voltages in a statistical way. The advantage of this method is that it can efficiently solve for a small number of node voltages in a large P/G network without solving the whole network. But the typical verification of a P/G circuit requires the analysis of the whole network or at least a portion of the network, not just a few nodes. The random walk method, however, is not very efficient for such tasks, as every node has to be simulated individually. Lowering accuracy may speedup the random walking process, but it may not be accepted for some applications requiring knowledge of accurate IR drops or voltage fluctuations.

In this paper, we propose a new random walk algorithm that combines direct solution methods like LU factorization method with the random walk concept. The idea is to partition a large P/G network into a number of smaller subcircuits and to solve each or a specific subcircuit in two steps. First we use the random walk process to solve for the boundary nodes of each subcircuit. Second we apply LU method to solve for the rest of nodes inside the subcircuit. Such process can be processed subcircuit by subcircuit or a hierarchical way if the solution of the whole network is required. Our experimental results show that the partial random walk can be one order of magnitude faster than the pure random walks specially when VDD/GND nodes are sparse and accuracy requirements are high.

This paper is organized as follows. In Section 2, we briefly review the random walk algorithm. In Section 3, we illustrate the new partial random work concept. The experimental results are shown in Section 4. Finally we conclude the paper in Section 5.

2. REVIEW OF RANDOM WALK ALGORITHM

The random-walk based approach to solving linear network exploits the fact that Kirchoff's current and voltage laws can be mapped to a random walking process such that KVL and KCL equations are equivalent to statistic formulas describing the random walking process [6]. Such a statistical process will become reasonable fast and accurate if the unknown node voltages can be sufficiently determined by visiting nearly by voltage-known nodes (like VDD/ GND nodes) that are not far away. As a result, the algorithm is suitable for P/G networks that have many VDD/GND nodes or pads (voltage-known nodes) evenly distributed inside a chip. Otherwise, it will take significant long walks before the walking process can stop at the voltage-known node. Also the random walk method can easily make the tradeoff between accuracy and runtime. But our experimental results show that the runtime can be significantly slowed down even with a very small increase in the accuracy requirement. Therefore, pure random walk process is inefficient for solving a large number of nodes with high accuracy, which is critical for the detailed signal integrity verification of large P/G networks.

3. PARTIAL RANDOM WALK

3.1 Basic Idea

The basic idea of the partial random walk concept is to allow the random walk process to solve boundary part of a subcircuit and then solve the rest of the nodes in the subcircuit by traditional direct methods like LU factorizations. The new algorithm combines the efficiency of both the random walk and LU factorization method to speed up the whole simulation process.

Specifically, we first partition a large P/G network into a number of small subcircuits such that LU method can solve them sufficiently fast. We notice that how a partitioning is performed or subcircuit is defined depends on the specific verification application. Then we apply the following two steps to solve each of the subcircuit: First we apply the random walk process to solve for all the boundary nodes of each subcircuit. Once the voltages of boundary nodes are known, we solve for voltages of the remaining nodes via LU method. Since the number of boundary nodes (which grows linearly with the size of a subcircuit) is typically smaller than the number of nodes inside each subcircuit (which grows quadratically with the

0-7803-8251-X/04/$17.00

size of a mesh-structured subcircuit), the runtime cost of solving those boundary nodes by random walk processes will be reduced significantly while the rest of nodes can be solved by the LU method efficiently.

Notice that if a number of subcircuits are to be solved, the order to solve those subcircuits should be arranged such that boundary node voltages, which are computed by random walk processes, should be reused as much as possible.

3.2 Equivalent Circuits For Voltage Sources

To speedup the simulation of subcircuits by LU method, we need to reduce the matrix sizes of the subcircuits by transforming the known boundary voltages to equivalent current sources. Specifically, let's have system-equation set Av = b. Suppose that we partition the circuit into two circuits, one is small subcircuit *I* and another is rest of the circuit *R*. In between, there are some boundary nodes connecting the two circuits. As a result, the circuit equation set can be rewritten in the following form:

$$\begin{bmatrix} A^{II} & A^{IB} & 0 \\ A^{BI} & A^{BB} & A^{BR} \\ 0 & A^{RB} & A^{RR} \end{bmatrix} \begin{bmatrix} v^I \\ v^B \\ v^R \end{bmatrix} = \begin{bmatrix} B^I \\ B^B \\ B^R \end{bmatrix} \quad (1)$$

where A^{II} is the internal matrix associated with internal variable vector v^I. A^{IB} and A^{BI} are the connection matrices between internal nodes v^I and boundary nodes v^B. Suppose that we obtain the node voltages of boundary node v^B by random walk processes. Then we have the following equations according to (1).

$$A^{II}v^I + A^{IB}v^B = B^I \quad (2)$$

then

$$A^{II}v^I = B^I - A^{IB}v^B \quad (3)$$

As a result, we can solve Eq (3) to obtain all the internal nodes voltages. Notice that $A^{IB}v^B$ is the vector of independent current sources computed from the known voltages of the boundary node vector v^B, which can be computed easily as shown in Fig. 3. The equivalent current sources are also used for VDD/GND nodes inside the subcircuits to reduce the MNA matrix size. Experimental results show using the equivalent current sources can significantly reduce the simulation time of subcircuits by LU method.

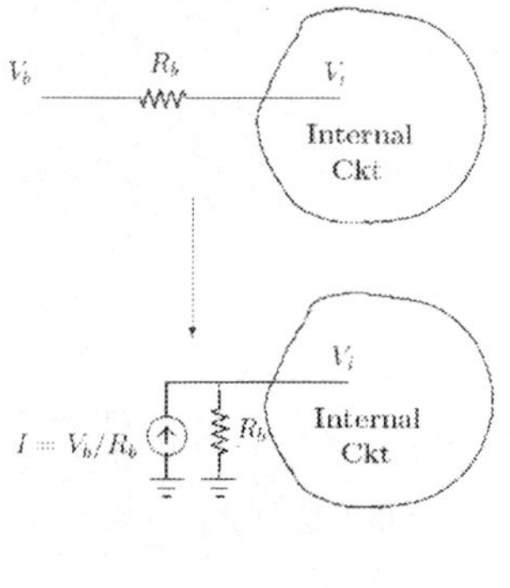

Figure 3. Equivalent current sources for voltage sources of the boundary nodes and internal VDD/GND nodes.

3.3 Extension to RLC network

In [6], P/G networks modeled as RC circuits are analyzed. In this subsection, we discuss how to extend this random walk to deal with RCL circuits. Here we consider self-inductor L.

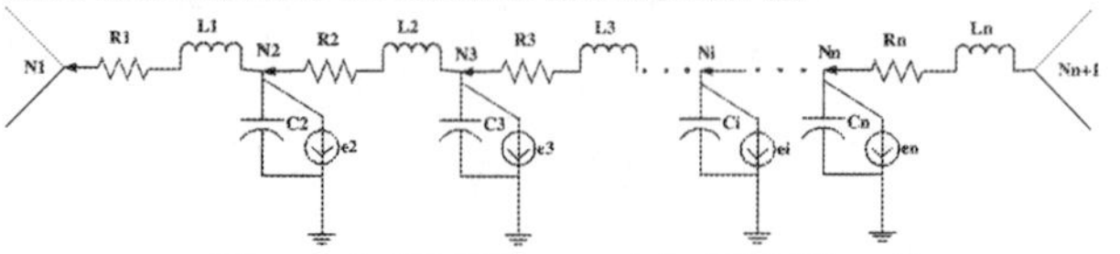

Figure 4. RLC Segments in P/G Circuits.

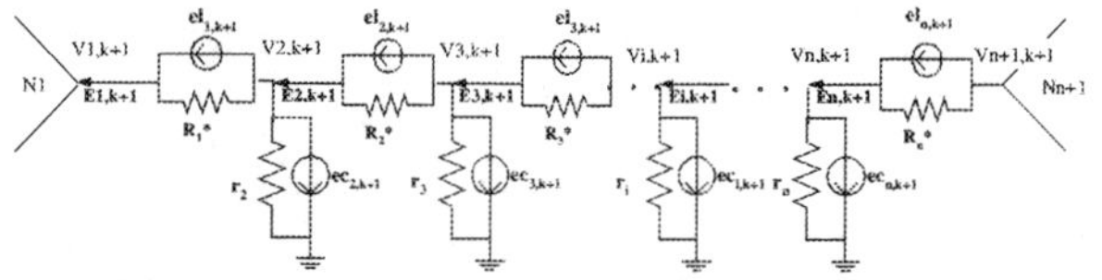

Figure 5. Simplified Discretized RLC Segments

For one RLC segment shown in Fig.4, if we apply trapezoidal integration with time step *h* and Norton equivalent circuit, we can transform the circuit in Fig.4 to the one in Fig.5. Specifically, during the transformation, a shunt capacitance branch *i* is transformed into an equivalent current and an equivalent resistor.

$$r_i = \frac{h}{2C_i}, \quad Ic_{i,k+1} = -\left(\frac{2C_i}{h}V_{i,k} + I_{i,k}\right), \quad (4)$$

where *k* is the time step index. Combine the current caused by devices and the equivalent current due to capacitor *i*, we obtain

$$ec_{i,k+1} = Is_{i,k+1} - \left(\frac{2C_i}{h}V_{i,k} + I_{i,k}\right). \quad (5)$$

Meanwhile, the floating R and L branches can be transformed into an equivalent current and an equivalent resistor connected in parallel.

$$R_i^* = 2L_i/h + R_i, \quad el_{i,k+1} = \left(\frac{h}{2L_i}V_{L,i,k} + I_{L,i,k}\right)\cdot\frac{2L_i/h}{2L_i/h + R_i}, \quad (6)$$

Let's define

$$g_{i-1,j,H} = \frac{1}{R^*_{i-1,j,H}}; \quad g_{i,j,H} = \frac{1}{R^*_{i,j,H}};$$

$$g_{i,j-1,V} = \frac{1}{R^*_{i,j-1,V}}; \quad g_{i,j,V} = \frac{1}{R^*_{i,j,V}};$$

$$g_{i,j} = g_{i-1,j,H} + g_{i,j,H} + g_{i,j-1,V} + g_{i,j,V} + \frac{1}{r_{i,j}},$$

where subscripts *H* and *V* mean horizontal and vertical direction. According to Fig.5, the KCL equation at node *(i,j)* can be formatted as the following

$$V_{i,j,k+1} = \frac{g_{i-1,j,H}}{g_{i,j}}V_{i-1,j,k+1} + \frac{g_{i,j,H}}{g_{i,j}}V_{i+1,j,k+1} + \frac{g_{i,j-1,V}}{g_{i,j}}V_{i,j-1,k+1} + \frac{g_{i,j,V}}{g_{i,j}}V_{i,j+1,k+1} \quad (7)$$
$$-\frac{1}{g_{i,j}}\left(Is_{i,j,k+1} - \left(\frac{2C_{i,j}}{h}V_{i,j,k} + I_{i,j,k}\right) + el_{i-1,j,H} - el_{i,j,H} - el_{i,j-1,V} + el_{i,j,V}\right)$$

Let's further define

$$I_{i,j,k+1} = \left(\frac{2C_{i,j}}{h} V_{i,j,k} + I_{i,j,k} \right) - el_{i-1,j,H} + el_{i,j,H} + el_{i,j-1,V} - el_{i,j,V}$$

Finally, we have

$$V_{i,j,k+1} = \frac{g_{i-1,j,H}}{g_{i,j}} V_{i-1,j,k+1} + \frac{g_{i,j,H}}{g_{i,j}} V_{i+1,j,k+1} + \frac{g_{i,j-1,V}}{g_{i,j}} V_{i,j-1,k+1} + \frac{g_{i,j,V}}{g_{i,j}} V_{i,j+1,k+1} + \frac{1}{g_{i,j}} I_{i,j,k+1} - \frac{1}{g_{i,j}} Is_{i,j,k+1} \quad (8)$$

Eq.(8) gives the probabilities of a random walking step from node (i,j) to its neighbor nodes (the coefficient of V_x is the probability to walking into each neighborhood node x) through resistor-inductance branches. Note that we will walk through the floating resistor and inductor, which are connected in series, in just one step.

4 EXPERIMENTAL RESULTS

The proposed algorithm has been implemented in C++. For the LU solver, we use SuperLU [9] to solve the linear equations from a subcircuit. The proposed algorithm is applied to analyze a RC P/G grid circuits with 63001 nodes and various ratios of VDD nodes. The performance of the proposed algorithm is also evaluated under different accuracy requirements. These performance results are compared with those of the pure random walk algorithm and SPICE.

Table 1 gives the CPU run times for the whole circuit, which is partitioned into 25 subcircuits, under different VDD percentages and accuracy in terms of delta, which is the absolute error margin for accuracy constraints. For instance, delta = 0.01v means that the error between the estimated one and real one is less than ± 0.01 volt with very high probability (99%)[6]. To obtain more accurate results, we also compare the two methods for simulating one particular subcircuit (50x50) of the circuit and the results are shown in Table 2. All the computations are carried out on a Linux workstation with dual 1.7GHz AMD processors and 2 GB memories.

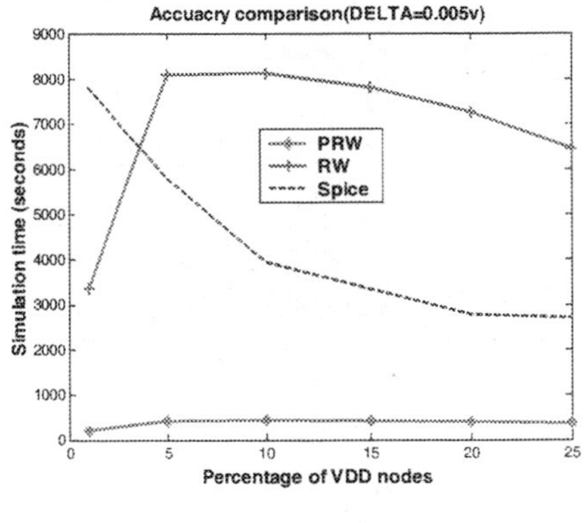

Figure 5. Runtime comparison of PRW, RW and SPICE on simulation of the whole circuit with delta = 0.005v

In each table, the first row shows the accuracy requirements in terms of different deltas. PRW and RW designate partial random walk and the pure random walk algorithms respectively. From Table 1 and Table 2, we can see that PRW algorithm is faster than other two methods for all the cases. The speedups over RW become significant (more than 10x) for high accuracy requirements (delta < 0.01v).

Figure 5 compares the runtimes of the PRW, the RW methods and SPICE under various VDD/GND node percentages in term of total number of nodes under constant accuracy requirement (delta =0.005v). For the random walk algorithm, theoretically if we decrease the number of VDD nodes, the random walk process will run slowly as it takes longer paths on average to find a home (node with known voltage). However, this is not always true, as the VDD nodes become very sparse, some node's voltages are small (we use 1.5volt as VDD value), as a result, the variations of the estimated voltages from true voltages become smaller, which in turn results in less number of random walks needed to meet the required accuracy as accuracy requirement is expressed in terms of absolute voltage values. So we may end up with small CPU runtimes as shown in Figure.5. But as the number of VDD node further increases, it eventually will help to save CPU time as shown in Figure. 5. For PRW method, since its random portion of run time is reduced, it becomes less sensitive to the changes of the VDD node percentages.

Figure 6 compares the runtime of the PRW and RW methods on simulating one particular subcircuit with 1% and 25% VDD node percentages respectively. The PRW method is faster than the RW method for all the cases. The speedup becomes more significant when the accuracy reaches 0.01 and beyond.

Figure 7 shows the impacts of the equivalent modeling on the CPU time for the whole circuit simulation with accuracy = 0.005 for different VDD node percentages. If no equivalent circuits are used, size of the circuit matrix will increase with the number of VDD nodes, as one VDD node introduces one extra node into the circuit matrix in MNA (modified nodal analysis) formulation, while the equivalent current of a voltage source reduces one node in circuit matrix, so there are two node difference in circuit matrix size for each VDD node added into a subcircuit. Since all the boundary nodes are voltage-known nodes, the MNA matrix size can be significantly reduced after we use equivalent modeling circuits for these boundary and VDD/GND nodes inside the subcircuits. As a result, the simulation time of LU method can be significantly reduced. On the contrary, as the matrix size increases, so does the CPU as shown in Figure 7 for simulation without using the equivalent circuits.

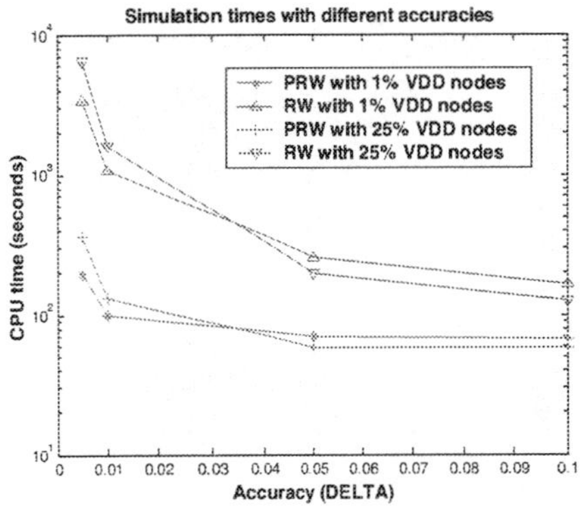

Figure 6. Runtime comparison of PRW and RW run on one subcircuit with different VDD nodes

5. CONCLUSION

In this paper, we have proposed a new circuit simulation algorithm, which combines the recent proposed statistical based random walk concept with LU factorization method to take advantage of both methods to speedup the simulation of large power distribution networks modeled RLC circuits. We also applied equivalent current circuits to speed up the simulation of subcircuits by LU method and extended the random talk method to deal RLC networks. Our experimental results show the significant speedup can be achieved over the pure random walk algorithm when VDD/GND nodes are spare and accuracy requirements are high.

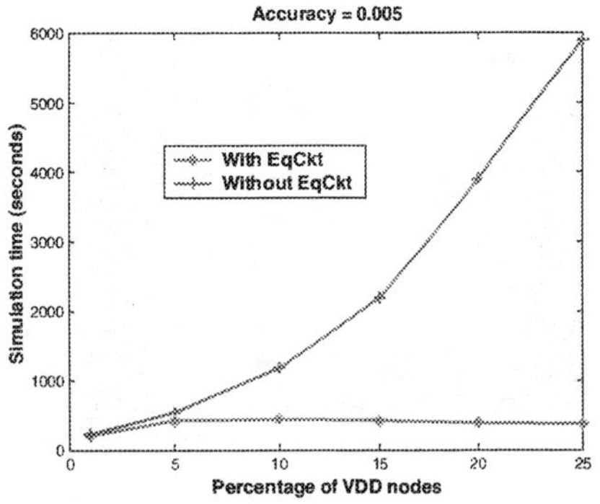

Figure 7. Comparison with and without equivalent current source modeling

Acknowledge

The authors would like to thank Haifeng Qian from the Univ of Minnesota for the discussion of the random walk algorithm.

REFERENCE

[1] G. Bai, S. Bobba and I. N. Hajj, "Simulation and optimization of the power distribution network in VLSI circuits", in *Proc. IEEE/ACM International Conf. on Computer-Aided Design*, pp. 481–486, 2000.

[2] T. Chen and C. C. Chen, "Efficient large-scale power grid analysis based on preconditioned Krylov-subspace iterative method", *Proc. ACM/IEEE Design Automation Conf.*, pp. 559–562, June, 2001.

[3] Y. Cao, Y. Lee, T. Chen and C. C. Chen, "HiPRIME: hierarchical and passivity reserved interconnect macromodeling engine for RLKC power delivery", *Proc. ACM/IEEE Design Automation Conf.*, pp. 379–384, June, 2002.

[4] H. H. Chen and D. D. Ling, "Power supply noise analysis methodology for deep-submicron VLSI chip design", *Proc. 34th ACM/IEEE Design Automation Conf.*, pp. 638–643, 1997.

[5] J.N. Kozhaya, S.R. Nassif, and F.N. Najm, "A multigrid-like technique for power grid analysis", *IEEE Trans. Computer-Aided Design*, vol. 21, no.10, pp. 1148-1160, Oct. 2002.

[6] H.-F. Qian, S.R. Nassif. S.S. Sapatnekar, "Random Walks in a Supply Network," *Proc. Design Automation Conference*, pp. 93-98, 2003.

[7] M. Zhao, R. V. Panda, S. S. Sapatnekar and D. Blaauw, "Hierarchical analysis of power distribution networks", *IEEE Trans. Computer-Aided Design*, vol. 9, no. 2, pp. 159–168, Apr. 1990.

[8] H.-H. Su, K.H. Gala, and S.S. Sapatnekar, "Fast analysis and optimization of power/ground networks", *Proc. IEEE/ACM International Conf. on Computer-Aided Design.*, pp. 477–482, 2000.

[9] SuperLU, http://crd.lbl.gov/~xiaoye/SuperLU/

Table 1. Runtime comparison of PRW, RW and SPICE for a whole circuit simulation (sec.)

Precision		0.005v		0.01v		0.05v		0.1v	
Vdd	SPICE	PRW	RW	PRW	RW	PRW	RW	PRW	RW
1%	7790	197	3363	99	1078	70	254	68	163
5%	5767	409	8103	147	2700	68	255	66	157
10%	3928	433	8129	152	2255	64	265	63	151
15%	3321	410	7798	144	2086	63	252	61	142
20%	2774	382	7248	137	1775	60	204	58	135
25%	2693	359	6457	132	1623	59	196	58	125

Table 2. Runtime comparison of PRW and RW for one subcircuit simulation (sec.)

Precision	0.001v		0.005v		0.01v		0.05v		0.1v	
Vdd	PRW	RW	PRW	RW	PRW	RW	PRW	RW	PRW	RW
1%	112	3474	7	145	3.75	37.43	2.85	7.4	2.82	6.85
5%	378	7904	18	327	6.06	82	2.77	7.88	2.73	6.63
10%	481	8611	21	338	7.02	84	2.70	7.68	2.62	6.25
15%	451	8037	20	318	6.65	79	2.6	7.39	2.53	5.93
20%	503	7196	22	286	7.16	74	2.54	7.12	2.44	5.73
25%	459	6689	20	268	6.70	66	2.42	6.71	2.34	5.3

MATRIX PENCIL BASED REALIZABLE REDUCTION FOR DISTRIBUTED INTERCONNECTS

Janet Wang
University of Arizona
wml@ece.arizona.edu

Omar Hafiz
University of Arizona
ohafiz@email.arizona.edu

ABSTRACT

In this paper, we propose a realizable parasitic reduction method for RLGC distributed interconnects . The proposed method generates a reduced order model based on a modified matrix pencil method. By using a set of analytic formulas, this method provides synthesied RLGC elements. This new model is applied to power grid and antena circuits involving triangular input waveforms, lossy transmission lines and discontinuities of interconnects. The results show better reduction ratio than the standard macromodels and good accuracy compared with the theoretical values.

1. INTRODUCTION

With increasing design complexity, the high capacity problem caused by the huge size of interconnect networks has pushed timing/noise analysis and transistor lever simulators to the limits. Model Order Reduction (MOR) based techniques are the only ways to solve this high capacity issue. Furthermore, the current design flows call for realizable reduction techniques so that incremental analysis, optimizations and ECO can be integrated into the design flows easily. However, recent work in MOR techniques, such as asymptotic waveform evaluation (AWE) [1], complex frequency hopping (CFH) [2], Pade approximation via Lanczos process (PVL) [3], and Passive Reduced-Order interconnect macromodeling algorithm (PRIMA) [4], have been focused on developing either passive or stable macromodels instead of realizable models that can be directly fed into general circuit simulators.

Cheng and Qin [12] proposed a generalized Y- Δ transformation based realizable reduction algorithm. In [13], Ismail and his co-authors developed a RLCK circuit crunching algorithm. Both of these methods work well for lumped RC/RLCK circuits, but are not suitable for distributed interconnects because they are generally described as macromodels to start with. In this paper, we propose a realizable reduction algorithm for distributed interconnects. First, a new modified matrix pencil technique is developed to provide the macromodel. This new technique introduces a phase shift model to reduce the number of terms needed in modeling the distributed interconnects. Then with a set of analytic formulas, the developed macromodel is realized as R,L,G,C lumped circuit elements. At this point, methods in [12] and [13] can be applied to further reduce the order.

This paper is organized as follows: the new phase shift model in discussed in Section II. A modified matrix pencil technique is introduced in Section III. A set of analytic equations is presented in Section IIII. Section V provides the implementation details. Section VI gives the experimental results. Finally, Section VII concludes the paper.

2. PHASE SHIFT MODEL

Standard MOR macromodels approximate the impulse response of high-speed interconnects using a series of complex exponentials in the time domain or complex poles in the frequency domain. The time-domain and frequency-domain fitting models can be expressed as:

$$f(t)=f_p(t)+f_{np}(t)=\sum_{\alpha=1}^{M}R_\alpha \exp(s_\alpha t)U(t)+f_{np}(t), \quad (1)$$

$$F(s)=F_p+F_{np}=\sum_{\alpha=1}^{M}\frac{R_\alpha}{s-s_\alpha}+F_{np}(s), \quad (2)$$

where $f(t)$ and $F(s)$ represent the observed responses in the time domain and frequency domain respectively, R_α are residues, s_α are poles, f_{np} and F_{np} stand for non-pole terms, and $U(t)$ denotes a unit step function. Such classes of techniques for extracting parameters for (1) (2) from frequency or time domain data can also be called Model-Based Parameter Estimation (MBPE) [11].

One of the difficulties with standard MOR macromodels is that they require a large number of terms to model interconnects that are electrically long. It is inefficient to model the portion of the waveform that occurs prior to the arrival of the waveform, i.e., a large number of points are required to model the portion of the waveform that is equal to zero because of causality. By incorporating phase shifts in the standard frequency-domain macromodel, macromodel can be expressed in the frequency domain as:

$$F(s)=\sum_{\alpha=1}^{M}\frac{R_\alpha \exp(-t_\alpha s)}{s-s_\alpha}+F_{np}(s), \quad (3)$$

where t_α model the phase shifts (time delays). This phase shift model is referred as Delay Reduced-Order Model (DROM) in the rest of this paper.

3. REDUCTION WITH THE MATRIX PENCIL TECHNIQUE

As stated above, in order to approximate interconnects response with fewer terms, DROM incorporates phase shifts (time delays) in addition to the poles. The time-domain expression for DROM is derived by taking the inverse Laplace transform of (3)

$$f(t)=\sum_{\alpha=1}^{M}R_\alpha \exp(s_\alpha(t-t_\alpha))U(t-t_\alpha)+f_{np}, \quad (4)$$

where $f(t)$ represents simulated or measured time-domain data. However, the matrix pencil method cannot be implemented directly for DROM because of the time delay terms. Therefore, (4) needs to be modified before applying the matrix pencil method. It should be noted that the time delays t_α are not necessarily distinct from each other. Assume that a transient waveform, which ranges from 0 to the maximum time $t_{\max}$, exhibits n different time delays. We partition (4) into n time segments so that we may utilize the matrix pencil method to deal with each time segment separately. Each time segment can be modeled as a sum of exponentials in the same way as the standard macromodels. The expression for each time segment and their summation can be written as:

$$f(t)=\sum_{j=1}^{n} f^{(j)} , \tag{5}$$

$$f^{(j)} = (\sum_{\alpha=1}^{M_j} R_\alpha \exp(s_\alpha (t-t_j)) + f_{np_j})[U(t-t_j)-U(t-t_{j+1})] . \tag{6}$$

Expression (6) represents the portion of the waveform for the time $t_j<t<t_{j+1}$, where M_j is the model order corresponding to the time segment. The two unit step functions guarantee $f^{(j)}$ is zero when $t<t_j$ or $t>t_{j+1}$. Since the matrix pencil technique requires uniformly sampled time-domain data, cubic spline interpolation can be applied to make the sampling interval uniform. In order to represent the time delay terms in the form required for the matrix pencil method, we replace $t-t_j$ with x and obtain

$$f^{(j)}(x) = (\sum_{\beta=1}^{M_j} R_\alpha \exp(s_\alpha x) + f_{np_j})[U(x)-U(x-(t_{j+1}-t_j))] \tag{7}$$

where $0 < x < t_{j+1} - t_j$. In the following paragraphs, we focus on extracting the poles and residues for the j_{th} term. The non-pole terms will be handled in Section V. Other terms in (5) can be obtained in the same way. First, sample the time-domain waveform and let $Z_\alpha = \exp(s_\alpha \delta_x)$, where δ_x is the sampling interval. This leads to

$$\begin{aligned} f^{(j)}(k) &= f^{(j)}(k\delta_x) \\ &= \sum R_\alpha z_\alpha^k \;; k = 0,1,2\cdots, N-1 \end{aligned} , \tag{8}$$

where N is the total number of samples. According to the matrix pencil method [10] and [11], two matrices $\bar{Y}_1$ and $\bar{Y}_2$ are defined as

$$\bar{Y}_1 = \begin{bmatrix} f^{(j)}(0) & \dots & f^{(j)}(L-1) \\ \dots & \dots & \dots \\ f^{(j)}(N-L-1) & \dots & f^{(j)}(N-2) \end{bmatrix}_{(N-L)\times L} , \tag{9}$$

$$\bar{Y}_2 = \begin{bmatrix} f^{(j)}(1) & \dots & f^{(j)}(L) \\ \dots & \dots & \dots \\ f^{(j)}(N-L) & \dots & f^{(j)}(N-1) \end{bmatrix}_{(N-L)\times L} , \tag{10}$$

where L is referred to as the pencil parameter. Then the matrix pencil is created as

$$\bar{Y}_2 - \lambda \bar{Y}_1 , \tag{11}$$

where λ is a scalar parameter. It is easy to prove that when $\lambda_1 = Z_\alpha$, where $\alpha = 1,2,...M_j$, the rank of the matrix (11) will be reduced. Hence, Z_α may be found as the generalized eigenvalues of the matrix pair $\{\bar{Y}_2;\bar{Y}_1\}$, which is an ill-conditioned *(N-L)×L* matrix.

Following [10] and [11], the singular value of the decomposition of $\bar{Y}_1$ is taken as follows:

$$\bar{Y}_1 = \sum_{i=1}^{M_j} \sigma_i u_i v_i^H = UDV^H . \tag{12}$$

Since the contaminated data $\bar{Y}_1$ has full rank, one can compute the truncated pseudo-inverse $\bar{Y}_1^{+} = VD^{-1}U^H$, where $U = [u_1, u_2, \cdots, u_{M_j}]$, $V = [v_1, v_2, \cdots, v_{M_j}]$ and

$$D = \begin{bmatrix} \sigma_1 & 0 & \cdots & 0 \\ 0 & \sigma_2 & \cdots & 0 \\ \cdot & & & \\ \cdot & & & \\ \cdot & & & \\ 0 & 0 & \cdots & \sigma_{M_j} \end{bmatrix} . \tag{13}$$

The model order M can be changed gradually to observe how well the macromodel matches with the simulation data. Another condition used to determine the model order M_j is $\frac{\sigma_{M_j}}{\sigma_1} = 10^{-p}$, where p is the number of significant decimal digits in the data. If p is chosen to be –3, then the singular values below 10^{-3} are assumed to be singular values associated with noise and are ignored. It is used as a condition in our computer code so that the model order can be determined automatically. It was proven in [11] that the solution requires the computation of the matrix eigenvalue problem:

$$(Z - z_\alpha I)\mathbf{z\alpha} = 0 , \tag{14}$$

where $Z = D^{-1}U^H Y_2 V$, z_α are eigenvalues and $\mathbf{z\alpha}$ are eigenvectors. The computation cost is reduced to the calculation of an eigenvalue problem for a $M_j \times M_j$ matrix.

Our code computes the eigenvalues using (14). Once M_j and z_α are known, the residues, R_α are obtained by solving the following least-squares problem

$$\begin{bmatrix} f^{(j)}(0) \\ f^{(j)}(1) \\ \cdot \\ \cdot \\ \cdot \\ f^{(j)}(N-1) \end{bmatrix} = \begin{bmatrix} 1 & 1 & \cdots & 1 \\ z_1 & z_2 & \cdots & z_{M_j} \\ \cdot & & & \\ \cdot & & & \\ \cdot & & & \\ z_1^{N-1} & z_2^{N-1} & \cdots & z_{M_j}^{N-1} \end{bmatrix} \begin{bmatrix} R_1 \\ R_2 \\ \cdot \\ \cdot \\ \cdot \\ R_{M_j} \end{bmatrix} . \tag{15}$$

3.1 Analytic Formulas for Realizable Reduction

DROM admittance parameters can be analytically derived as:

$$Y_{11} = \frac{1}{2Z_0} - \frac{1}{2Z_0} \cdot [\frac{(Z_1 - Z_0)(Z_0 + Z_2 + Z_3) + Z_3(Z_2 + Z_0)}{(Z_1 + Z_0)(Z_0 + Z_2 + Z_3) + Z_3(Z_2 + Z_0)}]\exp^{-2sTD_1} , \tag{16}$$

$$Y_{12} = Y_{21} = \frac{1}{2Z_0} - \frac{1}{2Z_0} \cdot [\frac{Z_3}{(Z_1+Z_0)(Z_0+Z_2+Z_3)+Z_3(Z_2+Z_0)}]\exp^{-s(TD_1+TD_2)}, \quad (17)$$

$$Y_{22} = \frac{1}{2Z_0} - \frac{1}{2Z_0} \cdot [\frac{(Z_1-Z_0)(Z_0+Z_2+Z_3)+Z_3(Z_2+Z_0)}{(Z_1+Z_0)(Z_0+Z_2+Z_3)+Z_3(Z_2+Z_0)}]\exp^{-2sTD_2}, \quad (18)$$

where TD_1 and TD_2 are the time delays of the two distributed lines, respectively. Matched loads Z_0, that are equal to the characteristic impedance of transmission line, are placed on both ends of the circuit to eliminate the multiple reflections. The elements Z_1, Z_2 and Z_3 can be resistors, inductors or capacitors. By choosing different elements it is possible to model various discontinuities within the interconnect . The equivalent circuit is depicted in Figure 1. Since in many cases such as power lines and antennas, the input signal is triangular. Here we use triangular waveform as our inputs to explain the realization procedure. Assume the triangular impulse response is denoted as TIR. If the rise time of the voltage source is very small, i.e., $\Delta t \to 0$, then there is an approximate relationship between the TIR_{ij} and Y_{ij},

$$Y_{ij}(s) = \frac{I_i(s)}{V_j(s)} \approx \frac{I_i(s)}{\Delta t} e^{s\Delta t} = \frac{TIR_{ij}(s)}{\Delta t}. \quad (19)$$

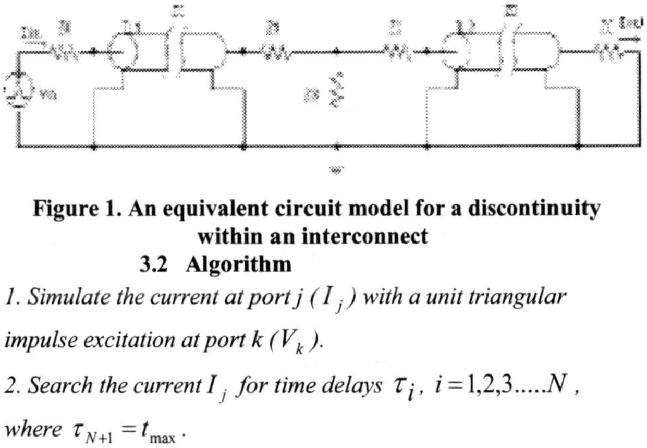

Figure 1. An equivalent circuit model for a discontinuity within an interconnect

3.2 Algorithm

1. Simulate the current at port j (I_j) with a unit triangular impulse excitation at port k (V_k).

2. Search the current I_j for time delays τ_i, $i = 1,2,3.....N$, where $\tau_{N+1} = t_{\max}$.

3. Use data interpolation to create a uniformly-spaced time record for I_j and let $TIR_{ik}(t) = I_j(t + \Delta t)$.

4. For i=1 to the number of time delays (N):

(a). Perform a discrete Fourier transform on the time segment from τ_i to τ_{i+1} and figure out its bandwidth.

(b). If the bandwidth $\geq 0.7 f_{\max}$, then the non-pole term np_i has an amplitude that is is equal to the value of the waveform at τ_i multiplied by Δt. Also set the actual starting approximation point $t_p = \tau + \Delta t$.

(c). If the bandwidth $< 0.7 f_{\max}$, let $t_p = \tau_i$. However, if τ_i is in the middle of a sharp slope, then try $t_p = \tau_i + \Delta t$ or use an even larger value for t_p.

(d). Approximate the waveform between t_p and τ_{i+1} using the matrix pencil method.

4. EXPERIMENT RESULTS

We now investigate a via A via in the transmission line. A via is modeled by setting $Z_1 = Z_2 = sL$ and $Z_3 = \infty$. Thus, substituting them into (16) to (18) yields the analytical admittance expressions as shown below:

$$Y_{11} = Y_{22} = \frac{1}{2Z_0} - \frac{sL}{2Z_0(sL+Z_0)}\exp(-2s\tau) \quad (21)$$

$$Y_{12} = Y_{21} = -\frac{1}{2(sL+Z_0)}\exp(-2s\tau) \quad (22)$$

When $L = 1nH$, the output currents are simulated using a group of triangular impulses with various rise times ($\Delta t = 1ps, 10ps, 50ps, 100ps$). These data together with the theoretical admittance Y_{21} are plotted in Figure 2.

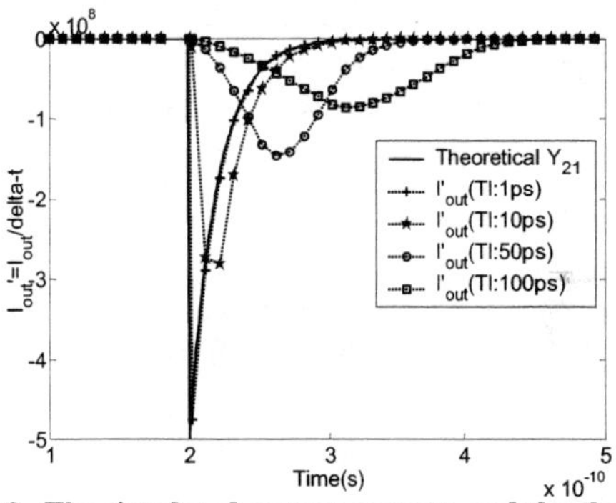

Figure 2. The simulated output currents and the theoretical Y_{21} for a via discontinuity

The poles and residues for DROM can be acquired by applying the matrix pencil method to this time segment. Table 1 lists the theoretical DROM admittance poles and DROM TIR poles for two triangular impulses with rise times of $\Delta t = 1ps$ and $\Delta t = 10ps$. As expected (see (6)) the pole locations associated with the short ($\Delta t = 1ps$) triangle impulse agree well with the theoretical value.

10ps (0.21ns-1ns)	1ps (0.201ns-1ns)	Ideal Impulse (0.2ns-1ns)
Real[t]: -0.1906	-0.5104	-0.5000
Imaginary: 0	0	0

Table 1: The theoretical DROM pole locations for the admittance parameters and DROM pole locations extracted from the simulated TIR_{21} for a via discontinuity

The transient response of DROM for the short TIR_{21} also agrees well with the original simulation TIR_{21}, as shown in Figure 3.

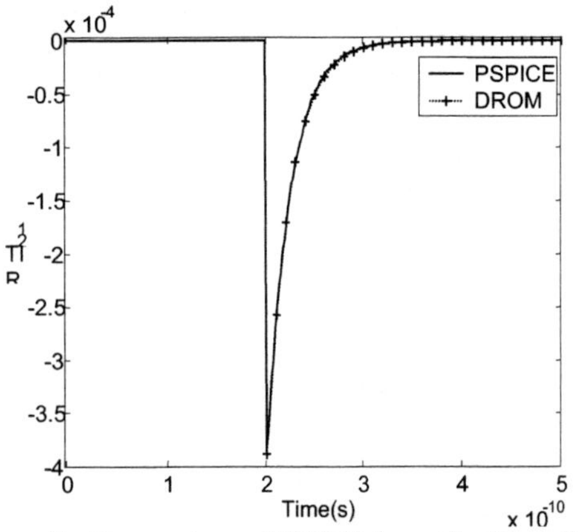

Figure 3. Comparison of DROM for a via discontinuity extracted from TIR_{21} simulation data (0.201ns-0.5ns) ($\Delta t = 1ps$)

10ps (0.21ns-1ns)	1ps (0.201ns-1ns)	Ideal Impulse (0.2ns-1ns)
1st: 0.01E-11	0.01E-12	0.01
2nd: -0.00788E-11	-0.009754E-12	-0.01

Table 2: The theoretical DROM non-pole amplitudes for the admittance parameters and the DROM non-pole amplitudes extracted from the simulated TIR_{11} for a via discontinuity

In Figure 4, we plot the model order variation of the standard macromodel in terms of the time delay. It is observed that the larger the time delay, the more terms are required to model the waveform using the standard macromodel. These results demonstrate that DROM greatly reduces the number of terms required to model interconnects, especially for electrically long lines.

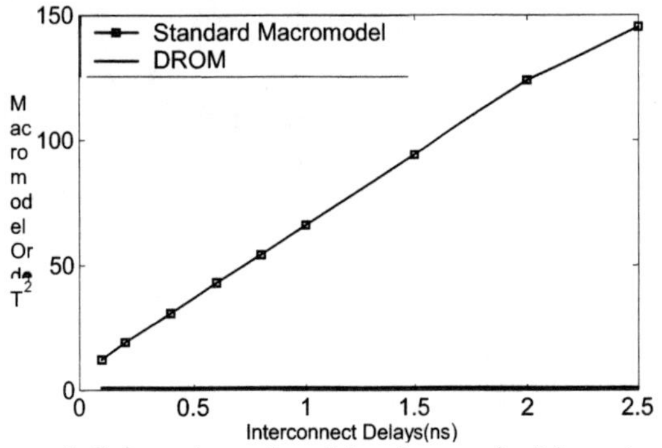

Figure 4. Comparison of model order required by a standard macromodel and DROM for a via discontinuity

5. CONCLUSION

We have described the algorithm for the matrix pencil technique based realizable parasitic reduction for distributed lines. A number of comparisons have shown that the new model greatly reduces the number of terms required by the standard macromodel.

References

[1] L. T. Pillage, R. A. Rohrer, "Asymptotic waveform evaluation for timing analysis," *IEEE Transactions on Computer-Aided Design*, vol. 14, pp. 639-649, May. 1990

[2] E. Chiprout, M. Nakhla, "Analysis of interconnect networks using complex frequency hopping," *IEEE Transactions on Computer-Aided Design*, VOL. 14, PP. 186-199, Feb. 1995

[3] P. Feldmann, R. W. Freund, "Efficient linear circuit analysis by Pade approximation via the Lanczos process," *IEEE Transactions on Computer-Aided Design*, vol. CAD-14, pp.639-649, 1995

[4] A. Odabasioglu, M. Celik, and L. T. Pileggi, "PRIMA: Passive Delay Reduced-Orderinterconnect macromodeling algorithm" *IEEE Transactions on Computer-Aided Design*, vol. 17, no. 8, pp.645-653, Aug. 1998

[5] R. Sanaie, E. Chiprout, M. S. Nakhla, "A fast method for frequency and time domain simulation of VLSI interconnects," *IEEE Transactions on Microwave Theory and Techniques*, vol. 42, no. 12, pp. 2562-2571, Dec. 1994

[6] R. Achar, M. S. Nakhla, "Efficient transient simulation of embedded subnetwords characterized by s-parameters in the presence of nonlinear elements," *IEEE Transaction on Microwave Theory and Techniques*, vol. 46, no. 12, p.2356-263, Dec. 1998

[7] W. T. Beyene, J. E. Schutt-Aine, "Efficient transient simulation of high-speed interconnects characterized by sampled data," *IEEE Transactions on Components, Packaging and Manufacturing Technology - Part B*, Vol. 21, no. 1, pp. 105-114, Feb. 1998

[8] R. Prony, "Essai Experimental et Analytique sur les Lois de la Dilatabilite de Fluides Elastiques et sur Celles del la Force Expansive de la Vapeur de L'alkool, a Differenctes Temperatures," J.l'Ecole Polytech. (Paris), 1, 1795, pp.24-76

[9] T. K. Sarkar, O. Pereira, "Using the Matrix Pencil Method to Estimate the Parameters of a Sum of Complex Exponentials", *IEEE Transactions on Antennas and Propagation Magazine,* Vol. 37 No. 1, pp. 48-55, Feb. 1995

[10] Yingbo Hua, T. K. Sarkar, "Generalized Pencil-of-Function Method for Extracting Poles of an EM System from Its Transient Response", *IEEE Transactions on Antennas and Propagation Magazine,* Vol. 37 No. 2, pp. 229-234, Feb. 1989

[11] E. K. Miller, "Model-Based Parameter Estimation in Electromagnetics: Part I. Background and Theoretical Development", *IEEE Transactions on Antennas and Propagation Magazine*, Vol. 40 No. 1, pp. 42-52, Feb. 1998

[12] Zhanhai Qin, and Chung-Kuan Cheng, " Realizable Parasitic Reduction Using Generalized Y- Δ Transformation", Proc. of Design Automation Conference, pp. 220-225, 2003.

[13] Chirayu Amin, Masud Chowdhury, Yehea Ismail, "Realizable RLCK Circuit Crunching", Proc. of Design Automation Conference, pp. 226-231, 2003.

Frequency Driven Repeater Insertion for Deep Submicron

N. Ahmed, M. H. Tehranipour, D. Zhou, M. Nourani
Center for Integrated Circuits & Systems
The University of Texas at Dallas, Richardson, TX 75083
{mht021000, nxa018600, zhoud, nourani}@utdallas.edu

Abstract—**Repeaters are now widely used to decrease delay and increase the performance of long interconnects. The maximum operating frequency is limited by the longest wire between two repeaters. The main goal of repeater insertion is to reduce the delay of the interconnects in a circuit. Increasing the number of repeaters on an interconnect increases the slew rate and the operating frequency but decreases the design performance in terms of area and power. This paper presents a new methodology of repeater insertion considering the circuit operating frequency to improve the slew rate and reduce the area overhead of the repeaters. We propose a mathematical treatment for finding the number and position of repeaters while keeping the same size of the optimal repeater size. Experimental results are shown for different frequencies and repeater sizes to indicate the improvements of slew rate and saving area overhead.**

I. INTRODUCTION

As very large scale integration (VLSI) circuits continue to be scaled aggresively past the 130-nm technology, performance of these IC's is being increasingly dominated by the interconnects. With technology scaling, more and more functionality is being integrated on-chip which results in an increase in die size inspite of the reduction in minimum feature size. As a result, the number of long wires and the length of these wires increases. The interconnect delay is also becoming a bottleneck in achieving high performance circuits. The power consumption of the buffers inserted in interconnects is another important factor in the interconnect design. Techniques which aim to reduce interconnect delay and power consumption are required for high speed circuit design.

Repeaters are now widely used to decrease the delay and increase the performance of long on-chip interconnections in CMOS VLSI. For large high-performance designs, the number of such repeaters can be very high and can take up significant fraction of active silicon and routing area. Additionally, as the operating frequency and leakage current increases while scaling, total chip power dissipation increases. A significant fraction of the total chip power dissipation arises due to long interconnect networks, especially in high-performance designs. For example, it has been reported that around 40%-70% of the total power consumption could be due to the clock distribution network [1].

High performance VLSI circuits are designed to work at a specific maximum clock frequency depending on their applications and the process technology used. The chip frequency increases as the technology scales down [2]. One of the constraints in achieving maximum clock frequency is clock rise time (t_r). The clock frequency is limited by the longest rise time in the clocking network [3]. The slew rate is inversely related to transition time and is defined as $(V_{90\%} - V_{10\%})/t_r$ as reducing t_r increases the slew rate. The clock rate limit in a buffered clock tree depends on the phase delay and transition time of each subtree instead of that of the whole clock tree. A pipeline scheme was presented to meet the GHz frequency challenge [4]. The distribution of clock signal on the chip must be done while minimizing the clock delay and clock skew and maximize the slew rate [5].

The operating frequency has a major impact on the number of repeaters that needs to be inserted. As the number of repeaters increases, the area and power dissipation of the drivers increases. In this paper, we propose a new repeater insertion methodology for improving the slew rate and reducing the area of the repeaters considering the operating frequency. This method is performed in two steps: First, the method finds the required number of repeaters on an interconnect to satisfy the operating frequency using an uniform repeater insertion. Second, the methodology proposes a new technique of repositioning the repeaters based on uniform shifting of the repeaters to the receiving end of the interconnect. For this purpose, we present a mathematical treatment for finding the number of repeaters and their positions. The second step of this methodology increases the slew rate and decreases the area compared to the first step without increasing the delay and power.

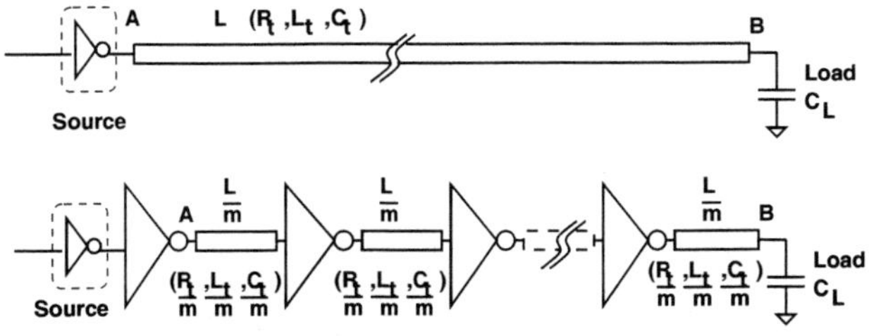

Figure 1. Uniform Repeater Insertion.

The rest of the paper is organized as follows. Section II describes the background about uniform repeater insertion, relation between operating frequency and number of repeaters and how delay, slew rate and power varies with position of the repeaters. The uniform shifting of the repeaters is discussed in Section III. Section IV discusses the experimental results. The paper ends with conclusion in Section V.

II. BACKGROUND

A. Uniform Repeater Insertion

Consider an interconnect of length, L and width, W between the source, i.e. the output buffer of the last stage, and a load as shown in Figure 1. The propagation delay of the interconnect increases quadratically with the line length. Repeaters are inserted to change the delay dependence on the length from quadratic to linear. There are m repeaters uniformly placed on the interconnect line, whose total resistance, inductance and capacitance are R_t, L_t and C_t, respectively. Each of the repeaters drives exactly identical interconnect length and is K times larger than the minimum size balanced inverter. The total capacitive load driven by each repeater is equal to the sum of the interconnect segment capacitance (C_t/m) and the intrinsic input capacitance of the repeater (C_r). C_r is proportional to the size of the repeater.

Many analytical delay models have been developed for different repeater and interconnect models [6][7]. To observe the effect of the repeater insertion on various parameters like phase delay and slew rate, we did SPICE simulations on an interconnect of length *L=1cm* and width *W= 1μm* for different values of m and K. All simulations are performed using $0.18\mu m$ technology. Each interconnect segment is modeled as a distributed RLC line. The R, L and C parameters are derived using Berkeley Predictive Technology Model (BPTM) [8]. The maximum voltage is 1.8V and the input rise time used is 100 ps. The input slew rate is calculated using $(V_{90\%} - V_{10\%})/(t_{90\%} - t_{10\%}) = 18V/ns$.

Figure 2 shows the delay and slew rate variation with different number of repeaters and for three different values of K. For this specific example, the optimum delay occurs at *m=1* and gradually increases as m

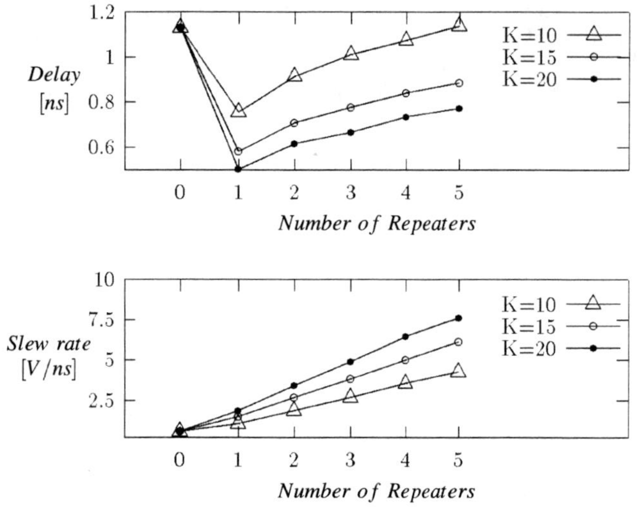

Figure 2. Delay and slew rate variation with number of repeaters.

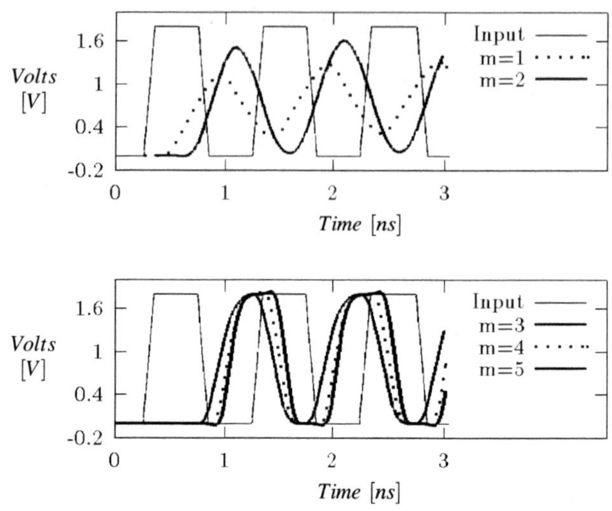

Figure 3. Output waveforms for an input signal frequency of 1GHz.

increases. This is because the interconnect segment delay is less than or equal to the repeater delay for $m > 2$. In such a situation, adding repeaters increases the overall delay. When the length of the interconnect increases, the number of repeaters required for optimal delay also increases. The slew rate corresponding to the optimal delay at $m = 1$ for *K=20* is 1.865 V/ns, i.e. a transition time of 900 ps. This severely limits the maximum frequency of data transfer across the interconnect. A higher transition time implies higher switching time which relates to more short circuit power dissipation. The transition time can be decreased by either increasing m or K. The number of repeaters are selected depending upon the delay and power specifications to obtain a better slew rate. In the next subsection, we discuss about repeater insertion with considering the frequency.

B. Relation Between Frequency and Number of Repeaters

The operating frequency also impacts on the number of repeaters to be inserted. We show a relation between them and develop a formulation to find the number of repeaters to be inserted for a given operating frequency. Figure 3 shows the output waveforms for different number of repeaters for an input signal frequency of 1GHz for an interconnect with $L = 1cm$ and $W = 1\mu m$. The size of the repeaters is $K = 20$. Even though $m = 1$ gives optimum delay, the output waveform does not reach the full voltage swing of 0 to 1.8V. As m increases, the slew rate improves and the output gets better in shape and pulse width. Depending on the slew rate constraint, the designer can choose $m \geq 3$ for the above design to work at 1GHz. It is clear that as the operating frequency increases, the number of repeaters also increases.

There is a limit to how long a wire can be in between two repeaters at a certain frequency. The maximum clock rate in a buffered clock tree running in pipeline scheme is formulated in [4]. Applying it to a single interconnect line, the clock rate limit is given by:

$$f_{pipe} = \frac{1}{2(\tau_{p_{max}} + \tau_{t_{max}} + \tau_h)}$$

$$\tau_{p_{max}} + \tau_{t_{max}} \leq \frac{T_{period}}{2} - \tau_h$$

Assuming a uniform placement of repeater insertion $\tau_{p_{max}}$, $\tau_{t_{max}}$ and τ_h are the phase delay, transition time ($V_{50\%}$ to $V_{90\%}$) and specified holding time of a sub-segment between any two repeaters, respectively. For a given slew rate or holding time constraint, the maximum length of wire between two repeaters can be found.

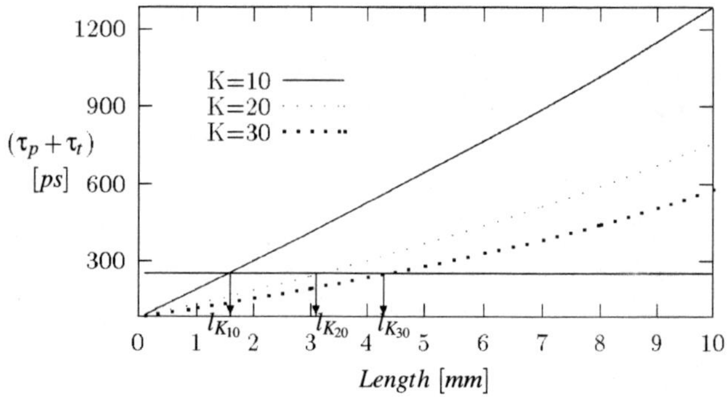

Figure 4. Variation of $(\tau_p + \tau_t)$ with length of the interconnect for different size factors K

Elmore delay model provides a good approximation for 50% delay but there is not much accuracy for 10% and 90% points of the output waveform. We perform SPICE simulations for a range of interconnect lengths between different inverter sizes. Figure 4 shows the variation of $(\tau_{p_{max}} + \tau_{t_{max}})$ with length for different K. For a holding time constraint of $25\% T_{period}$, the maximum length of wire between two repeaters of size K=10, K=20 and K=30 are 1.5mm, 3.1mm and 4.2mm, respectively. The number of repeaters of size K to be inserted in an interconnect of length L is determined by $m \geq \frac{L}{l_{K_{max}}}$, where $l_{K_{max}}$ is the maximum length of wire between two repeaters of size K. For example, for l=1cm and K=20, $m \geq 3.23$, meaning m=4 is the best choice to satisfy the operating frequency. Compared to the uniform repeater insertion, the required number of repeaters is increased when the frequency is considered. However, the main goal is to insert as less repeaters as possible to satisfy the frequency.

C. Relation Between Position of Repeaters and Slew Rate

The simulation results explained in the previous section for uniform repeater insertion suggest that in order to obtain a better slew rate either m or K is to be increased. In this section, we show a relationship between the position of the repeaters and the slew rate. This relationship

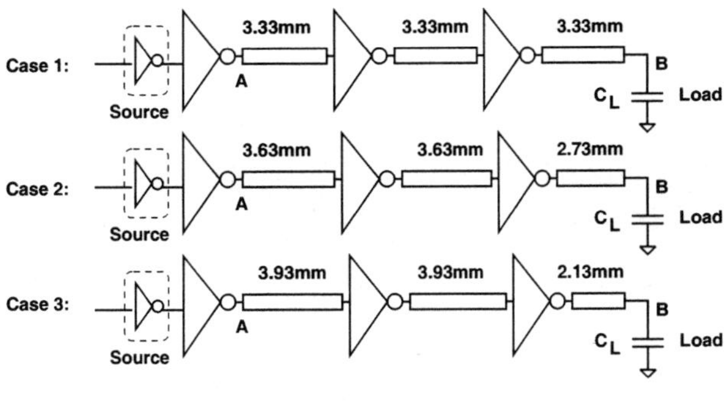

Figure 5. 3-stage repeaters with different positions with K=20.

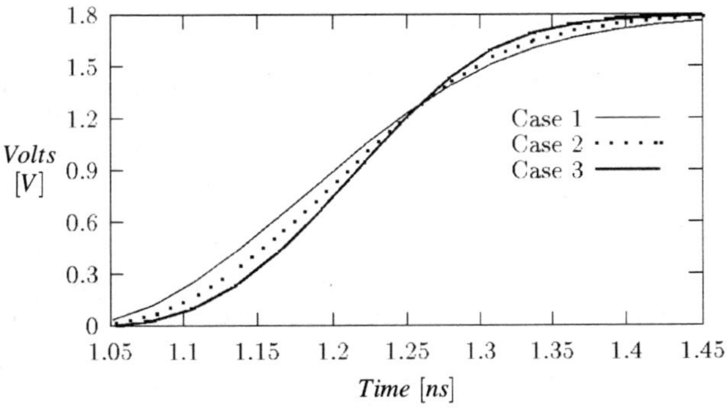

Figure 6. Output waveforms for different positions in a 3-stage repeater.

will be the foundation for further discussion on how to improve the slew rate by adjusting the position of the repeaters instead of increasing m or K. Figure 5 shows an example with three repeaters of size-factor $K=20$ inserted on an interconnect of length $L=1cm$ and width $W=1\mu m$. There are three different cases with repeaters placed in different positions. *Case1* shows the uniform repeater insertion. In *Case2* and *Case3* the second and third repeaters are equally shifted towards the load end by a fixed amount of *0.3mm* and *0.6mm*, respectively.

Figure 6 shows the output waveforms at B for the three different cases for f=1GHz. Clearly, the slew rate increases as the repeaters are shifted to the load end of the interconnect. By shifting the repeaters, the segment length driven by the first two repeaters is increased. As a result, the delay of these segments is increased. This increase in delay is compensated by the decrease in delay and transition time in the final segment. The overall delay in the shifted case is slightly greater than in uniform placement. With well defined buffer size and positioning, the slew rate can be improved.

Table I compares the three different cases in terms of phase delay, slew rate and power dissipation. The slew rate is improved by 33% in *Case3* compared to uniform repeater insertion at a cost of 2.5% increase in phase delay. The power dissipation is increased by 2.4%. As mentioned before, m=4 is the best choice for an interconnect of length $L=1cm$ and width $W=1\mu m$ operating at f=1GHz. The last row in the table shows the parameters for $m=4$. It is clearly seen that *Case3* (m=3) gives better performance in terms of delay, slew rate and power compared to the uniform placement with $m=4$. Based on these simulation results, a lower number of stages can improve the slew rate and power by adjusting the position of the repeaters, thereby saving area and power.

III. Uniform Shifting of Repeaters

Uniform repeater insertion is used to find the number of stages and repeater size to obtain optimal delay when driving interconnect load.

TABLE I
COMPARISON OF PHASE DELAY, SLEW RATE AND POWER FOR DIFFERENT POSITIONS OF REPEATERS.

	t_d[ps]	SR[V/ns]	P_{diss}[mW]
m=3, Case 1	653.2	5.7	4.412
m=3, Case 2	661.9	6.5	4.459
m=3, Case 3	669.4	7.6	4.517
m=4, Uniform	719.3	7.4	4.685

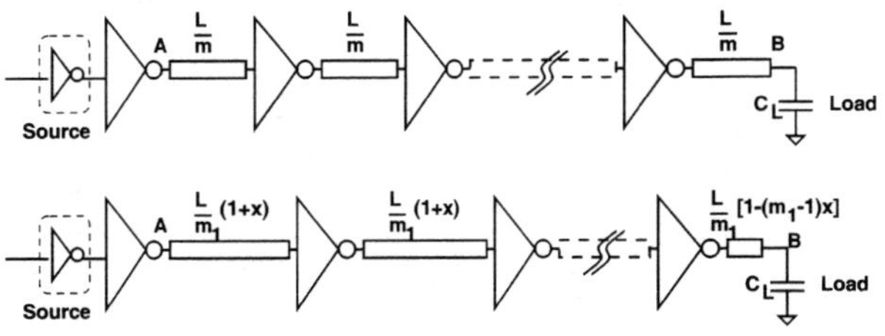

Figure 7. Uniform shifting of repeaters (size factor K).

Sometimes optimal delay may not satisfy the design specifications. Other parameters like transition time and power are also important. As seen in Section II-A, the corresponding transition time for optimal delay was high which limited the maximum operating frequency. Instead of increasing the number of repeaters, suitable position of repeaters can be determined to improve the performance. In this section, we propose a technique to find the number of repeaters lesser than uniform repeater insertion for a given operating frequency and their positions to obtain the same performance.

The first step is to perform uniformly placed repeater insertion to meet the design constraints. This can be done in two different methods. In the first method, the repeater size and the number of repeaters can be calculated by finding an analytical expression for $(\tau_p + \tau_t)$ and embedding it in the formulation discussed in Section II-B. The second method is to do a prior SPICE analysis and find the number of repeaters satisfying delay and power requirements, for a given frequency of operation. In our design methodology, we apply the latter method. In the next step, we determine the number of repeaters and the amount of shift to decrease the area.

Figure 7 shows two circuits with different repeater insertion methods for an interconnect of length L. In the first circuit, m repeaters with size factor K are uniformly placed along the interconnect. Noticing that the slew rate improves by shifting the repeaters to the receiving end, we select m_1 uniformly placed repeaters ($m_1 < m$) and shift them by a fixed factor x. The size factor of the repeaters in both circuits is the same.

A. Determination of parameter x

Each repeater stage is shifted by a fixed factor x towards the load end except the first repeater. The first $(m_1 - 1)$ segment lengths are increased by factor x and the last segment length is reduced by a factor of $(m_1 - 1) \times x$.

$$l_i = \begin{cases} \frac{L}{m_1}(1+x) & 1 \le i \le (m_1 - 1) \\ \frac{L}{m_1}[1-(m_1-1)x] & i = m_1 \end{cases}$$

Assuming the length of the last segment to be β times the segment length without shift, where $0 < \beta \le 1$. Equation 1 shows x interms of β and m_1. When $\beta = 1$, x becomes zero which implies uniform placement of repeaters.

$$l_{m_1} = \frac{L}{m_1}[1-(m_1-1)x] = \beta \times (\frac{L}{m_1})$$

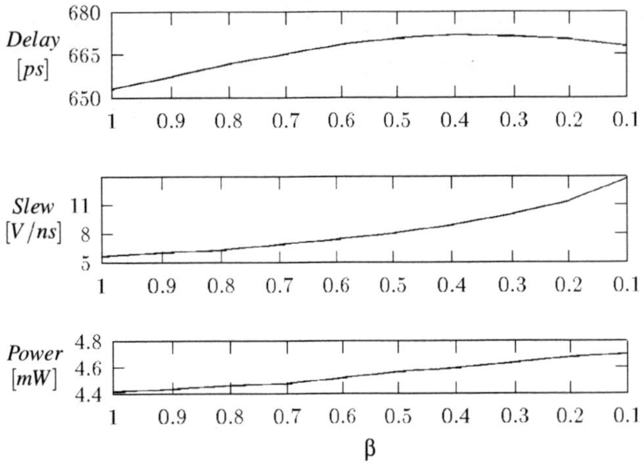

Figure 8. Variation of delay, slew rate and power with β.

$$x = \frac{(1-\beta)}{(m_1-1)} \quad (1)$$

Figure 8 shows the variation of delay and slew rate with β for three repeaters inserted on an interconnect of length $L = 1cm$ and width $W = 1\mu m$. When β is decreased, x increases and so the amount of shift. The delay of the first $(m_1 - 1)$ segments increases because the load driven by the repeaters increases. The delay reaches a maximum when $\beta = 0.4$ and gradually falls with decrease in β. This is because the decrease in delay in the last segment starts to dominate the sum of the increase in delays for the initial $(m_1 - 1)$ segments and the overall delay decreases. The slew rate and power continuously increase as the load driven by the last repeater decreases as β decreases. It is clear that in order to obtain a significant improvement in slew rate and meet the delay and power lower than the case of *m*=4 (as shown in Table I), the value of β must be chosen between 0.2 and 0.5 ($0.2 < \beta < 0.5$). From Equation 1, x can be expressed by the inequality $x > \frac{1}{2(m_1-1)}$ and then

$$x \simeq \frac{1}{m_1} \quad (2)$$

We choose $x = 1/m_1$ which satisfies this inequality.

B. Determination of parameter m_1

As seen in Figure 7, the repeater size is the same in both the circuits. Therefore, inorder to obtain a better slew rate, the length of the segment driven by the final repeater after applying shift should be less than the segment length driven in uniformly placed circuit. This can be expresssed as:

$$\frac{L}{m_1}[1-(m_1-1)x] < \frac{L}{m}$$

Substituting Equation 2, m_1 is determined by the inequality $m_1 > \sqrt{m}$. Due to shifting, the length of first $(m_1 - 1)$ segments increases. There is a limit on the length of interconnect ($l_{K_{max}}$) between two repeaters for a given operating frequency as explained in Section II-B. If the segment length exceeds $1.5 \times l_{K_{max}}$, $(\tau_p + \tau_t)$ is close to $\frac{T_{period}}{2}$, the output waveform degrades and has a traingular shape.

$$\frac{L}{m_1}(1+x) < 1.5 \times l_{K_{max}} \quad where\ x = \frac{1}{m_1}, m = \frac{L}{l_{K_{max}}}$$

Combining the two inequalites, the limit of m_1 can be expressed as:

$$max\{\sqrt{m}, \frac{m+\sqrt{m^2+6m}}{3}\} \leq m_1 < m$$

TABLE II

COMPARISON OF UNIFORM PLACEMENT AND UNIFORM SHIFT METHODS FOR OPERATING FREQUENCY F=1GHZ AND 2GHZ.

Frequency	*K*	*Repeaters*	*Area*	*Delay [ps]*	*SR [V/ns]*	*Power [mW]*
f= 1GHz	20	m=4, Uniform	80	719.3	6.55	3.73
	20	m=3, β = 0.4	60	673.7	6.85	3.54
f= 2GHz	27	m=6, Uniform	162	654.6	13.3	10.65
	27	m=5, β = 0.4	135	625.1	15.1	9.62

IV. Experimental Results

Table II compares the results of uniform placement of repeaters and fixed shifting of repeaters for two operating frequencies. The length of the interconnect $L = 1cm$ and width $W = 1\mu m$. As the operating frequency increses, the number of repeaters required and the size factor increases. The hold time (τ_h) constraint used is $25\% T_{period}$ which implies a $(\tau_p + \tau_t)$ period of 250 ps and 125 ps for operating frequency f = *1GHz* and f = *2GHz* respectively. The input rise time used is equal to $10\% \times T_{period}$. For the same performance, the number of repeaters for the uniform shift methodology is less compared to uniform placement of repeaters. The slew rate is improved by less number of repeaters, thereby saving area and power. As the technology shrinks, the operating frequency increases and the number of repeaters increases. The proposed methodology will be more effective interms of performance for higher operating frequencies.

V. Conclusion

The paper proposed a new methodology of repeater insertion to improve the slew rate and reduce the area/power. The method repositions the repeaters inserted during uniform repeater insertion by a fixed shifting factor. As reported in experimental results, the new repeater insertion methodology shows better results compared to uniform repeater insertion in terms of slew rate and area/power.

Acknowledgements

This work was supported in part by the National Science Foundation CAREER Award #CCR-0130513.

References

[1] T.Sakurai, "Design challenges for 0.1 *mu* m and beyond," in proc. *Design Automation Conference, Asia and South Pacific (ASP-DAC)*, pp. 553-558, 2000.

[2] J. Cong, L. He, K.Y. Khoo, C.-K. Koh, and Z. Pan, "Interconnect design for deep submicron ICs," in proc. *Int. Conf. on Computer Aided Design (ICCAD)*, pp. 478-485, 1997.

[3] M. Afghani and C. Svensson, "Performance of synchronous and asynchronous schemes for VLSI systems," *IEEE Trans. on Computers*, vol. 41, no. 7, pp. 858-872, 1992.

[4] X. Zeng, D. Zhou, "Design of GHz VLSI clock distribution circuit," in proc. *IEEE International Symposium on Circuits and Systems (ISCAS)*, pp. 391-394, 2001.

[5] G.E. Tellez and M. Sarrafzadeh, "Minimal buffer Insertion in Clock Trees with Skew and Slew Rate Constraints," *IEEE Trans. on Computer-Aidded Design of IC and Systems*, vol. 16, pp. 333-342, 1997.

[6] A. Nalamalpu and W. Burleson, "A Practical Approach to DSM Repeater Insertion: Satisfying Delay Constraints While Minimizing Area and Power," in proc. *ASIC/SOC Conference*, pp. 152-156, 2001.

[7] Y. I. Ismail and E. G. Friedman, "Repeater insertion in RLC lines for minimum propagation delay," in proc. *IEEE International Symposium on Circuits and Systems (ISCAS)*, pp. 404-407, 1999.

[8] *http://www-device.eecs.berkeley.edu/ ptm/interconnect.html*, Y. Cheng and C. Hu.

[9] *International Technology Roadmap for Semiconductors (ITRS)*, Semionductor Industry Association, San Jose, CA, 1999.

MODELING AND IMPLEMENTATION OF TWISTED DIFFERENTIAL ON-CHIP INTERCONNECTS FOR CROSSTALK NOISE REDUCTION

İ. Hatırnaz and Y. Leblebici

Swiss Federal Institute of Technology (EPFL)
Microelectronic Systems Laboratory (LSM)
CH-1015 Lausanne, Switzerland
Ilhan.Hatirnaz@epfl.ch, Yusuf.Leblebici@epfl.ch

ABSTRACT

A simple generic interconnect architecture is presented to allow effective cancellation of inductive and capacitive noise in high-speed on-chip interconnect lines. The approach is based on the principle of constructing periodically twisted differential line pairs for parallel interconnect segments in order to eliminate the mutual coupling influences. Detailed simulations show that the twisted-differential lines (TDL) provide high-speed and crosstalk-immnune interconnects, compared to single-ended and differential lines.

1. INTRODUCTION

With continued scaling of device features and interconnect dimensions down to deep-sub-micron and nanometer range, interconnects are becoming the limiting factor for performance and reliability in many system-on-chip (SoC) designs. Since the overall chip dimensions continue to increase with increasing system complexity, interconnects - especially the global connections between various system blocks on chip - tend to get longer. At the same time, wire width and wire separation continue to drop while their cross-sectional area is scaled down at a slower rate to prevent resistance values increase dramatically. This ongoing trend of controlling the RC delay, combined with the faster rise/fall times and longer wires, makes the inductive part of the wire impedance become comparable to its resistive part [1].

In this paper, we explore inductive coupling effects between neighboring parallel wires using a simple, physically-based equivalent circuit model, and we propose simple generic interconnect architecture to reduce cross-talk noise due to capacitive and inductive coupling between the interconnects and to reduce the interconnect delay, as well. This approach could prove to be a very suitable solution for the design of long high-speed bus lines that link various system sub-blocks on chip, achieving very low, predictable delays and noise levels even at very high switching speeds.

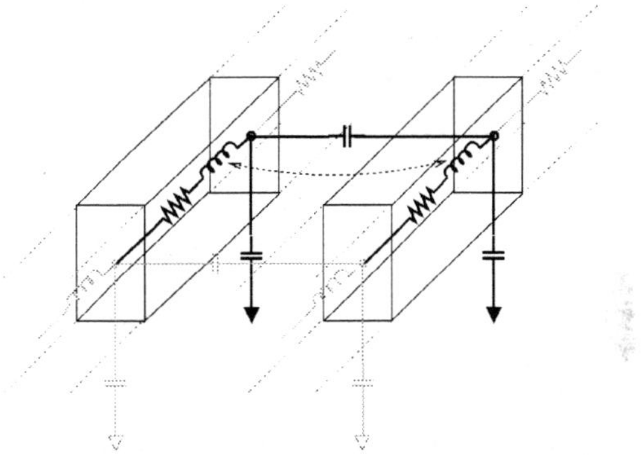

Fig. 1. RLCM model of an interconnect segment, showing inductive and capacitive coupling between two parallel lines.

2. ON-CHIP INTERCONNECTS

The classical approach for modelling on-chip interconnects is based on the assumption that the wire loads are mainly capacitive and lumped. In most cases, however, the load conditions imposed on the interconnection line are far from being simple. The line, itself a three-dimensional structure in metal (aluminium wires and tungsten vias), usually has a non-negligible resistance in addition to its capacitance. The (length/width) ratio of the wire usually dictates that the parameters are distributed, making the interconnect a true transmission line. Also, an interconnect is rarely isolated from other influences. In realistic conditions, the interconnection line is in very close proximity to a number of other lines, either on the same level or on different levels. The capacitive/inductive coupling and the signal interference between neighboring lines should also be taken into consideration for an accurate estimation of delay.

In general, if the time of flight across the interconnection line (as determined by the speed of light) is much shorter than the signal rise/fall times, then the wire can be modelled as a capacitive load, or as a lumped or distributed RC network. If the interconnection lines are sufficiently long and the rise times of the signal waveforms are comparable to the time of flight across the line, then the inductance also becomes important, and the interconnection lines must be modelled as transmission lines.

Figure 1 shows the simplified cross-section of two parallel interconnect lines, together with one segment of the distributed RLCM network that represents the resistive/capacitive/inductive loads as well as the capacitive and inductive coupling between the lines (also called the PEEC model, [2]).

The inductive effects mainly manifest themselves as the overshooting and undershooting of the signal edges, switching noise due to Ldi/dt voltage drop, and the long-range coupling. However, most of the techniques which have been used in order to reduce noise on wires, like shielding, increasing metal-to-metal spacing and etc., are more suitable for countering capacitive effects.

The capacitive cross-talk noise can be easily reduced by introducing a shield between the aggressor line and the victim line, because electric fields are terminated on the neighboring metallic nodes. However, the same is not necessarily true for the magnetic fields, which may extend well beyond the aggressor nodes. Therefore, the definition of the return path is very critical in determining the inductance of a wire. In the following, we demonstrate how the capacitive noise and the inductive noise can be suppressed significantly by applying a simple, repetitive interconnect pattern (structure) at the layout level.

3. INDUCTIVE COUPLING BASICS

Our first assumption is that we use two parallel traces for each signal line; driven in true differential mode: while one of the input nodes of the line is making a low-to-high transition, its complementary input node is making a high-to-low transition. Clearly, it requires gates (or line drivers) with two complementary outputs and also, differential receivers.

Using low-voltage swing differential signaling already offers a range of advantages: faster circuits, less crosstalk susceptibility, reduced power consumption and reduced electro-magnetic interference (EMI). All these benefits are mainly due to the fact that the differential driver needs to drive a load only to a few hundreds of millivolts, compared to a few volts depending on the technology used. Therefore, differential drivers are much smaller compared to single-ended drivers, which results in smaller change of current in time, followed by significant reduction of inductive noise [3]. Still, the use of full differential signaling is not capable of eliminating the inductive crosstalk between lines.

To allow a perfect cancellation of coupled magnetic and electric field components between two parallel adjacent lines, we consider using twisted differential line (TDL) [4]. The benefits of using twisted lines on printed circuit boards (PCB) are already studied and well-known [5]. Twisted line interconnect architectures have been proposed earlier for on-chip connections as well [5][6], but the systematic application of this structure together with full-differential signaling has not been studied or analyzed yet.

4. ANALYSIS OF THE TDL STRUCTURE

Now, assume that two adjacent differential signal lines are formed as shown in the Figure 2. In this figure, the length of lines between two twisting sections (twisting period) is assumed to be much larger than the distance between the differential pairs.

In the TDL structure the complementary signals of the differential pair are routed parallel to each other until to the twisting point, where, one signal of the differential pair changes its track on the same metal level; whereas, the other signal goes one level down through a via, crosses below its complementary signal and switches back to the initial metal layer. Hence, the signals exchange their routing tracks. This twisting of signals is repeated at equal intervals. It should be noted that the neighboring differential pairs have to be routed so that their twisting points are offset relative to them, as shown in Figure 2.

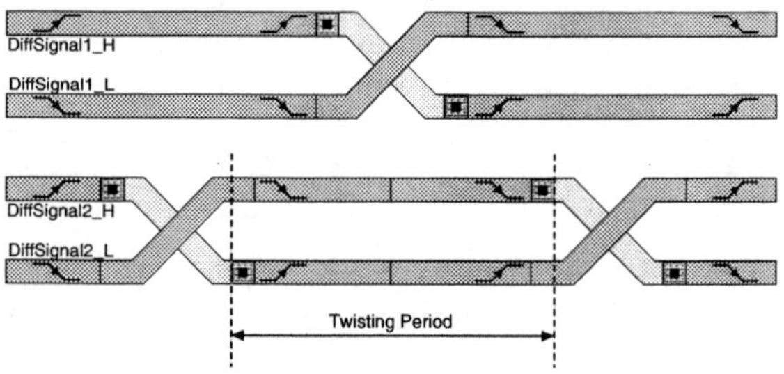

Fig. 2. Offset-twisted differential line arrangement at the layout level.

As far as the crosstalk issues are concerned, one of the differential pairs can be imagined as the aggressor and the other one as the victim line. It can be easily proved that the amount of induced voltages on the victim line will cancel out each other for any two adjacent twisted sections. Also note that in the proposed arrangement (offset twisted differential line), there is no fundamental difference between the aggressor and the victim line - i.e., the roles are completely reversible, and the cancelling effect would be observed in that case, as well.

Now, the phenomenon of inductive crosstalk noise cancellation can also be described by using the coupled PEEC models for the two differential lines. To simplify the view, only the partial inductance elements are shown. Each line segment is modeled by two equal partial inductances [2][3], as shown below in Figure 3. Note the current directions in each branch and the dots indicating the direction of inductive coupling.

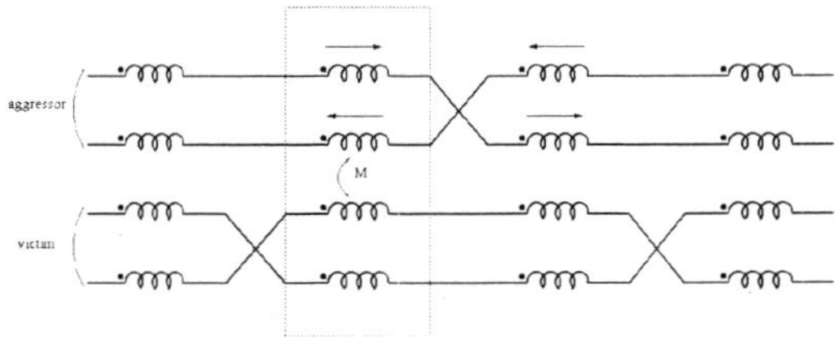

Fig. 3. Lumped circuit element model of two TDL segments, showing only inductive elements for simplicity.

Note that the polarity of the voltage induced by one inductor (L) on an adjacent inductor is determined by the relative location of the dots and by the current directions. At the same time, the magnitude of the induced voltage is determined by the amount of mutual inductance (M) between two adjacent line segments.

It can be seen that the voltage induced by each partial inductor will be cancelled out by the voltage (same magnitude, reverse polarity) induced on the neighboring segment. Also note that this is true for all line segments (even for those located further apart) and not just the closest ones. Furthermore, it can be shown easily that the offset TDL structure is similarly effective for the cancellation of capacitive coupling between the adjacent line pairs.

5. SIMULATION RESULTS

To demonstrate the effectiveness of the proposed approach, both delay and crosstalk simulations were made on the TDL segments, also, on the differential and single-ended straight lines for comparison purposes. In all the simulations the wires are located on the Metal-3 level, resulting in a total height of $3.750\mu m$ from the ground plane, with the minimum width ($0.6\mu m$), with a line separation of $0.6\mu m$ and metal thickness of $925nm$.

The line segments were modeled using the full PEEC model (as shown in Figure 1), including partial resistance, capacitance, inductance as well as coupling capacitance and mutual inductance values calculated by the 3-D extraction package OEA-NETAN (METAL/HENRY) ®[7].

Figure 4 shows the change in wire delay versus wire length for single-straight lines and differential lines. The delay is measured from the time the input voltage reaches 50% of the full-rail to the time the output reached the same voltage (propagation delay). Under these conditions (considering that there is no differential receivers) differential signals cause much more delay than the single-straight lines, simply because of the coupling capacitance between the differential signals and their opposite switching directions.

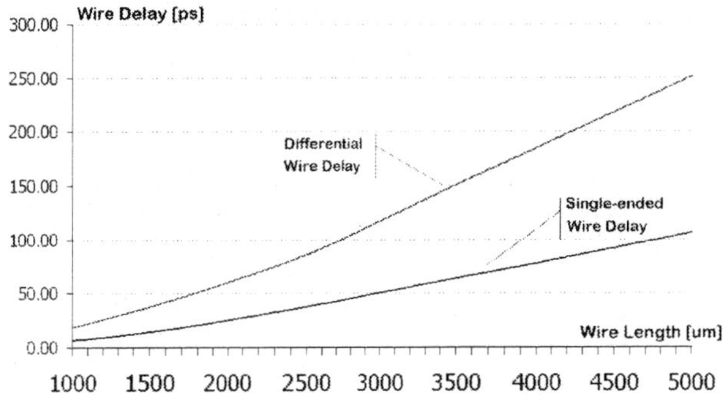

Fig. 4. The change of wire delay vs. wire length given for single-ended and differential lines.

The delay of the twisted-differential lines depend on a few parameters, which are the twisting period and the number of parallel vias used per layer change for twisting of signals. Therefore, the TDL delays of a 2mm long wire are given separately in Table 1. The segment length dictates the number of RLCM networks used to model the wire for one twisting period and only changes the granularity of our model. It is obvious from the table that for the same length of wire, there is a big change in delay depending on the twisting period and the number of vias (per layer change). The higher the twisting period, the smaller is the wire delay and the effect of the number of vias on the delay. As we shorten the twisting period, more twistings must be placed for routing the same length of wire, which results in a higher number of total vias and in a longer delay.

Segment Length[um]	Twisting	Number	Delay
100	400	1	13.82
		2	13.67
		3	13.61
10	100	1	21.72
		2	20.24
		3	19.74
	40	1	26.57
		2	22.65
		3	21.33
	20	1	24.47
		2	26.63
		3	23.94

Table 1. The delay of a 2mm long wire for different values of the twisting period, the segment length and the number of vias.

Table 2 compares the delay of a 2mm-long wire laid out with three different structures. For the TDL the delay for two different cases are provided, where, the 'worst' corresponds to the smallest twisting period and to the least number of vias (one via) per layer change (the value in the third-from-below cell in the right-most column in Table 1) and, the 'best' is exactly the opposite case among the considered ones, as given in the third-from-top cell in the same column. At this point it can be concluded that the TDLs, implemented with a combination of number of vias and of the twisting frequency, provide high-speed on-chip connections.

Straight-single wire	25.05ps
Differential wire	60.18ps
Twisted-differential wire (worst)	34.47ps
Twisted-differential wire (best)	13.61ps

Table 2. Delay of a 2mm wire implemented with different structures.

Crosstalk is the second issue to be considered. In all the crosstalk simulations, the aggressor is driven by a signal with rise/fall time of 100ps, while the victim line input is kept at a constant DC level (not left floating, i.e, modelling the case of static gate outputs). The voltage fluctuation at the output end of the victim line, terminated with a capacitive load, is measured as the crosstalk noise voltage.

Figure 5 shows the crosstalk noise voltage on a straight-differential victim line, Figure 5(a), subjected to the same conditions, compared to the noise on a TDL, Figure 5(b), with the same conditions and using the same geometry. The noise level on the TDL victim remains significantly lower than that on the straight-differential victim line. It can be seen clearly here as well that the TDL approach results in dramatically lower crosstalk noise. The maximum magnitude of the crosstalk noise signal on the victim line pair remains typically less than 2mV. It was also determined that the additional via resistance that are associated with the TDL structure do not significantly influence the results. With a typical via resistance of 1.2Ω per via, the magnitude of the noise signal increases to about 3mV, which is still much lower than the noise on the straight line.

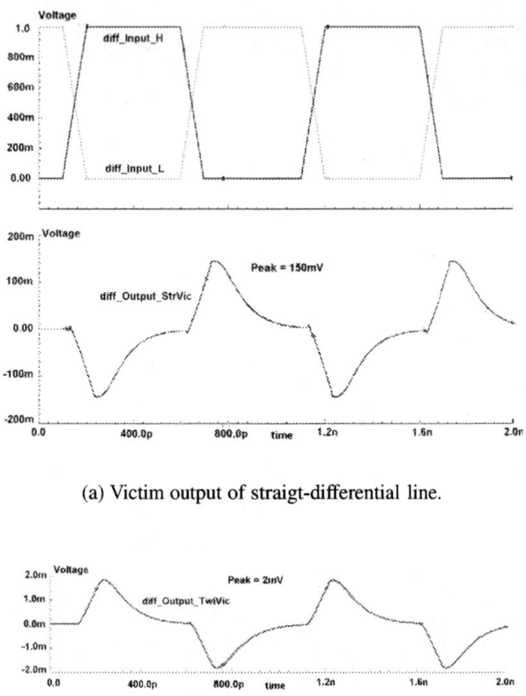

(a) Victim output of straigt-differential line.

(b) Victim output of twisted-differential line.

Fig. 5. Victim outputs of two lines with a length of 2000um. Note that the noise voltage amplitude is two orders of magnitude smaller in the twisted-differential case.

6. CONCLUSIONS

In this paper, we present a fully differential, offset twisted interconnect structure that is capable of reducing the crosstalk noise between adjacent line pairs significantly. The effectiveness of the proposed interconnect architecture is demonstrated with detailed simulation results. This approach could be applied very early and efficiently to construct highly noise-tolerant, on-chip high-speed bus structures for SoC.

7. REFERENCES

[1] H.B. Bakoglu, "Circuits, interconnections, and packaging for vlsi," *Addison Wesley*, 1990.

[2] A. Ruehli, "Inductance calculations in a complex integrated circuit environment," *IBM Journal of Research and Development*, vol. 16, pp. 470–481, Sept. 1972.

[3] Y. Massoud et al., "Managing on-chip inductive effects," *IEEE Transactions on VLSI Sytems*, vol. 10, pp. 799–805, December 2002.

[4] I. Hatirnaz Y. Leblebici, "Twisted differential on-chip interconnect architecture for inductive/capacitive crosstalk noise cancellation," *Int. Symposium on System-on-Chip*, 2003.

[5] J. Kim et al., "A novel twisted differential line for high-speed on-chip interconnections with reduced crosstalk," *Elec. Packaging Tech. Conf.*, 2002.

[6] K. Roy G. Zhong, C. Koh, "A twisted-bundle layout structure for minimizing inductive coupling noise," *IC-CAD*, pp. 406–411, 2000.

[7] E. Akcasu et al., "Net-an a full three dimensional parasitic interconnect distributed rcl extractor for large full-chip applications," *IEDM*, 1995.

DELAY BOUND BASED CMOS GATE SIZING TECHNIQUE

A. Verle, X. Michel, P. Maurine, N. Azémard, D. Auvergne

LIRMM, UMR CNRS/Université de Montpellier II, (C5506),
161 rue Ada, 34392 Montpellier, France

ABSTRACT

In this paper we address the problem of delay constraint distribution on CMOS combinatorial paths. We first define a way to determine on any path the reasonable bounds of delay characterizing the structure. Then we define two constraint distribution methods that we compare to the equal delay distribution and to an industrial tool based on Newton-Raphson like algorithms. Validation is obtained on a 0.25μm process by comparing the different constraint distribution techniques on various benchmarks.

1. INTRODUCTION

The goal of gate sizing is to determine optimum sizes for the gates in order that the path delay respects the constraints with the minimum area/power cost. Another parameter to consider is the feasibility of the constraint imposed on a given path. For that, indicators must be found to explore the design space and to select one solution among the available optimization alternatives such as sizing, buffering or technology remapping. The target of this paper is twofold: defining the delay bounds of a given path and determining a way for distributing a delay constraint on this path with the minimum area/power cost.

The problem of transistor sizing has been widely investigated using non linear programming techniques [1] or heuristics based on simple delay models [2]. Recently, in a pedagogical application [3] of the τ model, Sutherland [4], describing the gate delay as the product of electrical and logical efforts, proposed to minimize the delay on a path by imposing an equal effort that is a constant delay on all the elements of the path.

This way to select cell sizes can be proven mathematically exact [3] for a fanout-free path constituted of ideal gates (without parasitic capacitance or divergence branch). However this evenly budget distribution is far to be the optimal one with respect to delay and area for a real path, on which divergence branches and routing capacitance are not negligible. Starting from the definition of the design space in terms of minimum and maximum delay permissible on a given path, we propose in this paper a design space exploration method allowing an area/power efficient distribution of constraint on a combinatorial path.

The delay bound determination and the constraint distribution method are based on a realistic delay model [5] that is input slope dependent and able to distinguish between falling and rising signals. This model is shortly presented in part 2. In part 3 we give a method for defining delay bounds on a path. Different approaches for distributing a delay constraint are considered in part 4 and compared in part 5 on different benchmarks of increasing complexity. We finally conclude in part 6.

2. GATE DELAY MODELING

As previously mentioned sizing at the physical level imposes to use a realistic delay computation that must consider a finite value of the gate input transition time. As developed in [5] we introduce the input slope effect and the related input-to-output coupling in the model as:

$$t_{HL,LH}(i) = \frac{v_{TN,P}}{2}\tau_{INLH,LH}(i-1) + (1+\frac{2C_M}{C_M+C_L})t_{HL,LHstep}(i) \quad (1)$$

where $v_{TN,P}$ are the reduced value (V_T/V_{DD}) of the threshold voltage of the N,P transistors. $\tau_{INHL,LH}$ is the duration time of the input signal, taken to be twice the value of the step response of the controlling gate. C_M is the coupling capacitance between the input and output nodes. C_L is the output loading capacitance. Indexes (i), (i-1) specify the switching and the controlling gates, respectively.

Following [2], the step response of each edge is defined by the time interval necessary to load (unload) the gate output capacitance under the maximum current, I_{MAX}, available in the structure:

$$t_{HL,LHstep} = \frac{C_L \cdot \Delta V}{I_{MAX}} \quad (2)$$

Following the elegant model of [3] the evaluation of this step response on logic gates supplies a general expression given by:

$$t_{HL,LHstep} = \tau \cdot S_{HL,LH} \cdot \frac{C_L}{C_{IN}} \quad (3)$$

where τ is a time unit characterizing the process. C_{IN}, the gate input capacitance, is defined in terms of the P/N width ratio k. For simplicity, the S factors (logical effort of [3]) include all the current capability difference between the pull up (pull down) transistor equivalent to the corresponding serial array. These factors are configuration ratio dependent and characterize for each edge, the ratio of current available in an inverter and a gate of identical size.

Then considering an array of gates, the delay path can easily be obtained from (1) and (3) as a technology independent posynomial representation:

$$\frac{t_{HL,LH}}{\tau} = \theta = S'_1 \cdot \frac{C_2 + C_{P1}}{C_1} + \cdots + S'_{i-1} \cdot \frac{C_i + C_{Pi-1}}{C_{i-1}} + \ldots + S'_n \cdot \frac{C_L}{C_n} \quad (4)$$

where the S_i' include the logical effort and the input ramp effect, C_i represents the input capacitance of the gate and C_{pi} the output node total parasitic capacitance, including the interconnect and branching load.

3. DELAY BOUND DEFINITION

We consider realistic combinatorial paths on which two parameters are known and imposed:

- the output load capacitance of the last gate, that is determined by the input capacitance of the output register,
- the input capacitance of the first gate imposed by the loading conditions of the input register.

In that condition the path delay is bounded. These bounds can be determined, considering that the delay of a path (4) is a convex function of the gate input capacitance. This is illustrated in Fig.1 that gives the variation of the path delay with respect to the transistor sizing of a combinatorial path constituted of 13 gates. Note that the slope of the curve corresponds to the sensitivity of the path delay to the transistor sizing.

As shown the delay value decreases from a maximum value down to a minimum value that will be determined below. The maximum delay has been obtained imposing all the transistor sizes at the minimum allowed by the technology. This maximum value is a "reasonable" one but not the absolute maximum value. It is always possible to get a much greater value by loading minimum gates with an infinitely sized driver. This curve illustrates what we define by exploring the design space:

- near the maximum value, Θ_{Max}, of the delay the path sensitivity to the gate sizing is very important, a small variation of the gate input capacitance results in a large change in delay,
- at the contrary near the minimum Θ_{Min} the sensitivity is becoming very low and in that range any delay improvement is highly area/power expensive.

Evaluating the feasibility of a delay constraint Θ_c imposes to compare its value to the preceding bounds. If the Θ_c value is closed to the maximum Θ_{Max} the constraint satisfaction will be obtained at reasonable cost by transistor sizing otherwise it would be more profitable to reconfigure the logic or to insert buffers [6]. Let us define these bounds.

As previously mentioned we consider for Θ_{Max} the "reasonable" value obtained when all the gates are implemented with transistors of minimum size. For the minimum bounds we just use the posynomial property [7] of (6). Canceling the derivatives of (4) with respect to the gate input capacitances C_i we obtain a set of linked equations such as:

$$S'_{i-1} \cdot \frac{C_i}{C_{i-1}} - S'_i \cdot \frac{C_{i+1} + C_{Pi}}{C_i} = 0$$
$$S'_i \cdot \frac{C_{i+1}}{C_i} - S'_{i+1} \cdot \frac{C_{i+2} + C_{Pi+1}}{C_{i+1}} = 0 \ldots \quad (5)$$

Cell sizes can then be selected to match the minimum delay, by visiting all the gates in a topological order, starting from the output, such as:

$$C_i^2 = \frac{S'_i}{S'_{i-1}} \cdot C_{i-1} \cdot (C_{i+1} + C_{Pi}) \quad (6)$$

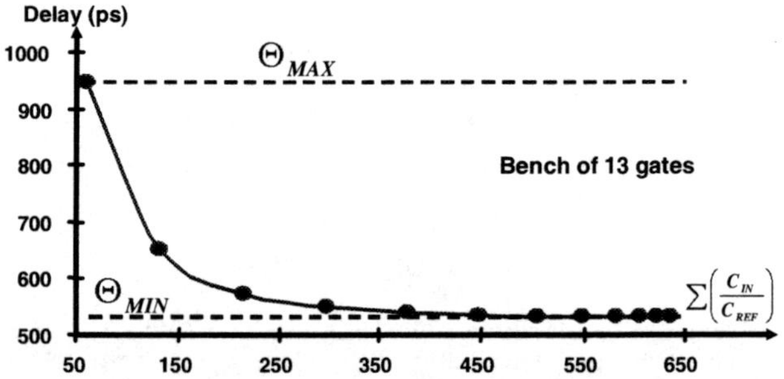

Fig.1. Illustration of the variation of the path delay with the gate sizing.

This results in a set of *n* linked equations that can be easily solved by iterations from an initial solution that considers C_{i-1} known and equal to a reference value C_{REF}. This reference can be set equal to the minimum value available (C_{MIN}) in the library or to any other one.

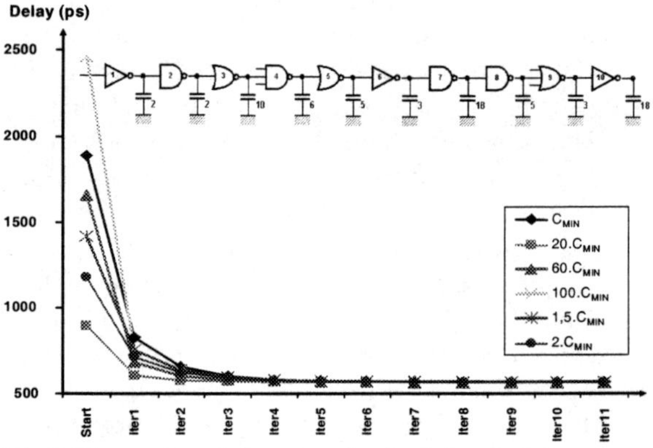

Fig.2. Illustration of the research of minimum delay on an array of ten gates for different values of the initial reference capacitance; the output load of each gate on the array is given in unit of C_{MIN}.

In Fig.2, we illustrate the variation of the calculation convergence with a choice of C_{REF}. As shown, whatever is the value of C_{REF} (C_{MIN} to100C_{MIN}), we always obtain a fast convergence to the minimum.

4. CONSTRAINT DISTRIBUTION

Determining the possible bounds of delay for a given path topology, the next step is to evaluate the feasibility of a constraint to be imposed on a path. The theory of constant effort or constant delay [4,8] provides an easy way to select the cell size for each stage but for real configuration it is far from the optimum and often results in oversized structures. For that we propose two techniques for the gate size selection in order to satisfy a constraint that we will compare in the next part to the constant delay method.

To define the first method we consider that imposing equal delay to the gates with an important value of the logical effort (S'_i), results in an important over sizing of these complex gates. The determination of the lowest delay bound directly provides the optimal delay distribution on the path that appears to be the fastest one. So we can use this distribution to define for each gate a weight or gain Θ_i relative to this distribution $\Theta_{Min}=\Sigma\Theta_{Mini}$. In that case we propose to distribute the delay constraint Θc on a path using a weight defined with respect to the minimum delay distribution as:

$$\theta_i = \theta_{Mini} \cdot \theta_c \Big/ \sum_i \theta_{Mini} \qquad (7)$$

This guarantees the conservation of the path delay distribution, obtained at the optimal solution, for any value of the constraint. Then processing backward from the output of the path, this directly gives, for each gate, the value of Θ_i that determines from (1,3) the size of the corresponding gate.

The second method of equal sensitivity is directly deduced from (5). Instead to search for the minimum we impose the same path delay sensitivity to the sizing, by solving:

$$S_{i-1} \cdot \frac{1}{C_{i-1}} - S_i \cdot \frac{C_{i+1} + C_{Pi}}{C_i^2} = a \ldots \qquad (8)$$

where "a" is a constant, representing the slope of the curve of Fig.3, that represents the variation of the delay between the bounds previously defined. Following the procedure used for the first method, the size of the gates is obtained from the iterated solution of (8) using as initial solution the sizing for the maximum delay value (all gates sized at C_{REF}). The different points on the curve of Fig.3 have been obtained from (8), by varying the value of "a", until "a" = 0 to get the minimum.

As expected, for a given value of the sensitivity factor "a", this curve represents the locus of the minimum delay solutions. No inferior solution can be found. Thus, varying the "a" value gives the possibility to explore the design space and to determine the minimum area sizing condition satisfying the delay constraint.

In Fig.3, we compare our approach to an industrial optimization tool (Amps from Synopsys). As shown the two methods give nearly equivalent design range exploration, however the equal sensitivity method results in a minimum area solution.

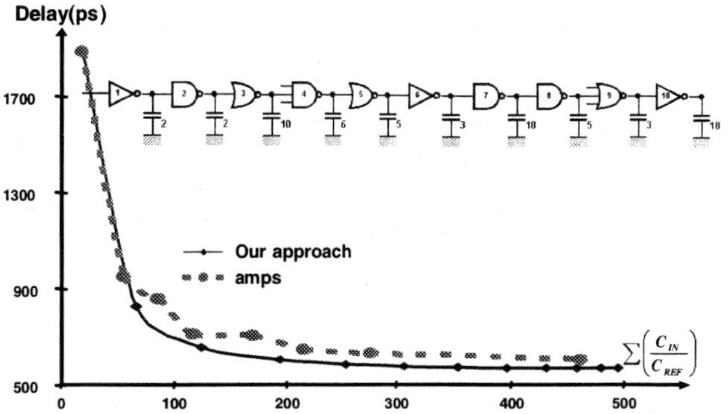

Fig.3. Design space exploration with constant sensitivity method.

5. VALIDATION

In order to validate these sizing and constraint distribution techniques we compare on different benchmarks the minimum delay value and the area obtained using the three investigated methods:

- equal distribution of delays (C), ($\Theta_i = \Theta_c/n$), [4] where *n* is the number of stages,
- weighted distribution (7), (B),
- equal gate sensitivity (8), (A),
- and using an industrial tool based on a Newton-Raphson based algorithm (D), [9] (Amps from Synopsys).

These benchmarks are constituted of array of gates (Nand, NOR, 2 and 3 inputs) with different loading conditions.

The comparison of the minimum delay values obtained with each technique is given in Table 1 for different paths. The targeted process is the STM 0.25µm with τ = 7.05ps. As shown the lowest minimum value of delay is obtained with both the weighted and the equal sensitivity techniques.

This ascertains the method used to determine the lowest bound of delay on a logical path. Note that around the minimum value of delay the area penalty is, of course, very large. This value of delay must be more considered as an indicator for the feasibility of the constraint than as a design target.

The next step is to compare for a given delay constraint the area of implementation obtained with the different distribution techniques. For that we impose on the different benchmarks a delay constraint defined between the bounds previously defined. Then we compare in Table 2 the area corresponding to the gate sizing allowing, with the different techniques, to match the constraint. We can observe that if for a weak constraint value the different techniques appear quite equivalent, for a tighter constraint the equal sensitivity distribution technique (A) allows a match with a much smaller area than the others. Note that all the values given in Tables 1 and 2 are obtained from Spice simulations (MM9 model) of the different benchmarks.

The weighted distribution (B) still results in a quite equivalent area but the equal delay distribution (C) and Amps (D) may result for quite complex paths in an important increase of area. For some constraint values they may fail to get a solution. Note that the equal sensitivity method is mathematically quasi-optimal and always gives slightly better results than the weighted distribution. However this last method, defined from a minimum delay solution obtained for a sensitivity value equal to zero, can be much more easily implemented.

Table 1

Gate Nb	Θ_{MAX} (ps)	Area µm	Siz. Tech.	Θ_{MIN} (ps)	Area µm
9	1874	42	A	620	987
			B	620	987
			C	676	391
			D	633	632
11	2085	46	A	698	1448
			B	698	1448
			C	777	440
			D	937	348
15	3479	3479	A	923	4337
			B	923	4337
			C	1023	1083
			D	960	3067
21	4583	94	A	1192	8419
			B	1192	8419
			C	1484	1039
			D	1693	1152
31	6560	138	A	1503	21578
			B	1503	21578
			C	1881	2226
			D	2426	1826

An illustration of these results is given in Fig.4 where we show for the path constituted of 31 gates, the complete exploration of the design space using the preceding constraint distribution methods. As shown for a delay constraint smaller than $\Theta_{max}/2$ the gain in area (power) using the equal sensitivity or the weighted distribution method is quite significant.

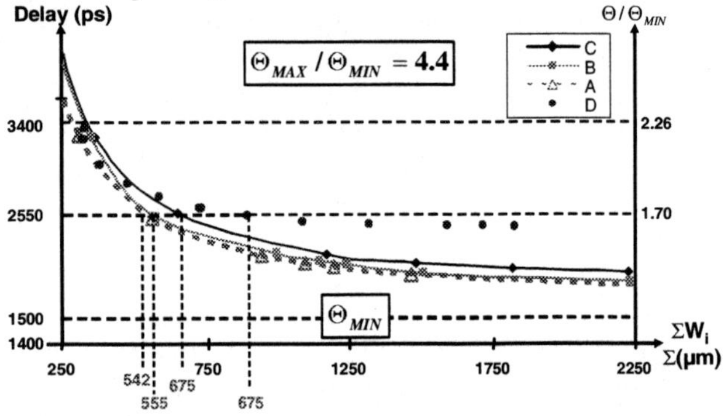

Fig.4. Illustration of the design space exploration, on the 31 gate path, using the different constraint distribution methods.

6. CONCLUSION

Based on a simple realistic delay model for gates, we have first determined an easy way to characterize the feasibility of a delay constraint imposed on a combinatorial path. We have defined reasonable maximum and real minimum delay bounds. Then we proposed two techniques to match a delay constraint on a path: the equal sensitivity and the weighted method that is a budgeting method. We have applied these methods on different benchmarks with various constraint conditions and compared the resulting implementation area with that obtained from an equal delay distribution and with an industrial tool. If for weak constraints the different methods are quite equivalent, for values near the minimum, the proposed methods always find a solution and result in an important area/power saving. Another point to be clarified further is to define at which distance of the minimum delay value it is area/power efficient to impose a constraint.

7. REFERENCES

[1] J. M. Shyu &Al , A. Dunlop, "Optimization-based transistor sizing" IEEE J. Solid State Circuits, vol.23, n°2, pp.400-409, 1988.

[2] J. Fishburn, A. Dunlop, "TILOS: a posynomial programming approach to transistor sizing" in Proc. Design Automation Conf. 1985,pp.326-328.

[3] C. Mead, M. Rem, "Minimum propagation delays in VLSI", ", IEEE J. Solid State Circuits, vol.SC17, n°4, pp.773-775, 1982.

[4] I. Sutherland, B. Sproull, D. Harris, "Logical Effort: Designing Fast CMOS Circuits", Morgan Kaufmann Publishers, INC., California, 1999.

[5] K. O. Jeppson, "Modeling the influence of the transistor gain ratio and the input-to-output coupling capacitance on the CMOS inverter delay", IEEE J. Solid State Circuits, vol.29, pp.646-654, 1994.

[6] S. Chakraborty, R. Murgai "Layout driven timing optimization by generalized DeMorgan transform" IWLS 2001, pp.53-59.

[7] M. Ketkar, K. Kasamsetty, S. Sapatnekar "Convex delay models for transistor sizing" Proc. of the 2000 Design Automation Conf. pp.655-660.

[8] J. Grodstein & al"A delay model for logic synthesis of continuously-sized networks", ICCAD 95, Nov 95.

[9] R. K. Brayton, R. Spence "Sensitivity and Optimization" Elsevier 1980

Table 2

Gate Nb.	Siz. Tech.	$\frac{\Theta_{MAX}}{\Theta_{MIN}}$	$\frac{\Theta_C}{\Theta_{MIN}}$	Area (µm)	$\frac{\Theta_C}{\Theta_{MIN}}$	Area µm
9	A	3	1.4	137	1.02	535
	B			147		560
	C			161		Fail
	D			144		632
11	A	11	2.15	66	1.1	310
	B			80		330
	C			94		440
	D			70		Fail
15	A	3.8	1.4	302	1.04	1333
	B			310		1407
	C			410		Fail
	D			324		3067
21	A	3.8	2.1	196	1.31	553
	B			214		558
	C			230		715
	D			198		1152
31	A	4.4	3.1	364	1.26	1275
	B			400		1361
	C			427		1970
	D			377		Fail

IMPROVING SYMBOLIC ANALYSIS IN CMOS ANALOG INTEGRATED CIRCUITS

Aguila-Meza J., Torres-Papaqui L., Tlelo-Cuautle E.

National Institute for Astrophysics, Optics and Electronics (INAOE)
Luis Enrique Erro No.1. Tonanzintla, Puebla. 72000, México
Tel/Fax: +52(222)247-0517. E-mail: <jaguila,lpapaqui>@inaoep.mx, e.tlelo@ieee.org

ABSTRACT

A symbolic-method to improve the computation of simple symbolic expressions (SEs), which represent the dominant behavior of a CMOS analog IC, is presented. It is demonstrated that the complexity in manipulating SEs, being minimized by modeling the behavior of the MOSFET using nullors, at different levels of abstraction. In this manner, to improve traditional symbolic-methods, the formulation of a compacted system of equations (CSEs) is computed herein by selecting the simplest model for the MOSFET, and by manipulating the interconnection-relationships of the circuit. The proposed method sets the guidelines to select the correct model, which minimizes computer-tasks by considering the bias and frequency operating conditions of every MOSFET.

Keywords: Analog Design Automation, Symbolic Analysis, Data Structures, Modeling, MOSFET, Nullor.

1. INTRODUCTION

Symbolic analysis is oriented to the analysis of circuits in which some or all their parameters are symbols [1]-[4]. The main objective is the computation of simple SEs which help to the designer to understand the way that a circuit works. Its application results in an inference guided towards an interactive design process through the generation of analytical design equations, which become to be quite useful for either, synthesis or optimization procedures [5]. The computation of the design equations like simple SEs requires of the application of approaching techniques [4]. In this manner, symbolic analysis methods oriented to provide analytical models represented by simple SEs, and representing the dominant behavior of a circuit, are very much needed.

Traditional symbolic analysis methods [4] compute SEs by applying simplification of symbolic terms before and during the formulation of the system of equations, and after the solution, and they generate only the dominant contribution of the solution. However, their range of application is limited to small circuits, which mainly depends on the number of nodes and on the complexity of the circuit-models, as well as on the connectivity, and on the error imposed by the matching of the design specifications, i.e. the approach. An alternative to improve the capability of symbolic analysis is to use models at different levels of abstraction [3]. Henceforth, an improved symbolic-method for CMOS analog ICs, whose behavior is modeled by nullors, is introduced. Basically, a hierarchical modeling approach using nullors, is given in section 2. A formulation approach of the CSEs is described in section 3. The proposed algorithm to select the correct model which minimizes the computational effort is summarized in section 4. Finally, the conclusions are given in section 5.

2. ANALOG MODELING

A very abstract model generates simple SEs, a very real one generates complex SEs. To minimize the complexity to compute SEs, it is recommended to generate models at different levels of abstraction, but avoiding the use of controlled sources, since they increase the computational effort and the order of the system of equations [2],[4]. To improve the formulation of a CSEs, the use of the nullor element is quite useful [2]-[3], since it generates uncoupled equations.

In this section, the behavior of the three terminals MOSFET is modeled using the nullor at different levels of abstraction, e.g. the generic MOSFET is shown in Fig. 1a. The transconductance, output conductance and gate-source capacitor can be easily added to Fig. 1a, as shown in Figs. 1b, 1c, and 1d. In the same manner, other parasitic elements can be added to model the behavior of the MOSFET according to the operating frequency ranges.

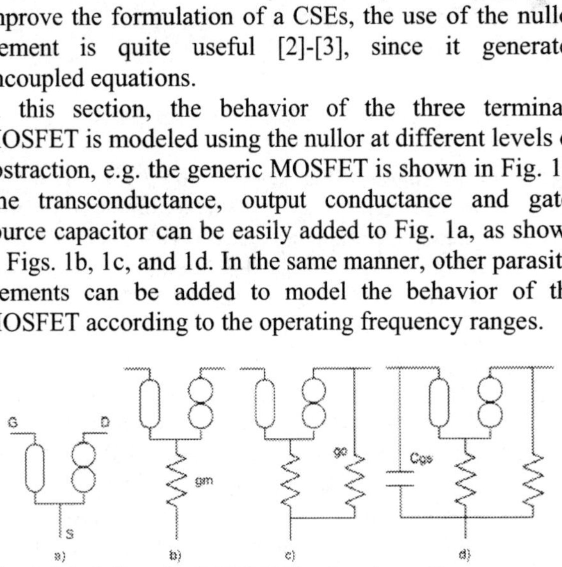

Fig. 1. Modeling the MOSFET using the nullor.

The use of the correct model which minimizes the computational effort in doing small-signal symbolic-analysis, depends on the biasing and operating frequency conditions, as shown in section 3. Furthermore, to cover the requirements imposed by analog design automation [5], intelligent systems could be applied to generate models that minimize the computational complexity in manipulating SEs.

The problem in selecting the appropriate model is reflected in the computational effort to generate SEs, e.g. transfer functions (TFs). For example: to compute the symbolic TF of the Miller OTA shown in Fig. 2, when using the model shown in Fig. 1d, to replace the eight MOSFETs, the resulting TF would have an order higher than two. On the other hand, by selecting the appropriate model to each MOSFET, as it is described in the following section, the resulting TF would be of order 1 or 2, good enough to characterize the OTA [6].

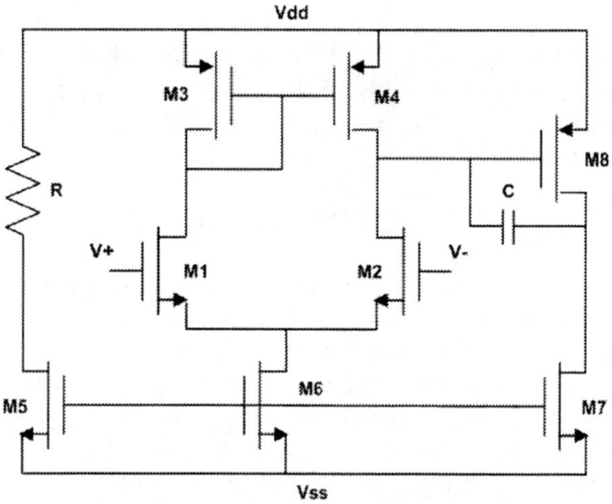

Fig. 2. Miller Operacional Transconductance Amplifier.

3. FORMULATION OF THE CSE

For a nullor circuit, the computation of a SE can be reduced to the manipulation of the nullator and norator interconnection relationships (IRs) [3]. The Cartesian product of the IRs results in the formulation of a CSEs, like by applying the nodal analysis method [2]. The order (***m***) of the CSEs is determined by the number of nodes (***n***), minus the numbers of nullors (N) [2]-[3].

In [2], a traditional symbolic-method to the formulation of a CSEs for nullor circuits, is given. The method consists in formulating an initial non-compacted system of equations (NCSEs), of order ***n***, expressed by (1). The NCSEs is reduced to a CSEs by applying the properties of nullators and norators, obtaining a CSEs of order ***m***=***n***-**N**, expressed by (2). Computing a CSEs from a NCSEs requires a hard computational effort to the manipulation of the IRs of the nullors, which are associated to the rows and columns of the Matrix in (1), until obtaining (2).

$$i_{NA} = Y_{NA} v_{NA} \quad (1)$$

$$i_{CNA} = Y_{CNA} v_{CNA} \quad (2)$$

An improved symbolic-method is given in [3], which avoids the formulation of the NCSEs by computing the CSEs directly. The computation of the elements of the Matrix in (2), are obtained by associating indexes row-column (R,C), from the Cartesian product of the IRs associated to the norators and nullators, respectively. The IRs of the admitances (Y) are associated according to (3) and (4), where i and j denote the nodes.

$$+Y_{ij} \forall (i = j) \quad (3)$$

$$-Y_{ij} \forall (i \neq j) \quad (4)$$

The indexes to compute the elements of the matrix in (2), are obtained as it is described below:

a) IRs of the norators: From the properties in (2), the associated indexes are structured by considering that the nodes ***i,j,*** where the norator is connected, are virtually short-circuited, and that indexes are associated to row variables (currents).
b) IRs of the nullators: From the properties in (2), the associated indexes are structured by considering that the nodes ***i,j,*** where the nullator is connected, are virtually short-circuited, and that indexes are associated to column variables (voltages).
c) The cartesian product of the IRs of the norators and nullators, generates the elements of the matrix in (2). Each element includes the sum of those admittances associated to each pair (R,C), which are searched in the IRs of the admittances.

Example 1: Lets consider the inverting amplifier shown in Fig. 3a. To compute the symbolic TF, the MOSFET can be modeled using the model in Fig. 1b. Also, to apply the nodal analysis method, the independent voltage source is transformed to a current source using one nullor [2]. The equivalent small-signal nullor circuit is shown in Fig. 3b.

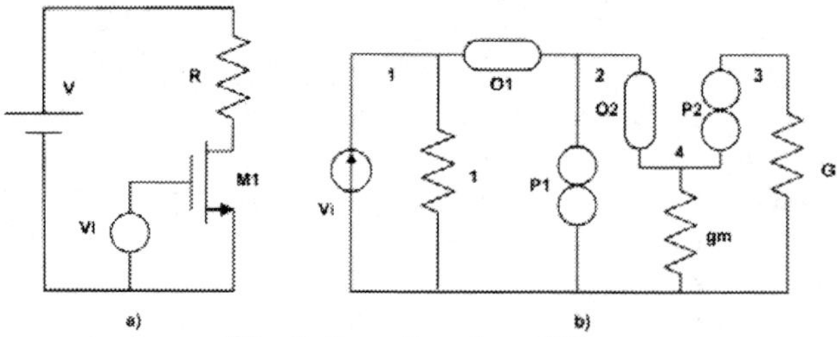

Fig. 3. Inverting Amplifier.

The order of the CSEs is equal to: ***m***=***n***-**N**=4-2=2. The matrix in (2) is computed as follows:

a) The IRs of the norators are shown in Table 1. The row-variables are: [(1), (3,4)]. Since norator P1 is connected between nodes 2,0 node 2 is eliminated [2].

b) The IRs of the nullators are shown in Table 2. The column-variables are: [(1,2,4), (3)].
c) The cartesian product of the IRs generates the 4 elements of the matrix in (2):

$$(Y_{11} \in Y_{CNA}) = (1,1) + (1,2) + (1,4)$$
$$(Y_{12} \in Y_{CNA}) = (1,3)$$
$$(Y_{21} \in Y_{CNA}) = (3,1) + (3,2) + (3,4) + (4,1) + (4,2) + (4,4)$$
$$(Y_{22} \in Y_{CNA}) = (3,3) + (4,3)$$

Searching the admittance-values associated to each pair (R,C) in Table 3, the final formulation of the CSEs is expressed by (5). The solution to (5), by applying Cramer's rule, with G=1/R, is given by (6).

TABLE 1.
IR OF THE NORATORS

Norator (P)	Associated Nodes
P1	2,0
P2	3,4

TABLE 2.
IR OF THE NULLATORS

Nullator (O)	Associated Nodes
O1	1,2
O2	2,4

TABLE 3.
IR OF THE ADMITANCES

Admittance	Associated Nodes
G	3,3
gm	4,4
1(vi)	1,1

$$\begin{bmatrix} v_i \\ 0 \end{bmatrix} = \begin{bmatrix} 1 & 0 \\ gm & G \end{bmatrix} \begin{bmatrix} v_{1,2,4} \\ v_3 \end{bmatrix} \qquad (5)$$

$$\frac{v_o}{v_i} = \frac{v_3}{v_i} = -\frac{g_m}{G} = -g_m R \qquad (6)$$

Example 2: Lets consider the inverting amplifier shown in Fig. 4a, in which the load resistance R has been replaced by an active load.

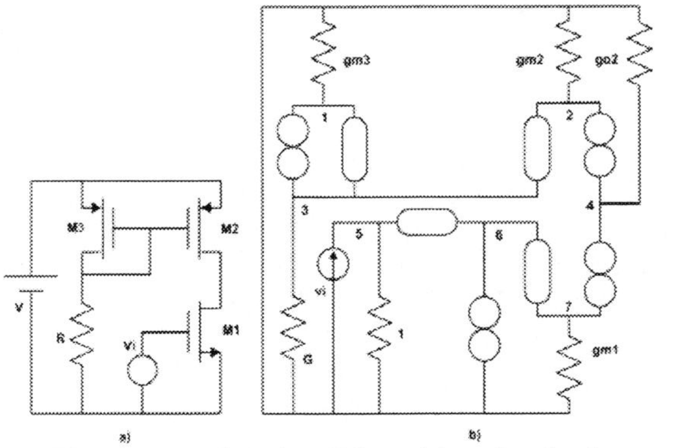

Fig. 4. Inverting Amplifier with active load.

To compute the symbolic TF, each MOSFET is modeled by considering its bias conditions. It is worth to mention that R in Fig. 3a, has been implemented by the output conductance of M2 in Fig. 4a, expressed by go2 in Fig. 4b. The order of the CSEs is equal to: ***m=n-*N**=7-4=3. The elements of the matrix in (2) are computed as follows:

a) The IRs of the norators are: (6,0), (4,7), (2,4), (1,3). The row-variables are: [(1,3), (2,4,7), (5)].
b) The IRs of the nullators are: (5,6), (6,7), (2,3), (1,3). The column-variables are: [(1,2,3), (4), (5,6,7)].

Searching the admittance-values to each pair (R,C) in Table 4, the final formulation is given by (7). The solution to (7) is expressed by (8), which is identical to (6), as it was expected using a current mirror as active load.

$$\begin{bmatrix} 0 \\ 0 \\ v_i \end{bmatrix} = \begin{bmatrix} g_{m3}+G & 0 & 0 \\ g_{m2} & g_{o2} & g_{m1} \\ 0 & 0 & 1 \end{bmatrix} \begin{bmatrix} v_{1,2,3} \\ v_4 \\ v_{5,6,7} \end{bmatrix} \qquad (7)$$

$$\frac{v_o}{v_i} = \frac{v_4}{v_i} = -\frac{g_{m1}}{g_{o2}} = -g_{m1} r_{o2} \qquad (8)$$

TABLE 4.
IR OF THE ADMITANCES

Admittance	Associated Nodes
G	3,3
1(vi)	5,5
gm1	7,7
gm2	2,2
go2	4,4
gm3	1,1

An important conclusion of this example, is that the elements biasing M2 are canceled, i.e. they do not appear in the final solution.

Example 3: The equivalent nullor circuit of the Miller OTA shown in Fig. 2, is shown in Fig. 5.

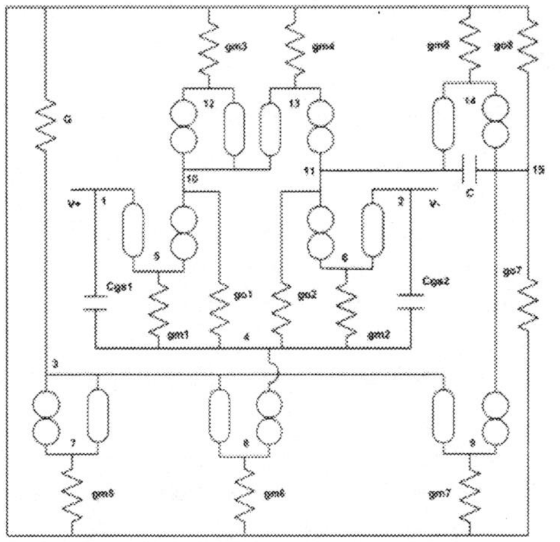

Fig. 5. Miller OTA using nullors

The order of the CSEs is equal to: ***m*=*n*-N**=15-8=7, by applying the MNA method [4], the order increases in one to include the differential voltage source between nodes 1 and 2. The CSEs is obtained as follows:

a) IRs of norators: (5,10), (6,11), (10,12), (11,13), (3,7), (4,8), (9,15), (14,15). The row-variables are: [(1),(2),(3,7),(4,8),(5,10,12),(6,11,13), (9,14,15)]
b) IRs of nullators: (1,5), (2,6), (10,12), (10,13), (3,7), (3,8), (3,9) and (11,14). The column-variables are: [(1,5),(2,6), (3,7,8,9), (4), (10,12,13), (11,14), (15)].

The Cartesian product generates 49 elements of the CSEs, and the stamp of the differential voltage source adds 15 elements, the final CSEs is expressed by (9).

$$\begin{bmatrix} 0 \\ 0 \\ 0 \\ 0 \\ 0 \\ 0 \\ 0 \\ v_d \end{bmatrix} = \begin{bmatrix} sC_1 & 0 & 0 & -sC_1 & 0 & 0 & 0 & 1 \\ 0 & sC_2 & 0 & -sC_2 & 0 & 0 & 0 & -1 \\ 0 & 0 & G+g_5 & 0 & 0 & 0 & 0 & 0 \\ -g_{m1}-sC_1 & -g_{m2}-sC_2 & g_6 & g_{m1}+g_{m2}+sC_1+sC_2+g_{o1}+g_{o2} & -g_{o1} & -g_{o2} & 0 & 0 \\ g_{m1} & 0 & 0 & -g_{m1}-g_{o1} & g_{o1}+g_3 & 0 & 0 & 0 \\ 0 & g_{m2} & 0 & -g_{m2}-g_{o2} & g_4 & sC+g_{o2} & -sC & 0 \\ 0 & 0 & g_7 & 0 & 0 & g_8-sC & sC+g_{o7}+g_{o8} & 0 \\ 1 & -1 & 0 & 0 & 0 & 0 & 0 & 0 \end{bmatrix} \begin{bmatrix} v_{1,5} \\ v_{2,6} \\ v_{3,7,8,9} \\ v_4 \\ v_{10,12,13} \\ v_{11,14} \\ v_{15} \\ i_{vd} \end{bmatrix} \quad (9)$$

As one sees, the MOSFETs M7-M8 of the Miller OTA has been modeled by using Fig. 1c, while M1-M2 has been modeled by using Fig. 1d. If four terminals MOSFETs were used, the nullor model would be more elaborated. The other MOSFETs has been modeled using Fig. 1b, since they are biasing M1-M2 and M8, mainly. The solution to (9) is computed by considering that: gm2=gm1, Cgs2=Cgs1, gm6=gm5, gm4=gm3, go2=go1, so that the simplified symbolic TF is expressed by (10).

$$TF = \frac{(-g_{m8}+sC)g_{m1}}{sC(g_{o7}+g_{o8}+g_{m8})} \quad (10)$$

4. PROPOSED ALGORITHM

The proposed algorithm to the computation of simple SEs representing the dominant behavior of a CMOS analog integrated circuit can be summarized as follows:

1. Obtain the Model of the behavior of each MOSFET by considering its bias and frequency operating conditions.
2. Compute the order (***m***) of the CSEs, which is determined by the number of nodes (***n***) minus the numbers of nullors (N): ***m*=*n*-N.**
3. Compute the CSEs as follows:
   a. Compute the IRs of the nullators which are associated to the column–variables (voltages).
   b. Compute the IRs of the norators which are associated to the row–variables (currents).
   c. Compute the IRs of the admittances.
   d. Compute the Cartesian product of the IRs of the norators and nullators to obtain the indexes (R,C) forming the elements of the matrix in (2).
   e. Compute the matrix in (2) by searching the admittance-values associated to each pair (R,C).
4. Compute the vectors i_{CNA} and v_{CNA} of (2).
5. Compute the desired transfer function by applying Cramer's rule or other method [4].

5. CONCLUSIONS

A novel symbolic-method to compute simple SEs in CMOS analog ICs, has been described. To improve the formulation of a CSEs, all MOSFETs has been modeled by nullors at different levels of abstraction. It was pointed out that the final SE must be simple and it should represent the dominant behavior of the circuit.
It was demonstrated that by modeling all MOSFETs using nullors, the computational complexity referred to the manipulation of SEs is minimized. Furthermore, the formulation has been carried out by computing the Cartesian product of the IRs of the norators and nullators. The important thing has been the insight that there are elements not appearing in the final solution of the CSEs, which has been calculated by applying Cramer´s rule. As a result, the proposed symbolic-method sets the guidelines to choose the correct model to minimize computational complexity, by considering the bias and frequency operating conditions of every MOSFET.

ACKNOWLEDGMENT

This work has been supported by CONACYT/MEXICO under project number J40321-Y.

6. REFERENCES

[1] Laurence P. Huelsman, "Applications of symbolic analysis to analog system design", IEEE ISCAS, pp. 1165-1168, 1992.
[2] Henrik Floberg, Symbolic analysis in analog integrated circuit design, Kluwer Academic Pub., 1997.
[3] Tlelo-Cuautle E., Sánchez-López C., Sandoval-Ibarra F., "Symbolic analysis: a formulation approach by manipulating data structures", IEEE ISCAS, vol. IV, pp. 640-643, Bangkok, Thailand, May 25-28, 2003.
[4] Fernández F. V., Rodríguez-Vázquez A., Huertas J. L. and Gielen G. E., "Symbolic Analysis Techniques", IEEE PRESS, New York, 1998.
[5] M. E. Schlarmann, R. L. Geiger, "Prototype implementation of a www based analog circuit design tool", IEEE ISCAS, vol. I, pp. 97-100, 2001.
[6] Randall L. G., Sanchez-Sinencio E., Active Filter Design Using Operational Transconductance Amplifiers: A Tutorial, IEEE Circuits&Devices Magazine, pp. 20-32, 1985.

A SYSTEMATIC APPROACH FOR ANALYZING FAST ADDITION ALGORITHMS USING COUNTER TREE DIAGRAMS

Naofumi Homma†*, *Jun Sakiyama*†, *Taihei Wakamatsu*†, *Takafumi Aoki*† *and Tatsuo Higuchi*‡

†Graduate School of Information Sciences, Tohoku University
Aoba-yama 05, Sendai 980-8579, Japan
E-mail: homma@aoki.ecei.tohoku.ac.jp
‡Department of Electronics, Tohoku Institute of Technology
Sendai 982-8577, Japan

ABSTRACT

This paper presents a unified representation of fast addition algorithms based on *Counter Tree Diagrams* (CTDs). By using CTDs, we can describe and analyze various adder architectures in a systematic way without using specific knowledge about underlying arithmetic algorithms. Examples of adder architectures that can be handled by CTDs include Redundant-Binary (RB) adders, Signed-Digit (SD) adders, Positive-Digit (PD) adders, carry-save adders, parallel counters (e.g., 3-2 counters and 4-2 counters) and networks of such basic adders/counters. In this paper, we focus on an application of CTDs to the analysis of two-operand RB adders with limited carry propagation. The analysis result shows that there exists possible two types of 3-stage CTDs for the RB adders. From this result, we can confirm that well-known RB adders are classified into one of the two types.

1. INTRODUCTION

Arithmetic circuits are of major importance in today's computing and signal processing systems. Numerous algorithms for arithmetic circuits have been developed and implemented since the early days of digital computers, and newer ones are still being proposed [1]. Most of the arithmetic circuits are designed by a designer who had trained in a particular way to understand the basic arithmetic algorithms. There are no systematic ways of deriving arithmetic circuits automatically due to the lack of unified theory for manipulating arithmetic circuit structures. Even the state-of-the-art logic synthesis tools can provide only limited capability to create arithmetic circuit structures.

Addressing this problem, this paper presents a unified representation of fast addition algorithms based on *Counter Tree Diagrams* (CTDs) [2]. In general, arithmetic circuits are composed of many primitive components; the most important components are adders and counters. However, various algorithms for fast addition have been investigated by many researchers in an ad hoc manner. There are no unified theoretical framework for bridging different addition algorithms. By using CTDs, one can describe and analyze arbitrary adder architectures in a systematic way without using any specific knowledge about underlying arithmetic algorithms. Examples of adder architectures that can be handled by CTDs include Redundant-Binary (RB) adders [3]–[5], Signed-Digit (SD) adders [6], carry-save adders, Positive-Digit (PD) adders [7], parallel counters (e.g., 3-2 counters and 4-2 counters [8]) and networks of such adders/counters.

*The author is also with PRESTO, JST, 4-1-8 Honcho Kawaguchi, Saitama, Japan.

In this paper, we focus on an application of CTDs to the analysis of fast addition algorithms with limited carry propagation. As an example, we give a systematic approach for analyzing two-operand RB adders, which are widely used in high-performance arithmetic circuits. The analysis result indicates that there exists possible two types of 3-stage CTDs for the RB adders. From this result, we reveal that the conventional RB adders which are designed independently by experts, can be classified into one of the two types under the adequate binary coding of CTD signals. The proposed analysis technique can be applied not only to two-operand RB adders but also to other types of adders.

2. COUNTER TREE DIAGRAMS

A *Counter Tree Diagram* (CTD) is defined as a digraph $G = (N(G), D(G))$, where $N(G)$ is a set of *counter nodes*, and $D(G)$ is a set of *directed edges* connecting counter nodes. The CTD is considered to be a data structure for representing networks of adders and counters. The counter node in CTDs is an abstraction of addition (or counting) function frequently appeared in arithmetic algorithms. The directed edge, on the other hand, represents operand dependency. Every directed edge is associated with a *weighted interval*, which represents the range of the corresponding operand. A weighted interval $w[a : b]$ is defined as a set of integers given by

$$w[a : b] \stackrel{\text{def}}{=} \{wa, w(a+1), \cdots, wb\}, \qquad (1)$$

where $a, b, w \in \mathbf{Z}$ and $a \leq b$. $\mathbf{Z}$ denotes the set of integers.

In general, addition of two weighted intervals are defined as

$$w_1[a_1 : b_1] + w_2[a_2 : b_2] \stackrel{\text{def}}{=} \{x + y \mid x \in w_1[a_1 : b_1] \text{ and } y \in w_2[a_2 : b_2]\}.$$

If $w_2 = cw_1$ ($c \in \mathbf{Z}$) and $|c| \leq |[a_1 : b_1]|$ (where $|c|$ is the absolute value of c and $|[a_1 : b_1]|$ is the number of elements in the set $[a_1 : b_1]$), then we can represent the result of addition as a *single weighted interval* as

$$w_1[a_1 : b_1] + w_2[a_2 : b_2] = w_1[a_1 + \min(ca_2, cb_2) : b_1 + \max(ca_2, cb_2)].$$

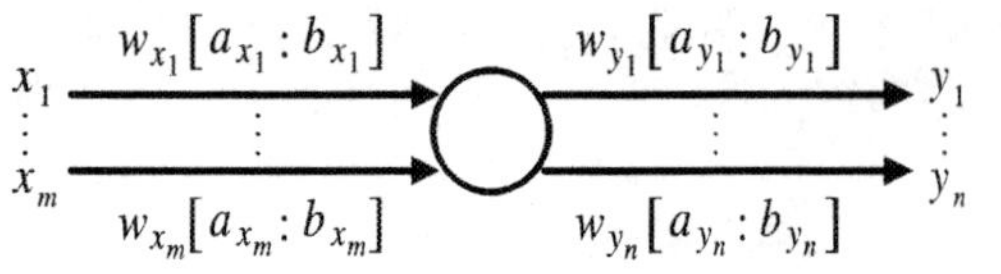

Fig. 1. Symbol of a counter node.

We shall define a *counter node* as an abstraction of addition (or counting) function. Figure 1 shows the symbol of the counter node. An m-input n-output *counter node* is defined as an element that converts m input variables to n output variables, where each variable is defined on a specific weighted interval. Let m input variables be denoted by $x_1, \cdots, x_m$ that are defined on the weighted intervals $w_{x_1}[a_{x_1} : b_{x_1}], \cdots, w_{x_m}[a_{x_m} : b_{x_m}]$, respectively. Let n output variables be denoted by $y_1, \cdots, y_n$ that are defined on the weighted intervals $w_{y_1}[a_{y_1} : b_{y_1}], \cdots, w_{y_n}[a_{y_n} : b_{y_n}]$, respectively. The function of the counter node is defined as

$$\sum_{i=1}^{m} x_i = \sum_{j=1}^{n} y_j, \tag{2}$$

where the weighted intervals associated with input/output variables must satisfy the following inclusion relation (called *counter node condition*):

$$\sum_{i=1}^{m} w_{x_i}[a_{x_i} : b_{x_i}] \subseteq \sum_{j=1}^{n} w_{y_j}[a_{y_j} : b_{y_j}]. \tag{3}$$

Equation (2) indicates that the function of the counter node is to convert a number of input variables to another number of output variables. Also, Equation (3) means that the counter node provides enough output dynamic range for representing all the input combinations.

The counter node can be used to represent arbitrary adder functions based on positional number systems including high-radix and redundant number systems. In order to describe addition algorithms, we need the network representation of counter nodes. For this purpose, we introduce *Counter Tree Diagrams* (CTDs).

For changing the level of abstraction in CTD representation, we define the basic transformations of CTDs: (i) decomposition/ composition of directed edges, and (ii) decomposition/composition of counter nodes. These manipulations preserve the function of the original CTD after transformation. When transforming a given CTD to equivalent CTDs of lower levels of abstraction, we need to examine the counter node condition (3) for every node in the transformed CTDs.

3. ANALYSIS OF TWO-OPERAND REDUNDANT-BINARY ADDERS

In this section, we show an application of CTDs to the analysis of two-operand Redundant-Binary (RB) addition algorithms with limited carry propagation. As a preliminary example, consider the CTD representation of a single-digit RB adder shown in Fig. 2 (a). This CTD does not suggest any information about the maximum length of carry propagation. Applying edge decomposition for carry signals and node decomposition, we have the structure shown in Fig. 2 (b). As for the right-hand counter node in this diagram, there still remains a possibility of $O(N)$ (N is the wordlength) carry propagation. Decomposing this counter node again, we can derive a 3-stage structure shown in Fig. 2 (c). This decomposition suggests an RB parallel adder architecture illustrated in Fig. 2 (d) that can perform constant-time two-operand addition. In this structure, the length of carry propagation is limited to 2 digit positions at most. The corresponding circuit implementation can be easily obtained by implementing every counter node with logic devices. Thus, CTD can be used to analyze RB adder structures without using specific knowledge about the circuit implementation.

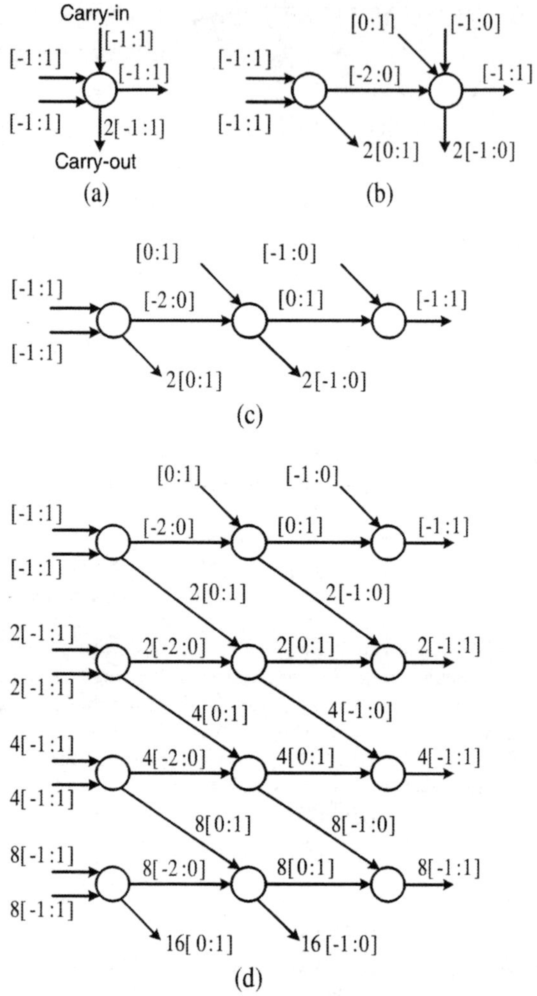

Fig. 2. CTD representation of an RB adder : (a) 1-digit RB adder, (b) 1-digit RB adder after edge decomposition and the first node decomposition, (c) 1-digit RB adder after the second node decomposition, and (d) overall architecture based on the 3-stage decomposition.

A major problem in the above discussion is that the decomposition is somewhat ad hoc and might be criticized for not being well-grounded. In the following, we give more systematic approach for the analysis of RB adders with limited carry propagation.

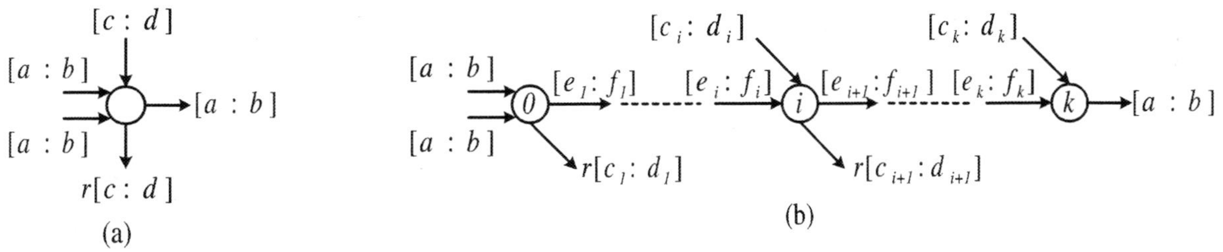

Fig. 3. CTD representation of an adder with the radix r and the digit set $[a : b]$: (a) 1-digit adder (general form), and (b) (k+1)-stage decomposition of the 1-digit adder.

Let us consider the positional number system defined by the radix r (≥ 2) and the integer digit set $[a : b] = \{a, a+1, a+2, \cdots, b\}$, where a and b are integers satisfying $a < b$. The single-digit two-operand adder using this number system can be represented by the CTD shown in Fig. 3 (a). Assume that this adder can be decomposed into the $(k+1)$-stage structure as shown in Fig. 3 (b), where the length of carry propagation is limited to k (≥ 1) digits at most. In the decomposed CTD, c_i, d_i, e_i, f_i ($i \in \{1, 2, \cdots, k\}$) are unknown integer parameters. These parameters must be selected so as to satisfy the counter node condition given by (3) for all the counter nodes in the CTD. Thus, we have

$$\begin{cases} [a:b]+[a:b] & \subseteq \; r\,[c_1:d_1]+[e_1:f_1], \\ [c_i:d_i]+[e_i:f_i] & \subseteq \; r\,[c_{i+1}:d_{i+1}]+[e_{i+1}:f_{i+1}], \\ & \quad\; (i \in \{1,2,\cdots,k-1\}) \\ [c_k:d_k]+[e_k:f_k] & \subseteq \; [a:b], \end{cases} \tag{4}$$

where the second inclusion relations must be omitted when $k = 1$.

For simplicity, we assume that the output of every counter node can be represented by a single weighted interval, which is reasonable assumption for practical design of parallel counters. This requires the following condition:

$$r - 1 \leq f_i - e_i, \tag{5}$$

for all i ($\in \{1, 2, \cdots, k\}$).

If the parameters, c_i, d_i, e_i, f_i ($i \in \{1, 2, \cdots, k\}$) do not have solutions, we can conclude that the $(k+1)$-stage decomposition given by Fig. 3 (b) is not possible for this adder. Thus, mathematical consideration of counter node conditions makes possible the systematic analysis of adder architectures for the given number system.

Assume that $r = 2$, $a = -1$ and $b = 1$. This corresponds to RB number system. In the case of $k = 2$, the above conditions (4) and (5) can be summarized as a set of inequalities:

$$\begin{cases} 2 & \leq & 2 \cdot d_1 + f_1, \\ -2 & \geq & 2 \cdot c_1 + e_1, \\ d_1 + f_1 & \leq & 2 \cdot d_2 + f_2, \\ c_1 + e_1 & \geq & 2 \cdot c_2 + e_2, \\ d_2 + f_2 & \leq & 1, \\ c_2 + e_2 & \geq & -1, \\ 1 & \leq & f_1 - e_1, \\ 1 & \leq & f_2 - e_2. \end{cases} \tag{6}$$

Solving the inequalities (6), we derive the two types of 3-stage architectures as shown in Fig. 4, where X, Y, Z, U, V, C_1 and C_2 denote CTD variables associated with the closest weighted intervals.

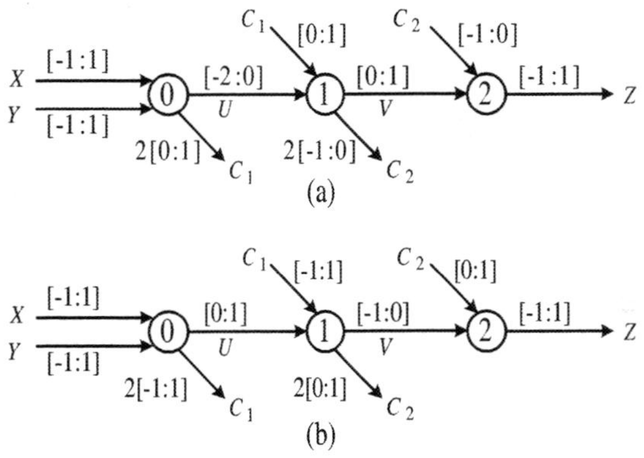

Fig. 4. Two types of 3-stage architectures for the RB adder: (a) Type I, and (b) Type II.

From the analysis result, we have confirmed that the well-known circuit implementations of RB adders[3]–[5] are classified into Type-I architecture under adequate binary coding of CTD variables. Figures 5–7 illustrate the RB adder implementations and their CTD interpretations, where x_i, y_i, z_i and u_i denote the ith bit of X, Y, Z and U, respectively, v, c_1 and c_2 denote the bit of V, C_1 and C_2, respectively. These three implementations are interpreted as Type-I architecture under the binary coding of CTD signals defined by Table 1–3, respectively. In Table 1 and 2, $*$ denotes either a 0 or a 1. Note that such RB adders are designed independently by expert designers. Thus, the use of CTDs provides a unified view of carry propagation free adders using RB number representation.

4. CONCLUSION

This paper has presented a unified representation of fast addition algorithms called Counter Tree Diagrams (CTDs) and its application to the analysis of two-operand Redundant-Binary adders with limited carry propagation. The analysis result shows that there are possible two types of 3-stage CTDs for RB adders. In addition, we revealed that the conventional implementations of two-operand RB adders are regarded as special instances within a specific type of CTDs. The CTD-based analysis will be also useful for other types of adder specifications.

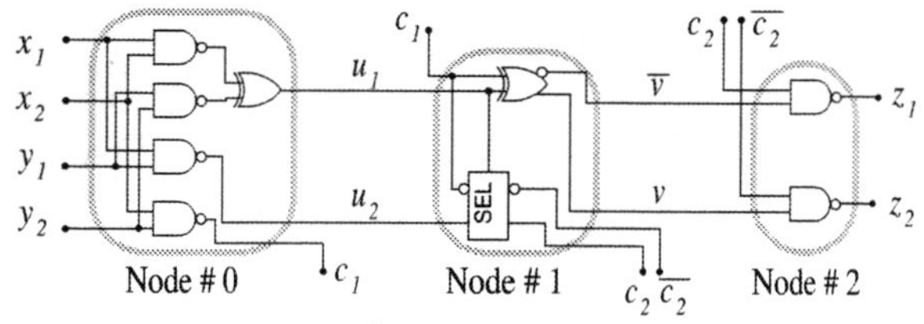

Fig. 5. RB adder implementation from [3].

Table 1. Binary coding of signals shown in Fig. 5.

$X/Y/Z$	$x_1x_2/y_1y_2/z_1z_2$
1	01
0	11
-1	10

U	u_1u_2
0	01
-1	1∗
-2	00

V	v
1	0
0	1

C_1	c_1
1	0
0	1

C_2	c_2
0	1
-1	0

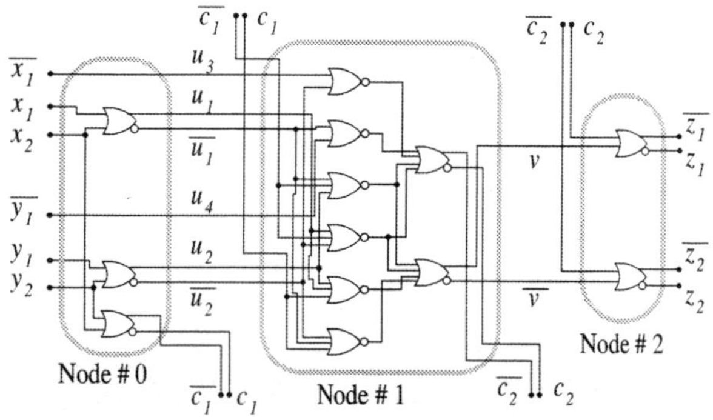

Fig. 6. RB adder implementation from [4].

Table 2. Binary coding of signals shown in Fig. 6.

$X/Y/Z$	$x_1x_2/y_1y_2/z_1z_2$
1	10
0	00
-1	01

U	$u_1u_2u_3u_4$
0	110∗ or 1110
-1	10∗1 or 011∗
-2	0011 or 1111

V	v
1	0
0	1

C_1	c_1
1	1
0	0

C_2	c_2
0	0
-1	1

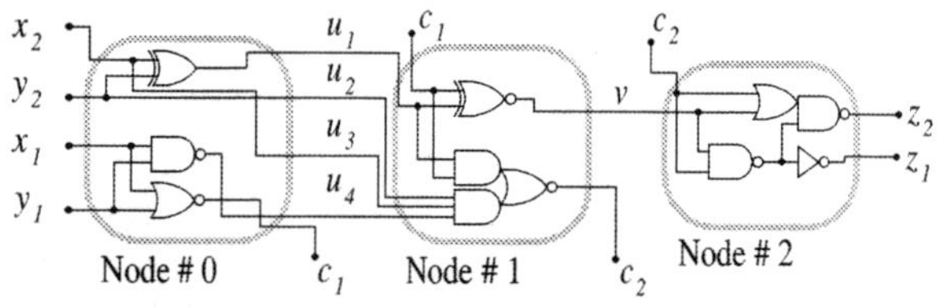

Fig. 7. RB adder implementation from [5].

Table 3. Binary coding of signals shown in Fig. 7.

$X/Y/Z$	$x_1x_2/y_1y_2/z_1z_2$
1	01
0	00
-1	11

U	$u_1u_2u_3u_4$
0	0111
-1	1101 or 1011
-2	0110 or 0001

V	v
1	0
0	1

C_1	c_1
1	1
0	0

C_2	c_2
0	0
-1	1

As described in this paper, a decomposed (low-level) CTD can be mapped onto physical logic devices directly. Assuming a specific circuit technology, we could optimize the performance of adders by changing the type of CTD decomposition as well as the way of encoding the CTD variables with binary signals. The CTD-based optimization of parallel adder circuits is being left for future work.

5. REFERENCES

[1] I. Koren, *Computer arithmetic algorithms 2nd Edition*, A K Peters, 2001.

[2] J. Sakiyama, N. Homma, T. Aoki, and T. Higuchi, "Counter Tree Diagrams: A unified framework for analyzing fast addition algorithms," *IEICE Trans. Fundamentals*, vol. E86-A, no. 12, Dec. 2003 (to be published).

[3] H. Makino, Y. Nakase, H. Suzuki, H. Morinaka, H. Shinohara, and K. Mashiko, "An 8.8-ns 54×54-bit multiplier with high speed redundant binary architecture," *IEEE J. Solid-State Circuits*, vol. 31, no. 6, pp. 773 – 783, June 1996.

[4] N. Takagi, H. Yasuura, and S. Yajima, "High-speed VLSI multiplication algorithm with a redundant binary addition tree," *IEEE Trans. Computers*, vol. 34, no. 9, pp. 789 – 796, Sept. 1985.

[5] S. Kuninobu, T. Nishiyama, and T. Taniguchi, "High speed mos multiplier and divider using redundant binary representation and their implementation in a microprocessor," *IEICE Trans. Electron.*, vol. E76-C, no. 3, pp. 436 – 445, Mar. 1993.

[6] S. Kawahito, M. Kameyama, T. Higuchi, and H. Yamada, "A 32×32-bit multiplier using multiple-valued MOS current-mode circuits," *IEEE J. Solid-State Circuits*, vol. 23, no. 1, pp. 124 – 132, Feb. 1988.

[7] S. Kawahito, M. Ishida, T. Nakamura, M. Kameyama, and T. Higuchi, "High-speed area-efficient multiplier design using multiple-valued current-mode circuits," *IEEE Trans. Computers*, vol. 43, no. 1, pp. 34 – 42, Jan. 1994.

[8] N. Ohkubo, M. Suzuki, T. Shinbo, T. Yamanaka, A. Shimizu, K. Sasaki, and Y. Nakagome, "A 4.4 ns CMOS 54×54-b multiplier using pass-transistor multiplexer," *IEEE J. Solid-State Circuits*, vol. 30, no. 3, pp. 251 – 257, Mar. 1995.

HWP: A NEW INSIGHT INTO CANONICAL SIGNED DIGIT

Fei Xu, Chip-Hong Chang and Ching-Chuen Jong

Centre for Integrated Circuits and Systems, Nanyang Technological University
School of EEE, Nanyang Avenue, Singapore 639798

ABSTRACT

A new Hamming Weight Pyramid (HWP) that resembles the Pascal triangle is proposed to succinctly compress the information about the distribution of the hamming Weight in Canonical Signed Digit (CSD) represented numbers in a visually appealing manner for analysis and synthesis. Many interesting properties are discovered in this regularly structured HWP. These properties lead to a novel and elegant way to convert decimal number to CSD representation. The proposed method avoids the need to convert the decimal numbers to their binary equivalence, which is an ineluctable intermediate process in the conventional decimal to CSD conversion algorithms.

1. INTRODUCTION

A signed digit representation of radix r is an infinite sequence of digits $a = (\ldots, a_2, a_1, a_0)$ with $a_i \in \{0, \pm1, \pm2, \pm(r-1)\}$. Signed digit representations of radix 2 are widely used for digital arithmetic operations, such as Booth coding and modified Booth coding [4, 8]. A signed digit representation is said to be minimum if it possesses the minimal hamming weight. Minimal signed digit representations [2] are crucial in reducing the hardware cost and power consumption in many complex arithmetic operations. The canonical signed digit (CSD) form, introduced in [8, 10], and the generalized non-adjacent form (GNAF) defined in [2, 3], are minimal signed digit representations for radix 2 and general radix $r>2$, respectively.

Any arbitrary N-bit 2's complement number can be represented in CSD form with no more than $(N+1)/2$ nonzero digits. It was shown in [10] that the expected number of nonzero digits in an N-digit CSD number tends asymptotically to $N/3 + 1/9$. On average, CSD uses 33% fewer nonzero digits than the binary number. To generate a CSD number, usually the decimal number needs to be converted to the binary form first. The traditional approach [2] to generate the CSD number from the binary form uses LUT.

CSD representation has not been widely used in hardware implementation of complex arithmetic operations due to the conversion overhead from the accustomed binary number system into CSD form. Recently, some new approaches based on look-ahead circuits [4, 7] and different derivations of CSD [1, 6] have been reported to circumvent the conversion problem. CSD representation also plays an important role in design automation algorithms leading to quality solutions to Multiple Constant Multiplication (MCM) problems and efficient digital filter design [5, 9]. As the constant coefficients are normally derived in decimal (integer) form, it would be useful to interpret the CSD number and its conversion directly from the decimal form.

In this paper, we provide a new insight into the formation of the CSD numbers. We derive a unique regular structure of the hamming weights of the CSD number sequence and name it as the Hamming Weight Pyramid (HWP). Many useful and attractive properties are extracted from HWP. A reduced HWP is also proposed. It can be constructed easily from a reduced sized lookup table. From the properties of the HWP, a new method to convert decimal number to its CSD format is derived with the computation complexity compatible to that of the conversion from decimal number to binary number. Direct conversion from decimal number to CSD form without the intermediate binary number makes it attractive as preprocessing step for many design automation algorithms that can be benefited from the substitution of integer operations by their CSD forms.

2. THE HAMMING WEIGHT PYRAMID (HWP)

To reduce the implementation cost of compound arithmetic operations composing of additions, subtractions and shifts in conventional binary computing machinery, the 2^{n+1} distinct integers, X in the range of $[-2^n, 2^n)$ are generally represented with signed digit (SD) representation as follows [10]:

$$X = \sum_{i=0}^{n} a_i 2^i, \quad a_i \in \{0, \pm1\} \tag{1}$$

The hamming weight of the SD is defined as the number of nonzero digits, a_i in (1). A canonical signed digit (CSD) representation is a unique SD representation that exhibits the property that no two consecutive digits are both nonzero [10], i.e,

$$a_i a_{i-1} = 0 \ \forall i \in [0, n] \tag{2}$$

Since negative integer in CSD can be trivially obtained from its positive counterpart by changing the signs of all the nonzero digits, it is sufficient to consider only the CSDs of natural numbers without loss of generality. An interesting symmetrical two-dimension ordered sequence is observed on the hamming weights of the CSD if we recursively decompose the CSD into a leading nonzero digit, a_L and a smaller CSD, as follows:

$$X = a_L + \sum_{i=0}^{L-1} a_i 2^i \tag{3}$$

Using first term of (3) as pivot and the second term as offset, the natural number sequence, 1, 2, 3, ... in CSD representation can be ordered in an interesting pyramid structure. Fig. 1 shows the pyramid of the hamming weights of CSD numbers, $2^i + j$, where i and j correspond to the row number and offset from the center column of the pyramid, respectively. The center column is numbered 0 and the column numbers are increasingly more negative (positive) towards its left (right). By reading the weights in a top-down, left-right order, they correspond to the CSDs of natural number sequence, 1, 2, 3, 4, ... All the corresponding CSD numbers in row i have the same leading '1' given by the corresponding number in the center column. For example, the corresponding CSD numbers in row 3 ($i = 3$) from left ($j = -2$) to right ($j = 2$) are $10\bar{1}0$, $100\bar{1}$, 1000, 1001 and 1010. All hamming weights in the same column correspond to the CSD represented numbers with their offset terms in (3) differ only in the leading nonzero digits, i.e., their offset terms upon further decomposition by (3) are identical. For example, the corresponding CSD numbers in column $j = -2$ begin from row 3 are $10\bar{1}0$, $100\bar{1}0$, $1000\bar{1}0$ and so on.

	-10	-9	-8	-7	-6	-5	-4	-3	-2	-1	0	1	2	3	4	5	6	7	8	9	10
0											1										
1											1										
2										2	1	2									
3									2	2	1	2	2								
4						3	2	3	2	2	1	2	2	3	2	3					
5	3	3	2	3	3	3	2	3	2	2	1	2	2	3	2	3	3	3	2	3	3
6	...	...	...	...	...	...	...	...	...	...	...	...	...	...	...	...	...	...	...	...	...

Figure 1. Hamming weight pyramid (HWP)

Since the least significant digits (LSDs) of the CSD numbers correspond to even numbered columns are identically zero, their hamming weights can be deduced from their corresponding odd CSD numbers obtained by shifting out the trailing zero digits. By omitting the even numbered columns, an Odd HWP (OHWP) for the hamming weights of CSD represented odd natural number sequence 3, 5, 7, 9 ... is obtained in Fig. 2.

2 2
2 2
3 3 2 2 3 3
3 3 3 3 2 2 3 3 3 3
4 4 3 3 4 4 3 3 3 3 2 2 3 3 3 3 4 4 3 3 4 4

Figure 2. The OHWP.

3. PROPERTIES OF HWP

The CSD represented numbers, 2^i with the minimum hamming weight of 1 form the middle pillar of the HWP from which equal number of entries are extended. The leftmost entry of any arbitrary row represents the hamming weight of a CSD number adjacent to one corresponding to the rightmost entry of the row above it. This formation has led to several interesting properties.

Property 1: The CSD numbers correspond to row i of HWP are the range of natural numbers that can be represented with i+1 digits and has a leading '1' in their CSD representation.

Property 2: The number of entries in row i ($i > 1$) of HWP, denoted as $N(i)$ can be described by any of the following recursive equations:

$$N(i) = N(i-1) + 2N(i-2) \tag{4}$$

$$\text{or} \quad N(i) = 2^i - N(i-1) \tag{5}$$

where $N(0) = N(1) = 1$. By unrolling the recursion, we have,

$$N(i) = \sum_{r=0}^{i} (-1)^r 2^{i-r} \tag{6}$$

Property 3: The total number of entries up to and including row i ($i \geq 0$), $S(i)$ is given by:

$$S(i) = \sum_{r=0}^{i} N(r) = \begin{cases} 2N(i) & \text{if } i \text{ is odd} \\ 2N(i) - 1 & \text{if } i \text{ is even} \end{cases} \tag{7}$$

Table 1 shows the values of $N(i)$ and $S(i)$ for the first 10 rows of the HWP.

i	0	1	2	3	4	5	6	7	8	9
$N(i)$	1	1	3	5	11	21	43	85	171	341
$S(i)$	1	2	5	10	21	42	85	170	341	682

Table 1: $N(i)$ and $S(i)$ of HWP

It is observed that the entries in row i of the HWP are entirely copied into the corresponding columns in row i+1. We define the core, $\boldsymbol{C}(i)$ of row i as a vector of elements that are copied from row i–1. The remaining elements to the left and right of $\boldsymbol{C}(i)$ are called the left and right wings, respectively. $\boldsymbol{W}_L(i)$ and $\boldsymbol{W}_R(i)$ are the vectors used to denote the order list of elements in the left and right wings, respectively. Based on the above nomenclatures, the following properties are stated:

Property 4: The cardinality of $\boldsymbol{C}(i)$ is given by:

$$|\boldsymbol{C}(i)| = N(i-1) = \sum_{r=0}^{i-1} (-1)^r 2^{i-r-1} \tag{8}$$

Property 5: The cardinality of $\boldsymbol{W}_L(i)$ or $\boldsymbol{W}_R(i)$ is given by:

$$|\boldsymbol{W}_L(i)| = |\boldsymbol{W}_R(i)| = \frac{N(i) - N(i-1)}{2} = N(i-2) = \sum_{r=0}^{i-2} (-1)^r 2^{i-r-2} \tag{9}$$

Property 6: For all $i \geq 2$, we have

$$\boldsymbol{W}_L(i) = \boldsymbol{W}_R(i) = \boldsymbol{C}(i-2) + \boldsymbol{I}_{|\boldsymbol{C}(i-2)|} \tag{10}$$

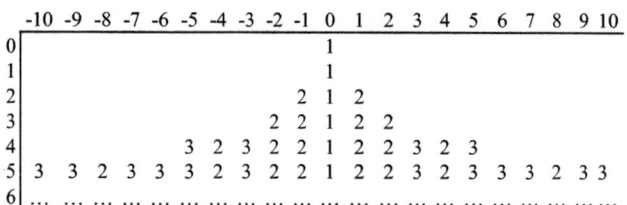

where I_n is a $n\times1$ vector whose elements are all '1's.

Based on the above properties, the rows of the HWP can be progressively generated after initializing the first two trivial rows to 1s. Fig. 3 illustrates the construction of row 4 of the HWP, where the cores and wings are represented by dotted and solid blocks, repectively.

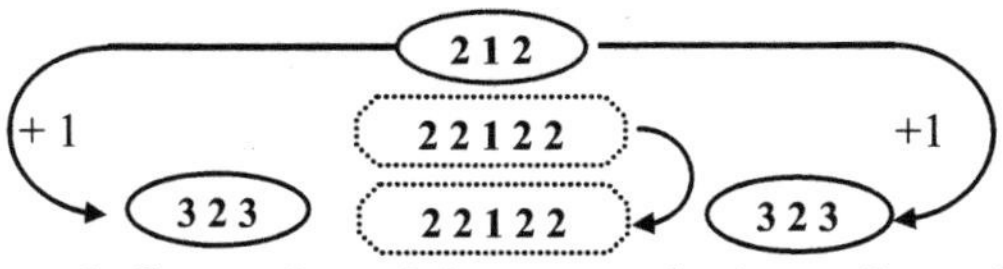

Figure 3. Generation of the core and wings of row 4.

Due to the symmetry, the number of entries towards either sides of the central element in a row is trivially given by $\frac{N(i)-1}{2}$.

Property 7: Let $h(i, j)$ denote the hamming weight appeared in row i, column j of the HWP, then

$$h(i,j)=\begin{cases} h(i-2,j)+1 & for\ i\geq 2\ and\ \frac{N(i-1)+1}{2}\leq |j|\leq \frac{N(i)-1}{2} \\ h(i-1,j) & for\ i\geq 2\ and\ 0\leq |j|\leq \frac{N(i-1)-1}{2} \end{cases}$$

and $h(0, 0) = h(1, 0) = 1$ (11)

By virtue of the above properties and the symmetry of the HWP, it is sufficient to store only the elements of $\boldsymbol{W_L}$ or $\boldsymbol{W_R}$. With proper indexing to the vectors, the information embedded in the HWP can be encoded succinctly. This reduces the storage of the HWP to less than a quarter of that required if the entries are sequentially stored along with the row and column indices. Table 2 illustrates the compressed format used to store a HWP up to row six. The elements in $\boldsymbol{W_L}(i)$ for i = 2, 3, ... are stored sequentially in the second row of Table 2. The numbers in the first row are the indices, i to the first element of $\boldsymbol{W_L}(i)$ listed in the second row. The third row is added to indicate the starting column numbers of $\boldsymbol{W_L}(i)$.

2	3	4			5					6										
2	2	3	2	3	3	3	2	3	3	4	3	4	3	3	2	3	3	4	3	4
-1	-2	-5			-10					-21										

Table 2. Memory organization of the HWP

Property 8: Let $csd(i, j)$ denote the CSD number corresponding to the hamming weight, $h(i, j)$ in the HWP. Then

$csd(i,j) << n = csd(i+n, 2^n\times j)$ (12)

where "<< n" means shifting the operand by n bits to the left.

Property 8 follows directly from multiplying both sides of (3) by 2^n. Multiplying the pivoting term by 2^n is the same as descending the entry in the HWP by n rows down the centre column while multiplying the offset term by 2^n is equivalent to shifting it outwards by 2^n times. Using Property 8, the CSD numbers differ only in the number of trailing zeros are readily identified. It is worth noting the traces of shifting the CSD numbers in the HWP. The sequences of shifts for ($10\bar{1}$, $10\bar{1}0$, $10\bar{1}00$, $10\bar{1}000$, $10\bar{1}0000$) and ($1010\bar{1}$, $1010\bar{1}0$, $1010\bar{1}00$) are marked in Fig. 4.

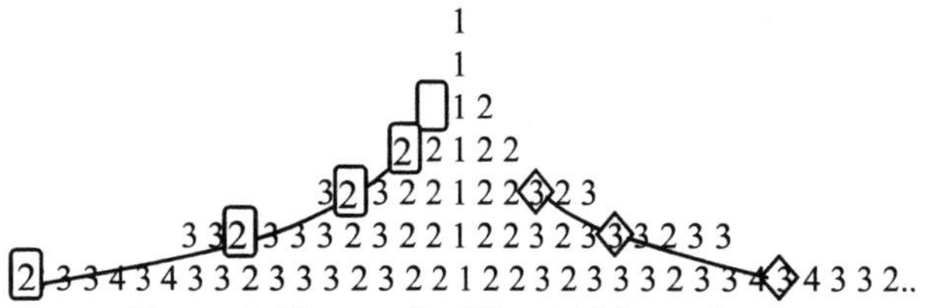
Figure 4. Trace of shifting CSD numbers.

Property 8 also provides the basis for the reduction of the HWP as the eliminated CSD entries in the reduced HWP can be retrieved from the shifting property. It should be noted that Properties 2 to 7 are still valid for the OHWP. In addition, the following property is unique to the OHWP.

Property 9: The entries in the OHWP appear in duplets of adjacent identical numbers. The corresponding CSDs of the duplet differ only in the signs of their least significant digits.

The duality property of the OHWP is highlighted in Fig. 5 by enclosing the duplets in brackets. The first pair of hamming weights corresponds to the CSD numbers $10\bar{1}$ and 101.

(2 2)
(2 2)
(3 3) (2 2) (3 3)
(3 3) (3 3) (2 2) (3 3) (3 3)
(4 4) (3 3) (4 4) (3 3) (3 3) (2 2) (3 3) (3 3) (4 4) (3 3) (4 4)

Figure 5. Duality property of OHWP.

If each duplet of hamming weights is replaced by a single number, the result is an interesting add-one HWP structure as shown in Fig. 6 which is identical to the original HWP with every entry incremented by one.

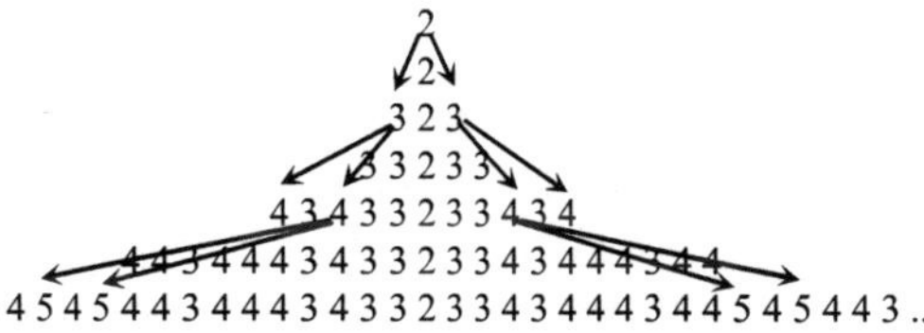
Figure 6. An add-one HWP generated from OHWP.

By subtracting one from every entries of Fig. 6, we obtained the structure of the original HWP. However, some information on the original CSD numbers is lost in this scale translation. In this case, the two least significant digits, $0\bar{1}$ and 01 are pruned. For example, when we replace the duplets of $10\bar{1}$ and 101 in Fig. 5 by a single hamming weight of 2 in the first entry of Fig. 6, we do not distinguish the difference between the least significant digits '01' and '$0\bar{1}$'. If these two digits are shifted out (hence reducing their hamming weights by 1), then the two CSD numbers are identically reduced to 1, which correspond to the first entry in the original HWP and has a hamming weight of 2–1 = 1. In the above example, $10\bar{1}$ and 101 are called the descendents and their shifted version, 1 is called the ancestor. Some ancestor-descendents relationships are marked on the add-one HWP in Fig. 6 with the arrows pointed from the ancestors to the descendents.

The above ancestor-descendents relationship is best expressed in the form of a ternary tree where each parent has three children. The left, middle and right children are obtained from their parent by appending to it a '$0\bar{1}$', '0' and '01', respectively. The ternary tree to generate the CSD numbers iteratively is shown in Fig. 7.

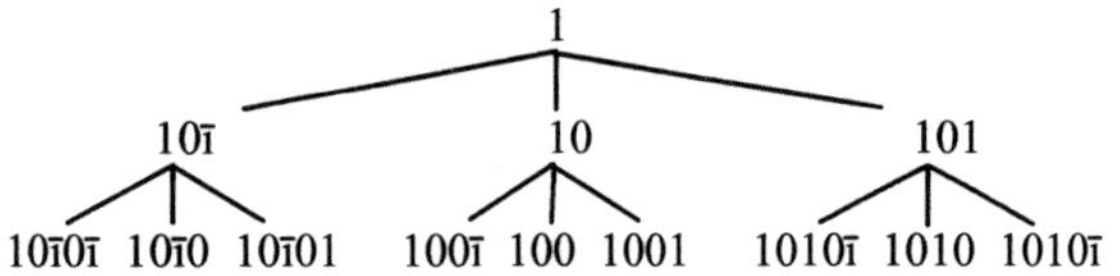

Figure 7. Ternary tree of CSD numbers

4. DECIMAL TO CSD CONVERSION

The HWP provides a new perspective to the synthesis of CSD numbers. With the help of Property 3 and (3), we derive the following algorithm for the conversion of a decimal number, d to its CSD representation, *csd*. The procedure is given as follows:

Step 1. Initialize all digits of *csd* to 0 and $sign = 1$.

Step 2. Search for the row index i such that $S(i-1) < d \leq S(i)$. Set the i-th digit of *csd* to *sign* and $d = d - 2^i$.

Step 3. If $d < 0$, set $d = -d$ and $sign = -sign$.

Step 4: Repeat Step 2 to 3 until $d = 0$.

The above computation requires h iterations to complete where h is the hamming weight of *csd*. Thus the average number of iterations for an N-digit CSD number is approximately $N/3$. If a lookup table is used to search for the row index in Step 2, the size of the table is proportional to the word length, as observed from Table 1 and Property 3. The search range is progressively narrow down as the offset term becomes smaller. The worst case computational complexity of the above conversion algorithm is thus bound by $O(N^2)$. The above algorithm can be easily extended to the conversion from 2's complement number and made amendable to hardware implementation.

As an example, the traces of converting the decimal number 45 to its CSD form $10\bar{1}0\bar{1}01$ are shown in Fig. 8. Let csd_i denote the i-th digit of the final CSD number. In the first iteration, $i = 6$, $csd_6 = sign = 1$ and $d = 45 - 64 = -19$. Since $d < 0$, $sign = -1$ and $d = 19$. In the second iteration, $i = 4$, $csd_4 = \bar{1}$ and $d = 19 - 16 = 3$. In the third iteration, $i = 2$, $csd_2 = \bar{1}$ and $d = 3 - 4 = -1$. Since $d < 0$, $sign = -sign = 1$ and $d = 1$. In the final iteration, $i = 0$, $csd_0 = 1$ and $d = 1 - 1 = 0$.

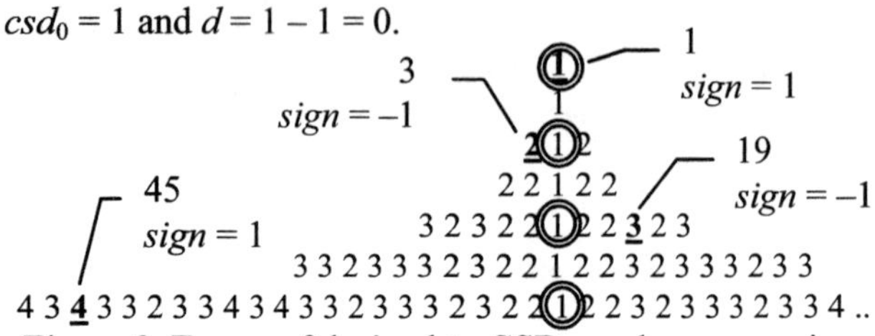

Figure 8. Traces of decimal to CSD number conversion.

5. CONCLUSION

A regularly structured Hamming Weight Pyramid is proposed which uncovers many interesting properties and essential relationship between natural numbers and their hamming weights in CSD representaion. A new method of converting decimal number to its CSD form is also derived. It is envisaged that applications that employ CSD form can be benefited from further exploitation of the derived properties. For example, efficient algorithms for maximizing the sharing of common subexpressions in multiple-constant-multiplication problems [5, 9] or ternary value finite automata [8] can be spawn from locating the constant coefficients and through direct manipulation on the proposed HWP.

6. REFERENCES

[1] D. A. Akopian, O. Vainio, S. S. Agaian, and J. T. Astola, "SBNR processor for stack filters," *IEEE Trans. on Circuits and Syst. – II*, vol. 44, no. 3, pp. 197-208, Mar. 1997.

[2] S. Arno and F. S. Wheeler, "Singed digit representations of minimal hamming weight," *IEEE Trans. on Computers*, vol. 42, no. 8, pp. 1007-1010, Aug. 1993

[3] W. E. Clark and J. J. Liang, "On arithmetic weight for a general radix representation of integers," *IEEE Trans. Information Theory*, vol. IT-19, no. 6, pp. 823-826, 1973.

[4] S. K. Das and M. C. Pinotti, "Fast VLSI circuits for CSD coding and GNAF coding," *Electronics Letters*, vol. 32, no. 7, pp. 632–634, Mar. 1996.

[5] R. I. Hartley, "Subexpression sharing in filters using canonic signed digit multipliers," *IEEE Trans. Circuits Syst. II*, vol. 43, no. 10, pp. 677-688, Oct. 1996.

[6] R. Hashemian, "A new method for conversion of a 2's complement to canonic signed digit number system and its representation," in *Proc. Int. Conf. On Signals, Systems and Computers*, vol. 2, pp. 904 -907 Nov. 1996.

[7] A. Herrfeld and S. Hentschke, "Look-ahead circuit for CSD-code carry determination," *Electronics Letters*, vol. 31, no. 6, pp. 434–435, Mar. 1995.

[8] K. Hwang, *Computer Arithmetic*, John Wiley & Sons, New York, 1979.

[9] M. Martínez-peiró, E. I. Boemo, and L. Wanhammar, "Design of high-Speed multiplierless filters using a nonrecursive signed common subexpression algorithm," *IEEE Trans. Circuit Syst.*, vol. 49, no. 3, pp. 196-203, Mar. 2002.

[10] R. W. Reitwiesner, "Binary arithmetic," in *Advances in Computers*, vol. 1, pp. 231-308, New York: Academic, 1966.

FAST MULTILEVEL FLOORPLANNING FOR LARGE SCALE MODULES

Ching-Chung Hu[1], De-Sheng Chen[2], Yi-Wen Wang[3]

Department of Information Engineering and Computer Science, Feng Chia University, Taichung, Taiwan
m9107876@knight.fcu.edu.tw[1], dschen@iecs.fcu.edu.tw[2], ywang@fcu.edu.tw[3]

ABSTRACT

With the advance of deep sub-micron, current methods are not effective to obtain acceptable layout for large scale modules. Hence, it is important to provide designers of SOC with a powerful floorplanner. In traditional approaches, it is common to simultaneously utilize clustering and declustering technologies, i.e. multiple phases to refine the solution quality. In this paper, we propose a top-down multilevel genetic floorplanning algorithm to handle the floorplanning and packing for large scale modules. The algorithm is simple and only needs the clustering phase. Experimental results show significantly better running time and promising solutions in comparison with other state-of-the-art research works.

1. INTRODUCTION

With the advance of VLSI technology, IC density and design complexity are growing at an incredible speed. Besides, the development of system on a chip (SOC) and IP reuse make number of modules in a design more numerous than previous times. Even though the difficulty of design is extremely augmented, the swifter and sterner time to market is still requested, in order to satisfy the demand of customers. Therefore, CAD tools which can efficiently deal with large scale design are awfully needed.

In the last twenty years, many floorplanners [1, 3, 4, 6] only coped with small designs or benchmarks. For example, the largest floorplan benchmark of MCNC provided is ami49 which only have forty nine modules and it is too fractional to compare with modern design. The execution time would sprout at an exponential speed when those floorplanners conduct large scale designs, hundreds or thousands modules for instance. Not only speed, but also the solution quality of those floorplanners would decrease and be unacceptable. In order to straighten out large scale problem, the concept of multilevel and hierarchical approach is used.

There were already many fields that manipulated multilevel or hierarchical approach to disentangle the large scale issue in VLSI years ago. For example, in graph partition, there were V-cycle multilevel partitioning in hMETIS[8] and top-down multilevel partitioning method in MLPR[9]. In cell placement, there were top-down hierarchical approach in Dragon [10] and V-cycle multilevel placement in mPL [7]. In floorplanning, a multilevel floorplanning algorithm extended from B*-tree [6], called MB*-tree [2], was recently proposed to tackle the problem of large scale designs. Thus, it is a sensible tendency to confront large scale issues with the notion of hierarchical or multilevel.

In this paper, a floorplanning algorithm with the concept of multilevel is proposed. It is based on a *genetic floorplan algorithm* (GFA) from [4] and thus called *multilevel genetic floorplan algorithm* (MLGFA). Unlike the V-cycle multilevel algorithms which usually have two phases, clustering and declustering, MLGFA only needs the clustering phase. Due to this one phase property, MLGFA is fast and easy to implement. Moreover, experimental results show that MLGFA scales very sell with the increasing circuit size, and the obtained result is comparable with the best known two-phase multilevel approach MB*-tree. Also, there is a significant runtime improvement over MB*-tree.

The rest of this paper is organized as follows. Section 2 gives the preliminaries. Section 3 introduces our multilevel genetic floorplan algorithm. Section 4 shows the experimental results and section 5 is the conclusion.

2. PRELIMINARIES

Consider a set of n rectangular modules $M=\{m_1,m_2,\ldots,m_n\}$. Each module $m_i \in M$ has two elements, (w_i,h_i) , where w_i and h_i denote the width and height of m_i. A floorplan $F = \{(x_i , y_i) \mid m_i \in M\}$is an assignment of n rectangular modules' coordinate locations and there is no overlap between any two modules. The objective of floorplanning is to minimize a particular cost function such as total area.

Genetic Algorithm is a stochastic optimization search algorithm, which is proposed by Holland [11], based on the mechanism of natural selection and evolutionary genetics. Genetic algorithm traverses whole solution space utilizing a *population* consisting of a set of *chromosomes*. A *generation* is a completion of *crossover*, *mutation*, and *selection*. In every new generation, a set of offsprings replaces the ancestors who are not suitable to survive.

3. THE MULTILEVEL GENETIC FLOORPLAN ALGORITHM

Each level of the proposed Multilevel Genetic Floorplan Algorithm (MLGFA) consists of three steps, clustering, topology determination, and super modules insertion. First of all, modules are selected from a sorted module list to form clusters. Then, slicing tree topology of the clusters, which correspond to super modules, are determined by the genetic floorplanner GFA. After that, the obtained super modules are inserted back to the module list. With the level of MLGFA lower and lower, the size of module list would rapidly be reduced. When it reaches a threshold, the final floorplan of the module list would be generated and the whole process is completed.

3.1 Useful Heuristics

Different from other multilevel framework [2, 7, 8], the concept of matching ratio from [5] is used here to choose only partial number of modules in the module list to form clusters. The usage of matching ratio had been proven usefully in top-down multilevel partitioning. Besides, a list of modules sorted by width is maintained during the whole process. A special property to the sorted list is that the width of each module is no less than its height. Therefore, if a module or a super module has height larger than its width, it has to be rotated before doing the insertion. With this heuristic, the selected clusters could be easily handled by the genetic floorplanner to get satisfactory solutions in a shot time.

3.2 The Algorithm

There are three parameters needed to be elaborated before explaining the pseudo code of MLGFA. First, nc is the *number of modules per cluster*. If nc is too large, running time of the floorplanner would rapidly increase. But if nc is too small, the solution space would be too small to obtain satisfying solutions. Hence, in order to reduce computational complexity and gain acceptable solutions, a suitable nc must be chosen for the floorplanner. The second is the *matching ratio* mr, $0 \leq mr \leq 1$. The number of modules to participate in clustering at each level would be the size of module list multiply by mr. According to [5], mr is suggested to be 1/3. The third is threshold T which is used to decide the entry condition of final level. When the size of module list is smaller than threshold value T, MLGFA would run the floorplanner for the last time and produce the final floorplan.

Figure 1 is the overall pseudo code of MLGFA. The input are all modules of the circuit and three parameters. The output is a floorplan. At line 1, MLGFA sorts all modules by their width. Line 5 is to decide the number of modules that would be kept in the module list at the end of level i, denoted as NLi. At line 7, select_cluster() is used to select nc modules to form a cluster CM and this function is described in more detail in Figure 2. At line 8, the floorplanner GFA is fed with CM and returns a sub-floorplan f_M. f_M would be added to a temporary module set S at line 9. When inner loop, line 6 to 10, terminates, modules in S are inserted into module list at line 11. In general, the final run of GFA at line 13 has a huge impact on the final floorplan generation, therefore it is worthy of spending more time on it.

Figure 2 shows the procedure of select_cluster(). The inputs are ML and nc, and the output is a set of modules CM. Each time the function is called, the size of ML would be reduced by nc. Line 1 and 2 indicate that the first nc modules from ML are selected and put into CM. Then, the selected modules are removed from ML and CM is returned.

```
Algorithm : MLGFA
Input  : M = All modules of this circuit
         T = A threshold to decide final level
         mr = Matching ratio
         nc = Number of modules per cluster
Output : F = A final floorplan
1. ML = a sorted modules list on M by their width
2. while |ML| > T
3. {
4.    S = φ;
5.    NLi = |ML|*(1 - mr);
6.    while |ML| > NLi {
7.       CM = select_cluster(ML,nc);
8.       f_M = GFA(CM);
9.       Add f_M to a module set S;
10.   }
11.   Insert modules of S into ML,
12. }
13. F = GFA(ML);
```

Figure 1: Pseudo code of MLGFA

```
Algorithm : select_cluster
Input  : ML = A module list, sorted by width
         nc = Number of modules per cluster
Output : CM = A set of modules
1. For i = 1 upto nc
2.    Add a module ML[i] to CM;
3. Remove CM from ML;
4. Return CM;
```

Figure 2 : Pseudo code of the select_cluster()

Although MLGFA has only one clustering phase, but as the experimental results will show, the obtained result is very competitive with other two-phase multilevel approaches. Also, because of this one phase property, there is a significant improvement in running time.

Figure 3 shows a simple example of running MLGFA on 17 modules. Let T = 6, mr = 1/3 and nc = 4. All 17 modules would be sorted by their width first. At level 1, NLi = 11 can be derived from the equation $NLi = |ML|*(1 - mr)$. According to the pseudo code in Figure 1, 8 modules are grouped into two clusters and two super modules are produced from these two clusters. Level 1 is completed by inserting these two new modules into the module list. The same procedure is repeated at level 2 and 3. At level 4, because the number of modules is smaller than threshold value T ($|M|$ = 5 < T = 6), the rest of modules would be collected and formed the final floorplan by running GFA for the last time.

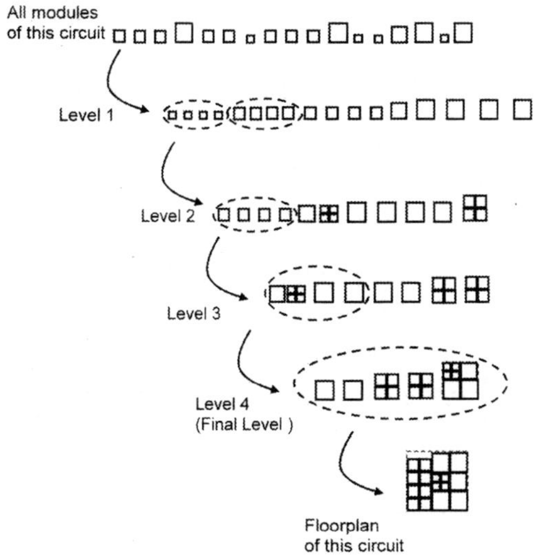

Figure 3 : A simple example of MLGFA

4. EXPERIMENTAL RESULTS

4.1 Data Sets

In order to compare with MB*-tree, we create benchmark circuits by duplicating the modules and nets of ami49, the largest MCNC benchmark circuit, by x times. x ranges from 1 to 200. Besides, n100, n200, and n300 are selected from GSRC block packing slot and these benchmarks are showed in Table I. Columns 1, 2 ,3 and 4 of Table I are names of benchmark circuits, number of modules, number of nets and total area of modules in the circuits, respectively.

4.2 Results

A comparison in time and dead area % is listed in Table II. Columns MB*-tree, B*-tree, Sequence Pair and O-tree of Table II are cited from [2]. The platform of [2] is SUN Ultra 60 workstation, CPU 450 Mhz and 2GB memory space. MLGFA is implemented in C++ language on SUN Ultra 10 workstation, CPU 440 Mhz and 384 MB memory space.

Column 1 and 2 of Table II are benchmark circuits and number of modules. Dead space % column of MLGFA shows best and average, repeated 10 times, results simultaneously (best / avg.). Denotation of NR in table II indicates that no result obtained within 5-hr CPU time on SUN Ultra 60. Table II shows that MLGFA can rapidly obtain acceptable and comparable solutions.

Figure 4 and 5 show the comparison between MLGFA and MB*-tree. Figure 4 shows the solution quality of MLGFA is comparable with MB*-tree. Figure 5 illustrates that the growth curve of running time of MLGFA is much more smooth and lower than the growth curve of MB*-tree. In particular, when the size of ami49 benchmark circuit scales up to 9800 modules, MLGFA took 21 times less running time than that required by MB*-tree.

Circuit	# modules	# nets	Total area(mm^2)
ami49x1	49	408	35.445
ami49x2	98	816	70.89
ami49x4	196	1632	141.78
ami49x10	490	4080	354.454
ami49x20	980	9160	708.908
ami49x40	1960	16320	1417.816
ami49x60	2940	24480	2126.724
ami49x80	3920	32640	2835.632
ami49x100	4900	40800	3544.54
ami49x150	7350	61200	5316.75
ami49x200	9800	81600	7089.08
n100	100	885	0.179501
n200	200	1585	0.175696
n300	300	1893	0.27317

Table I Benchmark characteristics

Table III is used to demonstrate the capability of MLGFA to handle more general large scale cases, n100, n200, and n300. Figure 6 shows the resulting floorplan of n300 constructed by MLGFA.

5. CONCLUSION

In this paper, a *Multilevel Genetic Floorplan Algorithm* MLGFA is proposed to handle the floorplanning and packing for large scale modules. Experimental results show that MLGFA scales well when the number of modules rapidly sprout. Currently, we are extending MLGFA to conduct both wirelength and area minimization, and some convincing result has been obtained.

6. REFERENCE

[1] P.-N. Guo, C.-K. Cheng, and T. Yoshimura," An O-tree Representation of Non-slicing Floorplan and Its Applications," *Proc. of DAC*, pp. 268–273, 1999.

[2] Hsun-Cheng Lee, Yao-Wen Chang, Jer-Ming Hsu, and Hannah H. Yang,"Multilevel Floorplanning/Placement for Large-Scale Modules Using B*-trees," *Proc. of DAC*, 2003.

[3] H. Murata, K. Fujiyoshi, S. Nakatake, and Y. Kajitani, "Rectangle-packing Based Module Placement," *Proc. of ICCAD*, pp. 472–479, 1995.

[4] Chang-Tzu Lin, De-Sheng Chen and Yi-Wen Wang, "An Efficient Genetic Algorithm for Slicing Floorplan Area Optimization," *Proc. of ISCAS*, pp. II-879 -II-882, 2002.

[5] Charles J. Alpert, Jen-Hsin Huang, and Andrew B. Kahng, "Multilevel Circuit Partitioning," *IEEE Trans. on CAD*, 1998.

[6] Yun-Chih Chang, Yao-Wen Chang, Guang-Ming Wu, and Shu-Wei Wu ,"B*-Trees: A New Representation for Non-Slicing Floorplans,"*Proc. of DAC*, 2000.

[7] T. F. Chan, J. Cong, T. Kong, J. R. Shinnerl, "Multilevel Optimization for Large-scale Circuit Placement," *Proc. of ICCAD*, pp. 171–176, 2000.

[8] G. Karypis and V. Kumar, "Multilevel k-way Hypergraph Partitioning," *Proc. of DAC*, pp. 343–348, 1999.

[9] Jason Cong and Chang Wu, "Global Clustering-Based Performance-Driven Circuit Partitioning," *Proc. of ISPD*, 2002.

[10] Maogang Wang Xiaojian Yang Majid Sarrafzadeh, "Dragon2000: Stander-cell Placement Tool for Large Industry Circuits", *Proc. of ICCAD*, 2000.

[11] J.H. Holland, "Outline for a logical theory of adaptive systems," *Journal of the ACM*, pp. 297–314, 1962.

Circuit	# modules	MB*-tree [2]		B*-tree[6]		Sequence Pair [3]		O-tree [1]		MLGFA	
		Dead space(%)	Time (min)	Dead space(%)	Time (min)	Dead space(%)	Time (min)	Dead space(%)	Time (min)	Dead space(%)	Time (min)
ami49x1	49	2.78	0.4	3.53	0.25	8.87	6.86	3.61	10.46	3.19 / 4.37	0.47
ami49x2	98	2.51	2.5	3.03	1.26	11.69	45.51	12.29	70.56	1.93 / 3.23	1.74
ami49x4	196	2.7	2.6	6.6	4.73	13	309	9.86	179.23	3.36 / 3.96	2.53
ami49x10	490	2.66	5.4	12.98	19.25	NR	NR	NR	NR	3.51 / 4.81	2.34
ami49x20	980	3.02	15.6	18.55	23.48	NR	NR	NR	NR	3.67 / 5.23	4.49
ami49x40	1960	3.72	24.8	27.33	53.21	NR	NR	NR	NR	3.80 / 5.70	4.39
ami49x60	2940	3.58	42.2	NR	NR	NR	NR	NR	NR	3.53 / 5.15	5.62
ami49x80	3920	3.67	57	NR	NR	NR	NR	NR	NR	3.37 / 4.37	6.08
ami49x100	4900	3.45	51.6	NR	NR	NR	NR	NR	NR	4.04 / 5.18	6.72
ami49x150	7350	3.42	142.2	NR	NR	NR	NR	NR	NR	3.56 / 4.18	9.68
ami49x200	9800	3.44	256.2	NR	NR	NR	NR	NR	NR	3.93 / 5.25	12.09

Table II Result comparison

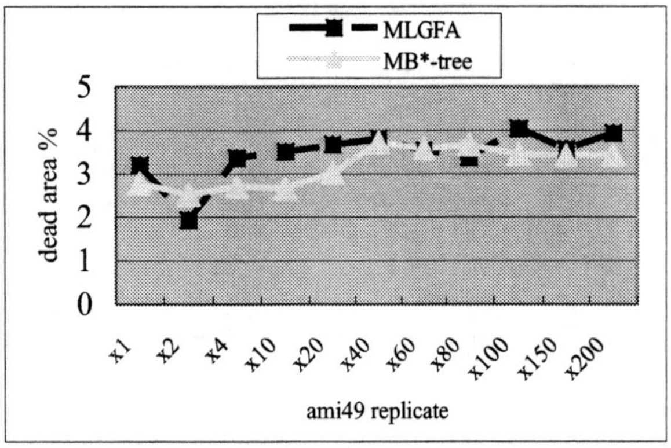

Figure 4 : Comparison of dead area % between MLGFA and MB*-tree

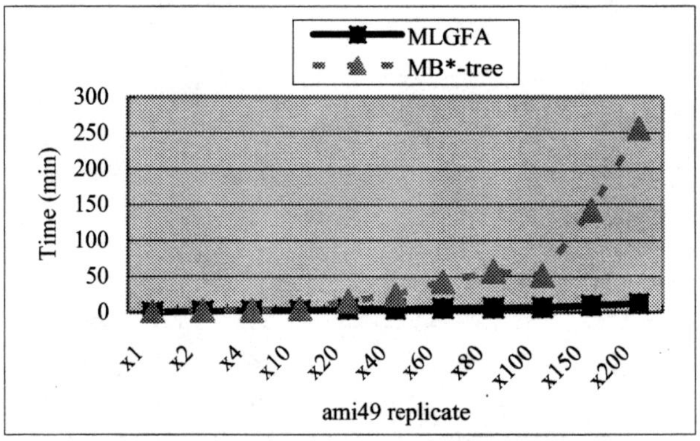

Figure 5 : Comparison of running time between MLGFA and MB*-tree

Circuit	# modules	MLGFA	
		Dead space (%)	Time (min)
n100	100	3.06 / 4.36	2.04
n200	200	3.25 / 4.51	1.78
n300	300	2.23 / 4.20	2.21

Table III Results for n100, n200 and n300

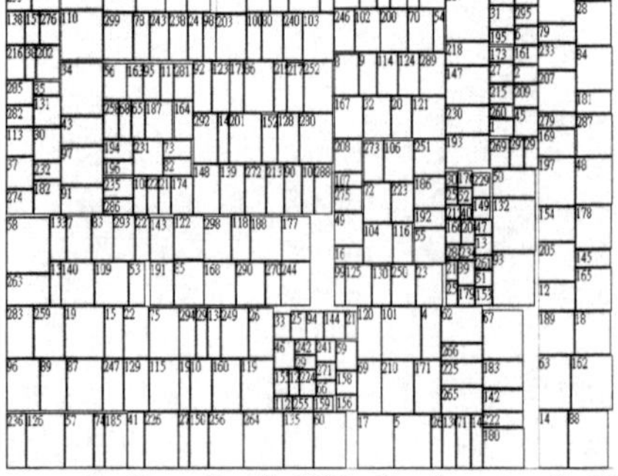

Figure 6 : Result of n300, dead space 3.68%

HOW MANY SOLUTIONS DOES A SAT INSTANCE HAVE?

Pushkin R. Pari, Lin Yuan, and Gang Qu
Electrical and Computer Engineering Department and Institute for Advanced Computer Studies
University of Maryland, College Park, MD 20742 USA

ABSTRACT

Our goal is to investigate the solution space of a given Boolean Satisfiability (SAT) instance. In particular, we are interested in determining the size of the solution space – the number of truth assignments that make the SAT instance true – and finding all such truth assignments, if possible. This apparently hard problem has both theoretical and practical values. We propose an exact algorithm based on exhaustive search that Solves the instance Once and Finds All Solutions (SOFAS) and several sampling techniques that estimate the size of the solution space. SOFAS works better for SAT instances of small size with a 5X-100X speed-up over the brute force search algorithm. The sampling techniques estimate the solution space reasonably well for standard SAT benchmarks.

1. INTRODUCTION

The Boolean satisfiability (SAT) problem seeks to decide, for a given formula, whether there is a truth assignment for its variables that makes the formula true. As the first computational task shown to be *NP-hard*, SAT plays the central role in theoretical computer science and finds numerous applications in various fields. Due to its discrete nature, SAT appears in many contexts in the field of VLSI CAD, such as automatic pattern generation, logic verification, timing analysis, delay fault testing, and channel routing[4].

Over the years, many SAT solvers have been developed based either on *local search* (e.g. GSAT, POSIT, SATO, Satz, WalkSAT, and Rel_SAT) or on *backtrack search* (e.g. GRASP, Chaff, and Zchaff). Links to these solvers can be found at the on-line SAT library [3]. Most of them focus on finding one truth assignment or proving that no such assignment exist. Although some solvers (e.g., GSAT[5]) do provide the option of finding multiple solutions, there is little discussion to the best of our knowledge, on how to obtain all the solutions to a given SAT instance or to determine the size of the solution space. Satometer [1] is the only similar work that estimates the percentage of the search space actually explored by a backtrack SAT solver. Nevertheless, these are important (and of course hard) problems that not only have theoretical value to unveil the structure of the SAT problem, but also can find real life applications, particularly in multiple objective optimization problems. For example, many VLSI CAD applications (such as logic optimization and channel routing) have their SAT formulation and knowing all the solutions gives designer freedom to optimize other design objectives simultaneously.

Naturally, there are two different approaches to finding all the solutions. One is by conducting a brute force search, which evaluates all the possible truth assignments for the variables. This guarantees that all solutions will be found, but the exponential growth of the solution space makes this method impractical. Another method is to repeatedly run a SAT solver until it fails to report any new solutions. The advantage of this approach is that it can find multiple solutions quickly. However, it may miss some solutions particularly, stochastic local search based solvers like GSAT and WalkSAT report unsatisfiable if no solutions are found within a given time.

We propose SOFAS (Solve Once and Find All Solutions) to exhaustively search for all the solutions. SOFAS reports all the solutions, unlike most solvers which tries to find one truth assignment. The basic idea is to scan the SAT instance clause by clause and prune the search space by deleting non-solutions. SOFAS speeds up the solution-pruning process significantly by renaming the variables, reordering the clauses, and carefully managing the solution candidates. The current version of SOFAS can only handle problems of moderate size, but it correctly finds all the solutions and is 5X-100X faster than the pure brute force search. We believe that its performance can be greatly enhanced by adding features such as the Davis Putnam procedure [2] and recursive learning [6, 7].

Enumerating all solutions eventually becomes an insurmountable task as the number of variables increases. Therefore, we propose a couple of sampling techniques to help the process by estimating the size of the solution space. The first method randomly assign values to a set of variables and then tries to determine the size of solution subspace with these variables fixed. The second method runs strategically different SAT solvers to find multiple solutions to the same instance and then compares these solutions to estimate the whole solution space.

We describe SOFAS in Section 2 and discuss the sampling techniques in Section 3. Section 4 reports our preliminary results and Section 5 concludes the paper.

2. SOFAS: SOLVE ONCE AND FIND ALL SOLUTIONS

Figure 1 gives the pseudo code of SOFAS. It reads the clauses one at a time, in a pre-determined order, and eliminates the variable assignment(s) that cannot satisfy this clause. When all the clauses are checked, the remaining assignment(s) are all solutions to the given SAT instance.

Figure 2 depicts the key steps of SOFAS by an example of the following 5-variable SAT formula: $\mathcal{F} = (x_1+x_4'+x_5)(x_3'+x_4+x_5)(x_2+x_4'+x_5')(x_2'+x_3'+x_4)(x_3+x_4'+x_5)$. We represent the solution space by a binary tree where the 2^n leaves, denoted by numbers $0, 1, 2, \cdots, 2^n - 1$ from left to right, correspond to all the possible assignments. At each non-leaf node, we pick an unassigned variable x_i and associate its left subtree with the all assignments with $x_i = 0$ and the right subtree with all assignments with $x_i = 1$. For example, the shaded node at the 4th level in Figure 2(a) represents $\{x_1 = 0, x_2 = 1, x_3 = 1\}$ while variables x_4 and x_5 remain unassigned.

We view each clause as a constraint that eliminates the assignments that fail to satisfy this clause. For instance, any assignment with $\{x_2 = 1, x_3 = 1, x_4 = 0\}$ will make clause $(x_2' + x_3' + x_4)$, and hence the formula $\mathcal{F}$, *FALSE*. Such assignments correspond to

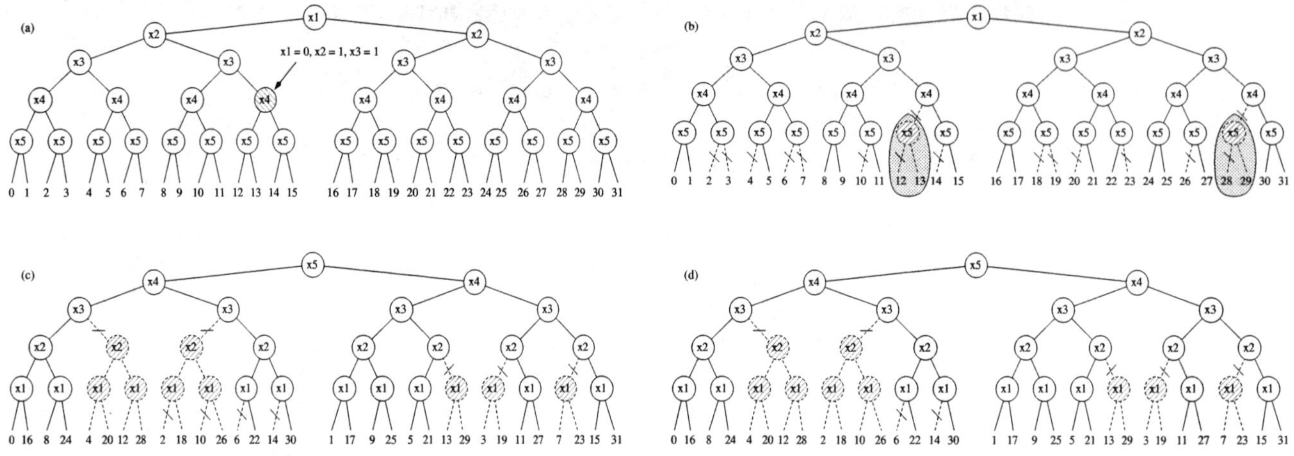

Fig. 2. Illustration of efficiency of SOFAS in finding all solutions to a small SAT instance. From left to right: (a) structure of the tree, (b) 16 original cuts, (c) 9 cuts after variable renaming, (d) 7 cuts after clause reordering.

Input: *a formula $\mathcal{F}$ over n Boolean variables $\{x_1, x_2, \cdots, x_n\}$.*
Output: *all the variable assignments that make $\mathcal{F}$ evaluate TRUE.*
Algorithm: Solve Once and Find All Solutions
/* Phase I: renaming the variables */
1. *compute the weighted occurrence of every variable x_i;*
2. *for $k = n, \cdots, 1$*
3. *rename the variable with the least weighted occurrence to y_k;*
4. *update the weighted occurrence of every remaining variable x_i;*

/* Phase II: reordering the clauses */
5. *sort the clauses by the largest indices of their variables in ascending order;*

/* Phase III: cut the non-solutions */
6. *mark all clauses unchecked;*
7. *let the solution space to be all the possible assignments;*
8. *while the solution space is non-empty and there are unchecked clauses;*
9. *cut all assignments that cannot satisfy the top ranked unchecked clause;*

/* Phase IV: report result */
10. *if the current solution space is empty*
11. *report $\mathcal{F}$ unsatisfiable;*
12. *else*
13. *report the solution space: all truth assignments to $\mathcal{F}$;*

Fig. 1. Pseudo code of SOFAS.

the two shaded subtrees in Figure 2(b), which will be pruned. A *cut* is the prune of a subtree and the clause $(x_2' + x_3' + x_4)$ results in two cuts. For the above formula $\mathcal{F}$, checking the five clauses in the order given in the definition of $\mathcal{F}$ results in 16 cuts as shown in Figure 2(b). Notice that there are 4 cuts associated with the last clause $(x_3 + x_4' + x_5)$, however, two of them have already been pruned by the first clause $(x_1 + x_4' + x_5)$. The 16 remaining leaves stand for all the truth assignment to $\mathcal{F}$.

Fact: Finding all truth assignments to a SAT formula is equivalent to eliminating all those that violate one or more clauses or to pruning the corresponding leaves from the binary decision tree.

SOFAS is based on this observation. It seeks to minimize the total number of cuts by renaming variables and reordering clause as one can see from the following lemmas.

Lemma 2.1 A clause with k variables in an n-variable formula makes 2^{n-k} truth assignments non-solutions. Furthermore, in the binary decision tree, these non-solutions correspond to $2^{i_k - k}$ subtrees each of size 2^{n-i_k}, where i_k is the largest index of the k variables.

In line 1, we define a clause with k literals to contribute 2^{-k} to the *weighted occurrence* of each of its variables. If a variable with a *weighted occurrence* ω is renamed to have the highest index n, then all the clauses that have this variable will result in $2^n \cdot \omega$ cuts based on Lemma 2.1. In light of this, we rename the variable with the smallest weighted occurrence to have the highest available index (line 3) and repeat this untill all the variables are renamed. This renaming procedure results in the following variable name convertion for our example formula $\mathcal{F}$: $\{y_1 = x_5, y_2 = x_4, y_3 = x_3, y_4 = x_2, y_5 = x_1\}$. We can then rewrite $\mathcal{F}$ as $(y_1 + y_2' + y_5')(y_1 + y_2 + y_3')(y_1' + y_2' + y_4)(y_2 + y_3' + y_4')(y_1 + y_2' + y_3)$. There will be only 9 cuts as depicted in Figure 2(c).

Some leaves of the binary tree may be pruned more than once when we consider the clauses one by one. For example, the leaves 2 and 10 in Figure 2(c) have been cut twice: first by the first clause $(y_1 + y_2 + y_5')$, then indirectly by the last clause $(y_1 + y_2' + y_3)$.

Lemma 2.2 For two clauses C and C', let x_k and $x_{k'} (k' \leq k)$ be the variable with the largest index and T and T' be any subtree pruned by C and C' respectively. If $k > k'$, then $T \cap T' = \phi$ or $T \subset T'$; if $k = k'$, then $T \cap T' = \phi$ or $T = T'$.

Lemma 2.2 suggests that if we check clause C' and make the corresponding cuts before we consider C, then no subtrees will be pruned twice. Phase II of the SOFAS algorithm (line 5 in Figure 1) enforces this by sorting the clauses by their largest indexed variables in ascending order. As a result, we need only 7 cuts for the rearranged formula $\mathcal{F} = (y_1 + y_2 + y_3')(y_1 + y_2' + y_3)(y_1' + y_2' + y_4)(y_2 + y_3' + y_4')(y_1 + y_2 + y_5')$ as shown in Figure 2(d).

After renaming variables and reordering clauses, SOFAS reads the clauses one at a time. For each clause, a set of non-overlapping intervals of the same length will be generated based on Lemma 2.3, to represent the non-solution subtrees. The union of all such intervals gives all the non-solutions and thus defines the solution space.

Lemma 2.3 For a k-literal clause with x_{i_k} as the highest indexed variable, the non-solution leaves can be represented by the union of intervals $[S+A, S+A+l-1]$, where $l = 2^{n-i_k}$ is the length of the interval, A is a clause-dependent constant, and S takes 2^{i_k-k} different values depending on the k literals in the clause.

In SOFAS we developed an efficient algorithm that (i) identifies a group of intervals that have already covered and skips them and (ii) merges consecutive intervals to keep the number of intervals minimal at all times. Details of this algorithm are omitted due to space limitation.

3. SOLUTION SPACE ESTIMATION BY SAMPLING

As the size of the SAT instance increases, both the search space and the potential solution space grow exponentially. Consequently any attempt in finding all the solutions will require an exponential run time. In this section, we present two efficient sampling techniques for the estimation of the size of the solution space.

3.1. Sampling over Smaller SAT Instances

This technique takes samples of solution space, by reducing the original SAT instance, which have a much smaller search space. It is based on the assumption that the average solution space size over a large number of smaller SAT instances generated from the original formula reflects the size of the original solution space.

In step 1, we create an unbiased estimation by eliminating all variables that have the same values over the entire solution space. We then create a smaller SAT formula in steps 2 and 3. If a selected variable x is assigned value '1', for example, we delete all the clauses with literal x and remove x' from all the remaining clauses. Note that this gives us a formula with k fewer variables and a much smaller solution space ($1/2^k$ of the original one). We then determine the solution space in step 4 where an unsatisfiable instance is considered to have zero solution. The repetition of steps 3 and 4 in steps 5 and 6 will help us to get a better estimation in step 7. Note that we do not assume a random distribution of the solution space. Instead, we take a large number of samples to estimate the average size of each solution subspace.

1. apply the Davis Putnam procedure [2] to determine the values of those variables that must have a constant value in all solutions. Let C be the list of c such variables.
2. randomly select k variables other than the c variables in C.
3. assign random values to these c variables
4. update the SAT formula and determine the number of solutions by solving for all solutions.
5. repeat steps 3 and 4 t times with different random assignments to the same set of k variables. Let $\{n_1, n_2, ..., n_t\}$ be the number of solutions in these t trials and $T = n_1 + \cdots + n_t$ be the total number of solutions.
6. repeat steps 3-5 K times and obtain the total number of solutions for each trial $\{T_1, T_2, ..., T_K\}$.
7. estimate the number of solutions for the original SAT formula to be $\frac{T_1+T_2+\cdots+T_K}{K*t} \cdot 2^k$

Fig. 3. Sampling over SAT instances of smaller size.

3.2. Sampling by (Strategically) Different Solvers

This is a variation of the following classical sampling technique: take 10 balls randomly from a blackbox, mark them and put them back into the box. Then take again 10 balls randomly, if 5 of them have been marked, then we estimate that there are around 20 balls in the box because half of the redrawn samples repeat. This relies on the fact that the sample drawing is conducted randomly.

However, when we apply a solver to a SAT instance, we have no guarantee that the solver will give us a random satisfying solution. When we repetitively solve the same problem with the same solver, it is not clear whether we will get the same solution or a different solution; and if different, whether the two solutions correlate with each other. In fact, many solvers have the tendency to find the same solution when solved repetitively.

To overcome these problems, we start with two solvers, $\mathcal{S}_1$ and $\mathcal{S}_2$, preferably strategically different solvers. We apply each solver to find a certain number of distinct solutions. To ensure that the solvers find different solutions each time, we append a new clause to the formula once a new solution is found. For example, if we have a solution $x_1 = 0, x_2 = 1$, and $x_3 = 0$ to a formula $\mathcal{F}$, we then add the clause $x_1 + x_2' + x_3$ to $\mathcal{F}$. Solving this new augmented instance guarantees us a new solution. Suppose that we have obtained k_1 and k_2 solutions by $\mathcal{S}_1$ and $\mathcal{S}_2$ respectively, where k solutions are reported by both solvers. We are able to estimate that the original instance has $\frac{k_1 \cdot k_2}{k}$ solutions.

This sampling technique solves the original SAT instance. However, it only looks for a certain number of solutions rather than finding all of the solutions. We argue that the run time to find a limited number of sample solutions will be much less than the time it takes to enumerate all the solutions. Our experiments also validate this argument. Finally, we mention that the two proposed methods – sampling over smaller SAT instances and sampling over different solvers – are orthogonal, and they can be combined for better run time efficiency.

4. EXPERIMENTAL RESULTS

We implement SOFAS and a naïve brute force search algorithm using the same data structure (for a fair comparison of their performance) and compared them on a set of uniform randomly generated 3-SAT formulas with 15 to 30 variables and a constant 4.3 clause-variable ratio. In the naïve brute force search algorithm, we exhaustively check for all the possible truth assignment one by one. An assignment becomes a solution if and only if all the clauses are satisfied. Otherwise, we move on to check the next truth assignment. SOFAS is 5X-100X times faster than this brute force search. We then compare the speed for various solvers, including SOFAS, Posit, Sato, and Satz, to find all the solutions for 3-SAT benchmarks. The 3-SAT instances include random formulas with 15, 20, and 25 variables as well as standard DIMACS and satlib benchmark instances [8, 3]. For solvers we find all the solutions by solving for one solution and then enforcing the solver to find a new solution until it fails to find one. Due to space constraints, we only mention that for these small- and medium-sized 3-SAT benchmarks, SOFAS is compatible with Posit, Sato, and Satz, particularly when there are many solutions. Furthermore, SOFAS always give the correct number of solutions, while other solvers cannot guarantee to find all the solutions occasionally.

We now evaluate the accuracy of the two solution space estimation methods. The SAT problems are the unforced uniform random 3SAT benchmarks from [3]. For space consideration, here we only report our results on two sets of benchmarks: 500 instances with 50 variables and 218 clauses, and 100 instances with 75 variables and 325 clauses. They are all satisfiable instances with the number of solutions ranging from one to a few thousand, which we obtain from repetitively running Zchaff.

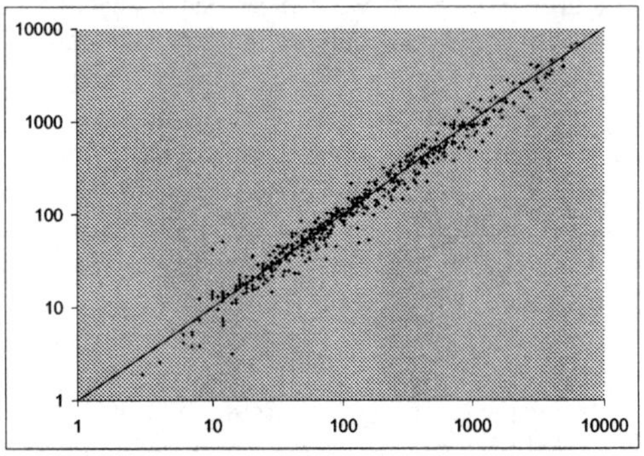

Fig. 4. Accuracy of sampling method I on 500 50-variable 3SAT instances. X axis: actual number of solution; Y axis: estimated number of solutions.

Figures 4 and 5 demonstrate the accuracy of the first sampling technique by plotting the actual and estimated number of solutions. The values of k,t and K are set to be 5, 10 and 10 respectively. That is, we randomly choose 10 sets of 5 variables and assign 10 different assignments for each set. We solve all the corresponding smaller sized SAT problems for all solutions by Zchaff and then estimate the number of solutions for the original problem by the formula given in Figure 3. The 45-degree line indicates the situation when the estimation meets exactly the actual number. Points above and below this line are the overestimated and underestimated cases respectively. Both figures show that our estimation is fairly close to the actual number of solutions. In fact, for the 500 50-variable benchmarks, the average error, measured by $\frac{1}{500}\sum_i \frac{estimation-actual\ number}{actual\ number}$, is only **0.2%** with most of the error comes from instances that have less than 20 solutions.

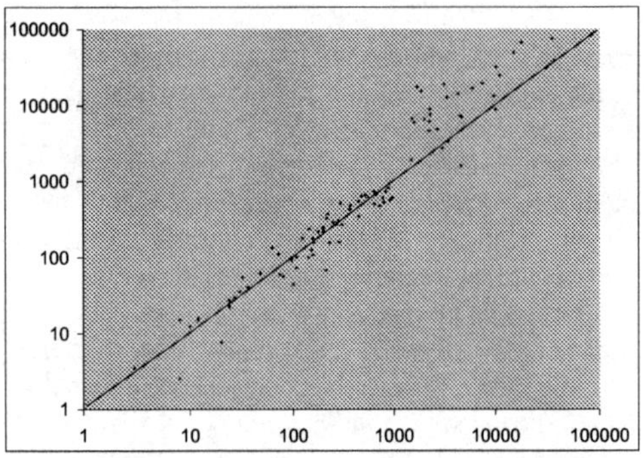

Fig. 5. Accuracy of sampling method I on 100 75-variable 3SAT instances. X axis: actual number of solution; Y axis: estimated number of solutions.

For the second method, we use Zchaff and Satz as the two strategically different solvers to obtain 100 and 250 (when applicable) solutions independently for each instance. We then compare these reported solutions and use the number of solutions found by both solvers to estimate the solution space of the original problem as we have discussed earlier. Figure 6 reports the result on 200 50-variable instances with at least 100 solutions. Although this method is faster than the previous sampling technique, one can see that it tends to underestimate the size of the solution space, particularly for those with large number of solutions. The reason is that the solutions, found by both solvers and those in the entire solution space, normally form groups rather than being randomly distributed. Therefore, instead of finding individual solutions that are in common, the two solvers usually report groups that have many solutions in common. This misleads us to underestimation. We expect to improve the accuracy by investigating on how to force solvers to find solutions that are far away to each other.

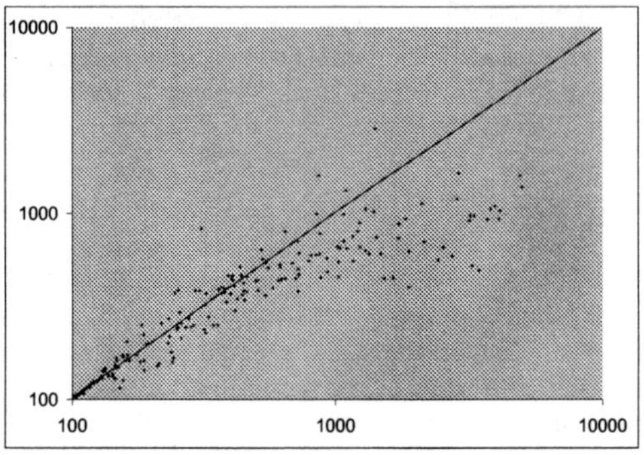

Fig. 6. Accuracy of sampling method II on 200 50-variable 3SAT instances. X axis: actual number of solution; Y axis: estimated number of solutions.

5. CONCLUSION

We present two different approaches to determine the solution space of a given SAT instance. The first one is an exact and effective algorithm based on pruning non-solutions targeting small to medium sized SAT problems. The second approach consists of two sampling techniques to estimate the size of the (large) solution space. We test these two estimation techniques on the SATLIB benchmark instances. The results show that they are fairly accurate. With these techniques, one can better understand the solution distribution and eventually the nature of the SAT problem. It also becomes possible to find better solutions for applications that require the knowledge of the entire solution space.

6. REFERENCES

[1] F.A. Aloul, B.D. Sierawski, and K.A. Sakallah. "Satometer: How Much Have We Searched?" *39th ACM/IEEE Design Automation Conference*, pp. 737-742, June 2002.

[2] M. Davis and H. Putnam. "A Computing Procedure for Quantification Theory", *Journal of the Association for Computing Machinery*, Vol. 7, No. 3, pp. 201-215, July 1960.

[3] H.H. Hoos and T. Stuzle. SATLIB: An Online Resource for Research on SAT. In *SAT 2000 (ed. I.P. Gent, H.V. Maaren, and T. Walsh*, pp. 283-292, IOS Press, 2000.

[4] J.P. Marques-Silva and K.A. Sakallah. "Boolean Satisfiability in Electronic Design Automation," *37th ACM/IEEE Design Automation Conference*, pp. 675-680, June 2000.

[5] B. Selman, H.A. Kautz, and B. Cohen. "Noise strategies for improving local search", *Proceedings of the 12th National Conference on Artificial Intelligence, AAAI'94*, pp. 337-343, 1994.

[6] J. P. M. Silva and K. A. Sakallah. "GRASP—A New Search Algorithm for Satisfiability", *Proceedings of the International Conference on Computer-Aided Design*, 1996.

[7] L. Zhang, C.F. Madigan, M.H. Moskewicz, and S. Malik. "Efficient Conflict Driven Learning in a Boolean Satisfiability Solver", *IEEE/ACM International Conference on Computer Aided Design*, pp. 279-285, November 2001.

[8] http://dimacs.rutgers.edu/

A New Generation of ISCAS Benchmarks from Formal Verification of High-Level Microprocessors

Miroslav N. Velev
http://www.ece.cmu.edu/~mvelev

***Abstract*—The paper presents a collection of 20 benchmark suites with a total of 1,132 ISCAS Boolean formulas from formal verification of high-level microprocessors, including pipelined, superscalar, and VLIW models with exceptions, multicycle functional units, branch prediction, instruction queues, and register renaming. These benchmarks can be used in research on testing, logic synthesis and optimization, equivalence verification, Decision Diagrams, and Boolean Satisfiability. The most complex formulas have more than 700,000 logic gates.**

1. Introduction

For better or for worse, benchmarks shape a field. Previous generations of ISCAS benchmarks [11][12] have been used in thousands of papers on testing, logic synthesis and optimization, equivalence verification, Decision Diagrams, and Boolean Satisfiability (SAT). Theorem-proving problems by Pelletier [49], and by Sutcliffe et al. [56] have also been studied widely. Boolean formulas in CNF format [34]—e.g., those generated in formal verification of high-level processors [73] with the tool flow employed in this paper [62]—have been used in dozens of SAT papers, including breakthrough ones such as those describing the SAT-solvers Chaff [46], BerkMin [25], and Siege [53], as well as in SAT competitions [39].

This paper presents a collection of 20 benchmark suites with a total of 1,132 ISCAS formulas from formal verification of pipelined, superscalar, and VLIW processors, including models with exceptions, multicycle functional units, branch prediction, instruction queues, and register renaming. The biggest benchmarks have more than 700,000 logic gates, and will further challenge tools for testing, logic synthesis and optimization, and equivalence verification, as well as Decision Diagrams, and SAT-solvers. These formulas were generated with a tool flow [62] that was used at Motorola [37] to formally verify a model of the M•CORE processor, and detected 3 bugs as well as corner cases that were not fully implemented. The tool flow was also used in an advanced computer architecture course [65], where students—having no prior knowledge of formal methods—designed and formally verified single-issue pipelined processors, as well as extensions with exceptions and branch prediction, and dual-issue superscalar implementations; a detailed description of the student bugs can be found in [67].

The benchmark suites contain both tautologies (formulas that are equivalent to the Boolean constant *true*, so that their negations will be unsatisfiable when evaluated by a SAT-solver), and non-tautologies (whose negations will be satisfiable). The former are from formal verification of correct processors, while the latter are from implementations with variations of actual bugs made when designing a correct model. That is, the variations introduced were not completely random, as done in other efforts to generate benchmark suites [28][32][33][45], but reflect realistic scenarios for errors that can be made when designing high-level microprocessors. The bugs were spread over the entire processor designs, and occurred either as single or multiple errors.

The benchmarks will facilitate research on circuit-based SAT-solvers [3][8][13][23][24][26][27][36][41][43][51], on variable ordering for SAT-solving [4][19][52][74], on preprocessing techniques for formula simplification [6][10][20][38][42][44][51], on deriving correlations between signals [5][40][41][47][48], and on efficient translation to CNF [34] format [7][9][21][22][35][50][57][68][70][71].

2. Background

The benchmarks are generated in formal verification of processors by *correspondence checking*—comparison of a pipelined implementation against a non-pipelined specification. *Flushing* [17]—feeding the pipeline with bubbles until all partially executed instructions are completed—is used to compute an *abstraction function*, *Abs*, mapping an implementation state to an equivalent specification state. *Controlled flushing* [18]—where the user overrides the processor stalling signals in order to eliminate the ambiguity of the instruction flow during flushing—results in simpler correctness formulas. The correctness criterion is expressed as a formula in the logic of EUFM [17]. The *safety property* (see Fig. 1) is a proof by induction, checking that all architectural state elements are updated in synchrony by either 0, or 1, or up to k instructions after each step of an implementation that starts from arbitrary initial state Q_{Impl}, possibly restricted by invariant constraints; k is the issue width of the implementation.

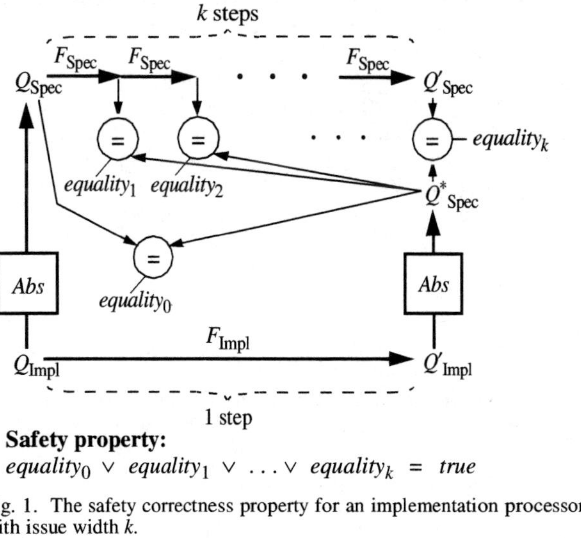

Safety property:
$equality_0 \vee equality_1 \vee \ldots \vee equality_k = true$

Fig. 1. The safety correctness property for an implementation processor with issue width k.

To prove *liveness*—that the processor will complete at least one new instruction after a finite number of steps, n—we can simulate the implementation symbolically for n steps and prove that:

$$equality_1 \vee equality_2 \vee \ldots \vee equality_{n \times k} = true,$$

omitting $equality_0$. Special abstractions and an indirect method for proving liveness—resulting in orders of magnitude speedup—are presented in [69]. The reader is referred to [1][2] for a discussion of microprocessor correctness criteria.

The syntax of EUFM [17] includes *terms* and *formulas*. Terms are used to abstract word-level values of data, register identifiers, memory addresses, as well as the entire states of memory arrays. A term can be an Uninterpreted Function (UF) applied on a list of argument terms, a term variable, or an *ITE* (for "if-then-else") operator selecting between two argument terms based on a controlling formula, such that *ITE(formula, term1, term2)* will evaluate to *term1* when *formula = true*, and to *term2* when *formula = false*. The syntax for terms can be extended to model memories by means of the functions *read* and *write* [17][61]. Formulas are used to model the control path of a microprocessor, as well as to express the correctness condition. A formula can be an Uninterpreted Predicate (UP) applied on a list of argument terms, a propositional variable, an *ITE* operator selecting between two argument formulas based on a controlling formula, or an equation (equality comparison) of two terms. Formulas can be negated and connected by Boolean connectives.

UFs and UPs are used to abstract the implementation details of functional units by replacing them with "black boxes" that satisfy no particular properties other than that of *functional consistency*. Namely, that the same combinations of values to the inputs of the UF (or UP) produce the same output value. Then, it no longer matters whether the original functional unit is an adder or a multiplier, etc., as long as the same UF (or UP) is used to replace it in both the implementation and the specification. Thus, we will prove a more general problem—that the processor is correct for any functionally consistent implementation of its functional units. However, this more general problem is easier to prove.

In our earlier work on applying EUFM to formal verification of pipelined and superscalar processors, we imposed some simple restrictions [58][59] on the modeling style for defining high-level processors, resulting in correctness formulas where most of the terms appear only in positive equations or as arguments to UFs or UPs. Such terms can be treated as distinct constants [15][72], thus significantly simplifying the correctness formulas, pruning the solution space, and resulting in orders of magnitude speedup of the formal verification; we call this property *Positive Equality*. The speedup is at least 5 orders of magnitude when formally verifying realistic dual-issue superscalar processors [72]. The speedup is increasing with the complexity of the implementation. Additional speedup of up to 2 orders of magnitude can be achieved by efficient translation to CNF [68][70][71].

In the extended version of the decision procedure EVC [62] that is used to generate the ISCAS benchmarks, the final Boolean formulas consists of AND, OR, NOT, and ITE (i.e., multiplexor) gates. Hashing [59] ensures that: there are no duplicate gates; merges an AND having another AND as input into a single AND having as inputs all the inputs of the two gates, except for the output of the merged AND, and similarly for an OR having another OR as input; eliminates duplicate inputs to AND and OR gates; and replaces an AND/OR with a constant if two inputs are complements of each other.

3. Description of the Benchmarks Suites

Suite 1: Variants of Single-Issue and Dual-Issue DLX Processors. Included are formulas from: a single-issue, 5-stage pipelined DLX [29] processor, `1dlx_c`; a dual-issue superscalar DLX with one complete pipeline (that can execute all instruction types) and one arithmetic pipeline, `2dlx_ca`; a dual-issue superscalar DLX with two complete pipelines, `2dlx_cc`; and intermediate dual-issue superscalar designs that were created when developing `2dlx_ca` and `2dlx_cc`. These benchmarks were used for the experiments in [59]. **Number of formulas in this suite:** 8 (all tautologies)

Suite 2: Single-Issue and Dual-Issue DLX Models with Exceptions, Multicycle ALUs, and Branch Prediction. The three main processors used in Suite 1 were extended with branch prediction; multicycle ALUs, multicycle Instruction Memory, and multicycle Data Memory; exceptions that could be raised from the ALUs, the Instruction Memory, and the Data Memory; and a return-from-exception instruction. Variants with both multicycle functional units and exceptions, as well as with all the features were also created. The most complex extensions of `2dlx_ca` and `2dlx_cc` are, respectively, `2dlx_ca_mc_ex_bp` and `2dlx_cc_mc_ex_bp`. These benchmarks are from the experiments in [60]. **Number of formulas in this suite:** 18 (all tautologies)

Suite 3: 100 Buggy Variants of the Most Complex Dual-Issue DLX. This suite contains formulas from 100 buggy versions of `2dlx_cc_mc_ex_bp` from Suite 2. The bugs are variants of actual errors made in creating the correct designs. These formulas were used in [64]. **Number of formulas in this suite:** 100 (all non-tautologies)

Suite 4: VLIW Processors with Speculative Execution. `9vliw` is the base processor; `9vliw_bp` is an extension with branch prediction; `9vliw_bp_mc` is a further extension with multicycle ALUs, instruction memory and data memory; and `9vliw_bp_mc_ex` is a version with both branch prediction, multicycle functional units, and exceptions. These models have 9 execution pipelines of 4 stages: 4 integer pipelines, two of which can perform both integer and floating-point memory accesses; 2 floating-point pipelines; and 3 branch-address computation pipelines. Data values are stored in 4 register files: integer, floating-point, predicate, and branch-address. Additionally, the architectural state includes a PC, a Data Memory, and two state elements from Intel's 64-bit architecture [31][55]—a Current Frame Marker used for register remapping, and an Advanced Load Address Table used to implement advanced loads. Every instruction is predicated with a qualifying predicate-register identifier that, if *true*, enables the instruction to modify architectural state. There can be up to 42 instructions in flight. These formulas were used for the experiments in [64]. **Number of formulas in this suite:** 14 (10 tautologies, and 4 non-tautologies)

Suite 5: 100 Buggy VLIW Processors. This suite contains formulas from 100 buggy versions of `9vliw_bp_mc` from Suite 4. The bugs were variants of actual errors made when creating the correct designs in that suite. These formulas were used for the experiments in [64]. **Number of formulas in this suite:** 100 (all non-tautologies)

Suite 6: Wide-Issue Out-of-Order Processors with Abstracted Reorder Buffer. Included are formulas from out-of-order processors with register-register ALU instructions. Newly fetched instructions are placed at the end of the Reorder Buffer (ROB) that is abstracted as described in [63]. The actual number of fetched instructions is determined by a Scheduler that is abstracted [63] with a generator of arbitrary Boolean values [60], thus modeling the operation of any scheduling mechanism. These designs do not have register renaming, and the operands are kept in the ROB entries, each associated with a computation slice that non-deterministically computes the result for that ROB entry as soon as both data operands become available. Included are benchmarks with up to 72 ROB entries, and issue/retire widths of up to 64 instructions. **Number of formulas in this suite:** 76 (all tautologies)

Suite 7: Liveness of Superscalar and VLIW Processors. The main benchmarks from Suite 2, and `9vliw` from Suite 4, as well as their extensions with exceptions and branch prediction, were checked for liveness (see Sect. 2). **Number of formulas in this suite:** 15 (all tautologies)

Suite 8: Wide-Issue Superscalar Processors. These benchmarks are from formal verification of wide-issue superscalar processors [64] that have 5 pipeline stages, can issue up to 15 instructions per cycle, and can execute ALU and load instructions, so that load-interlock stalling logic is implemented [29]. Included are formulas from designs with in-order issue and with out-of-order issue, and 6 buggy designs that were created inadvertently. **Number of formulas in this suite:** 23 (17 tautologies, and 6 non-tautologies)

Suite 9: Student Buggy Designs 1. These formulas are from student implementations of single-issue, 5-stage pipelined DLX processors [29], designed as a project in an advanced computer architecture course [65]. The models were created by 52 students (40 graduate and 12 undergraduate), working in 24 groups. The different coding styles of the many designers make these benchmarks ideal for developing simplification rules. The processors were implemented in a sequence of 6 steps, with both the correct and buggy implementations of each group for each step included. Furthermore, the abstraction function was computed with regular flushing, as opposed to controlled flushing [18], thus resulting in more complex formulas than the one from a single-issue pipelined DLX in Suite 1. For a full description of this project, and of the projects that resulted in the next two benchmark suites, the reader is referred to [65], and for a detailed discussion and classification of the bugs to [67]. **Number of formulas in this suite:** 294 (148 tautologies, and 146 non-tautologies)

Suite 10: Student Buggy Designs 2. These models were created in a sequel to the above project, where the same students extended their single-issue pipelined DLX processors with ALU exceptions, a return-from-exception instruction, and branch prediction. **Number of formulas in this suite:** 153 (61 tautologies, and 92 non-tautologies)

Suite 11: Student Buggy Designs 3. Benchmarks from another project in the same course—the students created dual-issue superscalar DLX models, where the first pipeline could execute ALU, load, and store instructions, and the second pipeline could execute ALU and branch instructions. Since controlled flushing [18] was not used, these formulas are significantly more complex than those from the dual-issue models in Suite 1. **Number of formulas in this suite**: 67 (25 tautologies, and 42 non-tautologies)

Suite 12: Single-Issue DLX Models with Instruction Queues 1. Extensions of `1dlx_c` from Suite 1 with instruction queues implemented as shift registers, like the instruction queue in the PowerPC 750 [30]. The queues have between 1 and 64 entries, for a total of 64 models. **Number of formulas in this suite:** 64 (all tautologies)

Suite 13: Single-Issue DLX Models with Instruction Queues 2. These models are extensions of `1dlx_c_mc_ex_bp` (a processor with multicycle functional units, exceptions, and branch prediction) from Suite 2 with instruction queues implemented as shift registers. The queues have between 1 and 64 entries, for a total of 64 benchmarks. Additionally, 10 buggy models of the processor with 64-entry instruction queue were created as variants of bugs made in implementing the correct models. **Number of formulas in this suite:** 74 (64 tautologies, and 10 non-tautologies)

Suite 14: VLIW Models with Instruction Queues 1. These are extensions of `9vliw` from Suite 4 with instruction queues of between 1 and 16 entries. 10 buggy variants of the processor with 6-entry instruction queue, and another 10 buggy variants of the processor with 8-entry instruction queue were created as variants of actual bugs made in developing the correct designs. **Number of formulas in this suite:** 36 (16 tautologies, and 20 non-tautologies)

Suite 15: VLIW Models with Instruction Queues 2. These are extensions of `9vliw_bp_mc_ex` from Suite 4 with instruction queues of between 1 and 16 entries. **Number of formulas in this suite:** 16 (all tautologies)

Suite 16: Dual-Issue DLX Models with Instruction Queues 1. These are extensions of `2dlx_ca` from Suite 1 with instruction queues of between 1 and 16 entries. **Number of formulas in this suite:** 16 (all tautologies)

Suite 17: Dual-Issue DLX Models with Instruction Queues 2. These are extensions of `2dlx_cc` from Suite 1 with instruction queues of between 1 and 16 entries. **Number of formulas in this suite:** 16 (all tautologies)

Suite 18: Dual-Issue DLX Models with Instruction Queues 3. Extensions of `2dlx_ca_mc_ex_bp` from Suite 2 with instruction queues of between 1 and 16 entries. **Number of formulas in this suite:** 16 (all tautologies)

Suite 19: Dual-Issue DLX Models with Instruction Queues 4. Extensions of `2dlx_cc_mc_ex_bp` from Suite 2 with instruction queues of between 1 and 16 entries. **Number of formulas in this suite:** 16 (all tautologies)

Suite 20: Out-of-Order Processors with Register Renaming and Completely Implemented Reorder Buffer and Reservation Stations. These processors [68] have completely implemented and instantiated Reorder Buffer and Reservation Stations. Up to 2 instructions are issued or retired per clock cycle. Register identifiers are renamed. The Reorder Buffer is implemented as a shift register like in the PowerPC 750 [30]. The suite includes formulas from 10 implementations with different Reorder Buffer sizes, and either deterministic or non-deterministic execution; some formulas contain case-splitting expressions to speed up the formal verification. **Number of formulas in this suite:** 10 (all tautologies)

4. Results

This section presents selected results for some of the challenging benchmarks in the above suites. The experiments were conducted on a Dell OptiPlex GX260 having a 3.06-GHz Intel Pentium 4 processor with a 512-KB on-chip L2-cache, 2 GB of physical memory, and running Red Hat Linux 9.0. The SAT-solver siege_v4 [53] was found to be extremely efficient on these benchmarks. From Suite 6, the formula from a model with 72 Reorder Buffer entries and issue/retire width of 4 instructions per cycle, was proved unsatisfiable in 556 seconds (no other SAT-solver could handle this formula). From Suite 8: a formula from a 14-wide model with in-order issue was proved unsatisfiable in 744 seconds after using efficient translation to CNF [68][70][71]. From Suite 13: the formula from the model with 40-entry instruction queue was proved unsatisfiable in 1,252 seconds. From Suite 14: the formula from a model with a 4-entry instruction queue was proved unsatisfiable in 784 seconds. And from Suite 20: the formula from a model with 6 Reorder Buffer entries and 6 Reservation Stations was proved unsatisfiable in 43 seconds.

5. Conclusions

The paper presented a collection of 20 benchmark suites with a total of 1,132 ISCAS Boolean formulas from formal verification of high-level microprocessors with many realistic features. The most complex formulas have more than 700,000 logic gates. These benchmarks can be used in research on testing, logic synthesis and optimization, equivalence verification, Decision Diagrams, and Boolean Satisfiability. Such ISCAS formulas will facilitate research on circuit-based SAT-solvers, and on SAT preprocessing techniques for formula simplification. The suites are available from [73].

References

[1] M.D. Aagaard, N.A. Day, and M. Lou, "Relating Multi-Step and Single-Step Microprocessor Correctness Statements," *FMCAD '02*, LNCS 2517, Springer-Verlag, November 2002.

[2] M.D. Aagaard, B. Cook, N.A. Day, and R.B. Jones, "A Framework for Superscalar Microprocessor Correctness Statements," STTT, Vol. 4, No. 3 (May 2003).

[3] M. Alekhnovich, and A.A. Razborov, "Satisfiability, Branch-Width and Tseitin Tautologies," *Symposium on Foundations of Computer Science (FOCS '02)*, November 2002.

[4] F.A. Aloul, I.L. Markov, and K.A. Sakallah, "Faster SAT and Smaller BDDs via Common Function Structure," *International Conference on Computer-Aided Design*, 2001.

[5] R. Arora, and M.S. Hsiao, "Enhancing SAT-Based Equivalence Checking with Static Logic Implications," *High Level Design Validation and Test Workshop (HLDVT '03)*, 2003.

[6] F. Bacchus, and J. Winter, "Effective Preprocessing with Hyper-Resolution and Equality Reduction," *Theory and Applications of Satisfiability Testing (SAT '03)*, 2003.

[7] M. Bauer, D. Brand, M. Fischer, A. Meyer, and M. Paterson, "A Note on Disjunctive Form Tautologies," SIGACT News, Vol. 4 (1973), pp. 17–20.

[8] J.D. Bingham, and A.J. Hu, "Semi-Formal Bounded Model Checking," *Computer-Aided Verification (CAV '02)*, LNCS 2404, Springer-Verlag, July 2002.

[9] T. Boy de la Tour, "An Optimality Result for Clause Form Translation," Journal of Symbolic Computation, Vol. 14 (1992), pp. 283–301.

[10] R.I. Brafman, "A Simplifier for Propositional Formulas with Many Binary Clauses," *Int'l. Joint Conference on Artificial Intelligence (IJCAI '01)*, 2001.

[11] F. Brglez, and H. Fujiwara, "A Neutral Netlist of 10 Combinational Benchmark Circuits," *International Symposium on Circuits and Systems (ISCAS '85)*, 1985.

[12] F. Brglez, D. Bryan, and K. Kozminski, "Combinational Profiles of Sequential Benchmark Circuits," *International Symposium on Circuits and Systems (ISCAS '89)*, 1989.

[13] E. Broering, and S.V. Lokam, "Width-Based Algorithms for SAT and CIRCUIT-SAT," *Theory and Applications of Satisfiability Testing (SAT '03)*, May 2003.

[14] R.E. Bryant, "Graph-Based Algorithms for Boolean Function Manipulation," IEEE Transactions on Computers, Vol. C-35, No. 8 (August 1986), pp. 677–691.

[15] R.E. Bryant, S. German, and M.N. Velev, "Processor Verification Using Efficient Reductions of the Logic of Uninterpreted Functions to Propositional Logic," ACM Transactions on Computational Logic (TOCL), Vol. 2, No. 1 (January 2001), pp. 93–134.

[16] R.E. Bryant, and M.N. Velev, "Boolean Satisfiability with Transitivity Constraints," ACM Transactions on Computational Logic (TOCL), Vol. 3, No. 4 (October 2002).

[17] J.R. Burch, and D.L. Dill, "Automated Verification of Pipelined Microprocessor Control," *Computer-Aided Verification (CAV '94)*, LNCS 818, Springer-Verlag, June 1994.

[18] J.R. Burch, "Techniques for Verifying Superscalar Microprocessors," *33rd Design Automation Conference (DAC '96)*, June 1996, pp. 552–557.

[19] E.M. Clarke, and O. Strichman, "A Failed Attempt to Optimize Variable Ordering with Tools for Constraints Solving," *Workshop on Constraints in Formal Verification*, 2002.

[20] J.M. Crawford, and L.D. Auton, "Experimental Results on the Crossover Point in Satisfiability Problems," *National Conference on Artificial Intelligence*, 1993.

[21] E. Eder, "An Implementation of a Theorem Prover Based on the Connection Method," *Artificial Intelligence: Methodology, Systems, Applications (AIMSA '84)*, 1985.

[22] U. Egly, and T. Rath, "On the Practical Value of Different Definitional Translations to Normal Form," *International Conference on Automated Deduction (CADE '96)*, 1996.

[23] J. Franco, M. Kouril, J. Schlipf, J. Ward, S. Weaver, M. Dransfield, and W.M. Vanfleet, "SBSAT: A State-Based, BDD-Based Satisfiability Solver," *6th International Conference on Theory and Applications of Satisfiability Testing (SAT '03)*, May 2003.

[24] M.K. Ganai, L. Zhang, P. Ashar, A. Gupta, S. Malik, "Combining Strengths of Circuit-Based and CNF-Based Algorithms for a High-Performance SAT Solver," *39th Design Automation Conference (DAC '02)*, June 2002.

[25] E. Goldberg, and Y. Novikov, "BerkMin: A Fast and Robust Sat-Solver," *Design, Automation, and Test in Europe (DATE '02)*, March 2002, pp. 142–149.

[26] S. Gopalakrishnan, V. Durairaj, and P. Kalla, "Integrating CNF and BDD Based SAT Solvers," *High-Level Design Validation and Test Workshop (HLDVT '03)*, November 2003.

[27] A. Gupta, A. Gupta, Z. Yang, and P. Ashar, "Dynamic Detection and Removal of Inactive Clauses in SAT with Application in Image Computation," *38th Design Automation Conference (DAC '01)*, June 2001, pp. 536–541.

[28] J.E. Harlow III, and F. Brglez, "Design of Experiments and Evaluation of BDD Ordering Heuristics," Software Tools for Technology Transfer (STTT), Vol. 3, No. 2 (2001).

[29] J.L. Hennessy, and D.A. Patterson, *Computer Architecture: A Quantitative Approach*, 3rd edition, Morgan Kaufmann, San Francisco, 2002.

[30] IBM, Inc., *PowerPC 740™/PowerPC 750™: RISC Microprocessor User's Manual*, 1999.

[31] Intel Corporation, *IA-64 Application Developer's Architecture Guide*, May 1999.

[32] K. Iwama, H. Abeta, and E. Miyano, "Random Generation of Satisfiable and Unsatisfiable CNF Predicates," Information Processing 92, Vol. 1: Algorithms, Software, Architecture, J. Van Leeuwen, *ed.*, Elsevier Science Publishers B.V., 1992, pp. 322–328.

[33] K. Iwama, and K. Hino, "Random Generation of Test Instances for Logic Optimizers," *31st Design Automation Conference (DAC '94)*, June 1994.

[34] D.S. Johnson, and M.A. Trick, *eds.*, *The Second DIMACS Implementation Challenge*, DIMACS Series in Discrete Mathematics and Theoretical Computer Science. http://dimacs.rutgers.edu/challenges

[35] T.A. Junttila, and I. Niemelä, "Towards and Efficient Tableau Method for Boolean Circuit Satisfiability Checking," *International Conference on Computational Logic (CL '00)*, LNAI 1861, Springer-Verlag, July 2000.

[36] A. Kuehlmann, M.K. Ganai, and V. Paruthi, "Circuit-Based Boolean Reasoning," *38th Design Automation Conference (DAC '01)*, June 2001, pp. 232–237.

[37] S. Lahiri, C. Pixley, and K. Albin, "Experience with Term Level Modeling and Verification of the M•CORE™ Microprocessor Core," *International Workshop on High Level Design, Validation and Test (HLDVT '01)*, November 2001.

[38] D. Le Berre, "Exploiting the Real Power of Unit Propagation Lookahead," *Workshop on Theory and Applications of Satisfiability Testing (SAT '01)*, June 2001.

[39] D. Le Berre, and L. Simon, "Results from the SAT'03 Solver Competition," *6th International Conference on Theory and Applications of Satisfiability Testing (SAT '03)*, 2003.

[40] C.M. Li, and Anbulagan, "Look-Ahead versus Look-Back for Satisfiability Problems," *Principles and Practice of Constraint Programming (CP '97)*, LNCS 1330, 1997.

[41] F. Lu, L.-C. Wang, K.-T. Cheng, J. Moondanos, and Z. Hanna, "A Signal Correlation Guided ATPG Solver and Its Applications for Solving Difficult Industrial Cases," *40th Design Automation Conference (DAC '03)*, June 2003.

[42] I. Lynce, and J.P. Marques-Silva, "Probing-Based Preprocessing Techniques for Propositional Satisfiability," *International Conference on Tools with Artificial Intelligence (ICTAI '03)*, November 2003.

[43] J.P. Marques-Silva, and L.G. e Silva, "Algorithms for Satisfiability in Combinational Circuits Based on Backtrack Search and Recursive Learning," *12th Symposium on Integrated Circuits and Systems Design (SBCCI '99)*, September–October 1999, pp. 192–195.

[44] J.P. Marques-Silva, "Algebraic Simplification Techniques for Propositional Satisfiability," *Principles and Practice of Constraint Programming (CP '00)*, September 2000.

[45] D. Mitchell, B. Selman, and H. Levesque, "Hard and Easy Distributions of SAT Problems," *10th National Conference on Artificial Intelligence (AAAI '92)*, July 1992, pp. 459–465.

[46] M.W. Moskewicz, C.F. Madigan, Y. Zhao, L. Zhang, and S. Malik, "Chaff: Engineering an Efficient SAT Solver," *38th Design Automation Conference (DAC '01)*, June 2001.

[47] Y. Novikov, "Local Search for Boolean Relations on the Basis of Unit Propagation," *Design, Automation and Test in Europe (DATE '03)*, March 2003.

[48] R. Ostrowski, E. Grégoire, B. Mazure, and L. Saïs, "Recovering and Exploiting Structural Knowledge from CNF Formulas," *Principles and Practice of Constraint Programming (CP '02)*, P. Van Hentenryck, *ed.*, LNCS 2470, Springer-Verlag, September 2002, pp. 185–199.

[49] F.J. Pelletier, "Seventy-Five Problems for Testing Automatic Theorem Provers," Journal of Automated Reasoning, Vol. 2, No. 2 (1986), pp. 191–216.

[50] D.A. Plaisted, and S. Greenbaum, "A Structure Preserving Clause Form Translation," Journal of Symbolic Computation (JSC), Vol. 2, 1985, pp. 293–304.

[51] D.K. Pradhan, "Logic Transformation and Coding Theory-Based Frameworks for Boolean Satisfiability," *High-Level Design Validation and Test Workshop (HLDVT '03)*, 2003.

[52] S. Reda, R. Drechsler, and A. Orailoglu, "On the Relation Between SAT and BDDs for Equivalence Checking," *Symposium on Quality of Electronic Design*, 2002.

[53] L. Ryan, Siege SAT Solver v.4. http://www.cs.sfu.ca/~loryan/personal/

[54] O. Shacham, and E. Zarpas, "Tuning the VSIDS Decision Heuristic for Bounded Model Checking," *Microprocessor Test and Verification (MTV '03)*, May 2003.

[55] H. Sharangpani, and K. Arora, "Itanium Processor Microarchitecture," IEEE Micro, Vol. 20, No. 5 (September–October 2000).

[56] G. Sutcliffe, C. Suttner, and T. Yemenis, "The TPTP Problem Library," *International Conference on Automated Deduction (CADE '94)*, LNAI 814, 1994, pp. 252–266. http://citeseer.ist.psu.edu/article/sutcliffe97tptp.html

[57] G.S. Tseitin, "On the Complexity of Derivation in Propositional Calculus," in *Studies in Constructive Mathematics and Mathematical Logic*, Part 2, 1968, pp. 115–125.

[58] M.N. Velev, and R.E. Bryant, "Exploiting Positive Equality and Partial Non-Consistency in the Formal Verification of Pipelined Microprocessors," *36th Design Automation Conference (DAC '99)*, June 1999, pp. 397–401.

[59] M.N. Velev, and R.E. Bryant, "Superscalar Processor Verification Using Efficient Reductions of the Logic of Equality with Uninterpreted Functions to Propositional Logic," *Correct Hardware Design and Verification Methods (CHARME '99)*, LNCS 1703, September 1999.

[60] M.N. Velev, and R.E. Bryant, "Formal Verification of Superscalar Microprocessors with Multicycle Functional Units, Exceptions, and Branch Prediction," *37th Design Automation Conference (DAC '00)*, June 2000, pp. 112–117.

[61] M.N. Velev, "Automatic Abstraction of Memories in the Formal Verification of Superscalar Microprocessors," *Tools and Algorithms for the Construction and Analysis of Systems (TACAS '01)*, Springer-Verlag, April 2001, pp. 252–267.

[62] M.N. Velev, and R.E. Bryant, "EVC: A Validity Checker for the Logic of Equality with Uninterpreted Functions and Memories, Exploiting Positive Equality and Conservative Transformations," *Computer-Aided Verification (CAV '01)*, LNCS 2102, July 2001.

[63] M.N. Velev, "Using Rewriting Rules and Positive Equality to Formally Verify Wide-Issue Out-Of-Order Microprocessors with a Reorder Buffer," *Design, Automation and Test in Europe (DATE '02)*, March 2002, pp. 28–35.

[64] M.N. Velev, and R.E. Bryant, "Effective Use of Boolean Satisfiability Procedures in the Formal Verification of Superscalar and VLIW Microprocessors," Journal of Symbolic Computation (JSC), Vol. 35, No. 2 (February 2003), pp. 73–106.

[65] M.N. Velev, "Integrating Formal Verification into an Advanced Computer Architecture Course," *ASEE Annual Conference & Exposition*, June 2003.

[66] M.N. Velev, "Automatic Abstraction of Equations in a Logic of Equality," *Automated Reasoning with Analytic Tableaux and Related Methods (TABLEAUX '03)*, 2003.

[67] M.N. Velev, "Collection of High-Level Microprocessor Bugs from Formal Verification of Pipelined and Superscalar Designs," *International Test Conference (ITC '03)*, October 2003.

[68] M.N. Velev, "Using Automatic Case Splits and Efficient CNF Translation to Guide a SAT-Solver When Formally Verifying Out-of-Order Processors," *Artificial Intelligence and Mathematics (AI&MATH '04)*, January 2004.

[69] M.N. Velev, "Using Positive Equality to Prove Liveness for Pipelined Microprocessors," *Asia and South Pacific Design Automation Conference (ASP-DAC '04)*, January 2004

[70] M.N. Velev, "Efficient Translation of Boolean Formulas to CNF in Formal Verification of Microprocessors," *Asia and South Pacific Design Automation Conference (ASP-DAC '04)*, January 2004.

[71] M.N. Velev, "Exploiting Signal Unobservability for Efficient Translation to CNF in Formal Verification of Microprocessors," *Design, Automation and Test in Europe (DATE '04)*, February 2004.

[72] M.N. Velev, "A New Correctness Proof for Positive Equality," submitted for publication.

[73] M.N. Velev, CNF and ISCAS Benchmark Suites. http://www.ece.cmu.edu/~mvelev

[74] D. Wang, E. Clarke, Y. Zhu, and J. Kukula, "Using Cutwidth to Improve Symbolic Simulation and Boolean Satisfiability," *IEEE International High Level Design Validation and Test Workshop (HLDVT '01)*, 2001.

ESDINSPECTOR: A NEW LAYOUT-LEVEL ESD PROTECTION CIRCUITRY DESIGN VERIFICATION TOOL USING A SMART-PARAMETRIC CHECKING MECHANISM

Rouying Zhan, Haigang Feng, Haolu Xie and Albert Wang
Department of Electrical and Computer Engineering, Illinois Institute of Technology
3301 S. Dearborn St., Chicago, IL 60616. Phone: (312) 567-6912, Email: awang@ece.iit.edu

ABSTRACT

On-chip ESD (electrostatic discharging) protection is a challenging IC design problem. New CAD tools are essential to ESD protection design prediction and verification at full chip level. This paper reports a novel smart parametric checking mechanism and the first intelligent CAD tool, entitled ESDInspector, developed for full-chip ESD protection circuitry design verification. Capability of the new tool is demonstrated using a practical design example in a 0.35μm BiCMOS.

1. INTRODUCTION

In order to avoid ESD-induced damages, ESD protection circuits are required for practical IC chips[1]. As IC technologies continue to migrate into the very-deep-sub-micron(VDSM) region, on-chip ESD protection circuit design emerges as a major design challenge to IC designers, particularly for mixed-signal and RF ICs. One of the major problems facing IC designers is the lack of practical CAD tools dealing with the complex ESD protection circuit design verification. As the result, the trial-and-error approach still dominates in ESD protection design practices.

The main challenge in developing ESD protection design CAD tools is to perform ESD protection design verification at whole chip level. The reasons follow: Firstly, success of an ESD protection design actually depends on the whole IC chip where mutual interactions between the ESD protection network and the core circuit being protected exist. Particularly, various parasitic ESD-type structures, within the ESD protection portion and/or inside the core circuit, have potentials of forming unexpected shunting paths. Since such parasitic discharge routes are normally not designed to handle ESD pulses, pre-mature ESD damages often occur as a consequence even though an individual ESD protection unit might work perfectly by itself. Secondly, numerous parasitic ESD devices may exist in a layout file, however only a handful of them might have a chance to become shunting paths under ESD stresses, other ones are just virtually insensitive to ESD stresses. Thus it is desirable for the verification tool to locate those life-threatening parasitic devices that designers should really worry about. Thirdly, even for all the designed ESD protection devices (i.e., intentional ESD elements), the verification tool should check whether they are designed properly in layout.

This work recognize that, unlike traditional IC checking operations, ESD protection design verification should include much more sophisticated checking aspects associated with ESD protection operation and consider all ESD devices in a chip as a network rather than just deal with them individually. In [2], an ESD/Latchup CAD tool is presented to locate parasitic npn transistors and pnpn paths (i.e. SCRs, silicon control rectifiers), which, however, offers only few basic layout-checking functions. Reference [3-4] discussed techniques to extract parasitic BJT transistors and provided a β-based method to locate life-threatening parasitic BJT transistors, which does not have the desired capacity to consider any type of ESD device. In [5] an ESD design rule checking program was presented. However, it only offered basic layout checking functions that check traditional layer spacing. In addition, all the above CAD tools only considered ESD devices individually when they tried to locate ESD devices in the layout.

This paper reports the first intelligent layout-level ESD protection circuitry design verification CAD tool, ESDInspector, developed based on a smart parametric checking mechanism. The paper is organized as follows. Section 2 reviews basic ESD protection operation. Section 3 introduces a new concept of ESD-critical parameters and presents the novel smart parametric checking mechanism. Section 4 describes the implementation of ESDInspector. Section 5 provides a practical ESD protection design example to demonstrate the capability of the ESDInspector.

2. ESD PROTECTION OPERATION

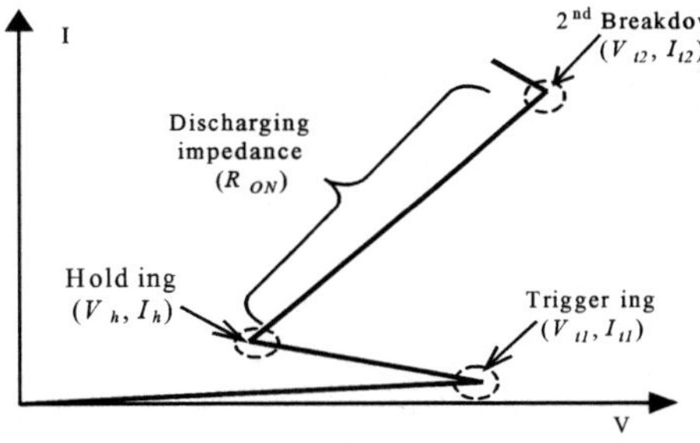

Fig. 1 Typical ESD snap-back protection I-V characteristics

Fig. 1 illustrates the typical I-V characteristics for ESD protection structures, where the ESD protection unit is triggered on under ESD stress, driven into a deep snapback region to form a low-impedance discharging path and to clamp the pad voltage to a low holding level. Apparently, the critical design parameters for an ESD protection structure include the triggering point (V_{t1}, I_{t1}), holding point (V_h, I_h), discharging resistance (R_{ON}) etc. Commonly used ESD protection structures include diode, BJT, MOSFET, SCR and other novel structures. Whatever ESD protection solutions to be used, these important parameters are universal to describe the behavior of them under ESD stresses.

3. SMART PARAMETRIC CHECKING MECHANISM

The goal of the layout-level ESD protection design verification is to locate critical ESD devices (critical ESD devices are those intentional or parasitic devices having a chance to be turned on under ESD stresses) in the layout and/or generate the input deck for schematic-level simulation. Some simple CAD tools were reported [2,5] to perform ESD design verification based on some basic layout checking functions, such as critical layer spacing checking and worst-case path resistance estimation. While such basic layout checking is important, it is only an over-simplified ESD DRC checking routine.

Upon the observation that ESD-critical parameters can be used to describe the behavior of all kinds of ESD devices, intentional or parasitic, this work propose to perform layout-level verification by quantitatively checking critical ESD parameters along with the connections between different ESD devices to locate critical ESD devices.

3.1. ESD-Critical Parameters

We first proposed the basic concept of *ESD-critical parameters*, which is a concise, but quantitative parameter matrix that describes the basic ESD protection operation. As discussed in the previous Section, these ESD-critical parameters should include important ESD operational features such as triggering point (V_{t1}, I_{t1}), holding point (V_h, I_h), discharging impedance (R_{ON}) etc. They determine where/when an ESD structure is turned on; how good the pad voltage clamping is; how efficient the ESD current may be discharged and how much ESD stress may be handled. In addition to these universal parameters, a few structure-specific parameters should also be included. For example, for BJT and MOSFET based ESD protection structures, the effective current gain, β, is an ESD-critical parameter because it determines how efficient the BJT is, which is the fundamental ESD mechanism in these structures. For SCR-type ESD protection structures, the β-product, i.e., $\beta_{NPN}\beta_{PNP}$, of their parasitic NPN and PNP BJTs is clearly an ESD-critical parameter because if it is less than 1, they will never have a chance to be turned on [1].

3.2. Smart Parametric Checking

Our layout-level verification procedure includes 2 steps. Step one, to extract all ESD devices, parasitic and intentional, from the layout. The task of the step two is then to locate those critical ESD devices and remove non-critical ones. In our implementation, step one is performed by an arbitrary ESD device extractor [6]. And we proposed a novel *smart parametric ESD checking mechanism* to perform the step two.

Smart parametric ESD checking mechanism is based upon the basic concept of ESD-critical parameters and practical ESD protection design flow. The mechanism can be understood by the following typical ESD protection design checking flow cases. In Case 1, one studies the extracted ESD netlist to see if all intentional ESD protection devices are there. If any of them missing, the layout has an error. In Case 2, assume two ESD protection devices extracted, A and B, are connected in parallel. If A has a sufficiently higher triggering voltage, V_{t1}, than B does, then, A should be removed since it would never have a chance to be turned on under an ESD pulse. In Case 3, if two extracted ESD devices, A and B, are in parallel connection and have comparable V_{t1} , one then proceeds to look into the discharging impedance R_{ON}. If R_{ON} of A is sufficiently larger than that of B, then the ESD current will discharge into device B, thus A can be removed. In Case 4, assume the ESD devices A and B have comparable V_{t1} and R_{ON}, one will look into their holding voltage V_h. The lower V_h will determine the pad voltage clamping behavior. These special Cases represent some common ESD protection design situations. Thus, implementation of the new smart parametric checking mechanism includes defining a group of ESD-critical parameter related checking criteria as to be discussed below.

A. β-product criterion

If device A is an SCR-type ESD protection device and its β-product is less than 1, A has no chance to be turned on by an ESD pulse and will be removed.

B. Triggering point criteria

B1. Maximum V_{t1} criterion

If V_{t1} of the device A is substantially higher than a given threshold voltage $V_{t1\text{-}max}$, pre-set for a specific ESD protection circuitry, then A cannot be turned on under an ESD pulse and will be removed. Typically, a "substantial" difference means a ~20%+ difference in the values upon our experience, which applies to all other value comparisons in the smart parametric checking procedures.

B2. V_{t1} calculation criterion

If ESD devices A and B are connected in series, the total triggering voltage of the combined A+B ESD network is the sum of V_{t1} of A and B.

B3. V_{t1} comparison criterion

If ESD devices A and B are connected in parallel and A has a substantially higher V_{t1} than B does, then A is a non-critical ESD device and will be removed.

C. Holding point criteria

C1. Maximum V_h criterion

If ESD device A has a holding voltage, V_h, substantially higher than a given threshold holding voltage $V_{h\text{-}max}$, then device A cannot provide the required sufficiently low pad voltage clamping under ESD stresses and will be removed.

C2. V_h comparison criterion

If ESD devices A and B are connected in parallel and A has a substantially higher V_h than B does, then A will be removed.

D. R_{ON} criteria

D1. R_{ON} comparison criterion

If ESD devices A and B, with comparable V_{t1}, are connected in parallel and device A has a substantially larger R_{ON} than B does, then A would not take the ESD current and will be removed.

More checking criteria can be added for ESD design verification as one understands more about the ESD protection behaviors and any novel ESD protection mechanisms are discovered. However, the smart parametric checking mechanism remains the same.

4. ESDINSPECTOR, A SMART CHECKER

A new CAD tool, entitled *ESDInspector*, has been developed based upon the novel smart parametric ESD checking mechanism. ESDInspector, along with the arbitrary ESD device extractor, ESDExtractor [6], aims to perform efficient and accurate full-chip ESD protection design verification at layout level. Both ESDExtractor and ESDInspector are parts of a complete CAD package (entitled ESDCat) for full-chip ESD design verification. Fig. 2 illustrates the whole layout-level verification solution provided by ESDCat.

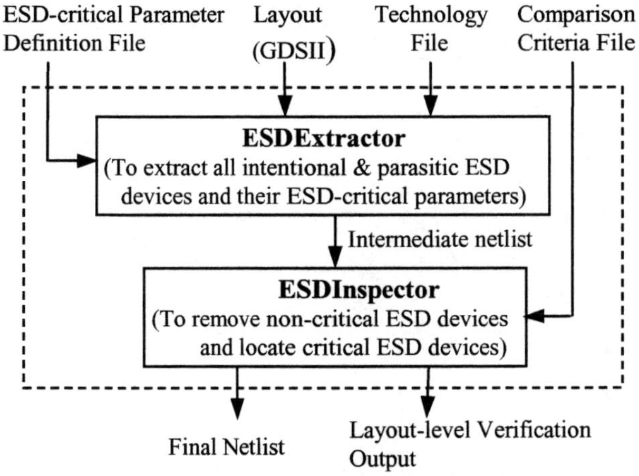

Fig. 2 The layout-level ESD protection design verification solution provided by ESDExtractor and ESDInspector.

4.1 ESDExtractor

The output of the ESDExtractor is an intermediate netlist including all intentional and parasitic ESD devices extracted from a layout file (in GDSII format). Extraction of arbitrary ESD devices are conducted based upon a novel subgraph isomorphism algorithm, discussed elsewhere [6].

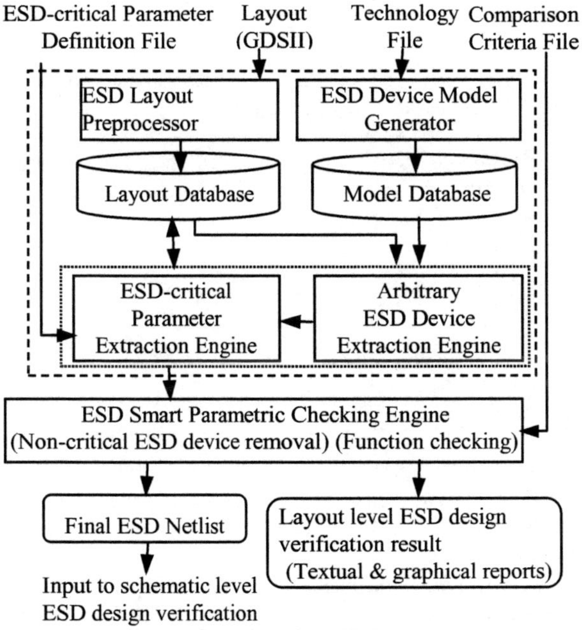

Fig. 3 A flow chart for ESDInspector

4.2 ESDInspector

After the auto extraction of ESD devices and related ESD-critical parameters, the smart parametric checking mechanism will be used to remove non-critical ESD devices and to locate critical ones. This is executed by the ESDInspector as shown in Fig.3, where the comparison engine will perform a comprehensive comparison routine upon the intermediate netlist using comparison criteria described previously. Both textual and graphical outputs are provided by the ESDInspector. A marker will flag where any critical ESD device is. In addition, a final ESD netlist is generated and that will be used for further ESD design verification at schematic level. Both ESDExtractor and ESDInspector are developed as technology-independent CAD tools written in C++ for both MS Windows and Unix operation systems.

5. APPLICATION EXAMPLE

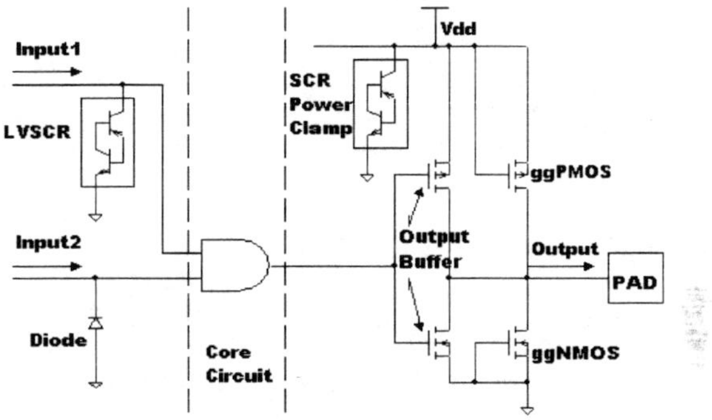

Fig.4 Schematic of an example ESD protection circuit

As an example, ESDExtractor and ESDInspector were applied to an IC chip with ESD protection circuitry shown in Fig. 4, where the core circuit being protected is a symbolic digital gate and an output CMOS buffer. Five types of ESD protection structures are used including a LVSCR (low voltage SCR) at input-1, a Diode at input-2, a complementary ggNMOS (ground-gate NMOS) and ggPMOS (grounded-gate PMOS) pair at output pad and a SCR as the power clamp. Fig. 5 shows the original chip layout in the GUI of ESDInspector. This chip was designed and fabricated in a commercial 0.35μm BiCMOS and tested for full function. Auto extraction of all ESD-type devices and the their parameters were conducted first by ESDExtractor, with the intermediate netlist generated listed below, where each extracted ESD device is presented by its name, terminal nodes and ESD-critical parameters in one line.

*complete netlist file created from D:/gdsii/demop6.gds
LVSCR1 Input1 Gnd $\beta_{NPN}\beta_{PNP}$=62.62758 Vt1=8.86156
It1=0.00504881 Vh=2.97375 Ih=0.0585548 Ron=8.36588
MVSCR1 Input1 Gnd $\beta_{NPN}\beta_{PNP}$=*56.84202 Vt1=9.78726*
It1=0.00572679 Vh=2.57119 Ih=0.0147246 Ron=8.36588
Diode-PW/NW1 Input2 Gnd Vt1=0.7 It1=0.00583679
Ron=187.36588
SCR1 Vdd Gnd $\beta_{NPN}\beta_{PNP}$=22.780656 Vt1=19.4619
It1=0.0238095 Vh=4.3142 Ih=0.100444 Ron=16
SCR2 Vdd Gnd $\beta_{NPN}\beta_{PNP}$=*11.1618 Vt1=19.5089 It1=0.01895*
Vh=4.37456 Ih=0.10233 Ron=17.09213
SCR3 Vdd Gnd $\beta_{NPN}\beta_{PNP}$ = *0.5588*
...
SCR6 Vdd Gnd $\beta_{NPN}\beta_{PNP}$ = *0.5588*
ggNMOS1 Output Gnd Gnd β=10.9947 Vt1=9.39176
It1=0.018315 Vh=7.60567 Ih=0.201368 Ron=19.66

ggPMOS1 Output Vdd Vdd β=3.27539 Vt1=9.66812 It1=0.0295047 Vh=7.6164116 Ih=0.1932792 Ron=14.375
ggPMOS2 Output Vdd Vdd β=3.27539 Vt1=9.66812 It1=0.0295047 Vh=7.6164116 Ih=0.1932792 Ron=14.375

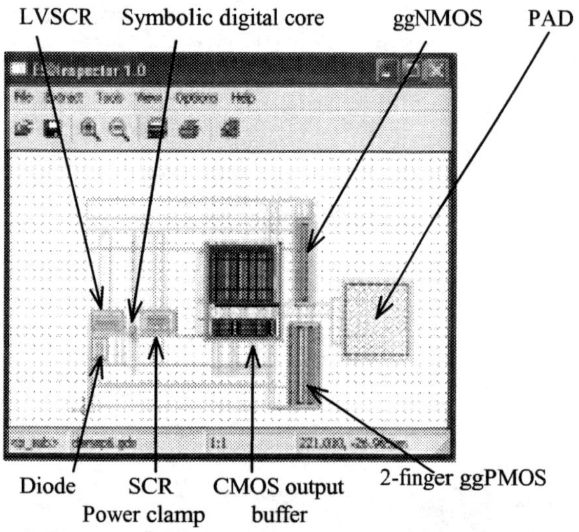

Fig. 5 Layout of the initial chip in GUI of ESDInspector.

This intermediate netlist correctly reports all intentional ESD protection devices, i.e., SCR1 for the SCR power clamp, LVSCR1 and Diode1 for the two input ESD protection devices, ggNMOS1 and ggPMOS1&2 for the output ESD protection devices. Note that two ggPMOS devices are extracted corresponding to the two-finger ggPMOS layout pattern. The extracted parasitic ESD-type devices include one middle-voltage SCR type device, MVSCR1 within the LVSCR1; and five extra SCR devices, i.e. SCR2 within the output buffer and SCR3/4/5/6 inside the core digital gate circuit. All parasitic ESD devices are printed in italics in the netlist.

Next, ESDInspector executes to eliminate any non-critical ESD devices extracted, using the smart parametric checking mechanism. In this example, the extracted parasitic MVSCR1 shows a much higher V_{t1} than LVSCR1 and SCR3-6 have β-products less than 1. As a result, they are non-critical ESD devices and should be removed. However, the parasitic SCR2 has a V_{t1} comparable to that of the intentional power clamp device, SCR1. Further, since SCR2 is in parallel with and has a discharging R_{ON} comparable to that of SCR1, potential risk exists for parasitic SCR2 to compete against the intentional SCR1 power clamp under ESD stresses. Hence, the parasitic SCR2 should be retained in the final ESD netlist, as listed below for further ESD design verification at layout or schematic level.

*final netlist file created from D:/gdsii/demop6.gds
LVSCR1 Input1 Gnd $\beta_{NPN}\beta_{PNP}$=62.62758 Vt1=8.86156 It1=0.00504881 Vh=2.97375 Ih=0.0585548 Ron=8.36588
Diode-PW/NW1 Input2 Gnd Vt1=0.7 It1=0.00583679 Ron=187.36588
SCR1 Vdd Gnd $\beta_{NPN}\beta_{PNP}$=22.780656 Vt1=19.4619 It1=0.0238095 Vh=4.3142 Ih=0.100444 Ron=16
SCR2 Vdd Gnd $\beta_{NPN}\beta_{PNP}$=11.1618 Vt1=19.5089 It1=0.01895 Vh=4.37456 Ih=0.10233 Ron=29.3
ggNMOS1 Output Gnd Gnd β=10.9947 Vt1=9.39176 It1=0.018315 Vh=7.60567 Ih=0.201368 Ron=19.66
ggPMOS1 Output Vdd Vdd β=3.27539 Vt1=9.66812 It1=0.0295047 Vh=7.6164116 Ih=0.1932792 Ron=14.375
ggPMOS2 Output Vdd Vdd β=3.27539 Vt1=9.66812 It1=0.0295047 Vh=7.6164116 Ih=0.1932792 Ron=14.375

Fig. 6 shows the graphical output for this circuit after the ESDInspector execution, with all devices in the final netlist shown and identified with frame markers for easy debugging. The run time of the whole execution by ESDExtractor and ESDInspector for this example circuit from prepossessing to GUI refresh is about 8.2s on Windows with 128MB memory.

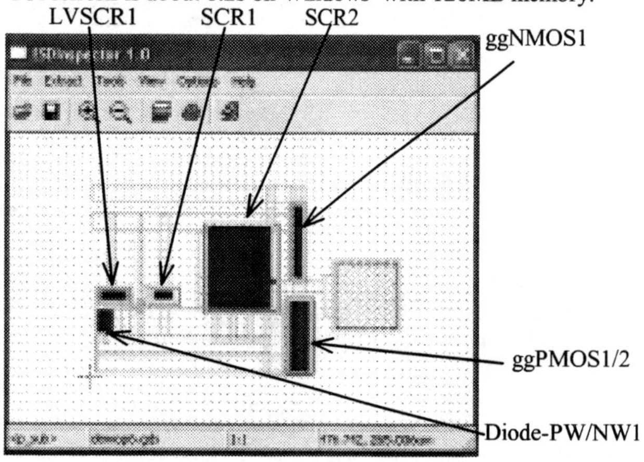

Fig. 6 Layout view after ESD device removal by ESDInspector identifies all critical ESD devices extracted as highlighted.

6. CONCLUSION

In conclusion, we report a new technology-independent CAD tool, ESDInspector, based on a new concept of *ESD-critical parameters* and a novel *smart parametric ESD checking mechanism*, developed to perform full-chip ESD protection design verification at layout level. A practical IC design example implemented in a 0.35µm BiCMOS is discussed to demonstrate the capacity of the ESDInspector

7. REFERENCES

[1] A.Wang, *On-Chip ESD Protection Design for ICs*, Klumer Academic, 2002, ISBN: 0-7923-7647-1.

[2] Ming-Dou Ker, *et al*, "Layout design and verification for cell library to improve ESD/Latchup reliability in deep-submicron CMOS technology", *IEEE CICC*, 1998, pp.537-540.

[3] T. Li, *et al*, "Automated extraction of parasitic BJTs for CMOS I/O circuits under ESD stress", *IEEE IRW*, 1997, pp.103-109.

[4] T. Li, *et al*, "Modeling, extraction and simulation of CMOS I/O circuits under ESD stress", *IEEE ISCAS,* 1998, pp.389-392.

[5] Q. Li, *et al*, "ESD design rule checker", *Proc. IEEE ISCAS*, 2001 pp499-502.

[6] R. Zhan, *et al,* "A technology-independent CAD tool for ESD protection device extraction" *IEEE ICCAD*, 2002, pp.510-513.

Fault Equivalence and Diagnostic Test Generation Using ATPG

Andreas Veneris, Robert Chang
University of Toronto
Dept ECE
Toronto, ON M5S 3G4
{veneris, rchang} @eecg.toronto.edu

Magdy S. Abadir
Motorola
7700 W. Parmer
Austin, TX 78729
m.abadir@motorola.com

Mandana Amiri
University of British Columbia
Dept ECE
Vancouver, BC V6T 1Z4
mandana@ece.ubc.ca

Abstract

Fault equivalence is an essential concept in digital design with significance in fault diagnosis, diagnostic test generation, testability analysis and logic synthesis. In this paper, an efficient algorithm to check whether two faults are equivalent is presented. If they are not equivalent, the algorithm returns a test vector that distinguishes them. The proposed approach is complete since for every pair of faults it either proves equivalence or it returns a distinguishing vector. This is performed with a simple hardware construction and a sequence of simulation/ATPG-based steps. Experiments on benchmark circuits demonstrate the competitiveness of the proposed method.

1 Introduction

Computing the complete set of fault equivalence classes in a combinational circuit is a classic problem in digital circuit design. Two faults are *functionally equivalent* (or *indistinguishable*) if no input test vector can distinguish them at primary outputs. Functional fault equivalence is a relation that allows faults in a circuit to be collapsed into disjoint sets of *equivalent fault classes*. Fault equivalence is essential in digital VLSI because it has significance in fault diagnosis [7], diagnostic test generation [5], testability analysis [7] nd logic synthesis [3] [7] [8] [9].

Methods to compute fault equivalence are classified as *structural* and *functional* [7]. Structural methods operate on the circuit graph to identify fault equivalence. These methods are fast but they have pessimistic results since they operate on fan-out free circuit regions only. Functional fault equivalence methods are more expensive but they typically identify more classes [1] [2]. These methods use logic implications and/or dominator information to prove equivalence. Since identifying logic implications is NP-hard [8], these methods do not utilize the complete set of logic implications and they may not return the complete set of fault equivalence classes.

In this paper, we present an efficient method that proves the equivalence of a pair of stuck-at faults. The method also returns a distinguishing vector if the faults are proven not to be equivalent. To simplify the discussion, we use the term "fault" to indicate a single stuck-at fault hereafter. To prove *fault equivalence* or perform *Diagnostic Automated Test Pattern Generation (DATPG)*, the method performs a sequence of simulation- and ATPG-based steps. It should be noted, ATPG is not exclusive to the method and other test generation and redundancy checking techniques (BDDs, SAT solvers etc) can be utilized.

The proposed methodology has a number of characteristics that make it attractive and practical. Unlike methods that alter existing ATPG tools [5], it uses conventional ATPG [4] [8] and a novel hardware construction to either prove equivalence of the fault pair or return a distinguishing vector. Therefore, it automatically benefits from advances in ATPG and remains straightforward to implement. It is also *complete* in the sense that for every fault pair it guarantees to prove equivalence or return a distinguishing vector, provided sufficient backtracking level in the ATPG engine. To the best of our knowledge, this is the first published results on the exact number of stuck-at fault equivalence classes for ISCAS'85 circuits.

The paper is organized as follows. Section 2 presents the two steps of the proposed functional fault collapsing and DATPG algorithm in terms of single stuck-at faults. Section 3 presents experiments and Section 4 concludes this work.

2 Fault Equivalence and DATPG

In this Section we present the *two* steps of the fault equivalence/DATPG method. The *first step* computes an approximation of fault equivalence classes using structural fault collapsing and input test vector simulation. Faults in the same class may or may not be equivalent, but faults in different classes are *guaranteed* to be not equivalent. The *second step* uses conventional ATPG and a novel hardware construction on pairs of faults in the same class as computed in Step 1 to either formally prove the faults are equivalent or perform DATPG. Therefore, more equivalent fault classes may be identified in this step.

2.1 Parallel Vector Simulation

The implementation starts with structural fault collapsing [7] to prove faults that are structurally proximal as equivalent so that only one representative fault from each such set needs be considered in later steps of the algorithm. Let set F be the complete set of representative faults. The faults in F are examined for equivalence by *Parallel Vector Simulation (PVS)*.

PVS is a simulation-based procedure that classifies these faults into potentially equivalence classes $F_1, F_2, \dots, F_n$ with respect to an input test vector set T. PVS identifies (maps) two faults f_A and f_B as *potentially equivalent* if and only if f_A and f_B give the exact same primary output responses for each test vector from the set T. If faults f_A and f_B are potentially equivalent, they are placed into the same class F_i. Faults in different classes are guaranteed not to

```
Parallel_Vector_Simulation(C, F, T)

(1) Simulate test vectors in T and create
    indexed bit-lists at every circuit line
(2) For every fault f s-a-v on line l do
(3)     fault_signature=0
(4)     set bit-list of l to value v
(5)     simulate at fan-out cone of l
(6)     update fault_signature
(7)     restore bit-lists at l fan-out cone
(8) Group faults with same signatures in
    same class F_i, i=1 ... n
```

Figure 1: Pseudocode for PVS (Step 1)

be equivalent since they already have different responses for some vector(s) in T. In fact, this vector(s) is also a distinguished vector for the faults in these classes.

Pseudocode for PVS is given in Fig. 1. The input to PVS is a circuit C, the collapsed set of faults F and a set of input test vectors T. In experiments, T is a relatively small set of 100-1000 test vectors with high stuck-at fault coverage [6]. The output of PVS is a set of fault classes $F_1, F_2, \ldots, F_n$ such that two faults f_a and f_b are in the same class F_i if and only if they have the exact same responses for all vectors in T.

At first, PVS simulates in parallel [7] all test vectors in T and stores the logic value of each line in an indexed bit-list [9] (Fig. 1, line (1)). Next, for every stuck-at v fault $f \in F$ on line l, value v is injected on l and simulated at the fan-out cone of l. The primary output bit-lists are treated as integer signatures for fault f and test set T (lines 2-6) before values are restored in line 7. This process is repeated for every fault in F and faults with the same signature are grouped together in line 8.

2.2 Fault Equivalence and DATPG

Given a pair of faults (f_A, f_B) from the same class F_i, Step 2 performs a hardware construction and employs conventional ATPG to *formally* prove their equivalence or return a test vector to distinguishe them (DATPG). It should be noted, formal fault equivalence or DATPG is performed in an *atomic step* for each pair of faults. We now outline the theory and implementation of this step in detail.

Assume, without loss of generality, faults f_A s-a-0 and f_B s-a-1 on lines l_A and l_B of the circuit. To examine their equivalence, the algorithm attaches two multiplexers, MUX_A and MUX_B, with common select line S. The 1-input of MUX_A is line l_A while the 0-input of the multiplexer is a constant logic value 0. Intuitively, a constant 0 indicates a s-a-0 fault on l_A. Similarly, the 0-input of MUX_B is the original line l_B and the 1-input is a constant 1.

This construction allows us to simulate the original circuit under presence of fault f_A when $S = 0$ and the original circuit under presence of f_B when $S = 1$. Therefore, if ATPG for select line S s-a-0 (ATPG for select line S s-a-1 produces similar results) exhausts the solution space to return no test vector and reports that the fault on S is redundant, then (f_A, f_B) is an equivalent fault pair [9]. In other words, the original circuit under the presense of each of the (f_A, f_B) faults behaves identically. Otherwise, the input test vector returned is a vector that distinguishes the two faults.

To illustrate this process, we can employ pair of values to indicate

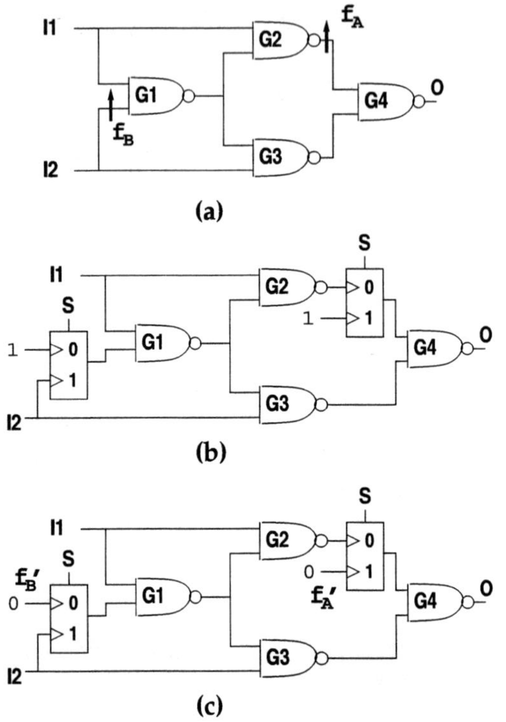

Figure 2: Circuits for Examples 1 and 2

simulation of *two* faulty circuits; one under the presence of f_A and the other under the presence of f_B. If the stuck-at fault on select line S is redundant, it implies that no $0/1$ and no $1/0$ value propagate(s) to any primary output. In other words, the two faulty circuits produce the same response for *all* input test vectors and the two faults are indistinguishable.

On the other hand, if a single $0/1$ $(1/0)$ difference is propagated to a primary output, then one of the two faults is *guaranteed* to be detected. Which fault is detected depends on the logic simulation value; if logic simulation gives 0 then f_B (f_A) is detected at the primary output since circuit under presence of f_A (f_B) and logic simulation have identical values. Both faults are detected if appropriate $0/1$ or $1/0$ simulation values propagate at different primary outputs. The examples that follow illustrate the above procedures.

```
Fault_Equivalence_DATPG(C, F_1, ..., F_n)

flag=0
for i=1 to n do
    randomly select f from F_i
    for every f' in Fi do
        perform the MUX construction
        if f' not equivalent to f do
            flag=1
            place f' in F_n+1
            store distinguishing vector
  if flag=1
     flag=0
     n=n+1
```

Figure 3: Fault Equivalence and DATPG

Table 1: Parallel Vector Simulation (Step 1)

ckt name	# of initial faults	faults after collaps.	# ATOM vectors	# of fault classes after PVS				CPU time (sec)
				ATOM vectors	ATOM and 500 random	500 random	1000 random	
c432	798	419	110	371	417	413	418	0.11
c499	2434	1314	127	901	1076	1027	1092	0.59
c880	1770	940	133	857	889	853	868	0.17
c1355	2412	1302	192	1046	1088	1010	1079	0.89
c1908	1802	975	210	714	748	684	767	0.50
c2670	3177	1627	242	1141	1178	1160	1184	0.42
c3540	4116	2143	264	1408	1548	1541	1580	10.9
c5315	7042	3743	216	2971	3404	3381	3415	2.06
c6288	14303	7479	64	6397	6597	6593	6597	0.99
c7552	10081	5321	393	4186	4280	4180	4273	5.98

Example 1: Consider the circuit in Fig. 2(a) and assume two faults from the same class F_i $f_A = G_2 \rightarrow G_4$ and $f_B = I_2 \rightarrow G_1$ both s-a-1. To test their equivalence, we can place two multiplexers, shown as boxes in Fig. 2(b), with common select line S. The 0-input to the first multiplexer is line G_2 and the 1-input of that multiplexer is a logic 1 to represent the presence of a s-a-1 fault. Similarly, the 0-input of the second multiplexer is a logic 1 while I_2 feeds the other input. In both cases, the output of each multiplexer connects to the original output in the circuit. The reader can verify that when $S = 0$, we operate on a circuit equivalent to the one in Fig. 2(a) under the presence of f_A and when $S = 1$ we operate on a circuit equivalent to the one in Fig. 2(a) under the presence of f_B. ATPG for select line S s-a-0 returns the fault is redundant. This indicates that the two circuits are functionally equivalent which, in turn, confirms that (f_A, f_B) is an equivalent fault pair.

Example 2: Consider again circuit in Fig. 2(a) and faults $f'_A = G_2 \rightarrow G_4$ and $f'_B = I_2 \rightarrow G_1$ this time both stuck at logic 0. A similar multiplexer construction as in Fig. 2(b) gives circuit shown in Fig. 2(c). The difference is that a logic 0 is placed on appropriate multiplexer inputs to indicate a stuck-at-0 fault. ATPG on common select line S s-a-0 returns test vector $(I_1, I_2) = (0, 0)$. This proves that the fault on S is not redundant and faults f'_A and f'_B are not equivalent. This is true since the test vector returned detects fault f'_A but does not even excite fault f'_B. In this case, the construction returns a distinguishing vector for fault pair (f'_A, f'_B).

Fig. 3 contains pseudocode for Step 2. For each class F_i ($i = 1 \ldots n$), a representative f is randomly selected. For each other member $f' \in F_i$, we perform the multiplexer construction to check whether f and f' are equivalent (lines 4-5). If they are not equivalent, f' (and all other such non-equivalent faults from F_i) is placed in new class F_{n+1} (lines 6-9) which will be examined later and the distinguishing vector is recorded. Observe, any such fault f' is guaranteed not to be equivalent with faults in any class $F_1, \ldots, F_{i-1}, F_{i+1}, \ldots, F_n$ by PVS. Faults placed in F_{n+1} may or may not be equivalent. Therefore, class F_{n+1} may get decomposed into new classes when it is examined later. The set of classes returned upon termination of the algorithm are also the exact fault equivalence classes for circuit C and fault set F.

3 Experiments

We implemented and ran the proposed method on an Ultra 5 SUN workstation with 128 Mb of memory. The details of the ATPG engine used can be found in [4][8]. We use a relatively low level 1 for recursive learning to provide a fair comparison with state-of-the-art diagnostic test generation tool DIATEST [5] and confirm the competitiveness of the approach. We evaluate the proposed techniques on ISCAS'85 combinational circuits optimized for area. Test vectors with high stuck-at fault coverage (ATOM vectors) are computed as in [6]. Run-times reported are in seconds.

Table 1 contains information about the first step of the algorithm. The first column of the table shows the circuit name and the second column contains the total number of stuck-at faults. This is roughly twice the number of lines, including branches. The third column shows the faults after structural fault collapsing [7].

We examine the performance of PVS with a set of random and stuck-at fault input test vectors. The number of ATOM vectors generated is shown in column 4. Columns 5–8 of Table 1 show the number of distinct fault classes upon termination of PVS for four different cases with respect to the test vector set T used: *(i)* ATOM vectors, *(ii)* ATOM vectors and 500 random vectors, *(iii)* 500 random vectors, and *(iv)* 1000 random vectors.

Intuitively, the more vectors we simulate the more accurate the results we expect in terms of number of classes, as discussed earlier. A study of the numbers indicates that a relatively small set of random vectors (case *(iv)*) gives sufficient resolution. In most cases, random simulation gives good resolution and there is little to gain with a pre-computed set of stuck-at fault test vectors.

This is also illustrated in Fig. 4 that depicts the number n of fault classes $F_1, F_2, \ldots, F_n$ versus the number of random vectors simulated. It is seen that the number of fault classes converges with a relatively small number of vectors. We use the classes from case *(iv)* as input to Step 2. The last column of the table contains the total run-time for Step 1.

Table 2 contains information that pertain to Step 2. Given fault pair (f_A, f_B), it tests whether the two faults are equivalent and returns a distinguishing vector if they are not. Columns 2-4 of Table 2 show values that pertain to the case when the faults are equivalent. The total number of fault pairs checked and the number of final (complete) fault classes are found in columns 2 and 3.

The relative error for PVS (Step 1), a simulation-based process, when compared to the formal engine of Step 2 is found in column 4. It is seen that in many cases the relative error is rather small (less than 10%). This suggests that simulation of random vectors provides in most cases sufficient resolution to compute fault equivalence. Therefore, the designer is presented with a relatively small trade-off between time and accuracy.

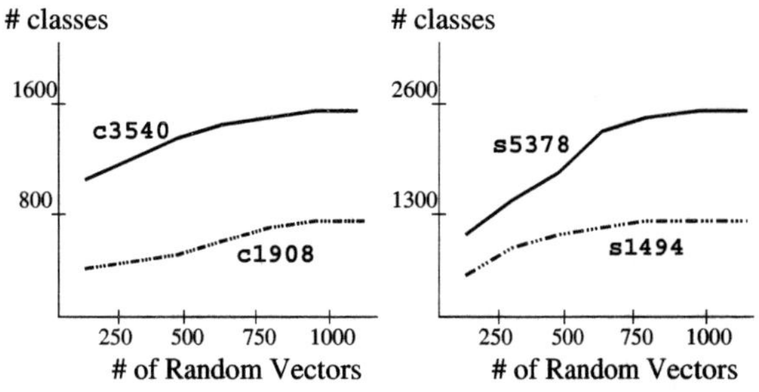

Figure 4: Performance of PVS

DATPG results are found in columns 5-7. Column 5 contains the number of distinguishable fault pairs checked. Columns 6 and 7 contain the *manner* these faults are detected. Recall from subsection 2.2, DATPG guarantees to detect one fault but it may detect both faults at *different* primary outputs. The numbers in these columns indicate that in as many as half of the cases, DATPG returns a test vector to distinguish *both* faults at different primary output (column 7). Column 8 of the Table 2 contains the average run-time for ATPG when the faults are equivalent (redundancy checking).

The last two columns in Table 2 provide comparison results with DIATEST [5] for the same list of fault pairs that are not equivalent. It is seen, that conventional ATPG and the hardware construction presented here provides an attractive alternative to a DATPG-specific engine. In fact, larger values of implication learning [8] will speed the ATPG tool and further improve performance of the method. This confirms the practicality of the approach that automatically benefits from advances in ATPG.

4 Conclusions

A method for fault equivalence and diagnostic test generation was presented. The method is practical since it uses conventional ATPG and a simple hardware construction to prove equivalence or return distinguishing test vectors. Experiments demonstrate its robustness, effectiveness and competitiveness.

Acknowledgments

The authors would like to thank Prof. W. K. Fuchs, Dr. E. Amyeen and S. Seyedi for their technical comments and support in this work.

References

[1] V. D. Agrawal, D. H. Baik, Y. C. Kim and K. K. Saluja, "Exclusive Test and its Applications to Fault Diagnosis," in *Proc. of IEEE Int'l Conf. on VLSI Design,* 2003.

[2] M. E. Amyeen, W. K. Fuchs, I. Pomeranz and V. Boppana, "Fault Equivalence identification using redundancy information and static and dynamic extraction," in *Proc. of IEEE VLSI Test Symposium, pp. 124-130,* 2001.

[3] S. C. Chang and M. Marek-Sadowska, "Perturb and Simplify: Multi-Level Boolean Network Optimizer," in *Proc. Int'l Conference on Computer-Aided Design, pp. 2-5,* 1994.

[4] H. Fujiwara and T. Shimono, "On the Acceleration of Test Generation Algorithms," in *IEEE Trans. on Computers, vol. C-32, no. 12,* December 1983.

[5] T. Gruning, U. Mahlstedt, and H. Koopmeiners, "DIATEST: A fast diagnostic test pattern generator for combinational circuits," in *Proc. Int'l Conf. on Computer-Aided Design, pp. 194-197,* 1991.

[6] I. Hamzaoglu and J. H. Patel, "New Techniques for Deterministic Test Pattern Generation," in *Proc. of VLSI Test Symposium, pp. 446-452,* 1998.

[7] N. Jha and S. Gupta, *Testing of Digital Systems*, Cambridge University Press, 2003.

[8] W. Kunz and D. K. Pradhan, "Recursive Learning: A New Implication Technique for Efficient Solutions to CAD Problems–Test, Verification, and Optimization," in *IEEE Trans. on Computer-Aided Design, vol. 13, no. 9, pp. 1143-1158* September 1994.

[9] A. Veneris and M. S. Abadir, "Design Rewiring Using ATPG," in *Proc. IEEE Trans. on Computer-Aided Design, vol. 21, no. 12, pp. 1469-1479,* December 2002.

Table 2: Fault Equivalence and DATPG (Step 2)

ckt	ATPG			DATPG			CPU (sec)	CPU comparison	
name	# pairs	# final	% err	# pairs	faults detected		fault	DATPG	DIATEST
	checked	classes	PVS	checked	one	both	equivalence	proposed	[7]
c432	1	419	0.2	111	70	41	0.03	0.00	0.00
c499	17	1106	1.3	84	28	56	0.14	0.02	0.08
c880	104	892	2.7	128	45	83	0.02	0.01	0.01
c1355	18	1094	1.4	80	25	55	0.22	0.07	0.05
c1908	335	830	7.6	129	39	90	0.06	0.05	0.01
c2670	6203	1443	18.0	127	73	54	0.07	0.05	0.05
c3540	8914	1839	14.1	116	62	54	0.10	0.09	0.09
c5315	184	3480	1.9	129	42	87	0.11	0.09	0.09
c6288	892	6973	5.4	94	77	17	0.42	0.05	0.09
c7552	8768	4737	9.8	130	46	84	0.50	0.16	0.13

A HYBRID-TYPE TEST PATTERN GENERATING MECHANISM

Chuen-Yau Chen and An-Chi Hsu

Department of Electrical Engineering, National Yunlin University of Science & Technology
Touliu City, Yunlin 64045, Taiwan, R.O.C.
Email: cychen@ieee.org

ABSTRACT

In this paper, we blend the weighted-random-pattern generator and the controllable-linear-feed-back-shift register to develop a hybrid-type test pattern generating mechanism. The whole testing is performed in two phases. The weighted-random-pattern generator drops some of the faults from the fault list during the first phase. Remain faults will be tested by the controllable-linear-feed-back-shift register during the second phase. We adopt controllable-linear-feed-back-shift register to generate the deterministic patterns instead of modifying the configuration of the weighted-random-pattern generator such that a better fault coverage can be achieved with a lower hardware penalty and a shorter test length.

1. INTRODUCTION

Buit-in self test (BIST) is the technique embedding the testbench in the circuits under test (CUT). This technique provides a suitable testing mechanism for the CUT, especially in the system-on-a-chip (SoC) era. BIST can be classified into three levels: the system level, the board level, and the chip level [1]. A chip-level BIST architecture includes three principal units: test pattern generation (TPG), output response analysis (ORA), and BIST controller. Although more and more TPG techniques have been developed, they are designed on the basis of the linear-feedback-shift register (LFSR) [1]-[10] and weighted-random-pattern generator (WRPG) [11]-[15]. The objectives of these techniques are the same: high speed, high fault coverage, and low hardware cost.

With regard to the LFSR-based test pattern generator, the testing efficiency heavily depends on the states of test patterns and test pattern length. These two factors are determined by the locations and the number of the exclusive-OR (XOR) gates in the LFSR. The advantage of an LFSR test pattern generator is that it can provide a complete set of test pattern for achieving a best fault coverage. However, because of the random (pseudo-random) sequence of the test patterns, getting an exact test pattern we need urgently in time is difficult. It takes much redundant time to generate many redundant test patterns. To avoid this problem, the LFSR-like test pattern generators called controllable linear feedback shift register (CLFSR) are developed [2]-[4].

Many researchers have involved themselves on developing the structures of WRPG [11]-[15] which are based on either structural analysis or deterministic test sets. For a structural-analysis-based WRPG, the weights are determined by testability measures or heuristic methods. The advantage is that it can generate enough weight sets. However, a higher fault coverage can not be achieved in this scheme. For a deterministic-test-sets-based one, a higher fault coverage can be achieved because the generated test patterns are composed of the weighted bits. However, once the weight sets are changed or the number of the test patterns is increased, the hardware should be reconfigured.

This work takes the concepts of CLFSR and WRPG to propose a hybrid-type test pattern generating (HTTPG) mechanism. WRPG is adopted to drop some faults in the original fault list before CLFSR taking this testing mission such that the duty of CLFSR can be alleviated. CFLSR is adopted to continue testing the faults that have not been detected by WRPG such that the WRPG can be implemented with a lower hardware complexity.

This work was supported by the Chip Implementation Center and the National Science Council, Republic of China, under Grant NSC-92-2218-E-224-007.

2. CONVENTIONAL TEST PATTERN GENERATORS

2.1. Linear Feedback Shift Register

An n-stage linear feedback shift register (LFSR) is composed of the delay elements (usually D flip-flops) and the exclusive-OR (XOR) gates that are configured as shown in Fig. 1. The operation of an LFSR can be characterized by [4]

$$\begin{bmatrix} S_0^+ \\ S_1^+ \\ \vdots \\ S_{n-3}^+ \\ S_{n-2}^+ \\ S_{n-1}^+ \end{bmatrix} = \begin{bmatrix} 0 & 1 & 0 & \cdots & 0 & 0 \\ 0 & 0 & 1 & \cdots & 0 & 0 \\ \vdots & \vdots & \vdots & \ddots & \vdots & \vdots \\ 0 & 0 & 0 & \cdots & 1 & 0 \\ 0 & 0 & 0 & \cdots & 0 & 1 \\ 1 & g_1 & g_2 & \cdots & g_{n-2} & g_{n-1} \end{bmatrix} \begin{bmatrix} S_0 \\ S_1 \\ \vdots \\ S_{n-3} \\ S_{n-2} \\ S_{n-1} \end{bmatrix} \quad (1)$$

where $\mathbf{S} = [S_0 \cdots S_{n-1}]^T$ stands for the present state of the LFSR, $\mathbf{S}^+ = [S_0^+ \cdots S_{n-1}^+]^T$ stands for the next state of the LFSR, and $g_i \in \{0, 1\} (i = 1, \cdots, n-1)$ specifies the existence of the feedback path from S_i. The combination of the feedback paths from S_i's determines the sequence of $\mathbf{S}^+$ appears. The LFSR can generate k distinct *nonzero* states where k has the maximum value of $2^n - 1$.

2.2. Weighted Random Pattern Generator (WRPG)

Fig. 2 shows the block diagram of the WRPG with three principle units: the LFSR with n bits, the weight generation logic composed of $m(= 2^n - 1)$ Boolean functions, and an array of m-to-n multiplexors. The LFSR generates n-bit random test patterns $\mathbf{E} = [e_1 \cdots e_n]$. The weight generator logic is designed to receive the pattern $\mathbf{E}$, and generate m bits of which the probability of the occurrence of '1' at $f_1, \cdots, f_m$ are $p_1, \cdots, p_m$, respectively. At the final stage, n of the m bits from f_i $(i = 1, \cdots, m)$ are selected by the array

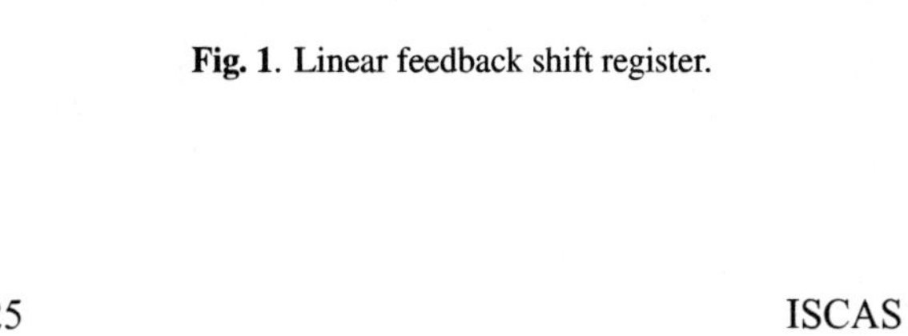

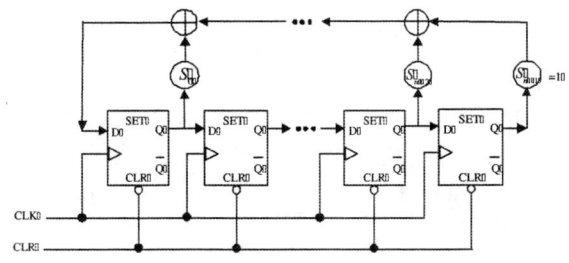

Fig. 1. Linear feedback shift register.

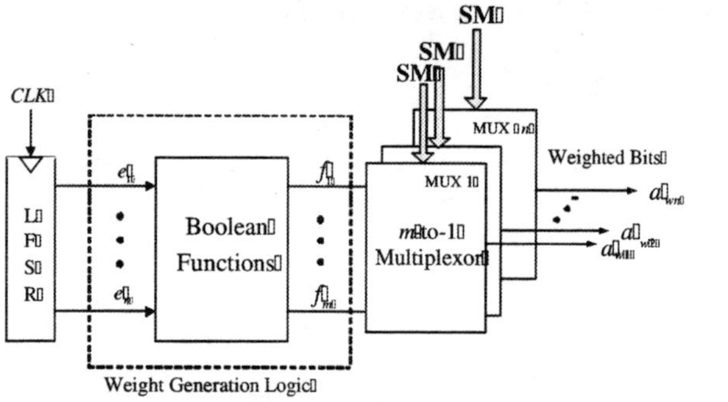

Fig. 2. Weighted random pattern generator.

of n parallel multiplexors controlled by the control signals $\mathbf{SM} = [SM_1 \cdots SM_n]$. These n selected bits are called the weighted bits $\mathbf{A_W} = [a_{w1} \cdots a_{wn}]$ that will be applied to the inputs of CUT for running the fault simulation.

3. HYBRID-TYPE TEST PATTERN GENERATING (HTTPG) MECHANISM

The flow chart in Fig. 3 shows the new test pattern generating mechanism. The job of the whole testing can be divided into two phases. They are PHASE 1, the weighted-random-pattern generator (WRPG) phase; PHASE 2, the controllable linear feedback shift register (CLFSR) phase. At the beginning, some detected faults f_d's are dropped from the original fault list L_{f_p} by running the fault simulations with the test patterns generated by automatic test pattern generator (ATPG). The remain faults f_{r0}'s are then transferred to PHASE 1 that will continue the fault simulations. In PHASE 1, ATPG determines the deterministic test patterns for the faults f_{r0}'s for calculating the weight sets $\mathbf{W}$'s of the CUT. Then, the WRPG will generate the weighted bits for testing the CUT. The faults that cannot be tested in this phase are called the Random Pattern Resistant (RPR) Faults f_{RPR}. These RPR faults are then sent to PHASE 2, and the CLFSR continues the fault simulations. Fig. 4 is the hardware architecture of the proposed CLFSR that shows the detail operations in PHASE 2. In PHASE 2, ATPG generates the corresponding deterministic test pattern $\mathbf{D} = [d_1 \cdots d_n]$ ($d_n \in \{0, 1\}$) for each RPR fault f_{RPR}. The control signal generator will calculate the required control signal $\mathbf{C} = [c_1 \cdots c_n]$ by

$$[c_1 \cdots c_n] = [d_1 \cdots d_n] \oplus [e_1 \cdots e_n] \qquad (2)$$

where $\mathbf{E} = [e_1 \cdots e_n]$ is the pattern set by the LFSR initially and '$\oplus$' is the bit-wise exclusive-OR (XOR) operator. The resultant control signal $\mathbf{C} = [c_1 \cdots c_n]$ are then applied to the Tuner for generating the test pattern $\mathbf{A} = [a_1 \cdots a_n]$ by

$$[a_1 \cdots a_n] = [c_1 \cdots c_n] \oplus [e_1 \cdots e_n] \qquad (3)$$

that is identical to the deterministic test pattern $\mathbf{D} = [d_1 \cdots d_n]$.

4. HARDWARE ARCHITECTURE OF THE HTTPG MECHANISM

Fig. 5 shows the hardware architecture of the new test pattern generator. This test pattern generator can be divided into two portions. They are the portion of WRPG and the portion of CLFSR. One LFSR is shared by WRPG and CLFSR. The patterns $\mathbf{E} = [e_1 \cdots e_n]$ generated by the LFSR are transferred to either the WRPG portion or the CLFSR portion according to the procedure control signal (PCS) that determines the operations of the demultiplexors. When $PCS = 0$, the WRPG is activated that will generate weighted bits $\mathbf{A_W} = [a_{w1} \cdots a_{wn}]$ to test the CUT. When $PCS = 1$, the CLFSR is activated. The patterns sent from LFSR will be tuned by the control signal $\mathbf{C}$. Then, the tuned test patterns $\mathbf{A} = [a_1 \cdots a_n]$ will be sent to test CUT.

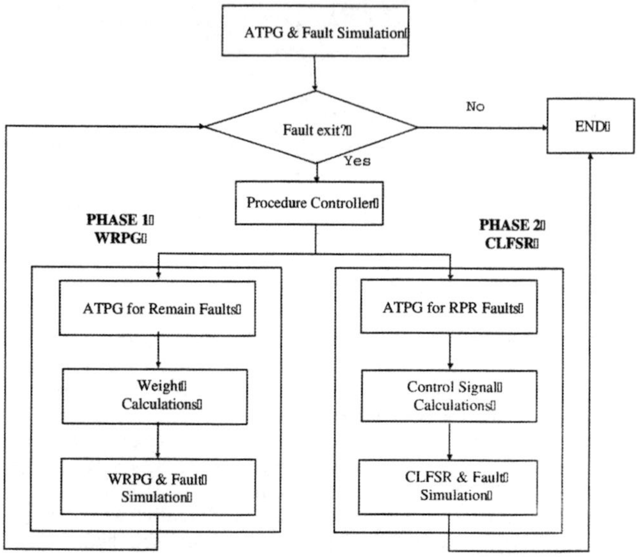

Fig. 3. Flow chart for the HTTPG mechanism.

5. SIMULATION RESULTS

The ISCAS-85 and ISCAS-89 circuit families are used as the benchmarks for verifying the function and performance of the proposed HTTPG mechanism. In this work, the width of the LFSR is chosen to be 3 bits that is actually $\mathbf{E} = [e_1 \cdots e_n]$ where $n = 3$. The pattern $\mathbf{E} = [e_1 \cdots e_3]$ is then sent to 3 parallel 1-to-2 demultiplexors (DeMUX) controlled by the procedure control signal PCS. In WRPG, the Boolean functions for the weight generating logic unit are designed as

$$f_1 = e_1 \cdot e_2 \cdot e_3 \qquad (4)$$

$$f_2 = e_2 \cdot e_3 \qquad (5)$$

$$f_3 = \overline{e_1 + e_2 \cdot e_3} \qquad (6)$$

$$f_4 = e_3 \qquad (7)$$

$$f_5 = e_1 + e_2 \cdot e_3 \qquad (8)$$

$$f_6 = \overline{e_2 \cdot e_3} \qquad (9)$$

$$f_7 = \overline{e_1 \cdot e_2 \cdot e_3} \qquad (10)$$

which provide the weight set $W = \{\frac{1}{8}, \frac{2}{8}, \frac{3}{8}, \frac{4}{8}, \frac{5}{8}, \frac{6}{8}, \frac{7}{8}\}$. The whole simulation is run by Turbo-Scan and Turbo-Fault developed by SynTest [16]. Turbo-Scan performs ATPG, and Turbo-Fault performs fault simulation.

5.1. Simulations on ISCAS-85 Circuit Family

With regards to the simulation results of ISCAS-85 circuit family, we compare this work with GLFSR [9] and WRPG [13]. Table 1 lists the comparisons of hardware complexity, fault coverage, and number of test patterns.

Consider the hardware complexity on the average. This work uses an 3-bit LFSR, so does the GLFSR scheme. Moreover, this work requires three extra XOR gates while the GLFSR requires four extra XOR gates. For the number of weight sets, only one weight set is sufficient for this work while the WRPG scheme requires about three weight sets.

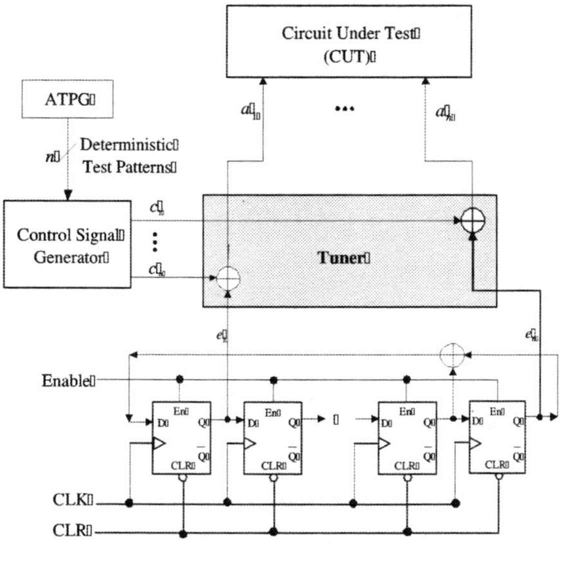

Fig. 4. Controllable linear feedback shift register.

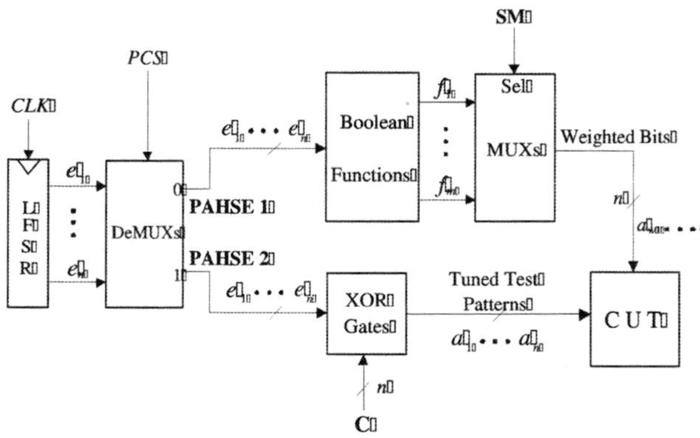

Fig. 5. Hardware architecture of HTTPG mechanism.

Consider the fault coverage on the average. This work performs a fault coverage of 97.6% that is higher than that performed by either GLFSR scheme or WRPG scheme.

Consider the number of test patterns on the average. This work requires about 740 test patterns which is much less than that required by the GLFSR scheme.

5.2. Simulations on ISCAS-89 Circuit Family

With regards to the simulation results of ISCAS-89 circuit family, we compare this work with WRPG [13]. Table 2 lists the comparisons of hardware complexity, fault coverage, and number of test patterns.

Consider the hardware complexity. This work requires an 3-bit LFSR and three extra XOR gates. For the number of weight sets, this work requires only one weight set while the WRPG scheme requires three weight sets on the average.

Consider the fault coverage on the average. This work performs an average fault-coverage of 98.54% which is higher than that performed by the WRPG scheme.

Consider the number of test patterns on the average. This work requires about 1213 test patterns which is much less than that required by the WRPG scheme.

6. CONCLUSION

We have proposed a hybrid-type test pattern generator based on WRPG and CLFSR in this paper. The WRPG and CLFSR cooperate with each other to finish the fault simulation by partitioning the whole testing job into two phases. WRPG perform the first-phase fault simulations to reduce the original fault list. CLFSR perform the second-phase fault simulations to test the RPR faults that have not been detected in the first phase and terminate the fault simulation process. Simulation results on the benchmarks show that this hybrid-type test pattern generator achieves high fault coverage of 97.6% for the ISCAS-85 circuit family and 98.54% for the ISCAS-89 circuit family with a reasonable number of test patterns and lower hardware cost.

7. REFERENCES

[1] M. Abramovici, M. A. Breuer, and A. D. Friedman, *Digital Systems Testing and Testable Design*, Computer Science Press, pp. 181-199, 432-448, 457-471.

[2] V. D. Agrawal, C. R. Kime, and K. K. Saluja, "A tutorial on built-in self-test part 1: principles", *IEEE Design & Test of Computers*, vol. 10, pp.73-82, Mar. 1993.

[3] V. D. Agrawal, C. R. Kime, and K. K. Saluja, "A tutorial on built-in self-test part 2: applications", *IEEE Design & Test of Computers*, vol. 10, pp. 69-77, June 1993.

[4] A. J. Al-Khalili and D. Al-Khalili, "A controlled probability random pulse generator suitable for VLSI implementation", *IEEE Trans. Instrum.*, vol. 39, pp. 168-174, Feb. 1990.

[5] B. Vasudevan, D. E. Ross, M. Gala, and K. L. Watson, "LFSR based deterministic hardware for at-speed BIST", in *Proc. 7th Annu. VLSI Test Symp.*, Apr. 1993, pp. 201-207.

[6] K. Douglas and M. Samiha, "Controllable LFSR for BIST", in *Proc. IEEE Instrumentation and Measurement Technology Conference*, May 2000, pp. 223-228.

[7] J. Savir and W. H. McAnney, "A multiple seed linear feedback shift register", *IEEE Trans. Comput.*, pp. 250-252, Feb. 1992.

[8] N. A. Touba and E. J. McCluskey, "Altering a pseudo-random bit sequence for scan-based BIST", in *Proc. Int. Test Conf.*, Oct. 1996, pp. 167-175.

[9] D. K. Pradhan and M. Chatterjee, "GLFSR-A new test pattern generator for bulit-in-self-test", *IEEE Trans. Computer-Aided Design*, vol. 18, pp. 238-247, Feb. 1999.

[10] L. T Wanf and E. T. McCluskey, "Hybrid designs generating maximum-length sequences", *IEEE Trans. Computer-Aided Design*, vol. 7, pp. 91-99, Jan. 1988.

[11] B. Reeb and H. J. Wunderlich, "Deterministic pattern generation for weighted random pattern testing", in *Proc. European Design and Test Conf.*, Mar. 1996, pp. 30-36.

[12] H. S. Kim, J. K. Lee, and S. Kang, "A new multiple weight set generation", in *Proc. Int. Computer Design Conf.*, Sept. 2001, pp. 513-514.

[13] M. Bershteym, "Calculation of multiple sets of weights for weighted random testing", in *Proc. Int. Test Conf.*, pp. 1031-1040, Oct. 1993.

[14] F. Brglez, G. Gloster, and G. Kedem, "Built-in self-test with weighted random pattern hardware", in *Proc. Int. Computer Design Conf.*, Sept. 1990, pp.161-166.

[15] H. S. Kim, J. K. Lee, and S. Kang, "A heuristic for multiple weight set generation", in *Proc. IEEE Int. ICCD. Conf.*, Sept. 2001, pp. 513-514.

[16] *SynTest Technologies*, Inc., Hsinchu, Taiwan.

Table 1. Performance factors comparisons. (Benchmark: ISCAS-85 circuit family)

ISCAS-85 Circuit Family	Number of bits		XORs		Weight Sets		Fault Coverage			Number of Test Patterns		
	This work	GLFSR [9]	This work	GLFSR [9]	This work	WRPG [13]	This work	GLFSR [9]	WRPG [13]	This work	GLFSR [9]	WRPG [13]
c17	3	–	3	–	1	–	100%	–	–	133	–	–
c432	3	–	3	–	1	1	98.28%	–	99.6%	166	–	332
c499	3	3	3	4	1	2	97.06%	95%	99.6%	198	256	376
c880	3	3	3	4	1	2	95.37%	95%	96.6%	289	210	376
c1335	3	3	3	4	1	1	99.23%	95%	98.6%	308	672	759
c1908	3	3	3	4	1	5	99.53%	95%	98.6%	350	880	1321
c2670	3	3	3	4	1	2	91.15%	90%	88.0%	1051	1280	708
c3540	3	3	3	4	1	3	95.12%	95%	98.2%	2830	420	1044
c5315	3	–	3	–	1	2	99.65%	–	99.8%	1043	–	1083
c6288	3	–	3	–	1	2	99.42%	–	100%	193	–	126
c7552	3	3	3	4	1	3	98.83%	90%	94.5%	1584	3008	1004
Average	3	3	3	4	1	3.1	97.6%	93.57%	97.35%	740.5	960.9	712.9

Table 2. Performance factors comparisons. (Benchmark: ISCAS-89 circuit family)

ISCAS-89 Circuit Family	LFSR bits	XORs	Weight Sets		Fault Coverage		Number of Test Patterns	
	This work	This work	This work	WRPG [13]	This work	WRPG [13]	This work	WRPG [13]
s27	3	3	1	–	100%	–	87	–
s298	3	3	1	1	99.78%	99.4%	107	222
s344	3	3	1	2	99.81%	100%	98	182
s349	3	3	1	2	99.35%	100%	97	182
s382	3	3	1	1	99.83%	99.2%	108	200
s386	3	3	1	5	99.82%	99.2%	389	681
s400	3	3	1	2	98.27%	99.8%	109	208
s420	3	3	1	3	95.86%	82.6%	154	269
s444	3	3	1	2	98.09%	100%	110	309
s510	3	3	1	2	99.86%	99.8%	439	516
s526	3	3	1	3	99.84%	99.5%	543	641
s641	3	3	1	2	99.91%	98.9%	815	762
s713	3	3	1	3	95.33%	97.1%	110	519
s820	3	3	1	2	99.91%	87.6%	285	999
s832	3	3	1	5	99.11%	95.7%	1087	1582
s838	3	3	1	3	99.93%	82.6%	1231	486
s953	3	3	1	5	99.29%	99.7%	1467	1359
s1196	3	3	1	3	99.94%	92.4%	1212	1208
s1238	3	3	1	4	97.32%	93.9%	928	1796
s1423	3	3	1	3	99.21%	97.7%	711	735
s1488	3	3	1	3	99.95%	99.5%	2209	1198
s1494	3	3	1	2	99.53%	99.3%	1209	1130
s5378	3	3	1	4	98.83%	96%	1198	1533
s9234	3	3	1	5	99.13%	84.5%	3228	2059
s13207	3	3	1	5	99.13%	96.5%	3210	3894
s15850	3	3	1	5	92.61%	94.5%	4121	3251
s35932	3	3	1	1	91.06%	100%	102	201
s38417	3	3	1	4	98.69%	96.8%	6938	6804
s38584	3	3	1	4	98.27%	97.8%	4081	4627
Average	3	3	1	3.1	98.54%	96.07%	1213.3	1412.6

PLACEMENT AND ROUTING OPTIMIZATION FOR CIRCUITS DERIVED FROM BDDS

Thomas Eschbach[1] *Rolf Drechsler*[2] *Bernd Becker*[1]

[1]Institute for Computer Science
Albert-Ludwigs-University
79110 Freiburg, Germany
{eschbach,becker}@informatik.uni-freiburg.de

[2]Institute for Computer Science
University of Bremen
28359 Bremen, Germany
drechsle@informatik.uni-bremen.de

ABSTRACT

The high complexity of circuits which currently consist of several millions of transistors, can only be managed using a concise design flow. Recently, the One-Pass Synthesis paradigm came up, i.e. to consider the whole design process as one flow instead of isolated steps. In this context, designing circuits based on the mapping of Binary Decision Diagrams (BDDs) shows several advantages.

While various BDD based approaches for logic minimization or design for testability have been proposed, in this paper we show that placement and routing of BDD circuits can be optimized at a high level of abstraction. Based on algorithms for reducing the number of nodes and edge crossings, we demonstrate on multiple benchmarks that significant improvements are possible in reasonable time.

1. INTRODUCTION

Technological advances in the last decades have allowed the production of large chips with millions of gates. Usually, when producing a chip, the design flow is split up into several individually optimized steps. In almost all cases, each one of these steps is supported by a powerful automated design tool, for example:

- high-level and logic synthesis
- mapping
- place and route

Significant effort has been spent trying to improve these individual steps, but in many cases, the optimization criterion of consecutive steps are different and their close interaction has not been considered sufficiently. Thus, the resulting designs were often suboptimal. E.g., a highly optimized technology independent netlist may produce a suboptimal final design when the mapping onto the target architecture is done. In many cases the logic optimization step does not use the underlying basic cell structure on the physical chip optimally. This can be avoided if the logic synthesis process takes the criteria of the final mapping and layout step into account.

Normally, several iterations of the complete design process have to be carried out to get reasonable results. This is a time consuming and expensive process. Therefore it makes sense to consider the final layout in earlier phases. The interaction between the synthesis, placement and routing phases are important since several quality related criteria are directly influenced. E.g. consideration of the interaction can result in smaller delays, less wastage area, lower power consumption, less crosstalk and better testability. As a promising solution to the resulting problems the One-Pass Synthesis methodology has been proposed. There are two main underlying ideas:

- to combine optimization steps that were split before
- to restrict the optimization in one level such that it fits better on the next

For a more detailed introduction see [1].

One very powerful approach in this context is based on circuits derived from a one-to-one mapping of BDDs [2]. The resulting circuits have nice properties regarding testability [3, 4] and power consumption [5]. To avoid crossings in the physical layout, redundant hardware (dummy nodes) can be inserted to obtain non-crossing ordered BDDs (NCOBDDs) as proposed in [6]. The dummy node insertion also allows a fine grained pipelining where every layer corresponds to a stage of a pipeline. Instead of decomposing the circuit in small macrocells, each containing a small number of BDD nodes (for example around twenty in [7]), and then laying out and routing the macrocells, the approach in [6] as well as the approach presented in this paper use the layered rooted tree structure of the BDD to generate a placement and routing for all nodes in one step.

In this paper, starting with a BDD representation whose size is optimized by using the sifting algorithm [8], we present a method for reducing wire crossings during BDD mapping without the addition of dummy nodes. We make use of the layered structure and the absence of cycles implied by the restricted order of BDDs. Based on an optimized BDD variable ordering, i.e. the variable ordering is fixed, we determine for each level an ordering of the nodes corresponding to multiplexer cells. Experiments show that the number of crossings can be considerably reduced.

2. PROBLEM DESCRIPTION

We give some basic properties of Binary Decision Diagrams (BDD), as far as they are important for the purposes of this paper. More details can be found e.g. in [9].

A reduced ordered Binary Decision Diagram (BDD) as introduced in [2] is a directed acyclic graph $G = (V, E)$ in which a Shannon decomposition is carried out in each node v that is not a sink:

$$f_v = \overline{x}_i f_{v_{x_i=0}} + x_i f_{v_{x_i=1}} \qquad (1)$$

x_i is called the decision variable in v. In a reduced ordered BDD each path from a source to a sink is consistent with a given ordering of the decision variables. (For an example see Figure 2.)

The size of the BDD is very sensitive to the chosen variable ordering, i.e. the BDD size may vary from linear to exponential in the number of variables for a given function f. In general, improving the variable ordering of BDDs is NP-complete [10]. However, efficient heuristic algorithms for improving the variable ordering are known, in particular the sifting method [8]. Currently, BDDs are commonly used for efficient representation and manipulation of Boolean functions, not only in the VLSI CAD community [11].

In the following, we consider synthesis approaches where circuits are derived from BDDs by a simple one-to-one mapping. This can easily be done by introducing a multiplexer for every node and corresponding wires for every edge of the BDD. A small example for the one-to-one mapping of one BDD node (a) into a multiplexer circuit (b) is given in Figure 1. Please note, that in the following BDD nodes are drawn as circles in contrast to derived multiplexers given as rectangles labeled with the corresponding decision (input) variable.

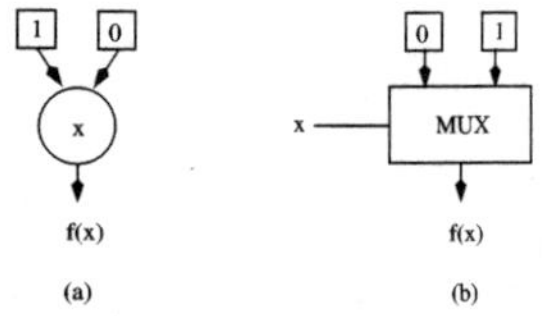

Fig. 1. Example transformation of a BDD (a) into a multiplexer based circuit (b).

Even though BDDs were not designed to produce the physical layout of a circuit, there were already very promising approaches to BDD-based circuit design. (For an overview see [1].) For totally symmetric functions, BDDs have a very regular structure that can be directly transferred into planar layouts. The nodes are locally connected to a maximum of four adjacent nodes. This leads to a layout with less crosstalk and short wires. For non-symmetric functions an approach to transfer the BDD into a lattice structures was proposed in [12]. It produces the same layout except for the fact that additional levels are needed. The main drawback of these methods is that the number of levels for some functions may become exponential in the number of variables. On the other hand, the NCOBDDs [6] only require one layer for every input variable at the cost of additional dummy nodes. Every BDD can be transformed into a NCOBDD by duplicating and reordering the nodes of the BDD to achieve a planar layout. These dummy nodes are then directly mapped onto the circuit using additional area and increase the energy consumption.

Due to this, in the following we consider "traditional BDDs", i.e. reduced ordered BDDs [2]. They provide a good representational compromise between regularity and compactness. To reduce the number of edge crossings of the given BDD in the layout process, a crossing reduction algorithm is applied. For an example see Figure 2.

Then it can be mapped onto a multiplexer based target technology. This step is an important part for the final layout of the circuit since unnecessary edge crossings complicate the process of routing. After this, the algorithm maps the nodes of the BDD to sub-circuits realizing the corresponding functions.

In the following we formulate the edge crossing reduction problem as a graph problem and introduce some common terms from this field: A directed graph $G = (V, E)$ is a *multi-layered* graph with d layers if the node set V is partitioned into disjoint subsets $V_1, V_2, \ldots, V_d$, i.e. $V_1 \cup V_2 \cup \ldots \cup V_d = V$ and $(\forall m \neq m')$

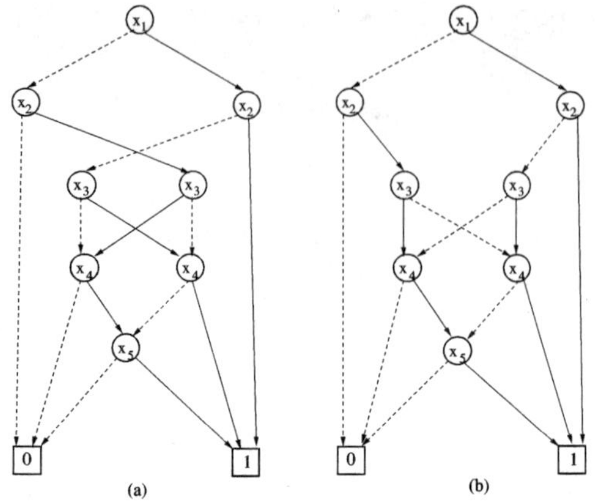

Fig. 2. This example shows a BDD before (a) and after (b) the crossing minimization.

$V_m \cap V_{m'} = \emptyset$, where V_m is called the *m-th layer* of the graph. All edges in E connect nodes in different layers. All nodes of the BDD which are labelled with the same variable can be assigned to the same layer to obtain a multi-layered graph.

A traditional approach to lay out a directed graph was introduced by [13]. It splits up the layout process into four steps:

- Cycle Removal: To obtain an acyclic graph, as few edges as possible are reversed.
- Layer Assignment: A proper layering will be computed by assigning every node to one layer.
- Crossing Reduction: The algorithm computes an ordering for all nodes on each layer which minimizes the total number of edge crossings ("multi-layer straight-line crossing minimization problem").
- X-coordinate assignment of all nodes.

This directed graph layout framework is well suited to lay out the BDD structure. Since BDDs are layered directed acyclic graphs, the goals of the first and the second step are already achieved.

To solve the (exact) multi-layer straight-line crossing minimization problem we have to determine an ordering ord_m for all layers containing all nodes in layer V_m so that the number of crossings is minimized. In the following, a set of orderings ord_m, $m \in \{1, \ldots, d\}$, is called an ordering for the graph. Unfortunately, even minimizing edge crossings in graphs with only two layers is NP-hard [14] and remains NP-hard if the ordering in one of the layers is fixed. Therefore the use of heuristic methods to solve this problem is justified. Many heuristic algorithms are known from literature, e.g. [15, 13, 16]. In this paper we use the popular *averaging* method first introduced by [13] which computes good solutions in a short time.
We then further reduce the number of crossings by post-processing it with the windows optimization heuristic method [17]. It decomposes the graph into smaller subgraphs which contain nodes form several layer and then computes an optimal ordering for each subproblem.

3. PLACEMENT AND CROSSING REDUCTION

In this work we first compute a BDD for a given circuit. The algorithm dynamically reorders the input variables to reduce the number of BDD nodes using sifting. This step directly influences the number of multiplexers used to implement the circuit, and thus on the properties of the resulting layout.

Then the averaging heuristic method is applied to reduce the number of edge crossings to obtain a good initial ordering which is then post-processed with the windows optimization algorithm. The averaging method helps to cut down overall run times, since a good starting point helps to avoid unnecessary calls of the locally optimal crossing reduction algorithm. Pseudo code for the complete flow is given below:

```
compute_ordering(circuit) {
  BDD = Compute_BDD_using_Sifting(circuit)
  rord = compute random ordering
  ordering = averaging(BDD, rord)
  ordering =windowsopt(BDD, ordering)
  return ordering
}
```

In the remaining part of this section the averaging and the windows optimization algorithm are briefly described.

3.1. Averaging heuristic method

The averaging heuristic method computes the position of node n on one layer with respect to the nodes on the layer above (below) which are directly connected to it. (Through the introduction of dummy nodes that are removed after the ordering has been computed we may assume that there are only edges between adjacent nodes.) It then sorts the nodes with respect to this pre-computed value. To compute a solution for the multi-layer straight-line crossing minimization problem the averaging technique makes use of the so called layer-by-layer sweep:

- Choose an initial ordering.
- Fix the positions of all nodes on the first layer.
- Compute the positions of all nodes on the second layer with respect to the fixed nodes located on the first layer.
- Fix the positions of the nodes on the second layer and compute the position of the nodes located on the third layer and so on.
- Start to sweep back, processing the second last layer considering only the nodes on the last layer until the ordering of the first layer is computed.
- Repeat the procedure until no further reduction in the number of crossings is achieved.

3.2. Windows Optimization

We only give a short description of the windows optimization heuristic method. Further details can be found in [17]. An initial ordering can be improved in the following way. A series of subsets of nodes with constant size, typically spreading over several layers, are extracted and processed with an exact approach with respect to their adjacent nodes. This approach is based on a dynamic programming method which makes use of a lower bound technique to reduce the search space. Only if the local solution induces a crossing reduction for the entire graph, is the new ordering used. The user has a fine grained control on the trade off between run time and solution quality by choosing the size of the window. The algorithm starts with a window width of four nodes per layer and a window depth of two layers. Then the algorithm increases the window depth to three and finally four layers dynamically to further reduce the number of crossings.

4. EXPERIMENTAL RESULTS

All algorithms are implemented in C. The experimental results are based on examples taken from benchmark circuits in [18] and [19]. The experiments were carried out on a 2 GHz personal computer with 1 GB main memory running the Linux OS. The run times are given in CPU seconds.

We utilized the $CUDD$ package [20] to compute a BDD for every circuit. $CUDD$ is a commonly accepted software tool to efficiently minimize and manipulate BDDs. Sifting [8] was used during the construction of the BDDs. The averaging heuristic method from section III.A was used to obtain a good initial embedding of the nodes. As observed in [21, 17], the averaging method computes high quality results in a short period of time compared to many other methods used in this field. Then windows optimization is applied.

In Table 1 the results of the implemented procedures are given. The second (third) column shows the number of multiplexers for each circuit without (with) using sifting. Next, in the fourth column we present the number of edge crossing produced by a random permutation of the multiplexers in each layer. The fifth and sixth columns provide the final results in terms of edge crossing (computed with the averaging heuristic method and after post-processing it with the windows optimization technique). The total run times for the averaging and the windows optimization method are given in the last two columns.

As already reported in [8] sifting reduces the number of nodes significantly. For completeness we give the precise numbers without using sifting in the second column. Compared to the averaging procedure, a random assignment of the nodes produces results which have on average nearly 200 times more crossings. Post-processing the results with the windows optimization technique further reduces the number of crossings on average by seven percent.

Table 2 is given to compare our results with the results utilizing the NCOBDDs published in [6]. In the second column the number of multiplexers required when using the NCOBDD structure is shown. Finally, the third column presents the number of multiplexers that are needed to obtain a circuit derived form the corresponding BDD. The additional dummy nodes in the NCOBDDs are necessary to allow a planar layout (and also pipelining). Of course, our approach does not compute crossing free layouts but it saves up to 90 % of the multiplexers compared to the NCOBDDs given in [6].

5. CONCLUSIONS

We have proposed a new approach to compute placement and routing for a circuit derived from the acyclic graph structure of the corresponding BDDs. This placement can be further improved by reordering the nodes within the layers to reduce the number of wire crossings. An algorithm was implemented that reduces the number of multiplexers and the number of wire crossings used. In this way, layout aspects can be considered at a high level of abstraction, and therefore in an early design phase. Experiments have shown the ef-

Table 1. Benchmark results of circuits

circuit	mplex	sift	rand	quality av	quality wo	time/s av	time/s wo
z4ml	33	16	157	3	3	0.27	1
cm138a	26	17	563	10	10	0.23	1
9sym	36	24	1328	8	8	0.49	3
cmb	53	28	2028	11	10	0.58	4
cu	75	31	1851	30	30	0.55	4
decod	41	40	1808	67	63	0.39	6
cm85a	50	36	1548	13	13	0.15	2
x2	51	36	1802	49	42	0.48	5
f51m	55	39	1512	51	48	0.43	3
cm162a	63	41	3477	23	23	0.71	3
pcle	64	41	4874	96	93	0.89	10
pm1	60	43	5587	37	35	0.88	5
i1	90	43	6923	50	50	1	8
cordic	78	49	3833	10	9	1	9
cc	85	60	5417	109	103	0.69	13
lal	128	75	14055	192	183	1	27
unreg	104	81	3379	140	128	0.49	12
count	119	81	43778	229	213	3	83
c8	112	81	18432	296	260	1	42
sct	124	82	9704	139	130	1	16
term1	424	96	18804	73	67	2	31
frg1	280	102	16084	257	244	2	35
pcler8	141	107	27243	336	326	2	30
b9	202	106	26314	181	177	2	36
ttt2	198	121	48158	548	509	3	95
i3	262	132	33421	0	0	3	28
cht	142	124	14889	621	560	0.97	38
comp	174	136	49102	143	135	4	219
alu2	187	168	17448	1578	1523	1	53
i2	1679	205	224872	69	69	17	154
i6	422	214	36755	4090	3806	1	99
apex7	1071	637	195568	1562	1521	7	287
i4	622	337	189885	129	121	8	169
i5	1053	322	193037	1130	1119	7	171
alu4	426	378	133126	5602	5309	4	371
i7	513	396	76898	5810	5273	2	128

ficiency of this approach. Current work focuses on the integration of the new algorithm into a complete layout environment.

Table 2. NCOBDDs vs. BDDs

circuit	multiplexers(NCOBDD)	multiplexers(BDD)
z4ml	77	16
cm138a	48	17
9sym	33	24
alu2	465	168
alu4	3482	378

6. REFERENCES

[1] R. Drechsler and W. Günther, *Towards One-Pass Synthesis*, Kluwer Academic Publishers, 2002.

[2] R.E. Bryant, "Graph - based algorithms for Boolean function manipulation," *TOC*, vol. 35, no. 8, pp. 677–691, 1986.

[3] P. Ashar, S. Devadas, and K. Keutzer, "Path-delay-fault testability properties of multiplexor-based networks," *INTEGRATION, the VLSI Jour.*, vol. 15, no. 1, pp. 1–23, 1993.

[4] B. Becker, "Testing with decision diagrams," *INTEGRATION, the VLSI Jour.*, vol. 26, pp. 5–20, 1998.

[5] P. Lindgren, M. Kerttu, M. Thornton, and R. Drechsler, "Low power optimization technique for BDD mapped circuits," in *Proceedings of Asia and South Pacific Design Automation Conference (ASP-DAC'01)*, 2001, pp. 615–621.

[6] A. Cao and C. K. Koh, "Non-Crossing Ordered BDD for Physical Synthesis of Regular Circuit Structure," in *Int'l Conf. on Comp. Design*, 2003.

[7] L. Macchiarulo, L. Benini, and E. Macii, "On-the-fly layout generation for ptl macrocells," *IEEE Design Automation and Test in Europe*, pp. 546–551, 2001.

[8] R. Rudell, "Dynamic variable ordering for ordered binary decision diagrams," *ICCAD*, pp. 42–47, 1993.

[9] R. Drechsler and B. Becker, *Binary Decision Diagrams – Theory and Implementation*, Kluwer Academic Publishers, 1998.

[10] B. Bollig and I. Wegener, "Improving the variable ordering of OBDDs is NP-complete," *IEEE Trans. on Comp.*, vol. 45, no. 9, pp. 993–1002, 1996.

[11] U. Wegener, *Branching programs and binary decision diagrams - theory and applications*, SIAM Monographs on Discrete Mathematics and Applications, 2000.

[12] M. Chrzanowska, Z. Wang, and Y. Xu, "A regular representation for mapping to fine-grain, locally-connected fpgas.," *In Proc. Midwest Symp. Circ. Syst*, pp. 2749–2752, 1997.

[13] K. Sugiyama, S.Tagawa, and M.Toda, "Methods for visual understanding of hierarchical system structures," *IEEE Transaction on Systems, Man and Cybernetics*, vol. 11, no. 2, pp. 109–125, 1981.

[14] M. R. Garey and D. S. Johnson, "Crossing number is NP-complete," *SIAM Journal on Algebraic and Discrete Methods*, vol. 4, pp. 312–316, 1983.

[15] P. Eades and D. Kelly, "Heuristics for reducing crossings in 2-layered networks," *Ars Combin.*, vol. 21, no. A, pp. 89–98, 1986.

[16] J. Warfield, "Crossing theory and hierarchy mapping.," *IEEE Transactions on System, Man, and Cybernetics*, vol. SMC-7, no. 7, pp. 505–523, 1977.

[17] T. Eschbach, W. Günther, R. Drechsler, and B. Becker, "Crossing Reduction by Windows Optimization," *Proceedings of the 10th International Symposium on Graph Drawing*, vol. LNCS 2528, pp. 285–294, 2002.

[18] F. Brglez, D. Bryan, and K. Kozminski, "Combinational profiles of sequential benchmark circuits," *Int'l Symp. Circ. and Systems*, pp. 1929–1934, 1989.

[19] K. McElvain, "Benchmark set: Version 4.0," *International Workshop on Logic Synthesis*, 1993.

[20] F. Somenzi, *CUDD: CU Decision Diagram Package Release 2.3.1*, University of Colorado at Boulder, 2001.

[21] M. Jünger and P. Mutzel, "2-layer straightline crosing minimization: Performance of exact and heuristic algorithms," *J. Graph Agorithms Appl.*, vol. 1, no. 1, pp. 1–25, 1997.

L-SIMULATOR: A MAGPEEC-BASED NEW CAD TOOL FOR SIMULATING MAGNETIC-ENHANCED IC INDUCTORS OF 3D ARBITRARY GEOMETRY

Haibo Long[1], Zhenghe Feng[1], Haigang Feng[2], Albert Wang[2] and Tianling Ren[3]

[1] State Key Lab on Microwave and Digital Communications, Dept. EE, Tsinghua University, Beijing, 100084, China, longhb00@mails.tsinghua.edu.cn, fzh-dee@mail.tsinghu.edu.cn

[2] Integrated Electronics Laboratory, Dept. ECE, Illinois Institute of Technology, Chicago, IL 60616, USA, fhg@ece.iit.edu, awang@ece.iit.edu

[3] Institute of Microelectronics, Tsinghua University, Beijing, 100084, China, rentl@tsinghua.edu.cn

ABSTRACT

This paper presents a new PEEC-based inductor simulation tool, entitled L-Simulator, which employs a novel magPEEC modeling algorithm and an existing FastCap modeling algorithm to address both magnetic and electrical coupling effects respectively, hence being capable to simulate 3D magnetic-enhanced RF IC inductors of arbitrary geometries. Applications on micromachined inductors in 0.2μm GaAs HEMT process and magnetic-cored micro inductors in 0.18μm CMOS technology are discussed.

1. INTRODCTION

Nowadays, magnetic materials have been widely used to make electro-magnetic (EM) structures such as MEMS (micro electromechanical systems) and on-chip spiral inductors for RF IC applications to improve inductances and quality factors. Inductors using magnetic films operated at GHz frequency have been reported [1].

Existing special EM CAD simulators for RF integrated inductors such as SISP [2], GEMCAP2 [3] and ASITIC [4] cannot deal with magnetic-enhanced inductor structures because of their inability to deal with magnetic materials involved. Existing commercial full-wave EM modeling software, such as HFSS, employs the finite element method (FEM) to carry out EM analysis that usually leads to a large number of unknowns associated with the global meshing required to cover all parts of the structure to be analyzed and its surrounding external space. Obviously, the corresponding equation-solving process requires excessive memory and CPU time.

New EM modeling technique and simulator are needed for accurate simulation and design of magnetic-enhanced RF IC micro inductors. We report a new EM simulation CAD tool, entitled L-simulator, which is developed to accurately simulate 3D EM structures of arbitrary geometries, including magnetic-enhanced RF IC micro inductors. The new L-simulator uses a novel magPEEC modeling algorithm we developed [5], which originated from a partial element equivalent circuit (PEEC) method initially proposed to model 3D multi-conductor systems in [6] and later extended to include dielectrics [7], to include the inductive effects of magnetic material. An existing FastCap electrical modeling technique [8] is adopted to include the capacitive effects of dielectrics. The magPEEC technique can address the impact of magnetic materials by accounting for fictitious magnetized currents on the surface of permeable materials, while the FastCap can evaluate the effects of dielectrics by using equivalent polarized charges on surface of dielectrics.

2. L-SIMULATOR: A NEW SIMULATOR FOR CONDUCTOR-MAGNET RF IC INDUCTORS

The new L-Simulator uses our new magPEEC modeling algorithm to enable accurately simulating RF IC inductors involving magnetic media of arbitrary 3D geometries. The flow chart of L-Simulator is shown in Fig. 1, where each ellipse denotes input or output data in the simulation flow and each rectangular represents a sub-program to process the input data and outputs results. The L-Simulator flow follows.

Firstly, geometrical parameters, including conductor structures, magnetic material entities and dielectric interfaces, and material parameters, including conductivity, dielectric constant and permeability, of all dielectrics, conductors and magnetic materials are provided as inputs to a meshing module, called Mesh-Pro.

Secondly, the Mesh-Pro partitions the spiral inductor into segments along the metal stripe to form corresponding equivalent circuits. It then discretizes each segment into *filament* cells, assuming uniform local distribution of conductive current, and magnetic material surface into *panel* cells, assuming uniform local distribution of fictitious magnetized current. Meanwhile,

0-7803-8251-X/04/$17.00 ©2004 IEEE

the dielectric-dielectric interfaces are meshed into panel cells, assuming uniform local distribution of polarized charges, and the dielectric-conductor interfaces are partitioned into panel cells, also assuming uniform local distribution of free charge and polarized charge. L-Cells (defined as inductive cells with current flowing on) composed of filament cells with their conductivities and panel cells with their permeabilities of both sides are prepared for processing by the integral-processing module, called magPEEC-Core, to calculate their magnetic couplings, which are expressed as LPEs (inductive partial elements). On the other hand, C-Cells (defined as capacitive cells with charges on) composed of panel cells with their conductivities and dielectric constants of both sides are processed by the integral-processing module, called FastCap-Core, to calculate their electrical couplings, which are expressed as CPEs (capacitive partial elements). The LPEs and CPEs, all based on cumbersome multi-fold integral over hexahedrons and quadrangles, are calculated one by one by using our fast and accurate integral programs, with accuracy level similar to the commercial Mathmatica software, however, much faster.

Thirdly, at given operating frequency, LPEs and CPEs are processed by matrix-processing modules, called LMat-Pro and CMat-Pro, to extract LMat (inductance matrix) and CMat (capacitance matrix), respectively. Since this step involves calculating large-scale matrices at given operating frequencies, it consumes the majority of the simulation time and requires the largest computer memory resources in the whole flow. A sparse matrix technique is proposed to accelerate the simulation procedure, which is currently under development.

Fourthly, LMat and CMat are used to form the equivalent circuit involving the metal segments mentioned above and are processed by the CirSyn-Pro module (a circuit synthesis program) to produce the VI-Dis (stands for potential and current distribution, i.e., the voltage across each segment and the current quantity in each segment at given operating frequency).

Next, using the VI-Dis data in a given frequency range, the PP-Pro module (a post-process program) can extract the two-port admittance matrix or s-matrix of the inductor. According to two-port microwave matrix, the VI-Dis data can then be used to extract the interest parameters such as quality factor, inductance, resistance and other capacitive parameters related to lumped-element equivalent circuit.

Finally, the LSta-Pro and Csta-Pro modules, based on static magnetic and static electrical methods, can be executed to extract the lumped inductive and capacitive parameters in low frequency range. These low-frequency parameters could be input into PP-Pro for post process. They are very helpful for analysis in some occasions.

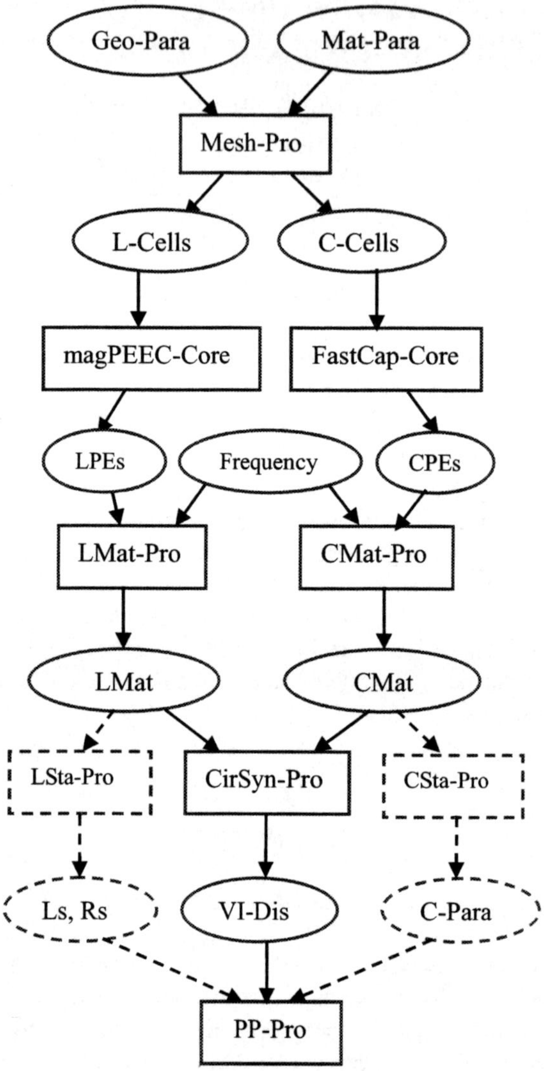

Fig. 1. Flow chart of the new simulator, L- Simulator.

3. APPLICATIONS FOR SUSPENDED INDUCTORS AND STACKED SPIRAL INDUCTORS WITH MAGNETIC CORES

Several application examples are discussed here, including simulating two reported inductors [9, 10] and our 3D super compact RF inductors designed and fabricated in a commercial 0.18μm, 6-metal, all copper CMOS technology [11] to verify various aspects of the L-Simulator.

In example-1, reported suspended inductors with different substrate etch-out depths fabricated in a 0.2μm GaAs HEMT D02AH process [9] was simulated using the new L-Simulator. The 4.8nH inductor shown in Fig. 2 (featuring 6 turns and 9μm strip width and 9μm strip spacing) was fabricated using wet etching to create hollow chambers of different depths in the GaAs substrate. The

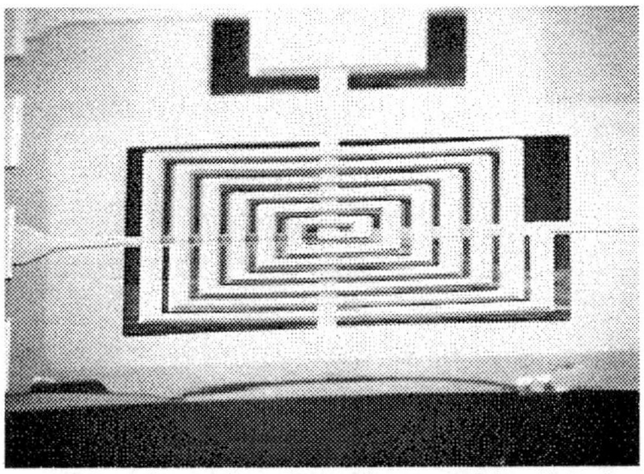

Fig. 2. Die photo for the micromachined inductor of 4.8nH with stripes suspended individually [9].

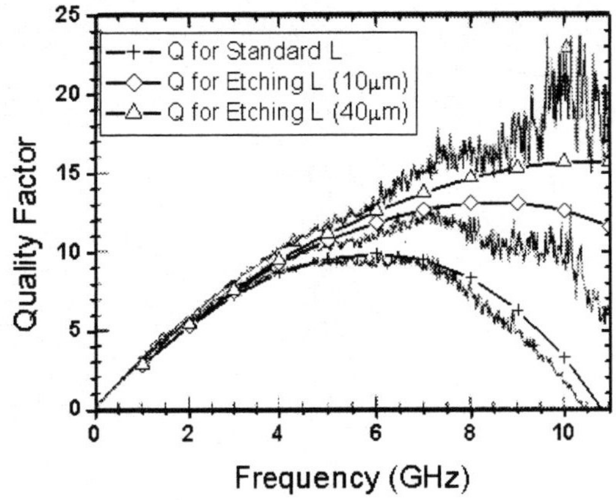

Fig. 3. Q-factor versus GaAs substrate etching depth of the 4.8 nH inductors: our simulation (solid line with symbols) versus measurement in [9].

simulated Q-factor data using our new L-Simulator are compared with reported measured data in Fig. 3. Fairly good agreement was observed for the Q-factor data across over a bandwidth of 10GHz. In addition, the simulated inductance of 4.85nH and resistance 10.58 Ω also agree well with the measured data of 4.8nH and 10.5 Ω, respectively. Since etching-created air chamber in GaAs substrate means to eliminate capacitive coupling effect, this example verifies the L-Simulator in dealing with the capacitive coupling effects.

In example-2, a reported infinite thin metal ring structure with magnetic layer underneath [10] was simulated using our new L-Simulator to verify the magPEEC algorithm. The simulation results by L-Simulator are compared with reported data of literature [10] in Fig. 4, where good agreements are observed.

Example-3 is a super compact 22μm x 23μm, six-layer stacked spiral cored inductor we designed and fabricated in a commercial 0.18μm 6-metal all copper CMOS technology [11] as shown in Fig. 5 for a structure with dummy M-core array, featuring four turns per spiral, 1μm metal line width and 0.5μm metal line spacing. The

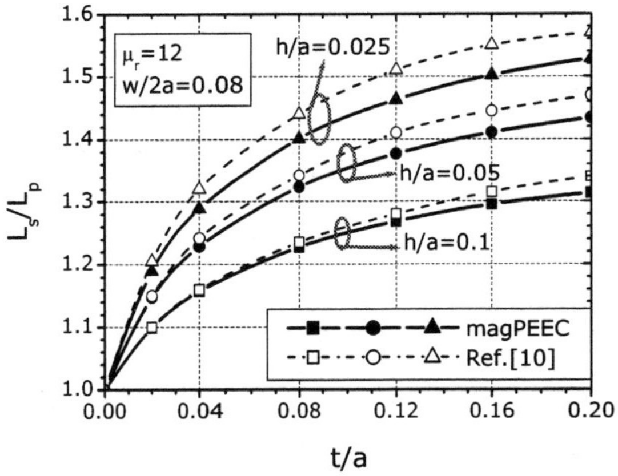

Fig. 4. Normalized inductances of a single metal ring over a magnetic layer versus the magnetic layer thickness with *h/a* as a parameter: our simulation (solid lines) versus reported data in Ref. [10].

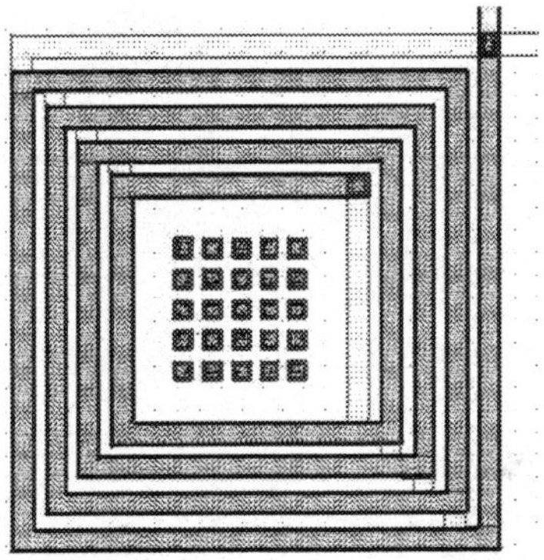

Fig. 5. A cross-section of a novel spiral inductor with stacked-via magnetic core arrays in 0.18μm CMOS [11].

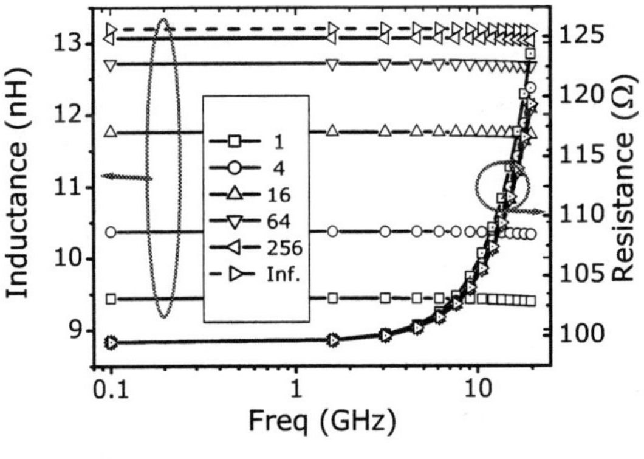

Fig. 6. Simulation results for the array-M-cored inductors with various relative permeability values as marked with legends.

measured inductance and resistance are 9.52nH and 98.1 Ω, respectively, for the inductor, which agrees well with the simulated 9.44nH and 99.50Ω shown in Fig. 6. Aiming to design transistor-sized super compact RF IC inductors with improved inductance and quality factor, we proposed a novel stacked-via magnetic-cored inductor structure for standard CMOS technologies [10]. The fabrication of these novel stacked-via-cored inductors will be fabricated in the 0.18μm CMOS. The new magPEEC-based L-Simulator was used to predict the performance of this novel design before running silicon. The layout versions were designed featuring a stacked-via core array of 5 x 5 cores with the total cross-section area of 12μm x 12μm, 7μm for core height H_{MC}, 1.12μm for core diameter D_{MC} and 1.6μm for core spacing SP_{MC}. The inductor metal structures are the same as the uncored one described above.

Figs .6 show the simulated inductance and resistance data for the inductors with M-core array. Firstly, the data show that inductance improvement by using magnetic cores saturates as the permeability increases to some threshold value (i.e., 100+ in this case). Secondly, the observation that the inductance~permeability improvement saturates as permeability approaches to 100+ suggests that one might not need to struggle to find the super-permeability magnetic materials for making super M-cored inductors. Instead, layout design shall optimize inductor performance. Thirdly, the array-M-cored structure may be a better choice in lifting up the upper limit of the inductance~permeability relationship observed by minimizing the core spacing SP_{MC} to increase the effective magnetic flux across the coils. Finally, the data suggest that the parasitic resistance is insensitive to the lossy nature of the magnetic materials used in array-M-cored inductors.

4. CONCLUSIONS

We report a new EM simulation tool, called L-Simulator, for simulating arbitrary 3D electro-magnetic RF IC inductors. L-Simulator uses a novel 3D electro-magnetic modeling algorithm, magPEEC, which extends existing PEEC approach to analyze arbitrary conductor-magnet structures by accounting for fictitious magnetic currents on magnetic material surface, and an existing capacitance extraction algorithm, FastCap, to deal with 3D structures with multiple dielectrics. The L-simulator can also be used to simulate other types of electro-magnetic structures of arbitrary 3D geometries. The L-Simulator was applied to several practical inductor designs for verification where good agreements between our simulation and measurements or other simulation are observed. Design prediction using the L-simulator was discussed for novel array-magnetic-cored super compact RF IC inductors.

5. ACKNOWLEDGEMENT

This paper is supported by National "973" R&D Project of China (NO. G1999033105), National Natural Science Foundation of China (NO. 60171015) and NSF of US (0302449)

5. REFENCES

[1] M. Yamaguchi, et al, "Sandwich-type ferromagnetic RF integrated inductor", *IEEE Trans. Microwave Theory and Tech., vol. MTT-49, no. 12*, pp. 2331-2335, Dec. 2001.

[2] Y. K. Koutsoyannopoulos, et al, "Systematic analysis and modeling of integrated inductors and transformers in RF IC design", *IEEE Trans. Circuits and Systems-II: Analog and Digital Signal Proceeding, vol. 47*, pp. 699-713, Aug. 2000.

[3] J. R. Long, et al, "The modeling, characterization, and design of monolithic inductors for Silicon RF IC's", *IEEE Journal of Solid-State Circuits*. vol. 32, pp. 357-369, Mar. 1997.

[4] A. M. Niknejad, et al, "Analysis, design, and optimization of spiral inductors and transformers for Si RF IC's", *IEEE Journal of Solid-State Circuits, vol. 33*, pp. 1470-1481, Oct. 1998.

[5] H. Long, et al, "magPEEC: extended PEEC modeling for 3D arbitrary electro-magnetic devices with application for M-Cored inductors", *2003 IEEE MTT-S Int. Microwave Symp. Dig.*, vol. 1, pp. 251-254, June 2003

[6] A. E. Ruehli, "Equivalent circuit models for three-dimensional multiconductor systems", *IEEE Trans. Microwave Theory and Tech., vol. MTT-22, no.3*, pp. 216-221, Mar. 1974.

[7] A. E. Ruehli, et al, "Circuit models for three-dimensional geometries including dielectrics", *IEEE Trans. Microwave Theory and Tech., vol. MTT-40, no. 7*, pp. 1507-1516, July 1992.

[8] K. Nabors, et al, "Multipole-accelerated capacitance extraction algorithms for 3-D structures with multiple dielectrics", *IEEE Trans. Circuits and Systems-I: Fundamental Theory and Applications, vol. 39*, pp. 946-954, Nov. 1992.

[9] R. P. Ribas, et al, "Micromachined microwave planar spiral inductors and transformers", *IEEE Trans. Microwave Theory and Tech., vol. MTT-48*, pp. 1326-1335, Nov. 2000.

[10] S. F. Mahnoud, et al, "Inductance and quality-factor evaluation of planar lumped inductors in a multilayer configuration", *IEEE Trans. Microwave Theory and Tech., vol. MTT-45*, pp. 918-923, Jun. 1997.

[11] H. Feng, et al, "Super compact RFIC inductors in 0.18 μm CMOS with copper interconnects", *Proc. IEEE MTT-S Int. Microwave Symp. Dig., vol. 1*, pp. 553-556, June 2002.

AN EFFICIENT APPROACH FOR HIERARCHICAL SUBMODULE EXTRACTION

Yi-Wei Lin and Jing-Yang Jou

Dept. of Electronics Engineering, National Chiao Tung University
Hsinchu, Taiwan, ROC

ABSTRACT

The growth of modern IC design complexity leads the consistency check and design verification during every level in design flow to be an important and challenged issue. In this paper, we propose an efficient approach to rebuild the hierarchical level from low level circuits. Our approach is based on the structure equivalent expansion algorithm to find repeated submodules in every circuit level to reconstruct circuit hierarchy. Without any addition library information, our approach is quite efficient in both time and space complexities by using only flatten netlists. The experiments on many real circuits containing combinational, sequential, and memory circuits show that our approach can rebuild most circuit hierarchical levels and also reduce the verification effort of the circuits.

1. INTRODUCTION

Design verification is a very important task at various levels of the design flow in the development of a System-on-a-Chip (SoC). The last step of design verification is to extract netlist from the layout and execute simulation to make sure the post-layout response meets all specifications. However, simulating the whole circuit directly on transistor level usually takes too much time and memory. Therefore, hierarchical circuit level extraction from flatten netlist is very helpful in speeding up the simulation and preventing the memory explosion problem.

To rebuild hierarchical circuit levels, the most useful and general method is to extract repeated subcircuits (submodules) from netlists. In nowadays IC design, there are often many repeated components to form the whole circuit, especially in memory circuits and datapath circuits of digital IC. If we can find these repeated components from the post-layout netlist, we can then conduct simulation or verification on only one of them. Therefore, we can greatly reduce the effort on simulating post-layout design and still achieve our original goal.

There are many relative researches on extracting repeated subcircuits in the past days. These previous approaches might have some deficiencies that make the extracted goals difficult or not efficient to achieve. Template library identification method [1,2,3] is similar to subcircuit isomorphism problem. It can be generally treated as finding a small circuit in a big circuit. It is quit efficient and useful in transistor-to-gate level transformation, but is not applicable in higher level transformations since it is rarely to have the higher level subcircuit libraries.

Templates generation and matching method [4,5] is improved from approach just mentioned. In order to extract subcircuits in multiple levels, this approach needs to generate all possible templates in the whole circuit. However, the generation is a very time-consuming and space-consuming task.

Structure equivalence expansion method [6,7] is to expand known subcircuits to new ones by structure equivalence. It starts at the smallest equivalent subcircuits, any transistor in netlist or any gate in gate-level, and then expands as big as possible by using all structure equivalence. However, the equivocal selection of start point and the lack of solution dealing with non-ideal expansion make this approach not good enough as expected.

In this paper, we propose an approach to rebuild design hierarchy efficiently in both time and space complexity without using any additional template information. We use a greedy strategy by selecting the subcircuit with the maximum number of repetitions to expand in every iteration. We update network information dynamically after extractions and avoid any equivocal expansion, so every expansion is distinct. Comparing to previous approach, we select the start point with global view and can deal with any type and any direction of non-ideal expansion that make our performance far better than previous approaches.

Our approach has been implemented and demonstrated on combinational, sequential, and memory circuits in wide range of circuit sizes. The experimental results show that our approach has rebuilt the hierarchy of the circuits and has reduced the verification effort of the circuits. The result also shows the near-linear performance in run time and memory usage.

This paper is organized as follows. In Section 2, the targeted problem is formulated mathematically. Then we propose our approach in Section 3. Section 4 shows the experimental results. At last, the conclusions are given in Section 5.

2. PROBLEM FORMULATION

Our formulation targets on the hierarchy situation and the effort for simulation and verification of a circuit. We use an example to illustrate our formulation.

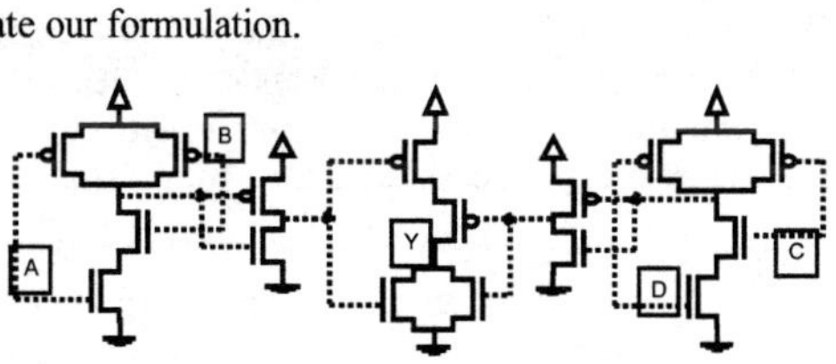

Figure 1 Transistor netlist Y=(A・B+C・D)'

Figure 1 is a small circuit performing function Y=(A・B+C・D)' in flatten netlist description. These 16 transistors are modeled by two types of module model, PMOS model and NMOS model. If we want to simulate the whole circuit in netlist

0-7803-8251-X/04/$17.00 ©2004 IEEE

format, then we will use the PMOS and NMOS transistor model to instantiate these 16 transistors. However, the effort to simulate the module model and the instantiated list of the module models is totally different. Therefore, we categorize the effort for simulation of the circuit into two classifications: ***Modeling cost*** represents the effort to simulate the module model. For example, the PMOS model and the NMOS model are the module models in figure 1. ***Instantiation cost*** represents the effort to simulate the list of the module models. The list of the module models is like the list of 8 PMOS and 8 NMOS in figure 1. The total efforts to simulate or verify the whole circuit, ***Verification cost***, can be considered as the sum of modeling cost and instantiation cost. These cost functions of figure 1 can be shown as the following:

$$MC_{flatten} = 2mc \quad (1)$$

$$IC_{flatten} = 16ic \quad (2)$$

$$VC_{flatten} = MC_{flatten} + IC_{flatten} = 2mc + 16ic \quad (3)$$

The terms MC, IC and VC are modeling cost, instantiation cost and verification cost respectively. mc and ic are the units of modeling cost and instantiation cost. These equations are based on the assumption that modeling cost of PMOS and NMOS model are equal. If we extract the AND gate as a new circuit module in figure 1, the cost functions will be following:

$$MC_{extracted} = 2mc_0 + mc_1 \quad (4)$$

$$IC_{extracted} = 4ic_0 + 2ic_1 \quad (5)$$

$$VC_{extracted} = MC_{extracted} + IC_{extracted} = 2mc_0 + mc_1 + 6ic \quad (6)$$

The term mc_0 is the modeling cost unit of MOS model (level 0), and mc_1 is the modeling cost unit of the NAND gate model (level 1). ic_0 and ic_1 are the units of instantiation cost in level 0 and level 1. ic_0 and ic_1 can be viewed as the same because the effort to simulate or verify every module instantiation is to repeat the submodule result only. If we treat the submodule, AND gate, as a small circuit, we will find $MC_{AND} = 6ic$ since there are 6 transistors composed a AND gate. The total verification cost reduction between before the extraction and after the extraction in the circuit of figure 1 can be summarized as:

$$VC' = VC_{flatten} - VC_{extracted} = 2mc_0 + 16ic - (2mc_0 + 12ic) = 4ic \quad (7)$$

The verification cost has been reduced after a new submodule is extracted. We can see the total verification cost reduction only depends on the instantiation cost. Therefore, we only need to emphasize the instantiation cost reduction during extraction.

However, if we extract the inverter and the NAND gate first before we extract the AND gate, the final verification cost is reduced only 2ic (from 16 to 14). Comparing the results of 2 different extractions, extracting AND gate directly has more verification cost reduction, but extracting three kinds of submodule step by step has a better hierarchy result.

However, the final goal of our extraction also contains the reconstruction of the hierarchy levels of the real circuits. For practical consideration, our target is to meet the real circuit hierarchy levels first. With the circuit hierarchy being well constructed, we then minimize the verification cost. Therefore, **the step by step extracting mechanism which extracts more hierarchical submodules is more suitable to our purpose.**

Then we analyze the verification cost function in general format to find the method to maximize the reduction.

$$VC_i = (S_i \times R_i + r_i)ic \quad (8)$$

$$VC_{i+1} = (S_i + R_i + r_i)ic \quad (9)$$

VC_i and VC_{i+1} are the verification cost of the circuit in level ***i*** and level ***i+1*** respectively. Equation (8) shows that the circuit has a submodule which is composed of S_i modules in level ***i*** with repetition R_i times. It also contains r_i modules of the rest of the circuit. Equation (9) indicates the submodule is extracted in level ***i+1***. After the extraction, VC_{i+1} is the summary of modeling cost S_i ic (the submodule is composed of S_i modules in level ***i***), and the instantiation cost ($R_i + r_i$)ic (containing R_i list of submodule and r_i rest modules). Therefore, the verification cost reduction after an extraction is computed in equation (10):

$$VC_i - VC_{i+1} = (S_i \times R_i - (S_i + R_i))dc \quad (10)$$

Our goal is to maximize the value of equation (10) between the two levels; that means our objective is to maximize the term $S_i \times R_i - S_i - R_i$ for every extracting selection. For hierarchical reason and implementation issue, the size of every submodule extracted, S_i, is the smaller the better. If we treat S_i as a constant value, our goal is to maximize a constant value times of R_i. Therefore, **our strategy is to extract the submodule with the maximum number of repetitions at every level to rebuild the hierarchy circuit.**

3. ALGORITHMS

The skeleton of our approach has been established after the previous formulation and analysis. Then we show the details of our approach.

3.1 Circuit Network Representation

For a circuit network *N*, the devices or components at every level in the circuits are represented as the set of modules, ***M[N]***. The set of edges, ***E[N]***, represents the one to one connection between two modules. For any module ***m*** in ***M[N]***, there are certain amount of terminals which are the ports to connect to other modules. These terminals for each module ***m*** in ***M[N]*** are represented as the set of terminals for a module ***T[m]***.

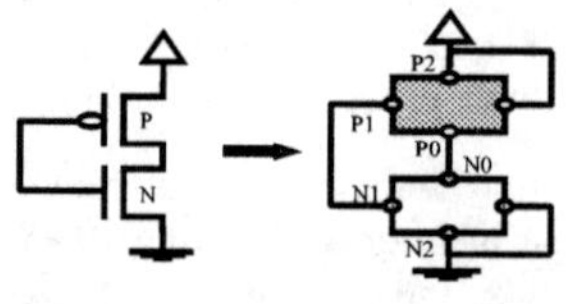

Figure 2 Circuit network representation

Figure 2 shows an inverter circuit represented in circuit network form. The grey square is the PMOS module and the white square is the NMOS module. P0, P1, P2, and P3 are the D, G, S, B terminal of the PMOS module respectively. So are N0~N3 of the NMOS module. The lines between these two modules are the edges that connect each other.

3.2 Submodule Extraction

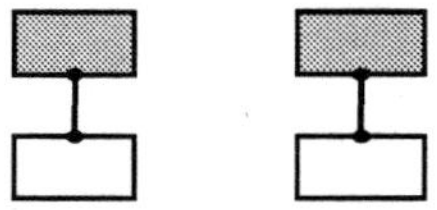

Figure 3 Two identical edges.

Figure 3 is an example showing two edges connected between one grey module and one white module. The dots on both sides of edges represent the terminals of the modules. As figure shows, the module types and the terminal types on each side of two edges are all identical. If we merge the modules on both sides of these two edges together, two identical submodules have been extracted. In other words, **merging the modules on both sides of identical edges can produce new submodules**. Therefore, our strategy which extracts submodules with maximum number of repetitions at every level in the circuit can be accomplished by merging the modules on both sides of the edge at every level in the circuit network.

However, this strategy has been limited by a common graph characteristic: **the overlap of the modules on both sides of the same type edges**.

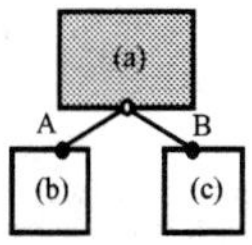

Figure 4 An example of modules overlap

Figure 4 shows a simple example of the overlap situation. Assume the type of edge ***A*** and edge ***B*** are the same. We can see that the upper sides of edge ***A*** and edge ***B*** are connected to the same module, module ***(a)***. If we merge both sides of the two edges, the two new submodules will be overlapped. Therefore, the original strategy must have some modification.

We use ***valid edges*** sieved out from the edges with the same type to ensure the submodules extracted with the valid edges will not overlap each other. For example, there is only one valid edge at most in figure 4 since only one non-overlap submodule can be extracted. We will classify the overlap situation into three classifications and describe the method finding valid edges for each classification.

3.2.1. Overlap in serious connection

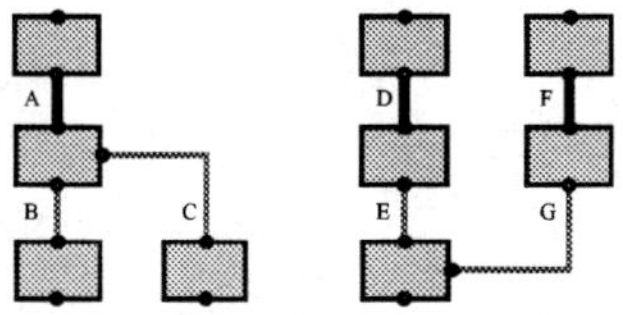

Figure 5 Overlap situation in series connection

Figure 5 is an example of the overlap situation in series connection. It shows many modules with the same type are connected in series. There are seven edges with the same type (***A~G***) and many modules on the sides of them are overlapped. We treat these edges connection as a small graph, use a depth-first search method to calculate the sequence of edge connections, and then use a bread-first search method to sieve the valid edges out. In figure 5, edge ***A***, ***D***, and ***F*** will be sieved out.

3.2.2. Overlap between parallel module I

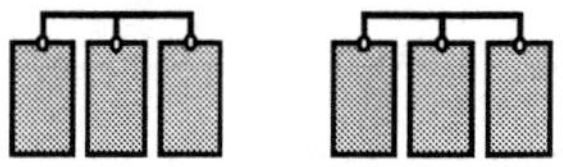

Figure 6 Overlap situation between parallel modules **I**

The edges which have the same types of modules and terminals on both sides are called ***parallel edges***. The modules connected by parallel edges are called ***parallel modules***. Our method dealing with the overlap parallel modules is to reduce all edges between a parallel module block to a special valid edge. For example, three edges of each three-parallel module block in figure 6 are reduced to a valid edge connecting three parallel modules. That means a parallel module block will be extracted to be a new submodule. However, the extraction based on the valid edge of different parallel module blocks may produce different type of submodules. Therefore, the parallel edges with the same type need to be handled specially.

3.2.3. Overlap between parallel module II

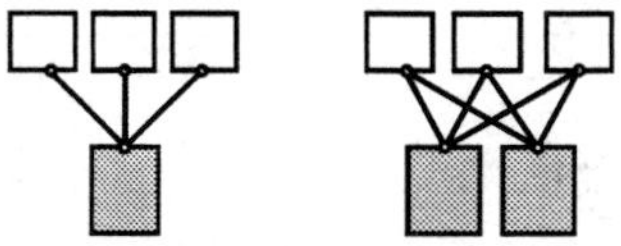

Figure 7 Overlap situation between parallel modules **II**

There is another overlap situation happening between parallel modules. Figure 7 shows many edges with the same types between two different types of parallel module blocks. We can see there are three edges with the same type whose bottom sides are all connected to the same module in the left part of figure 7. Theoretically, these three edges can reduce to a valid edge, but it cannot be confirmed without further network information since it shows nothing different to select any one of three. However, collecting further information may have lacking specific boundary issue and cause deficiency problems. We hold these edges until each of the modules on both sides are updated. Therefore, there are no valid edges sieved out in this situation.

After the overlap situation has been overcome, valid edges with the same type are gathered to be an edge group. We use ***G[N]*** to indicate the set of edge groups in the circuit network ***N***. Valid edges of each edge group correspond to the represented submodule which can be extracted. By sorting the valid edge number of each group in ***G[N]***, we can select the edge group with the maximum number of valid edges to extract the corresponding submodule at every level.

3.3 Complete Flow Chart

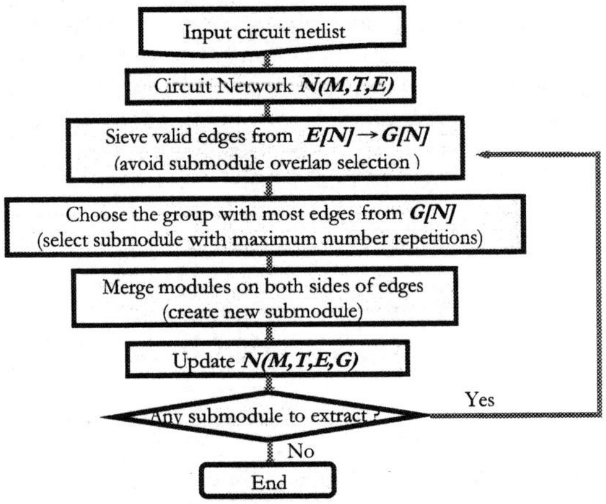

Figure 8 Flow Chart

Figure 8 summaries the algorithm steps described above. It starts from the flatten transistor netlist, and transforms the circuit to a circuit network. Then the greedy submodule extraction uses the property that the valid edges are the skeletons of the submodules. After the network information is updated, the extraction goes iteratively until all possible submodules are extracted.

4. EXPERIMEMT RESULTS

circuit	transistor number	FVC	VCR rate (1-fvc/tran#)	time	space (Byte)
Combinational Circuit					
8b-RA	226	51	77.44%	0.04s	0.672M
4*4 CSA	336	64	80.95%	0.09s	0.928M
C1908	1402	301	78.54%	0.10s	2.050M
C2670	2406	504	79.06%	0.17s	3.036M
C5315	5078	783	84.58%	0.57s	6.804M
C7552	6756	785	88.38%	1.05s	9.478M
Sequential Circuit					
S298	1396	269	80.73%	0.10s	2.02M
S382	1574	289	81.64%	0.11s	2.32M
S1238	2442	492	79.86%	0.20s	3.32M
S9234	12400	1151	90.72%	1.83s	19M
S15850	28556	1869	93.45%	5.41s	41M
S38417	82436	3631	95.60%	19.83s	106M
Memory Circuit					
32W2B	1628	337	79.30%	0.21s	3.034M
32W4B	2468	357	85.54%	0.43s	4.568M
32W8B	4148	377	90.91%	0.92s	8.072M
64W2B	2318	399	82.79%	0.39s	4.044M
64W4B	3542	417	88.23%	0.87s	6.654M
64W8B	5990	469	92.17%	1.52s	12.32M

Table 1 Experimental results

The combinational circuits experimented are some datapath circuits and some ISCAS-85 benchmark circuits. The sequential circuits are some ISCAS-89 benchmark circuits. The memory circuits with 32 and 64 words are extracted by post-layout netlists with the number of bits varies from 2~8 bits.

The first and second columns are the name of the circuit and the number of transistors respectively. Column 3 is the final verification cost (FVC), and column 4 is the verification cost reduction rate (VCR rate) compare to the verification cost of initial netlist. The final two columns are the time and memory space being consumed.

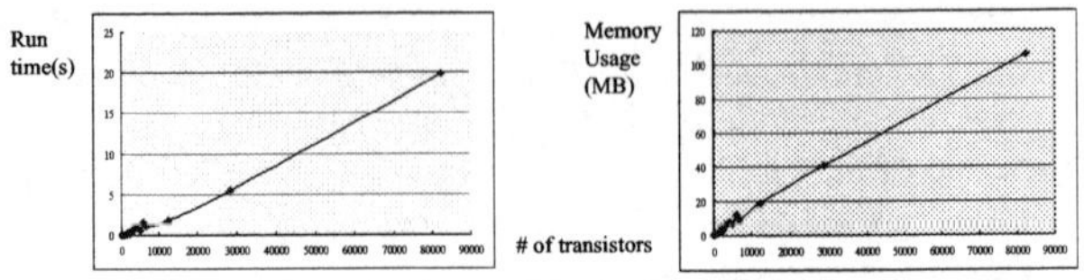

Figure 9 Time and Space Complexity

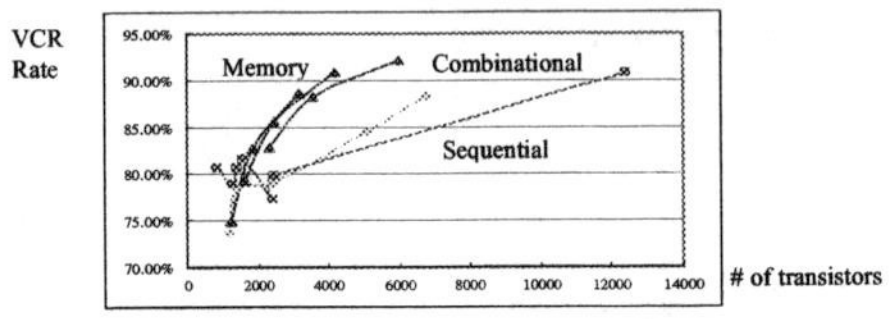

Figure 10 VCR rate vs. Circuit size

Figure 9 shows the near-linear complexity in both time and space consuming. Figure 10 shows the relationship between the circuit size and the VCR rate for three types of circuits. The VCR rate grows with the circuit size generally because there are more hierarchical levels in large circuits. For equal circuit size, memory circuits have the most VCR rate and sequential circuits have the least. It is because memory circuits often have more regular structures than others and sequential circuits conclude control units with almost no regularity.

5. CONCLUSIONS

In this paper, we propose an efficient approach for circuit hierarchy reconstruction and simulating effort reduction by extracting repeated submodules. Our approach is an iterative extraction mechanism using the transistor netlist as the only reference information. Because of the structure equivalence extraction, our approach can apply to circuits with any design types and circuits composed by any device. After the experiments, the results show the near-linear complexity for both time and memory usage.

6. REFERENCES

[1] M. Ohlrich, C. Ebeling, E. Ginting, and L. Sather, 'SubGemini: Identifying Subcircuits Using a Fast Subgraph Isomorphism Algorithm", Proceedings of Design Automation Conference, pp. 31-37, June 1993.

[2] Z. Ling and D. Y. Y. Yun, "An Efficient Subcircuit Extraction Algorithm by Resource Management", 2nd International Conference on ASIC, pp. 9-14, Oct 1996.

[3] W.-H. Chang, S.-D. Tzeng, and C.-Y. Lee, "A Novel Subcircuit Extraction Algorithm by Recursive Identification Scheme", 2001 IEEE International Symposium on Circuits and Systems, Vol.5, pp. 491 -494, 2001.

[4] A. Chowdhary, S. Kale, P. K. Saripella, N. K. Sehgal, and R. K. Gupta, "Extraction of Functional Regularity in Datapath Circuits", IEEE Transaction on Computer-Aided Design of Integrated Circuits and Systems, Vol.18, No.9, Sept 1999.

[5] D. Sreenivasa Rao, Fadi J. Kurdahi, "On Clustering for Maximal Regularity Extraction", IEEE Transaction on Computer-Aided Design of Integrated Circuits and Systems, Vol.12, No.8, Aug 1993.

[6] R.X.T. Nijssen and J.A.G. Jess, "Two-Dimensional Datapath Regularity Extraction", Proceedings of the 5th ACM/IEEE Physical Design Workshop, Reston, Virginia, 1996.

[7] Thomas Kutzschebauch, "Efficient Logic Optimization Using Regularity Extraction", Proceedings of the International Conference on Computer Design, pp. 487-493, 2000.

ALGORITHM FOR YIELD DRIVEN CORRECTION OF LAYOUT[1]

Yang Wang, Yici Cai, Xianlong Hong, Qiang Zhou

wangyang01@mails.tsinghua.edu.cn ,caiyc@tsinghua.edu.cn , hxl-dcs@tsinghua.edu.cn,zhouqiang@tsinghua.edu.cn

Abstract

As the development of VLSI technique, the critical dimension of IC has become smaller than the exposure wavelength. Due to the diffraction and interaction of optical waves, deformations between the image on wafer and the feature on layout are undeniable. This results in bad performance or even invalid circuits of the chips. OPC is critical compensation technique to correct the deformations on wafer images. This paper presents a layout correction and optimization algorithm called MOPC; it's a flexible and efficient core for the model-based OPC system. Since we divide the target features into different types before correction, the OPE between the target features and the environment features and the OPE between the neighboring segments of the inside feature are both considered during the correction.

Keywords: Layout, Optical Lithography, Model-Based OPC

1.Introduction

The chip yield is an important issue in IC manufacture all the time. Along with the development in IC technology, the deformations between mask and wafer become to affect the chip yield directly. With the shrinking of critical dimensions, Optical Proximity Effects (OPE) becomes an important factor to the performance of IC. Now the method to solve the problem that is widely adopted by industry is the Yield Driven Correction of Layout. Relative technology, Reticle Enhancement Technology (RET), mainly includes optical proximity correction (OPC), phase-shift mask (PSM) and assist and dummy features [1][2].

In today's deep submicron technology, the industry's adoption of OPCs means that the original design, the mask layout, and the structures printed on the wafer bear little resemble to each other, making the physical verification complex. The key verification challenge is to confirm that RET treatments on the photomask layout will accurately produce the desired result on the wafer, and to make sure that nothing is defective in the mask layout synthesis process. The improvement of the technology of ICs requires adopting OPC to enhance the performance of the circuits.

OPC reduces the deviation of photolithography features by changing the shapes of the layout features. Currently this method is divided into two sorts: rules-based approach [3] and model-based approach [4]. Rules-based approach needs to build the database of the enhancement rule first. Then in real application, one can get data from the database and use the data to enhance the features on the mask. Because of the limited data in the database and the approximation of the data values, this method is difficult to acquire high correction precision [5].

Model-based approach needs to choose the appropriate optical model in advance, the system uses the optical model to simulate the process of optical lithography and implement the enhancement of the features on the mask. This method is more time consuming, but due to its flexibility and higher precision, it is widely used. But current Model-Based OPC systems don't partition the models accurately, and they don't fully consider the OPE between the target features and the environment features, this effect and the OPE between the neighboring segments of the inside feature [6]. To improve the correction precision, we bring forward a high correction-precision, high flexibility kernel arithmetic of Model-Based OPC system.

This paper presents a layout correction and optimization algorithm for MOPC. We suggest an efficient classified operation for the complex features in layout. Some methods are introduced to ensure the high correction precision in MOPC. Pretreatment first we cut the edges of the correction target into small segments, each segment will be divided into different types according to its location on the basis of the Environment features. Compared with existing 2-D OPC algorithms[7], this method we proposed ensures that all types of the models can be included, at the same time simplifies the correction operation, and makes the system achieve high precision and efficiency. On basis of this, for further optimization we propose two kinds of optimization: single-loop and dual-loop, which are implemented and compared. The experimental results show the practicability and veracity of our MOPC.

The rest of the paper is organized as follows: in Section 2, we describe the Model-Based OPC system and preliminaries knowledge. In Section 3, we present the classified operation for the features and two optimization methods. Some experimental results are shown in Section 4. At last we give our conclusions in section 5.

2. Preliminaries knowledge

2.1 Model-based OPC system

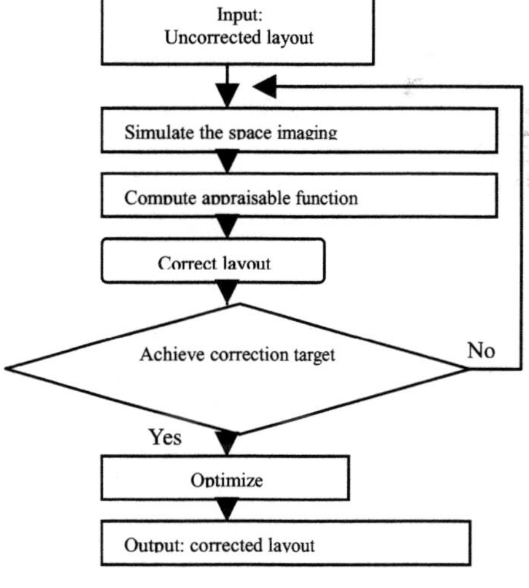

Fig.1 Model-Based OPC

The flow chart of Model-Based OPC is shown in Fig1.MOPC mainly includes: the simulation of photolithography, the computation of evaluation function, the correction of target

[1] This work is supported by the National Natural Science Foundation of China (NSFC) 90307017

feature, and the optimization of the corrected result.

The simulation of photolithography is the basic of the Model-Based OPC system [8][9], our MOPC uses KERNEL (the simulation program of photolithography) [8], to acquire the image on wafer corresponding to the features on the mask. Evaluation function is the important criterion of the coherence degree between the images and the features.

2.2 Environment features

The input data of MOPC is the correction target and the sequence of the environment features. Correction target is a whole polygon (convex polygon or concave polygon). Environment features are the environmental features that may bring optical proximity effects to the correction target. We use the variable AMBIT to express the OPE incidence. The value of AMBIT is related with the parameters of photolithography technology and the requirement of correction precision. (Fig2.)

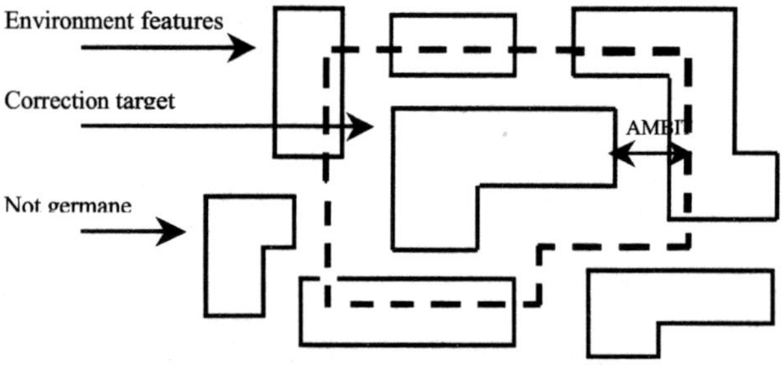

Fig2. Show the input correction target and the environment features. Mask features correction outputs the corrected features in the form of polygon.

Environment features are the features that can generate optical proximity effects to the target feature. During the process of the correction, the environment features must be taken into account to simulate optical lithography of the target features.

2.3 cutting the edges of the correction target

During the correction, we assume the change of light intensity is monotonefrom the outside to the inside of the feature, and the inside of the edge are exposed completely. So the targets of the correction operation are focused on the edges of the correction feature.

The length of the cutting segment is related with the type of the segment and the correction precision. Normally changes of the feature are great for line-end type segment and corner-type segment, the distortion can be serious, so the length of these two types is small. Edge-type segment is commonly a small line of the long edge, we can select longer cutting length for lower precision requirement; for higher precision requirement we can choose shorter cutting length or cut the Environment features.

At first, we cut the edges of the correction target into small segments; each segment will be marked with different types according to its location. There are three types: edge, line-end and corner, see Fig3. The segment is the basic unit of correction and different optimization strategy and evaluation functions are used

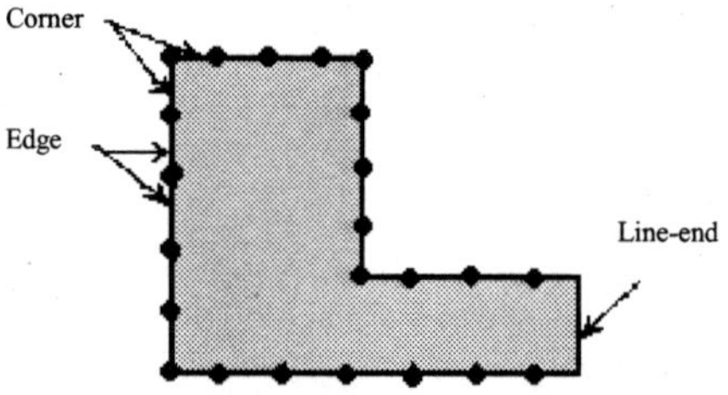

Fig3. Show the cutting of the correction target.

for different types of segments. This operation makes the correction target more specific and simple. The correction process deals with all the segments along the boundary of the correction target clockwise.

3. MOPC algorithm description

The arithmetic of MPOC uses some standard storage formats for the input layout data, these formats use Hierarchy Management to compress the storage space. During OPC operation, because the correction operation is related with the shape of the neighboring features, the Hierarchy Management is not applicable. The former layout data input file adopts grid-division; data is saved as integer format. Due to the small size of the additive correction feature, this method will increase the magnitude of the layout data greatly. For this reason, MPOC adopts scan line vertex representation method. This method flattens the hierarchical data structure and gets a simplex layout data of each layer. The features are saved in the coordinate order. During the correction the system uses the scan line to scan each layer of the layout following the top-down, left-right turn. The object of the correction is the edge of the feature, and the neighboring edges of the feature should also be considered.

Our MOPC cuts the edges of the correction target into correction segments . For different types of correction segments, we adopt different processing method. In this way we enhance the pertinence of OPC system, and improve the correction precision.

The single-loop optimization is implemented for the target feature and the dual-loop optimization is implemented between the target features and the environment features.

We use the width and the space of the feature as the criterion of selecting the target features. In order to find out all the target features that need correction we search all the layers of the layout and all the feature elements of each layer. Thus we set up two basic criterions: the corrected feature doesn't change the connection of the feature before correction; correction operation must be performed for all the features of small width or space that need correction. This is a great difference between our MOPC and practical 2-D OPC.

3.1 correcting the target feature

3.1.1 Correction strategy

(1) Correction strategy for the edge-type segment

Fig4. Shows the removal strategy of the edge-type segment on the dark mask. The segment will be moved along the vertical direction of the edge entirely.

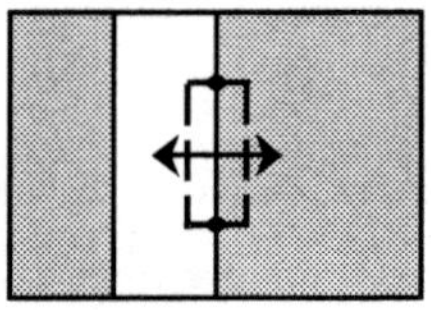

Fig.4 The edge-type segment

(2) Correction strategy for the corner-type segment

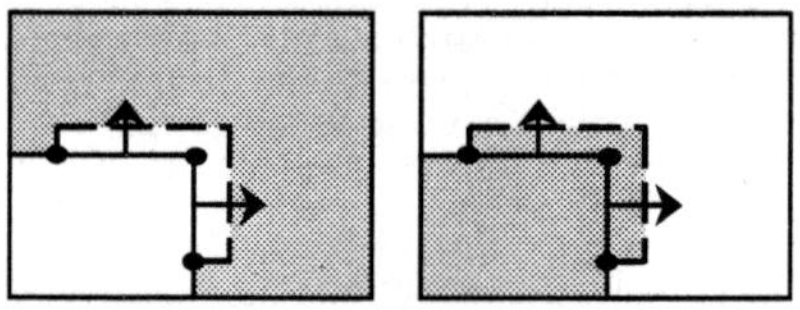

Fig5 The corner-type segment

Fig5. Shows the removal strategy of the inside/outside corner-type segment on the dark mask. The two corner segments will be moved inwards/outwards entirely.

(3) Correction strategy for the line-end-type segment

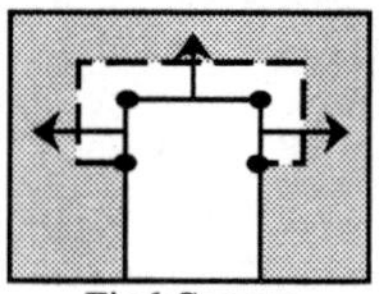

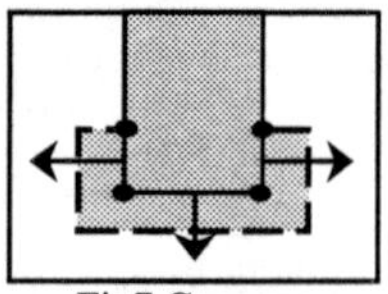

Fig6 Convex Fig7 Concave

The strategy for the convex/ concave line-end-type segment on the dark mask is shown in Fig6.7. The segment will be moved outwards/inwards together with two adjacent corner-type segments.

3.1.2 The evaluation function

The evaluation function as the criterion of correction denotes the difference between the image on wafer and the feature on mask. The objective is to minimize this difference.

Actually from the physical theory of the photolithography, the distortions between the images and features are inevitable in a certain extent. And our ultimate objective is to ensure the performance and yield. So we can control the objective value of evaluation functions according to the precision requirements.

(1) Evaluation for the correction of edge-type segment

For the edge-type segment we hope the image boundary is as close as possible to the feature boundary. So we choose the distance D_0, from the midpoint of segment to its corresponding imaging point, as the evaluation function. (Fig8)

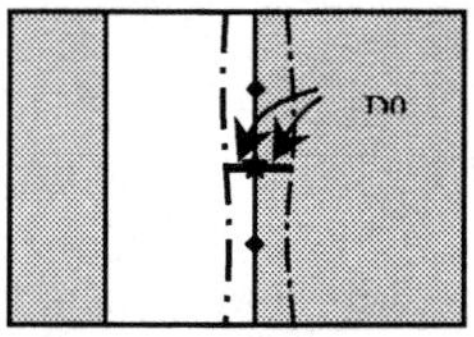

Fig8. Evaluation for the correction of edge-type segment

The evaluation function = D0

(2) Evaluation for the correction of corner-type segment

For corner-type segment, we hope its image can be sharper. After correction, the image boundary is closer to the vertex of the corner. But some parts have been over-corrected. Fig9 shows this situation. We use D_1 to represent the distance from image boundary to the vertex and D_2 to represent the over-corrected distance. Then we get the relation curve of these two parameters in

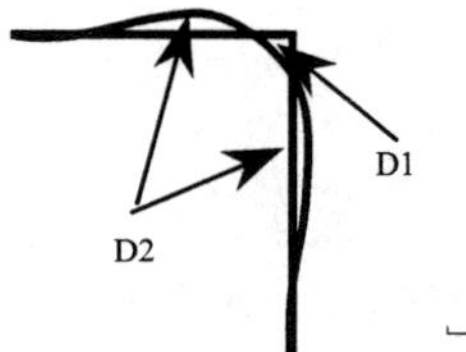

Fig9. Show the imaging features after corner correction and the expected features.

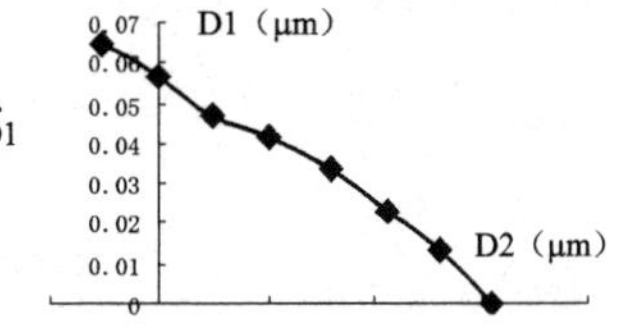

Fig10. Give the relation between two parameters.

Fig10.

So we have two choices for evaluation function: vertex distance and side distance. The Vertex distance (D_0) is the distance from the imaging point along the angle bisector to the vertex. The side distance is the sum of distances (D_1, D_2) from the midpoints of segments to their corresponding imaging points. (Fig11.)

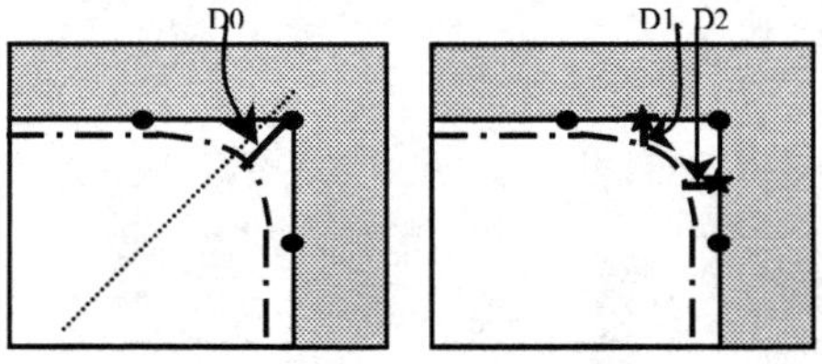

Fig11. The correction of the corner-type line

(3) Evaluation for the correction of line-end-type segment

For the line-end-type segment, our main objective is to adjust the line length and the round corners can be ignored in some extent. The evaluation function is the sum of three distances: D_0, D_1 and D_2. D_0 is from midpoint of line-end-type segment to its corresponding imaging point. D_1 and D_2 are from the imaging points along the angle bisector to the vertexes. (Fig12.)

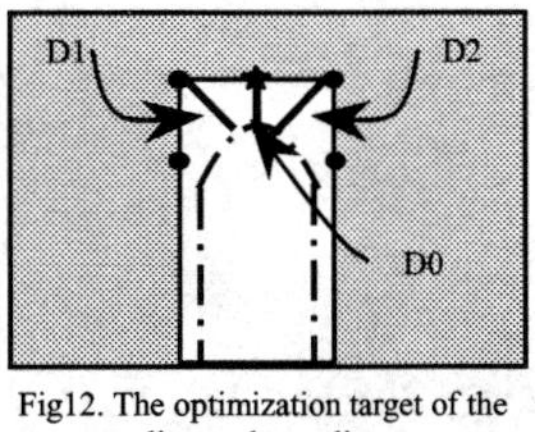

Fig12. The optimization target of the line-end-type line

The target function = D0+D1+D2

3.2 Optimizing the correction feature

3.2.1 single loop optimization for target feature

In the correction operation, the segments will be corrected one by one along the boundary. While one segment is under operation, all the others are kept in the uncorrected position. After all segments are corrected, the corrected segments will be connected to form a new polygon. But in practice, the corrections of segments will affect each other. So in the optimization step, we perform the correction iteratively to optimize the original result and the segments will be located in the last corrected positions.

We define different objective values for those evaluation functions in the iterations. First, the values can be relaxed and they will be constrained gradually in the iterations to meet the precision requirement. This method improves the correction result efficiently. We call it single loop optimization for target feature.

3.2.2 Dual loop optimization between target feature and environment Features

Environment features are the features that can generate optical proximity effects to the target feature. MOPC corrects the features on the mask one by one. During correction operation, the environment features are uncorrected. Actually, the correction of these environment features can affect the correction for the target feature, especially when they are very closer.

The further optimization takes the effect of corrected environment features into account. The target feature is combined with its environment features as a group. First we correct the target feature on condition of keeping the uncorrected environment features. Next, we correct the environment features one by one. At last, we correct the target feature again with the corrected environment. The correction results of target feature and its environment features are recorded and will be used in the following correction operations.

This optimization will improve the correction result further. We call it Dual loop optimization between target feature and environment features. In the complete correction and optimization operations, each feature will be corrected twice.

3.3 Post-treatment:

Smooth operation for the corrected features

After the correction, the number of the vertexes will increase several times. Fig13 is an example of the correction. The layout

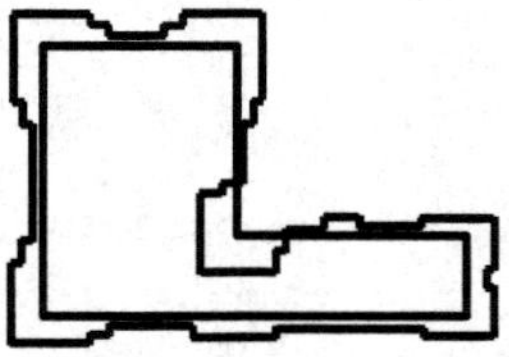

Fig 13 before smooth operation

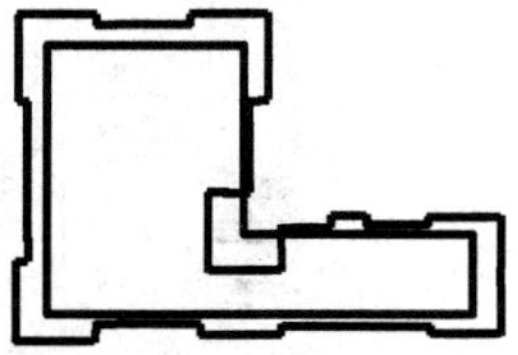

Fig14 after smooth operation

before correction has only 6 vertexes, but after correction it has 54 vertexes.

Obviously the correction operation increases the complexity of space, and it also brings a few difficulties in describing and producing the masks. In post-treatment, smooth operation is performed to remove small fluctuation that can be ignored. This operation doesn't affect the correction precision seriously. See Fig14, this target feature after smooth operation has 36 vertexes.

4.The Experimental Results

The Model-Based OPC system is implemented in C program on Sun Enterprise E450 with 4GB memory. The experimental results are showed in Fig. 15 and Fig. 16,17,18. The conditions of photolithography are as follows: wavelength = 250nm, numerical aperture = 0.5, defocus = 0.0, partial coherence factor = 0.7, minimum line width = 180nm.

The simulation of photolithography is the basic of the Model-Based OPC system; our MOPC uses KERNEL (the simulation program of photolithography) [10], to acquire the image on wafer corresponding to the features on the mask. All the test cases are adopted from the practical layout data, which ensure the strength of our experimental validation. To show the effectiveness of out method, we applied it to a set of random testing cases. Obviously, the Model-Based OPC system could achieve good results for every instance. Since we divide the target features into different types before correction, the OPE between the target features and the environment features and the OPE between the neighboring segments of the inside feature are both considered primal during the correction.

5. Conclusion

This paper presents a layout correction and optimization algorithm called MOPC; it's a flexible and efficient core for the model-based OPC system. The experimental results show the practicability and veracity of MOPC. In the future, Model-based OPC system will be improved further, and will be used more widely in IC fabrication.

Reference

[1] Jorg Thiele, Christoph Friedrich, Christoph Dolainsky, et al, "Integration of alternating phase shift mask technology into optical proximity correction," SPIE. vol.3679, pp.548-555, 1999

[2] A.B.Kahng, Shailesh Vaya, and Alexander Zelikovsky, "New Graph Bipartizations for Double-Exposure, Bright Field Alternating Phase-Shift Mask Layout," IEEE. pp.133-138, 2001

[3] Ryuji Takenouchi, Isao Ashia and Hiroichi Kawahira. "Development of a fast line width correction system," SPIE. vol. 4066 pp. 688-696, 2000

[4] Satomi Shioiri and Hiroyoshi Tanabe, "Fast Optical Proximity Correction: Analytical Method," SPIE vol. 2440 pp. 261-269, 1995

[5] Shi Rui, Cai Yici, Hong XianLong."The Selection and Creation of the Rules in Rules-Based Optical Proximity Correction." ASIC, pp.50-53, 2001

[6] Eiji Tsujimoto, Takahiro Watanabe, Kyoji Nakajo, "Automatic parallel optical proximity correction system for application with hierarchical data structure," SPIE Proceedings. Vol. 3679, pp.675-685, 1999

[7] H.Chuang,P.Gilbert,W.Grobman," Practical applications of 2-D optical proximity corrections for enhanced performance of 0.25 μm random logic devices" Electron Devices Meeting, vol.994, pp483-486,1997

[8] D. Dolainsky and W. Maurer, "Application of a Simple Resist Model to Fast Optical Proximity Correction," SPIE. vol.3051, pp.774-780, 1997

[9] Yong Liu, "Binary and Phase Shifting Mask Design for Optical Lithography," IEEE Transactions on Semiconductor Manufacturing, vol.5, NO.2, 1992

[10] Zhijin Chen and Zheng Shi, "A new method of 2D contour extraction for fast simulation of photolithographic process," <Chinese journal of semiconductors>, vol.23, No.7, pp766-7710, 2002

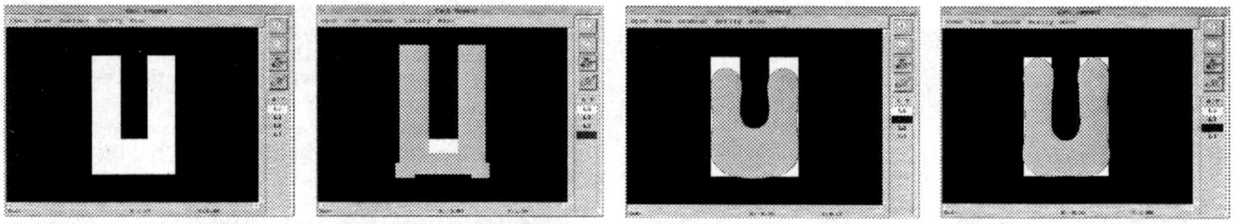

a) The target before correction b) The target after correction c) The imaging before correction d) The imaging after correction

Fig15. The Experimental Results for the line-end-type

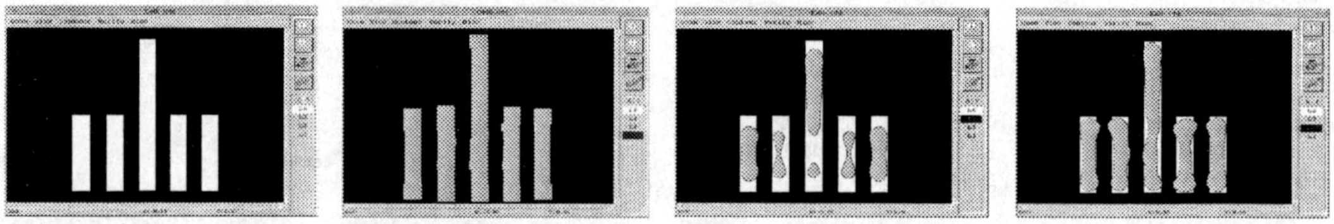

a) The target before correction b) The target after correction c) The imaging before correction d) The imaging after correction

Fig16. The Experimental Results for the edge-type

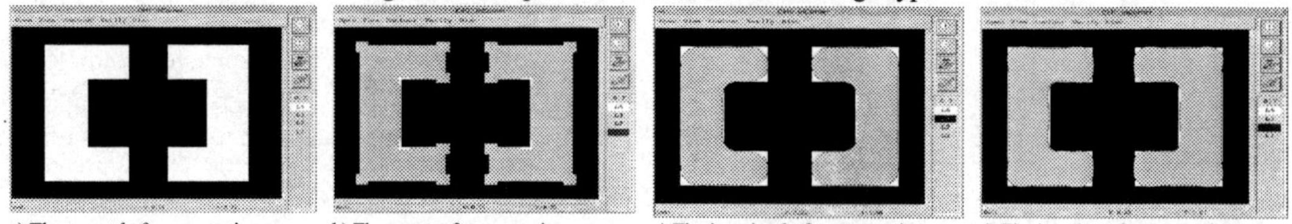

a) The target before correction b) The target after correction c) The imaging before correction d) The imaging after correction

Fig17. The Experimental Results for the corner-type

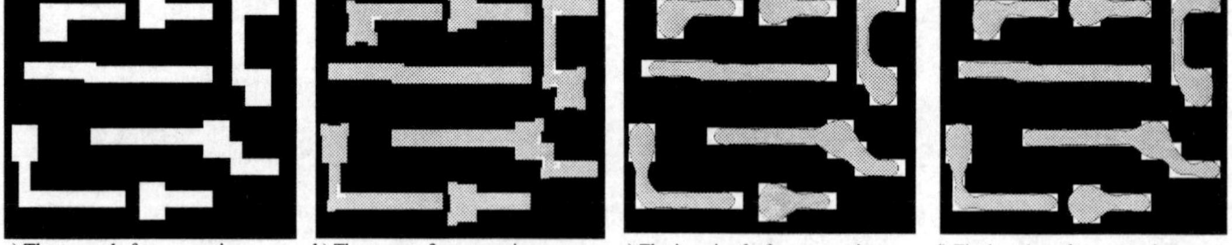

a) The target before correction b) The target after correction c) The imaging before correction d) The imaging after correction

Fig18. The Experimental Results for poly

SYMBOLIC NOISE ANALYSIS IN ANALOG INTEGRATED CIRCUITS

Sánchez-López C., Tlelo-Cuautle E.

National Institute of Astrophysics, Optics and Electronics (INAOE)
Av. Luis Enrique Erro No 1. Tonantzintla, Puebla. P.O.Box 51 & 216, 72000, MEXICO
Tel/Fax:+(52)(222) 247 05 17. E-mail: csanchez@inaoep.mx, e.tlelo@ieee.org

ABSTRACT

A novel symbolic analysis method to compute Noise Figure (NF) in analog circuits, is presented. The proposed method is based on manipulating data structures representing the interconnection relationships (IRs) of conductances, resistances and MOSFETs modeled by nullators and norators. It is demonstrated that the formulation of the system of equations being improved by computing the cartesian product of the indexes associated to the IR of norators and nullators. The proposed noise analysis method is highlighted by including several examples which show the feasibility of the technique. Finally, it is shown that the suitability is in good agreement compared to HSPICE noise simulations.

1. INTRODUCTION

Circuit analysis for analog ICs can be carried out using the commercially available numerical simulators such as HSPICE. However, symbolic simulators are preferred in analog design, since a symbolic expression helps to gain insight about the behavior of a circuit [3]. In this manner, a new technique to calculate noise in analog ICs by computing fully symbolic expressions of NF is proposed. The behavior of all active devices is modeled by nullor-circuits to improve symbolic computation as shown in [6], where a Compacted System of Equations (CSEs) is formulated by manipulating the indexes associated to the IRs of norators and nullators. In this paper, the symbolic-expression of NF for a differential pair and a CMOS Miller Operational Transconductance Amplifier (OTA) are derived and the total output noise-voltage is compared with HSPICE noise simulations. In doing symbolic noise analysis it is assumed that noise sources are uncorrelated, as it is used in most of the circuit simulators. A briefly description of the symbolic method given in [6], is described in Section 2. Section 3, is devoted to introduce the proposed method to compute NF in analog circuits. In Section 4, symbolic expressions of NF of a differential pair and a CMOS Miller OTA are computed by applying the proposed symbolic method. Finally, the conclusions are given in Section 5.

2. SYMBOLIC ANALYSIS OF NULLOR CIRCUITS

When MOSFETs are modeled with nullors, the resulting network contains a set of circuit-elements which simplifies the algorithms for doing symbolic circuit analysis [4, 6]. Furthermore, nullor circuits allows to compute a CSEs directly, which is expressed by:

$$i_{CNA} = Y_{CNA} v_{CNA}, \tag{1}$$

This work is supported by CONACYT/Mexico, under project number J40321-Y.

where Y_{CNA} represents the compacted linear admittance matrix, i_{CNA} the vector of stimulus and v_{CNA} the vector of nodal voltages variables. As one can see in [6], (1) is computed by manipulating the IRs associated to the rows and cols of the nullor-circuit. The symbolic technique can be summarized as follows:

1. Transform the analog circuit to its small-signal equivalent using nullors.
2. Compute the data-structure of the admittance-elements. To compute NF, the admittance-elements are squared [1, 2].
3. Compute the data-structure of nullators and norators IRs.
4. Compute the cartesian product of the combinations of pair of nodes associated to the elements in the matrix Y_{CNA}.
5. Search the resulting elements related to every pair of nodes and add them for every element of the reduced matrix.
6. Compute vectors i_{CNA} and v_{CNA}, in order to obtain a CSEs. To compute NF, these vectors are average expressions, since are modeled in terms of power spectral density [1, 2].
7. Solve for the desired output/input relationship.

3. PROPOSED METHOD

Although the MOSFET is a nonlinear device, when it is biased in the saturation region, its behavior can be modeled by a voltage controlled current source and this controlled source can be modeled using nullors, as shown in [4]. Using the properties of the nullor, a model for the MOSFET of three and four terminals can be obtained [5]. Since noise in integrated circuits is caused by some random physical phenomena which leads to small current and voltage fluctuations within circuit devices [1, 2], sources of noise associated to MOSFETs also can be added. In this manner, the proposed method to compute NF is based on modeling the active devices using nullors, where the parasitics elements and noise sources are added according to Fig. 4 and Fig. 5 for MOSFETs of three and four terminals, as shown in [5]. NF can be defined by (2), which is widely used as a measure of noise performance and it is usually expressed in decibels [1, 2].

$$NF = \frac{SNR_{in}}{SNR_{out}}, \tag{2}$$

Equation (2), can be rearranged to obtain the NF equation for circuits working in voltage-mode [2], and a similar analysis is carried out for current-mode circuits, where the NF equations becomes:

$$NF_v = \frac{\overline{V^2}_{n,out}}{A_v^2 N_{RS}}, \quad NF_i = \frac{\overline{I^2}_{n,out}}{A_i^2 N_{RS}}, \tag{3}$$

where $\overline{V^2}_{n,out}$= voltage of the total output-noise, $\overline{I^2}_{n,out}$= current of the total output-noise, A_v= voltage gain, A_i= current gain and N_{RS}= noise of the source resistance. In HSPICE a MOSFET is modeled with four noise current generators. Two of these represent the thermal noise associated with the parasitic drain and source series resistances. The other two are modeled as current sources from drain to source. One of them represents white shot noise and the other flicker noise. Equation (4) shows the models implemented in HSPICE for MOSFETs and resistances according to the noise level=0 [5].

$$\overline{I^2}_{n,R} = \frac{4kT}{R},\ \overline{I^2}_{n,M} = \frac{8}{3}kTg_M,\ \overline{I^2}_{n,M} = \frac{K_F I_{IDS}^{AF}}{C_{ox} L_{eff}^2 f}, \quad (4)$$

4. EXAMPLES

In the following two examples, the MOSFET is biased in the saturation region and the contribution of thermal noise is only considered in the computation of the symbolic NF.

4.1. CMOS Differential Pair

Lets consider the circuit shown in Fig. 1, which is a differential pair with noise sources connected between source-drain terminals. The representation of the equivalent circuit using nullors, is shown in Fig. 2, where the body effect has not been considered. The

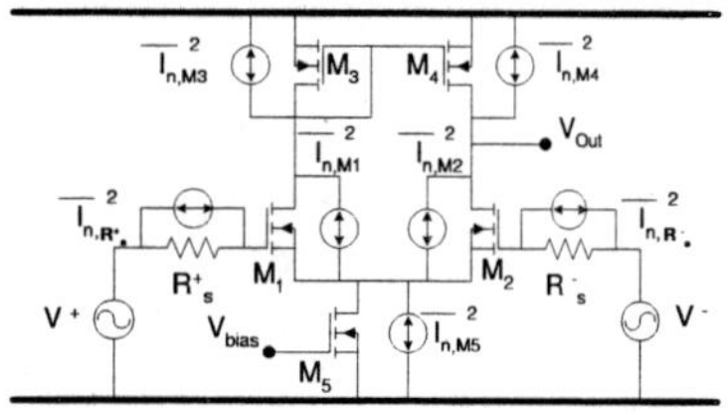

Fig. 1. Differential pair including noise sources.

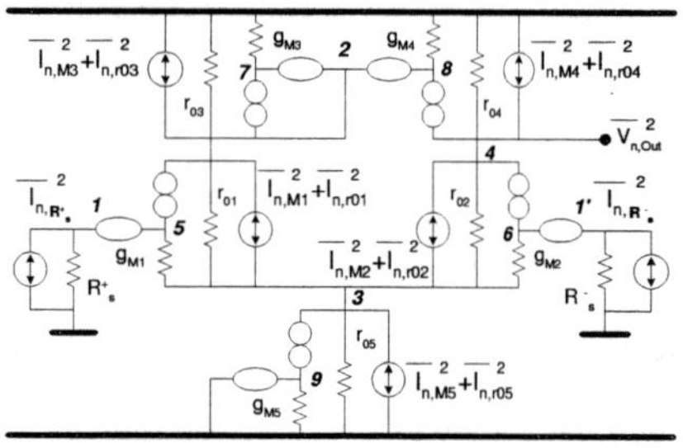

Fig. 2. Nullor circuit equivalent of Fig. 1.

next step consist in generate data structures of conductances, resistances and norator-nullator elements, where the IRs are shown in Table 1, and for the nullator-norator elements the IRs are shown in Table 2. The circuit shown in Fig. 2 has 9 nodes and 5 nullors, where the node $1'$=0 when the input signal V^+ is considered, so that the order of the CSEs will be 4×4. The voltage variables, according to Table 2 are $v_{1,5}$, $v_{2,7,8}$, v_3 and v_4 from the nullator properties, while the current variables are: i_1, $i_{2,5,7}$, $i_{3,9}$ and $i_{4,6,8}$ from the norator properties [4]-[5]. The resulting cartesian product of the IRs of norators and nullators becomes:

$$[(1),(2,5,7),(3,9),(4,6,8)] \times [(1,5),(2,7,8),(3),(4)], \quad (5)$$

Table 1. Data structure of conductances and resistances.

Conductance	Nodes	Resistance	Nodes
g_{M1}^2	$5,5$ or $3,3$	ro_1^2	$2,2$ or $3,3$
$-g_{M1}^2$	$5,3$ or $3,5$	$-ro_1^2$	$2,3$ or $3,2$
g_{M2}^2	$6,6$ or $3,3$	ro_2^2	$4,4$ or $3,3$
$-g_{M2}^2$	$6,3$ or $3,6$	$-ro_2^2$	$4,3$ or $3,4$
g_{M3}^2	$7,7$	ro_3^2	$2,2$
g_{M4}^2	$8,8$	ro_4^2	$4,4$
g_{M5}^2	$9,9$	ro_5^2	$3,3$
		R_s^{+2}	$1,1$
		R_s^{-2}	$1',1'$

Table 2. Data structure of nullators and norators.

Nullator	Nodes	Norator	Nodes
$O1$	$5,1$	$P1$	$2,5$
$O2$	$6,1'$	$P2$	$4,6$
$O3$	$7,2$	$P3$	$2,7$
$O4$	$8,2$	$P4$	$4,8$
$O5$	$9,0$	$P5$	$3,9$

$$\begin{aligned}&\mathbf{Y_{11}}=(1,1)+(1,5),\mathbf{Y_{12}}=(1,2)+(1,7)+(1,8),\mathbf{Y_{13}}=(1,3),\mathbf{Y_{14}}=(1,4),\\&\mathbf{Y_{21}}=(2,1)+(2,5)+(5,1)+(5,5)+(7,1)+(7,5),\mathbf{Y_{22}}=(2,2)+(2,7)+(2,8)\\&+(5,2)+(5,7)+(5,8)+(7,2)+(7,7)+(7,8),\mathbf{Y_{23}}=(2,3)+(5,3)+(7,3),\\&\mathbf{Y_{24}}=(2,4)+(5,4)+(7,4),\mathbf{Y_{31}}=(3,1)+(3,5)+(9,1)+(9,5),\mathbf{Y_{32}}=(3,2)\\&+(3,7)+(3,8)+(9,2)+(9,7)+(9,8),\mathbf{Y_{33}}=(3,3)+(9,3),\mathbf{Y_{34}}=(3,4)+(9,4),\\&\mathbf{Y_{41}}=(4,1)+(4,5)+(6,1)+(6,5)+(8,1)+(8,5),\mathbf{Y_{42}}=(4,2)+(4,7)+(4,8)\\&+(6,2)+(6,7)+(6,8)+(8,2)+(8,7)+(8,8),\mathbf{Y_{43}}=(4,3)+(6,3)+(8,3),\\&\mathbf{Y_{44}}=(4,4)+(6,4)+(8,4),\end{aligned} \quad (6)$$

Searching such elements from Table 1, the CSEs is shown by (7).

$$I = \begin{bmatrix} \frac{1}{R_s^{+2}} & 0 & 0 & 0 \\ g_{M1}^2 & g_{M3}^2 + \frac{1}{r_{o1}^2} + \frac{1}{r_{o3}^2} & -g_{M1}^2 - \frac{1}{r_{o1}^2} & 0 \\ -g_{M1}^2 & -\frac{1}{r_{o1}^2} & g_{M1}^2 + g_{M2}^2 + \frac{1}{r_{o1}^2} + \frac{1}{r_{o2}^2} + \frac{1}{r_{o5}^2} & -\frac{1}{r_{o2}^2} \\ 0 & g_{M4}^2 & -g_{M2}^2 - \frac{1}{r_{o2}^2} & \frac{1}{r_{o4}^2} + \frac{1}{r_{o2}^2} \end{bmatrix} V, \quad (7)$$

where

$$I = \begin{bmatrix} \overline{I^2}_{n,R_S} \\ \overline{I^2}_{n,M1} + \overline{I^2}_{n,M3} \\ \overline{I^2}_{n,M1} + \overline{I^2}_{n,M2} + \overline{I^2}_{n,M5} \\ \overline{I^2}_{n,M2} + \overline{I^2}_{n,M4} \end{bmatrix}, V = \begin{bmatrix} \overline{v^2}_{1,5} \\ \overline{v^2}_{2,7,8} \\ \overline{v^2}_{3} \\ \overline{v^2}_{4} \end{bmatrix}, \quad (8)$$

From (8), $\overline{V^2}_{n,out}=\overline{v^2}_4$. For, $g_{M2}=g_{M1}$, $g_{M4}=g_{M3}$, $r_{o2}=r_{o1}$ and $r_{o4}=r_{o3}$, the voltage gain is given by:

$$A_v^+ = \frac{g_{M1} g_{M3} r_{o1}^2 r_{o3}^2}{r_{o3}^2(g_{M3} r_{o1} + 1) + r_{o1}^2(g_{M3} r_{o3} + 1) + 2 r_{o1} r_{o3}}, \quad (9)$$

Finally, using (3), the symbolic computation of NF_v^+ is given by:

$$NF_v^+ = 1 + \frac{2}{3}\frac{1}{R_s^+ g_{M1} g_{M3}}\left[\frac{1}{r_{o1}^2} + \frac{1}{r_{o3}^2}\right]\left[\frac{1}{g_{M1}} + \frac{1}{g_{M3}}\right] + \frac{4}{3}\frac{1}{R_s^+ g_{M1}}\left[\frac{g_{M3}}{g_{M1}} + 1\right] \quad (10)$$

Now, considering the input signal V^-, the node 1=0 and the new voltage variable is $v_{1',6}$, the cartesian product is given by

$[(1'),(2,5,7),(3,9),(4,6,8)] \times [(1',6),(2,7,8),(3),(4)]$ and the modified elements in the Y_{CNA} matrix are:

$$\mathbf{Y_{11}}=(1',1')+(1',6), \mathbf{Y_{12}}=(1',2)+(1',7)+(1',8), \mathbf{Y_{13}}=(1',3),$$
$$\mathbf{Y_{14}}=(1',4), \mathbf{Y_{21}}=(2,1')+(2,6)+(5,1')+(5,6)+(7,1')+(7,6), \mathbf{Y_{31}}=(3,1')$$
$$+(3,6)+(9,1')+(9,6), \mathbf{Y_{41}}=(4,1')+(4,6)+(6,1')+(6,6)+(8,1')+(8,6), \tag{11}$$

the new CSEs is shown by (12) and the vectors I and V are given again by (8).

$$I = \begin{bmatrix} \frac{1}{R_s^{-2}} & 0 & 0 & 0 \\ 0 & g_{M3}^2 + \frac{1}{r_{o1}^2} + \frac{1}{r_{o3}^2} & -g_{M1}^2 - \frac{1}{r_{o1}^2} & 0 \\ -g_{M2}^2 & -\frac{1}{r_{o1}^2} & g_{M1}^2 + g_{M2}^2 + \frac{1}{r_{o1}^2} + \frac{1}{r_{o2}^2} + \frac{1}{r_{o5}^2} & -\frac{1}{r_{o2}^2} \\ g_{M2}^2 & g_{M4}^2 & -g_{M2}^2 - \frac{1}{r_{o2}^2} & \frac{1}{r_{o4}^2} + \frac{1}{r_{o2}^2} \end{bmatrix} V, \tag{12}$$

The voltage gain is $A_v^- = -g_{M1} r_{o1} || r_{o4}$ and the NF_v^- is given by:

$$NF_v^- = 1 + \frac{2}{3} \frac{1}{R_s^- g_{M1}} \left[\frac{g_{M3}}{g_{M1}} + 1 \right] \left[\frac{2 g_{M3}^2 r_{o3}^2 r_{o1}^2 + r_{o3}^2 + r_{o1}^2}{g_{M3}^2 r_{o3}^2 r_{o1}^2 + r_{o3}^2 + r_{o1}^2} \right] \tag{13}$$

Figure 3 shows a numerical comparison between $\overline{V^2}_{n,out}$ and HSPICE to both input signal V^+ and V^-. Noise models shown in (4) are considered to numerical simulations and as one can see, that simulations results in good agreement.

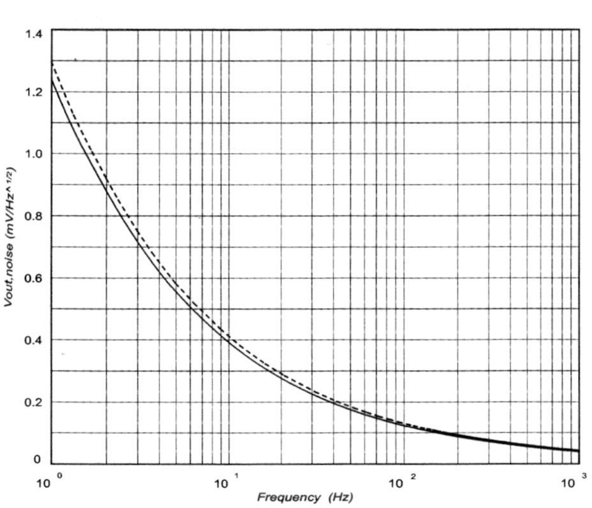

Fig. 3. Numerical comparison of the differential pair (Solid–HSPICE and Dashed–Symbolic).

4.2. CMOS Miller OTA

A CMOS OTA Miller with noise sources is shown in Fig. 4 and its equivalent circuit with nullors is shown in Fig. 5. Tables 3 and 4 show the IRs of conductances, resistances and nullator-norator elements. Considering only the nodes shown in Fig. 5, the order of the CSEs will be 5×5 where the node $1'=0$ for V^+. The voltage variables, according to Table 3 are $v_{1,6}$, $v_{2,8,9}$, v_3, $v_{4,10}$ and v_5 and the current variables are i_1, $i_{2,6,8}$, i_3, $i_{4,7,9}$ and $i_{5,10}$. The resulting cartesian product among variables is:

$$[(1),(2,6,8),(3),(4,7,9),(5,10)] \times [(1,6),(2,8,9),(3),(4,10),(5)], \tag{14}$$

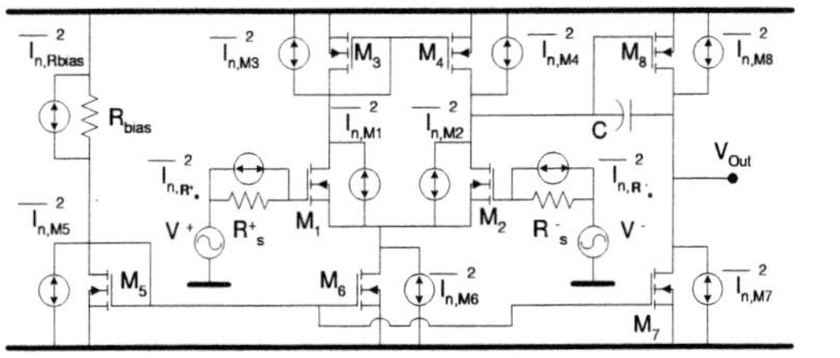

Fig. 4. CMOS OTA Miller with noise sources equivalent.

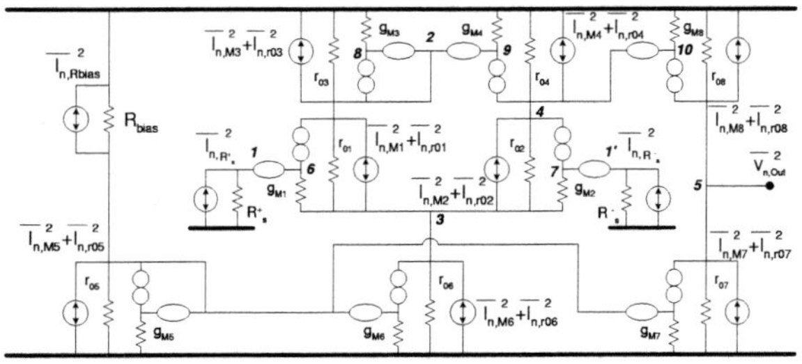

Fig. 5. Nullor circuit equivalent of Fig. 4.

Table 3. Data structure of conductances and resistances elements.

Conductance	Nodes	Resistance	Nodes
g_{M1}^2	$6,6$ or $3,3$	ro_1^2	$2,2$ or $3,3$
$-g_{M1}^2$	$6,3$ or $3,6$	$-ro_1^2$	$2,3$ or $3,2$
g_{M2}^2	$7,7$ or $3,3$	ro_2^2	$4,4$ or $3,3$
$-g_{M2}^2$	$7,3$ or $3,7$	$-ro_2^2$	$4,3$ or $3,4$
g_{M3}^2	$8,8$	ro_3^2	$2,2$
g_{M4}^2	$9,9$	ro_4^2	$4,4$
g_{M8}^2	$10,10$	ro_6^2	$3,3$
		ro_7^2	$5,5$
		ro_8^2	$5,5$
		R_s^{+2}	$1,1$
		R_s^{-2}	$1',1'$

Table 4. Data structure of nullators and norators.

Nullator	Nodes	Norator	Nodes
$O1$	$6,1$	$P1$	$2,6$
$O2$	$7,1'$	$P2$	$4,7$
$O3$	$8,2$	$P3$	$2,8$
$O4$	$9,2$	$P4$	$4,9$
$O8$	$10,4$	$P5$	$5,10$

$$\mathbf{Y_{11}}=(1,1)+(1,6), \mathbf{Y_{12}}=(1,2)+(1,8)+(1,9), \mathbf{Y_{13}}=(1,3), \mathbf{Y_{14}}=(1,4)$$
$$+(1,10), \mathbf{Y_{15}}=(1,5), \mathbf{Y_{21}}=(2,1)+(2,6)+(6,1)+(6,6)+(8,1)+(8,6),$$
$$\mathbf{Y_{22}}=(2,2)+(2,8)+(2,9)+(6,2)+(6,8)+(6,9)+(8,2)+(8,8)+(8,9),$$
$$\mathbf{Y_{23}}=(2,3)+(6,3)+(8,3), \mathbf{Y_{24}}=(2,4)+(2,10)+(6,4)+(6,10)+(8,4)$$
$$+(8,10), \mathbf{Y_{25}}=(2,5)+(6,5)+(8,5), \mathbf{Y_{31}}=(3,1)+(3,6), \mathbf{Y_{32}}=(3,2)$$
$$+(3,8)+(3,9), \mathbf{Y_{33}}=(3,3), \mathbf{Y_{34}}=(3,4)+(3,10), \mathbf{Y_{35}}=(3,5), \mathbf{Y_{41}}=(4,1)$$
$$+(4,6)+(7,1)+(7,6)+(9,1)+(9,6), \mathbf{Y_{42}}=(4,2)+(4,8)+(4,9)+(7,2)$$
$$+(7,8)+(7,9)+(9,2)+(9,8)+(9,9), \mathbf{Y_{43}}=(4,3)+(7,3)+(9,3),$$
$$\mathbf{Y_{44}}=(4,4)+(4,10)+(7,4)+(7,10)+(9,4)+(9,10), \mathbf{Y_{45}}=(4,5)+(7,5)$$
$$+(9,5), \mathbf{Y_{51}}=(5,1)+(5,6)+(10,1)+(10,6), \mathbf{Y_{52}}=(5,2)+(5,8)+(5,9)$$
$$+(10,2)+(10,8)+(10,9), \mathbf{Y_{53}}=(5,3)+(10,3), \mathbf{Y_{54}}=(5,4)+(5,10)$$
$$+(10,4)+(10,10), \mathbf{Y_{55}}=(5,5)+(10,5), \tag{15}$$

Again, searching the elements from Table 3, the matrix Y_{CNA} is given by (16).

$$\begin{bmatrix} \frac{1}{R^{+2}_{s}} & 0 & 0 & 0 & 0 \\ g^2_{M1} & g^2_{M3}+\frac{1}{r^2_{o1}}+\frac{1}{r^2_{o3}} & -g^2_{M1}-\frac{1}{r^2_{o1}} & 0 & 0 \\ -g^2_{M1} & -\frac{1}{r^2_{o1}} & g^2_{M1}+g^2_{M2}+\frac{1}{r^2_{o1}}+\frac{1}{r^2_{o2}}+\frac{1}{r^2_{o6}} & -\frac{1}{r^2_{o2}} & 0 \\ 0 & g^2_{M4} & -g^2_{M2}-\frac{1}{r^2_{o2}} & \frac{1}{r^2_{o4}}+\frac{1}{r^2_{o2}} & 0 \\ 0 & 0 & 0 & g^2_{M8} & \frac{1}{r^2_{o7}}+\frac{1}{r^2_{o8}} \end{bmatrix} \tag{16}$$

where vectors I and V are given by:

$$I = \begin{bmatrix} \overline{I^2}_{n,R_S} \\ \overline{I^2}_{n,M1}+\overline{I^2}_{n,M3} \\ \overline{I^2}_{n,M1}+\overline{I^2}_{n,M2}+\overline{I^2}_{n,M6} \\ \overline{I^2}_{n,M2}+\overline{I^2}_{n,M4} \\ \overline{I^2}_{n,M7}+\overline{I^2}_{n,M8} \end{bmatrix}, V = \begin{bmatrix} \overline{v^2}_{1,6} \\ \overline{v^2}_{2,8,9} \\ \overline{v^2}_{3} \\ \overline{v^2}_{4,10} \\ \overline{v^2}_{5} \end{bmatrix}, \tag{17}$$

Supposing $g_{M2}=g_{M1}$, $g_{M4}=g_{M3}$, $r_{o2}=r_{o1}$ and $r_{o4}=r_{o3}$, the voltage gain is given by: $A_v^+ = -g_{M1}g_{M3}g_{M8}(r_{o7}||r_{o8})(r_{o1}||r_{o4})^2$, where $\overline{V^2}_{n,out}=\overline{v^2}_5$. Using (3), the symbolic NF_v^+ is given by:

$$NF_v^+ = 1 + \frac{4}{3}\frac{1}{R_s^+ g_{M1}}\left[1+\frac{g_{M3}}{g_{M1}}\right] + \frac{2}{3}\frac{1}{R_s^+ g_{M1}g_{M3}g_{M8}} \left[\frac{1}{r^2_{o1}}+\frac{1}{r^2_{o3}}\right]\left[\frac{1}{g_{M1}g_{M3}}\left[\frac{1}{r^2_{o1}}+\frac{1}{r^2_{o3}}\right]+\frac{g_{M8}}{g_{M1}}+\frac{g_{M3}}{g_{M1}}+\frac{g_{M8}}{g_{M3}}\right], \tag{18}$$

If the input signal is V^-, then node 1=0 and the new voltage variable is $v_{1',7}$, the new cartesian product is given by:

$$[(1'),(2,6,8),(3),(4,7,9),(5,10)] \times [(1',7),(2,8,9),(3),(4,10),(5)], \tag{19}$$

and the modified elements in Y_{CNA} becomes:

$$\begin{aligned} &\mathbf{Y_{11}}=(1',1')+(1',7), \mathbf{Y_{12}}=(1',2)+(1',8)+(1',9), \mathbf{Y_{13}}=(1',3), \\ &\mathbf{Y_{14}}=(1',4)+(1',10), \mathbf{Y_{15}}=(1',5), \mathbf{Y_{21}}=(2,1')+(2,7)+(6,1') \\ &+(6,7)+(8,1')+(8,7), \mathbf{Y_{31}}=(3,1')+(3,7), \mathbf{Y_{41}}=(4,1')+(4,7) \\ &+(7,1')+(7,7)+(9,1')+(9,7), \mathbf{Y_{51}}=(5,1')+(5,7), \end{aligned} \tag{20}$$

The CSEs is shown by (21), where I and V are given by (17).

$$\begin{bmatrix} \frac{1}{R^{-2}_{s}} & 0 & 0 & 0 & 0 \\ 0 & g^2_{M3}+\frac{1}{r^2_{o1}}+\frac{1}{r^2_{o3}} & -g^2_{M1}-\frac{1}{r^2_{o1}} & 0 & 0 \\ -g^2_{M2} & -\frac{1}{r^2_{o1}} & g^2_{M1}+g^2_{M2}+\frac{1}{r^2_{o1}}+\frac{1}{r^2_{o2}}+\frac{1}{r^2_{o6}} & -\frac{1}{r^2_{o2}} & 0 \\ g^2_{M2} & g^2_{M4} & -g^2_{M2}-\frac{1}{r^2_{o2}} & \frac{1}{r^2_{o4}}+\frac{1}{r^2_{o2}} & 0 \\ 0 & 0 & 0 & g^2_{M8} & \frac{1}{r^2_{o7}}+\frac{1}{r^2_{o8}} \end{bmatrix} \tag{21}$$

The voltage gain is $A_v^- = g_{M1}g_{M8}(r_{o7}||r_{o8})(r_{o1}||r_{o4})$ and the NF_v^- is given by:

$$NF_v^- = 1 + \frac{2}{3}\frac{1}{R_s^- g^2_{M1}}\left[1+\frac{g_{M4}}{g_{M1}}\right]\left[\frac{2g^2_{M4}r^2_{o4}r^2_{o1}+r^2_{o4}+r^2_{o1}}{g^2_{M4}r^2_{o4}r^2_{o1}+r^2_{o4}+r^2_{o1}}\right] + \frac{2}{3}\frac{1}{R_s^- g^2_{M1}g_{M8}r^2_{o1}} + \frac{2}{3}\frac{1}{R_s^- g^2_{M1}g_{M8}}\left[\frac{g^2_{M4}r^2_{o1}+2}{g^2_{M4}r^2_{o4}r^2_{o1}+r^2_{o4}+r^2_{o1}}\right] \tag{22}$$

Finally, Fig. 6 shows a numerical comparison between $\overline{V^2}_{n,out}$ and HSPICE to both input signal V^+ and V^-. The solution to the CSEs has been made by applying Cramer's rule. The method given in [7] can be applied to improve the solution process, by evaluating determinant decision diagrams.

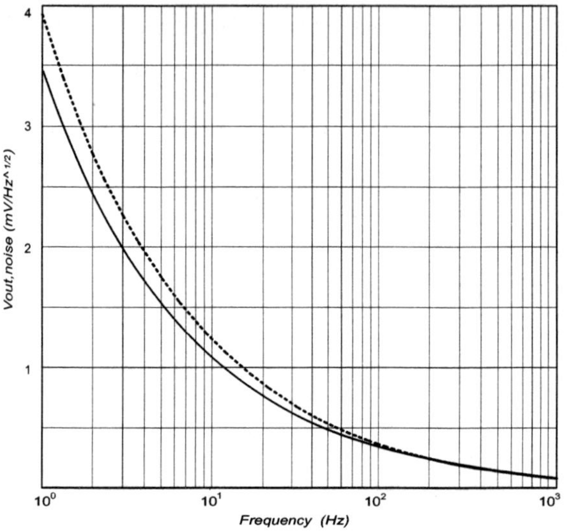

Fig. 6. Numerical comparison of the OTA Miller (Solid–HSPICE and Dashed–Symbolic).

5. CONCLUSIONS

A novel symbolic method to compute NF in analog circuits, has been described. It has been based on manipulating data structures related to the IRs of the circuit elements leading to an improvement and reducing the complexity in the formulation of CSEs. A comparison between HSPICE simulations and the proposed symbolic method has been presented, showing good approximation to calculate NF in analog circuits. In this manner, we conclude on the feasibility and suitability of the proposed method to be incorporated within a symbolic simulator.

6. REFERENCES

[1] A. Van Der-Ziel, *Noise, Characterization, Measurement* Prentice-Hall 1970.

[2] Behzad Razavi, *RF Microelectronics* Prentice-Hall, 1998.

[3] Georges Gilen and Willy Sansen. *Symbolic Analysis for Automated Design of Analog Integrated Circuits*, Kluwer Academic Publishers, 1991.

[4] Henrik Floberg, *Symbolic Analysis in Analog Integrated Circuit Design*, Kluwer Academic Publishers, 1997.

[5] Sánchez-López C., Tlelo-Cuautle E., Díaz-Sánchez A., *Computing the NF of MOST Circuits by Applying Symbolic Analysis*, IEEE MWSCAS, vol. III, pp. 429-432, 2002.

[6] Tlelo-Cuautle E., Sánchez-López C., Sandoval-Ibarra F., *Symbolic Analysis: A Formulation Approach by Manipulating Data Structures* IEEE ISCAS, vol. IV, pp. 640-643, 2003.

[7] X.-D. Tan and C.-J. Shi. *Symbolic Circuit-Noise Modeling and Analysis via Determinant Decision Diagrams*, Proc. Asia and South Pacific Design Automation Conference (ASP-DAC'00), Yokohama, Japan, pp. 283-287, January 2000.

AN EFFICIENT LOGIC EXTRACTION ALGORITHM USING PARTITIONING AND CIRCUIT ENCODING

Lily Huang, Tai-Ying Jiang and Jing-Yang Jou
Dept. of Electronics Engineering
National Chiao Tung University
Hsinchu, Taiwan, ROC
{*lily, giani, jyjou*} @eda.ee.nctu.edu.tw

Heng-Liang Huang
Legend Design Technology
hhuang@legenddesign.com

ABSTRACT

Nowadays, finding subcircuits in a larger circuit is primarily solved by using various heuristics based on graph isomorphism. These approaches are addressed on identifying one specific subcircuit at each time and may take numerous runs if many subcircuits need to be extracted. Therefore, they are not quite suitable for converting the whole circuit represented at transistor-level to a gate-level netlist. We present a logic extraction approach based on dc-connected component (DCC) partition and modified circuit-encoding algorithm to extract all kinds of subscircuits from the input circuit concurrently such that we can map each subcircuit represented as transistor level netlist to its corresponding logic gate. This mapping relation can be exploited to speed up the simulation of large circuits. Experiments on several real circuits, including sequential logic ones and combination logic ones, show the near-linear performance in run time and memory usage.

1. INTRODUCTION

Post-layout simulation is used to make sure the post-layout response meets the specification. However, to simulate the whole circuit represented at transistor level takes too much time. If the hierarchy relation between the originally flattened circuit and the gate level models can be obtained, many new spice-like simulators can exploit the hierarchical relation to speed up simulation or to prevent memory explosion problem while maintaining the accuracy of simulation results. *Logic extraction* is a technique by which subcircuits in the input circuit (such as elemental gates, flip-flops, etc.), whose transistor structure are the same as some specific logic gates in the given cell library, are recognized and mapped to its corresponding logic gate. In other words, *logic extraction* can rebuild the hierarchy of a flattened input circuit, whose hierarchy information is lost for some reasons, such that the simulation time and memory usage can be saved.

In this paper, we propose an approach to extract all kinds of subcircuits concurrently from the input circuit and to map them with their corresponding logic gates. As compared to other approaches, our approach is more suitable for *logic extraction.* The use of *circuit code* matching enable our approach significantly speed up the extraction of subcircuits. Furthermore, our approach need no pre-processing to find the candidates and can convert the whole input circuit with traversing the input circuit once only such that the computation time can be greatly saved.

This work was partially supported by Legend Design Technology Corporation.

The remainder of this paper is organized as follows. In section 2, some related works are also reviewed and summarized in section 2. Our approach is going to be introduced in section 3. In section 4, we will demonstrate our experimental results. Finally, we conclude this paper in section 5.

2. RELATED WORKS

In the literature, there are some previous works related to our problem [4,5,7]. The common objectives of them are to extract subcircuits from the large input circuit represented at transistor level.

In [4], Ohlrich proposed an approach mainly based on graph isomorphism. Possible candidates are first found and SubGemini is applied to further identify those candidates. Using graph isomorphism allows a general technology-independent solution. Nevertheless, this approach may lose the chance to exploit circuit properties and may become ineffective when input circuits are very large and with highly symmetric structures.

In [5], Ling proposed another approach based on subgraph isomorphism as well. The main difference between [4] and [5] is the construction of graphs. This approach can deal with the shorting-input problem and a near-linear time operational performance can be achieved while Constrained Resource Planning (CRP) is used. However, the set up time is needed for CRP engine and the edge complexity around high-degree nets is high. If an edge represents a global signal, such as clock and reset, verification requires strenuous effort.

DECIDE proposed in [7] solves subcircuit extraction problem based on a subgraph isomorphism algorithm as well. DECIDE uses a weighting function to generate a candidate set as starting nodes for identification. Identification adopts the recursive function acting like the breath-first search scheme. The branch-and-bound strategy used in recursive recognition greatly reduces the time of identification. However, since DECIDE uses the same bipartite circuit graph as SubGemini does, it cannot avoid the shorting-input problem, either.

These previous works aimed at identifying only one subcircuit at each time and not suitable for extracting many subcircuits such as the job of converting the whole circuit represented at transistor level to gate level. In this paper, we intend to develop an algorithm devoted to identify all kinds of subcircuits concurrently and to map them to their corresponding logic gates for rebuilding the hierarchy relation. Given a standard cell library as the source of the pattern circuits and an input circuit represented as transistor level netlist, whose hierarchy information is lost for some reasons, we first construct the Pattern Circuit Library (PCL) for each logic gate in the given cell library. By DCC partitioning, encoding each DCC, and *circuit code* matching, *logic extraction* is accomplished and need only one time of traversing the input circuit.

3. LOGIC EXTRACTION

Logic extraction is to convert the input circuit represented as the transistor level netlist to gate level one and to rebuild the hierarchy of an originally flattened input circuit.

In this section, graph representation is going to be introduced in section 3.1. DCC based partitioning method is discussed in section 3.2. *Circuit encoding* algorithm is demonstrated in section 3.3. At last, a complete flow is given in section 3.4.

3.1 Graph Representation

A circuit and its graph representation are shown in **Fig.1**. Our graph is a bipartite and undirected graph with devices, represented by circles, forming one kind of vertices, and nets (wires), represented by squares, forming the other kind. The global nets, 'vdd' and 'gnd', are treated as special nets. Edges in out graph connect two kinds of vertices, including circles and squares. Explicitly representing nets as vertices in the graph reduces the number of edges to represent the full interconnection of N device terminals from N(N-1)/2 to N. The edge reduction is tremendous when N is large for some high fan-out nets, such as clock, reset, and etc..

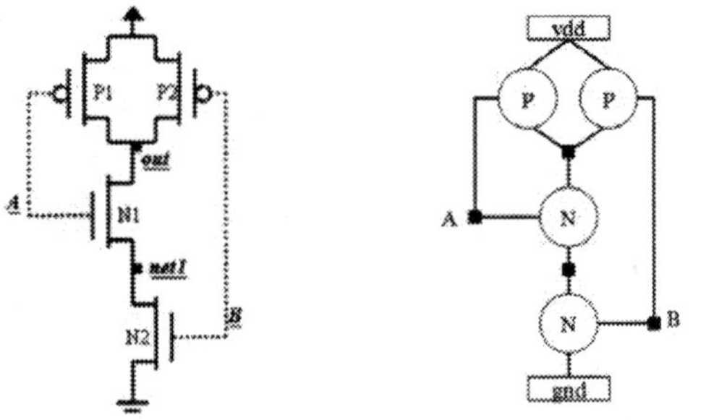

Fig.1 An example circuit and its corresponding graph

The device vertices can be distinguished by their types, i.e. pmos or nmos. An edge should be labeled according to the terminal type of its connecting device such that the circuit properties can be exploited. Because of the physical symmetry of MOSFET, the drain and the source have the same terminal properties. The edges are labeled by 's' if connecting to drain or source, and labeled by 'g' if connecting to gate. A net vertex is regarded as a "device set" which contains all the devices connected from it.

3.2 DC-Connected Component Partition Method

Most logic gates are dc-connected components (DCC) [8], except for transmission gate logics and flip-flops. A circuit is DCC if all of the transistors in the circuit can be collected by tracing the source-to-source connections between them. Therefore, we can extract each DCC by tracing the source-to-source connections. The circuit before performing DCC extraction and the circuit after performing DCC are shown in **Fig.2**. DCC extraction is suitable for partitioning because to partition circuits according to dc- connection needs tracing channels only and is easily conducted thereby.

To deal with transmission gates logics that DCC cannot handle, we exploit the unique property the transmission gate logic has – parallel pmos and nmos. With the unique property, transmission gate logics can be recognized and excluded from its previous DCC before circuit-encoding. Since a flip-flop contains two DCCs with their outputs connecting together to form a feedback loop, a flip-flop is not suitable to be partitioned by DCC extraction, either. Fortunately, the property of that their outputs connect together to form a feedback loop forms another characteristic to indicate the location of flip-flops.

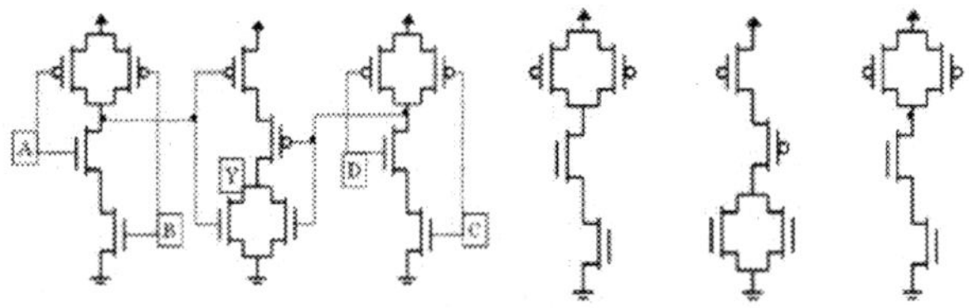

Fig.2 An DCC partitioning example

3.3 Circuit-Encoding Algorithm

We modify the circuit-encoding algorithm proposed in [1] to encode the characteristic graph of DCC and to provide a unique code independent of the name and the traversing order of the nodes. We set up *circuit matrix*, containing the connectivity between transistors and nets according to DCC graph. *Circuit matrix* is partitioned, regrouped, and reordered iteratively until it is the same as the one in the preceding step. That means all of the moss and nets are distinguishable and placed in their determinant positions, unless some groups contain symmetric elements that cannot be differentiated.

Take AOI4 as an example. The transistor level representation and its corresponding graph are shown in **Fig.3**. *Circuit matrix* is shown in **Table I**. The columns, A, B, C, and D, represent four nets and the column, V and G, represent Vdd and Gnd. Each row represents each transistor in the circuit. If a connection exists between a net and a transistor, we mark a '1' in the corresponding cross of the net column and the transistor row; otherwise we mark a '0'.

The details of partition and reordering are performed as the following steps:

I. Fill the column value by counting the number of elements in each row groups of a column. Group the columns with the same column values and reorder groups lexicographically from the highest to the lowest, except for the **V** column and **G** column. The result is shown in **Table II**. Notice that C and D are in the same group.

II. Fill the row value by counting the number of ones in each column group and group the rows by their own row value. Reorder lexicographically from the highest to the lowest. The result is shown in **Table III**. Repeat

Step I. The result of grouping column groups is the same as the table shown in **Table III**.

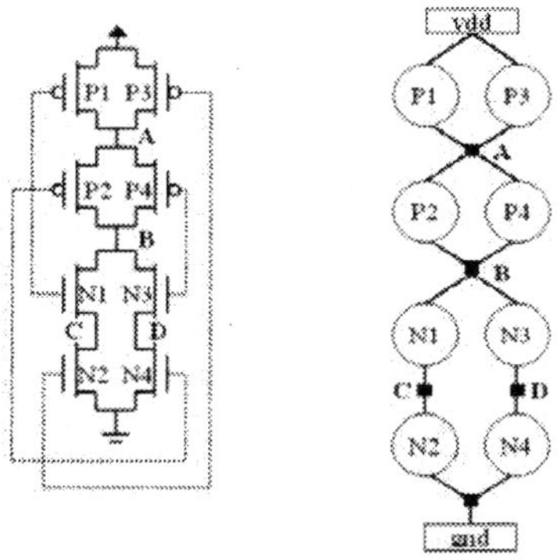

Fig.3 An AOI4 circuit and its bipartite graph

Table I. Initial *circuit matrix*

	A	B	C	D	V	G
P1	1	0	0	0	1	0
P2	1	1	0	0	0	0
P3	1	0	0	0	1	0
P4	1	1	0	0	0	0
N1	0	1	1	0	0	0
N2	0	0	1	0	0	1
N3	0	1	0	1	0	0
N4	0	0	0	1	0	1

Table II. *Circuit matrix* in Step I.

	A	B	C	D	V	G
P1	1	0	0	0	1	0
P2	1	1	0	0	0	0
P3	1	0	0	0	1	0
P4	1	1	0	0	0	0
N1	0	1	1	0	0	0
N2	0	0	1	0	0	1
N3	0	1	0	1	0	0
N4	0	0	0	1	0	1
column value	4 0	2 2	0 2	0 2	V	G

Table III. *Circuit matrix* in Step II.

	A	B	C	D	V	G	row value
P2	1	1	0	0	0	0	11000
P4	1	1	0	0	0	0	11000
P1	1	0	0	0	1	0	10010
P3	1	0	0	0	1	0	10010
N1	0	1	1	0	0	0	01100
N3	0	1	0	1	0	0	01100
N2	0	0	1	0	0	1	00101
N4	0	0	0	1	0	1	00101

III. Check the circuit symmetry before code generation. According to **Fig.4**, the symmetric elements in the DCC, (P2, P4), (P1, P3), (N1, N3), and (N2, N4) should get the same code since they are identical in the DCC structure. However, (N1, N3) and (N2, N4) do not have the same code due to the symmetric net pair (C,D). Since C and D still lie in the same column group in the final circuit matrix, the codes of (N1, N3) and (N2, N4) need to be modified by maximum code generation principle such that these symmetric devices have the same code. The codes of (N1, N3) and (N2, N4) are marked by the dotted circuit in **Table IV**. The final codes in decimal number are displayed in the last column.

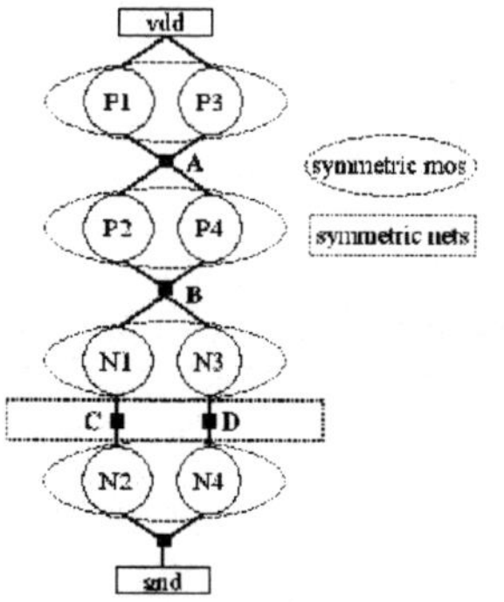

Fig.4 A graph to show the symmetric of AOI4

Table IV. *Circuit matrix* after symmetry modification

	A	B	C	D	V	G	(code)2	(code)
P2	1	1	0	0	0	0	110000	48
P4	1	1	0	0	0	0	110000	48
P1	1	0	0	0	1	0	100010	34
P3	1	0	0	0	1	0	100010	34
N1	0	1	1	0	0	0	011000	24
N3	0	1	1	0	0	0	011000	24
N2	0	0	1	0	0	1	001001	9
N4	0	0	1	0	0	1	001001	9

In the final *circuit matrix*, we get a *circuit code* including three properties used for subcircuit identification later:

1. *Matrix code*: [48, 48, 34, 34, 24, 24, 9, 9] used for the initial circuit recognition.
2. The gate-connected mos set: [(3, 5), (1, 8), (4, 7), (2, 6)] used to verify the g-connection after *matrix code* matching.
3. The index of the output net: [2]. The output net 'B' lies in the second column of the final matrix.

3.4 Logic Extraction

First, we use the elemental logic gates in the given standard cell library as the source of pattern circuits and build the Pattern Circuit Library (PCL). After PCL is ready, the input circuit is decomposed into DCCs and constructed into graphs. The identification process starts from PIs and encodes each DCC one by one along the signal flow. As to the multi-DCC circuits with single output, such as AND, OR, etc., their DCC sequences are recorded while decomposition is conducted, and appended to the circuit codes of the first DCC. After a DCC is encoded, we check whether the same *matrix code* exists in PCL. If so, the gate-connected mos set and the index of the output net need to be further checked to guarantee the subcircuit is identical to the one in PCL. If not, next encoded DCC is checked. The whole identification process can be done in a short time because matrix code matching can quickly filter out impossible candidates.

4. EXPERIMENTAL RESULTS

Our logic extraction approach is implemented in C++ language and has been successfully applied to several practical cases, including both combinational and sequential circuits. Preseving correctness, it shows effectiveness and efficiency as well. Experiments are taken on the following designs: No.1 ~ No.10 are ISCAS-85 benchmark circuits. No. 11 design is an arithmetic circuit that performs multiplication. No.12 ~ No.26 are ISCAS-89 benchmark circuits. Under our platform (Ultra Sparc II Workstation), the experimental results are given in **Table V**. We summarize the run time and memory usage versus 26 test circuits in **Fig.5**. For larger circuits, the two curves also prove our approach is near-linear in both time and memory.

To prove the correctness of *logic extraction*, we choose SVS (schematics vs. schematics) in Dracula to check the equivalence between the gate-level circuit converted by our approach from the given transistor-level circuit and the original gate-level description in the benchmark. We applied this verification method on the 26 test circuits and the correctness of the proposed approach has been proved.

5. CONCLUSION

A novel algorithm to convert the input circuit represented as the transistor level netlist to gate level one is proposed in this paper. The hierarchy relation between the transistor level netlist and the gate level one can be regained. Experimental results show that DCC partition is suitable for the searching boundary in an input circuit, the circuit-encoding algorithm can indeed provide an almost unique *matrix code* as the primary key for identification such that *logic extraction* has near-linear performance in both run time and memory usage.

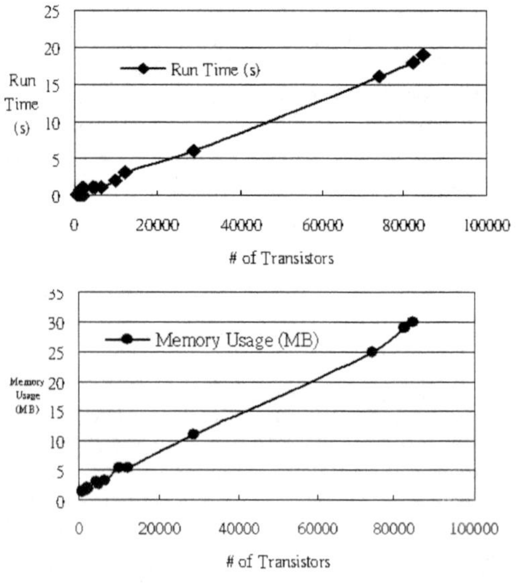

Fig.5 Performance

Table V. Experimental Results

No.	Circuit Name	#of Transistors	# of DCCs	Run Time (s)	Memory Usage (MB)
1	C432	900	238	0.09	1.46
2	C499	1704	438	0.10	1.82
3	C880	1176	262	0.12	1.46
4	C1355	2248	606	1.08	2.05
5	C1908	1402	317	0.07	1.52
6	C2670	2406	584	0.06	1.88
7	C3540	4568	1074	1.16	3.01
8	C5315	5078	1054	1.07	2.71
9	C6288	10136	2716	2.05	5.39
10	C7552	6756	1589	1.16	3.24
11	16b'	74408	18243	16.13	25.00
12	S298	1394	459	0.09	1.68
13	S344	840	251	0.05	1.38
14	S349	872	259	0.08	1.40
15	S382	1574	478	0.77	1.60
16	S526	1280	382	0.12	1.54
17	S641	1204	392	0.10	1.51
18	S731	1206	396	0.10	1.51
19	S838	2042	614	1.10	1.79
20	S953	2378	689	1.11	1.90
21	S1238	2442	640	0.89	1.89
22	S1488	2414	591	1.06	1.87
23	S9234	12400	3678	3.09	5.37
24	S15850	28556	7916	6.08	11.30
25	S38417	82436	22156	18.04	29.21
26	S38584	84802	22205	19.10	30.12

6. REFERENCE

[1] Kai-Ti Huang and David Overhauser, "A Novel Algo -rithm for Circuit Recognition", IEEE International Symposium on Circuit and Systems, pp. 1695-1698, 1995.

[2] C. Ebeling, "GeminiII: A Second Generation Layout Validation Tool", IEEE/ACM International Conference on Computer Aided Design, pp.610-615, Nov 1988.

[3] C. Ebeling and O. Zajicek, "Validating VLSI Circuit Layout by Wirelist Comparison", IEEE/ACM International Conference on Computer Aided Design, pp. 172-173, 1993.

[4] M. Ohlrich, C. Ebeling, E. Ginting, and L. Sather, "SubGemini: Identifying Subcircuits Using a Fast Subgraph Isomorphism Algorithm", IEEE/ACM Design Automation Conference, pp. 31-37, 1993.

[5] Z. Ling and D. Y. Yun, "An Efficient Subcircuit Extraction Algorithm by Resource Management", IEEE International Conference on ASIC, pp. 9-14, 1996.

[6] N. P. Keng and D. Y. Y. Yun, "A Planning/Scheduling Methodology for the Constrained Resource Problem", International Joint Conference on Artificial Intelligence, pp. 20-25, Aug 1989.

[7] Wei-Hsin Chang, Shuenn-Der Tzeng, and Chen-Yi Lee, "A Novel Subcircuit Extraction Algorithm by Recursive Identification Scheme", IEEE International Symposium on Circuits and Systems, pp. 491-494, 2001.

[8] D.C. Yuan, L.T. Pillage, and J.T. Rahmeh, "Evaluation by Parts of Mixed-Level DC-Connected Components in Logic Simulation", IEEE/ACM Design Automation Conference, 1993.

[9] Bruno T. Messmer and Horst Bunke, "Efficient Subgraph Isomorphism Detection: a Decomposition Approach", IEEE Transactions on Knowledge and Data Engineering, Mar/Apr 2000.

FREQUENCY-DOMAIN ERROR ANALYSIS OF LINEAR MULTISTEP METHODS

Giorgio Casinovi

School of Electrical and Computer Engineering
Georgia Institute of Technology
Atlanta, Georgia 30332, USA

Giuseppe M. Veca

Dipartimento di Ingegneria Elettrica
Università degli Studi di Roma "La Sapienza"
Via Eudossiana 18, 00184 Roma, Italy

ABSTRACT

This paper presents a comparative analysis of various numerical integration algorithms with respect to the error they introduce in the frequency spectrum of the numerical solution of a differential equation. It is shown that the most common linear multistep methods, often used for time-domain simulation, are ill-suited for frequency-domain analysis, and that much better accuracy can be obtained using another algorithm, recently developed for the specific purpose of performing continuous-spectrum frequency-domain simulation. Examples of numerical simulations that demonstrate this point are presented and discussed.

1. INTRODUCTION

In RF and microwave circuit design, frequency-domain simulators are often the preferred verification tools, because in most cases they are best suited to assess the performance of those particular types of systems. Algorithms based on the principle of harmonic balance [1, 2] are widely used for discrete-spectrum frequency-domain simulation. In many instances, however, it is necessary to verify the frequency-domain behavior of a system in the presence of signals that are best characterized as having continuous frequency spectra. In such cases, computing numerically the Fourier transforms of the signal waveforms obtained from time-domain simulations would appear to be a reliable method to evaluate the system's frequency-domain response.

This paper analyzes the error affecting the frequency spectra of time-domain solutions of linear ODE's yielded by linear multistep methods. It is shown that the most common time-domain simulation algorithms are ill-suited for frequency-domain analysis, and that much better accuracy can be obtained using an algorithm developed specifically for continuous-spectrum frequency-domain simulation [3, 4], which is described briefly in Section 2.

This work was supported in part by a Visiting Research Fellowship granted to the first author by the Università degli Studi di Roma "La Sapienza."

Throughout this paper, it will be assumed that the system being studied is linear and is described by the following equation:

$$\dot{\mathbf{x}}(t) = -\mathbf{A}\mathbf{x}(t) + \mathbf{B}\mathbf{u}(t). \tag{1}$$

In the frequency-domain, the solution of this differential equation is given by:

$$\mathbf{X}(j\omega) = (j\omega\mathbf{I} + \mathbf{A})^{-1}\mathbf{B}\mathbf{U}(j\omega). \tag{2}$$

The suitability of an algorithm for frequency-domain simulation will be assessed by comparing the spectrum of its numerical solution with the exact solution in (2). For this purpose, a relationship between the Fourier transform of a continuous-time signal and the z-transform of the sequence formed by its discretized samples is established in Section 3, while the actual error analysis is contained in Section 4. Finally, examples of numerical simulation results are given in Section 5.

2. FREQUENCY-DOMAIN SIMULATION ALGORITHM

This section describes briefly an algorithm devised for continuous-spectrum frequency-domain simulation of linear or nonlinear systems [3, 4].

It is assumed that both the input $\mathbf{u}(t)$ and the solution $\mathbf{x}(t)$ are band-limited; i.e., their frequency spectra have a finite upper bound ω_M. Let: $h = \pi/\omega_M$. According to the sampling theorem [5], $\mathbf{x}(t)$ can be represented by the following series expansion:

$$\mathbf{x}(t) = \sum_{k=-\infty}^{+\infty} \mathbf{x}_k \operatorname{sinc}[(t - t_k)/h], \tag{3}$$

where: $\mathbf{x}_k = \mathbf{x}(t_k)$, $t_k = kh$, and $\operatorname{sinc} x = \sin \pi x/(\pi x)$. Differentiating both sides of this equation yields:

$$\dot{\mathbf{x}}(t_n) = \frac{1}{h} \sum_{k=-\infty}^{+\infty} \mathbf{x}_k \operatorname{sinc}'[(t_n - t_k)/h].$$

But:

$$\begin{aligned}\mathrm{sinc}'[(t_n - t_k)/h] &= \mathrm{sinc}'(n-k) \\ &= \begin{cases} 0, & n = k \\ \dfrac{(-1)^{n-k}}{n-k}, & n \neq k. \end{cases}\end{aligned}$$

Therefore:

$$\dot{\mathbf{x}}(t_n) = \frac{1}{h} \sum_{\substack{k=-\infty \\ k\neq n}}^{+\infty} \mathbf{x}_k \frac{(-1)^{n-k}}{n-k} = \frac{1}{h} \sum_{\substack{k=-\infty \\ k\neq 0}}^{+\infty} \mathbf{x}_{n-k} \frac{(-1)^k}{k}. \quad (4)$$

Substituting this expression for $\dot{\mathbf{x}}(t_n)$ in (1) yields:

$$\frac{1}{h} \sum_{\substack{k=-\infty \\ k\neq 0}}^{+\infty} \mathbf{x}_{n-k} \frac{(-1)^k}{k} + \mathbf{A}\mathbf{x}_n = \mathbf{B}\mathbf{u}_n, \quad (5)$$

which is a set of infinitely many equations in the infinitely many unknowns $\mathbf{x}_n$. Choosing a finite number of points $t_{-N}, t_{-N+1}, \ldots, t_N$ (which is dependent upon the accuracy that one wishes to achieve), equation (5) becomes a set of $2N+1$ algebraic equations in the $2N+1$ unknowns $\mathbf{x}_{-N}$, $\mathbf{x}_{-N+1}, \ldots, \mathbf{x}_N$, which can be solved using standard methods. The spectrum of $\mathbf{x}(t)$ can then be computed by taking the Fourier transform of (3):

$$\mathbf{X}(j\omega) = \frac{\pi}{\omega_M} \sum_{k=-N}^{N} \mathbf{x}_k e^{-jk\pi\omega/\omega_M}. \quad (6)$$

This expression for $\mathbf{X}(j\omega)$ is valid for $|\omega| \leq \omega_M$.

3. DISCRETIZATION

In virtually all cases of practical interest, the steady-state response of a system can be computed only through numerical integration of (1), which requires discretizing continuous-time signals. Therefore the first step is to establish a relationship between the Fourier transforms of the continuous-time signals and the z-transforms of the sequences of their time samples. In principle, this relationship depends on the particular quadrature formula that is used to evaluate the Fourier integral numerically. For example:

$$\int_{-\infty}^{+\infty} \mathbf{u}(t) e^{-j\omega t}\, dt \approx h \sum_{n=-\infty}^{+\infty} \mathbf{u}_n e^{-j\omega h n},$$

where: $\mathbf{u}_n = \mathbf{u}(t_n)$. Therefore this particular quadrature formula yields the following relationship:

$$\mathbf{U}(j\omega) \approx h\mathbf{U}(z)\big|_{z=e^{j\hat{\omega}}},$$

where: $\hat{\omega} = \omega h$. Note that the same expression is obtained if the trapezoidal rule is used to evaluate the integral, because:

$$\begin{aligned}&\frac{h}{2} \sum_{n=-\infty}^{+\infty} \left(\mathbf{u}_n e^{-j\omega h n} + \mathbf{u}_{n+1} e^{-j\omega h(n+1)} \right) \\ &= \frac{h}{2} \sum_{n=-\infty}^{+\infty} \mathbf{u}_n e^{-j\omega h n} + \frac{h}{2} \sum_{n=-\infty}^{+\infty} \mathbf{u}_{n+1} e^{-j\omega h(n+1)} \\ &= h \sum_{n=-\infty}^{+\infty} \mathbf{u}_n e^{-j\omega h n}.\end{aligned}$$

These examples demonstrate that discretization establishes a mapping between ω and z which is given by[1]:

$$z = e^{j\hat{\omega}} = e^{j\omega h}. \quad (7)$$

In particular, the imaginary axis of the s plane is mapped onto the unit circle of the z-plane. It will be shown next that linear multistep numerical integration methods establish another relationship between s and z. *If this relationship is not consistent with (7), the numerical integration algorithm will introduce an additional error in the spectrum of the computed solution.*

4. ERROR ANALYSIS

Error analysis of numerical integration methods is the topic of a vast body of literature, but little or no attention has been paid to investigating how different algorithms affect the frequency spectrum of the computed solution. This section is devoted to a frequency-domain analysis of the error introduced in the numerical solution by linear multistep integration algorithms.

Linear multistep methods generate a numerical solution that satisfies a linear difference equation of the following type:

$$\sum_k a_k \mathbf{x}_{n-k} + \sum_k b_k \dot{\mathbf{x}}_{n-k} = 0. \quad (8)$$

Many common numerical integration algorithms (e.g. forward and backward Euler methods, trapezoidal method, backward differentiation formulae, etc.) belong to the class of linear multistep methods, as does the frequency-domain simulation algorithm described in Section 2.

Using (8) to solve (1) numerically yields the following difference equation:

$$\sum_k a_k \mathbf{x}_{n-k} + \sum_k b_k (-\mathbf{A}\mathbf{x}_{n-k} + \mathbf{B}\mathbf{u}_{n-k}) = 0,$$

whose z-transform is:

$$\begin{aligned}\left(\sum_k a_k z^{-k}\right) \mathbf{X}(z) - \left(\sum_k b_k z^{-k}\right) \mathbf{A}\mathbf{X}(z) & \\ + \left(\sum_k b_k z^{-k}\right) \mathbf{B}\mathbf{U}(z) &= 0.\end{aligned}$$

[1] This fact can be proven rigorously for a very wide class of numerical quadrature formulae that can be used for the evaluation of the Fourier integral.

Therefore $\mathbf{X}(z)$ is given by:

$$\mathbf{X}(z) = \left(-\frac{\sum_k a_k z^{-k}}{\sum_k b_k z^{-k}} \mathbf{I} + \mathbf{A} \right)^{-1} \mathbf{BU}(z). \qquad (9)$$

Thus the transfer function of the discretized equation is: $(s\mathbf{I}+\mathbf{A})^{-1}$, where:

$$s = -\frac{\sum_k a_k z^{-k}}{\sum_k b_k z^{-k}}. \qquad (10)$$

A comparison with (2) shows this transfer function matches that of the original differential equation only if $s = j\omega$ when $z = e^{j\hat{\omega}}$. Therefore the accuracy of the spectrum of the computed solution depends on how (10) maps the unit circle of the z-plane into the complex s-plane. To illustrate this point, three specific methods will be analyzed in detail.

The backward Euler method is defined by: $a_{-1} = -1$, $a_0 = 1$, $b_{-1} = h$, where h is the integration step. Therefore in this case (10) becomes:

$$s = -\frac{-z+1}{hz} = \frac{1}{h}(1 - z^{-1}).$$

It is readily verified that this equation maps the unit circle of the z-plane into a circle of radius $1/h$ centered at $s = 1/h$ and tangent to the imaginary axis of the s-plane [5]. This means that the spectrum of the computed solution is determined by the values that $(s\mathbf{I}+\mathbf{A})^{-1}$ takes on that circle, instead of the imaginary axis. As a consequence, a significant error may be introduced in the computed solution, unless the value of h is chosen so small that the frequency range of interest is mapped to a small arc around $s = 0$, where the circle is tangent to the imaginary axis.

If the trapezoidal method is used, the coefficient values are: $a_{-1} = -1$, $a_0 = 1$, $b_{-1} = b_0 = h/2$. The corresponding mapping from the z- to the s-plane is:

$$s = -\frac{-z+1}{h(z+1)/2} = \frac{2}{h}\frac{1-z^{-1}}{1+z^{-1}}.$$

If $z = e^{j\hat{\omega}}$, then:

$$s = \frac{2}{h}\frac{1-e^{-j\hat{\omega}}}{1+e^{-j\hat{\omega}}} = \frac{2}{h}\frac{e^{j\hat{\omega}/2}-e^{-j\hat{\omega}/2}}{e^{j\hat{\omega}/2}+e^{-j\hat{\omega}/2}} = j\frac{2}{h}\tan\frac{\hat{\omega}}{2}.$$

Although in this case the unit circle in the z-plane is mapped exactly on the imaginary axis in the s-plane, the nonlinearity of the tangent function warps the frequency scale of the transfer function. This, too, can have a significant impact on the accuracy of the spectrum of the computed solution.

Finally, the frequency-domain analysis algorithm outlined in Section 2 corresponds to the following choice of coefficients: $a_0 = 0$, $a_k = (-1)^k/k$ $(k \neq 0)$, $b_0 = -h$. The corresponding mapping between z and s is:

$$s = \frac{1}{h}\sum_{\substack{k=-\infty \\ k\neq 0}}^{+\infty} (-1)^k \frac{z^{-k}}{k}.$$

An analytical expression for the sum of this infinite series can be obtained from the following power series expansion:

$$\ln(1+z) = \sum_{k=1}^{+\infty} (-1)^{k-1}\frac{z^k}{k}.$$

Then:

$$\begin{aligned}
\sum_{\substack{k=-\infty \\ k\neq 0}}^{+\infty} (-1)^k \frac{z^{-k}}{k} &= \sum_{k=1}^{+\infty}(-1)^k\frac{z^{-k}}{k} + \sum_{k=-\infty}^{-1}(-1)^k\frac{z^{-k}}{k} \\
&= \sum_{k=1}^{+\infty}(-1)^k\frac{z^{-k}}{k} + \sum_{k=1}^{+\infty}(-1)^{-k}\frac{z^{k}}{(-k)} \\
&= -\sum_{k=1}^{+\infty}(-1)^{k-1}\frac{z^{-k}}{k} + \sum_{k=1}^{+\infty}(-1)^{k-1}\frac{z^{k}}{k} \\
&= -\ln(1+z^{-1}) + \ln(1+z) \\
&= \ln\left(\frac{1+z}{1+z^{-1}}\right) = \ln z.
\end{aligned}$$

Therefore, if $z = e^{j\hat{\omega}}$:

$$s = \frac{1}{h}\ln e^{j\hat{\omega}} = \frac{1}{h}j\hat{\omega} = j\omega.$$

This equation shows that, in this case, the unit circle of the z-plane is mapped on the imaginary axis of the s-plane in a way that preserves frequency scaling. This means that the spectrum of the numerical solution computed using this particular algorithm is exactly equal to the spectrum of the discretized input multiplied by the transfer function of the original continuous-time system. In practice, the truncation of the infinite series to a finite sum will affect the accuracy of the computed spectrum, but to a much lesser extent than in the two other cases examined previously.

5. NUMERICAL RESULTS

The error analysis carried out in the previous section was verified on the results of numerical simulations of a simple first-order low-pass *RC* circuit, with a 3-dB frequency of 128 kHz. The input was a signal with a uniform frequency spectrum between −512kHz and 512 kHz. Obviously, the frequency spectrum of the output signal should coincide with the filter's transfer function.

Simulations of this circuit were performed using the three methods previously analyzed. The same integration timestep ($h = 1/1024$ ms) and the same number of time points (257) were used in all cases. The spectrum of the numerical solution obtained from the backward Euler method is compared with the exact solution in Fig. 1 (solid and dashed lines, respectively). The error in the spectrum of the computed solution is apparent even at relatively low frequencies; in particular, the filter's bandwidth appears to be narrower than its actual value.

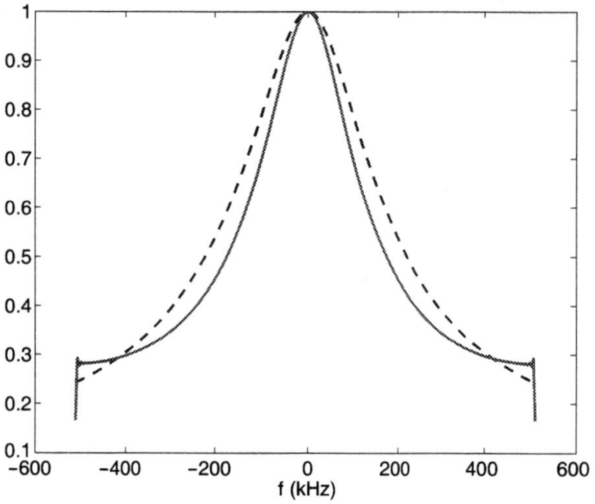

Fig. 1: Output spectrum, backward Euler method

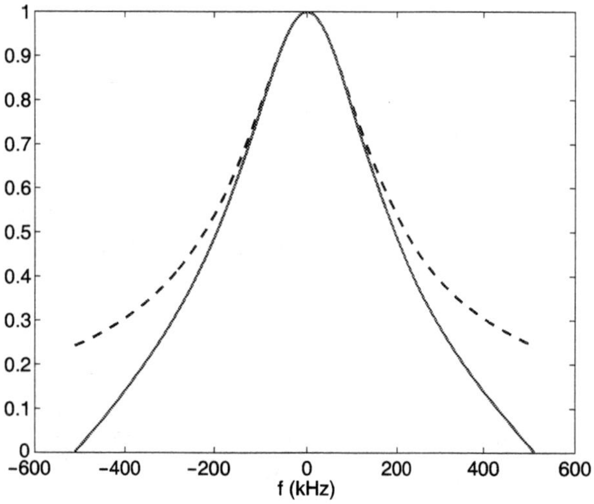

Fig. 2: Output spectrum, trapezoidal method

The results obtained from the trapezoidal method are shown in Fig. 2 (solid line), and are also compared with the exact solution (dashed line). While the two graphs essentially coincide at low frequencies, at high frequencies the error in the spectrum of the computed solution is significantly more pronounced than in the previous case. This fact illustrates a point that would seem counterintuitive on its face: the frequency-domain results obtained using the trapezoidal method can be less accurate than those obtained from the backward Euler method, despite the fact that, in the time domain, the opposite is generally true.

Finally, the results obtained from the frequency-domain simulation algorithm described in Section 2 are shown in Fig. 3: they are so close to the exact solution that the difference in the corresponding graphs is discernible only at the very high end of the frequency range.

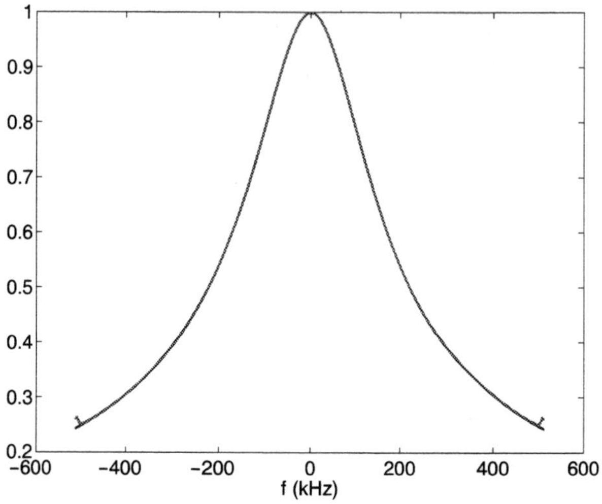

Fig. 3: Output spectrum, frequency-domain method

In conclusion, the theoretical analysis and numerical results presented in this paper have demonstrated that the choice of the numerical integration algorithm can affect the spectrum of the computed solution in a significant manner. In particular, algorithms that are reliable and accurate for time-domain simulation purposes do not necessarily generate equally accurate results in the frequency domain. Much better performance can be obtained from algorithms devised specifically for frequency-domain simulation purposes.

6. REFERENCES

[1] K. S. Kundert, J. White, and A. Sangiovanni-Vincentelli, *Steady-State Methods for Simulating Analog and Microwave Circuits*. Norwell, MA: Kluwer Academic Publishers, 1990.

[2] V. Rizzoli and A. Neri, "State of the art and present trends in nonlinear microwave CAD techniques," *IEEE Trans. Microwave Theory Tech.*, vol. 36, no. 2, pp. 343–365, Feb. 1988.

[3] G. Casinovi, "Continuous-spectrum frequency-domain simulation of nonlinear systems," in *Proceedings of the 2002 International Symposium on Circuits and Systems*, Scottsdale, AZ, May 2002, pp. II:57–60.

[4] ——, "An algorithm for frequency-domain noise analysis in nonlinear systems," in *Proceedings of the 39th Design Automation Conference*, New Orleans, LA, June 2002, pp. 514–517.

[5] A. V. Oppenheim and R. W. Schafer, *Discrete-Time Signal Processing*. Englewood Cliffs, NJ: Prentice Hall, 1989.

ENHANCING SCHEDULING SOLUTIONS THROUGH ANT COLONY OPTIMIZATION

Shekhar Kopuri and Nazanin Mansouri

Syracuse University
Department of Electrical Engineering and Computer Science
123 Link Hall, Syracuse, NY, 13244-1240
(315) 443-2483 (Ph) (315) 443-1122 (Fax)
skopuri, namansou@ecs.syr.edu, http://web.ecs.syr.edu/~namansou

ABSTRACT

In this paper a methodology based on Ant colony optimization is presented to generate optimal scheduling during high-level synthesis. The classical force equation of the Force-directed scheduling algorithm has been modified to accommodate the experiences accumulated by multiple agents in different iterations. In each iteration the obtained schedule is subjected to remaining steps of synthesis using standard techniques like clique partitioning for resource allocation and left edge algorithm. The results are used to improve scheduling in the next iteration.

1. INTRODUCTION

Ant-colony optimization(ACO) technique is a robust multi-agent population based approach to explore the solution space of a multi constraint optimization problem. First proposed by Dorigo et al in 1992, this technique is inspired by the ability of real ants to find the shortest path from the nest to the food source, collectively. ACO has been successfully applied to many scientific problems, especially those that are NP hard in nature. In this work, an Ants-based approach to solve the scheduling problem during high-level synthesis has been presented.

High-level synthesis can be defined as automated transformation of the behavioral representation of a hardware system into a structural design at the register transfer level (RTL). The design is synthesized using a library of components such that it conforms to a set of pre-defined constraints. The entire synthesis problem consists of many NP-complete multi-constraint optimization subtasks. The data-path and the controller are generated separately. The data-path synthesis is broadly divided into (1) *Scheduling*, the process of identifying individual operations, and assigning a time-step to each of these such that no dependencies are violated, (2) *Resource Allocation and Binding*, the process of identifying possible functional units that can perform each operation, and binding the operation to particular instances of these (3) *Register Allocation and Binding*, or the process of identifying the number and (optimized) assignment of various registers to store the result of each operation and (4) *Steering logic generation* or the process of generating the steering logic for the entire data-path (that can be performed differently depending upon whether the architecture has point to point, i.e. multiplexer-based or Bus-based interconnects).

This paper presents an ACO-based technique that generates a valid scheduling and optimizes it iteratively. The scheduling approach is inspired by the famous Force Directed Scheduling (FDS) algorithm proposed by Paulin and Knight in 1987 [1]. The force equation of the FDS has been modified by replacing the successor-predecessor force (PS-force) term with two new terms *viz. Trail* and *Global Experience*. These terms characterize the experience of one agent in generating a solution and the experience of the entire population, respectively. In each iteration a schedule is obtained which is subjected to the complete synthesis process. The results of the synthesis are used to update the experience terms in the next iteration. This iterative process yields a more realistic final schedule.

The remaining sections of this paper discuss this process in detail and are organized as follows. Section 2 is a review of the related research in ACO and scheduling in high-level synthesis. Section 3 presents the proposed methodology. The results are presented in Section 4. Finally, Section 5 draws conclusions and discusses the avenues for future work.

2. RELATED WORK

Ant colony optimization was first introduced by Dorigo et al [2, 3] to address combinatorial optimization problems. Since then it has been successfully applied to many multi-constraint practical problems[4, 5, 6, 7]. Dorigo et al proposed a multi-agent auto-catalytic methodology to solve multi-

This work is partly supported by a grant from MDC-NYSTAR, under award number 411903-G.

constraint optimization problems inspired by the ability of real ants to find the shortest path from nest to food source following the trials of a chemical substance called pheremone laid by other ants. The ants collectively strengthen the best trail over time while other less likely trails are weakened as pheremone is volatile. The first ants-based algorithm proposed was the AS algorithm(Ant System). Since then many modifications and versions suitable to specific problems have been proposed but all share the common features of (1) being a multi-agent population based optimization technique, (2) having a pheremone trail model that exhibits stigmergy (i.e. diminishing trial), and (3) taking decisions to change states probabilistically.

This paper attempts to solve the scheduling problem of high level synthesis using an Ant-based approach. Most scheduling algorithms used in High-level synthesis are derivatives of the Force-directed Scheduling algorithm presented by Paulin and Knight in 1987 [1]. Park and Kyung [8] proposed an iterative technique based on the graph-bisection problem by Kernighan and Lin. More recent research attempts combine different steps of synthesis: Heijeligers et al [9] and InSyn [10] use techniques like genetic algorithms and simulated evolution. One of the first attempts to use Ant-optimizations in High level synthesis is the work by Kienprasit and Chongstitvatana [11]. They proposed a system to do synthesis using dynamic ants. Their method uses a decision path graph which is initialy generated for each ant. Dynamic niche sharing is used to find the peaks which are updated within each cycle. This is iteratively improved over a population. They also suggested some heuristics that were generated faster results.

This paper presents an algorithm that tries to solve, specifically, the scheduling step of a high level synthesis process. Significantly, the algorithm has many parameters that can be determined empirically to make it adaptable to a typical synthesis framework. In other words, the methodology is less dependent on the type of architectures that are synthesized. It uses a modified force equation of the FDS algorithm. The experience of the entire population is based on the results of the synthesis process. The following section presents the the proposed algorithm.

3. METHODOLOGY

Force Directed Scheduling (FDS) algorithm has been widely used by many researchers in various forms since it was first proposed by Paulin and Knight [1]. It tries to minimize the overall concurrency under a fixed latency. At every time-step, the effect of scheduling an unscheduled operation is calculated, and the one with least worse effect is selected. This effect is termed as force, and comprises of two components: the self-force, S_{il}, and the predecessor-successor (PS) forces, PS_{il}, of all its predecessors and successors. Let an unscheduled operation i be scheduled at a time step

```
G_il = 0; i ∈ [1,N],  l ∈ [1,λ]
s^o = 0; No best schedule
for n=1 to n_iterations do
  M^n_ij = 0; i ∈ [1,N],  l ∈ [1,λ]
  n_ants = |ηN|; η < 1
  calculate S_ij  i ∈ [1,N],  l ∈ [1,λ]
  sort S in descending order;
  initialize n_ants with the least n_ants self-forces
  for m = 1 to n_ants do
    s = procedure A_FDS();
    calculate cost(s);
    update M^n_ij using equation 3
    if cost(s^o) > cost(s)
      s^o = s;
  update G_ij using equation 4
```

Table 1. Algorithm Ant_sched

l. Then, the self-force S_{il} represents the direct effect of this scheduling on the overall concurrency. It is given by:

$$S_{il} = \sum_{m=t_i^S}^{t_i^L} q_k(m)(\delta_{lm} - p_i(m)) \qquad (1)$$

where, t_i^S and t_i^L are the ASAP and ALAP times respectively, k is the type of operation, q_k is the type distribution for type k and δ_{lm} is the Kronecker delta function. This scheduling might cause the time frame of a predecessor or successor operation to change from $[t_i^S, t_i^L]$ to $[\tilde{t}_i^S, \tilde{t}_i^L]$. Then the ps-force exerted by this predecessor or successor is given by

$$PS_{il} = \frac{1}{\tilde{\mu}_i + 1} \sum_{m=\tilde{t}_i^S}^{\tilde{t}_i^L} q_k(m) - \frac{1}{\mu_i + 1} \sum_{m=t_i^S}^{t_i^L} q_k(m) \qquad (2)$$

where μ_i and $\tilde{\mu}_i$ are the mobility associated with the original and the changed time frames.

The FDS algorithm is of the order of O(n^3) time complexity. The results obtained are usually optimal but the method is time consuming, making it unsuitable for large graphs. Also, the basic FDS algorithm tries to reduce the number of functional units and memory units used. However, in the DSM realm the cost of the resources might not necessarily be dominant. The ACO-based technique presented here, that solves the scheduling problem for a *generic* circuit, has a modified force equation with the PS-force term replaced by two new terms the *Trail*, M_{il}^n and the *Global Experience*, G_{il}, representing the experience of one agent and that of the entire population, respectively. These experience terms are calculated based on the effectiveness of the schedule after complete synthesis. The process is captured algorithmically in the algorithm *Ant_sched*,

$n = 0$;
while $n \neq N$;
compute S_{ij} $i \in [1, N]$, $l \in [1, \lambda]$
compute F_{ij} $i \in [1, N], l \in [1, \lambda]$ using equation 5
choose an operation and time-step at random amongst x of those with least forces: $x = |\eta n|$, $\eta < 1$
schedule the operation and update the time frames
$n = n + 1$;

Table 2. Procedure A_FDS

and the procedure *A_FDS*. The subsections 3.1 and 3.2 interpret these algorithms elaborately. It is important to note that the self-force can be calculated in linear time where as the the time complexity of the ps-force is at least O(n^2)[1].

3.1. ACO-Based Scheduling Algorithm

Table 1 presents the ants-based scheduling algorithm. The experience of the ants within n-th iteration is maintained in a matrix $M^n(N, \lambda)$, where N is the number of operations to be scheduled and λ is the latency found by ASAP. Any element $M^n_{i,l}$, therefore, is the trail deposited by the n-th ant and represents their experience of scheduling operation i at the time-step l (i.e. the experience in the n-th iteration). It is initialized to a matrix of zeros. The initial self forces are calculated and sorted. A fixed fraction η of the N operations $s.t.(n_{ants} = |\eta N|)$ with the least self-forces are selected, and each is assigned to an ant that is initialized with corresponding operation and time-step as the first scheduled operation. Each ant then performs the rest of the scheduling process using the procedure A_FDS described in Table 2. The resulting schedule, s, is used in subsequent synthesis steps, and the process runs to completion. The cost of the synthesized design is calculated using the costs described in Table 3. It may be noted that the mechanism to calculate costs can be varied according to the type of circuits one is dealing with. Therefore, the effectiveness of the algorithm is dependent on how realistic is the cost calculation. Then, the trail matrix for the n-th cycle, M^n, is updated as follows:

$$M^n_{ij} = M^n_{ij} + \frac{cost(s) - cost(s^o)}{cost(s^o)} \quad \forall (i, j) \in s \qquad (3)$$

where, s^o is the best schedule thus far. If the schedule results in an improved design, s^o is changed to s. The trail laid by an ant is positive or negative depending on whether the improved synthesis results are obtained or not. The strength of the trail is proportional to the degree of improvement or worsening of the design. The trail evaporates by a factor ρ. A global matrix,$G(N, \lambda)$, to collect the trails laid by the entire population in all the cycles is maintained and is updated after every cycle in the following way

$$\forall (i, j) \quad G^{n+1}_{ij} = \rho G^n_{ij} + M^n_{ij} \qquad (4)$$

3.2. The Force Equation

The A_FDS uses following force equation to calculate force exerted when an operation i is scheduled at time-step l

$$F_{il} = \alpha \frac{S_{il}}{\sqrt{\frac{1}{N\lambda} \sum_{(i,j)=(1,1)}^{N,\lambda} S_{ij}}} - \beta M^n_{il} - \gamma G^n_{il} \qquad (5)$$

The first term is the normalized self-force component. The second term captures the experience accumulated in the current cycle and the third term captures the experiences accumulated over all the cycles. It should be noted that with increasing cycles the last two components increase in value, hence diminishing the importance of the the self-force over time. This is important as the self-force does not take the forces exerted by other operations into account. However, it can give a good starting point for the iterative process and eventually the algorithm *learns* this with experience. α, β, γ are constants determined by empirical means.

4. RESULTS

The algorithm Ant_sched was applied to four synthesis benchmarks viz. the Differential Equation, the Elliptical filter and two Discrete Cosine Transform benchmarks. Iterative improvement was seen in all the cases. Shown here in fig 1, fig 2 and fig 3 are the later three.Various values for the parameters $\eta, \lambda, \alpha, \beta, \gamma$ and ρ were simulated. Optimal schedules that led to minimal costs after synthesis were found in all the three cases within 20 iterations. Number of ants per iteration were 20. The values of η, α, β and γ were 0.5, 1, 2 and 2 respectively. ρ between 0.35 and 0.5 led to better solutions. As compared to the standard FDS algorithm the proposed algorithm generated better results (Table 4).

All the simulations were run on a Pentium 3, 800 MHz machine with 320Mb of RAM. None of the cases took more than few seconds to generate results. It is worthwhile to note that all the terms in the modified force equation can be calculated in linear time. The value of *n_iterations* is also low (<20 for the benchmarks). Also, the methodology is not bound by any specific allocation and binding technique nor limited to any specific cost estimation mechanism. It can be successfully implemented by varying the various parameters.

5. CONCLUSION

In this paper an ACO-based methodology to solve the scheduling problem in high-level synthesis for a generic design has

Hardware Unit	Cost
Single Cycle Multiplier	250
Single Cycle Adder/Subtractor/Comparator	50
Register	15
2-input Mux	15

Table 3. Cost Table

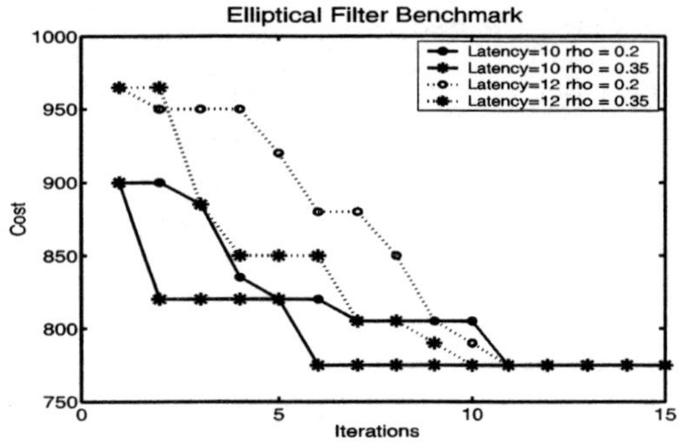

Fig. 1. Ant_sched on Elliptical filter Benchmark

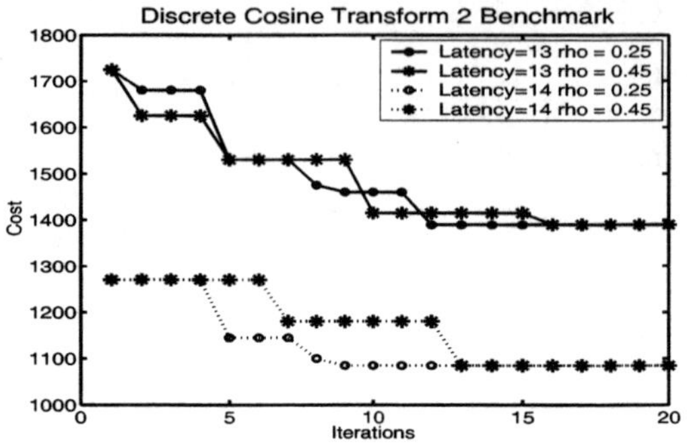

Fig. 3. Ant_sched on DCT2 Benchmark

been presented. Though iterative synthesis seems a time consuming step, we claim that this paper presents a new perspective to generating solutions to various individual steps of high level synthesis, each of which can be seen as a multi constraint optimization problem. Our method is independent of the architectures of the designs and tends to learn using the experience of a population instead of an individual. Another significant aspect is that we have used the classic FDS algorithm and modified it in a very simple form and generated solutions that developed iteratively. Our efforts in the future are targeted at modeling each of the individual problems of high level synthesis using Ant based optimization techniques and to perform extensive simulations to generate realistic values for the parameters in the algorithms. We also aim to tie them together to reduce the overall synthesis time.

6. REFERENCES

[1] John P. Knight Pierre G. Paulin, "Force-directed scheduling in automatic data path synthesis," *DAC*, pp. 195–202, 1987.

[2] M. Dorigo, *Optimization, Learning and Natural Algorithms*, PhD Thesis, Dipartimento di Elettronica, Politecnico di Milano, IT, 1992.

[3] M. Dorigo, V. Maneizzo, and A. Colorni, *Positive Feedback as a Search Strategy*, Technical Report 91-016, Dipartimento di Elettronica, Politecnico di Milano, IT, 1991.

[4] B.Bullnheimer, C.L. Mallows, and I.A. Wagner, *An Improved Ant System for Vehicle Routing Problem*, Technical Report POM-10/97 ,Institute of Management Science, University of Vienna, 1997.

[5] M. Dorigo, V. Maneizzo, and M. Trubian, "Ant system for job scheduling," *Belgian Journal of Operations Research, Statistics and Computer Science*, vol. 34, pp. 39–53, 1994.

[6] G.M. Gambardella and M. Dorigo, *HAS-SOP: A Hybrid AntSystem for Sequential Ordering Problem*, Technical Report 11-97, IDSIA, Lugano, CH, 1997.

[7] D. Costa and A. Hertz, "Ants can color graphs," *Journal of Operational Research Society*, vol. 48, pp. 295–305, 1997.

[8] I.C. Park and C.M. Kyung, "Fast and near optimal scehduling for data path synthesis," *DAC Proceedings*, pp. 650–685, June 1991.

[9] M.G.M. Heijligers, L.J.M. Cluitmans, and J.A.G. Jess, "High level sythesis, scheduling and allocation using genetic algorithms," *ASP-DAC*, pp. 61–66, 1995.

[10] A. Sharma and R.Jain, "Insyn : Integrated scheduling for dsp applications," *International Synposium on High level Sythesis*, pp. 96–103, 1994.

[11] R. Keinprasit and P. Chongstitvatana, "High level synthesis by dynamic ants," *International Journal of Intelligent systems*, 2003.

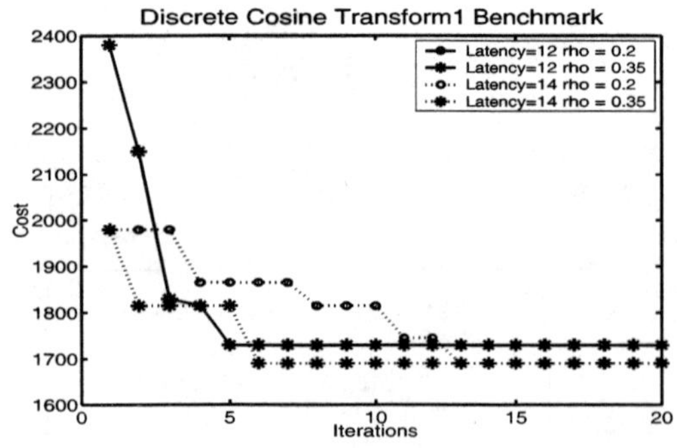

Fig. 2. Ant_sched on DCT1 Benchmark

Benchmark	Latency	FDS	A_FDS
Differential equation	4	875	875
	5	875	875
Elliptical filter	10	825	775
	12	775	775
Discrete Cosine Transform 1	12	1795	1730
	14	1795	1690
Discrete Cosine Transform 2	13	1210	1085
	14	1525	1390

Table 4. Cost of best solution for FDS and A_FDS procedures

SYNTHESIS SCHEME FOR LOW POWER DESIGNS WITH MULTIPLE SUPPLY VOLTAGES BY TABU SEARCH

Ling Wang   *Yingtao Jiang*   *Henry Selvaraj*

Department of Electrical & Computer Engineering
University of Nevada, Las Vegas
Las Vegas, NV 89154
USA

ABSTRACT

In this paper, a tabu-search-based behavior level synthesis scheme is proposed to minimize power consumption with resources operating at multiple voltages under the timing and the resource constraints. Unlike the conventional methods where only scheduling is considered, our synthesis scheme considers both scheduling and partitioning simultaneously to reduce power consumption due to the functional units as well as the interconnects among them. More importantly, our approach tends to efficiently address a few practical layout problems inherent to multiple voltage designs. In particular, we have configured our solutions as a three-tuple vector to account for both the timing and the partition. Cycling of the same solutions is prevented by applying a tabu list with an update mechanism enhanced with an aspiration function. In this way, the algorithm can search a large solution space with modest computation effort and fast convergence rate. Experiments with a number of DSP benchmarks show that the proposed algorithms achieve an average power reduction by 49.6%.

1. INTRODUCTION

With today's increasingly large and complex digital IC and system-on-chip (SoC) designs, power dissipation has emerged as a primary design consideration. Reduction of power consumption in VLSI designs can be achieved at various levels of the design hierarchy, ranging from the processing technology, circuit, logic, architectural and algorithmic (behavioral) levels, up to the system level. In this study, we focus on power minimization at the behavioral level.

As supply voltage is quadratically proportional to power consumption, the most effective way to reduce power consumption is to lower the supply voltage level for the circuit. However, reducing the supply voltage lead to circuit delay and the throughput penalties. One way of maintaining the throughput is to use the pipeline and explore the parallelism [1]. Another promising approach is to use multiple supply voltages on the chip [2][6]-[9][12]-[14][16], where nodes on the critical paths are assigned to the high-voltage resources to meet the timing constraints while nodes on the non-critical path are assigned to low-voltage resources to reduce power consumption.

Most designs using multiple voltage supplies at the behavioral level have a similar synthesis flow as shown in Fig. 1.a, with the emphasis is placed on scheduling the operations. In particular, R. Martin and J. Knight [9] developed a genetic search scheduling algorithm to minimize power consumption. Raje and Sarrafzaden [12] proposed a variable voltage scheduling algorithm for power minimization with a timing constraint. Chang and Pedram [2] proposed a dynamic programming scheduling technique for both non-pipelined and functionally pipelined data-paths. Johnson and Roy [6] developed an integer linear programming (ILP)-based scheduling algorithm with exponential time complexity. Manzak and Chakrabarti [8] presented a heuristic scheduling algorithm under timing and resource constraints with polynomial time complexity. In[14], Shiue and Chakrabarti presented a scheduling scheme to minimize the power consumption by considering the effect of the interconnect complexity and level shifters.

However, if only the scheduling is considered, multiple voltage synthesis design at the behavioral level can cause some physical layout problems. For instance, nodes with different voltages may be placed adjacent to each other in the physical layout, resulting in complex routing of interconnections and supply voltage lines. On the other hand, if we can partition a chip running at multiple voltages with fewer voltage regions and fewer signal passings between regions operating at different voltages, the above mentioned layout problems can be significantly alleviated. That is, partitioning the chip into different voltage islands shall be performed simultaneously with the operation scheduling.

Realizing the importance of partitioning at the behavioral level, we proposed the schemes to minimize power consumption under timing constraints [16], each scheme with a polynomial time complexity. In both schemes, the scheduling and partitioning are performed sequentially as shown in Fig.1.b. Since scheduling and partitioning are not independent from each other, it is more desirable to perform these two tasks simultaneously to achieve greater power reduction. If so, the synthesis flow works as shown in Fig.1.c.

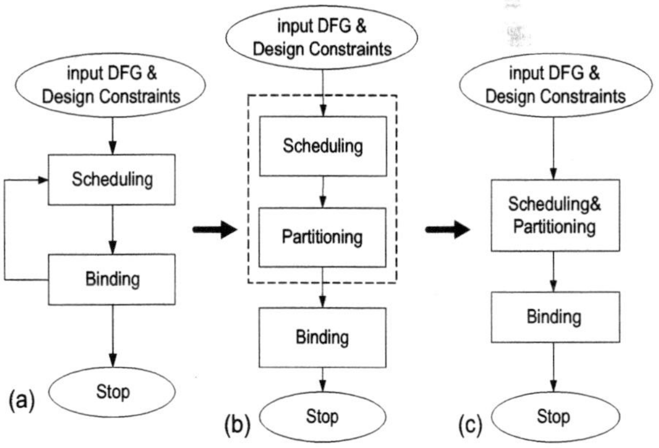

Fig.1 Flows of multiple supply voltage synthesis.

The scheduling and partitioning problems by nature are NP-hard problems [3]. Studies have shown that tabu search has been very effective in providing good solutions for such problems with a reasonable amount of time [5][11][15]. In this paper, we adopt the tabu search as the core algorithm and have developed a problem-specific modification for multiple voltage synthesis under timing and resource constraints.

2. PROBLEM STATEMENT

2.1. Notions

Definition 1 (Data Flow Graph). A data flow graph DFG is a directed acyclic graph $G=(V,E)$, where V is a set of nodes, and E is a set of edges between nodes. Here each node represents an operation, and a directed edge from node v_i to node v_j means execution of v_i must precede that of v_j.

Definition 2 (Timing Constraint). Timing constraint is the time available to execute the nodes in a DFG.

Definition 3 (Resource Constraint). Resource constraint is a set of allowed functional units that should run at specified supply voltages. Note that we could have several instances of the same functional unit running at different/same supply voltages.

Definition 4 (Spatial Voltage Cluster). A spatial voltage cluster is a set of nodes that operate at the same supply voltage. The resources assigned to the nodes in a spatial voltage cluster shall be placed near each other in the final layout to minimize the communication costs.

2.2. Problem Statement

The multiple voltage scheduling and partitioning (MVSP) problems under the timing and the resource constraints are defined as follows:

Given a DFG and timing and resource constraints, find a schedule and a partition that satisfy the given timing and resource constraints, with minimum power consumption.

A solution to above stated problem must satisfy the following requirements:

- each node in the DFG is assigned to a specific resource at its corresponding control step and
- all the nodes are partitioned into spatial voltage cluster(s).

Here, a solution S is defined as:

$$S=\{\langle v_1,r_1,c_1\rangle,\langle v_2,r_2,c_2\rangle,\ldots\langle v_i,r_i,c_i\rangle,\ldots\langle v_n,r_n,c_n\rangle\} \quad (1)$$

where (i) $v_i \in V$, $n=\|V\|$, (ii) r_i is the resource assigned to node v_i, and (iii) c_i is the cluster to which node v_i belongs and $c_i \in C$, C is the set of all spatial voltage clusters, $C=\{c_{i,j}(V_i)\}$, V_i is the voltage of cluster $c_{i,j}$.

The interconnection power consumption P_{int} with respect to a solution S is defined as

$$P_{\text{int}}(S)=\sum P_{\text{int}}^{(i,j)}(S) \quad (2)$$

where $P_{\text{int}}^{(i,j)}(S)$ is the power consumed by the interconnections between clusters c_i and c_j in the solution S, and

$$P_{\text{int}}^{(i,j)}(S)=0 \quad \textit{if } i=j \quad (3)$$

The resource power consumption P_{res} with respect to a solution S is defined as

$$P_{res}(S)=\sum_{i=1}^{n} P_{res}^{i}(S) \quad (4)$$

where n is the number of nodes.

Our goal is to find a solution S that can minimize the following objective function:

$$\sum P_{\text{int}}^{(i,j)}(S)+\sum_{i=1}^{n} P_{res}^{i}(S) \quad (5)$$

Subject to

$$T \le T_C \text{ and } R \le R_C \quad (6)$$

where (i) T is the total schedule time, (ii) T_C is the timing constraint, (iii) R is the total number of assigned resources at each control step, (iv) R_C is the number of the available resources (resource constraints) at each control step, and (v) n is the number of nodes.

Lemma [2]: Multiple voltage scheduling problem is NP-complete.

Theorem: Multiple voltage scheduling and partitioning problem is NP-complete.

Proof: It can be readily shown from the Lemma.

3. ALGORITHM DESCRIPTION

Tabu search is regarded as a meta-heuristic algorithm that takes a more aggressive approach than other search algorithms. It proceeds on the assumption that there is no value in choosing an inferior solution unless it is absolutely necessary, as in the case of getting out of a local optimum. Tabu search has been shown to be very useful in providing good solutions for many NP-hard problems, such as scheduling and graph partitioning, in a reasonable amount of time [5][11][15].

3.1. Scheduling and Partitioning by Tabu Search

The scheduling and the partitioning (Fig. 1.c) can be performed simultaneously by applying a tabu search algorithm detailed as following.

3.1.1 Representation of a Solution

A solution S is a three-tuple vector, as defined in Eq.(1). In Eq. (1), $r_i \in R_i$. R_i is the candidate resource list, which contains all the resources that could be potentially assigned to node v_i.

The delays of the resources in the candidate resource list R_i are required to satisfy the timing requirement, $d_i \le m + d^M$, where (i) d_i is the delay of any resource r_i in R_i, (ii) m is the mobility of node v_i, which is determined as the difference between its ASAP and ALAP schedules [16], and (iii) d^M is the delay of node v_i if it is assigned to the highest voltage resource in the design library.

The size of solution space is $(N \cdot M)^{\|V\|}$, where N is the voltage levels in the design library and M is the number of clusters.

3.1.2Creating a Starting Solution

A starting solution is generated as follows:

i) Generate a solution S by assigning the highest voltage resource to each node;
ii) Put all the nodes to one cluster;

3.1.3Objective Function

The objective function F with respect to a solution S has been defined in Eq. (5).

3.1.4 Mechanism of Generating a Neighborhood

Neighborhood of a solution S, denoted as N(S), contains all the solutions obtained by a move. A move from solution S to another S^* is represented by $(v_i,(r_i,c_l),(r_i^*,c_l^*))$, where i) v_i is the only node in S that its cluster and resource are to be modified, ii) resource r_i is assigned to v_i, node v_i belongs to cluster c_i in S, and iii) resource r_i^* is assigned to v_i in S^*, and node v_i belongs to a different cluster c_i^* in S^*. Note that resources r_i and r_i^* are of the same function type but operate at different supply voltages. As resource r_i^* operates at voltage V_j, c_i^* can be any of the clusters $c_{j,1}(V_j), c_{j,2}(V_j), ..., c_{j,m}(V_j)$. In this way, the search space includes all the clusters of interest.

Among all the solutions induced from neighborhood, the best solution will be the one that could lead to minimum resource power consumption and least interconnections among clusters. As a result, the power consumption due to the resources and interconnections can be maximally reduced. Depending on the assigned resources r_i^*, a feasible scheduling has also to be determined simultaneously to satisfy the timing and the resource constraints.

For N voltage levels in the design library and M clusters, the size of the neighborhood is $(N-1) \times M \times \|V\|$.

3.1.5 Structure of the Tabu List

Tabu list is constructed as a linear list L with length of k, which contains the k most recent moves. Going back to a solution S which was visited in the last k iterations is not allowed in order to prevent cycling on the same set of solutions.

Instead of keeping a whole move in the tabu list L, only the resource modification in the move is saved in L. That is, when (v_i, r_i, r_i^*) is stored in the tabu list, the move $M1$ $(v_i, (r_i, c_l), (r_i^*, c_{j,l}(V_j)))$ from S to S^* and the move $M2$ $(v_i, (r_i, c_l), (r_i^*, c_{j,k}(V_j)))$ from S to S^{**} are considered identical in the tabu list. Assume $M1$ is the current move and the move $M2$ is already in the tabu list, tabu condition can be overruled if F(S^*) has lower value than an *aspiration function* $A(M1)$, which is equal to F(S^{**}).

The update of the tabu list is given below:

```
if (Tabu condition is satisfied and  the current move is M
    from S*to S){
    if (A(M)>F(S)) A(M)=F(S);
    else{
        Set M at the end of L;
        Delete the oldest element in L;
    }
}
```

3.1.6 Scheduling & Partitioning by Tabu Search (Putting-Everything Together)

```
Input: DFG represented as G(V,E); // see Def. 1
Definitions:
    S: a feasible solution as Eq.(1);
    F: Objective function as Eq.(5);
    N(S): Neighborhood of S;
    L: tabu list with length k;
    A: Aspiration function;
    Max: number of iterations;
Initialization
    i=0;
    Generate an initial feasible solution S_i ;//Sec.3.1.2
    Initialize tabu list with length L and A;
    bests = S_i ; bestcost = F(bestS); besti = 0;
Body
    while (i < Max){
        i = i + 1;
        Generate neighborhood N( S_{i-1} )of S_{i-1} ; //Sec.3.1.4
        Select feasible solution S_i from N ( S_{i-1} );
        if ( S_i ∉ L or A( S_i ) > F( S_i )){
            if (F( S_i )<bestcost) {
                bests = S_i ; bestcost = F( S_i ); besti = i;
            }
        }
        Update tabu list L and aspiration function A;//Sec.3.1.5
    }
Output : the best solution bestS; // input for binding
```

3.2 Binding

After obtaining a satisfactory schedule and partition from the previous scheduling and portioning step, resource binding takes place (Fig. 1.c). As the resources assigned to the nodes with the same voltage supply will be placed close to each other in the physical layout, the interconnection power consumption between the clusters are significantly higher than that inside the clusters. Therefore, the binding algorithm proposed in [16] can be applied for each cluster to maximize the resource sharing among the nodes inside each cluster in order to reduce the interconnection cost.

4. EXPERIMENT RESULTS

In this section, we present the results obtained by applying our algorithm on some high-level synthesis benchmarks (Lattice, Differential Equation, Ellipf, Wdf7, Volterra, 7th order IIR filter, Wavelet and DCT). The number of nodes of each circuit is reported in column N of Table 1. The proposed algorithm is coded in C and runs on a Linux workstation with a 686 processor.

The design library, adopted from [2], consists of two types of functional units: adders and multipliers, and both can operate in any of the four voltage levels: 5.0V, 3.3V, 2.4V, and 1.5V. The delay and the power consumption of each function unit have been fully characterized.

The switching activity of the nodes is assumed to be 0.5. The timing constraint T_C (column Tc in Table 1) is set as 2T, where T is the delay obtained by ASAP schedule when all the sources operate at 5V (column Tc in Table 1). The resource constraint R is set as an integer. For instance, if the resource constraints R =2, it means that, at each control step, 2 multipliers or adders both operating at 5V, 3.3V, 2.4V, or 1.5V shall be allowed. In all the experiments, the number of iterations is set to 100, and the experimental results are tabulated in Table 1.

In Table 1, column E_{fun}^5 reports the energy consumption when all the functional units take the supply voltage of 5V. If the same circuits are synthesized using multiple supply voltages based on the proposed algorithm, the energy consumption is reported in column E_{fun}. It can be seen that, by average, multiple voltage designs consume 49.6% less power than their counterparts using single voltage supply. For the Wavelet circuit, for instance, 67.7% power reduction is obtained. Even for small circuit, Diffeq with only 20 nodes, we still see some power reduction.

Column *I* shows the interconnect cost (weighted sum of all interconnects) if the nodes of the DFG are partitioned into a number of clusters. For instance, the Lattice filter has a interconnect cost of 34, if the total number of clusters is 5 (shown in column Nc). Compared to the case when no partitioning is performed (i. e., only one cluster for its nodes as shown in column I_b), we can see the interconnect cost has been reduced by 38% (column R2%). On average, the interconnection reduction is 35% across all 9 benchmark circuits. Finally, the CPU times for synthesis runs are reported in seconds in the last column.

Fig 2 shows the convergence of the proposed algorithm for IIR filter. In all four cases ($R = 2,3,4,5$), the solutions converge after only 40 iterations. As a matter of fact, fast convergence has also been observed for all other benchmark circuits.

5. CONCLUSION

In this paper, we have presented a multiple voltage synthesis scheme based on tabu search to minimize power consumption under timing and resource constraints. Experiments with a number of DSP benchmarks show that the proposed algorithms can achieve significant power reduction.

6. REFERENCES

[1] P. Chandrakasan, S. Sheng, and R. W. Brodersen, "Low Power CMOS Digital Design," IEEE J. Solid-State Circuits. Vol. 27, pp.473-483, Apr. 1992.

[2] J. M. Chang, and M. Pedram, "Energy Minimization Using Multiple Supply Voltages," IEEE Trans. VLSI, Vol.5, pp.157-162, Dec. 1997.

[3] M. R. Garey and D. S. Johnson, Computers and Intractability: A guide to the Theory of NP-Completeness, Freeman, San Francisco, CA.

[4] F. Glover and C. Mcmillan, "The General Employee Scheduling Problem: an Integration of Management Science and Artificial Intelligence," Computers Oper. Res., Vol. 13, No. 3, pp.563-593, 1987.

[5] A. Hertz, and D. De Werra, "The Tabu Search metaheuristic: how we used it." Annals of Mathematics and Artificial Intelligence, Vol. 1, pp.111-121,1991.

[6] M. C. Johnson and K. Roy, "Datapath Scheduling with Multiple Supply Voltages and Level Converters," ACM Trans. Design Auto. Electronic Systems, Vol.2, pp. 227-248, July 1997.

[7] Y. R Lin and C. T. Hwang, "Scheduling Techniques for Variable Voltage Low Power Designs," ACM Trans. Design Auto. Electronic Systems, Vol.2, pp. 81-97, Apr. 1997.

[8] A. Manzak and C. Chakrabarti, "A Lower Power Scheduling Scheme with Resources Operating at Multiple Voltages," IEEE Tran. VLSI, Vol.10, pp. 6-14, Feb. 2002.

[9] R. Martin and J. Knight, "Power Profiler: Optimizing ASIC's Power Consumption at the Behavioral Level," Proc. IEEE/ACM Design Automat. Conf., pp. 42-47, 1995.

[10] R. Mehra, L. M. Guerra, and J. M. Rabaey, "Low Power Architectural Synthesis and the Impact of Exploiting Locality," J. VLSI. Signal Processing, Vol.13, pp.239-258,1996.

[11] E. L. Mooney and R. L. Rardin, "Tabu Search for a Class of Scheduling Problems," Ann. Oper. Res.,1993, Vol. 41, pp.253-278, 1993.

[12] S. Raje and M. Sarrafzadeh, "Variable Voltage Scheduling," Proc. Int. Symp. Low Power Design, pp.9-14, Apr 1995.

[13] S. Raje and M. Sarrafzadeh, "Scheduling with Multiple voltages," Integr. VLSI J., pp.37-60, Oct.1997.

[14] W. T. Shiue and C. Chaitali, "Low Power Scheduling with Resources Operating at Multiple Voltages," IEEE Trans. CASII, Vol.47, pp.536-543, June 2000.

[15] J. Skorin Kapov, "Tabu Search applied to the quadratic assignment problem", ORSA J. Computing, Vol.2, pp.33-45, 1989.

[16] L. Wang, Y. Jiang and H. Selvaraj, "A Synthesis Scheme for Low Power Designs with Multiple Voltages under Timing Constraints," Proc. NASA 11th VLSI Symp., 2003, Idaho.

Table 1 Experimental Results on benchmarks

Bench-mark	Nn	Tc	R	E^5_{fun} (pJ)	E_{fun} (pJ)	R1%	I_b	I	R2 %	Nc	CPU Time
Lattice	22	48	1	23598	13467.5	42.9	55	34	38	5	0.060
			2		10975.9	53.4		38	30	4	0.410
DIFFEQ	10	42	1	15614	15152.6	2..9	70	25	64	4	0.180
			2		12766.6	18.2		25	64	3	0.190
ellipf	37	58	1	23100	13463.7	41.7	119	51	57	5	0.24
			2		12235.4	47		56	52	4	1.140
			3		14046.7	39.2		57	52	4	1.15
			4		11447.5	50.4		58	51	4	1.23
Iir7	36	40	2	39212	20332.1	48.1	107	61	42	4	1.22
			3		15740.6	59.8		64	40	4	0.64
			4		14678.9	62.6		65	39	4	1.25
Wdf7	50	60	2	49464	24604	50.2	120	72	40	4	0.660
			3		19110.4	61.3		73	39	4	1.89
			4		19827.2	59.5		71	40	4	1.84
			5		18448	62.7		71	40	4	1.9
dct	42	34	4	43132	21977.8	49	132	79	40	4	1.650
			5		20165.7	53.2		72	45	4	1.650
volterra	32	40	3	43748	16814.8	61.5	81	55	32	4	0.370
			4		15122.7	65.4		54	33	5	0.86
wavelet	67	42	4	73180	28184.4	61.4	195	116	40	4	4.05
			5		23596.4	67.7		114	41	4	4.15

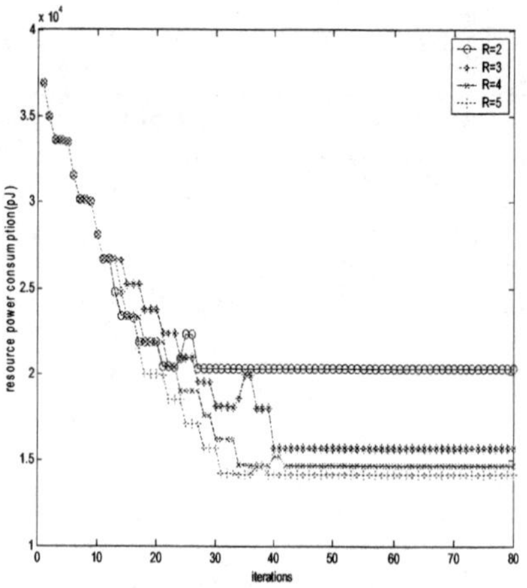

Fig. 2 IIR Circuit (power consumption vs. number of iterations) under different resource constraints.

GRAPHICS PROCESSOR UNIT (GPU) ACCELERATION OF FINITE-DIFFERENCE TIME-DOMAIN (FDTD) ALGORITHM

Sean E. Krakiwsky
University of Calgary, Canada
(403) 210-5460
krakiwsk@enel.ucalgary.ca

Laurence E. Turner
University of Calgary, Canada
(403) 220-5810
turner@enel.ucalgary.ca

Michal M. Okoniewski
University of Calgary, Canada
(403) 220-6175
michal@enel.ucalgary.ca

ABSTRACT

The Finite-Difference Time-Domain (FDTD) algorithm has become a tool of choice in many areas of RF and microwave engineering and optics. However, FDTD runs too slow for some simulations to be practical, even when carried out on supercomputers. The development of dedicated hardware to accelerate FDTD computations has been investigated. In this paper we demonstrate that off-the-shelf Graphics Processor Units (GPUs) can be successfully used to accelerate FDTD simulations. Using C++, OpenGL, and several OpenGL extensions, a modern GPU has been programmed to solve a simple two dimensional electromagnetic scattering problem. The GPU outperforms a Central Processing Unit (CPU) of comparable technology generation.

1. INTRODUCTION

The Finite-Difference Time-Domain (FDTD) numerical method is widely used for electromagnetic simulations [1], but the algorithm can be very slow when run in software. Recently, it has been proposed that Field Programmable Gate Array (FPGA) hardware is suitable for the acceleration of FDTD [2]-[6]. While significant acceleration has been demonstrated [3], these methods require the development of dedicated FPGA platforms, and will likely remain expensive. On the other hand, the Graphics Processor Unit (GPU) is a relatively inexpensive consumer oriented hardware product with massive computational power. Researchers have proposed the use of GPU products by such vendors as nVidia and ATI for the acceleration of solutions to physically-based problems such as fluid flow analysis, Fast Fourier Transforms (FFT), and crystal growth [7]-[9]. We investigate the suitability of the GPU for the acceleration of FDTD simulations.

2. BASICS OF FDTD METHOD

FDTD, first formulated by Yee [10] and further developed by Taflove and others [11], is a direct solution of Maxwell's time-dependent curl equations. Please refer to these references for a rigorous treatment of FDTD. This paper provides only a brief description of how to construct a 2-dimensional FDTD lattice. In 2-dimensional space, FDTD computations are performed in two, simple, staggered computational nodes as shown in Figure 1 and Figure 2. If the conductivity of the media of interest is zero, the electric field E_z and magnetic field H_z shown in Figure 1 and

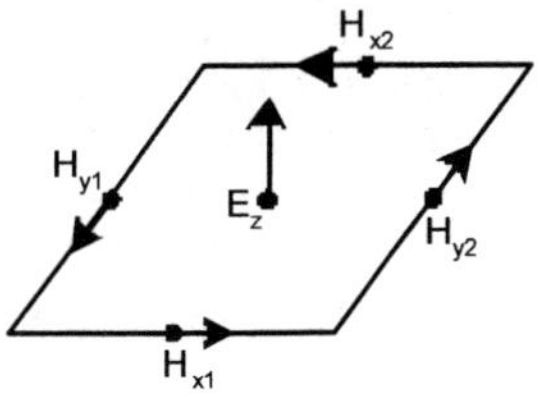

Figure 1: Electric Field E_z

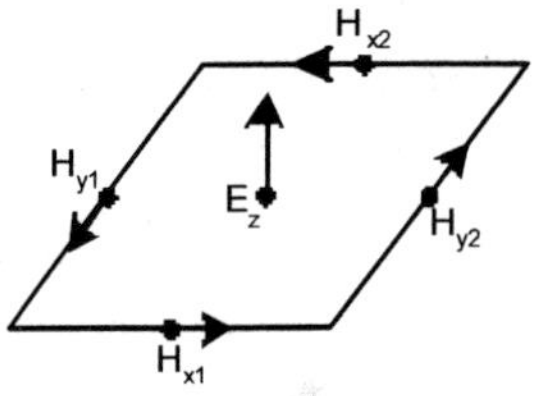

Figure 2: Magnetic Field H_z

Figure 2 respectively have the following two update equations:

$$E_z^{t+\Delta t} = E_z^t + \frac{\Delta t}{\varepsilon \cdot \Delta} \cdot \left[H_{x1}^{t+\frac{\Delta t}{2}} - H_{x2}^{t+\frac{\Delta t}{2}} + H_{y1}^{t+\frac{\Delta t}{2}} - H_{y2}^{t+\frac{\Delta t}{2}} \right] \qquad (1)$$

$$H_z^{t+\frac{\Delta t}{2}} = H_z^{t-\frac{\Delta t}{2}} + \frac{\Delta t}{\mu \cdot \Delta} \cdot \left[E_{x1}^t - E_{x2}^t + E_{y1}^t - E_{y2}^t \right] \qquad (2)$$

where, t is the simulation time in seconds;
Δt is the time step used in the simulation;
μ is the permeability at the H_z node;
ε is the permittivity at the E_z node; and
Δ is the length of a side of the quad in Figure 1.

It should be noted that in equations (1) and (2) the E_z and H_z fields are updated at different instances in absolute time (i.e. separated by half of a time step).

In many two dimensional problems (such as the simple scattering problem presented in Section 4) it typically suffices to consider only 3 out of 6 possible electric and magnetic field components. Furthermore, 3D problems can be assembled from appropriately constructed and coupled 2D slices. We consider a transverse electric field problem, where only H_y, H_x and E_z field components exist. Thus the $H_{x,y}$ update equation simplifies to:

$$H_{x,y}^{t+\frac{\Delta t}{2}} = H_{x,y}^{t-\frac{\Delta t}{2}} + \frac{\Delta t}{\mu \cdot \Delta} \cdot \left[E_{z1}^t - E_{z2}^t \right] \qquad (3)$$

Currently, we are using metal boundaries constituting Perfect Electric Walls (PEW). Therefore, no special computations are required in order to deal with the Dirichlet boundary conditions.

3. OVERVIEW OF GPU

Currently, the time between new GPU product releases is shorter than for CPUs and the latest consumer GPUs have more transistors that an Intel P4 CPU. It is the consensus in the graphics hardware community that this trend will continue for the next five years making GPUs much faster than CPUs for certain kinds of computations. While CPUs are optimized for generality, GPUs are optimized for vector arithmetic and parallelism. The main commercial driving force behind the increased performance of GPUs is the desire for real-time rendering of realistic or "cinematic quality" games and other entertainment related applications. The fast vector arithmetic, parallelism, and large memory bandwidth (27 Giga*bytes* per second in nVidia's FX 5900 chips) are attractive for various numerical methods such as FDTD. Furthermore, through interface standards such as OpenGL and Direct X, modern GPUs are programmable, making the implementation of novel applications possible. Therefore, we have at our disposal, a consumer priced PC-based product that is suitable for computationally intensive non-consumer oriented applications.

3.1. Graphics Pipeline

While exact GPU hardware diagrams are not in the public domain, enough of the architecture is conveyed to allow for programmability. In our context, the architecture is referred to as the "graphics pipeline" where the start of the pipeline is the CPU/GPU interface and the end of the pipeline is the render target, which is typically, but not always the frame buffer. The frame buffer is directly (or indirectly if using double buffering) displayed on the monitor. Figure 3 shows one view of the graphics pipeline that is relevant to our FDTD problem. For a more rigorous explanation of the graphics pipeline see [12].

We store the E_z, H_x, and H_x data arrays as one two dimensional texture by using standard OpenGL function calls. The E_z field corresponds to red, H_x to green, and H_x to blue - alpha is not used. The E_z and H_{xy} update equations are stored as assembly codes (called fragment programs) in program memory. The programmable Vertex Processor is used to perform geometric transformations and lighting. We do not make use of its programmability as we are only rendering one static quad per time step – leaving the Vertex Processor with very little work to do.

In between the programmable Vertex and Fragment Processors is the hardware that performs the primitive assembly, rasterization, and interpolation. This part of the hardware takes the vertices and vertex attributes from the vertex processor and converts them into triangle primitives and then fragments. A fragment may be thought of as a pixel-like entity that has the potential to become a pixel, but has not yet done so because it may still be modified downstream. Also, very important to our application, each fragment has an attribute list which includes texture coordinates. In the OpenGL glBegin-glEnd block of our application, the four vertices of the quad are matched with the appropriate four corners of a two dimensional texture. The texture coordinates for each fragment are calculated using interpolation. These coordinates act as data memory addresses and allow the Fragment Processor to execute the appropriate

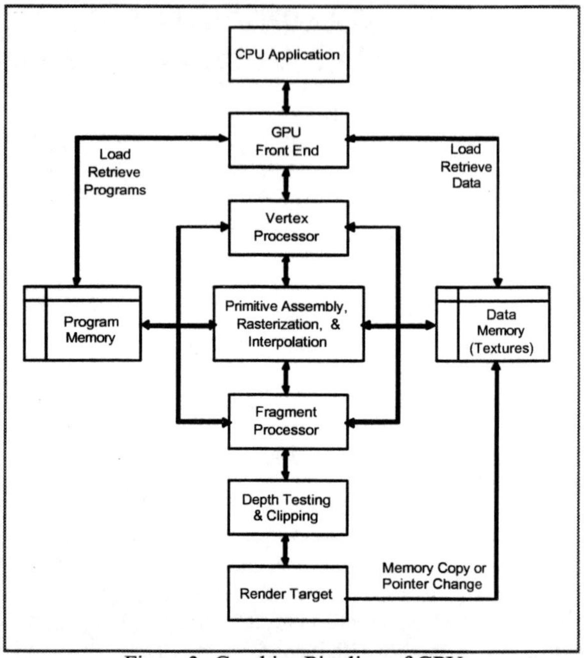

Figure 3: Graphics Pipeline of GPU

texture lookups thereby accessing the appropriate neighbors required in FDTD update equations. In addition to the texture lookups, the programmable Fragment Processor modifies each streaming fragment which in our case entails the execution of the FDTD update equations. This processor performs the vast majority of the work in our FDTD simulation.

Typically, the rendering target is the on screen buffer, called the frame buffer, which is displayed to the monitor. However, in our implementation the render target is an off screen buffer. In multi-pass rendering, the data in the render target is used as a source of data for subsequent passes. When the render target is the frame buffer, the OpenGL command glCopyTexSubImage2D can be used to copy the data from the frame buffer to texture memory. We use an off screen buffer to avoid the memory copy and simply bind the target as a texture and change the render target for the next pass. When a particular frame is to be displayed, the render target is changed to the frame buffer.

3.2. Precision, Range, and Storage

While older generations of GPUs severely limited precision, modern GPUs allow for full 32 bit IEEE floating point precision everywhere in the pipeline. The precision is provided because graphics programmers have demanded the support of sophisticated texture shading and rendering techniques applied in the middle of the pipeline, even though the final display on a typical monitor is only 8-bit. Several lower levels of precision alternatives exist, including 16-bit floating point and 8-bit fixed point. These other precision alternatives may improve performance, but not necessarily. For this paper, 8-bit fixed point arithmetic has been chosen. In future research, floating point precision will be investigated. With 8-bits of precision, only normalized FDTD simulations are possible. This means that the results of each update must be between zero and one. We implement normalization via impedance scaling and equations (1) and (3) become equations (4) and (5):

$$E_z^{t+\Delta t} = E_z^t + \frac{S_{fact}}{\varepsilon_r \sqrt{2}} \cdot \left[H_{x1}^{t+\frac{\Delta t}{2}} - H_{x2}^{t+\frac{\Delta t}{2}} + H_{y1}^{t+\frac{\Delta t}{2}} - H_{y2}^{t+\frac{\Delta t}{2}} \right] \quad (4)$$

$$H_{x,y}^{t+\frac{\Delta t}{2}} = H_{x,y}^{t-\frac{\Delta t}{2}} + \frac{S_{fact}}{\sqrt{2}} \cdot \left[E_{z1}^t - E_{z2}^t \right] \quad (5)$$

where, S_{fact} is the stability factor which we set to 0.9; and

ε_r is the relative permittivity at the E_z node.

If the real physical values are required, then de-normalization in the CPU is required.

Storage of data in system (non-GPU) memory is feasible and is initiated via simple OpenGL calls. The data is transferred from the GPU to system memory via the AGP bus which on the nVidia FX 5900 is AGP x8 (if the mother board supports it). Any significant usage of a memory copy, to system memory in this case, causes significant loss of performance. In some cases this performance degradation makes the use of the GPU for FDTD unfeasible. However, in many cases, it is not necessary to store data at every time step. Furthermore, it is typical that only fields at a few important points must be tracked, thereby drastically reducing the required bandwidth for storage purposes.

4. SIMPLE EM SCATTERING PROBLEM

Many practical 3-D electromagnetic problems reduce to 2-D problems, because of inherent symmetries. A simple 2-D electromagnetic scattering problem has been chosen for the speed comparisons of this paper and is described in [11]. Essentially, the problem consists of a dielectric cylinder, with relative permittivity equal to 4, of infinite length oriented in the z direction in free space as shown in Figure 4. A 2.6 GHz plane wave propagating in the y direction and allowing reflections to pass through it both ways is excited at a distance from the edge of the cylinder equal to the radius of the cylinder itself. In our context the absolute physical dimensions of the lattice are not important, but grid size is because it effects the speed of simulation. In our experimentation we vary the grid size from 128 x 128 to 2048 x 2048 as shown in the results section. The time steps of the simulation are 5 picoseconds.

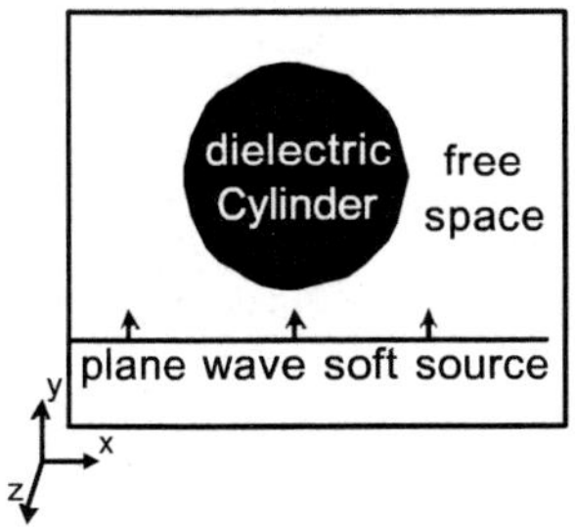

Figure 4: 2D Slice of Scattering Example

5. IMPLEMENTATION OF FDTD ON GPU

Update equations (4) and (5) are implemented on the GPU via the following steps.

A - Generate Two Texture Objects: One 2-dimensional texture is used for E_x, H_x, and H_y. In our implementation, real or physical zero corresponds to a color component of 0.5 in the graphics hardware. The plane wave source excitation is such that the peak/max and min/trough values for E and H never go outside one and zero. A second texture is created for the permittivity that varies across the lattice. Only the red component of the texture is used to store the permittivity, but other components could be used to store varying conductivity etc.

B - Load Fragment Programs for E_z and $H_{x,y}$: The update equations are loaded into program memory. The following assembly code constitutes the H_{xy} update equation:

```
!!FP1.0
DEFINE my_offset = {0.00390625, 0.0};
ADD R10, f[TEX0].xyxy, my_offset.yxxy;
TEX R0, f[TEX0], TEX0, 2D;
TEX R1.x, R10.xywz, TEX0, 2D;
TEX R2.x, R10.zwxy, TEX0, 2D;
ADD R3.y, R0.x, -R1.x;
ADD R3.z, R2.x, -R0.x;
MAD o[COLR].xyzw, {0.0, 0.17675, 0.17675, 0.0}.xyzw, R3.xyzw,
R0.xyzw;
END
```

This assembly code is written in compliance with the GL_NV_fragment_program specification detailed at the nVidia website [13]. The fragment program for the E_z update equation is similar in length and character. The plane wave source excitation is implemented by blending in a red line after the E_z pass. For this line only, the red component is set to the appropriate sinusoidal value with a call to glColor. Therefore, in order to simulate N time steps, 3N rendering passes are required (the 3rd one for the source is very fast).

C – Multi-Pass Rendering: In order to update both the E_z and $H_{x,y}$ fields, three passes are required, but on the third pass, which implements the plane wave source (i.e. the blended red line), the render target does not change. Figure 5 illustrates the approach of alternating rendering targets.

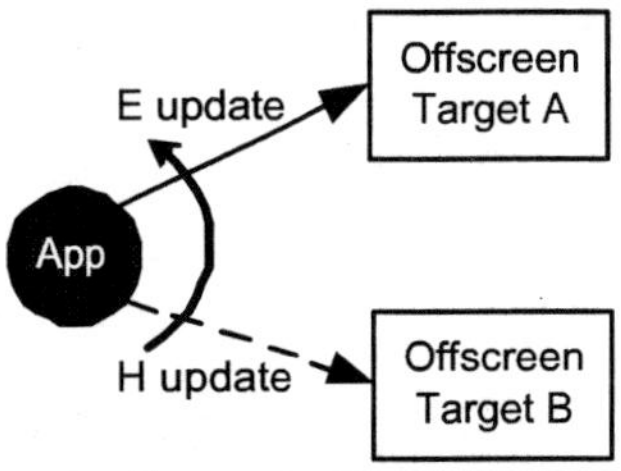

Figure 5: Alternating Rendering Targets

As the application updates the E_z field, rendering target A is pointed to. During this rendering pass, off screen rendering target B acts as a texture and is a source of data for the E_z update. The reverse is true when updating the $H_{x,y}$ fields. The vendor neutral WGL_ARB_render_texture extension [13] was used for this rendering approach.

6. GRAPHICAL DISPLAY

As previously stated, no display is necessary. However, it has been found that the inherent display capability of the GPU is convenient for debugging and may be useful to the user. Figure 7 and Figure 6 show the displays of the simulation after roughly 1500 time steps at grid sizes of 256 x 256.

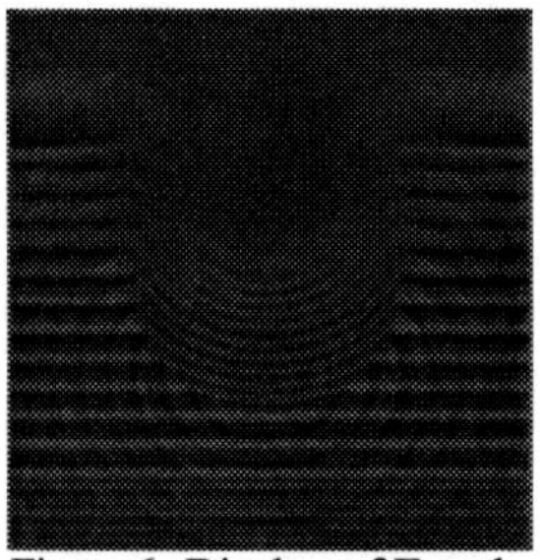

Figure 6: Display of E_z only

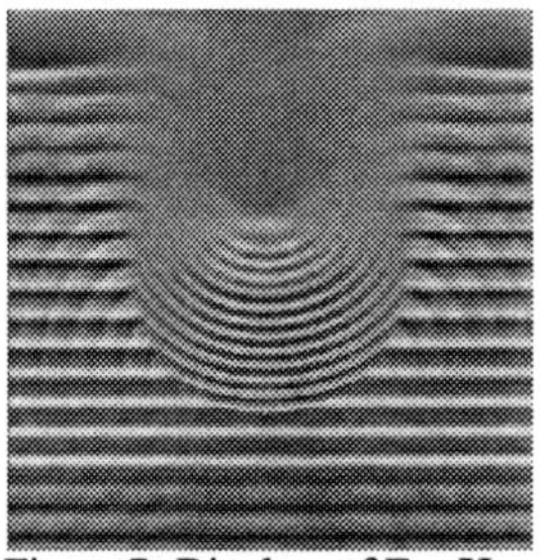

Figure 7: Display of E_z, $H_{x,y}$

Figure 6 shows only the E_z fields; peak E_z field values correspond to the bright red lines (white lines if viewing in black and white), while the dark lines corresponds to the trough field values represented by an absence of color. Figure 7 shows all three field components: E_z, H_x, and H_y correspond to the red, green, and blue components for each pixel. If the reader is viewing in color, the bright yellow lines correspond to peak E_z and H_x fields (i.e. red plus green equals yellow).

7. RESULTS

The CPU-based system used is an Intel P4 1.6 GHz machine with 256 MB RAM and AGP x4. The GPU used is the nVidia GeForce FX 5900 Ultra on a Gainward FX PowerPack board (with Golden Sample over-clocking). Table 1 contains a comparison of the execution time in seconds of the CPU relative to the GPU for 2000 time steps.

Table 1: Speedup Results for 2000 Time Steps

Grid Size	Time in Seconds		Speedup
	CPU	GPU	
128 x 128	1.98	1.47	1.41
256 x 256	10.6	1.77	6.00
512 x 512	63.5	6.24	10.2
1024 x 1024	255	24.6	10.4
2048 x 2048	1020	98.0	10.4

The execution time in seconds was measured by using the standard C time library exposed with the time.h file available in the MS Visual C++ 6.0 development environment. The GPU yielded a speedup of 1.41 and 6.0 for the 128 x 128 256 x 256 grid sizes respectively. At the higher grid sizes the speedup is roughly ten. There is a significant amount of overhead associated with changing the rendering target each pass. This overhead explains the low performance of the GPU at the smaller grid sizes whereby the FDTD computations constitute a smaller percentage of the work required to complete the simulation. However, at higher grid sizes the FDTD calculations outweigh this overhead. It is anticipated that as GPU architecture and driver improvements are made, the overhead in alternating rendering targets will be reduced significantly.

8. FUTURE WORK

The next steps in this vein of research are to (i) use higher levels of arithmetic precision; (ii) cultivate techniques to deal with more complex boundary conditions; (iii) develop the capability to execute 3-D FDTD simulations; (iv) build a more general framework allowing the user to easily modify simulation parameters; and (v) create a graphical interface that facilitates the inspection of particular nodes in real-time.

9. REFERENCES

[1] Taflove, Allen. "Advances in Computational Electrodynamics – The Finite Difference Time Domain Method". Norwood, MA: Artech House Inc., 1998.

[2] Marek, J. R., Mehalic, M. A. and Terzuoli A.J., "A dedicated VLSI architecture for Finite-Difference Time Domain calculations," in *Proc. of The 8th Annual Review of Progress in Applied Computational Electromagnetics*, Naval Postgraduate School, Monterey, CA, 1992.

[3] Schneider, R.N., Turner, L. E. and Okoniewski M.M.,"Application of FPGA technology to accelerate the Finite-Difference Time-Domain (FDTD) method," in *Proc. of TheTenth ACM International Symposium on Field-Programmable Gate Arrays*, Monterey, CA, 2002.

[4] Placidi, P., Verducci, L., Matrella, G., Roselli, L. and Ciampolini, P., "A custom VLSI architecture for the solution of FDTD equations," *IEICE Transactions on Electronics*, vol. E85-C, pp. 572-577, 2002.

[5] Verducci, L., Placidi, P., Matrella, G., Roselli, L., Alimenti, F., Ciampolini, P. and Scorzoni, A. , "A feasibility study about a custom hardware implementation of the FDTD algorithm," in *Proc. of The 27th General Assembly of theURSI*, Maastricht, Netherlands, 2002.

[6] Durbano, J. P., "Hardware implementation of a 1-dimensional Finite-Difference Time-Domain algorithm for the analysis of electromagnetic propagation," M.E.E. Thesis, Department of Electrical and ComputerEngineering, University of Delaware, Newark, USA, 2002.

[7] Harris, M. J., "Physically-Based Visual Simulation on Graphics Hardware", Graphics Hardware (2002), pp. 1-10

[8] Moreland, K., "The FFT on a GPU", Graphics Hardware (2003).

[9] Kim, T., "Visual Simulation of Ice Crystal Growth", Eurographics/SIGGRAPH Symposium on Computer Animation (2003).

[10] Yee, K.S., "Numerical Solution of initial boundary value problems solving Maxwell's equations inisotropic media," IEEE Trans. Antennas and Propagation, Vol. 14, 1966, pp.302-307

[11] Taflove, Allen, "Numerical solution of steady-state electromagnetic scattering problems usingthe time-dependent Maxwell's equations", IEEE Trans. Microwave Theory Technology. 23, 623 (1975).

[12] Fernando, R., "The Cg Tutorial," Addison-Wesley, April 2003.

[13] Kilgard, M. J, "NVIDIA OpenGL Extension Specifications", http://developer.nvidia.com, June 10, 2003.

A NEW MULTI-RAMP DRIVER MODEL WITH RLC INTERCONNECT LOAD

Lakshmi K. Vakati
University of Arizona
vakatil@ece.arizona.edu

Janet Wang
University of Arizona
wml@ece.arizona.edu

ABSTRACT

As the feature size is scaled down to 90 nm and below, fundamental modeling changes, such as the nonlinearity and higher frequencies of signals, require driver-load models to take into account propagation delay and slew rates. The conventional single Ceff (one-ramp) with lumped RC model is no longer accurate. In this paper we propose a new multi-ramp model with general RLC interconnects as loads. This new model accurately predicts both the 50% delay and the overall output waveform shape with inductance effects.

1. INTRODUCTION

The one-ramp single Ceff driver model [2][9] has been the cornerstone for driver delay and slope calculation. However, at 90 nm and below, it can generate delay errors and cause timing failures when used in conjunction with lumped RC load model. In order to capture the overall waveform shape at outputs, it is important to improve the accuracy of both the driver's model and the driven load's model.

Current work in driver-load modeling techniques have been focused on developing driver's model and output load model incoherently. Some important properties of nanometer designs, such as nonlinearity and higher frequency of the signals, have been neglected. As a result, both the inductance effect and the time of flight of propagated signals are not captured in the existing driver-load models. For example, in [10], driver load is modeled as RC circuit and a pi-model is synthesized by matching the first three moments of driving point admittance. However, the pi-model may no longer be synthesized if the inductive effects are significant. In [6], a new driving point model, similar in spirit to the RC pi-model, is proposed for on-chip interconnect wires with inductance. Yet, the model is based on matching the first four moments of driving point admittance. It may produce fake ringing and lose the time of flight effects. In [1], the authors proposed a two-ramp driver model to capture the time of flight effects. Then again, they model the driving point admittance by matching the first five moments. Their model may fail for interconnects with noticeable inductance effects.

In this paper, we present a new multi-ramp driver-load model. The driver's load is modeled as a distributed interconnect and is mapped to an effective capacitance value. Our model predicts both delay and slew accurately at no additional computation cost compared to [1].

The paper is organized as follows. We review the performance criteria to evaluate the importance of on-chip inductance and the concept of transmission line theory in the next section. Section 3 describes the proposed model including the implementation specifics and a summary of the modeling flow. In section 4 we present experimental results that show the effectiveness of the proposed model. And we conclude in section 5.

2. BACKGROUND

Deutsch et al. provided detailed conditional expressions to determine when distributed interconnect effects (usually modeled as transmission lines) are important for on-chip interconnects [3]. Consider a point to point net with length l, characteristic impedance Z_0, resistance, capacitance and inductance per unit length R, C and L, total capacitance C_T, and driven by a source impedance Z_{drv}. The line inductance needs to be included when

$$Z_{drv} < Z_0 \quad (1.a) \qquad \frac{Rl}{2Z_0} < 1 \quad (1.b)$$

$$T_r < 2T_f \quad (1.c) \qquad C_L << Cl, C_T \approx Cl \quad (1.d) \quad (1)$$

When the above criteria are satisfied inductance of the line can not be ignored. These criteria are satisfied in case of global interconnects whose resistance is less than 100 Ω/cm. The driver output waveform of one such interconnect is as shown in Figure 1.

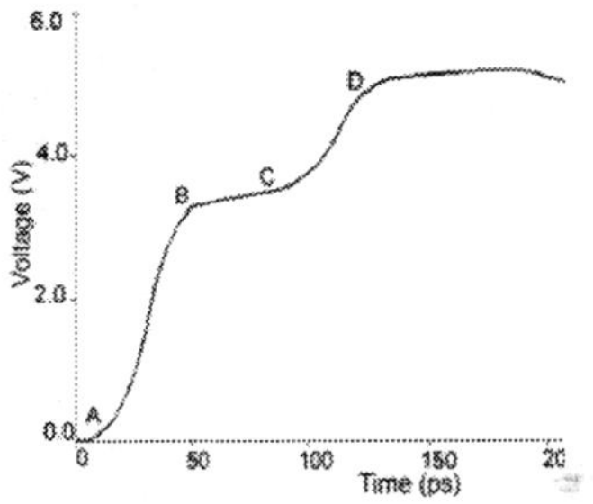

Figure 1. Driver output waveform of a 5mm RLC line driven by a 75X inverter. (R=70 ohm/cm, L=5.64 nH/cm, C=1.2 pF/cm)

As shown in Figure 1. the driver output waveform of an interconnect with significant inductance is not smooth as in RC cases and exhibits inflection points. The nature of the above waveform can be explained based on transmission line theory. First the output rises to an initial step (AB) and then, since the rise time is less or comparable to twice the time of flight delay, it waits for the reflections to return from the far end. The waiting period is $2T_f - T_r$ shown by the plateau (BC). Once the reflections return from the far end the output rises to another step (CD). The height of the initial step during the transition is given by,

$$\text{Height of initial step} = V_{DD} * f\text{, where } f = \frac{Z_0}{Z_0 + Z_{drv}} \quad (2)$$

Since Z_{drv} is less than Z_o the height of the initial step is greater than 50% of V_{DD}. The first reflection overshoots V_{DD} but the magnitude of the overshoot is in general negligible.

3. PROPOSED MODEL

From the previous discussion, it is clear that on-chip interconnects behave like transmission lines when the inductance criteria are satisfied. Hence we use admittance of a distributed lossy transmission line to model the driving point admittance. Ideally the driver output waveform should be modeled as a multipiecewise linear waveform or multi-ramp to capture the complete waveform including multiple reflections. But as the magnitude of the initial step is greater than 50% V_{DD}, only the first reflection is prominent. The overshoots and undershoots due to later reflections are negligible. As such we require a three-piece linear waveform to model the initial step, the plateau and the first reflection. The slope of the plateau depends on the equivalent capacitance of the transmission line. In most practical

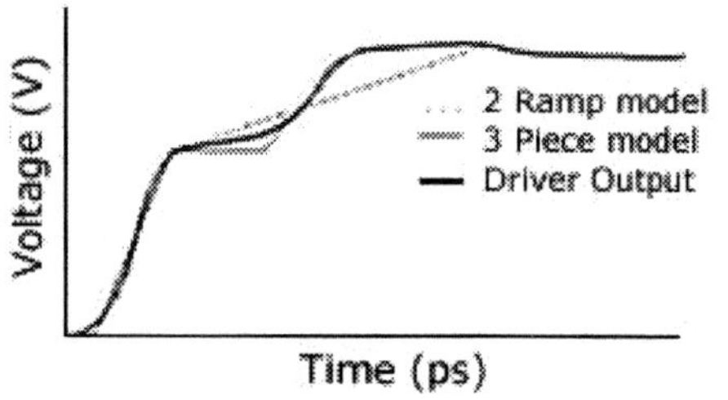

Figure 2. Driver output waveform, 3-piece and 2-ramp approximations

cases the plateau has a minor slope though not clear flat. So, as shown in Figure 2, representing the plateau by a flat step is more appropriate than the approach of finding a ramp that fits both the plateau and the first reflection [1]. The approach in [1] gives accurate delay prediction but may fail in accurate slew prediction. Though fitting the plateau with a linear piece rather than a flat step works better, it adds to the computational cost and it does not achieve noticeably better delay and slew accuracy. Hence we model the initial step by a ramp and the plateau by a flat step and the first reflection by another ramp. When inductive effects are not significant the entire transition can be modeled by one ramp. But if resistance shielding effects are predominant gate resistor model should be used to capture the exponential tail. Figure 3 shows our simplified driver output model. To model the driver output waveform as shown in Figure 3, we need to determine the slope of each ramp and the voltage breakpoint. Voltage breakpoint in our case is the voltage point the initial step rises to and is calculated by Equation 2.

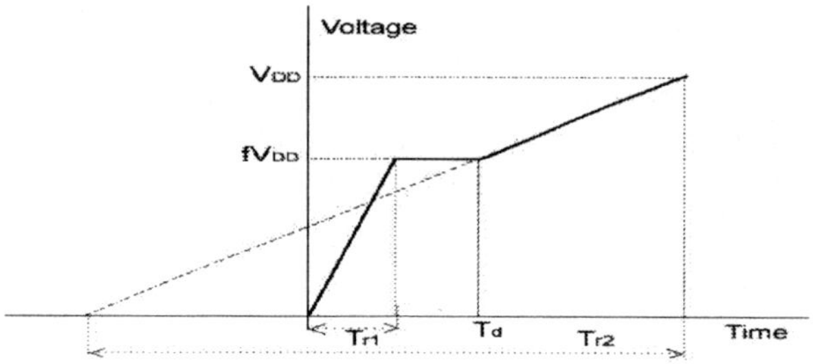

Figure 3. Simplified three-piece model of driver output waveform

The slope of the first ramp is (fV_{DD}/T_{r1}) and the slope of the second ramp is (V_{DD}/T_{r2}). The expressions are given by,

$$V(t) = fV_{DD}\frac{t}{T_{r1}} \qquad 0 < t < T_{r1}$$

$$V(t) = V_{DD}\frac{t}{T_{r2}} + \left(1 - \frac{T_d}{fT_{r2}}\right) fV_{DD} \qquad T_d < t < T_d + (1-f)T_{r2} \quad (3)$$

Here $T_d = 2T_f$, T_f is time of flight.

As we discussed in section 1, the interconnect model should be mapped to an effective capacitance value to resolve the incompatibility that exists between precharacterized look-up tables and RC/RLC loads. The approach to determine the effective capacitance value is to determine the capacitance load that has the same average current (therefore the same total charge transfer) as the transmission line model load. The charge averaging is for a finite period of time which we will refer to as the active region. The determination of this active region is critical to an accurate approximation. It was shown in [9] that equating the charge up to 50% point captures delay accurately, but fails in modeling the tail portion of the transition. Also equating the charge up to 100% point will not address this problem as this approach yields an averaged curve where both the delay and slew may be inaccurate [1]. A single effective capacitance cannot model the entire transition accurately. So we find two effective capacitances, where one capacitance models the first ramp and the second capacitance models the second ramp.

3.1 Implementation

The first effective capacitance is obtained by equating the charge transfer required by the admittance model and the capacitance value during the transition of the first ramp. The second capacitance is obtained by equating the charge transfer during the transition of the second ramp. We modeled the interconnect as a distributed lossy transmission line. At higher frequencies input impedance of the line is equal to the characteristic impedance. Hence we use characteristic admittance of the line in calculating Ceff. The characteristic admittance is given by

$$Y(s) = \sqrt{(sC+G)/(sL+R)} \quad (4)$$

$$Y(s) = \sqrt{C/L}\,\sqrt{[1+(G/C)y]/[1+(R/L)y]} \quad (5)$$

Where $y = \frac{1}{s}$. Y(s) can be expanded into a Maclaurin series of y around y = 0, or s = ∞. Therefore,

$$\sqrt{[1+(G/C)y]/[1+(R/L)y]} = 1 + m1y + m2y^2 + \cdots + mky^k + \cdot \quad (6)$$

where k-th moment is

$$m_k = \frac{\frac{d^k}{dy^k}\sqrt{[1+(G/C)y]/[1+(R/L)y]}}{k!} \quad (7)$$

we calculate the first five moments and use them directly instead of mapping them to a lower order model. And we experimentally found that the first three moments are sufficient to capture the time of flight effects. Several techniques related to stability problems in moment matching and realizable models have been explained in detail in [4]. For the first ramp,

$$V(s) = \frac{fVDD}{Tr1}\frac{1}{s^2} \quad (8)$$

The current delivered to the interconnect is given by

$$I(s) = V(s)Y(s) = \sqrt{\frac{C}{L}}\frac{fVDD1}{Tr1}\frac{1}{s^2}\left(\frac{s^5 + m1s^4 + m2s^3 + m3s^2 + m4s + m5}{s^5}\right) \quad (9)$$

The first effective capacitance C_{eff1} is the capacitance that requires the same charge transfer as the interconnect moments during the interval when the first ramp is in transition. Charge transferred to moments can be calculated by integrating I(t) from t = 0 to t = T_{r1}. The charge transfer associated with charging the effective capacitance for this interval is given by $C_{eff1}fV_{DD}$.

$$\int_0^{Tr1} I(t)dt = Ceff1 fV_{DD} \quad (10)$$

Solving the above equation for C_{eff1}

$$Ceff1 = \sqrt{\frac{C}{L}} \sum_{n=0}^{5} mn \frac{Tr1^{n+1}}{(n+2)!} \quad (11)$$

C_{eff1} can be obtained by iterating on T_{r1}. We start with an initial guess of C_{eff1} equal to the total capacitance and iteratively improve the effective capacitance until the value converges. T_{r1} at each step can be obtained from precharacterized cell information and the T_{r1} corresponding to the converged C_{eff1} is used to model the first ramp.

For the second ramp

$$V(s) = \frac{V_{DD}}{T_{r2}} \frac{1}{s^2} + \left(1 - \frac{T_d}{fT_{r2}}\right) fV_{DD} \frac{1}{s} \quad (12)$$

Charge transferred to moments can be calculated by integrating I(t) from t = T_d to t = T_d + (1-f)T_{r2}. The charge associated with charging the effective capacitance for this interval is given by $C_{eff2}(1-f)V_{DD}$.

$$\int_{Td}^{Td+(1-f)Tr2} I(t)dt = C_{eff2}(1-f)V_{DD} \quad (13)$$

C_{eff2} can be calculated by iterating on T_{r2}. we could not present all equations in detail due to space limitation. However T_{r2} calculated this way needs to be modified. The idea behind modeling the plateau by a flat step is that there is no charge transfer during this period ($T_d - T_{r1}$). Since effective capacitance is calculated by equating charge transfer, the T_{r2} corresponding to the converged C_{eff2}, should be modified to include this plateau time.

$$T_{r2n} = T_{r2} + T_d - T_{r1} \quad (14)$$

The modeling of driver output waveform requires calculating voltage breakpoint apart from determining slope of the two ramps. In order to model the voltage breakpoint we need to calculate on-resistance of the driver. We model on-resistance by a similar approach as adopted by thevinin models [11].

$$Rs = \frac{t_{90}(t_{in}, C_L) - t_{50}(t_{in}, C_L)}{\ln 5C_L} \quad (15)$$

Ideally one should find an effective capacitance and calculate on-resistance of the driver for this value of the load capacitance. However resistance does not change significantly by using total capacitance instead of effective capacitance [1].

3.2 Summary of modeling flow

Given the following information:

1. Line parasitics
2. precharacterized delay table for the driver

Steps for modeling driver output:

1. Find admittance moments m_1, m_2, m_3, m_4 and m_5 using (7)
2. Find driver on-resistance by (15) and calculate voltage breakpoint using (2).
3. Perform C_{eff1} iterations using (11) and compute T_{r1}
4. Check inductance criteria using (1).
   If inductance is significant
   - Model plateau by a flat step for a period of T_d - T_{r1}
   - Perform C_{eff2} iterations and compute T_{r2}
   - Modify T_{r2} to T_{r2n} using (14)
   - Model the driver output waveform using T_{r1}, T_{r2n} and voltage breakpoint.

   If inductance is not significant
   - Perform C_{eff} iterations using (11) with f=1 and compute T_r.
   - Model the output as a single ramp.

4. RESULTS

We tested our three-piece model by sweeping line lengths from 4mm to 10 mm and line widths from 1.2µm to 2.6µm. We swept driver strengths from 50X to 125X and input transition from 50ps to 150ps. We performed all experiments using 0.18µm CMOS technology. First we compare driving point waveforms obtained by our model with that of two-ramp model and SPICE simulations. We classified our test cases into low and high inductance nets. Table 1 and 2 show a set of cases with low and noticeable inductance effects. SPICE delay and slew numbers are compared with our new model, two-ramp and single ramp modeling results. From the table it is clear that for noticeable inductance effect nets, slew predictions of 2 ramp modeling exhibits substantial error and our multi-ramp model provided good results. In our model, the average error in delay was 9% and the average error in the slew rates was 2.2%.

Table 1. SPICE, new model, two-ramp[1] and one-ramp model comparison results for nets with inductance effects

Len/ Wid mm /µm	Line parasitics R(Ω) L(nH) C(pF)	Driver Size	Input Slew (ps)	Delay (ps)				Slew (ps)			
				SPICE	new model (%error)	2 ramp model [1] (% error)	1 ramp model (%error)	SPICE	new model (%error)	2 ramp model [1] (% error)	1 ramp model (% error)
4/1.2	75/5.9/1.42	100X	50	33.33	30.64 (-8.1%)	28.85 (-13.4%)	41.38 (24.2%)	112.16	102.16 (-8.9%)	138.36 (23.4%)	82.77 (26.2%)
4/1.4	67.7/5.3/1.52	100X	50	34.61	31.30 (-9.5%)	29.47 (-14.8%)	41.69 (20.45%)	113.51	103.46 (-8.8%)	141.69 (24.8%)	83.39 (-26.5%)

4/1.6	56.3/4.98/1.67	100X	50	36.03	32.58 (-9.6%)	30.67 (-14.9%)	43.68 (21.2%)	114.86	110.60 (-3.7%)	143.58 (25.0%)	84.36 (-26.6%)
5/1.2	81.5/6.44/1.57	100X	50	33.68	29.91 (-11.9%)	28.36 (-14.3%)	51.28 (52.2%)	140.09	137.03 (-2.2%)	172.91 (23.4%)	102.57 (-26.8%)
5/1.6	72/6.40/1.61	100X	50	34.12	30.32 (-11.1%)	29.78 (-12.72)	52.15 (52.84%)	143.69	140.04 (-2.5%)	180.25 (25.4%)	104.30 (-27.4%)
5/2.2	56/6.3/1.76	100X	50	34.69	30.86 (-11.0%)	29.92 (-13.7%)	53.56 (54.39%)	144.59	145.60 (0.7%)	183.98 (27.2%)	106.11 (-26.6%)
6/1.6	94/7.82/1.64	100X	50	32.88	29.69 (-9.7%)	28.65 (-12.8%)	62.95 (91.4%)	180.08	179.12 (-0.5%)	224.57 (24.7%)	125.91 (-30.1%)
6/2.2	77.4/7.28/1.80	100X	50	34.16	31.07 (-9.0%)	30.92 (-9.5%)	65.55 (91.9%)	185.14	185.97 (0.4%)	237.63 (28.3%)	131.11 (-29.2%)
6/2.4	62.8/6.74/1.92	100X	50	35.14	32.28 (-8.1%)	31.26 (-11.0%)	66.99 (90.63%)	189.18	190.33 (0.6%)	240.27 (27.0%)	134.00 (-29.2%)

Table 2. SPICE, new model and one-ramp model comparison results for low inductance nets

Len/Wid mm/µm	Line Parasitics R(Ω)/L(nH)/C(pF)	Driver Size	Input Slew (ps)	Delay(ps)			Slew(ps)		
				SPICE	new model (% error)	1 ramp model (% error)	SPICE	new model (% error)	1 ramp model (% error)
5/1.2	93.7/5.3/1	100X	50	31.62	28.26 (-10.6%)	39.45 (24.7%)	109.91	96.29 (-12.4%)	78.91 (-28.2%)
5/2.5	49.5/4.8/1.31	100X	50	34.60	30.65 (-11.4%)	44.79 (29.5%)	118.91	114.33 (-3.8%)	89.58 (-24.7%)
6/1.2	91/6.3/1.19	75X	50	35.92	32.53 (-9.4%)	56.12 (56.2%)	153.15	149.80 (-2.2%)	122.25 (-20.2%)
6/2	71.6/6/1.46	75X	100	74.32	65.38 (-12.0%)	95.63 (28.6%)	209.46	186.01 (-11.2%)	154.83 (-26.1%)
6/2.5	59.3/5.8/1.58	75X	100	76.57	67.45 (-11.9%)	98.46 (28.6)	211.71	198.49 (-6.2%)	161.27 (-23.8%)

5. CONCLUSIONS

In this paper we proposed a three-piece model based on transmission line theory that accurately predicts delay and slew for both low and high inductance nets. Results show that our three-piece model significantly reduces the error incurred due to the approach of fitting a single ramp for both plateau and first reflection.

6. REFERENCES

[1] Kanak Agarwal, Dennis Sylvester, and David Blaauw, "An effective capacitance based driver output model for on-chip RLC interconnects," *Design Automation Conference*, 2003, pp. 376-381

[2] R. Arunachalam, F. Dartu, and L.T. Pileggi, "CMOS gate delay models for general RLC loading," Int. Conf Computer Design, 1997, pp. 224-229

[3] A. Deutsch *et al.*, "When are transmission line effects important for on-chip interconnections?" *IEEE Trans. on Microwave Theory and Techniques*, 45, (Oct. 1997), 1836-1846.

[4] S. Lin and E. S. Kuh, "Pade approximation applied to lossy transmission line circuit simulation," *Int Symp. Circuits and Systems*, 1992, pp. 93-96.

[5] S. Lin, N. Chang, and O.S. Nakagawa, "Quick on-chip self and mutual inductance screen," *Int. Symp . Quality Electronic Design*, 2000, pp. 513-520.

[6] C.V. Kashyap and B.L. Krauter, "A realizable driving point model for on-chip interconnect with inductance," *Design Automation Conference*, 2000, pp. 190-195.

[7] B. Krauter, S. Mehrotra, and V. Chandramouli, "Including inductive effects in interconnect timing analysis," *Custom Integrated Circuits Conference*, 1999, pp. 445 -452.

[8] Y. Ismail, E. Friedman, and J. Neves, "Performance criteria for evaluating the importance of on-chip inductance," *Int Symp.Circuits and Systems*, 1998, pp. 244-247.

[9] J. Qian, S. Pullela, and L.T. Pillage, "Modeling the effective capacitance for the RC interconnect of CMOS gates," *IEEE Trans. CAD*, 13, (Dec 1994), pp. 1526-1535.

[10] P.R. O'Brien and T.L. Savarino, "Modeling the driving point characteristic of resistive interconnect for accurate delay estimation," *Int. Conf. Computer Aided Design*, 1989, pp. 512-515.

[11] F. Dartu, N. Menezes, and L.T. Pileggi, "Performance computation for pre-characterized CMOS gates with RC loads," *IEEE Trans. CAD*, 15, (May 1996), pp. 544-553.

[12] L.T. Pillage and R. Rohrer, "Asymptotic waveform evaluation for timing analysis," *IEEE Trans. CAD*, 9, (April 1990), pp. 352-366.

A COMPACT OPTIMIZATION METHODOLOGY FOR SINGLE-ENDED LNA

Gülin Tulunay and Sina Balkır

Department of Electrical Engineering, 209N WSEC,
University of Nebraska-Lincoln, Lincoln, NE 68588-0511, USA
e-mail: gulin@cruiser.unl.edu

ABSTRACT

An equation-based method for the optimization and design of single-ended low noise amplifiers (LNA) is presented. The performance metrics of the LNA such as gain, noise figure and input impedance are formulated in terms of the design variables. The parasitics are included early in the design process and the optimal LNA designs satisfying the required specifications are found by the optimizer. Optimal LNA designs based on a 0.35μm CMOS technology obtained by the presented approach are further verified by the Spectre simulator of the Cadence design environment. The specifications and performance of the designs are found to be in close agreement, validating the compact optimization technique presented in this work.

1. INTRODUCTION

Low noise amplifiers (LNA) are one of the key building blocks for RF receivers. They play a critical role in determining the overall system noise figure (NF) of the receiver. The main function of an LNA is to provide sufficient gain to overcome the noise of subsequent stages (e.g. mixers) while adding as little noise as possible.

Although several CMOS RF building blocks achieving tough requirements have been recently realized [1]-[4], there is no systematic design methodology that accounts for many specifications. An optimal design requires the simultaneous consideration of all specifications. Both active and passive devices with their parasitic elements should be taken into account during the design procedure so that stable designs can be achieved. Optimization of RF ICs has been addressed by researchers in the past [5,9]. However, results reported often employ compact but simplified expressions for the performance specifications and cannot directly match the simulation results of commercially available CAD tools. Hence, it is desirable to develop an optimization methodology that can provide an accurate starting point and consequently minimize the design time in a professional RF CAD environment.

In this paper, we use the simulated annealing algorithm in developing an equation-based optimization tool, that includes the effects of the parasitics during the optimization procedure. We use this methodology to optimize an inductively degenerated single-ended LNA topology. For inductor modeling, ASITIC [8] is used. The final design is further verified by the Spectre simulator of Cadence, and it is shown that the desired specifications are also met on a commercial grade platform.

In section 2, the single-ended inductively degenerated common-source LNA is analyzed and the important design goals that are aimed to be satisfied are outlined. In section 3, the inductor model used is presented and the equivalent circuit model including the inductor model and simple transistor models is shown. In section 4, the optimization flow and the overall design procedure are explained. In section 5, the simulation results of the designed LNA are given. Finally, section 6 concludes the paper.

2. LNA ARCHITECTURE

There are a number of different LNA topologies that have different advantages and disadvantages. Among these different topologies, the one with inductive source degeneration is chosen since it offers the possibility of achieving the best noise performance of any architecture[10]. The use of a cascode circuit can reduce the influence of gate-to-drain overlap capacitance C_{gd} and improve reverse isolation (S_{12}) of LNA. In Fig 1 the inductively degenerated common-source single-ended LNA is shown.

With such a source inductance, a real term in the input impedance can be generated without the need of real resistances which degrade the noise performance. However, resistive losses associated with spiral inductors are prohibitively large and for that reason, they also degrade the noise performance. Thus, the series resistances of these on-chip inductors should be incorporated into the optimization procedure.

The design of LNAs encompasses several goals. These include minimizing the noise figure of the amplifier, providing sufficient gain and providing a stable 50Ω input impedance to terminate a filter or a length of transmission line which delivers signal from antenna to the amplifier.

In the design of LNA, the channel width of the input

0-7803-8251-X/04/$17.00 ©2004 IEEE

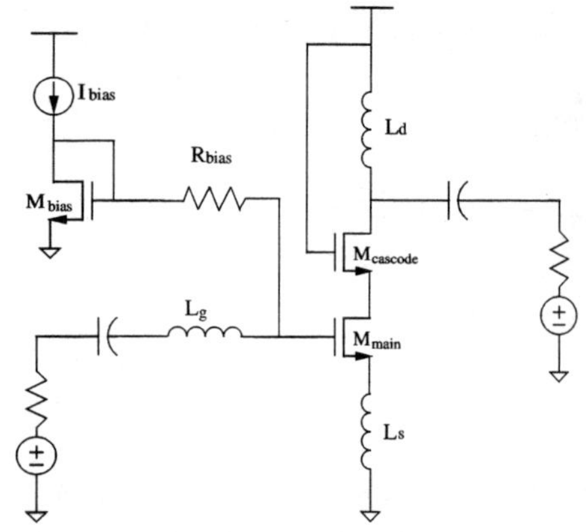

Fig. 1. The inductively source degenerated LNA

transistor is very critical, as it governs the LNA performance. It is important to minimize the noise contribution of the cascode device by minimizing the capacitance at its source. Merging the source of the cascode device with the drain of the input device proves to be an effective technique for reducing this capacitance[10]. Hence, the widths of the input transistor and the cascode transistor are chosen to be equal.

Width of the input transistor W_{main} together with the values of the load inductance L_d, the gate inductance L_g, the source inductance L_s and the bias current I_{bias} yields a total of five design variables.

3. INDUCTOR MODELING AND SMALL SIGNAL ANALYSIS

In this work, ASITIC is used for the modeling of inductors. From ASITIC, one can obtain the equivalent circuit model of an inductor including the value of the series resistance. This data obtained from ASITIC is then used to form look-up tables for each on-chip inductor. Since the gate inductance L_g is large, it will be an off-chip component.

The equivalent π circuit of a spiral is given in Fig 2.

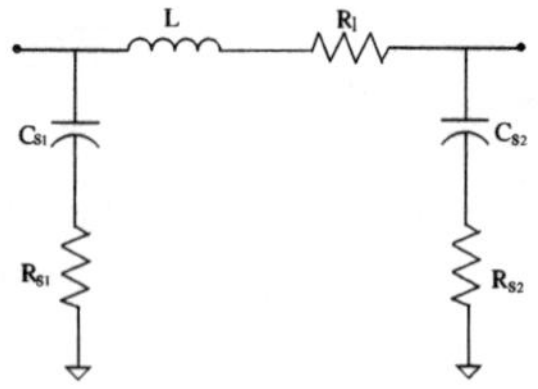

Fig. 2. The equivalent π circuit of a spiral inductor

Using ASITIC, two look-up tables are created for the load inductance L_d and the source inductance L_s. Each look-up table includes a range of different inductor values. For each individual inductor value the parasitics and the area of that inductor are also included in the look-up tables. Therefore, when the optimizer randomly picks an L_s and an L_d value from these look-up tables, it will also be able to get the information related to the parasitics and the areas of those chosen inductors.

Given this equivalent circuit model for a spiral inductor, and simple transistor models, the small signal equivalent circuit of the LNA will be as in Fig 3.

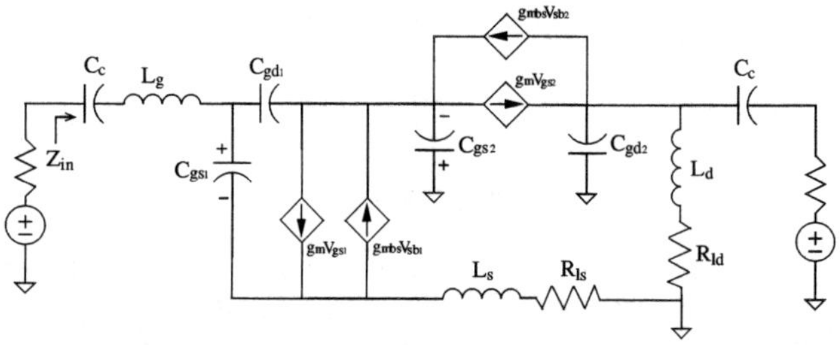

Fig. 3. The small signal equivalent circuit of the LNA

Using this equivalent circuit, the required equations for the optimizer can be obtained. In these equations the effects of C_{s1}, R_{s1}, C_{s2} and R_{s2} are ignored to avoid complex equations. Simulations using Spectre are carried out to observe the effects of these parasitics on the circuit performance. It is observed that the major parasitic element of an inductor that affects the performance of the LNA is its resistive loss due to substrate and metal layers used to form the inductor.

Small signal analysis of the circuit given in Fig 3, yields the equations for the noise figure (NF), input impedance (Z_{in}) and forward gain (S_{21}) in terms of the design variables as follows:

$$K_1 = \frac{(sL_s + R_{ls})(gm + sC_{gs})}{1 + gmbs(sL_s + R_{ls})} \quad (1)$$

$$K_2 = gm + gmbs + sC_{gs} \quad (2)$$

$$K_3 = \frac{gm + gmbs + s(C_{gs} + C_{gd})}{sC_{gd}} \quad (3)$$

$$K_4 = gm + sC_{gs} - K_1 gmbs + \left(1 + K_1 + \frac{C_{gs}}{C_{gd}}\right) K_2 \quad (4)$$

$$K_5 = (gm + gmbs)\left[\left(1 + K_1 + \frac{C_{gs}}{C_{gd}}\right)\frac{K_3}{K_4} - \frac{1}{sC_{gd}}\right] \quad (5)$$

$$Z_{in} = sL_g + \frac{1}{sC_c} + \frac{(1 + K_1)K_3}{K_4} \quad (6)$$

$$Z_{load} = \frac{\left[\left(R_s + \frac{1}{sC_c}\right) // (R_{ld} + sL_d)\right]}{1 + sC_{gd}\left[\left(R_s + \frac{1}{sC_c}\right) // (R_{ld} + sL_d)\right]} \quad (7)$$

$$|S_{21}| = \left| \frac{K_5 Z_{load}}{Z_{in} + R_s} \right| \quad (8)$$

$$NF = \frac{V_{os} + V_{od} + V_{ols} + V_{old} + V_{obias} + V_{ocas}}{V_{os}} \quad (9)$$

where V_{os} , V_{od}, V_{ols}, V_{old} , V_{obias} and V_{ocas} are the output referred thermal noise caused by the source resistance, main transistor, parasitic resistances of the inductors L_s and L_d, the bias resistance R_{bias} and cascode the transistor, respectively.

4. DESIGN-OPTIMIZATION FLOW

For optimization, the well-known simulated annealing algorithm is used. Annealing gives the system possibility to jump out of local minima with a reasonable probability. Hence, the probability of finding the global minima is high.

In this work, five design variables are chosen as explained in section 2. The optimization begins with the random selection of these five parameters. Then, a cost function value specific to this set of design variables, is computed. If the value of the cost function is acceptable, then the optimization process terminates. Otherwise, a new set of design variables is chosen, and a new cost function value associated with this new set of design variables is computed. As the number of iterations increases, the solution approaches to the optimum solution that minimizes the cost function most.

One of the most important issues in optimization is determining the cost function. The optimization of an LNA or in general, the optimization of any circuit is a constrained optimization problem, which can be written as follows [7]:

$$\min f(x) \text{ subject to } \begin{array}{ll} c_i(x) = 0 & i \in \mathcal{E} \\ c_i(x) \geq 0 & i \in \mathcal{I} \end{array} \quad (10)$$

where, $\mathcal{I}$ and $\mathcal{E}$ are the sets of inequality and equality constraints, respectively. f is the objective function to be minimized. In constrained optimization, we are confronted with the often conflicting goals of reducing the objective function and satisfying the constraints. Cost function is a measure of the balance between these two goals. In an unconstrained optimization problem, we can take the objective function as the cost function. However, if we have some constraints, and if we want the solution to be in the feasible region, then we have to define a different cost function to make the iterates satisfy the constraints.

A widely used cost function to assess the quality of the iterates is given below:

$$\phi(x;\mu) = f(x) + \frac{1}{\mu}\sum_{i\epsilon\mathcal{E}} |c_i(x)| + \frac{1}{\mu}\sum_{i\epsilon\mathcal{I}} [c_i(x)]^- \quad (11)$$

where, $x \,\epsilon\, \Re^n$ is a real vector with $n \geq 1$ design variables, $[c_i(x)]^- = max\{0, -x\}$ and the positive scalar μ is the penalty parameter determining the weight that we assign to constraint satisfaction relative to minimization of the objective.

In section 3, the design goals are formulated as an objective function in terms of the design variables. In this section, first the LNA design specifications will be formulated as equality and inequality constraints, and then, all of these LNA design objectives and constraints will be turned into a cost function, to be minimized by the simulated annealing algorithm.

The most important design goals of an LNA are considered as NF, S_{21} and S_{11}. Thus, we have three objective functions. The constraints are determined according to the LNA specifications. In this design, four of them are considered. The inequality constraints are given as follows:

- *Power*: If the power consumption of the bias circuitry is ignored, power can be approximated as

$$P = V_{dd} \cdot I_d \quad (12)$$

- *Area*: The area of the LNA on silicon can be approximated as the sum of transistor and inductor areas. This constraint can be used to impose an upper bound on the die area of the LNA.

$$Total\ Area = A(L_s) + A(L_d) + 2 \cdot W\ L \quad (13)$$

where, W and L are the gate width and length, respectively.

The equality constraints aim to satisfy input matching. Input matching requires two equality constraints to be satisfied at resonance frequency.

- The real part of the input impedance

$$\Re\{Z_{in}\} = 50\Omega \quad (14)$$

- The imaginary part of the input impedance

$$\Im\{Z_{in}\} = 0 \quad (15)$$

Finally, combining all of the objective functions and constraints, the cost function of the LNA will be as follows:

$$\phi = NF - |S_{21}| + \frac{1}{20}|c_1| + \frac{1}{2}|c_2| + \frac{1}{6}[c_3]^- + \frac{1}{6}[c_4]^- \quad (16)$$

where, c_1 and c_2 are the equality constraints related to input matching given as,

$$c_1 = \Im\{Z_{in}\} \quad c_2 = \Re\{Z_{in}\} - 50$$

and c_3 and c_4 are the inequality constraints related to area and power consumption, respectively, given as,

$$c_3 = \text{Upper bound on area} - \text{Total Area}$$

c_4 = Upper bound on power $- P$

where *Total Area* and P are as given in equations 13 and 12. The weights of the constraints in equation 16 are determined iteratively to get the best solution. The goal of the optimization algorithm is to minimize this cost function which in turn means to minimize NF and maximize S_{21} while satisfying the given constraints.

5. RESULTS

The optimal design obtained from the equation-based optimizer is shown in Table 1.

W_{main}	I_d	L_s	L_d	L_g
450μm	6mA	0.919nH	5.11nH	49nH

Table 1. The optimal design values

The upper bounds for power consumption and area are taken as $20mW$ and $0.07mm^2$, respectively. The performance of this optimal design is given in Table 2, which shows that all of the given constraints are satisfied.

$\vert S_{21}\vert$	13.96 dB
$\vert S_{11}\vert$	- 43.66 dB
NF	0.86 dB
Power	19.8 mW < 20 mW
Area	0.0675 mm^2 < 0.07 mm^2
$\Re\{Z_{in}\}$	50.13 $\approx$ 50 Ω
$\Im\{Z_{in}\}$	- 0.64 $\approx$ 0

Table 2. The performance of the optimal LNA

For a comparison of the values in Table 2, with the simulation results of Spectre circuit simulator, the circuit given in Fig 1 together with the optimal design values given in Table 1 and the resistive losses of the inductors L_s and L_d, is simulated by using the Spectre simulator of Cadence. The waveforms of $|S_{11}|$ and $|S_{21}|$ obtained from the simulator are given in Fig 4. At 900MHz, NF=1.02 dB, $|S_{21}|$=13.03 dB, $|S_{11}|$= -12.07 dB.

Due to the approximations in the equations, the simulated resonance frequency is slightly different than 900MHz. However, one can set the resonance frequency equal to the desired frequency, by fine tuning the off-chip component L_g, which is frequently done in practice.

6. CONCLUSION

In this paper, optimization of single-ended, inductively degenerated LNAs is demonstrated. The optimizer is based on very accurate and yet compact equations which are also in close agreement with the simulation results. Optimal LNA designs found satisfy the required specifications while achieving a reasonable gain and noise figure at the desired frequency. By tuning the off-chip inductor, the performance of the LNA circuit can be further improved without making any modification in the original design obtained from the optimizer. Consequently, this method cuts down the required design time significantly by minimizing the design iteration steps required in a commonly used CAD environment.

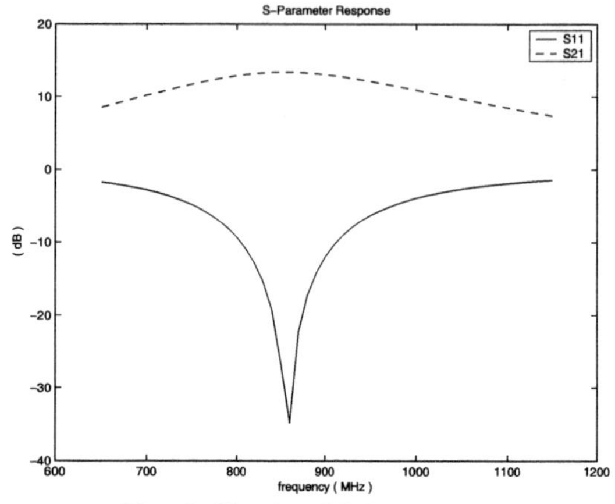

Fig. 4. Simulated S-Parameters

7. REFERENCES

[1] Gil J., Han K. and Shin H. "13 GHz 4.67 dB NF CMOS low-noise amplifier," *Elec. Letters*, vol.39, pp. 284-286, 2003.

[2] Taris T., Begueret J.B., Lapuyade H. and Deval Y. "A 1-V 2GHz VLSI CMOS Low Noise Amplifier," *RFIC Symposium*, pp. 527-530, 2003.

[3] Gatta F., Sacchi E., Suelto F., Vilmercati P. and Castello R., "A 2-dB noise figure 900-MHz differential CMOS LNA," *IEEE J. Solid-State Circuits*, vol. 36, pp. 1444-1452, 2001.

[4] Huang J.C., Weng R.M., Chang C.C., Hsu K. and Lin K.Y. , "A 2V 2.4GHz Fully Integrated CMOS LNA," *ISCAS 2001*, vol. 4, pp. 466-469, 2001.

[5] Gupta R. and Allstot D.J., "Parasitic-Aware Design and Optimization of CMOS RF Integrated Circuits," *IEEE MTT-S Int. Microwave Symposium*, vol. 3, pp. 1867-1870, June 1998.

[6] Ge J.Y. and Mayaram K. , "A Comparative Analysis of CMOS Low Noise Amplifiers for RF Applications," *ISCAS 1998*, vol. 4, pp. 349-352, 1998.

[7] Nocedal J. and Wright S.J. , *Numerical Optimization*, Springer-Verlag, New York, 1999.

[8] Niknejad A.M. and Meyer R.G., "Analysis, Design and Optimization of Spiral Inductors and Transformers for Si RF ICs," *IEEE J. Solid-State Circuits*, vol. 33, pp. 1470-1481, 1998.

[9] Vancorenland P., Ranter C.D., Steyaert M. and Gielen G. "Optimal RF Design using Smart Evolutionary Algorithms," *Design Automation Conf. 2000*, pp. 7-10, 2000.

[10] Derek K. Shaeffer, *The Design and Implementation of Low-Power CMOS Radio Receivers*, PhD Dissertation, Stanford University, 1998.

X RAY AND BLUE PRINT: TOOLS FOR MOSFET ANALOG CIRCUIT DESIGN ADDRESSING SHORT-CHANNEL EFFECTS

R. L. Oliveira Pinto[1], F. Maloberti[2]

[1]Texas A & M University, USA
[2]The University of Texas at Dallas, USA
rodrigo@ee.tamu.edu

ABSTRACT

An automatic engine built in the Java language designs transistors in few and easy steps. The results come from a combination of parameter extraction and simple analytical models that address the short-channel effects by referring them to the Early voltage. The outcome is precise output conductance, thermal noise coefficient and thermal noise from a friendly graphical user interface. Design and simulations of common-source amplifiers show the efficiency of the automation.

1. INTRODUCTION

Automation of analog circuits has been a hot topic for the past few years [1, 2], yet the analog designers still rely basically on circuit simulators and layout editors to accomplish their goals. There are solutions for complete sizing of MOSFET circuits, however, requiring manual modeling for further automation [3]. Circuit level design tools hide their methods from the analog designers who prefer to use their own design procedures [1]. Here we propose automation at transistor level by providing access to precise values of the main parameters of an amplifier to be used at the circuit level design.

Two design tools implemented in Java aid the designer to size MOSFET's for analog circuits. The first tool, which is called X Ray, extracts amplifier parameters mainly based on the inversion level [4, 5, 6, 7]. The second tool, which is called Blue Print, combines the data points from X Ray and analytical expressions [8] to size the transistors.

This paper describes the tools and the basis of the automation [8] as well as presents designs and simulation results to validate the method.

2. X RAY

X Ray is a program that extracts the Early voltage VA, thermal noise coefficient γ, and thermal noise i_d^2 based on a given range of inversion level i_f, drain-to-source-voltage V_{DS} and channel length L, for a specific technology, temperature T and channel type (N or P). The correct values of Early voltage are fundamental to design DC gain. Also, good prediction of the thermal noise is important to properly reduce its value in any amplifier design.

The main parameter extracted is the Early voltage for each point of each drain current because it reflects all the short channel effects modeled in the mobility and channel length [8], as equations (1) to (3) demonstrate. Equation (1) is the drain current I_D for hand design purposes. It is function of the gate-to-source voltage V_{GS} and drain-to-source voltage V_{DS} working with a constant Early voltage. V_{DSsat} is V_{DS} in the saturation point. μ is the mobility, C'_{ox} is the oxide capacitance per unit area, and V_T is the threshold voltage. The geometry is given by the channel width W and length L. Equation (2) is a simplified model of I_D for simulation purposes where μ and L are replaced by μ_{EFF} and L_{EFF}, respectively effective mobility and length, both containing short-channel effects. From (1) and (2) it is straightforward to find (3) and see that it is possible to refer the short-channel effects of μ_{EFF} and L_{EFF} to VA.

$$I_D = \frac{\mu C'_{OX}}{2}\frac{W}{L}(V_{GS}-V_T)^2\left(\frac{V_{DS}-VA}{V_{DSsat}-VA}\right) \quad (1)$$

$$I_D = \left(\frac{\mu_{EFF}}{L_{EFF}}\right)\frac{C'_{OX}}{2}W(V_{GS}-V_T)^2 \quad (2)$$

$$\frac{\mu_{eff}}{L_{eff}} = \frac{\mu}{L}\left(\frac{V_{DS}-VA}{V_{DSsat}-VA}\right) \quad (3)$$

The designer can work directly with points of VA during the hand design instead of complicated models of μ_{EFF} and L_{EFF}, however, the problem now is how to determine the space of extraction of VA that is applicable to as many designs as possible. The Early voltage depends on i_f, V_{DS}, L and T [9]. The inversion level and drain-to-source voltage have practical physical limits that naturally establish a space of extraction: i_f between 1 and 1000, which means from weak to strong inversion, and V_{DS} between 0V and the maximum voltage supply. We can normalize the channel length by the minimum length L_{MIN} and define a maximum normalized length for a particular design, for instance L/L_{MIN} equals to 9. The temperature is kept fixed. Once the points of VA are extracted they can be stored and used for several designs because i_f is

R. L. Oliveira Pinto is sponsored by CAPES, research agency of the Brazilian Ministries of Science and Technology and Education.

independent of geometry [4], and VA is not a function of the channel width W [9].

For design purposes, all the transistors that have the same i_f can share the same set of curves. The geometry is adjusted according to the drain current needed by using equations (4) and (5) [4], where I_S is the normalization current, n is the slope factor, and ϕ_t is the thermal voltage.

$$i_f = \frac{I_D}{I_S} \quad (4) \qquad I_S = \mu n C'_{ox} \frac{\phi_t^2}{2} \frac{W}{L} \quad (5)$$

Figure 1 presents the X Ray user interface showing its input and output areas.

The input area accepts steps of normalized channel length, as well as steps of inversion level. The minimum length for normalization has its respective field. V_{DS} receives a maximum range and resolution. The button "Load Model" opens a file chooser dialog box to assign a Spice transistor model to the system. An input field receives the temperature, and a radio button selects the channel type. Because i_f is independent of geometry, the system has a single transistor width as input.

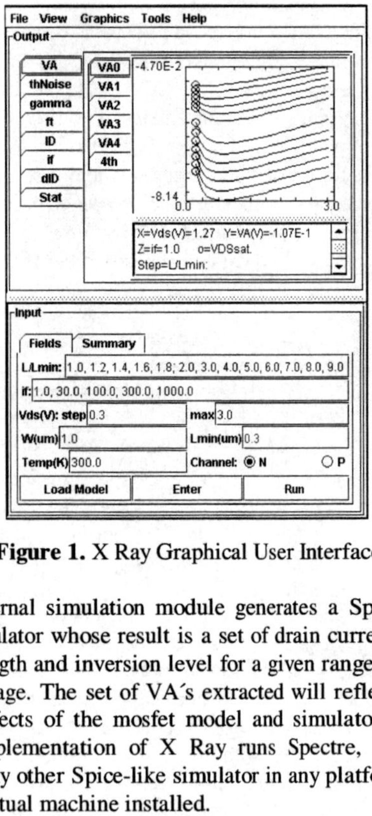

Figure 1. X Ray Graphical User Interface

An internal simulation module generates a Spice file and runs a simulator whose result is a set of drain currents for each channel length and inversion level for a given range of drain-to-source voltage. The set of VA´s extracted will reflect the short channel effects of the mosfet model and simulator used. The current implementation of X Ray runs Spectre, however, it could be any other Spice-like simulator in any platform that has the Java virtual machine installed.

The output area presents the results in sets of curves where each set is for one i_f. Figure 1 shows VA0 that is the Early voltage for if=1 where VA is in the y axis and V_{DS} is in the x axis. Each curve is for one step of L/L_{MIN}.

The thermal noise coefficient and thermal noise presented by X Ray follow expressions (6) and (7) [8], where q is the electronic charge and Δf is the bandwidth. The short-channel effects for γ and i_d^2 are taken into account by assigning VA for each i_f and V_{DS} in (6) and (7). Likewise γ and i_d^2, the intrinsic cut-off frequency f_T that accounts short-channel effects also comes from a mixture of an analytical model, equation (8) [8], and data points of VA.

$$\gamma_{sat} = \frac{1}{3}\left(2 + \frac{1}{i_f} - \frac{\sqrt{1+i_f}}{i_f}\right)\left(\frac{V_{DS} - VA}{V_{DSsat} - VA}\right) \quad (6)$$

$$\frac{i_d^2}{8qI_S\Delta f} = \gamma\left(\sqrt{1+i_f} - 1\right) \quad (7)$$

$$f_T = \frac{\mu\phi_t}{2\pi L^2} 2\left(\sqrt{1+i_f} - 1\right)\left(\frac{V_{DS} - VA}{V_{DSsat} - VA}\right) \quad (8)$$

For the hand design, the user can pick VA, γ and f_T from the graphics for a given i_f, V_{DS} and L [8]. Graphics for i_d^2 are also available, however, requiring adjustments according to the width needed. Another option is to size the transistor by means of Blue Print, as shown in the next section.

3. BLUE PRINT

Blue Print is an experimental calculator attached to X Ray that presents all the trade-offs involving a single transistor design by imitating the hand calculation process. Table 1 shows the expressions implemented by Blue Print, where some are analytical equations [4] while others are data points provided by X Ray. In both cases the variables involved are the ones required to design MOSFET amplifiers: the gate transconductance g_m, and the output resistance r_O, besides the ones already mentioned for noise, geometry, current, intrinsic cut-off frequency, and Early voltage. The analytical models used here work continuously from weak to strong inversion, hence, easy to be implemented for calculation purposes once there is no need of any interpolation function. The Blue Print graphical user interface shows only the design variables and hides the constants μ, n, C'_{ox}, ϕ_t. Figure 2 shows the interface, where each row represents a variable and each column one of the relations from Table 1.

The numbers given by Blue Print will be tailored for the simulator, technology and transistor model used by X Ray, even though the design equations [4] are different from the model used by the simulator. This approach is based on the assumption that the long channel transistor models should have similar behavior, both the ones that use any interpolation function to fill the gap between weak and strong inversion and those that work continuously, being the Early voltage the only parameter needed to be extracted in order to address short-channel effects for each particular model.

To design a transistor the user enters the desired value in the respective field and pushes one of the buttons in that row. If only one variable is left in that column, this variable is immediately determined. If one single variable is left in any other column as well, this one is also automatically calculated. This process continues until either more than one variable is left or all parameters are determined. One major advantage in the Blue Print is the fact that the numbers can be entered in any sequence, therefore providing to the designer freedom to choose what are the main parameters for that particular design, i.e., there is no preset design procedure. For instance, in the relation between V_{DSsat} and i_f, the designer can either enter first V_{DSsat} and then i_f or vice versa, the first one entered determines the other one. The designs are stored in the draft area under the calculator. Once the design is finished the user can reset the calculator and start another design.

The system supports several instances of both X Ray and Blue Print opened simultaneously, hence, allowing on the same screen the design of many transistors of N and P channels. This is a desirable feature while designing several transistors.

Table 1. Equations implemented in the Blue Print

Index	Expressions	Relations
1	$\frac{n\phi_t g_m}{I_D} = \frac{2}{1+\sqrt{1+i_f}}$	g_m, I_D, i_f
2	$i_f = \frac{I_D}{\mu n C'_{ox} \frac{\phi_t^2}{2} \frac{W}{L}}$	I_D, i_f, W, L
3 (data points)	f_T=f (i_f, V_{DS}, L)	i_f, L, f_T, V_{DS}
4	$\frac{V_{DSsat}}{\phi_t} = \left(\sqrt{1+i_f} - 1\right) + 4$	i_f, V_{DSsat}
5 (data points)	VA=f(i_f, V_{DS}, L)	i_f, L, V_{DS}, VA
6	$r_O = \frac{VA}{I_D}$	I_D, VA, r_O
7 (data points)	γ=f (i_f, V_{DS}, L)	i_f, L, V_{DS}, γ
8 (data points)	i_d^2=f (i_f, V_{DS}, L, W)	i_f, W, L, V_{DS}, i_d^2

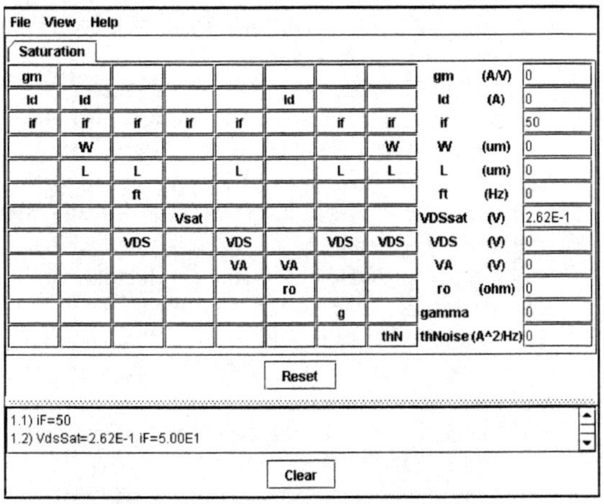

Figure 2. Blue Print Graphical User Interface

To access the data points the tool performs a simple linear interpolation between the points in the available range of values. If a certain value is out of range an error message is given. The precision, of course, will depend on the resolution of the points. This is why it is called Blue Print.

One can notice that depending on the sequence we can find more than one result for a certain parameter. It is also possible a different arrangement for the equations. For multiple results Blue Print takes the first one found. A better equation arrangement is a subject for another kind of study. The difficulties and constraints of a transistor level design, though, are clearly seen. The knowledge gained with this tool can help to create engines for circuit level design in the future.

In short, Blue Print automatically performs the hand calculation design that would require manipulation of several equations and manual reading of graphics.

4. DESIGNS AND SIMULATION RESULTS

This section compares results from X Ray / Blue Print with simulations using BSIM 3v3 of a common-source amplifier as the one seen in Figure 3. The technology used was TSMC25, L_{MIN}=0.3µm, run T21Q (MM_NON-EPI). We extracted the parameters for both N and P channels using i_f equals 1, 30, 100, 300 and 1000, L/L_{MIN} equals 1, 1.2, 1.4, 1.6, .1.8, 2, 3, 4, 5, 6, 7, 8 and 9, V_{DS} from 0 to 3V with 0.3V step, width equals 1µm, the minimum technology length and temperature of 300K.

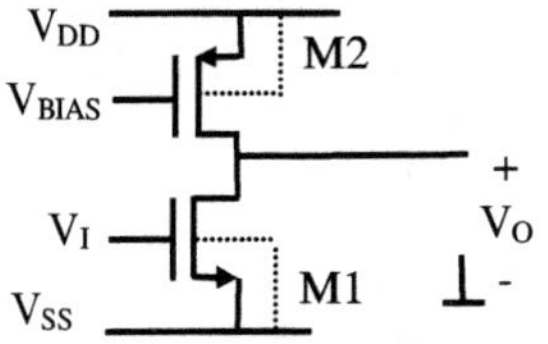

Figure 3. Common-Source (CS) Amplifier

Table 2 shows a design of a common source amplifier to compare the values given by Blue Print and the ones found in the simulations regarding the DC gain A_{vo} of the amplifier. The Blue Print column presents the steps used for the design, actually the ones input in the calculator, and the respective results. The first transistor designed was the NMOS, with i_f, W and L/L_{MIN} arbitrarily chosen. The V_{DS}, required in this method [8], is half of the voltage supply, V_{DD} to V_{SS}, of 1.5V. Among several other parameters, we primarily focus on g_m, and VA once they are directly related to the gain, as equations (9) and (10) [8] show, and I_D because it is needed to design the PMOS. W and VA are the main parameters for the P transistor, whose design also requires V_{DS} [8].

$$A_{VO} = \frac{g_{m1}}{g_{O1} + g_{O2}} \quad (9)$$

$$A_{VO} = \frac{1}{\phi_t n}\left(\frac{2}{\sqrt{1+i_{f1}}+1}\right)\left(\frac{1}{1/VA_1 + 1/VA_2}\right) \quad (10)$$

The design and simulations values agree fairly well. The Early voltages given by the simulations are very close to the ones provided by Blue Print, proving that the information of VA of a 1µm width transistor can be used for any width. The theoretical and simulated DC gains are, respectively, 25 and 37. Overall, the numbers quickly provided by Blue Print matched the ones found in the simulations.

Access to the short-channel effects of the thermal noise that are usually evaluated only during the simulations is desirable during the hand calculation design [8]. Table 3 compares the design and simulation of the CS amplifier for several voltage supplies. We chose this criterion due to the influence of V_{DS} on VA, and the correlation of VA with the thermal noise, as seen in equations (6) and (7) [8]. The parameters needed to calculate the noise are those given by Equation (11), which takes into account the short-channel effects based on data points of $i_d^2{}_1$, $i_d^2{}_2$, and VA_1, this last one to model those effects in g_{m1} [8]. Equation (11) also keeps the V_{DSsat}, which is normally ignored while calculating the influence of VA, because it presented better results for the noise.

The design for the NMOS was based on i_f equals 300, W of 12.525μm, minimum length, and the steps of V_{DS} presented in Table 3. The g_m for the NMOS is 2.51mA/V, whose $i_d^2{}_1$ and VA_1 are the ones seen in Table 3. The PMOS design also used i_f of 300, I_D equals 7.76μA, this one from the NMOS design, minimum length, and the same steps of V_{DS}. The final width for the PMOS was 35μm and the values for $i_d^2{}_2$ are also found in Table 3.

Table 2. CS Amplifier Design and Simulations for A_{vo}

Blue Print		Simulations	
N	P	N	P
Steps: 1: i_f=30 2: W=50μm 3: L=0.42μm 4: V_{DS}=0.75V	**Steps:** 5: i_f=30 6: I_D=221μA 7: L=0.42μm 8: V_{DS}=0.75V	 i_f=32.8 W=50.025μm L=0.45μm V_{DS}=0.51V	 i_f=32.8 I_D=227.2μA L=0.45μm V_{DS}=0.99V
Results: g_m=2.0mA/V I_D=221μA VA=-3.64V	**Results:** W=136μm VA=-10.7V	 g_m=2.81mA/V I_D=227.2μA VA=-3.95V	 W=140μm VA=-13.44V

Table 3. Noise: Blue Print X Simulations.

	Blue Print			Eq. (11)	Simul.
V_{DS} (V)	i_{d1}^2 $\left(\frac{A^2}{Hz}\right)$	i_{d2}^2 $\left(\frac{A^2}{Hz}\right)$	VA_1 (V)	$\sqrt{V_{N,in}^2}$ $\left(\frac{nV}{\sqrt{Hz}}\right)$	$\sqrt{V_{N,in}^2}$ $\left(\frac{nV}{\sqrt{Hz}}\right)$
0.9	3.68e-23	3.70E-23	-5.77	3.232	3.193
1.2	3.80e-23	3.68E-23	-6.75	3.155	3.172
1.5	3.91e-23	3.70E-23	-7.24	3.089	3.137
1.8	4.03e-23	3.73E-23	-7.46	3.028	3.115
2.1	4.16e-23	3.76E-23	-7.55	2.968	3.099
2.4	4.28e-23	3.79E-23	-7.56	2.907	3.089

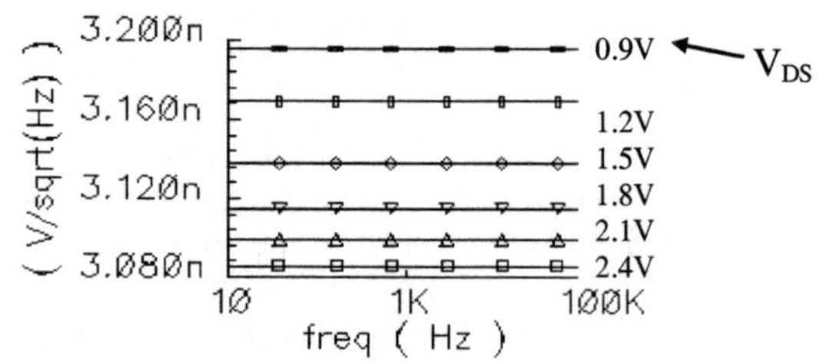

Figure 4. Noise for several values of V_{DS}.

$$V_{N,in}^2 = \left(i_{d1}^2 + i_{d2}^2\right)\frac{1}{\left(g_{m1}\left(\frac{V_{DS1} - VA_1}{V_{DSsat1} - VA_1}\right)\right)^2} \quad (11)$$

Calculations and simulations, again, are very close. Figure 4 shows the graphics of the noise for each V_{DS} presented in Table 3. The thermal noise using the long channel model would be 2.865nV/√Hz. The design model for the thermal noise with the short-channel effects is more precise than the long channel one.

5. CONCLUSIONS

This paper has presented an alternative approach to address short-channel effects for analog design by mixing analytical expression with data points. Short-channel effects were referred to the Early voltage. An engine for extraction of the Early voltage based in a well defined design space and calculation of thermal noise coefficient and thermal noise, as well as a tool for transistor level design were successfully implemented and tested. Theoretical and simulated results were compared and showed very good agreement, therefore, confirming the efficiency of the design method. The results presented here, for transistor level design, can be used as basis to build circuit level design tools.

6. REFERENCES

[1] B. Martin, "Automation Comes to Analog", IEEE Spectrum, pp. 70-75, June 2001.

[2] G. G. E. Gielen, R. A. Rutenbar, "Computer-Aided Design of Analog and Mixed-Signal Integrated Circuits", Proceedings of the IEEE, vol. 88, no. 12, December 2000.

[3] Hershenson, M.delM.; Boyd, S.P.; Lee, T.H, "Optimal design of a CMOS op-amp via geometric programming", Computer-Aided Design of Integrated Circuits and Systems, IEEE Trans. on , Volume: 20 Issue: 1 , Jan 2001.

[4] A.I.A. Cunha.; Schneider, M.C.; Galup-Montoro, "A MOS transistor model for analog circuit design", IEEE Journal of Solid-state Circuits, vol. 33, No. 10, October 1998.

[5] C. C. enz, F. Krummenacher and E. A. Vittoz, "An analytical MOS transitor model valid in all regions of operation and dedicated to low-voltage and low-current applications", Analog Integrated Circuits and Signal Processing, vol. 8, pp 83-114, July 1995.

[6] A.I.A. Cunha.; Schneider, M.C.; Galup-Montoro, "An explicit physical model for the long-channel MOS transistor including small-signal parameters", Solid-State Electronics, vol. 38, No. 11, November 1995.

[7] F. Silveira, D. Flandre, and P. G. A. Jespers, "A g_m/I_D Based Methodology for the Design of CMOS Analog Circuits and Its Application to the Synthesis of a Silicon-on-insulator Micropower OTA", IEEE Journal of Solid-State Circuits, vol. 31, no 9 September 1996.

[8] R. L. Oliveira Pinto, F. Maloberti, "Novel design Methodology for Short-Channel Mosfet Analog Circuits", IWSOC 2003, Calgary-Alberta, Canada, July 2003.

[9] Schneider, M.C.; Galup-Montero, C.; Filho, O.C.G.; Cunha, A.I.A. A single-piece charge-based model for the output conductance of MOS transistors Electronics, Circuits and Systems, 1998 IEEE International Conference on, Volume: 1, 1998.

[10] Shaeffer,D. K, Lee, T. H, "A 1.5 V 1.5 GHz CMOS Low Noise Amplifier," IEEE Journal of Solid State Circuits, Vol. 32, No. 5, pp .745-759, May 1997.

SEAMS - A SYSTEMC ENVIRONMENT WITH ANALOG AND MIXED-SIGNAL EXTENSIONS

H Aljunaid and T J Kazmierski

School of Electronics and Computer Science
University of Southampton
Southampton, SO17 1BJ, UK

ABSTRACT

We describe an efficient implementation of analog and mixed-signal extensions integrated with SystemC 2.0. SEAMS (SystemC Environment with Analog and Mixed-Signal extensions) uses a general-purpose analog solver to handle analog extensions and to provide modelling capabilities for general, mixed-mode systems with digital and non-linear analog behavior. We have extended the SystemC 2.0 kernel to invoke and synchronize our analog solver in each simulation cycle while maintaining compliance with the SystemC simulation cycle semantics. The operation of SEAMS is illustrated with the practical examples of a boost power converter and a 2GHz phase-lock loop frequency multiplier with noise and jitter models. Mixed-signal systems of this kind are known to be difficult to simulate as they exhibit disparate time scales which put most simulators in numerical difficulties. We hope that the practical experience of SEAMS might aid the recent efforts to standardize analog and mixed-signal extensions for SystemC.

1. INTRODUCTION

Several high level hardware description languages, such as VHDL and Verilog, have recently been extended to provide mixed-signal modelling capability and new standards for VHDL-AMS [1] and Verilog-AMS [2] are now widely available. In the light of the growing popularity of mixed, analog and digital ASICs, this is not surprising. Naturally, with the advent of SystemC, several research publications highlighted the possibility to extend the standard, digital modelling capabilities of SystemC [3, 4, 5, 6] to the analog domain. Bjornsen et al. [7] described an efficient software framework for rapid behavioral modelling and simulation of analog-to-digital converters based on SystemC. Their framework consists of three parts, the first is an analog extension representing analog signal classes, the second is a signal module library, which includes basic blocks required for A-D modelling, such as switched-capacitor integrators, operational amplifiers and a track-and-hold amplifier. The third part is a mixed-signal system test bench. Einwich *et al.* [6] presented a SystemC extension allowing an overall specification and verification of mixed-signal systems with linear analog models. They consider an ADSL line driver with analog linear filters as an example. This contribution outlines the concept of a general, mixed-signal SystemC simulator comprising an analog kernel with an underlying transient solver for non-linear, algebro-differential equations. We devote particular attention to the problem of synchronizing the analog kernel with the digital one. Our synchronization technique is compliant with the definition of SystemC simulation cycle semantics [4]. For this purpose, we link the analog kernel to the SystemC environment as a user module and synchronize it with the SystemC kernel via a lockstep synchronization algorithm. Operation of the extended, mixed-signal SystemC simulation platform is demonstrated using practical example of a boost power converter, in which analog behavior tightly interacts with a digital control loop. The results presented here might aid the current efforts to standardize analog extensions for SystemC [8].

2. ANALOG AND DIGITAL SOLVER SYNCHRONIZATION

Like in the case of most high-level hardware description languages, a SystemC model consists of a hierarchical network of parallel processes, which exchange messages under the control of the simulation kernel process and concurrently update the values of signals and variables. Signal assignment statements do not affect the target signals immediately, but the new values become effective in the next simulation cycle. The kernel process resumes when all the user defined processes become suspended either by executing a *wait* statement or upon reaching the last process statement. On resumption, the kernel updates the signals and variable and suspends again while the user processes resume. If the time of the next earliest event t_n is equal to the current simulation time t_c, the user processes execute a delta cycle.

To comply with the SystemC execution semantics, our

0-7803-8251-X/04/$17.00 ©2004 IEEE

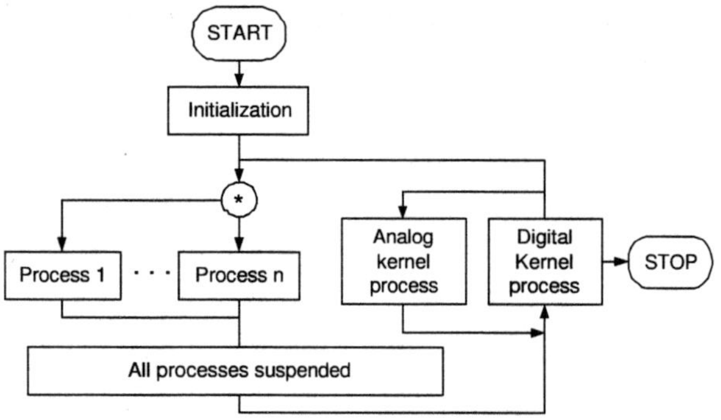

Fig. 1. Simulation cycle of a SystemC system with analog extensions.

mixed-signal SystemC simulator comprises an analog kernel, which runs as a SystemC process and drives the user defined analog modules. This scenario requires the analog solver to be able to handle delta cycles in a manner similar to that of digital processes, namely, the state of the analog solver may not be updated until after the SystemC kernel advances the simulation time ahead of the current simulation time t_c. In other words, the underlying analog solver not only must be able to execute delta cycles with the step size $h = 0$, but also must have a capability to backtrack to the state just before t_c (t_c^-), should new events at t_c change the analog stimili. The former requirement would be difficult to satisfy with most standard, SPICE-like simulators since they require a minimum step size greater than zero. Moreover, small step sizes can cause large round off errors and lead to inaccurate results. The analog model proposed in section 3 can be solved with arbitrary step size, including $h = 0$. Backtracking is achieved by retaining to the analog state (t_c^-) when required. The analog kernel repeatedly executes the simulation cycle shown in figure 1, which might involve delta cycles and backtracking. Analog simulators do not use events but instead employ an entirely different approach to time step control, namely, continuous step-size adjustment to minimize the errors caused by the numerical integration method used to solve differential equations in the circuit model. In order to minimize the error in the numerical integration, it may be necessary to repeatedly reject the circuit solution at a time point and to cut the time step. It is therefore necessary for the analog kernel in a SystemC environment not to advance past the current simulation time t_c unless a delta cycle occurs and re-evaluation of the current step is necessary. There are essentially two algorithms for the synchronization of two or more solvers running concurrently [9]. One, optimistic approach allows each simulator to progress in time until it runs out of internal events (internal activity has ceased) and then suspend. The alternative approach, adopted here, is the lock-step algorithm. The analog kernel advances until the current simulation time and, before suspending, schedules and event at the time equal to the current simulation time plus the next selected step size. This approach ensures that the SystemC kernel will make a step in time no larger that the analog kernel's step size. Since the analog kernel runs as a user process and is controlled by the SystemC kernel, no synchronization deadlock may happen. The only causes for deadlock can arise due to a failure to converge in the analog solver or due to unresolvable delta cycles. The signals that pass between the digital and analog domains are carried by global nets. The solution method adopted here allows each global net to be able to connect two partitions (the digital and analog simulator), one of which owns the net. This means that a general global net is divided into two smaller terminal nets, implemented as SystemC ports. The synchronization approach adopted in this paper is based on the lock-step principle to ensure that no results are thrown away and there is no need for backtracking. Most existing digital solvers cannot backtrack and therefore no fundamental changes are required if a mixed-signal system is integrated to the SystemC kernel. Our lock-step synchronization algorithm has been implemented as a modification to the digital kernel and can be described in the form of pseudo-code as follows:

```
time <- 0;
initialize both the analog and digital kernel;
while (time <= end_time) do

   while (immediate notifications are pending) do
      execute the analog kernel
      distribute notifications generated
        by the analog kernel on global nets

      while (there are active processes) do
         run a selected process
      end while (there are active processes)

      update signals

   // check if a delta cycle is necessary
   end while (immediate notifications are pending)

   advance time to the next timed notification
end while (time<= end_time)
```

In most cases, delta cycles caused by zero-delay paths are eventually resolved but, in general, sharing zero-delay paths between the two kernels should be avoided. If the maximum allowed number of delta cycles is exceeded, the algorithm treats this situations as deadlock and stops.

3. ANALOG MODEL

The description of the analog model relies on C++ classes that provide behavior for circuit nodes and primitive components. The component classes contain virtual methods that are invoked at matrix build time each time the analog solver requires to update the Jacobian according to modified nodal analysis stamps.

```
void analog::BuildCircuit(void){
 ...
  node v1("v1"),v2("v2"),vc("vc"),vout("vout"),vf("vf");
  node vramp("vramp");
  voltageS    E("E",v1,0,1.5);
  inductor    l1("l1",v1,v2,10e03,5.0);//10mH,5ohm
  resistor    ro("ro",vout,0,500.0);
  resistor    r1("r1",vout,vf,23e3);
  resistor    r2("r2",vf,0,10e3);
  MOS0        M1("M1",v2,vc,0);
  diode0      D1("D1",v2,vout);
  capacitor   Co("Co",Vout,0,1e-3);
  voltage_ramp vr("vr",vramp,0,1e-6);
 ...
} // analog::BuildCircuit() --------------------
```

The analog kernel invokes BuildCircuit()once prior to the simulation. Additional user-defined methods represent interfacing components that provide functionality for synchronization between SystemC ports and their corresponding values in the analog solver. The interfaces are bound at the build stage.

```
sc_signal<bool> ASig,S1;
sc_signal<double> S2;
sc_clock Clk("Clock", 0.1, SC_US,0.5);
  ...
    analog analog1("analog1");
    analog1.Vcontrol(S1);
    analog1.Vout(S2);
    analog1.clock1(ASig);

    digital digital1("digital1");
    digital1.Vout(S2);
    digital1.Vcontrol(S1);

    stim stim1("stim1");
    stim1.A(ASig);
    stim1.Clk(Clk);
  ...
```

Small step sizes including the case of a delta cycle where $h = 0$ are handled by smoothing the digital signal as illustrated in figure 2.

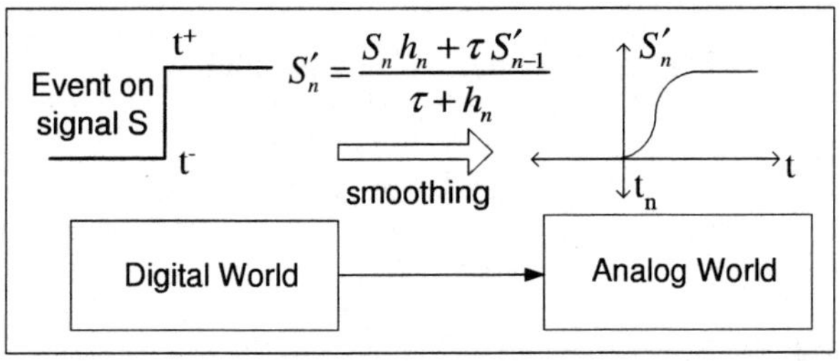

Fig. 2. Handling small step sizes.

4. PRACTICAL EXAMPLES

The analog extensions added to the language contain classes to handle electrical nodes and primitive analog components with user defined behavior necessary to build and solve the analog equation set. A boost 1.5V to 3.3V power converter and a 2GHz PLL-based frequency multiplier with noise and jitter have been used as examples of non-trivial analog and digital interaction. Systems of this kind usually put standard, SPICE-like simulators into difficulties because of the disparate time scales of their transients. In the case of a switched-mode power supply, the analog transient in the output circuit is four to five orders of magnitude slower than that of the fast switching waveforms in the digital controller. As a typical simulation in a system of this kind might require a hundred million time points, excessive CPU times often occur when the entire system is modelled on the circuit level The capacity of SystemC to enable system-level, mixed-signal modelling can vastly reduce simulation times where concepts need to be verified quickly and detailed, circuit-level modelling is not required. The power converter block structure is presented in figure 3. The controller's digital behavior in the example is modeled as a standard SystemC *SC_MODULE*. The testbench instantiates both the analog and digital module and provides global nets. Sample simulation results at steady-state are presented in figure 4.

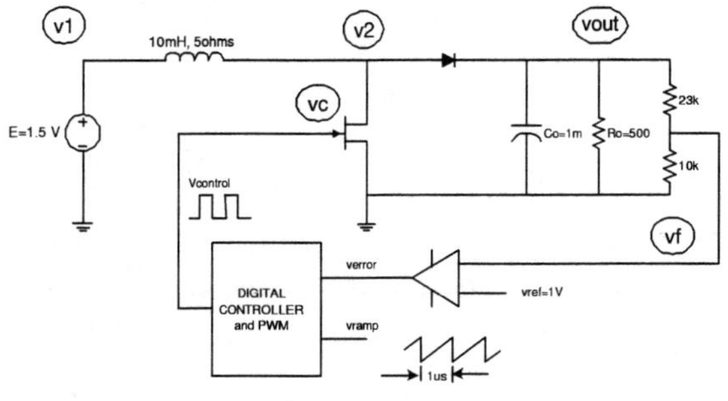

Fig. 3. Boost 1.5V to 3.3V switch-mode power supply with digital control.

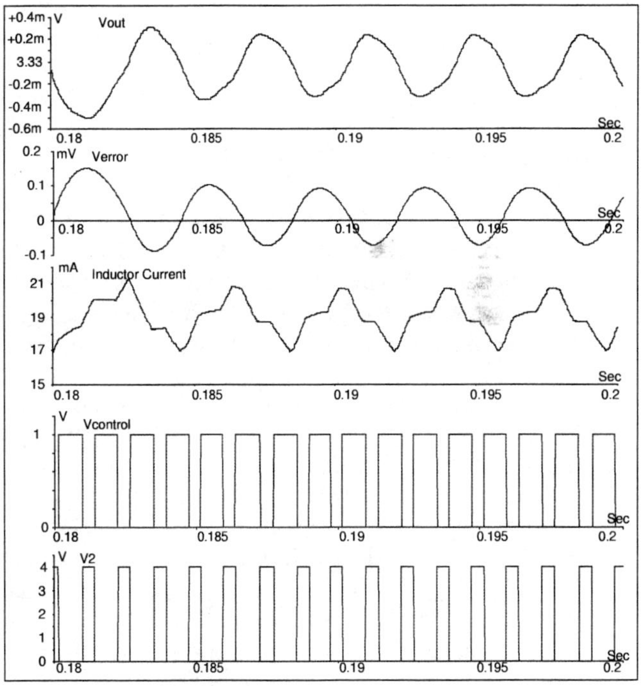

Fig. 4. SMPS simulation results for a 200ms time window in steady-state of the boost switch-mode power supply working in continuous mode.

The PLL multiplier simulation also requires millions of time steps to accurately reflect the effects of noise and jitter. The behavioral VCO model includes integration of the phase which converts the input voltage noise to a jitter in the output frequency. The system's block diagram is shown in figure 5 as well as the VCO jitter histogram calculated from the simulation results when the loop was in lock. Figure 6 shows different system values in the first micro seconds of the simulation.

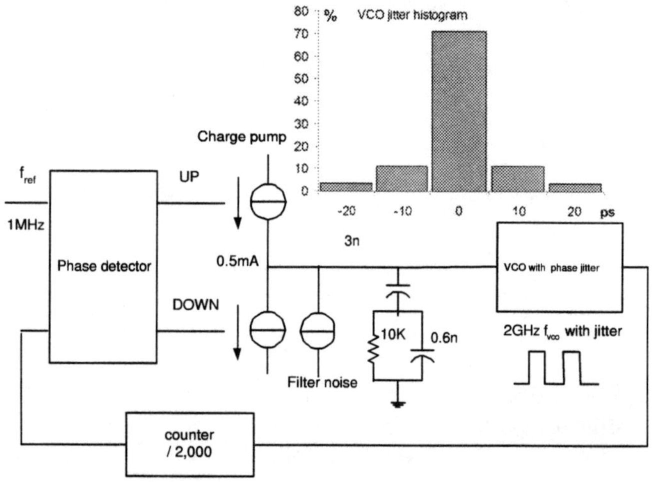

Fig. 5. 2GHz PLL multiplier with noise and jitter.

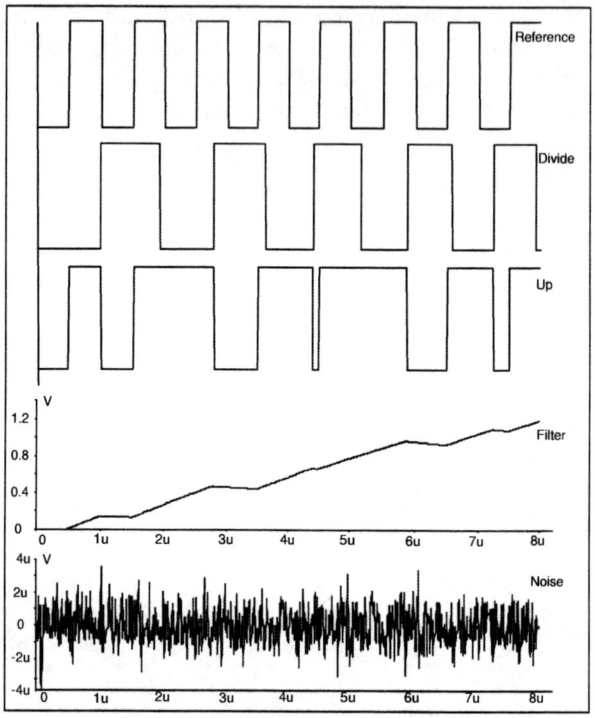

Fig. 6. 2GHz multiplier simulation results.

5. CONCLUSION

An extension to SystemC has been developed to simulate general, mixed-mode systems with digital and non-linear analog behavior. A novel analog solver has been developed coupled to the SystemC kernel via a lock-step synchronization algorithm. Its principle has been demonstrated using the practical examples of a mixed-signal boost power converter, in which analog behavior interacts with a digital control loop and a PLL frequency multiplier with noise and jitter.

6. REFERENCES

[1] E. Christen and K. Bakalar, "Vhdl-ams - a hardware description language for analog and mixed-signal application," *IEEE Trans CAS II*, vol. 46, no. 10, pp. 1263–1272, October 1999.

[2] *Verilog AMS Language Reference Manual, Accelera*, August 1998.

[3] J. Gerlach, "System level design using the systemc modeling platform," *Proc. SDL 2000*, 2000.

[4] W. Mueller, "The simulation semantics of systemC," *Proc. DATE 2001*, 2001.

[5] P. Panda, "Systemc - a modeling platform supporting multiple design abstractions," *Proc. ISSS 2001, Montreal, Canada*, 2001.

[6] K. Einwich, C. Clauss, G. Noessing, P. Schwarz, and H. Zojer, "SystemC extensions for mixed-signal system design," *Proc. FDL'2001, Lyon, France*, September 2001.

[7] T.E. Bonnerund, B. Hernes, and T. Ytterdal, "A mixed-signal, functional level simulation framework based on systemc for soc applications," *Proc. Custom IC Conf., CICC'2001, San Diego, California*, pp. 541–545, May 2001.

[8] A. Vachoux, C. Grimm, and K. Einwich, "Systemc-ams requirements, design objective and rationale," *Proc. DATE'2003, Munich, Germany*, pp. 388–393, 3-7 March 2003.

[9] M. Zwolinski, C. Garagate, Z. Mrcarica, T.J. Kazmierski, and A.D. Brown, "The anatomy of the simulation backplane," *IEE Proceedings on Computers and Digital Techniques*, vol. 142, no. 6, pp. 377–385, November 1995.

PROPERTIES OF FASTEST LINEARLY INDEPENDENT TRANSFORMS OVER GF(3)

Bogdan J. Falkowski and *Cheng Fu*

School of Electrical and Electronic Engineering
Nanyang Technological University,
Block S1, 50 Nanyang Avenue, Singapore 639798

ABSTRACT

Fastest linearly independent transforms over GF(3) have been investigated recently. This paper introduces new class of fastest transforms and discusses their various properties and relations. Experimental results for the fastest linearly independent transforms are compared with ternary Reed-Muller transform using ternary benchmark functions.

1. INTRODUCTION

Logic synthesis and testing using the Reed-Muller based spectral expansions were discussed by many authors [1, 2, 6, 9]. They are of great interest because the Reed-Muller based circuits have many desirable properties, such as simple testability vectors. The binary Reed-Muller transform itself is only one type of Linearly Independent (LI) transforms introduced in [7] and discussed in detail in [4] that are the special case of more general orthonormal expansions discussed in [1]. Further research showed that there are some LI transforms having the smallest computational cost that is lower than that of the binary Reed-Muller transform [8].

These authors recently introduced new fastest linearly independent transforms over GF(3) based on two classes [3]. In this article, another class of fastest linearly independent transforms over GF(3) is introduced and their properties and relations are compared with the known fastest transform over GF(3) as well as ternary Reed-Muller transform. The introduced fastest ternary transforms have the simplest butterflies of all possible LI transforms over GF(3) and due to its recursive equations they can be easily expanded to any dimension. In addition, some experimental results for them and ternary Reed-Muller transform for ternary benchmark functions have also been shown. The computational time shows advantage of the new fastest LI transform over calculation of the best ternary Reed-Muller transform for all ternary benchmark functions.

2. BASIC DEFINITIONS

Definition 1. Let M_n be a $N \times N$ $(N = 3^n)$ matrix with rows corresponding to minterms and columns corresponding to some switching ternary functions of n variables. If the sets of rows are linearly independent with respect to *ternary Galois Field*, then M_n has only one inverse in GF(3) and is said to be *Linearly Independent*. If M_n is Linearly Independent in GF(3), then M_n is a non-singular square matrix with respect to GF(3) and has a unique inverse M_n^{-1}.

Definition 2. The LI expansion for any n-variable ternary function $f(\overrightarrow{x_n})$ is

$$f(\overrightarrow{x_n}) = \sum_{j=0}^{3^n-1} A_j \overrightarrow{g_j}, \quad (1)$$

where g_j is any set of n-variable switching functions such that the matrix $M_n = \left[\overrightarrow{g_0}, \overrightarrow{g_1}, ..., \overrightarrow{g_{3^n-1}}\right]$, $\overrightarrow{g_j}$ represents the truth vector of the ternary functions, $0 \le j \le 3^n - 1$, A_j is the respective coefficient for the particular transform matrix M_n with modulo-3 inverse M_n^{-1}, and the symbol $\sum$ is the summation in GF(3).

Definition 3. The LI transform based on Definitions 1 and 2 can be described by the following general formula performed over GF(3):

$$\vec{F} = M_n \vec{A}, \quad (2)$$

and

$$\vec{A} = M_n^{-1} \vec{F}, \quad (3)$$

where $\vec{F} = \left[F_0, F_1, ..., F_{3^n-1}\right]^T$ is a column vector defining the truth vector of a ternary switching function $f(\overrightarrow{x_n})$ in a natural ternary ordering, M_n is a LI matrix of order $N = 3^n$ defined by any n-variable ternary switching function and $\vec{A} = \left[A_0, A_1, ..., A_{3^n-1}\right]^T$ is the coefficient column vector for the particular transform matrix M_n with the inverse matrix M_n^{-1} over GF(3).

3. PROPERTIES AND RELATIONS

There are six pairs of forward and inverse ternary fastest LI transforms presented in Table 1 denoted as Y1.1-Y1.6. All the presented transform matrices are square matrices with the dimension $3^n \times 3^n$, and they are constructed by submatrices M_{n-1}, O_{n-1}, I_{n-1}, J_{n-1} and Y_{n-1}, where O_{n-1} is a $3^{n-1} \times 3^{n-1}$ submatrix with all its elements equal to 0, I_{n-1} is identity submatrix and J_{n-1} is reverse-identity submatrix, Y_{n-1} is a $3^{n-1} \times 3^{n-1}$ submatrix with all its elements equal to 0 except one that is equal to 1 at the bottom left corner for transforms Y1.1-Y1.3 or at the top right corner for transforms Y1.4-Y1.6.

The following general matrix form of M_n will be used to present the properties of ternary fastest LI transforms.

$$M_n = \begin{bmatrix} M_{0,0} & M_{0,1} & \cdots & M_{0,3^n-1} \\ M_{1,0} & \ddots & & \vdots \\ \vdots & & \ddots & \vdots \\ M_{3^n-1,0} & \cdots & \cdots & M_{3^n-1,3^n-1} \end{bmatrix}$$

Property 1. The total numbers of non-zero elements in the forward and inverse transform matrices of Y1.1, Y1.2, Y1.4 and Y1.5 are equal to $3^n + 2^n - 1$.

There are 3^n '1's and $2^n - 1$ '2's in the inverse transform matrices while all non-zero elements are '1's in forward transform matrices.

Property 2. The total numbers of non-zero elements in the forward and inverse transform matrices of Y1.3 and Y1.6 are equal to $(3^{n+1} - 1)/2$.

All non-zero elements in the forward transform matrices are '1's and there are $(3^n - 1)/2$ '2's and 3^n '1's in the inverse transform matrices.

Property 3. The row i ($0 \le i \le 3^n - 1$) of the forward transform matrices of Y1.1 and Y1.2 has (1+*d*) '1' elements, where *d* is following the rules: *d* is the number of '2' elements behind the last 0 of the ternary representation of *i* if it does not contain '1', otherwise *d* is equal to zero.

The row *i* of their inverse transform matrices has one '1' element and *d* '2' elements.

If *D* is defined as {1, 2, ..., *d*}, then the '2's inside the row *i* are located at the columns: $i-(3^1-1)$, $i-(3^2-1)$, ..., $i-(3^d-1)$.

Property 4. $M_{(Y1.4)n} = [M_{(Y1.1)n}]^T$ and $M_{(Y1.5)n} = [M_{(Y1.2)n}]^T$, where the superscript T indicates matrix transpose operation.

Due to this property, similar approach to the position of non-zero elements in both forward and inverse transform matrices of Y1.4 and Y1.5 can be derived as Property 3.

Property 5. The row i ($0 \le i \le 3^n - 1$) of the forward transform matrices of Y1.3 and Y1.6 has (1+*d'*) '1' elements, where *d'* is following the rules: *d'* is the number of '0' elements behind the last non-zero element of the ternary representation of (*i*+1).

The row *i* of their inverse transform matrices has one '1' element and *d'* '2' elements.

If *D'* is defined as {1, 2, ..., *d'*}, then the '2's inside the row *i* are located at the columns: $i-(3^1-1)$, $i-(3^2-1)$, ..., $i-(3^{d'}-1)$.

Property 6. $M_{(Y1.6)n} = [M_{(Y1.3)n}]^T$, where the superscript T indicates matrix transpose operation.

Again, the position of non-zero elements in the forward and inverse transform matrices of Y1.6 can be described in a similar manner as Property 5.

Till now, the properties of the six pairs ternary fastest LI transform Y1.1-Y1.6 have been presented. Other six pairs of ternary fastest transforms that have similar properties to the transforms Y1.1-Y1.6 but require vertical or horizontal permutations also exist. They are denoted as Y2.1-Y2.6.

The following definitions will be used to derive the properties of the ternary fastest transforms.

The matrix M_n of ternary fastest LI transforms can be also represented in the following format,

$$M_n = \begin{bmatrix} M_{n-1}^{(1)} & O_{n-1} & M_{n-1}^{(2)} \\ O_{n-1} & M_{n-1}^{(5)} & O_{n-1} \\ M_{n-1}^{(3)} & O_{n-1} & M_{n-1}^{(4)} \end{bmatrix}$$

where all submatrices follow earlier notation.

Definition 4. The μ_{EH} operator on matrix M_n is defined as grouping the recursive equations in the submatrices vertically and interchanging them horizontally

$$\mu_{EH}[M_n] = \begin{bmatrix} M_{n-1}^{(2)} & O_{n-1} & M_{n-1}^{(1)} \\ O_{n-1} & M_{n-1}^{(5)} & O_{n-1} \\ M_{n-1}^{(4)} & O_{n-1} & M_{n-1}^{(3)} \end{bmatrix}.$$

Definition 5. The μ_{EV} operator on matrix M_n is defined as grouping the recursive equations in the submatrices horizontally and interchanging them vertically

$$\mu_{EV}[M_n] = \begin{bmatrix} M_{n-1}^{(3)} & O_{n-1} & M_{n-1}^{(4)} \\ O_{n-1} & M_{n-1}^{(5)} & O_{n-1} \\ M_{n-1}^{(1)} & O_{n-1} & M_{n-1}^{(2)} \end{bmatrix}.$$

Property 7. If M_a is a forward transform matrix in Y2.x, $x \in \{0,...,6\}$, and M_b is a forward transform matrix in Y1.y, $y \in \{0,...,6\}$, then when $x = y$,

$$M_a = \mu_{EH}[M_b]$$

Property 8. If M_a^{-1} is an inverse transform matrix in Y2.x, $x \in \{0,...,6\}$, and M_b^{-1} is an inverse transform matrix in Y1.y, $y \in \{0,...,6\}$, then when $x = y$,

$$M_a^{-1} = \mu_{EH}[M_b^{-1}]$$

Considering the matrix structure of Y2.x, $x \in \{0,...,6\}$ by Properties 5 and 6 the spectral coefficient vectors that are computed from the inverse transforms of Y2.x, $x \in \{0,...,6\}$ for any ternary function $f(\overrightarrow{x_n})$ are vertical permutation of the spectral coefficient vectors that are computed from the inverse transforms of Y2.y, $y \in \{0,...,6\}$, when $x = y$. Due to this fact, similar properties as Properties 1-4 can be obtained from the six ternary fastest transforms Y2.x, $x \in \{0,...,6\}$.

4. COMPUTATIONAL COSTS

The presented ternary fastest LI transforms have very regular structures with the introduction of submatrix Y_{n-1} in the recursive construction. The computational costs of transform depend on the way of the construction of the higher matrices from the basic submatrices and due to this reason, the ternary fastest LI transforms are the most efficient. A decomposition method for the transforms has been discussed in [3] and it will be used below.

Property 9. Based on the matrix decomposition method in [3], the number of additions required to compute the ternary fastest LI transforms Y1.1, Y1.2, Y1.4 and Y1.5 are $2^n - 1$ both for the forward and inverse transforms.

Property 9 also applies to Y2.1, Y2.2, Y2.4 and Y2.5.

Property 10. Based on the matrix decomposition method in [3], the number of additions required to compute the ternary fastest LI transforms Y1.3 and Y1.6 are $(3^n - 1)/2$ both for the forward and inverse transforms.

Property 10 applies to Y2.3 and Y2.6.

5. EXPERIMENTAL RESULTS

All the six ternary fastest LI transforms have been implemented using Microsoft Visual C++ and run on PIII 500 MHz computer with 128 MB RAM. They are run on several benchmark functions that have been modified to represent ternary functions instead of original MCNC and IWLS'93 binary benchmark functions. The translation from binary to ternary cases has been done by changing every 2 input (output) bits in binary files to an input (output) symbol in ternary files. If the number of input and/or output variables is odd, then a zero bit is added behind the binary cubes to make it even. For input (output), -- is taken as -, 00 is taken as 0, 01 is taken as 1 and 10 is taken as 2, whereas 11 is not used (taken as 0) for ternary case.

The numbers of non-zero spectral coefficients for all the six transform matrices Y1.1-Y1.6 are given in Table 2. The experimental results of best polarity and zero polarity ternary Reed-Muller transform are also given in the last two columns.

Comparing the results of the six ternary LI fastest transforms, it is clear that Y1.3 and its transpose transform matrix Y1.6 get some better results (less number of non-zero spectral coefficient resulting in simpler corresponding polynomial equation representing ternary function) than the results of other ternary LI fastest transforms, for example, Y1.3 has better results in alu4 for outputs 1, 2 and in inc for outputs 1, 2, 5, and Y1.6 has better results in alu4 for outputs 1, 3, 4, in apex4 for outputs 6, 7, 8, 10 and in inc for outputs 3, 5.

The results for the six ternary fastest transforms are better than the results of zero polarity ternary Reed-Muller transform for majority of cases. Even though ternary Reed-Muller transform gets smaller number of non-zero spectral coefficients than other transform matrices after searching all polarities, Table 2 shows that the fastest transforms can obtain less spectral coefficients in some cases, such as for ternary benchmark functions (TBF) apex4 and ex1010. It also should be noticed that in order to obtain the best polarity result for ternary Reed-Muller transform all 3^n spectra have to be calculated what take a lot of time. On the other hand, the transforms Y1.1-Y1.6 do not have polarity so the calculation needs to be performed only once for each of them. Table 3 gives the calculation time for Y1.1-Y1.3 and compares them with ternary Reed-Muller transform.

6. CONCLUSION

This article discusses six ternary fastest LI transforms that have the least computational complexity among all ternary LI transforms as the suitability of an LI transform in a given application depends not only on the choice of its basis function but also on the existence of efficient ways of its calculation as well as the complexity of its final polynomial expansion. Moreover similarly to known LI transforms for binary case that can be used for the development of word-level decision diagrams [10] the new fastest ternary transforms can have the same applications as well. In this article, it is also shown that the fastest transform are quite advantageous over well known ternary Reed-Muller transform in terms of the necessary computation and for some ternary functions the number of non-zero spectral coefficients of the polynomial expansion based on the fastest transform can be smaller. The ternary polynomial expansions based on the fastest transforms presented here can be easily implemented in terms of reversible logic gates [5].

7. REFERENCES

[1] F. Brown, *Boolean Reasoning*. Kluwer Academic, Boston, 1990.

[2] P. Davio, J.P. Deschamps and A. Thayse, *Discrete and Switching Functions*. George and McGraw-Hill, New York, 1978.

[3] B.J. Falkowski and C. Fu, "Polynomial Expansions over GF(3) based on Fastest Transformation", *Proc. IEEE 33rd Int. Symposium on Multiple-Valued Logic*, Tokyo, Japan, pp. 40-45, May 2003.

[4] B.J. Falkowski and S. Rahardja, "Classification and Properties of Fast Linearly Independent Logic Transformations", *IEEE Trans. on Circuits and Systems II: Analog and Digital Signal Processing,* vol. 44, no. 8, Aug. 1997, pp. 646-655.

[5] P. Kerntopf, "A comparison of logical efficiency of reversible and conventional gates", *Proc. 9th IEEE International Workshop on Logic Synthesis*, Dana Point, CA, May 2000, pp. 261-269.

[6] G.A. Kukharev, V.P. Shmerko and E.N. Zaitseva, *Multiple-Valued Data Processing Algorithms and Systolic Processors*. Science and Engineering, Minsk, 1990.

[7] M.A. Perkowski, "A Fundamental Theorem for EXOR Circuits", *Proc. IFIP WG 10.5 Workshop on Applications of the Reed-Muller expansion in Circuit Design,* U. Kebschull, E. Schubert, W. Rosenstiel, Eds., Hamburg, Germany, Sept. 1993, pp. 52-60.

[8] S. Rahardja and B.J. Falkowski, "Polynomial Expansions over GF(2) based on Fastest Transformation", *Proc. 35th IEEE International Symposium on Circuits and Systems*, Scottsdale, Arizona, vol. 3, May 2002, pp. 377-380.

[9] R.S. Stankovic, M. Stankovic, and D. Jankovic, *Spectral Transforms in Switching Theory, Definitions and Calculations.* IP Nauka, Belgrade, 1998.

[10] R.S. Stankovic and J.T. Astola, *Spectral Interpretation of Decision Diagram.* Springer-Verlag, New York, 2003.

TABLE 1. Fastest Transforms.

	Forward $M_n \Leftrightarrow M_n^{-1}$ Inverse
Y1.1	$\begin{bmatrix} M_{n-1} & O_{n-1} & O_{n-1} \\ O_{n-1} & I_{n-1} & O_{n-1} \\ Y_{n-1} & O_{n-1} & M_{n-1} \end{bmatrix} \Leftrightarrow \begin{bmatrix} M_{n-1}^{-1} & O_{n-1} & O_{n-1} \\ O_{n-1} & I_{n-1} & O_{n-1} \\ 2Y_{n-1} & O_{n-1} & M_{n-1}^{-1} \end{bmatrix}$
Y1.2	$\begin{bmatrix} M_{n-1} & O_{n-1} & O_{n-1} \\ O_{n-1} & J_{n-1} & O_{n-1} \\ Y_{n-1} & O_{n-1} & M_{n-1} \end{bmatrix} \Leftrightarrow \begin{bmatrix} M_{n-1}^{-1} & O_{n-1} & O_{n-1} \\ O_{n-1} & J_{n-1} & O_{n-1} \\ 2Y_{n-1} & O_{n-1} & M_{n-1}^{-1} \end{bmatrix}$
Y1.3	$\begin{bmatrix} M_{n-1} & O_{n-1} & O_{n-1} \\ O_{n-1} & M_{n-1} & O_{n-1} \\ Y_{n-1} & O_{n-1} & M_{n-1} \end{bmatrix} \Leftrightarrow \begin{bmatrix} M_{n-1}^{-1} & O_{n-1} & O_{n-1} \\ O_{n-1} & M_{n-1}^{-1} & O_{n-1} \\ 2Y_{n-1} & O_{n-1} & M_{n-1}^{-1} \end{bmatrix}$
Y1.4	$\begin{bmatrix} M_{n-1} & O_{n-1} & Y_{n-1} \\ O_{n-1} & I_{n-1} & O_{n-1} \\ O_{n-1} & O_{n-1} & M_{n-1} \end{bmatrix} \Leftrightarrow \begin{bmatrix} M_{n-1}^{-1} & O_{n-1} & 2Y_{n-1} \\ O_{n-1} & I_{n-1} & O_{n-1} \\ O_{n-1} & O_{n-1} & M_{n-1}^{-1} \end{bmatrix}$
Y1.5	$\begin{bmatrix} M_{n-1} & O_{n-1} & Y_{n-1} \\ O_{n-1} & J_{n-1} & O_{n-1} \\ O_{n-1} & O_{n-1} & M_{n-1} \end{bmatrix} \Leftrightarrow \begin{bmatrix} M_{n-1}^{-1} & O_{n-1} & 2Y_{n-1} \\ O_{n-1} & J_{n-1} & O_{n-1} \\ O_{n-1} & O_{n-1} & M_{n-1}^{-1} \end{bmatrix}$
Y1.6	$\begin{bmatrix} M_{n-1} & O_{n-1} & Y_{n-1} \\ O_{n-1} & M_{n-1} & O_{n-1} \\ O_{n-1} & O_{n-1} & M_{n-1} \end{bmatrix} \Leftrightarrow \begin{bmatrix} M_{n-1}^{-1} & O_{n-1} & 2Y_{n-1} \\ O_{n-1} & M_{n-1}^{-1} & O_{n-1} \\ O_{n-1} & O_{n-1} & M_{n-1}^{-1} \end{bmatrix}$

TABLE 2. Number of Non-zero Spectral Coefficients.

TBF	Output	Y1.1	Y1.2	Y1.3	Y1.4	Y1.5	Y1.6	RM^{-1} *Best polarity*	$RM_{<0>}^{-1}$ *Zero polarity*
alu4	O_1	1390	1390	1214	1393	1393	1214	442	1112
	O_2	2151	2151	1794	2150	2150	1816	382	1045
	O_3	1550	1550	1510	1557	1557	1459	592	1322
	O_4	1131	1131	1145	1130	1160	1111	702	1287
apex4	O_1	14	14	19	13	13	20	51	112
	O_2	74	74	81	70	70	69	78	128
	O_3	54	54	60	57	57	66	77	137
	O_4	52	52	57	57	57	66	79	116
	O_5	62	62	65	68	68	73	79	138
	O_6	106	106	95	103	103	89	76	146
	O_7	47	47	47	49	49	45	85	159
	O_8	60	60	62	59	59	55	82	160
	O_9	11	11	11	14	14	16	52	118
	O_{10}	49	49	41	46	46	35	51	118
ex1010	O_1	64	64	78	71	71	77	133	163
	O_2	70	70	80	66	66	77	134	160
	O_3	55	55	63	54	54	65	136	156
	O_4	60	60	73	61	61	72	127	152
	O_5	58	58	71	59	59	71	137	157
inc	O_1	32	32	24	33	33	30	17	47
	O_2	38	38	32	38	38	36	30	51
	O_3	31	31	27	32	32	25	22	54
	O_4	17	17	20	16	16	16	23	54
	O_5	18	18	12	18	18	12	2	6

TABLE 3. Calculation Time.

TBF	Y1.1	Y1.2	Y1.3	RM^{-1} *Bestl polarity*	$RM_{<0>}^{-1}$ *Zero polarity*
9sym	0.03	0.02	0.02	4.98	0.02
alu4	1.92	1.93	1.94	3750.94	2.08
apex4	0.10	0.09	0.10	27.53	0.15
clip	0.03	0.03	0.03	10.11	0.06
con1	0.01	0.01	0.01	2.38	0.01
ex5	0.35	0.34	0.35	27.72	0.50
ex1010	0.04	0.05	0.05	15.04	0.08
inc	0.01	0.02	0.01	5.79	0.02
misex1	0.01	0.02	0.02	5.08	0.02
rd84	0.01	<0.01	0.01	2.58	0.01
squar5	0.01	0.01	0.01	2.40	0.01
xor5	<0.01	<0.01	<0.01	0.68	0.03
z5xp1	0.01	0.02	0.02	5.72	0.06

MULTI-POLARITY HELIX TRANSFORM OVER GF(3)

Cheng Fu and *Bogdan J. Falkowski*

School of Electrical and Electronic Engineering
Nanyang Technological University,
Block S1, 50 Nanyang Avenue, Singapore 639798

ABSTRACT

In this paper, the new multi-polarity helix transform for ternary logic functions has been introduced. In addition, an extended dual polarity property that had been used to optimize Kronecker and quaternary Fixed-Polarity Reed-Muller (FPRM) expressions has been applied to calculate efficiently this new multi-polarity helix transform over GF(3). The experimental results for the new transform are compared with well known ternary Reed-Muller transform and it was found that the new transform is quite efficient in terms of non-zero spectral coefficients and corresponding memory storage.

1. INTRODUCTION

Reed-Muller (RM) transform that represent an important class of AND-EXOR expressions had been successfully applied in many areas such as signal processing [1], fault detection [5-7] and coding techniques, especially those concerned with group or block codes for error control [5, 6]. Fixed-Polarity RM (FPRM) is the RM polynomial expansion in which each variable has the same form [2, 6-8]. An n-variable ternary logic function can be expressed by 3^n different FPRM polynomial expansions, where each of them is canonical and can be differentiated from each other by its polarity number [2]. FPRM polynomial expansions with different polarity numbers generally possess different computational complexity, which is measured by the number of nonzero spectral coefficients or the number of literals in the FPRM polynomial expansion. The polarity number for which the number of the used computational complexity measure is smallest is called the optimal polarity number. A method that optimizes binary FPRM based on the relationship between two FPRMs with polarities that are dual was presented in [9]. This method was extended for the optimization of Kronecker expressions by introducing the term extended dual polarity in [3] and for the optimization of FPRM expressions over Galois Field (4) (GF(4)) in [4]. As mentioned in [3], Kronecker expansions are potentially better than FPRMs in optimization of logic functions if the criterion is the number of non-zero terms. Therefore it is interesting to find novel Kronecker based transforms that are efficient in final polynomial representations of logic functions, have nice properties as well as can be calculated in efficient way. In this paper, the new ternary multi-polarity transform based on Kronecker product is introduced that is named helix transform due to the symmetrical structure along the diagonal or reverse-diagonal in the transform matrices. Application of extended dual polarity for the efficient optimization of multi-polarity helix transform over GF(3) is also introduced. Experimental results that show big advantage in terms of non-zero spectral coefficients of the new multi-polarity helix transform when compared with ternary Reed-Muller transform are also presented in this paper.

2. BASIC DEFINITIONS

Definition 1. Let $\vec{F}=[F_0,F_1,...,F_{3^n-1}]^T$ represent a column vector defining the truth vector of a ternary function $f(\overrightarrow{x_n})$ in a natural ternary ordering. The helix transform H_n is an $N\times N$ $(N=3^n)$ matrix with rows corresponding to minterms and columns corresponding to some switching ternary functions of n variables. If the sets of rows are linearly independent with respect to *ternary Galois Field*, then H_n has only one inverse in GF(3). The truth vector can be obtained by the following equation,

$$\vec{F}=H_n\vec{A}, \tag{1}$$

where $\vec{A}=[A_0,A_1,...,A_{3^n-1}]^T$ is the spectral coefficient column vector for the particular transform matrix H_n. The spectral coefficient column vector can be reconstructed by the following equation,

$$\vec{A}=H_n^{-1}\vec{F}. \tag{2}$$

Addition and multiplication in GF(3) are given in Table 1 and Table 2, respectively [2].

0-7803-8251-X/04/$17.00 ©2004 IEEE

TABLE 1. Addition

+	0	1	2
0	0	1	2
1	1	2	0
2	2	0	1

TABLE 2. Multiplication

*	0	1	2
0	0	0	0
1	0	1	2
2	0	2	1

Definition 2. Let X be a matrix of order $m \times n$ and Y be a matrix of order $p \times q$, then their ternary Kronecker product, denoted by $Z = X \otimes Y$ that is executed over GF(3) is a partitioned matrix Z of order $(mp)\times(nq)$ defined as the following equation, where all the matrix elements belong to set $\{0,1,2\}$.

$$Z = X \otimes Y = \begin{bmatrix} x_{11}Y & x_{12}Y & \dots & x_{1n}Y \\ x_{21}Y & x_{22}Y & \dots & \dots \\ \dots & \dots & \dots & \dots \\ x_{m1}Y & \dots & \dots & x_{mn}Y \end{bmatrix} \qquad (3)$$

Definition 3. The zero polarity ternary helix transform matrix of size $N \times N$ $(N = 3^n)$ is created by applying *n-1* times ternary Kronecker product to basic helix transform of size 3×3 where such basic ternary helix transform matrices fulfill Definition 1.

$$H_n = \overset{n-1}{\otimes} H_1;$$

$$H_1 = \begin{bmatrix} 1 & 0 & 0 \\ 1 & 1 & 1 \\ 0 & 0 & 1 \end{bmatrix}, \text{ and } H_1^{-1} = \begin{bmatrix} 1 & 0 & 0 \\ 2 & 1 & 2 \\ 0 & 0 & 1 \end{bmatrix}.$$

Definition 4. For arbitrary *n*, the ternary Reed-Muller transform in zero polarity is based on the following forward and inverse equations [2]:

$$TRM_n = \overset{n-1}{\otimes} TRM_1 = \overset{n-1}{\otimes} \begin{bmatrix} 1 & 0 & 0 \\ 1 & 1 & 1 \\ 1 & 2 & 1 \end{bmatrix}, \qquad (4)$$

$$TRM_n^{-1} = \overset{n-1}{\otimes} TRM_n^{-1} = \overset{n-1}{\otimes} \begin{bmatrix} 1 & 0 & 0 \\ 0 & 2 & 1 \\ 2 & 2 & 2 \end{bmatrix}, \qquad (5)$$

where "$\overset{n-1}{\otimes}$" represents Kronecker product applied *n*-1 times to either the matrix TRM_1 in Equation 4 or TRM_1^{-1} in Equation 5 with additions and multiplications over GF(3).

3. MULTI-POLARITY HELIX TRANSFORM

In this section, the new ternary multi-polarity helix transform will be presented whose all matrices fulfill Definitions 1-3.

Let $H_n^{<k>}$ represent the *k*-th polarity of the transform H_n. For any number of variables, the generalized multi-polarity helix transform matrix $H_n^{<k>}$ is recursively defined as:

$$H_n^{<k>} = \overset{n-1}{\otimes} H_1^{<k>}.$$

In general,

$$H_n^{<k=k_0k_1\dots k_{n-1}>} = H_1^{<k_0>} \otimes H_1^{<k_1>} \otimes \dots \otimes H_1^{<k_{n-1}>}.$$

All the elements in $H_n^{<k>}$ belong to the set {0, 1, 2}. The ternary multi-polarity helix transform $H_n^{<k>}$ can be derived from $H_n^{<0>}$ using 3-adic shift in the columns of $H_n^{<0>}$.

Similarly, the inverse transform matrix $(H_n^{-1})^{<k>}$ can be obtained by

$$(H_n^{-1})^{<k>} = \overset{n-1}{\otimes}(H_1^{-1})^{<k>}.$$

In general,

$$(H_n^{-1})^{<k=k_0k_1\dots k_{n-1}>} = (H_1^{-1})^{<k_0>} \otimes (H_1^{-1})^{<k_1>} \otimes \dots \otimes (H_1^{-1})^{<k_{n-1}>}.$$

For any number of variables, $(H_n^{-1})^{<k>}$ can be obtained from $(H_n^{-1})^{<0>}$, also by using 3-adic shift in the rows of $(H_n^{-1})^{<0>}$. Table 3 gives all the polarities of forward and inverse multi-polarity helix transform matrix when *n*=1.

TABLE 3. Ternary Multi-Polarity Helix Transform

	Forward	Inverse
$x^{<0>} = \begin{bmatrix} 0 \\ 1 \\ 2 \end{bmatrix}$	$H^{<0>} = \begin{bmatrix} 1 & 0 & 0 \\ 1 & 1 & 1 \\ 0 & 0 & 1 \end{bmatrix}$	$(H^{-1})^{<0>} = \begin{bmatrix} 1 & 0 & 0 \\ 2 & 1 & 2 \\ 0 & 0 & 1 \end{bmatrix}$
$x^{<1>} = \begin{bmatrix} 1 \\ 2 \\ 0 \end{bmatrix}$	$H^{<1>} = \begin{bmatrix} 1 & 1 & 1 \\ 0 & 0 & 1 \\ 1 & 0 & 0 \end{bmatrix}$	$(H^{-1})^{<1>} = \begin{bmatrix} 0 & 0 & 1 \\ 1 & 2 & 2 \\ 0 & 1 & 0 \end{bmatrix}$
$x^{<2>} = \begin{bmatrix} 2 \\ 0 \\ 1 \end{bmatrix}$	$H^{<2>} = \begin{bmatrix} 0 & 0 & 1 \\ 1 & 0 & 0 \\ 1 & 1 & 1 \end{bmatrix}$	$(H^{-1})^{<2>} = \begin{bmatrix} 0 & 1 & 0 \\ 2 & 2 & 1 \\ 1 & 0 & 0 \end{bmatrix}$

4. EXTENDED DUAL POLARITY

Jankovic, Stankovic, and Moraga have introduced the notion of extended dual polarity property and used it for optimization of Fixed Polarity Reed-Muller expressions (FPRM) over GF(4) [4]. A generalization of this method was also proposed by the same authors to optimize Kronecker expressions [3]. In this section, the extended dual polarity property is also used to the new ternary multi-polarity helix transform. The relationships between two multi-polarity helix transforms of extended dual polarities are also derived.

***Definition** 5.* $P' = (p'_1,\ldots,p'_{i-1},p'_i,p'_{i+1},\ldots p'_n)$ is an extended dual polarity for the given polarity $P = (p_1,\ldots,p_{i-1},p_i,p_{i+1},\ldots p_n)$ if $p'_j = p_j$, $j \neq i$ and $p'_i \neq p_i$.

Example: For polarity P=(1, 2), the extended dual polarities are (0, 2) and (1, 0).

The relationship between spectra of two multi-polarity helix transforms is given by the following equation

$$\overrightarrow{A^{<p'>}} = (H_n^{-1})^{<p'>} \cdot \vec{F} = (H_n^{-1})^{<p'>} \cdot H_n^{<p>} \cdot \overrightarrow{A^{<p>}}.$$

If polarities p' and p are extended dual polarities as described in Definition 5, and $p'_i \neq p_i$, then their relationship can be presented as follows

$$\overrightarrow{A^{<p'>}} = (H_n^{-1})^{<p'>} \cdot \vec{F}$$

$$= \left(\left(\bigotimes_{j=1}^{i-1} (H_n^{-1})^{<p_j>} \right) \otimes (H_n^{-1})^{<p'_i>} \otimes \left(\bigotimes_{j=i+1}^{n} (H_n^{-1})^{<p_j>} \right) \right) \cdot$$

$$\left(\left(\bigotimes_{j=1}^{i-1} H_n^{<p_j>} \right) \otimes H_n^{<p'_i>} \otimes \left(\bigotimes_{j=i+1}^{n} H_n^{<p_j>} \right) \right) \cdot \overrightarrow{A^{<p>}}$$

Due to the properties of Kronecker product, the shown relation can be rewritten as

$$\overrightarrow{A^{<p'>}} = \left(I_i \otimes (H_n^{-1})^{<p'>} \cdot H_n^{<p>} \otimes I_{n-i} \right) \cdot \overrightarrow{A^{<p>}},$$

where I_k is identity matrix of order k.

It is clear that for any spectrum of polarity p, its dual polarity spectrum vector $\overrightarrow{A^{<p'>}}$ can be calculated very efficiently by only obtaining the results of multiplication between two helix transforms when n=1. Table 4 gives all possible matrix products for $(H_n^{-1})^{<p'>} \cdot H_n^{<p>}$.

TABLE 4. Relationships between Different Polarities

$(H^{-1})^{<0>} \cdot H^{<1>} = (H^{-1})^{<1>} \cdot H^{<2>} = (H^{-1})^{<2>} \cdot H^{<0>} = \begin{bmatrix} 1 & 1 & 1 \\ 1 & 2 & 0 \\ 1 & 0 & 0 \end{bmatrix}$
$(H^{-1})^{<0>} \cdot H^{<2>} = (H^{-1})^{<1>} \cdot H^{<0>} = (H^{-1})^{<2>} \cdot H^{<1>} = \begin{bmatrix} 0 & 0 & 1 \\ 0 & 2 & 1 \\ 1 & 1 & 1 \end{bmatrix}$

In order to perform the exhaustive search through all the polarities and to keep any two consecutive polarities as extended dual polarities, the following route will be used by traversing a 3-valued n-dimensional hypercube as shown in Fig. 1 for the case when n=3 [3, 4].

(000)-(001)-(002)-(012)-(011)-(010)-(020)-(021)-(022)-(122)-(121)-(120)-(110)-(111)-(112)-(102)-(101)-(100)-(200)-(201)-(202)-(212)-(211)-(210)-(220)-(221)-(222)

Figure 1. Searching Route

The properties of extended dual polarity for ternary multi-polarity helix transform provide an efficient way to get all the polarities of the helix transform in an order of extended dual polarities. Thus, the polynomial expansions based on multi-polarity helix transform of any ternary functions can be optimized using exhaustive search algorithm to get the best polarity spectrum.

5. EXPERIMENTAL RESULTS

In this section, the ternary multi-polarity helix transform has been implemented using Microsoft Visual C++ and run on PIII 500 MHz computer with 128 MB RAM. It is run on some benchmark functions that have been modified to represent ternary functions instead of original MCNC and IWLS'93 binary benchmark functions.

The translation from binary to ternary cases has been done by changing every 2 input (output) bits in binary files to an input (output) symbol in ternary files. If the number of input and/or output variables is odd, then a zero bit is added behind the binary cubes to make it even. For input (output), -- is taken as -, 00 is taken as 0, 01 is taken as 1 and 10 is taken as 2, whereas 11 is not used (taken as 0) for ternary case.

TABLE 5. Number of Non-Zero Spectral Coefficients

Ternary functions	Output	$H^{-1}_{<best>}$	$TRM^{-1}_{<best>}$	Decrease rate
apex4	O_1	22	51	56.9%
	O_2	63	78	19.2%
	O_3	63	77	18.2%
	O_4	62	79	21.5%
	O_5	59	79	25.3%
	O_6	77	76	-1.3%
	O_7	60	85	29.4%
	O_8	69	82	15.9%
	O_9	23	52	55.8%
	O_{10}	37	51	27.5%
clip	O_1	63	62	-1.6%
	O_2	88	75	-17.3%
	O_3	38	41	7.3%
ex1010	O_1	112	133	15.8%
	O_2	103	134	23.1%
	O_3	105	136	22.8%
	O_4	105	127	17.3%
	O_5	107	137	21.9%
inc	O_1	20	17	-17.6%
	O_2	28	30	6.7%
	O_3	16	22	27.3%
	O_4	21	23	8.7%
	O_5	7	2	-250%

Table 5 presents the experimental results on some ternary benchmark functions. The third column shows the number of non-zero spectral coefficients for the best polarity ternary helix transform. They are compared with the experimental results for the best polarity ternary Reed-Muller transform that are given in the fourth column of Table 5. The used ternary benchmark functions have more than one output. It can be seen that for the functions shown in the first column with their outputs in the second column, the ternary multi-polarity helix transform can obtain smaller number of non-zero spectral coefficients after searching out the best polarity for majority of cases when compared with the best polarity ternary Reed-Muller transform. The last column of this table also shows the decreasing and sometimes increasing rates in the number of non-zero spectral coefficients when the best polarity ternary Reed-Muller transform is compared with the best polarity helix transform.

6. CONCLUSION

In this article, novel Kronecker based transform called ternary multi-polarity helix transform has been introduced which has very regular structure and efficient calculation. Kronecker based extended dual polarity properties [3, 4] are revised and applied for efficient calculation of ternary multi-polarity helix transform. The presented properties and relationships give an efficient method to optimize the corresponding spectral polynomial expansions based on multi-polarity helix transform of any ternary function. The comparison of experimental results between ternary multi-polarity helix transform and ternary Reed-Muller transform are also discussed, and they show that for almost all the cases of ternary benchmark functions, our new ternary multi-polarity helix transform is more efficient than ternary Reed-Muller transform in terms of bigger number of zero spectral coefficients.

7. REFERENCES

[1] S. Agaian, J. Astola and K. Egiazarian, *Binary Polynomial Transforms and Nonlinear Digital Filters*. New York: Marcel Dekker, 1995.

[2] P. Davio, J.P. Deschamps and A. Thayes, *Discrete and Switching Functions*. New York: George and McGraw-Hill, 1978.

[3] D. Jankovic, R.S. Stankovic and C. Moraga, "Optimization of Kronecker expressions using the extended dual polarity property", *Proc. XXXVII Int. Scientific Conference on Information, Communication and Energy Systems and Technologies*, Nis, Yugoslavia, pp. 749-752, Oct 2002.

[4] D. Jankovic, R.S. Stankovic and C. Moraga, "Optimization of GF(4) expressions using the extended dual polarity property", *Proc. IEEE 33rd Int. Symposium on Multiple-Valued Logic*, Tokyo, Japan, pp. 50-55, May 2003.

[5] M.G. Karpovsky, ed., *Spectral Techniques and Fault Detection*. Orlando: Academic Press, 1985.

[6] K. Kinoshita, K. Asada and O. Karatsu, *Logic Design for VLSI*. Tokyo: Iwanami Shoten Publishers, 1985.

[7] T. Sasao, *Switching Theory for Logic Synthesis*. Boston: Kluwer Academic, 1999.

[8] R.S. Stankovic, M. Stankovic and D. Jankovic, *Spectral Transforms in Switching Theory, Definitions and Calculations*. Belgrade: IP Nauka, 1998.

[9] E.C. Tan, H. Yang, "Optimization of Fixed-Polarity Reed-Muller circuits using dual-polarity property", *Circuits, Systems and Signal Processing*, Vol. 19, No. 6, pp. 534-548, 2000.

A DESIGN FLOW FOR MULTIPLIERLESS LINEAR-PHASE FIR FILTERS: FROM SYSTEM SPECIFICATION TO VERILOG CODE

Kai-Yuan Jheng, Shyh-Jye Jou*, and An-Yeu Wu

Graduate Institute of Electronics Engineering, National Taiwan University, Taipei 106, Taiwan, ROC
*Department of Electrical Engineering, National Central University, Jhongli 320, Taiwan, ROC

ABSTRACT

This paper presents a design flow for the multiplierless linear-phase FIR filter synthesizer, which combines several research efforts. We propose a local search algorithm with variable filter order to reduce the number of adders further. In addition, several design techniques are adopted to reduce the hardware complexity of the system. By using this synthesizer, the system designers can design a filter efficiently and a chip can be successfully finished in a very short time.

1. INTRODUCTION

Recent rapid progress in very large scale integrated (VLSI) circuit technology has led to an emerging theme – "System-on-a-Chip" (SoC). With the increase in the density and complexity in VLSI circuit, the design costs for the development of a VLSI chip are also increased. It calls for rapid prototyping and design reuse of major silicon intellectual property (SIP) modules to alleviate the designer's effort and to speed up the design process. Therefore, computer aided design (CAD) tools play an important role in decreasing the design cycle time and accurately simulating the correctness of the circuit design.

The synthesizer we presented can automate the FIR filter design from the system specification to the corresponding synthesizable Verilog hardware description language (HDL) code. Because the synthesizer only requires the system-level specification, the synthesizer allows system designers, who are inexperienced in VLSI design, to design filters easily and concentrate on system design and performance evaluation. Therefore, by using this synthesizer, an efficient design of a chip can be successfully completed in a few minutes.

The rest of this paper is organized as follows. In Section 2, the design flow of the filter synthesizer and several hardware reduction methods are presented. An experimental result of a filter design example synthesized with our automatic design tool is then shown in Section 3. Finally, some conclusions will be given in Section 4.

This work was supported by MediaTek Inc., under NTU-MTK wireless research project.

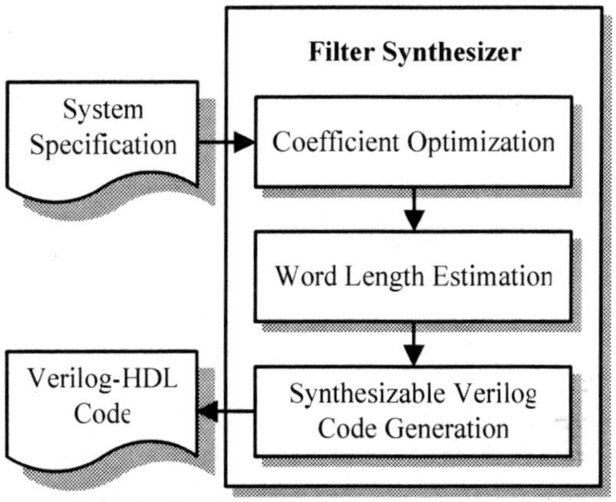

Fig. 1 Design flow of the synthesizer.

Table 1 Input data of system specification.

Filter Type (low-pass, high-pass, band-pass, band-stop)	
Normalized Passband and Stopband Edge Frequencies	ω_P , ω_S
Passband Ripple in Magnitude or dB	δ_P / A_P
Stopband Attenuation in Magnitude or dB	δ_S / A_S
Input and/or Output Data Word Length (bit)	W_{in} / W_{out}
Signal to Noise Ratio (dB)	SNR

2. SYNTHESIZER IMPLEMENTATION

The system configuration and dataflow of the synthesizer are shown in Fig. 1. The synthesizer consists of many subprograms. The main subprograms are the coefficient optimization, the word length estimation, and the synthesizable Verilog code generation. All programs are written in C++ language.

In this system, the input is the system-level specification, which is listed in Table 1. In addition, the architecture uses the symmetric transposed direct form filter structure with the MSB Fix technique [1], which is frequently adopted by high-speed designs.

2.1 Coefficient Optimization

- Coefficient Calculation

In this subprogram, we integrate the MATLAB engine [2] into our synthesis tool. The floating-point filter coefficient set is calculated by the generalized Reméz method [3] as given in the MATLAB gremez.m function.

● Optimization Algorithm

Numerous search algorithms for the design of multiplierless filters with canonic signed digit (CSD) or signed powers-of-two (SPT) coefficients have been proposed. However, they did not explore the possibility of further reduction of nonzero digits by taking the filter order as a variable parameter. In general, if a filter gradually increases the tap length N, its frequency response will become severer. Thus, we can allow more margins for coefficient quantization error by increasing the filter tap length. Besides, an observation [4] shows that one can start with a filter, which exceeds the given criteria that may involve acceptable level of increase in the filter order, but with much lesser total nonzero digits than the initial design. Therefore, we adopt a two-step local search algorithm proposed by Samueli [5] and exploit variable filter order to improve the method. The number of total nonzero digits typically decreases with N. However, there is a limit to N since the overhead increases with N.

2.2 Word Length Estimation

● Overflow Prevention

If the final output is within the range of the word length, overflow in partial sums are unimportant. This is a desirable property of 2's complement arithmetic. However, if the final output exceeds the range of the word length, the value of the output sample will be wrong and methods should be taken to prevent this. An approach is to avoid or allow limited overflow by scaling the coefficients. The coefficients h(k) may be scaled in the following way:

$$\hat{h}(k) = h(k) \cdot 2^{-R} \tag{1}$$

where

$$R = \left\lceil \log_2 \left(\sum_{k=0}^{N-1} h^2(k) \right)^{\frac{1}{2}} \right\rceil \tag{2a}$$

or

$$R = \left\lceil \log_2 \sum_{k=0}^{N-1} |h(k)| \right\rceil \tag{2b}$$

where R denotes right shift bit(s). The method given in (2a) probably lead to shorten internal word length than (2b) but this form of scaling will occasionally occur overflow which result in performance degradation. Therefore, the method in (2b) is adopted which never cause overflow because it is based on the worst-case conditions for overflow. Hence, the coefficient word length increases R bit(s) and the coefficients are then shifted right R bit(s) to prevent overflow.

● Internal Word Length Reduction

In digital signal processing, the finite word length has a strong effect on the system performance since it dominates the precision of the output signals. The increment of internal word length will lead to a better signal-to-noise ratio (SNR), but it would also increase the hardware complexity, consume more power, and slow down the system operation frequency. Therefore, it is a trade-off that the designer should take care of.

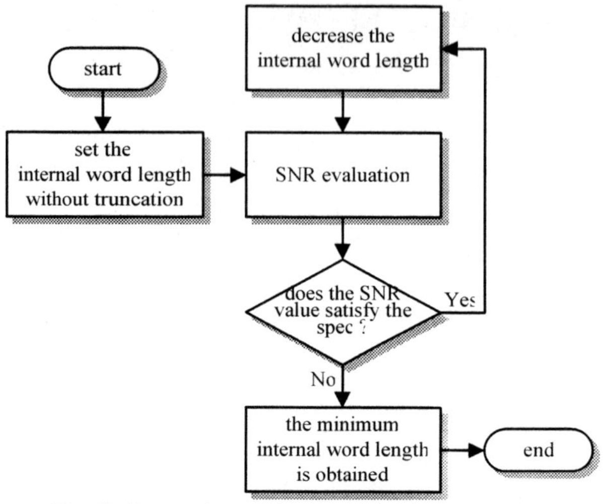

Fig. 2 Internal word length reduction flowchart.

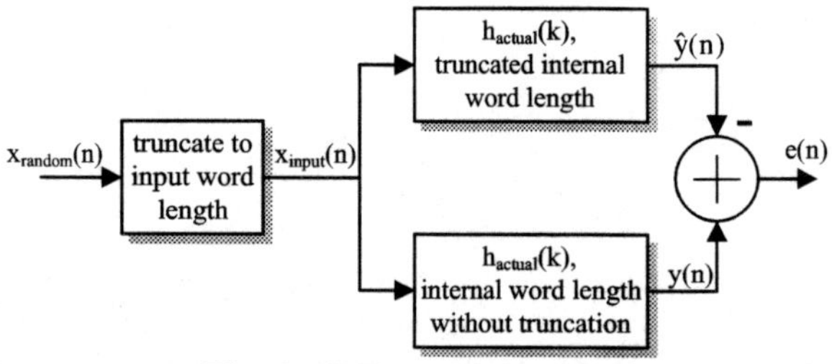

Fig. 3 SNR evaluation block.

It is observed that if designer is willing to accept some deviation from the given specifications, the decrement of internal word length enable a reduction of hardware complexity. In this subprogram, we involve a deviation index SNR that is defined in (3)

$$SNR = 10\log\left[\frac{E(y^2(n))}{E(e^2(n))}\right] = 10\log\left[\frac{E(y^2(n))}{E(y(n)-\hat{y}(n))^2}\right] \tag{3}$$

The internal word length reduction flowchart and SNR evaluation block are shown in Fig. 2 and Fig. 3 respectively. The initial internal word length will be evaluated for the result that does not introduce any error first. Then the internal word length will be decreased to the value that its SNR value still fits the specification. Finally, the minimum internal word length, which fulfills the specification, will be obtained.

2.3 Synthesizable Verilog Code Generation

Finally, we will generate three types of the symmetric transposed direct form FIR filters as shown in Fig. 4.

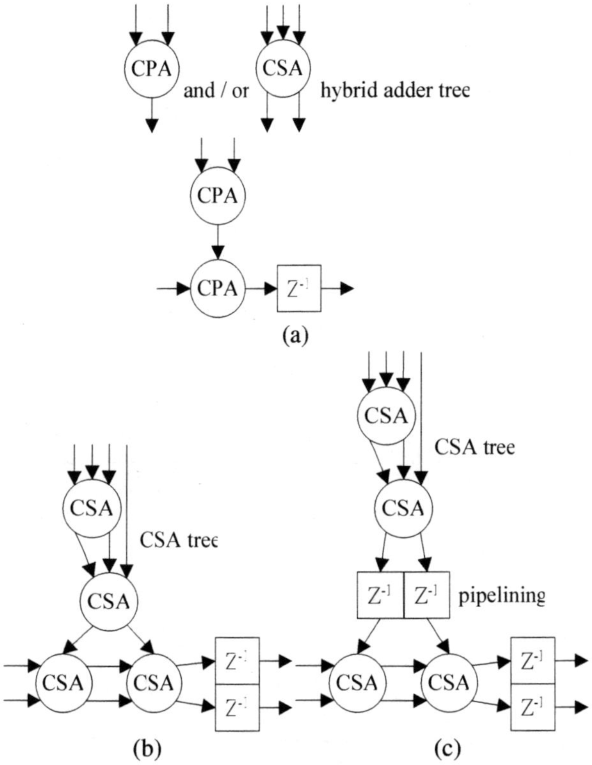

Fig. 4 Transposed direct form filter strcture written in (a) behavioral code, (b) DesignWare components and, (c) DesignWare components with pipelining.

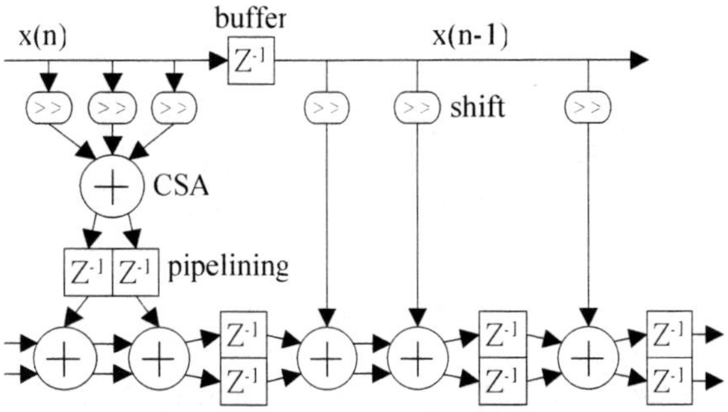

Fig. 5 The Structure C with an input buffer.

- Structure A: Fig. 4(a)

The transposed direct form filter structure is adopted and written in behavior level synthesizable Verilog-HDL code, which allows the synthesis tool to select the appropriate architecture for user's constraints.

- Structure B: Fig. 4(b)

The transposed direct form filter structure is utilized with carry save adders (CSA) written in DesignWare components [6] provided by Synopsys.

- Structure C: Fig. 4(c)

We exploit structure B with pipelining to achieve a two-CSA delay critical path. Moreover, the nonzero digits of most CSD coefficient sets is generally less than three so we can use a single input buffer rather than pipelining at each tap. Referring to Fig. 5, the input x(n) is for the taps whose nonzero digits are more than two and x(n-1) is for less then three.

Table 2

Minimum number of SPT terms required to attain -50dB NPR

Algorithm	#SPT	N
Max. SPT per coeff. = 4		
MILP [7]	68	28
Samueli [5]	66	28
Our Work #1	54	29
Our Work #2	52	30
Max. SPT per coeff. = 3		
MILP [7]	68	28
Samueli [5]	cannot reach -50 dB	
Our Work #3	57	29

Table 3 Word length estimation results.

Input Word Length	Coefficient Word Length	Internal Word Length	Output Word Length
14 bits	11 bits	14 bits	14 bits

3. DESIGN EXAMPLE

A linear-phase low-pass FIR filter is designed using our proposed method, the mixed integer linear programming (MILP) algorithm [7], and Samueli's local search algorithm [5]. The pass-band and stop-band edge frequencies are 0.3π and 0.5π, respectively. The normalized peak ripple (NPR) δ_{NPR}=-50dB. The word length of the input signal is assumed 14 bits.

The minimum number of SPT terms required by the various methods mentioned above is summarized in Table 2. The frequency responses and coefficients of the filter designed by our proposed method are shown in Fig. 6. When the maximum allowed number of SPT terms per coefficient is limited to four, the filter designed by our methods saves 22%(21%~24%) SPT terms and costs 5%(4%~7%) additional tap length. If the application requires us to limit the maximum number of SPT terms per coefficient to three, for a higher throughput rate, the filter designed using Samueli's algorithm failed to reach -50 dB NPR. However, using our proposed method can save 16% SPT terms and costs 4% additional tap length.

Secondly, the design results of the word length estimation are summarized in Table 3. In general, the SNR is set more than 40 dB for practical implementation.

Lastly, the design results are converted into three structures mentioned in Section 2.3. We then use the Synopsys Design Complier to synthesize the filters with TSMC 0.25μm process. The synthesis results of Work #1 are summarized in Table 4. The area is measured in equivalents of 2-input NAND gates.

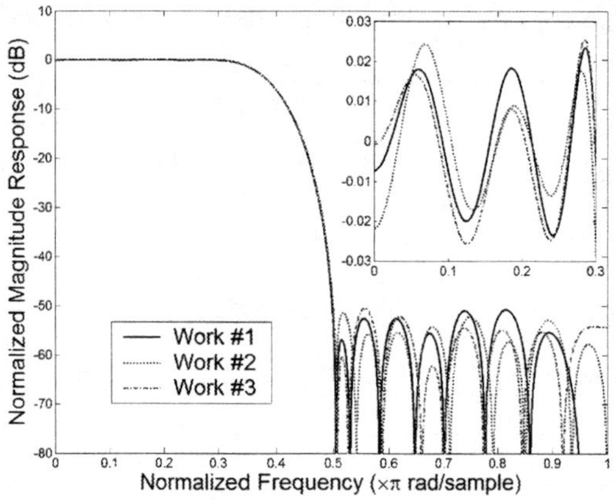

Work #1	
$h(0) = h(28) = -2^{-9} - 2^{-11}$	$h(8) = h(20) = 2^{-5} + 2^{-8} - 2^{-11}$
$h(1) = h(27) = -2^{-9} - 2^{-11}$	$h(9) = h(19) = 0$
$h(2) = h(26) = 2^{-8}$	$h(10) = h(18) = -2^{-4}$
$h(3) = h(25) = 2^{-7}$	$h(11) = h(17) = -2^{-4} + 2^{-7}$
$h(4) = h(24) = 0$	$h(12) = h(16) = 2^{-3} - 2^{-5} - 2^{-7}$
$h(5) = h(23) = -2^{-6} + 2^{-11}$	$h(13) = h(15) = 2^{-2} + 2^{-5} + 2^{-8}$
$h(6) = h(22) = -2^{-6} + 2^{-8} - 2^{-11}$	$h(14) = 2^{-1} - 2^{-3} + 2^{-8} + 2^{-10}$
$h(7) = h(21) = 2^{-6} + 2^{-10}$	

Work #2	
$h(0) = h(29) = -2^{-10}$	$h(8) = h(21) = 2^{-5} + 2^{-7} - 2^{-10}$
$h(1) = h(28) = -2^{-8}$	$h(9) = h(20) = 2^{-5}$
$h(2) = h(27) = 0$	$h(10) = h(19) = -2^{-4} + 2^{-6} + 2^{-8} + 2^{-10}$
$h(3) = h(26) = 2^{-7}$	$h(11) = h(18) = -2^{-3} + 2^{-5}$
$h(4) = h(25) = 2^{-7} - 2^{-10}$	$h(12) = h(17) = 0$
$h(5) = h(24) = -2^{-7} - 2^{-9}$	$h(13) = h(16) = 2^{-2} - 2^{-7} - 2^{-9}$
$h(6) = h(23) = -2^{-5} + 2^{-7} + 2^{-9}$	$h(14) = h(15) = 2^{-1} - 2^{-4} + 2^{-6}$
$h(7) = h(22) = 0$	

Work #3	
$h(0) = h(28) = -2^{-9} - 2^{-11}$	$h(8) = h(20) = 2^{-5} + 2^{-7} - 2^{-9}$
$h(1) = h(27) = -2^{-9} - 2^{-11}$	$h(9) = h(19) = 0$
$h(2) = h(26) = 2^{-8}$	$h(10) = h(18) = -2^{-4} - 2^{-8}$
$h(3) = h(25) = 2^{-7} + 2^{-11}$	$h(11) = h(17) = -2^{-4} + 2^{-8}$
$h(4) = h(24) = 0$	$h(12) = h(16) = 2^{-3} - 2^{-5} - 2^{-9}$
$h(5) = h(23) = -2^{-6} - 2^{-11}$	$h(13) = h(15) = 2^{-2} + 2^{-4} - 2^{-7}$
$h(6) = h(22) = -2^{-6} + 2^{-8} - 2^{-10}$	$h(14) = 2^{-1} - 2^{-3} + 2^{-5}$
$h(7) = h(21) = 2^{-6} + 2^{-9}$	

Fig. 6 Magnitude responses and coefficient sets of the filters.

The synthesis results show that structure A is suitable for the low-speed (133MHz) and area-efficient application; Structure B is suitable for the high-speed (400MHz) application; and Structure C is suitable for the very high-speed (800MHz) application. Therefore, our filter synthesizer can provide flexible hardware implementation for various applications.

Table 4 Synthesis results of Work #1.

(a) Timing Constraint: 7.50(ns)

Structure	A	B	C
Critical Path (ns)	7.46	4.65	4.65
Total Gate Count	5069	8103	9119
Combinational Area	2824	3613	3907
Noncombinational Area	2245	4490	5212

(b) Timing Constraint: 2.50(ns)

Structure	A	B	C
Critical Path (ns)	3.86	2.50	2.50
Total Gate Count	8338	8563	9225
Combinational Area	5799	4038	3973
Noncombinational Area	2539	4525	5252

(c) Timing Constraint: 1.25(ns)

Structure	A	B	C
Critical Path (ns)	3.86	1.57	1.25
Total Gate Count	8338	11520	12862
Combinational Area	5799	5595	5999
Noncombinational Area	2539	5925	6863

4. CONCLUSION

We have implemented a multiplierless FIR filter synthesizer written in C++ language and combined the MATLAB engine with our automatic design tool. We have also shown that the local search algorithm with variable filter order towards further reduction in the number of total nonzero digits. The variable filter order approach can be applied to other coefficient optimization algorithms.

Several design techniques are adopted to reduce the hardware complexity of the system. For flexible hardware implementation, we provide three structures that structure A is suitable for low-power applications and structures B, C are suitable for high-performance applications. We also find that the coefficient sets produced by our tool have many common terms, so common sub-expression elimination (CSE) techniques will be studied in the future.

5. REFERENCES

[1] B. C. Wong and H. Samueli, "A 200-MHz all-digital QAM modulator and demodulator in 1.2μm CMOS for digital radio applications," IEEE JSSC, pp. 1970-1979, Dec. 1991.

[2] "MATLAB external interfaces," MathWorks, Sep. 2003.

[3] D. J. Shpak and A. Antoniou, "A generalized Reméz method for the design of FIR digital filters," IEEE Trans. Circuits Syst., pp. 161-174, Feb. 1990.

[4] M. Bhattacharya and T. Saramaki, "Some observations on multiplierless implementation of linear phase FIR filters," IEEE ISCAS, pp. 193-196, May 2003.

[5] H. Samueli, "An improved search algorithm for the design of multiplierless FIR filters with powers-of-two coefficients," IEEE Trans. Circuits Syst., pp. 1044-1047, Jul. 1989.

[6] "DesignWare foundation library databook," Synopsys, Inc., Jan. 2002.

[7] Y. C. Lim and S. R. Parker, "FIR filter design over a discrete powers-of-two coefficient space," IEEE Trans. ASSP, pp. 583-591, Jun. 1983.

SHIELDING AREA OPTIMIZATION UNDER THE SOLUTION OF INTERCONNECT CROSSTALK[1]

Xin Zhao, Yici Cai, Qiang Zhou, Xianlong Hong

Dept. of CST, Tsinghua Univ.
Beijing 100084, P.R.China
Tel: +86-10-62785564
Fax: +85-10-62781489
e-mail: plantree99@mails.tsinghua.edu.cn

Lei He, Jinjun Xiong

Dept. of EE, UC, Los Angeles
Los Angeles, CA 90095-1594, USA
Tel: +1-310-206-2037
Fax: +1-310-206-4685
e-mail:lhe@ee.ucla.edu

ABSTRACT

As the technology advances into deep sub-micron era, crosstalk reduction is of paramount importance for signal integrity. Simultaneous shield insertion and net ordering (SINO) has been shown to be effective to reduce both capacitive and inductive coupling. Although shield insertion could reduce crosstalk efficiently, a large scale of unnecessary shields will bring the shielding area problem which is also critical for an efficient SINO algorithm. In this paper, we propose three novel algorithms using fewer shields to solve SINO problem: namely, net coloring (NC), efficient middle shield insertion (EMSI) and NC+EMSI. Compared to the corresponding algorithms in previous work [1], our algorithms can reduce shielding area largely with short runtime.

1. INTRODUCTION

It has been shown for years that interconnect crosstalk and delay have become bottle necks in determining circuit performance. Even though most current research on interconnect synthesis uses the RC model [2, 3, 4, 5, 6], it is evident that the RLC model becomes more appropriate as the on-chip inductive effect gains increasing prominence in gigahertz designs [7]. As the inductive has long-range effect in the sense that the mutual inductance between non-adjacent nets cannot be ignored when compared to the self inductance [1]. Therefore, shield insertion is needed to reduce inductive noise. However, as shielding area is directly decided by the number of shields, extra shields will waste the shielding area. It is of paramount important to find an efficient way to solve crosstalk problem. The more efficiently shields minimize the crosstalk, the fewer shields need to insert, and the shielding area could be much more optimized. Furthermore, as the increase of IC scale and the number of wires, the short routing resource and the routing congestion become more serious. It is crucial to develop algorithms to reduce the shielding area considering crosstalk at the same time.

Several previous studies have considered interconnect optimization under the RLC model. Assuming that current will return from the nearest shield, the loop inductance model is used in [1, 8, 9]. A table-based partial inductance model is adopted without pre-assuming any current return path [10, 11, 12]; additionally, a coupling inductance screening rule [9] is employed to decide the scope of the current return path. Also, a formula-based K_{eff} model is proposed in [8] as the figures of merit for inductive coupling. Although, the assumption of K_{eff} model is less intuitive to the designer [7], it is easy to compute and keeps a high fidelity versus the SPICE-computed RLC noise voltage for SINO solutions [8].

Based on K_{eff} model [8], Prof He provided greedy shield insertion, graph coloring and simulated annealing SINO algorithms to solve SINO/NB problem [1]. As overabundance shields are inserted in these algorithms, there is still much space for reducing the number of shields to economize the routing resources. In order to optimize the shielding area, in this paper, we develop three algorithms which are extremely efficient to reduce the shielding area while satisfying the crosstalk constraints. First, NC algorithms is proposed to color nets so that no sensitive wires adjacent to each other. Then, considering inserting shield in the key position, the EMSI algorithm is provided to reduce the inductive coupling. Finally, combining NC and EMSI algorithms into NC+EMSI algorithm, noise-bounded problems could be solved using fewer shields.

The remainder of this paper is organized as follows: Section 2 briefly introduces the SINO problem and mostly analyzes algorithms provided in this paper. Section 3 presents three SINO algorithms as Net Coloring algorithm, Efficient Middle Shield Insertion algorithm and NC+EMSI algorithm separately. Section 4 compares experimental results obtained by different algorithms to the existing algorithms [1]. Section 5 concludes the paper.

2. PROBLEM FORMULATIONS

2.1 Preliminaries

In this paper we consider coplanar interconnect structures [8] with inductive K_{eff} model [8]. Shielding area is directly decided by the number of shields with the same width as signal wires (denoted as s-wires). And the group of wires sandwiched between adjacent shields is called a block. Two net s_1 and s_2 are defined to be sensitive to each other if a switching signal on s_1 will cause s_2 to malfunction or vice-versa [1]. A sensitivity matrix S is used to represent the sensitivity of wires. In the rest of paper, formula-based K_{eff} model [8] is used to compute the inductive coupling coefficient K_{ij} between two wires.

[1] This paper is supported by NSFC(60176016) and High-Tech Research & Development (863) Program Of China 2002AAIZ1460 and Specialized Research Fund for the Doctoral Program of Higher Education:SRFDP-20020003008

2.2 Optimal SINO Problems

Considering capacitive and inductive coupling, SINO/NF and SINO/NB problems are defined in [1] to solve crosstalk minimization. For a given placement P, find a new placement P' by simultaneous shield insertion and net re-ordering such that the total area of P' is minimal and P' is noise free. This defines SION/NF problem. In the requirement that P' is capacitive noise free and the inductive coupling K_i satisfied $K_i \le K_{thresh}$ where K_{thresh} is a uniform thresh value, this forms SINO/NB problem.

2.3 Problem Analysis

Shields are used to reduce noise, while overabundance shields may waste the shielding area which is directly decided by the number of shields. Although, existing SINO algorithms [1] reduce both capacitive and inductive noise, there are still lots shields unnecessary. As greedy shield insertion operates [1] inefficiently not aiming at the key position influencing coupling most, we do lots of experiments to find the wires with max k value.

The coplanar interconnect structures containing 8, 16, 32, 64 s-wires with no shield are used to find the position of wires whose k value is the max in the block. We iterate the following sensitivity rates: 40%, 50% and 60% and run each condition under 50 random sensitivities. Table 1 shows the average position of wires having the max k value.

sensitivity rate	8 wires	16 wires	32 wires	64 wires
40%	3.10	6.05	15.42	31.24
50%	3.15	7.15	15.58	31.78
60%	3.10	7.45	15.60	31.68

Table 1. Average net position having the max value of k

The experimental results containing 64 s-wires under each 60%, 50%, 40% random sensitivities are provided in Figure 1. Here, x-axis indicates the experimental serial number of the 50 random sensitivities, and the y-axis indicates the wire order number from 0 to SIZE-1 (where SIZE is the number of wires).

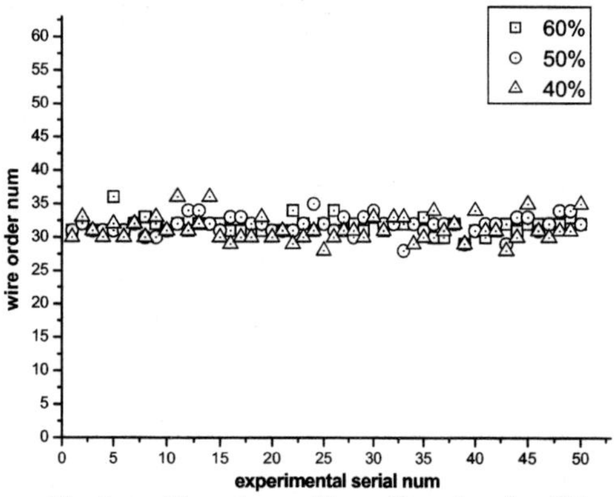

Fig. 1. The wire position of max k value (64 s-wires)

From the Figure 1, we find that k_max wire always appear from 30th to 32nd wire. So, the wire having the max k value mostly appears in the middle of the block, and by shielding these wires the k value in the block can be reduced efficiently. Finding this efficient way to reduce noise, fewer shields are needed and the shielding area can be optimized. The efficient middle shield insertion algorithm (EMSI algorithm) is developed based on this discussion.

3. ALGORITHM FOR OPTIMAL SINO PROBLEMS

In shielding area optimization, capacitive noise free is defined if wire s_i is not adjacent to any other wires sensitive to it, and the inductive noise free is defined if sensitive wires do not share a block. We first introduce a Net Coloring algorithm (NC algorithm) to solve noise-free problem, which makes no sensitive wires adjacent to each other. As the SINO/NF problem is over-constrained and may lead to over-designed solutions not according to realistic design constraints, which may need a large number of shields. Also, the EMSI algorithm depends on the initial placement. So we combined NC and EMSI algorithms into NC+EMSI algorithm to solve SINO/NB problem. The number of shields can be reduced by first running net coloring algorithm to reorder nets so that no sensitive nets are adjacent to each other, then invoking the EMSI algorithm could reduce inductive coupling noise using shield as few as possible.

3.1 Net Coloring Algorithm

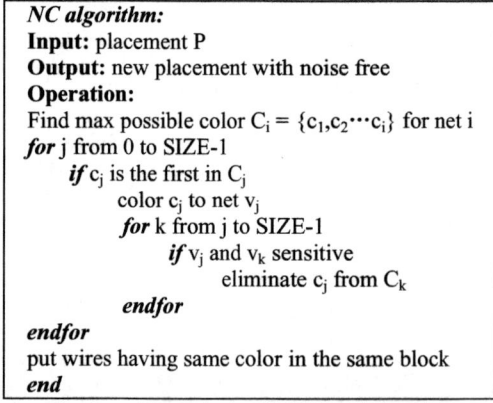

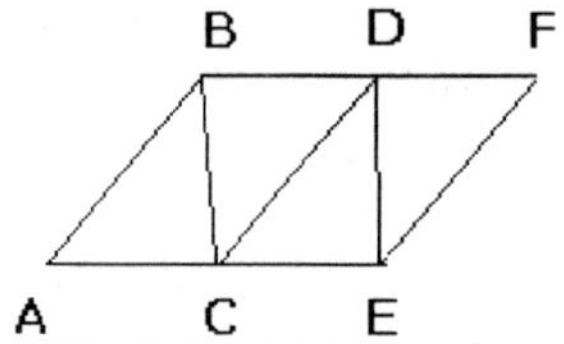

Fig 2. Sensitivity graph

The sensitivity of all wires could also be represented by graph shown in Figure 2, which is constructed as that a node corresponds to a wire and an edge exists if and only if the correspondent wires are sensitive to one another. By coloring the sensitivity graph, wires having the same color are not sensitive to each other.

NC algorithm:
Input: placement P
Output: new placement with noise free
Operation:
Find max possible color $C_i = \{c_1, c_2 \cdots c_i\}$ for net i
for j from 0 to SIZE-1
    if c_j is the first in C_j
        color c_j to net v_j
        for k from j to SIZE-1
            if v_j and v_k sensitive
                eliminate c_j from C_k
        endfor
endfor
put wires having same color in the same block
end

Fig. 3. Net Coloring SINO/NF Algorithm

In Figure 3, we present the net coloring based SINO algorithm to solve noise-free problem. It attempts to color all s-wires sequentially so that wires in the same block have a single color. Assuming that the set of colors is numbered by

$\{c_1, c_2 \ldots c_i\}$, where c_i represents the ith color. SINO/NC algorithm works in the following way: First, for each wire i enumerate the max available set of color C_i, where $C_i = \{c_1, c_2, \ldots c_i\}$. Then color the wires sequentially. Considering each wire, the first color in the set C_i of wire i is the final color to wire i. Meanwhile, in order to satisfy no wire sensitive to each other be colorred the same, we delete the color from the set of other pending wires which are sensitive to the wire i considered. Finally, put the wires having the same color in the same block, so that wires in the same block are not sensitive to each other.

3.2 Efficient Middle Shield Insertion Algorithm

Efficient middle shield insertion algorithm is mainly about how to insert the shield to solve SINO/NB problem. To optimize the shielding area, we should use as fewer shields as possible to reduce the inductive coupling noise.

> ***EMSI Algorithm:***
> **Input:** placement P, K_{thresh}
> **Output:** new placement with noise bounded
> **Operation:**
> Given a placement P
> ***for*** each block in P
>   ***if*** the k_{eff} value of block > K_{thresh}
>   insert a shield in the middle of the block
>   point to the new left block
>   ***else***
>   point to the next right block
> ***endfor***

Fig. 4. Efficient Middle Shield Insertion SINO/NB Algorithm

As discussed in section 2, the wire having the max k value mostly appears in the middle of the block, and by shielding these wires inductive coupling in the block can be reduced efficiently. In Figure 4, we present the efficient middle shield insertion algorithm for SINO problem (SINO/EMSI algorithm). This algorithm is a recursive procedure. Run through all the blocks in the given placement; calculate the max k value of the block. After inserting a shield in the middle of one block, the algorithm recursively checks the K_{eff} value of the two newly-generated blocks. If any of these is still greater than K_{thresh}, no matter whose k value is the max, an additional middle shield is inserted in the new block.

3.3 NC+EMSI Algorithm

As discussed in section 2, we combined NC and EMSI algorithms into NC+EMSI algorithm to solve noise-bounded problem. Net coloring operation is first used to eliminate capacitive noise. While, different from the NC algorithm which inserts shields between every pair of wires in different colors, net coloring before EMSI algorithm only inserts shields between the adjacent sensitive wires not having the same color. Then efficient middle shield insertion is operated to reduce inductive noise using as fewer shields as possible.

4. EXPERIMENTAL RESULTS

In this section, we apply these three algorithms on a large number of examples. And the results obtained by these algorithms are compared to the existing algorithms [1].

We have implemented all algorithms in the C++ programming language on a SUN E450. And we use the coplanar interconnect structures containing 32 and 64 s-wires to determine the performance of the algorithms for different combinations of K_{thresh} and uniform sensitivity rate. We use four different K_{thresh} values as 0.5, 1.0, 1.5, 2.0 respectively, here K_{thresh}=0.5 represents the total inductive coupling coefficient for each net is less than 0.5. And the sensitivity rate is changed from 40%, 50% to 60%. That the sensitivity rate is 40% represents each net is sensitive to 40% of all nets, and these sensitive nets are selected randomly. For each combination of K_{thresh} value and sensitivity rate, we present the resulting number of shielding wires for different algorithms. We run each algorithm on the same initial placement for twenty different random sensitivities. And the average of these twenty runs is shown in italic style in table2.

number of net	Kthresh	GC[1]	*NC*	SI[1]	*EMSI*	NO+SI[1]	*NC+EMSI*
	Net Sensitivity Rate :40%						
32	0.5	8.50	*3.94*	16.70	*13.94*	6.30	*6.01*
	1.0			15.90	*12.46*	5.30	*4.72*
	1.5			15.40	*8.68*	4.40	*3.99*
	2.0			13.80	*6.56*	4.00	*3.44*
	Net Sensitivity Rate :50%						
	0.5	11.00	*4.72*	18.90	*13.96*	8.40	*6.53*
	1.0			18.40	*12.46*	5.70	*4.97*
	1.5			18.00	*8.14*	4.80	*4.60*
	2.0			17.50	*6.44*	4.40	*3.87*
	Net Sensitivity Rate :60%						
	0.5	12.00	*5.02*	22.80	*14.08*	6.60	*6.72*
	1.0			22.10	*12.36*	6.00	*5.30*
	1.5			22.00	*8.22*	5.20	*4.60*
	2.0			21.50	*6.72*	4.50	*3.91*
64	Net Sensitivity Rate :40%						
	0.5	15.60	*3.92*	30.10	*28.92*	11.30	*7.76*
	1.0			28.50	*25.96*	9.20	*6.88*
	1.5			26.20	*18.32*	7.50	*5.82*
	2.0			23.50	*14.00*	6.80	*5.30*
	Net Sensitivity Rate :50%						
	0.5	19.30	*4.78*	32.10	*29.12*	14.30	*8.17*
	1.0			31.30	*25.88*	9.80	*7.29*
	1.5			30.60	*17.52*	8.60	*6.55*
	2.0			31.50	*13.96*	7.90	*5.71*
	Net Sensitivity Rate :60%						
	0.5	21.60	*5.11*	39.70	*28.72*	12.80	*8.21*
	1.0			39.20	*25.68*	11.40	*7.45*
	1.5			38.00	*17.36*	9.40	*6.73*
	2.0			36.80	*14.40*	8.10	*5.85*

Table 2. Number of shields inserted by SINO algorithms

For each combination of k and sensitivity rate, we compare our algorithms investigated above to the existing algorithms [1], the EMSI algorithm is compared to the greedy shield insertion algorithm [1], and the NC+EMSI algorithm is compared to the NO+SI algorithm [1], respectively here.

We can see from table 2, the greedy shield insertion algorithm is significantly worse than the EMSI algorithm. For example, in 32 wires with sensitivity rate (60%) and K_{thresh} changes from 0.5, 1.0, 1.5 to 2.0, EMSI algorithm reduces 27.25%, 30.44%, 43.06%, 46.19% shields respectively compared to SI algorithm [1]. Furthermore, as we expected, the EMSI algorithm has a higher efficiency to reduce the inductive coupling noise, along with the increase of K_{thresh}, the number of shields by EMSI algorithm reduces obviously, while the result of SI algorithm [1] reduces insignificantly. See 64 s-wires, sensitivity rate=50%, when K_{thresh} increase from 0.5, 1.0, 1.5 to 2.0 separately, the number of shields in EMSI algorithm to the given size (64 s-wires) decreases from 45.50%, 40.44%, 27.38% to 21.81% respectively with 23.69% reduction, but the results of

SI algorithm are 50.16%, 48.91%, 47.81%, 49.22% with at most 2.35% reduction. So, middle shield insertion in EMSI algorithm may efficiently reduce the inductive coupling noise, and the shielding area reduces significantly.

Compared to the graph coloring algorithm [1], the net coloring algorithm also needs 18.25%, 22.69% and 25.77% less shields separately than SINO/GC algorithm [1] in 64 s-wires with sensitivity rate from 40%, 50% to 60%.

Considering separated net ordering and shield insertion operation [1], NC+EMSI algorithm not only follows the requirement of net ordering operation about eliminate the capacitive coupling noise, but also gathers wires not sensitive to each other into the same block by the greatest extent, which is helpful to the following shield insertion operation to reduce the inductive coupling noise. From the experimental results, NC+EMSI algorithm also shows a good performance. Take 64 s-wires with sensitivity rate 60% and K_{thresh} value increases from 0.5, 1.0, 1.5 to 2.0 as example, the NO+SI algorithm [1] needs 20.00%, 17.81%, 14.69%, 12.66% shields, while relatively, the NC+EMSI algorithm needs 12.83%, 11.64%, 10.52%, 9.14% shields to the given size 64 , which means 7.17%, 6.17%, 4.17%, 3.52% less shielding area than the NO+SI algorithm[1].

	EMSI	NC	NC+EMSI
Runtime	0.013sec	0.00sec	0.004sec

Table 3. Approximate run times for SINO algorithms with sensitivity rate 60%

Finally, the runtime is showed in Table 3, where the times are for a single interconnect structures with a single run. SUN E450 is used to collect running time. From this table, all algorithms finished the examples in a few seconds. Therefore, large scale interconnect structures can be solved fast by algorithms we have proposed.

5. CONCLUSION AND DISCUSSIONS

As more and more high performance microprocessors and system-on-chip (SoC) operates in GHz+ scale, RLC crosstalk gains growing importance for signal integrity. In this paper, we have developed three efficient algorithms to solve the shielding area optimization. Extensive experiment results have shown that compared to the previous corresponding algorithms in [1] the proposed algorithms use less runtime while reduce the shielding area efficiently. SINO is a region based technique to reduce RLC crosstalk. Performing SINO within each region separately may introduce many dog-legs across regions. For full-chip routing optimization, we want to reduce the number of dog-legs as they deteriorate signal integrity [6]. We plan to study the row-based SINO problem with consideration of both shielding area and dog-leg minimization in the future, and develop efficient algorithms correspondingly.

6. REFERENCES

[1] L. He and K. M. Lepak, "Simultaneous shield insertion and net ordering for capacitive and inductive coupling minimization", in ISPD, pp. 55-60, 2000

[2] Gao, T. and Liu, C. L. "Minimum Crosstalk Channel Routing", in ICCAD, pp.692-696, 1993.

[3] Gao, T. and Liu, C. L. "Minimum crosstalk switchbox routing", in ICCAD, pp. 610-615, 1994.

[4] Yim, J. S. and Kyung, C. M. "Reducing Cross-Coupling amongn Interconnect Wires in Deep-Submicron Datapath Design", in DAC, pp.485-189, June, 1999.

[5] Xue, T. and Kuh, E. S. "Post global routing crosstalk synthesis", in IEEE Trans. CAD-16, pp.1418-1430, 1997.

[6] Chang, C.-C. and Cong, J. "Pseudo pin assignment with crosstalk noise control", in ISPD, pp. 41-47, April, 2000

[7] K. M. Lepak, I. Luwandi, and L. He, "Shield insertion and net ordering under explicit RLC noise constraint", in DAC, pp.199-202, June, 2001

[8] L. He and M. Xu, "Modeling and Layout Optimization for On-chip Inductive Coupling", U. of Wisconsin at Madison, Technical Report ECE-00-1

[9] S. Lin, N. Chang, and O. S. Nakagawa, "Quick on-chip self- and mutual-inductance screen", in International symposium on Quality of Electronic Design, pp.513-520, March, 2000.

[10] A. Ruehli, "Equivalent circuit models for three-dimensional multiconductor systems", IEEE Trans. On MIT, pp.216-221 , 1974.

[11] L. He, N. Chang, S. Lin, and O. S. Nakagawa, "An Efficient Inductance Modeling for On-Chip Interconnects", in IEEE CICC, pp.457-460, May, 1999.

[12] J. D. Ma and L. He, "Formulae and Applications of Interconnect Estimation Considering Shielding Insertion and Net Ordering", IEEE/ACM ICCAD, pp.327-332, November 2001.

PERFORMANCE METRICS FOR ASYNCHRONOUS DIGITAL CIRCUITS APPLICABLE TO COMPUTER-AIDED DESIGN

Rajani Parthasarathy and Ivan S. Kourtev

Department of Electrical Engineering
University of Pittsburgh
Pittsburgh, PA, USA, 15261

Abstract—Computer-Aided Design (CAD) circuit techniques for synchronous circuits generally improve the quality of a circuit based on previously specified performance metrics such as cycle-time, latency and throughput (often expressed in terms of the clock period). Limited techniques to minimize worst-case delays and physical area of *asynchronous* circuits have been developed. Optimization algorithms for these circuits must recognize their average-case speed and control performance by using quantifiable metrics specific to asynchronous circuits. The types of asynchronous circuits and the factors affecting their performance are surveyed in this paper. Techniques to compute and optimize the cycle-time, latency and throughput are analyzed to determine what these terms mean in the context of asynchronous circuits and the performance analysis of these circuits.

1. INTRODUCTION

The increasing complexity of integrated circuits (IC's) requires efficient CAD tools permitting designers to meet stringent design and performance requirements. Speed, area, and power, are generally viewed as the key performance metrics of synchronous digital IC's [1]—the speed of computations, for example, is expressed in terms of the clock period and is a quantifiable representation of the overall circuit performance. However, lack of CAD tools and optimization methods, including specific metrics to measure performance, limits progress in asynchronous design. Previous research in this area exists but includes relatively few discussions of performance analysis of asynchronous circuits. It is desirable to investigate what performance metrics are applicable to asynchronous circuits. In this paper, the factors affecting the performance of various types of asynchronous designs are discussed in Section 2, the performance metrics used in analyzing asynchronous circuits are addressed in Section 3, and conclusions are offered in Section 4.

2. ASYNCHRONOUS CIRCUITS

In a typical synchronous circuit, a global clock signal controls the timing of operations in the circuit and synchronizes all concurrent computations in the logic. Performance evaluation of synchronous circuits is in terms of the clock period length. In an asynchronous circuit, on the other hand, there is frequently no global synchronization and the sequencing of operations is achieved through local control in the logic circuit elements. In this section, different asynchronous circuits are analyzed to determine the factors affecting the performance of these asynchronous circuits.

Non-Pipelined Datapaths. Asynchronous circuits such as adders, multipliers and dividers have been demonstrated for arithmetic computations [2]. A simple adder implemented using a bounded-delay asynchronous circuit is shown in Figure 1(a). The performance of this circuit is evaluated by observing the data delay from the input to the output of the circuit. It takes 1 ns for the registers Reg A and Reg B to output the data to the combinational logic. The combinational logic then takes a maximum of 5 ns to compute the sum. Therefore, the total time taken by the bounded-delay adder to generate the sum is no longer than 6 ns. The delay element in Figure 1(a) models this worst-case delay of 6 ns.

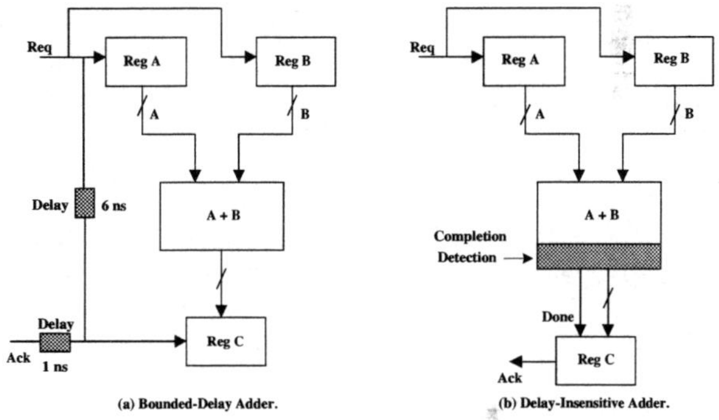

Fig. 1. Asynchronous adder: (a) Bounded-delay adder; (b) Delay-insensitive adder.

A delay-insensitive (DI) adder is shown in Figure 1(b) [note that the registers in the DI adder shown in Figure 1(b) are identical to the registers in Figure 1(a)]. Delay-insensitive circuits use a completion detection circuitry to indicate the completion of a process. The time taken by the combinational logic shown in Figure 1(b) to complete computation may vary depending on the data inputs to the adder. If, for example, the time taken by the combinational logic is, 4 ns for 50 % of all possible inputs and 5 ns for the remaining inputs, and if, the completion is detected in 1 ns, then the total computation time of the DI adder will be either 6 ns or 7 ns depending on the inputs. The computation time of an asynchronous circuit may, therefore, be a fixed worst-case value or vary over a range of data-dependent values [3]. In these cases, optimizing the asynchronous circuit is equivalent to minimizing the worst-case delay for bounded-delay circuits or the average delay for DI circuits.

Pipelined Datapaths. Pipelined asynchronous arithmetic circuits (adders and others) have been demonstrated. For example, an elastic asynchronous micropipeline [4] is illustrated in Figure 2(a) and uses a bundle of bounded-delay data wires. The delay-insensitive control logic uses a pair of a Request and Acknowledge handshake signals—a signal event on the Request wire indicates to the next stage that the data is ready and an event on the Acknowledge wire completes the data transfer. The total delay of a stage in a micropipeline is $t_{\text{bounded-delay}} = t_{\text{worst-case-comp}} + t_{\text{sync}}$ where $t_{\text{worst-case-comp}}$ and t_{sync} are the worst-case stage computation delay and the synchronization time during which the data and control signals are sent to the next stage, respectively.

In asynchronous pipelines with data-dependent delays, stage delays may vary. A DI pipeline with processing logic is illustrated in Figure 2(b). The total delay of a stage in a DI asynchronous pipeline is $t_{\text{DI}} = t_{\text{average-case-comp}} + t_{\text{completion}}$ where $t_{\text{average-case-comp}}$ and $t_{\text{completion}}$ are the average stage computation time and the time required by the completion detection circuitry [2], respectively.

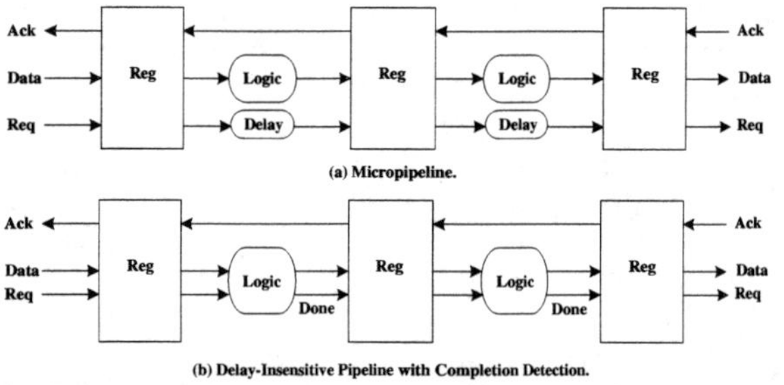

Fig. 2. Asynchronous Pipelines. (a) Micropipeline. (b) Delay-Insensitive Pipeline with Completion Detection.

In *unbalanced* pipelines (varying stage delays) the average-case performance of isolated stages can be matched by buffering these stages to obtain a deterministic throughput [1] (note, however, that buffering increases the area of the pipeline). Therefore, in pipelines with matched average delays (achieved by buffering), the area time product indicates an important area/speed tradeoff and is a more accurate measure of the circuit performance [1].

Counterflow pipelines [5] are asynchronous pipelines where instructions and data flow in opposite directions (*e.g.,* the 28-stage-long counterflow pipelined chip 'Zeke'). The throughput is a function of *(i),* the number of items in the pipelines and *(ii),* the instruction and data dependent interaction between the pipelines.

Asynchronous Memories. With the advent of asynchronous designs for processors and DSPs, asynchronous interfaces and peripherals for these asynchronous processors have been developed—asynchronous memory modules, for example, have been demonstrated for various processors. The self-timed approach is the most popular design technique used in the design of asynchronous memories [6]. The asynchronous static RAM (SRAM) in [6] has been designed using a self-timed approach and a handshaking communication protocol. In the self-timed asynchronous SRAM design in [6], the access times to different locations in the memory may differ. Completion detection circuits detect the completion of the read and write accesses to the memory, thereby improving the average speed of the memory. The memory performance depends on the capacity of the memory, read/write access times, and (in certain cases) on the locations of the memory that are being accessed.

Controller/Decoder Circuits. Asynchronous controller designs include cache controllers, various memory and disk drive controllers [7]. In *fundamental mode* asynchronous finite-state machine (FSM) designs [8], for example, only one input bit may change at a time and successive input changes must be delayed until all outputs stabilize [8]. The strict timing constraints in the fundamental mode FSM design, however, render this design unsuitable for large, complex cases. A less restrictive model, called the *burst-mode* design permits multiple input bit changes (input burst) to occur at the same time.

In synchronous circuits, the FSM changes state on the occurrence of a clock signal. The performance of synchronous sequential FSMs have been analyzed using the cycle time or clock period as the metric. In the asynchronous FSM using burst mode, state changes can occur only when a new input *burst* occurs. Therefore, there is no fixed cycle time for asynchronous FSMs that determines the performance of the circuit.

Mixed Interfaces. Problems in the distribution of clock signals across VLSI systems may be largely reduced by using Globally-Asynchronous, Locally-Synchronous systems (GALS) [9]. In a GALS system, synchronous modules using locally generated clocks communicate asynchronously with each other using a handshaking protocol. The local clocks may run at different speeds. A basic GALS module is shown in Figure 3 [10].

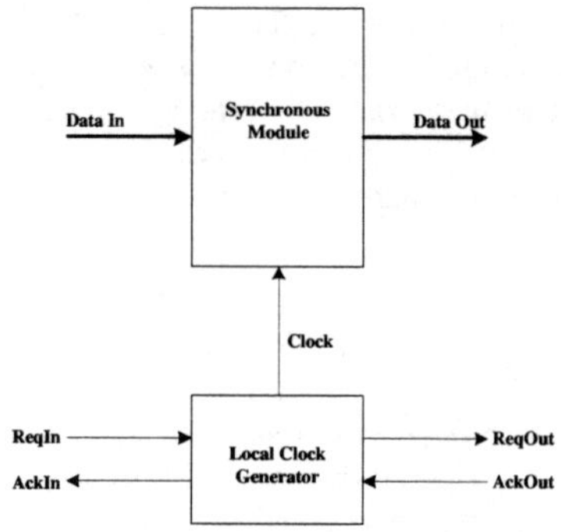

Fig. 3. Basic Module of Globally Asynchronous Locally Synchronous System [10].

It is difficult to determine the latency of a GALS system due to the existence of multiple clock domains. In [11], for example, the speed of a GALS processor is compared to the speed of a synchronous processor by means of simulations. The performance of a GALS system further depends on factors such as the specific algorithm, the type of synchronizer used between the synchronous and asynchronous modules, as well as the overall design partitioning [11]. It is also possible to use multiple local clocks in GALS with multiple supply voltages [11]. In this case, each synchronous module must be optimized for its own best clock period and supply voltage combination in order to yield optimal speed and power consumption of the overall system.

3. QUANTIFYING PERFORMANCE

The performance of asynchronous circuits has been analyzed using both simulation and analytical methods [2, 12]. While analytical methods may lead to arguably accurate results, these methods are difficult to formulate for complex asynchronous designs. Simulation results, on the other hand, may not always be accurate or may lead to non-deterministic performance.

In computational asynchronous circuits with data-dependent processing delays, for example, it may be impossible to compute the average delay of the circuit by simulation. A large set of random inputs must be used to accurately assess the average delay of such circuits [2]. Performance analysis based on average delays computed using random inputs is useful in research and experimental work where the same set of random inputs can be used to compare the performance of different designs [2] but cannot be used to obtain a deterministic performance measure of a circuit in general. Also, the set of random inputs may include inputs which do not occur in a real world application. In such cases the actual and simulated average delays may be different.

Analytical approaches modeling an asynchronous circuit using Petri Nets and other symbolic techniques using Binary Decision Diagrams (BDDs), have been proposed [12]. Generalized-Timed Petri Nets (GTPNs) [3] have been used to model different types of asynchronous circuits and measure the cycle time of asynchronous FSMs. Petri nets, however, exhibit a state explosion problem and are impractical for large designs. Also, the data-dependent delays used in the model [3] are based on the particular distribution of the data being used.

Symbolic techniques using BDDs and ADDs (Algebraic Decision Diagrams) have been developed to analyze the performance of asynchronous circuits [12]. These techniques use a discrete-time model of the asynchronous system specified by a set of nondeterministic finite state machines. In this model, a probability distribution of delays is assigned to all the FSM components with variable delays. These techniques permit analysis of larger systems compared to Petri-Nets. However, the accuracy of the analysis depends on the probability distribution of delays applied to the finite state machines.

Ideally, synchronous pipelines are optimized by increasing the throughput without increasing the latency. Because adding stages to a pipeline increases both the throughput and the latency, optimization at the micro-architecture level of synchronous pipelines involves choosing an optimum number of pipeline stages. In asynchronous pipelined circuits, however, it is possible to adjust the throughput and latency independent of each other by varying the average number of items in the pipeline [1]. Therefore, the optimization tools for asynchronous pipelines must consider the average number of data items in the pipeline while optimizing the throughput and latency of the pipeline. The optimum number of stages in the pipeline cannot be derived from a simple tradeoff between throughput and latency.

In pipelined circuits with matched gate delays in all stages, the pipeline throughput (computed by either using the gate delay values or by simulation) is used as a measure of the circuit performance. For pipelines with data-dependent processing times, analytical work is still in progress to develop methods for computing the circuit throughput—most results have been demonstrated through simulations [1] or modeling the circuit with Timed Petri Nets [3].

In synchronous circuits, reducing the stage delay—reducing the clock period without adding stages—generally increases the throughput. In asynchronous circuits, reducing the delay of a few stages in the pipeline to improve the performance of isolated stages does not necessarily improve the throughput of the pipeline [1]. An example of this situation is shown in Figure 4. The asynchronous pipeline shown in Figure 4(a) outputs 5 data items in 21 ns under steady-state conditions. If the delay of the first stage in this pipeline is reduced from 5 ns to 4 ns, then the resulting circuit [shown in Figure 4(b)] outputs 5 data items in 24 ns. The key to understanding this behavior is that a pipeline stage that is faster

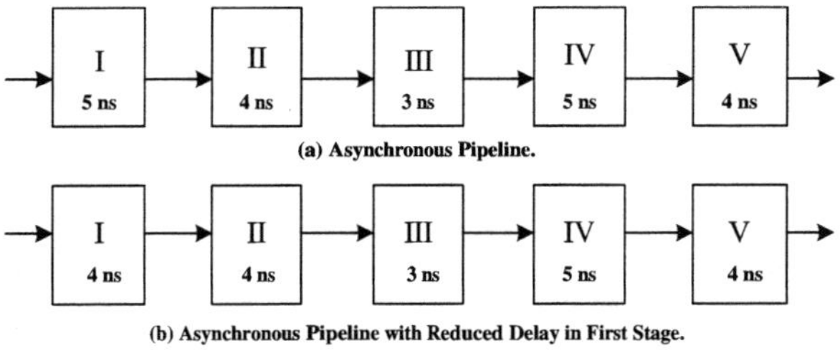

Fig. 4. Performance of Asynchronous Pipelines. (a) Asynchronous Pipeline. (b) Asynchronous Pipeline with Reduced Delay in First Stage.

than its preceding or succeeding stages must still wait for those stages to complete. It is possible to improve the performance of asynchronous pipelines by adding buffers between stages to match the stage delays but this approach increases the area of the circuit.

3.1. Unified Performance Metrics

Performance metrics for synchronous circuits primarily focus on the clock period (delay) and area of a circuit. The traditional performance metrics for synchronous digital circuits have been area A, delay T, area-delay product AT, and the area-squared delay product AT^2. However, the cost of heat removal and the requirements for portable systems make power consumption an increasingly important performance metric. Power-delay product (PDP) is often used to measure the performance of synchronous CMOS circuits. The power-delay product is a measure of the amount of average energy required to switch the output of a gate from low to high and from high to low. The PDP is a fairly accurate measure of the tradeoffs between power and delay in a circuit optimization problem. The PDP for a gate [13] may be computed using

$$PDP = 2(C_{load}V_{DD}^2 f_{max})\tau_p, \quad (1)$$

where f_{max} is the maximum operating frequency and τ_p is the average gate propagation delay. This expression evaluates to

$$PDP = C_{load}V_{DD}^2. \quad (2)$$

The result in Equation (1) may, however, be misleading if the energy per switching event is interpreted as a function of the operating frequency f_{max}. The energy is a function of the switching capacitance and the operating voltage and is independent of the operating frequency [13]. Therefore, the energy-delay product is used more often as a metric for synchronous circuits. The energy-delay product denoted by the notation Et is a good metric of the tradeoff between the power consumption and the speed of the circuit. The Et metric has been used in optimization problems to determine the optimal supply and threshold voltages and transistor sizing in synchronous circuits.

As the constraints on power consumption become more severe in low power applications, new metrics leading to both energy and delay efficient circuits have been proposed for both synchronous and asynchronous circuits. One such metric is Et^2 [14]. The metric Et^2 is independent of supply voltage variation for CMOS circuits. The tradeoff between energy and delay through voltage variation that exists for Et is, therefore, eliminated by Et^2. An example in [14] shows the computation of the optimal Et^2 for a pipeline which results in the optimal pipeline depth. The pipeline stages in this example, however, have equal delays. Methods to optimize unbalanced pipelines using the metric Et^2 must be explored.

A metric not limited to CMOS technology is Et^n where n is an optimization index [15]. It has been shown in [15] that a circuit composed of parallel or sequential components can be optimized using Et^n. The components of the circuit can be optimized independently and then recomposed to give the global optimum for the circuit.

Note that the energy and delay values must be estimated correctly for the metrics Et^2 and Et^n to be used in the optimization of asynchronous circuits. An energy estimation method has been described for asynchronous circuits in [16]. This method is, however, specific to quasi delay-insensitive (QDI) asynchronous circuits and is not accurate due to the dependence of the delays on the input data pattern. In QDI circuits, however, the dependence of the delays and energy consumption on the input pattern is weak. This method has been applied for the first time to an asynchronous MIPS R3000 microprocessor to estimate the energy consumption of different parts of the microprocessor [16]. Research is still in progress for using this energy estimation method in optimization problems involving the metric Et^2.

A new metric that has been recently proposed for synchronous circuits is the *hardware intensity* η [17]. Although the metric η has been proposed for synchronous circuits, it is possible to conceive methods to adapt η to asynchronous circuits as well. The hardware intensity defines a relation between the critical path delay, average energy dissipated per cycle and the circuit hardware. The metric η provides an architectural energy efficiency criterion for optimization. The metric η gives a measure of how much the logic has to be restructured to result in minimum energy and minimum delay for a given supply voltage. Simplified derivations of the metric η for a synchronous processor under different clocking conditions such as free running clock and partial clock gating are given in [17].

4. CONCLUSIONS

Performance analysis and optimization of asynchronous digital circuits is still an evolving area. Only a few performance analyses techniques and optimization tools are available and research in this area is just beginning.

In this paper, the various factors affecting the performance of asynchronous circuits have been discussed. Optimization tools must be aware of these factors for each type of asynchronous circuit. The performance measures used to compare and optimize different designs have also been reviewed. Traditional methods used to optimize synchronous circuits may not be sufficient to improve the performance of asynchronous circuits and different techniques must be developed. In a limited number of cases, the dependence of performance on the input data pattern further complicates the problem of optimization. There is a need for mature Computer-Aided Design techniques to analyze and optimize asynchronous circuits based on appropriate performance metrics.

5. REFERENCES

[1] D. Kearney and N. W. Bermann, "Performance Evaluation of Asynchronous Logic Pipelines with Data Dependant Processing Delays," *Asynchronous Design Methodologies*, pp. 4–13, IEEE Computer Society Press, May 1995.

[2] M. A. Franklin and T. Pan, "Performance Comparison of Asynchronous Adders," *Proc. International Symposium on Advanced Research in Asynchronous Circuits and Systems*, pp. 117–125, Nov. 1994.

[3] P. Kudva, G. Gopalakrishnan, and E. Brunvand, "Performance Analysis and Optimization for Asynchronous Circuits," *Proc. International Conf. Computer Design (ICCD)*, IEEE Computer Society Press, Oct. 1994.

[4] I. E. Sutherland, "Micropipelines," *Communications of the ACM*, Vol. 32, pp. 720–738, June 1989.

[5] B. Coates, J. Ebergen, J. Lexau, S. Fairbanks, I. Jones, A. Ridgway, D. Harris, and I. Sutherland, "A Counterflow Pipeline Experiment," *Proc. International Symposium on Advanced Research in Asynchronous Circuits and Systems*, pp. 161–172, Apr. 1999.

[6] V. W.-Y. Sit, C.-S. Choy, and C.-F. Chan, "A Four-Phase Handshaking Asynchronous Static RAM Design for Self-Timed Systems," *IEEE Journal of Solid-State Circuits*, Vol. 34, pp. 90–96, Jan. 1999.

[7] S. M. Nowick, M. E. Dean, D. L. Dill, and M. Horowitz, "The Design of a High-Performance Cache Controller: a Case Study in Asynchronous Synthesis," *Proc. Hawaii International Conf. System Sciences*, Vol. I, pp. 419–427, IEEE Computer Society Press, Jan. 1993.

[8] S. Hauck, "Asynchronous Design Methodologies: An Overview," Tech. Rep. TR 93-05-07, Department of Computer Science and Engineering, University of Washington, Seattle, 1993.

[9] E. G. Friedman, *Clock Distribution Networks in VLSI Circuits and Systems*. IEEE Press, 1995.

[10] C. Myers, *Asynchronous Circuit Design*. John Wiley & Sons, 2001.

[11] A. Iyer and D. Marculescu, "Power and Performance Evaluation of Globally Asynchronous Locally Synchronous Processors," *IEEE*, 2002.

[12] A. Xie and P. A. Beerel, "Symbolic Techniques for Performance Analysis of Timed Systems based on Average Time Separation of Events," *Proc. International Symposium on Advanced Research in Asynchronous Circuits and Systems*, pp. 64–75, IEEE Computer Society Press, Apr. 1997.

[13] S.-M. Kang and Y. Leblebici, *CMOS Digital Integrated Ciruits Analysis and Design*. McGraw-Hill, Inc., 1996.

[14] A. J. Martin, "Towards an energy complexity of computation," *Information Processing Letters*, Vol. 77, pp. 181–187, 2001.

[15] P. Pénzes and A. J. Martin, "Energy-Delay Efficiency of VLSI Computations," *Proc. of the Great Lakes Symposium on VLSI*, 2002.

[16] P. I. Pénzes and A. J. Martin, "An Energy Estimation Method for Asynchronous Circuits with Application to an Asynchronous Microprocessor," *Proc. Design, Automation and Test in Europe (DATE)*, pp. 640–647, Mar. 2002.

[17] V. Zyuban and P. Strenski, "Unified Methodology for Resolving Power-Performance Trade-offs at the Microarchitectural and Circuit Levels," *International Symposium on Low Power Electronics and Design*, pp. 161–171, Aug. 2002.

RTL/ISS CO-MODELING METHODOLOGY FOR EMBEDDED PROCESSOR USING SYSTEMC

Yoichi YUYAMA†, *Masao ARAMOTO*†, *Kazutoshi KOBAYASHI*†† *and Hidetoshi ONODERA*†

†Dept. of Comm. and Comp. Eng. , Graduate School of Informatics, Kyoto University, Kyoto 606-8501, Japan
††VLSI Design and Education Center, The University of Tokyo, Tokyo 113-8656, Japan

ABSTRACT

We propose ISS/RTL co-modeling methodology by describing both in common source file using SystemC. Our method enables rapid and easy generation/verification of RTL/ISS of customizable processor. We apply this method to processor "MiU–Processor". As a result code-sharing ratio is 67 %. For adding new instruction, we add only 12 lines to RTL/ISS shared part. Our ISS generation method is very effective for multi customizable processor SoC.

1. INTRODUCTION

Embedded systems typically consist of an embedded processor and several peripheral hard-wired logics. But the number of transistors that can be designed by a designer is limited, while the number of transistors on an SoC is increasing accroding to the Moore's law. Lots of processor-based SoC architectures are proposed [1–5] .

In these SoCs, a system consists of customized processors connected in parallel. Each processor is specialized for assigned function of system. Many customization approaches, for example, adding new feature to processor, varying bit width, and so force are proposed.

An instrcution set simulator (ISS) is required for each customized processsor. It is usually modeld with C/C++ languages in the behavioral level, while the RTL model of the processor is described with an HDL in the RTL level. But system level design language can describe hardware, software and system specification. This paper proposes ISS/RTL co-modeling methodology by describing both descriptions in a common source file. Our methodology enables rapid and easy design/verification of each model.

This paper is organized as follows. Section 2 describes customizable processors and its verification methodology. We explain ISS/RTL co-modeling methodology with SystemC [6] in Section 3. In Section 4, we apply the proposed method to our developed 機能特化型プロセッサ (MiU–Processor) and evaluate its code-sharing ratio, simulation speed and so on. Finally, we conclude this paper in Sect. 5.

2. CUSTOMIZABLE PROCESSOR AND ITS VERIFICATION FOR NEW SOCS

Recently, many embedded SoC architectures based on customizable processors are proposed. In this section, we explain conventional customizable processors and these verification flow.

2.1. Customizable Processor

As Application Specific Instruction set Processors (ASIP), PEAS–III [1, 2], Xtensa [3], etc. are proposed. Valen–C [4] is proposed as a variable-bit width processor. Almost all ASIPs consist of a processor core with built-in accelerators. These accelerators need to be designed in RTL.

MiU–Processor [7] , MeP(Media Embedded Processor) [5] and etc. are proposed as a customizable processor for multi-processor SoC. These methods divide whole system into some function blocks, which are distributed to customized processors optimized for assigned function. Many customization approaches, adding new instruction, deleting unnecessary instruction, co-processor and so on are proposed.

2.2. Verification of Customizable Processor

Verification time among the entire design flow is increasing according to the process minimization. We have to minimize the verification time for fast implementation. Generally speaking, customizable embedded processors are described in RTL. In the conventional approaches, RTL or ISS simulation is used for processor verification. An RTL verification simulates an actual circuit in a processor at RTL. It enables very accurate verification, but its simulation time becomes longer. To shorten the verification time, we have to use more abstract model than RTL for simulation.

On instruction-level simulations with ISS, each instruction on a target processor is executed virtually on a host computer. An ISS emulates the behavior of processor, for example, results of instruction execution, execution cycle, contents of a register file. An ISS model is always described with C/C++ languages. This model is designed separately with its RTL model. Figure 1-a shows this design style. This is not efficient for customizable processors, which need a different ISS for each customization. A simulation with the ISS model is much faster than the RTL model. But it is difficult to ensure equivalence of ISS and RTL. A bit-width-handling simulation is also very difficult in ISS described in C/C++.

3. RTL/ISS CO-MODELING USING SYSTEMC

In the conventional design flow, an RTL model is described with HDL, and an ISS model is described with C/C++. But many system level design languages have capabilities to describe entire system, hardware, software and system speci-

0-7803-8251-X/04/$17.00

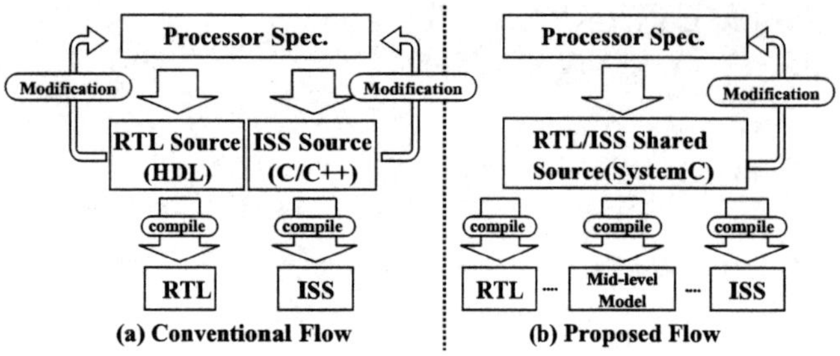

Fig. 1. Processor Design Flow with Conventional Method and Proposed RTL/ISS Code-Sharing Method

```
void function(void){
  // Feature common to RTL adn ISS model
  ....
#ifdef IN_RTL_DESCRIPTION
  // RTL model specific feature
  ....
#else
  // ISS model specific feature
  ....
#endif
  // Feature common to RTL adn ISS model
  ....
}
```

Fig. 2. Description Style of RTL/ISS co-modeling Using C Macro

fication. With SystemC, we can integrate both RTL and ISS models into a same source file to easily develop and verify both models. We apply the proposed method to our developed RISC processor "MiU–Processor". In this section, we explain the proposed co-modeling method in detail.

3.1. RTL/ISS Code-Sharing Methodology

An organization of a customizable processor varies according to an assigned function. It is very important to validate the equivalence of its RTL and ISS model. Our proposed method describes a customizable processor with SystemC. SystemC is implemented as a C++ class library, which requires no additional simulator. We can simulate the RTL model, by just compile it with a C++ compiler. This feature enables easy ISS generation from RTL and validation of ISS/RTL model. Figure 1-b shows a processor design flow using the proposed RTL/ISS code-sharing methodology.

However, there are different purposes between RTL and ISS models. Each model requires specific descriptions. We nullify a part of source file using C macro (e.g. `#ifdef`) to obtain RTL/ISS model from a single source file. This example is show in Figure 2.

RTL Model

The RTL model of a processor has to be logic synthesized into hardware. We have to give the first priority to logic synthesis. If the processor has pipeline stages, we divide each stage into module or processes. We have to set bit-width of intra-module variables and signals as wide as possible. Many signals and ports are necessary to make intra-module signals the same composition as those of an actual processor. In this case, we make assignment action for signals and ports as C macro to share more code between RTL and ISS.

ISS Model

Instruction-level simulations must be running as fast as possible, since it is used to validate the behaviors of the programs. In contrast with the RTL model, each stage does not have to be parallelized. It is only needed to emulate the processor behavior at the instruction level. We must obtain much information about programs on the processor to evaluate efficiency of customization and verification of its behavior. For example, we have to profile instruction statistics to customize its instruction set. Power consumption of the processor differs according to instruction sequences. We have to analyze instruction sequences to estimate power. To optimize the bit-width of data-path, we use statistics about values in registers. Update frequency of a register file can be obtained through instruction-level simulations, which is useful to optimize the size of the register file.

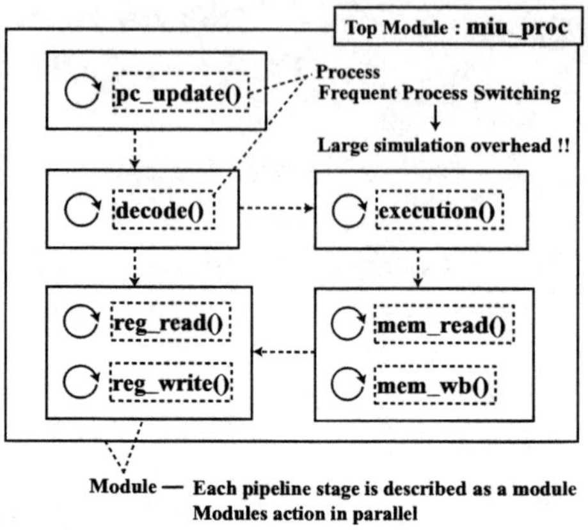

Fig. 3. RTL model of MiU–Processor

3.2. Overview and Organization of MiU–Processor

We apply the proposed method to a 5-stage pipelined RISC processor "MiU–Processor". Here, we explain an overview of the proposed ISS/RTL co-modeling methodology. and important point of modeling RTL/ISS model.

MiU–Processor has a conventional 5-stage pipelined Harvard architecture, which has a subset of the SH-2 [8] instruction set. Bit widths of the ALU and registers and number of registers are variable. 23 instructions can be removed to minimize the area. MiU–Processor is described in with SystemC RTL. Its file organization is shown in Fig. 3. In this model, each stage is described as a module separately. All the processes work in parallel.

In order to describe RTL and ISS models in a single SystemC source, we modify the original RTL model as follows.

Removal of pipeline

In SystemC, a process is executed with changes of signals. Each pipeline stage of MiU–Processor is described as a separate module, whose behavior is described in a `SC_METHOD` process. Every input signal activates process, which changes other output signals. In this structure, there happens many

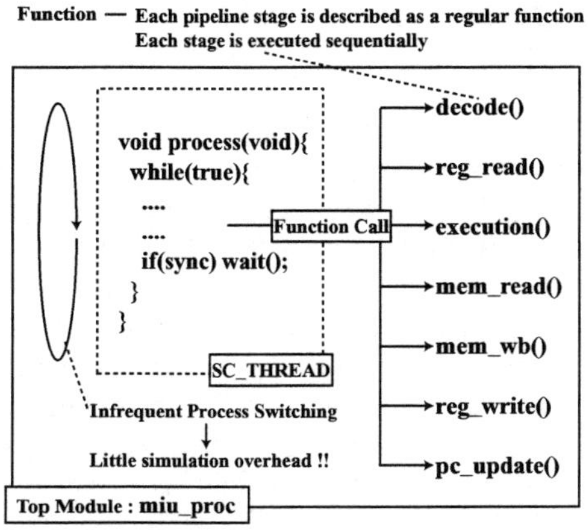

Fig. 4. ISS model of MiU–Processor

process switchings in one cycle. This causes a large simulation overhead(Figure 3). We remove pipelines from RTL for fast simulation(Figure 4). Each pipeline stage is described in a regular function, instead of a process. The ISS model executes only a single process that calls the regular function corresponding to each pipeline stage. It decreases process switching, and simulation time becomes shorter.

Log message constraint by log-level
Log messages like `printf()` consumes simulation time. We constraint display of log message according to its importance.

Switching clock sync/async mode
A process of SystemC is activated by changes of signals. A thread process is suspended by inserting `wait()`, but frequent process switching causes large overhead to lead to long simulation. Each stage of non-pipelined model is described as a single `SC_THREAD` process. This process call `wait()` to suspend for synchronization to the clock. In addition to this sync mode, we make async mode that dose not call `wait()` for high-speed simulation. `if(sync) wait();` in Figure 4 shows this mode. In sync mode, `wait()` is called to suspend process. But in async mode, `wait()` is not called and process is not suspended.

Substitution of SystemC specific function
SystemC-specific data types(e.g. `sc_int`, `sc_uint`) have its own methods, for example, `to_uint()`(conversion to `unsigned int`), `range()`(clipping partial bit range), concatenation of multiple `sc_int`/`sc_uint` variables and so on. These methods make large overhead. We substitute these methods with macro to speed up simulation. We use `int` and `unsigned int` for temporary variables instead of `sc_int` and `sc_uint`.

4. EVALUATION OF EQUIVALENCE AND SIMULATION SPEED OF RTL/ISS MODEL

Our proposed method makes it easy to generate and verify RTL/ISS models by describing in a single source code. This section explains evaluation of code-sharing ratio and simulation speed of each model.

```
// id: Instruction Decode stage
void miu_proc_id :: decode(void){
 switch(fetched_data){
  ....
  // Feature common to RTL adn ISS model
  // --- Beggining of additional part ---
  case CMP_STR:
   decode_set(D_CMP_STR,Rn,Rn,b2,b1,0,0);
   printd(DBG_ALL,"id:CMP/STR Rm,Rn\n");
   break;
  // --- End of additional part ---
   ....
 }
}

// ex: EXecution stage
void miu_proc_ex::execution(void){
 switch(decoded_instruction)
  ....
  // Feature common to RTL adn ISS model
  // --- Beggining of additional part ---
  case D_CMP_STR:
   T=(RANGE(Rmval,31,24)^RANGE(Rnval,31,24))
    ||(RANGE(Rmval,23,16)^RANGE(Rnval,23,16))
    ||(RANGE(Rmval,15, 8)^RANGE(Rnval,15, 8))
    ||(RANGE(Rmval, 7, 0)^RANGE(Rnval, 7, 0))
   ? 1 : 0;
   ex_dst = Rnval;
   break;
  // --- End of additional part ---
   ....
 }
}
```

Fig. 5. Changes by Adding Instruction "Comparing Strings" (Bold is added part)

4.1. Code-Sharing Rate of ISS/RTL Model

Table 1 shows the number of lines of RTL/ISS specific and shared descriptions in the source file of each stage. Total processor description is 2,800 lines, shared-part of which are 1,900 lines. Large part of entire description, 67% description, is shared. ISS specific part is only 88 lines, RTL specific part becomes 845 lines, about 30% of whole description.

Shared part of cpp files is 73%, and that of header files is 25%. A header file includes module information, ports, signals and processes. The RTL model has separate process each stage. But ISS model is non-pipelined model, whose stage has no information about module. So header files are not so shared as cpp files.

We evaluate code-sharing ratio of each file. It is propotional to the number of lines. For example, the shared part of "ifetch" stage with 38 lines is only 36%, while that of "ex" stage with 1199 lines is 70%.

To add new instructions, we mainly change decode and execution stage. By our proposed method, both stages are highly shared. We easily add/delete instructions in a shared-part not in a change ISS/RTL specific part. Figure 5 shows an example of an instruction addition(String Matching) of MiU–Processor. All we have to change is only 12 lines in an RTL/ISS shared part, out of 2,200 lines(total number of lines of "ex" and "id" stage).

4.2. Simulation Speed of ISS/RTL Model

We evaluate ISS simulation time on MiU–Processor with a PC/AT compatible computer(CPU: Celeron 2GHz, Mem-

Table 1. # of lines for each stage

Implementation file(.cpp)					
Source	Shared	ISS Only	RTL Only	Total	Shared Ratio
pc	36	10	42	88	41%
ifetch	13	4	21	38	34%
id	683	4	148	835	82%
regfile	43	25	61	129	33%
ex	912	21	266	1,199	76%
mem_wb	113	6	44	163	69%
Total	1,800	70	582	2,452	73%

Header file(.h)					
Source	Shared	ISS Only	RTL Only	Total	Shared Ratio
pc	15	3	26	44	34%
ifetch	15	3	21	39	38%
id	15	3	56	74	20%
regfile	15	3	27	45	33%
ex	19	3	101	123	15%
mem_wb	15	3	32	50	30%
Total	94	18	263	375	25%

ory: DDR-SDRAM 512MB, OS:Linux 2.4.24). We use GNU C Compiler(version 3.2) and SystemC(version 2.0.1). We measure number of cycle per second by 8×8 DCT program simulation. This result is shown in Table 2.

The RTL mode is able to simulate 20k cycle per second. Non-pipelined model simulates 75k cycle. To simulate a particular processor in an entire system, these accurate models are very useful.

To simulate an entire system, more abstract simulation is effective for speedup. Log message constraint makes simulation speed 500k cycle per second. By substitution of SystemC specific functions, 700k cycle per second simulation becomes possible. A clock async mode can simulate 2.5M cycle. When all high-speed options are enabled, simulation speed becomes 6.0M cycle per second. If using highly-abstract model for verification, equivalence between this model and actual processor is very important. We simulate DCT, FIR, sort and so on, and compare memory access of both models to evaluate equivalence. ISS in GDB [9] can simulate about 7.6M cycle per second when trace is on. Simulation speed of our proposed ISS is 6M cycle per second, about equal to this ISS.

Table 2. ISS Simulation Speed of MiU–Processor

	A	B	C	D	Speed
RTL					20k Cycle/s
	√				75k Cycle/s
↕	√	√			500k Cycle/s
	√	√		√	700k Cycle/s
	√	√	√		2,500k Cycle/s
Behavioral	√	√	√	√	6,000k Cycle/s

A : Removal of pipeline stages B : Log message constraint
C : Clock async mode
D : Substitution of SystemC specific features

It is very difficult to ensure equivalence of general ISS and actual processors. Our proposed ISS can simulate with the same speed as a general ISS. Furthermore, our method realizes bit-width-handling and multi-abstraction-level simulation. Our ISS generation method becomes very effective for customizable processor SoCs.

5. CONCLUSION

We propose an RTL/ISS code-sharing methodology with SystemC to generate and verificate both models easily. When compiling, C macro like `#ifdef` nullify unnecessary description for a requested model. This method makes up RTL/ISS model from single source file.

We apply our proposed method to customizable processor "MiU–Processor". We can share 67% of processor source file between RTL and ISS. For adding a string match instruction to the processor, we add only 12 lines in an RTL/ISS shared part of decode and execution stages.

Our proposed ISS can simulate as fast as a general one, and even enables bit-width-handling simulation. RTL/ISS code-sharing makes it easy to ensure equivalence of RTL and ISS. Our proposed ISS generation methodology is very efficient for a customizable-processor based SoC.

6. REFERENCES

[1] Makiko Itoh et. al., "PEAS–III: An ASIP Design Environment," in *ICCD 2000*, Sept. 2000, pp. 430–436.

[2] Jun Sato et. al., "An integrated design environment for application specific integrated processor," in *ICCD 1991*, Oct. 1991, pp. 414–417.

[3] Tensilica, Inc., *Xtensa*, http://www.tensilica.com.

[4] F. N. Eko et. al., "A soft-core processor architecture for embedded system design," in *APCHDL 1998*, July 1998, pp. 154–159.

[5] "MeP Media Embedded Processor," TOSHIBA, Inc., http://www.mepcore.com.

[6] Synopsys, Inc., Coware, Inc., Frontier Design, Inc., *SystemC Version 2.0 User's Guide*, 2001.

[7] Yoichi Yuyama et. al., "Heterogeneous Processor Architecture and Its Design Methodology to Shorten the Design Period of Embedded SoCs," in *SASIMI 2003*, Apr 2003, pp. 351–356.

[8] Hitachi, Inc., *SuperH*, http://www.superh.com/.

[9] Free Software Foundation, Inc., *GDB: The GNU Project Debugger*, http://www.gnu.org/software/gdb/gdb.html.

TEST VECTOR GENERATION AND CLASSIFICATION USING FSM TRAVERSALS

Ralph Marczynski, Mitchell A. Thornton, Stephen A. Szygenda
Department of Computer Science and Engineering
Southern Methodist University

ABSTRACT

Design correctness has become a bottleneck in the modern digital system design cycle. In an effort to improve current ad hoc simulation processes, this paper presents a method for the automated generation of simulation vectors using Symbolic FSM Traversal techniques. Generated vectors are classified into three categories, Forward Inter-Frontier, Reverse Inter-Frontier, and Intra-Frontier vectors; a classification based on a vector's ability to generate Forward-, Reverse-, and Inter-Frontier transitions in an FSM's state transition graph. Additionally, a State-Element Transition Relation (S-ETR) is introduced. This technique involves the construction of a Transition Relation (TR) for each state holding element and defining a smaller, incomplete, over-approximation of the TR. Combining the information present in the S-ETRs coupled with simulation is used to perform image computations.

1. INTRODUCTION

Functional verification is the process of determining some level of confidence that a design meets its specifications. One trend is the partial validation of designs through simulation [1], consisting of sending input to the design under test and observing the output. Often the design is judged by the equivalence found when comparing the output with that of some reference model or other abstraction of the circuit when it is exposed to the same stimulus. This approach is often used in the commercial environment; unfortunately this in not due to the method's effectiveness, but rather the lack of superior options.

The shortcoming of a simulation-based verification methodology is the limited functional coverage. Even for modest designs, the number of vectors needed to for an exhaustive test is too large to simulate in a feasible amount of time. In order to utilize simulation, one needs to choose the test vectors carefully as to gain the maximum amount of coverage and ensure tests for rare "corner-cases". Automated methods for such intelligent vector generation are the saving grace of simulation, but this area still requires further study to reach its full potential.

Formal methods provide alternatives that are potentially capable of overcoming the obstacles faced in simulation. A formal method is the application of mathematical methodologies to the specification and verification of systems [2]. The key advantage is the ability to exhaustively test the design with respect to its specification. To handle complex designs, symbolic BDD-based system representations [3,4,5] can be employed to allow for a compact representation of the state space and symbolic FSM traversal. With many mature to relatively mature technologies plagued with computational obstacles and the need for human intervention; a workable, practical, and effective verification flow appears to be rooted in the integration of multiple verification methods [6,7,8].

This paper addresses the integration and reuse of existing methods. Symbolic FSM traversal is used to generate test vectors, which are classified into three categories:

1.) *Forward Inter-Frontier Vectors:* vectors which will cause a transition from a state in frontier i to a state in frontier $i+1$.
2.) *Reverse Inter-Frontier Vectors:* vectors which will cause a transition from Frontier i to a state in frontier j, where $j < i$.
3.) *Intra-Frontier Vectors:* vectors that will cause transitions among states within a given frontier, including vectors causing self-loops.

Such a vector classification can be exploited during simulation. Vectors tracing a minimum length path between a given pair of a state's frontiers can be formed and statements about reachability can be made if no paths can be found. In the presence of temporal logic assertions, vectors with the greatest probability of cycle generation can be used to simulate for the failures of *eventuality* properties, as can Inter-Frontier transition vectors be used to simulate for the reachability of a fail state, necessary for validation of safety properties. In a guided search of the state space, Inter-Frontier communication complexity can also be used as heuristic for selecting further simulation origin states.

The remainder of this paper is organized as follows. Section 2 provides the necessary background concerning Symbolic FSM Traversal. Section 3 outlines the vector generation and grouping procedures, with Section 4 identifying its applications. Section 5 introduces *State-Element Transition Relations* (S-ETR) and their usefulness within a verification system integrating simulation with formal methods. Finally Section 6 provides concluding remarks.

2. BACKGROUND – FSM TRAVERSAL

FSM traversal is a traversal of a system's state transition graph (STG) in either a breath-first or depth-first manner. Starting at an initial state, typically the reset state, the search iterates through all nodes in the STG resulting in a set of all reachable states. The knowledge gained from such a procedure proves useful in many EDA-CAD settings.

An implicit STG representation is achieved through the construction of a *Transition Relation* (TR) [5], a function representing all possible transitions within a FSM. FSM

Traversal is then carried out as a series of *Image Computations*; a conjunction of the TR and all previously reached states with existential quantification of the resultant BDD over the present state variables [9]. The result of each iteration is then the set of states on the frontier (a set of next states), which are further re-labeled as present states and added to the set of reached states. The process is repeated until an evaluated frontier does not provide transitions into undiscovered states.

The BDD for the TR of complex systems can exceed memory limitations. The inputs to a system are not necessary for the image computation, therefore commonly used TRs differ from the one described in this paper in that inputs are removed through existential quantification, resulting in the *smoothed TR*. Smoothing can take place at different stages of TR construction, depending on the method [5,9]. Vector generation, however, requires the knowledge of input values with respect to state transitions; hence, the sub-optimal *monolithic* TR without smoothing is required. Discussion of a new representation of the TR, the S-ETR, is presented in Section 5, which significantly reduces the monolithic TR size through over-approximation, while retaining the input vector information.

3. TEST VECTOR GENERATION

At each image computation, test vectors for all transitions are determined and grouped into the three categories described in the Introduction. This classification enables the selection of vectors in order to meet specified coverage criteria (e.g. cycles, inter-frontier paths, heuristics based on complexity, etc.). In addition to the three classes of vectors, *State Entry Vectors* are determined; the only vectors which can possibly cause entry into a given state.

3.1. State Entry Vectors

For each reachable state, the TR can be manipulated in a straightforward manner to extract all the possible vectors that may lead to an entry into that state. The procedure is simply an existential quantification of the TR over the present state variables. The state entry vectors for the example FSM depicted in Figure 1 are presented in Table 1.

3.2. Forward Inter-Frontier Vectors

At each image computation, the vectors capable of causing entry into the new frontier are evaluated. The initial frontier R_0 is assumed to be the reset state. Table 2 presents the Forward Inter-Frontier Vectors for the example FSM. The procedure is implemented as follows:

$$R_{i+1} = TR \wedge R_i$$
$$S_{i+1} = \exists\,(PS \wedge IN)\, R_{i+1}$$
$$R_i Shifted = Re\text{-}Label\,(R_i,\, PS,\, IN)$$
$$R_{i+1} NewOnly = R_i Shifted \oplus S_{i+1}$$
$$Fwd\text{-}Inter\text{-}Frontier\ Vectors = R_{i+1} NewOnly \wedge R_{i+1}$$

3.3. Reverse Inter-Frontier Vectors

During a given image computation, the newly reached states on the frontier may contain states capable of a transition to a state

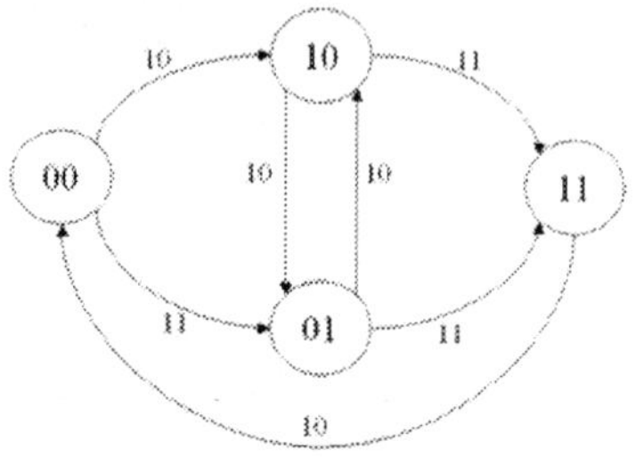

[illegible]	[illegible]	[illegible]	[illegible]	[illegible]	[illegible]
0	0	1	0	1	0
0	0	0	1	1	1
1	0	0	1	1	0
1	0	1	1	1	1
0	1	1	0	1	0
0	1	1	1	1	1
1	1	0	0	1	0

Figure 1: FSM and Corresponding Transition Relation

TABLE 1: STATE ENTRY VECTORS

State	Entry Vector(s)
00	0-
00	10
01	--
10	0-
10	10
11	0-
11	11

TABLE 2: FORWARD INTER-FRONTIER VECTORS

	Source State	Target State	Vector
Frontier 1	00	01	11
	00	10	10
Frontier 2	10	11	11
	01	11	11

lying in a previously reached frontier. Extracting such vectors (and classifying them accordingly) allows for additional procedures to generate "cycle" producing vectors. In order for this to be computed, we are forced to perform an image computation originating from only the newly reached states. Reusing the BDD from the Forward Inter-Frontier Vector procedure, the necessary steps are as follows:

$$R_{i+1} NewOnly\_Shifted = Re\text{-}Label\,(R_{i+1} NewOnly,\, NS, PS)$$
$$R\_fromNewOnly = R_{i+1} NewOnly\_Shifted \wedge TR$$
$$Rev\_\ Inter\text{-}Fronteir\_Vectors = R_i Shifted \wedge R\_fromNewOnly$$

The procedure computes Frontier $i+2$, and the union of Frontier $i+2$ with Frontier(s) j, with $j < i+1$, determines the Reverse Inter-Frontier Vectors. The only Inter-Frontier vector for the example FSM is (10) corresponding to source state (11) and target state (10).

3.4. Intra-Frontier Vectors

Finally, the set of vectors that cause transitions among states in a single frontier are evaluated. This classification allows for further generation of cyclic sequences and can be used in creating heuristics for guided state space traversal. Reusing the BDD from the previously defined procedures, the evaluation is:

$$Intra\text{-}Frontier\ Vectors = R\_fromNewOnly \wedge R_{i+1}\ NewOnly$$

The resultant set of vectors contains all vectors capable of initiating transitions within a single frontier, including vector resulting in self-loops. Figure 1 of the example FSM does not depict self-loop vectors; their presence is implied as the vectors that do not cause a state-pair transition. The Intra-Frontier Vectors for the example FSM are presented in Table 3.

TABLE 3: INTRA-FRONTIER VECTORS

	Source State	Target State	Vector
Frontier 1	01	01	1-
	10	01	10
	01	10	10
	10	10	0-
Frontier 2	11	11	0-
	11	11	11

4. APPLICATIONS

4.1. Minimum Path Vector Sequence

The grouping of vectors with respect to their role in FSM traversal allows for the efficient generation of vector sequences resulting in minimum paths between any given state pair. If any two states lie in two distinct frontiers, the minimum path between them can be found through manipulation of the Forward and Reverse Inter-Frontier Vectors found in the two state holding and possible intermediate frontiers.

A procedure has been implemented which determines minimum path vectors from reset to a specified reachable state. This involves finding the "deepest" frontier containing the target state (identified when the intersection between the states in the inspected Frontier intersected with the target state is not an empty set) and a traversal toward the reset state by selecting vectors creating the desired path during each iteration. This concept can be extended to any state pair by locating the source state, tracing either the Forward or Reverse Inter-Frontier vectors towards the target state.

If a pair of states lie in the same frontier, a similar procedure can be created considering Intra-Frontier vectors. Finally, if a Frontier containing the target state is not found, the target state can deemed unreachable

4.2. Heuristics for Guided State Space Exploration

Several techniques have been developed which limit the power of model checking to "find bugs" rather than exhaustively prove a system's correctness [10]. These methods rely on heuristics in order to target a portion of the state space that is most likely to contain flaws. Although these methods offer promise they are either dramatically inconsistent or require significant designer input. The information stored by the classes of vectors generated through the techniques presented in this paper can be used as heuristics for guided FSM searches.

The Intra-Frontier Vectors capture the amount of activity among states in a single frontier. If we suspect a frontier to contain an error state and is portrayed to be "highly-active", then that frontier may be marked as a candidate for more intense simulation, utilizing a variety of Intra-Frontier vector sequences. Continued traversal can then be focused on the "most-popular" state (a characteristic additionally obtainable from the Intra-Frontier Vectors) with a propagation of the observed properties into the next "partial" frontier.

5. STATE-ELEMENT TRANSITION RELATION

BDD explosion, in terms of the TR and/or the reachable state space, is one of the greatest obstacles when applying formal verification techniques incorporating FSM traversal. The techniques presented up to this point do little to help this situation. Complete FSM traversal using a sub-optimal TR is required to generate and group the vectors, further limiting the complexity of designs that can be investigated.

To address this limitation, a method for generating TRs local to each state-holding element is presented. Each *state-element TR (S-ETR)* is part of a partitioned TR providing information as to which input vectors cause a single state-holding element to change its value, essentially a bit relation with existential quantification of the present state variables. Such a TR can often be much simpler than a complete TR and our experiments show this to be the case. Table 5 compares the size of the S-ETR with the size of a monolithic TR without smoothing for selected circuits of the ISCAS89 benchmark suite.

Combining the S-ETRs produces an over-approximation of the reachable state space since complete state transition pairs are not enumerated, i.e. next states can be evaluated that do not actually occur in the FSM. Simulation can therefore be employed to determine which of the identified transitions are actually valid by simulating the vectors associated with the transitions. Using simulation to sample the behavior of the FSM has been shown to be efficient in other systems coupling simulation and formal verification methods [9]. The benefit of this approach is the reduction of the TR by incorporating simulation to "fill in" the relationships information missing in the S-ETR. The S-ETR, in turn, identifies the small portion of all possible input combinations that require simulation.

The technique is presented using the example FSM from Figure 1. We compute the S-ETR in a similar manner as the monolithic TR, only we do not perform the step of combining the bit relations to form the complete TR. Before merging wire relations, we trim all information that is not of local concern through the existential quantification of the present state variables. As a result, we obtain the vectors that generate local state element changes, presented in Table 5.

From the BDD representing the information in Table 5, we can extract all the *Inconsistencies* resulting from the over approximation. Clearly, a state exposed to a distinct input will transition to one and only one state. An *O(n)* procedure, where *n* is the number of cubes in the BDD, can be employed to identify the set of *Inconsistent* transitions, resulting in the set of present-state/next-state vector sets that use simulation to mark them valid or invalid. The procedure simply iterates over all the cubes, performs existential quantification of the next state variable and intersects each cube with the S-ETR. If the intersection results in a cube with more minterms than the original cube itself, the set of transitions is marked as inconsistent. Simulation can then separate which of the inconsistent set are valid transitions.

TABLE 4: S-ETR AND MONOLITH TR BDD SIZES FOR ISCAS89

Bench	S-ETR Size	Monolithic TR
s27	51	17
s208.1	29	40
s298	45	367
s344	612	415
s382	109	343
s386	13	131
s400	109	343
s420.1	53	80
s444	56	559
s510	16	182
s526	59	381
s526n	59	381
s641	1923	3990
s713	2302	4204
s820	138	252
s832	138	239
s953	1147	1023
s1488	8	310

TABLE 5: STATE ELEMENT CHANGE VECTORS

	State Element 0	State Element 1
Change	Vector	Vector
0→1	10	11
	11	10
1→1	01	00
	00	01
	11	11
0→0	00	00
	01	01
	11	10
1→0	10	10

One could imagine that an FSM can exist where each element is changed by any possible combination of vectors, leading to a simulation of all possible inputs to determine the next state. The results from experiments using the ISCAS89 benchmark suite show this to not be the typical case. In most cases, the number of possible vectors to cause a single element change was less than 20.

6. CONCLUSION

This paper presented several techniques for the manipulation of monolithic TRs in order to obtain vector sequences that can be used to traverse specified paths in an FSM during simulation. For the ease of such sequence generations, the vectors are grouped into classes, governed by a vector's role during FSM traversal.

Although the generated vectors may prove to be useful in some CAD settings; such as putting stress on assertions during simulation or creating heuristics for guided state searches, the techniques used rely on the ability to represent the TR and reachable state space during symbolic FSM traversal.

To loosen these restrictions, a procedure for deriving a State-Element Transition Relations is introduced. The S-ETR is built without the knowledge of the relationships among states, and it therefore only contains partial information. By coupling symbolic techniques for image computation with simulation, the missing information can be created on-the-fly by simulating state transition candidates produced by an analysis of the S-ETR and identifying them as valid or invalid.

7. REFERENCES

[1] International Technology Roadmap for Semiconductors, 2001 Edition, http://public.itrs.net/Files/2001ITRS/Home.htm.

[2] C. Kern and M. Greenstreet, "Formal Verification in Hardware Design: A Survey," *ACM Transactions on Design Automation of Electronic. Systems*, Vol. 4, April 1999, pp. 123-193.

[3] E.M. Clarke, O. Grumberg, and D. A. Peled, *Model Checking,* Cambridge: MIT Press, 1999.

[4] K. L. McMillan, *Symbolic Model Checking: An Approach to the State Explosion Problem*, Kluwer Academic Publishers, Boston/Dordrecht/ London, 1993.

[5] S.-Y. Huang and K.-T. Chen, *Formal Equivalence Checking and Design Debugging*, Kluwer Academic Publishers, Boston /Dordrecht /London,1998.

[6] D. Dill, "What's Between Simulation and Formal Verification?," slides from a presentation by Prof. Dill, Stanford University at DAC 1998.

[7] D. Dill and S. Tasiran, "Simulation meets Formal Verification," slides from a presentation at ICCAD 1999.

[8] E.M Clarke and J.M Wing, "Formal Methods: State of the Art and Future Directions," *Technical Report CMU-CS-96-178*, Carnegie Mellon University, 1996.

[9] Yang, C. H. and D. L. Dill: 1998, "Validation with Guided Search of the State Space", *Proc. of the Design Automation Conf.*

[10] Chen K.C "Memory verification needs fresh approach" http://www.eedesign.com/silicon/OEG20030428S0057, EEDesign, 2003

FORMAL VERIFICATION OF AN SOC PLATFORM PROTOCOL CONVERTER

Jounaïdi Ben Hassen and Sofiène Tahar

Concordia University
Electrical and Computer Engineering
Montreal, Quebec, Canada

ABSTRACT

In this paper we investigate the formal verification of the Memory Manager block of a System-on-a-Chip platform Protocol Converter using the FormalCheck tool of Cadence. The Memory Manager represents the main block of the protocol converter and is responsible for the reception of packets and their treatment for conversion. For the verification, we first extracted some constraints to define the environment for the Memory Manager. Then, we specified a number of relevant liveness and safety properties expressible in FormalCheck. Through extensive verification under the defined set of constraints, we have been able to find a few bugs in the design that were omitted by simulation. This experience demonstrates the usefulness of formal verification as complement to traditional verification by simulation.

1. INTRODUCTION

The increasing complexities of hardware designs have made verification and error detection on the critical path of the design process. Moreover, some of these errors may cause catastrophic loss of money, time, or even human life. This is in particular a serious problem for System-on-a-chip (SoC) designs which may contain processor cores, custom logic, memory and IP (Intellectual Property) blocks on a single chip. Traditionally, simulation is used to verify the "correctness" of a design. However, simulation is no longer able to keep pace with the increasing complexity of hardware designs. To overcome this difficulty, formal hardware verification methods [1] are now being deployed, and became useful tools for detection of functional design errors. By using formal verification alongside the design efforts, the overall design cycle can be reduced. This ensures a maximum design coverage while maintaining a high degree of confidence in the verification result. The objective of formal verification is to verify that the design model conforms to an abstract specification, consisting of a set of properties. Together, these properties describe (partially) the intended functionality of the design. Formal verification techniques have proven their efficiency by verifying industrial size systems.

This paper aims to explore the verification by model checking of the Memory Manager component of an SoC platform Protocol Converter System and to show that formal methods are strong enough for the verification of complex designs. The Protocol Converter system was designed by the "Groupe de Recherche en Micro-électronique" at the École Polytechnique de Montréal [2]. The architecture of the Memory Manager is described at the Register-Transfer Level coded in VHDL and represents the main block of the protocol converter system. The Memory Manager is composed of five modules: a Memory Manager Controller, an Address Counter Register, a Data Counter Register, a Packet Counter Register and a Packet Assembler. The model checking of the Memory Manager has been conducted using the FormalCheck tool of Cadence [3]. First, we extracted some constraints to define the environment for the Memory Manager. Then we specified a number of relevant liveness and safety properties expressible in FormalCheck. We successfully accomplished the properties verification under the defined set of constraints. This verification enabled us to find a number of bugs in the design that were not caught during the simulation process. This experience demonstrates how formal verification techniques can powerfuly complement traditional verification by simulation.

2. THE MEMORY MANAGER BLOCK

The Protocol Converter System at hand is based on an SoC platform developed at the École Polytechnique de Montréal [2]. The main advantage of such platform is its flexibility. In fact, its modularity allows the addition, the change or the drop of some modules without affecting the global architecture of the system. The Protocol Converter accepts incoming packets from a physical bus, converts their protocols and then sends them back through the physical interface. The Protocol Converter System is subdivided into three blocks as shown in Figure 1. The first block, composed of a Memory Manager and a Controller, is specialized in the reception of packets and preliminary treatments for the conversion. The second is responsible for the transmission of converted packets, while the third performs the

0-7803-8251-X/04/$17.00 ©2004 IEEE

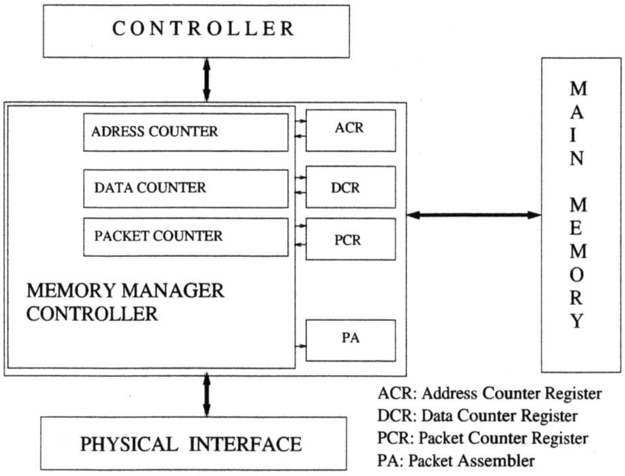

Fig. 1. The Memory Manager Architecture

conversion of packets coming from the first block and sends them back to the second block.

Packets come through a physical interface which communicates with the Memory Manager by transmitting data and some control signals. Upon reception, the Memory Manager stores incoming data in the Main Memory and transmits to the Controller the characteristics of the packet in transmission. These characteristics include the address of the packet in the Main Memory, the protocol of the packet and its size. The communication between the Memory Manager and the Controller is insured via the Memory Manager Controller. Once the protocol conversion is done, the Controller asks the Memory Manager, to remove the converted packet from the Main Memory. So, the address of this packet and some other control signals are transmitted to the Memory Manager. The functionality of the Memory Manager is insured by its different components. Besides the Memory Manager Controller, the Memory Manager has three registers and a Packet Assembler. Intuitively, the Memory Manager Controller is responsible for the communication between the Memory Manager and the Controller. The registers are used as counters of addresses and packets. The Packet Assembler is a module that concatenates packet's address and data.

3. ENVIRONMENT CONSTRAINTS

In FormalCheck, primary signals are assumed to be nondeterministic, meaning they could acquire any value within their range on any edge of the clock. However, in most cases correct design operation is allowed on a single edge of the clock. For this reason, properties should be observed using the appropriate clock edge. When we deal with formal verification by model checking, we should pay great attention to the set of defined constraints. Since constraints are generally used to simulate the environment on which the system operates, they are the most critical task. This set should be as complete as possible to describe the exact behavior of the environment. However, it should not contain more constraints than necessary to avoid obtaining an over-constrained system. In this work, we have established twelve constraints. A full list of all constraints is given in [4]. They include, for instance, a constraint to define the system clock, a constraint to define the reset of the system and some others to describe the duration and the synchronization of input signals (with respect to the clock system), as well as to describe the interaction between the external environment and the Memory Manager block. For example, the signal `Phyn_Sop` is activated during one cycle of the system's clock rising edge (`Sys_Clk=rising`) to indicate the start of packet submission. This can be expressed by the following *safety constraint*:

```
Constraint OneClockSop
  Type: Never
  Assume Never:(Phyn_Sop=1) and (WasSop=1)
```

This constraint means that the signal `Phyn_Sop` can never be activated during two clock cycles, where `WasSop` is a *state variable* that indicates if the signal `Phyn_Sop` is already activated:

```
WasSop: Range 0 to 1
  Initial: 0
  if (Phyn_Sop=1) and (Sys_Clk=0)
    then WasSop := 1;
    else WasSop := 0;
  end if;
```

The signal `Phyn_Sop` changes value only on the rising edge of the system's clock and can be expressed by the following safety constraint:

```
Constraint SyncSop
  Type: Always
  Assume Always: (Sys_Clk=rising) or
  (Phyn_Sop=stable)
```

meaning that the signal `Phyn_Sop` changes value only on the rising edge of the system's clock. In other words, if `Phyn_Sop` changes its value, forcibly the system's clock `Sys_Clk` is rising. Otherwise, `Phyn_Sop` remains stable.

4. PROPERTIES SPECIFICATION

After establishing the proper environment, and defining all needed constraints, we have specified sixteen *queries* (set of constraints and properties) of the Memory Manager, including liveness and safety properties. These properties have been defined based on the design textual documentation and the test benches given in [2]. In the following, we describe three samples for illustration purposes. A full list of all defined properties is given in [4]. Our main goals were to verify (1) the global reset of the Memory Manager, (2) the duration and the synchronization of output signals (with respect to the system's clock) and (3) the response of the Memory Manager according to received signals. In other terms, we wanted to verify that each time the Memory Manager receives a packet from the Physical Interface, it stores it in the Main Memory and informs the Controller. We wanted also to verify that the Controller is informed about all stored packets and vice-versa. For example, according to the Memory Manager documentation [2], when the Memory Manager begins the reception of a packet, it informs the Controller by activating the signal `Mgr_Sop` during one cycle. This activation should start at the falling edge of the system's clock defined by the signal `Sys_Clk`. In this query, we verify the duration of the signal `Mgr_Sop` (first property) and its synchronization with the clock (second property). In FormalCheck, this query is expressed as follows:

```
Query: Verif_Mgr_Sop
 Property: OneClkMgr_Sop
   Never: (Mgr_Sop=1) and (WasMgr_Sop=1)
 Property: SyncMgr_Sop
   Always: (Sys_Clk=falling) or
   (Mgr_Sop=stable)
 WasMgr_Sop: Range 0 to 1
  Initial: 0
  if (Mgr_Sop=1) and (Sys_Clk=1)
    then WasMgr_Sop := 1;
    else WasMgr_Sop := 0;
  end if;
```

where `WasMgr_Sop` is a state variable that indicates if the signal `Mgr_Sop` is already activated or not.

5. EXPERIMENTAL RESULTS

All verifications in this project were executed on an *Ultra 5* Sun workstation with 256 MB RAM and UNIX operating system. For all properties, we used *Symbolic (BDD)* as algorithm and *1-Step* as reduction technique [3]. We found these modes effective enough to conduct the verification process. The experimental results are shown in Table 1, including the status of the queries verification, the number of reached states (RS), the average state coverage (SC), the CPU time (real time) in seconds and the memory usage in MB. From Table 1, we can see that the system reset and signal duration properties are verified, but some others failed. For instance, when the Physical Interface starts a packet transmission, there is no guarantee that this packet will be detected by the Memory Manager and thereby will be stored in the Main Memory. In this case, the Controller will not be informed about this transmission and the packet will be lost. Intuitively, it seems logic that if the Memory Manager will not detect some packet transmissions, then it will not detect the end of these transmissions or the occurrence of some errors. In such cases, the Controller will not be informed by these activities performed by the Physical Interface. In fact, there is no direct dialogue between the Controller and the Physical Interface. Thus, all information concerning the packet transmission is transmitted to the Controller via the Memory Manager.

Table 1. Summary of Experimental Results

Query	Status	RS	SC (%)	CPU	Mem
Q1	Verified	2.68^{+7}	87.50	70	14.37
Q2	Verified	3.21^{+3}	96.77	10	2.32
Q3	**Failed**	5.09^{+3}	88.98	12	10.62
Q4	Verified	3.14^{+3}	96.77	9	2.32
Q5	**Failed**	5.25^{+3}	10	9	9.54
Q6	Verified	7.85^{+6}	97.06	9	2.32
Q7	**Failed**	8.89^{+6}	98.08	40	10.85
Q8	Verified	3.08^{+3}	96.77	8	2.32
Q9	Verified	4.37^{+5}	100	18	9.56
Q10	Verified	5.98^{+5}	96.97	15	9.66
Q11	Verified	4.38^{+5}	98.53	15	9.68
Q12	Verified	2.15^{+5}	100	18	9.66
Q13	Verified	2.15^{+5}	100	17	9.67
Q14	Verified	2.82^{+9}	98.41	12	2.34
Q15	Verified	2.82^{+9}	98.39	13	2.34
Q16	**Failed**	2.82^{+9}	98.44	192	16.63

In [2], there is an illustrated case in which packets are lost. When the Main Memory is full and no more space is available to store incoming packets, the Memory Manager will simply reject incoming packets without informing the Physical Interface. In such case, the Physical Interface will continue sending packets to the Memory Manager. However, according to [2], this situation was defined as the only case in which packets are lost. In our work, FormalCheck gives a counterexample and shows that in spite of memory space availability, incoming packets can still get lost. The failure of Q16 means that some packets are stored into the Main Memory but the Memory Manager does not inform the Controller about them once they are received. This means that some packets will remain in the Main Memory without conversion. Furthermore, the Controller will never ask for the suppression of these packets.

6. RELATED WORK

Related case studies, where model checking techniques were used to verify commercial products, are too numerous to be listed here. In the following, we elaborate on a few of them. For instance, in [5], FormalCheck was used to verify the implementation of a SCI-PHY (Saturn Compatible Interface for ATM-PHY devices) Level 2 protocol engine, commercialized by PMC-Sierra, Inc. From the set of eleven established properties, only one error has been detected. In comparison, the design considered in our project is more complex and presents a more generic behavior. In [6], the model checking tool VIS (Verification Interacting with Synthesis) [7], was adopted for the verification of an Asynchronous Transfer Mode (ATM) switch used for real applications in the Cambridge Fairisle network. This earlier work shows how VIS can partially verify large size circuit design by using specific reduction, abstraction and property division techniques. In [8], the same ATM switch fabric design is verified in FormalCheck and a comparison between the two hardware verification tools is given. From the experimental results in [8], it was shown that FormalCheck is faster than VIS and that the memory usage in FormalCheck is less than that in VIS for all verified properties. Moreover, no manual reduction or property composition were required in FormalCheck. This justifies again our choice of FormalCheck to verify our system. In [9], VIS was used for the verification of an ATM ring (ATMR) media access control (MAC) protocol. VIS was also used in [10] to formally verify a commercial product of PMC-Sierra, Inc. In this latter work, a design error which could lead the system to a deadlock state was detected. However, properties that involve the introduction of time delays were not verified because unlike FormalCheck, as used in our project, VIS does not support timed Verilog models. In [11], the MDG (Multiway Decision Graph) [12] tool was used in the formal verification of a Telecom System Block commercialized by PMC-Sierra, Inc., which processes a portion of the SONET (Synchronous Optical Network) line overhead of a received data stream. While the MDG tool possesses efficient features for data abstraction, it does not support neither Verilog nor VHDL to be considered in a complex project like ours.

7. CONCLUSION

In this study, we explored the formal verification by model checking of the Memory Manager Component of the Protocol Converter system. Our experimental results demonstrated the presence of many residual bugs in the design. These errors can lead to serious problems since they cause the non-conversion of many received packets and the non-liberation of the Main Memory. Despite the important effects of such situations on the functionalities of the Protocol Converter system, these errors were not detected by extensive post-design simulation efforts.

The main contribution of our work is the emphasis on the importance of formal methods for design verification. In our case, we were able to detect errors that were not detected by simulation. However, we should mention that formal techniques are not by themselves an alternative to verify if a design is correct with respect to a specification. They should be used with simulation as a complementary process to insure a maximum error detection.

In summary, our work joins other successful industrial-sized case studies in using model checking for hardware verification. We believe to have contributed fostering the evidence that model checking is now powerful enough to be widely used in industry to help in the verification of developed complex hardware designs.

8. REFERENCES

[1] C. Kern and M. Greenstreet, "Formal Verification in Hardware Design: A Survey," in *ACM Transactions on Design Automation of Electronic Systems*, Apr. 1999, vol. 4, pp. 123–193.

[2] S. Carniguian et al., "Intégration et Vérification d'un convertisseur de protocoles," Tech. report, Département de génie életrique, École Polytechnique de Montréal, Canada, 2002.

[3] Cadence, *Formal Verification Using Affirma FormalCheck, Version 2.4*, Aug. 1999.

[4] J. Ben Hassen and S. Tahar, "Formal Verification of a Protocol Converter Memory Manager using FormalCheck," Tech. report, Concordia University, Montreal, Canada, Apr. 2003.

[5] L. Barakatain and S. Tahar, "Functional Verification of a SCI-PHY Level 2 Protocol Engine," in *Proc. IEEE Int. Conf. on Information, Communications and Signal Processing*, Singapore, Oct. 2001.

[6] J. Lu et al., "Model Checking of a Real ATM Switch," in *Proc. IEEE Int. Conf. on Computer Design*, Austin, Texas, USA, Oct. 1998, pp. 195–198.

[7] R.K. Brayton et al., "VIS: A System for Verification and Synthesis," in *Proc. Conf. on Computer Aided Verification*, Rutgers, New York, USA, Jul. 1996, pp. 428–432.

[8] L. Barakatain and S. Tahar, "Model Checking of the Fairisle ATM Switch Fabric using FormalCheck," in *Proc. IEEE Canadian Conf. on Electrical and Computer Engineering*, Toronto, Canada, May 2001, pp. 907–912.

[9] H. Peng and S. Tahar, "Hardware Modeling and Verifiation of an ATM Ring MAC Protocol," in *Proc. IEEE Int. Conf. on Microelectronics*, Teheran, Iran, Nov. 2000, pp. 21–24.

[10] P. Murugesh and S. Tahar, "Formal Verification of the RCMP Egress Routing Logic," in *Proc. IEEE Int. Conf. on Microelectronics*, Kuwait City, Kuwait, Nov. 1999, pp. 89–92.

[11] M.H. Zobair and S. Tahar, "Formal Verification of a SONET Telecom System Block," in *Proc. Int. Conf. on Formal Engineering Methods*, Shanghai, China, Oct. 2002, pp. 447–458.

[12] F. Corella et al., "Multiway Decision Graphs for Automated Hardware Verification," in *Formal Methods in System Design*, Feb. 1997, vol. 10, pp. 7–46.

FPGA Based Accelerator for Functional Simulation

Mohamed N. Wageeh* Ayman M. Wahba* Ashraf M. Salem* Mohamed A. Sheirah**

* Mentor Graphics Egypt
Cairo, Egypt

** Computer & Systems Engineering Department
Faculty of Engineering- Ain Shams University
Cairo, Egypt

Abstract

In this paper, we introduce an FPGA-based approach to accelerate functional simulation. We achieve speedups between 5 and 100X over pure software simulation. This approach takes advantage of a simulator's Software Procedural Interface, provided by a commercial VHDL simulator. Our approach uses the Master-Slave co-simulation technique. The Master is the HDL simulator, which controls the time advance mechanism. The Slave is an FPGA board, where the DUT is synthesized in hardware. In the middle, we developed a communication library responsible for communicating the flow of events and values between both sides.

1. Introduction

VLSI designs are rapidly evolving, doubling in size with each generation. This causes dramatic increase of the simulation run time. Although the performance of software simulators has constantly improved, the rate of improvement falls far short of the rate at which design density is increasing [1]. Cycle-based simulation [2], formal verification [3], hardware emulation [4], and hardware acceleration [5] have become the state-of-the-art verification methodologies. One of the methodologies used for accelerating simulation is to optimize the simulator code. Simulation code could be generated with routines for unused VHDL features stripped off [5]. The VHDL simulator optimized in this way runs faster when the design is described mostly with simple constructs and expressions. But speedups reported for this methodology doesn't exceed 2X. Another software approach suggests C language-based design and verification methodology to enhance the simulation speed instead of the conventional HDL-based simulation [6][7]. But this methodology is most suitable for early design phases, rather than final verification before fabrication. Simulation assisted by special hardware is the best solution for speeding up the simulation of large design sections that have been tested and accepted by RTL simulations [8]. Although hardware emulation gives the best speedup, the cost is very expensive and it requires that the gate level design is already finished [6]. One interesting technology for acceleration relies on a VLIW-like virtual simulation processor mapped to an FPGA [9]. Companion to the processor is a compiler that compiles netlists into instructions to be streamed to the processor, along with the simulation vectors. Again, this technology applies to gate-level netlists only. The most efficient method of system level verification combines the benefits of HDL simulation and emulation is hardware acceleration. Essentially, the HDL designs can now be simulated in hardware and the designer can still use the software HDL simulator for debugging and to view the results. The benefits of this methodology are numerous; HDL code that represents the ASIC design (RTL synthesizable code) can be downloaded to the hardware accelerator and thus it runs at real hardware speed. The test-bench is executed in the HDL simulator, so it does not require any changes. Also, all the debugging capabilities of the HDL simulator will still be available.

Co-Simulation is the technique used for simulating a system having sub-components modeled in different types of models. Co-Simulation platforms can be divided into two main categories; Distributed co-simulation, and Master-Slave co-simulation. In Distributed co-simulation, two or more independent simulators are bound together with a backplane. Three basic approaches are widely used for the implementation of distributed co-simulation environments [10]. These approaches are the Sequential approach, the Standalone approach, and the Paired approach. The Sequential approach is a simple coupling of two simulators without support for feedback across the boundaries. The Standalone approach is a sort of one complex program that handles all types of models. It's also called "Native Mode" operation. In the Paired approach, also called Backplane approach, two or more simulators run in parallel. These simulators are synchronized through the backplane. On the other hand, Master-Slave co-simulation incorporates only one simulator (the master) augmented with some supporting component (the slave). Usually the slave is not an independent simulator and cannot operate in a standalone mode. The slave may be a software component or a piece of hardware.

This paper describes a verification platform that accelerates the simulation of hardware models written in synthesizable VHDL. The idea behind this platform is to put the "Design Under Test" (DUT) on a Field Programmable Gate Array (FPGA) chip as in figure 1. To

0-7803-8251-X/04/$17.00

verify the DUT, it is driven by stimulus coming from a test bench running on a VHDL simulator. A communication library is built to handle the flow of events between the simulator and the FPGA chip.

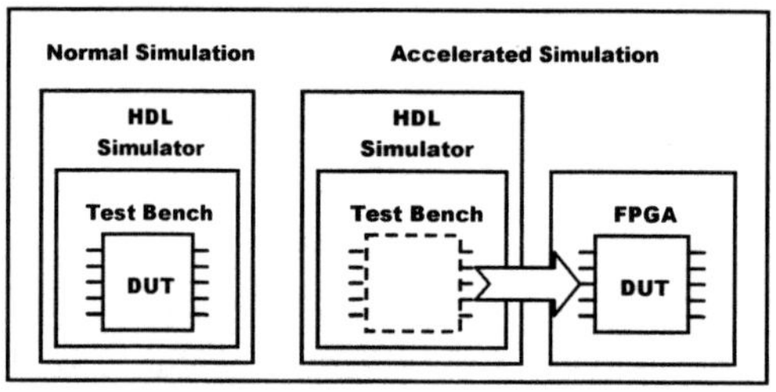

Figure 1: Accelerated Simulation

In section 2, we will give a brief description of the hardware part of the platform and its features and limitations. Next, in section 3, an overview and some background of the VHDL FLI will be presented. Then, in section 4 we will describe in detail our Master/Slave architecture. Finally, experimental results are presented followed by our comments and conclusions.

2. Accelerator Hardware

This platform is built around an FPGA chip mounted on a standard 32-bit 33MHz PCI board. A custom PCI controller is coupled with the FPGA to provide a fast, reliable, and flexible interface to a design programmed in the FPGA.

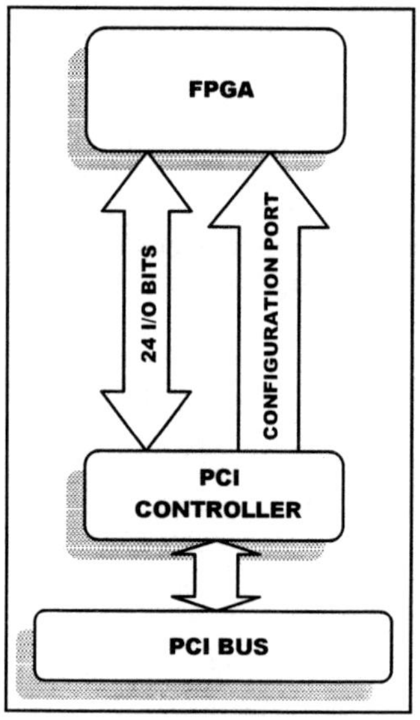

Figure 2: FPGA PCI-Board

The PCI Controller provides a 24-bit bidirectional I/O port between the controller and the FPGA. Each I/O bit can be individually enabled by host software as an input or output. This small number of software accessible pins is one of the limitations in our platform. The chip used is a Xilinx Spartan-II XC2S200-5PQ208C. This chip contains 1176 Configurable Logic Blocks, having 5292 Logic Cells, which makes it equivalent to 200,000 system gates [11].

3. Background and Basic Concepts

Digital simulators allow including components having simulation models written in C. This is usually done using Software Procedural Interface. In this article we use Modelsim's Foreign Language Interface (FLI)[12] to perform this task. FLI routines are C programming language functions that provide procedural access to information within the VHDL simulator. A user-written application can use these functions to traverse the hierarchy of a VHDL design, get information about and set the values of VHDL objects in the design, get information about a simulation, and control a simulation run. Now, we define some terminology within the context of HDL simulation. This terminology will help us understand how the proposed accelerator platform works.

A "**foreign architecture**" is a design unit that is instantiated in a design but that does not contain any VHDL code. Instead it is a link to a C model that can communicate to the rest of the design through the ports of the foreign architecture. Normally, a C model creates processes and reads and drives signal values; in essence, behaving in the same manner as VHDL code but with the advantage of the power of C and the ease of reading and writing files and communicating with other system processes. When an HDL simulator starts, it first goes through an **"elaboration"** phase during which the entire design is loaded and connected and initial values are set. During this phase all foreign shared-libraries are loaded and the initialization functions of all foreign architectures are executed. The **"simulation"** phase of the simulator begins when the first run command is executed and continues until a quit or restart command is executed. When a restart command is executed, the simulator goes through its elaboration phase again. A "**callback**" is a C function that is registered with the simulator for a specific reason. The registered function is called whenever the reason occurs. Callback functions generally perform special processing whenever certain simulation conditions occur. A "**process**" is a VHDL process that is created through the FLI. It can either be scheduled for a specific time or be made sensitive to one or more signals that trigger the process to run. The process is associated with a C function, and the C function is executed whenever the simulator runs the process.

4. Master/Slave Architecture

Figure 3 shows the software components of our platform. An FLI application should be compiled into a dynamic library, having an exported entry point. This entry point will be called by the host simulator, Modelsim, for initializing the application. The simulator loads the accelerator's library. The library utilizes the FLI to communicate with the simulator. The accelerator library is used to read ports information, create necessary processes/drivers during design elaboration, read new input values when events occur, and to schedule new events for output values read from the hardware. On the other side, the accelerator library uses APIs supplied with the FPGA-board to configure the hardware, send stimulus to the FPGA, and read outputs from the FPGA.

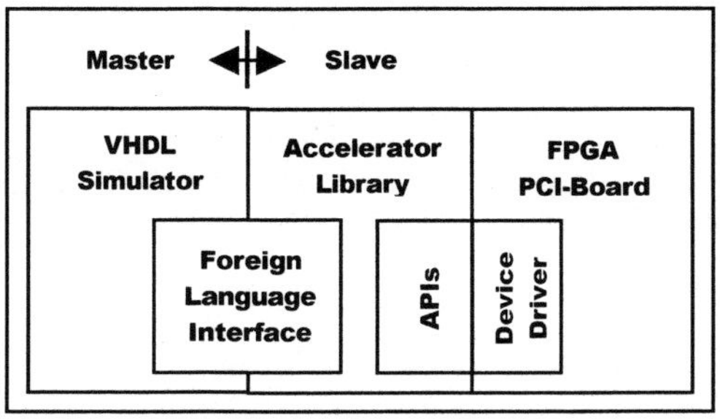

Figure 3: Master-Slave Architecture

Setting up the test environment consists of two major tasks. First, we need to transform the VHDL model of the DUT into hardware. Then, we need to write a skeleton for the DUT, and mark it as having a foreign architecture. This foreign architecture should point to the FLI application representing its model. To prepare the DUT to be put on the FPGA, we need to go through several steps. First, we synthesize the VHDL description of the DUT. In this step, we need to take care of the clock input. FPGA chips have special dedicated pins for receiving clock signals. These pins are connected internally to special clock buffers. These buffers provide an extra feature for clock signals, which is the elimination of clock skew. By default, synthesis tools detect and map clock signals to the primitive corresponding to these buffers.

In order to communicate with the chip we need to add constraints on DUT ports to be placed at pins accessible by software. Among the limitations in our hardware is that only 24 PFGA pins can be manipulated by software and this group of pins doesn't include any of the clock pins available in the chip. In order to avoid placement failure, we have to force the synthesis tool not to use clock buffers. In this way, we sacrifice the skew elimination feature of clock buffers, but this will not affect the results, since the hardware will run under the control of the simulation software, at a frequency much less than that of the final implementation. After that, comes the "Place and Route" (PAR) operation. In this stage we need to add the constraints mapping the DUT on accessible pins. The PAR tool uses these constraints, together with the output from the synthesis phase, to produce a floor-plan of the design on the chip. Finally, a bit file reflecting the design is generated to program the chip. Downloading the bit stream contained in this file is used to configure the logic cells and interconnect resources to behave exactly as the DUT. As the VHDL description of the DUT will no longer be available, the simulator needs to have a skeleton of the DUT specifying its interface (ports/generics). So we need to use the same exact entity declaration of the DUT. Along with the original entity declaration, we need to write an empty architecture, pointing to our FLI application. This is done by defining an attribute of type "string" in the architecture. This attribute should have the name "foreign", and its value should be composed of three parts. The following VHDL code gives an example.

```
ARCHITECTURE fpga OF DUT IS
  ATTRIBUTE foreign: string;
  ATTRIBUTE foreign OF fpga: ARCHITECTURE
  IS "fliAppInit fliApp.dll; parameter";
BEGIN
END fpga;
```

Where:

- fliAppInit: The name of the initialization function for this architecture (The Entry Point).
- fliApp.dll: The path to the dynamic library to load.
- parameter: A string that is passed to the initialization function. We use this part to specify the bit file, generated from the placement and routing, to be used in configuring the FPGA chip.

For the test-bench, we need to write a configuration telling the simulator to use the foreign architecture of the DUT, in case the original architecture resides in the same library.

```
CONFIGURATION cfg_tb OF testbench IS
  FOR tb_arch
    FOR dut_instance: DUT
      -- USE ENTITY work.DUT(HDL);
      USE ENTITY work.DUT(FPGA);
    END FOR;
  END FOR;
END cfg_tb;
```

FLI (Foreign Language Interface) applications should have an exported entry point for initialization. The declaration of the initialization function as specified by the simulator is as follows:

```
fliAppInit(
  mtiRegionIdT region,
  char *parameter,
  mtiInterfaceListT *generics,
  mtiInterfaceListT *ports
  )
```

The first parameter is a region ID that can be used to determine the location in the design for this instance. The second parameter is the last part of the string in the foreign attribute. The third parameter is a linked list of the generic values for this instance. The list will be NULL if there are no generics. The last parameter is a linked list of the ports for this instance.

During elaboration, when the simulator encounters an instance of a foreign architecture, it loads the dynamic library of the corresponding FLI application and calls its initialization entry point. Among the parameters passed to the entry point is a list of the instance ports. The proposed accelerator's initialization function does the following:

- Configure the FPGA chip.
- Traverse the ports list to collect needed information:
  - Save handles of the signals in the port list.
  - For each input/in-out port, a process is created and given the address of an I/O processing callback

defined in the library. This process is made sensitive to the input/in-out port, so that the associated callback is called whenever an event occurs at this port.

- For each out/in-out port, a driver is created to be used when scheduling events with values from the FPGA chip outputs.

- Construct a port mapping to map VHDL port names, to pin numbers of the FPGA chip.

- Register a cleanup callback to be called when stopping/restarting simulation to free any allocated memory.

During simulation, whenever an event happens on one of the DUT in/in-out ports, the process sensitive to this port (created during elaboration) is activated. Activating the process means that the corresponding callback function is called. At this time, the accelerator library receives control. The callback reads the new value from the simulator, and sends it to the FPGA-board to be applied on chip pins. Then, all outputs are read back from the chip. These values are used to schedule new events on the corresponding port drivers, to reflect the DUT response to input stimulus.

5. Experimental Results

We tested our accelerator on several designs and in all cases a great speedup in simulation was achieved. Table 1 presents the results obtained with a synthesizable model of PARWAN 8-bit CPU. PARWAN is a simple accumulator-based processor introduced in [13].

Simulation Time (μs)	Real Time (ms)		Speedup
	Normal	Accelerated	
20	2133	60	35.55
40	4977	70	71.10
60	7662	80	95.78
80	9844	90	109.38
100	12378	100	123.78

Table 1: Speedup Measurements for PARWAN

The simulation speedup is dependent on the following factors:

- Design Size: speedup is proportional to the gate count.
- Activity: high internal signal activity raises speedup; high signal activity on ports degrades speedup due to communication overhead.
- Simulation Time: as more events accumulate for processing, serial processing consumes a lot of CPU-time compared to concurrent processing in hardware due to parallelism.

6. Conclusion

In this article, we presented an FPGA-based accelerator for functional simulation of VHDL descriptions. The design unit under test (DUT) is synthesized on an FPGA chip, and the test-bench is kept in VHDL. The accelerator uses a VHDL simulator as the master of the simulation process, and the DUT emulated in hardware is the slave. The interface and the synchronization are implemented by the procedural interface of the simulator. The accelerator performance has been evaluated by simulating a simple processor. The performance gain is up to 125 times over the pure VHDL simulation. The factors affecting the gain are: the design size, the communication between the DUT and the test bench, and the length of the simulation time.

References

[1] J. Bauer, M. Bershteyn, I. Kaplan, P. Vyedin. "A Reconfigurable Logic Machine for Fast Event–Driven Simulation". In Design Automation Conference, June 1998.

[2] Z. Barzilai, J. Carter, B. Rosen, "HSS a high speed simulator," IEEE Trans. Computer Aided Design, vol. CAD-6, July, 1987.

[3] R.E Bryant, "Graph-based algorithms for Boolean function manipulation," IEEE Transaction on Computers, vol. C-25, Dec.1986

[4] D. Bittruf, Y. Tanurhan, "A survey for hardware emulators," Tec. Rep., ESPRIT Basic Research Project, No 8135, 1994.

[5] T. Ahn, K. Choi. "VHDL Simulation Acceleration Using Specialized Functions". In ISCAS, June 1997.

[6] J. Yim, Y. Hwang, C. Park, H. Choi, W. Yang. "A C-Based RTL Design Verification Methodology for Complex Microprocessor". In Design Automation Conference, 1997.

[7] S. Liao, S. Tjian, R. Gupta, "An Efficient Implementation of Reactivity for modeling hardware in the scenic Environment, In Design Automation Conference, 1997.

[8] R. Wisniewski, A. Bukowiec, M. Wegrzyn, "Benefits of Hardware Accelerated Simulation". In The International Workshop on Discrete-Event System Design, DESDes'01, June 2001.

[9] S. Cadambi, C. Mulpuri, P. Ashar. "A Fast, Inexpensive and Scalable Hardware Acceleration Technique for Functional Simulation". In Design Automation Conference, June 2002.

[10] D. Atef, A. M. Salem, H. Baraka. "An Architecture of Distributed Co-Simulation Backplane". In the 42nd Midwest Symposium on Circuits and Systems, Las Cruces, August 1999.

[11] Xilinx, "Xilinx Spartan-II FPGA Family Architecture," http://www.xilinx.com.

[12] Mentor Graphics, "ModelSim Foreign Language Interface," Version 5.5e, August 2001.

[13] Z. Navabi. "VHDL: Analysis and modeling of digital systems", McGraw-Hill, 1993.

FREQUENCY DOMAIN WAVELET METHOD WITH GMRES FOR LARGE-SCALE LINEAR CIRCUIT SIMULATION*

Jian Wang[1], *Xuan Zeng*[1], *Wei Cai*[2], *Charles Chiang*[3], *Jiarong Tong*[1] *and Dian Zhou*[4]

[1] ASIC & System State Key Lab., Microelectronics Dept., Fudan Univ., Shanghai 200433, P.R. China
[2] Mathematics Dept., University of North Carolina at Charlotte, Charlotte, NC 28269, USA
[3] Advanced Technology Group, Synopsys Inc., Mountain View, CA 94043, USA
[4] Electrical Engineering Dept., University of Texas at Dallas, Richardson, TX 75080, USA

ABSTRACT

In this paper, a *Frequency Domain Fast Wavelet Collocation Method with GMRES* (FFWCM-G) is proposed for the simulation of high-speed large-scale linear VLSI circuits. Taking advantages of wavelet expansion and adaptive scheme, the number of frequency points for calculating frequency response is minimized. Moreover, due to the efficiency of GMRES as the internal iterative solver, the proposed method can achieve nearly linear time complexity, which undoubtedly would be more promising to simulate large-scale integrated circuits.

1. INTRODUCTION

As the advance of high-speed deep-submicron VLSI technology, the more analog like circuit behaviors needs to be analyzed by circuit simulation for large scale complex systems. When dealing with circuit singularities represented as the sharp changing of waveforms, today's circuit simulators have been continuously challenged in both computation speed and simulation accuracy.

Time marching method, such as SPICE, is the most popular numerical approach in VLSI circuit simulation [1]. When handling with system singularities, time marching method has to reduce the time step sufficiently small to maintain the stability and accuracy of the marching scheme, thus slowing down the computation [1]. In the last decade, *Krylov* subspace based order-reduction techniques, such as AWE [2] and PVL [3], have achieved spectacular success in solving many practical VLSI interconnect problems. Such methods first reduce the scale of the original system to a relatively small order, and then obtain the response of the reduced-order system, as an approximation of that of the original system. However, the solution obtained by such methods is valid only at the neighborhood of the frequency expansion points, and there is no guarantee for the order reduction error bound within a wide frequency range.

A *Fast Wavelet Collocation Method* (FWCM) was first proposed in [4] for the initial boundary value problem of ODEs and PDEs, and later on applied in VLSI circuit simulation [5]. Compared with traditional approaches, FWCM can effectively deal with the time domain singularities and achieve uniform error distribution on the whole simulation interval in time domain. Unfortunately, this method suffers from the unbearable computation complexity since each of the N state variables is expanded by a number of M wavelets, which greatly restrains the method from simulating large-scale circuits. In order to accelerate the simulation speed of FWCM for large-scale circuit, a frequency domain wavelet method FFWCM was proposed in [6], where only the output variables are expanded by wavelets. Because the number of the output variables is much smaller compared with that of the state variables, FFWCM can significantly reduce the unknown variables and thus obtain simulation speed-up. However, in FFWCM, the frequency response of the system must be calculated at each collocation points in the frequency domain. Therefore FFWCM will lose the efficiency for the circuits that have many singularities in the frequency domain waveforms.

In this paper, we propose an approach to further accelerate the simulation speed of FFWCM by employing the *Krylov* subspace iterative method GMRES [7] as the internal solver for the large sparse linear system at each frequency point. With superior computational property of wavelets, the number of collocation points at which frequency response of the circuit is calculated, can be minimized. Furthermore, the GMRES method can lead to an almost linear time complexity for the large scale sparse liner systems.

The rest of the paper is organized as follows. In Section 2, we review the principles of frequency domain FFWCM and analyze its advantages and dominant factors for simulation time. Then in Section 3, we introduce GMRES algorithm and propose FFWCM with GMRES. To demonstrate the efficiency of the proposed method, a clock tree circuit is simulated by different methods in Section 4. Finally, we draw conclusions in Section 5.

* This research is supported partially by NSFC research projects 90307017, 60176017, 60076014 and 90207002, National 863 Program projects 2002AA1Z1340 and 2002AA1Z1460, Synopsys Inc., Science & Technology Key Project of Ministry of Education of China 02095, Cross-Century Outstanding Scholar's Fund of Ministry of Education of China, the Doctoral Program Foundation of Ministry of Education of China 2000024628, Shanghai Science and Technology Committee project 01JC14014, Shanghai AM R&D Fund 0107, and NSF grants CCR-0098275 and CCR-0306298.

2. FREQUENCY DOMAIN FAST WAVELET COLLOCATION METHOD

2.1. Principle of FFWCM

Without loss of generality, a linear circuit can be formulated as the following equations:

$$C\dot{x}(t) + Gx(t) = Bu(t) \tag{1}$$

$$y(t) = L^T x(t) \tag{2}$$

$$x(t)\big|_{t=0} = x_0 \tag{3}$$

where $x(t)$ is the N dimensional unknown vector and $u(t)$ is the K dimensional excitation vector. $y(t)$ is the unknown P dimensional output vector. And x_0 is the initial condition of the system. **C**, **G**, **B** and **L** are the coefficient matrixes, whose dimensions are $N \times N$, $N \times N$, $N \times K$ and $N \times P$, respectively. Without loss of generality, in the following we only consider the single output system, i.e., $P = 1$.

Performing the *Laplace* transform on (1) and (2), we have

$$X(s) = (sC + G)^{-1}[BU(s) + x_0] \tag{4}$$

$$Y(s) = L^T X(s) \tag{5}$$

where $X(s)$, $Y(s)$, $U(s)$ are the *Laplace* transforms of $x(t)$, $y(t)$, $u(t)$. The frequency response $Y(s)$ in (5) is actually defined in a two-dimensional complex plane of $s = \alpha + j\omega$, where α represents the real axis and $j\omega$ denotes the imaginary axis. We may reformulate $Y(s)$ as the sum of real part and imaginary part.

$$Y(s) = Y(\alpha + j\omega) = R(\alpha,\omega) + jI(\alpha,\omega) \tag{6}$$

By fixing α with A such that all the poles of the system are to the left of the line $x = A$, we obtain (7).

$$Y(s) = R(\omega) + jI(\omega) \tag{7}$$

There always exists a large Ω that the signal energy is small enough in the region $[\Omega, \infty)$. Then we can choose interval $[0,\Omega]$ as the approximation interval. Mapping the frequency interval $[0,\Omega]$ to the wavelet interval $[0, L]$ by $l = L \times \omega / \Omega$, we get

$$R(\omega) = R(\Omega \times l / L) = \tilde{R}(l) \tag{8}$$

$$I(\omega) = I(\Omega \times l / L) = \tilde{I}(l) \tag{9}$$

Then $\tilde{R}(l)$ and $\tilde{I}(l)$ can be expanded by wavelet basis functions.

$$\tilde{R}(l) = [e_{J,0}, e_{J,1} \cdots e_{J,M-1}][\varphi_{J,0}(l), \varphi_{J,1}(l) \cdots \varphi_{J,M-1}(l)]^T = E\varphi(l) \tag{10}$$

$$\tilde{I}(l) = [f_{J,0}, f_{J,1} \cdots f_{J,M-1}][\varphi_{J,0}(l), \varphi_{J,1}(l) \cdots \varphi_{J,M-1}(l)]^T = F\varphi(l) \tag{11}$$

where $E = [e_{J,0}, e_{J,1} \cdots e_{J,M-1}]$, $F = [f_{J,0}, f_{J,1} \cdots f_{J,M-1}]$, and $\varphi(l) = [\varphi_{J,0}(l), \varphi_{J,1}(l) \cdots \varphi_{J,M-1}(l)]^T$. $\varphi_{J,i}(l)$ (i = 0, 1, 2, ..., M–1) is the wavelet basis function in subspace V_{bJ} [1]. $e_{J,i}$ and $f_{J,i}$ (i = 0, 1, 2, ..., M–1) are the unknown wavelet coefficients. J is the wavelet order, and $M = 2^J \times L + 3$. Substituting (10) and (11) into (6), and discretize the equations at the collocation points [5] in subspace V_{bJ}, we have

$$E[\varphi(l_{J,0}), \varphi(l_{J,1}) \ldots \varphi(l_{J,M-1})]^T + jF[\varphi(l_{J,0}), \varphi(l_{J,1}) \ldots \varphi(l_{J,M-1})]^T$$
$$= [Y(A + j\omega_{J,0}), Y(A + j\omega_{J,1}) \ldots Y(A + j\omega_{J,M-1})]^T \tag{12}$$

where $\omega_{J,l} = l_{J,l} / L \times \Omega$ (l = 0, 1, 2, ..., M–1).

After solving the coefficient vectors E and F in (12), we will obtain the wavelet approximation of $Y(s)$.

2.2. Advantages and Limitations of FFWCM

2.2.1 Advantages

One of the major advantages of wavelet expansion is that there exists an adaptive scheme [4–6]. For a wavelet expansion system, the approximation space V_J can be decomposed into a set of orthogonal subspaces:

$$V_J = V_0 \oplus W_0 \oplus W_1 \oplus \ldots \oplus W_{J-1} \tag{13}$$

where J is the space level (order). The approximation accuracy depends upon the wavelet space level employed. The higher the space level J is, the less the error will be. In the adaptive scheme, wavelets will be automatically added to only those regions where the approximation doesn't reach specified accuracy. Thus, wavelet expansion can achieve a convergence rate of $O(h^4)$ with a minimal set of base functions in expansion space [4].

The number of the wavelet base functions is just the same with the number of the collocation points. Therefore, by adaptive scheme, the number of the collocation points in frequency domain, where equation (4) must be solved, can also be minimized. In addition, the error in frequency domain is uniformly distributed, with the error bound guaranteed by the adaptive scheme.

2.2.2 Limitations

In FFWCM, the computation is mainly composed of two parts. The first part is the calculation of frequency response $Y(A + j\omega_{J,i})$ (i = 0, 1, 2, ..., M–1) by solving equation (4) at the M frequency points. The second part comes from solving the wavelet coefficients in (12). Due to the compact support property [4] of wavelet, for each subspace W_J, the matrix in (12) is a tri-diagonal one. Therefore the wavelet coefficient vectors E and F in (12) can be solved in linear time [4–6].

So, for large scale circuit, computation time of FFWCM is mainly dominated by solving equation (4) at each frequency point, while the time for obtaining the wavelet coefficients is negligibly small.

3. FFWCM WITH GMRES

For solving large scale circuits, direct methods such as LU factorization are almost prohibitive due to their excessive computational cost. The iterative *Krylov* subspace methods, combined with proper preconditioning, are capable of solving large sparse linear system in almost linear time [7]. In this section, we propose to employ *Generalized Minimal RESidual method* (GMRES) [7] in FFWCM as the internal solver to further accelerate the simulation speed.

3.1. GMRES

Suppose a set of linear algebraic equations in the form of:

$$Ax = b \tag{14}$$

In iteration $n \geq 1$, GMRES picks the solution x_n from *Krylov* space $K_n(A, b) \equiv span\{b, Ab, \cdots, A^{n-1}b\}$ which solves the least square problem

$$\min_{z \in K_n} \|b - Az\| \tag{15}$$

in *Euclidean* norm. The least square problem is solved by constructing an orthonormal basis $\{v_1, v_1, \cdots, v_n\}$ for K_n using

Arnoldi method [7], starting with $v_1 = b/\|b\|$ as a basis for κ_1.

Collecting the basis vectors for κ_j in a matrix $\mathbf{V}_j = (v_1, \cdots, v_j)$, we get the decomposition associated with *Arnoldi* method:

$$\mathbf{A}\mathbf{V}_j = \mathbf{V}_{j+1}\mathbf{H}_j \tag{16}$$

where $\mathbf{H}_j$ is the upper *Hessenberg* matrix.

If $z \in \kappa_n$, it can be represented as $z = \mathbf{V}_n y$ for some *y*. So,

$$\mathbf{A}z = \mathbf{A}\mathbf{V}_n y = \mathbf{V}_{n+1}\mathbf{H}_n y \tag{17}$$

$$b = \|b\| v_1 = \|b\| \mathbf{V}_{n+1} e_1 \tag{18}$$

where e_1 is the first column of the identity matrix. The least square problem (15) is thus reduced to

$$\min_{z \in \kappa_n} \|b - \mathbf{A}z\| = \min_{y} \big\| \|b\| e_1 - \mathbf{H}_n y \big\| \tag{19}$$

By finding the solution y_n of the least square problem, we may obtain an approximate solution $x_n = \mathbf{V}_n y_n$ for equation (14). GMRES is terminated when norm $\|\mathbf{A}x_n - b\|$ is smaller than given error tolerance.

3.2. Applying GMRES to FFWCM

GMRES can serve as the internal solver for equation (4) in FFWCM. Several techniques should be considered when implementing GMRES in FFWCM.

Firstly, we should calculate the circuit response at the frequency points in the order required by the adaptive scheme. Specifically, we calculate the frequency response of the circuit at all collocation points in the order of subspace V_0, W_0, W_1, ...W_{J-1}. In each subspace, the frequency response is calculated in the ascending order of the collocation points.

Secondly, the initial value required by GMRES should be carefully chosen because the closer this initial value to real solution, the fewer iterations GMRES requires before convergence. Following are the principles for choosing the initial values.

Denote $\omega_{V0,i}$ (i = 0, 1, 2, ..., I_{V0}−1) the set of a number of I_{V0} collocation points in subspace V_0, which is sorted in an ascending order. Denote $\omega_{WJ,i}$ (i = 0, 1, 2, ..., I_{WJ}−1) the set of a number of I_{WJ} collocation points in the wavelet subspace W_J, which is also sorted in an ascending order.

a) At the first collocation point $\omega_{V0,0}$ in the subspace V_0, the initial value of the state variables can be chosen as the right-hand vector of the equation, or just an all-zero vector.

b) For the other collocation points $\omega_{V0,i}$ (i = 1, 2, ..., I_{V0}−1) in the V_0 subspace, we choose the circuit state variables solution at frequency point $\omega_{V0,i-1}$ to be the initial values of the state variables at frequency point $\omega_{V0,i}$.

c) For the collocation points in subspace W_J, the initial values of the state variables should be chosen by the solution of the equation at the most adjacent frequency point in the anterior subspace W_{J-1}, or the average of those at the two adjacent frequency points, if available.

Thirdly, preconditioning should be exploited to accelerate the convergence. Assuming that $u = \mathbf{P}x$, equation (14) becomes

$$\mathbf{A}\mathbf{P}^{-1}u = b \tag{20}$$

where **P**, named the preconditioner, should be an easily invertible approximation of **A**. The condition number of $\mathbf{A}\mathbf{P}^{-1}$ is much better than that of **A**, thus the convergence rate of GMRES can be improved by preconditioning. Experiments shows that incomplete LU factorization [8] of **A** will construct a good preconditioner for our problem.

In the process of GMRES, only matrix-vector productions, rather than the time-consuming matrix inversion, are involved, which readily exploits the sparsity and structure of the linear problem. The GMRES process is usually ended within only a few iterations for practical applications, which lead to an almost linear time cost [7]. Thus for the simulation of linear circuit with N unknown variables, if a number of M wavelets are employed, the time complexity for FFWCM with GMRES is approximately $O(MN)$. As M is only decided by the singularities of the waveform in frequency domain rather than the scale of the problem, the time complexity of FFWCM with GMRES can be considered as a nearly-linear one, i.e., $O(N)$.

4. NUMERICAL EXPERIMENTS

In this section, a clock tree circuit is employed to demonstrate the efficiency of the proposed method. As shown in Fig. 1, the clock tree circuit is modeled by RLC segments, which expands in a level-by-level scheme. So, the number of the elements in the tree will increase exponentially as the level of the tree goes up.

For all the experiments, frequency range of the simulation is set to [0, 15GHz], since at a frequency higher than that, the value of the transfer function is efficiently small. The result obtained by SPICE is set as the standard for calculating the error. Mean square value of the absolute error is chosen as the criteria for evaluating the result of each method. The transfer function of the RLC tree with 7 levels, obtained by HSPICE version 2001.2, is plotted is Fig. 2, which is similar to that of other circuit with a different level number.

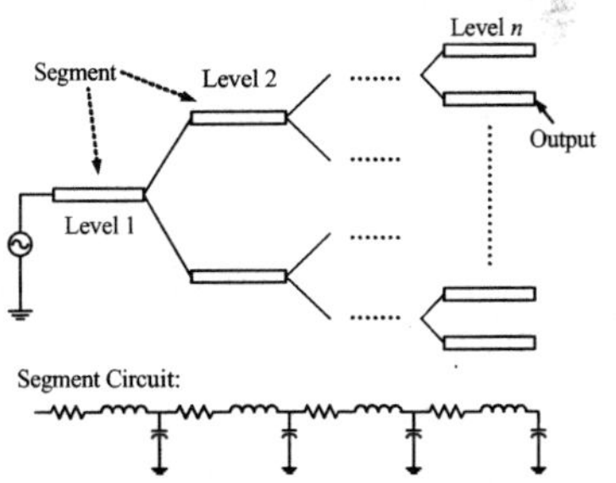

Fig. 1. A clock tree with n levels

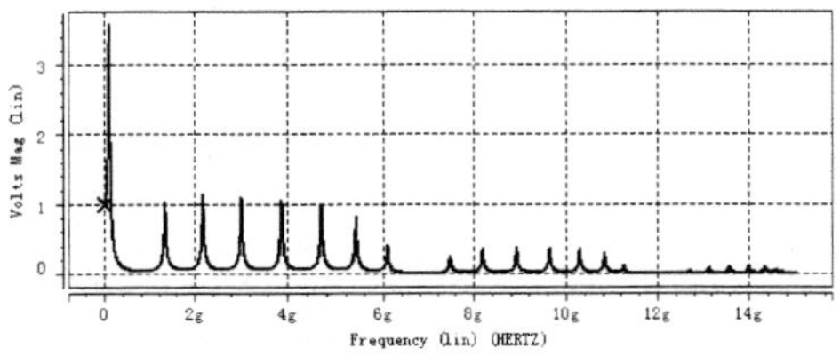

Fig. 2. Transfer function of the tree (n = 7)

4.1. Wavelet Collocation vs. Uniform Collocation

First, we test the efficiency of wavelet method to choose collocation points. For comparison, the method that uniformly chooses the points is also implemented. Here LU factorization method is used as internal solver for both approaches. The results are given in Table 1. Clearly, the wavelet collocation method can reach a higher accuracy with more than 30% reduction in the number of collocation points, compared with uniform collocation method. As discussed above, the whole computational cost is determined by solving equation (4) at each frequency point. Thus wavelet method can save much computation time by efficient reduction of collocation points.

Table 1. Collocation: wavelet vs. uniform

Tree Level n	Unknown Variables	Wavelet		Uniform	
		Points	Error	Points	Error
7	1,589	1015	0.00532	2100	0.00567
8	3,189	889	0.00343	2200	0.00354
9	6,389	809	0.00683	1200	0.00783
10	12,789	747	0.00421	1400	0.00422

4.2. GMRES vs. LU Factorization

The same examples are also used to demonstrate the efficiency of GMRES as the internal solver. LU factorization solver is employed for comparison. With both approaches, FFWCM is used for collocation and approximation. The error criteria for iteration convergence in GMRES is set to a sufficiently small number, actually, 10^{-7}. Under this criterion, the numerical errors between the results of these two methods are less than 0.1%. In addition, both methods employ the same number of wavelets, i.e., the same number of collocation points.

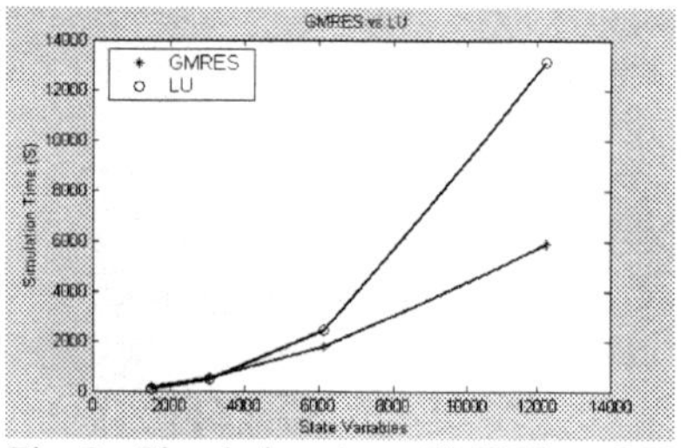

Fig. 3. Simulation time: GMRES vs. LU

As shown in Fig. 3, the time complexity for LU factorization and solving the resulting equations is over $O(n^2)$. However, the time complexity of the GMRES method, for the solution of the large-scale sparse algebraic equation generated in the simulation, is almost linear. In specific applications such as interconnect analysis, the number of unknown variables tends to become extremely large. GMRES is certainly more appropriate to serve as the internal solver. This also implies that, dealing with the ever increasing circuit scale, GMRES will behave even better over the traditional but widely used LU factorization method.

In addition, the error tolerance for judging GMRES convergence can be set to a relatively reasonable value in order to trade off the simulation accuracy and the simulation time.

4.3. FFWCM with GMRES

Finally, we make comparison between the proposed method – FFWCM with GMRES, and uniform collocation method with LU factorization as the solver. The latter is widely used in practical applications, such as the AC sweep analysis in SPICE. The number of collocation points and simulation error of both methods are the same as in Table I. In Fig. 4, simulation time is plotted, where it can be found that FFWCM with GMRES markedly surpasses the uniform collocation with LU factorization. Most importantly, the computation complexity of proposed method is almost linear, which will be more promising for the circuits with drastically increasing circuit scale.

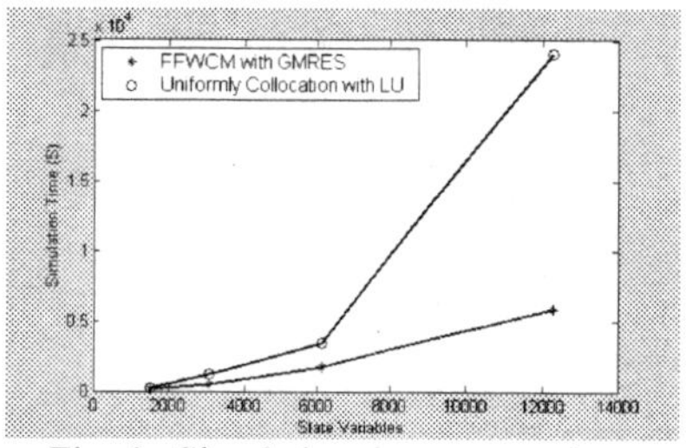

Fig. 4. Simulation time comparison

5. CONCLUSIONS

In this paper, we propose a frequency domain fast wavelet collocation method with GMRES, for the simulation of high-speed large-scale linear VLSI circuits. With an adaptive scheme of wavelet expansion, the collocation points at which frequency response of the circuit is calculated, can be reduced to a minimal number needed for guarantee the uniformly distributed error requirement. To efficiently solve the high-dimensional, sparse linear algebra equations, GMRES is employed as an iterative solver, which has almost linear time complexity. Several numerical examples are presented to demonstrate these properties of the proposed method.

FFWCM with GMRES has a nearly linear time complexity, which would become even more promising as the scale of the circuit increasingly growing. Further research may concern with how to adopt similar method in non-linear circuit simulation.

6. REFERENCES

[1] L. O. Chua, and P. M. Lin, *Computer Aided Analysis of Electronic Circuits: Algorithms, and Computational Techniques*, Prentice Hall, 1975.

[2] L. T. Pillage, and R. A. Rohrer, "Asymptotic waveform evaluation for timing analysis", *IEEE Trans. on CAD*, Vol. 9, No. 4, pp. 352–366, 1990.

[3] P. Feldmann, and R. W. Freund, "Efficient linear analysis by Padè approximation via Lanzos process", *IEEE Trans. on CAD*, Vol. 14, pp. 639–649, 1995.

[4] W. Cai, and J. Wang, "Adaptive multi-resolution collocation methods for initial boundary value problems of nonlinear PDEs", *SIAM J. Numerical Analysis*, Vol. 33, No. 3, pp. 937–970, 1996.

[5] D. Zhou, and W. Cai, "A fast wavelet collocation method for high-speed circuit simulation", *IEEE Trans. on CAS-I*, Vol. 46, No. 8, pp. 920–930, 1999.

[6] X. Zeng, S. Huang, J. Wang, and D. Zhou, "Frequency domain wavelet method for high-speed circuit simulation", *Proc. of IEEE ISCAS 2002*, part II, pp. 233–236, 2002.

[7] Y. Saad, *Iterative Methods for Sparse Linear Systems*, PWS Publishing Company, 1996.

[8] J. W. Demmel, *Applied Numerical Linear Algebra*, SIAM, 1997.

EVENT-DRIVEN DYNAMIC POWER MANAGEMENT BASED ON WAVELET FORECASTING THEORY

[1]*Ali Abbasian,* [1]*Safar Hatami,* [1]*Ali Afzali-Kusha,* [2]*Mehrdad Nourani,* [1]*Caro Lucas*
[1]Electrical and Computer Engineering Dept., University of Tehran, Tehran, Iran
[2]Department of Electrical Engineering, The University of Texas at Dallas
{abbasian, s.hatami}@ece.ut.ac.ir, afzali@ut.ac.ir, nourani@utdallas.edu, lucas@ipm.ir

ABSTRACT

The goal of dynamic power management is to reduce power dissipation in system level by putting system components into different states. This paper proposes a wavelet based approach that models the device behavior precisely. In spite of conventional stochastic models, this method eliminates some impractical assumptions which were shortcoming for previous methods. The stationary behavior of devices and the use of a memoryless distribution for device modeling are some of the aforementioned assumptions that have been relaxed in our model. Additionally, wavelet model can capture the local information accurately. Furthermore this algorithm is adaptive. This method has two additional benefits; firstly according to the device application a suitable wavelet basis can be used. Secondly, it has a sparse time-scale representation indicating that only a few coefficients in their wavelet representation have to be estimated. The simulation results show 95% accuracy in desktop Hard Disk Drive (HDD) states prediction and power saving by a factor of 2.

1. INTRODUCTION

Reducing power consumption in the electronic systems is one of the most important challenges in modern electronic system design due to the recent popularity of battery-operated portable devices and environmental concern related to desktops and servers. While the computational demands have drastically increased, the battery capacity has improved very slowly over the same time. So, in order to extend the battery service life time, reducing power dissipation is essential. Power reduction can be performed in different levels. Among them, reducing power consumption in system level is great of importance due to the fact that much of the power dissipation in portable electronic device comes from non digital component. Dynamic Power Management (DPM), which refers to a selective shutdown or slowdown of system components that are idle or underutilize, has proven to be a particularly effective technique for reducing power dissipation in such systems [2]. There are several methods for Power Management (PM); among them stochastically controls of PM are more efficient. These algorithms guarantee optimal results as long as the system that is power managed can be modeled well with memory-less distributions such as exponential distribution [1], [2], [3], [4]. Additionally the system components should have a stationary behavior [1], [5], [6], [7]. In the absence of these assumptions the optimal result will not be guarantied. In fact, system components not only have a nonstationary behavior, but also the memory less distributions can not model the system components behavior precisely.

In this paper, we introduce a novel method for power management at system level based on wavelet forecasting theory that enables very accurate modeling of system components with nonstationary behavior. In this method, the next event's time which is a nonstationary series is predicted by means of non-decimated wavelet. This method not only is an adaptive method but also is event-driven.

The remainder of this paper is organized as follow. Section 2 describes the desktop HDD behavior modeling. We present the wavelet forecasting theory for nonstationary process in section 3. Forecasting algorithm is presented in section 4. We show the simulation results for the HDD in section 5. Finally, we summarize our findings in section 6.

2. MODELING OF HDD BEHAVIOR

In this section a novel model for device behavior has been proposed. Figure 1(a) illustrates the states of a device without power management. As can be seen the device has two main states called *active* and *idle.* In active state there are some requests that must be serviced. When the requests were serviced, the device enrolls in idle state. In Figure 1(b) all events have been illustrated, upward arrows and downward arrows illustrate the events that the device state, changes from active to idle, and idle to active, respectively. In order to reduce power dissipation, it is essential to predict the idle length when a *to-idle* event is occurred. The *to-idle* events have been illustrated in Figure 1(c). The length of upward arrows determine the idle time, in this Figure horizontal axis is *to-idle* events numbers.

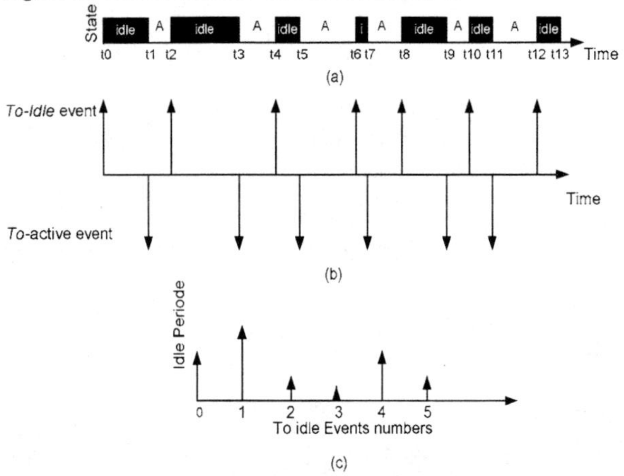

Figure 1. Device states (a), occurred event for the device (b), To idle events and their durations(c). i: idle, A: active

3. WAVELET FORECASTING THEORY FOR NON-STATIONARY PROCESS

The important question of how to model and forecast non-stationary process arises, and one of the main motivations behind using wavelets here, is that being well-localized in both time and frequency domain, they have the potential to naturally handle phenomenal whose spectral characteristics changes over time[16].

3.1. The class of locally stationary wavelet process

Suppose the stochastic array $X_{t,T}(t=0,...,T-1,T>0)$ is a representation of the idle length which is equal to the time between the end of former *active* state and the beginning of later *active* state. As mentioned before the values of $X_{t,T}$ are equal to the length of upward arrows in Figure1(c).

Deffinition1: This array is in the class of locally stationary wavelet (LSW) processes if there is a mean square representation of

$$X_{t,T} = \sum_{j=[\log_2 T]}^{-1} \sum_{k=-\infty}^{\infty} w_{j,k;T} \psi_{jk}(t) \xi_{jk} \tag{1}$$

where j and k are scale and location parameter, respectively, $w_{j,k;T}$ are real constants, $\{\psi_{j,k}(t)\}_{jk}$ is a non-decimated family of discrete compactly supported wavelets, ξ_{jk} is an orthonormal sequence of identically distributed zero mean random variables, and for each $j \le -1$, there exist a Lipschitz-continues function $W_j(z)$ on $[0,1)$ such that

- $\sum_{j=-\infty}^{-1} W_j(z)^2 < \infty$ uniformly on $z \in [0,1)$
- the Lipschitz constant L_j satisfy $\sum_{j=-\infty}^{-1} 2^{-j} L_j < \infty$
- there exist a sequence of constants C_j satisfy

$\sum_{j=-\infty}^{-1} C_j < \infty$, such that, for all T,

$$\sup_{k=0,...,T-1} \left| w_{j,k;T} - W_j(k/T) \right| C_j / T$$

the time-varying quantity $W_j(z)$ is defined in rescaled time $Z = t/T \in [0,1)$. As the non-decimated wavelet system is over complete, the amplitudes $w_{j,k;T}^2$ are not uniquely defined and therefore not identifiable. However, due to the regularity of $W_j(z)$ in the rescale time, the wavelet spectrum of $X_{t,T}$, defined by $S_j(z) = W_j(z)^2$ is unique. $S_j(z)$ measures the power of the process at a particular scale j and location z, and can be estimated by means of asymptotically unbiased multiscale estimators. The link between autocovariance function of a LSW process $X_{t,T}$, defined by $c_T(z,\tau) = Cov(X_{[zT],T}, X_{[zT]+\tau,T})$, and its wavelet spectrum $S_j(z)$ is shown in the next proposition. We first define the autocorrelation wavelets by $\Psi_j(\tau) = \sum_k \psi_{jk}(0)\psi_{jk}(\tau)$. They are symmetric with respect to τ and satisfy $\Psi_j(0)=0$ for all scales[12].

Proposition: under the assumptions of previous definition, $\|c_T - c\|_{L\infty} = O(T^{-1})$, where

$$c(z,\tau) = \sum_{j=-\infty}^{-1} S_j(z)\, \Psi_j(\tau). \tag{2}$$

Function $c(z,\tau)$ is called the local autocovariance function (LACV) of $X_{t,T}$. Formula (2) is a multiscale representation of the nonstationary autocovariance function $c(z,\tau)$. The representation is unique because the set $\{\Psi_j\}_j$ is linearly independent [13]. Hence, $c(z,\tau)$ can be estimated by the follow equation:

$$c\left(\frac{k}{T},\tau\right) = \sum_{j=-J}^{-1}\left(\sum_{l=-J}^{-1} A_{jl}^{-1}\Psi_l(\tau)\right) I_j\left(\frac{k}{T}\right),\ k=0,...,t-1, \tau \ne 0, \tag{3}$$

where the matrixes $A = (A_{jl})_{j,l<0}$ and $I_j\left(\frac{k}{T}\right)$ are defined by

$$A_{jl} = \langle \Psi_j, \Psi_l \rangle = \sum_\tau \Psi_j(\tau)\Psi_l(\tau) \text{ and } I_j\left(\frac{k}{T}\right) = \left(\sum_{s=0}^{t-1} X_{s,T}\psi_{jk}(s)\right)^2,$$

respectively, where $-\log(T) \le j \le -1, k = -1,...,t-1$.

3.2. Forecasting Theory

We now want to forecast an LSW process one step ahead, based on *t* observations $X_{0,T},..., X_{t-1,T}$ and we consider a linear predictor

$$\hat{X}_{t,T} = \sum_{s=0}^{t-1} b_{t-1-s;T}^{(t)} X_{s,T} \tag{4}$$

where coefficients $b_{t-1-s}^{(t)}$ minimize the mean square prediction error $E(\hat{X}_{t,T} - X_{t,T})^2$.

These coefficients can be obtained by solving the following equation:

$$\sum_{m=0}^{t-1} b_{t-1-m;T}^{(1)} \left\{ \sum_{j=-J}^{-1} S_j\left(\frac{n+m}{2T}\right)\Psi_j(m-n) \right\} = \sum_{j=-J}^{-1} S_j\left(\frac{t+n}{2T}\right)\Psi_j(t-n) \tag{5}$$

for all *n=0,..., t-1.*

Prediction equation can be obtained by using the relation between the wavelet spectrum and local autocovariance function as shown in (2). Furthermore, the prediction equation can also be written as

$$\sum_{m=0}^{t-1} b_{t-1-m;T}^{(1)} c\left(\frac{n+m}{2T}, m-n\right) = c\left(\frac{n+t}{2T}, t-n\right) \tag{6}$$

The above predictor is asymptotically unbiased but is not consistent and therefore has to be smoothed using e.g. a Gaussian kernel smoother with the bandwidth of g. it should be noted that in above equations only *t-1* values can be obtained for autocovariance from *t-1* previous observations. Consequently, for forecasting the next event, extending and smoothing the autocovariance function is unavoidable. In the next section we will consider the benefits of this new method and the relation between these concepts and previous formula, and The fifth section will discusses a data-driven method for choosing the smoothing parameter [12], [14].

4. FORECASTING ALGORITHM

We now address the question of how to estimate the unknown time-varying second order structure in the system of equation (5). In theory, the best linear predictor of X_t is given by (4), where $b_t = \left(b_{t-1-s;T}^{(t)}\right)_{s=0,\ldots,t-1}$ solves the prediction equations (5). In practice, each of the t components of vector b_t is estimated using our estimator of local autocovariance function based on observations $X_{0,T},\ldots,X_{t-1,T}$. Hence, we have to find a balance between the estimation error, potentially increasing with t, and the prediction error which is a decreasing function of t.

As a natural balancing rule which works well in practice, we suggest to choose a number of p such that the "clipped" predictor

$$\hat{X}_{t,T}^{(p)} = \sum_{s=t-p}^{t-1} b_{t-1-s;T}^{(t)} \qquad (7)$$

gives a good compromise between the theoretical prediction error and the estimation error.

We propose an automatic procedure for selecting the two parameters: the order p in (7) and the bandwidth g. we estimate the autocovariance $c(z,\tau)$ by smoothing over k/T to achieve consistency. For simplicity, we choose the same bandwidth g for all τ.

Also, we only incorporate the last p observations into the predictor. We select (g, p) using adaptive forecasting, i.e. we gradually update (g, p) according to the success of prediction. We first move backward by s observations and choose the initial parameters (g_0, p_0) for forecasting $X_{t-s,T}$. Next we forecast $X_{t-s,T}$ using not only (g_0, p_0) but also the eight neighboring pairs ($g_0+\delta\varepsilon_g$, $p_0+\varepsilon_p$), for ε_g, ε_p $\in$ {-1,0,1} and δ fixed . As we already know the true value of $X_{t-s,T}$, we compare the nine forecasts using a pre-selected criterion, and update (g, p) to be equal to the pair which gave the best forecast. We now use this updated pair, as well as its 8 neighbors, to forecast $X_{t-s+1,T}$ and continue in the same manner until we reach $X_{t-1,T}$. The updated pair (g_1, p_1) is used to perform the actual prediction, and can itself be updated later if we wish to forecast $X_{t,T}$, $X_{t+1,T}$,...Various criteria can be used to compare the quality of the pairs of parameters at each step. Denote by $\hat{X}_{t-i,T}(g,p)$ the predictor of $X_{t-i,T}$ computed using pair (g, p), and by $P_{t-i,T}(g, p)$ the length of the corresponding prediction interval. A suitable criterion is as follows: choose the pair which minimizes $\left|X_{t-i,T} - \hat{X}_{t-i,T}(g,p)\right| / P_{t-i,T}(g,p)$

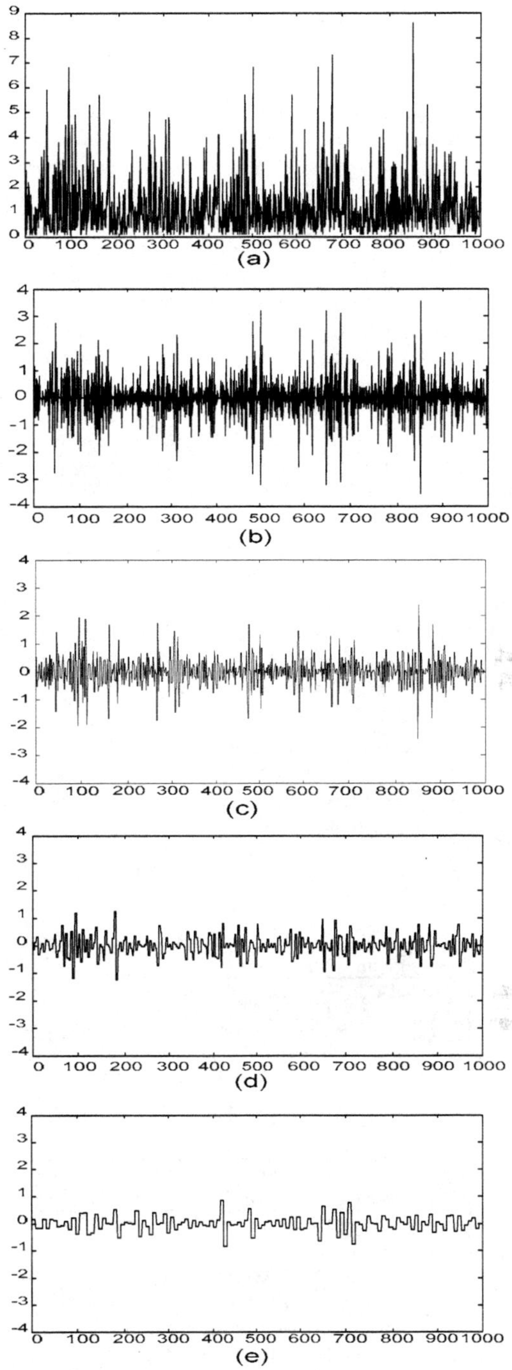

Figure 2. predicted series(a), first scale of estimated series(b), second scale of estimated series(c)third scale of estimated series(d), forth scale of estimated series(e)

5. SIMULATION RESULTS

In this simulation data has been collected from PC hard disk drive in an 8-h user request trace. The system was running Windows operating system with standard software (e.g., Word, Matlab, HSPICE, Visual .NET). It should be mentioned that all simulations have been done by using S-Plus 6.1 software.

At first, we obtain (g_1, p_1) using primery100 events and then the rest of data has been forecasted by (g_1, p_1). During this prediction the parameters (g_1, p_1) were adapted three times to satisfy the accuracy criterion. The Haar wavelet has been utilized as a basis to multiscale estimation. Figure 2 shows simulation results with respect to "lives" only at a limited number of scales where the vertical axis is idle time and the horizontal axis is event numbers. Particularly Figure 2(a) shows the predicted series and Figure 2(b)-2(e) represent the estimated series in scales one to four respectively. Note, since the coefficient of scale three and four and … are negligible compare to scales one and two, it is not essential to predict them. This decreases the computation drastically.

The accuracy of this method in prediction of the next event duration is about 95% compared to oracle. It should be mentioned that oracle is the ideal policy and its decision is based on a perfect knowledge about future events, and the goal in various model of DPM is to reach to this efficiency. In table I, *Pwr* is average power consumption (unit: *W*), N_{sd} is Number of shutdown, N_{wd} is Number of wrong shutdown causing a power overhead, and T_{ss} is average sleeping time per shutdown (unit: *sec*) [1]. It should be mentioned that *time out* policy is a policy that turn off the device after *t* seconds of idleness. As can be seen in Table I, this method reduces power dissipation by a factor of 2 compared to always on policy, and has minimum power consumption compared to other methods.

Table I: Desktop Hard Disk Drive simulation comparison

Algorithm	Pwr (W)	N_{sd}	N_{wd}	T_{ss}(s)
Oracle	1.64	164	0	166
proposed	1.74	160	8	150
TISMDP	1.92	156	25	147
Karlin's	1.94	160	15	142
Adaptive	1.97	168	26	134
30s timeout	2.05	147	18	142
120s timeout	2.52	55	3	235
DTMDP	2.60	105	39	130
Always on	3.48	0	0	0
CTMDP	3.90	326	318	4

6. CONCLUSION

In this paper we presented a wavelet based approach that models the device behavior precisely. In spite of the previous models, this method eliminates some impractical underlying assumptions which were critical for previous methods. Devices Stationary behavior is one of the aforementioned assumptions that have been in our model. Our model has several additional benefits compared to previous works. This new wavelet based model can capture the local information accurately. Furthermore this algorithm is adaptive meaning that if the device behavior changes this model will be adapted to it by adjusting only two parameters (g, p). Additionally, according to the device application a suitable wavelet basis can be used. Moreover, as illustrated in section 5 it has a sparse time-scale representation. The simulation results showed 95% accuracy in desktop HDD states prediction and power saving with a factor 2.

7. REFERENCES

[1] T. Simunic, L. Benini, P. Glynn, and G. D. Micheli, *"Event Driven Power Management," IEEE Trans. Computer Aided Design*, Vol. 20, No. 7. July 2001, pp. 840-856.

[2] Q. Qui, Q. Wu, and M. Pedram, *"Stochastic Modeling of a Power-Managed System-Construction and Optimization," IEEE Trans. Computer Aided Design,* Vol.20, No. 10, October 2001, pp. 1200-1217.

[3] D. Ramanathan, S. Irani, and R. K. Gupta, *"An Analysis of System Level Power Management Algorithms and Their Effects on Latency," IEEE Trans. Computer Aided Design,* Vol.21, No. 3, March 2002, pp. 291-305.

[4] S. Irani, S. Shukla, R. Gupta, *"Competitive Analysis of Dynamic Power Management Strategies for Systems with Multiple Power saving States," In Proceeding of DATE, 2002.*

[5] L. Benini, A. Bogliolo, G. D. Micheli, *"A Survey of Design Technique for System Level Dynamic Power Management," IEEE Trans. VLSI Sys,* Vol. 8, NO. 3, June 2000, pp. 299-316.

[6] M. Pedram, *"Power Optimization and Management in Embedded System" IEEE 2001.*

[7] Q. Qiu, and M. Pedram, *"Dynamic Power Management Based on Continuous Time Markov Decision Processes" In Proceeding of DAC 1999.*

[8] G. A. Paleologo, L. Benini, A. Bogliolo, G. D. Micheli, *"Policy Optimization for Dynamic Power Management" In proceeding of DAC 1998.*

[9] Q. Qiu, Q. Wu, and M. Pedram, *"Dynamic Power Management of Complex Systems using Generalized Stochastic Petri Nets," In Proceeding of DAC 2000.*

[10] E. Y. Chung, L. Benini, and G. D. Micheli, *"Dynamic Power Management Using Adaptive Learning Tree" IEEE 1999.*

[11] Q. Qiu, Q. Wu, and M. Pedram, *"Dynamic Power Management in a Mobile Multimedia System with Guaranteed Quality of service" In Proceeding of DAC 2001.*

[12] P. Fryzlewicz, S. Van Bellegem, and R. von Sachs, *"Forecasting non-stationary time series by wavelet process modeling" Department of Mathematics, Bristol, UK. February 25, 2002.*

[13] R. von Sachs, G. P. Nason, G. Kroisandt, *"Adaptive estimation of the evolutionary wavelet spectrum" University of Kaiserslautern, Germany, and University of Bristol, UK. October 15, 1997.*

[14] P. Fryzlewicz, *"Modeling and forecasting log-return as locally stationary wavelet process" July 31, 2002.*

[15] Mercurio, Danilo, Spokoiny, and Vladimir, *"Statistical inference for time-inhomogeneous volatility models" July 4, 2002.*

[16] C. S. Burrus, R. A. Gopinath, and H. Guo, *"Introduction to wavelets and wavelet transform,"* prentice-Hall, 1998.

A NOVEL ENCODING METHOD INTO SEQUENCE-PAIR

Chikaaki KODAMA[†*], *Kunihiro FUJIYOSHI*[†] *and Teppei KOGA*[‡]

[†]Dept. of Electrical and Electronic Eng., Tokyo University of Agri. & Tech., Tokyo 184-8588, Japan
[‡]School of Info. Science, Japan Advanced Inst. of Science and Tech., Ishikawa 923-1292, Japan

ABSTRACT

The sequence-pair was proposed to represent a rectangle packing and a placement, and is used to place modules automatically in VLSI layout design. Several decoding methods of sequence-pair were proposed. However, encoding methods are not found except the original one called "gridding". The gridding requires almost $O(n^3)$ time for a packing of n rectangular modules and it is hard to implement. Therefore, we propose a novel method to encode a given rectangle packing into a sequence-pair in $O(n \log n)$ time. We also propose a linear time method to obtain a sequence-pair from a given rectangular dissection represented by a Q-sequence, a recently proposed representation method of rectangular dissection. The proposed methods can be used for the compaction keeping topology, for example, in the post-process of the Force Directed Relaxation, a method used in module placement, and so on.

1. INTRODUCTION

The sequence-pair (seq-pair) [1] was proposed to represent a module placement and a packing. It is used for optimization of floorplan or module placement of VLSI physical design with stochastic algorithms such as simulated annealing or genetic algorithms. The merits of seq-pair are that it can represent an arbitrary rectangle packing and each seq-pair always has its corresponding packing(s). Since a seq-pair defines a relative position of all pairs of rectangles, it has wide applications. Various decoding methods of seq-pair have been proposed [2, 3, 4, 5]. On the other hand, studies about encoding methods are not found except for the original method called "gridding" [1].

The Force Directed Relaxation (FDR) [6, 7, 8] is another major approach to the module placement and has been studied for a long time. Since there are module overlappings in the placement, the operation to remove them is necessary in the post-process, almost keeping the topology of modules. Therefore, it is considered that the gridding can be used for the compaction keeping topology with a low temperature annealer in the post-process.

For the design reuse in renewed fabrication processes, we extract a set of macro device blocks, and block sizes shrink and shapes are changed. A method of packing these macro blocks without changing the topological relations was proposed as "topology constrained rectilinear block packing" in [9]. Gridding will also be able to use for topology constrained rectilinear block packing.

*E-mail: kodamada@fjlab.ei.tuat.ac.jp

However, the gridding requires almost $O(n^3)$ time for the placement of n rectangles. In addition to this disadvantageous big time complexity, it is difficult to implement.

In this paper, we propose (1) a fast encoding method to obtain a seq-pair from a given module placement in $O(n \log n)$ time. Also we propose (2) a linear time method to convert a given rectangular dissection represented by a Q-sequence (Q-seq) [10] to an equivalent seq-pair keeping topology.

By the proposed method (2) and a conventional method of obtaining a Q-seq from a given rectangle packing represented by a seq-pair [5], mutual conversion between a rectangle packing represented by a seq-pair and a rectangular dissection represented by a Q-seq becomes possible for the first time keeping the topology.

The organization of this paper is as follows. Sec. 2 defines preliminary terms, surveys gridding and presents a novel encoding method. Sec. 3 presents two procedures of converting a given rectangular dissection represented by a Q-seq to a seq-pair. Sec. 4 is for conclusions.

2. NOVEL METHOD TO ENCODE RECTANGLE PACKING INTO SEQ-PAIR

Before the explanation of the proposed encoding method from a given rectangle packing into a seq-pair, we introduce the definition of a seq-pair and the original encoding method called "gridding".

2.1. Sequence-Pair [1]

A **sequence-pair** (seq-pair) is an ordered pair of Γ_+ and Γ_-, where each of Γ_+ and Γ_- is a permutation of names of given n modules. For example, $(\Gamma_+; \Gamma_-) = (a\,b\,c\,d;\, b\,d\,a\,c)$ is a seq-pair of module set $\{a, b, c, d\}$. If module x is the i'th in Γ_+, we denote $\Gamma_+(i) = x$, as well as $\Gamma_+^{-1}(x) = i$. Similar notation is also used for Γ_-. To help intuitive understanding, we use a notation such as

$$(\Gamma_+; \Gamma_-) = (\cdot\cdot\ a \cdot\cdot b \cdot\cdot ; \cdot\cdot\ a \cdot\cdot b \cdot\cdot)$$

by which we mean

$$\Gamma_+^{-1}(a) < \Gamma_+^{-1}(b) \quad \text{and} \quad \Gamma_-^{-1}(a) < \Gamma_-^{-1}(b).$$

For every module pair $\{a, b\}$, a is in the left of b (equivalently, b is in the right of a) if

$$(\Gamma_+; \Gamma_-) = (\cdot\cdot\ a \cdot\cdot b \cdot\cdot ; \cdot\cdot\ a \cdot\cdot b \cdot\cdot).$$

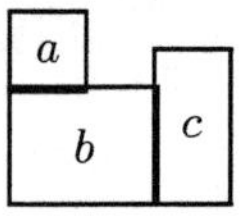

Fig. 1. A packing based on seq-pair $(abc; bac)$

Similarly, a is below b (equivalently, b is above a) if

$$(\Gamma_+;\Gamma_-) = (\cdots b \cdots a \cdots; \cdots a \cdots b \cdots).$$

A **Horizontal-Vertical-Relation-Set (HVRS)** for a set of modules is a set of horizontal (in the left/right of) or vertical (above/below) relations for all module pairs. For example, seq-pair $(a\,b\,c;\,b\,a\,c)$ corresponds to HVRS $\{a$ is above b, a is in the left of c, b is in the left of $c\}$ as shown in Fig.1. A seq-pair always corresponds to a realizable HVRS, and there is a seq-pair whose HVRS can lead to a minimum area placement.

From one seq-pair with n elements, a bottom left corner packing can be obtained in $O(n \log\log n)$ time [3].

2.2. Original Encoding Method "Gridding" [1]

The "gridding" is the original encoding method from a given rectangle packing into a sequence-pair. It is defined as follows.

For each rectangle x in a rectangle packing, pebble p is initially located at the upper right corner of x. We draw lines with pebble p which starts to move upward. It turns up and right alternatively until reaching the right upper corner of the boundary of the chip without crossing: i) boundaries of other rectangles, ii) lines drawn previously and iii) the boundary of the chip.

The drawn line is called "**up-right step-line**". It is always possible to draw such an up-right step-line for a rectangle. These lines refer to the corresponding rectangle names.

From the construction, no two up-right step-lines cross each other. It implies that these up-right step-lines can be linearly ordered, as well as the corresponding rectangles. Here we order the up-right step-lines from the left. Let Γ_+ be the rectangle name sequence in such order.

"**Up-left step-lines**" are drawn similarly as up-right step-lines. The difference is that an up-left step-line turns up and left alternatively. We order the up-left step-lines also from the left. Let Γ_- be the rectangle name sequence in such order. ■

Note that downward step-lines (down-left, down-right) defined in [1] are omitted because they are unnecessary.

The Basis of the Gridding

For rectangle a, let x coordinate of the left side be $x_l(a)$ and that of the right side $x_r(a)$. Similarly, let y coordinate of the lower side be $y_b(a)$ and that of the upper side $y_u(a)$. When the packing of rectangles a, b is given, the seq-pair $(\Gamma_+;\Gamma_-)$ corresponding to it must satisfy the following constraints.

Γ_+: if $x_l(a) < x_r(b)$ and $y_b(b) < y_u(a) \Rightarrow \Gamma_+^{-1}(a) < \Gamma_+^{-1}(b)$.

Γ_-: if $x_l(a) < x_r(b)$ and $y_b(a) < y_u(b) \Rightarrow \Gamma_-^{-1}(a) < \Gamma_-^{-1}(b)$.

For all pairs of rectangles, according to Γ_+ constraints, we obtain left-to-right relation of rectangle names in Γ_+ (${}_n\mathrm{C}_2$ combinations in all) and do topological sort to satisfy the whole relation. Then Γ_+ can be obtained. Similarly, Γ_- can be obtained. This method requires time in proportion to n^2, since we check all combinations of ${}_n\mathrm{C}_2$,

2.3. Fast-Gridding

Here we propose an efficient method called "Fast-gridding" to encode a rectangle packing into a seq-pair. This encoding method is possible in $O(n \log n)$ time (n is the number of rectangles).

Procedure Fast-gridding (Input: Rectangle packing / Output: Seq-pair $(\Gamma_+;\Gamma_-)$)

Step 1: Put a hypothetical rectangle S with null width and infinite height to the left of all rectangles.

Step 2: Obtain graph G as follows. Prepare a vertex corresponding to each rectangle including S. Make directed edges from one vertex corresponding to rectangle x to vertices corresponding to rectangles which can be lit by the beam coming straight from the right edge of x to rightward. G is clearly a planer graph and let vertical relation of each edge from one vertex be the same as the vertical relation of the corresponding beam.

Step 3: Initally, set permutation Γ_+ empty. In G, trace every vertex in depth first order starting from a vertex corresponding to S, giving the priority to the lower edge. While tracing each vertex, insert the vertex name (excluding the vertex corresponding to S) into immediately right of its parent vertex name on Γ_+. If the parent vertex name corresponds to S, insert it into the leftmost of Γ_+.

Step 4: Initally, set permutation Γ_- empty. In G, trace every vertex in depth first order starting from a vertex corresponding to S, giving the priority to the upper edge and insert vertex name similarly as in **Step 3** to obtain Γ_-. ■

In **Step 2**, vertical scan line scans all rectangles from left to right. Graph G can be obtained in $O(n \log n)$ time by computational geometry where names of rectangles giving out the beam to light the scan line are memorized as interval of y coordinate. As a result, the time complexity becomes $O(n \log n)$.

A seq-pair obtained by Fast-gridding imposes horizontal relation on any pair of elements only when the relation of the pair cannot be coded as vertical.

It is clearly understood that rectangle a is constrained to be left of rectangle b if and only if a path from vertex a to vertex b exists. In **Step 3**, since all vertices are traced in the depth first order, when visiting a certain vertex p, vertices corresponding to rectangles constrained to be above p are not yet visited. Each vertex corresponding to rectangles constrained to be left of rectangle p and already visited is the parent vertex name of p or exists left to it on Γ_+. Other names of vertices are right to it on Γ_+. Therefore in **Step 3**, according to Γ_+ conditions, all names of vertices determined to be left of p should be left to p and other vertices should be right to p on Γ_+. **Step 4** can be proved similarly.

An example of Fast-gridding is shown in Fig.2

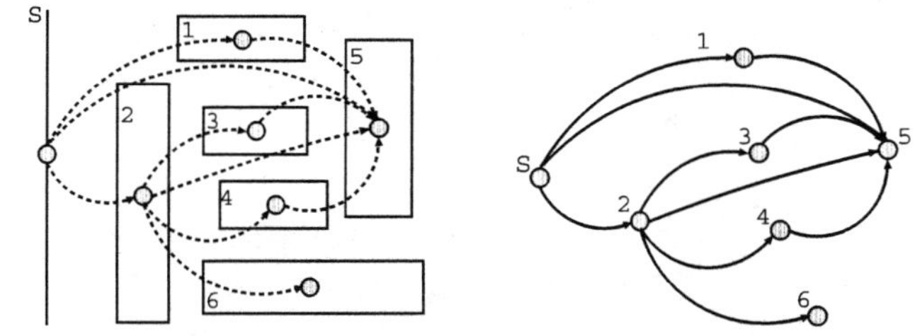

(a) An example of obtaining graph G (b) Graph G

Fig. 2. (a) Graph G obtained from a given placement with a hypothetical rectangle S. (b) In **Step 3**, all vertices of G are traced once in the order of "$S, 2, 6, 4, 5, 3, 1$" and $\Gamma_+ = (1\,2\,3\,4\,5\,6)$ is obtained. In **Step 4**, all vertices of G are traced once in the order of "$S, 1, 5, 2, 3, 4, 6$" and $\Gamma_- = (2\,6\,4\,3\,1\,5)$ is obtained.

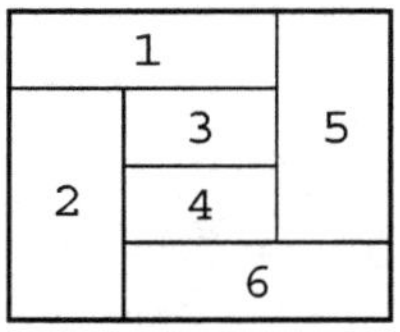

Fig. 3. Rectangular dissection represented by Q-seq $\mathcal{Q} = \mathcal{R}_2\mathcal{R}_1\mathcal{B}_5\mathcal{B}_1 1\mathcal{B}_3\mathcal{B}_2 2\mathcal{R}_6\mathcal{R}_4\mathcal{R}_3 3\mathcal{B}_4 4\mathcal{R}_5 5\mathcal{B}_6 6$

3. METHOD TO ENCODE RECTANGULAR DISSECTION REPRESENTED BY Q-SEQ INTO SEQ-PAIR

Here we propose two procedures of converting a given rectangular dissection represented by a Q-seq to a seq-pair of the same HVRS and size as the rectangular dissection. These conversion methods are possible in linear time, and the resultant seq-pair is a unique one since each seq-pair corresponds to separate HVRS.

Before the explanation of the proposed method, we introduce rectangular dissection and a Q-seq.

3.1. Rectangular Dissection

A **rectangular dissection** is a dissection of a rectangle into a set of rectangles called **rooms** with exclusive assignments of modules to rooms (no two modules share a single room.) Examples are shown in Fig.3. Only T-intersections are used to form the dissection except for the four corners of the outermost bounding rectangle [11]. (Two T-intersections on a point may look like a cross.) Each of the line segments and edges of the bounding rectangle is called a **seg**.

In this paper, as in the conventional floorplan or the topological placement, we only focus on the topology between rooms and segs in a rectangular dissection. Consequently, if the number of rooms is given, the variety of rectangular dissection becomes finite [12].

HVRS of rectangular dissection is defined as follows. If a vertical (horizontal) seg s adjoins right of (below) the room a and left of (above) room b, we say "room a is left of (above) room b via seg s."

3.2. Q-sequence [10]

Q-sequence (Q-seq) proposed by Sakanushi et al. is a method to represent a rectangular dissection of general structure with n rooms by letters of total length of $3n$. Each room x in a Q-seq has necessarily its corresponding $\mathcal{R}_x$ and $\mathcal{B}_x$. For example, rectangular dissection of Fig.3 is represented by a Q-seq
"$\mathcal{Q} = \mathcal{R}_2\mathcal{R}_1\mathcal{B}_5\mathcal{B}_1 1\mathcal{B}_3\mathcal{B}_2 2\mathcal{R}_6\mathcal{R}_4\mathcal{R}_3 3\mathcal{B}_4 4\mathcal{R}_5 5\mathcal{B}_6 6$."

Q-seq is made from **Q-state** defined as follows [13].

Definition 1 (Q-state) In a rectangular dissection, when the lower (right) end of a vertical (horizontal) seg s touches a horizontal (vertical) seg at the bottom right corner of room x in the shape of '$\perp$' ('$\dashv$'), Q-state of room x ($Q(x)$) has a room name followed by $\mathcal{R}$'s ($\mathcal{B}$'s) corresponding to rooms which are adjacent to the right (bottom) of seg s in the order from the bottom (right). ■

For example, a rectangular dissection of Fig.3 has a vertical and a horizontal seg touching in the shape of '$\perp$' at the bottom right corner of room 2 and, adjacent to the right of this vertical line, room 6, 4 and 3 stand in a column from the bottom. The Q-state of room 2 is "$Q(2) = 2\mathcal{R}_6\mathcal{R}_4\mathcal{R}_3$."

In Q-seq, "$\mathcal{R}_i$" ("$\mathcal{B}_i$") before the leftmost room name shows that room i is adjacent to the left seg (upper seg) of the bounding rectangle. We denote a sequence of $\mathcal{R}$'s ($\mathcal{B}$'s) with "$\mathcal{R}_i$" ("$\mathcal{B}_i$") as $Q(L)$ ($Q(U)$). For example, $Q(L)$ is the top "$\mathcal{R}_2\mathcal{R}_1$" in $\mathcal{Q}$ and it means room 2 and room 1 are adjacent to the left seg of the bounding rectangle. Similarly $Q(U)$ is "$\mathcal{B}_5\mathcal{B}_1$" and it means room 5 and room 1 are adjacent to the upper seg of the bounding rectangle.

Q-seq with n rooms is defined as the concatenation of $Q(L)$, $Q(U)$, and Q-states of rooms in Abe-order[1] such as $Q(L)\,Q(U)\,Q(1)\,Q(2)\cdots Q(n)$.

3.3. Simple Conversion Method from Q-seq to Seq-pair

The following procedure is a simple conversion method from a Q-seq to a seq-pair in $O(n)$ time where n is the number of rooms.

Procedure Qseq–SeqPair (Input: Q-seq $\mathcal{Q}$; Output: Seq-pair $(\Gamma_+;\Gamma_-)$)

Making Γ_+: Γ_+ is obtained by removing all positional symbols ($\mathcal{R}$'s and $\mathcal{B}$'s) from Q-seq $\mathcal{Q}$.

Making Γ_-: Get the rectangular dissection by decoding Q-seq $\mathcal{Q}$ and turn it clockwise by 90 degrees. Then get Q-seq $\mathcal{Q}'$ by encoding this. Γ_- is obtained by removing all positional symbols ($\mathcal{R}$'s and $\mathcal{B}$'s) from Q-seq $\mathcal{Q}'$ (shown in Fig.4). (**Procedure** Qseq–SeqPair **End**)

Since the length of $\mathcal{Q}$ is $3n$ where n is the number of rooms, Γ_+ can be obtained in $O(n)$ time. Also decoding and encoding of a Q-seq are possible in $O(n)$ time [10], so Γ_- is also obtained in $O(n)$ time. Therefore, procedure Qseq–SeqPair is carried out in linear time.

[1] In the process of encoding a Q-seq, the room of upper left corner is removed from the remaining rooms in order. This order is called "Abe-order" [10].

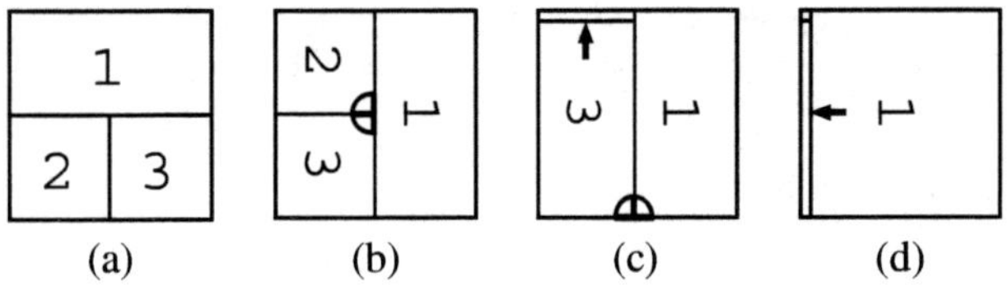

Fig. 4. An example of obtaining Γ_- by Qseq–SeqPair. (a) A rectangular dissection D corresponding to Q-seq "$\mathcal{R}_2\mathcal{R}_1\mathcal{B}_1 1\mathcal{B}_3\mathcal{B}_2 2\mathcal{R}_3 3$". (b) A rectangular dissection obtained by turning D clockwise by 90 degrees. (c) Room "2 ," the room of upper left corner (the prime room), was removed from the existing rooms. (d) Room "3," the prime room, was removed from the remaining rooms. Only room "1" remains, and a sequence "2, 3, 1" is obtained. This sequence is Γ_- of the seq-pair corresponding to D.

We can understand easily that the HVRS of the original Q-seq and that of the seq-pair obtained by conversion are equal because Abe-order [10] accords with Γ_+ of seq-pair.

3.4. Direct Conversion Method from Q-seq to Seq-pair

The following procedure makes it possible to obtain a seq-pair of the same HVRS as a given Q-seq in $O(n)$ time without decoding the given Q-seq.

Procedure Qseq–SeqPair (direct) $\left(\begin{array}{l}\text{Input: Q-seq } \mathcal{Q}\\ \text{Output: Seq-pair } (\Gamma_+;\Gamma_-)\end{array}\right)$

```
/*Making of Γ+*/
Obtain Γ+ by removing all positional symbols (R's and B's)
from Q;
/*Making of Γ-*/
Prepare a double linked list and insert nth room name in it;
for (i=n−1, n−2, ··· , 1){
  if (right adjacent symbol of ith room name in Q is R_k)
     Insert i to the immediate left of k in the list;
  else /*right adjacent symbol of ith room name in Q is B_k*/
     Insert i to the immediate right of k in the list;
}
The resultant list shows Γ-;
                    (Procedure Qseq–SeqPair (direct) End)
```

In this procedure, each element is inserted into the list (Γ_-) in the order from the last room of Q-seq, which is the same way as a decoding method of a Q-seq. Let the rectangular dissection after the insertion of ith room in $\mathcal{Q}$ on the decoding process be denoted as "$RD(i)$". When the ith room name in $\mathcal{Q}$ has inserted in Γ_- in the making, HVRS of $(\Gamma'_+;\Gamma_-)$ is the same as that of $RD(i)$, where Γ'_+ is a subsequence obtained from Γ_+ by eliminating the first $i-1$ elements.

If the next element to the ith room name on the Q-seq is $\mathcal{R}_x$, ith room should be placed immediately left of room x via a vertical seg. Since room x faces the left edge of the bounding box of $RD(i+1)$, elements on the left of x in Γ_- correspond to rooms below room x on $RD(i+1)$, and those on the right of x in Γ_- correspond to rooms right of or above x. Based on the decoding method of Q-seq, since ith room is inserted above the rooms below x and left of the other rooms, ith room name should be inserted immediately left of x on Γ_-.

If the next element to the ith room on the Q-seq is $\mathcal{B}_x$, this is similar.

4. CONCLUSIONS

We proposed (1) a novel encoding method from a given rectangle placement to a sequence-pair (seq-pair) in $O(n \log n)$ time and (2) two methods to obtain a seq-pair from a rectangular dissection represented by a Q-sequence (Q-seq) in linear time for the first time. By combining the proposed methods and the existing conversion method from a rectangle packing represented by a seq-pair to a Q-seq, we realized a mutual conversion between a rectangle packing (seq-pair) and a rectangular dissection (Q-seq). This conversion can be applied to a Corner Block List [14] (a rectangular dissection representation) as well as a Q-seq.

References

[1] H. Murata, K. Fujiyoshi, S. Nakatake, and Y. Kajitani: "Rectangle-packing-based module placement," in Proc. IEEE ICCAD, pp.472–479, 1995.

[2] T. Takahashi: "An algorithm for finding a maximum weight decreasing sequence in a permutation, motivated by rectangle packing problem," Technical Report IEICE, VLD96-30, pp.31–35, 1996 (in Japanese).

[3] X. Tang, R. Tian and D.F. Wong: "Fast evaluation of sequence pair in block placement by longest common subsequence computation," IEEE Trans. CAD, vol.20, no.12, pp.1406–1413, 2001.

[4] C. Kodama and K. Fujiyoshi: "An efficient decoding method of sequence-pair," in Proc. IEEE APCCAS, vol.2, pp.131–136, 2002.

[5] C. Kodama and K. Fujiyoshi: "Selected Sequence-Pair: An efficient decodable packing representation in linear time using sequence-pair," in Proc. IEEE ASP-DAC, pp.331–337, 2003.

[6] M. Sarrafzadeh and M. Wang: "NRG: Global and detailed placement," in Proc. IEEE ICCAD, pp.532–537, 1997.

[7] R. Forbes: "Heuristic acceleration of force-directed placement," in Proc. 24th DAC, pp.745–740, 1987.

[8] N. Quinn, and M. Breuer: "A force-directed component placement procedure for printed circuit boards," IEEE Trans. CAS, vol.CAS-26, no.6, pp.377-388, 1979.

[9] M. Z.-W. Kang, and W. W.-M. Dai: "Topology constrained rectilinear block packing for layout reuse," in Proc. ACM ISPD, pp.179–186, 1998.

[10] K. Sakanushi, Y. Kajitani and D. P. Mehta, "The quarter-state sequence floorplan representation," IEEE Trans. CAS–I, vol.50, no.3, pp.376–386, 2003.

[11] L. Stockmeyer: "Optimal orientations of cells in slicing floorplan designs," Info. and Control, vol.57, pp.91–101, 1983.

[12] T. Ohtsuki, N. Sugiyama and H. Hawanishi: "An optimization technique for integrated circuit layout design," in Proc. ICCST, pp.67–68, 1970.

[13] M. Tsuboi, C. Kodama, K. Sakanushi, K. Fujiyoshi and A. Takahashi "Linear time decodable rectangular dissection to represent arbitrary packing using Q-sequence," in Proc. SASIMI, pp.272–278, 2001.

[14] X. Hong, G. Huang, Y. Cai, J. Gu, S. Dong, C.-K. Cheng, and J. Gu, "Corner Block List: An efficient topological representation of non-slicing floorplan," in Proc. IEEE/ACM ICCAD, pp.8–12, 2000.

COMPUTING LARGE-CHANGE SENSITIVITY OF PERIODIC RESPONSES OF NONLINEAR CIRCUITS USING REDUCTION TECHNIQUES

P. Pai [†], *E. Gad* [‡], *R. Achar* [†], *R. Khazaka* [§], *M. Nakhla* [†]

[†] Dept. of Electronics, Carleton University, Ottawa, ON, Canada, K1S-5B6
[‡] School of Information Technology and Engineering (SITE), University of Ottawa, Ottawa, ON, Canada, K1N-6N5
[§] Dept. of Electrical and Computer Engineering, McGill University, Montreal, QC, Canada, H3A-2A7

ABSTRACT

This paper presents a new technique for computing Large-Change Sensitivity (LCS) of steady-state operating point in nonlinear circuits. The basic idea underlying the algorithm is the construction of a reduced system of nonlinear equations that preserves the derivatives of steady-state response with respect to the desired network parameters. Large change variations are then obtained by solving the reduced system instead of the original one.

1. INTRODUCTION

The rapid growth in the market of Silicon Radio Frequency Integrated Circuits (RFIC's) is placing new demands on current simulation tools for analog and mixed-signal nonlinear circuits. The key requirement that arises during various stages in the design process is the estimation of the effect of large variation in component or process parameters on the steady-state response of the circuit. Computing large-change sensitivity could arise in many situations such as in response to the need for estimating parasitic effects, parametric yield analysis, desensitizing a given design, tolerance assignment, fault and worst-case analysis. While techniques for computing small-change sensitivities could be used in finding sensitivity of desired output variables with respect to changes in circuit parameters, computation of large-change sensitivity is often needed since the parameter variation could be large.

The issue of large-change sensitivity has been tackled in a number of publications in the literature. Earlier approaches focused on the use of Householder formulas to derive efficient techniques for computing large change variation in response to changes in multiple circuit parameters [1]. These techniques, however, have been restricted to linear networks. Another approach tackled this problem in nonlinear circuits by using piecewise-linear representation for the nonlinear elements [2].

This paper presents a new approach for computing large-change sensitivities of the steady-state response in nonlinear circuits. A key advantage in the proposed approach is that it does not impose any restrictions on the form of nonlinearity that can be handled. Another advantage of the proposed algorithm stems from the fact that it allows computing sensitivity information with respect to all the desired parameters in the circuit without requiring further computational cost.

The main idea underlying the proposed algorithm is the construction of a reduced system of nonlinear equations. The key feature in this reduced system is that it shares with the original system a certain number of the leading derivatives of the steady-state response with respect to the desired circuit parameters. The variation of the steady-state response can then be obtained by incrementally changing the parameter and solving the reduced system. The key computational advantage of the proposed algorithm emanates from the fact that its size is dictated solely by the number of derivatives matched by the reduced system, which is typically several orders of magnitude smaller than the original system.

2. STEADY-STATE LARGE CHANGE SENSITIVITY

In computer-aided circuit design, it is often required to calculate the variation of network responses when a certain parameter or a set of parameters are changed. This problem is often referred to as large-change sensitivity. For the case where the network considered is nonlinear and the required response is the steady-state periodic response point of the network (given that the input excitation is periodic with period T), then this problem could be formulated using the Harmonic Balance (HB) approach as follows [3, 4],

$$\bar{\boldsymbol{G}}\bar{\boldsymbol{X}}(\lambda) + \jmath\boldsymbol{\Omega}\bar{\boldsymbol{C}}\bar{\boldsymbol{X}}(\lambda) + \bar{\boldsymbol{F}}\left(\bar{\boldsymbol{X}}(\boldsymbol{\lambda})\right) = \bar{\boldsymbol{B}} \qquad (1)$$

where $\bar{\boldsymbol{X}}(\lambda) \in \mathbb{C}^{m(2H+1)\times 1}$ is a vector that comprises the harmonics of all the circuit variable waveforms and λ is a given circuit parameter, with m representing the number of circuit variables and $(2H+1)$ is the number of harmonics used in representing the periodic waveform. The explicit dependency on λ in (1) indicates that $\bar{\boldsymbol{X}}$ is to be computed as λ is changed from a nominal value λ_0 to $\lambda_0 + \Delta\lambda$, where $\Delta\lambda$ is not necessarily small. The matrices $\bar{\boldsymbol{G}}$,$\bar{\boldsymbol{C}}$ and $\boldsymbol{\Omega} \in \mathbb{R}^{m(2H+1)\times m(2H+1)}$ are given by

$$\begin{aligned}
\bar{\boldsymbol{G}} &= BlockDiag\left[\cdots \boldsymbol{G} \quad \boldsymbol{G} \quad \boldsymbol{G}\cdots\right] \\
\bar{\boldsymbol{C}} &= BlockDiag\left[\cdots \boldsymbol{C} \quad \boldsymbol{C} \quad \boldsymbol{C}\cdots\right] \\
\boldsymbol{\Omega} &= BlockDiag[-\omega H \boldsymbol{I}_m \cdots \boldsymbol{0} \cdots \omega H \boldsymbol{I}_m] \qquad (2)
\end{aligned}$$

where $\boldsymbol{G}$ and $\boldsymbol{C} \in \mathbb{R}^{m\times m}$ are matrices representing the linear memoryless and memory elements in the network, respectively, $\boldsymbol{I}_m \in \mathbb{R}^{m\times m}$ is an identity matrix and $\omega = 2\pi/T$. Also in (1) $\bar{\boldsymbol{F}}(\bar{\boldsymbol{X}}) \in \mathbb{C}^{m(2H+1)\times 1}$ is the vector of harmonics contributed by the nonlinear elements in the network, and $\bar{\boldsymbol{B}}$ denotes the harmonics of the periodic independent sources in the circuit.

A direct approach of computing the changes in $\bar{\boldsymbol{X}}$ in response to variations in λ can be carried out by incrementally changing the value of λ, starting from λ_0, and solving the nonlinear system in (1) using iterative techniques such as Newton-Raphson method. This approach, however, incurs a large CPU cost for large integrated circuits with strong nonlinear behavior, as it requires repeated solution of a large coupled system of nonlinear equations with a dense Jacobian matrix.

3. PROPOSED ALGORITHM

The basic idea of the proposed algorithm is to track the variations in $\bar{\boldsymbol{X}}$ as λ is changed through solving a reduced system of nonlinear equations, instead of solving the original large system in (1).

3.1. Construction of the Reduced System

Constructing the reduced system of nonlinear equations is carried out by first computing the first, say q, scaled derivatives that appear in the Taylor series expansion of $\bar{\boldsymbol{X}}(\lambda)$ at $\lambda = \lambda_0$,

$$\bar{\boldsymbol{X}}(\lambda) = \sum_{i=0}^{\infty} \bar{\boldsymbol{A}}_i(\lambda_0)(\lambda - \lambda_0)^i \qquad (3)$$

where

$$\bar{\boldsymbol{A}}_i(\lambda_0) = \frac{1}{i!}\left.\frac{\partial^i \bar{\boldsymbol{X}}}{\partial \lambda^i}\right|_{\lambda=\lambda_0} \qquad (4)$$

The next step in constructing the reduced system is computing an orthonormal basis, $\boldsymbol{Q} \in \mathbb{C}^{m(2H+1)\times q}$, for the derivatives $\bar{\boldsymbol{A}}_i$, $0 \le i < q$ such that

$$\begin{aligned}
\text{ColSpan}[\boldsymbol{Q}] &= \text{ColSpan}\left[\bar{\boldsymbol{A}}_0 \cdots \bar{\boldsymbol{A}}_{q-1}\right] \\
\boldsymbol{Q}^T\boldsymbol{Q} &= \boldsymbol{I}_q \qquad (5)
\end{aligned}$$

where $\boldsymbol{I}_q$ is a $q \times q$ identity matrix. $\boldsymbol{Q}$ is then used in a change of variables in (1) through replacing $\bar{\boldsymbol{X}}$ with $\hat{\boldsymbol{x}}$ as follows

$$\bar{\boldsymbol{X}} \rightarrow \boldsymbol{Q}\hat{\boldsymbol{x}} \qquad (6)$$

The reduced system is then obtained by pre-multiplying the original system in (1) by $\boldsymbol{Q}^T$ yielding the following system

$$\hat{\bar{\boldsymbol{G}}}\hat{\boldsymbol{x}} + \hat{\bar{\boldsymbol{C}}}\hat{\boldsymbol{x}} + \hat{\bar{\boldsymbol{F}}}(\hat{\boldsymbol{x}}) = \hat{\bar{\boldsymbol{B}}} \qquad (7)$$

where

$$\begin{aligned}
\hat{\bar{\boldsymbol{G}}} &= \boldsymbol{Q}^T\bar{\boldsymbol{G}}\boldsymbol{Q} \\
\hat{\bar{\boldsymbol{C}}} &= \jmath\boldsymbol{Q}^T\boldsymbol{\Omega}\bar{\boldsymbol{C}}\boldsymbol{Q} \\
\hat{\bar{\boldsymbol{F}}}(\hat{\boldsymbol{x}}) &= \boldsymbol{Q}^T\bar{\boldsymbol{F}}(\boldsymbol{Q}\hat{\boldsymbol{x}}) \\
\hat{\bar{\boldsymbol{B}}} &= \boldsymbol{Q}^T\bar{\boldsymbol{B}} \qquad (8)
\end{aligned}$$

It is to be noted that the size of the reduced system in (7) is determined solely based on the number of derivatives chosen in computing the orthogonal basis $\boldsymbol{Q}$ in (5). This fact presents the main computational advantage in the proposed approach since tracing the variation of $\bar{\boldsymbol{X}}(\lambda)$ in response to the change in λ requires solving the system in (7) which is typically several orders-of-magnitudes smaller than the original system in (1). Details of tracing $\bar{\boldsymbol{X}}(\lambda)$ through solving the reduced system in (7) are presented next.

3.2. Solution of the Reduced System

The main observation from the presentation in the previous subsection is that the solutions $\hat{\boldsymbol{x}}$ of (7) and $\bar{\boldsymbol{X}}$ of (1) belong to different Euclidian spaces, namely, $\mathbb{C}^q$ and $\mathbb{C}^{m(2H+1)}$, respectively. An obvious consequence of this observation is that tracing $\bar{\boldsymbol{X}}(\lambda)$ via tracing $\hat{\boldsymbol{x}}$ requires a suitable mapping operator. This mapping is provided in the transformation matrix $\boldsymbol{Q}$. For example, if the solution at the nominal value λ_0 is denoted by $\bar{\boldsymbol{X}}_0$, then the corresponding solution in the reduced space is obtained from

$$\hat{\boldsymbol{x}}_0 = \boldsymbol{Q}^T\bar{\boldsymbol{X}}_0 \qquad (9)$$

To explain how to use (7) in computing changes in $\bar{\boldsymbol{X}}(\lambda)$, we use the following notations

$$\begin{aligned}
\bar{\boldsymbol{\Phi}}\left(\bar{\boldsymbol{X}}(\lambda), \lambda\right) &\equiv \bar{\boldsymbol{G}}\bar{\boldsymbol{X}} + \jmath\boldsymbol{\Omega}\bar{\boldsymbol{C}}\bar{\boldsymbol{X}} + \bar{\boldsymbol{F}}(\bar{\boldsymbol{X}}) - \bar{\boldsymbol{B}} \\
\hat{\bar{\boldsymbol{\Phi}}}\left(\hat{\boldsymbol{x}}(\lambda), \lambda\right) &\equiv \hat{\bar{\boldsymbol{G}}}\hat{\boldsymbol{x}} + \hat{\bar{\boldsymbol{C}}}\hat{\boldsymbol{x}} + \hat{\bar{\boldsymbol{F}}}(\hat{\boldsymbol{x}}) - \hat{\bar{\boldsymbol{B}}} \qquad (10)
\end{aligned}$$

to denote the dependencies of the solutions in both domains on λ. From (10), (7) and (6), the following can be deduced

$$\hat{\bar{\boldsymbol{\Phi}}}\left(\hat{\boldsymbol{x}}(\lambda), \lambda\right) = \boldsymbol{Q}^T\bar{\boldsymbol{\Phi}}\left(\boldsymbol{Q}\hat{\boldsymbol{x}}(\lambda), \lambda\right) \qquad (11)$$

Assuming that it is required to compute $\bar{\boldsymbol{X}}(\lambda)$ at $\lambda = \lambda_0 + \delta\lambda$, we proceed with solving the reduced system $\hat{\bar{\boldsymbol{\Phi}}}(\hat{\boldsymbol{x}}, \lambda_0 + \delta\lambda)$. This is a typical nonlinear system that can be solved using iterative techniques such as the Newton-Raphson (NR) method. This is usually done through iterating until convergence, in the following manner

$$\hat{\boldsymbol{x}}^{(i+1)} = \hat{\boldsymbol{x}}^{(i)} - \left[\hat{\bar{\boldsymbol{\Psi}}}\left(\hat{\boldsymbol{x}}^{(i)}, \lambda_0 + \delta\lambda\right)\right]^{-1} \hat{\bar{\boldsymbol{\Phi}}}\left(\hat{\boldsymbol{x}}^{(i)}, \lambda_0 + \delta\lambda\right) \quad (12)$$

where $\hat{\bar{\boldsymbol{\Phi}}}$ is obtained from (11) and $\hat{\bar{\boldsymbol{\Psi}}} \in \mathbb{C}^{q \times q}$ is the Jacobian matrix which is given by

$$\hat{\bar{\boldsymbol{\Psi}}} = \left[\hat{\boldsymbol{G}} + \hat{\boldsymbol{C}} + \boldsymbol{Q}^T \frac{\partial \bar{\boldsymbol{F}}(\bar{\boldsymbol{X}})}{\partial \bar{\boldsymbol{X}}} \boldsymbol{Q}\right]\Bigg|_{\lambda = \lambda_0 + \delta\lambda} \quad (13)$$

and $\hat{\boldsymbol{x}}^{(0)}$ is an initial guess solution taken here as $\hat{\boldsymbol{x}}_0$. Computing $\bar{\boldsymbol{X}}(\lambda_0 + \delta\lambda)$ is then achieved using the following mapping

$$\bar{\boldsymbol{X}}(\lambda_0 + \delta\lambda) = \boldsymbol{Q}\hat{\boldsymbol{x}}_\delta \quad (14)$$

where $\hat{\boldsymbol{x}}_\delta$ is assumed to be the solution of $\hat{\bar{\boldsymbol{\Phi}}}(\hat{\boldsymbol{x}}, \lambda_0 + \delta\lambda) = \boldsymbol{0}$, reached at the convergence of the NR iterations.

The key computational advantage of the proposed algorithm can be seen by noting that the main computational cost is the factorization of a $q \times q$ matrix in (13) and a series of q matrix-vector products, where $q \ll m(2H+1)$ the size of the original system.

4. COMPUTATION OF DERIVATIVES

This section addresses the issue of computing the derivatives $\bar{\boldsymbol{A}}_i, (i > 0)$ used in constructing the orthogonal basis $\boldsymbol{Q}$ in (5) and the associated computational complexity. It is to be noted here that $\bar{\boldsymbol{A}}_0$ represents the steady-state solution $\bar{\boldsymbol{X}}$ at the nominal point λ_0. This solution is assumed to be available through solving the nonlinear steady-state problem in (1). The approach adopted in computing the derivatives depends on whether the parameter λ belongs to the linear or nonlinear part of the network.

4.1. Derivatives w.r.t. linear parameters

Denoting the contribution of the linear part of the network by $\bar{\boldsymbol{Y}}$ where $\bar{\boldsymbol{Y}} = \bar{\boldsymbol{G}} + \jmath\boldsymbol{\Omega}\bar{\boldsymbol{C}}$, then computing $\bar{\boldsymbol{A}}_i$ when λ is assumed to be part of the linear network can be achieved using the following lemma.

Lemma 1 *Let $\bar{\boldsymbol{E}}_i \equiv 1/i!\partial^i\bar{\boldsymbol{Y}}/\partial\lambda^i$ and $\bar{\boldsymbol{D}}_k \equiv 1/k!\partial^k\bar{\boldsymbol{F}}/\partial\lambda^k$ computed at $\lambda = \lambda_0$, then $\bar{\boldsymbol{A}}_n$ can be computed as*

$$\left(\bar{\boldsymbol{E}}_0 + \bar{\boldsymbol{J}}_0\right)\bar{\boldsymbol{A}}_n = \sum_{j=1}^{n-1} \frac{j-n}{n} \bar{\boldsymbol{J}}_j \bar{\boldsymbol{A}}_{n-j} - \sum_{i=1}^{n} \bar{\boldsymbol{E}}_i \bar{\boldsymbol{A}}_{n-i} \quad (15)$$

where $\bar{\boldsymbol{J}}_i = 1/i!\partial^i\bar{\boldsymbol{J}}/\partial\lambda^i$ and $\bar{\boldsymbol{J}} = \partial\bar{\boldsymbol{F}}/\partial\bar{\boldsymbol{X}}$.

Proof: Substituting for $\bar{\boldsymbol{Y}}$ and $\bar{\boldsymbol{F}}$ using their Taylor series expansion and equating similar powers of λ yields

$$\bar{\boldsymbol{E}}_0\bar{\boldsymbol{A}}_n + \sum_{i=1}^{n} \bar{\boldsymbol{E}}_i\bar{\boldsymbol{A}}_{n-i} + \bar{\boldsymbol{D}}_n = \bar{\boldsymbol{0}} \quad (16)$$

Since $\frac{\partial\bar{\boldsymbol{F}}}{\partial\lambda} = \boldsymbol{J}(\bar{\boldsymbol{X}})\frac{\partial\bar{\boldsymbol{X}}}{\partial\lambda}$, then

$$\sum_{i=1}^{\infty} i\bar{\boldsymbol{D}}_i(\lambda-\lambda_0)^{(i-1)} = \sum_{i,j=0}^{\infty} j\boldsymbol{J}_i\bar{\boldsymbol{A}}_j(\lambda-\lambda_0)^{i+j-1} \quad (17)$$

Taking the n-th derivative w.r.t. λ and substituting $\lambda = \lambda_0$ results in

$$\bar{\boldsymbol{D}}_n = \bar{\boldsymbol{J}}_0\bar{\boldsymbol{A}}_n + \frac{1}{n}\sum_{i=1}^{n-1}(n-j)\bar{\boldsymbol{J}}_j\bar{\boldsymbol{A}}_{n-j} \quad (18)$$

Hence substituting from (18) into (16) proves the lemma. ■

4.2. Derivatives w.r.t. nonlinear parameters

In this case the system of nonlinear equations can be written as

$$\bar{\boldsymbol{Y}}\bar{\boldsymbol{X}}(\lambda) + \bar{\boldsymbol{F}}(\bar{\boldsymbol{X}}(\lambda), \lambda) = \bar{\boldsymbol{B}} \quad (19)$$

where the explicit dependency on λ appears in the nonlinear part only. We then have the following lemma for computing the derivatives $\bar{\boldsymbol{A}}_i$.

Lemma 2 *Let $\bar{\boldsymbol{U}} \equiv \partial\bar{\boldsymbol{F}}/\partial\lambda = \sum_i \bar{\boldsymbol{U}}_i(\lambda-\lambda_0)^i$, then*

$$\left(\bar{\boldsymbol{Y}} + \bar{\boldsymbol{J}}_0\right)\bar{\boldsymbol{A}}_n = -\frac{1}{n}\left(\sum_{j=1}^{n-1}(n-j)\bar{\boldsymbol{J}}_j\bar{\boldsymbol{A}}_{n-j} + \bar{\boldsymbol{U}}_{n-1}\right) \quad (20)$$

Proof: Taking the derivative w.r.t. λ in (19)

$$\bar{\boldsymbol{Y}}\frac{\partial\bar{\boldsymbol{X}}}{\partial\lambda} + \frac{\partial\bar{\boldsymbol{F}}}{\partial\lambda} + \frac{\partial\bar{\boldsymbol{F}}}{\partial\bar{\boldsymbol{X}}}\frac{\partial\bar{\boldsymbol{X}}}{\partial\lambda} = \boldsymbol{0} \quad (21)$$

Substituting from (3) into (21) and using $\partial\bar{\boldsymbol{F}}/\partial\bar{\boldsymbol{X}} = \bar{\boldsymbol{J}} = \sum_{i=0}\bar{\boldsymbol{J}}_i(\lambda-\lambda_0)^i$ and equating similar powers of λ proves the lemma. ■

5. NUMERICAL EXAMPLES

5.1. Example 1

To demonstrate the accuracy of the proposed technique, a nonlinear two-stage pin attenuator circuit containing 4 diodes was considered for LCS computation. The parameter considered for perturbation in this example was the threshold voltage of the diodes V_T. Exact LCS sensitivity was first computed by solving a number of nonlinear steady-state problems of size 368 corresponding to each value of V_T.

Using the proposed algorithm, on the other hand, required solving a system of size 16 only to track the harmonics of the steady-state solution as V_T is changed. Figure 1 shows a comparison of the harmonics obtained from solving the original system (1) at $V_T = 30$mV and those obtained from the reduced system constructed through matching the derivatives at $V_T = 25$mV.

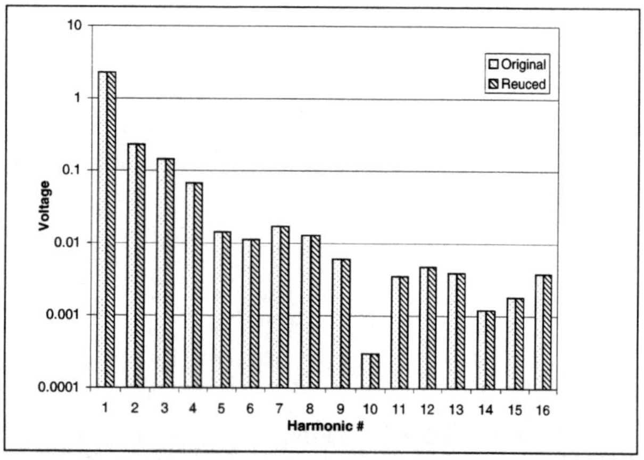

Fig. 1. Comparison for steady-state harmonics $V_T = 30$mv.

5.2. Example 2

A tuned amplifier circuit was considered in this example. Conventional HB analysis required 16 harmonics to approximate the steady-state solution. To track the output variation w.r.t. any parameter, a nonlinear system of equations of size 429 had to be solved repeatedly using NR method at each desired value for the parameter being traced. By contrast, using the proposed method, required applying NR method to solve a system of order 16 only. Figures 2 and 3 depict the variation of the 3rd harmonic at the output node as V_T and I_s parameters are changed. The graph also shows a comparison between the results obtained from solving the original system of order 429 using the HB technique and the reduced one whose order was 16.

6. CONCLUSION

A new accurate method for computing the LCS of steady-state operating point was presented. The proposed technique is based on generating a reduced system of equations to trace the solution as circuit parameters are varied through mapping the solution of the reduced system back to the space of the original system.

7. REFERENCES

[1] Jiri Vlach and Kishore Singhal, *Computer Methods for Circuit Analysis and Design*, Van Norstrand Reinolds, New York, 1983.

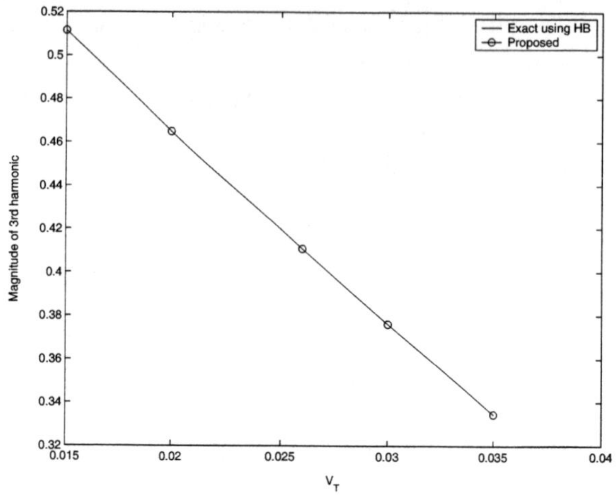

Fig. 2. Variation of the 3rd harmonic w.r.t. to V_T.

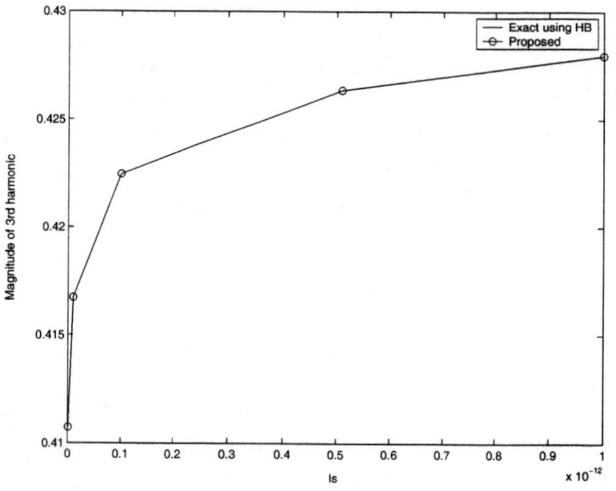

Fig. 3. Variation of the 3rd harmonic w.r.t. to I_S.

[2] M. T. Wong and M. Worsman, "Dc nonlinear circuit fault simulation with large change sensitivity," in *ATS Test Symposium*, 1998, pp. 366–371.

[3] Kenneth S. Kundert, Jacob K. White, and Alberto Sangiovanni-Vincentelli, *Steady-State Methods for Simulating Analog and Microwave Circuits*, Kluwer Academic, Boston, 1990.

[4] E. Gad, R. Khazaka, M. Nakhla, and R. Griffith, "A circuit reduction technique for finding the steady-state solution of nonlinear circuits," *IEEE Trans. Microwave Theory Tech.*, vol. 48, no. 12, pp. 2389–2396, Dec. 2000.

QUICK AND EFFECTIVE BUFFERED LEGITIMATE SKEW CLOCK ROUTING

Meng Zhao, Xinjie Wei, Yici Cai and Xianlong Hong

Department of Computer Science and Technology
Tsinghua University, Beijing 100084, China

ABSTRACT

In this paper, we propose a new quick and effective buffered legitimate skew clock routing algorithm BLST. We analyze the optimal buffer position in the clock path, and conclude the sufficient and heuristic condition for buffer insertion in clock net. During the routing process, this algorithm integrates topology generation, buffer insertion, and node merge together, and performs them parallel. Compared with the method [7] of buffer insertion after zero skew clock routing, BLST improves the maximal clock delay by at least 48%, and maximal clock skew by at least 22%. Compared with legitimate skew clock routing algorithm [3] with no buffer, this algorithm further decreases the total wire length and gets from 42% to 82% reduction on the maximal clock delay. The experimental results show our algorithm is quick and effective.

1. INTRODUCTION

The clock net is one of the most important parts of a synchronous VLSI, and clock skew is one of the most important factors for determining the performance and function of the clock net. With the development of clock routing, legitimate skew clock routing has been an active field. [1] combined DME and legitimate skew together, but it need an initial topology input. [2] presented synthesized topology generation and node merge together to complete legitimate skew driven clock routing, and [3] proved the feasibility of legitimate skew schedule.

Recent research has adopted many interconnect synthesis technique to optimize the interconnect performance, such as driver sizing, buffer insertion and wire sizing [4]. With the decrease of intrinsic delay of buffer and increase of chip size, it can be foreseen that buffer insertion is indispensable in high-performance VLSI design and production. [5][6] used buffer insertion in clock net routing to reduce the clock net power. [7] proposed a buffer insertion algorithm for clock delay and skew minimization.

This work was supported by NSFC60176016 and NSFC90307017.

Based on above problems, this paper presents a quick and effective buffered legitimate skew clock routing (BLST). We consider buffer insertion in the clock net topology design stage, and integrate clock tree topology design, sub-tree merging and buffer insertion together. Also, we continue using legitimate skew graph and matrix to compute the legitimate skew schedule [2][3], and ensure the feasibility of final result. The experimental results show that, BLST is superior to legitimate skew clock routing algorithm with no buffer [3] and can achieve savings on both clock delay and clock skew compared with the method of buffer insertion after zero skew clock routing [7].

The rest of this paper is organized as follows. In section 2, we introduce the delay model of interconnect and buffer, and deduce the sufficient condition and heuristic condition of buffer insertion. And we present the BLST algorithm and analyze the algorithm complexity of buffer insertion in section 3. Section 4 gives the experimental results.

2. CONDITIONS OF BUFFER INSERTION

2.1. Delay Model

For comparison and convenient computation, we continue to use Elmore [8] interconnect delay model as in figure 1(a). Where r_e and c_e are the unit wire resistance and the unit wire capacitance respectively.

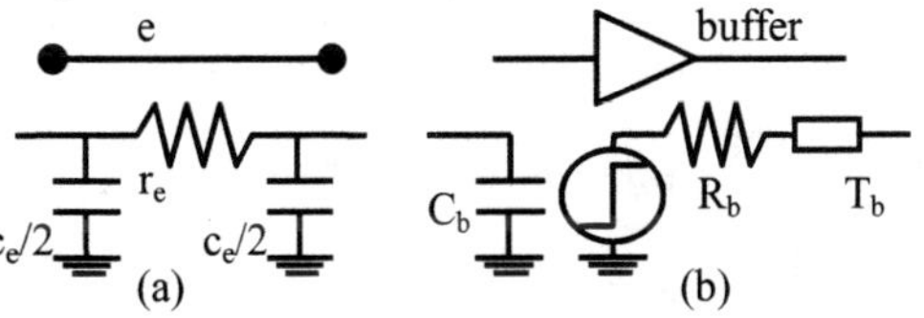

Figure 1. (a) Interconnect Model; (b) Buffer Delay Model

Figure 1(b) gives the buffer delay model, where T_b, R_b, and C_b are intrinsic delay, input capacity and output resistance of buffer, respectively.

2.2. Buffer Position

Firstly we discuss the optimal buffer position in an edge. According to Elmore delay model, figure 2 gives the distributed RC delay model of a buffered wire from s_f to s_l. Where C_l is the

0-7803-8251-X/04/$17.00 ©2004 IEEE

load capacity of subtree $Tree_{s1}$ whose root is s_1, T_1 is the delay from subtree $Tree_{s1}$'s root to clock sink, L is the distance from s_f to s_1, $L*a_1$ is the distance between s_f and buffer position, r and c are the unit wire resistance and the unit wire capacitance respectively.

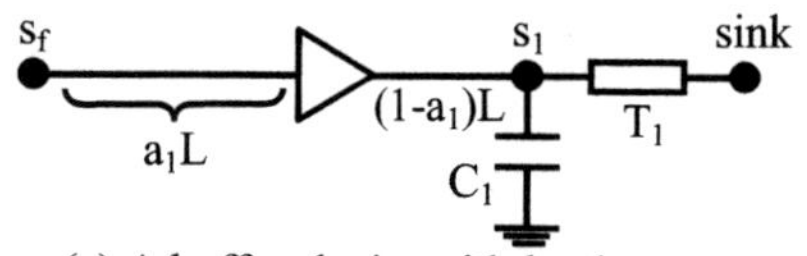

(a) A buffered wire with load

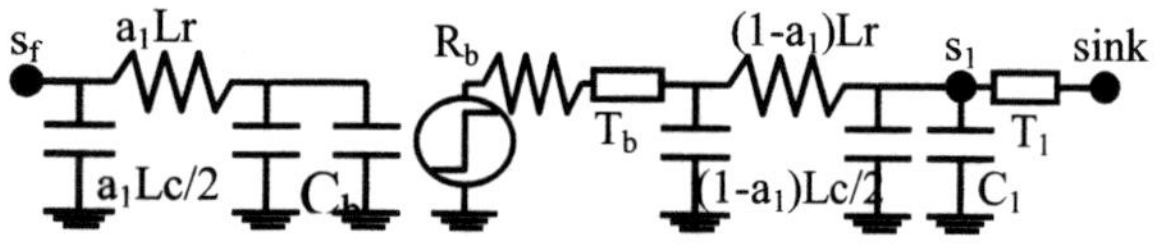

(b) Distributed RC delay model
Figure 2.

According to figure 2(b), we can get the delay from s_f to clock sink as expression (1).

$$Delay_1 = L\cdot a_1\cdot r\cdot(\frac{L\cdot a_1\cdot c}{2}+C_b)+R_b\cdot(L\cdot(1-a_1)\cdot c+C_1) \qquad (1)$$
$$+L\cdot(1-a_1)\cdot r\cdot(\frac{L\cdot(1-a_1)\cdot c}{2}+C_1)+T_b+T_1$$

Based on expression (1), we can deduce that $Delay_1$ is minimized when $a_1=\frac{1}{2}+\frac{(C_1-C_b)}{2\cdot c\cdot L}+\frac{R_b}{2\cdot r\cdot L}$. According to the benchmark parameter value in [2][3][7]: the load capacity of clock sink C_{sink} is larger than Cb; we can conclude as follows:

$$\text{If } C_{sink}\geq C_b,\ L\leq\frac{R_b}{r}\text{, then } a_1\geq 1 \qquad (2)$$

Because buffer position is between s_f to s_1, the range of a_1 should be [0, 1]. Therefore, based on expression (2), the condition of delay minimization from s_f to s_1 is $a_1=1$ when $L\leq R_b/r$, that means buffer position is at subtree's root.

We tested on different benchmarks in [3][7] and found that $L\leq R_b/r$. Therefore, in order to achieve delay minimization on this paper's benchmarks, we should set $a_1=1$, i.e., we should put buffer at the position of child.

2.3. Sufficient Condition for Buffer Insertion

When a buffer is positioned at a subtree's root, other two buffers should be inserted at its brother node for maintaining the balance of clock tree. Figure 3 shows buffer insertion at two brother subtrees' roots s_1 and s_2. L is the distance between s_1 and s_2, L*mid is the distance between their father node s_f and s_1, and the value range of mid is [0,1].

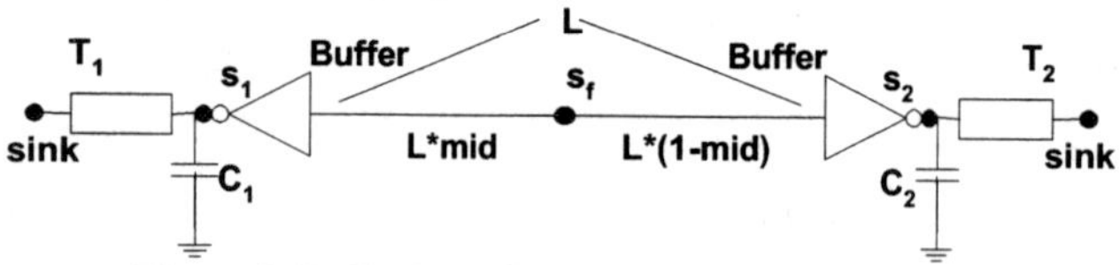

Figure 3. Buffer Insertion at Two Subtrees' Roots

Let $Delay_1$ denote the delay from s_f to s_1 without buffer insertion, and $Delay_1$' denote the delay from s_f to s_1 after buffer insertion, then we get the expression (3) and (4).

$$Delay_1 = L\cdot mid\cdot r\cdot(\frac{L\cdot mid\cdot c}{2}+C_1)+T_1 \qquad (3)$$

$$Delay_1' = L\cdot mid\cdot r\cdot(\frac{L\cdot mid\cdot c}{2}+C_b)+R_b\cdot C_1+T_b+T_1 \qquad (4)$$

If we can immediately get optimization after inserting buffer at the position of s_1 or s_2, delay from s_f to s_1 or s_2 should be reduced, then we obtain the sufficient condition for buffer insertion as follows:

$$Delay_1 < Dealy_1' \Rightarrow mid > \frac{R_bC_1+T_b}{rL(C_1-C_b)} \qquad (5)$$

$$Delay_2 < Dealy_2' \Rightarrow mid < 1-\frac{R_bC_2+T_b}{rL(C_2-C_b)} \qquad (6)$$

According to (5) and (6), we continue to deduce as follows:

$$\Rightarrow(\frac{R_b\cdot C_1+T_b}{C_1-C_b}+\frac{R_b\cdot C_2+T_b}{C_2-C_b})/r<L \qquad (7)$$

Therefore, expression (7) is the sufficient condition for buffer insertion. And because buffer reduces path delay by shielding the large load of subtree, the necessary condition which needs to be met is that the capacity of buffer should be smaller than the load capacity of subtree:

$$C_1>C_b \text{ and } C_2>C_b \qquad (8)$$

Further, according expression (5) and (6), we can deduce that: under sufficient condition, the value range of mid should be shrunk into a sub-range as in expression (9) after buffer insertion, in order to ensure delay decrease.

$$mid\in[\frac{R_b\cdot C_1+T_b}{r\cdot C_1\cdot L-r\cdot C_b\cdot L},\ 1-\frac{R_b\cdot C_2+T_b}{r\cdot C_2\cdot L-r\cdot C_b\cdot L}] \qquad (9)$$

2.4. Heuristic Condition for Buffer Insertion

We test on benchmarks in [3], and found that: during the process of clock routing, the sufficient condition for buffer insertion can not be met. Because of circuit parameter, the clock net with a small physical size can not meet the expression (7) all the time. Consequently, for some small physical size circuits, heuristic condition for buffer insertion should be considered.

We have known that inserting buffer in clock tree can reduce the path delay, but buffer also brings side effect since buffer has intrinsic delay and increases the circuit's power. So, we should synthetically consider these above factors in algorithm of this paper, and find a tradeoff point of clock delay and buffer number. In the clock tree construction algorithm of this paper, the subtrees of any level are merged because of many similar aspects such as load capacity, load delay (the delay to sink) and their distance, so the clock subtrees' roots of the same level have the approximate load. According to this character, we should insert the same buffers at the clock nodes of the same level.

To mitigate power dissipation, we do not insert buffer at every level of clock tree. For a clock tree, in which only one level is inserted buffer, the middle level is the optimal level for decreasing delay and power dissipation [9]. Therefore, buffer is firstly inserted in the middle level of clock tree. Then, buffer is inserted in the middle level between clock sink level and clock middle level because RC of clock high level affect more on delay than other lower levels. The above process continues recursively, until buffer is inserted at the child node of clock source.

For a clock net with **n** sinks, the clock tree has totally **h** levels: $h=\lceil\log_2 n\rceil$. And according to the above heuristic buffer insertion method, the number of clock tree levels with buffer ***m*** should meet the following expression:

$$m \le (\frac{h}{2} + \frac{h}{2^2} + \frac{h}{2^4} + \ldots + 2^2 + 2) = h - 2 \quad (10)$$

Also we set that the clock sink level is level **0**, and the clock source level is the level **h**. Then we insert buffer at level **mb** as follows:

$$mb = h - \left\lfloor \frac{h}{2^i} \right\rfloor, (i = 1, 2, \ldots; mb \le h - 1) \quad (11)$$

So expression (11) is the ***heuristic*** condition for buffer insertion in term of level. Correspondingly, the necessary conditions for buffer insertion (8) should still be met.

3. BLST ALGORITHM

3.1. Overview

Our quick and effective buffered legitimate skew clock routing approach (BLST) includes three major steps:

(i) Generating an appropriate clock tree topology,

(ii) Examining the conditions of buffer insertion and inserting buffer accordingly,

(iii) Merging two brother nodes iteratively.

Figure 4 gives an outline description of our algorithm, where M is the shortest distance matrix of legitimate skew range graph of sinks. For detail of M, please refer to [3].

Input: the set of clock sinks: **S** ={s1, s2, …, sn}
Clock source: s_0
Shortest path matrix of legitimate skew graph: **M**

Output: clock routing tree: T

Procedure BLST
While (|**S**| > 1) {
    Topology(S);
    Select two brother nodes s_l and s_r;
    BufferInsertion;
    Merge-Node;
    Update **S**;
    Update **M**;
}
Optimize(T);

Figure 4. BLST Algorithm Overview

The main idea is as follows. First, we use topology generation approach **Topology** in [3][10] to produce clock tree topology of one clock tree level. Then, we select two brother nodes, and do **BufferInsertion**: judging whether they meet the condition of buffer insertion, and inserting buffer appropriately if yes. Afterwards, **Merge-Node** [3] utilizes the legitimate skew ranges between sinks (denoted as **M**) to merge the two brother nodes and tries to minimize the wire length and delay to sink, also updates **S**. We repeat the above steps until there is only one node in **S**. Finally, **Optimize** [2] improves the clock tree structure.

3.2. Buffer Insertion

During buffer insertion process: **BufferInsertion**, we firstly set the value range of mid: [0, 1]. Then we judge if buffer insertion condition (expression (8) and (7) or (11)) is met. If the ***sufficient*** condition for buffer insertion is met, operation of buffer insertion is done and the value range of mid is updated as expression (9). If only the ***heuristic*** condition for buffer insertion is met, operation of buffer insertion is also done, but no need to update the value range of mid. If no buffer insertion condition is met, no buffers need to be inserted. Figure 5 gives the pseudo code of **BufferInsertion**.

Procedure BufferInsertion
BI=0;
[Min(mid), Max(mid)]=[0,1];
If ($C_1 > C_b$ & $C_2 > C_b$) {
    If ($(\frac{R_b \cdot C_1 + T_b}{C_1 - C_b} + \frac{R_b \cdot C_2 + T_b}{C_2 - C_b})/r < L$) {
        BI=1;
        [Min(mid), Max(mid)]= $[\frac{R_b \cdot C_1 + T_b}{r \cdot C_1 \cdot L - r \cdot C_b \cdot L}, 1 - \frac{R_b \cdot C_2 + T_b}{r \cdot C_2 \cdot L - r \cdot C_b \cdot L}]$;
    }
    Else If (the level of s_{father} is $h - \left\lfloor \frac{h}{2^i} \right\rfloor (i = 1, 2, \ldots)$)
        BI=1;
}
If (BI=1) Update the load of s_l and s_r;

Figure 5. Process of Buffer Insertion

3.3. Complexity Analysis on Buffer Insertion Problem

In this section, we discuss about the complexity of buffer insertion problem based on the above routing method, without routing complexity considered. Based on the routing algorithm in section 3.1 of this paper, we can insert buffer at the child nodes of every non-sink node. As in section 2.4 of this paper, for a clock net with **n** sinks, the clock tree has totally **h** levels: $h = \log_2 n$, and has totally N=n-1 non-sink nodes. If the total buffer number is not limited, every non-sink node has the possibility to be inserted buffer at its child nodes; then the complexity of optimal buffer insertion problem is 2^N. However, if we adopt the buffer insertion method in figure 5, the complexity of clock routing algorithm is not increased, then the speed of clock tree construction is very quick and the buffer insertion result is near optimal.

4. EXPERIMENTAL RESULTS

The proposed **BLST** algorithm has been implemented in C and tested 6 benchmark circuits on a Sun E450. The benchmark of s1423, s5378 and s15850 are in [2], and the benchmark of n50, n100, n300 and n500 are in [7].

Table I gives the comparison result between LST in [3] and **BLST** of this paper. The values in parenthesis are the percentage of reduction over **LST**'s results. The parameter of buffer to be inserted in s1423, s5378, s15850 is as follows: Rb=40Ω, Cb=60fF, Tb=2.0ps. The unit wire resistance is 0.004Ω/μm, and the unit wire capacitance is 20pF/μm^2.

The difference between LST and BLST is only the buffer insertion. We can see from table I, compared with the results of [3], the maximal clock delay is evidently decreased at least 42%. Because of the mask of buffer in the subtrees, the effect on topology from discrepancy of capacity and delay is mitigated, the effect of distance works mainly, and this is consistent with the object of total wire length minimization. The total clock wire length is further decreased in the BLST algorithm. Also, with the tight restriction of legitimate skew, the maximal clock skew is close to zero. Furthermore, the execution of BLST algorithm is very quick, and the run time of BLST is in the same magnitude as that of LST. This shows that our buffer insertion approach does not increase the computation complexity of clock tree construction as the analysis in section 3.3.

Table I. Comparison between [3] and BLST

		LST	BLST
S1423	*L*	99636	94212(-5.44%)
	D	4.53e-11	2.60e-11(-42.6%)
	S	4.93e-12	1.20e-11
	T	0.11	0.07
S5378	*L*	164591	164417(-0.11%)
	D	1.02e-10	5.43e-11(-46.8%)
	S	1.24e-11	5.01e-11
	T	0.56	0.87
S15850	*L*	433736	417388(-3.77%)
	D	3.71e-10	6.38e-11(-82.8%)
	S	3.89e-12	4.08e-11
	T	3.37	7.4

L: total wire length (um), *D*: max clock delay (s), *S*: max clock skew (s), *T*: CPU time (s).

Table II. Comparison between [7] and BLST

		[7]	[10]	BLST
n100	*D*	674.31	123.0(-81.7%)	66.5(-90.1%)
	S	52.27	0.44	30.4(-41.8%)
	B	4 levels		9
n300	*D*	710.10	403.4(-43.1%)	116.3(-83.6%)
	S	48.82	0.94	37.0(-24.2%)
	B	5 levels		14
n500	*D*	660.70	491.6(-25.6%)	338.0(-48.8%)
	S	72.32	1.38	30.0(-58.5%)
	B	4 levels		15

D: max clock delay (s), *S*: max clock skew (s), *B*: buffer number.

Table II gives the result comparison between [7], [10] (zero skew clock routing algorithm which integrates topology generation and node merge together), and BLST. The parameter of buffer to be inserted in benchmark circuits of [7] is as follows: Rb=100Ω, Cb=10fF, Tb=30ps. The unit wire resistance is 0.014Ω/μm, and the unit wire capacitance is 0.86fF/μm^2.

The values in parenthesis are the percentage of reduction over results in [7]. We limit the buffer number of BLST as no more than 15. [7] inserts buffer after completing routing, and puts buffer insertion as a post-process step of clock routing. Whereas BLST inserts buffer during clock routing, and integrates topology generation, buffer insertion, and nodes merging together.

We can see from table II, because of the appropriate topology generation method, the zero skew clock routing algorithm [10] outperforms [7] from 25% to 81.7% on maximal clock delay. And with the same number of buffers, the max clock delay of BLST outperforms that of [7] at least 48%. Compared with [10], the further optimization of max clock delay of BLST arises from two reasons: first, we interlude buffer insertion in the process of merging node; second, [7] constructs zero-skew clock net, while we utilize legitimate skew range to further optimize routing result during clock tree construction.

5. CONCLUSIONS

A new quick and effective buffered legitimate skew clock routing algorithm **BLST** has been proposed in this paper. In the process of clock routing, we integrate clock topology generation, buffer insertion and merging brother nodes together, expedite the computation speed, increase the execution efficiency and achieve a better result than [3] and [7]. In addition, we utilize legitimate skew ranges when constructing clock tree, and make post-process improvement. The experimental results show our **BLST** algorithm is quick and effective.

6. REFERENCES

[1] Tsao, C.-W. A., Koh, C.-K , "UST/DME: A Clock Tree Router for General Skew Constraints," IEEE/ACM International Conference on Computer-Aided Design, Digest of Technical Papers Nov 5-Nov 9 2000: 1092-3152

[2] Zhao Meng, Cai Yici, Hong Xianlong, Liu Yi, "Legitimate Skew Driven Clock Tree Routing and Optimization", Chinese Journal of Semiconductors, 2003, Vol.24 No.4: 438-444 (in Chinese)

[3] Meng Zhao, Yi Liu, Yici Cai, Xianlong Hong, "Effective Legitimate Skew Driven Clock Tree Routing," ASIC, 2003. Proceedings. 5th International Conference

[4] J.Cong, L.He, K.-Y.Khoo, C.-K.Koh, and D.Z.Pan, "Interconnect design for deep submicron ICs," in Proc. Int. Conf. on Computer Aided Design, 1997: 478–485

[5] Ashok Vittal, Malgorzata Marek-Sadowska, "Low-Power Buffered Clock Tree Design," IEEE Transactions on Computer Aided Design of Integrated Circuits and Systems, Vol. 16, No. 9, September 1997

[6] Jatuchai Pangjun and Sachin S. Sapatnekar, "Low Power Clock Distribution using Multiple Voltages and Reduced Swings," IEEE Transactions on VLSI Systems, Vol. 10, No. 3, June 2002: 309 - 318

[7] X. Zeng, D. Zhou and W. Li, "Buffer Insertion for Clock Delay and Skew Minimization," Proceedings of the ACM International Symposium on Physical Design, 1999: 36-41

[8] W C Elmore, "The transient response of damped linear network with particular regard to wideband amplifier", Applied Physics, 1948, 19: 55-63

[9] Zhiyan Li, Xiaolang Yan, Ning Zheng, "A Multi-Staged Zero Skew Clock Net Routing," Acta Electronica Sinica, Vol. 26, No. 2, Feb. 1998: 95-98 (in Chinese)

[10] Yi Liu, Meng Zhao, Xianlong Hong, Yici Cai, Weimin Wu, "A clustering-based algorithm for zero-skew clock routing with buffer insertion", ASIC, Proceedings. 4th International Conference on, 2001: 183-186

THEORY OF T-JUNCTION FLOORPLANS IN TERMS OF SINGLE-SEQUENCE

*Xuliang Zhang**

SII EDA Technologies Inc.
2-5, Hibikino, Wakamatsu-ku, Kitakyushu
Fukuoka, 808-0135 Japan
xuliang@sii.co.jp

Yoji Kajitani

The University of Kitakyushu
1-1, Hibikino, Wakamatsu-ku, Kitakyushu
Fukuoka, 808-0135 Japan
kajitani@env.kitakyu-u.ac.jp

ABSTRACT

In a placement, two rectangles on a plane are non-overlapping if and only if the relationship between them is one of *ABLR* (*above, below, left-of, right-of*)-relations. From the standpoint that a system of ABLR-relations is specified by the designer to confine the placement, the first concern is its consistency, i.e. if there is a corresponding placement. The second is an efficient way to construct a corresponding placement. The third is a handy coding of the ABLR-system all the way. In this paper, the first concern is solved by an existence condition of the primal- and dual-orders of rectangles. The second is answered by a linear time construction algorithm of a T-junction floorplan. This is new in its speed and generality with respect to the number of rooms. The third is by a new coding *Single-Sequence* SS. Its unique suitability to handle the T-junction floorplan is remarkable. Using the merit that SS can control the distribution of empty rooms, a novel application is suggested to *space-planning*, a recent trend in VLSI physical design to budget the space for congestion relief and interference separation (though the detail is not contained here for the space).

1. INTRODUCTION

Physical design of recent system chips is facing difficulties due to increasing complexities and requests for timing integrity and routability issues. Among them, placement is the key technology in the design flow. Let us reflect back to the fundamentals.

A placement of n rectangles is said *consistent* if any two rectangles are *non-overlapping* with each other. Two rectangles x and y are non-overlapping if and only if the relationship between x and y is one of *ABLR-relations*, i.e. x is *above, below, left-of,* or *right-of* y. The set of all ABLR-relations between two rectangles is called the *ABLR-system*. As long as area minimization is involved, an ABLR-system defines a placement leaving not much freedom. Hence it is still believed that it is essential in optimization to list all consistent ABLR-systems "without duplication". Then the problem turns to develop an ABLR-generator. The terrifying project was first challenged by Onodera et.al.[4] but even a fast computer fails to search them of over 7 or 8 rectangles. From this point, Polish expression [2] for slicing structured placements was a preceding ABLR-system generator. After Onodera et.al.[4], BSG[6, 7] in 1994 is considered the first success. Various ABLR-system generators followed: SP[5] and TCG[13] for general topological structures, O-tree[9] and B*-tree[12] for topology and physical dimension mixed structures, Q-seq[10, 20] and CBL[11] and Prime-graph[15] for T-junction floorplans.

While if an ABLR-system is a constraint specified by a designer, the first concern is its consistency, i.e., if there is a corresponding placement. One reason why the way by Onodera et. al. does not work well is because they have no efficient way to filter the consistent ABLR-systems.

In this paper, this is solved by an elegant way. It associates with a smart characterization of consistent ABLR-systems in terms of a permutation of 1, 2, ..., n, called *Single-Sequence* and denoted by SS[18, 21, 22]. SS may look like a normalized SP[19]. But it is different from SP since it is an independent idea based on Abe-labelling of the T-junction floorplan[1, 20]. One surfaced consequence is in the consistency check as mentioned above. Another is a linear time construction of the corresponding floorplan. They are impossible by any ABLR-system generator without using Abe-labelling.

Generally, there are two approaches to get a placement from the ABLR-system. One is to use a pair of horizontal and vertical constraint graphs to apply the 1-dimensional compaction. The other is to use the T-junction floorplan in which n rooms satisfy the ABLR-system and rectangles embedded in them inheriting the ABLR-relations. Because of the merit that floorplanning secures a space for free use of individual elements, it is becoming popular. This is also our framework.

BSG, SP, O-tree, B*-tree, and TCG are developed for the former framework and not suitable for us in spite of efforts[16, 19] to convert. The Q-sequence, CBL, Prime-graph were invented to handle T-junction floorplans. But they focus on the rooms instead of contents, so not suitable for our framework though [17] made some efforts using CBL.

Our proposing algorithm outputs the T-junction floor-

*He is currently a visitor researcher in The university of Kitakyushu plan configuration on a special topological grid in linear time of the number of rooms. It has the capability to create empty rooms intentionally. This suggests an idea of latent topological white spaces to be distributed for congestion relief.

In the following, Section 2 will discuss the consistency of ABLR-systems, and other sections will be devoted to coding a floorplan to SS and decoding an SS to a T-junction floorplan. Proofs are all omitted for the space.

2. CONSISTENT ABLR-SYSTEM

Matrix $ABLR$ is defined as follows: Rows and columns denote the labels of rectangles arranged in the same order. Entry (x, y) is determined as $(x, y) = A$ (or B, L, R) if x is above y (or x is below y, x is left-of y, x is right-of y, respectively), which we denote as xAy (xBy, xLy, xRy, respectively). As an example, an ABLR-system of rectangles a, b, c, d (n=4) is given below(left).

$$ABLR = \begin{array}{c|cccc} & a & b & c & d \\ \hline a & * & L & A & L \\ b & R & * & A & A \\ c & B & B & * & L \\ d & R & B & R & * \end{array}, \quad AL = \begin{array}{c|cccc} & a & b & c & d \\ \hline a & * & L & A & L \\ b & - & * & A & A \\ c & - & - & * & L \\ d & - & - & - & * \end{array}$$

The diagonal entry (*) is not defined. Since xBy and xRy are equivalent to yAx and yLx, respectively, we replace each B and R with (-) in the matrix without losing any information. The resultant matrix is called the AL-matrix. See above right. BL-, AR-, BR-matrices are similarly defined. We observe that this AL-matrix is upper-triangulated. This is not incidental.

Theorem 1 *An ABLR-system is consistent if and only if there are two orders of rectangles, called the prime-order and dual-order, such that the former makes the AL-matrix upper-triangulated and the latter does the BL-matrix.*

In our example, (a b c d) is the prime-order. Before getting BL-matrix, we normalize the labels of rectangles to integers according to the prime-order. In our example, we let $a \to 1, b \to 2, c \to 3, d \to 4$. Then, by exchanging elements $(x, y) = A$ with $(y, x) = B$, a BL-matrix is obtained as shown below (left).

$$BL' = \begin{array}{c|cccc} & 1 & 2 & 3 & 4 \\ \hline 1 & * & L & - & L \\ 2 & - & * & - & - \\ 3 & B & B & * & L \\ 4 & - & B & - & * \end{array}, BL = \begin{array}{c|cccc} & 3 & 1 & 4 & 2 \\ \hline 3 & * & B & L & B \\ 1 & - & * & L & L \\ 4 & - & - & * & B \\ 2 & - & - & - & * \end{array}$$

By rearranging the order, we have an upper-triangulated BL-matrix as shown in the right. Then, Theorem 1 guarantees the ABLR-system being consistent. Order (3 1 4 2) is the dual-order. In fact, a placement is realized as shown in Fig.1(a) or (b) by T-junction floorplans of five or six rooms, each room embedded with a rectangle or left empty. Check that all ABLR-relations among embedded rectangles are correctly realized in both, and also that the ABLR-system is not realizable by any floorplan of four rooms or less.

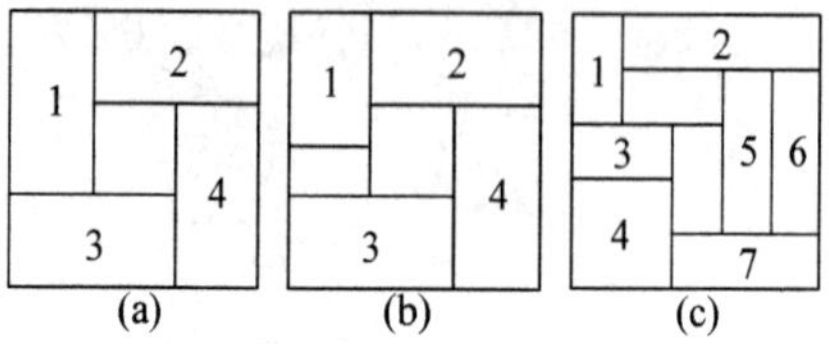

Fig. 1. (a), (b) T-junction floorplans with the same ABLR-system by dual-order S=(3 1 4 2). (c) S=(4 3 7 1 5 6 2).

If the normalized labels feature the property of numbers, a smart property is observed.

Theorem 2 *Entry (x,y) in the upper-triangle of the upper-triangulated BL-matrix is L if $x < y$, and B otherwise.*

Thus we can forget the matrix, just remember the dual-order, from which we can recover the matrix.

By the way, all permutations of n integers count $n!$ varieties. A natural question is : Can any permutation be the dual-order of some consistent ABLR-system? The answer is positive.

Theorem 3 *Any permutation can be the dual-order. More precisely, any permutation π of n labels corresponds to a T-junction floorplan in which n of rooms contain rectangles and they satisfy the ABLR-system with π the dual-order.*

3. T-JUNCTION FLOORPLAN AND CODING

For Abe-labelling and related properties, we refer to [20]. An important feature of Abe-labelling is that every room is visited exactly once and ends at the bottom-right room so that every room has a unique Abe-label. This is an inborn feature of the T-junction floorplan: an integer k ($n \geq k \geq 1$) uniquely designates one room. One surfaced property from this is that the dual-order, which characterizes the ABLR system, is obtained by a way dual to Abe-labelling.

Coding:(Dual-order from a floorplan) Prepare a null queue S. Let the bottom-left room be the current room. Append its Abe-label to S. Notice the seg, say s, that ends at the right-top corner of the current room. If it is vertical (or horizontal), let the bottommost (or leftmost) room that is adjacent to the seg on the other side be the new current room. Append its Abe-label to S. Repeat the process until the current room contains the top-right corner. (This room is the last.)

See a floorplan in Fig.2 where SS is the dual-order.

Now we have observed that even a consistent ABLR-system imposed on n rectangles may not be satisfied by any n room floorplan. In this case, we have only to use a larger floorplan in which not all rooms are called. Then a rectangle is given an integer label which is different from the Abe-label of the room in which the rectangle is assigned. To avoid any confusion from such double labelling, we introduce a framework: A room is focused only if it is assigned with a rectangle and given the label of the integer which is the order of such rooms along Abe-labelling. Each rectangle inherits the label of the room to which it is assigned.

Now our problem is: Given SS(permutation of 1,2,.., n), find a T-junction floorplan that its n rooms are assigned with rectangles that satisfy the ABLR-system derived from SS by the rule: for two rectangles x and y contained in this order, xLy if (x, y) is incremental and xBy otherwise.

4. EMPTY-ROOM FILLING

Constructing a T-junction floorplan allowing empty-rooms, the result is not unique. To avoid this occurrence, we first characterize the floorplans with the least empty rooms.

A room r has four boundaries. If a boundary is a seg s that terminates its both ends at the midpoints of other boundary segs, s and r are called the *shunt-seg* and *shunt-room*, respectively. A room that has no shunt-seg is called a *spiral-room*. In Fig.1(b), room 1 and the empty room below are both shunt-rooms, and the room with no label in Fig.1(a) is a spiral-room. An empty shunt-room does not affect the ABLR-relations if it is reduced by shifting the shunt-seg inward to crush it. Hence the empty rooms contained in our target floorplan are all spiral rooms.

The next problem is to unveil spiral empty-rooms in SS. There may exist more than one implicit empty-rooms. Therefore, our proposed algorithm discovers the first one in Abe-order and identifies it by its previous room in SS. Then, a virtual rectangle is filled there to obtain new SS whose empty-rooms are reduced by one. The process is repeated if empty-rooms remain. A similar problem has been handled in [16] and [19]. But their standpoint is to characterize the sequence-pair that will not produce a spiral pattern of four rectangles in the placement.

For arbitrary two integers j and k, define $< j : k >$ as the set of integers between j and k excluding j and k. For example, $< 3 : 7 > = < 7 : 3 > = \{4, 5, 6\}, < 6, 7 > = \emptyset$.

Consider a permutation S of 1, 2, .., n. A pair (x, y) of consecutive integers is said *monotonic separator* if either (1) $x < y$ and any integer of $< x : y >$ that is before x in S is smaller than any one of $< x : y >$ after y, or (2) $x > y$ and any integer of $< x : y >$ that is before x in S is larger than any one of $< x : y >$ after y. Otherwise, it is a *non-monotonic separator.*

Given SS={4 3 7 1 5 6 2} (its corresponding floorplan is shown in Fig.1(c)), monotonic separators are (4,3), (6,2), (5,6), and (3,7). While (7,1) and (1,5) are not.

Theorem 4 *Given SS, if pair (p, q) is the first non-monotonic separator, there is a spiral empty-room in the next (along Abe-order) of room r.*

The spiral empty room is determined as follows. If $p < q$ $(p > q)$, r and s are integers of $< p : q >$, where $r < s$, and s is before p and r after q in SS(r before p and s after and smaller q). The closest of such integers is r. If a virtual rectangle is filled in the empty room next to r along Abe-order, SS is renewed as follows: Give the virtual rectangle with label $r + 1$ and put it between p and q. Relabel the current integers $r + 1$, ..., n to r+2, ..., n+1, respectively.

The detection and filling are possible in $O(n + e)$, n is the length of the original $SS$and e is the maximal number of empty rooms[14, 22].

5. DECODING: FLOORPLAN FROM SS

Floorplans hereinafter are assumed to contain no empty-room. The prime-seg of room r is s_r. The right and bottom walls of the chip are both considered as the prime-segs of room n and labelled as s_n. Similarly, the topmost and left-most walls are the prime-segs of room 0 with label s_0.

We image a topological grid G on the xy-coordinate system. The rows and columns are defined as the extensions of segs. Since every seg of the floorplan is a prime-seg of a unique room and each room has its Abe-label, let rows and columns of G inherit these labels. If it is horizontal (vertical), its y-coordinate (x-coordinate) is y_r (x_r). Let V and H denote sets of suffices (actually room labels) of x_k's (y_k's). Then, $V \cup H = \{0, 1, ..., n\}$ and $V \cap H = \{0, n\}$. As an example, the floorplan shown in Fig.2 defines $V = \{0, 1, 4, 5, 6, 9\}$ and $H = \{0, 2, 3, 7, 8, 9\}$.

Then, G consists of extended prime-segs which are arranged in increasing order of suffices horizontally and vertically. Precisely, G consists of columns at x_j for $j \in X$ and rows at y_j for $j \in Y$. They are arranged so that $x_j < x_k$ if $j < k$, and $y_j < y_k$ if $j > k$. See Fig.2.

To fix the output of floorplan, we determine each room r by calculating the x- or y-grids of its four boundaries.

Definition 1:(mbs number)In SS, $mbs(r)$ is the maximum integer before r but smaller than r. The default value for $mbs(r)$ is 0.

Definition 2:(mas number)In SS, $mas(r)$ is the maximum integer after r but smaller than r. The default value for $mas(r)$ is 0.

Definition 3:(mal number)In SS, $mal(r)$ is the minimum integer after r but larger than r. The default value for $mal(r)$ is n+1, where n is the largest integer in SS.

Definition 4:(mbl number)In SS, $mbl(r)$ is the minimum integer before r but larger than r. The default value for $mbl(r)$ is n+1, where n is the largest integer in SS.

For example, for SS=(4 9 5 1 3 8 6 2 7), mas(9)=8,mas(5)=3, mas(1)=0, mbs(9)=0, mbs(4)=0, mbs(8)=5, mal(4)=5, mal(9)=10, mal(7)=10, mbl(3)=4, mbl((9)=10,mbl(4)=10.

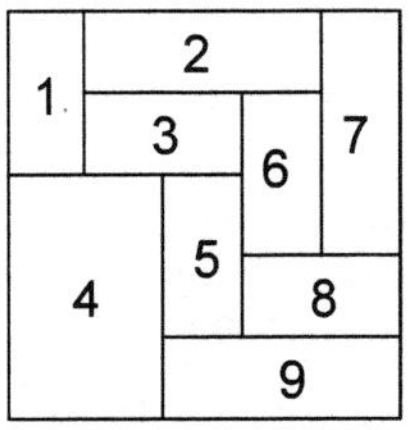

Fig. 2. Floorplan on grid G for SS=(4 9 5 1 3 8 6 2 7)

To help calculate mbs, mas, mal, and mbl, we introduce a basic data structure *double linked list* L with functions $Prev(s)$ and $Next(s)$ for an element s, which return the element previous to s and the one next to s, respectively. Function $Del(s)$ deletes s from L. They work in a constant time by using an additional data structure *address array* AA, which stores the addresses of elements in L. When we calculate $Next(s)$ and $Prev(s)$ or execute $Del(s)$, we first derive the address of s in AA, which stores the addresses of elements in L for directly access L.

It will take two steps to calculate mbs, mas, mal, and mbl. The aided data structure L stores the same elements for the two steps, i.e., from n to 1 in decreasing order.

In the first step, $mas(r)$ and $mbl(r)$ (r=1,2,...,n) can be calculated one by one from the left of SS to the right, where $mas(r)$=$Next(r)$, $mbl(r)$=$Prev(r)$. After r is handled, it is deleted from L.

In the second step, $mbs(r)$ and $mal(r)$ (r=1,2,...,n) can be calculated one by one from the right of SS to the left, where $mbs(r)$=$Next(r)$, and $mal(r)$=$Prev(r)$. After r is handled, it is deleted from L.

The total time complexity of calculating mbs, mas, mal, mbl is $O(n)$.

Lemma 1 *For a room r, $mbs(r)$, $mbl(r)$-1, $mal(r)$-1, and $mas(r)$ are the xy-grid numbers of the left-, bottom-, right-, and top-boundaries, respectively.*

A floorplan sketch obtained this way is shown in Fig.2.

6. WHITE SPACE PLANNING IN VLSI LAYOUT

In layout design of nano-scale systems on a chip, the key is space budgeting. The authors believe that this is successfully solved only if we are able to construct a structure on the space such that every dissected part has its identity in the whole area. Abe-labelling defined on the T-junction floorplan would support this idea. We have developed some techniques by Single-Sequence SS that control the empty rooms: Given a specification of a consistent ABLR-system, insert a series of empty rooms connecting two designated rooms without disturbing existing ABLR-relations. This will be useful for practical applications. But for the space, we only refer to Zhang, et. al.[22].

7. CONCLUSION

Standing on the point that the fundamental constraint on the rectangle placement is the ABLR-system, we first discussed the consistency issues. It revealed that the consistent ABLR-system is completely characterized by a sequence of integers from 1 to n, where n is the number of rooms to which n rectangles are assigned. Coding and decoding problems between T-junction floorplans and the sequences are solved elegantly.

8. REFERENCES

[1] M. Abe, "Covering The Square by Squares without Overlapping", Journal of Japan Mathematical Physics, Vol.4, No. 4, pp. 359-366, 1930 (in Japanese)

[2] D. F. Wong, and C. L. Liu, "A New Algorithm for Floorplan Designs," Proc. 23rd DAC, pp.101-107, 1986.

[3] S. Wimer, I. Koren, I. Cederbaum, "Floorplans, Planar Graphs, and Layouts", IEEE Trans. CAS, Vol. 35, No. 3, pp.267-278, 1988.

[4] H. Onodera and Y. taniguchi and K. Tamaru, "Branch-and-Bound Placement for Building Block Layout", Proc. 28th DAC, pp.433-439, 1991

[5] H. Murata, K. Fujiyoshi, S. Nakatake, and Y. Kajitani, "VLSI Module Placement Based on Rectangle-packing by the Sequence-Pair," IEEE Trans. CAD, Vol. 15, No. 12, pp. 1518–1524, 1996.

[6] S. Nakatake, K. Fujiyoshi, H. Murata and Y. Kajitani, "Module Placement on BSG-Structure and IC Layout Applications", Proc. ICCAD, pp.484-451, Nov. 1996.

[7] S. Nakatake, K. Fujiyoshi, H. Murata, Y. Kajitani, "Module placement on BSG-structure and IC layout applications", Proc. ICCAD, pp.484-451, Nov. 1996.

[8] H.Murata, K. Fujiyoshi, T. Watanabe, Y. Kajitani, "A Mapping from Sequence-Pair to Rectangular Dissection", Proc. ASP-DAC, pp.625-634, 1997

[9] P. N. Guo, C. K. Cheng, and T. Yoshimura, "An O-Tree Representation of Non-Slicing Floorplan and Its Applications," Proc. 36th DAC, pp.286-291, Jun., 1999.

[10] K. Sakanushi, K. Midorikawa, and Y. Kajitani, "A General and Fast Floorplaning by Reduct-Seq Representation," Technical Report of IEICE, (VLD2000-24), Vol.100, No.120, pp.109-116, 2000.

[11] X. Hong, S. Dong, Y. Ma, Y. Cai, C. K. Cheng, and J. Gu, "Corner Block List: An Efficient Topological Representation of Non-Slicing Floorplan", Proc. ICCAD 2000, pp. 8-12, Nov., 2000.

[12] Y.-C. Chang, Y.-W. Chang, G.-M. Wu, and S.-W. Wu, "B*-trees: A New Representation for Non-Slicing Floorplans," Proc. 37th DAC, pp. 458-463, LA, CA, June 2000.

[13] J.-M Lin and Y.-W Chang, "TCG: A Transitive Closure Graph-Based Representation for Non-Slicing Floorplans", Proc. DAC, pp.764-769, 2001

[14] M. Tsuboi, C. Kodama, K. Sakanushi, K. Fujiyoshi, and A. Takahashi "A mapping from sequence-pair to rectangular dissection", Proc. SASIMI pp.625-633, 2001

[15] C. W. Zhuang, K. Sakanushi, L. Y. Jin and Y. Kajitani, "An Enhanced Q-Sequence Augmented with Essential Empty Room Insertions and Parenthesis Trees", Proc. DATE, pp.61-68, Mar., 2002

[16] C. Kodama and K. Fujiyoshi, "An Efficient Decoding Method of Sequence-pair with Reduced Redundancy", IEICE Trans. Fundamentals, Vol. E85-A, No.12, pp.2785-2794, Dec., 2002.

[17] Y. Ma, X. Hong, S. Dong, Y. Cai, C.K. Cheng, J. Gu, "Stairway Compaction using Corner Block List and Its Applications with Rectilinear Blocks", Proc. ASP-DAC 2002, pp.387-392

[18] Y. Kajitani, "The Single Sequence That Unifies Placement and Floorplanning", presented at the presession meeting of ASP-DAC, 2003, "Asian Semi-Conductor University Cooperations", January 2003.

[19] C. Kodama, K. Fujiyoshi, "Selected Sequence-Pair: An Efficient Decode Packing Representation in Linear Time Using Sequence-Pair", Proc. ASP-DAC 2003,pp.331-337.

[20] K. Sakanushi, Y. Kajitani, and D. P. Mehta, "The Quarter-State-Sequence Floorplan Representation", IEEE Trans. CAS-I: Fundamental Theory and Applications, Vol. 50, No. 3, March 2003

[21] X. Zhang, X. Zhu, Y. Kajitani, "Layer Based Area Partition Using Single-Sequence for Preferable Routes",Proc. ASICON 2003

[22] X. Zhang, Y. Kajitani, "Space-Planning: Placement of Modules with Controlled Empty Area by Single-Sequence" Proc. ASP-DAC Jan. 2004, pp. 25-30.

GENERATING RANDOM BENCHMARK CIRCUITS FOR FLOORPLANNING

Tao Wan and Malgorzata Chrzanowska-Jeske
Department of Electrical and Computer Engineering
Portland State University
Portland, Oregon 97207 USA
Email: {want, jeske}@ece.pdx.edu

ABSTRACT

Benchmark files are widely used to evaluate and compare new algorithms and tools. In this paper, a fairly simple although reliable random floorplanning benchmark generator, BGen, based on Rent's rule and a novel net-degree distribution model, is presented. Through direct and indirect validation, it is shown that BGen generates realistic distribution of module sizes and netlist information.

1. INTRODUCTION

Benchmarks are essential to measure the performance, capability and limitation of newly designed algorithms and tools, especially in VLSI design automation (DA) area. The DA community heavily relies on circuit benchmarks to test and compare their algorithms [2]. For example, the floorplanning tool ELF-SP [14], which is a general floorplanning tool that uses sequence-pair representation, was tested against five MCNC [1] benchmarks and achieved encouraging results. Unfortunately, these benchmarks, which were released from 1985 to 1993, are now obsolete and may not adequately represent the complexity of modern design [2]. Thus, research on developing new benchmarking tools for floorplanning is needed.

The goal of this work is to generate floorplanning benchmarks with a larger number of modules and nets, and with information about power dissipation. These benchmarks are designed for general floorplanning tools and were tested using ELF-SP. We developed a tool, called BGen, to generate benchmarks that consist of many modules (more than 100) with various distributions of module sizes and aspect ratios. They include interconnect information that is randomly generated based on a novel interconnect net-degree distribution model from [13]. Power dissipation information is also integrated to allow testing of floorplanning tools in optimizing temperature gradient and power distribution. Power consumption of different types of modules, such as digital, analog and RF, that might be included in a system-on-chip, are considered. To simplify the problem and user interface, we intend to use the minimum number of input parameters. Furthermore, other unspecified metrics of generated benchmark circuits should fall into the reasonable ranges seen in real circuits.

Previous approaches to benchmark generation were presented in [3], [4], [5], [9], [11], and [12]. Darnauer and Dai's rmc [3] was the first tool to generate random benchmark circuits for routability measurement. Its input parameters include the number of circuit inputs and outputs, the number of LUTs, average fan-in and Rent's exponent. In [4] and [5], Hutton et al. defined important characteristics of combinational and sequential circuits and developed a tool CIRC to extract them. Then a tool GEN was applied to generate random circuits parameterized by those characteristics. Pistorius et al. [9] presented another tool, PartGen, to generate netlists for partitioning to multiple FPGAs. PartGen integrated five subgenerators for different types of netlists: regular combinational logic, irregular combinational logic, memory blocks, combinational and sequential logic and interconnections. Tool gnl [11] is based on a bottom-up clustering approach according to Rent's rule [7]. At last, [12] reviewed existing benchmark generation methods and discussed the advantages and drawbacks of different methods through direct validation.

Most of the above methods were designed for partitioning and routability measurement of FPGAs. Circuits generated by them lack module size (area) information, which is the most important information for floorplanning tools. Furthermore, some input metrics from the above generator tools are unnecessary for floorplanning. BGen uses the number of modules, the number of primary inputs/outputs, module size distribution, power variance, internal net fraction factor, average number of pads and Rent's exponent as input constraints. The number of nets and fan-out distributions are not included, since they have implicit relationship with Rent's rule and can be derived from it (see detail in Section 4). Delay, which is related to timing, is omitted in current version and will be included in future work. Other metrics, such as edge length distribution, circuit shape and sequential shape described in [4] and [5], are not considered here since they are not related to floorplanning. The generated benchmark data sets include three different files:

- .dat file: the number of modules, module sizes and aspect ratios
- .nets file: the number of I/O pads, reference I/O pads positions, the number of nets and their interconnections
- .pwr file: power number for each module

The organization of this paper is as follows: Section 2 describes our method to generate module sizes with different distributions. Procedures to generate power numbers and netlist information are presented in Section 3 and Section 4, respectively. Experimental results and validations are presented in Section 5 and Section 6 presents the conclusion.

2. MODULE SIZE DISTRIBUTIONS

At first, module size distributions of MCNC benchmarks were studied. We found that ami49 benchmark could be approximated with Exponential distribution while apte benchmark approximated with Gaussian distribution.

Based on a random number generator urand[0,1], our tool uses four common distributions to generate benchmarks: Uniform distribution, Gaussian distribution, Exponential distribution and Poisson distribution. Aspect ratios of modules are uniformly distributed based on urand[0.5, 2], or set as fixed user inputs. The number of modules generated by BGen is limited only by the available computer memory. Users can specify the number of modules, their distribution as well as lower and upper bounds of aspect ratio. In the early stage of this work, data samples with 100, 200, 500 and 1000 modules had been tested using ELF-SP. Results showed that ELF-SP achieved 95% or above effective area ratio (soft-module packing) for benchmarks within 200 modules regardless of their distributions. As the number of modules increases, the result tends to get worse, especially for hard-module placement. The effective area ratio could be as low as 20%. But soft-module floorplanning can still reach the effective area ratio of 80%~90%. It was also shown that ELF-SP gives similar packing efficiency for different probability distributions.

3. INTEGRATING POWER NUMBERS

The format of .pwr file is fairly simple. It consists of one column of power numbers with the same order as in .dat file. According to Liu et al.'s [8] experiments, power number of each circuit module is roughly proportional to its area. More specifically, the power demand P_i for each block is

$$P_i = C(1+e_i)A_i\,, \tag{1}$$

where C is a constant representing the average power density, e_i is a random number between $-eMax$ and $eMax$ ($0<eMax<1$) that represents the dynamic nature of power dissipation for each module and A_i is the module size. Users are allowed to assign different C ($C1$, $C2$ ...) to digital, analog and RF parts of an ASIC design or system-on-chip.

4. NETLIST INFORMATION

Netlist is a list of all the nets in the circuit and defines module connectivity. For our benchmarks, netlist information is stored in .nets file. The procedure of generating .net file is rather complex. Basically, it can be divided into three tasks.

4.1. Pin positions

Initially, a reference chip size, a reference chip width and a reference chip height are computed to determine positions of the I/O pads. The reference chip size is assumed as the summation of sizes of all modules.

$$A_{ref} = \sum_{i=1}^{M} A_i\,, \tag{2}$$

where A_i is the size of module i.

The reference chip width W_{ref} and the reference chip height H_{ref} are computed as

$$W_{ref} = H_{ref} = \sqrt{A_{ref}}\,. \tag{3}$$

According to Rent's rule [7], which is the simple power-law relationship between the number of pads, T, and the number of modules in a circuit, M.

$$T = kM^{p}\,, \tag{4}$$

where k is the average number of pads per module and p is the Rent's exponent. T can also be assigned by users,

$$T = nPI + nPO\,, \tag{5}$$

where nPI and nPO are the number of primary input pads and number of primary output pads, respectively.

I/O pads are uniformly distributed along four sides of the reference chip. Each pad is assigned a pad number.

4.2. Net-degree distribution

To randomly generate appropriate netlist, we have to estimate the net-degree distribution.

Two major net-degree distribution models have been developed in recent years. Zarkesh-Ha et al. [15] established a closed-form expression for the fan-out distribution. Stroobandt [10] used polynomials to model the net generating process.

To achieve accurate estimation of net-degree distribution, we derived a weighted exponential model [13]. In this paper, the net-degree distribution for a circuit consisting of m modules is represented as an m+1-element vector $N(i)|_m$. The internal net-degree distribution and external net-degree distribution are denoted as $N_i(i)|_m$ and $N_e(i)|_m$ respectively. Their normalized representations are denoted as $D(i)\big|_m$, $D_i(i)\big|_m$ and $D_e(i)\big|_m$. Considering a circuit growing process from m modules to $2m$ modules, we recorded the changes in $D_i(i)\big|_m$ and $D_e(i)\big|_m$ during the net generating process.

For $2 \le n \le m$,

$$\Delta D_i(n)\big|_{2m} = f \cdot PP \cdot W(n) \sum_{i=2}^{n} \frac{D_e(i)\big|_m \cdot D_e(n+2-i)\big|_m}{\sum_{j=2}^{m+1}(W(i+j-2)\cdot D_e(j)\big|_m)} \tag{6}$$

where PP is the unknown probability with which external nets will be chosen to combine with other external nets, f is the fraction of new internal nets to all newly connected nets [10] and $W(n)$ is a (m+1)-element weight vector that models how nets with higher net degree have higher probability to be combined. In this paper, $W(n)$ is simply defined as n for n-terminal internal nets and (n+1)-terminal external nets.

For $m+1 \le n \le 2m$,

$$\Delta D_i(n)\big|_{2m} = f \cdot PP \cdot W(n) \sum_{i=n-m+1}^{m+1} \frac{D_e(i)\big|_m \cdot D_e(n+2-i)\big|_m}{\sum_{j=2}^{m+1}(W(i+j-2)\cdot D_e(j)\big|_m)} \tag{7}$$

For $3 \le n \le m+1$,

$$\Delta D_e(n)\big|_{2m} = (1-f)\cdot PP \cdot W(n-1) \sum_{i=2}^{n-1} \frac{D_e(i)\big|_m \cdot D_e(n+1-i)\big|_m}{\sum_{j=2}^{m+1}(W(i+j-2)\cdot D_e(j)\big|_m)} - 2\cdot PP\cdot D_e(n)\big|_m \tag{8}$$

For $m+2 \le n \le 2m+1$,

$$\Delta D_e(n)\big|_{2m} = (1-f)\cdot PP \cdot W(n-1) \sum_{i=n-m}^{m+1} \frac{D_e(i)\big|_m \cdot D_e(n+1-i)\big|_m}{\sum_{j=2}^{m+1}(W(i+j-2)\cdot D_e(j)\big|_m)} \tag{9}$$

For n=2,

$$\Delta D_e(2)\big|_{2m} = -2 \cdot PP \cdot D_e(2)\big|_m \tag{10}$$

From Equation (4), (8), (9) and (10), it can be derived

$$PP = k\cdot\frac{(2m)^{p-1}-m^{p-1}}{sum1+sum2+sum3}$$

$$sum1 = -2\sum_{n=2}^{m+1} D_e(n)\Big|_m$$

$$sum2 = \sum_{n=3}^{m+1}\sum_{i=2}^{n-1}\frac{(1-f)\cdot W(n-1)\cdot D_e(i)\big|_m\cdot D_e(n+1-i)\big|_m}{\sum_{j=2}^{m+1}(W(i+j-2)\cdot D_e(j)\big|_m)} \quad (11)$$

$$sum3 = \sum_{n=m+2}^{2m+1}\sum_{i=n-m}^{m+1}\frac{(1-f)\cdot W(n-1)\cdot D_e(i)\big|_m\cdot D_e(n+1-i)\big|_m}{\sum_{j=2}^{m+1}(W(i+j-2)\cdot D_e(j)\big|_m)}$$

With the value of PP, Equation (6), (7), (8), (9) and (10) are applied to update the distribution of $D_i(i)\big|_m$ and $D_e(i)\big|_m$. The above equations represent a single iteration from m modules to $2m$ modules. To find out the net-distribution of a circuit with M modules, starting from a single module, this process has to be repeated until $2m$ is equal to or greater than M. Figure 1 describes the algorithm.

```
Algorithm net-degree-distribution(M)
Input: M, f, p, k
Output: net-degree distribution N_i and N_e
begin
initialize vectors: D_i, D_e, N_i and N_e;
D_e(2)=k;(*D_e distribution for a single module*)
H=ceil(log_2 M);(*Determine the number of iterations*)
for h=0 to H-1 do
    m1=2^h;
    m2=2^(h+1);
    compute PP;
    update D_i and D_e;
end for
for i=2 to M do
    N_i(i)=round(D_i(i)*M);
    N_e(i)=round(D_e(i)*M);
end for
end.
```

Figure 1: Algorithm of computing net-degree distribution

4.3. Random netlist generation

After knowing the internal and external net-degree distributions, we use a random number generator to generate the netlist. Figure 2 shows the whole process.

```
Algorithm netlist()
Input: N_i, N_e, M, T
Output: netlist
begin
determine N_iMax; (*max net degree for internal nets*)
determine N_eMax; (*max net degree for external nets*)
for i=2 to N_iMax do
    for j=1 to N_i(i) do
        randomly select i numbers from (1,2,...M);
    end for
end for
for i=2 to N_eMax do
    for j=1 to N_e(i) do
        randomly select i-1 numbers from (1,2,...M);
        randomly select 1 number from (1,2,...T);
    end for
end for
end.
```

Figure 2: Algorithm of generating netlist

5. RESULTS AND VALIDATIONS

An important issue of benchmark generation is the quality of randomly generated circuits. These circuits should be realistic, which means they should exhibit the similar properties and characteristics as real circuits [11]. Papers [4] and [11] proposed two different approaches to verify the quality of generated circuits: direct validation and indirect validation.

5.1. Direct validation

Direct validation implies direct comparison of important characteristics of generated benchmarks to those of real circuits. At first, a partitioning tool hMetis [6] was applied to measure the Rent characteristics of MCNC benchmark circuits. Then those parameters were used in BGen to generate corresponding MCNC clones. We used Uniform, Exponential, Gaussian, Poisson and Gaussian distributions for MCNC ami33, ami49, apte, hp and xerox respectively. Total area, total number of nets and average net degree were compared as shown in Table 1, Table 2 and Table 3 respectively. Numbers in column "BGen" are the averages over 6 generated clones. The experiments were carried on a SUN UltraSPARC 10 workstation.

Table 1 shows that BGen well captures the area information except for ami49. Considering the fact that the numbers of modules in MCNC benchmarks are limited and the module size distributions in these benchmarks are only approximated to one of the well-known distributions, the results are acceptable.

In order to validate the accuracy of net-degree distribution model incorporated in BGen, we include Zarkesh-Ha et al.'s model and Stroobandt's model for comparison. In evaluating the total number of nets, BGen achieves on average 6.00% difference comparing to the original benchmark examples, while Zarkesh-Ha et al.'s gets 48.51% and Stroobandt's 20.79%. For average net degree, the average difference for BGen is 16.21%, while for Zarkesh-ha et al.'s is 16.04% and for Stroobandt's is 15.46%.

5.2. Indirect validation

Using indirect validation method, shown in Figure 3, we tested the clone circuits as well as the original circuits with floorplanning tool ELF-SP and compared the results (soft-module packing), as given in Table 4 and Table 5. For each MCNC circuit, we generated 3 clones. For each test circuit, we run ELF-SP floorplanner 5 times. The number of iterations was limited to 4000, and β, that represents the weight associated with wires in the floorplanner cost function ($cost=Area/10^{scale}+\beta Wirelength$) [14], was set to 1000.

It can be observed that effective area ratio results for MCNC benchmarks and their clones are similar. For total wire length, BGen performs slightly better than Zarkesh-Ha et al.'s. For ami49 and xerox, the errors for BGen are rather large due to the overestimation of the total area. Zarkesh-Ha's model largely underestimates the total number of nets and thus compensates this overestimation to some degree. A more important reason is that in ami49 and xerox there are many nets connecting the same modules. ELF-SP is thus able to place these modules very close. On the contrary, in the process of netlist generation in clones by BGen, these nets are randomly distributed to different modules, and therefore resulting in ELF-SP cannot place these modules really close.

6. CONCLUSIONS

In this paper, we described a method to generate parameterized random benchmark circuits for floorplanning. Our tool, BGen, is fairy simple, fast and reliable. The validation results prove that BGen is able to generate realistic circuits for floorplanning tools.

The benchmark generation uses the novel net-degree distribution model, that results in more realistic interconnect density. Other circuit parameters, such as net switching activity, delay and more complex power variance, are being considered for integration into BGen.

Table 1: Total areas

name	num. of modules	actual	BGen	diff%
ami33	33	1156449	1323312	14.43
ami49	49	35445419	58163758	64.09
apte	9	46561628	45535604	-2.20
hp	11	8830584	7306334	-17.26
xerox	10	19350296	22182419	14.64

Table 2: Total number of nets

name	actual	Zarkesh-Ha et al.	Stroobandt	BGen
ami33	123	53	144	115
ami49	408	222	449	429
apte	97	43	131	93
hp	83	42	96	81
xerox	203	132	256	227

Table 3: Average net degree

name	actual	Zarkesh-Ha et al.	Stroobandt	BGen
ami33	4.15	2.87	2.26	2.25
ami49	2.26	2.95	2.04	2.05
apte	2.89	2.86	2.87	3.15
hp	2.89	2.83	2.62	2.78
xerox	2.37	2.74	2.07	2.07

Table 4: Eff. area ratio (%)

name	actual	Zarkesh-Ha et al.	BGen
ami33	94.82	97.13	96.26
ami49	97.85	98.20	97.34
apte	96.07	95.46	97.26
hp	92.20	98.06	96.61
xerox	92.71	98.20	97.39

Table 5: Total wire length

name	actual	Zarkesh-Ha et al.		BGen	
		length	diff%	length	diff%
ami33	54038	37060	-31.42	70878	31.16
ami49	831220	1268200	52.57	1613666	94.13
apte	410360	230193	-43.90	562233	37.01
hp	140580	80282	-42.89	168553	19.90
xerox	294540	553533	87.93	650580	120.88

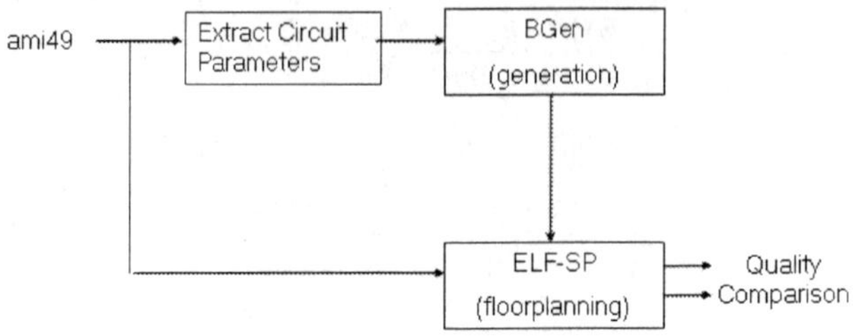

Figure 3: Indirection validation method

7. REFERENCES

[1] Computer-aided Design Benchmarking Laboratory. Available at: http://www.cbl.ncsu.edu/benchmarks/.

[2] C. J. Alpert, "The ISPD98 Circuit Benchmark Suite," In *Proc. of ISPD*, ACM Press, pp. 80-85, April 1998.

[3] J. Darnauer and W. W. Dai, "A Method for Generating Random Circuits and its Application to Routability Measurement," In *Proc. of Intl. Symp. on FPGAs*, pp. 66-72, Feb 1996.

[4] M. D. Hutton, J. P. Grossman, J. S. Rose and D. Corneil, "Characterization and Parameterized Random Generation of Digital Circuits," *IEEE Trans. CAD*, Vol. 17, No. 10, pp. 955-966, Oct 1998.

[5] M. D. Hutton, J. S. Rose and D. Corneil, "Automatic Generation of Synthetic Sequential Benchmark Circuits," *IEEE Trans. CAD*, Vol. 21, No. 8, pp. 928-940, Aug 2002.

[6] G. Karypis and V. Kumar, "hMetis: A Hypergraph Partitioning Package," Nov 1998. Available at: http://www-users.cs.umn.edu/~karypis/metis/hmetis/main.shtml.

[7] B. S. Landman and R. L. Russo, "On a Pin versus Block Relationship for Partitions of Logic Graphs," *IEEE Trans. on Compute.*, pp. 1469-1479, 1971.

[8] I. Liu, H. Chen, T. Chou, A. Aziz and D. F. Wong, "Integrated Power Supply Planning and Floorplanning," In *Proc. of ASP Design Automation Conference*, Jan 2001.

[9] J. Pistorius, E. Legai and M. Minoux, "PartGen: A Generator of Very Large Circuits to Benchmark the Partitioning of FPGAs," *IEEE Trans. CAD,* Vol. 19, No. 11, Nov 2000.

[10] D. Stroobandt, *Analytical Methods for a priori Wire Length Estimates in Computer Systems*, PhD Dissertation, Gent University, Faculty of Applied Sciences, Belgium, 1998.

[11] D. Stroobandt, P. Verplaetse and J. V. Campenhout, "Generating Synthetic Benchmark Circuits for Evaluating CAD Tools," *IEEE Trans. CAD*, Vol. 19, No. 9, pp. 1011-1022, Sep 2000.

[12] P. Verplaetse, J. V. Campenhout and D. Stroobandt, "On Synthetic Benchmark Generation Methods," In *Proc. of ISCAS*, Vol. IV, pp. 213-216, May 2000.

[13] T. Wan and M. Chrzanowska-Jeske, "Prediction of Interconnect Net-Degree Distribution Based on Rent's Rule," In *Proc. of SLIP*, February 2004.

[14] B. Wang, M. Chrzanowska-Jeske, and G. Greenwood, "ELF-SP: Evolutionary Algorithm for Non-Slicing Floorplans with Soft Modules," In *Proc. of ICECS*, pp. 681-684, 2002.

[15] P. Zarkesh-Ha, J. A. Davis, W. Loh, and J. D. Meindl, "Prediction of Interconnect Fan-Out Distribution Using Rent's Rule," In *Proc. of SLIP*, pp. 107-112, April 2000.

A PROGRAMMABLE ARRAY OF SILICON NEURONS FOR THE CONTROL OF LEGGED LOCOMOTION

Francesco Tenore[1], *Ralph Etienne-Cummings*[1,2], *M. Anthony Lewis*[3]

[1] Electrical and Computer Engineering, Johns Hopkins University, Baltimore MD 21218
[2] Institute of Systems Research, University of Maryland, College Park, MD 20742
[3] Iguana Robotics, Inc., P.O. Box 625, Urbana, IL 61803

{fra, retienne}@jhu.edu; tlewis@iguana-robotics.com

ABSTRACT

The biological foundation of most natural locomotory systems is the Central Pattern Generator (CPG). The CPG is a set of neural circuits found in the spinal cord, arranged to produce oscillatory periodic waveforms that activate muscles in a coordinated manner. A 2nd generation VLSI CPG emulator chip — with more and improved neurons, enhanced flexibility, and a higher degree of programmability — has been developed to synchronize oscillators with different frequencies and phases, also produced by the chip, through the coupling of integrate-and-fire (IF) silicon neurons. These oscillators are then used to control the movement of a robot's limbs by using the IF neurons to set a specific phase difference between the oscillators. The chip's architecture is examined in detail, and the construction and implementation of the artificial neural networks that produce the waveforms required for locomotion is described.

1. INTRODUCTION

One of the most fundamental activities performed in nature by living beings is locomotion. An essential element of most periodic biological systems is the Central Pattern Generator (CPG). A CPG in a locomotory system is produced using neurons coupled together to achieve phasic relationships necessary for coordinated, gait-type movements. A programmable neural array was built in silicon to be able to attain these phasic relationships in order to apply them to the limbs of a biped robot.

In silico neural networks are nothing new. In building these networks, two extreme philosophies are often used. On the one hand, silicon chips are characterized by discrete time events, communication within the chip realized through the Address Event Representation (AER) protocol [1], and massive numbers of spiking neurons, in which general principles of synaptic operation are implemented [2,3]. On the other hand, silicon chips that rely on exceptionally detailed models such as the Hodgkin-Huxley neuron model or the Morris-Lecar [4,5,6,7], which allow a dramatically reduced number of neurons and are characterized by a very large parameter space.

Somewhere in between is the approach taken for the chip described here, which builds on work on a 1st generation chip [8,9], in which the neuron model is not as detailed as the ones based on the Hodgkin-Huxley or the Morris-Lecar models, but it can produce continuous time signals as opposed to the ones that require an AER protocol for communication. With locomotory systems in mind, the chip described in this paper can also accept analog input voltages, which correspond to sensory feedback signals that allow modulation of the gait.

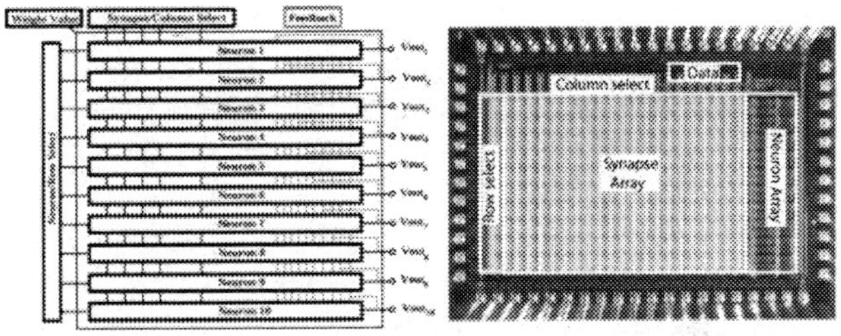

Fig. 1. CPG emulator chip high level architecture and micrograph (3.3x2.1 mm^2)

The main differences from the 1st generation chip are in the higher number of neurons and their greater versatility, as they can be used as continuous envelope-type neurons or as spiking neurons. The synapses are built such that any one of them can be programmed to be excitatory or inhibitory. The current chip also has 10 feedback synapses which allow all the neurons to be interconnected, and the rest of the synapses allow for external signals such as sensory feedback to modulate the charging and discharging of the membrane capacitance. Also, the charging and discharging of this capacitance is now an exponential function of time, as opposed to the linear function that the previous chip exhibited. Thus, it becomes harder, after a certain point, for excitatory or inhibitory input spikes to change the membrane voltage. This allows for better coupling between CPGs.

2. ARCHITECTURE

The 2nd generation chip described here is an array of 10 neurons fabricated in silicon using a 0.5 μm CMOS process. Each neuron was developed to have a silicon equivalent of a membrane capacitance, 22 synapses, and an axon hillock. The chip can also implement a refractory period. Figure 1 shows a micrograph and high-level architecture of the chip. Additionally, the chip contains digital memories, used to modify synaptic weights and to modulate the membrane conductance. The chip's neurons are fully interconnected, i.e. each neuron's output can be fed as a

0-7803-8251-X/04/$17.00 ©2004 IEEE

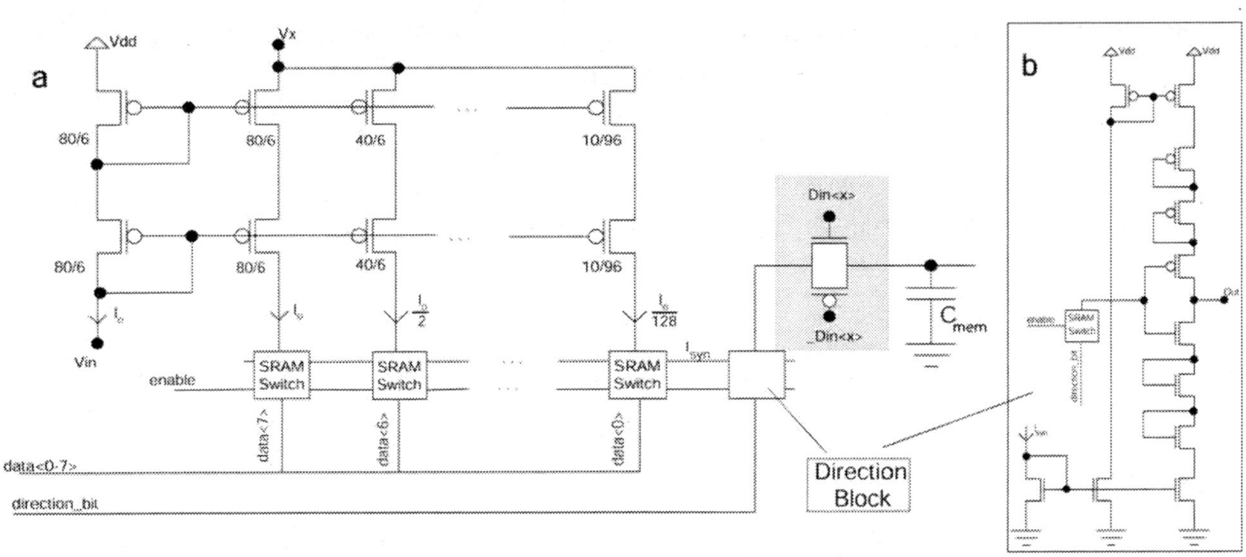

Fig. 2. a) Synapse schematic. Transistor sizes expressed in feature size dimensions, λ (1λ=0.35μm). The shaded pass transistor part is not present in the 4 analog synapses. b) Detailed architecture of the direction block.

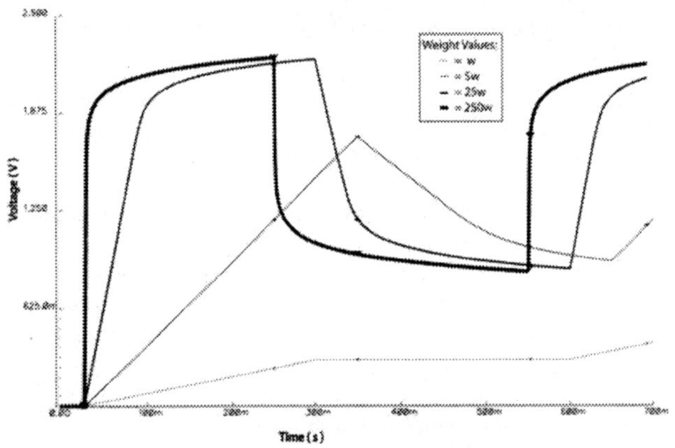

Fig. 3. Membrane voltage charging and discharging speeds for different weight values. (Simulation result).

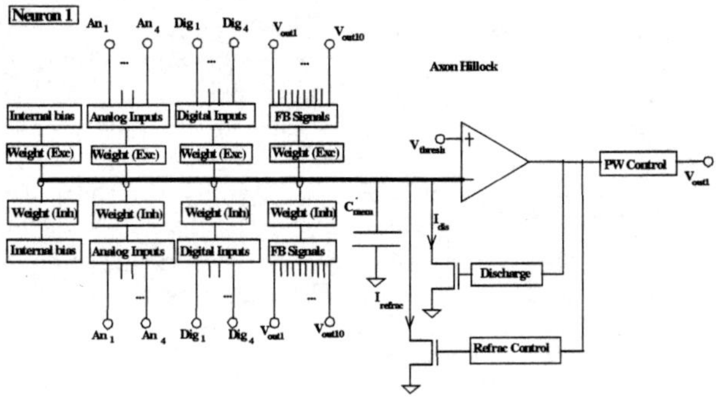

Fig. 4. Block diagram of a single neuron.

synaptic input to itself or any other neuron. Addressing a particular neuron/synapse pair for weight modifications occurs through row and column registers. The weight to be stored in the chip's memory is contained in a 9-bit data register. The data register's first bit selects whether the input is excitatory or inhibitory, and the remaining 8 bits set the synaptic weight.

2.1. Synapses

The chip contains 4 different types of synapses, depending on the source of the input. There are 8 external synapses (4 analog, 4 digital), 10 feedback synapses (one for each neuron) and a test synapse for testing purposes.

The architecture of a synapse is shown in Figure 2. Each synapse contains an 8-bit digital-to-analog converter (DAC) to program the synaptic weights into the chip. The weights are produced by summing eight scaled copies of a bias current, where each scaled copy is twice the size of the previous one. These biasing weights control which current copy is summed to the final synapse output. SRAM memory cells, which store the value of the synaptic weight, act as switches that either allow the current to flow to the direction block or prevent it from flowing. The voltage V_x is either an input analog voltage (making the synapse analog) or V_{dd}. A digital synapse has an external digital voltage that connects to the inputs of the shaded switch box shown in Figure 2a (the switch is shaded as it is not found in analog synapses). A feedback synapse is identical to a digital synapse except that the signal that controls the switch is a neuron output. Each synapse also has a direction block, shown in Figure 2b, that allows the current to charge or discharge the membrane capacitance, thus making the synapse excitatory or inhibitory respectively. The charging and discharging speeds of the membrane capacitance are shown in Figure 3 for different weight values. It is important to note that in both the charging and the discharging phases, after an initial time-interval of linearity, the capacitor charges and discharges more and more slowly as it approaches the voltage drop across the two diode-connected pMOS and nMOS transistors shown in Figure 2b. This feature acts as an automatic gain control, and allows a wider range of coupling weights between neurons to be used without affecting the entrainment (a linear charge up profile limits the coupling weights to a narrow range of small values).

2.2. Neurons

The chip's 10 neurons are all identical and completely interconnected. The output of any given neuron can be fed back to any one or all 10 neurons. Each neuron also has a self-feedback synapse. Figure 4 shows a block diagram of a single neuron with its synapses acting upon the membrane potential. The output of a hysteretic comparator, which emulates a neuron's axon hillock, switches from low to high when its upper threshold is reached by the voltage on the membrane capacitor. Similarly, it resets when the voltage decreases to the hysteretic comparator's lower threshold value.

The strength with which the membrane capacitor is discharged can be tuned by the discharge block. The duration of a spike or an oscillation is then set through a pulse-width control block. This block is necessary in order to be able to modify spike durations independently of synaptic inputs. The pulse-width control block is essentially an RS flip-flop, which is *set* when the comparator output goes high. At this point, a synapse-like structure with tunable strength charges up another capacitor, which *resets* the output signal (and itself) when the threshold of the PW control block comparator is reached. Therefore, the speed with which this charge up occurs can also be easily tuned. The modulation of the pulse width determines the type of neuron that is created. A slow *reset* of the flip-flop is typically accompanied by a slow charge up of the membrane potential, thereby creating slow oscillations characteristic of envelope neurons. A fast charging of the membrane capacitance and a fast *reset* of the flip-flop, on the other hand, creates spiking neurons. Hence, the chip is able to create two different types of neurons, and its versatility lies in the fact that any neuron can be of either type.

Finally, the "refrac control" block shown in Figure 4 allows the neurons to also have a refractory period. The duration of this period is set in the same way the pulse width control block is reset — the refractory control block output stays high (thus preventing the membrane capacitor to charge up) until its own capacitor's voltage reaches a certain threshold which resets its flip-flop and allows the membrane capacitor to charge up once again. Other characteristics of the neuron are covered in [10].

3. NEURAL NETWORKS

With the two different types of neurons that the chip can output, neural networks can be created such as the one shown in Figure 5a. The figure shows an envelope neuron and a spiking neuron connected to each other, such that the envelope neuron's output can influence the charging of the membrane capacitor of the spiking neuron. With no feedback, the envelope neuron continues oscillating at its fixed frequency, whereas the spiking neuron, not having anything to charge its membrane capacitance, would stay low. In the scenario depicted in the figure, the spiking neuron starts spiking when the envelope neuron's output is high, charging up the other neuron's membrane capacitance, thus allowing it to start spiking. When the envelope neuron's output goes back low, nothing charges the spiking neuron's membrane capacitance so it stays low as well.

This spiking output can be used in conjunction with a complementary one, as shown in Figure 5b, to create two 180° out-of-phase spiking neurons. These signals can then be used to actuate a robot's hip. This complementary situation is similar to

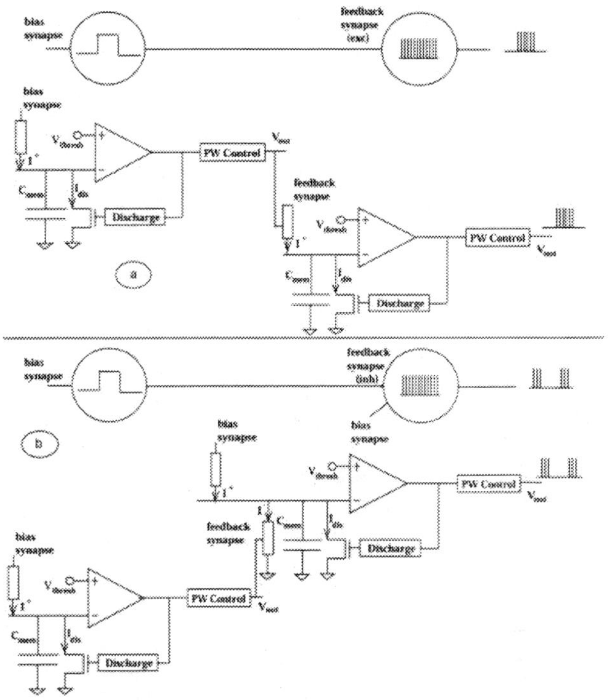

Fig. 5. a) An envelope neuron exciting a spiking neuron. b) An envelope neuron inhibiting a spiking neuron. Note that charging, in this case, occurs through a bias synapse.

the one described above but essentially differs for two elements. First, the use of an excitatory synapse ("bias synapse" in Figure 5b) with constant charging of the capacitor is required. Second, the feedback synapse from the envelope neuron is now an inhibitory one. The output of the motor neuron is in the form of spikes if the input from the feedback synapse is low and if only the excitatory synapse has an effect on the membrane. Similarly, the output is low if the envelope neuron is high, since its inhibition strength is set to be greater than the excitation of the charging synapse. These enveloped spikes can then be used to drive motors on a robot's hips.

4. RESULTS

Antagonistic to the scenario just described is a situation in which a single spike is used to modify the period of an envelope waveform. In this case, the spike occurs either when the capacitor is charging (i.e. the output is low) or when the capacitor is discharging (i.e. the output is high). The effect of adding quanta of charge as the envelope neuron is charging is a decrease of its period. Adding charge as the neuron is trying to remove charge (discharging) makes the period increase. The longer the duration of the spike and the stronger the synaptic weight, the more readily these situations can be observed. Figures 6a and 6b present these situations. In particular, Figure 6a shows the charging and discharging of a neuron's membrane potential through external excitatory and inhibitory synaptic inputs. In this figure, 3 different situations can be identified. 1) As excitatory spikes charge the membrane capacitance, the voltage across it grows until the hysteretic comparator's upper threshold is reached and the neuron fires, thereby resetting the voltage on the membrane capacitance through the discharge

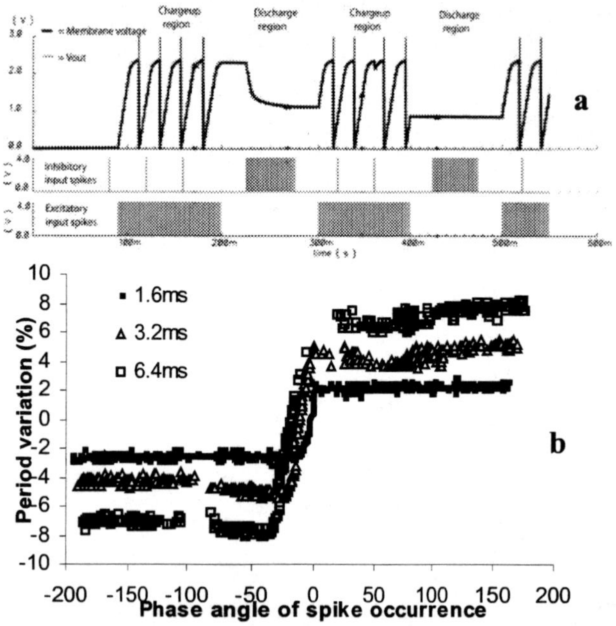

Fig. 6. a) Effect of spike charging and discharging on membrane capacitance and neuron output. b) Effect of spikes of various durations on the period of an envelope neuron.

block. The discharge weight is large to ensure complete resetting of the membrane voltage such that charging restarts from the same voltage level. 2) As this happens, inhibitory spikes slow down the charging and as a result, the frequency of the output spikes decreases. These spikes have a stronger effect if they occur as the membrane potential is close to the hysteretic comparator's upper threshold, and a smaller effect if the capacitance has been recently discharged. The second stream of membrane charge up shows this more clearly. 3) Inhibitory spikes then proceed to bring the voltage down, first linearly, as was seen, then exponentially. Finally, excitatory spikes bring the voltage back up until the neuron fires and the process starts anew. (As said previously, data from the membrane voltage are simulated, as there is no external access to the membrane potential.)

Figure 6b presents results collected from the chip and shows the effect of a spike on the period of an envelope neuron. In the figure, 0° (±180°, respectively) corresponds to the occurrence of a spike as the envelope neuron has just reached its hysteretic comparator's lower (higher) threshold and, therefore, starts charging up (discharging). As can be seen, the envelope neuron's period decreases if the spike occurs during charge up (-180° - 0°), and increases if it occurs during discharge (0° - +180°). The three curves differ for spike duration, and the figure also shows that a longer lasting spike has a larger effect on the envelope neuron's period.

The spikes being fed to an envelope neuron are used in applications to create entrainment. If two envelope neurons have similar but not identical frequencies, a master oscillator can trigger a spike entrainer neuron with its same frequency but at a different phase to synchronize the two envelopes. Thus, the two signals will have the same frequency, but the phase is determined by the spike entrainer. Since the spike can be generated at any phase [10], the two signals can be synchronized and be out-of-phase by any amount. This phase difference allows the generation of different locomotory gaits [10].

5. CONCLUSIONS

An array of programmable silicon neurons was implemented in a VLSI chip to generate the required waveforms for emulating a biological Central Pattern Generator. Continuous and spiking outputs can be obtained and modulated as desired. Furthermore, by appropriately coupling the various neurons it is possible to create the waveforms necessary to control a robot's limbs. The chip builds on a previous chip and expands its versatility to get closer to a self-contained locomotion controller for walking robots.

Acknowledgement

The authors would like to thank Zi Rong Xu for his invaluable contributions to the previous versions of this chip.

6. REFERENCES

[1] M. Mahowald, *An analog VLSI system for stereoscopic vision*, Kluwer Academic Publisher, Boston, MA, 1994.

[2] D.H. Goldberg, G. Cauwenberghs, and A.G. Andreou, "Probabilistic synaptic weighting in a reconfigurable network of VLSI integrate-and-fire neurons," *Neural Networks*, vol. 14, no. 6-7, pp. 781-793, 2001.

[3] G.Indiveri, A.M. Whatley, and J. Kramer, "A reconfigurable neuromorphic VLSI multi-chip system applied to visual motion computation," *Proc. MicroNeuro '99*, April 1999.

[4] C. Rasche, R. Douglas, M. Mahowald, "Characterization of a pyramidal silicon neuron," *Neuromorphic Systems: Engineering silicon from neurobiology*, L. S. Smith and A. Hamilton, eds, World Scientific, 1st edition, 1998.

[5] M. Simoni, S. DeWeerth, "Adaptation in an aVLSI model of a neuron," *IEEE Transactions on circuits and systems II: Analog and digital signal processing*. 46(7):967-970, 1999.

[6] A.L. Hodgkin, A.F. Huxley. "A quantitative description of ion currents and its applications to conduction and excitation in nerve membranes," *Journal of Physiology (Lond.)*, 117:500-544, 1952.

[7] C. Morris, H. Lecar, "Voltage oscillations in the barnacle giant muscle fiber," *Biophysics J.*, vol. 35, pp. 193-213, 1981.

[8] M.A. Lewis, R. Etienne-Cummings, A.H. Cohen, M. Hartmann, "Toward biomorphic control using custom aVLSI chips," *Proceedings of the International conference on robotics and automation*, San Francisco, CA, 2000.

[9] M.A. Lewis, R. Etienne-Cummings, M.J. Hartmann, A.H. Cohen, Z.R. Xu, "An *in silico* central pattern generator: silicon oscillator, coupling, entrainment, and physical computation," *Biological Cybernetics*, 88, 2, 2003, pp. 137-151.

[10] F. Tenore, R. Etienne-Cummings, M.A. Lewis, "Entrainment of silicon central pattern generators for legged locomotory control," *Advances in Neural Information Processing Systems 16*, Pending Publication, 2004.

SIGNAL RECONSTRUCTION FROM SPIKING NEURON MODELS

Dazhi Wei and John G. Harris

Computational NeuroEngineering Lab
University of Florida
Gainesville, FL 32611

ABSTRACT

We describe a method for perfect signal reconstruction from spiking neuron models such as integrate-and-fire or leaky integrate-and-fire neurons. These neural models encode a single analog signal in the timing of asynchronous digital pulses. We show that using only the output firing times of these neurons, we can recover a bandlimited input signal to within machine precision. A major application of this work is for a replacement of conventional analog-to-digital converters in some applications where simpler analog hardware is traded off for more complex reconstruction on the part of the subsequent digital processor. Realistic SPICE simulations of CMOS spiking neurons show that accurate reconstruction with more than 12-bit precision can be achieved. The effects of frequency aliasing, noise, and temporal quantization are considered.

1. INTRODUCTION

Numerous models of spiking biological neurons have been introduced in the literature, see Gerstner for a recent review [1]. These models span a spectrum from complex, biologically realistic models to simpler, less realistic models. These models map a set of continuous synaptic inputs $x_k(t)$ to a discrete number of firing events t_i. We concentrate on a subset of these models that linearly filter a single synaptic input $x(t)$ to produce a continuously varying cell potential. When the cell potential rises above a fixed cell threshold, a discrete action potential is created. This class of models includes the popular integrate-and-fire (IF) and leaky IF (LIF) neurons. We show how to mathematically reconstruct the continuous input based on the firing pattern. This work builds on the pioneering efforts of Lazar and Toth who realized a link between mathematical frame theory and reconstruction for the case of asynchronous sigma-delta converters [2].

This material is based upon work supported by NSF under Award ID: 0135946 and by NASA under award no. NCC 2-1363.

2. RECONSTRUCTION FROM INTEGRATE-AND-FIRE NEURONS

We first consider the reconstruction from the simplest spiking model: the integrate-and-fire neuron without any refractory period. In this model, the synaptic input $x(t)$ is integrated until its value reaches the cell threshold θ. At this time (t_i) an action potential is created and the integration is reset to zero. The firing times must satisfy:

$$\int_{t_i}^{t_{i+1}} x(t)dt = \theta \,, \forall i \tag{1}$$

Let us assume that $x(t)$ is bandlimited to $[-\Omega_s, \Omega_s]$, and $t_i, i \in Z$ be a timing sequence with maximum adjacent interval $(t_{i+1}-t_i) < T$, where the Nyquist period $T = \pi/\Omega_s$. Any bandlimited signal can be expressed as a low-pass filtered version of an appropriately weighted sum of delayed impulse functions [3, 4].

$$\begin{aligned} x(t) &= h(t) * \sum_j w_j \delta(t - s_j) \\ &= \sum_j w_j h(t - s_j) \end{aligned} \tag{2}$$

where w_j are scalar weights, $s_j = (t_j + t_{j+1})/2$, $h(t)$ is the impulse response of the low-pass filter and $*$ denotes the convolution operator. The purpose of creating another timing sequence s_j is to improve the reconstruction efficiency. The impulse response of the ideal low-pass filter is given by:

$$h(t) = \sin(\Omega_s t)/(\Omega_s t) \tag{3}$$

Now the signal recovery problem is simplified as how to calculate the appropriate weights. If $s_j = jT$ is a uniform sampling sequence, standard sampling theory can be used to show that the impulse weight reduces to $w_j = x(s_j)$.

But generally the weights need to be reconstructed using the encoding information. Substituting equation 2 into equation 1, we obtain

$$\theta = \int_{t_i}^{t_{i+1}} x(t)dt \quad = \quad \int_{t_i}^{t_{i+1}} \sum_j w_j h(t - s_j)dt$$

$$= \sum_j w_j \int_{t_i}^{t_{i+1}} h(t - s_j)dt$$
$$= \sum_j w_j c_{ij} \quad (4)$$

where c_{ij} are constants that can be numerically computed with:

$$c_{ij} = \int_{t_i}^{t_{i+1}} h(t - s_j)dt \quad (5)$$

The resulting set of linear equations is given by $\underline{\mathbf{C}}\,\underline{\mathbf{W}} = \underline{\theta}$ in matrix form. Unfortunately, $\underline{\mathbf{C}}$ is usually ill-conditioned necessitating the use of a SVD-based pseudo-inverse conditioning technique (pinv in Matlab) to calculate the weight vector $\underline{\mathbf{W}}$:

$$\underline{\mathbf{W}} = \underline{\mathbf{C}}^+\,\underline{\theta} \quad (6)$$

Given appropriate weights, eq. 2 can be used to numerically calculate $x(t)$ to within machine precision. A similar derivation can be derived for any spiking neuron model that uses simple integration and/or linear operators. We have successfully reconstructed continuous waveforms from spike trains for leaky integrate-and-fire neurons (see Section 3) and for adaptive neurons whose effective threshold changes based on its recent spike history[5].

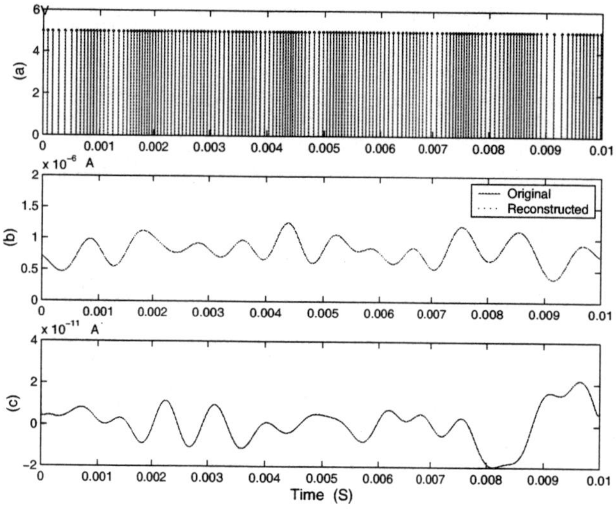

Fig. 1. Reconstruction results from IF neuron (a) the spike train, (b) the original and reconstructed signals, and (c) the error between the original and the reconstructed signals

Figure 1 shows a reconstructed signal using this method for an integrate-and-fire neuron. The input signal is a Gaussian random noise signal bandlimited to $[-3000\pi, 3000\pi]$, and the corresponding Nyquist period $T = 1/3$mS. Since the maximum adjacent spike interval (0.142mS) is less than T, this method can be used to reconstruct the input signal.

The simulation results show the effective signal to noise ratio (SNR) of the reconstruction is 103.3dB. SNR is computed as the power of the input signal divided by the power of the error between the original and the reconstructed signals.

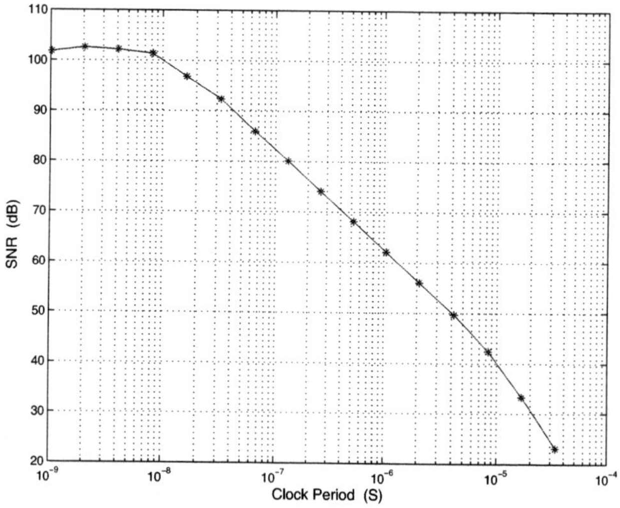

Fig. 2. Plot of SNR vs. clock period used in reconstruction

For most practical applications, the spike train will be synchronized to a fast clock on a digital microprocessor or a DSP. This synchronization will necessarily introduce some time jitter to the recorded arrival time of each spike. Figure 2 shows a plot of SNR vs. the simulated clock period for the same random noise input used in Figure 1. This plot indicates how fast the digital synchronization clock must run in order to achieve particular levels of performance. For example, if an SNR of 60dB was desired then a clock period of 1μS or less would be required, meaning a frequency of at least 1MHz. The plot also gives an idea of how much jitter can be allowed in the timing signal in the electronics and in the transmission process. Uniform delays in all of the spikes results in a simple delay in the reconstructed signal so it is the variation among the delays that causes distortion.

Since it is assumed that the input signal is bandlimited to some frequency, an anti-aliasing pre-filter is strictly required just as it is for standard Nyquist rate sampling. It is well known that for standard Nyquist rate sampling, higher frequencies are simply mapped to lower frequencies preserving the amount of energy. On the other hand, the integration step of the integrate-and-fire neuron provides an attenuation of these higher frequencies so that they cause less corruption of the reconstructed signal. Figure 3 shows a plot of SNR vs. the frequency of an added high-frequency sine wave. As the frequency of the added sine wave increases, its detrimental effect is reduced.

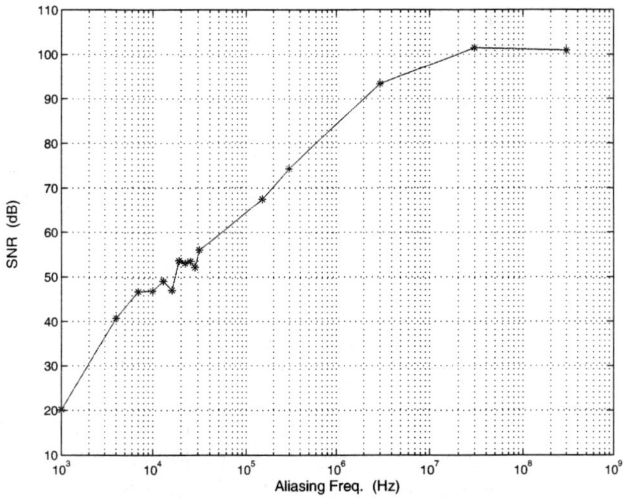

Fig. 3. Plot of SNR vs. aliasing frequency

3. HARDWARE

The possibility of accurate reconstruction from spiking neuron circuits has many implications. One promising line of research is that a spiking neuron circuit could be used as a substitute for an analog-to-digital converter (ADC) for remote sensing, implanted biomedical devices and other power-limited applications. In these scenarios, a continuous waveform is transformed to a spike train using extremely simple and low-power hardware. The asynchronous digital spikes can then be efficient transmitted without incurring the usual noise and crosstalk effects of analog signals. Depending on the transmission system, the spikes may or may not need to be synchronized to a digital clock before transmission. The continuous signal can then be reconstructed at a time and location where power dissipation and circuit size are not an issue. In effect, we have traded off for simpler, lower-power hardware at the transmitter for a more complex digital reconstruction process at the receiver.

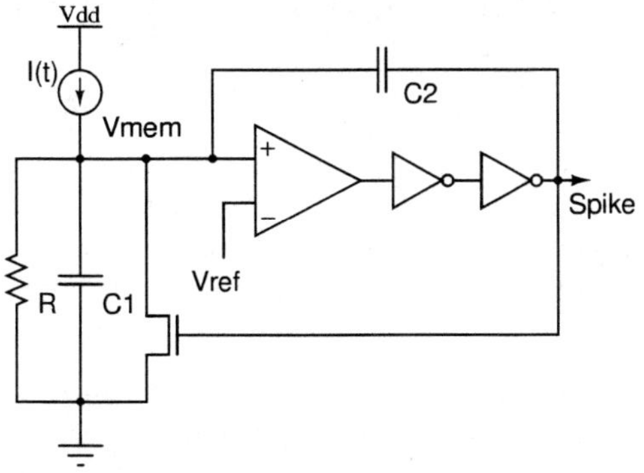

Fig. 4. Original spiking neuron from Mead

This type of conversion, in effect, delays quantization CMOS spiking models have a rich history. One of the earliest silicon neurons discussed in Mead's textbook [6] is shown in Figure 4. The threshold voltage of the neuron V_{ref} is selected to let V_{mem} be ground in the reset state. This is for the purpose of simplifying the calibration of the neuron in later calculations.

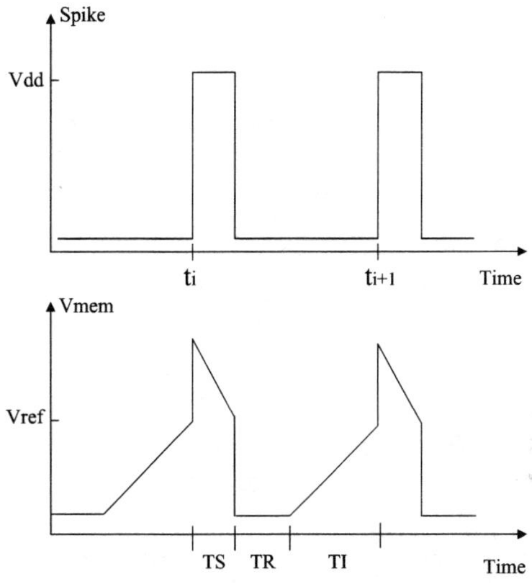

Fig. 5. Waveforms from spiking neural circuit

Figure 5 shows the shape of the capacitor voltage V_{mem} and the output spike of the neuron. The time of the spike (t_i) is defined to be the rising edge of the neuron output. In the figure, T_I represents the integration period, while T_S the spike width and T_R the refractory period. Obviously V_{mem} is related to $x(t)$ only during the T_I period. We can derive the following equations:

$$C\frac{dV_{mem}}{dt} + \frac{V_{mem}}{R} = x(t) \tag{7}$$

$$V_{mem}(t_i + T_S + T_R) = 0 \tag{8}$$

$$V_{mem}(t_{i+1}) = V_{ref} \tag{9}$$

In fact equations 7, 8 and 9 describe the encoding operation of the LIF neuron on the input current. They are used to generate a linear system of equation in a similar fashion to the ideal integrate-and-fire neuron discussed in section 2, except

$$c_{ij} = \int_{t_i+T_s+T_R}^{t_{i+1}} h(t - s_j)\, e^{(t-t_{i+1})/(RC)} dt \tag{10}$$

which is the leaky integration of $h(t - s_j)$ over the time period $[t_i + T_s + T_R,\, t_{i+1}]$.

Because process variations can cause the circuit parameters to deviate from the designed parameters, the neuron may need to be calibrated. We input two constant currents I_1 and I_2 to the neuron, and measure the integration periods T_{I1} and T_{I2} of the output spike train, respectively. We can get the parameters from equations below:

$$V_{ref} = RI_1(1 - e^{-T_{I1}/(RC)}) \quad (11)$$

$$V_{ref} = RI_2(1 - e^{-T_{I2}/(RC)}) \quad (12)$$

In our simulation, the input current to the LIF neuron is

$$\begin{aligned} x(t) = 200(2.5 &+ 0.5\sin(2\pi 1000t) + 0.3\sin(2\pi 600t) \\ &+ 0.2\sin(2\pi 130t))\, nA \quad (13) \end{aligned}$$

and $V_{ref} = 3$V, $R = 15$MΩ, and $C = 18$pF

We can use equation 6 to calculate the weights for each impulse at s_j, then use equation 2 to reconstruct the signal, which is shown in Figure 6. The signal to noise ratio (SNR) is 80.2dB, and the effective number of bits (ENOB) precision is $(SNR - 1.76)/6.02 \approx 13$ bits.

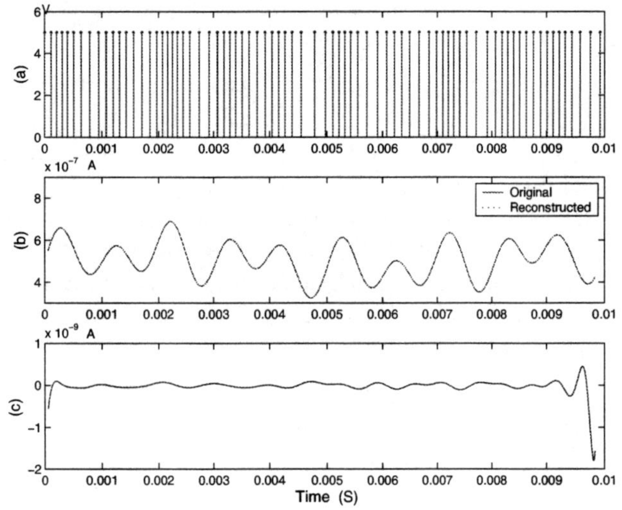

Fig. 6. Reconstruction results from LIF neuron (a) the spike train, (b) the original and reconstructed signals, and (c) the error between the original and the reconstructed signals

4. CONCLUSIONS AND FUTURE WORK

We have described a method for perfect signal reconstruction from spiking neuron models. We show that using only the output firing times of these neurons, we can recover a bandlimited input signal to within machine precision. A major application of this work is for a replacement of conventional analog-to-digital converters in some applications where simpler analog hardware is traded off for more complex reconstruction on the part of the subsequent digital processor. Realistic SPICE simulations of CMOS spiking neurons show that accurate reconstruction with more than 12-bit precision can be achieved.

There is much more work to be done in the reconstruction process. We are seeking more computationally efficient approximations for $h(t)$ than the truncated sinc function in use now. This will lead to simpler estimation of the constants c_{ij}. In order to reconstruct the signal in real-time a window based methodology is necessary where finite width windows are used to reconstruct the signals which can then be pieced together.

These reconstruction algorithms can be immediately useful for existing neuromorphic chips that use spiking neurons to generate chip outputs. For instance, many imager chips have been built that use spikes to generate outputs using an address-event style readout. For these chips, a continuous bandlimited signal can be reconstructed from the chip output. The bandlimited frequency of each pixel would naturaly be determined by the intensity, and hence SNR, of each pixel.

Finally, the reconstruction algorithm also provides an "interpretation" for actual recorded spike trains from real biological neurons. Though this interpretation is limited by the inherent assumptions of the underlying model, it provides an alternative to the simple binning that is commonly performed to derive continuous waveforms from spike trains in applications such as brain-machine interfaces.

5. REFERENCES

[1] W. Gerstner and W. Kistler, *Spiking Neuron Models Single Neurons, Populations, Plasticity*, Cambridge University Press, 2002.

[2] A. Lazar and L. Toth, "Time encoding and perfect recovery of bandlimited signals," *ICASSP*, vol. 6, pp. 709–712, April 2003.

[3] R. Duffin and A. Schaeffer, "A class of nonharmonic fourier series," *Transactions of the American Mathematical Society*, vol. 72, pp. 341–366, 1952.

[4] H. Feichtinger and K. Gröchenig, "Theory and practice of irregular sampling," in *J.J. Benedetto and M.W. Frazier, editors,Wavelets: Mathematics and Applications*, Boca Raton, FL, 1994, pp. 305–363.

[5] M. Chacron, K. Pakdaman, and A. Longtin, "Interspike interval correlations...," *Neural Computation*, vol. 14, no. 2, pp. 253–278, February 2003.

[6] C. Mead, *Analog VLSI and neural systems*, Addison-Wesley, Reading, Massachusetts, 1989.

AN EVENT–BASED VLSI NETWORK OF INTEGRATE–AND–FIRE NEURONS

Elisabetta Chicca, Giacomo Indiveri and Rodney J. Douglas

Institute of Neuroinformatics, University of Zurich and ETH Zurich
Winterthurerstrasse 190, CH–8057 Zurich, Switzerland

ABSTRACT

The growing interest in pulse–based neural networks is encouraging the development of hardware implementations of massively parallel, distributed networks of Integrate–and–Fire (I&F) neurons. We have developed a mixed–mode (analog/digital) VLSI device that comprises a reconfigurable network of I&F neurons and adaptive synapses. The synapses receive input spikes and the neurons transmit output spikes (events) using an asynchronous Address–Event Representation (AER). In this paper we describe the network architecture, present experimental data demonstrating the characteristics of the single elements on the chip, and show that a competitive network configuration has Winner–Take–All (WTA) behavior and produces spike synchronization.

1. INTRODUCTION

Networks of I&F neurons have been shown to exhibit a wide range of useful computational properties, including feature binding, segmentation, pattern recognition, onset detection, input prediction, *etc.* [1]. These types of networks are very well suited for VLSI implementation. Recent and growing interest in pulse–based neural networks, together with the emergence of a standard that allows VLSI neurons to communicate using asynchronous pulse–frequency modulated events (spikes), have led to the development of a large number of VLSI implementations of networks of I&F neurons (see the ISCAS04 invited session on spiking neural networks). The asynchronous communication protocol is based on the Address–Event Representation [2, 3]. In this representation input and output signals are real–time, digital events that carry analog information in their temporal structure (interspike intervals). Each event is represented by a binary word encoding the address of the sending node. On–chip arbitration schemes are used to handle event "collisions" (cases in which sending nodes attempt to transmit their addresses at exactly the same time). Systems containing more than two AER chips can be assembled using additional off–chip arbitration. These off–chip arbiters can also use lookup–tables and processing elements to remap, time–stamp and perform digital operations on address–events [2, 4].

In this paper we present an AER chip comprising a network of I&F neurons and dynamic synapses. The I&F neuron circuit is described and fully characterized in [5]. The circuits implementing the synapses are of two types [6, 7]. Both types integrate input spikes, producing biologically plausible dynamics: one type is compact and exhibits short–term depression, while the other is larger, but can exhibit either short–term depression or facilitation, depending on its parameter settings. The parameters that control the synaptic dynamics and their strength are global and can be set by external voltage references.

The AER input synapses and AER output neurons offer the possibility of implementing networks of arbitrary topology when the device is interfaced to a dedicated PCI–AER board [4], able to log, monitor, map and generate address–events. In addition to externally addressable AER synapses, we included synaptic circuits with hard–wired on–chip connectivity to implement a competitive network topology. The circuits used on this chip are to a large extent technology independent and the network could be scaled up to arbitrary size.

2. NETWORK ARCHITECTURE

The architecture of the VLSI network of I&F neurons is shown in Fig. 1(a). It is a two–dimensional array containing a row of 32 neurons, each connected to a column of afferent synaptic circuits. Each column contains 14 AER excitatory synapses, 2 AER inhibitory synapses and 6 locally connected (hard–wired) synapses. When an address–event is received, the synapse with the corresponding row and column address is stimulated. If the address–events routed to the neuron with the corresponding column address integrate up to the neuron's voltage threshold for spiking, then that neuron generates an address–event which is transmitted off–chip. Arbitrary network architectures can be implemented using off–chip look–up tables and routing the chip's output address–events to one or more AER input synapses. The synapse address can belong to a different chip, therefore,

This work was supported by the EU grant ALAVLSI (IST–2001–38099).

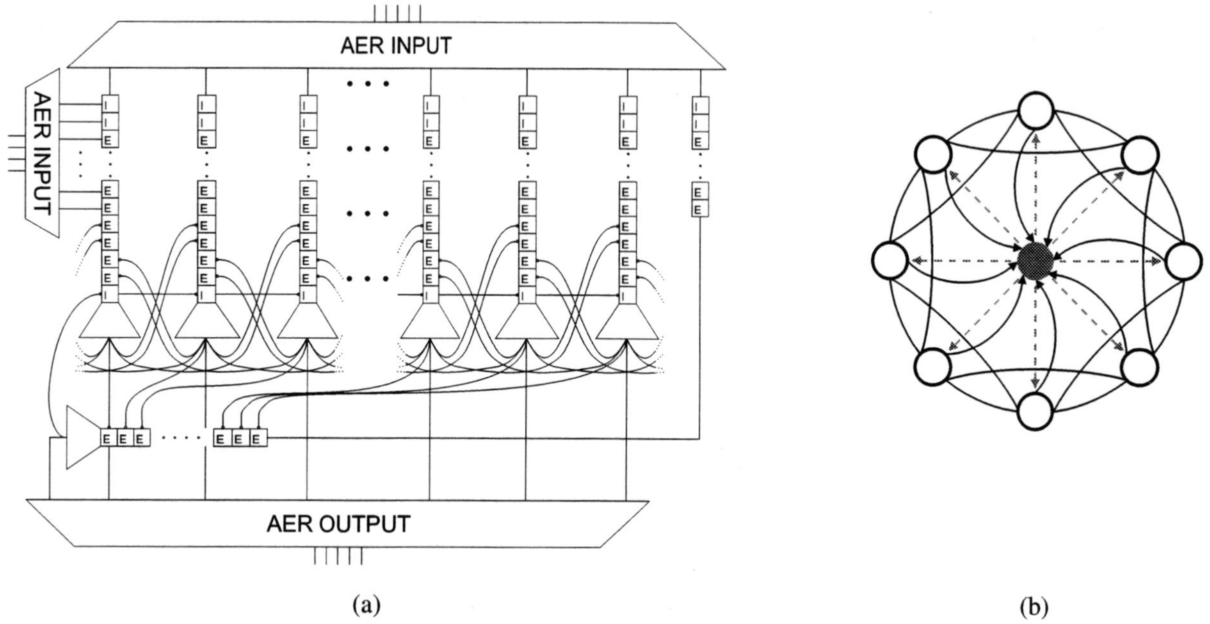

Fig. 1. (a) Chip architecture. Squares represent excitatory (E) and inhibitory (I) synapses, trapezoids represent I&F neurons. The I&F neurons can transmit their spikes off–chip and/or to locally connected synapses (see text for details). (b) Schematic representation of the connectivity pattern implemented by the internal hard–wired connections (closed boundary condition). Empty circles represent excitatory neurons and the filled circle represents the global inhibitory neuron. Solid/dashed lines represent excitatory/inhibitory connections. Connections with arrowheads are monodirectional, all the others are bidirectional.

arbitrary multi–chip architectures can be implemented.

Synapses with local hard–wired connectivity are used to realize a competitive (WTA) network with nearest neighbor and second nearest neighbor interactions (see Fig. 1): 31 neurons of the array send their spikes to 31 local excitatory synapses on the global inhibitory neuron; the inhibitory neuron, in turn, stimulates the local inhibitory synapses of the 31 excitatory neurons; each excitatory neuron stimulates its first and second neighbors on both sides using two sets of locally connected synapses. The first and second neighbor connections of the neurons at the edges of the array are connected to pads. This allows us to leave the network open, or implement closed boundary conditions (to form a ring of neurons [8]), using off–chip jumpers.

All of the synapses on the chip can be switched off. This allows us to inactivate either the local synaptic connections, or the AER ones, or to use local synapses in conjunction with the AER ones. In addition, a uniform constant DC current can be injected to all the neurons in the array. The amplitude of this current can be set through a global bias voltage.

The chip was implemented using a standard $0.8\ \mu m$ CMOS technology. The layout of the whole array, including the AER input and output sections covers an area of about $1.1 \times 1.9\ mm^2$. The layout of one column of the array, including the I&F neuron, the 16 AER synapses and the 6 local ones covers an area of about $31 \times 1500\ \mu m^2$. Only about 6% of this area is occupied by the neuron ($31 \times 86\ \mu m^2$).

In theory this network can scale up to any arbitrary size, both in terms of the number of neurons, and the number of synapses. In practice the network size is limited by the AER bandwidth available. If we consider a network of neurons configured via the PCI–AER board with 30% connectivity (a typical figure used in modeling studies), in which (typically) only 10% of the neurons fire at a mean rate of 100Hz, the speed of the (non–optimized) AER circuits implemented on the current chip limits the maximum number of possible neurons to approximately 1000. Using the same $0.8\ \mu m$ CMOS technology used for the current device, an array of 1000×300 I&F neurons and synapses would require a silicon area of approximately $31 \times 20\ mm^2$.

3. EXPERIMENTAL RESULTS

To verify the correct behavior of the circuits on the chip we injected the same DC current to all the neurons in the array and measured the network's response properties as a function of different configuration parameters (such as the strengths of different synaptic weights). In these experiments we did not stimulate the neurons via the AER synapses (that have been tested previously and shown to function correctly).

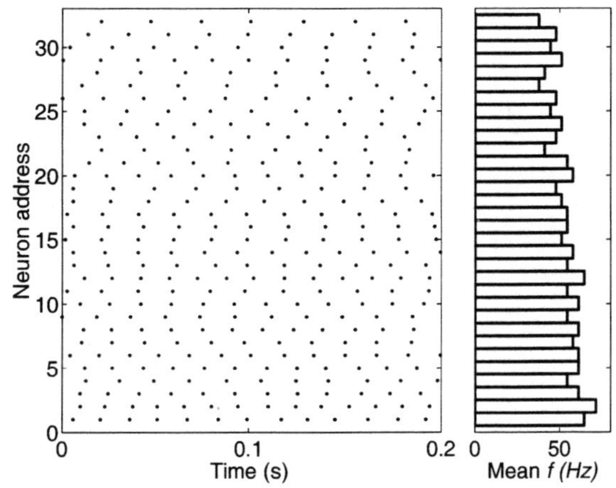

Fig. 2. Network response to homogeneous constant input current with all synaptic connections disabled. Left panel: raster plot of the network activity. Right panel: mean output frequencies. The differences in mean output frequency are due to device mismatch effects both in the input transistors and in the I&F neuron circuit elements.

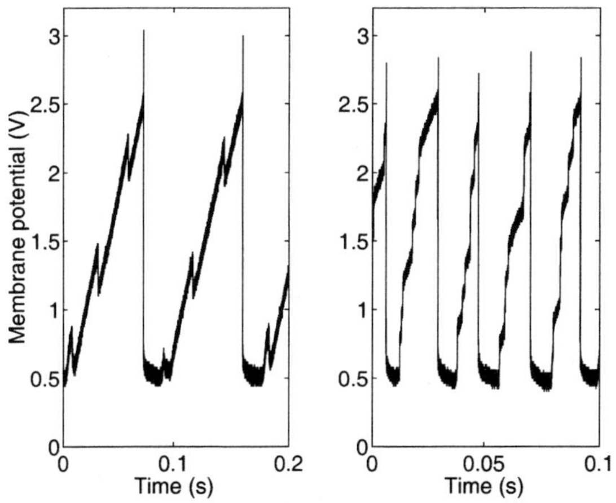

Fig. 3. Membrane potentials. Left panel: membrane potential of one of the excitatory I&F neurons in the network. The neuron integrates a constant DC current while receiving inhibitory spikes from the global inhibitory neuron. Right panel: membrane potential of the global inhibitory neuron. This neuron integrates the same constant DC current while receiving excitatory inputs from all the active excitatory neurons in the array.

3.1. Basic experiments

Initially we performed a set of basic experiments to test the functionality of the main building blocks of the chip: the neurons; the synapses; and the AER sections.

In a first experiment we switched off all the local hard–wired connections, injected a constant DC current to all the neurons and monitored their spiking activity using the PCI–AER board. In Fig. 2 we show a raster plot of the expected regular firing observed. The differences in mean firing rate are due to device mismatch effects both in the input transistors and in the I&F neuron circuit elements.

In a second experiment, we tested the competitive network topology (without lateral interactions) by switching on the connections in both directions between the excitatory neurons and the global inhibitory neuron. In this case, in addition to the constant DC current, the excitatory neurons integrate inhibitory inputs that tend to decrease their output firing rates, while the global inhibitory neuron integrates its excitatory inputs that increase its mean firing rate. The membrane potential of all the neurons in the array can be measured through an on–chip voltage scanner, which allows either all the neurons in parallel to be probed (multiplexed in time) or only one neuron at a time. In Fig. 3 we show the membrane potential of one of the excitatory I&F neurons in the network, next to the membrane potential of the global inhibitory neuron.

3.2. Network behavior

In these sets of experiments we activated the network's hard–wired connections to implement two different types of competitive networks with lateral connections. In both cases we activated the hard–wired connections from the excitatory neurons to the inhibitory one and the connections from the inhibitory neuron to the excitatory ones, stimulated the network by injecting a constant DC input current to all the neurons, and used the PCI–AER board to monitor the network spiking activity.

In the first experiment, symmetric nearest neighbor lateral connections were activated. Even in this extremely simplified case, with constant homogeneous inputs and symmetric connectivity, the network was able to produce a classical WTA behavior. Although all neurons should receive the same input current, due to device–mismatch effects, one neuron wins the competition and suppresses, through the inhibitory neuron, all other ones, while exciting nearest neighbors (see Fig. 4). As the coupling between neurons was set to be relatively strong, the excitatory and inhibitory neurons synchronized their spiking activity.

In the second experiment we activated both first and second neighbor excitatory connections. When the strength of these connections is asymmetric and global inhibition is strong enough, the network generates a traveling wave of activity, as shown in Fig. 5. Global inhibition allows the winning neurons to suppress all the others and the asymmetric lateral excitation propagates the activity in one direction. The neurons at the edge of the array were connected to form a ring [8], so that the wave could propagate cyclically through the array.

In both experiments the spiking activity of the neurons

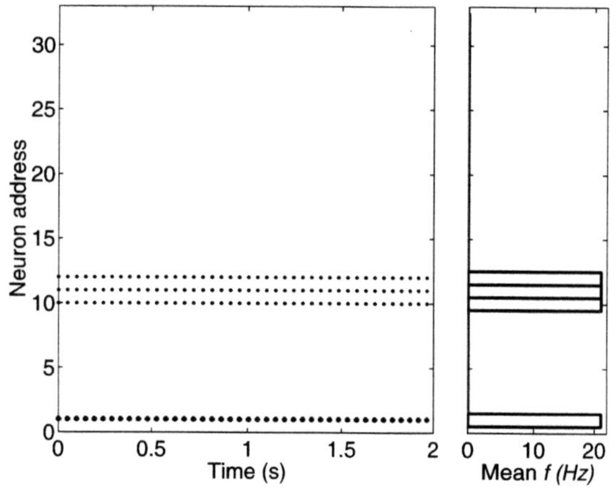

Fig. 4. WTA behavior. Left panel: raster plot of the network activity in response to a constant DC input current with lateral excitatory (first neighbor) connections, excitatory to inhibitory connections and global inhibition activated. The neuron with address 1 is the global inhibitory neuron. Right panel: mean output frequencies.

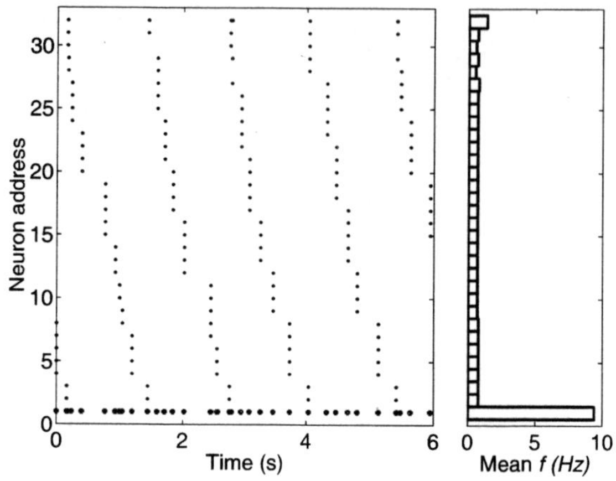

Fig. 5. Traveling wave. Left panel: raster plot of the network activity in response to a constant DC input current with asymmetric excitatory first and second nearest neighbor connections and with global inhibition. The neuron with address 1 is the global inhibitory neuron. Right panel: mean output frequencies.

is highly synchronized. This is a consequence of the parameters used in these experiments. These are extreme cases, used to characterize the architecture with its hard–wired competitive network topology, in which the input is a simple homogeneous constant current, and the strength of the connections is set to relatively high values, to amplify the small differences in neuronal activity due to mismatch parameters.

4. DISCUSSIONS AND FUTURE WORK

In this paper we have presented a reconfigurable VLSI array of AER neurons and synapses with additional on–chip connectivity that implements a competitive network topology. We demonstrated the correct behavior of the main blocks present on the device and showed how, even using the simplest possible input stimulus, the local hard–wired competitive network can give rise to interesting behaviors such as WTA functionality, spike synchronization and traveling wave generation. So far, we have used a simple constant DC current as the input to the network. In future work, we will take advantage of the PCI–AER board to stimulate the network with Poisson distributed spike trains, or with address–events generated by neuromorphic vision [9] or auditory sensors. Having multiple instances of the same synaptic circuit for each neuron will allow us to explore the effect of adaptation to several competing stimuli. We will use this device to implement real–time models of selective attention systems [5]; we will also study the network's ability to generate traveling waves, persistent activity (even after the input stimulus is removed), binding by synchronization, and other behaviors observed in cortical circuits.

5. REFERENCES

[1] W. Maass and C. M. Bishop, Eds., *Pulsed Neural Networks*, The MIT Press, 1999.

[2] S. R. Deiss, R. J. Douglas, and A. M. Whatley, "A pulse-coded communications infrastructure for neuromorphic systems," in *Pulsed Neural Networks*, W. Maass and C. M. Bishop, Eds., chapter 6, pp. 157–178. MIT Press, 1998.

[3] K. A. Boahen, "Communicating neuronal ensembles between neuromorphic chips," in *Neuromorphic Systems Engineering*, T. S. Lande, Ed., pp. 229–259. Kluwer Academic, Norwell, MA, 1998.

[4] V. Dante and P. Del Giudice, "The PCI–AER interface board," in *Report on the 2001 Workshop on Neuromorphic Engineering*, 2001, pp. 99–103.

[5] G. Indiveri, "Modeling selective attention using a neuromorphic analog VLSI device," *Neural Computation*, vol. 12, no. 12, pp. 2857–2880, Dec 2000.

[6] E. Chicca, G. Indiveri, and R. J. Douglas, "An adaptive silicon synapse," in *Proc. IEEE International Symposium on Circuits and Systems*. IEEE, May 2003.

[7] C. Rasche and R. Hahnloser, "Silicon synaptic depression," *Biological Cybernetics*, vol. 84, no. 1, pp. 57–62, 2001.

[8] R. Hahnloser, R. Sarpeshkar, M. Mahowald, R. J. Douglas, and S. Seung, "Digital selection and analog amplifi cation coexist in an electronic circuit inspired by neocortex," *Nature*, vol. 405, pp. 947–951, 2000.

[9] J. Kramer, "An integrated optical transient sensor," *IEEE Trans. on Circuits and Systems II*, vol. 49, no. 9, pp. 612–628, Sep 2002.

A TIME DOMAIN WINNER-TAKE-ALL NETWORK OF INTEGRATE-AND-FIRE NEURONS

J. P. Abrahamsen, P. Häfliger, and T. S. Lande

University of Oslo , Norway
e-mail: jenspa@ifi.uio.no

ABSTRACT

A time domain winner-take-all circuit based on simple self-resetting integrate-and-fire neurons is presented in this paper. Integrate-and-Fire (I&F) neurons can translate the intensity of a current input into a time domain signal: Strong input current will lead to an early spike output and weak input to a late output. By making the self-reset line global for all neurons, only the first spiking neuron, which is the neuron with the strongest input, will ever spike, and thus, win over the others.

This WTA circuit was conceived as part of an imager chip to process current input from a motion detection array, thus detecting the row and column of maximum change of illumination. The fact that this WTA processes analog input and produces spike output is most convenient for the address event interface (AER) that conveys the WTA output off-chip.

We verified the WTA functionality with experiments of an AMS 0.6μm CMOS implementation. Some suggestions on how to achieve additional functions by simple extensions of the circuit are discussed.

1. INTRODUCTION

1.1. WTA circuits

A winner-take-all circuit finds the maximum among a set of inputs, shutting off the other inputs. This is a useful function in many sensory processing tasks that try to focus attention on the most interesting/salient of an array of sensory inputs. Solving this task on-chip with dedicated parallel computing elements is much faster than processing the raw sensor data serially on a computer.

WTA implementations are often implemented as neural network like structures where model neurons exert negative feedback on their neighbours, such that only the strongest can maintain its activity and suppresses the other neurons. Maybe the best known example of a WTA circuit is the one introduced by Lazzaro et. al. [1]. It is a really space conservative implementation with only two transistors per input/output element in its most basic form, plus one additional transistor operated as a common bias current source for all elements.

The result can be read out as a current that is equal to a bias current for the winning cell and zero for the losers, or as a voltage V_m which logarithmically encodes the magnitude of the input current of the winner and is close to zero for the losers. A single additional transistor operating in subthreshold with V_m as its gate voltage can turn V_m back into a current that for the winning WTA cell is proportional to its input and close to zero for the others.

1.2. Integrate-and-fire neurons

Integrate-and-fire (I&F) neurons are among the simplest neuronal models that produce pulse-event output, just like their biological analogue. They simply integrate a current input on a capacitance (referred to as 'soma voltage' in the following) until a threshold is reached. This triggers a pulse-event output (spike, action potential), after which they reset themselves. Probably the most popular implementation in analog electronics is C. Mead's self-resetting neuron [3], which is similar to the time-domain WTA cell, shown in figure 1. In Mead's self-resetting neuron $Vspike$ is connected to the transistor above the $VspikeLen$ transistor, and the part labeled "common for all neurons" is not present.

The input current I is integrated as 'soma voltage' V_{soma} on the capacitor C_{soma} until the digital switching threshold V_θ of the amplifier (two inverters in series) is reached, at which point the output voltage goes high. A charge is fed back through $C_{feedback}$, stabilizing V_{soma} somewhat above the threshold.

When the output goes high, the neuron starts to discharge slowly. The discharge current is controlled by $V_{spikeLen}$. When V_{soma} has fallen to the threshold again, the amplifier output goes low, and the borrowed charge from the feedback capacitor is returned. The cycle is complete and can start again.

The function of an I&F neuron can be described as translating an input current into a latency signal [4, 2]: After an I&F has been reset, it will fire after an interval T that is inversely proportional to its input current I. On the physical device the proportionality constant is given by the threshold voltage V_θ and the integrating capacitance C (actually the total input capacitance $C_{soma} + C_{feedback}$).

$$T = \frac{V_\theta\, C}{I} \tag{1}$$

If the I&F neuron resets itself instantaneously with every output spike (i.e. if we neglect that the output pulses are not infinitely short and that there is a certain refractory period), its output pulse frequency ν will be inversely proportional to that interval and thus, proportional to the input current.

$$\nu = \frac{1}{T} = \frac{I}{V_\theta\, C} \tag{2}$$

This work was supported by the EU 5th Framework Programme IST project CAVIAR.

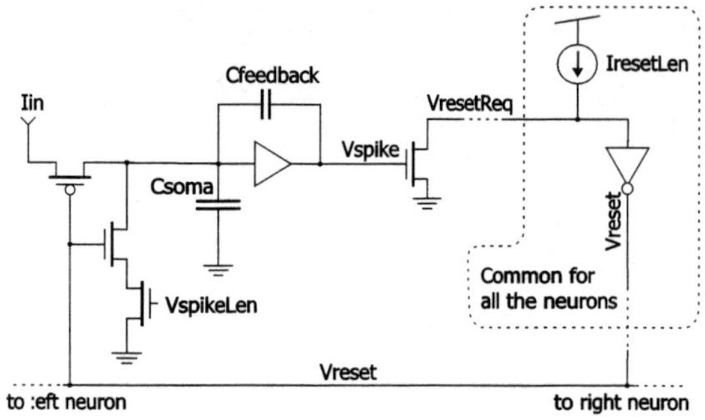

Figure 1: A neuro-WTA cell shown together with the current source and inverter common for all cells.

1.3. Address Event Representation

The Address event representation (AER) is a communications protocol emulating point-to-point connections for spike signals. It multiplexes spikes from different sources (e.g. I&F neurons) on a shared digital bus. Thus, a transmission of a digital number on the bus identifies the source of a spike being transmitted. At the receiver end, this 'sender-address' is decoded and a pulse is reconstructed at the appropriate location (e.g. a synapse).

This protocol can for instance be used to send a number of analog signals encoded in spike frequency from a sensor chip with only a limited number of pins. As mentioned, I&F neurons conveniently turn analog signals into spike frequency, and can thus serve as the first stage of an interface between analog signals and the AER protocol.

2. IMPLEMENTATION

We designed our WTA circuit to process current input from an array of motion sensitive optic sensors. It should determine the strongest input and convey it as frequency encoded pulse signals to an AER interface. The obvious solution would have been to simply put the classic WTA and an I&F neuron in sequence. But since a WTA is usually realized as a neural network structure we thought of a way of letting the I&F neurons perform both tasks, i.e. the translation of current into pulse events and the WTA function.

We used the I&F property of encoding input strength into the time domain as a latency: if all neurons in an array are reset at the same time, then the next neuron to fire will be the one with the strongest input. We now simply use that first neuron's output pulse to reset all neurons instead of just resetting itself.

The modified I&F neuron is shown in figure 1 together with the components that are global for all neurons: a current-source and an inverter. When a neuron's output goes high, the global reset-request line, which is otherwise kept at Vdd by a weak pull-up transistor, is pulled down. This is in effect a wired NOR. The strong inverter then pulls up the global reset line, starting the discharge process for all the neurons. The pull-up transistor provides only a small current, ensuring a fast onset and a delayed reset of the common reset-pulse. See section 2.1 for an explanation on why this is desirable.

In many applications, we are not interested in having a winner if the inputs are all smaller than a certain level. For example on our imager chip we are not interested in finding a winner when the perceived motion is unclear or if there are no moving objects at all. We want to define a minimum threshold for the winning input for the WTA to make a decision.

This function could be implemented in current mode as a leakage from each neuron's C_{soma}. The neuron's will only be able to spike when the input current is larger than the leakage current. This would require one additional transistor per neuron and it would slow down the response time.

We chose to implement that computation in time domain instead. This is done by simply adding an extra neuron to the WTA array that receives a constant current input. We call it a 'timeout neuron'. This neuron will fire first and thus, win, if no other receives a larger current. Its output will set the minimal output frequency for winners. Compared to the current mode solution we save the extra transistor per WTA neuron and eliminate mismatch problems (in the leakage currents). Additionally, the timeout neuron assures us that the WTA circuit is still operating, by sending out events from the timeout neuron.

2.1. Properties and limitations

The time-domain WTA will spike with a frequency proportional to the winner's input current. Thus, it does not only determine the winner, but also provides information about this winners maximal input current, at no extra cost in layout space.

Two general measures that characterize the quality of a WTA circuit are its absolute accuracy and its resolution of discrimination. The absolute accuracy can be expressed as the minimal difference in input current of the winning cell to the other cells, in order for the WTA to reliably choose the correct winner. It is mainly dependent on device mismatch and can be improved by using larger device geometries. Both, the classical WTA and our time domain WTA suffer from these mismatches. In our implementation the mismatch in soma capacitance of the neurons is the one that leads to errors in the conversion of the analog currents into time domain signals and thus, to errors in the WTA computation. Capacitance mismatch is normally significantly smaller (roughly estimated no more than 5% for our capacitor sizes) than transistor mismatch (especially for subthreshold currents). So if a WTA processes currents from sensors, the expected mismatch error in the sensor signals is usually more significant. Also on our test chip, we could only supply inputs to the WTA either from transistors connected as current sources or from optical sensors. Thus, we could not measure the absolute accuracy of the WTA with well controlled current inputs.

The resolution of discrimination expresses the WTA's ability to reach a clear decision: If two input currents cause the WTA to issue an inconclusive output (in our case two spiking neurons), within what range can one change one of the two currents before the WTA reaches a clear decision again. This measure is independent of the absolute accuracy and can be estimated theoretically, even when assuming perfect device matching: two neurons can win if the delay between the latency T_1 of the (correct) first spike and the hypothetical la-

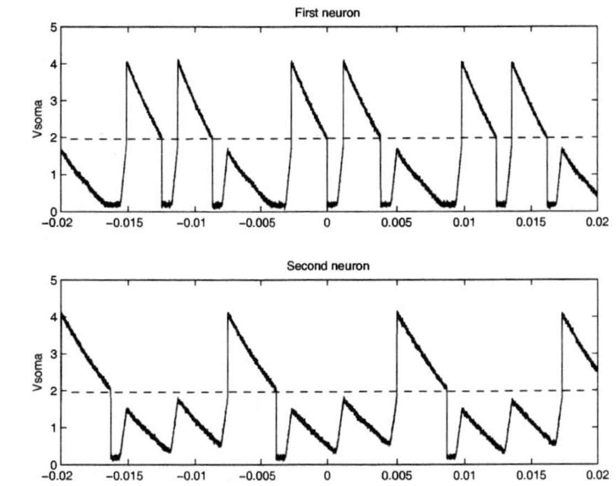

Figure 2: The head-start effect: A neuron is not completely discharged when a new cycle begins, giving it a head-start.

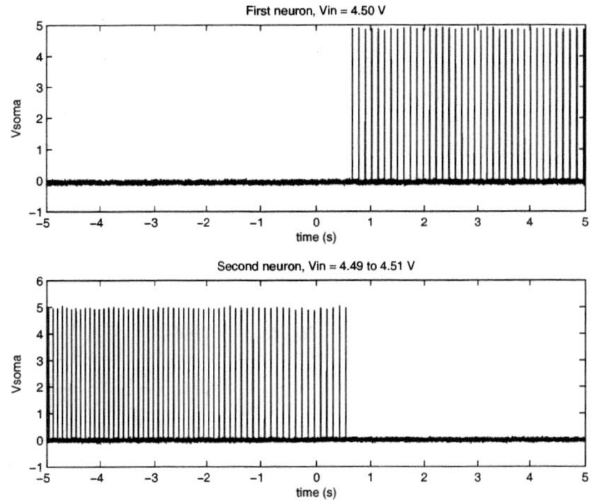

Figure 3: Two neurons in competition, where the lower neurons input is gradually decreased.

tency of the first runner-up T_2 is shorter than the delay t_d between T_1 and the global reset. Using equation (1) this leads to an expression for the minimal difference of the inverse input currents that the circuit can distinguish.

$$(\frac{1}{I_2} - \frac{1}{I_1}) > \frac{t_d}{V_\theta\, C} \tag{3}$$

Expressing the difference as a percentage $x = 1 - \frac{I_2}{I_1}$ of the winning current we get:

$$x > \frac{k}{1+k}, k = I_1 \frac{t_d}{V_\theta\, C} \tag{4}$$

In our implementation $C = 1.56$pF and $V_\theta \approx 1.9$V. We estimate the reset delay $t_d \approx$1ns. It should be clear though that it is dependent on the capacitive load on the common wired OR line, and thus, dependent on the number of WTA neurons. For arrays that are substantially larger, precautions must be taken to keep this delay small. We can set the maximum current from the photo array by a bias current in our target application and we typically chose currents below 10nA, i.e. well in the subthreshold domain. For 10nA currents, the WTA is capable to discriminate differences of 0.0007%. Thus, in our intended operating range we were not able to observe ambiguous WTA output at all during our experiments.

To reach a discrimination worse than 2%, currents above 30μA are necessary. The classic WTA implementation offers clear relative discrimination (output voltage more than 50% different) at about 2% difference in the input currents. This quality is about one order of magnitude worse than ours in our intended operating range, but it is independent of the absolute input current for a huge range of currents. Thus, our solution offers sharper discrimination only if the range of input currents is limited.

Formula (4) also tells us, that the delay t_d of the onset of the global reset can influence the discrimination. For this reason, it was clear that the parameter current $IresetLen$ (figure1) should be chosen small. But there is an even better reason for this: Less obvious, and more severe, for the WTA to operate properly the delay of the *falling* edge of Vreset requires to be of a minimal length too. If that delay is too short, the effect in figure 2 is observed. The figure shows the soma voltage of two neurons, which both received constant current input. The top neuron had a slightly larger input and should therefore win. The current difference was large enough to discriminate correctly according to equation (4). Still, the lower neuron wins every third spike.

This is because the duration of the reset (the width of the Vreset pulse) is too short for all neurons to discharge completely. As figure 2 shows, when the lower neuron looses, it is not completely reset, whereas the winning top neuron is. That is partially due to device mismatch and also due to the Early effect. The top neuron is depleted more quickly back to the switching threshold and thus, its output pulse ends, while the second neuron's soma voltage is still larger than 0V. Thus, for every consecutive competition the lower neuron has an increasing head-start until it ultimately wins a competition. Figure 2 shows an example where the lower neuron unjustly wins every third competition. Therefore, it is important to choose $IresetLen$ small enough for the reset pulse to extend the output pulse of the winner by some percentage, to ensure that all neurons are properly reset.

3. RESULTS

We implemented the time domain WTA circuit on an 0.6μm AMS CMOS chip, together with an 48x48 array of optical motion sensors. On the chip there is a separate 4 neuron WTA for test purposes with pFET transistors connected as current sources delivering input currents. Two 48 neuron WTA are processing the total current from the imager columns and rows respectively.

Figure 3 shows measurements of two neurons in the four neuron WTA, where the input current to the initially winning neuron (lower trace) is gradually decreased (note the slight decrease in the spike frequency), letting another neuron take over (upper trace). The input currents were about 27pA. With

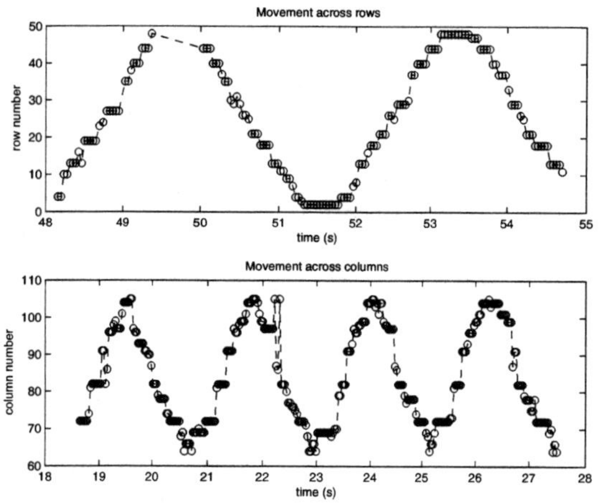

Figure 4: Output from our motion tracker circuit in response to a bright spot moving across it's visual field.

very careful tweaking we were able to make two neurons fire alternately for currents around $6nA$. We ruled out the 'head start' effect as a cause, since we could not affect this behaviour by adjusting $I_{resetLen}$ and because there was not one single spike of one neuron every n spikes of the other (as we would expect from the 'head start' effect). We observed that the alternating winners won 10-20 consecutive times, indicating that noise is determining the winner. Two *simultaneous* winners could only rarely be observed. Only in simulation, could we have two neurons fire fire constantly together. This was for example the case for input currents $\approx 8,4nA$ with a relative difference less than 0.0004 %. For a simulated 0.003 % difference the circuit discriminated correctly again. These numbers are in agreement with the theoretical formula (4).

Figure 4 shows the output of the WTA for our motion tracker circuit, i.e. with real sensor input. We were moving a bright spot accross the chips visual field vertically (across the imager rows, upper plot) and horizontally (columns, lower plot).

4. EXTENSIONS

We briefly present two simple useful extensions of the WTA circuit, providing additional functions.

4.1. All neurons stronger than a threshold win

The first variant selects not only one winner but all neurons with input larger than a threshold. This is achieved by making all but the timeout neuron both self resetting and globally resetable. In order to achieve this, two additional transistors, as depicted in figure 5 are added, and the wired OR transistor is removed. Additionally only the timeout neurons output is connected to the global reset line (block A in the figure).

4.2. The n strongest neurons win

In the second variant this timeout neuron's constant current input is replaced by a switchable constant current source (block

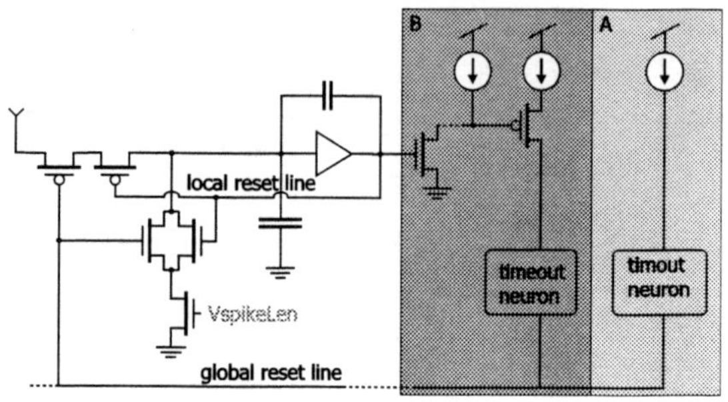

Figure 5: Extension cell with two optional blocks.

B in figure 5). It is turned on by the output pulses of the WTA neurons and thus, it counts the pulses until it reaches its threshold. The outputs are connected to the wired OR like the basic WTA. The strength of the current source determines how many inputs are needed. This allows a fixed number of the strongest neurons to spike.

5. CONCLUSION

By looking at the output of I&F neurons as time domain signals that encode analog input currents in spike latency, a novel kind of WTA circuit was conceived. Determining the strongest of several analog signals in time domain does no longer require analog circuits, but merely asynchronous logic.

Furthermore the time domain WTA's property of turning current inputs into spike outputs makes it suitable as interface to the AER protocol, which in turn is well suited to convey a large number of analog values off chip with a limited number of pins.

The winning neuron does not merely convey the fact that its input current is the strongest, but it also conveys that strength as spike frequency. In our target application, this strength can be used as a measure of the confidence in the signal, and this information is therefore useful for further processing steps.

All these properties make the time domain WTA well suited to our intended application of determining the strongest output of a motion sensor array and transmitting its output offchip by the AER protocol. Other target applications can be any kind of sensory-array VLSI chips.

6. REFERENCES

[1] J. Lazzaro, S. Ryckebusch, M.A. Mahowald, and C.A. Mead. Winner-take-all networks of o(n) complexity. *NIPS*, 1, 1988.

[2] W. Maass and T. Natschläger. Networks of spiking neurons can emulate arbitrary hopfield nets in temporal coding. *Network: Comp. in Neural Sys.*, 8:355–372, 1997.

[3] C. Mead. *Analog VLSI and Neural Systems*. Addison Wesley, 1989.

[4] S. Thorpe, F. Fize, and C. Marlot. Speed of processing in the human visual system. *Nature*, 381:520–522, June 1996.

MULTI-BAND OFDM: A NEW APPROACH FOR UWB

Anuj Batra, Jaiganesh Balakrishnan, and Anand Dabak
DSPS R&D Center, Texas Instruments, 12500 TI Blvd, MS 8649, Dallas, TX 75206
{batra, jai, dabak}@ti.com

ABSTRACT

In this paper, a multi-band OFDM (MB-OFDM) system for ultra-wideband (UWB) communication is described. In this system, the UWB spectrum is divided into bands that are 528 MHz wide and the data is transmitted across the bands using a time-frequency code. Within each sub-band, an OFDM modulation scheme is used to transmit the information. An overview of the MB-OFDM system and various systems design considerations are discussed. In addition, the performance results in realistic multi-path environments are provided for the MB-OFDM system.

1. INTRODUCTION

Over the last year and half, UWB communication systems have received significant attention from both the industry and as well as academia. The primary reason for the increased attention is due to the landmark ruling by the Federal Communications Commission (FCC). In February 2002, the FCC allocated 7,500 MHz of spectrum (from 3.1 GHz to 10.6 GHz) for use by UWB devices. This ruling has helped to create new standardization efforts, like IEEE 802.15.3a, that focus on developing high-speed wireless communication systems for personal area network.

Another reason for the excitement about UWB is because this technology promises to deliver data rates that can scale from a data rate of 110 Mb/s at a distance of 10 meters up to a data rate of 480 Mb/s at a distance of 2 meters in realistic multi-path environments all while consuming very little power and silicon area. It is expected that UWB devices will be able to provide low cost solutions that can satisfy the consumers need for data rates as well as enable new consumer market segments.

The paper is organized as follows. Section 2 provides a description of various UWB approaches and motivates the design of a MB-OFDM system. Section 3 describes the system parameters for the MB-OFDM system. Some of the design considerations for a MB-OFDM solution are discussed in section 4. Section 5 provides performance results of the MB-OFDM system based on detailed link level simulations and section 6 concludes.

2. OVERVIEW OF PREVIOUS UWB SYSTEMS

The traditional design approach for UWB communication system involved using narrow time-domain pulses that occupy a very wide spectrum [1]. The main disadvantage of these systems is the challenge of building RF and analog circuits with large bandwidths, high speed analog-to-digital converters (ADCs) to process this extremely wideband signal, and the significant digital complexity required to capture the multi-path energy in dense multi-path environments.

An approach based on pulsed multi-band [2] alleviated the need to process the signal over a large bandwidth. This approach divides the spectrum into several smaller sub-bands, whose bandwidth is approximately 500 MHz, and single-carrier modulation techniques are used to transmit the information on each of the sub-bands. By interleaving the symbols across sub-bands, the UWB system can still maintain the same transmit power as if it were using the entire bandwidth. By using smaller sub-bands to transmit the information, the processing bandwidth for the receiver can be reduced, thereby lowering the complexity of the design, reducing the power consumption of the radio, lowering the overall solution cost, and improving spectral flexibility and worldwide compliance.

The primary disadvantage of a pulsed multi-band system is the difficulty in collecting significant multi-path energy when using a single RF chain. The amount of multi-path energy collected is limited by the dwell time on each sub-band. In addition, when a small number of RAKE fingers are employed, the system performance is very sensitive to group delay variations introduced by the analog front-end components. Finally, the pulsed multi-band system places very stringent band-switching timing requirements (< 100 ps) at both the transmitter and receiver, leading to complicated synthesizer designs.

Multi-path energy collection is an important issue because it fundamentally determines the range and robustness of a communications system. By combining orthogonal frequency division multiplexing (OFDM) system with multi-banding, the strengths of the pulsed multi-band system can be retained while still addressing the issue of multi-path energy capture. This new system, MB-OFDM [3,4], has several nice properties, including the ability to efficiently capture multi-path energy with a single RF chain, insensitivity to group delay variations, and the ability to deal with narrowband interferers at the receiver without having to sacrifice either sub-bands or data rate. In addition, the MB-OFDM system offers simplified synthesizer architectures and relaxes the band-switching timing requirements. The only potential drawback of this MB-OFDM approach is that the transmitter is more complex because of the IFFT and slightly higher peak-to-average ratio than a pulse-based multi-band system.

3. SYSTEM OVERVIEW OF MULTI-BAND OFDM

For the sake of simplicity, a MB-OFDM system consisting only three bands is described herein. Figure 1 illustrates how the OFDM symbols are transmitted in a MB-OFDM system. In this example, it has been implicitly assumed that the time-frequency coding (TFC) is performed across just three OFDM symbols; however, in practice, the TFC pattern can have a much longer periodicity.

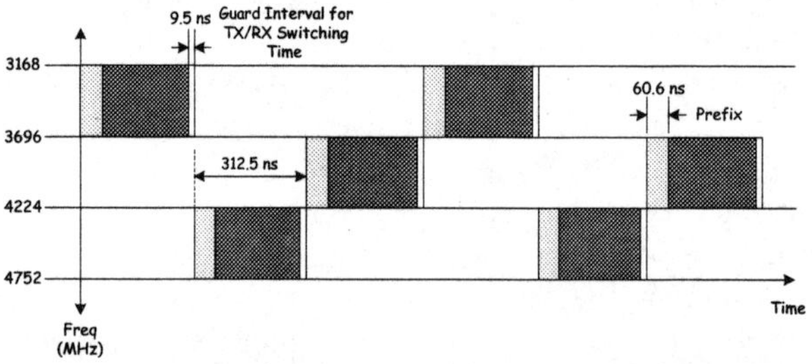

Figure 1 – Example of TF coding for an MB-OFDM system.

In a MB-OFDM system, a guard interval (9.5 nanoseconds) is appended to each OFDM symbol and a zero-padded prefix (60.6 nanoseconds) is inserted at the beginning of each OFDM symbol. The guard interval ensures that there is sufficient time for the transmitter and receiver to switch to the next carrier frequency. A zero-padded prefix provides both robustness against multi-path and eliminates the need for power back-off at the transmitter. More details about the zero-padded prefix will be described in a later section.

An example of a block diagram for a transmitter architecture that implements the MB-OFDM system is shown in Figure 2. The structure of the MB-OFDM solution is very similar to that of a conventional wireless OFDM physical layer, except that the carrier frequency is changed based on the time-frequency code. In addition, other modifications have been made to reduce the complexity. For example, the constellation size has been limited to QPSK in order to reduce the internal precision of the digital logic, specifically the IFFT and FFT, and to reduce the required precision of the ADC and DAC. The larger spacing between the sub-carriers, as compared to narrow-band OFDM systems such as IEEE 802.11a, also contributes to the simplicity of the MB-OFDM implementation. The large spacing relaxes the phase noise requirements on the carrier synthesis circuitry and improves robustness to synchronization errors.

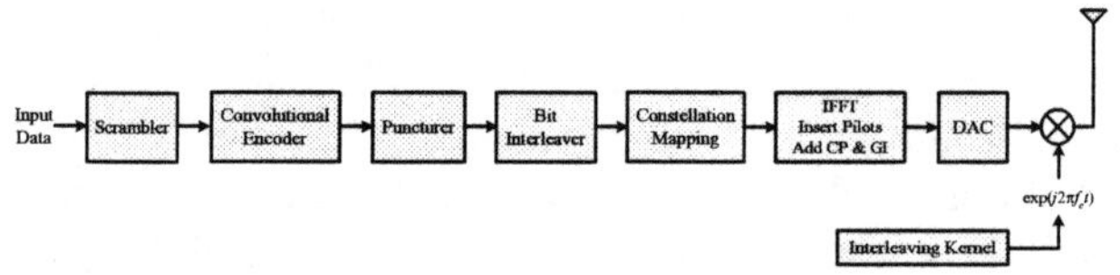

Figure 2 – Example architecture for a MB-OFDM system.

3.1. System Parameters

The system parameters for the 110 Mb/s, 200 Mb/s, and 480 Mb/s modes of MB-OFDM solution are given in Table 1. Note that this system is also capable of transmitting additional data rates of 55, 80, 160, and 320 Mb/s. This system employs an orthogonal frequency division multiplexing modulation schemes with a total of 128 sub-carriers, only 122 of which carry energy. Of the 122 sub-carriers that carry energy, 100 are devoted to data, 12 are assigned to pilot tones, and the remaining 10 are used as guard tones.

Table 1 – MB-OFDM System Parameters

Data Rate	110 Mb/s	200 Mb/s	480 Mb/s
Constellation	QPSK	QPSK	QPSK
FFT Size	128	128	128
Coding Rate	11/32	5/8	¾
F-D Spreading	No	No	No
T-D Spreading	Yes	Yes	No
Symbol Length	312.5 ns	312.5 ns	312.5 ns

Forward error correction coding in conjunction with either frequency-domain spreading or time-domain spreading is used to vary the data rate of the system. The different coding rates are generated by puncturing an industry standard R = 1/3, K = 7 convolutional code with generator polynomial $[133, 145, 175]_8$. The exact puncturing patterns to generate the various coding rates are specified in [3]. The frequency-domain spreading is obtained by forcing the input data into the IFFT to be conjugate symmetric. An advantage of using this frequency-domain spreading is that output will always be real; implying that only the real portion of the transmitter needs to be implemented. Time-domain spreading is obtained by transmitting the OFDM symbol followed a permutation of that OFDM symbol.

4. DESIGN CONSIDERATIONS

In this section, some of the design considerations, including the frequency synthesizer architecture and the length of the zero-padded prefix, for a MB-OFDM system are discussed.

4.1. Frequency Planning and Synthesizer Design

The frequency planning shown in Figure 1 was chosen for two specific reasons: to allow sufficient guard band on the lower side of band 1 and the upper side of band 3, which simplifies the design of a pre-select filter; and to simplify the design of the synthesizer and ensure that the system can switch between the center frequencies within a few nanoseconds.

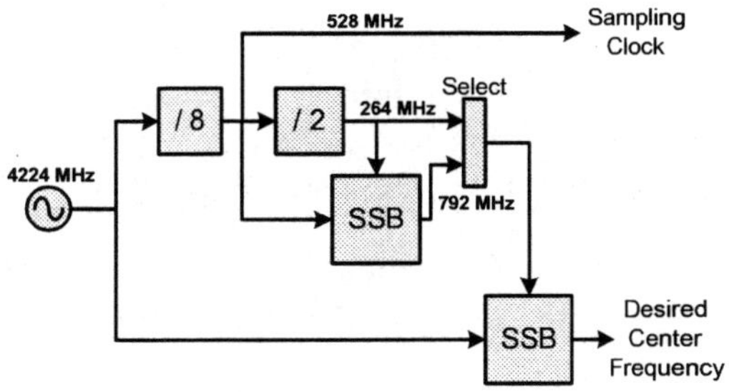

Figure 3 – Example of a MB-OFDM synthesizer architecture.

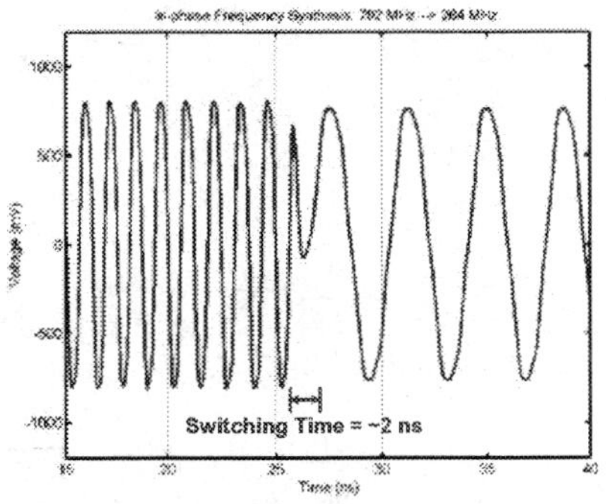

Figure 4 – Circuit-level simulations of the in-phase portion of the frequency-switching architecture

An example of a fast frequency-synthesizer circuit for the MB-OFDM solution is shown in Figure 3. This synthesizer exploits the relationship between the center frequencies for each of the bands and the oscillator frequency. The basic idea is that each of the center frequencies is generated using a single-sideband beat product of the oscillator frequency with another frequency derived from the oscillator. This other frequency is obtained by using a combination of frequency dividers and single-sideband mixers. For example, the center frequency for band 1 is generated by mixing 4224 MHz with 792 MHz to obtain a frequency 3432 MHz. The 792 MHz signal is generated by mixing 528 MHz (= 4224 / 8) and 264 MHz (= 4224 / 16). Similarly, the center frequencies for bands 2 and 3 can be obtained by mixing 4224 MHz with 264 MHz (= 4224 MHz / 16).

The advantage of this architecture is that all of the center frequencies are generated from a single PLL. Since these frequencies are available at all times, switching between the different sub-bands can be accomplished within a few nanoseconds. The exact time will be limited by the response time of the multiplexer. Figure 4 depicts the plot from a circuit-level simulation of the in-phase portion of frequency-switching architecture shown in Figure 3. From this plot, the nominal switching time is around 2 ns. The result is similar for the quadrature portion of the frequency-switching architecture.

4.2. Zero-padded Prefix

An OFDM system offers inherent robustness to multi-path dispersion with a low-complexity receiver. This property is a result of adding a cyclic prefix (CP). A receiver using a CP transforms the linear convolution with the channel impulse response into a circular convolution. Note that a circular convolution in the time domain is equivalent to multiplication in the Discrete Fourier Transform (DFT) domain. Hence, a single-tap frequency domain equalizer is sufficient to undo the effects of the multi-path channel on the transmitted signal.

The length of the CP determines the amount of multi-path energy captured. Multi-path energy not captured during the CP window results in inter-carrier-interference (ICI). Therefore, the CP length needs to be chosen to minimize the impact due to ICI and maximize the multi-path energy collected, while keeping the overhead due to the CP small.

The UWB channel models are highly dispersive; a 4-10 m, non-line-of-sight (NLOS) channel environment has an RMS delay spread of 14 ns [5]. The average captured energy for this channel environment and ICI are illustrated as a function of the CP length in [4]. To capture sufficient multi-path energy and minimize the impact of ICI for all channel environments, the CP length is chosen as 60.6 ns.

OFDM systems typically use a CP to provide robustness against multi-path. However, the same robustness can be obtained by using a zero-padded prefix (ZPP) instead of the CP [6]. The only modification required at the receiver is to collect additional samples corresponding to the length of the prefix and to use an overlap-and-add method to obtain the circular convolution property.

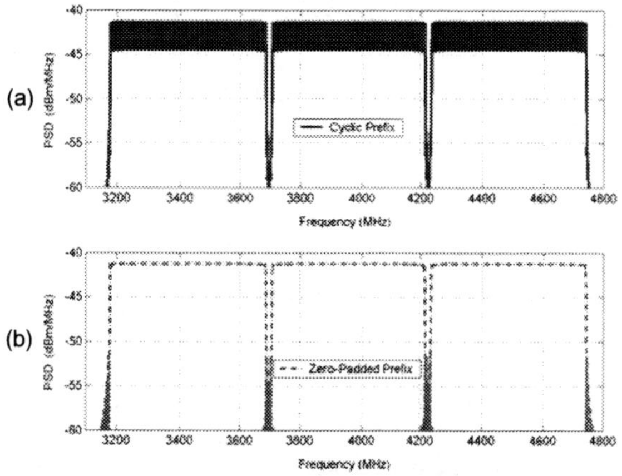

Figure 5 – Power spectral density plots for an MB-OFDM system using: (a) a CP, and (b) a ZPP.

The advantage of using a ZPP is that the power back off at the transmitter can be avoided. When a CP is used, structure is introduced into the transmitted signal, which leads to ripples in the power spectral density. Since the emissions in an UWB system is limited by the FCC, any ripples in the power spectral density (PSD) will result in transmit power back off. In fact, the power back off could be as large as 1.5 dB, which results in a lower range. With a ZPP, the structure in the transmitted signal is eliminated and therefore, there are no ripples in the PSD. Figure 5 illustrates the ripples in the power spectral density for an MB-OFDM system that uses a CP and a ZPP. So when a ZPP is used, no power back off is required at the transmitter; resulting in a system with the maximum achievable range.

5. SYSTEM PERFORMANCE

The performance of a MB-OFDM system is evaluated for various data rates and channel environments via simulations. A description of the channel models for three specific multi-path environments: line-of-sight (LOS) 0-4 m, NLOS 0-4 m, and NLOS 4-10 m, and the 100 channel realizations for each channel model are provided in [5]. Note that these channel realizations include a log-normal shadowing component with a standard deviation of 3 dB. The simulations incorporated losses due to front-end filtering, clipping at the DAC, ADC degradation (4-bits for 110/200 Mbps and 5-bits for 480 Mbps), packet acquisition, channel estimation, clock frequency mismatch (± 20 ppm at the TX and RX), carrier offset recovery, carrier tracking, etc.

Table 2 shows the range, as a function of data rate, for which the MB-OFDM system can achieve a link success probability of 90% for a PER of 8%. The link success probability is defined as the percentage of channel realizations for which the system can successfully acquire and demodulate a packet with a PER of less than 8%. These results demonstrate that the MB-OFDM system can attain a distance of approximately 11 meters in all channel environments for a data rate of 110 Mbps.

Table 2 – 90% link success probability distance for a MB-OFDM system.

Data Rate	110 Mb/s	200 Mb/s	480 Mb/s
AWGN	20.5 m	14.1 m	8.9 m
LOS: 0-4 m	11.4 m	6.9 m	2.9 m
NLOS: 0-4 m	10.7 m	6.3 m	2.6 m
NLOS: 4–10 m	11.5 m	6.8 m	—

6. SUMMARY

In this paper, a MB-OFDM system for wireless UWB communication was described. In addition, several design considerations were discussed. Performance results show that this system is able to support a data rate of 110 Mbps at distances of over 10 m.

7. ACKNOWLEDGEMENTS

The authors would like to thank Ranjit Gharpurey for helping to develop the MB-OFDM proposal and for generating the frequency synthesis circuit simulations.

8. REFERENCES

[1] M. Z. Win and R. A. Scholtz, "On the Robustness of Ultrawide Band Signals in Dense Multipath Environments," *IEEE Comm. Letters*, Vol. 2, No. 2, February 1998.

[2] V. S. Somayazulu et al., "Design Challenges for Very High Data Rate UWB Systems," *Proc. Asilomar Conf. On Systems, Signals and Comp.*, November 2002.

[3] A. Batra et al., "Multi-band OFDM Physical Layer Proposal for IEEE 802.15 Task Group 3a," *IEEE P802.15-03/268r1-TG3a*, September 2003.

[4] J. Balakrishnan et al., "A Multi-band OFDM System for UWB Communication," *Proc. IEEE Conf. on Ultra Wideband Systems and Technologies*, Reston, VA, November 2003.

[5] J. R. Foerster, et al., "Channel Modeling Sub-committee Report Final," *IEEE P802.15-02/490r1-SG3a*, November 2002.

[6] B. Muquet et al., "Cyclic Prefix or Zero Padding for Wireless Multicarrier Transmission?," *IEEE Transactions on Communications*, vol. 50, no. 12, December 2002.

Pulsed OFDM modulation for Ultra wideband Communications

Ebrahim Saberinia, Jun Tang, Ahmed H. Tewfik and Keshab K. Parhi
Electrical and Computer Engineering Department
University of Minnesota
Minneapolis, USA

***Abstract*— In this paper, we describe a new orthogonal frequency division multiplexing (OFDM) scheme, named Pulsed-OFDM, for ultra wideband (UWB) communications. Pulsed-OFDM modulation uses pulsed sinusoids instead of continuous sinusoids to send information in parallel over different sub-carriers. Pulsating OFDM symbols spreads the spectrum of the modulated signals in the frequency domain leading to a spreading gain equal to the inverse of the duty cycle of the pulsed sub-carriers. The spreading gain provided by this system leads to an enhanced performance in multipath fading environments. We also show that at the receiver part of the multipath diversity can be exploited by pulsed-OFDM system leading to a low complexity and low power consumption transceiver structure. Easy implementation and frequency spreading of pulsed-OFDM modulation make it a proper modulation for UWB communications where the ratio of bandwidth to the data rate is large but power spectral density is limited. We design a low complexity and low power consumption pulsed-OFDM system for the IEEE 802.15.3a wireless personal area networks and compare it with the normal OFDM system has been proposed for this standard. We also provide realistic simulation results for the measured indoor propagation channels provided by the IEEE 802.15.3a standard to demonstrate the advantages of the pulsed OFDM modulation for UWB communications.**

1. INTRODUCTION

UWB communication systems use signal with a fractional bandwidths that is lager than 20% of the center frequency or more than 500 MHz [1]. UWB communication systems offer several potential advantages. Prior research indicates that the wide bandwidth of such systems makes them more robust to multipath interference [2]. Recognizing the potential benefits of UWB systems, the FCC has opened up the 3.1-10.6 GHz to indoor UWB transmission subject to power limitations consistent with part 15 of commission's rules [1].

Several UWB systems been proposed in the literature. Most of this system uses single-carrier modulations of very short mono-pulses. These systems can be divided in two different categories, namely time modulated ultra wideband (TM-UWB) and direct sequence phase coded UWB (DSC-UWB). TM-UWB sends low duty cycle pulses with a large bandwidth of several GHz [2-4]. Pulse position modulation (PPM) or pulse amplitude modulation (PAM) is used to modulate these pulses with information symbols. DSC-UWB on the other hand uses high duty cycle pulse trains that are modulated by pseudo random sequences to spread the signal [4-5].

Using multi-carrier modulations like orthogonal frequency division multiplexing (OFDM) for UWB communications is also studied in last two years [6][7][8]. Since OFDM is a spectrum efficient modulation technique that designed for band-limited systems while UWB communication is a power limited system with very large bandwidth, OFDM should be combined with a frequency spreading technique to be suitable for UWB communications. Forward error correction codes with interleaving are used with OFDM to exploit the frequency diversity and combat with frequency selective channels which could lead to additional complexity and latency.

In a UWB system, because of the large available bandwidth, we can acquire the diversity gain with simple tools. In this paper we describe another slow frequency hopping MC-UWB system with a new OFDM scheme that uses orthogonal pulsed sinusoids instead of continuous sinusoids that are used in normal OFDM. Pulsating OFDM symbol spreads its spectrum in the frequency leading to a spreading gain equal to the inverse of the duty cycle of pulse train. The system exploits the fact that pulsed-OFDM produces additional multipath diversity that can be easily captured using one or more traditional OFDM receivers depending on desired performance goals, thereby reducing or eliminating frequency selective fading. This modulation can be used with slow frequency hopping to achieve a low complexity high performance MC-UWB system. In order to show the advantages of pulsed-OFDM modulation for UWB communications in a realistic scenario, we design a pulsed-OFDM system for the IEEE 802.15.3a wireless personal area networks which seems to be the first commercial system works with UWB. We describe how to design a pulsed OFDM system with low complexity and compare its performance and power consumption with the Multi-band OFDM system which is the leading proposal for the IEEE 802.15.3a standard. The simulation results show that the pulsed OFDM system outperforms the normal OFDM system with considerable lower power consumption and complexity which is critical to the IEEE 802.15.3a devices.

2. NON-PULSED OFDM[]

In normal OFDM modulation a block of N data symbols is transmitted over N orthogonal sub-carriers in parallel. An OFDM block has the following form:

$$x(t)=\sum_{n=0}^{N-1} b_n e^{j2\pi n f_0 t} \qquad 0 \le t \le NT \tag{1}$$

where b_n is the data symbol that is transmitted over nth sub-carrier. If $f_0 = 1/NT$, then the N sub-carriers are orthogonal in the sense that:

$$\frac{1}{NT}\int_0^{NT}\left(e^{j2\pi n f_0 t}\right)\left(e^{j2\pi m f_0 t}\right)^* dt = \delta(n-m), \tag{2}$$

where $\delta(n)=1$ if $n=0$ and $\delta(n)=0$ otherwise. The receiver uses this orthogonally to demodulate each sub-carrier data independent of others.

This research was supported in part by Army Research Office under grantn umber DA/DAAG19-01-1-0705 and NSF funding number CCR-0313224.

The OFD signa can be easi y generated and demodu ated digita y using fast Fourier transforms FFT and FFT . n a typica OFD transcei er b oc a b oc of N data symbo s $\mathbf{b} = b_0 \ldots b_{N-1}{}^T$ modu ate N digita orthogona sub-carriers using an FFT operation

$$\mathbf{x} = \mathbf{F}^{-1}\mathbf{b} \qquad 3$$

The output of the ana og to digita con erter at the recei er is e ua to

$$y\ n = \sum_{m=0}^{L} h_t(mT)x\ n-m\ + w\ nT \qquad 4$$

where $h_t\ t$ is oera channe impu se response with duration T_{ch} and L is determined by the ength of channe impu se response as $L = \left\lceil \frac{T_{ch}}{T} \right\rceil$ and $w\ t$ is the fi tered additi e white Gaussian noise A GN . uation 4 can be e pressed in the ector-matri form as

$$\mathbf{y} = \mathbf{H}\mathbf{F}^{-1}\mathbf{b} + \mathbf{n} \qquad 5$$

where matri $\mathbf{H}$ is a Toep it channe matri .

Non-pu sed OFD systems suffer from the fre uency se ecti ity of the fading channe s. f the channe fre uency response has a nu or is ery sma on a sub-carrier center fre uency then the data that was sent o er that sub-carrier cannot be reco ered. App ying forward error correction codes and inter ea ing the coded data before OFD modu ation e p oits fre uency di ersity to combat with fre uency se ecti e channe s. For dense mu tipath fading channe s hea y coding is re uired in order to achie e re iab e communication o er a sub-carriers which add to the comp e ity and atency of the system and decrease the o era bit rate. n the ne t section we describe how the pu sed OFD system pro ides additiona di ersity to combat with fre uency se ecti e channe s without adding to the comp e ity of the system.

3. PULSED-OFDM MODULATION

The pu sed-OFD system uses orthogona pu sed sinusoids instead of continues sinusoids as para e sub-carriers to modu ate data. The pu sed-OFD can be generated by upsamp ing the output of the FFT before sending it into DAC in the transmitter The upsamp ing is done by inserting *K-1* eroes between samp es of the signa . The resu ting pu sed OFD signa is then a pu se train with duty cyc e e ua to *1/K*. e wi refer to K as the processing gain of the pu sed-OFD system. Let $\mathbf{x} = \mathbf{F}^{-1}\mathbf{b}$ be the output of the FFT b oc . The output signa is upsamp ed to $\mathbf{x}_u$ before it is fed into a DAC and passes pu se shaping fi ter channe and recei er fi ter with o era impu se response $h_t\ t$. After samp ing the output of the recei er fi ter with rate $1/T$ the output is e ua to

$$y_u\ n = \sum_{m=0}^{L} h_t(mT)x_u\ n-m\ + n_u\ n$$

where L is determined by the ength of the channe impu se response as $L = \left\lceil \frac{T_{ch}}{T} \right\rceil$. f we group the output signa into K subgroups as fo ows

$$y_i\ n = \sum_{m=0}^{L'} h_i(m)x\ n-m\ + n_i\ n \qquad 7$$

where $h_i(m) = h_t(\ i + mK\ T)$. uation 7 shows that when we send a pu sed OFD o er a channe with ength L the recei ed signa is e ui a ent to K norma OFD recei ed signa s on K different channe s with ength $L' = \left\lceil \frac{L}{K} \right\rceil$. n matri - ector form the e uation 7 can be written as fo ow

$$\mathbf{y}_i = \mathbf{H}_i\mathbf{F}^{-1}\mathbf{b} + \mathbf{n}_i \qquad i = 0 \ldots K-1$$

Ta ing the FFT of the ectors y_i s diagona i es each irtua channe and pro ides recei ed ector of the fo owing form

$$\mathbf{z}_i = \mathbf{D}_i\mathbf{b} + \tilde{\mathbf{n}}_i \qquad i = 0 \ldots K-1 \qquad 9$$

$$\mathbf{D}_i \square \mathbf{F}\mathbf{H}_i\mathbf{F}^{-1} = diag\ H_i\ 0\ \ H_i\ f_0\ \ldots\ H_i\ \ N-1\ f_0$$

where $\boldsymbol{H}_i\ f$ is the Fourier transform of the ith irtua digita channe impu se response. The output of the FFT then is e ua to

$$\mathbf{z}_i = \mathbf{D}_i\mathbf{b} + \tilde{\mathbf{n}}_i \qquad i = 0 \ldots K-1. \qquad 10$$

According to e uation 10 the u sed-OFD modu ation pro ides K di ersity branches for each data channe . This di ersity is e ui a ent to transmission o er K irtua different channe . Another way of presenting the additiona di ersity achie ed by the pu sed OFD system is in fre uency domain. As is we - nown the upsamp ing process spreads the fre uency of the signa o er a band K times arger than the origina by repeating the origina signa spectrum in fre uency domain. Since the signa s in the fre uency domain are repeated K times and each fre uency e periences different f at fading we ha e di ersity branches for each sub-carrier.

The di ersity branches can be combined with each other to combat the fre uency se ecti ity of the fading channe . f the distance between the fre uencies of sub-carriers carrying the same data is arger than the fre uency spread of the fading channe then each di ersity branch e periences independent f at fading and the di ersity gain is ma imi ed. The optimum way of combining di ersity branches is ma ima ratio combining 9 . sing this method the pu sed-OFD signa is demodu ated as

$$\hat{b}\ n = \frac{\sum_{i=0}^{K-1} \hat{d}_i\ n\ z_i\ n}{\sum_{i=0}^{K-1} \left|\hat{d}_i\ n\right|^2}. \qquad 11$$

4. PULSED OFDM VERSUS NON-PULSED OFDM IN UWB COMMUNICATIONS

n order to compare pu sed and non-pu sed OFD modu ation in a rea istic situation we use the 02.15.3a standard physica ayer as a framewor . The 02.15.3a high bit rate wire ess persona area networ standard is targeting data transmission at rates of 110 b/s o er 10 meters 220 b/s o er 4 meters and 40 b/s o er 1 meter. The eading proposa for this standard is mu ti-band OFD system that supported by a group of companies and uni ersities. The mu ti-band OFD system is a system that uses norma OFD modu ation. n that proposa the who e a ai ab e u tra wideband spectrum between 3.1-10. G is di ided into se era sub-bands with sma er bandwidth of 52

. n each sub-band a norma OFD modu ated signa with $N=128$ sub-carriers and Q S modu ation is used. The main difference between the mu ti-band OFD system and other narrowband OFD systems is the way that different sub-bands are used in the system. The transmission is not done continua y on a sub-bands. Rather it is time mu tip e ed between different bands band hopping in order to use a sing e hardware for communications o er different sub-bands.

The 02.15.3a transcei ers wi be used in portab e de ices such as camcorders CD p ayers and aptops as we as on fi ed de ices such as TVs and des tops. The comp e ity and power consumption of the transcei ers is therefore a ery important issue for this standard to be successfu .

4.1. Pulsed-OFDM system for IEEE 802.15.3a WPAN

To meet these re uirements of the 02.15.3a standard the mu ti-band OFD proposa transmits a norma OFD symbo in each sub-band e ery 242.5 nsec. The number of sub-carriers are N =12 and the bandwidth of each sub-band is e ua to 52 . Con o utiona error correction codes and fre uency spreading are used to dea with fre uency se ecti e fading channe s. Different bit rates are achie ed by using different channe coding and fre uency spreading rates. For 110 b/s data transmission rate a code with rate 11/32 is used.

n order to reduce comp e ity in the design of mu ti-band pu sed-OFD system for 02.15.3a standard we use N=32 sub-carriers. A con o utiona error correcting code of fi ed rate e ua to 2/3 is used for a bit rates.

oth the pu sed-OFD and non-pu sed-OFD systems ha e the same RF front end. Thus in the ana og domain e cept for the D/A and A/D parts the two systems ha e simi ar comp e ity and power consumption. The transmitter and recei er structure of the pu sed-OFD in the base band digita domain is i ustrated in Figure-2. n the fo owing subsections we describe how to FFT and FFT b oc s and coder and decoder b oc s.

4.1.1. FFT and IFFT modules

y reducing the number of sub-carriers on y a 32-point FFT is re uired at the transmitter side which is much simp er than a 12 -point FFT in the mu ti-band OFD system. t is nown that the comp e ity of an FFT or FFT processor is proportiona to N og$N$ where N is the si e of the data. At the recei er side the pu sed-OFD re uires four 32-point FFT processors whi e the non-pu sed-OFD system on y re uires a sing e 12 -point FFT. owe er e en though the computationa comp e ity of four 32-point FFT processors is sti sma er than that of a 12 -point FFT four para e FFT structures wi occupy significant y arger area than a sing e 12 -point FFT processor.

Fortunate y by digging into the hardware structure of the FFT processor we ha e found that it is possib e to reduce four para e FFT structures down to one without introducing significant e tra comp e ities. n Figure 3 a wide y used hardware imp ementation structure of an FFT processor is i ustrated. This structure is ca ed radi -4 mu ti-path de ay commutator R4 DC structure. n the figure the b oc mar ed with C4 represents a 4-input-4-output commutator the b oc with F4 is a 4-point butterf y structure and the symbo ⊗ represents a mu tip ier. The b oc s with numbers represent the de ay e ements in the path. One important wea ness of the hardware structure is the ow hardware uti i ation. A of the hardware e ements in the FFT processor such as commuters butterf y structures and mu tip ier are on y be uti i ed 25 which means most of the time the hardware e ements are not used.

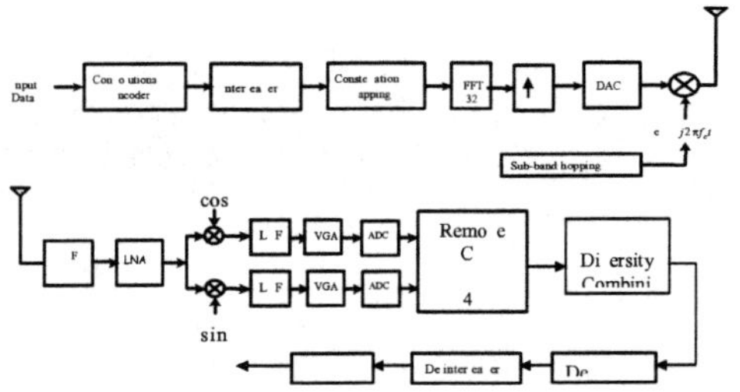

Figure 2. Transmitter and recei er structure for mu ti-band u sed OFD

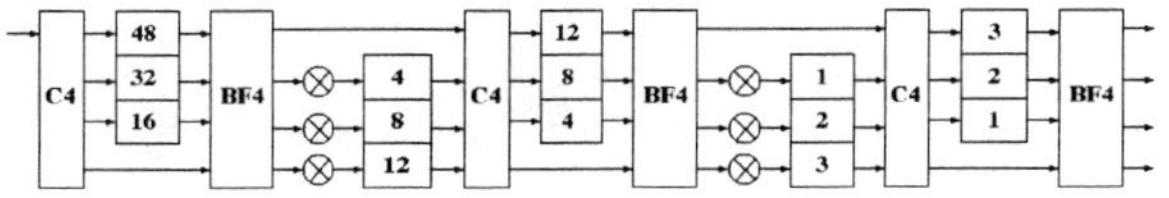

Figure 3. A 4-point R4 C FFT imp ementation structure 10

There e ist other more efficient imp ementations of the FFT processor. owe er this structure is the most wide y used one due to its regu ar and simp e contro circuits. n pu sed-OFD the ow hardware efficiency of the structure ma es it possib e to combine mu tip e para e FFT processors into the same structure by uti i e time mu tip e . As stated in the 10 four para e inputs such as in the recei er of the pu sed-OFD can be combined using mu tip e ers and buffers into the same hardware structure. Thus the hardware efficiency can be increased up to 100 . A of the e ements inside the structure are fu y uti i ed. Compared with non-pu sed-OFD which uses a sing e 12 -point FFT structure the 4-para e 32-point FFT processor not on y has ower computationa and hardware comp e ity but a so has higher hardware efficiency.

4.1.2. Coder-Decoder

The mu ti-band pu sed-OFD system uses con o utiona channe coding of rate 2/3 whi e the mu ti-band OFD scheme uses codes with rates 11/32 and 11/1 both punctured from rate 1/3 codes . en though the coding rate of pu sed-OFD system is greater than non-pu sed system the performance of pu sed-OFD e ceeds non-pu sed-OFD as is shown in the ne t section. That is because of the mu tipath di ersity achie ed by pu sed-OFD . owe er direct imp ementation of the Viterbi decoder for the rate-2/3 systematic con o utiona code has much higher comp e ity than that of a rate-1/3 con o ution codes. This is due to the fact that each point in the tre is representation of a rate-1/3 con o utiona code has on y two inputs whi e a point in the tre is representation of a rate-2/3 con o utiona code has four inputs. The increase in the number of inputs wi significant y increase the comp e ity of the add-compare-se ect ACS unit. The on y choice is then to use punctured codes. Figure-4 i ustrates the performance difference between a systematic rate-2/3 codes and a punctured rate-2/3 code from a rate-1/2 code in A GN channe . t can be seen that the performance difference between the optima rate-2/3 code and the punctured code is ess than 0.5d at a R = 10^{-5}. Therefore we choose the use of punctured codes in the mu ti-band pu sed-OFD system for 02.15.3a.

The introduction of punctured code wi increase the c oc rate and the power consumption of the Viterbi decoder due to the insertion of nu symbo s. owe er as we wi see in section the decoder structure for the punctured code in pu sed-OFD sti has considerab y ower power consumption than that of the non-pu sed-OFD because of the c oc rate that they operate.

4.2. Performance

n order to compare the performance of the pu sed and non-pu sed systems we ran a comp ete simu ation of both systems o er the channe mode s described in the 02.15.3a channe mode ing report. n the report C 4 are used to mode channe s at 10 meters. Figure-5 shows the resu ts of the simu ation of both the non-pu sed and pu sed-OFD systems when operating at 110 b/s o er C 4 channe . n this figure the recei ed bit error rate R is p otted ersus distance for both systems. The R must be ess than 10^{-5} as re uired by the standard. Figure 5 shows that the pu sed-OFD system can operate at a range of 9.7 meters on C -4 type channe s whi e the range of norma OFD is on y . meters. ence we conc ude that pu sed-OFD outperforms non-pu sed-OFD in dense mu tipath channe under identica bit rate and bandwidth conditions.

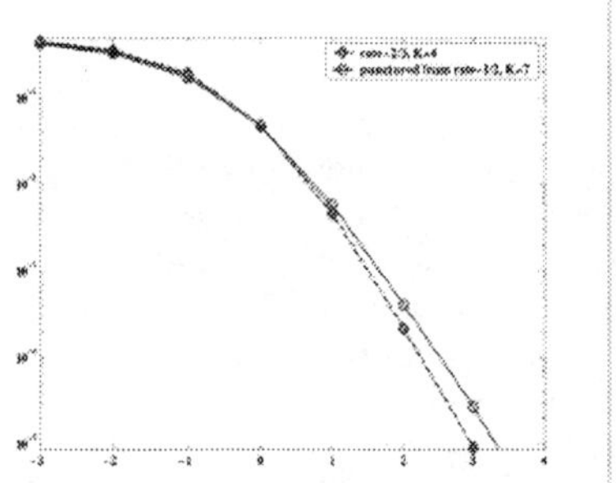

Figure 4. erformance comparison of rate-2/3 code and punctured rate-1/2 code with the same number of states in additi e white Gaussian noise AGN channe.

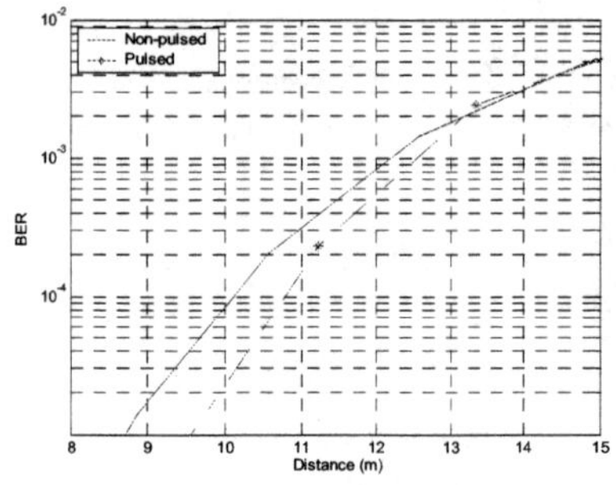

Figure 5. it error rate ersus distance for pu sed and nun-pu sed-OFD systems in channe C -4

4.3. Power Consumption Comparison

esides the soupier performance and ower comp e ity the pu sed-OFD system has a big ad antage o er non-pu sed OFD in terms of power consumption. The power consumption of a VLS chip is determined by its c oc rate the supp y o tage and the capacitance of the circuit and can be rough y computed with

$$P = C_{tota} V_0^2 f \qquad 12$$

where C_{tota} denotes the tota capacitance of the circuit V_0 is the supp y o tage and f is the c oc fre uency. According to the ana ysis of the pre ious sections we conc ude that the pu sed-OFD re uires ower c oc fre uency and eads to ower power consumption. This fact is i ustrated in Tab e-1 where the c oc rates of different parts of transmitter and recei er are isted for both pu sed and non-pu sed systems. Actua y due to the simp icity of the circuits in pu sed-OFD such as FFT processors and Viterbi decoders the C_{tota} in pu sed-OFD is sma er than the one in non-pu sed-OFD which wi ead to further reduction of the power consumption.

Tab e 1 C oc rate in different parts of transcei er for non-pu sed and pu sed OFD

	nput data	After Coding with uncturing	Output of FFT	nput to FFT	nput to Decoder
Non-pu sed	110	330	320 sps	320 sps	320 bps
u sed	110	220	0 sps	320 sps	1 0 bps

5. CONCLUSION

u sed-OFD modu ation is a simp e fre uency spreading techni ue for mu ti-carrier systems that pro ide additiona di ersity in the mu tipath fading channe . n the rea istic situations i e the 02.15.3a wire ess persona area networ s using pu sed OFD modu ation eads to a superior performance compared to a the norma OFD system whi e the comp e ity and power consumption of the transmitter and recei er is much ower.

R F R NC S

1 pected performance and attribute criteria appro ed for 02.15.3a A t Y Seection Criteria http //grouper.ieee.org/groups/ 02/15/pub/2003/ an03/03031 r 02-15_TG3a- Y-Se ection-Criteria.doc

2 Summary of the app ication presentations from the Study Group 02.15.3a ca for appications http //grouper.ieee.org/groups/ 02/15/pub/2003/ an03/03030 r0 02-15_TG3a-Technica -Re uirements.doc

3 Anu atra et a . " u ti-band OFD hysica ayer proposa " merged proposa for 02.15.3a http //ieee 02.org/15/pub/Down oad.htm u y 2003.

4 FCC Notice of roposed Ru e aing "Reision of art 15 of the Commission s Ru es Regarding tra- ideband Transmission Systems " T-Doc et 9 -153.

5 A. . Tewfi and . Saberinia " igh it Rate tra-ideband OFD ", *IEEE GLOBECOM 2002.*

Zhendao ang Gianna is G. . " ire ess mu ticarrier communications," *IEEE Signal Processing Magazine* o . 17 pp. 29-4 ay 2000.

7 A. V. Oppenheim and R. Schafer "Digita signa processing" rentice a 1st edition an. 1975.

eff Foerster "Channe mode ing sub-committee report fina " http//ieee02.org/15 /pub/Down oad.htm No . 2002.

9 G. L. Stuber *Principales of Mobaile Communications* uwer Academic ub ishers 2nd edition February 2001.

10 L. R. Rabiner and . Go d *Theory and Application of Digital Signal Processing* rentice- a nc. 1975.

INTERFERENCE RESILIENT TRANSMISSION SCHEME FOR MULTIBAND OFDM SYSTEM IN UWB CHANNELS

** Seung Young Park, **Gadi Shor, and *Yong Suk Kim*
* Samsung Advanced Institute of Technology, Suwon, Korea
** Wisair, Tel-Aviv, Israel
Email: young@ieee.org

ABSTRACT

In this paper, we propose an interference resilient transmission scheme for multiband orthogonal frequency division multiplexing (MB-OFDM) system in ultra wideband (UWB) channels. In our proposed scheme, the interleaver is combined with the OFDM symbol repeater to reduce the effect of cochannel interference (CCI) from single simultaneously operating piconet (SOP). When a collision occurs, the interference is mitigated by increasing the effective coding rate through the use of interleaver and by reducing a symbol collision probability through the use of symbol repeater. Our simulation results demonstrate that the proposed scheme can effectively reduce the CCI from a single SOP.

1. INTRODUCTION

The Federal Communications Commission (FCC) has allocated 7,500 MHz of spectrum for unlicensed use of ultra-wideband (UWB) devices in the frequency band of 3.1 to 10.6 GHz, provided that the occupied bandwidth is at least 500MHz and the radiated power is less than –41.3 dBm/MHz within the signal band.

UWB is an emerging technology as a solution for the IEEE 802.15.3a (TG3a) standard [1], of which the purpose is to provide specifications for low-complexity, low-cost, low-power consumption, and high data-rate wireless connectivity among devices within the personal operating space. The data rate must be high enough (greater than 110 Mbps) to satisfy a set of consumer multimedia industry needs for wireless personal-area networks (WPAN) communications.

Recently, multiband orthogonal frequency division multiplexing (MB-OFDM) has been proposed as a candidate technique for IEEE 802.15.3a standard [2], where the system uses a total of 122 sub-carriers which are modulated using quadrature phase shift keying (QPSK) and forward error correction coding (e.g., convolutional coding) is used. Each OFDM symbol is transmitted through different frequency slots according to time-frequency interleaving (TFI) patterns.

MB-OFDM devices (DEVs) can communicate each other and establish a piconet under the control of piconet coordinator (PNC). Since the DEVs in a piconet share the same TFI pattern, PNC can allocate a specific time slot for each connection between DEVs, or employ carrier sense multiple access with collision avoidance (CSMA/CA) to avoid mutual interference. However, when geographically overlapped multiple piconets are considered, there is no coordination of transmissions among different piconets. That is, the MB-OFDM must possess robustness to interference from other simultaneously operating piconets (SOPs), which is achievable by employing a combination of coding and frequency diversity.

To further provide robustness to interference, we propose an interference resilient transmission scheme, which combines the interleaver and the OFDM symbol repeater. The main idea of our proposed scheme is that when a collision occurs, the interference is mitigated by increasing the effective coding rate through the use of interleaver and by reducing a symbol collision probability through the use of symbol repeater, simultaneously. Our simulation results demonstrate that the proposed scheme can effectively mitigate the interference from other SOPs.

The remainder of the paper is organized as follows. A basic configuration of the MB-OFDM system is discussed in Section II. Section III describes the proposed transmission scheme including the collisions between two geographically overlapped piconets. In Section IV, the proposed scheme is evaluated with computer simulations. The concluding remarks are finally given in Section V.

2. SYSTEM DESCRIPTION

In this section, we briefly describe the MB-OFDM system model.

A. System Model

The transmitter of the MB-OFDM system considered in this paper is shown in Figure 1. A finite sequence of binary information bits is convolutionally encoded with a coding rate of 1/3 in 110Mbps mode as in [2]. Each coded bit is tone- and symbol-interleaved as described in [2], mapped into one of qudrature phase shift keying (QPSK) constellation points. After pilot tones are inserted in proper positions for the phase tracking [2], the QPSK symbols are represented by $\{X_{l,k}\}$, where l and k represent the OFDM symbol and subcarrier indicies, respectively. The common way to implement the inverse Fourier transform is by an inverse Fast Fourier Transform (IFFT) algorithm. When a 128-point IFFT is used, the coefficients 1 to 61 are mapped to the same numbered IFFT inputs. And then, the same coefficients are complex-conjugated and mapped into IFFT inputs 67 to 127. The rest of the inputs, 27 to 37 and the 0 (DC) input, are set to zero [2]. By using this technique, the IFFT output signal consists of real number only, so the transmitter needs only I-channel in the RF part. The lth OFDM symbol consists of 50 QPSK symbols and 22 pilot tones, which is given by

$$x_{l,n} = \sum_{\substack{k=-61 \\ k\neq 0}}^{61} X_{l,k} e^{j\frac{2\pi kn}{128}}, \tag{1}$$

where n denotes a discrete-time index and $X_{l,k} = X^*_{l,-k}$ for $1 \le k \le 61$. After performing the IFFT, the output is cyclically extended and a guard interval is added to generate an output with the desired length.

A transmission of the MB-OFDM system is shown in Figure 2, where each OFDM symbol is transmitted at different frequency slot based on the specific TFI pattern in each piconet.

B. Collision Characteristics

In this paper, we assume that there is only one SOP. Figure 3 (a) shows the collision example where the reference piconet uses the TFI pattern {f1, f2, f3, f1, f2, f3, repeats}, whereas the interfering piconet uses the TFI pattern {f1, f3, f2, f1, f3, f2, repeats}. We note that every third received OFDM symbol is completely received without interference. In other words, only 200 coded bits out of 600 coded bits can be received without interferecnce. Even though the channel coding with lowest coding rate of 1/3 is employed [2], it is difficult to decode this signal correctly. Therefore, we need to devise a new interleaving/transmission scheme, which is robust against interference.

3. PROPOSED SCHEME

In this section, we describe the proposed transmission scheme. To mitigate the effect of interference, the simplest way is to increase a redundancy in the transmitted siganl at the cost of the loss of spectrum efficiency. In this paper, we provide a redundancy in two ways. One is the same symbol repetition without any data loss compared to that of the original one, the other is the interleaver modification to improve the decoding performance through the effective coding rate reduction.

A. Symbol Repetition

In MB-OFDM system, RF part of the transmitter needs only I-channel, since an OFDM signal has only real components as discussed in Section II-A. As a consequence, we can achieve frequency diviersity by combining two subcarriers containing the same data in the receiver. When we use I- and Q-channel in the RF part at the transmitter, however, we can tansmit different symbols on all the subcarriers. It means that we can transmit the same OFDM signals twice without any data rate loss. Furthermore, it also achieves frequency diversity by combining two subcarriers on the same positions in the different frequency slots. In sequel, we can decrease the collision probability so as to mitigate the effect of interference. Figure 3 (b) shows the collision example for the repeated transmission. In this case, we note that 4 OFDM symbols out of 6 OFDM symbols are received without interference, which means that these 4 OFDM symbols contain 600 coded bits, and hence it may be decoded correctly without any informaiton loss due to collision.

In actual environment, however, the interference signal and the desired signal are not time-aligned due to no coordination of transmissions among different piconets. It produces more collisions than that shown in Figure 3 (b), so that we may not decode this signal correctly. This motivates us to find an additional method to mitigate the effect of the collisions. In the following section, we propose more effective interleaving scheme to mitigate this effect.

B. Transmitter Description of the Proposed Scheme

Figure 4 shows a transmitter structure of the proposed scheme. For example, 200 information bits are convolutionally encoded with a coding rate of 1/3. Each 200 coded-bit output of each generator polynomial of the encoder is tone-interleaved, mapped into one of QPSK constellation points and OFDM-modulated by 128-points IFFT. A transmission of the proposed scheme is described in Figure 5, where each OFDM symbol is transmitted twice at different frequency slot based on the specific TFI pattern.

C. Receiver Processing based on Effective Coding Rate

For example, when collision occurs at 1st, 2nd and 5th OFDM symbol, 200 coded bits from second polynomial ouput can be received from coherently combining 3rd and 4th OFDM symbols and 200 coded bits from third polynomial ouput can be received from 6th OFDM symbols. Therfore, equivalent coding rate is 1/2 as shown in Figure 6. In this case, 4 out of 6 OFDM symbols are collided, i.e., 400 coded bits from two polynomails is available, then, we can decode the desired signal correctly.

In the proposed scheme, therefore, the effect of interference can be mitigated by increasing the effective coding rate.

4. SIMULATION RESULTS

In this section, we provide simulation results demonstrating performance enhancement of the proposed system in single SOP.

The performance simulations incorporate losses due to front-end 5th order butter-worth filter, clipping at the DAC, ADC degradation (5-bits), multi-path, packet acquisition, channel estimation, carrier phase tracking, etc. As specified in the test, the shadowing component was removed from both the reference and interfering links by normalizing each channel realization to unity multi-path energy. In this paper, only channel model 3 of [3] is considered and its realizations used for the reference link and the interfering link are the same as in [2]. The arrival times of these two links are assumed to be not time-aligned. And this difference is assumed to be a uniformly-distributed random variable.

A 1/3-rate convolutional code with generator polynomials $[117\ 155\ 127]_8$ is used. The test link is established such that the reference link is set at a distance of $d_{ref} = 0.707$ of the 90% link success probability distance [2]. The distance separation at which an interfering piconets can be tolerated is obtained by averaging the performance over all combinations of the reference link and interferer link channel realizations for each channel environment.

Although there are various methods to detect the collided symbols in the receiver, in this paper, we use collision detection algorithm based on the received signal power. By comparing the received signal powers on the same frequency slot and the averaged received signal power, we can determine which symbol is collided.

To get an insight into average behavior, all simulations herein were performed for 2000 1024 byte packets at each channel realization.

Figure 7 (a) and (b) shows average packet error rate (PER) for 3- and 7-band system of 110Mbps mode (1/3-rate convolutinal code is used) in two SOPs [2]. As expected, it is shown that the PER performance degrades as a distance ratio of the interference link and the reference link, d_{int}/d_{ref}, decreases. For 3-band system, SOP performance can be dramatically improved compared to that of the scheme described in [2]. For 7-band system, on the other hand, SOP performance of the scheme in [2] shows better performance compared to that of 3-band system, since the collision probability is much lower than that of 3-band. Yet, SOP performance of the proposed scheme can be also improved.

5. CONCLUSIONS

In this paper, we propose interference resilient transmission scheme for MB-OFDM system combining the interleaver and the symbol repeater. When collision occurs, SOP performance can be improved by increasing the effective coding rate. Our simulation results indicate that the proposed scheme can effectively mitigate the interference from other SOPs.

ACKNOWLEDGMENT

The authors would like to thank Dr. A. Batra, Dr. J. Balakrishnan from Texas Instruments and Mr. H. J. Kim from Samsung Advanced Institute of Technology for their insightful and constructive comments.

REFERENCES

[1] IEEE 802.15WPAN High Rate Alternative PHY Task Group 3a (TG3a). Available at http://www.ieee802.org/15/pub/TG3a.html

[2] A. Batra et al, "Multi-band OFDM physical layer proposal for IEEE 802.15 task group 3a," available at http://grouper.ieee.org/groups/802/15/pub/2003/Jul03/03268r0P802-15_TG3a-Multi-band-CFP- document.doc.

[3] J. Foerster et al, "Channel modeling sub-committee report," available at http://grouper.ieee.org/groups/802/15/pub/2002/Nov02/

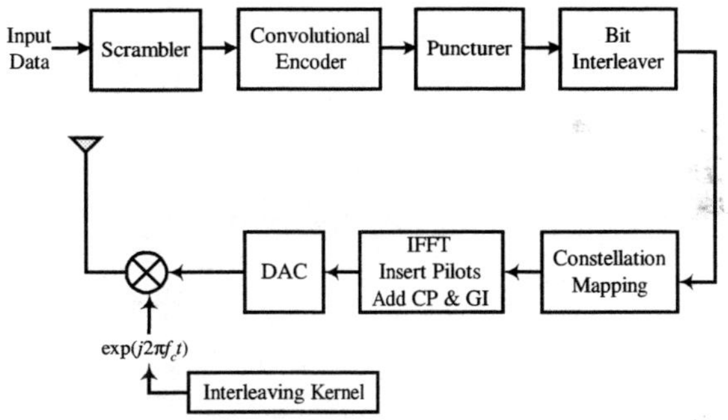

Fig. 1 Transmitter structure of the MB-OFDM system.

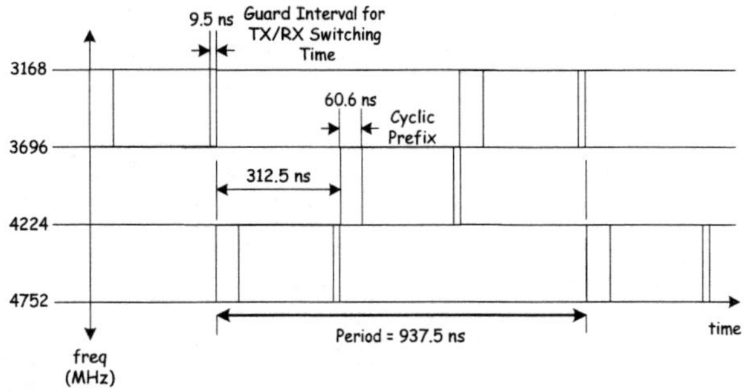

Fig. 2 Transmission of MB-OFDM system

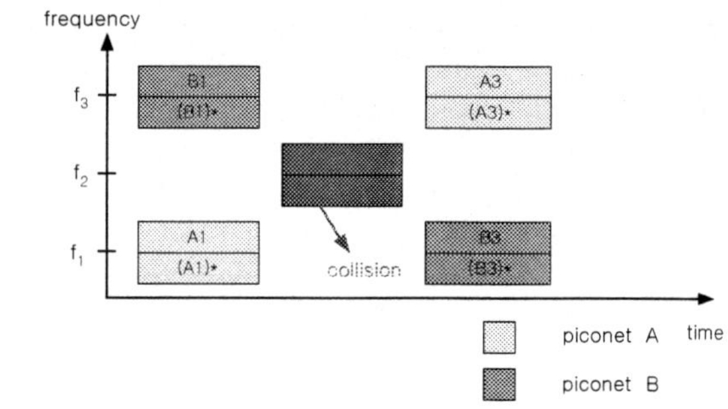

(a) Scheme in [2]

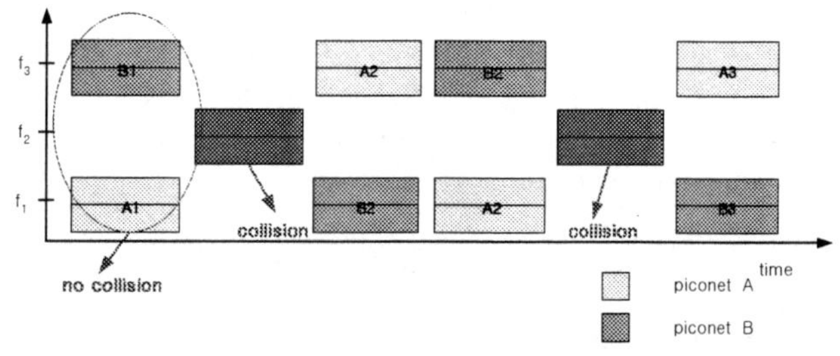

(b) Proposed scheme

Fig. 3 Collision example

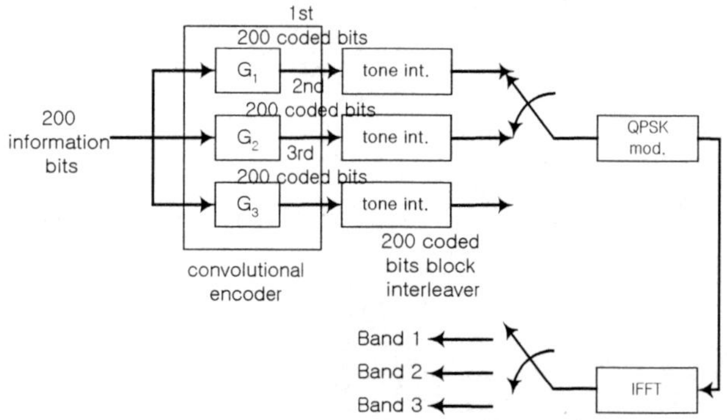

Fig. 4 Transmitter structure for the proposed scheme

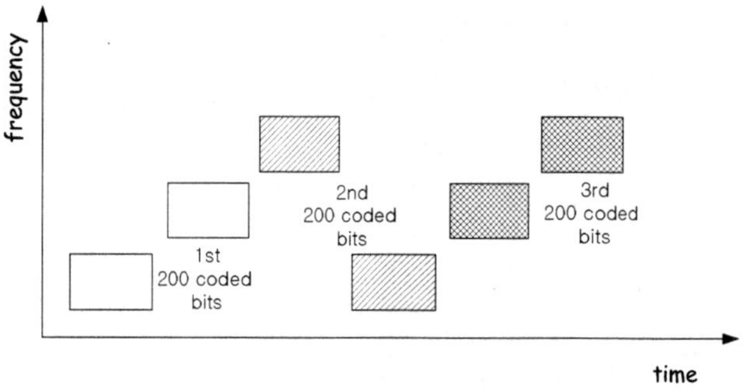

Fig. 5 Transmission example

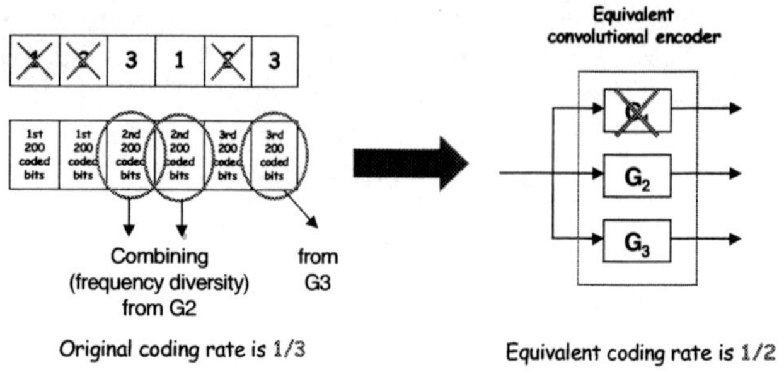

Fig. 6 Decoding example

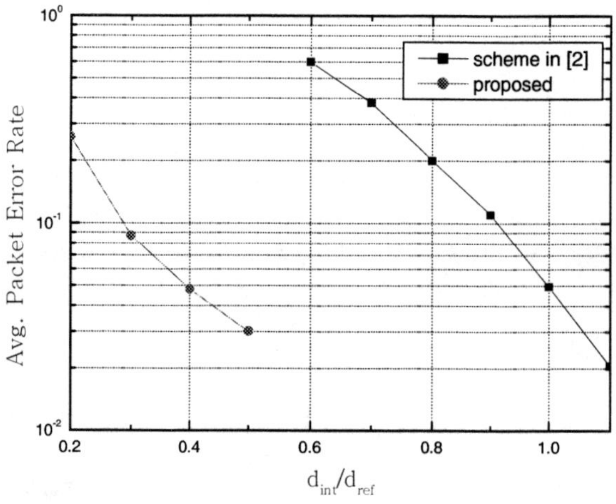

(a) 3 band system

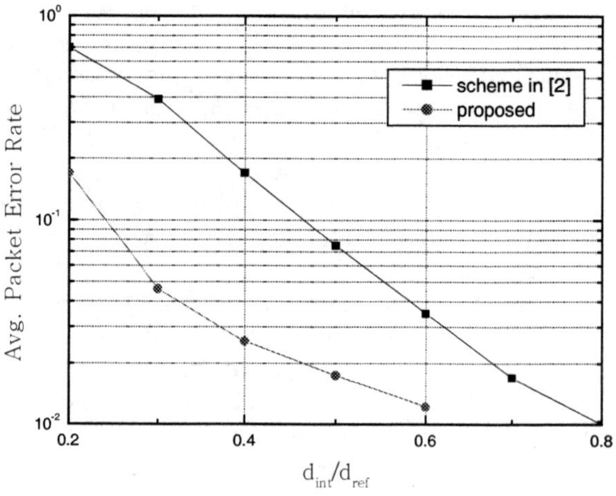

(b) 7 band system

Fig. 7 PER performance versus $d_{\text{int}} / d_{\text{ref}}$

High Performance Solution for Interfering UWB Piconets with Reduced Complexity Sphere Decoding

Jun Tang, Ahmed H. Tewfik and Keshab K. Parhi
Dept. of Electrical and Computer Engineering
University of Minnesota
email: {tangjun, tewfik, parhi}@ece.umn.edu

***Abstract*— The sphere decoding (SD) algorithm has been widely recognized as an important algorithm to solve the maximum likelihood detection (MLD) problem, given that symbols can only be selected from a set with a finite alphabet. The complexity of the sphere decoding algorithm is much lower than the directly implemented MLD method, which needs to search through all possible candidates before making a decision. However, in high-dimensional and low signal-to-noise ratio (SNR) cases, the complexity of sphere decoding is still prohibitively high for practical applications. In this paper, a simplified SD algorithm, which combines the K-best algorithm and SD algorithm, is proposed. With carefully selected parameters, the new SD algorithm, called SD-KB algorithm, can achieve very low complexity with acceptable performance degradation compared with the traditional SD algorithm. The low complexity of the new SD-KB algorithm makes it applicable to the simultaneously operating piconets (SOP) problem of the multi-band orthogonal frequency division multiplex (MB-OFDM) scheme for the high-speed wireless personal area network (WPAN). We show in particular that the proposed algorithm provides over 4dB gain in bit error rate (BER) performance over the baseline MB-OFDM scheme when several piconets interfere with each other. The SD-KB algorithm can provide pseudo-MLD solutions, which have significant performance gain over the baseline method, especially when the signal-to-interference ratio (SIR) is low. The cost of performance improvement is higher complexity. However, the new SD algorithm has predictable computation complexity even in the worst scenario.**

I. Introduction

The sphere decoding (SD) algorithm was first studied in computer science and has been recently used in the communication area, especially in solving multiple-input-multiple-output (MIMO) or multi-path-related demodulation and decoding problems, [1]–[4]. The SD algorithm tries to solve the problem of finding a lattice point $\mathbf{u}$ in the space $\mathcal{S}^N$, whose transformed point, $\mathbf{Mu}$ in a subspace of $\mathbb{R}^M$, has the minimum Euclidean distance to a noisy observation vector $\mathbf{x}$, i.e.,

$$\mathbf{u} = \arg\min_{\mathbf{u}\in\mathcal{S}^N} \|\mathbf{x} - \mathbf{Mu}\|^2, \tag{1}$$

where $\mathbf{u}$ is an N-dimensional array and each element in $\mathbf{u}$ belongs to a set $\mathcal{S}$ that is a finite alphabet. The number of elements in $\mathcal{S}$ is N_s. The lattice matrix $\mathbf{M}$ is an M-by-N matrix, which maps the space $\mathcal{S}^N$ to a subspace of $\mathbb{R}^M$. The array $\mathbf{x}$ is an M-dimensional array, containing the observation signal corrupted by white noise $\mathbf{w}$, i.e.,

$$\mathbf{x} = \mathbf{Mu}_0 + \mathbf{w}. \tag{2}$$

It has been proven that this problem is NP-hard [5]. However, in low-dimensional cases, the sphere decoding (SD) algorithm, which can be used to solve the problem and has much lower complexity than direct searching. The sphere decoding algorithm first defines a region (a subspace of $\mathcal{S}^N$), where the optimal solution must lie within, and then searches for the optimal solution in this region. The size of the search space (sphere) can be reduced as the search progresses. The expected complexity of the sphere decoding algorithm is $\mathcal{O}(N^2)$ to $\mathcal{O}(N^3)$ when the signal-to-noise (SNR) ratio is high [2]. However, the average complexity of SD algorithm is much higher when the SNR is low and the complexity in the worst case still grows exponentially as N increases linearly. The empirically obtained average complexity of the SD algorithm over SNR and the number of dimensions N is shown in Figure 1. In the figure, the lattice matrix $\mathbf{M}$ is randomly generated, such that the norm of each row in $\mathbf{M}$ was normalized to 1 and the rows in the matrix were independent from each other. The figure shows that the average computation time increases exponentially as the number of dimensions increases linearly, when the SNR is low (e.g. lower than 10dB in this figure).

This research was supported in part by Army Research Office under grant number DA/DAAG19-01-1-0705.

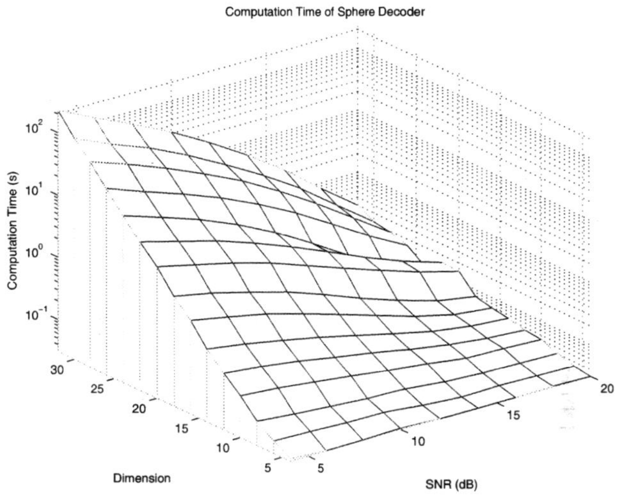

Fig. 1. The experimental results of the complexity of the SD over SNR and dimension N.

The high complexity of the SD in low SNR and high-dimensional cases makes it necessary to develop a sub-optimal algorithm, which can speed up the computation significantly with minimum performance loss.

A alternative method method, called the K-best method, has been proposed to solve (1) by limiting the search scope in each dimension [6] and thus reducing the computational complexity. In the K-best method, in every dimension, at most K layers with minimum distances to the received signal are retained for the computation of the next dimension. The K-best method has predictable complexity even in the worst case. In the worst case, the complexity of the K-best method is $\mathcal{O}(KNN_s)$. Unlike the SD algorithm, the K-best method does not include any iteration, thus it is easy to implement with high-speed hardware structures by utilizing pipelining techniques [6]. However, the K-best method is not an optimal algorithm in the sense that it cannot guarantee performance, especially when the number of dimensions is large.

In this paper, a new SD algorithm, called the SD-KB algorithm, is proposed. The new algorithm reduces the computational complexity by combining the K-best method into the SD algorithm.

II. THE NEW SD-KB ALGORITHM

A. Algorithm

The SD algorithm as explained in Section I is a depth-first algorithm, which iteratively searches along paths through all of the dimensions. The K-best method, on the contrary, searches the K most likely layers in each dimension before it goes to the next dimension. Here, we discuss how to combine the two, such that we can have both predictable complexity and a near optimal solution by tuning the parameters of the algorithm. The new sphere decoding with K-best (SD-KB) algorithm divides the N dimensions of the lattice into two parts. The first part, $\mathcal{P}_1$, contains L dimensions. In these dimensions, the depth-first sphere decoding algorithm is used. The other part, $\mathcal{P}_2$, contains the remaining $N-L$ dimensions. In these dimensions, the K-best method is introduced to speed up the computation. Before applying the SD-KB algorithm, some preprocessing is required on the matrix $\mathbf{M}$ to reduce the performance loss due to the introduction of the K-best method. This preprocessing is similar to the preprocessing in the sphere decoding algorithm [7], which changes the ordering of the searching on the dimensions to increase the computational speed. In the SD-KB algorithm, the same preprocessing not only increases the computational speed, but also *improves the detection performance*.

Figure 2 shows a simple example of the SD-KB algorithm. In this example, $N=5$, $K=2$, $L=2$ and $\mathcal{S}=\{1,-1\}$ ($N_s=2$). This figure shows the trellis representation of the MLD problem. From the top to the bottom there are 5 dimensions defined by the lattice matrix M. In the first two dimensions, i.e., *Dim. 5* and *4*, the sphere decoding iteration is used. In the figure, one step in the iteration is marked with the solid line from *Dim. 5* to *Dim. 4*. When the sphere decoding iteration reaches the last dimension in $\mathcal{P}_1$, i.e., *Dim. 4* in the example, the detection procedure goes to the first dimension in $\mathcal{P}_2$, i.e., *Dim. 3* in the example. In $\mathcal{P}_2$, at most K layers in each dimension, those that have the minimum distance to the received signal, are kept for the computation of the next dimension. In Figure 2, the layers kept are marked with shadowed circles. In the next dimension, the candidate layers are computed based on the K layers kept in the previous dimension. The maximum number of candidate layers required to compute in each dimension is KN_s. In the last dimension of $\mathcal{P}_2$, i.e., *Dim. 1*, only the best solution is kept. In the example, after the step in the sphere-decoding iteration shown with the solid line and the followed 2-best computation, the best solution which is kept until the last dimension in $\mathcal{P}_2$ is the one marked with a dashed line. With $\oplus$ and $\ominus$ representing "+1" and "−1", respectively, the dashed line solution corresponds to $\{+1,+1,+1,-1,-1\}$.

If the distance of this solution to the received signal is smaller than the radius of current search sphere, the recorded shortest distance and the radius of the sphere are then updated to the distance of the newly obtained solution and this solution is saved to memory. After reaching the last dimension in $\mathcal{P}_2$, the algorithm will trace back to the last dimension in $\mathcal{P}_1$, i.e., *Dim. 4* in the example, for the next step in the sphere decoding iteration. The same procedure will be repeated until the last step of the sphere decoding iteration. In every dimension, no matter whether it is in $\mathcal{P}_1$ and $\mathcal{P}_2$, only those layers that have smaller distances to the received signal than the radius of the search sphere are computed. If in a dimension in $\mathcal{P}_2$, the distances between all of the layers and the received signal are larger than the radius of the searching sphere, the K-best computation will stop at this dimension with no output and immediately traces back to the last dimension in $\mathcal{P}_1$ for the next step of the SD iteration.

In the SD-KB algorithm, L is a critical parameter which affects

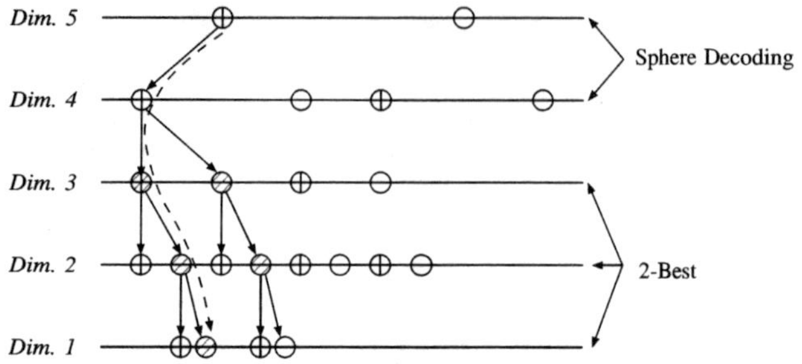

Fig. 2. An example of modified sphere decoding algorithm. $N=5$, $L=2$ and $K=2$

the performance and computational speed. When the value of L increases to N, the SD-KB algorithm reduces to the traditional sphere decoding algorithm and has optimal performance but slowest computational speed. When the value of L decreases to 0, the SD-KB algorithm becomes the K-best method and has the fastest computational speed but the worst performance degradation. The choice of the value of L is balanced by the requirements on performance and computational speed for a specific application. In the worst case, the complexity of the SD-KB algorithm is $\mathcal{O}(K(N-L)N_s^{L+1})$. The predictable worst case complexity of SD-KB makes it feasible to implement it with hardware technologies such as ASIC. Thus, high-speed implementation is possible compared with the relatively low-speed software implementations.

B. Simulation Result

A simulation of the performance and computational speed of the original SD algorithm and the SD-KB algorithm was performed with a 800MHz Sun Blade 1000 system. The simulation results are shown in Figure 3. In this figure, when L is smaller than or equal to 8, the computational time of the problem is at least 10 times smaller than the original sphere decoding algorithm. Furthermore, even with $L=4$, the performance is still within 1dB of the original algorithm. We can conclude that with carefully chosen K and L, it is possible to speed up the computation significantly without severely degrading performance. The computational time increases exponentially when the value of SNR decreases as the value of L is large and increases in a polynomial manner as the value of L is small.

It is also worth noting that when the SNR is very high, such as above 18dB in Figure 3(a), the computational time of the proposed algorithm with small value of L takes longer than the traditional SD algorithm. This phenomenon is due to the very fast convergence speed of the SD algorithm when the SNR is high.

III. APPLICATION TO SOP PROBLEM IN MB-OFDM SCHEME FOR HIGH-SPEED WPAN

The IEEE 802.15.3a is a task group (TG) focused on an alternative physical layer technique for the short distance high speed wireless personal area communications (WPAN) based on the ultra wide-band (UWB) techniques. After the down-selection procedure at the standards meeting in July, 2003, the multi-band orthogonal frequency division multiplex (MB-OFDM) scheme has emerged as the leading proposal due to its simplicity and ease of implementation [8]. However, the MB-OFDM has poor support for simultaneously operating piconets (SOP) due to its slow-hopping pattern, OFDM technology and the limited number of frequency bands available. In WPAN systems, multiple uncoordinated piconets may broadcast in the same area at the same time. The interference from other piconets

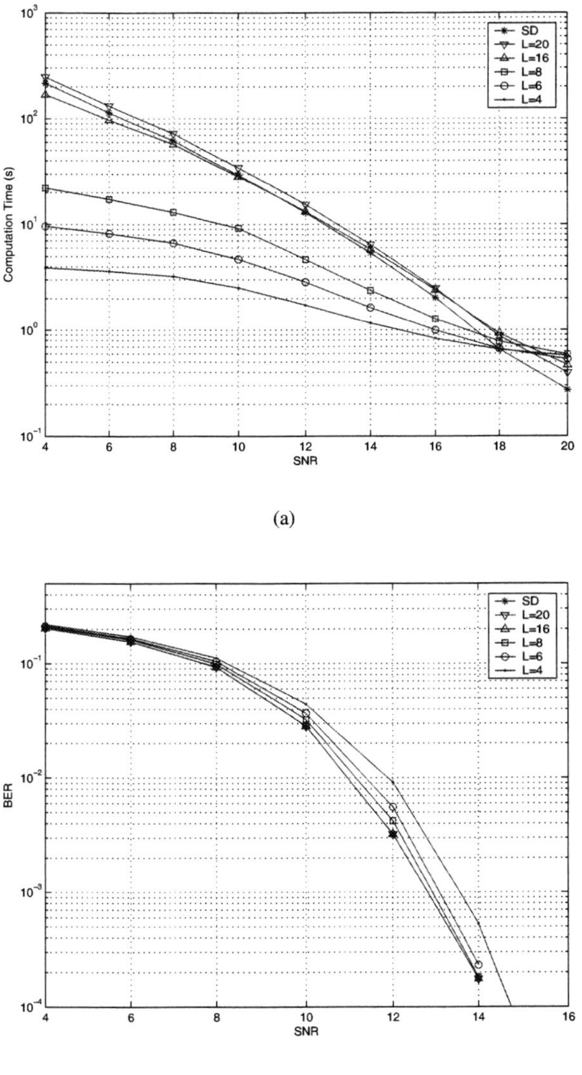

Fig. 3. Comparison of the computational time and performance of the original sphere decoding and SD-KB algorithm: (a) the comparison of the average computational time and (b) the comparison of the performance (bit error rate).

will severely degrade performance. The most challenging difficulty in the SOP problem is the uncoordinated nature of the piconets, i.e., the piconets are not synchronized and no multiple access scheme is available between different piconets. Several techniques have been proposed to mitigate the problem of SOP. Some of them try to reduce the chance of the collisions between piconets. These techniques include half pulse repetition frequency (PRF) scheme [9] and a time-spreading scheme. These techniques require some modifications of the baseline MB-OFDM scheme. For example, the half-PRF technique requires some form of coordination between the piconets.

The MLD technique can be used to mitigate the performance degradation when collision happens and it only requires modification of the receiver. It can also be combined with other techniques for further performance improvement.

A. Problem Statement and Solution

When collision happens, at the receiver side, the received signal $\mathbf{r}$ is:

$$\mathbf{r} = \alpha \mathbf{HPF}^H \mathbf{b} + \beta \mathbf{OH}_i \mathbf{PF}^H \mathbf{i} + \mathbf{w}, \tag{3}$$

where $\mathbf{F}^H$ is the IFFT matrix applied at the transmitter, $\mathbf{H}$ is the channel matrix of the desired signal $\mathbf{b}$, $\mathbf{H}_i$ is the channel matrix of the interference signal $\mathbf{i}$, $\mathbf{P}$ represents the operation of adding cyclic prefix (CP) at the transmitter, $\mathbf{O}$ is the shift matrix, which represents the time asynchrony between the desired signal and the interference signal, and $\mathbf{w}$ is the white Gaussian noise with variance σ^2. Array $\mathbf{b}$ contains the QPSK symbols in one OFDM symbol transmitted by the transmitter from the same piconet of the receiver, while $\mathbf{i}$ contains the QPSK symbols transmitted by the interferer, which is in another piconet not synchronized to the receiver's piconet. Parameters α and β are the attenuations of the reference and interference signal, respectively, due to the propagation path loss and the shadowing effect. In this paper, we only consider the SOP problem of two piconets. However, the conclusion can be extended to the SOP problem of multiple piconets. The FFT solution leads to the optimal solution in the absence of the interference $\mathbf{i}$. When the interference is present, the traditional solution becomes:

$$\begin{aligned} \mathbf{u} &= \alpha \mathbf{FP}^{-1}\mathbf{HPF}^H \mathbf{b} + \beta \mathbf{FP}^{-1}\mathbf{OH}_i \mathbf{PF}^H \mathbf{i} + \mathbf{FP}^{-1}\mathbf{w} \\ &= \Lambda \mathbf{b} + \mathbf{Gi} + \hat{\mathbf{w}}, \end{aligned} \tag{4}$$

where $\mathbf{P}^{-1}$ represents the operation of removing cyclic prefix, $\mathbf{F}$ is the FFT matrix; $\Lambda = \alpha \mathbf{FP}^{-1}\mathbf{HPF}^H$ is a diagonal matrix due to the properties of the circular matrix, and $\hat{\mathbf{w}} = \mathbf{FP}^{-1}\mathbf{w}$ is still white Gaussian noise with variance σ^2. However, unlike $\mathbf{\Lambda}$, $\mathbf{G} = \beta \mathbf{FP}^{-1}\mathbf{OH}_i\mathbf{PF}^H$ is not a diagonal matrix, generally, due to the shift matrix $\mathbf{O}$. In this scenario, the traditional solution will no longer be optimal. The MLD solution of the problem is:

$$\mathbf{u}_{ML} = \arg\min_{\mathbf{u}\in \mathcal{S}^N} \left(\mathbf{FP}^{-1}\mathbf{r} - \mathbf{\Lambda u}\right)^H \mathbf{R}^{-1} \left(\mathbf{FP}^{-1}\mathbf{r} - \mathbf{\Lambda u}\right), \tag{5}$$

where $\mathbf{R} = E\left[(\mathbf{Gi}+\mathbf{w})(\mathbf{Gi}+\mathbf{w})^H\right]$ is the correlation matrix of the interference and the noise. Under the assumption that the elements in $\mathbf{i}$ follow an i.i.d. distribution and are independent from the noise $\mathbf{w}$, $\mathbf{R} = \mathbf{GG}^H + \sigma^2\mathbf{I}$. After whitening the noise, the MLD problem is equivalent to:

$$\mathbf{u}_{ML} = \arg\min_{\mathbf{u}\in \mathcal{S}^N} \left\|\mathbf{U}(\mathbf{FP}^{-1}\mathbf{r} - \mathbf{\Lambda u})\right\|^2, \tag{6}$$

where $\mathbf{U}$ is the upper triangular matrix in the Cholesky decomposition of matrix $\mathbf{R}^{-1}$, i.e., $\mathbf{R}^{-1} = \mathbf{U}^H\mathbf{U}$. If we denote $\mathbf{x} = \mathbf{UFP}^{-1}\mathbf{r}$ and $\mathbf{M} = \mathbf{U\Lambda}$, the problem in (6) reduces to problem (1) and can be solved by the SD or SD-KB algorithm (with some performance loss) as explained in the previous sections.

Although the MLD can provide the optimal performance, the complexity of the MLD, even with sphere decoding, is too high in the high-dimensional and low SNR scenario. In the SOP problem, the number of dimensions is equal to the number binary information bits transmitted over an OFDM symbol. Even with a relatively low data rate (110Mbps or 200Mbps) and a spreading factor of two[1], the number of dimensions N is equal to 100, which is prohibitively complex with direct implementation of the MLD or sphere decoding algorithms. However, the complexity of the SD-KB algorithm is still in an acceptable scope with carefully chosen values for K and L. At the same time, it can provide significant performance improvement compared to the baseline demodulation method. It is worth noting

[1]The spreading is implemented in a conjugate manner. See [8] for detail.

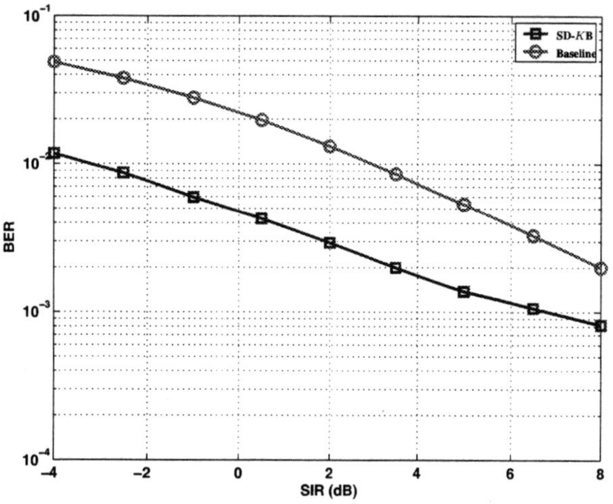

Fig. 4. Simulation results of the application of SD-KB algorithm in the SOP problem of MB-OFDM systems. The curve marked with squares is the result of SD-KB algorithm and the one marked with cycle is the result of the baseline techniques. No MLD or sphere decoding results are provided in the figure due to the prohibitively high complexity of the MLD problem in the low SNR scenarios.

that noise power in SOP also includes the power of the interference signal, which may be equal to or even larger than the power of the desired signal, i.e., the transmitter in the interfering piconet is closer to the receiver than the transmitter in the reference piconet. In most cases, when collision occurs, the SNR, after whitening the interference signal, will be low.

B. Simulation Results

A simplified software simulation was performed to compare the performance of the SD-KB algorithm and the baseline technique. In the simulation, for simplicity, the following assumptions were made: both the channels of the reference signal and the interference signal can be precisely estimated, the receiver is perfectly synchronized to desired transmitter, the front end A/D converter has infinite precision and both the channels of the reference signal and the interference signal are static during the transmission of a packet.

The simulation results are shown in Figure 4. From the figure, the SD-KB algorithm provides over 4dB gain in BER performance compared with the baseline method. When the SIR is very low, the performance gain is even larger. The values of K and L are 16 and 20 in the simulation of the SD-KB algorithm, respectively, which are chosen to balance computational complexity and performance degradation (compared with the sphere decoding algorithm).

Compared with the baseline demodulation, frequency domain equalizer and diversity combining techniques, the complexity of the SD-KB algorithm is higher. This increase in complexity is the cost of better performance. Note that we also have to separately estimate the channels of the reference signal and the interference signal, which makes the channel estimation function unit more complex than that of the baseline technique. However, note that we do not need to precisely estimate the interferer's channel. We only need the correlation matrix $\mathbf{R}$, or its Cholesky decomposition $\mathbf{U}$, instead. This makes the channel estimate in the MLD feasible with some blind channel estimate techniques or with known pilot tones transmitted in the reference OFDM symbols since no training sequence can be obtained from the interference signal.

IV. Conclusion

In this paper, a new sphere decoding algorithm, SD-KB algorithm, has been presented. By combining K-best algorithm into the SD algorithm, the complexity of the SD algorithm can be significantly reduced. At the same time, the performance degradation is still in an acceptable level with carefully chosen values of K and L. The larger the values of K and L, the better the performance and the higher the computational complexity. The smaller the values of K and L, the worse the performance and the lower the complexity. The SD-KB algorithm is specifically useful for the high-dimensional and low SNR cases. The complexity of the algorithm in the worst case is $\mathcal{O}(K(N-L)N_s^{L+1})$. The predictable worst case complexity makes it feasible to implement the algorithm with VLSI circuit chips.

The application of SD-KB algorithm in the SOP problem of MB-OFDM scheme for high-speed WPAN is presented as an example. Compared with baseline techniques, the new algorithm can provide over 4dB performance gain in two simultaneously operating piconets. The performance gain is more significant when the SIR is very low. The SD-KB algorithm only modifies the structure of the receiver in the digital signal processing (DSP) part. The communication protocol and the transmitter remain unchanged. It can also be combined with other proposed techniques, which reduces the chance of collisions, to further improve the performance. However, it also requires more complex channel estimation techniques for both the reference signal and the interference signal.

Acknowledgement

The authors sincerely thank Wanlun Zhao for his help on the software implementation of sphere decoding algorithm.

References

[1] A. M. Chan and I. Lee, "A new reduced-complexity sphere decoder for multiple antenna systems," in *IEEE International Conference on Communications*, Apr. 2002, pp. 460–464.

[2] H. Vikalo and B. Hassibi, "Maximum-likelihood sequence detection of multiple antenna systems over dispersive channels via sphere decoding," *EURASIP Journal on Applied Signal Processing*, no. 5, pp. 525–531, 2002.

[3] H. Vikalo, B. Hassibi, and U. Mitra, "Sphere-constrained ML detection for frequency-selective channels," in *IEEE International Conference on Acoustics, Speech, and Signal Processing (ICASSP)*, vol. 4, Apr. 2003, pp. IV–1 – IV–4.

[4] E. Viterbo and J. Boutros, "A universal lattice code decoder for fading channels," *IEEE Transactions on Information Theory*, vol. 45, pp. 1639–1642, July 1999.

[5] D. Micciancio, "The hardness of the closest vector problem with preprocessing," *IEEE Transactions on Information Theory*, vol. 47, pp. 1212–1215, Mar. 2001.

[6] K.-W. Wong, C.-Y. Tsui, R. S.-K. Cheng, and W.-H. Mow, "A VLSI architecture of a K-best lattice decoding algorithm for MIMO channels," in *IEEE International Symposium on Circuits and Systems (ISCAS)*, vol. 3, May 2002, pp. III–273 – III–276.

[7] E. Agrell, T. Eriksson, A. Vardy, and K. Zeger, "Closest point search in lattices," *IEEE Transactions on Information Theory*, vol. 48, no. 8, pp. 2201–2214, Aug. 2002.

[8] A. Batra, J. Balakrishnan, *et al.*, "Multi-band OFDM physical layer proposal for IEEE 802.15 task group 3a," IEEE P802.15-SG3a-03/268r0, IEEE P802.15 Working Group for Wireless Personal Area Networks (WPANs), July 2003.

[9] E. Ojard and J. Karaoguz, "Reduced duty cycle multiband OFDM," IEEE P802.15-SG3a-03/273r0, IEEE P802.15 Working Group for Wireless Personal Area Networks (WPANs), July 2003.

CHANNEL ESTIMATION AND SYNCHRONIZATION WITH SUB-NYQUIST SAMPLING AND APPLICATION TO ULTRA-WIDEBAND SYSTEMS

Irena Maravić[†], *Martin Vetterli*[†‡] *and Kannan Ramchandran*[‡]

[†]Audio-Visual Communications Laboratory
Swiss Federal Institute of Technology in Lausanne, CH-1015 Lausanne, Switzerland

[‡]Department of EECS, University of California at Berkeley, Berkeley CA 94720, USA

ABSTRACT

We consider the problem of low-complexity channel estimation and timing in digital ultra-wideband receivers. We extend some of our recent results on sampling of certain classes of parametric non-bandlimited signals and develop a frequency domain framework that yields high-resolution estimates of all relevant channel parameters by sampling a received signal below the traditional Nyquist rate. In particular, we show that the minimum required sampling rate in an UWB receiver is determined by the so-called *innovation rate*, which corresponds to the number of degrees of freedom of the received UWB signal. Our framework allows for faster acquisition compared to current digital solutions and potentially reduces power consumption and complexity of digital UWB receivers significantly. It is particularly suitable in applications such as ranging or positioning and can also be used for identification of more realistic UWB channel models, where different propagation paths undergo different frequency-selective mitigation.

1. INTRODUCTION

Ultra-wideband (UWB) technology has recently received much attention as a potential solution to a variety of short-range problems, such as accurate ranging and positioning as well as multipath fading mitigation in indoor wireless networks [1] [4]. UWB systems use trains of pulses of very short duration, typically on the order of a nanosecond, thus spreading the signal energy from near DC to a few gigahertz. Although a possibility of using extremely short pulses has been investigated for at least two decades, primarily in the radar community, there still remains a lot to be done for this technology to become pervasive. Some of the important issues include low-cost and low-power designs and efficient algorithmic solutions suitable for digital implementation.

The nature of UWB signaling brings new challenges both in the analysis and practice of reliable systems. One of the main design challenges is rapid synchronization, as its accuracy and complexity directly affect the system performance [4]. This problem becomes even more involved with a current trend to minimize the number of analog components needed, and perform as much processing digitally as possible [2]. Namely, due to the extreme bandwidths involved, digital implementation may lead to prohibitively high costs in terms of power consumption and receiver complexity. For example, conventional techniques based on sliding correlators would require very fast and expensive A/D converters (operating in the gigahertz range) and, therefore, high power consumption. Furthermore, implementation of such techniques in digital systems would have almost unaffordable complexity as well as slow convergence time, since one has to perform exhaustive search over thousands of fine bins, each at the nanosecond level. In order to improve the acquisition speed, several modified timing recovery schemes have been proposed, such as a bit reversal search [3], or the correlator-type approach which exploits properties of beacon sequences [4]. Even though some of these methods have already been in use in certain analog systems, the need for very high sampling rates, along with the search-based nature of such methods, makes them less attractive for digital implementation.

In this paper, we present a new approach to channel estimation and timing in digital UWB receivers, which allows for sub-Nyquist sampling rates and reduces the receiver complexity, while retaining a desirable performance. The idea is based on our recent results on sampling of certain classes of parametric non-bandlimited signals that have a finite number of degrees of freedom per unit of time, or a *finite rate of innovation*. We show that the minimum required sampling rate in UWB systems is determined by the innovation rate of the received UWB signal [5] [6], rather than the Nyquist rate. Our approach allows for lower sampling rates and reduced complexity and power consumption compared to other digital techniques [2] [3]. It is particularly suitable in applications such as ranging or positioning, but can also be used for estimation of more general UWB channels [1].

2. LOW-SAMPLING RATE PARAMETER ESTIMATION

2.1. Problem Statement

A number of propagation studies for UWB signals have been done, which take into account temporal properties of a channel or characterize a spatio-temporal channel response [1]. A typical model for the impulse response of a multipath fading channel is given by

$$h(t) = \sum_{l=1}^{L} a_l \delta(t - t_l), \tag{1}$$

where t_l denotes a signal delay along the l-th path, while a_l is a complex propagation coefficient. Equation (1) implies that a received signal $y(t)$ is made up of a sum of attenuated and delayed replicas of a transmitted signal $s(t)$, i.e.

$$y(t) = \sum_{l=1}^{L} a_l s(t - t_l) + \eta(t), \tag{2}$$

where $\eta(t)$ denotes receiver noise. Let $Y(\omega)$ denote the Fourier transform of the received signal

$$Y(\omega) = \sum_{l=1}^{L} a_l S(\omega) e^{-j\omega t_l} + \mathcal{N}(\omega), \qquad (3)$$

where $S(\omega)$ and $\mathcal{N}(\omega)$ are Fourier transforms of $s(t)$ and $\eta(t)$ respectively. Clearly, spectral components are given by a sum of complex exponentials, thus we can convert the problem of estimating the unknown channel parameters $\{t_l\}_{l=0}^{L-1}$ and $\{a_l\}_{l=0}^{L-1}$ into the classic spectral estimation problem. High-resolution spectral estimation is well-studied: there exists a rich body of literature on efficient algorithms and achievable performance bounds [7]. In the following, we will adopt a model-based approach and show how it can be used to obtain high-resolution estimates of all the relevant parameters from a subsampled version of the received signal.

2.2. Polynomial Realization of the Model-Based Estimator

Suppose that the received signal $y(t)$ is filtered with an ideal bandpass filter $H_b = \text{rect}(\omega_1, \omega_2)$, and sampled uniformly at a rate $R_s \geq \frac{\omega_2 - \omega_1}{2\pi}$. Suppose next that N uniformly spaced frequency samples of $Y(\omega)$ are available, that is,

$$Y[n] = Y(\omega_n) = Y(\omega_1 + (n-1)\omega_0), \quad n = 1, ..., N. \qquad (4)$$

where $\omega_0 = \frac{\omega_2 - \omega_1}{N-1}$ and let $Y_s[n] = Y[n]/S[n]$, where $S[n]$ are the samples of the Fourier transform $S(\omega_n)$ of the transmitted UWB pulse. Assuming that in the considered frequency band the above division is not ill-conditioned, the samples $Y_s[n]$ can be expressed as a sum of complex exponentials (3), i.e.,

$$Y_s[n] = \sum_{l=1}^{L} a_l e^{-j\omega_n t_l} + \mathcal{N}(\omega_n). \qquad (5)$$

The polynomial approach exploits the fact that in the absence of noise, each exponential $\{e^{-jn\omega_0 t_l}\}_{n \in Z}$ can be "nulled out" or annihilated by a first order FIR filter $H_l(z) = (1 - e^{-j\omega_0 t_l} z^{-1})$, that is,

$$e^{-jn\omega_0 t_l} * [1, -e^{-j\omega_0 t_l}] = 0. \qquad (6)$$

Consider thus an L-th order FIR filter $H(z) = \sum_{m=0}^{L} H[m] z^{-m}$, having L zeros at $z_l = e^{-j\omega_0 t_l}$,

$$H(z) = \prod_{l=1}^{L} (1 - e^{-j\omega_0 t_l} z^{-1}). \qquad (7)$$

Note that $H[m]$ is the convolution of L elementary filters with coefficients $[1, -e^{-j\omega_0 t_l}]$, $l = 1, ...L$. Since $Y_s[n]$ is the sum of complex exponentials, each will be annihilated by one of the roots of $H(z)$, thus we have:

$$(H * Y_s)[n] = \sum_{k=0}^{L} H[k] Y_s[n-k] = 0, \quad n = L+1, ..., N. \qquad (8)$$

In matrix form, the system (8) is equivalent to

$$\begin{pmatrix} Y_s[L+1] & Y_s[L] & \cdots & Y_s[1] \\ Y_s[L+2] & Y_s[L+1] & \cdots & Y_s[2] \\ \vdots & \vdots & \ddots & \\ Y_s[2L] & Y_s[2L-1] & \cdots & Y_s[L] \\ \vdots & \vdots & & \end{pmatrix} \cdot \begin{pmatrix} H[0] \\ H[1] \\ \vdots \\ H[L] \end{pmatrix} = \mathbf{0}. \qquad (9)$$

By setting $H[0] = 1$, one can uniquely solve for $H[m]$, $m = 1, ..., L$, from only $2L$ equations in (9). Therefore, the information about the delays t_l can be extracted from the roots of $H(z)$, while the corresponding coefficients a_l are then estimated by solving the system of linear equations (5), by fitting L exponentials $e^{-j\omega_n t_l}$ to the data set $Y_s[n]$.

The above result can be interpreted in the following way: the signal $y(t)$ is projected onto a low-dimensional subspace corresponding to its bandpass version. This projection is a unique representation of the signal as long as the dimension of the subspace is greater than or equal to the number of degrees of freedom. That is, since $y(t)$ has $2L$ degrees of freedom, $\{t_l\}_{l=0}^{L-1}$ and $\{a_l\}_{l=0}^{L-1}$, it suffices to use only $2L$ adjacent coefficients $Y_s[n]$.

2.3. Algorithm in the Presence of Noise: Subspace Approach

In the case of noiseless data, any subspace of sufficient dimension can be used for estimation. In practice, noise will be present, and this can be dealt with by oversampling and using standard techniques from noisy spectral estimation, such as the singular value decomposition (SVD). Besides, in the presence of noise, it is desirable to estimate the channel from a frequency band where the signal-to-noise ratio (SNR) is highest. This brings us to a more practical version of the model-based approach, the so-called subspace or SVD-based estimator, which avoids the root finding step and relies only on a right deployment of matrix manipulations [6].

Namely, the key is to observe two properties of the matrix $\mathbf{Y_s}$ in (9). The first one is that in the case of noiseless data, $\mathbf{Y_s}$ has rank L. This allows us to reduce the noise level by approximating a noisy data matrix with a rank L matrix, by computing its SVD, i.e, $\mathbf{Y_s} = \mathbf{U_s}\mathbf{\Lambda_s}\mathbf{V_s}^H + \mathbf{U_n}\mathbf{\Lambda_n}\mathbf{V_n}^H$, where $\mathbf{U_s}$ and $\mathbf{V_s}$ contain principal left and right singular vectors of $\mathbf{Y_s}$. The second property is the Vandermonde structure of $\mathbf{U_s}$ and $\mathbf{V_s}$, that is, they both satisfy the so-called shift-invariant subspace property [6],

$$\overline{\mathbf{U_s}}^d = \underline{\mathbf{U_s}}_d \cdot \mathbf{\Phi} \text{ and } \overline{\mathbf{V_s}}^d = \underline{\mathbf{V_s}}_d \cdot \mathbf{\Phi}^H \qquad (10)$$

where $\mathbf{\Phi}$ is a diagonal matrix having elements $e^{j\omega_0 d t_l} = e^{j\omega_0 \hat{t}_l}$ along the main diagonal, while $\overline{(\cdot)}^d$ and $\underline{(\cdot)}_d$ denote the operations of omitting the first d rows and the last d rows of $(\cdot)$ respectively. Therefore, the time delays $\{t_l\}_{l=1}^{L}$ can be uniquely determined from the eigenvalues λ_l of the operator that maps $\underline{\mathbf{U_s}}_d$ onto $\overline{\mathbf{U_s}}^d$ (or $\underline{\mathbf{V_s}}_d$ onto $\overline{\mathbf{V_s}}^d$), that is, $t_l = \frac{N\angle\lambda_l}{2\pi d}$. The advantage of using values of d larger than $d = 1$, is that the separation between the estimated time delays $\hat{t}_l$ is effectively increased d times, which is of particular interest in the case of estimating closely spaced dominant components in a low SNR regime [6]. Once the time delays have been estimated, the coefficients a_l can be found from the Vandermonde system (5). Finally, we note that the number of dominant components is estimated as the number of dominant singular values of $\mathbf{Y_s}$.

A major computational requirement of the above method is associated with the SVD step, which is an iterative algorithm with the computational order of $\mathcal{O}(N^3)$ per iteration. In [6], we suggested alternative methods of lower complexity, such as the *Power method* and the method of *Orthogonal iteration*. These methods have computational requirements on the order of $\mathcal{O}(N^2)$ per iteration, and they generally converge in less than 10 iterations.

3. EXTENSIONS AND SOME APPLICATIONS

3.1. Estimating More Realistic UWB Channel Models

Our frequency-domain framework can be extended to handle the more complex case of an UWB channel that takes into account certain bandwidth-dependent properties [1]. Namely, as a result of the very large bandwidth of UWB signals, components propagating along different propagation paths undergo different frequency selective distortion and a more realistic UWB channel model is of the form

$$h(t) = \sum_{l=1}^{L} a_l p_l(t - t_l), \tag{11}$$

where $p_l(t)$ are different pulse shapes that correspond to different paths. In this case, the spectral coefficients of the received signal are given by

$$Y[n] = S[n] \sum_{l=1}^{L} P_l[n] a_l e^{-j\omega_n t_l} + \mathcal{N}[n], \tag{12}$$

where $P_l[n]$ are now unknown coefficients. One possible way to obtain a closed form solution is to approximate the coefficients $P_l[m]$ with polynomials of degree $D \leq R - 1$, that is,

$$P_l[n] = \sum_{r=0}^{R-1} p_{l,r} n^r. \tag{13}$$

Equation (12) now becomes

$$Y[n] = S[n] \sum_{l=1}^{L} a_l \sum_{r=0}^{R-1} p_{l,r} n^r e^{-j\omega_n t_l} + \mathcal{N}[n]. \tag{14}$$

By denoting $c_{l,r} = a_l p_{l,r}$ and $Y_s[n] = Y[n]/S[n]$, we obtain

$$Y_s[n] = \sum_{l=1}^{L} \sum_{r=0}^{R-1} c_{l,r} n^r e^{-j\omega_n t_l} + \mathcal{N}[n]. \tag{15}$$

In this case, the shift-invariance property no longer holds, however, one can prove that a filter with multiple roots at $z_l = e^{-j\omega_0 t_l}$, i.e.

$$H(z) = \prod_{l=1}^{L} (1 - e^{-j\omega_0 t_l} z^{-1})^R = \sum_{k=0}^{RL} H[k] z^{-k} \tag{16}$$

will be the annihilating filter for $Y_s[n]$ [6]. Therefore, the information about the time delays t_l can be extracted from the roots of the filter $H(z)$. The corresponding pulse shapes are then estimated by solving for the coefficients $c_{l,r}$ from the system (15).

In previous work [1], Cramer, Win and Scholtz used an array of sensors to spatially separate the multipath components, which is then followed by identification of each path using the so-called Sensor-CLEAN algorithm. Note that our approach does not require an antenna array, and takes advantage of the fact that the multipath components will have independent delays at the reception [6].

3.2. Low-Complexity Rapid Acquisition in UWB localizers

One of the most appealing applications of our framework can be found in ultra-wideband systems used for precise position location. Such UWB transceivers use low duty-cycle periodic transmission of a coded sequence of impulses to insure low-power operation [4]. Yet, rapid synchronization presents a major design challenge, and current systems still rely on exhaustive search through all possible code positions. In order to show how our approach can be used in this framework, note that the received signal $y(t)$ can be modeled as a convolution of L, possibly different, impulses with a known coding sequence $g(t)$, that is,

$$y(t) = \sum_{l=1}^{L} a_l p_l(t - t_l) * g(t). \tag{17}$$

The DFT coefficients of $y(t)$ are now given by

$$Y[n] = \sum_{l=1}^{L} a_l P_l[n] G[n] e^{-jn\omega_c t_l}, \tag{18}$$

where $\omega_c = 2\pi/T_c$, while T_c denotes the cycle time. If we use the polynomial approximation of coefficients $P_l[n]$, the total number of degrees of freedom per cycle is $2RL$, which determines the minimum sampling rate.

Another advantage of our solution is that it allows for a "multiresolution" or two-step approach, that is, one can first obtain a rough estimate of the sequence timing, by sampling the signal at a low rate over the entire cycle. Later, precise delay estimation can be carried out by increasing the sampling rate, yet sampling the received signal only within a narrow time window where it is present. The rationale for using the two-step approach is that a sequence duration T_s typically spans a small fraction of the cycle time T_c (e.g. less than 20%), where search-based methods [2] [4], result in a long acquisition time and apparently "waste" power in processing time slots where the signal is not present. In the following, we will demonstrate the effectiveness of the two-step approach with a numerical example.

4. SIMULATION RESULTS

In this section, we show some simulation results that illustrate the performances of the proposed algorithms. We consider an UWB localization system, where a coded sequence of 127 UWB impulses is periodically transmitted, while the sequence duration spans approximately 20% of the cycle time T_c, as illustrated in Figure 1(a). We first consider the channel model (1), assuming $L = 6$ paths with one dominant (containing 70% of the total power). In Figure 1(b), we show the root mean square error (RMSE) of time delay estimation for the dominant component. The results indicate that the method yields highly accurate estimates for a wide range of SNR's, and this with sub-Nyquist sampling rates. Note that the achievable RMSE with the approach using digital matched filters [2], is RMSE=1 (sample).

The two-step delay estimation method is analyzed next. During the coarse synchronization phase, the signal is sampled uniformly over the entire cycle at a low rate $N_l = 0.05N_n$ (N_n is the Nyquist rate). For low values of SNR (that is, less than -5dB), the samples are averaged over multiple cycles in order to increase the effective SNR. The second step is fine synchronization, where the signal is sampled at a higher rate $N_h = 0.5N_n$, yet within a narrow time window. RMSE of the two-step approach is shown in Figure 2(a). The error is compared to the RMSE obtained when the signal is sampled uniformly at a rate $0.5N_n$ over the entire cycle. Clearly, the two solutions yield a similar performance, however, with the two-step approach, the computational requirements are reduced by a factor of 20, while the power consumption reduces roughly by a factor of 3.3. In Figure 2(b), we plot the RMSE of delay estimation for different combinations of N_l and N_h, assuming that the total number of cycles is 5. The number of cycles during each phase is chosen such that the power consumption remains constant. Clearly, the performance improves as N_h increases, yet at the expense of increased computational requirements.

We next consider the case of a channel model given by (1), assuming that a received signal is made up of $L = 70$ pulses and 8 closely spaced dominant components, as shown in Fig. 3(a). In Fig. 3(b), we illustrate the effects of quantization on the estimation performance, specifically, we consider 4-7 bit architectures. Clearly, as the number of bits increases, the overall performance improves. Generally, the 5-bit architecture already yields a very good performance. Also note that when $n_b \geq 5$ and the value of SNR is low (e.g. $SNR < 0dB$), quantization has almost no impact on RMSE. However, as SNR increases, quantization noise becomes dominant and determines the overall performance.

5. CONCLUSIONS

We presented a frequency-domain framework for high-resolution parameter estimation in digital UWB receivers, which is based on our recent sampling results for certain parametric non-bandlimited signals. Our approach requires lower sampling rate and, therefore, lower complexity and power consumption compared to existing digital solutions. Besides, it leads to faster acquisition and allows for identification of more realistic channel models without resorting to complex algorithms. Our approach is particularly suitable in applications such as positioning or ranging, but can be used in other UWB applications as well, primarily for synchronization and channel characterization purposes.

6. REFERENCES

[1] R. J. Cramer, R. A. Scholtz and M. Z. Win, "Evaluation of an Ultra-Wideband Propagation Channel", *IEEE Trans. on Antennas and Propagation*, Vol. 50, No. 5, pp 561-570, May 2002.

[2] I. O' Donnell, M. Chen, S. Wang and R. Brodersen, "An Integrated, Low-Power, UWB Transceiver Architecture for Low-Rate Indoor Wireless System", IEEE CAS Workshop on Wireless Communications and Networking, Sept. 2002.

[3] E. Homier and R. Scholtz, "Rapid Acquisition of UWB Signals in a Dense Multipath Channel", *in Proc. IEEE Conf. on UWB Systems and Technologies*, May 2002.

[4] R. Fleming, C. Kushner, G. Roberts and U. Nandiwada, "Rapid Acquisition for Ultra-Wideband Localizers", *in Proc. IEEE Conf. on UWB Systems and Technologies*, May 2002.

[5] I. Maravic, M. Vetterli and K. Ramchandran, "High-resolution acquisition methods for wideband communication systems", *In Proc. of ICASSP*, April 2003.

[6] I. Maravic, M. Vetterli and K. Ramchandran, "High-resolution methods for channel estimation and synchronization in UWB systems", *IEEE Trans. on Signal Processing*, submitted, 2003.

[7] B. D. Rao and K. S. Arun, "Model based processing of signals: A state space approach", *Proceedings of the IEEE*, Vol. 80, No. 2, pp. 283-309, February 1992.

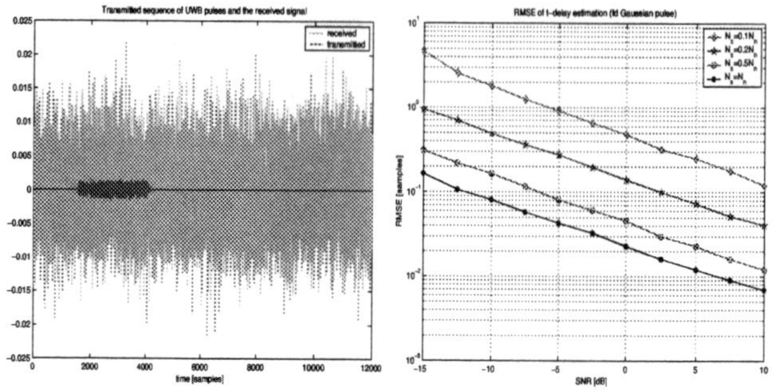

(a) (b)

Fig. 1. Timing recovery in UWB systems (a) *Received UWB sequence (red) and received signal during one cycle (blue). The transmitted pulse is an ideal first-derivative Gaussian impulse with the duration of (approximately) $T_p = 5$ samples, while the spacing between the transmitted pulses is 20 samples.* **(b)** *Root-mean square error (RMSE) of delay estimation vs. SNR (defined as the ratio between the power of the sequence and noise power in one cycle), and this for different sampling rates N_s. N_n denotes the Nyquist rate.*

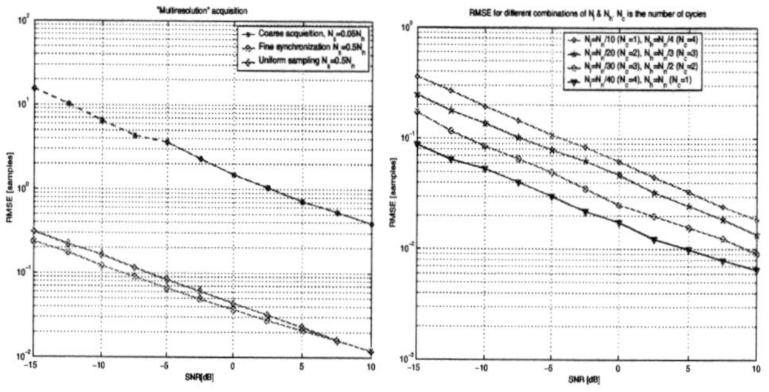

(a) (b)

Fig. 2. Two-step delay estimation (a) *RMSE of two-step time delay estimation is compared to the RMSE obtained with high-rate uniform sampling over the entire cycle.* **(b)** *RMSE of delay estimation for different combinations of N_l (the sampling rate for coarse synchronization) and N_h (the sampling rate for fine synchronization). N_c denotes the number of averaging cycles during each phase.*

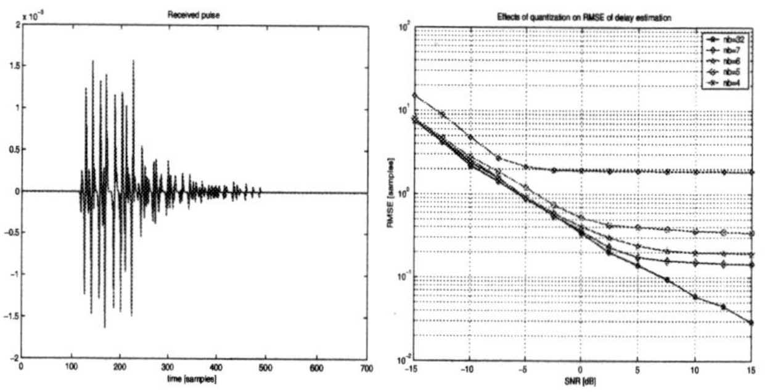

(a) (b)

Fig. 3. Estimating higher-rank channel models (a) *Received UWB signal made up of 70 pulses, with 8 components being dominant (containing approximately 85% of the total power). Average spacing between the received dominant components is $2T_p$, where T_p denotes the pulse duration.* **(b)** *Effects of quantization on the RMSE of delay estimation (dominant components only) for 4-7 bit receiver architectures. The results are compared to the case when the number of bits is $n_b = 32$. The sampling rate is one fourth the Nyquist rate ($N_s = N_n/4$).*

SILICON SPIKE-BASED SYNAPTIC ARRAY AND ADDRESS-EVENT TRANSCEIVER

R. Jacob Vogelstein[1], *Udayan Mallik*[2], *and Gert Cauwenberghs*[2]

[1]Department of Biomedical Engineering
[2]Department of Electrical and Computer Engineering
Johns Hopkins University, Baltimore, Maryland 21218

ABSTRACT

An integrated array of 2,400 spiking silicon neurons, with reconfigurable synaptic connectivity and adjustable neural spike-based dynamics, is presented. At the system level, the chip serves as an address-event transceiver, with incoming and outgoing spikes communicated over an asynchronous event-driven bus. Internally, every cell implements a spiking neuron that models general principles of synaptic operation as observed in biological membranes. Synaptic conductance and synaptic reversal potential can be dynamically modulated for each event. The implementation employs mixed-signal charge-based circuits to facilitate digital control of system parameters and minimize variability due to transistor mismatch. In addition to describing the structure of the silicon neurons, we present experimental data characterizing the operation of the 3mm × 3mm chip fabricated in 0.5μm CMOS technology.

1. INTRODUCTION

An increasing number of experimental microchip designs are inspired by biological structures, particularly those found in the vertebrate and invertebrate nervous systems. Neuromorphic systems engineering [1, 2] emulates both function and structure of biological neural systems in silicon, and correspondingly achieves high levels of efficiency in the implementation of sensory systems for vision [3] and audition [4]. The complexity of neural computation, beyond sensory perception, requires a multi-chip approach and a proper communication protocol between chips to implement higher levels of processing and cognition.

The common language of neuromorphic chips is the "Address-Event Representation" (AER) communication protocol [5], which uses time-multiplexing to emulate extensive connectivity between neurons. In its original formulation, AER implements a one-to-one connection topology; to create more complex neural circuits, convergent and divergent connectivity is required. The AER framework has proved essential in enhancing the connectivity between multi-chip neuromorphic modules [6, 7, 8, 9, 10, 11]. AER "transceivers" [6, 10, 11] call for a memory-based projective field mapping that enables routing an address-event to multiple receiver locations. Accordingly, the chip described in this paper contains 2,400 neurons but no hardwired connections between cells, rather depending on an external infrastructure to route events to their appropriate targets.

There are a few examples of reconfigurable neural array transceivers in the literature [10, 11], but the one presented here differs in some important aspects. This paper concentrates on the design of the neural cell, which has two novel attributes. First, the silicon neuron implements a discrete-time version of the single compartment, conductance-based *membrane equation*—a standard model describing current flux through biological neural membranes—which enables more sophisticated simulations than a standard integrate-and-fire model allows. Second, this design permits an unlimited number of connections between neurons, with independent control of connection strength and synaptic reversal potential on a per-connection basis.

This work is partially funded by NSF Award #0120369 and ONR Award #N00014-99-1-0612. RJV is supported by an NSF Graduate Research Fellowship.

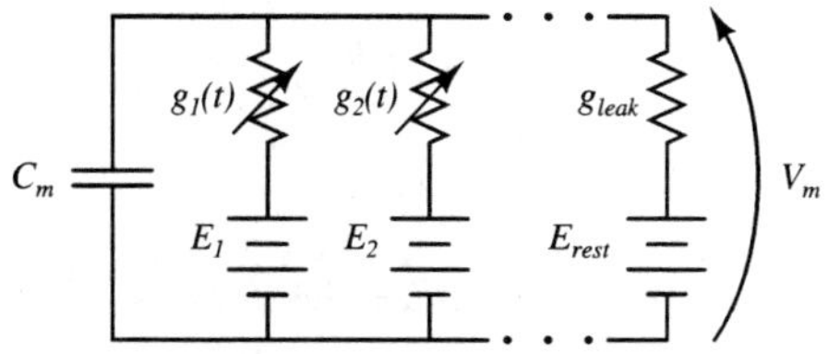

Fig. 1. Single-compartment model of a biological neuron with multiple synapses and a static leak conductance.

2. NEURAL MODEL

A number of silicon neurons have been presented in the literature with varying degrees of biological accuracy. The most detailed and accurate silicon models feature many parameters and are very flexible [12], but occupy a large on-chip area and therefore limit the number of cells that can be fabricated on a single chip. The simplest models contain only a few transistors and are well-suited for implementation in a large-scale network, but deviate significantly from the biology and have few adjustable parameters. Many applications would benefit from a balance between these two extremes: a more biologically accurate neural model allows for more sophisticated simulations of cognitive functions, but only in the context of a sophisticated network architecture. We have therefore designed a small-footprint, highly configurable, "general-purpose" silicon neuron that implements a standard model of biological neural membranes.

A popular model used in computational neuroscience to describe the current flux through biological neural membranes is the single-compartment model (Fig. 1). The model is specified by the membrane equation:

$$C_m \frac{dV_m}{dt} = g_1(t) \cdot (E_1 - V_m) + g_2(t) \cdot (E_2 - V_m) + \cdots \quad (1)$$
$$+ g_{leak} \cdot (E_{rest} - V_m)$$

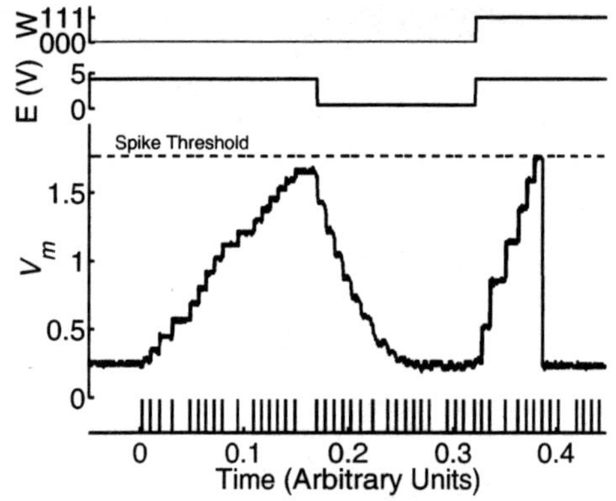

Fig. 2. Data captured from an oscilloscope during operation of the chip. The lower trace illustrates the membrane potential (V_m) of a single neuron in the array as a series of events are sent at times marked at the bottom of the figure. The synaptic reversal potential (E) and synaptic weight (W) are drawn in the top two traces.

where C_m is a large membrane capacitance, V_m is the membrane potential, $g_i(t)$ is a time-varying synaptic conductance, and E_i is a synaptic reversal potential. Although biology operates in continuous time, most neural interactions occur on the millisecond time scale, so it is possible to simulate the internal dynamics of a neuron using fast, discrete time steps. Similarly, while multiple synaptic inputs can be active simultaneously in a real neuron, it is essentially equivalent to activate a group of synapses in rapid succession due to biology's low precision in the time domain. We exploit both of these observations in the design of the silicon neuron.

The neural cell schematic is shown in Figure 3, along with event generation circuitry to trigger and communicate output spikes (see Sec. 3). The cell size, including the event generation circuitry, is 40μm $\times$ 60μm. Using a simple switched-capacitor architecture, this circuit implements a discrete-time version of the membrane equation:

$$C_m \frac{\Delta V_m}{\Delta T} = g(T) \cdot (E - V_m) \tag{2}$$

By sequentially activating switches X1 and X2, a packet of charge proportional to the externally supplied and dynamically modulated reversal potential E is added to (or subtracted from) the membrane capacitor C_m. The amount of charge transferred is also dependent on which of the three binary-sized "synaptic weight" capacitors (C0-C2) are enabled: these elements are dynamically switched on and off by applying voltage to the gates of transistors M1-M3. When sufficient charge has been integrated on C_m to cause V_m to exceed a "spike threshold", the neuron generates an event (see Sec. 3). Figure 2 illustrates the functionality of one neuron in the array as it receives a sequence of events with both the synaptic reversal potential and the synaptic weight dynamically varied.

There are a few advantages of this architecture over previous AER transceiver designs [11]. First, it allows for simulation of an unlimited number of synapses on every cell, as each incoming event can be assigned a unique weight and reversal potential. Second, it simulates biologically realistic conductance-based synapses. The use of conductance-based synapses in a neural model can have important implications: unlike standard integrate-and-fire models, the order of events in a conductance-based model is an essential factor in determining the neural output (Fig. 2). Third, charge-based circuits exploit better matching between capacitors than between MOS transistors due to threshold variations, which results in greater uniformity of operation across the chip. Finally, there is very little charge leakage off the membrane capacitor, allowing for large dynamic range in the implementation of neural dynamics on various time scales. Since neural integration is discrete-time, it is also possible to decouple event timing from emulated time, and dynamically warp the time axis [13].

3. EVENT GENERATION

Information encoded by neurons in the array is represented by the time between successive events. Therefore, event generation and communication is an essential element of the design. The event generation circuitry of [11] is embedded in every cell (Fig. 3, right). An event—signaled by a low voltage on $\overline{R_{req}}$—should be generated each time a neuron's membrane potential exceeds the spike threshold. In the circuit, charge integrated on the membrane capacitor (C_m) of a cell (see Sec. 2) results in an increase in potential applied to the gate of M4, the input terminal of a comparator. Since V_m can rise very slowly, the comparator is implemented as a current-starved inverter, with M5 biased in weak inversion, for reduced power dissipation. The spike threshold is set by V_{thresh}, the voltage applied to the source of M4; this value is shared by all cells in the array and is externally controlled. The corresponding input-referred threshold is approximately equal to $V_{thresh} + V_{T_n}$, where V_{T_n} is the threshold voltage of M4. When V_m exceeds this value, a positive-feedback loop implemented by transistors M6-M8 is activated, triggering a spike event by driving V_m to the positive rail and V_{comp} to ground.

A high voltage on M15 activates $\overline{R_{req}}$, the output node of a row-wise wired-NAND, and indicates to the row arbitration circuitry that a cell in that row has generated an event and needs to be serviced. Until this occurs, the row and column acknowledge signals R_{ack} and C_{ack} will remain low, maintaining the positive feedback loop and preventing any further inputs from affecting the cell. The row arbitration circuity indicates it has selected a row by driving one pair of R_{scan} and R_{ack} signals high. All cells in that row with pending events will then pull their $\overline{C_{req}}$ signals low, indicating to the column arbitration circuitry that they have generated events and need to be serviced. Finally, the column arbitration circuitry indicates which column it has selected by driving one column's C_{req} signal high. At that point, both R_{ack} and C_{ack} are asserted (for one cell only) so the positive-feedback loop is inactivated and the reset circuit implemented by nMOS transistors M9 and M10 causes V_m to become V_{reset} (like V_{thresh}, V_{reset} is shared by all cells in the array and is externally controlled). As V_m drops below the comparator's threshold voltage, V_{comp} is pulled high by M5 and the column and row requests ($\overline{C_{req}}$ and $\overline{R_{req}}$) are removed. This completes the handshaking sequence between a cell and the arbitration circuitry.

4. EXPERIMENTAL RESULTS

Every incoming event is routed to one or more neurons and is associated with a particular binary weight and an analog reversal potential (E). Additionally, a spike threshold voltage (V_{thresh}) and resting potential (V_{reset}) are set globally for the entire chip. To quantify neurons' dependence on these parameters, we have performed three experiments. First, to determine the effect of the

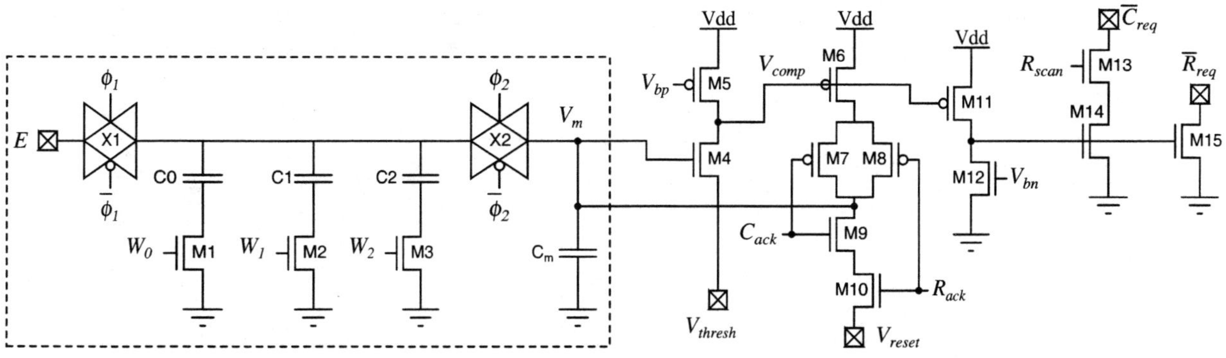

Fig. 3. Silicon single-compartment neuron (inside dashed box), with event generation circuitry (shown right, [11]).

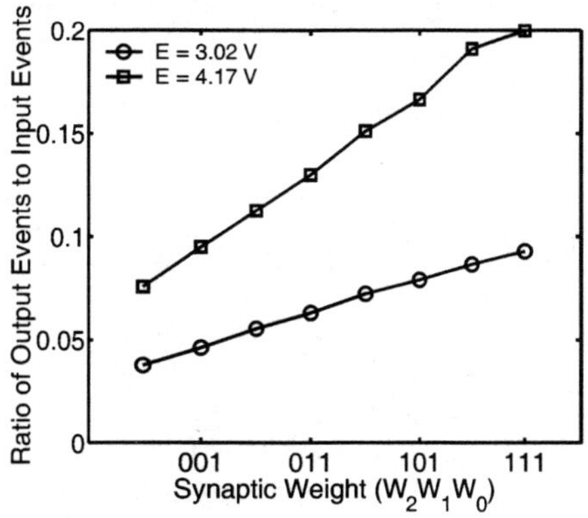

Fig. 4. Average ratio of output events to input events versus synaptic weight. Data are averaged across 10 trials per cell at each value of synaptic weight, and then averaged over all 2,400 cells in the array.

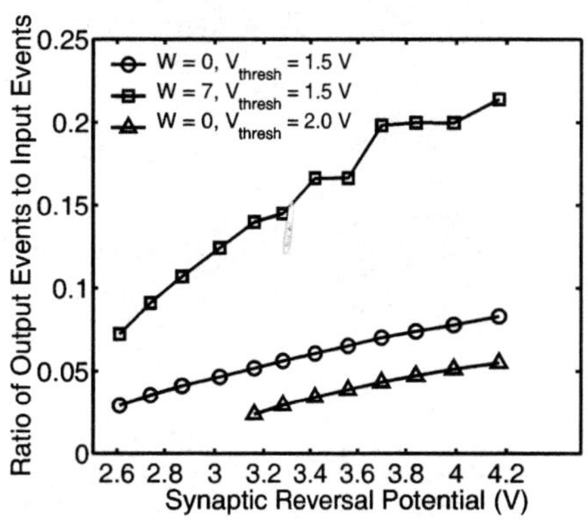

Fig. 5. Average ratio of output events to input events versus synaptic reversal potential (E). Data are averaged across 10 trials per cell at each value of synaptic weight, and then averaged across all 2,400 cells in the array.

weight capacitors C0-C2 (see Fig. 3), each neuron's membrane potential was reset to a fixed voltage and a series of excitatory events at a fixed reversal potential and a given synaptic weight were sent to the cell. The number of events required to elicit a spike was recorded and after ten measurements at each synaptic weight, another cell in the array was tested. The results over all 2,400 cells were summarized by plotting the average ratio of output events to input events versus synaptic weight (Fig. 4). The design called for a greater slope for the lines in Figure 4, but this was limited by a large parasitic capacitance when a weight capacitor was in the "off" state.

The second experiment was designed to quantify the effect of the synaptic reversal potential (E). Here, each neuron's membrane potential was reset to a fixed voltage and a series of events at a given excitatory synaptic reversal potential and a fixed synaptic weight were sent to the cell. Again, the number of events required to elicit a spike was recorded. However, instead of varying the weight (as in the first experiment), the same cell was then re-tested with a different value of E. The results over all 2,400 cells were summarized by plotting the average ratio of output events to input events versus synaptic reversal potential (Fig. 5). Although in normal operation the spike threshold voltage V_{thresh} is likely to be fixed, in some cases it may be desirable to dynamically vary this parameter. Therefore, we repeated the experiment described above using two different values of V_{thresh} (Fig. 5).

As discussed in Section 2, one of the advantages of implementing the neurons in charge mode is that it minimizes variability across the chip. To quantify the mismatch between neurons, we conducted a third experiment wherein all of the event parameters (synaptic weight, synaptic reversal potential, spike threshold voltage, and resting potential) were held constant at values that were barely supra-threshold. The average ratio of the number of output events to input events was measured as before, and the distribution was plotted as a histogram (Fig. 6a). The variability is small, with a standard deviation of 0.0017 around the mean of 0.0210. To see if there was any systematic variation, the number of input events required to elicit an output event was converted into a normalized gray-scale value and the value for each neuron was plotted as a single pixel in a 60×40 bitmap, where darker pixels correspond to a larger number of output events per input events (Fig. 6b). Evidently, there is a gradient toward lower response rates in the upper-right quadrant of the array.

5. CONCLUSION

We have presented a novel neural array transceiver consisting of 2,400 spiking neurons that each implement a discrete-time version of the biological membrane equation and include a single "general-

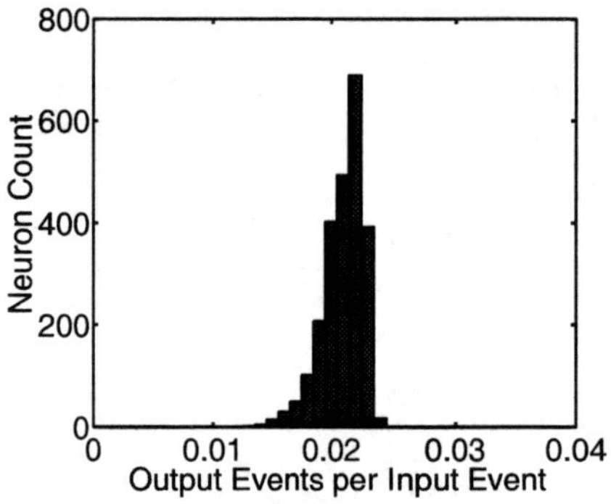

(a)

(b)

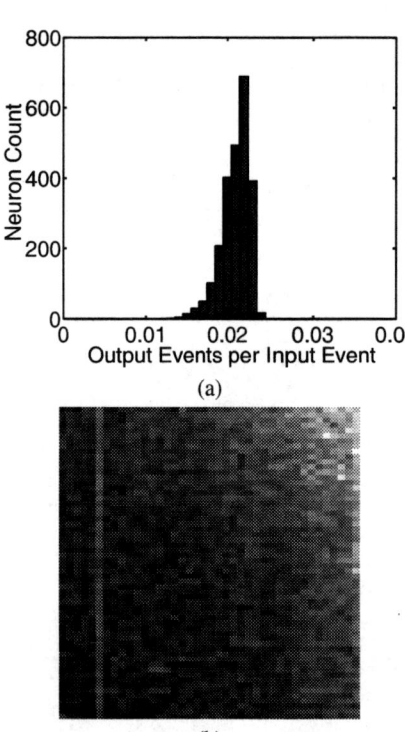

Fig. 6. (a) Histogram showing distribution of the average ratio of output events to input events. (b) Bitmap illustrating spatial trends in the variation of the average ratio of output events to input events (see text for details).

purpose" synapse with dynamically configurable weight and reversal potential. The data shown verify the functionality of the chip. Future work will focus on embedding multiple chips in a large system containing digital memory to store connection patterns and synaptic parameters, a microcontroller to manage a shared AER bus, and a high-speed interface with a neuromorphic sensor or personal computer. This system will allow rapid prototyping and emulation of large-scale networks of realistic neurons and, ultimately, interface with real biological neurons in computer-controlled wetware experiments. An interesting extension of the functionality of these networks is to incorporate silicon models of spike-based learning [14] in the address-domain [13].

Fig. 7. Chip micrograph. Center: 60×40 neuron array. Periphery: row and column address-event decoders and arbitrating encoders.

6. REFERENCES

[1] C. Mead, *Analog VLSI and Neural Systems*, Addison-Wesley, Reading, MA, 1989.

[2] T.S. Lande, Ed., *Neuromorphic Systems Engineering— Neural Networks in Silicon*, Kluwer Academic, Norwell, MA, 1998.

[3] C. Koch and H. Li, Eds., *Vision Chips: Implementing Vision Algorithms with Analog VLSI Circuits*, IEEE Computer Press, 1995.

[4] A. van Schaik, E. Fragniere, and E. Vittoz, "Improved silicon cochlea using compatible lateral bipolar transistors," in *Adv. Neural Info. Proc. Sys. 8*, D.S. Touretzky, M.C. Mozer, and M.E. Hasselmo, Eds., pp. 671–677. MIT Press, Cambridge, MA, 1996.

[5] M. Mahowald, *An analog VLSI system for stereoscopic vision*, Kluwer Academic Publishers, Boston, MA, 1994.

[6] K. A. Boahen, "Point-to-point connectivity between neuromorphic chips using address events," *IEEE Trans. Circ. Sys.-II*, vol. 47, no. 5, pp. 416–434, 2000.

[7] C. M. Higgins and C. Koch, "Multi-chip neuromorphic motion processing," in *Proc. 20th Ann. Conf. Adv. Res. VLSI*, D. S. Wills and S. P. DeWeerth, Eds., Los Alamitos, CA, 1999, pp. 309–323, IEEE Computer Society.

[8] S.-C. Liu, J. Kramer, G. Indiveri, T. Delbrück, and R. Douglas, "Orientation-selective aVLSI spiking neurons," in *Adv. Neural Info. Proc. Sys. 14*, T. G. Dietterich, S. Becker, and Z. Ghahramani, Eds. MIT Press, Cambridge, MA, 2002.

[9] N. Kumar, W. Himmelbauer, G. Cauwenberghs, and A. Andreou, "An analog VLSI chip with asynchronous interface for auditory feature extraction," *IEEE Trans. Circ. Sys.-II*, vol. 45, no. 5, pp. 600–606, 1998.

[10] G. Indiveri, A. M. Whatley, and J. Kramer, "A reconfigurable neuromorphic VLSI multi-chip system applied to visual motion computation," *Proc. MicroNeuro'99*, Apr. 1999.

[11] D. H. Goldberg, G. Cauwenberghs, and A. G. Andreou, "Probabilistic synaptic weighting in a reconfigurable network of VLSI integrate-and-fire neurons," *Neural Networks*, vol. 14, no. 6-7, pp. 781–793, 2001.

[12] M. Mahowald and R. Douglas, "A silicon neuron," *Nature*, vol. 354, pp. 515–518, 1991.

[13] R. J. Vogelstein, F. Tenore, R. Philipp, M. S. Adlerstein, D. H. Goldberg, and G. Cauwenberghs, "Spike timing-dependent plasticity in the address domain," in *Adv. Neural Info. Proc. Sys. 15*, S. Becker, S. Thrun, and K. Obermayer, Eds. MIT Press, Cambridge, MA, 2003.

[14] P. Hafliger and M. Mahowald, "Spike based normalizing Hebbian learning in an analog VLSI artificial neuron," in *Learning On Silicon*, G. Cauwenberghs and M. A. Bayoumi, Eds., pp. 131–142. Kluwer Academic, Norwell, MA, 1999.

BIOLOGICALLY INSPIRED ARTIFICIAL NEURAL NETWORK ALGORITHM WHICH IMPLEMENTS LOCAL LEARNING RULES

Ausra Saudargiene[1,2], *Bernd Porr*[1] *and Florentin Wörgötter*[1]

[1]Department of Psychology, University of Stirling, Stirling FK9 4LA, Scotland
[2]Department of Informatics, Vytautas Magnus University, Kaunas, Lithuania
<ausra,bp1,worgott>@cn.stir.ac.uk

ABSTRACT

Artificial neural networks (ANNs) are usually homogenous in respect to the used learning algorithms. On the other hand, recent physiological observations suggest that in biological neurons synapses undergo changes according to local learning rules. In this study we present a biophysically motivated learning rule which is influenced by the *shape* of the correlated signals and results in a learning characteristic which depends on the dendritic site. We investigate this rule in a biophysical model as well as in the equivalent artificial neural network model. As a consequence of our local rule we observe that transitions from differential Hebbian to plain Hebbian learning can coexist at the same neuron. Thus, such a rule could be used in an ANN to create synapses with entirely different learning properties at the same network unit in a controlled way.

1. INTRODUCTION

Learning rules used to update the weights in artificial neural network algorithms are the same for all inputs and units. However, recent physiological experiments suggest that in biological neurons synaptic modifications depend on the location of the synapse (1) i.e. synaptic strength is regulated by local learning rules.

The same synapse may be strengthened and weakened depending on the temporal order of the pre- and postsynaptic activity. The weight grows if the presynaptic signal precedes the postsynaptic signal, and shrinks if the temporal order is reversed. Such form of synaptic modifications is called spike-timing-dependent plasticity (STDP) (2). However, not only the timing of the pre- and postsynaptic activity, but also the the shapes of the signals may define the properties of synaptic plasticity. This claim is supported by the fact that strong depolarization, necessary to induce synaptic changes, has a different origin and a varying shape along the dendritic tree. Close to the soma learning is driven by steep and short back-propagating spikes which become more shallow and longer in duration while back-propagating into the dendritic tree (3). In distal parts, where back- propagating spikes fail to invade, slow and wide local Na^+- and Ca^{2+} channel-dependent dendritic spikes provide the necessary depolarization (1). These observations suggest that synaptic modifications are location-dependent.

In this paper we present a biophysical model of STDP which captures the dependence of synaptic changes on the membrane potential shape. The model uses a differential Hebbian rule to correlate the NMDA synaptic conductance and the derivative of the membrane potential at a synapse. We will show that the model reproduces the STDP weight change curve in a generic way and is sensitive to the different shapes of the membrane potential. The model predicts that learning depends on the synapse location on the dendritic tree. Then we will describe the equivalent circuit diagram and discuss the model referring to system-theory, presenting it in the context of filter transfer functions at the end of this article.

2. BIOPHYSICAL MODEL

The model represents a dendritic compartment with a single NMDA synapse (Fig. 1 A). The NMDA channels are essential in inducing synaptic plasticity as their blockade to a large degree prevents STDP (1). It is believed that NMDA channel-mediated Ca^{2+} influx triggers the chain reactions involving CaMKII, calmodulin, calcineurin and in this way affects the synaptic strength (4). The NMDA synaptic conductance, regarded as a presynaptic signal, is given by :

$$g(t) = \bar{g}\hat{g}(t) = \bar{g}\frac{e^{-t/\tau_1} - e^{-t/\tau_2}}{1 + \kappa e^{-\gamma V(t)}} \quad (1)$$

where V is the membrane potential, $\bar{g}_N = 4\ nS$ peak conductance, $\bar{g}_N = 4\ nS$ $\tau_1 = 40\ ms$, $\tau_2 = 0.33\ ms$ time constants and $\eta = 0.33/mM$, $[Mg^{2+}] = 1\ mM$, $\gamma = 0.06/mV$ (5). The membrane potential is expressed as:

$$C\frac{dV(t)}{dt} = \rho\, g(t)[E - v(t)] + i_{dep}(t) + \frac{V_{rest} - V(t)}{R}, \quad (2)$$

where ρ is the synaptic weight of the NMDA-channel, g its conductance, $E = 0\ mV$ its equilibrium potential. The current i_{dep} is used to account for the depolarizion caused by other sources than synaptic inputs, such as back-propagating spikes or local dendritic regenarative potentials. The last term represents the leakage current. The resting potential $V_{rest} = -70\ mV$, membrane capacitance $C = 50\ pF$ and the membrane resistance to $R = 100\ M\Omega$.

The differential Hebbian learning rule for the synaptic change is defined as:

$$\frac{d\rho}{dt} = \hat{g}(t)V'(t), \qquad (3)$$

where $\hat{g}$ is the normalized conductance function of the NMDA channel, the pre-synaptic influence quantity, and V' is the derivative of the postsynaptic membrane potential.

The depolarizing membrane potentials, which trigger synaptic plasticity, vary along the dendritic tree. We use a short and steep back-propagating action potential to model the synaptic changes close to the soma, and long and shallow dendritic spike to account for synaptic modifications in the distal parts. The back-propagating spike and the dendritic spike, measured $210\mu m$ and $860\mu m$ from the soma, respectively, are presented in Fig. 1 B and have been taken from (6; 7). The depolarization coming from these spikes is very strong, therefore we may neglect the contribution of the NMDA synaptic input. Instead of using Eq. 2 we calculate the change of the membrane potential using the given shape of the spike and then substitute its derivative in the learning rule (Eq. 3).

We obtain an asymmetrical weight change curve if the depolarization is provided by a steep back-propagating spike (Fig. 1 C). The synapse is weekened if $T < 0$ and strengthened if $T > 0$, where T is the temporal difference between the presynaptic activity and the postsynaptic activity. $T > 0$ means that the postsynaptic signal follows the presynaptic signal at the NMDA channel and vice versa. However, we observe a shifted curve if the depolarization comes from the shallow dendritic spike. The synaptic weight grows even for negative values of $T > -20ms$. Thus, we get plain Hebbian learning between $-20ms$ and ∞.

The model reproduces the STDP curve in a generic way. The shape of the weight change curve is strongly influenced by the shape of the depolarizing membrane potential, which induces plasticity. The slow rising flank of this signal is the essential factor of the transition from an asymmetrical to a symmetrical weight change characteristic. As the depolarizing potentials vary in different parts of the dendritic tree, these results suggest that learning rules are local and depend on the location of the synapse in biological neurons.

The electrical circuit equivalent to the model described above is presented in Fig. 2. Elements R_1, C_1 define the shape of the presynaptic signal g. R_3 corresponds to the in-

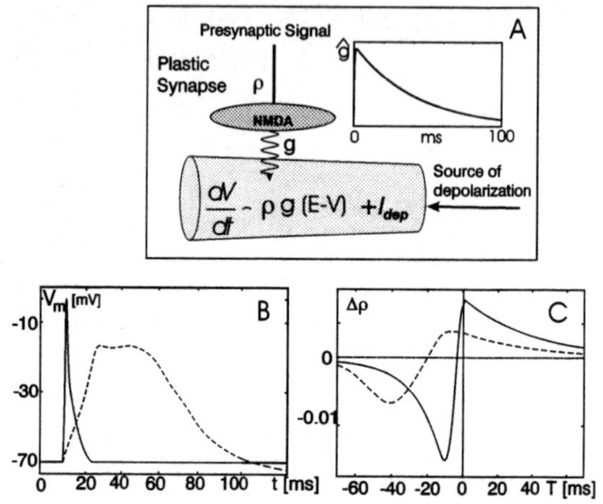

Fig. 1: *Schematic diagram of the model. A) Components of the membrane model. The inset shows the NMDA synaptic conductance function. B) Depolarizing membrane potentials: steep back-propagating spike and shallow dendritic spike $210\mu m$ and $860\mu m$ from the soma, respectively C) The resulting weight change curves. The shallow depolarizing potential leads to potentiation even for negative values $0 < T < -20ms$ values.*

tracellular resistance, R_2 and C_2 describe the passive membrane properties and alltogether determine the shape of the postsynaptic signal v. The derivative of v, obtained after the filtering in the last R_2 and C_2 circuit, is multiplied by g. The resulting so called weight change is fed to a gain-controlled amplifier and influences the postsynaptic signal v. Various shapes of the postsynaptic sigal v may be obtained by adjusting the values of R_2 , R_3 and C_2 and would lead to different learning characteristics.

3. BIOLOGICALLY INSPIRED SITE-SPECIFIC LEARNING ALGORITHM

We represent a further step of abstraction in Fig. 3. This block-diagram is not directly equivalent to the circuit in Fig. 2 but it captures the main observation emerging from the biophysical model. Namely that learning depends on the location of the synapse, i.e. is driven by the derivative of a postsynaptic signal specific at a given site. In an artificial neural network system this would mean that output signal undergoes a transformation specific for each input and only then its derivative is applied to update the weight of a given input. The diagram of such an algorithm is presented in Fig. 3. We can still roughly associate the NMDA characteristic to the pathways $x_{1,\ldots,n}$ representing many (possibly different) inputs and the source of depolarization (e.g., the back-propagating spike) to the pathway x_0. Hence this pathway enters the summation node with an unchangable weight ρ_0. This circuit is a modified version of the ISO learning circuit (8). ISO learning is a drive-reinforcement

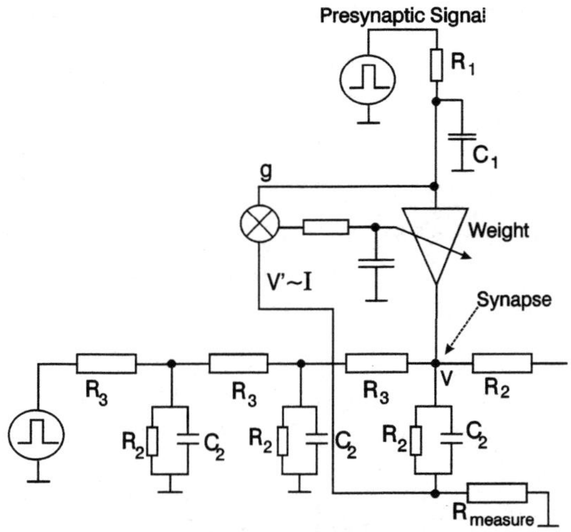

Fig. 2: *Equivalent electrical circuit of the learning algorithm. Postsynaptic signal v is differentiated by R_2 C_2 circuit and multiplied by the presynaptic signal g to obtain the weight which influences the postsynaptic signal v via a gain-controlled amplifier.*

algorithm for temporal sequence learning where the weights change according to the relative timing of the input signals. All inputs x_0, x_1,..., x_n are filtered using bandpass filters h_0, h_1,..., h_n, weighted by ρ_0, ρ_1,..., ρ_n and summed to produce the output v: $v = \rho_0 u_0 + \sum_{i=1} \rho_i u_i$, where $u = x * h$. Different from the ISO learning, here the output is also filtered with the filters h_{11},...,h_{nn}, and only then the derivatives of the obtained signals v_1',...,v_n' are used to change the weights of the corresponding inputs:

$$\frac{d}{dt}\rho_i = \mu u_i v_i' \quad \text{where } v_i = v * h_{ii},\ \mu \ll 1. \qquad (4)$$

We assume that the input x_0 is dominating the output and its weight ρ_0 is fixed. We apply the analytical solution derived for the ISO learning (8) to calculate the weight change curve for different shapes of the filtered output signal (for details see Appendix). For a steep output signal entering the learning rule we obtain differential Hebbian learning, and for a shallow one we get a curve similar to plain Hebbian learning (Fig. 3B,C). The parameters of the filters h_{11},...,h_{nn} which transforms the output signal determines this transition.

4. DISCUSSION

The biophysical model of STDP inspired an artificial neural network algorithm with site-specific learning rules. The biophysical model is based on a differential Hebbian learning rule which correlates the NMDA synaptic conductance with the derivative of the membrane potential. The results show that the weight change curve strongly depends on the

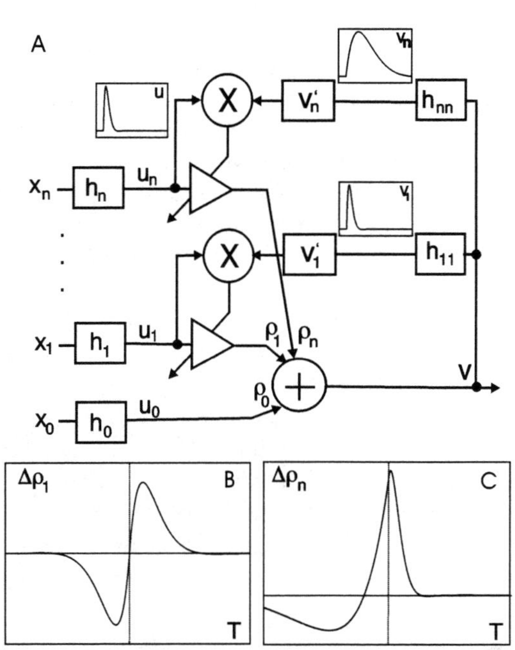

Fig. 3: *A) Algorithm for site-specific learning. Transfer functions are denoted as h, changing weights ρ as an aplifier. All inputs are filtered. Weight ρ_0 is fixed. Weights ρ_1,..., ρ_n are updated using the derivatives v_1',...,v_n' of the filtered output v. Filter functions h_{11} ,...,h_{nn} differ for each input. B) Analytically calculated weight change curve if the filtered output has a steep rising flank C) Analytically calculated weight change curve if the filtered output has a shallow rising flank.*

shapes of the depolarizing membrane potential at the location of the synapse. This signal changes its shape along the dendrite and may be provided by different mechanisms such as back-propagating spikes close to the soma and dendritic spikes in the distal parts. Therefore we predict that learning rules are location-dependent. Close the soma, where learning is driven by short back-propagating spikes, the synaptic modifications are bidirectional, described by an asymmetrical STDP curve. In the distal parts, where synaptic changes are induced mainly by long-lasting dendritic spikes, synapses undergo potentiation even for negative values of T. The same learning rule leads to different synaptic modifications and it is self-adjusting following the shapes of the depolarization source in different locations of the dendritic tree.

The typical approach to model STDP is to assume a certain weight change curve which does not depend on the local properties of the cell, e.g. (9). A few more detailed models take into consideration the postsynaptic signal which is associated with the membrane potential, e.g. (10; 11; 12) and observe that its shape influences the shape of the weight change curve. These models differ from our as the rule of (10) is based on TD learning, while (11; 12) rely on the absolute Ca^{2+} concentration in the weight updating algorithm.

Our algorithm offers the possibility to easily define a parameter-controlled learning rule in an artificial neural network. We have just now started trying to solve an instrumental conditioning problem, where the actions of the learner influence its inputs and hence the learning with such an architecture. A small network of sub-comparmentalized neurons is linked to a simple agent that reacts to stimulus presentation with an orienting behaviour following the stimulation of the right neuronal subset. The goal is to train it with one unconditioned stimulus (US) and several conditioning stimuli (CS) only one of which is correlated to the unconditioned stimulus. The US will always trigger the correct output neurons to elicit the orienting response. Conversely, each CS elicit a response in many input neurons, some of which are better correlated with *each other* than others. Hebbian learning *between* these CS inputs will "extract" and strenghten the better-correlated neurons. This leads, after learning, to a drive from *all* CS regardless of their correlation with the US. Now we get reliable (but mostly *wrong*) behavioural reactions. Since all but one CS are temporally uncorrelated to the US, differential Hebbian learning will lead to a weakening of all "wrong" CS. By the end, the system has learned to drive a small subset of only a few neurons in a feed-forward way, eliciting a response that will lead to the desired behaviour. This is work in progress an no results exist so far. Nevertheless, it clearly shows how such sub-compartmentalized learning rules could be used for behavioral control.

5. APPENDIX

The weight change curves are calculated using the analytical solution obtained for ISO learning (8). We assume that the output is dominated by x_0 and the contribution of other inputs is negligible ($\rho_i \mid_{t=0} = 0, i > 0$). Then the pairs of the filter functions h_0 and h_{11}, h_0 and h_{22}, etc., h_0 and h_{nn} can be considered as single filter functions h_{01},..., h_{0n}. These filters are specific for each input x_1,...,x_n pathway and shape the output signal v_i whose derivative enters the learning rule. The filters h are described by: $h(t) = \frac{1}{b} e^{at} \sin(bt)$ with $a := -\pi f / Q$ and $b := \sqrt{(2\pi f^2 - a^2}$, where f is the center frequency and Q is the damping. Then the cumulative weight change at the $i - th$ pathway is given by: for $T \geq 0$

$$\rho_i(T) = \mu \frac{b_i M_i \cos(b_i T) + (a_i P_i + 2a_{0i} |p_i|^2) \sin(b_i T)}{b_i (P_i + 2a_i a_{0i} + 2b_i b_{0i})(P_i + 2a_i a_{0i} - 2b_i b_{0i})} e^{-T a_i}$$

and for $T < 0$

$$\rho_i(T) = \mu \frac{b_{0i} M_i \cos(b_{0i} T) + (a_{0i} P + 2a_i |p_{0i}|^2) \sin(b_{0i} T)}{b_{0i} (P_i + 2a_{0i} a_i + 2b_{0i} b_i)(P_i + 2a_{0i} a_i - 2b_{0i} b_i)} e^{-T a_i},$$

where $M_i = |p_i|^2 - |p_{i0}|^2$, $P_i = |p_i|^2 + |p_{i0}|^2$ and $p_i = |a_i|^2 - |b_i|^2$,$p_{0i} = |a_{0i}|^2 - |b_{0i}|^2$, $i > 0$. The parameters for the weight change curves presented in Fig. 3 are: $f_{01} = 0.01$, $Q_{01} = 0.6$, $f_{0n} = 0.002$, $Q_{0n} = 0.6$, $f_1 = f_n = 0.01$, $Q_1 = Q_n = 0.6$.

6. REFERENCES

[1] N. L. Golding, P. N. Staff, and N. Spurston, "Dendritic spikes as a mechanism for cooperative long-term potentiation," *Nature*, vol. 418, pp. 326–331, 2002.

[2] G-Q. Bi and M. Poo, "Synaptic modification by correlated activity: Hebb's postulate revisited," *Annu. Rev. Neurosci.*, vol. 24, pp. 139–166, 2001.

[3] J. C. Magee and D. Johnston, "A synaptically controlled, associative signal for Hebbian plasticity in hippocampal neurons," *Science*, vol. 275, pp. 209–213, 1997.

[4] G. Q. Bi, "Spatiotemporal specificity of synaptic plasticity: cellular rules and mechanisms," *Biol. Cybern.*, vol. 87, pp. 319–332, 2002.

[5] C. Koch, *Biophysics of Computation*, Oxford University Press, 1999.

[6] M. E. Larkum, J. J. Zhu, and B. Sakmann, "Dendritic mechanisms underlying the coupling of the dendritic with the axonal action potential initiation zone of adult rat layer 5 pyramidal neurons," *J. Physiol. (Lond.)*, vol. 533, pp. 447–466, 2001.

[7] G. Stuart, N. Spruston, B. Sakmann, and M. Häusser, "Action potential initiation and backpropagation in neurons of the mammalian central nervous system," *Trends Neurosci.*, vol. 20, pp. 125–131, 1997.

[8] B. Porr and F. Wörgötter, "Isotropic sequence order learning," *Neural Comp.*, vol. 15, pp. 831–864, 2003.

[9] S. Song, K. D. Miller, and L. F. Abbott, "Competitive Hebbian Learning through spike-timing-dependent synaptic plasticity," *Nature Neurosci.*, vol. 3, pp. 919–926, 2000.

[10] R. P. N. Rao and T. J. Sejnowski, "Spike-timing-dependent Hebbian plasticity as temporal difference learning," *Neural Comp.*, vol. 13, pp. 2221–2237, 2001.

[11] G. C. Castellani, E. M. Quinlan, L. N. Cooper, and H. Z. Shouval, "A biophysical model of bidirectional synaptic plasticity: Dependence on AMPA and NMDA receptors," *Proc. Natl. Acad. Sci. (USA)*, vol. 98, no. 22, pp. 12772–12777, October 23 2001.

[12] H. Z. Shouval, M. F. Bear, and L. N. Cooper, "A unified model of NMDA receptor-dependent bidirectional synaptic plasticity," *Proc. Natl. Acad. Sci. (USA)*, vol. 99, no. 16, pp. 10831–10836, 2002.

SPIKE BASED LEARNING WITH WEAK MULTI-LEVEL STATIC MEMORY

H. Kolle Riis and P. Häfliger

Institute of Informatics, University of Oslo, Norway
e-mail: haavarkr@ifi.uio.no, hafliger@ifi.uio.no

ABSTRACT

In this paper we present a VLSI implementation of a learning synapse that uses a spike based learning rule to adjust its weight. The weight is stored on a recently presented weak multi-level static memory cell (MLSM) [1]. This memory cell stores a voltage on a capacitance and that voltage is weakly driven to the closest of several stable levels. We verified the suitability of this memory for this task in a VLSI chip implementation. An array of integrate and fire neurons with four of these learning synapses each was implemented on a 0.6μm AMS CMOS chip. The learning capability of these neurons was tested in simple spike and rate based pattern recognition tasks in a two neuron network. Cross-inhibition between them lead to improved decorrelation of the output spikes, inducing a tendency in the neurons to specialize on different patterns.

1. INTRODUCTION

Neural models that use short pulse-events or spikes for communication have become ever more popular over the last years. Neurophysiological experiments indicate that this kind of signal representation is essential for processing in certain parts of the nervous system [2, 3]. Mathematical models try to measure and use the information capacity of spike trains and also electronic devices start to apply such asynchronous digital pulse signals to advantage. The circuit proposed in this paper makes use of temporal patterns of spike signals to define an unsupervised learning behavior.

A central problem of neuromorphic aVLSI learning circuits is the storage of the learning state variables. A suitable memory should be space conservative and distributed, always accessible to the neuromorphic local learning mechanisms in the artificial synapses. It should preferably not require global digital control signals, that add noise to the analog computations and require space for routing. It should be able to maintain the states over long periods of time (ideally years), since learning is a slow process. Some form of analog storage would be ideal for the analog update mechanisms.

Different methods have been used in the past. For example, digital static memory in combination with AD/DA conversion, either local in a synapse [4] or global with refreshing a local capacitor. Their major disadvantage is the required digital control signals, most severe in the case of central storage. In contrast, analog floating gate storage (analog flash ROM) with Fowler Nordheim tunneling and hot electron injection for the writing operations, seems to offer all desired properties [5–7]. It is truly analog and retains stored values for years. And indeed, we have used this kind of storage for several previous designs. The major drawback is that tunneling and hot electron injection structures are badly characterized and subject to severe device mismatches, at least in the affordable standard CMOS processes that research institutes like ours have access to. Therefore we introduced a novel weak multi-level static memory cell [1] as an alternative to present synaptic memory. In this paper we present a first application that makes use of it.

2. METHODS

Each synapse is attached to an I&F neuron and its state is expressed by the synaptic weight w. w expresses the increase of the 'soma voltage' of the I&F neuron as a result of a single input spike to the synapse. An I&F neuron simply accumulates all the synaptic inputs from all its synapses until it reaches its firing threshold. Then it produces an output spike (or action potential (AP)) and resets its soma voltage. Simultaneously, the learning mechanism changes the weight by Δw with every output spike, where Δw is:

$$\Delta w = w(\alpha c - \beta w) \tag{1}$$

w is the synaptic weight which in this paper is expressed in units of the neuron's threshold voltage. Thus, an input spike to a synapse with weight bigger than 1 will immediately trigger an AP. c is a variable we call 'correlation signal'. It is incremented by one with every input spike, it decays over time with fixed decay rate, and it is reset with every weight update. Thus, when the weights are updated (i.e. when the neuron produces an AP) c has a tendency to be big if there has been recent input activity and it is small if there has not.

Two conditions must be met for the weight to settle. First, the neuron's 'weight vector' $\vec{w}$, consisting of all the weights of the synapses attached to the neuron, must points in the same direction as the vector of the correlation signals $\vec{c}$ (sampled as the neuron fires). Secondly, the weight vector's length $\|\vec{w}\|$ has to be equal to $\sqrt{\frac{\alpha}{\beta}}$. The first condition lets $\vec{w}$ follow the pattern of recent input activity $\vec{c}$, if this input is correlated with output spikes. That makes it a 'Hebbian' learning rule. Note that I&F neurons will produce more output spikes the closer $\vec{w}$'s direction matches the direction of the vector of recent input activity. As a consequence, the learning accelerates

This work was supported by the EU 5th Framework Programme IST project CAVIAR.

the closer the weight vector gets to its attractive point. If it faces different input patterns, this self-reinforcing mechanism makes the neuron pick the input pattern that is initially the closest to $\vec{w}$. The second condition says that this learning rule implicitly normalizes the weight vector. Thus, it keeps the growth of the weights in check at all times and keeps them from saturating, which is essential for aVLSI implementations. Please refer to [8] for a deduction of these properties.

We have extended on an earlier implementation [8] and equipped it with a weak MLSM [1] for weight storage. The MLSM is a capacitive storage cell and injecting to or sinking current from it, changes the stored voltage V_w. In addition, weak currents slowly drive the voltage to the closest of five stable weight levels. The topography of the MLSM in conjunction with the learning mechanism restricts us from using an increased number of stable weight levels, since the stored voltage should remain in the subthreshold domain while the minimum spacing between stable weight levels is ~250mV. The stored voltage is applied to the gate of an nMOS transistor, which serves as a current source that is opened by an input spike to the synapse and sinks a current from the neuron during that input pulse. We define the theoretical weight w as the increase in the soma voltage caused by an input pulse to the synapse normalized by the fire threshold V_{fire}. Since we operate this nMOS transistor in subthreshold we can write:

$$w = I_0 e^{\frac{V_w}{U_T}} \frac{t_{spike}}{C_{soma} V_{fire}} \qquad (2)$$

I_0 summarizes device, process and physical constants, U_T is the thermal voltage and t_{spike} the length of the incoming spike. The transistor's source voltage is assumed to be at 0V. The weight updates are computed in two circuit blocks, one that computes the positive term $\alpha c w$, that increases the weight, and another computing a negative term βw^2, that decreases the weight. True to the learning rule they are triggered by the neuron's action potential. Their output is a digital voltage pulse, the length of which is modulated to express the magnitude of the increase and decrease respectively. Those pulses open current sources to move the voltage in the MLSM up or down.

The 'learn up' circuit (figure 1) represents the variable c as a voltage V_c which is proportional to c with proportionality constant C_c/Q_c (Q_c being the charge injected to the capacitance C_c for one input spike). It is incremented through transistors M1 and M2. M2 operates as a current source and is switched on by the incoming spike. The constant decay is achieved through transistor M3. Transistors M6 to M8 produce the circuits output pulse. Its length is proportional to V_c. The AP triggers the process by opening transistors M8 (not conducting). If V_c is bigger than V_{dd} minus the bias voltage on M7, then the output voltage goes low. This bias voltage is kept very close to V_{dd} and thus, we make the simplifying assumption that V_c only needs to be bigger than 0 to initiate that low output voltage. The output remains low as long as V_c remains bigger than 0. It does not do so for long though, since the AP also closes M4 and thus a constant current depletes V_c. That means the length of the active low output pulse (and thus, the change of V_w) is proportional to the initial value of

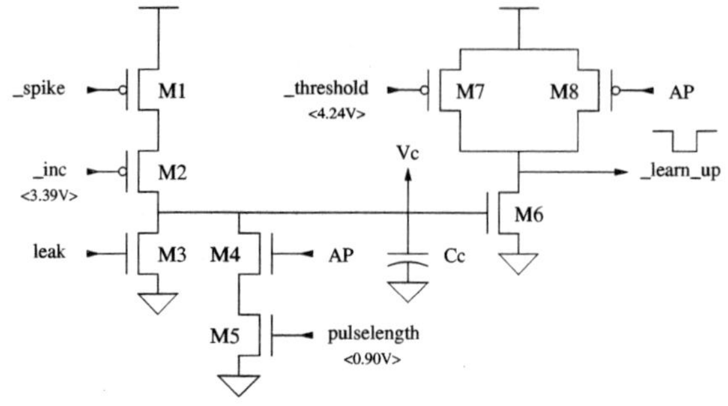

Fig. 1. *The learn up circuit. Modulates the length of the* ***_learn_up pulse*** *which opens a current source that injects a current to the MLSM capacitance.* V_c *is the voltage on* C_c.

V_c. And it follows that the change in w is proportional to the product of w and V_c.

$$\Delta V_w = U_T \alpha c \rightarrow \Delta w \approx \frac{\delta w}{\delta V_w} \Delta V_w = w \alpha c \qquad (3)$$

$$\alpha = \frac{Q_c}{(I_{leak} + I_{pulselength}) U_T} \cdot \frac{I_{up}}{C_w} \qquad (4)$$

I_{up} is the current injected to the memory capacitance C_w while I_{leak} and $I_{pulselength}$ the current through transistor M3 and M5 respectively.

The 'learn down' output pulse also depends on the discharge time of a capacitance through a current source (figure 2). The capacitance C_{down} is charged up to V_{dd} while the AP is low. As the AP goes high, the output of the NAND gate goes low as its second input V_{down} stays high for a while and starts to discharge only slowly through transistor M2. The current through that transistors is given by the gate to source voltage, where the gate voltage is kept constant and the source voltage is V_w. The active low output pulse finishes when the switching threshold of the NAND gate is reached. Thus, the length of the output pulse and therefore the decrease of V_w, is proportional to $I_0 e^{\frac{V_w}{U_T}}$ (We neglect effects of the slope factor and thus, this term is proportional to w, see equation 2). And finally, this lets us state that the change of w is proportional to w^2.

$$\Delta V_w = U_T \beta w \rightarrow \Delta w \approx \frac{\delta w}{\delta V_w} \Delta V_w = \beta w^2 \qquad (5)$$

$$\beta = \frac{(V_{dd} - 2.1) C_{down} C_{soma} V_{fire}}{I_0{}^2 e^{\frac{V_{w\_max}}{U_T}} U_T t_{spike}} \cdot \frac{I_{down}}{C_w} \qquad (6)$$

I_{down} is the current sinked from the memory capacitance C_w while 2.1V denotes the switching threshold of the NAND-gate.

3. RESULTS

Experiments have been conducted on a 0.6μm AMS CMOS chip. It contained an array of 32 test neurons, each consisting of four learning synapses, one inhibitory synapse, one excitatory synapse and the soma. During measurements, the MLSM

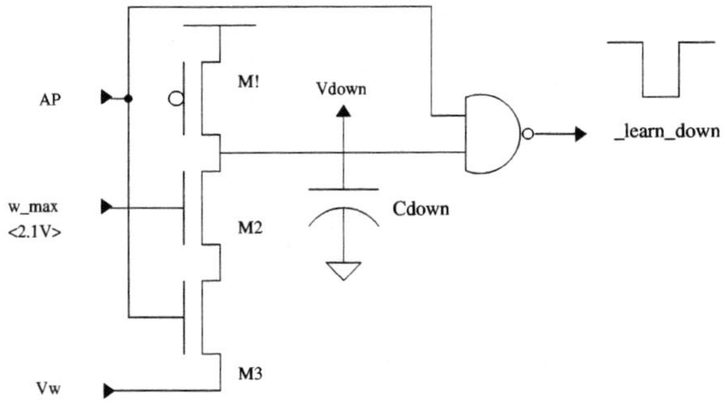

Fig. 2. *The learn down circuit. Modulates the length of the* ***learn_down pulse*** *which opens a current source that draws a current from the MLSM capacitance.*

	No inhibition		Cross-inhibition	
Neuron	1	2	1	2
Pattern 1	96.7%	73.3%	90.0%	56.7%
Pattern 2	3.33%	26.7%	10.0%	43.3%
Complete decorrelation	**30.0%**		**46.7%**	

Table 1. *Table lists the percentage of patterns learned for each neuron during 30 spike-pattern learning test runs with and without inhibition.*

holds five stable weight levels equally spaced between 1.20V and 180mV.

As mentioned before, the learning rule will make the neuron adapt to a spatial pattern of recent input activity. What the synapses considers to be 'recent' is determined by the time constant of the leakage of the correlation signal. Thus, by tuning this leakage, one can obtain different learning behaviors. If that leakage time constant is short relative to the input spike frequency, only synapses that received spike inputs that were almost coincident with the output spike get rewarded. Thus, the neuron will adapt to a spatio-temporal *pattern of coincident spikes*. On the other hand, if the correlation signal leakage time constant is long, the average activity over some time will determine the reward at a synapse and the neuron will adapt to *patterns of coincident average activity*.

To illustrate these two different behaviors we conducted an experiment where we stimulated two synapses of one neuron (no graph shown). Each synapse received an independent Poisson distributed spike train of 10Hz average frequency. If the correlation signal time constant was short (**_leak**=0.8V), the learning mechanism would perceive the input as two different spatial spike patterns: one where only synapse 1 receives a spike and another where only synapse 2 receives a spike. Thus, the neuron adapted to one of the two patterns by either raising the weight of synapse 1 or the weight of synapse 2 to the maximum and decreasing the other to a minimum.

If, however, we increased the correlation signal time constant (**_leak**=0.6V), the learning mechanism could no longer differentiate two different temporal patterns, but was only seeing one rate pattern of 10Hz input to both synapses. The weights would both settle at an intermediate strength.

For future applications we are interested to try the learning mechanism on unsupervised classification tasks. Therefore, we conducted an experiment with competitive learning in a two neuron network. Cross inhibition (connecting the outputs to the other neuron's inhibitory synapse) should decorrelate the neuron's outputs and force them to adapt to different input patterns. Again we did this with a strong leakage (**_leak**=0.8V, learning spike patterns) of the correlation signal first. The same 50Hz Poisson spike train was sent to synapses 1 and 2 of both neurons (pattern 1) and another 50Hz Poisson spike train stimulated synapses 3 and 4 in both neurons (pattern 2). We initialized all weights to be in a strong ($V_w \approx 1.2V$) state first. Figure 3 shows the evolution of the weights of synapses 1 of both neurons for one such experiment. The first neuron adapted to pattern 1 (upper trace) and the other to pattern 2 (lower trace). This was also confirmed by looking at the other 6 weights. Note also that over the 10 seconds that the experiments lasted, the weight update steps are smaller than the resolution of the weak multi-level memory cell (250mV) and the discretization does not exert any noticeable effect on the analog computation.

This particular figure shows a successful decorrelation experiment of the two neurons. However, this was not the case in all experiments. In about 50% of the experiments the neurons would adapt to the same spike pattern. This happened because the competition between them was not effective enough: When the weights were strong it took but a single spike input to trigger a spike output. The inhibition was usually not fast enough to prevent the other neuron from firing, if it was triggered by the same spike. Still the decorrelation did actually not fail completely. That became obvious as we repeated the experiment without the cross inhibition. Then, both neurons showed an affinity to input pattern 1 (as they actually did as well with cross inhibition). That is because the synapse pair of synapse 1 and 2 tended to be stronger due to device mismatches (coincidently for both chips used in the experiment). The neurons would fall for that same input pattern together in more than 70% of the experiments. Thus, in the experiments with cross inhibition the 'stronger' of the two neurons, neuron 1, was able to force neuron 2 to choose the other pattern more often. This was the case since the current from the initially stronger synapses of neuron 1 did charge the soma capacitance slightly faster and this sometimes did prevent the slightly slower neuron 2 from firing. We intend to increase that effect by adding a resistance to the soma input in future designs. Table 1 lists the percentages of patterns picked by the neurons in 30 test runs, with and without cross inhibition.

The results were similar for a decorrelation experiment with low correlation leakage (**_leak=0.6V**, learning rate patterns). We stimulated learning synapse 1 in both neurons for one second with Poisson distributed input frequency of 100Hz (pattern 1), then learning synapse 2 in both neurons, also for a second (pattern 2). This time we initialized all weights to be low ($V_w \approx 0.45V$). We repeated the two patterns 9 times. The graph in figure 4 shows the output frequencies of the two

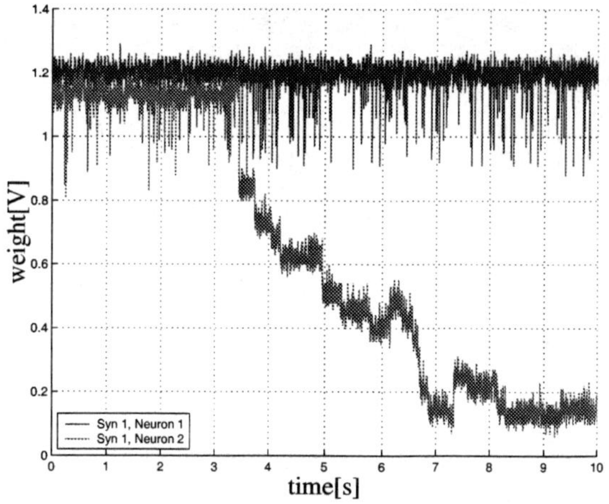

Fig. 3. *Trace of synaptic weight one for both neurons in the neural network. Neuron one maintains weight value and learns input pattern one. Neuron two depresses weight and learns input pattern two.*

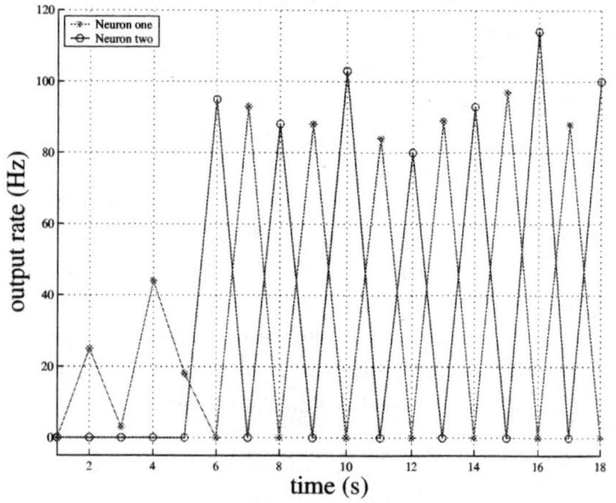

Fig. 4. *Output frequency of the two neurons.*

neurons averaged over one second for one test run. Neuron one learns pattern 1 (presented at uneven time steps) while neuron two remains inactive at start before it reacts to pattern 2 (presented at even time steps). Again, we observed a successful decorrelation in 50% of the trials. And again, in more than 95% of the failures both neurons adapted to pattern 1, thus confirming that an unsuccessful outcome in only 50% of the experiments actually constitutes an improvement.

4. CONCLUSION

We adapted a VLSI implementation of a spike timing dependent learning algorithm to operate with a weak multi-level memory cell for weight storage. This storage cell allows the weight updates to be computed by an analog circuit. Thus, if a weight changes into the attractive area of another discrete state, that is the accumulative result of analog changes. Only over long intervals (of up to several seconds) is the analog resolution lost and the state discretized.

In a number of on-chip experiments we confirmed the implementations ability to learn simple pattern classification tasks. Cross-inhibition lead to competitive learning and improved the decorrelation of the neurons. We suggest as small change for future designs to make the cross-inhibition more effective and thus, to achieve still better decorrelation.

With appropriate parameter settings the circuit can change its sensitivity from patterns of coincident spikes to patterns of coincident average activity. Thus, it can express both spike timing dependent learning or more classical rate dependent Hebbian learning. Observing the weight evolution in those experiments one can nicely observe the small analog update steps. The fast analog computation is not affected by the memory's weak and slow discretization.

In the classification tasks applied in this work also the final reduction of the weights to discrete states (when the neuron was idle for some time after the learning) did not compromise the result. It remains, however, to be seen, how the final discretization affects the learning of more complex (i.e. non-binary) spatio-temporal patterns.

5. REFERENCES

[1] P. Häfliger and H. Kolle Riis. A multi-level static memory cell. In *Proc. of IEEE ISCAS*, volume 1, pages 22–25, Bangkok, Thailand, May 2003.

[2] S. Thorpe, F. Fize, and C. Marlot. Speed of processing in the human visual system. *Nature*, 381:520–522, June 1996.

[3] H. Markram, J. Lübke, M. Frotscher, and B. Sakmann. Regulation of synaptic efficacy by coincidence of postsynaptic APs and EPSPs. *Science*, 275:213–215, 1997.

[4] P. Heim and M. A. Jabri. Long-term CMOS static storage cell performing AD/DA conversion for analogue neural network implementations. *Electronic Letters*, 30(25), December 1994.

[5] M. Holler, S. Tam, H. Castro, and R. Benson. An electrically trainable artificial neural network (ETANN) with 10240 'floating gate' synapses. *International Joint Conference on Neural Networks*, (II):191–196, June 1989.

[6] C. Diorio, P. Hasler, B. A. Minch, and C. Mead. A floating-gate MOS learning array with locally computed weight updates. *IEEE Transactions on Electron Devices*, 44(12):2281–2289, December 1997.

[7] P. Häfliger and C. Rasche. Floating gate analog memory for parameter and variable storage in a learning silicon neuron. In *Proc. of IEEE ISCAS*, volume II, pages 416–419, Orlando, 1999.

[8] P. Häfliger. *A spike based learning rule and its implementation in analog hardware*. PhD thesis, ETH Zürich, Switzerland, 2000. http://www.ifi.uio.no/~hafliger.

SPIKE SYNCHRONIZATION IN A NETWORK OF SILICON INTEGRATE-AND-FIRE NEURONS

Shih-Chii Liu and Rodney Douglas

Institute of Neuroinformatics, University of Zurich and ETH Zurich
Winterthurerstrasse 190, CH-8057 Zurich, Switzerland
e-mail: shih,rjd@ini.phys.ethz.ch

ABSTRACT

Spike-event based computational processing by neuronal networks depends on a variety of spatio-temporal mechanisms that are distributed across neuronal synapses, dendrites, and somata. The overall processing of spatio-temporal patterns of spikes requires global temporal coherence between all these processes both within a neuron, and across all neurons of the network. We are evaluating the use of networks of hybrid analog-digital integrate-and-fire neurons, fabricated using VLSI, for studying real-time event-based processing. These hardware networks have the advantages of real-time operation, and inherent global synchronization by their analog implementation. However, one of the technical challenges in the operation of such networks is the unavoidable fabrication related variance between the different neurons of the network. Here we show that despite this variance, an important behavior of neuronal networks, spike synchronization, can be obtained from a simple network of VLSI neurons interconnected by lateral excitation and global inhibition.

1. INTRODUCTION

Event-based processing by neurons and their networks involves various mechanisms such as synapses whose efficacy depends on the temporal structure of their input spike train [1, 2]; spatial interactions between synapses distributed across a neuron's dendrites; voltage- and action potential ('spike') dependent adaptations; and predictive learning based on the phase difference between pre- and post-synaptic events [3].The overall computation performed by a network assumes a global temporal coherence between all these processes both within a neuron, and across all neurons of the network.

It is difficult to explore the real-time behavior of such systems using conventional simulation techniques because of the computational cost of simulating a distributed system on a serial processor, and also because the encoding of the problem onto an inherently serial binary processor necessarily destroys the continuous spatio-temporal nature of the original network.

An alternative to conventional simulation is to construct these networks directly in hardware, using analog VLSI technology. In this case the neurons are encoded directly into physical analogs on a chip, and so their behavior is inherently synchronized by the real-time physics of the hardware. Fabrication in hardware also opens the route to applying these experimental networks directly in systems that include real sensors and effectors. With these advantages in mind, a variety of silicon spike-based multi-neuron chips with different circuit and synaptic models are currently being developed using hybrid analog/digital VLSI [4, 5, 6, 7, 8].

The construction and operation of such large-scale networks of spiking neurons presents significant technical challenges. For example, providing a flexible infrastructure for event-based communication in a medium where point-to-point wires similar to axons are not generally practical; noise problems consequent on having digital and analog circuitry intertwined on the same chip; and the variance between neurons that arises out of inherent non-idealities in the fabrication of chips by the silicon foundry. As a step towards characterizing the performance of such hardware networks, we have evaluated whether the neurons which are interconnected by lateral excitation and global inhibition are able to synchronize their spike generation, despite the considerable variance in their individual properties.

2. MULTI-NEURON NETWORK

The analog VLSI (aVLSI) multi-neuron chip in this work comprises 64 leaky integrate-and-fire neurons, each of which can be stimulated through different types of dynamic synapses. The circuit was fabricated in a 0.8 μm CMOS process. The architecture of the system is shown in Fig. 1. The chip comprises a set of generic neuronal and synaptic elements that can be reconfigured into a desired network architecture. The particular configuration is determined by setting the coupling parameters, and by routing spike events by an asynchronous spike transmission protocol called the address-event representation (AER) protocol. This mechanism routes spike events between neurons and synapses that are labelled by

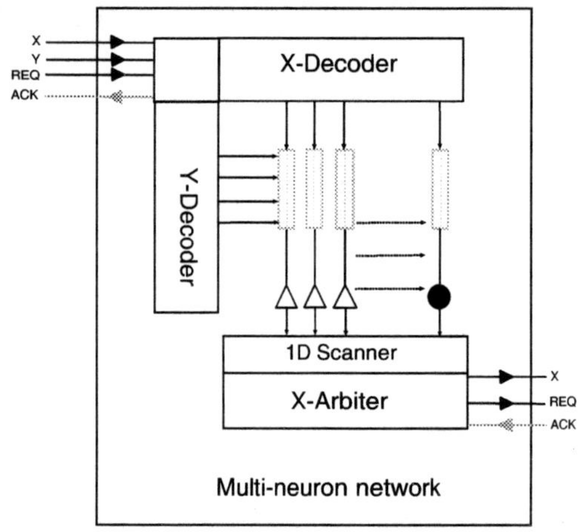

Fig. 1. Block diagram of chip architecture. The neurons and a set of synapses are labelled by unique address labels. REQ and ACK signals are used to ensure correct transmission of the addresses between chips. The X and Y decoder blocks decode the incoming addresses for the selected synapse and neuron. The X arbiter communicates the addresses of the active neurons to another chip. The open triangular symbols represent the somas of the 63 excitatory neurons and the solid black circle represent the soma of the global inhibitory neuron which is driven by all the excitatory neurons.

unique address bits [9, 4, 5]. This protocol permits neurons to be virtually connected on, or across chips.

2.1. Chip Architecture

There are two different sets of synapses on-chip. The first set of synapses can be individually addressed from an external device using the AER interface (gray boxes, Fig. 1). They consist of non-depressing excitatory and inhibitory synapses and excitatory short-term dynamic synapses. These synapses are used to make long-range, less structured connections, between neurons.

The second set of synapses consist of non-depressing excitatory and inhibitory synapses and short-term dynamic synapses [10]. These synapses can be individually activated, but their characteristics are controlled by global bias parameters, and so the same pattern of connection is applied to all neurons in that layer. For example, all the neurons can be connected to their nearest neighbors through excitatory synapses by setting the appropriate global parameters.

On this chip, 63 out of 64 neurons are excitatory neurons and the remaining neuron is an inhibitory neuron. This ratio of excitatory neurons to inhibitory neurons while much greater than the ratio found in the cortex allows for the exploration of properties of different cortical network architectures. The excitatory neurons can drive the inhibitory neuron who, in return, can inhibit all the excitatory neurons by choosing certain global parameters. The spiking activity of the neurons can be monitored through the asynchronous transmission bus (through the address labels) while scanning circuits allow us to monitor the membrane potentials of the neurons.

2.2. Neuron and Synapse Circuits

The neuron circuit implements an integrate-and-fire model with a constant leak current [6]. The normal parameters in this model are represented in the neuron circuit: the threshold voltage, the refractory period, the pulse width of the spike, and the leak current. The neurons can also adapt their firing rate to a constant input similar to the cortical pyramidal cells.

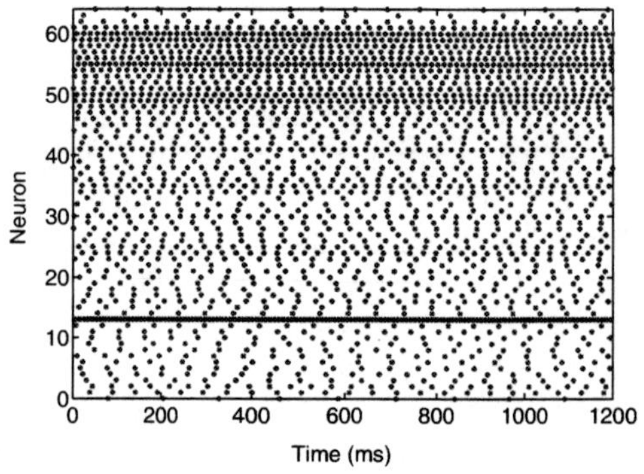

Fig. 2. Measurements of the spiking activity from the multi-neuron array in response to constant input currents. Raster plot of spikes from the neurons identified by their address label on the y-axis in response to a constant injected current. Notice that neurons numbered 50–60 have a much higher spike rate than most of the other neurons on the chip.

The excitatory or inhibitory synaptic circuits are derived from a simple synaptic model. The synaptic current increases immediately a spike arrives, but the time constant of the decay of this current is adjustable. The synapses have two types of short-term dynamics: Depressing, and facilitating. Both of these synapses implement theoretical models similar to that of Abbott and colleagues [2].

3. CHARACTERIZATION

Even though the VLSI circuits of neurons and synapses are identical across the chip, fabrication and layout non-idealities lead to differing spike rates from the neurons with the same injected current. Figure 2 shows a raster plot of a sample distribution of the spikes obtained from the neurons on a particular chip in which we used, primarily, minimum length transistors. The individual points represent spikes of the neurons which are labelled by their addresses on the y-axis. Almost all neurons respond to the constant current input by generating a regular spike train. But the average rate of these action potentials differs widely across the chip.

This regional variation in the performance of neurons could be reduced by using modern fabrication processes, and very large dimension transistors. However, large transistor dimensions come at the cost of reducing the number of neurons and synapses per chip area.

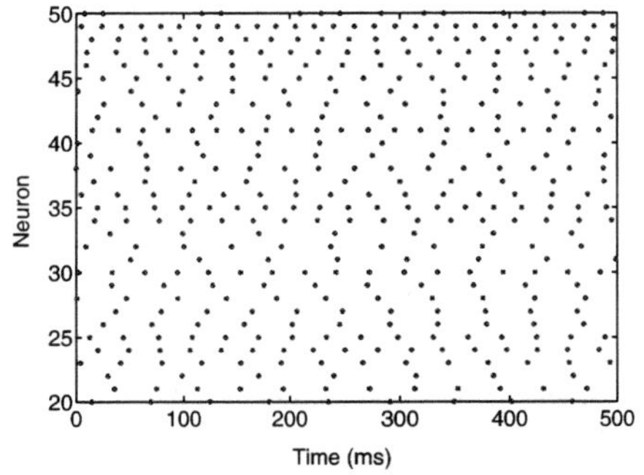

(a)

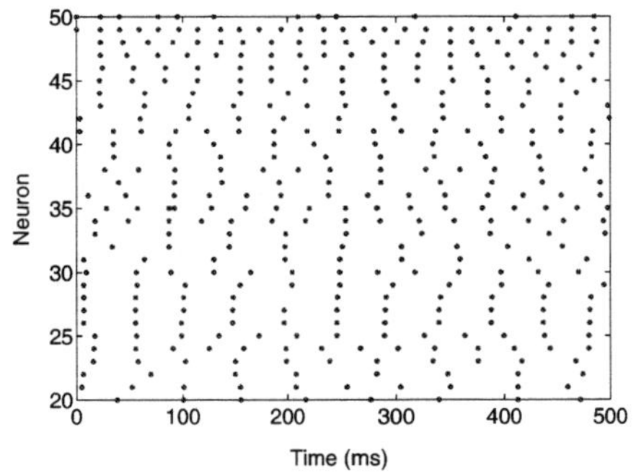

(b)

Fig. 3. Raster plots of spike activity of a set of neurons in response to a constant input current. (a) The spiking activity is regular for the individual neurons but their spikes are not synchronized. (b) With nearest-neighbor coupling, the spikes become synchronized between neighboring neurons.

4. SYNCHRONIZATION

Synchronization between neurons of their spike generation has been widely studied in the neuroscience, computational, and physiology community [11, 3]. It is a powerful mechanism to entrain groups of neurons to drive their common postsynaptic targets more effectively. Conventional simulation studies of synchronization have usually been carried out with networks of homogeneous neurons that have local excitatory or inhibitory connections, or global inhibition. We have taken these configurations as the starting point for our evaluations. As a first step, we have considered only homogenous nearest neighbor lateral excitatory connections.

4.1. Excitatory coupling

We expected that, with lateral coupling activated, the variance in the firing rates of the neurons might increase further, due to the additional variances contributed by the synapses themselves, and so prevent synchronization.

We measured the spiking patterns of neurons in response to a constant input current, and compared their correlation when uncoupled, to their correlation in the presence of nearest neighbour connections. Figure 3 shows that the neurons synchronize with one another, especially when their firing rates are similar. We measured the degree of synchronization by computing the averaged cross-correlogram between a single neuron and all other neurons, and compared the results for both cases (Fig. 4). The curve (dashed dotted line) for the nearest-neighbor coupling shows an increase in correlation for all neurons, as observed previously in the modeling literature.

4.2. Global inhibition

Inhibitory coupling is known to be another effective mechanism for synchronizing neurons[12]. We investigated whether global inhibition together with local excitatory coupling could improve the degree of synchronization in the hardware network. We introduced global inhibition by activating the connections to and from the sole inhibitory neuron that is driven by all the excitatory neurons, as in a winner-take-all architecture.

The results (solid line) plotted in Fig. 4 show that under conditions where the network does not implement a hard winner-take-all function, the average correlation of some neurons in this network configuration increased significantly.

5. CONCLUSION

There is large variance in the characteristics of the 64 integrate-and-fire neurons on our chip. For example, their individual responses to constant injected currents (or gain, measured as spike rate per nA) varied over a range of approximately 3. The variance was correlated systematically with location on the chip, which is commonly observed in the fabricated chips. Analog circuits, like ours, that operate in the subthreshold domain are particularly sensitive to these anomalies because of the exponential characteristics of the current-voltage relationship of transistors. We hope to reduce this

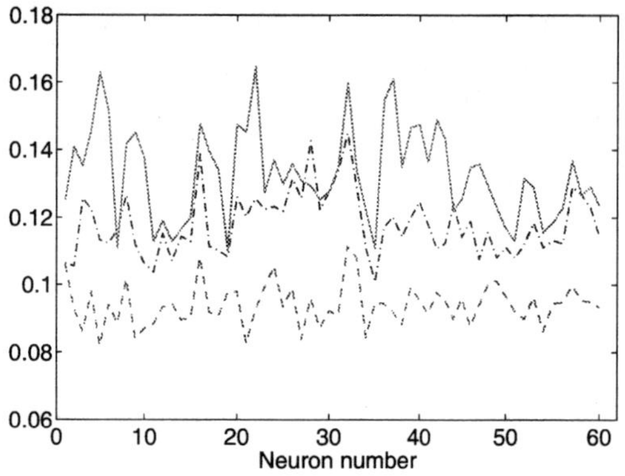

Fig. 4. Normalized correlation averaged across all other neurons for each neuron. The dashed line corresponds to the case without any coupling; the dashed dotted line corresponds to the case where nearest-neighbor coupling is present, and the solid line corresponds to the case where global inhibition is also introduced.

variance in average firing rate in future iterations of this circuit, by incorporating suitable somatic adaptive conductances that draw the individual neurons toward a similar output rate.

The spontaneous activity of the neurons was uncorrelated. This is a pleasing result because we anticipated that the power supply coupling or digital switching transients would cause artefactual synchronization of the neurons. Introduction of lateral coupling led to good synchronization of the nearest neighbors. When the global inhibition was incorporated into the circuit, the synchronization extended across more distant neighbours. Of course, these results are expected on first principles. However, the important point here is that spike-based processing can be usefully studied in hardware neuronal networks, despite fabrication-dependent variances.

6. ACKNOWLEDGMENTS

We acknowledge members of the INI hardware group for the development of the software and hardware infrastructure; in particular Giacomo Indiveri, Adrian Whatley and Tobi Delbrück. We acknowledge Kevan Martin for the discussions on the chip architecture. This work is partially funded by EU-grant IST-2001-34124.

7. REFERENCES

[1] M. Tsodyks, K. Pawelzik, and H. Markram, "Neural networks with dynamic synapses.," *Neural Comput*, vol. 10, no. 4, pp. 821–35, May 15 1998.

[2] F. S. Chance, S. B. Nelson, and L. F. Abbott, "Synaptic depression and the temporal response characteristics of V1 cells.," *J Neurosci*, vol. 18, no. 12, pp. 4785–99, Jun 15 1998.

[3] W. Maass and C.M. Bishop, Eds., *Pulsed Neural Networks*, MIT Press, Boston, MA, 1999.

[4] S. R. Deiss, R. J. Douglas, and A. M. Whatley, "A pulse-coded communications infrastructure for neuromorphic systems," in *Pulsed Neural Networks*, W. Maass and C. M. Bishop, Eds., chapter 6, pp. 157–178. The MIT Press, 1998.

[5] K. A. Boahen, "Communicating neuronal ensembles between neuromorphic chips," in *Neuromorphic Systems Engineering*, T.S. Lande, Ed., chapter 11, pp. 229–259. Kluwer Academic Publishers, Boston, MA, 1998.

[6] S.-C. Liu, J. Kramer, G. Indiveri, T. Delbrück, T. Burg, and R. Douglas, "Orientation-selective aVLSI spiking neurons," *Neural Networks: Special Issue on Spiking Neurons in Neuroscience and Technology*, vol. 14, no. 6/7, pp. 629–643, 2001.

[7] G. Indiveri, T. Horiuchi, E. Niebur, and R. Douglas, "A competitive network of spiking VLSI neurons," in *World Congress on Neuroinformatics*, F. Rattay, Ed., Vienna, Austria, Sept 24–29 2001, ARGESIM/ASIM Verlag, ARGESIM Reports.

[8] A. van Schaik, "Building blocks for electronic spiking neural networks," *Neural Networks: Special Issue on Spiking Neurons in Neuroscience and Technology*, vol. 14, no. 6/7, pp. 617–628, 2001.

[9] M. Mahowald, *VLSI analogs of neuronal visual processing: a synthesis of form and function*, Ph.D. thesis, California Institute of Technology, Pasadena, CA, 1994, Computation in Neural Systems Dept.

[10] M. Boegerhausen, P. Suter, and S.-C. Liu, "Modeling short-term synaptic depression in silicon," *Neural Computation*, vol. 15, no. 2, pp. 331–348, 2003.

[11] R.E. Mirollo and S. H. Strogatz, "Synchronization of pulse-coupled biological oscillators," *SIAM J. Appl. Math*, vol. 50, pp. 1645–1662, 1990.

[12] C. van Vreeswijk, L.F. Abbott, and G.B. Ermentrout, "When inhibition not excitation synchronizes neural firing," *J Comput Neurosci*, vol. 1, no. 4, pp. 313–321, 1994.

A SPIKE-BASED ANALOGUE CIRCUIT THAT EMPHASISES TRANSIENTS IN AUDITORY STIMULI.

Natasha Chia and Steve Collins

Department of Engineering Science
University of Oxford
Parks Road, Oxford
England OX1 3PJ
email:steve.collins@eng.ox.ac.uk

ABSTRACT

There is increasing evidence that onsets in an auditory stimulus can serve as important cues for useful functions including grouping components of the same sound and sound localisation. Motivated by this observation an analogue circuit has been designed to extract temporal information from an auditory input. The first component of the proposed system is a bank of subthreshold filters. The half-wave rectified output from each filter is then converted to a series of spikes by a set of parallel comparators, each of which detects when the power in a specific frequency interval exceeds a specific threshold. This system therefore creates a representation that emphasises the fine temporal structure in an auditory stimulus that can be used to separate different sound sources.

1. INTRODUCTION

Unlike existing artificial systems, humans and higher animals can separate sounds from different sources with apparent ease. This glaring contrast between the performance of natural and artificial systems has lead several researchers to investigate the advantages that may arise if artificial systems reproduce the behaviour of natural systems more closely.

One approach to the separation of auditory signals that has been developed emphasises two features of the natural auditory systems. One of these features is simply the existence of two ears. The other feature is that neurons in both the auditory nerve and the cochlear nucleus show a strong response to any onset, that is any sudden increase in the strength of an auditory stimulus[1]. These features of the auditory system have be replicated in a system developed by Smith [2, 3, 4]. In this system binaural recordings are first filtered in a gammatone filterbank. The output from each filter is then converted to a phase-lock spike code to provide a phase-locked onset detector within each frequency band. Near coincident of onsets in different frequency bands can then be used to group together the different components of a sound. Finally, the onset times for each group are used to compute the relative delay in the detection of the same onset in each ear, the interaural time difference (ITD). Since the first onset from any sound arrives at each ear via the shortest path the ITDs can be used to detect the direction of a sound, even when only one onset is detected. This system is therefore capable of grouping together different parts of same sound as detected by each ear, i.e. monaurally. The ITDs between two ears can then be used to localise sound sources. Although this system can be used to separate sounds from different sources it is too computationally expensive to be executed in real time. However, the simplicity of the model makes it suitable for implementation in analogue VLSI. Unlike previous similar work, which has concentrated upon neuromorphic systems design[5], our aim is to create a biologically inspired system that provides sound localisation information for binaural signal processing.

A description of the bank of subthreshold filters required to separate an auditory stimulus into its constituent parts is presented in section 2. Using subthreshold circuits to implement these filters has several advantages. However, the response of the resulting filters will be sensitive to changes in operating temperature unless they are carefully designed. A bias circuit has therefore been used that ensures that the transconductance of a subthreshold MOSFET is constant. Using these circuits the response of the filters can be made robust to temperature variations. The 'spike' generating circuits used to convert the response of each filter into a series of 'spikes' is described in section 3. The first part of this circuit is a half-wave rectifier. The output from this circuit is then converted to 'spikes' by a comparator that generates an output spike whenever the acoustic power within the associated frequency band rises above a threshold value. This circuit therefore extracts useful temporal information from the acoustic stimuli. However, a system containing only

This work was funded by EPSRC grant GR/R74581.

one comparator per filter will fail to preserve the information contained in the relative amplitudes in each frequency band. The system that has been designed therefore has several comparators at the output of each half-wave rectifier. These comparators create a 'spike' representation of a stimulus that retains both its temporal and amplitude features and hence preserves the information that could be critical in separating sounds from different sources.

2. BANK OF FILTERS

One concern when facing the task of implementing the system developed by Smith is that for biological plausibility a gammatone filterbank has been used. However, simulations have shown that this unusual type of filter can be replaced by a simple second order band-pass filter without a significant degradation in system performance. The circuit diagram for the band-pass filter, that will be used in a prototype system is shown in Figure 1. Analysis of this second order filter circuit shows that its transfer function is

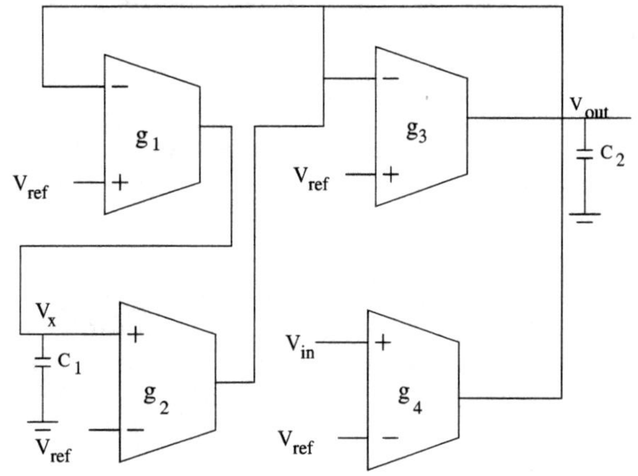

Fig. 1. *A schematic diagram of the type of second order biquad filter used in the filterbank.*

$$H_{BPF} = \frac{s\frac{g_4}{C_2}}{s^2 + s\frac{g_3}{C_2} + \frac{g_1 g_2}{C_1 C_2}} \tag{1}$$

To create a bank of filters and leave space for other circuits each filter must be relatively compact. It is therefore desirable to use small capacitance values, typically 1pF, in each filter. With these capacitance values audio frequency filters can be created if the transconductance element is implemented using MOSFETs operating in subthreshold. In this operating regime the output current from a simple five transistor transconductance amplifier (TA), shown to the left of figure 2 is[6]

$$I_{out} = I_{bias} \tanh(q(V_{in1} - V_{in2})/2nkT) \tag{2}$$

In its linear regime the transconductance of this circuit is therefore

$$g_{out} = \frac{qI_{bias}}{2nkT} \tag{3}$$

One characteristic of subthreshold TAs, highlighted by equation 3, is that their transconductance is dependent upon temperature. The problem is that from equation 1 it can be shown that this will create a temperature dependence of the frequency response of the filter. In fact simulations show that with a constant bias voltage the centre frequency of a typical filter can vary from 280Hz at 273K to 1513 Hz at 328K. This is an unacceptably large part of the auditory frequency range. In order to make the filter response more robust to changes in operating temperature a bias circuit is required that ensures that transconductance of the TAs is independent of temperature. To achieve this the circuit shown to the right hand side of figure 2 has been used to generate the gate bias voltage for the current source within each TA.

In the bias generating circuit transistors M6-M9 form a current mirror that ensures that the current through the two branches of the circuit is identical. Analysis of the circuit[7] shows that if $g_{M11}R > 1$ then $g_{M10} = R^{-1}$. The transconductance of transistor M10 is therefore controlled by the resistance R. The temperature dependence of the filter response will then be determined by the temperature dependences of this resistor and the capacitors within the filter. In fact simulations show that with this bias circuit the centre frequency of the same typical filter varies from 679Hz at 273K to 717Hz at 328K.

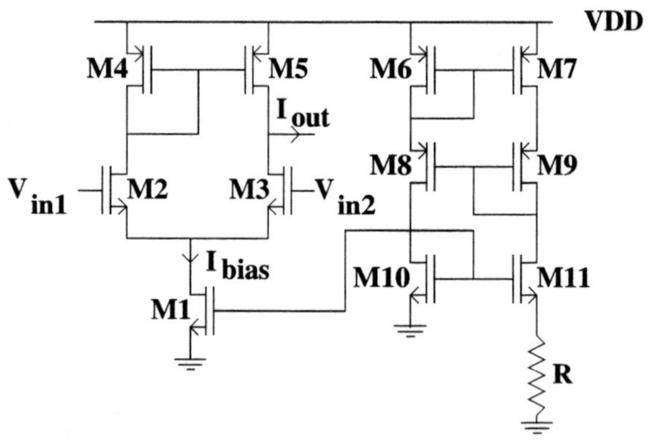

Fig. 2. *A schematic circuit diagram showing the simple transconductance element used in the filters and the circuit used to generate the bias voltage for the transconductance amplifier.*

The architecture developed by Smith consists of a bank of 32 filters covering the frequency range from 100Hz to 7.5kHz. The two possible approaches to varying the values of g_1 and g_2 within each filter in order to cover this frequency range are either to change the bias voltage on transistor M1 within each transconductance amplifier or to change its geometry. Since the first approach would require a bias circuit for each filter, the second approach has been adopted. Furthermore, to ensure that the quality factor and the gain of all filters are the same the values of g_3 and g_4 in each filter are proportional to the values of g_1 and g_2.

The final stage in designing the filter is to determine the area of each of the transistor required to avoid too much mismatch between the devices in each input pair and current mirror. Since the quality factor of each filter is 10 a maximum tolerable variation in centre frequency of 1% was chosen. Conservative Monte Carlo simulations, allowing no correlations between devices, were then used to determine the size of transistors needed to match this specification. As expected these simulations showed that large devices are needed and the final circuit layout required an area of $370\mu m$ by $180\mu m$. Although this is relatively large, there is plenty of space available on a die to implement the other functions.

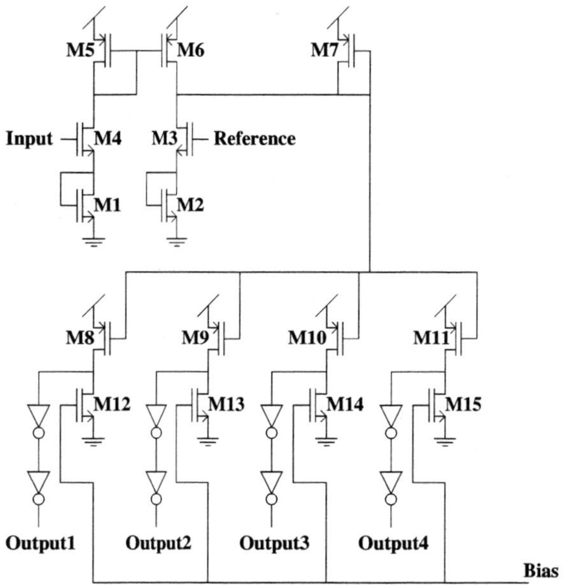

Fig. 3. *A half-wave rectifier circuit and a set of parallel comparators used to create a 'spike' representation of the circuit input signal.*

3. SPIKE GENERATING CIRCUITS

The first stage in creating a 'spike' representation of the response of each filter is to half-wave rectify the filter response. This function is performed by the circuit at the top of figure 3 which compares the output voltage from the filter with a reference value. The d.c. component of the output voltage from each filter is typically 1 V. For this circuit to operate in subthreshold each of the two input devices, M3 and M4, is therefore placed in series with a diode connected device M1 and M2. When the output voltage from the filter is more than the reference value then transistor M3 sinks less current than that sourced by transistor M6. Under these conditions there will be no current flowing through transistor M7. In contrast when the filter output voltage falls below the reference level, then the current through transistor M6 is less than that flowing through transistor M3 and current is drawn through transistor M7. Thus current only flows through this transistor when the input voltage, supplied by the filter output, is below the reference value. This current therefore represents a half-wave rectified version of the filter output.

The final stage of creating a 'spike' representation of the filter response is to compare the output from the half-wave rectifier with a threshold value. Using a single comparator it is possible to create an output 'spike' whenever the power in the corresponding filter exceeds a threshold value. However, this is only a coarse representation of the signal in a particular frequency range. To create a better representation of the input signal several comparators are therefore required. In the example circuit shown in figure 3 there are four current mode comparators. The input to each of these current comparators is a copy of the current representing the output of the half-wave rectifier circuit. Within each comparator the current sourced by this p-channel device is compared to the current flowing through an n-channel device. In the circuit shown in figure 3 the width of the n-channel device within each comparator has been varied, so that the reference current in the different comparators are I_{ref}, $2I_{ref}$, $3I_{ref}$, and $4I_{ref}$. Ideally, the current through a MOSFET operating in subthreshold saturates when its drain-source voltage is greater than approximately 100 mV. The comparator output voltage should therefore be very close to one of the two supply voltages. However, simulations show that the source-drain dependence of the subthreshold current through the MOSFETs is sufficient to prevent this ideal situation. To avoid the possible ambiguity that this could cause the output voltage from each comparator is passed through two inverters.

The response of a 'spiking' circuit with four outputs is shown in figure 4. This figure shows the response of the system to the decaying sinusoid input at the top of the figure. The other four signals in the figure then correspond to the output voltages of four comparators, after the inverters. Of these four signals Output1 has the lowest threshold value and Output4 the highest. The figure clearly shows that the leading edge of each spike follows a minimum in the input signal. As required the spikes are therefore phase-locked to the input. Except for the first cycle of the input, during which there is an onset transient in the filter response, these results also show that the amplitude of the input signal at each minimum is represented by the pattern of activity across the four outputs. In particular these outputs form a thermometer encoding of the amplitude of the signal minimum. Since the amplitude of each signal minimum is individually encoded this system is able to represent the signal power with a very high temporal resolution.

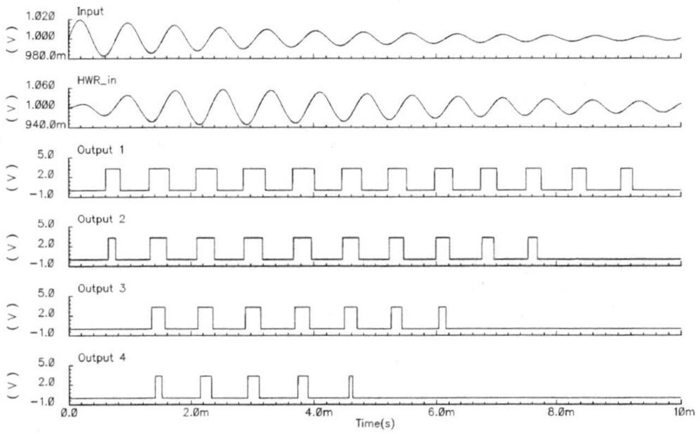

Fig. 4. *The response of a system with four comparators to the onset of an exponentially decaying sinusoidal input. The input is shown at the top of the figure together with the output from the filter. The other four signals are the outputs from the four comparators, each with a different threshold.*

4. CONCLUSIONS

Smith has shown that using a biologically inspired spike representation of an auditory input signal it is possible to determine the onset time of a stimulus very accurately. The onset times within different frequency bands can then be used monaurally to group together the different frequency components of a sound. In a binaural system the onset of each sound can also be used to calculate an interaural time difference which is related to the location of a sound source. This system therefore creates two useful cues that can be used to separate the sounds from different sources.

The simplicity of the system developed by Smith means that it is suitable for implementation in analogue VLSI. As a first stage in implementing this system a bank of filters has been designed. By using transconductance elements operating in subthreshold it is easy to cover the range of frequencies required by this system. However, there are problems when designing circuits based on subthreshold devices in particular they can be very sensitive to changes in operating temperature and variations between 'identical' devices can be large. Simulations results have been presented which show that the first of these problems can be avoided by using a bias circuit that ensures that the transconductance of the transconductance amplifiers is independent of temperature. Variations between devices are then limited to an acceptable level by correct sizing of individual transistors. The result is a filter with a centre frequency that is defined to within 1% which occupies a relatively small area. To capture the fine temporal structure of each component of a stimulus the output from each filter is half-wave rectified. The amplitude of each minimum in the the filter output signal is then converted to series of 'spikes' by a group of comparators operating in parallel. These spikes are phase-locked to the input and form a thermometer type coding of the amplitude of each minimum. This simple circuit is therefore capable of extracting data from a stimulus that is useful for two functions; monaural grouping of the components of a single sound and binaural direction finding. By using this simple component circuit future artificial systems may at last be able to match the performance of natural systems.

5. REFERENCES

[1] E.M. Rouiller, "Functional organization of the auditory pathways," in *The Central Auditory System*, G. Ehret and R. Romand, Eds. Oxford University Press, 1997.

[2] L.S. Smith, "Using depressing synapses for phase locked auditory onset detection," in *Artificial Neural Networks: ICANN 2001*, G. Dorffner, H. Bischof, and K. Hornik, Eds. 2001, vol. 2130 of *LNCS*, pp. 1103–1108, Springer.

[3] L.S. Smith, "Phase-locked onset detectors for monaural sound grouping and binaural direction finding," *Journal of the Acoustical Society of America*, vol. 111, no. 5, pp. 2467, 2002.

[4] L.S. Smith, "Biologically inspired robust onset detection," *Journal of the Acoustic Society of America*, vol. 114, no. 4, pp. 2198, 2003.

[5] Tor Sverre Lande, *Neuromorphic Systems Engineering*, Kluwer Academic Publishers, 1998.

[6] C.A. Mead, *Analog VLSI and Neural Systems*, Addison-Wesley, 1989.

[7] T. H. Lee, *The Design of CMOS Radio-Frequency Integrated Circuits*, Cambridge University Press, 1998.

Transmultiplexers as precoders in modern digital communication: a tutorial review

P. P. Vaidyanathan and B. Vrcelj
Dept. Electrical Engineering, California Institute of Technology
Pasadena, CA 91125, USA
ppvnath@systems.caltech.edu bojan@systems.caltech.edu

Abstract.[1] In this paper we review the recent impact of transmultiplexers in digital communications. Filter bank precoders, conditions for equalization, and multiuser interference cancellation are reviewed. The idea behind blind channel identification is reviewed as well. The emphasis is mostly on the theoretical infrastructure, and the list of references provide a wealth of related information.

I. INTRODUCTION

Transmultiplexers have been known in digital communications for many years. Historically the transmultiplexer has been viewed as a system that converts from time multiplexed components of a signal to a frequency multiplexed version, and back [1], [2], [18]. The mathematical theory of transmultiplexers however allows more general interpretations and, therefore, applications. Some of these include channel equalization, channel identification and so forth. The role of transmultiplexers in digital communications has gained new importance because of many recent results in filter bank precoders. The works of Giannakis et al., Xia, and Lin and Phoong have had a particular impact in this area. The pioneering work of Giannakis et al. shows that such precoders not only allow equalization of linear finite-spread channels, but also blind equalization based only on second order statistics. Furthermore, cancellation of multiuser interference in CDMA channels without knowledge of channel coefficients has been shown to be possible.

In this paper we review the fundamental infrastructure behind these methods. In Sec. II, the transmultiplexer system is introduced and the mathematical framework deveoped. In Sec. III we present the theory behind the cancellation of multiuser interference. To the best of our knowledge the approach used here is new. It gives rise to several well known systems as special cases, including the single user DMT system [16], [8], [21], and the multiuser Amour system [4]. Section IV describes the meaning of bandwidth expansion and then derives the Amour system of [4] which was introduced in the literature in a different context, namely, the cancellation of multiuser interference. In Sec. V we make brief remarks on handling the effects of channel noise. In Sec. VI the fundamentals behind blind identification of channels using transmultiplexers is reviewed. Some of the important topics that are not covered in detail are briefly mentioned in Sec. VII. Important references are cited for further reading, and there are excellent papers in this special session covering related topics.

[1]Work supported in part by the ONR grant N00014-99-1-1002, USA.

Notations. Standard multirate notations from [18] will be freely used. For example the notation $[H(z)]_{\downarrow M}$ represents the z-transform of the decimated version $h(Mn)$. Lower and upper case notations such as $h_k(n)$ and $H_k(z)$ will be consistently used to denote sequences and their z-transforms. The "communications notations" used here are quite standard, though some readers might benefit by reading [20] for this.

II. THE TRANSMULTIPLEXER SYSTEM

Most of our discussions in this paper will center around the structure shown in Fig. 1 (and its generalization in Fig. 2) called a transmultiplexer. The signals $s_k(n)$ are symbol streams (such as PAM or QAM signals, [11], [20]). These could be symbols generated by different users who wish to transmit messages over the channel. Or they could be different independent parts of the signals generated by one user [20]. The distinction between these two scenarios is not required for our discussions here. The symbol streams $s_k(n)$ are passed through the interpolation filters or transmitter filters $F_k(z)$ to produce the signals

$$x_k(n) = \sum_i s_k(i) f_k(n - iP)$$

The filters $F_k(z)$ are also called pulse shaping filters because they take each sample of $s_k(n)$ and "put a pulse $f_k(n)$ around it". The sum $x(n)$ of the signals $x_k(n)$ is then transmitted over a common channel. The channel is described by a linear time invariant filter $C(z)$ followed by additive noise. At the receiver end, the filters $H_k(z)$ have the task of separating the signals and reducing them to the original rates by P-fold decimation.

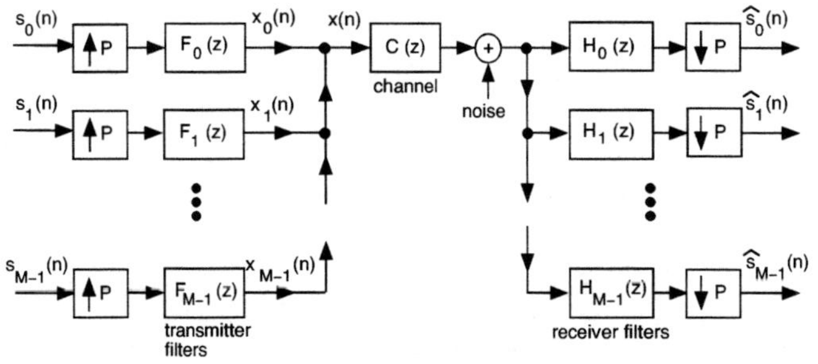

Fig. 1. The M-user transmultiplexer.

II.1. Preliminary remarks

Notice that since the M signals are mutiplexed into one channel, it is necessary to have $P \geq M$. When $P > M$ we have a *redundant transmultiplexer*, whereas $P = M$ corresponds to a *minimal* transmultiplexer.

The received signals $\widehat{s}_k(n)$ in general are different from $s_k(n)$ for several reasons. First, there is multiuser interference or **MUI**. This means that $\widehat{s}_k(n)$ is affected not only by $s_k(n)$ but also by $s_m(n), m \neq k$. Second,the channel $C(z)$ introduces a linear distortion called intersymbol interference or **ISI** (even when $M = 1$), and finally there is additive noise. The task at the receiver is therefore to minimize the effects of these distortions so that the transmitted symbols $s_k(n)$ can be detected from $\widehat{s}_k(n)$ with acceptably low probabilities of error. In absence of noise it is possible to compensate or **equalize** the effect of the channel completely and obtain perfect symbol reconstruction, that is $\widehat{s}_k(n) = s_k(n)$. The theory of perfect-reconstruction transmultiplexers for the case $P = M$ was developed in [22] and [6]. The case $P > M$ is more useful, as it eliminates some of the difficulties associated with practical filter design. For example when $P > M$ it is possible to equalize an FIR channel with the help of FIR filters $H_k(z), F_m(z)$ alone [8,13].

A special case of the redundant transmultiplexer has been widely used in DSL systems which use the discrete multitone modulation or **DMT** techniques in the transceiver [5], [8], [20]. In this system the filters are chosen from a simple uniform DFT filter bank (this appears at the end of this section). The more general theory of redundant transmultiplexers is fairly recent. Noteworthy here are the fundamental contributions from Lin and Phoong and early contributions from Xia (see references at the end). Pioneering developments on redundant transmultiplexers, also called *filter bank precoders* came from the group of Giannakis, et al., who showed both equalizability and blind identifiability (these terms will become clear as we proceed further).

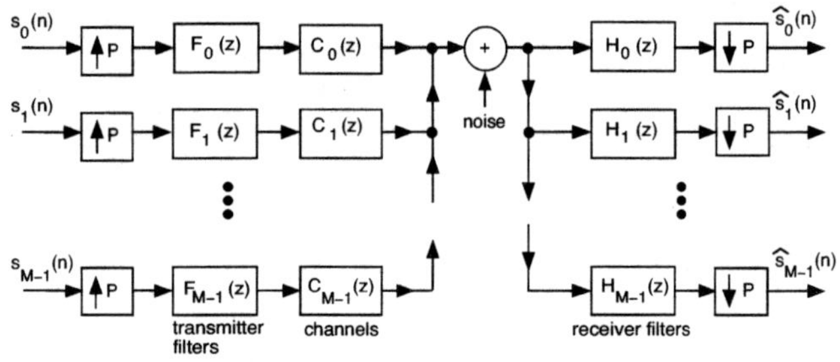

Fig. 2. Generalization of the M-user transmultiplexer.

A generalization of the transmultiplexer system is shown in Fig. 2. Again there are M users transmitting the symbol streams $s_m(n)$, but over M *different channels* $C_m(z)$. A noisy superimposition of these is received by the M receivers, which are expected to separate out the components $s_m(n)$. If the channel transfer functions $C_m(z)$ are identical for all m this reduces to the traditional transmultiplexer of Fig. 1. This generalization has become important in the context of multiuser communications over wireless channels. A technique for cancellation of MUI in these systems was first developed by Giannakis, et al. [4].

II.2. Mathematical analysis

Even though Fig. 2 is more general than Fig. 1, it is just as easy to analyze. So we will start with this figure. An important result in the theory of multirate systems is called the polyphase identity [18]. This states that if a linear time invariant (LTI) filter $G(z)$ is sandwiched between an expander and decimator as shown in Fig. 3 then the result is equivalent to an LTI system with impulse response $d(n) = g(Pn)$ (decimated version). That is, the transfer function of the overall system is $D(z) = [G(z)]_{\downarrow P}$. Returning to Fig. 2 the transfer function from $s_m(n)$ to $\widehat{s}_k(n)$ is therefore given by

$$T_{km}(z) = [H_k(z)C_m(z)F_m(z)]_{\downarrow P} \qquad (1)$$

If we choose the filters $\{F_m(z)\}$ and $\{H_k(z)\}$ such that

$$[H_k(z)C_m(z)F_m(z)]_{\downarrow P} = \delta(k - m), \qquad (2)$$

then multiuser interference is cancelled and there is perfect recovery of symbols, i.e., the **PR property**.

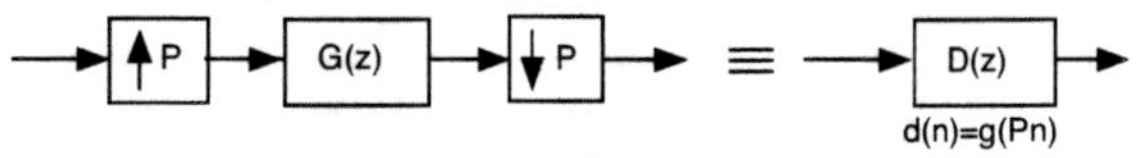

Fig. 3. The polyphase identity.

It is possible to satisfy (2) even when $P = M$. Readers familiar with the idea of *biorthogonal filter banks* will recognize that with $G_m(z) \triangleq C_m(z)F_m(z)$, the sets $\{H_k(z)\}$ and $\{G_m(z)\}$ satisfying (2) form a biorthogonal filter bank. However, even when the channels $C_m(z)$ are known and FIR, it is not true that there exist FIR solutions or stable IIR solutions $F_m(z), H_k(z)$ satisfying (2). For example when $P = M = 1$, the condition is like $H(z)C(z)F(z) = 1$ and cannot be satisfied when all three transfer functions are FIR.

If we allow an interpolation factor $P > M$ (redundant transmultiplexer) then there are many advantages. For example assume the channels are FIR:

$$C_m(z) = \sum_{n=0}^{L} c_m(n)z^{-n} \qquad (3)$$

Then if we choose $P \geq M + L$, there is a clever way to obtain perfect symbol recovery with FIR filters $F_m(z)$ and $H_k(z)$, as we shall see.

II.3. Eliminating interblock interference

Figure 4 shows the mth transmitter and kth receiver redrawn in polyphase form [18]

$$F_m(z) = \sum_{i=0}^{P-1} z^{-i}R_{i,m}(z^P), \; H_k(z) = \sum_{i=0}^{P-1} z^{i}E_{k,i}(z^P) \qquad (4)$$

With noise ignored, the path between the mth transmitter and kth receiver can be described in terms of a matrix $\mathbf{C}_m(z)$ whose elements are

$$[\mathbf{C}_m(z)]_{\ell,i} = [z^{\ell-i}C_m(z)]_{\downarrow P}$$

For example when $L = 2$ and $P = 5$, $\mathbf{C}_m(z)$ is

$$\begin{array}{c} \quad P-L \quad\quad\quad\quad\quad\quad | \quad\quad L \\ \left(\begin{array}{ccc|cc} c_m(0) & 0 & 0 & z^{-1}c_m(2) & z^{-1}c_m(1) \\ c_m(1) & c_m(0) & 0 & 0 & z^{-1}c_m(2) \\ c_m(2) & c_m(1) & c_m(0) & 0 & 0 \\ 0 & c_m(2) & c_m(1) & c_m(0) & 0 \\ 0 & 0 & c_m(2) & c_m(1) & c_m(0) \end{array}\right) \end{array}$$

This is called the **blocked version** of $C_m(z)$.

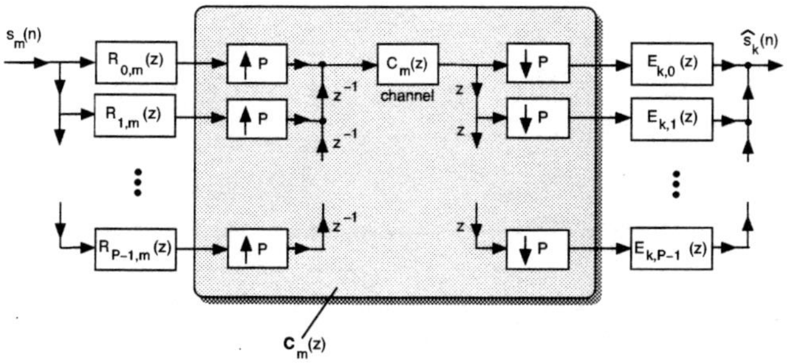

Fig. 4. The mth transmitter and kth receiver in polyphase form.

The preceding matrix is called a *pseudocirculant* and the interested reader can find out more about it in [18] and references therein. The first $P-L$ columns have constant entries and the last L columns have z^{-1} in them. Similarly the last $P-L$ rows have constant entries. Thus the blocked version can be partitioned in two ways:

$$\mathbf{C}_m(z) = P \overset{\quad P-L \qquad L}{\begin{pmatrix} \mathbf{A}_m & \mathbf{B}(z) \end{pmatrix}} = \begin{matrix} L \\ \\ P-L \end{matrix} \overset{P}{\begin{pmatrix} \mathbf{C}(z) \\ \\ \mathbf{D} \end{pmatrix}} \qquad (5)$$

where $\mathbf{A}_m$ is a $P \times (P-L)$ matrix and $\mathbf{D}$ is a $(P-L) \times P$ matrix (both constant Toeplitz matrices). In Fig. 4 the transfer function from $s_m(n)$ to $\widehat{s}_k(n)$ is given by

$$[E_{k,0}(z) \quad E_{k,1}(z) \quad \ldots \quad E_{k,P-1}(z)]\,\mathbf{C}_m(z) \begin{bmatrix} R_{0,m}(z) \\ R_{1,m}(z) \\ \vdots \\ R_{P-1,m}(z) \end{bmatrix}$$

Using (5) this can be written in either one of the two forms

$$[E_{k,0}(z)\, E_{k,1}(z) \ldots E_{k,P-1}(z)] \begin{pmatrix} \mathbf{A}_m & \mathbf{B}(z) \end{pmatrix} \begin{bmatrix} R_{0,m}(z) \\ R_{1,m}(z) \\ \vdots \\ R_{P-1,m}(z) \end{bmatrix}$$

or

$$[E_{k,0}(z) \quad E_{k,1}(z) \ldots E_{k,P-1}(z)] \begin{pmatrix} \mathbf{C}(z) \\ \\ \mathbf{D} \end{pmatrix} \begin{bmatrix} R_{0,m}(z) \\ R_{1,m}(z) \\ \vdots \\ R_{P-1,m}(z) \end{bmatrix}$$

Since $\mathbf{B}(z)$ and $\mathbf{C}(z)$ have z^{-1} in them, they represent interference between input vectors occuring at different times at the input of $\mathbf{C}_m(z)$. This is called inter-block interference or **IBI**, and it is convenient to eliminate it. We will describe two ways to construct the transmitter and receiver filters in such a way that IBI is eliminated.

1. ***Zero-padding***. In the first method we eliminate $\mathbf{B}(z)$ by setting

$$R_{P-L,m}(z) = \ldots = R_{P-1,m}(z) = 0.$$

   That is, we insert a block of L zeros at the end of each block of $P-L$ symbols (Fig. 5(a)). This zero-padding method is also called *zero-prefixing*.

2. ***Zero-jamming***. In the second method we eliminate $\mathbf{C}(z)$ by choosing

$$E_{k,0}(z) = \ldots = E_{k,L-1}(z) = 0.$$

   That is, we replace a block of L samples with zeros, at the beginning of each block of P successive *received* symbols (Fig. 5(b)).

Our discussions in this paper will be restricted to the zero padding case. In the zero-padding method the transfer function $T_{km}(z)$ from $s_m(n)$ to $\widehat{s}_k(n)$ is given by

$$[E_{k,0}(z)\, E_{k,1}(z) \ \ldots \ E_{k,P-1}(z)]\,\mathbf{A}_m \begin{bmatrix} R_{0,m}(z) \\ R_{1,m}(z) \\ \vdots \\ R_{P-L-1,m}(z) \end{bmatrix} \qquad (6)$$

where the $P \times (P-L)$ matrix $\mathbf{A}_m$ has the form

$$\mathbf{A}_m = \begin{bmatrix} c_m(0) & 0 & \ldots & 0 \\ c_m(1) & c_m(0) & \ldots & 0 \\ \vdots & \vdots & \ddots & \vdots \\ c_m(L) & & & \\ 0 & c_m(L) & & \\ \vdots & & \ddots & \vdots \\ 0 & 0 & \ldots & c_m(L) \end{bmatrix} \qquad (7)$$

This represents the effects of the mth channel completely.

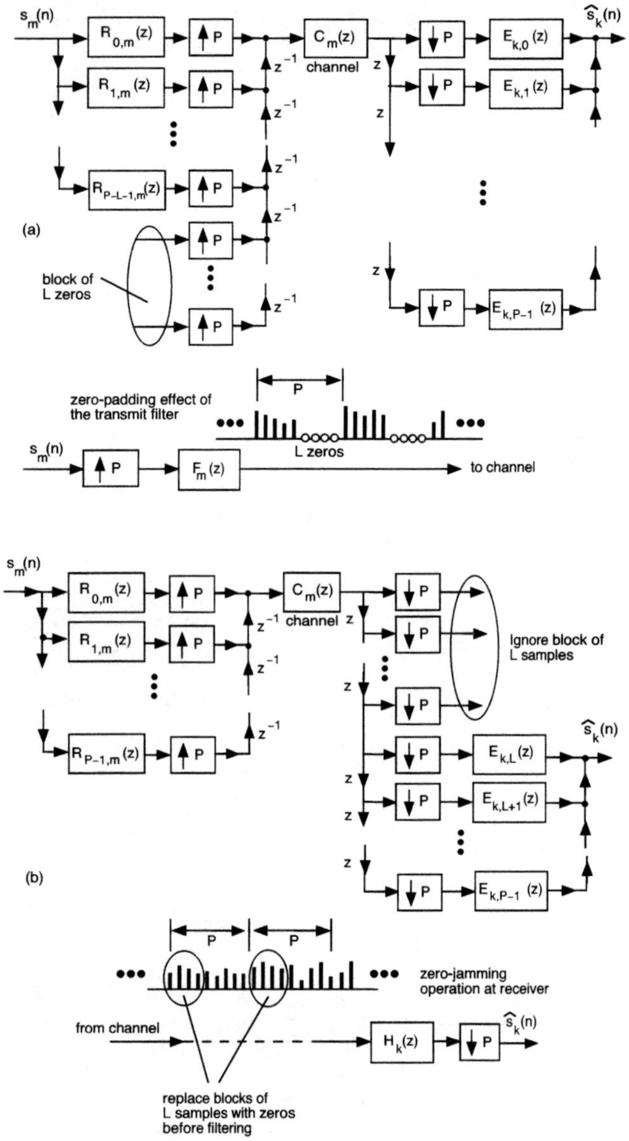

Fig. 5. Two ways to force the effective channel matrix to be a constant. (a) Zero-padding, and (b) zero-jamming.

III. CANCELLING MULTIUSER INTERFERENCE

The matrix $\mathbf{A}_m$ given above is a *full-banded Toeplitz matrix.* Because of this it has a very beautiful property: given any nonzero number ρ_k, the matrix satisfies the equation

$$[1\ \rho_k^{-1}\ \ldots\ \rho_k^{-(P-1)}]\,\mathbf{A}_m = C_m(\rho_k)\,[1\ \rho_k^{-1}\ \ldots\ \rho_k^{-(P-L-1)}] \quad (8)$$

where $C_m(z) = \sum_n c_m(n) z^{-n}$ as usual. This is quite readilty verified directly.[2] We exploit this in the choice of our receiver and transmitter filters, and show how MUI can be cancelled. Consider the transfer function $T_{km}(z)$ given by (6). Suppose we choose

$$H_k(z) = a_k(1 + \rho_k^{-1} z + \rho_k^{-2} z^2 + \ldots + \rho_k^{-(P-1)} z^{(P-1)}) \quad (9)$$

$$F_m(z) = r_{0,m} + r_{1,m} z^{-1} + \ldots + r_{P-L-1,m} z^{-(P-L-1)} \quad (10)$$

where ρ_k are distinct for $0 \le k \le M-1$. Using the identity (8), Eq. (6) representing $T_{km}(z)$ can be rearranged as

$$\begin{aligned} & a_k\,[1\ \rho_k^{-1}\ \ldots\ \rho_k^{-(P-1)}]\,\mathbf{A}_m \begin{bmatrix} r_{0,m} \\ r_{1,m} \\ \vdots \\ r_{P-L-1,m} \end{bmatrix} \\ &= a_k C_m(\rho_k)\,[1\ \rho_k^{-1}\ \ldots\ \rho_k^{-(P-L-1)}] \begin{bmatrix} r_{0,m} \\ r_{1,m} \\ \vdots \\ r_{P-L-1,m} \end{bmatrix} \\ &= a_k C_m(\rho_k) F_m(\rho_k) \end{aligned}$$

Thus the transfer function from $s_m(n)$ to $\widehat{s}_k(n)$ is

$$T_{km}(z) = a_k C_m(\rho_k) F_m(\rho_k) \quad (11)$$

which is a constant independent of z. Assume further that the multipliers a_k are chosen as

$$a_k = \frac{1}{C_k(\rho_k)} \quad (12)$$

Then the perfect symbol recovery condition $T_{km}(z) = \delta(k-m)$ becomes

$$F_m(\rho_k) = \delta(k-m), \quad 0 \le k, m \le M-1 \quad (13)$$

Even with ρ_k chosen as arbitrary (but distinct) numbers, these can be satisfied as long as $F_m(z)$ have M degrees of freedom (FIR with order $\ge M-1$). This gives the condition $P \ge M + L$ which we shall replace with

$$P = M + L \quad (14)$$

Then the preceding conclusions are valid as long as the channel order $\le L$. The multipliers a_k in Eq. (12) which are part of the receiver filters (9) can be regarded as z-domain equalizers. If $\rho_k = e^{j\omega_k}$ these become *frequency domain equalizers.*

A number of conclusions can be made here. The condition (13) guaratees that multiuser interference is cancelled *even if the channels are unknown.* Only their order L needs to be known. The knowledge of the channels is required only to design the equalizers (12) for each receiver. Since FIR channels have only finite number of zeros, we can always choose ρ_k so that $C_k(\rho_k) \neq 0$. *So, channel equalization is always possible!*

[2]Since Toeplitz matrices represent convolution (LTI filtering) when appropriately used, this property can be regarded as a manifestation of the elementary fact that LTI systems reproduce exponentials [10].

Example. *The DFT filter bank.* Assume $C_m(z) = C(z)$ for all m, $P = M+L$, and $\rho_k = W^{-k}$, $W \stackrel{\Delta}{=} e^{-j2\pi/M}$. Then

$$H_k(z) = \frac{1}{C(W^{-k})} \sum_{n=0}^{P-1} z^n W^{nk} \qquad (15)$$

The PR condition (13) yields the transmit filters $F_m(z) = \sum_{n=0}^{M-1} W^{-mn} z^{-n}/M$.

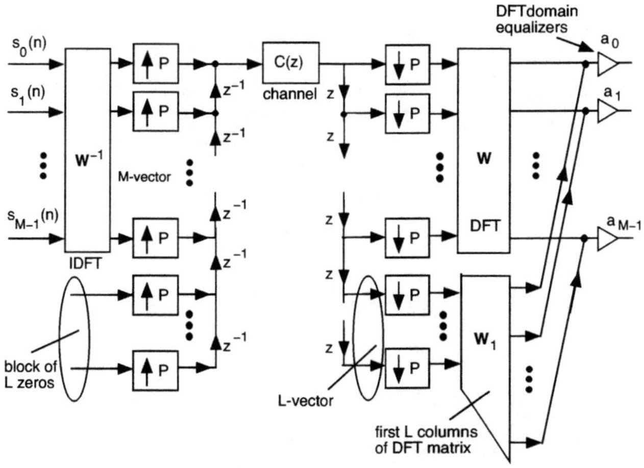

Fig. 6. Zero-padding at the transmitter and cyclic-prefix-like equalizers at the receiver.

If all users are in one place[3] the system can be drawn as shown in Fig. 6 where $\mathbf{W}$ is the $M \times M$ DFT matrix and $\mathbf{W}_1$ is the submatrix of $\mathbf{W}$ obtained by retaining the first $L = (P - M)$ columns. The equalizers

$$a_k = 1/C(W^{-k}) = 1/C(e^{j2\pi k/M}) \qquad (16)$$

invert the channel frequency response sampled on the DFT grid.

If we start with the *zero-jamming* approach instead of zero-padding, we can develop a similar example. After going through the details we will discover the transceiver shown in Fig. 7. Readers familiar with the **cyclic prefix** system [16] will readily recognize it in this figure! Thus the cyclic prefix system is a special case of the equalizer system developed in this section and therefore has the perfect symbol recovery property in absence of noise. Our analysis above shows that the equalizers a_k appearing in cyclic prefix systems also appear naturally in a *zero padding system.*

[3]In the multiuser scenario, the symbol streams $s_m(n)$ are separate independent users. On the other hand, in a context like the DMT channel, $s_m(n)$ are obtained from one stream $s(n)$ by parsing [20].

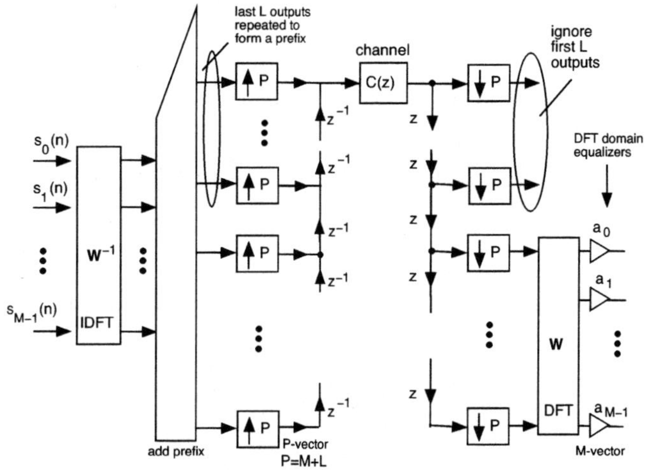

Fig. 7. The cyclic prefix system used in discrete multitone modulation.

IV. CONTROLLING BANDWIDTH EFFICIENCY

The transmitter system of Fig. 2 makes a linear combination of M independent symbol streams. If these have to be separated successfully later, then the spacing between the samples of $s_k(n)$ must be at least M times larger than the spacing T between the symbols entering the channels $C_k(z)$. But the actual spacing between samples of $s_k(n)$ is PT secs. See Fig. 8(a) which demonstrates the idea with $T = 1$. The excess space, measured by the ratio

$$\gamma = \frac{P}{M} = \frac{M+L}{M},$$

is called the *bandwidth expansion factor.* It is a price paid, in terms of the redundancy, which allows equalization of FIR channels of order L.

A slight variation of the redundant transmultiplexer system of Fig. 2 results if we make each user look like K users. This can be done by blocking the mth user K-fold as in Fig. 9. The K "subusers" $s_{m,i}(n)$ are merely substreams of $s_m(n)$. The system is mathematically equivalent to MK users, though the MK channels are not all different. The perfect recovery requirement is now $\widehat{P} \geq MK + L$ which we replace with

$$\widehat{P} = MK + L \qquad (17)$$

The advantage of this system is that the banwidth expansion factor is now

$$\widehat{\gamma} = \frac{\widehat{P}}{MK} = \frac{MK+L}{MK} \qquad (18)$$

which can be made arbitrarily close to unity by making K large. See Fig. 8(b). MUI cancellation and equalization can be achieved exactly as in the previous section by appropriate design of the transmit and receiver filters. This is precisely the **Amour system** developed by Giannakis et al. [4] for MUI cancellation in CDMA systems wherein, each of the $F_{m,i}(z)$ is regarded as a CDMA code (each user has

K codes). There are K different coefficients ρ_k used by each user, and these are called signature points in [4].

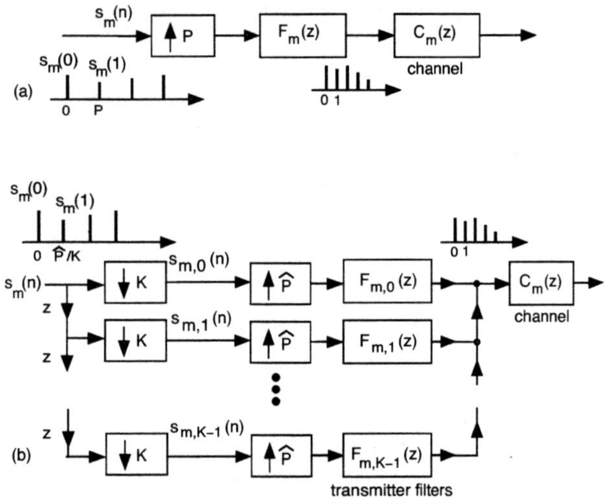

Fig. 8. Explaining bandwidth expansion.

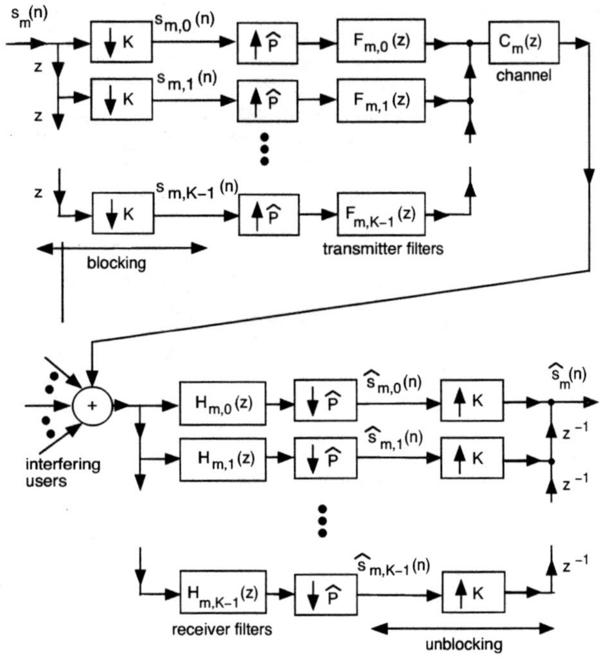

Fig. 9. The mth user viewed as K separate users.

V. OPTIMIZING FOR NOISE

If the statistics of the channel noise is known then it is possible to optimize the receiver to reduce its effect. To give a flavor for this we describe a scenario developed by Lin and Phoong [8]. Consider again the transmultiplexer of Fig. 1. With the transmitter and receiver filter banks expressed in polyphase matrix form [18] we can draw the transceiver as in Fig. 10. Assume as before that the filters have order $< P$ so $\mathbf{E}(z)$ and $\mathbf{R}(z)$ are constants. In the zero-padding scheme

$$\mathbf{R}(z) = \begin{bmatrix} \mathbf{R}_0 \\ \mathbf{0} \end{bmatrix} \qquad (19)$$

where $\mathbf{R}_0$ is $M \times M$. The transfer function from $\mathbf{s}(n)$ to $\widehat{\mathbf{s}}(n)$ is the product $\mathbf{T}(z) = \mathbf{E}(z)\mathbf{C}_b(z)\mathbf{R}(z)$. where $\mathbf{C}_b(z)$ is the blocked version of the channel (like $\mathbf{C}_m(z)$ in Fig. 4). With $\mathbf{R}(z)$ restricted as above, this becomes $\mathbf{EAR}_0$, where $\mathbf{A}$ is as in Eq. (7). The condition for perfect recovery in absence of noise is therefore

$$\mathbf{EAR}_0 = \mathbf{I} \qquad (20)$$

With $\mathbf{R}_0$ and $\mathbf{A}$ given, the $M \times P$ matrix $\mathbf{E}$ has to be chosen as a left inverse of $\mathbf{AR}_0$. Since $P > M$ the left inverse is not unique. The left inverse which minimizes the channel noise at the receiver input is worked out in [8].

Consider again Fig. 1 and assume no redundancy, that is, $P = M$. If we assume that we are allowed to have $1/C(e^{j\omega})$ at the receiver and restrict the filter bank $\{F_m(z), H_k(z)\}$ to the class of orthonormal filter banks [18], then an interesting result can be proved. Namely the optimum filter bank which minimizes the probability of error (for fixed bit rate and transmitted power) is the so-called principal component filter bank or **PCFB** for the effective power spectrum

$$S_{eff}(e^{j\omega}) \stackrel{\Delta}{=} S_{ee}(e^{j\omega})/|C(e^{j\omega})|^2,$$

where $S_{ee}(e^{j\omega})$ is the power spectrum of the additive noise $e(n)$. Further details on this result can be found in [21].

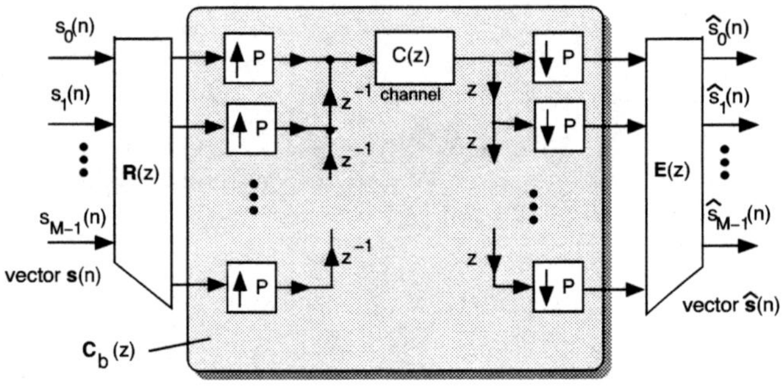

Fig. 10. Transmultiplexer of Fig. 1 in polyphase form.

All systems described in the preceding paragraphs fall under the category of *zero-forcing equalizers*. i.e., equalizers which tend to cancel the effect of $C(z)$ completely without taking into account the effects of noise. It turns out that this is not necessarily the best thing to do (*intuition*: if $C(e^{j\omega})$ is very small for some frequencies, then $1/C(e^{j\omega})$ can amplify the channel noise severely). There exist equalizers based on minimizing the mean squared error between $s_k(n)$ and $\widehat{s}_k(n)$. These are called MMSE equalizers. Some details can be found in [13] and [4].

VI. BLIND IDENTIFICATION

In many practical situations as in mobile communications, the channel transfer functions are unknown, and have to be estimated before they can be equalized. Such estimation can be done either with the help of training signals or by blind identification methods. It is well known that with non redundant systems ($P = M$) blind identification is not possible unless we use fourth order moments such as the Kurtosis of the data [15], [17]. An important result in recent years is the observation that blind identification is indeed possible without the use of fourth order moments, if we use *redundant* transmultiplexers or filter bank precoders [3], [14].

Consider again the single channel system of Fig. 10. Assume as in Sec. V that the channel is FIR with order $\leq L$, that the receiver filters have order $\leq P-1$, and that the transmitting filters have order $\leq M-1$. In particular therefore $\mathbf{R}(z)$ is as in Eq. (19). Figure 11 shows the path from the transmitted symbols to the channel output $y(n)$. For convenience we consider the blocked version $\mathbf{y}(n)$ as indicated. With the vector $\mathbf{s}(n)$ as defined in the figure, we then have

$$\mathbf{y}(n) = \mathbf{A}\mathbf{R}_1\mathbf{s}(n)$$

where $\mathbf{A}$ is as in (7) and reproduced below:

$$\mathbf{A} = \begin{bmatrix} c(0) & 0 & \dots & 0 \\ c(1) & c(0) & \dots & 0 \\ \vdots & \vdots & \ddots & \vdots \\ c(L) & & & \\ 0 & c(L) & & \\ \vdots & & \ddots & \vdots \\ 0 & 0 & \dots & c(L) \end{bmatrix}$$

This is a full-banded Toeplitz matrix representing the FIR channel of order $\leq L$.

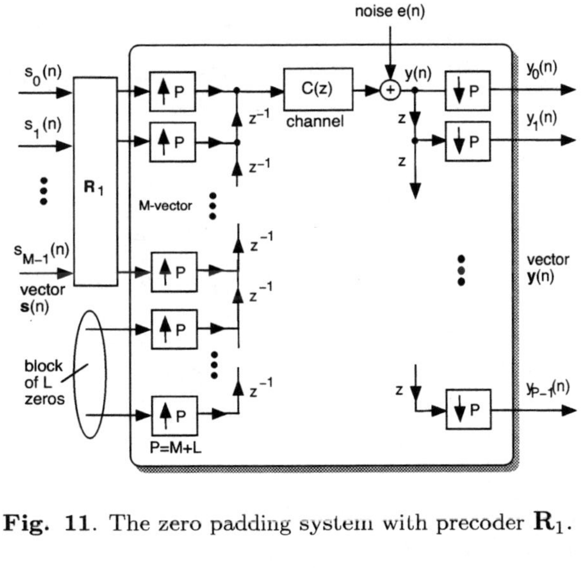

Fig. 11. The zero padding system with precoder $\mathbf{R}_1$.

Assume the channel $c(n)$ is unknown. We now argue that the observation of $\mathbf{y}(n)$ can be used to identify the channel $c(n)$ upto a scale-factor ambiguity. This is called *blind identification* because the input stream $\mathbf{s}(n)$ is unknown, unlike in training-based channel identification. Briefly, imagine we observe the output vector $\mathbf{y}(n)$ for a certain duration, say $0 \leq n \leq J-1$, and write the equation

$$\underbrace{[\mathbf{y}(0) \;\; \mathbf{y}(1) \;\; \dots \;\; \mathbf{y}(J-1)]}_{\mathbf{Y} \text{ matrix; size } P\times J}$$

$$= \underbrace{\mathbf{A}}_{P\times M} \underbrace{\mathbf{R}_1}_{M\times M} \underbrace{[\mathbf{s}(0) \;\; \mathbf{s}(1) \;\; \dots \;\; \mathbf{s}(J-1)]}_{\mathbf{S} \text{ matrix; size } M\times J} \qquad (21)$$

At this point we assume that the symbol stream $\mathbf{s}(n)$ is *rich,* that is, there exists a J such that $\mathbf{S}$ has full rank M. Since $\mathbf{A}$ and $\mathbf{R}_1$ have rank M, the product on the right hand side of Eq. (21) has rank M. So the $P \times J$ data matrix $\mathbf{Y}$ has rank M, and there are $P-M$ or L linearly independent vectors orthogonal to all the columns in $\mathbf{Y}$. That is, there is a $L \times P$ matrix $\mathbf{V}$ with L independent rows such that

$$\mathbf{VY} = \mathbf{VAR}_1\mathbf{S} = \mathbf{0} \qquad (22)$$

Since $\mathbf{R}_1\mathbf{S}$ has rank M, this implies

$$\mathbf{VA} = \mathbf{0} \qquad (23)$$

As $\mathbf{V}$ is $L \times P$ with rank L, there are $P - L = M$ independent *columns* which annihilate $\mathbf{V}$ from the right. But the M columns of the lower triangular matrix $\mathbf{A}$ not only annihilate $\mathbf{V}$, they are linearly independent as well. So *any annihilator* of $\mathbf{V}$ is in the column space of $\mathbf{A}$. In particular consider nonzero vectors of the form $\begin{pmatrix} \times \\ \mathbf{0} \end{pmatrix}$ where $\times$ has length $L+1$. The only vector of this form which annihilates $\mathbf{V}$ from the right is the 0th column of $\mathbf{A}$. This column (hence $c(n)$) can therefore be identified upto scale.

Does the method work when the data is contaminated by noise? In this case we have $\mathbf{y}(n) = \mathbf{AR}_1\mathbf{s}(n) + \mathbf{e}(n)$, where $\mathbf{e}(n)$ is the blocked version of channel noise $e(n)$. Assume $\mathbf{s}(n)$ and $\mathbf{e}(n)$ are jointly wide-sense stationary and uncorrelated. Denoting the autocorrelation of $\mathbf{y}(n)$ by $\mathbf{R}_y$, and so forth, we then have

$$\mathbf{R}_y - \mathbf{R}_e = \mathbf{AR}_1\mathbf{R}_s\mathbf{R}_1^\dagger\mathbf{A}^\dagger.$$

By repeating the preceding arguments about ranks and annihilators, it is readily verified that the matrix on the right has precisely L independent left-annihilators. Since the left hand side can be estimated from the received signal and noise statistics (assumed known), we can therefore estimate the L annihilators. These are also the annihilators of $\mathbf{A}$. So the channel can be identified as before.

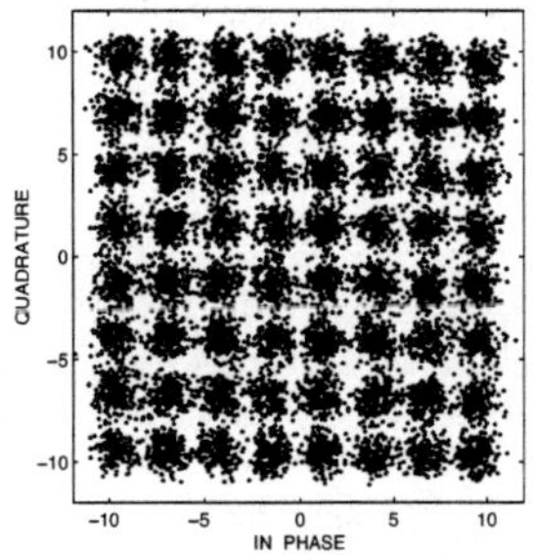

Figure 12. Result of equalization after blind identification.

Example. To demonstrate the idea with an example, we consider a simple 4th order FIR channel ($L = 4$) with $C(z)$ given by $-0.7684-0.8655z^{-1}+0.4305z^{-2}-0.3204z^{-3}+0.4992z^{-4}$. We choose a transmitter using a 64-QAM constellation, and assume that $\mathbf{s}(n)$ is a blocked version with $M = 12$ so there are 12 subusers $s_k(n)$. Therefore $P = M + L = 16$. Assuming the noise $e(n)$ is white and the SNR at the channel output is 25 dB, we estimate the channel using the above method. Once the channel is estimated, it can be used in equalization. The scatter diagram for the equalized signal is shown in Fig. 12.

VII. CONCLUDING REMARKS

One problem which was not addressed in this review is the minimization of redundancy in the precoder.[4] First observe that in Fig. 1 if we are allowed to have infinite length equalizers then we can always use $1/C(e^{j\omega})$ to equalize the channel, as long as $C(e^{j\omega})$ does not have unit circle zeros. In this case there is no need for redundancy, in fact no need for the filter bank precoder at all. It is more interesting and desirable to have FIR equalizers and we found that the use of redundancy L (where L is the channel order) is sufficient. But is it possible to have smaller redundancy? The answer is indeed yes, under some conditions. The results are rather involved, and can be found in fairly recent papers [23], [13], [9], [12].

Another interesting aspect that was not addressed is resistance to channel nulls. To explain this idea recall that the inverse of the channel response $C(\rho_k)$ is often involved in the equalization, for example Eq. (12) and Eq. (16). If $C(\rho_k)$ is very small or zero then we have a problem. A simple way to overcome this difficulty is to add extra redundancy of L samples in the precoder (e.g., instead of adding L zeros in Fig. 6 add $2L$ zeros). In this way, the coefficients $C(\rho_k)$ are involved for $M+L$ different values of ρ_k. As shown in [7], this allows us a choice at the receiver to choose any M out of the $M + L$ values of $C(\rho_k)$ for equalization. Since $C(z)$ has order L, at most L of these $C(\rho_k)$'s could be zero and we can always pick the largest M coefficients $C(\rho_k)$ for equalization. Such designs are called null-resistant designs. Many details on this can be found in [4], [7].

Research on the topic of filter bank precoders is still growing. For example there are four papers on filter bank precoders in the Sept. 2003 issue of the *IEEE Transactions on Signal Processing*! In [19] we introduce a new *frequency domain method for blind identification*, which offers some practical advantages. There remain many interesting problems in this area as the reader will learn from the other papers in this special session.

VIII. REFERENCES

[1] Akansu, A. N., Duhamel, P., Lin, X., and Courville, M. de. "Orthogonal transmultiplexers in communications: a review," IEEE Trans. SP, April 1998.

[2] Bellanger, M., "On computational complexity in digital transmultiplexer filters", IEEE Trans. on Comm., vol. 30, pp. 1461–1465, July 1982.

[3] Giannakis, G. B. "Filter banks for blind channel identification and equalization," IEEE Signal Processing Letters, vol. 4, no. 6, pp. 184–187, June 1997.

[4] Giannakis, G. B., Wang, Z., Scaglione, A., and Barbarossa, S., "Amour — generalized multicarrier transceivers for blind CDMA regardless of multipath," IEEE Trans. Comm., pp 2064–2076, Dec. 2000.

[5] Kalet, I. "The multitone channel", IEEE Trans. Comm., pp. 119–124, Feb. 1989.

[6] Koilpillai, R. D., Nguyen, T. Q., and Vaidyanathan, P. P. "Some results in the theory of cross-talk free transmultiplexers," IEEE Trans. on Acoustics, Speech and Signal Processing, vol. 39, pp. 2174-2183, Oct. 1991.

[7] Liang, J., and Tran, T. D., "A partial DFT based OFDM system resistant to channel nulls", Proc. IEEE Asilomar conference on Signals, Systems, and Computers, Nov. 2002.

[8] Lin, Y.-P., and Phoong, S.-M., "Perfect discrete multitone modulation with optimal transceivers," IEEE Trans. SP, vol. 48, pp. 1702–1712, June. 2000.

[9] Lin, Y.-P., and Phoong, S.-M., "Minimum redundancy for ISI free FIR filter bank transceivers," IEEE Trans. SP, vol. 50, pp. 842–853, April 2002.

[10] Oppenheim, A. V., and Schafer, R. W. *Discrete-time signal processing,* Prentice Hall, 1999.

[11] Proakis, J. G. *Digital communications,* McGraw Hill 1995.

[12] Ribeiro, C. B., de Campos, M. L. R, and Diniz, P. S. R., "FIR equalizers with minimum redundancy", Proc. of the ICASSP, v. III, pp. 2673-2676, May 2002.

[13] Scaglione, A., Giannakis, G. B., and Barbarossa, S. "Redundant filter bank precoders and equalizers Part I: Unification and optimal designs", IEEE Trans. Signal Processing, vol. 47, pp. 1988-2006, July 1999.

[14] Scaglione, A., Giannakis, G. B., and Barbarossa, S. "Redundant filter bank precoders and equalizers Part II: Synchronization and direct equalization", IEEE Trans. Signal Processing, vol. 47, pp. 2007-2022, July 1999.

[15] Shalvi, O., and Weinstein, E. "New criteria for blind deconvolution of nonminimum phase systems (channels)", IEEE Trans. Info. Th., vol. 36, pp. 312-321, Mar. 1990.

[16] Starr, T., Cioffi, J. M., and Silverman, P. J. *Understanding DSL technology,* Prentice Hall, Inc., 1999.

[17] Tkacenko, A., and Vaidyanathan, P. P. "Generalized Kurtosis and applications in blind equalization of MIMO channels,", Proc. IEEE Asilomar Conference on Signals, Systems, and Computers, Nov. 2001.

[18] Vaidyanathan, P. P. *Multirate systems and filter banks,* Prentice Hall, Inc., 1993.

[19] Vaidyanathan, P. P., and Vrcelj, B. "A frequency domain approach for blind identification with filter bank precoders," IEEE ISCAS 2004, submitted.

[20] Vaidyanathan, P. P., "Filter banks in digital communications," IEEE CAS Mag., vol. 1, pp. 4–25, 2001.

[21] Vaidyanathan, P. P., Lin, Y.-P., Akkarakaran, S., and Phoong, S.-M., "Discrete multitone modulation with principal component filter banks," IEEE Trans. Circuits and Systems-I, vol. 49, pp. 1397–1412, Oct. 2002.

[22] Vetterli, M. "Perfect transmultiplexers," Proc. IEEE Int. Conf. ASSP, pp. 2567–2570, Japan, April 1986.

[23] Xia, X-G. "New precoding for intersymbol interference cancellation using nonmaximally decimated multirate filter banks with ideal FIR equalizers," IEEE Trans. Signal Processing, vol. 45, no. 10, pp. 2431–2441, Oct. 1997.

[24] Xia, X-G. "Modulated coding for intersymbol interference channels," Marcel Dekker, Inc., 2001.

[4] For large M the bandwidth expansion factor is small even for moderate L and this is not a serious issue.

ZERO-FORCING EQUALIZATION FOR TIME-VARYING SYSTEMS WITH MEMORY

Cássio B. Ribeiro, Marcello L. R. de Campos, and Paulo S. R. Diniz

Electrical Engineering Program and Department of Electronics and Computer Engineering
COPPE/Poli/Federal University of Rio de Janeiro
P.O. Box 68504, 21945-970, Rio de Janeiro, RJ, Brazil

ABSTRACT

In this article we derive conditions for existence of zero-forcing multiuser detectors which achieve perfect reconstruction of the transmitted symbols. Compared to similar works in the literature, the derived conditions allow a reduction in the length of the equalizer filters, while imposing no constraints on the channel order and on the order of the precoder filters. Since the analysis is carried out in time domain, time-varying filters are also considered.

1. INTRODUCTION

The forthcoming release of new wireless communications systems motivates research focusing on the development of receivers which are capable of coping with the severe mobile environment, overcoming its limitations and providing reliable communications at high data rate.

In this work we derive theoretical conditions for the existence of symbol-level multiuser detectors which achieve perfect reconstruction of the transmitted symbols. The conditions are derived for a general communications system that could be interpreted, e.g., as a wireless communications system.

For the derivation of the theoretical conditions the system is modeled as a time-varying precoder with memory, which is an extension from previous works [1, 2, 3], where only memory-less precoders were considered, and also from [4], where only time-invariant precoders were considered. With this extension the results presented in this article can be applied directly to filterbank-based systems, like Discrete Wavelet MultiTone.

2. SYSTEM DESCRIPTION

In this section we describe the communications system considered throughout this article. The system model is shown in Figure 1, where $s_m(n)$, and $\hat{s}_m(n)$ are the transmitted and received symbols for user m, respectively, $m = 0, \ldots, M-1$. The channel is modeled by $h(k)$, and the transmit filters are given by $f_m(k,n)$, where the index n indicates the time-varying nature of the transmit filters.

The output of the synthesis filter bank (precoder) is given by

$$u(k) = \sum_{m=0}^{M-1} \sum_{n=-\infty}^{\infty} s_m(n) f_m(k-nN, n), \tag{1}$$

where $s_m(n)$ is the symbol transmitted by the m-th user at time instant n, and $f_m(k,n)$ is the filter corresponding to user m. In this article we will not impose any constraints on the maximum length of the precoder filters. As a consequence, equation (1) introduces intersymbol interference (ISI) that must be removed in the receiver.

The signal is transmitted through a linear time-invariant (LTI) channel $h(k)$, and is received together with additive Gaussian noise with zero mean $v(k)$, i.e.,

$$y(k) = x(k) + v(k) = \sum_{l=-\infty}^{\infty} h(l)u(k-l) + v(k). \tag{2}$$

The receiver estimates the transmitted symbols for the m-th user by filtering the signal $y(k)$ by $g_m(k)$ and then decimating the output by N, as shown in Figure 1. This process is described by

$$\hat{s}_m(n) = \sum_{j=-\infty}^{\infty} y(nN-j) g_m(j). \tag{3}$$

In order to obtain a compact representation in matrix form for the process, we will define the $M \times 1$ vectors

$$\mathbf{s}(n) = \begin{bmatrix} s_0(n) & s_1(n) & \cdots & s_{M-1}(n) \end{bmatrix}^T \tag{4}$$

$$\hat{\mathbf{s}}(n) = \begin{bmatrix} \hat{s}_0(n) & \hat{s}_1(n) & \cdots & \hat{s}_{M-1}(n) \end{bmatrix}^T, \tag{5}$$

and the $N \times 1$ vectors

$$\mathbf{u}(n) = \begin{bmatrix} u(nN) & \cdots & u(nN+N-1) \end{bmatrix}^T \tag{6}$$

$$\mathbf{v}(n) = \begin{bmatrix} v(nN) & \cdots & v(nN+N-1) \end{bmatrix}^T \tag{7}$$

$$\mathbf{y}(n) = \begin{bmatrix} y(nN) & \cdots & y(nN+N-1) \end{bmatrix}^T. \tag{8}$$

Equations (1) and (3) can be rewritten as

$$\mathbf{u}(n) = \sum_{i=-\infty}^{\infty} \mathbf{F}_i(n)\mathbf{s}(n-i) \tag{9}$$

$$\hat{\mathbf{s}}(n) = \sum_{j=-\infty}^{\infty} \mathbf{G}_j \mathbf{y}(n-j), \tag{10}$$

where the elements of the $N \times M$ matrix $\mathbf{F}_i(n)$ and of the $M \times N$ matrix $\mathbf{G}_i$ are

$$\{\mathbf{F}_i(n)\}_{l,m} = f_m(iN+l, n), \; \{\mathbf{G}_i\}_{m,l} = g_m(iN-l)$$
$$m = 0, \ldots, M-1, \quad l = 0, \ldots, N-1. \tag{11}$$

0-7803-8251-X/04/$17.00 ©2004 IEEE

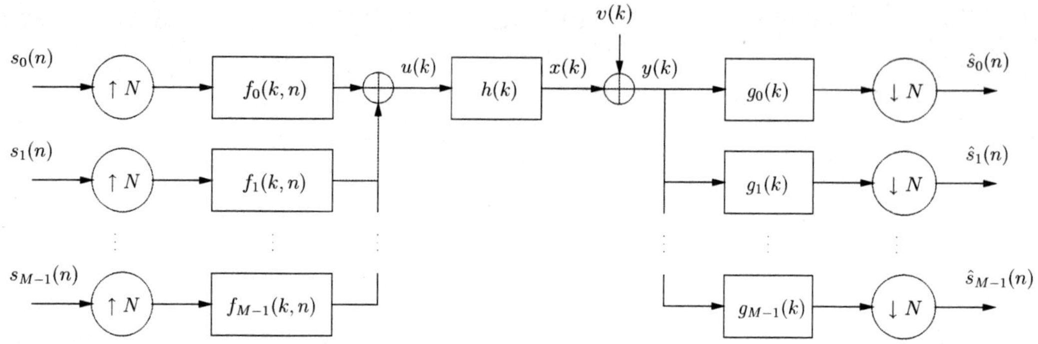

Fig. 1. Communications system in time domain as a MIMO system.

Define the $N \times N$ matrix $\mathbf{H}_l$ as

$$\mathbf{H}_l = \begin{bmatrix} h(lN) & h(lN-1) & \cdots & h(lN-N+1) \\ h(lN+1) & h(lN) & \cdots & h(lN-N) \\ \vdots & \vdots & \ddots & \vdots \\ h(lN+N-1) & h(lN+N-2) & \cdots & h(lN) \end{bmatrix}. \quad (12)$$

The received signal is then given by

$$\mathbf{y}(n) = \sum_{l=-\infty}^{\infty} \mathbf{H}_l \mathbf{u}(n-l) + \mathbf{v}(n). \quad (13)$$

In order to simplify the notation we will introduce now the assumptions that will be considered in the following sections for equalization:

(a) the channel is modeled as an L-th order FIR filter, with $h(0), h(L) \neq 0$;

(b) the spreading factor (or block length) is greater than or equal to the number of symbols, i.e., $N \geq M$;

(c) $f_m(k,n)$ are causal length-TN FIR filters, and $g_m(k)$ are causal length-QN FIR filters, T and Q integers.

Assumption (c) implies that matrices $\mathbf{F}_i(n)$ are zero, except for $i = 0, \ldots, T-1$; matrices $\mathbf{G}_j$ are zero, except for $j = 0, \ldots, Q-1$. Assumption (a) implies that matrices $\mathbf{H}_l$ are zero, except for $l = 0, \ldots, B$, with $B = \lceil \frac{L}{N} \rceil$, where $\lceil \cdot \rceil$ denotes the smallest integer greater than or equal to $(\cdot)$.

Based on these assumptions, we can write

$$\mathbf{u}(n) = \sum_{i=0}^{T-1} \mathbf{F}_i(n)\mathbf{s}(n-i) \quad (14)$$

$$\mathbf{y}(n) = \sum_{l=0}^{B} \mathbf{H}_l \mathbf{u}(n-l) + \mathbf{v}(n) \quad (15)$$

Let us define the $(B+Q)N \times (B+Q+T-1)M$ matrix $\boldsymbol{\mathcal{F}}$ as

$$\boldsymbol{\mathcal{F}} = \begin{bmatrix} \mathbf{F}_{T-1}(n-Q-B+1) & \cdots & \mathbf{F}_0(n-Q-B+1) & \cdots & \mathbf{0} \\ \vdots & \ddots & & \ddots & \vdots \\ \mathbf{0} & \cdots & \mathbf{F}_{T-1}(n) & \cdots & \mathbf{F}_0(n) \end{bmatrix}, \quad (16)$$

the $QN \times (B+Q)N$ matrix $\boldsymbol{\mathcal{H}}$ as

$$\boldsymbol{\mathcal{H}} = \begin{bmatrix} \mathbf{H}_B & \cdots & \mathbf{H}_0 & \mathbf{0} & \cdots & \mathbf{0} \\ \mathbf{0} & \mathbf{H}_B & \cdots & \mathbf{H}_0 & \cdots & \mathbf{0} \\ \vdots & \vdots & & & & \vdots \\ \mathbf{0} & \mathbf{0} & \cdots & \mathbf{H}_B & \cdots & \mathbf{H}_0 \end{bmatrix} \quad (17)$$

and the $(B+Q+T-1)M \times 1$ vector $\bar{\mathbf{s}}(n)$ as

$$\bar{\mathbf{s}}(n) = \begin{bmatrix} \mathbf{s}^T(n-T-Q-B+1) & \cdots & \mathbf{s}^T(n) \end{bmatrix}^T \quad (18)$$

From (14) and (15), we can write the input-output relation of the communications system in Figure 1 as

$$\hat{\mathbf{s}}(n) = \boldsymbol{\mathcal{G}}\boldsymbol{\mathcal{H}}\boldsymbol{\mathcal{F}}\,\bar{\mathbf{s}}(n) + \boldsymbol{\mathcal{G}}\bar{\mathbf{v}}(n), \quad (19)$$

where the $M \times QN$ matrix $\boldsymbol{\mathcal{G}}$ is defined as

$$\boldsymbol{\mathcal{G}} = \begin{bmatrix} \mathbf{G}_{Q-1} & \cdots & \mathbf{G}_0 \end{bmatrix}. \quad (20)$$

3. ZF MULTIUSER DETECTION

In this section we examine the system described in the previous section in order to derive necessary conditions for ZF equalization. From equation (19), ZF equalization is obtained if $v(k) = 0$ and

$$\boldsymbol{\mathcal{G}}\boldsymbol{\mathcal{H}}\boldsymbol{\mathcal{F}} = \begin{bmatrix} \mathbf{0}_{M \times (Q+B-1)M} & \mathbf{I}_M \end{bmatrix}. \quad (21)$$

The above condition is satisfied if and only if the right side of (21) belongs to the space spanned by the columns of $\boldsymbol{\mathcal{H}}\boldsymbol{\mathcal{F}}$, i.e.,

$$\begin{bmatrix} \mathbf{0}_{M \times (Q+B-1)M} & \mathbf{I}_M \end{bmatrix}^T \in \mathcal{R}\{\boldsymbol{\mathcal{F}}^T \boldsymbol{\mathcal{H}}^T\}, \quad (22)$$

where $\mathcal{R}\{\cdot\}$ denotes the space spanned by the columns of $\{\cdot\}$.

Let us express the above relation in terms of the null space of $\boldsymbol{\mathcal{H}}$, denoted as $\mathcal{N}\{\boldsymbol{\mathcal{H}}\}$. Let us define the $r \times 1$ vector $\mathbf{e}_r(i)$ as the i-th column of the $r \times r$ identity matrix. Thus, we can say that if there is a $(Q+B)N \times 1$ vector $\boldsymbol{\alpha}$ such that

$$\boldsymbol{\mathcal{F}}^T \boldsymbol{\alpha} = \mathbf{e}_{(Q+B)M}(i + (Q+B-1)M), \quad (23)$$

then the condition expressed in equation (22) is satisfied if and only if $\boldsymbol{\alpha} \in \mathcal{R}(\mathcal{H}^T)$.

Instead of dealing with $\mathcal{R}(\mathcal{H}^T)$, that is not easy to characterize, we will work with the nullspace of $\mathcal{H}$, that has a straightforward characterization as a function of the zeros of $h(k)$. Without impairing the analysis, let us consider $BN - L = 0$, i.e., the channel length is a multiple of the decimation/interpolation ratio N, and then we can write the null space of $\boldsymbol{\mathcal{H}}$ as

$$\mathcal{N}(\boldsymbol{\mathcal{H}}) = \text{span}\{\left[1 \ v_l \ \cdots \ v_l^{(Q+B)N-1}\right]^T, l=0,\ldots,L-1\}, \quad (24)$$

where v_l are the roots of the polynomial $h(n)$.

Since the ZF condition can be generalized to $\hat{\mathbf{s}}(n) = \mathbf{s}(n-\Delta)$, we can use a general $(B+Q)N$ vector $\boldsymbol{\gamma}$ instead of e in (23), and then, from the definition of $\boldsymbol{\mathcal{F}}$ in (16), we can rewrite (23) as

$$\begin{bmatrix} \mathbf{0} & \cdots & \cdots & \mathbf{0} \\ \vdots & & & \vdots \\ \mathbf{0} & \cdots & \cdots & \mathbf{0} \\ \mathbf{F}_0^T(n-Q-B+1) & \cdots & \mathbf{F}_{T-1}^T(n-1) & \mathbf{0} \\ \mathbf{0} & \ddots & & \mathbf{F}_{T-1}^T(n) \\ \vdots & & \ddots & \vdots \\ \mathbf{0} & \cdots & \mathbf{0} & \mathbf{F}_0^T(n) \end{bmatrix} \boldsymbol{\alpha} + \\ + \begin{bmatrix} \mathbf{F}_{T-1}^T(n-Q-B+1) & \mathbf{0} & \cdots & \mathbf{0} \\ \vdots & \ddots & & \vdots \\ \mathbf{F}_1^T(n-Q-B+1) & \cdots & \mathbf{F}_{T-1}^T(n-2) & \mathbf{0} \\ \mathbf{0} & \cdots & \cdots & \mathbf{0} \\ \vdots & & & \vdots \\ \mathbf{0} & \cdots & \cdots & \mathbf{0} \end{bmatrix} \boldsymbol{\alpha} = \boldsymbol{\gamma} \quad (25)$$

where it was considered that $T \leq B+Q$, that will be shown later to be always satisfied.Since (25) must have a solution for any vector $\boldsymbol{\gamma}$, after some algebraical manipulations we conclude that (25) is equivalent to

$$\text{diag}(\mathbf{F}_0^T(n-Q-B+1),\ldots,\mathbf{F}_0^T(n))\,\boldsymbol{\alpha} = \boldsymbol{\gamma}' \quad (26)$$

$$\tilde{\mathbf{F}}\boldsymbol{\alpha} = \boldsymbol{\gamma}'' \quad (27)$$

where

$$\tilde{\mathbf{F}} = \begin{bmatrix} \mathbf{F}_{T-1}^T(n-Q-B+1) & \cdots & \mathbf{0} & \mathbf{0} \\ \vdots & \ddots & \vdots & \vdots \\ \mathbf{F}_1^T(n-Q-B+1) & \cdots & \mathbf{F}_{T-1}^T(n-2) & \mathbf{0} \end{bmatrix}. \quad (28)$$

The problem in equation (23) is now split in two different problems, and we must find conditions for the existence of a vector $\boldsymbol{\alpha}$ that simultaneously solves equations (26) and (27). The solution to (26) can be found in [2], hence our task is to find which restrictions should be added to the solution to (26) in order for it to solve also (27).

If we assume $\mathbf{F}_0(n)$ has full column rank, we can write without loss of generality $\mathbf{F}_0(n) = [\mathbf{I}_M \ \ \boldsymbol{\Phi}_0^T(n)]^T\mathbf{F}(n)$, where $\mathbf{F}(n)$ is a full-rank $M \times M$ matrix. Hence we can rewrite (26) as

$$\text{diag}([\mathbf{I}_M \ \boldsymbol{\Phi}_0^T(n-Q-B+1)],\ldots,[\mathbf{I}_M \ \boldsymbol{\Phi}_0^T(n)])\boldsymbol{\alpha} = \overline{\boldsymbol{\gamma}}' \quad (29)$$

where $\overline{\boldsymbol{\gamma}}' = \text{diag}(\mathbf{F}^T(n-Q-B+1),\ldots,\mathbf{F}^T(n))^{-1}\boldsymbol{\gamma}'$. The main objective now is to parameterize the solutions of (29) that satisfy the ZF condition, and then show how to guarantee that these solutions will also follow (27).

We will divide the vector $\boldsymbol{\alpha}$ in length-M and length-$(N-M)$ blocks, such as

$$\boldsymbol{\alpha} = \left[\tilde{\boldsymbol{\alpha}}_{Q+B-1}^T \ \ \hat{\boldsymbol{\alpha}}_{Q+B-1}^T \ \ \cdots \ \ \tilde{\boldsymbol{\alpha}}_0^T \ \ \hat{\boldsymbol{\alpha}}_0^T\right]^T \quad (30)$$

where $\tilde{\boldsymbol{\alpha}}_q$ and $\hat{\boldsymbol{\alpha}}_q$ are $M \times 1$ and $(N-M) \times 1$, respectively, for $q = 0,\ldots,Q+B-1$. From equations (29) and (30), we have

$$\tilde{\alpha}_q + \boldsymbol{\Phi}_0^T(n-q)\,\hat{\alpha}_q = \overline{\boldsymbol{\gamma}}'_{qM:(q+1)M-1} \quad (31)$$

Let us also define the $(Q+B)(N-M) \times 1$ vector $\hat{\boldsymbol{\alpha}}$ as

$$\hat{\boldsymbol{\alpha}} = \left[\hat{\boldsymbol{\alpha}}_{Q+B-1}^T \ \ \cdots \ \ \hat{\boldsymbol{\alpha}}_0^T\right]^T \quad (32)$$

Using (31) and (32), we can write $\boldsymbol{\alpha} = \boldsymbol{\Theta}\,\hat{\boldsymbol{\alpha}} + \overline{\boldsymbol{\gamma}}'$, where the $(Q+B)N \times (Q+B)(N-M)$ matrix $\boldsymbol{\Theta}$ is given by

$$\boldsymbol{\Theta} = \text{diag}\left\{[-\boldsymbol{\Phi}(n-Q-B+1) \ \ \mathbf{I}_{N-M}]^T,\cdots,\ [-\boldsymbol{\Phi}(n) \ \ \mathbf{I}_{N-M}]^T\right\} \quad (33)$$

If $\boldsymbol{\alpha} \in \mathcal{R}(\mathcal{H}^T)$ then $\boldsymbol{\alpha}$ is orthogonal to all vectors in $\mathcal{N}(\boldsymbol{\mathcal{H}})$. Then, we need to establish restrictions on the values of Q, $N-M$ and L that guarantee orthogonality between $\boldsymbol{\alpha}$ and the vectors in the set $\mathcal{N}(\boldsymbol{\mathcal{H}})$. Let $\mathbf{V}$ be the $(Q+B)N \times L$ Vandermonde matrix generated using the zeros of the channel $h(n)$, v_l, $l = 0,\ldots,L-1$), as in equation (24). From equations (30), (31), and (33), we can write

$$\mathbf{V}^T\boldsymbol{\Theta}\,\hat{\boldsymbol{\alpha}} = -\mathbf{V}^T\overline{\boldsymbol{\gamma}}', \quad (34)$$

Equation (34) admits solution for any $\overline{\boldsymbol{\gamma}}'$ if and only if $\text{rank}(\mathbf{V}^T\boldsymbol{\Theta}) = L$. Then, a necessary condition for existence of solution is

$$(Q+B)(N-M) \geq L \ \Rightarrow \ Q \geq \left\lceil \frac{L}{N-M} \right\rceil - B. \quad (35)$$

In [2, 4], the authors show that if the precoder is time-invariant, another condition that must be satisfied is $N - M \geq \mu$, where μ is the number of congruous zeros in the largest set of congruous zeros [4], i.e., zeros such that $v_l^N = \rho$, $l = 0,\ldots,\mu-1$. For time-variant precoders there is still a condition for ZF equalization that is dependent on the zeros of the channel, but that can be made independent of the number of congruous zeros. As shown in [3], defining δ as the number of different sets of precoders present in $\mathcal{F}$, ZF equalization is possible if $(N-M)\delta \geq \mu$. In order to avoid the evaluation of sets of congruous zeros, it suffices to consider the worst case, $\mu = L$, and in that case δ is such that $(N-M)\delta \geq L \Rightarrow \delta \geq \frac{L}{N-M}$

Once we have found the solution to (26), we have to guarantee that this solution has enough degrees of freedom so that it can also be a solution to (27). The difference between the number of columns and the number of rows of $(\mathbf{V}^T\boldsymbol{\Theta})$ is the number of degrees of freedom in the solution of (34). The vector $\hat{\boldsymbol{\alpha}}$ that satisfies (34) can be rewritten as

$$\hat{\boldsymbol{\alpha}} = \hat{\boldsymbol{\alpha}}^* + \overline{\boldsymbol{K}}\,\boldsymbol{\epsilon}, \tag{36}$$

where $\hat{\boldsymbol{\alpha}}^*$ is one solution for (34), the $(Q+B)(N-M) \times (Q+B)(N-M) - L$ matrix $\overline{\boldsymbol{K}}$ spans the nullspace of $\mathbf{V}^T\boldsymbol{\Theta}$, and $\boldsymbol{\epsilon}$ is a $(Q+B)(N-M) - L \times 1$ vector.

We are now able to verify the conditions the solution of (26) must satisfy in order to be also a solution of (27). Substituting (36) in (27) and using $\boldsymbol{\alpha} = \boldsymbol{\Theta}\,\hat{\boldsymbol{\alpha}} + \overline{\boldsymbol{\gamma}}'$, we have

$$\tilde{\mathbf{F}}\left(\boldsymbol{\Theta}\left(\hat{\boldsymbol{\alpha}}^* + \overline{\boldsymbol{K}}\boldsymbol{\epsilon}\right) + \overline{\boldsymbol{\gamma}}'\right) = \boldsymbol{\gamma}'' \tag{37}$$

that can be rewritten as

$$\left(\tilde{\mathbf{F}}\boldsymbol{\Theta}\overline{\boldsymbol{K}}\right)\boldsymbol{\epsilon} = \boldsymbol{\gamma}'' - \tilde{\mathbf{F}}\left(\boldsymbol{\Theta}\hat{\boldsymbol{\alpha}}^* + \overline{\boldsymbol{\gamma}}'\right) \tag{38}$$

For the equation above to admit solution for any $\hat{\boldsymbol{\alpha}}^*$, $\boldsymbol{\gamma}''$ and $\boldsymbol{F}_t^T$, $t = 0, \ldots, T-1$, the number of columns of the matrix multiplying $\boldsymbol{\epsilon}$ in the equation above must be greater than or equal to the number of rows, i.e.,

$$\begin{aligned} (Q+B)(N-M) - L \geq (T-1)M \;\Rightarrow \\ \Rightarrow\; Q \geq \left\lceil \frac{L + (T-1)M}{N-M} \right\rceil - B \end{aligned} \tag{39}$$

These results show that for time-variant precoders the amount of redundancy, $N-M$, does not depend on the number of congruous zeros, allowing the system design to be independent of the channel realization. If we use precoders with memory, the conditions for ZF equalization are the same previously derived [3], except for the minimum order of the equalizer, that is now given by (39).

4. EXPERIMENTAL RESULTS

In this section we present some experimental results obtained by computer simulations. The results consist in an average of 50 transmissions, each one comprising approximately 10000 BPSK symbols for each $s_m(n)$, $m = 0, \ldots, M-1$. The performance is measured as the mean squared error (MSE) between the transmitted and received sequences. The precoder filters were randomly generated such that the constraint on the full column rank of $\mathbf{F}_0(n)$ is respected.

Figure 2 shows the variation of the MSE with the redundancy $N-M$. The simulation was carried out for $N = 10$,

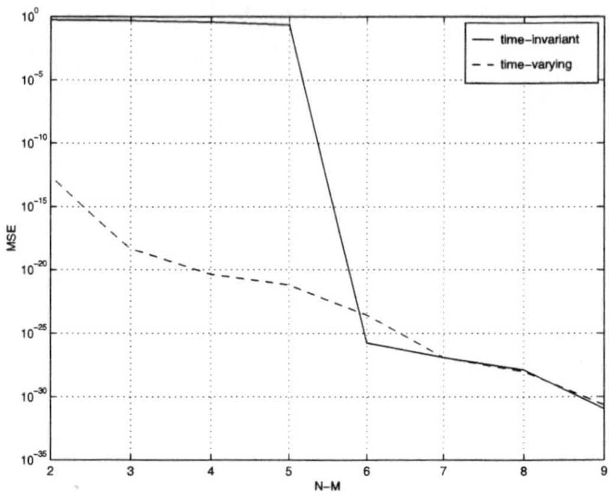

Fig. 2. MSE as a function of $N-M$ for a channel with a set of congruous zeros with 6 elements.

$N-M = \{2,3,4,5,6,7,8,9\}$, $D = 6$, where D is the period of the time-varying $\mathbf{F}_0(n)$. The channel model $h(k)$ is a 13-coefficient FIR filter, such that for $N = 10$, there is a set of congruous zeros with six elements. Hence, for time-invariant precoder ZF equalization will be possible for $N-M \geq 6$, what can be verified from Figure 2. For time-varying precoders ZF equalization is possible for any value of $N-M \geq 1$, provided that there are enough different precoder filters. This can be also verified from Figure 2.

5. CONCLUSION

In this work we derived theoretical conditions for the existence of symbol-level multiuser detectors which achieve perfect reconstruction of the transmitted symbols. The extension of the results in [3] for time-varying precoders with memory allows the application of the techniques for ZF equalization to MUD in systems using long filters in the precoders, as DWMT. Compared to related works, e.g. [1], the obtained relations allow a reduction in the length of the equalizer filters, and also allows transmission through channels with impulse responses longer than the block length.

6. REFERENCES

[1] Anna Scaglione, Georgios B. Giannakis, and Sergio Barbarossa, "Redundant filterbank precoders and equalizers part I: Unification and optimal designs," *IEEE Transactions on Signal Processing*, vol. 47, no. 7, pp. 1988–2006, July 1999.

[2] Cássio B. Ribeiro, Marcello L. R. de Campos, and Paulo S. R. Diniz, "FIR equalizers with minimum redundancy," in *Proc. ICASSP*, April 2002, vol. 3.

[3] Cássio B. Ribeiro, Marcello L. R. de Campos, and Paulo S. R. Diniz, "Zero-forcing multiuser detection in CDMA systems using long codes," Proc. of GLOBECOM, Dec. 2003.

[4] Yuan-Pei Lin and See-May Phoong, "Minimum redundancy for ISI free FIR DMT transceivers," *IEEE Transactions on Signal Processing*, vol. 50, no. 4, pp. 842–859, May 2002.

ON PILOT PATTERN DESIGN FOR PSAM-OFDM SYSTEM

Wei Zhang[†]*, Xiang-Gen Xia*[‡]*, and P. C. Ching*[†]

[†]Department of Electronic Engineering, The Chinese University of Hong Kong
Shatin, N. T., Hong Kong. Email: {wzhang, pcching}@ee.cuhk.edu.hk
[‡]Department of Electrical and Computer Engineering, University of Delaware
Newark, DE, 19716, USA. Email: xxia@ee.udel.edu

ABSTRACT

Pilot symbol assisted modulation (PSAM) of subcarriers is an attractive scheme for high data-rate OFDM transmission systems and has been widely used in broadcasting systems like DVB-T. In this paper, a new and simple pilot pattern for PSAM-OFDM system operated in the mobile communication environment is proposed and analyzed in terms of its bit error rate. Analytical and simulation results show that the proposed clustered pilot pattern for OFDM channel estimation gives a better performance than the conventional equally spaced pilot pattern in fading channel, yet without sacrificing the bandwidth efficiency.

1. INTRODUCTION

Orthogonal frequency division multiplexing (OFDM) is a promising technique for high speed transmissions in a frequency selective fading environment. It has been adopted as a standard for European digital video broadcasting terrestrial transmission (DVB-T) [1] and wireless local area network (W-LAN) [2].

To achieve high data rate with good performance in current OFDM systems, we use coherent detection [3]. The coherent detection relies on the knowledge of the channel state information (CSI). One method to estimate the CSI distorted by the fading process is to send pilot symbols among the transmitted data symbols. The locations of these pilot symbols, i.e. the pilot pattern, can be block type or comb type [4]. The former allocates a pilot symbol to each subcarrier at one OFDM symbol, whereas the latter inserts pilots to a subset of subcarriers only. Upon receiving the corrupted pilot symbols at the receiver, the characteristics of the pilot subchannel can be estimated using the Least Square (LS) or Minimum Mean Square Error (MMSE) method [5]. Subsequently, the CSI of the data subchannel can be derived by interpolation.

This work was partially supported by a research grant awarded by the Hong Kong Research Grant Council.

Typically, for combating the fading channel the pilot symbols in DVB-T are placed among the transmitted data symbols with equal space. In this paper, a new pilot pattern is proposed in which two pilots are clustered as a group while a series of these pilot groups are equally positioned along the frequency axis. The new pilot pattern can reduce half of the noise power of the pilot subchannel estimate.

The rest of this paper is organized as follows. In Section 2, a brief description of PSAM-OFDM system is presented. A new pilot pattern is introduced in Section 3. The BER formula in multipath fading channel is given in Section 4. Results and discussions are shown in Section 5. Finally, in Section 6, we draw the conclusion.

2. PSAM-OFDM

An OFDM baseband system considered in this paper is reviewed as follows. Firstly, the binary information data are grouped and mapped into QAM or QPSK signal. After pilot insertion, the modulated sequences are converted into a time domain signal. Following the inverse DFT, guard interval is appended to prevent inter-symbol interference (ISI) [3]. Then the transmitted signal is sent to the multipath fading channel. After removing the guard interval and performing DFT and channel estimation at the receiver, the demodulated signal of the kth subcarrier can be represented as

$$Y(k) = X(k)H(k) + W(k) \tag{1}$$

for $0 \leq k \leq N-1$, where N is the number of subcarriers in an OFDM symbol. $H(k)$ and $W(k)$ represent the frequency transfer function of the channel and additive white Gaussian noise (AWGN) with zero mean and variance of σ_w^2, respectively.

Various OFDM channel estimation algorithms with periodic pilot sequences have been proposed to compensate the deleterious effects of multipath fading channels [4, 5, 6]. The typical comb type pilot-based OFDM channel estimation is illustrated in Fig. 1.

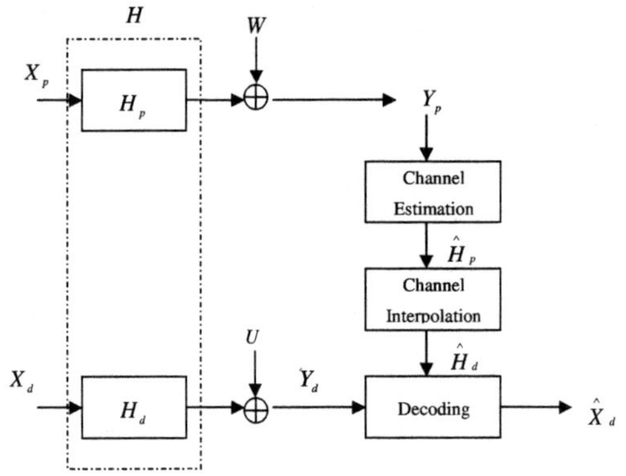

Fig. 1. Block diagram of the typical pilot-based OFDM channel estimation

The received pilot symbol and data symbol can be expressed as

$$Y_p = X_p H_p + W \tag{2}$$

and

$$Y_d = X_d H_d + U \tag{3}$$

respectively, where H_p denotes the pilot subchannel and H_d represents the data subchannel, W and U are the noise terms in the pilot and data subchannel respectively. The pilot subchannel can be estimated from the received pilot symbol and the original transmitted pilot symbol. The LS channel estimation [5] is given by

$$\hat{H}_p = \frac{Y_p}{X_p} \tag{4}$$

The data subchannel information can be estimated by taking interpolation between the pilot subchannel estimates. Here, linear interpolation is used and the data subchannel estimate can be written as

$$\hat{H}_d(k) = (1 - \frac{l}{L})\hat{H}_p(m) + \frac{l}{L}\hat{H}_p(m+1) \tag{5}$$

where L is the pilot spacing. The variable m denotes the location of pilot symbol and l is the distance between the mth pilot subchannel and the kth data subchannel.

The transmitted data symbol can be retrieved by

$$\hat{X}_d = \frac{Y_d}{\hat{H}_d} \tag{6}$$

3. CLUSTERED PILOT PATTERN

Since the number of pilot symbols is a tradeoff between bandwidth efficiency and channel estimation accuracy, the placing of a limited number of pilot symbols among the transmitted data symbols is important to the performance of channel estimation. In other words, given a fixed number of pilots, it is desirable to consider an effective scheme of assigning their positions so as to give a better channel estimate. The problem is very intriguing. In DVB-T standard, an equally spaced pilot pattern is adopted. The pilot symbols are placed in every 12 subcarriers in one OFDM symbol.

In this paper, a new pilot pattern is proposed, called clustered pilot pattern. In this pattern, every two neighboring pilots are grouped into one cluster, and these clustered pilot groups are equally spaced. The merit of the clustered pilot pattern is that it can reduce one half of the noise power of the channel estimate. At the receiver, before the interpolation manipulation, the two pilot subchannel estimates in each cluster are averaged so that the noise power at the pilot subchannel estimate can be reduced. By substituting the average of the two pilot subchannel estimates into (5), a more accurate data subchannel estimate will be obtained. Usually, the improvement of system performance in terms of bit error rate (BER) is a major concern and will be discussed in the following sections.

Note that the number of pilot symbols in the proposed pilot pattern is the same as the conventional equally spaced pilots. Hence the accuracy of channel estimate is increased without the loss of bandwidth efficiency.

4. ANALYTICAL BER

In the following, we assume the channel is a wide sense stationary uncorrelated scattering (WSSUS) Rayleigh fading channel. The average bit error rate for an OFDM-QPSK system is given by [7]

$$P_b(E) = \frac{1}{2}[1 - \frac{1}{2}\frac{\frac{\rho_1+\rho_2}{\sqrt{2}}}{\sqrt{1+\frac{1}{2\bar{\gamma}_b}-\frac{(\rho_1-\rho_2)^2}{2}}} - \frac{1}{2}\frac{\frac{\rho_1-\rho_2}{\sqrt{2}}}{\sqrt{1+\frac{1}{2\bar{\gamma}_b}-\frac{(\rho_1+\rho_2)^2}{2}}}] \tag{7}$$

where $\bar{\gamma}_b$ denotes the average SNR per bit, and

$$\rho_1 = \frac{\mu_1}{\sigma_1\sigma_2}, \rho_2 = \frac{\mu_2}{\sigma_1\sigma_2} \tag{8}$$

$$\mu_1 + j\mu_2 = \frac{E\{\hat{H}_d H_d^*\}}{2} \tag{9}$$

$$\sigma_2^2 = \frac{E\{|\hat{H}_d|^2\}}{2} \tag{10}$$

In our analysis, linear interpolation is used. Hence (9) and (10) can be easily obtained from (5). To evaluate (9) and (10), we also need to know the channel's frequency correlation function (FCF). Assume a 6-tap typical urban (TU) channel model [8], it has

$$h(t) = \sum_{k=1}^{6} c_k r_k(t) \delta(t - \tau_k) \quad (11)$$

where c_k and τ_k denote the kth path amplitude and propagation delay respectively, and $r_k(t)$ are independent stationary complex zero mean Gaussian process with the unit variance for the kth path. Thus, the resulting channel's FCF in one OFDM symbol is given by

$$E\{H(m)H^*(n)\} = \sum_{k=1}^{6} c_k^2 \exp \frac{-j2\pi(m-n)\tau_k}{T_s} \quad (12)$$

where m and n denote the index of the mth and nth frequency subchannel respectively and the asterisk represents the conjugate.

Considering the Doppler spread induced by the mobile receiver, the resultant time-varying channel is approximated as a time-invariant channel and the approximation error is moved to the noise [9].

5. SIMULATION RESULTS

Extensive computer simulations have been carried out to compare and evaluate the performance of the conventional pilot pattern and the proposed clustered pilot pattern. The OFDM parameters in the simulation are from DVB-T 2K mode [1]. The duration of one OFDM symbol is $224\mu s$ (guard interval excluded). The sampling interval is $7/64\mu s$. The bandwidth is $8MHz$ and the center carrier frequency is assumed to be $470MHz$. The length of cyclic prefix is $1/16$ of 2048, the number of carriers.

The TU channel model is considered in our simulation and the uncorrelated multipath fadings with a Jakes Doppler power spectrum are simulated [10]. Modulation of the subcarrier is assumed to be QPSK. To evaluate the efficacy of the proposed method, the total number of pilots is kept unchanged for all simulations. Thus, if the pilot spacing of the equally spaced pilot pattern is L, that of the clustered pilot pattern should be $2L$.

Fig. 2 gives the BER verse Eb/N0 for PSAM-OFDM system in a multipath Rayleigh fading channel. It can be seen that the analytical BER yields a good approximation to the simulation BER when the speed of the receiver is 100 km/h and 200 km/h respectively. This provides us with a convenient tool to analyze the performance of PSAM-OFDM system. Careful examination reveals that the BER curve of equally spaced pilots crossed with the BER curve of the clustered pilots at 24 dB and 26 dB, here we called these crossing SNR values (CSV), at 100 km/h and 200 km/h respectively. This clearly illustrates that the clustered pilots gives better BER performance only when SNR is lower

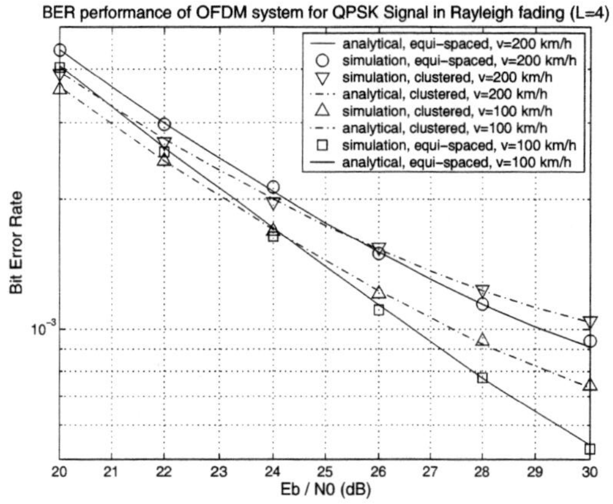

Fig. 2. BER Performance of PSAM-OFDM system in Rayleigh fading channel with various pilot patterns and speed

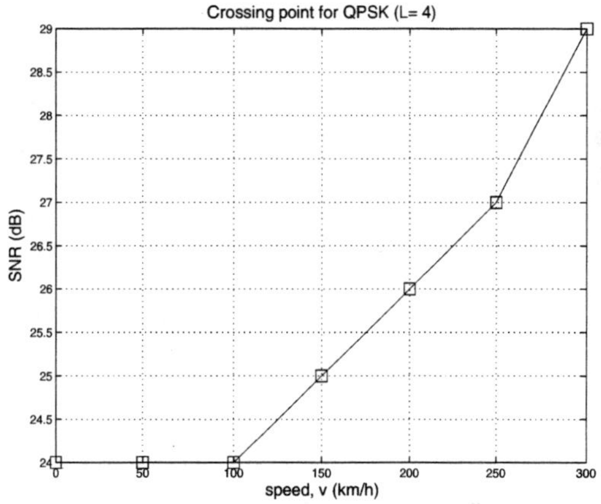

Fig. 3. SNR vs. speed (km/h) for the crossing point

than the CSV. This can be explained that the deleterious effect for BER is mainly from the fading and the de-noising effect of the clustered pilots is almost negligible at high SNR. Moreover, the larger the pilot spacing, the less accurate the data channel estimate. Therefore, the clustered pilots with double pilot spacing of that of the equally spaced pilots yields the worse BER performance in fading environment. In summary, the pilot pattern selection is a tradeoff between fighting noise and combating fading.

In Fig. 3, the CSV is plotted against various travelling speeds in km/h. It shows that the faster the speed, the higher the CSV, when speed is above 100 km/h. This indicates a larger workspace of the clustered pilot pattern as the speed is increasing.

Fig. 4 gives a practical scheme of pilot pattern selection. The crossing SNR value and crossing BER value against various pilot spacing are plotted by using (7), which is verified as a good approximation to the BER simulation in Fig. 2. When the operating environment of the PSAM-OFDM system is in region 'A' marked in the diagram, the clustered pilot pattern is recommended. Correspondingly, in

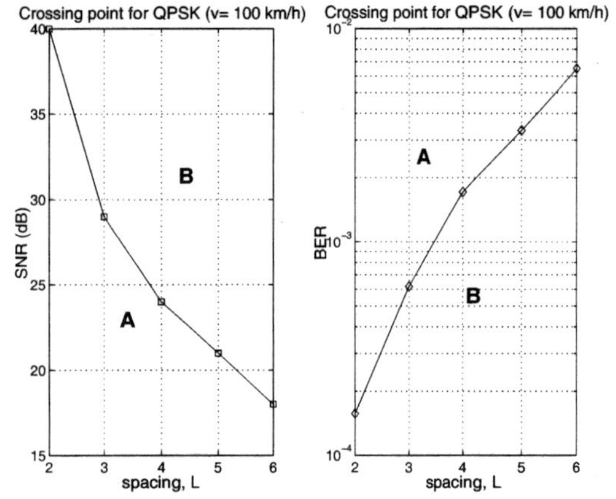

Fig. 4. SNR and BER vs. pilot spacing for the crossing point (A- clustered pilot pattern recommended, B- equally spaced pilot pattern recommended)

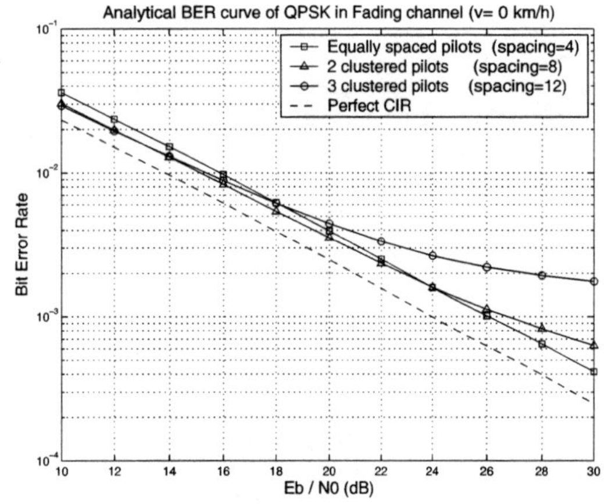

Fig. 5. Analytical BER performance comparison of different pilot patterns (equally spaced pilots, two clustered pilots and three clustered pilots)

region 'B', the conventional pilot pattern should be used. This scheme provides system designer an easy way of selecting a suitable pilot pattern for a fixed number of pilots that will give a better BER performance under different operating conditions.

One pilot pattern which clusters three pilots together is also considered and its BER performance compared with the other two pilot patterns is shown in Fig. 5. It is easily seen that the three clustered pilots cannot give better BER results than the two clustered pilots as the larger spacing results in a worse channel estimates in fading environment.

6. CONCLUSION

In this paper, a new and simple pilot pattern which clusters two pilots together is proposed for the PSAM-OFDM system. Compared with the conventional pilot pattern, it gives a better performance in fading channel under some condition without sacrificing the bandwidth efficiency. By deriving the analytical BER in mobile fading channel, a practical scheme for pilot pattern selection is formulated.

7. REFERENCES

[1] *Digital Video Broadcasting: Framing Structure, Channel coding, and Modulation for Digital Terrestrial Television* , Geneva, Switzerland, Jan. 1999. Draft EN300 744 V1.2.1.

[2] *Wireless LAN Medium Access Control (MAC) and Physical Layer (PHY) Specifications: High-Speed Physical Layer in the 5GHz Band* , Piscataway, NJ: IEEE Std. 802.11a, Sep. 1999.

[3] A. R. S. Bahai and B. R. Saltzberg, *Multi-Carrier Digital Communications: Theory and Application of OFDM*, Kluwer Academic/Plenum, 1999.

[4] S. Coleri, M. Ergen, A. Puri and A. Bahai, "Channel estimation techniques based on pilot arrangement in OFDM systems," *IEEE Trans. Broadcasting*, vol. 48, pp. 223-229, Sept. 2002.

[5] M. Morelli and U. Mengali, "A comparison of pilot-aided channel estimation methods for OFDM systems," *IEEE Trans. Signal Processing*,vol. 49, pp. 3065-3073, Dec, 2001.

[6] O. Edfors, M. Sandell, J. -J. van de Beek, S. K. Wilson and P. O. Borjesson, "OFDM channel estimation by singular value decomposition," *IEEE Trans. Commun.*,vol. 46, pp. 931-939, July 1998.

[7] M. X. Chang and Y. T. Su, "Performance analysis of equalized OFDM systems in Rayleigh fading," *IEEE Trans. Wireless Commun.*, vol. 1, pp. 721-732, Oct. 2002.

[8] Commission of the European Communities, *COST 207: Digital Land Mobile Radio Communications. Luxembourg: Final Report*, Office for Official Publications of the European Communities, 1989.

[9] X.-G. Xia, *Modulated Coding for Intersymbol Interference Channels*, New York: Marcel Dekker, Oct. 2000.

[10] C.-X. Wang and M. Pätzold, "Methods of generating multiple uncorrelated Rayleigh fading processes," in *Proc. 57th IEEE Vehicular Technology Conf.*, Jeju, Korea, pp. 510-514, April 2003.

COMPLETE CHARACTERIZATION OF CHANNEL INDEPENDENT GENERAL DMT SYSTEMS WITH CYCLIC PREFIX

Soura Dasgupta[1], *and Ashish Pandharipande*[2]

[1]Department of Electrical & Computer Engineering
The University of Iowa
Iowa City, IA-52242, USA.
dasgupta@engineering.uiowa.edu

[2]Department of Electrical & Computer Engineering
University of Florida
Gainesville, FL 32611-6130, USA.
ashish@dsp.ufl.edu

ABSTRACT

The following fact is well known about Discrete Multitone Transmission (DMT) systems: In the special case of Orthogonal Frequency Division Multiplexing (OFDM) when the input and output transforms are the IDFT and DFT matrices respectively, and the length of cyclic prefix is longer thant the channel length, ISI free transmission is possible for almost all channel parameter values. In this paper, we ask whether more general DMT systems with cyclic prefix enjoy similar *channel resistance*? We show that among all possible FIR transmitting and receiving filters, of arbitrary order, channel resistant ISI free transmission requires (a) that the receive filters be matched to the transmit filters, and (b) that to within a scaling and delay, the transmit and receive filters it must have IDFT and DFT coefficients. Thus we prove that, should cyclic prefix be applied, then trvial variations of OFDM are the *only* channel resistant DMT system.

1. INTRODUCTION

A general form of Discrete multitone (DMT) modulation, is depicted in fig. 1. An M-point block transformation, $A(z)$ is applied to M-parallel data streams, followed by a parallel to serial conversion (block P to S). For an FIR channel with transfer function,

$$H(z) = \sum_{i=0}^{\kappa} h_i z^{-i}, \tag{1.1}$$

a redundancy of length κ is added by the CPI block, and is removed by the CPR block. Thus each M-block of $v(n)$ is converted to an N-block of $s(n)$,

$$N = \kappa + M. \tag{1.2}$$

Supported by ARO contract DAAD19-00-1-0534 and NSF grants ECS-9970105, ECS-0225432 and CCR-9973133. Pandharipande was with the Department of Electrical and Computer Engineering, the University of Iowa, Iowa City, IA-52242, USA when this work was partially completed.

After CPR, one performs serial to parallel conversion (StoP block), and the inverse block transformation, $B(z)$. The overall system has the equivalent description of fig. 2, where $F_i(z)$ and $G_i(z)$ serve as transmit and receive filters respectively.

In its most popular version of Orthogonal Frequency Division Multiplexing (OFDM) the general DMT is specialized to select

$$A(z) = W^H,$$

and

$$B(z) = W$$

where W is the M-point DFT matrix having ik-th element

$$[W]_{ik} = e^{-j\frac{2\pi(i-1)(k-1)}{M}}. \tag{1.3}$$

Further, the form of redundancy employed in OFDM is *cyclic prefix*, where each M-block of $v(n)$ is converted to an N-block of $s(n)$, by prepending the last κ samples of the block. OFDM is a standard in many wireline and wireless applications, [1].

In OFDM $F_i(z)$ and $G_i(z)$ are mutually matched, respectively causal and anticausal of degree $M-1$ with coefficients of $F_i(z)$ being the cefficients of the i-th column of W^H. This leads to considerable spectral overlap between the subchannels. At the same time, this selection of the input/output transforms, ensures that together with cyclic prefix and (1.3) ensure the attractive feature of *channel resistance.* In particular Inter Symbol Interference (ISI) free transmission results for almost all channels with degree less than M. The only channels excepted from this rule are those having unit circle zeros at the M-th roots of unity.

More specifically, with

$$X(n) = [x_0(n), \cdots, x_{M-1}(n)]^T, \tag{1.4}$$

and

$$\hat{X}(n) = [\hat{x}_0(n), \cdots, \hat{x}_{M-1}(n)]^T \tag{1.5}$$

one has

$$\hat{X}(n) = \Delta(h_0, \cdots, h_\kappa)X(n) \tag{1.6}$$

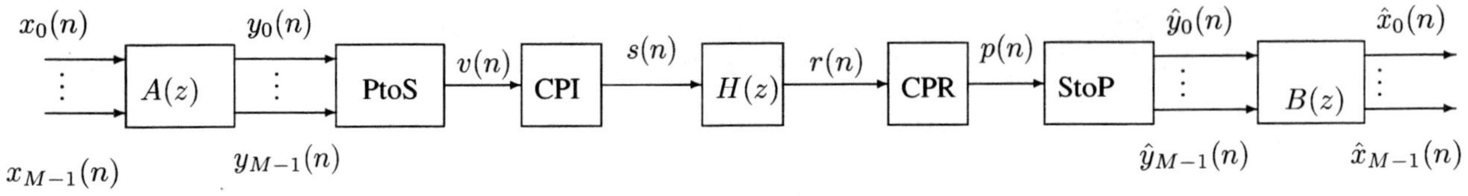

Fig. 1. The DMT system.

with diagonal, $\Delta(h_0, \cdots, h_\kappa)$ nonsingular for almost all h_i barring those for which $H(z)$ has a unit circle zero at an M-th root of unity. Thus, for almost all channels of order no greater than κ, an one tap equalizer at each subchannel output ensures ISI removal. We call such a system *Channel Resistant DMT.*

Improved spectral separation is advocated in [7]. Such separation is achieved using an alternative to OFDM, referred to as *overlapped discrete multitone modulation* or *discrete wavelet multitone* (DWMT) modulation. The basic structure of a DWMT system is the same as that of a DMT system. The only difference lies in the fact that in a DWMT system, the modulating and demodulating operations are implemented using an inverse fast wavelet transform (IFWT) and a forward fast wavelet transform (FWT) respectively. The filters used in DWMT systems also violate (A). Consequently, the DWMT system is more susceptible to Inter Block Interference (IBI) than the DMT system [7]. DWMT systems are however more effective against Inter Carrier Interference (ICI) than DMT systems, when perfect synchronization is lacking and/or the channel length is underestimated. More general orthogonal block transforms, with zero padding redundancy, and precoding are treated in [2]-[6].

Given the prevalence of cyclic prefix redundancy in practical systems, the issues raised in [7], and the obvious attraction of channel resistant trasmission, in this paper we pose the following question. Are there longer length $F_i(z)$ and $G_i(z)$ that together with cyclic prefix redundancy, ensure channel resistance? We thus characterize all $M \times M$ FIR

$$A(z) = \sum_{i=-p_1}^{p_2} A_i z^{-i} \text{ and } B(z) = \sum_{i=-q_1}^{q_2} B_i z^{-i}, \quad (1.7)$$

where p_i, q_i are arbitrary nonnegative integers, such that with cyclic prefix redundancy, and $H(z)$ as in (1.1), (1.6) holds with diagonal $\Delta(h_0, \cdots, h_\kappa)$ nonsingular for almost all h_i. We show that all such $A(z)$, $B(z)$ yield $F_i(z)$ $G_i(z)$, that are mutually matched to within complex scaling constants, and are identical to their counterparts yielded by the conventional DMT, to within scaling constants and delays. Thus channel resistance with cyclic prefix is incompatible with greater spectral separation.

2. FORMULATION

In fig. 1 observe that with,

$$\begin{aligned} \mathbf{v}(n) &= [v(Mn), \cdots, v(Mn - M + 1)]^T \\ &= [y_0(n), \cdots, y_{M-1}(n)]^T \\ &= A(z)X(n). \end{aligned}$$

At the same time, the addition of the cyclic prefix redundancy is characterized by:

$$\begin{aligned} \mathbf{s}(n) &= [s(Nn), \cdots, s(Nn - N + 1)]^T \\ &= \begin{bmatrix} 0 & I_\kappa \\ I_M & \end{bmatrix} \mathbf{v}(n) \end{aligned}$$

Further, with

$$\mathbf{r}(n) = [r(Nn), \cdots, r(Nn - N + 1)]^T$$

one has

$$\mathbf{r}(n) = \begin{bmatrix} h_0 & z^{-1}h_{N-1} & z^{-1}h_{N-2} & \cdots & z^{-1}h_1 \\ h_1 & h_0 & z^{-1}h_{N-1} & \cdots & z^{-1}h_2 \\ \vdots & \vdots & \vdots & \vdots & \vdots \\ h_{N-1} & h_{N-2} & \cdots & h_1 & h_0 \end{bmatrix} \mathbf{s}(n).$$

The CPI block is characterized by:

$$\mathbf{p}(n) = [p(Mn), \cdots, p(Mn - M + 1)]^T$$

and

$$\mathbf{p}(n) = \begin{bmatrix} 0 & I_M \end{bmatrix} \mathbf{r}(n), \quad (2.8)$$

Finally one has

$$\hat{X}(n) = B(z)\mathbf{p}(n). \quad (2.9)$$

Define circulant(η) to be the square circulant matrix with first row η. Then, [2], one has

$$\mathcal{H} = \text{circulant} \begin{bmatrix} h_0 & 0 & \ldots & 0 & h_\kappa & \ldots & h_1 \end{bmatrix}, \quad (2.10)$$

and

$$\hat{X}(n) = B(z)\mathcal{H}A(z)X(n). \quad (2.11)$$

with $\mathcal{H}$ $M \times M$. Further we note, that because of (2.8) and (2.9), fig. 1 is equivalent to fig. 2 with

$$[F_0(z), \cdots, F_{M-1}(z)] = \left[1, z^{-1}, \ldots, z^{-(M-1)}\right] A(z^M) \quad (2.12)$$

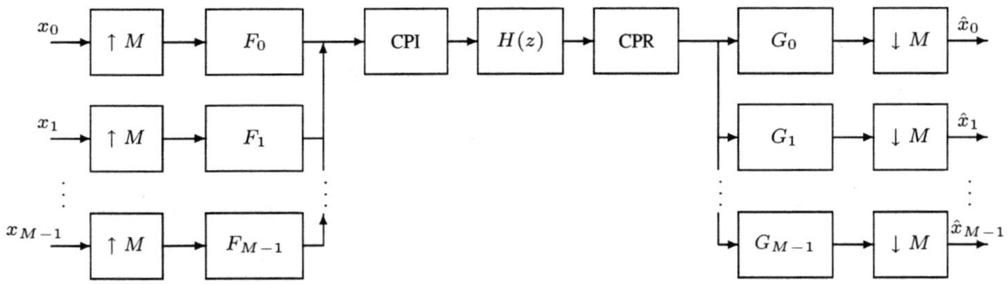

Fig. 2. A multicarrier system with cyclic prefix.

and

$$[G_0(z), \cdots, G_{M-1}(z)] = \left[1, z, \cdots, z^{(M-1)}\right] B^T(z^M). \tag{2.13}$$

Channel resistance then requires that with diagonal,

$$\Delta(h_0, \cdots, h_\kappa)$$

nonsingular for almost all h_i,

$$B(z)\mathcal{H}A(z) = \Delta(h_0, \cdots, h_\kappa). \tag{2.14}$$

Observe with the $M \times M$ circulant shift matrix

$$J = \text{circulant}\begin{bmatrix} 0 & 1 & \dots & 0 \end{bmatrix}, \quad \mathcal{H} = h_0 I + \sum_{i=1}^{\kappa} J^{M-i} h_i. \tag{2.15}$$

Thus (2.14) requires that with diagonal Δ_i and at least one Δ_i nonsingular

$$B(z)A(z) = \Delta_0, \text{ and } \forall 1 \leq i \leq \kappa \quad B(z)J^{M-i}A(z) = \Delta_i. \tag{2.16}$$

3. THE MAIN RESULT

The following Lemma is useful.

Lemma 3.1 *Suppose two $M \times M$ nonsingular matrices $\hat{B}$ and $\hat{A}$ are such that both $\hat{B}\hat{A}$ and $\hat{B}J^{M-1}\hat{A}$ are diagonal. Then for some permutation matrix P and diagonal matrices D_B and D_A,*

$$\hat{B} = D_B P W^H$$

and

$$\hat{A} = W P^T D_A.$$

Proof: As $\hat{B}\hat{A}$ is diagonal, the expression for $\hat{B}$ ensures that for $\hat{A}$. Call $\bar{A}_i$ the $M \times (M-1)$ matrix comprising all but the i-th column of $\hat{A}$ and $b_i = [b_{i1}, \cdots, b_{iM}] \neq 0$ the i-th row of $\hat{B}$. As

$$\text{rank}(\bar{A}_i) = M - 1,$$

$$\begin{bmatrix} b_i \\ b_i J^{M-1} \end{bmatrix} \bar{A}_i = 0 \Rightarrow \text{rank}\left(\begin{bmatrix} b_i \\ b_i J^{M-1} \end{bmatrix}\right) = 1.$$

Thus for some $\alpha_i \neq 0$,

$$[b_{i1}, \cdots, b_{iM}] = \alpha_i [b_{i2}, \cdots, b_{iM}, b_{i1}].$$

Thus, for $1 \leq l, i < M$,

$$b_{il} = \alpha_i b_{i,l+1}$$

and

$$b_{iM} = \alpha_i b_{i1}.$$

Thus

$$\alpha_i^M = 1$$

and

$$b_i = b_{i1}[1, \alpha_i \cdots, \alpha_i^{M-1}].$$

■

In (2.16) as at least one Δ_i is nonsingular, $B(z)$, $A(z)$ are nonsigular. Thus from Lemma 3.1, for almost all z,

$$B(z) = D_B(z)P(z)W^H$$

and

$$A(z) = WP^T(z)D_A(z),$$

with $D_B(z)$, $D_A(z)$ diagonal and $P(z)$ a permutation matrix at almost all z.

Observe, that for D_i nonzero, diagonal, P_i, P permutation matrices, and D a diagonal matrix,

$$D_1 P_1 + D_2 P_2 = DP$$

iff

$$P_1 = P_2.$$

Thus $P(z) = P$ is constant. Further, from (1.3),

$$D_A(z) = \sum_{i=-p_1}^{p_2} D_{Ai} z^{-i}$$

and

$$D_B(z) = \sum_{i=-q_1}^{q_2} D_{Bi} z^{-i}.$$

Further

$$D_B(z)D_A(z) = \Delta_0.$$

It is readily seen that the product of two-sided scalar polynomials $a(z)$ and $b(z)$ is a constant iff for some l,

$$a(z) = a_l z^l$$

and

$$b(z) = b_{-l} z^{-l}.$$

Thus, as $D_B(z)$ and $D_A(z)$ are diagonal, for some constant nonsingular, diagonal Λ_A and Λ_B and

$$\mathcal{Z} = \text{diag}\ \{z^{i_1}, \cdots, z^{i_M}\}$$

$$B(z) = \Lambda_B \mathcal{Z} P W^H \text{ and } A(z) = W P^T \mathcal{Z}^{-1} \Lambda_A. \quad (3.17)$$

As $W^H \mathcal{H} W$ is always diagonal, this is also sufficient for (2.14), with

$$\Delta(h_0, \cdots, h_\kappa)$$

nonsingular unless $H(z)$ has a zero at an M-th root of unity. Call the the transmit and receive filters of OFDM $\hat{F}_i(z)$ and $\hat{G}_i(z)$ respectively, i.e.

$$\left[\hat{F}_0(z), \cdots, \hat{F}_{M-1}(z)\right] = \left[1, z^{-1}, \ldots, z^{-(M-1)}\right] W^H, \quad (3.18)$$

and

$$\left[\hat{G}_0(z), \cdots, \hat{G}_{M-1}(z)\right] = \left[1, z, \cdots, z^{(M-1)}\right] W^T. \quad (3.19)$$

Then, because of (2.12), (2.13) and (3.17), we have the following main result.

Theorem 3.1 *Under (1.7), (2.14) holds for all h_i with*

$$\Delta(h_0, \cdots, h_\kappa)$$

nonsingular unless $H(z)$ has a zero at an M-th root of unity, iff the following two conditions hold.

(i) With $\hat{F}_i(z)$ as in (3.18), for some complex $\alpha_i \neq 0$, integer k_i and permutation P, $[F_0(z), \cdots, F_{M-1}(z)]$ equals

$$[\alpha_0 z^{Mk_0} \hat{F}_0(z), \cdots, \alpha_{M-1} z^{Mk_{M-1}} \hat{F}_{M-1}(z)] P;$$

(ii) for some complex

$$\beta_i \neq 0,$$

$$G_i(z) = \beta_i F_i^*(1/z).$$

The conditions of Theorem 3.1 can be interpreted as follows. To within a scaling each $\hat{F}_i(z) = \hat{G}_i^*(z^{-1})$. To within a scaling and a delay the coefficients of each $\hat{F}_i$ are either M-point DFT coefficients or M-point IDFT coefficients. The permutation matrix P simply rearranges the filters among the channels. Thus essentially one has the traditional OFDM, with the same level of specral overlap between the subchannels.

4. CONCLUSIONS

We have shown that the only channel resistant DMT systems are trivial variants of OFDM, as long as redundancy in the form of cyclic prefix is employed. This stands in contrast to the results of [3] where general filter bank systems under precoding are shown to be channel resistant even with judiciously chosen non-DFT based filters. Evidently the tight coupling between DFT based systems and channel resistance demonstrated in this paper, is attributable to the use of cyclic prefix.

5. REFERENCES

[1] J.S. Chow, J.C. Tu and J.M. Cioffi, "A discrete multitone transceiver system for HDSL applications", *IEEE Journal on Selected Areas in Communications*, pp 895-908, Aug 1991.

[2] Y. Lin, and S. Phoong, "Perfect discrete multitone modulation with optimal transceivers", *IEEE Transactions on Signal Processing*, pp 1702 -1711, June 2000.

[3] A. Scaglione, G. B. Giannakis, and S. Barbarossa, "Redundant Filterbank Precoders and Equalizers", *IEEE Transactions on Signal Processing*, SP-47:1988-2022, July 1999.

[4] A. Pandharipande and S. Dasgupta, "Optimum DMT based transceivers for multiuser communications", in *Proc. ICASSP*, 2001, Salt Lake City, Utah.

[5] P.P. Vaidyanathan, "Filter Banks in Digital Communications", *IEEE Ccts and Sys Mag*, Mar. 2001.

[6] X.-G. Xia, "New precoding for intersymbol interference cancellation using nonmaximally decimated Multirate Filterbanks with ideal FIR equalizers", *IEEE Trans. Signal Processing*, pp 2431-2441, 1997.

[7] S.D. Sandberg and M.A. Tzannes, "Overlapped discrete multitone modulation for high speed copper wire communications", *IEEE Journal on Selected Areas in Communications*, pp 1571-1585, Dec 1995.

ANTIPODAL PARAUNITARY PRECODING FOR OFDM APPLICATION

See-May Phoong, Kai-Yen Chang

Dept. of EE & Grad. Inst. of Comm. Engr.
National Taiwan Univ.
Taipei, Taiwan, ROC

Yuan-Pei Lin

Dept. Electrical and Control Engr.
National Chiao Tung Univ.
Hsinchu, Taiwan, ROC

ABSTRACT

Paraunitary (PU) matrices have found many applications. In this paper, a special class of PU matrices, namely the antipodal PU (APU) matrices, is used as precoding matrices for OFDM systems. Both the zero-forcing and MMSE receivers will be derived for precoded OFDM systems with APU precoding matrices. The performance of such precoded OFDM systems will be analyzed. We will show that using a APU precoding matrix, we are able to average the noise variance in both the time and frequency domains, and this obtains time and frequency diversity. Experiments show that precoded OFDM systems with MMSE receivers have a much better bit error rate performance than the conventional OFDM system.

1. INTRODUCTION

Multirate systems and filter banks have played an important role in various areas of signal processing [1]. Of particular interest is the class of paraunitary (PU) matrices. One attractive feature of these matrices is their energy conservation property which can avoid the noise or error amplification problem. In the past, the design and complete parameterization of PU matrices have been successfully derived. In this paper, we are going to study a special class of PU matrices, namely the antipodal paraunitary (APU) matrices. An $M \times M$ polynomial matrix $\mathbf{T}(z) = \sum_{i=0}^{N-1} \mathbf{T}_i z^{-i}$ is APU if all the entries of $\mathbf{T}_i$ are ± 1 and it satisfies[1]

$$\widetilde{\mathbf{T}}(z)\mathbf{T}(z) = MN\mathbf{I}.$$

The tilde notation denotes $\widetilde{\mathbf{T}}(z) = \mathbf{T}^H(1/z^*)$, where ^{H} is transpose-conjugation and * is the complex conjugation. For the special case of constant (memoryless) matrices, APU matrices reduce to scaled Hadamard matrices. Various methods have been proposed for the construction of APU matrices [2] [3]. The application of APU matrices in synchronous spread spectrum communications has been explored [2] and promising results have been demonstrated.

In this paper, we will apply APU matrices to linearly precoded OFDM systems. Linearly precoded OFDM systems have been studied by a number of researchers [4] [5] [6]. When the OFDM system has a DFT precoding matrix, it was shown to be the same as the so-called the single carrier with frequency domain equalizer (SC-FDE) system, which was first introduced in [7]. In [4], it was shown that the SC-DFE system has the maximum diversity gain among all linearly precoded OFDM systems. In [5] [6], BER minimized precoder for OFDM system was considered. For high SNR transmission, the SC-FDE system is optimal in the sense that it minimizes the bit error rate among OFDM systems with any orthogonal precoding matrix. In these studies, the precoders are constant matrices and the resulting precoded OFDM systems belong to the class of block transmission systems.

OFDM systems with APU precoding matrices are overlapped block transmission systems; a block of data symbols is transmitted over several blocks of transmitted signals. By doing so, we are able to average the noise variance in both the time and frequency domains and this achieves time and frequency diversity. Both the zero-forcing and MMSE receivers will be derived. Performances of the proposed systems will be analyzed and compared with the conventional OFDM system.

2. PRECODED OFDM SYSTEMS

Fig. 1 shows the block diagram of a precoded OFDM system. In a precoded OFDM transmitter, the kth input block $\mathbf{s}(k)$ consisting of M modulation symbols, e.g. QAM symbols, are first passed through an M by M precoding matrix $\mathbf{T}(z)$. The output of $\mathbf{T}(z)$ is given by

$$\mathbf{u}(k) = \sum_{i=0}^{N-1} \mathbf{T}_i \mathbf{s}(k-i).$$

In this paper, we consider only APU precoding matrices. APU precoding matrices enjoy two main advantages. Firstly, they have very low complexity. Their implementation involves only additions and there exists an efficient butterfly structure for a broad class of APU matrices [3]. Secondly, as we will show later, APU matrices have the ability to average the noise variance over both the time and frequency domains. We assume that the APU precoding matrix $\mathbf{T}(z)$ is *normalized* so that $\widetilde{\mathbf{T}}(z)\mathbf{T}(z) = \mathbf{I}_M$. In other words, all the entries of $\mathbf{T}_i$ are $\pm 1/\sqrt{MN}$. After taking the M-point IDFT of $\mathbf{u}(k)$, we get:

$$\mathbf{x}(k) = \mathbf{W}^H \mathbf{u}(k),$$

where $\mathbf{W}$ is the $M \times M$ DFT matrix with its klth entry given by $[\mathbf{W}]_{kl} = 1/\sqrt{M} exp(-j2\pi kl/M)$. Before $\mathbf{x}(k)$ is transmitted, a cyclic prefix (CP) of length L is added. Note that unlike the conventional block transmission system, the transmitted block $\mathbf{x}(k)$ now contains information of N blocks of input vectors $\mathbf{s}(k-i)$

This work was supported in parts by National Science Council, Taiwan, ROC, under NSC 92-2219-E-002-015 and 92-2213-E-009-022, Ministry of Education, Taiwan, ROC, under Grant # 89E-FA06-2-4, and the Lee and MTI Center for Networking Research.

[1] One can generalize the definition of APU matrices to include complex matrices. In this case, all the entries of the coefficient matrices $\mathbf{T}_i$ will have equal magnitude.

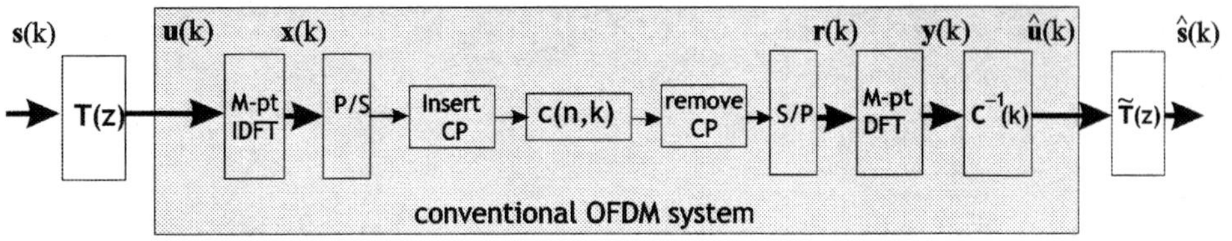

Fig. 1. An OFDM system with APU precoding matrix $\mathbf{T}(z)$.

for $0 \leq i < N$. In this paper, we assume that the channel is slowly varying so that for each OFDM block, the channel response does not vary. We model the combined effect of DAC, transmit filter, channel, receive filter and ADC as an equivalent discrete time system with $c(n,k)$ denoting the nth tap of the impulse response when the kth block is sent. We also assume that the CP length L is large enough so that for all k, $c(n,k) = 0$ whenever $n > L+1$. The channel noise $\nu(n)$ is assumed to be an AWGN (complex) with variance N_0.

At the receiver end, the first L samples of the received block that correspond to the CP are discarded to remove the inter block interference. We obtain the $M \times 1$ vector $\mathbf{r}(k)$. Taking the DFT of $\mathbf{r}(k)$, we get

$$\mathbf{y}(k) = \mathbf{W}\mathbf{r} = \mathbf{C}(k)\mathbf{u}(k) + \boldsymbol{\nu}(k),$$

where $\mathbf{C}(k)$ is an $M \times M$ diagonal matrix whose (ℓ, ℓ)th entry is given by the DFT coefficient of $c(n,k)$:

$$C_\ell(k) = \sum_{n=0}^{L+1} c(n,k) e^{-j2\pi n\ell/M}. \quad (1)$$

The noise vector $\boldsymbol{\nu}(k)$ is an AWGN vector with autocorrelation matrix $N_0\mathbf{I}_M$. Assume that the channel does not have spectral null so that $\mathbf{C}(k)$ is invertible. After multiplying the diagonal matrix $\mathbf{C}^{-1}(k)$, we get

$$\widehat{\mathbf{u}}(k) = \mathbf{u}(k) + \mathbf{C}^{-1}(k)\boldsymbol{\nu}(k). \quad (2)$$

In the absence of channel noise, the vector $\hat{\mathbf{u}}(k) = \mathbf{u}(k)$ for all k. When the precoding matrix $\mathbf{T}(z)$ is PU, we can get a zero forcing receiver by taking $\widetilde{\mathbf{T}}(z)$ as the decoding matrix, as indicated in Fig. 1. Note that when we take $\mathbf{T}(z) = \widetilde{\mathbf{T}}(z) = \mathbf{I}_M$, the system in Fig. 1 reduces to the conventional OFDM system. It should be emphasized that even though the precoded OFDM system has an overlapping-block transmitter, the channel impulse response $c(n,k)$ can be different for different block number k and the system in Fig. 1 still has the zero-forcing property.

2.1. Noise Analysis

Define the noise vector in the kth block as

$$\boldsymbol{\beta}(k) = \widehat{\mathbf{u}}(k) - \mathbf{u}(k) = \mathbf{C}^{-1}(k)\boldsymbol{\nu}(k).$$

The autocorrelation matrices of $\boldsymbol{\beta}(k)$ are given by

$$\mathcal{R}_\beta(k,\ell) = \mathcal{E}[\boldsymbol{\beta}(k)\boldsymbol{\beta}^H(k-\ell)] = N_0\delta(\ell)\mathbf{C}^{-1}(k)\mathbf{C}^{-H}(k). \quad (3)$$

Because $\mathbf{C}^{-1}(k)$ is a diagonal matrix, we see from the above equation that $\boldsymbol{\beta}(k)$ is also an AWGN vector but each entry has a different variance.

Define the output noise vector $\mathbf{e}(k) = \widehat{\mathbf{s}}(k) - \mathbf{s}(k)$. Then it can be viewed as the output of $\widetilde{\mathbf{T}}(z)$ with the input vector $\boldsymbol{\beta}(k)$. Therefore, we can write

$$\mathbf{e}(k) = \sum_{\ell=0}^{N-1} \mathbf{T}_\ell^H \boldsymbol{\beta}(k+\ell).$$

Using the facts that $\boldsymbol{\beta}(k)$ is an AWGN vector and $\widetilde{\mathbf{T}}(z)$ is a normalized PU matrix, one can verify that its zeroth autocorrelation matrix at the kth block is given by

$$\mathcal{R}_e(k,0) = \mathcal{E}[\mathbf{e}(k)\mathbf{e}^H(k)] = \sum_{\ell=0}^{N-1} \mathbf{T}_\ell^H \mathcal{R}_\beta(k+\ell, 0)\mathbf{T}_\ell,$$

where $\mathcal{R}_\beta(i,0)$ is the zeroth autocorrelation matrix of $\boldsymbol{\beta}(i)$ given in (3). Note that $\mathcal{R}_\beta(i,0)$ is a diagonal matrix. Looking at the ith diagonal term of $\mathcal{R}_e(k,0)$, we can write the noise variance at ith subchannel (when the kth block is being processed) as

$$\sigma_{i,\mathbf{T}}^2(k) = \frac{1}{N} \sum_{\ell=0}^{N-1} \left[\frac{1}{M} \sum_{n=0}^{M-1} \frac{N_0}{|C_n(k+\ell)|^2} \right], \quad (4)$$

where we have used (3) and the fact that all the entries of $\mathbf{T}_\ell$ have magnitude equal to $1/\sqrt{MN}$. The quantity $\sigma_{i,\mathbf{T}}^2(k)$ is independent of i; all subchannels have the same noise variance! Moreover the decoding matrix $\widetilde{\mathbf{T}}(z)$ has an averaging effect on the channel gains over a time period of N blocks. Note that we do not make any assumption about the APU matrix $\mathbf{T}(z)$. Any APU precoding matrix can achieve (4). From (4), we also see that the performance of the precoded OFDM with a zero-forcing receiver degrades significantly when some of the channel gains are small. The noise variances in all subchannels will be very large over a period of N blocks. To solve this problem, an MMSE receiver is needed and will be derived in the next section.

2.2. Comparisons with Other Systems

When we take $\mathbf{T}(z) = \mathbf{I}_M$, the system in Fig. 1 becomes the conventional OFDM system. In this case $\mathbf{u}(k) = \mathbf{s}(k)$. Thus, for the conventional OFDM system, we can obtain from (2) the output noise variance at the ith subchannel as

$$\sigma_{i,ofdm}^2(k) = \frac{N_0}{|C_i(k)|^2}. \quad (5)$$

The variance $\sigma_{i,ofdm}^2(k)$ depends on the block index k as well as the frequency index i, and it is inversely proportional to $|C_i(k)|^2$. For highly frequency selective channels, some of the gains $|C_i(k)|$ can be small and the performance of the OFDM system will be affected by these spectral nulls.

If we generalize the definition of APU matrices to include complex matrices, then the DFT matrix $\mathbf{W}$ is APU. When we take $\mathbf{T}(z) = \mathbf{W}$, i.e., the DFT matrix, the system in Fig. 1 becomes the SC-FDE system [7] [5]. By carrying out the same derivation, one can show that the noise variance of the SC-DFE system can be obtained by simply setting $N = 1$ in (4). The noise variance at the ith subchannel when the kth block is sent is given by

$$\sigma_{i,sc}^2(k) = \frac{1}{M} \sum_{n=0}^{M-1} \frac{N_0}{|C_n(k)|^2}. \tag{6}$$

Observe from the above expression that $\sigma_{i,sc}^2(k)$ is independent of the frequency index i. All the subchannels have the same noise variance and they are equal to the average noise variance of the conventional OFDM system.

We can clearly see the difference between the conventional OFDM, the SC-FDE and the precoded OFDM systems from the three expressions in (5), (6) and (4). Because the decoding matrix $\widetilde{\mathbf{T}}(z)$ is PU, it has the energy (or power) conservation property [1]. The average output noise variance for the three systems is the same. However they distribute these noise variances to the subchannels differently. For the conventional OFDM system, each subchannel can have a very different noise variance, especially when the channel is highly frequency selective. From (5), we see that subchannels having small $|C_n(k)|$ will suffer from large noise variances. On the other hand, the SC-FDE system has an average effect in frequency domain; it averages over all subchannels. For fast fading channel, $\sum_{n=0}^{M-1} 1/C_n(k)$ can vary from block to block and the performance of SC-DFE system will have a large variation with respect to k. From (6), we see that if $\sum_i |C_i(k)|^2$ is small for some k, the whole kth block will be severely affected by noise amplification problem. The precoded OFDM system with precoder $\mathbf{T}(z)$ has an averaging effect in both frequency and time-domain; it averages over all subchannels and over N OFDM blocks.

3. MMSE RECEIVER FOR PRECODED OFDM SYSTEMS

As we have mentioned earlier, in the presence of spectral nulls, precoded OFDM systems with zero-forcing receivers suffer from serious performance degradation. To avoid this problem, an MMSE receiver is needed. In the derivation of the MMSE receiver, we assume that the transmitted signals $\mathbf{s}(k)$ satisfy

$$\mathcal{E}\{\mathbf{s}(k)\mathbf{s}^H(k-\ell)\} = E_s\delta(\ell)\mathbf{I}_M.$$

In other words, the symbols are uncorrelated and have equal signal power. The fact that $\mathbf{T}(z)$ is normalized PU implies that $\mathbf{u}(k)$ also satisfies $\mathcal{E}\{\mathbf{u}(k)\mathbf{u}^H(k-\ell)\} = E_s\delta(\ell)\mathbf{I}_M$.

We assume that the receiver removes the first L samples corresponding to the CP so that there is no inter block interference. Given the received vector $\mathbf{r}(k)$, we want to design an MMSE receiver. As the DFT matrix $\mathbf{W}$ is invertible, there is no loss of generality if we consider the vector $\mathbf{y}(k) = \mathbf{W}\mathbf{r}(k)$. Given the vector $\mathbf{y}(k)$, we want to design an MMSE receiver. Consider an MMSE receiver (possibly time-varying) with N coefficient matrices $\mathbf{Q}(k, \ell)$ for $0 \le \ell \le N-1$. Given the input vector $\mathbf{y}(k)$, the output of the MMSE receiver can be described as:

$$\widehat{\mathbf{s}}(k) = \sum_{\ell=0}^{N-1} \mathbf{Q}(k,\ell)\mathbf{y}(k+\ell),$$

where $\mathbf{Q}(k,\ell)$ are $M \times M$ matrices. Our goal is to find $\mathbf{Q}(k,\ell)$ so that the following mean squared error is minimized.

$$\mathcal{E}\{(\widehat{\mathbf{s}}(k) - \mathbf{s}(k))^H(\widehat{\mathbf{s}}(k) - \mathbf{s}(k))\}.$$

Applying the orthogonality principle, one can verify that the MMSE solution is given by:

$$\mathbf{Q}(k,\ell) = \mathbf{T}_\ell^H \mathbf{\Lambda}(k+\ell),$$

where the diagonal matrix $\mathbf{\Lambda}(k)$ is given by

$$\mathbf{\Lambda}(k) = E_s\mathbf{C}^H(k)\Big(E_s\mathbf{C}(k)\mathbf{C}^H(k) + N_0\mathbf{I}_M\Big)^{-1}.$$

The nth diagonal entry of $\mathbf{\Lambda}(k)$ is given by

$$\lambda_n(k) = \frac{C_n^*(k)}{|C_n(k)|^2 + N_0/E_s}.$$

From the expression of $\mathbf{Q}(k,\ell)$, we see that the MMSE receiver can be decomposed into a time-varying diagonal matrix $\mathbf{\Lambda}(k)$ and the time-invariant matrix $\widetilde{\mathbf{T}}(z)$. Therefore, we can implement the MMSE receiver as Fig. 2. Comparing the zero-forcing and MMSE receivers in Fig. 1 and Fig. 2 respectively, one immediately sees that their only difference is the one-tap equalizer and they have the same implementational complexity. When there is no noise, i.e. $N_0 = 0$, the MMSE receiver reduces to the zero-forcing receiver.

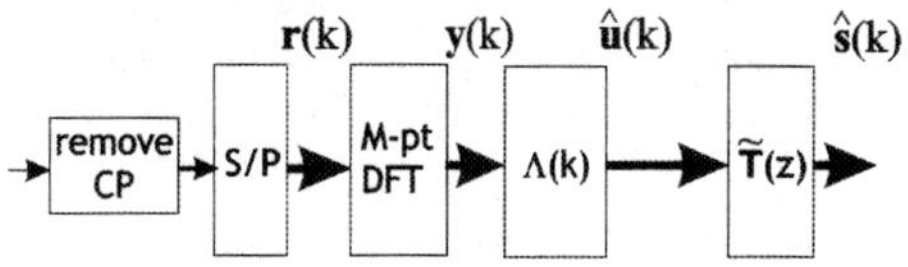

Fig. 2. An MMSE receiver for the precoded OFDM system.

One can verify that for the precoded OFDM system with an MMSE receiver, all the subchannels also have the same error variance and it is given by

$$\sigma_{mmse}^2(k) = \frac{1}{N}\sum_{\ell=0}^{N-1}\Big(\frac{1}{M}\sum_{n=0}^{M-1}\frac{N_0}{|C_n(k+\ell)|^2 + N_0/E_s}\Big).$$

From the above expression, it is clear that the decoding matrix $\widetilde{\mathbf{T}}(z)$ has an averaging effect in both frequency and time-domain; it averages over all subchannels and over N OFDM blocks. Moreover when some of the channel gains $|C_n(k)|$ approach zero, the error variance $\sigma_{mmse}^2(k)$ does not goes to infinity. In fact, the error variance is upper bounded by E_s. As we will see in the next section that by using a MMSE receiver, the performance of the precoded OFDM system is improved significantly.

4. SIMULATION

In this section, we carry out Monte-Carlo experiments to verify the performance of precoded OFDM systems with different precoders. The transmission channels are the modified Jakes fading channels described in [8]. In the experiments, we will use channel models with two different ratios of doppler frequency and transmission rate. A larger value of r indicates that the channel is changing faster. The ratio $r = 0.0001$ corresponds to a slowly varying channel whereas $r = 0.001$ corresponds to a channel that varies 10 times faster. The number of taps of the channels is 16. The channel noise $\nu(n)$ is AWGN with variance N_0. In our simulation, we assume that the receiver knows the exact channel response. The DFT size is $M = 64$ and the length of cyclic prefix is $L = 16$.

APU matrices of different length N will be used as the precoding matrices. When $N = 1$, the APU matrix reduces to the Hadamard matrix. It is known [5] that the OFDM system with a Hadamard precoding matrix has the same bit error rate performance as the SC-DFE system. The input vector $\mathbf{s}(n)$ consists of QPSK symbols with power equal to E_s. We plot the bit error rate curves versus SNR (signal to noise ratio), which is equal to E_s/N_0. In the simulation, we do not consider MMSE receiver for the conventional OFDM system because the bit error rate performance of OFDM systems with MMSE receivers is identical to that of OFDM systems with zero-forcing receivers.

The results for $r = 0.0001$ are shown in Fig. 3. From the figure, we see that the performance of precoded OFDM system with a zero-forcing receiver is worse than that of the OFDM system. This is because when the transmission encounters deep fading at some frequency bins, all the outputs of precoded OFDM receiver will be seriously affected by channel noise. On the other hand, for OFDM system, only a portion of the outputs will be seriously affected. However when an MMSE receiver is employed, the precoded OFDM systems have a much better performance than the OFDM system. If we compare the performance of precoded OFDM systems with different precoders, we see that when the channel is slowly varying, using a longer precoding matrix does not provide much gain in performance. This is because when the channel variation in the time domain is small, averaging the performance in the time domain has little effect on the performance.

For channel that is varying 10 times faster with $r = 0.001$, the results are shown in Fig. 4. Again we see that precoded OFDM system with a zero-forcing receiver does not perform well and using an MMSE receiver can greatly improve the performance of precoded OFDM systems. Also note that the performance improves as N (the length of the precoding matrix) increases. As the channel is fast varying, averaging in the time domain can provide additional gain. If we compare the cases of $N = 1$ and $N = 8$, averaging over 8 blocks can yield an additional gain of more than 2 dB when the bit error rate is 10^{-5}.

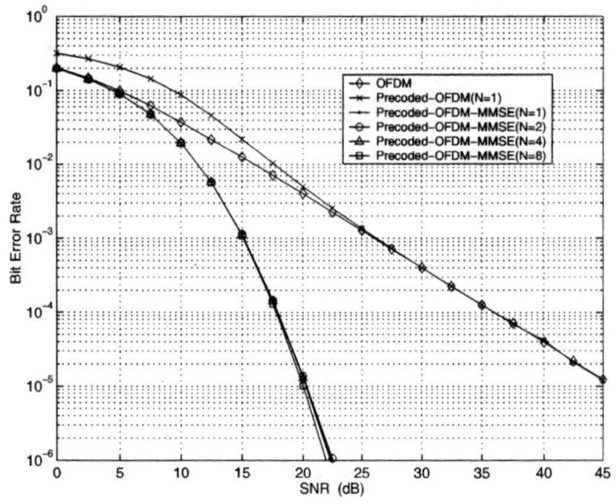

Fig. 3. Bit error rate performance for slowly varying channels

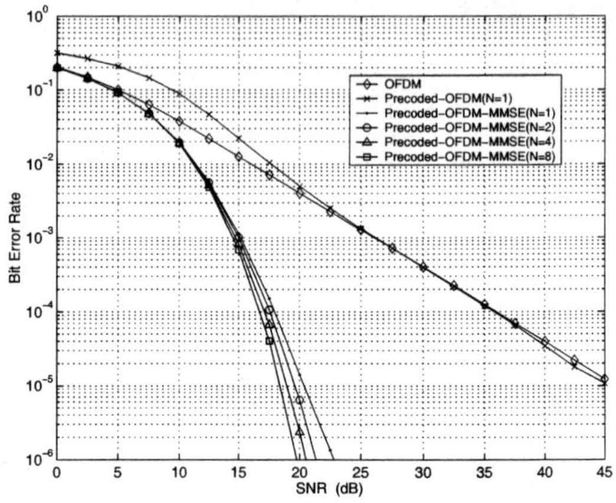

Fig. 4. Bit error rate performance for fast varying channels

5. CONCLUSIONS

In this paper, we have studied OFDM systems with APU precoding matrices. Using an APU precoding matrix, we can average the noise variances in both the time and frequency domains. We have derived MMSE receivers for precoded OFDM systems. Experiments show that precoded OFDM systems with MMSE receivers have a much better bit error rate performance that the conventional OFDM system.

6. REFERENCES

[1] P. P. Vaidyanathan, *Multirate systems and filter banks*, Prentice-Hall, 1993.

[2] G. W. Wornell, "Emerging applications of mutirate signal processing and wavelets in digital communications," *Proceedings of the IEEE*, vol. 84, pp. 586–1187, Aug. 1996.

[3] S. M. Phoong and Y. P. Lin, "Lapped Hadamard Transforms and Filter Banks," *Proc. IEEE Int. Conf. Acout., Speech and Signal Proc.*, pp. VI-509–512, Hong Kong, April 2003.

[4] Z. Wang and G. B. Giannakis, "Linearly precoded or coded OFDM against wireless channel fades?," in *Proc. Third IEEE Workshop Signal Process. Adv. Wireless Commun.*, Taoyuan, Taiwan, Mar. 2001.

[5] Y. P. Lin and S. M. Phoong, "BER minimized OFDM systems with channel independent precoders," *IEEE Trans. Signal Proc.*, pp. 2369–2380, Sep. 2003.

[6] Y. Ding, T N. Davidson, Z.-Q. Luo and K. M. Wong, "Minimum BER block precoders for zero-forcing equalization," *IEEE Trans. Signal Proc.*, pp. 2410–2423, Sep. 2003.

[7] H. Sari, G. Karam, and I. Jeanclaude, "Frequency-Domain Equalization of Mobile Radio and Terrestrial Broadcast Channels," Globecom, San Francisco, CA, 1994.

[8] P. Dent, G. E. Bottomley and T. Croft, "Jakes fading model revisited," *Electronic Letters*, pp. 1162-1163, June 1993.

NONLINEAR DIGITAL FILTERS FOR BEAUTIFYING FACIAL IMAGES IN MULTI-MEDIA SYSTEMS

Kaoru Arakawa

Dept. of Computer Science, Meiji University

ABSTRACT

Nonlinear digital filters for new application in multi-media systems, that is beautification of face images, are proposed. These filters can remove undesirable skin components such as wrinkles and spots on face images, and accordingly make the skin look smooth and clear without affecting other principal components in the images. Three types of nonlinear filters are proposed; one is quite a simple one which is effective for removing roughness and small spots on skin and can be easily realized by simple hardware, one is an extended one which is a little more complicated but can smooth the roughness and small spots more naturally, and the other more complicated one which can preserve the natural roughness on skin and can remove relatively large spots on skin. Computer simulations verify the high performance of these filters.

1. INTRODUCTION

In multimedia systems, such as TV phone and video conferences, we have a lot of opportunities to transmit face images. Conventionally, information systems have been developed to transmit and display the original information as correctly as possible, however, this principle may not be valid in recent multi-media systems, because multi-media information appeals to human sense.

In this paper, new types of nonlinear digital filters are proposed for such multi-media systems considering human sense in the way that human face images are automatically beautified on the display. These filters perform signal separation based on the signal amplitude as well as the frequency, thus they can filter out small-amplitude random fluctuation in face images corresponding to the wrinkles, spots, rash and so on, and which are unnecessary components on skin and make the skin look bad. Using the filters proposed here, these rough components on skin can be automatically removed, and the face images come to look beautified.

These filters have nonlinear structures, since linear filters cannot remove such small-amplitude fluctuation without blurring the edges in the images. Three types of nonlinear filters are proposed here. The first one is quite a simple one which is effective for smoothing the skin completely, removing roughness and small spots on skin. This filter can be easily realized by simple hardware. The second is an extended form of the first, which is a little more complicated but can smooth the skin more naturally. The last is more complicated one but can preserve the natural roughness on skin and can remove relatively large spots on skin. Finally in this paper, computer simulations for actual face images are shown to verify their high performance.

2. ε–FILTER: A SIMPLE METHOD TO BEAUTIFY SKIN IN FACE IMAGES

The ε-filter is a nonlinear digital filter proposed by the author's group before for noise reduction of nonstationary signals such as images[1]. The input-output relationship of the ε-filter is represented as

$$y(n) = x(n) + \sum_{i=-N}^{N} a_i F(x(n+i) - x(n)) \qquad (1)$$

where $x(n)$ and $y(n)$ are the input and the output signal at time n respectively. $x(n)$ is supposed to be the sum of an original signal and small-amplitude random noise. a_i is the filter coefficient of a linear non-recursive low-pass filter, satisfying the following condition to keep the DC level unchanged.

$$\sum_{i=-N}^{N} a_i = 1 \qquad (2)$$

F is a nonlinear function which takes a form as Fig.1, bounded within a certain value ε as follows.

$$|F(p)| \leq \varepsilon \qquad ; -\infty \leq p \leq \infty \qquad (3)$$

When $|p| \leq \varepsilon$, $F(p)=p$ and this filter works as a linear nonrecursive low-pass filter. Thus, if the amplitude of the added noise is less than $\varepsilon/2$, this filter can smooth the noisy input signal in the part where the original signal does not change much. On the other hand, in the part where the original signal largely changes, this filter preserves the changes in the original signal, since the difference between the input and the output of this filter is limited to a certain value ε' as follows.

$$F.|y(n)-x(n)| = \left|\sum_{i=-N}^{N} a_i F(x(n+i)-x(n))\right| \le \sum_{i=-N}^{N} |a_i| \varepsilon \equiv \varepsilon' \qquad (4)$$

Thus, this filter smoothes out small-amplitude high-frequency noises, while preserving large-amplitude changes in the original signal, even if these changes contain high-frequency abrupt components. This filter is named as an ε-filter and can be easily implemented by adding the simple nonlinear function to the conventional nonrecursive linear low-pass filter. The ε-filter equals to a linear nonrecursive low-pass filter, when the value ε is large enough.

In this explanation, signals are expressed in a one-dimensional form, but when the filter is applied to images, the filter is modified into a two-dimensional form by replacing the time point n with the pixel (n,m), a_i with a_{ij}, and the summation with a two-dimensional one.

3. COMPONENT-SEPARATING ε-FILTER: AN EXTENDED FORM OF THE ε-FILTER

The ε-filter is powerful to reduce small-amplitude random noise, while preserving the abrupt jumps in the input signals. However, the performance of the ε-filter is not satisfactory enough, when the signal contains continuous changes, such as a slope and a sinusoidal hump, since the ε-filer is designed to be ideal when the signal contains just flat components and discontinuous jumps. When the signal contains continuous changes, noise remaining and signal distortion occur. In order to solve this problem, a component separating ε-filter (CS filter for short) is proposed[3].

The CS filter consists of three ε- filters H, E, E' as shown in Fig. 2. In this figure, first, the filter H extracts a rough signal component s(n) from the input signal x(n). Next, the residual t(n), that is x(n)-s(n), is processed by the filter E, and s(n) is processed by E'. Finally, the output y(n) is obtained by combining these outputs v(n) and z(n). Here, H adopts a large value for ε, which means that H is close to a linear low-pass filter in order to extract the rough structure s(n). On the other hand, E and E' are strongly nonlinear ε-filters, which remove small-amplitude

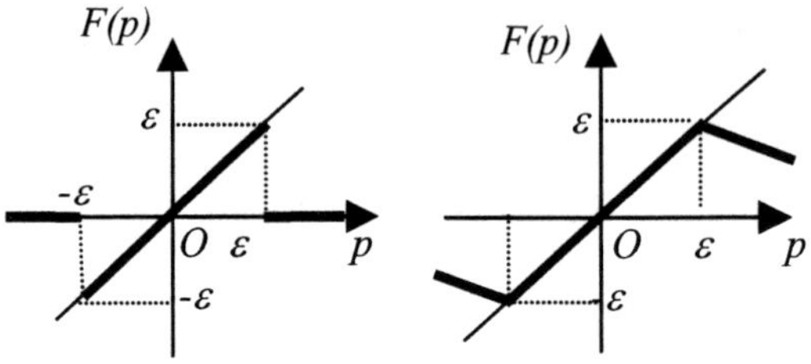

(a) (b)

Fig. 1 Examples of nonlinear functions F.

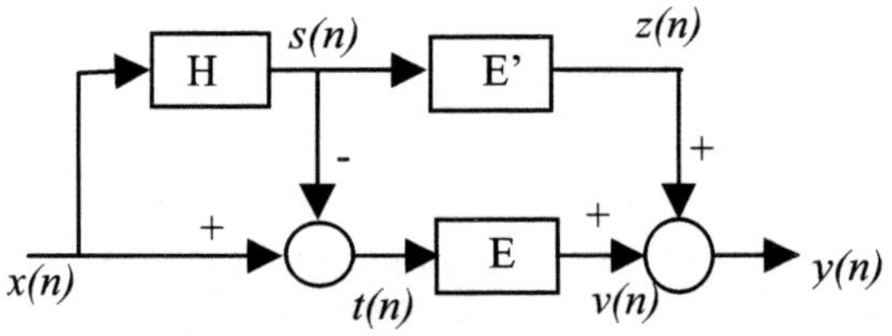

Fig.2 Schematic diagram of CS filter.

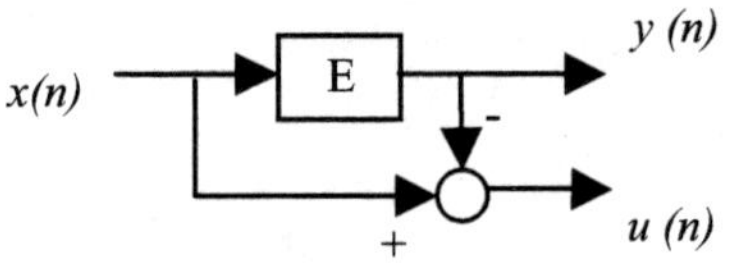

Fig.3 Signal separation using the ε-filter.

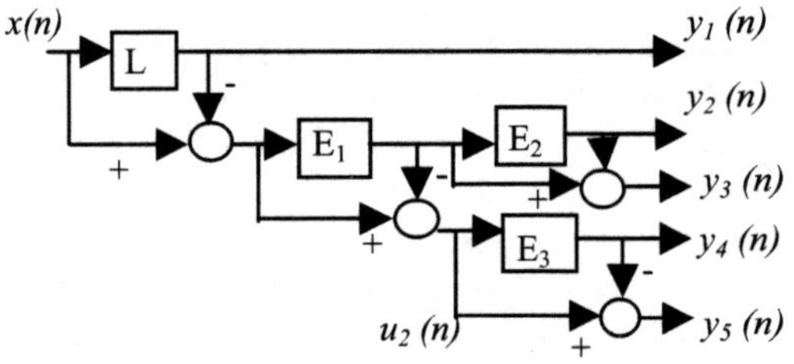

Fig.4 Schematic diagram of an ε-nonlinear filter bank.

high-frequency noise while preserving the large-amplitude abrupt changes in the input. Since continuous changes in the input is subtracted before ε-filtering, noise remaining and signal distortion are avoided. Suppose that the filter coefficients in E and E' are denoted as a_i and a_i', the window sizes are as N and N', and the nonlinear functions are as F_E and F'_E respectively, y(n) is expressed as

$$y(n) = x(n) + \sum_{i=-N}^{N} a_i F_E(t(n-i)-t(n)) + \sum_{i=-N'}^{N'} a_i' F_E'(s(n-i)-s(n)) \qquad (5)$$

Since the amplitude of F_E and F_E' are limited, the difference between the input and the output $|y(n)-x(n)|$ are also limited to a certain value, thus this filter can preserve the large-amplitude abrupt changes while removing the small amplitude random noises in the same way as the ε-filter.

4. FACE IMAGE PROCESSING USING ε-NONLINEAR FILTER BANK

Since the ε-filter removes noise considering its amplitude as well as the frequency, signal separation considering both the signal amplitude and frequency can be realized using the ε-filter as shown in Fig.3. Here, the input $x(n)$ is separated into $y(n)$ and $u(n)$, where $y(n)$ is the output of the ε-filter, composed of large-amplitude or low-frequency components and $u(n)$ of small-amplitude and high-frequency ones. Since all the values a_i's can set to be positive, the amplitude of $u(n)$ can be limited to ε from eq(4). Using this idea, a filter bank, named as a ε-nonlinear filter bank (ε-filter bank for short), as shown in Fig.4 can be realized[4]. Here, L denotes a linear low-pass filter, and $E_1 \sim E_3$ denote ε-filters. First, the filter L separates the input to a low-frequency component and a high-frequency one. Here, the low-frequency components corresponds to a rough signal structure of face image, and the high corresponds to the fine structure of the face. In the fine structure, the outputs $y_2(n)$ and $y_3(n)$ correspond to a relatively low-frequency component compared with $y_4(n)$ and $y_5(n)$, and also $y_3(n)$ and $y_5(n)$ correspond to small-amplitude components compared with $y_2(n)$ and $y_4(n)$ respectively. When we think of our actual face, wrinkles are usually represented as small-amplitude and high-frequency signals, while spots are small-amplitude and intermediate-frequency ones. Thus, wrinkles can be extracted in the signal $y_4(n)$ and spots in $y_3(n)$. Natural roughness of skin, represented as quite small-amplitude and high-frequency signals are obtained in $y_5(n)$. Such roughness of skin should be preserved in the output image in order to show the face as a natural one. The human face beautifying system proposed here first separates the face image components into $y_k(n)$ $(1 \leq k \leq 5)$ by the ε-nonlinear filter bank and then subtract $y_3(n)$ and $y_4(i,j)$ from the input. Undesirable wrinkles and spots can be removed from the input image automatically.

However, if this system is applied to actual face images, some small-amplitude edges around weak components such as lips are also degraded, since such small edges contain some components close to $y_3(n)$ and $y_4(n)$. Therefore, this system must be modified so that $y_3(n)$ and $y_4(n)$ are subtracted only when the pixel is not around the edge. Moreover, in order to avoid the output image from blurring, the subtracted value should be limited within a certain small value. Finally, the output image $z(n)$ of this system is described as a rule-based system using the ε-filter bank as follows.

IF $|med(n)-x(n)|<\delta_2$
THEN $y'_3(n) = 0$, ELSE $y'_3(n) = y_3(n)$
IF $|u_2(n)|> \delta_1$ AND $|med(n)-x(n)|< \delta_2$
THEN $y'_4(n) = 0$, ELSE $y'_4(n) = y_4(n)$
$z(n) = x(n)-G(y'_3(n)+y'_4(n))$

Here, $med(n)$ denotes the output of a median filter around the pixel n, G a nonlinear function as Fig.1(b), and δ_1 and δ_2 are positive parameters; these parameters and the value ε in G are set experimentally. Pixels on edges are detected from the difference between the input and the median, because the difference is smaller on the edges. In order not to suppress small-amplitude wrinkles, the condition $|u_2(n)| > \delta_1$ is added for $y'_4(n)$.

5. COMPUTER SIMULATIONS

Fig. 5 shows the performance of the ε-filter, the simplest one, compared with that of a conventional linear smoothing filter. Here, the window sizes are 5x5 for both. We can see that the ε-filter can smooth the skin without blurring the entire image, and make the face look young and beautified. Fig.6 shows the comparison of the performance of the three nonlinear filters proposed here. We can see that the boundary between the light area and a dark one is too sharply emphasized in the case of the simple ε-filter, while that is softer in the CS filter. In the case of the ε-filter bank, the boundary especially at the noise is more smooth and natural. Moreover, the outputs of the ε-filter and the CS filter are too much smoothed, but the output using the ε-filter bank contains small amount of roughness which make the face look more natural. Fig. 7 shows the performance for relatively large spots on skin. We can see that such large spots can be also removed naturally as well as wrinkles in the case of the ε-filter bank.

6. CONCLUSIONS

Three types of nonlinear filters for beautifying face image in multimedia systems are proposed. They can remove undesirable roughness on skin, such as wrinkles and spots, in order to make the skin look better, and are based on the idea of signal separation considering both the frequency and the amplitude of the signal. These filters are different in the complexity in implementation and the quality of output image; the more the system is complicated, the more natural the output face image becomes. Which one should be used depends on each situation. Computer simulations for actual face images show the high performance of these filters.

In these filters, the way to design them optimally is already proposed for the ε-filter and the CS-filter[3]. As to the ε- nonlinear filter bank, the parameters are still set experimentally. If this system is applied only to face images, such experimental setting is usually valid, since the quantitative characteristics of the signal are almost the same for every face image. But considering the change of the brightness and the size of the face image, some database of the filter parameters for various circumstances must be prepared. Details on this subject is for further research.

7. REFERENCES

[1] H. Harashima, K. Odajima (currently K. Arakawa), Y. Shishikui, H. Miyakawa, "ε-Separating Nonlinear Digital Filter and Its Applications", *Trans. IEICE*, vol.J-66-A, no.4, pp.297-304, April 1982 (in Japanese).

[2] H. Watabe, Y. Arakawa, and K. Arakawa, "Nonlinear Filters for Multimedia Applications", *Proc. IEEE ICIP'99*, 27AO3.6, pp.174-179, Oct.1999.

[3] K. Arakawa and H. Naito, "Extended Component-Separating Nonlinear Filter and Its Application to Face Image Beautification", *19-th Fuzzy System Symposium*, pp.411-414, Sept. 2003.

[4] T.Okada, S. Miyazaki, H. Watabe, K. Arakawa, and Y. Arakawa, " Nonlinear Filter Bank Using e-Filters and Its Application to Face Image Processing", *Proc. IEEE ISPACS 2002*, B3-2, Nov. 2002.

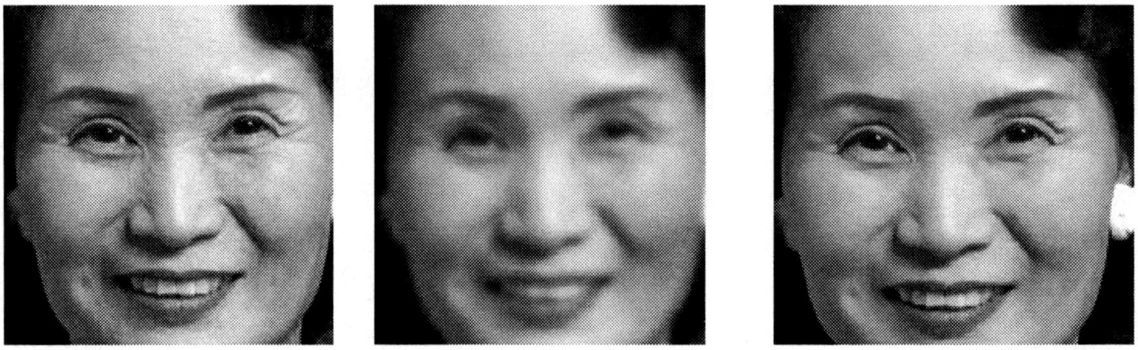

(a) Input face image. (b) Output of a linear smoothing filter. (c) Output of the ε-filter.

Fig.5 Results of processing an face image by a conventional linear smoothing filter and the ε-filter.

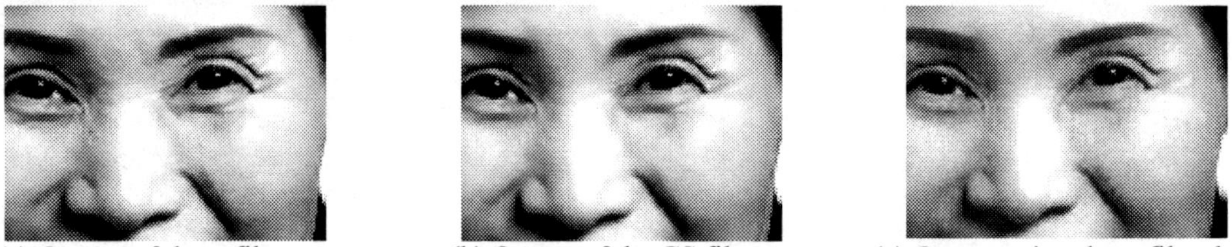

(a) Output of the ε-filter. (b) Output of the CS filter. (c) Output using the ε-filter bank.

Fig.6 Comparison of the performance of the three nonlinear filters proposed here.

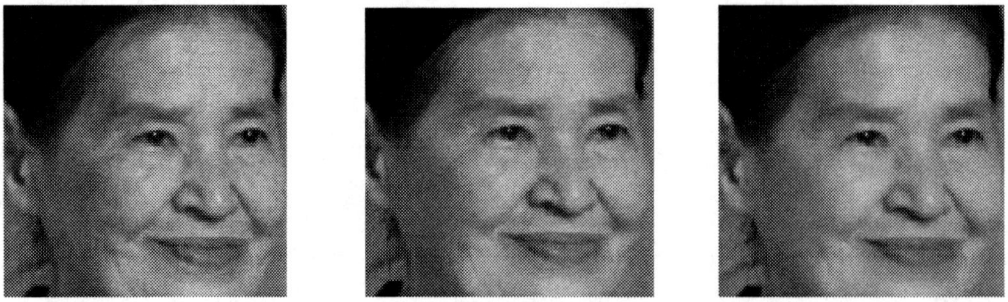

(a) Input face image. (b) Output of the ε-filter. (c) Output using the ε-filter bank.

Fig.7 Performance of the ε-filter bank for spots on a face compared with that of the ε–filter.

Video Error Concealment Techniques using Progressive Interpolation and Boundary Matching Algorithm

Tsung Han Tsai, Yu Xuan Lee and Yu Fong Lin
Department of Electrical Engineering
National Central University, Taiwan
E-mail：han@ee.ncu.edu.tw

Abstract

Recently the transmission of the multimedia data under internet environment has been wildly used in many applications. However, the video data is very sensitive to the transmission error caused by packet lost. This induces decoded video data with error propagation and makes the video quality very poor. In this paper we proposed a low complexity error concealment algorithm using progressive interpolation and boundary matching for object-based video coding. By use of the bidirectional frame as the referenced frame, error propagation in inter frame coding mode can be reduced largely. Our approach has the advantage of better video quality within a small search range and low complexity requirement.

1. Introduction

Based on the progress in communicational techniques, the transmission of multimedia data becomes more and more important especially in digital image video signal. However the realistic communication channels are error prone-environments. Encoded video bitstream is very sensitive to the transmission error suchlike the loss of video packets of data or the lost of synchronization. Because the video compressed bitstream is made of variable length code (VLC) that every symbol is coded by different number of bits for increasing of coding efficiency. These symbols are described for video information of the intra frames and inter frames. Therefore video packet loss will induce the error propagation in the decoded video data and decay the video quality seriously.

Error concealment intends to conceal the effects induced by error propagation during video compressed data under transmission. It utilizes the received video data without packet loss to achieve recovery of the lost data. Temporal domain error concealment algorithms are usually used in interframes [1]. The video data of the inter frames are composed of motion vector and the DCT (Discrete Cosine Transform) coefficients of prediction error. If the motion vector is received correctly, the missing Macroblock (MB) data is set to their corresponding motion compensated MB by the motion vector. Unfortunately under packet switch network transmission environment, the video packet loss would induce the lost of motion vector and the DCT coefficients of the prediction error. Under this situation, the collocated MB in the reference frame would be chosen. It limits the performance of this algorithm. Due to the lost of motion vector, many algorithms are proposed to achieve recovery of motion vector [2], [3]. In this situation the search range dominates the degree of complexity of the algorithm. Spatial interpolation algorithms are also proposed which use the available video data around the missing MB and interpolation formulation to achieve recovery of the missing video data within the MB. Some examples had been proposed including averaging, linear interpolation, multidirectional interpolation, and motion field interpolation [4], [5], [6]. These approaches assume the pixel values around the missing MB are very smooth, but in the most video sequence this assumption would not be always true. If this condition is not satisfied, the blurring effect would be caused by the interpolation especially in the missing MB with edge presentation.

In this paper we propose a low complexity error concealment algorithm using progressive interpolation and boundary matching for object-based video coding. By use of the bidirectional frame as the referenced frame, error propagation in inter frame coding mode can be reduced largely. This paper is organized as follows. In Section 2, the proposed algorithm based on BMA and progressive interpolation is described. In Section 3 we describe the low complexity approach in detail and explain how to utilize B frame to be a referenced frame. In Section 4 the simulation result compared to other traditional techniques are presented. Finally, Section 5 concludes this paper.

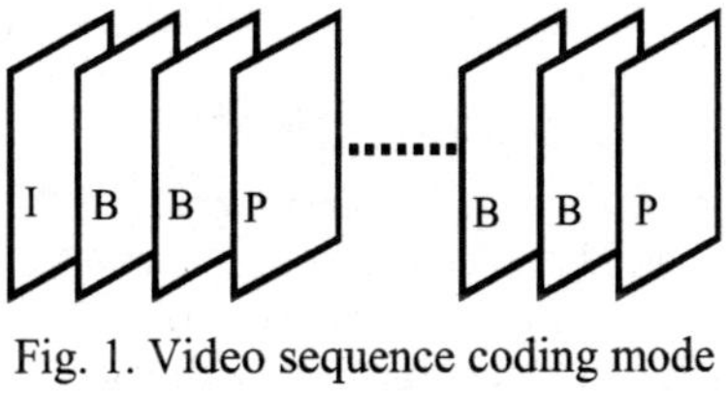

Fig. 1. Video sequence coding mode

2. Proposed Algorithm

2.1. Progressive Interpolation

Prediction frame coding is the essential techniques in modern video compression standard. As shown in Fig. 1, three kinds of coding mode, I, B, P, are applied. However, each kind of coding mode suffers from different tolerance in the lost of the data. Among many cases, data lost in P frame will induce higher degree of quality decayed due to its data dependence and propagation to the other P and B frame coding.

The progressive interpolation is based on the linear interpolation formulation, but under the video packet lost environment the available video data around the missing MB may not be received correctly. In this case we use the top side and the bottom side pixel value around the missing MB to interpolate the left side pixel value first. Then we use the three sides of pixel, top, bottom, and left value around the missing MB to interpolate the missing pixel value by linear interpolation using (1) and (2). The statement is illustrated in Fig. 2.

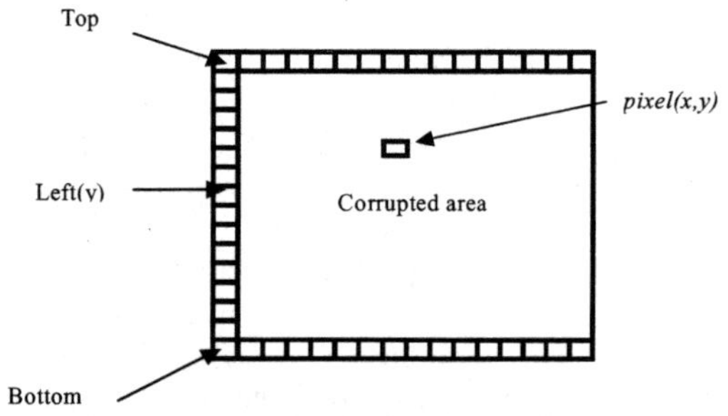

Fig. 2. Progressive interpolation diagram

$$left(y) = \frac{Top * (15 - y) + Bottom * y}{15} \quad (1)$$

$$pixel(x,y) = \frac{Top(x)*(15-y) + Bottom(x)*y + Left(y)*(15-x)}{15+(15-x)} \quad (2)$$

2.2. Boundary Match Algorithm

The Boundary Matching Algorithm (BMA) is proposed in 1993 [2]. It is used to estimate the lost motion vector in the reference frame, depending on the boundary difference (top, bottom and left) with the finite search range. The operation is described in (3).

$$\hat{f}_R(x,y,n) = \hat{f}_R(x+\hat{d}_x, y+\hat{d}_y, n-1) + f_D(x,y,n) \quad (3)$$

$\hat{f}_R(x,y,n)$: reconstructed pixel value.

$\hat{f}_D(x,y,n)$: texture data (differential pixel value, DFD).

$\hat{d}_x, \hat{d}_y$: search position (limited by the search range).

By use of the sum of square error in the left, top and bottom to be the cost function, it measures the degree of the difference by (4), (5), (6) and (7). In the above equations, assume that the texture data is received correctly in the decoder and the motion data is corrupted. Under this environment condition, BMA utilizes the texture data in the current frame and reconstructed pixel value in previous frame to reconstruct the corrupted pixel data in the current frame.

$$C_A = \sum_{x=x_0}^{x_0+N-1} (\hat{f}_R(x, y_0, n) - f_R(x, y_0 - 1, n))^2 \quad (4)$$

$$C_L = \sum_{y=y_0}^{y_0+N-1} (\hat{f}_R(x_0, y, n) - f_R(x_0 - 1, y, n))^2 \quad (5)$$

$$C_B = \sum_{x=x_0}^{x_0+N-1} (\hat{f}_R(x, y_0 + N - 1, n) - f_R(x, y_0 + N, n))^2 \quad (6)$$

$$C = C_A + C_L + C_B \quad (7)$$

Where x_0 and y_0 are the top-left pixel coordinate, N is the MB size, n is the time unit of the video frame and C is the total difference that is sum of top, bottom and left difference. This will lower the computational complexity. In (3), the displaced frame difference (DFD) is taken into account with packet loss situation. But in the packet lost situation, the DFD and motion vector of the video packet are both lost. At this situation, [2] also proposed another algorithm to resolve this problem. It used the DFD from the top of missing MB, then calculated the latest estimated motion vector by BMA. This algorithm is called Extended Boundary Matching Algorithm (EBMA). In our proposed approach, we only use the difference of boundary pixel to measure the search result directly without consideration of DFD.

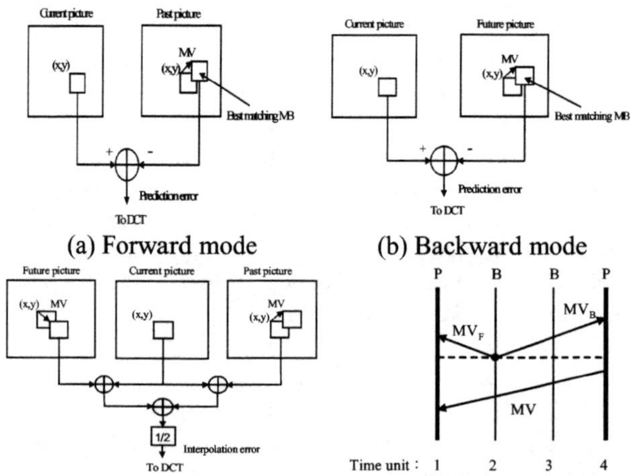

(a) Forward mode (b) Backward mode

(c) Interpolated mode (d) Direct mode

Fig. 3. Four kinds of coding mode of the B frame.

3. Low Complexity Approach

Motion compensation is used in general video coding standard to reduce temporal redundancy like MPEG1, MPEG2, MPEG4 and H.263.....etc. It means temporal correlation between current frame and reference frame is established by use of motion vector provided by motion estimation.

In this paper we propose an efficient video error concealment algorithm with low complexity issue. This algorithm utilizes the B frame as reference frame for video error concealment. As shown in Fig. 3 there are four kinds of video coding mode used in the B frame. In each coding mode there exists different coding procedure. The first one is backward coding mode. In this mode the MB is coding with prediction of the past P frame. It likes general coding mode in the P frame. The second is forward coding mode. In this mode the MB is coding with prediction of the future P frame. It presents property of the bidirectional prediction in the B frame. The third is interpolated coding mode. In this mode the MB is coding with bidirectional property concurrently and averages prediction error then sends it to DCT processing. It means that two times of motion estimation are executed, one is for past P frame and the other one is for future P frame. The fourth is direct coding mode. In this coding mode the MB is coding with bidirectional property. However, the motion vector in this coding mode is a linear relationship between past P frame and future P frame. By this linear relationship there is no any motion estimation procedure executed in this coding mode. From the view point of the coding complexity, the complexity of the backward coding mode and forward coding mode are the same. The most complicated coding mode in the B frame is interpolated coding mode because there are two times of motion estimation procedure. The lowest complexity of these four video coding modes is direct mode because no any motion estimation is executed in this coding mode.

The principle of this proposed algorithm is executed for the packet lost in the P frame then identifies corrupted MBs. Mapping these corrupted MBs to the B frame next to the corrupted P frame by collocated MB in the B frame. Then identify which kind of video coding mode of this collocated MB is performed. After this procedure each collocated MB would be determined in one of four kinds of video coding mode. In each kind of coding mode there exists a video error concealment strategy corresponding to each kind of coding mode of the B frame. The whole algorithm flow chart is shown in the Fig. 4.

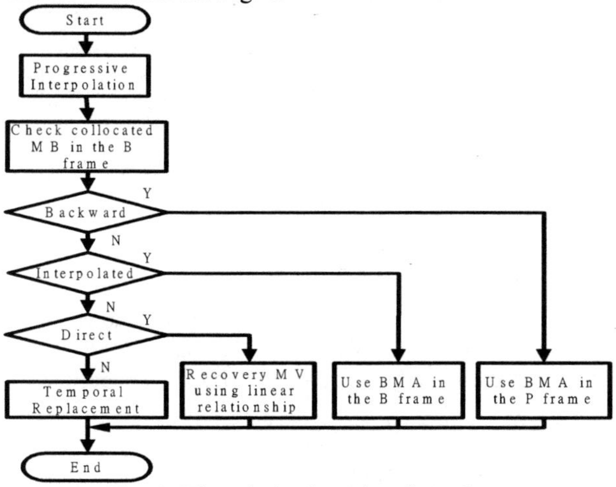

Fig. 4. The whole algorithm flow chart.

A. *Backward mode*

In this mode the collocated MB in the B frame is referenced to the future P frame or I frame such that the degree of error propagation in this mode is very obvious. By this reason the video quality of the collocated MB is not good enough to be a referenced frame for using of BMA. Therefore the original BMA that references from previous P frame or I frame is chosen in the backward mode.

B. *Interpolated mode*

In this coding mode the B frame is located after the erroneous P frame and could be regarded as the reference frame. Because the B frame is much similar to the erroneous P frame than the P frame referenced by the erroneous P frame, the erroneous P frame will propagate the erroneous video data to the referenced B frame. At first we need to use the progressive interpolation to conceal the erroneous P frame then apply the BMA in the B frame. As shown in Fig.5, after the erroneous P frame is chosen to be the referenced frame, since that B frame is much similar than the P frame referenced by the erroneous P frame, the precision for block matching calculation can be enhanced.

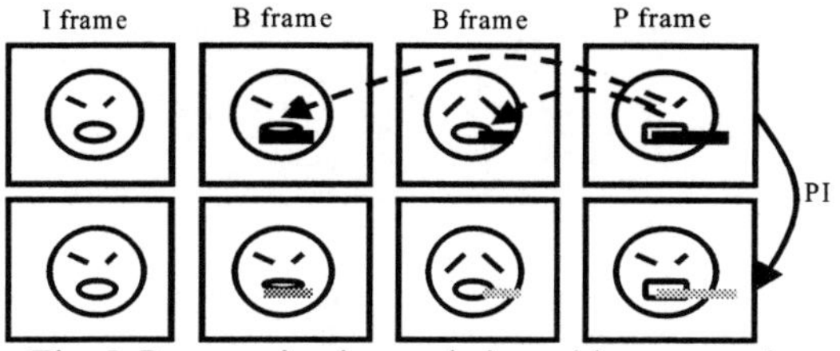

Fig. 5. Progressive interpolation with propagation procedure diagram.

C. *Direct mode*

In the direct mode the motion vector of the collocated MB is achieved by linear scaling to the corresponding MB of the future P frame or I frame such that the motion vector in this mode is obtained with low computation requirement. If the collocated MB in the B frame is received correctly, then the bidirectional motion vector can be decoded without any error. The motion vector of the corrupted MB in the P frame can be estimated with low computation requirement by linear scaling relationship.

D. *Forward mode*

In this mode the collocated MB in the B frame is referenced to the past P frame or I frame such that the influence of the error propagation at this mode is very small. As result, there is a very cleanly referenced MB. This collocated MB is very similar to the corrupted MB in the P frame such that the temporal replacement is chosen for this mode with a little texture data distortion.

4. Simulation Result and Analysis

We use the Microsoft MPEG-4 reference software to construct the whole simulation environment and assume that the video packet lost is in the P frame only. The P frame will make error propagation. Thus it will generate the random packet lost pattern that referenced by the decoder to introduce the video packet lost situation. The test video sequence is encoded with the bit rate of the 1Mbps. The frame rate is 30Hz and the sequence format is CIF with object base type. Because this experiment emphasis that our proposed algorithm is more efficient to smaller search range result, so we take the object based format into account to avoid the experiment result influenced by the still background.

Fig. 6 – 8 show the average PSNR quality for different concealment methods based on the several video sequences. The packet lost rate in the P frame is 1%, 3% and 5% respectively. 50 simulation results are taken and computed. The temporal replacement is replaced by the collocated MB of the referenced frame [1], the EBMA is proposed in [2]. The search range is used in ±4 pixel. From these Figures, t is obvious that our proposed algorithm does not suffer from high packet lost very much compared to other methods. At each packet lost situation, our algorithm also maintain the best quality comparing to others.

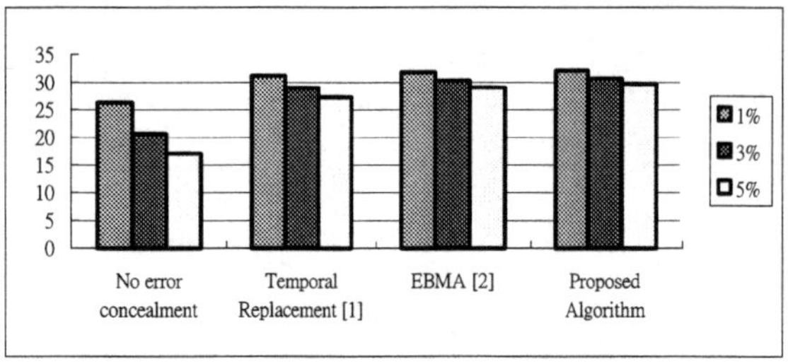

Fig. 6. The average PSNR (dB) evaluation of Bream.

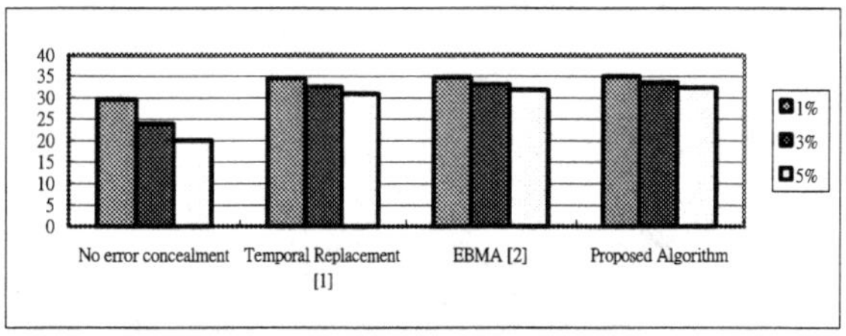

Fig. 7. The average PSNR (dB) evaluation of Foreman.

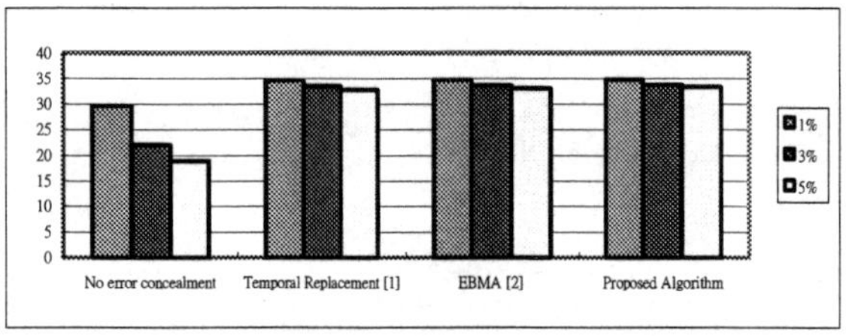

Fig. 8. The average PSNR (dB) evaluation of Weather.

The error concealment strategy is using temporal replacement for forward mode and linear scaling for direct mode to achieve recovery of the motion vector in the corrupted MB of the P frame. The search operation of EBMA or BMA needs much computation requirement of video decoder. For every corrupted MB, the EBMA is executed for each one of them such that search operation spends large amount of computation requirement. In our proposed approach, by identified the collocated MB of the corrupted MB, the BMA is executed only for the collocated MB which is backward mode or interpolated mode, such that the search operation is much less than it of EBMA. In a long term period of the error concealment, the computation requirement can be reduced. The reduction ratio is defined in (8) and presented in Fig. 9.

$$\text{ReductionRatio} = \frac{\#\text{of search operation in proposed algorithm}}{\#\text{of serach operation in EBMA}} \quad (8)$$

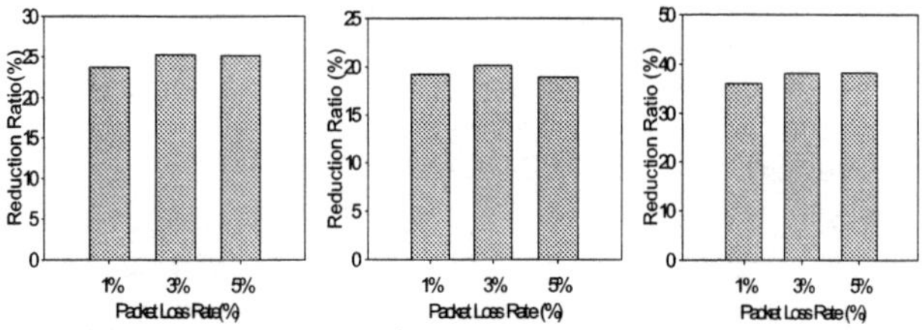

(a) Bream (b) Foreman (c) Weather

Fig. 9.Reduction Ratio.

The video error concealment result of subjective view is presented by frame view in Fig. 10. These sequences include bream, foreman and weather respectively. The packet lost ratio of presented result is 3% and search range of EBMA and proposed algorithm is ±4 pixels.

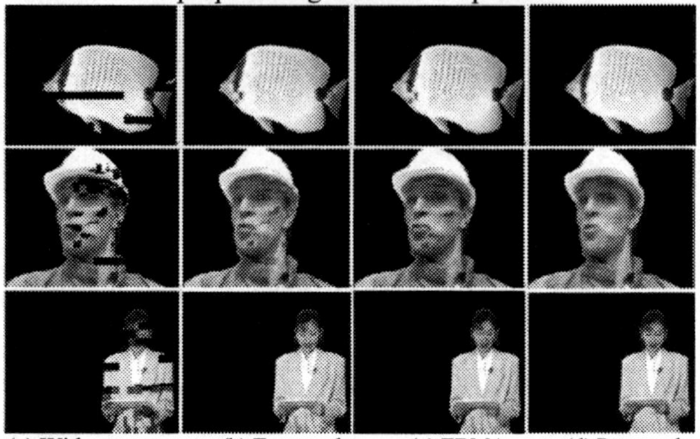

(a) Without error concealment (b) Temporal replacement (c) EBMA (d) Proposed algorithm

Fig. 10. Error concealment result by different algorithms

5. Conclusions

In this paper, a low complexity error concealment approach is proposed. In this approach, the B frame is used as the referenced frame. The coding mode of collocated MB is identified into four kinds of error concealment strategy respectively. The search operation is executed only for two kinds of error concealment strategy. With the proposed error concealment approach, it achieves better video quality and the overall computation requirement can be reduced in a long term period.

References

[1] S.Aign & K. Fazel, "Error detection & concealment measures in MPEG-2 video decoder," in Proc. Of the International Workshop on HDTV'94, Toorino, Oct. 1994.

[2] W-M. Lam, A.R Reibman & B. Liu, "Recovery of lost or erroneously received motion vector" in Proc. ICASSP, Vol. 5, April 1993, pp V417 – V420.

[3] Choog Soo Park; Jongchul Ye; Sang Uk Lee "Lost motion vector recovery algorithm;" Circuits and Systems, 1994. ISCAS '94, 1994 IEEE International Symposium on, Volume: 3, 30 May-2 Jun 1994 Page(s): 229 -232 vol.3.

[4] Kwok, W.; Huifang Sun "Multi-directional Interpolation For Spatial Error Concealment", Consumer Electronics, 1993. Digest of Technical Papers. ICCE, IEEE1993International Conference on, 8-10 Jun 1993 Page(s): 220 –221.

[5] S.Aign and K. Fazel "Temporal and Spatial Error Concealment Techniques for Hierarchical MPEG-2 Video Codec" IEEE International Conference in Communication, Vol.3, pp. 1778-1783, 1995.

[6] ME Al-Mualla, N Canagarajah, DR Bull, "Error concealment if lost motion vector using motion field interpolation", Proceeding of ICIP, 1998, pp 512-516.

JPEG BASED IMAGE COMPRESSION WITH ADAPTIVE RESOLUTION CONVERSION SYSTEM

Kazuhiro Shimauchi[*], *Masahiro Ogawa*[*] *and Akira Taguchi*[**]

Department of Electrical and Electronic Engineering, Musashi Institute of Technology
1-28-1 Tamazutsumi Setagaya-ku Tokyo 158-8557 Japan
[*]{shimauchi, ogawa}@sp.ee.musashi-tech.ac.jp, [**] ataguchi@eng.musashi-tech.ac.jp

ABSTRACT

It is well known that the standard JPEG algorithm causes visually disturbing blocking effects when we obtain a high compression ratio image. In order to obtain a better quality image with high compression ratio, we propose a new image compression method, which is combined with the standard JPEG and an adaptive resolution conversion. Since the resolution conversion can reduce the image data, the proposed hybrid method is realized smaller quantization parameter than the standard JPEG, and we obtain a high quality decoded image. In the proposed method, compression is performed by two factors as the quantization and the resolution conversion. We study how to adjust two factors. The proposed method archives high quality image compression only adding the small system to the JPEG.

1. INTRODUCTION

Transform-based data compression is by far the most popular choice in both image and video coding applications. Due to its near-optimal energy compaction property and the availability of fast algorithms and hardware implementations, the block-based discrete cosine transform (DCT) is the used in most of current image and video compression standards, such as JPEG[1] and MPEG[2]. The block-based DCT based coding can successfully compress images with nearly no perceptible artifacts. However, at low bit rates, a major problem associated with the block-based DCT compression is that the decoded images manifest visually objectionable artifacts. One of the well-known artifacts in low bit rate transform-coded images is the blocking effect, which is noticeable in the form of undesired visible block boundaries.

The bit rate control in the block-based DCT coding is achieved by only changing quantization steps. It should make quantization steps large if we want to obtain high compression ratio. The result of this condition, quantization error becomes large, it causes the blocking artifact.

In order to obtain a better compression performance, we consider about image compression method, which is combined with the standard JPEG and resolution conversion. The image compression by the decimation is effective for the flat areas of an image, because that the area does not have high frequency component, sufficiently. And, at the high compression ratio, the high frequency coefficients of DCT are quantized by the large quantization steps. The result of quantization, almost high frequency coefficients are zero. This means that the number of the decimate-able blocks is large and compression by decimation becomes more effective in the case of high compression. If it is possible to achieve effective compression by a resolution conversion, it is expected that the JPEG codec makes quantization error and blocking artifact small. The JPEG-based image coding using resolution conversion[3] compresses an image by uniform decimation and the JPEG encoder. This method realizes a better compression performance than the standard JPEG, but, in some bit rate cases, its performance is inferior to the JPEG because that the method decimates an image uniformly without consideration for frequency characteristic.

In this paper, we propose a new image compression method, which is combined with the standard JPEG and the adaptive resolution conversion based on local frequency characteristic to realize the targeted compression ratio. In the proposed method, an image is divided into 16x16-sized blocks and decimate-able blocks are decimated to 8x8-sized. After the adaptive resolution conversion, each block is coded by JPEG encoder. We prove that it is effective to add compression by the decimation to the JPEG. Furthermore, in the proposed method, the controlling bit rate is performed by two factors as adaptive decimation and quantization. Thus, we study how to adjust two factors in order to realize user's operation to ease (only one factor) as same as the JPEG.

2. JPEG BASED IMAGE COMPRESSION USING ADAPTIVE RESOLUTION CONVERSION

2.1. The standard JPEG algorithm

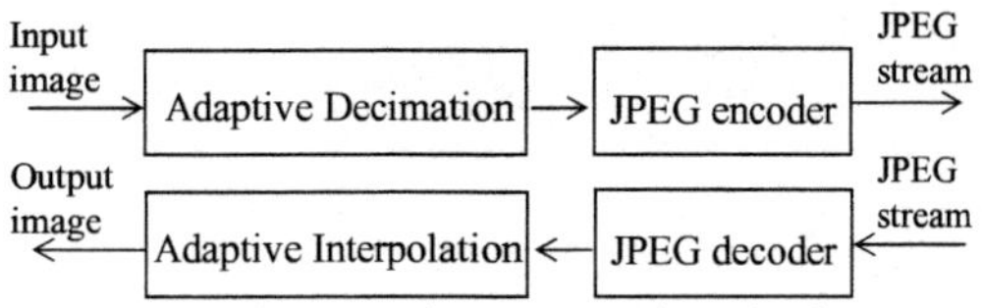

Fig. 1 The proposed method.

Firstly, in order to explain principle of the blocking artifact, we show the standard JPEG algorithm.

In the standard JPEG, an image is divided into the 8x8-sized blocks, and each block is compressed by DCT, quantization and Huffman coding. In these processes, the compression ratio is controlled by the quantization process. To achieve controlling compression ratio, a parameter that is called Quantization Parameter (*QP*) is implemented. The quantization steps $q(i, j)$ for the 8x8-point DCT coefficient is determined by

$$q(i,j) = QP \times Q(i,j) \qquad for \;\; i,j = 0,1,2\cdots 7 \qquad (1)$$

where, $Q(i, j)$ is the quantization table which is defined by Ref.[1]. Since the compression ratio is controlled by only *QP*, it should make *QP* large if we want to obtain high compression ratio. The result of this condition, quantization error becomes large, and it causes the blocking artifact.

2.2. Over view of the proposed method

In this paper, we propose a new image compression method which is combined the standard JPEG and an adaptive resolution conversion. In order to realize this system, we add adaptive resolution conversion blocks to the JPEG encoder and decoder as shown in Fig.1. In encoding process, an image is divided into small blocks. These blocks are decimated based on the local frequency characteristic and desired compression ratio. After the adaptive decimation, every block is coded by the JPEG encoder. On the other hand, in decoding process, the coded bit stream is decoded by the JPEG decoder. And the decoded blocks are interpolated based on result of adaptive decimation.

2.3. Adaptive Decimation

In the proposed method, an image is divided into 16x16-sized blocks. Every block is checked if it contains edges or fine texture. If the current block has no edges or fine texture, the 16x16-sized block is decimated to 8x8-sized block. In another case, the size of block is unchanged.

Figure 2 shows the block diagram of the adaptive decimator. Each block is judged whether the block can be decimated or not, by the detector which is shown in dashed line part of Fig. 2. The detector estimates high frequency component that is lost by low pass filtering in decimation process. In order to evaluate the lost high frequency component of 16x16-sized block signal,

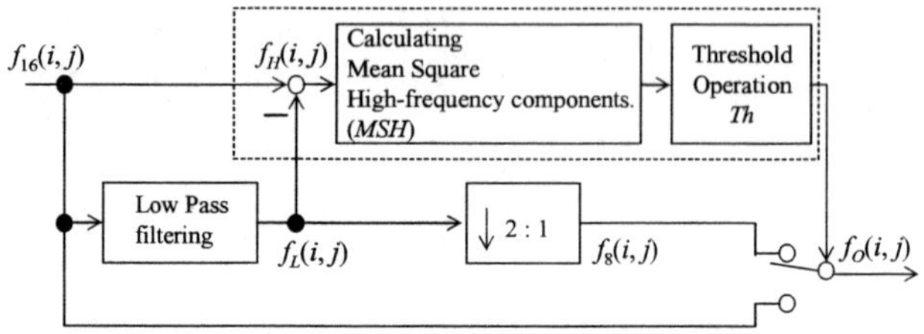

Fig. 2 The adaptive decimator.

we define *MSH* (Mean Square High frequency component) as

$$MSH = \frac{1}{16 \times 16} \sum_{i=0}^{15} \sum_{j=0}^{15} (f_H(i,j))^2 \qquad (2)$$

$$f_H(i,j) = f_{16}(i,j) - f_L(i,j) \qquad (3)$$

where, $f_{16}(i, j)$ shows original 16x16-sized block signal. $f_L(i, j)$ is obtained by low pass filtering of $f_{16}(i, j)$. The low pass filter of the proposed method is the Gaussian filter[4]. Since, we use interpolation which is proposed by Ref.[5] at the adaptive interpolation section shown in Fig.1.

The judgment of the detector is shown as

$$f_O(i,j) = \begin{cases} f_8(i,j) & if \;\; MSH < Th \\ f_{16}(i,j) & othrewise \end{cases} \qquad (4)$$

where, $f_8(i, j)$ is down sampled signal(8x8-size),

$$f_8(i,j) = f_L(2i, 2j) \qquad for \;\; i,j = 0,1,2,\cdots,7 \quad (5)$$

In the proposed method, the value of threshold parameter *Th* is depend on with the compression ratio. There is very important relationship between *Th* and quantization factor *QP*. We will discuss about it in section 4.

2.4. Sorting blocks after the adaptive decimation

The output of the adaptive decimation $f_O(i, j)$s are different sized blocks, respectively. The 8x8-sized blocks and the 16x16-sized blocks are mixed in the output signal. It is necessary that the stored data is sent with every block size, and the total data size is reduced as much as possible under decodable conditions.

We proposed sorting blocks method as shown in Fig. 3. This method packs continuing 8x8-sized blocks into an original 16x16-sized block. An 16x16-sized block can pack four 8x8-sized blocks, but we determine the largest number is three. The fourth 8x8-sized block is filled with value of 128. The decoder can decipher that current 16x16-sized block is decimated or not by reading filled 128s. And the number of the 128s block explains how many 8x8-sized blocks are continuing.

The proposed sorting blocks method achieves compacting data and satisfies the decodable conditions. Furthermore, Fig. 3 is compressed by the JPEG encoder.

Fig. 3 Sorting blocks after the adaptive decimation.

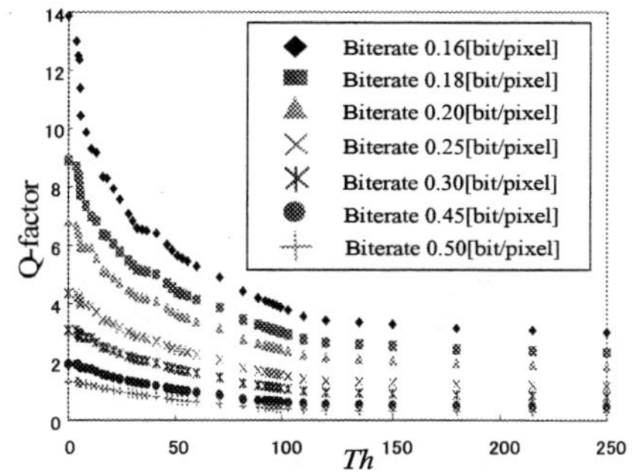

Fig. 4 The relationship between *Th* and *QP*.

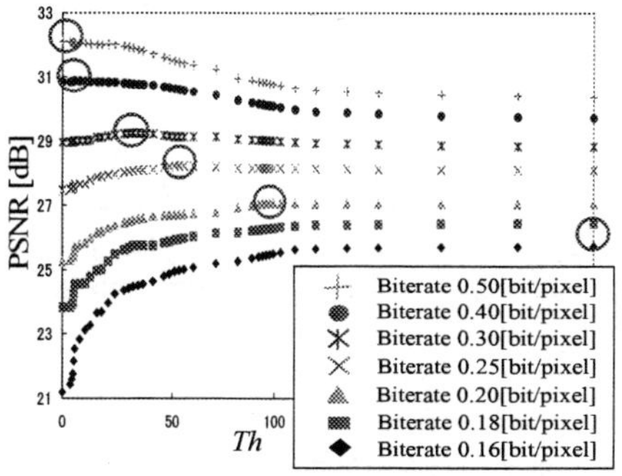

Fig. 5 The relationship between *Th* and PSNR with same bit rate.

2.5. Adaptive Interpolation (Decoding Process)

In the decoding process, a bit stream is decoded by the JPEG decoder. As we have described in the preceding section, the decoder of the proposed method can decipher the result of the adaptive decimation. The decoder reads the filled 128s, and the decimated blocks are reconstructed by the interpolation. The proposed method applies the interpolator of Ref.[5] to the decoder, it realizes compensation for the lost high frequency component to some degree.

3. THE BIT RATE CONTROL BY THE QUANTIZATION AND THE RESOLUTION CONVERSION

3.1. Reducing the complexity of the rate control

As we described in 2.1, the bit rate is controlled by the quantization parameter *QP* of the JPEG codec. On the other hand, degree of the adaptive decimation is operated by the parameter *Th* as we have described in 2.3. The proposed method combines the standard JPEG and the adaptive resolution conversion; therefore a user has to operate 2 parameters of *QP* and *Th*. It is very complex. It is necessary that *Th* is determined automatically by *QP*, which is inputted by a user. We study how to adjust the two parameters from computer simulation.

3.2. Derivation of the relationship between *QP* and *Th*

To investigate the relationship between *QP* and *Th,* firstly, we simulate image compression using the proposed method for the "Girl"(8[bit] grayscale, 256x256[pixels]) [6]. Figure 4 shows the relationship between *Th* and *QP* in several bit rates.

The proposed method does not use decimation when the $Th = 0$. In this case, the performance of the proposed method is equal to the standard JPEG. As shown in Fig. 4, if the *Th* is larger, it is possible to make *QP* smaller Therefore, it is expected that it is effective to add compression by the decimation to the standard JPEG for decreasing blocking noise. However, if the decimation is applied to larger area, the decoded image blurred. We want to find the combination of *QP* and *Th* which can obtain the highest quality of a decoded image. Quality of an image is evaluated by PSNR(Peak Signal to Noise Ratio):

$$PSNR = 10\log_{10}\left(\frac{255^2}{MSE}\right)\ [dB] \qquad (6)$$

$$MSE = \frac{1}{M \cdot N}\sum_{i=0}^{M-1}\sum_{j=0}^{N-1}\left(X(i,j) - S(i,j)\right)^2 \qquad (7)$$

where $S(i, j)$ and $X(i, j)$ is the original and reconstructed images, respectively, and M, N is the horizontal and vertical sizes of the image.

We add PSNR-axis to Fig. 4, and consider the combination in 3-demantional function. Figure 5 shows *Th*-PSNR plane of it. In 0.50[bit/pixel], it turns out that it is better not to carry out the adaptive decimation. On the other hand, it becomes clear that all blocks should be decimated to obtain highest quality decoded image in 0.16[bit/pixel]. Figure 5 also shows that the appropriate *Th* is depending on the bit rate. We could find the combination of *QP* and *Th* that it obtains the highest quality of a decoded image for the "Girl" from Fig. 4 and Fig. 5. If there are similar relationships between *QP* and *Th* for general natural images, it is possible to apply the relation-ship for the other images. We simulate image compression using the proposed method for the six images shown in Fig. 6(8[bit] grayscale, 256x256[pixels])[6]. Figure 6 shows that the relationship between *QP* and *Th* for the six images. A tendency can be found out in the relation of the two parameters. The relationship between *Th* and *QP* is derived as

$$Th = 9.6QP^2 + 0.2QP - 25.3 \qquad (8)$$

The equation realizes that appropriate *Th* is determined automatically if *QP* is inputted by user.

4. EXTENSION TO COLOR IMAGES OF THE PROPOSED METHOD

Since continue-tone color images are used in many applications, extension to color images of the codec is very important. We apply the proposed method to color images.

In general coding, each RGB color image is transformed into a YCbCr representation using the standard equations because that the color space is available for continue-tone color image compression. We adopt the color space and apply the proposed method to the Y (Luminance) component. Other components Cb and Cr are decimated depending to the results of the Y component, since, Cb and Cr have edges at the same positions as Y component in many cases of continue-tone color images.

5. SIMULATION RESULTS

To evaluate the performance of the proposed method, we simulate image compression using the proposed method. The averaged MSE of three RGB color components is used in the PSNR(*eq*.(6)), we define this PSNR as $PSNR_{color}$. Figure 7 shows the $PSNR_{color}$ of the decoded image in several bit rates for the "Lenna" (24[bit] RGB,

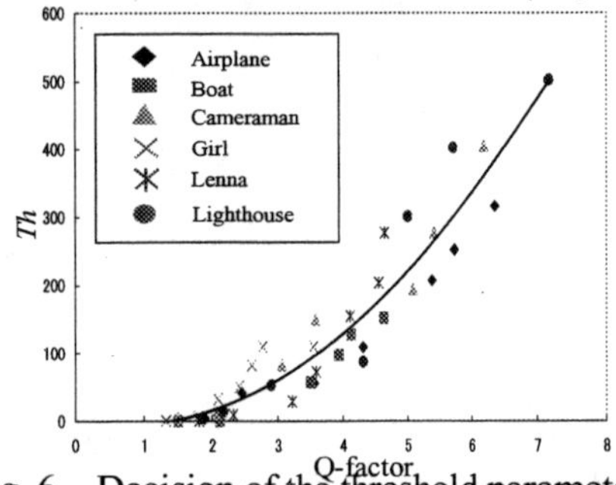

Fig. 6 Decision of the threshold parameter *Th*.

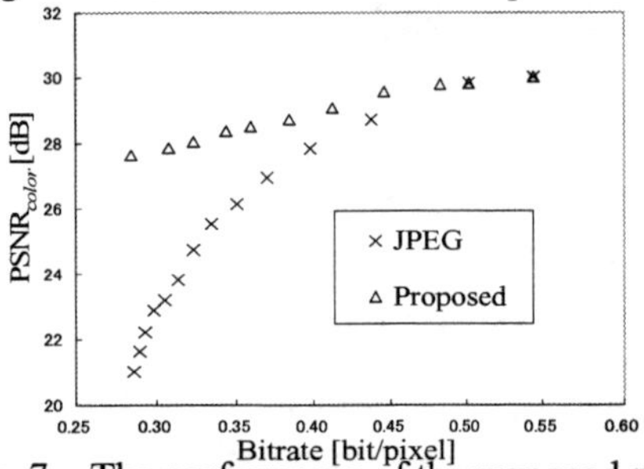

Fig. 7 The performance of the proposed method.

512x512[pixels])[6]. In the bit rate is under 0.50[bit/pixel], it turns out that the proposed method is superior to the standard JPEG. On the other hand, in the bit rate is higher than 0.50[bit/pixel], the performances of the proposed method are same as the standard JPEG. It means that there are no decimate-able blocks in the conditions. Figure 8 shows the decoded images of the proposed, JPEG and new standard JPEG2000 at the same compression ratio of 0.33[bit/pixel]. Clearly, the decoded image of our method indicates a better quality than that of the standard JPEG. And, the proposed method archives high quality image compression closing to new standard JPEG2000, with only adding the small system to JPEG.

6. CONCLUSION

In this paper, we proposed a new image compression method which is combined the standard JPEG and an adaptive resolution conversion based on the local frequency characteristic. The controlling the compression ratio in the proposed method was adjusted with the relationship between decimation and quantization. The results of simulation showed that it is effective to add the resolution conversion to the standard JPEG encoder-decoder at the high compression ratio.

REFERENCES

[1] *Digital Compression and Coding of Continuous-tone still images*, ISO 10918-1, 1991.
[2] *Generic Coding of Moving Pictures and Associated Audio Information*, ISO/IEC 13818-2, 1996.
[3] T. K. Truong, L. J. Wang, I. S. Reed, and W. S. Hsieh, "Image Data Compression Using Cubic Convolution Spline Interpolation," *IEEE Trans. Image Processing*, vol.9, no.11, pp.1988-1995, Nov. 2000.
[4] P. J. Burt and E. H. Adelson, "The Laplacian Pyramid as a compact image code," *IEEE Trans. Communications*, vol.COM-31, no.4, pp.532-540, April 1983.
[5] Y. Takahashi, A. Taguchi, "An Arbitrary Scale Image Enlargement Method with the Prediction of High-Frequency Components," *IEICE Trans. (Japanese Edition)*, vol.J84-A, no.9, pp.1192-1201, Sept. 2001.
[6] *The USC-SIPI Image Database*, http://sipi.usc.edu

(a) JPEG ($PSNR_{color}$ 25.55[dB]) (b) Proposed ($PSNR_{color}$ 28.35[dB]) (c) JPEG2000 ($PSNR_{color}$ 31.03[dB])

Fig. 8 Decoded images (bitrate 0.33[bit/pixel])

PERCEPTUAL CODING OF DIGITAL COLOUR IMAGES BASED ON A VISION MODEL

C. S. Tan, D. M. Tan, and H. R. Wu
School of Computer Science & Software Engineering
Monash University, Victoria 3800, Australia
{ctan,dmt,hrw}@csse.monash.edu.au

ABSTRACT

This paper presents a perceptual coder for Y-Cr-Cb (YUV) colour spaces. The approach extends from the single colour channel model of [1], [2] and the RGB colour domain model of [3] that mimics the human visual system (HVS) in exploiting its intra-band and inter-orientational masking properties. Although the proposed perceptual coder retains most of the Embedded Block Coding with Optimized Truncation (EBCOT) coding features in [4] and is fully bit-stream compliant with JPEG2000 standard [5], the simulation results have revealed comparable or better visual performance over both the MSE and visual masking of JPEG2000 Verification Model 8.0 coder.

1. INTRODUCTION

Digital image coding and compression constitutes an integral part of multimedia systems with applications to learning and education, visual communications, tele-medicine and etc. The JPEG2000 image coding standard [5] recently introduced several advancement and many new features such as wavelet transform and region of interest (ROI) coding. The result of this is an improvement in the quality of coded images. Other enhancements such as the adoption of block coding paradigm based on the concept of EBCOT [4] also ensures superior performance over the other wavelet based coders such as reported in [6] and [7]. The block coding paradigm of EBCOT [4] and hence JPEG2000 [5] generates independent bit-streams for each block which are packed into quality layers [8].

Central to delivering the optimized bit stream in JPEG2000 [5] is the rate-distortion (R-D) function and the context arithmetic coding. The distortion measure used in the R-D optimization is the Mean Squared Error (MSE) or visually weighted MSE. However, it is believed that methods relying mainly on MSE or its derivatives such as Signal-to-Noise Ratio (SNR) or Peak Signal-to-Noise Ration (PSNR) are inaccurate measurement for perceptual quality [9]. As the HVS based coding is attuned towards visual quality as opposed to MSE, which is purely statistical in nature and that does not take into account the psycho-visual redundancy that exists in the HVS, it is important to have a perceptual distortion metric that is attuned to aspects of the HVS.

Current literary understanding of the HVS is far from complete and while there is no general agreement as to what model best represents the HVS, several models [10] and [11] have emerged in recent year which have gained popular usage.

2. VISION MODELLING

Human vision can be modelled by two separate stages: optical and cortical. The former accounts for the limitation of the eyes' sensitivity to vision relative to background luminance level and spatio-temporal frequencies. This visual sensitivity as described in [12] is determined by contrast, the level of intensity between two patterns. The contrast is often described by an attenuation function in objective metric as the contrast sensitivity function (CSF). In the proposed vision model, the CSF is applied as uniform frequency-specific sensitivity weights as in [13]. The latter accounts for the masking phenomenon of the HVS where the ability to identify a visual pattern can be weakened or diminished in the presence of other visual patterns. The masking function which includes the excitation and inhibition of neuron responses can be classified into spatial and temporal model. There exist four aspects of masking in the spatial model: intra-band, inter-band, inter-orientational and colour maskings. For coding of still images, temporal masking that concerns with masking along the temporal axis of spatial information is not relevant and hence is not implemented.

3. PROPOSED MASKING MODEL

The masking model used in this paper follows that of Tan et al. [2], [1], [3], which can be traced back to the work of Watson [14], and Teo and Heeger [15]. Teo and Heeger [15] used a generalised contrast gain control model (CGC). The model can be adapted to serve all four aspects of spatial masking. The model, which approximates the behaviour of neural signals from the visual cortex, comprises of several stages such as linear transform, normalisation, and detection.

In their earlier work, Teo and Heeger [15] utilized a quadrature mirror filter (QMF) in the linear transform, but steerable pyramid was used subsequently [16] to decompose images into different frequencies and orientation bands. The steerable pyramid was chosen over QMF to avoid aliaising. The transformation of signals into its frequency domain is necessary to suit the neuron responses of the eyes.

The normalisation stage describes the response at orientation θ, R_θ, of the visual system subjected to the process of inhibition and excitation in the visual pathway as represented by (1),

$$R_\theta = k_o \frac{(A_\theta)^{p_o}}{\sum_\phi (A_\phi)^{q_o} + \sigma_o^2} \quad (1)$$

where A_θ is the coefficient of the transform at orientation θ, A_ϕ are the coefficients at the same spatial location as A_θ spanning across all orientations (inclusive of A_θ), σ_o is the saturation coefficient, k_o is the scaling factor, p_o and q_o are respectively the excitation and inhibition exponents. In our model, q_o is set to 2. Both σ_o and k_o are chosen to avoid rapid saturation of the response. As in Watson [14], a condition for $p_o > q_o > 0$ is imposed.

A simple l_2 norm function is used to determine the distortion between the original and processed images as defined in (2)

$$D = \left| R_o - R_p \right|^2 \tag{2}$$

where R_o and R_p are the responses of the original and processed images respectively. The overall inter-orientational distortion, D_o, between the two images is determined by pooling equation (2) for all coefficients spanning all frequencies.

In the case of the intra-band masking, the response, R_I, is defined by (3),

$$\boldsymbol{R_I} = \boldsymbol{K_I} \frac{(\boldsymbol{A_y})^{\boldsymbol{P_I}}}{(\boldsymbol{A_y})^{q_I} + \frac{8}{\boldsymbol{N}} \sum_{\boldsymbol{z}} (\boldsymbol{A_z})^{q_I} + \boldsymbol{\sigma}_{\text{var}}^2 + \boldsymbol{\sigma_I^2}} \tag{3}$$

where $q_I = 2$, A_y is the coefficient at spatial frequency location y, $\sum_z (A_z)^2$ is the sum of all squared coefficients around the neighbourhood of location y, N is the number of neighbourhood coefficients, σ_{var}^2 is the variance of the neighbourhood coefficients, k_I and σ_I^2 are the scaling and saturation coefficients respectively, with condition being imposed on $p_I > q_I > 2$. The variance, σ_{var}^2 takes care of local texture masking [17]. High variances are expected for region with higher texture contents.

For intra-band masking, the distortion measure, D_I, between the original and processed images is determined by pooling equation (2) across all frequencies.

The final distortion measure, D_T, encompassing both the intra-band and inter-orientational masking is then computed by (4),

$$D_T = g_I D_I + g_o D_o \tag{4}$$

where g_I and g_o are the proportional contributing factors for both intra-band and inter-orientational masking, respectively.

4. PROPOSED CODER

The proposed coder as illustrated in Figure 1 is based on the EBCOT coding structure [4] as adopted by JPEG2000. The distortion measurement in the EBCOT is replaced by the proposed HVS based distortion measure consisting of CSF filtering, intra-band and inter-orientational masking function.

The transform kernel used in the proposed coder is Daubuchies 9/7 wavelet filters [18]. The coefficients are critically sampled. The lower implementation complexity and lower computational cost of this filters justifies its used in the proposed coder over the aliaising-free steerable pyramid filters.

Refer to Figure 1, the input image is block-based frequency decomposed into raw coefficients with Daubuchies 9/7 wavelet filters. The raw coefficients are passed on to the perceptual distortion measure and the bit-plane coder. The output of the bit plane coder is the quantised coefficients which are passed on to the perceptual distortion measure. The CSF filtering is applied to both the raw and quantized coefficients, and their outputs are fed to the masking function to determine their intra-band and inter-orientational masking responses. The pooling and detection stage computes the final distortion measure that is passed on to R-D optimiser along with the rate generated by the bit plane coder.

5. PARAMETERISATION

For the proposed coder to operate on colour components Y-Cr-Cb, the CSF and the parameters of the masking function are optimised for each colour component to obtain an approximation to the behaviour of the HVS. It must be emphasised that these parameters in Table 1 only represent sub-optimal values as there are 42 parameters need to be optimised.

Model Parameters			
	Y	U	V
CSF-LL	0.95	1.03	1.28
CSF-1	1.15	1.23	1.35
CSF-2	1.33	1.39	1.40
CSF-3	1.41	1.34	1.35
CSF-4	1.30	1.10	1.13
CSF-5	1.02	0.65	0.85
k_o	0.9876	0.9800	0.9300
σ_o	6.925	15.02	10.11
p_o	2.145	2.04	2.215
g_o	0.35	0.501	0.35
k_I	1.09	1.11	0.98
σ_I	2.505	11.00	1.505
p_I	2.153	2.17	2.30
g_I	0.37	0.85	0.402

Table 1: Sub-optimal parameters. One set of CSF coefficients for each resolution level for the 5-level wavelet transform with Daubuchies 9/7 filters.

6. RESULTS AND CONCLUSIONS

The performance of the proposed coder has been evaluated with comparative force-choice subjective tests [19], [20]. With each original image coded at 1.0 bpp, 0.5 bpp, 0.25 bpp and 0.125 bpp, a total of 20 images were generated from 5 different images. The images are assessed on a 21-inch, 0.25mm dot pitch Sun Monitor with display resolution of 1280 x 1024 pixels. The testing involves 6 participants assessing the performance of the images generated by the proposed coder against that of MSE and CVIS (simple vision modelling) of JPEG2000 VM8 coder. The CVIS images are generated with masking gain, g = 0.5. The presentations of the 20 images are pseudo randomized to left or right and 1 to 20. The test results presented in Table 2 suggest that the proposed coder produces images with noticeable superior visual quality at low to medium bitrate between 0.125 and 0.25 bpp. The performance improvement of the proposed coder is due to its ability to apportion higher bit budget to code region of higher visual importance.

With small sample size taken in comparative the subjective test, the current result is preliminary but promising. Further investigation and testing is on going.

Image	Bitrate (bpp)	Scores (%)			
		Test 1		Test 2	
		A	B	A	C
goldhill	1.0	0	100	67	33
	0.5	83	17	83	17
	0.25	67	33	100	0
	0.125	83	17	83	17
sail	1.0	67	33	83	17
	0.5	83	17	100	0
	0.25	83	17	100	0
	0.125	100	0	100	0
pepper	1.0	17	83	33	67
	0.5	83	17	83	17
	0.25	50	50	83	17
	0.125	67	33	67	33
lena	1.0	67	33	50	50
	0.5	67	33	50	50
	0.25	83	17	100	0
	0.125	83	17	100	0
tulip	1.0	33	87	67	33
	0.5	67	33	83	17
	0.25	33	87	100	0
	0.125	83	17	83	17

Table 2: Comparative Forced-Choice Subjective Results (A – proposed coder, B – MSE of JPEG2000, C – CVIS of JPEG2000)

7. REFERENCES

[1] D. M. Tan, H. R. Wu, and Z. Yu, "Perceptual Coding of Digital Monochrome Images," *IEEE Sinal Processing Letters*, to appear in 2004.

[2] D. M. Tan, H. R. Wu, and Z. Yu, "Vision Model Based Perceptual Coding of Digital Images," Proc. of International Symposium on Intelligent Multimedia, Video and Speech Processing, Hong Kong, pp. 87-91, 2-4 May 2001.

[3] D. M. Tan, H. R. Wu, and Z. Yu, "Perceptual Coding of Digital Colour Images," Proc. of IEEE International Symposium on Intelligent Signal Processing and Communication Systems, Nashville, Tennessee, USA, 20-21 Nov 2001.

[4] D. Taubman, "High Performance Scalable Image Compression with EBCOT," *IEEE Transactions on Image Processing*, vol. 9, no. 7, pp. 1158-1170, July 2000.

[5] M. Boliek, C. Christopoulos, and E. Majani, "JPEG2000 Part I Final Committee Draft Version 1.0," *Technical Rep, ISO/IEC JTC1/S9/WG1 N1646*, 2000.

[6] J. M. Shapiro, "Embedded Image Coding Using Zerotrees of Wavelet Coefficients," *IEEE Transactions on Image Processing*, vol. 41, pp. 3445-3462, 1993.

[7] A. Said and W. A. Pearlman, "A New Fast and Efficient Image Codec based on Set Partitioning in Hierarchical Trees," *IEEE Transactions on Circuits and Systems for Video Technology*, vol. 6, issue 3, pp. 243-250, June 1996.

[8] M. Rabbani and R. Joshi, "An Overview of the JPEG 2000 Still Image Compression Standard," *Signal Processing: Image Communication*, vol. 17, issue 1, pp. 3-48, Jan, 2002.

[9] B. Girod, "What's Wrong with Mean-Squared Error?," in *Digital Images and Human Vision*, A. B. Watson, Ed. MIT: MIT Press, pp. 207-220, 1993.

[10] T. Chen and H. R. Wu, "Objective Quality Assessment for Digital Video Coding based on Human Vision Model," Proceedings of the 3rd World Multiconference on Systemics, Cybernetics and Informatics (SCI '99) and the 5th International Conference on Information Systems Analysis and Synthesis (ISAS '99), vol. 6, pp. 36-41, Orlando, USA, Aug 1999.

[11] C. J. Van Den Branden Lambrecht, "Testing Digitial Video System and Quality Metrics based on Perceptual Models and Architecture, CH -1015." Lausanne, Switzerland: EPFL, 1996.

[12] F. L. Van Nes and M. A. Bournan, "Spatial Modulation Transfer Function in the Human Eye," *Journal of the Optical Society of America*, vol. 57, no. 3, pp. 401-406, 1967.

[13] Z. Yu, H. R. Wu, S. Winkler, and T. Chen, "Vision Model Based Impairment Metric to Evaulate Blocking Artifacts in Digital Video," *Proceeding of IEEE*, vol. 90, pp. 154-169, Jan 2002.

[14] A. B. Watson and J. A. Solomon, "A model of Visual Contrast Gain Control and Pattern Masking," *Journal of the Optical Society of America*, pp. 2379-2391, 1997.

[15] P. C. Teo and D. J. Heeger, "Perceptual Image Distortion," *Proceedings of IEEE International Conference on Image Processing*, Austin, Texas, pp. 982-986, Nov 1994.

[16] E. P. Simoncelli, W. T. Freeman, E. H. Adelson, and D. J. Heeger, "Shiftable Multiscale Transform," *IEEE Transactions on Information Theory*, vol. 38, pp. 587-607, 1992.

[17] R. Safranek and J. Johnston, "A Perceptually Tuned Sub-band Image Coder with Image Dependent Quantization and Post-quantization Data Compression," *Proceeding of IEEE ICASSP*, pp. 1945-1948, 1989.

[18] M. Antonini, M. Barlaud, P. Mathieu, and I. Daubechies, "Image Coding Using Wavelet Transform," *IEEE Transactions on Image Processing*, vol. 1, no. 2, pp. 205-220, 1992.

[19] R. M. Kaplan and D. P. Saccuzzo, *Psychological Testing: Principles Applications and Issues.*, 5th. ed., Thomson Learning Inc., 2001.

[20] L. Friedenberg, *Psychological Testing: Design, Analysis and Use.* Allyn & Bacon, A Simon & Shcuster Company, 1995.

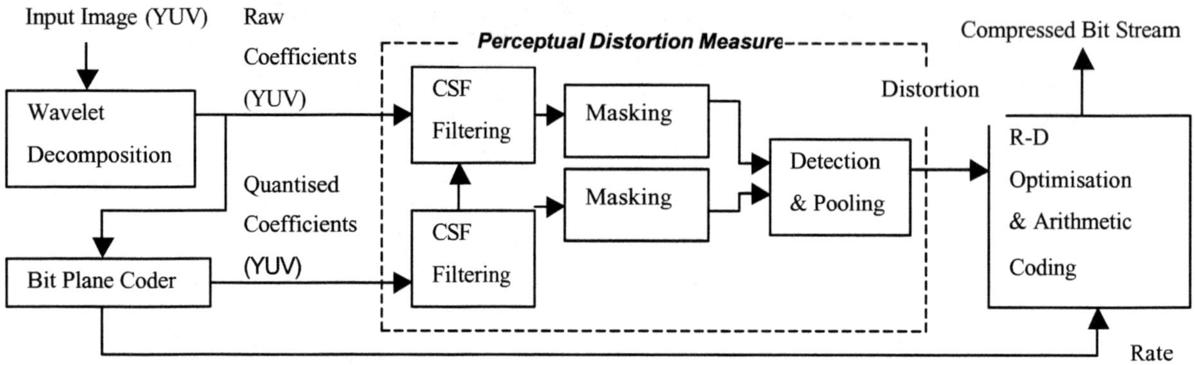

Figure 1: Perceptual Distortion Metric embedded in EBCOT structure

Figure 2(a) Image coded to 0.125 bpp by the proposed coder

Figure 2(b) Image coded to 0.125 bpp by JPEG2000 with MSE

Figure 2(c) Image coded to 0.125 bpp by JPEG2000 with CVIS

Figure 2(d) Original image

SMART NOISE REDUCTION SYSTEM BASED ON ALE AND NOISE RECONSTRUCTION SYSTEM

Naoto Sasaoka[1], Yoshio Itoh[1], Kensaku Fujii[2] and Yutaka Fukui[1]

[1]Faculty of Engineering, Tottori University
4-101 Koyama-minami, Tottori 680-8552, Japan
[2]Faculty of Engineering, Himeji Institute of Technology
2167 Shosha, Himeji 671-2201, Japan

ABSTRACT

A noise reduction technique to reduce wideband and sinusoidal noise in noisy speech is proposed. We have proposed the noise reconstruction system (NRS) using the linear prediction error filter (LPEF) and system identification model. However, since the sinusoidal noise has the characteristic just like the vowel, the conventional system cannot reduce the sinusoidal noise. In this paper, we introduce the adaptive line enhancer (ALE) to reduce the sinusoidal noise. The ALE estimates a current signal correlated with a delayed signal. The speech signal is non-stationary signal in long time interval. On the other hand, the sinusoidal noise is stationary noise. For example, ventilating fan and engine noise. So the autocorrelation of the sinusoidal noise is maintained although that of the speech signal fades. Thus the ALE can estimate only the sinusoidal noise. Therefore, the noise reduction system based on the ALE and the NRS reduces not only the wideband noise but also sinusoidal noise.

1. INTRODUCTION

A large variety of noise reduction techniques have been proposed for reducing background noise in noisy speech with microphone array [1], spectral subtraction (SS) [2] and so on. The microphone array can be considered as a directional microphone with a blind spot in the arrival bearing of noise. In case many noise sources exist, the microphone array cannot avoid increasing the number of microphones. On the other hand, SS method is known to use one microphone. In the case of the SS method, the musical tones arise from residual noise, and the processing delay occurs. Additionally, the SS method requires prior estimation of a noise spectrum. It means that the SS method requires the speech/non-speech section detector under the noisy environments.

In order to solve these problems, we have proposed the noise reconstruction system (NRS) based on linear prediction and system identification [3], [4]. The NRS uses a linear prediction error filter (LPEF) and an adaptive digital filter (ADF). First, the noisy speech is whitened by LPEF. Assuming that the background noise is generated by exciting a linear system with the white signal, it can be reconstructed from the whitened noise by estimating the transfer function of the linear system.

However, since the sinusoidal noise has the characteristic just like the vowel, the sinusoidal noise cannot be reduced by the NRS. In real environments, the background noise includes not only the wideband noise but also the sinusoidal noise. Therefore, it is important to reduce both the wideband noise and the sinusoidal noise.

In this paper, we introduce the adaptive line enhancer (ALE) for reducing the sinusoidal noise [5]. Generally, the ALE is used to enhance the speech signal or sinusoidal signal superposed on wideband noise. However, in this paper, ALE is used to reduce the sinusoidal noise superposed on the speech. The ALE can estimate only the current input signal $x(n)$ correlated with a delayed signal $x(n-\tau)$, where τ denotes delay. Since the phoneme of speech changes, the speech signal is non-stationary signal in long time interval. On the other hand, the sinusoidal noise is the stationary signal. For example, ventilating fan and engine noise. As the delay τ increases, the correlation of the sinusoidal noise is maintained although that of the speech signal fades. Thus, the ALE can estimate only the sinusoidal noise.

2. THE PRINCIPLE OF NOISE REDUCTION

2.1. Wideband noise reduction based on Noise Reconstruction System

Figure 1 shows the noise reconstruction system to reduce the wideband noise [3], [4], where $x(n)$ is the noisy speech. The $x(n)$ is represented as

$$x(n) = s(n) + \xi(n) \tag{1}$$

where $s(n)$ and $\xi(n)$ are clean speech and wideband noise, respectively. $w(n)$, $\hat{\xi}(n)$ and $\hat{s}(n)$ are output of LPEF, the reconstructed noise and the enhanced speech signal, respectively. $H_{LPEF}(z)$ and $H_{NRF}(z)$ represent the transfer function of the LPEF and NRF (Noise Reconstruction Filter), respectively. NRF is a transversal type filter.

First, the prediction error signal $w(n)$ is obtained by LPEF. The $w(n)$ is represented as

$$w(n) = x(n) - \sum_{k=1}^{L} h'_k(n)x(n-k) \qquad (2)$$

where $h'_k(n)$ represent the tap coefficients of the LPEF. The coefficients $h'_k(n)$ converge such that the prediction error signal whitens [6]. Since the speech signal can be represented as the stationary signal in short time interval, most of the speech signals will be predicted by the linear predictor. On the other hand, assuming that the wideband noise is generated by exciting a linear system with the white signal, the noise becomes the white signal.

Next, we consider the reconstruction of the wideband noise. Assuming that the wideband noise is generated by the linear system $H_N(z)$ with the white signal, the wideband noise can be reconstructed from the whitened signal $w(n)$ by estimating the transfer function of the noise generating system. This estimation is performed by the NRF which is used as system identification model, where $\xi(n)$, $s(n)$ and $\hat{s}(n)$ are a desired signal, disturbance and a estimation error signal, respectively. Finally, the enhanced speech signal is obtained by subtracting the reconstructed noise from $x(n)$.

Since the sinusoidal noise has the characteristic just like the vowel, the NRS cannot reduce the sinusoidal noise. In this paper, we introduce the sinusoidal noise reduction system based on ALE.

2.2. Sinusoidal noise reduction based on ALE

2.2.1. The principle of sinusoidal noise reduction

Let the sinusoidal noise be the stationary periodic signal. For example, ventilating fan and engine noise. On the other hand, the phoneme of speech changes, so the speech signal is the non-stationary signal.

The noisy speech $x(n)$ is given as follows:

$$x(n) = s(n) + \eta(n) \qquad (3)$$

where $s(n)$ and $\eta(n)$ represent clean speech and sinusoidal noise. The delayed signal is represented as

$$x(n-\tau) = s(n-\tau) + \eta(n-\tau) \qquad (4)$$

where τ is delay. Since characteristics of speech change, the correlation of speech fades as delay τ increases. On the other hand, the delayed sinusoidal noise $\eta(n-\tau)$ is correlated with $\eta(n)$.

The proposed sinusoidal noise reduction system is shown in Fig. 2 [5]. The ALE is composed of delay and

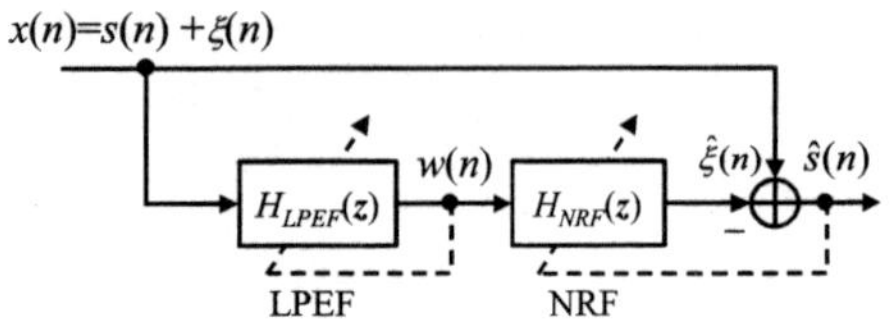

Fig. 1 Noise Reconstruction System (NRS).

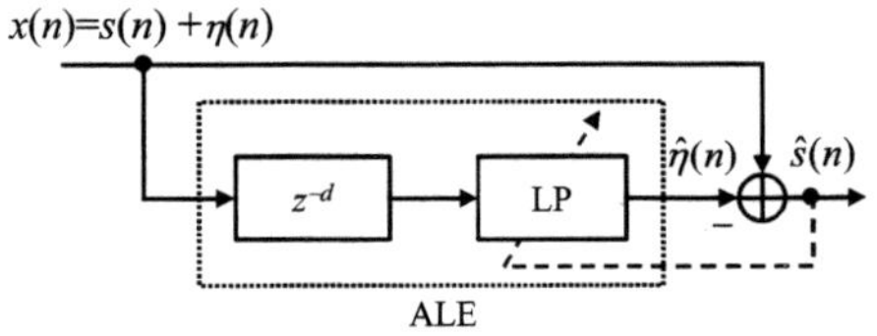

Fig. 2 Adaptive Line Enhancer (ALE).

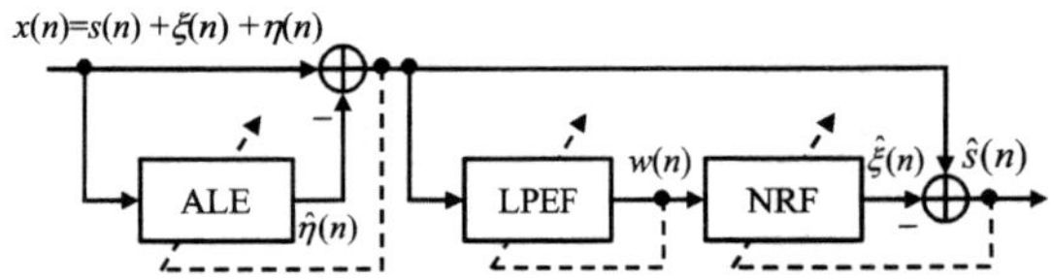

Fig. 3 Proposed Noise Reduction System.

Linear Predictor. The output of ALE is given by

$$\hat{\eta}(n) = \sum_{k=1}^{M} h''_k(n)x(n-d-k) \qquad (5)$$

where $h''_k(n)$ and d are tap coefficients and decorrelation parameter, respectively.
The ALE can estimate only the current signal correlated with the delayed signal. Assuming that there is enough decorrelation parameter d to fade correlation of speech out, the ALE can estimate only the sinusoidal noise. Finally, the enhanced speech is obtained by subtracting the estimated sinusoidal noise from noisy speech $x(n)$.

2.2.2. The step size control

In case that a fixed step size for updating the tap coefficients is used, it is difficult to carry out reducing noise with the high quality of enhanced speech. Thus, we introduce the step size control to ALE. In a speech section, the small step size is used not to estimate the speech. In non-speech section, the large step size is used to track the sinusoidal noise.

The relationship equation between quantity of decreasing noise and step size is used to control the step size [7]. The relationship is given as follows:

$$\mu = \frac{2E_0 P_S}{P_N + E_0 P_S} \quad (\mu_{\min} \le \mu \le \mu_{\max}) \qquad (6)$$

where μ, $\mu_{\min}$ and $\mu_{\max}$ are step size, minimum and maximum of step size, respectively. The $\mu_{\min}$ is used to prevent step size from being too small to track the background noise. Additionally, the $\mu_{\max}$ is used to prevent step size from being so large that the estimation accuracy of the background noise is degraded. P_S and P_N represent power of the desired signal and disturbance, respectively. P_S and P_N are represented as

$$P_S(n) = \hat{\eta}(n)^2 + \rho P_S(n-1) \quad (7)$$

$$P_N(n) = \hat{s}(n)^2 + \rho P_N(n-1) \quad (8)$$

where ρ represents the forgetting factor.

2.3. Sinusoidal and Wideband noise reduction

The noise reduction system based on the NRS and the ALE is proposed to reduce the wideband and the sinusoidal noise.

Figure 3 shows the noise reduction system to reduce the sinusoidal and the wideband noise. The proposed system uses the ALE in first stage and the NRS in second stage. The noisy speech is represented as follows:

$$x(n) = s(n) + \xi(n) + \eta(n) \quad (9)$$

where $s(n)$, $\xi(n)$ and $\eta(n)$ represent clean speech, wideband noise and sinusoidal noise, respectively.

First, the sinusoidal noise is reduced by ALE. Since the characteristic of the sinusoidal and the wideband noise is different, the ALE estimates only the sinusoidal noise. Thus, the wideband noise superposed on the speech is obtained. Next, the wideband noise is reduced by the NRS. Finally, the enhanced speech is obtained.

3. SIMULATION RESULTS

The performance of the proposed noise reduction system was evaluated. All sound data prepared in simulations were sampled by 8kHz with 16bit resolution. The speech signal spoken by the male talker was used. As the background noise, the colored noise and the sinusoidal noise were used. The colored noise was generated by passing white noise through the second order infinite impulse response (IIR) filter. The filter coefficients of second order IIR filter are given as follows:

$$C_1 = -2r\cos\theta \quad (10)$$

$$C_2 = r^2 \quad (11)$$

where C_1 and C_2 are filter coefficients. In this simulation, r and θ were set to 0.8 and 0.79, respectively. Additionally, the sinusoidal noise is given as follows:

$$\eta(n) = \sum_{k=1}^{10} \cos(2\pi k f_0 nT + 2\pi r(k)) \quad (12)$$

where f_0, T and $r(k)$ ($0 \le r(k) < 1$) represent pitch frequency of the sinusoidal noise, sampling period and uniform random numbers, respectively.

The adaptive algorithms for updating tap coefficients are LMS (Least Mean Square) algorithm and NLMS (Normalized Least Mean Square) algorithm [6]. Table 1 shows the each parameter in this simulation.

Table 1 Parameters used in this simulation

ALE (NLMS)	Number of tap coefficients	200
	Decorrelation parameter d	199
	Quantity of decreasing noise E_0	0.0019
	Forgetting factor ρ	0.995
	Minimum of step size	0.003
	Maximum of step size	0.023
LPEF (LMS)	Number of tap coefficients	128
	Step size	0.1
NRF (NLMS)	Number of tap coefficients	128
	Step size	0.01

The SNR (Signal to Noise Ratio) was used to evaluate the noise reduction ability. The SNR_{in} and SNR_{out} represent the input and output SNR, respectively. These indices are defined as follows:

$$SNR_{in} = 10\log_{10}\frac{\sum_{j=1}^{N} s^2(j)}{\sum_{j=1}^{N} \delta^2(j)} \quad (13)$$

$$SNR_{out} = 10\log_{10}\frac{\sum_{j=1}^{N} y_S^2(j)}{\sum_{j=1}^{N} y_\delta^2(j)} \quad (14)$$

where N is the number of samples. $s(j)$ and $\delta(j)$ are clean speech and background noise, respectively. $y_s(j)$ and $y_\delta(j)$ represent speech and noise components included in the enhanced speech, respectively.

3.1. Results of sinusoidal and wideband noise reduction

The noise reduction ability of the proposed system was evaluated. In this simulation, the sinusoidal noise superposed on the wideband noise, whose power ratio was set to 0dB, was used as background noise. Figure 4 shows the waveform of results. From the simulation results, the SNR increases to about 9.9dB. It can be seen that the proposed system has the potential for reducing the sinusoidal and wideband noise, simultaneously.

3.2. Results of sinusoidal noise reduction

The sinusoidal noise reduction ability of the proposed system was evaluated because there was possibility that the NRS degrades the sinusoidal noise reduction ability. The simulation results were shown in Table 2, where ALE represents a result of noise reduction using only the ALE. From the simulation results, since the result of the proposed system approximately equals that of only the

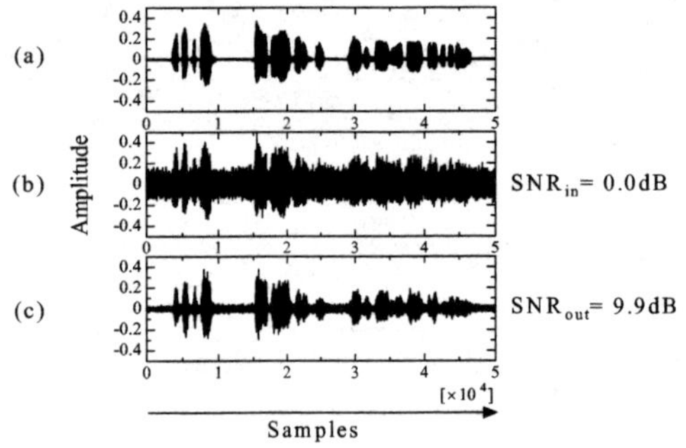

Fig. 4 Results of sinusoidal and wideband noise reduction. (a) Original Speech (b) Noisy Speech (c) Enhanced Speech

Table 2 Sinusoidal noise reduction. (SNR_{in}=0.0[dB])

	SNR_{out}[dB]
ALE	18.1
Proposed system	18.5

Table 3 Wideband noise reduction. (SNR_{in}=0.0[dB])

	SNR_{out}[dB]
NRS	8.1
Proposed system	7.8

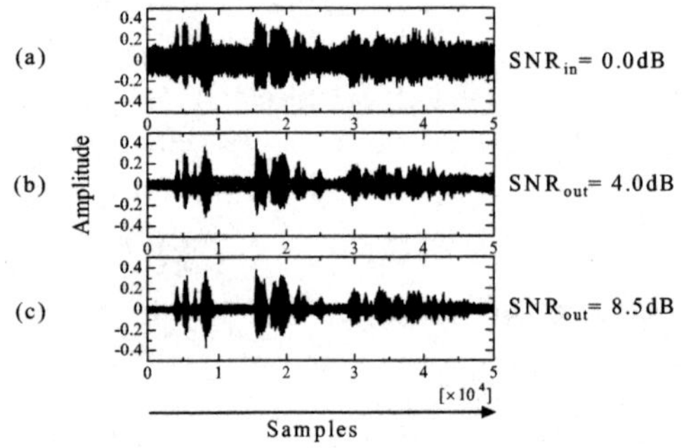

Fig. 5 Results of actual background noise reduction.
(a) Noisy Speech (b) Speech enhanced by only NRS
(c) Speech enhanced by the proposed system

ALE, it is considered that the noise reconstruction system does not influence the sinusoidal noise reduction.

3.3. Results of wideband noise reduction

The wideband noise reduction ability of the proposed system was evaluated because there was possibility that the ALE degrades the wideband noise reduction ability. The simulation results were shown in Table 3, where NRS represents a result of noise reduction using only the NRS. From the simulation results, since the result of the proposed system approximately equals that of only the NRS, it is considered that the ALE does not influence the wideband noise reduction.

3.4. Results of actual background noise reduction

The actual background noise reduction ability was evaluated. In this simulation, the tunnel noise, which includes the wideband, ventilating fan and engine noise, is used as background noise. Figure 5 shows the waveform of results. In Fig. 5(b), the result of only NRS is indicated. In Fig. 5(c), the result of the proposed system is illustrated. Compare the Fig. 5(c) with Fig. 5(b), the SNR_{out} is improved by about 4.5dB. It indicates that the proposed system has the potential for reducing the actual background noise.

4. CONCLUSION

A noise reduction technique to reduce the wideband and the sinusoidal noise in noisy speech has been proposed. We have proposed the NRS using the linear prediction error filter and system identification model. However, since the sinusoidal noise has the characteristic just like the vowel, the NRS cannot reduce the sinusoidal noise. In this paper, we have introduced the ALE to reduce the sinusoidal noise. From the simulation results, we have verified that the proposed noise reduction system reduces both the wideband noise and the sinusoidal noise. Further research involves the more improvement of noise reduction ability.

REFERENCES

[1] J. F. Flanagan, J. D. Johnston, R. Zahn and G. W. Elko, "Computer-steered microphone arrays for sound transaction in large rooms," J. Acoust. Soc. Am., 78, pp.1508-1518, Nov. 1985.

[2] S. F. Boll, "Suppression of acoustic noise in speech using spectral subtraction," IEEE Trans. Acoust., Speech, Signal Processing, vol. ASSP-27, no.2, pp.113-120, April 1979.

[3] A. Kawamura, K. Fujii, Y. Itoh and Y. Fukui, "A new noise reduction method using estimated noise spectrum," IEICE Trans. Fundamentals, vol.E85-A, no.4, pp.784-789, April 2002.

[4] A. Kawamura, K. Fujii, Y. Itoh and Y. Fukui, "A new noise reduction method using linear prediction error filter and adaptive digital filter," Proc. ISCAS2002, vol.III, pp.488-491, May 2002.

[5] Y. Sugihara, K. Fujii, A. Kawamura, Y. Itoh and Y. Fukui, "A study on methods to reduce sinusoidal noises superposing on speech signal," Technical report of IEICE. US2002-98, Jan. 2003.

[6] S. Haykin, Introduction to adaptive filters, Macmillan publishing company, New York, 1984.

[7] H. Itakura and Y. Nishikawa, "On some characteristics of an echo canceller using a learning identification algorithm," IEICE Trans. Fundamentals, vol.J60-A, no.11, pp.1015-1022, Nov. 1977.

A 39x48 GENERAL-PURPOSE FOCAL-PLANE PROCESSOR ARRAY INTEGRATED CIRCUIT

Piotr Dudek

Department of Electrical Engineering and Electronics
University of Manchester Institute of Science and Technology (UMIST)
PO Box 88, Manchester M60 1QD, United Kingdom
p.dudek@umist.ac.uk

ABSTRACT

This paper presents the implementation of a general-purpose programmable vision chip, with a 39×48 SIMD processor-per-pixel array, fabricated in a 0.35μm CMOS technology. The chip employs Analogue Processing Elements to achieve cell density of 410 cells/mm^2. The array operates at 1.25MHz with power consumption of 12μW/cell and executes low-level image-processing algorithms in real-time. Chip architecture, circuit and layout design issues are discussed. Experimental results are presented.

1. INTRODUCTION

In a retina-like smart sensor thousands of tiny processors have to be placed, together with photodetectors, on a single integrated circuit. The trade-offs between speed, accuracy, functionality, processor area and power consumption have to be very carefully resolved. The choices of a suitable architecture and efficient circuit techniques are of critical importance. Digital vision chips, based on bit-serial SIMD (Single Instruction Multiple Data) processor arrays have been reported in the literature [1-2]. Compact digital processing elements can be built, but due to limited local memory present in these devices they are suitable mostly for binary image processing. However, some gray-scale image processing is usually required and techniques based on analogue circuits seem to be more suitable for this purpose than digital solutions (at least in present-day CMOS technologies). The analogue approaches appear to offer better efficiency in terms of speed, area and power consumption [3-4].

Previously, we proposed an approach to the design of a general-purpose programmable vision chip, which is essentially a combination of a "conventional" SIMD architecture and an "unconventional" processing element circuitry [4]. A new silicon implementation of our vision chip concept (i.e. the SCAMP-2 chip) has been recently fabricated. In the new chip the design of the Analogue Processing Element (APE) has been further simplified and refined [5], while the Flexible Global Readout Architecture [6] has been added, to enhance chip's ability to perform data reduction at the sensor level. In this paper the chip design is overviewed and test results are presented.

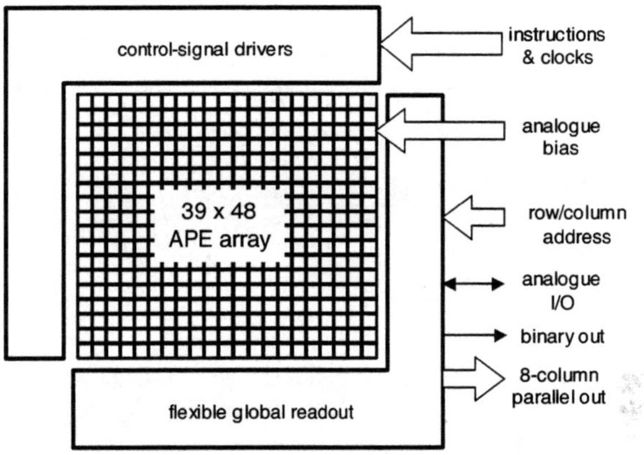

Fig. 1. Architecture of the SCAMP-2 vision chip

2. CHIP ARCHITECTURE

The processing core of the chip is formed by a 2-D mesh-connected processor-per-pixel array, shown in Figure 1. The processing elements in the array (APEs) are implemented as "analogue microprocessors" (they process data according to a software program, but achieve this using analogue circuits, and store data in a form of analogue sampled signals). A single controller issues instructions, which are broadcast to each APE in the array, and the APEs execute instructions in parallel, according to the SIMD paradigm, each operating on their local data. The pixel-parallel arrangement is perfectly suited to execute early vision (low-level image processing) algorithms.

The design of the APE in the SCAMP-2 chip has been previously reported [5]. Its architecture is shown in Figure 2. Briefly, each APE contains a photodetector, nine "registers" (each capable of storing analogue data, e.g. a gray-scale pixel value or another scalar variable), a 4-neighbour local communication network, I/O circuits for external data transfers, and a local-autonomy activity-flag circuit.

The overall principle, when designing the APE, was that of a "minimum hardware" processor. It was motivated by the desire to reduce the size of the APE, so that large resolutions of pixel-per-processor arrays are feasible. At the same time, though, the accuracy of processing had to be maintained at an acceptable level. A further consideration was the reduction in power consumption.

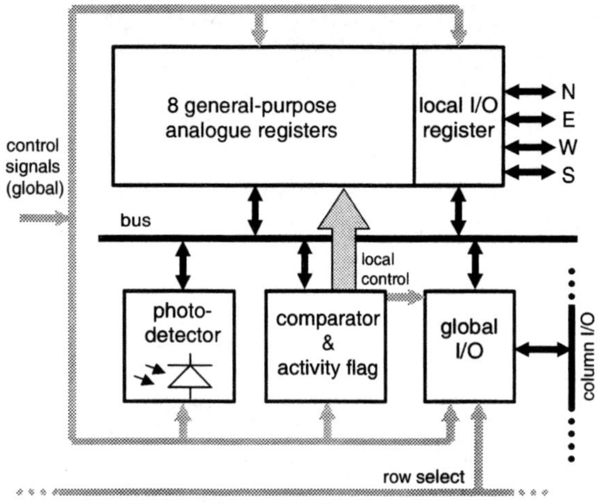

Fig.2. Architecture of the APE

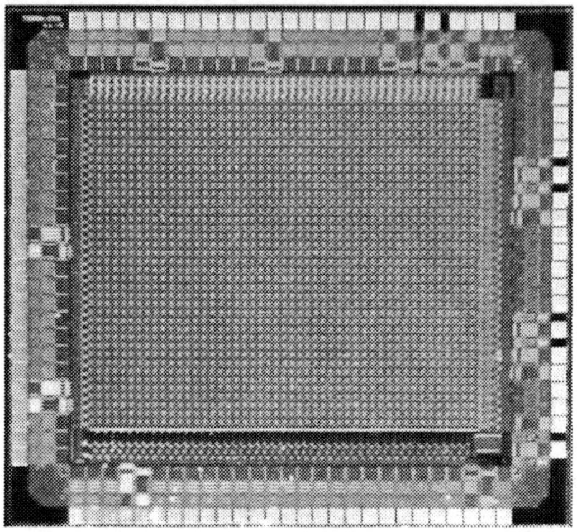

Fig.3. Chip microphotograph

2.1. ALU-free design

A primary function of any algorithmically-programmable processor (i.e. a Universal Turing Machine) is to perform a programmed sequence of arithmetic and logic operations on data stored in the memory. On each machine instruction cycle the arguments of an operation are taken from the memory (e.g. processor's registers), presented to inputs of an appropriately configured computational circuit (known to microprocessor designers as ALU - Arithmetic Logic Unit), and the result of the operation is stored back into the memory.

The design of the APE follows this principle. However, it is notable that no dedicated hardware exists to implement arithmetic operations. Instead, the basic operations of addition, inversion and division are executed directly in the registers/analogue bus system, by means of switched-current signal processing techniques [5]. This results in significant silicon area savings (no need for ALU hardware). Consequently, it can also improve the accuracy of processing - increasing the size of transistors in the registers (in the present implementation, the analogue registers occupy almost 70% of the APE area) reduces the errors associated with analogue storage, and at the same time improves the transistor matching required, for example, to achieve accurate multiplication coefficients for convolution kernels, etc.

It has to be noted, that in this "minimum hardware" APE, the silicon area and accuracy improvements are achieved at a cost of a simplified instruction set and therefore somewhat reduced processing speeds. However, since the typical low-level image processing algorithms are rather simple, this massively parallel system easily achieves performance adequate for real-time computer vision.

3. IMPLEMENTATION

The chip has been implemented in a 0.35μm single-poly 3-metal layer CMOS technology. The chip microphotograph is shown in Figure 3. The 39×48 processor array is surrounded by the control-signal distribution network and the read-out circuitry. The total chip area, including I/O pads, is equal to $10mm^2$. (The chip is a scaled-down version of a planned 128×128 array). The APE cell size is equal to 49.35μm×49.35μm. The array operates at 1.25MHz clock, with 2.5V supply voltage and maximum power consumption of 12μW per cell.

3.1. Control signals

Control signals, which are used to configure APEs in the array, are derived from instruction-code-words that have to be provided to the chip on each clock cycle. This means, that the controller (sequencer), issuing instructions to the SIMD array, is located off-chip. This is a preferred solution for a proof-of-concept research design, as a suitable controller can be easily implemented on a programmable-logic device (or replaced by lab instruments as required). Similarly, all analogue bias voltages are provided externally.

Control signals are routed horizontally and vertically, over the processor array. A complete set of control-signal drivers exists for each row/column of the array. Global analogue biasing voltages (all of them driving high-impedance nodes) are routed along other control signals. Great care has been taken during the layout stage to ensure that the coupling between digital control signals and sensitive analogue circuit nodes is minimized where necessary. It has to be said, that using a 3-metal layer technology is somewhat restrictive, especially when it comes to providing good (low impedance) power supply rails, and while the results have been satisfactory for this implementation it is expected that the availability of another metal layer would further improve the noise performance of the circuits (especially if the array size is scaled up).

3.2. Read-out

The peripherial circuitry includes the Flexible Global Readout Architecture, described in more detail in [6]. It permits addressing individual APEs in the array (for random-access readout) as well as groups of the APEs, to facilitate certain global operations. It is expected that in typical high-frame-rate applications reading-out of full gray-scale images will not be necessary. Instead, some image "descriptors" or features will be obtained from the sensor/processor array system (e.g. coordinates of targets, counts of objects, etc.). Nevertheless, a mechanism for column-parallel read-out of binary images (i.e. processing results, such as an edge map, etc.) is also provided, via an 8-bit output port. An analogue read-out of gray-scale images (or data arrays corresponding to processing results) is also possible. The chip also accepts an analogue input data. There are no on-chip A/D or D/A converters in the present implementation. Instead, off-the-shelf components are used in the test system. Again, this option has been chosen to simplify the prototype and allowed to concentrate the chip design effort on the more innovative parts of the system.

3.3. Image Sensor

An image sensor is embedded within the processor array on a pixel-per-processor basis. An n-diffusion diode is used as a photodetector. The pixel fill-factor is approximately 5.6%. The design of the image sensor is fairly simple and has not been optimized for optical performance. However, care has been taken to minimize the mismatch by appropriately scaling the transistors. The measured fixed-pattern noise is equal to 0.4% (rms). The photodetector can operate in the integration mode – the integration capacitor is formed by the inherent capacitance of the photodiode and the gate capacitance of an n-MOS transistor, which is used (in a cascode configuration) to provide voltage-to-current conversion. The output current values are matched to the analogue register range (0 to 1.7μA). The sensitivity of the photodetector can be regulated, by changing the reset voltage and the transconductance of the transistor stage. The light-to-current characteristic can be close-to-linear for dark pixels and compressed (quadratically) for bright pixels. Additionally, the photodetector can operate in a logarithmic-compression, continuous mode.

The metal layers used for routing of control-signals and power supply voltages are also used as an optical shield, to reduce the incident light in the processing part of the APE, while providing an opening over the photodiode area to maximise the light sensitivity of the photosensor. As a result, the measured light-induced leakage in the analogue registers is 50 times below the sensitivity of the photodetector (i.e. if full-scale pixel brightness value is obtained after 10 ms integration time, for instance, then during this time the analogue value held in a register will change by 2% of the maximum value). This is an acceptable value, particularly since usually the integration time is much longer than the required retention time of an analogue value. However, the leakage error has to be taken into account if data is to be retained for longer periods (e.g. a few video frames). Furthermore, it has to be noted that just like any other focal-plane processor array, the chip could be "blinded" by a strong light source, leading to unpredictable processing results. This fact should be detected by the vision chip system, if it is required to work in an uncontrolled environment.

4. ACCURACY

The registers in the APE are implemented as S^2I memory cells (with additional features allowing to conditionally disable current-storage operation as well as switch-off biasing currents to save power when the registers are not being accessed). The storage of data in analogue memories, however, is not error-free. Apart from the fact that leakage currents lead to a change in the value of the stored sample, as described above, there are also errors associated with the sampling process (clock feedthrough and charge injection from MOS switches, output conductance effects) as well as noise. The arithmetic operations performed using analogue circuits are not perfectly accurate either. Furthermore, these errors are accumulated as the data is repeatedly sampled and processed throughout the execution of an algorithm. It is therefore of paramount importance, that the analogue errors are kept at an acceptable level.

The measured error of the single transfer operation (i.e. loading one register with the analogue value stored in the other register) reveals a signal-independent error of 56nA, which amounts to 3.3% of the maximum signal level of 1.7μA. However, it can be shown [4] that this error can be easily compensated algorithmically (each storage performs signal inversion, thus after two operations the offset error cancels out). More important is the signal-dependent error of the transfer operation. Its magnitude is equal to 6nA (i.e. 0.35%). Register-to-registers variations of the error, which are the most important as they will determine the fixed-pattern noise in the processing result, are equal to 0.042% (rms). Finally, there is a random noise associated with each transfer operation, and this was measured to be equal to 0.092% (rms).

It is not sufficient, though, to consider the accuracy of analogue memories only. The accuracy of arithmetic operations is also very important. An addition operation relies on current summation and is accurate, apart from a small error (similar to the transfer operation error), which can be attributed to the limited accuracy of the analogue memory cells. The division operation, on the other hand, is inaccurate, as it relies on matching between the transistors in two or more registers. In the APE, the fixed-pattern noise of a straightforward division-by-two instruction (performed by dividing a current between two memory cells) has been measured to be equal to 2.32%. However, this can be significantly reduced, to below 0.2%, with a five-step algorithm based on an error compensation technique described in [7]. Overall, it is expected that the achieved levels of accuracy should be sufficient for robust implementation of a majority of low-level image processing algorithms. It also should be noted, that due to the "dividing in registers" concept, the mismatch problems associated with our approach are inherently smaller than ones present in CNN-based processors [3] (which require 9 multipliers, in addition to memory cells, to be fitted into the limited processor area).

5. IMAGE PROCESSING EXAMPLES

The SCAMP-2 architecture is that of a software-programmable general-purpose processor array for which a variety of image processing algorithms can be developed. Figure 4 includes examples of two simple algorithms, which have been implemented and executed on the SCAMP chip. The edge detection result is the sum of the absolute values of two images, each obtained by convolving the input image with a 3×3 Sobel convolution kernel (horizontal and vertical). The median filtering is performed in a 3×3 neighbourhood. For comparison, results of the same algorithms in an ideal case (calculated from the input images using numerical computation) are shown in Figure 4. The overall error is due to the accumulation of analogue processing errors and noise. The total rms difference between experimental and ideal results (excluding borders) is 2.1% for edge detection and 1.2% for median filtering. The images were obtained at 25 frames per second (fps) with processor array operating with a 1.25MHz clock. The processing times for the two algorithms are as follows: 30.3μs for edge detection and 157μs for median filtering. This, potentially, enables operation at several thousands fps. It is also worth noting, that the total power consumption in the processor array operating at 25 fps is only 73nW per pixel for the edge detection algorithm and 380nW per pixel for the median filter algorithm. To the best of our knowledge these power consumption figures are lower for our chip than can be achieved by any other processor.

6. CONCLUSIONS

A general-purpose programmable vision chip with a 39×48 SIMD processor-per-pixel array has been designed, fabricated, and tested. The concept of a "minimum hardware analogue microprocessor" allowed the size of the APE to be reduced so that cell density of 410 cells/mm^2 was achieved. At the same time, good accuracy of processing (particularly in terms of fixed-pattern-noise) has been achieved. The speed of the processing is appropriate to execute early vision algorithms in high frame-rate applications, while the extremely low power consumption makes the chip suitable for low power applications.

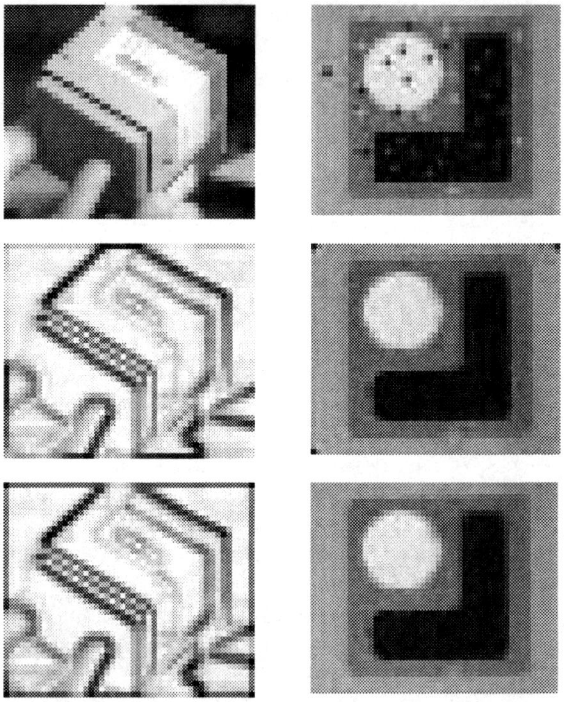

Figure 4. Results of image processing on SCAMP-2: left: edge detection, right: median filter. top: original images; middle: results obtained on chip, bottom: comparison with ideal (numerical) processing results.

ACKNOWLEDGEMENT

This work has been supported by the EPSRC under grant no. GR/R52688/01

REFERENCES

[1] M.Ishikawa et. al. "A CMOS Vision Chip with SIMD Processing Element Array for 1ms Image Processing", Proc. ISSCC'99, TP 12.2, 1999.

[2] F. Paillet, et. al., "Making the most of 15kλ^2 silicon area for a digital retina", Proc. SPIE, vol. 3410, AFPAEC'98. 1998.

[3] G. Liñán-Cembrano et al. "A Processing Element Architecture for High-Density Focal-Plane Analog Programmable Array Processor Arrays", Proc. ISCAS 2002, vol.III, pp.341-344, May 2002.

[4] P.Dudek and P.J.Hicks, "An Analogue SIMD Focal Plane Processor Array", IEEE International Symposium on Circuits and Systems, ISCAS 2001, vol.IV, pp.490-493, May 2001.

[5] P.Dudek, "A Processing Element for an Analogue SIMD Vision Chip", Proc. European Conference on Circuit Theory and Design, ECCTD'03, vol.III, pp.221-224, September 2003.

[6] P.Dudek, "A Flexible Global Readout Architecture for an Analogue SIMD Vision Chip", Proc. ISCAS 2003, May 2003.

[7] J.-S. Wang and C.-L. Wey, "Accurate CMOS Switched-Current Divider Circuits", Proc. ISCAS'98, vol I, pp.53-56, May 1998.

AN IMPROVED ANALOG COMPUTATION CELL FOR PARIS II, A PROGRAMMABLE VISION CHIP

Sébastien Moutault, Hervé Mathias, Jacques-Olivier Klein, Antoine Dupret

Institut d'Electronique Fondamentale, Bâtiment 220, Université de Paris 11, 91405 Orsay Cedex

Abstract

An improved analog programmable computation cell is addressed. Based on an original methodology, the analog processor features high computational performance along with a reduced power consumption and die area. Simulations validate the concept. Comparisons with state of the art digital processors are given.

1. Introduction

The present trend in vision chips is to make them programmable and to simplify their interface so as to integrate them into complete systems [1-4]. This dramatically increases their complexity with regard to the retinas of first generation [5]. Mainly intended for embedded applications, these SoCs have to face constraints of both reduced power consumption and processing time.

For low-level image processing, the accuracy of the calculations is rarely crucial, typically 7 to 8 bits. Therefore, analog based computational systems may be used. They offer high compactness and low power consumption at the expense of a more tedious design [1-4] than their purely digital counter parts.

Based on the use of a programmable analog computation cell, our approach drastically reduces the number of functional units [6] present in a vision chip. Our methodology may ultimately lead to the automation of the analog processing circuit design [7].

Using this methodology, we designed a programmable vision chip: PARIS II, which will be described at first. We shall then detail the key element of this architecture: the programmable computation cell. We shall end by the presentation of its performance and their comparison with digital, state of the art, systems.

2. Global architecture

PARIS II is a SIMD machine. It includes a NxM photosensor's array to which an array of Nx(k M) memory points is associated [8], where k is the number of memory elements per pixel. The so formed (NxM+Nx(k M)) matrix is bordered, on one side, by a vector of N processors. A column of multiplexers selects the column of pixels or of memories to be used by the processor (cf. figure 1). A sequencer, implemented for example by a digital IP CPU, delivers the successive processors' instructions.

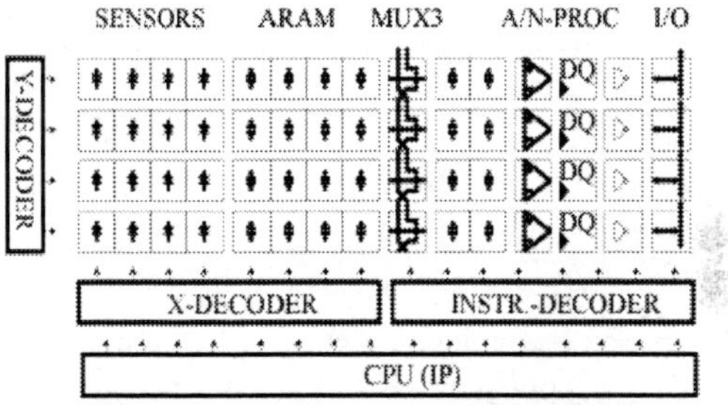

Figure 1.: Architecture of PARIS II

The processor is a switched capacitor analog computing unit. A first version has been implemented in PARIS I, a particularly compact programmable vision chip [2, 9]. This circuit is able to implement the main algorithms encountered in low-level image processing. Its basic operators are the conditional Multiplication ACcumulation (MAC) and the comparison [9]. Because of the small area of PARIS I's processor, no error compensation has been incorporated. A precision of 6 to 7 bits per calculation is achieved. It is sufficient to perform operations such as spatial filtering by a linear convolution kernel [9]. On the other hand, errors in calculation become critical for highly iterative calculations (e.g. stochastic relaxation, implementation of CNNs...). Besides, PARIS I requires (2N+1) clock cycles to complete a N-bit MAC.

Derived from PARIS I processor, we so designed a new computation unit that speeds up calculations while increasing their accuracy. Higher performance is obtained by increasing the clock frequency and reducing the number of phases required for a MAC. Precision, mainly lessened by charge injections, clock feedthrough and amplifiers offsets, is improved by an original structure that we now detail.

0-7803-8251-X/04/$17.00 ©2004 IEEE

3. New computation cell

In a switched capacitor structure, the ultimate calculation precision is limited by the achievable accuracy on the capacitor ratios (in the order of 0.1 %). This constitutes the higher error level limit for the rest of the circuit. The design of our cell, and especially the switches topology, has been aimed at finding a compromise between accuracy, compactness and speed. The structure of our new switched capacitor processor is given figure 2.

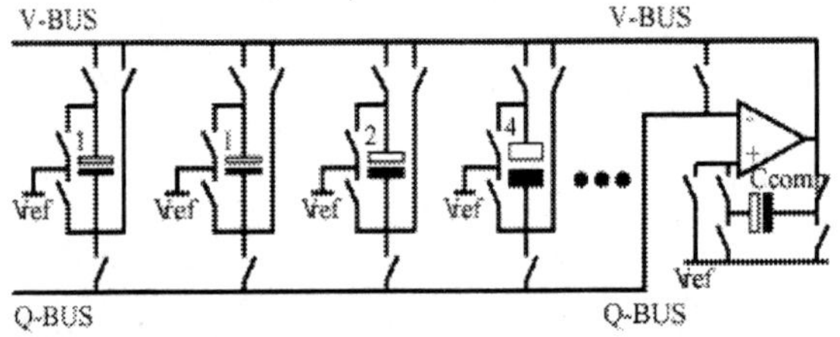

Figure 2.: Simplified schematic of the processing unit

Table 1 summarizes the various switches configurations corresponding to the available operations. Each processor instruction fixes at the same time the switches configurations for the OTA and for the associated analog registers. The MAC of an analog value by a 4-bit constant now only requires 4 clock cycles with the following sequence of operations: writing on the left-most capacitor, clearing the other (weighted) capacitors and setting them into the feedback loop of the OTA, accordingly to the coefficient of the MAC.

For a division by 2, as described in the example given section 4, only the capacitor with a weight of 1 is put in the feedback loop. This configuration results in charge balancing between the 2 capacitors, which carries out the division.

Label	Operation name	o_swXCm[4:0]
ARnop	No Operation	00000
ARfdb	Feedback	11000
ARwtp	Write Positive	01001
ARwtn	Write Negative	00110
ARtrf	Transfer	10010
ARcmp	Compare	10010
ARrst	Reset	00011
Label	Operation name	o_swXOTA[5:0]
OTAnop	No Operation	100101
OTAcmpMode	OTA in compare mode	100100
OTAsOffset	OTA offset sample	011011
OTAcOffset	OTA offset compensation	100101

Table 1 : switches configuration for the processors

An additional clock cycle is required to store the obtained value in an analog register.

A new strategy has been developed to lower the impact of parasitic capacitances found on buses and POLY/POLY capacitors. Indeed, the classical solutions [10, 11] cannot be used directly here since our capacitors may be charged either positively or negatively.

To reduce the influence of the POLY/BULK parasitic capacitances, every writing operation (negative copy, positive copy, transfer) performed on the bottom plate of a computing capacitor is made using a low impedance source. Moreover, this plate is never used to transfer charges to an other capacitor (transfer +/- operations) but is stuck at a constant voltage during such operations.

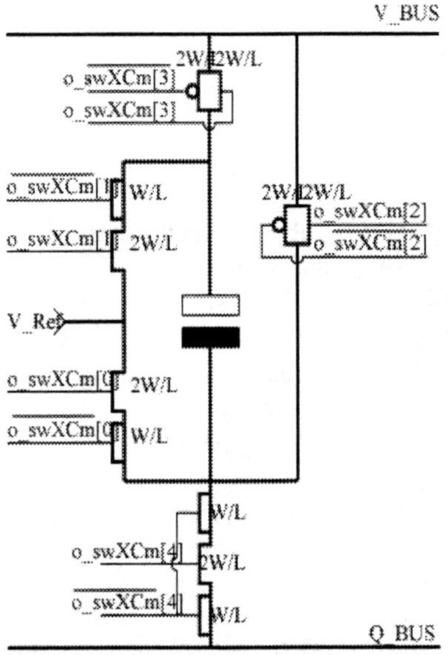

Figure 3.: Analog Register and its associated switches

The non-inverting input of the amplifier is also maintained at a constant voltage, which ensures that the inverting input voltage experiences only small variations due to the finite gain amplifier. The corresponding parasitic capacitance interference is thus reduced.

To perform clock feedthrough compensation, active switches on both sides of the computing capacitors are driven as often as possible by the same clock signal. In this way, the parasitic charges injected on each plate are really close. Hence the stored voltage is not modified.

Charge injection compensation is obtained either by using dummy transistors for NMOS switches [10] or by a correct sizing of the complementary transistors in CMOS switches. For example, switch o_swXOTA[5] is a NMOS switch with only one dummy transistor connected to the high impedance net. The other side of the switch is directly connected to the low impedance voltage reference source.

All the switches are implemented with minimum size transistors. An optimal value of 200fF has been found for the computation capacitors in order to lessen the effect of the uncompensated clock feedthrough and so achieve the requested precision performance. The corresponding kT/C noise, lower than 150µV, is negligible.

An OTA's offset compensation phase has also been included (figure 4 – operations "OTAsOffset" followed by "OTAcOffset"): the amplifier is at first configured as a voltage follower and its offset is stored in its load capacitor "Ccomp". This capacitor is then reversed and connected to the non-inverting input to perform the offset compensation. This phase can be performed only once.

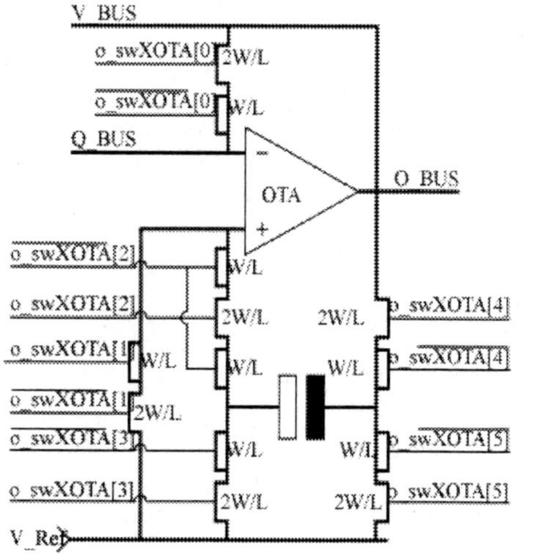

Figure 4.: OTA offset sampling and compensation

The amplifier has been designed as a compromise between area and performance: given the reserved area for the whole analog computing cell, the pixels size and pitch, the amplifier has to fit in a 50µm x 50µm square. The aimed analog computation performance is 40MIPS with a 0.1% precision for a 0-4V voltage range on a 2pF load. The amplifier characteristics are given Table 2.

Static gain	64 dB
Slew-Rate	130 V µs^{-1}, 220 V µs^{-1}
Gain Bandwidth Product	81 MHz on a 2 pF load
Phase Margin	69° on 2 pF load
Output Swing	0-4 V
Power	30 µA, 5 V supply

Table 2 : main characteristics of the OTA

4. Simulation results and performance

The whole architecture has been designed in a 0.6µm CMOS technology. Simulations, with typical mean conditions, were made on the extracted layout of the computing cell to estimate the performance in terms of speed and precision. Figure 5 shows a MAC performed on an input varying from 0 to 4V, by steps of 250mV, by a coefficient set to 0.5 by program.

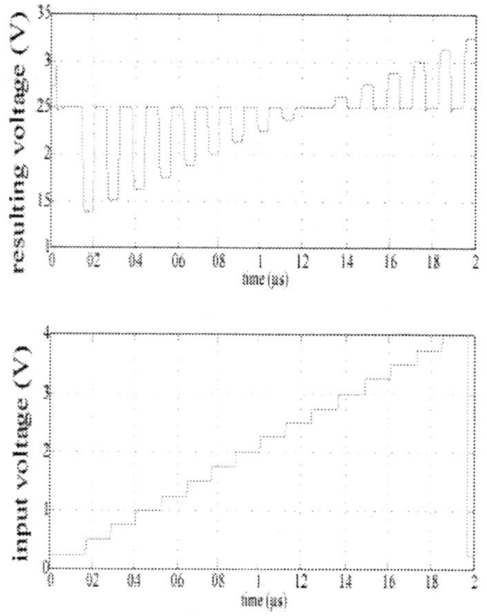

Figure 5.: Input and output simulation results for a MAC operation. A 0.5V coefficient is applied to the input voltage

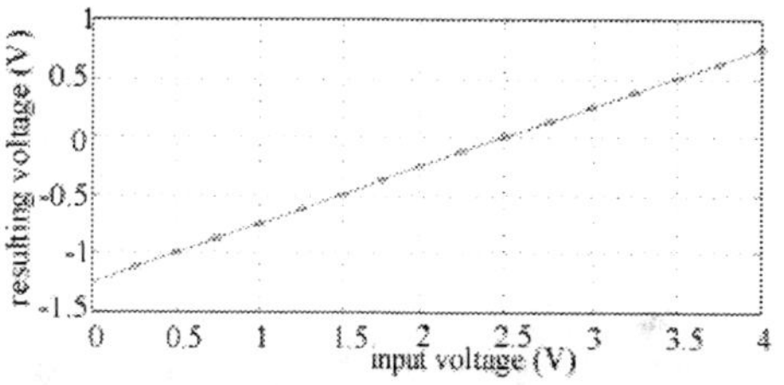

Figure 6.: Simulated output voltages (referenced to 2.5V) and the corresponding linear regression curve

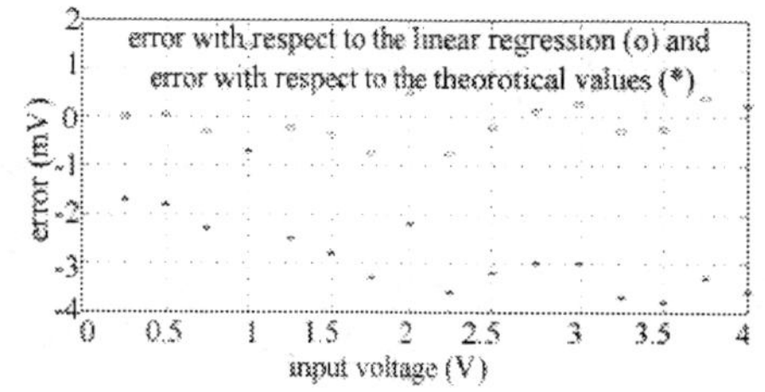

Figure 7.: Errors obtained on the simulated results

The curve, given figure 5, shows the measured points and the corresponding linear regression. The linearity error and the absolute error are given figure 6. The latter is less than 4mV. Our dynamic range being in a 0 to 4V range, the obtained precision is thus 10 bits. The last curve (figure 8) represents the difference between the two calculated errors. It shows that the error found on our results comes from two different sources: an input dependent error due to charge injection and a constant offset mainly caused by clock feedthrough for certain configurations of switches.

The simulations also show that a MAC is performed in 4 clock cycles, for a 40MHz clock. Since PARIS II is composed of 64 analog computing cells working in SIMD mode, the global performance of our architecture is 640MMAC·s^{-1}. The simulations have also shown that the power consumption is of 0.2W.

The global layout of PARIS II is given figure 9. Its area is 1.6mm², with k=3, N=64 and M=32.

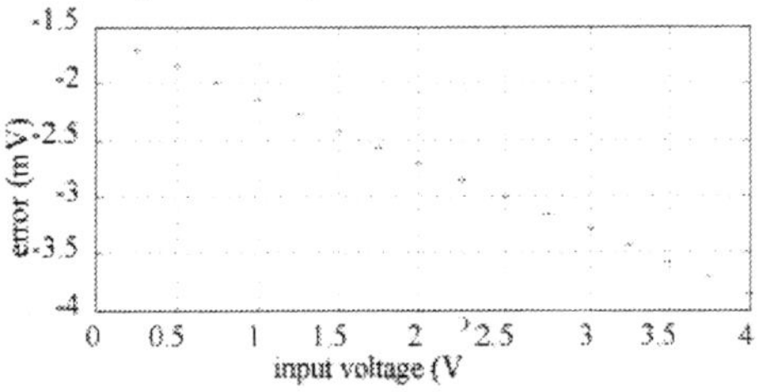

Figure 8.: Measured difference between the 2 precedent errors

Table 8. compares PARIS II with other computing architectures. PARIS II obtains the best performance at a low power cost but with a precision of 10 bits (compared to the 32 bits of PIV and NM6403 which, on the other hand, have a much higher power consumption and larger die area). Last generation DSPs (e.g.: TMS320C6415) have slightly higher performance (1200MMACs^{-1}) but a power consumption of 2W and are designed in 0.13μm CMOS technology.

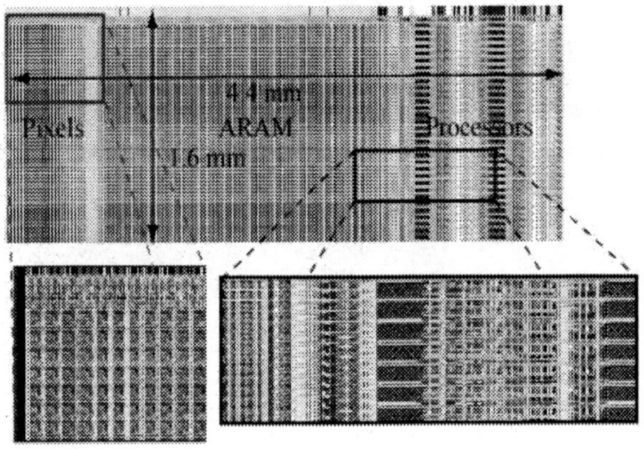

Figure 9.: Layout of PARIS II

5. Conclusion

A new switched capacitor analog computing cell has been designed to improve the performance of our low-level image processing architecture in terms of speed and precision. The results obtained in simulation show that the global architecture constitutes a good trade-off between cost, power consumption and computational performance. It is particularly well suited for embedded vision applications even when highly recursive algorithms are implemented.

	P. IV	68HC12	NeuroMatrix NM6403	PARIS II
Power (W)	58	0.25	1,6	0.2+0.2
Clock (GHz)	2.4	0.025	0.04	0.04
Area (mm²)	131	~3	~50	1,6+6
MMAC s^{-1}	350	0.69	40	640
Techno.(μm)	0.13	0.25	0.5	0.6

Table 3 : main characteristics of some conventional processors versus PARIS II

6. References

[1] Dudek, P. and J. Hicks. A CMOS General-Purpose Sampled-Data Analogue Microprocessor. in Pro. of the 2000 IEEE International Symposium on Circuits and Systems. 2000. Genève, Suisse.

[2] Dupret, A., J.-O. Klein, and A. Nshare. *A programmable vision chip for CNN based algorithms.* in *CNNA*. 2000. Catania: IEEE 00TH8509.

[3] Etienne-Cummings, R., Z. Kalayijan, and D. Cai, *A programmable focal-plane MIMD image.* IEEE Journal of Solid State Circuits, 2001. **36**(1): p. 64-73.

[4] Linan, G., et al. A 0.5μm CMOA 10^6 Transistors Analog Programmable Array Processor for Read-Time Image Processing. in Proc. of 1999 Europeen Solid-State Circuis Conference. 1999.

[5] Moini, A., *Vision Chips*. Kluwer Academic Publishers, ed. I. 0-7923-8664-7. 2000.

[6] Klein, J.-O., et al. A Universal Switched Capacitor Computation Cell

Applied to a Programmable Vision Chip. in ECCTD'03. 2003. Cracow: EALiE AGH-UST.

[7] Moutault, S., J.-O. Klein, and A. Dupret. A universal switched capacitors operator for the automatic synthesis of analog computation circuits. in IEEE-CAMP2003. 2003. New Orleans, LA: IEEE press.

[8] Nshare, A., J.O. Klein, and A. Dupret. Improved ARAM for PARIS, an Original Programmable Vision Chip. in CNNA2002. 2002. Frankfurt.

[9] Dupret, A., J.-O. Klein, and A. Nshare, *A DSP-like analog processing unit for smart image sensors.* International Journal of Circuit Theory and Applications, 2002. **30**: p. 595-609.

[10] Eichenberger, C. and W. Guggenbuhl, *On Charge Injection in Analog MOS Switches and Dummy Switch Compensation Techniques.* IEEE Transactions on Circuits and Systems, 1990. **37**(2): p. 256-264.

[11] Wegmann, G., E. Vittoz, and F. Rahali, *Charge Injection in Analog MOS Switches.* IEEE Journal of Solid State Circuits, 1987. **22**(6): p. 1091-1096.

A CNN-DRIVEN LOCALLY ADAPTIVE CMOS IMAGE SENSOR

R. Carmona, C. M. Domínguez-Matas, J. Cuadri, F. Jiménez-Garrido and A. Rodríguez-Vázquez

Instituto de Microelectrónica de Sevilla-CNM-CSIC
Campus de la Universidad de Sevilla. Avda. Reina Mercedes s/n,
41012-Sevilla, Spain. E-mail: rcarmona@imse.cnm.es

ABSTRACT

A bioinspired model for mixed-signal array mimics the way in which images are processed in the visual pathway. Focal-plane processing of images permits local adaptation of photoreceptor structures in silicon. Beyond simple resistive grid filtering, nonlinear and anisotropic diffusion can be programmed in this CNN chip. This paper presents the local circuitry for sensors adaptation based on the mixed-signal VLSI parallel processing infrastructure in CMOS.

1. INTRODUCTION

The retina operates on the captured visual stimuli at early stages in the process of vision. Complex spatio-temporal processing encodes visual information into a reduced set of channels [1] to be delivered to the brain by the optic nerve. This model inspires a feasible alternative to conventional digital image processing. We are interested , in particular, in local monitoring and control of the photosensing devices for contrast enhancement. This capability improves the perceived sensation by extracting the reflectance information from the acquired luminance matrix [2]. This task is gracefully implemented in the biological retina. Concurrent sensing and massively parallel processing provides enough computing power to realize these tasks in mixed-signal VLSI. This paper presents a method for locally adapting integrating photosensors exposure time to the illumination conditions that can be implemented in an already functioning CNN chip architecture based on the mammalian retina.

2. BIOINSPIRED CNN PROCESSOR

Visual stimuli trigger the formation of patterns of activation in the retina. These patterns are processed as they advance towards the optic nerve. Contrarily to the spike-like coding of neural information found elsewhere, they are continuous-time analog waves [3]. The biological motivation is the lack of bandwidth offered by the spike-like neural impulses to handle the data contained in the visual stimulus. The captured signals are promediated and the high-gain characteristics of the cones and the bipolar cells are shifted to adapt to light conditions. These operations have a local scope and depend on the recent history of the cells. Once adaptation is achieved, patterns of activity are formed dynamically.

A CNN model has been developed that approximates the observed behaviour of different parts of the retina [4]. This model has been implemented in a 2-layer CNN chip, designed and fabricated in a standard 0.5μm CMOS technology [5]. It contains a central array of $32 \times 32 \times 2$ processing nodes arranged in a 2-layer structure. The evolution of each node, $C(i, j, k)$, in layer k, is described by:

$$\tau_k \frac{dx_{n_{ijk}}}{dt} = -g[x_{ijk}(t)] + b_{00k}u_{ijk} + z_{ijk} + \\ + \sum_{l=-r_1}^{r_1} \sum_{m=-r_1}^{r_1} \sum_{l=-r_1}^{r_1} a_{lmn} f[x_{i+l,j+m,k+n}(t)] \quad (1)$$

where $f()$ and $g()$ are nonlinear functions of the state variable, $x_{ijk}(t)$. Each layer incorporates intra- and interlayer feedback and feedforward connections, a_{lmn} and b_{lmn}, a bias term z_{ijk}, and its own time constant τ_k. Programming different dynamics in this CNN model is possible by adjusting these parameters. Different reaction-diffusion equations can be mapped into this architecture, resulting in propagative and wave-like phenomena, similar to those found at the biological retina. Fig. 1 shows an effect observed in the outer plexiform layer (OPL) of the retina and programmed in the chip —the detection of spatio-temporal edges followed by de-activation of the patterns—, sampled at different points in the evolution of the network dynamics [5].

Apart from the two different programmable CNN nodes, the basic processing element contains local analog and logic memories, for the storage of intermediate results, and a programmable local logic unit. The basic processing element occupies 188μm $\times$ 186μm, resulting in a cell density of 29.24cells/mm^2 (not considering the control circuit overhead). The power consumption of the whole chip has been estimated in 300mW. The fastest time constant is designed to be under 100ns. The chip can handle analog data with an

Partially funded by ONR Project N-000140210884, CE Project IST-1999-19007 (DICTAM) and the Spanish MCyT Project TIC1999-0826.

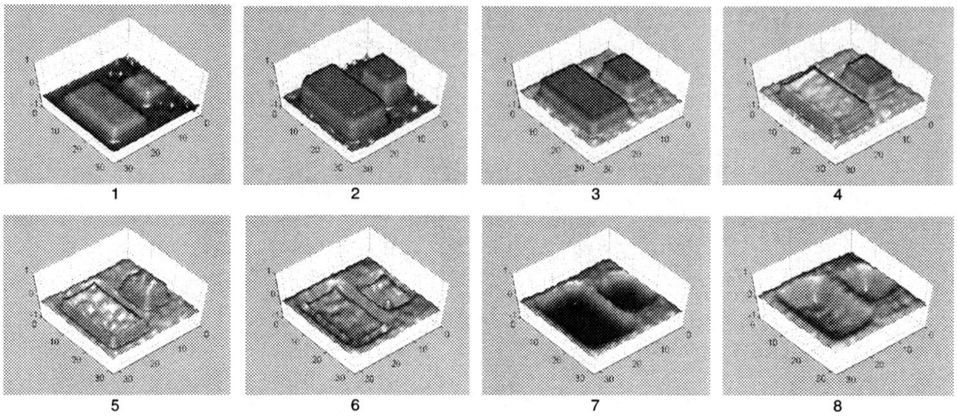

Fig. 1. Spatio-temporal edge detection and de-activation (fast layer).

equivalent resolution of 7.5bits (measured). In future versions we plan to incorporate adaptive control of the sensors based on focal-plane processing feedback.

3. ADAPTIVE OPTO-ELECTRONIC INTERFACE

Capturing light in CMOS technology counts on the separation of photogenerated electron-hole pairs by effect of an electric field, normally provided by a reverse-biased pn junction. Through a linear resistive load, the photocurrent, I_{ph}, generates an instantaneous voltage, proportional to the incident light power. Noise can be filtered out by integrating this current in the diode's parasitic capacitor C_p. Thus, for a given integration time t_{int}, there is a linear relation between the average incident light power density over the sensor, through I_{ph}, and the pixel voltage V_{ph}:

$$V_{ph}(t_{\text{int}}) = V_{\text{ref2}} - \frac{I_{ph}}{C_p} t_{\text{int}} \tag{2}$$

While CMOS photosensors have a maximum dynamic range of 5 to 7 decades, light intensity on natural scenes can vary over up to 14 decades. Thus, linear sensors usually produce images with over-exposed and under-exposed regions. To accommodate a larger light intensity range within the photodiode dynamic range, voltage compression is required.

The perceptual quality of an image is closely related with the ability of separating the irradiance $E(x,y)$, and the reflectance $\rho(x,y)$ —where most of the relevant information is—, both contained in the luminance signal $I(x,y)$. This task is continuously performed in the retina. As a consequence of Weber's law, the perception gain is inversely proportional to the local average of brightness, $\tilde{I}(x,y)$. If $E(x,y)$ is the main responsible of $\tilde{I}(x,y)$, then [2]:

$$\rho(x,y) \propto \frac{I(x,y)}{\tilde{I}(x,y)} \tag{3}$$

Based on the intra- and inter-frame correlation found in natural scenes, the local average voltage of a previous frame, $\tilde{V}_{ph}(n-1)$, can be employed to control the integration time of the next frame, $t_{\text{int}}(n)$, so as to establish a shorter integration time for strongly illuminated areas and a larger one for the areas lying in the dark. The required local information is provided by the CNN core circuitry in tens of nanoseconds. Fig. 2 depicts the schematic of the local adaptation circuit. Photogenerated currents are integrated in the sense capacitor. M_{reset} works as the electronic

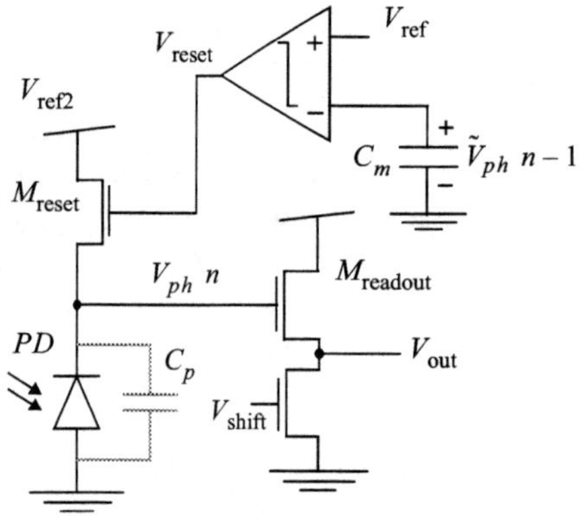

Fig. 2. In-pixel adaptation circuit.

shutter. While the local stored average value, $\tilde{V}_{ph}(n-1)$, remains below a global reference inverted ramp delivered to every pixel, V_{ref}, the photosensor voltage is shorted to V_{ref2}. When the ramp crosses the value of the stored average, the photodiode starts discharging C_p. The larger $\tilde{V}_{ph}(n-1)$, the sooner the inverse ramp crosses it, and the longer the integration time. Thus, image areas with a stronger average illumination, will have a shorter exposure time, while image areas lying in the dark will have a longer $t_{\text{int}}(n)$, following:

$$t_{\text{int}}(n) = t_{\min} + \frac{t_{\max} - t_{\min}}{V_{\text{ref2}} - V_{\min}} \tilde{V}_{ph}(n-1) \tag{4}$$

where $t_{\max}$ and $t_{\min}$ are the maximum and minimum integration times for each frame. Before the ramp restarts, V_{out} is sampled and $\tilde{V}_{ph}(n)$ is computed by the CNN array.

A high dynamic range image has been built with 11 snapshots of the same scene taken at different integration times. Fig. 4(a) shows a 256-level representation of this HDR image. Information is truncated by the limited DR. If it is re-captured with a simulated pixel array with local adaptation to the average illumination, some information is recovered (Fig. 4(b)). This result can be compared with logarithmic compression (Fig. 4(c)). The average brightness employed is obtained by linear diffusion, but it can be the result of a more involved computation: anisotropic diffusion, diffusion with controlled contours, etc [6].

Stability concerns arise if $\Delta t_{\text{int}} = t_{\max} - t_{\min}$ grows, leading to non-convergence of the series defined by Eq. 4. If the first term, $t_{\min}$ is made dependent on V_{ph}, this series converges for a larger range of Δt_{int}. The global adaptation circuit in Fig. 5 provides an average integration time that is a function of the total image average brightness $\overline{V}_{ph}(n-1)$. Then, if the difference between the maximum and the minimum is fixed to Δt_{int}, Eq. 4 can be rewritten:

$$t_{\text{int}}(n) = \bar{t}_{\text{int}}(n-1) - \frac{\Delta t_{\text{int}}}{2}[1 - \frac{2\tilde{V}_{ph}(n-1)}{V_{\text{ref2}} - V_{\min}}] \quad (5)$$

This is accomplished by a circuit providing a ramp signal whose period is proportional to $\overline{V}_{ph}(n-1)$, i. e. inversely proportional to the history of the average illumination. Fig. 6 shows several frames of an artificially distorted sequence. After a few frames, $\bar{t}_{\text{int}}(n)$ converges and a locally adapted capture takes place. Observing the evolution of $\bar{t}_{\text{int}}(n)$, it dinamically adapts to the average brightness, increasing when dark elementes enter the scene and decreasing if the image gains in brightness. When the image remains still $\bar{t}_{\text{int}}(n)$ converges to an optimum. Fig. 3 displays how tracking a fixed gray level evolves to a stable integration time, and to the same medium gray in all cases.

4. CONCLUSIONS

A simple but precise model of the real biological retina renders a feasible efficient implementation of an artificial vision device. The results of CNN processing can be employed to locally adapt the photosensors operation. The application of this adaptation method, based on the local average brightness of previous image frames, result in a perceptually enhanced image capture.

5. REFERENCES

[1] B. Roska and F.S. Werblin, "Vertical interactions across ten parallel, stacked representations in the mammalian retina," *Nature*, vol. 410, pp. 583–587, Mar. 2001.

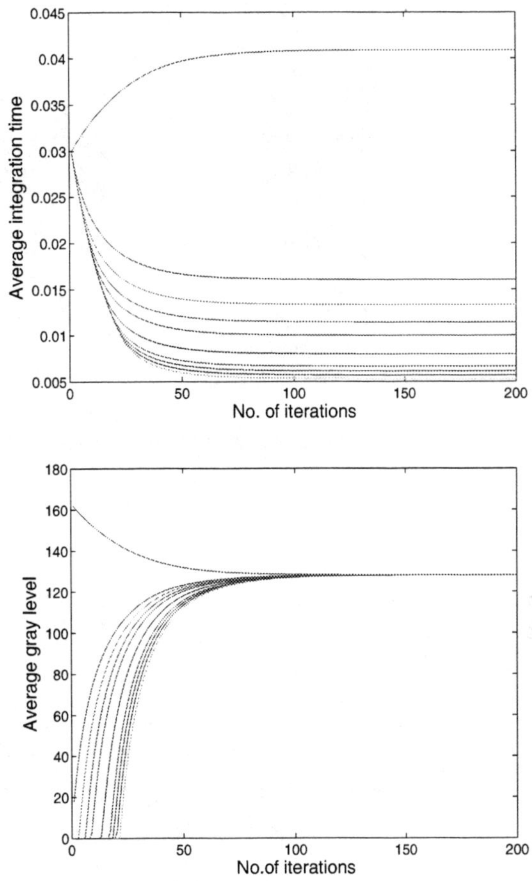

Fig. 3. Simulated capture of a plain gray picture.

[2] V. Brajovic, "A model for reflectance perception in vision," in *Bioengineered and Bioinspired Systems, Proceedings of SPIE*, May 2003, vol. 5119, pp. 307–315.

[3] F. Werblin, "Synaptic connections, receptive fields and patterns of activity in the tiger salamander retina," *Investigative Ophthalmology and Visual Science*, vol. 32, no. 3, pp. 459–483, Mar. 1991.

[4] F. Werblin; T. Roska and L. O. Chua, "The analogic cellular neural network as a bionic eye," *International Journal of Circuit Theory and Applications*, vol. 23, no. 6, pp. 541–569, Nov. 1995.

[5] R. Carmona et al., "A bio-inspired two-layer mixed-signal flexible programmable chip for early vision," *IEEE Transactions on Neural Networks*, vol. 14, no. 5, pp. 1313–1336, Sept. 2003.

[6] Cs. Rekeczky; T. Roska and A. Ushida, "Cnn-based difference-controlled adaptive nonlinear image filters," *International Journal of Circuit Theory and Applications*, vol. 26, pp. 375–423, July 1998.

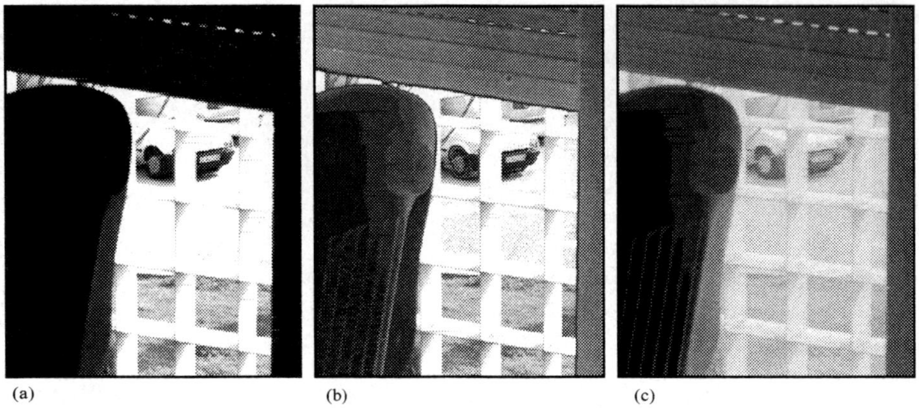

Fig. 4. Simulated capture of a still picture with local adaptation (center).

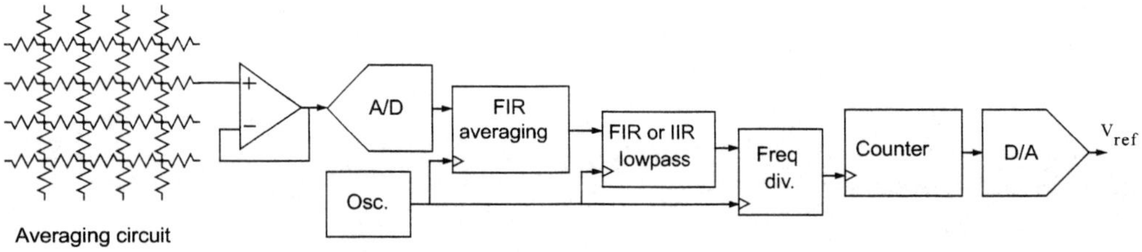

Fig. 5. Conceptual schematic of the global adaptation circuit.

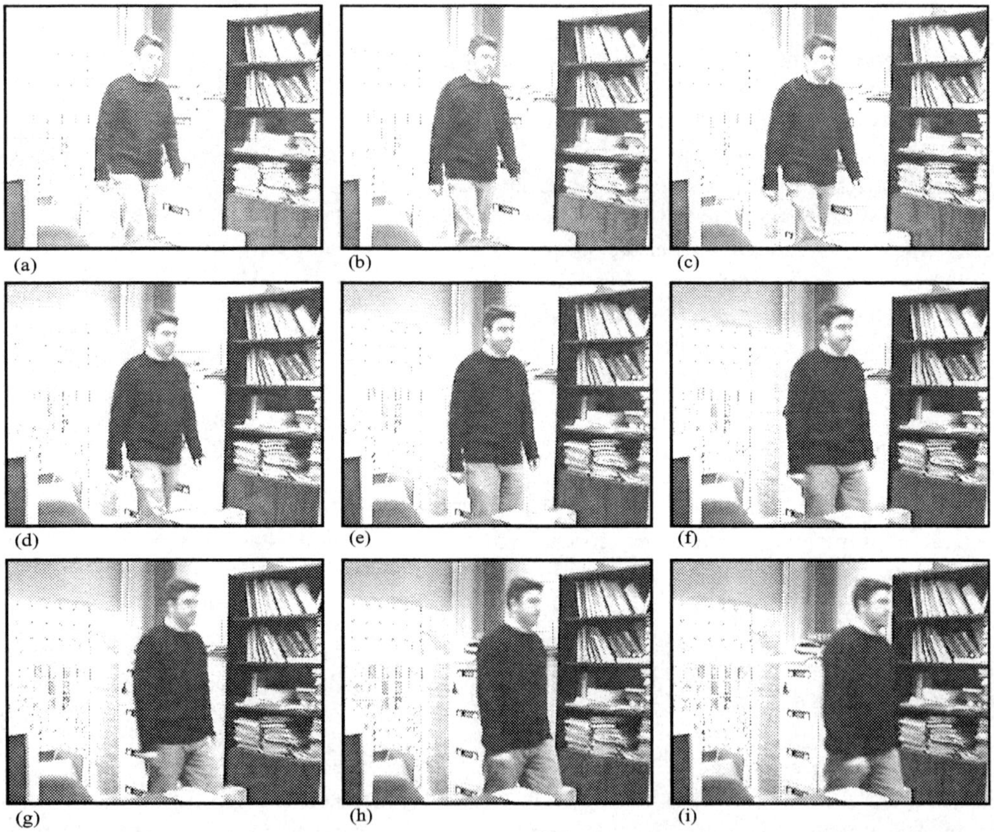

Fig. 6. Simulated adaptation in a sequence taken at 25fps, 1 out of each 3 frames shown.

N $\times$ 16 CELLULAR TEST CHIPS FOR LOW-PASS FILTERING LARGE IMAGES

A. Kananen, M. Laiho, K. Halonen

Electronic Circuit Design Laboratory
Helsinki University of Technology
Otakaari 5 A, 02150 ESPOO, Finland
Asko.Kananen@ecdl.hut.fi

A. Paasio

Laboratory of Microelectronics
University of Turku
Lemminkaisenkatu 14-18 B
20520 TURKU, Finland
arjupa@utu.fi

ABSTRACT

In this paper, measurement results of a test chip realizing low-pass filtering parallel processor are given and some of the observed problems are discussed. The chip was designed to test the implementation of a network processor where the size of the network is smaller than the actual image size. In this processor it is possible to write in, calculate the output and read out the results simultaneously. In this way savings can be achieved in silicon size, power consumption and processing time. Measurement results indicate that the system is functional in principle and is capable of processing thousands of images in a second. But, also several problems were found. A new design is introduced where the problems found in the first design are taken into account and a larger grid size chip was implemented. The goal of the new chip is to investigate the problems in designing this kind of a system in a larger scale. Some preliminary measurement results were also available at the time of writing and they were included in the article.

1. INTRODUCTION

The ever increasing need for more calculation power from processors, especially in the field of image processing, has lead to several new processor architectures. One of these architectures is the cellular nonlinear network (CNN) paradigm [1]. In [2] a CNN universal machine (CNN-UM) was introduced, where memory and local logic were included inside the cells of the locally connected network. So far the largest implemented silicon realization has been a chip [3] where the processor grid consisted of 128×128 parallel processors. However, the reached grid sizes have been relatively small when compared to the used resolutions, for instance in video processing, where the smallest standardized image size is QCIF with 176×144 pixels. In many applications some of the features of a CNN-UM are not needed and the programmability can also be reduced. Therefore, in many cases an application specific array processor/processors could lead to an area-saving result where there would also be room for the additional digital circuitry to link the processor to the outside world and the array processor would only do the tasks that require the enormous processing power of the parallel system. When designing application specific processors everything can be optimized for the application and in this way savings can be achieved also in power consumption. This kind of an approach was introduced in [4].

This work was supported by the Academy of Finland

In this paper measured results are given for a test chip that was manufactured to test the methods introduced in [4] [5] on silicon. After briefly showing some of the measurement results and problems with the first test chip, a new test chip is introduced where the goal is in exploring the problems in large arrays.

2. BASIC ASSUMPTIONS OF THE PROCESSING TASKS

In the paper [4] problems in designing large array processors were discussed and a method to reduce the number of cell rows in the array was proposed. With the method the speed of the processing was not lost but the power consumption and the silicon area were reduced. However, the method is not generally compatible with all the processing tasks. It is valid, however, every time for at least two different cases: first, if there is feedback and the processing is linear, and second, if there is no feedback and the network is connected in a feedforward manner only. In this section the basic idea of the method will be presented and the system for linear low-pass filtering is also shown.

The method is based on the fact that for the above mentioned cases the information of the whole image to be processed is not needed all the time to get the desired output. In the linear case with feedback the network has only one energy minimum and the influence of the neighboring cells reduces the further they are from the cell that is monitored. And for the feedforward case, only the neighboring cells have any influence on the output of the monitored cell.

If the size of the processing network is reduced in the manner that was described in [4], in order to process images larger than the network, there has to be a possibility to control the connectivity of the first and the last physical cell row. This basically means that when the evaluation phase is at the top or at the bottom of the image, the first and last rows have to be connected to the border cells, respectively. In the phase of evaluating image in the middle rows the physical top and bottom rows are connected to each other forming a circularly connected array. In this way it is possible to give the first and the last rows similar neighborhood as they would have in the full image size network, for example zero-flux. And in the middle of the image these same cell rows are connected as if they were neighboring cell rows.

3. LOW-PASS FILTER REALIZATION

Low-pass filtering is used in many image processing tasks, for example in reducing the errors in transmission or in reducing the

0-7803-8251-X/04/$17.00

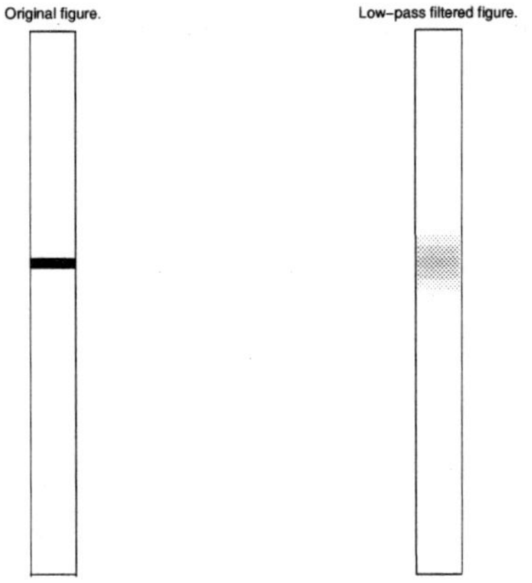

Fig. 1. Simulated sphere of influence in one dimension.

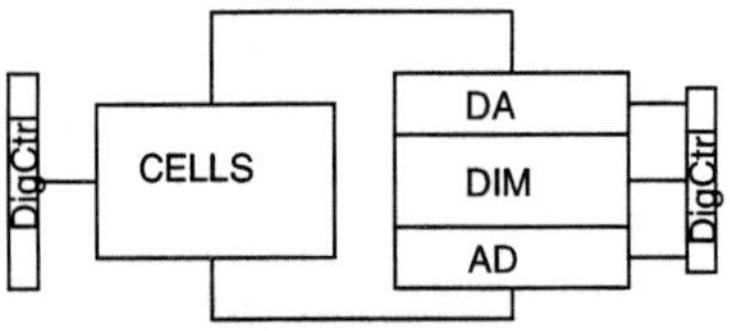

Fig. 2. Block diagram of the realized chip.

bitrate of video compression by smoothing the image and therefore reducing the number of high frequency components in DCT. There are several ways to do some type of low-pass filtering, but in this case the chosen way was the resistive grid network that can be also described with Cellular Nonlinear Network templates [6]. In this case the implemented templates were introduced in [7] and they are slightly modified from the direct conversion from the resistive network to CNN because the center term in the B-template, that should be the resistor ratio, is spread to all the components of the template. In this way, an additional smoothing operation is achieved. The implemented templates are shown in Eq. 1.

$$A = \begin{bmatrix} 0 & 1 & 0 \\ 1 & -4 & 1 \\ 0 & 1 & 0 \end{bmatrix} \quad B = \begin{bmatrix} 0.1 & 0.1 & 0.1 \\ 0.1 & 0.2 & 0.1 \\ 0.1 & 0.1 & 0.1 \end{bmatrix} \quad (1)$$

Since the template realizes, in principle, a resistive network filtering, the operation is inherently linear and the reduction in cell rows can be applied. The sphere of influence was simulated to be 5 rows in both directions from the row to be calculated. That is illustrated in Fig.1 where the input was all white with one row black in the middle. When this requirement was taken into account, and with all the needed row operations, a 16 row network was chosen to be implemented. It's control can be handeled using a 4 bit control circuit that generates all the control signals.

3.1. Realization of the system

In Fig.2 the block diagram of the realization is shown. It consists of a digital image memory (DIM), digital-to-analog converters for each column (DA), the cell grid (CELLS), digital controls (DigCtrl) and the column analog-to-digital converters (AD).

The input image is first loaded in digital format into the DIM using the digital I/O of the chip. After loading the whole image

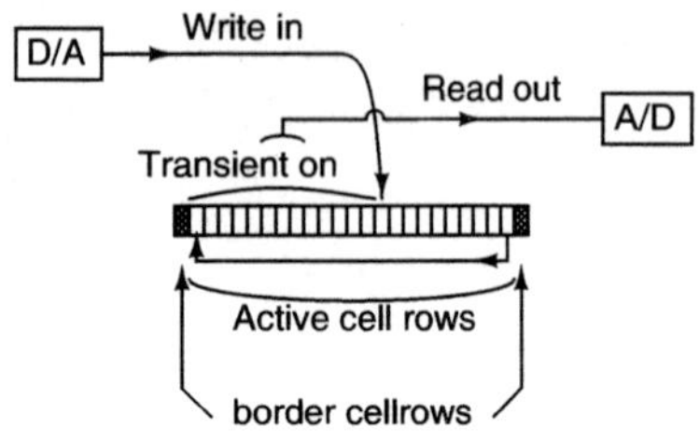

Fig. 3. Processing flow in the network.

into DIM, the processing of the image can start. The processing flow is shown in Fig.3. In this figure the network is shown from the side where all the cell rows can be seen. First, a D/A-conversion is performed row-wise on the image and the analog value is fed into the network. This is shown in the figure as the Write in -operation. Then the Write in -operation continues in a row-by-row manner with constant clocking frequency. When 5 rows of the image have been loaded in the network, the first row has the needed neighboring information for the final evalution of the output. During the Write in for the 6th row the result of the 1st row should become ready and simultaniously with the 7th Write in -opereation the result of the first row can be read out to the A/D-converter from where the digital result is read back to the DIM. Also here the operation is done row-by-row. In this way the processing result of the network is ready in the digital memory only after 6 clock cycles after the last input row is read to the network. If the network was full size, the same amount of time would have been needed for the Write in -operation, then one cycle would have been needed for the evalution and finally the same amount of cycles would be required for reading out teh result as with the writing in. Therefore it can be stated that the processing flow is faster here than with the full size network.

3.2. Low-pass cell

The cell itself is relatively simple as can be seen in the block diagram in Fig.4. The reading in is controlled by a switch that connects the input of the cell to the output of the DA-converter. The current mode input is read to the analog memory of the cell and the evaluation starts immediately after the reading in. The B-template block is separate from the A-template calculation block. When the input is read to the cell, the current corresponding to the center element of the B-template is fed to the summing node along with the influence from the neighboring cells. From the summing node the calculation of the A-template starts. The output is read after the summing node has settled to its final value and the output switch then connects the output node to the input of the AD-converter.

4. TEST CHIPS

Two test chips have been manufactured using the previously presented realization. The number of the cell rows was optimized to 16 to provide the required neighborhood for all the cells and to fulfill the input requirements of the following gradient calculation block, where the outputs of three consecutive cell rows was needed. The first chip was designed to test the functionality on system level and therefore the width of the cell rows was chosen to be 4 for easier testability. The measured results are presented in the first subsection.

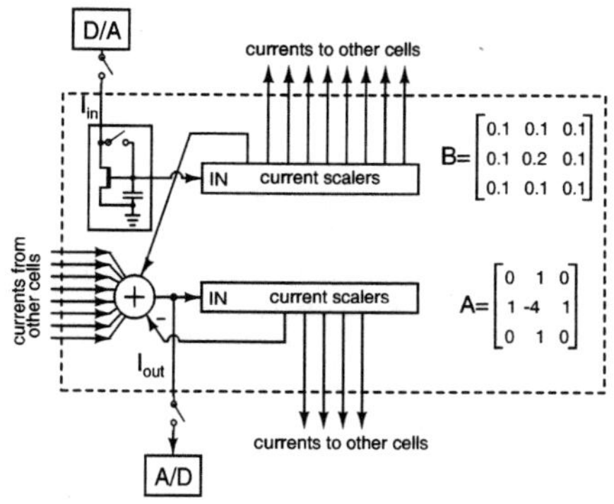

Fig. 4. Cell in principle.

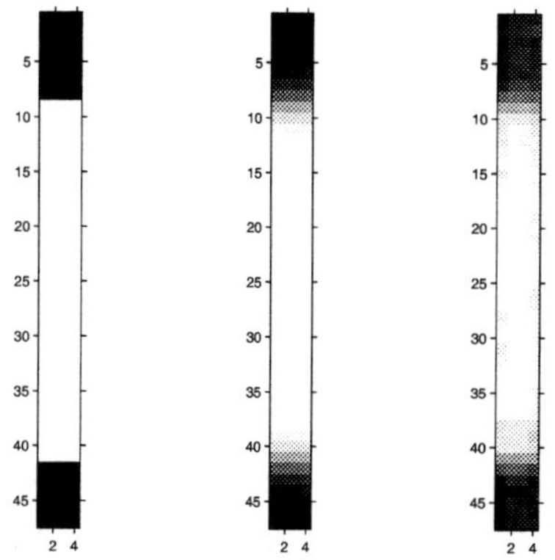

Fig. 5. Original image, ideal low-pass filtered image and the measured output.

Using the information gained from the first chip, the second version of the system was designed. It was aimed to exploit countermeasures for the problems of large arrays, like the bias-distribution and mismatch between cells in a large scale. Measurement results are not yet available, but the modifications and the used methods to prevent mismatch are discussed in the second subsection.

4.1. Measurements of the 4×16 test chip

The test chip was mainly designed for system evaluation and so it was decided that for easier testability and handling of the different parts the size of the cell grid was 4 columns. To be able to test the system the image memory consisted of 48 rows. The resolution of the converters was 6 bits. The used process was standard digital CMOS process with 6 metals and line width of $0.25\mu m$. Due to the dynamic realization of the analogue memories in the cells, there is not a way directly to monitor the progression of processing. Therefore the evaluation of the converters was made possible individually by either connecting them as standalone devices or by connecting them to an A/D/A-converter loop.

The Fig.5 shows a measurement result where the system level operation of the network was tested. The figure shows first the input image, next to it is the ideal result and then the third is the measured digital output. Image sizes are in the figure 4×47 because one memory row was not functional. In the figure it can be seen that there is column-to-column level differencies. This is mainly due to differencies in the bias-currents of the column AD-converters. The error was measured also for the DA-converters but since the processor does low-pass filtering they are not so visible in the output. Anyhow, the output shows that the system of reduced cell rows itself does not cause errors.

In the case of Fig.6 the input to the network was constant for all the cells to test the differences between the cells. The output of all the four columns is pictured in its own window. In the figure it is visible that there are differences between the cells, especially in the first and the last column. Also the level shift between the columns is visible in the figure.

Fig.7 reveals the most severe problem with the chip. In the figure there is pictured the output of one cell of each column when the measurement is repeated 2500 times. There are large differences between the outputs in different measurement rounds. When this was later investigated, the source of the problem turned out to be the output buffer of the AD-converter that was too weak to push the memory cell up every time.

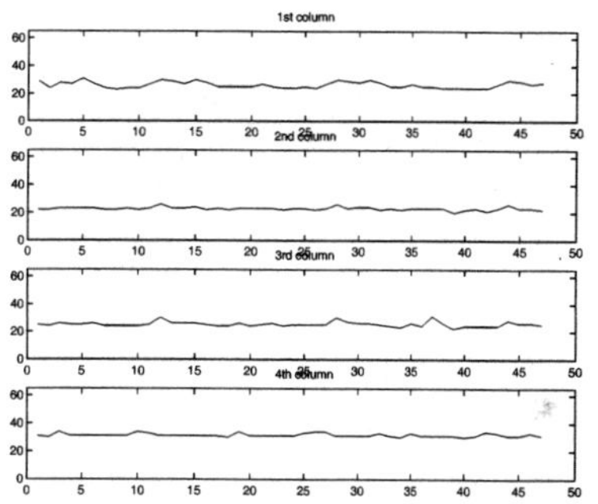

Fig. 6. Column outputs for a processed uniform image.

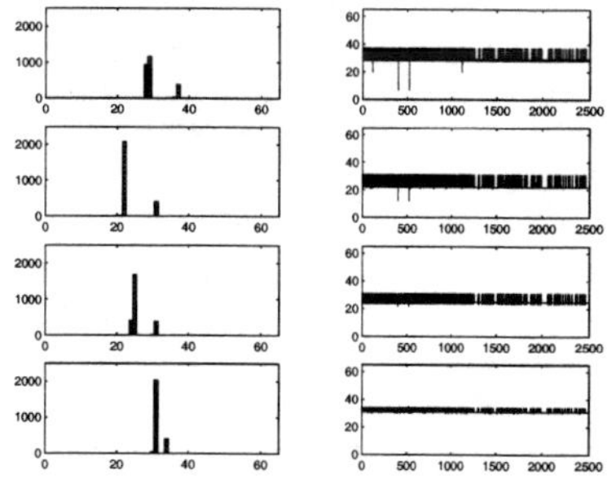

Fig. 7. Measured outputs of the cells of one cell row when the measurement is repeated 2500 times (right) and the distribution of the outputs (left).

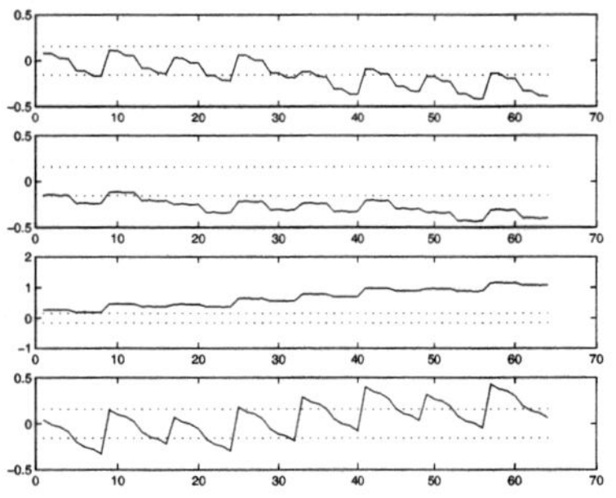

Fig. 8. Outputs of the converters relative to the ideal conversion.

The problem with the buffer caused that it was not possible to measure the maximum intended operation speed. Anyhow, the shown measurements were made with $350ns$ processing time for each cell row. This leads to $52.5\mu s$ processing time for a QCIF size image, if the network was 176 columns wide. The power consumption was 2.5 mW per column during the processing.

4.2. The 64×16 test chip

With the results of the first test chip the second version of the chip was designed. The number of columns in the design was chosen to be 64 and the image memory was 56×64×8 bits. In the design process the main effort was put on to improve the matching between the current biases of the converters and cells in the network. Also the manufacturing process was changed to $0.18\mu m$ CMOS process with the same 6 metals for routing. In the following, the improvements for the new chip are briefly discussed.

4.2.1. Modifications in the design

In the first test chip the biases to converters were simply delivered by using current mirrors. In Fig.8 the outputs of the first test chips DA-converters are shown. The figures show the difference of the ideal conversion and the output. The dotted lines are 0.5 LSB away from the ideal. Obviously the errors are intolerable large, especially if anywhere close to 8 bit accuracy is to be achieved. Each converter needs two bias currents, one to provide the offset of 5μA and the other for the actual conversion bias. In the measurements it turned out that the differences between different converters could be as large as 2.5 % in the offset only. Another problem with the bias distribution is that when the grid size grows, the more currents are needed and the further away the extremes are. In this case when 64 bias and offset currents were needed for both AD- and DA-converters the distribution was made by first mirroring the current from outside the chip into 8 currents. These 8 currents were distributed to the next stage of the bias distribution where each of the currents were mirrored to eight new currents. As a result the needed 64 currents were obtained.

These bias problems were simply solved by introducing dummy transistors to bias distribution circuitry and making the mirror transistors considerably larger. This way the matching between transistors was improved and also the preliminary measurement results show that the offset errors were reduced to maximum 0.1 % from the nominal value.

The cell basically remained the same except for the the size of the mirroring transistors grew relatively to the minimum line width. To decrease the level shift effect between the columns the border cells were similar to the actual cells in layout, only the unwanted connections were cut.

The resolution of the converters in the new design is 8 bits in both DA- and AD-conversion. The AD-converter type was also changed to asynchronous SAR-type converter [8] in which there is no need for fast clocking signals that could generate noise in the chip. In this system that is especially beneficial since the conversion is done simultaneously with the processing of the filter.

The processor part of the layout was 950μ m × 450μm and with the converters the total size is 0.65mm^2. Even for 176 columns wide system the size would be less than 2mm^2.

5. CONCLUSIONS

Measurement results of a test chip to realize low-pass filtering parallel processor with reduced number of cell rows were shown and the problems were discussed. In the system level the method of reduced cell rows was shown to be functional. Then a new design where the same system was implemented in a larger scale was introduced and the modifications that were made after the measurements of the first chip were briefly described. The array size was still relatively small and therefore this kind of a calculation block is possible to be attached on any processor chip. The future work includes also an addition of programmability of the resistive network resistor ratio.

6. REFERENCES

[1] L.O.Chua and L.Yang, 'Cellular Neural Networks: Theory', *IEEE Transactions on Circuits and Systems*, Vol. 35, 1257–1272 (1988)

[2] T. Roska, L. O. Chua, "The CNN Universal Machine: An Analogic Array Computer", *IEEE Transactions on Circuits and Systems-II*, vol. 40, pp.163–146, 1993.

[3] G. Linan, R. Dominguez-Castro, S. Espejo, A. Rodriguez-Vazquez," ACE16k: an Advanced Focal-Plane Analog Programmable Array Processor", *European Solid-State Circuits Conference*, Villach, Austria, pp. 216-219, September 2001.

[4] A. Kananen, A. Paasio, K. Halonen, "CNN Applications from Hardware Point of view: Video Sequence Segmentation", *Int. J. Circ. Theor. Appl.* pp.117–137 ,2002.

[5] A.Kananen, A.Paasio, K.Halonen, 'Overlapping Issues in Designing Large CNNs', *the 6th International Workshop on Cellular Neural Networks and their Applications*, Catania, Italy, pp-321-3325, 2000.

[6] B.E.Shi, L.O.Chua, 'Resistive Grid Image Filtering: Input/Output Analysis via the CNN Framework', *IEEE Transactions on Circuits and Systems - part I*, Vol. 39, 531–548, (1992)

[7] A.Stoffels, T.Roska, L.O.Chua, 'Object-Oriented Image Analysis for Very-Low-Bitrate Video-Coding Systems Using the CNN Universal Machine', *International Journal of Circuit Theory and Applications*, Vol. 25, 235–258, (1997)

[8] M.Laiho, A.Paasio, 'An Analogue-To-Digital Converter, a Current Scaler and a Method for Controlling a Function of the Current Scaler', *patent application WO03069782*.

A MIXED-SIGNAL CMOS DTCNN CHIP FOR PIXEL-LEVEL SNAKES

V.M. Brea, D.L. Vilariño, D. Cabello

Department of Electronics and Computer Science
University of Santiago de Compostela
Santiago de Compostela, Spain
Phone:+34981563100, Ext. 13572. Fax:+34981528012. Email:victor@dec.usc.es

ABSTRACT

This paper introduces the processing core of a full-custom mixed-signal CMOS chip intended for Pixel-Level Snakes. Among the different parameters to optimize on the top-down design flow our methodology is focused on area. This approach results in a single-instruction-multiple-data chip implemented by a DTCNN with a correspondence between pixel and processing element. This is the first prototype for pixel-level snakes; an integrated circuit with a 9×9 resolution manufactured in a $0.25\mu m$ CMOS STMicroelectronics technology process.

1. INTRODUCTION

Active contours are a particular case of the multidimensional deformable models where an elastic curve embedded in an image evolves toward some salient features after minimizing a potential function involving internal and external forces [1]. The main advantages of the active-contour-based techniques lie in the processing of local and global information extracted from the image and in their ability to combine low-level data extracted from the image with a priori knowledge (high-level data) on the domain of application. Their main drawbacks are related to their mathematical formulation which makes the processing speed strongly dependent on the flexibility of the contour evolution. Therefore the use of these techniques in applications needing fast time responses might be seriously restricted. The Pixel-level Snakes (PLS) make up an active-contour-based technique which overcomes these drawbacks due the pixel-level discretization of the curve, which also makes this approach suitable for its implementation onto a single instruction multiple data architecture consisting of a simple control section communicating with an array of processing elements with a pixel to processing element correspondence [2].

DTCNNs [3] have shown to be a very efficient architecture to host the PLS algorithm. Their inherent robustness and ease of control irrespective of the mathematical domain of the signal processing, analog or digital, facilitate their hardware implementation. This work deals with the design and implementation of the first PLS chip mapped onto a 9×9 DTCNN architecture with the CNN array operating in the analog domain, while the control circuitry works in the digital domain [4]. The data extracted from device-level HSPICE simulations and layout areas forecast the feasibility of the upscaling of the CNN array for a future DTCNN visual chip for real-time image processing intended for video traffic monitoring.

The paper is organized as follows. Section 2 briefly introduces the domain for which the chip is intended: the PLS technique in moving-object-segmentation-based applications. Section 3 provides the mathematical model onto which the PLS algorithm is projected: a DTCNN. Section 4 collects some of the hardware issues on the design, as well as some chip data. Finally, Section 5 gathers the main conclusions drawn from the work.

2. APPLICATION: THE PLS TECHNIQUE

Moving object segmentation is a real-time task where the PLS technique has shown a good performance [2, 5]. In this kind of applications, the initial contours of every frame are set from the delimitation of the objects of interest in the previous one. In visual systems provided with a fast acquisition stage the moving objects are slightly shifted/deformed among successive frames of the sequence, and thus the initial contours are very close to their final shape and position making it easier to delimit the objects of interest by using only local information to guide the contour evolution [1]. Hence, a PLS on-chip implementation is a good approach to meet the required high processing speed. It should be noted, however, that this way of setting up the initial contours cannot be extended to the first frame in the sequence where different approaches should be applied depending on every application. Particularly, for video surveillance tasks the setup of the initial contour can be based on the capability of managing topologic transformations by PLS [2, 5]. This consists of defining a region of control with an initial closed contour. Once the moving object is completely inside the control area, there is a topologic transformation which pro-

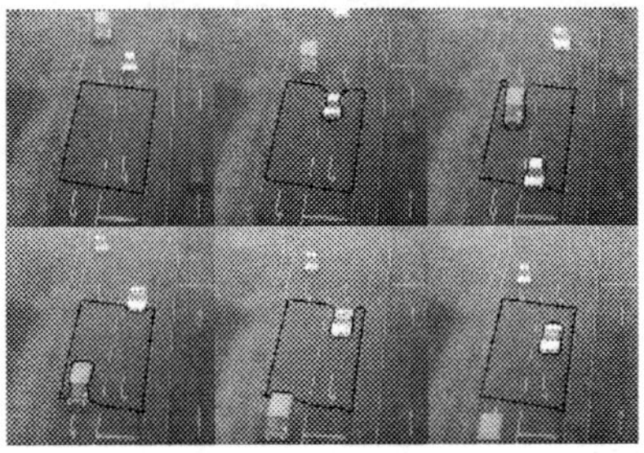

Fig. 1. Video traffic monitoring with PLS.

duces two new contours. One of them restores the initial control area. The other delimits the moving object inside the control area. Fig. 1 shows an example of a video traffic monitoring sequence following such an approach with a square shape bounding the area of control[1]. The initial square contour is deformed to be adapted to the entering vehicles, leading to the splitting of the initial contour into two new ones: one of them restoring the initial control area, and the other one bounding and tracking the moving vehicle.

3. THE ARCHITECTURE: DTCNN

The PLS algorithm is fed with two input images: the so-called active contour image, a B/W image where the contours are represented as black 8-connected pixels, and the external potential, a real-valued array previously extracted from the original image, which along with the internal potential image, a real-valued array derived from the contours themselves, guide their evolution [2].

The PLS algorithm is projected onto a multilayer 2D-DTCNN network with cyclic time-variant templates controled by a clock signal ck. As the algorithm operations are successive over time, every building layer is translated into a processing step, at most, fed by the scaled outputs of two former layers. As the outputs from previous steps are binary values settled during the ck cycle, both are equally computed. This implies that there is no difference between the feedback and the control templates, and that is the reason why we label the templates A and A^+, they are interchangeable. In our application the $A - A^+$ selection is made by minimizing the control circuitry in the DTCNN PLS cell [4].

Fig. 2 displays the so-called generic DTCNN PLS cell, a schematic representation within the framework of the CNN Universal Machine (CNNUM) [6] irrespective of the subsequent mathematical domain implementation, either analog or digital. Since the PLS algorithm processes both: B/W and gray-scale images, the cell can be viewed as two sub-

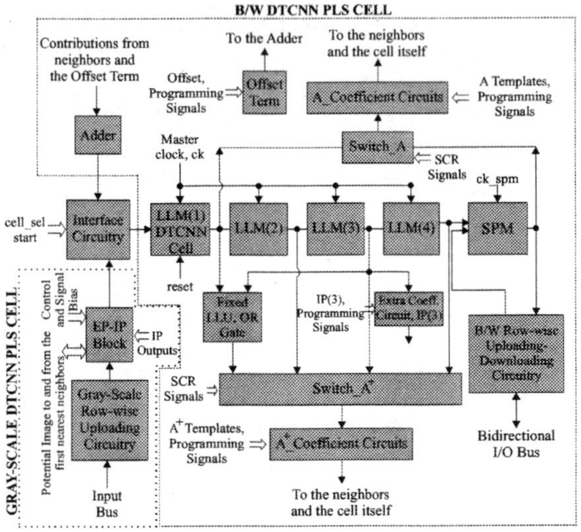

Fig. 2. The generic DTCNN PLS cell.

cells communicating with each other through the *Interface Circuitry*. The so-called local logic memories (LLM) [6] inside the B/W sub-cell make up the core of the DTCNN PLS cell, a shift register which keeps the active contours from one DTCNN operation to another. Every LLM is implemented with a dynamic memory. Nevertheless, the special memory (SPM) shown in Fig. 2 is implemented with a static memory. The reason is that this is the device in which the initial contours are stored.

The shift register is the best option for the core of the B/W cell circuitry independently of the mathematical domain of the PLS on-chip implementation. The B/W contour image is dynamically changed according to the DTCNN equations, whereas the static external potential image is stored in the $EP - IP$ block (External Potential-Internal Potential), from which the B/W guiding information image is provided. The A and A^+ coefficient circuits steer their outputs to the cells within the first order of neighborhood and to the cell itself. This collects the self-feedback and the neighboring contributions by the *adder circuit*. The connectivity with the neighbors is completed with the flux of the gray-scale potential image to and from the first four neighbors along the four cardinal directions. The *interface circuitry* selects which input drives the core of the cell: the one coming from the *coefficient circuits* and the *offset term*, or that one coming from the $EP - IP$ block. In the first case, a conventional B/W processing step is being processed, while in the second one, the guiding information image is being introduced into the shift register.

4. CIRCUIT DESIGN ISSUES

The constraints imposed on the top-down design flow of the DTCNN PLS chip depend on the specifications to optimize.

[1] *Copyright 1998 by H.H. Nagel. Institut Für Algorithmen und Kognitive Systeme. Fakultät für Informatik, Universität Karlsruche (TH). Postfach 6980, D-76128 Karlsruhe, Germany. http://i21www.ira.uka.de/*

Area consumption, power dissipation and processing speed are among the main goals to meet in an integrated circuit. Nevertheless, for a given CMOS technology process there is a trade-off between these parameters, leading to a maximum value expressed in the $performance$ equation [4, 7]:

$$Performance = Speed/(Power \times Area) \qquad (1)$$

In our design it is apparent that the capability of processing large images will improve the performance of the final system. Hence, the principles of design behind our circuits are focused on area. Furthermore, due to the parallelism of the SIMD architecture, a lesser cell area increases the processing speed as well. Concerning the power dissipation, this is set by the minimum SNR imposed by the application. If the accuracy needs are not so tight, the physics of computation of the analog circuits leads to an efficient approach, which will likely outperform the digital solution. In the latter, assuming negligible the static power dissipation, the power consumption is proportional to the clock frequency. Nevertheless, the inherent parallelism of the DTCNN architecture weakens the processing speed requirements, avoiding the need for a high frequency in a digital implementation.

The parameters listed in Eq. 1 depend on the mathematical domain of the processing core. This is determined by the precision requirements of the PLS algorithm in the moving object segmentation domain. These, like any other early-vision-based task, are under the border line of 7-8 bits of accuracy, making the analog approach superior to the digital one [11, 12]. More specifically, the accuracy of an analog implementation is limited by two types of random errors: inherent noise and mismatch effects. The low accuracy of the PLS application, inferior to 5 bits, along with the relatively high voltage for the analog circuitry, 3.3 V in a $0.25\mu m$ technology process, makes the mismatch be the dominant error on the final performance of the circuit. The magnitude of mismatch errors depends on the technology used, the transistor area, the distance between transistors, and on the circuit biasing conditions. Nevertheless, irrespective of the design procedure at circuit level, many of the constraints on area can be introduced upper on the top-down design flow, when projecting the PLS algorithm onto a DTCNN.

4.1. Coefficient Circuits

The key to achieve low-area coefficient circuits is the robustness [8]. This states how large the deviations on each coefficient are allowed in order to keep a correct output. The higher the robustness, the larger the mismatch errors can be, and thus the looser the hardware requirements.

In the PLS algorithm, the lowest relative robustness is around 5% ($< 5bits$), so that, the analog solution outperforms the digital one. Current steering D/A converters with 5 bits of scalability (4 bits + sign) seem to be the best option for the coefficient circuit from the area consumption point of view. In these circuits, high currents along with low (W/L) ratios for the current mirrors produce lower mismatch errors. The sign of the coefficients, however, also has an effect on the magnitude of the mismatch errors. These are higher on the negative coefficients, implemented with NMOS current mirrors, due to their lower V_{gs} compared to that of the PMOS counterpart [9]. This leads us to design only PMOS current mirros for the PLS coefficient circuits with dimensions of $(0.8 \times 2.1)\mu m^2$ [4].

4.2. Shift Register and Round About Switches

The shift register stores the B/W contour image in the $LLMs$ of Fig. 2. As the area consumption is the main goal to optimize, a dynamic implementation for the shift register is preferable to the static version, leading to a lower power consumption as well. The drawbacks are related to its more complex design caused by dynamic effects like charge injection and clock feed-through errors, which could yield misprocessed data.

Both the charge injection and feed-through errors are minimized with large capacitive nodes, leading to large inverters in the shift register, amounting to an area of $(3.25 \times 1)\mu m^2$, much greater than that of the minimum size transistor in the $0.25\mu m$ CMOS technology, $(0.8 \times 0.25)\mu m^2$. The passing of '1' weak logic levels in NMOS-pass-transistor gates is performed with a high enough overdrive voltage, 3.3 V, while the shift register is supplied with 2.5 V. The '0' weak logic levels in the PMOS-pass-transistor-gates are restored in the $LLMs$ [4].

4.3. Gray-scale Cell

The Gray-scale or $EP - IP$ block makes a comparison among neighboring potentials along the current processing direction guiding the contours toward the maximum of the potential image. The internal potential is calculated with the coefficient circuits of the cell, while the external potential image is uploaded from outside and stored in a current memory inside the $EP - IP$ block. Later, both the external and internal potential values are summed obeying the KCL to produce the potential image itself. The local potential image values are sent to the first neighboring cells along the current processing direction to make the aforementioned comparison for guiding the contours.

The $EP - IP$ block is an analog circuit supplied with 3.3 V basically made up of a current memory for the potential values and a subsequent comparator. As for most of the PLS-based applications 5 bits of precision are sufficient for guiding the contours, the accuracy requirements are weak enough for an SI current memory to process the external potential image with an acceptable level of error. Here the principles of design are focused on minimizing mismatching by means of low W/L ratios in the two transistors of the basic current mirror of the SI cell, up to $(0.8/3)$, as well as on maintaining a large geometry to deal with feed-through and charge injection errors. Finally, the neighboring potential levels are compared by a differential pair assisted by a positive feedback circuit with a low impedance node to collect the currents and transform them into voltages, similarly to a sense amplifier in a DRAM. Here the principles of design are focused on drawing a strong logic level to the shift register in the B/W cell [4].

4.4. Chip Data

Fig. 3 displays a microphotograph of the DTCNN PLS chip, a 9×9 prototype, in a $0.25\mu m$ CMOS technology from STMicroelectronics, where the main blocks have been appropriately labeled. The PLS templates are looked up and delivered to the 9×9 analog DTCNN array from an SRAM and through a bank of 5 bits current steering D/A converters with their corresponding buffers, OAs in voltage-follower configuration, to lower the fan-out seen by the SRAM, and thus to speed up the template processing.

The chip has been designed in a full-custom style with an area of $4.84mm^2$ in a PGA68 (minimum required for the foundry with this package), and two power supplies: 3.3 V for the most sensitive circuitry, and 2.5 V for the digital circuits. Every cell occupies around $(83 \times 45)\mu m^2$ which amounts to a cell density close to $270cells/mm^2$, similar to some of the state-of-the-art CMOS CNN implementations [10]. The power dissipation is high, $\sim 0.5mW/cell$, since this has not been the main specification to optimize. Future releases of the DTCNN PLS chip, however, should focus on the power consumption. Concerning processing speed, the goal is to process images at video rate, $30 frames/s$. A frequency of ~ 1 MHz for the execution of the PLS algorithm itself is high enough for these applications [4]. Nevertheless, the pin count must be appropriately upscaled on a larger CNN array in order to achieve video rate processing.

5. CONCLUSION

The synergy of the DTCNN architecture and the PLS technique is a promising tool for moving-object-segmentation-based applications. This has been proven in this paper with the first 9×9 CMOS PLS prototype in a $0.25\mu m$ technology with full-custom style and with the area as the main specification to optimize. Future releases will include the power consumption as another parameter to look at, as well as the incorporation of the acquisition and preprocessing stages.

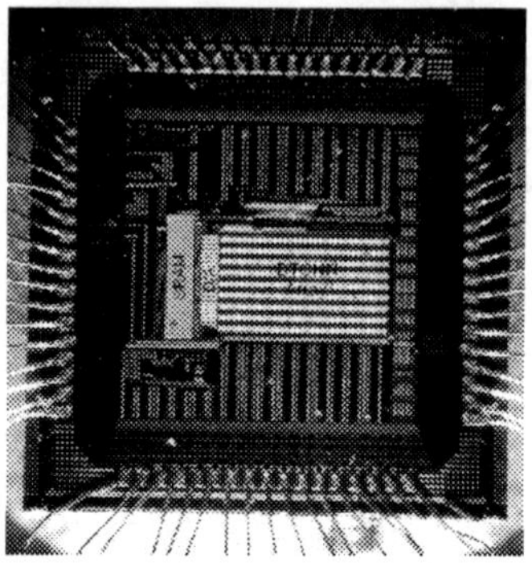

Fig. 3. Microphotograph of the PLS chip.

6. ACKNOWLEDGMENTS

This work was funded by Ministerio de Ciencia y Tecnología (Spain) under the Project TIC2003-09521.

7. REFERENCES

[1] M. Kass et al. "Snakes: Active Contours Models", International Journal of Computer Vision, vol. 1, pp. 321–331, 1988.

[2] D.L. Vilariño et al. "Cellular Neural Networks and Active Contours: A Tool for Image Segmentation", Image and Vision Computing, vol. 21, no. 2, pp. 189–204, 2003.

[3] H. Harrer et al. "Discrete-Time Cellular Neural Networks", Int. J. Circ. Theor. Appl., vol. 20, pp. 453–467, 1992.

[4] V.M. Brea, D.L. Vilariño, A. Paasio, D. Cabello "Design of the Processing Core of a Mixed-Signal CMOS DTCNN Chip for Pixel-Level Snakes", IEEE TCAS-I (Accepted).

[5] D.L. Vilariño, D. Cabello, V.M. Brea "An Analogic CNN-Algorithm of Pixel Level Snakes for Tracking and Surveillance Tasks", CNNA2002, pp. 84–91, 2002.

[6] T. Roska and L.O. Chua "The CNN Universal Machine: An Analogic Array Computer", IEEE TCAS-II, vol. 40, no. 3, pp. 163–173, 1993.

[7] P. Kinget et al. "Analog VLSI Integration of Massive Parallel Processing Systems", Kluwer Academic Publishers, 1997.

[8] A. Paasio et al. "CNN Template Robustness with Different Output Nonlinearities", Int. J. Circ. Theor. Appl., vol. 27, n. 1, pp. 87–102, 1999.

[9] P.G. Drennan et al. "Understanding MOSFET Mismatch for Analog Design", IEEE J. on Solid-State Circuits, vol. 38, pp. 450–456, 2003.

[10] G. Liñán et al. "ACE16K: A 128x128 Focal Plane Analog Processor with Digital I/O", CNNA2002, pp. 132–139, 2002.

[11] A. Paasio, K. Halonen, "A New Cell Output Nonlinearity for Dense Cellular Nonlinear Network Integration", IEEE TCAS-I, vol. 48, n. 3, pp. 272–280, 2001.

[12] V.M. Brea et al. "On the Mathematical Domain of a CMOS Discrete-Time Cellular Non-linear Network Cell", ECCTD'03, vol. 3, pp. 361–364, 2003.

AN OVERVIEW AND COMPARISON OF ANALYTICAL TCP MODELS

Inas Khalifa and Ljiljana Trajković

School of Engineering Science, Simon Fraser University
Vancouver, British Columbia, Canada
{ikhalifa, ljilja}@cs.sfu.ca

ABSTRACT

Modeling TCP performance is an important issue that attracted research attention over the past decade. In this paper, we present an overview of models used to capture the TCP behavior. We compare several existing analytical models with respect to modeled attributes, modeling assumptions, and validation techniques. We also identify features that new TCP models should possess. Finally, we address the importance of devising common validation techniques and performance evaluation metrics for TCP models.

1. INTRODUCTION

Recent traffic measurements in several large campus networks indicated that over 90% of network traffic is transported by the Transmission Control Protocol (TCP) [3]. This ubiquity has encouraged research on modeling TCP behavior. Several analytical TCP models have recently been proposed [2], [3], [5], [6], [7], [9]. TCP models may be used to improve the existing and evaluate new congestion control algorithms and TCP implementations. For example, it has been demonstrated that TCP SACK outperforms TCP Tahoe when segment losses are independent, while Tahoe outperforms SACK when segment losses are correlated [9]. They may also be used to define TCP-friendly behavior or investigate the interaction between TCP and queue management algorithms [4].

In this paper, we examine TCP models and present a framework for model comparison with respect to the modeling assumptions and validation techniques. We also address the lack of common validation approaches and evaluation metrics. In Section 2, we present an overview of TCP. Model classification is discussed in Section 3. Section 4 describes existing analytical TCP models, while Section 5 compares the modeling assumptions and model validation techniques. Conclusions are given in Section 6.

2. OVERVIEW OF TCP

TCP provides a reliable connection-oriented data service in packet-switched networks. This is achieved by employing acknowledgments (ACKs), sequence numbers, and timers. TCP employs window-based congestion control mechanisms to adjust its congestion window size ($cwnd$). $cwnd$ is the maximum amount of data that the sender can transmit before receiving an ACK. The receiver also advertises a limit ($rwnd$) on the amount of outstanding data. Data transmission is always governed by the window size $W_m = \min(cwnd, rwnd)$.

A client initiates a TCP connection by sending a SYN segment to the server. The connection is established using a three-way handshake. The sender then uses the *slow start* algorithm to detect the available bandwidth by incrementing $cwnd$ by one segment upon receipt of each ACK. When $cwnd$ reaches the slow start threshold ($ssthresh$), the sender enters *congestion avoidance* phase and $cwnd$ is incremented by one segment per round trip time (RTT).

If a segment is not acknowledged within the retransmission timeout interval (RTO), the sender infers that the segment is lost (TO loss). It retransmits the lost segment, doubles the RTO (exponential backoff), and switches to slow start. Segment loss may also be detected using triple duplicate ACKs (TD losses). When TD losses occur, the sender uses the *fast retransmit* algorithm: it halves the $cwnd$ and retransmits the missing segment without waiting for RTO to expire. The *fast recovery* algorithm is then used to control data transmission until a non-duplicate ACK is received [1].

A TCP receiver uses the delayed acknowledgment algorithm. It may acknowledge b segments with one ACK, where $b \geq 1$. ACKs are generated: (i) for at least every second full-sized segment of new data, (ii) within 500 ms upon the arrival of an unacknowledged segment, (iii) for out-of-order segments, to trigger fast retransmit, and (iv) for segments that fill gaps in the sequence number space.

3. MODEL CLASSIFICATION

An IP-based communication network employs routers with queue management mechanisms to enable data transfers. TCP delivers byte streams between pairs of hosts. TCP models investigate the network from three different perspectives. Models in the first class consider queue management

0-7803-8251-X/04/$17.00 ©2004 IEEE

and they characterize TCP flows by their average throughput [4]. Models in the second class consider the interaction between TCP flows and queue management mechanisms. They typically assume simple topologies, number of flows, and direction of data flow [8]. Models in the third class deal with TCP dynamics. They are the main focus of this paper. These models consider the network from the perspective of the TCP layer and characterize it by parameters such as loss, average drop probability p, and average RTT [5], [7].

Analytical TCP models can also be classified according to the length of the TCP transfer, which determines the congestion control algorithms and loss detection mechanisms that need to be incorporated into the model. In case of *short-lived* transfers, TCP performance is strongly affected by the connection establishment and slow start phases, with segment losses mostly being TO losses. Models for *long-lived* transfers capture the steady-state performance of TCP, which is dominated by the congestion avoidance phase, and losses may be TD as well as TO. There are also models for TCP transfers of *arbitrary* length.

4. ANALYTICAL TCP MODELS

We illustrate the evolution of modeling techniques by surveying six analytical TCP models. We define three TCP performance attributes: *throughput* T as the total amount of data sent by a TCP source per unit time (including retransmissions), *goodput* G as the total amount of data correctly received per unit time, and *latency* L as the time required to successfully complete a TCP transfer.

4.1. Models for Long-Lived Transfers

Mathis, et al., [5] developed a model that predicts the steady-state throughput of long-lived TCP transfers in the presence of light to moderate segment losses. The model considers the congestion avoidance phase and assumes no TO losses. The segment loss process is periodic, with a constant probability p. In other words, every segment loss is followed by the successful delivery of $\frac{1}{p}$ segments. Therefore, the evolution of $cwnd$ follows a periodic sawtooth pattern during equilibrium. The number of segments transmitted during each period is $\frac{1}{p}$. Given a maximum window size of W_m, the minimum value of $cwnd$ is $\frac{W_m}{2}$. Moreover, if the receiver acknowledges every segment, then $cwnd$ is incremented by one segment every RTT. Hence, the duration of each period is $\frac{W_m}{2} \times RTT$. The area under the sawtooth, $(\frac{W_m}{2})^2 + \frac{1}{2}(\frac{W_m}{2})^2$, is also equal to $\frac{1}{p}$. Thus, the throughput T is:

$$T = \frac{MSS}{RTT}\frac{K}{\sqrt{p}}, \tag{1}$$

where MSS is the maximum segment size and K is a constant that depends on the loss model and the acknowledgment strategy. For example, $K = \sqrt{3/2}$ when loss is periodic and delayed acknowledgment is not employed.

Padhye, et al., [7] developed a stochastic model for the steady-state throughput of long-lived bulk transfer TCP flows. The behavior of TCP congestion control is modeled in terms of *rounds*. A round starts with the transmission of $cwnd$ segments and ends when the first ACK is received. Hence, the duration of a round is equal to RTT. It is assumed to be independent of $cwnd$. The model considers both TO and TD losses and $cwnd$ limitation by W_m. Moreover, it assumes a correlated (*bursty*) loss model: if a segment is lost, all remaining segments in the same round are also lost. Stochastic systems techniques are used to determine the expected values of the number of segments transmitted in a round and the duration of the round in terms of the loss probability p. The throughput is approximated as:

$$T = \min\left(\frac{W_m}{RTT}, \frac{1}{RTT\sqrt{\frac{2bp}{3}}+RTO_0\min(1,3\sqrt{\frac{3bp}{8}})p(1+32p^2)}\right), \tag{2}$$

where RTO_0 is the initial value of RTO. The model can be extended to capture the goodput G of the transfer by computing the number of received, instead of transmitted, segments [7].

Altman, et al., [2] proposed a model for the throughput of long-lived TCP transfers, subjected to a stationary ergodic loss process. The dynamics of TCP are modeled by observing the instantaneous transmission rate $X(t)$, defined as the number of packets in the network (window size) divided by RTT at time t. The instants Y_n when loss events occur are modeled by a stationary ergodic point process $\{Y_n\}_{n=-\infty}^{+\infty}$. The inter-loss duration S_n is equal to $(Y_{n+1} - Y_n)$ when loss is TD, and $(Y_{n+1} - Y_n - E[t_{TO}])$ when loss is TO, where $E[t_{TO}]$ is the average duration of the timeout period. The throughput T, computed as the time average of the process $X(t)$, is:

$$\begin{aligned} T = \; & \frac{1}{RTT\sqrt{bp}}(1 - \lambda_{TO}E[t_{TO}]) \times \\ & \sqrt{\frac{1+\nu}{2(1-\nu)} + \frac{1}{2}\hat{C}(0) + \sum_{k=0}^{\infty}\frac{1}{2^k}\hat{C}(k)}\,, \end{aligned} \tag{3}$$

where $\hat{C}(k) = (E[S_nS_{n+k}] - E[S_n]^2)/E[S_n]^2$ is the normalized covariance, ν is the factor used to reduce $cwnd$ when loss occurs, and λ_{TO} is the number of TO losses per unit time. When all losses are TD and random, (3) becomes identical to (1). When $W_m \neq \infty$, an explicit expression for T cannot not be obtained. Expressions for the lower and upper bounds on T are derived, instead.

4.2. Models for Transfers of Arbitrary Length

Cardwell, et al., [3] developed a model for the latency of TCP transfers of arbitrary length by extending [7] to include connection establishment and initial slow start. The expected latency $E[L_{CE}]$ of the three-way handshake used during the connection establishment (CE) phase is:

$$E[L_{CE}] = RTT + RTO_0\left(\frac{1-p_r}{1-2p_r} + \frac{1-p_f}{1-2p_f} - 2\right), \quad (4)$$

where p_f is the segment loss rate in the forward path from the server to the client, and p_r is the loss rate in the reverse path. The data transfer latency $E[L]$ required to complete a transfer of size N segments is computed as the sum of four components:

$$E[L] = E[L_{ss}] + E[L_{loss}] + E[L_{ca}] + E[L_{delack}], \quad (5)$$

where L_{ss} is the latency of the initial slow start, E_{loss} is the expected cost of TO losses or fast recovery that occurs at the end of the initial slow start, L_{ca} is the expected time required to transfer the remaining $(N - E[d_{ss}])$ segments, and L_{delack} is the cost of the first delayed ACK if $cwnd_0$ is equal to 1. Both $E[L_{ss}]$ and $E[L_{loss}]$ are functions of RTT, b, W_m, p, N, $E[t_{TO}]$, and the initial window size $cwnd_0$. $E[L_{ca}] = (N - E[d_{ss}])/T$, where T is the throughput [7]. $E[d_{ss}]$ is the expected number of segments sent before a loss is encountered during the initial slow start. Finally, $E[L_{delack}]$ is equal to 150 ms for Windows platforms or 100 ms for BSD UNIX.

Sikdar, et al., [9] modeled the latencies of arbitrary length transfers of TCP Tahoe, Reno, and SACK by estimating the transfer time given that the transfer experiences no loss, a single loss, and multiple loss indications. The expected latency to transfer N segments is:

$$\begin{aligned} E[L(N)] &= E[L_{CE}] + E[L_{delack}] + E[L_{ml}(N)] \\ &+ (1-p)^N E[L_{nl}(N)] + p(1-p)^{N-1} E[L_{sl}(N)], \end{aligned} \quad (6)$$

where $E[L_{CE}]$ is given in (4) with $p_f = p_r = p$, and $E[L_{delack}]$ is as given in [3]. $E[L_{ml}(N)]$ is the expected latency with M loss indications. It is a function of N, RTT, W_m, $E[t_{TO}]$, $cwnd_0$, and M. $E[L_{nl}(N)]$ is the expected latency when no losses occur. It is a function of N, RTT, and W_m. $L_{sl}(N)$ is the expected latency when there is a single loss and it is a function of N, RTT, W_m, $E[t_{TO}]$, and $cwnd_0$. The last three terms in (6) are computed for TCP Tahoe, Reno, and SACK. The steady-state throughput of long-lived transfers is:

$$T = \frac{d \times MSS}{E[L_{ss}(p)]}, \quad (7)$$

where $d = \frac{1}{p}$ is the average number of segments transmitted between two loss indications during steady-state. $E[L_{ss}(p)]$ is the expected time to transfer d segments. It is calculated by averaging the expected time to transmit d segments in the presence of multiple losses over all possible values of $cwnd$ and all possible positions of loss indications within the window.

4.3. A Model for Short-Lived Transfers

Mellia, et al., [6] devised a recursive analytical model for the average latency of short-lived TCP transfers during the slow start phase by exhaustively enumerating loss scenarios and their probabilities. The connection establishment latency $E[L_{CE}]$ is calculated using (4) with $p_f = p_r = p_s$, where p_s is the SYN segment dropping probability. The latency is computed using L_m^w, defined as the average time spent to successfully send m segments with an initial $cwnd$ of w. L_n^1 is the average time required to transfer n segments with $cwnd_0 = 1$. L_n^1 is equal to $(L_1^1 + L_{n-1}^2)$. It is derived *recursively* as a function of p, RTO, RTT, and $L_{m'}^{w'}$ where $m' < m$. L_m^m is the average time required to transfer m segments that belong to the same window. For example,

$$\begin{aligned} L_1^1 &= RTT + RTO\tfrac{p}{1-2p} \\ L_2^2 &= RTTq^2 + qp(RTO + RTT + L_1^1) \\ &\quad + pq(RTO + L_1^1) + p^2(RTO + L_2^1) \end{aligned} \quad (8)$$

where $q = 1 - p$, and p is uniformly distributed. L_m^w is computed recursively up to $m = 9$, where the data transfer time is $L_9^1 = L_1^1 + L_{9-1}^2$. The procedure may be repeated further. However, the complexity grows exponentially since all loss scenarios need to be considered.

5. ASSUMPTIONS AND MODEL VALIDATION

The modeling assumptions for the surveyed models are summarized in Table 1. They are classified in three categories [3].

Data Transfer length and congestion control algorithms affect TCP performance. Table 1 shows that neither slow start after TO losses nor fast retransmit are considered, although TO losses are common [7]. ISS denotes the initial slow start performed after connection establishment.

End Points assumptions deal with the TCP implementation, the number of segments b acknowledged by a single ACK, and whether or not the window limitation W_m is taken into account. They also include the loss detection mechanism and the duration of RTO. Note that b alone is not sufficient to describe the complex delayed ACK mechanism. Other common assumptions are the use of full-sized segments, *greedy* sources that generate segments as fast as $cwnd$ allows, and that both Nagle algorithm and silly window syndrome are neglected.

Network assumptions deal with the loss process seen by data and ACK segments. No model assumes a specific topology, a data flow direction, or a queue management algorithm. All models neglect transmission and processing delays and assume that RTT is independent of $cwnd$.

Validation is the process of evaluating how accurately a model reflects the real-world phenomena that it tries to capture. Models can be validated by comparing their performance with results from: (i) simulations, (ii) controlled

Table 1. Summary of the modeling assumptions. The TCP algorithms are Connection Establishment (CE), Initial Slow Start (ISS), Congestion Avoidance (CA), and Fast Recovery (FRC). "Exp." indicates that exponential backoff is employed.

Model	**Length**	**TCP algorithms**	b	W_m	**Loss detection**	RTO	**Data/ACK loss**
Mathis [5]	Long-lived	CA	1, 2	Yes	TD	None	Periodic/None
Padhye [7]	Long-lived	CA	Any	Yes	TO, TD	Exp.	Bursty/None
Altman [2]	Long-lived	CA	Any	No	TO, TD	Exp.	Stationary/None
Cardwell [3]	Arbitrary	CE, ISS, CA	2	Yes	TO, TD	Exp.	Bursty/SYN only
Sikdar [9]	Arbitrary	CE, ISS, CA, FRC	2	Yes	TO, TD	Exp.	Bursty/SYN only
Mellia [6]	Short-lived	CE, ISS, FRC	1	No	TO, TD	Exp.	Uniform/SYN only

Table 2. Model validation techniques. The subscripts m and t denote model prediction and trace measurement, respectively.

Model	**Simulations**	**Controlled meas.**	**Live meas.**	**Comparison**	**Evaluation Metric**
Mathis [5]	Yes	Yes	No	None	Least mean squared fit error
Padhye [7]	Yes	Yes	No	None	$(T_m - T_t)/T_t$
Altman [2]	No	Yes	No	[7]	None
Cardwell [3]	Yes	Yes	Yes	[5], [7]	$(L_m - L_t)/L_t$, $(L_m - L_t)/RTT$
Sikdar [9]	Yes	No	Yes	[3], [7]	$(L_m - L_t)/L_t$
Mellia [6]	Yes	No	No	None	None

measurements under conditions that reflect the modeling assumptions, and (iii) live measurements from the Internet.

Table 2 shows that most models were neither validated using live measurements, nor compared to other models. Several models use evaluation metrics to measure how they deviate from the results used for validation. These metrics only represent the average relative error between the model prediction and the traffic measurement. With such a diversity of validation methods, traffic measurements, and evaluation metrics, it is rather difficult to compare the models.

6. CONCLUSIONS

In this paper, we presented an overview of TCP models. We surveyed a number of analytical TCP models and showed that in all models, T is inversely and L is directly proportional to $RTT, p, RTO, E[t_{TO}]$. We compared models with respect to the modeling approaches, assumptions, and validation techniques. The comparison indicated that new models should include: slow start after TO losses, fast recovery, and more accurate capture of delayed acknowledgment. In order to demonstrate their merits, new models need to be compared to existing models based on a common set of traffic measurements. Such quantitative comparisons could be even more valuable if a comprehensive set of evaluation metrics were identified. These metrics need to reflect model predictions over a range of network variables and parameters, transfer lengths, and geographical network span.

7. REFERENCES

[1] M. Allman, V. Paxson, and W. Steven, "TCP congestion control," RFC 2581, Apr. 1999.

[2] E. Altman, K. Avrachenkov, and C. Barakat, "A stochastic model of TCP/IP with stationary random losses," *ACM Computer Communication Review*, vol. 30, no. 4, pp. 231–242, Oct. 2000.

[3] N. Cardwell, S. Savage, and T. Anderson, "Modeling TCP latency," in *Proc. INFOCOM*, Mar. 2000, pp. 1742–1751.

[4] V. Firoiu and M. Borden, "A study of active queue management for congestion control," in *Proc. INFOCOM*, Mar. 2000, pp. 1435–1444.

[5] M. Mathis, J. Semke, Jamshid Mahdavi, and T. Ott, "The macroscopic behavior of the TCP congestion avoidance algorithm," *ACM Computer Communication Review*, vol. 27, no. 3, pp. 67–82, Jul. 1997.

[6] M. Mellia, I. Stoica, and H. Zhang, "TCP model for short lived flows," *IEEE Communications Letters*, vol. 6, no. 2, pp. 85–87, Feb. 2002.

[7] J. Padhye, V. Firoiu, D. Towsley, and J. Kurose, "Modeling TCP reno performance: A simple model and its imperical validation," *IEEE/ACM Transactions on Networking*, vol. 8, no. 2, pp. 133–145, Apr. 2000.

[8] R. Roy, R. C. Mudumbai, and S. S. Panwar, "Analysis of TCP congestion control using a fluid model," in *Proc. ICC*, Mar. 2001, pp. 2396–2403.

[9] B. Sikdar, S. Kalyanaraman, and K. S. Vastola, "Analytic models and comparative study of the latency and steady-state throughput of TCP Tahoe, Reno and SACK," in *Proc. GLOBECOM*, Nov. 2001, pp. 1781–1787.

SELF-SIMILARITY IN MAX/AVERAGE AGGREGATED PROCESSES

G. Mazzini[1], *R. Rovatti*[2], *G. Setti*[1]

[1]DI - University of Ferrara, via Saragat 1, 44100, Ferrara - ITALY and
ARCES - University of Bologna, via Toffano 2/2, 40125, Bologna - ITALY,
[2] DEIS - University of Bologna, viale Risorgimento 2, 40136, Bologna - ITALY and
ARCES - University of Bologna, via Toffano 2/2, 40125, Bologna - ITALY
g.mazzini@ieee.org, rrovatti@arces.unibo.it, gsetti@ing.unife.it

ABSTRACT

Second-order self-similar processes are fully statistically charachterized by their activity factor and Hurst parameter, which are usually extracted from the computation of the autocovariance function of the process at different aggregation levels. Unfortunately, such an extraction procedure is difficult to be performed on experimental data or tested in analytical investigation. We here first aim at solving this problem by proposing a criterion to verify the self-similarity directly from the original process. Such a criterion is then applied to evaluate the self-similar features of the process obtained by averaging or maximizing the output of several independent sources. For both such cases we show that the resulting process is also self-similar with an higher Hurst parameter with respect to the original ones.

1. INTRODUCTION

The seminal papers [1] [2] [3] experimentally highlighted the bursty-like behavior of ethernet LAN traffic, stemming from both single or multiple, intranet and internet data sources, across a wide range of time scales. This "fractal-like" behavior, is usually referred to as *self-similarity* [4] [5] [6]. As it is well known, conventional Markov or Poisson processes cannot capture such a burstiness in a satisfactory way [7] since they becomes smoother (i.e., less bursty) as the observation time scale increases or as the number of traffic source increases. On the contrary, second-order self-similar processes are nowadays one of the most credited model for such bursty traffic.

The fundamental statistical features of a self-similar process can be highlighted by referring to the autocovariance function of the so-called aggregated processes, namely those obtained by averaging non-overlapping blocks of the original samples.

More specifically, let $x(i)$ be the i-th sample of a generic stationary process X and define the i-th sample $x^{[m]}(i)$ of the aggregated process $X^{[m]}$ as

$$x^{[m]}(i) = (1/m) \sum_{j=0}^{m-1} x(mi+j)$$

where $x^{[1]}(i) = x(i)$ and m is called the aggregation factor.

Let us first indicate with $p^{[m]}(k) = E[x^{[m]}(i)x^{[m]}(i+k)]$, where $p^{[1]}(k) = p(k)$, and $A = E[x^{[m]}(i)] = E[x^{[1]}(i)]$ the second-order moment and the activity of the process, respectively, which are independent of i thanks to the process stationarity. With this, the autocovariance function of the aggregate process can be expressed as

$$\begin{aligned} C^{[m]}(k) &= E[x^{[m]}(i)x^{[m]}(i+k)] - E^2[x^{[m]}(i)] \\ &= p^{[m]}(k) - A^2 \end{aligned}$$

A process is said to be asymptotically second-order self-similar if there constants $H \in]0.5, 1[$, $\beta > 0$ and $m' > 0$ exist such that if $m >> m'$ and k is large enough:

$$\frac{C^{[m]}(0)}{C^{[m']}(0)} \sim \left(\frac{m}{m'}\right)^{2H-2} \tag{1}$$

$$\frac{C^{[m]}(k)}{C^{[m]}(0)} \sim \frac{C^{[m']}(k)}{C^{[m']}(0)} \sim \beta k^{2H-2} \tag{2}$$

where H, usually called Hurst parameter, compounds the degree of the source burstiness. In other terms, the closer to 1 is the H the more self-similar is the process, while if H is close to 0.5 the process shows a Poisson behavior [1].

With these definitions, the self-similar process is completely defined by means of the Hurst parameter H and the activity factor A.

Many recent studies [8][9][10][11] focus their attention on the generation of self-similar traffic with the aim to estimate network systems performance in different traffic conditions. We here aim at obtaining a twofold goal. First we introduce an easy criterion for extracting the self-similar features of a process. Then, we use it to show that a self-similar behavior is maintained when some sort of post-processing is applied on the output of self-similar sources. More

This work has been partially supported by MIUR under the framework of the FIRB program specifically, we consider as set of independent self-similar sources and study the process obtained by averaging their maximum or by considering the maximum of the source activity. It is worthwhile to stress that such processing has no relationship with the construction of the aggregation process entailed in the definition of the notion of self-similarity.

Let us indicate with $x_j(i)$ the output of the j-th self-similar source, with $j = 1, \ldots, N$, at time step i. We assume that the sources are binary, so that $x_j(i) = 0$ or 1 which represents the "idle" or "busy" status of the channel. This kind of sources are usually called ON-OFF and it has been shown to be modeled by a suitably defined chaotic Markov map with an infinite number of branches [11]. It is worthwhile to notice that for this process the activity factor of the j-th source can obviously be directly computed from the second-order moments as $p_j(0) = E[x_j^2(i)] = E[x_j(i)] = A_j$. We will also assume that each source is characterized by an activity factor A_j and Hurst parameter H_j.

We are also interested in the study of the following two process

$$y_a(i) = \frac{1}{N}\sum_{j=1}^{N} x_j(i) \tag{3}$$

$$y_{\max}(i) = \max_j [x_j(i)] \tag{4}$$

which have been obtained from the original one by applying an averaging operator or by computing the maximum over the value of all possible sources. The first case is interesting when the system performance depend on the normalized occupation of the channel, while the second one is relevant one needs to know if activity is present.

Note also that $y_a(i)$ is a real process with $0 \leq y_a(i) \leq 1$ while $y_{\max}(i)$ is a discrete process characterized by $y_{\max}(i) \in \{0, 1\}$ as the original sources.

The paper is structured as follows: in Section 2 a single condition to test the self-similarity of a process is stated in order to give an useful tool to verify this behavior after the processing application. Section 3 discusses the source averaging processing while section 4 studies the source maximum processing. Some conclusions are finally drawn in Section 5.

2. AN EFFICIENT SELF-SIMILARITY CRITERION

We here need to suitably recast conditions (1) and (2) to develop an efficient criterion for testing the self-similarity features of a process. The main idea is to avoid performing an investigation on all the the aggregated processes for several values of m, and to develop a self-similarity criterion for the original, unaggregated process.

This procedure is formally grounded by the following theorem.

Theorem 1. *A process X is asymptotically self-similar if*

$$C(k) = C^{[1]}(k) \sim k^{2H-2} \tag{5}$$

Proof. Condition (5) may be explicitly recast as $C(k) = E[x(i)x(i+k)] - A^2$, so that, by recalling the definitions of the aggregated process and of $C^{[m]}(k)$, we have

$$\begin{aligned} C^{[m]}(k) &= \frac{1}{m^2}\sum_{n=0}^{m-1}\sum_{j=0}^{m-1} E[x(mi+n)x(mi+mk+j)] - A^2 \\ &\sim \frac{1}{m^2}\sum_{n=0}^{m-1}\sum_{j=0}^{m-1} (mk+j-n)^{2H-2} \end{aligned}$$

where the asymptotic equivalence holds by assuming m and k large enough, so that $mk+j-n > 0$. To proceed further, let us exploit the definition of Riemann upper and lower sums to state the asymptotic equivalence of this double sum to the corresponding integral, namely

$$\begin{aligned} C^{[m]}(k) &\sim \frac{1}{m^2}\int_1^m\int_1^m (mk+j-n)^{2H-2}\,dj\,dn \\ &\sim \frac{m^{2H-2}}{(2H-1)2H}\left[(k-1)^{2H} + (k+1)^{2H} - 2k^{2H}\right] \\ &\sim m^{2H-2}k^{2H-2} \end{aligned} \tag{6}$$

where asymptotic conditions $2H(2H-1)k^{2H-2} \sim (k-1)^{2H} + (k+1)^{2H} - 2k^{2H}$ has also been exploited.

Let us now compute $C^{[m]}(0)$ from the very definition of $C^{[m]}(k)$, i.e.

$$\begin{aligned} C^{[m]}(0) &= \frac{2}{m^2}\sum_{n=0}^{m-2}\sum_{j=n+1}^{m-1} E[x(mi+n)x(mi+j)] \\ &+ \frac{E[x^2(0)]}{m} - A^2 \end{aligned}$$

where the term $E[x^2(0)]/m$, corresponding to the case $j = n$, vanishes for m large enough. With this, and by exploiting (5), we easily get

$$\begin{aligned} C^{(m)}[0] &\sim \frac{2}{m^2}\sum_{n=1}^{m-1}(m-n)n^{2H-2} \\ &\sim \frac{2}{m^2}\int_1^m (m-n)n^{2H-2}\,dn \\ &\sim \frac{1}{(2H-1)H}m^{2H-2} \end{aligned} \tag{7}$$

By using (6) and (7), we easily have that the self-similarity condition (2) holds with $\beta = (2H-1)H$. □

Let us finally stress that Theorem 1 establishes a link between the autocovariance function of the aggregate process and of the original one, which can be exploited to easily verify and characterize self-similarity features of the process.

3. SOURCE AVERAGING PROCESSING

Theorem 2. *The process generated by a source obtained by applying the averaging operator* (3) *to* N *self-similar sources with Hurst coefficients* H_j *and activity* $A_j = p_j(0)$ $j = 1, \ldots, N$ *is again self-similar and characterized by the following parameters*

$$\begin{aligned} A_a &= \frac{1}{N}\sum_{j=1}^{N} p_j(0) \\ C_a(k) &\sim \frac{1}{N^2}k^{2H_\alpha-2} \end{aligned}$$

where $H_\alpha \geq H_j$ *for each* $j = 1, \ldots, N$.

Proof. The computation of A_a trivial derives from its very definition. The proof of the autocovariance trend rely on the use of Theorem 1 for the non aggregate process. By using the very definition of autocovariance and by using the source independence condition, we have

$$C_a(k) = \left[\frac{1}{N^2}\sum_{j=1}^{N} p_j(k) + \frac{1}{N^2}\sum_{j=1}^{N}\sum_{\substack{q=1\\q\neq j}}^{N} p_j(0)p_q(0)\right] - \left[\frac{1}{N}\sum_{j=1}^{N} p_j(0)\right]^2$$

$$= \frac{1}{N^2}\sum_{j=1}^{N}[p_j(k) - p_j^2(0)] = \frac{1}{N^2}\sum_{j=1}^{N} C_j(k)$$

Note that by applying Theorem 1, we have that $C_j(k) = k^{2H_j-2}$ and when k is large enough the trend is regulated by the source characterized to have the larger exponent, i.e., by the source with the larger Hurst parameter. □

In the case of independent sources with $p_j(0) = p(0)$ and $H_j = H$ for $j = 1, \ldots, N$ we have the obvious result $A_a = p(0)$ and $C_a(k) = C(k)/N$.

4. SOURCE MAXIMUM PROCESSING

Theorem 3. *The process generated by a source obtained by applying the maximum operator* (4) *to* N *self-similar sources with Hurst coefficients* H_j *and activity* $A_j = p_j(0)$ $j = 1, \ldots, N$ *is again self-similar and characterized by the following parameters*

$$\begin{aligned} A_{\max} &= 1 - \prod_{j=1}^{N}[1 - p_j(0)] \\ C_{\max}(k) &\sim k^{2H_\alpha-2}\prod_{\substack{j=1\\j\neq\alpha}}^{N}[1 - p_j(0)]^2 \end{aligned}$$

where $H_\alpha \geq H_j$ *for each* $j = 1, \ldots, N$.

Proof. To evaluate the activity of this processing we have to compute $A_{\max} = E[y_j(i)]$. To do so, observe that $y_j(i)$ is in the same domain of the original sources and that $E[y_j(i)] = P\{\max_j[x_j(i)] = 1\}$. Now, by defining the commutative and associative $+$ operator such that $0 + 0 = 0$, $1 + 0 = 1$ and $1 + 1 = 1$, we have:

$$\begin{aligned} A_{\max} &= P\left\{\overset{N}{\underset{j=1}{+}} x_j(i) = 1\right\} \\ &= 1 - P\left\{\overset{N}{\underset{j=1}{+}} x_j(i) = 0\right\} \\ &= 1 - \prod_{j=1}^{N} P\{x_j(i) = 0\} \\ &= 1 - \prod_{j=1}^{N}[1 - p_j(0)] \end{aligned}$$

where the only way to obtain $\overset{N}{\underset{j=1}{+}} x_j(i) = 0$ is that all the $x_j(i) = 0$ for any j and due to the source independence, this may be computed by taking the product of the probability $P\{x_j(i) = 0\} = 1 - p_j(0)$.

As far as the autocovariance function is concerned, let us notice first that

$$\begin{aligned} &P\left\{\overset{N}{\underset{j=1}{+}} x_j(i+k) = 1 \text{ and } \overset{N}{\underset{j=1}{+}} x_j(i) = 1\right\} \\ &= P\left\{\overset{N}{\underset{j=1}{+}} x_j(i+k) = 1 \middle| \overset{N}{\underset{j=1}{+}} x_j(i) = 1\right\} \\ &\quad \times P\left\{\overset{N}{\underset{j=1}{+}} x_j(i) = 1\right\} \\ &= \left[1 - P\left\{\overset{N}{\underset{j=1}{+}} x_j(i+k) = 0 \middle| \overset{N}{\underset{j=1}{+}} x_j(i) = 1\right\}\right] A_{\max} \\ &= \left[1 - \frac{P\left\{\overset{N}{\underset{j=1}{+}} x_j(i+k) = 0 \text{ and } \overset{N}{\underset{j=1}{+}} x_j(i) = 1\right\}}{A_{\max}}\right] A_{\max} \\ &= A_{\max} - P\left\{\overset{N}{\underset{j=1}{+}} x_j(i+k) = 0 \text{ and } \overset{N}{\underset{j=1}{+}} x_j(i) = 1\right\} \end{aligned}$$

where the term $\overset{N}{\underset{j=1}{+}} x_j(i+k) = 1$ has been substituted with the more tractable $\overset{N}{\underset{j=1}{+}} x_j(i+k) = 0$ that holds if and only if all the $x_j(i) = 0$ for each j.

The same approach may be used to remove the left event $\overset{N}{\underset{j=1}{+}} x_j(i) = 1$, which leads to

$$\begin{aligned} &P\left\{\overset{N}{\underset{j=1}{+}} x_j(i) = 1 \text{ and } \overset{N}{\underset{j=1}{+}} x_j(i+k) = 0\right\} \\ &= 1 - A_{\max} - P\left\{\overset{N}{\underset{j=1}{+}} x_j(i+k) = 0 \text{ and } \overset{N}{\underset{j=1}{+}} x_j(i) = 0\right\} \end{aligned}$$

By merging the last two results we get

$$P\left\{\overset{N}{\underset{j=1}{+}} x_j(i+k)=1 \text{ and } \overset{N}{\underset{j=1}{+}} x_j(i)=1\right\}$$
$$=2A_{\max}-1+P\left\{\overset{N}{\underset{j=1}{+}} x_j(i+k)=0 \text{ and } \overset{N}{\underset{j=1}{+}} x_j(i)=0\right\}$$
$$=2A_{\max}-1+\prod_{j=1}^{N} P\left\{x_j(i+k)=0 \text{ and } x_j(i)=0\right\}$$

where the last equation is a straightforward consequence of sources independence. We need now to compute the term $P\{x_j(i+k)=0 \text{ and } x_j(i)=0\}$ which depends on a single source. By its very definition we have

$$P\{x_j(i+k)=0 \text{ and } x_j(i)=0\}$$
$$=[1-P\{x_j(i+k)=1|x_j(i)=0\}]\,P\{x_j(i)=0\}$$
$$=\left[1-\frac{P\{x_j(i)=0 \text{ and } x_j(i+k)=1\}}{1-p_j(0)}\right][1-p_j(0)]$$
$$=1-p_j(0)-P\{x_j(i)=0|x_j(i+k)=1\}\,p_j(0)$$
$$=1-p_j(0)-\left[1-\frac{P\{x_j(i)=1 \text{ and } x_j(i+k)=1\}}{p_j(0)}\right]p_j(0)$$
$$=1-2p_j(0)+p_j(k)$$

By recalling the autocovariance definition and by using the above result we further obtain

$$\begin{aligned}
C_{\max}(k) &= P\left\{\overset{N}{\underset{j=1}{+}} x_j(i+k)=1 \text{ and } \overset{N}{\underset{j=1}{+}} x_j(i)=1\right\}-A_{\max}^2\\
&= 2A_{\max}-1+\prod_{j=1}^{N}[1-2p_j(0)+p_j(k)]-A_{\max}^2\\
&= \prod_{j=1}^{N}[[1-p_j(0)]^2+p_j(k)-p_j^2(0)]-\prod_{j=1}^{N}[1-p_j(0)]^2\\
&= \prod_{j=1}^{N}[[1-p_j(0)]^2+C_j(k)]-\prod_{j=1}^{N}[1-p_j(0)]^2
\end{aligned}$$

Note that when k is large enough $C_j(k)$ tend to vanish, so the first product may be recast as

$$\prod_{j=1}^{N}[[1-p_j(0)]^2+C_j(k)]$$
$$\sim \prod_{j=1}^{N}[1-p_j(0)]^2+\sum_{j=1}^{N} C_j(k)\prod_{\substack{q=1\\q\neq j}}^{N}[1-p_j(0)]^2$$

which complete the proof by exploiting Theorem 1 and by following the same consideration reported in the proof of Theorem 2. □

In the case of independent sources with $p_j(0)=p(0)$ and $H_j=H$ for $j=1,\ldots,N$, we have $A_{\max}=1-[1-p(0)]^N$ and $C_{\max}(k)\sim k^{2H-2}[1-p(0)]^{2(N-1)}$. If the activity $p(0)$ of the single source tends to zero, then $A_{\max}\sim Np(0)$ and $C_{\max}(k)\sim k^{2H-2}[1-2Np(0)]$.

5. CONCLUSIONS

We here report some results on the analytical characterization of self-similar sources. Theorem 1 allows to introduce a simple criterion that can be used to verify the self-similarity feature of a process without computing its aggregation. Such a Theorem is the key point to evaluate the self-similar features of the process obtained by averaging (Theorem 2) or maximizing (Theorem 3) the output of several independent sources. While the activity factor of the resulting source strongly depends on the particular process, in both cases we show that the resulting process is also self-similar with an higher Hurst parameter with respect to the original ones. It is also worthwhile to stress that the last two results also hold when one or more sources are not self-similar. In fact, all non-self-similar sources have an autocovariance decay faster with respect to the self-similar ones, so that their contribution result to be vanishing in the same way as a source with $H=0.5$, i.e. the lower bound for the Hurst parameter range. The unique important condition for this kind of extension is obviously that at least one source is self-similar. If this property holds for different kinds of sources post-processing still remains an open question which need to be addressed with the goal of realizing flexible artificial traffic sources to be employed for off-line testing of network apparatus.

6. REFERENCES

[1] W.E. Leland, M.S. Taqqu, W. Willinger, D.V. Wilson, "On the Self-Similar Nature of Ethernet Traffic," IEEE Trans. on Net., Vol.2, pp. 1-15, 1994.

[2] W.E. Leland, D.V. Wilson, "High Time-resolution Measurement and Analysis of LAN Traffic: Implications for LAN Interconnection," Proc. INFOCOM'91, 1991.

[3] A. Erramilli, P. Pruthi, W. Willinger, "Self-Similarity in High-Speed Network Traffic Measurements: Fact or Artifact?, " 12th Nordic Teletraffic Seminar-NTS 12 (VTT Symposium 154), 1995.

[4] A. Erramilli, P. Pruthi, W. Willinger, "Recent Developments in Fractal Traffic Modeling," Proc. ITC '95, 1995.

[5] I. Norros, "The Management of Large Flows of Connectionless Traffic on the Basis of Self-Similar Modeling," Proc. ICC'95, 1995.

[6] I. Norros, "On the Use of Fractional Brownian Motion in the Theory of Connectionless Networks," IEEE JSAC, Vol.13, pp.953-962, 1995.

[7] V. Paxson, S. Floyd, "Wide Area Traffic: The Failure of Poisson Modeling," IEEE/ACM Trans. on Networking, Vol. 3, pp. 226-244, 1995.

[8] E. Erramilli, P. Singh, "An application of deterministic chaotic maps to model packet traffic," Queueing Systems, vol. 20, pp. 171-206, 1995.

[9] Z.Fan, P.Mars, "Self-Similar Generation and Parameter Estimation Using Wavelet Transform," Proc. GLOBECOM'97, 1997.

[10] A. Giovanardi, G. Mazzini, R. Rovatti, G. Setti, "Features of Chaotic Maps with Self-Similar Trajectories," Proc. NOLTA'98, 1998.

[11] R. Rovatti, G. Mazzini, G. Setti, A. Giovanardi, "Statistical modeling and design of discrete-time chaotic processes: advanced finite-dimensional tools and applications," Proceedings of the IEEE, vol 90, n. 5, May 2002, pp. 820 -841.

MODELLING OF TCP PACKET TRAFFIC IN A LARGE INTERACTIVE GROWTH NETWORK

D.K. Arrowsmith and M. Woolf

Mathematical Sciences, Queen Mary, University of London
London E1 4NS, UK

ABSTRACT

We simulate packet traffic on a large interactive growth network designed to reflect the topology of the real internet. A closed-loop algorithm (TCP) is used to define packet transfer. We compare long range dependent and Poisson sources, and the response of the network to changes in routing queue service rate. By limiting queue length we simulate the effects of packet loss. We vary the pattern of hosts in the network and compare results from different networks generated using the same parameters. The two source types produced very different results highlighting the importance of LRD source models. Results also showed robustness with respect to changes in network and host pattern. This demonstrates the validity of the interactive growth method for characterising real scale-free networks.

1. INTRODUCTION

Simulation of packet traffic is a very useful tool in understanding the behaviour of networks. In this paper we attempt to move closer to simulation of real networks without using nearly exact modelling offered by network simulators such as NS [1] and Opnet [2]. These are limited in the size of network they can handle, and the complexity of their models makes extraction of useful information more difficult.

Simulated packet traffic in regular networks [3, 4] was compared for long range dependent (LRD) [5] sources with Poisson or short range dependent (SRD) sources. These were open-loop simulations. On the internet transmission control protocol (TCP), which is closed loop, dominates. Erramilli *et al* [6] showed that TCP feedback modifies self-similarity in offered traffic without creating or removing the self similarity itself. For these reasons, we have moved to simulations that model in a slightly simplified manner the most commonly used version of TCP.

A further aspect of our original simulations was that regular networks were modelled (Manhattan, hexagonal and triangular networks). Clearly this is not close to the real situation. For this paper we have changed to scale-free graphs that match the topology of the internet as closely as possible.

Studies of the topology of the internet [7, 8] have shown that it is a scale-free network (Barabasi). Faloutsos *et al* [9] found that the node degree has a power law probability distribution, $P(k) \sim Ck^{-2.22}$, where k is the node connectivity. The interactive growth (IG) networks [10] are created by mirroring the way in which the internet is generated in reality. Shi and Mondragon [11] have shown that these graphs model both the degree distribution of the actual network and the tier structure. An important feature is the 'rich-club' phenomenon in which important ('rich') nodes preferentially attach to one another and are also more likely to be connected to by new nodes. The move away from regular networks brought up the problem that some nodes were very highly connected and using the same service rate for each node was no longer appropriate. For this reason we have arranged for the service rate to be linked to the connectivity of each node.

2. NETWORK TRAFFIC MODEL

The network topology used was a 1024 node IG network with a power law index of 2.22. Each node is designated as either a host or a router. Routers have a single routing queue that receives packets in transit across the network, and releases them back onto the network at a rate governed by the connectivity of the node. The simulation is of the fixed-increment time advance type rather than next-event time advance [12]. This allows the routing queue service rate to be set as $0.25n^a$ packets per time tick of the simulation, where n is the degree. The index a has been chosen to be between 1 and 2 in this paper. Hosts have identical routing queues and function in the same way, but additionally act as sources. They have transmit buffers that hold packets generated by LRD and SRD traffic sources until they have been acknowledged. The exact mechanism for this is described below.

A simplified version of TCP Reno [13] is used as the network protocol. This is the predominate protocol used on the internet at present. This version is derived from that described in Erramilli *et al* [6]. It concentrates on the slow start mechanism because in real networks this both affects all connections, and is the dominant effect for most connections. It also has a more marked effect on dynamics than congestion avoidance.

As in our previous work, a double intermittency map [14] is used as the basis for each LRD traffic source; a uniform random number generator for each SRD source. Specifically, we use the family of *Erramilli* interval maps, [14], $f = f_{(m_1,m_2,d)} : [0,1] \to [0,1]$, where

$$f(x) = \begin{cases} x + (1-d)\,(x/d)^{m_1}, & x \in [0,d], \\ x - d\,((1-x)/(1-d))^{m_2}, & x \in (d,1], \end{cases} \quad (1)$$

Here $d \in (0, 1)$, and the parameters $m_1, m_2 \in [3/2, 2]$ induce *map intermittency.* The map f is used to produce an *orbit* of real numbers $x_n \in [0, 1]$ which is converted to binary sequence by associating the symbols '0' or '1' to the intervals $[0, d]$, $(d, 1]$ respectively. It is this binary output that is seen to represent the packets output of a host.

However, they are used in a slightly different way. One sojourn period in the 'on' side represents a whole file which is then *windowed* using the TCP *slow-start* algorithm: *at the start of the file the window size is set to 1. Only a single packet is sent at that time tick.* When a packet reaches its destination an acknowledgement is returned to the source: *once this packet has been acknowledged, the window size is doubled and the next two packets are sent.* When both these packets have been acknowledged: *the window size is doubled again and a new window of packets is sent. The doubling process is repeated until the end of the file or the maximum window size is reached which is constant until the end of the file is reached.* (Congestion avoidance is not modelled for the reason given above.) Thus the dynamics now involves not only a dynamical packet traffic mechanism but also *window size* dynamics.

This dynamical output is augmented by a window variable on the same time base. At a given node i of the network, and time $t = n$, there is a current state $x_i(n)$, and also a current window size, $w_i(n)$, for the number of packets that can be sent at time $t = n$. There is also a *residual file size* $s_i(n)$, at node i which is given by the number of iterates of f such that $f^{s_i(n)}(x_i(n)) > d$, and $f^{s_i(n)+1}(x_i(n)) < d$. The source will send $p_i(n) = \min\{w_i(n), s_i(n)\}$ packets. The full dynamics therefore takes the form, see [6]:

For $x_i(n) < d$, i.e. no packet mode -
$w_i(n+1) = 0$, and $x_i(n+1) = f(x_i(n))$.
For $x_i(n) > d$, i.e. packet mode -

$$w_i(n+1) = \begin{cases} 1, & \text{if} \quad x_i(n-1) < d, \\ \min\{2w_i(n), w_{max}\}, & \text{otherwise,} \end{cases} \tag{2}$$

and $x_i(n+1) = f^{p_i(n)}(x_i(n))$.

The above assumes all packets in a window are acknowledged within the retransmission timeout (RTO). If not so the window is sent again with the RTO doubled and window size reset to 1. During the 'off' periods of the map the window size is set to zero and no packets are sent. The initial RTO size is calculated using the standard exponential average method [15].

The chosen routing algorithm is similar to that used previously for regular networks [3, 4]. Before the simulation starts an all-pairs shortest path algorithm is used to calculate shortest paths between all pairs of hosts in the network (all links between nodes are assumed to have unit length). This is read in as a look-up table and consulted whenever a packet is to be moved from a node. At each time step packets are forwarded from the head of each routing queue. Any acknowledgements reaching their destination trigger the release of the next window of packets from that host.

The number forwarded is governed by the service rate formula [1]. The rules for forwarding packets are as follows: *A neighbouring node with the shortest path to the destination is always selected.* Often more than one neighbour shares this minimum distance, in which case the link that has had fewest packets forwarded across it is selected. *If these counts are equal a random selection is made.* If the selected neighbour has reached is routing queue length limit the packet is dropped.

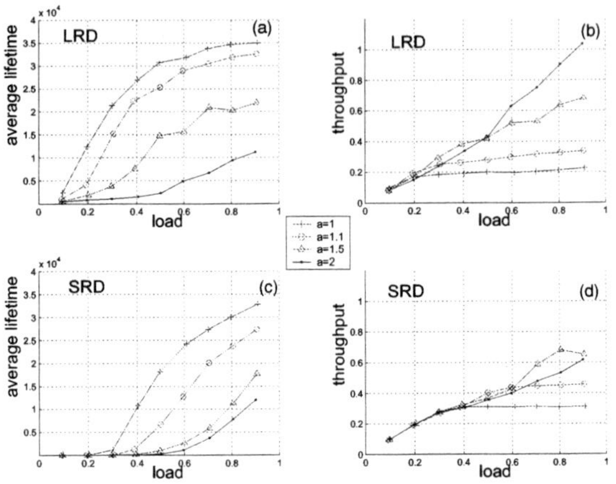

Figure 1. Comparisons between the open-loop simulations of [3] and the closed-loop simulations on scale-free IG networks of the same size and host density.

3. SIMULATION RESULTS

In Fig. 1 hosts were selected randomly. Clearly the IG networks with TCP become congested much more rapidly and have much lower throughputs. The reason for this becomes clear when you consider the TCP. The requirement that packets be acknowledged before the next window is sent is quite conservative. It means that hosts cannot send new windows more frequently than the round trip times. Any congestion is reacted to immediately because this increases RTT's and effectively backs off the sources. Since file sizes are not that big and window sizes are often reset to 1 because of the RTO limit, throughput can never be that high. The IG network with TCP therefore has much lower capacity than the open loop networks. In fact if TCP were used with a regular network this would perform even more poorly because of the longer average path length in the network. For this reason more detailed comparisons between these simulations and our earlier work are not possible.

Note that average lifetimes includes the waiting time in transmit buffers. Real systems would not include this. In addition loads quoted are those offered to the transmit buffers, rather than the load exiting theses buffers. We have done this to allow the end users experience to be modelled.

In fig. 2 we compare different sources and routing queue service rates. As before a 1024 node IG network is used. All nodes with degree 1 or 2 (589 out of the 1024) are set as hosts, the rest as routers. There is a marked difference between LRD and SRD sources. Comparing throughputs for the two source types at the lower server index values of 1 and 1.1, behaviour is similar. Throughput matches load up to a threshold, and then levels out. However, this threshold is more than 50% higher for SRD sources. For $a = 1.5$,

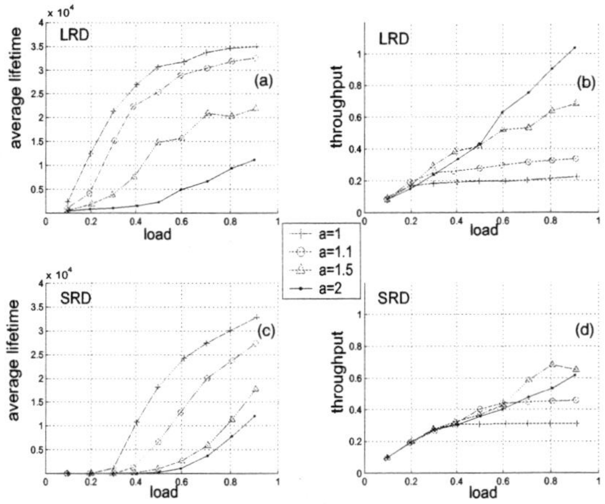

Figure 2. The number of packets that can be served at each time instant is increased according to a power law n^a, where n is the degree of the node and $a = 1, 1.1, 1.5, 2$.

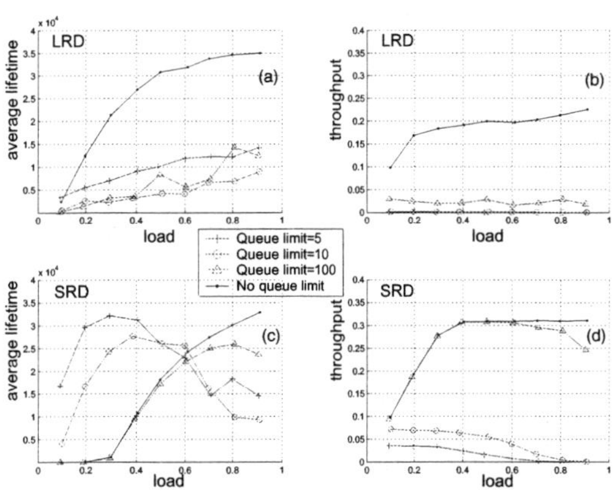

Figure 4. Effect of varying queue length limit on average packet lifetime and throughput for LRD and SRD sources.

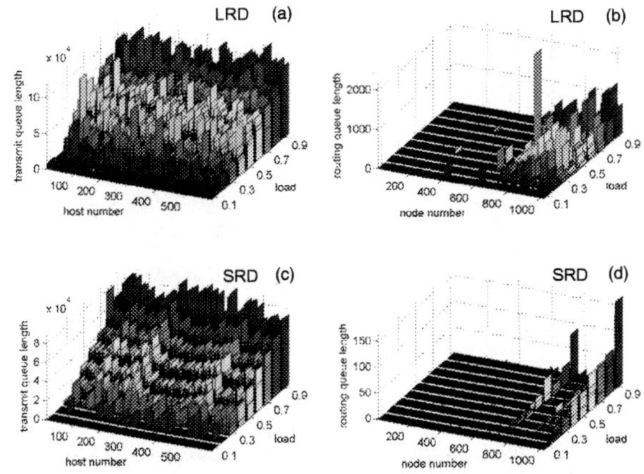

Figure 3. Queue lengths for host nodes of an IG network are shown as load increases, for both SRD and LRD sources.

throughputs are similar, with no discernable threshold. For the highest server strength the situation is reversed and the network with LRD sources has a higher throughput, able to handle the maximum load applied to it without becoming overloaded.

Average lifetimes behave similarly to open loop simulations in that, for low loads lifetimes for SRD sources are much less than for LRD. In SRD sources we see the same transition seen in the open-loop case from free to congested. As the service rate is increased this transition becomes less and less sharp. This is because the increasing network capacity slows the onset of congestion. Fig. 3 illustrates the behaviour of transmit queue lengths and routing queue lengths. Transmit queue lengths reflect the behaviour seen in the average lifetimes for both types of source. At low loads they are very small, as loads increase they rise rapidly. As routing queue service rates increase maximum queue lengths do not change greatly, but the number of long queues reduces. The reason for this is a subject for further investigation.

Routing queues are much shorter than transmit queues. For the LRD sources there is no well defined pattern to the queues, except that they all occur in the higher degree half of the plots (plots are arranged in order of degree). The behaviour for SRD sources is a bit different. Queue lengths are much shorter. There is a clear progression in size as the load increases, and the same nodes produce queues at different loads. The fact that certain nodes become congested whatever the load suggests that the distribution of nodes is important. At higher server strengths the most highly connected nodes become dominant (and produce the longest queues). The much longer queues seen in the transmit buffers show that these queues are the dominant factor contributing to average lifetimes.

We simulate the loss of packets in real networks by limiting the length of the routing queues. Fig. 4 shows data for the same network and host pattern for four different queue limits. The server index value is 1. Very severe packet loss is simulated here - this does not equate to the level of packet loss in real networks. In the LRD case decreasing the queue limit reduces the average lifetime considerably. Limited queues means packets get dropped at the routers and are therefore re-sent more frequently, resulting in shorter waits in the transmit buffer. The effect is more pronounced for the very low queue length limit.

Average lifetimes for SRD sources behave quite differently. For lifetimes peak at a load of 0.3 for the queue limit of 5. As the queue limit is raised this peak shifts to higher load values. With an infinite queue the lifetime curve has the familiar 's' shape.

Throughputs for LRD sources are greatly reduced when a queue limit is imposed. For the smaller queue limits throughput is very low indeed.

For SRD sources throughput is similar for the queue limit of 100, but again greatly reduced for the smaller queue limits. Even so, these throughputs are still much greater than for the LRD sources. This is linked to the shorter

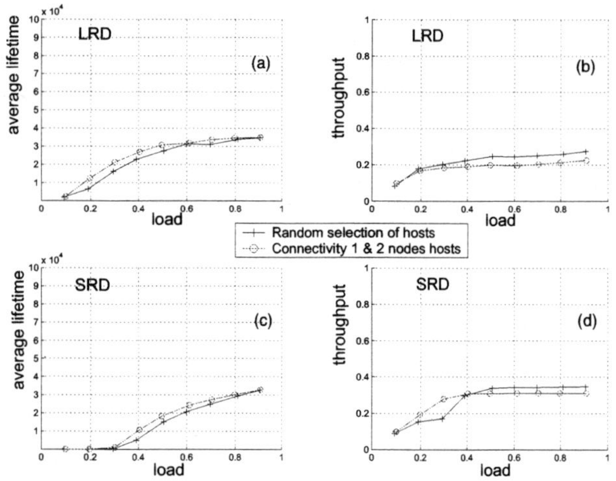

Figure 5. Average lifetime and throughput characteristics for different topological spreads of hosts in an IG network.

queues for SRD sources. Most are less than a 100 so this limit has little effect.

Fig. 5 shows results from the same IG network, but with the host pattern changed so that hosts are selected randomly, but with the same density (589 out of 1024). Results are similar, but the network with randomly placed hosts always performs better than the one with hosts placed at the low degree hosts. Average lifetimes are lower and throughput slightly higher.

The results have been shown to be robust by taking different random selections of host, and different IG networks with the same parameters.

4. SUMMARY

We have extended our earlier work with regular lattices and open-loop packet traffic algorithms. The use of large IG networks and the TCP protocol have brought us much closer to simulating the internet and other scale-free networks.

The robustness of the results with respect to changes in IG network, and changes in host pattern validate the IG method as a way of characterising real scale-free networks such as the internet. They also show the same kind of repeatability seen in our previous work.

Scale-free networks clearly have very different properties to regular networks. The fact that a small number of nodes have very high degree is probably the major point of difference. For this reason we have only made rough comparisons between the two types of graph.

The most outstanding differences are found when comparing throughput for LRD and SRD sources. Clearly the *bursty* property of LRD sources can have a very pronounced effect in terms of queue build up when several highly connected nodes are adjacent to one another. Very high server rates can overcome the problems caused by LRD sources, and even improve throughput over SRD sources in extreme cases.

We have simulated very severe packet loss. This has a dramatic effect on throughput, but should not be equated with the real world situation.

Our results show what the user experiences when the network is presented with a much higher load than it can easily handle. At very low loads the TCP mechanism would prevent the network ever getting congested. TCP is a very effective control mechanism in this sense. Even larger networks are preferable which will be addressed in future work.

5. REFERENCES

[1] The Network Simulator 2, ns2 - http://www.isi.edu/nsnam/ns/.

[2] Opnet, Opnet Tech. Inc.-http://www.opnet.com.

[3] M. Woolf, D.K. Arrowsmith, R.J. Mondragon, J.M. Pitts, Optimization and phase transitions in chaotic model of data traffic, *Phys. Rev E*, 66, 046106 (2002).

[4] D.K. Arrowsmith, R.J. Mondragon, J.M. Pitts, M. Woolf, Internet packet traffic congestion, *Proc Int. Sym. Cir. and Sys.*, 2003, ISCAS '03., **3**, 2003, 746 -749.

[5] W.E. Leland, M.S. Taqqu, W. Willinger and D.V. Wilson, On the Self-Similar Nature of Ethernet Traffic. *Proc. ACM SIGCOMM* 93, 1993.

[6] A. Erramilli, M. Roughan, D. Veitch and W. Willinger, Self-Similar Traffic and Network Dynamics, *Proc. IEEE*, **90**, 2002, 800-819.

[7] R.Albert and A-L Barbasi, Statistical Mechanics of Complex Networks, *Rev. Mod. Phys.* **74**, 2002, 47-97.

[8] L. Subramanian, S. Agarwal, J. Rexford and R. H. Katz, Characterizing the Internet Hierarchy, *Proc. of INFOCOM 2002*, June, 2002.

[9] M. Faloutsos, P. Faloutsos, and C. Faloutsos, On Power-Law Relationships of the Internet Topology, *Proc. ACM/SIGCOMM, Comput. Commun. Rev.*, **29**, 251-262, 1999.

[10] S. Zhou and R. J. Mondragon, Towards modelling the Internet topology - the Interactive Growth model, *Proc. of the 18th Int. Teletraffic Cong., ITC18*, Berlin, 2003.

[11] S. Zhou and R. J. Mondragon, The rich-club phenomenon in the Internet topology, *to appear IEEE Communications Letters* , 2003.

[12] A.M. Law and W.D. Kelton, Simulation Modeling and Analysis, 3rd Edition, *McGraw-Hill*, 2000.

[13] W. R. Stevens, TCP/IP Illustrated, Volume 1: The Protocols, *Addison-Wesley*, 1994.

[14] A Erramilli, R.P. Singh and P.Pruthi, *Proc. of the 14th Int. Teletraffic Conf.* 1994, edited by James W. Roberts (Elsevier, Amsterdam, 1994), 329-38.

[15] W. Stallings, Data and Communications, *Prentice Hall Int. Inc.*,2000.

ON THE CORRELATION OF TCP TRAFFIC IN BACKBONE NETWORKS

H.X. Nguyen *, *P.Thiran*

EPFL: Swiss Federal Institute of Technology
LCA-ISC-I&C (Institute of Communication Systems)
CH-1015 Lausanne, Switzerland

C. Barakat †

INRIA - Planete group
06902 Sophia Antipolis
France

ABSTRACT

In this paper, we study the second order statistics of traffic in an Internet backbone. We model the traffic at the flow level by a Poisson shot noise process. This model is quite parasimonious, and is driven only by variables that can be easily obtained from measurements, namely flow sizes, durations and arrival rate. We consider the auto-correlation of TCP traffic where the loss process of each TCP connection is assumed to be Poisson. Using a stochastic differential equation, we are able to provide an upper bound on the auto-covariance function of the aggregated TCP traffic whose tightness is shown by simulations with the network simulator-*ns*.

1. INTRODUCTION

Second order statistics of network traffic, namely the auto-correlation, plays an important role in network performance evaluation (e.g., [1]) and traffic modelling (e.g., [2]). There is a huge amount of work in the literature which investigates the second order characteristics of network traffic (see [3] and the references therein). A recent trend in studying network traffic is to model traffic as fluid flows where each flow is a stream of packets which have common source and/or destination [4, 5]. Modelling traffic at the packet level is very difficult, since traffic on a link is the result of a high level of multiplexing of numerous flows whose behavior is strongly influenced by the transport protocol and by the application. From a simplicity stand point, it is much easier to monitor flows than to monitor packets in routers. The instantaneous rate of a flow depends on the network and transport protocol dynamics. Transmission Control Protocol (TCP) flows, which account for about 95% of the Internet traffic, can have a highly varying instantaneous rate due to the complexities in the mechanisms that TCP employs. However, the majority of work on auto-correlation of network traffic in the literature concentrates on either studying the correlation of a single TCP connection [3], or of the aggregated traffic in which flows have simple shapes such as rectangle.

The objective of this work is to study the auto-correlation of traffic in an Internet backbone where links are usually not congested. To this end, we use a Poisson shot noise process where a flow is a generic notion which can be a TCP flow or a UDP stream. We will concentrate only on the auto-correlation of traffic where all flows are TCP connections. Instead of using Markovian models for studying auto-correlation of TCP traffic (e.g., [3]), our approach uses a stochastic differential equation to obtain a tight bound on the auto-covariance function of TCP flows. We find an upper bound on the auto-correlation of the aggregated TCP traffic. This upper bound is a function of only a few flow parameters such as flow arrival rate, flow sizes and flow durations. These parameters can be easily computed by a router (e.g., using a tool such as NetFlow, which provides flow information in Cisco routers). Such a bound on the auto-covariance function of backbone traffic is useful for backbone operators who currently have only very basic information about the traffic.

The rest of the paper is organized as follows. Section 2 gives a summary of the shot noise model that was developed in previous work. Section 3 presents the auto-correlation of TCP traffic. Section 4 contains the simulation results and Section 5 concludes the paper.

2. SHOT NOISE MODEL

Our traffic model in this paper is the shot-noise model which was developed by Barakat et.al. in [6, 4]. In [6], one definition of flow is a stream of packets having the same source and destination IP addresses, same source and destination port numbers, and the same protocol number. Alternatively, a flow can also be defined as a stream of packets having the same /24 destination address prefix (i.e., only the 24 most significant bits).

Let T_n, $n \in \mathbf{Z}$, denote the arrival time of the n-th flow to the backbone link under consideration. Let S_n and D_n denote the size and duration of the n-th flow. The size of a

*H.X. Nguyen is supported by grant DICS 1830 of the Hasler Foundation, Bern, Switzerland.

†C. Barakat performed the work while at EPFL

flow is the volume of data it transports during its lifetime. Let $X_n(t-T_n)$ denote the rate of the n-th flow at time t (e.g., in bits/s), with $X_n(t-T_n)$ equals to zero for $t < T_n$ and for $t > T_n + D_n$. We call $X_n(.)$ the flow rate function or shot. The total rate of data through the backbone link, which we denote by $R(t)$, is the result of multiplexing of all shots: $R(t) = \sum_{n \in \mathbf{Z}} X_n(t-T_n)$. Flow rate functions $\{X_n(.)\}$ are assumed to be independent and identically distributed. These assumptions hold relatively well on a backbone link [6]. The total data rate $R(t)$ can be seen as a shot-noise process [7], where the term "shot" is synonymous of the "flow rate function".

Assume that traffic flows arrive at the backbone link as a homogeneous Poisson process of finite rate. Using elements from theory of Poisson shot-noise process, the authors in [6] found that the auto-covariance function of the aggregated traffic can be expressed as

$$\begin{aligned} C_R(\tau) &= \mathbf{E}[R(t-\tau)R(t)] - \mathbf{E}[R^2(t)] \\ &= \lambda \mathbf{E}\left[\mathbf{1}_{\{D_n > |\tau|\}} \int_0^{D_n - |\tau|} X_n(u) X_n(u+|\tau|) du \right], \quad (1) \end{aligned}$$

Where $\mathbf{1}_A$ is the indicator function of event A.

3. CORRELATION STRUCTURE OF TCP TRAFFIC

The auto-correlation of the traffic strongly depends on the shot shape we consider, and thus on the dynamics of the flow rate, which in turn depends on many factors such as the definition of flows, the transport mechanism, the application nature, etc. In some important cases, we can make use of the protocol information to derive the shot function $X_n(.)$. The most typical example is TCP, whose dynamics shape the flows and can be captured by analytical models. TCP is a window-based flow control protocol that provides reliable end-to-end communication in data networks. It is designed to adapt to the various traffic conditions of the network: a TCP connection progressively increases its transmission rate until it receives some indication that the capacity along its path is almost fully utilized. On the other hand, when the network cannot accommodate the traffic, the data rate of the connections is reduced. More specifically, the transmission rate of a TCP connection is governed by the additive-increase multiplicative-decrease (AIMD) mechanism which is as follows. Between congestion events (we also call them loss events, since they are usually the times at which a packet loss is detected at the sender), the rate of a TCP connection increases linearly with a slope α, which is inversely proportional to the square of the round-trip time (RTT) of the connection. At the congestion events, the rate of a TCP connection is reduced by half. Precisely, α is related to the RTT by: $\alpha = 1/bRTT^2$ where b is the acknowledgement factor which indicates how many packets are included in one acknowledgement [8] (typically, $b = 2$).

In this work, we assume that congestion events have a Poisson distribution with intensity λ_ℓ. This assumption may not hold in practice, but is needed for the following analytical derivation.

Assume that all traffic flows are long-lived TCP flows, i.e. the congestion avoidance phase is dominant over the life time of the flows. Rewrite equation (1), we get

$$\begin{aligned} C_R(\tau) &= \lambda \mathbf{P}(D_n > |\tau|) \\ &\quad \mathbf{E}\left[\int_0^{D_n - |\tau|} X_n(u) X_n(u+|\tau|) \mathrm{d}u \,\middle|\, D_n > |\tau| \right]. \quad (2) \end{aligned}$$

Now if we condition on $D_n = d$ with $d > |\tau|$ then

$$\begin{aligned} &\mathbf{E}\left[\int_0^{D_n - |\tau|} X_n(u) X_n(u+|\tau|) \mathrm{d}u \,\middle|\, D_n = d, d > |\tau| \right] \\ &= \int_0^{d-|\tau|} \mathbf{E}_d \left[X_n(u) X_n(u+|\tau|) | \, d > |\tau| \right] \mathrm{d}u \\ &= (d - |\tau|) \mathbf{E}_d \left[X_n(t) X_n(t+|\tau|) | \, d > |\tau| \right]. \end{aligned}$$

Notation $\mathbf{E}_d$ is used to indicate the expected value is calculated under the condition $D_n = d$. The last equality is derived under the assumption that the TCP rate is in its stationary regime at time 0 so that $\mathbf{E}_d [X_n(t) X_n(t+|\tau|)]$ does not depend on time t.

From here on, we do the calculation for $\tau > 0$. For $\tau < 0$, the calculation is similar. Let $r_{d|\tau}(\tau) = \mathbf{E}_{d|\tau}[X_n(t) X_n(t+\tau)]$, where the subscript $d \mid \tau$ indicates the expected value is calculated under the conditions $D_n = d$, and $d > \tau$. Let

$$\begin{aligned} \mathrm{d}r_{d|\tau}(\tau) &= r_{d|\tau}(\tau + \mathrm{d}\tau) - r_{d|\tau}(\tau) \\ &= \mathbf{E}_{d|\tau}[X_n(t)(X_n(t+\tau+\mathrm{d}\tau) - X_n(t+\tau))]. \end{aligned}$$

We now consider the condition of occurrence of loss in $[t+\tau, t+\tau+\mathrm{d}\tau]$. When a loss appears in $[t+\tau, t+\tau+\mathrm{d}\tau]$, the rate is divided by 2, so

$$X_n(t+\tau+\mathrm{d}\tau) - X_n(t+\tau) = -\frac{X_n(t+\tau)}{2} + \alpha \mathrm{d}\tau,$$

Whereas when there is no loss in $[t+\tau, t+\tau+\mathrm{d}\tau]$, we have

$$X_n(t+\tau+\mathrm{d}\tau) - X_n(t+\tau) = \alpha \mathrm{d}\tau.$$

Since the loss process is Poisson with rate λ_ℓ, the probability that a loss appears in $[t+\tau, t+\tau+\mathrm{d}\tau]$ is independent of $X_n(.)$, and is equal to $\lambda_\ell \mathrm{d}\tau$. If all the flows have the same RTT, and thus the same α, it follows that

$$\begin{aligned} \mathrm{d}r_{d|\tau}(\tau) &= -\frac{\mathbf{E}_{d|\tau}[X_n(t) X_n(t+\tau)]}{2} \lambda_\ell \mathrm{d}\tau \\ &\quad + \mathbf{E}_{d|\tau}[X_n(t)] \alpha \mathrm{d}\tau. \quad (3) \end{aligned}$$

From [8], we have for a Poisson loss process: $\mathbf{E}_{d|\tau}[X_n(t)] = 2\alpha/\lambda_\ell$.

Inserting this value in (3), we obtain the following ordinary differential equation

$$\frac{\mathrm{d}r_{d|\tau}(\tau)}{\mathrm{d}\tau} + \frac{\lambda_\ell}{2} r_{d|\tau}(\tau) = \frac{2\alpha^2}{\lambda_\ell}.$$

From [4], we have the initial condition for Poisson losses $r_{d|\tau}(0) = \mathbf{E}_{d|\tau}[X_n^2(t)] = 4\mathbf{E}_{d|\tau}[S_n]^2/3d^2$.

Solving the above equation and simplifying the result using $\mathbf{E}_{d|\tau}[X_n(t)] = \mathbf{E}_{d|\tau}[S_n]/d$ ([4]), we obtain:

$$r_{d|\tau}(\tau) = \frac{1}{3}\frac{\mathbf{E}_{d|\tau}[S_n]^2}{d^2} e^{-\alpha\tau\frac{d}{\mathbf{E}_{d|\tau}[S_n]}} + \frac{\mathbf{E}_{d|\tau}[S_n]^2}{d^2}.$$

Let us define $f_{d|\tau}$ as the probability density function (pdf) of the random variables D_n whose probability space is $[\tau, \infty]$. Substituting the expression of $r_{d|\tau}$ in (2), we get

$$C_R(\tau) = \lambda \mathbf{P}(D_n > \tau) \int_\tau^\infty (d-\tau) \left(\frac{1}{3}\frac{\mathbf{E}_{d|\tau}[S_n]^2}{d^2} e^{-\alpha\tau\frac{d}{\mathbf{E}_{d|\tau}[S_n]}} + \frac{\mathbf{E}_{d|\tau}[S_n]^2}{d^2}\right) f_{d|\tau}\mathrm{d}d.$$

Since the function $f(x) = x^2 e^{-\alpha\tau d/x}/3d^2$ is convex for all $x > 0$, we have:

$$\frac{1}{3}\frac{\mathbf{E}_{d|\tau}[S_n^2]}{d^2} e^{-\alpha\tau\frac{d}{\mathbf{E}_{d|\tau}[S_n]}} \leq \mathbf{E}_{d|\tau}[\frac{1}{3}\frac{S_n^2}{d^2} e^{-\alpha\tau\frac{d}{S_n}}].$$

Thus,

$$\begin{aligned} C_R(\tau) &\leq \lambda \mathbf{P}(D_n > \tau) \\ &\int_\tau^\infty (d-\tau) \left(\mathbf{E}_{d|\tau}[\frac{1}{3}\frac{S_n^2}{d^2} e^{-\alpha\tau\frac{d}{S_n}}] + \frac{\mathbf{E}_d[S_n^2]}{d^2}\right) f_{d|\tau}\mathrm{d}d \\ &\leq \lambda \mathbf{P}(D_n > \tau) \left(\frac{1}{3}\mathbf{E}\left[\frac{S_n^2(D_n-\tau)}{D_n^2} e^{-\alpha\tau\frac{D_n}{S_n}} \middle| D_n > \tau\right]\right. \\ &\quad + \left.\mathbf{E}\left[\frac{S_n^2(D_n-\tau)}{D_n^2}\middle| D_n > \tau\right]\right). \end{aligned}$$

As $C_R(\tau)$ is an even function, we have the general result:

$$\begin{aligned} C_R(\tau) &\leq \lambda \mathbf{P}(D_n > |\tau|) \\ &\left(\frac{1}{3}\mathbf{E}\left[\frac{S_n^2(D_n-|\tau|)}{D_n^2} e^{-\alpha|\tau|\frac{D_n}{S_n}} \middle| D_n > |\tau|\right]\right. \\ &+ \left.\mathbf{E}\left[\frac{S_n^2(D_n-|\tau|)}{D_n^2}\middle| D_n > |\tau|\right]\right). \end{aligned} \quad (4)$$

There are two separate terms in the bound. The first term is an exponential function of τ, which vanishes quickly for large values of τ. The second term is a linear function of τ, which decreases slowly as τ increases. As a result, the bound has two distinct behaviors: one for small values of τ, when the exponential term dominates and one for large values of τ, when the linear term dominates. Furthermore, the loss rate λ_l does not figure in the bound and the round-trip time only appears in the exponential term via α.

4. SIMULATION RESULTS

We present a validation of the bound on the auto-covariance function of TCP traffic by simulation. We use the *ns* simulator to study two different scenarios. In the first scenarios, all flows have the same size but different durations. In the second scenario, both sizes and durations of flows change during each simulation.

4.1. Simulation scenario

In our simulations, a set of TCP Newreno flows transmit files over a 10 Mbps link which corresponds to the backbone link. Each flow transmits one file and all flows cross the backbone link in the same direction. The duration of each simulation is equal to 1000 seconds. Delayed acknowledgement option of TCP is enabled and each packet has a size of 500 bytes. Before arriving on the backbone link, all flows experience some packet losses with a probability of 3% (to introduce randomness in the durations of flows). TCP flows are generated according to a Poisson process. The rate of the Poisson process and the file sizes are chosen such that the 10Mbps backbone link always remains under-utilized. The round trip time of all TCP flows is set to 80ms. We compute the rate with which data cross the 10Mpbs link and store the variation of this rate as a function of time. This rate is used to calculate the auto-covariance of TCP traffic. We also measure the size and duration of each flow, which produces samples for the processes $\{D_n\}$ and $\{S_n\}$. The instantaneous rate $R(t)$ is measured by averaging the number of bytes that cross the 10Mpbs link over the interval of 100ms. In each simulation, we plot the auto-covariance of the real simulation traffic, and the upper bound of the auto-covariance for TCP traffic (4). In our simulations, the RTT was set to 80ms. α will then be: $\alpha = 1/(bRTT^2) = 500 * 8/(2 * 0.08^2)(bits/sec^2)$.

4.2. Constant-size flows

We set the arrival rate of TCP flows to 2 flows per second and we give file sizes constant values equal to 25Kbytes, 50Kbytes, 100Kbytes, 250Kbytes and 500Kbytes. We run a set of 10 simulations for each value of the file size. In each simulation, all files have the same size. The average of the 10 values obtained from simulations are plotted. The 95% confidence intervals for the actual traffic are also plotted, while the 95% confidence interval for the approximations are omitted because they are too narrow. The results are plotted in Figures 1(a) to (e).

4.3. Variable-size flows

We repeat the previous simulations with variable file sizes. To generate variable sizes, for each flow we pick a real num-

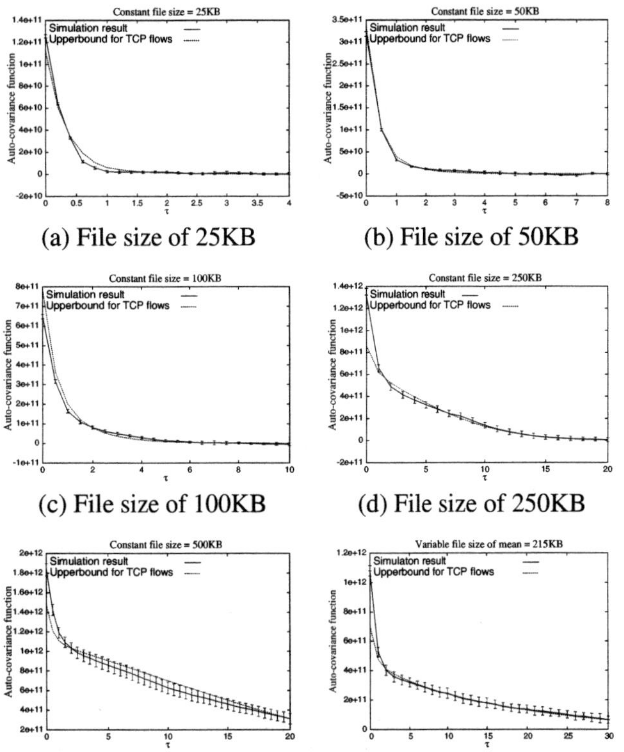

(a) File size of 25KB (b) File size of 50KB

(c) File size of 100KB (d) File size of 250KB

(e) File size of 500KB (f) Variable file size

Fig. 1. Simulation results showing the auto-covariance function $C_R(\tau)$ versus time lag τ.

ber randomly between 1 and 3 with a uniform distribution. The size of the flow in Kbytes is 10 to the power of the selected number. This way we get an average file size of 215KB. The results are plotted in Figure 1(f).

4.4. Observations

From the simulation results, we can observe that for all values of τ, the gap between the upper bound in (4) and the real auto-covariance of the traffic is very small. For large values of τ, both the upper bound and the auto-covariance of the simulated traffic decrease slowly as a function of τ with a rate close to linear. This is consistent with our finding in (4), where for large value of τ the exponential term vanishes and only the linear term contributes to the bound. For small values of τ, both the auto-covariance of the simulated traffic and the upper bound decrease exponentially fast as a function of τ. This can be explained by the domination of the exponential term in the bound. Note here that the bound (4) is obtained under some strong assumptions (i.e., losses are Poisson, and other approximations detailed in [6]), which may not hold in practice. This explains the instances in our simulation results where the bound (4) is not respected.

5. CONCLUSION

In this paper, we use a Poisson shot noise model to study the auto-correlation of Internet traffic in non-congested backbone links. We provide an upper bound on the correlation of aggregated TCP traffic where all flows are long-lived TCP flows. The upper bound is a function of only three flow parameters (λ: arrival rate, D_n flow duration, and S_n flow size) which can be obtained from passive measurements quite easily [4]. Such a bound can be used in network dimensioning and management to study the impact of flow arrival, flow sizes and durations on the auto-correlation of the traffic and hence on dimensioning the backbone.

6. REFERENCES

[1] F.Park, G.Kim, and M.Crovella, "On the effect of traffic self-similarity on network performance," *Proc. of SPIE International Conference on the Performance and Control of Netwokk systems*, Nov 1997.

[2] M.Krunz and A.Makowski, "Modelling video traffic using M/G/Infinity input processes: A compromise between Markovian and LRD models," *IEEE JSAC*, vol. 16, pp. 733–748, Jun 1998.

[3] D. R. Figueiredo, B. Liu, V. Misra, and D. Towsley, "On the autocorrelation structure of TCP traffic," *'Computer Networks Journal' Special Issue on 'Advances in Modeling and Engineering of Long-Range Dependent Traffic'*, 2002.

[4] C. Barakat, P. Thiran, G. Iannaccone, C. Diot, and P. Owezarski, "Modeling Internet backbone traffic at the flow level," *IEEE Transactions on Signal processing*, vol. 51, no. 8, Aug 2003.

[5] N. Hohn, D. Veitch, and P. Abry, "Cluster processes: a natural language for network traffic," *IEEE Transactions on Signal processing*, vol. 51, no. 8, Aug 2003.

[6] C. Barakat, P. Thiran, G. Iannaccone, C. Diot, and P. Owezarski, "A flow-based model for Internet backbone traffic," *ACM SIGCOMM Internet Measurement Workshop*, Nov 2002.

[7] D.Daley and D.Vere-Jones, *An introduction to the theory of point processes*, New York: Springer-Verlag, 1988.

[8] E. Altman, K. Avrachenkov, and C. Barakat, "A stochastic model for TCP/IP with stationary random losses," *ACM SIGCOMM*, Sep. 2000.

A ROBUST APPROACH TO ACTIVE QUEUE MANAGEMENT CONTROL IN NETWORKS

Sabato Manfredi, Franco Garofalo

Faculty of Engineering
University of Naples Federico II
Via Claudio 21, Napoli 80100, Italy.
Email: {smanfred, fgarogal}@unina.it

Mario di Bernardo

Faculty of Engineering
University of Sannio
Benevento 82100, Italy.
Email: dibernardo@unisannio.it

ABSTRACT

This paper is concerned with the design of improved AQM control schemes by means of an appropriate reduction method for time-delay systems. The aim is to stabilize the delay differential equations model of TCP (Transmission Control Protocol) behavior. A robust observer is introduced to estimate online the transmission window resulting in a window-based congestion control scheme. The robust stability and dynamic behavior in the presence of network parameter uncertainties is validated and tested through numerical simulations on a single bottleneck.

1. INTRODUCTION

In the past few years, networks have become an essential part of many engineering applications. A particularly cumbersome task when dealing with networks, for example with Internet, is to model, analyze and control the transient and steady-state behaviour of traffic flows particularly at congested links. Recently, models have been proposed which approximate the packet transmission as a continuous flow with the aim of correctly estimating average steady-state quantities such as transmission rates [1, 2, 3]. Their main feature is that of capturing the behaviour of each transmitting source and its interaction with the control signals due to congestion originated from the link (or router) to which that source is connected. The control mechanism often used to prevent the congestion phenomenon is the Transmission Control Protocol (TCP). During a TCP communication, the receiver sends back to the sender an acknowledgment signal for each packet which has been received. In practice, the TCP sender transmits w packets (w is called window size) and waits for their respective acknowledgments. If packets are acknowledged, the sender then increases w, while if a packet is dropped (i.e. not acknoweledged by the receiver), w is halved (multiplicative decrement). This mechanism is termed *drop-tail*. A more refined flow control strategy, the so-called RED (Random Early Drop) control, is achieved through a feedback mechanism based on marking (or dropping) packets according to the average queue length. This information when acknowledged by the receiver, allows the transmitter to increase or reduce its transmission rate in better accordance with the actual queue usage. The main advantages of RED control schemes are the elimination of flow-synchronization problems and the attenuation of traffic outbursts through the control of the average queue length. The principal disadvantage is the difficulty of tuning the control characteristic parameters. Much ongoing research effort has been spent to analyse RED schemes [4] and facilitate their design [5, 6]. It has been shown that control theory can offer an invaluable set of tools to improve the performance of existing AQM schemes which can be seen as particular types of feedback control systems [2, 5, 7, 8, 9]. This paper is concerned with the design of improved AQM control schemes to cope with unwanted variations of such characteristic parameters as the average round-trip time and load. In particular:

- an appropriate reduction method for time-delay systems is proposed to cope with round-trip time and load variations and hence used to design a suitable AQM control scheme;
- a robust observer is used in the control loop to avoid direct measurement of the transmission window, as this would be unpractical in applications.
- the resulting control law is validated and tested through numerical simulations.

2. BACKGROUND

2.1. A fluid model of TCP behavior

A fluid model of TCP dynamical behaviour was derived in [1] using the theory of stochastic differential equations. The model describes the evolution of the average characteristic variables on the network such as the average TCP window size and the average queue length. Extensive simulations in *Network-Simulator (NS)*-2 have shown that the model captures indeed the qualitative behaviour of TCP traffic flows. Hence, it is particularly useful for the design of innovative Active Queue Management control schemes for TCP-controlled flows using a control theory approach. Under the assumption of neglecting the TCP timeout, the model is described by the following set of nonlinear coupled ODEs:

$$\dot{W}(t) = \frac{1}{R(t)} - \frac{W(t)}{2}\frac{W(t-R(t))}{R(t-R(t))}p(t-R(t))$$

$$\dot{q} = \begin{cases} -C + \frac{N(t)}{R(t)}W(t), & \text{q>0} \\ \max\{0, -C + \frac{N(t)}{R(t)}W(t)\}, & \text{q=0} \end{cases} \tag{1}$$

where W is the average TCP window size (packets), q the average queue length (packets), T_p the propagation delay, R the transmission round-trip time ($R = \frac{q}{C} + T_p$), C the queue capacity (packets/sec), N the number of TCP sessions and p the probability of a packet being marked. If we assume $N(t) = N$, $R(t) = R_0$ to be the nominal values of R and N, we can then linearize the dynamic model (1) about the operating point (W_0, q_0, R_0) (see [7]

0-7803-8251-X/04/$17.00 ©2004 IEEE

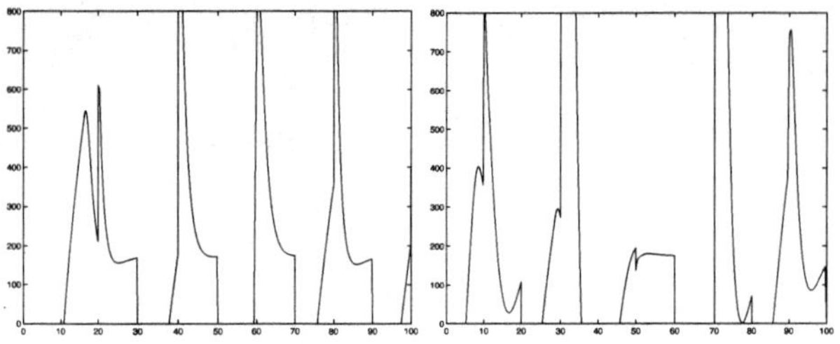

(a) (b)

Fig. 1. Time evolution of the queue length for the PI-based AQM scheme applied to the nonlinear model for the cases A (a) and B (b).

for further details) obtaining the following set of delay differential equations (DDEs):

$$\begin{aligned}\delta\dot{W}(t) &= -\frac{N}{R_0^2 C}(\delta W(t) + \delta W(t - R_0)) \\ &- \frac{1}{R_0^2 C}(\delta q(t) - \delta q(t - R_0)) - \frac{R_0 C^2}{2N^2}\delta p(t - R_0) \\ \delta\dot{q}(t) &= \frac{N}{R_0}\delta W(t) - \frac{1}{R_0}\delta q(t) \end{aligned} \quad (2)$$

where $\delta W \doteq W - W_0$, $\delta q \doteq q - q_0$, $\delta p \doteq p - p_0$. In what follows we will consider the same parameter values used in [7], i.e. $C = 3750$ pkt/s,$R_0 = 0.246$ s, $N = 60$, $q_{max} = 800$ pkt corresponding to the steady-state operating regime $W_0 = 15$ pkt, $q_0 = 175$ pkt, $p_0 = 0.008$.

2.2. AQM Control

Using this model, novel congestion control strategies have been proposed to improve the performance of standard TCP-RED algorithms (see for example [5] and references therein). The goal is to keep the queue length q within some desired bounds in order to avoid wild fluctuations and packet losses. Typically, the control approach consists of designing a linear feedback control law using the marking probability δp as the control input. Namely, the marking probability is varied in order to keep δq in the model above as close as possible to zero (so that the queue length q stays as close as possible to its nominal value). For example, to deal with the disadvantages of a classical RED controller, a PI (proportional-integral) controller is proposed in [7] to obtain zero steady-state error, better responsiveness and robustness (see [7] for further details). The controller is successful in achieving a better regulation of the queue length when compared with classical methods. Nevertheless, it performs poorly in the presence of round-trip time or load variations which are bound to occur in more realistic networks. In fact, the round-trip time R often varies with respect to its nominal value R_0 because of the occurrence of congestion phenomena (varying queue delays) and/or the presence of sessions number N variations. Variations in the average round-trip time can cause the onset of unstable dynamics. Thus, the ultimate aim of AQM control is that of making the system robust against variations of the load N and the roud trip time R. To test our controllers, we will consider two cases: **Case A** where the round trip time R is white noise variable in the interval[1] $(0.2, 0.5)$ ms; **Case B** where larger variations of the round-trip propagation delay ($R \in (0.1, 1)$) and load variations are considered (N is also a white noise random variable $\in (40, 270)$). Round trip time variations are implemented by varying independently the round trip propagation delay T_p. Fig. 1(a) shows the time evolution of the queue length for the PI-controlled nonlinear system for Case A. The limited robustness properties of the PI control action can be observed leading to several packet losses (corresponding to queue overflows $q \geq q_{max} = 800$). The queue dynamics for Case B are shown in Fig. 1(b). Again, we note a performance degradation with the occurrence of several packet losses due to buffer overflows. Thus, in the presence of realistic variations of round trip time and load about their nominal values, classical PI-AQM controllers perform poorly both in terms of their stability margins and unwanted packet losses due to buffer overflows. We shall therefore seek to synthesise an improved robust AQM control strategy to cope with unexpected variations of the load and the round-trip time.

[1]We note that the above interval corresponds to a typical uncertainty interval about the nominal value of the round trip time in a communication network, also in the absence of strong congestion [10].

3. CONTROL DESIGN

In what follows we will carry out the design of an output feedback control scheme for active queue management (AQM) which is robust to unwanted variations of R and N. In so doing we will proceed in two stages:

- Firstly, we will synthesise a full state feedback controller based on an appropriate reduction method, recently proposed in [11, 12] which allows to cope with parameter uncertainties;
- Then, we will design a robust observer to estimate the transmission window online, avoiding the requirement for full state availability. Such an observer will then be used to synthesise a robust AQM scheme.

To address the former point we will use the Reduction Method for time-delay systems presented in [11, 12, 13].

4. A REDUCTION BASED AQM CONTROL

The reduction transformation method presented in [11, 12] can be briefly outlined as follows (see [11, 12] for further details). Given a time-delay system of the form

$$\begin{aligned}\dot{\mathbf{x}}(t) &= A_0\mathbf{x}(t) + A_1\mathbf{x}(t-\tau) + B_0 u(t) + B_1 u(t-\tau) \\ y &= Cx(t)\end{aligned} \quad (3)$$

with known and fixed nominal delay τ, initial conditions $x(0) = \Psi(t), t \in [-\tau, 0]$ and $u(t) = \Phi(t)$ for t$\in [-\tau, 0]$, the idea is to compensate the effects of the delay by considering the following linear transformation of the system states:

$$z(t) \doteq x(t) + \int_{t-\tau}^{t} e^{A(t-\tau-\Theta)} A_1 x(\Theta) d\theta + \int_{t-\tau}^{t} e^{A(t-\tau-\eta)} B_1 u(\eta) d\eta. \quad (4)$$

Under this transformation, it can be shown (see [11, 12]) that system (3) can be recast in terms of the new state variables z as the delay-free set of ODEs:

$$\dot{\mathbf{z}}(t) = A\mathbf{z}(t) + Bu(t) \quad (5)$$

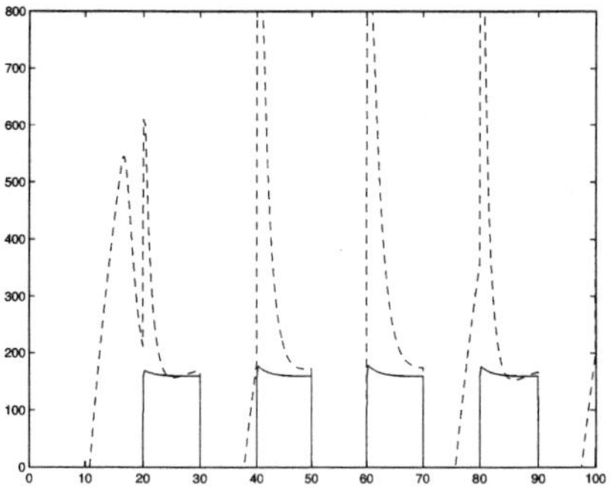

Fig. 2. Time evolution of the queue length for the PI and OFS method based on reduction transformation methods for Case A.

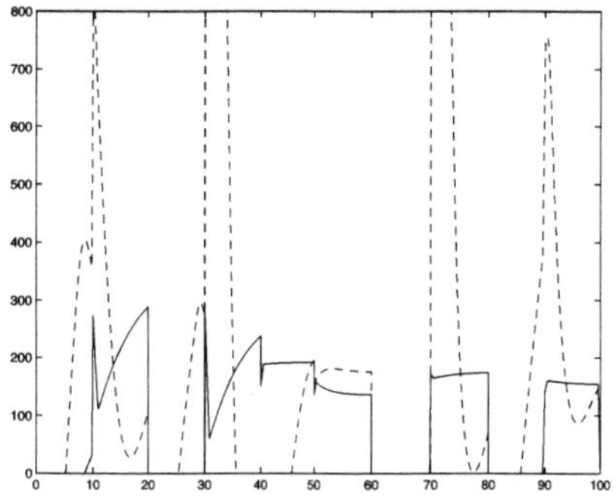

Fig. 3. Time evolution of the queue length for the PI and OFS method based on reduction transformation methods for Case B.

where the matrices A and B are solutions of the following matrix equations: $A = A_0 + e^{-\tau A}A_1$, $B = B_0 + e^{-\tau A}B_1$. Under the assumption that (A, B) is stabilizeable, system (3) can be stabilized using a classical state-feedback designed on the reduced model (5) given by $u(t) = Kz(t)$. Consequently by means of this reduction method approach, we can derive a stabilizing control for delayed system (3) designing a stabilizing controller for the ordinary system (5) with control methods derive for finite-dimensional systems. The matrix K can be chosen according to one of the many available methods for delay-free dynamical systems. This approach will now be used to obtain an improved AQM control scheme. Firstly, notice that the linearized time-delay model of TCP-IP given by (2) is equivalent to (3) with $x = (\delta w\ \delta q)^T$, $u = \delta p$, $\tau = R_0$ and:

$$A_0 = \begin{pmatrix} -\frac{N}{R_0^2 C} & -\frac{1}{R_0^2 C} \\ \frac{N}{R_0} & -\frac{1}{R_0} \end{pmatrix}, \quad A_1 = \begin{pmatrix} -\frac{N}{R_0^2 C} & \frac{1}{R_0^2 C} \\ 0 & 0 \end{pmatrix}$$

$$B_0 = \begin{pmatrix} 0 \\ 0 \end{pmatrix}, \quad B_1 = \begin{pmatrix} -\frac{R_0 C^2}{2N^2} \\ 0 \end{pmatrix}. \qquad (6)$$

Thus applying the reduction transformation method, a control was chosen of the form $u(t) = Kz$ with z given by (4) and K tuned using an optimal LQ approach on the reduced model The resulting evolution of the closed-loop nonlinear system are shown in Figg. 2, 3 for Case A and Case B respectively. The performance of the AQM management scheme based on reduction transformation methods is compared with that of a PI controller. It is observed that the novel strategy, which directly accounts for the presence of delays in the TCP-IP model, is more robust to variations in N and R. It guarantees an acceptable transient and steady-state performance while reducing the packet loss due to buffer overflows ($q \geq q_{max} = 800$). A disadvantage of this control scheme is that it requires the window size W for feedback purposes as this is typically hard to measure on line. To overcome this problem, we will now introduce a robust observer to estimate W online.

4.1. A robust observer

In this section we use a reduced robust observer to estimate the windows size variations δW for the linear time-delay system (2). (Note that W can be obtained by adding δW to the nominal value W_0.) The design uses the general linear matrix inequality approach presented in [14] in order to guarantee the stability of the closed-loop system and reduce the effect of uncertainties on the estimates. In particular, given a time-delay system of the form (3), let $y(t) = Cx(t) = [1\ 0]$ be the measured output vector and $\delta\hat{W}(t)$ the quantity to be estimated. Assuming D to be an estimate of the upper bound of the uncertainties acting on the system, according to the approach presented in [14], given a positive scalar γ and a matrix E, if there exists a matrix X and two symmetric positive definite matrices P and Q such that:

$$\begin{bmatrix} M & PA_1 & XC & 0 & XD \\ A_1^T & -A_1 & 0 & 0 & 0 \\ C^T X^T & 0 & -\bar{Q} & 0 & 0 \\ 0\bar{B}^T P & 0 & 0 & -I & 0 \\ DX^T & 0 & 0 & 0 & -I \end{bmatrix} < 0 \qquad (7)$$

where $M = A_0^T P + PA_0 + Q + \frac{1}{\gamma^2}E^T E$, $\bar{B} = [B_0, B_1]$, $\bar{Q} = \frac{1}{2}Q$ then the system, expressed using Laplace transforms as:

$$\begin{aligned} \hat{W}(s) &= E\left(sI_n - A(z) + LC(z)\right)^{-1}\bar{B}\bar{u}(s) \\ &+ E\left(sI_n - A(z) + LC(z)\right)^{-1} Ly(s) \end{aligned} \qquad (8)$$

with $L = P^{-1}X$, $\bar{u}(t) = [u(t)\ u(t-\tau)]^T$, generates a robust estimate of $\hat{W}(s)$ so that the time evolution of the estimate $\hat{W}(t)$ satisfies one of the following conditions:

1. $\lim_{t\longrightarrow\infty}(W(t) - \hat{W}(t)) = 0$ if $R = R_0$, $N = N_0$,
2. $\|W(t) - \hat{W}(t)\|_2$ is bounded under variations of R and N.

(See [14] for further details and the proofs of the result stated above.)

4.2. Application to AQM

Using the observer design strategy presented above, we design a novel output feedback stabilization scheme for AQM considering the reduction method synthetized in the previous section complemented by a reduced robust observer. This estimates online the transmission window W which is then fed back together with the measured queue length q. Considering the nominal linear time-delay system (2),we solved the LMI problem (7) via algebraic manipulation software (the details of the computations are omitted here for the sake of brevity). The estimation error on the window size W in the presence of variations on both N and R (Case B) is depicted in Fig. 4 showing good estimation dynamics. Using the estimated W in the control scheme derived in section 4, we then obtained a robust AQM scheme whose performance is shown in Fig. 5 where the performance of a simpler PI controller is also depicted. Firstly, we would like to emphasise that the observer-based

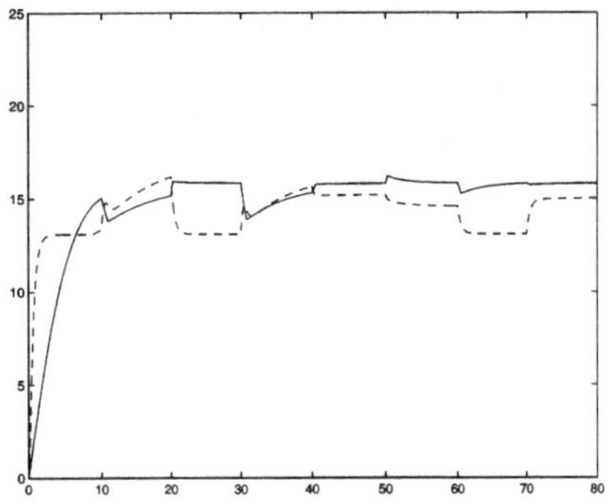

Fig. 4. Time evolution of the window size (solid line) and window size estimated (dashed line) for case B on nonlinear system.

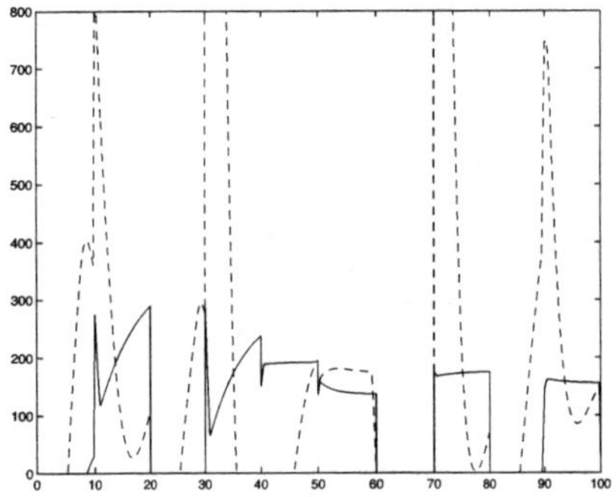

Fig. 5. Time evolution of the queue length for the PI and OFS method based on reduction transformation methods with observer for case B.

control performance is comparable with that of the state feedback control scheme based on reduction methods shown in Fig. 3. This confirms that the observer provides an efficient online estimation of W without altering the qualitative performance of the control system. The evolution of the queue length for the robust AQM scheme and the PI controller, depicted in Fig. 5 clearly shows that the robust AQM scheme presents an improved transient and steady-state performance. Namely, the PI controller shows consistent variations of the queue length and regions where packets are lost ($q \geq 800$). The robust AQM, instead, guarantees an acceptable steady-state response and eliminates all packet losses over the time range of interest.

5. CONCLUSIONS

We have shown that it is possible to improve the performance of classical control AQM control strategies under variations of the round trip time and the load. As such variations are bound to occur in practical applications, the need was outlined for more sophisticated robust controllers. In particular, a robust output feedback scheme for AQM was proposed. The design proceeded in two stages. Firstly, a set of transformations were considered to render the system robust against round-trip time and load variations. Then the resulting feedback controller, relying on full availability of the states, was modified by inserting an appropriately designed robust observer. The observer allowed the online estimation of the transmission window W. Thus, a robust AQM was synthesised and tested numerically. The numerical simulations confirmed that the new scheme exhibits a much improved performance when compared to the available PI AQM control strategy. In particular, the queue length dynamics was shown to exhibit less fluctuations and no packet losses in the presence of parameter variations and uncertainties. Current work is aimed at testing the strategy presented in this paper both numerically (via NS-2 simulations) and experimentally.

6. REFERENCES

[1] V. Misra, W.B. Gong, and D. Towsley, "Fluid-based analysis of a network of aqm routers supporting tcp flows with an application to red," *ACM SIGCOMM*, 2000.

[2] S. Low, F. Paganini, and J.C. Doyle, "Internet congestion control," *IEEE Control Systems Magazine*, pp. 28–43, 2002.

[3] S.Manfredi, M. di Bernardo, and F. Garofalo, "A nonlinear competitive model of traffic flows at congested links in networks," *ISCAS*, 2003.

[4] M. May, T. Bonald, and J.C. Bollot, "Analytic evaluation of red performance," *IEEE INFOCOM 2000.*

[5] C. V. Hollot, V. Misra, D. Towsley, and W. Gong, "A control theoretic analisys of red.," *IEEE INFOCOM 2001.*

[6] T. J. Ott, T. V. Lakshman, and L. H. Wong, "Sred:stabilized red," *IEEE INFOCOM 1999.*

[7] C. V. Hollot, V. Misra, D. Towsley, and W. Gong, "Analysis and design of controllers for aqm routers supporting tcp flows," *IEEE ACM Trans on Automatic Control*, pp. Vol 47, no. 6, 2002.

[8] Aweya, Ouelette, and D.Y. Montuno, "A control theoretic approach to active queue management.," *Computer Newtworks*, pp. 36, 203–235, 2001.

[9] Y. Chait, C. V. Hollot, V. Misra, S. Oldak, D. Towsley, and W. Gong, "Fixed and adaptive model-based controllers for active queue management," *American Control Conference*, pp. pp 2981–2986, 2001.

[10] Larry, L. Peterson, and Bruce S. Davie, "Computer networks: a system approach," *Morgan Kaufmann, 2000, 2nd Edition.*

[11] W.H. Know and A.E. Pearson, "Feedback stabilization of linear systems with delayed control," *IEEE Trans. Automatic Control*, pp. 266–269, 1980.

[12] Young, "Reduction transformation methods for input delayed systems:a survey," *10th CISL Winter Workshop*, pp. 19–30, 1997.

[13] Kojima, Uchida, Shimemura, and Ishijima, "Robust stabilization of a systems with delays in control," *IEEE Trans. Automatic Control*, pp. 1694–1698, 1994.

[14] A. Fattouh, O. Sename, and J. M. Dion, "A lmi approach to robust observer design for linear time-delay systems," *IEEE Conference on Decision Control 2000.*

ON DYNAMIC BEHAVIOR OF WEAKLY CONNECTED CELLULAR NEURAL NETWORKS

Marco Gilli and Fernando Corinto

Department of Electronics, Politecnico di Torino, Torino - Italy.

ABSTRACT

It was recently shown that weakly connected cellular neural/nonlinear networks (consisting of locally coupled oscillators) represent a suitable architecture for modelling biological neuro-computers. Such networks are described by large systems of nonlinear differential equations and may exhibit a rich dynamics, including chaos and complex bifurcation phenomena. We focus on space invariant cellular nonlinear networks and show that their dynamic behavior can be investigated through a spectral method, based on the application of the describing function technique. For a generic coupling, the spectral approach yields some approximate analytical conditions, that are useful for estimating some important network features and in particular for distinguishing between stationary (stable) and non-stationary behavior. In case of weak coupling the spectral method allows one to estimate the whole set of stable and unstable periodic limit cycles.

1. INTRODUCTION

Recent studies on thalamo-cortical systems have shown that networks composed by weakly connected oscillators represent a good architecture for a neurocomputer [1, 2].

Such structures can be adequately modelled by cellular neural/nonlinear networks (CNNs), a new paradigm of analog dynamic processor arrays, that was introduced some years ago in the electrical engineering community [3, 4, 5, 6, 7]. CNNs are described as a 2 or n-dimensional array of identical nonlinear dynamical systems (called cells), that are locally interconnected. In most applications the connections are specified through space-invariant templates (that consist of small sets of parameters identical for all the cells) [6]. The local connectivity has allowed the realization of several high-speed VLSI chips [8]: this is the main advantage of CNNs with respect to general neural models.

The mathematical model of a CNN consists of a large system of locally coupled nonlinear ordinary differential equations (ODEs), that may exhibit a rich spatio-temporal dynamics, including several attractors and bifurcation phenomena [9]. A complete study of their dynamics would require to classify, for given sets of parameters, all the attractors and possibly to estimate their domains of attraction.

Time domain numerical simulation has allowed one to discover several spatio-temporal dynamic phenomena in CNNs [9], but it is not suitable for discovering and classifying all the attractors of a high-dimensional network. In fact the global dynamic analysis, through the sole numerical simulation, would require to identify for each choice of network parameters all sets of initial conditions that converge to different attractors. This would be a formidable, and practically impossible task.

Recently, some harmonic balance (HB) based techniques have been applied to the study of the dynamic behavior of space-invariant CNNs (see [10]-[18]). In particular in [16, 18] the describing function technique was exploited for predicting and classifying some significant spatio-temporal phenomena. In [19, 20] a spectral technique was applied to the study of periodic oscillations and their bifurcations in one-dimensional arrays of Chua's circuits.

In this paper we focus on weakly connected CNNs of oscillators and we show that their global dynamic behavior can be studied through a spectral method based on a suitable extension of the describing function technique. In case of generic coupling, the proposed spectral approach provides some simple approximate analytical conditions for estimating some important network features and in particular for distinguishing between stationary and non-stationary behavior. In case of weak coupling the spectral method is also suitable for investigating the set of all stable and unstable limit cycles.

2. CELLULAR NEURAL NETWORK MODELS

According to [21] we assume that a cellular nonlinear network be described by the following system of $N \times M$ nonlinear ordinary differential equations:

This work was supported in part by *Ministero dell'Istruzione, dell'Università e della Ricerca*, under the FIRB project no. RBNE012NSW. An extended version of this manuscript is under publication in the IEEE, Transactions on Circuits and Systems, I.

$$L(D_t)x_{ij}(t) = \sum_{(k,l)\in N_r(i,j)} T^A_{ij,kl}(x_{ij},x_{kl})\, f(x_{kl})$$

$$+ \sum_{(k,l)\in N_r(i,j)} T^B_{ij,kl}(u_{ij},u_{kl})\, u_{kl}$$

$$+ \sum_{(k,l)\in N_r(i,j)} T^C_{ij,kl}(x_{ij},x_{kl})x_{kl} \ + \ z_{ij} \quad (1)$$

The state variables x_{ij} are assumed to be arranged on a regular rectangular grid and are denoted by two indexes ($1 \leq i \leq N, 1 \leq j \leq M$); $f(\cdot)$ is a Lipschitz nonlinear function and represents the output; u_{ij} is the input; z_{ij} is a constant bias term; $L(D_t)$ is a rational function of the differential operator $D_t = d/dt$; $T^A_{ij,kl}(x_{ij},x_{kl})$ and $T^C_{ij,kl}(x_{ij},x_{kl})$ are the output and the state feedback templates, that in general might be space-variant nonlinear functions of the state variables x_{ij} and x_{kl}; $T^B_{ij,kl}(u_{ij},u_{kl})$ is the input template; $N_r(i,j)$ denotes the neighborhood of interaction of each state-variable x_{ij}. The model is completed by specifying the *initial conditions*, i.e. $x_{ij}(0)$, and the space *boundary conditions*. Note that equations (1) may contain higher order time derivatives, whereas multi-layer templates are not allowed.

The above equations simplify if templates T^A and T^C are linear and space-invariant and the inputs and the bias terms are null.

$$L(D_t)x_{ij}(t) = \sum_{|n|\leq r,|m|\leq r} T^A_{nm}\, f(x_{i+n,j+m})$$

$$+ \sum_{|n|\leq r,|m|\leq r} T^C_{nm}\, x_{i+n,j+m} \quad (2)$$

3. PERIODIC OSCILLATIONS IN WEAKLY CONNECTED CNNS

We assume that in absence of coupling each cell presents an oscillating behavior, i.e. that it exhibits at least one stable asymptotic limit cycle.

In order to illustrate our method, we assume that the input and the bias term be null and that the CNN be described as a one-dimensional array with neighborhood radius $r = 1$. State equations (2) simplify as follows:

$$L(D_t)x_i(t) = T^A_{-1}f(x_{i-1}) + T^A_0 f(x_i) + T^A_1 f(x_{i+1})$$

$$+ \ T^C_{-1}\, x_{i-1} + T^C_0\, x_i + T^C_1\, x_{i+1} \quad (3)$$

We also suppose that the boundary conditions be either of Dirichlet type (i.e. $x_0(t) = x_{N+1}(t) = 0$) or of Neumann type (i.e. $x_0(t) = x_1(t), x_N(t) = x_{N+1}(t)$).

According to the assumptions of the describing function technique, the state $x_i(t)$ is represented through a bias term and a single temporal harmonic with suitable amplitude, phase and frequency:

$$x_i(t) \approx \hat{x}_i(t) = P_i + Q_i \sin\left(\hat{\omega}t + \sum_{k=1}^{i-1} \eta_k\right) \quad (4)$$

where P_i denotes the bias, Q_i the amplitude of the first harmonic, $\hat{\omega}$ is the frequency and η_k indicates the phase shift between $\hat{x}_{i+1}(t)$ and $\hat{x}_i(t)$.

We remark that, according to the above representation (4), each cell state may have a different amplitude, bias and phase, whereas the frequency is identic for all cells. We also point out that the single harmonic approximation (4) can be effectively exploited for investigating the existence of periodic limit cycles, even if the cell state waveforms are not quasi-sinusoidal and contain several harmonics [22].

The output of the nonlinear function $f(\cdot)$, when the input is (4), admits of the following first harmonic representation (that in several cases can be expressed through a closed analytical form):

$$f(\hat{x}_i(t)) \approx F^P_i(P_i,Q_i) + F^Q_i(P_i,Q_i)\sin\left(\hat{\omega}t + \sum_{k=1}^{i-1}\eta_k\right) \quad (5)$$

where:

$$F^P_i(P_i,Q_i) = \frac{1}{2\pi}\int_{-\pi}^{\pi} f\left[P_i + Q_i\sin(\theta)\right] d\theta$$

$$F^Q_i(P_i,Q_i) = \frac{1}{\pi}\int_{-\pi}^{\pi} f\left[P_i + Q_i\sin(\theta)\right]\ \sin(\theta)\ d\theta$$

We proceed as follows. By substituting in (3) the approximate expressions (4) and (5) for $\hat{x}_i(t)$ and $f(\hat{x}_i(t))$, we derive a set of $3N$ nonlinear equations in the $3N$ unknowns, $P_1,...,P_N,Q_1,...,Q_N,\eta_1,...,\eta_{N-1},\hat{\omega}$. The set of $3N$ equations is reported below ($1 \leq i \leq N$):

$$L(0)\ P_i = T^A_{-1}F^P_{i-1} + T^A_0 F^P_i + T^A_1 F^P_{i+1}$$
$$+ \ T^C_{-1}P_{i-1} + T^C_0 P_i + T^C_1 P_{i+1} \quad (6)$$

$$\mathbf{Re}[L(j\hat{\omega})]\ Q_i = (T^A_{-1}F^Q_{i-1} + T^C_{-1}Q_{i-1})\ \cos(\eta_{i-1})$$
$$+ \ (T^A_0 F^Q_i + T^C_0 Q_i)$$
$$+ \ (T^A_1 F^Q_{i+1} + T^C_1 Q_{i+1})\ \cos(\eta_i) \quad (7)$$

$$\mathbf{Im}[L(j\hat{\omega})]\ Q_i = -(T^A_{-1}F^Q_{i-1} + T^C_{-1}Q_{i-1})\ \sin(\eta_{i-1})$$
$$+ \ (T^A_1 F^Q_{i+1} + T^C_1 Q_{i+1})\ \sin(\eta_i) \quad (8)$$

The above equations should be completed by the boundary conditions, that for Dirichlet type give rise to $P_0 = Q_0 = P_{N+1} = Q_{N+1} = 0$, whereas for Neumann type can be expressed as $P_0 = P_1$, $Q_0 = Q_1$, $P_N = P_{N+1}$, $Q_N = Q_{N+1}$, and $\eta_0 = \eta_N = 0$.

The space invariance structure of equation (8) allows one to derive that $\sin(\eta_i)$ and $\sin(\eta_{i-1})$ are related by a difference equation with non-constant coefficients. For both Dirichlet and Neumann boundary conditions, we have:

$$\begin{aligned} \sin(\eta_i) &= \left(\frac{T_{-1}^A F_{i-1}^Q + T_{-1}^C Q_{i-1}}{T_1^A F_{i+1}^Q + T_1^C Q_{i+1}} \right) \sin(\eta_{i-1}) \\ &+ \frac{\mathbf{Im}[L(j\hat{\omega})] Q_i}{T_1^A F_{i+1}^Q + T_1^C Q_{i+1}} \quad (2 \le i \le N-1) \\ \sin(\eta_1) &= \frac{\mathbf{Im}[L(j\hat{\omega})] Q_1}{T_1^A F_2^Q + T_1^C Q_2} \end{aligned} \quad (9)$$

The above equation can be explicitly solved by exploiting standard techniques for scalar difference equations (see [23]). Through some non-trivial algebraic manipulations we obtain the following expression for $\sin(\eta_i)$ $(2 \le i \le N-1)$:

$$\begin{aligned} \sin(\eta_i) &= \left[\sum_{k=1}^{i-1} \frac{T_{-1}^A h_k + T_{-1}^C}{T_1^A h_i + T_1^C} Q_k^2 \left(\prod_{j=k+1}^{i-1} \frac{T_{-1}^A h_j + T_{-1}^C}{T_1^A h_j + T_1^C} \right) \right. \\ &\left. + Q_i^2 \right] \cdot \frac{\mathbf{Im}[L(j\hat{\omega})]}{Q_i Q_{i+1} (T_1^A h_{i+1} + T_1^C)} \end{aligned} \quad (10)$$

where $h_i = F_i^Q / Q_i$.

Then by substituting the expression of $\sin(\eta_{N-1})$ into equation (8) (with $i = N$ and $Q_{N+1} = F_{N+1}^Q = 0$ in case of Dirichlet boundary conditions and $\eta_N = 0$ in case of Neumann boundary conditions) the following equation is obtained:

$$\mathbf{Im}[L(j\hat{\omega})] \left[\sum_{k=1}^{N-1} \left(\prod_{j=k}^{N-1} \frac{T_{-1}^A h_j + T_{-1}^C}{T_1^A h_{j+1} + T_1^C} \right) Q_k^2 + Q_N^2 \right] = 0 \quad (11)$$

The above equation (11) is exact only in the framework of the describing function approximation. However in most significant cases [22, 24] it can be exploited for estimating the conditions under which periodic oscillations occur. The following Proposition holds for generic coupling coefficients.

Proposition 1: If at least one of the coefficients Q_k and one of the coefficients (T_1^A, T_1^C) are different from zero and in addition the following set of conditions is satisfied:

$$T_{-1}^A \ge 0,\ T_1^A \ge 0,\ T_{-1}^C \ge 0,\ T_1^C \ge 0,\ h_j > 0\ (j = 1, ...N) \quad (12)$$

then the describing function system (6)-(8) admits of a solution if and only if there exists $\hat{\omega}$ such that

$$\mathbf{Im}[L(j\hat{\omega})] = 0 \quad (13)$$

If the conditions of Proposition 1 are satisfied, then the solution of system (6)-(8) can be significantly simplified. In fact, if $\mathbf{Im}[L(j\hat{\omega})] = 0$, from equation (10) we derive that only two values of phase shifts η_i are admissible:

$$\eta_i = \begin{cases} 0 \\ \pi \end{cases} \quad 1 \le i \le N-1 \quad (14)$$

We also derive that for each $\hat{\omega}$ that satisfies (13) and for each one of the 2^{N-1} admissible phase shifts $\{\eta_1, ..., \eta_{N-1}\}$, $\eta_i \in \{0, \pi\}$, the total number of unknowns of system (6)-(8) can be reduced to $2N$, i.e. $P_1, ..., P_N$, $Q_1, ..., Q_N$.

In case of weak coupling, the above considerations allows one to estimate the whole set of stable and unstable limit cycles (see [24] for details).

Proposition 2: If the template coupling coefficients ($T_{\pm 1}^A$ and $T_{\pm 1}^C$) are sufficiently small, then for each $\hat{\omega}$ satisfying (13), the describing function system (6)-(8) admits of 2^{N-1} solutions (corresponding to the 2^{N-1} admissible phase shifts $\{\eta_1, ..., \eta_{N-1}\}$, $\eta_i \in \{0, \pi\}$).

It is shown in [24] that Proposition 2 provides an accurate prediction of the actual number of limit cycles for weakly connected CNNs described by equations (3). A spectral technique for investigating the stability properties of each limit cycle is also presented in [24].

4. CONCLUSIONS

We have shown that the dynamic behavior of weakly connected CNNs of oscillators can be studied by exploiting the describing function technique. In case of generic coupling we have yielded a set of simple approximate analytical conditions for estimating the existence of periodic oscillations (Proposition 1). In case of weak coupling we have also provided an accurate prediction of the total number of stable and unstable limit cycles (Proposition 2).

5. REFERENCES

[1] F. C. Hoppensteadt and E. M. Izhikevitch, "Oscillatory neurocomputers with dynamic connectivity," *Physical Review Letters*, vol. 82, no. 14, pp. 2983-2986, 1999.

[2] F. C. Hoppensteadt and E. M. Izhikevitch, *Weakly connected neural networks*, Springer-Verlag, NY, 1997.

[3] L. O. Chua and L. Yang,"Cellular neural networks: Theory," *IEEE Transactions on Circuits and Systems*, vol. 35, pp. 1257-1272, 1988.

[4] L. O. Chua and L. Yang, "Cellular neural networks: Applications," *IEEE Transactions on Circuits and Systems*, vol. 35, pp. 1273-1290, 1988.

[5] L. O. Chua and T. Roska, "The CNN paradigm," *IEEE Transactions on Circuits and Systems: Part I*, vol. 40, pp. 147-156, 1993.

[6] L. O. Chua and T. Roska, *Cellular neural networks and visual computing*, Cambridge University Press, U.K., 2002.

[7] L. O. Chua and T. Roska, "The CNN universal machine: an analogic array computer," *IEEE Transactions on Circuits and Systems: Part I*, vol. 40, pp. 163-173, 1993.

[8] A. R. Vàzquez, M. Delgado-Restituto, E. Roca, G. Linan, R. Carmona, S. Espejo and R. Dominguez-Castro, "CMOS Analogue Design Primitives," in *Towards the visual microprocessor - VLSI design and use of Cellular Network Universal Machines*, J. Wiley, Chichester., pp. 87-131, 2002.

[9] *IEEE Transactions on Circuits and Systems: Part I. Special issue on nonlinear waves, patterns and spatio-temporal chaos in dynamic arrays*, vol. 42, Oct. 1995.

[10] K. R. Crounse and L. O. Chua, "Methods for image processing and pattern formation in cellular neural network: a tutorial," *IEEE Transactions on Circuits and Systems: Part I*, vol. 42, pp. 583-601, 1995.

[11] G. Setti, P. Thiran and C. Serpico, "An approach to information processing in 1-dimensional cellular neural networks - Part II: Global propagation," *IEEE Transactions on Circuits and Systems: Part I*, vol. 45, 1998.

[12] L. Goras and L. O. Chua, "Turing patterns in CNNs: Equations and behaviors," *IEEE Transactions on Circuits and Systems: Part I*, vol. 42, pp. 612-626, 1995.

[13] M. Gilli, "Investigation of chaos in large arrays of Chua's circuits via a spectral technique," *IEEE Transactions on Circuits and Systems: Part I*, vol. 42, pp. 802-806, 1995.

[14] P. P. Civalleri and M. Gilli, "A spectral approach to the study of propagation phenomena in CNNs," *International Journal of Circuit Theory and Applications*, vol. 24, pp. 37-47, 1996.

[15] M. Gilli, "Analysis of periodic oscillations in finite-dimensional CNNs, through a spatio-temporal harmonic balance technique", *International Journal of Circuit Theory and Applications*, vol. 25, pp. 279-288, 1997.

[16] P. P. Civalleri and M. Gilli, "Analysis of periodic and chaotic oscillations in one-dimensional arrays of Chua's circuits," *1997 European Conference on Circuit Theory and Design*, pp. 353-358, Budapest, 1997.

[17] P. P. Civalleri and M. Gilli, "A harmonic balance technique for the analysis of periodic attractors and their bifurcations in cellular neural networks," *IEEE Fifth International Workshop on Cellular Neural Networks and their Applications*, pp. 106-111, London, 1998.

[18] M. Gilli and P. P Civalleri, "A HB technique for the classification of periodic and chaotic attractors in one-dimensional arrays of Chua's circuits," *IEEE International Symposium on Nonlinear Theory and its Applications*, Dresda, 2000.

[19] , M. Gilli, P. Checco, and F. Corinto,"Periodic orbits and bifurcations in one-dimensional arrays of Chua's circuits," *International Symposium on Circuits and Systems*, vol. III, pp. 781-784, Bangkok (Thailand), May 2003.

[20] M. Gilli, P. Checco, and F. Corinto,"A spectral technique for the analysis of nonlinear dynamic arrays," *IEEE Eleventh International Workshop on Nonlinear Dynamics of Electronics Systems*, pp. 85-88, Schuls (Switzerland), May 2003.

[21] M. Gilli, T. Roska, L. O. Chua and P. P. Civalleri, "CNN Dynamics represents a broader class than PDEs," *International Journal of Bifurcation and Chaos*, vol. 12, pp. 2051-2068, 2002.

[22] A. I. Mees, *Dynamics of feedback systems*, John Wiley, New York, 1981.

[23] H. Levy and F. Lessman, *Finite difference equations*, Dover Publication, Inc., New York, 1992.

[24] M. Gilli, F. Corinto, and P. Checco, "Periodic oscillations and bifurcations in cellular nonlinear networks," *Internal report, July 2003*, under publication in the *IEEE Transactions on Circuits and Systems - I*.

TOWARDS A BIO-INSPIRED MIXED-SIGNAL RETINAL PROCESSOR

Timothy G Constandinou[1], Julius Georgiou[1,3] and Chris Toumazou[1,2]

[1] EEE Dept, Imperial College of Science, Technology and Medicine, London, SW7 2BT, UK.
[2] Toumaz Technology Limited, Culham Science Centre, Abington, Oxfordshire, OX14 3DB, UK.
[3] Geosilicon Limited, 7 Thessalonikis Avenue, Strovolos, Nicosia 2020, Cyprus.

ABSTRACT

A robust distributed architecture for real-time object-based processing is presented for tasks such as object size, centre and count determination. This approach uses the input image to enclose a feedback loop to realize a data-driven pulsating action. Outlined is the top level design for hardware implementation in a standard CMOS technology.

1. INTRODUCTION

A modern advanced image processing system uses an external camera to stream the image data to the processor, executing a software algorithm. Such a modular scheme demands huge bandwidth requirements for the video transmission and therefore heavy power requirements. Several early filtering applications can benefit from combining the phototransduction and processing at the pixel level. A new breed of vision chips have recently emerged that strive to achieve precisely this. A generic reconfigurable architecture to provide such pixel-level processing is the cellular neural network processor [1]. Other systems have been inspired by the unparalleled computational efficiency of living organisms in solving complex image processing tasks. These biologically-inspired (or retinomorphic [2]) systems have been realized to perform tasks such as image enhancement and feature extraction. Object-based processing is a fundamental task for many early vision applications. The segmentation of various objects in an image has been traditionally implemented in software using techniques such as the snake algorithm [3]. It has not been till recently, that dedicated hardware has been developed for such tasks as object-based attention selection [4] and contour length measurement [5]. This paper proposes a novel scheme [6] suitable for such object-based computation based on a distributed processing architecture. Although several of the features have been biologically inspired, the algorithm is fundamentally synthetic. By using this hybrid approach, a realistically hardware implementable system can be developed benefiting from increased computational efficiency provided by the bio-inspired analogue processing elements. The reduced power consumption enables realization of mobile diagnosis devices which would otherwise be technically unachievable.

The target application for hardware implementation is microscopic cellular population analysis as a microelectronic alternative to haemocytometry. The developed system (ORASIS) is required to provide cellular count and size information on microscopic images such as those shown in figure 1.

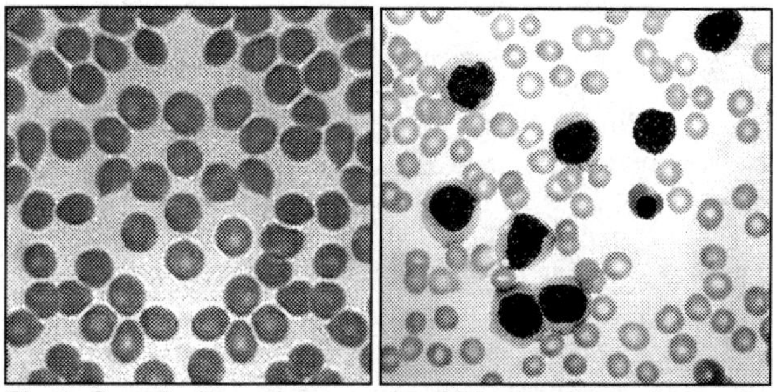

Fig 1. Sample input images of red-blood cell specimens for application in microscopic cellular population analysis

2. ALGORITHM [6]

A continuous-time edge-detection technique is used to form the contours and trigger the data-driven processing. On detection of an object boundary, the initial state for the signal flow is set. By propagating an inward fill, the contour can be reduced until this converges to the centre. The central point is detected by utilizing spatiotemporal integration; i.e. a summation of the cells set within the receptive field within a certain time window. On centroid detection, the object is reset and output transmitted, thus realizing an inward pulsating action. The frequency of pulsation determines the size, i.e. radius of this object. Figure 2 illustrates this interaction graphically through computer simulations.

This scheme can be applied in two different modes of operation; either as a single shot "capture and process" mode or using the above described continuous pulsating mode. The trade-off between these two modes of operation is accuracy (due to averaging) versus power consumption (due to increased duty cycle.)

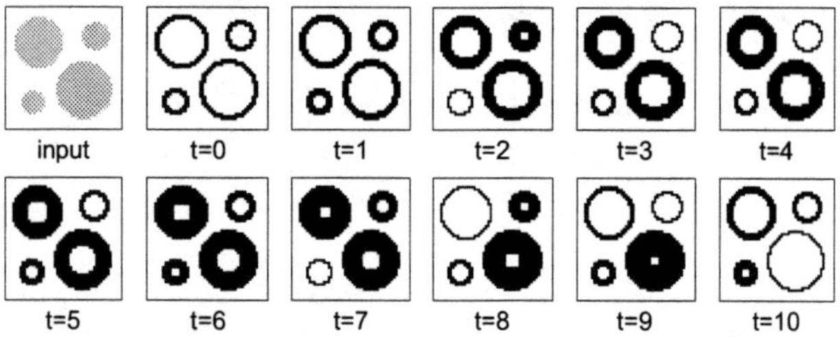

Fig. 2 Computer simulation results of the bio-pulsating contour reduction algorithm, with snapshots taken at time intervals at the propagation delay of the processing.

3. METHOD

Objects are defined as regions in the image with intensity below (or above) the average level of the input image. The edges are detected by computing the difference in neighbouring cell intensities and contours are formed if a continuous edge is found; i.e. at the nodes which have two edges leading to them. The contour reduction is facilitated by setting a cells state if any of its adjacent cells have been set in addition to the object criterion being satisfied. The rate of the contour reduction is preset by introducing a delay element in the propagation cycle. The unusual feature of this method is the absence of any pre-defined synchronisation signal, for example, a clock. The only synchronisation is obtained through the data-driven object reset scheme but on a local, rather than a global basis. The reset is generated on detection of an object centre. As previously mentioned, this detection involves counting that all local pixel-cells have been set within a certain time period.

4. BIOLOGICALLY INSPIRED APPROACH

Many of the implemented circuits and functions have in fact been biologically inspired. As in the mammalian retina, the front-end circuitry includes continuous time logarithmic photo detection, in addition to localised smoothing (averaging) and adaptive edge detection for the signal conditioning. Furthermore, the signal propagation based on localised interaction works in a similar way to the orientation-selective V1 cells in the primary visual cortex. The centroid determination is implemented using a pseudo centre-surround receptive field technique; very similar to the functional organisation of the retina. This has been implemented using delay and propagate, integrate and fire type neuronal circuits; producing a truly spike-domain output as in the case of ON/OFF ganglion cells.

5. HARDWARE IMPLEMENTATION

The presented architecture is currently being realized into circuit blocks for implementation in a standard 0.18μm CMOS process provided by UMC. The circuit topology is a unique combination of both weak-inversion analogue providing micropower operation with asynchronous logic for robustness and stability.

The complete top level system architecture is shown in Figure 3. This contains an X*Y array of smart pixels; containing both the photodetecting devices and local processing circuitry. At the column and row headers are address encoders which relay the data received through a digital bus for off-chip communication. Such an encoding scheme is often referred to as address event representation (AER.) This approach is possible due to the very low output bandwidth requirement that avoids the polling of all pixels.

Containing several current-mode circuits; each pixel requires a bias current reference. A current-mode distribution scheme is adopted implementing a tree-like hierarchy. The PTAT master reference supplies the bias currents to the four corners of the array. These corner currents are then duplicated for each row and subsequently for each column, resulting in each pixel receiving an individual bias. This vastly reduces errors arising from bias current variations; a major headache when using voltage-mode current distribution. The improved current matching is due to using low-proximity current-copying circuits thus minimizing any mismatch errors; discussed in further detail in section 6.

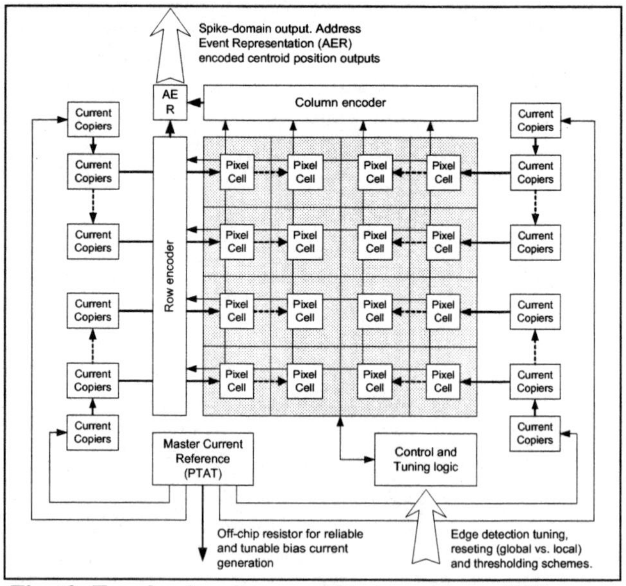

Fig. 3 Top level system architecture of an X*Y array illustrating the bias distribution and output readout schemes.

In order to facilitate the contour computation, the processing must occur at the pixel corners, as illustrated in the pixel-cell architecture shown in Figure 4.

The required functional (pixel-level) blocks; all continuous time topologies, are given below:

a. Light detection: Active photodiode (continuous time) utilizing n+ implant p-substrate junction diodes.
b. Edge detection: Discrete output using thresholding technique [7] utilizing differential current-mode hysteresis for computation of object contours.
c. Local averaging: Narrow-field for input image smoothing and wide-field for object detection; using current-mode circuitry for thresholding.
d. Local resetting: Dynamic switching regulated with local average current-mode thresholding [7] for object segmentation, to provide localised (object) resetting.
e. Neuromorphic logic: performing delay-and-propagate computation for signal flow and centre-surround-like computation for centroid determination.
f. Memory: Basic 1-bit memory implemented using digital (asynchronous) RS flip-flop for storage of present cellular state.

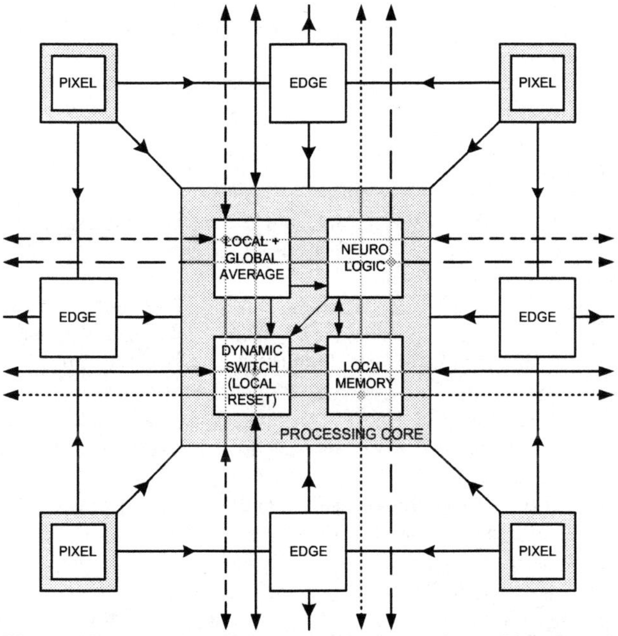

Fig. 4 Proposed cellular architecture for object-based processing illustrating organisation and connectivity of functional blocks within a quad-pixel arrangement.

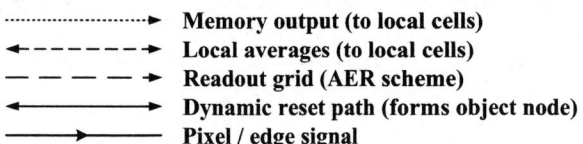

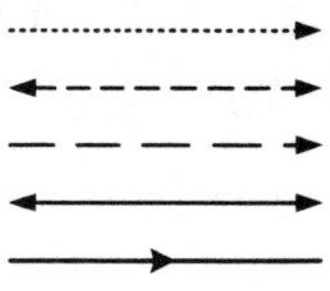

Memory output (to local cells)
Local averages (to local cells)
Readout grid (AER scheme)
Dynamic reset path (forms object node)
Pixel / edge signal

6. DEVICE MISMATCH

A fundamental design issue for ensuring circuits operating in weak inversion will work is device matching. The device mismatch arises from process parameter variations mainly in gate oxide thickness and doping concentrations, resulting in device threshold voltage and drain current variations. Since the gm/I ratio is at a maximum for devices operating in weak inversion, this signifies that subthreshold circuits are those most affected by device mismatch [8].

In designing a system requiring image acquisition capabilities in standard CMOS technology, careful consideration must be taken into such sources of error. Non-uniformities in the pixel array; referred to as fixed pattern noise (FPN) are mainly due to offset and gain mismatches between the in-pixel amplifiers. This error, if uncompensated for, would normally render a processing algorithm unusable; however, the method presented has proved robust. Through computer simulations, the inherent immunity to both FPN and physical defects has been verified; discussed in section 7. For both this reason and the high lighting conditions present in the target application, the required dynamic range is limited and therefore the FPN will not pose a serious problem.

However, mismatch errors are not limited to FPN. All circuit blocks requiring critically matched device pairs or groups are susceptible to such errors, for example differential pairs or current mirrors. Subsequently all such circuits require additional attention from schematic design through layout. Specialist simulation techniques such as Monte-Carlo analysis in additional to careful layout [9] can minimize these mismatch errors to both improve performance characteristics and production yields.

7. SIMULATED RESULTS

The proposed system has been simulated at all levels; from the top-level distributed algorithm, to the bottom-level photodiode device physics. These have been facilitated using a selection of simulation tools including the Cadence Spectre Simulator and MATLAB in addition to custom developed code. For the scope of this paper, only the high-level algorithmic simulations shall be discussed.

By using the above mentioned mathematical tools, the distributed algorithm has been simulated with a wide variety of input images. Artificial fixed-pattern noise (FPN) and process defects have been introduced to demonstrate the inherent robustness and fault-tolerant properties of the contour-reduction algorithm. Through successive simulation using randomly generated noise and defect errors, statistical data has been compiled to illustrate a trend for the robustness and stability, given in Figure 5.

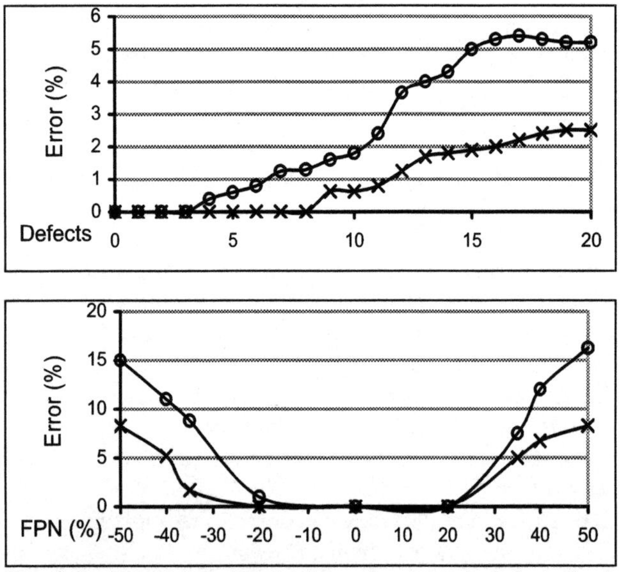

Fig. 5 Statistical data illustrating robustness to defects (top) and FPN (bottom,) acquired through successive computer simulation of the bio-pulsating contour reduction algorithm.

—o— **Object size computation**
—×— **Object count computation**

8. TARGET SPECIFICATIONS

The target design specifications for the ORASIS chip are listed in table 1.

Technology		UMC 0.18µm1P6M CMOS
Supply voltage		1.8V core (3.3V I/O)
Dynamic Range		from 50mWm^{-2} to 5kWm^{-2}
Responsivity		0.2AW^{-1}m^{-2} @ λ=500nm
Maximum tolerable FPN		+/- 15%
Cellular	**Cell Area**	90µm x 90µm
	Active fill factor	11%
	Pixel power	18nW (typical)
	Edge power	20nW (maximum)
	Averaging power	95nW (typical)
	Logic power	5nW (maximum)
	Total online power	138nW
System	**Chip area**	25mm^2
	Array resolution	40 x 40 cells
	Total array power	345µW (typical)
	Total periphery power	100µW (maximum)
	Total online power	345µW
	Duty cycle (online)	10%
	Total effective power	44.5µW

Table 1 Target design specifications for ORASIS cell- and system level hardware implementation

9. CONCLUSION

This paper outlines the top-level hardware implementation of the bio-pulsating contour reduction algorithm [6]. This is a parallel, distributed algorithm performing asynchronous object recognition breaking the bottleneck of traditional, sequential von Neuman based computational paradigm. The globally asynchronous scheme is regulated by employing data-generated local synchronisation, reducing power consumption and improving the signal-to-noise ratio. By incorporating the processing in the front end, the bandwidth requirements have been reduced by at least four orders of magnitude. Both the functionality and robustness have been verified through extensive computer simulation and by implementing an explicit architecture; the hardware realisation has been targeted for micropower operation, realising a retinal vision processor.

10. ACKNOWLEDGEMENTS

The authors would like to acknowledge the Basic Technology grant (UKRC GR/R87642/02) and the AMx technology grant (EPSRC GR/R96583/01,) in addition to Toumaz Technology Limited for sponsoring this research.

11. REFERENCES

[1] T. Roska, A. Rodriguez-Vazguez, "Towards visual microprocessors," Proc. IEEE, Vol. 90, pp. 1244-1257, 2002.

[2] K.A. Boahen, "A retinomorphic vision system," IEEE Micro, Vol. 16, pp. 30-39, 1996

[3] M. Kass, A. Witkin and D. Terzopoulos, "Snakes: active contour models," Int. J. Comput. Vision, Vol. 1, pp. 321-331, 1988

[4] T.G. Morris, T.K. Horiuchi and S.P. Deweerth, "Object-based selection within an analog VLSI visual attention system," IEEE Trans. Circuits Syst. 2, Vol. 45, pp. 1564-1572, 1998.

[5] S Liu and J.G Harris, "Dynamic wires: an analog VLSI model for object processing," Int. J. Comput. Vision, Vol. 8, pp. 221-239, 1992.

[6] T.G. Constandinou, T.S. Lande, C. Toumazou, "Bio-pulsating architecture for object-based processing in next generation vision systems," IEE Electronics Letters, Vol. 39 (16,) pp. 1169-1170, 2003.

[7] T.G Constandinou, J. Georgiou and C. Toumazou, "A nanopower mixed-signal edge-detection circuit for pixel-level processing in next generation vision systems," IEE Electronics Letters, Vol. 39 (25,) pp. 1774-1775, 2003.

[8] A. Papasovic, A.G. Andreou, C.R. Westgate, "Characterisation of Subthreshold MOS Mismatch in Transistors for VLSI Systems," Analog IC's & Signal Proc., Vol. 6, pp. 75-85, 1994.

[9] R.A. Hastings, "The Art of Analog Layout," Prentice Hall, 2000.

BRAIN-MACHINE INTERFACES USING THIN-FILM SILICON MICROELECTRODE ARRAYS

Daryl R. Kipke, PhD.

Department of Biomedical Engineering and Department of Electrical Engineering and Computer Science, University of Michigan, Ann Arbor MI, USA 48109-2125

ABSTRACT

The development of next-generation neuroprosthetic systems will be propelled by improved microelectrode front-ends for providing reliable and selective neural recordings. Silicon, planar micromachined microelectrode arrays have the design space for a diverse set of neuroprosthetic applications. Recent results in which these microelectrodes have been shown to consistently and reliably record high-quality neuronal spikes over several months suggest that this technology provides an appropriate foundation for ongoing neuroprobe development and refinement.

1. INTRODUCTION

The scientific and medical foundation for restoring movement and function in paralyzed patients through controlled stimulation of the peripheral nerves is firmly established. While state-of-the-art neuroprostheses can provide selective activation of muscle groups throughout the arm and hand, the overall performance of these neuroprostheses could be greatly enhanced by increasing the number and quality of the volitional signals produced by the patients to control their devices. Several recent studies in monkeys and humans suggest that neuronal recordings from motor or pre-motor areas of the cerebral cortex are likely sources for these types of control signals. This is especially intriguing because recent advances in functional neurosurgery and multi-channel microelectrode technologies bring into sharp focus the possibility of obtaining appropriate chronic neuronal recordings from motor areas in the cerebral cortex of a paralyzed patient. One of the primary unmet challenges of today is to develop implantable penetrating probes that are safe and effective for use in humans and can also be integrated into advanced neuroprostheses controlled directly with cortical signals.

Extracellular recording of spikes from individual or small clusters of cortical neurons for extended periods of time (many months) is now a relatively generic neurophysiological method that has been addressed using various types of probes, including bundles of microwires, microwires embedded in a neurotrophic assembly, polymer substrate probes, and several types of silicon-substrate probes. This paper reports on the use of a particular type of silicon micromachined probe—the so-called "Michigan" probe system—for chronic unit recording in the cerebral cortex. The various components of this probe system have been developed over the past decade [1, 2] (recently reviewed in [3]) and have several favorable attributes including batch fabrication, high reproducibility of geometrical and electrical characteristics, easy customization of recording site placement and substrate shape, small size, the ability to combine it with a silicon ribbon cable and the ability to include on-chip electronics for signal conditioning [4]. In addition, although the devices are planar, they can be microassembled into three-dimensional arrays [5] and combined with a flexible polymer interconnect cable [6].

This paper reviews progress towards characterizing the recording performance of Michigan silicon probes in animals. Some of the results were previously presented [7, 8]. The probes were found to consistently and reliably provide high-quality spike recordings during this time period. These findings support subsequent development and refinement of this probe technology for neuroscience and medical applications.

2. METHODS AND RESULTS

The Michigan silicon probe process begins with selective diffusion of boron into a silicon wafer to define the shape and thickness of the device [5]. Lower dielectric layers of silicon dioxide and silicon nitride are next deposited using chemical vapor deposition. These layers provide insulation from the conductive boron-doped substrate. The conductive interconnect material (polysilicon in most cases) is deposited, patterned, and then insulated with upper dielectrics as just described. In order to establish recording sites, selective access through the upper dielectrics to the polysilicon layer is achieved using a combination of wet and dry etching. Metal is next deposited and patterned to form the electrode sites (typically iridium) and bond pads (gold) using a lift-off technique. Dielectrics in the field area are removed using a dry etch. After the wafer is thinned, it is placed in ethylenediamine-pyrocatechol which selectively etches away the undoped silicon. This process uses eight photolithographic masks, and enables the design of a wide

0-7803-8251-X/04/$17.00

variety of planar probes, with specific site configuration and physical layout customized to the targeted neural structure.

The typical probes used in rodent studies consisted of four penetrating shanks, each with four uniformly spaced recordings sites of the same size (Fig. 1A and B). The probes were provided by the Center for Neural Communication Technology at the University of Michigan, a NIH National Resource Center (www.engin.umich.edu/facility/cnct/). All animal procedures complied with the United States Department of Agriculture guidelines for the care and use of laboratory animals and were approved by the University of Michigan Animal Care and Use Committee.

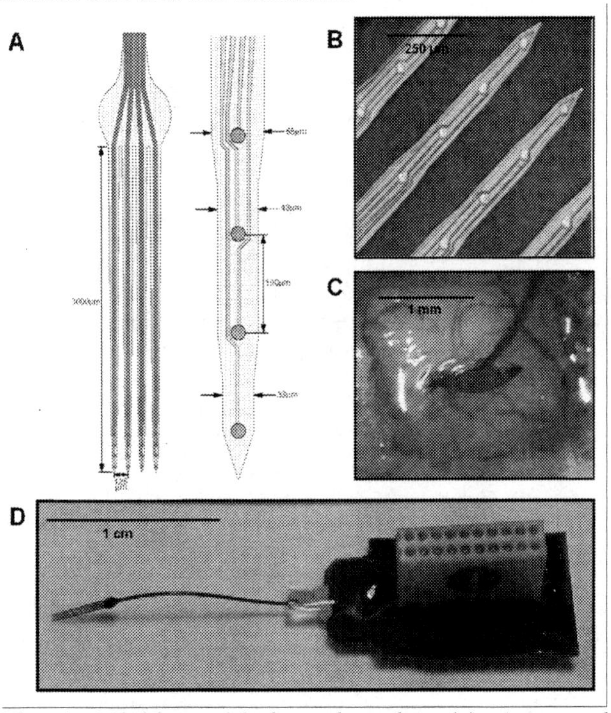

Fig. 1. Michigan 16-channel probe. (a) A general schematic of the four-shank, sixteen channel probe used in this experiment. Some probes contained a 20 µm by 60 µm "well" between the first recording site and the tip. (b) Four shank, sixteen channel probe with 20 µm diameter recording sites. (c) Implanted probe in auditory cortex. (d) Complete probe assembly. The implantable shanks of the probe are integrated with a 1 cm long, 5 µm thick ribbon cable, which is then integrated with a silicon substrate containing the bond pads (approximately 1.5 mm by 0.8 mm). This is bonded to the Omnetics ® connector, which interfaces to the instrumentation system. Silicone rubber is applied to the probe/cable junction to provide a grasping region.

The most complete dataset to date includes results from 14 silicon probes implanted into a consecutive series of 10 rats. Seven of the probes were implanted in the auditory cortex and seven were implanted in the motor cortex. Neural recordings and probe impedance measurements were obtained from each subject daily for the first four post-operative weeks and at least once (typically 5 times) weekly thereafter. The longest implant in this series was 127 days post operation. With one exception, each case was terminated due to factors other than probe failure (i.e., no discriminable units on any probe). These factors included an adverse reaction to anesthesia (one case) and failure of the acrylic head-cap (4 cases). Five subjects were terminated with functioning probes in order to obtain endpoint histology.

The silicon probes proved to be effective for recording discriminable single and multi-unit spike activity. The spike amplitudes ranged from 50-800 µV peak-to-peak with background noise typically below 20 µV peak-to-peak. A brief recording segment across all of the sites in one subject illustrates the typical range and extent of signals (Fig. 2). The recorded spike activity consists of moderate to high amplitude spikes associated with typical multi- and single-unit spike activity, respectively. In the example of Fig. 2, 15 of the 16 sites were found to have discriminable spikes, resulting in a recording yield of 15/16 = 94%. As described in the next section, unit activity was discriminated objectively using a fixed and well-described algorithm (see methods).

Recording yield was consistently high in each of the 10 subjects in this study during the entire assessment period (Fig. 3). The recording yield varied between 88-94%, with an average of 92% calculated over all of the subjects (Fig. 3). The recording yield was also fairly consistent from day to day, as indicated by the associated relatively small average standard error (variation) of 3.5% for each recording day.

3. DISCUSSION AND CONCLUSIONS

The results of this study establish the basic functionality of Michigan silicon micromachined probes for chronic, long-term neural recording in cerebral cortex. In particular, the probes were found to provide reliable, high-quality multichannel spike recordings during the four-month assessment periods used in this study. In summary, 207 of the 224 (92.4%) probes included in this study recorded discriminable spike activity. The day to day recording quality was expressed in terms of an estimated signal-to-noise ratio (SNR) computed from the peak-to-peak voltage amplitudes of spikes relative to the peak-to-peak amplitudes of noise.

4. ACKNOWLEDGEMENTS

The probes were provided by the University of Michigan Center for Neural Communication Technology sponsored by NIH/NCRR grant P41 RR09754. The authors acknowledge the contributions of P. Rousche and members of the Neural Engineering Lab, most notably D. Pellinen and K. Otto. The authors would like to thank Matthew Holecko for assistance in the histological analysis. They thank Ning Gulari for fabrication of the probes and Rania Oweiss for packaging.

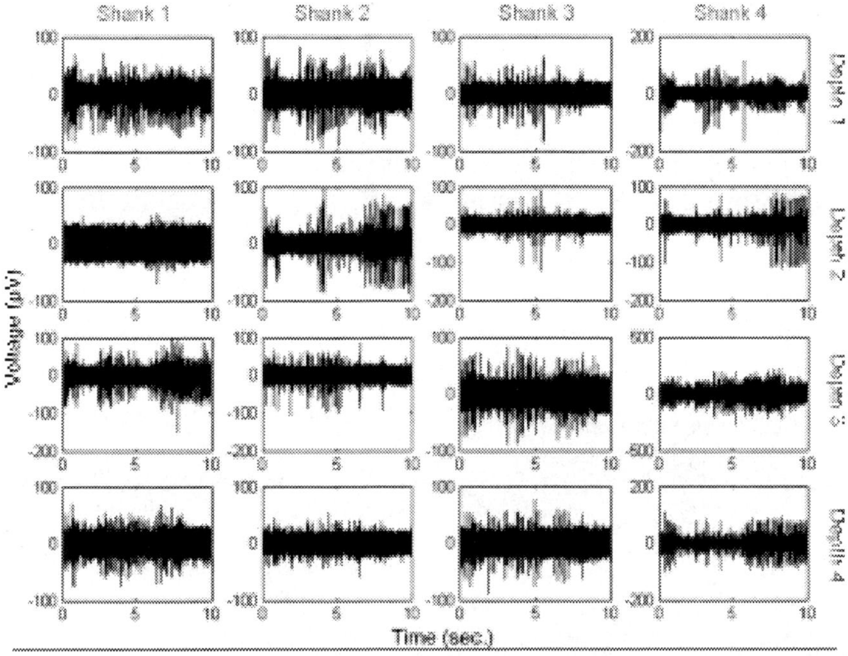

Fig. 2. A five second snapshot of data collected simultaneously on 16-channels implanted in motor cortex on day 113 from animal BMI-4. Fifteen of sixteen channels have sortable single units present.

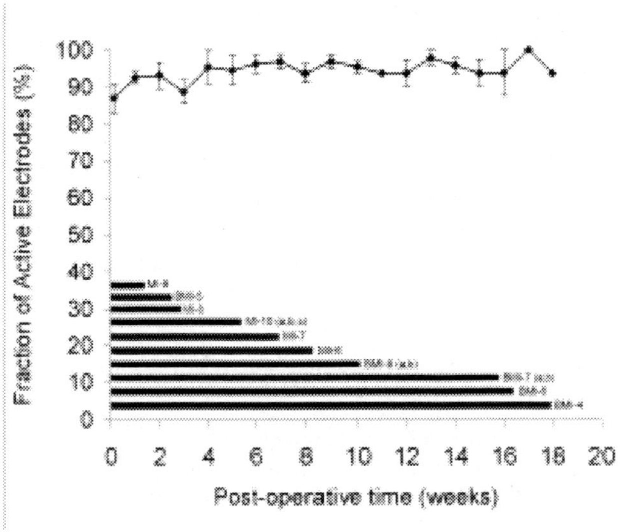

Fig. 3. The average percent of active electrodes on all implanted arrays. Time bars at the bottom of the figure represent each animal's contribution to mean results.

5. REFERENCES

[1] K. L. Drake, K. D. Wise, J. Farraye, D. J. Anderson, and S. L. BeMent, "Performance of planar multisite microprobes in recording extracellular single-unit intracortical activity," *IEEE Trans Biomed Eng*, vol. 35, pp. 719-32, 1988.

[2] J. F. Hetke, J. L. Lund, K. Najafi, K. D. Wise, and D. J. Anderson, "Silicon ribbon cables for chronically implantable microelectrode arrays," *IEEE Trans Biomed Eng*, vol. 41, pp. 314-21, 1994.

[3] J. F. Hetke and D. J. Anderson, "Silicon microelectrodes for extracellular recording," in *Handbook of Neuroprosthetic Methods*, W. E. Finn and P. G. LoPresti, Eds. Boca Raton, FL: CRC Press, 2002.

[4] Q. Bai and K. D. Wise, "Single-unit neural recording with active microelectrode arrays," *IEEE Trans Biomed Eng*, vol. 48, pp. 911-20, 2001.

[5] Q. Bai, K. D. Wise, and D. J. Anderson, "A high-yield microassembly structure for three-dimensional microelectrode arrays," *IEEE Trans Biomed Eng*, vol. 47, pp. 281-9, 2000.

[6] J. F. Hetke, J. C. Williams, D. S. Pellinen, R. J. Vetter, and D. R. Kipke, "3-D silicon probe array with hybrid polymer interconnect for chronic cortical recording," presented at 1st International IEEE EMBS Conf. on Neural Engin., Capri, Italy, 2003.

[7] D. R. Kipke, R. J. Vetter, J. C. Williams, and J. F. Hetke, "Silicon-substrate intracortical microelectrode arrays for long-term recording of neuronal spike activity in cerebral cortex," *IEEE Trans Neural Systems and Rehab. Engin*, vol. 11, pp. 151-155, 2003.

[8] R. J. Vetter, J. C. Williams, J. F. Hetke, E. A. Nunamaker, and D. R. Kipke, "Spike recording performance of implanted chronic silicon-substrate microelectrode arrays in cerebral cortex," *IEEE Trans Biomed Eng*, Submitted (invited).

CNN WAVE BASED COMPUTATION FOR ROBOT NAVIGATION PLANNING

Paolo Arena, Adriano Basile, Luigi Fortuna, Mattia Frasca

Dipartimento di Ingegneria Elettrica Elettronica e dei Sistemi
Universitá degli Studi di Catania - Viale A. Doria, 6 - 95100 Catania, Italy
E-mail: mfrasca@diees.unict.it

ABSTRACT

In this work a methodology for real-time robot navigation in a complex, dynamically changing environment, based on wave computation and implemented by Cellular Neural Networks (CNNs) is introduced.

The key point of the approach is to consider the environment in which robot moves as an excitable medium. Obstacles and targets represent the source of autowave generation. The wavefronts propagating in the CNN medium provide to the robot all the information to achieve an adaptive motion avoiding the obstacles and directed to the target.

In particular the paradigm of reaction-diffusion (RD) equations are used to implement a CNN-based wave computation for the navigation control. Experimental results validating the approach are shown.

1. INTRODUCTION

Wave computing is an emerging paradigm of nonlinear science. In this paper wave computing is applied to the problem of robot navigation in a complex environment with obstacles. The idea underlying the approach is the use of a reaction-diffusion medium as an analog processor for the control of the robot.

A very common approach to the robot navigation problem is the use of artificial potential fields, where the whole experimental arena is mapped into the computational architecture of the robot navigator: obstacles generate repulsive fields and targets generate attractive fields. However the computational resources required by this approach are quite large, and therefore real-time adaptation to moving obstacles is difficult to achieve.

On the other hand there exist experimental implementations of chemical controllers that calculate the optimal path for a robot moving in an obstacle-filled environment. These controllers are analog processors mostly based on the excitation wave dynamics in the Belousov-Zhabotinsky (BZ) medium [1, 2]. Their intrinsic parallel capabilities allow to overcome the limits of the potential field approach, but their main drawback is the limited speed of diffusion of the wavefronts in chemical reactions.

Starting from these considerations, in this work CNNs [3, 4] are applied to generate autowaves for the navigation control of a robot, allowing to overcome the limits of chemical processors by keeping the peculiarities of the paradigm. The basic idea is the use of reaction-diffusion (RD) equations that can be implemented in CNNs and reflect the nature of the BZ medium.

The robot arena is mapped on a CNN: its role is to select obstacles (or the target) and identify them as sources of repulsive (or attractive) autowaves. Subsequently a RD-CNN generates autowave fronts. The robot is allowed to freely move within the arena. A simple AND operation reveals the wavefront collisions with the robot, and a steering command is sent to the robot for the generation of highly adaptive trajectories. This methodology allows the selection of the optimal robot trajectory in dynamically changing environment in extremely reduced times: under this perspective such an algorithm, that solves a complex task, succeeds in being even faster than, in several cases, the robot typical reaction times. Experimental results confirming the suitability of the approach are briefly discussed.

2. RD-CNN FOR ROBOT NAVIGATION CONTROL

RD phenomena occur in many systems [5], but they obey the same laws that can be represented under the following general RD equations:

$$\frac{d\mathbf{u}}{dt} = f(\mathbf{u}) + \nabla^2\mathbf{u} \tag{1}$$

in which $\mathbf{u}$ is a vector of at least two elements. For instance $\mathbf{u}$ can represent the dynamics of an activator/inhibitor system diffusing in a two dimensional medium and showing pattern formation. In this paper we are interested in nonlinear media, represented by equation (1) showing wave propagation. These waves are very important for the navigation control, since the trajectory computation can be eas-

This work was supported by the Italian "Ministero dell'Istruzione, dell'Universitá e della Ricerca" (MIUR) under the PRIN project "Innovative Bio-Inspired Strategies for the Control of Motion Systems" ily realized through their generation, propagation and interaction. Under this perspective nonlinear media represent a class of parallel computers with unique features: parallel input of data, parallel information processing and parallel outputs [7].

In our approach the reaction-diffusion medium is devoted to control the trajectory of a robot in an environment with obstacles, where a target point for the robot to reach can also be fixed. Both obstacles and target are mapped onto the nonlinear medium as autowave sources, stimulating different regions or cells. A key characteristic of autowaves is that they annihilate when they collide [8], in such a way they intrinsically determine the path of points equidistant from the obstacles.

CNNs can map RD equations [10] if the following equations for the single CNN generic cell C_{ij} are taken into account:

$$\begin{cases} \dot{x}_{ij;1} = -x_{ij;1} + p y_{ij;1} - s y_{ij;2} + D_1 (y_{i+1,j;1} + \\ \qquad + y_{i-1,j;1} + y_{i,j+1;1} + y_{i,j-1;1} - 4 y_{i,j;1}) \\ \dot{x}_{ij;2} = -x_{ij;2} + p y_{ij;2} + s y_{ij;1} + D_2 (y_{i+1,j;2} + \\ \qquad + y_{i-1,j;2} + y_{i,j+1;2} + y_{i,j-1;2} - 4 y_{i,j;2}) \end{cases} \tag{2}$$

considering also, as the output nonlinearity, the classical saturation one.

Taking into account that the same model for the RD-CNN was also successfully used as the basic neuron model for bio-inspired locomotion control [6], it derives that the class of RD-CNN discussed in this paper represents a unifying paradigm for the solution of locomotion control, from the low level of the motion pattern generation, to the very high level of the trajectory planning.

The CNN framework constitutes a real-time implementation of the RD equations for navigation control. Moreover, thanks to the universal computing properties of a CNN chip, a huge number of additional algorithms can be used, making them able to perform additional complex image filtering routines, that are needed to have a complete and really working algorithm for robot trajectory planning. In fact, a first, essential step that has to be performed a priori, lies in the target/obstacle recognition. In our case, the input is provided by a video camera making a motion picture of the environment where the robot moves. By taking into account the peculiar characteristics of the obstacles (i.e. their shape), several algorithms, already implemented and tested on VLSI CNN chips, can be adapted to the environment features and applied to provide an image containing only the obstacles surrounding the robot, whose position is assumed to be known at any given time. This image is processed by the RD-CNN (the analog processor) in which autowaves propagate.

In particular, the CNN cells corresponding to the position of obstacles are stimulated by setting initial conditions for the first layer of the CNN to $x_{ij} = +1$. Waves, generated at the obstacle positions, propagate until they collide with the robot. Depending on the kind of source generating the wave (obstacle or target) and on the particular compass direction the wavefront comes from, the new direction for the robot can be directly determined. Thus the trajectory planning is not performed before the robot starts moving, but occurs step by step while the robot moves, soon after each wave-robot collision. In particular, after the completion of a robot movement, one could establish to process another frame or not, i.e. one can decide either to reset the RD-CNN and start with a new configuration of obstacles and target, or to continue with the previously acquired configuration. This allows to account also for moving obstacles or to speed-up the algorithm. The details of the algorithm used and experimental results are presented in the following Section.

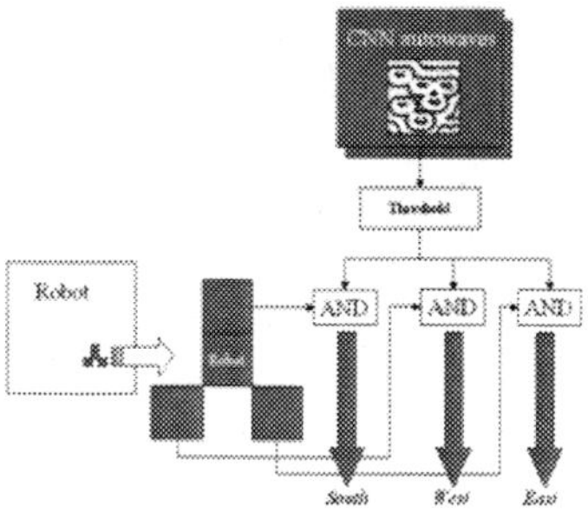

Fig. 1. Flow diagram of the CNN algorithm (the case of autowaves generated by obstacles is shown).

3. THE CNN ALGORITHM

Two complementary RD-CNNs are used: one is the medium for propagation of waves generated by obstacles, while the second deals with the target. The flow diagram of the algorithm as regards the case of autowaves generated by obstacles is shown in Fig. 1. The robot is viewed as a four active pixel object according to Fig. 1 with a given orientation. After a threshold operation an AND operation is performed between the snapshot of wave propagation, represented by the state of the first layer of the RD-CNN, and the image containing the robot. Depending on which of the robot pixels is first reached by the wavefront, a particular motion instruction is given to the robot. The parameters of the RD-CNN are chosen according to $p = 1.7$, $s = 1$, $D_1 = 0.1$, $D_2 = 0.1$.

Simulation results have been obtained by using a dedicated framework, written in C++, for the simulation of the robot and of the CNN algorithm. A 50×50 CNN has been simulated. Fig. 2 shows several examples of simulation results. In this case, instead of a target to reach, the robot

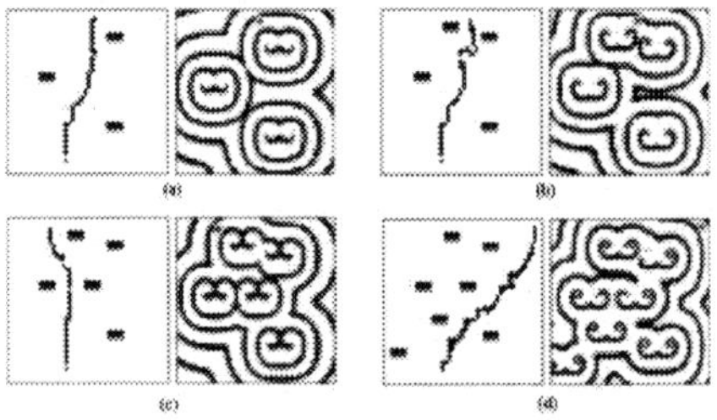

Fig. 2. Simulation results: case with only obstacles.

focuses to proceed in a pre-specified direction (in this case north). Similar results have been obtained in the case in which also the target is taken into account.

As it can be seen from the simulation results, the algorithm presented is really effective. However, there can be cases in which the robot could be trapped into a local minimum, for example when target and obstacles are in particular symmetrical points with respect to the robot. In such cases the robot moves alternatively between the same two positions. To overcome such a condition the strategy, a random perturbation on the robot trajectory is added. Moreover, in view of a real implementation, it is not necessary that the video feedback is done on the whole robot arena, but it can be assumed that the processing takes place only on a certain region surrounding the robot. Due to the adaptive characteristics of the methodology, the environment focused by the camera can vary as the robot moves. So for example, at the beginning, when the target could be out of the captured area, the robot is assigned to move only in the direction of the target. So its first task could be to avoid obstacles while moving in a particular direction. As the robot position approaches the target, the final focus could be taken into consideration. Moreover, taking limited, next-to-robot frames improves the computational efficiency, since distant obstacles do not generate wavefronts; so the local trajectory is influenced only by obstacles in the immediate vicinity of the robot. These aspects are really important, in view of a real implementation on the RD-CNN chip.

4. RESULTS ON A ROVING ROBOT

The algorithm discussed previously is quite general and can be implemented with two different installations of the camera: on the ceiling of the laboratory (as it has been assumed in the simulations) or on board, taking a picture of a small planar space in front of the robot. Thus, a *world-centered perception*, or a *robot-centered perception* can be implemented by using the general paradigm of reaction-diffusion navigation control. The experimental set-up is provided by a real environment, where a roving robot is required to move in a pre-specified direction, in this case south: so, no particular target positions are given. The camera is connected to a personal computer simulating the RD-CNN algorithm.

Fig. 3. Example of a typical trajectory of the robot while avoiding an obstacle. The camera is placed on the ceiling of the laboratory.

In the first experiment the *world-centered perception* case is reproduced. At each step the robot camera takes a picture of the environment: obstacles are used as initial conditions for the RD-CNN. When the wavefronts reach the robot, depending on the direction the wavefront comes from, the robot executes the proper command. The robot succeeds in going in the pre-specified direction avoiding two obstacles, as it can be appreciated in Fig. 3. Here the positions successively occupied by the robot are reported through red points. From the analysis of the figures it emerges that the robot is indeed able to avoid both the obstacles, but it passes much closer to the one than the other. This potential problem can be explained as follows. Since autowave fronts annihilate while colliding, if the robot moves in a direction where a wavefront has just been annihilated, it will "sense" the subsequent front when the robot will be next to its source, i.e. the obstacle. A suitable synchronization among the robot motion and the wave generation should avoid the problem.

The second experiment refers to the *robot-centered perception* strategy: the camera is placed on the robot via a vertical pole whose height is about $1m$. The camera is positioned in such a way that the focal plane is almost parallel to the ground and the robot is situated in the central bottom position within the frames. In this condition the frames refer to the environment in the forward motion direction of the robot. The whole experiment is reported in Fig. 4. Here some snapshots are illustrated: as it can be noticed, the robot is able to avoid both the obstacles, while proceeding in the pre-specified direction.

5. CONCLUSIONS

A novel paradigm of computing based on reaction-diffusion CNN equations has been applied to real-time robot naviga-

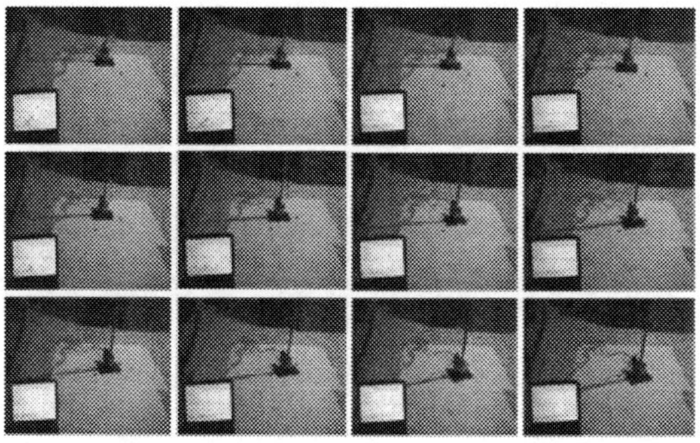

Fig. 4. Example of robot-centered navigation control based on RD-CNN. Several snapshots of a video are shown.

tion in complex, dynamically changing environment. The environment in which the robot moves is mapped onto a RD-CNN (the excitable medium) where obstacles and the target generate autowaves. The waves generated by obstacles are interpreted by the robot as repulsive stimuli while wave generated by the target as attractive stimuli. The robot optimizes its trajectory employing spatio-temporal dynamics of 'repulsive' and 'attractive' events. Each element of the excitable medium acts as a very primitive computing device. The medium's sites take on ideally continuous values and the site interacts locally; therefore the path planning problem is solved through an analog and massively parallel computation, allowing also for a real-time control of the robot's trajectory, also when high motion speeds are required to the robotic structure.

With respect to traditional schemes for the computation of the path planning, the approach presented here is totally innovative since it uses massively parallel computing. For instance the core of the traditional potential field approach is the calculation, for each environment state, of the potential function for each point of the robot arena. In case of moving obstacles on dynamically changing situations, the advantages of the approach based on an analog processor are apparent: in our case each obstacle/target is a wave source independently on its position. Another advantage of the approach is the adaptability of the parallel computation to a vast class of optimization problems.

Future developments will concern the implementation on VLSI CNN chip of the approach and the possibility of applying the proposed methodology to cooperating robots. Each of the robot is driven by its own analog processors with its local vision field (as in the case of robot-centered perception) and may interact with the other robots to find a global optimal solution.

Acknowledgment

The authors would like to thank A. Adamatzky and A. Rodriguez-Vazquez for their helpful comments and suggestions.

6. REFERENCES

[1] K. Agladze, N. Magome, R. Aliev, T. Yamaguchi and K. Yoshikawa, "Finding the optimal path with the aid of chemical wave", *Physica D*, 106 , pp. 247–254, 1997.

[2] A. Adamatzky and B. De Lacy Costello, "Collision-free path planning in the Belousov-Zhabotinsky medium assisted by a cellular automaton", *Naturwissenschaften 89*, 10, pp. 474-478, 2002.

[3] L. O. Chua, L. Yang, "Cellular Neural Networks: Theory and Applications", *IEEE Transactions on Circuits and Systems I*, 35, pp. 1257-1290, 1988.

[4] L. O. Chua, T. Roska, "The CNN paradigm", *IEEE Transactions on Circuits and Systems*, 40, pp. 147-156, 1993.

[5] J. D. Murray, *Mathematical Biology*, Springer-Verlag, Berlin, 1989.

[6] P. Arena, L. Fortuna, M. Branciforte. "Reaction-diffusion CNN algorithms to generate and control artificial locomotion", *IEEE Transactions on Circuits and Systems I*, 46(2):253–260, 1999.

[7] A. Adamatzky, *Computing in Nonlinear Media and Automata Collectives* (IoP Publishing, Bristol and Philadelphia, 2001).

[8] V.I.Krinsky (editor), *Self-Organization: Autowaves and Structures Far from Equilibrium*, Springer-Verlag, Berlin, 1984

[9] P. Arena, S. Baglio, L. Fortuna, G. Manganaro, "Self-Organization in a Two-Layer CNN", *IEEE Trans.on Circuits and Systems - Part I*, vol. 45, no.2, 157-162, 1998.

[10] L. O. Chua, "CNN: A Paradigm for complexity", in *Visions of Nonlinear Science in the 21st Century*, (J. L. Huertas, W.-K. Chen, R. N. Madan, eds.) World Scientific Series on Nonlinear Science, Series A, vol. 26, 1999.

[11] P. Julian, A. Desages, B. D'Amico, "Orthonormal high-level canonical PWL functions with applications to model reduction", *IEEE Trans on Circuits and Systems I*, Vol. 47, No. 5, pp. 702 -712, 2000.

[12] P. Arena, L. Fortuna, L. Occhipinti, "A CNN Algorithm for Real Time analysis of DNA microarrays", *IEEE Transaction on Circuits and Systems, I: Fundam. Theory and Applic.*, Vol. 49, no. 3, 2002.

[13] T. Roska, L. Kèk, L. Nemes, A. Zaràndy, M. Brendel, and P. Szolgay (Eds.) "CNN Software Library (Template and Algorithms), Version 7.2", DNS-1-1998, (CADET-15), Computer and Automation Institute, Hungarian Academy of Sciences, Budapest, 1998.

[14] P. Arena, A. Basile, M. Bucolo, L. Fortuna, A. Virzì, "A bio-inspired Visual Feedback locomotion control based on CNN Universal Machine", *Journal of Circuits, Systems, and Computers*, in press.

FINITE ITERATION DT-CNN – NEW DESIGN AND OPERATING PRINCIPLES

C. Merkwirth

Computational Biology and
Applied Algorithmics Group
Max-Planck-Institut für Informatik
66123 Saarbrücken, Germany
e-mail: cmerk@mpi-sb.mpg.de

J. Bröcker, M. Ogorzałek, J. Wichard

Department of Electrical Engineering
AGH University of Science and Technology
al. Mickiewicza 30
30-059 Kraków, Poland
e-mail: maciej@agh.edu.pl

ABSTRACT

In this paper we propose to use the Discrete-Time Cellular Neural Network (DT-CNN) in a finite iterate mode. In such a mode of operation no special requirements on template stability properties are needed. We propose a constructive back propagation based algorithm for template design. For a given number of iterations we can find optimal sequence of templates for a given problem to be solved. Our novel approach is demonstrated by design of a digit recognition DT-CNN.

1. INTRODUCTION

When first introduced by Chua and Yang [1] Cellular Neural Networks were considered as a large array of simple processing elements – continuous time dynamical systems – characterized by only local connectivity. The local connections defined by so called templates were fixed throughout the network and kept constant when performing a single operation. Further a discrete time version of the CNN principle has been introduced (DT-CNN) [2]. In both proposed types of networks the underlying principle is that computation is done by system dynamics, i.e. the response of the network to given stimuli is defined by the limit behavior in time – convergence of all responses to some fixed point [3]. The achieved steady state is predefined by the connection templates and by the the network input and initial conditions.

Here we propose to use the DT-CNN as a "finite iteration" computing device. The network will perform a fixed number of iterations in contrast to the standard convergence approach. We propose also to use a different template design strategy the main feature of which is to use for each iteration a different template. It should be stressed that in the finite iterate approach we can drop the requirement that templates guarantee convergence of the network to a fixed point – no requirements on stability, symmetry or any other template properties are needed [4].

We propose also a constructive approach to design the templates – a learning strategy. For the finite iterate DT-CNN this is based on optimization and use of advanced back propagation techniques. Fixing the number of iterates as a design parameter we can run the optimization algorithm which will find an optimal sequence of templates minimizing a suitably chosen cost function of the considered problem (eg. pattern recognition).

Parts of this work are supported by the Research Training Network *COSYC of SENS* No. HPRN-CT-2000-00158 within the 5th Framework Program of the EU and by the Deutsche Forschungsgemeinschaft (DFG) grant LE 491/11-1. We would like to thank the people of the AGH University of Science and Technology for their support.

2. THE CNN TOPOLOGY

The states of the DT-CNN cells y_{ij}^t evolve for iterations $t = 0, \ldots, T-1$ according to:

$$\begin{aligned} x_{ij}^{t+1} &= \sum_{l,m=-\frac{K-1}{2}}^{\frac{K-1}{2}} \left(A_{l,m}^t y_{i+l,j+m}^t \right. \\ &\qquad \left. + B_{l,m}^t u_{i+l,j+m}\right) + b^t \\ y_{ij}^{t+1} &= \sigma(x_{ij}^{t+1}), \end{aligned} \tag{1}$$

wherein the u_{ij} denote the input pattern and $\sigma(x_{ij})$ is the activation function, further discussed in subsection 2.1. If we use different template weights (A^t, B^t and offsets b^t) in each iteration t, we can massively improve the performance of the DT-CNN in classification tasks. In this case, the total number P of adjustable scalar parameters can be calculated by $P = T(2K^2+1) - K^2$, where T is the number of *template layers*, and K is the number of rows of the quadratic template matrices. For eg $K = 5$, this results in 51 weights per template layer. A^0 is not used since $y_{ij}^0 = 0$ by initialization.

2.1. The activation function

The activation function $\sigma(x)$ depicted in Figure 1 is a piecewise linear function that does not fully saturate for input values x far from the origin [5]. This feature ensures a still nonvanishing derivative of the training error with respect to very large template entries. A training algorithm exploiting derivative information can thus escape from such usually quite suboptimal solutions. In contrast, networks having a saturating sigmoid activation function may cause the training algorithm to get stuck at such places, for the gradient almost vanishes for large arguments. Since this feature is only relevant for the training, the slope far from the origin can be set to converge to zero during the backpropagation epochs.

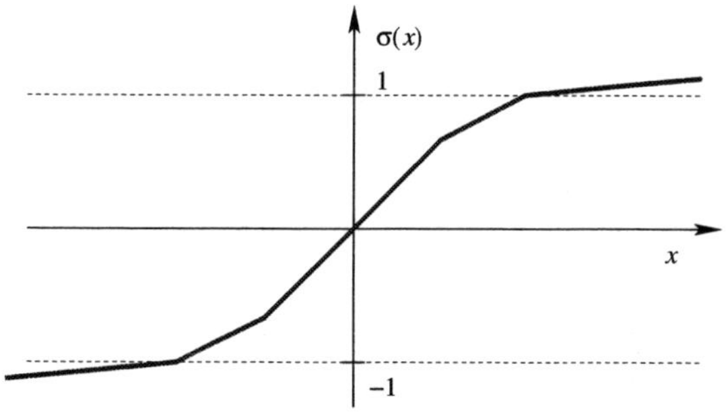

Fig. 1. A piecewise linear activation function is used as nonlinearity. Note that the function does not saturate at ± 1, but the slope decreases to 0.01. The singularity of the derivative at isolated points does not adversely affect a gradient based training algorithm.

3. THE DT-CNN CLASSIFIER

This paper is concerned with using a DT-CNN as a classifier. In a binary decision problem we present an input pattern u to the receptive field of the DT-CNN, we want to get a decision whether this input pattern belongs to a specific class or not. Since the DT-CNN is a spatial processing engine, we have to convert the spatial output y_{ij}^T into one output value suitable as decision variable. This is done by simply averaging these 256 states y_{ij}^T and comparing the result to a certain threshold (usually zero).

In the case of *Handwritten Digit Recognition* the task is to identify handwritten digits (from zero to nine) from a 16 by 16 pixel grayscale image. This obviously is a classification problem with *ten* instead of two classes. This multiclass classification problem can be converted into 10 binary classification problems to be solved independently as follows. The first classifier has to distinguish the zeros from the rest, the second the ones from the rest etc. All ten classifiers are trained on the same input patterns. The desired outputs presented to the first classifier are $+1$ for input patterns showing a zero, otherwise -1. For the second classifier the desired outputs are $+1$ for observations for input patterns showing a one, and so on.

4. TRAINING OF DT-CNN CLASSIFIERS

The training of DT-CNN classifiers is based on minimization of a function henceforth called the *training error*. The training error depends on the training data and the network templates and is minimized with respect to the latter. More specifically, let $(z^i, u^i), i = 1 \ldots N$ be the training data consisting of input output pairs. Further, denote by $f(u^i; A, B, b)$ the output of DT-CNN classifier, where A, B and b stand for the entity of the templates. The training error then has the form

$$E(A, B, b) := \sum_{i=1}^{N} \lambda(z_i - f(u^i; A, B, b)). \quad (2)$$

The *loss function* λ measures the deviation of the CNN output from the desired value z^i. In optimization usually a quadratic loss function is used, basically due to the simplicity of the resulting derivatives. We however advocate the use of an *ϵ–insensitive absolute loss function.* The advantages of this strategy are subject of the next subsection. Further subjections are concerned with the details of the optimization procedure. The training of networks with shared weights requires an augmented error function. Although a shared weight network may be initialized with equal templates in each layer and trained as such, we found this procedure to be quite unstable. A better way to train such networks is to initialize them with different weights and include the following penalty term

$$D(A, B, b) := \sum_{t=2}^{T} \lambda(A^{t-1} - A^t) + \lambda(B^{t-1} - B^t) + \lambda(b^{t-1} - b^t)$$

into the training error. Here λ is to be understood as applied componentwise and then summed. The total error to be minimized is the convex sum

$$TE(A, B, b) := (1 - \gamma)E(A, B, b) + \gamma D(A, B, b).$$

4.1. The ϵ–insensitive loss function

We use an *ϵ–insensitive absolute loss function* $\lambda(\xi)$ to calculate the training error as plotted in Figure 2. The output of the DT-CNN has zero loss and gradient if it lies inside the ϵ–margin of the desired output. This forces the algorithm to learn the misclassified training patterns instead of adjusting the weights with gradient steps of already correctly classified patterns. Training patterns that cannot be correctly classified have only a linear contribution to the loss. This provides an appropriate tradeoff between tolerating outliers and penalizing classification errors. In our case, we use $\epsilon = 0.9$ which seems to be very high, but a smaller ϵ degrades the classification performance. The use of ϵ–insensitive loss functions in the context of learning problems is described in [6].

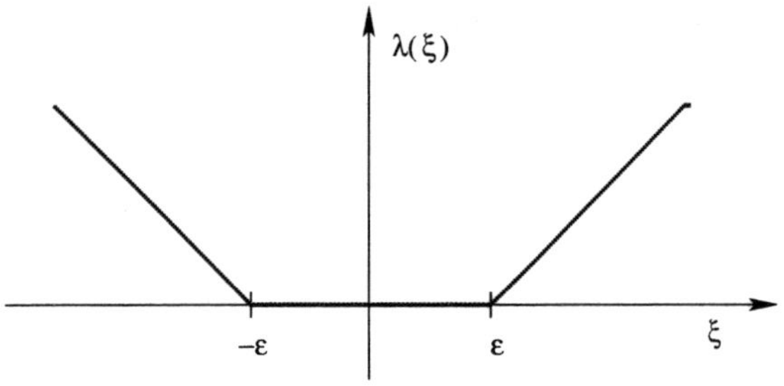

Fig. 2. The ϵ-insensitive loss function used for the training of the DT-CNNs. Though the derivative of the loss function is not continuous, it can be used in the backpropagation training without problems.

4.2. Stochastic gradient descent and back propagation

The method of training a Neural Network considered in this paper is known as *stochastic gradient descent* or *online learning*. The templates are updated recursively according to the rule

$$\begin{aligned} A_n &= A_{n-1} - \delta A_n, \\ B_n &= B_{n-1} - \delta B_n, \\ b_n &= b_{n-1} - \delta b_n, \end{aligned} \quad (3)$$

where $n = 1 \ldots N$, the number of training samples. The update δA_n depends on the nth training sample only. Namely, we set

$$\begin{aligned} \delta A_n &= \mu \frac{\partial}{\partial A} \lambda(z_n - f(u^n; A_{n-1}, B_{n-1}, b_{n-1})), \\ \delta B_n &= \mu \frac{\partial}{\partial B} \lambda(z_n - f(u^n; A_{n-1}, B_{n-1}, b_{n-1})), \\ \delta b_n &= \mu \frac{\partial}{\partial b} \lambda(z_n - f(u^n; A_{n-1}, B_{n-1}, b_{n-1})), \end{aligned}$$

Thus, in contrast to batch learning, where gradients are accumulated over all samples of the data set for the *same template setting* before a gradient step is performed, a tiny gradient step is done here after each observation has been processed. A sweep through the whole data set is called *epoch.* The global stepsize $\mu = 0.0007$ is exponentially decreased every five epochs by a factor of 0.8. For every epoch the order in which the samples of the training data set are processed is randomized. In all simulations below, the number of training epochs is fixed to 50.

Note that, in contrast to batch learning, the algorithm will not terminate even in a global minimum. Nonetheless, it may provide good solutions: The parameters classifying the first sample correctly usually provide a reasonable guess for the second sample, and, after being updated again, will provide a good guess for the third sample and so on. In a linear problem, online learning provides the optimal solution in one epoch.[1] The algorithm does not terminate as well, but more epochs will yield spurious results (overfitting). In a nonlinear setup more epochs but with smaller steps are required. The stochastic gradient descent furthermore does not suffer from getting stuck in local minima.

A detailed discussion of the advantages of stochastic gradient descent over batch learning can be found in [7].

In the training of shared weight networks, the second term in the error function has to be taken into account. The update rule for the templates in online learning in this case is

$$\begin{aligned} \delta A_n &= \mu(1-\gamma) \frac{\partial}{\partial A} \lambda(z_n - f(u^n; A_{n-1}, B_{n-1}, b_{n-1})) \\ &\quad + \gamma \frac{1}{P} \frac{\partial}{\partial A} D(A, B, b), \end{aligned}$$

and similarly for B and b, where P is the number of parameters.

4.3. Shrinking

A reduction in training time can be achieved by using a heuristics called *shrinking*. The evaluation of training samples that are easy to classify is postponed, while the algorithm concentrates on samples that are harder to classify. In detail this means that if a training observation repeatedly yields a vanishing classification error (i.e. is inside the ϵ–interval of the loss function), it is disclosed from the training set for a certain number of epochs determined by the following rules. For each sample, store two counters C_1 and C_2. Initially, both counters are equal to one. At the beginning of any epoch, decrement C_1 by one. If C_1 becomes zero, evaluate the sample. If the output for this sample turns out to be inside the ϵ-insensitive margin, increase C_2 by one and then reset C_1 to the value of C_2. However, if an observation is outside the ϵ-insensitive margin, reset both C_1 and C_2 to 1. Then start the new epoch.

4.4. Building classifier ensembles

A common way to improve the performance of classifiers is ensemble building (see [8] and the references therein). This is done by training several (6-16) DT-CNNs on randomly chosen subsets of the training data where the training starts with random weight initializations. To compute the output of the ensemble for one input pattern, the output decision variables of all DT-CNNs belonging to that ensemble are averaged. The averaging is not done on the template weights. Please note that we have to construct one ensemble for each of the ten binary classification tasks.

[1]This however requires a more sophisticated choice of the stepsize μ, namely the *Kalman gain*

5. DIGIT RECOGNITION – RESULTS

The ZIP Code Data Set[2] consists of normalized handwritten digits, automatically scanned from envelopes by the U.S. Postal Service. The images, originally binary and of different sizes and orientations, have been deslanted and size normalized, resulting in 16 by 16 grayscale images (see [7]). There are 7291 training observations and 2007 test observations, distributed as follows: The 2009 samples of the ZIP

Digit	0	1	2	3
Train	1194	1005	731	658
Test	359	264	198	166

Digit	4	5	6	7	8	9
Train	652	556	664	645	542	644
Test	200	160	170	147	166	177

Table 1. Distribution of classes in the ZIP Code data sets

Code test set are neither used for training nor to derive stopping criteria nor for model selection for classifier ensembling. The patterns are presented to the training algorithm as input-output pairs together with the desired the class label out of 0,1,. . . ,9 (supervised learning setting).

5.1. Classification Performance

Computational demand did not allow for a systematic check of all possible combinations of the template size K and the number of layers T. So $K = 5$ was used for all simulations (see table 2). With standard nearest-neighbor coupling $K = 3$ we could not achieve an adequate classification performance, for template size $K = 7$ we observed training and test errors starting to disagree (overfitting).
From Table 2 one can observe that varying T from one to ten layers, the classification rate increases significantly. For more template layers the rate saturates around 96.5%. For comparison, we trained a polynomial SVM classifier (see [9]) of degree 4 with $C = 100$ on the ZIP Code training set and applied it to the test set.

T	1	2	3	4	5	6
Rate [%]	79.7	91.6	94.9	95.9	96.0	96.1

T	8	10	14	18	24	SVM
Rate [%]	95.6	96.5	96.6	96.2	96.6	95.4

Table 2. Classification rates of final classifiers on the ZIP Code test set versus number of template layers T. Note that uniform random guessing yields a trivial classification rate of 10%.

[2]The ZIP Code data set can be obtained from
www-stat-class.stanford.edu/~tibs/ElemStatLearn

6. CONCLUSIONS

This article discusses a new class of Discrete-Time Cellular Neural Networks which operate in the finite iterate regime. This new approach does not rely on convergence and stability of the network dynamics but uses the CNN as a one-step computing engine. The templates are designed using a minimization approach. As an application example we have confirmed excellent performance of a DT-CNN classifier for digit recognition. The results on a database of handwritten digits outperform a polynomial Support Vector classifier. A main drawback of the method is the higher computational effort for training ensembles of DT-CNN classifiers than for SVM classifiers. Nonetheless, the authors hope that a high speed hardware implementation can be realized based on the templates determined by the proposed method.

7. REFERENCES

[1] L.O. Chua and L.B Yang, "Cellular neural networks: Theory," *IEEE Trans. on Circuits and Systems*, vol. 35, pp. 1257–1272, 1988.

[2] H. Harrer and J.A. Nossek, "Discrete-time cellular neural networks," *Int. J. Circuit Theory and Applications*, vol. 20, pp. 453–467, 1992.

[3] A. Kilinc S. Arik and F. Acar Savaci, "Global asymtotic stability of discrete-time cellular neural networks," in *Proc. of IEEE Int. Workshop on Cellular Neural Networks and their Appl.*, 1998, pp. 52–55.

[4] Á. Zárandy, "The art of CNN template design," *Int. J. Circuit Theory and Applications - Special Issue: Theory, Design and Applications of Cellular Neural Networks*, vol. 17, pp. 5–24, 1999.

[5] Y. LeCun et al, "Efficient BackProp," in *Neural Networks: Tricks of the trade*, G. Orr and K. Mller, Eds., vol. 1524 of *Lecture Notes in Computer Science*, pp. 9–50. Springer Verlag, 1998.

[6] V. Vapnik, *The Nature of Statistical Learning Theory*, Springer Verlag, New York, 1999.

[7] Y. LeCun et al, "Gradient-based learning applied to document recognition," *Proceedings of the IEEE*, vol. 86, no. 11, pp. 2278–2324, 1998.

[8] C. Merkwirth, M. Ogorzałek, and J.D. Wichard, "Stochastic gradient descent training of ensembles of DT-CNN classifiers for digit recognition," in *Proc. of the ECCTD*. Kraków, Poland, 2003, pp. 337–341.

[9] C. C. Chang and C.J. Lin, "Libsvm - A library for support vector machines," 2001. www.csie.ntu.edu.tw/~cjlin/libsvm

A NUMERICAL DESIGN APPROACH FOR HIGH SPEED, DIFFERENTIAL, RESISTOR-LOADED, CMOS AMPLIFIERS

Ethan Crain, Michael Perrott

Microsystems Technology Laboratory, MIT

ABSTRACT

A simple numerical procedure is introduced to allow straightforward design of high speed, resistor loaded, differential amplifiers in modern CMOS processes whose device characteristics dramatically depart from traditional square law characterisitics. The analytical form of the procedure allows for an intuitive perspective of the varying gain-bandwidth product for such amplifiers. Calculations based on the method are compared to Hspice simulated results based on a 0.18u CMOS process. Its application to the design of high speed, source-coupled logic (SCL) gates and latches is also discussed.

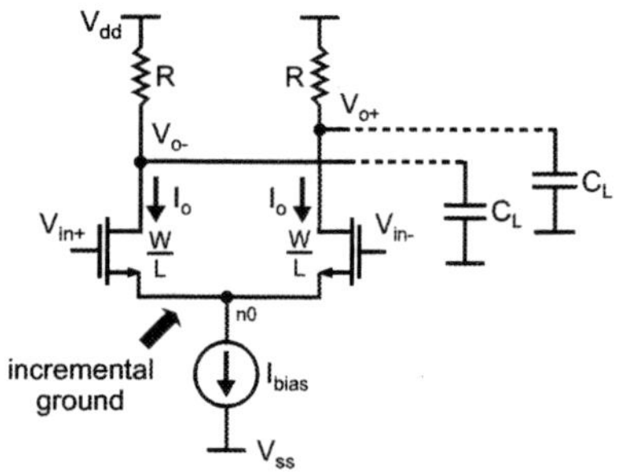

Fig. 1. Differential amplifier used in calculations.

1. INTRODUCTION

CMOS analog design techniques have traditionally assumed square law characteritics for device I-V curves when calculating the impact of device properties on circuit performance. However, the square law assumption is quickly becoming highly inaccurate with the introduction of finer line width processes due to nonideal effects such as velocity saturation. As a result, the accuracy of traditional design equations is steadily degrading, and analog designers are in need of alternate approaches to such formulations.

Thus far, there have been two responses to dealing with changing device characteristics in the analog design community. The first has been to assume square law I-V characteristics in calculations, and then rely on a simulator such as SPICE to tweak in final device parameter adjustments. Unfortunately, the square law is rapidly becoming inaccurate to the point that the analytical calculations are practically useless — all design time is then spent on SPICE simulations. Such an approach removes intuition from the designer's grasp, leads to a lengthy design process (since many tweaks are required), and often leads to suboptimal performance. The second approach is to completely automate the analog design process — the user simply specifies performance specifications and some possible topologies, and customized software takes care of the rest [1]. Unfortunately, while very useful for the design of standard analog blocks, such an approach removes creativity from the designer's grasp and offers little intuition for the creation of new circuit topologies.

We propose an alternate approach to this issue — develop numerical procedures for designing specific classes of circuits which resemble hand analysis, but use simulated device characteristics in place of analytical expressions. By sticking with procedures similar to hand analysis, much intution can be gained about design tradeoffs. By using simulated device characteristics, the results are made accurate so that little or no tweaking is required in SPICE.

This paper applies the above philosophy to the design of high speed, resistor-loaded, differential amplifiers. These structures are tremendously useful in circuit applications whose speed requirements exceed the abilities of full-swing logic circuits. Implications for the design of SCL latches, registers, and gates are also discussed.

2. BACKGROUND

Figure 1 displays a resistor-loaded, differential amplifier used in high speed applications. The resistors are often realized within a reasonably small area using unsilicided polysilicon, and introduce less capacitance than other loads such as triode PMOS devices or diode-connected NMOS devices. Further increases in bandwidth can be achieved at the expense of chip area by introducing inductors into the loads [2].

Design of resistor-loaded amplifiers entails choosing appropriate device sizes and resistance values given three design specifications:

- Allowable power dissipation: I_{bias}
- Desired voltage swing: V_{sw}
- Desired DC voltage gain: $|A_v|$

An additional specification for the amplifier is its bandwidth — its value is constrained by choice of the above three specifications as well as the load that the amplifier is required to drive (assumed capacitive). We define intrinsic bandwidth (BW) as the amplifier bandwidth that results when the amplifier drives an identical stage without additional wiring capacitance. Since actual circuits contain wiring capacitance, the intrinsic bandwidth offers only an upper bound on achievable performance, but is still a very useful metric. Note that, to achieve the maximum bandwidth, the transistor length, L, will always be assumed to be set to its minimum value for the discussion to follow.

Figure 1 allows us to relate the first two design specifications to other circuit parameters. When zero differential input voltage is applied to the amplifier, the bias current through each transistor is observed to be

$$I_o = I_{bias}/2.$$

As the input differential voltage is varied, the current through each resistor ranges between 0 and I_{bias}. Therefore, the maximum single-ended voltage swing at the amplifier output is

$$V_{sw} = I_{bias}R = 2I_oR \tag{1}$$

The third design specification, DC gain, is derived about the bias point of zero differential input voltage using the small signal transistor model shown in Figure 2. Here we have assumed that node $n0$ in Figure 1 is at incremental ground as the differential voltage is varied. Ignoring capacitance for this DC calculation, we write

$$|A_v| = g_m(R||\frac{1}{g_{ds}}) \Rightarrow g_m = |A_v|/R + |A_v|g_{ds} \tag{2}$$

Unfortunately, evaluation of the above equation requires calculation of g_m and g_{ds} as a function of the device bias current and size. As pointed out earlier, hand calculations assuming square law I-V characteristics prove inaccurate for this task. Our proposed method of addressing this issue is described in the following section.

3. PROPOSED APPROACH

We will now show that we can create a design framework in which all design calculations revolve around the solution of just one key variable given the three constraints described earlier. This key variable is current density, and is defined as

$$I_{den} = \frac{I_o}{W},$$

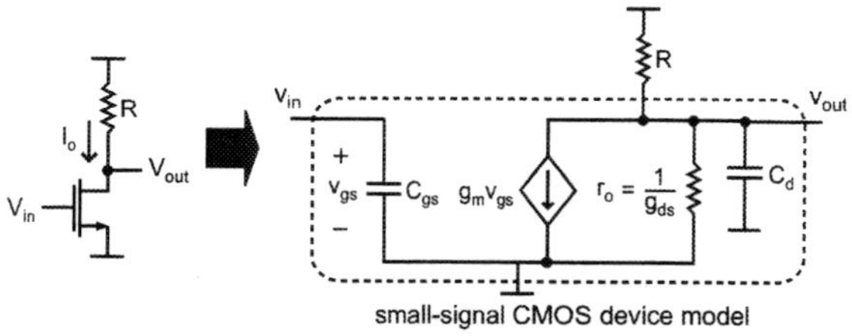

Fig. 2. Small signal model for amplifier.

where W is the width of the transistor as indicated in Figure 1.

Two key relationships involving current density will now be derived. The first is a gain/swing constraint formulation that will set the value of I_{den}. The second is a gain-bandwidth product expression that incorporates the impact of I_{den}.

3.1. Derivation of Gain/Swing Constraint Formulation

Given a fixed transistor length, $L = L_{min}$, both the g_m and g_{ds} values of a CMOS device are dependant primarily on the transistor width, W, and bias current I_o. Given a fixed value for I_o, as set by power dissipation requirements, it is straightfoward to sweep W of the device in SPICE to obtain simulated plots of $g_m(I_o, W)$ and $g_{ds}(I_o, W)$. We then define $g_{m0}(I_{den})$ and $g_{ds0}(I_{den})$ as

$$g_{m0}(I_{den}) = g_m(I_o, W)/W, \ \ g_{ds0}(I_{den}) = g_{ds}(I_o, W)/W \tag{3}$$

Let us now revisit the swing and gain constraints discussed in the Background section. Combining Equation 1 and Equation 2, we obtain

$$g_m = \frac{2|A_v|}{V_{sw}}I_o + |A_v|g_{ds}.$$

We relate g_m and g_{ds} to the simulated characteristics defined in (3) as

$$Wg_{m0}(I_{den}) = \frac{2|A_v|}{V_{sw}}I_o + |A_v|Wg_{ds0}(I_{den}).$$

Dividing through by W, we obtain the key gain/swing constraint formulation as a function of current density:

$$g_{m0}(I_{den}) = \frac{2|A_v|}{V_{sw}}I_{den} + |A_v|g_{ds0}(I_{den}) \tag{4}$$

The above expression states that current density is completely set by the choice of gain, swing, and the simulated g_m, g_{ds} curves.

3.2. Derivation of Gain-Bandwidth Tradeoff

To examine the tradeoff between gain and intrinsic bandwidth, we first note that the capacitive load can be approximately related to the amplifier device size as

$$C_L = WC_{L0} = W(C_{gs0} + C_{d0}), \tag{5}$$

where C_{L0} is the simulated capacitive load normalized to an effective W equal to one. Justification for the above expression follows from the fact that the amplifier is driving an identical structure for its load and that both C_{gs} and C_d scale linearly with the device width, W.

Calculation of the intrinsic bandwidth is computed as

$$BW(rad/s) = \frac{1}{RC_L} = \frac{2I_{den}}{V_{sw}C_{L0}}. \tag{6}$$

The amplifier gain is found through algebraic manipulation of Equation 4:

$$|A_v| = \frac{g_{m0}(I_{den})}{(2/V_{sw})I_{den} + g_{ds0}(I_{den})}. \tag{7}$$

The gain-bandwidth product is then found by combining the above two expressions:

$$|A_v| \cdot BW = \frac{g_{m0}(I_{den})}{C_{L0}} \frac{1}{1 + V_{sw}g_{ds0}(I_{den})/(2I_{den})}. \tag{8}$$

Given g_{ds0} is negligibly small, the above expression reverts to the classic g_m/C expression familiar to analog designers. However, one must note that g_m is a function of current density — the implications of this point will be brought home in the following section.

4. INTUITIVE INSIGHTS FROM METHOD

The first useful insight of the proposed method is that it provides an intuitive picture of the dependance of gain-bandwidth product on current density. Figure 3 displays a gain-bandwidth plot for a 0.18u NMOS device according to Equation 8. Each curve utilized a $g_{m0}(I_{den})$ curve and estimate of C_{L0} generated in Hspice from a SPICE model file for the 0.18u CMOS process. The top curve assumes $g_{ds} = 0$, while the bottom one includes its effect based on $g_{ds0}(I_{den})$ generated from Hspice. In either case, we see that gain-bandwidth product is increased as current density is increased, so that high current density is desirable in high speed applications.

The second useful insight of the proposed method is that it reveals that current density is not a free variable — it is determined by the gain and swing requirements of the amplifier as well as the $g_{m0}(I_{den})$ and $g_{ds0}(I_{den})$ characteristics of the device. Figure 4 displays a graphical interpretation of Equation 4 in setting the current density. Ignoring the influence of $g_{ds0}(I_{den})$, the current density is determined as the intersection of the $g_{m0}(I_{den})$ curve for the CMOS process with a straight line whose slope is $2|A_v|/V_{sw}$. As gain is increased relative to a given voltage swing, the line slope is increased and I_{den} must be reduced. Combining

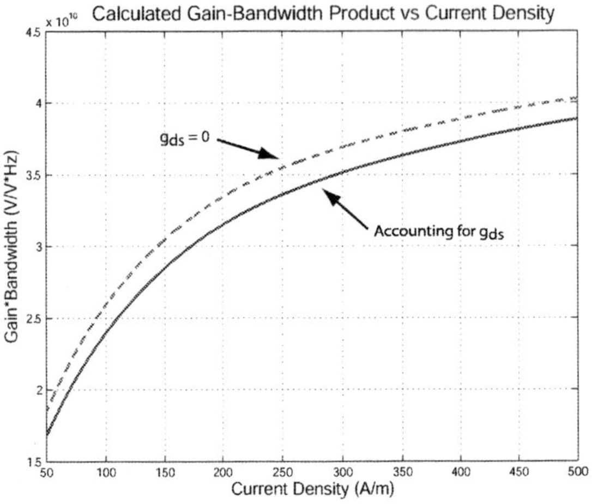

Fig. 3. Calculated Gain-Bandwidth product vs I_{den}.

this observation with Figure 3, we see that higher gains lead to reduced gain-bandwidth products.

Note that the impact of finite output conductance, $g_{ds0}(I_{den})$, is to add to the straight line whose slope is $2|A_v|/V_{sw}$, which leads to further reduction of the resulting current density setting. Therefore, finite output conductance degrades the achievable gain-bandwidth product of the differential amplifier structure.

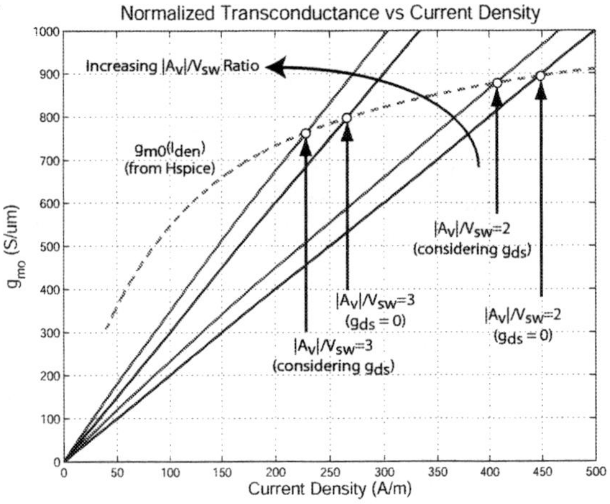

Fig. 4. Current density settings versus gain/swing.

5. RESULTS

The proposed procedure was used to design several differential amplifiers in a 0.18u CMOS process (only NMOS devices were used) with varying gain values. The swing and power dissipation were held constant at $V_{sw} = 1$V and $I_{bias} = 2$mA, respectively, and the bandwidth was calculated based on Equation 6. Table 1 displays a comparison of the calculated gain and bandwidth values to the Hspice simulation results. In the Hspice simulation, the amplifier has the same topology as shown in Figure 1 and is loaded by an identical amplifier stage whose output is set to a constant voltage in order to eliminate Miller effect on the capacitive load it presents.

Target Gain	Calculated BW (GHz)	Simulated Gain	Simulated BW (GHz)
2.00	14.45	2.03	13.74
3.00	8.30	3.02	8.17
4.00	5.18	4.00	5.31
5.00	3.27	4.98	3.48
6.00	1.99	5.97	2.19

Table 1. Calculated vs simulated amplifier performance.

Table 1 reveals that the proposed design procedure is quite accurate with respect to achieving the desired gain for the amplifier. The calculated versus simulated bandwidth values are not as accurate, but are still within $\pm$ 10 % of each other. The discrepancy in bandwidth is likely due to the fact that the capacitive load is not strictly a linear function of W as assumed in Equation 5.

6. APPLICATION TO SCL DIGITAL CIRCUITS

It is interesting to note that high speed digital structures also make use of such differential amplifier structures. Figure 5 illustrates a high speed SCL latch and a NAND/AND gate. The differential amplifiers embedded in such structures are turned on or off based on other differential pairs below them. When turned on, their behavior corresponds to that of the basic differential amplifier structure.

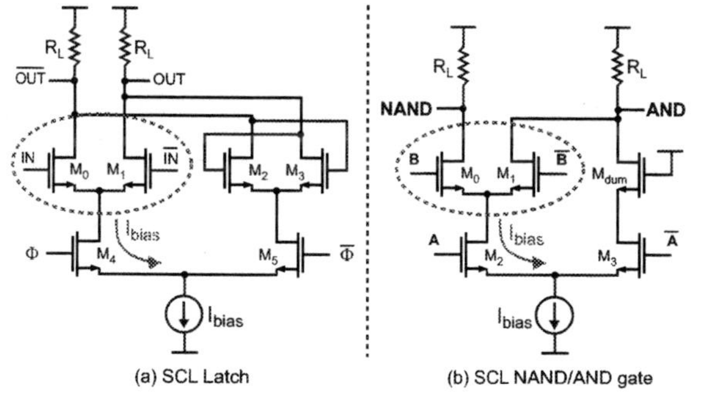

Fig. 5. Digital high speed circuits.

We have found that the following heuristic design method works well for such circuit structures:

1. Use the proposed method to design the differential amplifier portion of the structure with given gain, swing, and bias current requirements. We have found that a choice of gain in the range of 1.25 to 1.75 works well (the swing and bias current values depend on the application). In the latch example of Figure 5 (a), this step yields sizes for M_0 and M_1.

2. Choose identical sizes for transistors that feed off the same diff pair as the differential amplifier above. In the latch example, this would lead to M_2 and M_3 having the same sizes as M_0 and M_1.

3. Choose sizes that are roughly 20 % larger for the transistors that feed the above differential pairs. In the latch example, the widths of M_4 and M_5 would then be set to be 20% higher than the widths of M_0 and M_1 (L should be minimum in all cases). Note that this progressive scaling technique is commonly applied in digital design (see page 298 of [3]) — the value of 20% is not necessarily optimal but has worked well for us in practice.

7. CONCLUSIONS

This paper presented a simple numerical technique to design high speed differential amplifiers with resistor loads without relying on square law assumptions for the CMOS devices. By combining hand analysis with SPICE generated data, intuition of such issues as gain-bandwidth product properties is achieved while still obtaining highly accurate design calculations. Calculations from the method were compared to Hspice simulations, and reveal that the formulations are highly accurate with respect to achieving desired gain, and reasonably accurate for bandwidth estimation. A heuristic extension of the method can be applied to high speed SCL logic gates and latches.

8. REFERENCES

[1] H. Liu, A. Singhee, R.A. Rutenbar, and L.R. Carley, "Remembrance of Circuits Past: Macromodeling by Data Mining in Large Analog Design Spaces," in *Design Automation Conference*, June 2002, pp. 437–442.

[2] T.H. Lee, *The Design of CMOS Radio-Frequency Integrated Circuits*, Cambridge University Press, 1998.

[3] Jan M. Rabaey, *Digital Integrated Circuits, A Design Perspective*, Prentice Hall, 1996.

Analysis Techniques for Obtaining the Steady-State Solution of MOS LC Oscillators

Makram M. Mansour[†‡§], Amit Mehrotra[†§], William W. Walker[‡], Amit Narayan[§]
[†] University of Illinois at Urbana-Champaign, Urbana, Illinois 61801
[‡] Fujitsu Laboratories of America, Sunnyvale, California 94085
[§] Berkeley Design Automation, Santa Clara, California 95054

Abstract— Accurate analysis techniques for estimating the periodic steady-state solution of MOS LC oscillators using short-channel device equations are presented. These techniques allow us to make quantitative estimates of the oscillator steady-state performance without the need for time-consuming transient simulations using simulators such as SPICE. Further, our engineering techniques provide insight and quantitative understanding on the design of current-day, deep-submicron MOS LC oscillators and serve as a starting point in a design methodology that includes complete phase noise/timing jitter analysis and optimization. Our analytical results for a cross-coupled LC oscillator that was previously fabricated and tested are in good agreement with SPICE simulations.

I. Introduction

Estimating the periodic steady-state (PSS) solution for oscillators is an essential step for any CAD tool as well as analytic approach aimed at obtaining accurate estimates for oscillator phase-noise and timing jitter performance. One obvious way for obtaining the PSS solution is by running SPICE transient simulations *long enough* until all initial transients die out and the oscillator reaches steady-state. However, this technique is impractical and as an example, consider the simulation of a high-Q oscillator. This requires very long transient simulations: a Q of 1,000 suggests that the turn-on transient time starting from a zero initial state will be of the order of 1,000 cycles of the oscillation frequency. Advanced RF circuit simulators such as SpectreRF compute the PSS solution by first running transient analysis to obtain good circuit initial conditions and then running PSS analysis. Further, the PSS analysis requires multiple transient simulation runs at each iteration for it to converge. Efficient numerical techniques, both in the time-domain and frequency-domain, have been implemented in such CAD tools which are based on harmonic balance and shooting methods. Not until an accurate periodic steady-state solution is reached can phase noise and timing jitter analysis on these oscillators be performed.

From the analytic point of view of obtaining the oscillator PSS solution, few significant attempts were made [4]–[6]. Most of these attempts use long-channel equations for modeling the MOS transistor I-V characteristics. However, these equations do not take into account short-channel effects, such as velocity saturation and mobility degradation and hence, are inaccurate for current-day oscillator designs. Therefore, there is a need for new accurate analytical methods which use the more accurate short-channel MOS equations [7] to obtain the oscillator PSS solution.

In this paper, we develop analytically using short-channel MOS device equations the periodic steady-state expressions of the most common MOS LC oscillator topology: *the cross-coupled LC oscillator*. These equations serve as starting point for an oscillator design methodology that allows us to derive analytically closed-form expressions for the oscillator perturbation projection vector (PPV) [1]–[3]. This, in turn, leads to the calculation of a single scalar c, which completely characterizes the phase noise and timing jitter performance of an oscillator. Hence, such design methodology based on analytical expressions allows us to accurately estimate the oscillator phase noise and timing jitter from its constituent circuit parameters using hand analysis. This can further lead to a systematic and quantitative approach for analytically optimizing oscillators with respect to phase noise. Theoretical predictions of the oscillator's steady-state performance are compared with results of circuit simulations using SPICE for a previously fabricated oscillator [8] for which simulations sufficiently represent measured results.

II. Short-Channel MOS Equations

The square-law (long-channel) characteristic equations provide moderate accuracies for MOS devices with minimum channel lengths greater than $4\,\mu$m. However, in the deep submicron regime, below $0.5\,\mu$m, these equations are inapplicable due to higher-order short-channel effects which necessitate using more complex models, such as the *short-channel MOS device equations* [7]. This is a sufficiently accurate, yet compact model for short-channel devices (in the saturation region, i.e., $V_{DS} \geq V_{DS,sat}$) and is given by

$$I_D = W C_{ox} v_{sat} \frac{(V_{GS} - V_t)^2}{V_{GS} - V_t + E_{sat} L}, \qquad (1)$$

where

$$V_{DS,sat} = \frac{(V_{GS} - V_t) E_{sat} L}{V_{GS} - V_t + E_{sat} L}, \qquad (2)$$

$$E_{sat} = \frac{2 v_{sat}}{\mu_{eff}}, \quad \mu_{eff} = \frac{\mu_0}{1 + \Theta(V_{GS} - V_t)},$$

Θ is a parameter that is inversely proportional to the oxide thickness usually in the range from $0.1\,\mathrm{V}^{-1}$ to $0.4\,\mathrm{V}^{-1}$, C_{ox} is the gate oxide capacitance per unit area, μ_0 is the *low-field* mobility of the carrier, W is the device width, L is the device length, V_{GS} is the device gate-source voltage, V_t is the threshold voltage, and v_{sat} is the velocity saturation index.

III. Cross-Coupled LC Oscillator

The most common and convenient LC oscillator configuration for realization in MOS technology is the cross-coupled LC oscillator shown in Fig. 1. In monolithic form, transistors M1 and M2 are biased with a tail current source I_{SS}, connected at their sources. The inductors are usually on-chip spiral inductors with relatively low-Q (less than 10) and the resistors R_L represent the inductor parasitics. The load capacitors C_L represent chip parasitics plus any added load capacitances. These capacitors are essential for the operation of the oscillator and in the simplest form they are composed purely of chip parasitics. Tuning of the oscillator frequency can also be achieved by using MOS or diode varactors.

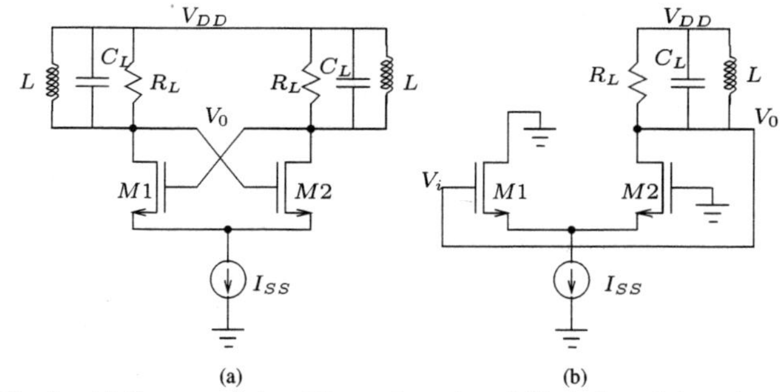

Fig. 1. (a) Cross-coupled oscillator schematic and (b) single-ended version.

A. Periodic Steady-State Analysis

The following analysis applies equally to a corresponding p-channel cross-coupled pair with appropriate sign changes. The PSS analysis begins by assuming that M1 and M2 are perfectly matched. We will do the analysis for the single-ended version of the oscillator (shown in Fig. 1(b)), knowing that we will have the same frequency and distortion characteristics as the fully differential form. The only difference is the magnitude of the output waveforms which is half of that of the fully differential form.

For this PSS analysis, we assume that the output resistance of the tail current source $R_{SS} \to \infty$ and that the output resistance of each transistor $r_0 \to \infty$ since these assumptions do not strongly affect the large-signal behavior of the circuit. We also assume that the drain resistors are small enough that neither transistor operates in the triode region if $V_i \leq V_{DD}$. Applying short-channel equations to transistor M1,

$$I_1 = WC_{ox}v_{sat}\frac{(V_{GS_1}-V_t)^2}{V_{GS_1}-V_t+E_{sat}L}, \tag{3}$$

and defining the parameters V_a and b as

$$V_a = \frac{I_{SS}}{WC_{ox}v_{sat}}, \quad b = \frac{E_{sat}L}{V_a},$$

we can normalize the voltages and currents in (3) as

$$\begin{aligned}\frac{I_1}{I_{SS}} &= \frac{WC_{ox}v_{sat}}{I_{SS}}\frac{(V_{GS_1}-V_t)^2}{V_{GS_1}-V_t+E_{sat}L}\\ &= \frac{1}{V_a}\frac{(V_{GS_1}-V_t)^2}{V_{GS_1}-V_t+E_{sat}L}.\end{aligned}$$

A similar derivation can be performed for transistor M2. Thus, the normalized drain current equations for both transistors are

$$i_1 = \frac{v_1^2}{v_1+b}, \tag{4}$$

$$i_2 = \frac{v_2^2}{v_2+b}, \tag{5}$$

where, by definition

$$i_1 = \frac{I_1}{I_{SS}}, \qquad v_1 = \frac{V_{GS_1}-V_t}{V_a},$$
$$i_2 = \frac{I_2}{I_{SS}}, \qquad v_2 = \frac{V_{GS_2}-V_t}{V_a}.$$

From KCL at the sources of M1 and M2 we have

$$i_1 + i_2 = 1. \tag{6}$$

Using (4), (5), and (6) to eliminate i_2 gives

$$v_1 = \frac{1}{2}\left(i_1 + \sqrt{i_1^2 + 4i_1 b}\right), \tag{7}$$

$$v_2 = \frac{1}{2}\left(1 - i_1 + \sqrt{1 + i_1^2 - 2i_1 + 4b - 4i_1 b}\right). \tag{8}$$

From KVL around the input loop, the input voltage can be written as

$$V_i = V_{GS_1} - V_{GS_2},$$

which is conveniently normalized as

$$v_i = \frac{V_i}{V_a} = \frac{(V_{GS_1}-V_t)-(V_{GS_2}-V_t)}{V_a} = v_1 - v_2,$$

to finally get

$$v_i = \frac{1}{2}\left(2i_1 - 1 + \sqrt{i_1^2+4i_1 b} - \sqrt{1+i_1^2-2i_1+4b-4i_1 b}\right). \tag{9}$$

Equation (9) can be solved for i_1. Note that (9) is only valid for $i_1 \in [0,1]$. This implies the solution is valid when

$$v_i \in \left[-\frac{1}{2}\left(1+\sqrt{1+4b}\right), \frac{1}{2}\left(1+\sqrt{1+4b}\right)\right], \tag{10}$$

and when $v_i > \frac{1}{2}\left(1+\sqrt{1+4b}\right)$, $i_1 = 1$. On the other hand, $i_1 = 0$ when $v_i < -\frac{1}{2}\left(1+\sqrt{1+4b}\right)$. Furthermore, equation (9) was derived assuming that the transistors are in saturation region. Note that if the transistor goes into linear region when the current has completely switched to one of the branches, the above derivation still holds. The above derivation becomes invalid only if $V_{DS} < V_{DS,sat}$ for some transistor with v_i in the range given in (10).

Consider the case when $v_i = \frac{1}{2}\left(1+\sqrt{1+4b}\right)$ and $i_1 = 1$. We need to determine the conditions when M1 is in saturation for this case. Since $i_2 = 0$, $V_{GS_2} = V_t$ and therefore $V_S = -V_t$. Hence, $V_{DS_1} = V_t$ and

$$V_{GS_1} = V_t + V_a v_i,$$

and from (2) we have

$$V_{DS,sat_1} = \frac{(V_{GS_1}-V_t)E_{sat}L}{V_{GS_1}-V_t+E_{sat}L}, \tag{11}$$

$$= V_a\frac{\frac{1}{2}\left(1+\sqrt{1+4b}\right)b}{\frac{1}{2}\left(1+\sqrt{1+4b}\right)+b}. \tag{12}$$

Therefore $V_{DS} \geq V_{DS,sat}$ translates to

$$\frac{V_t}{V_a} \geq \frac{\frac{1}{2}\left(1+\sqrt{1+4b}\right)b}{\frac{1}{2}\left(1+\sqrt{1+4b}\right)+b}.$$

Usually this is not a very stringent condition and can be easily achieved in designs. Note that V_t is larger than V_{t_0} due to the body effect.

B. Oscillator Universal Design Curves

Solving equation (9) results in three possible solutions for i_1. Only one of these solutions behaves properly (i.e., for $v_i > \frac{1}{2}(1+\sqrt{1+4b})$, $i_1 = 1$ and $i_1 = 0$ for $v_i < \frac{1}{2}(1+\sqrt{1+4b})$). This behavior is shown in the plots of i_1 versus v_i for $b = 0, \ldots, 5$ in Fig. 2. Notice that $-\frac{1}{2}(1+\sqrt{1+4b}) \leq v_i \leq \frac{1}{2}(1+\sqrt{1+4b})$.

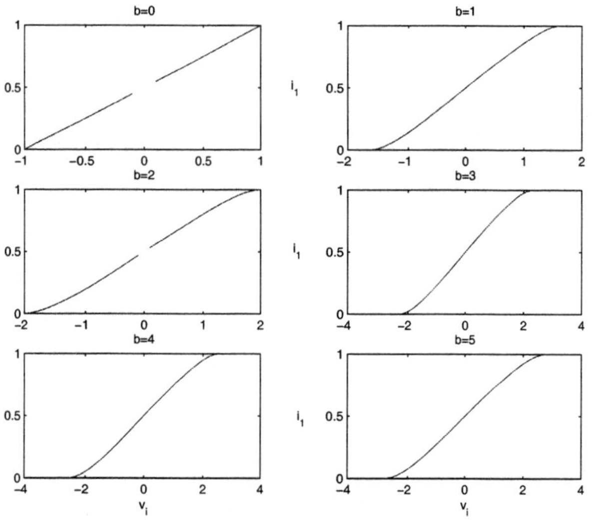

Fig. 2. Normalized drain current i_1 versus normalized signal voltage v_i for different values of b.

The normalized current i_1 depends on the normalized signal voltage v_i and the parameter b, and hence numerical solution of the waveforms can be used to produce universal curves for oscillator design. These curves can be used to determine the Fourier coefficients in the Fourier series expansion of i_1, i.e.,

$$i_1(t) = i_{1,1}\cos\omega t + i_{1,3}\cos 3\omega t + i_{1,5}\cos 5\omega t + \ldots \quad (13)$$

In order to calculate the harmonic components of the current, it is assumed that the voltage waveforms are almost sinusoidal. This, in general, is a valid assumption if the Q of the circuit is moderate, which we will assume. The results of a computer solution for the normalized drain current waveforms are shown in Figs. 3(a)-3(d). In these plots the odd harmonic currents are plotted as a function of v_i for various values of b. (Note that even harmonic currents are absent). The curves show a monotonic rise and represents the case where the device is always saturated. For the other curves in Figs. 3(c) and 3(d), the point where i_1 begins to fall at low v_i coincides with partial operation in the triode region.

C. Comparison of Analysis versus Simulations

To confirm the validity of the proposed characterization technique we compare the analysis results to SPICE simulations for a 4.7 GHz oscillator with on-chip inductor [8] shown in Fig. 4. This oscillator is fabricated using 0.35 μm CMOS process and uses transistor parasitics as capacitors in the oscillator.

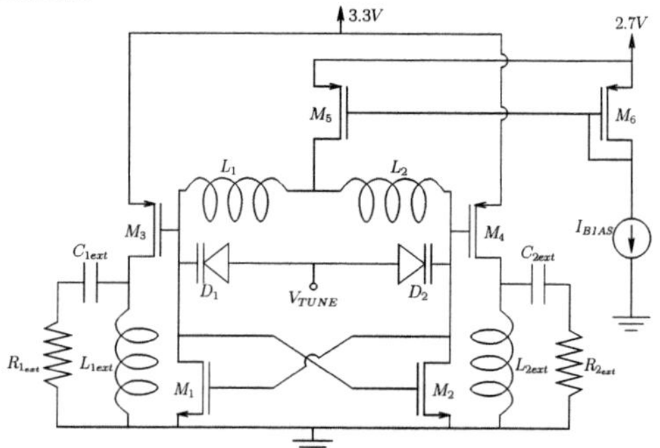

Fig. 4. 4.7 GHz oscillator with on-chip inductor.

Short-channel device model parameters were extracted for the specific technology by running test circuits to give: $v_{sat} = 86301.58\,\mathrm{m/s}$, $E_{sat} = 9.234\,\mathrm{V}/\mu$, and $V_t = 0.43\,\mathrm{V}$. All loading effects are conveniently reflected to the output and derived as an effective resistance R_L, seen at the drain of M1. The output voltage amplitude was calculated using the first harmonic component of I_1 as $V_m = I_{1,1}R_L = 0.2\,\mathrm{V}$. The drain current I_1 can then be derived using the oscillator universal curves and plotted as seen in Fig. 5.

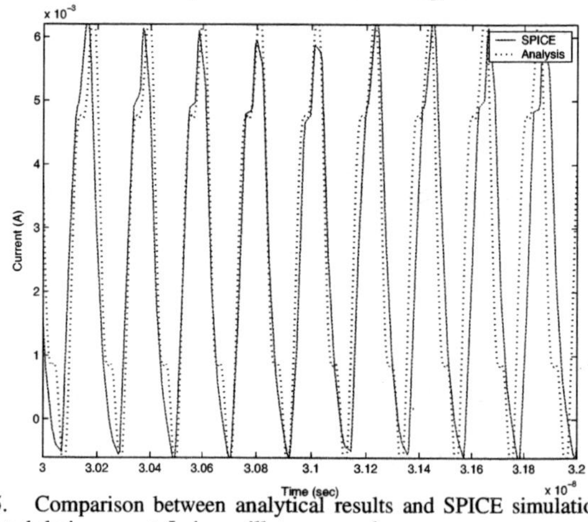

Fig. 5. Comparison between analytical results and SPICE simulations for estimated drain current I_1 in oscillator example.

Waveform distortion in the oscillator voltage may be of importance in some applications and can be calculated as follows. Drain current in transistor M1 in Fig. 4 produces third, fifth, and seventh harmonics as given in Figs. 3(a)-3(d). Once these are known, the harmonic voltage in V_1 is readily calculated assuming that C_L is the dominant loading at the drain of M1 (the inductors are assumed an open circuit at all harmonics of the oscillation frequency). Table I lists the resulting normalized harmonic components. The nth harmonic distortion in V_1 is given by

$$\mathrm{HD}_n = \frac{I_{1,n}}{n\omega_0 C_L V_m}, \quad n = 3, 5, 7. \quad (14)$$

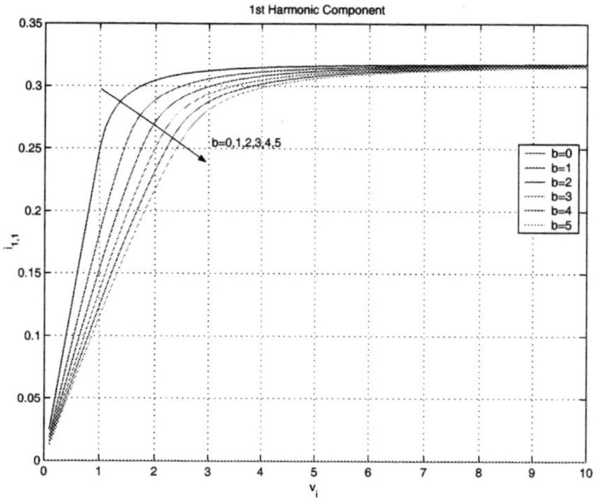

(a) 1st harmonic component

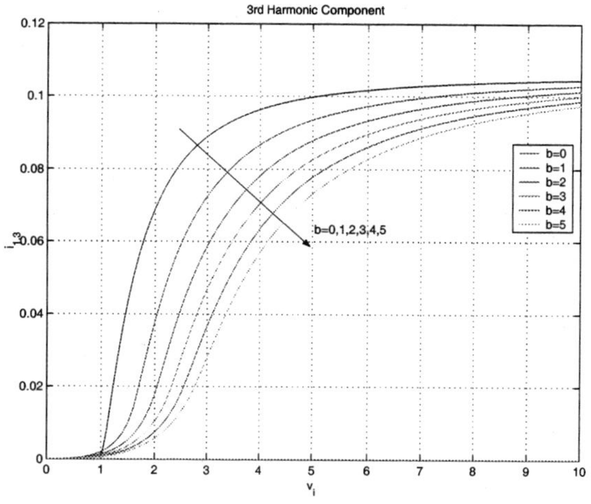

(b) 3rd harmonic component

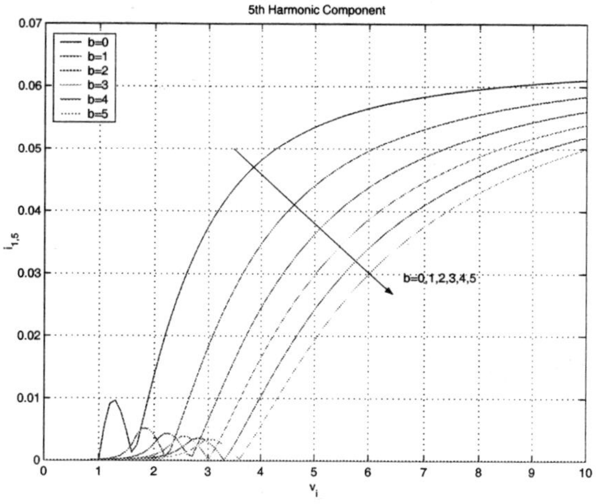

(c) 5th harmonic component

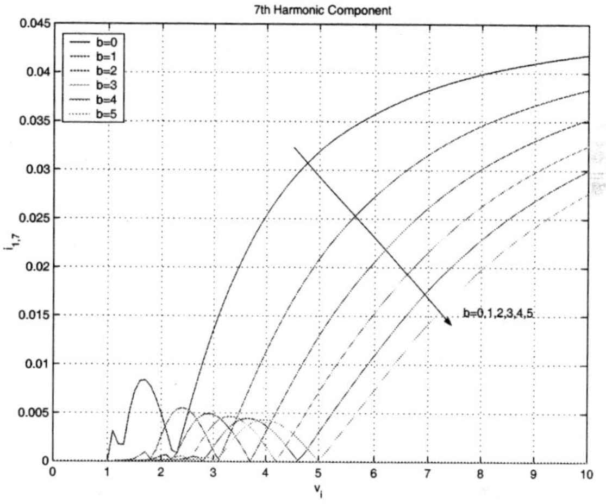

(d) 7th harmonic component

Fig. 3. Normalized harmonic currents versus normalized signal voltage and parameter b.

IV. Conclusion

Accurate analytic expressions for the cross-coupled LC oscillator PSS solution have been derived. The derivation was based on short-channel MOS device equations assuming the transistors operate in the saturation region. The techniques presented for the cross-coupled oscillator are intuitive and allow for similar derivations for other oscillator topologies. The validity of the proposed analysis has been tested by comparing our analysis results with SPICE simulations of an actual fabricated and measured oscillator, where we find that our analysis is in good agreement with simulations as well as measured results.

TABLE I
NORMALIZED HARMONIC COMPONENTS.

Freq. [GHz]	SPICE†	Analysis
4.7847	1	1
9.5694	4.97m	0
14.3541	29.82m	26.92m
19.1388	2.51m	0
23.9235	5.91m	4.97m
28.7081	0.62m	0
33.4928	1.77m	1.5m

† Even harmonics are present due to nonlinear capacitive parasitics that were not accounted for in the analysis.

References

[1] A. Demir, A. Mehrotra, and J. Roychowdhury, "Phase noise in oscillators: A unifying theory and numerical methods for characterization," *IEEE TCAS*, vol. 47, no. 5, pp. 655-674, May 2000.

[2] A. Demir, A. Mehrotra, and J. Roychowdhury, "Phase noise and timing jitter in oscillators," in *Proceedings of IEEE CICC*, pp. 45-48, 1998.

[3] A. Demir, D. Long, and J. Roychowdhury, "Computing phase noise eigenfunctions directly from steady-state Jacobian matrices," in *Proceedings of IEEE/ACM ICCAD*, pp. 283-289, 2000.

[4] R. Meyer, *EECS 242 Course Notes*, University of California at Berkeley, 1999.

[5] A. Mehrotra, *ECE 383 Course Notes*, University of Illinois at Urbana-Champaign, 2001.

[6] B. Razavi, *Design of Analog CMOS Integrated Circuits.* McGraw Hill, 2000.

[7] P. Ko, "Approaches to Scaling," *Chapter 1 in VLSI Electronics*, vol. 18, Academic Press, 1989.

[8] P. Kinget, "A fully integrated 2.7 V 0.35 μm CMOS VCO for 5 GHz wireless applications," in *Digest of Technical Papers IEEE ISSCC*, pp. 226-227, 1998.

Macromodeling of Digital Libraries for Substrate Noise Analysis

Zhe Wang [1] zhewang@ece.umn.edu
Rajeev Murgai [2] murgai@fla.fujitsu.com
Jaijeet Roychowdhury [1] jr@ece.umn.edu

ABSTRACT

The demand for low cost, low power, and small area electronic devices calls for system-on-a-chip (SoC) designs. Integration of complex digital blocks and high performance analog functions onto single SoCs induces signal integrity between noisy digital circuits and sensitive analog sections. Such signal integrity degrades the performance of analog circuits and even causes functional failures of victim circuits. In order to account for this interference in circuit design phases, substrate noise analysis becomes particularly important, especially in deep submicron digital and mixed-signal circuits. To this end, it is critical to estimate efficiently and accurately the noise injection from the digital circuit with tens of millions of transistors. In this work, we develop *techniques that automatically extract* low-complexity time-varying macromodels for digital blocks, at the cell library building phase. Tailored for substrate noise analysis, the extracted macromodel includes three major noise injection mechanisms. The efficacy and accuracy of our macromodel are confirmed in the simulation results. Thanks to the linear time-varying (LTV) model reduction based on the Time-Varying Padé (TVP) method, our macromodel extraction features high accuracy with affordable complexity. Equally attractive is the accurate-by-construction substrate noise model generation by merging the macromodel extraction into cell library building phase.

I. INTRODUCTION

The demand for electronic devices with high performance, low cost and low power consumption, together with continuously advancing silicon technology, results in an increasing number of system-on-a-chip (SoC) designs. Such designs give rise to *signal integrity problems* between noisy digital circuits and sensitive analog sections. A major cause for signal integrity problems is *substrate coupling*, defined as any voltage deviation in the bulk node of a device caused by currents propagating through a substrate [8]. When thousands of transistors in the digital circuit switch, they inject considerable current into the common substrate. This injected current travels through the substrate and eventually interferes with analog circuits on the same die. As feature sizes decrease and clock frequencies increase, the substrate noise created by digital switching increases dramatically. Therefore, in nanometer process (0.25μm and below), substrate noise analysis becomes increasingly important.

Substrate noise analysis consists of the following components: i) calculation of the amount of currents injected by the digital circuitry into the substrate; ii) modeling of the path that leads the injected currents to the analog circuitry, i.e., extraction of the substrate model; iii) evaluation of substrate noise impact on the analog circuitry. In this paper, we will concentrate on the first component.

The most accurate means to quantify the substrate interference would be the full SPICE model. However, its feasibility is questionable due to the large digital circuit size. Not to mention the inclusion of package model, substrate model, and power supply model. In the literature, several approaches have been proposed to speed up the computation by replacing the SPICE model of digital blocks with their corresponding macromodels. These approaches include: a macromodel consisting of a current (noise) source that is obtained by running the SPICE model for each gate [12]; a model utilizes capacitors controlled by ideal switches [3]; and a model that stores in a cell library independent time varying current sources for individual cells [4].

Nevertheless, to reduce complexity, these approaches sacrifice accuracy. This is because all of them rely on manually derived models, which yields an accuracy that is heavily dependent on the individual researcher's understanding of the physical nature of the digital circuits. Reminiscent of existing macromodels, we also start from the SPICE-level circuit descriptions. But different from all existing works, in this paper we provide a new perspective by introducing an *automatically* generated macromodel for substrate noise analysis.

The automatic feature of our approach stems from the fact that the low-complexity models are generated using algorithms. More specifically, the digital circuit is first partitioned into digital cells each containing one or more nonlinear devices. We then convert each of these digital cells into a LTV macromodel, and collect all of them in a cell library. Notice that the conversion/extraction and collection can be carried out off-line, and thereby does not hinder the real-time substrate noise computation. The latter can be implemented by "dragging-and-dropping" corresponding macromodels from the library into the chip-level substrate noise analysis representation.

The simulation results confirm that: compared to SPICE model, our macromodel approach is up to *160* times as fast, with only 3.21% peak noise error. Summarizing, by slightly modifying cell characterization methodology, our macromodel extraction is capable of generating bottom-up accurate-by-construction models for full-chip substrate noise analysis.

The rest of the paper is organized as follows. In Section II, we will present three major noise injection mechanisms. In Section III, the development of our macromodel will be presented in conjunction with a brief review of TVP. Macromodel examples, together with simulations and comparisons, will be presented in Section IV. Finally, summarizing remarks will be given in Section V.

II. NOISE INJECTION MECHANISMS

Since all transistors and contacts are connected to the substrate directly or through reverse biased junctions, noise can be injected to the substrate through several mechanisms. Among all noise injection mechanisms, we will briefly review three major ones in this section.

A. Noise Injection through Contacts

When digital circuits operate, transistors switch on and off at the same rhythm. Consequently, current spikes are generated in power supply lines. Flowing through bond wires and package lead frames, the current spikes induce noise in the power lines due to the impedance of the package and wires by $L(dI/dt)$ and IR, as shown in Figure 1. Such noise is known as simultaneous

[1] Department of Electrical and Computer Engineering, University of Minnesota, Minneapolis, Minnesota, USA.
[2] Fujitsu Laboratories of America, Inc, Sunnyvale, CA, USA.

0-7803-8251-X/04/$17.00 ©2004 IEEE

switching noise (SSN), or, delta-I noise. As a result, power supply lines in digital circuits are contaminated by SSN. With tens of millions of transistors on a single chip, the SSN can easily reach hundreds of millivolts.

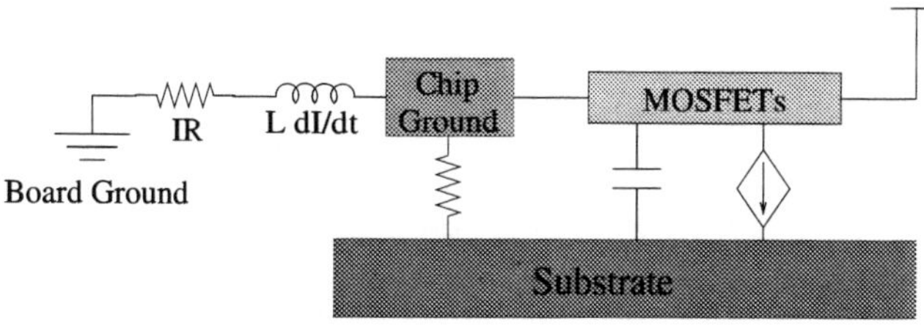

Fig. 1. Current spikes flow through bond wires and package lead frames cause fluctuations in the internal supply voltage. During switches, MOSFETs also inject currents to substrate through capacitive coupling and impact ionizations

Current spikes can also be generated when the load capacitance is charged or discharged. The latter occurs due to the state change at the IC outputs. As the current spikes flow through the pins and bond wires, voltage fluctuations can be observed.

Both SSN and the voltage fluctuations on the I/O pad will induce current injection into the substrate through resistive P^+ contacts.

B. *Noise Injection through Capacitive Junctions*

In digital circuits, routine MOSFET operations implicate capacitance-alike behavior of the junctions between transistors and the substrate. This is because they are reversely biased. This capacitance plays the role of connecting bridge for current leak from switching MOSFETs. The magnitude of this current leakage is proportional to the transient speed and the junction capacitance. As a result, noise injection becomes increasingly important in high speed digital circuits.

C. *Impact Ionization*

It is well known that strong electric field exists between drain and source in submicron transistors. The presence of such a strong electric field enables electrons (in NMOS devices) to acquire sufficient energy to become "hot." These hot electrons impact the drain, produce ionization and dislodging holes that are swept to the substrate, which appear as substrate currents [1]. This impact ionization current density is typically in the order of mA/μm, and thereby dominates junction leakage. As technology advances, the channel length keeps shrinking, but the power supply voltage decreases at a slower pace. Consequently, the electric field increases, and can lead to intensive electron-hole pair generation [6]. Different from NMOS transistors, hot carrier induced substrate currents are smaller in PMOS transistors, due to the lower mobility of holes.

III. Macromodel for Noise Injection

Based on these three noise injection mechanisms, we will next develop a macromodel that is extracted from a SPICE-level description. Any circuit connected to a noisy power supply can be represented by modeling the power supply noise as a *small* system input $u(t)$, in addition to the *large* signal vector containing its logic inputs $\boldsymbol{b}_l(t)$[1]. The resultant nonlinear system driven by both inputs are given by:

$$\frac{d\boldsymbol{q}(\boldsymbol{y}(t))}{dt}+\boldsymbol{f}(\boldsymbol{y}(t))=\boldsymbol{b}_l(t)+\boldsymbol{b}u(t),\ \boldsymbol{z}_t(t)=\boldsymbol{d}^T\boldsymbol{y}(t), \qquad (1)$$

[1]For notational simplicity but without loss of generality, we do not consider the effects of ground bounce here, since the latter can be included in the same manner as the power supply noise.

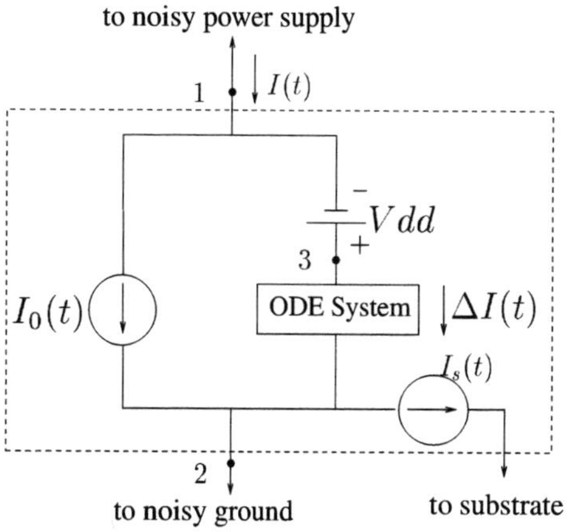

Fig. 2. Proposed macromodel of a digital cell.

where $\boldsymbol{b}$ is the vector that links the small-signal input (that is, the power supply noise) to the rest of the system, $\boldsymbol{y}(t)$ is a $m\times 1$ vector containing a total of m unknown node voltages and branch currents; $\boldsymbol{q}(\cdot)$ and $\boldsymbol{f}(\cdot)$ are nonlinear functions describing the charge/flux and resistive terms in the cell, respectively; the system outputs $\boldsymbol{z}_t(t)$ consist of current flows to ground and current leaks to substrate; and $\boldsymbol{d}$ is the vector that link the output to the rest of the system. Notice that one of the system outputs is current drawn from power network induced not only by the logical input of the digital cell $\boldsymbol{b}_l(t)$, but also by the power supply noise captured by $u(t)$. In contrast, the macromodel in [4] can be interpreted as a solution of 1 with $u(t) = 0$ over a time period (usually a clock cycle). In other words, [4] assumes perfect power supply voltage, which implies that no interaction between the device current and the supply voltage is captured. It has been recently shown that when the power supply voltage drops 10% from (the ideal value) $2V$ in a $0.25\mu m$ technology, the peak current through an inverter changes more than 30%. Unfortunately, such an error is inherited to any SSN estimation methods utilizing the model as in [4].

In order to establish a small-sized macromodel that can be generated automatically from SPICE-level circuit descriptions, we adopt a general method called TVP [9]. This method was developed originally for mixed-signal/RF/analog circuits, and was applied to mixers and switched-capacitor filters. However, if appropriately adapted, TVP can be readily applied to substrate noise analysis by reducing large digital logic blocks for *SSN and IR drop prediction purposes*, as we develop and demonstrate here.

Separating the time scales of the small input $u(t)$ and the logic input $\boldsymbol{b}_l(t)$, (1) can be re-expressed in MPDE form as:

$$\begin{gathered}\frac{\partial\boldsymbol{q}(\hat{\boldsymbol{y}})}{\partial t_1}+\frac{\partial\boldsymbol{q}(\hat{\boldsymbol{y}})}{\partial t_2}+\boldsymbol{f}(\hat{\boldsymbol{y}}(t_1,t_2))=\boldsymbol{b}_l(t_1)+\boldsymbol{b}u(t_2)\\ \hat{\boldsymbol{z}}_t(t_1,t_2)=\boldsymbol{d}^T\hat{\boldsymbol{y}}(t_1,t_2)\qquad \boldsymbol{z}_t(t)=\hat{\boldsymbol{z}}_t(t,t),\end{gathered} \qquad (2)$$

where the hatted variables are bivariate (i.e., two-time scales) forms of the corresponding variables in (1). In fact, it has been proved in, e.g., [10], that any solution of (1) generates a solution of (2).

Solving (2) when $u(t_2) = 0$, and linearizing around this point, the outputs linear in the input $u(t)$ can be obtained. With the solution denoted by $\hat{\boldsymbol{y}}^*(t_1)$, the outputs are given by $\boldsymbol{z}_t(t) = \hat{\boldsymbol{z}}_t(t,t) = \boldsymbol{d}^T\hat{\boldsymbol{y}}^*(t)$. Recalling that the outputs are two currents, we represent $\boldsymbol{z}_t(t)$ as two current sources $I_0(t)$ and $I_s(t)$, as shown in Fig. 2. Evidently, $I_0(t)$ is identical to the one resulted by applying the macromodel in [4]. Although $I_0(t)$ and $I_s(t)$ are generally time-varying, they are 'fixed' in the sense that they are uniquely determined by the circuit (digital cell/block) itself, regardless of the power supply variation. Since $I_s(t)$ is only a small part of the total substrate noise, we can ignore the variation of $I_s(t)$ introduced by the fluctuation of power supply voltage without losing much accuracy, but the resulted macromodel will be much simpler. Doing so, the output vector z in (2) becomes a scalar z thereafter. In order to capture the variation of current power supply induced by the fluctuation, one also needs to solve the following *linear* MPDE:

$$\frac{\partial(C(t_1)\hat{x})}{\partial t_1} + \frac{\partial(C(t_1)\hat{x})}{\partial t_2} + G(t_1)\hat{x} = bu(t_2) \qquad (3)$$
$$\hat{z}(t_1,t_2) = d^T\hat{x}(t_1,t_2) \qquad z(t) = \hat{z}(t,t),$$

where vectors $\hat{x}$, $\hat{z}$, and z are the small-signal versions of $\hat{y}$, $\hat{z}_t$ and z_t, respectively; $C(t_1) = (\partial q(\hat{y})/\partial\hat{y})|_{\hat{y}^*(t_1)}$ and $G(t_1) = (\partial f(\hat{y})/\partial\hat{y})|_{\hat{y}^*(t_1)}$ are time-varying matrices. Eq. (3) reveals a linear relationship between the bivariate output $\hat{z}(t_1,t_2)$ and the small input signal $u(t_2)$. But this linear relationship is time-varying in the system time scale t_1. To obtain the time-varying transfer function from $u(t_2)$ to $\hat{z}(t_1,t_2)$, let us carry out the Laplace transform of (3) with respect to t_2, and collect observations at a total of $N+1$ instances $\{t_{1,n}\}_{n=0}^N$ with $t_{1,0}=0$, and $t_{1,N}=T_1$. In the following, we will consider the case where the system is periodic in t_1, and take T_1 to be one period of the system[2]. With s denoting the Laplace variable along the t_2 time axis, and capital symbols denoting transformed variables, it can be readily verified that the time-varying transfer function is given by:

$$\begin{aligned} H(s) &= [H(t_{1,1},s),\ldots,H(t_{1,N},s)]^T \\ &= \mathcal{D}^T[s\mathcal{C}+\mathcal{G}+\Delta\mathcal{C}]^{-1}\bar{B}, \end{aligned} \qquad (4)$$

such that $H(s)U(s) = \hat{Z}(s)$ with definition $\hat{Z}(s) = [\hat{Z}^T(t_{1,1},s),\ldots,\hat{Z}^T(t_{1,N},s)]^T$. In establishing (4), we also used the following notation: $\bar{B} = \mathbf{1}_{N,1}\otimes b$, where $\mathbf{1}_{N,1}$ is a N by 1 all-one vector, and $\otimes$ denotes Kronecker product; $\mathcal{D} = I_N \otimes d$, and $\Delta = (\text{dial}\{1/\delta_1,\ldots,1/\delta_N\}(I_N - J_N))\otimes I_m$, where I_N stands for a N by N identity matrix, and J_N a N by N circulant matrix with first column $[0,1,0,\ldots,0]^T$, and first row $[0,\ldots,0,1]$; $mN\times mN$ matrices $\mathcal{C}$ and $\mathcal{G}$ consist of $C(t_{1,n})$ and $G(t_{1,n})$, $\forall n\in[1,N]$, respectively. With m and N being the number of system states (i.e., node voltages and branch currents) and the number of samples in t_1, respectively, the product mN could be very large. The latter then poses prohibitive computational complexity. Therefore, we apply model order reduction techniques on Eq. (4). Along the lines of [9], a model of reduced order $q < mN$ can be obtained by casting (4) into the standard form $H(s) = \mathcal{D}^T[I_{mN} - s\mathcal{A}]^{-1}\mathcal{R}$ with definitions $\mathcal{A} = -[\mathcal{G}+\Delta\mathcal{C}]^{-1}\mathcal{C}$ and $\mathcal{R} = [\mathcal{G}+\Delta\mathcal{C}]^{-1}\bar{B}$, and applying Krylov subspace methods [7, 11]. With block Arnoldi algorithm, the resultant qth order transfer function that approximates $H(s)$ in (4) is given by [2]:

$$H_q(s) = L_q^T[I_q - sT_q]^{-1}R_q, \qquad (5)$$

where $L_q = V_q^T\mathcal{D}$ is a $q\times N$ matrix, T_q is a $q\times q$ block-Hessenberg matrix, $R_q = V_q^T\mathcal{R}$ is a $q\times 1$ vector, and V_q is the $mN\times q$ matrix consisting of the q orthogonal bases generated by applying block Arnoldi algorithm to $\mathcal{A}$ and $\mathcal{R}$.

Transforming (5) into time domain, we have the following system representation:

$$-T_q\frac{dx}{dt} + x = R_q u(t) \qquad z(t) = l_q(t)x(t), \qquad (6)$$

where x is a vector of size q, $z(t)$ is the output, and $l_q(t)$ is the $q\times 1$ time-varying vector that relates the system (states) to the output. To link $l_q(t)$ with the $q\times N$ matrix L_q in (5), we notice that the nth column of L_q is nothing but $l_q(t_{1,n})$, $\forall n\in[1,N]$. Eq. (6) corresponds to an ODE system, which translates the noise in power supply grids $u(t)$ to its corresponding current change $z(t)$. As a supplement to the current $I_0(t)$, we denote $z(t)$ as $\Delta I(t)$.

As shown in Fig. 2, the resultant LTV macromodel consists of three major components: a current source $I_0(t)$, a ODE system generating the current $\Delta I(t)$ and a current source I_s injecting current into substrate. The former ($I_0(t)$) is the current that the digital cell consistently draws, assuming perfect power supply and ground. The ODE system in our macromodel turns out to be LTV, which acts as a current source and the current $\Delta I(t)$ is determined by the voltages at node 2 and node 3, i.e., voltage variations at ground and power supply, respectively. Benefited from applying model reduction techniques, the LTV ODE system contains only $1\sim 10$ nodes, which corresponds to a marked reduction in comparison with hundreds of nodes in the original digital cell. Notice that the ODE system parameters captured in T_q, R_q, and L_q do not depend on the voltage variation, and can thus be computed off-line, and stored together with $I_0(t)$ and $I_s(t)$ in a cell library. The proposed macromodel enjoys high-accuracy and low-complexity, when included in a complete substrate noise analysis circuit that contains the package model, on-chip power networks, and substrate model, and is thus readily applicable to large scale circuits.

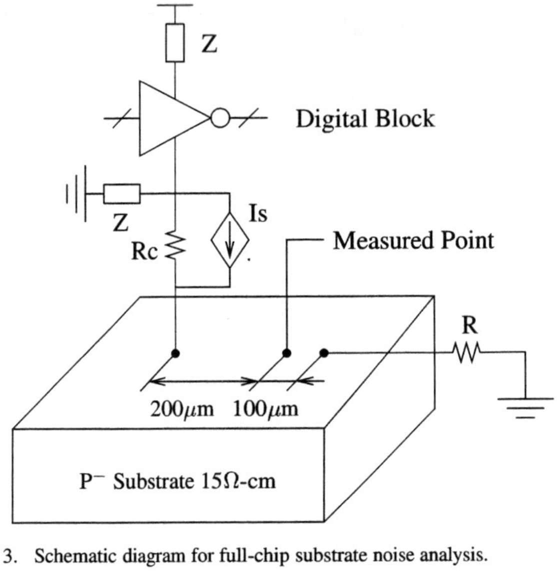

Fig. 3. Schematic diagram for full-chip substrate noise analysis.

IV. Simulation Results

To demonstrate the performance of our macromodel, we apply the extraction method to a full-chip substrate noise analysis example. As depicted in Fig. 3, the system consists of

[2]For more general cases, and frequency domain treatments, the reader is referred to [9].

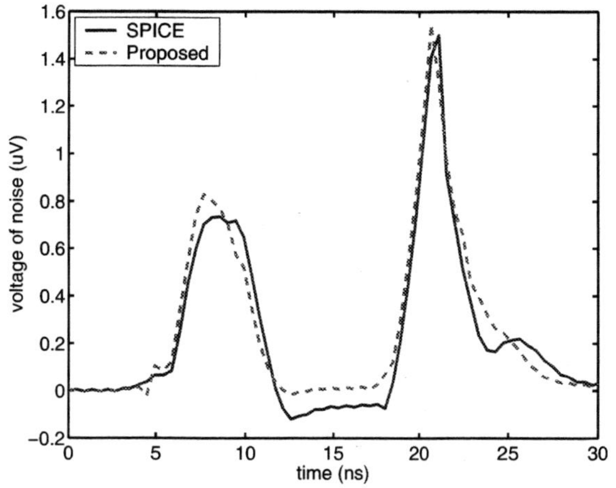

Fig. 4. The time-domain comparison of substrate noise induced by an 100-inverter-block using SPICE MOSFET model and our proposed macromodel.

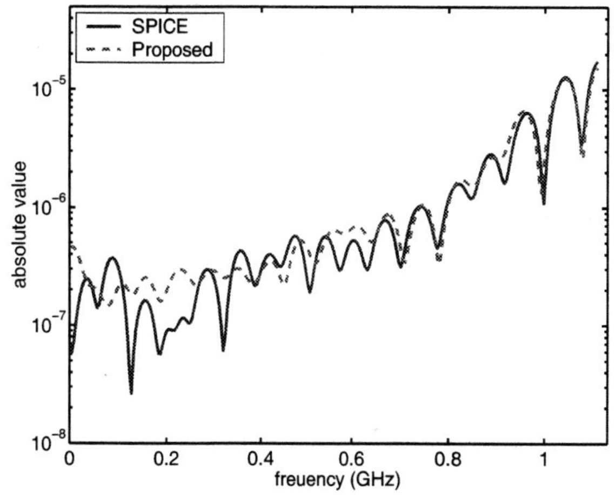

Fig. 5. The frequency-domain omparison of substrate noise induced by an 100-inverter-block using SPICE MOSFET model and our proposed macromodel.

a power supply network, package, a digital block, and a substrate. The digital block is constructed by stacking a total of 50 inverter chains, each containing 2 inverters. The logic input of the digital block has a period 30ns. In our simulations, the package and power supply network are modeled as an impedance $Z = 2 + j2 \times 10^{-9}\Omega$ in Fig. 3; the resistance of ground contact is $R_c = 2\Omega$; and the resistance of ground bias network is $R = 2\Omega$. The substrate is P$^-$ type with dimension $width \times length \times depth = 1000 \times 1000 \times 300\mu\text{m}^3$, and a sheet resistance 15Ω−cm. Discretisizing the substrate into $50 \times 50 \times 50\mu\text{m}^3$ cubes, we model it as a resistive mesh. The digital block is located 200μm from the measuring point, which is 100μm from the ground bias point.

For the digital block, the MOSFET is simulated using Schichman-Hodges model with $\mu_n C_{ox}\frac{W}{L} = 1.6 \times 10^{-4}$A/V^2 for both NMOS and PMOS, load $C = 0.5$pF (see e.g., [5]). Although not mandatory, the Schichman-Hodges model (instead of more comprehensive models, such as BSIM3) is adopted here for simplicity. In order to verify the capability of our macromodel in reflecting the power supply variation, we use a power supply voltage 10% below the ideal value of 2V.

We choose uniform step size of $t_{1,n} = 450n$ ps, $\forall n \in [0, N]$, and the ODE system order $q = 2$. The latter implies a computational complexity similar to the macromodel in [4]. For comparison purpose, we use both SPICE-level description and our macromodel extraction to represent the digital block, and carry out the substrate noise analysis. With SPICE-level description, it takes 291.3s; whereas with our macromodel, it only takes 1.81s, which is 160 times faster. As evident in Fig. 4, the discrepancy between their peak values is 0.048μV, which corresponds to an error of 3.21%. It is also worth mentioning that the resultant noise using our macromodel closely matches the true not only in peak value, but also in shape, which is corroborated by the comparison in frequency domain by taking Fourier Transform of their corresponding results, as depicted in Fig. 5.

V. Conclusions

Efficient and accurate computation of noise injection from the digital circuits consists of a major challenge in substrate noise analysis. In this paper, using TVP and model reduction techniques, we established a macromodel for noise injection computation with low complexity and high accuracy. A key attraction of this macromodel is the inclusion of three major noise mechanisms. Thanks to its low complexity, the macromodel enables system-level noise analysis, even in large-scale circuits. The accuracy renders our digital cell macromodel interact with power grid voltage variations just as the original cell does, thus provides reliable results even with noisy power supply/ground. As we presented in section IV, our proposed model significantly speeds up the computation (160 times faster) while still offering excellent accuracy (3.21% peak noise error).

VI. Acknowledgment

This work was supported by NSF under Grant No.s CCR-0204278 and CCR-0312079, by DARPA under Grant No. SA0302103, and by SRC CADT program.

References

[1] X. Aragones, J. Gonzalez, and Rubio. *Analysis and Solutions for Switching Noise Coupling in Mixed-Signal ICs*. Kluwer Academic Publishers, Boston, 1 edition, 1999.

[2] D. L. Boley. Krylov space methods on state-space control models. *Circuits, Systems and Signal Processing*, 13:733–758, 1994.

[3] H. H. Chen and D. D. Ling. Power supply noise analysis methodology for deep-submicron VLSI design. In *Proc. of IEEE DAC*, pages 638–643, Anaheim, CA, June, 1997.

[4] A. Dharchowdhury, R. Panda, D. Blaauw, R. Vaidyanathan, B. Tutuianu, and D. Bearden. Design and analysis of power distribution networks in PowerPC^{TM} microprocessors. In *Proc. of IEEE DAC*, pages 738–743, Anaheim, CA, June 15-19, 1998.

[5] D. A. Johns and K. Martin. *Analog Integrated Circuit Design*. John Wiely & Sons Inc, 1996.

[6] P. Miliozzi, L. Carloni, E. Charbon, and A. Sangiovanni-Vincentelli. SUBWAVE: a methodology for modeling digital substrate noise injection in mixed-signal ICs. In *Proc. of IEEE on Custom Integrated Circuits Conference*, pages 385–388, May 5-8, 1996.

[7] A. Odabasioglu, M. Celik, and L.T. Pileggi. Prima: passive reduced-order interconnect macromodeling algorithm. In *IEEE/ACM International Conference on Computer-Aided Design*, pages 257–260, San Jose, CA, Nov 9-13, 1997.

[8] S. Ponnapalli, N. Verghese, W. K. Chu, and G. Coram. Preventing a noisequake. *Circuits and Devices Magazine*, 17:19–28, November 2001.

[9] J. Roychowdhury. Reduced-order modeling of time-varying systems. *IEEE Trans. on Circuits and Systems II: Analog and Digital Signal Processing*, 46(10):1273–1288, October 1999.

[10] J. Roychowdhury. Analysing circuits with widely separated time scales using numerical PDE methods. *IEEE Trans. on Circuits and Systems I: Fundamental Theory and Applications*, 48(5):578–594, May 2001.

[11] Y. Saad. *Iterative Methods for Sparse Linear Systems*. PWS-Kent, Boston, 1996.

[12] M. v. Heijningen, M. Badaroglu, S. Donnay, M. Engels, and I. Bolsens. High-level simulation of substrate noise generation including power supply noise coupling. In *Proc. of IEEE DAC*, pages 738–743, Los Angeles, CA, June 5-9, 2000.

CONTINUATION METHOD IN MULTITONE HARMONIC BALANCE

Suihua Lu, Amit Narayan, Amit Mehrotra

Berkeley Design Automation
2902 Stender Way, Santa Clara, CA 95054

ABSTRACT

Design of communication circuits often requires computing quasi-periodic steady state response to multiple inputs of different frequencies that may not always be harmonically related. Multitone Harmonic Balance method can be used to provide an accurate solution if the circuit is mildly non-linear. However, the convergence of Harmonic Balance methods is not guaranteed if the initial guess is far from the actual solution. We present a novel approach to apply continuation method in multitone Harmonic Balance to significantly improve the convergence properties of this technique. In addition to being more robust, we show that the continuation based method can be significantly more efficient in terms of run-time compared to traditional fixed step source-stepping methods. Simulation results on strongly nonlinear circuits are used to illustrate the effectiveness of the algorithm.

1. INTRODUCTION

The exploding demand for low cost and high performance wireless products has increased the need for more efficient and accurate simulation techniques for RF ICs. Designers are often interested in measurements such as intermodulation distortion, gain compression and intercept point, that require accurate computation of steady state response of a circuit under incommensurate multitone excitation.

Harmonic Balance [1, 2, 3] is a well-established frequency domain technique for periodic and quasi-periodic steady state analysis of mildly nonlinear circuits. Harmonic Balance method has the advantage of being very accurate and fast when the (quasi-)periodic steady state solution can be accurately represented by a small number of Fourier coefficients. Frequency dependent distributed elements like transmission lines and S-parameter measured data can be handled naturally and the simulation run time is relatively insensitive to the numerical values of excitation frequencies.

In Harmonic Balance the circuit differential equations are represented in terms of the Fourier coefficients and differentiation in time domain is replaced by algebraic multiplication in frequency domain. This results in a large system of algebraic equations. The system is typically solved using Newton Raphson method. Typically, a Krylov subspace method such as GMRES [4, 5] is used to compute the inverse of the large Jacobian matrix which results from the linearization of the nonlinear equations.

Newton Raphson method converges quadratically once the initial guess is near the solution. However, if the initial guess is not good, Newton Raphson algorithm may fail to converge. Thus, a good initial guess of the solution is critical to the convergence of the problem.

Finding a good initial guess for the Multitone Harmonic Balance problem is a challenging problem and several techniques can been applied. A DC solution can been used as an initial guess. Alternatively, if the frequencies commensurate, an initial transient analysis can be carried out, and the resulting waveforms can be transformed to frequency domain to be the initial guess. However, these techniques might work only for a limited number of scenarios.

If the circuit is responding in a strongly nonlinear fashion to one large tone (e.g. the LO tone in the mixer) and weakly nonlinear fashion fashion to other moderate/small tones (e.g the RF tone), one can turn off the moderate/small tones and run a single tone Harmonic Balance analysis first, and use the solution as initial guess for the multitone problem.

For problems where a circuit is responding nonlinearly to multiple excitations, the DC solution as well as the solution from one tone problem can be very far away from the actual solution causing Newton to fail.

Simple source stepping scheme can be applied here to aid the convergence where the power of the independent source is ramped-up to its final value in small incremental steps. Since the solution from a previous step usually provides a good guess for the Newton Raphson, it has a higher chance of convergence. However, to ensure convergence the step size has be small and the method can become very slow requiring enormous computational time. If the step size is increased, Newton-Raphson may not converge.

In this paper, we propose to use the "predictor-corrector" continuation method [6] to address the convergence problem in multitone Harmonic Balance. The tangent predictor at the previous solution is used to predict the new solution, while Newton Raphson is used as the corrector. We show that this method is very robust and converges on very nonlinear problems while providing significant improvement in run-time over traditional source-stepping algorithms.

2. BACKGROUND

2.1. Multitone Harmonic Balance Formulation

Consider a circuit described by a set of nonlinear differential-algebraic equations (DAEs):

$$\frac{dq(v(t)}{dt} + f(v(t)) + b(t) = 0 \quad (1)$$

where $v(t) \in \mathbf{R}^N$ is the vector of node voltages, $b(t) \in \mathbf{R}^N$ is the vector of inputs, $f(v(t)) \in \mathbf{R}^N$ is the vector of sums of resistive currents at each node, $q(v(t)) \in \mathbf{R}^N$ is the vector of node charges (or fluxes), and N is the number of circuit nodes.

Let the circuit be driven by multiple periodic input sources $b_1(t), b_2(t), \cdots, b_M(t)$ with frequency $f_1, f_2, \cdots, f_M$ respectively, denote $B^j_{k_j}$ to be the Fourier coefficients of $b_j(t)$:

$$b(t) = b_1(t)+b_2(t)+\cdots+b_M(t) = \sum_{k_1=-\infty}^{\infty} B^1_{k_1} e^{j2\pi k_1 f_1 t} + \sum_{k_2=-\infty}^{\infty} B^2_{k_2} e^{j2\pi k_2 f_2 t} + \cdots + \sum_{k_M=-\infty}^{\infty} B^M_{k_M} e^{j2\pi k_M f_M t} \quad (2)$$

The circuit possesses a quasi-periodic steady state response. The solution vector $v(t)$ can be approximated with truncated Fourier series.

$$v(t) = \sum_{k_1=-K_1}^{K_1} \cdots \sum_{k_M=-K_M}^{K_M} V_{k_1 \cdots k_M} e^{j2\pi(k_1 f_1 + \cdots + k_M f_M)t} \quad (3)$$

Then circuit equation (1) can be expressed in frequency domain as

$$\sum_{k_1=-K_1}^{K_1} \sum_{k_2=-K_2}^{K_2} \cdots \sum_{k_M=-K_M}^{K_M} e^{j2\pi(k_1 f_1 + \cdots + k_M f_M)t} [j2\pi(k_1 f_1 + \cdots + k_M f_M) Q_{k_1 \cdots k_M} + F_{k_1 \cdots k_M} + B_{k_1 \cdots k_M}] = 0$$
$$\text{where} \quad B^j_{k_j} = B_{0,\cdots,0,k_j,0,\cdots,0} \quad (4)$$

There are two approaches to solve equation (4): multi-dimensional DFT and one dimensional DFT using artificial frequency mapping. The multidimensional DFT approach is restricted to the box truncation of the harmonics [1], and the artificial frequency mapping approach is generally faster and more flexible.

Multirate methods [7, 8, 9, 10] that constructs a partial differential equation can be served as a novel view point to illustrate the artificial frequency mapping. Without lose of generality, a two tone example is used to illustrate the box truncation. Equation (1) can now be written as

$$\frac{\partial q(\hat{v}(t_1,t_2)}{\partial t_1} + \frac{\partial q(\hat{v}(t_1,t_2)}{\partial t_2} + f(\hat{v}(t_1,t_2)) + b(t_1) + b(t_2) = 0 \quad (5)$$

where the original circuit has a solution $v(t) = \hat{v}(t_1 = t, t_2 = t)$.

Similarly, Equation (4) in frequency domain becomes

$$\sum_{k_1=-K_1}^{K_1} \sum_{k_2=-K_2}^{K_2} e^{j2\pi k_1 f_1 t_1 + j2\pi k_2 f_2 t_2} [j2\pi(k_1 f_1 + k_2 f_2) Q_{k_1 k_2} + F_{k_1 k_2} + B_{k_1 k_2}] = 0 \quad (6)$$

Evaluating the solution along $(t_1 = t, t_2 = \frac{f_1}{f_2} \frac{1}{2K_2+1} t)$, equation (6) becomes

$$\sum_{k_0=-K_0}^{K_0} e^{j2\pi k_0 f_0 t} [j2\pi(k_1 f_1 + k_2 f_2) Q_{k_0} + F_{k_0} + B_{k_0}] = 0 \quad (7)$$

where $k_0 = k_1(2K_2+1) + k_2$, $K_0 = (2K_1+1)(2K_2+1)$ and $f_0 = (2K_2+1)f_1$. In other words, there exists a linear one to one map that maps between (k_1, k_2) to k_0, such that $k_0 f_0$ is equally spaced in the artificial frequency axis. A compact matrix form of the above equation is

$$\Omega \mathbf{Q}(\mathbf{V}) + \mathbf{F}(\mathbf{V}) + \mathbf{B} = 0 \quad (8)$$

where $\mathbf{Q} = [Q^t_{-K_0}, \cdots, Q^t_{K_0}]^t$, $\mathbf{F} = [F^t_{-K_0}, \cdots, F^t_{K_0}]^t$, $\mathbf{B} = [B^t_{-K_0}, \cdots, B^t_{K_0}]^t$ and $\Omega = diag(-j(k_1(-K_0)2\pi f_1 + k_2(-K_0)2\pi f_2), \cdots, -j(k_1(K_0)2\pi f_1 + k_2(K_0)2\pi f_2))$. A standard Harmonic Balance solver can be used to solve equation (8).

2.2. Continuation Method

To solve the nonlinear equation

$$f(x) = 0 \quad f, x \in \mathbf{R}^N \quad (9)$$

the most popular method is Newton Raphson (NR) which sets up the following iterative procedure

$$x^{(k+1)} = x^{(k)} - \left[J\left(x^{(k)}\right)\right]^{-1} f\left(x^{(k)}\right) \quad (10)$$

where $J_{ij} = \frac{\partial f_i}{\partial x_j}$. This procedure converges quadratically to the solution if the initial guess $x^{(0)}$ is sufficiently close to the solution. In many cases this procedure fails to converge to the solution.

Consider the following generalization of (9)

$$f(x,\lambda)=0 \quad f,x\in \mathbf{R}^N \quad \lambda\in\mathbf{R}^p \tag{11}$$

We can view the solution process of (9) as a particular case of solution of (11) for a given λ.

Assume that the solution of equation (11) at $\lambda=\lambda_0$ is known or can be easily computed. The continuation method amounts to calculating solutions on the solution manifold

$$(\mathbf{y}^0,\lambda_0),(\mathbf{y}^1,\lambda_1),\cdots,$$

until one reaches a target point, say at $\lambda=\lambda_T$.

With the predictor-corrector methods, the jth continuation step starts from a solution $(\mathbf{y}^j,\lambda_j)$ and attempts to calculate the solution $(\mathbf{y}^{j+1},\lambda_{j+1})$ for the "next" λ, namely, λ_{j+1} in two steps:

$$(\mathbf{y}^j,\lambda_j)\overset{\text{predictor}}{\longrightarrow}(\bar{\mathbf{y}}^{j+1},\bar{\lambda}_{j+1})\overset{\text{corrector}}{\longrightarrow}(\mathbf{y}^{j+1},\lambda_{j+1})$$

In many circuits, the solution manifold may fold around itself. Consider the point λ_{SNB}, such that there are no solutions for $\lambda>\lambda_{SNB}$ and for $\lambda<\lambda_{SNB}$ there are two solutions. Therefore there is no close neighborhood around λ_{SNB} where a unique solution exists. Such points are call *saddle node bifurcation (SNB)* points. The fact that the Jacobian is singular at such point causes many simple continuation methods to fail. Parameterization techniques are used to resolve this problem by adding an additional equation. (Pseudo-)Arclength technique that uses the arclength as the new parameter is widely used in computing DC operating points [11].

3. CONTINUATION METHOD IN MULTITONE HARMONIC BALANCE

The frequency domain circuit equation (8) can be written as

$$\mathbf{G}\triangleq\mathbf{YV}+\Omega\mathbf{Q}(\mathbf{V})+\mathbf{F}(\mathbf{V})+\mathbf{B_1}+\lambda\mathbf{B_2}=0 \tag{12}$$

where $\mathbf{YV}$ term is added to explicitly denote the distributed elements, $\mathbf{B_2}$ is the vector of Fourier coefficients of of periodic excitations that needs to be ramped up, and $\mathbf{B_1}$ is the vector of Fourier coefficients of DC sources and other periodic excitations.

For $\lambda=0$, the solution $\mathbf{V}=[V^t_{-K_0},\cdots,V^t_{K_0}]^t$ is the Fourier coefficients of the DC solution or the solution to the one tone harmonic balance. And we are interested in solution at $\lambda=1$.

3.1. Tangent Predictor

Taking differential of both sides of equation (12), one obtains:

$$\begin{aligned}0=d\mathbf{G}&=\frac{\partial\mathbf{G}}{\partial\mathbf{V}}d\mathbf{V}+\frac{\partial\mathbf{G}}{\partial\lambda}d\lambda\\&=\mathbf{J}_{HB}d\mathbf{V}+\mathbf{B_p}d\lambda\end{aligned} \tag{13}$$

where $\mathbf{J}_{HB}$ is the Jacobian of Harmonic Balance. Therefore,

$$d\mathbf{V}^{j+1}=-(\mathbf{J}_{HB}|_{V^j})^{-1}\mathbf{B_p}d\lambda^{j+1} \tag{14}$$

Krylov subspace methods like GMRES [4, 5] are used to compute $(\mathbf{J}_{HB}|_{V^j})^{-1}\mathbf{B_p}$ efficiently.

3.2. Secant Predictor

Polynomial extrapolation method can be used as a predictor as well. A polynomial in λ of degree p that passes through $(p+1)$ points $(\mathbf{V}^{j-p},\lambda_{j-p}),\cdots(\mathbf{V}^{j-1},\lambda_{j-1}),(\mathbf{V}^j,\lambda_j)$ can be constructed. The predictor is formed by evaluating the polynomial at $\lambda=\lambda_{j+1}$.

The trivial predictor is the zeroth-order polynomial, when the predictor is the previous solution, $\bar{\mathbf{V}}^{j+1}=\mathbf{V}^j$. Thus, the source stepping method can be regarded as a special case of polynomial predictor.

A polynomial of the first order, the secant, is an alternative to the tangent predictor described in Section 3.1:

$$d\mathbf{V}^{j+1}=\frac{\mathbf{V}^j-\mathbf{V}^{j-1}}{\lambda_j-\lambda_{j-1}}d\lambda^{j+1} \tag{15}$$

The secant predictor (equation (15)) is computationally cheaper than the tangent predictor (equation (14)), since there is no inversion of the Jacobian involved. However, the predictor itself is not as good as the tangent predictor.

3.3. Arclength

A more numerically stable continuation method that can follow the solution manifold around a bifurcation point is the so-called (pseudo)-Arclength method [6, 1]. Both $\mathbf{V}$ and λ are parameterized using the arclength parameter s. Let the augmented states be $\tilde{\mathbf{V}}=[\mathbf{V}^T,\lambda]^T$. The additional equation can be written as:

$$(d\tilde{\mathbf{V}}^{j+1})^T(\tilde{\mathbf{V}}^{j+1}-\tilde{\mathbf{V}}^j)=s^{j+1}-s^j \tag{16}$$

In other words, it forces the new solution $\tilde{\mathbf{V}}^{j+1}$ to lie on a hyperplane perpendicular to the tangent prediction vector $d\tilde{\mathbf{V}}^{j+1}$.

3.4. Corrector And Step Control

Let μ denote the general parameter, which can be λ or s. The general algorithm is the following:

1. Initialization: let $\mu_i = 0$ and $\mu_f = \Delta$ and the known solution at μ_i be $x(\lambda(\mu_i))$
2. Predictor step: use the solution at μ_i to predict the solution at μ_f
3. Corrector step: use the predicted value of the solution at μ_f to obtain the actual solution at μ_f
4. Step control: if the corrector step failed, set $\mu_f = 0.5(\mu_i + \mu_f)$ and repeat; if the corrector step succeeded and $\mu_f = 1$, then done; else set $\mu_i = \mu_f$, $\mu_f = 3\mu_f - 2\mu_i$ and repeat.

Newton Raphson method is used in the correction step. We observed that in each correction step, the inversion of the Jacobian which is computed using Krylov subspace methods is the most expensive. Thus, for efficient step control, we don't want to be overly aggressive. Relatively small but cheap steps are sometimes preferred as the GMRES run time can be lower.

4. SIMULATION RESULTS AND CONCLUSIONS

A rectifier driven by two large input tones is used as an example. The circuit responses in a strongly nonlinear manner to both tones. Using the solution from DC, or a one tone harmonic balance result in non-convergence. The traditional source stepping using equally spaced steps works, however, the number steps needs to be around 15 to achieve convergence with an average number of 5.33 NR iterations per step. Using tangent following allows the simulator to dynamically change the steps size, thus convergence can be achieved in 13 steps with an average of 4.3 NR iterations. A 1.5X run time speed up is achieved as well.

A single balanced mixer driven by an large LO tone and a RF tone is used as another example. The RF tone is set to be strong enough to drive the circuit close to compression. The multitone analysis converges only with the use of continuation method. Using tangent following method to ramp up both tones takes 11 steps, with an average 4.4 NR iterations.

We conclude that continuation method applied to multitone Harmonic Balance can significantly improve the convergence properties of this technique. In addition to being more robust, the continuation based method can be significantly more efficient in terms of run-time compared to traditional fixed step source-stepping methods.

5. REFERENCES

[1] Kenneth S Kundert, Jacob K White, and Alberto L Sangiovanni-Vincentelli, *Steady-state methods for simulating analog and microwave circuits*, Kluwer Academic Publishers, Boston ; Dordrecht, 1990.

[2] Kenneth S Kundert, "Introduction to RF simulation and its application," *IEEE Journal of Solid state circuits*, vol. 44, no. 9, 1999.

[3] Kartikeya Mayaram, David C. Lee, and Shahriar Moinian, "Computer-aided circuit analysis tools for RFIC simulation: algorithms, features and limitations," *IEEE Transactions on circuits and Systems-II: Analog and digital signal processing*, vol. 47, no. 2, 2000.

[4] Y. Saad, *Iterative Methods for Sparse Linear Systems*, Boston: PWS Publishing Company, 1996.

[5] Ricardo Telichevsky, Kenneth S Kundert, and Jacob K White, "Efficient steady-state analysis based on matrix-free Krylov-subspace methods," in *Proceedings on the Design Automation Conference*, June 1995, pp. 480–484.

[6] R. Seydel, *Practical Bifurcation and Stability Analysis: from equilibrium to shaos*, Springer-Verlag, New York, NY, 1994.

[7] Hans-Georg Brachtendorf, Günther Welsch, and Rainer Laur, "A novel time-frequency method for the simulation of the steady state of circuits driven by multi-tone signals," in *Proceedings of the 1997 IEEE International Symposium on Circuits and Systems*, 1997, vol. 3, pp. 1508–1511.

[8] Jaijeet S Roychowdhury, "Analyzing circuits with widely separated time scales using numerical pde methods," *IEEE Transactions on circuits and Systems-I: Fundamental theory and applications*, vol. 48, no. 5, pp. 578–594, 2001.

[9] Jaijeet S Roychowdhury, "Efficient methods for simulating highly nonlinear multi-rate circuits," in *Proceedings of the 34th Design Automation Conference*, 1997, pp. 269–274.

[10] Amit Mehrotra, *Simulation and Modelling Techniques for Noise in Radio Frequency Integrated Circuits*, Ph.D. thesis, University of California at Berkeley, 1999.

[11] Jaijeet S Roychowdhury and R.C. Melville, "Homotopy techniques for obtaining a dc solution of large-scale MOS circuits," *ACM*, 1996.

MILLIMETER-WAVE CMOS DEVICE MODELING AND SIMULATION

Chinh H. Doan, Sohrab Emami, Ali M. Niknejad, and Robert W. Brodersen

Berkeley Wireless Research Center
Dept. of EECS, University of California, Berkeley

ABSTRACT

Challenges for modeling and simulating active and passive 130-nm CMOS devices at mm-wave frequencies (>30 GHz) are discussed. Small-signal lumped circuits with appropriate parasitic elements are used to model the active transistor devices with excellent broadband accuracy. Passive element transmission lines are discussed as generic scalable reactive elements suitable for forming resonant circuits with intrinsic transistor capacitance. The trade-offs between physical and electrical circuit models for the transmission lines are presented. Our approach demonstrates that relatively simple models can be used to accurately predict the small-signal performance of CMOS active and passive devices from dc up to mm-wave frequencies.

1. INTRODUCTION

The rapid scaling of CMOS to shorter channel lengths has enabled circuits to operate well into the millimeter-wave (mm-wave) frequency range [1][2]. The design of circuits at frequencies near the limits of any given technology requires accurate modeling of both the active and passive components. Millimeter-wave modeling of CMOS devices presents several unique challenges due to the low-resistivity silicon substrate, parasitic source, drain, and gate resistances, and the multi-layer dielectric with its relatively high loss tangent. In order to properly account for such limitations, new methodologies must be developed which involve careful modeling at several levels of abstraction, allowing trade-offs between accuracy, flexibility, and simulation speed. The methodology that we have developed to model and simulate both active and passive devices up to 65 GHz will be described along with experimental verification.

2. TYPICAL CMOS PROCESS

One of the most important differences between silicon and compound semiconductor technologies is the lossy nature of the substrate. The silicon substrate of a modern CMOS process has a resistivity of around 10 Ω-cm, whereas a III-V semi-insulating substrate has a resistivity of around 10^7–10^9 Ω-cm and can effectively be treated as an ideal dielectric. The lossy substrate poses problems for both active and passive devices. For transistors, signals which couple to the substrate through the source and drain junction capacitors incur significant losses at mm-wave frequencies. For passive components—inductors, capacitors, and transmission lines—the close proximity of the lossy substrate to the metal conductors lowers the attainable quality-factor (Q) of the passives.

Another important characteristic of a CMOS process is the metallization stack. The gate material for a silicon MOSFET is silicided polysilicon, which has a much higher sheet resistance (~10 Ω/□) than that of the metal used for the gates of GaAs FETs. Modern CMOS technologies have also migrated towards all-copper interconnects. While copper possesses the benefit of a higher conductance than aluminum or gold, the fabrication steps require a uniform density of all metal layers. Thus, floating dummy fill metal must be added to empty areas to increase the local density. Conversely, large areas of metal (e.g., for ground planes) are forced to have slots in order to reduce the density. The slots result in additional conductor loss, while the floating dummy metal can create unwanted and unpredictable coupling if exclusion areas are not defined.

3. TRANSISTOR MODELING

The traditional microwave approach uses measured S-parameter models for the design of small-signal circuits. Although S-parameter models are quite accurate and useful, a lumped small-signal circuit model provides several additional benefits. The small-signal circuit model is supported by all simulators and provides smooth data which results in improved convergence. The circuit model also allows extrapolation to frequencies beyond the measurement capabilities of the test equipment.

The physical layout of a single finger of an NMOS device is shown in Fig. 1, which illustrates the significant high-frequency loss mechanisms. A mm-wave transistor model must account for the series source and drain resistances (R_S, R_D), polysilicon gate resistance (R_G), non-

0-7803-8251-X/04/$17.00 ©2004 IEEE

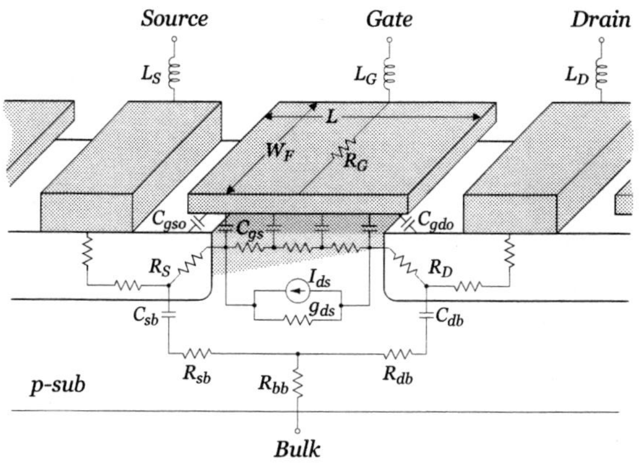

Fig. 1. Physical model for one finger of an NMOS transistor.

quasi-static (NQS) channel resistance (r_{nqs}), and resistive substrate network (R_{sb}, R_{db}, and R_{bb}) [3].

The series source and drain resistances are dominated by the intrinsic spreading resistance in the source-drain extension (SDE) region near the channel. The NQS channel resistance models the effective increase in gate resistance due to the finite channel charging time, and can be shown to be inversely proportional to g_m. The gate resistance, R_G, accounts for the distributed RC nature of the polysilicon gate, and can be approximated using a single lumped resistor. Additionally, series inductors must be added to all terminals—L_G, L_D, L_S—to properly model the delay effects associated with interconnect wiring.

A test chip consisting of transistors in common-source configuration with W/L ranging from 40/0.13 to 320/0.13 was fabricated in a 130-nm CMOS process. On-wafer mm-wave characterization from 0.04 to 65 GHz was performed on a Cascade Microtech probe station using an Anritsu 37397C VNA and Cascade Microtech GSG coplanar probes. At each bias point, a small-signal lumped model with frequency-independent component values was extracted using a hybrid optimization algorithm in Agilent IC-CAP.

S-parameters for the simulated model and measured data up to 65 GHz are shown in Figs. 2 and 3 for a 100/0.13 NMOS transistor biased at V_{GS}=0.65 V and V_{DS}=1.2 V. The excellent broadband accuracy of the simulation compared to the measured data is representative of all of our models and verifies that the topology of our small-signal model is correct.

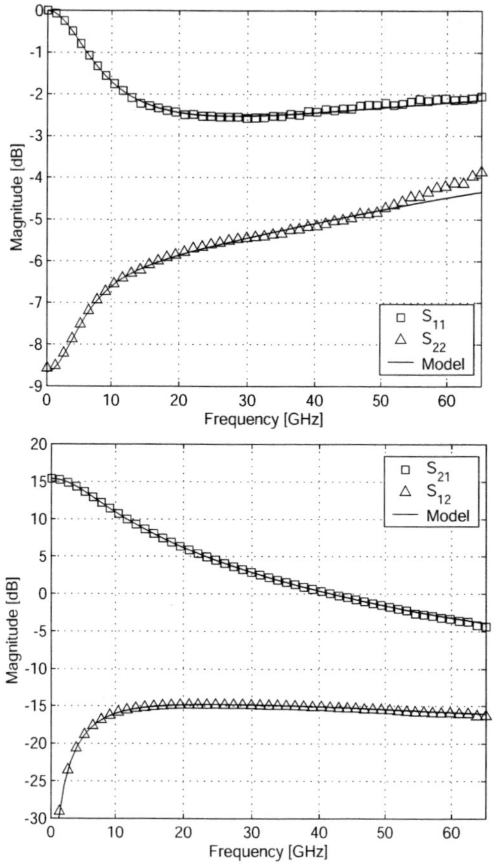

Fig. 2. Measured and modeled S-parameters (magnitude) for a 100/0.13 NMOS transistor (V_{GS}=0.65 V, V_{DS}=1.2 V).

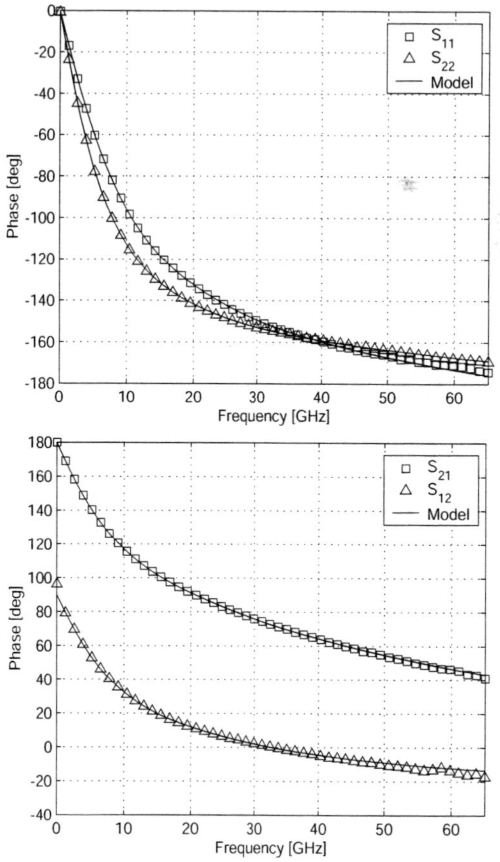

Fig. 3. Measured and modeled S-parameters (phase) for a 100/0.13 NMOS transistor (V_{GS}=0.65 V, V_{DS}=1.2 V).

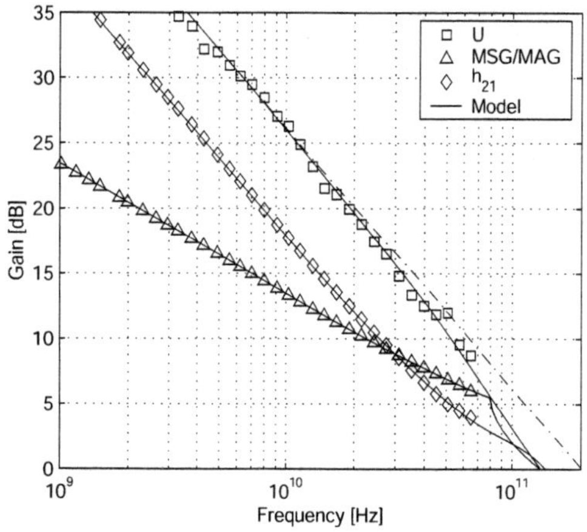

Fig. 4. Measured and modeled gains for a 100/0.13 NMOS transistor (V_{GS}=0.65 V, V_{DS}=1.2 V).

Furthermore, it also demonstrates that distributed effects and frequency-dependent losses caused by the skin effect can be adequately accounted for using only lumped components with frequency-independent values.

The transistor gains—Mason's unilateral gain, maximum stable gain (MSG), maximum available gain (MAG), and current gain—for this device are plotted in Fig. 4. The accurate modeling of the unilateral gain is particularly important. Unlike the MSG and current gain, Mason's unilateral gain is a very strong function of all resistive losses. Therefore, accurately fitting the unilateral gain validates that the important loss mechanisms have been properly modeled. Using the small-signal model, the extrapolated f_{max} is 135 GHz, which is much smaller than the value of 200 GHz if a 20 dB/decade slope is assumed.

4. TRANSMISSION LINE MODELING

For mm-wave circuit design, matching networks and resonators are often realized using transmission lines.

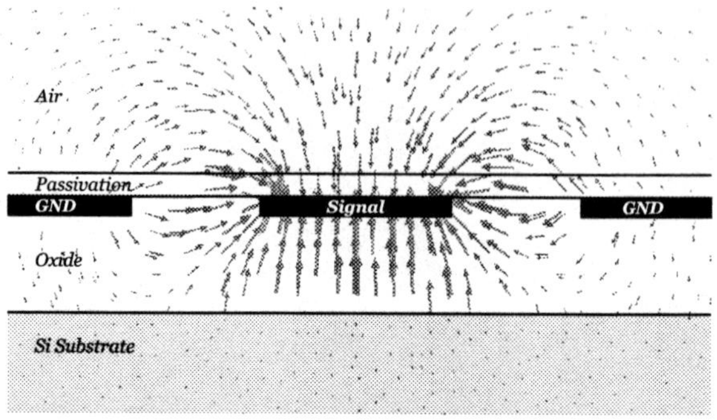

Fig. 5. Cross-section of the electric field distributions from a 3-D EM simulation (HFSS) of a CPW transmission line.

Transmission lines are inherently scalable in length, making them easier to model than lumped passives. Another benefit of using transmission lines is that the well-defined ground return path confines the magnetic and electric fields and significantly reduces any coupling to adjacent structures. Therefore, the lines can be modeled individually without concern that nearby structures will affect the performance characteristics of the line when used in more complex circuits.

The fabricated CMOS test chip also included 1-mm long coplanar waveguide (CPW) transmission lines of different cross-sectional dimensions, producing a large range of characteristic impedances. Equation-based models derived for microwave transmission lines are not applicable for lines integrated on silicon substrates. The models typically assume thin conductors, ground planes which are far from the signal lines, and high-quality dielectrics. The lossy silicon substrate is very close to the signal lines (~5 μm) and the metal thickness is on the order of the dimensions of the conductor width and signal-to-ground spacing.

Electromagnetic simulations of the passives based on the physical layout were performed using Ansoft HFSS (Fig. 5). Several simplifications were used to improve simulation time with a small degradation in accuracy. The

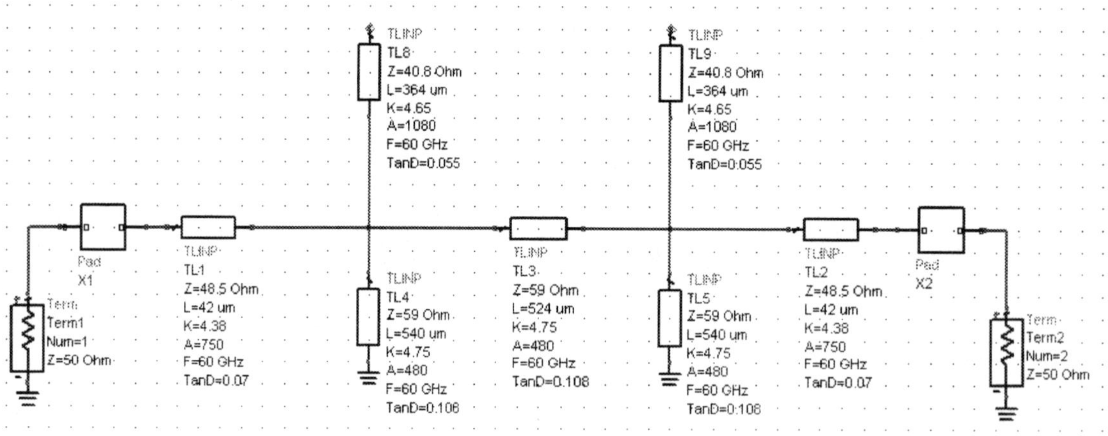

Fig. 6. Schematic of the 30-GHz CPW bandpass filter.

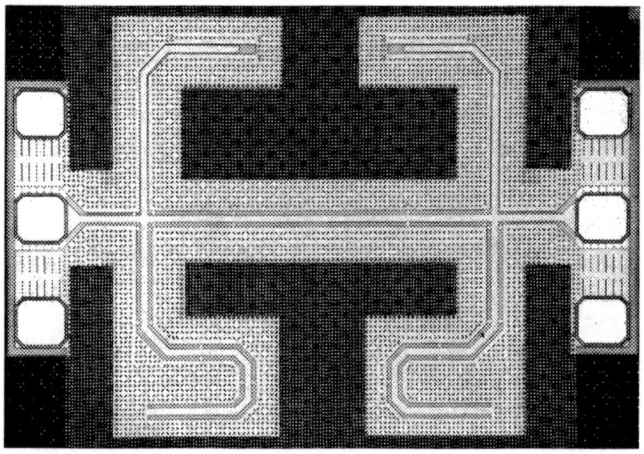

Fig. 7. Chip micrograph of the 30-GHz CPW filter.

silicon substrate is modeled as uniformly doped, and the multilayer dielectrics are modeled with two layers: the oxide and passivation with effective dielectric constants of 4 and 6.3, respectively. Additionally, the dielectric loss tangent, which degrades with frequency, is assumed to be constant. Although 3-D EM simulations enable the characterization of arbitrary passive devices, they suffer from long simulation times (many hours) and the inability to include several poorly characterized effects (substrate doping profile, conductor surface roughness, etc.). Physical EM simulations, therefore, serve as a good verification tool, but are too slow to be used during the design phase.

Another approach is to use electrical models in ADS which capture the high-level behavior of the lines. The model parameters are relatively easy to extract from measured data since only four parameters are required to model the broadband performance of each transmission line: characteristic impedance, effective dielectric constant, attenuation constant, and loss tangent. A first-order frequency-dependent loss model is used. The model assumes that the conductor loss is only caused by the skin effect losses, and the shunt loss is due to a constant loss tangent.

Using simple electrical models has many advantages. The simulation time is very fast, and the models can be easily integrated into commonly-used simulators such as Agilent ADS. Additionally, scalable (in length) models enable the use of powerful optimizers to tune the circuit performance. A broadband pad model was also extracted which consists of a shunt capacitance and series inductance. From our experience, it was not necessary to model the second-order effects such as bends, junctions, end-effects, radiation, and discontinuities to have sufficiently accurate circuit characterization.

To validate our passive models, a 30-GHz bandpass filter, which was composed of series and shunt stubs of the modeled CPW lines, was designed using ADS (Fig. 6). The optimizer was used to determine the line lengths, and

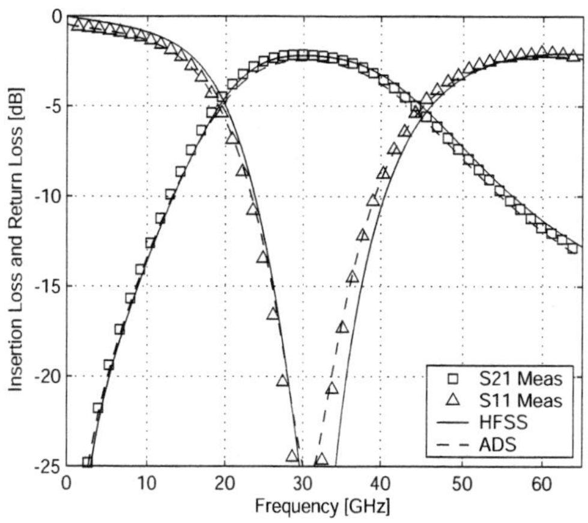

Fig. 8. Measured and simulated results for the 30-GHz CPW filter.

the pads were also included as part of the filter. A die photo of the filter is shown in Fig. 7, and the measured and simulated results for the 30-GHz filter are plotted in Fig. 8, demonstrating excellent broadband agreement for both the physical EM model as well as the simple, scalable electrical models.

5. CONCLUSIONS

We have presented small-signal CMOS transistor models and scalable electrical transmission line models which accurately match measured data up to 65 GHz. Our approach demonstrates that relatively simple models can be used to predict the small-signal performance of CMOS active and passive devices up to mm-wave frequencies. With the available small-signal models, it is now possible to design CMOS circuit blocks, such as amplifiers and oscillators, for mm-wave transceivers.

6. ACKNOWLEDGMENTS

The authors thank STMicroelectronics for chip fabrication. This work was funded by the DARPA TEAM project and the industrial members of the Berkeley Wireless Research Center.

7. REFERENCES

[1] H. Wang, "A 50GHz VCO in 0.25μm CMOS," in *IEEE Int. Solid-State Circuits Conf. Dig. Tech. Papers*, Feb. 2001, pp. 372–373.

[2] M. Tiebout, H.-D. Wohlmuth, and W. Simbürger, "A 1V 51GHz fully-integrated VCO in 0.12μm CMOS," in *IEEE Int. Solid-State Circuits Conf. Dig. Tech. Papers*, Feb. 2002, pp. 238–239.

[3] C. Enz and Y. Cheng, "MOS transistor modeling for RF IC design," *IEEE J. Solid-State Circuits*, vol. 35, pp. 186–201, Feb. 2000.

IMPROVED DESIGN OF FREQUENCY-RESPONSE-MASKING FILTERS USING ENHANCED SEQUENTIAL QUADRATIC PROGRAMMING

Wu-Sheng Lu

Dept. of Elec. and Comp. Engineering
University of Victoria
Victoria, BC, Canada V8W 3P6

Takao Hinamoto

Graduate School of Engineering
Hiroshima University
Higashi-Hiroshima 739-8527, Japan

ABSTRACT

Sequential quadratic programming (SQP) algorithms are widely recognized to be among the most successful algorithms for nonconvex optimization. This paper attempts to develop an SQP-based method for frequency-response-masking (FRM) filters. We explain how the complementarity conditions in the SQP algorithm help reduce the amount of computation required to update the Lagrange multipliers in a significant manner. Simulation results are presented to demonstrate the algorithm's performance that compares favorably with several existing design methods.

1. INTRODUCTION

The frequency-response-masking (FRM) technique originated in [1] has proved to be effective for the design of digital filters with narrow transition bands. Several design methods for linear-phase and low-group-delay, FIR and IIR, basic and multistage FRM filters have been investigated in the past, see [1]–[13] and the references cited there. Among others, available design methods include joint optimization of all subfilters using semidefinite programming (SDP) [11] and second-order cone programming (SOCP) [12][13]. Although these methods work well in general, a problem with them is the large number of constraints that inevitably effects design efficiency and, in the case of high-order FRM filters, may cause numerical difficulties.

In this paper, the joint optimization of subfilters is approached in a rather different way, namely, via an enhanced sequential quadratic programming (SQP) technique. Although, to the best knowledge of the authors, it appears to be the first attempt to use SQP for the design of FRM filters, SQP algorithms are widely recognized to be among the most successful algorithms for nonconvex constrained optimization problems [14]. Since the minimax design of an FRM filter can be formulated as a nonconvex constrained minimization problem, SQP is a natural candidate tool for the design. However, our primary reason to develop an SQP-based design methodology is that the complementarity conditions in an SQP formulation are found to be effective in reducing the number of constraints that actually participate in the optimization. Our design method is rather general in the sense that it is applicable to both basic and multistage FRM filters with linear phase response or low group delay. Because of space limitation and for illustration clarity, however, our attention here is focused on the class of basic, linear phase FIR FRM filters. Technical details of the proposed method are given in Secs. 2 and 3. Design examples with performance comparisons are presented in Sec. 4.

2. PROBLEM FORMULATION

Following [1], the reader is referred to the structure in Fig. 1 where all filters are assumed to have linear-phase responses, and the lengths of the masking filters are either both even or both odd. The transfer functions of the prototype and masking filters are respectively denoted by

$$H_a(z) = \sum_{k=0}^{N-1} h_k z^{-k},\ H_{ma}(z) = \sum_{k=0}^{N_a-1} h_k^{(a)} z^{-k},$$
$$H_{mc}(z) = \sum_{k=0}^{N_c-1} h_k^{(c)} z^{-k}$$

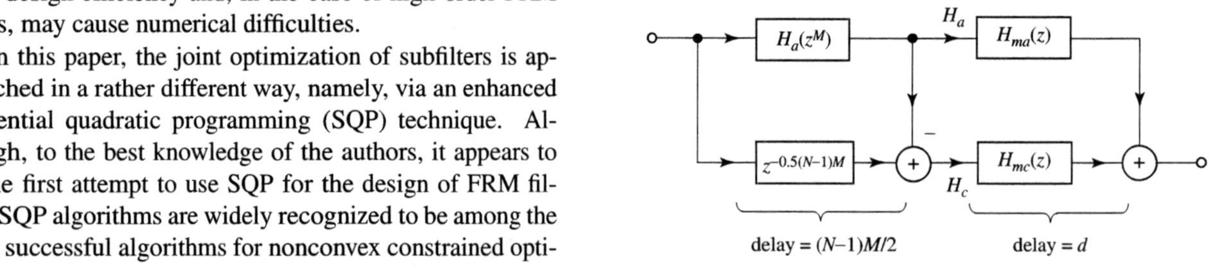

Fig. 1. A basic FRM filter structure.

In what follows, H_a, H_{ma} and H_{mc} are referred to as *subfilters*. Without loss of generality, the FRM filter can be treated as a zero-phase FIR filter and all subfilters are assumed to be of odd length. The frequency response of the FRM filter is then given by

$$H(\omega, \boldsymbol{h}) = [\boldsymbol{a}^T \boldsymbol{c}(\omega)][\boldsymbol{a}_a^T \boldsymbol{c}_a(\omega) - \boldsymbol{a}_c^T \boldsymbol{c}_c(\omega)] + \boldsymbol{a}_c^T \boldsymbol{c}_c(\omega) \quad (1)$$

where

$$\begin{aligned}
\boldsymbol{a} &= [h_{(N-1)/2}\ 0.5h_{(N+1)/2}\ \cdots\ 0.5h_{N-1}]^T \\
\boldsymbol{c}(\omega) &= [1\ \cos M\omega\ \cdots\ \cos[(N-1)M\omega/2]]^T \\
\boldsymbol{a}_a &= [h^{(a)}_{(N_a-1)/2}\ 0.5h^{(a)}_{(N_a+1)/2}\ \cdots\ 0.5h^{(a)}_{N_a-1}]^T \\
\boldsymbol{c}_a(\omega) &= [1\ \cos\omega\ \cdots\ \cos[(N_a-1)\omega/2]]^T \\
\boldsymbol{a}_c &= [h^{(c)}_{(N_c-1)/2}\ 0.5h^{(c)}_{(N_c+1)/2}\ \cdots\ 0.5h^{(c)}_{N_c-1}]^T \\
\boldsymbol{c}_c(\omega) &= [1\ \cos\omega\ \cdots\ \cos[(N_c-1)\omega/2]]^T
\end{aligned}$$

and $\boldsymbol{h} = [\boldsymbol{a}^T\ \boldsymbol{a}_a^T\ \boldsymbol{a}_c^T]^T$. The minimax design of the FRM filter amounts to finding a vector $\boldsymbol{h}$ that solves the minimax optimization problem

$$\underset{\boldsymbol{h}}{\text{minimize}}\{\underset{\omega\in\Omega}{\text{maximize}}\ W(\omega)|H(\omega,\boldsymbol{h}) - H_d(\omega)|\} \quad (2)$$

where $H_d(\omega)$ is a real-valued desired frequency response, $W(\omega) \geq 0$ is a weighting function, and $\Omega = \{\omega : 0 \leq \omega \leq \pi\}$.

Let β be an upper bound of $W((\omega)|H(\omega,\boldsymbol{h}) - H_d(\omega)|$ on Ω. As the first step of the optimization we convert the problem in (2) into a constrained minimization problem

$$\text{minimize} \quad \beta \quad (3a)$$

$$\text{subject to: } W(\omega)|H(\omega,\boldsymbol{h}) - H_d(\omega)| \leq \beta \text{ for } \omega \in \Omega \quad (3b)$$

For practical exercise of optimization techniques, the constraint in (3b) is imposed on a dense grid of frequencies $\Omega_d = \{0 \leq \omega_1 \leq \cdots \leq \omega_K \leq \pi\}$ and the problem in (3) becomes

$$\text{minimize} \quad \beta \quad (4a)$$

$$\text{subject to: } W(\omega_i)|H(\omega_i,\boldsymbol{h}) - H_d(\omega_i)| \leq \beta \text{ for } \omega_i \in \Omega_d \quad (4b)$$

3. DESIGN METHOD

3.1. An SQP-based algorithm

The constraints in (4b) can be made more specific as

$$\begin{aligned}
&p_i(\boldsymbol{h},\beta) = \\
&\beta + W(\omega_i)[H(\omega_i,\boldsymbol{h}) - H_d(\omega_i)] \geq 0,\ 1 \leq i \leq K \quad (5a)\\
&p_{K+i}(\boldsymbol{h},\beta) = \\
&\beta - W(\omega_i)[H(\omega_i,\boldsymbol{h}) - H_d(\omega_i)] \geq 0,\ 1 \leq i \leq K \quad (5b)
\end{aligned}$$

Defining $\boldsymbol{x} = [\beta\ \boldsymbol{h}^T]^T$ and $\boldsymbol{e} = [1\ 0\ \cdots\ 0]^T$, (4) can be expressed as

$$\text{minimize} \quad \boldsymbol{e}^T\boldsymbol{x} \quad (6a)$$

$$\text{subject to: } p_i(\boldsymbol{x}) \geq 0 \quad 1 \leq i \leq 2K \quad (6b)$$

Since $-p_i(\boldsymbol{x})$ are not convex functions, (6) is a nonconvex problem.

The Lagrangian of (6) is defined by

$$L(\boldsymbol{x},\boldsymbol{\mu}) = \boldsymbol{e}^T\boldsymbol{x} - \sum_{i=1}^{2K}\mu_i p_i(\boldsymbol{x})$$

where μ_i for $1 \leq i \leq 2K$ are the Lagrange multipliers. The solution of problem (6) must satisfy the Karush-Kuhn-Tucker (KKT) conditions [15]

$$\begin{aligned}
\nabla L(\boldsymbol{x},\boldsymbol{\mu}) &= \boldsymbol{0} && (7a)\\
p_i(\boldsymbol{x}) &\geq 0 \quad 1 \leq i \leq 2K && (7b)\\
\mu_i &\geq 0 \quad 1 \leq i \leq 2K && (7c)\\
\mu_i p_i(\boldsymbol{x}) &= 0 \quad 1 \leq i \leq 2K && (7d)
\end{aligned}$$

It is the KKT conditions that form the basis of our design algorithm and a subsequent analysis of the algorithm. Suppose one starts with a reasonable initial point $\boldsymbol{x}_0$ (which may be produced using the method in [1]) and an initial $\boldsymbol{\mu}_0 = \boldsymbol{0}$. In the kth iteration, $\{\boldsymbol{x}_k, \boldsymbol{\mu}_k\}$ is updated to $\{\boldsymbol{x}_{k+1}, \boldsymbol{\mu}_{k+1}\} = \{\boldsymbol{x}_k, \boldsymbol{\mu}_k\} + \{\boldsymbol{\delta}_x, \boldsymbol{\delta}_\mu\}$ such that (7a), (7b), and (7c) are approximately satisfied up to the first order, and (7c) is precisely satisfied. This first-order approximation leads to

$$\begin{aligned}
\boldsymbol{Y}_k\boldsymbol{\delta}_x + \boldsymbol{e} - \boldsymbol{A}_k^T\boldsymbol{\mu}_{k+1} &= \boldsymbol{0} && (8a)\\
\boldsymbol{A}_k\boldsymbol{\delta}_x &\geq -\boldsymbol{c}_k && (8b)\\
\boldsymbol{\mu}_{k+1} &\geq \boldsymbol{0} && (8c)\\
(\boldsymbol{\mu}_{k+i})_i(\boldsymbol{A}_k\boldsymbol{\delta}_k + \boldsymbol{c}_k)_i &= 0 \quad 1 \leq i \leq 2K && (8d)
\end{aligned}$$

where $\boldsymbol{Y}_k = \nabla^2 L(\boldsymbol{x}_k, \boldsymbol{\mu}_k)$, $\boldsymbol{c}_k = [p_1(\boldsymbol{x}_k)\ \cdots\ p_{2K}(\boldsymbol{x}_k)]^T$, and

$$\boldsymbol{A}_k = \begin{bmatrix} \nabla^T p_1(\boldsymbol{x}_k) \\ \vdots \\ \nabla^T p_{2K}(\boldsymbol{x}_k) \end{bmatrix} \quad (9)$$

Equations (8a)–(8d) turn out to be the *exact* KKT conditions for the quadratic programming (QP) problem

$$\text{minimize} \quad \frac{1}{2}\boldsymbol{\delta}^T\boldsymbol{Y}_k\boldsymbol{\delta} + \boldsymbol{\delta}^T\boldsymbol{e} \quad (10a)$$

$$\text{subject to: } \boldsymbol{A}_k\boldsymbol{\delta} \geq -\boldsymbol{c}_k \quad (10b)$$

Let the solution of (10) be denoted by $\boldsymbol{\delta}_x$, the Lagrange multiplier $\boldsymbol{\mu}_{k+1}$ can then be determined by (8a) and (8d) as follows. First, the $2K$ components of $\boldsymbol{A}_k\boldsymbol{\delta}_x + \boldsymbol{c}_k$ are examined. For the component indices with $(\boldsymbol{A}_k\boldsymbol{\delta}_x + \boldsymbol{c}_k)_i > 0$, the complementarity conditions in (8d) imply that $(\boldsymbol{\mu}_{k+1})_i = 0$. Since $\boldsymbol{\delta}_x$ satisfies (10b), the rest of indices are those where $(\boldsymbol{A}_k\boldsymbol{\delta}_x + \boldsymbol{c}_k)_i = 0$ and the complementarity conditions are satisfied regardless of the values of $(\boldsymbol{\mu}_{k+1})_i$. These possibly nonzero Lagrange multipliers can be determined using (8a) as

$$\hat{\boldsymbol{\mu}}_{k+1} = (\boldsymbol{A}_{ak}\boldsymbol{A}_{ak}^T)^{-1}\boldsymbol{A}_{ak}(\boldsymbol{Y}_k\boldsymbol{\delta}_x + \boldsymbol{e}) \tag{11}$$

where the rows of $\boldsymbol{A}_{ak}$ are those rows of $\boldsymbol{A}_k$ satisfying $(\boldsymbol{A}_k\boldsymbol{\delta}_x + \boldsymbol{c}_k)_i = 0$ and $\hat{\boldsymbol{\mu}}_{k+1}$ denotes the associated Lagrange multiplier. Having computed $\hat{\boldsymbol{\mu}}_{k+1}$, vector $\boldsymbol{\mu}_{k+1}$ can be obtained by inserting zeros wherever necessary in $\hat{\boldsymbol{\mu}}_{k+1}$. It should be stressed that typically the number of nonzero Lagrange multiplies, say $\hat{K}$, is much smaller than the number of constraints imposed in (4b), K (usually $\hat{K} < 0.1K$). Consequently, computing $\hat{\boldsymbol{\mu}}_{k+1}$ using (11) which involves inversion of an $\hat{K} \times \hat{K}$ matrix does not impose a computational burden. Moreover, since in the $2K$ linear constraints in (10b) only a small fraction of them are *active*, solving the QP problem in (10) can be carried out efficiently when an active-set type algorithm [15] is utilized.

Having obtained $\boldsymbol{\delta}_x$ and $\boldsymbol{\mu}_{k+1}$, point $\boldsymbol{x}_k$ is then updated to $\boldsymbol{x}_{k+1} = \boldsymbol{x}_k + \boldsymbol{\delta}_x$, and $\boldsymbol{Y}_k$, $\boldsymbol{c}_k$, and $\boldsymbol{A}_k$ are updated to $\boldsymbol{Y}_{k+1}$, $\boldsymbol{c}_{k+1}$, and $\boldsymbol{A}_{k+1}$, respectively. The iteration continues until a convergence criterion in terms of the progress made, i.e., $\|\boldsymbol{\delta}_x\|_2$, or the total number of iterations is met. The coefficients of the optimized subfilters can be found in the solution vector $\boldsymbol{x}^*$ as $\boldsymbol{h}^* = \boldsymbol{x}^*(2 : \text{end})$.

3.2. Convex relaxation of problem (10)

The Hessian matrix of the Lagrangian, $\boldsymbol{Y}_k$, is in general not positive definite, hence problem (10) is not a convex QP problem. A convex relaxation of problem (10) can be made by replacing the Hessian matrix $\boldsymbol{Y}_k$ in (10a) with a positive definite matrix, still denoted by $\boldsymbol{Y}_k$, with $\boldsymbol{Y}_0 = \boldsymbol{I}$ using the Broyden-Fletcher-Goldfarb-Shanno (BFGS) recusion [15] that updates $\boldsymbol{Y}_k$ to

$$\boldsymbol{Y}_{k+1} = \boldsymbol{Y}_k + \frac{\boldsymbol{\eta}_k\boldsymbol{\eta}_k^T}{\boldsymbol{\delta}_x^T\boldsymbol{\eta}_k} - \frac{\boldsymbol{v}_k\boldsymbol{v}_k^T}{\boldsymbol{\delta}_x^T\boldsymbol{v}_k} \tag{12a}$$

where

$$\boldsymbol{v}_k = \boldsymbol{Y}_k\boldsymbol{\delta}_x$$

$$\boldsymbol{\eta}_k = \theta\boldsymbol{\gamma}_k + (1-\theta)\boldsymbol{v}_k \tag{12b}$$

$$\boldsymbol{\gamma}_k = -(\boldsymbol{A}_{k+1} - \boldsymbol{A}_k)^T\boldsymbol{\mu}_{k+1} \tag{12c}$$

$$\theta = \begin{cases} 1 & \text{if } \boldsymbol{\delta}_x^T(\boldsymbol{\gamma}_k - 0.2\boldsymbol{v}_k) \geq 0 \\ \frac{0.8\boldsymbol{\delta}_x^T\boldsymbol{v}_k}{\boldsymbol{\delta}_x^T(\boldsymbol{v}_k - \boldsymbol{\gamma}_k)} & \text{otherwise} \end{cases} \tag{12d}$$

In this way, (10) becomes a convex QP problem which possesses a unique global minimizer that can be obtained using an efficient algorithm such as an active-set algorithm. A desirable feature of the BFGS update is that if $\boldsymbol{Y}_k$ is positive definite, then $\boldsymbol{Y}_{k+1}$ is also positive definite. With $\boldsymbol{Y}_0 = \boldsymbol{I}$, therefore, the QP subproblems involved in the entire design process are all guaranteed to be convex QP problems.

3.3. Implementation

Initial subfilters can be obtained using the method proposed in [1]. For given $H_d(\omega)$, weighting function $W(\omega)$, a grid of frequencies Ω_d, and an initial $\boldsymbol{h}_0$, the value of β_0 can be calculated as

$$\beta_0 = \max_{\Omega_d} W(\omega_i)|H(\omega_i, \boldsymbol{h}_0) - H_d(\omega_i)|$$

The SQP-based algorithm starts with initial point $\boldsymbol{x}_0 = [\beta_0 \ \ \boldsymbol{h}_0^T]^T$ and $\boldsymbol{Y}_0 = \boldsymbol{I}$. For the design of basic FIR FRM filters, the matrix $\boldsymbol{A}_k$ in (9) is a $2K \times (N+N_a+N_c+5)/2$ matrix whose ith and $(K+i)$th rows for $1 \leq i \leq K$ are given by $[1 \ \ W(\boldsymbol{a}_a^T\boldsymbol{c}_a - \boldsymbol{a}_c^T\boldsymbol{c}_c)\boldsymbol{c} \ \ W(\boldsymbol{a}^T\boldsymbol{c})\boldsymbol{c}_a \ \ W(1-\boldsymbol{a}^T\boldsymbol{c})\boldsymbol{c}_c]$ and $-[-1 \ \ W(\boldsymbol{a}_a^T\boldsymbol{c}_a - \boldsymbol{a}_c^T\boldsymbol{c}_c)\boldsymbol{c} \ \ W(\boldsymbol{a}^T\boldsymbol{c})\boldsymbol{c}_a \ \ W(1-\boldsymbol{a}^T\boldsymbol{c})\boldsymbol{c}_c]$ respectively, where the frequency-dependence for W, $\boldsymbol{c}$, $\boldsymbol{c}_a$, and $\boldsymbol{c}_c$ have been omitted. Reliable convex QP solvers are available, for example, in MATLAB Optimization Toolbox: `quadprog` uses an interior-point method while `qp` adopts an active-set method.

4. DESIGN EXAMPLES

The method described in Sections 2 and 3 was applied to design two one-stage linear-phase FRM filters that were addressed in the literature [1][6][11].

Example 1: The design parameters were $N = 45$, $N_a = 41$, $N_c = 33$, $M = 9$, $\omega_p = 0.6\pi$, $\omega_a = 0.61\pi$, $W(\omega) \equiv 1$ for $\omega \in [0, \ \omega_p]\bigcup[\omega_a, \ \pi]$ and $K = 950$. It took the algorithm 110 iterations to converge to a solution FRM filter whose amplitude response and passband ripple are shown in Fig. 2a and b, respectively. It is interesting to note that among the $2K = 1900$ inequality constraints (see (6b)), the average number of active constraints in the entire design process was only 27. In other words, the average size of the matrix $\boldsymbol{A}_{ak}\boldsymbol{A}_{ak}^T$ in (11) was 27×27. The maximum passband ripple and minimum stopband attenuation were 0.0667 dB and 42.38 dB, respectively, which compare favorably with the design of the same FRM filter in [1] (with passband ripple = 0.0896 dB and stopband attenuation = 40.96 dB) and in [11] (with passband ripple = 0.0674 dB and stopband attenuation = 42.25 dB).

Example 2: The design parameters were $N = 123$, $N_a = 56$, $N_c = 78$, $M = 21$, $\omega_p = 0.4\pi$, $\omega_a = 0.61\pi$, and $K = 1100$. The weighting function $W(\omega)$ was piecewise constant with $W(\omega) \equiv 1$ in the passband and $W(\omega) \equiv 12$ in the stopband. It took the proposed algorithm 150 iterations to converge to a solution FRM filter. The average number of active constraints was 67. The amplitude response and passband ripple of the FRM filter are depicted in Fig. 3(a) and (b), respectively. The maximum passband ripple and minimum stopband attenuation are 0.0898 dB and 61.66 dB. For comparison, the maximum ripple and minimum stopband

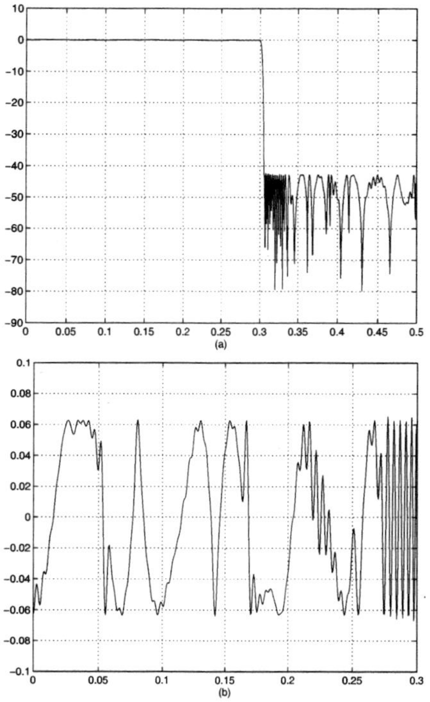

Fig. 2. (a) Amplitude response and (b) passband ripple of the FRM filter in Example 1, all in dB.

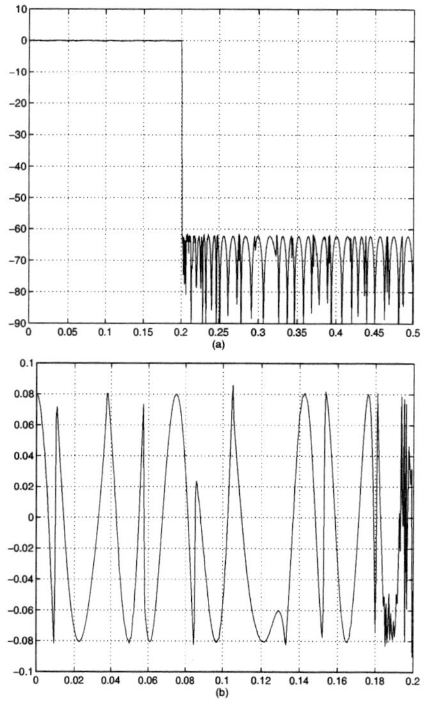

Fig. 3. (a) Amplitude response and (b) passband ripple of the FRM filter in Example 2, all in dB.

attenuation are 0.0864 dB and 60 dB in [6] and are 0.0855 dB and 60.93 dB in [11].

5. REFERENCES

[1] Y. C. Lim, "Frequency-response masking approach for the synthesis of sharp linear phase digital filters," *IEEE Trans. Circuits Syst.*, vol. 33, pp. 357-364, April 1986.

[2] T. Saramaki and A. T. Fam, "Subfilter approach for designing efficient FIR filters," *Proc. 1988 ISCAS*, pp. 2903-2915.

[3] Y. C. Lim and Y. Lian, "The optimum design of one- and two-dimensional FIR filters using the frequency response masking technique," *IEEE Trans. Circuits Syst. II*, vol. 40, pp. 88-95, Feb. 1993.

[4] Y. C. Lim and Y. Lian, "Frequency-response masking approach for digital filter design: complexity reduction via masking filter factorization," *IEEE Trans. Circuits Syst. II*, vol. 41, pp. 518-525, Aug. 1994.

[5] T. Saramaki, Y. C. Lim, and R. Yang, "The synthesis of half-band filter using frequency-response marking technique,' *IEEE Trans. Circuits Syst. II*, vol. 42, pp. 58-60, Jan. 1995.

[6] T. Saramaki and H. Johansson, "Optimization of FIR filters using frequency-response masking approach," *Proc. 2001 ISCAS*, vol. II, pp. 177-180.

[7] L. Svensson and H. Johansson, "Frequency-response masking FIR filters with short delay," *Proc. 2002 ISCAS*, May 2002.

[8] T. Saramaki and Y. C. Lim, "Use of the Remez algorithm for designing FRM based FIR filters," *Circuits, Systems, Signal Processing*, vol. 22, no. 2, pp. 77-97, 2003.

[9] L. D. R. de Barcellos, S. L. Netto, and P. S. R. Diniz, "Optimization of FRM filters using the WLS-Chebyahev approach," *Circuits, Systems, Signal Processing,* vol. 22, no. 2, pp. 99-113, 2003.

[10] Y. Lian, "Complexity reduction for FRM-based FIR filters using prefilter-equalizer technique," *Circuits, Systems, Signal Processing,* vol. 22, no. 2, pp. 137-155, 2003.

[11] W.-S. Lu and T. Hinamoto, "Optimal design of frequency-response-masking filters using semidefinite programming," *IEEE Trans. Circuits Syst.*, I, vol. 50, pp. 557-568, April 2003.

[12] W.-S. Lu and T. Hinamoto, "Optimal design of FIR frequency-response-masking filters using second-order cone programming," *Proc. ISCAS'2003*, vol. 3, pp. 878-881, Bangkok, May 2003.

[13] W.-S. Lu and T. Hinamoto, "Optimal design of IIR frequency-response-masking filters using second-order cone programming," to appear in *IEEE Trans. Circuits Syst., I.*

[14] P. T. Boggs and J. W. Tolle, "Sequential quadratic programming," *Acta Numerica*, vol. 4, pp. 1-51, 1995.

[15] R. Fletcher, *Practical Methods of Optimization*, 2nd ed., Wiley, 1987.

FREQUENCY-RESPONSE MASKING TECHNIQUE INCORPORATING EXTRAPOLATED IMPULSE RESPONSE BAND-EDGE SHAPING FILTER

Ya Jun Yu[1,2], *Yong Ching Lim*[3], *Kok Lay Teo*[2] *and Guohui Zhao*[2]

[1]Temasek Laboratories, Research TechnoPlaza, Nanyang Technological University, Singapore 639798

[2]Applied Mathematics Department, The Hong Kong Polytechnic University, Hong Kong

[3]Electrical & Electronic Engineering, Nanyang Technological University, Singapore 639798

ABSTRACT

Techniques based on the frequency-response masking (FRM) approach are known to be highly efficient for achieving drastic reduction of the number of multipliers and adders in narrow transition band linear-phase finite-impulse response (FIR) filters. Basic FRM structure comprises a band-edge shaping filter and two masking filters. It is observed that the impulse response of the band-edge shaping filter of FRM shows quasi-periodic property. Previous study demonstrated that the quasi-periodic property of the impulse responses of lowpass or highpass digital FIR filters can be exploited to reduce the complexity of the filter by using extrapolated impulse response technique. Extrapolating the idea, the band-edge shaping filter in FRM may be designed by the extrapolated impulse response technique. In this paper, an FRM technique incorporating extrapolated impulse response band-edge shaping filter is introduced. An iterative optimization procedure is proposed to design the FRM subfilters. An example taken from the literature is used to illustrate that the number of multipliers produced by our new technique is less than 80 percent of earlier results.

1. INTRODUCTION

Frequency-response masking (FRM) technique [2–5] produces a filter network, as shown in Fig.1, which comprises a band-edge shaping filter, $F(z^L)$, and two masking filters, $G_1(z)$ and $G_2(z)$, to synthesize narrow transition band linear phase digital FIR filters. $F(z^L)$ is obtained by replacing each delay element of a prototype filter, $F(z)$, by L delay elements and therefore has very sharp transition bands and very low arithmetic complexity. However, if the desired FIR filter has an extremely narrow transition band, the order of the band-edge shaping filter may still be high. In such cases, the complexity of the band-edge shaping filter may be reduced by the two-stage FRM approach [2, 6, 7] or prefilter-equalizer structures [8].

It is interesting to note that the impulse response of the band-edge shaping filter, $f(n)$ for $|n| \leq \frac{N_F-1}{2}$, shows quasi-periodic property, where N_F is the filter length. An example of an impulse response sequence $f(14)$ through $f(61)$ is shown in Fig. 2. Based on this observation, an FRM technique incorporating extrapolated impulse response band-edge shaping filter is introduced in this paper.

The extrapolated impulse response approach [1] was proposed to synthesize lowpass and highpass filters with reduced complexity by making use of the correlation between blocks of impulse response samples. In this approach, smaller side lobes of impulse response samples are approximated as scaled versions of larger ones so as to extrapolate the filter length. Thus, the hardware complexity of the filter is reduced.

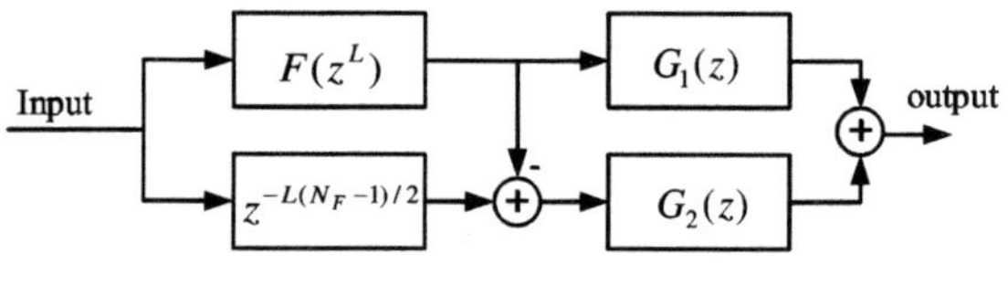

Fig. 1: A filter synthesized using the frequency-response masking technique.

In this paper, the extrapolated impulse response approach is incorporated into the FRM technique to reduce the complexity of the band-edge shaping filter. An optimization procedure is proposed to iteratively optimize the extrapolated band-edge shaping filter and the masking filters. It is shown, by means of an example, that the numbers of adders and multipliers of the resulting filter are less than 80 percent of those of the filter obtained earlier [5].

2. THE EXTRAPOLATED IMPULSE RESPONSE

This section reviews the extrapolated impulse response approach for synthesizing digital filters.

Assume that a linear phase FIR filter has a zero-phase transfer function of

$$H(z) = h(0) + \sum_{n=1}^{N} h(n)(z^n + z^{-n}), \tag{1}$$

where $h(n)$ is the impulse response of the digital filter. The impulse response may be partitioned into a central segment and several side segments in such a way that the side segments resemble each other fairly closely but their magnitudes diminish with increasing distance from the central segment. In the extrapolated impulse response technique, the largest side segment is selected as the primary side segment and the remaining secondary side segments are expressed as scaled versions of this primary side segment. An example of the side segments is shown in Fig. 2.

Consider the case where the impulse response may be partitioned into a central segment and $R + 1$ quasi-periodic side segments. Assume that the durations of side segments are equal to d. Assume further that the primary segment (called the zeroth side segment) ranges from $n = M + 1$ through $n = M + d$, and the rth secondary side segment ranges from $n = M + rd + 1$ through

0-7803-8251-X/04/$17.00 ©2004 IEEE

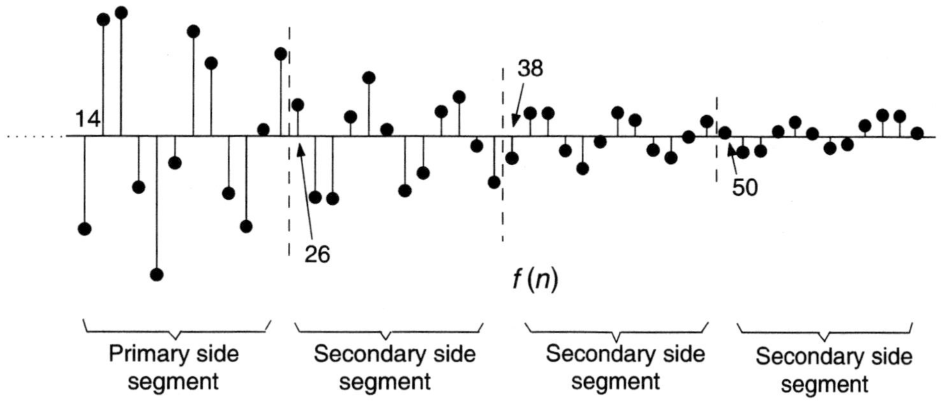

Fig. 2: The impulse response sequence $f(14)$ through $f(61)$ of the band-edge shaping filter $F(z)$, obtained by using the iterative optimization approach [5].

$n = M + (r+1)d$ for $r = 1, \cdots, R$, $H(z)$ may be rewritten as

$$H(z) = h(0) + \sum_{n=1}^{M} h(n)(z^n + z^{-n}) + \sum_{r=0}^{R}\sum_{m=1}^{d} h(M+m+rd)\left(z^{M+m+rd} + z^{-(M+m+rd)}\right). \tag{2}$$

If the segments for $r \geq 1$ are approximated as scaled versions of the segment for $r = 0$, $H(z)$ can be approximated by

$$H(z) \approx \hat{H}(z) = h(0) + \sum_{n=1}^{M} h(n)(z^n + z^{-n}) + \sum_{m=1}^{d} h(M+m)\sum_{r=0}^{R}\alpha_r\left(z^{M+m+rd} + z^{-(M+m+rd)}\right), \tag{3}$$

where α_r is the rth scaling factor and $\alpha_0 = 1$.

The frequency response $\hat{H}(e^{j\omega})$ of the extrapolated impulse response filter is given by

$$\hat{H}(e^{j\omega}) = h(0) + 2\sum_{n=1}^{M} h(n)\cos(n\omega) + 2\sum_{m=1}^{d} h(M+m)\sum_{r=0}^{R}\alpha_r \cos\left((M+m+rd)\omega\right). \tag{4}$$

In [1], the values of $h(M+m)$ for $m = 1, \cdots, d$ are fixed *apriori* using the values of the minimax optimum design. Thus, the optimization problem is simplified to be a linear problem of determining the filter coefficients $h(n)$ for $n = 0, \cdots, M$, and scaling factors α_r for $r = 0, \cdots, R$; the problem was solved using a linear programming algorithm. In the paper, we adopt an iterative technique to optimize $h(n)$ and α_r for all n and r. Further details will be discussed in Section 3.

3. AN ITERATIVE OPTIMIZATION PROCEDURE

Let $F_0(z)$ be the overall transfer function of the filter shown in Fig. 1.

$$F_0(z) = F(z^L)G_1(z) + \left[1 - F(z^L)\right] G_2(z). \tag{5}$$

Let ω_p and ω_s be the passband edge and stopband edge, respectively, for the filter $F_0(z)$. In the FRM technique, for a given L, one of the following two cases (but not both) may yield a set of θ and ϕ satisfying $0 \leq \theta < \phi \leq \pi$, where θ and ϕ are the passband and stopband edges for the prototype filter $F(z)$. In the first case, referred to as Case A, the parameters are given by [2, 3, 6]

$$l = \lfloor L\omega_p/(2\pi)\rfloor,\ \theta = L\omega_p - 2l\pi,\ \phi = L\omega_s - 2l\pi \tag{6}$$

and in the second case, denoted by Case B, by

$$l = \lceil L\omega_s/(2\pi)\rceil,\ \theta = 2l\pi - L\omega_s,\ \phi = 2l\pi - L\omega_p. \tag{7}$$

Here, $\lfloor x \rfloor$ is the largest integer less than or equal to x, and $\lceil x \rceil$ is the smallest integer larger than or equal to x.

Our technique starts from an initial design, $F^{(0)}(z)$, $G_1^{(0)}(z)$ and $G_2^{(0)}(z)$, obtained using the approach proposed in [5], in which, the band-edge shaping filter, $F(z^L)$, provides the desired overall filter performance on $[\Omega_{p1},\ \Omega_{p2}] \cup [\Omega_{s1},\ \Omega_{s2}]$, whereas the masking filters, $G_1(z)$ and $G_2(z)$, provide the overall filter performance on $[0,\ \Omega_{p1}] \cup [\Omega_{s2},\ \pi]$, where

$$\begin{array}{llll} & \text{Case A Design} & \text{Case B Design} & \\ \Omega_{p1} & = 2l\pi/L & = (2l-1)\pi/L & \\ \Omega_{p2} & = (2l\pi+\theta)/L & = (2l\pi-\phi)/L & = \omega_p \\ \Omega_{s1} & = (2l\pi+\phi)/L & = (2l\pi-\theta)/L & = \omega_s \\ \Omega_{s2} & = (2l+1)\pi/L & = 2l\pi/L. & \end{array} \tag{8}$$

When the band-edge shaping filter is synthesized using the extrapolated impulse response technique, the optimization of the sub-filters of the FRM technique can be carried out by embedding another iterative procedure for the design of the band-edge shaping filter. In the embedded iterative procedure, the extrapolated coefficients and scaling factors are iteratively optimized; their effects are most significant in the frequency band $[\Omega_{p1},\ \Omega_{p2}] \cup [\Omega_{s1},\ \Omega_{s2}]$. Assume that the impulse response of $F(z)$ is quasi-periodic from sample $f(M+1)$ through $f(M+(R+1)d)$, and each period has duration d. Furthermore, the zero-phase frequency response of $F(z^L)$, $G_1(z)$ and $G_2(z)$ is denoted as $F(L\omega)$, $G_1(\omega)$ and $G_2(\omega)$. The proposed optimization procedure for the design of a lowpass filter, $F_0(z)$, with passpand and stopband ripples of δ_p and δ_s is as follows:

***Step* 0**: Set $p = 0, \epsilon_G^{(p)} = 0$.

***Step* 1**: Set $p = p + 1$. Determine the parameters of $G_1^{(p)}(\omega)$ and $G_2^{(p)}(\omega)$ that minimize $\epsilon_G^{(p)}$.

$$\epsilon_G^{(p)} = \max_{\omega \in [0, \Omega_{p1}] \cup [\Omega_{s2}, \pi]} |W(\omega)| \left[H_G^{(p)}(\omega) - D(\omega) \right]. \quad (9)$$

In (9), the first band is the passpand, where $W(\omega)$ and $D(\omega)$ are equal to unity. The second band is the stopband, where $W(\omega)$ is equal to δ_p/δ_s and $D(\omega)$ is equal to zero. In the sequel, the same desired and weighting functions are used. Furthermore, in (9),

$$H_G^{(p)}(\omega) = F^{(p-1)}(L\omega) G_1^{(p)}(\omega) + \left[1 - F^{(p-1)}(L\omega)\right] G_2^{(p)}(\omega). \quad (10)$$

***Step* 2**: Set

$$\begin{aligned} &q = 0, \epsilon_\alpha^{(-1)} = 0, \epsilon_F^{(0)} = 0, \\ &\boldsymbol{X}_1^{(q)} = \left[f^{(p-1)}(0), \cdots, f^{(p-1)}(M)\right]^T, \\ &\boldsymbol{X}_2^{(q)} = \left[f^{(p-1)}(M+1), \cdots, f^{(p-1)}(M+d)\right]^T, \\ &\boldsymbol{X}_3^{(q)} = \left[\alpha_0^{(p-1)}, \cdots, \alpha_R^{(p-1)}\right]^T. \end{aligned} \quad (11)$$

***Step* 3**: Set $q = q + 1$. Determine the values of $\boldsymbol{X}_1^{(q)}$ and $\boldsymbol{X}_3^{(q)}$ that minimize $\epsilon_\alpha^{(q)}$.

$$\epsilon_\alpha^{(q)} = \max_{\omega \in [\Omega_{p1}, \Omega_{p2}] \cup [\Omega_{s1}, \Omega_{s2}]} |W(\omega)| \left[H_\alpha^{(q)}(\omega) - D(\omega) \right], \quad (12)$$

where

$$H_\alpha^{(q)}(\omega) = F^{(q)}(L\omega) G_1^{(p)}(\omega) + \left[1 - F^{(q)}(L\omega)\right] G_2^{(p)}(\omega). \quad (13)$$

In (13),

$$F^{(q)}(L\omega) = \boldsymbol{K}_1^{(q-1)}(L\omega)^T \begin{bmatrix} \boldsymbol{X}_1^{(q)} \\ \boldsymbol{X}_3^{(q)} \end{bmatrix}, \quad (14)$$

where

$$\begin{aligned} \boldsymbol{K}_1^{(q-1)}(w) = [&1, 2\cos(\omega), \cdots, 2\cos(M\omega), \\ &2\sum_{m=1}^{d} f^{(q-1)}(M+m)\cos((M+m)\omega), \cdots, \\ &2\sum_{m=1}^{d} f^{(q-1)}(M+m)\cos((M+m+Rd)\omega)\Big]^T. \end{aligned} \quad (15)$$

***Step* 4**: Set $q = q + 1$. Determine the values of $\boldsymbol{X}_1^{(q)}$ and $\boldsymbol{X}_2^{(q)}$ that minimize $\epsilon_F^{(q)}$.

$$\epsilon_F^{(q)} = \max_{\omega \in [\Omega_{p1}, \Omega_{p2}] \cup [\Omega_{s1}, \Omega_{s2}]} |W(\omega)| \left[H_F^{(q)}(\omega) - D(\omega) \right], \quad (16)$$

where

$$H_F^{(q)}(\omega) = F^{(q)}(L\omega) G_1^{(p)}(\omega) + \left[1 - F^{(q)}(L\omega)\right] G_2^{(p)}(\omega). \quad (17)$$

In (17),

$$F^{(q)}(L\omega) = \boldsymbol{K}_2^{(q-1)}(L\omega)^T \begin{bmatrix} \boldsymbol{X}_1^{(q)} \\ \boldsymbol{X}_2^{(q)} \end{bmatrix}, \quad (18)$$

where

$$\begin{aligned} \boldsymbol{K}_2^{(q-1)}(\omega) = [&1, 2\cos(\omega), \cdots, 2\cos(M\omega), \\ &2\sum_{r=0}^{R} \alpha_r^{(q-1)} \cos((M+rd+1)\omega), \cdots, \\ &2\sum_{r=0}^{R} \alpha_r^{(q-1)} \cos((M+rd+d)\omega)\Big]^T. \end{aligned} \quad (19)$$

***Step* 5**: If $\left|\epsilon_\alpha^{(q-1)} - \epsilon_\alpha^{(q-3)}\right| \leq \Delta$ and $\left|\epsilon_F^{(q)} - \epsilon_F^{(q-2)}\right| \leq \Delta$, where Δ is a prescribed tolerance, then set

$$F^{(p)}(L\omega) = \boldsymbol{K}_2^{(q-1)}(L\omega)^T \begin{bmatrix} \boldsymbol{X}_1^{(q)} \\ \boldsymbol{X}_2^{(q)} \end{bmatrix}, \quad (20)$$

and go to Step 6. Otherwise, go to Step 3.

***Step* 6**: If $\left|\epsilon_G^{(p)} - \epsilon_G^{(p-1)}\right| \leq \Delta$, then stop. Otherwise, go to Step 1.

Steps 1, 3 and 4 can be accomplished using linear programming. In the above algorithm, the two masking filters are optimized in Step 1 and they focus mainly on generating the desired response on $[0,\ \Omega_{p1}] \cup [\Omega_{s2},\ \pi]$. The extrapolated band-edge shaping filter is optimized by iteratively adjusting the extrapolated coefficients and scaling factors, i.e., $[\boldsymbol{X}_1^T, \boldsymbol{X}_3^T]^T$ and $[\boldsymbol{X}_1^T, \boldsymbol{X}_2^T]^T$, respectively, in Step 2 through 5 and it focuses on generating the desired response on $[\Omega_{p1},\ \Omega_{p2}] \cup [\Omega_{s1},\ \Omega_{s2}]$. The sharing of the frequency-response-shaping responsibility was first proposed by Saramäki and Johansson [5].

4. EXAMPLES

We choose the same example used in [5, 6] to illustrate the advantage of our proposed technique.

The specifications of [5, 6] are reproduced as follows: $\omega_p = 0.4\pi$, $\omega_s = 0.402\pi$, $\delta_p = 0.01$, and $\delta_s = 0.001$. For the optimum conventional direct-form FIR filter design, the minimum order to meet the given criteria is 2541, requiring 2541 adders and 1271 multipliers when the coefficient symmetry is exploited.

Using iterative optimization approach [5], the overall number of multipliers was minimized by choosing $L = 21$. The best solution is obtained for $N_F = 123$, $N_1 = 56$ and $N_2 = 78$, where N_1 and N_2 are the filter lengths of $G_1(z)$ and $G_2(z)$, respectively. The filter requires 129 multipliers and 254 adders.

Notice that the impulse response of the band-edge shaping filter $F(z)$, obtained using the iterative optimization approach [5], is quasi-periodic from $f(14)$ through $f(61)$ with a period of 12, as shown by the stem plot in Fig. 2. Choose M to be 13 and d to be 12, i.e., $f(n)$ for $|n| \leq 25$ are realized accurately and the period of the impulse response segments are 12 samples. Hence, the impulse response sequences $f(26)$ through $f(37)$, $f(38)$ through $f(49)$ and $f(50)$ through $f(61)$ are approximated as scaled versions of $f(14)$ through $f(25)$. Using out technique, the band-edge shaping filter with length 123 can be extrapolated from a filter with length of 51. However, in this case, the passband and stopband ripples of the obtained overall filter are 0.018 and 0.0018, respectively.

To meet the original specifications, more coefficients in the band-edge shaping filter must be optimized accurately. In this example, an additional 8 coefficients, $f(62)$ through $f(65)$ and $f(-62)$ through $f(-65)$, were appended to the filter and optimized together with the coefficients in the central segment; this reduces the passband and stopband ripples to 0.01 and 0.001, respectively, i.e., besides $f(n)$ for $|n| \leq 25$, $f(62)$ through $f(65)$ are also realized accurately, whereas the impulse response sequences $f(26)$ through $f(61)$ are approximated as scaled versions of $f(14)$ through $f(25)$.

The frequency response of the extrapolated band-edge shaping filter with length 131 is shown in Fig. 3, which is similar as the one presented in [5]. Thirty-six multipliers and 62 adders are needed for the realization of this band-edge shaping filter when the coefficient symmetry is exploited [9]. They are approximately 60% and 50% of those used in the band-edge shaping filter of the earlier design [5]. The overall multipliers and adders needed are 103 and 194, which are 79.8% and 76.4% respectively of the earlier design [5]. The price to pay for is an increase of 6 percent in the filter order. The overall frequency response is depicted in Fig. 4.

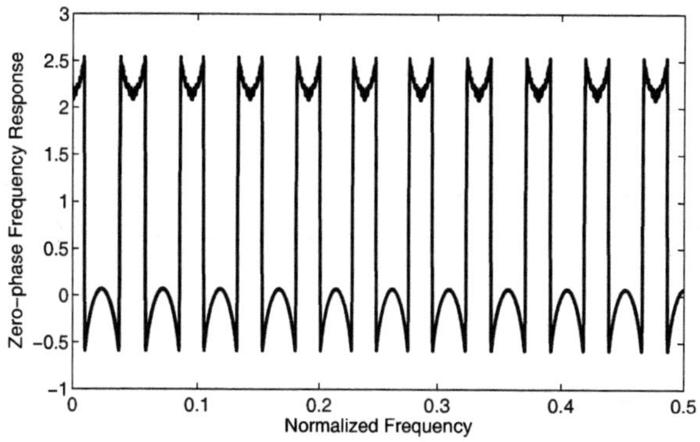

Fig. 3: Frequency response of the band-edge shaping filter, $F(z)$, with length 131, extrapolated from a filter with length 51.

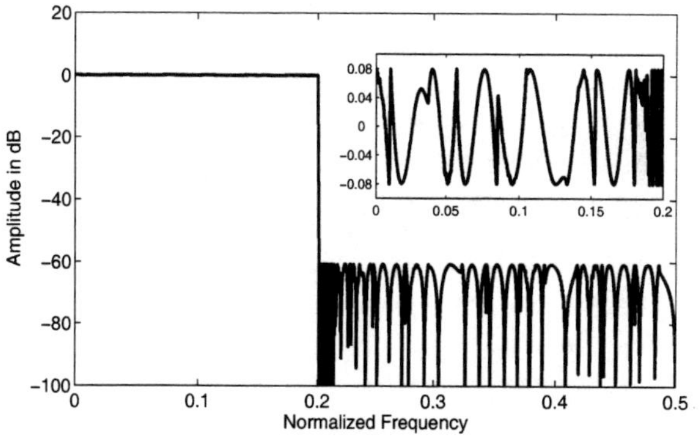

Fig. 4: Frequency response of the overall filter, in which the band-edge shaping filter with length 131 is extrapolated from a filter with length 51.

5. CONCLUSION

The impulse response of the band-edge shaping filter in FRM technique shows quasi-periodic property. The arithmetic complexity of the FIR filters using FRM approach has been reduced by incorporating the extrapolated impulse response band-edge shaping filters.

6. REFERENCES

[1] Y.C. Lim, "Extrapolated impulse response FIR filters", *IEEE Trans. Circuits, Syst.*, vol. 37, pp. 1548-1551, Dec. 1990.

[2] ——, "Frequency-response masking approach for the synthesis of sharp linear phase digital filters," *IEEE Trans. Circuits Syst.*, vol. CAS-33, pp.357-364, April. 1986.

[3] Y. C. Lim and Y. Lian, "The optimum design of one- and two-dimensional FIR filters using the frequency-response masking technique," *IEEE Trans. Circuits Syst II.*, vol. 40, pp. 88-95, Feb. 1993.

[4] Y.C. Lim, Y.J. Yu, H.Q. ZHeng and S.W. Foo, "FPGA implementation of digital filters synthesized using the frequency-response masking technique," *Proc. IEEE Int. Conf. Circuits Syst.*, Sydney, Australia, vol. II, pp173-176, 2001.

[5] T. Saramäki and H. Johansson, "Optimization of FIR filters using the frequency-response masking technique," in *Proc. IEEE Int. Conf. Circuits Syst.*, Sydney, Australia, vol. II, pp. 177-180, 2001.

[6] T. Saramäki and Y. C. Lim, "Use of the Remez algorithm for designing FIR filters utilizing the frequency-response masking approach," in *Proc. IEEE Int. Conf. Circuits Syst.*, Orlando, FL, vol. III, pp. 449-455, 1999.

[7] Y.J. Yu, T. Saramäki and Y.C. Lim, "An iterative method for optimizing FIR filters synthesized using the two-stage frequency-response masking technique," in *Proc. IEEE Int. Conf. Circuits Syst.*, Bangkok, vol. III, pp. 874-877, 2003.

[8] Y. Lian, "A new frequency-response masking structure with reduced complexity for FIR digital design," in *Proc. IEEE Int. Conf. Circuits Syst.*, Sydney, Australia, vol. II, pp. 609-612, 2001.

[9] Y.J. Yu, G.H. Zhao, K.L. Teo and Y.C. Lim, "Optimization and implementation of extrapolated impulse response filters," in *Proc. IEEE Int. Conf. Neural networds & Signal Processing*, vol. I, pp. 744-747, Nanjing, Dec. 2003.

FREQUENCY-RESPONSE MASKING BASED FILTERS WITH THE EVEN-LENGTH BANDEDGE SHAPING FILTER

Jianghong Yu, and Yong Lian

Department of Electrical and Computer Engineering
National University of Singapore
10 Kent Ridge Crescent, Singapore, 119260
email: eleliany@nus.edu.sg

ABSTRACT

This paper presents the design of Frequency Response Masking (FRM) filter using an even-length filter as bandedge shaping filter. The optimization of each subfilter is carried out with the help of the Sequential Quadratic Programming (SQP) technique. It is shown, by means of examples, that the proposed FRM filters with even-length bandedge shaping filter leads to designs comparable to the original FRM filters.

1. INTRODUCTION

The Frequency Response Masking (FRM) technique was first introduced in [1]. It is one of the most computationally efficient ways for the design of FIR filters with arbitrary bandwidth and very narrow transition bandwidth. Fig.1 shows a basic FRM structure which includes three subfilters and a delay block. The subfilter $H_a(z)$ is called bandedge shaping filter whose role is to form the narrow transition band of the overall filter through interpolation. The arbitrary bandwidth is produced by the interpolated bandedge shaping filter and its complement. The subfilters $H_{Ma}(z)$ and $H_{Mc}(z)$ are called masking filters which are used to remove the passband images caused by the interpolated bandedge shaping filter. The frequency responses of each subfilter can be found in Fig.2. Much effort has been made to improve the FRM technique [2-10]. These works were focused on further reduction of complexity for FRM by introducing either modified FRM structure [2-6] or nonlinear optimization techniques to optimize all subfilters simultaneously [7-10]. Such effort has made FRM based filters more efficient. However, all reported FRM structures employ an odd-length filter as the bandedge shaping filter. The usage of even-length filter as the bandedge shaping filter in FRM has never been exploited. In this paper, we present the detailed analysis on why an even-length filter is not as good as an odd-length filter for bandedge shaping filter and provide a solution to utilize even-length filter as bandedge shaping filter.

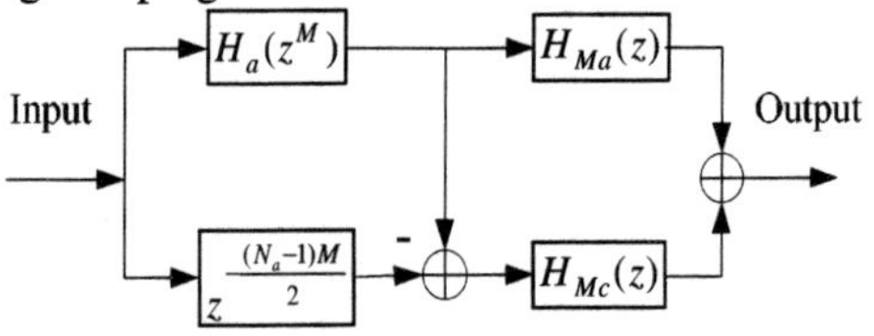

Fig. 1 A basic structure of FRM.

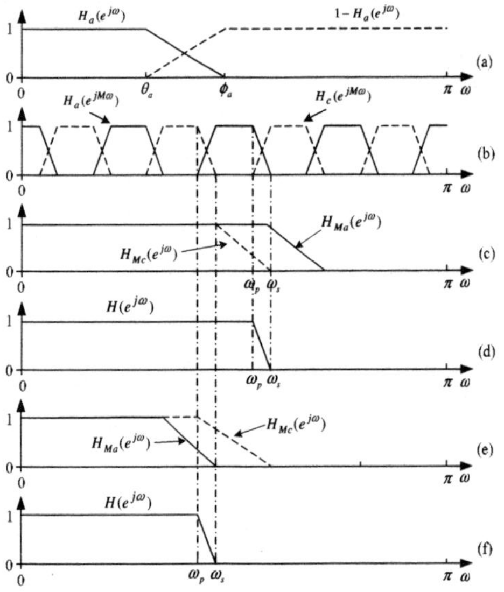

Fig. 2 Frequency responses of each subfilter in FRM.

The paper is organized as follows. In section 2, we analyze the ripple of an FRM filter with even-length bandedge shaping filter and compare it with the original FRM filter. In Section 3, a design procedure is given for FRM filters with even-length bandedge shaping filter. Design examples are given in Section 4.

2. RIPPLE ANALYSIS OF FRM USING EVEN-LENGTH BANDEDGE SHAPING FILTER

It is a well known fact that the passband gain of a lowpass even-length symmetric FIR filter can be either 1 or -1. When such a filter is used as bandedge shaping filter in an FRM structure, it will produce a complementary bandedge shaping filter whose gain in the passband will be 1 or 2. The z-transform transfer function of an FRM filter is given by:

$$H(z)=H_a(z^M)H_{Ma}(z)+\left[z^{-\frac{(N_a-1)M}{2}}-H_a(z^M)\right]H_{Mc}(z) \quad (1)$$

Note that the interpolation factor M should be an even number if an even-length bandedge shaping filter is used. The frequency response of the overall system can be written as:

$$H(e^{j\omega})=H_a(e^{jM\omega})H_{Ma}(e^{j\omega})+[1-H_a(e^{jM\omega})]H_{Mc}(e^{j\omega}) \quad (2)$$

The frequency responses of each subfilter in an FRM filter with even-length bandedge shaping filter are shown in Fig.3. It is clear that there are total 3 transition bands that can be used to form the transition band of the overall filter as indicated in Figs. 3c, 3e, and 3g, respectively. The cases show in Figs. 3c and 3e correspond to Case A and B, respectively, in the original FRM approach. The case in Fig. 3g exists only under even-length bandedge shaping filter approach. Let's denote it as Case C. This case seems to be not practical as it requires a masking filter having the same transition bandwidth as the overall filter, as shown in Fig. 3f. It is shown in Section 4 that Case C is usable if a proper optimization scheme is utilized in optimize all subfilters jointly.

To design the FRM filter based on the even-length bandedge shaping filter, it is interested to know how the overall filter ripple is affected under such an approach. Compared with the original FRM, the difference of the proposed approach occurs in the protuberant shape of $H_c(z^M)$ in frequency regions II and III as indication on Fig. 3a. We shall analyze the ripple effect in frequency regions II and III for Case A. Same procedures can be applied to Case B and C. For other frequency regions, the analysis can be found in [1].

Let $G(\omega)$ and $\delta(\omega)$ be the desired value and deviation of the overall filter, $G_p(\omega)$ and $\delta_p(\omega)$ be the desired value and deviation of the interpolated bandedge shaping filter. $G_{ma}(\omega)$, $G_{mc}(\omega)$, $\delta_{ma}(\omega)$, and $\delta_{mc}(\omega)$ are the desired value and deviation of $H_{Ma}(z)$ and $H_{Mc}(z)$, respectively. In frequency region II, according to (2), we have

$$\begin{aligned}G(\omega)+\delta(\omega)&=\left[G_p(\omega)+\delta_p(\omega)\right][G_{ma}(\omega)+\delta_{ma}(\omega)]\\&+[1-G_p(\omega)-\delta_p(\omega)][G_{mc}(\omega)+\delta_{mc}(\omega)]\end{aligned} \quad (3)$$

In this region $G(\omega)=G_{ma}(\omega)=G_{mc}(\omega)=1$. Substituting the condition into (3) and ignoring the second-order terms, we have

$$\delta(\omega)=G_p(\omega)\delta_{ma}(\omega)+[1-G_p(\omega)]\delta_{mc}(\omega) \quad (4)$$

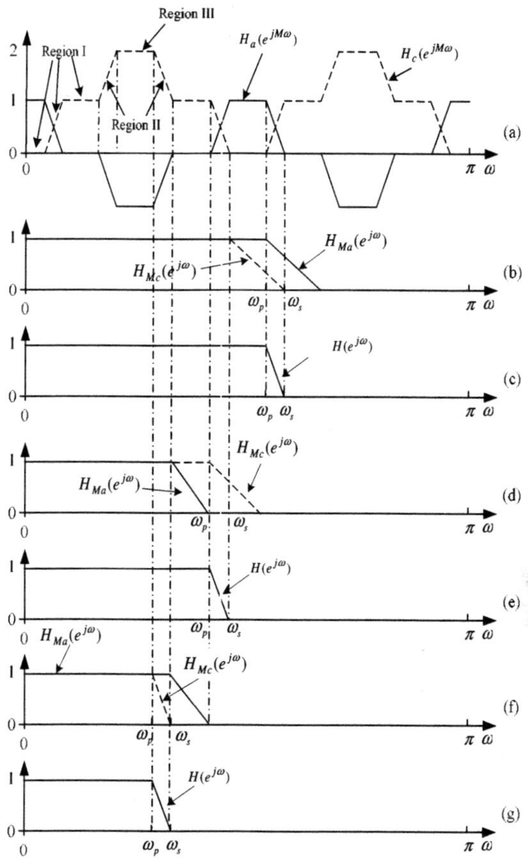

Fig. 3 Frequency response of each subfilter in FRM with even-length bandedge shaping filter.

As $-1< G_p(\omega) <0$ in this region, (4) can be expressed as

$$|\delta_{mc}(\omega)|\le|\delta(\omega)|\le|\delta_{ma}(\omega)|+2|\delta_{mc}(\omega)| \quad (5)$$

Similarly, in frequency region III, we can find the ripple of the overall filter by substituting $G_p(\omega)=-1$ into (4),

$$\delta(\omega)=2\delta_{mc}(\omega)-\delta_{ma}(\omega) \quad (6)$$

or in the range

$$\big||\delta_{ma}(\omega)|-2|\delta_{mc}(\omega)|\big|\le|\delta(\omega)|\le|\delta_{ma}(\omega)|+2|\delta_{mc}(\omega)| \quad (7)$$

From (5) and (7), we can see that the ripples in frequency regions II and III can be much larger than the ripples of any one of masking filters. If traditional design methods such as linear programming or Remez method are used to design each subfilter iteratively as in [1] or [11], the ripples of two masking filters must be at most half of given specifications for overall filter. The reduced ripple of masking filters result the increase of complexity of masking filters. From the above analysis, we can conclude that the even-length bandedge shaping filter is not good choice for FRM if an iterative synthesis approach is used.

However, the jointly optimization techniques introduced in [7-10] prompt the possible ways to minimize the ripple in regions II and III while keep the complexities of all subfilters in check. It is shown in the next two sections that the even-length based FRM filter can produce a design comparable to the original FRM if Sequential Quadratic Programming (SQP) is used.

3. THE DESIGN METHOD

To compensate the ripple in the regions II and III, a design method that is capable of optimizing all subfilters jointly is highly desired. Suppose the desired frequency response is $H_d(\omega)$, which equals to unity in the passband, and zero in the stopband, we can define the error function as

$$E(\omega) = \left| H(e^{j\omega}) - H_d(\omega) \right| \quad (8)$$

$H(e^{j\omega})$ is the frequency response given in (2). The optimization is carried out over a set of dense frequency grid $\omega_k \in \Omega = [0\ \omega_p] \cup [\omega_s\ 0.5]$ such that all values of $E(\omega_k)$ satisfy the given specifications over Ω. This is a multi-objective optimization problem. If a weighting function *weight*(ω_k) is introduced, the multi-objective problem is transferred into a single objective optimization problem. The new single objective function is

$$f = \sum_{\omega_k \in \Omega} weight(\omega_k) E(\omega_k) \quad (9)$$

The optimization goal is to minimize f in (9). The initial *weight*(ω_k) equals to unity in the passband, and equals to the ratio of passband ripple and stopband ripple, δ_p/δ_s, in the stopband. If a proper weighting function is selected, an equal ripple solution can be achieved. Although methods in [7-10] are possible for such a task, we use SQP function, *fgoalattain*, available in Matlab [12-14]. The design can be done in following steps.

Step 1: Determine the initial solution as described in [1],[2] or [11].
Step 2: Call *fgoalattain* to determine parameters to minimize the objective function in (9).
Step 3: If passband ripple or stopband ripple is larger than the given specification, update the weight vector by using the algorithm in [15].
Step 4: Repeat step 2 until an equal ripple solution is reached or the change in the solution is smaller than predefined tolerance ε, which was set to 10^{-8}.

4. DESIGN EXAMPLES

Let's illustrate the proposed method by designing a lowpass filter taking from [1] with normalized passband and stopband edges at 0.3, and 0.305, respectively. The passband and stopband ripples are 0.01. The minimum filter length is 389 for a minimax design. To satisfy the given specification, the filter designed with odd-length bandedge shaping filter requires an interpolation factor of 6, and the lengths of subfilters are N_a=67, N_{Ma}=21, and N_{Mc}=13. 52 multipliers are needed.

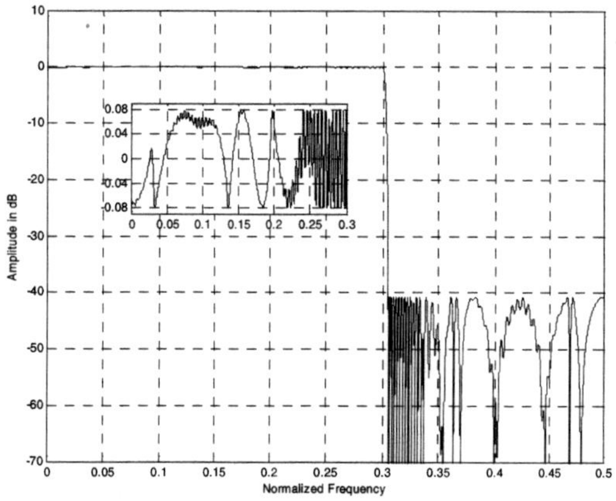

Fig. 4 Frequency response of overall filter.

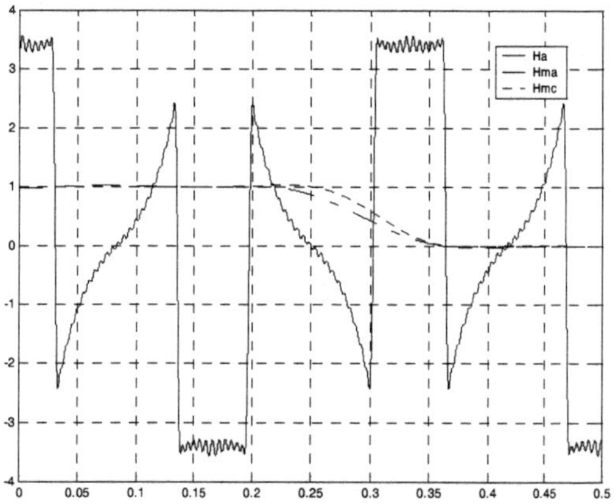

Fig. 5 Frequency responses of each subfilter.

If even-length bandedge shaping filter is used, the lengths of subfilters are N_a=68, N_{Ma}=13, N_{Mc}=19. 51 multipliers are needed, which is slightly better than the original FRM. The frequency responses of the overall filter and all subfilters for an even-length based FRM are shown in Figs. 4 and 5.

To illustrate the Case C design, we design a lowpass filter with normalized passband and stopband edges at 0.17 and 0.175, respectively. The passband and stopband ripples are 0.01. The design requires an interpolation factor of 8, the lengths of subfilters are N_a=52, N_{Ma}=26, N_{Mc}=20. 49 multipliers are needed. Using the original FRM, the subfilter lengths are N_a=49, N_{Ma}=28, N_{Mc}=12, which is 4 multipliers less than the one with even-length design. This shows that for Case C one of the masking filters is not necessary as sharp as the overall filter when using SQP to optimize all subfilters. The frequency responses of the overall filter and all subfilters are shown in Figs.6 and 7, respectively.

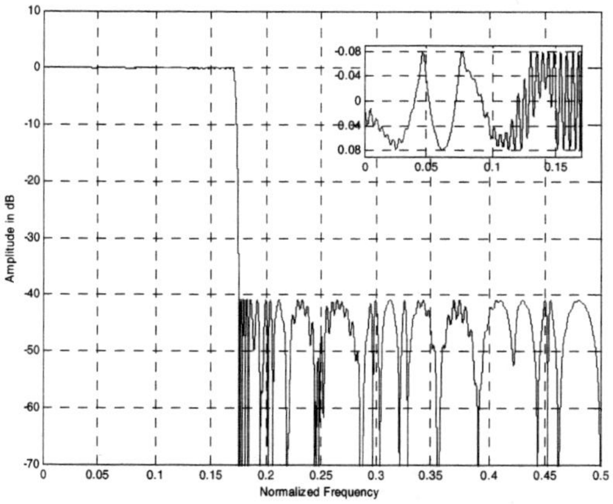

Fig. 6 Overall system frequency response

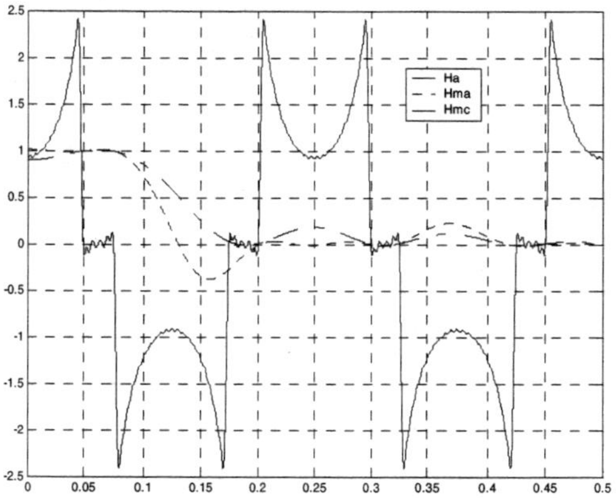

Fig 7 Frequency responses of all subfilters.

5. CONCLUSION

We have proposed a method to synthesis FRM filters based on the even-length bandedge shaping filter. The difficulties faced by using the even length filter as the prototype in FRM are analyzed. Solutions are provided to address the issues in even-length based FRM filter. It seems that the SQP optimization technique is a good candidate to jointly optimize all subfilters in the proposed method. The design examples show that even-length based FRM achieves a similar performance as the original FRM.

6. ACKNOWLEDGEMENT

This work was supported by the National University of Singapore Academic Research Fund R263-000-160-112.

7. REFERENCE

[1]. Y. C. Lim, "Frequency-response masking approach for the synthesis of sharp linear phase digital filters," *IEEE Trans. Circuits Syst.*, vol. 33, pp. 357-364, Apr. 1986.

[2]. Y. C. Lim and Y. Lian, "The optimal design of one- and two- dimensional FIR filters using the frequency response masking technique," *IEEE Trans. Circuits Syst., Part 2.*, vol. 40, pp. 88-95, Feb. 1993.

[3]. Y. C. Lim and Y. Lian, "Frequency-response masking approach for digital filter design: complexity reduction via masking filter factorization," *IEEE Trans. Circuits Syst. II*, vol. 41, pp. 518-525, Aug. 1994.

[4]. Y. Lian, L. Zhang, and C.C. Ko, "An improved frequency response masking approach for designing sharp FIR filters," *Signal Processing*, 81, pp. 2573-2581. 2001.

[5]. Y. Lian, and C.Z. Yang, "Complexity reduction by decoupling the masking filters from bandedge shaping filter in frequency-response masking technique," *Circuits, Systems and Signal Processing,* vol.22, No.2, pp.115-135, Mar. 2003.

[6]. Y. Lian, "Complexity reduction for frequency-response masking based FIR filters via prefilter-equalizer technique," *Circuits, Systems and Signal Processing*, vol. 22, No.2, pp.137-155, Mar. 2003.

[7]. T. Saramäki, and H. Johansson, "Optimization of FIR filters using the frequency-response masking approach," in *Proc. 2001 ISCAS,* vol. II, pp.177-180, May 2001.

[8]. Y. J. Yu, and Y. C. Lim, "FRM based FIR filter design – the WLS approach," in *Proc. 2002, ISCAS*, vol III, pp.221-224, May 2002.

[9]. W. S. Lu, T. Hinamoto, "Optimal design of frequency-response-masking filters using semidefinite programming," *IEEE Trans. Circuits and Syst.*, vol. 50, pp. 557-568, Apr. 2003.

[10]. W. S. Lu, and T. Hinamoto, "Optimal design of FIR frequency-response-masking filters using second-order cone programming," *Proc. ISCAS'2003,* vol. 3, pp.878-881, Bangkok, May 2003.

[11]. T. Saramäki, and Y.C. Lim, "Use of Remez algorithm for designing FRM based FIR filters," *Circuits, Systems and Signal Processing*, vol.22, No.2, pp.77-97, Mar. 2003.

[12]. S. P. Han, "A global convergent method for nonlinear programming," *J. Optimization Theory and Applications*, vol. 22 p297, 1977.

[13]. M. J. D. Powell, "The convergence of variable metric methods for nonlinear constrained optimization calculations," *Nonlinear Programming 3*, (O.L. Mangasarian, R. R. Meyer, and S. M. Robinson, eds.), Academic Press, 1978.

[14]. T. Coleman, M. A. Branch, and A. Grace, *User's Guide of Optimization Toolbox.* Natick, MA: The MathWorks, Inc, Jan. 1999.

[15]. Y. C. Lim, J. H. Lee, C. K. Chen, and R. H. Yang, "A weighted least squares algorithm for quasi-equiripple FIR and IIR digital filter design," *IEEE Trans. Signal Processing*, vol. 40, pp.551-558, Mar. 1992.

AN EFFICIENT ALGORITHM FOR THE OPTIMIZATION OF FIR FILTERS SYNTHESIZED USING THE MULTISTAGE FREQUENCY-RESPONSE MASKING APPROACH

Juha Yli-Kaakinen and Tapio Saramäki

Institute of Signal Processing
Tampere University of Technology
P. O. Box 553, FIN-33101 Tampere, Finland

Ya Jun Yu

Dept. of Electrical and Computer Engineering
National University of Singapore
Singapore 119260

ABSTRACT

A very efficient technique to drastically reduce the number of multipliers and adders in narrow transition-band linear-phase finite-impulse response digital filters is to use the one-stage or multistage frequency-response masking (FRM) approach originally introduced by Lim. In the original synthesis techniques developed by Lim and Lian, the subfilters in the overall implementation are separately designed. As shown earlier by the first two authors of this contribution as well as Johansson, the arithmetic complexity in the one-stage FRM filter designs can be considerably reduced by using the following two-stage technique for simultaneously optimizing all the subfilters. First, a suboptimal solution is found by using a simple design scheme. Second, this solution is used as a start-up solution for further optimization being carried out by an efficient unconstrained nonlinear optimization algorithm. This paper exploits this approach for synthesizing multistage FRM filters. An example taken from the literature illustrates that both the number of multipliers and the number of adders for the resulting optimized FRM filters are approximately 70 percent compared with those of the filters synthesized using the original multistage FRM filter design schemes.

1. INTRODUCTION

ONE of the most efficient techniques for synthesizing narrow transition-band linear-phase finite-impulse response (FIR) digital filters with a drastically reduced number of multipliers and adders compared with the conventional direct-form implementation is the one-stage or multistage frequency-response masking (FRM) approach [1]–[4]. The price to be paid for these reductions is a slight increase in the overall filter order. A drawback of the original multistage FRM synthesis techniques [1]–[4] is that the subfilters have been separately designed. In [5], it has been shown by the first two authors of this paper as well as by Johansson how the arithmetic complexity can be further decreased in the one-stage FRM approach by simultaneously designing the subfilters. The purpose of this contribution is to introduce a two-step technique for reducing the overall arithmetic complexity in multistage FRM approach by simultaneously designing all the subfilters. First, a straightforward design scheme is used for estimating the subfilter orders and other design parameters as well as for generating a start-up solution for further optimization. Second, this solution is improved with the aid of an efficient iterative unconstrained nonlinear optimization algorithm utilizing properly selected nonuniformly spaced frequency grid points.

An example taken from the literature shows that the arithmetic complexity of the resulting optimized FRM filters is approximately 70 percent compared with those of the filters synthesized using the original multistage FRM design schemes. In addition, this example shows that the estimated subfilter orders are in the close vicinity of the actual values minimizing the arithmetic complexity of the overall filter. This fact considerably reduces the computational workload to arrive at the solution with the lowest arithmetic complexity.

This work was supported by the Academy of Finland, project No. 44876 (Finnish Centre of Excellence Program (2000–2005)). Juha Yli-Kaakinen was also financed by a postdoctoral research grant from the Academy of Finland, project No. 75492. This work was partially carried out while Ya Jun Yu was visiting the Institute of Signal Processing, Tampere University of Technology, Finland.

2. MULTISTAGE FRM APPROACH

This section briefly reviews the basic principle of using the multistage FRM approach for designing linear-phase FIR filters. In addition, it is discussed how to accurately estimate the design parameters minimizing the arithmetic complexity of the multistage FRM filters resulting when using the proposed two-step design technique to be described in Section 3.

2.1. Filter transfer function

In the multistage FRM approach with R stages, the linear-phase FIR filter transfer function $H(z)$ is recursively constructed as follows:

$$H(z) \equiv F^{(0)}(z) = F^{(1)}(z^{L_1})G_1^{(1)}(z) + \left[z^{-L_1 N_F^{(1)}/2} - F^{(1)}(z^{L_1})\right]G_2^{(1)}(z) \quad \text{(1a)}$$

$$F^{(1)}(z) = F^{(2)}(z^{L_2})G_1^{(2)}(z) + \left[z^{-L_2 N_F^{(2)}/2} - F^{(2)}(z^{L_2})\right]G_2^{(2)}(z) \quad \text{(1b)}$$

$$\vdots \qquad \vdots \qquad \vdots$$

$$F^{(R-1)}(z) = F^{(R)}(z^{L_R})G_1^{(R)}(z) + \left[z^{-L_R N_F^{(R)}/2} - F^{(R)}(z^{L_R})\right]G_2^{(R)}(z) \quad \text{(1c)}$$

where

$$F(z^{L_R}) = \sum_{n=0}^{N_F^{(R)}} f^{(R)}(n)z^{-nL_R}, \quad \text{and} \quad G_k^{(r)}(z) = \sum_{n=0}^{N_k^{(r)}} g_k^{(r)}(n)z^{-n}, \quad \text{(1d)}$$

for $k = 1, 2$ and $r = 1, 2, \ldots, R$ are the filters to be designed. Here, the impulse-response coefficients $f^{(R)}(n)$, $g_1^{(r)}(n)$, and $g_2^{(r)}(n)$ for $r = 1, 2, \ldots, R$ possess an even symmetry. It can be shown that in order to arrive at a desired overall solution with a linear-phase response, the $N_1^{(r)}$'s and $N_2^{(r)}$'s for $r = 2, 3, \ldots, R$ as well as $N_F^{(R)}$ have to be even. Furthermore, $N_1^{(1)}$ and $N_2^{(1)}$ should be both either even or odd. Finally, it should be pointed out that if $N_1^{(r)}$ and $N_2^{(r)}$ are different, then, instead of $G_2^{(r)}(z)$ $[G_1^{(r)}(z)]$, $z^{-(N_1^{(r)}-N_2^{(r)})/2}G_2^{(r)}(z)$ $[z^{-(N_2^{(r)}-N_1^{(r)})/2}G_1^{(r)}(z)]$ has to be used for $N_1^{(r)} > N_2^{(r)}$ $[N_1^{(r)} < N_2^{(r)}]$. These selections guarantee that the delays of both of the terms of $F^{(r)}(z)$'s are equal, thereby ensuring that the overall transfer function $H(z)$ is a linear-phase filter with an even symmetry.

An efficient implementation for $H(z)$ using the minimum number of delay elements has been given in [3], [4]. Figure 1 shows this implementation for the three-stage case ($R = 3$). In this figure, $\widehat{L}_0 = 1$, $\widehat{L}_1 = L_1$, $\widehat{L}_2 = L_1 L_2$, and $\widehat{L}_3 = L_1 L_2 L_3$. How to determine the remaining parameters can be found in [3], [4].

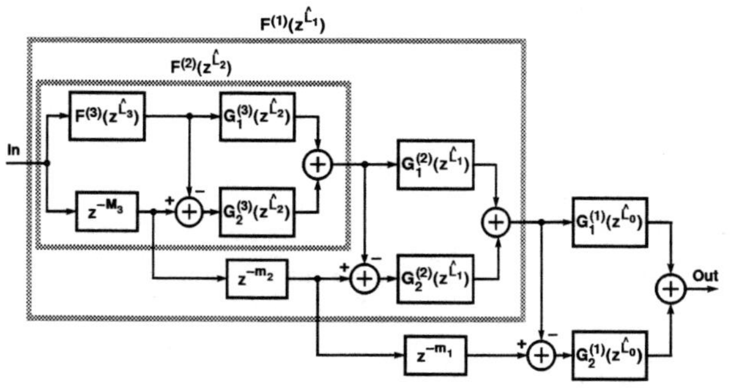

Fig. 1 An efficient implementation for a filter synthesized using the three-stage FRM approach.

2.2. Estimation of the design parameters for the proposed synthesis scheme

Given R, the number of stages in the FRM approach, the passband and stopband edges of $H(z)$, denoted by ω_p and ω_s, respectively, and the passband and stopband ripples, denoted by δ_p and δ_s, respectively, the problem is to determine the L_r's and the subfilter orders to meet the specifications and to minimize the number of multipliers as given by[1]

$$N_F^{(R)}/2+1+\sum_{k=1}^{2}\Big[\lfloor (N_k^{(1)}+2)/2\rfloor+\sum_{r=2}^{R}(N_k^{(r)}/2+1)\Big]. \quad (2)$$

Here, it is assumed that the coefficient symmetries are exploited in the overall implementation.

In the original synthesis schemes and in the proposed one to be considered in the next section, after fixing the L_{r+1}'s, for each $F^{(r)}(z)$ for $r=0,1,\ldots,R-1$, it is crucial to check whether it is a so-called Case A or Case B design (see, e.g., [3], [4]) and to determine the parameters l_{r+1}, θ_{r+1}, and ϕ_{r+1} as follows:

Step 1: Set $r=1$.

Step 2: For the given value of L_r, $F^{(r-1)}(z)$ is either a Case A design or a Case B design (not both). A Case A design is applicable if l_r, θ_r, and ϕ_r are determined as

$$l_r=\lfloor L_r\omega_p/(2\pi)\rfloor,\ \theta_r=L_r\omega_p-2l_r\pi,\ \phi_r=L_r\omega_s-2l_r\pi, \quad (3a)$$

and the resulting θ_r and ϕ_r satisfy $0\le\theta_r<\phi_r\le\pi$. A Case B design, in turn, can be used if l_r, θ_r, and ϕ_r are determined as[2]

$$l_r=\lceil L_r\omega_s/(2\pi)\rceil,\ \theta_r=2l_r\pi-L_r\omega_s,\ \phi_r=2l_r\pi-L_r\omega_p, \quad (3b)$$

and the resulting θ_r and ϕ_r satisfy $0\le\theta_r<\phi_r\le\pi$.

Step 3: If $r=R$, then stop. Otherwise, set $r=r+1$, $\omega_p=\theta_{r-1}$, $\omega_s=\phi_{r-1}$, and go to Step 2.

For a Case A design, the passband and stopband edges of $G_1^{(r)}(z)$ and $G_2^{(r)}(z)$ are given by

$$\omega_p^{(G_1^{(r)})}=[2l_r\pi+\theta_r]/L_r,\quad \omega_s^{(G_1^{(r)})}=[2(l_r+1)\pi-\phi_r]/L_r. \quad (4a)$$

$$\omega_p^{(G_2^{(r)})}=[2l_r\pi-\theta_r]/L_r,\quad \omega_s^{(G_2^{(r)})}=[2l_r\pi_r+\phi_r]/L_r. \quad (4b)$$

For a Case B design, the corresponding edges are given by

$$\omega_p^{(G_1^{(r)})}=[2(l_r-1)\pi+\phi_r]/L_r,\quad \omega_s^{(G_1^{(r)})}=[2l_r\pi-\theta_r]/L_r. \quad (5a)$$

$$\omega_p^{(G_2^{(r)})}=[2l_r\pi-\phi_r]/L_r,\quad \omega_s^{(G_2^{(r)})}=[2l_r\pi+\theta_r]/L_r. \quad (5b)$$

For $F^{(R)}(z)$, in turn, the passband and stopband edges are located at $\omega=\theta_R$ and $\omega=\phi_R$, respectively.

There exist the following important observations concerning the original and the proposed design schemes. First, the order of $F^{(R)}(z)$ is very close to the order of an FIR filter having the above-mentioned passband and stopband edges and the passband and stopband ripples of the overall design. The same is true for the $G_1^{(r)}(z)$'s and $G_2^{(r)}(z)$'s in the original design schemes, whereas the corresponding orders for the proposed technique are only 60 percent of those of the original synthesis schemes. Therefore, the orders of all the subfilters can be accurately estimated by using the formula given in [6]. Second, the number of multipliers, as given by Eq. (2), typically achieves the minimum value when the transition bandwidths of each pair $G_1^{(r)}(z)$ and $G_2^{(r)}(z)$ for $r=1,2,\ldots,R$ are approximately the same.

There are the following consequences of the above observations. First, if the transition bandwidths of each pair $G_1^{(r)}(z)$ and $G_2^{(r)}(z)$ are the same, then

$$L_1=L_2=\cdots=L_R=\Big[2\beta(\omega_s-\omega_p)/\pi\Big]^{\frac{-1}{R+1}}, \quad (6)$$

where $\beta=1$ for the original design techniques and $\beta=0.6$ for the proposed design scheme, minimizes the arithmetic complexity. Second, because the orders of all the subfilters can be estimated very accurately for all selections of the L_r's, it is beneficial to find, in terms of the estimated orders, the values of L_r's in the vicinity of the above values to minimize the number of multipliers.[3] Third, most importantly, as will be seen in connection with examples in Section 4, the resulting estimated minimum subfilter orders $N_1^{(r)}$'s and $N_2^{(r)}$'s and $N_F^{(R)}$ as well as the estimated values of L_r's are in the very close vicinity of the actual values minimizing the arithmetic complexity of the overall FRM filter, that is, the number of multipliers and adders required to meet the given overall criteria. This fact considerably reduces the computational workload to arrive at the solution with the lowest arithmetic complexity.

3. PROPOSED TWO-STEP DESIGN SCHEME FOR MULTISTAGE FRM APPROACH

This section describes the proposed two-step technique for simultaneously designing all the subfilters in the multistage FRM filter.

3.1. Algorithm for finding an initial filter

Given R and the specifications of $H(z)$, the initial values for the $G_1^{(r)}(z)$'s and $G_2^{(r)}(z)$'s as well as for $F^{(R)}(z)$ can be found in the following steps:

Step 1: Determine the L_r's and the orders of the subfilters, in terms of the estimation formulas, as described in Subsection 2.2. In addition, determine their passband and stopband edges as well as their orders according to the discussion of Subsection 2.2.

Step 2: Design the subfilters with the estimated orders using the Remez multiple exchange algorithm so that the weights in the passband and stopband regions are $1/\delta_p$ and $1/\delta_s$, respectively.

A more sophisticated method for designing initial filters for a two-stage FRM approach has been proposed in [7].

3.2. Further optimization

The initial solution obtained by using the above simple algorithm can be improved with the aid of an efficient unconstrained nonlinear optimization technique. The optimization problem is to find the adjustable parameter vector Φ as given by

$$\Phi=\Big[g_1^{(1)}(0),\ldots,g_1^{(1)}\lfloor N_1^{(1)}/2\rfloor,g_2^{(1)}(0),\ldots,g_2^{(1)}\lfloor N_2^{(1)}/2\rfloor,\ldots,$$
$$g_1^{(R)}(0),\ldots,g_1^{(R)}(N_1^{(R)}/2),g_2^{(R)}(0),\ldots,g_2^{(R)}(N_2^{(R)}/2),$$
$$f^{(R)}(0),\ldots,f^{(R)}(N_F^{(R)}/2)\Big] \quad (7a)$$

to minimize

$$\max_{\omega\in[0,\,\omega_p]\cup[\omega_s,\,\pi]}|E(\Phi,\,\omega)|, \quad (7b)$$

[1] $\lfloor x\rfloor$ stands for the largest integer that is smaller than or equal to x.

[2] $\lceil x\rceil$ stands for the smallest integer that is larger than or equal to x.

[3] For this purpose, a MATLAB routine has been generated.

where

$$E(\Phi,\,\omega) = W(\omega)[H(\Phi,\,\omega) - D(\omega)], \tag{7c}$$

$$D(\omega) = \begin{cases} 1, & \omega \in [0,\,\omega_p] \\ 0, & \omega \in [\omega_s,\,\pi], \end{cases} \quad W(\omega) = \begin{cases} 1/\delta_p, & \omega \in [0,\,\omega_p] \\ 1/\delta_s, & \omega \in [\omega_s,\,\pi], \end{cases} \tag{7d}$$

and $H(\Phi,\,\omega)$ is the zero-phase frequency response of the overall transfer function $H(z)$.

In order to solve this problem, the passband and stopband regions are discretized into the frequency points $\omega_j \in [0,\,\omega_p]$ for $j = 1, 2, \ldots, J_p$ and $\omega_j \in [\omega_s,\,\pi]$ for $j = J_p + 1, J_p + 2, \ldots, J_p + J_s$. The resulting discrete minimax problem is to find Φ to minimize

$$\epsilon = \max_{1 \le j \le J_p + J_s} |E(\Phi,\,\omega_j)|. \tag{8}$$

This problem can be solved using an effective unconstrained nonlinear optimization algorithm. For this purpose, the function `fminimax` from the optimization toolbox provided by MathWorks, Inc. [8] has been used. When using this function, the user has to provide a function which evaluates the objective function, that is, the error function to be minimized at the given frequency points as well as the gradients of the objective function with respect to the adjustable parameters at these points. The solution meeting the given criteria is obtained when ϵ becomes less than or equal to unity.

In order to arrive at the solution meeting the given criteria with the minimum number of multipliers, the above two algorithms are carried out for various order combinations in the close vicinity of the estimated orders. Furthermore, if there exist more sets of the L_r's resulting practically the same minimum number of multipliers, all these sets are tried and, finally, the set resulting in the minimized complexity is selected.

3.3. Practical filter synthesis

In order to achieve good results with the nonlinear optimization technique of Subsection 3.2, the frequency spacing of the error function $E(\Phi,\,\omega)$ must be dense enough. Otherwise, the resulting error function may have spikes in the intervals between the sampling points after the minimization. This problem can be overcame by using fairly large number of sampling points. It has been experimentally observed that the proper selection for this number, denoted for later use by I, is of the order of 3–6 times the order of $F^{(R)}(z)$ times the product of the L_r's. The more grid points are used, the more accurate is the final solution. However, at the same time the convergence rate of the optimization algorithm becomes slower.

It has turned out that the convergence to a very accurate solution can be significantly speeded up by performing the optimization in the following steps:

Step 1: Set $k = 1$ and select a dense set of uniformly spaced grid points $\Omega_{\text{dense}} = \{\omega_1, \omega_2, \ldots, \omega_I\}$ in $[0,\,\omega_p] \cup [\omega_s,\,\pi]$ as well as a sparse initial set of uniformly spaced grid points $\bar{\Omega}_0 = \{\bar{\omega}_1, \bar{\omega}_2, \ldots, \bar{\omega}_J\}$ with $J = \lfloor I/100 \rfloor$ in $[0,\,\omega_p] \cup [\omega_s,\,\pi]$.

Step 2: Find in Ω_{dense} the abscissas of the local maxima of $|E(\Phi,\,\omega_j)|$, that is, those ω_j's in Ω_{dense} for which

$$|E(\Phi,\,\omega_{j-1})| < |E(\Phi,\,\omega_j)| > |E(\Phi,\,\omega_{j+1})|, \tag{9}$$

where $|E(\Phi, \omega_0)| = |E(\Phi, \omega_{J+1})| = 0$. Store these abscissas of the extrema into Ω_{new}. Set $\bar{\Omega}_k = \bar{\Omega}_{k-1} \cup \Omega_{\text{new}}$.

Step 3: Solve the discrete minimax problem of Eq. (8) using $\bar{\Omega}_k$ as the set of the frequency points.

Step 4: If $|\bar{\Omega}_{k-1}| = |\bar{\Omega}_k|$, then stop.[4] Otherwise, set $k = k + 1$ and go to Step 2.

In the above algorithm, the extrema of $|E(\Phi,\,\omega)|$ at Step 2 are located by evaluating $|E(\Phi,\,\omega)|$ over a dense set of frequencies spanning the approximation regions. Then, these new extremal points are included into the frequency points of the previous iteration and the optimization is performed using the updated frequency points. This process is continued until the number of frequency points remain the same between two successive optimizations.

[4] Here, $|\Omega|$ denotes the size of the set Ω.

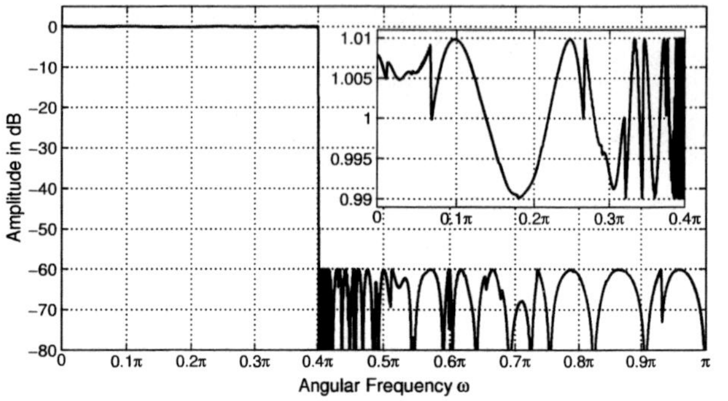

Fig. 2 Magnitude response for the best proposed two-stage overall filter for $L_1 = L_2 = 6$ and $\sum_{n=0}^{N_F} f(n) = 2.5$.

Table 1 Specifications for the separate stages as well as the estimated orders of the subfilters for the three-stage FRM filter.

	r	L_r	Case	l_r	θ_r	ϕ_r	$N_F^{(R)}$	$N_1^{(r)}$	$N_2^{(r)}$
$F^{(0)}(z) \equiv H(z)$	1	4	B	1	0.392π	0.400π	–	10	15
$F^{(1)}(z)$	2	4	B	1	0.400π	0.432π	–	10	14
$F^{(2)}(z)$	3	4	B	1	0.272π	0.400π	39	8	18

4. NUMERICAL EXAMPLE

Consider the specifications [3]–[5]: $\omega_p = 0.4\pi$, $\omega_s = 0.402\pi$, $\delta_p = 0.01$, and $\delta_s = 0.001$. For the conventional direct-form FIR filter, the minimum order to meet the given criteria is 2541, requiring 1271 multipliers and 2541 adders when the coefficient symmetry is exploited.

For the optimized one-stage FRM approach [4], $L = 21$ minimizes the number of multipliers required in the implementation. This filter is a Case A design with $l = 4$, $\theta = 0.4\pi$, and $\phi = 0.442\pi$. The minimum orders for $G_1(z)$, $G_2(z)$, and $F(z)$ to meet the given specifications are $N_1 = 55$, $N_2 = 77$, and $N_F = 122$, respectively. The overall number of multipliers and adders for this design are 129 and 256 [5], respectively, that are ten percent of those required by an equivalent conventional direct-form design (1271 and 2541). The overall filter order is 2639 that is only four percent higher than that of the direct-form design (2541).

For the two-stage FRM approach, the arithmetic complexity of the overall filter is minimized by $L_1 = L_2 = 6$. For these selections, $F^{(0)}(z) \equiv H(z)$ and $F^{(1)}(z)$ are Case A designs ($l = 1$) with $\theta_1 = 0.4\pi$ and $\phi_1 = 0.412\pi$; $\theta_2 = 0.4\pi$ and $\phi_2 = 0.472\pi$, respectively. The best solution resulting when using the proposed synthesis scheme is obtained by $N_F^{(2)} = 70$, $N_1^{(2)} = 14$, $N_2^{(2)} = 22$, $N_1^{(1)} = 17$, and $N_2^{(1)} = 21$. As the measure of goodness the overall number of multipliers, as given by Eq. (2), was used. However, if there exist several solutions requiring the same minimum number of multipliers, then the solution with the shortest overall delay was selected. For the above filter, the number of multipliers and adders are 76 and 148, respectively, that are approximately 60 percent of those of the corresponding one-stage design at the expense of a seven percent increase in the overall filter order (from 2689 to 2673). Furthermore, the number of multipliers and adders are 71 percent of those of the original multistage FRM synthesis schemes [3], [4] and only six percent of those required by the conventional direct-form design.

According to the discussion of Section 2.2, the minimum estimated orders for $F^{(2)}(z)$, $G_1^{(2)}(z)$, $G_2^{(2)}(z)$, $G_1^{(1)}(z)$, and $G_2^{(1)}(z)$ are $N_F^{(2)} = 71$, $N_1^{(2)} = 16$, $N_2^{(2)} = 21$, $N_1^{(1)} = 15$, and $N_2^{(1)} = 22$, respectively. Since the orders of both $G_1^{(1)}(z)$ and $G_2^{(1)}(z)$ should be even or odd and the orders of all $G_1^{(2)}(z)$, $G_2^{(2)}(z)$, and $F^{(2)}(z)$ should be even, it was tried several combinations in the vicinity of the estimated orders.

[5] The two additional adders required to combine the subfilters are included in this figure. In the sequel, the same will be done in the case of multistage FRM filters.

Table 2 Summary of FRM filter designs in the example under consideration.

Method	R	$L_1 \ldots L_R$	$N_1^{(3)}$	$N_2^{(3)}$	$N_1^{(2)}$	$N_2^{(2)}$	$N_1^{(1)}$	$N_2^{(1)}$	$N_F^{(R)}$	N_{mult}	N_{ove}	CPU
Conventional direct-form filter		1	–	–	–	–	–	–	2541	1271	2541	–
Original one-stage FRM approach [3]	1	16	–	–	–	–	70	98	162	168	2690	–
Optimized one-stage FRM approach [5]		21	–	–	–	–	55	77	122	129	2639	–
Original two-stage FRM approach [3]			–	–	28	36	26	40	74	107	2920	–
Iterative two-stage FRM approach [7]	2	6	–	–	16	20	17	23	74	79	2807	–
Proposed			–	–	14	22	17	21	70	76	2673	9 min
Original three-stage FRM approach [3]	3	4	16	32	18	24	16	28	40	94	3196	–
Proposed			6	16	8	16	11	15	38	61	2763	4 min

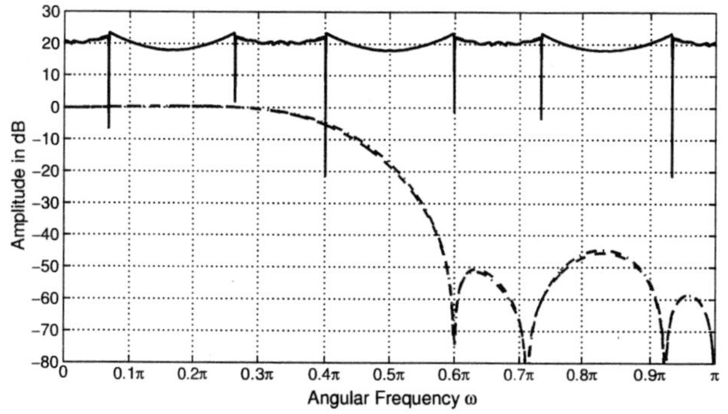

Fig. 3 Magnitude responses for $G_1^{(1)}(z)$ (dashed-line), $G_2^{(1)}(z)$ (dot-dashed line) and $F^{(1)}(z^6)$ (solid line) for the best proposed two-stage filter for $L_1 = L_2 = 6$ and $\sum_{n=0}^{N_F} f(n) = 2.5$.

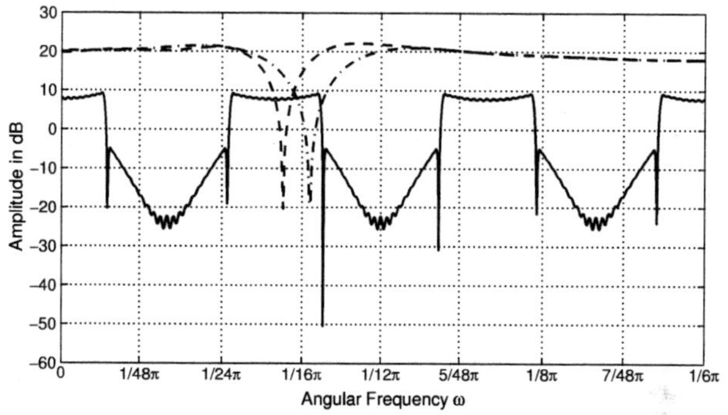

Fig. 4 Magnitude responses for $G_1^{(2)}(z^6)$ (dashed-line), $G_2^{(2)}(z^6)$ (dot-dashed line) and $F^{(2)}(z^{36})$ (solid line) for the best proposed two-stage filter for $L_1 = L_2 = 6$ and $\sum_{n=0}^{N_F} f(n) = 2.5$.

For the optimal design, the orders of $G_1^{(1)}(z)$ and $G_1^{(2)}(z)$ differ only by two and the orders of $G_2^{(1)}(z)$, $G_2^{(2)}(z)$, and $F^{(2)}(z)$ only by one from the estimated ones. For this design, $\delta_p = 10\delta_s = 0.009\,79$ when using the limitation $\sum_{n=0}^{N_F} f(n) = 2.5$. With this limitation the maximum amplitude value out of the amplitude responses for the five subfilters is minimized. The magnitude response as well as the passband details for the overall filter are shown in Fig. 2. Figures 3 and 4 show the magnitude response for $G_1^{(1)}(z)$, $G_2^{(1)}(z)$, and $F^{(1)}(z^6)$ as well as for $G_1^{(2)}(z^6)$, $G_2^{(2)}(z^6)$, and $F^{(2)}(z^{36})$, respectively. It is interesting to observe from Fig. 3 the similarity of the amplitude responses of $G_1^{(1)}(z)$ and $G_2^{(1)}(z)$.

There are the following two important properties of the proposed FRM filters to be pointed out. First, the subfilter orders estimated according to the discussion of Section 2.2 are very close to the orders minimizing the arithmetic complexity, implying that the two-step optimization technique described in Section 3 has to be carried out a very few times. Second, there are numerous solutions giving approximately the same result. These solutions differ from each other in the sense that independent of the maximum value of the zero-phase frequency response of $F^{(R)}(z)$, practically the same passband and stopband ripples for the overall FRM filter are achieved. Future work is devoted to studying this property in more detail.

For a three-stage design, the overall number of multipliers is minimized by $L_1 = L_2 = L_3 = 4$ (the value obtained using the procedure of Subsection 2.2). Table 1 in the previous page shows the specifications for $F^{(0)}(z)$, $F^{(1)}(z)$, and $F^{(2)}(z)$ as well as the estimated orders $N_1^{(r)}$ and $N_2^{(r)}$ for $r = 1, 2, 3$ and $N_F^{(3)}$. The best solution resulting when using the proposed synthesis scheme is obtained by $N_F^{(3)} = 38$, $N_1^{(3)} = 6$, $N_2^{(3)} = 16$, $N_1^{(2)} = 8$, $N_2^{(2)} = 16$, $N_1^{(1)} = 11$, and $N_2^{(1)} = 15$. For this filter, the number of multipliers and adders are 61 and 116, respectively, that are 65 percent of those of the original multistage FRM filter synthesis techniques [3], [4] and less than five percent of those required by the conventional direct-form design.

Some of the characteristics for the filters designed in this section using various algorithms are summarized in Table 2. In this table, N_{mult} denotes the number of multipliers required to implement the overall filter, N_{ove} denotes the order of the overall filter, and CPU denotes the time required to achieve the optimum solution from the initial solution generated using the algorithm of Subsection 3.1 with $I = 10^4$ on a 500 MHz DEC DS20 AlphaStation running MATLAB 6.5.

REFERENCES

[1] Y. C. Lim, "Frequency-response masking approach for the synthesis of sharp linear phase digital filters," *IEEE Trans. Circuits Syst.*, vol. CAS-33, pp. 357–364, Apr. 1986.

[2] Y. C. Lim and Y. Lian, "The optimum design of one- and two-dimensional FIR filters using the frequency response masking technique," *IEEE Trans. Circuits Syst. II*, vol. 40, pp. 88–95, Feb. 1993.

[3] T. Saramäki, "Design of computationally efficient FIR filters using periodic subfilters as building blocks," in *The Circuits and Filters Handbook*, W.-K. Chen, Ed. CRC Press, Inc., 1995, pp. 2578–2601.

[4] T. Saramäki and Y. C. Lim, "Use of Remez algorithm for designing FRM based FIR filters," *J. Circuits, Syst., Signal Processing*, vol. 22, no. 2, pp. 77–97, Mar./Apr. 2003.

[5] T. Saramäki, J. Yli-Kaakinen, and H. Johansson, "Optimization of frequency-response-masking based FIR filters," *J. Circuits, Syst., Computers*, vol. 12, no. 5, Oct. 2003.

[6] O. Herrmann, L. R. Rabiner, and D. S. Chan, "Practical design rules for optimum finite impulse response lowpass digital filters," *Bell Syst. Tech. J*, vol. 52, no. 6, pp. 769–799, July/Aug. 1973.

[7] Y. J. Yu, T. Saramäki, and Y. C. Lim, "An iterative method for optimizing FIR filters synthesized using the two-stage frequency-response masking technique," in *Proc. IEEE Int. Symp. Circuits Syst.*, Bangkok, Thailand, May 25–28 2003.

[8] T. Coleman, M. A. Branch, and A. Grace, *Optimization Toolbox User's Guide*, The MathWorks, Inc., Jan. 1999, Version 2.

A CONSTANT-Q SPECTRAL TRANSFORMATION WITH IMPROVED FREQUENCY RESPONSE

Danillo B. Graziosi, Cristiano N. dos Santos, Sergio L. Netto, and Luiz W. P. Biscainho

PEE/COPPE & DEL/POLI, UFRJ
POBox 68504, Rio de Janeiro, RJ, 21945-970, BRAZIL
{danillo, csantos, sergioln, wagner}@lps.ufrj.br

ABSTRACT

This paper introduces a new transform intended for audio processing. The proposed transform exhibits two interesting features in the frequency domain, namely: a constant-Q characteristic and a steep response for each output channel. Constant Q implies that the spectral description of the transformed signal is performed along a log-like scale, as opposed to the linear scale of standard transforms, such as the discrete Fourier transform (DFT). The consequent variable resolution makes the new transform especially suited for the analysis of musical audio signals. The improved frequency response, as compared also to the DFT, is achieved by its implementation as a bank of very selective filters based on the frequency-response masking (FRM) approach. Such selectivity results in higher isolation between adjacent channels of the overall transform. Application of the new transform to the analysis of musical signals is illustrated through a computer experiment.

I. INTRODUCTION

The discrete Fourier transform (DFT) is a powerful tool for signal analysis, constituting a true mathematical bridge between time and frequency domains [1]. The DFT-based analysis, however, can be shown to present significant interference between the outputs of adjacent channels. In addition, considering the way the occidental musical scales were historically built, one can deduce that for musical audio signals, a logarithmic frequency representation would be more natural than the linear-frequency scale inherent to the DFT. This paper then proposes a new transform which attempts to overcome these two drawbacks related to the DFT. The new transform, the so-called constant-Q fast filter bank (CQFFB), is generated by the combination of the constant-Q transform (CQT) introduced in [2], [3] with the improved response characteristic of the fast filter bank (FFB), based on the frequency-response masking (FRM) approach and presented in [4], [5], [6].

The remaining of this paper is organized as follows: In Section II a brief description of the CQT is provided, focusing on the positive aspects of having a log-like frequency scale. In Section III, the sliding fast Fourier transform (sFFT) is interpreted under a filter-bank perspective for implementing the DFT. The FFB is then described in Section IV as a generalization of the sFFT, whose prototype filters can be replaced by more selective ones to achieve an improved frequency response. Section V presents the novel CQFFB, combining the positive issues of the CQT (log-like frequency scale) and the FFB (selective frequency response). Finally, computer experiments are included in Section VI to illustrate the capability of the CQFFB to analyze musical audio signals. Computational complexity issues will be addressed in a future paper.

II. THE CONSTANT-Q SPECTRUM TRANSFORMATION

When using the DFT to analyze musical audio signals, the resulting linear frequency scale yields a badly balanced signal description, since it concentrates too much information in the high-frequency region.

In [2], a constant-Q method, which allows the description of the frequency domain in a logarithmic scale, is presented as a tool for the analysis of musical audio signals. A strong motivation behind a constant-Q transform (CQT) is to allow harmonic frequencies to be represented in equal intervals in the transform domain. In that manner, any fundamental frequency together with its associate harmonics define a linear pattern which can be easily identified.

For the DFT, the frequency resolution $(\delta f)_{\mathrm{DFT}}$ is a constant value, given by the sampling frequency f_s divided by the total number of samples N being transformed:

$$(\delta f)_{\mathrm{DFT}} = \frac{f_s}{N}. \tag{1}$$

The N frequencies directly represented are

$$f_k = \frac{k}{N} f_s, \tag{2}$$

for $k = 0, 1, \ldots, (N-1)$.

The standard CQT decomposes the signal into components given by the following frequencies:

$$f_k = \left(2^{1/12}\right)^{\alpha k} f_{\min}, \tag{3}$$

for $k = 0, 1, \ldots, (N-1)$, where α defines the frequency resolution in fractions of a semitone. Then, $\alpha = \frac{1}{2}$ corresponds to a quarter-tone resolution, which suffices for many applications. Such resolution corresponds to a selectivity factor

$$Q = \frac{f_k}{(\delta f)_{\mathrm{CQT}}} = \frac{f_k}{\left(2^{1/24}-1\right) f_k} \approx \frac{1}{0.0293} \approx 34. \quad (4)$$

To achieve a constant Q, one should use a variable number of points

$$N_k = \frac{f_s}{(\delta f)_{\mathrm{CQT}}} = \frac{f_s Q}{f_k} \quad (5)$$

to determine each transform sample. Noticing that now

$$f_k = \frac{Q}{N_k} f_s, \quad (6)$$

we get that for the CQT the index k equals the selectivity factor Q. We then define the CQT of a signal $x(n)$ based on the definition of the DFT, but with the number of points given by equation (5), that is,

$$X_{\mathrm{CQT}}(k) = \frac{1}{N_k} \sum_{n=0}^{N_k-1} \mathcal{W}(n,k) x(n) e^{-j\frac{2\pi}{N_k} Qn} \quad (7)$$

for $k = 0, 1, \ldots, (N-1)$, where $\mathcal{W}(n,k)$ is a windowing function used to reduce blocking effects. A Table listing the characteristics of a 156-channel CQT for a sampling rate of $f_s = 32$ kHz can be found in [2].

III. THE SLIDING FFT

The N-point sliding FFT (sFFT) can be seen as the operation [4]

$$\begin{aligned} X(k) &= \sum_{i=0}^{N-1} x(n+i) W_N^{ki} \\ &= \left(\sum_{i=0}^{N-1} q^{-i} W_N^{ki} \right) \{x(n)\}, \end{aligned} \quad (8)$$

where q is the delay operator, such that $q^i\{x(n)\} = x(n-i)$ and $W_N = e^{-j\frac{2\pi}{N}}$. Hence, in the z domain, the sFFT operator can be expressed as

$$\mathrm{sFFT}(z) = \sum_{i=0}^{N-1} z^i W_N^{ki} = \prod_{i=0}^{L-1} \left[1 + \left(z W_N^k\right)^{2^i} \right], \quad (9)$$

where $L = log_2 N$. The FFB, to be reviewed in the next section, is a generalization of the sFFT which results from describing the sFFT(z) as indicated in equation (9).

Example 1: Using equation (9), the transfer function of channel 34 (linked to the quarter-tone CQT) of a 256-point sFFT is given by

$$\begin{aligned} H_{34}(z) &= \prod_{i=0}^{7} \left[1 + \left(z W_{256}^{34}\right)^{2^i} \right] \\ &= G_a^{0,34} G_a^{1,68} G_a^{2,136} G_a^{3,16} G_a^{4,32} G_a^{5,64} G_a^{6,128} G_a^{7,0}, \end{aligned} \quad (10)$$

where

$$\begin{cases} G_a^{0,34} = 1+z^1 W_{256}^{34}; & G_a^{1,68} = 1+z^2 W_{256}^{68} \\ G_a^{2,136} = 1+z^4 W_{256}^{136}; & G_a^{3,16} = 1+z^8 W_{256}^{16} \\ G_a^{4,32} = 1+z^{16} W_{256}^{32}; & G_a^{5,64} = 1+z^{32} W_{256}^{64} \\ G_a^{6,128} = 1+z^{64} W_{256}^{128}; & G_a^{7,0} = 1+z^{128} W_{256}^{0}, \end{cases} \quad (11)$$

such that $G_a^{i,j} = (1 + z^{2^i} W_N^j)$, with $j = [(34 \times 2^i) \bmod N]$. The corresponding magnitude response of channel 34 is depicted in Figure 1, where one can readily see that the first sidelobes are 13 dB below the channel passband. In order to normalize the sFFT response to the 0 dB level, the channel transfer function should be scaled by a factor of $c = 20 \log_{10} N = 48.2$ dB.

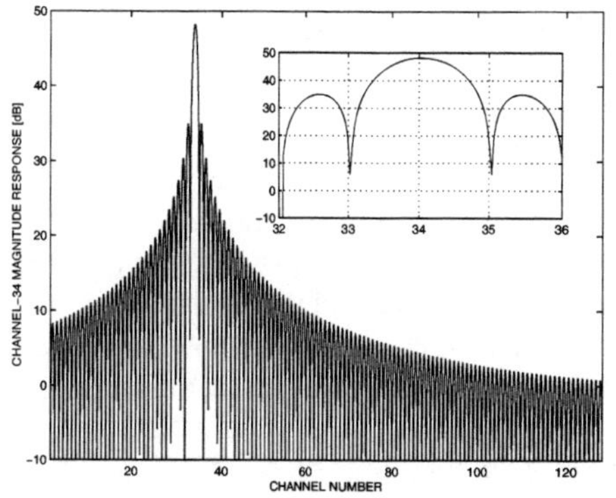

Fig. 1. Example 1: Magnitude response of channel 34 of a 256-point sFFT.

■

IV. THE FAST FILTER BANK

In equation (9), following the description in [4], the factor

$$G_a^{0,k}(z) = 1 + z W_N^k \quad (12)$$

can be seen as the kernel filter of the sFFT(z), since all other factors can be derived from it following a frequency scaling operation

$$z W_N^k \to (z W_N^k)^i. \quad (13)$$

A general fast filter bank (FFB) can be then generated from the sFFT by substituting the first-order kernel filter $G_a^{0,k}(z)$ by any other filter. Of course, higher order filters can yield improved frequency responses [4].

Example 2: Figure 2 depicts the magnitude response of channel 34 of a 256-point FFB using the filters given in [4].

Another interpretation for the FFB, based on the frequency-response masking (FRM) approach [7], is given in [6], where alternative FFB subfilters are also provided. Applications of the FFB include a programmable filter [6] and automatic music transcription [8].

■

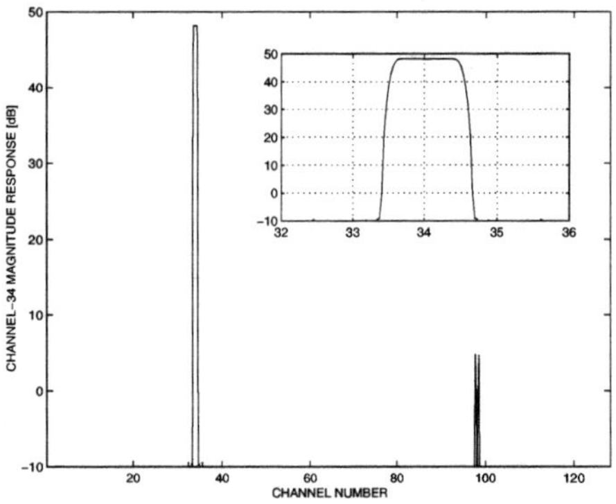

Fig. 2. Example 2: Magnitude response of channel 34 of a 256-point FFB.

V. AN IMPROVED CONSTANT-Q TRANSFORMATION

The so-called constant-Q fast filter bank (CQFFB) was devised as a blend of the CQT and FFB techniques. In that manner, we combine the constant-Q behavior of the CQT with the improved frequency response of the FFB. To achieve such symbiotic combination, however, we must first overcome a discrepancy between these two techniques. In fact, while the CQT calls for a variable number of time-domain samples to compute each frequency-domain sample, as given in equation (5), the FFB requires the underlying DFT to be computed from a power-of-two number of time-domain samples. The adopted strategy was to modify the CQT in the following way: instead of varying the number of input samples employed to compute each output sample in order to guarantee constant Q, we keep the number of input samples fixed as a power-of-two integer and change accordingly the sampling frequency, which becomes

$$f_s(k) = f_k \frac{N}{Q}, \qquad (14)$$

with f_k defined by equation (3), as before. After such modification, considering the quarter-tone resolution (Q=34), the Nyquist criterion requires

$$2f_k < f_s(k) \Rightarrow N > 68 \Rightarrow N = 128, 256, \ldots. \qquad (15)$$

Naturally, resampling of the input signal may need to include some kind of anti-aliasing filtering. In a preliminary version of the CQFFB we used the MATLAB®command `resample`.

Table I lists the characteristics of the 156-channel CQT, which uses a fixed sampling rate f_s = 44.1 kHz, and CQFFB, which uses a fixed number of N = 256 samples in each channel.

TABLE I

CHARACTERISTICS OF THE 156-CHANNEL CQT (WITH $f_s = 44.1$ KHZ) AND CQFFB (WITH $N = 256$ SAMPLES), BOTH WITH A SELECTIVITY FACTOR $Q = 34$.

Channel	Midinote	f_k (Hz)	N_k [CQT] (samples)	$f_s(k)$ [CQFFB] (Hz)	Time (ms)
0	53	175	8568	1318	194.3
6	56	208	7209	1566	163.5
12	59	247	6070	1860	137.6
18	62	294	5100	2214	115.6
24	65	349	4296	2628	97.4
30	68	415	3613	3125	81.9
36	71	494	3035	3720	68.8
42	74	587	2554	4420	57.9
48	77	699	2145	5263	48.6
54	80	831	1804	6257	40.9
60	83	988	1518	7439	34.4
66	86	1175	1276	8847	28.9
72	89	1398	1073	10526	24.3
78	92	1664	901	12529	20.4
84	95	1978	758	14893	17.2
90	98	2350	638	17649	14.5
96	101	2797	536	21060	12.2
102	104	3327	451	25050	10.2
108	107	3956	379	28536	8.6
114	110	4710	318	35464	7.2
120	113	5608	267	42225	6.1
126	116	6675	225	50259	5.1
132	119	7942	189	59799	4.3
138	122	9461	158	71236	3.6
144	125	11216	134	84450	3.0
150	128	13432	112	101135	2.5

VI. COMPUTER EXPERIMENT

Example 3: We formed a test signal $x(n)$ composed by six sinusoids of different frequencies 185.0, 196.0, 587.3, 622.3, 1046.5, and 1108.7 Hz (corresponding to F#, G, D, D#, C, and C#, respectively), sampled at a rate of $f_s = 44.1$ kHz. The entire signal has a total of 44100 samples, or 1 s. This signal was processed by the sFFT, FFB, CQT, and CQFFB tools, all with 100 frequency bands between 130.8 and 2282.4 Hz (corresponding to C and halfway between C# and D, respectively), for the sake of uniformity. Hence, the frequency resolution for the sFFT and FFB, which perform a linear frequency sampling, was

$$(\delta f)_{\text{sFFT}} = (\delta f)_{\text{FFB}} = \frac{2282.4 - 130.8}{100} = 21.5 \text{ Hz}. \qquad (16)$$

Accordingly, for the CQT and CQFFB we had $Q = 34$, corresponding to a quarter-tone frequency resolution, as suggested before.

The output magnitudes obtained by all four analysis tools for the input signal $x(n)$ are depicted in Figure 3. The sFFT response can be seen in Figure 3(a), from which one may notice that the sFFT yielded a nonzero spectrum for all frequency components and it was unable to resolve properly the low-frequency signal components. The FFB response is shown in Figure 3(b), where one may easily see that although the FFB was able to point out the purely sinusoidal characteristic of the input signal, it was not able, however, to separate properly the two low-frequency components. In Figure 3(c), the response of the CQT indicates that it succeeds in analyzing all sinusoidal components clearly, while presenting some sort of noisy behavior throughout the spectrum. Finally, Figure 3(d) depicts the magnitude response of the proposed CQFFB, which is able to show clearly the sinusoidal nature of the input signal, due to its selective response, with all six components clearly visualized, due to its logarithmic frequency scale.

■

VII. CONCLUSION

In this paper, two techniques previously known in the literature were reviewed, namely: the constant-Q transform (CQT), presented in [2], and the fast filter bank (FFB), introduced in [4]. A novel transform was then proposed, exhibiting a constant-Q characteristic, as opposed to the linear frequency resolution of the FFB (including the traditional sFFT), and improved frequency response, if compared to the standard sFFT-like response of the CQT. These two features combined together make the so-called constant-Q fast filter bank (CQFFB) especially suitable for audio applications, including analysis, coding, and transcription, as indicated by a computer experiment.

VIII. REFERENCES

[1] P. S. R. Diniz, E. A. da Silva, and S. L. Netto, *Digital Signal Processing; System Analysis and Design*, Cambridge, Cambridge, UK, 2002.

[2] J. C. Brown, "Calculation of a constant Q spectral transform," *J. Acoustical Society of America*, vol. 89, no. 1, pp. 425–434, Jan. 1991.

[3] J. C. Brown and M. S. Puckette, "An efficient algorithm for the calculation of a constant Q transform," *J. Acoustical Society of America*, vol. 92, no. 5, pp. 2698–2701, Nov. 1992.

[4] Y. C. Lim and B. Farhang-Boroujeny, "Fast filter bank (FFB)," *IEEE Trans. Circuits and Systems–II: Analog and Digital Signal Processing*, vol. 39, no. 5, pp. 316–318, May 1992.

[5] B. Farhang-Boroujeny and Y. C. Lim, "A comment on the computational complexity of sliding FFT," *IEEE Trans. Circuits and Systems–II: Analog and Digital Signal Processing*, vol. 39, no. 12, pp. 875–876, Dec. 1992.

[6] Y. C. Lim and B. Farhang-Boroujeny, "Analysis and optimum design of the FFB," *Proc. IEEE Int. Symp. Circuits and Systems*, pp. II.509–II.512, London, UK, 1994.

[7] Y. C. Lim, "Frequency-response masking approach for the synthesis of sharp linear phase digital filters," *IEEE Trans. Circuits and Systems*, vol. CAS-33, pp. 357–364, Apr. 1986.

[8] S. W. Foo and W. T. Lee, "Transcription of polyphonic signals using fast filter bank," *Proc. IEEE Int. Symp. Circuits and Systems*, pp. III.241–III.244, Scottsdale, USA, May 2002.

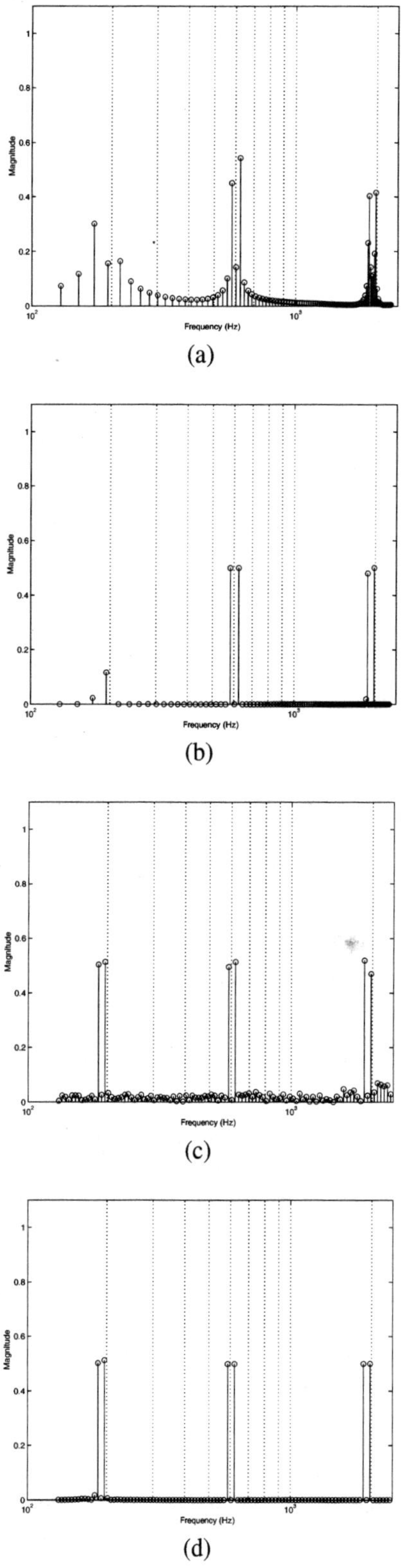

Fig. 3. Example 3: Magnitude responses of 100-point transformations of an input signal composed by six sinusoidal components: (a) sFFT; (b) FFB; (c) CQT; (d) CQFFB.

A ZERO-VOLTAGE AND ZERO-CURRENT SWITCHING THREE-LEVEL DC-DC CONVERTER WITH SECONDARY-ASSISTED REGENERATIVE PASSIVE SNUBBER

*Tingting Song *, Nianci Huang *, and Adrian Ioinovici **, Senior Member ,IEEE*

*Department of Electrical Engineering, Sichuan University, China, E-mail: songtt@21cn.com
** Holon Academic Institute of Technology, Israel, E-mail: adrian@hait.ac.il

ABSTRACT

An energy-recovery snubber, developed previously for full-bridge PWM phase-shift converters, is inserted in the secondary side of the transformer of a three-level converter. As a result, ZVS is obtained for the outer power switches and ZCS for the inner switches. The voltage on the rectifier diodes is clamped at a reasonable voltage, parasitic secondary ringing being avoided .The energy of the snubber is fully transferred to the load. Using a three-level converter, the voltage stresses on both power switches and rectifier diodes are reduced at half of that on their counterparts in full-bridge converters. Experimental results prove the expected high-efficiency of the proposed converter.

1. INTRODUCTION

AC-DC power factor correctors operating with a 380 V or 440V line are suitable as input stages for high-power FB-DC-DC converters. For complying with regulations as the IEC 61000-3-2, the DC-link voltage of the corrector has to be substantially increased, as compared to the line voltage, thus increasing the voltage stress (equal to the link voltage) on the switches of the full-bridge converter. If the full-bridge converter is replaced by a three-level converter, the voltage stress across its power switches is reduced to half of that on their counterparts in the full-bridge. Thus the rating of the switches is considerably reduced (approximately to half), and consequently, the on-resistances of the switches and the conduction losses are reduced. A phase-shift control can be adopted in order to realize a robust and simple PWM control.

The switching losses can be reduced by turning on/off all the switches under a ZVS condition. This solution is not suitable for the inner switches at light load, when only the energy of the (small) leakage inductance is available to realize ZVS, what is not sufficient. It is also not suitable for high power applications, where, due to the high voltages, IGBTs have to be used. For IGBT switches, ZCS would be more desirable than ZVS, due to their high turn-off switching loss caused by the tail current.

In these cases, it is preferable to turn-on/off the leading (outer) switches with ZVS, and the lagging switches with ZCS (ZVZCS three-level converters [1]-[4]). In order to realize a ZCS condition, the primary current has to be kept at zero during the freewheeling stage. The papers cited above make use of different primary-assisted solutions in this purpose: either an additional saturable inductor, or a storage capacitor is added into the primary circuit, increasing the losses. Even the use of extra diodes, or an additional transformer in the primary do not solve the problem of severe parasitic ringing in the secondary side of the transformer and the overvoltage on the rectifier diodes. Generally, the primary-assisted ZVZCS converters are suitable up to power levels of 5 kW. These problems can be solved [4] by using an active snubber in the transformer secondary, but this would increase cost and control complexity.

The realization of ZVZCS has been previously studied extensively for FB-PWM phase-shift controlled converters. New ideas, by using secondary-assisted solutions to create the ZCS condition for the inner legs, have been developed for these converters. It was proposed to insert a regenerative passive snubber in the transformer secondary [5]-[11]. The main idea of all these papers is to use a snubber formed by 1-2 capacitors and 1-3 diodes, inserted between the rectifier's diodes and the output filter. The snubber's capacitor(s) is charged during the energy transfer stage, then it clamps the voltage on the rectifier diodes at a certain level, and assures a zero-primary-current during the transition of the lagging switches, when it discharges on the load, thus recuperating the energy.

The aim of this paper is to use the idea of a regenerative snubber for a three-level converter. As the snubbers cited above operate almost in the same manner, only one of them will be analyzed here.

The proposed converter is presented in Section 2, together with a description of its operation in steady state. Simulation and experimental results are given in Section 3, with Conclusions in the last section.

0-7803-8251-X/04/$17.00 ©2004 IEEE

2. PROPOSED THREE-LEVEL CONVERTER WITH PWM PHASE-SHIFT CONTROL AND SECONDARY-ASSISTED PASSIVE REGENERATIVE SNUBBER

The DC/DC circuit is given in Fig.1. The capacitors C_{s1}, C_{s2} are used to limit the voltage on each switch in off-state to half of the input voltage. The outer power switches M1 and M4 are the leading ones, and the inner power switches M2, M3 are the lagging ones. L_{lk} represents the leakage inductance of the high-frequency transformer. C_A, D_{A1} and D_{A2} form the regenerative snubber. $D_{R1} - D_{R4}$ are the rectifier diodes. V_{ab} and I_P represent the primary voltage and current. V_{rec} is the secondary(rectifier) voltage . Let m denotes the ratio n_1/n_2 .

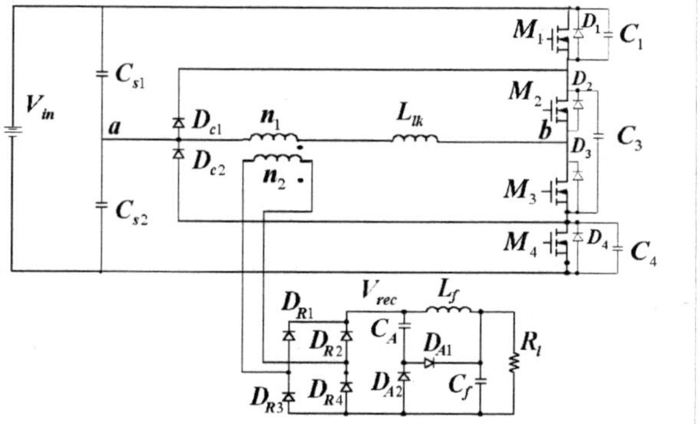

Fig. 1 Three-level converter with secondary energy recovery snubber

In a steady-state switching cycle ,the converter goes through eight modes. The theoretical switching diagram and main waveforms are given in Fig.2.

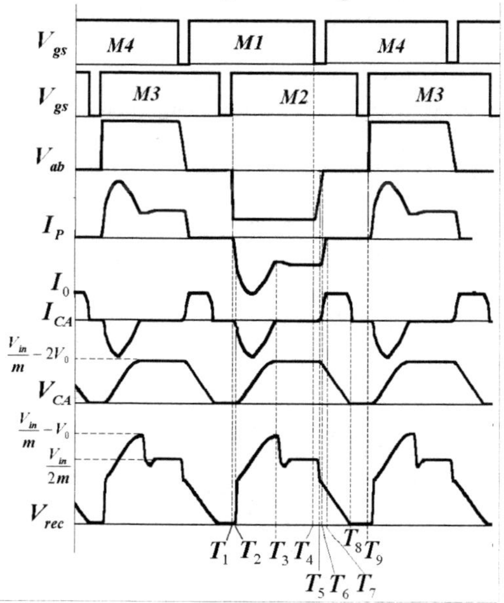

Fig.2 Switching diagram

MODE 1 [$t_1 \sim t_2$]: At t_1, when M1 was in conduction, and the primary current was zero, M2 is turned-on with ZCS due to the presence of the inductance L_{lk}. I_p in absolute value increases linearly until it reaches the reflected output current I_0/m The circuit in this stage is given in Fig. 3a ,and its equivalent ,as reflected to the primary , in Fig 3b.

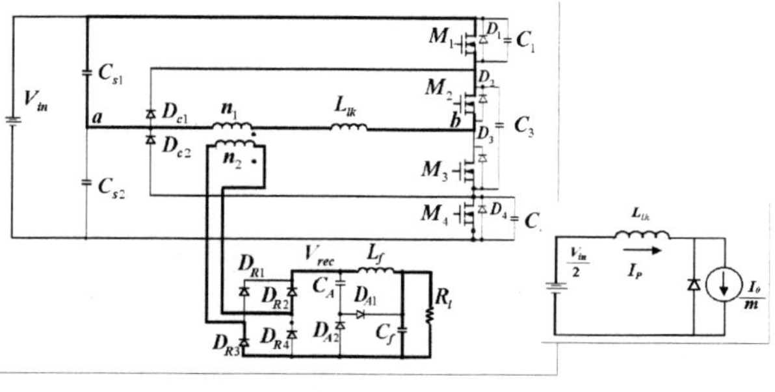

(a) **(b)**

Fig.3 .Equivalent switching topology for MODE 1; a) full circuit; b)simplified circuit reflected to the primary

Accordingly, $I_P(t) = ((V_{in}/2)/L_{lk}) \times t$ (1)

MODE 2 [$t_2 \sim t_3$]: At t_2, I_p in absolute value reaches the reflected load current, D_{R1}, D_{R4} turn-off (ZCS ,due to L_{lk}'s presence in Mode 2), and the line-to-load energy transfer process starts (Fig 4 a,b). Concomitantly, the snubber's C_A begins charging through D_{A1} (D_{A1} turns-on with ZVS and ZCS).

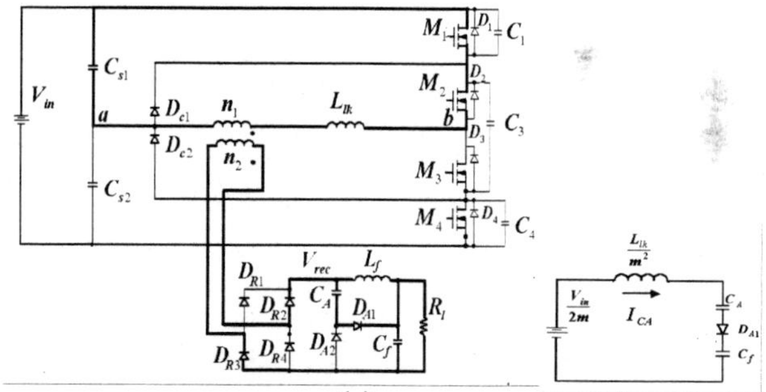

(a) (b)

Fig 4. Equivalent switching topology for MODE 2; a)full circuit; b)simplified circuit reflected to secondary

Accordingly:

$$I_{CA} = \frac{V_{in}/2 - mV_0}{\sqrt{L_{lk}/C_A}} \times Sin(\omega(t - t_1)) \, , \, \omega = \frac{1}{\sqrt{L_{lk}C_A/m^2}} \tag{2}$$

$$V_{CA} = \frac{V_{in}}{2m} - V_0 - (\frac{V_{in}}{2m} - V_0) \times Cos(\omega(t - t_1)) \tag{3}$$

$$V_{rec} = V_{CA} + V_0 = \frac{V_{in}}{2m} - (\frac{V_{in}}{2m} - V_0) \times Cos(\omega(t - t_1)) \tag{4}$$

MODE 3 [$t_3 \sim t_4$]: The current in D_{A1}, which was evolving sinusoidally , reaches the value 0 at t_3 , D_{A1} turns-off (ZCS), C_A remains charged at its maximum value attained at the end of MODE 2 : $V_{in}/m-2V_0$. The equivalent switching topology is given in Fig.5.

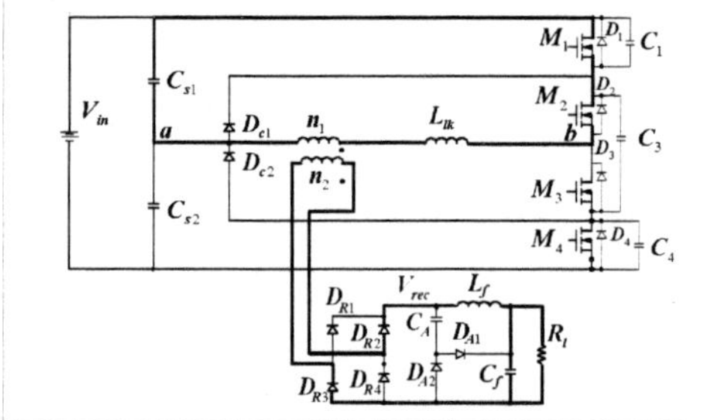

Fig.5. Equivalent switching circuit for MODE 3

MODE 4 [$t_4 \sim t_5$]: At t_4, the leading switch M1 is turned off with ZVS due to its parallel capacitor C_1. The primary current charges C_1 and discharges C_3. As usually, C_3 is assumed large enough, such that its voltage remains constant: $V_{in}/2$.

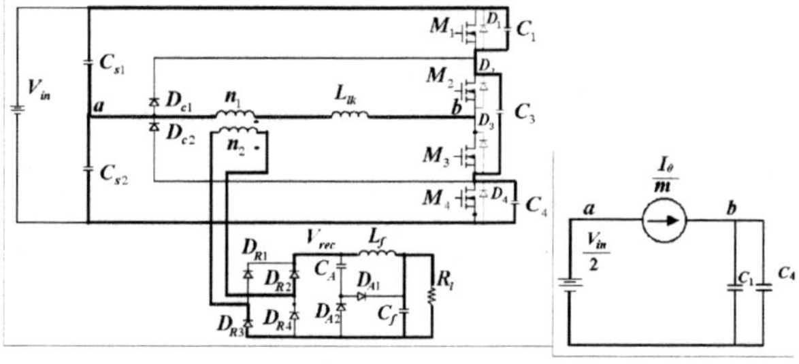

(a) (b)

Fig.6 Equivalent switching topology for MODE 4
a) full circuit; b)simplified circuit reflected to the primary

according to Fig.6, $I_0 = mI_p$, $V_{ab} = mV_{rec}$. V_{rec} decrease:

$$V_{rec} = \frac{1}{m}\left(\frac{V_{in}}{2} - \frac{I_0/m}{C_1+C_4}t\right) \quad (5)$$

MODE 5 [$t_5 \sim t_6$]: At t_5, V_{rec} reaches the C_A's voltage ($V_{in}/m-2V_0$). D_{A2} turns-on (ZVS) ,and C_A discharges on the load, thus recuperating the accumulated energy in MODE 2. The rectifier voltage, due to V_{CA} decreases much slower than V_{ab}, which continues to decrease at the approximate same rate as in the precedent mode. The difference between V_{ab} and mV_{rec} is applied to the leakage inductance, as a result, I_p starts decreasing. The equivalent circuit is given in Fig.7. accordingly

$$I_p = \frac{1}{m}[I_0(1-\frac{C_p}{C_p+C_A})Cos\omega t + \frac{C_p}{C_p+C_A}I_0],\ \omega = m\sqrt{\frac{C_p+C_A}{C_pC_AL_{lk}}} \quad (6)$$

$$V_{CA} = \frac{V_{in}}{m} - 2V_0 - \frac{1}{(C_p+C_A)\omega}I_0 Sin\omega t + \left(\frac{1}{C_A} + \frac{C_p}{(C_p+C_A)C_A}\right)I_0 t \quad (7)$$

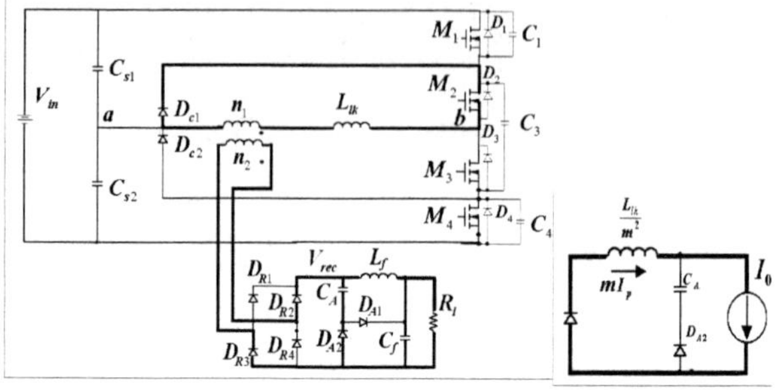

(a) (b)

Fig.7 Equivalent switching topology for MODE 5;
a) full circuit; b)simplified circuit reflected to secondary

MODE 6 [$t_6 \sim t_7$]: At t_6, C_4 was completely discharged, D_{c1} turns-on (clamping V_{C4} to zero), thus allowing for ZVS turn-on of M4. The switching configuration is given in Fig. 8. $V_{ab}=0$, V_{rec} is applied across L_{lk} ,so that I_p drops quickly.

$$I_p = \frac{1}{m}\left[(I_p(t_6)m - I_0)Cos\omega t - \frac{V_{rec}(t_6)}{Z_0}Sin\omega t + I_0\right] \quad (8)$$

$$V_{CA} = Z_0(I_p(t_6)m - I_0)Sin(\omega t) + V_{rec}(t_6)Cos(\omega t) \quad (9)$$

$$\omega = \sqrt{\frac{1}{L_{lk}C_A/m^2}},\quad Z_0 = \sqrt{\frac{L_{lk}}{m^2C_A}}$$

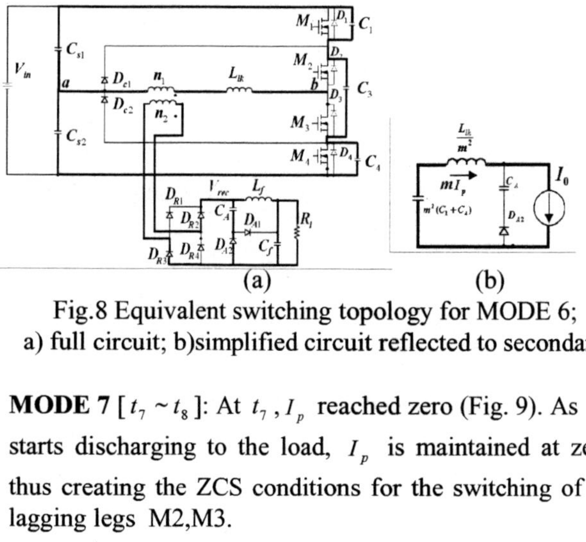

(a) (b)

Fig.8 Equivalent switching topology for MODE 6;
a) full circuit; b)simplified circuit reflected to secondary

MODE 7 [$t_7 \sim t_8$]: At t_7, I_p reached zero (Fig. 9). As C_A starts discharging to the load, I_p is maintained at zero, thus creating the ZCS conditions for the switching of the lagging legs M2,M3.

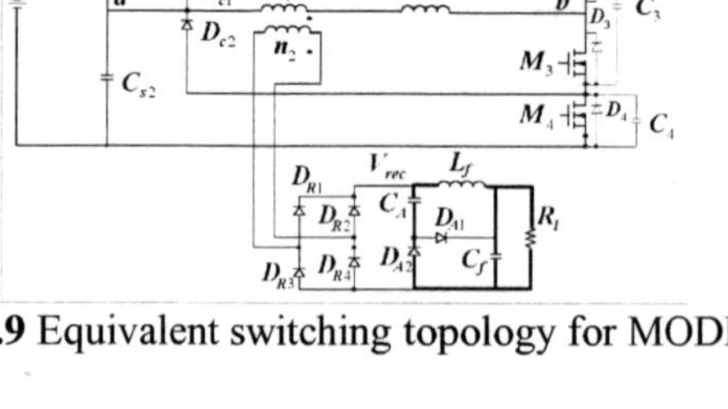

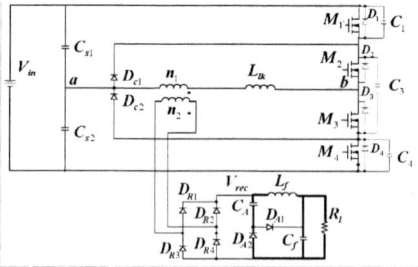

Fig.9 Equivalent switching topology for MODE 7

$$V_{CA} = V_0 + (V_{rec}(t_7) - V_0)Cos\omega t - \sqrt{\frac{L_f}{C_A}} I_0 Sin\omega t , \omega = \frac{1}{\sqrt{L_f C_A}} \quad (10)$$

$$I_{CA} = I_0 Cos\omega t + \frac{(V_{rec}(t_7) - V_0)}{\sqrt{L_f / C_A}} Sin\varpi t \quad (11)$$

MODE 8 [$t_8 \sim t_9$] At t_8, C_A is completely discharged, thus all its energy is recuperated to load. The rectifier diodes turn-on with ZVS, and the load current freewheels through the rectifier (Fig. 10). I_p remains zero, preparing the ZCS turn-on of M3.

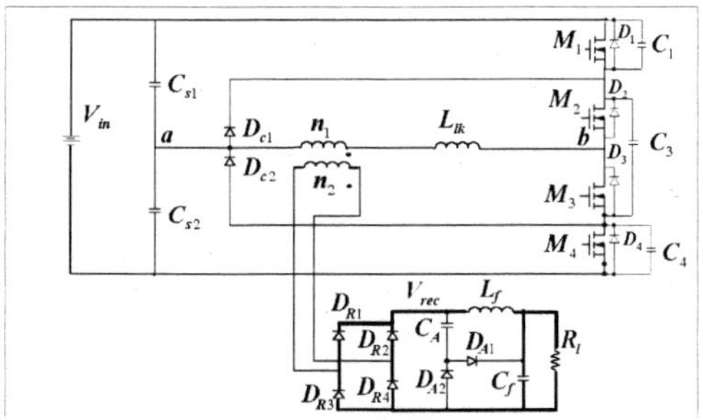

Fig.10 Equivalent switching topology for MODE 8

3.SIMULATION AND EXPERIMENTAL RESULTS

A prototype of 4.5kW (150V, 30A), operating at 40kHz was realized in the laboratory. $n_1 = 20$, $n_2 = 20$, $L_{lk} = 5uH$ $C_A = 330n$, $L_f = 1.5mH$, $C_f = 47u$, $C_3 = 10n$, M1-M4 are switches of type SGH80N60UFD, all the diodes in the converter are FFA60U60DN. The main simulated waveforms are shown in Fig. 11.

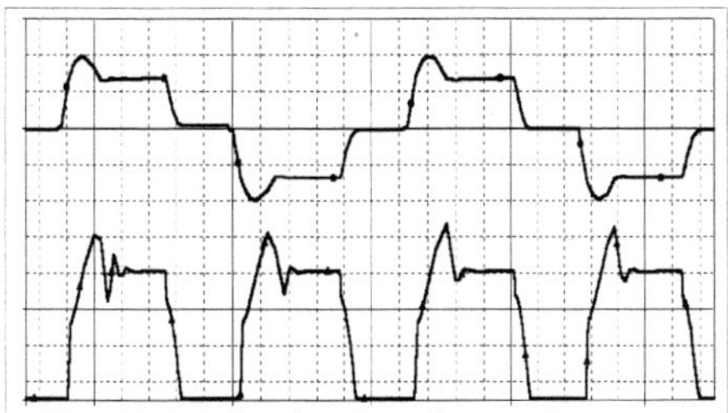

Fig. 11. Main simulated waveforms

The experimental I_p and V_{rec} are shown in Fig.12.

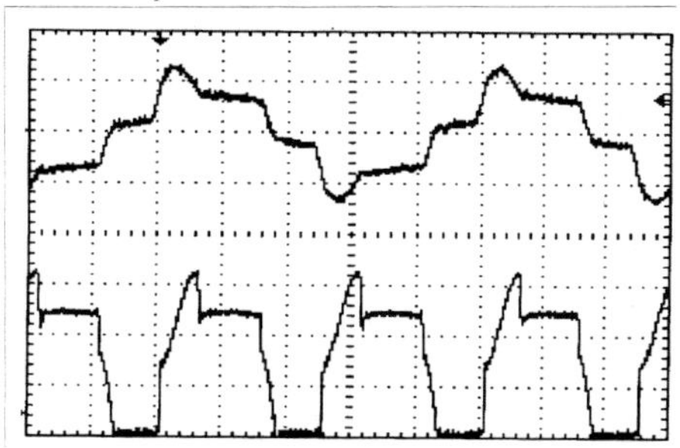

Time: 5us/div

Fig.12. Main experimental waveforms

The measured efficiency at full load is 93.7%.

4.CONCULSION

Both the simulated and experimental results show that the voltage stress on the power switches is reduced to Vin/2. The inner switches turn on/off with ZCS ,as the primary current was reset during this period. There is no redundant circulation of primary current in the freewheeling stage. The rectifier voltage stress is maintained at a low value . No secondary ringing is present .All these advantages gave a high efficiency

11. REFERENCES

[1] X.Ruan, L.Zhou, and Y.Yan, “A novel zero-voltage and zero-current-switching PWM three-level converter,” in Proc.IEEE PESC Rec, pp. 1075-1079, 2001.

[2] S.J.Leon, F.Canales, P.M.Barbosa, and F.C.Lee, “A primary-side-assisted zero-voltage and zero-current switching three-level DC-DC converter with phase-shift control,” in Proc.IEEE APEC Rec,pp.641-647,2002.

[3] X Ruan and B Li, “Zero-voltage and zero-current-switching PWM hybrid full-bridge three-level converter,” in Proc. IEEE PESC Rec., 2002.

[4] F.Canales, P.M.Barbosa, and F.Lee, “A zero voltage and zero-current switching three level DC/DC converter,” in Proc.IEEE APEC, pp.314-320, 2000.

[5] E.S.Kim, K.Y.Joe, M.H.Kye, Y.H.Kim, and B.D.Yoon, “An improved soft-switching PWM FB DC/DC converter for reducing conduction losses,” IEEE Trans.Power Electron.,vol.14,no.2,pp.258-264,March 1999.

[6] J.G.Cho, J.W.Baek, C.Y.Jeong, and G.H.Rim, “Novel zero-voltage and zero-current-switching full-bridge PWM converter using a simple auxiliary circuit,” IEEE Trans. Ind. Aplic, Vol.35, no. 1,pp.15-20, Jan./Feb. 1999.

[7] J.G.Cho,J.W.Baek, D.W.Yoo, H.S.Lee, and H.Rim, “Novel zero-voltage and zero-current-switching (ZVZCS) full bridge PWM converter using a simple auxiliary circuit,” in Proc.IEEE APEC Rec,pp.834-839, 1998.

[8] E.S.Kim and Y.H.Kim, “A ZVZCS PWM FB DC/DC converter using a modified energy-recovery snubber,” IEEE Trans.Ind.Electron, vol. 49,no.5, pp.1120-1127,Oct.2002.

[9] H.S.Choi,J.W.Kim,B.H.Cho, “Novel zero-voltage and zero-current switching (ZVZCS) full-bridge PWM converter using coupled output inductor,” IEEE Trans.Power Electron.,vol.17,no.5,pp.641-648,Sept.2002.

[10] H.S.Choi ,J.H.Lee,B.H.Cho,and J.W.Kim, “Analysis and design considerations of zero-voltage and zero-current switching (ZVZCS) full-bridge PWM converters ,” in Proc. IEEE PESC Rec,pp.1-6, 2002.

[11] T.T.Song and N. Huang, “Analysis and modeling of a novel ZVSZCS FB PWM converter,” Journal of Circuits, Systems, and Computers ,(special jubilee issue on Power Electronics),Vol.13,No.3,June 2004.

HYSTERETIC CONTROLLER FOR CMOS ON-CHIP SWITCHING POWER CONVERTERS

G. Villar[1], E. Alarcón[1], H. Martínez[1], E. Vidal[1], F. Guinjoan[1], S. Porta[2], A. Poveda[1]

[1] Department of Electronic Engineering.
Universitat Politècnica Catalunya
Campus Nord – Mòdul C4.
C/. Gran Capità s/n
08034 Barcelona, Spain.
e-mail: ealarcon@eel.upc.es

[2] Department of Electrical and
Electronic Engineering
Universidad Pública de Navarra
31006 Pamplona, Spain.

ABSTRACT

The design and implementation of a CMOS analog integrated circuit that provides versatile sliding-mode control laws for high-frequency switching power converters is described. The controller circuit implements general-purpose linear-surface state control laws incorporating compensating dynamics. The analog controller integrated circuit considers current-mode circuit techniques, hence exhibiting good performance as far as operation speed, power consumption, suitability to poorly regulated power supplies and robustness in front of interferences are concerned. The circuit allows modular connection, being composed of externally linearized transconductors based on current conveyors, current-mode amplifiers and filtering stages, and a transimpedance high-speed hysteretic comparator. The IC includes power MOSFETs and their drivers so that all active elements for implementing a switching power converter are included, only remaining as external components the power reactive elements. Layout details for a a CMOS 0.35μm technology implementation are discussed. The circuit may be used either as a standalone IC controller or as controller circuit that is technology-compatible with on-chip switching power converters and on-chip loads for future powered Systems on Chip

1. INTRODUCTION

1.1 CMOS on-chip efficient power management

In telecommunications and computing systems, the continuous trend in miniaturizing power processing subsystems, which are in charge of guaranteeing high efficiency in the use of supply energy, stands from the global system-level impact of those subsystems in terms of volume and weight, and thus on portability. This statement is particularly true for actual and future generation systems-on-chip (SOC), a trend which provides a line of convergence for the implementation of current and future systems for portable, mobile and autonomous applications, in which power management is one of the key performance limiting factors as regards to ergonomics and operability time. The ultimate step consequently consists in the fully monolithic integration of the power converter together with the same circuits that constitute its load within either the same substrate or chip package, yielding a complete Powered System on a Chip (PSOC). Despite the notable research efforts in the field of integrated switching power converters, paradigm of energy-efficient processing circuits, further investigations are required to achieve further miniaturization and better efficiency whilst retaining compatibility with an integrated technology, in particular with digital CMOS technology, the standard technology for signal processing and computation and the target for RF IC implementations. In the investigation towards integration of switching power converters, research both on optimized converter topologies [22], [23], [24], [42], active devices [25], integrable passive elements [26], [27], [28] and control methods [29], [30] is required. As opposite to previous works on most of those areas, the circuit integration of high performance control methods has not been hitherto fully investigated as regard their implementation difficulties, although some very strategic stringent applications require such high performance [31], [33].

Switching power converter circuits are characterized as nonlinear dynamic systems. This aspect justifies the lack of standard methods for modeling and controlling those systems. In this sense, their conventional control methods are based on the application of classical techniques of linear feedback, after a previous linearization around an equilibrium point of the nonlinear dynamic equations which model the converter equivalent average behavior. This approximation notably restricts the dynamic capabilities of those switching power conversion systems.

In front of that situation, there exist several control methods which exhibit remarkable dynamic performance due to the use of fast-dynamics state variables of the converter, being possible to point out, among others, current-mode control, one-cycle control and sliding-mode control. This work is devoted to the implementation of this latter high-performance control method, which is introduced in the following paragraph.

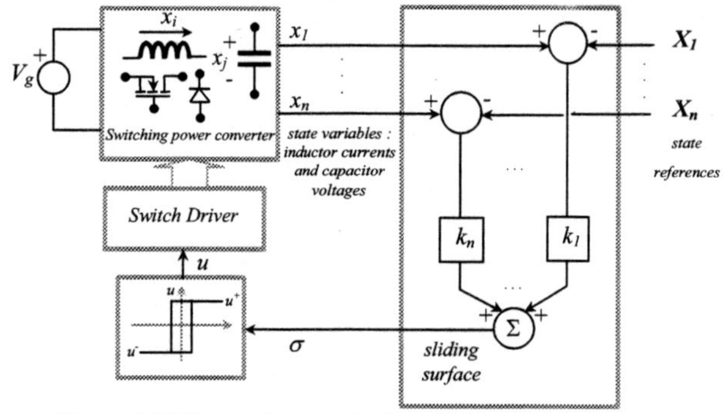

Figure 1 Sliding-mode control of a switching power converter

1.2 Sliding-mode control techniques for switching power converters

Sliding-Mode control techniques are considered a plausible alternative to PWM control strategies in the area of switching DC-DC power regulators, since they improve both the dynamic response and the robustness of these power conversion circuits against parameter variations and external perturbations, both input voltage and load variations. Thorough theoretical foundations for the application of these sliding-mode techniques to general systems exist in the literature ([1], [2]), and its application to the control of switching DC-DC power converters is a topic of current interest [3-6]. In this field, a general-purpose control law [1, 4,13], consisting in a linear sliding surface σ, is described by:

$$\sigma \equiv k_1(x_1 - \mathbf{X}_1) + \ldots + k_n(x_n - \mathbf{X}_n) = 0$$
$$\sigma > 0 \Rightarrow switch\ ON \qquad \sigma < 0 \Rightarrow switch\ OFF \qquad (1)$$

where $x_1,\ldots,x_n$ are the state variables of the power converter, $\mathbf{X}_1,\ldots,\mathbf{X}_n$ are their associated references and $k_1,\ldots,k_n$ are the sliding surface coefficients. In order to avoid sliding regimes with infinite switching frequencies, a practical approach to sliding-mode control requires the addition of a certain hysteresis level to the comparator that is in charge of continuously evaluating the σ combination to provide the ON/OFF control signal. To illustrate the technique, figure 1 represents a generic sliding-mode controller providing the control action to a switching power converter.

1.3 Implementation of Sliding-Mode Controller

In spite of the sliding-mode control advantages, and concerning the circuit implementation of a sliding-mode controller, not much work exists in the area, to the best knowledge of the authors. Note that a high-speed implementation is required, since sliding-mode control is an instantaneous time-domain technique that handles state variables switching ripple, being hence affect by speed limitations. This fact is worsened by the continuous trend in increasing switching frequencies due to the resulting miniaurization of the power circuit.

As regards previous implementation attempts, note that, being the sliding control an instantaneous control technique with multiple inputs which does not require excessive processing complexity, digital signal processing presents the disadvantages of the associated computation time and high power consumption, together with the necessity of including several A/D converters. Nevertheless, there exist digital implementations based on look-up-tables mapped on EPROMs [3], or by means of microcontrollers [5] that have shown to be effective for unpractically low switching frequencies (*i.e.* tens of kHz). On the other hand, the few existing discrete analog implementations of sliding-mode controllers [6], [14] make use of classical voltage-mode processing, this is, they use high-gain operational amplifiers operated in closed-loop negative-feedback configurations, which results in severe bandwidth limitations, both small-signal and large signal -slew rate-.

In front of these limitations, current-mode circuit techniques [7], [18] are considered in this work. The use of current-mode circuit techniques to implement high-performance controllers for switching power converters is discussed in detail in [16]. It is well known that, in a low impedance circuit environment, the use of currents as information-carrying signals yields reduced voltage variations, and thus, in a reduced impact of parasitic capacitances, representing an increment in the maximum operation speed of these circuits. Apart from this key effect, other advantages of current-mode processing are the local open-loop high-bandwidth processing and the capability of directly implementing signal aggregations by virtue of Kirchoff's current law, as well as other functions in a compact manner, both linear (amplifiers, filters) and nonlinear (polynomial, PWL). As opposite to this, some disadvantages are related to current-mode processing, namely, accuracy errors due both to the loss of precision associated to the loss of negative feedback, and the need to use mismatched current replicas when copying signals. It is worth noting, however, that this loss of precision reduces its impact in controller implementation applications, compared to the design of open-loop processors (*e.g.* continuous-time filters), since in the former the global feedback lowers the precision requirements. Last but not least, a standing feature of current-mode operation is that, due to the operation associated to low-impedance nodes, it is more robust to interferences or crosstalks in a mixed signal or switching microelectronic environment, as it is the case of an ON/OFF controller intended to control switching converters, possibly within a SOC environment.

In the work presented herein, the use of current-mode circuit techniques is proposed to synthesize, design and implement a high-performance microelectronic analog sliding-mode controller. The use of current-mode circuit techniques results in an idealization (in the sense of minimum delay) of the sliding-mode controller operation in terms of speed. It is well known in sliding-mode theory that the high-gain effect of this technique requires high switching frequencies (ideally infinite). The circuit proposal described in this paper operates accurately at frequencies (≈2 MHz) more than one decade over previous designs, both digital [5], and analog [6], [14], [19].

In what follows, this paper will describe design details of the sliding mode analog controller in section 2, whereas section 3 will present the IC controller implementation and layout. The paper is summarized and conclusions are drawn in section 4.

2. ANALOG CMOS DESIGN DETAILS

2.1 Analog sliding-mode controller architecture

Concerning the architecture of the controller, shown in figure 2, it should be emphasized that, due to the current-mode representation, the circuit naturally allows modular operation by simple connection at the summation node (labeled *C* in the schematic), where the linear combination which defines the switching hyperplane (1) is conformed. The designed circuit restricts the general case described by (1) to a two-variable case. This notwithstanding, the architecture and the controller circuit is of general purpose due to its modular capabilities, since it can provide sliding-mode control to a wide variety of switching DC-DC converters such as Buck, Boost, Buck-Boost or higher order converters such as Cuk [4] and SEPIC [12]. Note that, when the b_{MASTER} switch is ON and the $b_{MASTER/SLAVE}$ switch is OFF, the current-mode version of the sliding surface, i_σ, is continuously compared with the zero level by the next comparator stage so as to force the control action described by equation (1). By turning the $b_{MASTER/SLAVE}$ control switch ON, this common node allows the addition of other parallel processing stages, and allows also the insertion of a current-mode synchronization ramp for the purpose of controlling steady-state switching frequency [17].

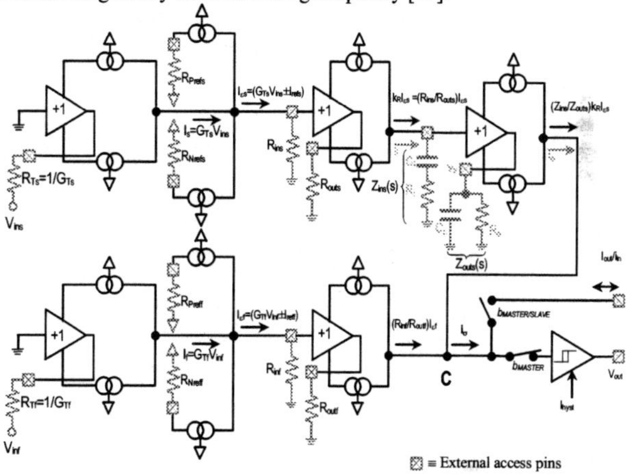

Figure 2 Architecture of of the sliding-mode IC controller

As regards the controller input interface, although state variables in the switching power converter are either capacitor (floating) voltages or inductor currents, the latter are sensed either by means of shunt resistors or Hall-effect sensors, hence appearing as differential floating voltages for the controller.

Note that since the controller prototype is designed as a standalone IC controller, the user flexibility is allowed by selection of external components that define the controller parameters. In the case of a fixed controller for on–chip power controllers, either passive components or their active emulation should be integrated on–chip.

The presented architecture features several extensions from previous designs [36][37][39].

The input stage considers and external resistor as the input of the controller so as to capture dynamic ranges that are out of th IC supply margins. Since voltages within the architecture are refered to the mid point between the supply voltages, this medium point is generated on-chip by means of a buffered resistor divider. Finally, the IC includes a driver and power MOSFET section so that all active elements (both signal and power processing) for implementing a switching power converter are included, and only power reactive elements remain external.

Most of the signal processing within the circuit, as it has been introduced, is performed in current-mode. With this purpose, the main building block used in the controller is the current conveyor (CCII), a building-block [7] that bases its operation both on voltage copying or buffering (between high-impedance node Y and low-impedance node X) and current copying or buffering (between low-impedance node X and high-impedance node Z), both open-loop-wise. Details of the circuit design of the current-conveyor block are described in [36][37][39]. The CCII has been improved as regards impedance at node X, which is critical for the proper operation of the controller, by using local feedback [40].

It is well known [3-6] that actual implementations of sliding-mode control require the addition of a certain hysteresis level to the output comparator which is in charge of continuously evaluating the i_σ combination -the current-mode version of equation (1)-, so as to avoid sliding regimes with infinite switching frequencies, although slow implementations rely on uncontrollable circuit implementation delays [6, 14]. In order to achieve that hysteresis effect, it is used the hysteretic current-mode comparator described in [36][37][39], based on a current-input voltage-output comparator which includes a CMOS inverter that confers it zero offset [10], [21]. Note that the hysteresis level is electronically adjustable, this allowing an increased controllability, as some theoretical works reveal [20], [32].

2.2 Behavioural modelling of the controller

In order to properly design the electrical parameters of the CCII building blocks, the architecture of the sliding-mode controller was reproduced SIMULINK with behavioral models of the CCII (including, for all of the three ports, resistance value, impedance bandwidth, saturation levels, as well as voltage and current copying accuracy). Several simulations were performed in order to study the effect of the different CCII nonidealities on different controller operation indexes. Figure 3 shows as an example the current at the input of the hysteretic comparator as a function of CCII current copying bandwidth.

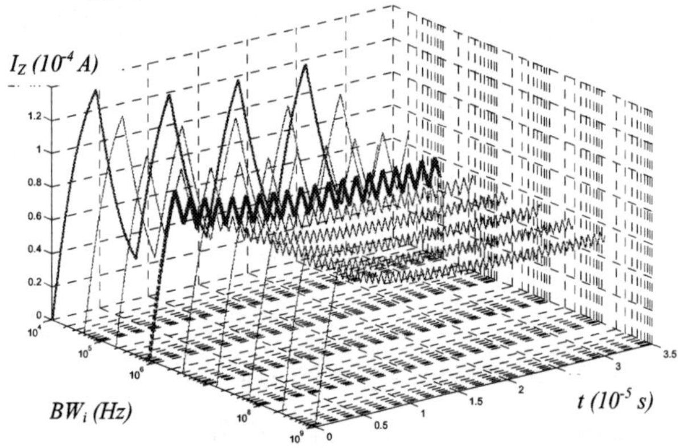

Figure 3 Closed-loop converter control. Behavioral simulations

2.3 Power MOSFETs and drivers

The designed integrated circuit encompasses a power section that includes two power MOSFETs in standard CMOS technology, their correspondent drivers, as well as a nonoverlapping circuit to guarantee complementary operation.

Power transistors have been designed to allow a conduction current of *500 mA*, targeting either on-chip converter low power applications or as a means of driving extenal driver for medium to high power converters. The switch scheme implements synchronous rectification.

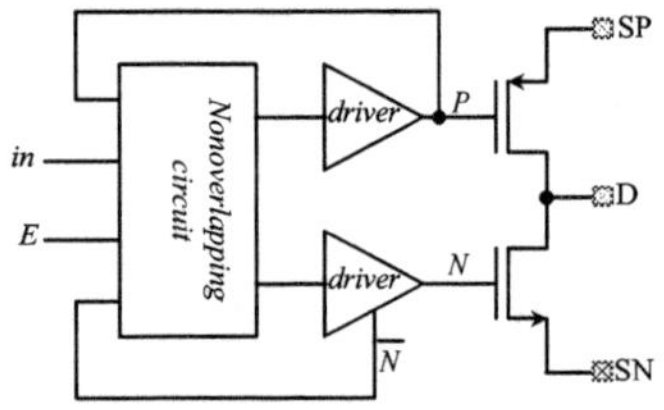

Figure 4 Schematic of the power section

The driver stage has been designed as a chain of tapered buffers with a scaling factor of 5, a reasonable tradeoff of circuit complexity *versus* power consumption and area ocupation.

Both power transistors have independent driving to allow the inclusion of the nonoverlapping circuit. An additional logic signal (enable, *E*) has been included for shut down the power section.

3 CONTROLLER IMPLEMENTATION AND LAYOUT

A prototype controller using the aforementioned standard 0.35μm CMOS technology was integrated. Figure 5 shows the full-custom layout of the complete general-purpose sliding-mode analog controller previously described, whereas figure 6 shows a detail of the power section including drivers and power transistors. In order to reduce supply interference, it can be observed the separation of the pad ring as well as the consideration of dedicated supply pads for the controller, the driver stage and the power MOSFETs, these last ones requiring several pads to guarantee current handling capability.

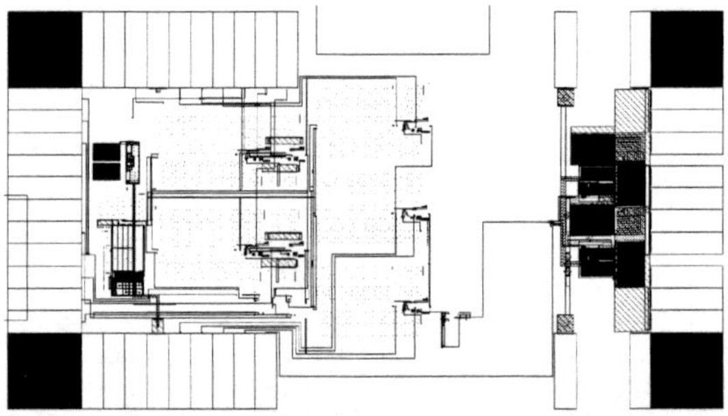

Figure 5 Layout of the complete sliding-mode controller

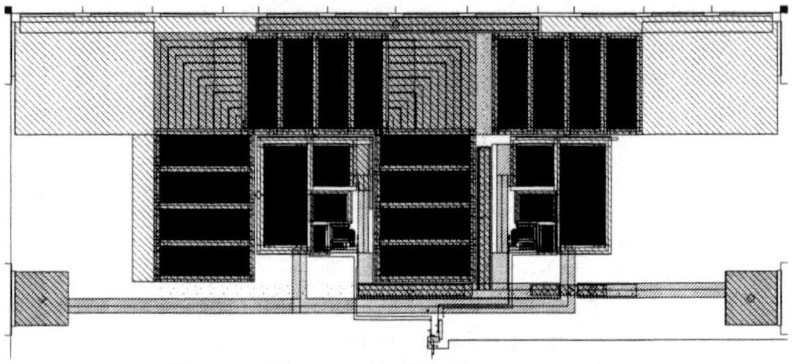

Figure 6 Layout detail of power section

4. CONCLUSIONS

This work presents a CMOS microelectronic implementation of a sliding-mode controller with linear sliding surfaces for switching power converters. The IC controller includes a power stage with complementary power MOSFETs to implement synchronous rectification and their correponding drivers. The design of the analog building blocks considered behavioural modelling to study the impact of circuit nonidealities over controller operation. The circuit features low power consumption, high speed and small occupied area, mainly due to the use of current-mode circuit techniques.

Acknowledgements. This work has been partially funded by project TIC-2001-2157-C02-01 from the Spanish MCYT and EU FEDER fund.

REFERENCES

1. H. Sira-Ramirez, “Sliding Motions in Bilinear Switched Networks”, *IEEE Transactions on Circuits and Systems*, pp. 919-933, August 1987.
2. V.I. Utkin, “Sliding Mode and their Applications in Variable Structure Systems”, *Mir Editorial,* Moscow, 1978.
3. D. Biel, E. Fossas, F. Guinjoan, E. Alarcón and A. Poveda, “Application of sliding-mode control to the design of a buck-based sinusoidal generator”, *IEEE Transactions on Industrial Electronics,* Vol. 48, No. 3, June 2001, pp. 563-571.
4. L. Martínez-Salamero, J. Calvente, R. Giral, A. Poveda and E. Fossas, “Analysis of a Bidirectional Coupled-Inductor Cuk Converter

Operating in Sliding Mode", *IEEE Transactions on Circuits and Systems-I*, vol.45, n°4, pp. 355-363, April 1998.
5. M.Oppenheimer, I. Husain, M. Elbuluk and J. De Abreu, "Sliding Mode Control of the Cuk Converter", *proceedings of the IEEE Power Electronics Specialists Conference (PESC96), pp. 1519-1526.*
6. M. Carpita, "Sliding Mode Controlled Inverter with Switching Optimization", *proceedings of the EPE Journal*, Vol 4. N°3, Sept. 98.
7. C. Toumazou, J. Lidgey and D. Haigh, Editors, "Analogue IC design: the Current-mode approach", *IEE Circuits and Systems Series 2*, Peter Peregrinus Editorial, 1991.
8. M. H.L. Chow, K.W. Siu, C.K.Tse and Y.S. Lee, "A Novel Method for Elimination of Harmonics in Single-Stage PFC Switching Regulators", *IEEE Trans. on Power Electronics*, Vol.13, N°1, Jan. 98, pp. 75.
9. A. Payne and C. Toumazou , "Practical Integrated Current Conveyors", 11.2 Chapter in *IEEE ISCAS'94 Tutorials,* Edited by C. Toumazou, N. Battersby and S. Porta.
10. A. Rodríguez Vázquez, R. Domínguez Castro, F. Medeiro and M. Delgado Restituto, "High Resolution CMOS Current Comparators: Design and Applications to Current-Mode Function Generation", *Analog Integrated Circuits and Signal Processing*, 7, pp-149-165,1995
11. E. Alarcón, D. Biel, F. Guinjoan, E. Fossas, E. Vidal and A. Poveda, "Current-Mode BiCMOS Sliding-Mode Controller Circuit for AC Signal Generation in Switching Power DC-DC Converters", *IEEE Midwest Symposium on Circuits and Systems, MWSCAS'99*, NMSU, New Mexico, August 1999.
12. J. Hernanz, L. Martínez-Salamero, A. Poveda and E. Fossas, "Analysis of a Sliding-Mode Controlled SEPIC Converter", *The Transactions of The Institute of Electrical Engineers of Japan*, Vol. 116-D, N°11, Nov. 1996.
13. R. Giral, L. Martínez-Salamero, J. Hernanz, J. Calvente, F. Guinjoan, A. Poveda and R. Leyva, "Compensating Networks for Sliding-Mode Control", *Proc. IEEE International Conference on Circuits and Systems, ISCAS'95*, pp. 2055-2058.
14. H. Chiacchiarini, P. Mandolesi and A. Oliva, "Nonlinear Analog Controller for a Buck Converter: Theory and Experimental Results", *Proc. IEEE International Symposium on Industrial Electronics, ISIE'99*, pp. 601.
15. K. Koli and K. Halonen, "CMRR enhancement techniques for current-mode instrumentation amplifiers", *IEEE Transactions on Circuits and Systems-I : Fundamental theory and applications*, vol 47, n°5, May 2000, pp 622-632.
16. E. Alarcón, "Microelectronic design of controllers for switching power converters", *PhD thesis*, Technical University of Catalunya (UPC), Barcelona, Spain, 1999.
17. P. Mattavelli, L. Rossetto and G. Spiazzi, "General-purpose sliding-mode controller for DC-DC converter applications", *proceedings of the IEEE Power Electronics Specialists Conference (PESC93)*, pp. 609.
18. M. Ismail and T. Fiez, Analog VLSI: Signal and Information Processing, McGraw-Hill, N.Y., 1994.
19. G. Escobar, R. Ortega, H.Sira-Ramirez, J.P. Vialin and I. Zein, "An experimental comparison of several nonlinear controllers for power converters", *IEEE Control Systems Magazine*, Volume 19, Issue 1, Feb 1999, pp. 66 -82.
20. V. M. Nguyen and C.Q. Lee, "Tracking Control of buck converter using sliding-mode with adaptive hysteresis", *proceedings of the IEEE Power Electronics Specialists Conference 95 (PESC95)*, pp. 1086
21. Traff, H., "Novel approach to high speed CMOS Current Comparators", Electronics Letters, Volume: 28, Issue: 3, 30 Jan. 1992, Pages: 310 -312.
22. S. Sakiyama, J. Kajiwara, M. Kinoshita, K. Satomi, K. Ohtani, A. Matsuzawa, "An On-Chip High-Efficiency and Low-Noise DC/DC Converter Using Divided Switches with Current Control Technique", *IEEE International Solid-State Circuits Conference*, pp. 156-157, 1999.
23. X. Zhou, P-L Wong, P. Xu, F.C. Lee and A. Q. Huang, "Investigation of candidate VRM topologies for future microprocessors", *IEEE Transactions on Power Electronics*, pp. 1172-1181, November 2000.
24. S. Reynolds, "A DC-DC converter for short-channel CMOS technologies", *IEEE Journal of Solid-State Circuits*, Vol.32, N°1, Jan 97.
25. S. Ajram, G. Salmer, "Ultrahigh Frequency DC-to-DC Converters Using GaAs Power Switches", *IEEE Transactions on Power Electronics*, pp. 594-602. September 2001.
26. S. Iyengar, T. M. Liakopoulos, C. H. Ahn, "A DC/DC Boost Converter Toward Fully On-Chip Integration Using New Micromachined planar Inductors", *30th Annual IEEE Power Electronics Specialists Conference*, 1999. PESC 99, Volume: 1, 1999, pp. 72-76.
27. L. Daniel, C. R. Sullivan and S. R. Sanders, "Design of microfabricated inductors", *IEEE Transactions on Power Electronics*, vol 14, n°4, pp. 709-723, July 1999.
28. C. H. Ahn and M. G. Allen, "A comparison of two micromachined inductors (bar- and meander-type) for fully integrated boost DC/DC power converters", *IEEE Transactions on Power Electronics*, vol 11, n°2, pp. 239-245, March 1996.
29. W. Lau and S. R. Sanders, "An integrated controller for a high frequency buck converter", *1997 IEEE PESC Record*, pp 246-254.
30. A.P. Dancy and A.P. Chandrakasan, "Ultra Low Power Control Circuits for PWM Converters", *proceedings of the IEEE Power Electronics Specialists Conference 97 (PESC'97)*, pp. 21.
31. G. Hanington, P.F. Chen, P. Asbeck and L.E. Larson, "High-efficiency power amplifier using dynamic power-supply voltage for CDMA applications", *IEEE Transactions on microwave theory and techniques*, vol 47, N°8, pp. 1471-1476, August 1999.
32 T. Szepesi, "Stabilizing the frequency of hysteretic current-mode DC/DC converters", *IEEE Transactions on Power Electronics*, vol PE-2, n°4, pp. 302-312, October 1987.
33. J. Kim and M. A. Horowitz, "An efficient digital sliding controller for adaptive power-supply regulation", *IEEE Journal of solid-state circuits*, vol 37, n°5, pp. 639-647, May 2002.
34. B. Arbetter, R. Erickson, D. Maksimovic, "DC-DC Converter Design for Battery-Operated Systems", *26th Annual IEEE Power Electronics Specialists Conference, 1995*, Volume 1, pp. 103-109.
35. R. W. Erickson, D. Maksimovic, "Fundamentals of Power Electronics", *Ed. Kluwer Academic Publsihers*, 2nd edition, 1999
36. E. Alarcón, A. Romero, A. Poveda, S. Porta and L. Martínez-Salamero, 'Current-mode analogue integrated circuit for sliding-mode control of switching power converterts', *IEE Electronics Letters*, 3rd February 2002, Vol. 38, No.3.
37. E. Alarcón, A. Romero, A. Poveda, S. Porta and L. Martínez-Salamero, 'General-purpose Sliding-Mode Controller Analog Integrated Circuit for Bidirectional Switching DC-DC Power Converters', *Analog Integrated Circuits and Signal Processing,* vol 38, Feb. 04, pp 203-214.
38. Alarcón, G. Villar, E. Vidal, H. Martínez and A. Poveda, 'General-purpose one-cycle feedforward controller for switching power converters: A high-speed current-mode CMOS VLSI implementation', *proceedings of the 44th IEEE Midwest Symposium on Circuits and Systems (MWSCAS01)*, Dayton, Ohio, USA, August 2001, pp. 290-293.
39. E. Alarcón, A. Romero, A. Poveda, S. Porta and L. Martínez-Salamero, 'Sliding-mode control analog integrated circuit for switching DC-DC power converters', *proceedings of the 2001 IEEE International Symposium on Circuits and Systems - ISCAS'01*, Sydney, Australia, May 2001, pp. 500-503.
41 F. Seguin and A. Fabre, "New second generation current conveyor with reduced parasitic resistance and bandpass filter application", *IEEE Trans. on Circuits and Systems I*, June 2001, Vol.48, No.6, pp 781-785.
42. G. Villar, E. Alarcón, F. Guinjoan and A. Poveda, 'A design space exploration for integrated switching power converters', *proceedings of the 2003 IEEE International Symposium on Circuits and Systems - ISCAS'03*, Bangkok, Thailand, May 2003, pp. 304-307.

State Trajectory Prediction Control for Boost Converters

Kelvin K.S. Leung and Henry S.H. Chung†

Department of Electronic Engineering
City University of Hong Kong
83 Tat Chee Avenue, Kowloon Tong, Kowloon, Hong Kong
†Fax.: (852) 2788 7791
†Email: eeshc@cityu.edu.hk

Abstract **– Boost and flyback converters have a right-half-plane zero in their control-to-output transfer function. This property makes the controller difficult to be designed with the classical frequency-domain approach to ensure good output regulation and fast response over wide frequency bandwidth of input voltage and output load perturbations. Instead of a small-signal design approach, this paper presents a cycle-by-cycle state trajectory prediction (STP) control method for boost converters. The method is based on predicting the output voltage after a hypothesized switching action. The output can revert to steady state in two switching actions under large-signal input voltage and output load disturbances. Theoretical predictions are verified with experimental results of a 120W 18/24V prototype.**

I. Introduction

Over the last three decades much effort have been made to research new control schemes for switching converters to achieve good output regulation and dynamic response. As switching converters are inherently nonlinear, most of the design strategies are based on small-signal techniques [1]. However, some converters, like boost and flyback converters, have a right-half-plane zero in their control-to-output transfer function. This property makes the controller difficult to be designed with the classical frequency-domain approach for ensuring good output regulation and fast response over wide bandwidth of input voltage and output load perturbations [2].

There has been considerable work along large-signal modeling and control of switching converters. A direct and viable approach of designing the control scheme is to apply some nonlinear system control techniques, in which the nonlinear power stage is represented by state-space averaged models [3]. Typical design strategy is to use the Lyapunov asymptotic stability theory to derive a control law that ensures a global stability region and optimizes both state trajectories and control energy [4]. Other approaches use robust nonlinear control algorithms for the small-signal models that achieves global or semi-global stability [5].

Another approach is based on cycle-by-cycle control schemes, in which the controlled switch is dictated by the instantaneous values of the circuit variables. Examples of these are the one-cycle control of [6], the sliding mode control schemes of [7], the bang-bang control [8], and the digital control [9]. However, the control using one-cycle control has no information on the output load disturbances and is for buck converters only. In the sliding-mode control, the trajectory is restricted along the sliding surface and will converge to the operating point after many switching cycles. Bang-bang or hysteresis control can provide a fast dynamic response and tracking of the control variables. However, the hysteresis control does not work with all types of systems. If a control of this type, based on the output voltage, is tested with a boost converter, the results are disastrous. Paper [8] proposes the use of state-trajectories control that the converter can achieve steady-state operation for a step change in input voltage or output current in one on/off control, but the control requires either sophisticated digital processor or analog computation.

Concluding the above control methods, the best solution is the one that can achieve the ultimate goals proposed in [8], but with a simple implementation. This paper presents a cycle-by-cycle STP control method for boost converters. The method is based on predicting the output voltage after a hypothesized switching action. Several switching criteria are derived to dictate the state of the main switch. Theoretical predictions are verified with experimental results of a 120W 18/24V converter prototype.

II. Principles of Operations

A. Steady-state operation

Fig. 1 shows the schematic of the boost converter, which has three possible topologies. In Topology 1, the switch *S* is on and the diode *D* is off. The inductor *L* is charging up from the input source v_i. The output load *R* is supplied from the output capacitor *C*. The output voltage is v_o. Thus,

$$\frac{di_L}{dt} = \frac{1}{L} v_i \tag{1}$$

and

$$\frac{dv_o}{dt} = \frac{dv_C}{dt} = \frac{1}{C} i_C \tag{2}$$

where i_L is the inductor current, v_C is the capacitor voltage, and i_C is the capacitor current.

In Topology 2, *S* is off and *D* is on. The energy stored in *L* will release to the load, together with the input source. Thus,

$$\frac{di_L}{dt} = \frac{1}{L}(v_i - v_o) \quad (3)$$

and

$$\frac{dv_o}{dt} = \frac{dv_C}{dt} = \frac{1}{C} i_C \quad (4)$$

In Topology 3, S and D are off. No energy remains in L. The output load is supplied from C.

$$\frac{di_L}{dt} = 0 \quad (5)$$

and

$$\frac{dv_o}{dt} = \frac{dv_C}{dt} = \frac{1}{C} i_C \quad (6)$$

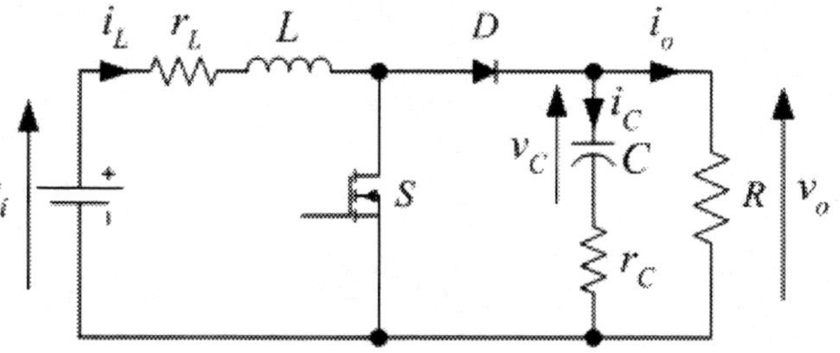

Fig. 1 Schematic of boost converter.

If the output ripple voltage is much smaller than the average output voltage in the steady state, it can be assumed that the output current i_o is a constant. Fig. 2 shows the typical waveforms of the v_o and i_C varies between $v_{o,\max}$ and $v_{o,\min}$. The state of S is determined by predicting the area under i_C with a hypothesized switching action till $i_C = 0$ and comparing the area with a fixed ratio of the output error at that instant. The control scheme is depicted in Fig. 3.

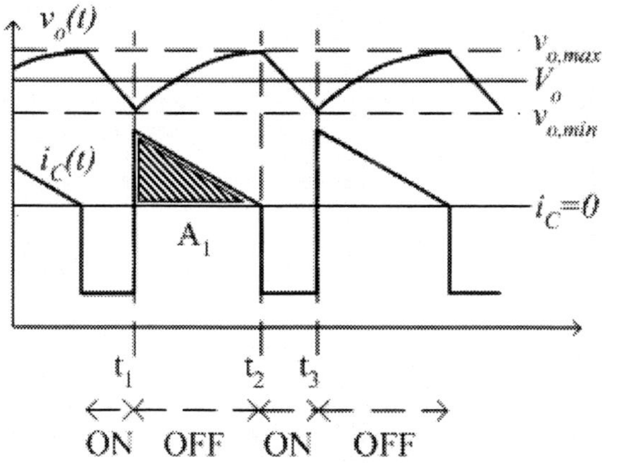

Fig 2 Typical waveforms of v_o and i_C.

v_{gate} is derived from the output 'Q' of an RS flip-flop. The input 'R' is v_{g_off}, which commands S off. The input 'S' is v_{g_on}, which commands S on. When v_{g_off} and v_{g_on} are in logic 'HIGH', the state of S will be changed.

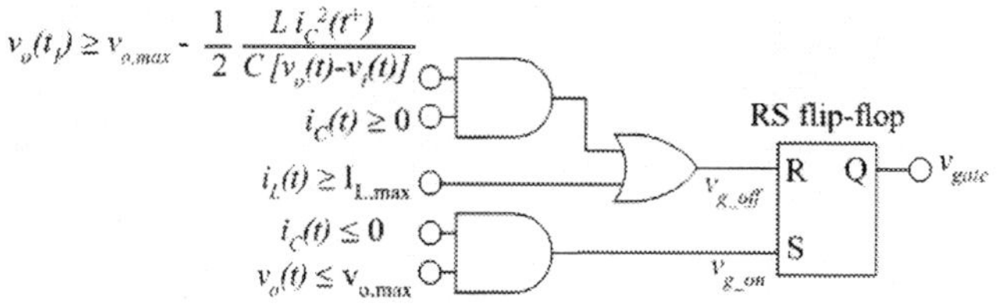

Fig. 3 Control Circuit of STP boost converter.

1) *Generation of v_{g_off} for switching S off*

As shown in Fig. 2, S is originally in the on state and is switched off at t_1. The objective is to determine t_1, so that v_o equals $v_{o,\max}$ at t_2 (at which $i_C = 0$). The shaded area A_1 under i_C is integrated from t_1 to t_2. Thus,

$$\Delta v_o = v_o(t_2) - v_o(t_1) = v_{o,\max} - v_o(t_1) = \frac{1}{C}\int_{t_1}^{t_2} i_C \, dt \quad (7)$$

If A_1 is approximated by a triangle, it can be shown that

$$\int_{t_1}^{t_2} i_C \, dt = \frac{1}{2}\frac{L \cdot i_C^2(t_1^+)}{[v_o(t_1) - v_i(t_1)]} \quad (8)$$

where $i_C(t_1^+)$ is the value of i_C after S is switched off. i_C is inconsistent before and after switching S off at t_1,

$$i_C(t_1^-) \neq i_C(t_1^+) \quad (9)$$

$i_C(t_1^+)$ is obtained by subtracting $i_o(t_1^-)$ from $i_L(t_1^-)$,

$$i_C(t_1^+) \cong i_L(t_1^-) - i_o(t_1^-) \quad (10)$$

S has to be switched off, in order to ensure that v_o will not exceed $v_{o,\max}$ in the subsequent topology. Thus, by substituting (8) into (7), the criteria for switching S off are

$$v_o(t_1) \geq v_{o,\max} - \frac{1}{2}\frac{L \cdot i_C^2(t_1^+)}{C \cdot [v_o(t_1) - v_i(t_1)]} \quad (11)$$

and

$$i_C(t_1^+) \geq 0 \quad (12)$$

For the sake of safety, S will also be switched off if i_L is larger than I_{L_max}. That is,

$$i_L(t) \geq I_{L\_\max} \tag{13}$$

Fig. 3 shows how v_{g_off} is generated from (11)-(13). The second term of the right-hand-side in (11) is realized by using an analog computational unit [8] that perform the function of

$$v_{off} = v_y \left(\frac{v_z}{v_x}\right)^m \tag{14}$$

where $m = 2$, $v_x = 1$, $v_y = \frac{1}{2} \frac{L}{C \cdot [v_o(t_1) - v_i(t_1)]}$, and $v_Z = i_C(t_1^+)$.

v_{off} is then subtracted from $v_{o,\max}$ and compared with v_o to implement (11).

2) Generation of v_{g_on} for switching S on

As shown in Fig. 2, S can be switched on at t_2 when $v_o = v_{o,\max}$ and $i_C = 0$. Thus,

$$i_C(t_2) \leq 0 \tag{15}$$

and

$$v_o(t_2) \leq v_{o,\max} \tag{16}$$

Fig. 3 shows how v_{g_on} is generated from (15) and (16).

B. Transient response

The criteria of (11)-(16) are applicable for both steady-state operation and large-signal disturbances. Followings illustrate the transient responses of the converter under a load change with different initial switching condition. Fig. 4(a) shows the corresponding waveforms, when i_o is suddenly decreased and S is on initially. Since i_L cannot change suddenly, i_C will increase significantly. Eqs. (11) and (12) are satisfied. v_{g_off} is generated. S is then switched off. The converter may go into the discontinuous conduction mode, showing that the control method is independent on the mode. Fig. 4(b) shows the corresponding waveforms when i_o is decreased suddenly and S is initially off. i_C increases. Eqs. (11) and (12) are satisfied. Thus, S will keep on in the off state, until v_{g_on} commands S to on. Fig. 4(c) shows the corresponding waveforms when i_o is increased suddenly and S is initially on. i_C decreases instantaneously. Eqs. (11) and (12) are not satisfied. Eq. (15) will be satisfied and (16) may be satisfied. S will keep on in the on state. Fig. 4(d) shows the corresponding waveforms when i_o is increased suddenly and S is initially off. i_C decreases instantaneously. Eq. (15) will be satisfied. S will then be on when (16) is satisfied.

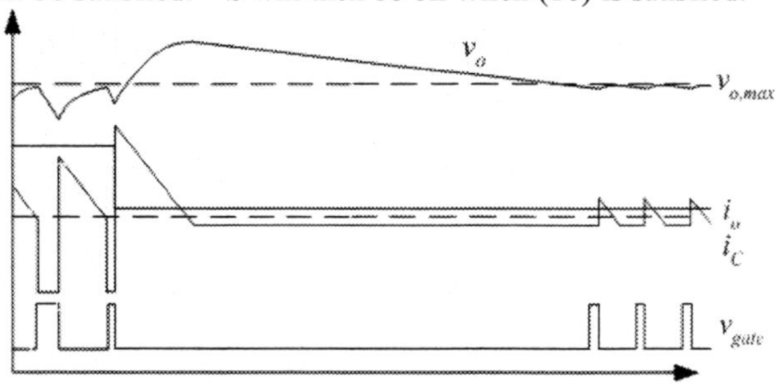

(a) i_o decreases when S is ON.

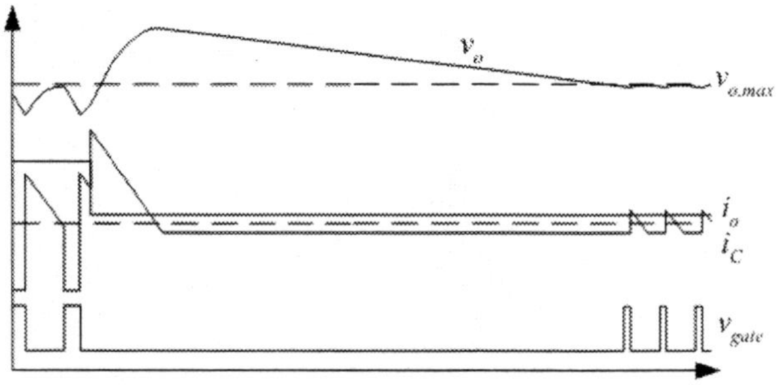

(b) i_o decreases when S is OFF.

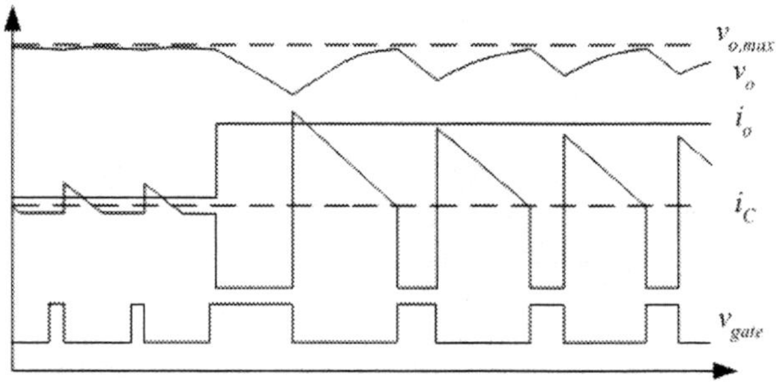

(c) i_o increases when S is ON.

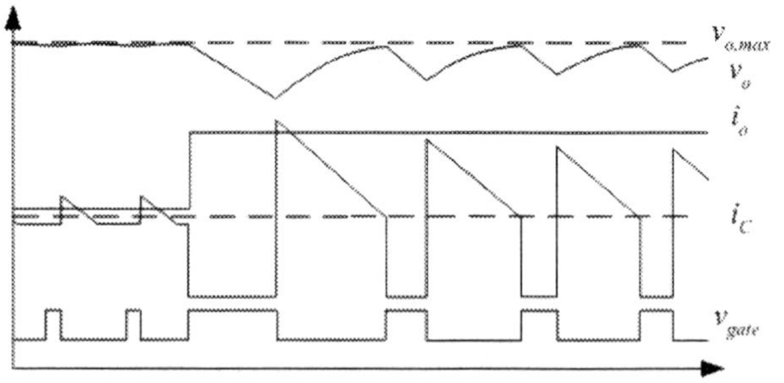

(d) i_o increases when S is OFF.

Fig. 4 Waveform of v_o, i_o, i_C and v_{gate} under load change.

III. Experimental Prototype

A 120W prototype has been built. The component values are: L = 12μH and C = 300 μF. v_o is regulated at 24V and the nominal input voltage is 18V. Fig. 5(a) shows the transient responses of the converter when i_o is changed from 0.5A (12W) to 5A (120W). Fig. 5(b) shows the transient responses when i_o is changed from 5A (120W) to 0.5A (12W).

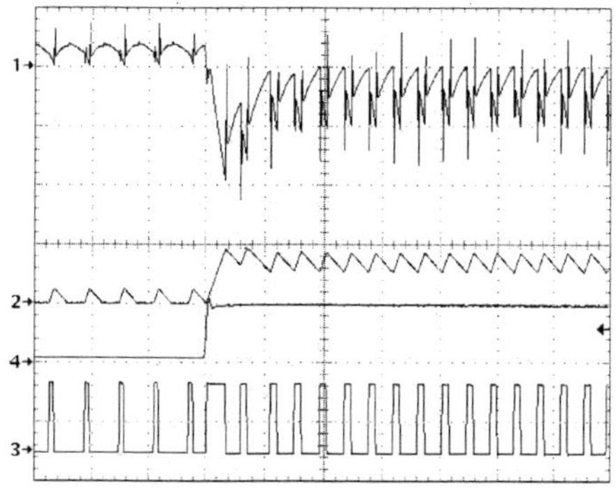

(a) i_o is changed from 0.5A(12W) to 5A(120W). [Ch1: v_{out} (50mV/div)].

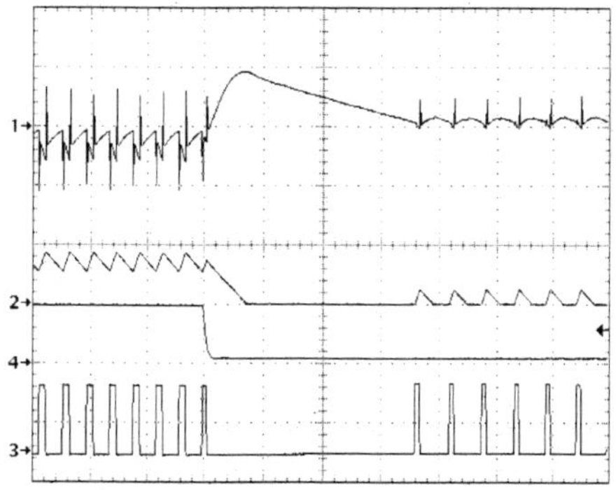

(b) i_o is changed from 5A(120W) to 0.5A(12W). [Ch1: v_{out} (100mV/div)]

Fig. 5 Transient response under output current change [Ch2: i_L (10A/div); Ch3: v_{gate} (5V/div); Ch4: i_o (5A/div)]

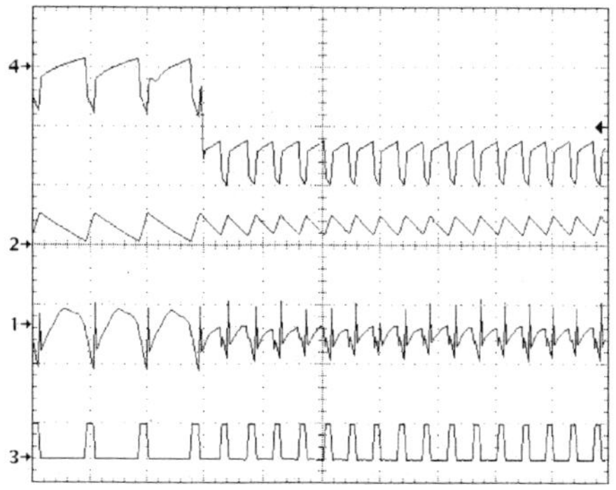

(a) v_i is changed from 20V to 18V.

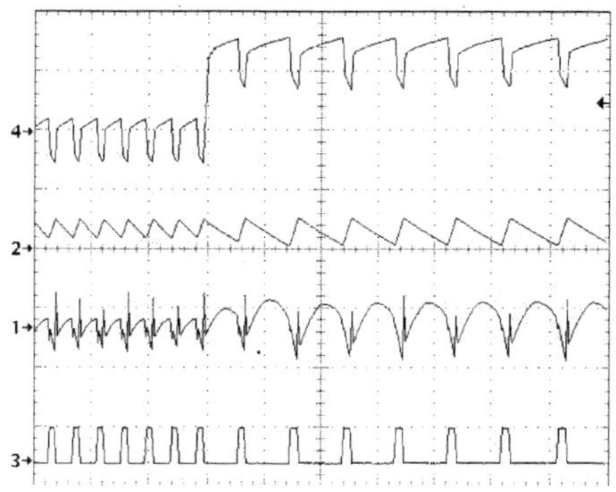

(b) v_i is changed from 18V to 20V.

Fig. 6 Transient responses under input voltage change. [Ch1: v_{out} (50mV/div); Ch2: i_L (10A/div); Ch3: v_{gate} (10V/div); Ch4: v_i (2V/div)]

Fig. 6(a) shows the transient responses of the converter when v_i is changed from 20V to 18V. Fig. 6(b) shows the transient responses of the converter when v_i is changed from 18V to 20V.

IV. Conclusions

An STP technique that is applied to achieve fast transient response of the boost converter is presented. The output voltage can revert to steady state within two switching actions when it is subject to large-signal disturbances. The STP performances have been studied experimentally with a 120W 18/24V converter prototype.

References

[1] K.K. Sum, *Switch Mode Power Conversion: Basic Theory and Design*, Marcel Dekker, 1984.

[2] L. H. Dixon, "Closing the feedback loop," in *Switching Regulated Power Supply Design Seminar Manual*. Lexington, KY: Unitrode, 1986.

[3] Y.F. Liu and P.C. Sen, "A general unified large signal model for current programmed dc-to-dc converters," *IEEE Trans. Power Electron.*, vol. 9, no. 4, pp. 414-424, July 1994.

[4] N. Kawasaki, H. Nomura, and M. Masuhiro, "A new control law of bilinear dc-dc converters developed by direct application of Lyapunov," *IEEE Trans. Power Electron.*, vol. 10, no. 3, pp. 318-325, May 1995.

[5] R. Naim, G. Weiss, and S. Ben-Yaakov, "H^{∞} control applied to boost power converters," *IEEE Trans. Power Electron.*, vol. 12, no. 4, pp. 677-683, July 1997.

[6] K. Smedley and S. Ćuk, "One-cycle control of switching converters," *IEEE Trans. Power Electron.*, vol. 10, no. 6, pp. 625-633, Nov. 1995.

[7] L. Rossetto, G. Spiazzi, P. Tenti, B. Fabiano, and C. Licitra, "Fast-response high-quality rectifier with sliding mode control," *IEEE Trans. Power Electron.*, vol. 9, no. 2, pp. 146-152, Mar. 1994.

[8] W. Burns and T. Wilson, "State trajectories used to observe and control the behavior of a voltage step-up dc-to-dc converter," *IEEE Trans. Aerosp. Electron.*, vol. 12, no. 6, pp. 706-717, Nov. 1976.

[9] C. C. Fang, "Sampled-data modeling and analysis of one-cycle control and charge control," *IEEE Trans. Power Electron.*, vol. 16, no. 3, pp. 345-350, May 2001.

DESIGN EQUATIONS FOR SUB-OPTIMUM OPERATION OF CLASS E AMPLIFIER WITH NONLINEAR SHUNT CAPACITANCE

T. Suetsugu[1], Senior Member, IEEE, *and M. K. Kazimierczuk*[2], Senior Member, IEEE

[1] Dept. of Electronics and Computer Science, Fukuoka University
8-19-1, Nanakuma, Johnan, Fukuoka 814-0180, Japan
[2] Dept. of Electrical Eng., Wright State University
Dayton, OH 45345 USA

ABSTRACT

Design equations for satisfying sub-optimum operating conditions of Class E amplifier with a nonlinear shunt capacitance with a grading coefficient of 0.5 at the duty cycle $D = 0.5$ are derived. By exploiting sub-optimum class E operation, various amplifier parameters such as input voltage, operating frequency, output power, and load resistance can be set as design specifications. An example of a design procedure of the class E amplifier is given. The theoretical results were verified with PSpice simulation.

1. INTRODUCTION

The shunt capacitance of the class E amplifier consists of only the transistor output parasitic capacitance at very high operating frequencies. The MOSFET output capacitance is nonlinear; it varies with the drain-to-source voltage. An analysis of the class E amplifier with a nonlinear shunt capacitance of a grading coefficient 0.5 was presented by Chudobiak for optimum operation at the duty cycle $D = 0.5$ [4]. But this result still does not show design equations for sub-optimum operation of the class E amplifier. Although the optimum operation at $D = 0.5$ is the most important case because it offers the maximum output power capability, sub-optimum operating conditions have also valuable features, such as lower peak switch voltage and lower peak switch current.

In this paper, design equations are derived for sub-optimum operation for a class E amplifier with a nonlinear shunt capacitance of grading coefficient 0.5. In this analysis, expressions for all elements, peak switch voltage, peak switch currents, and output power capability are derived. A class E amplifier circuit was designed and its operation was verified with PSpice using Level 3 MOSFET model.

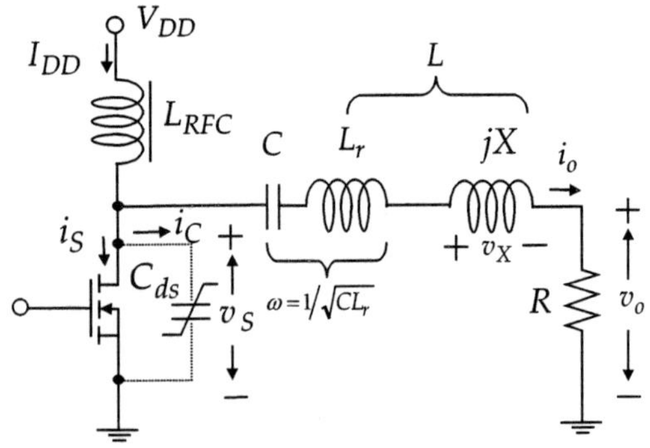

Fig. 1. Basic circuit of a class E amplifier.

2. WAVEFORMS OF THE CLASS E AMPLIFIER WITH NONLINEAR CAPACITANCE

The circuit analyzed in this paper is a class E amplifier shown in Fig. 1. The shunt capacitance of the amplifier consists only of the MOSFET output capacitance. The derivations of design equations are carried out under the following assumptions:

1. The inductance of the choke coil L_{RFC} is large enough to neglect its current ripple.
2. The internal resistance of the choke coil is zero; therefore, the DC voltage drop across the choke is zero.
3. The loaded quality factor Q of the output resonance circuit is high enough so that the output current can be considered a sine wave.
4. The load resistance includes parasitic resistances of the series resonance circuit, i.e., the resonance circuit is considered to be a pure reactance.
5. The MOSFET on-resistance transistor is zero.
6. The MOSFET turns on and off instantly.
7. The shunt capacitance C_{ds} is entirely comprised of the MOSFET output capacitance.
8. The grading coefficient of the MOSFET output capacitance is 0.5. The shunt capacitance is described by

$$C_{ds} = \frac{C_{j0}}{\sqrt{1 + \frac{v_S}{V_{bi}}}} \qquad (1)$$

where C_{j0} is the shunt capacitance at the drain-to-switch voltage $v_S = 0$ and V_{bi} is the built-in potential of the MOSFET body diode.

It is also assumed that the switch voltage satisfies the ZVS condition at the switch turn-on and the switch voltage is not negative before the switch turn-on.

When the switch is OFF, i.e., for $0 < \theta \le \pi$, the current through the shunt capacitance is

$$i_C(\theta) = I_{DD} - I_m \sin(\theta + \phi) \qquad (2)$$

where I_{DD} is the dc input current, I_m is the output current amplitude, and $\theta = \omega t$. The switch voltage v_S is the integral of the shunt capacitor current i_C given by

$$i_C = C_1 \frac{dv_S}{d\theta}. \tag{3}$$

Integrating both side of this equation with respect of θ gives

$$\omega \int C_1 dv_S = \int i_C d\theta. \tag{4}$$

Solving this equation, the switch voltage v_S for $0 < \theta \le \pi$ can be derived as

$$v_S(\theta) = V_{bi}\left\langle \left\{ \frac{I_{DD}\theta + I_m[\cos(\theta+\phi) - \cos\phi]}{2V_{bi}\omega C_{j0}} + 1 \right\}^2 - 1 \right\rangle. \tag{5}$$

3. Circuit Parameters for ZVS Operation at $D = 0.5$

Imposing the ZVS condition given by

$$v_S(\pi) = 0 \tag{6}$$

on equation (5) yields

$$\frac{\pi I_{DD} - 2I_m \cos\phi}{2\omega C_{j0} V_{bi}} + 1 = \pm 1. \tag{7}$$

A solution that satisfies condition with a non-negative switch voltage and ZVS condition can be obtained only when the right side of (7) is equal to 1. This case leads to

$$I_{DD} = \frac{2I_m}{\pi}\cos\phi \tag{8}$$

Since the power loss in the circuit is zero, the output power is equal to the input power, i.e., $I_{DD}V_{DD} = I_m^2 R/2$, I_m can be obtained as

$$I_m = \frac{4V_{DD}}{\pi R}\cos\phi. \tag{9}$$

I_{DD} can be obtained as

$$I_{DD} = \frac{8V_{DD}}{\pi^2 R}\cos^2\phi. \tag{10}$$

Substituting (9) and (10) into (2), the shunt capacitance current at the switch turn-on $i_C(\pi)$ is obtained as

$$i_C(\pi) = \frac{4V_{DD}\cos\phi}{\pi R}\left(\frac{2}{\pi}\cos\phi + \sin\phi\right). \tag{11}$$

Note that $i_C(\pi) = 0$ at $\phi = \arctan(-2/\pi) = -0.567$ rad. The output power P_o is obtained as

$$P_o = \frac{8V_{DD}^2}{\pi^2 R}\cos^2\phi. \tag{12}$$

Fig. 2 plots the normalized output current amplitude $I_m R / V_{DD}$, the normalized output power $P_o R / V_{DD}^2$, and the normalized shunt capacitance current at the switch turn-on $Ri_C(\pi)/V_{DD}$ as functions of ϕ. Since $Ri_C(\pi)/V_{DD} < 0$, the operating range is $-\pi/2 < \phi < -0.567$ rad. The output power P_o is zero at $\phi = -\pi/2$. In this case, the dc input current I_{DD} is zero and the switch voltage waveform is a part of a sinusoidal

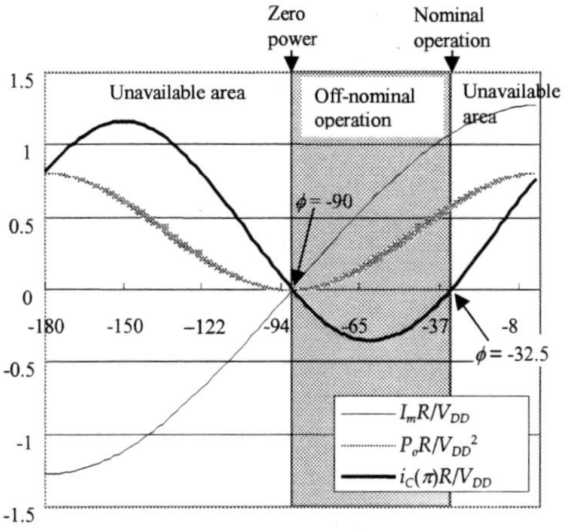

Fig. 2. Normalized output current amplitude $I_m R/V_{DD}$, normalized output power $P_o R/V_{DD}^2$, and normalized shunt capacitance current at switch turning-on $i_C(\pi)R/V_{DD}$ as functions of ϕ.

waveform. The switch voltage slope at the switch turn-on is zero when $\phi = -0.567$ rad. In this case, the amplifier achieves the optimum operation and the product of the output power P_o and the load resistance R reaches the maximum value given by

$$P_o R / V_{DD}^2 = 8/(\pi^2 + 4). \tag{13}$$

Since the dc component of the voltage drop across the choke inductor L_{RFC} is zero

$$V_{DD} = \frac{1}{2\pi}\int_0^{2\pi} v_S(\theta) d\theta$$

$$= \frac{1}{2\pi}\left\{ \frac{V_{DD}^2 \cos^2\phi}{(\pi\omega C_{j0}R)^2 V_{bi}}\left[\frac{4\pi}{3} + \left(\pi - \frac{24}{\pi}\right)\cos^2\phi\right] - \frac{2V_{DD}\sin 2\phi}{\pi\omega C_{j0}R} \right\}. \tag{14}$$

Equation (14) is a quadratic of $\omega C_{j0}R$, which has two possible solutions. One solution is not suitable for this circuit because $\omega C_{j0}R < 0$ when $-\pi/2 < \phi < -0.567$ rad and the shunt inductance short circuits the dc supply voltage. The other solution of (14) for $\omega C_{j0}R$ is

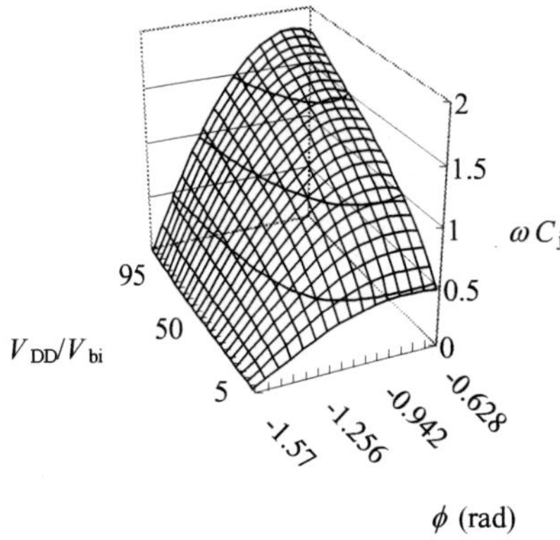

Fig. 3. $\omega C_{j0}R$ versus ϕ and V_{DD}/V_{bi}.

$$\omega C_{j0}R = \frac{\cos\phi}{\pi^2}\left[-2\sin\phi + \sqrt{4\sin^2\phi + \left[\left(\frac{2}{3}\pi^2 - 16\right)\cos^2\phi + \pi^2\right]\frac{V_{DD}}{V_{bi}}}\right]. \quad (19)$$

Fig. 3 plots $\omega C_{j0}R$ as a function of ϕ and V_{DD}/V_{bi}. It can be seen that $\omega C_{j0}R$ is large when V_{DD}/V_{bi} is large. $\omega C_{j0}R$ is about 1.8 when $V_{DD}/V_{bi} = 120$. $\omega C_{j0}R$ is zero when $\phi = -\pi/2$ and it reaches the maximum value when $\phi = -0.8$ rad. The sum of the output voltage $v_o(\theta)$ and the fundamental component of the voltage $v_X(\theta)$ across the reactance X can be expressed as

$$v_1(\theta) = V_o\sin(\theta+\phi) + \frac{XV_o}{R}\cos(\theta+\phi) = V_1\sin(\theta+\phi_1) \quad (20)$$

where

$$V_1 = V_o\sqrt{1+\left(\frac{X}{R}\right)^2} \quad (21)$$

is the amplitude of v_1, and

$$\phi_1 = \phi + \arctan\left(\frac{X}{R}\right). \quad (22)$$

Because the reactance of the resonant circuit $C - L_r$ is zero at the operating frequency,

$$0 = \frac{1}{\pi}\int_0^{2\pi} v_S(\theta)\cos(\theta+\phi_1)d\theta. \quad (23)$$

Using this equation, ϕ_1 is derived as

$$\phi_1 = \arctan\left\langle\left\{\frac{V_{DD}}{V_{bi}}\frac{1}{3\pi}\cos^2\phi\sin\phi + \omega C_{j0}R\left(\frac{4}{\pi} - \frac{\pi}{2}\right)\right\} \Big/ \left\{\frac{V_{DD}}{V_{bi}}\left[\left(\frac{16}{\pi^2} - \frac{1}{3}\right)\cos^2\phi - \frac{4}{3}\right]\frac{\cos\phi}{\pi} + \omega C_{j0}R\frac{\pi}{2}\sin\phi\right\}\right\rangle. \quad (24)$$

The reactance X is given by

$$X = R\tan(\phi - \phi_1). \quad (25)$$

Fig. 4 plots X/R as a function of ϕ and V_{DD}/V_{bi}. It can be

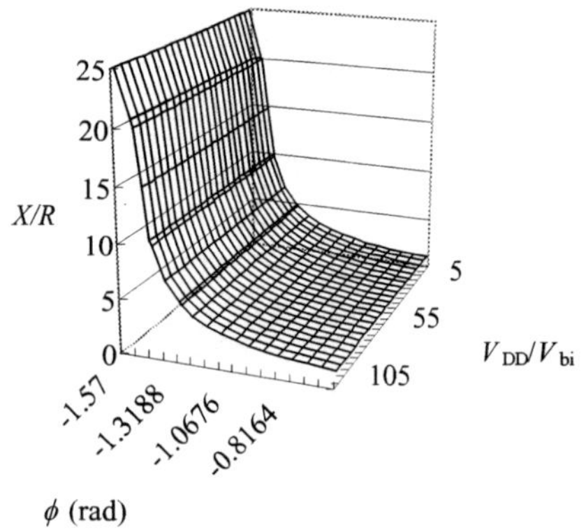

Fig. 4. X/R versus ϕ and V_{DD}/V_{bi}.

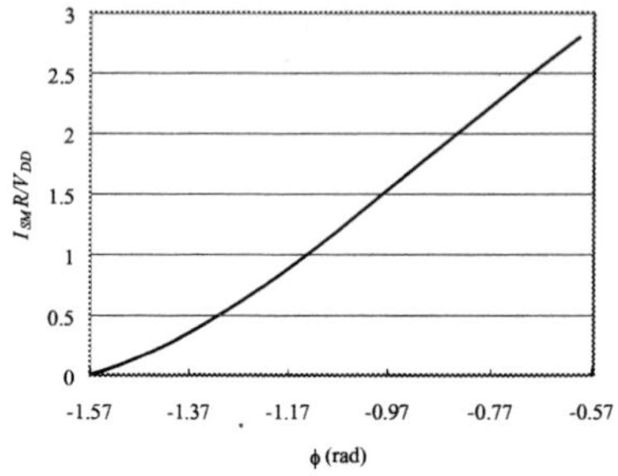

Fig. 5. Normalized maximum switch current $I_{SM}R/V_{DD}$ versus ϕ.

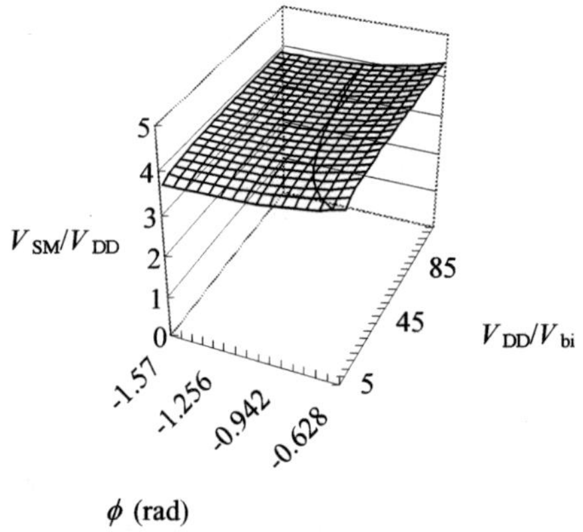

Fig. 6. Normalized V_{SM}/V_{DD} versus ϕ and V_{DD}/V_{bi}.

seen that X/R increases as ϕ increases. X/R is about 20 for $\phi = -\pi$ and it is about 1 for $\phi = -0.567$ rad. There is a little change of X/R with variations of V_{DD}/V_{bi}. The maximum switch current I_{SM} is calculated as a sum of the input current I_{DD} and the output current amplitude I_m. It is derived as

$$I_{SM} = I_{DD} + I_m = \frac{4V_{DD}}{\pi R}\cos\phi\left(\frac{2}{\pi}\cos\phi + 1\right). \quad (26)$$

Fig. 5 shows normalized maximum switch current $I_{SM}R/V_{DD}$ as a function of ϕ. It can be seen that the normalized maximum switch current $I_{SM}R/V_{DD}$ as ϕ increases from –1.57 rad to

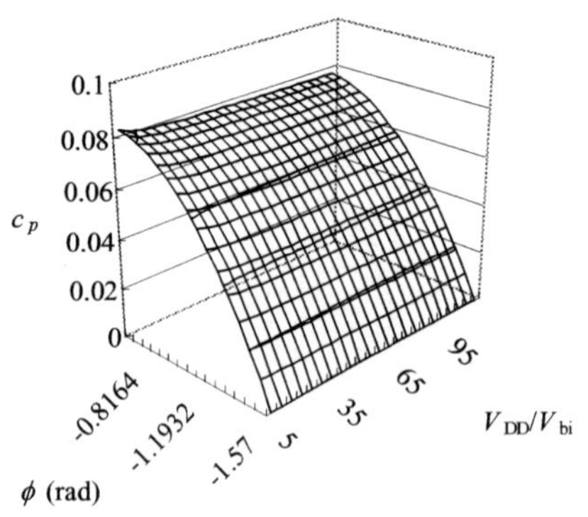

Fig. 7. c_p versus ϕ and V_{DD}/V_{bi}.

−0.57 rad. The maximum switch voltage V_{SM} occurs when the shunt capacitance current i_C is zero. The maximum switch voltage V_{SM} is equal to the switch voltage at $\theta = \theta_1$, i.e., $V_{SM} = v_S(\theta_1)$, where

$$\theta_1 = -\phi + \sin^{-1}\left(\frac{2}{\pi}\cos\phi\right). \qquad (27)$$

Fig. 6 shows normalized maximum switch voltage V_{SM}/V_{DD} as a function of ϕ and V_{DD}/V_{bi}. It can be seen that V_{SM}/V_{DD} increases with increasing ϕ. V_{SM}/V_{DD} is less than about 4.5 and greater than 3.5 when V_{DD}/V_{bi} is less than 120. The output power capability c_p is defined as

$$c_p = \frac{P_o}{I_{SM}V_{SM}} = \frac{I_{DD}V_{DD}}{I_{SM}V_{SM}}. \qquad (28)$$

Fig. 7 shows the output power capability c_p as a function of ϕ and V_{DD}/V_{bi}. It can be seen that c_p decreases as ϕ increases. c_p reaches the maximum value when ϕ is approximately −0.6 rad.

4. Design Example

A design procedure is shown for an example circuit whose operating frequency is $f = 4$ MHz, $V_{DD} = 20$ V, $P_o = 2.6$ W, $R = 40\ \Omega$, $V_{bi} = 0.7$ V, and $Q = 10$. From (11)

$$\cos\phi = \frac{\pi}{V_{DD}}\sqrt{\frac{PR}{8}} = \frac{\pi}{20}\sqrt{\frac{2.6\cdot 40}{8}} = 0.566 \qquad (28)$$

from which $\phi = 0.969$ rad. From (17),

$$C_{j0} = \frac{\cos\phi}{\pi^2\omega R}\left[-2\sin\phi + \sqrt{4\sin^2\phi + \left[\left(\frac{2}{3}\pi^2 - 16\right)\cos^2\phi + \pi^2\right]\frac{V_{DD}}{V_{bi}}}\right]$$

$$= 898\ \text{pF}, \qquad (29)$$

$$\phi_1 = \arctan\left\langle\left\{\frac{V_{DD}}{V_{bi}}\frac{1}{3\pi}\cos^2\phi\sin\phi + \omega C_{j0}R\left(\frac{4}{\pi} - \frac{\pi}{2}\right)\cos\phi\right\}\right.$$

$$\left./\left\{\frac{V_{DD}}{V_{bi}}\left[\left(\frac{16}{\pi^2} - \frac{1}{3}\right)\cos^2\phi - \frac{4}{3}\right]\frac{\cos\phi}{\pi} + \omega C_{j0}R\frac{\pi}{2}\sin\phi\right\}\right\rangle$$

$$= 0.16\ \text{rad}, \qquad (30)$$

$$X = R\tan(\phi_1 - \phi) = 40\cdot\tan(0.16 + 0.969) = 84.5\ \Omega, (31)$$

$$L = \frac{RQ}{\omega} = \frac{40\cdot 10}{2\pi\cdot 4\cdot 10^6} = 15.9\ \mu\text{H}, \qquad (32)$$

$$C = \frac{1}{\omega(RQ - X)} = \frac{1}{2\pi\cdot 4\cdot 10^6(40\cdot 10 - 84.5)} = 126\ \text{pF}, (33)$$

$$V_m = RI_m = \frac{4V_{DD}}{\pi}\cos\phi = \frac{4\times 20}{\pi}0.566 = 14.4\ \text{V}. \quad (34)$$

Then, V_{SM} can be obtained using (27) and (5),

$$\theta_0 = -\phi + \sin^{-1}\left(\frac{2}{\pi}\cos\phi\right) = -0.6\ \text{rad}, \qquad (35)$$

$$V_{SM} = V_{bi}\left\langle\left\{\frac{I_{DD}\theta_0 + I_m[\cos(\theta_0 + \phi) - \cos\phi]}{2V_{bi}\omega C_{j0}} + 1\right\}^2 - 1\right\rangle = 79.2$$

V (36)

I_{SM} can be obtained from (26),

$$I_{SM} = \frac{4V_{DD}}{\pi R}\cos\phi\left(\frac{2}{\pi}\cos\phi + 1\right) = \frac{4\cdot 20}{\pi\cdot 40}\cdot 0.566\cdot\left(\frac{2}{\pi}\cdot 0.566 + 1\right) = 0.49\ \text{A}. (37)$$

5. Simulation Results

The example circuit designed in Section VI was simulated with PSpice. The SPICE MOSFET MODEL Level 3 was used for the MOSFET. In accordance with the calculations, we set CBD = 898 pF, PB = 0.7 V, and MJ = 0.5. The simulated waveforms of switch voltage v_S and the output voltage v_o are shown in Fig. 8. The switch voltage v_S became negative just before the switch turns on. The ZVS operation was achieved. The peak switch voltage V_{SM} was 92 V. It was about 10 V different from the calculated value. The output voltage amplitude V_m was 14 V. It was very close to the calculated value.

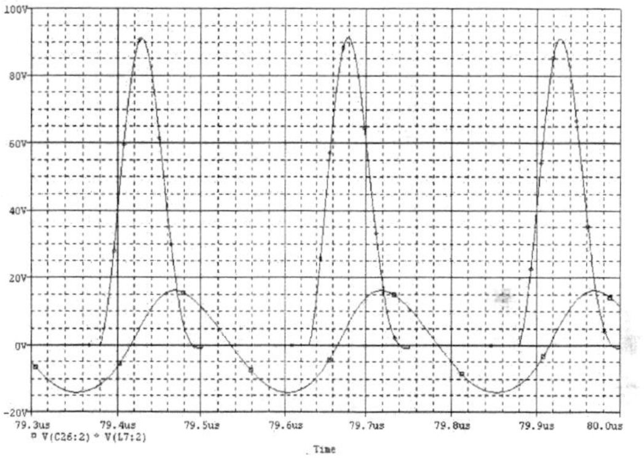

Fig. 8. Simulated waveforms of the switch voltage v_S and the output voltage v_o.

6. Conclusions

In this paper, the class E amplifier with a nonlinear shunt capacitance for the sub-optimum operation was analyzed and the design equations were derived. The operation was verified with PSpice simulation. By exploiting sub-optimum operation, designer can consider various operating conditions

References

[1] N. O. Sokal and A. D. Sokal, "Class E - A new class of high-efficiency tuned single-ended switching power amplifiers," *IEEE J. Solid-State Circuits*, vol. SC-10, pp. 168-176, June 1975.

[2] M. J. Chudobiak, "The use of parasitic nonlinear capacitors in class E amplifiers," *IEEE Trans. Circuits Syst. I*, vol. 41, no. 12, pp. 941-944, Apr. 1994.

[3] M. K. Kazimierczuk and D. Czarkowski, *Resonant Power Converters*, New York: John Wiley & Sons, Inc., 1995, Chapter 13.

TOWARD THE INTEGRATION OF MICROSYSTEMS SUPPLY

Bruno ESTIBALS [1], Corinne ALONSO [1], Alain SALLES [1], Angel CID-PASTOR [1], Henri CAMON [1], Luis MARTINEZ-SALAMERO [2]

1 LAAS-CNRS
Laboratoire d'Analyse et d'Architecture des Systèmes
7, avenue du Colonel Roche – F-31077 Toulouse Cedex 04, France
Phone : (33) +5 61 33 64 43 – E-Mail : bruno.estibals@laas.fr

2 ETSE-URV
Escola Tècnica Superior d'Enginyeria, Universitat Rovira I Virgili
Avda. Països Catalans 26, Campus Sescelades, 43007 Tarragona, Spain

ABSTRACT

This paper presents a global methodology for the complete design of an integrated inductor. After describing the different existing topologies, a method providing the geometrical dimensions from the desired inductance value is reported. Then, the frequency dependency of the inductance and its parasitic capacitance and resistance are analyzed by means of specific CAD Tools. A specific analysis of thermal failure mechanisms is presented next. Temperature and current density of different shapes of connection pad are subsequently studied.

1. INTRODUCTION

Microsystems or MEMS (*for Micro Electro Mechanical Systems*) are implemented by combining individual devices, either monolithically or by means of hybrid packaging technology, with the aim of creating a complete system performing specific functions. These functions are mainly environment perception, data processing, communications and action/reaction on the environment. The electrical architecture of a microsystem includes electronic and/or optical circuits, signal generators and receivers, microsensors, microactuators and microgenerators. The autonomy of the microsystem is guaranteed by microgenerators which supply the electrical energy at a very small scale. However, current solutions for microgeneration are still based on electrochemical storage elements which are heavy, bulky, polluting and require frequent and long recharging. On the other hand, many efforts are being carried out in this field to improve the performances of present microgenerators by developing new accumulators, fuel cells and hybrid structures.

Moreover, the problem of microsystem supply turns into a power processing problem at a very small scale, *i.e.* the design of an interface between a microgenerator and the different elements of a microsystem ensuring a stable level of electric energy with specified regulation indexes.

The design of such supplies is constrained of severe limitations ranging from the electric requirements, *i.e.* high voltage (*around 10V*) and small current (*electrostatic levels*), small voltage (*less than 3V*) and high current (*several Amperes*), to the need of specific components as switches, capacitors and inductors as well as microsystem actuators (*electrostatic, magnetic or thermal*).

The main bottle neck in the integration of power supplies is the microelectronic integration of inductors [1]. Although a great amount of research has been devoted to this field in the last years, so far no technology has reached a real degree of maturity to announce immediate dramatic changes in the supply integration. This is due to the complex interdisciplinary nature of the magnetic integration process where a correct establishment of a global methodology concerning all parts involved in the process can be a determinant factor of success.

This paper presents a global methodology for the complete design of an integrated inductor. After describing the different existing topologies, a method providing the geometrical dimensions from the desired inductance value is reported. Then, the frequency dependency of the inductance and its parasitic capacitance and resistance are analyzed by means of specific CAD Tools. A specific analysis of thermal failure mechanisms is presented next. Temperature and current density are subsequently studied and different shapes of connection pads are proposed.

2. GEOMETRICAL INDUCTANCE DESIGN

2.1. The different inductance topologies.

A considerable effort has been devoted to modeling and design of integrated inductors whose only practical options are bond wires and planar spiral geometries. Although bond wires permit to achieve a high quality factor *Q*, with typical *Q*'s in the 20-50 range, their inductance values are constrained and can be rather sensitive to production fluctuations. On the other hand, planar spiral inductors have limited *Q*'s, but their inductances are well-defined over a broad range of process variations. Thus, planar spiral inductors have become essential elements of communication circuit blocks or integrated converters.

Square spirals are popular because of the ease of their layout. Squares are easily generated even in simple Manhattan style layout tools. However, other polygonal spirals have also been used in circuit design. Some designers prefer polygons with more than fours sides to improve performance. Among these, hexagonal and octagonal inductors are used widely. Figure 1 shows the layout for square, octagonal and circular inductors respectively.

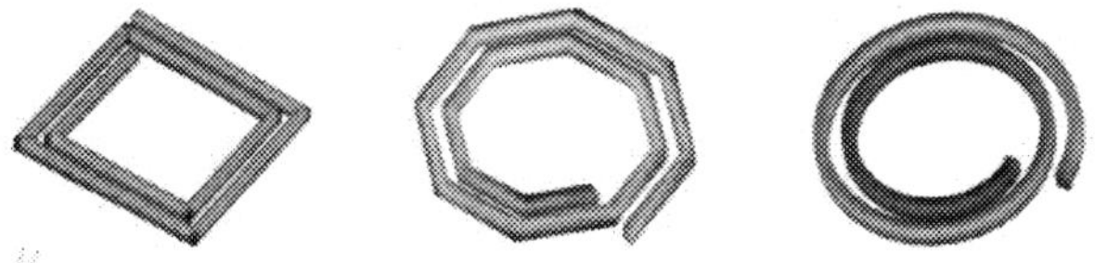

a. b. c.

Figure 1. 3D Model of the Simulated Integrated Inductances obtained with Coventor Solution.
a. Square – b. Octagonal – c. Circular.

2.2. Inductance evaluation using an analytical model.

For a given shape, an inductor is completely specified by the number of turns *n*, the turn width *w*, the turn spacing *s* and any of the following parameters : the outer diameter d_{out}, the inner diameter d_{in}, the average diameter $d_{avg} = 0.5.(d_{out} + d_{in})$ or the fill ratio defined as $\rho = \frac{d_{out} - d_{in}}{d_{out} + d_{in}}$. The thickness of the inductor has only a very small effect on inductance.

To facilitate the design of such components, significant work has gone into modeling spiral inductors using lumped circuit models. The parasitic resistors and capacitors in this model have physically intuitive simple expressions, but the inductance value lacks a simple accurate expression. This inductance can be computed exactly by solving Maxwell's equations. Another technique is based on the Greenhouse method to compute the inductance [2-3]. It offers sufficient accuracy and adequate speed, but cannot provide an inductor design directly from specifications and is cumbersome for initial design. More, only inductance of square design can be calculated.

In order to evaluate shape influence on inductance value, another simple and accurate expression for the inductance of a planar spiral can be obtained by approximating the sides of the spirals by symmetrical current sheets of equivalent current densities. For example, in the case of the square, four identical current sheets can be obtained. The current sheets on opposite sides are parallel to one another, whereas the adjacent ones are orthogonal. Using symmetry and the fact that sheets with orthogonal current have zero mutual inductance, the computation of the inductance is now reduced to evaluating the self-inductance of one sheet and the mutual inductance between opposite current sheets. These self and mutual inductances are evaluated using the concepts of geometric mean distance (*GMD*), arithmetic mean distance (*AMD*), and arithmetic mean square distance (*ASMD*). The resulting expression is [3] :

$$L_{gmd} = \frac{\mu . n^2 . d_{avg} . c_1}{2} . \left(\ln\left(\frac{c_2}{\rho}\right) + c_3 . \rho + c_4 . \rho^2 \right) \quad (1)$$

where the coefficients c_i are layout dependent and are shown in table 1, ρ is the fill ratio defined previously and μ the permeability. Although the accuracy of this expression worsens as the ratio $\frac{s}{w}$ becomes large, it exhibits a maximum error of 8% for $s \le 3w$. Note that typical practical integrated spiral inductors are built with $s \le w$. The reason is that a smaller spacing improves the inter-winding magnetic coupling and reduces the area consumed by the spiral. A large spacing is only desired to reduce the inter-winding capacitance. In practice, this is not a concern as this capacitance is dwarfed by the underpass capacitance.

Layout	c_1	c_2	c_3	c_4
Square	1.27	2.07	0.18	0.13
Hexagonal	1.09	2.23	0.00	0.17
Octogonal	1.07	2.29	0.00	0.19
Circle	1.00	2.46	0.00	0.20

Table 1. Coefficients for current sheet expression [16].

Figures 2 and 3 show the inductance dependency on inner diameter and number of turns respectively for different shapes of inductance : square, hexagonal, orthogonal and circle. The corresponding dimensions for both width *w* and space *s* is 1µm whereas the thickness *t* is equal to 3µm. in figure 2, turn number *n* is equal to 20 ; in figure 3, outer diameter d_{out} is equal to 700µm. note that the square spiral topology provides the biggest inductance value irrespective of the variant parameter.

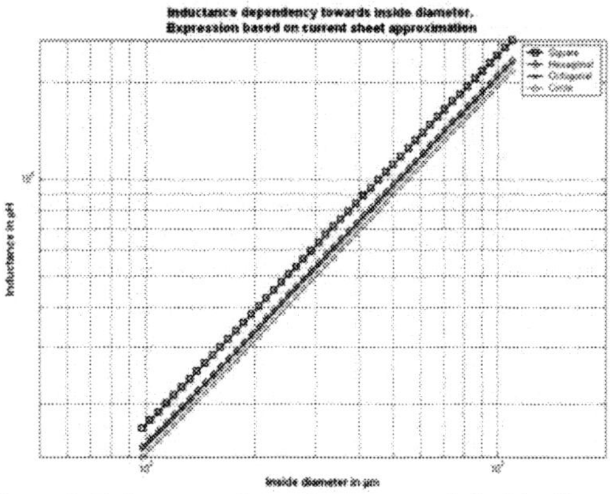

Figure 2. Inductance dependency towards inside diameter.
w = s = 1µm ; t = 3µm ; n = 20.

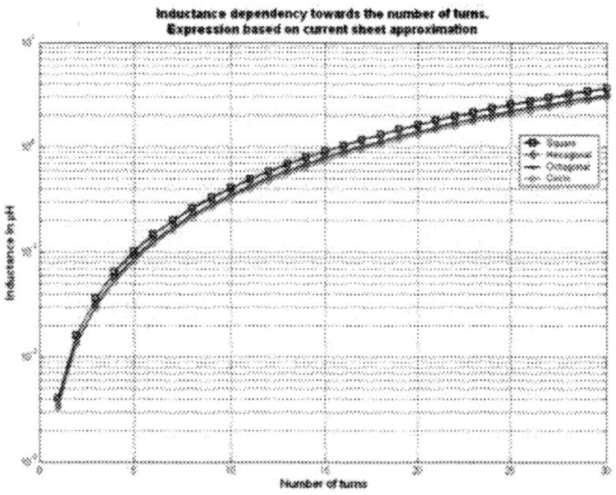

Figure 3. Inductance dependency towards number of turns
w = s = 1µm ; t = 3µm ; d_{out} = 700µm

There are two major difficulties that must be overcome in order to improve the usefulness of this geometry. First, closed magnetic circuits must be completed using a thick magnetic material with high permeability to reduce magnetic reluctance and to minimize magnetic field interference (*making the adhesion of the thick magnetic layer a concern*). Second, the spiral conductor must have as small resistance as possible to reduce power consumption in the conductors.

MemHenry Module of Coventorware solution uses an integral equation approach that requires a mesh only in the interior of conductors, thus eliminating the need for the user to mesh all free space [4]. Combining this approach with a multipole-accelerated solution algorithm, MemHenry can rapidly complete the extraction of resistance and inductance frequency dependence for complicated 3D structures. Using the same model, the designer can analyze the dependence of inductance with respect to varying position, dimension, or even mechanical deformation.

Our first task has been validating the MemHenry module. Therefore, several simulations have been performed to obtain the frequency dependence of inductance and resistance. Figure 4 and 5 show the corresponding results for all shapes depicted in figure 1.

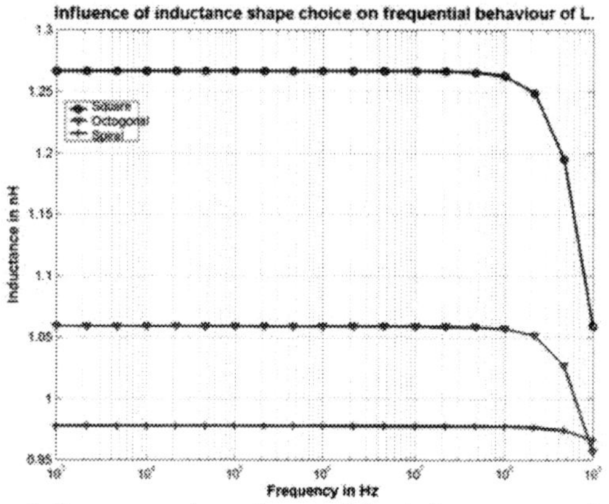

Figure 4. Frequency dependence of the inductance as a function of its geometrical shape.
$w = s =$ 10µm ; $t =$ 3µm ; $n = 2$.

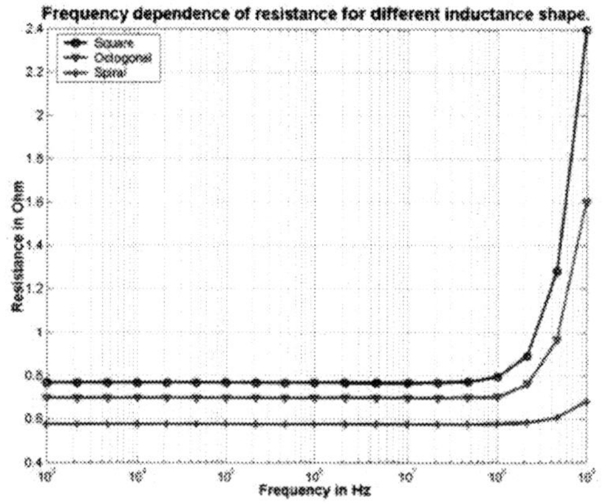

Figure 5. Frequency dependence of the resistance as a function of its geometrical shape.
$w = s =$ 10µm ; $t =$ 3µm ; d_{out} = 100µm

Figures 4 and 5 show clearly that, using same geometrical parameters, square design give systematically the highest inductance and serial resistance values.

An important figure of merit for inductors is the quality factor Q. Q factors are estimated by :

$$Q = \frac{\mathrm{Im}[y_{11}]}{\mathrm{Re}[y_{11}]} = \frac{L\omega}{R} \qquad (2)$$

Where L, R and ω are inductance, resistance and angular frequency, respectively. Quality factors of inductors measured in this way increase linearly with frequency until one peak. For higher frequencies, Q-factors decrease with frequency. This decrease is caused by a capacitive coupling of signals through a parasitic metal to substrate capacitor, C_{p1}. In general, this Q-estimation technique is valid for frequencies below the frequency where the peak of Q occurs. If the frequency is increased sufficiently,

$\frac{\text{Im}[y_{11}]}{\text{Re}[y_{11}]}$ becomes zero. This point is the frequency at which the average magnetic energy storage is approximately equal to the average electric energy storage, *ie*. the self resonant frequency of the inductor. Figure 6 shows the *Q*-factor dependency for different inductance shapes. Note that circular shapes exhibit better *Q*. Therefore, shapes are chosen for RF circuits and systems working at 5-6 GHz [5].

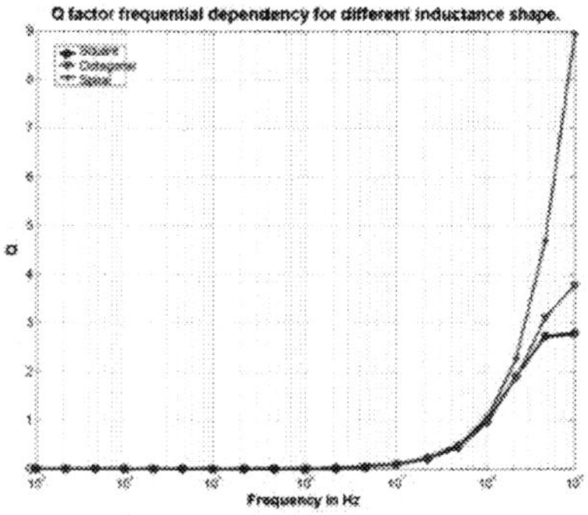

Figure 6. *Q* factor frequential dependency for different inductance shape.
$w = s = 10\mu m$; $t = 3\mu m$; $d_{out} = 100\mu m$

A straightforward way to increase *Q*-factors is to reduce series resistance by thickening or widening conductors, although widening them increase the substrate parasitic capacitance. This increase can limit the operating frequency range for the inductors. Therefore, we decide to limit wideness and to try to realize conductors as narrow as possible.

3. ELECTRO-THERMAL STUDY OF ENERGY-STORAGE IN NANO-INDUCTORS

Inductance with sub-micronic dimensions are approaching the mean-free path of the conduction electrons (*typically a few 10nm*). Therefore, it is becoming increasingly important to characterize and understand the electrical transport properties and failure mechanisms of such thin conductors.

The first part of our work has consisted in validating MemEtherm for such dimensions. With this aim, we have compared our results with those obtained by Durkan in [6] for a same shape. Then, we have focused our work on via connection. Indeed, when connecting the component, vias from 50µm to few hundreds micrometers large are employed and present an important geometrical discontinuity using the 0.3µm inductor's wire. Therefore, conductors failures would appear in this point because of the thermal mechanical stress.

We have used MemEtherm toolbox of CoventorWare, which is a specialized solver used to analyze structural deformations of MEMS devices due to Joule heating. By applying a voltage or current flow as a boundary condition, the solver computes the resulting temperatures.

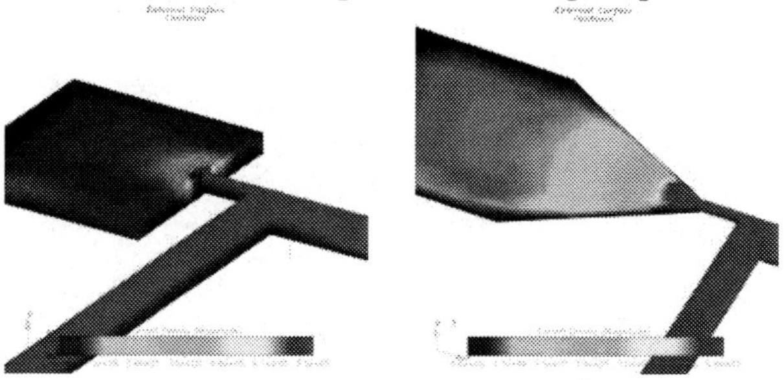

a. b.

Figure 7. Current density repartition towards contact shape.
a. Classical square – b. Funnel shape.

Two different connection topologies have been designed and simulated. Figure 7a. shows the classical square shape connected with the end of a nanowire, which presents huge current density discontinuity. Current density in the contact is evaluated to be hundred times smaller than in the wire, implying electromigration weakening the contact. Figure 7b. shows that the transition between the wire and the funnel-shape contact is softened. As the current density gradient is smaller, electromigration effects will be significantly reduced.

4. CONCLUSION

A global methodology for the complete design of an integrated inductor has been presented. Analytical expressions and Coventorware simulations have permitted to conclude that square shape is the most adapted design to obtain high inductance values. Electro-thermal simulations for nano-inductors have been performed leading to the conclusion that progressive funnel shapes must be used to connect electrical pads to nano-wire.

5. REFERENCES

[1] B. Estibals, C. Alonso, "Design of Power Converters for New Low Power Applications", *Proc. Of EPE-PEMC 2002*, Dubrovnick, Cavtat, paper n° ssls_04.pdf, 2002.

[2] H-M. Greenhouse, "Design of Planar Rectangular Microelectronic Inductors", *IEEE Trans. Parts, Hybrid Packaging*, vol. PHP-10, n° 2, 1974.

[3] C-P. Yue, C. Ryu, J. Lau, T-H. Lee, S-S. Wong, "A physical model for planar spiral inductors on silicon", *Proc. Of IEEE Intern. Electron Devices Meeting*, San Fransisco, CA, 1996.

[4] Web site : http://www.coventor.com/

[5] R-B. Merrill, T-W. Lee, H. You, R. Rasmussen, L-AS. Moberly, "Optimization of high Q integrated inductors for multilevel metal CMOS", *Proc. Of IEEE Int. Electron Devices Meeting, Washington DC*, pp. 38.7.1-38.7.4, 1995.

[6] C. Durkan and M. E. Welland, "Analysis of failure mechanisms in electrically stressed gold nanowires", *Ultramicroscopy*, Elsevier Science, pp. 125-133, 2000.

ENHANCED MULTIPLE HUFFMAN TABLE (MHT) ENCRYPTION SCHEME USING KEY HOPPING

Dahua Xie and C.-C. Jay Kuo

Integrated Media Systems Center and Department of Electrical Engineering
University of Southern California, Los Angeles, CA 90089-2564, USA
E-mails: dahuaxie@usc.edu and cckuo@sipi.usc.edu

ABSTRACT

Efficient multimedia encryption algorithms play a key role in multimedia security protection. One algorithm known as the MHT (Multiple Huffman Tables) method was recently developed by Wu and Kuo in [4] and [5]. Although MHT has many desirable advantages, it is vulnerable under the chosen-plaintext attack (CPA). In this work, an enhanced MHT encryption scheme is proposed to remedy this drawback using the idea of key hopping. It is demonstrated by mathematical analysis that our proposed scheme is secure against chosen plaintext attack.

1. INTRODUCTION

The rapid development of multimedia compression and processing techniques (e.g. MPEG4, H.264, G.273, AMR) has brought about a proliferation of novel multimedia applications, such as on-line music streaming, real-time video monitoring. As a consequence, the problem of multimedia data security is receiving an increasing amount of attention nowadays in both academia and industry. Among various multimedia security topics, a cryptographically secure multimedia encryption scheme is always of fundamental importance.

Traditional crypto algorithms, such as the private-key 3DES and the public-key RSA, work perfectly well with regular security applications. However, they fail to give satisfactory performance when it comes to multimedia content encryption. The reason is two-fold. First, the size of multimedia content is often much larger than that of the traditional text data, which makes encrypting the whole content impractical. Second, in many multimedia applications, there is some associated timing constraint, especially for real-time on-line applications. The speed of traditional cryptographic ciphers may be too slow to catch upon the real-time requirements. In summary, a good multimedia encryption scheme has to achieve a high level of content security while keeping a low computation cost at the same time. Designing such a scheme is therefore a challenging task.

A promising direction of research in this field, pioneered by Wu and Kuo [4], [5], is to combine encryption with entropy coding by using multiple statistical models in the entropy codec. Entropy coding is the last stage of many multimedia compression systems, where a symbol stream is converted into a bit-stream via a statistical model. The major advantage of this scheme is that encryption is carried out at the same time of entropy coding, thus demanding only a negligible amount of computation. The high semantic security is guaranteed in the sense that, without knowing the key, the bit-stream cannot be correctly decoded. The entropy coder considered by Wu and Kuo was the Huffman codec and the QM-coder. Even though the multiple Huffman tables (MHT) method proposed by Wu and Kuo has many desirable properties, it is vulnerable to the chosen-plaintext attack (CPA). An enhanced MHT encryption scheme is proposed in this work to overcome this drawback. It is proved mathematically that the proposed algorithm is secure against the chosen plaintext attack.

The rest of the paper is organized as follows. Previous work is reviewed in Section 2 with emphasis on the security analysis of the Multiple Huffman Table (MHT) encryption scheme. We introduce the concept of chosen-plaintext attack (CPA) and show that the basic MHT scheme is vulnerable under CPA. An enhanced algorithm is presented in Section 3 to overcome this drawback. It is proved by crypto-analysis that the enhanced MHT encryption scheme is CPA-secure. Finally, concluding remarks and future research directions are given in Section 4.

2. SECURITY ANALYSIS OF MHT

In this section, we briefly introduce several existing approaches and discuss their pros and cons. In particular, we will focus on the basic MHT scheme and its security analysis under chosen plaintext attack (CPA).

2.1. Selective Encryption Methods

In *selective encryption* schemes, only a particular portion of the bit stream is selected and encrypted using a

0-7803-8251-X/04/$17.00

cryptographic cipher with the remaining part unchanged, in hope that the selected portion is crucial to understanding the multimedia content. The success of a selective scheme thus counts heavily on choosing the right part that contains the richest content meaning of an image/video.

Several existing selective encryption schemes are primarily based on DCT coefficient and motion vector encryption/scrambling [1],[2],[3]. However, later research indicates that all these schemes either do not achieve high semantic security or incur a significant computational overhead. This is mainly due to the following two reasons. First, DCT is known to capture the image energy concentration but NOT the intelligibility concentration, which tends to be scattered among all frequency domains. Second, selective encryption before entropy coding impairs the compression ratio, while performing encryption after entropy coding results in a prohibitively high computational overhead. Currently, there seems no good selective encryption scheme that can meet the criteria mentioned in the introduction.

2.2. Encryption Using Multiple Statistical Models

In the entropy coding stage, symbol streams are converted to binary sequences according to some predefined code table (Huffman table), which suggested the possibility of integrating encryption with entropy coding. Wu and Kuo ([4], page 5) proposed to use an adaptive entropy coder based on multiple Huffman tables, which is called the Multiple Huffman Table (MHT) encryption scheme. The choice of particular Huffman tables and the order in which they are used are kept secret as the key of the encryption algorithm. The basic MHT scheme can be briefly described as follows:

Basic MHT Encryption Scheme

1. Generate 2^k Huffman tables, numbered 0 to $2^k - 1$.
2. Generate a random vector $P = (P_1\ P_2 \ldots P_n)$, where each P_i is a k-bit number from 0 to $2^k - 1$.
3. For the i-th symbol in the data stream, use table $P_{(i \bmod n)}$ to encode it.

Rather than directly encrypting the input data, MHT conceals Huffman tables in a secret manner such that, without knowing the particular tables and the order they are used, it is almost impossible to correctly decode the bit stream and get the original symbol stream. The biggest advantage is that no additional computational overhead is required. What needed is a little bit more memory space to store multiple Huffman tables and the (computationally almost negligible) operation to index these tables during encoding.

Despite its many advantages, the basic MHT scheme is only secure under the ciphertext-only attack, which is unfortunately not strong enough in modern multimedia applications. In particular, we will show that MHT succumbs to a more harsh attack mode, *i.e.* the chosen-plaintext attack (CPA).

2.3. Chosen-Plaintext Attack (CPA)

Chosen-plaintext attack (CPA) is a common attack mode to be considered carefully in modern cryptography. In the CPA scenario, an attacker has the ability to choose any plaintext at his/her will and observes the corresponding ciphertext. In other words, the attacker has unlimited access to an (maybe public) encryption oracle. It is generally required in modern cryptography that a good encryption scheme should be at least CPA-secure since it is prudent to assume the attacker may obtain, by whatever means, unlimited access to an encryption oracle.

Some notations are given below. Assume that there are totally S different symbols, and the symbol alphabet is denoted by $A = \{ a_1, a_2, \ldots, a_S \}$. A Huffman table T is a transform that converts each symbol a_i in A to a variable-length binary sequence $T(a_i)$, called a_i's code. The index of Huffman table used to encode symbol a_i is denoted by $\#(a_i)$. The vector $P = (P_1\ P_2 \ldots P_n)$ is called the *key hopping sequence,* which represents the successive occurrence of the index of Huffman tables that are used to encode incoming symbols. In other words, the key (*i.e.* its associated Huffman table) is hopping around according to this sequence. By some simple analysis, we can reach the conclusion that the basic MHT is vulnerable under CPA as specified in the following proposition.

Proposition I: Under CPA, the basic MHT scheme can be broken in nS times of encryption oracle access, where S is the size of the symbol alphabet and n is the length of the key hopping sequence.

Proof:

It is clear that for a given Huffman table T, we can effectively recover T if all its symbol-code entries (a_i, $T(a_i)$) are known for all $a_i \in A$ which is the symbol alphabet. Now, it is easy to see that, in the basic MHT scheme, if $i \equiv j \pmod n$, then the i-th symbol and the j-symbol are encoded by the same Huffman table $P_{(i \bmod n)}$. In other words, the key hopping sequence is simply a repetition of the length-n random vector $P = (P_1\ P_2 \ldots P_n)$. This observation quickly leads to a successful CPA attack that recovers all n Huffman tables. The attack method is described below:

Step 1. Input length-n symbol stream $a_1\ a_1 \ldots a_1$ (n successive a_1) to the encryption oracle and obtain the output sequence $c_1\ c_2 \ldots c_n$. Now, the attacker knows that for Huffman table P_1, $T_{P1}(a_1) = c_1$, for Huffman table P_2, $T_{P2}(a_1) = c_2$, all the way up to $T_{Pn}(a_1) = c_n$, for Huffman table P_n. In other words, the attacker has successfully found a_1's code in all n Huffman tables. This takes n times of oracle access.

Step 2. This time, the attacker inputs the symbol stream $a_2\ a_2\ ...a_2$ (n successive a_2) to the encryption oracle. By the same token, the attacker find a_2's code in all n Huffman tables by observing the output. This takes another n times of oracle access.

Continuing in this fashion, in Step S (where S is the size of the symbol alphabet) a_S's code in all n Huffman tables are recovered. At this point of time, the attacker has successfully accumulated all nS symbol-code pairs (a_i, $T(a_i)$) for a_i being any one of S symbols and T any one of the n Huffman tables. The attacker is now able to construct all n Huffman tables and break the basic MHT scheme. The total computational effort is nS times of the encryption oracle access. The proof is completed.

The above attack strategy indicates that the design of the key hopping sequence lies at heart of the security of the MHT scheme. If the key hopping sequence has some specific property, an attacker can accordingly construct some particular plaintexts and launch a CPA. In the basic MHT scheme, a length-n repetition pattern of the key hopping sequence enables the attacker to recover all Huffman tables by exploiting particular length-n symbol streams.

3. ENHANCED MHT ENCRYPTION SCHEME

In this section, we propose an enhanced MHT scheme to overcome the CPA weakness. Then, it is shown by crypto-analysis that the proposed scheme is secure under CPA.

3.1. Enhanced MHT Encryption Scheme

Now, let us ask the following question. What are the desirable properties that a key hopping sequence should have to thwart a CPA? First, it should appear statistically "identical" to the attacker. Given two sequences, a CPA attacker should not be able tell them apart by statistical properties such as the mean and the variance. Second, it must be statistically "random" enough so that the attacker cannot gather any useful information about which key is used to encrypt which symbol. Suppose that there is a mechanism which, on a given input, will produce a key hopping sequence $k_1, k_2, \dots k_n \dots$. According to the above discussion, we would require the following two properties:

1. For a different initial input, the ensemble of key hopping sequence $\{ k_1, k_2, \dots k_n \dots\}$ forms a probability indistinguishable distribution.
2. If some successive occurrence of k_n is known to an observer who does not know the generating mechanism, it is computationally infeasible for this observer to compute the next occurrence.

It is thus natural to come up with the idea of using a cryptographically secure pseudo-random number generator (PRNG) to generate a key hopping sequence since the output sequence of a PRNG satisfies the above properties. Such a mechanism is termed a *key hopper*. It is worthwhile to point out that the situation here is similar to that of a frequency hopping spread-spectrum communication system, where the frequency hops according to a (preset) pseudo-random sequence (hence our name of key hopping sequence and key hopper).

In our proposed scheme, we use a one-way hash function $h(\bullet)$ to emulate a key hopper by first selecting a random seed s, and then outputing the values $h(s)$, $h(s+1)$, $h(s+2)$, … .([6], page 173). Although this ad-hoc method has not been proven to be cryptographically secure, it appears sufficient for most applications. The detailed algorithm is described below.

Enhanced MHT Encryption Scheme

1. Generate $n = 2^k$ different Huffman coding tables, numbered from 0 to n-1
2. Select a cryptographic good m-bit hash function $h(\bullet)$ (For example, 160-bit SHA-1 hash function) and generate a random m-bit seed s. Calculate $r = \lfloor m/k \rfloor$
3. Compute $h(s)$ and truncate the hash value into k-bit blocks. Write $h(s) = t_1 \| t_2 \| \dots \| t_r \| rem$ with each t_i representing a number from 0 to n-1 and *rem* the remaining bits if m is not a multiple of k
4. for $i = 1$ to r
   Use Huffman table t_i to encode one symbol
5. After encoding r symbols in Step 4, set $s = s+1$ and go to Step 3.

3.2. Crypto-Analysis of Enhanced MHT Scheme

In this subsection, we perform the crypto-analysis of our enhanced MHT scheme under the CPA scenario. It is argued that, in the enhance MHT scheme, the former attacking strategy of learning all entries and recovering a Huffman table is no longer valid. In short, we can prove the following result for the enhanced MHT scheme under the CPA.

Proposition II: For the enhanced MHT scheme, a CPA attacker's strategy is no better than an exhaustive key search. The size of the exhaustive key search space is 2^m, where m is the output length of the key hopper.

Proof:

The key of the enhanced MHT scheme consists of (1) the n Huffman tables and (2) the random seed s used to produce the key hopping sequence. The goal of a CPA attacker is to recover both part of the key. Suppose in an attack, the encryption oracle has seen the following access record: the plaintext $m_1m_2...m_n...$and the corresponding ciphertext $c_1c_2...c_n...$. We will reorganize this plaintext/ciphertext pair by writing them into the form of a 3-tuple (m_i, c_i, $\#(m_i)$), where $\#(m_i)$ is the key hopping sequence (*i.e.* the index of a Huffman table used to encode

m_i). We also assume that the attacker has reconstructed n Huffman tables $T_1 \dots T_n$ at the end of the attack.

First, we argue that if the attacker does not know the key hopping sequence $\#(m_i)$ generated by s, then there is no way of verifying whether his attack was successful or not. This is because, in the 3-tuple (m_i, c_i, $\#(m_i)$), c_i is always a possible legitimate code of m_i. If the attacker does not know the value of $\#(m_i)$, he/she cannot determine whether the recovered Huffman tables $T_1 \dots T_n$ were correct or not based solely on the access record (m_i, c_i).

Our next claim is that the attacker has to take a guess on the random seed s in his/her attack. Otherwise, he/she will not be able to know the key hopping sequence generated by s. As mentioned above, there is no way to determine whether the Huffman tables were correct or not in this case.

Now, the attacker picks a random number s', which is a guess of the true seed s. From the value of s', the attacker is able to calculate a candidate key hopping sequence. Using this sequence as the value of $\#(m_i)$ in the oracle access record (m_i, c_i, $\#(m_i)$), the attacker can gradually build up entries in Huffman tables. For example, if the candidate key hopping sequence begins with 3, 45, 17, then the attacker can fill c_1 in the entry for m_1 in Huffman table 3, c_2 in the entry for m_2 in Huffman table 45, and c_3 in the entry for m_3 in Huffman table 17. Generally speaking, if the oracle access record is sufficiently long and contains enough occurrences of each symbol in the alphabet, at some point of time all entries of n Huffman tables will be filled up.

After this happens, the attacker encodes the plaintext using the recovered Huffman tables and compares the result to the oracle access record (m_i, c_i, $\#(m_i)$). If $s' = s$, then the candidate key hopping sequence is the same as the true key hopping sequence. In this case, the attacker will end up with the exactly same n Huffman tables as those used by the encryption oracle. The encoding results using these recovered Huffman tables therefore always agree with the oracle access record. Otherwise, there must be a conflict. The attacker will eventually find at some point of time that the encoding result using the recovered Huffman tables is different from the oracle access record, which indicates that $s' \neq s$.

Since there is apparently no clue about how to guess the value of s, the attacker's strategy is no better than an exhaustive search. He/she randomly picks a seed and executes the above operation to see whether a conflict occurs or not. The random seed s is an m-bit binary sequence so the size of key search space is 2^m. On the average, an attack would expect to execute 2^{m-1} such operations to find the correct seed s and recover the Huffman tables.

After the attacker has successfully reconstructed the Huffman tables, there still remains the random seed, the other half of the key associated with a particular encrypted bit-stream. Along the same line of reasoning, the strategy of attacker is still no better than wild guessing. In this case, he/she randomly picks a seed, generates the key hopping sequence, decodes the bit-stream using the correct Huffman tables, and checks whether the decoded content is semantically meaningful. Again, the key space is 2^m and on average, another 2^{m-1} operations are mandated (pick a seed, decode and check) to sift out the correct seed associated with the particular ciphertext which the attacker tries to break.

In summary, the average search space to construct the correct Huffman tables is 2^{m-1} and, after that, the attacker needs another 2^{m-1} search operations to find out the key hopping sequence. Therefore, the total search space is 2^m. This completes the proof.

4. CONCLUSIONS AND FUTURE WORK

In this research, we presented an improvement of a multimedia encryption scheme called the Enhanced MHT Encryption Scheme that is secure under the chosen-plaintext attack (CPA). The basic idea is to use a pseudo-random number generating mechanism to generate a key hopping sequence, and choose different keys according to this sequence. The randomness and unpredictability of this sequence prevent a CPA attacker gaining additional information from the oracle access and confine the attack strategy to exhaustive search.

The advantage of this methodology is that it does not rely on the security strength of any underlying cryptographic cipher. Using a stronger cipher as the building block, this approach can be easily extended to produce more cryptographically secure encryption schemes.

5. REFERENCES

[1] L.Tang, "Methods for encryption and decrypting MPEG video data efficiently", *Proc. Of the 4th ACM international conference on multimedia,* Boston, Nov 1996

[2] C. Shi and B.Bhargava, "A fast MPEG video encryption algorithm", *Proc. Of the sixth ACM international conference on multimedia,* Bristol, UK, Sep 1998

[3] L. Qiao and K.Nahrstedt, "A new algorithm for MPEG video encryption", *Proc. of the first international conference on imaging science, systems, and technology,* July 1997, Las Vegas.

[4] C. Wu and C.-C. J. Kuo, "Efficient multimedia encryption via entropy codec design", *SPIE international symposium on electronic imaging,* San Jose, Jan 2001

[5] C. Wu and C.-C. J. Kuo, "Fast encryption methods for audiovisual data confidentiality", *SPIE Photonics East - Symposium on Voice, Video, and Data Communications,* Boston, Nov 2000

[6] A.J.Menezes, P.C.van Oorschot and S.A.Vanstone, "Handbook of Applied Cryptography", *CRC Press,* 1996

VISUAL CRYPTOGRAPHY FOR PRINT AND SCAN APPLICATIONS

Wei-Qi Yan, Duo Jin, Mohan S Kankanhalli

School of Computing
National University of Singapore
Singapore 117543

ABSTRACT

Visual cryptography is not much in use in spite of possessing several advantages. One of the reasons for this is the difficulty of use in practice. The shares of visual cryptography are printed on transparencies which need to be superimposed. However, it is not very easy to do precise superposition due to the fine resolution as well as printing noise. Furthermore, many visual cryptography applications need to print shares on paper in which case scanning of the share is necessary. The print and scan process can introduce noise as well which can make the alignment difficult. In this paper, we consider the problem of precise alignment of printed and scanned visual cryptography shares. Due to the vulnerabilities in the spatial domain, we have developed a frequency domain alignment scheme. We employ the Walsh transform to embed marks in both of the shares so as to find the alignment position of these shares. Our experimental results show that our technique can be useful in print and scan applications.

1. INTRODUCTION

Visual cryptography (VC) is basically a secret sharing scheme extended for images [1] and its distinguishing characteristic is the ability of secret restoration without the use of computation. Figure 1 is an example of the use of visual cryptography.

The two top images figure 1(a) and figure 1(b), share 1 and share2 are two randomly generated images which carry the secret information. If we print the two shares on transparencies and superimpose them, we can clearly see the secret as shown in figure 1(c).

Briefly speaking, the VC technique technique is for binary images where α is the secret image, γ is a randomly generated share while β is the other share such that:

$$\alpha_i + \beta_i = \gamma_i, \;\; i = 0, 1, 2, \ldots, n$$

Thus without β and γ, α cannot be deduced at all [2]. This scheme provides perfect security with simplicity [1]. Visual cryptography possesses these characteristics:

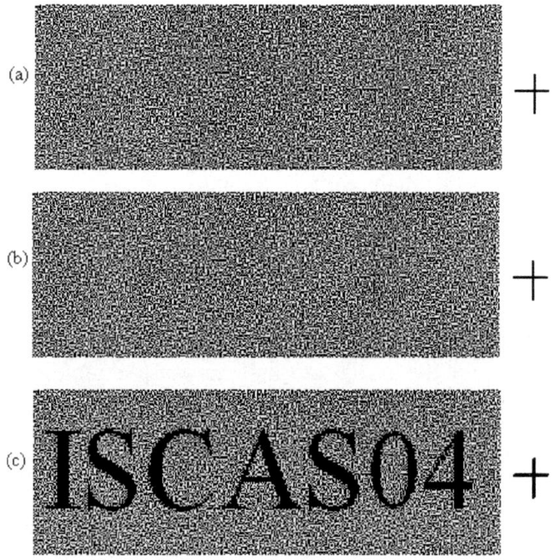

Fig. 1. Cross Alignment for Visual Cryptography

- Perfect security
- Decryption (secret restoration) without the aid of a computing device
- Robustness against lossy compression and distortion due to its binary attribute.

However, the shortcomings of visual cryptography are as salient as its merits. There are three main drawbacks in visual cryptography:

- It results in a loss of resolution. The restored secret image has a resolution lower than that of the original secret image.
- Its original formulation is restricted to binary images. For color images, some additional processing such as halftoning and color-separation are required.
- The superposition of two shares is not easy to perform unless some special alignment marks are provided.

0-7803-8251-X/04/$17.00 ©2004 IEEE

The manual alignment procedure can be tedious especially for high resolution images.

The first two shortcomings have been discussed in literature earlier [3] [4] [5] [6]. We will focus on the third problem in this paper. The shares of VC printed on transparencies are very difficult to be overlapped with proper alignment even if we ignore the printing errors. A wide variety of applications of visual cryptography would require the printing of the shares on paper like that of documents, checks, tickets or cards [7] [8]. In such cases, scanning of the printed shares is inevitable for restoring the secret. The scanned shares (with printing, handling and scanning errors) have to be superimposed in order to reconstruct the secret image which could be some photo, code or other such important information.

In this paper, we concentrate on the print and scan applications of visual cryptography. i.e. to obtain the precise position of scanned shares which requires rotation and alignment correction. Putting alignment marks in the spatial domain is extremely vulnerable to cropping and editing. Therefore, we use the Walsh transform domain to embed perceptually invisible alignment marks. We show that the Walsh transform helps in recovering the marks inspite of noise and we can precisely align the scanned shares to recover the secret.

2. BACKGROUND

In order to carry out the superposition, initially a spatial tag is marked beside the shares. In figure 1, we put a cross beside each share. For restoring the secret, the two crosses need to be precisely overlapped. If this is done, the secret image will be revealed. Another solution to this problem is by utilizing the extended visual cryptography scheme [9]. This scheme shares a secret by using two protection images **B** and **C**. The procedure of visual cryptography is performed as: **A** = **B'** + **C'** where the secret **A** is divided into two shares **B'** and **C'**. On these shares **B'** and **C'**, images **B** and **C** are also visible. During restoration, images **B** and **C** are aligned to make them disappear (by cancelling) revealing the secret in the process. An example of this technique is shown in figure 2, figure 2(a) and figure 2(b) are shares, figure 2(c) is the reconstructed image.

Actually, figure 1 and figure 2 belong to the same class of techniques since they both work in the spatial domain. The problem with this class is that the alignment marks are visible to an attacker and thus can be easily removed by cropping or localized image alteration. We therefore explore the alternative idea of using marks in the frequency domain. In particular, we consider the use of the discrete Walsh transform, which is useful for pulse signals and is distinct from the discrete Fourier transform **(DFT)**, discrete cosine transform **(DCT)** and discrete wavelet transform **(DWT)**. Walsh functions are a complete set of orthogonal functions

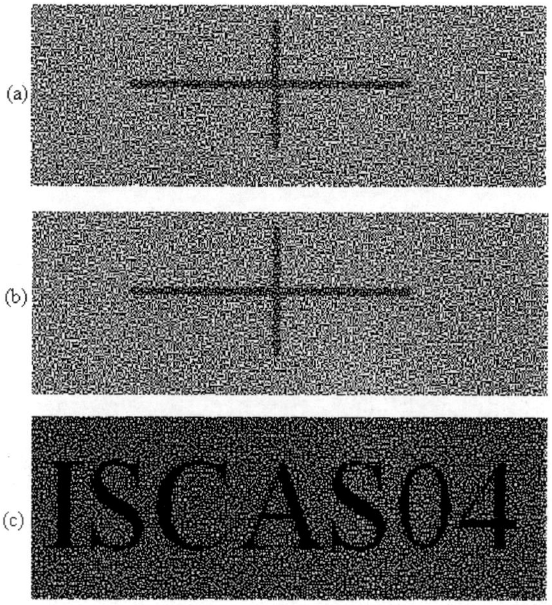

Fig. 2. Cross Alignment by Using Extended Visual Cryptography

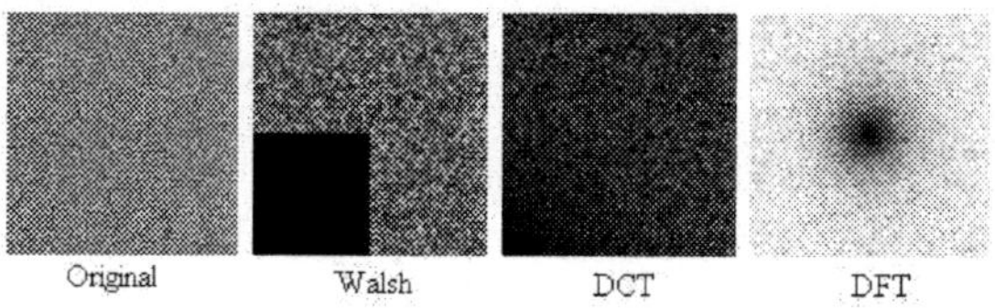

Fig. 3. The original shares and their transformations

with the the value being only -1 and 1. We use the 2D discrete Walsh transform:

$$W_{xy}(u,v) = \frac{1}{N_x}\frac{1}{N_y}\sum_{y=0}^{N_y-1}\sum_{x=0}^{N_x-1} f(x,y)\cdot(-1)^{\alpha} \qquad (1)$$

The inverse transform is given by:

$$f(x,y) = \sum_{v=0}^{N_y-1}\sum_{u=0}^{N_x-1} W_{xy}(u,v)\cdot(-1)^{\alpha} \qquad (2)$$

where $\alpha = \sum_{r=0}^{p_x-1} x_r u_r + \sum_{s=0}^{p_y-1} y_s v_s$, $f(x,y)$ is a pixel of the image, (x,y) is its position. $W_{xy}(u,v)$ represents the transform coefficients, $N_x = 2^{P_x}$, $N_y = 2^{P_y}$, (P_x and P_y are positive integers), x_r, u_r, y_s and v_s are either 0 or 1 (i.e. one bit of x, u, y and v respectively).

Unlike the Walsh transform, transforms like **DFT**, **DCT** and **DWT** are mainly used for continuous tone color images. The results of applying these three transformations to a VC share is shown figure 3. In figure 3, the left image is a VC share. The subsequent images show the result of applying

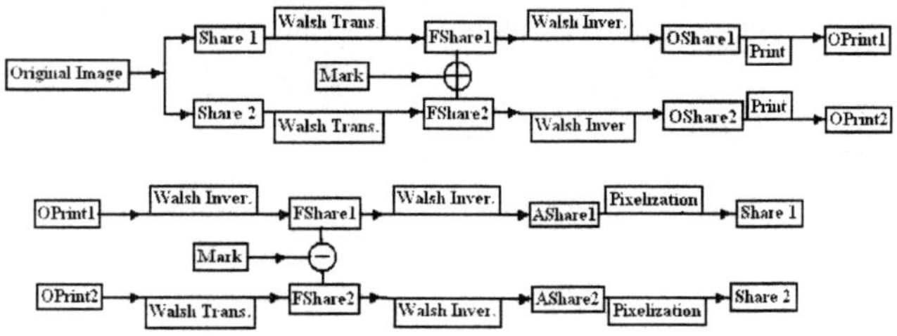

Fig. 4. Flowchart of Process for Print and Scan VC

the Walsh, **DCT** and the **DFT** transforms. The differences are quite apparent. Note that the bottom-left rectangle of the image for the Walsh transform is totally dark. This information can exploited in removing noise by filtering the coefficients in this quadrant. The overview of our print and scan scheme for the superposition of VC shares is shown in the flowchart of figure 4. The basic idea is to introduce some alignment marks in the Walsh transform domain.

3. OUR WORK

In this section, we will describe our contributions. During encryption, as shown in figure 4, we apply the Walsh transform on the shares. Then we embed marks in the high frequency coefficients of the transform. Then the inverse transform is applied to obtain the new shares with hidden marks that are printed on paper to be transmitted via public channels.

During the process of decryption, we scan the paper image and extract the marks by performing the Walsh transform to obtain the approximate alignment for shares superimposition. We then fine-tune the alignment by performing rotation and translation. The rotation is done by using:

$$\begin{bmatrix} x\prime \\ y\prime \end{bmatrix} = \begin{bmatrix} \cos(\alpha) & -\sin(\alpha) \\ \sin(\alpha) & \cos(\alpha) \end{bmatrix} \cdot \begin{bmatrix} x \\ y \end{bmatrix} \quad (3)$$

The rotation adjustment in increments of angle α is done as shown in figure 5. The translation adjustment by $\triangle x$ and $\triangle y$ is done as shown in figure 6. The criteria for finding the best alignment position is that the superimposed image should have the least number of black pixels if we perform the XOR operation between them. This is because the XOR operation allows for perfect restoration of the secret image [3]. Our algorithm can thus be described as follows:

input : Scanned Shares
output : Revealed Secret

1: Scan the two printed shares;
2: Do initial alignment by aligning Walsh transform domain marks;
3: Do final alignment by minimizing black pixels (after XOR) using shifts & rotates;
4: Reveal the secret;

Algorithm 1: Alignment of the VC Shares

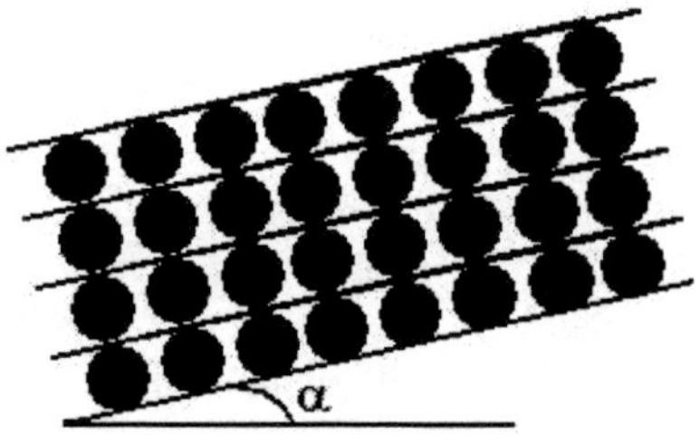

Fig. 5. Adjustment of Visual Cryptography Shares

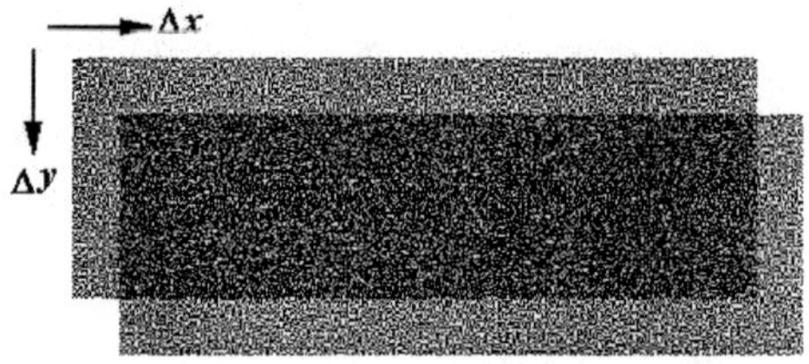

Fig. 6. The Shift Operation to the Overlapping Shares

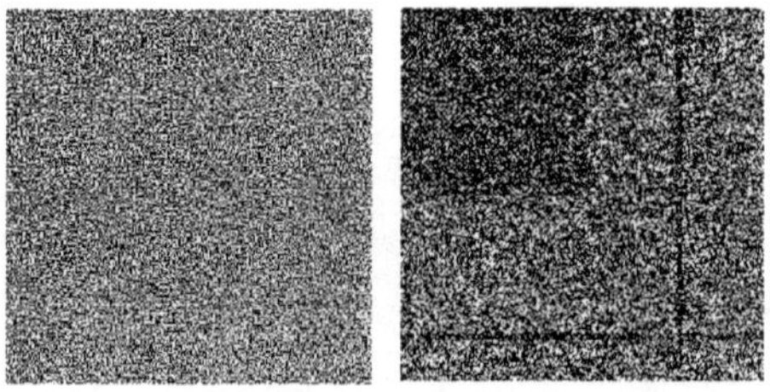

Fig. 7. Marked VC Share in Walsh Transform Domain

4. RESULTS

In this section, we will provide the results for visual cryptography. Figure 7 shows a share and the mark in the Walsh

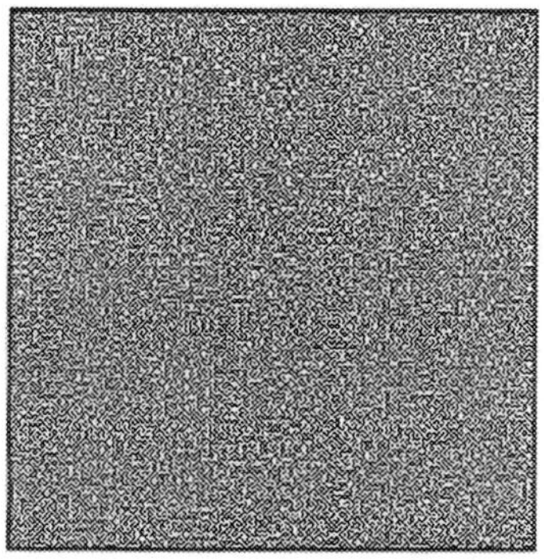

Fig. 8. The Scanned Watermarked VC Shares

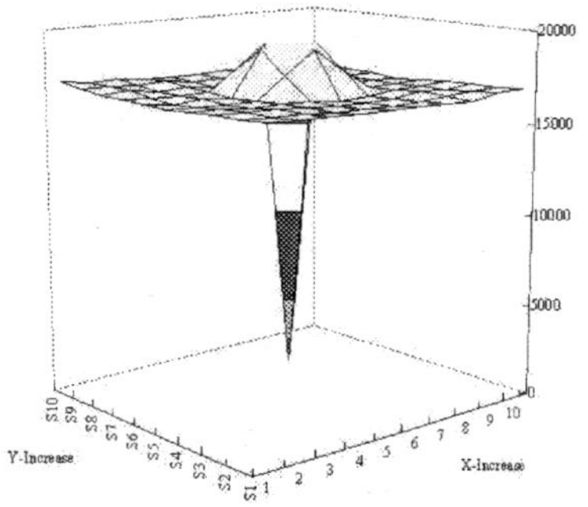

Fig. 9. Number of Black Pixels at Various Alignments

transform domain. The mark is in the form of a cross. In the implementation, we print the image by using the default setting of the Lexmark T622 PS3 printer at 1200dpi and scan the share image at 300dpi resolution. Figure 8 is an example of a scanned marked share. Figure 9 shows the minimization of black pixels when the correct alignment is obtained. The revealed secret after performing the XOR operation is shown in figure 10.

5. CONCLUSION

In this paper, we have tried to solve the practical problem associated with the use of visual cryptography. Many VC applications involve printing the share on the paper channel. If this is the case, then a print and scan technique needs to be developed for recovering the secret image. However, precise alignment at high resolutions is then a problem. We therefore propose the use of the Walsh transform to embed alignment marks in the transform domain. These marks are used as guides to precisely align the shares automatically. Our experimental results point to the viability of the use of VC for print and scan applications. Our future work will focus on the applications of visual cryptography on the 2D bar code.

Acknowledgement: This work has been supported in part by a fellowship from the Singapore Millennium Foundation (SMF).

Fig. 10. Secret Image Revealed by the XOR operation

6. REFERENCES

[1] M. Naor and A. Shamir, "Visual cryptography," in *Proc. of Advances in Cryptology*. 1995, vol. 950, pp. 1–12, Springer-Verlag.

[2] A. Shamir, "How to share a secret," *Communications of the ACM*, vol. 22, no. 11, pp. 612–613, Nov 1979.

[3] D. Jin, "Progressive color visual cryptography," Masters degree thesis, School of Computing, National University of Singapore, Singapore, July 2003.

[4] Y.C. Hou, C.Y. Chang, and F. Lin, "Visual cryptography for color images based on color decomposition," in *Proc. of 5th Conference on Information Management*, Taipei, Nov 1999, pp. 584–591.

[5] Y.C. Hou, C.Y. Chang, and SF Tu, "Visual cryptography for color images based on halftone technology," in *Proc. of International Conference on Information Systems, Analysis and Synthesis, World Multiconference on Systemics, Cybernetics and Informatics*, 2001.

[6] Young-Chang Hou, "Visual cryptography for color images," *Pattern Recognition*, vol. 36, pp. 1619–1629, 2003.

[7] B.S. Zhu, J.K. Wu, and M.S. Kankanhalli, "Print signatures for document authentication," in *Proc. of ACM Conference on Computer and Communications Security*, October 2003, pp. 145–153.

[8] N. Degara-Quintela and F. Perez-Gonzalez, "Visible encryption: using paper as a secure channel, security and watermarking of multimedia contents," in *Proc. of SPIE'03*, 2003, vol. 5020.

[9] G. Ateniese, C. Blundo, A. De Santis, and D. Stinson, "Extended schemes for visual cryptography," *Theoretical Computer Science*, vol. 250, pp. 143–161, 2001.

Web Search Steganalysis: Some Challenges and Approaches

R. Chandramouli
Multimedia Systems, Networking, and Communications (MSyNC) Laboratory
Department of Electrical and Computer Engineering
Stevens Institute of Technology
Email:mouli@stevens-tech.edu

Abstract—**This paper presents some important issues in searching the Internet for covert messages. Several related theoretical and practical problems are discussed. Web search is formulated as a mathematical optimization problem and solutions are proposed. Two algorithms: *coordinated search* and *random search* are discussed within this optimization framework along with their pros and cons.**

I. Introduction

A covert channel is defined as: *any communication channel that can be exploited by a process to transfer information in a manner that violates the systems security policy* [1]. It seems that the Internet has opened up new avenues for covert communication. Some of them are the following:

- **Digital data as covert channel:** digital data such as image, video, and audio are easily created, manipulated, and distributed across the Internet with the use of current technologies and devices. These are good channels for hiding information with medium to high message carrying capacities. There are several techniques available today that can embed messages within these data types without causing significant perceptual distortion. For example, the least significant bits of a digital image can be replaced by message bits. The resultant image can then be placed on a website for a receiving party to download and extract the hidden message. More about this topic can be found in [2].
- **TCP/IP protocol as covert channel:** Some features of the TCP/IP protocol suite can be used to send covert messages as discussed in [3]. Encrypted or non-encrypted information can be encapsulated within otherwise normal TCP/IP packets. The TCP/IP header information can also be modified to encode secret messages. There are some fields in the packet header that are not used by the current communication networks. These fields can be used as message carriers.
- **Timing channel:** An user of a time-shared computing server can transmit covertly by varying the rate at which it sends jobs for processing. Since the response time of the computing server depends on its instantaneous load, other users can get a *noisy* version of the covert information by measuring the response time to their own jobs. One of the earliest work on this topic is by Lipner [4].

These covert channels are an immense cause of security concern because they can be used to pass malicious messages. These messages could be in the form of computer viruses, spy programs, terrorist messages, etc. Therefore, detecting these covert channels is an important issue that needs to be addressed.

Information hiding/embedding also known as *steganography* is a popular research area currently. Hiding information in digital medium such as a digital image has been receiving a lot of attention these days. Steganalysis is the counterpart of steganography that deals with detecting data that could contain hidden messages. Current steganalysis algorithms use statistical methods to determine the presence/absence of a hidden message. In this paper we assume image steganography without loss of generality. Also, we focus on detecting digital medium as covert channels in the Internet. We believe that the techniques presented here could be applied in detecting some other types of covert channels also. A work closely related to this paper is by Bloom [5].

First, we remark that every website and every digital image on these sites is a potential covert channel. However, it is clear that we cannot search every website for covert messages. In addition to the sheer problem of the large number of websites, we also face the following issues:

- Some covert message carrying websites may not have public links.
- Websites are created, moved, and destroyed on a daily basis, perhaps even randomly.
- Websites carrying covert messages may not be found by current search engines because of the web search metrics used by them. For example, some search engines show webpages that are linked to by other webpages. We have no reason to believe that a webpage containing images with covert messages will be have several links pointing to it. This beats the purpose of covert communication!
- A webpage like e-bay could contain thousands of images thus making it computationally very difficult to detect hidden messages.

Since exhaustive search of the Internet is infeasible, some natural follow-up questions are the following:

- How can the search efficiency be improved?
- How can *a priori* information about the Internet sites be exploited?
- What are the options for an efficient resource allocation for covert message search?

In this paper, we attempt to address some of these questions and provided a mathematical formulation of this problem. The solution to this mathematical formulation gives us some insights on how to design algorithms for web search to detect covert communication. The paper is organized as follows. Web search steganalysis is posed as an optimization problem in Section II followed by concluding remarks in Section III.

II. Web Search Steganalysis as an Optimization Problem

Consider the following web search steganalysis problem formulation. Let us say that a covert message X may be present in one of M webpages with a priori probabilities p_j, $j = 1, 2, \ldots, M$. Note that the set of M webpages of interest can be chosen in a variety of ways, such as:

- An external intelligence information such as email trace, tapping phone conversations, etc. could raise suspicion about certain websites.
- As discussed in [5], some websites can be safely eliminated before the search begins due its security level. For instance, there is little reason to believe that government websites (.mil, .gov) may contain images with hidden information.
- Websites of organizations with radical political or religious views may be a good candidate for search. Such a suspicion could also be strongly supported by the text content of the website.
- Tracking http requests in the backbone network [5], we can find out about sites that do not have a publicly available link that may be a cause for concern.

Once some kind of a criterion or side information is used to short list the candidate websites for searching, the next issue is to compute the probabilities p_j, $j = 1, 2, \ldots, M$. There are three possible scenarios:

- **(a) complete information case**: the probabilities $\{p_j\}$ are completely known.
- **(b) partial information case**: only an ordering of these probabilities are known, i.e., it is known that $p_1 \geq p_2 \geq \ldots \geq p_M$, but not their exact values.
- **(c) no information case**: $\{p_j\}$ completely unknown.

Each of these three cases occur could occur in practice. Complete information about $\{p_j\}$ may be available to government intelligence and law & order agencies. Partial information about the probabilities may be obtained with the help of some side information from law enforcement agencies, monitoring some suspect web sites, Internet chat rooms, tracing http requests, etc. The last case of course occurs when a blind web search is conducted for stego information. Nevertheless, in this paper we provide detailed analysis of case (b) only. We however note that this analysis is extendable to the other two cases also.

If J denotes the set of websites in which the message X may be found, then, the message location distribution on J is $p : J \rightarrow [0, 1]$ and $\sum_{j \in J} p_j = 1$. Since the websearch is limited by total search time/effort due to a variety of physical and logical limitations, it may be necessary to locate X's website with the minimal amount of time/effort. Let $b : J \times [0, \infty) \rightarrow [0, 1]$ be the *location function* such that $b(j, z)$ denotes the conditional probability of locating X with the amount of search effort spent in site j, $z \geq 0$, given that X is in site j. Note that the value of b is affected by a number of factors such as the quality of the steganalysis algorithms being used to analyze the images in a website, computational resources available, quality and amount of side information about the embedding algorithm, etc. Then the total probability of locating the message X is given by,

$$\sum_{j \in J} p_j b(j, f(j)), \tag{1}$$

where $\sum_{j \in J} f(j)$ is the total effort. A cost function on J is a function $c : J \times [0, \infty) \rightarrow [0, \infty)$ such that $c(j, z)$ gives the cost of applying search effort z in site j. The function $f : J \rightarrow [0, \infty)$ gives the amount of effort spent in each site, called an *allocation on* J. Then,

$$P[f] = \sum_{j \in J} p_j b(j, f(j)), \quad C[f] = \sum_{j \in J} c(j, f(j)) \tag{2}$$

denotes the probability of locating the covert message and the cost resulting due to the allocation f, respectively. As an example, if $f(j) = T_j$ denotes the amount of search time spent to localize X then $c(j, .) = T_j$ denotes the amount of time spent in searching site j. Note that, other allocation functions can also be used depending on the resource constraint. Suppose F denotes the allocations and T is the upper bound on the total time cost in the localization search process, then, the proposed search problem is to find $f^* \in F$ by solving the following constrained optimization problem,

$$C[f^*] \leq T, \quad \text{and} \quad P[f^*] = \max\{P[f] : f \in F \text{ and } C[f] \leq T\}. \tag{3}$$

Then f^* is the optimal message localization search allocation for total time cost T. Note the resemblance of this formulation with optimal search theory [6].

Suppose $b(j, T_j) = 1 - e^{-T_j}$, $j = 1, 2, \ldots, M$, that is, the probability of locating a message in site j increases exponentially with the amount of time spent in searching that site. At present, we only have an intuitive justification for this detection probability model. However, we note that this model may not be far from a practical model. In practice, suppose there are L (L large) number of steganalysis algorithms that we wish to run on all the images downloaded from a website. It is reasonable to assume that as the number of steganalysis algorithms run on the images increases, the probability of the covert message not being detected by any one of these will decrease. Of course, here we assume that the individual steganalysis algorithms are such that the rate of decrease of the miss probability is exponential. The proposed approach can also be extended to other models for $b(.,.)$.

Let $0 < \sum_j T_j \leq T$ be the total time constraint for web-search steganalysis. Then, following an analysis presented in [6] we show here without proof that the probability of locating the message with optimal search effort allocation satisfies:

$$1 - e^{-T/M} \leq P[f^*] \leq 1 - T\left[\Pi_{j=1}^{M} p_j\right]^{\frac{1}{M}} e^{-T/M}, \tag{4}$$

and, if $\sum_j T_j = T$ then the optimal search time that must be allocated for searching site j is given by,

$$T_j = max(0, ln\frac{p_j}{\lambda}), \; j = 1, 2, \ldots, M \tag{5}$$

$$\lambda = \left[\Pi_{j=1}^{M} p_j\right]^{\frac{1}{M}} e^{-T/M}. \tag{6}$$

We observe that, from Eq. (4) it is possible to estimate the total time bound T for a desired message localization accuracy.

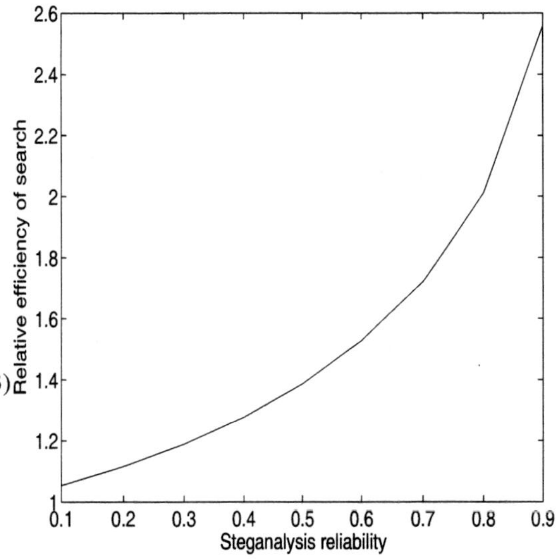

Fig. 1. Relative efficiency of random search w.r.t. coordinated search. Higher value of relative efficiency implies coordinated search strategy is more efficient.

Next, we consider a second model for web search. Let us say a particular website is being searched to locate a message. Due to practical constraints such as computational constraints, time constraints, modelling errors, and reliability of steganalysis algorithms being used, it may only be possible to detect the presence of the covert message with probability q in each search attempt independent of previous searches. Also, assume that false alarms are negligible. This is justified because most of the steganalysis algorithms allow us to put an upper bound on the false alarm probability. Now, there are two possible ways to search the website, namely, *coordinated search* and *random search*. In a coordinated search, the images within a site that were analyzed previously for a covert message and did not yield a positive result are completely stored within the searching algorithm's/computer's memory. These images are avoided in future searches. On the other hand, in a random search, the search algorithm does not maintain the list of previously searched images thus eliminating the need for large built-in memory requirement. Instead, it simply searches the entire website in a random fashion. Clearly, the coordinated search seems to be more effective than random search; however, its complexity and memory requirement are higher. The basic question we ask is: when is choosing the random search better than coordinated search, if at all? The answer to this question will give us valuable information in trading off efficiency for cost and simplicity. If N_c is the total number of times the website is searched, then the probability of detecting the covert message is $d_c = 1-(1-q)^{N_c}$. For the random search strategy, by adopting the technique given in [6], we can see that the detection probability is $d_r = 1 - e^{qN_r}$. Suppose we want $d_c = d_r$, then we see that it produces the following relative efficiency factor of the random search message detection w.r.t. coordinated search,

$$\Lambda = \frac{N_r}{N_c} = -ln(1-q)/q, \; 0 < q < 1. \tag{7}$$

From Fig. 1 we see that if the steganalysis accuracy is not high, then random search performs almost as good as coordinated search. In addition, no extra memory cost or coordination is necessary. Λ in Eq. (7) also serves as a useful measure to compare the performance of any other web search algorithm w.r.t. the coordinated search which is optimal.

III. Conclusion

From the analysis presented here, we conclude that it is possible to design efficient web search algorithms to detect covert messages. The proposed mathematical web search model admits a wide variety of resource constraints. Depending on the application, implementation, hardware, and steganalysis probability of error constraints, a suitable resource model can be used to derive an optimal web search strategy using the proposed technique. Depending on the reliability of the steganalysis algorithms employed and the storage constraint one of two strategies, namely, coordinated search or random search can be chosen. It is seen that for a certain range of steganalysis reliability, both these methods give comparable performance.

Acknowledgement

This material is based on research sponsored by Air Force Research Laboratory under agreement number F306020-02-2-0193 and NSF DAS 0242417. The U.S. Government is authorized to reproduce and distribute reprints for Governmental purposed notwithstanding any copyright notation thereon. The views and conclusions contained herein are those of the authors and should not be interpreted as necessarily representing the official policies or endorsements, either expressed or implied, of Air Force Research Laboratory or the U.S. Government.

References

[1] U. S. D. O. D. 1985., "Trusted computer system evaluation criteria."

[2] I. Cox, J. Bloom, and M. Miller, *Digital Watermarking: Principles & Practice*. Morgan Kaufmann, 2001.

[3] C. Rowland, "http://www.firstmonday.dk/issues/issue2_5/rowland/."

[4] S. Lipner, "A comment on the confinement problem," *Fifth symposium on Operating systems principles*, pp. 192–197, Nov. 1975.

[5] J. Bloom, "Smartsearch steganalysis," *SPIE Conf. on Security and Watermarking of Multimedia Contents*, vol. 5020, pp. 167–172, 2003.

[6] L. Stone, *Theory of optimal search*. Academic Press, 1975.

ROBUST VQ-BASED DIGITAL WATERMARKING FOR MEMORYLESS BINARY SYMMETRIC CHANNEL

Jeng-Shyang Pan[§], *Min-Tsang Sung*[§], *Hsiang-Cheh Huang*[★], *and Bin-Yih Liao*[§]
[§]Dept. Electronic Eng., Nat'l Kaohsiung Univ. of Applied Sciences, Kaohsiung, Taiwan, R.O.C.
[★]Dept. of Electronics Engineering, National Chiao Tung University, Hsinchu, Taiwan, R.O.C.
jspan@cc.kuas.edu.tw

ABSTRACT

Vector quantization (VQ) has been distinguished for its high compression rate in lossy data compression applications. And VQ-based watermarking plays a newly developed branch in digital watermarking research fields. In this paper, we propose optimized schemes for VQ-based image watermarking. We overcome the VQ index assignment problem with genetic algorithm, which is suitable for transmitting the watermarked image over noisy channels. We obtain better robustness of the watermarking algorithm against the effects caused by channel noise in our simulations after inspecting the results from several test images. In addition, to compare with existing schemes in literature, the watermarked image quality in our scheme has approximately the same quality, with better performance in robustness, to the schemes proposed by other researchers. This also proves the effectiveness of our proposed schemes in VQ-based image watermarking for copyright protection.

1 INTRODUCTION

Applications of digital watermarking become more and more prevailing recently. With the widespread use of Internet, people nowadays easily retrieve digital multimedia items, especially the digital images, through the World Wide Webs. Ironically, because of the digital nature of multimedia, including easy transferring and flexible editing, they suffer from infringing upon the intellectual properties of original owners of such digital contents. Therefore, in addition to conventional schemes such as cryptographic methods, digital watermarking offers another useful solution for copyright protection.

The aim of digital watermarking is to insert a random number sequence, copyright messages, ownership identifiers, or control signals, called the *watermark*, into the digital multimedia and/or into the associated secret keys. After inserting or embedding the watermark by specific algorithms, the original media will be slightly modified, and the modified media is called the watermarked media. There might be no or little perceptible differences between the original media content and the watermarked one. After transmitting the watermarked media via the Internet or the communication channels, the embedded watermarks can later be extracted or detected from the watermarked multimedia and/or the secret keys for authentication or identification.

In this paper, we focus on digital image watermarking with vector quantization (VQ). Among the watermarking researches in literature, in addition to the traditional schemes to watermark into the spatial domain pixels or the transform domain coefficients, e.g., DCT coefficients, VQ based watermarking [1][2][3][4][5] is a newly developed branch that can be further explored for researches and applications. In this paper, we propose an optimized, VQ-based image watermarking scheme with genetic algorithm (GA) [6] suitable for transmitting watermarked images over memoryless binary symmetric channels. We ameliorate our previous works and the schemes proposed by other researchers. The results with our algorithm have both better performance in watermark robustness and better quality in watermarked image comparing with conventional VQ-based watermarking schemes.

This paper is organized as follows. In Section 2, we explain the backgrounds for VQ and the importance of index assignment. In Section 3, we make use of genetic algorithm for optimizing VQ codebook index assignment and VQ-based image watermarking. In Section 4, we depict the effects caused by the memoryless binary symmetric channel. Simulation results are presented in Section 5, and we also compare those with the results from existing algorithms in literature. And we conclude this paper in Section 6.

2 BACKGROUNDS FOR VECTOR QUANTIZATION AND CODEBOOK INDEX ASSIGNMENT

To reduce the space requirement for storage and the bandwidth requirement for communication, a wide variety of compression techniques had been developed [7]. For multimedia applications, less significant information can be sacrificed for higher compression rate, since human sensory system is less sensitive to detailed information. In this kind of applications, VQ had received considerable attention for its high compression rate and its essential role in various compression applications. A vector can be fixed numbers of consecutive samples of audio data or a small block of image and video data, for example, the grey-level values of a 4×4 pixel image block form a 16-dimentional

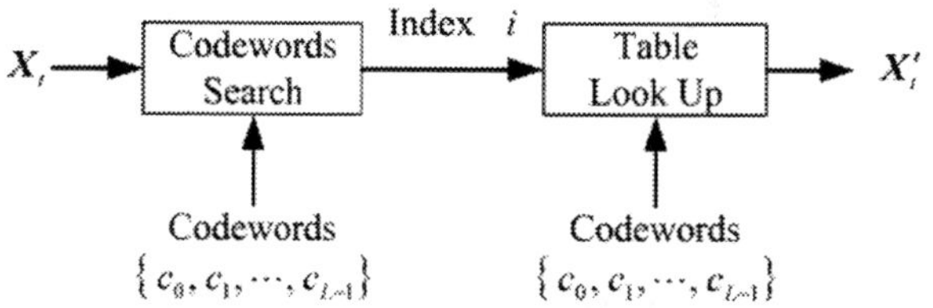

Figure 1. A block diagram for vector quantization.

codevector. Figure 1 gives a block diagram illustration of the operation of VQ compression.

The original image X is composed of the combination of all input vectors, X_t, $\forall t$. For practical implementations, people use small blocks in the original image X to represent the input vector X_t. In the sender end, the codeword search process looks for a "nearest" codeword, c_i, from the codebook for the given input vector X_t. Euclidean distance is employed in the search process to measure the distance between the codeword c_i and the input vector X_t, as indicated in Eq. (1),

$$i = \arg\min_j D(X_t, c_j), \quad j = 1, 2, \cdots, L-1 \tag{1}$$

where $D(\bullet, \bullet)$ denotes the Euclidean distance.

The index of selected codeword is then transmitted to the receiver end. With the same codebook, the decompression process can easily reconstruct vector X'_t by simple table look-up, as depicted in Figure 1. There will be distortion introduced by the compression-decompression process, on the one hand, since X'_t is only an approximated version of the original X_t. If we work on 8-bit grey-level image, using a block size of 4×4 pixel and a codebook of 256 codewords, then the compression ratio is up to $\frac{4\times4\times8}{\log_2 256} = 16$. All the reconstructed vectors, X'_t, $\forall t$, make up the reconstructed image, X'.

On the other hand, the channel noise will also induce channel errors during transmission. The effect of channel errors is to cause errors in the received indices, which makes the reconstructed image deteriorated. Thus, distortions are introduced in the decoding step. Distortion due to an imperfect channel, such as the memoryless binary symmetric channel to be discussed in Section 4, can be reduced by assigning suitable indices to codevectors. If the number of codevectors is L, the possible combination of indices to codevectors is $L!$. To test all the $L!$ assignments is an NP-hard problem [8].

The codebook plays an essential role in VQ. The codebook size, or the number of codewords in a codebook, is a trade-off between the reconstructed image quality and the compression rate. In addition, the VQ index assignment is also an important issue for codebook design. The codewords in the codebook decide the resultant compression distortion. A dedicated procedure is required for the generation of appropriate codebook. Among other alternatives, LBG algorithm [9] is widely used in various applications.

3 OPTIMIZATION OF VQ INDEX ASSIGNMENT AND VQ-BASED WATERMARKING

As we described in Section 2, the optimization of index assignment for vector quantizer is computationally intractable for large codebook size even if very powerful computer is used, because there exist $L!$ possible ways to arrange the indices of codevectors for L codevectors [8][9][10][11]. Hence, in this paper, we employ genetic algorithm (GA) to optimize the codevector index assignment that is suitable for VQ-based watermarking transmitting over noisy channels.

Assuming that the input image is X, which has size $M \times N$ for generating the VQ codebook. We employ the LBG algorithm [9] with binary splitting to train the codebook for X, and we obtain the codebook with length L, $C = \{c_0, c_1, \cdots, c_{L-1}\}$. Next, we employ GA to split C into two groups, namely, $C' = \{c'_0, c'_1, \cdots, c'_{\frac{L}{2}-1}\}$ and $C'' = \{c''_0, c''_1, \cdots, c''_{\frac{L}{2}-1}\}$. For one index in C', it has a one-to-one corresponding counterpart in C''. We can see that $C' \cup C'' = C$ and $C' \cap C'' = \phi$. Before GA training, the codebook indices in C are randomly split into two sub-codebooks with equal number of indices. We define a fitness function in GA by

$$\text{WM\_err} = \sum_{i=0}^{\frac{L}{2}-1} (c'_i - c''_i)^2 \tag{2}$$

The goal of GA is to minimize the fitness value in Eq. (2). We perform the crossover and mutation procedures in GA for 200 iterations [6][8], and obtain the minimized fitness value, WM_err, and acquire the optimized assignment of VQ indices.

Assuming that the watermark W has size $M_W \times N_W$. And the original image is X, with size $M \times N$. In the watermark embedding process, we need to adopt relationships between the binary watermark, which has bit '0' and '1', and the split codebooks C' and C''. Suppose we find the index of the current codevector c_i in C.

Embedding procedures are:

- If the embedded bit is '0', and the index of the current codevector c_i is in C', keep the index unchanged.
- If the embedded bit is '1', and the index of the current codevector c_i is in C'', keep the index unchanged.
- If the embedded bit is '0', and the index of the current codevector c_i is in C'', replace c_i by choosing a corresponding index in C'.
- If the embedded bit is '1', and the index of the current codevector c_i is in C', replace c_i by choosing a corresponding index in C''.

By doing the procedures above, we perform GA training with fitness function in Eq. (2) to obtain the optimized index assignment with the minimized fitness value. After training, the watermarked image X' is transmitted to the receiver.

In extracting the watermark, because the watermarked image $\boldsymbol{X}'$ may experience channel errors during transmission, the receiver may obtain the possibly corrupted image, $\boldsymbol{X}''$. We will deal with the channel error problem in Section 4. In the receiver side, we use the same sub-codebooks $\boldsymbol{C}'$ and $\boldsymbol{C}''$ to extract the watermark. We use table look-up to find the VQ indices of the received image $\boldsymbol{X}''$. If the index in one codevector belongs to $\boldsymbol{C}'$, we determine the extracted bit to be '0'; if not, the extracted bit is '1'. By gathering all the extracted watermark bits, we obtain the extracted watermark $\boldsymbol{W}'$.

4 VQ INDEX TRANSMISSION OVER THE BINARY SYMMETRIC CHANNEL

In a memoryless binary symmetric channel (BSC), the input and output alphabet sets consist of the binary elements, '0' and '1'. The transition or conditional probabilities are symmetric:

$$P(0|1) = P(1|0) = \varepsilon, \tag{3-a}$$

and

$$P(1|1) = P(0|0) = 1-\varepsilon. \tag{3-b}$$

Eqs. (3-a) and (3-b) express the channel transition probabilities. Given a channel symbol was transmitted, the probability that it is received in error is ε.

We define a channel model for memoryless BSC with bit error probability ε,

$$P(c_j \mid c_i) = (1-\varepsilon)^{m-H(c_i,c_j)} \cdot \varepsilon^{H(c_i,c_j)} \tag{4}$$

where $m = \log_2 L$ is the number of bits to represent the indices, L stands for the length of codebook, $H(c_i, c_j)$ denotes the Hamming distance between the two indices c_i and c_j. Suppose that $m \cdot \varepsilon \ll 1$, we can further simplify Eq. (4) by [11]

$$P(c_j \mid c_i) = \begin{cases} \varepsilon, & H(c_i,c_j)=1; \\ 1-m\cdot\varepsilon, & H(c_i,c_j)=0; \\ 0, & H(c_i,c_j)>1. \end{cases} \tag{5}$$

With the channel model in Eq. (4) or the simplified channel model in Eq. (5), we propose a method to calculate "channel watermarking distortion" by

$$E_{\mathrm{cw}} = \sum_{i=0}^{L-1}\sum_{j=0}^{L-1} P(c_i)\cdot P(c_j \mid c_i)\cdot \delta(c_i, c_j) \tag{6}$$

where

$$\delta(c_i, c_j) = \begin{cases} 0, & c_i \text{ and } c_j \text{ belong to the same group;} \\ 1, & c_i \text{ and } c_j \text{ belong to different groups.} \end{cases} \tag{7}$$

E_{cw} is a measure of expected bit error rate (BER) of the extracted watermark from transmitting watermarked image over memoryless BSC. The smaller the E_{cw} value, the less distortion the watermarked image, hence the better robustness the watermark.

By considering the combined effects of watermark embedding and the channel noise, we can employ GA again by changing the fitness function in Eq. (2) with

$$f_n = \mathrm{PSNR}_n + \lambda \cdot \frac{1}{E_{\mathrm{cw},n}} \tag{8}$$

where f_n, PSNR_n, $E_{\mathrm{cw},n}$ denote the fitness value, PSNR value, and expected BER value of the nth iteration in GA, respectively. λ is the weighting factor to balance the effects caused by both effects of image quality and the robustness.

5 SIMULATION RESULTS

We employ several well-known test images, including Lena, F-16, fishing boat, pepper, and baboon, to test the effectiveness and superiority of our proposed algorithms. The size of the test images employed are all 512×512. The watermark to be embedded, $\boldsymbol{W}$, has size 128×128, also shown in Figure 2(a). Next, we perform VQ operation with block size $\frac{512\times512}{128\times128} = 4\times4$. The codebook size is $L = 512$ in our simulations. We employ the schemes described in Section 3 and Section 4 to train the assignment of indices with GA. For simulating the watermarked image transmission over the memoryless binary symmetric channel, we set the transition probability $\varepsilon = \frac{1}{36} \cong 0.028$.

We evaluate the performances of our algorithm by checking the watermarked image quality and the robustness of the extracted watermark. On the one hand, we examine the watermarked image quality by calculating its Peak Signal-to-Noise Ratio (PSNR) value. On the other hand, robustness denotes the capability for the watermark to combat with the errors induced by the channel noise. We measure the robustness with the Bit Error Rate (BER), by calculating the frequency of bit errors when detecting a multi-bit watermark message, between the embedded watermark $\boldsymbol{W}$ and extracted one $\boldsymbol{W}'$.

Simulation results with the existing scheme and our schemes are presented in Table I after training 200 iterations in GA. The weighting factor λ in Eq. (8) influences the effects from watermarked PSNR and extracted BER. The PSNR values between conventional VQ-based watermarking scheme and optimized index assignment for VQ-based watermarking are approximately the same for all the test images from the objective point of view. There is no difference between the two watermarked images if we view them subjectively. After optimization, our results have higher PSNR values. In addition, we have the lower BER in the extracted watermark.

Table I. Comparisons of the existing scheme and the schemes proposed in this paper, with the fitness function demonstrated in Eq. (8).

Scheme	Watermarked PSNR	Bit Error Rate in extracted watermark
The scheme in [1]	30.16 dB	0.1197
With weighting factor $\lambda = 0.25$ in Eq. (8)	30.32 dB	0.0824
With weighting factor $\lambda = 0.5$ in Eq. (8)	30.23 dB	0.0695

Table II. The comparisons of the actual bit error rate (BER) with or without GA optimization. BER greatly reduced after GA optimization in VQ index assignment.

test image	existing scheme	optimized	BER reduction ratio
Lena	0.1197	0.0543	54.64%
F-16	0.1140	0.0556	51.28%
fishing boat	0.1136	0.0623	45.16%
pepper	0.1189	0.0633	46.76%
baboon	0.1130	0.0754	33.27%

In Table II, we show comparisons of watermark robustness between the results of existing scheme and our scheme. We measure watermark robustness by calculating the actual BER values after training with the fitness function in Eq. (6). The results from different test images with our schemes reveal better robustness, which outperform those in literature. BER values greatly reduced after GA optimization in VQ index assignment. Moreover, from the BER reduction ratio, we can see the effectiveness of our proposed schemes.

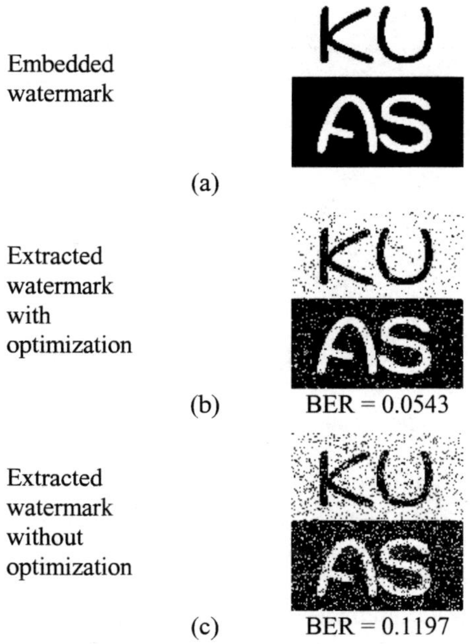

Figure 2. Comparisons of the embedded watermark and extracted ones. (a) Embedded watermark, with equal numbers of 0's and 1's. (b) Extracted watermark with GA-training. (c) Extracted watermark without GA-training.

We demonstrate the extracted watermarks in Figure 2(b) and Figure 2(c), respectively. The extracted watermark looks clearer in Figure 2(b) than the one in Figure 2(c). Hence, by training index assignment with GA and by properly choosing the GA fitness function, we can obtain the better robustness with our proposed scheme. Summing up, our proposed algorithm has better robustness to compare with other results in literature.

6 CONCLUSION

In this paper, we proposed an optimized scheme for VQ-based image watermarking, which is suitable for transmitting over noisy channels. We addressed fundamental concepts, and the importance and complexity for index assignment in VQ. By splitting the codebook and assigning different indices for embedding the watermark, the proposed algorithm outperforms those in literature. Simulation results also presented the better robustness of our watermarking algorithm, with comparable quality in the watermarked images. With GA, we can explore the optimized index assignment for watermark embedding and the effect caused by channel noise.

In the future work, other optimization techniques will be developed into our system. We can further optimize our system by considering the combined effects from VQ codebook design, index assignment, and errors from memoryless BSC, to obtain the better watermarked image quality and the better robustness of the algorithm.

ACKNOWLEDGEMENT

This work was supported by National Science Council (Taiwan, ROC) under Grant No. NSC91-2219-E-151-002.

REFERENCES

[1] Z. M. Lu and S. H. Sun, "Digital image watermarking technique based on vector quantisation," *Electronics Letters*, vol. 36, no. 4, pp. 303–305, Feb. 2000.

[2] Z. M. Lu, J. S. Pan, and S. H. Sun, "VQ-based digital image watermarking method," *Electronics Letters*, vol. 36, no. 14, pp. 1201–1202, Jul. 2000.

[3] H.-C. Huang, F. H. Wang and J. S. Pan, "Efficient and robust watermarking algorithm with vector quantisation," *Electronics Letters*, vol. 37, no. 13, pp. 826–828, Jun. 2001.

[4] H.-C. Huang, F. H. Wang and J. S. Pan, "A VQ-based robust multi-watermarking algorithm," *IEICE Trans. Fundamentals*, vol. 85-A, no. 7, pp. 1719–1726, Jul. 2002.

[5] M. Jo and H. D. Kim, "A digital image watermarking scheme based on vector quantisation," *IEICE Trans. Information and Systems*, vol. E85-D, no. 6, pp. 1054–1056, Jun. 2002.

[6] D. E. Goldberg, *Genetic Algorithm in Search, Optimization, and Machine Learning*, Addison-Wesley: Reading, MA, 1989.

[7] K. Sayood, *Introduction to Data Compression*, 2nd Ed., Morgan Kaufmann: San Francisco, CA, 2000.

[8] J. S. Pan, F. R. McInnes, and M. A. Jack, "Application of parallel genetic algorithm and property of multiple global optima to VQ codevetor index assignment for noisy channels," *Electronics Letters*, vol. 32, no. 4, pp. 296–297, Feb. 1996.

[9] Y. Linde, A. Buzo, and R. M. Gray, "An algorithm for vector quantizer design," *IEEE Trans.Communication*, vol. 28, no. 1, pp. 84–95, Jan. 1980.

[10] J. S. Pan and S. C. Chu, "Non-redundant VQ channel coding using tabu search strategy," *Electronics Letters*, vol. 32, no. 17, pp. 1545–1546, Aug. 1996.

[11] N. Farvardin, "A study of vector quantization for noisy channels," *IEEE Trans. Information Theory*, vol. 36, no. 4, pp. 799–809, Jul. 1990.

USING INVISIBLE WATERMARKS TO PROTECT VISIBLY WATERMARKED IMAGES

Yongjian Hu [1 2 3]*, *Sam Kwong* [3] *and Jiwu Huang* [1]

[1] School of Info. Sci. and Tech.,Sun Yat-Sen Univ., Guangzhou, PRC
[2] Dept. of Automatic Control Engg. South China Univ. of Tech., Guangzhou, PRC
[3] Dept. of Computer Science, City Univ. of Hong Kong, Kowloon, Hong Kong
E-mail: eeyjhu@scut.edu.cn

ABSTRACT

Physical visible watermarks have been widely used for centuries. Now digital visible watermarks such as electronic logos find their applications in digital library, video broadcasting, and other multimedia services. Several visible watermarking techniques have been proposed in the literature, and meanwhile, some problems with visible watermarks are also under investigation. Among these problems, watermark removal and unauthorized insertion are two major concerns. In this paper, we propose using an invisible watermark in visibly watermarked images to overcome these problems. When a visibly watermarked image is in question, the invisible watermark can provide appropriate ownership information. We first investigate what kind of invisible watermark is needed, and then, focus on the details of the invisible watermarking technique. The experiments have shown that the proposed algorithm can provide a very effective protection for visibly watermarked images.

1. INTRODUCTION

With the rapid spread of computer networks and the wide use of multimedia technologies, many watermarking techniques are now under investigation for protecting owner's intellectual property rights. Which protections are appropriate relies on the kind of data and environments. Visible watermarks are useful for protecting online images because they discourage unauthorized copying [1]. Several visible watermarking techniques have been proposed in the literature (e.g. [2]-[4]). In particular, the technique developed by the IBM has now been used in many projects including the Vatican Library and the National Gallery of Art [1][2]. Generally, visible watermarking requires that each watermark should be easily visible, unobtrusive, and hard to remove. From the viewpoint of commercial use, however, the last requirement is very important. Although we can use a number of methods to make the visible watermark difficult to remove, we have to admit that removal is not impossible. In fact, some researchers have attempted to remove the embedded watermark by using image inpainting techniques [5]. Besides robustness, there exists another problem for visible watermarks. As pointed out in [3], a visible watermark bearing a certain logo does not constitute a proof of ownership. That is, some one can insert the logo of others within an image and claim that the resulting image comes from them. Due to the existence of such threats, it is necessary to develop a new mechanism to protect visibly watermarked images.

In this paper, we propose an invisible watermarking technique in the dual watermarking system to provide additional protection for visibly watermarked images. We first describe the properties and requirements of such an invisible watermark, and then, give the details of the watermarking algorithm. The experimental results have shown that the proposed algorithm is very effective to verify the ownership of the visibly watermarked image.

* This work was supported by City University research grant 7001488, NSFC (60325208, 60172067, 60133020), "863" Program (2002AA144060), NSF of Guangdong (013164) and Funding of Ministry of Education, China.

2. NEW WATERMARKING TECHNIQUE

2.1. Problem analysis

Generally, there are mainly two problems associated with visible watermarking. The first is that a visible watermark must be difficult to remove; the second is that it must be able to withstand the impersonator problem [3]. Most efforts in the literature focus on the first problem. However, to the best of our knowledge, there are few papers available for the second problem. The researchers in [3] and [4] proposed an invisible fragile watermarking scheme to solve this problem. They determined whether the embedded visible watermark is genuine or unauthentic by detecting the alteration to the image.

In [4], Mohanty *et al.* presented a dual watermarking technique which attempts to establish the owner's right to the image and detect the intentional and unintentional tampering of the image. However, this early research is simply

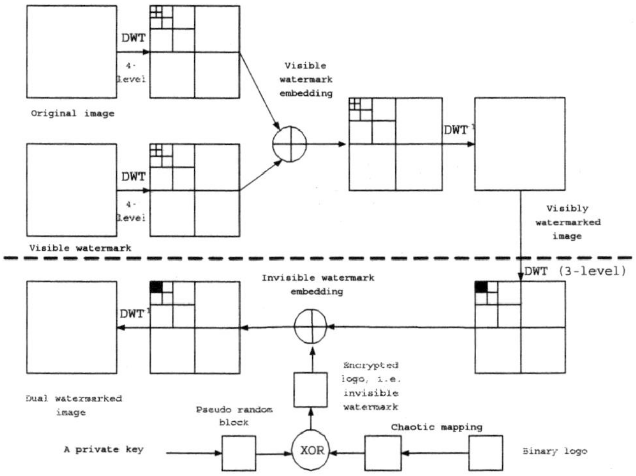

Figure 1: A dual watermarking system. The upper part is for 4-level DWT domain visible watermarking. The lower part is for invisible watermark embedding where the black area represents the embedding region.

a combination of visible and invisible watermarking algorithms. It first used a block-DCT based visible watermarking algorithm to embed a gray-level watermark image, and then, regarded the resulting image as a new image to carry out invisible watermarking. Invisible watermarking is performed in spatial domain. The fragile watermark consisting of pseudo-random binary sequence $\{0,1\}$ is EX-ORed with the kth bit-plane of the image. All bit-planes (Ex ORed and non-Ex ORed) of the image are merged to obtain the final watermarked image. They claimed that if anybody tries to tamper the visible watermark intentionally, they can know the extent of tampering with the help of invisible watermark detection algorithm. However, they obviously did not distinguish between the tampering of the visibly watermarked image and that of the embedded visible watermark. Their fragile watermark can detect whether there is any change in the visibly watermarked image, but it is unable to tell whether those changes are targeted at the embedded visible watermark or the visibly watermarked image.

In [3], Wong and Memon used an invisible authentication watermark to ensure the identity of a visibly watermarked image. Any modification to the visible watermark would be reflected in a corresponding error in the fragile watermark. However, their scheme is too sensitive to be used in most practical applications.

Some apparent disadvantages of using fragile watermarks for this purpose can be described as follows:

1) Common image processing such as compression, filtering, noise addition or geometric distortion can not hinder the embedded visible watermark from indicating ownership, but they can destroy the fragile watermark easily.

2) When the owner's visible watermark is visually removed or tampered, or replaced by another unauthorized visible watermark, one can not identify the right owner to the image with the fragile watermarking scheme.

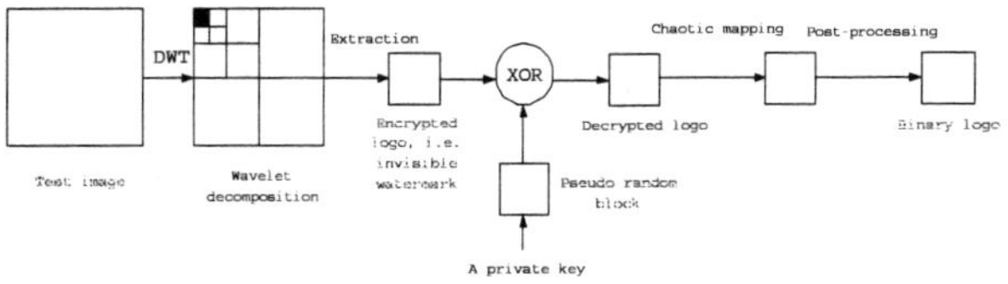

Figure 2: Invisible watermark extraction.

Based on the above analysis, we conclude that the fragile watermarking scheme is not the most suitable one for this application. We need an invisible watermark that can enhance ownership protection of the visibly watermarked image. Ownership assertion watermarks are typically robust. So we need a robust instead of fragile invisible watermarking scheme in the dual watermarking system. We present the properties and associated requirements of such robust watermarks as follows:

1) The watermark should be invisible and has no apparent interference with the visibly watermarked image.

2) The watermark is desirously extracted without resorting to the visibly watermarked image.

3) The watermark must be difficult to remove and can resist non-malicious changes such as image compression and malicious attacks such as image inpainting/replacement. It is desirable for the invisible watermark to survive most attacks that the visible watermark can survive. Visibly watermarked images are usually compressed for online use. So malicious attacks may be targeted at the compressed-decompressed versions of images. The most challenge to such a watermarking technique is to make the invisible watermark robust against operations like compression and image inpainting.

For better understanding of the terminologies employed, the diagram of a dual watermarking system is shown in Figs.1 and 2. Below we describe the details of the proposed invisible watermarking scheme.

2.2. Robust invisible watermarking algorithm

As stated in subsection 2.1, resistance to compression and image inpainting is the most important challenge to the design of such an invisible watermarking algorithm. Since content-preserving image processing such as compression has low-pass nature, it is reasonable to choose the low-pass component for watermark embedding. In this paper, we propose to embed the invisible watermark in the low-pass subband of a three-level wavelet decomposition image. Image inpainting/replacement will affect all the frequency components of any watermarking scheme based on a global transform, whereas the DWT-based scheme produces watermarks with local spatial support. The selection of three-level decomposition is to make the number of low-pass coefficients large enough for watermark embedding.

The embedding and extraction strategy is based on the method in [7]. Let $X = [x_k]$ denote the low-pass coefficients of the visibly watermarked image, and let the watermark, $M = [m_k]$, be the binary sequence. Here $m_k = 0$ or 1 and $k(= 1, 2, ..., K)$ is the single index of watermark sequence. The detailed steps of embedding can be referred to [7].

In the process of watermark detection, we use the inverse manner of embedding for extracting the watermark.

2.3. Constructing invisible watermarks

Many forms of invisible watermarks can be used in this application. In this paper, we choose the watermark in the form of the binary image of the embedded visible watermark so that the extracted logo can indicate the ownership without additional computing. To increase the security of the invisible watermark, the invisible watermark image is shuffled with some techniques like chaotic mapping before embedding. Here we use the Arnold's cat map to transform the binary logo.

Let X= $\begin{bmatrix} x \\ y \end{bmatrix}$ be a $n \times n$ matrix of watermark image, Arnold's cat map is the transformation described in [8]:

$$\Gamma \begin{bmatrix} x \\ y \end{bmatrix} \rightarrow \begin{bmatrix} x+y \\ x+2y \end{bmatrix} \bmod n.$$

where mod is the modulo of the $\begin{bmatrix} x+y \\ x+2y \end{bmatrix}$ and n. In this paper, $n = 64$ so that the periodicity is 96.

After transformation, a binary block which follows uniform distribution and has the same size as the binary logo is EX-ORed with the binary logo to create the encrypted watermark (see Fig.1). The seed of the pseudo-random number generator is used as the private key. In watermark detection, without the private key, one can only obtain a meaningless binary sequence. However, legal users can recreate the binary logo by EX-ORing the extracted block with the pseudo-random block generated with the private key.

2.4. Post-processing extracted invisible watermarks

From experiments we notice that image operations like compression and inpainting/replacement have different effects on the extracted binary image. The former produces random noise-like pixels across the entire extracted image whereas the latter only ruins parts of the image. This is because compression could alter values of some low-pass coefficients and thus cause parts of embedded information bits not to be extracted correctly. With higher compression ratio, more noise-like pixels would appear. On the other side, inpainting/replacement, which is often a kind of spatial operation,

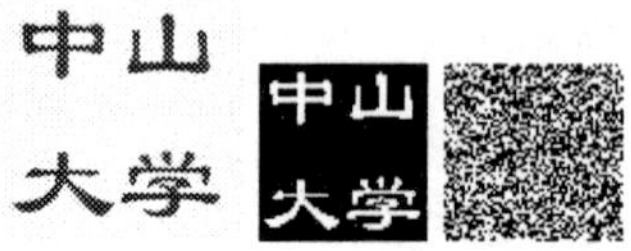

Figure 3: The visible watermark ($512 \times 512 \times 8bits$) (left), the binary logo ($64 \times 64 \times 1bits$)(center), and its encrypted one (right).

only causes information bits embedded in targeted areas to be lost. To remove noise-like pixels resulting from compression, we use a median filter to post-process the extracted binary image. The structure elements of the filter is defined as $S = [0\ 1\ 0; 1\ 0\ 1; 0\ 1\ 0]$. After such a non-linear filtering, most isolated noise-like pixels can be removed.

3. EXPERIMENTAL RESULTS

We have performed many experiments to demonstrate the effectiveness of the proposed algorithm. The experimental results on $Lena(512 \times 512 \times 8bits)$ are reported here. The wavelet employed is $sym8$. $Q = 18$. The gray-level watermark image, i.e., the visible watermark, and its binary logo and the encrypted version of the logo are shown in Fig.3.

Content-adaptive visible watermarking is crucial for visibility as well as robustness. In this paper, we adopt an image-adaptive visible watermarking scheme based on our previous work in [6]. The only change to that algorithm is that we scale the gray value of the watermark image into the range of the host image before wavelet decomposition instead of during subband-by-subband watermark addition. The diagram of the visible watermarking algorithm is shown in the upper part in Fig.1. The detailed steps for embedding can be referred to [6].

After performing visible watermarking, we obtain the visibly watermarked image in Fig.4(left). This image is then processed with the proposed invisible watermarking scheme to produce the dual watermarked image in Fig.4(right). Compared to the original visibly watermarked image, the dual watermarked image has a very high PSNR (peak signal to noise ratio) (45.86dB). Therefore, the insertion of invisible watermark has very small impact on the image quality. If there is no interference on this dual watermarked image, the invisible watermark, i.e., the binary logo, can be losslessly extracted. The results are shown in Fig.5. When the image suffers from unintentional processing such as compression, the logo can be reconstructed with some watermark bits lost. The results under JPEG compression with quality factor 50 and 30 are exhibited in Fig.6. It can be observed that the proposed post-processing is very effective to filter isolated noise-like pixels. When the image suffers from intentional attacks such as image replacement, the results in

Figure 4: The visibly watermarked image (left) and the dual watermarked image (right).

Figure 5: The extracted encrypted logo (left) and the decrypted one (right).

Fig.7 show that the proposed structure of the invisible watermark is strongly against this type of attack. The reason is that the useful information bits of the binary logo have been efficiently diffused by chaotic mapping before embedding and, on the other side, image replacement can only destroy the information bits in the targeted area. This explanation is also applicable to more fatal and complex attacks like image inpainting where most parts of the image content remain unchanged. Due to the use of low-pass embedding strategy, the proposed algorithm also shows strong resistance against attacks such as additive noise, smoothing and scaling. Because of paper length limitation, we can not show the experimental results.

4. CONCLUSION

We have proposed a new way for protecting visibly watermarked images. We first pointed out that a robust invisible watermark should be used, and then, give details of designing a robust invisible watermarking algorithm including the construction of the watermark, embedding, extraction and post-processing. Future efforts will focus on how to enhance robustness of the embedded invisible watermark, for example, to make the watermark against geometrical attacks which challenge all existing invisible watermarks. How to embed more watermark information is another major concern. The research on dual watermarking system is very significant for the practical application of visible watermarks.

Figure 6: The results under JPEG compression with quality factor 50 and 30, respectively. From left to right, the decrypted logos (the first and the third) and their corresponding post-processed ones (the second and the fourth).

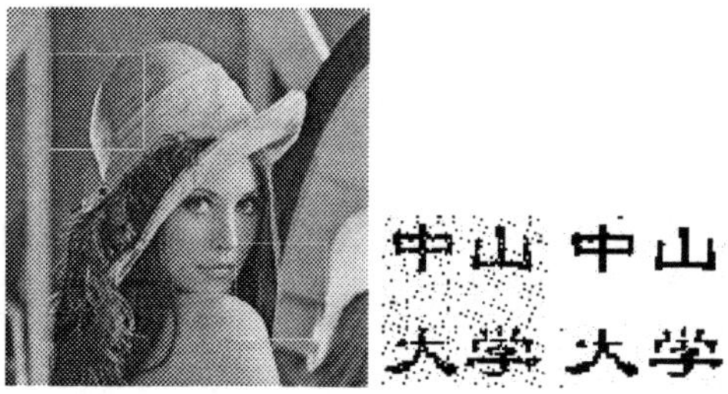

Figure 7: The partly replaced test image (left), the extracted decrypted logo (center) and its post-processed one (right).

5. REFERENCES

[1] F. Mintzer, "Developing digital libraries of cultural content for Internet access ," *IEEE Communications Magazine,* Vol. 37, Jan. 1999, pp. 72 -78

[2] G.W. Braudaway, K.A. Margerlein, and F.C. Mintzer, "Protecting public-available images with a visible image watermark," *Proc. SPIE Conf. on Optical Security and Counterfeit Deterrence Techniques,* vol. SPIE 2659, Feb. 1996, pp. 126-132

[3] P.W. Wong and N. Memon, "Secret and public key image watermarking schemes for image authentication and ownership verification", *IEEE Trans. on Image Processing,* vol.10, Oct. 2001, pp. 1593-1601

[4] S.P. Mohanty, K.R. Ramakrishnan, M.S. Kankanhalli, "A dual watermarking technique for images," *ACM Multimedia (2),* 1999, pp. 49-51

[5] C.H. Huang, and J.L. Wu, "Inpainting attacks against visible watermarking schemes," *Proc. SPIE Conf. on Security And Watermarking Of Multimedia Contents,* vol. SPIE 4314, 2001, pp. 376-384

[6] Y. Hu, and S. Kwong, "Wavelet domain adaptive visible watermarking," *Electronics Letters,* vol. 37, Sep. 2001, pp. 1219-1220

[7] H. Inoue, A. Miyazaki, and T. Katsura, "Wavelet-based watermarking for tampering proofing of still images," *IEEE Int. Conf. on Image Processing,* Vol. 2 , Sep. 2000, pp. 88 -91

[8] G. Peterson, "Arnold's Cat Map," http://online. redwoods.cc.ca.us/instruct/darnold/maw/catmap.htm

A 2.5 MILLIWATT SOS CMOS RECEIVER FOR OPTICAL INTERCONNECT

A. Apsel, Z. Fu

Cornell University
Ithaca, NY, 14853

ABSTRACT

We demonstrate a low power, high bit rate, cross coupled differential receiver in Silicon on Sapphire (SOS) CMOS to be used as part of an inter-chip optical interconnect. The internal amplifier of the transimpedance first stage provides high gain without requiring large, capacitive input gates. The resulting transimpedance stage extends the bandwidth of the differential receiver when small photodetectors are used. We fabricated this receiver in an SOS CMOS process to simplify the packaging of chip-to-chip interconnects with CMOS processors. The total measured power consumption of this receiver is 2.5mW at gigabit rates, in a 0.5 μm UTSi^{TM} SOS CMOS process.

1. INTRODUCTION

As processor speeds continue to increase, the speed of inter-chip communication has become the bottleneck of high performance systems. Optical interconnects designed for low power consumption at high bit rates offer an excellent means of overcoming data rate limitations imposed by conventional wiring between chips [1]. As optical interconnects reduce in power and size, their usefulness at the box and board level of a computer will increase dramatically.

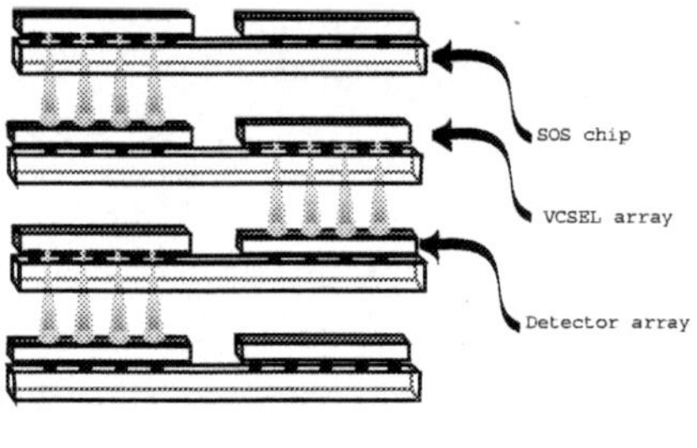

Figure 1: A stacked array of VCSELs, Detectors, and SOS CMOS circuitry compose a 3-D interconnect.

CMOS compatibility is essential in the production of inexpensive optical communication channels between CMOS chips. Silicon on sapphire (SOS) processes pair CMOS technology with a sapphire substrate. Sapphire substrates provide advantages, in addition to electrical isolation, of optical transparency at wavelengths from 300nm to 6000nm. Using SOS electronics hybridized to photonic elements, we can build through substrate interconnects at different levels in a vertically stacked array of microchips, as shown in figure 1. Furthermore, bulk-less CMOS SOS processes dramatically reduce parasitic capacitances and improve device isolation, enabling the design of low power, high-speed CMOS circuits [2].

In this paper we present an innovative cross-coupled differential amplifier for use in an input stage of a differential optical receiver. We demonstrate the receiver in a silicon on sapphire process (SOS) for through wafer chip-to-chip interconnects. In the discussion to follow we show how a cross coupled amplifier structure fabricated in an SOS process can be used to improve the bandwidth of a transimpdedance amplifier while maintaining low power consumption and packaging simplification. We also show experimental results for such an amplifier fabricated in an SOS CMOS process and tested as part of a through wafer optical interconnect.

2. DIFFERENTIAL TRANSIMPEDANCE AMPLIFIERS

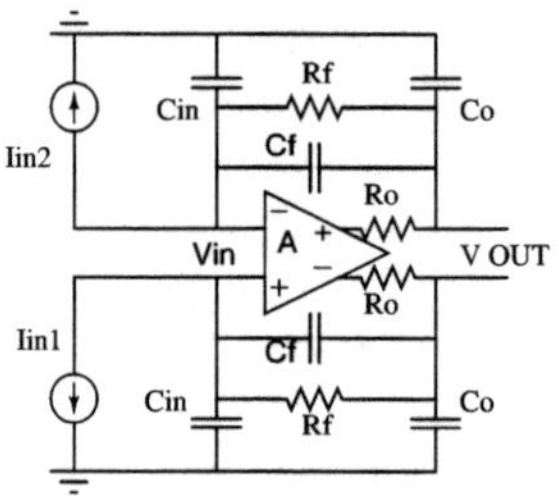

Figure 2: Diagram of a differential transimpedance amplifier.

We begin the examination of this transimpedance amplifier with a brief discussion of general differential transimpedance amplifiers as shown in figure 2. For more detailed analysis see [3]. When a transimpedance amplifier is designed for high bandwidth performance, the bandwidth of the internal differential amplifier may be only 2 to 3 times the bandwidth of the transimpedance design. In these cases the transmipedance, Z_{out}, is a second order function

$$Z_{out} = \frac{-AR_f}{1+A} \frac{1}{(1 + s(\frac{1}{1+A}R_fC_{in} + \frac{B}{1+A}) + s^2\frac{BR_fC_{in}}{1+A})} \tag{1}$$

where $\frac{1}{B}$ is the bandwidth of the differential amplifier, A refers to the low frequency gain of the internal amplifier, R_f is the feedback resistance, C_{in} is the combined photodetector and input gate capacitance of the amplifier, and C_o and R_o are the next stage load capacitance and output resistance of the amplifier. The pass-band of the filter ends at $\omega = \omega_0$ when the slope of the frequency response drops by 40 dB/decade. We find the following expressions for ω_0.

$$\omega_0 = \sqrt{\frac{1+A}{BR_fC_{in}}} \tag{2}$$

We note that since $1/B$ is proportional to ω_0,

$$\omega_0 \propto \frac{1+A}{R_f C_{in}} \tag{3}$$

indicating that ω_0 is maximized when A is maximized and C_{in} is minimized while maintaining the necessary ratio between $R_f C_{in}$ and $\frac{1}{B}$ for stability and limited peaking. These points are crucial to the design of an efficient optoelectronic receiver. The following section compares the cross coupled transimpedance amplifier to an amplifier without cross coupled loads.

3. CROSS COUPLED DIFFERENTIAL AMPLIFIER

3.1. Theory

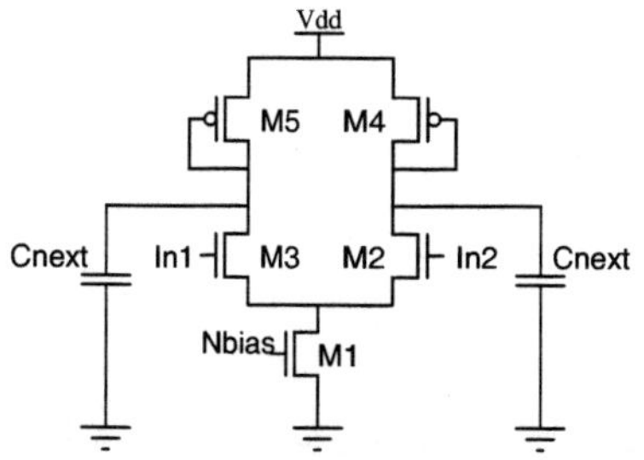

Figure 3: Diagram of diode load amplifier.

Figure 3 shows a differential amplifier with self-biasing diode connected current sources. This is a simple amplifier which can be used in a low power differential transimpedance amplifier circuit with resistive feedback. It provides the power supply noise and interference rejection desired of a differential receiver circuit. From the small signal model, assuming the circuit to be symmetric, we can write the transfer function

$$\frac{\Delta V_o}{\Delta V_i} = \frac{V_{gp1} - V_{gp2}}{V_1 - V_2} = \frac{g_{mn}}{g_{mp} + g_{dp} + g_{dn}} \frac{1 - s\frac{C_{gdn}}{g_{mn}}}{1 + s\frac{C_N + C_{gsp} + C_{gdn}}{g_{mp} + g_{dp} + g_{dn}}} \tag{4}$$

using the approximation that $C_{gsn} >> C_{gdn}$.

The gain bandwidth product of this amplifier is linearly related to g_m and $\frac{1}{C_N + C_{gsp} + C_{gdn}}$. The gain of the differential amplifier, defined as $\frac{g_{mn}}{g_{mp} + g_{dp} + g_{dn}}$, is adjusted by sizing the input nFet transistors relative to the diode connected pFETs. There is a limit on how small we can make the diode connected pFETs and still drive subsequent stages. Due to this limitation, the g_m and therefore size of the input nFETs must can be made large to acheive high gain. Unfortunately, a high gain amplifier is designed at the expense of a large input gate, adding capacitance to the input of the transimpedance amplifier, and degrading overall bandwidth. This is not a significant factor when photodetectors are large and have high capacitance values. However, the capacitance of the input transistors can be significant when small photodetectors with small capacitances (below 200fF) are used. Furthermore, aggressive bandwidth improvement of the transimpedance amplifier as a whole cannot occur if we require very large input photodetectors.

Figure 4 shows a modification to the diode connected amplifier in figure 3. The modification maintains the transimpedance amplifier's compatiblitity with gain stages and comparators which may be used in the design of a complete receiver. The transimpedance amplifier shown in figure 4 has positive feedback, and will not work as a stable differential amplifier if the stabilizing resistors are removed and large signal inputs are applied without control of the DC bias point of the inputs. For this reason it is best to use diode load amplifiers as self-biasing gain stages in a multistage receiver design.

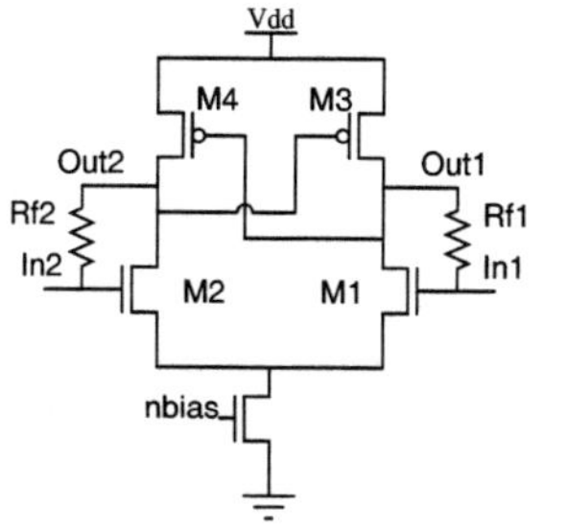

Figure 4: Cross coupled transimpedance amplifier used as alternative receiver first stage.

As in the diode load differential transimpedance amplifier, the cross coupled amplifier provides offset and interference rejection since the circuit is symmetric to both inputs and outputs. Furthermore, the common source architecture provides power supply noise rejection. The cross-coupled transimpedance amplifier has the advantage of generating gain without using large input transistors which cut the bandwidth of the transimpedance stage as detailed in the previous section. Analysis of the small signal model is necessary to understand the benefits of this configuration.

Again, we assume that the circuit is symmetric for this analysis. From the small signal model we can compute the full transfer function.

$$\frac{V_{g1} - V_{g2}}{V_1 - V_2} = \frac{-g_{mn}}{g_{dp} + g_{dn} - g_{mp}} \frac{1 - s\frac{C_{gdn}}{g_{mn}}}{1 + s\frac{C_{gdn} + C_{gsp} + C_n}{g_{dp} + g_{dn} - g_{mp}}} \tag{5}$$

As long as the amplifier is designed for high gain, the pole of this function will occur before the zero. In practice we are able to design a cross coupled amplifier with a small C_{gdp} compared to C_n and C_{gdn} such that the gain-bandwidth products of the two circuits are approximately equal. The subtle advantage of the cross coupled transimpedance amplifier can be seen via the expression for DC gain, A.

$$A = \frac{g_{mn}}{g_{dp} + g_{dn} - g_{mp}} \tag{6}$$

We can improve this figure in two ways. First, we can increase g_{mn} by designing large input transistors as in the previous case, with an eventual cost in the bandwidth $\approx \frac{1+A}{R_f C_{in}}$ of the transimpedance amplifier. Second, we have the additional freedom of sizing g_{mp} relative to g_{dp} and g_{dn}. The latter option allows us to use additional freedom to size input transistors small while maintaining high gain. This is not much of a factor when input photodiodes are large (capacitances on the order of .5 pF). However, in order to achieve the best performance from a transimpedance receiver, we would like to minimize C_{in} by using a small photodiode. In this case, any added capacitance at the input of the amplifier degrades receiver performance.

As we examine the cross-coupled circuit we notice that it utilizes positive feedback to achieve high gain. It is important to note that positive feedback can be used to improve gain as we have demonstrated, but this circuit will only be stable when used with stabilizing resistors and while g_{mp} is not greater than $g_{dp} + g_{dn}$. If we consider the qualitative behavior of the circuit, we will recognize why this is true. Assume a current is sunk through $M2$ through feedback resistor R_{f2} to a photodiode. At the same time, no current is sourced through R_{f1}. V_{out2} will rise to accommodate the voltage drop, pulling the gate of $M3$ high and the voltage on node $Out1$ low. That voltage will pull the gate of $M4$ low, providing additional gain (with a cost in delay) and allowing it to source more current. If the pFETs and g_{mp} are chosen and sized correctly, the additional current will be sourced primarily through the feedback resistor, R_{f2}, allowing node $Out2$ to settle. If g_{mp} is too large, the current sourced through $M4$ in response to a change in V_{out1} will be very large. In this case, a large current will be pushed through input $M2$, causing node $Out1$ to rise more. As this node increases, the feedback cycle continues until both outputs have hit the supply rails. Clearly this configuration is unstable, however, the circuit configuration shown here with balanced feedback under the sizing constraint that $g_{mp} < g_{dp} + g_{nd}$ enables the design of a stable, high performance receiver.

3.2. Silicon on Sapphire

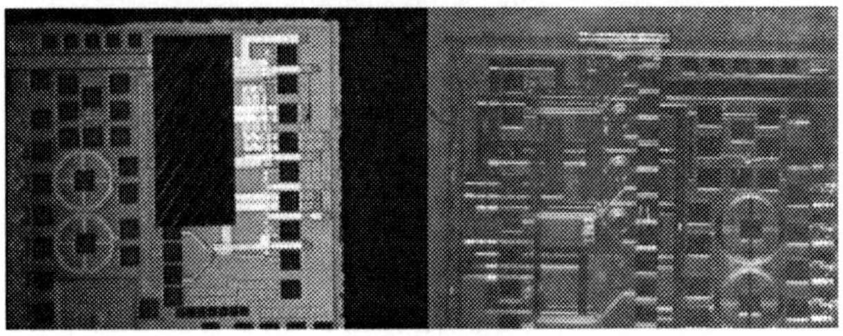

Figure 5: Front and back of SOS receiver array with flip bonded photodetectors. On the right hand side figure an array of PIN photodiodes can be seen through the SOS CMOS substrate. Layout of each cell is 30 × 30 μm in .5 μm SOS process.

Packaging of optoelectronic VLSI systems is simplified when signals transmit directly through electronic substrates [1, 4]. This enables flip bonding of any front side contact optical elements such as VCSEL's or passive nano-photonics. Standard bulk and SOI CMOS wafers are opaque over the wavelengths of commercially available VCSEL sources (850nm, 980nm). Even in upper wavelength ranges of 1550nm, transmission in silicon is well below 100 percent, requiring significant thinning of the substrate. Working systems have been produced at 1310nm and 1550nm using this technique, [4], however the complexity of this approach as well as the optical loss incurred adds cost and reduces yield significantly.

Unlike silicon wafers, SOS wafers are transparent to wavelengths ranging from 300nm to well over 5 μm. Transmission of light at visible and IR wavelengths in sapphire approaches 90 percent for 1mm thick samples, while transmission through the same thickness of silicon is far more lossy [5]. Figure 5 shows top and bottom views an array of cross coupled receivers fabricated in SOS CMOS and packaged with an array of GaAs PIN photodiodes. We can easily see the PIN photodiodes bonded to the receiver through the transparent substrate.

Sapphire substrates, in addition to providing excellent optical characteristics, facilitate design and fabrication of high-speed

Amplifier	Input nFETs	pFETs	nFET current source
Diode Load	100/0.5	4/0.5	15/0.5
Cross Coupled	30/0.5	4/0.5	20/0.5

Table 1: Transistor sizes in W/L for amplifier circuits used receivers.

and low power electronics. Devices produced in SOS processes are electrically isolated from each other, reducing parasitic capacitances, latch-up, cross talk, and ultimately power consumption. The CMOS electronics reported in this paper were fabricated in a 0.5μm UTSi^{TM} Peregrine SOS process available through MOSIS. Using this process, we were able to produce receivers for 750Mbps data links consuming less than 2.5mW per channel with bit error rates of better than 10^{-12}.

3.3. Results

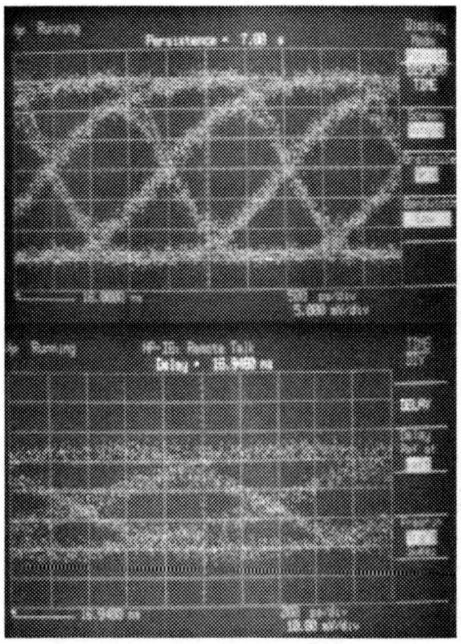

Figure 6: Eye diagrams from receiver tested optically through the substrate with flip-bonded photodiodes. (Top) Response at 700Mbps. (Bottom) Response at 1Gbps.

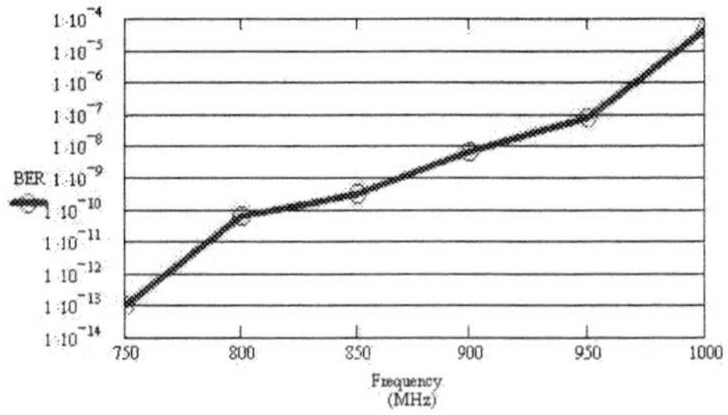

Figure 7: Measured results of bit error rate tests on the cross-coupled receiver tested as a part of an optical interconnect using a $2^{31} - 1$ psuedo-random bit sequence.

We have designed, simulated, fabricated, and tested the cross-coupled amplifier shown in figure 5. The layout of the transimpedance

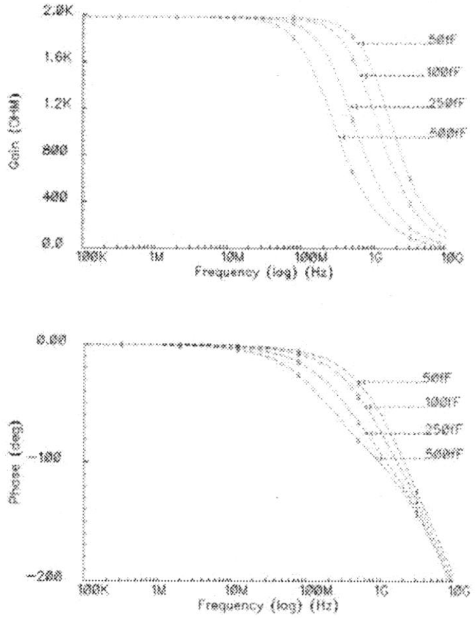

Figure 8: Simulation of Cross-coupled SOS receiver with 500 fF, 250 fF, 100 fF, and 50fF input capacitance. Bandwidth at 50fF is approximately 1.4 GHz. Data rate is 2Gbps.

first stage is extremely compact and works for low gain applications, without secondary gain stages other than a comparator. Table 1 shows the transistor sizes used in this layout and simulation. We can see from this table that the same gain can be acheived in the cross-coupled amplifier with input gates less than 1/3 the size of the input gates used in a diode load amplifier. The small number of stages allows us to use the entire bandwidth of the first stage without degrading phase and latency of the receiver. In order to maintain the high bandwidth for testing, we add weak output driver capable of driving the signal to only 0.35V on 50 Ωs, but adding only a few degrees of phase shift.

We fabricated an array of 3 cross-coupled receivers shown in figure 5, in a 0.5 μm UTSi^{TM} CMOS process. Figure 6 shows measured eye diagrams representing the large signal behavior of the full receiver with 250fF input photodetectors flip bonded to the SOS die. We tested the receivers using a directly modulated VCSEL with a 10dB extinction ratio at 1.25Gbps controlled by a high speed PRBS generator with a bitstream output of length $2^{23} - 1$ as the optical input. We measured total power consumption of these receivers as 2.5mW per channel at data rates up to 1 Gbps with 40μA sensitivity for digital output levels.

Noise in our measured results reflects noise of the optical channel, VCSEL, detector, and receiver. Figure 7 shows the measured BER's (bit error rates) for the receiver as part of an optical interconnect using a PRBS bit sequence of length $2^{31} - 1$. BER's for this receiver in this configuration were measured as low as 10^{-13} for data rates of 750Mbps. This low BER indicates that our low noise, low power receiver in SOS is practical for chip-to-chip interface applications such a busses where BER's must approach 10^{-12} or better to rival high speed electrical signaling.

Simulation results shown in figure 8 indicate that reducing the capacitance of input photodetectors will further improve the performance of this receiver, as indicated in the theory section of this paper. Given that simulation results for C_D=250fF agree well with our measured results, we expect 2Gbps performance of the receiver when C_D=50fF at the same power level based on an estimated bandwidth of 1.4 GHz in simulation. A receiver with a resistive or diode load differential amplifier of the same gain and bandwidth will not show a marked overall bandwidth improvement when 50fF capacitance photodiodes are used. The bandwidth in a resistive load configuration with 50fF detectors is approximately 700 MHz in simulation. We expect the increase in bandwidth for this cross-coupled configuration to degrade the noise floor somewhat, but calculations indicate that degradation with be less than an order or magnitude. Sustainable BER's are estimated to be 10^{-12}, a range acceptable for interconnects within computer networks.

While we have demonstrated a low power, high speed receiver which can easily be packaged for chip-to-chip optical interconnects using a 0.5μm process, we believe that with scaling, the performance of this receiver will continue to improve dramatically. Based on our calculations for SOS and SOI technologies, we believe that data rates of 10 Gbps can be acheived in 0.13μm technology for power consumptions of less than 0.5 mW per channel, assuming 25fF input capacitance detectors.

4. CONCLUSIONS

We have shown that a cross-coupled SOS CMOS transimpedance amplifier can be employed to dramatically improve the bandwidth of low power opto-electronic interconnects while maintaining a low noise floor. The use of SOS CMOS technology in conjunction with this receiver configuration enable the power consumption to remain low at higher bit rates while facilitating simplified packaging of 3-D interconnects. We have demonstrated such a receiver configuration as part of a through wafer optical interconnect and have achieved low noise and low power operation at bit rates approaching 1Gbps. We have also shown that the use of low capacitance photodiodes would allow us to operate this receiver at rates above 2Gbps at the same power level. This type of low-power, CMOS compatible solution has the potential to solve many of the problems that arise due to scaling of wired interconnects.

5. REFERENCES

[1] A. Andreou, Z. Kalayjian, A. Apsel, P. O. Pouliquen, R. A. Athale, G. Simonis, and R. Reedy, "Hybrid integration of surface emitting vcsel's with ultra-thin silicon on sapphire (sos) cmos vlsi circuits," *Circuits and Systems Magazine*, Aug. 2001.

[2] J. B. Kuo and Ker-Wei Su, CMOS VLSI *Engineering Silicon-on-Insulator*, Kluwer Accademic Publishers, 2000.

[3] A. Apsel and A. G. Andreou, "A 10 milliwatt 2 gbps cmos optical receiver for optoelectronic interconnect," *Proceedings of the International Symposium on Circuits and Systems*, May 2003.

[4] O. Vendier, S. Bond, M. Lee, S. Jung, M. Brooke, N. M. Jokerst, and R. P. Leavitt, "Stacked si CMOS circuits with a 40-Mb/s through-silicon optical interconnect," *IEEE Photonics Technology Letters*, vol. 10, no. 4, pp. 606–608, Apr. 1998.

[5] J. Jamieson, R. McFee, G. Plass, R. Grube, and R. Richards, *Infrared Physics and Engineering*, McGraw-Hill, 1963.

ACTIVE SUBSTRATES FOR OPTOELECTRONIC INTERCONNECT

Donald Chiarulli, Steven Levitan, Jason Bakos
Departments of Computer Science and Electrical Engineering
University of Pittsburgh

Charlie Kuznia
Peregrine Semiconductor

ABSTRACT

We present the design of an intelligent optoelectronic chip carrier (IOCC). This is an active package that is the basis for short haul, PCB[1] and MCM[2,3] level, optical interconnect. Our goal is a new solution to one of the most difficult problems associated with the packaging of chip-level optical interconnections; the dense and spatially interleaved integration of optical signaling with electrical signals, power and ground. Our approach is based on an "active substrate" using Peregrine UTSi silicon on sapphire technology and the adaptation of laser drilling techniques to create vias through the sapphire. The result is an optoelectronic package that supports full CMOS performance, is mechanically and electrically compatible with current ball grid array (BGA) technology for electronic interconnect, and provides windows for active side optical I/O and substrate-side thermal extraction paths.

OVERVIEW

The I/O pad set for most contemporary VLSI components is a large array of area pads in which electrical I/O pins are distributed among a larger number, often as many as three or four times as many, power and ground pins. This spatial distribution of power and ground pins over the chip area and the close proximity between I/O and supply pins cannot be changed without impacting the I/O performance of the CMOS circuit. Any technique for augmenting these interconnects with an optical interconnection architecture must provide both a device interface and a connection structure that does not disturb this relationship.

Most of the optoelectronic VLSI packages and interconnection architectures proposed to date take one of two approaches. In one approach [fouad] the optical and electrical connections are segregated, with signal pins connected to optoelectronic devices by flip-chipping to the center of the device and power and ground connections drawn from the edges of the chip. In the other approach [], the optoelectronic and electronic device are mounted separately on a common substrate such as PCB or MCM material. The former is simply unrealistic since restricting power and ground connections to the edges of the chip severely limit device performance. In the latter, the electrical PCB/MCM substrate is a performance-limiting bottleneck.

In this paper we present and alternative design, an intelligent optoelectonic chip carrier (IOCC). The IOCC is mechanically compatible with existing ball grid array (BGA) technology and can be directly mounted onto an optoelectrontic PCB or similar MCM substrate. The only mechanical difference in the IOCC is that optical windows are embedded in the solder bump array for arrays of incoming and outgoing optical channels. Electrically the IOCC is quite different from a conventional BGA chip carrier. It is an active device, not a passive interconnect. It is fabricated on a Peregrine UTSi, Silicon on Sapphire (SoS), substrateand thus can include driver and receiver circuits as well as serialization, channel coding, and multiplexing logic in the package. By locating optical drivers directly below the contact pins for the CMOS devices, we can achieve the tightest integration and best performance possible with current technology.

0-7803-8251-X/04/$17.00 ©2004 IEEE

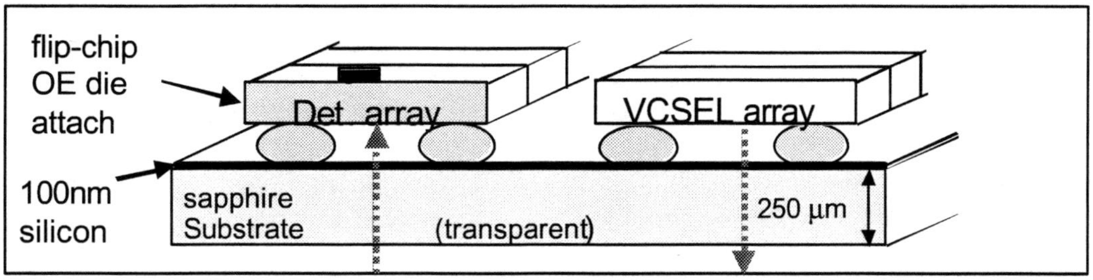

Figure 1: Example of UTSi O/E device structure. Optical signals enter and exit through the transparent substrate to flip-chip bonded OE devices

ACTIVE SUBSTRATE TECHNOLOGY

The active substrate of the IOCC is a Silicon-on-Sapphire (SoS) mixed signal device fabricated using the Peregrine UTSi process. VCSEL and detector arrays for optoelectronic interfaces are flip chip bonded to the top surface as shown in Figure 1 with emitters and detectors pointed down,

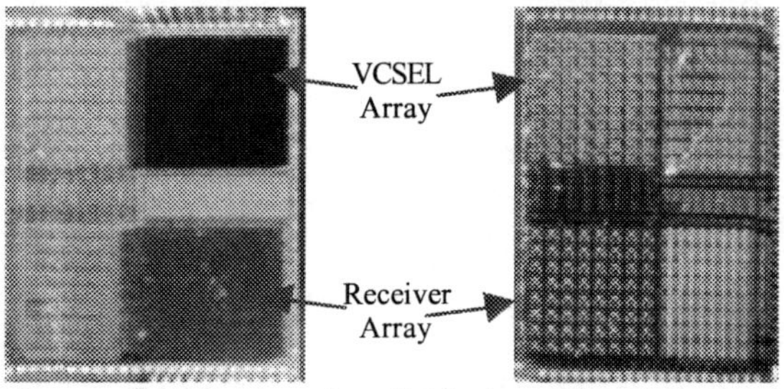

Figure 2: Photographs of 64 channel switch O/E switch designed at the University of Pittsburgh and fabricated by Peregrine The left image shows the top side, the right image shows the bottom side (flipped).

into the substrate. Since the sapphire is transparent, the optical channels traverse this material and enter or exit the chip from the bottom surface. This orientation has the key advantage over other optoelectronic architectures that the top it retains a thermal extraction path through the top of the device. SoS technology can host both digital and analog circuitry in the substrate. Unlike conventional silicon CMOS, the insulating sapphire material eliminates both substrate crosstalk and parasitic capacitance to the substrate. This enables the dense integration of multiple channels of analog drivers with receivers (TIAs and limiting amplifiers) and other sensitive analog circuitry with high speed digital circuitry serialization, encoding/decoding, ECC, and switching.

Figure 2 shows top and bottom view photographs of a 64 channel fiber optic switch using a 2D, 8x8, VCSEL array. The top view shows the transparent UTSi device and bonded OE arrays. On the left side of the chip is the CMOS circuitry for routing

Figure 3: Bottom view looking through transparent substrate. Flip-chip bonded photodectors can be seen through windows opened in the SoS circuitry

signals from the receiver array (top right) to the transmitter array (bottom right). The bottom view shows windows left in the SoS circuitry for optical signals. Figure 3 is a close-up of the chip, where

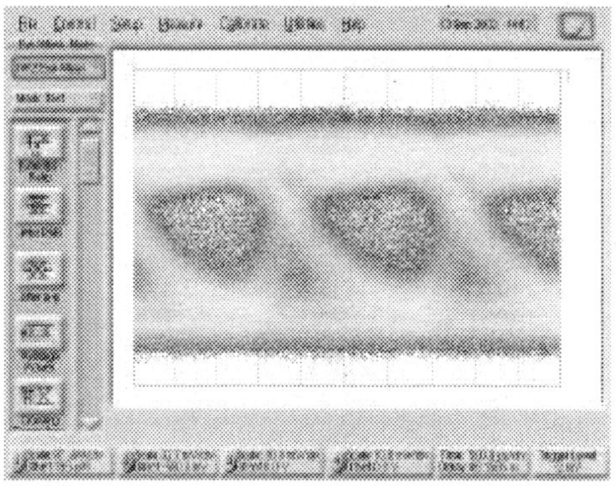

Figure 4: 500Mhz eye diagram for probed operation of the circuit in Figure 9. This shows receiver output for an optical signal that was coupled out of a FIG

the array of PIN detectors is clearly visible through the substrate windows. Receiver operation at 500Mhz is shown in the eye-diagram in Figure 12.

CHIP CARRIER DESIGN

Although the SoS substrate supports digital CMOS circuitry, as in the system above, it is not possible to match the density and performance of bulk CMOS in this process. Thus the IOCC is designed to have one or more CMOS dies flip-bonded to the substrate along with the VCSEL and detector arrays. In order to support the electrical I/O requirements discussed above, as well as providing the stable, low noise power and ground supply connections required for high-speed CMOS, the IOCC provides area pad connections on its bottom surface using laser-drilled vias through the sapphire substrate.

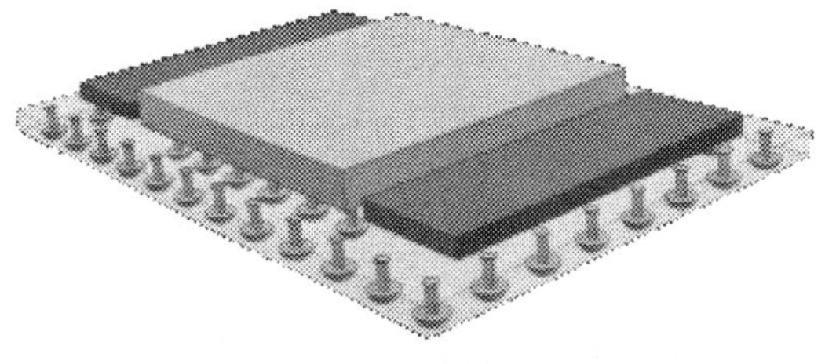

Figure 5: Top view of IOCC package

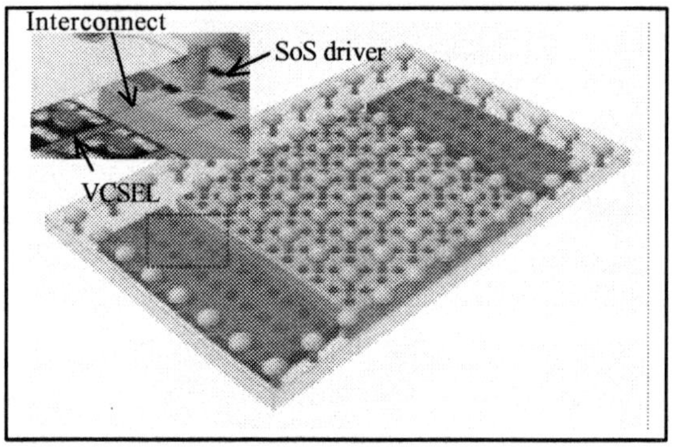

Figure 6: Bottom view of IOCC package

Figures 5 and 6 are conceptual drawings of the design. Figure 5 shows the top side with one CMOS die and two opto-electronic chips flip-chip bonded to the logic side of the SoS die. Figure 6 is a view from the bottom side. The transparent SoS substrate allows us to see in this view: the solder-ball array on the bottom side of the die, the vias that have been laser drilled and plated through the SoS substrate, and the bond pad arrays for the CMOS and optolelectronics interleaved with the via contacts on the logic side. The inset shows a close-up view that illustrates a CMOS bond pad with driver logic and interconnect routing to the VSCEL in the SoS. This arrangement, drawn to scale with a 500um pitch for the solder-ball/via array pitch and and 250um pitch for the flip-chip bond pads, leaves ample room on the logic side of the substrate for driver and receiver logic and interconnect through the metal layers of the SoS device.

There are two key features of the resulting structure. The first is that the substrate itself supports active devices, not passive routing. Thus, on the SoS substrate we can not only provide driver and receiver circuitry for the optoelectronic pathways, we can also provide buffering, bus control, level shifting, and driving among all of the chips on the substrate.

The second key feature of the IOCC is the implementation of electrical vias through the sapphire substrate to connect electrical signal, power and ground connections to solder balls on the bottom surface of the substrate. Sapphire is an extremely hard material and mechanical drilling techniques cannot be used. Our approach uses focused laser drilling to create vias using a 355nm laser housed in an ESI model 5300 laser drilling machine. This machine is capable of 2-4um positioning accuracy with can provide the necessary hole diameters of 60-80um in approximately five seconds of drilling time per hole. This process is the first application of large scale vias for packaging applications and the first use of post processing steps for SoS designed to provide plated metal contacts through these holes.

In our first demonstration device, we are designing an IOCC which will support a CMOS microcontroller die along with VCSEL and receiver arrays. The active substrate will provide level conversion, driving, receiving as well as multiplexing of signals between the microcontroller and the optoelectronics. The design will provide 36 to 48 optical channels that will be allocated to the major I/O bus of the microcontroller. The package itself will be mounted on a transparent substrate to allow it to optically communicate with a second IOCC which will provide memory and system I/O.

SUMMARY

The need for ever increasing I/O bandwidth for VLSI systems drives us to consider optical I/O. However, to date most attempts to integrate optics with state-of-the-art VLSI integration levels incur severe penalties in the electrical I/O performance. Either both optical and electronic I/O must be segregated on the chip, to support flip-chip integration; or else multi-chip carriers must be used, degrading high speed signal integrity. Our solution is an optically transparent, electrically active substrate that enables tight integration of multiple chips, both standard CMOS devices with area pads, as well as optoelectonic devices. Electrical interconnects among devices are mediated by active devices, while optical interconnect between devices is through the transparent substrate. All I/O is preformed on the same common face of the substrate, allowing access to the backside of the devices for thermal management.

References

1) Donald Chiarulli, and Steven Levitan, "Chip-to-Chip Multi-point Optoelectronic Interconnections" in *Optics in Computing*, OSA Technical Digest, (Optical Society of America, Washington DC, 2003) pp 111-113.
2) Jason Bakos, Donald Chiarulli, and Steven Levitan, "Optoelectronic Multi-Chip Module Demonstrator System" in *Optics in Computing*, OSA Technical Digest, (Optical Society of America, Washington DC, 2003) pp 117-119.
3) Leo Selavo, Donald Chiarulli, and Steven Levitan, "Smart Optical Transceiver Architecture with Dynamic Channel Encoding" in *Optics in Computing*, OSA Technical Digest, (Optical Society of America, Washington DC, 2003) pp 47-49.
4) Fouad Kiamilev....
5) Philippe Marchand ....

INTEGRATED RADIOACTIVE THIN FILMS FOR SENSING SYSTEMS

Amit Lal, Hui Li and Hang Guo

*Sonic*MEMS Laboratory, School of Electrical and Computer Engineering
Cornell University, 402 Phillips Hall, Ithaca, NY 14853, USA, email: lal@ece.cornell.edu

ABSTRACT

Hybrid electromechanical systems with integrated power supplies are made possible with radioisotope thin films. A self-powered pressure transducer with pulsed radio-frequency (RF) output and a nanopower betavoltaic microbattery suitable for low-power CMOS are demonstrated in this paper. The pressure transducer collects charges emitted from the radioisotope with a dielectric cantilever to realize a self-reciprocating process, whose period is modulated by the ambient pressure. A RF pulse is generated at the end of each cycle by impulse excitation of the dielectric waveguide. The pressure can be determined by externally detecting the period between RF pulses.

1. INTRODUCTION

We previously presented a self-reciprocating radioisotope-powered cantilever [1]. Self-reciprocation provides the possibility of self-powered MEMS for autonomous applications. Furthermore, such a self-powered system is potentially capable of working for at least the half-life time of the radioisotope, which is 100.2 years for ^{63}Ni.

MEMS vacuum sensors have been developed [2-5], which either use thermal conductivity measurement, or utilize diaphragm deflection. In this paper we implement a micro ionization-based vacuum sensor. In macro ionization-based vacuum gauges, electrodes are heated to boil electrons, and electrons are accelerated to ionize residual gases [6]. We utilize radioactive-decay-generated electrons, which naturally have high energy (average electron energy is 17.3 keV for ^{63}Ni), to ionize the residual gas. In addition, self-powered RF transmission containing ionization history is demonstrated via the self-reciprocating cantilever.

MEMS technology integrates functionality in microscale volumes, but it is difficult to apply similar scaling to power sources. Recently, researches have been done to explore microcombustion based energy generation [7,8], micro-fuel-cells [9], micro chemical batteries (lithium batteries) [10] and micro solar cells. Compared with those technologies, micro betavoltaic batteries offer high energy density; long lifetime and most importantly, radioisotopes can be formed in thin films, allowing integration directly into MEMS with little additional volume requirement.

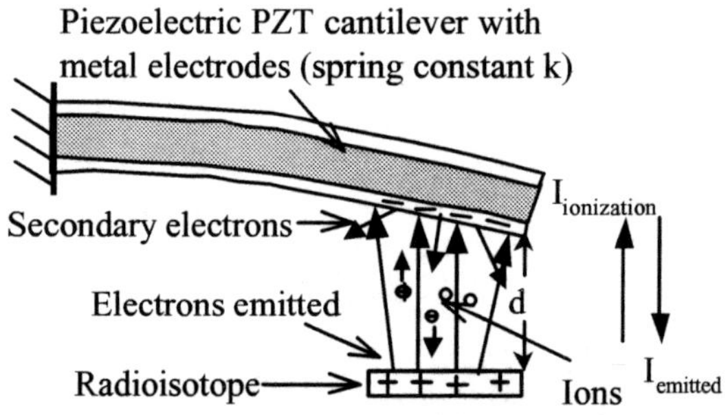

Figure 1. A sketch of the cantilever powered by a radioisotope. The emitted electrons are collected by the cantilever.

2. PRINCIPLE OF OPERATION

2.1 Self-reciprocating cantilever

Electrons from a radioisotope film are collected by a piezoelectric PZT cantilever as illustrated in Figure 1. Positive charges are left on the radioisotope source because of charge conservation. The electrostatic force pulls the cantilever toward the source. The cantilever eventually touches the source. At this moment the charges on the cantilever and the source are neutralized and the cantilever is released to begin a new cycle. An electromechanical model has been developed to describe the movement of the cantilever driven by a radioisotope [1]. The model shows the period of the reciprocation cycle can be determined as:

$$T = (\sqrt{d_0} - \beta)\sqrt{\varepsilon_0 kA} / \alpha I \qquad (1)$$

where d_0 is the initial distance between the cantilever and the radioisotope, ε_0 is the vacuum permittivity, k is the spring constant of the cantilever, A is the area of the cantilever exposed to the radioisotope and I is the current from the radioisotope. α, β are two empirical constants: α gives what portion of the electrons that are actually collected on the cantilever and β is related to the amount

0-7803-8251-X/04/$17.00 ©2004 IEEE

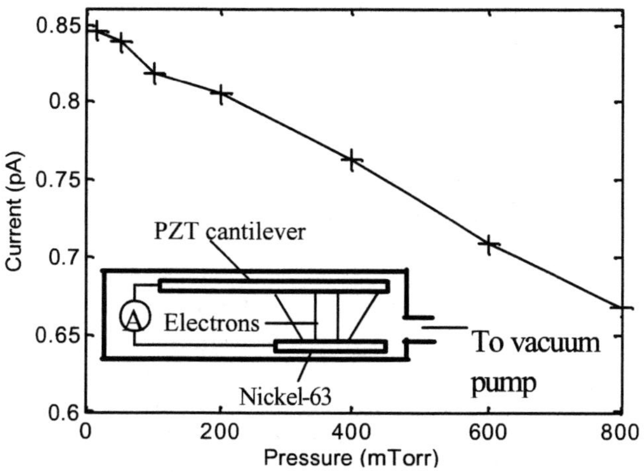

Figure 2. The current provided by the ^{63}Ni source varies with the pressure.

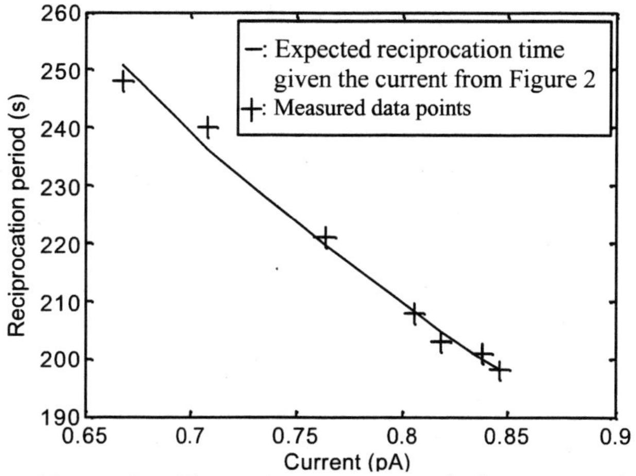

Figure 3. The reciprocation period changes as the current from the radioisotope changes, as a function of pressure, enabling pressure measurement.

of residual charges left in the cantilever when the cantilever is released. It can be seen that the reciprocation cycle time decreases as the current from the radioisotope increases.

2.2 Gas ionization

A typical measured current versus pressure relationship is shown in Figure 2. Here a 1 milliCurie ^{63}Ni source is expected to produce 5.92 pA (3.7×10^7 electrons/second $\times1.6\times10^{-19}$ ampere/electron). The reasons for much lower values measured could be: first, the ions already in the ambient gas neutralize some of the electrons emitted. Secondly, the current measurement imposes a short circuit electrical boundary condition on the two electrodes. Any ions generated eventually recombine. In this process the emitted electrons are absorbed by an amount proportional to particular density or pressure between the electrodes.

As the current varies in response to pressure, so will the cantilever reciprocation time. The measured reciprocation times shown in Figure 3 verify this.

2.3 RF pulse generation

In addition to the self-reciprocation, a RF pulse is also generated at the time of cantilever release. As the electrode of the PZT cantilever which is facing the source collects the electrons, an electrical field is built up across the PZT. At the moment the cantilever touches the source the sudden shorting of the charges on one electrode introduces a sudden change of the electric field and hence the voltage across the cantilever. This results in the impulse excitation of the PZT dielectric waveguide that generates the RF pulse. Figure 4 illustrates this process. Figure 5 shows a waveform measured at an oscilloscope connected across the PZT cantilever electrodes. Because this RF pulse is generated only at the end of a reciprocation cycle, if we use a receiver to detect the RF signal, and count the time interval between signals, we can determine the pressure from the time interval.

3. EXPERIMENTAL RESULTS OF THE PRESSURE TRANSDUCER

In order to implement a self-powered RF transmitter in an IC package format, a PZT cantilever, a ^{63}Ni source and an antenna were integrated inside a DIP package as shown in Figure 6. The cantilever has dimensions of 8 mm × 3 mm ×

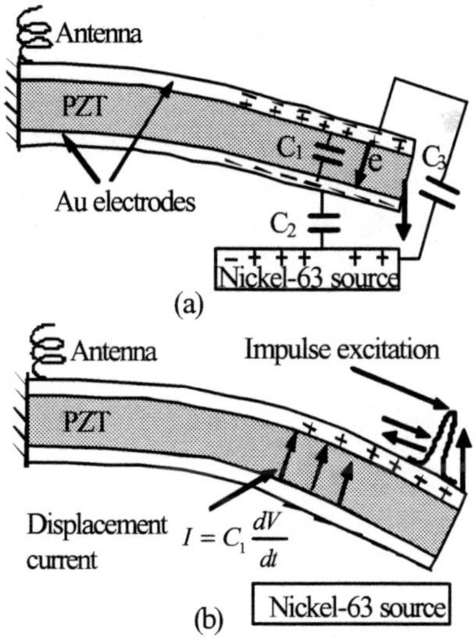

Figure 4. (a) The capacitance of the piezoelectric cantilever builds up an electric field as the charges are built on the two electrodes. (b) The sudden shorting of the charges on one side results in a sudden release of the electric field and hence the voltage across the cantilever. This results in a current $I = C_1\, dV/dt$ that excites the dielectric RF mode of the PZT.

100 μm. A hand-wound coil is soldered to it as an

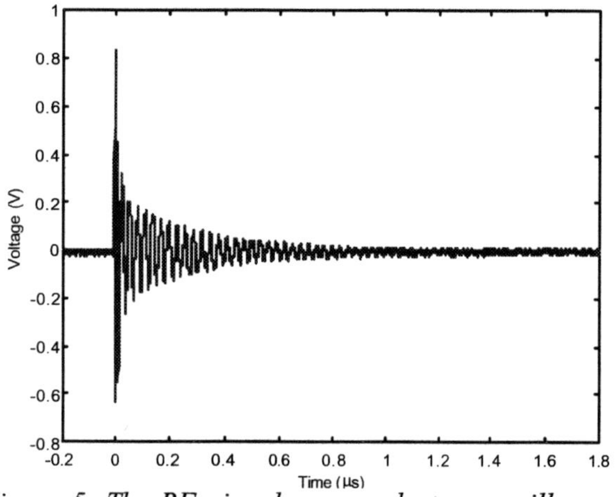

Figure 5. The RF signal measured at an oscilloscope connected across the PZT cantilever electrodes. The signal has a frequency of 33 MHz and a peak-to-peak voltage of 1.47 V.

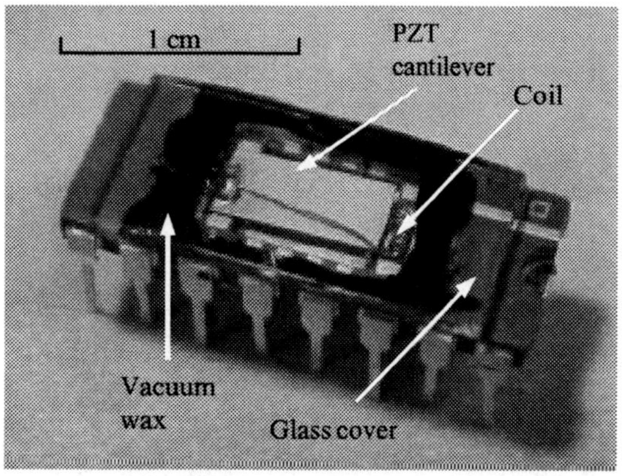

Figure 6. A PZT cantilever is mounted inside a chip carrier. A self made coil is soldered to it. The glass cover is glued to the package with a high molecular weight vacuum wax that can provide good sealing for the vacuum needed. An inlet on the backside provides connection to a vacuum system.

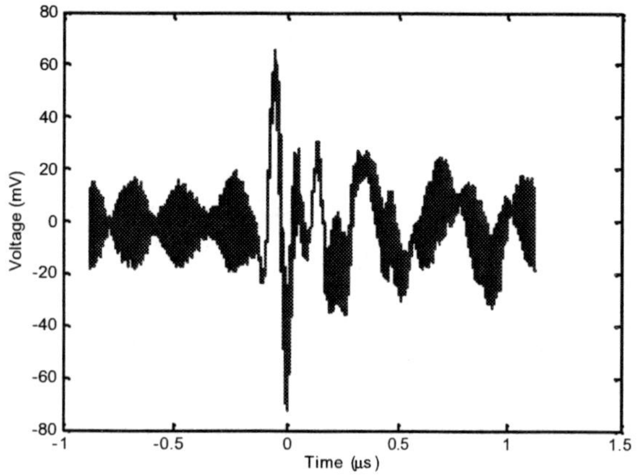

Figure 7. A typical pulse detected by the coil placed 0.1 m away from the DIP package. The frequency is 10 MHz. The peak-to-peak voltage is 138 mV.

antenna. Underneath the cantilever is the ^{63}Ni beta source. An inlet is drilled on the back of the chip carrier to be connected to a vacuum system. The chip carrier is sealed with a glass cover by high-molecular weight vacuum wax. The pressure inside the chip carrier can be maintained between 15 mTorr to 1 Torr. A coil (100 turns, ϕ 6 mm) with 25 μH inductance placed at 0.1 m to the chip carrier connected to an oscilloscope is used to pick up the signal from the PZT cantilever. The pulse could be detected at a distance as large as 0.2 m but with lower probability.

Figure 7 shows a typical signal obtained on the oscilloscope. By timing the interval between received signals the reciprocation cycle time can be determined. The detected wireless signal has a different frequency than the the PZT-oscilloscope frequency due to the different loads across the PZT. Figure 8 shows the cycle time variation versus pressure. A linear relation is observed with a sensitivity of 66 millisecond/mTorr.

The vacuum sensing is done by measuring the reciprocation cycle of the cantilever. According to Equation 1, for faster sensing, one might use a higher activity source or make the initial distance between the cantilever and the radioisotope thin film smaller.

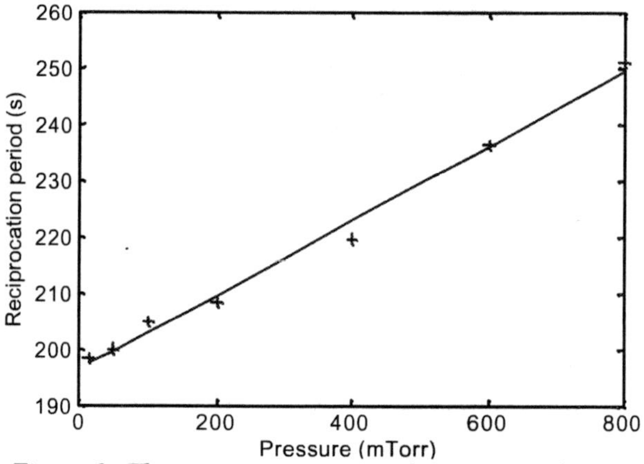

Figure 8. The reciprocation period is measured versus the pressure. A linear relationship is found with a change rate of 66 millisecond/mTorr.

4. BETAVOLTAIC MICROBATTERY

The betavoltaic effect pertains to generation of electron-hole pairs (EHPs) under beta particle incidence on a semiconductor. When EHPs diffuse into the depletion region of the semiconductor *pn*-junction, the electrical field of the depletion region will sweep them into the external circuit. Because the resulting current is from *n*- to *p*-type semiconductor, net power can be extracted.

The maximum kinetic energy of the electrons emitted from ^{63}Ni, 66.7 keV, is less than 200~250 keV, the threshold energy necessary to cause substantial permanent damage to the silicon crystalline matrix [11]. The *I-V* curve for a betavoltaic microbattery with a load R_L can be derived as:

$$I = I_p - I_0\left[\exp\left(\frac{eV}{kT}\right) - 1\right] \quad (2)$$

and open circuit voltage V_{oc}, short circuit current I_s and maximum output power P_{max} as:

$$V_{oc} = \frac{kT}{e}\ln\left(1 + \frac{I_p}{I_0}\right) \quad (3)$$

where I_p is the current generated by radioisotope, I_0 the leakage current of *pn*-junction device. The actual incident current has loss due to the reflected electrons and secondary electrons when the emitted electrons from radioisotopes incident on the surface of the *pn* diode.

From the circuit model, we can see that a very small leakage current of *pn*-junction is required for larger V_{oc} in the microbattery. For photovoltaic cells and high-power betavoltaic battery, the value of I_p is usually on the order of *mA* to 100*mA*. So the leakage current I_0 is not important and could be on the order of *nA* or even μA. However, for the microbattery using very low radioactivities (1-100mCi), I_p is on the order of *pA* or *nA*. If the leakage current I_0 is on the scale of *nA* or μA, there will be minimal power output.

Bulk micromachining is used to fabricate the *pn*-junction device with the inverted pyramid array as reservoirs for the liquid $^{63}NiCl/HCl$ solution, which helps to increase the pn junction area for increased power as used in the past for photovoltaic cells. The area magnification due to the pyramids is 1.85. The activity of liquid $^{63}NiCl/HCl$ solution is 8μCi/μl, and 125μl is used for 1mCi. The short circuit current for this device is 2.86 *nA*, the open circuit volt is 128m*V*, and maximum output power is 0.32 *nW*. Figure 9 shows the tested results of betavoltaic microbatteries in DIP packages, using 0.25mCi and 1mCi of electroplated ^{63}Ni sources, respectively.

On-chip (/integrated) series cascade of smaller diodes is being developed to achieve higher voltages necessary to drive electronic circuits and MEMS devices. For the microbatteries presented in this paper, the source density of current ^{63}Ni thin film is 0.0625mCi/mm^2 (=1mCi/16mm^2). It can be expected that the output power of 100~200nW will

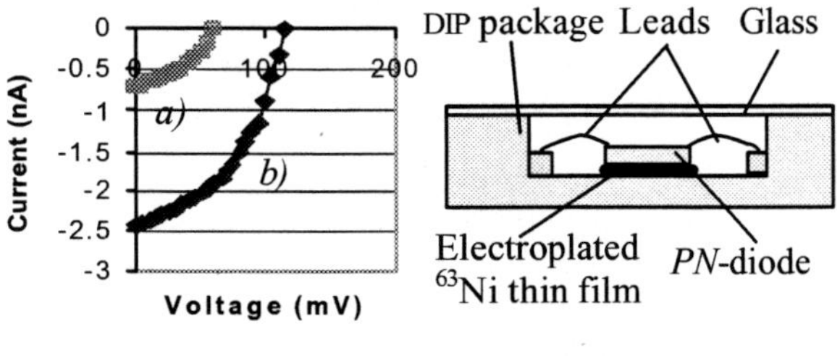

Figure 9. Tested results of packaged betavoltaic microbatteries Left: tested results with a) 0.25mCi of electroplated ed ^{63}Ni b) 1mCi of ^{63}Ni Right: betavoltaic microbatttery in DIP package

be achieved when the source density increases to 1mCi/mm^2, with the output voltage up to 20~30V.

5. CONCLUSION

A ^{63}Ni radioisotope powered cantilever for vacuuming sensing with RF transmission and experimental betavoltaic batteries, which utilized very low levels of activities (1 milliCurie or less) using ^{63}Ni source to produce sub-nanowatt power are demonstrated. An antenna outside can receive a RF pulse to determine the reciprocation period and the pressure. This vacuum sensing system can be used for long-term pressure sensing applications in building walls, in implantable devices, etc. There also exists the chance that the sensing system can be used for gas sensing purpose. As the sensing mechanism is ionization based, how it works with different gas species will have to be investigated first. Nanopower betavoltaic batteries can be used for nanopower electronics to enable circuits for long-term operation integrated with physical sensing.

6. REFERENCES

[1] H. Li, A. Lal, J. Blanchard and D. Henderson, "Self-reciprocating Radioisotope-powered Cantilever," *J. Appl. Phys.*, **92,** pp 1122-27, 2002.

[2] T. M. Berlicki, "Thermal Vacuum Sensor with Compensation of Heat Transfer," *Sensors and Actuator A*, **93,** pp 27-32, 2001.

[3] M. Esashi, S.Sugiyama, etc. al, "Vacuum-sealed Silicon Micormachined Pressure Sensors," *Proceedings of the IEEE*, **86**, pp 627-39, 1998.

[4] H. Miyashita and M. Esashi, "Wide Dynamic Range Silicon Diaphragm Vacuum Sensor by Electrostatic Servo System," *J. Vac. Sci. Technol. B*, **18**, pp 2692-7, 2000.

[5] E. H. Klaassen and G. T. A. Kovacs, "Integrated thermal conductivity vacuum sensor," *Technical Digest, Solid State Sensor and Actuator Workshop*, pp 249-52, June, 1996.

[6] A. Roth, *Vacuum Technology*, Sec. Edition, North-Holland, Section 6.7, pp 310-21, 1982.

[7] S.B.Schaevitz, A.J.Franz, K.F.Jensen, and M.A.Schmit, "A Combustion-Beased MEMS Thermoelectric Power Generator," *Dig. Tech., Transducers '01*, pp 30-3, 2001.

[8] Chunbo Zhang, K. Najafi, L.P.Bernal, and P.D.Washabaugh, "An Integrated Combustor-Thermoelectric Micro Power Generator," *Dig. Tech., Transducers '01*, pp 34-7, 2001.

[9] WooYoung Sun, Geun Kim, and Sang Sik Yang "Fabrication of Micro Power Source (MPS) Using a Micro Direct methanol fuel cell (μDMFC) for the medical application," *Tech. Dig. 14th MEMS*, pp 341-4, 2001.

[10] J.B.Lee, Z.Chen, M.G.Allen, A.Rohatgi, and R. Arya, "A miniaturized high-voltage solar cell array as an electrostatic MEMS power supply," *IEEE J.MEMS*, 4(3), pp 102-8, 1995.

[11] P. Rappaport and J. J. Loferski, "Thresholds for Electrons Bombardment Induced Lattice Displacements in Si and Ge", Phys. Rev., **100**, pp 1261-6, 1995.

WIDENING THE DYNAMIC RANGE OF THE READOUT INTEGRATION CIRCUIT FOR UNCOOLED MICROBOLOMETER INFRARED SENSORS

Alexander Belenky, Alexander Fish, Shy Hamami, Vadim Milrud and Orly Yadid-Pecht

The VLSI Systems Center, Ben-Gurion University, Beer-Sheva, Israel, e-mail: oyp@ee.bgu.ac.il

ABSTRACT

A simple and robust technique for widening the dynamic range (DR) of the readout circuits for uncooled microbolometer arrays is presented. The proposed method allows reducing the number of bits required for microbolometers non-uniformities compensation, significantly reducing the complexity of the pre-integration compensation circuitry. Applying the proposed technique results in reduced memory size and smaller area occupied by the circuits. In this proposed solution regular readout circuits can be used, and additional hardware is placed in the periphery, without affecting spatial and temporal resolution. System architecture and operation are discussed and an analytic analysis is presented.

1. INTRODUCTION

Over the past several years, uncooled infrared (IR) detectors have been rapidly developed into large size focal plane arrays (FPA) [1]. IR detectors have been an important technology for both civilian and military applications, including IR search and track, medical examination, astronomy, forward-looking infrared (FLIR) systems, missile guidance, and other strategic armaments.

The most common thermal uncooled detector is a resistance microbolometer device. The resistance microbolometer for detection of thermal radiation is a micromachined device with a resistance which changes in correspondence with the amount of absorbed IR radiation [2]. With each microbolometer functioning as a pixel, a two dimensional (2-D) image or picture representation of the incident IR radiation can be generated by translating the changes in resistance of each microbolometer into a time-multiplexed electrical signal that can be displayed on a monitor or stored in a computer. The circuitry used to perform this translation is usually known as the readout integrated circuit (ROIC), and is fabricated as an integrated circuit in the silicon substrate [3]. Usually the changes in bolometer resistance value caused by IR radiation are translated to a signal current and integrated using the capacitive transimpedance amplifier (CTIA) for the integration time (Tint period) [4,5]. The key ROIC design challenges are to integrate the required detector interface circuitry into the desired pixel pitch area, provide low noise and dynamic range (DR) with low power dissipation.

Designing of an advanced ROIC for microbolometer FPA should take in account non-uniformity in bolometer parameters, like the temperature coefficient of resistance (TCR), resistance, IR radiation observation coefficient, heat capacity, and thermal conductivity. This spatial non-uniformity is a natural result of a microbolometer fabrication process. The bolometer response magnitude due to these non-uniformities can be much larger than its response to the incident IR radiation, resulting in requirement of very wide dynamic range (WDR) from the ROIC. Also, large scene temperature excursions require large DR with local area contrast enhancement [6]. In order to relax ROIC WDR constrains, different compensation techniques have been proposed in the literature [3,7]. For example, in [3] each bolometer is biased to a unique programmable potential using an on-chip 14 bit DACs to compensate the detector spatial non-uniformity magnitude response. These programmable potentials are determined during the calibration phase and are then stored in a digital memory.

This paper presents a novel technique that widens the ROIC DR, allowing compensation circuitry simplification. This simplification manifests itself in reduced DAC accuracy, resulting in smaller memory size and reduced area occupied by the circuits. In this suggested solution regular detector arrays and readout circuits can be used and additional hardware is placed in the periphery, without affecting spatial and temporal resolution. Simplification of the compensation circuitry, by applying the presented method, results in a slight increase in system noise equivalent temperature difference (NETD), making the proposed technique more suitable for civilian applications where low-cost and simplicity are the main demands.

Section 2 presents the overview of the existing architectures of the microbolometer FPA and describes possible compensation methods. Operation description and analytic analysis of the proposed method for WDR are presented in Section 3. Section 4 concludes the paper.

2. BACKGROUND

Figure *1* shows the principle architecture of a commonly used microbolometer array (FPA) including the basic ROIC. The FPA includes (a) an NxM array of sensing microbolometers with row selection switches, (b) a column shared thermally shorted to the substrate (blind) bolometers, (c) a CTIA stage, (d) sample-and-hold (S/H) circuitry and (e) multiplexer for signal serial readout. The microbolometer current signals flow down by the columns. It then integrated (by the integrators at the base of each column) during the readout for a Tint period. In the bottom of each column a reference blind bolometer is used to compensate the variation in temperature of the corresponding sensing bolometers.

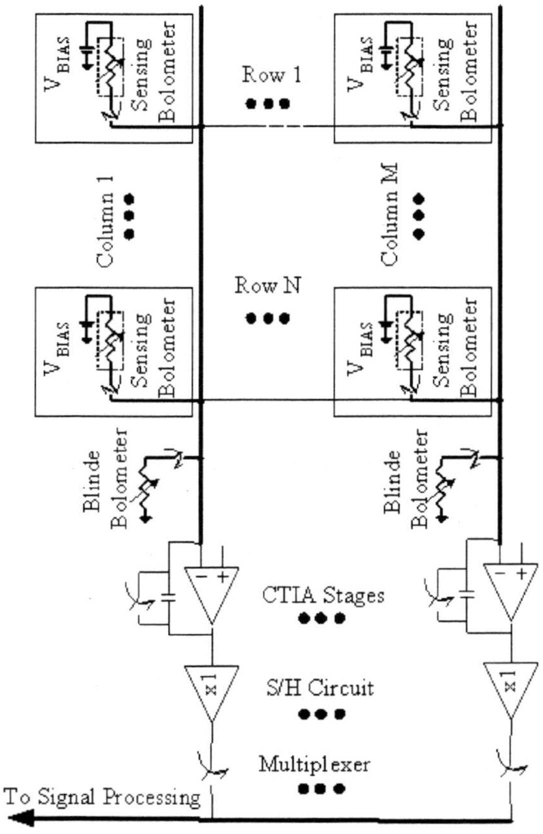

Figure 1: Principle architecture of a commonly used microbolometer array.

As mentioned above, the signal current that integrated as a result of the sensing bolometer response to the incident IR radiation can be much smaller than the one, caused by the bolometer spatial non-uniformity. These non-uniformities in the microbolometer response characteristics lead to a very wide DR ROIC constrain, which is almost impossible for implementation, especially in advanced technologies with low power supply. In order to achieve an output signal dominated by the level of incident IR radiation, compensation for detector non-uniformity is required. An example of an existing non-uniformity compensation technique is shown in Figure 2, as presented by Indigo and Sofradid companies.

According to this method, the compensation is performed using two MOS transistors in a common gate configuration. These transistors bias the sensing and the load blind bolometers to unique programmable potentials using on-chip 14 bit Digital-to-Analog-Converters (DACs). Digital coefficients for the bias and offset DACs are stored in memory for each bolometer during the calibration phase. The drain current of the n-channel MOSFET is used to compensate the sensing bolometer current such that only temperature difference induced currents are integrated. Even so, a large portion of the output range of the amplifier is dedicated to cover the residual offset range. According to Figure 2, the ROIC operation is given by:

$$I_{INT} = I_{BS} - I_{BB} = I_{OFF} + I_{SIG} \quad (1)$$

$$Vout = Vout_{OFF} + Vout_{SIG} = I_{INT} \cdot T_{INT} / C_{INT} \quad (2)$$

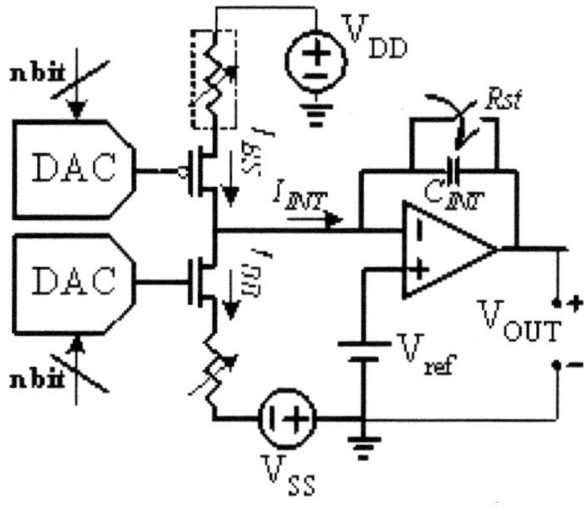

Figure 2: Indigo and Sofradir compensation circuit example.

$$Vout = (I_{OFF} + I_{SIG}) \cdot T_{INT} / C_{INT} \quad (3)$$

where I_{INT} is the subtracted current, I_{OFF} is the offset portion, I_{SIG} is the signal current, I_{BS} is the bias current flowing through the sensing bolometer, I_{BB} is the bias current flowing through the blind bolometer, C_{INT} is the integration capacitance, T_{INT} is the integration time and V_{OUT} is the CTIA output voltage.

In addition, a common "two points non-uniformity calibration" (NUC) procedure is carried out in order to achieve a uniform background picture. This NUC is performed next to the CTIA, yielding offset and gain correction.

3. WDR METHOD DESCRIPTION

3.1. Operation Overview

In order to simplify the compensation technique, the number of compensation bits (i.e. DAC precision) should be reduced. This simplification results in smaller memory size needed for the compensation coefficients storage, reduced DAC area and simplifying system complexity. However, according to (3), the reduced DAC accuracy increases the integrated offset current I_{OFF} and thus results in larger CTIA output voltage V_{OUT}. Large I_{OFF} value causes to saturated ROIC system, limited by the CTIA DR. According to (3), for a given I_{OFF} value, the V_{OUT} can be reduced or by increasing the integration capacitor C_{INT} or by reducing the integration time T_{INT}. Since the first solution is a non-practical for implementation in the modern VLSI designs due to very large area occupied by the capacitor, the proposed technique is based on an adaptive control of integration time in each column of FPA.

Fig. 3 shows the proposed method for WDR enhancement applied on a common used ROIC. The proposed system consists of (a) a sensing microbolometer, (b) Sofradir and Indigo companies ROIC, (c) CTIA circuit that is implemented as a switched capacitor to allow integration, (d) an analog comparator that compares the output voltage with an appropriate threshold as will be defined below, (e) a control unit that controls the reading operation, the timing of the comparisons and provides corresponding resets for each integration capacitor, (f) a digital memory and (g) a clock generator.

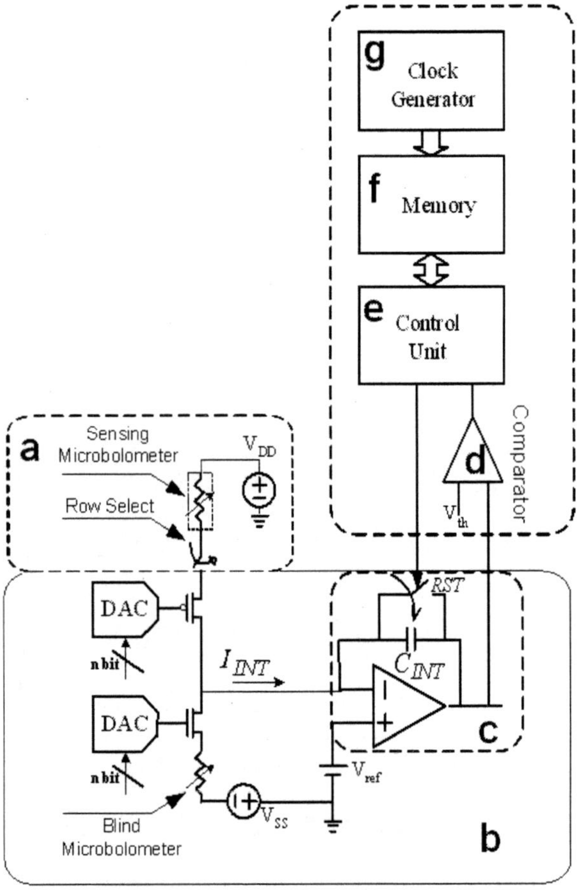

Figure 3: The proposed technique for WDR enhancement applied on a common used ROIC.

In this proposed algorithm, the required expansion of the DR is determined by a series of W-bits [8]. The total integration time is subdivided into several integration times, which are progressively shorter, according to the down-going series:

$$T_{INT}/X^1,\ T_{INT}/X^2, \ldots, T_{INT}/X^W \qquad (4)$$

where X>1 and T_{INT} represents the full integration time. The outputs of a selected row are compared with an appropriate threshold, at certain time points:

$$T_{INT} - T_{INT}/X^1,\ T_{INT} - T_{INT}/X^2, \ldots, T_{INT} - T_{INT}/X^W \qquad (5)$$

The appropriate threshold voltage is determined by:

$$V_{th} = V_{\max\_swing} \cdot \left(1 - \frac{1}{X}\right) \qquad (6)$$

where V_{th} is the threshold value, V_{max_swing} is the CTIA maximal voltage swing.

If any of the checks determines that the voltage on the integration capacitor will pass the V_{max_swing} voltage at the end of the current integration time, then the integration capacitor is reset and is allowed to integrate once more, but for a shorter period of time. A reset pulse is applied to that capacitor only if it was reset in the preceding comparison time point. The number of resets during the full integration time T_{INT} is counted and is saved in SRAM memory storage and output in due time. Once the row is read at time Tint, we retrieve this additional information from the memory. An analog CTIA output voltage value is readout at the end of the integration (at time point T_{INT}). This enables the voltage representation of the detector integrated current value to be described as a floating-point representation. In this floating-point representation, the exponent will describe the scaling factor for the actual integration time, while the mantissa will be the regular output at time T_{INT}. This way, the actual voltage representation of the detector integrated current value would be:

$$Value = Man \cdot \left(\frac{T_{INT}}{T_{INT}/X^{EXP}} \right) = Man \cdot X^{EXP} \qquad (7)$$

where Value is the actual voltage representation of the detector integrated current value, Man (Mantissa) is the analog or digitized CTIA voltage value that has been read out at the time point T_{INT}, X is a chosen constant (X>1) and EXP is the number of resets that have been applied during the integration process.

Fig. 4 describes this method via a time domain diagram. W_1, $W_2 \ldots W_W$, represent the exponent bits, i.e. W_1 represents the point of decision, $T_{INT} - T_{INT}/X_1$, whether reset or not for the first time is given, W_2 for the next point and so forth.

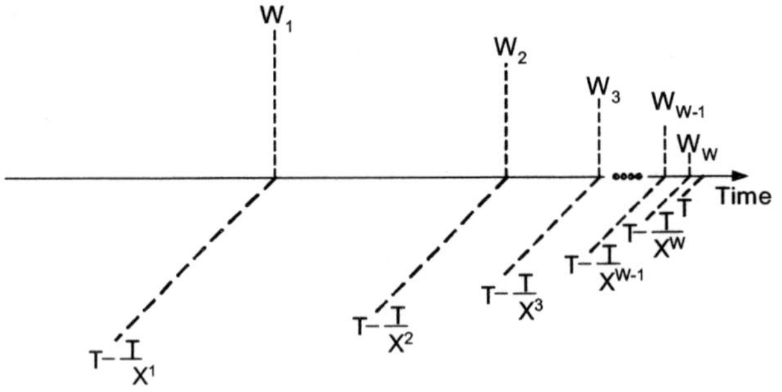

Figure 4: Time domain diagram.

It should be noted that the comparator role is to avoid the ROIC system saturation (i.e. exceeding Vmax_swing) in order to enable a proper restoring of the voltage representation of the detector integrated current value. The proposed method provides system immunity to the comparator offset voltage allowing simple, low power and low area comparator implementation. This immunity can be ensured using the following constrain on the appropriate threshold voltage from (6):

$$V_{th} = \left[V_{\max\_swing} - 2 \cdot V_{offset} \right] \cdot \left(1 - \frac{1}{X}\right) \tag{8}$$

3.2. Analytic Analysis

In order to examine the influence of the proposed method on the existing ROIC performances, the analytic analysis has been performed. A brief summary of this analysis is presented herein.

Some basic assumptions have been made, while examining the proposed method: (a) the sensing microbolometer resistance, ΔR(T), can be expressed as:

$$\Delta R(T) = R_0(1 - e^{-\alpha(T-T_0)}) \approx R_0 \frac{\alpha}{\gamma}(T_0 - T) \tag{9}$$

where R_0 is the microbolometer resistance at standard temperature T_0, α is the TCR at T_0, γ is defined as $\Delta T_a/\Delta T_b$ which is the ambience to microbolometer temperature change ratio, and T is the substrate temperature. (b) MOS transistors have been designed with $1/g_m \ll R_0$, where g_m is transistor transconduction. (c) Johnson noise was taken in account as the dominant noise component. (d) *n* is a number of resets that were performed during T_{INT} period and *k* is the number of reduced compensation bits.

The technique influences the CTIA output signal voltage to ΔT as given by:

$$\frac{v_{out}}{\Delta T} \square I_{BS} \cdot \frac{\alpha}{\gamma} \cdot \frac{T_{INT}}{C_{INT}} \cdot \frac{1}{2^n} \left[\frac{V}{{}^0 K} \right] \tag{10}$$

Johnson noise output contribution of the two branches is given by:

$$V_{no}{}_{total} \square \sqrt{\frac{4kT_b}{R_0} \cdot T_{INT}} \Big/ C_{INT} \tag{11}$$

The new NETD can be calculated using (10) and (11), and it is expressed by:

$$NETD = \frac{V_{no\,total}}{\left(v_{out} / \Delta T \right)} = \frac{\gamma}{\alpha} \cdot \frac{1}{I_{BS}} \cdot \sqrt{\frac{4kT_b}{R_0}} \cdot \sqrt{\frac{2^n}{T_{INT}}} \quad [{}^o K] \tag{12}$$

The relation between the reduced number of compensation bits *k* and number of executed resets *n* is determined by:

$$I_{OFF} 2^k \le I_{INT\_MAX} 2^n - I_{SIG} \tag{13}$$

where I_{OFF} and I_{SIG} are as represented in (1) before the reduction in number of compensation bits, and I_{INT_MAX} is the maximum integrated current that allows operation in the permitted DR of the CITA. Rewriting (13) leads to a straight forward connection between *k* and *n*, as follows:

$$k \le \log_2 \left(\frac{I_{INT\_MAX} 2^n - I_{SIG}}{I_{OFF}} \right) \tag{14}$$

According to (14), *k* is always larger than *n* for given I_{INT_MAX}, I_{SIG} and I_{OFF}, thus requiring smaller system memory storage. Note that the proposed system performance improves with increasing I_{SIG}/I_{OFF} ratio. On the other hand, applying the proposed technique results in an increased NETD (only as square root of 2^n), as can been seen from (12).

4. CONCLUSIONS

A simple technique for widening the dynamic range (DR) of the readout circuits for uncooled microbolometers arrays was presented. The fact, that in this proposed solution regular readout circuits can be used and additional hardware is placed in the periphery, without affecting spatial and temporal resolution, makes it very useful for different ROIC architectures. The applying of the method on the existing advanced ROIC system was described. The summary of analytic analysis, examining the proposed technique influence on the DR, NETD and compensation system was shown. It was found that this method allows reducing the number of bits required for microbolometers non-uniformities compensation, while increasing the NETD system value. The novel system properties enable relaxing the complexity of the pre-integration compensation circuitry and thus it is found most suitable for civilian applications, where low-cost and simplicity are the main demands. The future work includes fabrication of system employing the proposed method.

5. REFERENCES

[1] C. Chen, X. Yi, J. Zhangn, and B. Xiong, "Micromachined uncooled IR bolometer linear array using VO_2 thin films," *International Journal of Infrared and Millimeter Waves*, vol. 22, no. 1, 2001.

[2] C. Jansson, U. Ringh, and K. Liddiard, "Theoretical analysis of pulse bias heating of resistance bolometer infrared detectors and effectiveness of bias compensation," *Proc. SPIE*, vol. 2552, pp. 644-652, 1995.

[3] W. J. Parrish et al, "Methods and circuitry for correcting temperature-induced errors in microbolometer focal plane array", *U.S. Patent 6 028 309*, Feb. 10, 1998.

[4] C. C. Hsieh, C. Y. Wu, F. W. Jih, and T. P. Sun, " Focal-Plane-Arrays and CMOS Readout Techniques of Infrared Imaging Systems," *IEEE Transactions on Circuits and Systems for Video Technology*, vol. 7, no. 4, August 1997.

[5] L. Kozlowski, S. Cabelli, R. Kezer, and W. Kleinhans, "10x132 CMOS/CCD readout with 25μ pitch and on-chip signal processing including CDS and TDI," *in Infrared Readout Electronics, Proc. SPIE*, vol. 1684, pp. 222–230, 1992.

[6] DARPA web site :http://www.darpa.mil/darpatech99/ Presentations/scripts/mto/MTOImagingSensorsScript.txt

[7] M. A. Wand et al, "Ambient temperature micro-bolometer control, calibration, and operation", *U.S. Patent 6 267 501*, July 31, 2001.

[8] O. Yadid-Pecht, A. Belenky " In-Pixel Autoexposure CMOS APS. "IEEE Journal of Solid-State Circuits, Vol. 38, No. 8, pp. 1-4, August 2003.

A 16 × 16 PIXEL SILICON ON SAPPHIRE CMOS PHOTOSENSOR ARRAY WITH A DIGITAL INTERFACE FOR ADAPTIVE WAVEFRONT CORRECTION

Eugenio Culurciello and Andreas G. Andreou

Electrical & Computer Engineering, Johns Hopkins University

ABSTRACT

We report on a 16 × 16 pixel CMOS photosensor array fabricated in silicon on sapphire CMOS technology. The transparency of the substrate allows imaging from both the back and front side and opens new applications for CMOS active pixel sensor arrays. A digital asynchronous interface is employed to minimize the design complexity and the the power consumption. The analog pixel value is encoded as a pulse density stream of address events. The multiple threshold MOS transistors available in this technology enable circuit optimizations for low power and low voltage operation. We characterize the pixel sensitivity and show experimental data from both back and front side imaging.

1. INTRODUCTION

Silicon on Sapphire CMOS (SOS-CMOS) available from Peregrine semiconductors [1] is a promising technology for hybrid optoelectronic microsystems [2]. The transparency of the substrate to wavelengths from infra-red to ultra-violet, opens opportunities for applications in high speed free space interconnects and 3D integration. An example of such an application is adaptive optical wavefront correction [3]. In addition to the transparent substrate, this particular SOI process offers MOS transistors with three different thresholds thus enabling the optimization of both analog and digital circuits for low power.

In this paper we report on a 16 by 16 pixels SOS-CMOS photosensor array fabricated in the Peregrine SOS-CMOS $0.5\mu m$ process. To the best of our knowledge the first standard CMOS photosensor array capable of transducing light simultaneously from both sides of the die.

2. PHOTODIODE DESIGN AND CHARACTERIZATION

The fabricated SOS-CMOS imager employs a native PIN photodiode as photosensitive element. Spectral and temporal characterization of such structures have been reported previously in [4, 5] and more recently in [6].

Using the ultra thin silicon photodiodes has advantages and disadvantages. Photon absorbtion in the ultra thin (100nm) silicon layer is small thus severely degrading the quantum efficiency to blue and UV wavelengths. This is in contrast to bulk CMOS photo-detectors that are sensitive to red and infra-red and have weak response to blue. Blue and ultraviolet sensitivity requires ultra-shallow junctions that are hard to achieve in standard bulk CMOS processes. Using a PIN photodiodes decreases junction capacitance and thus yields devices with bandwidth in excess of 5GHz [5]. The photodiode used in our pixel has a horizontal structure $100nm$ thick and $16.4\mu m$ long, with a $1.2\mu m$ intrinsic silicon layer between anode and cathode. The photocurrent is integrated on a $250fF$ capacitor. In our application, of an adaptive wavefront correction system, the poor absorbtion in the PIN photodiodes is an advantage as we are only employing the photosensor array in the feedback loop and the PIN photodiode is present just to sample the light from the laser beam.

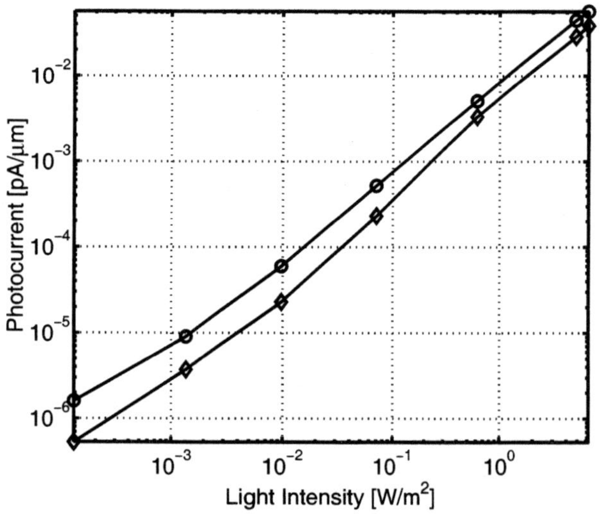

Fig. 1. Front (diamond) and back (circle) side photocurrent at $555nm$

When the PIN photodiode is reverse-biased with the nominal supply voltage of $3.3V$, we measured high levels of dark current ($17pA$) (approximately $1pA/\mu m$ of diode length). This imposes a lower bound on the lower limit on the sensor's dynamic range. Assuming a conservative threshold of $0.5V$ for triggering an event, in the dark a pixel takes $250fF \cdot 0.5V/17pA = 7ms$ to integrate enough light to produce an event. This is clearly a small time constant compared to other image sensor based on the same architecture [7], and ultimately limits the quality of the images. Integrating for longer period of time reduces the noise in integrating pixels. High level of dark current result in decreased image contrast. Events from the sensor arise with time constant of $27\mu s$ ($7ms/256$, where 256 is the number of cells). This is the minimum event rate at the output of the image sensor. Ambient light generates a photocurrent of $30 - 40pA$, therefore the image sensors SNR is only 2-3. A low-power ($5mW$) laser pointer light generates $100pA$ and an integration time of $1.25ms$. A high intensity illumination lamp generates a maximum of $600pA$ with an integration time of $0.2ms$. The problem with excess leakage currents in the the PIN photodiodes can be solved by re-designing the photosensitive structure.

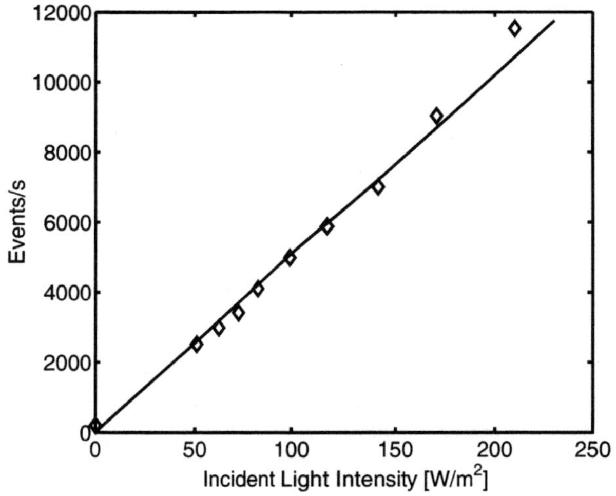

Fig. 2. Event rate versus light intensity

Figure 1 shows a plot of the photocurrent per unit length of the SOS PIN photodiode from both front and back side of the die. Light intensity was measured with a calibrated photometer and a variable high intensity source. Note that light integrated from the back side generates higher photocurrents than the front side. This occurs because the front side of the die is covered by metal and SiO_2 layers that filter some of the incident light. Figure 2 reports the event frequency versus incident light intensity at $555nm$ for a single pixel in the array. Using backside illumination a 100% fill factor can be achieved. For the data in this figure, the light was focused on a single pixel on the array using a lens, and the light intensity from the laser was varied using neutral density filters. The event frequency f_{ev} is linear with light intensity I_{in}. The relationship is given in formula 1 where the parameter L_s equals to 51.

$$f_{ev}[Hz] = L_s \cdot I_{in}[W/m^2]) \qquad (1)$$

The output event frequency spans approximately 2 orders of magnitude (200 to 11500Hz) and thus the array is capable of encoding data with 4 to 6 bits of precision, which is sufficient for the intended application.

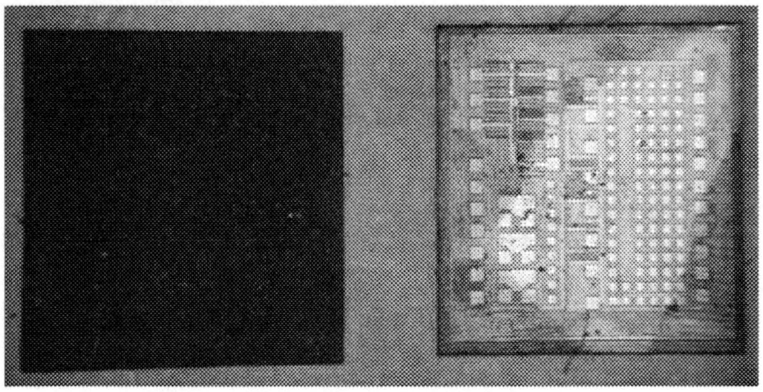

Fig. 3. An SOS-CMOS test-die before and after mechanical polishing.

3. SYSTEM ARCHITECTURE

Conventional optical image sensor arrays convert light into a time varying voltages that are buffered at the pixel level using a voltage follower and subsequently synchronously serially scanned to the output. The read-out circuit thus allocates equal portion of the channel bandwidth to all pixels independently of activity and continuously dissipates power as the scanner is always active. The system presented in this paper employs pulse frequency modulation at the pixel level converting the analog data into asynchronous pulses at the pixels' output. Such digital representation of light intensity yields devices capable of large dynamic range imaging and low-power dissipation [8, 9, 7]. Pulse modulated output has also been reported for biomimetic image gradient sensors [10].

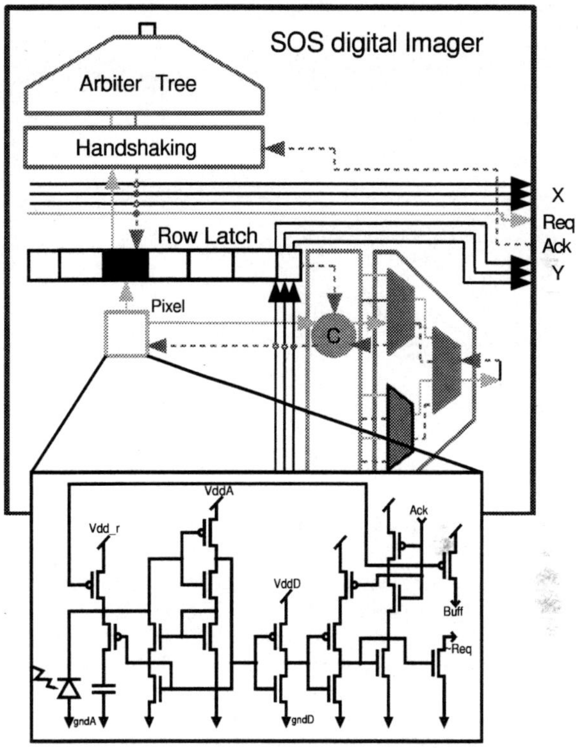

Fig. 4. SOS-CMOS Photosensor Array Architecture

We now describe the pulse frequency modulation scheme implemented at the pixel. Individual pixels integrate light on a local capacitor and when a threshold is reached, they request access to the output bus. Their address (X, Y locations) appears at the output after arbitration (refer to Arbiter Tree in Figure 4) in the form of an event. The value of the light intensity is inversely proportional to the inter-event interval. The image sensor read-out is initiated by individual pixels, therefore the available output bandwidth is allocated according to pixels demand. A detail analysis and comparison of synchronous and asynchronous readout schemes is beyond the scope of this paper and can be found in [11]. The pixel circuit implementing the pulse density modulation is shown as insert in Figure 4. The layout of the pixel is shown in Figure 5.

Arbitration is executed in two steps: rows are arbitrated first then arbitration occurs between active pixel in the selected row [7]. Once a row has been selected and row events are stored in the

Row Latch of Figure 4, other rows of pixels can proceed with light integration and competition for the output bus. This organization increases the sensor's throughput. An Address bus as well as a Request and an Acknowledge signal interface the system to the receiver which reconstructs the analog signal. Reconstruction of the images can be performed by counting the number of events in a given window of time (*histogram imaging*), or by computing the inter-event time between successive events from the same pixel (*inter-event imaging*) [7]. An analysis of optimal receiver design for such class of imagers is discussed in the paper by Apsel and Andreou [12].

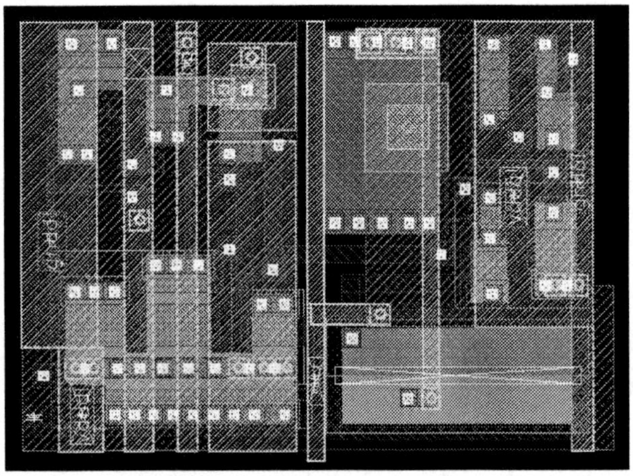

Fig. 5. Pixel Layout

4. EXPERIMENTAL RESULTS

The die area for the sensor array is $0.66cm^2$ without output pads, and $1.23cm^2$ with pads. Pixel size is 29.6 by $42\mu m$. A micrograph of the die is shown in Figure 8. The SOS wafer was polished only on the front side after fabrication. To obtain a clear die on both sides, we have polished the back side of the die using a mechanical lapping machine. Lapping was performed up to a surface roughness of $1\mu m$. The mechanical polishing resulted in an optically clear die on both side. Alternatively index matching fluid can be employed to fill in the asperities of the backside [2]. The effectiveness of the mechanical polishing is evident in figure 3.

Sample image sequences given from the sensor are given in Figure 6 and 7. Figure 6 is obtained by illuminating on the front side, Figure 7 on the back side of the die (BSI). In both Figure a moving spot is sampled 4 times and the output event rate was $0.48MSamples/s$. These images were reconstructed by creating a normalized histogram after collecting 10000 events. Noisiness of the images is due to errors in the collections of addresses and to the limited number of pixels.

Power consumption of the analog array is: $0.60\mu W$. Since there are 256 pixels in the array, this amounts to a consumption of about $2.32nW$ per pixel. Low power operation is a result of employing the pulse frequency modulation pixel [7] as well as optimized circuit design employing MOS transistors with three different thresholds. The energy consumption for each event is on the order of $4.83fJ$. Digital power consumption was measured to be $1.1mW$. This includes the use of 3 pseudo-CMOS logic signals in

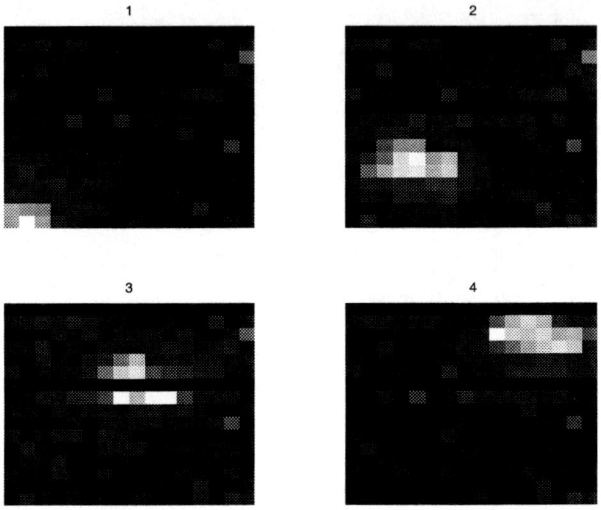

Fig. 6. Output from a light spot moving up-right impinging on the front side of the sensor array

the array, that account for most of the power consumption. In the next version of the chip, the wired-or will be replaced with a tree based fully CMOS design, thus reducing dramatically the power dissipation. The circuit design both for the pixel and periphery employs the available zero threshold transistors [2] to reduce the complexity, minimizing the number of transistors and wires necessary for bias circuits.

Both analog and digital power consumption measurements were conducted using a supply voltage of $3.3V$ and with an output event rate of $0.48Mevents/s$. The average amount of energy spent by the array to communicate an event to an external receiver is $2.3\mu J$. This figure takes into account the switching of 9 pads (4 + 4 addresses and a *request* signal) and the power consumption of the pad drivers, that in this technology is 40pJ/event (from simulation data).

5. SUMMARY

We have designed, fabricated and tested a 16 by 16 pixel SOS-CMOS photoreceptor array with a digital interface. We characterize the individual components of the array and demonstrate functionality for both front and back side imaging.

6. ACKNOWLEDGEMENTS

The authors would like to kindly thank Dan Gianola (Mechanical Engineering Department) for the assistance with the lapping machinery and Francisco Tejada for his help with bonding and packaging of the dies. Fabrication was provided by MOSIS through a DARPA fabrication grant, Dr. Ravi Athale point of contact. This work was supported by NSF grant *Microscale Adaptive Optical Wavefront Correction, ECS-0010026*.

7. REFERENCES

[1] Peregrine Semiconductors Inc., "0.5um FC design manual (52/0005)," URL: http://www.peregrine-semi.com/, March

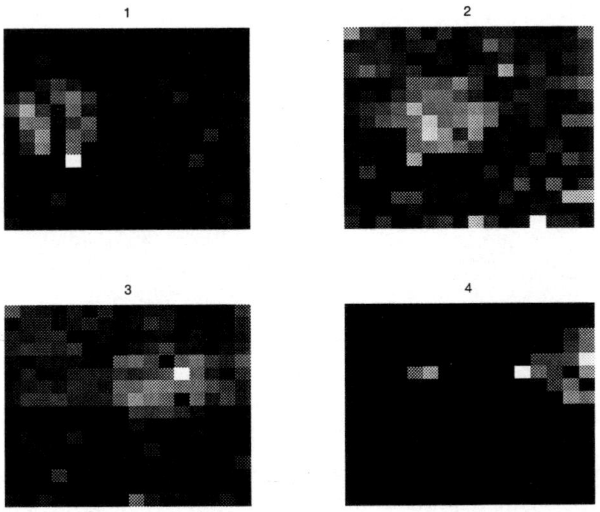

Fig. 7. Output from a light spot moving up-right impinging on the back side of the sensor array

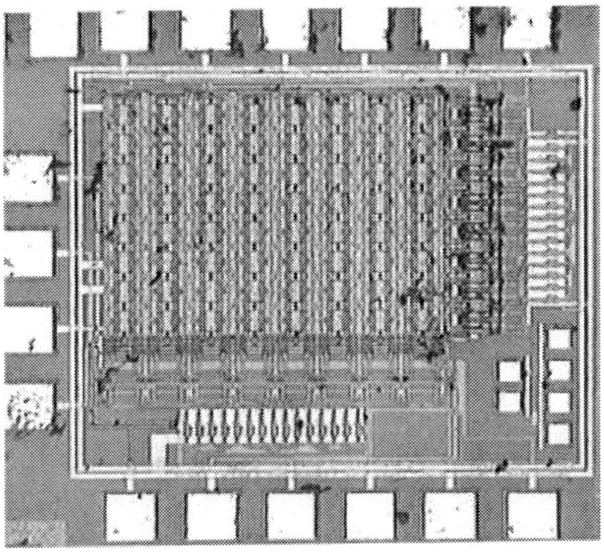

Fig. 8. Die micrograph for the 16x16 sensor array.

2003.

[2] A. G. Andreou, Z.K. Kalayjian, A. Apsel, P.O. Pouliquen, R.A. Athale, G. Simonis, and R. Reedy, "Silicon on sapphire CMOS for optoelectronic microsystems," *Circuits and Systems Magazine, IEEE*, vol. 1, no. 3, pp. 22–30, 2001.

[3] M. Cohen, R.T. Edwards, G. Cauwenberghs, M.A. Vorontsov, and G.W. Carhart, "AdOpt: Analog VLSI stochastic optimization for adaptive optics," in *Proc. Int. Joint Conf. Neural Networks (IJCNN'99)*, Washington DC, 1999, vol. 4, pp. 2343–2346.

[4] P. Abshire, *Sensory Information Processing Under Physical Constraints*, Ph.D. thesis, Johns Hopkins University, Baltimore, MD, october 2001.

[5] A. Apsel, E. Culurciello, A.G. Andreou, and K. Aliberti, "Thin film pin photodiodes for optoelectronic silicon on sapphire CMOS," in *ISCAS '03. Proceedings of the 2003 International Symposium on Circuits and Systems*, May 2003, vol. 4, pp. 908 – 911.

[6] A. Uehara, K. Kagawa, T. Tokuda, J. Ohta, and M. Nunoshita, "Back-illuminated pulse-frequency modulated photosensor using silicon-on-sapphire technology developed for use as epi-retinal prosthesis device," *Electronics Letters*, , no. 15, pp. 1102–1104, July 2003.

[7] K. A. Boahen E. Culurciello, R. Etienne-Cummings, "A biomorphic digital image sensor," *IEEE Journal of Solid-State Circuits*, vol. 38, no. 2, pp. 281 –294, February 2003.

[8] W. Yang, "A wide-dynamic range, low-power photosensor array," in *IEEE International Solid-State Circuits Conference, 1994. Digest of Technical Papers. ISSCC. 1994*, San Francisco, CA, February 1994, pp. 230 – 231.

[9] L.G. McIlrath, "A low-power low-noise ultrawide-dynamic-range CMOS imager with pixel-parallel A/D conversion," *Solid-State Circuits, IEEE Journal of*, vol. 36, pp. 846 – 853, May 2001, Issue 5.

[10] P.-Y.; Mortara A.; Nussbaum P.; Heitger F. Barbaro, M.; Burgi, "A 100 x 100 pixel silicon retina for gradient extraction with steering filter capabilities and temporal output coding," *Solid-State Circuits, IEEE Journal of*, vol. 37, pp. 160 – 172, February 2002, Issue 2.

[11] E. Culurciello and A.G. Andreou, "A comparative study of access topologies for chip-level address-event communication channels," *IEEE Transactions On Neural Networks*, September 2003, Special Issue On Hardware Implementations.

[12] A. Apsel and A.G. Andreou, "Quality of data reconstruction using stochastic encoding and an integrating receiver," in *Proceedings of the 43th Midwest Symposium on Circuits and Systems*, Ames, MI, August 2000, pp. 183–186, Best Student Paper Award.

COMPENSATION FOR CLOCK SKEW IN VOICE OVER PACKET NETWORKS BY SPEECH INTERPOLATION

Tõnu Trump

Ericsson AB
Stockholm, Sweden
tonu.trump@ericsson.com

ABSTRACT

In voice over packet networks, where the clocks controlling sampling rates at end-points are not always synchronized to each other, small differences in local clock frequencies accumulate over time and result in systematic buffer over/under-runs. This paper describes some signal processing techniques that can be used to compensate for this phenomenon often referred to as "clock skew". The studied algorithms effectively change the number of samples in the receiver play-out buffer dependent on estimated clock skew at the receiver, thereby avoiding systematic over/under-runs of the buffer and allowing shorter de-jittering buffer lengths which, in turn, result in lower end-to-end delay and improved speech quality. The proposed compensation algorithms are based on interpolation techniques.

1. INTRODUCTION

There is a growing interest in merging the existing telecommunication and computer networks into one multimedia communication network, which were able to provide all kinds of accesses including the highly interactive applications like telephony and teleconferencing. This development calls for solution of several new signal processing problems. This paper addresses some synchronization issues occurring in telephony over the packet switched networks e.g. Internet telephony.

In a traditional circuit switched telephony network a telephone exchange receives a synchronization clock from a switch located higher up in the hierarchy and distributes it down to all relevant pieces of equipment. This type of "global" synchronization is not always feasible in packet switched systems that do not have any permanent connection between the parties. This means that each receiver / transmitter operating on a packet switched network has to rely upon its own local clock and the clock frequencies of the parties may differ from each other in some extent. The difference in clock frequencies (clock skew) will cause problems in the receivers if proper measures are not taken to compensate for it. For instance, if the clock frequency in the transmitter is higher than that of the receiver, the receiver will systematically receive more speech samples than it is able to replay according to its own clock which leads to systematic overfill of the receiver's buffers. Similarly, if the clock frequency of the transmitter is lower than that of the receiver, the receiver will systematically find itself in a lack of speech samples to replay.

As discussed above, the clock skew is caused by differing sampling frequencies in the transmitter and receiver. As such, the skew can be considered as a deterministic phenomenon. In a packet switched network the information packets are transmitted from router to router towards the destination. On their way they are temporarily stored in the buffers of the routers. Durations of such storages are dependent on the current network load. As a result, the packets are delayed by different amounts of time in the network which appear as a delay jitter at the receiver. This jitter is typically compensated for by so called jitter absorbing buffer [1, 2], adding a variable extra delay before a packet is played out so that the overall delay time is kept constant (fixed jitter buffer) or varied slowly to minimize the overall delay at some given packet loss rate dependent on the current network conditions (adaptive jitter buffer). The delay jitter can be modeled as a random process as opposed to the deterministic behavior of the clock skew.

The clock skew problem has been recognized in several recent studies [3, 4, 5, 6, 7, 8, 9, 10, 11]. Some measurement results quantifying the skew can be found in [4] and [5]. In [4], a variation of clock rate up to $\pm 0.5\%$ between nominally similar clocks in workstations and personal computers, is reported. In [5] the clock skew was measured between several commercially available Internet phones and software clients. Average skew up to 55 parts per million (ppm) was measured for the IP phones and up to 300 ppm for software clients.

0-7803-8251-X/04/$17.00 ©2004 IEEE

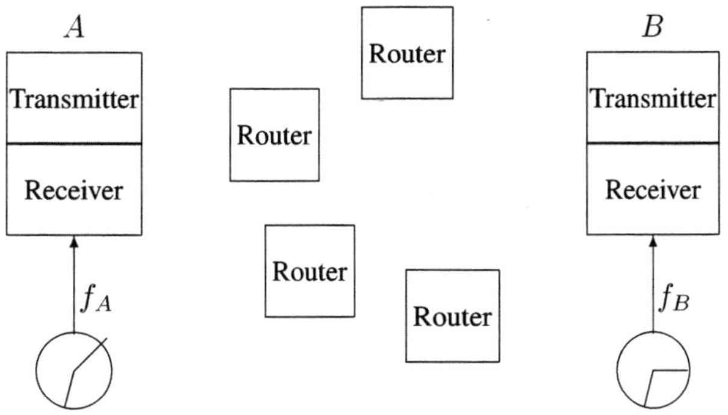

Figure 1: Simplified transmission system.

2. SKEW ESTIMATION

The investigated scenario is depicted in Figure 1, where two terminals, A and B, communicate over a packet switched network shown by a number of routers. Let us assume that the clock frequencies of A and B differ, so that the signal is sampled at rate f_A in the transmitter A and replayed with sampling frequency f_B in the receiver B. Also assume, that each transmitted packet consists of M samples (or of a compressed representation of these). Then the transmitter in A sends packets at time instants

$$t_A = \frac{M}{f_A}t, \qquad t = 1, 2, \ldots, \tag{1}$$

while the receiver in B expects the packets at time instants

$$t_B = \frac{M}{f_B}t + d_B, \qquad t = 1, 2, \ldots, \tag{2}$$

where d_B is some transmission delay expected by B. The actual packet arrival times are given by

$$t_R = \frac{M}{f_A}t + d_R + v(t), \qquad t = 1, 2, \ldots, \tag{3}$$

where d_R is the minimum transmission delay from A to B over the network and $v(t)$ is a random variable characterizing the extra delay added by the network (delay jitter) at time t. Let us now compute the difference between the actual and expected arrival times of the packets

$$\begin{aligned} y(t) = t_R - t_B &= \frac{M(f_B - f_A)}{f_A f_B}t + (d_R - d_B) + v(t) \\ &= \tau t + a + v(t). \end{aligned} \tag{4}$$

The latter is the linear trend model, where the parameters a and τ are to be estimated. The parameter a can be interpreted as a correction to the initial guess about the minimum transportation delay. The parameter $\tau = \frac{M}{f_A} - \frac{M}{f_B}$ is the difference of packet periods in A and B. The parameter τ is the clock skew.

There are several algorithms available for estimation of the clock skew from the linear trend model above. In [9] an algorithm based on examination of cumulative minima of the packet arrival times is developed to detect the skew. In [7] the least squares algorithm is used to estimate the model parameters. In [8] an algorithm based on linear programming is proposed and it is shown that it outperforms the least squares algorithm as well as the Paxson's algorithm [9]. In [10] it is shown that the linear programming algorithm is in fact maximum likelihood trend estimator for exponential noise (jitter) model. It is further argued that the good properties of linear programming approach result from the underlying (shifted) exponential distribution being a better model for one way packet delays through the network than Gaussian distribution. In particular it should be noted that the delay that forms the noise process is always positive valued. In addition, a recursive algorithm suitable for implementation in real time systems is suggested in [10]. In [11] the convex-hull approach is used to derive some skew estimation algorithms. The resulting algorithms are similar to those of [8] and [10]. Additionally, the scenario where clock resets can occur during the call is discussed in [11].

3. SKEW COMPENSATION

In this section we discuss several approaches to compensate for the clock skew, once detected.

3.1. Clock adjustment approach

Several papers suggest to avoid the problem by making the involved clocks precise enough by making use of the global positioning system (GPS) or by some other means see e.g. [6, 12]. The others propose to compensate for the clock skew, once detected, by appropriately changing the clock rate of one of the clocks see e.g. [3]. The underlying assumption in both cases is that the involved clocks can be adjusted. This certainly true for some equipment but may not be possible for some others. For example, it may be difficult in the case of software clients running on PC-s, as the equipment may use independent clock on audio card to control the sampling rate. Another complication occurs if one of the parties is a gateway interconnecting IP network and Public Switched Telephone Network (PSTN). Playout rate in such a gateway is typically synchronized to the PSTN clock and cannot be arbitrarily changed for individual calls. To overcome the difficulties mentioned above this paper concentrates to another possibility for clock skew compensation, namely changing the sampling rate by signal processing techniques.

3.2. Signal processing approach

The easiest way to compensate for the skew would be to duplicate a sample value in the playout buffer each time the receiver clock has gained one sample time relative to the transmitter clock and correspondingly to remove a sample each time the receiver clock has lost one sample time. This simple method would, however, lead to some degradation of speech quality by reasons discussed later in this paper. Better results can be obtained if addition/deletion of the samples is done during silence periods in the speech signal e.g. in conjunction with silence suppression. The latter would give satisfactory result for speech signals in silent background but may be problematic with a more general content.

In [4] it is suggested to locate repetitive signal segments and add or remove one such a segment from the speech signals when the need arises. This approach is similar to error concealment by waveform substitution. As noted in [4], the technique works best if the signal has many (pseudo) repetitive segments like voiced speech but may lead to problems with signals having rich content.

We now turn to some more general techniques that can be used to change sampling rate. The classical system for changing sampling period by the factor K/L as described in [13] is depicted in Figure 2. It first performs up-sampling by an integral factor L by means of filling $L-1$ zeros between each of the original signal samples. This is followed by filtering through a low pass filter (LPF) with cutoff frequency $f_0 = \frac{1}{2L}$ and gain L. The result of this operation is low pass filtered through a filter with cutoff frequency $f_1 = \frac{1}{2K}$ and gain 1 to avoid aliasing and then down-sampled by the factor K. Obviously the two low pass filters can be combined into a single one.

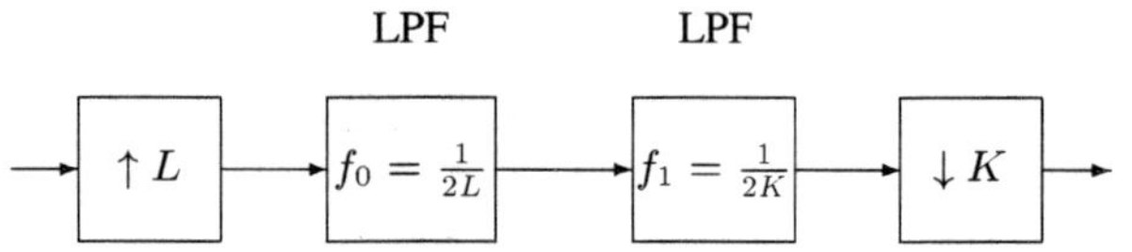

Figure 2: System changing sampling period by factor K/L.

The system is capable to perform any desired change of sampling rate by a rational factor. The complexity of the scheme grows, however, with L and can be impractical for the application at hand, where only a very small change in sampling rate, $\frac{K}{L} = 1 + \frac{\tau f_B}{M}$, is required. We therefore consider some lower complexity alternatives provided by spline interpolation [14, 15]. Splines are piecewise polynomials with pieces that smoothly connect together. A zero order B-spline, $\beta^0(t)$, is a rectangular pulse and n-th order B-spline is $n+1$ times convolution of $\beta^0(t)$ with itself.

It can be shown that spline interpolation can be represented as convolution of the signal samples with so called cardinal splines

$$\hat{x}(t) = \sum_{k=-\infty}^{\infty} x(n)\eta^n(t-k), \tag{5}$$

where $x(n)$ is the original discrete signal and $\hat{x}(t)$ is its continuous time representation via cardinal spline basis functions. The (continuous time) cardinal spline of degree n is given by

$$\eta^n(t) = \sum_{k=-\infty}^{\infty} \sum_{j=1}^{[n/2]} \alpha_j z_j^{|k|} \beta^n(t-k), \tag{6}$$

where

$$\alpha_j = \frac{1}{b_1^n(n/2)(z_j - z_j^{-1}) \prod_{i=1;i\neq j}^{[n/2]} (z_j + z_j^{-1} - z_i - z_i^{-1})}, \tag{7}$$

b_1^n is the discrete spline obtained from the continuous one by sampling at integral time instants, z_j are the $[n/2]$ smallest roots of the z transform of b_1^n and $[\cdot]$ represents rounding down to the nearest integer. This representation is convenient for investigating aliasing effects related to spline interpolation. The spectrum of the cardinal spline is

$$H^n(f) = \frac{(\sin(\pi f)/\pi f)^{n+1}}{b_1^n(0) + \sum_{k=1}^{[n/2]} 2b_1^n(k)\cos(2\pi f k)}. \tag{8}$$

The telephone channel is band-limited between 300 and 3400 Hz with sampling rate 8 kHz. Approximating the signal spectrum with constant inside the telephone band and with zero outside the band and keeping in mind that spectrum of a time-discrete signal is periodic we can plot the spectrum of spline-interpolated telephone band signal as in Figure 3. The rectangular spectrum (solid line) corresponds to the ideal interpolator, shown in Figure 2, with ideal low pass filters. The three other lines correspond to the zero (dotted), first (dashed-dotted) and third (dashed) order spline interpolators. Note, that the zero order spline interpolation is in fact equivalent to the simple method of duplicating or deleting samples, discussed in the beginning of this chapter. The first order spline-interpolation is the same as linear interpolation between signal samples. It can be observed from Figure 3 that the undesired aliasing components for zero– and first–order interpolators are rather strong as compared to the signal components in base–band. This must be seen as a warning for audible distortion if this type of interpolation is used for skew compensation. It can also be seen that the aliasing power decreases rapidly with increasing order of spline interpolation. Note that the spectra correspond to the equivalent continuous time signals given by (5) that are actually never explicitly computed. Discrete signals at

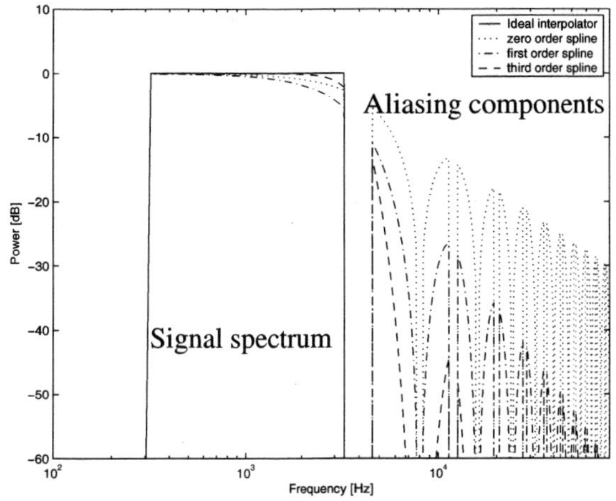

Figure 3: Equivalent spectra of ideal and some spline interpolators for 8000 Hz sampling rate.

the output of spline interpolator correspond to sampling of those at the new rate resulting in folding the components above half the new sampling rate to the baseband.

4. RESULTS AND DISCUSSION

In our listening experiments we compared the result of the simple sample duplication / removal approach (zero order spline), with linear interpolation (first order spline) and cubic spline interpolation techniques. The effects of sampling rate change by 0.1%, 0.5% and 1% was investigated. Speech and music samples were used as the listening material. Aliasing distortion for 0.1% sampling rate change with zero order interpolation was noticeable with speech signals and clearly audible with music samples. With skew 0.5 % and larger, the distortion became clearly audible with speech signals as well. The linear interpolation scheme resulted in noticeable distortion from 0.5 % sampling rate change for speech signals and from 0.1 % for music signals. The distortion was often experienced as a kind of "vibrato" effect. Aliasing distortion from cubic spline interpolation was not noticed with speech signals in our listening experiments. With music, some distortion was observed with skew of 0.5 % and higher.

The general conclusion from the experiments is that the higher is the skew rate, the higher order spline interpolation is required to make the compensation without introducing audible distortion. A possible explanation to this is that the dominating aliasing components, when folded back to the base-band after a small change in sampling rate, appear close to the original signal components and are partly masked by them. The smaller the skew, the smaller will be the difference in frequency between the original and the first folded bands and, hence, the stronger is the masking effect.

We also observed, that aliasing was subjectively more disturbing for music samples than for speech for the same skew rate and interpolation scheme.

5. REFERENCES

[1] P. DeLeon and C. J. Sreenan, "An Adaptive Predictor for Media Playout Buffering," *Proc. IEEE International Conference on Acoustics, Speech, and Signal Processing,* Phoenix, vol.6, pages 3097–3100, March, 1999.

[2] J. F. Kurose and K. W. Ross *Computer Networking: A Top-Down Approach Featuring the Internet,* Addison Wesley, 2001.

[3] D. Fober, S. Letz, Y. Orlarey, "Clock Skew Compensation over a High Latency Network," *Proc. International Computer Music Conference* Gothenburg, Sweden, 2002, pp.548–552

[4] O. Hodson, C. Perkins and V. Hardman "Skew detection and compensation for internet audio applications" *Proc. IEEE International Conf. on Multimedia and Expo,* New York, July 2000, Vol. 3, pp. 1687–1690

[5] Wenyu Jiang, K. Koguchi and H. Schulzrinne "QoS Evaluation of VoIP End-points" *Proc. IEEE International Conference on Communications,* Anchorage, Alaska, May 2003, pp. 1917–1921

[6] H. Melvin and L. Murphy "An Investigation into the use of Synchronised Time to Improve VoIP Service," *Proc. First Joint IEI/IEEE Symposium on Telecommunications Systems Research* Dublin, Ireland, Nov. 2001.

[7] R. Noro and J.-P. Hubaux "Clock Synchronization of MPEG-2 Services Over Packet Networks" *Telecommunication Systems,* vol. 11, No. 1,2, pp.3-16, March 1999

[8] S. B. Moon, P. Skelly and D. Towsley, "Estimation and Removal of Clock Skew from Network Delay Measurements," *Proc. IEEE Conf. Computer Communications* New York, Mar. 1999, pp. 227-234.

[9] V. Paxson, "On Calibrating Measurements of Packet Transit Times," *Proc. Joint International Conference on Measurement and Modeling of Computer Systems,* Madison, WI, USA, June 1998, pp. 11-21.

[10] T. Trump "Maximum Likelihood Trend Estimation in Exponential Noise," *IEEE Trans. on Signal Processing* Vol. 49, Sept. 2001, pp.2087–2095

[11] Li Zhang, Z. Liu and C. H. Xia "Clock Synchronization Algorityms for Network Measurements" *Proc. IEEE Infocom,* June 2002, Vol 1, pp.160–169

[12] A.Pasztor and D. Veitch, "PC based Precision Timing without GPS," *Proc. ACM Sigmetrics* Marina Del Rey, CA, USA, June 2002.

[13] A. Oppenheim and R. Schafer, *"Discrete Time Signal Processing,"* Prentice Hall, 1989.

[14] M. Unser, A. Aldroubi and M. Eden, "B-Spline Signal Precessing: Part I – Theory," *IEEE Trans. on Signal Processing* Vol. 41, Feb. 1993, pp.821–833

[15] M. Unser, "Splines – A Perfect Fit for Signal and Image Precessing," *IEEE Signal Processing Magazine* Vol. 16, Nov. 1999, pp.22–38.

FAST ADAPTIVE COMPONENT WEIGHTED CEPSTRUM POLE FILTERING FOR SPEAKER IDENTIFICATION

Arthur L. Swanson[1], *Ravi P. Ramachandran*[2] *and Steven H. Chin*[3]

1. L-3 Communications, arthur.l.swanson@l-3com.com
2. Rowan University, ravi@rowan.edu
3. Rowan University, chin@rowan.edu

ABSTRACT

Mismatched training and testing conditions for speaker identification exist when speech is subjected to a different channel for the two cases. This results in diminished speaker identification performance. Finding features that show little variability to the filtering effect of different channels will make speaker identification systems more robust thereby achieving a better performance. It has been shown that subtracting the mean of the pole filtered linear predictive (LP) cepstrum from the actual LP cepstrum results in a robust feature. This feature is known as the pole filtered mean removed LP cepstrum. Another robust feature is the adaptive component weighted (ACW) cepstrum particularly with mean removal. In this paper, we combine the ACW cepstrum with the pole filtering concept to configure a more robust new feature, namely, the pole filtered mean removed ACW cepstrum. This new method is fast and shows a higher performance then the pole filtered mean removed LP cepstrum and the mean removed ACW cepstrum. Experimental results are given for the TIMIT database involving a variety of mismatched conditions.

1. INTRODUCTION

Speaker recognition refers to the concept of recognizing a speaker by his/her voice or speech samples [1][2][3]. Some of the important applications of speaker recognition include customer verification for bank transactions, access to bank accounts through telephones, control on the use of credit cards, and for security purposes in the army, navy and airforce. The two tasks within speaker recognition are *speaker identification* and *speaker verification*. Speaker identification (ID) deals with a situation where the person has to be identified as being one among a set of persons by using his/her voice samples. The objective of speaker verification is to verify the claimed identity of that speaker based on the voice samples of that speaker alone. A claimant speaker is either accepted or rejected by the system.

The speaker ID problem may further be subdivided into *closed set* and *open set*. The closed set speaker ID problem refers to a case where the speaker is known *a priori* to belong to a set of M speakers. In the open set case, the speaker may be out of the set and hence, a "none of the above" category is necessary. Another distinguishing aspect of speaker recognition systems is that they can either be text-independent or text-dependent depending on the application. In the text-independent case, there is no restriction on the sentence or phrase to be spoken, whereas in the text-dependent case, the input sentence or phrase is fixed for each speaker. The focus of this paper is on text-independent, closed set speaker identification.

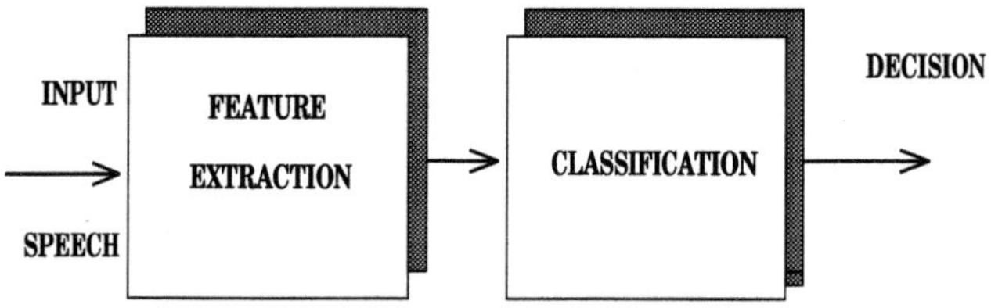

Figure 1: A general diagram of a recognition system

Speaker recognition consists of two stages, namely, *Feature Extraction* and *Classification* as shown in Fig. 1. Feature extraction is associated with obtaining the characteristic patterns of the signal that are representative of the speaker in question. The parameters or features used in speaker recognition are a transformation of the speech signal into a compact acoustic representation that contains information useful for the identification of the speaker. This is often done using short-time linear predictive (LP) [4] analysis which leads to an all-pole LP vocal tract model. The LP coefficients are converted to the LP cepstrum [4] which in turn, is the feature vector. The classifier uses the features to render a decision as to the speaker identity or verifies the claimed identity of the speaker.

The recognition task is highly successful if the environmental conditions for training and testing are the same (known as matched conditions). Studies have shown that recognition performance degrades when the training and testing conditions are not the same (known as mismatched conditions) [5][6][7]. This occurs when the speaker is trained on one type of telephone (handset, cordless or speakerphone) and during the testing phase, a different type of telephone is used. In this particular case, channel mismatch is encountered and this contributes to the degradation in the performance. Channels have a filtering effect on the speech and alter the overall spectral envelope of the speech signal. Assuming that the speech and channel spectra are well approximated by the all-pole LP model, it is observed that a channel influence on the speech leads to an additive component on the LP cepstrum. Estimating and removing this additive channel component will mitigate the channel effect and make speaker recognition systems more robust. One method of estimating the additive channel component is to take the mean of the LP cepstrum vectors over an utterance [8]. It has been shown that a better estimate is obtained by taking the mean of the pole filtered LP cepstrum [9][10][11]. Removal of the mean of the pole filtered LP cepstrum from the LP cepstrum vectors results in a more robust feature and is known as the pole filtered mean removed cepstrum (PFMRC). Another channel estimate is based on the mean of the adaptive component weighted (ACW) cepstrum [12][13][14]. Removal of the mean of the ACW cepstrum from the ACW cepstrum vectors results in another robust feature and is known as the mean removed ACW cepstrum (MRACW).

In this paper, we combine the concept of pole filtering to the ACW cepstrum to configure a new channel estimate and a new feature. The channel estimate is the pole filtered mean of the ACW cepstrum. Removal of the pole filtered mean of the ACW cepstrum from the ACW cepstrum vectors results in the new robust feature and is known as the pole filtered mean removed ACW cepstrum (PFMRACW). This method is computationally efficient just like its PFMRC and MRACW counterparts. It also gives a better performance than the PFMRC and MRACW approaches.

2. LINEAR PREDICTIVE CEPSTRUM

The autoregressive LP model for speech is given by the difference equation [4]

$$s(n) = \sum_{i=1}^{p} a_i s(n-i) + e(n) \quad (1)$$

where $s(n)$ is the speech signal, $e(n)$ is the prediction error and a_i are the predictor coefficients. It can be noted that $s(n)$ is predicted as a linear combination of the previous p samples. The all-pole LP transfer function is given by

$$H(z) = \frac{S(z)}{E(z)} = \frac{1}{A(z)} = \frac{1}{1 - \sum_{i=1}^{p} a_i z^{-i}}. \quad (2)$$

where $S(z)$ and $E(z)$ are the z-transforms of $s(n)$ and $e(n)$ respectively. In practice, the predictor coefficients a_i are computed over short intervals (typically 10 ms to 30 ms) called frames during which the vocal tract configuration is assumed to be stationary. This is done using the autocorrelation method [4][15] which guarantees that $H(z)$ is a stable function.

The predictor coefficients a_i are converted to the LP cepstrum $clp(n)$ $(n \geq 1)$ by an efficient recursive relation given as [4][15]

$$clp(n) = a_n + \sum_{i=1}^{n-1} (\frac{i}{n}) clp(i) a_{n-i} \quad (3)$$

Since $clp(n)$ is of infinite duration, the feature vector of dimension p consists of the components $clp(1)$ to $clp(p)$ which are the most significant due to the decay of the sequence with increasing n.

2.1. Pole Filtered Mean Removed Cepstrum (PFMRC)

As mentioned earlier, when speech is subjected to channel interference, an additive component due to the channel manifests itself on the LP cepstrum. To compensate for the channel effect, this component is estimated as the mean or the pole filtered mean of the LP cepstrum and removed by subtraction. For simple mean subtraction, the feature vector is

$$cmrc(n) = clp(n) - E[clp(n)] \quad (4)$$

where the expectation is taken over an utterance consisting of a number of frames.

The LP poles with narrow bandwidths that lie close to the unit circle usually represent the formants and are less sensitive to channel and noise effects. Hence, these poles do not contribute to the channel estimate as they contain much speech information. In contrast, the broad bandwidth poles model the spectral tilt, sub-glottal variation and the channel effects. These poles offer a better estimate of the channel. Pole filtering modifies the LP poles so as to broaden the bandwidth of the formant poles [9][10][11]. Broadening the bandwidth of the formant poles is performed by transforming the LP polynomial so as to weight the predictor coefficients as given by

$$H(z/\gamma) = \frac{1}{A(z/\gamma)} = \frac{1}{1 - \sum_{i=1}^{p} a_i \gamma^i z^{-i}}. \quad (5)$$

where $0 < \gamma \leq 1$. Given the original LP poles p_i, the new set of poles are γp_i. The cepstrum formed from these modified poles (denoted as $cpflp(n)$) is related to the LP cepstrum as [9][10][11].

$$cpflp(n) = \gamma^n clp(n) \quad (6)$$

The feature vector $cpfmrc(n)$ is known as the pole filtered mean removed cepstrum (PFMRC) and is computed as given below.

- Select the parameter γ.
- For each frame of speech, calculate $clp(n)$ and $cpflp(n)$.
- Find the channel estimate $E[cpflp(n)]$ where the expectation is taken over all speech frames in an utterance.
- Find the feature vector $cpfmrc(n) = clp(n) - E[cpflp(n)]$.

3. ADAPTIVE COMPONENT WEIGHTED CEPSTRUM

The first step in developing the ACW cepstrum [12] is to perform a partial fraction expansion of the LP function $H(z) = 1/A(z)$ to get

$$\begin{aligned} \frac{1}{A(z)} &= \sum_{k=1}^{p} \frac{\lim_{z \to p_k} \left[\left(1 - p_k z^{-1}\right) / A(z) \right]}{1 - p_k z^{-1}} \\ &= \sum_{k=1}^{p} \frac{r_k}{1 - p_k z^{-1}} \end{aligned} \quad (7)$$

The experiments in [12] reveal that the residues r_k show considerable variations especially for nonformant poles when the speech is degraded. Therefore, the variations in r_k were removed by forcing r_k to be $constant = 1$ for every k. Hence, the resulting transfer function is a pole-zero type of the form

$$\begin{aligned} H_{acw}(z) &= \frac{N(z)}{A(z)} \\ &= \sum_{k=1}^{p} \frac{1}{1 - p_k z^{-1}} \\ &= \frac{1}{A(z)} \sum_{k=1}^{p} \prod_{i=1 \neq k}^{p} (1 - p_i z^{-1}) \\ &= p \frac{1 - \sum_{k=1}^{p-1} b_k z^{-k}}{1 - \sum_{k=1}^{p} a_k z^{-k}} \end{aligned} \quad (8)$$

It has been shown in [14] that $N(z)$ is the derivative of $A(z)$ with respect to z and hence, the coefficients b_k are easily found from a_k. Applying the recursion in Eq. (3) to b_k and a_k results in two cepstrum sequences $cb(n)$ and $clp(n)$ respectively. The ACW cepstrum is $cacw(n) = clp(n) - cb(n)$. For simple mean subtraction (MRACW method), the feature vector is

$$cmracw(n) = cacw(n) - E[cacw(n)] \tag{9}$$

where the expectation is taken over an utterance consisting of a number of frames.

3.1. Pole Filtered Mean Removed Adaptive Component Weighted Cepstrum (PFMRACW)

The contribution of this paper is to combine the pole filtering concept to the ACW cepstrum to get a better channel estimate and a more robust feature vector. The first step is to choose a value of γ between 0 and 1 and perform a partial fraction expansion of the pole filtered LP function $1/A(z/\gamma)$ to get

$$\frac{1}{A(z/\gamma)} = \sum_{k=1}^{p} \frac{s_k}{1 - q_k z^{-1}} \tag{10}$$

Setting $s_k = 1$ for every k gives a transfer function

$$\begin{aligned} H_{pfacw}(z) &= \frac{M(z)}{A(z/\gamma)} \\ &= \sum_{k=1}^{p} \frac{1}{1 - q_k z^{-1}} \\ &= p\frac{1 - \sum_{k=1}^{p-1} m_k z^{-k}}{1 - \sum_{k=1}^{p} \gamma^k a_k z^{-k}}, \end{aligned} \tag{11}$$

Again, $M(z)$ is the derivative of $A(z/\gamma)$ with respect to z and hence, the coefficients m_k are easily found from $\gamma^k a_k$. Applying the recursion in Eq. (3) to m_k results in the cepstrum sequence $cm(n)$. The cepstrum corresponding to the denominator of $H_{pfacw}(z)$ is $cpflp(n)$ (see Eq. (6). The pole filtered ACW cepstrum is expressed as $cpfacw(n) = cpflp(n) - cm(n)$. The feature vector (denoted as $cpfmracw(n)$) is known as the pole filtered mean removed ACW cepstrum (PFMRACW) and is computed as given below.

- Select the parameter γ.
- For each frame of speech, calculate $cacw(n)$ and $cpfacw(n)$.
- Find the channel estimate $E[cpfacw(n)]$ where the expectation is taken over all speech frames in an utterance.
- Find the feature vector $cpfmracw(n) = cacw(n) - E[cpfacw(n)]$.

4. EXPERIMENTAL RESULTS

Closed set, text-independent speaker identification experiments are carried out using the TIMIT database. Thirty eight speakers from the New England dialect are considered. The speech is downsampled from 16 kHz to 8 kHz. For each speaker, there are 10 sentences. The first five are used for training a vector quantizer (VQ) classifier using the Linde-Buzo-Gray (LBG) method [16] and the squared Euclidean distance as the distortion measure. A VQ codebook is designed for each of the 38 speakers. The training conditions include clean speech and speech subjected to representative

Training Condition	Testing Condition	PFMRC γ, ISR	MRACW ISR	PFMRACW γ, ISR
Clean	CMV	0.90, 58.4	60.0	0.95, 65.8
Clean	CPV	0.90, 59.5	56.3	0.90, 67.4
CMV	CPV	0.85, 66.3	70.0	0.80, 74.2
CPV	CMV	0.80, 66.8	65.8	0.90, 73.7
Clean	EMV	0.70, 67.9	65.3	0.80, 79.5
Clean	EPV	0.75, 70.5	68.4	0.70, 78.4
EMV	EPV	0.80, 74.2	70.0	0.85, 81.6
EPV	EMV	0.70, 71.6	70.5	0.90, 78.4
CMV	EMV	0.80, 64.2	62.6	0.95, 68.4
CMV	EPV	0.95, 62.6	62.6	0.95, 66.3
CPV	EMV	0.85, 58.9	55.8	0.75, 62.1
CPV	EPV	0.85, 61.6	57.4	0.95, 58.9

Table 1: Speaker identification success rate (ISR) as a percent. The best value of γ is shown. The acronym PFMRC is for pole filtered cepstral mean subtraction. The acronym MRACW is for adaptive component weighted cepstral mean subtraction. The acronym PFMRACW is for pole filtered adaptive component weighted cepstral mean subtraction.

bandpass telephone channels [17]: (1) the Continental Mid Voice (CMV) channel, (2) the Continental Poor Voice (CPV) channel, (3) the European Mid Voice (EMV) channel and (4) the European Poor Voice (EPV) channel. The remaining five sentences are individually used for testing thereby giving 190 test utterances.

The testing conditions correspond to channel corrupted speech. Consider a particular test feature vector. This is quantized by each of the 38 codebooks. The quantized vector is that which is closest (according to the squared Euclidean distance) to the test feature vector. Hence, 38 different distances are recorded, one for each codebook. This process is repeated for every test feature vector. The distances are accumulated over the entire set of feature vectors. The codebook which renders the smallest accumulated distance identifies the speaker. The identification success rate (ISR) is the number of utterances for which the speaker is identified correctly divided by the total number of utterances tested. The VQ codebook size is 64. For both training and testing, silent or low-energy frames are discarded by energy thresholding. Also, a 12th order LP analysis is used with 30 ms frames having an overlap of 20 ms. All the feature vectors have dimension 12.

Table 1 shows the results for speaker identification. Different values of γ were tried. The best values of γ are from 0.70 to 0.95. In Table 1, the result for the best value of γ is given for each training/testing combination. The performance of our new PFMRACW method is better than the both the PFMRC and MRACW approaches (except for only one case when the CPV channel is used for training and the EPV channel is used for testing).

For both the PFMRC and PFMRACW methods, the best value of γ depends on the training and testing conditions. However, results show that the ISR varies very little for values of γ between 0.70 and 0.95. Decreasing γ below 0.70 does result in significant performance loss and hence, these values should not be used. Figure 2 shows the ISR versus γ for the PFMRC and PFMRACW methods for the case when training is done on the CMV channel and testing is done on the CPV channel. The question of what γ to use can more easily be answered since the variation in the ISR is relatively low for values of $0.70 \leq \gamma \leq 0.95$. By examining the results for

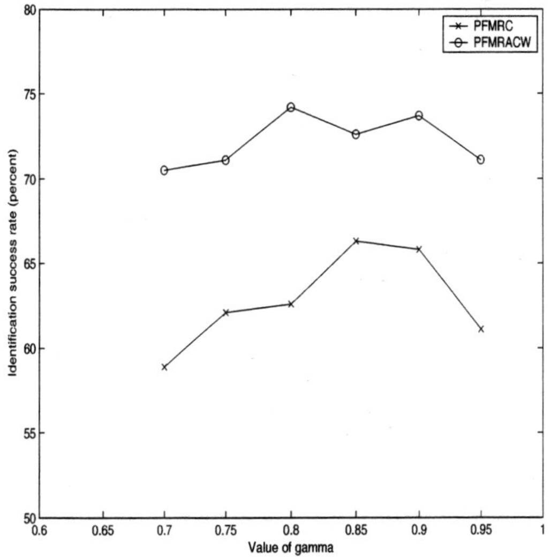

Figure 2: The ISR versus γ for the PFMRC and PFMRACW methods for the case when training is done on the CMV channel and testing is done on the CPV channel.

Training Condition	Testing Condition	PFMRC ISR	MRACW ISR	PFMRACW ISR
Clean	CMV	56.3	60.0	56.3
Clean	CPV	57.4	56.3	60.5
CMV	CPV	66.3	70.0	72.6
CPV	CMV	63.2	65.8	73.7
Clean	EMV	62.6	65.3	77.9
Clean	EPV	68.9	68.4	77.9
EMV	EPV	71.6	70.0	81.6
EPV	EMV	70.5	70.5	76.8
CMV	EMV	56.8	62.6	62.6
CMV	EPV	61.1	62.6	51.1
CPV	EMV	58.9	55.8	59.5
CPV	EPV	61.6	57.4	57.4

Table 2: Speaker identification success rate (ISR) as a percent for $\gamma = 0.85$ is shown. The acronym PFMRC is for pole filtered cepstral mean subtraction. The acronym MRACW is for adaptive component weighted cepstral mean subtraction. The acronym PFMRACW is for pole filtered adaptive component weighted cepstral mean subtraction.

all the training and testing conditions attempted, it is beneficial to fix γ at 0.85. Table 2 shows the results for $\gamma = 0.85$. It is clear that the ISR for both the PFMRC and PFMRACW are generally slightly below the best possible and that the PFMRACW method is almost always the best (the exception being two cases). Table 2 also shows the results for the MRACW method for the sake of completeness.

5. REFERENCES

1. G. R. Doddington, "Speaker recognition - identifying people by their voices" *Proc. IEEE*, vol. 73, pp. 1651–1664, November 1985.
2. A. E. Rosenberg, "Automatic speaker verification: A review", *Proc. IEEE*, vol. 64, pp. 475–487, April 1976.
3. J. P. Campbell, "Speaker recognition: A tutorial", *Proc. IEEE*, vol. 85, pp. 1437–1462, September 1997.
4. L. R. Rabiner and R. W. Schafer, *Digital Processing of Speech Signals*, Prentice-Hall, 1978.
5. A. E. Rosenberg and F. K. Soong, "Recent research in automatic speaker recognition", in *Advances in Speech Signal Processing*, edited by S. Furui and M. M. Sondhi, Marcel Dekker, pp. 701–738, 1991.
6. R. J. Mammone, X. Zhang and R. P. Ramachandran, "Robust speaker recognition - A feature based approach", *IEEE Signal Proc. Mag.*, vol. 13, pp. 58–71, September 1996.
7. R. P. Ramachandran, K. R. Farrell, Roopashri Ramachandran and R. J. Mammone, "Robust Speaker Recognition - General Classifier Approaches and Data Fusion Methods", *Pattern Recognition*, Vol. 35, No. 12, pp. 2801–2821, December 2002.
8. B. S. Atal, ""Effectiveness of linear prediction characteristics of the speech wave for automatic speaker identification and verification", *Jour. of the Acoust. Soc. of Amer.*, vol. 55, pp. 1304–1312, June 1974.
9. D. Naik and R. J. Mammone, "Channel normalization using pole filtered cepstral mean subtraction", *SPIE Int. Symp. on Optics, Imaging and Instrumentation*, vol. 2277, San Diego, California, pp. 99–110, July 1994.
10. D. Naik and R. J. Mammone, "Pole filtered cepstral mean subtraction", *IEEE Int. Conf. on Acoust., Speech and Sig. Proc.*, Detroit, Michigan, pp. 157–160, May 1995.
11. R. P. Ramachandran and K. R. Farrell, "Fast Pole Filtering for Speaker Recognition", *IEEE Int. Symp. on Circuits and Systems*, Geneva, Switzerland, pp. V-49–V-52, May 28–31, 2000.
12. K. T. Assaleh and R. J. Mammone, "New LP-derived features for speaker identification", *IEEE Trans. on Speech and Audio Proc.*, vol. 2, pp. 630–638, October 1994.
13. M. S. Zilovic, R. P. Ramachandran and R. J. Mammone, "Speaker identification based on the use of robust cepstral features obtained from pole-zero transfer functions", *IEEE Trans. on Speech and Audio Proc..* vol. 6, pp. 260–267, May 1998.
14. M. S. Zilovic, R. P. Ramachandran and R. J. Mammone, "A Fast Algorithm for Finding the Adaptive Component Weighted Cepstrum for Speaker Recognition", *IEEE Trans. on Speech and Audio Processing*, Vol. 5, No. 1, pp. 84–86, January 1997.
15. L. R. Rabiner and B.-H. Juang, *Fundamentals of Speech Recognition*, Prentice-Hall, 1993.
16. Y. Linde, A. Buzo and R. M. Gray, "An algorithm for vector quantizer design", *IEEE Trans. on Comm.*, vol. COM-28, pp. 84–95, Jan. 1980.
17. J. Kupin, "A wireless simulator (software)," CCR-P, April 1993.

A LOUDNESS ENHANCEMENT TECHNIQUE FOR SPEECH

Marc A. Boillot

Computational Neuro-Engineering Lab
University of Florida, Gainesville, FL
emb034@email.mot.com

*John G. Harris**

Computational Neuro-Engineering Lab
University of Florida, Gainesville, FL
harris@cnel.ufl.edu

ABSTRACT

We propose a novel application of the vocoder postfilter to increase perceived loudness of clean speech without increasing signal energy or degrading intelligibility. The critical band concept in auditory theory states that perceived loudness of a narrow-band signal will increase when the bandwidth of that signal increases beyond a critical band, even though the energy remains constant. Our post-filter technique applies formant bandwidth expansion to the vowel regions of speech without changing the vowel power to elevate perceived loudness. Vowels are known to contain the highest energy, have a smooth spectral envelope, long temporal sustenance, and for this reason are suitable candidates to target for a loudness enhancement technique. ISO-532B loudness analysis patterns and listening tests are provided to demonstrate a perceptual loudness improvement corresponding to a 2dB power gain.

1. INTRODUCTION

It is known that when the bandwidth of noise at fixed intensity increases beyond a critical band, loudness will increase [1]. In this paper a bandwidth expansion technique which preserves energy is employed to increase loudness in vowels without sacrificing intelligibility or naturalness. Consonants provide the acoustic cues for intelligibility and are typically the precipitated target for speech enhancement methods. Vowels are high energy, resonant, long in duration, and spectrally smooth in nature. They are good candidates for this procedure and studies have shown that vowel bandwidths can be widened without degrading intelligibility [2]. It has also been suggested that the overall loudness level of continuous speech is closer to the average peak loudness than the average loudness [1].

The proposed loudness filter is an extension of the adaptive post-filter and noise spectral shaping filter used to improve the subjective speech quality of vocoded speech [3].

Funding was provided by the iDEN Technology and the DSP Product Development Group of Motorola (UF# 10483) (* contact author)

We incorporate the ISO-532B analysis method to objectively evaluate the improvement in loudness [4]. The ISO-532B defines a quantitative procedure for calculating the loudness of steady-state tones, and can be used to precisely determine the specific loudness and loudness patterns. We transcribed the ISO-532B code in [5], into a Matlab equivalent function for our analysis. However, no measure is known to exist which calculates the comprehensive loudness of continuously varying complex sounds such as speech. We use the ISO-532B analysis to quantify the loudness improvement on an accepted loudness scale. Loudness patterns are generated for comparison. Our technique focuses on vowels which are more steady-state and longer in duration than consonants. We assume that the ISO-532B loudness in vowel regions corresponds to a subjective perception of loudness in those regions.

Speech enhancement methods are employed as techniques which attempt to emphasize the salient characteristics of speech. These techniques focus on emphasizing certain regions of speech which have been experimentally determined to contain important acoustic or auditory cues. In most cases no well defined measure exists which truly maps a subjective quality to an objective quantity. In many cases speech enhancement implies some form of noise suppression where the defined target is noise. Noise is the contaminant signal, and the approach is to develop estimation functions which maximize a signal to noise ratio. Certain spectral subtraction techniques have been extended with psychoacoustics in mind such as that of [6][7], which incorporate perceptual masking knowledge in the noise reduction process. For these methods a well defined problem and tractable solution usually exists: to suppress the audible noise.

Other speech enhancement methods typically operate on clean speech with the intent of improving intelligibility. By virtue, they are highly subjective and usually require exhaustive listening tests. It is extremely difficult to define tangible variables which correlate a measure of quality or intelligibility. For this reason speech enhancement techniques for improving intelligibility are heuristic by nature and are difficult to quantify since the target response is highly subjective. In this study we instantiate the auditory percept of loudness to be our target variable, much as noise is to spectral subtraction methods. We utilize the ISO-532B loudness analysis to quantify the increase of loudness, and we propose a speech enhancement method with a simple and well defined purpose: To increase the perceptual loudness of speech without adding energy.

Speech processing by nature distorts the signal in a prescribed sense. For the reasons previously stated, some measure of intelligibility is required for this task. Our technique has been applied to the point at which a change in intelligibility is noticeable but still acceptable. We have included a DRT listening test with our loudness test to evaluate the effects on intelligibility.

2. LPC BANDWIDTH EXPANSION

LPC is a well understood method used to generate an all-pole model of a speech signal [8]. LPC analysis inherently provides a way to increase or decrease vowel bandwidths. Pole resonance and bandwidth are a direct function of the pole distance from the unit circle. As the pole moves closer to the unit circle the resonance increases and the bandwidth decreases. Hence, it is possible to change the pole bandwidth by simply changing the evaluation radius of the unit circle. The LPC technique used to alter pole bandwidth is based on McCandless procedure [9] and shown in Eq (1).

This technique provides a way to evaluate the z transform on a circle with radius r greater than or less than the unit circle $r = 1$ [10]. For $0 < r < 1$ the evaluation is on a circle closer to the poles and the contribution of the poles has effectively increased, thus sharpening the pole resonance. Stability is of concern since $1/A(\tilde{z})$ is no longer an analytic expression within the unit circle. For $r > 1$ (bandwidth expansion) the evaluation is on a circle farther away from the poles and thus the pole resonance peaks decrease and the pole bandwidths are widened. The poles are always inside the unit circle and $1/A(\tilde{z})$ is stable. The bandwidth adjustment technique simply requires a scaling of the LPC coefficients by a power series of r, Eq (1).

$$A(\tilde{z})|_{\tilde{z}=re^{jw}} = \sum_{k=0}^{p}(a_k r^{-k})\, e^{-jwk} \tag{1}$$

If the poles are well separated [8], the bandwidth can be related to the pole radius k by

$$B \approx -\ln(k/r) f_s/\pi \tag{2}$$

where $r = 1.0$ defines evaluation on the unit circle. When the evaluation is on a circle other than the unit circle, the bandwidth increase is related to the new evaluation circle with radius $\hat{r} > 1$ by

$$\triangle B \approx -[\ln(k/\hat{r}) - \ln(k/r)] f_s/\pi$$

$$\triangle B \approx \ln(\hat{r}) f_s/\pi \tag{3}$$

The bandwidth adjustment technique has been used as a weighting function for spectral distortion measures in speech recognition [11], and to sharpen formant bandwidths as a speech enhancement method to improve speech intelligibility in noise. A bandwidth factor is chosen to narrow the formant bandwidths which effectively raise the formant peaks above the noise. It is also common to speech coding and has been used as a compensation filter for the bandwidth underestimation problem and as a postfilter to enhance the relative quality of vocoded speech due to quantization [3].

The transfer function of the bandwith expanded filter is defined in Eq (4) for which α and β denote the bandwidth expansion terms and are reciprocals of an evaluation radius, $1/r$.

$$H(z) = \frac{A(z/\alpha)}{A(z/\beta)} \tag{4}$$

The selection depends on the values of the α and β parameters. For $0 < \alpha < \beta < 1$, the filter provides a sharpening of the formants, or a narrowing of the formant bandwidth. Noise spectral shaping uses the perceptual weighting filter of Eq (4) to generate a noise error function used to weight the importance of certain frequency regions. The noise error function defines a reference for bit-allocation which retains the formant pole locations and adjusts only the pole bandwidths. This technique is also used to compensate for the bandwidth underestimation problem due to quantization. Quantization generates audible noise in the reconstructed speech. The filter of Eq (4) has been applied as a postfilter to improve the overall quality of the synthesized speech and alleviate these effects [12]. The quantization process tends to sharpen formants and a motivation behind the noise shaping technique is to slightly increase the bandwidth prior to encoding. For $0 < \beta < \alpha < 1$, the filter provides a broadening of the formants and is a bandwidth expansion filter.

3. METHODS/RESULTS

None of the previously mentioned applications of bandwidth alteration have been applied for the loudness enhancement of speech. Based on our understanding of the critical band concept of hearing, we have reason to investigate the filter of Eq (4) for increasing speech loudness. For the following tests, a constant bandwidth expansion factor was applied to simply evaluate the subjective loudness and the ISO-532B loudness. In later experiments the bandwidth factor was made a function of tonality, using the Spectral Flatness Measure (SFM) [6] for bandwidth control, a compressive nonlinearity function was used to smooth the change of radius over time, and an approximation to the

ISO-532B was included. The approximation can be used with the tonality measure to control the level of bandwidth expansion such that regions are scaled to similar loudness.

The technique was applied on a real-time (frame by frame) basis. We used 10th-order LPC analysis with 32ms frame sizes, 50% frame overlap, per frame energy normalization, and overlap and add reconstruction. Filter states were preserved from each frame to the next and no sub-frame interpolation of coefficients was applied. Durbin's method with a Hamming window was used for the auto-correlation LPC analysis. All speech examples were bandlimited from 100Hz to 16KHz. Each frame was passed through the filter of Eq (4) with $\alpha = 0.8$ and $\beta = 0.4$ and reconstructed with the overlap and add method of Hamming windows. The bandwidth has been expanded for loudness enhancement to the point at which a change in intelligibility is noticeable but still acceptable.

4. LOUDNESS TEST

A subjective listening test of random words were selected from the TI-46 for presentation to the listener. The test consisted of 240 utterances ($fs = 10KHz$) at a comfortable listening level. Each subject listened to the speech utterances through Sony MDR-V200 padded headphones. The test took about 15 minutes for each of the 13 participants who were untrained in audiology.

The listening test was a GUI which presented the listener an option to select which of two words (of equal energy) is louder. One word was the original, the other, the modified with bandwidth expansion. To determine the potential dB gain improvement, a dB scaling of the modified words was transparently included in the test. The modified words were randomly scaled between -1 and -3dB (no weighting), and the user was not given information as to which word was modified or how much it was scaled. The results of these choices roughly determine by how many dB the bandwidth expansion technique can perceptually improve loudness. Fig (1) shows the crossover point (-2dB) at which, on average, a modified word could be scaled down to sound as loud as the original. We are perceptible to intensity variations greater than 1dB which corresponds to the just noticeable degree of noise modulation in the higher critical bands of human hearing [1]. The loudness listening tests reveal a perceived loudness gain of 2dB, with 95% confidence intervals.

A simple sensitivity screening test was also intermittently included to evaluate the listener's loudness resolution. Every so often two identical words were presented one of which was gain scaled to determine their just noticeable difference in loudness. Fig (1) also shows the average loudness resolution of the listeners. With headphones, results indicate they begin to detect a 0.5dB difference which increases as gain separation increases (at 0dB difference the results are 50% as expected).

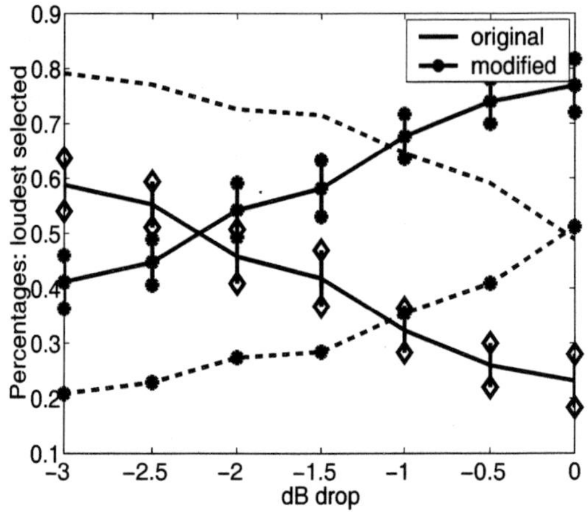

Fig. 1. Average dB crossover gain for 13 listeners in the subjective loudness listening test using the bandwidth expansion filter in Eq (4). Solid line shows the 2dB crossover for equal subjective loudness, and dotted line shows the subject sensitivity screening.

Fig (2) shows the ISO-532B loudness analysis of a TIMIT test sentence. Results are shown only for a constant bandwidth expansion and as a demonstration to show the change in perceptual loudness across the entire utterance. The energy level has been normalized on a per frame basis for both sentences which have equal power and extremely close ITU P.56 levels. We intentionally restricted loudness analysis to be between 100Hz to 4KHz. This range is typical of vocoders. The resulting loudness pattern demonstrates an improvement in the sone level of speech especially in the vowel regions using the basic technique. Similar ISO-532B loudness pattern improvements were found for most TIMIT sentences applied with technique.

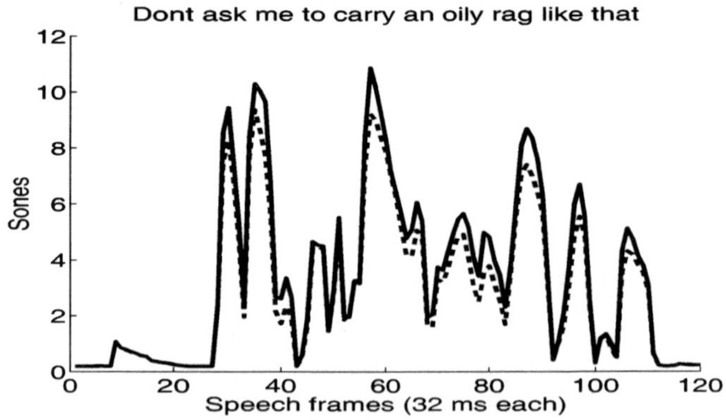

Fig. 2. ISO-532B loudness pattern of a TIMIT speech sentence. Original (dotted) and bandwidth expanded (solid).

5. INTELLIGIBILITY TEST

To address the question of intelligibility, we administered a variant of the Diagnostic Rhyme Test [13]. The DRT provides a way to measure the change in intelligibility between different speech processing systems. The DRT is a paired comparison test which isolates subjective differences in the processing of unvoiced sounds.

Table 1. Vocabulary of words used for Rhyming Test of Intelligibility, subdivided into confusable setsI-III.

I:	F, S, YES, X
II:	8, A, H, K
III:	B, C, D, E, G, P, T, V, Z, 3

An utterance in Table 1 is randomly selected along with another word from the same confusable set and either the enhancement algorithm is applied or not. To obscure the utterance, white noise is added to the processed or unprocessed utterance. The test consisted of 60 utterances at 0 dB SNR. The listener listened to the speech utterances through headphones at a comfortable listening level. The test took about 15 minutes for each of the 10 participants. Table 2 presents the results of the rhyme test in Table 1 for the combination of confusable sets I-III for 0 dB SNR. Results are presented as a recognition percentage and the $\pm$ bars represent a 95% confidence interval of the mean across listeners for each confusion set. Table 2 show slightly better intelligibility results with the bandwidth expansion speech enhancement procedure.

Table 2. Average intelligibility results of the rhyme test for 10 listeners hearing 60 words at 0 dB SNR. Table results are displayed as the percent correct P_N population mean

	Unprocessed	Processed	difference
I	86.4 ± 10.6	90.2 ± 6.9	$+3.8$
II	92.0 ± 7.3	91.2 ± 7.7	-0.8
III	82.4 ± 6.1	86.1 ± 8.5	$+3.7$
All	85.8 ± 3.7	88.1 ± 5.5	$+2.3$

6. CONCLUSIONS

We have shown that simple constant linear bandwidth expansion elevates loudness for clean speech without adding signal energy and without degrading intelligibility. This is supported by our informal subjective listening tests and ISO-532B analysis. Intelligibility results indicate that the technique does not compromise signal integrity in consonantal regions for clean conditions. Additionally, a tonality measure should be employed to target only vowel regions of speech for the bandwidth expansion procedure. Continued investigation is encouraged to reveal the relationship between loudness and intelligibility, and to what degree loudness can increase before intelligibility is sacrificed.

7. REFERENCES

[1] E. Zwicker and H. Fastl, *Psychoacoustics*, Springer Series, 1998.

[2] J.R. Dubno and M.F. Forman, "Effects of spectral flattening on vowel identification," *J. Acoustic. Sob. Am*, vol. 82(5), pp. 1503–1511, Nov. 1987.

[3] J.H. Chen and A. Gersho, "Adaptive postfiltering for quality enhancement of coded speech," *IEEE Trans. on Speech and Audio Proc.*, vol. 3(1), pp. 59–71, 1995.

[4] ISO-532, "Acoustics - method for calculating loudness level," *ISO Geneva, Switzerland*, 1975.

[5] E. Zwicker, H. Fastl, and C. Dallmayr, "Basic-program for calculating the loudness of sounds from their 1/3-oct band spectra according to iso 532 b," *Acustica, Letters to the editors*, vol. 55, pp. 63–67, 1984.

[6] D. Tsoukalas, J. Mourjopoulos, and G. Kokkinakis, "Speech enhancement based on audible noise suppression," *IEEE Trans. on Acoustic., Speech, and Signal Proc.*, vol. 5(6), pp. 497–514, Nov 1997.

[7] N. Virag, "Single channel speech enhancement based on masking properties of the human auditory system," *IEEE Trans. on Speech and Audio Proc.*, vol. 7(2), March 1999.

[8] L.R. Rabiner and B.H. Juang, *Fundamentals of Speech Recognition*, Prentice-Hall, Englewood Cliffs, NJ, 1993.

[9] S. McCandless, "An Algorithm for Automatic Formant Extraction using linear predictive spectra," *IEEE Trans. on Acoustic., Speech, and Signal Proc.*, vol. ASSP-22, pp. 135–141, 1974.

[10] J.D. Markel and A.H.Gray, *Linear Prediction of Speech*, Springer-Verlag, 1976.

[11] F.K. Soong and M.M. Sondhi, "A frequency weighted itakura spectral distortion measure and its application to speech recognition in noise," *IEEE Trans. on Acoustic., Speech, and Signal Proc.*, vol. 36(1), pp. 41–48, Jan 1988.

[12] R. Salami and et al, "Design and description of cs-acelp: A toll quality 8 kb/s speech coder," *IEEE Trans. on Speech and Audio Proc.*, vol. 6(2), pp. 116–130, 1998.

[13] W.D. Voiers, *Ch.34 Diagnostic Evaluation of Speech Intelligibility*, Dowden, Hutchinson, and Ross, Inc., 1977.

USABLE SPEECH DETECTION USING A CONTEXT DEPENDENT GAUSSIAN MIXTURE MODEL CLASSIFIER

Robert E. Yantorno, Brett Y. Smolenski, Ananth N. Iyer, Jashmin K. Shah
Temple University/ECE Dept. 12th & Norris Streets, Philadelphia, PA 19122-6077, USA
robert.yantorno@temple.edu, bsmolens@temple.edu, aniyer@temple.edu, shah@temple.edu
http://www.temple.edu/speech_lab

ABSTRACT

Speech that is corrupted by nonstationary interference, but contains segments that are still usable for applications such as speaker identification or speech recognition, is referred to as "usable" speech. A common example of nonstationary interference occurs when there is more than one person talking at the same time, which is known as co-channel speech. In general the above speech processing applications do not work in co-channel environments; however, they can work on the extracted usable segments. Unfortunately, currently available usable speech measures only detect about 75% of the total available usable speech. The first reason for this high error stems from the fact that no single feature can accurately identify all the usable speech characteristics. This situation can be resolved by using a Gaussian Mixture Model (GMM) based classifier to combine several usable speech features. A second source of error stems from the fact that the current usable speech measures treat each frame of co-channel data independently of the decisions made on adjacent frames. This situation can be resolved when a Hidden Markov Model (HMM) is used to incorporate any context dependent information in adjacent frames. Using this approach we were able to obtain 84% detection of usable speech with a 16% false alarm rate.

1. INTRODUCTION

In the field of audio restoration, such as the removing of clicks in gramophone recordings, it is common practice to first detect and remove the damaged portions of the signal and then interpolate the removed sections [1]. This approach has the advantage of only processing the damaged portions of the signal. It is interesting to note that this approach to audio restoration is commonly used when dealing with nonstationary interference, but has not yet been applied to the nonstationary interference in co-channel speech.

In addition, even for speech having little interference there exists several common situations like coughing, yawning, laughing, etc., that lie outside the training sets of typical speech processing applications. Further, most speech processing systems use feature vectors derived from physical models of the speech production process, such as the Linear Prediction Coefficients (LPC), which assume an all-pole model of the vocal tract [2]. However, for some classes of phonemes, such as nasals, it is known that the underlying production mechanisms are best described using pole-zero models [3]. It is unlikely that the above mentioned kinds of data would be very useful when input to a speech processing application, and hence, one would not want to processes these segments. The situation is similar to when the statistician identifies and removes outliers or when one chooses not to process low energy and unvoiced speech frames

The traditional approach to processing highly corrupted speech has been to enhance the speech while attenuating the interference [4]. However, recently a novel approach to co-channel speech processing has been proposed [5]. Like the audio restoration approach to click removal, the portions of the co-channel speech that are highly corrupted are first detected and removed. Within a co-channel utterance, where both speakers are contributing the same overall energy, there exist several segments of speech where one of the speakers is 15 dB or more above the other speaker [5]. It has been shown that when the target speaker is at least 15 dB greater than the interfering speaker, 80% reliable identification of the target speaker can be obtained [6]. Hence, these segments with a high Target-to-Interferer Ratio (TIR) may be considered usable with respect to speaker identification. The TIR was computed by taking the value, in dB, of the ratio of signal power to interferer power. Since for speaker identification it is not necessary to make a decision on every frame of data, the system can be implemented in a co-channel environment by extracting out and processing only the usable segments. Fortunately, current research has shown that about 35% of a co-channel utterance is usable speech [6].

Recent advances in co-channel speech processing have produced several usable speech measures, which yield some indication of the TIR [7] [8] [9] [10]. Such measures are necessary to determine usability in an operational environment, since *a priori* knowledge of the TIR would not be available. Unfortunately, currently available usable speech measures only detect about 75% of the total available usable speech. One reason for this high detection error stems from the fact that the measures treat each frame of co-channel data independently of the decisions made on adjacent frames. Another source of error stems from the fact that no single usable speech measure is capable of identifying all of the characteristics of usable speech [11]. It is the goal of this research to increase the performance of usable speech identification by combining several measures and making the classification process context dependant.

The system in Figure 1 (below) illustrates the approach taken. The features used in this research were Linear Prediction Coefficients (LPC) along with a linear discriminant (LD) based feature derived from the LPC residual. The features were then orthogonalized, to make them independent, and passed through a GMM classifier. Previously proposed approaches used the much less sophisticated classification techniques of nonlinear estimation and Quadratic Discriminant Analysis (QDA) on only two usable speech measures, which did not make use of contextual information [11] [12]. The decisions of the GMM are then passed trough a Maximum Likelihood Sequence Detector to determine the most probable sequence of usable and unusable states given the output of the output of the GMM.

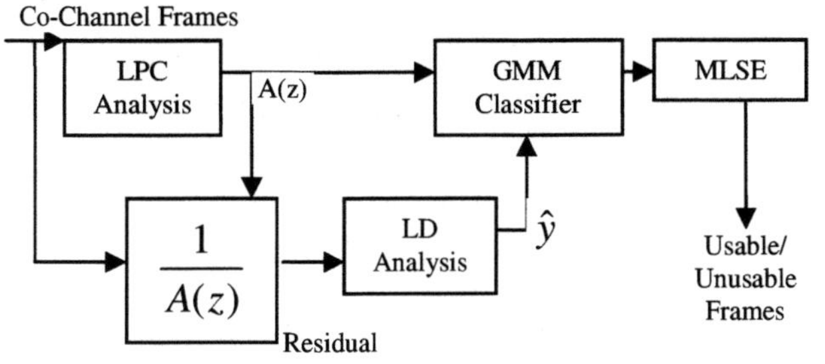

Figure 1: Block diagram of context dependent usable speech classifier.

Using a GMM type classifier is used, the desirable features are those that have a distribution well modeled by a mixture of Gaussians. Although the actual distribution of the LPC's of a speaker for a particular phoneme may not be Gaussian distributed, the estimate of them is [13]. When one includes the estimates of the LPC's across several phonemes, a mixture of Gaussians should result regardless of the orthogonalization stage. In addition, Linear Discriminant Analysis (LDA) is used on the LPC residual to yield an additional novel feature that incorporates any remaining useful information. Further research using additional nonlinear features having the above desirable properties is currently ongoing.

2. BACKGROUND

2.1 Linear Discriminant Analysis

Linear Discriminant Analysis (LDA) was used in an attempt to capture all the remaining information left in the LPC residual using one additional feature. The goal of linear discriminant analysis is to use a linear transformation to project the set of raw testing data vectors onto a vector space of lower dimension such that some metric of class discrimination is maximized [14]. The metric most often used is the ratio of the between class scatter (variance) to the within class scatter:

$$trace\{S_w^{-1} S_b\} \tag{1}$$

The result of this minimization for the two class (usable or unusable) problem is the following linear transformation (matrix equation):

$$\hat{y} = (\mu_1 - \mu_2)^T S_w^{-1} x \tag{2}$$

where the μ are the two mean vectors of the two class's data vectors and $\boldsymbol{x}$ is the data test vector [14]. The mean vectors and within-class-scatter matrix were estimated using the sample mean and sample variance of the training vectors.

This transformation produces the 1-dimensional feature $\hat{y}$ from the LPC-residual data frames, which for this research where 80 samples (10msec frame at 8kHz sampling rate) in length. Hence, the transformation is from $\Re^{80}$ to $\Re$. Since the feature generated by this approach is a linear combination of a large number of independent identically distributed random variables, the feature's probability distribution will be highly Gaussian regardless of the distribution of $\boldsymbol{x}$ [15]. Exploring other metrics as well as nonlinear transformations is currently ongoing.

2.2 Gaussian Mixture Model

Formally a random vector $\boldsymbol{x}$ that is described by a mixture of M Gaussians has a probability density function of the form:

$$f(x) = \sum_{i=1}^{M} \lambda_i N(\mu_i, \Sigma_i) \tag{3}$$

where the $N(\mu_i, \Sigma_i)$ are multivariate Gaussian distributions having mean vector μ_i and covariance matrix Σ_I [16]. The λ_i sum to one and indicate the relative weight of each Gaussian component in the mixture. It can be shown that any distribution can be approximated with arbitrary precision using a mixture of enough Gaussians [14].

To obtain the parameters λ_i, μ_i, and Σ_i, the Expectation Maximization (EM) algorithm was used, which is an iterative implementation of maximum likelihood estimation using incomplete information about the underlying probability distributions [16]. Eight mixture components (M=64) were used, since this amount produced the lowest detection error.

In general, each covariance matrix in the mixture contains N^2 elements (where N is the dimension of the feature vector) that need to be estimated. If the features are chosen such that they are independent, than all the off-diagonal elements of the covariance matrices will be zero [15]. Hence, one would like to use independent features, since only N parameters would need to be estimated.

2.3 Hidden Markov Model

In order to make the classifier context dependent, it would be helpful to use a statistical model that exploits as much *a priori* information about the TIR as possible [14]. One challenge regarding this is that the segmental TIR process is a nonstationary process. To model the non-stationary aspects of the TIR, the following HMM is proposed, Figure 2 (below).

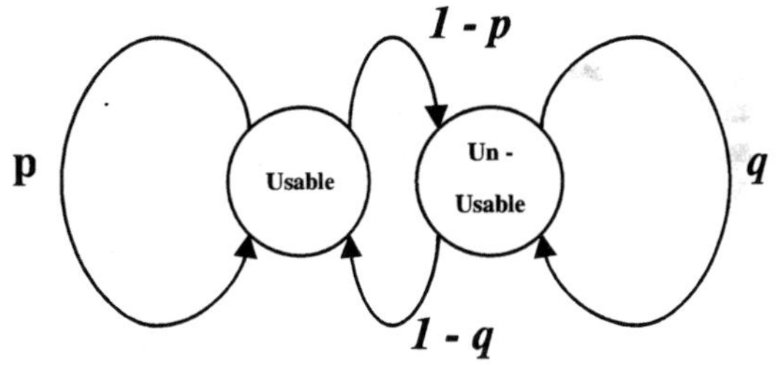

Figure 2: State diagram of the HMM process of co-channel speech frames.

We say the model is 'hidden' because one cannot observe the actual states, just the statistical characteristics of the signal for a particular state [17]. For the usable state, one person is talking with little interference. For the unusable state, both talkers are contributing about equal energy. Hence, the transition probabilities of this process are related to the statistics of the silent portions of speech. Each state corresponds to a 40ms frame of the co-channel signal and, hence, the signal is quasi-stationary in this time interval [2].

The state transition matrix $\boldsymbol{T}$ of this process is:

$$T = \begin{bmatrix} p & 1-p \\ 1-q & q \end{bmatrix} \tag{4}$$

where p is the probability of the next frame being usable given that the current frame is usable, and q is the probability of the next frame being unusable given that the current frame is unusable. These probabilities were estimated using the measure's training data. One can notice that this model will only make use of dependence between adjacent frames. Fortunately, current research has shown that little dependence exists between anything but adjacent frames [5]. Using the state transition matrix in conjunction with the celebrated Forward-Backward algorithm it is possible to determine the maximum likelihood sequence of states given the output of the GMM classifier [17].

3. METHODS

For this research 10 single speaker utterances, 5 male and 5 female, were taken from the TIMIT speech corpus. These 10 utterances were used to form a co-channel speech database of 45 co-channel files (10 choose 2 = 45). The files were down sampled to 8kHz and the longer file in each pair was truncated to make both files the same length. The files were then combined at 0 dB overall TIR to form the co-channel utterance. To control the variability and eliminate any bias between the dialect regions, only one dialect region was used (region 1 of the TIMIT data base).

It should be noted that in an operational environment it is highly unlikely that two speakers would be talking over each other during the entire utterance. In addition, each utterance would not have exactly the same length or have the same energy. The reason for using this approach was to capture the worst possible scenario, with respect to both speakers, that one could expect in a co-channel environment.

Once the co-channel utterance was formed it was broken down into 40 ms frames with no overlap, since it has been demonstrated that speaker identification reliability has little dependence on overlap [6]. For each frame, the values for the features, TIR, signal energy, and spectral flatness were computed. Signal energy and spectral flatness were necessary in order to exclude silence and unvoiced frames, since usable speech measures would not be used with these frames. Usable speech measures are designed to work with only voiced speech, since unvoiced frames provide little information useful for speaker identification [6].

Training data was used for obtaining the parameters of the GMM classifier and MLSE detector. Once these models were obtained, it was possible to use testing data to classify what frames of the co-channel speech were usable ($|TIR| > 15dB$) and those frames that were unusable ($|TIR| < 15dB$). The absolute value is necessary, since usable speech can come from both speakers. Half of the 45 co-channel speech files (22) were randomly selected to train the system. The remaining half (23 co-channel files) were used for testing.

4. RESULTS

Figure 3 (below) shows the classification results for the APPC measure alone, quadratic discriminant analysis (QDA) classifier using APPC and SAPVR-residual measures as features, context independent GMM classifier using the LPC-based features, and the context dependent GMM classifier. Since the minimum probability of error criterion is used in determining the decision boundary surface of the classifier, the percent of Misses (%Misses = 100% - %Hits) equals the percentage of false alarms. However, one can easily choose to weight the false alarms differently than the misses and obtain a different decision boundary surface. The context dependent GMM classifier was able to obtain 84% detection of usable speech with a 16% false alarm rate. This amounts to a 38% reduction in total detection error over the APPC measure alone.

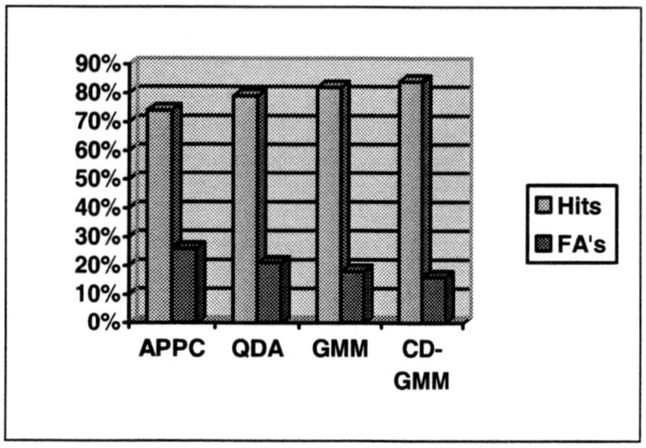

Figure 3: Percent hits and False Alarms (FA) for APPC measure, QDA classification using APPC and SAPVR-residual measures, context independent GMM classifier using LPC-based features, and context dependent (CD) GMM classifier.

5. FURTHER RESEARCH

More usable features need to be developed. Some current candidates include using pole-zero parameters, sinusoidal model parameters, as well as, nonlinear features of the vocal tract such as those derived from the Teager energy operator [18]. Further, parameters derived from the glottis such as the Liljencrantz-Fant model of the glottal flow derivative, may give an indication of the quality of the speech [19]. In addition, the use of other classification techniques such as Support Vector Machines (SVM) has yet to be explored.

The current approach to usable speech segmentation is to partition the signal into short fixed length frames with no overlap. Segmentation is always necessary, since the speech signal is nonstationary [2]. However, a more intelligent approach to segmenting the speech signal would be to identify the stationary regions in the speech signal and process those entire segments. Iterative feature extraction and sequential usable speech detection should improve the resolution capabilities of the classifier as well as make it context dependent by default.

Also, usable speech can be defined with respect to the intended application, as opposed to the TIR value, by studying what types of frames work with the system. In addition to improving speaker identification systems, several other applications of usable speech are currently under development including a speaker count and speaker separation system [20].

ACKNOWLEDGEMENT

This effort was sponsored by the Air Force Research Laboratory, Air Force Material Command, and USAF, under agreement number F30602-02-2-0501. The U.S. Government is authorized to reproduce and distribute reprints for Government purposes notwithstanding any copyright annotation thereon.

DISCLAIMER

The views and conclusions contained herein are those of the authors and should not be interpreted as necessarily the official policies or endorsements, either expressed or implied, of the Air Force Research Laboratory, or the U.S. Government.

6. REFERENCES

[1] S. J. Godsill and P. J. W. Rayner, *Digital Audio Restoration: A Statistical Model Based Approach*, New York: Springer, 1998.

[2] L. R. Rabiner and R. W. Schafer, *Digital Processing of Speech Signals*, Englewood Cliffs, NJ: Prentice-Hall, 1978.

[3] D. O'Shaughnessy, *Speech Communications: Human and Machine,* New York: Institute of Electrical and Electronics Engineers, 2000.

[4] J. S. Lim, ed., *Speech Enhancement*, Englewood Cliffs, NJ: Prentice-Hall, 1983.

[5] R. E. Yantorno, "Co-Channel Speech Study", Final Report for Summer Research Faculty, Sponsored by AFRL/IF Laboratory, Rome, NY. 1999.

[6] J. Lovekin, R. E. Yantorno, D. S. Benincasa, S. J. Wenndt, and M. Huggins, "Developing Usable Speech Criteria for Speaker Identification", ICASSP 2001, pp. 421-424, May 2001.

[7] K. R. Krishnamachari, R. E. Yantorno, D. S. Benincasa, and S. J. Wenndt, "Spectral Autocorrelation Ratio as a Usability Measure of Speech Segments Under Co-channel Conditions", IEEE International Symposium on Intelligent. Signal Processing and Communication Systems, November 2000.

[8] J. Lovekin, K. R. Krishnamachari, R. E. Yantorno, D. S. Benincasa, and S. J. Wenndt, "Adjacent Pitch Period Comparison (APPC) as a Usability Measure of Speech Segments Under Co-channel Conditions", IEEE International Symposium on Intelligent Signal Processing and Communication Systems, November 2001.

[9] A. R. Kizhanatham, R. E. Yantorno, S. J. Wenndt, "Co-channel Speech Detection Approaches Using Cyclostationarity or Wavelet Transform", 4th IASTED International Conference on Signal and Image Processing, July 2002.

[10] N. Chandra, R. E. Yantorno, D. S. Benincasa, and S. J. Wenndt, "Usable Speech Detection Using the Modified Spectral Autocorrelation Peak-to-Valley Ratio Using the LPC residual", 4th IASTED International Conference on Signal and Image Processing, July 2002.

[11] B. Y. Smolenski, R. E. Yantorno, and S. J. Wenndt, "Fusion of Co-Channel Speech Measures Using Independent Components and Nonlinear Estimation", IEEE International Symposium on Intelligent Signal Processing and Communication Systems, November 2002.

[12] B. Y. Smolenski, R. E. Yantorno, and S. J. Wenndt, "Fusion of Usable Speech Measures Quadratic Discriminant Analysis", IEEE International Symposium on Intelligent Signal Processing and Communication Systems, December 2003.

[13] S. M. Kay, *Fundamentals of Statistical Signal Processing*, Englewood Cliffs, NJ: Prentice-Hall, 1998.

[14] S. Theodoridis and K. Koutroumbas, *Pattern Recognition*, San Diego, CA: Academic Press, 1999.

[15] H. Stark and J. W. Woods, *Probability, Random processes, and Estimation Theory for Engineers*, Englewood Cliffs, NJ: Prentice-Hall, 1994.

[16] G. J. McLachlan and K. E. Basford, *Mixture Models: Inference and Applications to Clustering*, New York, NY: M. Dekker, 1988.

[17] X. D. Huang, Y. Ariki, and J. A. Mervyn, *Hidden Markov Models for Speech Recognition*, Edinburgh: Edinburgh University Press, 1990.

[18] T. F. Quatieri, *Discrete-time Speech Signal Processing: Principles and Practice*, Upper Saddle River, NJ: Prentice-Hall, 2002.

[19] D. G. Childers, *Speech Processing and Synthesis Toolboxes*, New York: John Wiley, 2000.

[20] B. Y. Smolenski, R. E. Yantorno, D. S. Benincasa, and S. I. Wenndt, "Co-Channel Speaker Segment Separation", ICASSP, May 2002.

DECOMPOSITION AND RECOGNITION OF A MULTI-CHANNEL AUDIO SOURCE USING MATCHING PURSUIT ALGORITHM

David B. Bjornberg, Sedig Agili and Aldo Morales

Electrical Engineering Program
Penn State University at Harrisburg
Middletown, PA 17057

ABSTRACT

Performing real time note detection of multiple audio sources is a difficult task due to the complex nature of audio signals. Typically notes are close in pitch and the harmonic overtones of each note are strongly interlaced, thus preventing standard filtering techniques. In this paper, we develop signal-optimized harmonic atom matching pursuit algorithm, which will be used to decompose an audio signal in terms of elementary waveforms called harmonic atoms. Matching pursuit is an iterative algorithm that correlates the signal against a dictionary set of Gabor atoms and searches for the atoms that exhibit the highest level of correlation. The resulting matched atoms allow for accurate and efficient detection of a single note or multiple notes played simultaneously. In this paper prior knowledge of the signal is used to calculate a parameter set to avoid a costly search over the complete dictionary. The parameter set defined for our proposed algorithm includes time-location, decay rate, frequency, scale and phase, which are calculated at the onset of each note played. The onset (transient) is detected by using a fourth order digital difference then an appropriate interval and optimal window atom is selected. This optimized algorithm is demonstrated through synthesized analysis example. The application of this algorithm in multi-source audio signal can also be demonstrated.

1. INTRODUCTION

Audio signals are composed of two superimposed structures, transients and stationary parts. Transients are rapid changes in the audio signal and represent the dynamics of the signal. Stationary parts are audio signals that remain constant for long durations. Stationary parts are ideal for analysis, while transient structures add to the complexity of the analysis. For example, a guitar audio signal, which is governed by the vibration on a finite length string, is not stationary since the vibration is not of infinite duration. In addition, the signal is composed of excitation transients, which are caused by the initial deflection of the string.

In order to gain insight on signal characteristics a guitar signal was recorded and then analyzed using MATLAB. The signal of a single plucked guitar note is shown in Figure 1. Figure 1a illustrates the start of the signal at approximately 0.35 seconds and shows that it is roughly sinusoidal. Figure 1b shows that the envelope decay is exponential in nature. In addition, Figure 1a illustrates the excitation transient, where this transient represents the onset of the signal after string deflection. This rapid magnitude change does provide a unique feature that can be used later to initiate the optimized matching algorithm.

It is commonly known that musical instruments have a unique and complex harmonic distribution. For example if two close notes are played simultaneously and if a standard filtering technique were used to identify which signals were present, the filter would need to be able to discriminate down to the tens of Hertz, which is not practical.

2. THE MATCHING PURSUIT ALGORITHM

Matching pursuit [1-3] is an iterative algorithm for deriving signal decompositions in terms of elementary functions. In order to decompose complex signals, the matching pursuit algorithm first implements a specialized form of an elementary function called the Gabor atom. The advantage of using Gabor atoms is that they are ideal for representing complex non-stationary signals. The generic Gabor atom is introduced below in equation (1).

$$g_{(s,u,\xi)}(t)=\frac{1}{\sqrt{s}}w\left(\frac{t-u}{s}\right)e^{i2\pi\xi(t-u)} \tag{1}$$

A set of Gabor atoms is commonly referred to as the dictionary. Thus, the generic Gabor atom in equation (1) is indexed through the dictionary by a specific parameter set (s-scale, u-time shift, ξ-frequency). The windowing function $w\left(\frac{t-u}{s}\right)$ shown in equation (1), is used to take into account the time dependence of the audio signal.

The concept of the matching pursuit algorithm is to correlate an input signal with a dictionary set of expansion functions, which are based on Gabor atoms. Once the ideal window is selected, the matching pursuit algorithm identifies the Gabor atom that exhibit the highest degree of correlation with the input signal. When the optimal match atom is identified, that atom will be subtracted from the original signal producing a new signal called the residual. The residual is then again correlated to the remaining dictionary atoms and this process continues until the amplitude of the residual signal is below a specified threshold. This process can be expressed by the following equation.

$$s(t) = \sum_{m=1}^{M} \lambda_m g_m(t) + R_m(t) \quad (2)$$

In (2) the input signal $s(t)$ can be expressed as a summation of λ_m scaled of M Gabor atoms $g_m(t)$ with the addition of a residual term $R_M(t)$. In order to account for a full range of expansion functions, the dictionary must be defined for all the possible Gabor atoms. Thus, one of the problems of this algorithm is the intense computational demands. Typically the dictionaries are over-complete since a generic matching pursuit algorithm assumes no previously known information about the signal as described in [1-2].

3. HARMONIC ATOM BASED MATCHING PURSUIT METHOD

A specialized algorithm that offers increased computational efficiency is the harmonic atom based matching pursuit algorithm [3]. This method takes advantage of the harmonic structure of a given audio signal. As described previously each note will correspond to a specific set of harmonics. With this concept in mind, the Gabor set of atoms can be used to create a new set of harmonic atoms defined as follows.

$$h(t) = \sum_{k=1}^{K} c_k g_{(s,u,\xi_k)}(t) \quad (3)$$

As shown in equation (3) the harmonic atom is an extension of the Gabor atom. The harmonic atom $h(t)$ represents a K length series of Gabor atoms. This series summation exhibits an increasing frequency component ξ_k. The harmonic structure is approximately $\xi_k \approx k\ \xi_o$, where k represents the k'th harmonic of the signal. Since the damping coefficient is equal to the decay rate of the signal, this value can be easily found for a given input signal. Once this value is determined our proposed algorithm is able to optimize the harmonic atom's structure to match the effects of various decay (damping) levels.

In essence, this method uses the new harmonic atom structure to subdivide the standard Gabor atom based dictionary into a smaller harmonic sub-dictionary. The algorithm operation is very similar to the standard Gabor atom based matching pursuit, except that now the dictionary is based on harmonic atoms. Harmonic decomposition results in the following signal expansion [3]:

$$s(t) = \sum_{m=1}^{M} \lambda_m h_m(t) + R_M(t) = \sum_{m=1}^{M}\sum_{k=1}^{K} c_m g_{(m,k)}(t) + R_M(t) \quad (4)$$

As shown in equation (4) the signal is decomposed in terms of harmonic atoms plus a residual, which can also be expressed as a double summation using Gabor atoms. The advantage of this technique is an increase in computational efficiency; since the recognition of a single fundamental frequency will also, by association, recognize the corresponding harmonics. Thus, as stated in [3], this method is ideal for real time audio note detection.

4. IMPLEMENTATION

We optimized the harmonic atom based matching pursuit algorithm for the specific case of a guitar audio signal. As discussed guitar audio signals are not time stationary, therefore, for efficiency purposes, an exponential windowing function was used to damp a series of sinusoids. The next important signal parameter is the location of the transient samples. It was found that by using fourth order digital difference the transient points were quickly identifiable. Now that the transient locations are identified, the algorithm continues by deriving the optimum windowing function for each interval. This can be a difficult process since the signal decay rate is nonlinear. The technique implemented starts by taking the absolute value of the signal and then applying a digital low-pass filter. The low-pass filtering operation was implemented using MATLAB and it is based on a Butterworth filter design. The algorithm then utilizes an iterative approach at determining the appropriate windowing functions. Since iterative calculations are computationally demanding, it is advantageous to minimize the number of calculations. This can be done by introducing a minimized interval processing. Recall that the transient samples have been located, thus for the most accurate windowing function derivation one would use the full interval between transient samples to derive the windowing function. However, it is not necessary to use the full interval between transients; it was found that accurate results could be achieved by using a smaller interval. For this research an interval size of five thousand samples was used which corresponds to a 0.227 second time interval, with a sampling frequency of 22,050 Hz.

The next concept is to pick the appropriate frequency atom. A Short Time Fourier Transform (STFT) is performed on this interval. Fortunately, audio signals do not exhibit sub-harmonics; this implies that the first spectral peak from the STFT analysis is always a fundamental frequency. This concept is profound since the algorithm does not need to perform an iterative approach to decide the appropriate frequency atom, thus increasing the algorithm's efficiency. It was decided to use three Gabor atoms to form the harmonic atom set, since beyond the second harmonic the frequency domain magnitude was greatly diminished. The algorithm continues by determining the appropriate scale. Since STFT was performed, the algorithm takes note of the relationship in magnitude between the fundamental frequency and two of its corresponding harmonics. The approach used is iterative in nature and takes into account the magnitude relationship between the harmonic atoms. Again the iterative algorithm compares a list of possible amplitude coefficients with the signal and decides the best fit. The resolution of the amplitude coefficient is very important, since it has a direct effect on all three Gabor atoms. Fortunately, it is possible to reduce the size of the possible coefficients by understanding that the coefficient only needs to be positive, since the algorithm will derive the necessary phase component at a later time. In addition, the analyses of real guitar signals that are recorded in a wav format have a limited range of ± 1; furthermore this range could be reduced more by the recording process. The next issue is deriving the appropriate phase coefficient for each Gabor atom of the harmonic atom set. Because of the complex nature of the standing waves set upon a string there is no guarantee that the various frequency signal components be in phase. Thus, it is necessary to derive the optimum phase coefficient for each Gabor atom of the harmonic atom set. Again an iterative approach was used to determine the proper phase coefficient. It was found that determining the phase of the fundamental frequency is simpler than determining the phase of the harmonics. Thus, as the phase coefficients are found they are used to optimize the pursuit of the harmonic's phase coefficient. While computationally demanding the phase coefficient has a major effect on the optimization of the harmonic atom set. The effect of an improper phase coefficient is very apparent in the decomposition process, since it becomes difficult to remove the desired signal components. The above described algorithm is shown in a flow chart in Figure 4.

5. SYNTHESIZED SIGNAL ANALYSIS EXAMPLE

To test the proposed algorithm, we generated a synthesized signal. The sinusoids used for synthesis were of various amplitudes, composed of three fundamental frequencies and two corresponding overtones and of various decay rates. As shown in Figure 2 the synthesized signal consists of three distinct intervals, notice the sharp excitation transients and the various decay rates. In order to visualize the decomposition process, Figure 3 depicts the first pass of the decomposition performed on the third note of Figure 2. The interval processing of the third transient of figure 2 is shown in Figure 3a along with its associated STFT spectrum in Figure 3b. The algorithm yields the first optimized harmonic atom as shown in Figures 3c and 3d. The decomposition continues by subtracting the harmonic atom from the original signal, which yields the residual signal as shown in Figures 3e and 3f. Time domain comparisons of Figures 3a and 3e illustrate a reduction in amplitude but little more information. The decomposition process is more accurately monitored from the STFT analysis, which is illustrated when comparing Figures 3b and 3f. The harmonic atom selected has a fundamental frequency of 550 Hz, and corresponding harmonics at 1100 Hz and 1650 Hz as shown in Figure 3d. Notice the correlation in magnitude between the harmonic atom and the selected frequencies of the original signal. The residual signal STFT shown in Figure 3f illustrates the significant reduction in magnitude at the fundamental frequency of 550 Hz. The reduction in magnitude is approximately one third of the original signal. Further reduction can be achieved by increasing the resolution of the scale and phase derivation, with an undesired trade-off of efficiency.

6. CONCLUSIONS

While matching pursuit algorithms are very computationally demanding, they allow for accurate and compact signal decomposition of complex time-dependant signals. We proposed an efficient implementation of harmonic based matching pursuit algorithm, first transient recognition which yielded the interval processing technique was obtained. Then the derivation process of the harmonic atom parameter set was achieved. The concept of algorithm resolution was a key issue during the parameter derivation. The resolution can be used to balance the algorithm's accuracy versus efficiency. The algorithm was demonstrated through synthesized signals. However, the application of this algorithm in multi-source audio signals can also be demonstrated.

7. REFERENCES

[1] Goodwin, M., "Matching pursuit with damped sinusoids", *Acoustics, Speech, and Signal Processing IEEE International Conference,* April 1997.

[2] Goodwin, M.; Vetterli, M., "Atomic decomposition of audio signals", *Applications of Signal Processing to Audio and Acoustics, IEEE ASSP Workshop,* October 1997.

[3] Gribonval, R.; Bacry, E., "Harmonic decomposition of audio signals with matching pursuit", *Signal Processing, IEEE Transactions on , Volume: 51 Issue: 1*, Jan 2003.

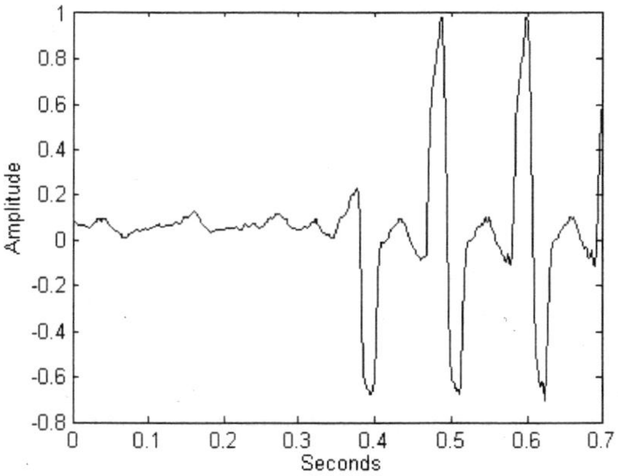

Figure 1a Guitar signal transient

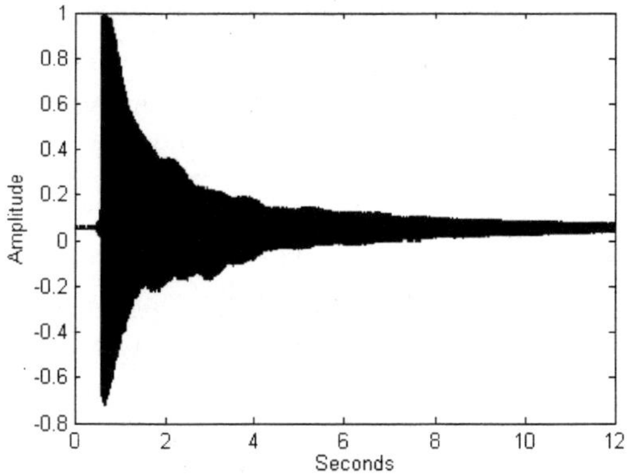

Figure 1b Guitar signal envelope

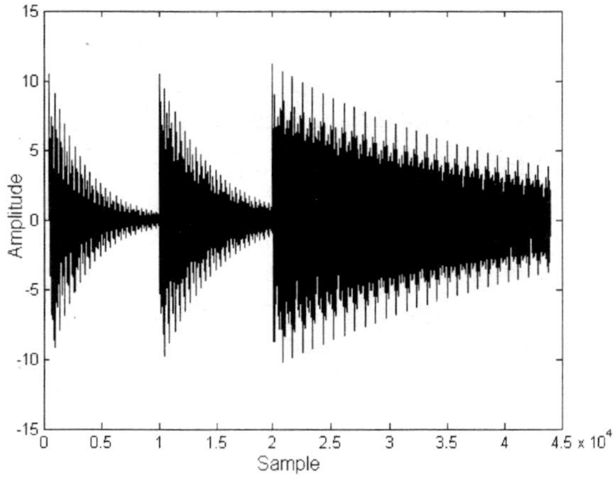

Figure 2 Synthesized Signal Envelope

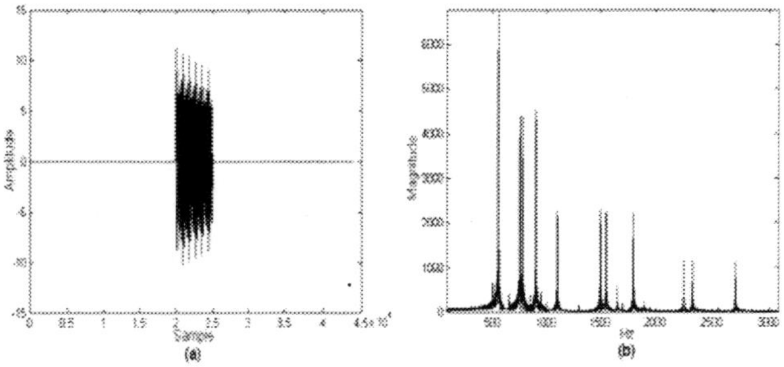

Figure 3: (a) Original Signal, (b) Original Signal STFT

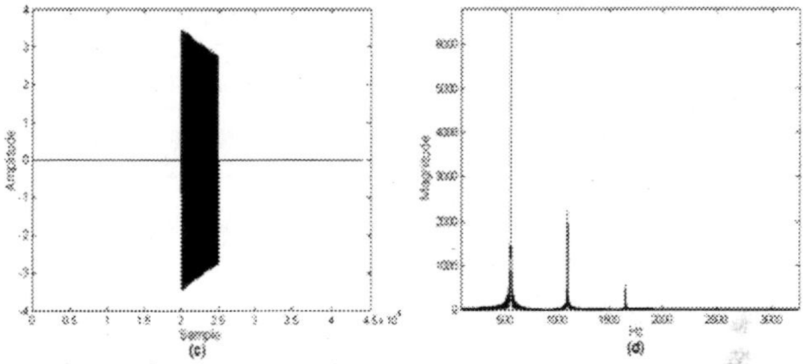

(c) Harmonic Atom, (d) Harmonic Atom STFT

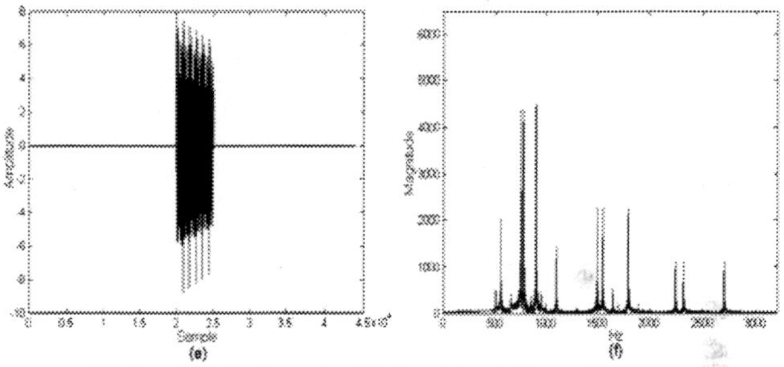

(e) Residual Signal, (f) Residual Signal STFT

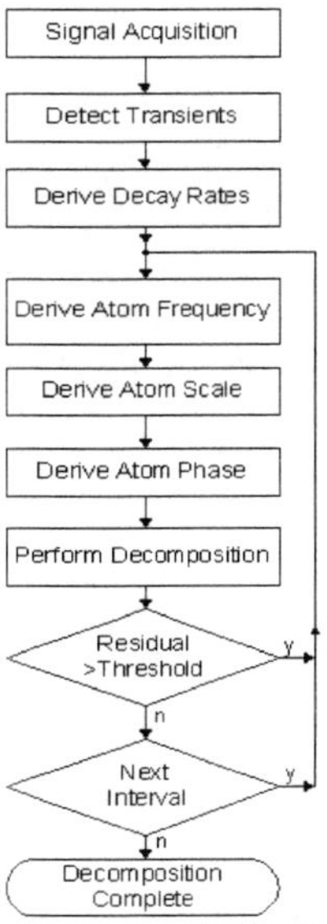

Figure 4: Flow chart for the optimized algorithm

MIXED SIGNAL ASPECTS OF BEHAVIORAL MODELING AND SIMULATION

Gabriel Popescu and Leonid B. Goldgeisser

Synopsys Inc
2025 NW Cornelius Pass Rd.
Hillsboro, OR 97124

ABSTRACT

System level designs incorporate digital analog and considerable mixed signal components. With systems growing in size and complexity, AMS HDLs become increasingly important for modeling and simulation of entire mixed signal, mixed technology systems. The size of those systems often becomes prohibitively large for analog simulation. Moreover, strong nonlinear behavior shown in most complex systems makes their accurate modeling and efficient simulation a skillful art.

In this paper we analyze some important aspects of mixed signal simulation and present possible resolutions for issues raised. The focus is on the effect of HDL definitions on simulation behavior.

1. INTRODUCTION

The recent development of analog and mixed signal (AMS) HDLs in the area of system level designs enabled modeling of entire systems using a single language with subsequent simulation in a single simulator. This integration of two different domains (Analog and Digital) gives the user a powerful and flexible tool to verify complex nonlinear behavior of vast systems [1], [2]. Skillfully modeled and simulated, AMS can combine the advantages of both domains: its efficiency can reach the performance of digital simulation while its accuracy is comparable in areas of interest to analog simulation. However, simulating an entire sophisticated system using a complex tool is quite challenging and requires a keen understanding of hardware description language features, capabilities and limitations. In recent years there were several publications dedicated to AMS HDLs and associated issues ([3], [4] and [6] to name just a few). Some papers discuss issues described in the current paper (most notably the AMS initialization issues discussed in [5]) however our main interest is to show what problems one can expect from combining working (in digital context) digital models and working (in analog context) analog models into a single system (and suddenly discovering that the AMS simulation results are not that was expected). Some of the discussion remains valid in multilingual context. We show what the issues are and propose our approach for resolving them.

2. MIXED SIGNAL SYSTEM EXAMPE

Figure 1 represents a typical example of circuits commonly dealt with by an AMS simulator. Such circuits generally include a considerable digital part, usually IC as well as extensive analog part. The analog part could be implemented in electrical, mechanical, thermal (or in a combination thereof) or in any other technology.

Figure 1 shows an audio system model implemented in VHDL-AMS. The audio system includes mixed-signal (analog and digital) IC's for clock generation and A-to-D conversion, a DSP algorithm section for response leveling and sound effects, board level analog electronics for filtering and amplification, as well as a mixed-technology loudspeaker with mechanical non-linearities and resonance characteristics. Using this example one can verify the entire system performance. There are several challenges one must overcome with this model. Most of those challenges result from combining analog and digital parts together. In the next section we will discuss some of those issues.

3. MIXED SIGNAL ISSUES

Major difficulties in combining the digital and the analog part in one model arise from the following three differences:

- differences in signal representation (i.e. discrete vs. continuous)
- separate time point selection and synchronization,
- differences in the simulation paradigm.

3.1. Digital and Analog signal representation.

Digital parts of the circuit use discrete variables, while analog parts use continuous variables. This creates a requirement to convert discrete signals into continuous signals (e.g. voltages and currents) and vice versa when connecting those parts together. The HDL definition specifies a set of basic tools to accomplish this conversion.

3.2. Independent Time Steps.

For performance reasons, analog and digital time points should not be required to be identical since it is too expensive to force an analog solution for each digital step. The analog solution should be found only when necessary. However, at certain points digital and analog parts have to exchange information.

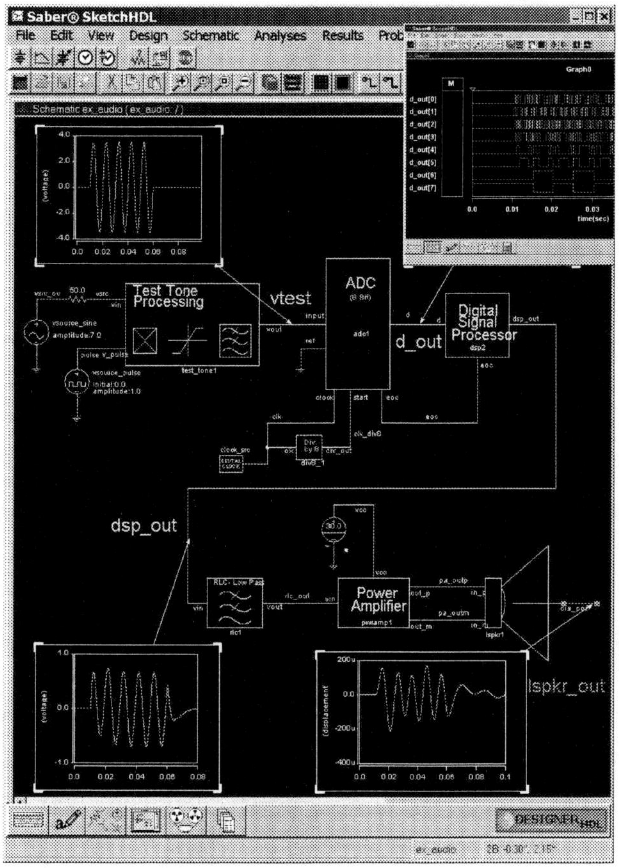

Figure 1: Mixed Signal Audio Example.

3.3. Different simulation paradigms.

There are differences in design and verification approaches for analog and digital domains. One of the examples of such differences is the Operating Point or DC analysis. It is a very well established concept in the analog world while on the digital side the concept of DC does not even exist. In many cases, the designer would like to obtain the operating point of the simulated circuit, but inadequate modeling of digital components might prevent this. In the next several sections we are going to use a relatively simple example to demonstrate the issues described above.

4. MIXED SIGNAL LANGUAGE CONCEPTS.

4.1. VHDL-AMS simulation cycle

Figure 2 illustrates a simplified concept of a VHDL-AMS simulation cycle [1]. The inner loop is called Delta *Cycle* and it corresponds to digital signal propagation and/or finding an analog solution at the point. (This is different from analog iterations, such as Newton-Raphson iterations. Such analog iterations are part of the analog solution). Delta cycles are necessary because processes can update signals that affect other processes or the analog part. New analog solution can in turn trigger an execution of a sensitive process. The outer loop represents the time advancing loop.

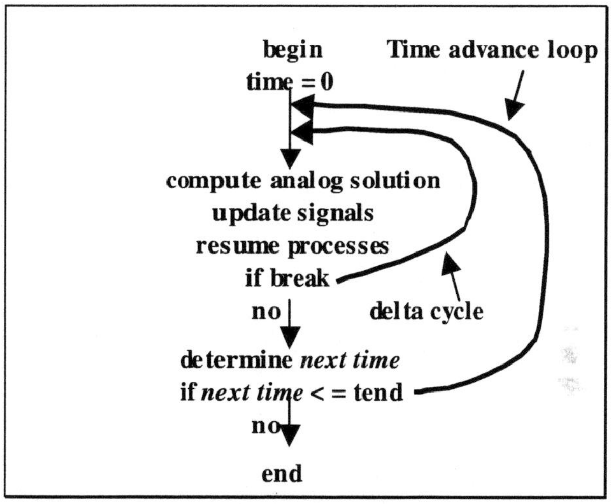

Figure 2 VHDL-AMS Simplified Simulation Cycle.

To distinguish between Operating Point, Transient and Frequency Analyses, VHDL-AMS introduces a special signal DOMAIN. Other HDL's use similar approaches. The value of this signal for those analyses is QUIESCENT_DOMAIN, TIME_DOMAIN and FREQUENCY_DOMAIN respectively.

For a pure analog circuit the quiescent solution is identical with the DC solution. All the independent source values are fixed for time equal zero. All derivatives with respect to time in the model's equations are set to zero.

Since digital designers usually are not interested in DC, they often even do not incorporate an accurate DC model (or even any model at all). The result might be that in a mixed signal circuit the quiescent solution is ill-defined.

Additional, yet similar problems appear in the case of multilingual designs where the models have to be carefully designed with respect to the simulation cycle used for the analysis in order to cover the desired behavior. In such cases a language specific simulation cycle or a unified simulation cycle can be used. The final solution should be the result of a compromise between the generality of the models and the specifics of the simulation scenario. The analysis of this solution is out of the scope of this paper.

4.2. Mixed Signal Interfaces.

Most issues discussed in this paper are caused by analog to digital (A/D) and digital to analog (D/A) interfaces.

The purpose of those interfaces is to accommodate the continuous time equations with the discrete time equations.

The function of the D/A interface is two-fold:

- It has to transform a discrete digital value into a continuous analog one.
- It has to manifest the discontinuity to the analog solver. This means controlling the analog time step and the integration algorithm at the time of discontinuity.

The A/D interface is essentially a threshold crossing involving a waveform and a reference. Its major responsibility is to detect the time of the crossing within acceptable accuracy and manifest proper signal sensitivity to the digital part.

Even though all AMS HDLs offer similar capabilities in terms of A/D and D/A interfaces, their implementation and the level of control they offer to the user is quite different. In VHDL-AMS the D/A interface is performed via the *break* statement, which indicates discontinuity for the analog solver. The *break* statement is the most fundamental D/A interface and it can include the initial conditions specification to facilitate convergence. *'ramp* and *'slew* signal attributes defined in VHDL-AMS use the *break* statement and are discussed later. The definition of the D/A and A/D interfaces is illustrated in **Figure 3.** Graph a of **Figure 3** shows a digital signal **D**, that has an event at time t1. The analog quantity **A** in **Figure 3.b** depends on the signal **D**. Therefore, in order to obtain an accurate waveform of **A**, the analog solver must find solution at **X** and **Y** (i.e. t1- and t1+). The event on the digital side creates the *possibility* of a discontinuity on the analog side. Since it is way to expensive to (re)calculate the analog solution every time any digital signal changes its value, it is a responsibility of the model writer to indicate the discontinuity via *break.* Missing a discontinuity might result in an erroneous solution as shown by the dashed line in ***Figure 3*.b**. An example of a *'ramp* attribute is shown in **Figure 3.c**. Here, analog value **R** follows the instant change in **D** with finite slope. *Break, 'ramp* and *'slew* are only basic D/A tools offered by VHDL-AMS. The user can use them to build much more elaborate D/A interfaces addressing arbitrary requirements.

For the A/D interface the language feature is the attribute *'above*, illustrated in **Figure 3.d** and e. In this case the comparison between the values of the quantities X and ref schedules transactions on the signal S of type Boolean.

In addition various tools can perform automatic insertion of mixed signal interfaces, thus making them transparent to the user [2], [3]. User defined mixed signal interfaces can be implemented as separate objects, referred to here as "interface models", or as a part of a different model.

In the second approach, A/D and D/A interfaces are included in the models they allow an accurate modeling of inherent inertial behavior in the real world. In this approach, physical devices can have several models depending on where the model is used. This approach is illustrated in the following case study.

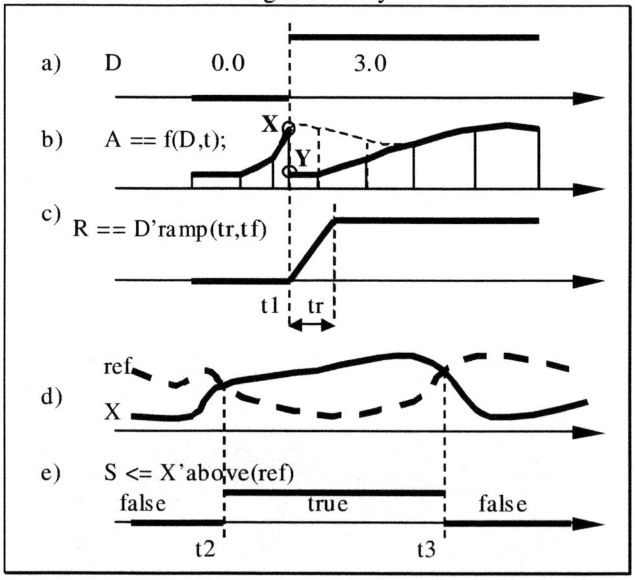

Figure 3 Mixed Signal Interfaces

5. CASE STUDY

The example discussed in detail is a simple inverter chain. For the purpose of the discussion the inverter I3 in **Figure 4** has an analog model at the transistor level while inverters I1 and I5 are modeled as pure digital gates. The inverted I2 contains the D/A interface and the inverter I4 contains the A/D interface.

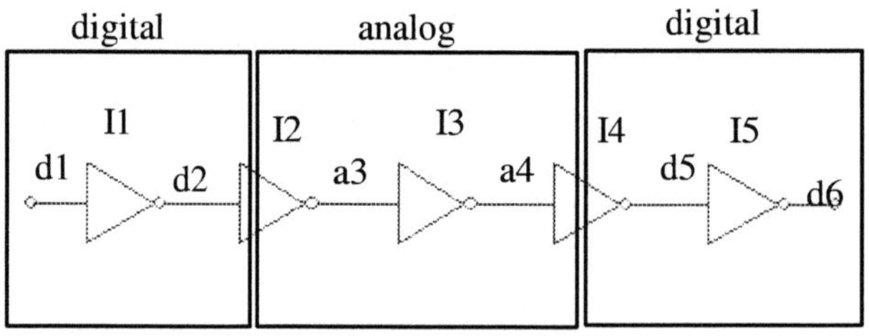

Figure 4 Case study: Inverter Chain

5.1. Simplistic models case

Suppose the type of the signals in the digital world in **Figure 4** is bit with the values '0' and '1'. The value of the signal d1 is set to '1'. The inverters are modeled as follows:

1. The digital inverters are modeled by the relation:
   output <= not input after tp;

2. The D/A interface inverters are modeled by:

process begin
if input = '0' then s <= analog_1 after tp;
else s <= analog_0 after tp;
end if;
wait on input;
end process;
output == s'ramp(tp,tp);

3. The analog inverters are transistor level models.
4. The A/D interface inverters are modeled by:

process begin
if input'above(threshold) then output <= '0' after tp;
else output <= '1' after tp;
end if;
wait on input'above(threshold);
end process;

For the quiescent solution all the derivatives with respect to time on the analog side are set to zero. All the signals on the digital side have the value equal to the initial value, '0' in this case. During the initialization phase all processes are executed until they suspend. Transactions are scheduled for the signals on the left side of the assignment operator. After the initialization phase the analog (DC) solution is computed. The values of the signals are the initial values. Crossed threshold of the A/D inverter I4 resulted in an event, which in turn caused signals update and processes execution.

This schedules another transaction of the output of this inverter. The quiescent solution is presented in **Table 1**. S.V. referred to the Simulated Value (i.e. the value obtained by the simulator) while C.V. corresponds to the consistent DC solution (i.e. value expected to be measured if the entire circuit was implemented using only analog models).

Table 1 Case study; Inverter Chain; simple models.

Net	D1	d2	a3	a4	d5	d6
S.V.	'1'	'0'	0.0	3.0	'0'	'0'
C.V.	'1'	'0'	3.0	0.0	'1'	'0'

5.2. AMS enhanced models

In order to get the consistent DC solution during the quiescent domain simulation the digital activity has to propagate until there are no more transactions scheduled. For VHDL-AMS, during this analysis the only events propagated are delta events (i.e. zero delay events). To achieve this the models on the digital side have to contain zero delayed functionality specified for the quiescent domain. This results in a domain dependent model:

process begin
if (domain = quiescent_domain)
then output <= not input;
else output <= not input after tp;
end if;
wait on input;
end process;

At the D/A in this particular case the signal's s value has to be available directly on the analog side. Also the analog solution has to be computed for all the changes of this signal. This results again in a domain dependent model:

if domain = quiescent_domain use output == s;
else output == s'ramp(tp,tp);--this works only in TR
end use;
break on s; –necessary indicate a discontinuity at DC;

At the A/D interface the only domain dependent aspect is the delays. In this case the simulation cycle will consist of a set of analog DC solution computations followed by signal updating and process resumption until all transactions are exhausted. The transitions are presented in **Table 2**.

Table 2 Case study; Inverter Chain; enhanced models.

Net	d1	d2	A3	A4	d5	d6
1	'1'	'0'	0.0	3.0	'0'	'0'
2	'1'	'0'	3.0	0.0	'0'	'0'
3(C.V.)	'1'	'0'	3.0	0.0	'1'	'0'

The enhanced behavior was achieved using analysis dependent models. The combination simulation cycle / analysis dependent model for VHDL is similar to the basic approach in the MAST simulation cycle [3].

6. CONCLUSIONS

In order to get the desired enhanced AMS behavior the A/D and D/A interfaces as well as the analysis specific models have to be carefully considered in the context of the specific simulation cycle used for analysis. The paper discusses some of these aspects indicating patterns to solve the inherent problems.

7. REFERENCES

[1] "IEEE Standard VHDL Analog and Mixed-Signal Extensions", IEEE Std. 1076.1-1999.

[2] "Verilog-AMS Language Reference Manual", IEEE Std. 1364-2001

[3] R. Scott Cooper, "The Designer's Guide To Analog & Mixed-Signal Modeling Illustrated with AHDL-AMS and MAST", Avant! Corp. 2001.

[4] Peter J. Ashenden, Gregory D. Peterson and Darell A. Teegarden, "The System Designer's Guide to VHDL-AMS"

[5] *Ruan, K.G,* "Initialization of mixed-signal systems in VHDL-AMS", BMAS 2001

[6] *Christen E, Bakalar K,* "VHDL-AMS-a hardware description language for analog and mixed-signal applications", IEEE Transactions on Circuits and Systems II: Analog and Digital Signal Processing, Vol. 46, Issue: 10, Oct. 1999 Pp: 1263 - 1272

BEHAVIOURAL MODELLING OF ANALOGUE FAULTS IN VHDL-AMS – A CASE STUDY

Mark Zwolinski, Andrew D. Brown
Electronic Systems Design Group
School of Electronics and Computer Science
University of Southampton
Southampton SO17 1BJ, UK

ABSTRACT

Analogue fault simulation is needed to evaluate the quality of tests, but is very computationally intensive. Behavioural simulation is more abstract and thus faster than fault simulation. Using a phase-locked loop as a case study, we show how behavioural fault models can be derived from transistor-level fault simulations and that faulty behaviour can be accurately modeled.

1. INTRODUCTION

In the digital domain, the single stuck fault model has been established as a useful representation of fabrication defects for some 30 years [1]. This model has been supplemented by other fault models, such as bridging faults and delay faults. These models are used to develop test patterns and cover of the faults is a useful metric of the effectiveness of the a particular set of test patterns.

In the analogue world, there are two main schools of thought. On the one hand, the quality of manufactured circuits is assessed against performance specifications. This approach is difficult to apply in the context of integrated circuits as it implies that detailed measurements must be made of the performance of a circuit. More recently, attention has turned to the idea of circuit-level fault models, such as opens and shorts [2]. Tests can be developed for such faults in a manner analogous to that used for digital faults [3]. Such structural testing has not, however, been widely adopted, partly because there is often no clear relationship between structural faults and overall performance, and partly because validating test patterns requires vast numbers of simulations.

Analogue fault simulation generally requires the repeated insertion of short or open faults into a circuit netlist and the (serial) simulation of these perturbed netlists. Methods for concurrent fault simulation of circuits have been proposed [4], but, unlike in the digital world, differences between good and faulty simulations tend to fall in a continuous range, rather than in terms of discrete values. Moreover, to gain useful knowledge about the behaviour of a circuit under fault conditions, it is necessary to simulate a circuit under a range of parametric variations. This can lead to the need for thousands of fault simulations, requiring vast computing resources.

Behavioural modelling of a circuit under fault conditions would lead to a reduction both in the simulation time and in the number of simulations. Behavioural simulation tends to be faster than transistor level simulation. Different faults can tend to result in similar faulty behaviours. The difficulty is that the overall faulty behaviour is not known until a set of transistor level fault simulations has been completed [5]. Knowing, however, the possible range of faulty behaviours and knowing which faults may be serious, but difficult to detect, can guide the circuit designer, both to minimize the likelihood of particular faults by careful layout and to include suitable test structures to facilitate testing.

The motivation for this paper is to show how a behavioural model of a complex circuit under fault conditions can be developed. Very often the fault model can be derived by simply modifying circuit parameters. This suggests that it may be possible to develop a link between performance and structural faults. It should also be noted that the recent availability of reliable VHDL-AMS and Verilog-AMS simulators has made this work timely.

This paper takes the form of a case study. A phase-locked loop is modelled at the transistor level and faults inserted into the netlist. These faults are then classified according to the observed behaviour of the circuit. From this a set of behavioural models can be developed.

2. TRANSISTOR LEVEL FAULT SIMULATIONS

For this case study, a commercial phase-locked lop (PLL) design has been used. For reasons of confidentiality and space, we will only summaries the circuit description here. The structure of the circuit is shown in figure 1.

0-7803-8251-X/04/$17.00 ©2004 IEEE

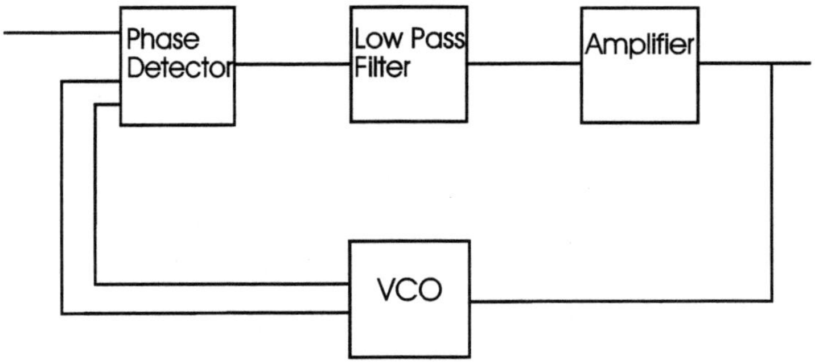

Figure 1 PLL Structure

The circuit is partitioned into five blocks: a phase detector, a low-pass filter, an amplifier, a voltage controlled oscillator (VCO) and a voltage reference block (not shown). In total, there are 95 MOS transistors in the circuit. To simulate the circuit over 12μs, with an input stimulus of 1.5MHz (18 cycles) takes about 11 seconds of CPU time on a 1.7GHz Pentium 4 processor.

The fault model is based on open and short circuits around each transistor [3]. Each transistor can have its gate shorted to the drain or source or the drain and source shorted together. Additionally, the drain, source or gate connections can be open-circuit. The short and open circuits are approximated by 1Ω and 100MΩ, respectively. Thus, there are 6 faults per transistor. In total, therefore 95x6=570 faults may be simulated, although in practice, the total is less as some transistors would already be configured with the gate and source connected and some short circuits effectively occur twice, on different transistors. Thus the total simulation time needed for all faults would be nearly 2 hours. (This assumes that each fault simulation takes about the same time – in practice some simulations will take longer and some may fail.)

Furthermore, each faulty circuit is simulated using Monte Carlo analysis to model the parametric variation that might be expected because of process variations. It is usual to perform 30 Monte Carlo simulations for each model, increasing the total number of simulations to 17100 and the expected CPU time to 60 hours.

Because of the computational effort involved, we limited our study to faults in the VCO. This has 33 transistors and, after equivalent and redundant faults are removed, 155 distinct possible faults. In practice, the VCO is perhaps the most critical block in the PLL.

Inserting faults by hand into a netlist is both tedious and error-prone. Therefore we have used a software tool, ANAFINS [6], to automate the process. A list was made of faults and from this, a series of perturbed netlists was generated and transistor-level simulations were performed. It would, of course, also be tedious to examine simulated waveforms by hand in order to identify similar behaviours. Therefore the simulation files were set up to automatically measure characteristics of the simulation. Six characteristics were extracted in the period between 4 μs and 12 μs:

- Oscillation frequency
- RMS output voltage
- Average output voltage
- Output offset voltage
- Mid-point of the output voltage swing
- RMS supply current.

By extracting just six figures from each simulation, it was also possible to save a significant amount of disk space. These figures, gathered from 30 Monte Carlo simulations of the fault-free circuit and of each faulty circuit, were used in the next step of the modelling procedure.

3. FAULT CLUSTERING

The intention in this work was to model a circuit under fault conditions by changing the circuit parameters of a behavioural model of the fault-free circuit. From all the simulations, listed above, we had 4680 sets of results. By including the parametric variations, it is possible to determine which of the groupings corresponding to the original 155 faults (and the fault-free case) overlap and can thus be collapsed. In practice, it is impossible to evaluate these results manually. Here, we used a Robust Heteroscedastic Probabilistic Neural Network (RHPNN) to classify and group the results.

The RHPNN is explained in full elsewhere [7]. The original design objective of the RHPNN was to classify data in several dimensions (in this case circuit simulation responses) as pass or fail. Here its use is extended to determine clusters of faulty circuit behaviours. The output from the RHPNN is sorted to find the mean values in each cluster and hence values for the behavioural models.

The RHPNN requires two phases: a training phase and a testing phase. The training phase requires a small set of data, but can take a significant amount of time to run – in this case about an hour. The testing phase uses the full data set (4680 results), but runs in a few seconds. The RHPNN collapsed the full range of faults into 51 distinct sets (with one fault-free set). Part of the classification is shown in Table 1.

Table 1 Extract of RHPNN fault clustering output

Cluster	Fault	VCO_OFFSET	VCO_AMP
51	0	-0.17452	2.2326
42	0	-0.20668	2.0135
86	0	-0.36199	2.8823
1	0	-0.30682	2.3777
0	0	-0.34792	1.9224
119	1	2.0306	0.000087
81	2	-1.5208	0.000066
146	2	-1.3282	0.000065
109	3	-0.71632	0.000002
22	4	-0.20466	2.0172
91	4	-0.3615	2.8834

Table2 Parameters for behavioural VCO fault model

	AMP	FOSC	OFFSET
Fault-free	2.2	1.54e6	-0.23
m14_mdop1	0.0	1.54e6	-2.42
m15_mdss1	0.0	1.54e6	-1.52
m30_mgds1	1.0	1.69e6	-1.12

4. BEHAVIOURAL FAULT MODELLING

Figure 2 shows a VHDL-AMS model of the VCO. As can be seen, there are four generic parameters that can be set by the instantiating model. By choosing appropriate values for these generics, as determined by the RHPNN grouping, different fault conditions can be modeled. Figure 3 shows the full PLL model. The structure of the behavioural model is a little different to that of the netlist. The filter and the amplifier are combined into one block. The other two circuit blocks, the phase detector and the low pass filter take one and two generic parameters respectively. Notice that this models the full PLL, not the configuration used for characterization.

Table 2 shows the circuit parameters used to model four of the fault conditions. Figure 4 shows the transient analysis results of these four fault conditions modeled both behaviourally and structurally. As can be seen, the behavioural fault model and the structural fault model produce very similar simulation results. The differences occur in the start-up transients. This is to be expected as the parameters for the behavioural models were derived by measuring circuit characteristics between 4 μs and 12 μs. Moreover, in reality any decision about the quality of a manufactured circuit would be taken once initial transients had died away.

The transient analysis of the behavioural model takes about 1 s of CPU time – a speed-up of an order of magnitude over the transistor-level model.

5. CONCLUSIONS

We have demonstrated that it is possible to derive behavioural models that reproduce the effects of transistor level faults in an analogue circuit. In order to derive the parameters for these models it is necessary to perform full transistor-level simulations and to analyze the results in detail. All these steps can be automated. The behavioural models could be used to explore the overall testability of an analogue or mixed-signal circuit.

6. ACKNOWLEDGEMENTS

The author acknowledges the contributions of the following people, without whom this paper would not have been possible: I.M Bell and S.J Spinks for developing the ANAFINS software; A.J. Perkins for developing the transistor-level models; Z.R. Yang for writing the RHPNN program and C.H. Tan for developing the fault clustering software and an earlier version of the behavioural models.

7. REFERENCES

[1] Abramovici M., M. A. Breuer, A. D. Friedman, *Digital Systems Testing and Testable Design*, IEEE Press 1990.

[2] C. Sebeke, J.P. Teixeira and M.J. Ohletz, "Automatic Fault Extraction and Simulation of Layout Realistic Faults for Integrated Analogue Circuits", *Proc. European Design and Test Conf.*, pp. 464-468, 1994.

[3] S.J.Spinks, C.D.Chalk, I.M.Bell, M.Zwolinski, "Gen-eration and Verification of Tests for Analog Circuits Subject to Process Parameter Deviations", *JETTA*, 20(1), pp. 1-13, 2004.

[4] Y. Kilic, M. Zwolinski, "Concurrent Transient Fault Simulation of Nonlinear Analogue Circuits", *6th IEEE International Mixed Signal Testing Workshop*, 2000.

[5] C. Chalk and M. Zwolinski, "Macromodel of CMOS Operational Amplifier Including Supply Current Variation", *Electronics Letters*, 171(31), pp. 1398-1400, 1995.

[6] I.M.Bell, K.R.Eckersall, S.J.Spinks, G.E.Taylor, "Fault orientated test and fault simulation of mixed signal integrated circuits", *Proc. ISCAS*, pp. 389-392, 1995.

[7] Z. R Yang, M. Zwolinski, C. D Chalk and A. C. Williams, "Applying A Robust Heteroscedastic Probabilistic Neural Network to Analog Fault Detection and Classification", *IEEE Trans. CAD.* 19(1), pp. 142-151, 2000.

```
library IEEE;
use IEEE.math_real.all;
use IEEE.electrical_systems.all;
entity vco is
  generic (amp: REAL := 2.2;
           c_fr: REAL := 1.5e6;
           vco_gain: REAL := 1.0e6;
           offset: REAL := -0.2);
  port (terminal ain, aout1, aout2 :
        electrical);
end entity vco;

architecture analog of vco is
  quantity vco_out : voltage;
  quantity vin across ain;
  quantity vout1 across iout1 through
           aout1;
  quantity vout2 across iout2 through
           aout2;
  constant wc: real:=MATH_2_PI*c_fr;

begin
  vco_out==sin(wc*NOW-trunc(wc*NOW/
    MATH_2_PI) * MATH_2_PI + MATH_2_PI *
    vco_gain * vin'INTEG);
  if vco_out>0.0 use
    vout1 == -amp + offset;
    vout2 == amp + offset;
  else
    vout1 == amp + offset;
    vout2 == -amp + offset;
  end use;
end architecture analog;
```

Figure 2 VHDL-AMS model of VCO

```
library IEEE;
use IEEE.electrical_systems.all;
entity pll is
end entity pll;

architecture netlist of pll is
  constant PD_GAIN:REAL :=  4.0;
  constant LP_BK_FREQ :REAL :=  99.7e3;
  constant LP_GAIN:REAL := 17.84;
  constant VCO_AMP:REAL := 2.2;
  constant VCO_FOSC:REAL := 1.54e6;
  constant VCO_GAIN:REAL := 1.28e6;
  constant VCO_OFFSET:REAL := -0.23;
  terminal vin_rf, vlocal_osc, vout_ph_det,
     vout, vout1 :electrical;

begin
pd: entity WORK.phase_detector
    generic map(PD_GAIN)
    port map
(vlocal_osc,vin_rf,vout_ph_det);
lpf: entity WORK.pll_lpf
    generic map(LP_BK_FREQ, LP_GAIN)
    port map (vout_ph_det,vout);
osc: entity WORK.vco generic map(VCO_AMP,
     VCO_FOSC, VCO_GAIN, VCO_OFFSET)
     port map (vout, vout1, vlocal_osc);
vsource:  entity WORK.vsin
     generic map(0.0, 1.5e6, 3.0)
     port map (vin_rf, ground);
end architecture netlist;
```

Figure 3 VHDL-AMS model of PLL

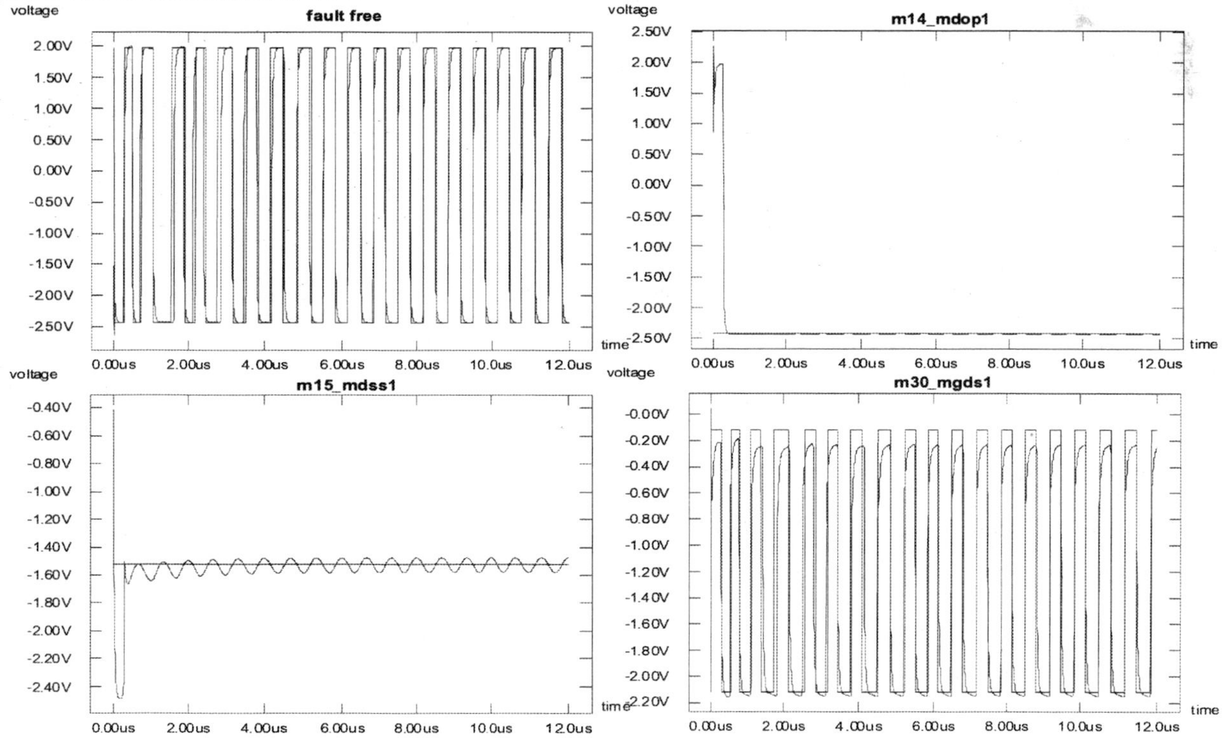

Figure 4 Comparison of behavioural and transistor level fault simulations

A NEW MODEL ARCHITECTURE FOR CUSTOMER SOFTWARE INTEGRATION

Ken G. Ruan

Synopsys Inc. 2025 NW Cornelius Pass Rd., Hillsboro, OR 97124
TEL: (503) 547-6399, Email: ken.ruan@synopsys.com

Abstract – Mixed-Signal Mixed-Technology simulation increasingly being used in complex system design verification. Now, it has been adopted by many mixed-technology industries, including automotive and aerospace industries, for designing an Electronic-Control-Unit (ECU). Micro-controllers are often used in ECUs for better features and performance. Design verification of micro-controller based ECUs is a significant challenge to simulation technology.

A new model architecture is proposed in this paper. An embedded control system co-simulator, Saber(tm)MC, is under development based on this architecture. Simulation of an entire ECU can be performed with a set of intuitive commands. Software developer may use a debugger to debug the customer code while an entire ECU is being simulated. A brushless DC motor controller is used to illustrate the proposed model architecture and demonstrate that it fits in well in ECU design verification.

I. INTRODUCTION

In the automobile industry and other mixed-technology industries, micro-controllers are often used in Electronic-Control-Unit (ECU) for ignition control, window/door operation and other functions. It is common to find 50-60 micro-controllers in newly designed automobiles to provide better features and performance.

Simulation programs in non-electronic areas are often developed in programming languages, such as C/C++, not in hardware description languages, such as MAST [1] and VHDL-AMS [2]. There are strong demands from mixed-technology industries that electronic and non-electronic models work together in simulating entire ECUs.

There are two major reasons why simulating micro-controller based ECUs is a significant challenge:

1. Micro-controllers vary from one to another. There are so many different micro-controllers available. It is not practical to develop micro-controller models in hardware description language;
2. Control software is normally written in programming languages that vary from ECU to ECU, even when the same micro-controller is used. Integration of customer software in ECU simulation must be addressed appropriately.

In recent years, instruction set description languages [3]-[4] have been developed for micro-controller modeling. Advances in instruction set simulation technology have made it possible to run control software on a virtual micro-controller. However, there is still a significant coupling gap between control software and the control object, the peripheral of the micro-controller.

In a hardware description language [1]-[2], a foreign function may be used to evaluate model quantities. The limitation of this approach is that the foreign function is a fixed portion of the model. It must be precompiled and included in an object library that is accessible by the simulator. This restriction becomes impractical for ECU simulation since customer software may vary from design to design. It is not cost effective to develop a new model whenever there is change in control software.

A new model architecture is proposed to remedy the limitation of fixed links with customer software. In this architecture, a mixed-technology, mixed-signal simulator, Saber(tm), is used to simulate blocks in the MAST hardware description language. Special interface models are developed in MAST for the I/O ports of micro-controllers. Due to the fact that I/O ports of micro-controllers are designed with common essential functionality, only a small set of interface models are needed. Each interface model consists of an interface template in MAST and a partner object called a "port class" that is written in C++. For each interface template, there is a set of signals that are defined as foreign to the Saber(tm) simulator. Coupling between these foreign signals and their counter parts in the port class is accomplished using mixed-mode interface [6]. The interface template model is generic in the sense that both interface template and port class are independent of any particular micro-controller. Any micro-controller part specific functionality is supported through an adapter.

An adapter is developed for a micro-controller part and may be shared by other micro-controller parts within the related micro-controller family. Customer programs and interface models are coupled with the port class via application interface functions. Interface functions are automatically generated based on brief specifications on the design configuration that are provided by the ECU designer and are dynamically loaded into the simulator during simulation. In this approach, the connections between customer software and the interface models are not fixed.

In addition, customers may use a debugger to debug their code as software developers do, performing such operations as setting or removing a breakpoint, stepping over to next statement, stepping into a function, viewing execution results and others.

With this architecture, simulation and debugging of an entire ECU with embedded micro-controller can be done.

In the next section, the proposed model architecture is described. The Saber(tm)MC simulator that is based on this architecture is described in Section III. A simplified brushless DC motor controller with a control algorithm in the C language is described in Section IV. Simulation results are shown in Section V. Finally, the Conclusion section draws together the key points of this paper.

II. A MODEL ARCHITECTURE

The proposed model architecture is illustrated in Figure 1. An interface model consists of four portions: a MAST template, a port class, an interface module and customer software.

Block 1 is the MAST template portion of an interface model. An example template for a pwm port is shown in APPENDIX A.

The pwm template has four pins - d2, d1, d0 and pwmclock - that are defined for netlist connections. Each connection pin is the output of a smc_d2a hypermodel [5] where a foreign state is defined. The state of a connection pin is determined by the corresponding foreign state.

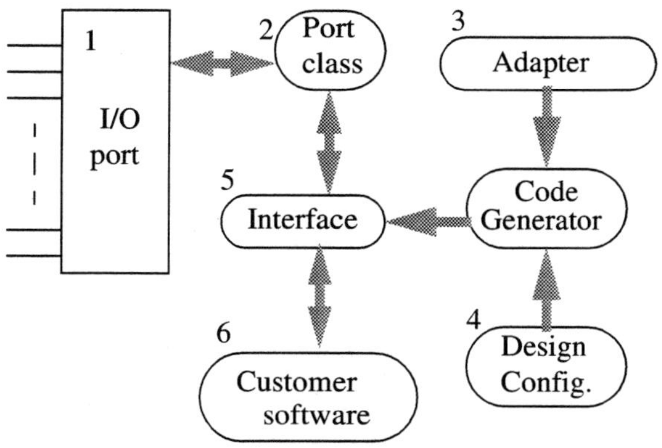

1. MAST template for an I/O port
2. Port class in C++
3. An adapter, designed for a micro-controller
4. Configuration of a system to be simulated
5. Interface functions to customer software that is generated by the code generator. See next section
6. Control software provided by the user

Figure 1. Block diagram of model architecture

Block 2 represents the port class portion of an interface model in C++. For each foreign state of the interface template, there is a corresponding member state in the port class. Both foreign states and member states are registered with the Saber(tm) simulator during initialization. Each foreign state in an interface template must match its member state in the port class. This is achieved by using the Mixed-Signal Interface [6].

Block 3 is an adapter that is developed for a micro-controller family such as the c166 family [7]. An adapter is a collection of functions, including data types, in C++ that represent I/O port behavior for a micro-controller family [8]. The adapter functions and customer functions, represented by Block 6, are coupled with port classes through application interface functions that are automatically generated based on customer's design configurations (See block 4). Since descriptions on I/O ports normally apply to micro-controllers within the same family, it is reasonable to have a single adapter for a family.

Block 4 is a design configuration file for the system under simulation. It specifies how the customer software will be linked with the interface models in the design. An example of a configuration file is given in APPENDIX B.

Customer functions may be in either source code or object code. Adapter functions, automatically generated application interface functions and customer functions are compiled and linked to create a shared object library. Functions in this library are dynamically loaded by the Saber(tm)MC simulator during simulation. Since member functions in a port class are only the interface to customer program, the behavior of the interface model is determined by the adapter functions and the customer software.

The interface functions, block 5, couple the adapter functions and the customer functions with member functions of interface port classes. As mentioned earlier, these interface functions are generated based on the contents of a design configuration file provided by the customer.

An example of a design configuration file is shown in APPENDIX B. Specifications in the configuration file include the Saber(tm) netlist file name (design name), the micro-controller part name, the name of the micro-controller family, the contents of special function registers used by the I/O ports, the customer function invoking format, and other details. As one can see, the format is very intuitive.

A make file may be created as an option. As long as there is no change in the design configuration specifications, there is no need to change the adapter interface functions. This allows the user to modify the customer code as needed without re-generating interface functions.

An embedded system can be represented by three portions: the micro-controller, customer software and control object.

The Micro-controller is represented by interface templates that are modeled in the proposed model architecture. The control object is described in a netlist and/or a schematic that consists of MAST models and the interface templates mentioned earlier. Connections between control object and micro-controller can be made by links between template pins, accordingly. The customer software is designed to be executed by either the host computer or an instruction set simulator (ISS). The embedded system can then be simulated as an entity.

The architecture of Saber(tm)MC simulator for embedded system simulation is described in next section.

III. THE Saber(tm)MC SIMULATOR

The Saber(tm)MC simulator is currently under development. The architecture of Saber(tm)MC is illustrated in Figure 2.

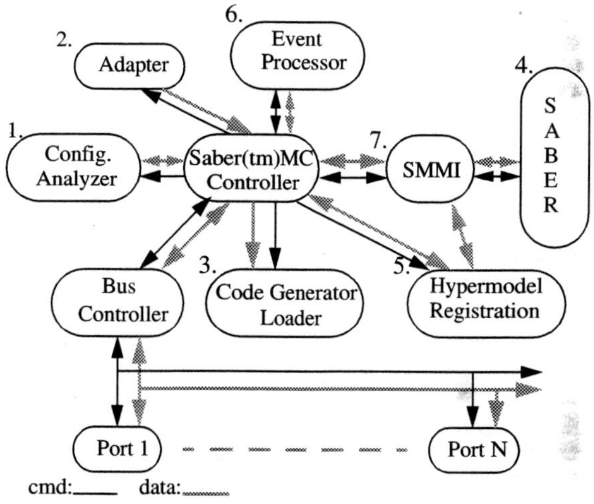

Figure 2. Architecture of Saber(tm)MC

The configuration analyzer (Block 1) reads the design configuration file first. A design instance is initiated that consists of all I/O port class instances specified in the configuration file (Block Port 1 - N). Based on the specified micro-controller part, an adapter (Block 2) for the part is loaded into Saber(tm)MC. Based on the customer's design configuration and the adapter, interface functions in C++ are generated by the Code Generator (Block 3). These source files, together with customer functions, are compiled to create a dynamic link library.tmOnce when the library is created, the Saber(tm) simulator (Block 4) is activated. Saber(tm) micro-controller simulation is a co-simulation. The events in all port class instances are handled by the event processor (Block 6). Communication between port classes and interface templates are achieved using SMMI (Block 7), the Saber(tm) Mixed-Mode Interface [6].

Port member states in all port class instances are registered against all foreign states in the interface template instances in the Saber(tm) netlist. This procedure is called "hypermodel registration" (Block 5). Finally, the simulation command is carried out until the specified end time is reached.

The simulation flow of Saber(tm)MC is illustrated in Figure 3.

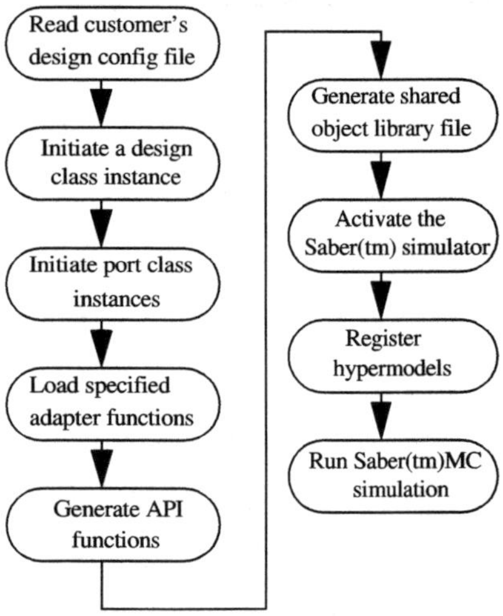

Figure 3. Saber(tm)MC simulation flow

IV. BRUSHLESS DC MOTOR CONTROLLER

A simplified brushless DC motor controller is used to illustrate how Saber(tm)MC simulation is accomplished. Its schematic block diagram is shown in Figure 4.

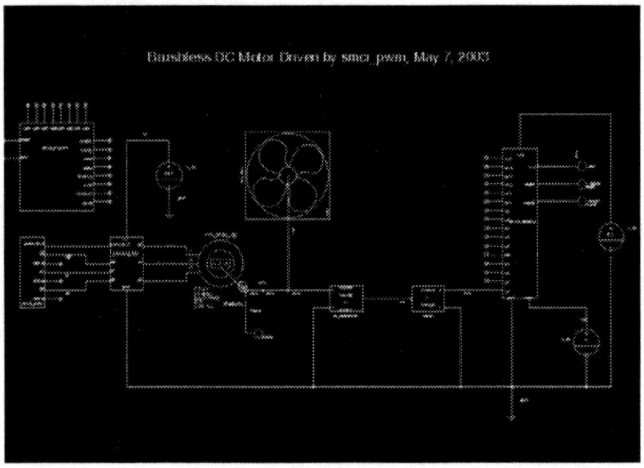

Figure 4. Schematic of a DC motor controller

Two I/O ports are used in this design. The smci_pwm interface template is used for the pwm port of a micro-controller c166. Its outputs are three pulse width modulated logic signals with phase shifted by $2/3\ \pi$ from each other. These pwm signals are applied at the inputs of the pwmavg_3ph template. Three analog outputs are generated from the logic signals. These analog signals are proportional to the duty cycle of the pwm signals, correspondingly. When the duty cycle of a pwm signal varies between 0 to 1, the output voltage varies between ground and power supply.

The analog outputs of pwmavg_3ph drive a brushless DC motor that is loaded with an electric fan. The rotation speed of the motor is sensed and converted into an analog signal, vrm. vrm is applied at the input of smci_analog template used as an A/D converter port for the micro-controller.

The analog input of the smci_analog is converted into a binary number that is stored in the c166_adcdat special function register of the a/d converter port, named result. Refer to APPENDIX C.

The contents of the result register is used by the rmctrl function to calculate the instance rotational speed, freq, of the brushless DC motor. The pwm3ph function evaluates the corresponding duty cycle of each phase that is stored in the compare register of the pwm port. The pwm port class then generates pwm signals that are sent to smci_pwm template in the form of events via SMMI.

In this example, an entire micro-controller based control unit is represented, including the control object hardware (represented in Saber(tm) netlist), the micro-controller (represented by the interface templates smci_pwm, smci_analog and its supporting I/O port classes) and the customer software (represented by rmctrl and pwm3ph functions). Simulation may be carried out using a simple script containing AIM [9] commands.

Simulation results are discussed in the next section.

V. SIMULATION RESULTS

At the beginning, the rotational speed of the DC motor is zero. It reaches its steady speed after a transition period. See the waveform, wrm, in Figure 5.

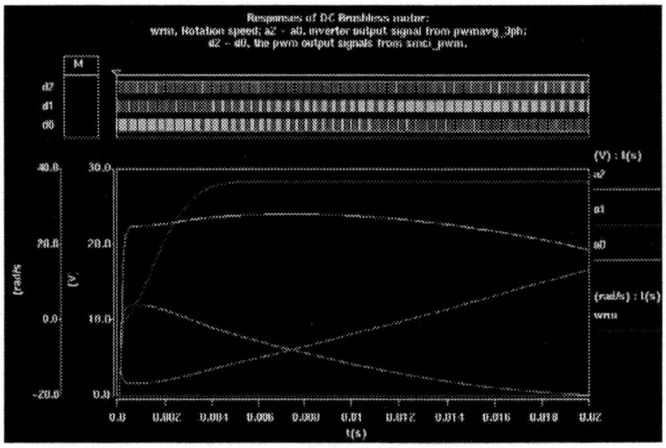

Figure 5. Simulation results

The pwm signals, d2, d1 and d0 from the smci_pwm template are logic signals shown in the upper portion in Figure 5. These signals are pulses with duty cycle controlled by the variables drva, drvb and drvc calculated by the customer function adv3ph. The PWM signals are converted into analog signals a2, a1 and a0 respectively by the pwmavg_3ph template and are shown in the analog signal region in Figure 5.

The customer may use a debugger to debug the customer code by setting breakpoints, stepping over statements, stepping into a function and viewing variable values, and other operations while simulation is in progress.

VI. CONCLUSION

A new model architecture was described. In this model architecture, customer software can be dynamically integrated with Saber(tm) templates based on customer specifications.

Saber(tm)MC, an embedded system simulator based on the new model architecture, is under development. A brushless DC motor controller design was described to demonstrate that Saber(tm)MC is capable of integrating customer software with Saber(tm) designs and that entire micro-controller based ECU can be simulated as an entity. The user may use a programming debugger to debug customer software during simulation just as a software developer would.

This provides a powerful tool for embedded control system design verification.

REFERENCES

[1] "Mast Language Reference Manual", Synopsys, Inc., 1994

[2] "IEEE Standard VHDL Language Reference Manual (Integrated with VHDL-AMS Changes)", Design Automation Standard Committee of the IEEE Computer Society, 1997

[3] R. Cmelik, D. Keppel: "Shade: A Fast Instruction Set Simulator for Execution Profiling," Proc. SIGMETRICS, ACM, Nashville, TN, 1994, pp 128-137

[4] G. Hadjiyiannis, S Hanono, S. Davadas: "ISDL: An Instruction Set Description Language for Retargetability," Technical report, MIT, 1996

[5] "Template Library, Model Fundamentals", Chapter 11, Synopsys, Inc. 1994

[6] "Mixed-Mode Interface", Synopsys, Inc., 1989.

[7] "C166/ST10 v7.5 C Cross-Compiler User's Guide," Altium Inc., 2001

[8] "C161 peripharal manual", Altium Inc., 2001

[9] "AIM Fundamentals: Customer SaberDesigner(tm) Using AIM," Synopsys, Inc., March 1998

APPENDIX A. A PWM TEMPLATE

A PWM (pulse width modulator) interface template is shown here. The name of this template is smci_pwm.

The template is for a three-phase PWM driver. d2, d1 and d0 are the three logic outs with independent phase and duty cycle. All three outputs are synchronized with the pwm control signal pwmclock.

```
template smci_pwm d2 d1 d0 pwmclock

state logic_4 d2,d1,d0,pwmclock
{
smc_d2a.d2a d2
smc_d2a.d1a d1
smc_d2a.d0a d0
# The pwm control signals
smc_d2a.pwmclock pwmclock
}
```

APPENDIX B. DESIGN CONFIGULATION FILE

The design configuration file for the brushless DC motor controller example is listed below.

```
Design Name : ctrl1_c166 ;
Part Name : c166 ;
Master : partner ;
Program file : rmctrl.c rmctrl.h;
Virtual Mode : host ;
Clock Config : noClock;
ebc mode: c166_ebcMode0  ebcmod0 ;
ebc mode: c166_ebcMode1  ebcmod1 ;
Channel Config : internal ;
SMCI Interface :  smci_pwm ;
Arguments : c166_cc6con cc6con;
Arguments : c166_cmp13 cmp13;
Arguments : c166_t12of offset;
Arguments : c166_cc6ic cc6_eic;
Arguments : double freq;
Arguments : int maxValue;
Arguments : c166_cc60 drva ;
Arguments : c166_cc61 drvb ;
Arguments : c166_cc62 drvc ;
Arguments : c166_cout60 cout60 ;
Arguments : c166_cout61 cout61 ;
Arguments : c166_cout62 cout62 ;
Arguments : c166_cout63 clk ;
Registers : logic_4 cout60 cout61 cout62 ;
Function  : irc : drv3ph(smcTime, maxValue, freq) : pw-
            mic ;
Config Ends : ;
Channel Config : internal ;
SMCI Interface :  smci_analog ;
Arguments : c166_adcon adcon ;
Arguments : c166_adcon1 adcon1 ;
Arguments : c166_adctr0 adctr0 ;
Arguments : c166_adctr2 adctr2 ;
Arguments : c166_adctr2in adctr2in ;
Arguments : c166_adcdat2 inject ;
Arguments : c166_adcdat result ;
Arguments : c166_adceic adc_eic;
Arguments : c166_adccic adc_cic;
Arguments : c166_adceoc eoc ;
Registers : int rm;
Function  : irc : rmctrl(smcTime, adcon, adcon1) : eoc ;
Function  : irc : adccic(&adcon, &adcon1) : overrun ;
Config Ends : ;
Part Ends : ;
dctrcmd : dctr (pf tr, ts 1n, te 20m) ;
```

APPENDIX C. A CUSTOMER FUNCTION

The customer functions, rmctrl and drv3ph, used in this example are shown below. The rmctrl function is used by the smci_analog interface template for frequency evaluation and drv3ph is used by smci_pwm for pulse width calculation.

```
#include "c166.h"
#include "rmctrl.h"

c166_adcon adcon = {2,2,0,0,0,0,1,0,3,0};
c166_adcon1 adcon1 = {1,0,0,0,3,4};
c166_adctr2 adctr2 = {0,0,16,16};
c166_adcdat result = {0,0};
......
c166_cc6con cc6con = {2,2,0,1,0,7};
c166_cmp13 cmp13 = 50000;
c166_t12of offset = 0;
c166_cc6ic cc6_eic = {0,1,1,0,14,2};
int maxValue = 65535;
int drva, drvb, drvc;

int rmctrl(double t,  c166_adcon con,  c166_adcon1 con1)
{
   int siglIdx = result.chnr;
   int fullScale = 1023;
   if (con1.res == 1) fullScale = 255;
   switch (siglIdx) {
      case 0:{
         rm = result.adres;
         freq = (double) wrmMax*rm/fullScale;
         break;
      }
   }
   return 0;
}
void drv3ph(double t, int scl, double f)
{
   // Calculate the duty cycle in number of clock cycles.
   double theta = xp/2* f *t;
   double dPhase = MATH_2PI/3.0;
   drva = 0.5*scl*(1.0+sin(theta));
   drvb = 0.5*scl*(1.0+sin(theta + dPhase));
   drvc = 0.5*scl*(1.0+sin(theta + 2*dPhase));
   return;
}
```

PROGRAMMING INTERFACE REQUIREMENTS FOR AN AMS SIMULATOR

Martin Vlach

Lyngument, Inc.

ABSTRACT

Application Programming Interfaces (APIs) have been successfully used in digital simulators for years. A similar approach is proposed for Analog and Mixed Signal (AMS) simulators. Potential applications are examined, and requirements for an API are derived.

1. INTRODUCTION

The concept of using standard Application Programming Interfaces (APIs) to control simulation, insert external models, and implement innovative applications has been successfully used in the digital simulation – examples include VPI of Verilog[1], FLI of Modelsim[2] and the emerging standard VHPI (VHDL Programming Interface) for VHDL[3].

Analog and Mixed Signal (AMS) simulators based on standard Hardware Description Languages (HDLs) such as MAST®[4], VHDL-AMS and Verilog-AMS have become a viable technology for design modeling. However, standard APIs for Analog and Mixed Signal simulators do not exist. A viable API for an AMS simulator would provide a path to a *standard* method of interfacing new applications with the various AMS simulators. Such an approach would provide benefits to the simulator user who needs novel simulator controls, simulation meta-analyses, or simulation result analyses.

The benefits of standard APIs are similar to those that Hardware Description Languages now provide to the modeler: the ability to focus on the *solution* without having to spend time on the implementation details of that solution in any specific simulator.

2. APPLICATIONS

There are a large number of potential applications that would benefit from having a well-defined *standard* interface to AMS simulators. Some of these are:

- Design traversals, netlisters, connectivity extractors. These applications examine the elaborated structure of the design, without actual simulation, in order to answer questions about the design.
- Behavioral model extraction tools. These tools could examine the static design, automatically derive a higher-level model or choose one from a library, and use simulation to adjust details of the behavioral model.
- Debugging environments. See the following subsection.
- Simulation testbench and verification tools ease the task of verifying whether a design conforms to specifications as it is being refined.
- Code profilers and coverage tools may help in determining whether all regimes of a model are exercised by a test suite.
- Power estimation tools could be developed based on design traversal, with or without simulation.
- Simulation history recording and display would help designers as they explore the design space.
- Meta-simulation procedures, such as large change sensitivity, optimization, and large signal steady state, can be implemented in external and 3rd party tools.
- Design space exploration, such as parameter sweeps or Monte Carlo can be implemented by external tools.
- Detection of unsafe operating regions, for example by calculating conditions "on-the-fly" or from the recorded simulation history.
- Fault insertion, such as device failures due to radiation effects may be enabled when applied together with appropriate models.
- Co-simulation, backplane interfaces. At this time there are no established methods of co-simulation in the analog domain. However, an API may help spur research in this area.

2.1. Debugging Environments

Debugging tools are those applications that help designers isolate problem areas in their design. In the AMS domain there are three debugging environments:

2.1.1. Design Debugging

Design debugging assumes that

- both individual component models are correct, and that

- the simulator provides a good approximation to the solution of the mathematical model.

Design debugging tool helps the designer by simplifying tasks such as changing topology and circuit parameters.

2.1.2. Model Debugging

Model debugging helps the engineer to uncover problems with individual models, such as incorrect equations, or operating out of the model's domain of definition.

2.1.3. Simulation Debugging

Simulation debugging includes issues such as convergence and accuracy problems. This kind of debugging would be possible for simulators that employ standard Hardware Description Languages (HDLs) for the description of the design, since the semantics is well defined, and external tools may verify the correctness of simulator implementation.

From a practical point of view, there might be a need to make available algorithm-specific functionality. This kind of debugging would likely be tightly coupled to a specific simulator.

3. API COMPONENTS

Since standard APIs already exist for digital HDLs, and AMS HDLs are generally supersets of the digital languages, the AMS APIs should also be designed as supersets of the underlying digial API.

There are several areas that contribute to a well-designed AMS simulator API. They are described in the following sections. The terminology follows that developed for VHPI[3] and VHPI-AMS[5], an extension of VHPI to the AMS domain.

3.1. Utilities

The utility functions include tasks such as printing, displaying and comparing values, and error handling mechanisms. The extensions in this area are obvious and easily defined.

3.2. Static Design Data

Static design data access includes access to the structure of the design and details of the HDL models. It includes all information that is available prior to the start of the simulation. Extensions in this area are also quite obvious, and follow the corresponding extension of the HDL to its AMS version.

3.3. Access to And Modification of Objects

The AMS extensions fall in several categories:

- Access to values of the new objects introduced by AMS extensions to the HDLs. These include read-only access of simulation quantities such as voltages and currents. The extensions are obvious and not hard to implement. They enable tools that can monitor progress of simulation.
- Modification of static design values, such as model parameters and component values. These changes are also relatively easy to implement. They enable tools for design exploration – parameter sweeps, Monte Carlo.
- Modification of simulation quantities. Since AMS simulation requires simultaneous soulution of nonlinear ordinary differential equations, modification of values such as instantaneous voltages or currents by assigning new values is not acceptable, and new methods must be devised (see below).

Instead of modifying the instantaneous values of analog quantities, changes to topology must be introduced. For example, there are some requirements imposed by a debugger application that may fall into this category:

- Forcing a value of an across quantity (e.g. voltage) onto a node. This is equivalent to inserting a source into the topolgy
- Forcing two values to be equal - a "short" in the topology.
- Decoupling two values - an "open" in the topology.

Topology changes in the AMS domain are more complex to implement than similar changes in the digital domain, due to the simultaneous (instead of sequential) nature of the simulation cycle.

If an AMS simulator implements the ability to make these kinds of changes through an API (rather than by much costlier change to source, recompilation, and reelaboration), an additional capabilitiy is enabled – the ability to insert or delete elements. This ability enables applications that may wish to change simulation models, for example replacing a model with a failed variant.

3.4. Simulation Interaction And Control

The API calls in this group query the state of the analog kernel, and control it by modifying its configuration parameters.

Some AMS language extensions, notably VHDL-AMS, intentionally disregard possible implementation strategies. While this is possible in the case of model definition, using a mathematical description, it makes much less sense at the API level. To make them practical, AMS APIs should take into consideration known solution algorithms. At issue are aspects such as

- Ease of implementation of the interaction in a kernel. If a particular functionality is required by an AMS API, but would require major kernel algorithm modifications, then those changes will not happen.

- Usefulness and usability for the application developer. It will likely become necessary to develop certain aspects of applications (e.g. convergence debuggers) that are targeted to specific simulators, or at least to specific solution algorithms.
- Speed. Simulation speed is one of the more important aspects of simulation. Introducing an API to a simulator must not slow the simulator when the API is not in use.

Examples of commonly used solution algorithms include:

- Full matrix methods,
- Relaxation methods,
- Combination of the two, such as subdivision of the system into multiple independently solved subsystems.

Examples of simulation interaction include:

- Access to intermediate iteration values during nonlinear system analysis.
- Access to intermediate steps during numerical solution.
- Participation in determining convergence criteria.
- Participation in determining truncation error.
- Participation in determining analog time step.

3.5. Foreign Model Instantiation And Communication

There are few widely adopted methods for using analog "foreign models". The methods that have been used are always of necessity tightly coupled to the particular analog kernel. Most of the published methods assume SPICE-like algorithms and Modified Nodal Analysis (MNA) formulation.

The interface should provide support for alternative algorithms implemented by an analog kernel. The following issues must be addressed:

- Type of formulation – e.g. Nodal, MNA, Tableaux, etc, with suitable defaults. This will have the effect of defining the dependent and independent variables.
- Require that the application model return the values of dependent variables for given independent variable values.
- Optionally require that the application model return Jacobian information for the dependent variables with respect to the independent variables.

In principle it would be possible to define an interface that is purely declarative, in which the characteristic equations (as a function of parameters and ports) are conveyed to the elaborator via API calls. However, this has little practical value.

4. ADDITIONAL CONSIDERATIONS

4.1. Data Representation Issues

One of the problems encountered by VHPI is the need to convert internal simulation representations of the numerous data types into a simulator-independent form. In digital simulation, this is a major obstacle, since it impacts simulation speed. Luckily analog kernels seem to be exclusively based on using doubles, and data representation conversion is less likely to cause a problem.

An area where data conversion may be a problem is in the representation of time. In AMS simulation, the use of double to represent time may not be sufficiently accurate, and simulators may implement representations with higher resolution. An AMS API should define calls that manipulate time by the simulator, without the need for data conversion.

4.2. Restricted Access to Protected Data

The access to protected models should be restricted to the information that can be found in a model vendor library and what is necessary for interfacing the library cells in the design.

This may have the side effect of not being able to perform some analog debugging functionality at all. This is due to the simultaneous solution nature of analog algorithms.

4.3. Resetting Simulation State

Some of the proposed applications suggest additional mechanism to reset the simulation state.

For the AMS extensions the analog simulation kernel concept of **analog system state** is required. Analog system state allows the setting of the initial conditions for the differential equations and quantity values to a state other than the state at time zero. Such a concept is likely going to be simulator specific.

4.4. Support for Additional Simulation Capabilities

Digital simulator APIs concentrate on transient simulation in time domain. AMS simulators have additional capabilities that must be addressed in the API. They include

- dc operating point
- small signal frequency domain (ac)
- noise in the frequency domain

4.5. Multiple Languages

Practical simulation environments use simulators that are capable of using multiple representations, including SPICE, Verilog-AMS, VHDL-AMS, C language models.

One of the challenges of AMS APIs will be to enable access to such simulators in a language independent way.

5. BENEFITS

Numerous useful applications have been suggested, but none of these have been implemented as industrial solutions. One of the reasons is that in order to investigate new approaches to solving industrial-sized problems, the analog engine must be capable of solving industrial-sized systems. The application algorithms can be tried relatively easily on small systems using special-purpose engines, but incorporating them into existing simulators is usually not feasible.

With a standardized access to simulation engines, many different approaches could be easily investigated on true designs. This would lead to a broad experimentation with AMS add-on applications. The initial investment on the part of simulator developers would be returned many times by a broader acceptance of simulation.

Another aspect to consider is that often the algorithms are likely to be quite costly. However, with an API they would be implemented in code external to the engine. The engine would not be burdened when the cost is not warranted, but the application would be easily attached when needed.

6. REFERENCES

[1] "IEEE Verilog 2001," IEEE Std 1364-2001

[2] "Model Technology, a Mentor Graphics Company", http://www.model.com/

[3] "IEEE DASC .VHDL PLI Task Force," http://www.model.com/

[4] "MAST Hardware Description Language", http://www.synopsys.com/products/mixedsignal/saber/mast_ds.html

[5] "VHPI-AMS Subcommittee of the 1076.1 Working Group," http://www.lynguent.com/VHPI-AMS/index.html

MULTIPLE DOMAIN BEHAVIORAL MODELING USING VHDL-AMS

Peter R. Wilson, J. Neil Ross, Andrew D. Brown and Andrew Rushton

University of Southampton, United Kingdom

ABSTRACT

With the increasingly high level of system integration it is becoming necessary to model not only electronic behavior of systems, but also interfaces to "real world" applications and the detailed physical behavior of elements of the system in question. The emergence of standard languages such as VHDL-AMS has made it possible to now describe a variety of physical systems using a single design approach and simulate a complete system. Application areas where this is becoming increasingly important include mixed-signal electronics, electro-magnetic interfaces, integrated thermal modeling, electro-mechanical and mechanical systems (including MEMS), fluidics (including hydraulics and micro-fluidics), power electronics with digital control and sensors of various kinds.

In this paper, we will show how the behavioral modeling of multiple energy domains is achieved using VHDL-AMS, demonstrating with the use of examples how the interactions between domains takes place, and provide an insight into design techniques for a variety of these disciplines. The basic framework is described, showing how standard packages can define a coherent basis for a wide range of models, and specific examples used to illustrate the practical details of such an approach. Examples such as integrated simulation of power electronics systems including electrical, magnetic and thermal effects, mixed-domain electronics and mechanical systems are presented to demonstrate the key concepts involved in multiple energy domain behavioral modeling.

1. INTRODUCTION

The emergence of the new modeling language VHDL-AMS (defined by IEEE Std 1076.1 in 1999 [1]) has proved extremely useful in defining and modeling mixed-technology systems – of which electro-magnetic components are a good example. The modeling language allows a mixture of so-called domains (such as electrical, magnetic, thermal and so on) as well as providing analogue and digital modeling capabilities. The application of a mixed-technology approach to modeling multiple-domain components has been described in some detail by classic papers such as those by Cherry [2], Laithwaite [3] and Carpenter [4]. In these papers, the key concept of energy conservation throughout the complete system is emphasized, with the use of equivalent circuit elements to represent behavior in different energy domains. Examples of this include using a circuit model of a resistor in the thermal domain to denote "thermal resistance" or an "inductance" in the magnetic domain to provide a loss term.

The basic approach for modeling devices in VHDL-AMS is to define a model *entity* and *architecture*(s). The model entity defines the interface of the model to the system and includes connection points and parameters. A number of architectures can be associated with an entity to describe the model behavior such as a behavioral or physical level description. A complete model consists of a single entity combined with a single architecture. The *domain* or technology type of the model is defined by the type of terminal used in the entity declaration of the ports. The proposed IEEE Std 1076.1.1 [5] defines standard types for multiple energy domains including electrical, thermal, magnetic, mechanical and radiant systems. Within the architecture of the model, each energy domain type has a defined set of *through* and *across* variables (in the electrical domain these are *voltage* and *current* respectively) that can be used to define the relationship between the model interface pins and the internal behavior of the model. Christen, et. al. [6],[7] and Vachoux [8] each provide an overview of the rationale behind the VHDL-AMS language especially with regard to the specific extensions addressed by the new set of packages, namely natures, quantities and terminals.

In the "conventional" electronics arena, the nature of the VHDL-AMS language is designed to support "mixed-signal" systems (containing digital elements, analog elements and the boundary between them) with a focus on IC design. Typical examples of this kind of application are described in [9]-[15]. There are of course many other similar applications that can be found in the literature in this area. Where the strengths of the VHDL-AMS language have really become apparent however, is in the multi-disciplinary areas of mechatronic and micro electro mechanical systems (MEMS) [16-20]. In this paper, we have highlighted several interesting examples that illustrate the strengths of this modeling approach, with emphasis on multiple-domain simulations.

2. ELECTRO-MAGNETIC-THERMAL MODELING

As an example of how the packages can be used in practice a simple electro-magnetic transformer is used to illustrate some of the key concepts involved. If a simple two winding transformer (schematic shown in figure 1) is implemented using a mixed–technology approach, the simple structure consists of two winding models interfacing between the electrical and magnetic domains and a magnetic model of the core.

Using this approach, structural models of magnetic components can be built up that encompass the multiple domain capability in the VHDL-AMS language.

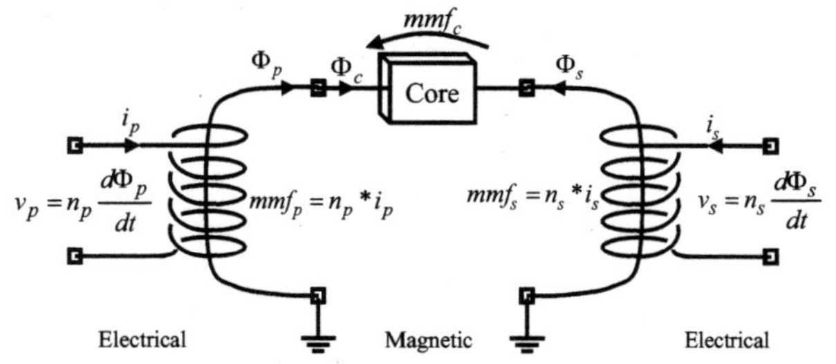

Figure 1: Electro-Magnetic model of transformer

Using this approach, the winding model is defined using the listing below:

```
1  use work.electrical_systems.all;
2
3  entity winding is
4     generic (r : real := 0.0;
5              n : real := 1.0);
6     port (
7       terminal ep,em : electrical;
8       terminal mp,mm : magnetic
9     );
10 end entity winding;
11
12 architecture simple of winding is
13  quantity h across f through mp to mm;
14  quantity v across i through ep to em;
15 begin
16   h == i*n;
17   v == - n*f'dot + i*r;
18 end architecture simple;
```

Note that in order to use the electrical and magnetic definitions, the electrical_systems package must be referenced using the use statement in line 1. Note that the terminals are defined with electrical or magnetic types and the resulting quantities (v & i in the electrical domain and h & f in the magnetic domains) have the correct units, symbols and types defined by the referenced package. The same approach is used for the magnetic core model as shown below:

```
use work.energy_systems.all;
use work.electrical_systems.all;

entity core_linear is
   generic (ur   : real := 1.0;
            len  : real := 1.0e-2;
            area : real := 1.0e-4
            );
   port (terminal p,m : magnetic);
end entity core_linear;

architecture simple of core_linear is
constant mg : real
      := PHYS_MU0*ur*area/len;
quantity mmf across f through p to m;
begin  -- architecture linear
   assert len /= 0.0
      report "len should not be 0!"
     severity error;
   f == mg * mmf;
end architecture simple;
```

Using this basic approach a non-linear magnetic model such as the Jiles-Atherton (JA) model [21]-[23] can be built up that can be simulated to estimate power loss, distortion and circuit behavior accurately. A typical BH curve using a VHDL-AMS model is given in figure 2 below.

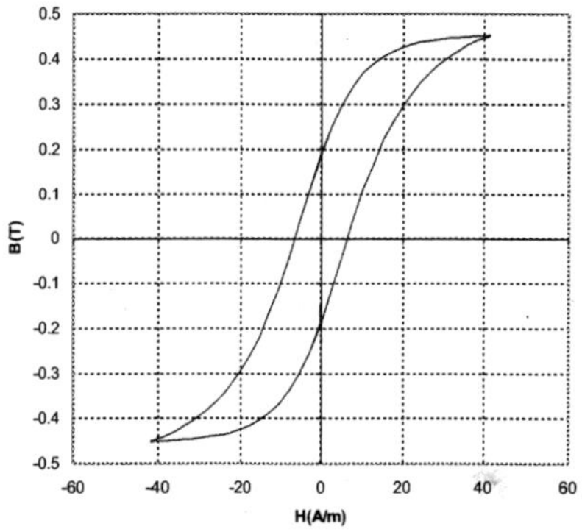

Figure 2: JA Model BH Curve

While this is useful, the real strength of the multiple-domain approach comes into play when the complete system energy balance is considered. The practical meaning of the BH loop is energy lost (given by the area contained within the loop) that is practically emitted as heat. The model can therefore be practically extended to include thermal effects. The modeling of thermal effects in a multiple-domain simulation has been applied for electronic devices by Hefner [24]-[25], and for magnetic devices by various authors [26]-[30]. Using similar techniques to the analogous models in the magnetic domain, elemental models can be devised for resistance, capacitance, convection, conduction and emission in the thermal domain. Building these models into a circuit-like implementation using VHDL-AMS allows the accurate dynamic simulation of thermal behavior and its impact on other domains (in this case magnetic and electrical). A typical thermal circuit is shown in figure 3 and the effect on the BH loops as the temperature increases is shown in figure 4.

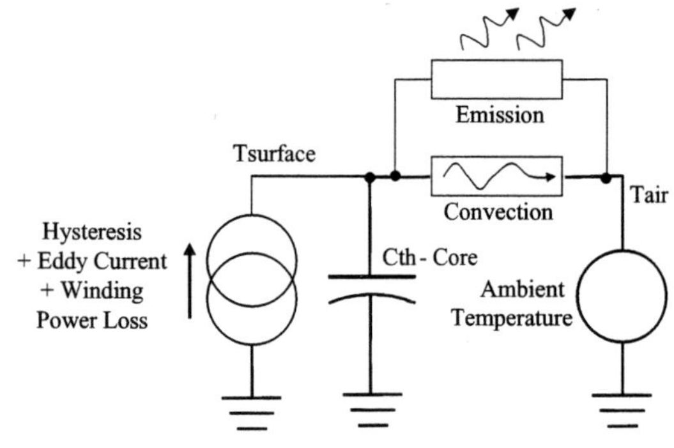

Figure 3: Thermal Circuit for Electric-Magnetic Model

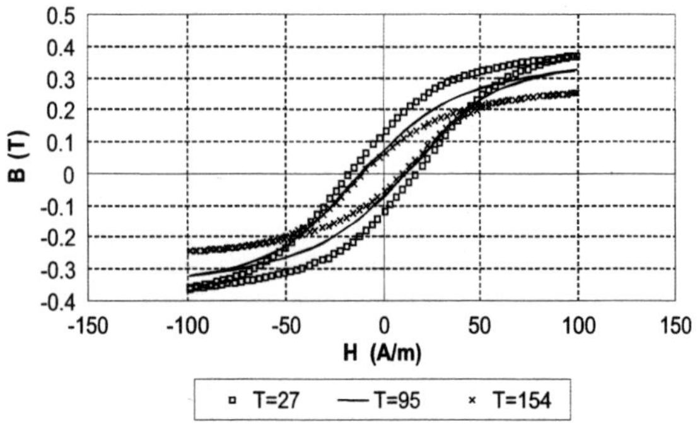

Figure 4: Thermally dependent BH curves

To illustrate the similarity between the domain models, the thermal capacitance model in VHDL-AMS is given below. Note the use of the thermal pin definitions, requiring the thermal_systems package.

```
use work.thermal_systems.all;
use work.ai_standard.all;
entity ctherm is
  generic (cth:real := 0.0);
  port (terminal th,tl : thermal);
end entity ctherm;

architecture simple of ctherm is
  quantity tc across heatfl through th to tl;
begin  -- simple architecture
  assert cth /= inf and cth /= undef
  report "cth specified incorrectly"
  severity error;

  heatfl == cth * tc'dot;
 end architecture simple;
```

3. MIXED-SIGNAL MODELING

With the extension of VHDL (digital) to the analog domain with VHDL-AMS, it is possible to interface digital logic signals directly to analog electrical signals. There are two mechanisms required for this, digital to analog and analog to digital. An example of a simple analog to digital "conversion" model is a simple comparator with an electrical voltage input, and a digital output. In VHDL-AMS this is easily modeling using the threshold function **'above**. This an example of VHDL-AMS code to implement a basic comparator function is given below, where **vin** is an analog quantity and **out** is a logic output.

```
if vin'above(vh) then
  out := '1';
elsif not vin'above(vl) then
  out := '0';
end if;
```

Conversely, a digital to analog interface is implemented using the **'ramp** function that specifies a dynamic analog output change, depending on a discrete signal. An example of a basic analog quantity (**vout**) being driven by a discrete signal (**vsig**) is given by the code below where **tt** is the transition or ramp time.

```
vout == vsig'ramp(tt)
```

4. ELECTRO-MECHANICAL MODELING

A final significant application area for VHDL-AMS has been the modeling of electro-mechanical systems, particularly micro-machines (or MEMS) [16]-[17]. Exactly the same principles are used for these devices, with the mechanical domain models defined as required for the mechanical equations. It is worth noting that the mechanical models are divided into rotational (angular velocity and torque) and translational (force and distance) types. A typical simple example of a mixed domain system is a motor, in this case a simple DC motor. Taking the standard motor equations as shown in (1) & (2), it can be seen that the parameter *ke* links the rotor speed to the electrical domain (back emf) and the parameter *kt* links the current to the torque.

$$V = L\frac{di}{dt} + iR + Ke\omega \quad (1)$$

$$T = Kti - J\frac{d\omega}{dt} - D\omega \quad (2)$$

This is implemented using the VHDL-AMS model shown below:

```
use work.electrical_systems.all;
use work.mechanical_systems.all;

entity dc_motor is
    generic (kt : real;
             j  : real;
             r  : real;
             ke : real;
             d  : real;
             l  : real);
    port (terminal p, m   : electrical;
          terminal rotor  : rotational_v );
end entity dc_motor;

architecture behav of dc_motor is
    quantity w across t through rotor
      to rotational_v_ref;
```

```
    quantity v across i through p to m;
begin
  v == l*i'DOT + i*r + ke*w;
  t == i*kt - j*w'DOT - d*w;
end architecture behav;
```

5. CONCLUSIONS

It has become crucial for effective design of integrated systems, whether on a macro- or microscopic scale, to accurately predict the behavior of such systems prior to manufacture. Whether it is ensuring that sensors or actuators operate correctly, or integrated components such as magnetics also operate correctly, or analyzing the effect of parasitics and non-ideal effects such as temperature, losses and non-linearities, the requirement for multiple-domain modeling has never been greater. Now languages such as VHDL-AMS offer an effective and efficient route for engineers to describe these systems and effects, with the added benefit of standardization leading to interoperability and model exchange. The challenge for the EDA industry is to provide adequate simulation and particularly modeling tools to support engineering design.

1. REFERENCES

[1] IEEE Std 1076.1, IEEE, Rochester, N.Y., USA

[2] E. C. Cherry, "The duality between inter-linked electric and magnetic circuits", *Proceedings of the Physics Society*, Volume 62, 1949 pp101-111

[3] E. R. Laithwaite, "Magnetic Equivalent Circuits for electrical machines", *PROC. IEE*, Vol. 114, November 1967, pp1805-1809

[4] C.J. Carpenter, "Magnetic Equivalent Circuits", *PROC. IEE.*,Vol 115,No 10, October 1968, pp1503-1511

[5] Proposed IEEE Std 1076.1, IEEE, Rochester, N.Y., USA

[6] Christen, E. and Bakalar, K., "VHDL-AMS-a hardware description language for analog and mixed-signal applications", *IEEE Transactions on Circuits and Systems II: Analog and Digital Signal Processing*, Volume 46, Issue 10 , Oct. 1999, pp 1263 -1272

[7] Christen, E. and Bakalar, K., "VHDL 1076.1-analog and mixed-signal extensions to VHDL", *Design Automation Conference*, 1996, Proceedings EURO-DAC '96, 16-20 Sept. 1996, pp 556 -561

[8] Vachoux A, "Analog and mixed-signal extensions to VHDL", *ANALOG INTEGRATED CIRCUITS AND SIGNAL PROCESSING* 1998, Vol 16, Iss 2, pp 185-200

[9] Sabiro, SG, "Mixed-mode system design: VHDL-AMS", *Microelectronic Engineering*, 2000, Vol 54, Iss 1-2, pp 171-180

[10] Godambe NJ and Shi CJR, "Behavioral level noise modeling and jitter simulation of phase-locked loops with faults using VHDL-AMS", *JOURNAL OF ELECTRONIC TESTING-THEORY AND APPLICATIONS* 1998, Vol 13, Iss 1, pp 7-17

[11] Perkins AJ; Zwolinski M; Chalk CD; Wilkins BR,"Fault modeling and simulation using VHDL-AMS",*ANALOG INTEGRATED CIRCUITS AND SIGNAL PROCESSING* 1998, Vol 16, Iss 2, pp 141-155

[12] Boccuni I; Gulino R; Palumbo G, "Behavioral model of analog circuits for nonvolatile memories with VHDL-AMS", *ANALOG INTEGRATED CIRCUITS AND SIGNAL PROCESSING* 2002, Vol 33, Iss 1, pp 19-28

[13] Mongellaz B; Marc F; Milet-Lewis N; Danto Y, "Contribution to ageing simulation of complex analogue circuit using VHDL-AMS behavioural modelling language", *MICROELECTRONICS RELIABILITY* 2002, Vol 42, Iss 9-11, pp 1353-1358

[14] Sida M; Ahola R; Wallner D, "Bluetooth transceiver design and simulation with VHDL-AMS", *IEEE CIRCUITS & DEVICES* 2003, Vol 19, Iss 2, pp 11-14

[15] Vogels, M.; De Smedt, B.; Gielen, G, "Modeling and simulation of a sigma-delta digital to analog converter using VHDL-AMS", *2000 IEEE/ACM International Workshop on Behavioral Modeling and Simulation*, (2000), p 5-9

[16] Endemano A; Desmulliez MPY; Dunnigan M,"System level simulation of a double stator wobble electrostatic micromotor ", *SENSORS AND ACTUATORS A-PHYSICAL* 2002, Vol 99, Iss 3, pp 312-320

[17] Gibson D; Carter H; Purdy C, "The use of hardware description languages in the development of microelectromechanical systems ", *ANALOG INTEGRATED CIRCUITS AND SIGNAL PROCESSING* 2001, Vol 28, Iss 2, pp 173-180

[18] Bontoux P; O'Connor I; Gaffiot F; Letartre X; Jacquemod G, "Behavioral modeling and simulation of optical integrated devices", *ANALOG INTEGRATED CIRCUITS AND SIGNAL PROCESSING* 2001, Vol 29, Iss 1-2, pp 37-47

[19] Zhang TH; Cao F; Dewey AM; Fair RB; Chakrabarty K, "Performance analysis of microelectrofluidic systems using hierarchical modeling and simulation", *IEEE TRANSACTIONS ON CIRCUITS AND SYSTEMS II-ANALOG AND DIGITAL SIGNAL PROCESSING* 2001, Vol 48, Iss 5, pp 482-491

[20] Voigt P; Schrag G; Wachutka G, "Microfluidic system modeling using VHDL-AMS and circuit simulation", *MICROELECTRONICS JOURNAL* 1998, Vol 29, Iss 11, pp791-797

[21] D.C. Jiles & D.L. Atherton, "Theory of Ferromagnetic Hysteresis (invited)", *Journal of Applied Physics*, Vol. 55 , No 6, Mar 1984, pp2115-2120

[22] D.C. Jiles & D.L.Atherton, "Theory of Ferromagnetic Hysteresis", *Journal of Magnetism and Magnetic Materials*, 61, 1986, pp48-60

[23] D.C. Jiles & D.L. Atherton, "Ferromagnetic Hysteresis", *IEEE Transactions on Magnetics*, Vol 19, No. 5, Sep 1983, pp2183-2185

[24] A. Hefner, "A Dynamic Electro-thermal model for the IGBT", *IEEE IAS Conference Record*, p1094, 1992

[25] A.R. Hefner and D.L. Blackburn, "Simulating the dynamic Electro-thermal behaviour of Power Electronic Circuits and Systems", *Conf. Rec. IEEE Workshop on computers in Power Electronics*, p143, 1992.

[26] P. Tenant, J.J. Rousseau, L. Zegadi, "Hysteresis modeling taking into account the temperature", *European Power Electronics Conference proceedings*, 1995, Volume 1, p1.001-1.006.

[27] A. Maxim, D. Andreu & J. Boucher, "New spice behavioral macromodeling method of magnetic components including the self-heating process", *PESC Record - IEEE Annual Power Electronics Specialists Conference*, Vol. 2, 1999, pp 735-740

[28] Hsu, J.T & Vu-Quoc, L., "A rational formulation of thermal circuit models for electro-thermal simulation –Part I: Finite Element Method", IEEE Transactions on Circuits and Systems, Vol 43, No. 9, pp 721-732, Sep. 1996

[29] Hsu, J.T & Vu-Quoc, L., "A rational formulation of thermal circuit models for electro-thermal simulation –Part II: Model Reduction Techniques", IEEE Transactions on Circuits and Systems, Vol 43, No. 9, pp 733-744, Sep. 1996

[30] P. R. Wilson, J. N. Ross & A. D. Brown, "Simulation of Magnetic Component Models in Electric Circuits including Dynamic Thermal Effects", *IEEE Transactions on Power Electronics*, Vol. 17, No. 1, Jan. 2002, pp55-65

STATISTICAL PROPERTIES OF CHAOTIC SEQUENCES GENERATED BY JACOBIAN ELLIPTIC CHEBYSHEV RATIONAL MAPS

Tohru Kohda

Kyushu University,
Dept. Computer Science and Communication Engg,
Fukuoka 812, Japan

ABSTRACT

Jacobian elliptic Chebyshev map is introduced as a rational version of Chebyshev polynomial, whose invariant measure is defined by homeomorphism of tent map. Such an elliptic function and its derivative, another elliptic function, induce Jacobian elliptic Chebyshev curve. Real-valued orbits on such a curve provide uniform distribution and give sequences of i.i.d. 2-dimensional binary random variables.

1. INTRODUCTION

Sequences of independent and identically distributed (i.i.d.) binary random variables [1] have found sigificant applications in modern digital communication systems, *e.g.*, spread spectrum (SS) systems and cryptosystems as well as in computational applications requiring random numbers [2]. As pointed out by Ulam and Von Neumann [3], the logistic map is a strong candidate for pseudorandom number generation without using linear feedback shift register (LFSR) sequences. A chaos-based generator of random bits has been proposed [4, 5]. Motivated by Ulam-Neumann's sophisticated statement, we have recently shown that a class of ergodic maps with *equidistributivity property* (EDP) can generate sequences of i.i.d. binary random variables if their associated binary function satisfy the *constant summation property* (CSP) [6]–[7]. Furthermore, we have defined Jacobian elliptic Chebyshev rational map as a rational version of Chebyshev polynomial and shown that it can generate a sequence of i.i.d. binary random variables [8]. Introducing derivative of Jacobian elliptic Chebyshev rational map defines an elliptic curve [9]. Real-valued orbits on the curve are shown to produce a sequence of i.i.d. 2-dimensional binary random variables.

This work was supported in part by Grant-in-Aid for Scientific Research of Ministry of Education, Science and Culture of Japan under grants No.15017271(2003) and No.15360206(2003).

2. INVERSE FUNCTION OF HOMEOMORPHISM AND INVARIANT MEASURE

Let us begin by reviewing a close relationship between inverse function of homeomorphism and an ergodic map $\tau : I = [d, e] \rightarrow I$ with its unique absolutely continuous invariant (ACI) measure, denoted by $f^*(\omega)d\omega$ [10]. Several maps are known as follows:

Bernoulli map with its ACI measure $f^*(\omega)d\omega = d\omega$:

$$\tau_B(\omega) = 2\omega (\bmod 1) = \begin{cases} 2\omega, & 0 < \omega < \frac{1}{2}, \\ 2\omega - 1, & \frac{1}{2} \leqslant \omega < 1. \end{cases} \quad (1)$$

Logistic map with ACI measure $f^*(\omega)d\omega = \dfrac{d\omega}{\pi\sqrt{\omega(1-\omega)}}$:

$$L_2(\omega) = 4\omega(1-\omega),\ \omega \in [0, 1]. \quad (2)$$

Piecewise linear map of p branches with $f^*(\omega)d\omega = d\omega$ [11]:

$$N_p(\omega) = (-1)^{\lfloor p\omega \rfloor} p\omega (\bmod p),\ \omega \in [0, 1]. \quad (3)$$

Chebyshev polynomial maps of degree p ($p = 2, 3, \cdots$) with $f^*(\omega)d\omega = \dfrac{d\omega}{\pi\sqrt{1-\omega^2}}$ [11]–[12]:

$$T_p(\omega) = \cos(p\cos^{-1}\omega),\ \omega \in [-1, 1]. \quad (4)$$

We have

Definition 1 *(Topological Conjugation) [10] Two transformations $\overline{\tau} : \overline{I} \rightarrow \overline{I}$ and $\tau : I \rightarrow I$ on intervals $\overline{I}$ and I are called topologically conjugate if there exists a homeomorphism h: $\overline{I} \stackrel{\text{onto}}{\longrightarrow} I$, such that*

$$\tau(\omega) = h \circ \overline{\tau} \circ h^{-1}(\omega). \quad (5)$$

Suppose $\tau(\cdot)$ and $\overline{\tau}(\cdot)$ have their ACI measures, denoted by $f^*(\omega)d\omega$ and $\overline{f}^*(\overline{\omega})d\overline{\omega}$, respectively. Then, these ACI measures have the relation

$$f^*(\omega) = \left|\frac{dh^{-1}(\omega)}{d\omega}\right| \overline{f}^*(h^{-1}(\omega)) \quad (6)$$

holds.

Remark 1: If we take the tent map $N_2(\omega)$ as $\bar{\tau}(\cdot)$, then $f^*(\omega)$ is simply represented by the derivative of inverse function of homeomorphism $h(\omega)$. For example, sin function is defined in the inverse function form:

$$\omega = \int_0^{\sin\omega} \frac{du}{\sqrt{1-u^2}}, \quad \sin^{-1}\omega = \int_0^{\omega} \frac{du}{\sqrt{1-u^2}}. \tag{7}$$

In fact, Ulam and von Neumann [3] got $L_2(\omega)$ via the inverse of homeomorphism $h^{-1}(\omega) = \frac{2}{\pi}\sin^{-1}\sqrt{\omega}$.

3. CONSTANT SUMMATION PROPERTY IMPLIES STATISTICAL INDEPENDENCE OF CHAOTIC SEQUENCES

Next let us review the elementary theory of coin tossing [1]. Consider the Bernoulli map $\tau_B(\omega)$ for an initial point ω whose binary expansion is given by

$$\left.\begin{array}{l} \omega = 0.d_1(\omega)d_2(\omega)\cdots d_n(\omega);\ d_k(\omega)\in\{0,1\}, \\ \tau_B(\omega) = 0.d_2(\omega)d_3(\omega)\cdots. \end{array}\right\} \tag{8}$$

The functions $d_k(\cdot)$, called Rademacher functions, furnish us with a model of independent tosses of a 'fair' coin. It is well known that for almost every ω, $\{d_k(\omega)\}_{k=1}^{\infty}$ can imitate coin tossing. Hence, $\{d_k(\omega)\}_{k=1}^{\infty}$ is considered as a theoretical model producing a sequence of i.i.d. binary random variables (BRVs). Obviously, the important question arises, can any other map and its associated pulse function generate a sequence of i.i.d. BRVs? We have got an affirmative answer to this question [6, 7], which is firstly, the map should satisfy EDP and secondly, the binary function should satisfy CSP [1].

Let us show that EDP and CSP guarantee statistical independence of a sequence of i.i.d. random variables and that these correspond to trivial and non-trivial partition of the interval. To evaluate statistical properties of chaotic sequence, we introduce Perron-Frobenius (PF) operator P_τ acting on a function of bounded variation $H(\omega) \in L^\infty$ for $\tau(\omega)$ which can be defined in several forms.

Definition 2 *[10] The simplest one is defined by*

$$\int_I P_\tau H(\omega)G(\omega)d\omega = \int_I H(\omega)G(\tau(\omega))d\omega \tag{9}$$

as its adjoint opertator $G(\tau(\omega))$, called Koopman operator. The second one is represented by

$$P_\tau H(\omega) = \frac{d}{d\omega}\int_{\tau^{-1}([d,\omega])} H(y)dy = \sum_{i=1}^{N_\tau} |g_i'(\omega)| H(g_i(\omega)), \tag{10}$$

where $g_i(\omega) = \tau_i^{-1}(\omega)$ is the $i-$th preimage of ω and N_τ denotes the number of preimages.

[1] Fortunately, EDP is satisfied by many well-known maps [7]. Furthermore, EDP is proven to be invariant under topological conjugation.

ACI measure $f^*(\omega)d\omega$ satisfies the PF equation:

$$P_\tau f^*(\omega) = f^*(\omega). \tag{11}$$

The PF operator of map $x = \tau(y)$ is a transfer operator of input distribution $f_y(y)dy$. The resulting output one $f_x(x)dx$ which comes from two input distributions gives *conservation law of probability measure*, defined by

$$f^*(x) = \frac{f^*(y_1)}{|\tau'(y_1)|} + \frac{f^*(y_2)}{|\tau'(y_2)|}. \tag{12}$$

Replacing $(x, y_i, |\tau'(y_i)|^{-1})$ by $(\omega, g_i(\omega), |g_i'(\omega)|)$ gives

$$f^*(\omega) = |g_1'(\omega)|f^*(g_1(\omega)) + |g_2'(\omega)|f^*(g_2(\omega)) \stackrel{\text{def}}{=} P_\tau f^*(\omega) \tag{13}$$

whose generalized version gives eq.(10). Hence we get

Definition 3 *[6] If a piecewise-monotone onto map $\tau(\omega)$ satisfies*

$$|g_i'(\omega)|f^*(g_i(\omega)) = \frac{1}{N_\tau} f^*(\omega), \quad 0 \leqslant i \leqslant N_\tau - 1, \tag{14}$$

then it is said to satisfy EDP.

EDP implies 2 facts. Firstly, infinitesimal measure $f^*(\omega)d\omega$ is equicontributed. Secondly, the occurrence of symbols s_0 and s_1 generated by a binary function $\sigma(\omega)$, defined by

$$\sigma(\omega) = \begin{cases} s_0, & \text{for} \quad \omega \in J_0 \\ s_1, & \text{for} \quad \omega \in J_1 \end{cases} \tag{15}$$

is equiprobable with respect to trivial partition of the interval, $I = J_{s_0} \cup J_{s_1}$, $J_{s_0} \cap J_{s_1} = \phi$. That is, the map $\tau(\cdot)$ with EDP can generate a sequence of i.i.d. binary random variables s_0 and s_1.

Now let us consider a stationary real-valued sequence $\{H(X_n)\}_{n=0}^{\infty}$, where $X_n = \tau^n(\omega)$. The ensemble average $\mathbf{E}[H(X_n)]$ is defined by

$$\mathbf{E}[H(X_n)] = \int_I H(\tau^n(\omega))f^*(\omega)d\omega. \tag{16}$$

From stationarity of the process, we denote $\mathbf{E}[H(X_n)]$ by $\mathbf{E}[H(X)]$.

Definition 4 *[6, 7] For a class of piecewise monotone onto maps with EDP, if its associated function $H(\cdot)$ satisfies*

$$\frac{1}{N_\tau}\sum_{i=0}^{N_\tau-1} H(g_i(\omega)) = \mathbf{E}[H(X)], \tag{17}$$

then $H(\cdot)$ is said to satisfy CSP.

Theorem 1: [6, 7] *For a class of piecewise-monotone onto maps with EDP, following three symmetric properties:*

S1: *the symmetric binary function* $C_T(\omega)$ *with* $T = \{t_r\}_{r=0}^{2M}$, *defined as*

$$C_T(\omega) = \bigoplus_{r=0}^{2M} \Theta_{t_r}(\omega) \quad (18)$$

$$\Theta_{t_r}(\omega) = \begin{cases} 0, & \textit{for} \quad \omega < t_r \\ 1, & \textit{for} \quad \omega \geq t_r \end{cases} \quad (19)$$

satisfying $t_r + t_{2M-r} = d + e$, $r = 0, 1, \cdots, M$;

S2: *the symmetric ACI measure, defined as* $f^*(d + e - \omega) = f^*(\omega)$;

S3: *the symmetric map, defined as* $\tau(d + e - \omega) = \tau(\omega), \quad \omega \in I$

give

$$P_\tau\{C_T(\omega) f^*(\omega)\} = \mathbf{E}[C_T] f^*(\omega). \quad (20)$$

Relation (20) implies CSP of $C_T(\omega)$ holds, which guarantees $\{C_T(\tau^n(\omega)\}_{n=0}^{\infty}\}$ is a sequence of i.i.d. BRVs with probability $\mathbf{E}[C_T]$. Hence, $C_T(\omega)$ is expected to hold great promise for generating running-key sequences in a stream cipher system and a lot of sequences of i.i.d. binary random variables in Monte-Carlo simulations if the system implemented in floating-point environment can overcome inevitable round off errors.

4. JACOBIAN ELLPTIC CHEBYSHEV RATIONAL MAP

Katsura and Fukuda [13] gave a rational function version of the logistic map $L_2(\omega)$ with parameter k using a homeomorphism $h^{-1}(\omega) = \frac{1}{K(k)} \mathrm{sn}^{-1}(\sqrt{\omega}, k)$, defined by

$$R_2^{\mathrm{sn}^2}(\omega, k) = \frac{4\omega(1-\omega)(1-k^2\omega)}{(1-k^2\omega^2)^2}, \quad (21)$$

$$f^*(\omega, k) d\omega = \frac{d\omega}{2K(k)\sqrt{\omega(1-\omega)(1-k^2\omega)}}, \quad (22)$$

where $\mathrm{sn}(\omega, k)$ is the inverse function of the elliptic integral with modulus k ($|k| < 1$) and $K(k)$ is the complete elliptic integral, each of which is defined respectively as

$$\omega = \int_0^{\mathrm{sn}(\omega,k)} \frac{dv}{\sqrt{(1-v^2)(1-k^2v^2)}}, \quad (23)$$

$$K(k) = \int_0^{\frac{\pi}{2}} \frac{d\theta}{\sqrt{1-k^2\sin^2\theta}}. \quad (24)$$

Kohda and Fujisaki [8] have recently introduced the *Jacobian elliptic Chebyshev rational map* with modulus k using homeomorphism $h^{-1}(\omega, k) = \frac{\mathrm{cn}^{-1}(\omega,k)}{2K(k)}$ of $N_p(\overline{\omega})$, $p = 2, 3, 4, \cdots$, defined as

$$R_p^{\mathrm{cn}}(\omega, k) = \mathrm{cn}(p\,\mathrm{cn}^{-1}(\omega, k), k), \; \omega \in [-1, 1], \quad (25)$$

$$f^*(\omega, k) d\omega = \frac{d\omega}{2K(k)\sqrt{(1-\omega^2)\{(1-k^2)+k^2\omega^2\}}}, \quad (26)$$

where $\mathrm{cn}(\omega, k)$ is the inverse function of the elliptic integral with modulus k, *i.e.*,

$$\omega = \int_{\mathrm{cn}(\omega,k)}^{1} \frac{dv}{\sqrt{(1-v^2)\{(1-k^2)+k^2v^2\}}}. \quad (27)$$

We know that $R_p^{\mathrm{cn}}(\omega, k)$ satisfies the semi-group property $R_r^{\mathrm{cn}}(R_s^{\mathrm{cn}}(\omega, k), k) = R_{rs}^{\mathrm{cn}}(\omega, k)$ and when $p = 2$, gives

$$R_2^{\mathrm{cn}}(\omega, k) = \frac{1 - 2(1-\omega^2) + k^2(1-\omega^2)^2}{1 - k^2(1-\omega^2)^2}. \quad (28)$$

Using the trivial relation $K(0) = \frac{\pi}{2}$, we get

$$R_p^{\mathrm{cn}}(\omega, 0) = T_p(\omega), \; f^*(\omega, 0) d\omega = \frac{d\omega}{\pi\sqrt{1-\omega^2}}. \quad (29)$$

5. DYNAMICS OF REAL-VALUED ORBIT ON JACOBIAN ELLIPTIC CURVE

The Weierstrass elliptic fuction $\wp(z)$ ($z \in \mathcal{C} \bigcup \{\infty\}$), as the inverse function of the elliptic integral in Weierstrass normal form, defined as

$$z = \int_{\wp(z)}^{\infty} \frac{dt}{\sqrt{4t^3 - g_2t - g_3}} \quad (30)$$

has the duplication formula

$$\wp(2z) = \frac{1}{4}\{\frac{\wp''(z)}{\wp'(z)}\}^2 - 2\wp(z) \quad (31)$$

which gives

$$\wp(2z) = R_2^{\wp}(z) = \frac{z^4 + g_2(z^2/2) + 2g_3z + (g_2/4)^2}{4z^3 - g_2z - g_3} \quad (32)$$

and satisfies the "algebraic relation", called elliptic curve

$$(\wp')^2 = 4\wp^3 - g_2\wp - g_3, \quad (33)$$

where g_2 and g_3 are known quantities defined in terms of lattice Λ subject to the restriction $g_2^3 - 27g_3^2 \neq 0$. Furthermore, duplication formulae for $\wp(2z)$ and $\wp'(2z)$ give 2-dimensional maps [14].

Another elliptic function is introduced as a consequence of such investigation. [9] The duplication formula for $x_n = \mathrm{cn}\,\omega$, $y_n = (\mathrm{cn}\,\omega)'$ leads us to get the 2-d map $(x_{n+1}, y_{n+1}) = (\phi_2^{\mathrm{cn}}(x_n, y_n), \psi_2^{\mathrm{cn}}(x_n, y_n)) = (\mathrm{cn}\,2\omega, \frac{(\mathrm{cn}\,2\omega)'}{2})$ defined as:

$$x_{n+1} = \frac{x_n^2 - y_n^2}{x_n^2 + y_n^2}, \quad (34)$$

$$y_{n+1} = \frac{2x_ny_n(1-k^2+k^2x_n^4)}{(x_n^2+y_n^2)^2} \quad (35)$$

which has a *stable* invariant curve, defined by $Y^2 = (1 - X^2)(1 - k^2 + k^2X^2)$.The adjective *stable* means the curve behaves as an *attractive* curve as shown in Fig. 1. On

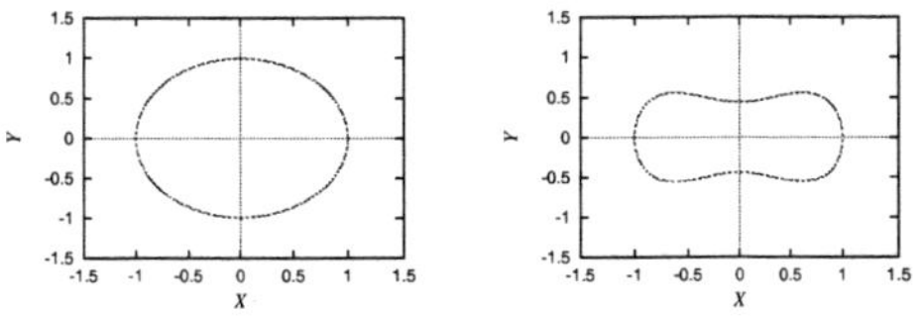

(a) $k = 0$ (b) $k = 0.9$

Fig. 1. Stable invariant curves.

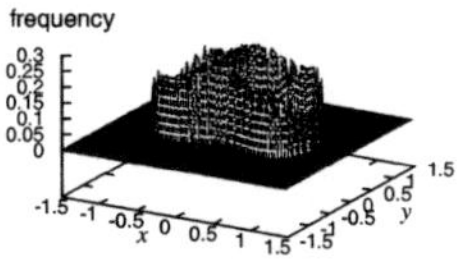

Fig. 2. Joint distributions with $k = 0.9$.

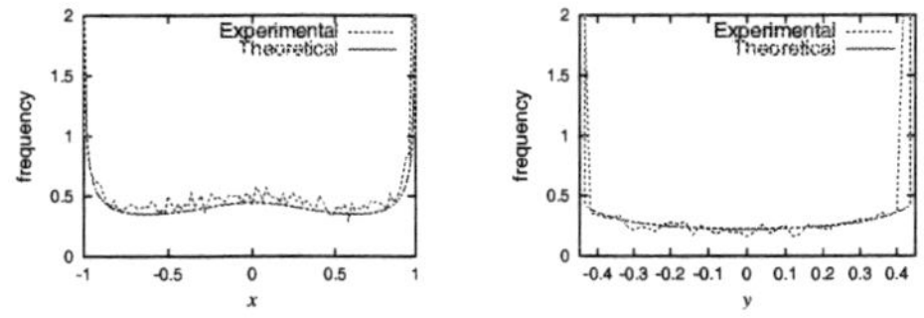

(a) (b)

Fig. 3. Marginal distributions (a) μ_X and (b) μ_Y with $k = 0.9$.

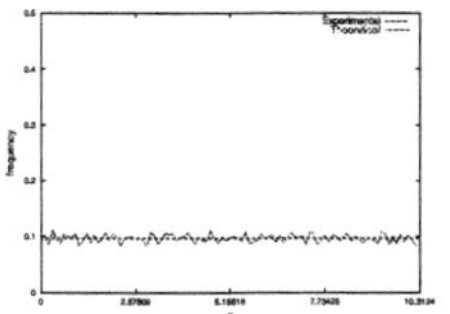

Fig. 4. Distribution μ_u along the invariant curve with $k = 0.9$.

the contrary, invariant curve $Y^2 = 4X(1-X)(1-k^2X)$ for the 2-dimensional map $(\phi_2^{\mathrm{sn}^2}(x_n, y_n), \psi_2^{\mathrm{sn}^2}(x_n, y_n)) = (\mathrm{sn}^2 2\omega, \frac{(\mathrm{sn}^2 2\omega)'}{2})$ of $R_2^{\mathrm{sn}^2}(\omega)$ with $x_n = \mathrm{sn}^2\,\omega,\ y_n = (\mathrm{sn}^2\,\omega)'$, defined as:

$$x_{n+1} = \frac{y_n^2}{(1-k^2x_n^2)^2}, \tag{36}$$

$$y_{n+1} = \frac{2y_n(1-2x_n+k^2x_n^2)(1-2k^2x_n+k^2x_n^2)}{(1-k^2x_n^2)^3} \tag{37}$$

is *unstable*. Observed measure of real-valued orbits, generated by 2-d map $(\phi_2^{\mathrm{cn}}(x_n, y_n), \psi_2^{\mathrm{cn}}(x_n, y_n))$ as shown in Fig. 2 supports that there exist ACI measures on the curve. Figs. 3 (a) and (b) show its marginal distributions of y and x, respectively defined by $\mu_X = \int_{-1}^{1} f^*(x,y)dy$ and $\mu_Y = \int_{-1}^{1} f^*(x,y)dx$ as well as their theoretical ones which are given by integrands of elliptic integrals for inverse functions $(\mathrm{cn}\,u)^{-1}$ and $\left(\frac{d\,\mathrm{cn}\,u}{du}\right)^{-1} = -(\mathrm{sn}\,u\,\mathrm{dn}\,u)^{-1}$, respectively, *e.g.*, μ_X is identical to $f^*(\omega,k)d\omega$ (eq. (26)) [9]. This implies derivatives of inverse functions of homeomorphisms to the tent map $N_2(u)$ with $\theta = \mathrm{am}(u,k)$ are represented by a kind of elliptic functions, where $\mathrm{am}(u,k)$ is Jacobi's amplitude function, defined by

$$u = \int_0^\theta \frac{d\theta}{\sqrt{1-k^2\sin\theta}}. \tag{38}$$

Fig. 4 shows distribution of the u-axis along the invariant curve, denoted by μ_u, *i.e.*, which is identical to the uniform ACI measure of $N_2(u)$, defined as $f^*(u)du = du$.

6. CONCLUDING REMARKS

Jacobian elliptic Chebyshev rational map and its associated binary function can generate a sequence of i.i.d. 2-dimensional p-ary random variables. Measure of real-valued orbits on elliptic curve is numerically and theoretically observed.

7. REFERENCES

[1] P. Billingsley, *Probability and Measure* Third Edition, John Wiley & Sons, Inc., 1995.

[2] D. Knuth, "The Art of Computer Programming 2, Seminumerical Algorithms," 2nd ed., Addison-Wesley, Reading, Mass, 1981.

[3] S. M. Ulam and J.Von Neumann, "On combination of stochastic and deterministic processes", *Bull. Math. Soc.*. **53**, p.1120, 1947.

[4] T.Stojanovski and L.Kocarev, "Chaos based random number generators PartI:Analysis," *IEEE Trans, Crcuits Syst.I,* vol.43, no.3,pp.281-288, 2001.

[5] L.Kocarev and Goce Jakimoski, "Pseudorandom bits-generated by chaotic maps," *IEEE Trans, Crcuits Syst.I,* vol.50, no,1, pp.123-126, 2003.

[6] T. Kohda and A. Tsuneda, "Statistics of Chaotic Binary Sequences", IEEE Trans. Information Theory, Vol.43, No.1,pp.104–112, 1997.

[7] T. Kohda, "Information Sources using chaotic dynamics", *Proc. IEEE*, **90**, no.5, 641–661, 2002.

[8] T.Kohda and H.Fujisaki, "Jacobian Elliptic Chebyshev Rational Maps", *Physica D*, **148**, pp.242-254, 2001.

[9] T. Kohda, A.Katoh, Y.Hattori and T.Yoshimura, "2-d dynamics of Jacobian Elliptic Chebyshev rational maps," *Proceedings of the 11th workshop on Nonlinear Dynamics of Electronic System*, pp.129–132, 2003.

[10] A. Lasota and M. C. Mackey, "*Chaos, Fractals and Noise*", Springer-Verlag, 1994.

[11] S. Grossmann and S. Thomae, "Invariant distributions and stationary correlation functions of one-dimensional discrete processes", *Z. Naturforsch.*, **32a**, 1353–1363, 1977.

[12] T. J. Rivlin, "*Chebyshev Polynomials*", John Wiley & Sons, Inc., 1990.

[13] S. Katsura, W. Fukuda, "Exactly Solvable Models Showing Chaotic Behaviour", *Physica A*, **130**,pp.597-605, 1985.

[14] R.Brown and L.O.Chua, "Clarifying Chaos:Examples and Counterexamples", *Journal of Chaos and Bifurcation*, **6**-2, pp.219–249, 1996.

OUTAGE IN CHAOS COMMUNICATION

A. J. Lawrance and *G. Ohama*

University of Birmingham, UK
Email: A.J.Lawrance@bham.ac.uk

Shiga University, Japan
Email: ohama@biwako.shiga-u.ac.uk

ABSTRACT

The paper investigates the notion of outage, as an assessment of the likelihood of a bit being in error, for chaos-based communication systems. It is seen as a complementary measure of performance to bit error rate. For antipodal chaos shift keying (CSK) a theoretically exact formula is derived which is then used in specific cases to obtain useful qualitative and quantitative information. Illustrations are with independent Gaussian spreading as an upper benchmark and with the logistic map spreading as typical of effective chaos; a uniform spreading distribution is also used. Comparisons are obtained between these spreading distributions and between different extents of their spreading, calibrated against per bit signal to noise ratio. The distribution of bit error probability over spreading sequences is also obtained, being the fundamental quantity for both outage and bit error rate. The form of this distribution illustrates the degree to which bit error rate is a stable representation of performance.

1. INTRODUCTION

The assessment of performance in chaos-based communication modeling has in the past mainly been limited to bit error rates. Another general idea of efficiency in communication systems is *outage*; this concept tries to capture the rate of transmission for which a bit error is likely, according to some quantification of 'likely', rather than the rate at which it actually occurs. The aim of this paper is to sketch an exact approach to the calculation of outage and to give exemplification using correlation decoders in simple CSK systems. The work follows on from our recent approach [1] which has produced mathematically exact bit error rates for some detectors in CSK models, in contrast to earlier work which rested on analytical approximation or simulation.

2. OUTAGE AND CSK SYSTEMS

The quantification of the likelihood of error of an individual bit is that the probability of bit error (*PBE*) is greater than α, say, a *severity* parameter; *PBE* has range $(0,\frac{1}{2})$ and so also $0<\alpha<\frac{1}{2}$; the smaller the value of α the more severe or demanding the definition of outage. Typical values of α are 10^{-3} for voice transmissions and 10^{-4} for data transmissions. In CSK systems one imagines being able to calculate the *PBE*'s for an individual bit as a function of its known spreading segment, and hence determining a distribution of *PBE* values over all possible spreading segments. *Outage rate* is then the probability from this distribution for which $PBE>\alpha$. The indicator function $I^\alpha \equiv I(PBE>\alpha)$ is used to obtain the proportion of the spreading segments with $I^\alpha=1$; a reliable bit gives $I^\alpha=0$. A poor communication system will have a high theoretical probability of $I^\alpha=1$. This is an aspect of the distribution of *PBE* values which will be studied; the usually calculated theoretical bit error probability is the mean of this distribution. Thus, both outage probability and the traditional bit error probability are different aspects of a bit error probability distribution.

An antipodal coherent CSK system is considered in which the chaotic spreading sequence $\{X_i\}$ is generated by the chaotic map $\tau_X(z),\, c<z<d$ which has mean μ_X, variance σ_X^2 and a symmetric invariant distribution. Assume AWGN $\{\varepsilon_t\}$ of variance σ^2 in the channel. A spreading factor N is used in transmitting each binary bit b. A +1 bit value is signified by transmitting the spreading sequence values unaltered, while for a −1 the spreading values are reflected about their mean. The

received bit sequence $R=(R_1,R_2,...,R_N)$ is thus of the form

$$R_i=\mu_X+b(X_i-\mu_X)+\varepsilon_i, i=1,2,\ldots,N, b=\pm 1. \quad (2.1)$$

At the receiver a synchronized exact copy of the spreading segment $(X_1,X_2,...,X_N)$ is assumed known, since the system is coherent, and a correlation decoder

$$C(X,R)\equiv\sum_{i=1}^{N}(X_i-\mu_X)(R_i-\mu_X) \quad (2.2)$$

is used to decide bit type, positive values indicating a $+1$ and negative values a -1.

The probability of a $+1$ bit error as a function of its own spreading segment is given by

$$\begin{aligned}BER(X_1,X_2,...,X_N)&=P\{C(X,R)<0\,|\,b=1\}\\&=P\left\{\sum_{i=1}^{N}(X_i-\mu_X)^2+\sum_{i=1}^{N}\varepsilon_i(X_i-\mu_X)<0\right\}\end{aligned} \quad (2.3)$$

which usefully is the same as that for a -1 bit error in this case. The probability (2.3) can be further evaluated by noting that, conditional on $(X_1,X_2,...,X_N)$, the second summation in the argument of the right-hand side is a Gaussian random variable. Then, with $\Phi(\cdot)$ denoting the distribution of a standardized Gaussian variable, there is the result

$$BER(X_1,X_2,...,X_N)=\Phi\left\{-\sqrt{\sum_{i=1}^{N}(X_i-\mu_X)^2}\Big/\sigma\right\} \quad (2.4)$$

which takes values in the range $(0,\frac{1}{2})$. The outage indicator variable is defined as

$$I^{\alpha}(X_1,X_2,...,X_N)=\begin{cases}1 \ \text{if} \ BER(X_1,X_2,...,X_N)>\alpha\\0 \ \text{if} \ BER(X_1,X_2,...,X_N)\le\alpha.\end{cases} \quad (2.5)$$

The outage probability $O(\alpha)$ is the expectation of (2.5) where as the bit error rate is the expectation of (2.4). As with (2.4) from our previous work, [1], (2.5) can be evaluated exactly. From (2.5), the outage rate $O(\alpha)$ becomes

$$O(\alpha)=\int_{x:\,BER(x)>\alpha}f(x)dx, \quad (2.6)$$

where $f(x)$ is the $N-$dimensional joint density function referring to the values $x=(x_1,x_2,...,x_N)$ of a one-bit spreading segment. The region of the integration in (2.6) can be expressed as

$$x:\sum_{i=1}^{N}(x_i-\mu_X)^2<\left[\sigma\Phi^{-1}(1-\alpha)\right]^2 \quad (2.7)$$

and so from (2.4), (2.6) becomes

$$O(\alpha)=\int_{x:\,\sum_{i=1}^{N}(x_i-\mu_X)^2<\left[\sigma\Phi^{-1}(1-\alpha)\right]^2}f(x)dx. \quad (2.8)$$

This key expression allows further specific results to be obtained.

First consider the case of no spreading, chaotic or not, that is $N=1$, applying to any distribution of a single spreading variable X; the inequality in (2.7) can be written

$$(x-\mu_X)^2<\left[\sigma\Phi^{-1}(1-\alpha)\right]^2 \quad (2.9)$$

and expressed using the signal to noise ratio $(\sigma_X/\sigma)^2$, as

$$\mu_X-\frac{\Phi^{-1}(1-\alpha)}{\sigma_X/\sigma}\sigma_X<x<\mu_X+\frac{\Phi^{-1}(1-\alpha)}{\sigma_X/\sigma}\sigma_X$$

with (2.6) becoming

$$O(\alpha)=\int_{\mu_X-\frac{\Phi^{-1}(1-\alpha)}{\sigma_X/\sigma}\sigma_X}^{\mu_X+\frac{\Phi^{-1}(1-\alpha)}{\sigma_X/\sigma}\sigma_X}f(x)dx. \quad (2.10)$$

This shows, as expected, reducing outage as the signal to noise ratio increases and that when the ratio is less than $\sigma_X\Phi^{-1}(1-\alpha)/\{\frac{1}{2}(d-c)\}$ that the outage probability is 1. Note also the reducing outage as α increases, that is when the definition of outage becomes less severe. It shows why outage will be large for symmetric unimodal spreading distributions and small for distributions with large amounts of tail probability, as for the $beta(\frac{1}{2},\frac{1}{2})$ distribution. Explicit results will be given for Gaussian, and $beta(\frac{1}{2},\frac{1}{2})$ distributions, as will follow.

3. GAUSSIAN DISTRIBUTION OF SPREADING

While interest is primarily in chaotic spreading sequences, independent Gaussian ones usually give a poor performance, they do provide an upper bench mark for the use of chaotic spreading, and are very tractable. In this situation, (2.4) is the function $\Phi\left\{-(\sigma_X/\sigma)\sqrt{\chi_N^2}\right\}$ of a χ_N^2 variable with N degrees of freedom, following from $\sum_{i=1}^{N}(X_i-\mu_X)^2 \big/ \sigma_X^2$ having a χ_N^2 distribution. Thus, the outage probability (2.6) can be expressed in terms of the distribution function, $F_{\chi_N^2}(x)$, of a χ_N^2 as

$$O(\alpha) = F_{\chi_N^2}\left(\left[\frac{\Phi^{-1}(1-\alpha)}{\sigma_X/\sigma}\right]^2\right). \tag{3.1}$$

This will usually be considered and plotted as a function of the *per bit signal to noise ratio*, *pbSNR*, defined here as $N\sigma_X^2/\sigma^2$, not in decibel form. The result shows that the larger the signal to noise ratio, the nearer the outage probability $O(\alpha)$ is to zero, as it should be. As a function of α, outage decreases from 1 at $\alpha=0$ to 0 at $\alpha=\frac{1}{2}$. The plots in Figure 1 illustrate the decrease in outage probability for increasing N as the *pbSNR* increases from unity, and the corresponding increase in outage probability as the *pbSNR* decreases from unity to zero.

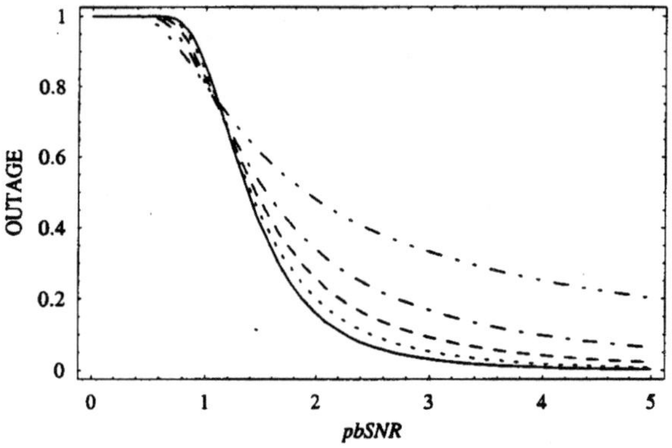

Figure 1. Plot of outage probability $O(\alpha)$ against pbSNR for $\alpha = 0.1$ and independent Gaussian spreading sequence when $N=1$ (dot-dot-dashed line), $N=2$ (dot-dashed line), $N=3$ (dashed line), $N=4$ (dotted line) and $N=5$ (solid line).

4. CHAOTIC SPREADING

Turning to $O(\alpha)$ for chaotic spreading, the region of integration in (2.8) is considered. The bit error probability (2.4) now needs to be expressed as

$$BER(x) = \Phi\left\{-\sqrt{\sum_{i=1}^{N}\left(\tau^{i-1}(x)-\mu_X\right)^2} \Big/ \sigma\right\} \tag{4.1}$$

where $\tau^{i-1}(x)$ is the $(i-1)^{\text{th}}$ iteration of the map starting with initial value x, and which itself follows the invariant distribution of the map. The region of integration is thus some segments of the interval $[c,d]$ given by

$$x : \Phi\left\{-\sqrt{\sum_{i=1}^{N}\left(\tau^{i-1}(x)-\mu_X\right)^2} \Big/ \sigma\right\} > \alpha \tag{4.2}$$

and in general not easy to determine. For a logistic map, $N=2$ is at the limit of analytical explicitness, but still too involved to report here. Illustrations from using (4.1) and (4.2) are given in Figure 2 for a range of severity parameter values and in particular show the sensitivity of the outage rate to *pbSNR* and the severity parameter.

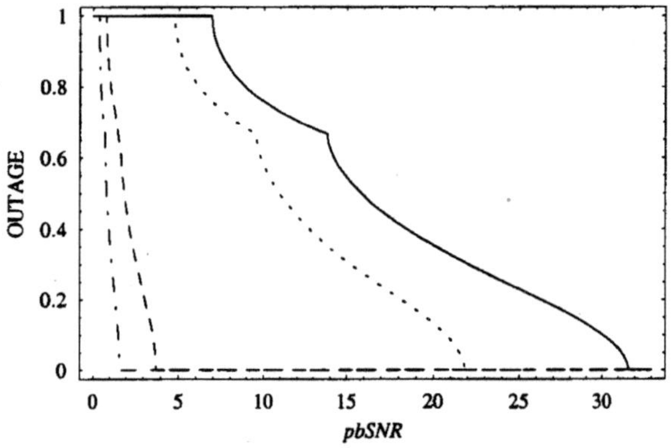

Figure 2. Plot of outage probability $O(\alpha)$ against pbSNR for logistic map spreading spreading $(N=2)$ and selected α, $\alpha=0.0001$ (solid line), $\alpha=0.001$ (dotted line) and $\alpha=0.01$ (dashed line) and $\alpha=0.1$ (dot-dashed line).

5. DISTRIBUTION OF BIT ERROR PROBABILITY AND OUTAGE

The distribution of bit error probability (2.4) over all possible spreading sequences is not usually considered but it extends the understanding of bit error and provides a further way of assessing a communications system; clearly, there is advantage in having a fairly constant bit error robability so that its expectation provides a good description of bit by bit behaviour. This distributional aspect might also be useful when comparing systems.

For general chaotic spreading (2.4) gives

$$BER(x) = \Phi\left\{-\frac{\sigma_X}{\sigma}\sqrt{N}S_\tau(x)\right\},$$

$$S_\tau^2(x) = \tfrac{1}{N}\sum_{i=1}^{N}\left(\tau_X^{i-1}(x)-\mu_X\right)^2 \Big/ \sigma_X^2, \tag{5.1}$$

and the distribution function of BER as

$$P(BER < z) = 1 - P\left\{S_\tau^2(X) > \left(\frac{\Phi^{-1}(1-z)}{(\sigma_X/\sigma)\sqrt{N}}\right)^2\right\}. \tag{5.2}$$

The distribution of $S_\tau^2(X)$ is the key aspect here, and in general, not available explicitly without approximation, except when $N=1$. In this case

$$P(BER < z) =$$

$$1 - F_X\{\mu_X + \sigma\Phi^{-1}(1-z)\} + F_X\{\mu_X - \sigma\Phi^{-1}(1-z)\}$$

$$\text{for } \Phi\left(\frac{d-c}{2\sigma}\right) < z < \tfrac{1}{2} \tag{5.3}$$

and is otherwise zero. In density terms this gives the required result

$$f_{BER}(z) = 2\sigma\Phi^{-1\prime}(z) f_X\{\mu_X + \sigma\Phi^{-1}(z)\},$$

$$\text{for } \Phi\left(\frac{d-c}{2\sigma}\right) < z < \tfrac{1}{2}. \tag{5.4}$$

This $N=1$ result is tractable for Gaussian, uniform and beta spreading distributions; it becomes most explicit in the Gaussian case as

$$f_{BER}(z) = 2\frac{\Phi^{-1\prime}(z)}{\sigma_X/\sigma}\phi\left(\frac{\Phi^{-1}(z)}{\sigma_X/\sigma}\right),$$

$$\Phi\left(\frac{d-c}{2\sigma}\right) < z < \tfrac{1}{2}. \tag{5.5}$$

When $\sigma_X = \sigma$, which is equivalently $pbSNR$=1 since $N=1$, (5.5) gives a uniform distribution over $(0,\frac{1}{2})$. In such situations there is no preferred or modal value of bit error, and thus the distribution average or bit error rate is unsatisfactory, to a degree. A comparison of these spreading distributions in this case is given in Figure 3.

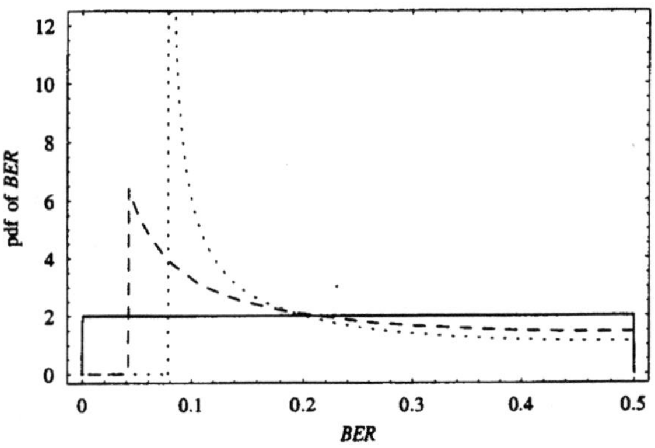

Figure 3. Distributions of the bit error probability when $N=1$ and pbSNR=1 for the spreading distributions $beta(\frac{1}{2},\frac{1}{2})$ (dotted line), uniform (dashed line) and Gaussian (solid line).

Superiority of the $beta(\frac{1}{2},\frac{1}{2})$ distribution is again evident but long and fairly constant tails are a feature of all three distributions.

6. RELATION BETWEEN BIT ERROR AND OUTAGE

Approximating the distribution of $B(X)$ from (5.1) over a chosen number K (≥ 1) intervals $\left(\frac{1}{2}+\frac{1}{2}\frac{k-1}{K}, \frac{1}{2}+\frac{1}{2}\frac{k}{K}\right)$, $k=1,2,\ldots,K$, covering $(\frac{1}{2},1)$, the bit error rate becomes

$$BER \simeq \tfrac{1}{4K} + \tfrac{1}{2K}\sum_{k=1}^{K-1} O\left\{\tfrac{1}{2}\left(1-\tfrac{k}{K}\right)\right\}. \tag{6.1}$$

The result shows that bit error rate and outage probabilities are complementary measures of bit error, to an extent. Whereas bit error can be expressed in terms of outage, outage cannot be expressed in terms of bit error. Distinctly, outage includes a severity parameter which offers the advantage of adjustment to circumstances in applications.

6. REFERENCE

[1] A. J. Lawrance & G. Ohama (2003). "Exact calculation of bit error rates in communication systems with chaotic modulation," *IEEE Transactions on Circuits and Systems –I: Fundamental Theory and Applications, vol.* 50, *No.* 11, pp.1391-1400, 2003.

ACKNOWLEDGMENT

The authors are grateful to Professor G Mazzini (University of Ferrara) for stimulating this investigation.

ANALYSIS OF HYBRID SYSTEMS BY MEANS OF EMBEDDED RETURN MAPS

J. Krupar, A. Mögel and W. Schwarz

Dresden University of Technology
Department of Electrical Engineering
Mommsenstrasse 13, 01062 Dresden, Germany

ABSTRACT

The behavioural analysis of hybrid systems is to a large extend an open problem. In this paper it is shown how the construction of an embedded map can be used to determine the characteristics of the output signal of a hybrid system. The method is demonstrated with two DC-DC converter examples.

1. INTRODUCTION

Hybrid systems often occur in modern electronics. They consist of continuous subsystems and digital subsystems. Fig. 1 depicts a signal of a hybrid system. Typical hybrid systems are clock generators, DC-DC converters, charge pumps, switched mechanical actors etc. Analysis of these systems often is more or less complicated and a variety of tools exists [1], [2], [3] for this purpose. However, conditions under that one of the tools is applicable are always special and need to be checked in each application.

The use of return maps unifies the analysis for most of these systems. Depending on the special application the embedded map often can be simplified or reduced in the number of state variables.

In Section 2 we summarize the derivation of the return map of a hybrid system. Section 3 derives the two-dimensional map for a DC-DC boost converter operating in the current programmed mode (CPM). Transient simulation and models for the control loop are discussed. The map is simplified into a one dimensional one in section 3.2. To this map statistical analysis is applied. In section 4 parametric optimization of given maps are done considering the practical aspects EMI and switching loss.

2. RETURN MAP OF A HYBRID SYSTEM

The hybrid system switches between different parameters when a switching event occurs. This is the case when either a state condition is satisfied or a time event (clock signal) occurs. At each event the parameters of the system change. Between the events the parameters are constant.

Knowing the solutions of the continuous-time subsystem the characteristics of the continuous-time output signal are calculated from the return map g. When the system has no memory Eq. (1) or (2) can be derived.

$$A_{n+1} = g(A_n) \tag{1}$$

$$t_{n+1} = g(t_n) \tag{2}$$

If there is a memory then a subsequence of events must be taken into account. The occurence times of the switching events are:

$$\tau_{n+1} = \tau_n + t_{n+1} \tag{3}$$

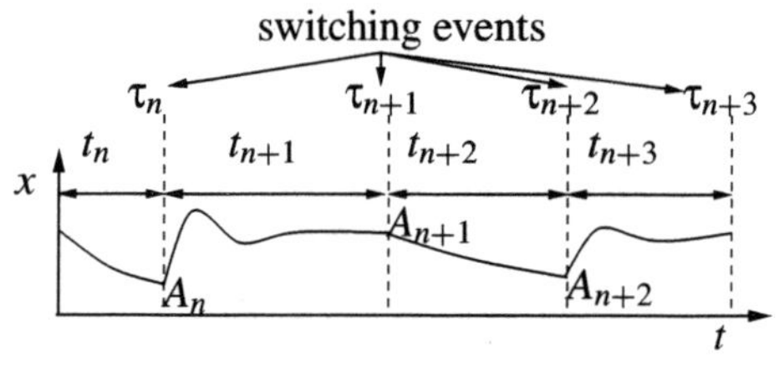

Fig. 1. Signal produced by a hybrid system.

Based on a map $\tau_{n+1} = f(\tau_n)$ analysis tools for systems having one piecewise linear in time varying state and periodic varying upper and lower boundaries are proposed in [4].

In some cases the signal can be modeled by a pulse process with linear filtering. Then its power density spectrum (PDS) can be calculated directly from the map:

$$S(\omega) = \sum_{n=1}^{\infty} c_n \Theta_n(\omega) \tag{4}$$

Θ_n is the characteristic function of τ_n and c_n is the correlation sequence of the pulses A_n. This approach is proposed in [5].

3. ANALYSIS EXAMPLE

Fig.2 depicts a DC-DC converter in current programmed mode (CPM). The switch S is controlled by the output of a RS flip-flop. The flip-flop is set by a usually periodic clock signal clk and reset when the inductor current i_l reaches the reference current i_{ref}. The switch-off event is a state condition and the switch-on event is a time event.

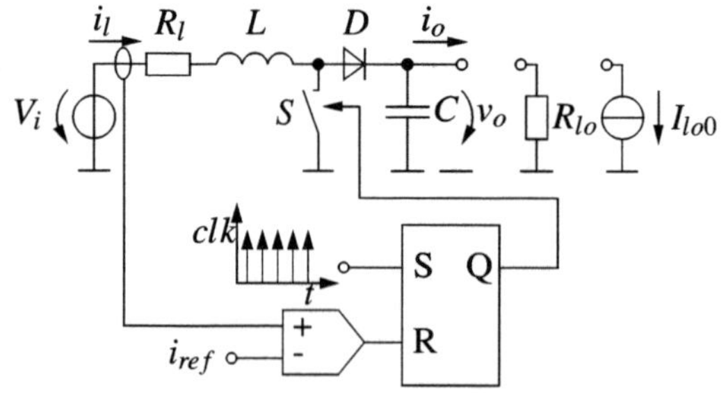

Fig. 2. CPM boost converter.

3.1. Two-dimensional analysis

From the state equations the embedded map of the state variables of subsequent events can be derived:

$$\begin{pmatrix} v_{o,n+1} \\ i_{l,n+1} \end{pmatrix} = g\begin{pmatrix} v_{o,n} \\ i_{l,n} \end{pmatrix} \quad (5)$$

Fig. 3 shows the map of the first state variable and Fig. 4 that of

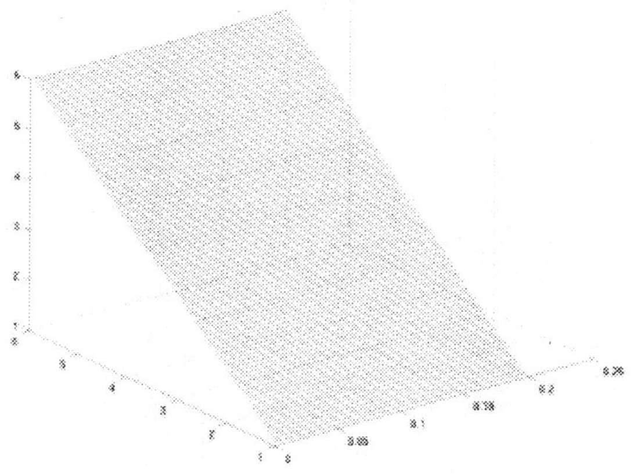

Fig. 3. CPM boost converter v_o map.

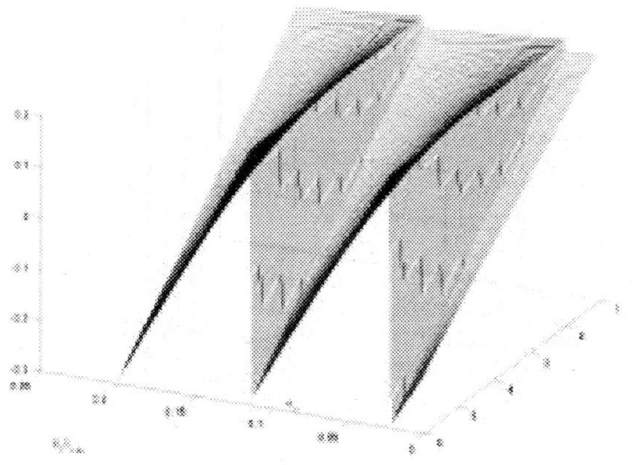

Fig. 4. CPM boost converter i_l map.

the second one. The exact twodimensional map is rather large but can be easily derived and hence it is not proposed here.

Using these maps the step response of the open loop converter is calculated by iteration of the map. The so calculated response is depicted in Fig. 5. It of course fits very good with the spice simulation but only takes a small fraction of the time of the spice simulation. Note that the converter operates chaotically and hence the mean, maximum and minimum values are chaotic too [6]. In the example the value of C was set to a very small one to show the capability of the map of calculating ripple values.

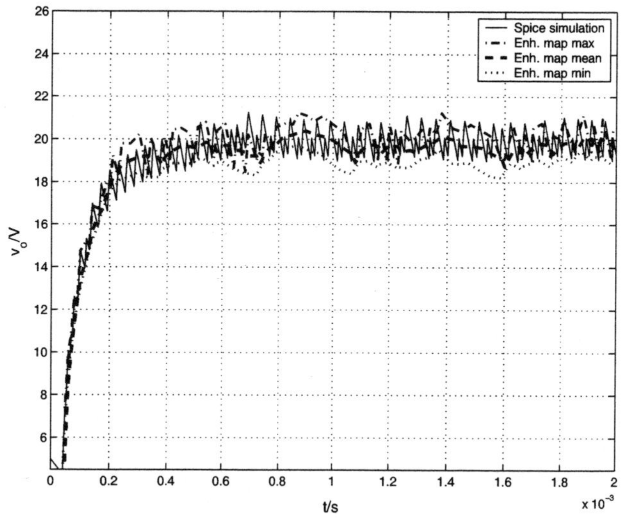

Fig. 5. Open loop step response of CPM DC-DC converter.

Simulation of the closed loop response is also possible using the embedded map. The purpose of the control loop is to keep the output voltage at a constant level independent from load and input voltage variations. The control can be a sampled or a continuous time one. Modelling the control loop is done by a separate map that derives the control variable (here i_{ref}) from the states, [7].

$$c_{n+1} = g_c(x_n) \quad (6)$$

In case of a sampled control this map usually is a FIR or IIR filter. A proportional feedback of v_{o2} e.g. is modeled by:

$$i_{ref,n+1} = k_{cont} v_{o,n} \quad (7)$$

The event-discrete modelling of a continuous-time control loop is possible approximately only. One approach is taking the average over one cycle and applying an appropriated FIR or IIR filter to the average to model P-,PD-,PI- and PID-controllers. As the solutions of the continuous part are known the average can be determined from the states at the switching event by a function g_{mean}.

The model of e.g. a continuous proportional control loop is:

$$i_{ref,n+1} = k_{cont} g_{mean}(v_{o,n}) \quad (8)$$

Using the discrete event model statistical analysis of the open loop or closed loop system (e.g. mean value, ripple and PDS calculation) is possible. However statistical analysis of the two-dimensional map would over-extend this paper. The analysis is proposed for a one-dimensional case here.

3.2. One-dimensional analysis

If the output voltage is not of prime interest in the analysis it can be set to a constant value when the output capacitor C is large (small v_o ripple condition) and the controller produces a constant output. A second simplification is neglecting the voltage drop over R_l. In a well designed converter both conditions will be full-filled. Then the inductor current i_l increases and decreases nearly linearly in time. Fig. 6 depicts the inductor current waveform of the CPM

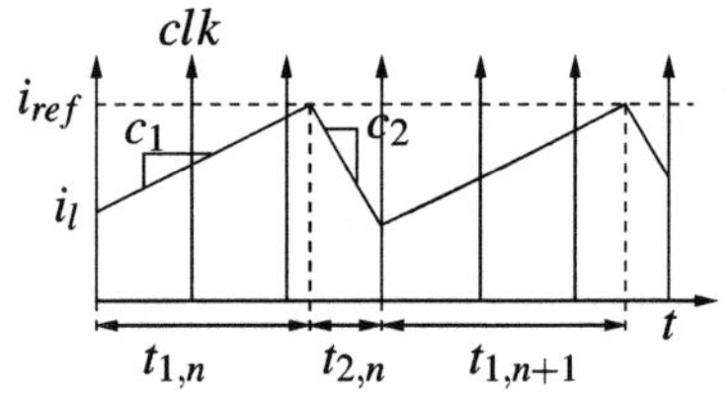

Fig. 6. CPM boost converter i_l map, one-dimensional model.

converter. The model becomes one-dimensional. Eq. (5) simplifies to:

$$i_{l,n+1} = i_{ref} - c_2 \left(T - \frac{i_{ref} - i_{l,n}}{c_1} \text{mod} T \right) \tag{9}$$

The map of subsequent time intervals also can be derived:

$$\begin{aligned} t_{2,n} &= T - t_{1,n} \text{mod} T \\ t_{1,n+1} &= -\frac{c_2}{c_1} t_{2,n} = \alpha t_{2,n} \end{aligned} \tag{10}$$

These maps now are used to apply **statistical analysis**. First the density of the continuous variable i_l is calculated from the density of the map knowing the piecewise continous solution of i_l. Using Eq. (9) we obtain the minimum inductor current $i_{l1} = i_{ref} - c_2 T$. The inductor current density is different from zero for values $i_{l1} \leq i_l \leq i_{ref}$. The density f_{i_l} of inductor current follows directly from the density f_{t_1}, Eq. (10).

$$f_{i_l} = \frac{\int_{i_{l1}}^{i} f_{t_1}\left(\frac{1}{c_1}(i_{ref} - \tilde{i})\right) d\tilde{i}}{\int_{i_{l1}}^{i_{ref}} \int_{i_{l1}}^{\hat{i}} f_{t_1}\left(\frac{1}{c_1}(i_{ref} - \tilde{i})\right) d\tilde{i} d\hat{i}} \tag{11}$$

f_{t_1} is determinable by the frobenius-perron operator. At integer values of α f_{t_1} is identically distributed.

$$f_{t_1} = \frac{1}{\alpha T} \tag{12}$$

Then Eq. (11) is determinable analytically.

$$f_{i_l} = 2 \frac{i_l - i_{l1}}{\left(i_{ref} - i_{l1}\right)^2}, \quad i_{l1} \leq i_l \leq i_{ref} \tag{13}$$

The mean value is determined from f_{i_L}.

$$\bar{i_l} = \frac{1}{i_{ref} - i_{l1}} \int_{i_{l1}}^{i_{ref}} i_l f_{i_l} di \tag{14}$$

For integer α Eq. (14) simplifies to:

$$\bar{i_l} = \frac{i_{l1} i_{ref}^2 + \frac{1}{3}\left(i_{l1}^3 + 2 i_{ref}^3\right)}{\left(i_{ref} - i_{l1}\right)^2} \tag{15}$$

The maximum current ripple is calculated directly from the return map. In our case we have:

$$i_{l,rip}^{max} = -\frac{1}{2} c_2 T \tag{16}$$

The mean ripple is obtainable e.g. from the expectation value of t_2.

$$\overline{i_{l,rip}} = -\frac{1}{2} c_2 E(t_2) \tag{17}$$

Again for integer α analytical expressions are obtained.

$$\overline{i_{l,rip}} - \frac{1}{4} c_2 T \tag{18}$$

The PDS of i_l is obtainable by statistical analysis, too. One possibility is a time series analysis using Eq. (10), [8], [9]. In [5] an analytical approach is proposed to calculate the PDS directly from the map using characteristic functions.

4. DESIGN EXAMPLE

In [9] is shown that the CPM boost converter can be modeled by a system (Fig. 2) that generates the set-pulses of the RS flip-flop by a second comparator that compares i_l by a second reference value instead of the use of a clock signal. The sequence of reference values is produced by return map Eq. (9). However when the con-

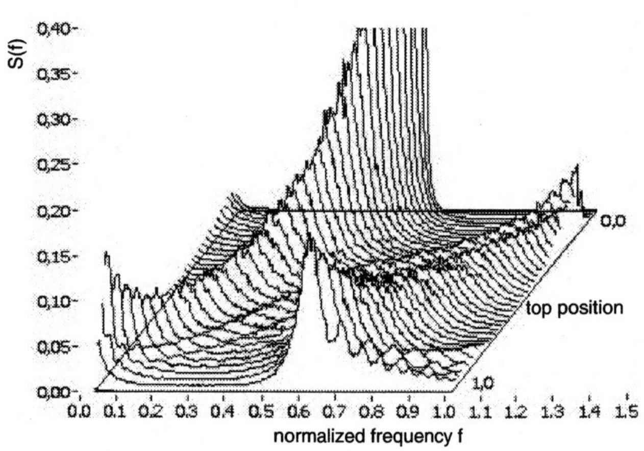

Fig. 7. PDS, tent map, top moved.

version ratio $\frac{V_o}{V_i}$ is smaller than two the converter operates periodically and hence produces large harmonic PDS components. If we build the model above instead of the CPM realisation we can use any map for the generation of lower boundary in order to improve EMI. In the following examples $\frac{V_o}{V_i} = 1.8$ is used. First we apply a tent map and move the top position between 0 and 1. Fig. 7 shows the PDS. The PDS is calculated by time series analysis of the embedded map. The best EMI is achieved at a top position arround 0.7. Fig. 8 depicts the results for bernoulli map with parameter first minumum min_1. It is defined by $min_1 = \frac{1}{number\ of\ segments\ of\ map}$. Fig. 9 depicts the same results for renyi map using another scale. The tendency is in principle the same as in Fig. 8. For $min_1 \geq \frac{1}{1.57}$ subharmonics appear, where 1.57 is the so-called golden mean. Both figures show that the number of segments should be large to achieve good EMI.

The CPM converter as well as the schemes used in Figs. 7 to 9 are able to produce time intervals $t_{1,n}, t_{2,n} \to 0$ and hence $f \to \infty$. Then large switching loss is produced by the switching device. To overcome this problem limitation of the minimum time intervals is required. One approach is downscaling of the range of output values of the map. The result is depicted in Fig. 10, where the gain value is scaled between 0 and 1. Gain is defined by $gain = \frac{y_{max}}{i_{ref}} = \frac{x_{max}}{i_{ref}}$, $y_{min} = x_{min} = 0$, where x and y are the map input and

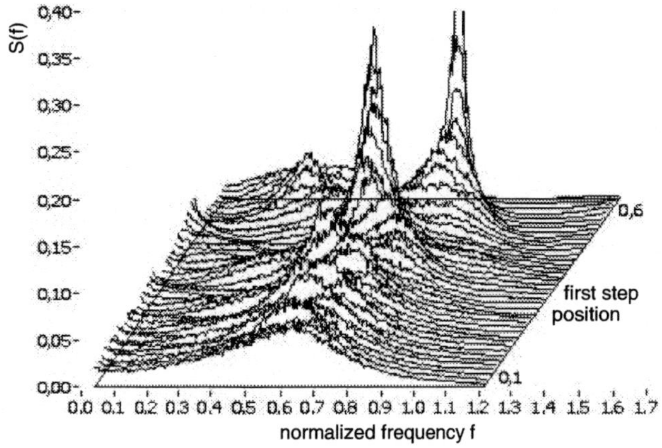

Fig. 8. PDS, bernoulli map, first step moved.

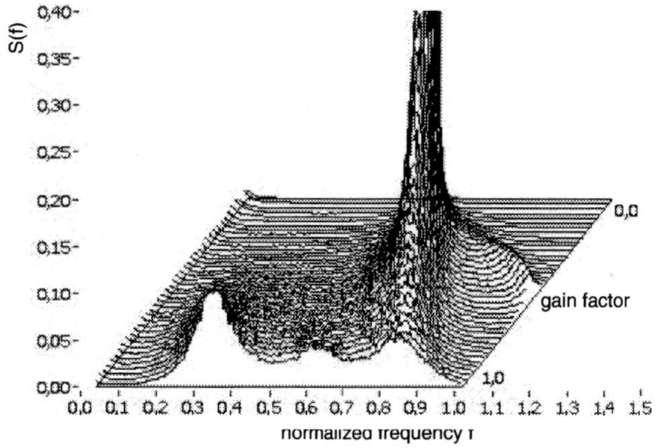

Fig. 10. PDS, tent map, gain scaled.

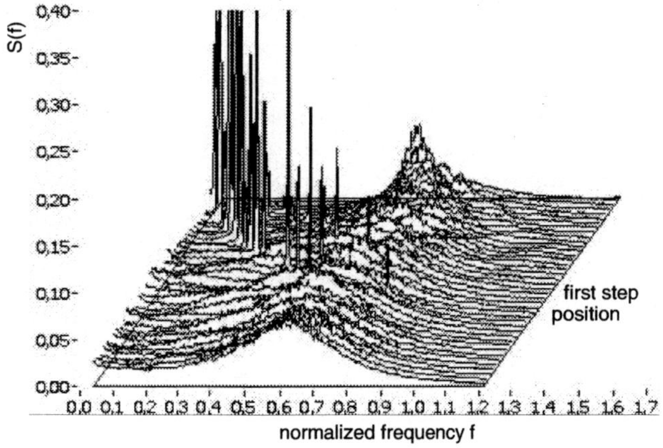

Fig. 9. PDS, renyi map, first step moved.

output values respectively. By reduction of the gain the maximum frequency is reduced but Fig. 10 clearly indicates the increasing maximum power density. This figure together with the power loss calculation may be used to make a good compromise. Power loss calculation may be done by statistical analysis of the map, too.

5. CONCLUSIONS

Modelling of hybrid systems is done by use of embedded return maps. The paper proposes two examples. Time series analysis as well as statistical analysis is done by this approach. Map models are capable of unifying the analysis of a large class of hybrid systems.

In the design example we were able to design the map and fit it to practical requirements respecting trade-offs.

Further improvements can be achieved including the analysis of switching losses, conduction losses etc. into the statistical analysis of the embedded map.

6. REFERENCES

[1] B. Lehman and R. M. Bass, "Switching frequency dependent averaged models for pwm dc-dc converters," *IEEE Transactions on Power Electronics*, vol. 11, no. 1, pp. 89–98, January 1996.

[2] B. Lehman and R. M. Bass, "Extensions of averaging theory for power electronic systems," *IEEE Transactions on Power Electronics*, vol. 11, no. 4, pp. 542–553, July 1996.

[3] V. A. Caliskan, G. C. Verghese, and A. M. Stanković, "Multi-frequency averaging of dc-dc converters," *IEEE Transactions on Power Electronics*, vol. 14, no. 1, pp. 124–133, January 1999.

[4] H. Torikai T. Saito and W. Schwarz, "Switched dynamical systems with double periodic inputs: An analysis tool and its application to the buck-boost converter," *IEEE Transactions on Circuits and Systems–I*, vol. 47, no. 7, pp. 1038–1046, July 2000.

[5] A. L. Baranovksi and W. Schwarz, "On spectral analysis of impulse process," in *Proc. NDES'02*, Izmir, Turkey, June 2002, pp. 37–40.

[6] O. Woywode, J. Weber, H. Güldner, A. L. Baranovksi, and W. Schwarz, "Statistical analysis of chaotic dc-dc converters," in *Proc. EAIT'01*, Kharagpur, India, December 2001.

[7] J. Krupar and W. Schwarz, "Discrete time and continuous time modelling of a dc-dc converter – a comparison," in *Proc. NDES'02*, Scuol, Switzerland, May 2003, pp. 145–148.

[8] J. Krupar and W. Schwarz, "Spread spectrum clock generation – chaotic and periodic modulation schemes," in *Proc. ECCTD'03*, Krakow, Poland, September 2003, pp. 235–238.

[9] A. L. Baranovksi, A. Mögel, W. Schwarz, and O. Woywode, "Chaotic control of a dc-dc-converter," in *ISCAS'00*, Geneva, Switzerland, May 2000, vol. II, pp. 108–112.

ANDRONOV-HOPF BIFURCATION OF SINUSOIDAL OSCILLATORS UNDER THE INFLUENCE OF NOISE

Wolfgang Mathis

Department of Electrical Engineering, University of Hannover, Germany
mathis@tet.uni-hannover.de

ABSTRACT

In this paper, we consider Andronov Hopf bifurcation in sinusoidal electrical oscillators under noisy disturbance. For this purpose the deterministic description of oscillators must be generalized using the concept of so-called invariant measures. It turns out that in noisy oscillators two inequivalent bifurcation approaches are available. By means of the van der Pol oscillator these noisy Andronov Hopf bifurcation concepts are illustrated.

1. INTRODUCTION

Oscillators belong to the earliest electronic circuits. In 1913 Meissner developed a first tube oscillator for radio transmitter applications; see Mathis [18] for details of the history of electrical oscillators. Of course, tubes are replaced in almost all modern oscillator circuits by transistors but the functionality of these circuits does not changed. The oscillatory behavior of these circuits can only be obtained if nonlinearity is included. This leads to nonlinear differential equations for the description of oscillator circuits and was a big obstacle for a successful circuit analysis. Furthermore a complete and systematic design process for the class of oscillatory circuits including nonlinearities is missing until now although some progress is made during the last two decades. Recent efforts are related to the development of design approaches using computer algebra systems where the nonlinear describing equations can be solved at least in the case of polynomial nonlinearity models. These approaches apply nonlinear transformation and equivalence principles (see e.g. Mathis [20]) as well as perturbation series (averaging [17], [21], Volterra series [8]) and the Andronov-Hopf bifurcation theorem ([15], [1], [11], [22]) in order to deal with nonlinearities. Although the main ideas of this concept are well-known further research is needed in order to develop a reliable design system for oscillators which can be used by circuit designers.

Whereas the deterministic approach proceeds the problem of noise in oscillator circuits came in the focus of research during the last few years. Since electrical oscillators based on nonlinear devices we have to consider noise in nonlinear circuits. Some interesting results were published by Weiss, Mathis [28] as well as Wyatt, Coram [30] but physical descriptions of thermodynamical noise are restricted to reciprocal circuits with positive resistors. Weiss and Mathis used ideas from Stratonovich [23]. Therefore a noise analysis of oscillator circuits from first principles which includes nonreciprocal devices (e.g. transistors; equivalent to negative resistors using feedback principle) is not possible until now. Only the noise behavior of certain noise aspects of active devices can be studied by thermodynamic concepts (see e.g. [29]). If we are interested in noise properties of entire oscillator circuits nonreciprocal effects have to be considered. With respect to noise properties of electronic oscillator circuits many research was done in the area of phase noise; see Lee, Hajimiri [14] and Goldberg [9] for an overview. Since a sophisticated basis of phase noise in nonreciprocal circuits is missing most authors use phenomenological approach. In our paper we study noise aspects of oscillator circuits and its impact to the deterministic Andronov-Hopf bifurcation phenomenon.

2. DETERMINISTIC CIRCUIT DESCRIPTION

Although the describing equations of oscillatory circuits are generally of the type of so-called differential algebraic equations (DAEs [16]) we will consider only those oscillator models which can be described by explicit ordinary differential or state-space equations

$$\dot{x} = F(x), \qquad F : \mathbb{R}^n \to \mathbb{R}^n \tag{1}$$

In contrast to transient analysis problems we have to consider (stable) asymptotic solutions of (1) and especially limit cycles in oscillator circuits. Since in higher dimensional cases no systematic methods for calculating limit cycles are available Papalexi, Mandelstam and Andronov developed the bifurcation approach for electrical oscillators using ideas from Poincaré's results (see [18] and in particular the papers of Bissell [7] and of Aubin, Dalmedico [6]). A first overview on this subject was published in 1935 and later on results were included in the 1937 monograph of Andronov, Witt and Chaikin (English version: 1966 [1]). The main idea of Papalexi, Mandelstam and Andronov was the embedding of (1) into a μ-parametrized family of differential equations

$$\dot{x} = F(x, \mu), \qquad F : I\!R^n \times I\!R \to I\!R^n \tag{2}$$

and searching for a qualitative changing of asymptotic solutions within this family. In sinusoidal oscillators is a changing from a stable equilibrium to a stable limit cycle of particular interest and therefore Mandelstam et al. [15] studied this case intensively; today this phenomenon is called Andronov-Hopf bifurcation. Unfortunately their fundamental results were not noticed by other researchers in electrical engineering. It lasted more than 40 years until this subject was considered by Mees and Chua [22] in 1979. Afterwards Andronov-Hopf bifurcation became an essential subject in the theory of electrical circuits (e.g. Mathis [16]). It should be remarked that in higher dimensional systems ($d > 2$) additional methods (e.g. the center manifold theorem) are needed (see e.g. Guckenheimer, Holmes [11]).

Whereas many aspects of deterministic systems can be studied efficiently in "time-domain" differential equations (1) or (2) there is an equivalent description that emphasized the statistical point of view. It is known from classical mechanics (e.g. Goldstein [10]) that differential equations of form (1) or (2) can be formulated as Frobenius-Perron evolution equation or as generalized Liouville equation. In these cases we are interested in the dynamics of a suitable class of density functions $f : I\!R \to I\!R$ an equation with an associated Frobenius-Perron-Operator P^t can be written down

$$f(x,t) = P^t\left(f(x)\right). \tag{3}$$

This equation is closely related to the generalized Liouville equation (see Tuckerman et al. [24]

$$\frac{\partial f}{\partial t} = -\text{div}\left\{f \cdot F\right\} = -\sum_{i=1}^{n} \frac{\partial (f \cdot F_i)}{\partial x_i}. \tag{4}$$

where P^t is related to the Liouville operator in the energy-preserving case. The changing from equations (1) or (2) to equations (3) or (4) can be interpreted in the following manner: instead considering the dynamics of single initial points we consider weighted whole sets of initial points where the density function f is the weighting function. An advantage of this representation is that it can be generalized to the more general class of noisy systems.

3. STOCHASTIC CIRCUIT DESCRIPTION

It was already mentioned that in the case of non-reciprocal circuits a physical derivation of dynamical equations for noisy electrical networks from first principle is not available. Therefore we need a heuristic approach.

Using the so-called Langevin approach of stochastic systems we start with a deterministic description of circuits and add a stochastic process ξ

$$\frac{dx}{dt} = F(x) + \sigma(x)\,\xi, \tag{5}$$

where the coefficient $\sigma(x)$ characterizes the coupling of the noise source. The first term should be interpreted as dissipation term where as the second term corresponds a fluctuation term.

Using the concept of stochastic differential equations ξ has to be interpreted as a generalized white noise process but in order to solve these equations a more generalized concept of integration (e.g. Kunita [13]) is needed. In essential there are two concepts of stochastic integration which are due to Ito and Stratonovich, respectively, and associated types of stochastic differential equations

$$dx = F(x)dt + \sigma(x)\,dW \tag{6}$$

where W is the so-called Wiener process. Both concepts are mathematical equivalent to a Fokker-Planck partial differential equation which generalizes in some sense the concept of Liouville equations (see Arnold [3], section 4.2)

$$\frac{\partial f}{\partial t} = -\sum_{i=1}^{n} \frac{\partial (f \cdot F_i)}{\partial x_i} + \sum_{i,j=1}^{n} \frac{\partial^2 (\sigma^2 f)}{\partial x_i \partial x_j}. \tag{7}$$

In the case of linear stochastic differential equations - the original subject of Langevin - there is no difference between Ito's and Stratonovich's type. Unfortunately stochastic differential equations (of Ito or Stratonovich type) are sound concepts only from a mathematical point of view if we consider nonlinear Langevin equations. The reason is that each type corresponds to a certain interpretation rule; otherwise its meaning is not well defined. It is interesting to see that for nonlinear Langevin equations in contrast to linear ones the deterministic equation (without noise) does not correspond to the averaged equation (see van Kampen's paper for further details [25])

$$\left\langle \frac{dx}{dt} \right\rangle = \frac{d\left\langle x \right\rangle}{dt} = \left\langle F(x) \right\rangle + \left\langle \sigma(x)\,\xi \right\rangle. \tag{8}$$

Even if σ is constant we note that the function F and the average operator $\langle\cdot\rangle$ does not commute. Only if $\langle F(x)\rangle = F(\langle x\rangle)$ is valid (that is in the linear case) the averaged equation (first moment of the stochastic process x) fulfills the deterministic equation $\dot{x} = F(x)$. Therefore with van Kampen [26], [27] we come to the conclusion that there is no good reason why the dissipation term should be identical to the vector field of the deterministic equation. However only in the reciprocal case a sound physical concept is available (e.g. Weiss [28]).

4. BIFURCATION CONCEPTS IN NOISY CIRCUITS AND SYSTEMS

In the last section we consider two related concepts for describing general stochastic or noisy circuits and systems if a deterministic description of the form $\dot{x} = F(x)$ is assumed. If we consider non-reciprocal (nonlinear) circuits we have to apply the heuristic Langevin approach where an associated stochastic differential equation can be derived. From a systems theoretical point of view stochastic differential equations belong to the class of state space equations which are formulated in time-domain. An alternative concept of describing noisy circuits uses a probability density function f which satisfies a Fokker-Planck-type equation.

In this section we are concerned with parametrized families of stochastic dynamical systems in the Langevin form $\dot{x} = F(x, \mu) + G(x)\xi$ and its associated Fokker-Planck equation. Although it is known that both concepts are equivalent from a mathematical point of view it turns out that there are different concepts of stochastic bifurcation. The earlier stochastic bifurcation concept based on a Fokker-Planck description and was founded in physical applications by Horsthemke and Lefever [12]. In this approach qualitative changing of the stationary solution within the family of Fokker-Planck equations is studied. Although it is a suggestive concept which can be illustrated easily there is no time dependence and therefore it is rather a static concept to bifurcation. In the mathematical literature it is called "P-bifurcation" (e.g. Arnold [5]).

A dynamical concept of stochastic bifurcation is based on the stochastic differential equation. In contrast to the P-bifurcation concept where we look for qualitative changes of the asymptotic probability density function the dynamical (or D-) bifurcation concepts is concerned with qualitative changes of certain properties within the family of stochastic differential equations. For this purpose suitable analogues for equilibrium points of deterministic differential equations is needed. It turns out (see Arnold [5]) that so-called invariant measures of stochastic flows are adequate analogues for deterministic equilibrium points. In doing so we assume that like in the deterministic case a stochastic differential equation is replaced by a "stochastic flow" or so-called cocycle (Arnold [5]).

Note that if x_0 is a deterministic equilibrium point of a cocycle $\varphi(t, \omega, x_0) = x_0$ then the Dirac measure δ_{x_0} is stationary and invariant. Therefore there is a close relationship of deterministic equilibrium points and invariant measures.

Therefore the fundamental question of D-bifurcation is "Are there other invariant measures than Dirac measures?". It turns out that a necessary condition for qualitative changing in the sense of D-bifurcation is the vanishing of a Lyapunov exponent. It should be mentioned that there is no general relation between P-bifurcation and D-bifurcation.

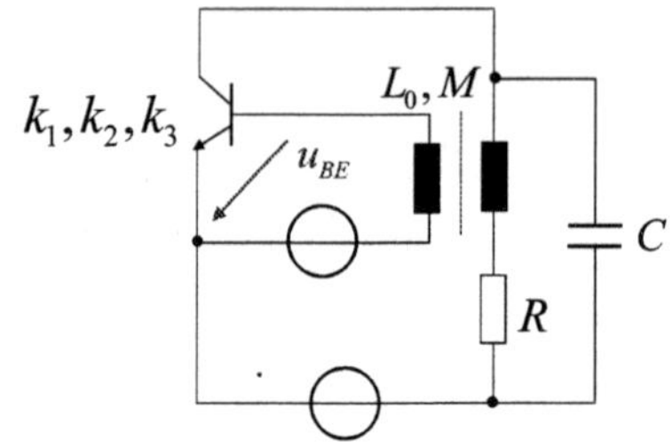

Fig. 1. Transistor Meissner Oscillator

For illustrating these bifurcation concepts we restrict us to 2-dimensional systems. For higher dimensional systems stochastic concepts for normal forms and/or center manifolds are needed (see Arnold [5]). We consider a Meissner oscillator circuit in fig. 1. If $k_2 = 0$ the following circuit equation for the voltage between basis and emitter can be derived ($\omega_0^2 = 1/(LC)$)

$$\ddot{u}_{BE} + \left(\frac{R}{L} - \omega_0^2 M(k_1 - 3k_3 u_{BE}^2)\right)\dot{u}_{BE} + \omega_0^2 u_{BE} = 0. \tag{9}$$

Equation (9) can be normalized in the standard van der Pol form $\ddot{x} - (\mu - \gamma x^2)\dot{x} + x = 0$. Now we assume with Ariaratnam [2] that we have a noisy resistor which results in an additive decomposition of

$$\ddot{x} - (\mu_0 + \sigma\xi - \gamma x^2)\dot{x} + x = 0. \tag{10}$$

If (10) is converted into first order equations and polar coordinate transformations are applied to the system we obtain after a stochastic averaging the following stochastic differential equation for the amplitude process $a(t)$

$$da = \frac{1}{2}\left(\mu_0 + \frac{5}{8}\sigma^2 - \frac{1}{4}\gamma a^2\right) a dt + \left(\frac{3}{8}\right)^{\frac{1}{2}} a\, dW_a. \tag{11}$$

For the analysis of D-bifurcations we have to determine the stability of stationary solutions $a_s(t)$ by means of associated Lyapunov exponents. If small variations $r(t)$ of $a_s(t)$ are considered the following linearized stochastic differential equation for amplitude process can derived

$$dr = \frac{1}{2}\left(\mu_0 + \frac{5}{8}\sigma^2 - \frac{3}{4}\gamma a_s^2\right) r dt + \left(\frac{3}{8}\right)^{\frac{1}{2}} \sigma r\, dW_a. \tag{12}$$

For the bifurcation analysis the zeros of the Lyapunov exponents have to be found

$$\lambda = \frac{1}{2}\left(\mu_0 + \frac{1}{4}\sigma^2 - \frac{3}{4}\gamma\langle a_s^2\rangle\right). \tag{13}$$

Obviously it results a zero for the trivial solution $a_s = 0$ that at $\mu_0 = -\sigma^2/4$. Furthermore for the same value of μ_0 we have a zero for $a_s \neq 0$, too, such that we have a D-bifurcation.

For studying P-bifurcation we need a solution of the stationary Fokker-Planck equation associated to the noisy van der Pol equation. It turns out that a first P-bifurcation occurs at $\mu_0 = \sigma^2/8$ where the peak of the probability density function shifts as $a_p := 2((\mu_0 - \sigma^2/8)/\gamma)^{1/2}$. Another changing occur at $\mu_0 = \sigma^2/2$ where the uni-modal density centered at the origin changes to a bi-modal density possessing a ring of peaks; see Ariaratnam [2]. Note that μ_0-values for D- and P-bifurcation differ substantially. Additional examples can be found in Arnold [5].

5. CONCLUSION

In our paper similarities and differences of describing methods for deterministic and stochastic circuits and systems are discussed. Especially we discussed some difficulties with respect to a sound physical interpretation of describing equations if non-reciprocal and nonlinear circuits are considered. Moreover we discovered the basic ideas of two concepts for the analysis of bifurcations in noisy nonlinear circuits and systems. By means of a van der Pol equation derived from a Meissner oscillators these bifurcation concepts are illustrated.

6. REFERENCES

[1] Andronov, A.A., A.A. Witt, S.E. Khaikin, Theory of Oscillators. Pergamon, Oxford, 1966 (first published in Russian in 1937)

[2] Ariaratnam, S.T.: Stochastic Bifrucation in Hereditary Systems. Proc. 8th ASCE Spec. Conference on Prob. Mech. Struct. Reliability. Notre Dame University, 23-26 July 2000, USA

[3] Arnold, L., The unfoldings of dynamics in stochastic analysis. Comput. Appl. Math. 16, 1997, 3-25

[4] Arnold, L.; P. Imkeller, Normal forms for stochastic differential equations. Probab. Theory Relat. Fields 110, 1998, 559-588

[5] Arnold, L., Random dynamical systems. Berlin-Heidelberg-New York 1998

[6] Aubin, D.; A.D. Dalmedico: Writing the History of Dynamical Systems and Chaos: Longue Durée and Revolution, Disciplines and Cultures. Historia Mathematica 29, 2002, pp. 273-339

[7] Bissell, C.: A.A. Andronov and the Development of Soviet Control Engineering. IEEE Control Systems Magazine 18, 1998, pp. 56-62

[8] Chua, L.O.; Y.S. Tang: Nonlinear oscillation via Volterra Series. IEEE Trans. Circuits Syst. CAS-29, 1982, 150-168

[9] Goldberg, B.-G.: Oscillator phase noise revisited – a Heuristic review. RF Design, January 2002, pp. 52-64

[10] Goldstein, H.; C. Poole; J Safko: Classical Mechanics (3rd Edition). Addison Wesley, San Francisco 2001

[11] Guckenheimer, J.; P. Holmes, Nonlinear oscillations, dynamical systems, and bifurcation of vector fields. Springer-Verlag, Berlin-Heidelberg 1983

[12] Horsthemke, W.; R. Lefever: Noise-Induced Transitions. Springer-Verlag, Berlin-Heidelberg 1984

[13] Kunita, H., Stochastic flows and stochastic differential equations. Cambridge University Press 1990

[14] Lee, T.H.; A. Hajimiri: Oscillator Phase Noise: A Tutorial. IEEE Journ. Solid-State Circuits SC-35, 2000, pp. 326-336

[15] L. Mandelstam, N. Papalexi, A.A. Andronov, S. Chaikin, A. Witt. Exposé des Recherches Récentes, sur les Oscillations Non Linâires. *Zeitschr. f. Techn. Physik* 4 (1935) 81-134.

[16] Mathis, W., Theory of Nonlinear Circuits (in German) Springer-Verlag, Berlin-Heidelberg, 1987

[17] Mathis, W.; C. Keidies: Application of Center Manifolds to Oscillator Analysis. Proc.12th European Conference on Circuit Theory and Design (ECCTD'95), Istanbul, Trkei, 27.-31. August 1995

[18] Mathis, W., Historical remarks to the history of electrical oscillators (invited). In: Proc. MTNS-98 Symposium, July 1998, IL POLIGRAFO, Padova 1998, 309-312.

[19] Mathis, W., Nonlinear stochastic circuits and systems - A geometric approach, Proc. 4th MATHMOD 2003, Vienna, February 5-7, 2003

[20] Mathis, W., Transformation and Equivalence. In: W.-K. Chen (Ed.): The Circuits and Filters Handbook. CRC Press & IEEE Press, Boca Raton 2003

[21] Prochaska, M.; W. Mathis: Symbolic Analysis Methods for Harmonic Oscillators. Proc. 12th Intern. Symp. Theor. Electrical Eng (ISTET03), Warschawa 2003, Poland

[22] Mees, A.I, L.O. Chua, The Hopf Bifurcation and its Application to Nonlinear Oscillations in Circuits and Systems. IEEE Trans. Circuits and Systems, vol. 26, 1979, 235-254

[23] Stratonovich, R.L.: Nonlinear Thermodynamics I. Springer-Verlag, Berlin-Heidelberg, 1992

[24] Tuckerman, M.E.; C.J. Mundy; G.J. Martyna: On the Classical Statistical Mechanics of Non-Hamiltonian Systems. Europhysics Lett. 45, 1999, pp. 149

[25] van Kampen, N.G., Thermal fluctuations in nonlinear systems. Journ. Math. Phys. 4, 1963, 190-194

[26] van Kampen, N.G., The Validity of Nonlinear Langevin Equations. Journ. Statist. Physics 25, 1981, 431-442

[27] van Kampen, N.G., Ito versus Stratonovich. Journ. Statist. Physics 24, 1981, 175-187

[28] Weiss, L.; W. Mathis, A Thermodynamical Approach to Noise in Nonlinear Networks. Int. Journ. Circ. Theor. Appl. 26, 1998, 147-165

[29] Weiss, L.; W. Mathis: A thermodynamic noise model for nonlinear resistors. IEEE Trans. Electron Dev. Lett. 20, 1999, pp. 402-404

[30] Wyatt, J.L.; G.J. Coram: Nonlinear Device Noise Models: Satisfying the Thermodynamic Requirements. IEEE Trans. Electron Dev. 46, 1999, pp. 184-193

COMPUTATION BY NATURAL SYSTEMS DEFINED

Ruedi Stoop and Norbert Stoop

Institute of Neuroinformatics, ETH and University of Zurich
Winterthurerstr. 190, Zürich 8057

ABSTRACT

An explicit measure of the computation performed by general systems (electronic circuits, neurons, mechanical devices, etc.) is defined. We propose that the deeper nature of computation, and thus of any measure of computation, is in its reduction of complexity. The latter we understand as the "obstruction against prediction", experienced by an observer. We demonstrate the applicability and usefulness of this concept in different examples, which include some of the most studied families of dynamical systems. The measure can also be computed for higher-dimensional and experimental systems.

1. INTRODUCTION

Real world (="natural") systems have recently received increased scientific attention, because of their obvious efficiency properties. A simple bee, e.g., shows that biological systems can perform perception tasks in a way yet unchallenged by artificial (digital) systems. Often, in this context, it is argued that these natural systems "perform computations". However, this notion of computation generally remains undefined. It may seem that in artificial digital systems, the notion of computation has an intrinsic meaning. Only recently, the fundamentals of computation have been investigated in more details. One line of approach was to define computation in terms of the difficulty of solving classes of fundamental problems, under (mostly: time) constraints. Another line of approach was the extension from intrinsically rational to real weights, as is advocated by biologically motivated neural networks. Depending on the starting point, widely diverging concepts emerged [1, 2, 3]. All that these concepts, however, offer no measure of computation.

The approach that we pursue in this contribution is radically different. It is our understanding that before making extended use of a notion, it first should be defined. In physics, this is generally done by conveying a means for measuring the property (this measure can be as general as verification, which is restricted to the value set $\{0, 1\}$). This is why we are going to define a measure of computation for natural systems. The measure will be applicable to most classes of dynamical systems.

Such a measure of computation cannot be tied to a particular problem to be solved. Rather, it should be defined as an average over all possible problems possible to solve by the device. As the most natural and general enough starting point of computation, we first focus on the initial condition (in the context of artificial digital computation often called operands). Among all possible inputs, low- as well as high-complexity input strings have to be expected, with no particular distribution. The only possible way of implementing this requirement is by taking averages over the input space. Next, a sufficiently general notion of a computing system should be defined, capable of processing the initial condition. In accordance with the intuitive understanding of "computation", we let a general map represent the computational system. This can be motivated by insights from neural network theory, where the simplest tractable representation of a computational unit is a map (general neural networks are universal approximators of functions). More general space-time systems may pose conceptual difficulties. However, many specific aspects will nevertheless be describable in terms of maps. The initial condition to this map (e.g., a single input for one-valued operators, or a string composed from two numbers in the case of a two-valued operator) is then iterated by the map; this is what we take as the representation of the computation. The simplest way is to look upon input as binary digit expansions that are fed into the map. Without restriction of generality, the map hence can be assumed to operate on some subset of the unit interval. The computation halts, when the transient behavior of the map has been passed. (This criterium demands, in principle, a measure of accuracy by the observer. Since in our approach we shall operate on sets of unspecified precision, this assumption will automatically be incorporated.) Because of the general formulation, various results of computation, in particular, periodic, quasiperiodic, as well as chaotic behavior are consistent with the approach. In order to obtain an overall characterization of the computation performed, we will have to average (possibly in a generalized sense) over all initial conditions.

Real systems can be defined as systems that are noisy, with unspecified values on small digits. To define computation in a way applicable to these systems, it makes no sense to define computation by means of single, isolated, initial conditions. Instead, it implies to operate on the level of sets of neighborhoods, where the linear approximation Df, or f' in the case of dimension one, is the appropriate representation of the map f. This replaces the classical view of computation based on a mapping between points, involving questions of precisions. To arrive at a measure of computation, it appears that the most natural way would be to consider the number of iterations needed, starting from any initial condition, to arrive at any of the possible results. If a non resolution-dependent characterization is to be given, this implies taking the asymptotic limit $n \to \infty$. In this way, as the simplest measure of computation, the quantity $\lim_{n\to\infty} \frac{1}{n} \int \rho(x) \log \mid f^{n'}(x) \mid dx$ emerges, where $\rho(x)$ denotes the natural density. The obtained measure, however, coincides with the Lyapunov exponent of the system, implying that by the notion of computation, no novel quality would be obtained. This is in stark contrast to our intuitive notion, that does not foresee such a coincidence.

In the quest for a more suitable concept, the question emerges what the deeper notion of computation could be. Using a generally accepted perception of computation as the starting point, our answer is, that it should be understood as the reduction of the difficulty of prediction in the statistical sense. This quantity should be averaged over the input space, working on sets of neighboring trajectories because of the finite accuracy required (in this sense, our notion of computation indeed will be vaguely related with the concept of Lyapunov exponents). Computation thus should reduce complexity, when the latter is understood as the obstruction against predictability. In this way, the following rough picture emerges: Consider an arbitrary input, represented as a binary expansion from the unit interval. To this input, already a complexity can be attached, as it may have emerged as the result of some iterated mapping. We shall intrinsically assume that the input complexity can be of any possible complexity (in the Kolmogorov sense [4, 5] or in the sense of a complexity of prediction as defined below). By means of the system that we expect to perform the "computation", this input complexity is transformed into an output complexity (measured, again, by a complexity of prediction). The reduction of the input complexity by means of the map defines our measure of computation.

2. C_S COMPLEXITY

Our outlined concept of a measure of computation thus prerequisites the definition of an appropriate measure of complexity. This measure should express the difficulty of making predictions for real systems based on past observations. As such, it needs to be based on neighborhoods of trajectories in real number space, rather than on individual trajectory in the space of rational numbers. As a consequence, this measure of complexity will be different from the Kolmogorov complexity.

Recently, such a measure of complexity has been worked out [6, 7]. Its main features will be outlined below. Let us consider a dynamical system with discrete time, defined by a map f on some set M in the Euclidean space R^n. Pick an arbitrary point x_0 in the phase space, take some neighborhood $U = U(x_0)$ and consider the orbits $\{f^{(n)}U\}$, $n \in N$. We are interested in observables that relate to measures that are multiplicative along the orbit, i.e., for which the n-step average is evaluated as $(\prod_{k=0}^{n-1} \mu(f^{(k)}(x)))^{1/n}$, where $x \in U$ is the initial state of a particular orbit. Examples of such measures are derivatives, probabilities, etc. Take such a measure $\mu(x)$ and define our observable ν as $\mu(x) =: \exp(\nu(x))$. Our goal is to study the problem of prediction of the next values $\nu(f^{(r)}(x)), r > n$, along the orbits. For the decay of the probability P of retaining a particular measurement value of the observable during a system evolution of n steps, we employ the large deviation ansatz [8] $P(\nu, n)d\nu \sim e^{-ng(\nu)}d\nu$. The thermodynamic formalism implies [8] that

$$g(\nu) = \nu - S(\nu), \tag{1}$$

where $S(\nu)$ is an entropy function. In more detail, the thermodynamic formalism departs from a partition function $Z(n, \beta, \nu)$, where n is the level or depth of the partition and β can be viewed as an inverse temperature. With $Z(n, \beta, \nu)$, a free energy $F(\beta) = \lim_{n\to\infty} \frac{1}{n} \log(Z(n, \beta, \nu))$ is associated, where in $F(\beta)$ we suppressed the dependence on the observable ν. In the absence of phase transitions, an entropy function is obtained by means of the Legendre transform $S(\nu) = \nu\beta - F(\beta)$. Entropy functions $S(\nu)$ have the property of strict convexity, with infinite derivatives at the two end-points of the curve (in the absence of phase transition effects).

A suitable complexity measure is defined as the difficulty of prediction of the observable ν, averaged over all system behaviors. Equation (1) implies that the probability for observing trajectories with a specific value of ν, as a function of n behaves as

$$P(\nu, n)d\nu \sim e^{-n(\nu - S(\nu))}d\nu. \tag{2}$$

As $\nu \geq S(\nu)$, the smaller $\nu - S(\nu)$, the better the prediction based on the past of the orbits will be. Orbits with $\nu = S(\nu)$ will yield perfect long-time prediction. Indeed, this situation characterizes the long-time average of the natural invariant measure (depending on the observable, Lyapunov exponent, information dimension, e.g. [6, 7]). As the complexity is the *difficulty* of making correct predictions, over all length scales, the average

$$\int \frac{S(\nu)}{\nu} d\nu \tag{3}$$

is defined as the measure of complexity.

In order to facilitate the comparison of systems with different topological entropies and to extend the range of applications, the above concept can be refined. ν and $S(\nu)$ can be rescaled as $\tilde{\nu} = \nu/\nu_0$ and $\tilde{S}(\tilde{\nu}) = S(\nu)/\nu_0$, where ν_0 is the topological length scale exponent. Geometrically, this corresponds to a similarity transformation of the entropy function's graph at $(0,0)$, which maps the topological length scale exponent ν_0 to unity. In this case, our complexity measure assumes the form $C_s(\nu) := {\nu_0}^2 \int \frac{\tilde{S}(\tilde{\nu})}{\tilde{\nu}} d\tilde{\nu}$. where on the left-hand side, ν refers to the chosen observable. To obtain a fined-tuned distinction of dynamical systems according to their complexity, we may exponentiate the front factor and the integrand independently. Then the most general form of our measure is obtained as

$$C_s(\gamma, \beta)(\nu) := {\nu_0}^{2\beta} \frac{\nu_1}{\nu_1 - \kappa} \int \left(\frac{\tilde{S}(\tilde{\nu})}{\tilde{\nu}} \right)^{\gamma} d\tilde{\nu}, \qquad (4)$$

where γ and β are weightening exponents. To avoid divergence, we require $\gamma > -1$. Most relevant for our purpose, however, will be $C_s(1,0)$, as this characterizes the natural measure of the difficulty of prediction under suppression of lengthscale aspects.

3. COMPUTATION MEASURED

As the desired measure of computation should reflect the ability of the system to reduce the difficulty of prediction, the most natural and straightforward way to define it is to define it as the quantity

$$CO_s = 1/(C_s(1,0) + 1), \qquad (5)$$

where $C_s(1,0)$ is the $(1,0)$-complexity measure as defined above and where 1 has been added in the denominator in order to prevent a possible singularity caused by $C_s(1,0) = 0$. This measure of computation is statistical in nature, as it is extracted by means of the thermodynamic formalism. It does not require explicit hierarchical analysis and is non-divergent by construction. If the maximal scaling index ν_{max} is finite, the measure itself will be finite, bounded from below by zero and from above by the value $1 + \log(\nu_{max})$. $CO_s = 1$ indicates a system that performs optimal (decisive) computation, whereas $CO_s \leq \frac{1}{2}$ indicates that the system performs no notable computation. It is worth noting that in this way the measure of computation will be finite, in all realistic cases.

Examples: To get some idea for the implications by this definition, we focus on the the natural partition of the phase space generated by the iteration of a map f and denote the associated observable by $\varepsilon = log\frac{1}{|f'|}$ [6, 7]. According to their computational measure, systems can be divided

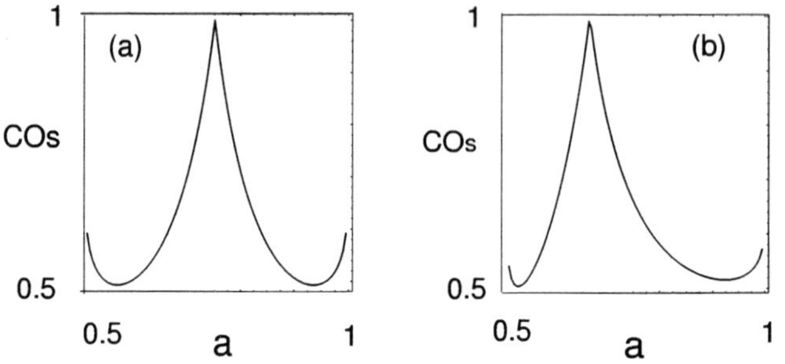

Fig. 1. Computation CO_s. a) Hyperbolic tent map paradigm, parameter a: location of peak. b) Nonhyperbolic bungalow tent map paradigm, parameter a: location of right diagonal point.

into classes of increased computation: I ($CO_s < 0.5$), II ($0.5 < CO_s < 1$), III ($CO_s = 1$). The first class is given by systems that almost perform no computation. In this class, we have the intermittent systems, as their complexity when measured by means of $C_s(1,0)$, is highest among the prototypical classes of dynamical systems [6, 7]. This is in agreement with observations that biological complexity is highest at the border between order and chaos [9, 10]. Two of the most prominent 1-d dynamical system classes are the (skewed) tent map family, as the standard representative for hard hyperbolic chaotic systems, and the bungalow tent map as the standard representative for nonhyperbolic chaotic systems. Both systems allow for a simple analytic determination of the computation measure. As a function of the family parameter, they mostly belong to the intermediate, second, class, as is shown in Fig. 1. For the parabola family $y = ax(1-x)$, $a \in [2,4]$, computation can be calculated analytically only for the isolated parameter values. This family belongs to a mixture between the second and the third class, depending on the parameter value. For $a = 4$, as $C_s(1,0)(\varepsilon) = 1/2$, the computation is $2/3$. The third class, with maximal computation $C_s(1,0) = 1$, is characterized by the most simple emergent behavior. As outlined in the introduction, stable periodic orbits are one possible example. This is why in the periodic windows, the parabola falls into this class. However, also classical arithmetic operations are members. In the case of addition where one operand is fixed (e.g., to the value $x = a$), the map $f : y = x + a$ mod 1 can be taken as a representation. Since the complexity of this map is zero, maximal computation $CO_s = 1$ is obtained. The identical result is obtained, if multiplication by a is represented by $f : y = ax$ mod 1.

4. DISCUSSION

In this paper, we constructed an observable-dependent measure of computation. The range of applications of our method extends to a very broad class of systems. The evaluation is particularly simple, if a generating or approximate generating partition is available. More generally, our measure can be calculated, whenever an entropy function of scaling exponents can be evaluated. These cases include experimental time series (see, e.g.,[8]).

Our measure provides three main insights. First, the numerical results for 1-d maps point out that nonhyperbolicity *per se* does not have a strong influence on computation. Second, when we compare the fully developed parabola with the symmetric tent map, these systems can be transformed by means of a conjugacy, preserving two points of $S(\varepsilon)$ (the natural measure and the topological measure). For an observer making predictions, they appear as distinct instead. This is captured by our measure, which yields computation $CO_s = \frac{2}{3}$ and $CO_s = 1$, respectively. Third, for intermittent systems, that are at the border between chaos and order, the computation emerges to be smallest. Their apparent job therefore is not to compute, but to provide sufficient complexity, on which computational elements then can do efficient computation.

Our concept of computation also can be interpreted in the context of periodic orbits of dynamical systems. It has recently been shown that unstable orbits can easily be stabilized [11]. Stabilized orbits are among the simplest computational results in our approach. If the stabilization is achieved by means of simple limiter control, it has a simple interpretation in terms of inhibitory neuronal connections. In our framework, this control process is interpreted as a process with maximal computation. We also evaluated our complexity measure for experimental neuron data (unpublished). In an in vivo experiment of cat visual cortex V1, inter-spike intervals ("ISI") between firing events were recorded and analyzed, for two distinct stimulation paradigms [12] (stimulation by noisy patterns moving into the neuron's preferred direction and square-wave stimulation as the optimal stimulus). Using our measure of computation, we obtained the insight that optimal stimulation of the neuron led to larger computation if compared to non-optimal stimulation. Thus, optimal stimulation leads to an improved computation. The vast amount of information arriving at different stages of cortical computation may render the assessment of the computation hardly feasible, if the system is divided into small computational steps. In this case, we propose to replace the whole cascade process by one single map, for which the computation can be evaluated along the lines outlined.

What is the advantage of such a viewpoint, apart from putting computation on the fundament of measureability? Computation in natural systems appears in different stages. Each one taking particular properties into account, ruling out some complexity, but retaining complexity to be combined with results from other computations. With the current approach, the computation performed by individual units in the process as well as the computation performed by the whole can be evaluated and compared. It can be expected that from this comparison, more insight into the relation between computation performed by parts and done by the whole system can be obtained. This knowledge will be required in order to exploit computation by natural, in particular biological, systems for technical purposes.

The authors strongly benefited from discussions with W.-H. Steeb on the nature of computation. R. S. acknowledges the support by a KTI-contract with Phonak AG hearing systems.

5. REFERENCES

[1] L. Blum, F. Cucker, M. Shub, and S. Smale, *Complexity ans Real Computation*, Springer, 1997.

[2] H. Siegelmann, *Neural Networks and Analog Computation*, Birkhäuser, 1998.

[3] Y. Hardy and W.-H. Steeb, *Classical and Quantum Computing*, Birkhäuser, 2001.

[4] A.N. Kolmogorov, "Three approaches to the quantitive definition of information," *Probl. Inf. Theory*, vol. 1, no. 3, 1965.

[5] M. Li and P. Vitányi, *An Introduction into Kolmogorov Complexity and its Applications*, Springer, 1997.

[6] R. Stoop and N. Stoop, "An integral measure of complexity," in *Control of Oscillations, Proceedings of the Conference Progress in Nonlinear Science, in Honor of 100 Birthday of A.A. Andronov*, V. Shalfeev, Ed., 2002, vol. III, pp. 290–296.

[7] R. Stoop, N. Stoop, and L.A Bunimovich, "Complexity of dynamics as variability of prediction," *J. Stat. Phys.*, p. in press, 2003.

[8] J. Peinke, Parisi J., Roessler O., and Stoop R., *Encounter with Chaos*, Springer, 1992.

[9] C.G. Langton, "Studying artificial life with cellular automata," *Physica D*, vol. 22, pp. 120–140, 1986.

[10] S.A. Kaufmann, *The origin of order*, Oxford University Press, 1993.

[11] R. Stoop and C. Wagner, "Scaling properties of simple limiter control," *Phys. Rev. Letters*, vol. 90, pp. 154101–1–154101–4, 2003.

[12] R. Stoop, D. Blank, A. Kern, J.-J. v.d.Vyver, S. Lecchini, and C. Wagner, "Collective bursting in layer iv," *Cog. Brain Research*, vol. 13, pp. 293, 2002.

AUDIO SOURCE SEPARATION
BASED ON INDEPENDENT COMPONENT ANALYSIS

Shoji Makino † *Shoko Araki* † *Ryo Mukai* † *Hiroshi Sawada* †

† NTT Communication Science Laboratories, NTT Corporation
2-4 Hikaridai, Seika-cho, Soraku-gun, Kyoto 619-0237, Japan
{maki, shoko, ryo, sawada}@cslab.kecl.ntt.co.jp

ABSTRACT

This paper introduces the blind source separation (BSS) of convolutive mixtures of acoustic signals, especially speech. A statistical and computational technique, called independent component analysis (ICA), is examined. By achieving nonlinear decorrelation, nonstationary decorrelation, or time-delayed decorrelation, we can find source signals only from observed mixed signals. Particular attention is paid to the physical interpretation of BSS from the acoustical signal processing point of view. Frequency-domain BSS is shown to be equivalent to two sets of frequency domain adaptive microphone arrays, i.e., adaptive beamformers (ABFs). Although BSS can reduce reverberant sounds to some extent in the same way as ABF, it mainly removes the sounds from the jammer direction. This is why BSS has difficulties with long reverberation in the real world. If sources are not "independent," the dependence results in bias noise when obtaining the correct unmixing filter coefficients. Therefore, the performance of BSS is limited by that of ABF. Although BSS is upper bounded by ABF, BSS has a strong advantage over ABF. BSS can be regarded as an intelligent version of ABF in the sense that it can adapt without any information on the array manifold or the target direction, and sources can be simultaneously active in BSS.

1. INTRODUCTION

Speech recognition is a fundamental technology for communication with computers, but with existing computers, the recognition rate drops rapidly when more than one person is speaking or when there is background noise. On the other hand, humans can engage in comprehensible conversations at a noisy cocktail party. This is the well known cocktail-party effect, where the individual speech waveforms are found from the mixtures. The aim of audio source separation is to provide computers with this cocktail party ability, thus making it possible for computers to understand what a person is saying at a noisy cocktail party.

Blind source separation (BSS) is an emerging technique, which enables the extraction of target speech from observed mixed speeches without the need for source positioning, spectral construction, or a mixing system. To achieve this, attention has focused on a method based on independent component analysis (ICA). ICA extracts independent sounds from among mixed sounds. This paper considers ICA in a wide sense, namely nonlinear decorrelation together with nonstationary decorrelation and time-delayed decorrelation. These three methods are discussed in a unified manner [1, 2]. There are a number of applications for

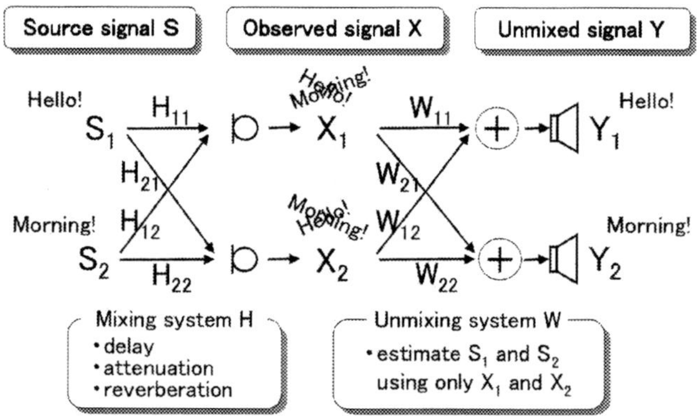

Figure 1: BSS system configuration.

the BSS of mixed speech signals in the real world [3], but the separation performance is still not good enough [4, 5].

Since ICA is a purely statistical process, the separation mechanism has not been clearly understood in the sense of acoustic signal processing, and it has been difficult to know which components were separated, and to what degree. Recently, the ICA method has been investigated in detail, and its mechanisms have been gradually uncovered by using theoretical analysis from the perspective of acoustic signal processing [6] as well as experimental analysis based on impulse response [7]. The mechanism of BSS based on ICA has been shown to be equivalent to that of an adaptive microphone array system, i.e., N sets of adaptive beamformers (ABFs) with an adaptive null directivity aimed in the direction of unnecessary sounds.

From the equivalence between BSS and ABF, it becomes clear that the physical behavior of BSS reduces the jammer signal by making a spatial null towards the jammer. BSS can further be regarded as an intelligent version of ABF in the sense that it can adapt without any information on the source positions or period of source existence/absence.

2. WHAT IS BSS?

Blind source separation (BSS) is an approach for estimating source signals $s_i(n)$ using only the information of mixed signals $x_j(n)$ observed at each input channel. Typical examples of such source signals include mixtures of simultaneous speech signals that have been picked up by several microphones, brain waves recorded by multiple sensors, and interfering radio signals arriving at a mobile station.

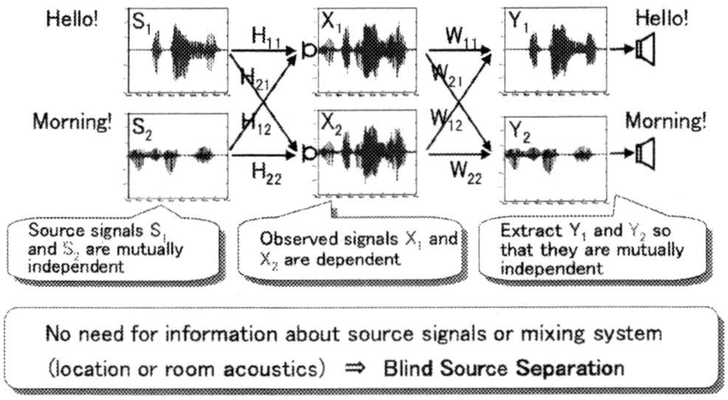

Figure 2: Task of blind source separation of speech signals.

2.1. Mixed Signal Model for Speech Signals in a Room

In the case of audio source separation, several sensor microphones are placed in different positions so that each records a mixture of the original source signals at a slightly different time and level. In the real world where the source signals are speech and the mixing system is a room, the signals that are picked up by the microphones are affected by reverberation. Therefore, the N signals recorded by M microphones are modeled as

$$x_j(n) = \sum_{i=1}^{N}\sum_{p=1}^{P} h_{ji}(p)s_i(n-p+1) \ (j=1,\cdots,M), \quad (1)$$

where s_i is the source signal from a source i, x_j is the signal received by a microphone j, and h_{ji} is the P-taps impulse response from source i to microphone j.

This paper focuses on speech signals as sources that are nongaussian, nonstationary, nonwhite, and that have a zero mean.

2.2. Unmixed Signal Model

To obtain unmixed signals, unmixing filters $w_{ij}(k)$ of Q-taps are estimated, and the unmixed signals are obtained as

$$y_i(n) = \sum_{j=1}^{M}\sum_{q=1}^{Q} w_{ij}(q)x_j(n-q+1) \ (i=1,\cdots,N). \quad (2)$$

The unmixing filters are estimated so that the unmixed signals become mutually independent. This paper considers a two-input, two-output convolutive BSS problem, i.e., $N = M = 2$ (Fig. 1) without a loss of generality.

2.3. Task of Blind Source Separation of Speech Signals

It is assumed that the source signals s_1 and s_2 are mutually independent. This assumption usually holds for sounds in the real world. There are two microphones which pick up the mixed speech. Only the observed signals x_1 and x_2 are available and they are dependent. The goal is to adapt the unmixing systems w_{ij}, and extract y_1 and y_2 so that they are mutually independent. With this operation, we can obtain s_1 and s_2 in the output y_1 and y_2. No information is needed on the source positions or period of source existence/absence. Nor is any information required on the mixing systems h_{ji}. Thus, this task is called *blind* source separation (Fig. 2).

Note that the unmixing systems w_{ij} can at best be obtained up to a scaling and a permutation, and thus cannot itself solve the dereverberation/deconvolution problem [8]. A robust and precise method for solving the permutation problem of frequency-domain BSS was proposed in [9], and a minimal distortion principle for solving the scaling problem was proposed in [10].

3. WHAT IS ICA?

Independent component analysis (ICA) is a statistical method that was originally introduced in the context of neural network modeling [11]. Recently, this method has been used for the BSS of sounds, fMRI and EEG signals of biomedical applications, wireless communication signals, images, and other applications. ICA thus became an exciting new topic in the fields of signal processing, artificial neural networks, advanced statistics, information theory, and various application fields.

Very general statistical properties are used in ICA theory, namely information on statistical independence. In a source separation problem, the source signals are the "independent components" of the data set. In brief, BSS poses the problem of finding a linear representation in which the components are mutually independent. ICA consists of estimating both the unmixing matrix $\mathbf{W}(\omega)$ and sources s_i, when we only have the observed signals x_j.

The unmixing matrix $\mathbf{W}(\omega)$ is determined so that one output contains as much information on the data as possible. The value of any one of the components gives no information on the values of the other components. If the unmixed signals are mutually independent, then they are equal to the source signals.

4. HOW SPEECH SIGNALS CAN BE SEPARATED?

This paper attempts a simple and comprehensive (rather than accurate) exploration from the acoustical signal processing perspective in the frequency domain. With the ICA-based BSS framework, how can we separate speech signals?

The simple answer is to diagonalize $\mathbf{R}_Y$ in each frequency bin, where $\mathbf{R}_Y$ is a (2×2) matrix:

$$\mathbf{R}_Y = \begin{bmatrix} \langle\Phi(Y_1)Y_1\rangle & \langle\Phi(Y_1)Y_2\rangle \\ \langle\Phi(Y_2)Y_1\rangle & \langle\Phi(Y_2)Y_2\rangle \end{bmatrix}. \quad (3)$$

The function $\Phi(\cdot)$ is a nonlinear function. The operation $\langle\cdot\rangle$ is the averaging operation used to obtain statistical information. We want to minimize the off-diagonal components, while at the same time, constraining the diagonal components to proper constants.

The components of the matrix $\mathbf{R}_Y$ correspond to the mutual information between Y_i and Y_j. At the convergence point, the off-diagonal components, which are the mutual information between Y_1 and Y_2, become zero:

$$\langle\Phi(Y_1)Y_2\rangle = 0, \quad \langle\Phi(Y_2)Y_1\rangle = 0. \quad (4)$$

While at the same time, the diagonal components, which only control the amplitude scaling of the output Y_1 and Y_2, are constrained to proper constants:

$$\langle \Phi(Y_1)Y_1 \rangle = c_1, \quad \langle \Phi(Y_2)Y_2 \rangle = c_2. \tag{5}$$

To achieve this convergence, we use the recursive learning rule [12, 13].

$$\mathbf{W}_{i+1} = \mathbf{W}_i + \eta \Delta \mathbf{W}_i, \tag{6}$$

$$\Delta \mathbf{W}_i = \begin{bmatrix} c_1 - \langle \Phi(Y_1)Y_1 \rangle & \langle \Phi(Y_1)Y_2 \rangle \\ \langle \Phi(Y_2)Y_1 \rangle & c_2 - \langle \Phi(Y_2)Y_2 \rangle \end{bmatrix} \mathbf{W}_i. \tag{7}$$

When $\mathbf{R}_Y$ is diagonalized, $\Delta \mathbf{W}$ converges to zero.

4.1. Second Order Statistics (SOS) Approach

If $\Phi(Y_1) = Y_1$, we have the simple decorrelation:

$$\langle \Phi(Y_1)Y_2 \rangle = \langle Y_1 Y_2 \rangle = 0. \tag{8}$$

This is not sufficient to achieve independence, therefore, we cannot solve the problem. This can be understood in a comprehensive way in that we have four unknown parameters W_{ij} in each frequency bin, but only three equations in (4) and (5) since $Y_1 Y_2 = Y_2 Y_1$ when $\Phi(Y_i) = Y_i$, that is, the simultaneous equations become underdetermined. Accordingly the simultaneous equations cannot be solved.

However, when the sources are nonstationary, the second order statistics is different in each time block. As a result, more equations are available and the simultaneous equations can be solved. This is the *nonstationary decorrelation* approach [14].

Similarly, when the sources are nonwhite, we have a delayed correlation for a multiple time delay:

$$\langle \Phi(Y_1)Y_2 \rangle = \langle Y_1(m) Y_2(m + \tau_i) \rangle = 0, \tag{9}$$

The second order statistics is different in each time delay, thus more equations are available and the simultaneous equations can be solved. This is the *time-delayed decorrelation* (TDD) approach [15].

These are the approaches of *second order statistics* (SOS).

4.2. Higher Order Statistics (HOS) Approach

On the other hand if, for example, $\Phi(Y_1) = \tanh(Y_1)$, we have:

$$\langle \Phi(Y_1)Y_2 \rangle = \langle \tanh(Y_1)Y_2 \rangle = 0. \tag{10}$$

With a Tailor expansion of $\tanh(\cdot)$, (10) can be expressed as

$$\langle (Y_1 - \frac{Y_1^3}{3} + \frac{2Y_1^5}{15} - \frac{17Y_1^7}{315} \ldots)\, Y_2 \rangle = 0, \tag{11}$$

thus we have higher order or *nonlinear decorrelation*, then we can solve the problem. Or more simply, we could say that we have four equations in (4) and (5) for four unknown parameters W_{ij} in each frequency bin. Accordingly the simultaneous equations can be solved.

This is the approach of *higher order statistics* (HOS) [16].

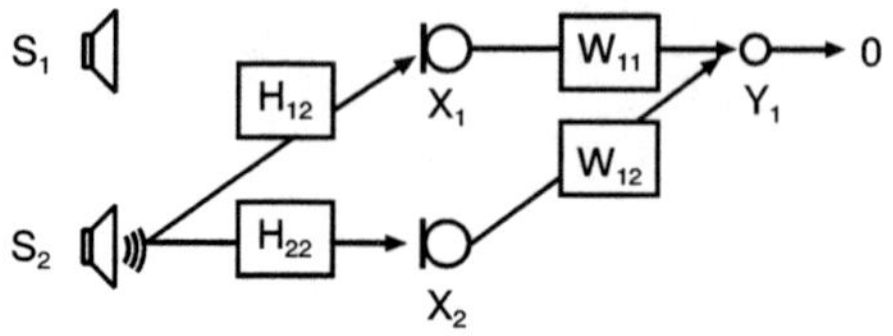

(a) ABF for target S_1 and jammer S_2.

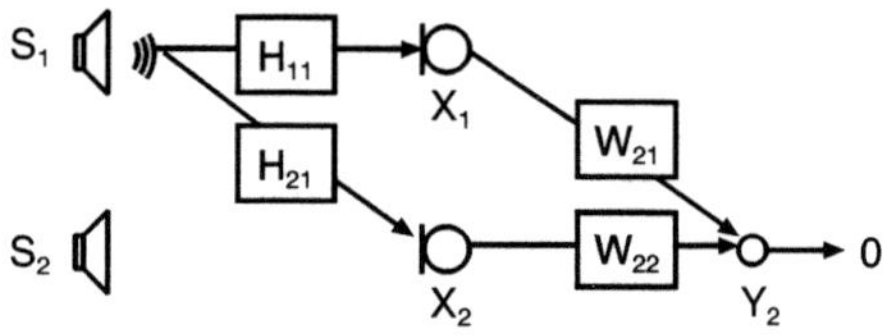

(b) ABF for target S_2 and jammer S_1.

Figure 3: Two sets of ABF-system configurations.

5. SEPARATION MECHANISM OF BSS

BSS is a statistical, or mathematical method, so the physical behavior of BSS is not obvious. We are simply attempting to make the two output signals Y_1 and Y_2 independent. Then, what is the physical interpretation of BSS?

We can understand the behavior of BSS as two sets of ABFs [6]. An ABF can create only one null towards the jammer when two microphones are used. BSS and ABFs form an adaptive spatial null in the jammer direction, and extract the target.

5.1. Frequency-Domain Adaptive Beamformer (ABF)

Here, we consider the frequency-domain adaptive beamformer (ABF), that can adaptively remove a jammer signal. Since the aim is to separate two signals S_1 and S_2 with two microphones, two sets of ABFs are used (see Fig. 3). That is, an ABF that forms a null directivity pattern towards source S_2 by using filter coefficients W_{11} and W_{12}, and an ABF that forms a null directivity pattern towards source S_1 by using filter coefficients W_{21} and W_{22}. Note that the direction of the target or the impulse responses from the target to the microphones should be known, and that the ABF can adapt only when a jammer is active but a target is silent.

The separation performance of BSS is compared with that of ABF. Figure 4 shows the directivity patterns obtained by BSS and ABF. In Fig. 4, (a) and (b) show directivity patterns by $\mathbf{W}$ obtained by BSS, and (c) and (d) show directivity patterns by $\mathbf{W}$ obtained by ABF. When $T_R = 0$, a sharp spatial null is obtained by both BSS and ABF [see Figs. 4(a) and (c)]. When $T_R = 300$ ms, the directivity pattern becomes duller for both BSS and ABF [see Figs. 4(b) and (d)].

6. DISCUSSIONS

BSS was interpreted from the physical standpoint showing the equivalence between frequency-domain BSS and two sets of microphone array systems, i.e., two sets of adaptive beamformers (ABFs) [6]. Convolutive BSS can be understood as multiple ABFs that generate statistically inde-

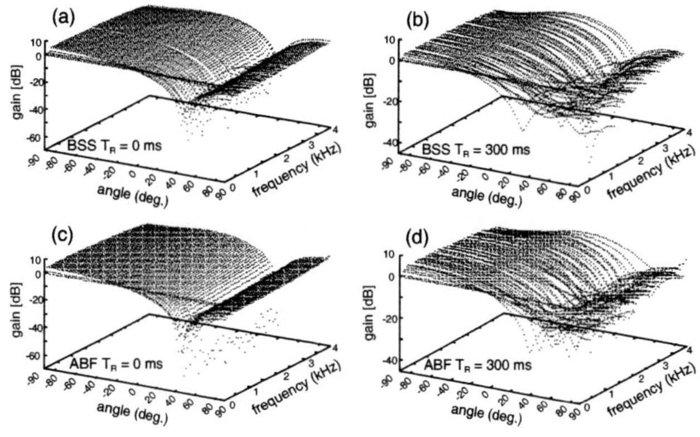

Figure 4: Directivity patterns (a) obtained by BSS (T_{R}=0 ms), (b) obtained by BSS (T_{R}=300 ms), (c) obtained by ABF (T_{R}=0 ms), and (d) obtained by ABF (T_{R}=300 ms).

pendent output, or more simply, an output with minimal crosstalk.

Because ABF and BSS mainly deal with sound from the jammer direction by making a null towards the jammer, the separation performance is fundamentally limited [5]. This understanding clearly explains the poor performance of BSS in the real world with long reverberation. If the sources are not "independent," their dependency results in bias noise to obtain the correct unmixing filter coefficients. Therefore, the BSS performance is upper bounded by that of the ABF.

However, in contrast to the ABF, no assumptions regarding array geometry or source location need to be made in BSS. BSS can adapt without any information on the source positions or period of source existence/absence. This is because, instead of adopting power minimization criterion that adapt the jammer signal out of the target signal in ABF, a cross-power minimization criterion is adopted that decorrelates the jammer signal from the target signal in BSS. It was shown that the least squares criterion of ABF is equivalent to the decorrelation criterion of the output in BSS. The error minimization was shown to be completely equivalent to a zero search in the cross-correlation.

Although the performance of the BSS is limited by that of the ABF, BSS has a major advantage over ABF. A strict one-channel power criterion has a serious crosstalk or leakage problem in ABF, whereas sources can be simultaneously active in BSS. Also, ABF needs to know the array manifold and the target direction. Thus, BSS can be regarded as an intelligent version of ABF.

7. CONCLUSIONS

The blind source separation (BSS) of convolved mixtures of acoustic signals, especially speech, was examined. Source signals can be extracted only from observed mixed signals, by achieving nonlinear, nonstationary, or time-delayed decorrelation. The statistical technique of independent component analysis (ICA) was studied from the acoustic signal processing point of view.

ACKNOWLEDGMENTS

We thank Dr. Hiroshi Saruwatari for detailed discussion.

REFERENCES

[1] S. Makino, "Blind Source Separation of Convolutive Mixtures of Speech," in *Adaptive Signal Processing: Applications to Real-World Problems*, J. Benesty and Y. Huang, Eds., Springer, Berlin, Jan. 2003.

[2] J. F. Cardoso, "The three easy routes to independent component analysis; contrasts and geometry," in *Proc. ICA*, Dec. 2001, pp. 1–6.

[3] T. W. Lee, A. J. Bell, and R. Orglmeister, "Blind source separation of real world signals," *Neural Networks*, vol. 4, pp. 2129–2134, 1997.

[4] M. Z. Ikram and D. R. Morgan, "Exploring permutation inconsistency in blind separation of speech signals in a reverberant environment," in *Proc. ICASSP2000*, June 2000, pp. 1041–1044.

[5] S. Araki, R. Mukai, S. Makino, T. Nishikawa, and H. Saruwatari, "The fundamental limitation of frequency domain blind source separation for convolutive mixtures of speech," *IEEE Trans. Speech Audio Processing*, vol. 11, no. 2, pp. 109–116, Mar. 2003.

[6] S. Araki, S. Makino, Y. Hinamoto, R. Mukai, T. Nishikawa, and H. Saruwatari, "Equivalence between frequency domain blind source separation and frequency domain adaptive beamforming for convolutive mixtures," *EURASIP Journal on Applied Signal Processing*, accepted.

[7] R. Mukai, S. Araki, H. Sawada, and S. Makino, "Separation and dereverberation performance evaluation of frequency domain blind source separation," *Journal on Acoust. Sci. & Tech.*, accepted.

[8] E. Weinstein, M. Feder, and A. V. Oppenheim, "Multichannel signal separation by decorrelation," *IEEE Trans. Speech Audio Processing*, vol. 1, no. 4, pp. 405–413, Oct. 1993.

[9] H. Sawada, R. Mukai, S. Araki, and S. Makino, "A robust approach to the permutation problem of frequency-domain blind source separation," in *Proc. ICASSP*, Apr. 2003, pp. 381–384.

[10] K. Matsuoka and S. Nakashima, " Minimal distortion principle for blind source separation," in *Proc. ICA*, Dec. 2001, pp. 722–727.

[11] J. Herault and C. Jutten, "Space or time adaptive signal processing by neural network models," in *Neural networks for computing: AIP conference proceedings 151*, New York J. S. Denker, ed., American Institute of Physics, Ed., 1986.

[12] S. Amari, A. Cichocki, and H. Yang, "A new learning algorithm for blind source separation," in *Advances in Neural Information Processing Systems 8*, pp. 757–763, MIT Press, 1996.

[13] A. Cichocki, R. Unbehauen, and E. Rummert, "Robust learning algorithm for blind separation of signals," *Electronics Letters*, vol. 30, no. 17, pp. 1386–1387, 1994.

[14] K. Matsuoka, M. Ohya, and M. Kawamoto, "A neural net for blind separation of nonstationary signals," *Neural Networks*, vol. 8, no. 3, pp. 411–419, 1995.

[15] L. Molgedey and H. G. Schuster, "Separation of a mixure of independent signals using time delayed correlations," *Physical Review Letters*, vol. 72, no. 23, pp. 3634–3636, 1994.

[16] A. Hyvarinen, H. Karhunen, and E. Oja, *Independent component analysis*, John Wiley & Sons, 2001.

ICAR: INDEPENDENT COMPONENT ANALYSIS USING REDUNDANCIES

Laurent Albera[1,2], *Anne Ferréol*[2], *Pascal Chevalier*[2], *Pierre Comon*[1]

(1) I3S, Algorithmes-Euclide-B, BP 121, F-06903 Sophia-Antipolis Cedex , France
(2) THALES Communications, 146 Boulevard de Valmy, BP 82, F-92704 Colombes, France

{albera,comon}@i3s.unice.fr {anne.ferreol,pascal.chevalier}@fr.thalesgroup.com

ABSTRACT

A new Blind Source Separation (BSS) algorithm, called ICAR and using only Fourth Order (FO) statistics of the data, is proposed. The latter method is compared by computer experiments with the well-known methods COM1, COM2, JADE, FastICA and SOBI. Since ICAR has given very good convergence results and has performed the source separation in the presence of a Gaussian noise with unknown spatial correlation, it appears as being one of the most attracting BSS algorithms.

1. INTRODUCTION

Blind Source separation (BSS) or Independent Component Analysis (ICA) have lately raised great interest. These problems find their place in numerous applications including telecommunications, speech processing, or biomedical engineering. For instance, in antenna array processing, if several sources impinge of an array of discrete sensors, and if the channel time spread associated with every source and every sensor is negligible compared to the symbol period, then the signals received can be modeled as a static mixture of the transmitted sources. BSS aims at extracting the sources from the sole observation of the mixtures received on the array. Several techniques are available, depending on the assumptions made. In this paper, it will be assumed that sources are mutually statistically independent. Contrary to Principal Component Analysis (PCA), which exploits only statistics of order 2, ICA resorts to statistics of higher order, and is thus able to impose some stronger statistical independence than just a mere decorrelation. This is made possible if sources are not Gaussian, and made easier if there are at least as many sensors as sources.

Since the early works of Jutten [11], the concept of ICA has evolved, and most of the material has been presented in the seminal paper of Comon [7], where it is proposed to maximize contrast functionals; an algorithm is proposed there, which shall be referred to as COM2, for Contrast-based Maximization of squared fourth order cumulants. Some years later, he also proposed a simple algorithm [8], which will be called COM1 in the sequel, maximizing a squares-free criterion [13]. On the other hand, Cardoso and Souloumiac have proposed [4] slightly later an efficient algorithm performing Joint Approximate Diagonalization of Eigen cumulant matrices (JADE). Later, Hyvarinen and others propose the so-called FastICA fixed point algorithm, which extracts one source at a time, first for real signals [12], then for complex [3]. All these methods can be sensitive to the presence of additive Gaussian noise, of unknown spatial coherence; such a noise is frequent in applications such as HF radiocommunications. In fact, they perform as a preliminary stage an exact second-order decorrelation by a preliminary "spatial whitening"; this operation is sub-optimal in several respects, because it puts too much emphasis on second order statistics.

The solution (also sub-optimal though) proposed in this paper consists of getting rid of this whitening stage, and of using exclusively higher order statistics, namely fourth order cumulants. More precisely, the redundancy theoretically present in the quadricovariance of the observations is exploited. This concept can be extended to statistics of order strictly higher than 4, allowing for instance to address the case of *underdetermined mixtures*, in which the number of sources present exceeds in permanence the number of sensors. Such extensions to order 6, or more generally to order $m{=}2q$ ($q{\geq}2$), have been recently proposed by the authors under the names of BIRTH [2] and BIOME [1].

2. NOTATIONS AND ASSUMPTIONS

Let N sensors be available, and denote $x_n(k)$ the noisy observations received on these sensors, $1 \leq n \leq N$. The vector of observations $\boldsymbol{x}(k) \stackrel{\text{def}}{=} [x_1(k)\; x_2(k)\; \cdots\; x_N(k)]^{\mathsf{T}}$ can be modeled in the form below:

$$\boldsymbol{x}(k) = \boldsymbol{A}\,\boldsymbol{s}(k) + \boldsymbol{\nu}(k) = \sum_{p=1}^{P} \boldsymbol{a}_p\, s_p(k) + \boldsymbol{\nu}(k) \qquad (1)$$

where $s_p(k)$ denotes the unknown sources, and $\boldsymbol{A} \stackrel{\text{def}}{=} [\boldsymbol{a}_1\; \boldsymbol{a}_2\; \cdots\; \boldsymbol{a}_P]$, $\boldsymbol{s}(k) \stackrel{\text{def}}{=} [s_1(k)\; s_2(k)\; \cdots\; s_P(k)]^{\mathsf{T}}$, $\boldsymbol{\nu}(k) \stackrel{\text{def}}{=} [\nu_1(k)\; \nu_2(k)\; \cdots\; \nu_N(k)]^{\mathsf{T}}$ denote respectively the $N \times P$ mixing matrix, and the source and noise random vectors, assumed to be independent. Also define for very index k the entries of the 4-th order cumulant tensor, $\mathcal{C}_{x}$, of a random process $x(k)$, stationary and ergodic up to order 4:

$$C_{de,x}^{fg} \stackrel{\text{def}}{=} \mathsf{Cum}\{x_d(k),\ x_e(k),\ x_f(k)^*,\ x_g(k)^*\} \quad (2)$$

It is then possible to store every entry of tensor $\mathcal{C}_x$ in a matrix Q_x, sometimes called the *quadricovariance*:

$$\forall 1 \le d, e, f, g \le N,\ Q_x(N(d-1)+g, N(f-1)+e) = C_{de,x}^{fg} \quad (3)$$

where $Q_x(r, q)$ corresponds to the (r, q)-th component of Q_x. Note that this arrangement is of course not unique, but this is the one that yields the best performances in terms of identifiability [6]. Matrix Q_x can be denoted more generally as $\mathcal{C}_{4,x}^{1}$ [1].

Let us add that the expression of 4-th order cumulants of zero-mean complex variables as a function of moments appears in many papers; see for instance [10] for non zero mean variables. We also assume the following:

H1. For every index k, the source vector $s(k)$ is stationary and ergodic (extension to cyclostationary and cycloergodic is straightforward), with complex values, mutually independent at order 4 (i.e. all cross cumulants of order 4 are null);

H2. For every k, the noise vector $\nu(k)$ is stationary and ergodic;

H3. For every k, $s(k)$ and $\nu(k)$ are statistically independent;

H4. Source *kurtoses* (standardized autocumulants of order 4), $\kappa_{ppp,s}^{ppp} = C_{ppp,s}^{ppp}/\pi_p^2$, are non zero and have all the same sign ($\pi_p \stackrel{\text{def}}{=} \mathsf{E}[|s_p(k)|^2]$ denotes the power of the p-th source);

H5. The number of sources is smaller than the number of sensors: $P \le N$. The mixture is then referred to as *overdetermined*;

H6. The mixing matrix A is full rank and does not contain any null entry.

The goal is to determine a separating matrix, W, such that

$$y(n) \stackrel{\text{def}}{=} W^{\mathsf{H}} x(n) \quad (4)$$

is an estimate of vector $s(k)$ up to a multiplicative *trivial* matrix (i.e. of the form $\Lambda\Pi$ where Λ is invertible diagonal and Π is a permutation). In ICAR, as in numerous other BSS algorithms, the construction of W requires the identification of mixture A.

3. THE CORE OF THE ICAR METHOD

3.1. Properties of the quadricovariance

The multilinearity property enjoyed by cumulants allows to write

$$Q_x = [A \otimes A^*]\, Q_s\, [A \otimes A^*]^{\mathsf{H}} \quad (5)$$

where Q_x and Q_s denote quadricovariance matrices of $x(k)$ and $s(k)$, of size $N^2 \times N^2$ and $P^2 \times P^2$, respectively. Because sources are independent, Q_s is diagonal. However, it is not full rank. Another decomposition,

$$Q_x = \mathcal{A}_Q\, \mathcal{Q}_s\, \mathcal{A}_Q^{\mathsf{H}} \quad (6)$$

exhibits two new matrices. The first one, $\mathcal{Q}_s \stackrel{\text{def}}{=} \text{diag}(\left[C_{11,s}^{11}\ C_{22,s}^{22}\ \cdots\ C_{PP,s}^{PP}\right])$, is diagonal of size $P \times P$, and invertible from (**H4**). The second one, $\mathcal{A}_Q$, of size $N^2 \times P$, is also of full rank from (**H6**) and [1, prop. 1], can be written as:

$$\begin{aligned} \mathcal{A}_Q &\stackrel{\text{def}}{=} [a_1 \otimes a_1^*\ a_2 \otimes a_2^*\ \cdots\ a_P \otimes a_P^*] \\ &= [[A^* \Phi_1]^{\mathsf{T}}\ [A^* \Phi_2]^{\mathsf{T}}\ \cdots\ [A^* \Phi_N]^{\mathsf{T}}]^{\mathsf{T}} \end{aligned} \quad (7)$$

where

$$\Phi_n \stackrel{\text{def}}{=} \text{diag}([A(n,1)\ A(n,2)\ \cdots\ A(n,P)]) \quad (8)$$

Entries $A(n, \cdot)$ of the diagonal matrix Φ_n, of size $P \times P$, form the n-th line of A. (Matrices $\mathcal{Q}_s$ and $\mathcal{A}_Q$ are referred to as $\zeta_{4,s}$ et $\mathcal{A}_2^1$ in [1]).

3.2. Principle of ICAR

the principle is similar to that described in [2] with the difference that statistics of order 4 are utilized instead of 6. This difference is important because 4-th order statistics do not allow any more to address *underdetermined mixtures*, at least in the present framework. Since sources have the same kurtosis sign, it is possible to determine a unique real square root matrix of Q_x or $-Q_x$. Without restricting the generality, assume source kurtoses are all positive. The square root of Q_x can be computed by an Eigen Value Decomposition (EVD), from (6):

$$Q_x^{1/2} = E_s\, L_s^{1/2} = \mathcal{A}_Q\, \mathcal{Q}_s^{1/2}\, V^{\mathsf{H}} \quad (9)$$

where L_s ($L_s^{1/2}$ denotes a square root of L_s) is the $P \times P$ diagonal matrix containing the P non-zero eigenvalues of Q_x, and E_s is the $N^2 \times P$ matrix of associated normalized eigenvectors. Because $\mathcal{A}_Q$ is full rank, it can be shown that (**H4**) amounts to assuming that diagonal elements of L_s are non zero and have the same sign [1, prop. 2], here positive. Moreover, (9) establishes a link between $Q_x^{1/2}$ and $\mathcal{A}_Q$, where V is a unitary matrix, uniquely defined once L_s and E_s are fixed. Next, (9) and (7) relate $Q_x^{1/2}$ and A such that:

$$\begin{aligned} Q_x^{1/2} &= \left[[A^* \Phi_1 \mathcal{Q}_s^{1/2} V^{\mathsf{H}}]^{\mathsf{T}} \cdots [A^* \Phi_N \mathcal{Q}_s^{1/2} V^{\mathsf{H}}]^{\mathsf{T}}\right]^{\mathsf{T}} \\ &\stackrel{\text{def}}{=} [\,\Gamma_1{}^{\mathsf{T}}\ \Gamma_2{}^{\mathsf{T}}\ \cdots\ \Gamma_N{}^{\mathsf{T}}\,]^{\mathsf{T}} \end{aligned} \quad (10)$$

where $\boldsymbol{\Gamma}_n = \boldsymbol{A}^* \boldsymbol{\Phi}_n \boldsymbol{\mathcal{Q}}_s^{1/2} \boldsymbol{V}^{\mathsf{H}}$ is the n-th matrix block of $\boldsymbol{Q}_x^{1/2}$, of size $N \times P$. Consequently, matrix $\boldsymbol{V}$ diagonalizes the $N(N-1)$ matrices $\boldsymbol{\Theta}_{n_1,n_2}$ given by

$$\forall 1 \le n_1 \ne n_2 \le N, \qquad \boldsymbol{\Theta}_{n_1,n_2} = \boldsymbol{\Gamma}_{n_1}^{\sharp} \boldsymbol{\Gamma}_{n_2} \tag{11}$$

where $\sharp$ denotes pseudo-inversion. In fact, from the expression of $\boldsymbol{\Gamma}_n$ yielded by (10), and under assumptions **(H5)**-**(H6)**, the pseudo-inverse of $\boldsymbol{\Gamma}_n$ can be written as $\boldsymbol{\Gamma}_n^{\sharp} \stackrel{\text{def}}{=} (\boldsymbol{\Gamma}_n{}^{\mathsf{H}} \boldsymbol{\Gamma}_n)^{-1} \boldsymbol{\Gamma}_n{}^{\mathsf{H}}$. As a consequence, matrices $\boldsymbol{\Theta}_{n_1,n_2}$ can be rewritten in the form below:

$$\forall 1 \le n_1 \ne n_2 \le N, \qquad \boldsymbol{\Theta}_{n_1,n_2} = \boldsymbol{V}\, D_{n_1,n_2}\, \boldsymbol{V}^{\mathsf{H}} \tag{12}$$

where matrices $D_{n_1,n_2} \stackrel{\text{def}}{=} \boldsymbol{\Phi}_{n_1}^{-1} \boldsymbol{\Phi}_{n_2}$ are diagonal. Denote $\boldsymbol{V}_{sol}$ the unitary matrix that jointly diagonalizes matrices $\boldsymbol{\Theta}_{n_1,n_2}$. Then $\boldsymbol{V}_{sol} = \boldsymbol{V}\boldsymbol{\mathcal{T}}$ with $\boldsymbol{\mathcal{T}}$ unitary. Thus, from (9), $\boldsymbol{V}_{sol}$ allows to identify $\boldsymbol{\mathcal{A}}_Q$ up to a unitary matrix:

$$\boldsymbol{Q}_x^{1/2}\, \boldsymbol{V}_{sol} = \boldsymbol{\mathcal{A}}_Q\, \boldsymbol{\mathcal{Q}}_s^{1/2}\, \boldsymbol{\mathcal{T}} \stackrel{\text{def}}{=} \widehat{\boldsymbol{\mathcal{A}}_Q} \tag{13}$$

On the other hand, from assumptions **(H5)**-**(H6)**, for every pair $(p_1, p_2)|_{p_1 \ne p_2}$ belonging to $\{1,2,\ldots,P\}^2$, there exists at least a pair $(n_1, n_2)|_{n_1 \ne n_2}$ belonging to $\{1,2,\ldots,N\}^2$ such that $D_{n_1,n_2}(p_1,p_1) \ne D_{n_1,n_2}(p_2,p_2)$. This implies that $\boldsymbol{\mathcal{T}}$ is trivial. Then from (13), matrix $\widehat{\boldsymbol{\mathcal{A}}_Q}$ is consequently an estimate of $\boldsymbol{\mathcal{A}}_Q$, up to a trivial matrix.

Next from (10), equation (13) can be rewritten as

$$\begin{aligned} \boldsymbol{Q}_x^{1/2}\, \boldsymbol{V}_{sol} &= \left[[\boldsymbol{A}^* \boldsymbol{\Phi}_1 \boldsymbol{\mathcal{Q}}_s^{1/2} \boldsymbol{\mathcal{T}}]^{\mathsf{T}} \cdots [\boldsymbol{A}^* \boldsymbol{\Phi}_N \boldsymbol{\mathcal{Q}}_s^{1/2} \boldsymbol{\mathcal{T}}]^{\mathsf{T}} \right]^{\mathsf{T}} \\ &\stackrel{\text{def}}{=} [\boldsymbol{\Sigma}_1{}^{\mathsf{T}}\, \boldsymbol{\Sigma}_2{}^{\mathsf{T}} \cdots \boldsymbol{\Sigma}_{N^2}{}^{\mathsf{T}}]^{\mathsf{T}} = \widehat{\boldsymbol{\mathcal{A}}_Q} \end{aligned} \tag{14}$$

The matrix block $\boldsymbol{\Sigma}_1$, formed of the N first rows of $\boldsymbol{Q}_x^{1/2}$ $\boldsymbol{V}_{sol}$, corresponds to $\boldsymbol{A}^*$, up to a trivial matrix:

$$\boldsymbol{\Sigma}_1 = \boldsymbol{A}^*\, \boldsymbol{\Phi}_1\, \boldsymbol{\mathcal{Q}}_s^{1/2}\, \boldsymbol{\mathcal{T}} \tag{15}$$

where $\boldsymbol{\mathcal{Q}}_s^{1/2}$ and $\boldsymbol{\Phi}_n$ are diagonal for every n, $1 \le n \le N$. This method is named ICAR1. One could also think of several improvements, for instance by averaging the N blocks $\boldsymbol{\Sigma}_n$ for estimating $\boldsymbol{A}^*$, giving rise to ICAR2.

3.3. Refinement of the method

In order to fully exploit the information contained in matrix $\widehat{\boldsymbol{\mathcal{A}}_Q}$ (14), namely its redundancies of $\boldsymbol{\mathcal{A}}_Q$ in (7), it is possible to mimic the last step of FOBIUM [9]. Indeed, from (13) and (7):

$$\begin{aligned} \widehat{\boldsymbol{\mathcal{A}}_Q} &= \left[\lambda_{\xi(1)} \left[\boldsymbol{a}_{\xi(1)} \otimes \boldsymbol{a}_{\xi(1)}^* \right] \cdots \lambda_{\xi(P)} \left[\boldsymbol{a}_{\xi(P)} \otimes \boldsymbol{a}_{\xi(P)}^* \right] \right] \\ &\stackrel{\text{def}}{=} \left[\boldsymbol{b}_{\xi(1)} \cdots \boldsymbol{b}_{\xi(P)} \right] \end{aligned} \tag{16}$$

where $|\lambda_p| = |C_{ppp,\boldsymbol{s}}^{ppp}|^{1/2}$, and ξ is a permutation mapping on $\{1, 2, \ldots, P\}$. It is then possible to associate every vector $\boldsymbol{b}_{\xi(p)}$ of size $N^2 \times 1$ with a matrix $\boldsymbol{B}_{\xi(p)}$ of size $N \times N$, whose columns are precisely the N successive vectors $\boldsymbol{b}_{\xi(p)}$ formed of N values:

$$\boldsymbol{B}_{\xi(p)} = \lambda_{\xi(p)} \left[\boldsymbol{a}_{\xi(p)} \boldsymbol{a}_{\xi(p)}^{\mathsf{H}} \right]^* \tag{17}$$

A mere diagonalization of matrices $\boldsymbol{B}_{\xi(p)}^*$ allows to yield the P directional vectors $\boldsymbol{a}_p$, by retaining each time the eigenvector associated with the dominant eigenvalue, and up to a permutation and a scalar multiplicative factor. The method with the latter improvement is named ICAR3.

4. SIMULATIONS

Two computer experiments show the performances of ICAR and some efficient BSS techniques (COM1, JADE, FastICA, SOBI). $P = 4$ statistically independent sources, i.e. 2 BPSK and 2 QPSK, all with a raised cosine pulse shape of roll-off equal to 0.25, are received by a UCA of $N = 4$ identical sensors of radius R such that $R/\lambda = 0.55$ (λ: wavelength). The four sources, assumed synchronized, have the same input SNR (Signal to Noise Ratio) of 20dB and the noise is spatially and temporally white Gaussian. The symbol period T_1 associated with the first BPSK is equal to three times the sample period T_e. The other sources have a symbol period equal to twice the sample period. The directions of arrival of the sources are such that the source steering vectors are orthogonal and the associated carrier residus are such that $f_{c1}\, T_e = 0$, $f_{c2}\, T_e = 0.3$, $f_{c3}\, T_e = 0.2$ and $f_{c4}\, T_e = 0.1$. The performance measure assumed to evaluate the quality of the extraction of source p is the maximal signal to interference plus noise ratio associated with source p, denoted $SINRM_p$ [5]. It can be compared to the optimal $SINRM_p$ computed using the exact mixing matrix instead of the estimated one and denoted by *Optimum SMF*. These are precisely the comparisons that are drawn now. Figure 1 displays the SINRM of source 3 averaged over 200 realizations, associated with ICAR3, and BSS techniques JADE, COM1, FastICA and SOBI, as a function of the number of samples. Figure 1 shows the good performance of the ICAR3 algorithm facing the other methods. Note that the SOBI method gives in this simulation good results since sources have been chosen with different spectral densities,

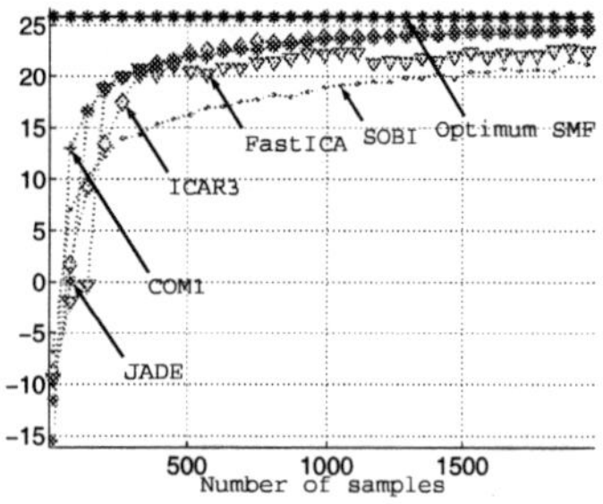

Fig. 1. $SINRM_3$ for a SNR of 20 dB

especially taking different carrier residus. Similar results are observed for the other sources, and are not reported here.

The influence of the noise spatial coherence coefficient ρ is next studied. Now we have 3 sources, i.e. 2 BPSK and 1 QPSK, all with a raised cosine pulse shape of roll-off equal to 0.25, are assumed to be received by a UCA of $N=5$ identical sensors of radius R such that $R/\lambda = 0.55$. Their symbol periods are equal to $T_1 = 2T_e$, $T_2 = 3T_e$ and $T_3 = 4T_e$ respectively. Their carrier residus are chosen equal to zero. The source steering vectors are built orthogonal, the SNR is 0 dB and 1500 samples are used for separation in this scenario. We apply the COM1, COM2, JADE, SOBI and ICAR3 methods, and the SINRM associated with each source is computed and averaged still over 200 realizations. The Gaussian noise is modeled as a sum of two noises, $\boldsymbol{\nu_{in}}(k)$ and $\boldsymbol{\nu_{out}}(k)$, of covariance $\boldsymbol{R_{\nu}^{in}}$ and $\boldsymbol{R_{\nu}^{out}}$ respectively, such that:

$$\boldsymbol{R_{\nu}^{in}}(r,q) \overset{\text{def}}{=} \sigma^2\delta(r-q)/2 \quad \boldsymbol{R_{\nu}^{out}}(r,q) \overset{\text{def}}{=} \sigma^2\rho^{|r-q|}/2 \quad (18)$$

where σ^2 is the global noise variance per sensor. Note that $\boldsymbol{R_{\nu}}(r,q) \overset{\text{def}}{=} \boldsymbol{R_{\nu}^{in}}(r,q) + \boldsymbol{R_{\nu}^{out}}(r,q)$ is the (r,q)-th component of the global noise covariance. Contrary to COM1, COM2, JADE and SOBI, algorithm ICAR3 is totally insensitive to the increase in coefficient ρ. In fact, the classical methods such as COM1, COM2, JADE, FastICA and SOBI require a prior spatial whitening based on second order moments. This stage theoretically needs the perfect knowledge of the noise covariance. If this is not the case, a whitening of the observed data is performed instead, which is biased. ICAR3 does not suffer from this drawback, since it uses only 4-th order cumulants, which are (asymptotically) insensitive to Gaussian noise, regardless of its space/time color. Again, similar results have been observed for sources 1 and 2.

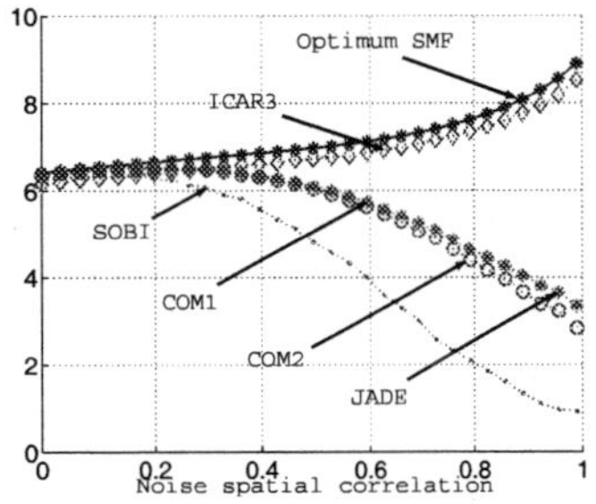

Fig. 2. *SINRM*$_3$ for a SNR of 0 dB

5. CONCLUSION

The new algorithm proposed in this paper, referred to as ICAR, utilizes only fourth order statistics of observations, and seems to be much more attractive than previous (classical) BSS techniques needing prior second order decorrelation, according to our computer simulations.We currently work on another solution to the problem presented in section 3.3, in other words, on another technique to extract the mixing matrix $\boldsymbol{A}$ contained in $\boldsymbol{\mathcal{A}}_Q$ (7).

6. REFERENCES

[1] L. ALBERA, A. FERREOL, P. COMON, P. CHEVALIER, "Blind Identification of Overdetermined and underdetermined Mixtures of sources (BIOME)", *Linear Algebra Applications*, april 2003, submitted.

[2] L. ALBERA, A. FERREOL, P. COMON, P. CHEVALIER, "Sixth order blind identification of underdetermined mixtures (BIRTH) of sources", in *ICA 03, Fourth International Symposium on Independent Component Analysis and Blind Signal Separation*, Nara, Japan, April 1-4 2003, pp. 909–914.

[3] E. BINGHAM, H. HYVARINEN, "A fast fixed-point algorithm for independent component analysis of complex valued signals", *Int. J. of Neural Systems*, vol. 10, no. 1, pp. 1–8, 2000.

[4] J.-F. CARDOSO, A. SOULOUMIAC, "Blind beamforming for non-gaussian signals", *IEE Proceedings-F*, vol. 140, no. 6, pp. 362–370, December 1993.

[5] P. CHEVALIER, "Optimal separation of independent narrow-band sources : Concept and Performances", *Signal Processing, Elsevier*, vol. 73, pp. 27–47, 1999.

[6] P. CHEVALIER, L. ALBERA, A. FERREOL, P. COMON, "On the virtual array concept for higher order array processing", *IEEE Transactions On Signal Processing*, 2003, submitted.

[7] P. COMON, "Independent Component Analysis, a new concept ?", *Signal Processing, Elsevier*, vol. 36, no. 3, pp. 287–314, April 1994.

[8] P. COMON, "From source separation to blind equalization, contrast-based approaches", in *ICISP 01, Int. Conf. on Image and Signal Processing*, Agadir, Morocco, May 3-5 2001, pp. 20–32.

[9] A. FERREOL, L. ALBERA, P. CHEVALIER, "Fourth order blind identification of underdetermined mixtures of sources (FOBIUM)", in *ICASSP 03, 2003 IEEE International Conference on Acoustics Speech and Signal Processing*, Hong Kong, China, April 6-10 2003, pp. 41–44.

[10] A. FERREOL, P. CHEVALIER, L. ALBERA, "Higher order blind separation of non zero-mean cyclostationary sources", in *EUSIPCO 02, XI European Signal Processing Conference*, Toulouse, France, September 3-6 2002, vol. 5, pp. 103–106.

[11] J. HERAULT, C. JUTTEN, B. ANS, "Détection de grandeurs primitives dans un message composite par une architecture de calcul neuromimétique en apprentissage non supervisé", in *GRETSI 85, Dixième colloque sur le Traitement du Signal et des Images*, Nice, France, Septembre 1985, pp. 1017–1022.

[12] H. HYVARINEN, E. OJA, "A fast fixed-point algorithm for independent component analysis", *Neural Computation*, vol. 9, no. 7, pp. 1483–1492, 1997.

[13] N. THIRION, E. MOREAU, "New criteria for blind signal separation", in *IEEE Workshop on Statistical Signal and Array Processing*, Pennsylvania, US, August 2000, pp. 344–348.

BLIND SIMO CHANNEL ESTIMATION FOR CPM USING THE LAURENT APPROXIMATION

Shawn P. Neugebauer and Zhi Ding

University of California, Davis
Department of Electrical and Computer Engineering
Davis, CA 95616–5294

ABSTRACT

We demonstrate that by employing the main-pulse Laurent approximation, blind second-order-statistics-based channel identification can be performed for partial-response CPM signals having modulation indices other than one-half. We use the approximation to formulate the problem in the single-input multiple-output model, permitting fractional sampling of the channel output, and we compare the estimation performance, for several multipath-distorted CPM signals, of several blind channel identification methods. Simulation results demonstrate the efficacy of existing techniques for general CPM signals and highlight the necessity of exploiting the known source statistics and pulses.

1. INTRODUCTION

Most of the existing blind channel identification methods developed for CPM signals are either not applicable to most CPM signals or are computationally complex. Nearly all such work has been done for GMSK as used in the GSM system, e.g., [1, 2], and most of that work has taken advantage (even if indirectly) of particular simplifying characteristics in the system configuration. The results that are applicable to the class of CPM signals generally involve very computationally-complex algorithms. Even though there exist high-quality linear approximations for many other CPM signals, second-order statistical methods for blind channel identification have not been applied to multipath-distorted CPM signals.

Cirpan and Tsatsanis [3] were among the first to recognize the apparent unsuitability of the classical methods to CPM and they proposed the blind stochastic maximum likelihood (SML) method. Although the technique exploits intrinsic properties of CPM, the method is very computationally intensive. Nguyen and Levy [4] were the first to demonstrate a more computationally-efficient algorithm for blind channel estimation for CPM signals. Their expectation-maximization Viterbi algorithm (EMVA) can simultaneously identify the channel and detect the input, even for the nonlinear finite-state machine in CPM, given equally-likely information symbols. The EMVA method gives excellent performance for multipath-distorted CPM signals, but, being based on the Viterbi algorithm, it still has high computational complexity.

The seminal work of Tong *et al.* [5] (the TXK algorithm) was the first to demonstrate blind channel identification using only the second-order statistics (SOS) of the channel output. The TXK algorithm showed that with the assumption of white input symbols, single-input multiple-output (SIMO) channels can be blindly estimated (to within a scaling factor). Afkhamie and Luo [6] proposed a modification of the TXK algorithm to specifically accommodate temporally-correlated sources. López-Valcarce and Dasgupta [7] recently developed an extension of the TXK algorithm that accommodates correlated symbols (we refer to the result as the LVD algorithm). The LVD algorithm uses the same autocorrelation matrices required by the TXK algorithm, except it is more computationally efficient than the approach of Afkhamie and Luo. Of the two, we focus on the LVD algorithm herein.

Another class of SOS-exploiting blind channel-identification methods is the so-called subspace methods, initiated by Moulines *et al.* [8] (we refer to this approach as the MDCM algorithm). Unlike the TXK and LVD algorithms, the MDCM algorithm does not specifically exploit knowledge of the statistics of the input symbols, but it can accommodate correlated input symbols. Furthermore, the MDCM algorithm can be modified to incorporate prior knowledge of a known pulse-shaping function. Schell *et al.* [9] were the first to demonstrate this for a second-order blind algorithm, and Ding *et al.* [10] first modified the MDCM algorithm (we refer to this as MDCM with prior knowledge—MDCM-WPK). A comprehensive comparison of the various SIMO methods can be found in [11].

Laurent [12] showed that a binary CPM signal can be represented using a sum of pulse amplitude-modulated (PAM) signals. When the frequency-shaping pulse has finite length, the Laurent representation (LR) is exact. Even if strictly violated (e.g., for GMSK), the approximation error using the LR is very low. The first-order term in the LR, a PAM signal, has been shown to provide, in many cases, a good approximation of the CPM signal; we refer to this as the single-pulse Laurent approximation (SPLA). When the modulation index is one-half, the SPLA minimizes the approximation error (this simple, high-quality approximation has been thoroughly exploited for GMSK, see, e.g., [1, 2]). For other modulation indices, Laurent's "main" pulse, a weighted sum of time-shifted versions of the pulse in the first-order term of the LR, minimizes the approximation error. We refer to the resulting PAM signal as the main-pulse Laurent approximation (MPLA).

We demonstrate that by employing the MPLA, blind SOS-based channel identification can be performed for partial-response CPM signals, including signals having modulation index other than one-half. After formulating the problem in the SIMO model, we compare the channel-estimation performance, for several multipath-distorted CPM signals, of the TXK, LVD, MDCM, and MDCM-WPK methods, all of which use the second-order statistics of the channel output. The results demonstrate the existence of techniques for effective, computationally-efficient blind channel identification for general binary CPM signals.

2. SIGNAL MODEL AND APPROXIMATION

CPM signals result from driving a frequency modulator with a PAM signal. The FM signal produced has continuous phase, and so avoids characteristics that lead to poor spectral properties. We denote by $s(t, \boldsymbol{\alpha})$ the complex envelope of a binary CPM signal. Then, following [13],

$$s(t, \boldsymbol{\alpha}) = \exp\left[j\pi h \psi(t, \boldsymbol{\alpha})\right] \tag{1}$$

with time-varying phase

$$\psi(t, \boldsymbol{\alpha}) = \sum_n \alpha_n q(t - nT) \tag{2}$$

and where $\boldsymbol{\alpha} = \{\alpha_n = \pm 1\}$ are the information symbols, T is the symbol period, $h = \frac{2k}{p}$ is the modulation index (k, p relatively prime), and $q(t)$ is the phase-shaping pulse, which is related to the frequency-shaping pulse $g(t)$ by

$$q(t) = \int_{-\infty}^{t} g(\zeta) d\zeta. \tag{3}$$

Different choices of $g(t)$ (non-zero over $t \in [0, LT]$, in practice) and h control the bandwidth of and the intersymbol interference (ISI) in the baseband signal (e.g., GMSK uses a Gaussian pulse shape and $h = \frac{1}{2}$ to provide good bandwidth properties at the expense of ISI); an excellent reference is [14].

Several researchers have confirmed the utility of the Laurent representation and have then employed it to develop simplified receivers (e.g., [15]) or to analyze CPM signals (e.g., [16, 17]. Our informal experiments have confirmed the utility of the SPLA and the MPLA for a wide variety of binary CPM signals with various pulse shapes, truncation lengths, and modulation indices. Although the SPLA (using the pulse $C_0(t)$ in the first PAM component of the LR) is the best approximation (in the MSE sense) when $h = \frac{1}{2}$, the MPLA is optimum for other $h \in [0, 1]$. Laurent's main pulse, $P(t)$, has been shown to provide an approximation, linear in special "pseudosymbols," that minimizes the MSE between the approximation and the CPM signal. The pulse $P(t)$ is a weighted sum of time-shifted $C_0(t)$ pulses, and the two pulses have the same length. For $h = \frac{1}{2}$, $P(t) = C_0(t)$. Our focus herein is on other modulation indices, therefore we use as an approximation to (1)

$$\tilde{s}(t, \boldsymbol{\alpha}) = \sum_{n=-\infty}^{\infty} b_{0,n} P(t - nT) \tag{4}$$

with complex-valued pseudosymbols

$$b_{0,n} = \exp\left[j\pi h \sum_{i=-\infty}^{n} \alpha_i\right] \tag{5}$$

and $P(t)$ the main pulse of width $(L+1) \cdot T$; the other parameters are as before.

The pseudosymbols are nonlinear in the information symbols, reflecting the inherent nonlinearity in CPM, but the approximation is linear in the pseudosymbols, enabling the use of many classical techniques. Generally, the pseudosymbols are correlated, even if the information symbols are not [16]. With uncorrelated $\boldsymbol{\alpha}$, the pseudosymbol correlations are given by

$$R_b(k) = E\left[b_{0,i} b_{0,i+k}^*\right] = \cos(\pi h)^{|k|}. \tag{6}$$

For the often analyzed case of $h = \frac{1}{2}$, we take $R_b(k) = \delta(k)$, i.e., the pseudosymbols are uncorrelated. We see that $R_b(k)$ is symmetric about $h = \frac{1}{2}$, it is slowly decaying as h approaches 0 (1) from above (below), it is impulse-like for $h \approx \frac{1}{2}$, and it is independent of pulse shape. We refer to [16] for additional detail on the LR (e.g., the general definition of $C_0(t)$—which depends on $q(t)$, h, and L).

3. BLIND CHANNEL ESTIMATION FOR CPM

Following [7], we consider the FIR SIMO model

$$\overline{\mathbf{y}}_n = \sum_{i=0}^{l} \mathbf{h}_i a_{n-i} + \overline{\mathbf{w}}_n \tag{7}$$

where $\{a_n\}$ is the zero-mean W.S.S. sequence of transmitted symbols, $\{\overline{\mathbf{y}}_n\}$ are $p \times 1$ vectors of channel outputs, $\{\overline{\mathbf{w}}_n\}$ are $p \times 1$ vectors of white noise, and $\{\mathbf{h}_i\}$ are $p \times 1$ vectors constructed from the finite support, linear, causal channel impulse response $h(t)$ (includes pulse-shaping and/or propagation effects). The problem can be reformulated as

$$\mathbf{y}_n = \mathcal{H}\mathbf{a}_n + \mathbf{w}_n \tag{8}$$

where $\mathbf{y}_n$ and $\mathbf{v}_n$ are $mp \times 1$ vectors constructed from $\{\overline{\mathbf{y}}_n\}$ and $\{\overline{\mathbf{w}}_n\}$, $\mathcal{H}$ is an $mp \times (m + l)$ Toeplitz matrix constructed from $\{\mathbf{h}_i\}$, $\mathbf{a}_n$ is the $d \times 1$ vector $[a_n \; a_{n-1} \; \cdots \; a_{n-d+1}]^T$, and $d = m + l$. The SIMO model is applicable to situations where multiple sensors receive the same signal, or when the continuous-time channel has excess bandwidth and a single-sensor output can be fractionally sampled. The development herein supports both situations.

The TXK algorithm uses $\mathcal{R}_y(0) = E\left[\mathbf{y}_n \mathbf{y}_n^H\right]$ and $\mathcal{R}_y(1) = E\left[\mathbf{y}_n \mathbf{y}_{n-1}^H\right]$ to estimate $\hat{\mathcal{H}}$, assuming (as with most SOS-based methods) that $\mathcal{H}$ has full rank and $\mathbf{a}_n$ is white. The LVD algorithm relaxes the requirement on $\mathbf{a}_n$ but requires $\mathcal{R}_a(0) = E\left[\mathbf{a}_n \mathbf{a}_n^H\right]$ and $\mathcal{R}_a(1) = E\left[\mathbf{a}_n \mathbf{a}_{n-1}^H\right]$ (with a minor rank condition). In contrast to the TXK-based methods, the MDCM algorithm exploits the Toeplitz structure of $\mathcal{H}$, and in doing so does not require $\mathbf{a}_n$ be white (although $\mathcal{R}_a(0)$ must have full rank). The MDCM-WPK extends the MDCM algorithm by decomposing the composite channel into a known pulse and an unknown propagation channel, resulting in a simpler minimum-eigenvector problem. Refer to [11] for detailed development of the TXK, MDCM, and MDCM-WPK algorithms; refer to [7] for detailed development of the LVD algorithm.

We desire a SIMO formulation of the multipath-distorted CPM signal

$$r(t) = s(t, \boldsymbol{\alpha}) \star c(t) \tag{9}$$

with $c(t)$ the propagation-channel impulse response. Due to the nonlinearity of $s(t, \boldsymbol{\alpha})$, it is clear that such a formulation is not possible directly in terms of the information symbols $\boldsymbol{\alpha}$ (i.e., we cannot simply use $a_n \triangleq \alpha_n$). Although [18] demonstrates the applicability of the multi-sensor SIMO model when using $a_n \triangleq s(nT)$, fractionally-sampled CPM is not (directly) amenable to this approach. However, by adopting the MPLA, we can approximate the multipath-distorted signal as

$$r(t) \approx \tilde{r}(t) = \tilde{s}(t, \boldsymbol{\alpha}) \star c(t) \tag{10}$$

which implies

$$h(t) \approx \tilde{h}(t) = P(t) \star c(t). \tag{11}$$

Therefore, because $\tilde{s}(t, \boldsymbol{\alpha})$ is linear in the pseudosymbols $b_{0,n}$ we take $a_n \triangleq b_{0,n}$ to obtain a SIMO approximation of multipath-distorted CPM. We leave for future work analysis of the effect of approximation error on identifiability of the channel or the quality of the channel estimates.

We discuss the applicability of several SOS-based methods to blind estimation of CPM signals in this model. We ultimately desire an estimate of $c(t)$, however, an estimate of $\tilde{h}(t)$ would also be useful since $P(t)$ could be deconvolved from the estimate to obtain an estimate of $c(t)$. When $h = \frac{1}{2}$, the pseudosymbols are uncorrelated, and $\tilde{h}(t)$ can be blindly estimated using the standard approaches (e.g., TXK). When $h \neq \frac{1}{2}$, the pseudosymbols are temporally correlated, and the MDCM method, which is tolerant of correlated input, and the LVD method, which specifically exploits correlated input can both estimate the composite channel. Furthermore, with prior knowledge of $P(t)$, we can use MDCM-WPK to obtain better estimates of $c(t)$. However, as noted in [9], it is essential that at least the pulse be identifiable; we leave as future work analytical study of the identifiability of Laurent pulses or the composite channel.

4. SIMULATION RESULTS

To illustrate the utility of the methods described, we use them to blindly estimate channels for several multipath-distorted CPM signals. To fairly evaluate the methods, we compare the estimates from all but the MDCM-WPK method with the approximate composite channel $\tilde{h}(t)$, after gain and phase adjustment of the estimate; for the MDCM-WPK method, we first compute $\hat{\tilde{h}}(t) = \hat{c}(t) \star P(t)$. In all cases, 100 iterations are performed per SNR value. The SNR is defined as in [7]. The configuration of the methods is as follows. All methods are given knowledge of the channel length; methods for estimating it are available. LVD uses exact $\mathcal{R}_a(0)$ and $\mathcal{R}_a(1)$, of the required size, constructed from $R_b(k)$. Both MDCM and MDCM-WPK use the noise subspace approach; MDCM-WPK is given the main pulse and the exact length of the propagation channel. For comparison purposes, the MMSE estimate of the composite FIR channel is estimated from the exact pseudosymbols. All the blind methods estimate the noise variance as the smallest eigenvalue of $\mathcal{R}_y(0)$ and subtract it from the autocorrelation matrices to obtain de-noised estimates.

Figure 1 demonstrates the NRMSE performance of the methods for GMSK having $L = 2$, $BT = 0.3$, and $h = 0.3$ and received in the multipath channel $[1 \; -0.7 + 0.5j]$. We use $T = 4$ (corresponding to 4 sub-channels), $m = 8$ & $l = 2$ ($d = 10$), and 1000 symbols per iteration. The TXK method fails completely because it cannot handle correlated input. However, the LVD method, when given the correlation characteristics of the pseudosymbols, works well. Perhaps surprisingly, the MDCM method out-performs the LVD method. This might be because the autocorrelation function of the pseudosymbol sequence does not have finite support. Not surprisingly, the MDCM-WPK out-performs the other blind methods, although not substantially. This is probably due primarily to the approximation error of the main pulse. The MMSE method out-performs the blind methods, providing a floor on the methods that estimate the composite channel.

Figure 2 demonstrates the performance of the methods for a generalized tamed frequency modulation (GTFM) [19] signal having $L = 2$, $B = 0.62$, $r = 0.36$, and $h = 0.7$ and received in the three-ray multipath channel $[0.804 \; -0.581e^{j\frac{\pi}{6}} \; -0.124e^{j\frac{\pi}{4}}]$. The channel introduces severe frequency-selective distortion. We use $T = 4$, $m = 6$ & $l = 2$ ($d = 8$), and 500 symbols per iteration—half the symbols used in the previous example despite the more difficult channel. Again, the TXK method fails completely because of the temporal correlation of the pseudosymbols. The LVD and MDCM methods both perform well, producing usable channel estimates for all values of SNR shown. The MDCM-WPK again out-performs the other methods. Figure 3 compares the main $P(t)$ pulse, as used in MDCM-WPK, with the $C_0(t)$ pulse.

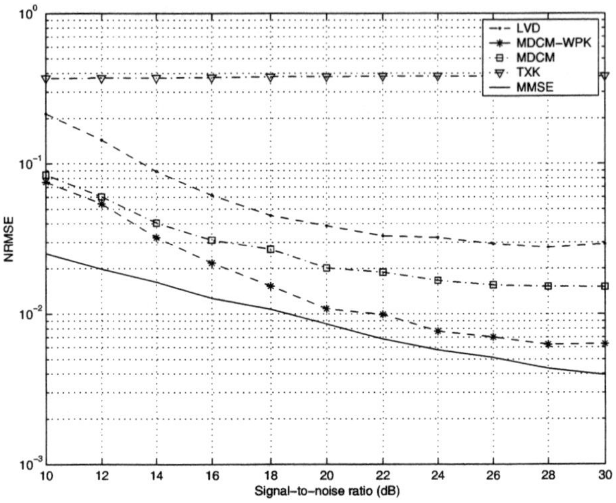

Fig. 1. Comparison of various SOS-exploiting methods for blind channel identification for GMSK ($L = 2$, $BT = 0.3$, $h = 0.3$) received in two-ray multipath channel.

5. CONCLUSION

We demonstrated that by employing the main-pulse Laurent approximation, blind SOS-based channel identification can be performed for partial-response CPM signals, include those having non-standard modulation indices. We formulated the problem in terms of the single-input multiple-output model, and compared the channel-estimation performance of several well-known SOS-based blind SIMO estimation algorithms. We demonstrated that accommodation of source-symbol statistics is essential, and exploitation is preferable. Although we used binary CPM herein, the approach *may* be applicable to M-ary CPM through the use of the approximation of Mengali *et al.* [16]. Future work will involve analytical study of the identifiability of the approximate SIMO channels, evaluation of the effects of approximation error on quality of estimation, and investigation of tolerance to channel-length over-/under-estimation.

6. REFERENCES

[1] Zhi Ding and Ge Li, "Single-channel blind equalization for GSM cellular systems," *IEEE J. Select. Areas Comm.*, vol. 16, no. 8, pp. 1493–1505, Oct. 1998.

[2] Dieter Boss, Karl-Dirk Kammeyer, and Thorsten Petermann, "Is blind channel estimation feasible in mobile communication systems? A study based on GSM," *IEEE J. Select. Areas Comm.*, vol. 16, no. 8, pp. 1479–1492, Oct. 1998.

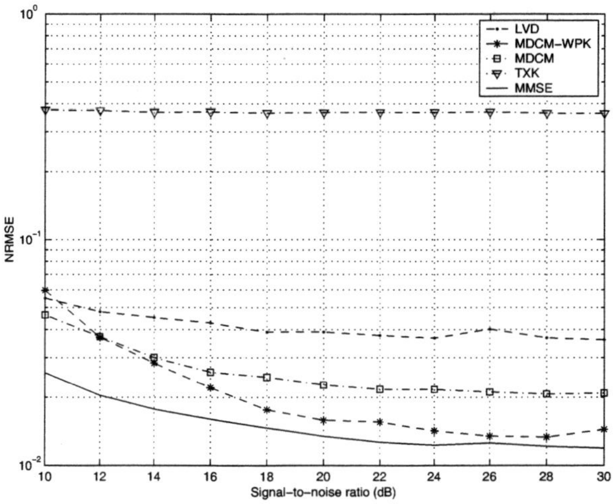

Fig. 2. Comparison of various SOS-exploiting methods for blind channel identification for GTFM ($L = 2$, $B = 0.62$, $r = 0.36$, $h = 0.7$) received in severe three-ray multipath channel.

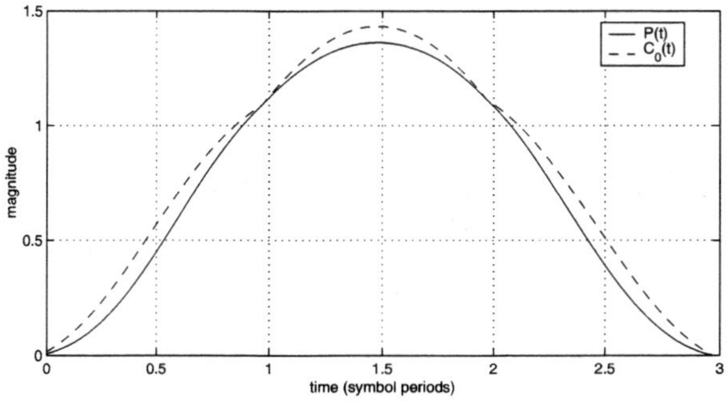

Fig. 3. Comparison of $C_0(t)$ and $P(t)$ pulses for GTFM ($L = 2$, $B = 0.62$, $r = 0.36$, $h = 0.7$).

[3] H. A. Cirpan and M. K. Tsatsanis, "Blind receivers for nonlinearly modulated signals in multipath," *IEEE Trans. Sig. Proc.*, vol. 47, no. 2, pp. 583–586, Feb. 1999.

[4] Hoang Nguyen and Bernard C. Levy, "Blind ML detection of CPM signals via the EMV algorithm," in *Proc. IEEE Int'l. Conf. Acoustics, Speech, Signal Processing*, Orlando, Florida, May 2002, IEEE, vol. 3, pp. 2457–2460.

[5] Lang Tong, Guanghan Xu, and Thomas Kailath, "Blind identification and equalization based on second-order statistics: A time domain approach," *IEEE Trans. Inform. Theory*, vol. 40, no. 2, pp. 340–349, Mar. 1994.

[6] Kaywan H. Afkhamie and Zhi-Quan Luo, "Blind identification of FIR systems driven by Markov-like input signals," *IEEE Trans. Sig. Proc.*, vol. 48, no. 6, pp. 1726–1736, June 2000.

[7] Roberto López-Valcarce and Soura Dasgoupta, "Blind channel equalization with colored sources based on second-order statistics: A linear prediction approach," *IEEE Trans. Sig. Proc.*, vol. 49, no. 9, pp. 2050–2059, Sept. 2001.

[8] Eric Moulines, Pierre Duhamel, Jean-François Cardoso, and Sylvie Mayrargue, "Subspace methods for the blind identification of multichannel FIR filters," *IEEE Trans. Sig. Proc.*, vol. 43, no. 2, pp. 516–525, Feb. 1995.

[9] Stephan V. Schell, Dale L. Smith, and Sumit Roy, "Blind channel identification using subchannel response matching," in *Conference on Information Science and Systems, Proceedings*, Princeton, NJ, Mar. 1994, pp. 858–862, 28th annual meeting.

[10] Zhi Ding and Z. Mao, "Knowledge based identification of fractionally sampled channels," in *Proc. IEEE Int'l. Conf. Acoustics, Speech, Signal Processing*, Detroit, USA, May 1995, vol. 3, pp. 1996–1999.

[11] Zhi Ding and Ye (Geoffrey) Li, *Blind equalization and identification*, Marcel Dekker, New York, NY, 2001.

[12] Pierre A. Laurent, "Exact and approximate construction of digital phase modulations by superposition of amplitude modulated pulses (AMP)," *IEEE Trans. Comm.*, vol. COM-34, no. 2, pp. 150–160, Feb. 1986.

[13] John G. Proakis, *Digital Communications*, McGraw-Hill, second edition, 1989.

[14] John B. Anderson, Tor Aulin, and Carl-Erik W. Sundberg, *Digital phase modulation*, Applications of Communications Theory. Plenum Press, New York, New York, 1986.

[15] Ghassan Kawas Kaleh, "Simple coherent receivers for partial response continuous phase modulation," *IEEE J. Select. Areas Comm.*, vol. 7, no. 9, pp. 1427–1436, Dec. 1989.

[16] Umberto Mengali and Michele Morelli, "Decomposition of M-ary CPM signals into PAM waveforms," *IEEE Trans. Inform. Theory*, vol. 41, no. 5, pp. 1265–1275, Sept. 1995.

[17] A. Napolitano and C.M. Spooner, "Cyclic spectral analysis of continuous-phase modulated signals," *IEEE Trans. Sig. Proc.*, vol. 49, no. 1, pp. 30–44, Jan. 2001.

[18] Shawn P. Neugebauer, "Blind SIMO channel estimation for CPM signals," in *Conference Record of the Thirty-Sixth Asilomar Conf. on Signals, Systems and Computers*, Pacific Grove, CA, 9–12 Nov. 2002, vol. 1, pp. 654–658.

[19] Kah-Seng Chung, "Generalized tamed frequency modulation and its application for mobile radio communications," *J-SAC*, vol. SAC-2, no. 4, pp. 487–497, July 1984.

ADAPTIVE SUPER-EXPONENTIAL ALGORITHMS FOR BLIND DECONVOLUTION OF MIMO SYSTEMS

Kiyotaka Kohno[1], *Yujiro Inouye*[2], *Mitsuru Kawamoto*[3] *and Tetsuya Okamoto*[4]

Dept. of Electronic and Control Systems Engineering, Shimane University
1060 Nishikawatsu, Matsue, Shimane 690-8504, Japan
[3]Bio-Mimetic Control Research Center, RIKEN, Moriyama, Nagoya 463-003, Japan
[1]kohno@yonago-k.ac.jp, [2]inouye@riko.shimane-u.ac.jp, [3]kawa@ecs.shimane-u.ac.jp

ABSTRACT

Multichannel blind deconvolution of finite-impulse response (FIR) or infinite-impulse response (IIR) systems is investigated using the multichannel super-exponential method. First, some properties are shown for the rank of the correlation matrices relevant to the multichannel super-exponential method. Then, the matrix inversion lemma is extended to the degenerate rank case. Based on these results, two types of adaptive multichannel super-exponential algorithms are presented, that is, the one in covariance form and the other in QR-factorization form.

1. INTRODUCTION

Multichannel blind deconvolution has recently received attention in such field as digital communications, image processing and neural information processing.

Recently, Shalvi and Weinstein proposed an attractive approach to single-channel blind deconvolution called the *super-exponential method* (SEM) [1]. Extensions of their idea to multichannel deconvolution were presented by Inouye and Tanebe [2], Martone [3], [4], and Yeung and Yau [5]. In particular, Martone [3] proposed an adaptive version of the SEM based on low-rank processing [6], but the Martone algorithm require a rank-revealing technique, while the present paper presents an explicit formula for revealing the rank of relevant correlation matrices in the absent of noise.

In the present paper, we show some properties of the rank of the relevant correlation matrices, and present a matrix pseudo-inversion lemma. Based on these results, we propose two type of adaptive multichannel super-exponential algorithms (AMSEA's), the one in covariance (correlation or Kalman-filter) form and the other in QR-factorization form.

The present paper uses the following notation: Let Z denote the set of all integers. Let $\boldsymbol{C}^{m\times n}$ denote the set of all $m\times n$ matrices with complex components. The superscripts T, $*$, H and $\dagger$ denote, respectively, the transpose, the complex conjugate, the complex conjugate transpose (Hermitian) and the (Moore-Penrose) pseudoinverse operations of a matrix. Let $i=\overline{1,n}$ stand for $i=1,2,\cdots,n$.

2. ASSUMPTIONS AND PRELIMINARIES

We consider an MIMO channel system with n inputs and m outputs as described by

$$\boldsymbol{y}(t)=\sum_{k=-\infty}^{\infty}\boldsymbol{H}^{(k)}\boldsymbol{s}(t-k),\qquad t\in Z,\tag{1}$$

where

$\boldsymbol{s}(t)$ n-column vector of input (or source) signals,
$\boldsymbol{y}(t)$ m-column vector of channel outputs,
$\boldsymbol{H}^{(k)}$ $m\times n$ matrix of impulse responses.

The transfer function of the channel system is defined by

$$\boldsymbol{H}(z)=\sum_{k=-\infty}^{\infty}\boldsymbol{H}^{(k)}z^k,\qquad z\in C.\tag{2}$$

For the time being, it is assumed for theoretical analysis that noise is absent in (1).

To recover the source signals, we process the output signals by an $n\times m$ equalizer (or deconvolver) $\boldsymbol{W}(z)$ described by

$$\boldsymbol{z}(t)=\sum_{k=-\infty}^{\infty}\boldsymbol{W}^{(k)}\boldsymbol{y}(t-k),\qquad t\in Z.\tag{3}$$

The objective of multichannel blind deconvolution is to construct an equalizer that recovers the original source signals only from the measurements of the corresponding outputs.

We put the following assumptions on the systems and the source signals.

A1) The transfer function $\boldsymbol{H}(z)$ is stable and has full column rank on the unit circle $|z|=1$ [this implies that the unknown system has less inputs than outputs, i.e., $n\leq m$, and there exists a left stable inverse of the unknown system].

A2) The input sequence $\{\boldsymbol{s}(t)\}$ is a complex, zero-mean, non-Gaussian random vector process with element processes $\{s_i(t)\}$, $i=\overline{1,n}$ being mutually independent. Moreover, each element process $\{s_i(t)\}$ is an i.i.d. process with a nonzero variance σ_i^2 and a nonzero fourth-order cumulant γ_i. The variances σ_i^2's and the

0-7803-8251-X/04/$17.00 ©2004 IEEE

fourth-order cumulants γ_i's are unknown.

A3) The equalizer $\boldsymbol{W}(z)$ is an FIR system of sufficient length L so that the truncation effect can be ignored.

Remark 1: As to A1), if the channel system $\boldsymbol{H}(z)$ is FIR, then a condition of the existence of an FIR equalizer is rank$\boldsymbol{H}(z) = n$ for all nonzero $z \in C$ [7]. Moreover, if $\boldsymbol{H}(z)$ is irreducible, then there exists an equalizer $\boldsymbol{W}(z)$ of length $L \leq nK$, where K is the length of the channel system [7]. Besides, it is shown that there exists generically (or except for pathological cases) an equalizer $\boldsymbol{W}(z)$ of length $L \leq \lceil \frac{nK}{m-n} \rceil$, where $\lceil x \rceil$ stands for the smallest integer that is greater than equal to x.

For now, there are two approaches to multichannel (or MIMO) blind deconvolution, a *concurrent blind deconvolution approach* and a *deflationary blind deconvolution approach.* The former is to deconvolve (or recover) concurrently all the source signals, while the latter is to deconvolve sequentially (or iteratively with respect to sources) the source signals one by one. The former approach is employed in the present paper and the latter approach will be developed in a forthcoming paper.

Let us consider an FIR equalizer with the transfer function $\boldsymbol{W}(z)$ given by

$$\boldsymbol{W}(z) = \sum_{k=L_1}^{L_2} \boldsymbol{W}^{(k)} z^k, \tag{4}$$

where the length $L := L_2 - L_1 + 1$ is taken to be sufficiently large. Let $\tilde{\boldsymbol{w}}_i$ be the Lm-column vector consisting of the tap coefficient (corresponding to the ith output) of the equalizer defined by

$$\tilde{\boldsymbol{w}}_i := \left[\boldsymbol{w}_{i,1}^T, \boldsymbol{w}_{i,2}^T, \cdots, \boldsymbol{w}_{i,m}^T\right]^T \in \boldsymbol{C}^{mL}, \tag{5}$$

$$\boldsymbol{w}_{i,j} = \left[w_{i,j}{}^{(L_1)}, w_{i,j}{}^{(L_1+1)}, \cdots, w_{i,j}{}^{(L_2)}\right]^T \in \boldsymbol{C}^L, \tag{6}$$

where $w_{i,j}{}^{(k)}$ is the (i, j)th element of matrix $\boldsymbol{W}^{(k)}$.

Inouye and Tanebe [2] proposed the *multichannel super-exponential algorithm* for finding the tap coefficient vectors $\tilde{\boldsymbol{w}}_i$'s of the equalizer $\boldsymbol{W}(z)$, of which each iteration consists of the following two steps:

$$\tilde{\boldsymbol{w}}_i^{[1]} = \tilde{\boldsymbol{R}}_L^{\dagger} \tilde{\boldsymbol{d}}_i \qquad for\ i = \overline{1, n}, \tag{7}$$

$$\tilde{\boldsymbol{w}}_i^{[2]} = \frac{\tilde{\boldsymbol{w}}_i^{[1]}}{\sqrt{\tilde{\boldsymbol{w}}_i^{[1]H} \tilde{\boldsymbol{R}}_L \tilde{\boldsymbol{w}}_i^{[1]}}} \qquad for\ i = \overline{1, n}, \tag{8}$$

where $(\cdot)^{[1]}$ and $(\cdot)^{[2]}$ stand respectively for the result of the first step and the result of the second steps. Let $\tilde{\boldsymbol{y}}(t)$ be the Lm-column vector consisting of the L consecutive inputs of the equalizer define by

$$\tilde{\boldsymbol{y}}(t) := \left[\bar{\boldsymbol{y}}_1(t)^T, \bar{\boldsymbol{y}}_2(t)^T, \cdots, \bar{\boldsymbol{y}}_m(t)^T\right]^T \in \boldsymbol{C}^{mL}, \tag{9}$$

$$\bar{\boldsymbol{y}}_i(t) := [y_i(t - L_1), y_i(t - L_1 - 1), \cdots, y_i(t - L_2)]^T \in \boldsymbol{C}^L, \tag{10}$$

where $y_i(t)$ is the ith element of the output vector $\boldsymbol{y}(t)$ of the channel system in (1). Then the correlation matrix $\tilde{\boldsymbol{R}}_L$ is represented as

$$\tilde{\boldsymbol{R}}_L = E\left[\tilde{\boldsymbol{y}}^*(t) \tilde{\boldsymbol{y}}^T(t)\right] \in \boldsymbol{C}^{mL \times mL}, \tag{11}$$

and the forth-order cumulant vector $\tilde{\boldsymbol{d}}_i$ is represented as

$$\begin{aligned} \tilde{\boldsymbol{d}}_i &= E\left[|z_i(t)|^2 z_i(t) \tilde{\boldsymbol{y}}^*(t)\right] \\ &\quad -2E\left[|z_i(t)|^2\right] E\left[z_i(t) \tilde{\boldsymbol{y}}^*(t)\right] \\ &\quad -E\left[z_i{}^2(t)\right] E\left[z_i{}^*(t) \tilde{\boldsymbol{y}}^*(t)\right], \end{aligned} \tag{12}$$

where E$[x]$ denotes the expectation of a random variable x. We note that the last term can be ignored in case of E$[s_i{}^2(t)]$=0 for all $i = \overline{1, n}$, in which case E$[z_i{}^2(t)]$=0 for all $i = \overline{1, n}$.

3. PROPERTIES OF CORRELATION MATRICES AND MATRIX PSEUDO-INVERSION

We consider the rank deficiency problem of the correlation matrix $\tilde{\boldsymbol{R}}_L$ of $\{\tilde{\boldsymbol{y}}(t)\}$ in (11) with respect to the length $L = L_2 - L_1 + 1$ of the equalizer $\boldsymbol{W}(z)$. This problem is very important for solving the equation (7) and also a fundamental issue in low-rank adaptive signal processing [6].

Theorem 1: Let $\tilde{\boldsymbol{R}}_L \in \boldsymbol{C}^{mL \times mL}$ be the correlation matrix defined by (11) for the channel system $\boldsymbol{H}(z)$ with n inputs and m outputs satisfying A1) and A2), where $L = L_2 - L_1 + 1$ is the length of the equalizer $\boldsymbol{W}(z)$. Then the following statements hold true:
1) If $m = n$, the $\tilde{\boldsymbol{R}}_L$ is nonsingular for $L = 1, 2, \cdots$.
2) If $m > n$ and $\boldsymbol{H}(z)$ is the transfer function of an FIR system of length K, then the sequence $\{\tilde{\boldsymbol{R}}_L\}$ decreases monotonically as L increases, and

$$\text{rank}\tilde{\boldsymbol{R}}_L = nL,\ for\ L \geq nK. \tag{13}$$

3) If $m > n$ and if $\boldsymbol{H}(z)$ is the rational transfer function of an IIR system, then the sequence $\{\text{rank}\tilde{\boldsymbol{R}}_L\}$ decreases as L increases, and

$$\lim_{L \to \infty} \frac{1}{L} \text{rank}\tilde{\boldsymbol{R}}_L = n. \tag{14}$$

In order to develop an adaptive version of the multichannel super-exponential algorithm, the matrix inversion lemma [9] should be extended to the rank-degenerate case. The following lemma gives an explicit formula of the pseudoinverse for a positive semidefinite Hermitian matrix $\boldsymbol{A}$ added to a general rank-one matrix $\boldsymbol{b}\boldsymbol{b}^H$.

Lemma 1: Let $\boldsymbol{A} \in \boldsymbol{C}^{n \times n}$ be a positive semidefinite Hermitian matrix, and $\boldsymbol{b} \in \boldsymbol{C}^n$ be a nonzero vector. Let the linear vector space $\boldsymbol{C}^n$ be uniquely decomposed as $\boldsymbol{C}^n$=Im$\boldsymbol{A} \oplus$(Im$\boldsymbol{A}$)$^\perp$, where Im$\boldsymbol{A}$ denotes the image space of $\boldsymbol{A}$ and (Im$\boldsymbol{A}$)$^\perp$ denotes the orthogonal complement of Im$\boldsymbol{A}$. Let $\boldsymbol{b} \in \boldsymbol{C}^n$ be decomposed uniquely as

$$\boldsymbol{b} = \boldsymbol{b}_1 \oplus \boldsymbol{b}_2\ with\ \boldsymbol{b}_1 \in Im\boldsymbol{A}\ and\ \boldsymbol{b}_2 \in (Im\boldsymbol{A})^\perp. \tag{15}$$

Let $\boldsymbol{Q}$ be defined as

$$\boldsymbol{Q} = \boldsymbol{A} + \boldsymbol{b}\boldsymbol{b}^H \in \boldsymbol{C}^{n \times n}. \tag{16}$$

Then the pseudoinverse $\boldsymbol{Q}^\dagger$ of matrix $\boldsymbol{Q}$ is explicitly expressed, depending on the values of vectors $\boldsymbol{b}_1$ and $\boldsymbol{b}_2$ and matrix $\boldsymbol{A}$, as follows:

1) If $\boldsymbol{b}_2 = 0$, then

$$\boldsymbol{Q}^\dagger = \boldsymbol{A}^\dagger - \frac{\boldsymbol{A}^\dagger \boldsymbol{b}_1 \boldsymbol{b}_1^H \boldsymbol{A}^\dagger}{1 + \boldsymbol{b}_1^H \boldsymbol{A}^\dagger \boldsymbol{b}_1}. \tag{17}$$

2) If $\boldsymbol{b}_2 \neq 0$ and $\boldsymbol{b}_1 = 0$, then

$$\boldsymbol{Q}^\dagger = \boldsymbol{A}^\dagger + \frac{\boldsymbol{b}_2 \boldsymbol{b}_2^H}{(\boldsymbol{b}_2^H \boldsymbol{b}_2)^2}. \tag{18}$$

3) Let l be a non-negative number defined by

$$l := |1 + \boldsymbol{b}_1^H \boldsymbol{Q}_b^\dagger \boldsymbol{b}_2|^2 - \boldsymbol{b}_1^H \boldsymbol{Q}_b^\dagger \boldsymbol{b}_1 \boldsymbol{b}_2^H \boldsymbol{Q}_b^\dagger \boldsymbol{b}_2, \tag{19}$$

where $\boldsymbol{Q}_b^\dagger$ is defined by

$$\boldsymbol{Q}_b^\dagger := \boldsymbol{A}^\dagger - \frac{\boldsymbol{A}^\dagger \boldsymbol{b}_1 \boldsymbol{b}_1^H \boldsymbol{A}^\dagger}{1 + \boldsymbol{b}_1^H \boldsymbol{A}^\dagger \boldsymbol{b}_1} + \frac{\boldsymbol{b}_2 \boldsymbol{b}_2^H}{(\boldsymbol{b}_2^H \boldsymbol{b}_2)^2}. \tag{20}$$

Then in the case when $\boldsymbol{b}_1 \neq 0$, $\boldsymbol{b}_2 \neq 0$ and $l \neq 0$,

$$\boldsymbol{Q}^\dagger = \boldsymbol{Q}_b^\dagger - \boldsymbol{Q}_b^\dagger [\boldsymbol{b}_1, \boldsymbol{b}_2] \boldsymbol{Q}_d [\boldsymbol{b}_1, \boldsymbol{b}_2]^H \boldsymbol{Q}_b^\dagger, \tag{21}$$

where

$$\boldsymbol{Q}_d := \frac{1}{l} \left[\begin{array}{c|c} -\boldsymbol{b}_2^H \boldsymbol{Q}_b^\dagger \boldsymbol{b}_2 & 1 + \boldsymbol{b}_1^H \boldsymbol{Q}_b^\dagger \boldsymbol{b}_2 \\ \hline 1 + \boldsymbol{b}_2^H \boldsymbol{Q}_b^\dagger \boldsymbol{b}_1 & -\boldsymbol{b}_1^H \boldsymbol{Q}_b^\dagger \boldsymbol{b}_1 \end{array} \right] \in \boldsymbol{C}^{2\times 2}. \tag{22}$$

Remark 2: We can show $0 < l \leq 1$ under the assumptions.

4. ADAPTIVE SUPER-EXPONENTIAL ALGORITHMS

Except for the case when the number of outputs equals the number of inputs, i.e., $m = n$, the correlation matrix $\tilde{\boldsymbol{R}}_L$ is not of full rank. Situations with the number of independent sources (or inputs) being strictly less than the number of sensors (or outputs) are often encountered in various applications such as digital communication, image processing and neural information processing. Moreover, if the underlying channel system exhibits slow changes in time, processing all the available data jointly is not desirable, even if we can accommodate the computational and storage loads of the batch algorithm in (7) and (8), because different data segments correspond to different channel responses. In such a case, we want to have an adaptive algorithm which is capable of tracking the varying characteristics of the channel system. In the following, we propose two types of AMSEA's, that is, the one in covariance (correlation or Kalman-filter) form and the other in QR-factorization form.

Consider the batch algorithm in (7) and (8). The equation (8) constraints the length of vector $\tilde{\boldsymbol{w}}_i$ to equal one, and thus we assume this constraint is always satisfied using a normalization or an automatic gain control (AGC) of $\tilde{\boldsymbol{w}}_i$ at each time t. To develop an adaptive version of (7), we must specify the dependency of each time t and rewrite (7) as

$$\tilde{\boldsymbol{w}}_i(t) = \tilde{\boldsymbol{R}}_L^\dagger(t) \tilde{\boldsymbol{d}}_i(t) \quad , \quad i = \overline{1,n}. \tag{23}$$

Here the subscript L of $\tilde{\boldsymbol{R}}_L(t)$ is omitted for simplicity hereafter. The recursions for time-updating of matrix $\tilde{\boldsymbol{R}}(t)$ and vector $\tilde{\boldsymbol{d}}_i(t)$ in (23) are given as

$$\tilde{\boldsymbol{R}}(t) = \alpha \tilde{\boldsymbol{R}}(t-1) + (1-\alpha) \tilde{\boldsymbol{y}}^*(t) \tilde{\boldsymbol{y}}^T(t), \tag{24}$$

$$\tilde{\boldsymbol{d}}_i(t) = \alpha \tilde{\boldsymbol{d}}_i(t-1) + (1-\alpha) \tilde{\boldsymbol{y}}^*(t) \tilde{z}_i(t), \tag{25}$$

where

$$\tilde{z}_i(t) := (|z_i(t)|^2 - 2 < |z_i(t)|^2 >) z_i(t) - < z_i^2(t) > z_i^*(t). \tag{26}$$

Here $< |z_i(t)|^2 >$ and $< z_i^2(t) >$ denote respectively the estimates of $E\left[|z_i(t)|^2\right]$ and $E\left[z_i(t)^2\right]$ at time t, α is a positive constant close to, but less than one, which accounts for some exponential weighting factor or forgetting factor [9].

By applying Lemma 1 for calculating the pseudoinverse of $\tilde{\boldsymbol{R}}(t)$, we obtain the following theorem which determines $\tilde{\boldsymbol{w}}_i(t)$ from $\tilde{\boldsymbol{w}}_i(t-1)$, $\tilde{\boldsymbol{y}}(t)$ and $z_i(t)$.

Theorem 2: The recursion for $\tilde{\boldsymbol{w}}_i(t)$ is

$$\tilde{\boldsymbol{w}}_i(t) = \boldsymbol{P}(t) \tilde{\boldsymbol{R}}(t) \tilde{\boldsymbol{w}}_i(t-1) + \boldsymbol{k}(t) \left[\tilde{z}_i(t) - \tilde{\boldsymbol{y}}^T(t) \tilde{\boldsymbol{w}}_i(t-1) \right], \tag{27}$$

where

$$\boldsymbol{k}(t) := (1-\alpha) \boldsymbol{P}(t) \tilde{\boldsymbol{y}}^*(t), \tag{28}$$

$$\tilde{z}_i(t) := (|z_i(t)|^2 - 2 < |z_i(t)|^2 >) z_i(t) - < z_i^2(t) > z_i^*(t), \tag{29}$$

$$< |z_i(t)|^2 > := \beta < |z_i(t-1)|^2 > + (1-\beta) |z_i(t)|^2, \tag{30}$$

$$< z_i^2(t) > := \beta < z_i^2(t-1) > + (1-\beta) z_i^2(t), \tag{31}$$

and the formula of the recursion for $\boldsymbol{P}(t)$ from $\boldsymbol{P}(t-1)$ by using Lemma 1 is very lengthy and is omitted for page limit. Here β is a positive constant less than α. These equations are initialized by their values appropriately selected or calculated by the batch algorithm in (7) and (8) at initial time t_0 and used for $t = t_0 + 1$, $t_0 + 2, \cdots$.

Before presenting another type of adaptive algorithms, we mention the following lemma on the so-called QR-factorization of a general matrix $\boldsymbol{A}$.

Lemma 2 [8],[10]**:** Given an $n \times n$ Hermitian $\boldsymbol{A} \in \boldsymbol{C}^{n\times n}$. Let r be a chosen integer satisfying $|\lambda_r| > |\lambda_{r+1}|$, where the eigenvalues $\lambda_1, \lambda_2, \cdots, \lambda_n$ of $\boldsymbol{A}$ are arranged in decreasing order of magnitude. Given an $n\times r$ matrix $\boldsymbol{Q}_0$ with orthonormal columns and generate a sequence of matrices $\{\boldsymbol{Q}_k\} \subset \boldsymbol{C}^{n\times r}$ as follows:

$$\boldsymbol{Z}_k = \boldsymbol{A} \boldsymbol{Q}_{k-1}, \tag{32}$$

$$\boldsymbol{Q}_k \boldsymbol{R}_k = \boldsymbol{Z}_k \quad : \textit{QR-factorization}, \tag{33}$$

where $\boldsymbol{Q}_k \in \boldsymbol{C}^{n\times r}$ is a matrix with orthonormal columns and $\boldsymbol{R}_k \in \boldsymbol{C}^{r\times r}$ is an upper triangular matrix. If $\boldsymbol{Q}_0$ is not unfortunately chosen, then the sequence $\{\boldsymbol{Q}_k\}$ converges to a matrix of r dominant eigenvectors, and the upper triangular sequence $\{\boldsymbol{R}_k\}$ converges the diagonal matrix of r dominant eigenvalues.

By applying Lemma 2 for calculating the pseudoinverse of $\tilde{\boldsymbol{R}}(t)$, we have the following theorem which gives an adaptive solution $\tilde{\boldsymbol{w}}_i(t)$ of (23) from $\boldsymbol{Q}_r(t-1)$, $\boldsymbol{Q}_r(t-2)$, $\tilde{\boldsymbol{d}}_i(t-1)$, $\tilde{\boldsymbol{y}}(t)$ and $z_i(t)$ (where, for example, $\boldsymbol{Q}_r(t-1) \in \boldsymbol{C}^{mL\times r}$ represents approximately r dominants eigenvectors of $mL \times mL$ matrix $\tilde{\boldsymbol{R}}(t-1)$).

Theorem 3: Let r be fixed as $r = nL$, where n is

the number of the inputs of the channel system in (1) and L is the length of the equalizer in (4). Then an adaptive solution $\tilde{\boldsymbol{w}}_i(t)$ of (23) is

$$\tilde{\boldsymbol{w}}_i(t) = \boldsymbol{Q}_r(t-1)\boldsymbol{R}_r^{-1}(t)\boldsymbol{Q}_r^H(t)\tilde{\boldsymbol{d}}_i(t), \quad (34)$$

where $\boldsymbol{Q}_r(t)$ and $\boldsymbol{R}_r(t)$ is obtained by the QR decomposition of matrix $\boldsymbol{Z}(t)$ defined by $\boldsymbol{Z}(t) := \tilde{\boldsymbol{R}}(t)\,\boldsymbol{Q}_r(t-1)$, which is decomposed as

$$\begin{aligned}\boldsymbol{Z}(t) &= \boldsymbol{Q}_r(t)\boldsymbol{R}_r(t) \in \boldsymbol{C}^{mL\times r},\\ &\boldsymbol{Q}_r(t) \in \boldsymbol{C}^{mL\times r},\ \boldsymbol{R}_r(t) \in \boldsymbol{C}^{r\times r}, \quad (35)\end{aligned}$$

and the update of $\boldsymbol{Z}(t)$ is

$$\begin{aligned}\boldsymbol{Z}(t) &= \alpha\boldsymbol{Z}(t-1)\boldsymbol{Q}_r^H(t-2)\boldsymbol{Q}_r(t-1)\\ &\quad + (1-\alpha)\tilde{\boldsymbol{y}}^*(t)\tilde{\boldsymbol{y}}^T(t)\boldsymbol{Q}_r(t-1). \quad (36)\end{aligned}$$

The update of $\tilde{\boldsymbol{d}}_i(t)$ is

$$\tilde{\boldsymbol{d}}_i(t) = \alpha\tilde{\boldsymbol{d}}_i(t-1) + (1-\alpha)\tilde{\boldsymbol{y}}^*(t)\tilde{z}_i(t), \quad (37)$$

where

$$\tilde{z}_i(t) := (|z_i(t)|^2 - 2 < |z_i(t)|^2 >)z_i(t) - < z_i^2(t) > z_i^*(t), \quad (38)$$

$$< |z_i(t)|^2 > = \beta < |z_i(t-1)|^2 > + (1-\beta)|z_i(t)|^2, \quad (39)$$

$$< z_i^2(t) > = \beta < z_i^2(t-1) > + (1-\beta)z_i^2(t). \quad (40)$$

These equations are initialized by their values appropriately selected or calculated by the batch algorithm in (7) and (8) at an initial time t_0 and used for $t = t_0+1$, $t_0+2, \cdots$.

Remark 3: If the number n of inputs varies dynamically, we should estimate the number of the inputs before using Theorem 3.

5. SIMULATIONS

For page limit, we show only one of the simulation results in Figure 1 by using the AMSEA in covariance form (23) - (31). We considered an MIMO channel system with two inputs and three outputs, and assumed that the length of the channel is three ($K = 3$), the length of the equalizer is six ($L = 6$), and two source signals are the 4-PSK and the 8-PSK signals. As a measure of performance, we use the multichannel intersymbol interference (MISI). The last matrix $\boldsymbol{H}^{(2)}$ of the impulse response of the channel was varied by approximately 3 times at discrete time t=10,000. The values of α and β were chosen as α=0.999 and β=0.05, respectively. Figure 1 shows the result for the time-variant system obtained by using 50,000 data samples. The details of the simulation results will be shown in the conference.

6. CONCLUSIONS

We have investigated multichannel blind deconvolution of FIR or IIR systems using the multichannel super-exponential method. We have shown some properties of the correlation matrices relevant to the multichannel super-exponential method and then presented a pseudo-inversion lemma. Based on these results, we have proposed two types of adaptive multichannel super-

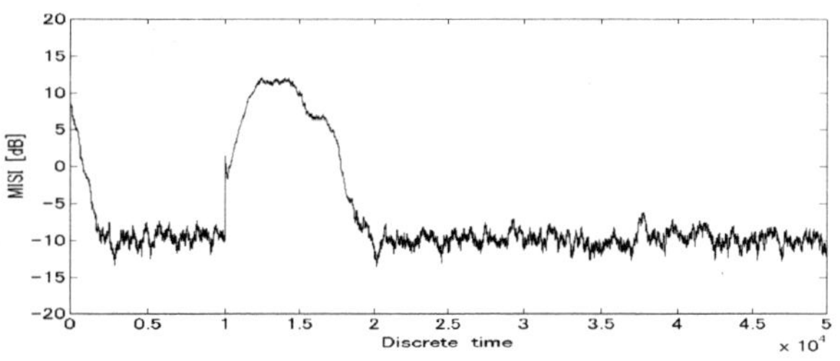

Figure 1: Performance of AMSEA in covariance form.

exponential algorithms (AMSEA's), the one in covariance form and the other in QR-factorization form.

7. REFERENCES

[1] O. Shalvi and E. Weinstein, "Super-exponential methods for blind deconvolution," *IEEE Trans. Information Theory*, vol. 39, no. 2, pp. 504-519, Mar. 1993.

[2] Y. Inouye and K. Tanebe, "Super-exponential algorithms for multichannel blind deconvolution," *IEEE Trans. Signal Processing*, vol. 48, no. 3, pp. 881-888, Mar. 2000.

[3] M. Martone, "An adaptive algorithm for antenna array low-rank processing in cellular TDMA base stations," *IEEE Trans. Communications*, vol. 46, no. 5, pp. 627-643, May 1998.

[4] M, Martone, "Fast adaptive super-exponential multistage beamforming for cellular base-station transceivers with antenna arrays," *IEEE Trans. Vehicular Tech.*, vol. 48, no. 4, Jul. 1999.

[5] K. L. Yeung and S. F. Yau, "A cumulant-based super-exponential algorithm for blind deconvolution of multi-input multi-output systems," *Signal Process.*, vol. 67, pp. 141-162, 1998.

[6] P. Strobach, "Low-rank adaptive filters," *IEEE Trans. Signal Process.*, vol. 44, no. 12, Dec. 1996.

[7] Y. Inouye and R-W. Liu, "A system-theoretic foundation for blind equalization of an FIR MIMO channel system," *IEEE Trans. Circuits and Systems – I, Fundam. Theory Appl.*, vol. 49, no. 4, pp. 425-436, Apr. 2002.

[8] G. H. Groub and C. F. Van Loan. *Matrix Computations, 2nd Ed.*, Baltimore, MD: The Johns Hopkins University Press, 1989

[9] S. Haykin. *Adaptive Filter Theory, 3rd Ed.*, Upper Saddle River, NJ: Prentice-Hall, 1996.

[10] G. W. Stewart, "Methods of simultaneous iteration for calculating eigenvectors of matrices," in *Topics in Numerical Analysis* II, J. H. Miller, Ed. New York: Academic, pp. 169-185, 1975.

BEYOND ICA: ROBUST SPARSE SIGNAL REPRESENTATIONS

Andrzej CICHOCKI, Yuanqing LI, Pando GEORGIEV and Shun-ichi AMARI

Riken, Brain Science Institute, Saitama 351-0198, Wako-shi, JAPAN
cia@brain.riken.jp

ABSTRACT

In many applications it is necessary to perform some decomposition of observed signals or data in such a way that components have some special properties or structures such as statistical independence, sparsity, smoothness, non-negativity, prescribed statistical distributions and/or specific temporal structure. In this paper we discuss cost functions whose minimization solve such problems and we present new properties that characterize optimal solutions for sparse representations. Especially, we discuss robust cost functions in order to find sparse representation of noisy signals. Furthermore, we discuss sub-band decomposition preprocessing to relax independence conditions for source signals.

1. INTRODUCTION

Data decomposition and representation are widely used in signal processing and neural computing. Recently, several novel methods and approaches have been proposed for decomposition and representations of signals and images, especially, Independent Component Analysis (ICA), Sparse Component Analysis (SCA) and Non-negative Matrix Factorization (NMF) [1, 4, 6, 11, 12]. All these methods can be expressed algebraically as some specific problems of matrix factorization: Given observation (often called sensor or data) matrix $\mathbf{X} \in \mathbb{R}^{m \times N}$ perform the matrix factorization

$$\mathbf{X} = \mathbf{AS} + \mathbf{E}, \qquad (1)$$

where $\mathbf{A} \in \mathbb{R}^{m \times n}$ represents basis data matrix or mixing matrix (depending on application), n is the number of sources, m the number of observations, N the number of samples, $\mathbf{E} \in \mathbb{R}^{m \times N}$ is a matrix representing errors or noise and matrix $\mathbf{S} \in \mathbb{R}^{n \times N}$ contains the corresponding hidden components that give the contribution of each basis vector. Often these components represent unknown source signals with specific temporal structures, features or properties. For example, the rows of matrix $\mathbf{S}$ should be as sparse as possible for SCA or as independent as possible for ICA or take only nonnegative values for NMF or values with specific constraints [6, 9, 10]. It is important to note that the statistical independence and sparsity are different criteria or concepts. For example, for mixture of two uniform distributed sources, independent components and sparse components are quite different. However, for generalized Gaussian distributed random signals with pdf: $p(y, \beta) = ce^{-|y/\sigma|^{\beta}}$ both sparsity and normalized kurtosis (or entropy) are consistent in the sense that the independence and sparsity criteria exactly coincide for super-Gaussian signals with positive kurtosis. However, in general sparsity and independence are completely different criteria and may give quite different results and interpretations.

Although some decompositions or matrix factorizations provide an exact reconstruction data (i.e., $\mathbf{X} = \mathbf{AS}$), we shall consider here decompositions which are approximative in nature, however they should be robust to noise and enforce some desirable constraints. Different cost functions and imposed constraints may lead to different types of matrix factorizations. In blind source separation (BSS) problem the data matrix $\mathbf{X} = [\mathbf{x}(1), \mathbf{x}(2) \ldots, \mathbf{x}(N)]$ can be represented by vectors $\mathbf{x}(k)$ $(k = 1, 2, \ldots, N)$ for many time instants as multiple measurements or recordings, thus the compact aggregated matrix equation (1) can be written in a vector form as the system of linear equations: $\mathbf{x}(k) = \mathbf{A}\,\mathbf{s}(k) + \mathbf{e}(k)$, where $\mathbf{x}(k) = [x_1(k), \ldots, x_m(k)]^T$ is the vector of the observed signals at the discrete time instant k while $\mathbf{s}(k) = [s_1(k), \ldots, s_n(k)]^T$ is the vector of components at the same time instant. The above formulated problems are related closely to linear inverse problem or more generally, to solving a large ill-conditioned system of linear equations (overdetermined or underdetermined depending on applications) where it is necessary to estimate reliably not only vectors $\mathbf{s}(k)$ but also to identify a matrix $\mathbf{A}$ for noisy data[1]. It can be shown that using optimization approach we can extend these techniques to other useful concepts such Smooth Component Analysis (SmoCA), Dense Component Analysis (DenCA) and other representations with specific features or constraints.

2. ROBUST SPARSE SIGNAL REPRESENTATIONS

Sparse Component Analysis (SCA) and sparse signals representations (SSR) arise in many scientific problems, especially, where we wish to represent signals of interest by using a small (or sparse) number of basis signals from a much larger set of signals [3, 4, 9, 10]. In fact, finding a sparse (or in many cases the sparsest) solution can be viewed equivalently as the problem of selecting very few columns $\mathbf{a}_j \in \mathbb{R}^m$ (called atoms) of the matrix $\mathbf{A} = [\mathbf{a}_1, \mathbf{a}_2, \ldots, \mathbf{a}_n]$ (called dictionary) to represent the observation vector $\mathbf{x}$ which is referred as the subset selection problem. We can state the subset selection problem as follows: Find an optimal subset of $r << n$ columns from the matrix $\mathbf{A}$, which we de-

[1]In this contribution, we assume that we employ two stage procedure in which in the first stage a mixing matrix $\mathbf{A}$ is estimated, e.g., by employing sparsification procedure in time frequency domain (e.g., wavelets package) and next clustering approach.

note by $\mathbf{A}_r \in \mathbb{R}^{m\times r}$ such that $\mathbf{A}_r\mathbf{s}_{r*} \cong \mathbf{x}$, or equivalently $\mathbf{A}_r\mathbf{s}_{r*} + \mathbf{e}_r = \mathbf{x}$, where $\mathbf{s}_{r*} \in \mathbb{R}^r$ is a vector consisting non zero elements of the vector $\mathbf{s}_* \in \mathbb{R}^n$ and $\mathbf{e}_r \in \mathbb{R}^m$ represents some residual error vector which norm should be as small as possible or below of some threshold. Usually, we have interest in sparsest and unique representation, i.e., it is necessary to find solution having the smallest possible number of non-zero-components. The problem can be reformulated as the following optimization problem:

$$(P_\rho) \quad J_\rho(\mathbf{s}) = \|\mathbf{s}\|_\rho = \sum_{j=1}^{n} \rho(s_j) \quad \text{s. t.} \qquad \mathbf{A}\,\mathbf{s} = \mathbf{x}, \qquad (2)$$

where $\mathbf{A} \in \mathbb{R}^{m\times n}$, (usually with $n >> m$) and $\|\mathbf{s}\|_\rho$ suitably chosen function which measures the sparsity of the vector $\mathbf{s}$. It should be noted the sparsity measure does not need be necessary a norm, although we use such notation. For example, we can apply Shannon, Gauss or Renyi entropy or normalized kurtosis as measure of the (anti-)sparsity [7]. In the standard form, we use l_p-norm with $0 \le p \le 1$. Especially, l_0 quasi-norm attract a lot of attention since it ensures sparsest representation [3, 4, 9, 10]. Unfortunately, such formulated problem (2) for l_p-norm with $p < 1$ is rather very difficult, especially for $p = 0$ it is NP-hard, so for a large scale problem it is numerically untractable. For this reason, we often use Basis Pursuit (BP) or standard Linear Programming (LP) for $\|\mathbf{s}\|_\rho = \|\mathbf{s}\|_1$, with $\rho = p = 1$.

In practice, due to noise and other uncertainty (e.g. measurement errors) the system of linear underdetermined equations should not be satisfied precisely but with some prescribed tolerance (i.e., $\mathbf{A}\,\mathbf{s} \cong \mathbf{x}$ in the sense that $\|\mathbf{x} - \mathbf{A}\,\mathbf{s}\|_q \le \varepsilon$). From the practical point of view as well as from a statistical point of view, it is convenient and quite natural to replace the exact constraints $\mathbf{x} = \mathbf{A}\,\mathbf{s}$ by the constraint $\|\mathbf{x} - \mathbf{A}\,\mathbf{s}\|_q \le \varepsilon$, where choice of l_q-norm depends on distribution of noise and specific applications. Especially, the l_2-norm constraints for Gaussian distributed noise leads minimization of the standard cost function

$$(BPD) \qquad J_{2,1}(\mathbf{s}) = \|\mathbf{x} - \mathbf{A}\,\mathbf{s}\|_2^2 + \alpha\,\|\mathbf{s}\|_1, \qquad (3)$$

which is referred as Basis Pursuit Denoising (BPD), where $\alpha > 0$ is a regularization parameter which controls the trade-off between approximation error $\varepsilon(\alpha)$ and sparsity. In some applications, we need to impose a specific sparsity index r, expressed by required maximum number on nonzero elements, with $r < m$ of the vector $\mathbf{s}_*$ [3, 9]. For noisy and uncertain data we propose to use a more flexible and robust cost function (in comparison to the standard (P_ρ) problem) which will be referred as Extended BPD ($EBPD$):

$$(EBPD) \qquad J_{q,\rho}(\mathbf{s}) = \|\mathbf{x} - \mathbf{A}\,\mathbf{s}\|_q^q + \alpha\,\|\mathbf{s}\|_\rho, \qquad (4)$$

There are several possible basic choices for l_q and sparsity criteria ($\|\mathbf{s}\|_\rho = \|\mathbf{s}\|_p$) For example, for the uniform (Laplacian) distributed noise we should choose l_∞-Chebyshev norm (l_1-norm). Some basic choices of ρ (for $l_q = 2$) are $\rho = 0$ (minimum l_0 quasi norm or atomic decomposition related with the matching pursuit (MP) and FOCUSS algorithm), $\rho = 1$ (basis pursuit denoising) and $\rho = 2$ (ridge regression). These three regularizers have different characteristics and applications. Both atomic decomposition and basis pursuit denoising enforce a sparse solution whilst ridge regression will encourage all entries of the vector $\mathbf{s}$ to be non-zero. Atomic decomposition is computationally very demanding due to its combinatorial nature. Both ridge regression and basis pursuit denoising can be converted to the standard quadratic (QP) optimization problem. The optimal choice of ρ norms depends on distribution of noise in sparse components. For example, for noisy components, we can use robust norms such as Huber function defined as $\|\mathbf{s}\|_{\rho_H} = \sum_i \rho_H(s_i)$, where $\rho_H(s_i) = s_i^2/2$ if $|s_i| \le \beta$ and $\rho_H(s_i) = \beta\,|s_i| - \beta^2/2$ if $|s_i| > \beta$, and/or epsilon norm defined as $\|\mathbf{s}\|_\varepsilon = \sum_j |s_j|_\varepsilon$ where $|s_j|_\varepsilon = \max\{0,\ (|s_j| - \varepsilon)\}$ (see Fig. 1).

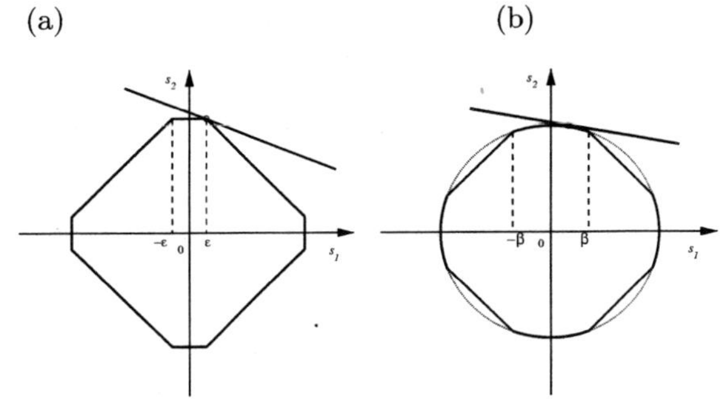

Figure 1: Plots illustrating robust measures of sparsity for 2-D problem: $\min J_\rho(\mathbf{s})$ subject to the constraint $a_1s_1 + a_1s_1 = x$: (a) epsilon norm, (b) Huber function.

2.1. Tuning the parameter α

It should be noted that if the regularization parameter α is too large or too small the optimal solution of the $EBPD$ can be useless. By increasing α, we increase sparsity of $\mathbf{s}_*$ till all entries will be zero for $\alpha \ge \|\mathbf{A}^T\,\mathbf{x}\|_q^*$ (where $\|\cdot\|_q^*$ denotes dual norm to $\|\cdot\|_q$). More precisely, for $0 < \alpha \le \|\mathbf{A}^T\,\mathbf{x}\|_q^*$ the solution $\mathbf{s}_*$ is a piecewise linear function of α, under condition that the solution is unique. In the general case, however, the number of non-zero elements of $\mathbf{s}_*$ is not necessarily a monotonic function of α. In the special case, when the basis matrix $\mathbf{A}$ is orthogonal, i.e., $\mathbf{A}\mathbf{A}^T = \mathbf{A}^T\mathbf{A} = \mathbf{I}_n$ with $m = n$ the $EBPD$ problem for $q = 2$ and $\rho = p = 1$ has an explicit solution given by $\mathbf{s}_* = \text{sign}(\mathbf{s}_0)\ .*\ [|\mathbf{s}_0| - \alpha\mathbf{1}]_+$ or in scalar form $s_{j*} = \text{sign}(s_{j0})[|s_{j0}| - \alpha]_+$, where $\mathbf{s}_0 = \mathbf{A}^T\,\mathbf{x}$ is the minimum norm solution to $\mathbf{A}\,\mathbf{s} = \mathbf{x}$ without any constraints and $[x]_+ = \max\{0,\ x\}$ [3].

If a specific sparsity profile is imposed $r = \text{card}(\mathbf{s}_*) < m$ (where card means the number of nonzero elements) a good heuristic is to solve the problem $EBPD$ for different values of α, finding approximately the smallest value of α that satisfy desired constraint.

The practical importance of the $EBPD$ approach in comparison to the standard LP or BP approach is that the $EPBD$ allows for treating the presence of noise or errors due to mismodeling. Moreover, using the $EBPD$ approach, we have possibility to adjust the sparsity profile (i.e., adjust the number of non-zero components) by tuning the parameter α. In contrast, by using the LP approach we do not have such option. Furthermore, the method can be applied both for overcomplete and undercomplete models.

2.2. Fundamental Properties - Uniqueness and Optimal Solution for Sparse Representations

One fundamental question that is actually investigated by many researchers is to find sufficient conditions for a vector $\mathbf{x}$ to have a unique possibly sparsest and optimal representation as a linear combination of columns of the matrix $\mathbf{A}$ and to find such condition that the various heuristic and greedy algorithms such as BP, BPD, FOCUSS or MP (Matching Pursuit) ensure optimal or close to optimal representations [3, 4, 7, 9, 10]. Another important issue is to establish equivalence conditions for various criteria and algorithms in the sense that obtained solutions have non-zero components at the same locations and with the same signs [10].

The following theorem provide fundamental properties and conditions which characterize optimal and unique solution of sparse representation of signals.

Theorem 1 *Let us consider the underdetermined system* $\mathbf{A}\,\mathbf{s} = \mathbf{x}$ *with* $\mathbf{A} \in \mathbb{R}^{m \times n}$ *and* $m < n$ *for which specific model selection algorithms (BP, DBP, MP, FOCUSS) returned the sparse representation in the form* $\mathbf{x} \cong \mathbf{A}_1 \mathbf{s}_{1*} = \sum_{j \in I} \mathbf{a}_j s_{1j*}$*, where* $\mathbf{A}_r = \mathbf{A}_1 \in \mathbb{R}^{m \times r}$ *designates the* $r \leq m$ *columns of* $\mathbf{A}$ *associated with* r *non-zero elements* s_{1j*} *of the estimated sparse vector* $\mathbf{s}_*$*. Furthermore, let a submatrix* $\mathbf{A}_2 \in \mathbb{R}^{m \times (n-r)}$ *designate the* $(n-r)$ *columns of the matrix* $\mathbf{A}$ *which are associated with the elements of the vector* $\mathbf{s}_*$ *equal to zero. This sparse representation* $\mathbf{s}_*$ *is unique (in the sense of uniqueness of matrix* $\mathbf{A}_1$*) and optimal (in the sense of minimal error* $\|\mathbf{e}_r\|_2$*) if and only if all the components of the vector*

$$\mathbf{g}_1 = \mathbf{A}_2^T \, [\mathbf{A}_1^+]^T \, \mathrm{sign}(\mathbf{s}_{1*}), \qquad (\mathbf{A}_1^+ = [\mathbf{A}_1^T \mathbf{A}_1]^{-1} \mathbf{A}_1^T) \quad (5)$$

have magnitudes strictly less than one (i.e., $\|\mathbf{g}_1\|_\infty < 1$*). However, if the elements of the vector* $\mathbf{g}_1$ *have magnitudes of less than or equal one the solution is optimal but not unique. In the case, when* $\|\mathbf{g}_1\|_\infty > 1$ *the solution is not optimal, that is the specific algorithm did not recover the sparse representation.*

It should be noted that the vector $\mathbf{g}_1$ in (5) is rather insensitive to values of entries of $\mathbf{s}_{1*}$ since it depends only on their signs. Since we unlikely to know the optimal solution *a priori* the above theorem may look useless. However, using Theorem 1, we can easily check whether the LP/BP or any heuristic algorithm, provides an optimal desired solution or not. In fact, the above Theorem has quite general nature and the condition (5) can be applied for many heuristic and greedy algorithms for subset selection such as BP, BPD, MP, and FOCUSS algorithm.

The above basic results can be extended and generalized for arbitrary l_p-norm or any norm non necessary differentiable one, if we apply the concepts of the subgradient and subdifferential.

2.3. Correspondence between Underdetermined and Overdetermined Systems of Linear Equations

Recently, we have discovered that there exists an interesting correspondence (or duality) between the fundamental properties and also between some class of algorithms for overdetermined and underdetermined equations, especially for the minimum l_1-norm problems. These basic properties will be illustrated only by one theorem which is dual to our Theorem 1.

Theorem 2 *[2] Let the* $\mathbf{s}_* \in \mathbb{R}^n$ *will be a solution of inconsistent system of overdetermined equations* $\mathbf{x} \cong \mathbf{A}\,\mathbf{s}$ *in the sense that its associated residual error vector* $\mathbf{e}(\mathbf{s}) = \mathbf{x} - \mathbf{A}\,\mathbf{s}_*$ *has* $r \geq n$ *zero elements. Furthermore, let the reduced submatrix* $\mathbf{A}_r = \mathbf{A}_1 \in \mathbb{R}^{r \times n}$ *designates the* r *rows of the matrix* $\mathbf{A}$ *associated with* r *zero elements of the residual error vector* $\mathbf{e}(\mathbf{s}_*)$ *and* $\mathbf{A}_2 \in \mathbb{R}^{(m-r) \times n}$ *corresponds to the remaining subset of* $(m-r)$ *inconsistent equations which have nonzero residuals:* $\mathbf{e}_2(\mathbf{s}_*) = \mathbf{x}_2 - \mathbf{A}_2 \mathbf{s}_* \neq \mathbf{0}$*. For such formulated problem the vector* $\mathbf{s}_*$ *is a unique and optimal solution (in the sense of minimum* $\|\mathbf{x} - \mathbf{A}\,\mathbf{s}\|_1$*) if and only if all the components of the vectors*

$$\mathbf{g} = [\mathbf{A}_1^+]^T \, \mathbf{A}_2^T \, \mathrm{sign}[\mathbf{e}_2(\mathbf{s}_*)] \qquad (6)$$

have magnitudes less than one (i.e., $\|\mathbf{g}\|_\infty < 1$*). However, if the elements of the vector* $\mathbf{g}$ *have magnitudes of less than or equal one the solution* $\mathbf{s}_*$ *is optimal but not unique. In the case when* $\|\mathbf{g}\|_\infty > 1$ *the solution is not optimal.*

The above correspondence and duality properties can be extended and generalized for arbitrary l_p-norm problems. Furthermore, we found that the family of FOCUSS algorithms [7] for underdetermined case ($m < n$) is dual to the family of IRLS (Iteratively Reweighted Least Squares) algorithms for overdetermined case ($m > n$) [1].
The main merits and advantages of the presented correspondences between overdetermined and underdetermined problems are that, the results obtained for overdetermined problems can be transformed for underdetermined problems and vice-versa. In fact, many such results and efficient algorithms exist for the overdetermined systems of linear equations but relatively few are available for the underdetermined case. We believe, that the discovered correspondence, duality and direct analogies between underdetermined and overdetermined systems enable us to reduce this gap considerably.

3. BEYOND ICA - MULTIRESOLUTION SUB-BAND DECOMPOSITION ICA

Despite the success of using standard ICA in many applications, the basic assumptions of ICA may not hold for some source signals hence some caution should be taken when using standard ICA to analyze real world problems, especially in biomedical signal processing. In fact, by definition, the standard ICA algorithms are not able to estimate statistically dependent original sources, that is, when the independence assumption is violated. In this section, we will present a natural extension and generalization of ICA called Multiresolution Sub-band Decomposition ICA (MSD-ICA) which relaxes considerably the assumption regarding mutual independence of primarily sources [5]. The key idea in this approach is the assumption that the wide-band source signals can be dependent, however only some of their narrow band subcomponents are independent. In other words, we assume that each unknown source can be modelled or represented as a sum of narrow-band sub-signals (sub-components): $s_i(k) = s_{i1}(k) + s_{i2}(k) + \cdots + s_{iK}(k)$. Let us assume that only a certain set of sub-components are independent. Provided that for some of the frequency sub-bands (at least one, say j) all sub-components, say $\{s_{ij}(k)\}_{i=1}^{n}$, are mutually independent or temporally decorrelated, then we can easily estimate the mixing or separating system under condition that these sub-bands can be identified by some *a priori* knowledge or detected by some self-adaptive process.

For this purpose, we simply apply any suitable standard ICA/BSS algorithm, however not for all available raw sensor data but only for suitably pre-processed (e.g., sub-band filtered) sensor signals. The basic concept in the MSD-ICA is to divide the sensor signal spectra into their subspectra or sub-bands, and then to treat those subspectra individually for the purpose at hand. The sub-band signals can be ranked and further processed independently. By applying any standard ICA/BSS algorithm for specific sub-bands and raw sensor data, we obtain sequence of separating matrices $\mathbf{W}_0, \mathbf{W}_1, \ldots, \mathbf{W}_K$, where $\mathbf{W}_0$ is the separation matrix from the original data $\mathbf{x}$ and $\mathbf{W}_j$ is the separating matrix from preprocessing sensor data $\mathbf{x}_j$ in j-th sub-band.

In order to identify for which sub-bands corresponding source subcomponents are independent, we propose to compute the matrices $\mathbf{G}_{jq} = \mathbf{W}_j \mathbf{W}_q^{-1}$, $\forall j \neq q$, where $\mathbf{W}_q$ is estimating separating matrix for q-th sub-band. If subcomponents are mutually independent for at least two sub-bands, say for the sub-band j and sub-band q, then the global matrix $\mathbf{W}_j \mathbf{W}_q^{-1} = \mathbf{P}_{jq}$ will be generalized permutation matrix with only one nonzero (or dominated) element in each row and each column. This follows from the simple observation that in such case the both matrices $\mathbf{W}_j$ and $\mathbf{W}_q$ represent inverses (for $m = n$) of the same mixing matrix $\mathbf{A}$ (neglecting nonessential scaling and permutation ambiguities). In this way, we can blindly identify essential information for which frequency sub-bands the source subcomponents are independent and we can easily identify correctly the mixing matrix. Furthermore, the same concept can be used to estimate blindly the performance index and to compare performance of various ICA algorithms, especially for large scale problems.

In the preprocessing stage we can use any linear transforms, especially, more sophisticated methods, such as block transforms, multirate sub-band filter bank or wavelet transforms, can be applied. We can extend and generalize further this concept by performing the decomposition of sensor signals in a composite time-frequency domain rather than in frequency sub-bands as such. This naturally leads to the concept of wavelets packets (sub-band hierarchical trees) and to block transform packets [1, 5, 11]. Such preprocessing techniques has been implemented in ICALAB [8].

4. CONCLUSIONS

In this paper we have proposed several extensions and modifications of the optimization algorithms for robust sparse component analysis and independent component analysis where various constraints or preprocessing are imposed in order to satisfy physical conditions. The proposed cost functions are quite flexible and general in the sense that can be applied for noisy data with various distributions of noise. The optimal solutions can be obtained by converting the optimization problems to the standard procedures: LP, QP or SOCP (Second Order Cone Program) and solved by powerful and efficient algorithms. Moreover, we proposed a new procedure for estimation of performance of blind source separation using the concept of multiresolution sub-band decomposition ICA which also relaxes considerably the condition on independence of original sources and allows us to find optimal sub-bands.

5. REFERENCES

[1] A. Cichocki and S. Amari, *Adaptive Blind Signal and Image Processing*, John Wiley, Chichester, May 2003 (Revised and corrected edition).

[2] J. A. Cadzow, "Minimum l_1, l_2 and l_∞ norm approximate solutions to an overdetermined system of linear equations", Digital Signal Processing 12, pp. 524-560, 2002.

[3] J.J. Fuchs. On sparse representations in arbitrary redundant bases, Technical report, IRISA, Dec. 2002. submitted to IEEE Trans. Inform. Theory.

[4] R. Gribonval and M. Nielson, "Highly sparse representations from dictionaries are unique and independent of the sparseness measure", (submitted) 2003.

[5] A. Cichocki and P. Georgiev P, "Blind source separation algorithms with matrix constraints", IEICE Transactions on Information and Systems, vol. E86-A, No.3, March 2003, pp.522-531.

[6] D.D. Lee and H. S. Seung, "Learning of the parts of objects by non-negative matrix factorization", Nature, Vol. 401, Oct. 1999, pp.788-791.

[7] K. Kreutz-Delgado, J.F. Murray, B.D. Rao, K. Engan, T.-W. Lee and T.J. Sejnowski, "Dictionary learning algorithms for sparse representation", Neural Computation, vol.15, No. 2, pp. 349-396, 2003.

[8] A. Cichocki, S. Amari, K. Siwek, T. Tanaka et al. (2002) "ICALAB Toolboxes for Signal and Image Processing", *www.bsp.brain.riken.go.jp*

[9] S. Chen, D.L. Donoho, M.A. Saunders, "Atomic decomposition by basis pursuit", SIAM Journal on Scientific Computing, vol.20, pp.33-61, 1999.

[10] D.L. Donoho and M. Elad, "Maximal sparsity representation via l_1 minimization", The Proc. Nat. Aca. Sci., Vol. 100, pp. 2197-2202, March 2003.

[11] M. Zibulevsky, P. Kisilev, Y.Y. Zeevi, and B.A. Pearlmutter, "Blind source separation via multinode sparse representation", In Advances in Neural Information Processing Systems, (NIPS2001) Vol. 14. Morgan Kaufmann, pp. 185–191.

[12] Y.Li, A. Cichocki and S. Amari, "Analysis of sparse representation and blind source separation", Neural Computation, vol.16, pp. 1-42, 2004.

BLIND DETECTION OF PHOTOMONTAGE USING HIGHER ORDER STATISTICS

Tian-Tsong Ng, Shih-Fu Chang
Department of Electrical Engineering
Columbia University, New York, NY
{ttng,sfchang}@ee.columbia.edu

Qibin Sun
Institute of Infocomm Research
Singapore
qibin@i2r.a-star.edu.sg

ABSTRACT

In this paper, we investigate the prospect of using bicoherence features for blind image splicing detection. Image splicing is an essential operation for digital photomontaging, which in turn is a technique for creating image forgery. We examine the properties of bicoherence features on a data set, which contains image blocks of diverse image properties. We then demonstrate the limitation of the baseline bicoherence features for image splicing detection. Our investigation has led to two suggestions for improving the performance of the bicoherence features, i.e., estimating the bicoherence features of the authentic counterpart and incorporating features that characterize the variance of the feature performance. The features derived from the suggestions are evaluated with Support Vector Machine (SVM) classification and shown to improve the image splicing detection accuracy from 62% to about 70%.

1. INTRODUCTION

Photomontage refers to a paste-up produced by sticking together photographic images. In olden days, creating a good composite photograph required sophisticated skills of darkroom masking or multiple exposures of a photograph negative. In today's digital age, however, the creation of photomontage is made simple by the cut-and-paste tools of the popular image processing software such as Photoshop. With such an ease of creating good digital photomontages, we could no longer take image authenticity for granted especially when it comes to legal photographic evidence [1] and electronic financial documents. Therefore, we need a reliable and objective way to examine image authenticity.

Lack of internal consistency, such as inconsistencies in object perspective, in an image is sometimes a telltale sign of photomontage [1]. However, unless the inconsistencies are obvious, this technique can be subjective. Furthermore, forgers can always take heed of any possible internal inconsistencies.

Although image acquisition device with digital watermarking features could be a boon for image authentication, presently there still is not a fully secured authentication watermarking algorithm, which can defy all forms of hacking, and the hardware system has to secure from unauthorized watermark embedding. Equally important are the issues such as the need for both the watermark embedder and detector to use a common algorithm and the consequence of digital watermarks degrading image quality.

On the premise that human speech signal is highly Gaussian in nature [2], a passive approach was proposed [3] to detect the high level of non-gaussianity in spliced human speech using bicoherence features. Unlike human speech signal, the premise of high guassianity does not hold for image signal. It was shown [4] that bispectrum and trispectrum of natural images have a concentration of high values in regions where frequency components are aligned in orientation, due to image features of zero or one intrinsic dimensionality such as uniform planes or straight edges. As images originally have high value in higher order spectrum, detecting image splicing based on the same principle of increased non-gaussianity would be a very low signal-to-noise problem, not to mention the possible complex interaction between splicing and the non-linear image features.

Recently, a new system for detecting image manipulation based on a statistical model for 'natural' images in the wavelet domain is reported [5]. Image splicing is one kind of image tampering the system takes on; however, no further detail about the technical approach is provided in the article.

Image splicing is defined as a simple joining of image regions. We currently do not address the combined effects of image splicing and other post-processing operations. Creation of digital photomontage always involves image splicing although users could apply post-processing such as airbrush style edge softening, which can potentially be detected by other techniques [5]. In fact, photomontages with merely image splicing, as in Figure 1, can look deceivingly authentic and each of them only took a professional graphic designer 10-15 minutes to produce.

Figure 1: Spliced images that look authentic subjectively

In this paper, we pursue the prospect of grayscale image splicing detection using bicoherence features. We first examine the properties of the proposed bicoherence features [3] in relation to image splicing and demonstrate the insufficiency of the features. We then propose two new methods on improving the performance of the bicoherence features for image splicing detection. Lastly, we evaluate the methods using SVM classification experiments over a diverse data set of 1845 image blocks. More details about this work are included in [6].

2. BICOHERENCE

Bicoherence is a normalized bispectrum, i.e., the third order correlation of three harmonically related Fourier frequencies of a signal, $X(\omega)$ [7]:

$$b(\omega_1,\omega_2)=\frac{E[X(\omega_1)X(\omega_2)X^*(\omega_1+\omega_2)]}{\sqrt{E[|X(\omega_1)X(\omega_2)|^2]E[|X(\omega_1+\omega_2)|^2]}}=|b(\omega_1,\omega_2)|e^{j\Phi(b(\omega_1,\omega_2))}$$

When the harmonically related frequencies and their phase are of the same type of relation, i.e., when there exists (ω_1, ϕ_1), (ω_2, ϕ_2) and $(\omega_1+\omega_2, \phi_1+\phi_2)$ for $X(\omega)$, $b(\omega_1,\omega_2)$ will have a high magnitude value and we call such phenomena *quadratic phase coupling* (QPC). As such, the average bicoherence magnitude would increase as the amount of QPC grows. Besides that, bicoherence is insensitive to signal gaussianity and bispectrum is often used as a measure of signal non-gaussianity [8].

2.1. Bicoherence Features

Motivated by the effectiveness of the bicoherence features used for human-speech splicing detection [3], similar features are extracted from a bicoherence with

- Mean of magnitude: $M = |\Omega|^{-1}\Sigma_\Omega|b(\omega_1, \omega_2)|$
- Negative phase entropy: $P=\Sigma_n p(\Psi_n)\log p(\Psi_n)$

where

$\Omega=\{(\omega_1, \omega_2)|\ \omega_1=(2\pi m_1)/M,\ \omega_2=(2\pi m_2)/M,\ m_1, m_2= 0,\dots,.M-1\}$

$p(\Psi_n)= |\Omega|^{-1}\Sigma_\Omega 1(\Phi(b(\omega_1, \omega_2))\in \Psi_n)$, $1(\cdot)$=*indicator function*

$\Psi_n=\{\phi|-\pi+(2\pi n)/N\leq\phi< -\pi+2\pi(n+1)/N\},\ n=0,\dots, N-1$

2.2. Estimation of Bicoherence Features

With limited data sample size, instead of computing 2-D bicoherence features from an image, 1-D bicoherence features can be computed from N_v vertical and N_h horizontal image slices of an image and then combined as follows:

$$fM=\sqrt{\left(\tfrac{1}{N_h}\sum_i M_i^{Horizontal}\right)^2+\left(\tfrac{1}{N_v}\sum_i M_i^{Vertical}\right)^2}$$

$$fP=\sqrt{\left(\tfrac{1}{N_h}\sum_i P_i^{Horizontal}\right)^2+\left(\tfrac{1}{N_v}\sum_i P_i^{Vertical}\right)^2}$$

In order to reduce the estimation variance, the 1-D bicoherence of an image slice is computed by averaging segment estimates:

$$\hat{b}(\omega_1,\omega_2)=\frac{\frac{1}{k}\sum_k X_k(\omega_1)X_k(\omega_2)X_k^*(\omega_1+\omega_2)}{\sqrt{\left(\frac{1}{k}\sum_k|X_k(\omega_1)X_k(\omega_2)|^2\right)\left(\frac{1}{k}\sum_k|X_k(\omega_1+\omega_2)|^2\right)}}$$

We use segments of 64 samples in length with an overlap of 32 samples with adjacent segments. For lesser frequency leakage and better frequency resolution, each segment of length 64 is multiplied with a Hanning window and then zero-padded from the end before computing 128-point DFT of the segment.

In Fackrell et al. [9], it is suggested that N data segments should be used in the averaging procedure for estimating a N-point DFT bispectrum of a stochastic signal. Overall, we use 768 segments to generate features for a 128x128-pixel image block.

3. IMAGE DATA SET

Our data set [10] is collected with sample diversity in mind. It has 933 authentic and 912 spliced image blocks of size 128 x 128 pixels. The image blocks are extracted from images in CalPhotos image set [11]. As the images are contributions from photographers, in our case, they can be considered as authentic i.e., not digital photomontages.

The authentic category consists of image blocks of an entirely homogenous textured or smooth region and those having an object boundary separating two textured regions, two smooth regions, or a textured regions and a smooth region. The location and the orientation of the boundaries are random.

The spliced category has the same subcategories as the authentic one. For the spliced subcategories with object boundaries, image blocks are obtained from images with spliced objects; hence, the splicing region interface coincides with an arbitrary-shape object boundary. Whereas for the spliced subcategories with an entirely homogenous texture or smooth region, image blocks are obtained from those in the corresponding authentic subcategories by copying a vertical or a horizontal strip of 20 pixels wide from one location to another location within a same image.

4. PROPERTIES OF BICOHERENCE FEATURES

We are interested in investigating the performance of bicoherence features in detecting spliced images on the three object interface types for which such performance varies over, i.e. smooth-smooth, textured-textured, and smooth-textured. Figure 2 shows the scatter plot of the bicoherence magnitude feature (*fM*) of the authentic and spliced image blocks with a particular object interface type. The plots also show how well the edge percentage (y-axis) captures the characteristics of different interface types. The edge pixels are obtained using Canny edge detector. The edge percentage is computed by counting the edge pixels within each block. As the plots for bicoherence phase feature (*fP*) are qualitatively similar, they are omitted due to space constraints.

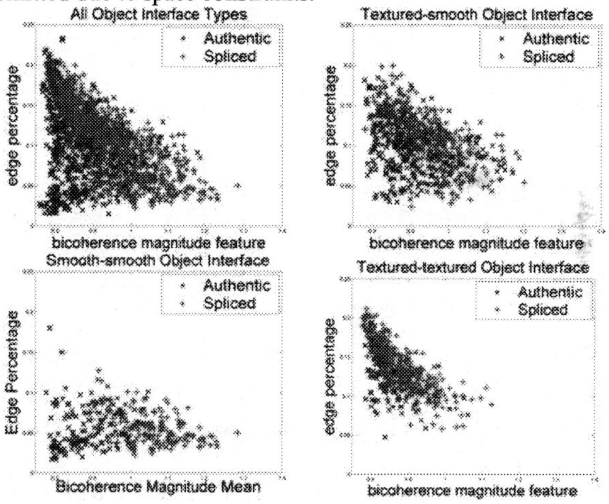

Figure 2: Bicoherence magnitude feature for different object interface types

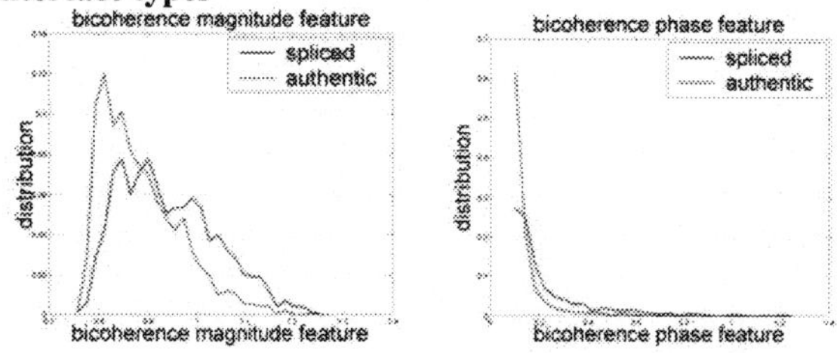

Figure 3: Distribution of the bicoherence magnitude feature, *fM*, (left) and the phase feature, *fP* (right)

We observe that the performance of the bicoherence feature in distinguishing spliced images varies for different object interface types, with textured-textured object interface type being the worst case. Figure 3 shows the distribution of the features for the authentic and spliced image categories. We can observe that the distributions of the two image block categories are greatly overlapped, although there are noticeable differences in the peak locations and the heavy tails. Hence, we would expect poor classification between the two categories if the features were to be used directly.

5. METHODS FOR IMPROVING THE PERFORMANCE OF BICOHERENCE FEATURES

Our investigation on the properties of bicoherence features for images leads to two methods for augmenting the performance of the bicoherence features in detecting image splicing:

1. By estimating the bicoherence features of authentic images.
2. By incorporating image features that capture the image characteristics on which the performance of the bicoherence features varies, e.g., edge pixel percentage feature (*fE*) capture the characteristics of different object interface.

5.1. Estimating Authentic Counterpart Bicoherence Features

Assume that for every spliced image, there is a corresponding authentic counterpart, which is similar to the spliced image except that it is authentic. The rationale of the approach, formulated as below, is that if the bicoherence features of the authentic counterpart can be estimated well, the elevation in the bicoherence features due to splicing could be more detectable.

$$\begin{aligned} f_{Bic} &= g(\Lambda_I(image), \Lambda_S(image, s), s) + \varepsilon \\ &\approx g_1(\Lambda_I(image)) + g_2(\Lambda_S(image, s), s) + \varepsilon \\ &\approx f_{Authentic} + \Delta f_{Splicing} + \varepsilon \end{aligned}$$

where Λ_I is a set of splicing-invariant features while Λ_S is a set of features induced by splicing, s is a splicing indicator and ε is the estimation error. In this formulation, g_1 corresponds to an estimate of the bicoherence feature of the authentic counterpart, denoted as $f_{Authentic}$ and g_2 corresponds to the elevation of the bicoherence feature induced by splicing, denoted as $\Delta f_{Splicing}$. With $\Delta f_{Splicing}$, splicing would be more detectable after the significant interference from the splicing-invariant component, g_1, is removed. $\Delta f_{Splicing}$ can be estimated with $f_{Bic} - f_{Authentic}$, which we call *prediction discrepancy*. The $f_{Authentic}$ estimation performance would be determined by two factors, i.e., how much we capture the splicing-invariant features and how well we map the splicing-invariant features to the bicoherence features.

A direct way to arrive at a good estimator is through an approximation of the authentic counterpart obtained by depriving an image of the splicing effect. As a means of 'cleaning' an image of its splicing effect, we have chosen the texture decomposition method based on functional minimization [12], which has a good edge preserving property, for we have observed the sensitivity of the bicoherence features to edges.

5.2. Texture Decomposition with Total Variation Minimization and a Model of Oscillating Function

In functional representation, an image, f defined in $\Omega \subset R^2$, can be decomposed into two functions, u and v, within a total variation minimization framework with a formulation [12]:

$$\inf_u \left\{ E(u) = \int_\Omega |\nabla u| + \lambda \|v\|_*, f = u + v \right\}$$

where the u component, a structure component of the image, is modeled as a function of bounded variation while the v component, representing the fine texture or noise component of the image, is modeled as an oscillation function. $\|\cdot\|_*$ is the norm of the oscillating function space and λ is a weight parameter for trading off variation regularization and image fidelity.

The minimization problem can be reduced to a set of partial differential equations known as Euler-Lagrange equations and solved numerically with finite difference technique. As the structure component could contain arbitrarily high frequencies, conventional image decomposition by filtering could not attain such desired results. In this case, the structure component will serve as an approximation for the authentic counterpart, hence, the estimator for $fM_{Authentic}$ and $fP_{Authentic}$ are respectively $\hat{f}M_{Authentic} = fM_{structure}$ and $\hat{f}P_{Authentic} = fP_{structure}$.

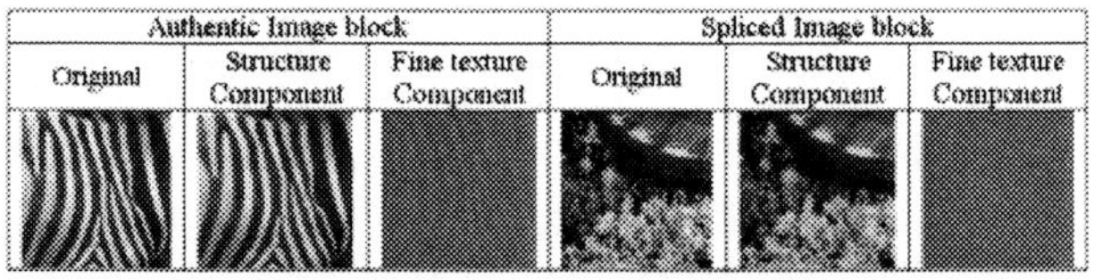

Figure 4: Examples of texture decomposition

For the linear prediction discrepancies between the bicoherence features of an image and those of its authentic counterpart, i.e., $\Delta fM = fM - \alpha \hat{f}M_{Authentic}$ and $\Delta fP = fP - \beta \hat{f}P_{Authentic}$, the parameters α and β, without being assumed to be unity, are learnt by Fisher Linear Discriminant Analysis in the 2-D space (fM, $\hat{f}M_{Authentic}$) and (fP, $\hat{f}P_{Authentic}$) respectively, to obtain the subspace projection where the between-class variance is maximized relative to the within-class variance, for the authentic and spliced categories.

We evaluate effectiveness of the estimator, as shown in Figure 5 using the prediction discrepancy for the magnitude and phase features. Our objective is to show that the new features (ΔfM, ΔfP) have a stronger discrimination power between authentic and spliced compared to the original features (fM, fP). This objective is partially supported by observing the difference between Figure 5 and Figure 3 (In Figure 5, two distributions are more separable)

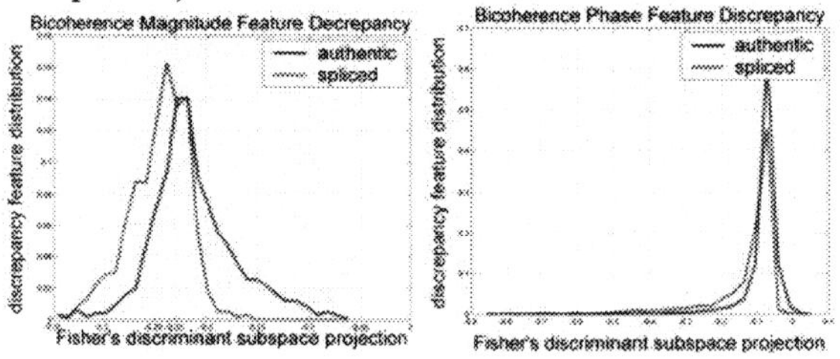

Figure 5: Distribution of prediction discrepancy

6. SVM CLASSIFICATION EXPERIMENTS

We herein evaluate the effectiveness of the features, which are derived from the proposed method, i.e., prediction discrepancy and edge percentage using our data set. SVM classifications with RBF kernel are performed with parameters chosen for ensuring no overfitting as verified by 10-fold cross-validation. Three statistics obtained from 100 runs of classification are used to evaluate the performance of feature sets:

- Accuracy mean: $M_{accuracy} = \frac{1}{100}\sum_i (N^i_{S|S} + N^i_{A|A}) / (N^i_{\bullet|S} + N^i_{\bullet|A})$
- Average precision: $M_{precision} = \frac{1}{100}\sum_i N^i_{S|S} / N^i_{S|\bullet}$
- Average recall: $M_{recall} = \frac{1}{100}\sum_i N^i_{S|S} / N^i_{\bullet|S}$

where S and A represents Spliced and Authentic respectively and $N^i_{A|B}$ denotes the number of samples B detected as A in the ith run. The results of the experiment are shown below:

Feature Label	Feature Name
Orig	magnitude and phase features $\{fM, fP\}$
Delta	Prediction discrepancy $\{\Delta fM, \Delta fP\}$
Edge	Edge percentage fE

Feature Set	$M_{accuracy}$	$M_{precision}$	M_{recall}
Orig	0.6259	0.6354	0.5921
Delta	0.6876 (+6.2 %)	0.6685	0.7477
Orig+Delta	0.7028 (+7.7 %)	0.6725	0.7925
Orig+Edge	0.7005 (+7.5 %)	0.6780	0.7667
Delta+Edge	0.6885 (+6.3 %)	0.6431	0.8517
Orig+Delta+Edge	0.7148 (+8.9 %)	0.6814	0.8098

Note: Statistical t-tests for classification results using feature set $\{fM, fP\}$ against all other results are performed. The null hypothesis (i.e., the mean of the two results are the same) is rejected at a 0.05 significance level for all tests.

Below are the observations from the classification results:

1. Prediction discrepancy features alone obtain 6.2 % improvement in $M_{accuracy}$ over the original bicoherence features.
2. Edge percentage improves the performance of the bicoherence features by 7.5 % in $M_{accuracy}$.
3. Prediction discrepancy and edge percentage are redundant with respect to each other.
4. The best performance (last row) obtained by incorporating all the proposed features is 71 % in $M_{accuracy}$, which is 8.9 % better than the baseline method (first row).

The results are encouraging as it shows the initial promise of the authentic counterpart estimation. The third observation may be an indication that the prediction discrepancy features are less affected by image texturedness. Hence, if the estimation of the authentic counterpart bicoherence features can be further improved, it may help in the classification of the toughest case where the object interface type is textured-textured.

The block level detection results can be combined in different ways to make global decision about the authenticity of a whole image or its sub-regions. For example, Figure 6 illustrates the idea of localizing the suspected splicing boundary.

7. CONCLUSIONS

In this paper, we have shown the difficulties of image splicing detection using bicoherence features, despite the technique being effective on human speech signals. We have also empirically shown how the performances of the bicoherence features depending on the different object interface types. Two methods are proposed for improving the capability of the bicoherence features in detecting image splicing. The first exploits the dependence of the bicoherence features on the image content such as edge pixel density and the second offsets the splicing-invariant component from bicoherence and thereby obtains better discriminative features. Finally, we observe improvements in SVM classification after the derived features are incorporated.

Figure 6: Spliced image blocks (marked with a red box)

This is the first step of our effort in using bicoherence features for image splicing detection. We will next seek a model to get an insight on why bicoherence is sensitive to splicing, from which other effective features can be derived.

8. ACKNOWLEDGEMENTS

Gratitude to A*STAR, Singapore for sponsoring the first author, to Lexing Xie, Shahram Ebadollahi and Anita Huang for their discussions and helps, and to Huei-Sim Tang for the professionally produced photomontages.

9. REFERENCES

[1] W. J. Mitchell, "When Is Seeing Believing?", *Scientific American*, pp. 44-49, February 1994.

[2] J.W.A. Fackrell and S. McLaughlin. "Detecting nonlinearities in speech sounds using the bicoherence", *Proc. of the Institute of Acoustics*, 18(9), pp. 123–130, 1996.

[3] H. Farid, "Detecting Digital Forgeries Using Bispectral Analysis", *Technical Report*, AIM-1657, MIT AI Memo, 1999.

[4] Krieger, G., Zetzsche, C. and Barth, E., "Higher-order statistics of natural images and their exploitation by operators selective to intrinsic dimensionality", *Proc. of IEEE Signal Processing Workshop on HOS*, pp. 147-151, 21-23 July 1997.

[5] H. Farid, "A Picture Tells a Thousand Lies", *New Scientist*, 179(2411), pp. 38-41, Sept. 6, 2003.

[6] T.-T. Ng and S.-F. Chang, "Blind Image Splicing and Photomontage Detection Using Higher Order Statistics", *ADVENT Technical Report*, #201-2004-1, Columbia University, http://www.ee.columbia.edu/dvmm/, Jan 2004.

[7] Y. C. Kim and E. J. Powers, "Digital Bispectral Analysis and its Applications to Nonlinear Wave Interactions", *IEEE Trans. on Plasma Science,* vol. PS-7, No.2, pp. 120-131, June 1979.

[8] M. Santos et al., "An estimate of the cosmological bispectrum from the MAXIMA-1 CMB map", Physical Review Letters, 88, 241302, 2002

[9] J. W. A. Fackrell, P. R. White, J. K. Hammond, R. J. Pinnington and A. T. Parsons, "The interpretation of the bispectra of vibration signals-I. Theory", *Mechanical Systems and Signal Processing*, Vol. 9(3), pp. 257-266, 1995.

[10] Data set of authentic and spliced image blocks, DVMM, Columbia Univ., http://www.ee.columbia.edu/dvmm/researchProjects/AuthenticationWatermarking/BlindImageVideoForensic/

[11] CalPhotos: A database of photos of plants, animals, habitats and other natural history subjects. Digital Library Project, University of California, Berkeley.

[12] L. A. Vese and S. J. Osher, "Modeling Textures with Total Variation Minimization and Oscillating Patterns in Image Processing", *UCLA C.A.M. Report,* 02-19, May 2002.

DATA HIDING FOR BI-LEVEL DOCUMENTS USING SMOOTHING TECHNIQUE

Huijuan Yang and Alex C. Kot

School of Electrical & Electronic Engineering
Nanyang Technological University, Singapore 639798
Email: ehjyang@ntu.edu.sg.

ABSTRACT

In this paper, a novel method for watermarking on binary images to authenticate the owner or authorized user is proposed. The proposed method makes use of noise suppression patterns which takes into account the smoothness in 5x5 neighborhood. It has a moderate capacity of the watermark while preserving very good invisibility to human eyes. The resulted watermarked binary image is an enhancement to the original image in terms of smoothness. The fast computational speed and time complexity makes the method suitable for real-time authentication. Retrieval of the watermark doesn't require the presence of the original binary document image. However the secret key and the location map information are needed for the retrieval. The watermark used in the algorithm is the Gold-like sequence. Discussions on issues in binary image watermarking have also been presented. Experimental results validate the arguments made in the paper.

1. INTRODUCTION

The wide proliferation of digital media such as digital images, audio, video and text images (documents) has aroused great interest for copyright protection, annotation and authentication. Watermarking has been regarded as one of the means for this copyright protection and authentication. Authentication of digital documents is of utmost importance because of its wide usage in today's eOffice world. Yet it is a very challenging job to embed watermark in binary images, as it is easily to be perceived and noticeable by changing bits in binary images.

A method for embedding data in binary images by employing shuffling technique is proposed in [1]. The same number of bits has been embedded in each block based on the flippability score of the pixel within the block. The score is measured by the change in smoothness and connectivity. A shuffling table or a key is needed for correctly extracting the watermark bits. A look up table is provided for a 3×3 pattern, the watermarking effects depend very much on how the table is defined. Therefore it is very subjective and difficult to judge. Furthermore, the table size increases exponentially when the size of neighborhood increases.

In [2], a data-hiding scheme is proposed, in which the bits flipped in the host image must be adjacent to another bit that has a value equal to the former's new value. Thus the hiding effect is not quite visible. In other words, they look for the boundary pixel to flip such that the hiding effects are not easily to be perceived by human eyes. However, it may lose connectivity and results in noisy images in some situations, as the connectivity issue has not been taken into consideration during the embedding process. The contribution of this method is its weight matrix generation scheme and the large capacity, which can hide as many as $\lfloor \log_2(mn+1) \rfloor - 1$ bits of data for each image block.

Another watermarking method catered for text document is proposed in [3]. In this scheme, the data is embedded in the 8-connected boundary of a character. Pairs of patterns that are dual to each other are used for watermark embedding and extraction. The available patterns that can be used in the watermarking depend on the resolution of the document images, thus the capacity of the watermark is not as high as expected. An investigation of watermarking for binary images on frequency domain, in particular, DC components of DCT is presented in [4]. The quality of the binary images is not so good due to the reason that a post-embedding binarization process is involved. However, it may provide certain degree of robustness to common image processing. An objective distortion measure for binary images is proposed in [5]. This method is based on the reciprocal distance calculation, which matches well with subjective evaluation. A binary image authentication method based on the distortion measure discussed in [5] is proposed in [6].

2. THE PROPOSED METHOD

2.1. Gold-like Sequence Generation

Gold-like sequence [7] is used as the watermark in the present approach. The reason why Gold-like sequence is chosen as the watermark lies in: the linear span of the Gold-like sequence is very large compared with its period whilst retaining the desirable properties of m sequences. The Gold-like sequence $I(a,b)$ is generated by generating the m sequences with polynomials $h(x)$ and $\hat{h}(x)$ separately, and then add the two m sequences chip by chip by synchronous clocking. It is defined blow:

$$I(a,b) = \{a, b, a \oplus b, a \oplus Tb, a \oplus T^2 b, .., a \oplus T^{N-1} b\} \qquad (1)$$

where, $\{a_n\}$ and $\{b_n\}$ are reciprocal m-sequences generated from $h(x)$ and $\hat{h}(x)$. T is a shift operator.

2.2. Some Definitions

The denoise patterns considered in the proposed method are based on a 5x5 neighborhood. The designations of pixels in the neighborhood block are defined in Figure 1.

w_{21}	w_{22}	w_{23}	w_{24}	w_{9}
w_{20}	w_{6}	w_{7}	w_{8}	w_{10}
w_{19}	w_{5}	p	w_{1}	w_{11}
w_{18}	w_{4}	w_{3}	w_{2}	w_{12}
w_{17}	w_{16}	w_{15}	w_{14}	w_{13}

Figure 1. Designations of pixels in 5x5 neighborhood. Pixels in grey are the ones which need to be checked further.

It is assumed that the eight neighbors of the point (i, j) (p) are: $(i-1, j-1)$ (W_6), $(i-1, j)$ (W_7), $(i-1, j+1)$ (W_8), $(i, j-1)$ (W_5), $(i, j+1)$ (W_1), $(i+1, j-1)$ (W_4), $(i+1, j)$ (W_3), $(i+1, j+1)$ (W_2), where W_1, W_3, W_5 and W_7 are 4-neighbours of pixel p, while W_2, W_4, W_6 and W_8 are 8-neighbour of pixel p. The flippability of the center pixel within the block depends mainly on the smoothness of the center pixel and its neighbors in the 5x5 neighborhood.

2.3. The Denoise Mechanism

Usually 3x3 neighborhood is particularly desirable in binary image processing, i.e., thinning. However, it has difficulty in preserving the connectivity and smoothness of the shape of the original image [8]. If the original image pattern is noisy, it is necessary to consider the 5x5 neighborhood. The same rules hold for data hiding in binary images. The consistency and smoothness of the resulted watermarked image is of utmost concern. The topology and smoothness of the resulted watermarked image are taken into consideration in the present approach. The watermarking process works as a denoising process, which handles the boundary noise, erosion. It does not create any new erosion, change the shape and topology of the original image. The denoise patterns considered can be illustrated in Figure 2.

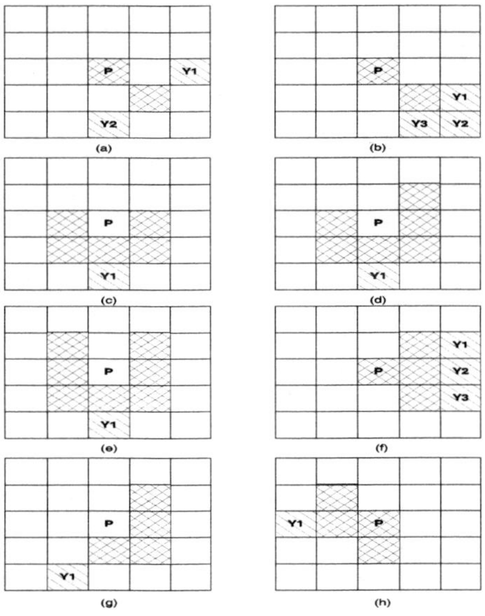

Figure 2. Denoise patterns used in the watermark embedding process excluding symmetric cases of rotation, mirroring and complement.

The rules for checking of the flippability based on these denoise patterns are illustrated as follows, assume the object pixel is black (represented by 1) and the background pixel is white (represented by 0).

For the limb pattern illustrated in Figure 2 (a), further checking on pixels Y_1 (w_{11}) and Y_2 (w_{15}) is required. If both pixels are of the object color, then the center pixel p can be flipped. In this case, the center pixel is treated as a noise protrusion pixel. The condition is defined below.

$$(p \wedge w_2) \wedge (\overline{w}_1 \wedge \overline{w}_3 \wedge \overline{w}_4 \wedge \overline{w}_5 \wedge \overline{w}_6 \wedge \overline{w}_7 \wedge \overline{w}_8) \wedge (w_{11} \wedge w_{15}) = 1 \quad (2)$$

where, $\overline{w}$ is the complement of w.

For the limb pattern in Figure 2(b), if all three pixels Y_1 (w_{12}), Y_2 (w_{13}) and Y_3 (w_{14}) are of object color, the center pixel p is most likely a noisy pixel, then it can be flipped. The condition is defined below.

$$(p \wedge w_2) \wedge (\overline{w}_1 \wedge \overline{w}_3 \wedge \overline{w}_4 \wedge \overline{w}_5 \wedge \overline{w}_6 \wedge \overline{w}_7 \wedge \overline{w}_8) \wedge (w_{12} \wedge w_{13} \wedge w_{14}) = 1 \quad (3)$$

For the "U" pattern in Figure 2(c), (d) and (e), if the pixel Y_1 (w_{15}) is of the background color, then the hole in the center pixel p can be filled. The conditions are defined below accordingly.

$$(\overline{p} \wedge w_1 \wedge w_2 \wedge w_3 \wedge w_4 \wedge w_5) \wedge (\overline{w}_6 \wedge \overline{w}_7 \wedge \overline{w}_8) \wedge (\overline{w}_{15}) = 1 \quad (4)$$

$$(\overline{p} \wedge w_1 \wedge w_2 \wedge w_3 \wedge w_4 \wedge w_5 \wedge w_8) \wedge (\overline{w}_6 \wedge \overline{w}_7) \wedge (\overline{w}_{15}) = 1 \quad (5)$$

$$(\overline{p} \wedge w_1 \wedge w_2 \wedge w_3 \wedge w_4 \wedge w_5 \wedge w_6 \wedge w_8) \wedge (\overline{w}_7) \wedge (\overline{w}_{15}) = 1 \quad (6)$$

For the "T" pattern in Figure 2(f), if all three pixels Y_1 (w_{10}), Y_2 (w_{11}) and Y_3 (w_{12}) are of the object color, then the center pixel p can be flipped. In such a scenario, the center pixel is most likely a protrusion. The condition is defined below.

$$(p \wedge w_1 \wedge w_2 \wedge w_8) \wedge (\overline{w}_3 \wedge \overline{w}_4 \wedge \overline{w}_5 \wedge \overline{w}_6 \wedge \overline{w}_7) \wedge (w_{10} \wedge w_{11} \wedge w_{12}) = 1 \quad (7)$$

For the "L" pattern in Figure 2(g), if the pixel Y_1 (w_{16}) has the same the background color, then the center pixel p (which can be in either object or background color) can be flipped. Otherwise, flip the center pixel will most likely create a triangle noise in the center. The condition is defined below.

$$(\overline{p} \wedge w_1 \wedge w_2 \wedge w_3 \wedge w_8) \wedge (\overline{w}_4 \wedge \overline{w}_5 \wedge \overline{w}_6 \wedge \overline{w}_7) \wedge (\overline{w}_{16}) = 1 \quad (8)$$

$$(p \wedge w_1 \wedge w_2 \wedge w_3 \wedge w_8) \wedge (\overline{w}_4 \wedge \overline{w}_5 \wedge \overline{w}_6 \wedge \overline{w}_7) \wedge (\overline{w}_{16}) = 1 \quad (9)$$

For the "Staircase" pattern in Figure 2(h), if the pixel Y_1 (w_{19}) is of the object color, the center pixel can be flipped. In such a case, the pattern is no longer a staircase pattern. The condition is defined below.

$$(p \wedge w_3 \wedge w_5 \wedge w_6) \wedge (\overline{w}_1 \wedge \overline{w}_2 \wedge \overline{w}_4 \wedge \overline{w}_7 \wedge \overline{w}_8) \wedge (w_{19}) = 1 \quad (10)$$

2.4. The Embedding and Verification Mechanism

The watermark embedding process given in Figure 3 consists of the following steps:

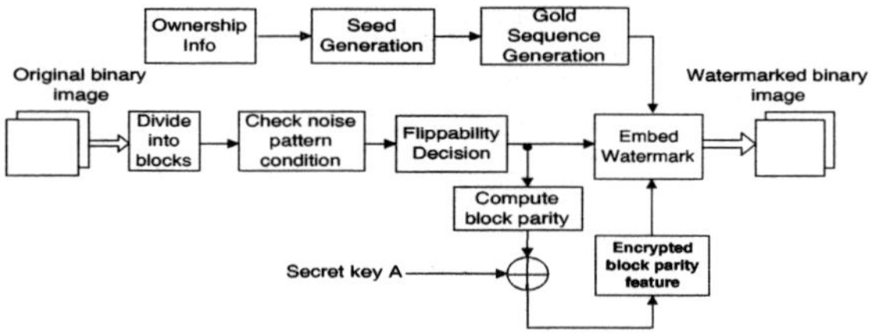

Figure 3. Block diagram of watermark embedding process

Step 1. Generate the watermark. The seed which is generated from the authorized user or owner information is fed to the watermark generator to generate the watermark.

Step 2. Divide the original binary image into blocks. An odd number (e.g., 7x7) is selected as the block size in the current study. Other sizes may also be considered. However, large block size may results in decrease in the capacity.

Step 3. Determine the flippability of the block based on the denoise pattern condition checking as discussed in Section 2.3. Within one block, the flippability checking will be performed on 3x3 neighbourhood which is centered on all pixels except the block boundary ones. Further checking is performed in one or two more pixels in 5x5 neighborhood, which may or may not be in the same block. Take as an example illustrated in Figure 1, the moving window is centered at pixels, i.e., p and w_1 to w_8 within the block. However, the boundary pixels, i.e., w_9 to w_{24} are excluded from being centered.

Step 4. Compute the encrypted odd-even feature of the block. A secret key A is used to be XORed with the odd-even features of each block of the original image. The purpose of using the secret key is to ensure the integrity of the content. If the embedding is based on the plain odd-even features of the block, the adversaries are able to change the values of other pixels in the same block while keeping the odd-even features of the block the same. In such a case, the forged document can still be verified even if the content has been changed.

Step 5. Embed the watermark based on the encrypted odd-even features of the block. One bit is flipped within one 7x7 block.

The encrypted odd-even feature of the block will be employed to embed the watermark bits, i.e., to embed "0", the XORed number of black pixels inside the block is an even number. To embed "1", the XORed number of black pixels inside the block is an odd number. The location map which corresponds to one bit per block is then XORed with the secret key B for later watermark retrieval. The watermark verification process can be illustrated in block diagram shown in Figure 4.

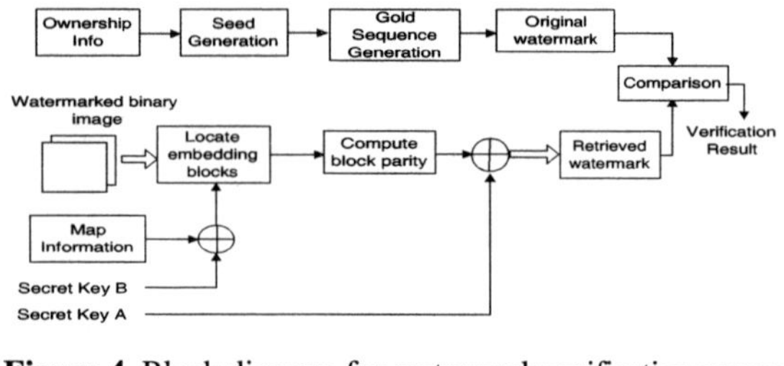

Figure 4. Block diagram for watermark verification process

The length of key A equals to the number of flippable blocks in the image. While the length of key B equals to the number of blocks in the image, which is of the same length as the map information. One simple way for the generation of key A and key B can be using the Gold sequence generator. In such a case, only the secret seed information needs to be sent to the receiver side, e.g., 6 bits of the seed value. Thus, the encrypted map information and the seed value of the secret key A and key B needs to be sent over to the receiver side. In order to achieve this goal, the PKI technology can be employed as a wrapper for this secure exchange between the sender and the receiver, i.e., the sender may use his/her secret key to encrypt the secret information, at the receiver side, the receiver may use the sender's public key to recover the secret information.

Once the secret and map information have been recovered, the secret key A and key B thus can be generated. The secret key B is firstly XORed with the map information bits to get back the embedding location. The secret key A is then used to retrieve back the parity feature of the block. Thereafter the watermark bits can be retrieved.

3. EXPERIMENTAL RESULTS

Several observations have been made on watermarking of the binary images. The document image that is captured through a paper document, thereafter scanned and binarized by means of thresholding, this process usually resulted in a degraded and noisy document. Noise removing is a very important preprocessing step to facilitate other subsequent operations to the document images. Watermarking on digital document images is usually done by introducing noise and degradation, which resulted in a poor quality document image. If the process can work towards enhancement of the quality of the images, it will not only serve the authentication or copyright protection purposes but also facilitate the subsequent OCR process.

So far in the existing literature, no standard for evaluating the binary image watermarking algorithm has been established. However, it can always be accepted by most of the researchers that the basic requirements should be satisfied for a good watermarking method. The resulted watermarked binary should be topologically equivalent to the original binary image. It means that the flipping of the pixels in the image doesn't loss the object shape. The flipping of the pixels in the binary images should preserve connectivity for both foreground (object pixel) and background. The watermarking process should prevent excessive erosion and should not completely delete the feature pattern of the binary images, therefore can maintain the topology of the binary images. The number of pixels flipped in certain region should be balanced so that the number of pixels flipped from white to black roughly should equal to the number of pixels flipped from black to white, e.g., the resulted watermarked image is still of the same intensity to human eyes.

The proposed method has been implemented and some tests are conducted. The block size selected is 7x7. One handwritten text document and one pure text document have been used as the text images. The results are illustrated in Figure 5 and 6 respectively. It can be observed from the results that the proposed method is quite effective in removing noise. The resulted watermarked image is of good quality compared with the original one. Another observation can be made is that the capacity of the watermark depends on the noisy nature of the original image. The pure text document has more information and much noisier than the handwritten document. Therefore, its capacity is much larger than that of handwritten document, even if the two images are of similar size.

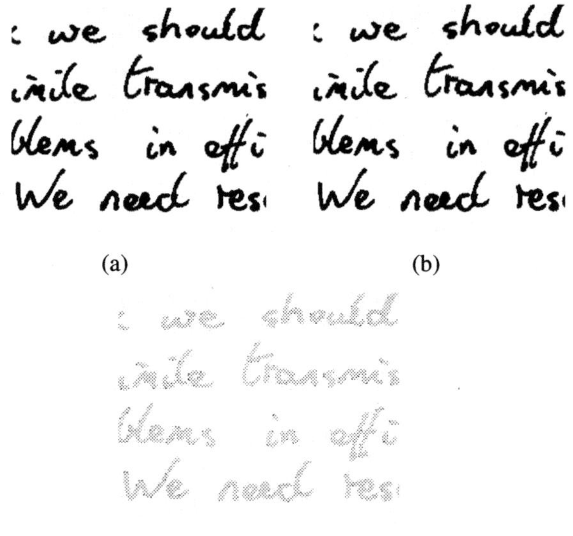

(c)

Figure 5. Hiding effect on a handwritten document: (a) the original image, which is of size 272x272. (b) 354 bits are embedded by the proposed approach. (c) the difference image, the black dot is the pixel that has been flipped.

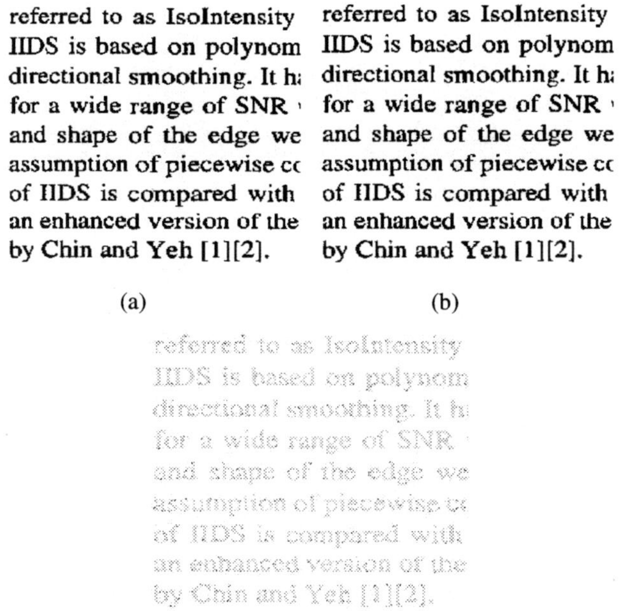

(c)

Figure 6. Hiding effect on a pure text document: (a) the original image, which is of size 280x280. (b) hide 590 bits by our proposed approach. (c) the difference image, the black dot is the pixel that has been flipped.

4. CONCLUSIONS

In this paper, a novel method for embedding watermark in the binary document images for the purpose of document authentication is proposed. The computation load and time complexity of the proposed method are rather low due to the reason that only the condition checking is required for the watermark embedding. Thus it is suitable for real-time on-line embedding and verification. The quality of the watermarked document is comparable if not better than the original document due to the utilizing of the noise suppression pattern. The proposed approach didn't introduce noise or create erosion to the original binary image. The use of large span of Gold-like sequence makes the brute force attack more difficult. The denoising approach adopted in this study put forward a new concept in the watermarking field, that is: watermarking can not only work towards degrade the quality of original document, but also it can work towards enhance the quality of the original document, which therefore will facilitate subsequent OCR process. Our future direction will be focused on improve the embedding process to make it blind for the convenience of end users.

REFERENCES

[1] M. Wu, E. Tang and B. Liu, "Data Hiding in Digital Binary Image", *IEEE International Conference on Multimedia and Expo.*, 2000, pp. 393-396.

[2] Y.-C.Tseng and H.-K. Pan, "Data Hiding in 2-Color Images", *IEEE Transactions on computers*, Vol. 51, No. 7, July 2002.

[3] Q. Mei, E. K. Wong, and N. Memon, "Data Hiding in Binary Text Documents", *Proceedings of SPIE* Vol. 4314, pp. 369-375, 2001.

[4] H. Lu, X. Shi, Shi, Y. Q., Kot A. C. and L. Chen, "Watermark embedding in DC Components of DCT for binary images", 2002 IEEE Workshop on Multimedia Signal Processing, pp. 300-303, 9-11, Dec. 2002.

[5] H. Lu, A. C. Kot and Y.Q. Shi, "Distance-Reciprocal Distortion Measure for Binary Document Images", to appear in *IEEE Signal Processing Letters*, Feb. 2004.

[6] H. Lu, Kot A. C. and J. Cheng, "Secure data hiding in binary images for authentication", Proceedings of the 2003 International Symposium on Circuits and Systems, ISCAS'03, Vol. 3, pp. III-806---III-809, 25-28 May 2003.

[7] H. Yang, "Watermarking on Digital Audio Signals", *M.Eng. Thesis*, Nanyang Technological University, February 2001.

[8] A. Rosenfeld, "Connectivity in digital pictures", J. ACM, 17, (1), 1970, pp. 146-160.

A ROBUST AUDIO WATERMARKING SCHEME

Hyoung Joong Kim[*], *Taehoon Kim*[*], *In-Kwon Yeo*[+]

*Department of Control/Instrumentation Engineering
Kangwon National University
Chunchon 200-701, Korea

+Divison of Mathematics & Statistical Informatics
Chonbuk National University
Jeonju 561-756, Korea

ABSTRACT

Spread-spectrum watermarking scheme adds random noise ***w*** into the host signal ***x*** to generate watermarked signal **s**. Blind detector computes inner product <*w*, *s*> or its variants (i.e., correlation) and decides if the host signal contains embedded bits or not by its sign of the inner product. Such detection method may not be sufficiently robust in some applications due to high false-positive or false-negative rate. This paper introduces a novel method to reduce detection errors considerably by excluding outliers in the signal. First, this paper identifies the usefulness of the outliers. Second, this paper identifies the problem of outliers. Due to the vague definition of outlier, it was impossible to draw unique criterion for outlier. Thus, third, stable and robust definition of outlier is adopted. Informed watermarking scheme for spread-spectrum is exploited to reduce false-positive and false-negative errors. Such detectors can also enhance detection rate in other watermarking schemes. This scheme can enhance robustness of image watermarking and audio watermarking as well..

1. INTRODUCTION

Watermarking schemes embed watermark signal ***w*** into the host signal ***x***. The watermark signal ***w*** can be embedded with or without taking ***x*** into consideration. (Some researchers call the latter as blind embedding and the former as informed embedding [5].) The host signal ***x*** contains important information that may be exploited to achieve better watermarking. If the valuable information is not exploited, good watermark signal ***w*** should be designed to mitigate the adverse effects due to the host signal. Thus, there is no reason to exploit the host signal as side information.

So-called informed embedding schemes [2][4][5] [6] exploit the characteristics of host signal in the embedding phase. It is obviously clear that the host signal ***x*** can be used for embedding since it is available in the embedding phase. Thus, informed embedding is a logical concept.

However, ***x*** is not available to the detector. Thus, one may think detection with side information is impossible. In blind schemes, detection is achieved without recourse to the host signal ***x***. Note that imperceptibility requirement of watermarking scheme assumes $\boldsymbol{s} \approx \boldsymbol{x}$, which implies that the watermarked signal ***s*** is very much similar to ***x***. Thus, the detector can exploit ***s*** as side information of ***x***. This simple fact is a starting point in the design of an efficient informed detector.

Informed detection has been widely used especially in quantization-based watermarking schemes [2][4][6]. However, informed watermarking concept can be applicable to any watermarking schemes including spread-spectrum schemes, patchwork schemes, and so on. This paper introduces examples of informed watermarking for spread-spectrum schemes and shows how the error of spread-spectrum can be reduced.

Spread-spectrum watermarking scheme is widely used in audio, image, and video watermarking. This scheme is easy to implement. Spread-spectrum is robust against most attacks except jitter attacks. This scheme is the most common form of blind watermarking. A pseudo-random signal, ***w***, is modulated on the host signal and detection relies on a hypothesis test on the result of a correlation of the watermarked signal ***s*** with ***w***. Hence, the watermark signal ***w*** can be considered as a private key, which must be available to both the embedding and detecting modules.

Blind detector computes inner product <*w*, *s*> or its variants and decides embedded bits by the sign of the inner product. Such detection method may not be robust due to high false-positive or false-negative error. The error comes mostly from the outliers in ***s***, and consequently, ***x***. Thus, it is better to exclude outliers from correlation computing. This is the main topic of this paper.

In general, outliers are defined to be the sample values greater than 3σ or -3σ, where σ is the standard deviation of the sample values. However, due to the undesirable distribution, i.e., far from normal distribution, large standard deviation, high skewed distribution, new definition of outlier is needed. This paper describes the importance of outlier exclusion during both embedding and detection. In addition, this paper introduces good statistic for robust and stable definition of outlier. The scheme can be applied to image watermarking and audio watermarking as well to enhance robustness.

2. SPREAD_SPECTRUM WATERMARKING

Spread-spectrum scheme embeds $\boldsymbol{w}$ into the host signal $\boldsymbol{x}$. Embedding schemes include additive and multiplicative schemes. Additive embedding scheme embeds watermark signal $\boldsymbol{w}$ into host signal $\boldsymbol{x}$ to obtain watermarked signal $\boldsymbol{s}$ as follows:

$$s(n) = x(n) + \alpha w(n). \tag{1}$$

General assumption required for good detection is that the host signal $\boldsymbol{x}$ is flat and has a small variation [3]. The small variation means that audio samples are highly correlated with each other among neighbors. It also means that audio signal is rich in low-frequency components. However, small variance is not always the case. Some audio samples are rich in mid-frequency or high-frequency components. Even though there is no malicious attack, the correlation-based detection can be erroneous due to this characteristic of the host signal. These errors are called false-positive or false-negative errors. At least the error probability should be as low as 10^{-8}. Thus, efficient detection schemes are needed to reduce such errors. It is assumed that the watermarked signal $\boldsymbol{s}$ is contaminated by noise or malicious attack. The received signal is denoted by $\boldsymbol{r}$. Once the watermarked signal is modified, it is natural to assume that the watermark is also modified as $\hat{\boldsymbol{w}}$.

One possible solution for reducing false-positive or false-negative solution is by using denoising scheme. The denoising approach is to remove host signal as much as possible so that we compute $<\boldsymbol{w}, \hat{\boldsymbol{w}}>$ rather than $<\boldsymbol{w}, \boldsymbol{s}>$. Denoising schemes include low-pass filtering or Wiener filtering.

Non-blind scheme extracts $\hat{\boldsymbol{w}}$ by subtracting $\boldsymbol{x}$ from $\boldsymbol{r}$. The non-blind scheme computes correlation between watermark signal $\boldsymbol{w}$ and the extracted watermark signal $\hat{\boldsymbol{w}}$. Cox et al. [3] suggest that the entry of $\boldsymbol{w}$ is chosen independently according to $N(0,1)$ and the normalized correlation value should be larger than 6 (or six sigma in standard deviation) to make sure that $\boldsymbol{w}$ is similar to $\hat{\boldsymbol{w}}$. Statistically speaking, if $\hat{\boldsymbol{w}}$ is equivalent to $\boldsymbol{w}$, the inner product $<\boldsymbol{w}, \hat{\boldsymbol{w}}>$ is distributed as a chi-square distribution with the degree of freedom m where m is the number of entries of $\boldsymbol{w}$ and its expected value and variance are m and $2m$, respectively. Thus, the normalized correlation will be around $m^{1/2}$. It is extremely unlikely to have such high value unless they are almost the same signal because of the independence of $\boldsymbol{w}$.

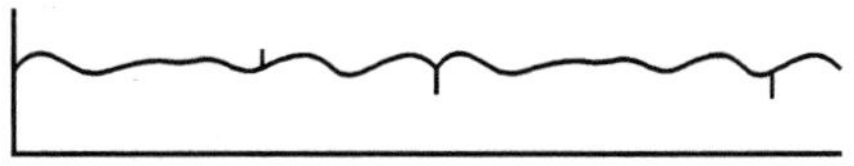

(a) Almost uniformly distributed signal model

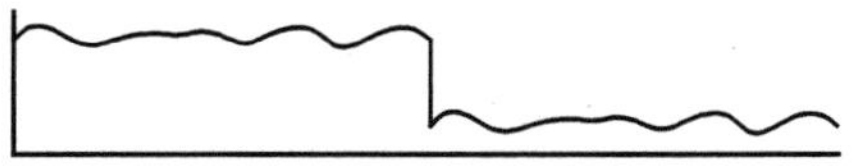

(b) Almost bimodal signal model

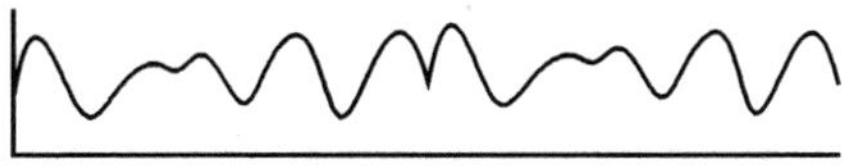

(c) Real signal model

Figure 1. A typical 1-D signal model

Blind scheme computes correlation between watermark signal $\boldsymbol{w}$ and watermarked signal $\boldsymbol{r}$ since detector does not have $\boldsymbol{x}$. Hartung and Girod [7] multiply $\boldsymbol{r}$ by $\boldsymbol{w}$ such that $<\boldsymbol{r}, \boldsymbol{w}> = \sum [x(n) + \alpha\hat{w}(n)]w(n)$ and detect watermarks by computing the sign of $<\boldsymbol{r}, \boldsymbol{w}>$. This scheme works well if $<\boldsymbol{x}, \boldsymbol{w}>$ vanishes. However, this is not always the case.

Therefore, denoising scheme, a method to remove $\boldsymbol{x}$ as much as possible has been introduced. One such denoising scheme is filtering out of $\boldsymbol{x}$ from $\boldsymbol{s}$ or $\boldsymbol{r}$ by low-pass filter [8]. Most of the filtering schemes exploit highly correlated neighbors. In other words, they assume that $x(n) \approx x(n+1)$ or $x(n) \approx x(n-1)$. Then, we can get

$$\begin{aligned} r(n) - r(n+1) &= [x(n) + \alpha\hat{w}(n)] - [x(n+1) + \alpha\hat{w}(n+1)] \\ &\approx \alpha\hat{w}(n) - \alpha\hat{w}(n+1). \end{aligned} \tag{2}$$

So long as Equation (2) holds, assuming that $\Sigma\hat{w}(n+1)w(n) \approx 0$, we can have

$$\begin{aligned} \Sigma[r(n) - r(n+1)]w(n) &= \Sigma[\alpha\hat{w}(n) - \alpha\hat{w}(n+1)]\, w(n) \\ &\approx \alpha\, \Sigma\hat{w}(n)w(n). \end{aligned}$$

Figure 1 shows some example signals. The signals can be modeled in various ways. Almost uniformly distributed signal model (a) has small variance. It is believed that this type of signal quite fits well into the model without any preprocessing. Our experiment shows that uniformly distributed model is more susceptible to errors. However, this model is most suited with low-pass filtered detection. The high error comes from high value of mean. The error can be reduced considerably by making its mean zero (see Figure 2). Mean zero and small standard deviation are the requirements of spread-spectrum watermarking.

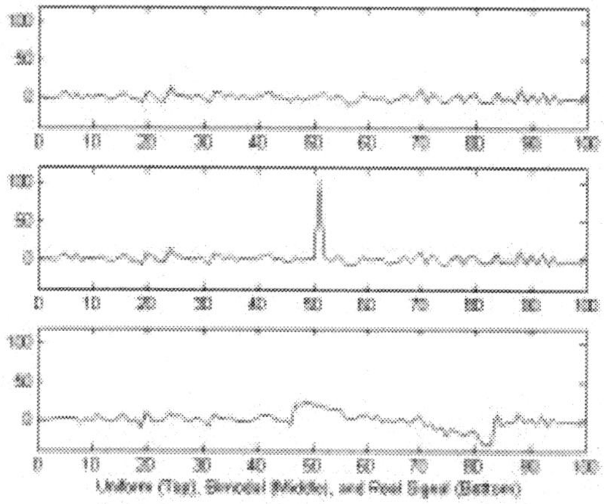

Figure 2. Low-pass filtered version of Figure 1

Note that bimodal model is better than the uniform case when there is no preprocessing. However, uniform case is the best when it is preprocessed. Interestingly enough, real signal model is better than bimodal case in the experiments. Our experiments give us two insights. First, low-pass filtered model is far better than blind detection without preprocessing. That is why denoising is required. Second, lengthy sequences are better, which is a well-known axiom in spread-spectrum watermarking. Since the expected value of the normalized correlation is around $m^{1/2}$ when $\hat{w}$ is similar to w, the larger m leads the larger probability that the normalized correlation is bigger than 6. Figure 2 is very much promising in terms of error rates with very short pseudo-random sequences. Even with a sequence with length 100, the error rates are less than 5% even with α=2.

3. INFORMED WATERMARKING WITHOUT OUTLIER

The performance of low-pass filtered spread-spectrum watermarking is very promising with short sequences. People want to hide more information into the same channel. The channel capacity is inversely proportional to the length of the sequence. The shorter the sequence, the larger the capacity. Our experiment shows that low-pass filtered sequences are not thoroughly free from errors. If possible, even shorter sequences will be preferred for higher capacity.

The solution to reducing errors to as low a level as possible can be found from informed watermarking. This paper delves into the informed watermarking scheme, which is more correctly described as communication with side information at the decoder. Decoder assumes $s \approx x$ in time-domain or $S \approx X$ in transform-domain, which is the side information itself. Figure 2 shows the low-pass filtered signal of **s**. The figure at the top shows the low-pass filtered signal of almost uniformly distributed signal. Its sample standard deviation is 4.96. The middle figure is that of bimodal signal. Its sample standard deviation is 11.50. The sample standard deviation of the real signal is 14.73. We see that the larger the standard deviation, the worse the detection rate. When the low-pass filtered signal is plain, we can say that $x(n) \approx x(n+1)$ and $\Sigma[r(n) - r(n+1)]w(n) \approx \alpha\, \Sigma\hat{w}(n)w(n)$. In two bottom plots of Figure 2, the variation of signal is big so that the effect of $\Sigma[x(n) - x(n+1)]w(n)$ still remains in calculating the correlation. That is the reason why the real signal is worse than bimodal case in terms of detection rate.

Note that the errors due to large variations can be remedied by excluding outliers. In case of bimodal model, the outliers are clearly identifiable: these are the samples lying between 50 and 53 (see Figure 2). Since the values are so dominantly large that they can affect the correlation too much, it is better to neglect them in correlation computing. Then, the errors in our experiment can be reduced to a level very close to zero in low-pass filtered model.

However, the real signal model is worse than the bimodal one in terms of error detection with excluded outliers. As is shown in Figure 1, there is no dominant pinnacle as in the bimodal case. Thus, there is no way to find an apparent cause of errors. However, we can define outliers which may cause the error. In this paper, the outlier is a sample whose value is larger than $k\sigma$, where k is a constant and σ is the standard deviation. Mean of the low-pass filtered sequence is assumed to be zero.

Our experiments show that the errors can be reduced considerably. Bimodal case is a simple one. Since there is just one pinnacle, by neglecting that sample or neglecting the outlier, we can get perfect detection. But real signals are rather complex. By neglecting samples larger than $k\sigma$, (k = 1, 2, 3), the detection error can be reduced. Spread-spectrum in time-domain is not robust. Thus, it is advisable to embed watermark signal in transform-domain. Embedding and detection in the transform-domain is different from that in the time-domain. During the embedding phase, we compute $\sigma(\boldsymbol{X})$, where $\boldsymbol{X}$ is the host signal in the transform-domain. We add watermark signal $\boldsymbol{W}$ in the transform-domain as follows:

$$\boldsymbol{S} = \boldsymbol{X} + \alpha\, \boldsymbol{W}. \qquad (3)$$

If the value of $|X(n)| > k\sigma$, then we assume that $W(n) = 0$ so that watermark signal is not added. During detection phase, we also compute $\sigma(\boldsymbol{S})$, where $\boldsymbol{S}$ is the watermarked signal in the transform-domain. If the value of $|S(n)| > k\sigma$, then we assume that $W(n) = 0$ so that correlation is not computed.

In the transform-domain, low-pass filtering is not feasible in general since the samples in the transform-domain have little correlation among neighbor samples. Note that the standard deviation of signal in frequency domain in our experiment is 154.30, which is quite large compared to that in the time-domain. Moreover, the Probability Density Function (PDF) is highly skewed. This scheme works well in the highly skewed model. Our experiment shows that detection error with 0.1σ is around 0.002 and is smallest out of the results. The reason why a nonsense quantity 0.1σ is chosen as a criterion of determining outliers is that the standard deviation is so large and the probability density function (PDF) is highly skewed. From our experiments we can conclude that outliers depend on host signals. It means that it is not easy to define optimal criterion for outliers. Thus, the outliers should be defined very carefully.

Table 1. Detection error with noise signal length 100

Strength	Outlier	Time-Domain Filtered Signal	Transform-Domain Signal
α=2	$\|M_i\|>3$	0.005	0.010

Let μ and σ be a sample mean and a sample standard deviation, respectively. Then, an outlier is usually defined as the value being more than 3σ far from μ. However, since both the mean and the standard deviation are not robust and sensitive to outliers, sometimes it makes an unexpected result and gives a difficulty in interpretation to use a measurement based on those quantities when extreme values exist in the data. Indeed, several statistics based on robust quantities such as median are found for testing outliers [1]. We consider the following statistic as a measurement of detecting outliers

$$M_i = \frac{2(x_i - x_{50\%})}{x_{84\%} - x_{16\%}}, \qquad (3)$$

where $x_{50\%}$ stands for the sample median and $x_{84\%}$ and $x_{16\%}$ denote the 84 percentile and the 16 percentile, respectively. We define the observation having $|M_i| \geq 3$ as an outlier. When the data are drawn from a normal distribution, we see that $x_{84\%}-x_{16\%} \approx 2\sigma$, $x_{50\%} \approx \mu$, and $M_i \approx (x_i-\mu)/\sigma$, which is a widely-used statistic for detecting an outlier. Table 1 shows that the definition of outlier is effective and robust. In other words, the outlier definition is independent of the values of variance and mean of the samples. Moreover, Table 1 shows that outlier extraction based on the statistic in Equation (3) can give very high chance of reducing errors. The errors should be reduced by other means such as introducing error-correcting codes or more novel watermarking schemes.

4. CONCLUSION

Spread-spectrum watermarking scheme adds random noise $\boldsymbol{w}$ into the host signal $\boldsymbol{x}$ to generate watermarked signal $\boldsymbol{s}$. Blind detector computes inner product $<\boldsymbol{w}, \boldsymbol{s}>$ and decides embedded bits by its sign of the inner product. Such detection method is not robust due to high false-positive or false-negative rate. This paper introduces a novel method to reduce detection errors considerably by excluding outliers in the signal. Stable and robust definition of outlier is adopted. Informed watermarking scheme for spread-spectrum is exploited to reduce false-positive and false-negative errors. Effectiveness of the proposed scheme is demonstrated by thorough experimentation. This scheme works well in time-domain and frequency-domain as well. Such detectors can also enhance detection rate in other watermarking schemes.

Our main concern is focused on the simple question: How can we reduce detection errors with simple statistic? The answer is quite simple: Exclude outliers. Statistic in Equation (3) is effective since it enables us to define outliers automatically no matter what sample variance and sample distribution. Thus, robustness against various attacks is beyond the scope of this paper.

5. REFERENCES

[1]. V. Barnett, and T. Lewis, *Outliers in Statistical Data*, John Wiley & Sons, New York, 1994.

[2]. B. Chen, and G.W. Wornell, "Achievable performance of digital watermarking systems," *Proc. IEEE Int. Conf. Multimedia Computing and Systems*, pp. 13-18, 1999.

[3]. I.J. Cox, J. Kilian, T. Leighton, and T. Shamoon, "Secure spread spectrum satermarking for multimedia," *IEEE Tr. on Image Proc.* vol. 6, pp. 1673-1687, 1997.

[4]. I. J. Cox, M. L. Miller, and A. L. McKellips, "Watermarking as communications with side information," *Proc. of the IEEE*, vol. 87, pp. 1127-1141, 1999.

[5]. I. J. Cox, M. L. Miller, and J. Bloom, *Digital Watermarking*, Morgan Kaufmann Publishers, 2001.

[6]. J. Eggers, and B. Girod, *Informed Watermarking*, Kluwer Academic Publishers, 2003.

[7]. F. Hartung, and B. Girod, "Watermarking of uncompressed and compressed video," *Signal Proc.*, vol. 66, no. 3, pp. 283-301, 1998.

[8]. E.T. Lin, C.I. Podilchuk, and E. J. Delp, "Detection of image alterations using semi-fragile watermarks," *Proc. SPIE Int. Conf. Security and Watermarking of Multimedia Contents*, vol. 3971, 2000.

SLANT TRANSFORM WATERMARKING FOR TEXTURED IMAGES

Anthony T. S. Ho, Xunzhan Zhu, Yong Liang Guan and Pina Marziliano

Nanyang Technological University
School of Electrical and Electronic Engineering
Singapore, 639798

ABSTRACT

In this paper, we propose a digital watermarking algorithm based on the Slant transform for the copyright protection of complex textured images. A lot of images in practical applications, such as remote sensing and medical archiving, contain complex textures. With more AC energies concentrating in the high frequency components, images with complex textures usually suffer more from common image processing operations than smooth images do. Therefore, watermarking robustness is more an issue for textured images though it is commonly recognized that highly textured images provide more capacity for data hiding without introducing too much degradation to the visual quality of original images. Experimental results showed that our Slant transform watermarking provides significant advantage for watermark insertion and retrieval for images with complex textures such as satellite images. An analytical comparative study on the performance of the Slant transform adapting our earlier watermarking schemes for fast Hadamard transform will be performed based on its robustness against various Stirmark and Checkmark attacks.

1. INTRODUCTION

Copyright protection and authentication of digital multimedia contents is increasingly becoming more and more important in parallel with the tremendous development of the Internet. To counter this growing information security problem of illegal distribution and counterfeiting, much research effort has been invested in digital watermarking in the past few years. A lot of images in practical applications, such as remote sensing and medical archiving, contain complex textures. Although it is commonly recognized that highly textured images provide more capacity for data hiding without introducing too much degradation to the visual quality of original images, watermarking robustness is more an issue for these images for the reason that usually textured images suffer more from common image processing operations than smooth images.

In this paper, we propose an improved digital watermarking algorithm based on the Slant transform for the copyright protection of textured images. The Slant transform has been applied to many image processing applications, such as transform coding and image restoration [1, 2, 3] along with many fast algorithms have been proposed for it [4]. The Slant transform has many similar properties to the Walsh-Hadamard transform. Designed to match basis vectors to areas of constant luminance slope, the Slant transform is good for compact energy in "smooth" images. However, in terms of watermarking, it works better for textured images, for the energy spread becomes a significant advantage in watermarking and offers a good spread of middle to higher frequencies with significant energies for robust information hiding. This has brought us to consider protecting the copyright of textured images using the Slant transform watermarking.

The proposed algorithm will be evaluated based on its robustness against various Stirmark [5] and Checkmark [6] attacks using test images with different textures. The performance comparison results of the Slant transform for complex textured image watermarking against Hadamard transform will also be presented. The rest of this paper is organized as follows. In Section 2, the Slant transform is briefly reviewed. Then, our proposed watermarking method is elaborated in Section 3. Simulation results are presented in Section 4, these include comparison of the Slant transform scheme to the Hadamard transform scheme in [7, 8]. Finally, the conclusions are given in Section 5.

2. THE SLANT TRANSFORM

Let $\mathbf{U}$ be the original image of size $N \times N$, its 2-D Slant transform is given by

$$\mathbf{V} = \mathbf{S}_N \mathbf{U}\, \mathbf{S}_N^T \tag{1}$$

where $\mathbf{S}_N$ is the $N \times N$ unitary Slant matrix. The inverse transformation to recover $\mathbf{U}$ from the transform components matrix $\mathbf{V}$ is given by

$$\mathbf{U} = \mathbf{S}_N^T \mathbf{V}\, \mathbf{S}_N. \tag{2}$$

The Slant transform is a member of the orthogonal transforms. It has a constant function for the first row, and has a

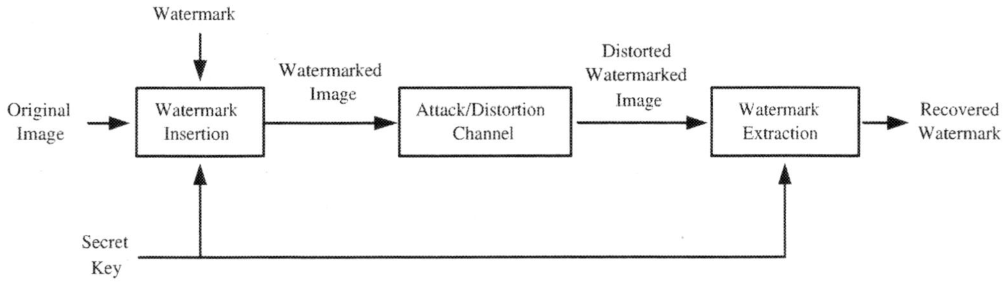

Fig. 1. Generic block diagram of a blind watermarking system.

second row which is a linear (slant) function of the column index. The matrices are formed by an iterative construction that exhibits the matrices as products of sparse matrices, which in turn leads to a fast transform algorithm.

The Slant transform matrix of order two is given by

$$\mathbf{S}_2 = \frac{1}{\sqrt{2}} \begin{bmatrix} 1 & 1 \\ 1 & -1 \end{bmatrix}. \tag{3}$$

The Slant matrix of order four is obtained by the operation

$$\mathbf{S}_4 = \frac{1}{\sqrt{2}} \left[\begin{array}{cc|cc} 1 & 0 & 1 & 0 \\ a_4 & b_4 & -a_4 & b_4 \\ \hline 0 & 1 & 0 & -1 \\ -b_4 & a_4 & b_4 & a_4 \end{array}\right] \bullet \left[\begin{array}{c|c} \mathbf{S}_2 & \mathbf{0} \\ \hline \mathbf{0} & \mathbf{S}_2 \end{array}\right] \tag{4}$$

where a_4 and b_4 are scaling constants. The orthonormality conditions lead to:

$$a_4 = 2b_4 \text{ and } b_4 = \frac{1}{\sqrt{5}}.$$

Equation (4) can be generalized to give the Slant matrix of order N in terms of the Slant matrix of order $N/2$ by the following recursive relation:

$$\mathbf{S}_N = \frac{1}{\sqrt{2}} \left[\begin{array}{cc|c|cc|c} 1 & 0 & \mathbf{0} & 1 & 0 & \mathbf{0} \\ a_N & b_N & & -a_N & b_N & \\ \hline \multicolumn{2}{c|}{\mathbf{0}} & \mathbf{I}_{\frac{N}{2}-2} & \multicolumn{2}{c|}{\mathbf{0}} & \mathbf{I}_{\frac{N}{2}-2} \\ \hline 0 & 1 & \mathbf{0} & 0 & -1 & \mathbf{0} \\ -b_N & a_N & & b_N & a_N & \\ \hline \multicolumn{2}{c|}{\mathbf{0}} & \mathbf{I}_{\frac{N}{2}-2} & \multicolumn{2}{c|}{\mathbf{0}} & -\mathbf{I}_{\frac{N}{2}-2} \end{array}\right] \bullet \left[\begin{array}{c|c} \mathbf{S}_{\frac{N}{2}} & \mathbf{0} \\ \hline \mathbf{0} & \mathbf{S}_{\frac{N}{2}} \end{array}\right]. \tag{5}$$

The matrix $\mathbf{I}_{\frac{N}{2}-2}$ is the identity matrix of dimension $\frac{N}{2} - 2$. The constants a_N, b_N, may be computed by the formulas:

$$a_{2N} = \left(\frac{3N^2}{4N^2-1}\right)^{1/2} \text{ and } b_{2N} = \left(\frac{N^2-1}{4N^2-1}\right)^{1/2}. \tag{6}$$

3. WATERMARKING IN SLANT TRANSFORM DOMAIN

A generic block diagram of a blind watermarking system [?] is shown in Figure 1. The original image is not necessary at the watermark extraction stage. This refers to a "blind" watermarking process. A visually recognizable pattern is embedded by modifying the Slant transform coefficients of relevant sub-blocks of the host image. The detailed image-embedded watermark insertion and extraction algorithm are discussed in this section.

Copyright information in the form of a trademark or logo can be created as a pattern for watermarking. In our experiment, a grayscale image of size 64×64 is used as the watermark. The watermark embedding process is shown in Figure 2. We adapt a similar image-embedded watermark insertion algorithm as in [7], while using the Slant transform domain instead of the Hadamard transform domain. The algorithm can be described as follows.

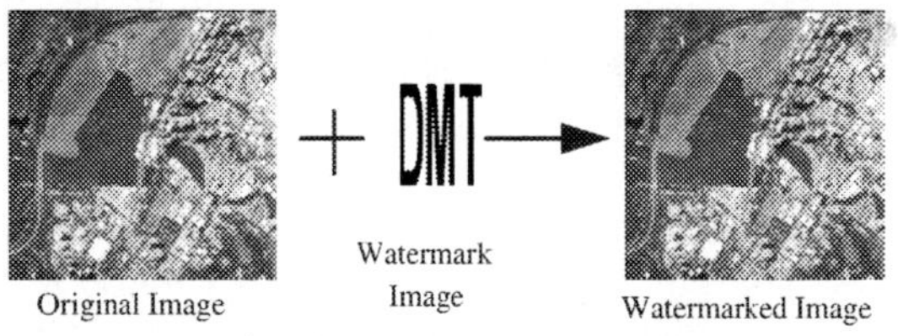

Fig. 2. Image-embedded watermark embedding process.

The watermark image, $\mathbf{W}$, is first transformed into a set of Slant transform coefficients by Equation (1). A Slant transform matrix of order 64 is applied to this image, and then a Slant transform components matrix is obtained. The DC component is stored in the secret key file and the AC components are used for watermark embedding.

Let the original image be $\mathbf{X}$. Similar to the algorithm used in [7, 8], it is decomposed into a set of non-overlapped sub-blocks. An pseudo-random number generator is used to select a certain number of sub-blocks for watermark embedding, whose initial seed is also contained in the secret key file. In every selected sub-block, sixteen middle and high frequency coefficients are used for consequent modulation. The coefficient selection affects the performance of the watermarking scheme significantly. The high frequency components are relatively vulnerable to compression operations, while the low frequency components must be retained for visual quality of the watermarked image. Therefore, most existing watermarking schemes choose to embed the watermark into the middle frequency band. In our scheme, embedding locations as shown in Figure 3 are adopted. These locations are selected based on a tradeoff analysis between robustness and data integrity.

Let the watermark Slant transform coefficients denoted by w_i. The AC coefficients of Slant transformed original image sub-blocks, before and after inserting watermark are denoted by x_i and x_i^*, respectively, where $i \in (0, n]$ and n is the number of the watermark coefficients to be embedded into every sub-block, which is set to 16 in our experiment. The embedding formula is given as follows

$$x_i^* = \alpha w_i \tag{7}$$

where α is the watermark strength factor that controls the tradeoff between visual quality of the watermarked image and robustness of the watermarking scheme.

After embedding, the original coefficient x_i is replaced by x_i^* and a new 8×8 matrix of Slant transform components of image sub-block is obtained. The inverse Slant transform is then applied to the matrix using Equation (2) to obtain the luminance matrix of the watermarked image sub-block. After performing the watermark insertion for all the selected sub-blocks of the original image, a watermarked image, $\hat{\mathbf{X}}$, is obtained. At the same time, as indicated earlier, the secret key file has been generated for subsequent decoding.

In watermark detection, the positions of the sub-blocks with watermark embedded are computed using the initial state number of the pseudo-random number generator that is contained in the key file. All the selected sub-blocks are Slant transformed. Let these coefficients denoted by $\hat{x_i^*}$ and the retrieved watermark Slant transform coefficients by $\hat{w}_i$, where $i \in (0, n]$ and n is the number of the watermarked coefficients in every sub-block. The watermark extraction formula is given by

$$\hat{w}_i = \frac{\hat{x_i^*}}{\alpha}. \tag{8}$$

The above extracted AC coefficients and the DC component stored in the key file are rearranged into a Slant transform components matrix. The extracted watermark image, $\hat{\mathbf{W}}$, is then obtained by an inverse Slant transform using Equation (2).

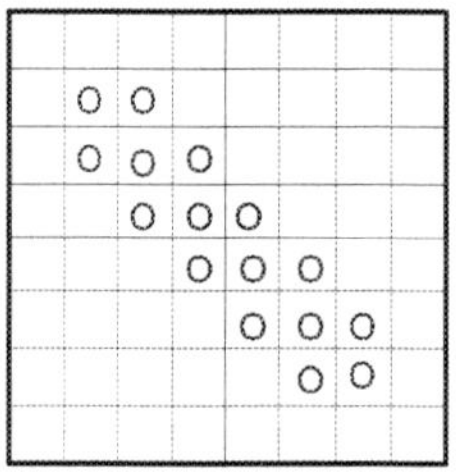

Fig. 3. Embedding locations in one selected Slant transform components block.

4. SIMULATION RESULTS

We use two 512×512 gray-scale satellite images with different distinct textures to test our algorithm. The original and watermarked images are shown in Figure 4. Results show that there are no perceptually visible degradations on the watermarked images with a PSNR of 37.43 dB for *Singapore* and 40.65 dB for *Dolomites*.

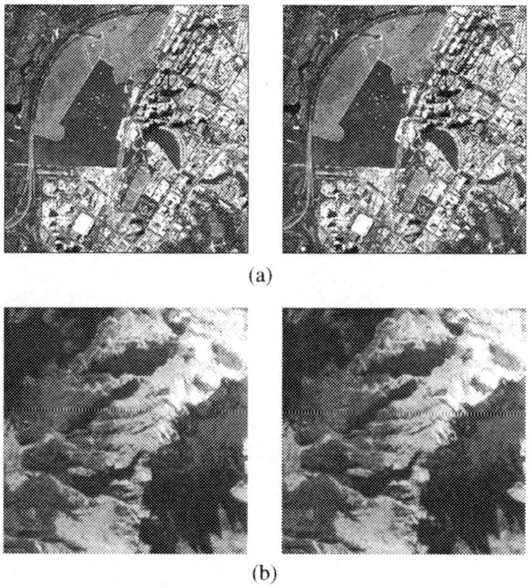

Fig. 4. Original images (left) and watermarked images (right): (a) *Singapore*; (b) *Dolomites*.

The Slant transform watermarking scheme is benchmarked using Stirmark 4 [5] and Checkmark [6]. Sample results of the *Singapore* are reported in Table 1. We can see that the Slant transform watermarking is robust to additive noise and common image processing techniques.

We also compare the robustness of the Slant scheme to the Hadamard scheme in [7, 8] against the typical JPEG compression attack. The results are shown in Figures 5. We can see that, for the two typical textured images, the Slant transform survive JPEG compression far better than the Hadamard transform. Since the Slant transform is designed to match basis vectors to areas of constant lumi-

Benchmarking Operations	Extracted Watermark	Correlation Coefficient
Gaussion filtering	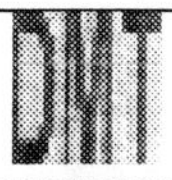	0.9640
Wiener filtering	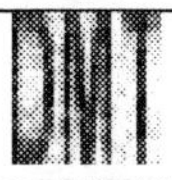	0.9500
JPEG compression	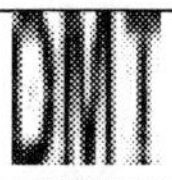	0.9273
Wavelet compression	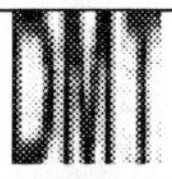	0.9446
Sharpening		0.9529
Additive noise		0.9514

Table 1. Sample benchmarking results for *Singapore* image.

nance slope, it is good for compact energy in "smooth" images. However, in terms of watermarking, it works better for textured images, for there would be more Slant transform bands remained intact due to the energy spread to more middle or low frequency bands. It follow therefore that the Slant transform is very robust when applied for watermarking textured images, which usually suffer in common image processing operations such as compression.

5. CONCLUSIONS

This paper has presented a robust watermarking technique for copyright protection of textured images based on the Slant transform. The embedding and extraction processes have been described in details. The Slant transform has more useful middle and high frequency bands than several high gain transforms, especially for images with complex textures such as satellite images, which provides a good "hidden" space for digital watermarking. This research is based on our previous work on the Hadamard transform [7, 8]. Comparison results between the Slant transform and the Hadamard transform showed the significant advantage of the Slant transform for robust watermarking of complex textured images.

6. REFERENCES

[1] R. J. Clarke, *Transform Coding of Image*, Academic Press, New York, 1985.

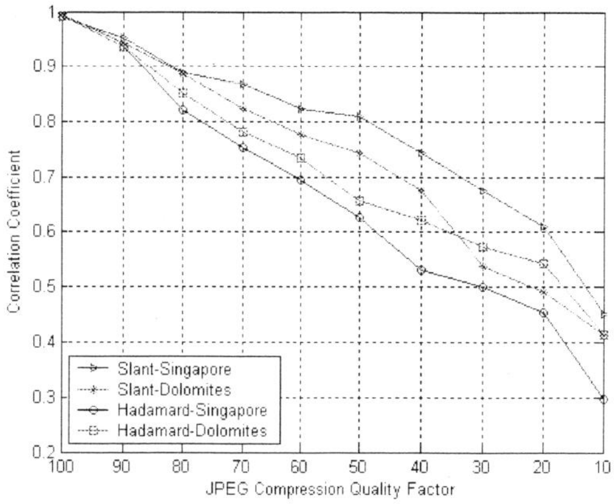

Fig. 5. Performance comparisons between Slant transform and Hadamard transform.

[2] W. K. Pratt, L. R. Welch, and W. H. Chen, "Slant transform for image coding," *Proc. Symp. Appl. Walsh Function*, Mar. 1972.

[3] W. K. Pratt, W. H. Chen, and L. R. Welch, "Slant transform coding," *IEEE Trans. Commun.*, vol. COM-22, Aug. 1974.

[4] Z. D. Wang, "New algorithm for the slant transform," *IEEE Trans. Pattern Anal. Machine Intell.*, vol. pami-4, pp. 551–555, Dec. 1982.

[5] http://www.cl.cam.ac.uk/ fapp2/watermarking/stirmark/.

[6] http://watermarking.unige.ch/Checkmark/index.html.

[7] A. T. S. Ho, J. Shen, and S. H. Tan, "Digital image-in-image watermarking technique for copyright protection of satellite images using the fast Hadamard transform," in *IGARSS'02*, Toronto,Canada, June 2002.

[8] A. T. S. Ho, J. Shen, and S. H. Tan, "A robust digital image-in-image watermarking technique for satellite images," in *ICICS'01*, Singapore, Oct. 2001.

A REVIEW OF VIDEO REGISTRATION METHODS FOR WATERMARK DETECTION IN DIGITAL CINEMA APPLICATIONS

Hui Cheng

Sarnoff Corporation
201 Washington Road, Princeton, NJ 08540
hcheng@sarnoff.com

ABSTRACT

To deter content theft in digital cinema applications, forensic watermarking has been proposed to track a digital content and to determine the time and the place where the theft occurs. However, to be effective in digital cinema applications, a forensic watermark must be invisible and secure. The detection of the forensic watermark must also be robust to attacks and processing, such as camcorder capture, low bit rate compression, D/A/D conversion, cropping, resizing and frame rate conversion. To achieve the required robustness of detection, many proposed forensic watermarking algorithms adopt reference-based (informed or non blind) watermark detection where the original video is used to assist the detection. To use the original video, one must first register it to the candidate video, where the forensic watermark needs to be extracted. In this paper, we will review a number of video registration algorithms proposed for forensic watermark detection in digital cinema applications. Among all the proposed registration algorithms, we will discuss and compare the following three algorithms: (1) A semi-automatic temporal, spatial and volumetric video registration algorithm proposed by IRISA; (2) A key frame based temporal alignment algorithm proposed by Universite Catholique de Louvain and (3) A spatial, temporal and volumetric video registration algorithm proposed by Sarnoff.

1. INTRODUCTION

Content theft due to piracy is a major concern of distribution of high-value copyrighted materials, e.g. movies, digitally. Although companies, such as Boeing [1], has established an end-to-end secure content delivery system where digital movie files maintain compressed and encrypted from the post-production facility to the moment of presentation in movie theaters. Possible theft of the digital content can still occur, either by the theft of film prints directly from studios and by camcorder capture in the theater. Most DRM (Digital Rights Management) approaches have little control over such pirating. Therefore, a content owner needs forensic tools to determine where and when the theft occurs and to identify the party who licensed material and was responsible for preventing such pirating activities.

As stated by SMPTE DC28.4 [2], the requirements for such *download watermark* and *exhibition watermark* are (1) high fidelity, (2) robust to the combination of capture and compression, (3) secure against unauthorized removal and embedding, (4) low latency for embedding. The above four requirements pose significant challenges to watermark design, embedding and detection. However, the detection of forensic watermarks can rely on the use of a reference video (i.e. the original, unmarked video) and it can be an expensive non-real-time process. In a survey conducted by Boeing [1], all four vendors for forensic watermarking solutions agree that the use of originals during watermark detection are necessary for serious geometric distortion and can improve the security and reliability of a forensic watermarking system.

In order to use the original for watermark detection, one must first register the original video (also referred as the reference video) and the candidate video (the video under the investigation and from which the watermark needs to be extracted). A candidate video is a processed version of the watermarked reference video. It is sometimes generated by capturing a watermarked video displayed in a theater, and then enhancing and/or compressing it. Therefore, significant misalignments exist between the reference and the candidate videos. These misalignments are, in general, a combination of spatial, temporal and valuemetric (e.g. histogram) misalignments. Spatial misalignment is the result of spatial manipulation of a video sequence, such as warping, cropping and resizing (e.g. capturing a movie with 2.35:1 aspect ratio using a camcorder with 4:3 aspect ratio). The main causes of temporal misalignment are the change of temporal resolution, such as frame rate conversion (e.q. 3-2 pull down), and frame dropping or frame repeat used by video compression (i.e. MPEG-4). Video capturing is another cause, because displaying and capturing generally are not synchronized and operate at different frame rates. In addition, reference and candidate videos have different color histograms. This is the result of video processing, such as compression, filtering or gamma change. It can also be the result of white balance or automatic gain control (AGC) applied in camcorder capture. To correct these misalignments, spatial, temporal and histogram registrations are needed. Spatial [3][4], temporal [5][6] and histogram [7] registrations have been studied separately. However, these approaches are generally not designed for watermark detection and there is not an integrated approach for spatial, temporal and valuemetric video registration.

In this paper, we will review and compare three video registration algorithms proposed for watermark detection in digital cinema application. They are (1) the spatial and temporal luminance correction proposed by IRISA [8]; (2) the key frame based temporal alignment proposed by Universite Catholique de Louvian (UCL) [9] and (3) the spatial, temporal and histogram video registration algorithm [10] proposed by Sarnoff in their iTrace system [11].

2. IRISA VIDEO REGISTRATION ALGORITHM

The video registration proposed by IRISA [8] first registers the reference and the candidate sequences temporally by comparing the average luminance variation[1]. The two sequences registered temporally are then spatially aligned using a semi-automatic spatial registration. Finally, spatial and temporal luminance change is corrected by jointly estimating the spatial and temporal luminance transfer function.

The spatial registration proposed by IRISA is semi-automatic. First, feature points are extracted automatically using the Harris detector. The response of Harris detector at the pixel (x,y), $E(x,y)$ is defined as

$$E(x,y)=\sum_{u,v}\omega_{u,v}\left|I(x+u,y+v)-I(x,y)\right|^2,$$

where $I(x,y)$ is the luminance at the pixel (x,y) and $\omega_{u,v}$ is a Gaussian filter. The feature points are selected as those local maxima of $E(x,y)$ whose values are larger than a given threshold. Then, an operator selects a subset of the feature points and establishes the correspondence among them. Finally, the spatial transformation is computed through the set of feature points and their correspondence determined by the operator.

IRISA also proposed an approach for compensating spatial and temporal luminance variation due to projection and AGC. Denote the t-th reference frame as $I(x,y,t)$, the t-th candidate frame as $J(x,y,t)$ and the watermark as W. The luminance change caused by display and camcorder AGC is modeled as

$$J(x,y,t)=\alpha(t)*\lambda(x,y)*\left[I(x,y,t)+W\right],$$

where $\alpha(t)$ represents the temporal-varying gain resulting from AGC and $\lambda(x,y)$ is the spatial-varying gain resulting from the display, e.g. the lens, the screen and the ambient condition. Let $q(x,y,t)=J(x,y,t)/I(x,y,t)\approx\alpha(t)*\lambda(x,y)$ and define

$$\rho(x.y)\equiv\sum_{t}q(x,y,t)=\sum_{t}\left[\alpha(t)*\lambda(x,y)\right]=M_\alpha*\lambda(x,y)$$

$$\sigma(x.y)\equiv\sum_{x,y}q(x,y,t)=\sum_{x,y}\left[\alpha(t)*\lambda(x,y)\right]=M_\lambda*\alpha(t)$$

$$\sum_{x,y,t}q(x,y,t)=\sum_{x,y,t}\left[\alpha(t)*\lambda(x,y)\right]=M_\alpha*M_\lambda$$

Then, the estimate of $I(x,y,t)+W$, $\hat{J}(x,y,t)$, is computed as

$$\hat{J}(x,y,t)=\frac{J(x,y,t)*M_\alpha*M_\lambda}{\rho(x,y)*\sigma(t)}.$$

3. UCL KEY FRAME BASED TEMPORAL REGISTRATION

The temporal video registration algorithm proposed by Universite Catholique de Louvain, Belgium [9] aligns two spatially and temporally misaligned video sequences temporally by establishing a correspondence among key-frames of the two sequences. The correspondence among frames that are not key frames is computed by interpolation of the temporal correspondence of nearest key-frames.

In UCL's approach, a frame is declared as a key frame if the histogram of its luminance component is sufficiently different from the luminance histogram of the previous frame. Let $H(t,j)$ be the value of the j-th bin of a 64-bins luminance histogram of frame t. Then, the different between $H(t)$ and $H(t-1)$ is defined as $d(t,t-1)$.

$$d(t,t-1)=\sum_{j=1}^{64}\left|H(t,j)-H(t-1,j)\right|$$

Frame t is a key frame, if $d(t,t-1)$ is larger than a threshold T. The threshold used in the implement is adaptive both locally and globally to achieve reliable and robust key frame detection.

After key frames are extracted from both sequences, the indices of the key frames are matched for temporal alignment. However, since the key frame detection does not guarantee to find the same key frames in both sequences, the algorithm will only match a subset of the key frames detected. The number of key frames that can be omitted is a parameter set by the user and it is denoted as N_o. Let $s_o(i,k)$ and $s_c(i,k)$ be the frame index of the i-th key frame of the k-th processing window of the original and the candidate sequence, respectively. The temporal misalignment is modeled as a frame shift plus a frame rate change. That is

$$s_Q(i,k)=\beta_1(k)s_c(j)+\beta_0(k).$$

The two parameters $\beta_0(k)$ and $\beta_1(k)$ are assumed to be constant over each processing window and slow varying as a function of k. Denote the l-th possible match from the candidate sequence to the original sequence among the key frames of the k-th processing window as $\alpha_{l,k}(i)$. That is the original frame i is matched to the candidate frame $\alpha_{l,k}(i)$. Then, the temporal registration is formulated as a minimization problem:

$$\alpha_{l^*,k}(i)=\arg\min_{l}E_1(l,k)+E_2(l,k)$$

where $E_1(l,k)$ is the model error and $E_2(l,k)$ penalizes large changes of $\beta_0(k)$ and $\beta_1(k)$.

$$E_1(l,k)=\min_{\beta_0(k),\beta_1(k)}\sum_{i=1}^{K-K_o}\left[s_o(i,k)-\beta_1(k)s_c(\alpha_{l,k}(i))-\beta_0(k)\right]^2$$

$$E_2(l,k)=\left[\beta_0(k)-\beta_0(k-1)\right]^2+\left[\beta_1(k)-\beta_1(k-1)\right]^2$$

When generating all possible matches, UCL assumes that the orders of frames or shots are the same among original and candidate sequences. In addition, a match is only allowed if the signs of luminance changes of the two key frames, one from the original sequence and the other from the candidate sequence are the same. But the image contents are not used as part of the temporal matching cost. UCL also has a special treatment for detecting frame drop and correcting it on the fly.

4. SARNOFF SPATIAL TEMPORAL AND HISTOGRAM (STH) VIDEO REGISTRATION

Sarnoff also developed a spatial, temporal and histogram (STH) registration algorithm [10] in their forensic watermarking system

[1] No detailed information of temporal registration is given in the paper.

for digital cinema applications[2] [11]. They first model spatial, temporal and histogram misalignments in videos, from which forensic watermarks need to be extracted. Denote the i-th reference frames (or field) as I_i and the j-th candidate frames as J_j. Then, the three types of misalignments are modeled as follows.

4.1 Sarnoff Temporal Misalignment Model

When the rate of capture is larger than half the rate of display, temporal misalignment can be modeled using a 2-frame integration model, $\varphi_t(\cdot)$. That is,

$$J_j \equiv \varphi_t(I_i, I_{i-1}; \lambda_{j,i}) = \lambda_{j,\alpha(j)} \cdot I_{\alpha(j)} + (1-\lambda_{j,\alpha(j)}) \cdot I_{\alpha(j)-1}, \quad (1)$$

where $\alpha(j)$ is the matching index that maps a candidate video frame to a reference frame. $\lambda_{j,\alpha(j)}$, $0<\lambda_{j,\alpha(j)} \le 1$, is the weight of $I_{\alpha(j)}$, and it can be estimated efficiently using a closed-form formula [10].

The above model represents many widely used temporal operations, such as frame drop/repeat in MPEG-4 or frame rate conversion (e.g. 3-2 pull-down). It is also a good model for video capture. Since capture and display are not synchronized, most captured frames are a linear combination of two consecutive displayed frames. In this case, $\lambda_{j,i}$ is the percentage of exposure of I_i during the capture of J_j.

4.2 Model of Histogram Misalignment

Modern camcorders are sophisticated video recording systems. In order to produce visually pleasing videos, a set of algorithms is used to improve the appearance of a captured video. Some of these algorithms will alter the RGB histograms of a capture video. The most important one is Automatic Gain Control (AGC). Depending on the average luminance, AGC will adjust the sensitivity of the camcorder by applying a gain to the captured video signal. In addition, White Balance that determines what color is white also affects a histogram. Finally, the end-to-end gamma from display to capture may not be unity. In this paper, we will only correct the histogram of the luminance because the forensic watermark only modifies the luminance component. We represent histogram transformation using a table look-up, denoted as $\varphi_h(\cdot)$. For each frame, $\varphi_h(\cdot)$ has 256 parameters, which map gray levels in a candidate video frame, to those in the reference frame.

4.3 Model of Spatial Misalignment

Many factors can lead to spatial distortion in captured video. First, the camcorder may not be placed on the optical axis of the projector. In addition, each capture may use different zoom or crop parameters. Finally, the camcorder may not be completely stationary during the capturing process. Therefore, a captured video is usually a perspective projection of the original video. For this application, we further assume that the transformation is plane perspective, and eliminate the dependency on the depth.

4.4 STH Video Registration

Let $\varphi_s(\cdot)$, $\varphi_t(\cdot)$ and $\varphi_h(\cdot)$ be temporal, histogram and spatial transformations, respectively. The combined transformation, denoted as $\varphi(\cdot)$, is $\varphi_t(\varphi_h(\varphi_s(\cdot)))$. Although the above composition depends on the order of how these three transformations are combined, the difference among different compositions are small and can be ignored. For captured frame j, we denote the parameters of the histogram transformation $\varphi_h(\cdot)$ as γ_j^h and the parameters of the plane perspective transformation $\varphi_s(\cdot)$ as γ_j^s. Also, for simplicity, we denote all parameters of $\varphi(\cdot)$, except the mapping index $\alpha(j)$, as γ_j. That is, $\gamma_j = [\lambda_{j,\alpha(j)}, \gamma_j^h, \gamma_j^s]$.

Using the above model, the registration can be formulated as a minimization of a matching cost. That is, given the reference and the candidate videos, estimate the matching indices, $\alpha(j)$, and model parameters, γ_j, that minimize the distortion between J_j and the model prediction of J_j. However, the minimization is subject to a causal constraint on $\alpha(j)$. That is, for any given j_1 and j_2, if $j_1 < j_2$, then $\alpha(j_1) \le \alpha(j_2)$. The causal constraint requires the same temporal ordering among the candidate video as the reference one. This is enforced by most video processing algorithms. In the case of video capture, it means that no frames displayed in the past can be captured in the future. In addition, in this paper, we adopt the mean squared error as the distortion measure. Therefore, the registration parameters $[\gamma^*(j), \alpha^*(j)]$ are computed as

$$[\gamma^*(j), \alpha^*(j)] = \underset{\substack{[\gamma(0),\cdots,\gamma(M),\\ \alpha(0)\le\cdots\le\alpha(M)]}}{\arg\min} \sum_{j=0}^{M} \left\| J_j - \varphi[I_{\alpha(j)-1}, I_{\alpha(j)}; \gamma_j] \right\|^2 \quad (2)$$

where M is the number of captured frames, and $\gamma(j)$ is γ_j. Since $\gamma(j)$ for $j = 0,\cdots,M$ can be optimized independently of each other, and there is only causal dependency among $\alpha(j)$, the optimization defined in (2) can be solved using dynamic programming [12].

To solve (2) using dynamic programming, we first partition the minimization into *stages* according to the index of the candidate frame, j. The *state* for each stage is the reference frame index, denoted as i. In a grid defined by stages and states, $\alpha(j)$ defines a mapping from stages to states. We call this mapping a path from one stage to another. Therefore, the solution of (2) is a path from stage 0 to stage M that has the minimal accumulated mean squared error. However, because of the causal constraint, $\alpha(0) \le \alpha(1) \le \cdots \le \alpha(M)$, the solution to (2) can only be a feasible path, if it is monotonically non-decreasing in value. Therefore, the solution to (2) is a monotonically non-decreasing path from stage 0 to stage M that has the minimal accumulated mean squared error (MSE).

[2] This work was performed under the support of the U.S. Department of Commerce, National Institute of Standards and Technology, Advanced Technology Program, Cooperative Agreement Number 70NANB1H3036.

Denote the minimal accumulated MSE over a feasible path from stage 0 to stage j as $\delta(j)$. Then, the dynamic programming algorithm contains the following three steps:

(1) Local minimization at each node (j,i).

$$\gamma^*(j \mid \alpha(j)=i) = \arg\min_{\lambda(j)} \left\| J_j - \varphi\left[I_{i-1}, I_i; \gamma_j\right] \right\|^2 \quad (4)$$

(2) Recursively, compute $\delta(j)$, for $j = 0,1,\ldots,M$.

(3) After $\delta(M)$, the minimal accumulated MSE for the last stage is calculated, back trace to get $\left[\gamma^*(j), \alpha^*(j)\right]$.

4.5 Minimization of Local Prediction Error

As shown above, to use dynamic programming, we need to minimize the local prediction error defined in (4). Since there is no closed-form solution to (4), we will use an iterative approach called ICM (Iterative Conditional Mode) to solve the minimization. The basic idea of ICM is to fix a set of parameters and optimize on the rest. In our case, we will iteratively fixed two of the three sets of parameters and optimize on the other one. Using ICM, the optimization of local prediction becomes three optimizations: optimization of temporal frame integration, spatial registration and histogram registration.

The optimization of temporal frame integration can be computed using in closed-form [10]. Histogram registration, also known as histogram shaping [7], maps one histogram to another. In this paper, we use a non-parametric histogram shaping algorithm that is similar to the one discussed in [7]. However, it is modified to give a smaller MSE than the original algorithm. We adopt the algorithm proposed in [3] as the spatial registration algorithm used in the proposed video registration. This algorithm minimizes the MSE matching cost given the temporal and the histogram transformations using the gradient decent algorithm and the 8-parameter plane perspective transformation.

4.6 Contextual Constraints

Video registration is an ill-posed inverse problem. For a given original video and a processed video, there may exist more than one solution. However, due to the nature of the prior knowledge of the application, the solutions to the same problem may have significantly different probabilities. For example, frame repeat or frame drop is usually used infrequently, and they are seldom repeated more than once. When there are a large number of similar frames, they are more likely cause by a scene of little motion than caused by consecutive uses of frame repeat. In addition, contextual constraints are needed for histogram registration. Without them, histogram registration can fail in special cases. For example, when a reference video contains uniform frames (e.g. black frames), any frames in the candidate video can be mapped to the uniform frame by histogram registration with a MSE of 0, which obviously is not a desired solution. Therefore, contextual constraints, the prior knowledge of what a solution must satisfy, can be used to reduce the solution space, to improve the accuracy and the robustness against noises. Using contextual constraints in the form of cost functions, $C(\gamma(0), \alpha(0), \cdots, \gamma(M), \alpha(M))$, the optimization problem in (2) can be extended to

$$\left[\gamma^*(j), \alpha^*(j)\right] = \arg\min_{\substack{[\gamma(0),\cdots,\gamma(M),\\ \alpha(0)\le\cdots\le\alpha(M)]}} \sum_{j=0}^{M} \left\| J_j - \varphi\left[I_{\alpha(j)-1}, I_{\alpha(j)}; \gamma(j)\right] \right\|^2 + C(\gamma(0), \alpha(0), \cdots, \gamma(M), \alpha(M)).$$

REFERENCES

[1] J.P. Lixvar, "Watermarking requirements for Boeing digital cinema," *Proc. of SPIE/IS&T Electronic Imaging*, Santa Clara, CA, January 2003, vol. 5020, pp. 546-552.

[2] M. Watson, "Conditional access study group watermark recommendations," *Proc. of SMPTE DC28.4 Watermarking Workshop*, March 15, 2001.

[3] R. Kumar, H.S. Sawhney, J. C. Asmuth, A. Pope and S. Hsu, "Registration of video to geo-referenced imagery," *Proc. ICPR 98*, Brisbane. Australia, Aug. 1998.

[4] S. Baudry, P. Nguyen, H. Maitre, "Estimation of geometric distortions in digital watermarking," *Proc. of IEEE Int'l Conf. on Image Processing*, Rochester, NY, Sept. 2002.

[5] Y. Caspi and M. Irani, "Alignment of non-overlapping sequences," *Proc. of IEEE Int'l Conf. on Computer Vision*, Vancouver, BC, Canada, July 2001.

[6] H. Cheng, "Temporal registration of video sequences," *Proc. ICASSP'2003*, Hong Kong, China, April 2003.

[7] A.K. Jain, *Fundamentals of Digital Image Processing*, Prentice Hall, 1989.

[8] P. Nguyen, R. Balter, N. Montfor and S. Baudry, "Registration methods for non blind watermark detection in digital cinema applications," *Proc. of SPIE/IS&T Electronic Imaging*, Santa Clara, CA, January 2003, vol. 5020, pp. 553-562.

[9] D. Delannay, C de Roover and B. Macq, "Temporal alignement of video sequences for watermarking systems," *Proc. of SPIE/IS&T Electronic Imaging*, Santa Clara, CA, January 2003, vol. 5020, pp. 481-492.

[10] H. Cheng and M.A. Isnardi, "Spatial, temporal and histogram video registration for digital watermark detection," *Proc. of Int'l Conf. on Image Processing*, Barcelona, Spain, September 2003.

[11] J. Lubin, J.A. Bloom and H. Cheng, "Robust second-generation watermarking for tracking in digital cinema," *Proc. of SPIE/IS&T Electronic Imaging*, Santa Clara, CA, January 2003, vol. 5020, pp. 536-545.

[12] A.E. Bryson, *Dynamic Optimization*, Addison-Wesley, '98.

3-NEIGHBORHOOD MOTION ESTIMATION IN CNN SILICON ARCHITECTURES

Lauri Koskinen, Ari Paasio, Kari Halonen

Electronic Circuit Design Laboratory
Helsinki University of Technology
P.O.Box 3000, FIN-02015 HUT, Finland
email: lauri.koskinen@ecdl.hut.fi

ABSTRACT

A novel 3-neighborhood connection for CNN silicon architectures is presented. The 3-neighborhood connection is ideal for computing the matching criterion, such as MSE or SAD, needed in the motion estimation of current video coding standards such as MPEG-4 or H.264. The 3-neighborhood connection can achieved with eight additional connections. An integrated chip with two 9X9 CNN test arrays capable of MSE and SAD calculation was implemented in 0.18 CMOS. Simulations in MPEG-4 show comparable performance to fully digital solutions. In worst case SAD accuracy conditions there is only under a 0.2dB drop in PSNR and under a 1.3% increase in coded bits. With the MSE the figures are 0.1dB and 0.71%

1. INTRODUCTION

With the advent of 2.5G and 3G mobile phones incorporating video cameras low power video encoding increases to an even greater importance. In most video coding implementations, such as MPEG-4 [1] which is used in 3G networks, the most important object of power consumption reduction is block-based motion estimation. With the multiple reference frames of the forthcoming video coding standard H.264 [2] the proportion of motion estimation increases even further.

Presented here is a novel 3-neighborhood method of connecting Cellular Nonlinear Network (CNN) cells to enable the block-based analog calculation of the matching criterion such as minimum square error (MSE) or sum of absolute differences (SAD) used in motion estimation. One possible implementation of the analog calculation of SAD is in local processing in connection with sensor arrays such as [3]. Another possible implementation is in MPEG-4 Core profile shape segmentation applications such as [4]. Calculating the SAD in connection with a sensor array cannot use a DCT transformed and quantized reference frame but the analog SAD can still be used as a starting point for conventional reduced complexity motion estimation algorithms. Matching criterion calculation with CNN has been previously suggested in [5] where the calculation is only done on a pixel-by-pixel basis and not block-based as is suggested here.

This work is supported by the Academy of Finland (#1168301/00) and (#1366361/99)

2. THE NOVEL 3-NEIGHBORHOOD

In a Cellular Nonlinear Network (CNN) the computation is divided into simple parallel operations thereby achieving large computational power. In an original CNN model [6] and in the analog calculation core of it's extension, the CNN Universal Machine (CNN-UM) [7], the state of a cell depends dynamically on the states of the cells connected to it. This state can be expressed with

$$\frac{dx_{i,j}}{dt} = -x_{i,j} + \sum_{C_{k,l}\in N_r(i,j)} A(i,j;k,l)y_{k,l} + \sum_{C_{k,l}\in N_r(i,j)} B(i,j;k,l)u_{k,l} + z \quad (1)$$

where $x_{i,j}$ is the cell state, $u_{i,j}$ is the cell input and $y_{i,j}$ is the cell output. A and B are the space invariant feedback and feed-forward coefficients, respectively. These coefficients represent the weights in the connections between the cell $C_{k,l}$ and the cell $C_{i,j}$. $N_r(i,j)$ is the neighborhood of the cell $C_{i,j}$ and z sets the operating point of the cell. The modification of the coefficient weights A, B and z achieve the different operations of the network. These weights are called templates. The equivalent of a A-template is a spatial IIR-filter and the equivalent of a B-template is a spatial FIR-filter. In Fig.1 the B-template implementation of a CNN cell is shown. To implement the 3rd neighborhood two such circuits are needed in succession.

Motion estimation (ME) achieves most of the bit-rate reduction in current video coding standards. In ME a macroblock of pixels in the current frame is compared to same size blocks within a search area of one or more reference frames. Only H.264 uses multiple reference frames. To find the best match a matching criterion is used. The best match is the reference macroblock. The reference and the current

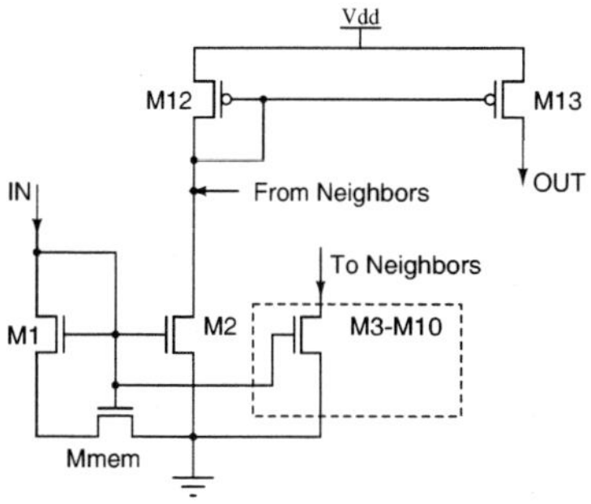

Fig. 1. Linear non-propagative cell

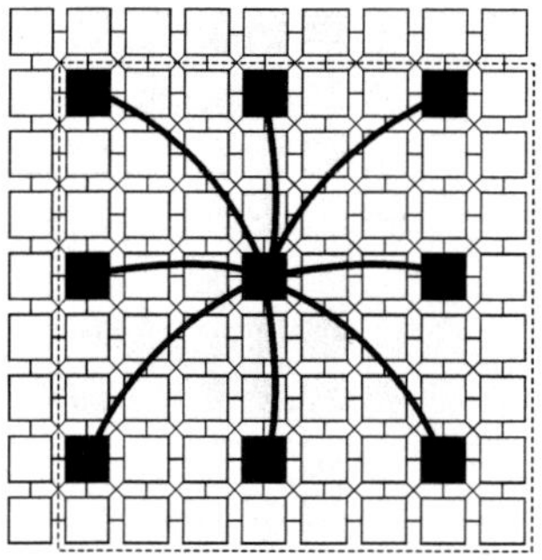

Fig. 3. The 3-neighborhood for analog matching criterion calculation. A normal 8X8 macroblock is also outlined.

macroblock are the subtracted and the remainder (prediction error) is further coded along with the motion vector which points to the reference macroblock. Due to its efficient digital implementation the most often used matching criterion in block-based ME is the SAD

$$SAD_N(dx,dy) = \sum_{i=x,j=y}^{x+N,y+N} |ref(i+dx,j+dy) - cur(i,j)| \quad (2)$$

where N is the block size, x,y the coordinates of the top left hand corner of the block and dx,dy the current candidate motion vector. The macroblock sizes used both in MPEG-4 and H.264 are 16X16 and 8X8. H.264 also uses a 4X4 block size. A superior option to the SAD is the MSE

$$MSE_N(dx,dy) = \frac{1}{N^2} \sum_{i=x,j=y}^{x+N,y+N} (ref(i+dx,j+dy) - cur(i,j))^2 \quad (3)$$

which is less efficient to implement in digital solutions.

In all previous 2-dimensional CNN implementations the cells are connected to it's four vertical and horizontal neighboring cells or eight of it's nearest neighbors. In Fig.2a the cell shaded in black is connected to it's neighbors shaded in grey. Such a connection is defined as the cell's 1-neighborhood. In Fig.2b the black cell is connected in a 2-neighborhood which increases the number of needed connections to 24.

To implement an averaging operation with B-templates within a 8X8 macroblock would require a 4-neighborhood.

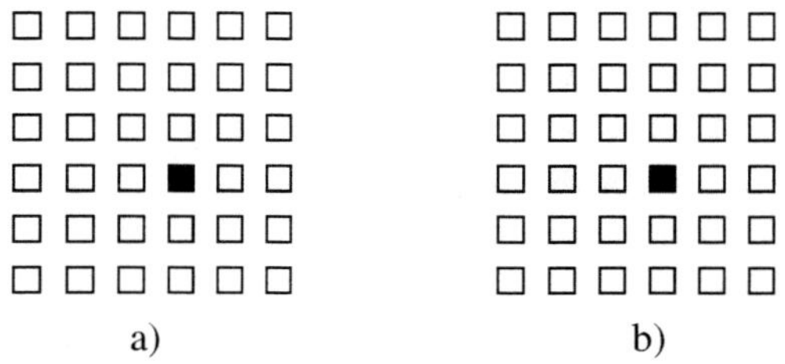

Fig. 2. Different conventional neighborhoods: a) 1-neighborhood b) 2-neighborhood

The number of connections per cell needed for a 4-neighborhood is 80 which makes a VLSI realization extremely unefficient. Another option would be to implement the averaging operation with an A-template. A-template implementations suffer from accumulating analog inaccuracies and, as with IIR filters, the effect boundary cannot be defined exactly.

To reduce the number of connections while enabling an accurate calculation of the matching criterion the novel 3-neighborhood shown in Fig.3 is developed. In Fig.3 the 3rd neighborhood connections are shown with the thick lines and the thin lines represent the 1st neighborhood connections. The matching criterion of a 9X9 block can be calculated efficiently with this type of connection. With the proposed connection the calculation of the SAD is done as follows: The absolute value or the square of the difference between the current pixel and the reference pixel is calculated in the individual cells. The difference values are then averaged by multiplying by 1/9 and summing. Thereby all the connections shown in Fig.3 reflect the 1/9 multiplying operation. The combined effect is the sum of the values in a 9X9 block as can be seen in Fig.3 where the pixels shaded with gray belong to the analog SAD block. The SAD value can then be read from the center cell. To calculate the 4X4 MVs required by H.264 with CNN architectures would require irregular connections and thereby space variant templates. The basic 16X16 MVs defined by multiple video coding standards would require a 4-neighborhood corresponding to the 3-neighborhood presented here increasing the number of connections so that a silicon implementation would be inefficient

The B-templates of an CNN cell can be realized with the multiplying current mirror structure shown in Fig.1. For the SAD operation an absolute value circuit of Fig.4 [8] was implemented in conjunction with the cell of Fig.1. For the MSE operation an efficient implementation was achieved by implementing the absolute value operation and the quadratic operation [9] separately. The quadratic circuit shown in Fig.5 needs a positive input which is why the absolute value

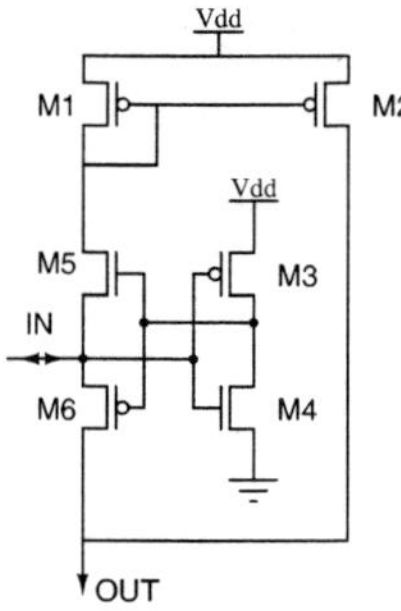

Fig. 4. Absolute value circuit

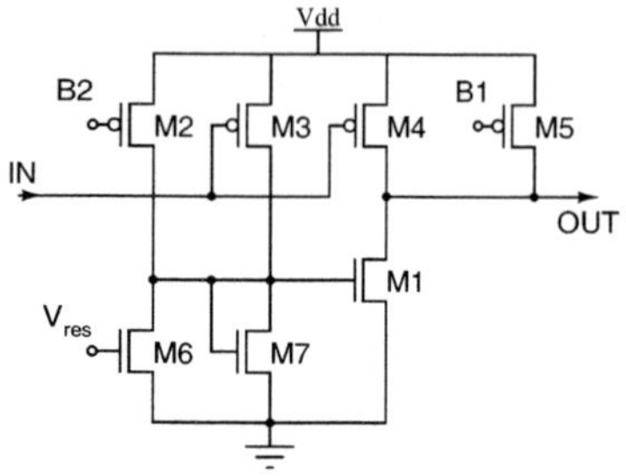

Fig. 5. Quadratic circuit

circuit was also needed. The output scaling needed for the MSE can be implemented with the output current mirror. In the digital domain using these analog SAD values the motion estimation is reduced to minimum value operation within a block. The sizes of the implemented SAD and MSE cells, including the absolute value and quadratic circuits, were 25.3x34.0 and 45.2x33.6 μm, respectively.

To reduce the effect of the difference between the 9X9 block size of the analog SAD and the 8X8 block size of the macroblock the simulations were also calculated with the third neighborhood template

$$B_{3n_2} = \begin{bmatrix} \frac{4}{64} & \frac{6}{64} & \frac{6}{64} \\ \frac{6}{64} & \frac{9}{64} & \frac{9}{64} \\ \frac{6}{64} & \frac{9}{64} & \frac{9}{64} \end{bmatrix}. \quad (4)$$

This template reduces the proportionate effect of the 3X3 blocks which include pixels that do not belong to the macroblock. These pixels are the ones outside the outlined macroblock in Fig.3. An ideal solution to the oversize analog SAD block would require space variant templates in the first neighborhood.

3. SIMULATION RESULTS

The simulated transistor level SAD and MSE values were input into a simulated digital MPEG-4 architecture using the full search motion estimation algorithm. No output buffer was implemented so the quantization parameter Q_p was varied to control the number of coded bits. Due to the 9X9 block size of the analog matching criterion calculation MPEG-4 advanced prediction mode was chosen to be used in motion estimation. Advanced prediction means that the 9X9 analog SAD values and the motion estimation difference resulting from these values was compared to conventional motion estimation with a 8X8 block size. Due to computational power issues in digital encoders the conventional method in motion estimation is to first calculate 16X16 block size motion vectors and then do a refined 8X8 block size search around the 16X16 values. To have an exact comparison between the analog 9X9 SAD values and the conventional 8X8 SAD values here the full search was done solely in 8X8 block size. Full pixel resolution was used in the motion vectors.

In analog operations accuracy considerations must be taken into account. When considering a B-template operation the accuracy errors are due to the mismatch of transistors in the current mirrors that achieve the multiplying operations needed in calculating the output value of a cell. Such mismatch effects are throughly analyzed in [10]. The inaccuracy does not affect the summing operation but the values of the templates. Inaccuracy in the templates gives incorrect weights to different pixels. Transistor sizes of the circuits were calculated with Monte Carlo simulations so that the deterioration in rate-distortion values was negligible. In calculating the MSE both the template inaccuracy and inaccuracy of square circuit have to be taken onto account. The square circuit can be designed at a high precision but the size requirements of the cell cell limit the accuracy. Both the SAD and MSE were simulated with averaging templates used in both neighborhoods and the template from Eq.4 in the second neighborhood.

The results for the sequences foreman and news are collected in table.1. With the SAD using averaging templates in both neighborhoods showed better results. The calculation of the MSE showed better results when using the the Eq.4 template in the second neighborhood. Ideally the best results were shown in the sequence foreman by using MSE and the template from Eq.4 in the second neighborhood. Then the results show a 0.061dB drop in PSNR and a 0.17% drop in the number of coded bits. This can be thought of as equivalent coding performance. With the MSE the worst case results were 0.092dB drop in PSNR and a 0.17% increase in the number of coded bits with the sequence foreman. With the SAD the ideal results with the sequence news showed a 0.05dB drop in PSNR and a 0.77% increase in the number of coded bits. The worst case results showed a 0.128dB drop in PSNR and a 1.26% increase in the number of coded bits. The chip has been sent to the Foundry. The Layout of the designed chip is shown in Fig.6. The cell arrays are labeled to separate them from the logic and DA and AD-converter arrays.

	Foreman		News	
	PSNR(dB)	BR(%)	PSNR(dB)	BR(%)
Ideal SAD 1	-0.123	1.14	-0.050	0.77
Ideal SAD 2	-0.907	8.00	-0.284	2.59
Worst Case SAD 1	-0.128	1.26	-0.054	0.93
Worst Case SAD 2	-0.917	8.24	-0.316	2.65
Ideal MSE 1	-0.617	2.47	-0.271	2.11
Ideal MSE 2	-0.061	-0.17	-0.040	0.05
Worst Case MSE 1	-0.653	2.96	-0.368	2.25
Worst Case MSE 2	-0.092	0.17	-0.061	0.34

Table 1. Drop in PSNR and percentual increase in the number of coded bits for the sequences foreman and news. The number 1 refers to averaging templates in both neighborhoods. The number 2 refers to Eq.4 template used in the third neighborhood.

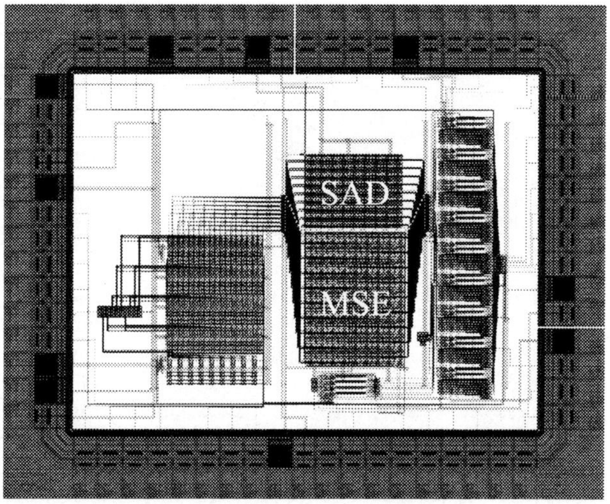

Fig. 6. Layout of the designed chip with the cell arrays outlined

4. CONCLUSIONS

A novel 3-neighborhood connection for CNN architectures has been presented. The 3-neighborhood connection is intended for video coding applications using parallel processing such as CNN shape segmentation or sensor arrays with local processing. The connection enables the efficient calculation of block-based motion estimation matching criterion. Due to the type of connections in the analog calculation the block size is 9X9 in contrast to to the macroblock size of 8X8 as used in current video coding standards. While using the SAD the 9X9 block causes only a -0.123dB penalty in PSNR and a 1.14% increase in coded bits. The worst case values calculated with Monte Carlo simulations show that the penalty is still only -0.128dB and 1.26%. With simple changes in the cell logic the MSE can be also implemented. With the MSE comparable performance to conventional digital full search using SAD can be achieved.

5. REFERENCES

[1] Fugunaka S.; et al. (ed.), *MPEG-4 Video Verification Model version 16.0.* ISO/IEC JTC1/SC29/WG11 Coding of Moving Pictures and Associated Audio, Noordwijkerhout, Netherlands N3312, March 2000.

[2] Wiegand T. (ed.), *Text of Final Committee Draft of Joint Video Specification (ITU-T Rec. H.264 | ISO/IEC 14496-10 AVC).* ISO/IEC JTC1/SC29/WG11 Coding of Moving Pictures and Associated Audio, Klagenfurt, Austria N4920, July 2002.

[3] Dudek, P.; Hicks, P.J.; An analogue SIMD focal-plane processor array. *IEEE International Symposium on Circuits and Systems, 2001. ISCAS 2001.* Vol. 4 , 2001 pp. 490 -493

[4] Koskinen, L.; et. al. Effect of CNN shape segmentation on MPEG-4 shape bit-rate *IEEE International Symposium on Circuits and Systems, 2002. ISCAS 2002.* Vol. 4 , 2002 pp. 552 -555

[5] Czuni, L.; Sziranyi, T.; Motion segmentation and tracking optimization with edge relaxation in the cellular nonlinear network architecture. *Proceedings of the 2000 6th IEEE International Workshop on Cellular Neural Networks and Their Applications, 2000. (CNNA 2000).* pp. 51 -56

[6] Chua L.O., Yang L. Cellular Neural Networks: Theory. *IEEE Transactions on Circuits and Systems.* 1988, Vol. 35, pp. 1257-1272

[7] Roska T., Chua L. O. The CNN Universal Machine: An Analogic Computer. *IEEE Transactions on Circuits and Systems-II.* 1993, Vol. 40, pp. 163-146

[8] Baturone I., et. al. Implementation of CMOS Fuzzy Controllers as Mixed-Signal Integrated Circuits. *IEEE Transactions on Fuzzy Systems.* 1997, Vol. 5, pp. 1-19

[9] Mika Laiho, Ari Paasio, Asko Kananen, Kari Halonen, 'Realization of Couplings in a Polynomial Type Mixed-Mode Cellular Neural Network' To be published in the *International Journal of Neural Systems.*

[10] Lakshmikumar K. R.; Hadaway R. A.; Copeland M. A.; Characterization and Modeling of Mismatch in MOS Transistors for Precision Analog Desing. *Int. J. Solid-State Circ.,* Vol. sc-21, pp. 1057-1066, 1986.

Self-Synchronized Audio Watermark in DWT Domain

Shaoquan Wu[1], *Jiwu Huang*[1](contacting author. isshjw@zsu.edu.cn), *Daren Huang*[1], *Yun Q. Shi*[2]

1. School of Information Science and Technology, Sun Yat-Sen University, Guangzhou 510275, P. R. China.

2. Dept. of ECE, New Jersey Institute of Technology, NJ 07102, USA.

Abstract

Robust audio watermarks present some challenges. In this paper, we propose a DWT-based audio watermarking algorithm to resist shifting and cropping. Synchronization codes are embedded into audio with informative watermark, and thus the watermark generated has the ability of self-synchronization. To achieve good robustness performance, we embed the synchronization codes and the watermark into the low frequency coefficients in DWT domain. By exploiting the time-frequency localization capability of DWT to decrease the computation load for searching synchronization codes effectively, we resolve the conflict between robustness and computation complexity of synchronization code. In addition, we analyze the performance of the proposed algorithm in terms of data payload, error probability of synchronization codes, *SNR*, and *BER*. The experimental results show the watermark is robust against common signal processing manipulations and attacks, such as Gaussian noise corruption, resampling, requantization, cropping and MP3 compression.

1. Introduction

Compared with image watermarking, audio watermarking receives less attentions. According to IFPI (International Federation of the Phonographic Industry) [1], audio watermarking should meet the following requirements: 1) The watermark should not degrade perception of audio. 2) The algorithm can offer more than 20 dB *SNR* for watermarked audio and 20 bps (bit per-second) data payload for watermark. 3) The watermark should resist most common audio processing operations and attacks, such as D/A and A/D conversions, temporal scaling (stretch by ±10%), additive and multiplicative noise, MP3 compression. 4) The watermark should prevent unauthorized detection, removal and embedding, unless the quality of audio become very poor. These requirements present many challenges to robust audio watermarking.

Most of the recent audio watermarking algorithms can be grouped into two categories: algorithms in time domain [2,3] and algorithms in frequency domain [4,5,6]. The main weaknesses of the existing algorithms include low payload for data hiding and low robustness performance of watermark, especially low performance in resisting shifting and cropping [7,8].

Since the watermarked audio is likely suffered from shifting and cropping (such as editing, IP packet loss, etc.), it may be necessary to introduce the synchronized watermarking, where the synchronization codes are embedded into the original audio with the watermark [6,9]. The synchronization codes are exploited to locate the positions where the watermark is embedded. Huang et al. [6] hid the Bark code in time domain as the synchronization code while embedding the watermark into DCT coefficients. Due to the nature of time domain algorithm and the limitation of code length, the synchronization code is not robust enough.

This paper presents an audio watermarking algorithm in DWT (Discrete Wavelet Transform) domain. The features of the algorithm include: 1) Synchronization codes are embedded into audio with the informative watermark, and so the watermark has the ability of self-synchronization. 2) Synchronization code and watermark are embedded into the low frequency coefficients in DWT domain to achieve good robustness performance against to common signal processing procedures and noise corruption. 3) By exploiting the time-frequency localization capability of DWT to decrease the computation load for searching synchronization codes effectively, we resolve the conflict between robustness and computation complexity of synchronization code. 4) We provide analytical formula to estimate *SNR* based on embedding strength S. We also estimate the *BER* of watermark after Gaussian noise corruption. The correctness of these formulas is fully demonstrated by the experiment results.

The experimental results show the watermark is robust against common signal processing manipulations and attacks, such as Gaussian noise corruption, resampling, requantization, cropping and MP3 compression.

2. The Proposed Algorithm

2.1 Synchronization Code

The algorithm exploits a PN-sequence as synchronization code. Suppose $\{a_i\}$ is an original synchronization code and $\{b_i\}$ is an unknown sequence with the same length. If the number of different corresponding bits between $\{a_i\}$ and $\{b_i\}$ are less than and equal to a predefined threshold, e, the $\{b_i\}$ will be determined as the synchronization code. The analysis of error probability in searching synchronization codes is given in Section 3.2.

2.2 Synchronization Codes in DWT Domain

Synchronization codes can be embedded into time domain or frequency domain. The advantage of embedding in time domain is low cost for searching computation. However, the robustness is low due to the limitation of embedding strength. If the synchronization codes are embedded into frequency domain, such as DFT and DCT domain, the robustness will be improved but the searching cost will increase greatly.

Since the robustness of synchronization codes is very important to watermark extraction, we embed the synchronization code into DWT domain to achieve high robustness. To decrease the searching cost, we apply the time-frequency localization feature of DWT to improve the efficiency of searching synchronization codes.

Assume that $f(i)$ is original audio signal. Without loss of generality, we construct the following two audio sections (A and B) with M samples:

$$A=[f_0, f_1, \cdots, f_{M-1}] \text{ and } B=[f_{2^K}, f_{2^K+1}, \cdots, f_{2^K+M-1}]$$

where K is the decomposition level of DWT.

Let $AL(i)$ and $BL(i)$ denote the coefficients of low-frequency sub-band of section A and B under K-level DWT, respectively. Except a few boundary coefficients, we have

$$AL(i+1) = BL(i) \quad (1)$$

For example, if $K=1$, we have

$$AL(i+1)=\sum_{j=0}^{L-1} h_j \cdot f_{2(i+1)+j} = \sum_{j=0}^{L-1} h_j \cdot f_{2^K+(2i+j)} = BL(i)$$

where L is the length of wavelet filters and $\{h_i\}$ denotes the low pass filter of wavelet.

It can be found that the Equation (1) is valid for K=1. If K>1, according to the definition of K-level DWT and mathematical induction, it is easy to prove that Equation (1) is also valid except for some the boundary coefficients.

The number of those boundary coefficients that are not satisfy Equation (1) is less than L+2 by a careful estimation.

In searching synchronization code sample by sample, we can use Equation (1) to save computation cost. Suppose that we had calculated the $AL(i)$ for section A. When the shift number of samples is 2^K, like that B is a shift of A with 2^K samples, for section B we just need to calculate L+2 coefficients at most by Equation (1).

Assume that there is one synchronization code in an audio section with M samples. For DWT, it just needs 2^K times K-level DWT transform to find the synchronization code at most, comparing with M times for DCT/DFT transform. Moreover, the computation cost for DWT is lower than that for DCT/DFT. Thus, DWT will greatly save the searching cost compared with DFT and DCT.

2.3 Data Embedding

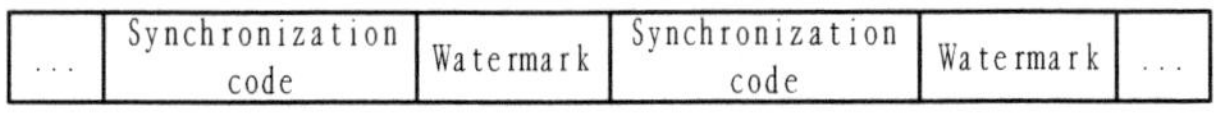

Fig. 1 Data structure of $\{m_i\}$.

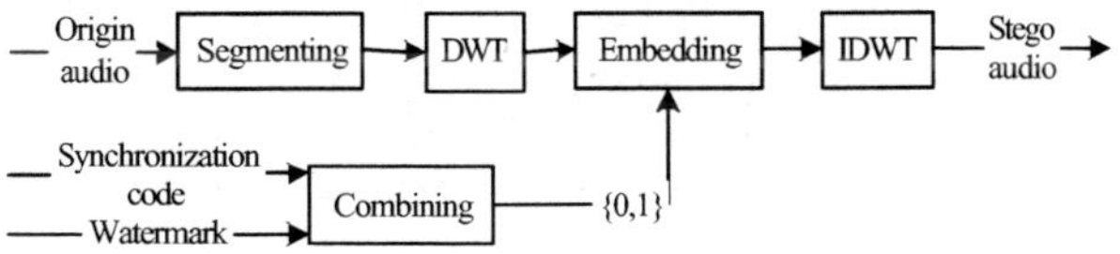

Fig. 2 Watermark embedding.

Before embedding, the synchronization codes and watermark signal should be converted into a binary sequence $\{m_i\}$. The data structure of $\{m_i\}$ is showed in Fig. 1.

Then split the original audio to proper sections and perform DWT on every section. The sequence $\{m_i\}$ is embedded into the low-frequency sub-band of each section. The length of audio section depends on the embedding data amount and DWT decomposition level.

The rules for embedding $\{m_i\}$ are as follows [10]:

$$c_i^* = \begin{cases} c_i - Mod(c_i, S) + 3S/4 & if \quad m_i = 1 \\ c_i - Mod(c_i, S) + S/4 & if \quad m_i = 0 \end{cases} \quad (2)$$

$$Mod(c_i, S) = \begin{cases} c_i - \lfloor (c_i/S) \rfloor * S & if (c_i \geq 0) \\ c_i + (\lfloor (-c_i/S) \rfloor + 1) * S & if (c_i < 0) \end{cases}$$

where c_i and c_i^* are the coefficients of the low-frequency sub-band of original audio and watermarked audio, respectively. $\lfloor \ \rfloor$ is the floor function. S denotes the embedding strength. The value of S should be as large as possible under the imperceptible constraint.

2.4 Data Extracting

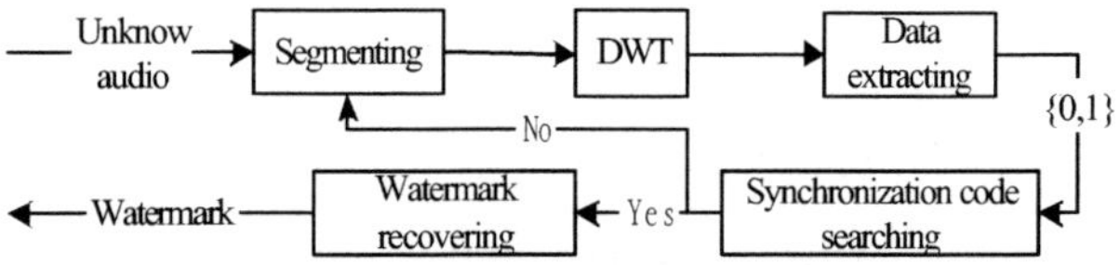

Fig. 3 Watermark extraction

When extracting the hidden data, we can split the test audio into sections (at least one synchronization code should be included in a section) and then perform DWT with the proper levels on every section. Let c_i^* denote the coefficients of low-frequency sub-band of each section, we can extract m_i^* from c_i^* by using the following rule [10]:

$$m_i^* = \begin{cases} 1 & if \quad Mod(c_i^*, S) \geq S/2 \\ 0 & if \quad Mod(c_i^*, S) < S/2 \end{cases} \quad (3)$$

Before extracting the informative watermark, we need to search the synchronization code in sequence $\{m_i^*\}$. With the synchronization codes, the embedding locations of the watermark can be determined. The watermark extraction is shown in Fig. 3.

3. Performance Analysis

We use BER (Bit Error Rate) and SNR (Signal to Noise Ratio) to analyze the performance of the proposed algorithm. In this section, we will evaluate the performance of the proposed algorithm in terms of data payload of watermark, error probability of synchronization code, SNR with embedding strength S, and BER of watermark under Gaussian noise corruption.

3.1. Data Payload

The *data payload* refers to the number of bits that are embedded into origin audio within a unit of time, measured with *bps* (bit per-second). Suppose that the sampling rate of audio is R (Hz) per-second and the wavelet decomposition level is K. The data payload B of the algorithm is defined as:

$$B = R/2^K \ bps \quad (4)$$

3.2. False Positive Error and False Negative Error

There are two types of errors in searching synchronization codes, false positive error and false negative error. A false positive error occurs when a synchronization code is detected in the location where no synchronization code is embedded while a false negative error occurs when an existing synchronization code is missed. In the case of a false positive error occurs, the bits after the locations of the false synchronization code will be regard as the watermark bits. In the later, some watermark bits will be lost.

The false positive probability P_1 and false negative probability P_2 can be calculated as follows:

$$P_1 = \frac{1}{2^p} \cdot \sum_{k=p-e}^{p} C_p^k \quad (5)$$

$$P_2 = \sum_{k=e+1}^{p} C_p^k \cdot (BER_{sync})^k \cdot (1 - BER_{sync})^{p-k} \quad (6)$$

where p is the length of a synchronization code and e is the threshold introduced in Section 2.1. Equation (5) indicates that the false positive probability P_1 is independent on watermark attack.

3.3. Estimation Formulas

We give two estimation formulas based on orthogonal wavelet for the proposed algorithm. The results are similar with bi- and no-orthogonal wavelets.

Based on the observations made on dozens of different types of audio signal, we can assume that the random variable $\xi = c_i - \lfloor c_i / S \rfloor \cdot S$ obeys uniform distribution on [0,*S*], where c_i is a DWT coefficient of the low frequency subband used to embed data and S the embedding strength as introduced in Section 2.3. By this assume and the Parseval Theorem, we can calculate that the average change of the low subband of original audio is $7 \cdot S^2 / 48$. Finally, we give the function that links the *SNR* after embedding, original audio F, number of audio samples N, DWT level K, and embedding strength S.

$$SNR = -10 \log_{10}\left(\frac{N}{2^K} \cdot \frac{7 \cdot S^2}{48 \cdot \|F\|_2^2} \right) \quad (7)$$

With Equation (7) we can decide the embedding strength S directly according to *SNR* without actual embedding attempt.

It is easy to prove that the low subband of noise with Gaussian distribution $N(0, \delta^2)$ still obeys Gaussian distribution $N(0, \delta^2)$ under K^{th}-DWT. By the embedding and extraction rules in Section 2.3 and Section 2.4, we know that if the noise strength at the low subband of watermarked audio is more than $S/4$, the bit error will occur. So we have the *BER* of watermark under Gaussian Noise $N(0, \delta^2)$ corruption:

$$BER = \frac{2}{\sqrt{2\pi}\delta} \int_{S/4}^{+\infty} e^{-\frac{x^2}{2\delta^2}} dx \quad (8)$$

Equation (8) estimates the algorithm performance and indicates that the *BER* of the watermark is independent on the original audio F under the corruption of additive Gaussian noise.

The correctness of these two formulas is completely demonstrated by the simulation results.

4. Experimental Results

We test our algorithm on two 16-bit signed mono audio signals sampled at 44.1 kHz with the length of about 14 seconds in WAVE format. The two audio signals are with the different proprieties (march music and light music), denoted as *march.wav* and *light.wav*.

The synchronization codes compose of m-sequence with period 63 and the watermark is a 64*64 binary image. The threshold e in Section 2.1 is set to be 21. Haar wavelet is applied with the decomposition level of 5.

With Equation (4), we can estimate the data payload B, 1367 bps. It needs an audio about 3 seconds to embed a synchronization code and a watermark. The false positive probability P_1 of

Table 1 Gaussian noise corruption

march.wav

δ	*SNR (db)*	*BER(%)*	P_2(%)
1000	19.21	2.12	$3.44*10^{-19}$
1300	16.90	7.76	$8.52*10^{-8}$
1600	15.09	16.08	$2.06*10^{-2}$
1900	13.58	23.52	2.73
2200	12.31	29.89	22.88

light.wav

δ	*SNR(%)*	*BER(%)*	P_2(%)
200	20.61	0.15	$3.72*10^{-44}$
300	17.38	3.22	$2.18*10^{-15}$
400	14.87	11.41	$8.63*10^{-5}$
500	12.94	20.95	$7.28*10^{-1}$
600	11.32	28.47	15.97

Table 2 Resampling

march.wav			light.wav		
Resample Rate (Hz)	*BER* (%)	P_2(%)	Resample Rate (Hz)	*BER* (%)	P_2(%)
22050	0	0	22050	0	0
11025	0	0	11025	0	0
8000	0.12	$2.78*10^{-46}$	8000	1.89	$3.02*10^{-20}$
6000	1.65	$1.67*10^{-21}$	6000	9.16	$1.81*10^{-6}$
4000	27.12	10.73	4000	19.34	$2.67*10^{-1}$

Table 3 Requantization

march.wav			light.wav		
Quantization Bits	*BER* (%)	P_2 (%)	Quantization Bits	*BER* (%)	P_2 (%)
24	0	0	24	0	0
8	0	0	8	0	0

Table 4 MP3 compression

march.wav			light.wav		
Bit Rate (*kbps*)	*BER* (%)	P_2(%)	Bit Rate (*kbps*)	*BER* (%)	P_2(%)
128	0	0	128	0	0
96	0	0	96	0.78	$1.64*10^{-28}$
80	0.39	$4.56*10^{-35}$	80	3.76	$5.31*10^{-14}$
56	2.34	2.77×10^{-18}	56	7.54	$4.96*10^{-8}$
48	9.94	7.83×10^{-6}	48	9.50	$3.49*10^{-6}$
32	21.78	1.15	32	11.23	$6.58*10^{-5}$

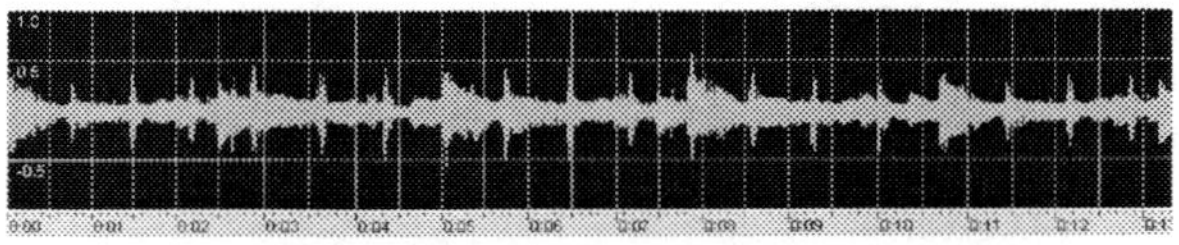

(a) Watermarked Audio (*light.wav*)

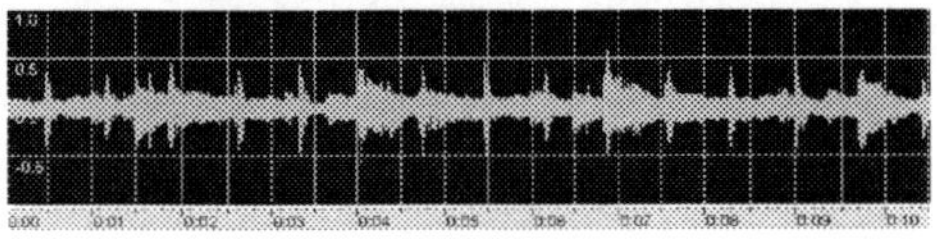

(b) Cropped Audio

Fig. 4 Anti-cropping of Synchronization Code

synchronization code is 0.56% based on Equation (5).

The values of embedding strength *S* are 9000 for march.wav and 2500 for ligth.wav, respectively. The *SNR*s between the origin audios and the watermarked audios are 23.52 dB for march.wav and 23.13 dB for ligth.wav. Table 1~4 show the results for Gaussian noise corruption, resampling, requantization, and Mp3 compression attacks.

To test the ability of resisting to cropping, we cut an audio section from the watermarked audio, shown in Fig. 4. It is obvious that there are two synchronization codes and two watermarks in the cropped audio. Using the extracting algorithm, we can extract the two synchronization code and watermarks.

Through Table 1~4, we can see that the watermarks and synchronization codes are very robust to various attacks. For example, the *BER* of extracted data are 21.78% and 11.23% under MP3 compression with the lowest bit rate (32 *kbps*) for 44.1 kHz audio. Accordingly, the false negative probabilities P_2 are 1.15% and $6.58*10^{-5}$%. The watermarks can easily be distinguished.

The results of resmapling attack indicate that the proposed algorithm is robust to time-scaling if the scaling factor is found.

We also test the performance of proposed algorithm with different wavelet bases. Many wavelet bases, such as Daubechies wavelets, Coiflets wavelets, and Symlets wavelets, are applied in our work. The conclusion is that the choice of wavelet base has little effect to the performance of the algorithm. Thus, we exploit the simplest wavelet base, Haar wavelet.

5. Conclusions

In this paper, we propose a self-synchronized audio watermarking algorithm in DWT domain. The main contributions are as follows.

1) Watermark is embedded with synchronization codes and thus the self-synchronized watermark has the ability to resist shifting and cropping.

2) The synchronization codes and watermarks are embedded into low-frequency sub-band in DWT domain and thus achieve good robustness performance against to common signal processing procedure and noise corruption.

3) The time-frequency localization feature of DWT is exploited to improve the efficiency for searching synchronization codes.

4) We provide analytical formula to estimate *SNR* based on embedding strength *S*. We also estimate the *BER* of watermark after Gaussian noise corruption. The correctness of these formulas is fully demonstrated by the experiment results.

After conducting and comparing several groups of experiments, we find that the embedding strength *S* is greatly depended on the type and magnitudes of the original audio signals. It is not the best choice to use a fixed *S* for an audio. Our future works will include that how to apply the adaptive embedding [11]. It may be another consideration to apply error correct coding technique [12] to improve the robustness.

Acknowledgments. Supported by NSF of China (60325208, 60172067, 60133020), "863" Program (2002AA144060), NSF of Guangdong (013164); New Jersey Commission of Science and Technology via NJWINS.

References

[1] Stefan Katzenbeisser, Fabien A.P. Petitcolas, editors, Information Hiding Techniques for Steganography and Digital Watermarking, ARTECH HOUSE, INC., 2000.

[2] Michael A., Gerzon and Peter G. Graven., "A high-rate buried-data channel for audio CD", *Journal of the Audio Engineering Society*, 43(1/2):3--22, January--February 1995.

[3] D. Gruhl, A.Lu and W. Bender, "Echo Hiding", *Proc. of 1st Information Hiding Workshop*, LNCS vol. 1174, Berlin, Germany: Springer-Verlag, pp. 295-315, 1996.

[4] W. Bender, D.Gruhl, and N. Morimoto, "Techniques for data hiding", *IBM Systems Journal*, 1996, 35(3/4): 131-336.

[5] M.Cooperman and S.Moskowitz, "Steganographic method and device", *U. S. Patent* 5, 613,004, Mar., 1997.

[6] J. Huang, Y. Wang, and Y.Q. Shi, "A blind audio watermarking algorithm with self-synchronization", *Proc. of IEEE, Int. Sym. on Circuits and Systems*, vol. 3, pp. 627-630, 2002.

[7] Bassia, P., Pitas, I., Nikolaidis N., "Robust audio watermarking in the time domain", *IEEE Trans. on Multimedia*, 3(2):232-241, 2001.

[8] Byeong-Seob Ko, Nishimura R., Suzuki Y., "Time-spread echo method for digital audio watermarking using pn sequences", *Proc. of IEEE Int. Conf. on Acoustics, Speech, and Signal Processing*, vol. 2, pp. 2001-2004, 2002.

[9] W.-N. Lie and L.-C. Chang, "Robust and high-quality time-domain audio watermarking subject to psychoacoustic masking", *Proc. of IEEE, Int. Sym. on Circuits and Systems*, vol. 2, pp. 45-48, 2002.

[10] J. Liu, The application of wavelet in image compression and digital watermarking, Doctoral Dissertation of Zhejiang University, 2001.4.

[11] J. Huang, Y. Q. Shi, and Y. Shi, "Embedding image watermarks in DC components," *IEEE Trans. on Circuits and Systems for Video Technology*. 10(6): 974-979, 2000.

[12] J. Huang and Y. Q. Shi, "Reliable Information Bit Hiding," *IEEE Trans. on Circuits and Systems for Video Technology*. 2002, 12(10): 916-920.

FEATURE DIFFERENCE ANALYSIS IN VIDEO AUTHENTICATION SYSTEM

Dajun He[1], Zhiyong Huang[2], Ruihua Ma[1] and Qibin Sun[1]

[1]Institute for Infocomm Research (I2R)
21 Heng Mui Keng Terrace, Singapore, 119613
Email: {djhe, ruihua, qibin}@i2r.a-star.edu.sg

[2]School of Computing
National University of Singapore, Singapore, 117543
Email: huangzy@comp.nus.edu.sg

ABSTRACT

In most video authentication systems, the difference between features of the original and received videos is often used to decide the authenticity. In this paper, by employing mutual information to represent the similarity between the original and received video frames, we theoretically analyze the relationship between the feature difference and the video distortion. The relationship we derived is applied to estimate the maximum allowable feature difference in a video authentication system and to show how the feature difference varies with quantization step. Experimental results demonstrate that the derived relationship is reasonable and helpful for designing a robust video authentication system.

1. INTRODUCTION

In most video applications, original video may undergo various processing before reaching final users. So a distortion between the original video and the received video may exist. This may be either incidental distortion, which is introduced by normal video processing such as compression, resolution conversion and geometric transformation, or intentional distortion, which is introduced by malicious attack. A robust video authentication system should, hence, tolerate the incidental distortion while being capable of detecting the intentional distortion. Many researchers have designed robust image/video authentication algorithms to meet above requirements based on watermarking strategy, which is sometimes termed "self-embedding" authentication system [1] as shown in Figure 1. In this system, a robust and important feature of the image/video is extracted and embedded into the image/video at the sending site; the detector retrieves this original feature from the watermark and compares it with the feature extracted from the received image/video to determine the authenticity of the image/video. If the difference exceeds a threshold, the received image/video will be claimed as an un-authentic image/video. This threshold, which should be determined before an authentication system is designed, refers to the maximum feature difference between the original video and the video that has undergone various normal video processing.

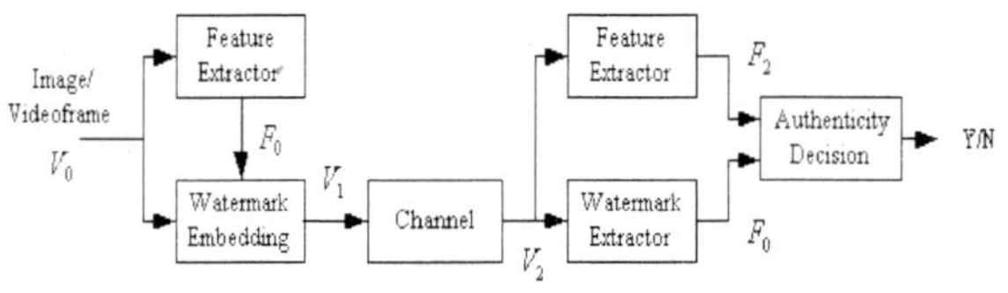

Figure 1 Block-diagram of a "self embedding" authentication system

Dittmann [2] and Queluz [3] used the edge of image as the feature to generate a digital signature in their authentication systems. Although they claimed that this feature is robust to high quality compression and scaling, a threshold is still used to improve the robustness. Nevertheless, the authors did not mention how this threshold is obtained. In our previous works [4, 5], a semi-fragile watermark instead of digital signature is used in a video authentication system. The Error Correction Coding (ECC) scheme and cryptographic hash scheme were employed to improve the system's robustness and security. However, a threshold was also needed to decide the authenticity of video objects. And this threshold was determined in an empirical way. It requires a lot of computation; and the effectiveness of the threshold itself also poses a problem.

In this paper, we will adopt a theoretical approach to derive the threshold based on analyzing the relationship between the feature difference and the video distortion. Moreover, using the derived relationship, we will also show how the feature difference varies with the quantization step in video compression. The paper is organized as follows: the relationship between the feature difference and the video distortion is analyzed in Section 2. Experimental results are presented in Section 3. We conclude the paper in Section 4.

2. FEATURE DIFFERENCE ANALYSIS

2.1 Mutual information and feature difference

Mutual information is a basic concept in information theory. It measures the amount of the information that one random variable contains about another random variable [6]. The definition is as follows:

$$I(X;Y)=\sum_{x\in\widetilde{X}}\sum_{y\in\widetilde{Y}}p(x,y)\log\frac{p(x,y)}{p(x)p(y)} \qquad (1)$$

where X is a discrete random variable with set $\widetilde{X}$ and probability density function $p(x)$; Y is a discrete random variable with set $\widetilde{Y}$ and probability density function $p(y)$; and $p(x,y)$ is the joint probability density function of X and Y . To compute mutual information, the probability density functions must be computed or estimated. It is worth noting that in real applications, the estimation is crucial but not always trivial.

Mutual information has been widely used to measure the similarity between two images, especially in image registration. For example, two images are considered geometrically aligned if the mutual information of image intensity values is maximized [7].

As mentioned in the Introduction Section, in video authentication, we are concerned about the feature difference between the original video frame and the distorted video frame to determine whether the distorted one is still authentic. When the feature is well selected to represent the video frame, we should be able to express the feature difference between two video frames in terms of mutual information of these two video frames in the following linear form:

$$d(F_0,F_2)=L_f-\alpha * I(V_0;V_2) \qquad (2)$$

where V_0 and F_0 represent the original video frame and its feature respectively; V_2 and F_2 represent the distorted video frame and its feature respectively; and $I(V_0;V_2)$ is the mutual information between the original video frame and the distorted video frame. L_f is the size or length of the selected feature.

Intuitively, the relationship between the feature difference and the video distortion should meet following two requirements:

a. The difference should be zero if the original video frame and its distorted video frame are identical.
b. The difference should be maximized if the original video frame and the distorted video frame are independent.

Thus, we have

$$\alpha=\frac{L_f}{I(V_0;V_0)}=\frac{L_f}{H(V_0)} \qquad (3)$$

where $H(V_0)$ represents the entropy of the original video frame [6].

Equation (3) can also be acquired from the angle of information theory because the entropy of an image can be used to represent the complexity of this image. Given a video frame, more complexity it is, the larger its entropy is. On the other hand, the size of feature extracted from a video frame increases with the complexity of this video frame. Thus, a relationship between entropy of a video frame and the size of feature, which is similar to equation (3), must exist.

Substituting equation (3) into (2), equation (2) can be rewritten as

$$d(F_0,F_2)=L_f-L_f*\frac{I(V_0;V_2)}{H(V_0)} \qquad (4)$$

This is the relationship between the feature difference and the video distortion. From this equation, we can know that the feature difference between two video frames only depend on the mutual information of these two video frames because $H(V_0)$ will be fixed once a video frame is given. Therefore, the computation of the feature difference becomes that of the mutual information. Again, mutual information computation is a process of the distribution estimation for video frames.

Now let's take a look back at L_f . In [4], features are all converted into Quasi-Gray binary code, called feature code, to ensure that one-bit change in feature code only represents one unit modification on the feature of video content so that the difference between two features can be measured by just calculating the Hamming distance. So, in this paper, we will assume that all features are converted into Quasi-Gray binary code. Thus, the two terms "feature difference" and "feature distance" are interchangeable in this paper.

In the next subsection, two important applications of the derived relationship will be introduced: one is to estimate the maximum difference between feature of the original video frame and that of the processed video frame if the video processing is acceptable; the other is to show how the feature difference varies with the quantization step in video compression.

2.2 Two applications

2.2.1 Maximum allowable feature difference

As shown in Figure 1, if the video only undergoes normal video processing, the distortion introduced by watermark embedding and video processing must be imperceptible. In other words, this distortion should be limited. Let D represents the maximum allowable distortion between

V_0 and V_2, the maximum allowable feature difference can be calculated as follows:

$$\max(d(F_0;F_2)) = \max_{E(d(V_0;V_2))\le D} (L_f - \alpha * I(V_0;V_2)) \quad (5)$$

$$= L_f - \alpha * \min_{E(d(V_0;V_2))\le D} (I(V_0;V_2)) \quad (6)$$

$$= L_f - \alpha * R(D) \quad (7)$$

The second term of equation (7) is the definition of Rate Distortion function. Since $R(D)$ is a non-increasing convex function of D [6], the maximum feature difference could be calculated if the maximum allowable distortion D is known. For a given video frame, we consider its Just Noticeable Difference (JND) as the maximum acceptable distortion in the video authentication system. Thus, the maximum feature difference can be obtained on the JND.

2.2.2 Feature difference and video compression

In video compression, video quality degradation mainly comes from quantization. So we will look for the relationship between the feature difference and the quantization step since the video compression is considered as a normal processing in video authentication. Let C_i be the original DCT coefficient and q_i the quantization step. Then, the reconstructed DCT coefficient ($C_i^{'}$) is

$$C_i^{'} = C_i + \Delta(q_i) \quad (8)$$

where $\Delta(q_i)$ is considered as an additive uniformly distributed noise [8]; and the probability density function of $\Delta(q_i)$ is given by

$$p(\Delta(q_i)) = \begin{cases} 1/q_i & if \quad |\Delta(q_i)| < q_i/2 \\ 0 & Others \end{cases} \quad (9)$$

Using the same notation in Section 2.1, the difference between features of the original video frame and the reconstructed video frame can be calculated as

$$d(F_0;F_2) = L_f - a * I(V_0;V_2)$$

$$= \alpha * (H(V_0) - I(V_0;V_2)) \quad (10)$$

According to properties of entropy and mutual information,

$$H(V_0) - I(V_0;V_2) = H(V_0 / V_2) \quad (11)$$

$$= H((V_0 - V_2)/V_2) \quad (12)$$

Following that conditioning reduces the entropy, we further get

$$H(V_0) - I(V_0;V_2) \le H(V_0 - V_2) \quad (13)$$

$$= H(\sum_i (C_i - C_i^{'})) \quad (14)$$

Using the theorem, termed as independence bound on entropy, in information theory, we have

$$H(\sum_i (C_i - C_i^{'})) \le \sum_i H(C_i - C_i^{'}) \quad (15)$$

$$= \sum_i H(\Delta(q_i)) \quad (16)$$

$$= \sum_i \log(q_i) \quad (17)$$

Equation (17) is the entropy of variable $\Delta(q_i)$ with a uniform distribution. Therefore, the upper bound for feature difference between the original video frame and the reconstructed video frame can be finally expressed as

$$d_{\max}(F_0;F_2) = \alpha * \sum_i \log(q_i) \quad (18)$$

If all the quantization steps are identical to be q, equation (18) can be further written as

$$d_{\max}(F_0;F_2) = \beta * \log(q) \quad (19)$$

This relationship clearly indicates how compression affects the feature difference between the original video frame and its compressed version.

3. EXPERIMENATAL RESULTS

In this section, we use the feature selected in [4] for evaluation. It is a 44 bits binary data. Please refer to [4] for more detail on feature selection. During evaluation, video "Akiyo" is used as the testing video.

Firstly, we compute the maximum allowable feature difference based on the JND given by Watson [9]. During computation, the DCT coefficients are classified into 64 independent channels by placing the coefficients in the same position in the DCT blocks into the same channel. These 64 channels are scanned in Zig-Zag order before the first 30 channels are selected for computation. This is in line with the fact that features selected in video authentication system usually only represent the low and middle frequency information due to the requirement of robustness. We also assume that the channels are Gaussian channels except that the DC channel is assumed to be a Laplacian channel. The upper bound of the maximum allowable feature difference is shown in Figure 2. From this figure, we can see that the maximum allowable feature differences are quite stable within the whole video sequence. Similar results have also been obtained in evaluating other testing videos. This indicates that the maximum allowable feature difference is a value almost independent of video content. Thereafter, the maximum allowable feature difference could be calculated before a robust video authentication system is designed. Note that, however, the calculated value "6" is different from the value "3" that we obtained in experiments [4].

This is due to two factors: one is that the value in Figure 2 is an upper bound; the other is that the selected feature only partially reflects the information in video frames and is not as sensitive as expected. That is, we need to understand that there is always a trade-off between robustness and sensitivity for feature selection in video authentication.

Secondly, we evaluate the feature difference between the original video frame and the reconstructed video frame. The relationship between the feature difference and quantization step is shown in dashed line in Figure 3. For comparison, we also test the theoretical relationship based on equation (19), shown in solid line in Figure 3. The experimental results are not very close to the analytical results derived in Section 2 but agree in terms of tendency.

4. CONCLUSIONS

In this paper, we have derived an analytical relationship between the feature difference and the video distortion based on Mutual Information for video authentication. To evaluate its validity, we applied it to estimate the maximum allowable feature difference, which is an important parameter in designing a robust video authentication system. In addition, we also showed how feature difference varies with the quantization step in video compression. Experimental results have confirmed the validity of our analytical results and the usefulness of the derived relationship in the design of a robust video authentication system.

The same approach can also be applied to relate the feature difference to the geometrical manipulations such as rotation and scaling. In the future, we will investigate how to select/combine different features according to the theoretical results in the design of a robust video authentication system.

11. REFERENCES

[1] Martinian, E.; Wornell, G.W.; and Chen, B., "Authentication with Distortion Criteria", Submitted to *IEEE Trans. Inform. Theory*.

[2] Ditmann, J.; Steinmetz, A.; and Steinmetz, R., "Content-based digital signature for motion pictures authentication and content-fragile watermarking", *Multimedia Computing and Systems*, 1999. IEEE International Conference on, Volume: 2, 1999, Page(s): 209 -213 vol.2.

[3] Queluz, M.P., "Towards robust, content based techniques for image authentication", *Multimedia Signal Processing*, 1998, IEEE Second Workshop on, 1998, Page(s): 297 –302.

[4] Dajun He; Qibin Sun; and Qi Tian, "A Semi-fragile Object Based Video Authentication System", *Circuits and Systems*, 2003, ISCAS '03, Proceedings of the 2003 International Symposium on, Volume: 3, May 25-28 2003 Page(s): 814 -817.

[5] Qibin Sun; Shih-Fu Change; Maeno, K; and Suto,M , " A New Semi-fragile Image Authentication Framework Combining ECC and PKI Infrastructures", IEEE International Symposium on *Circuits and Systems*, 2002, ISCAS 2002, Volume: 2, Page(s): 440-443.

[6] T. M. Cover, J.A. Thomas, *Elements of Information Theory*, New York: John Wiley & Sons, 1991.

[7] Maes, F.; Collignon, A.; Vandermeulen, D.; Marchal, G.; and Suetens, P., "Multi-modality image registration by maximization of mutual information", *Mathematical Methods in Biomedical Image Analysis*, 1996., Proceedings of the Workshop on , 21-22 June 1996, Page(s): 14 -22.

[8] Kundur, D., "Implications for high capacity data hiding in the presence of lossy compression", *Information Technology: Coding and Computing*, 2000, Proceedings. International Conference on, 27-29 March 2000, Page(s): 16 -21

[9] Watson, A.B., "Visually optimal DCT quantization matrices for individual images", Data Compression Conference, 1993. DCC '93, 30 March-2 April 1993, Page(s): 178 -187.

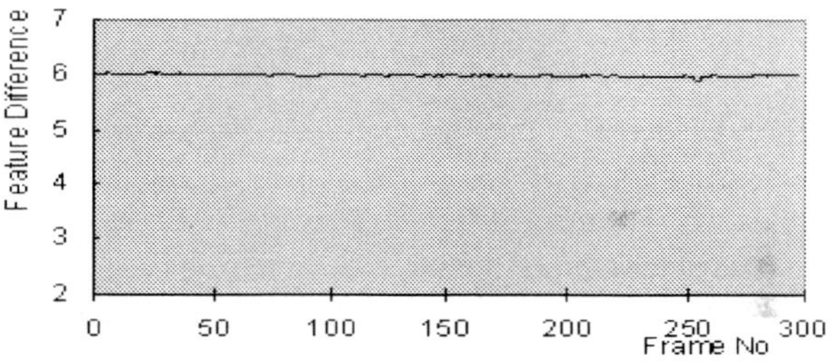

Figure 2 Maximum allowable feature differences for video "Akiyo". The horizontal axis represents frame number. From the figure, we can find that all differences are around 6.

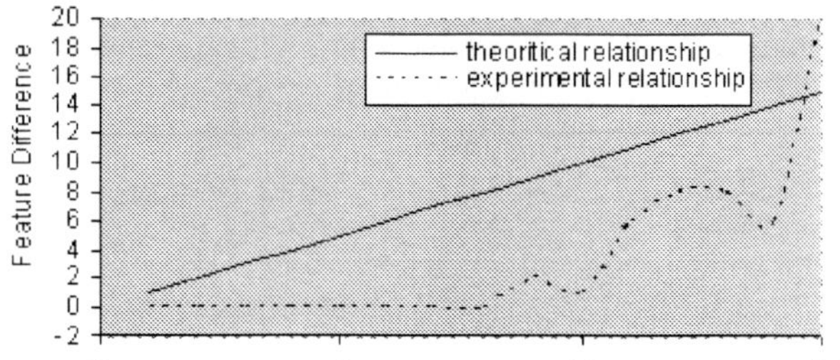

Figure 3 The relationship between feature difference and quantization step. The solid line represents the relationship based on theory; the dashed line is the experimental result. Two results agree in terms of tendency.

PREDICTION OF BRAIN ELECTRICAL ACTIVITY IN EPILEPSY USING A HIGHER–DIMENSIONAL PREDICTION ALGORITHM FOR DISCRETE TIME CELLULAR NEURAL NETWORKS (DTCNN)

F. Gollas, C. Niederhöfer, R. Tetzlaff

Johann Wolfgang Goethe University, Frankfurt am Main, Germany
Institute of Applied Physics
Robert-Mayer-straße 2-4
f.gollas@iap.uni-frankfurt.de

ABSTRACT

Several investigations have shown, that a higher–dimensional nonlinear signal analysis can contribute to the problem of detecting precursors for impending epileptic seizures in electroencephalographic recordings. In previous work we analyzed brain electrical activity using Volterra-Systems [1] and CNN [2]. The outline of this paper is to propose a new higher-dimensional DTCNN prediction algorithm. First results will be given for a long term recording of brain electrical activity.

1. INTRODUCTION

Epilepsy is the most common chronical disorder of the nervous system [3], and affects approximately 1% of the worlds population. The mechanisms responsible for the generation of seizures in common are not well understood.

Although, a bulk of literature has been published, finding precursors of impending seizures is an outstanding problem. Recent investigations have shown that nonlinear time series analysis of invasive electroencephalographic (EEG) recordings can contribute to a better understanding of neural behavior in epilepsy.

Linear prediction algorithms are well understood and commonly used since a long time in a wide range of applications e.g. in the field of speech analysis and coding. The general form for a predictor of the order p, providing an estimator for the signal $\mathbf{x}(n)$ out of p previous samples, is given by

$$\hat{\mathbf{x}}(n) = \mathcal{P}(\mathbf{x}(n-1), \mathbf{x}(n-2), \ldots, \mathbf{x}(n-p)). \quad (1)$$

By minimiziation of the power of the prediction error

$$\Xi(n) = \mathbf{x}(n) - \hat{\mathbf{x}}(n)). \quad (2)$$

an approximation of $x(n)$ is obtained, exceedingly in regard to the short time behavior of the signal generating system. In previous investigations we studied the prediction of brain

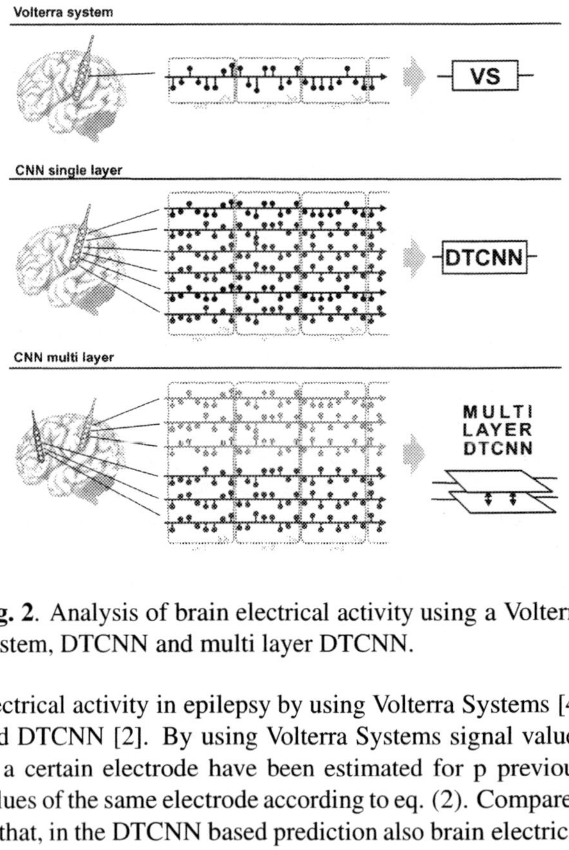

Fig. 2. Analysis of brain electrical activity using a Volterra System, DTCNN and multi layer DTCNN.

electrical activity in epilepsy by using Volterra Systems [4] and DTCNN [2]. By using Volterra Systems signal values of a certain electrode have been estimated for p previous values of the same electrode according to eq. (2). Compared to that, in the DTCNN based prediction also brain electrical activity from neighboring electrodes of the same electrode array have been considered, i.e.

$$\begin{aligned} x_j(n) \;=\; & \mathcal{P}[x_{j-1}(n-1), \ldots, x_{j-1}(n-p)| \\ & x_j(n-1), \ldots, x_j(n-p)| \\ & x_{j+1}(n-1), \ldots, x_{j+1}(n-p)] \end{aligned} \quad (3)$$

where j denote the electrode. By taking the assumption that

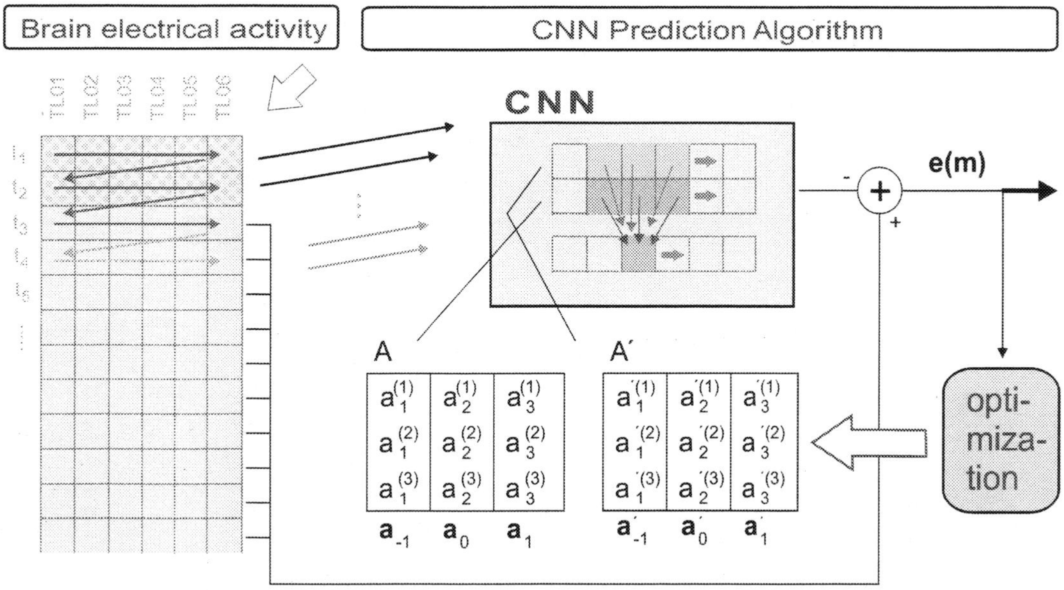

Fig. 1. Schematic flowchart of the below described single layer algorithm.

the interaction between different brain regions change significantly before the onset of an epileptic seizure we propose in this contribution a generalization of the DTCNN algorithm to two different electrode arrays $l, l^\dagger$. Thereby the prediction is performed according to

$$\begin{aligned} x_j^l(n) \;=\; & \mathcal{P}[x_{j-1}^l(n-1),\ldots,x_{j-1}^l(n-p)| \\ & x_j^l(n-1),\ldots,x_j^l(n-p)| \\ & x_{j+1}^l(n-1),\ldots,x_{j+1}^l(n-p)| \\ & x_{j-1}^{l^\dagger}(n-1),\ldots,x_{j-1}^{l^\dagger}(n-p)| \\ & x_j^{l^\dagger}(n-1),\ldots,x_j^{l^\dagger}(n-p)| \\ & x_{j+1}^{l^\dagger}(n-1),\ldots,x_{j+1}^{l^\dagger}(n-p)]. \end{aligned} \tag{4}$$

Apparently two types of coupling, the couplings within one electrode array and also the couplings between different arrays, are considered. The different prediction schemes are given in Fig. 2.

2. PREDICTION BY DTCNN USING NONLINEAR WEIGHT FUNCTIONS

In this paper autonomous delay type DTCNN with nonlinear polynomial type weight functions are considered for a prediction of brain electrical activity, which can be described by state equations according to

$$\begin{aligned} x_i(t_{n+1}) = \; & -x_i(t_n) + \\ & \sum_{\tau=0}^{T} \sum_{k=1}^{K} \sum_{j \in \mathcal{S}_i(r)} a_j^{(k)(\tau)} x_j^k(t_{n-\tau}) \end{aligned} \tag{5}$$

where $\mathcal{S}_i(r)$ denotes the sphere of influence of cell S_i and $id(x)$ is chosen as output function.

Following the considerations in [5, 6] 30 s of signal data are assumed to be quasi stationary. The data acquired from depth electrodes during presurgical diagnostics in the University of Bonn is sampled at a rate of 200 Hz. According to this ascertainments, brain electrical activity is segmented into segments of 6000 values. Signals from electrodes located in the region of the epileptic focus and electrodes in another area of the brain are used for the following investigations.

With a direct neighborhood interaction, layers l and delay order one the state equation eq. (5) can be regarded as predictor given by

$$\begin{aligned} \hat{x}_i^l(t_{n+1}) = \; & -x_i^l(t_n) + \\ & \sum_{l^\dagger=1}^{2} \sum_{j=-1}^{i+1} a_j^{l^\dagger l}(x_{i+j}^{l^\dagger}(t_n)) + \\ & \sum_{l^\dagger=1}^{2} \sum_{j=-1}^{i+1} a'^{l^\dagger l}_j(x_{i+j}^{l^\dagger}(t_{n-1})) \end{aligned} \tag{6}$$

where $a_j^{l^\dagger l}$ and $a'^{l^\dagger l}_j$ denote the couplings between layer $l^\dagger$ and layer l and the cell coupling is given by third order polynomials

$$\begin{aligned} a_j^{l^\dagger l}(x_i(t_n)) &= \sum_{k=1}^{3} a_j^{(k)\,l^\dagger l}\, {x_i^{l^\dagger}}^k(t_n) \\ a'^{l^\dagger l}_j(x_i(t_{n-1})) &= \sum_{k=1}^{3} a'^{(k)\,l^\dagger l}_j\, {x_i^{l^\dagger}}^k(t_{n-1}). \end{aligned} \tag{7}$$

The dynamics of the underlying DTCNN depends on a gene [2] of 72 parameters. During the optimization procedure the

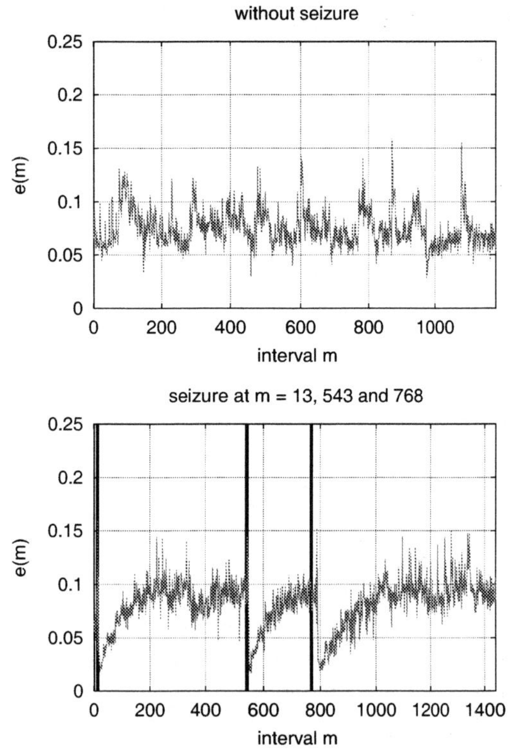

Fig. 3. Single Layer DTCNN - **Up** : Segment of brain electrical activity not containing any seizures. No distinct changes can be observed. **Down** : Segment of brain electrical activity containing 3 seizures. During and after a seizure the error drops down.

relative mean square error for a data segment m defined by

$$e(m) = \sum_{l} \sum_{n} \sum_{i} \frac{(x_i^l(t_n) - \hat{x}_i^l(t_n))^2}{x_i^{l\,2}(t_n)} \qquad (8)$$

has to by minimized. During optimization the Powell [7] method is deployed.

In our investigations, the estimation of the next value is not based on the previously estimated values, but on the original brain electrical activity. Therefore after the calculation of the error in each timestep the cellstate is reinitialized with recorded EEG data.

3. RESULTS

In Fig. 1 the calculation process of the above described algorithm from the data acquisition to the result for a single layer is depicted. The used long term data originated from a male patient of 19 years. It lasts more than 6 days of continuous measurement and contains 10 seizures. The sampling rate is 200 Hz and 48 channels were recorded.

A data segment is taken as the state of a DTCNN, afterwards the randomly initialized templates are applied and the

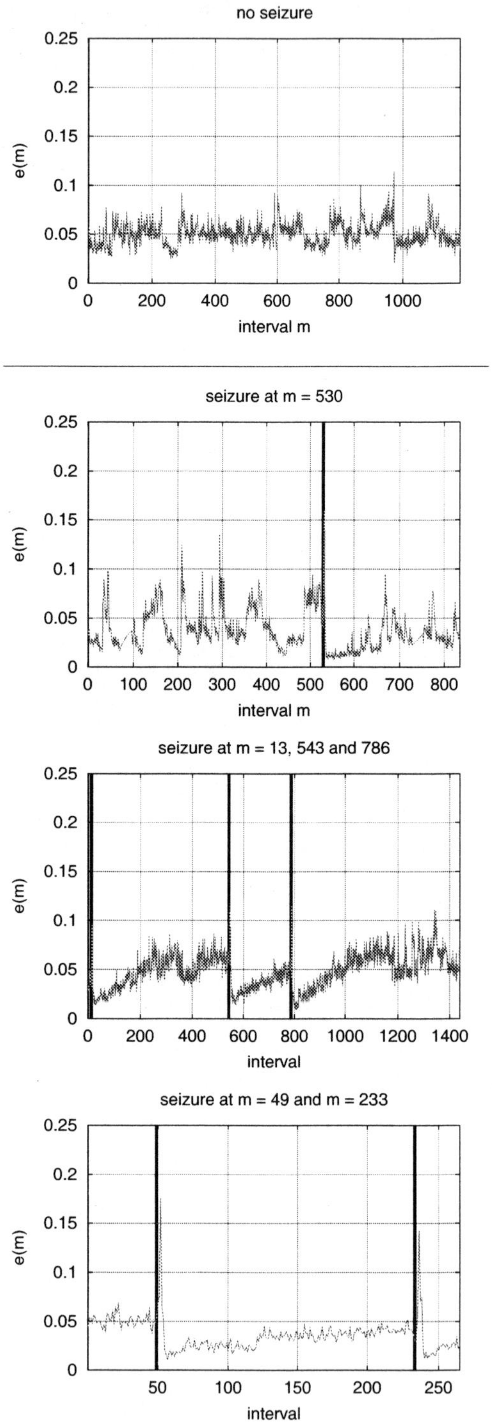

Fig. 4. Multi Layer DTCNN - **Up** : Segment of brain electrical activity not containing any seizures. No distinct changes can be observed. **Down** : Segment of brain electrical activity containing 3 seizures. Each time during and after a seizure the error drops down.

prediction error is calculated for each timestep. The parameters of the system are updated when the error for the whole segment is calculated. This is done by using the SCNN simulation system [8, 9]. The results obtained in previous investigations [4, 10] are based on short time recordings of several patients suffering from epileptic seizures. The following results were obtained by the analysis of the above mentioned long term recordings.

Fig. 3 shows the obtained prediction errors of one segment of brain electrical activity which does not contain any seizures and one segment of brain electrical activity containing 3 seizures for the single layer DTCNN. Regarding this result the observed value of the prediction error $e(m)$ does not show any distinct changes in the upper figure but during and after a seizure in the down figure there is a distinct change in the behavior of the observed error. Every time a seizure takes place the value of the prediction error drops down from its normal mean of about 0.1. After a seizure a slow but continuous recovery to the normal mean can be seen.

In Fig. 4 the results for the multilayer calculations are depicted. In the upper part the data segment without a seizure can be seen and there is no distinct change in the behavior of the prediction error. In the lower part of the figure three different datasets can be seen, all containing one or more seizures and showing distinct changes during and after the seizure. As in the single layer results the prediction error drops down from its former mean and slowly recovers after a seizure.

4. CONCLUSION

In this paper, new results for the prediction of brain electrical activity in epilepsy are discussed. Thereby the observed prediction error e_m shows distinct changes during an epileptic seizure. These first results indicate that the proposed prediction procedure may be useful for analysis of brain electrical activity in epilepsy.

5. ACKNOWLEDGMENTS

The authors would like to thank K. Lehnertz and C. E. Elger from the University Clinic of Epileptology in Bonn, Germany, who provided the invasive recorded brain electrical activity.

6. REFERENCES

[1] M. Schetzen, *The Volterra and Wiener Theories of Nonlinear Systems*, Wiley, 1980.

[2] Leon O. Chua, *CNN: A Paradigm for Complexity*, vol. 31 of *A*, World Scientific Series on Nonlinear Science, University of California, Berkeley, 1998.

[3] Jr. M.D. Ph.D. Jerome Engel, *Seizures and Epilepsy*, Contemporary neurology series; 31, 1989.

[4] C. Niederhöfer, S. Suna, and R. Tetzlaff, "Nonlinear prediction of brain electrical activity in epilepsy with a volterra rls algorithm," in *2002 IEEE International Symposium on Circuits and Systems*, 2002.

[5] T. H. Bullok, J. Z. Achimowicz, R. B. Duckrow, S. S. Spencer, and V. J. Iragui-Madoz, "Biocoherence of intracranial ""eeg"" in sleep, wakefulness and seizures," *Elsevier, Electroencephalography and clinical Neurophysiology*, , no. 103, pp. 661–678, 1997.

[6] G. Dumermuth, P.J. Huber, B. Kleiner, and Th. Gasser, "Analysis of the interrelations between frequency bands of the ""eeg"" by means of the bisprectrum. a preliminary study.," *Elsevier, Electroencephalography and Clinical Neurophysiology*, , no. 31, pp. 137–148, 1971.

[7] W. H. Press, S. A. Teukolsky, W. T. Vetterling, and B. P. Flannery, *Numerical Recipes in C: The Art of Scientific Computing*, Cambridge University Press, New York, 1992.

[8] C. Ames, D. Feiden, G. Geis, R. Kunz, A. Loncar, F. Puffer, R. Schoenmeyer, R. Tetzlaff, and D. Weiß, *SCNN 2000 Dokumentation*, Institut für Angewandte Physik, Johann Wolfgang Gothe Universität Frankfurt am Main, 2000.

[9] A. Loncar, R. Kunz, and R. Tetzlaff, "Scnn 2000 - part i: Basic structure and features of the simulation system for cellular neural networks," Catania, 2000.

[10] D. Weiß, "Nichtlineare Prädiktion hirnelektrischer Aktivität mit Zellularen Nichtlinearen Netzwerken bei Epilepsie," Diploma Thesis, Johann Wolfang Goethe University, Frankfurt am Main, Germany, 2001.

Fuzzy Neural Network Classification Design Using Support Vector Machine

Chin-Teng Lin, Chang-Moun Yeh, and Chun-Fei Hsu
Department of Electrical and Control Engineering
National Chiao-Tung University, Hsinchu, Taiwan

Abstract

Fuzzy neural networks (FNNs) for pattern classification usually use the backpropagation or C-cluster type learning algorithms to learn the parameters of the fuzzy rules and membership functions from the training data. However, such kinds of learning algorithms usually cannot minimize the empirical risk (training error) and expected risk (testing error) simultaneously, and thus cannot reach a good classification performance in the testing phase. To tackle this drawback, a support-vector-based fuzzy neural network classification (SVFNNC) is proposed. The SVFNNC combines the superior classification power of support vector machine (SVM) in high dimensional data spaces and the efficient human-like reasoning of FNN in handling uncertainty information. The learning algorithm consists of two learning phases. In the phase 1, the fuzzy rules and membership functions are automatically determined by the clustering principle. In the phase 2, the parameters of FNN are calculated by the SVM with the proposed adaptive fuzzy kernel function. To investigate the effectiveness of the proposed SVFNNC, it is applied to the Iris, Vehicle and Dna datasets. Experimental results show that the proposed SVFNNC can achieve good classification performance with drastically reduced number of fuzzy kernel functions.

Index Terms – Fuzzy neural network, fuzzy kernel function, support vector machine.

I. Introduction

Much research has been done on fuzzy neural networks (FNNs), which combine the capability of fuzzy reasoning in handling uncertain information [1] and the capability of neural networks in learning from processes [2]. They have been successfully applied to classification, identification, control, pattern recognition, and image processing, *etc* [3]. In particular, many learning algorithms of fuzzy (neural) classifiers have been presented and applied in pattern classification and decision-making systems [4]. Conventionally, the selection of fuzzy if-then rules often relies on a substantial amount of heuristic observation to express proper strategy's knowledge. Obviously, it is difficult for human experts to examine all the input-output data to find a number of proper rules for the fuzzy system. Most pre-researches used the backpropagation (BP) and/or C-cluster type learning algorithms to train parameters of fuzzy rules and membership functions from the training data [5], [6]. However, such learning only aims at minimizing the classification error in the training phase, and it cannot guarantee the lowest error rate in the testing phase. In statistical learning theory, the support vector machine (SVM) [7] has been developed for solving this bottleneck. The SVM performs structural risk minimization and creates a classifier with minimized VC dimension. As the VC dimension is low, the expected probability of error is low to ensure a good generalization. The SVM keeps the training error fixed while minimizing the confidence interval. So, the SVM has good generalization ability and can simultaneously minimize the empirical risk and the expected risk for pattern classification problems. However, the optimal solutions of SVM rely heavily on the property of selected [8] kernel functions, whose parameters are always fixed and are chosen solely based on heuristics or trial-and-error nowadays.

In this paper, we develop a support-vector-based fuzzy neural network classification (SVFNNC), which combines the superior classification power of SVM in high dimensional data spaces and the high efficient human-like reasoning power of FNN in handling uncertainty information. The SVFNNC is the realization of a new idea for the adaptive kernel functions used in the SVM. The use of the proposed fuzzy kernels provides the SVM with adaptive local representation power, and thus brings the advantages of FNN (such as adaptive learning and economic network structure) into the SVM directly. Finally, the experimental results on four datasets (Iris, Vehicle, Dna) from the UCI Repository and Statlog collection show that the proposed SVFNNC classification method can automatically generate the fuzzy rules, improve the accuracy of classification, reduce the number of required kernel functions, and increase the speed of classification.

II. Construction of Initial Fuzzy Neural Network

A four-layered fuzzy neural network (FNN) is shown in Fig 1, which is comprised of the input, membership function, rule, and output layers. This four-layered network realizes the following form of fuzzy rules:

Rule R_j: If x_1 is A_{j1} and .. and x_m is A_{jm},
Then y is d_j, $j=1, 2, \cdots, n$, (1)

where A_{jm} are the fuzzy sets of the input variables x_i, and d_j are the consequent parameter of y. For the ease of analysis, a fuzzy rule 0 is added as:

Rule 0 : If x_1 is A_{01} and .. and x_m is A_{0m}
Then y is d_0 (2)

where $A_{0k}=1$ for $k=1, 2, \cdots, m$ are the fuzzy sets of the input variables, and d_0 is the consequent parameter of y in the fuzzy rule 0. Thus, the overall output as the summation of all input signals:

$$O = \sum_{j=1}^{n} d_j \times a_j + d_0, \tag{3}$$

where the connecting weight d_j is the output action strength the scalar d_0 is a bias and a_j, $j = 1, 2, \cdots, m$ are the input variables of the FNN. Thus the FNN mapping can be rewritten in the following input-output form:

$$O = \sum_{j=1}^{n} d_j \times a_j + d_0 = \sum_{j=1}^{n} d_j \prod_i u_i^j(x_i) + d_0. \tag{4}$$

Moreover, for constructing the initial fuzzy rules of the FNN, the fuzzy clustering method is used to partition a set of data into a number of overlapping clusters based on the distance in a metric space between the data points and the cluster prototypes. Each cluster in the product space of the input-output data represents a rule in the rule base. The goal is to establish the fuzzy preconditions in the rules. In this work, we use an aligned clustering-based approach proposed in [9]. This method produces a partition result as shown in Fig. 2.

A rule corresponds to a cluster in the input space, with $\mathbf{m}_i$ and $\mathbf{D}_i$ representing the center and variance of that cluster. For each incoming pattern **x**, the strength a rule is fired can be interpreted as the degree the incoming pattern belongs to the corresponding cluster. For computational efficiency, we can use the firing strength derived in (3) directly as this degree measure

$$F^i(\mathbf{x}) = \prod_i a_i = e^{-[\mathbf{D}_i(\mathbf{x}-\mathbf{m}_i)]^T[\mathbf{D}_i(\mathbf{x}-\mathbf{m}_i)]} \in [0,1] \tag{5}$$

where $\mathbf{x}=[x_1, x_2, x_3, \cdots, x_m]^T$ is the FNN input vector, and $F^i(\mathbf{x}) \in [0, 1]$. In the above equation the term $[\mathbf{D}_i(\mathbf{x}-\mathbf{m}_i)]^T[\mathbf{D}_i(\mathbf{x}-\mathbf{m}_i)]$ is the distance between **x** and the center of cluster i. Using this measure, we can obtain the following criterion for the generation of a new fuzzy rule. Let **x** be the newly incoming pattern. Find

$$J = \arg \max_{1 \le j \le c(t)} F^j(\mathbf{x}), \tag{6}$$

where $c(t)$ is the number of existing rules at time t. If $F^J \le \bar{F}(t)$, then a new rule is generated, where $\bar{F}(t) \in (0, 1)$ is a prespecified threshold that decays during the learning process. Once a new rule is generated, the next step is to assign initial centers and widths of the corresponding membership functions. Since our goal is to minimize an objective function and the centers and widths are all adjustable later in the following learning phases, it is of little sense to spend much time on the assignment of centers and widths for finding a perfect cluster. Hence we can simply set

$$\mathbf{m}_{[c(t)+1]} = \mathbf{x}, \tag{7}$$

$$\mathbf{D}_{[c(t)+1]} = \frac{-1}{\beta} \cdot diag\left[\frac{1}{\ln(F^J)} \cdots \frac{1}{\ln(F^J)}\right] \tag{8}$$

according to the first-nearest-neighbor heuristic, where $\beta \ge 0$ decides the overlap degree between two clusters.

III. Fuzzy Kernel

The proposed fuzzy kernel $K(\hat{\mathbf{x}}, \hat{\mathbf{z}})$ in this paper is defined as

$$K(\hat{\mathbf{x}}, \hat{\mathbf{z}}) = \begin{cases} \prod_{i=1}^{n} u_j(x_i) \cdot u_j(z_i), & \text{if } x_i \text{ and } z_i \text{ are both in the } j\text{-th cluster} \\ 0, & \text{otherwise,} \end{cases} \tag{9}$$

where $\hat{\mathbf{x}} = [x_1, x_2, x_3, \cdots, x_n] \in R^n$ and $\hat{\mathbf{z}} = [z_1, z_2, z_3, \cdots, z_n] \in R^n$ are any two training samples, and $u_j(x_i)$ is the membership function of the j-th cluster. Assume the training samples $\mathbf{s}=\{(x_1, y_1), (x_2, y_2), \cdots, (x_n, y_n)\} \in X$ are partitioned into m classes of clusters through fuzzy clustering in Section II and n is the total number of training samples. We can perform the following permutation of training samples

$$\begin{aligned} &cluster\ 1 = \{(\boldsymbol{x}_1^1, y_1^1), \ldots, (\boldsymbol{x}_{k_1}^1, y_{k_1}^1)\} \\ &cluster\ 2 = \{(\boldsymbol{x}_1^2, y_1^2), \ldots, (\boldsymbol{x}_{k_2}^2, y_{k_2}^2)\} \\ &\vdots \\ &cluster\ m = \{(\boldsymbol{x}_1^m, y_1^m), \ldots, (\boldsymbol{x}_{k_m}^m, y_{k_m}^m)\}, \end{aligned} \tag{10}$$

where k_i is the number of points belonging to the i-th cluster, so that we have $\sum_{i=1}^{m} k_i = n$. Then the fuzzy kernel can be calculated by using the training set in (10), and the obtained kernel matrix **K** can be rewritten as the following form

$$\mathbf{K} = \begin{bmatrix} \mathbf{K}_1 & \mathbf{0} & \cdots & \mathbf{0} \\ \mathbf{0} & \mathbf{K}_2 & \ddots & \vdots \\ \vdots & \ddots & \ddots & \mathbf{0} \\ \mathbf{0} & \cdots & \mathbf{0} & \mathbf{K}_m \end{bmatrix} \in R^{n \times n} \tag{11}$$

where $\mathbf{K}_i$ is defined as

$$\mathbf{K}_i = \begin{bmatrix} K(x_1^i, x_1^i) & K(x_1^i, x_2^i) & \cdots & K(x_1^i, x_{k_i}^i) \\ K(x_2^i, x_1^i) & K(x_2^i, x_2^i) & \ddots & \vdots \\ \vdots & \ddots & \ddots & K(x_{k_i-1}^i, x_{k_i}^i) \\ K(x_{k_i}^i, x_1^i) & \cdots & K(x_{k_i}^i, x_{k_i-1}^i) & K(x_{k_i}^i, x_{k_i}^i) \end{bmatrix} \in R^{k_i \times k_i} \tag{12}$$

In order that the fuzzy kernel function defined by (11) is suitable for application in SVM, we must prove that the fuzzy kernel function is symmetric and positive-definite Gram Matrices.

Theorem 1 : For the fuzzy kernel defined by (11), if the membership functions $u(x_i): R \to [0, 1]$, $i = 1, 2, \cdots, n$, are positive-definite functions, then the fuzzy kernel is a Mercer kernel.

IV. Learning Algorithm of SVFNNC

The learning algorithm of the SVFNNC consists of two phases. The first phase establishes initial fuzzy rules. In the second phase, the optimal parameters of SVFNNC are calculated by SVM technique. The details are given below:

Learning Phase 1 – Establishing initial fuzzy rules

The whole algorithm for the generation of new fuzzy rules as well as fuzzy sets in each input variable is as follows. Suppose no rules are existent initially. IF **x** is the first incoming input pattern THEN do

PART 1. {

Generate a new rule with center $\mathbf{m}_1 = \mathbf{x}$ and width $\mathbf{D}_1 = diag\left(\frac{1}{\sigma_{init}}, \cdots, \frac{1}{\sigma_{init}}\right)$, where σ_{init} is a prespecified constant. After decomposition, we have one-dimensional membership functions, with $m_{1i} = x_i$ and $\sigma_{1i} = \sigma_{init}$, $i = 1, \cdots, n$. In addition, after we determine the precondition part of fuzzy rule, we also need to properly assign the consequence part of fuzzy rule. Here we define two output nodes for doing two-cluster recognition. If output node 1 obtains bigger exciting value, we know this input-output pattern belongs to class 1. Hence, initially, we should assign the proper weight w_{Con-1} for the consequence part of fuzzy rule as follows:

{ IF the output pattern $\mathbf{y}$ belongs to class 1 (namely, $\mathbf{y} = [1 \ \ 0]$), $w_{Con-1} = [1 \ \ 0]$ for indicating output node 1 been excited,

ELSE

$w_{Con-1} = [0 \ \ 1]$ for indicating output node 2 been excited. } }

ELSE for each newly incoming input $\mathbf{x}$, do

PART 2. {

Find $J = \arg \max_{1 \le j \le c(t)} F^j(\mathbf{x})$, as defined in (5). We should check if the newly incoming output pattern $\mathbf{y}$ is different from the maximal excited rule:

{ IF $w_{Con-J} \neq \mathbf{y}$,{ set $c(t+1) = c(t) + 1$ and generate a new fuzzy rule, with $\mathbf{m}_{c(t+1)} = \mathbf{x}$, $\mathbf{D}_{c(t+1)} = \frac{-1}{\beta} diag\left(\frac{1}{\ln(F^J)}, \cdots, \frac{1}{\ln(F^J)}\right)$ and $w_{Con-c(t+1)} = \mathbf{y}$,

β decides the overlap degree between two clusters. In addition, after decomposition, we have $m_{new-i} = x_i$, $\sigma_{new-i} = -\beta \times \ln(F^J)$, $i = 1, \cdots, n$. Do the following fuzzy measure for each input variable i:

{ $Degree(i, t) \equiv \max_{1 \le j \le k_i} E\left[\mu(m_{new-i}, \sigma_{new-i}), \mu(m_{ji}, \sigma_{ji})\right]$,

where k_i is the number of partitions of the ith input variable, and $E(\cdot)$ is defined in (10).

IF $Degree(i, t) \le \rho(t)$ THEN adopt this new membership function, and set $k_i = k_i + 1$, ELSE set the projected membership function as the closest one.} }} ELSE

{ Continue to check if $F^J \ge \bar{F}_{in}(t)$. If answer is YES, we do nothing. Otherwise, we also generate a new fuzzy rule with $\mathbf{m}_{c(t+1)} = \mathbf{x}$, $\mathbf{D}_{c(t+1)} = \frac{-1}{\beta} diag\left(\frac{1}{\ln(F^J)}, \cdots, \frac{1}{\ln(F^J)}\right)$, and the respective consequent weight $w_{Con-a(t+1)} = \mathbf{y}$. In addition, we also need to do the fuzzy measure for each input variable i. }}

In the above algorithm, $c(t)$ is the rule number at time t, β decides the overlap degree between two clusters, We shall see later in **Learning Phase 2** how we can use the results from the SVM method to determine these weights.

Learning Phase 2 - Calculating the parameters of SVFNNC

Through learning phase (1), the initial structure of SVFNNC is established and we can then use SVM to find the optimal parameters of SVFNNC based on the proposed fuzzy kernels. The dual quadratic optimization of SVM is solved in order to obtain an optimal hyperplane for any linear or nonlinear space:

maximize $L(\vec{\alpha}) = \sum_{i=1}^{l} \alpha_i - \frac{1}{2} \sum_{i,j=1}^{l} y_i y_j \alpha_i \alpha_j K(\mathbf{x}_i, \mathbf{x}_j)$

subject to $0 \le \alpha_i \le C$, $i = 1, 2, \cdots, l$, and $\sum_{i=1}^{l} y_i \alpha_i = 0$,

where $K(\mathbf{x}_i, \mathbf{x}_j)$ is the fuzzy kernel in (11), $\mathbf{x}_i$'s are training data, and C is a user-specified positive parameter to control the tradeoff between complexity of the SVM and the number of nonseparable points. This quadratic optimization problem can be solved and a solution $\vec{\alpha}_0 = (\alpha_1^0, \alpha_2^0, \ldots\ldots, \alpha_{nsv}^0)$ can be obtained, where α_i^0 are Lagrange coefficients, and nsv is the number of support vectors. The corresponding support vectors $\mathbf{sv} = [\mathbf{sx}_1, \mathbf{sx}_2, \cdots, \mathbf{sx}_i, \cdots, \mathbf{sx}_{nsv}]$ can be obtained, and the constant (threshold) d_0 in (3) is

$$d_0 = \frac{1}{2}\left[\left(w_0 \cdot x^*(1)\right) + \left(w_0 \cdot x^*(-1)\right)\right] \text{ with } w_0 = \sum_{i=1}^{nsv} \alpha_i y_i x_i, \quad (13)$$

where nsv is the number of fuzzy rules (support vectors); the support vector $x^*(1)$ belongs to the first class and support vector $x^*(-1)$ belongs to the second class. Hence, the parameters in the fuzzy rules of SVFNNC can be calculated by $d_j = y_j \alpha_j$ in (3) and $\mathbf{m}_j = \mathbf{sx}_j$, j=1, 2, $\cdots$, nsv, where d_j is the coefficient corresponding to $\mathbf{m}_j = \mathbf{sx}_j$.

V. Experimental Results

The classification performance of the proposed SVFNNC is evaluated on four well-known benchmark datasets. From the UCI Repository, we choose one dataset: Iris dataset. From Statlog collection we choose two datasets: Vehicle, and Dna datasets. These three datasets will be used to verify the effectiveness of the proposed SVFNNC. We scale all training data to be in [-1, 1] and also accordingly adjust testing data to be in [-1, 1]. It is noted that in the original Iris and Vehicle datasets, testing data are not available, so we repeat the 2-fold cross-validation 100 times using different partitions of the datasets and report the best cross-validation rate here.

These experimental results show that the proposed SVFNNC is good at maintaining the good generalization ability. Moreover, performance comparisons among the existing fuzzy neural network classifiers [6], the RBF-kernel-based SVM [8], and the proposed SVFNNC are made in Table I. This table shows that the SVFNNC produces lower testing error rates as compared to FNN classifiers [10], [11], and uses less support vectors as compared to the normal SVM using fixed-width RBF kernels. In summary, the proposed SVFNNC exhibits better generalization ability on the testing data and use much smaller number of fuzzy rules.

VI. CONCLUSIONS

(3) This paper proposed a support-vector-based fuzzy neural network classifier (SVFNNC), which combines the superior classification power of support vector machine (SVM) in high dimensional data spaces and the efficient human-like reasoning of fuzzy neural network (FNN) in handling uncertainty information. The SVFNNC is the realization of a new idea for the adaptive kernel functions used in the SVM. The use of the proposed fuzzy kernels provides the SVM with adaptive local representation power, and thus brings the advantages of FNN into the SVM directly.

REFERENCES

[1] K. Tanaka and H. O. Wang, *Fuzzy Control Systems Design and Analysis*, New York: Wiley, 2001.

[2] B. Kosko, *Neural Networks and Fuzzy Systems*, Englewood Cliffs, NJ: Prentice-Hall, 1992.

[3] C. T. Lin and C. S. G. Lee, *Neural Fuzzy System: A Neural-Fuzzy Synergism to Intelligent System*, Englewood Cliffs, NJ: Prentice-Hall, 1996.

[4] L. Y. Cai and H. K. Kwan, "Fuzzy classifications using fuzzy inference networks," *IEEE Trans. Syst., Man, Cybern. Pt B*, vol. 28, pp. 334-347, June. 1998.

[5] B. Gabrys and A. Bargiela "General fuzzy min-max neural network for clustering and classification," *IEEE Trans. Neural Networks,* vol. 11, pp. 769-783, May 2000.

[6] K. Nozaki, H. Ishibuchi, and H. Tanaka, "Adaptive fuzzy rule-based classification system" *IEEE Trans. Fuzzy Syst.*, vol. 4, pp. 238-250, Aug. 1996.

[7] V. Vapnik, *Statistical Learning Theory,* New York: Wiley, 1998.

[8] C. W. Hsu and C. J. Lin, "A comparison of methods for multiclass support vector machines," *IEEE Trans. Neural Networks,* vol. 13, pp. 415-525. March 2002.

[9] C. F. Juang and C. T. Lin, "An on-line self-constructing neural fuzzy inference network and its applications," *IEEE Trans. Fuzzy Syst.*, vol. 6, pp. 12-32, Feb. 1998.

[10] H. M. Lee, C. M. Chen, J. M. Chen, and Y. L. Jou "An efficient fuzzy classifier with feature selection based on fuzzy entropy," *IEEE Trans. Syst., Man, Cybern. Pt B,* vol. 31, pp. 426-432, June 2001.

[11] M. R. Berthold and Jay Diamond, "Constructive training of probabilistic neural networks", *Neurocomputing,* vol. 19, pp. 167-183, 1998

TABLE I. Classification error rate comparisons among FNN classifier, SVM classifier, and SVFNNC, where NA means "not available".

Datasets	FNN [10, 11]		SVM [8]		SVFNNC	
	Number of rules	Error rate	Number of support vectors	Error rate	Number of rules	Error rate
Iris	NA	4.3%	16	3.3%	7	4%
Vehicle	NA	29.9%	343	1%	69	1.42%
Dna	NA	16.6%	1152	4.21%	304	5.4%

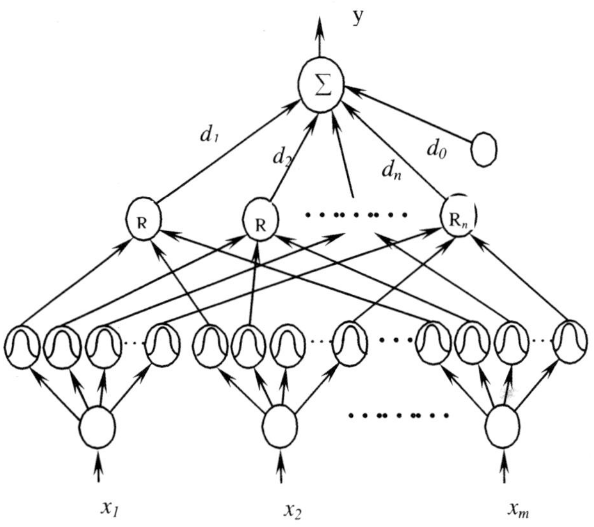

Fig. 1. The structure of the four-layered fuzzy neural network.

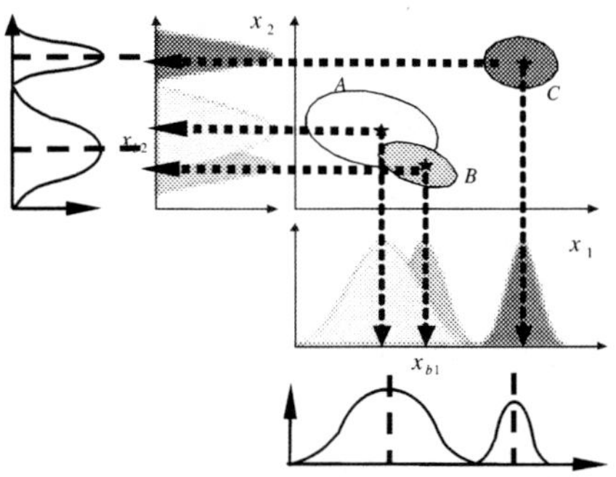

Fig. 2. The aligned clustering-based partition method giving both less number of clusters as well as less number of membership functions.

GLOBAL CONVERGENCE ANALYSIS OF DECOMPOSITION METHODS FOR SUPPORT VECTOR MACHINES

Norikazu Takahashi and Tetsuo Nishi

Department of Computer Science and Communication Engineering,
Kyushu University
6-10-1 Hakozaki, Higashi-ku, Fukuoka 812-8581 Japan

ABSTRACT

Convergence property of decomposition methods for solving quadratic optimization problems arising in support vector machines is studied. It is shown that under a mild condition any decomposition method converges to an optimal solution.

1. INTRODUCTION

Support vector machines (SVMs) have recently attracted great attention in various fields such as pattern recognition, machine learning, and signal processing because of their high generalization performance (see [1] for example).

Learning of an SVM leads to a quadratic programming (QP) problem. Since a large number of training examples are dealt with for SVM learning, the QP problem has a very large size in general. Although many techniques are available for solving QP problems, these conventional methods cannot be directly applied to the learning of SVM because large matrix computations are required. To overcome this difficulty, decomposition methods such as sequential minimal optimization (SMO) algorithm [2] and SVM^{light} [3] have been proposed.

A decomposition method tries to find an optimal solution by updating the values of variables iteratively. In each step, a small number of variables are selected based on some criterion and a QP problem with respect to the selected variables is solved. Since the QP problem to be solved in each iteration is small, large matrix computations can be avoided. It has been shown via computer simulations that decomposition methods are useful for solving large QP problems within reasonable computation time [2],[3],[4]. On the other hand, however, convergence properties of decomposition methods have not been fully understood so far.

This research was partly supported by Japan Society for the Promotion of Science, Grant-in-Aid for Encouragement of Young Scientists, 15760268, 2004 and the 21st Century COE Program 'Reconstruction of Social Infrastructure Related to Information Science and Electrical Engineering'.

In this paper, the convergence of a general form of decomposition methods is studied. We will first show that under a certain condition any decomposition method converges to an optimal solution of the QP problem. Our proof is a direct application of the well-known global convergence theorem [5] for more general optimization methods. We will next show that under the same condition any decomposition method with the stopping criterion stops in a finite number of iterations. The main result of this paper can be considered as a generalization of some existing results [6], [7], [8], [9].

2. PRELIMINARIES

For later discussions we give here some definitions and a key theorem which plays an important role in our proof. All of them can be found in Reference [5].

Definition 1 *Let $\Gamma \subset X$ be a given solution set and let $\boldsymbol{A}$ be a point-to-set mapping on X. A continuous real-valued function Z on X is said to be a descent function for Γ and $\boldsymbol{A}$ if it satisfies the following two conditions.*

i) If $\boldsymbol{x} \notin \Gamma$ and $\boldsymbol{y} \in \boldsymbol{A}(\boldsymbol{x})$ then $Z(\boldsymbol{y}) < Z(\boldsymbol{x})$.

ii) If $\boldsymbol{x} \in \Gamma$ and $\boldsymbol{y} \in \boldsymbol{A}(\boldsymbol{x})$ then $Z(\boldsymbol{y}) \leq Z(\boldsymbol{x})$.

Definition 2 *A point-to-set mapping $\boldsymbol{A}$ from X to Y is said to be closed at $\boldsymbol{x} \in X$ if the assumptions i) $\boldsymbol{x}(k)$ converges to $\boldsymbol{x}$ while belonging to X and ii) $\boldsymbol{y}(k)$ converges to $\boldsymbol{y}$ while belonging to $\boldsymbol{A}(\boldsymbol{x}(k))$ imply $\boldsymbol{y} \in \boldsymbol{A}(\boldsymbol{x})$. The point-to-set mapping $\boldsymbol{A}$ is said to be closed on X if it is closed at each point of X.*

Theorem 1 *Let $\boldsymbol{A}$ be a point-to-set mapping on X, and suppose that, given $\boldsymbol{x}(0)$ the sequence $\{\boldsymbol{x}(k)\}_{k=0}^{\infty}$ is generated satisfying $\boldsymbol{x}(k+1) \in \boldsymbol{A}(\boldsymbol{x}(k))$. Let a solution set $\Gamma \subset X$ be given, and suppose that*

i) all points $\boldsymbol{x}(k)$ are contained in a compact set $S \subseteq X$,

ii) there is a descent function Z for Γ and $\boldsymbol{A}$, and

0-7803-8251-X/04/$17.00 ©2004 IEEE

iii) the mapping A is closed at points outside Γ.

Then the limit of any convergent subsequence of $\{x(k)\}$ is a solution.

3. DECOMPOSITION METHOD

Let $\{(x_i, y_i)\}_{i=1}^{l}$ be a set of training examples, where $x_i \in \mathbb{R}^n$ is the i-th input pattern and $y_i \in \{1, -1\}$ the label of the class to which x_i belongs. Let $q_{ij} = y_i y_j K(x_i, x_j)$ where $K(\cdot,\cdot)$ denotes the kernel function used by a SVM. Then, learning of a SVM for the set of training examples $\{(x_i, y_i)\}_{i=1}^{l}$ leads to the following optimization problem:

Problem 1 *Find $\alpha = [\alpha_1, \alpha_2, \ldots, \alpha_l]^T$ which minimizes*

$$W(\alpha) = \frac{1}{2}\sum_{i=1}^{l}\sum_{j=1}^{l} q_{ij}\alpha_i\alpha_j - \sum_{i=1}^{l}\alpha_i$$

under the constraints

$$\sum_{i=1}^{l} y_i\alpha_i = 0 \quad (1)$$

$$0 \le \alpha_i \le C \quad (i = 1, 2, \ldots, l) \quad (2)$$

where C is a positive constant.

For widely used kernel functions such as a radial-basis function, $Q = [q_{ij}] \in \mathbb{R}^{l \times l}$ becomes positive semi-definite [10]. We therefore assume throughout this paper that Q is positive semi-definite. Under this assumption, Karush-Kuhn-Tucker condition is the necessary and sufficient condition for α to be an optimal solution of Problem 1.

Let Ω be the set of all α satisfying (1) and (2). The set Ω is called the feasible region for Problem 1. Let us define $I_{up}(\alpha)$ and $I_{low}(\alpha)$ as follows:

$$\begin{aligned} I_{up}(\alpha) &= I_0(\alpha) \cup I_1(\alpha) \cup I_2(\alpha) \\ I_{low}(\alpha) &= I_0(\alpha) \cup I_3(\alpha) \cup I_4(\alpha) \end{aligned}$$

where

$$\begin{aligned} I_0(\alpha) &= \{i \mid 0 < \alpha_i < C\} \\ I_1(\alpha) &= \{i \mid y_i = 1 \text{ and } \alpha_i = 0\} \\ I_2(\alpha) &= \{i \mid y_i = -1 \text{ and } \alpha_i = C\} \\ I_3(\alpha) &= \{i \mid y_i = 1 \text{ and } \alpha_i = C\} \\ I_4(\alpha) &= \{i \mid y_i = -1 \text{ and } \alpha_i = 0\} \end{aligned}$$

As shown in Reference [11], the necessary and sufficient condition for $\alpha \in \Omega$ to be an optimal solution of Problem 1 can be expressed as follows:

$$\min_{i \in I_{up}(\alpha)} F_i(\alpha) \ge \max_{i \in I_{low}(\alpha)} F_i(\alpha) \quad (3)$$

where $F_i(\alpha)$ is defined by

$$F_i(\alpha) = y_i\left(\sum_{j=1}^{l} q_{ij}\alpha_j - 1\right)$$

It is obvious from (3) that if there exists a pair of indices (i, j) such that

$$i \in I_{up}(\alpha),\ j \in I_{low}(\alpha),\ F_i(\alpha) < F_j(\alpha)$$

then α is not an optimal solution. Such a pair of indices is called a violating pair at α.

Since Problem 1 is very large in general, conventional techniques cannot be applied due to large matrix computations. To overcome this difficulty, decomposition methods such as SMO algorithm [2] and SVM^{light} [3] have been proposed. A general form of decomposition methods is expressed as follows:

Algorithm 1 (Decomposition method) *Given a set of training examples $\{(x_i, y_i)\}_{i=1}^{l}$, kernel function $K(\cdot,\cdot)$ and a positive constant C, execute the following procedures.*

1) *Set $\alpha(0) = \mathbf{0}$ and $k = 0$.*
2) *Choose q ($\le l$) indices $i_1, i_2, \ldots, i_q$ from $\{1, 2, \ldots, l\}$. Let $B = \{i_1, i_2, \ldots, i_q\}$ and $N = \{1, 2, \ldots, l\} \setminus B$.*
3) *Solve Problem 1 under the additional constraint that $\alpha_i = \alpha_i(k)$ for all $i \in N$. Let the solution of this optimization problem be denoted by α^*.*
4) *Set $\alpha(k+1) = \alpha^*$. Add 1 to k and go to Step 2).*

It is obvious from Step 3) that $\alpha(k) \in \Omega$ for all k. Since each optimization problem in Step 3) has q variables where q is much smaller than l in general, it is easier to find a solution of this problem than to find a solution of the original problem, i.e., Problem 1. Especially, in the case where $q = 2$, an optimal solution can be explicitly expressed in terms of y_i, q_{ij}, and $\alpha_i(k)$, $i \in N$ [2]. Note that $\alpha \in \Omega$ is an optimal solution of the subproblem in Step 3) if and only if

$$\min_{i \in I_{up}(\alpha) \cap B} F_i(\alpha) \ge \max_{i \in I_{low}(\alpha) \cap B} F_i(\alpha)$$

$$\alpha_i = \alpha_i(k), \quad \forall i \in N$$

4. CONVERGENCE OF DECOMPOSITION METHODS

In what follows, we will denote by Γ the set of optimal solutions of Problem 1. Formally, Γ is defined as

$$\Gamma = \{\alpha \in \Omega \mid \min_{i \in I_{up}(\alpha)} F_i(\alpha) \ge \max_{i \in I_{low}(\alpha)} F_i(\alpha)\}$$

Also, for a given $M \subset \{1, 2, \ldots, l\}$ and an $\alpha \in \Omega$, we define $\Gamma_M(\alpha)$ as follows:

$$\Gamma_M(\boldsymbol{\alpha}) = \{\boldsymbol{\beta} \in \Omega \mid \beta_i = \alpha_i \ \forall i \in N, \min_{i \in I_{up}(\boldsymbol{\beta}) \cap M} F_i(\boldsymbol{\beta}) \geq \max_{i \in I_{low}(\boldsymbol{\beta}) \cap M} F_i(\boldsymbol{\beta})\}$$

Using this notation, we can express the set of solution of the subproblem in Step 3) as $\Gamma_B(\boldsymbol{\alpha}(k))$.

Lemma 1 *For any $M \subset \{1, 2, \ldots, l\}$ and any $\boldsymbol{\alpha} \in \Omega$, the point-to-set mapping $\Gamma_M(\boldsymbol{\alpha})$ is closed.*

Proof Let $\{\boldsymbol{\beta}(n)\}_{n=1}^{\infty}$ be any sequence such that

$$\boldsymbol{\beta}(n) \in \Gamma_M(\boldsymbol{\alpha}), \quad \forall n \tag{4}$$

and $\lim_{n\to\infty} \boldsymbol{\beta}(n) = \bar{\boldsymbol{\beta}}$. In order to show closedness of $\Gamma_M(\boldsymbol{\alpha})$, we only have to show that $\bar{\boldsymbol{\beta}} \in \Gamma_M(\boldsymbol{\alpha})$. Let us suppose that $\bar{\boldsymbol{\beta}} \notin \Gamma_M(\boldsymbol{\alpha})$. Then there exists at least one pair of indices (i, j) such that

$$i \in I_{up}(\bar{\boldsymbol{\beta}}) \cap M, \ j \in I_{low}(\bar{\boldsymbol{\beta}}) \cap M, \ F_i(\bar{\boldsymbol{\beta}}) < F_j(\bar{\boldsymbol{\beta}}) \tag{5}$$

Since $\boldsymbol{\beta}(n)$ converges to $\bar{\boldsymbol{\beta}}$, there is an n_0 such that

$$I_{up}(\boldsymbol{\beta}(n)) = I_{up}(\bar{\boldsymbol{\beta}}), I_{low}(\boldsymbol{\beta}(n)) = I_{low}(\bar{\boldsymbol{\beta}}) \quad \forall n \geq n_0 \tag{6}$$

It follows from (5) and (6) that there is an n_1 ($\geq n_0$) such that

$$i \in I_{up}(\boldsymbol{\beta}(n)) \cap M, \ j \in I_{low}(\boldsymbol{\beta}(n)) \cap M, \ F_i(\boldsymbol{\beta}(n)) < F_j(\boldsymbol{\beta}(n)), \ \forall n \geq n_1$$

which contradicts (4). Thus we have $\bar{\boldsymbol{\beta}} \in \Gamma_M(\boldsymbol{\alpha}(k))$. ∎

Let us now define the point-to-set mapping $\boldsymbol{A}(\boldsymbol{\alpha})$ as follows:

$$\boldsymbol{A}(\boldsymbol{\alpha}) = \begin{cases} \cup_{M \in S_q(\boldsymbol{\alpha})} \Gamma_M(\boldsymbol{\alpha}), & \text{if } \boldsymbol{\alpha} \notin \Gamma \\ \Gamma, & \text{if } \boldsymbol{\alpha} \in \Gamma \end{cases}$$

where $S_q(\boldsymbol{\alpha})$ is the family of all sets $M \subset \{1, 2, \ldots, l\}$ such that i) the cardinality of M is q and ii) M contains at least one violating pair at $\boldsymbol{\alpha}$.

Lemma 2 *For any $\boldsymbol{\alpha} \in \Omega$, the point-to-set mapping $\boldsymbol{A}(\boldsymbol{\alpha})$ is closed at points outside Γ.*

Proof As shown in Lemma 1, the set $\Gamma_M(\boldsymbol{\alpha})$ is closed for any M and $\boldsymbol{\alpha}$. Since the cardinality of $S_q(\boldsymbol{\alpha})$ is finite and a finite union of closed sets is again closed, $\cup_{M \in S_q(\boldsymbol{\alpha})} \Gamma_M(\boldsymbol{\alpha})$ is closed. ∎

Lemma 3 *The function $W(\boldsymbol{\alpha})$ is a descent function for the solution set Γ and the point-to-set mapping $\boldsymbol{A}(\boldsymbol{\alpha})$.*

Proof Let $\boldsymbol{\beta}$ be any point belonging to $\boldsymbol{A}(\boldsymbol{\alpha})$. In the case where $\boldsymbol{\alpha} \in \Gamma$, it is obvious that $W(\boldsymbol{\beta}) = W(\boldsymbol{\alpha})$ because $\boldsymbol{\beta} \in \Gamma$. On the other hand, in the case where $\boldsymbol{\alpha} \notin \Gamma$, we easily see that $\boldsymbol{\alpha} \notin \Gamma_M(\boldsymbol{\alpha})$ and $\boldsymbol{\beta} \in \Gamma_M(\boldsymbol{\alpha})$ for all $M \in S_q(\boldsymbol{\alpha})$. This means $W(\boldsymbol{\beta}) < W(\boldsymbol{\alpha})$. Thus $W(\boldsymbol{\alpha})$ is a descent function. ∎

Theorem 2 *Suppose that B in Step 2) of Algorithm 1 contains at least one violating pair unless $\boldsymbol{\alpha}(k)$ is an optimal solution. Then the limit of any convergent subsequence of $\{\boldsymbol{\alpha}(k)\}_{k=0}^{\infty}$ is a solution of Problem 1.*

Proof From the definition of $\boldsymbol{A}(\boldsymbol{\alpha})$ and the assumption of the theorem, we easily see that the sequence $\{\boldsymbol{\alpha}(k)\}_{k=0}^{\infty}$ satisfies $\boldsymbol{\alpha}(k+1) \in \boldsymbol{A}(\boldsymbol{\alpha}(k))$ for all k. Obviously all points $\boldsymbol{\alpha}(k)$ are contained in the compact set Ω. From Lemma 2 the mapping $\boldsymbol{A}(\boldsymbol{\alpha})$ is closed at points outside Γ. Moreover, from Lemma 3 there is a descent function for Γ and $\boldsymbol{A}(\boldsymbol{\alpha})$. Thus we can conclude from Theorem 1 that the limit of any convergent subsequence of $\{\boldsymbol{\alpha}(k)\}_{k=0}^{\infty}$ is a solution. ∎

5. TERMINATION OF DECOMPOSITION METHODS

Many decomposition methods use

$$\min_{i \in I_{up}(\boldsymbol{\alpha})} F_i(\boldsymbol{\alpha}) \geq \max_{i \in I_{low}(\boldsymbol{\alpha})} F_i(\boldsymbol{\alpha}) - \tau \tag{7}$$

as a stopping criterion for practical reason instead of (3), where τ is a positive tolerance parameter [3],[4]. In this section, we consider the general form of decomposition methods with the stopping criterion (7) as shown below. We will show that for any value of τ it stops in a finite number of iterations.

Algorithm 2 *Given a set of training examples $\{(\boldsymbol{x}_i, y_i)\}_{i=1}^{l}$ and a positive constant C, execute the following steps.*

1) *Set $\boldsymbol{\alpha}(0) = \boldsymbol{0}$ and $k = 0$.*
2) *If $\boldsymbol{\alpha} = \boldsymbol{\alpha}(k)$ satisfies (7) then stop.*
3) *Choose q ($\leq l$) indices $i_1, i_2, \ldots, i_q$ from $\{1, 2, \ldots, l\}$. Let $B = \{i_1, i_2, \ldots, i_q\}$ and $N = \{1, 2, \ldots, l\} \setminus B$.*
4) *Solve Problem 1 under the additional constraint that $\alpha_i = \alpha_i(k)$ for all $i \in N$. Let the solution of this optimization problem be denoted by $\boldsymbol{\alpha}^*$.*
5) *Set $\boldsymbol{\alpha}(k+1) = \boldsymbol{\alpha}^*$. Add 1 to k and go to Step 2).*

Theorem 3 *Suppose that B in Step 3) of Algorithm 2 contains at least one violating pair unless $\boldsymbol{\alpha}(k)$ is an optimal solution. Then Algorithm 2 stops in a finite number of iterations for any value of τ.*

Proof Suppose Algorithm 2 does not stop. Let $\{\boldsymbol{\alpha}(k_t)\}_{t=0}^{\infty}$ be any convergent subsequence of $\{\boldsymbol{\alpha}(k)\}_{k=0}^{\infty}$ and $\bar{\boldsymbol{\alpha}}$ be the limit of the subsequence. Then we have

$$\min_{i \in I_{up}(\boldsymbol{\alpha}(k_t))} F_i(\boldsymbol{\alpha}(k_t)) < \max_{i \in I_{low}(\boldsymbol{\alpha}(k_t))} F_i(\boldsymbol{\alpha}(k_t)) - \tau, \quad \forall t$$

Since there exists a t_0 such that

$$I_{up}(\alpha(k_t)) = I_{up}(\bar{\alpha}),\ I_{low}(\alpha(k_t)) = I_{up}(\bar{\alpha}) \quad \forall t \geq t_0$$

we have

$$\min_{i \in I_{up}(\bar{\alpha})} F_i(\alpha(k_t)) < \max_{i \in I_{low}(\bar{\alpha})} F_i(\alpha(k_t)) - \tau, \quad \forall t \geq t_0$$

Letting $t \to \infty$ we have

$$\min_{i \in I_{up}(\bar{\alpha})} F_i(\bar{\alpha}) \leq \max_{i \in I_{low}(\bar{\alpha})} F_i(\bar{\alpha}) - \tau \tag{8}$$

On the other hand, it follows from Theorem 2 that $\bar{\alpha}$ is a solution of Problem 1, that is, $\bar{\alpha}$ satisfies

$$\min_{i \in I_{up}(\bar{\alpha})} F_i(\bar{\alpha}) \geq \max_{i \in I_{low}(\bar{\alpha})} F_i(\bar{\alpha})$$

This contradicts (8). Therefore Algorithm 2 stops in a finite number of iterations. ∎

We finally consider termination property of some typical decomposition methods. In the generalized SMO algorithm [11], B consists of two indices i and j such that

$$i \in I_{up}(\alpha(k)),\ j \in I_{low}(\alpha(k)),\ F_i(\alpha(k)) < F_j(\alpha(k)) - \tau$$

for each k, and (7) is employed as the stopping criterion. Since such a pair (i, j) is obviously a violating pair at $\alpha(k)$, from Theorem 3 we can conclude that the generalized SMO algorithm stops in a finite number of iterations. In SVM ^{light} [3], B is determined by solving a certain optimization problem. As one can easily see, the resulting B contains at least one violating pair (see [3] or [6] for more details). Therefore, as well as SMO algorithm, we can conclude from Theorem 3 that SVM^{light} algorithm stops in a finite number of iterations.

6. CONCLUDING REMARKS

Convergence property of decomposition methods for SVM learning was studied in this paper. Making use of a well-known global convergence theorem, we first proved that under a very mild condition on the selection of B any decomposition method converges to an optimal solution. We next proved that under the same condition any decomposition method with a stopping criterion (7) stops in a finite number of iterations.

Concerning this problem, similar results have recently presented by Lin [6], [7], [8]. However, these results were derived under the assumption that any principal submatrix of Q whose size is less than or equal to q is positive definite. Moreover, the method of selecting B must satisfy some special conditions. In the present paper, on the other hand, only positive semi-definiteness of Q is assumed, and only requirement for B is that it contains at least one violating pair. Therefore, we can conclude that our results are much more general than those given by Lin.

7. REFERENCES

[1] B. Schölkopf, C. J. C. Burges, and A. J. Smola, Eds., *Advances in Kernel Methods: Support Vector Learning*, MIT Press, Cambridge, Massachusetts, 1999.

[2] J. C. Platt, "Fast training of support vector machines using sequential minimal optimization," in *Advances in Kernel Methods: Support Vector Machines*, B. Schölkopf, C. Burges, and A. Smola, Eds. MIT Press, Cambridge, MA, 1998.

[3] T. Joachims, "Making large-scale support vector machine learning practical," in *Advances in Kernel Methods: Support Vector Machines*, B. Schölkopf, C. Burges, and A. Smola, Eds. MIT Press, Cambridge, MA, 1998.

[4] S. S. Keerth, S. K. Shevade, C. Bhattacharyya, and K. R. K. Murthy, "Improvements to Platt's SMO algorithm for SVM classifier design," *Neural Computation*, vol. 13, pp. 637–649, 2001.

[5] D.G. Luenberger, *Linear and Nonlinear Programming*, Addison-Wesley Publishing, Reading, MA, 1989.

[6] C.-J. Lin, "On the convergence of the decomposition method for support vector machines," *IEEE Transactions on Neural Networks*, vol. 12, pp. 1288–1298, Nov. 2001.

[7] C.-J. Lin, "Asymptotic convergence of an SMO algorithm without any assumption," *IEEE Transactions on Neural Networks*, vol. 13, pp. 248–250, Jan. 2002.

[8] C.-J. Lin, "A formal analysis of stopping criteria of decomposition methods for support vector machines," *IEEE Transactions on Neural Networks*, vol. 13, pp. 1045–1052, Sept. 2002.

[9] N. Takahashi and T. Nishi, "On termination of SMO algorithm for support vector machines," in *Proceedings of ISEE2003*, Nov. 2003.

[10] S. Haykin, *Neural Networks*, Prentice Hall, Upper Saddle River, NJ, 1999.

[11] S. S. Keerthi and E. G. Gilbert, "Convergence of a generalized SMO algorithm for SVM classifier design," *Machine Learning*, vol. 46, pp. 351–360, 2002.

SELF-ORGANIZING TOPOLOGICAL TREE

Pengfei Xu and Chip-Hong Chang
Centre for High Performance Embedded Systems, Nanyang Technological University
Nanyang Avenue, Singapore 639798

ABSTRACT

The Self-organizing Maps (SOM) introduced by Kohonen implement two important operations: vector quantization (VQ) and a topology-preserving mapping. In this paper, a tree structured SOM algorithm named Self-organizing Topological Tree (SOTT) is proposed. Unlike the conventional tree structured SOMs [3,4,5], every layer of the proposed SOTT is organized simultaneously instead of layer by layer. A reduction of computational complexity from $O(N)$ to $O(\log N)$ is achieved by the tree structured search in both the training phase and the test phase. Furthermore, such topological tree that maintains both intra-layer and inter-layer topologies during and after learning can be seen as endowing the SOM topological map with the progressive decoding and multi-resolution capabilities.

1. INTRODUCTION

Self-organizing Maps (SOM), and its variants, are the most popular artificial neural network algorithms in the unsupervised learning category [2, 3]. They are widely used in many engineering applications, e.g., clustering of high-dimensional data, classification of images and acoustic patterns, data visualization, data mining and so on [3]. SOM is essentially a projection from higher dimension to lower dimension with topology preservation property, whereby data that are in close proximity in the higher dimension input space are mapped to the neighboring outputs in the lower dimension. This topology preservation property is the most important feature of SOM and it is realized by the unique neighborhood adaptation function.

Some tree structured variants of SOMs have been proposed before to reduce the computational complexity by using tree structured search rather than full search for the winning neuron [3, 4, 5]. In these algorithms, the layers of the tree are organized layer by layer. Thus, the training data need to be input several times, which is not desirable in online, real-time and large database applications. To alleviate the bias brought by tree structured search, a lateral search is usually used in these algorithms. However, this lateral search will increase the computational complexity, and more severely, destroy the inter-layer topology between the parent neuron and children neurons.

In this paper, a novel Self-organizing Topological Tree (SOTT) is proposed. The aim is to update all layers of neurons simultaneously during training, so that the training set only needs to be fed once. We are also interested in preserving both the intra-layer topology between neighboring neurons within the same layer and inter-layer topology between the parent and children neurons across layers during learning. To design such an algorithm, two main problems need to be addressed: (1) the bias brought by the tree structured search, and (2) how to maintain both the inter-layer and intra-layer topologies while updating all layers of neurons simultaneously. In the proposed SOTT, a multi-path search is used to reduce the bias subjected to the tree structured constraint and some specific update rules are designed to preserve the topologies between the intra-layer and inter-layer neurons.

The rest of this paper is organized as follows. In Section 2, the proposed learning algorithm of the SOTT is described. Vector quantization of images is used as a vehicle to test the learning ability of the proposed SOTT. The results are compared with other algorithms in Section 3. The paper is concluded in Section 4.

2. SOTT STRUCTURE AND TRAINING ALGORITHM

A SOTT can be viewed as a multi-layer SOM, with fixed depth and breadth. A SOTT realizes the mapping of an n-dimensional Euclidean space R^n onto a finite codebook space W:

$$\Lambda : \mathbf{R}^n \rightarrow W \quad (1)$$

where the input vector $\boldsymbol{x} = (x_1, x_2, \ldots, x_n) \in R^n$ and $W = \{\boldsymbol{w}_{ij} \in R^n \ \forall i = 1, 2, \ldots, L \text{ and } j = 1, 2, \ldots, N_i\}$ define a finite set of codevectors, $\boldsymbol{w}_{ij}$ which are points in R^n.

The structure of the SOTT network is characterized by L, the number of layers and N_i is the number of neurons at the i-th layer for each $i = 1, 2, \ldots, L$. A 2-D Quad-tree and a 1-D string structured SOTT with $L = 3$, $N_i = 4^{i-1}$ are shown in Fig. 1.

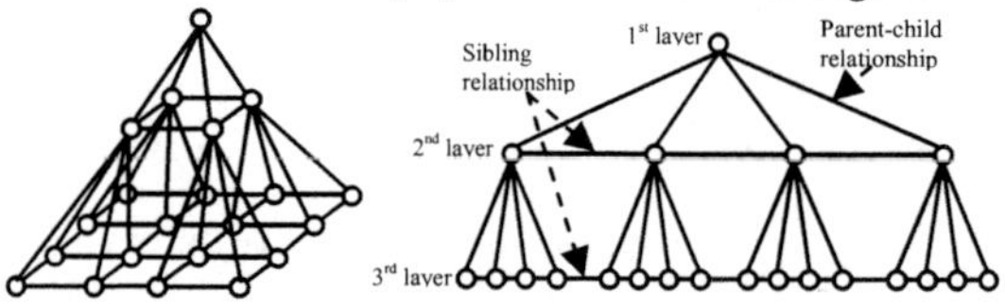

(a) (b)

Figure 1. Structure of SOTT (a) 2-D quad tree (b) 1-D string

Two kinds of relationships exist in a SOTT: (1) the intra-layer neighborhood relationship, (2) the inter-layer parent-child relationship. Let G_i be a fully connected graph where its vertices and edges are defined respectively by the neurons and their interconnections at the i-th layer. From the formulation of a near-neighbor mesh of a finite lattice, we define the k-neighborhood $\mathcal{N}_v$ of the neuron $v \in G_i$ as the spatial support of the position of v in G_i. A neuron, u is said to be in the k-distance neighborhood of the neuron v if there is a connected path from u to v and their minimal Euclidean distance, $\|u-v\| \le k$. Furthermore, for each non-leaf neuron $v \in G_i$, a subtree H_v of neurons rooted at v with a branch factor of B can be found. A neuron u is said to be a child of v iff $v \in G_i$ and $u \in G_{i+1} \cap H_v$. The neurons of $G_{i+1} \cap H_v$ that have the same parent neuron v, are called the siblings.

Tree structured search can be used in tree structured SOMs to speedup the learning and the later mapping process. Unlike the TS-SOM [4, 5], which uses a fixed neighborhood for the adaptation, the neighborhood, $\mathcal{N}_v(t)$ of a winning neuron, v for each layer of our proposed SOTT is monotonically decreasing with the training epoch time, t. Using a fixed neighborhood results in a map which is not globally ordered and consists of small patches of local ordering between which the ordering direction changes discontinuously. To reduce the bias brought by the tree structured search with neighborhood taper, we use a multipath search to find the winning path. Every neuron on this winning path is taken as the winning neuron on its layer. These winning neurons drive the learning of their neighbors simultaneously and independently on the layers they resided. The main steps involved in the training algorithm are network initialization, input data fetching, winning path search, update of winning neurons and their neighborhoods, and convergence check. In what follows, some of these key training steps that make our SOTT unique are discussed.

2.1. Initialization

As opposed to other clustering algorithms, e.g., GLA [6], initialization is not that crucial in SOMs and its variants, so is SOTT thanks to the topological neighborhood. A better initialization scheme may enable the network to converge faster but it usually requires a priori knowledge of the input data distribution. In image processing applications, we initialize the neurons of the SOTT with uniformly distributed weight vectors on the major diagonal of the input space.

$$w_{ij} = \left[\frac{255\left(j-\frac{1}{2}\right)}{N_i}, \ldots, \frac{255\left(j-\frac{1}{2}\right)}{N_i} \right] \forall i = 1, 2, \ldots, L \text{ and } j = 1, 2, \ldots, N_i \tag{2}$$

This simple initialization method provides a good neighborhood and parent-child relationship among all neurons on the tree.

2.2. Input permutation

Competitive learning has the advantage of being online over the batch algorithms like GLA [6]. However, on-line learning has the order dependent problem, i.e., the final weight vectors are affected by the order of the input sequence. For example, if the pixels of an image are input in a raster scanning order, the pixels towards the bottom-right of the trained image will not be well learned due to the diminishing learning rate. Thus we propose a novel global butterfly jumping sequence to permute the images into quasi-repetitive patterns so that the online learning can also be fast and effective. It is defined by a mapping, $\pi: Z \rightarrow Z^n$ of an input order number $r \in Z$ to a n-dimensional coordinate system, $(x_1, x_2, \ldots, x_n) \in Z^n$ and $x_i \in [0, N)$,where Z is a finite integer space. The butterfly permutation π is given by:

$$\pi(r)=\left\{(x_1,x_2,\ldots,x_n)\in Z^n \mid x_j=\sum_{i=0}^{\lceil \log_2 N\rceil-1} 2^{\lceil \log_2 N\rceil-1-i}\left(r_{ni}\oplus r_{ni+j}\right)\forall j\neq n, x_n=\sum_{i=0}^{\lceil \log_2 N\rceil-1} 2^{\lceil \log_2 N\rceil-1-i} r_{ni}\right\} \tag{3}$$

where r_i is the i-th bit of the binary representation of the decimal number r, and r_0 is the least significant bit.

This global butterfly jumping sequence is a generalization of the 2-D localized butterfly jumping sequence proposed by [7] to span the entire input space without any constraint on the dimensionality of the input vectors. The difference in the permutation effect of the sequences proposed by [7] and ours for a 2-D image is shown in Fig. 2. In Fig. 2, the permuted data are read back in the raster scan order to form a 2-D image to make the spatial characteristics visually perceivable. The training data permuted by the global jumping sequence are more evenly distributed in the 1-D sequence. Thus, the main features of the image are learnt at the early possible encounters while the learning rate is still high and similar patterns are learnt repeatedly to enhance competitive learning.

(a) (b) (c)

Figure 2. (a) the original image and the effects of (b) localized butterfly permutation (c) global butterfly permutation

2.3. Winning path search

Searching for winning neurons has always been the most time consuming bottleneck task in SOMs and its variants. The search strategy also influences the quality of learning. A multi-path tree structured search is proposed here for our SOTT to achieve an impressive speedup with only a small degradation in the output quality comparing with the exhaustive search in classical SOMs. The pseudo codes of the algorithm are given in Fig. 3.

Find_win_path($x(t)$, *root*)
  -- $x(t)$: input at time t; *root*: root neuron of SOTT;
begin
  initialize κ; -- initialize path search factor
  for (i = 1 to L) initialize λ_i; -- initialize the emphasis at each layer
  v = root; $e_v = \| x(t) - w_v \|$; -- compute error at root
  $V_1 = (v, e_v)$; --set of neurons and errors
  (v, e_v) = ***Find_win_child***($x(t)$, 1, V_1); -- find winning children
  win_leaf = v; -- recursively until a winning leaf is found
  win_path = ***Trace_parent***(v); -- trace the winning path from the leaf
end

Find-win-child($x(t)$, i, V_i)
-- $x(t)$: input at t; i: layer number; V_i: neuron-error pairs, (v, e_v) at i-th layer;
begin
  if ($i = L$) **return** first element of V_i; -- return winning leaf neuron
  Initialize $V_{i+1} = \phi$; –Initialize an empty set for layer *(i+1)*
  for (each $(v, e_v) \in V_i$) $V_{i+1} = V_{i+1} \cup$ ***Compute_child_error***($x(t)$, i, v, e_v);
  -- compute the errors for all children of node v and put them in the set.
  Sort the elements of V_{i+1} in ascending order of error;
  Delete all except the first κ neuron-error pairs from V_{i+1};
  return ***Find_win_child***($x(t)$, i+1, V_{i+1});
end

Compute_child_error($x(t)$, i, v, e_v)
-- $x(t)$: input at time t; i: layer number; v: neuron; e_v: error of v;
begin
  compute $e_{u_j} = \left\| x(t) - w_{u_j}(t) \right\| + \lambda_i e_v \quad \forall u_j \in child(v)$;
  return $\bigcup_{j=1}^{B} \left(u_j, e_{u_j}\right)$;
end

Figure 3. Algorithm for the win path search

In Fig. 3, the subroutine ***Compute_child_error*** computes the distortion between the input $x(t)$ and the B children of the parent neuron, v at level i. The fractional constant, called the leakage ratio λ_i is introduced so that adjacent layers, instead of just one layer, are considered in finding the best matching neuron at the leaf layer. The value of λ_i is fixed at each layer but varying across layers to facilitate the control of performance trade-off at different layers. By tuning the leakage ratio λ_i, some intermediate layers can be emphasized according to the specific requirement of the application. This can be useful in some applications like progressive compression, progressive transmission, and color classification.

Another key parameter, κ is introduced in the subroutine ***Find_win_child*** to produce an equivalent effect of a κ-path search algorithm. ***Find_win_child*** is a recursive routine that calls ***Compute_child_error*** to calculate the errors of the children of a set of input neuron-error pairs. The κ best matching neurons are used as input to search for the κ best matching successor neurons at the next layer until the leaf layer is reached. Searching for the best matching leaf neuron from among the children of a single best matching neuron at the upper-layer is suboptimal. This is because the weight vector of the final winning neuron at the leaf layer for a single path search is most probably not the closest to the input vector. The parameter κ provides the control necessary to minimize the bias caused by branching at the higher level of the hierarchy. The value of κ can be tuned to trade computational complexity for the overall performance. From the simulation result, SOTT performs much better with $\kappa = 2$ than with $\kappa = 1$. The performance of the SOTT with $\kappa = 2$ is almost the same as that of the full search but its computational complexity is much smaller and is only a little higher than that of $\kappa = 1$. If κ is made different for some layers, we have a more complicated but interesting dynamic multi-path search.

The recursive procedure ***Find_win_child*** is called at the main routine ***Find_win_path*** with the root neuron and its error equal to the Euclidean distance from the input vector. Once the winning leaf neuron is found, the function ***Trace_parent*** traverses the winning path bottom-up via the parent neuron of the winning leaf neuron until the root neuron is reached.

2.4. Update Neurons

Once the winning path is found, each neuron on this path is deemed as the winning neuron of the layer it resided. These neurons will drive the learning of their neighbors within their own layer, as in SOM. The learning of SOTT is composed of several SOM learning running parallelly and independently.

The classical SOM learning function is given by

$$w_i(t+1) = w_i(t) + \alpha(m)h(i,c,m)\left[x(t) - w_i(t)\right] \tag{4}$$

where c is the winning neuron, and $w_i(t)$ denotes the weight of neuron i at time t. $\alpha(t)$ is so-called the learning rate function, which is usually designed as a monotonic decreasing function of the sweep number m. A sweep is a period during which the updating parameters remain constant. In our simulations, we set the number of input data in one sweep to be 6 times the number of neurons on the leaf layer. $\alpha_i(m) = \alpha_i(0)k_1^m$ is one of the most widely used learning rate functions, e.g., in [7] (Unless explicitly specified, the common SOM parameters set in this paper are benchmarked according to [7]). The initial learning rate, $\alpha(0)$ is usually set to a value less than 1. The learning rate for the neighborhood of the winning neuron is scaled by the taper function $h(i,c,m)$. This neighborhood taper is usually defined as a decreasing Gaussian function of the spatial distance between the winning neuron and its neighbors, i.e.,

$$h(i,c,m) = e^{-\frac{\|C_i - C_c\|^2}{\sigma(m)^2}} \tag{5}$$

C_i is the coordinate vector of neuron i. Clearly, the neuron closer to the win neuron will get a higher value of $h(i,c,m)$. $\sigma_i(m) = \sigma_i(0)k_2^m$ is the neighborhood width at sweep number m. Only the neurons within $\sigma(m)$ are neighbor neurons that need to be updated by (4). $\sigma(0)$, the initial neighborhood width, is usually set to a value between 1 and $N/2$, where N is the network size. Both the learning rate, $\sigma(m)$ and the neighborhood width, $\sigma(m)$ decrease with m. The constants, k_1 and k_2, are set to a value typically between 0.5 and 1.

By letting every layer of SOTT learns like a SOM simultaneously, it is hoped that it will perform as good as a SOM in terms of clustering performance and topological order. However, there is no provision to preserve the correlation between the parent and children neurons until now. During updating, the neurons are likely to be dragged far away from their parents in the learning process as mentioned in [4]. This has a detrimental effect of destroying the hierarchical topology, and enlarges the distortion and the already problematic bias of tree structured network.

Maintaining both the intra-layer and the inter-layer relationships well is instrumental to achieving good performance of the proposed SOTT. To keep the neurons of the same family close during the entire training process, the following updating rules are imposed. (1) The initialization of neighborhood widths, $\sigma(0)$ of each layer is made to be proportional to the number of neurons on that layer, i.e., $\sigma_i(0)/N_i$ is constant. (2) The children neurons will be updated if and only if (a) their parent neuron is also updated and (b) they are sufficiently close to the winning neuron of their layer. Rules (1) and (2a) are designed to keep and inter-layer relationship, and (2b) is designed to neglect trivial updates with little or no effect in order to reduce the computational complexity.

3. Simulation Result

We use the SOTT to perform a mapping from the 3-D RGB space to 1-D as in typical color quantization, and then visualize the weights of the neurons in RGB space. The result of color quantization of Lena is displayed in Fig. 4. In Fig. 4, the squares represent 4 neurons on the first layer, the circles represent 16 neurons on the second layer, and the triangles represent 64 neurons on the third layer. The tightly coupled parent-child topographical relationship is evident from Fig. 4. Furthermore, every layer in the SOTT has its own topological order comparable to that obtained by a single sheet SOM. The quality of the topological map of the third layer of SOTT in Fig. 4 is compared with a classical 64 neuron 1-D string SOM objectively using the metric J [3], which is defined as:

$$J = \sum_{i=2}^{N} \|w_i - w_{i-1}\| - \|w_N - w_1\| \tag{6}$$

where N is the number of neurons and w_i is the weight vector of the i-th neuron. Lower value of J represents better topological order. The result plotted in Fig. 5 shows that the quality of the topological order in both cases are comparable.

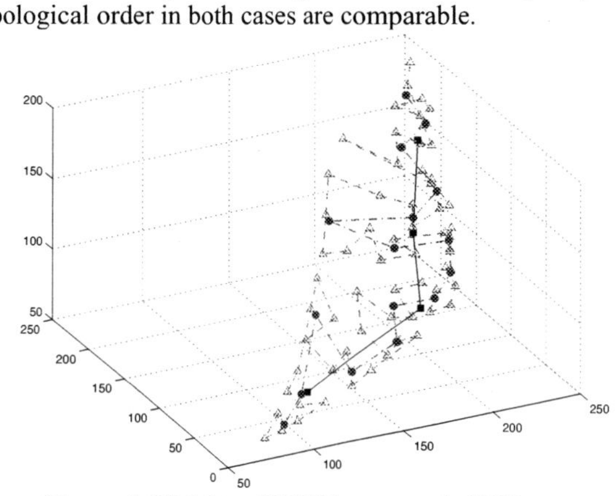

Figure 4. Weights of SOTT neurons in RGB space

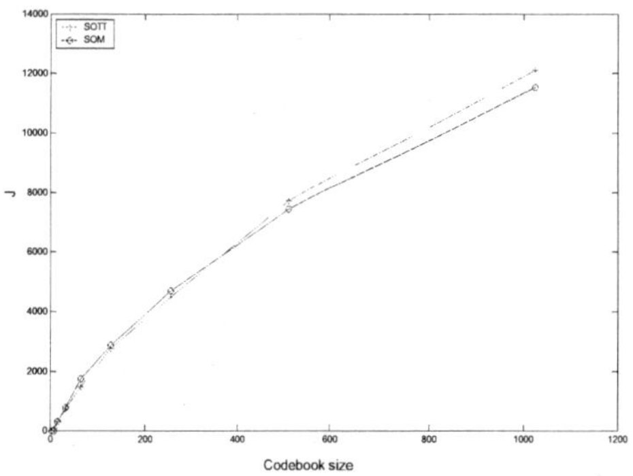

Figure 5. Topological order quality J of SOM and SOTT

Next, we compare the quantization performance of SOTT with the classical SOM [7], GLA [6] and GLA-TSVQ [1] on image clustering. GLA is a classical vector quantization algorithm, also known as LBG or *k*-mean clustering. GLA-TSVQ is a tree structured vector quantizer based on GLA. The results on the training of image Lena are shown in Table 1 (run on Pentium VI 2.1G, 512 MB memory PC).

Table 1: Comparison of PSNR (dB) and training time (s)

N	SOM		SOTT		GLA		TSVQ	
	PSNR	Time	PSNR	Time	PSNR	Time	PSNR	Time
4	24.95	**0.08**	24.95	**0.08**	**25.00**	2.5	**25.00**	2.5
8	28.48	**0.13**	28.49	0.18	**28.69**	10.5	28.16	4.6
16	**30.60**	**0.27**	30.25	0.31	30.58	44.3	30.48	5.3
32	32.38	0.86	32.23	**0.76**	**32.40**	59.1	32.22	11.8
64	**33.91**	2.53	33.70	**1.62**	33.85	124.5	33.58	13.3
128	**35.40**	9.44	35.22	**3.89**	35.35	184.9	34.92	12.9
256	**36.94**	35.80	36.55	**9.00**	36.80	555.2	36.22	16.3
512	**38.39**	147.20	37.96	22.01	38.14	1062.6	37.72	**19.9**
1024	**39.95**	604.13	39.45	64.73	39.50	1745.5	39.06	**19.0**

Other figures of merits are also compared and summarized in Table 2. SOM and SOTT are online, while GLA and GLA-TSVQ are batch algorithms. Unlike batch algorithms, online algorithms do not have to wait for all the training data to be stored in memory before processing can begin. SOTT and GLA-TSVQ have multiresolution property by virtue of their tree structure. By using tree structured search, SOTT and GLA-TSVQ lower the computation time at the cost of a slight degradation in quality. SOM and SOTT possess intra-layer topology due to the neighborhood adaptation. Both SOTT and GLA-TSVQ have ordered inter-layer topology as a result of SOTT's specific learning rules and the top-down generation of GLA-TSVQ. From Table 2, SOTT ranks the highest in its scores in all features but the output quality. However, the degradation in PSNR is very small (less than 0.5 dB). Overall, the quantization performance of SOTT is still competitive. It is worthwhile to trade such tiny degradation of quality for the dramatic increase in speed and many attractive features that would otherwise not be possible to possess simultaneously through other clustering methods.

Table 2: Overall comparison of different algorithms

Properties	SOM	SOTT	GLA	GLA-TSVQ
Type of algorithm	Online	online	batch	batch
Multiresolution	No	Yes	No	Yes
Rank in quality	1	3	1	4
Rank in speed	3	1	4	1
Intra-layer topology	Yes	Yes	No	No
Inter-layer topology	No	Yes	No	Yes

4. CONCLUSION

This paper presents a new Self-organizing Topological Tree (SOTT) to overcome the computational intensive search in the training and mapping processes of single sheet SOM. Multipath search and specific update rules are designed to reduce the bias of tree structured search while keeping the inter-layer and intra-layer topologies intact. Experimental results show that the performance of the SOTT is comparable to the full search SOM and its computational efficiency is higher for large network size. The SOTT features the graceful topological ordering, mutli-resolution maps and on-line learning capability, which are very useful for developing efficient and elegant new solutions to data clustering and pattern classification applications.

5. REFERENCES

[1] A. Buzo, A. H. Gray, Jr., R. M. Gray, and J. D. Markel, "Speech coding based upon vector quantization," *IEEE Trans. Acoust., Speech, Signal Processing,* vol. ASSP-92, pp. 562-574, Oct. 1980.

[2] S. Haykin, *Neural Networks: A Comprehensive Foundation.* Prentice Hall, 1998

[3] T. Kohonen, *Self-organizing Maps*, vol. 30 of *Springer Series in Information Sciences*, Springer, Berlin,Heidelberg,New York, third extended edition edition, 2001.

[4] P. Koikkalainnen and E. Oja, "Self-organizing hierarchical feature maps," in *Proceeding of International Joint Conference on Neural Networks(IJCNN'90)*, San Diego, CA, 1990, vol. 2, pp. 279–284.

[5] P. Koikkalainnen, "Progress with the tree-structured self-organizing map," in *11th European Conference on Artificial Intelligence (ECAI'94)*, San Diego, CA, 1994, pp. 211–215.

[6] Y. Linde, A. Buzo, and R. M. Gray, "An algorithm for vector quantizer design," *IEEE Trans. Commun.*, vol. COM-28, pp. 84–95, 1980.

[7] S. O. Pei and Y. S. Lo, "Color image compression and limited display using Self-organization Kohonen Map," *IEEE Transaction on Circuits and System for Video Technology, vol. 8, no.* 2, pp. 191–205, April 1998.

SENSITIVITY ANALYSIS OF LOW-COMPLEXITY VECTOR QUANTIZERS FOR FOCAL-PLANE IMAGE COMPRESSION

José Gabriel R. C. Gomes and Sanjit K. Mitra

Department of Electrical and Computer Engineering, University of California
Santa Barbara, CA 93106 USA. E-mail: (gabriel,mitra)@ece.ucsb.edu

ABSTRACT

Most high-performance block-coding systems for image compression, such as JPEG, have been designed for software or dedicated digital hardware implementations where the data are already assumed to be available in digital format. In modern CMOS photosensors, smart-pixel technologies have allowed the realization of basic signal processing tasks at the pixel level, in analog format before analog-to-digital (A/D) conversion. The elimination of A/D converters and implementation of block-coding directly over analog blocks of pixels in such sensors can be attractive both in terms of area savings and power consumption. The design of block encoders, under the strong hardware constraints that derive from the A/D converter removal, has been investigated in this paper. We present a comparison of three systems in terms of rate, distortion and complexity, and present a numerical simulation analysis of their sensitivity to implementation errors. The conclusion of the analysis is that linear-transform coding vector quantizers outperform full-search vector quantizers and warping hyperbolic-tangent neural networks, in terms of performance, complexity and robustness, for a CMOS imaging sensor implementation.

1. INTRODUCTION

In the CMOS photo-sensors of modern imaging sensors, the raw data are stored on a pixel-by-pixel basis inside capacitors distributed over a pixel array. The data read-out can be implemented by random access systems [1]. The sensor design research is currently focusing on enhancing the pixel cell (the *smart-pixel* shown in Fig. 1 has analog signal processing capabilities embedded in it), so that more processing power is available at the pixel level. As a result, signal processing tasks that were traditionally placed outside the sensor chip are being done inside it. A number of signal processing algorithms implemented at the pixel level have been recently advanced, including deblurring [2], edge detection [3] and motion detection [4].

One particularly interesting task to be applied at the lowest possible level, taking advantage of local processing resources of each pixel (or of a block of pixels, as shown in the right-hand part of Fig. 1) is image compression before data read-out. Some low-precision analog block-processing scheme followed by comparators simultaneously accomplishes compression and data-conversion, so all the data-conversion hardware can be removed. The external buffering requirements are relaxed and a higher frame-capture rate is allowed, for the same power consumption. Besides

This work was supported in part by CAPES/Brazil, by a University of California MICRO grant with matching support from Philips Research Laboratory, and in part by Microsoft Corporation.

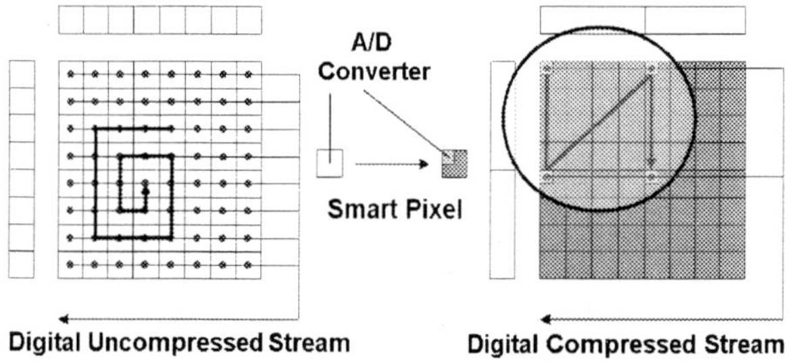

Fig. 1. The CMOS digital-pixel sensor (left) has one A/D converter for each pixel and can be addressed as a random-access memory. In *smart-pixel* technologies, analog signal processing can also be performed inside the pixel and before A/D conversion (represented by a dark shade of gray inside the pixels in the figure). A simple extension of the *smart-pixel* concept to block processing, using the same analog processing techniques, would allow for data read-out already in compressed, block-coded, format (right).

this main advantage, there is the bonus of possible size reduction [5], because replacing analog-to-digital (A/D) converters by analog processing at the pixel level might make the cells smaller (higher resolution for the same silicon area).

The principal element of the image compression at the focal plane is the mapping of rectangular blocks of analog samples directly into compressed binary words, that are parts of a longer bit-stream describing the whole image. The idea is reproduced in Fig. 2, and has been studied in terms of computational complexity requirements in [6] and [7], from the point of view of neural network design[1].

In this work, we study specifically the robustness of three systems that are possible candidates for mixed-signal implementation of block encoders: (1) the full-search vector quantizer or VQ (in this paper *VQ* is used interchangeably to mean both *vector quantizer* and *vector quantization*), that is sometimes also denoted as a self-organizing map (SOM) [8]; (2) the linear transform coder; (3) the non-linear (warping) transform coder. These last two can also be modelled as multilayer perceptrons (MLPs) [9]. A brief description of these systems and their main features is given in Sections 2 and 3. Numerical simulations of the three systems and conclusions are presented in Section 4.

[1]The motivation for modelling these block coders as neural networks derives from the fact that the input is analog and the output is digital, and the system behavior is completely specified by pairs of inputs and target outputs. The pairs are grouped in training sets from which the design can be done.

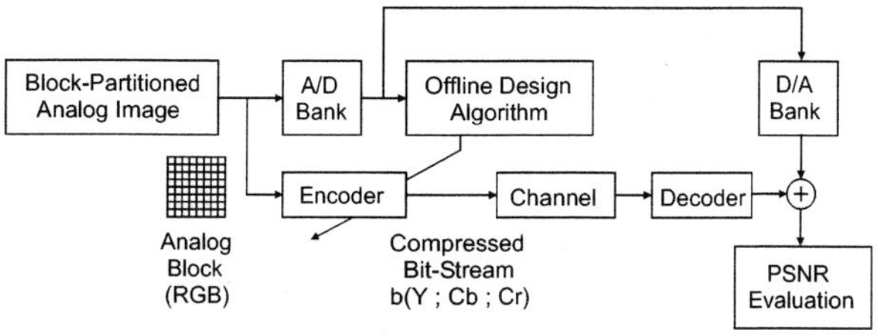

Fig. 2. Direct block coding of analog vectors at the pixel level using a mixed-signal encoder. We would like $b(Y; C_b; C_r)$ to reproduce the stream generated by a digital compression system.

2. LOW-COMPLEXITY VECTOR QUANTIZATION

Most high-performance block coders, such as JPEG [10], have been designed for software or dedicaded digital hardware implementations in which the data is already assumed to be available in digital format. In the case of image compression performed directly over raw data present at the CMOS photo-sensor, the elimination of A/D converters can be done by mapping analog vectors directly into block codes using neural networks. That brings us back to the problem of evaluation of the complexity of block encoders, but with stronger instruction and hardware constraints than those faced by digital implementations. For example, it is reasonable to assume that entropy coding is not available, and most of the complexity gains come from the removal of the A/D converters and digital hardware for image compression. Under these constraints, we investigate three block coding techniques that are amenable to mixed-signal implementation and compare their computational complexity and sensitivity properties.

Linear Transform Coding is used to indicate the VQ for which the residual vectors (or pixel-blocks) of DPCM coding of DC levels are pre-multiplied by a linear transform, and scalar quantizers are designed and then used to process the highest-energy coefficients in the transform domain. The design is done in the following way: we denote each transform domain vector by X, and specify a vector T with thresholds for scalar quantization of each component of X. After all vectors X are given a code according to T, the centroids of each codebook cell are computed and used as reconstruction vectors. Then, all vectors X are assigned, according to their codes, to a reproduction vector $\hat{X}$, and so a distortion and entropy performance ((H, D) pair) can be computed for the VQ. Given a training set X, the pair (H, D) is a function only of T, and so T can be designed to minimize a given cost function $J = D + \lambda H$. From the similarity with MLPs, we also call these VQs Linear MLPs (or simply LMLPs).

Non-Linear Transform Coding is similar to the previous idea, except that one nonlinear step is included between the input data X and the data Y that is effectively scalar-quantized (in other words, applied to T) [7]. Here also we denote an input vector by X, but before quantization we multiply it by a gain g and add a bias b, and apply a nonlinear warping:

$$Z = tanh\left(g\begin{bmatrix} 0 \\ X \end{bmatrix} + b\right) \qquad (1)$$

Next we pre-multiply Z by W and add a bias B, where the pair of parameters (W, B) is designed to provide projections along the principal components of the subspace spanned by the *lattice data* [7] inside a very high-dimensional feature space:

$$Y = W(g, b)Z + B(g, b) \qquad (2)$$

All vectors Y are given binary codes according to the scalar quantizers described by vector T, and the same codes are used for the corresponding vectors X. Then the centroids of cells of vectors X with the same code can be computed and used as VQ reconstruction vectors. Again, all vectors X are encoded by $\hat{X}$ according to their codes generated from Y and T, and a (H, D) pair can be computed for the VQ. Given a training set X, the pair (H, D) is a function of T, g and b, and these parameters are computed to minimize $J = D + \lambda H$. From the use of a regular lattice to represent the shape of the input data density, we also call these systems Kernel Lattice VQs (KLVQs) [7].

Full-search VQ, also known as learning VQ (LVQ), is the usual way to encode the input vectors X directly, using an unstructured vector codebook with reproduction vectors computed by the LBG algorithm [11], in order to minimize exclusively the MSE, in other words using $\lambda = 0$ in the cost expressions previously mentioned. Since its computational complexity grows exponentially with the number of bits in the code, this scheme is not amenable to simple mixed-signal implementation.

Computational complexity assessments for all three schemes, in a different range of bit allocations for two components, are provided in Table 1. The complexity problem of the full-search VQ becomes clear in this table. Also, we have only taken into account the complexity of radial computation of distances from an input vector X to all codevectors in the codebook. The complexity of selecting the winner and encoding the decision results, which are mainly logical tasks, has not been taken into account, so the complexity of full-search VQ is larger than what appears in Table 1.

Table 1. Comparison of Linear Transform Coding, Non-Linear Transform Coding and Full-Search VQ in terms of computational complexity. By comparing these numbers with the infinite-precision performance in Figs. 3 and 4, note that MLPs have much less complexity than full-search VQ for approximately equivalent rate-distortion performance.

Type	Allocation	Complexity			
	LinearMLP	Mpp	App	Tpp	bpp
1	[0; 1]	0	0.063	0.063	0.063
2	[1; 1]	0	0.125	0.125	0.125
3	[1; 2]	0	0.313	0.250	0.188
4	[2; 2]	0	0.500	0.375	0.250
5	[2; 3]	0	0.938	0.625	0.313
6	[3; 3]	0	1.375	0.875	0.375
	KLVQ	Mpp	App	Tpp	bpp
7	[0; 1]	0.375	0.483	0.063	0.063
8	[1; 0]	0.375	0.483	0.063	0.063
9	[1; 1]	0.375	0.500	0.125	0.125
10	[2; 0]	0.375	0.625	0.188	0.125
11	[1; 2]	0.375	0.688	0.250	0.188
12	[2; 2]	0.375	0.875	0.375	0.250
13	[2; 3]	0.375	1.313	0.625	0.313
14	[3; 3]	0.375	1.750	0.875	0.375
	VQ (FullSearch)	Mpp	App	Tpp	bpp
15	1-bit	0.250	0.375	0.063	0.063
16	2-bit	0.500	0.625	0.125	0.125
17	3-bit	1.000	1.125	0.188	0.188
18	4-bit	2.000	2.125	0.250	0.250
19	5-bit	4.000	4.125	0.313	0.313
20	6-bit	8.000	8.125	0.375	0.375
21	7-bit	16.000	16.125	0.438	0.438

(1) Mpp, App, Tpp and bpp: Number of multiplications, additions, comparisons (thresholding) and bits per pixel.
(2) Overheads for color conversion, DPCM and linear transform: 6.125 Mpp, 6.688 App, 1.063 Tpp and 0.375 bpp for all systems.
(3) Numbers 1 to 21 relate to systems in Figs. 3 and 4.

3. KERNEL PCA SENSITIVITY

Kernel PCA as proposed in Eqs. (1) and (2) faces a problem of sensitivity to hardware implementation errors. That is because solutions of low J (on the lower convex hull of the (H, D) points) usually have a very high bias b. That means most of the information passes from the first layer of the MLP to its second layer in the form of 7th or 8th digits at the highly saturated results of the hyperbolic tangents. As a consequence, the W matrix of the second layer has very large values (on the order of 10^5 for X dynamic ranges normalized to 1.0), which makes an analog hardware implementation extremely sensitive to errors is W. So, for the time being we provide a complete sensitivity analysis of the other systems (LMLP and full search VQ), but restrict the KLVQ performance evaluation to errors on numerical operations and quantization of the thresholds T. The parameters of W are represented with infinite precision, so that the results in Sec. 4 are meaningful, but that also indicates that high-bias nonlinear transform coding schemes should not be used for analog-hardware block coding. However we display the current results as an indication that a second level of linear transformation might apply, and a complexity theory can be developed for these systems as well. It could be possible to increase the robustness of KLVQ systems by choice of low-bias solutions and use of different Kernel functions, such as polynomials or radial-basis expressions.

4. SIMULATION RESULTS

In this section we present results for an example of compression of 4×4 pixel-blocks using DPCM and two transform coefficients (2-D transform-domain data). A *training database* is composed using 21 images. The DC level of all pixel-blocks is encoded by DPCM and removed from each block. The integer-valued transform proposed in [12] is applied to the residual of each block, and only the first two frequency components are retained (2-D transform-domain data X), to keep the simulations light. The method of design for LMLPs, KLVQs and full-search VQs is described next.

Linear encoders (LMLPs) are designed with bit allocations varying from $[0; 1]$ to $[3; 3]$. That is done by partitioning the 2D transform domain data with one scalar quantizer per dimension. The VQ codebook is formed by computation of the centroids of the space-domain data according to the partitions of the transform domain. The thresholds T are computed by a nonlinear unconstrained optimization routine (Nelder-Mead simplex method) so that the reconstruction mean-squared-error (MSE), denoted by D in Section 2, is minimized.

Nonlinear encoders (KLVQs) are designed with bit allocations also varying from $[0; 1]$ to $[3; 3]$. The 2-D transform-domain data is analyzed through Kernel PCA by $tanh(x)$ warping and decorrelation using the second layer of the sigmoidal MLP. This generates a second set of 2-D vectors, that describes projections of the transform domain into a 2-D subspace of the Kernel PCA feature space [7]. This second set is partitioned using one scalar quantizer per dimension, and the VQ codebook is formed by computation of the centroids of the space-domain data according to *the partitions of the Kernel PCA subspace*. The thresholds T are again computed by the same numerical optimization routine, so that the reconstruction MSE is minimized.

The special property of the KLVQ map is that the lattice used for description of the statistics of the input-space data X is adjustable. By keeping the lattice canonical, as was done in [7], we can generate a family of nonlinear MLPs for VQ by changing the gain g and bias b, which are the same scalar values for all units at the KLVQ first layer. In this example, we chose values of gain varying from 0.005 to 10 and bias varying from -1.5 to 10. For each (g, b) pair, distortion minimization was performed from $[0; 1]$ to $[3; 3]$ bit allocations and the (H, D) values of the best encoders were applied to a program for computation of lower convex hull points. The lower convex hull is described by nine points that are shown in the sensitivity analysis of Figs. 3 and 4.

Full-search VQ encoders (SOMs) are designed directly for the 2-D transform domain data using a standard form of LBG [11]. Reconstruction of the transform domain coefficients is then performed directly by look-up-table decoding, and from the decoded vectors the MSE is also directly computed. Full-search VQs with 1 to 7 bits of indexing are designed.

After the encoders are designed, numerical simulations for sensitivity evaluation are performed: first of all, infinite-precision assessments of rate-distortion performance are obtained. That gives the lower-bound curves presented in the figures, that can be used as a reference for performance degradation for each system. The effects of the two types of perturbations have to be simulated:

1. *The performance degradation due to the implementation of static thresholds, matrices and vectors with finite precision.* The term *static* is used here to indicate values that have to be implemented as fixed-hardware parts, by means of capacitance ratios or transconductance amplifiers. The precision available today for such analog hardware implementation is in the range of 6 to 8 bits depending on the integration process and application of careful layout techniques [13]. We simulate this effect by performing scalar quantization of the thresholds designed (in the case of transform coders, linear and nonlinear), and by performing scalar quantization of each coefficient in the codebook memory (in the case of full-search vector coders).
2. *The performance degradation due to operation imprecision* (for example, the implementation of an analog multiplication is noisy both in the representation of its operands and in the effectuation of the product). Again, the precision available today for the analog hardware implementation of multiplications and additions (including vector accumulation and inner products) is in the range of 6 to 8 bits [13]. We simulate this effect by adding independent random processes, described by a sequence of i.i.d. random variables uniformly distributed in the range $[0, a]$, to the operands and results of each analog operation that is simulated. The upper limit a is equal to the signal dynamic range divided by the maximum number of bins allowed by the given precision, for example $a = 1/64$ for transform coefficients with 1.0 dynamic range and 6-bit precision.

In each plot, and for each family of encoders, there is a lower compression bound given by infinite precision, and a degraded-performance curve given by the combination of both the effects mentioned above. The test set used for evaluation is the well-known 'peppers' image, which is not an element of the 21-image training set. The following observations can be made about the data presented in Figs. 3 and 4, in connection with the complexity data from Table 1:

- For an ideal implementation (infinite precision) and very low complexity, KLVQs have the best rate-distortion performance (the lowest curves are the solid ones). However,

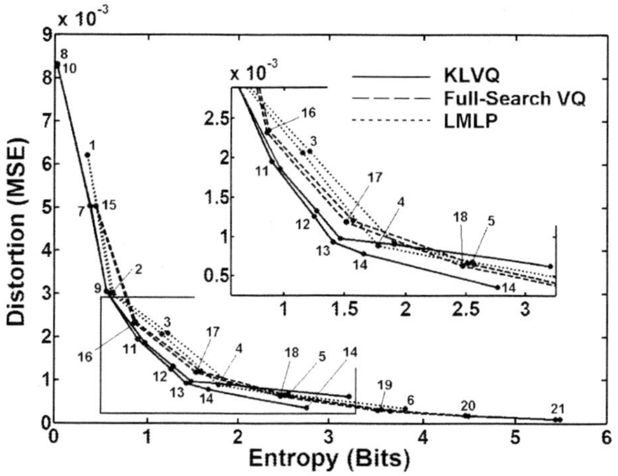

Fig. 3. Comparison of KLVQ, full-search VQ and Linear Transform Coding in terms of compression performance; for each pair of lines, the lower one corresponds to an infinite precision implementation, and the upper one to 7-bit precision. The numbers 1 to 21 refer to the complexity entry values in Table 1.

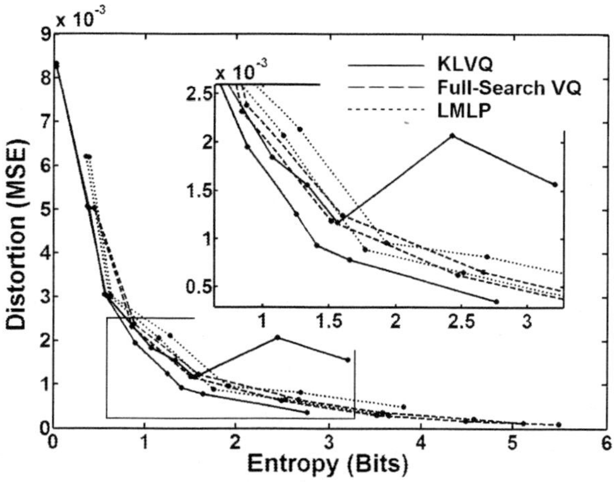

Fig. 4. Comparison of KLVQ, full-search VQ and Linear Transform Coding in terms of compression performance; for each pair of lines, the lower one corresponds to an infinite precision implementation, and the upper one to 6-bit precision.

it is affected severely by implementation errors. The degradation is still reasonable for 7-bit precision (not taking into account quantization of W coefficients), but it gets clearly unacceptable for 6-bit precision, and that happens particularly for allocations more complex than 5 bits.

- Full-search VQ has the best robustness to implementation errors (deviation between ideal and finite-precision curves), and also a good rate-distortion performance. However, we see from Table 1 that its complexity is too high for implementation over a block of analog pixels in a sensor.
- The linear transform coders (LMLPs) achieve the best agreements between performance, complexity and robustness. They have a distortion that is similar to that of full-search VQs and KLVQs, especially for systems more complex than 4 bits; and at the same time they have very low complexity (from Table 1) and are robust to analog implementations with at least 6 bits of precision.

All three systems are able to perform a direct mapping of analog data vectors into a compressed bit-stream; however, our numerical simulation has shown that Linear MLPs are the only encoders that simultaneously achieve a good rate-distortion performance and robustness to implementation errors, with a remarkably low computational complexity. This analysis was performed over a range of different rates and types of implementation errors.

5. REFERENCES

[1] D. X. D. Yang, B. Fowler, and A. El Gamal, "A Nyquist-rate pixel-level ADC for CMOS image sensors," *IEEE J. of Solid-State Circuits*, vol. 34, no. 3, March 1999, pp. 348–356.

[2] X. Liu and A. El Gamal, "Simultaneous image formation and motion blur restoration via multiple capture," in *Proc. IEEE Int. Conf. Acoustics, Speech and Signal Processing*, Salt Lake City, Utah, May 2001, vol. 3, pp. 1841–1844.

[3] G. Cauwenberghs and J. Waskiewicz, "Analog VLSI cellular implementation of the boundary contour system," in *Proc. Neural Information Processing Systems Conf.*, Denver, CO, November 1998, pp. 657–663.

[4] E. Funatsu, "Artificial retina large scale integration with on-sensor projection function for high-speed motion detection," *Optical Engineering*, vol. 41, no. 11, November 2002, pp. 2709–2718.

[5] A. Rodríguez-Vázquez, S. Espejo, R. Domínguez-Castro, R. Carmona, and G. Liñán, "CMOS design of focal plane programmable array processors," in *Proc. European Symp. on Artificial Neural Networks*, Bruges, Belgium, April 2001, pp. 57–62.

[6] J. G. R. C. Gomes and S. K. Mitra, "Low-complexity image compression without A/D conversion using analog multilayer perceptron," in *Proc. European Conf. on Circuit Theory and Design*, Krakow, Poland, September 2003, pp. III.281–III.284.

[7] J. G. R. C. Gomes and S. K. Mitra, "Kernel PCA for quantization of analog vectors on a pyramid," in *Proc. IEEE Int. Workshop on Neural Networks for Signal Processing*, Toulouse, France, Sep 2003, pp. 579–606.

[8] T. Kohonen, *Self-Organizing Maps*, New York: Springer-Verlag, 2001.

[9] S. Haykin, *Neural Networks - A Comprehensive Foundation*, New Jersey: Prentice-Hall, 1999.

[10] W. B. Pennebaker and J. L. Mitchell, *JPEG: Still Image Data Compression Standard*, New York: Van Nostrand Reinhold, 1993.

[11] Y. Linde, A. Buzo, and R. M. Gray, "An algorithm for vector quantizer design," *IEEE Trans. on Communications*, vol. COM-28, January 1980, pp. 84–95.

[12] H. Malvar, A. Hallapuro, M. Karczewicz, and L. Kerofsky, "Low-complexity transform and quantization with 16-bit arithmetic for H.26L," in *Proc. IEEE Int. Conf. on Image Processing*, Rochester, NY, USA, September 2002, pp. II.489–II.492.

[13] G. Liñán, *Diseño de Chips Programables de Señal Mixta con Bajo Consumo de Potencia para Sistemas de Visión en Tiempo Real*, Ph.D. Dissertation, University of Seville, Spain, June 2002.

CHAOTIC COMMUNICATIONS USING NONLINEAR TRANSFORM-PAIRS

W. P. Tang and H. K. Kwan

Department of Electrical and Computer Engineering
University of Windsor
401 Sunset Avenue, Windsor, Ontario, Canada N9B 3P4
kwan1@uwindsor.ca

ABSTRACT

In this paper, we describe a general chaotic communication system with nonlinear transform-pairs and saw-tooth functions. Some chaotic communication systems can be deduced from this general model. It provides a novel approach for encryption and decryption by setting nonlinear transform-pairs as the keys in the sender and the receiver. Simulation results of several such chaotic communication systems with direct and parameter modulations are presented. Both modulation methods can be used to transmit signals in noise disturbed Gaussian channel. The parameter modulation method provides the feasibility to transmit multiple signals simultaneously over a single channel with a 40dB Gaussian noise in which the extended Kalman filter (EKF) is used.

1. INTRODUCTION

Recently, chaotic communication system has received growing interests. Among previous chaotic communication systems, transform-pair with a modulus operation is often used for chaos synchronization and encryption [1]-[6]. In [1], a nonlinear digital filter and its inverse filter with finite precision are used for chaos digital encoding and decoding. The transform-pairs used in [1] are linear digital filters with left shift operation. An improvement of [1] is given by [2] with large filter parameters and it uses number theory to work for a classic encryption. The structure of the encoder [2] is that of a digital random-number generator. Using the property that chaotic systems are sensitive to system parameters, an adaptive chaotic digital encoder with variable parameters is proposed [3], and higher complexity and much better correlation properties can be achieved for increased security. Digital filter pairs have been researched in [4], [5]. The digital filter pair is shown to have an ability to hide signals in chaotic carrier. In [4], the coder should be a maximally disturbed and invertible channel, a general structure containing a static nonlinearity and a dynamical subsystem is derived and the system characteristics are specified. The proposed method in [5] uses a discrete time nonlinear filter with a saw-tooth nonlinearity induced by the two's complement overflow in digital filters, as a chaotic generator. Direct chaos modulation and parameter modulation are used. Adaptive digital filter in the receiver has been adopted to recover the signals in an autonomous way [3], [5]. In [6], a collection of research results during the last ten years on the conjunction of chaos and cryptography is presented. In this paper, we present a general chaotic communication system consisting of one or more nonlinear transform-pairs and saw-tooth functions in cascade, from which some chaotic communication systems can be deduced. It provides a method for encryption and decryption by negotiating one or more nonlinear transform-pairs as the keys in the sender and the receiver, which is different from the classic encryption method [7] that uses parameters as the key. In Section 2, the proposed general chaotic communication system is described, three theorems are derived, and several examples of the general system are given. In Sections 3-4, direct modulation method and parameter modulation method are applied to show the use of such transform-pairs for secure communications, and simulation results of seven examples are presented. A summary is given in Section 5.

2. GENERAL CHAOTIC COMMUNICATION SYSTEM

The general chaotic communication system is shown in Fig. 1 with two cascaded stages. The sender consists of $\varphi_1(\cdot)$ and $\varphi_2(\cdot)$ transform-pairs each with a saw-tooth non-linearity in the feedback path. The receiver performs the complementary operations of the sender. It consists of the complementary $\widetilde{\varphi}_2(\cdot)$ and $\widetilde{\varphi}_1(\cdot)$ transform-pairs each with a saw-tooth non-linearity. It can be proved that the general cascaded system has the ability to transmit signal without loss.

The non-linearity in the system has a general saw-tooth shape, which is defined by:

$$f_1(x) = x - 2c * round(\frac{x}{2c}) \quad c > 0 \,,\ |x| < c \qquad (1a)$$

$$f_2(x) = x \bmod(c) \qquad 0 < x < c \qquad (1b)$$

where c is a positive real or integer value.

To explain chaotic synchronization of the general cascaded system, we firstly consider the simple case that has a single $\varphi(\cdot)$ in the sender and a single complementary $\widetilde{\varphi}(\cdot)$ in the receiver. The Theorem 1 below refers to this configuration.

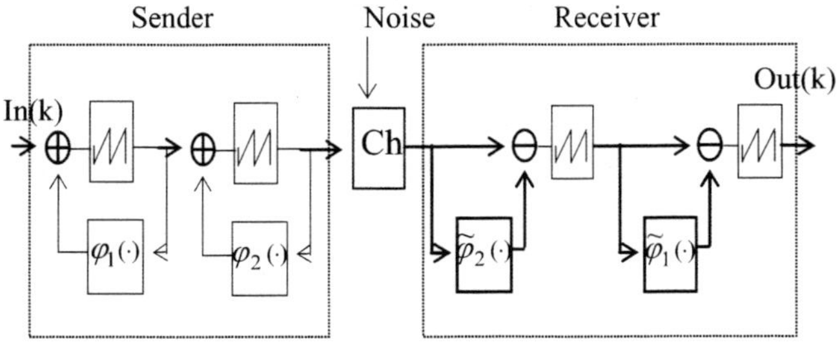

Fig. 1. General cascaded chaotic system (cascaded stage = 2).

Theorem 1: Let the input and output of a single transform-pair chaotic communication system without channel noise be in(k) and out(k). For a positive constant c, if |in(k)|<c then the saw-tooth non-linearity function $f(x) = f_1(x)$; or if 0<in(k)<c then $f(x) = f_2(x)$. The transform-pair in the sender and receiver system is $\varphi(\cdot)$ and $\widetilde{\varphi}(\cdot)$ respectively. If $\varphi(\cdot)$ is the same as $\widetilde{\varphi}(\cdot)$, then $\varepsilon(k) = out(k) - in(k) = 0$.

From the Theorem 1, it can be deduced that the output of the receiver is still equal to the input of the sender when more transform-pairs each with a saw-tooth non-linearity are inserted into the sender and the receiver respectively. We shall summarize this property in Theorem 2.

Theorem 2: The general cascaded system with saw-tooth non-linearity operations and transform-pairs $\varphi_1(\cdot) - \widetilde{\varphi}_1(\cdot), \varphi_2(\cdot) - \widetilde{\varphi}_2(\cdot), \cdots\cdots, \varphi_L(\cdot) - \widetilde{\varphi}_L(\cdot)$ has the ability to transmit signal without a loss when all the transform-pairs between the sender and the receiver are perfectly matched.

The sender of the general cascaded system can exhibit a chaotic behavior if one of its transform-pairs functions is not stable. Without a loss of generality, let us consider the case of the single transform-pair as in the Theorem 1. This property is stated in Theorem 3.

Theorem 3: Let in(k) and s(k) be independent input and output signal of the sender, respectively.
$s(k) = f(in(k) + \varphi(s(k-1)))$ where $f(x)$ is defined by (1a) or (1b). If $|\varphi'(s(k))| > 1$ where $\varphi'(s)$ denotes the derivative of $\varphi(s)$, and $\left|\frac{in(k) + \varphi(s(k-1))}{c}\right| \notin Z$ where Z is the set of all integers, for $k = 1,2,\ldots,\infty$, then s(k) is chaotic.

From the Theorem 2 and the Theorem 3, other chaotic systems such as [1]-[6] can be deduced from the general system. Table 1 gives some examples of chaotic communications with a single transform-pair.

Table 1. Transform-pairs examples.

Transform-pairs	φ(s(k)) with c = 1, f(s)=s mod(c)	φ'(s(k))
1	$s(k-1)+s(k-1)^2$, $\varphi(s(k)) \notin Z$	1+s(k)
2	$a^{s(k-1)}$, $\|a\| \geq e$, $a \notin Z$	$\ln(a)*a^{s(k)}$
3	ln(s(k-1)), 0<s(k)<1	1/s(k)
4	1/s(k-1), $s(k) \neq 0$, $\varphi(s(k)) \notin Z$	$-1/(s(k))^2$
5	$e^{2s(k-1)}$	$2e^{2s(k)}$
6	$3s(k-1) - s(k-1)^2$, $\varphi(s(k)) \notin Z$	3-2*s(k)

3. DIRECT MODULATION METHOD

Two modulation methods can be used for the general chaotic communication system: Direct modulation method and parameter modulation method. The direct modulation uses a modulated signal as the input of the chaotic system. For the direct modulation method, the system can be synchronous and asynchronous, which means the receiver can recover the signal at any time, not necessary from the time when the sender sends the signal. This is because the signal is recovered based on the input values of the receiver and does not matter how much channel delay is involved.

Given the three transform-pair examples listed in Table 2, the simulation results are shown in Fig. 2. In the simulations, the following input is used: $in(k) = 0.2\sin(2\pi(k - 15.3846)/61.5385) + 0.5$ and Gaussian noise is added to channel of 40dB for the case of Fig. 2(a) and of 30-60dB for the case of Fig. 2(b). From Fig. 2, we can observe that each transform-pair has different transmission ability in the presence of channel noise. The ranking of SNR performance from the highest to the lowest is the polynomial, exponent, and logarithmic transform-pairs.

Table 2. Simulation examples of direct modulation.

Transform-pairs	φ(s(k)) with c=1, f(s)=s mod(c)
Polynomial	$s(k-1)+s(k-1)^2$, $\varphi(s(k)) \notin Z$
Exponent	$3.0^{s(k-1)}$
Logarithmic	ln(s(k-1)), 0<s(k)<1

4. PARAMETER MODULATION METHOD

The parameter modulation method uses modulated signal(s) as the parameter(s) in the sender of the chaotic system, and an adaptive algorithm with a real-time tracking ability is used in the receiver to recover the parameter(s). The adaptive algorithm to be used can be Kalman filter, NLMS filter, RLS filter, and others.

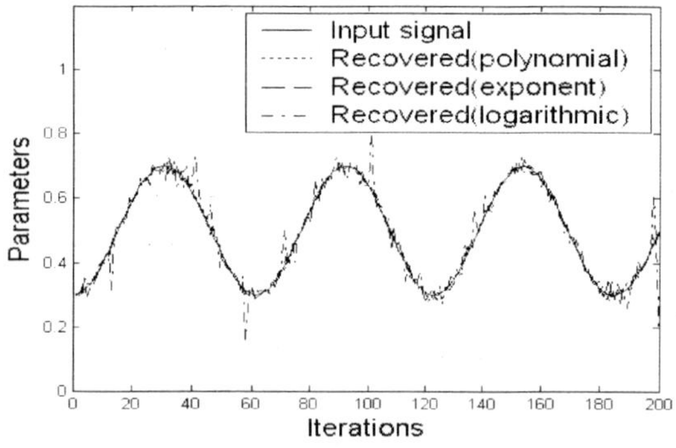

(a)

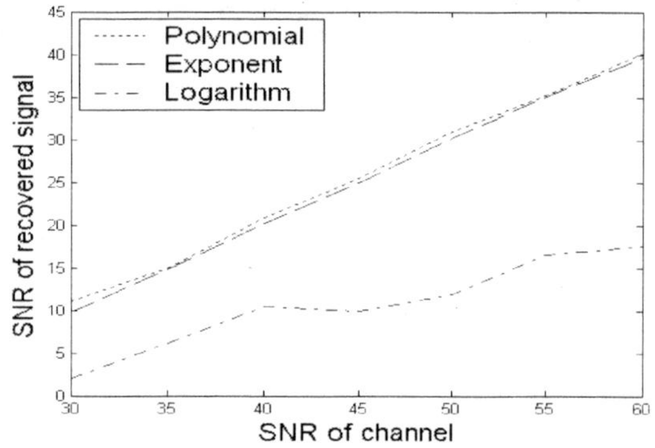

(b)

Fig. 2. Simulations results of direct modulation. (a) Recovered signals (SNR of channel is 40dB). (b) SNR of recovered signals (in dB) versus SNR of channel (in dB).

A well-known adaptive algorithm for such a nonlinear system is the extended Kalman filter (EKF) [8]-[9], which is summarized as follows: Let **a**' and $\mathbf{a}'^{-}_{k}$ denote the filter parameters of the receiver and the *k*th component of a, the partial derivative of the nonlinear measurement matrix φ(a, s(k)) with respect to $\mathbf{a}'^{-}_{k}$ is given by:

$$H(k) = \left. \frac{\partial \varphi(\mathbf{a}', s(k))}{\partial \mathbf{a}'} \right|_{\mathbf{a}' = \mathbf{a}'^{-}_{k}}$$

Let P(k) and Q_m denote, respectively, the error covariance propagation and the measurement noise covariance, the Kalman gain matrix is given by:

$$G(k) = P(k)^{-} \cdot H(k) \cdot \left[H(k) P(k)^{-} H(k)^{T} + Q_m \right]^{-1}$$

Let ε(k) denotes the error of the receiver, the state estimate update for the filter parameters of the receiver is given by:

$$a'(k) = a'(k)^{-} + G(k) \cdot \varepsilon(k)$$

Let Q_p denotes the process noise covariance, the error covariance propagation is given by:

$$P(k) = (I - G(k)H(k)) \cdot P(k)^{-} + Q_p$$

Given the three transform-pair examples listed in Table 3, the simulation results are shown in Fig. 3. In the simulations, three sinusoid signals are used. The input in(k) is defined in Section 3. M_2(k)=0.2sin(0.0247k) and M_3(k) =0.2sin(0.0503k+1.5708) are used to modulate the parameters a_1 and a_2 respectively. Gaussian noise is added to the channel of 40dB for the cases of Fig. 3(a)-(c), and of 30-60dB for the case of Fig. 3(d). As observed from Fig. 3, the EKF has been successfully used to track modulated parameter signals. Comparing Fig. 2(b) and Fig. 3(d), the results indicate that the parameter modulation has a better performance when the SNR of the channel is between 30dB to 35dB over a noise-disturbed Gaussian channel. From Fig. 3, we can see that the logarithmic transform-pair has the best SNR performance in the parameter modulation approach. Note that it has the worst SNR performance in the direct modulation case.

Table 3. Simulation examples of parameter modulation.

Transform-pairs	φ(s(k)) with c=1, f(s)=s mod(c)	H(k)
Polynomial	a_1s(k-1)+a_2s(k-2), φ(s(k)) ∉ Z	[s(k-1) s(k-2)]
Exponent	$a_1^{s(k-1)}$	s(k)*$a_1^{s(k)-1}$
Logarithmic	ln(a_1*s(k-1)), 0<s(k)<1	1/a_1

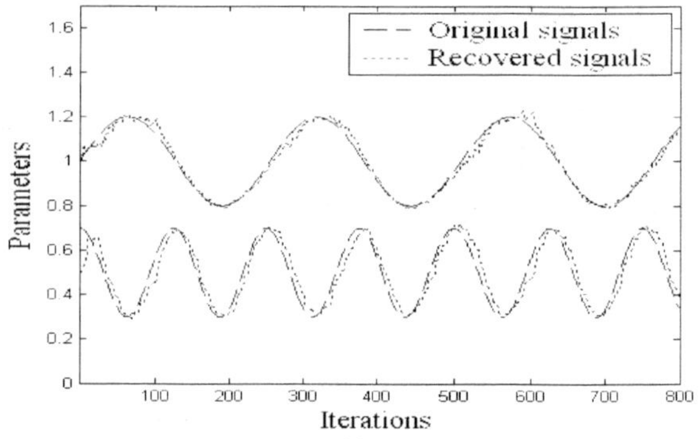

(a)

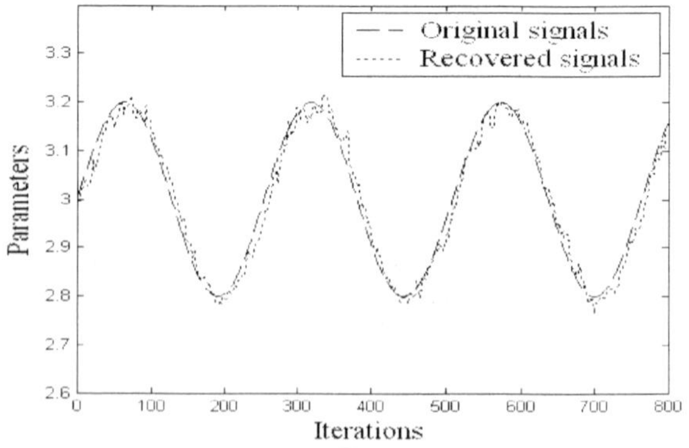

(b)

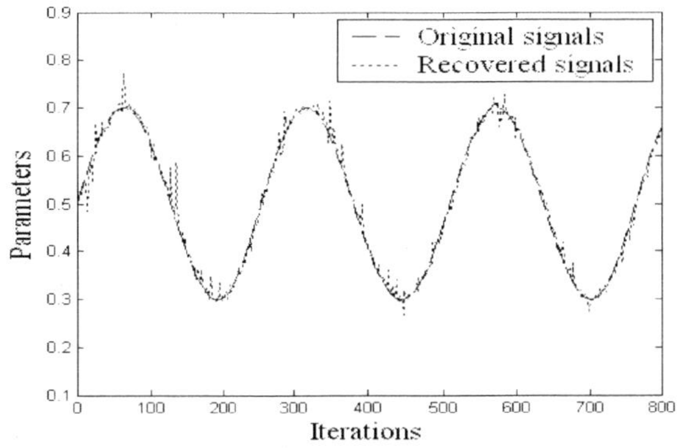

(c)

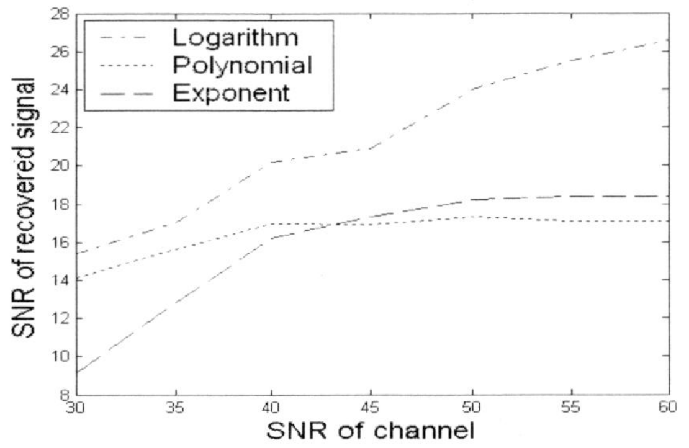

(d)

Fig. 3. Simulations results of parameter modulation (SNR of channel is 40dB). (a) Polynomial. (b) Exponent. (c) Logarithmic. (d) SNR of recovered signals (in dB) versus SNR of channel (in dB).

In practice, transmitting more than one modulated signal over a single noise-disturbed Gaussian channel is possible using the parameter modulation with the polynomial transform-pair, which has been shown in Fig. 3(a). Fig. 4 shows the results of the case of transmitting multiple (in this case, 15) signals over a single noise-disturbed Gaussian channel using the parameter modulation with the polynomial transform-pair.

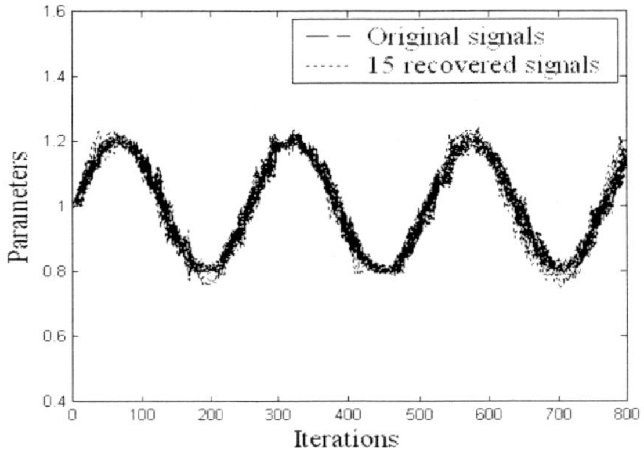

Fig. 4. Transmitting 15 signals simultaneously using parameter modulation. SNR of channel is 40dB.

In Fig. 4, the polynomial transform-pair is defined as:

$$\varphi(s(k)) = a_1 \cdot s(k-1) + \sum_{i=2}^{15} a_i \cdot s(k-i)^2 \ , \ a_j = \overline{a}_j + M(k)$$

$$\overline{a}_j = 1, j = 1,...,15$$

$$M(k) = 0.2\sin(0.0247k)$$

5. SUMMARY

In this paper, several examples of a general chaotic communication system with nonlinear transform-pairs and saw-tooth functions have been constructed and simulated using the direct and parameter modulations. The results of simulations indicate that the proposed systems can transmit signals in noise-disturbed Gaussian channel. Feasibility of transmitting multiple signals simultaneously over a single channel with simulated additive Gaussian noise has also been validated using the parameter modulation.

6. REFERENCES

[1] D. R. Frey, "Chaotic Digital Encoding: An Approach to Secure Communication," *IEEE Trans. Circuits Syst. II*, vol. 40, pp. 660–666, Oct. 1993.

[2] Michael J. Werter. "An Improved Chaotic Digital Encoder," *IEEE Trans. Circuits Syst. II*, vol. 45, pp. 227-229, Feb. 1998.

[3] J. Zhang, X. Dai, and P. Xu. "Improvement to Frey's Adaptive Chaotic Encoder," *Proceedings of the 3rd World Congress on Intelligent Control and Automation*, vol. 4, pp. 2479-2483, 2000.

[4] M. Gotz, K. Kelber, and W. Schwarz, "Discrete-Time Chaotic Encryption Systems - Part I: Statistical Design Approach," *IEEE Trans. Circuits Syst. I*, vol. 44, pp. 963-970, Oct. 1997.

[5] Adrian Leuciuc, "Information Transmission using Chaotic Discrete-Time Filter," *IEEE Trans. Circuits Syst. I*, vol. 47, pp.82-88, Jan. 2000.

[6] F. Dachselt and W. Schwarz, "Chaos and Cryptography," *IEEE Trans. Circuits Syst. I*, vol. 48, pp. 1498-1509, Dec. 2001.

[7] W. E. Burr, "Selecting the Advanced Encryption Standard," *IEEE Security & Privacy Magazine,* vol. 1, pp. 43 -52, March-April 2003.

[8] H. Leung and Z. Zhu, "Performance Evaluation of EKF-Based Chaotic Synchronization," *IEEE Trans. Circuits Syst. I*, vol. 48, pp. 1118 -1125, Sept. 2001.

[9] M. Han, J. Xi, and S. Xu. "Application of Kalman Filter to Chaotic Prediction of Sunspots," *Proceedings of the 4th World Congress on Intelligent Control and Automation*, vol. 1, pp. 406-409, 2002.

A New Charge-Packet Driven Mismatch-Calibrated Integrate-and-Fire Neuron for processing Positive and Negative Signals in AER based Systems

Bernabé Linares-Barranco, Teresa Serrano-Gotarredona, Rafael Serrano-Gotarredona, and Jesús Costas-Santos

Instituto de Microelectrónica de Sevilla, Ed. CICA, Av. Reina Mercedes s/n, 41012 Sevilla, SPAIN. Phone: 95-505-6666, Fax: 94-505-6686, E-mail: bernabe@imse.cnm.es

Abstract

We present the design and experimental measurements of an integrate-and-fire pixel for Address-Event-Representation (AER) transceiver chips such that (a) input events can be weighted according to a digital word, (b) this weight includes a sign bit, (c) the incoming event is acompanied by a sign bit, and (d) the pixel can be calibrated to compensate for mismatch in large arrays of these pixels. A prototype has been fabricated in the AMS 0.35μm CMOS process, whose experimental measurement results are provided.

1. Introduction

Address-Event-Representation (AER) technique is emerging as a strong candidate for providing a powerful framework in bio-inspired large-signal multi-layer hierarchically-structured scalable neural processing systems for vision and auditory applications [1]. Originally proposed by Silvilotti [2] for replicating the state of an array of pixels (neurons or cells) from one (emitter) chip onto another (receiver) chip, the technique can be used for spreading the state of one chip onto many processing chips [3]. Extra digital processing can easily be added while data is transferred between chips [4]. The approach is modularly expansible by assembling chips in a matrix fashion so that larger 2D images can be processed at the same speed. Also, since recently, ways are being proposed to include real-time programmable convolution processing on-chip by simply taking advantage of the AER nature [5].

In a traditional point-to-point AER communication [6], the state of an array of pixels in one chip is replicated on another chip. Fig. 1 shows an emitter chip composed of an array of neurons (pixels or cells) whose state evolves in time with a fairly slow time constant[1].

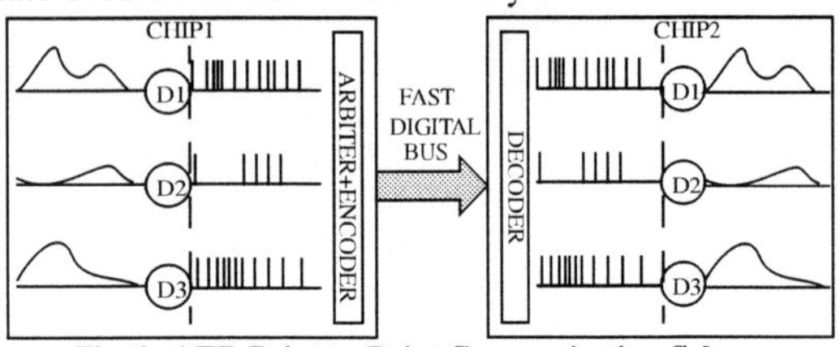

Fig. 1: AER Point-to-Point Communication Scheme

This continuous-time state feeds an integrate-and-fire oscillator that transforms the incoming continuous-time signal (like the photo current of a photo diode in a silicon retina pixel) into a Pulse-Frequency-Modulated (PFM) digital signal. This spiking signal is such that pulses should be very narrow (in the order of *nano-seconds*), while inter-spike interval should be much slower (typically, in the range of *mili-seconds*). This allows powerful time multiplexing, as explained next.

Each time a pixel produces a spike, it communicates asynchronously with a peripheral arbiter which will take the pixel coordinate and communicate it through a high-speed digital inter-chip bus to one or multiple receiver chips, using asynchronous communication. The high-speed inter-chip bus will show a flow of pixel addresses or coordinates. Those pixels that are more active and produce higher spike rates will access the inter-chip bus more frequently, while those less active will access the bus less times. The receiver chips will be reading the addresses on the bus, direct them to the pixel with the same coordinate, which will integrate them and recover the original continuous-time activity of the same pixel in the emitter chip.

This point-to-point communication can be extended easily to a multi-receiver scheme [3], so that for example a retina emitter chip can send its output to many post-processing chips in parallel. Also, using more complicated schemes, multiple emitter chips can merge their outputs to a smaller set of processing chips. Furthermore, operations in the address space (such as image translations, rotations, and generic transformations) are easily added by inserting look-up tables (PROMs) in the interchip buses [4]. The approach is also extensible to multi-layer hierarchically structured neural systems.

Presently, we are developing an AER convolution chip. In this chip, when an input pulse is received, it is sent to a projection field around the present address. Each pixel in the projection field receives a current

1. In conventional frame rate video transmission, pixels are sampled at $25Hz$, which implies a bandwidth below $12.5Hz$ for a pixel state variation signal.

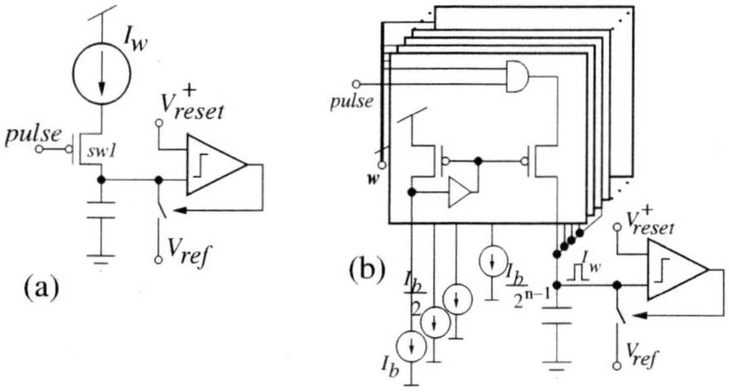

Fig. 2: (a) Fixed-Weight Single-Sign Integrate-and-Fire Neuron, (b) Programmable-Weight Single-Sign Integrate-and-Fire Neuron

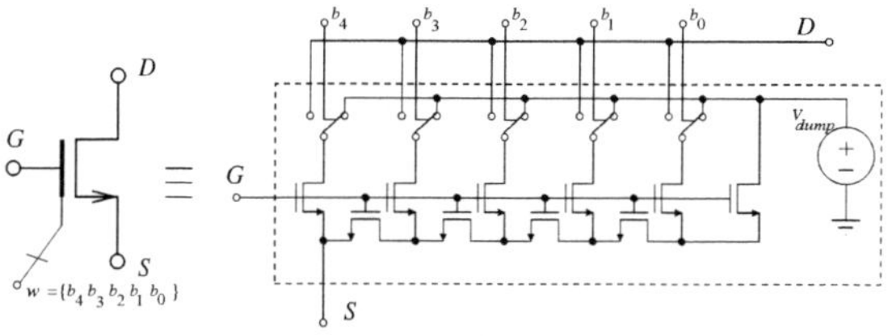

Fig. 3: Ladder-Based Current Splitter for Binary Weighted Current Generation

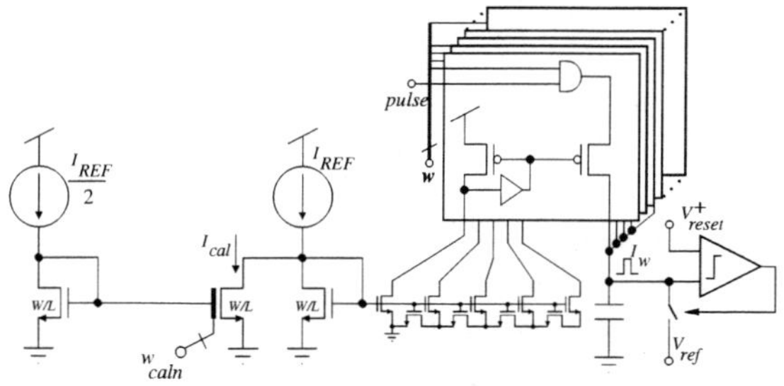

Fig. 4: Mismatch-Calibrated Programmable-Weighted Single-Sign Integrate-and-Fire Neuron.

pulse (charge packet) which is added onto an integrate-and-fire neuron. The current pulse of each pixel is weighted according to a digital word which represents the kernel weight for that pixel.

2. Pixel Circuit

Depending on the application, the pixel of an AER receiver (or transceiver) chip can be simple or very complicated. For example, in the case of plain reconstruction without further AER output generation, a simple diode-capacitor integrator might be sufficient [8]. However, in the case of a transceiver chip that receives AER input signals and generates AER output signals, the natural way is to use an integrate-and-fire neuron to collect the incoming pulses and generate the corresponding output pulses as well. This is the case we will consider from now on. Several situations can be distinguished, depending on whether or not we would like to weight the incoming pulses before integration, or if we would like to assign a sign to pulses and/or weights.

A. Traditional Fixed-Weight Single-Sign Integrate-and -Fire Neuron

The simplest situation is when a pixel receives pulses with the same weight and sign. In this situation each incoming pulse should produce a fixed charge package, which will be added to the integration capacitor of an integrate-and-fire neuron. A possible implementation is shown in Fig. 2(a). The incoming pulse activates switch $sw1$ which will inject a fixed charge, assuming the pulse is always of fixed duration (for example, if produced by a monostable at the periphery of the integrator array [9]). This circuit suffers from high mismatch between pixels, caused by the mismatch in the charge packages. Note that the charge package depends on parasitic feedthrough capacitances, the switch channel charge, and the precise timing of the pulse.

B. Programmable-Weight Single-Sign Integrate-and-Fire Neuron

The circuit in Fig. 2(a) is not directly extensible to the case where one needs to weight the charge package according to a digital word **w** stored in a register. In this case, if the weight is of n bits, we need to have n circuits each capable of providing a proper binary weighted charge. Then, depending on the value in the weight register **w**, a given combination of those circuits will be activated for an incoming pulse. In order to achieve precise binary weighting, we need to generate a charge package that can be precisely controlled by, for example, a precise bias current $I_b/2^j$. Source switching is a good technique to achieve this [10]. The circuit in Fig. 2(b) exploits this technique. A set of n binary weighted bias currents $I_b/2^j$ feed n current mirrors. The mirrors output transistor sources are switched or not depending on the bits of the weight stored in the weight register **w**. The charge produced by each branch is proportional to its bias current $I_b/2^j$, and the resulting pulse current I_w is properly weighted by the digital word **w**. The buffer at the mirror input is included to minimize parasitic feedthrough effects.

C. Programmable-Weight Mismatch-Calibration Sin-gle -Sign Integrate-and-Fire Neuron

The binary weighted bias currents $I_b/2^j$ in Fig. 2(b) need to be available in each pixel. These currents suffer from great mismatch from pixel to pixel in case we need to implement large arrays. Even if each pixel contains a complete ladder-based current splitter, as shown in Fig. 3, the mismatch of these ladders can be as high as 50% for arrays spread over about $2mm$ distances [11]. Fortunately, there is a way of calibrating these splitters and improve mismatch by over a factor of 10 [11]. The solution involves combining two of those ladders, as is shown in Fig. 4.

D. Programmable-Weight Mismatch-Calibration Dou-ble-Sign Integrate-and-Fire Neuron

When processing generic convolutions with programmable kernels, there is a need to implement kernels that may take values with either positive and negative signs [5]. For example, in the case of Gabor filters or orientation sensitive kernels, half of the kernel shows a positive sign while the other half is symmetric with negative sign. Consequently, a generic convolution chip needs to be prepared for programming kernel weights of both signs. Furthermore, the output image generated by convolving an input intensity or contrast image[2] will have pixels with either positive or negative signs. Consequently, the output AER stream of such processing includes an extra sign bit. If this output is fed again to a further convolution processing chip, then the input image AER stream comes also with its sign bit. Consequently, a generic programmable weight convolution chip for use in a generic multilayer system has to be able to receive AER input streams with sign bit, process kernels that include a sign bit, and produce output AER streams that include a sign bit as well. Therefore, the integration pixel of such a convolution chip will receive an event pulse with sign, which needs to be added or subtracted to the integration capacitor, depending on the signs of the input pulse and the stored weight.

A valid solution to avoid the implementation of double sign circuitry is to have two paths: the excitatory path and the inhibitory path with excitatory and inhibitory connections and two integrators (one for representing excitatory states and the other for inhibitory states). This is a very biologically inspired solution, but presents the following problem in AER systems: the state of a pixel is represented by the frequency difference of the excitatory and inhibitory integrators. Consequently, the activity of each integrator can be very high even if the net difference is small. However, each integrator will be consuming power as well as AER communication bandwidth. In order to avoid this problem, we propose to implement a double-sign integrate-and-fire neuron with only one integration capacitor.

The proposed solution consists of adding a top/down symmetrically replicated[3] version of the circuit in Fig. 4, to itself. All elements should be replicated, except for the capacitor. Also, the 2-input AND gates are replaced by 3-input ones, because an extra input is needed for the event sign. The resulting circuit has two opposite sign charge package generation circuits with their respective programmable weights ***w*** and calibration circuits. They both feed a unique integration capacitor. This capacitor is connected to two complementary comparator/reset circuits that detect whether the accumulated net (positive and negative) charge has reached either the positive V^+_{reset} or the negative V^-_{reset} threshold. When one of the thresholds is

2. Intensity or contrast images coming from an imager or retina have all pixels with positive values.

3. Changing sign of currents, nmos transistors to pmos ones, and pmos ones to nmos ones.

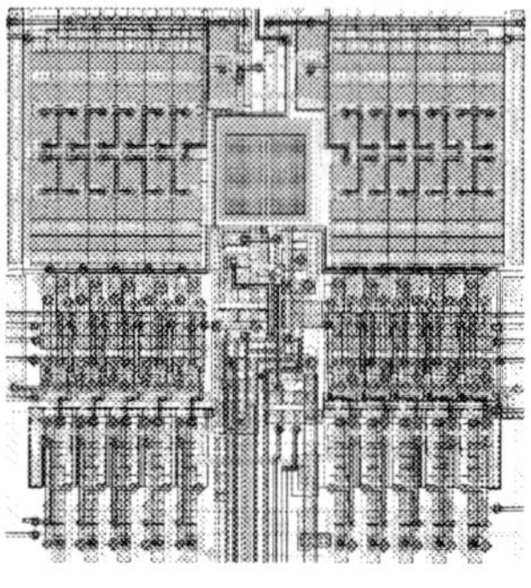

Fig. 5: Layout of fabricated Double-Sign Mismatch-Calibrated Integrate-and-Fire Neuron

reached a pulse is produced for the corresponding sign and the capacitor is reset to the central voltage between the two thresholds V_{ref}. Positive charge is added to the integration capacitor when both input pulse and stored weight have the same sign, while a negative charge is added when they have opposite sign.

3. Experimental Results

A test integration pixel was fabricated in the AMS $0.35\mu m$ CMOS process. In this particular case, both positive and negative programming weights and both calibration weights were implemented with a five bit resolution. The area of the pixel is $70 \times 70\mu m^2$. The first thing to verify is that the charge packages are precisely and linearly controlled by the current I_w programmed in the weights ***w***. If excessive parasitic charge is present for each pulse (like capacitive feedthrough) then there would be no linearity between the integration capacitor charging rate and the controlling current I_w or weight ***w***. Fig. 6(a) shows the measured integration capacitor voltage when pulses arrive at a fixed frequency with a positive sign and the stored weight is positive. Each slope corresponds to a different weight. As can be observed, all slopes stay fairly linear with time. The staircases are actual capacitor voltage, while the superimposed straight lines are the computed best fit. Fig. 6(b) shows the resulting average capacitor voltage slopes as a function of weight current I_w. The maximum value for I_w was $100nA$. As can be seen, the dependence is fairly linear, which means that the charge packages produced are reasonably well controlled by the programming weights. In the case of Fig. 6, input pulses had a duration of $T_h = 100ns$ with a periodicity of $T = 10\mu s$, and the voltages were set to $V_{ref} = 1.5V$, $V^+_{reset} = 2.5V$, and $V^-_{reset} = 0.5V$.

In order to test the double sign feature of the integrate-and-fire neuron, we obtained the measurements shown in Fig. 7. They were obtained by applying input spikes of width $T_h = 100ns$, period $T = 10\mu s$, while adding a sign signal of period $1ms$ but with different duty cycles. In Fig. 7 both programmable weights w were set to their maximum values ($I_w = 100nA$). Fig. 7(a) shows the capacitor

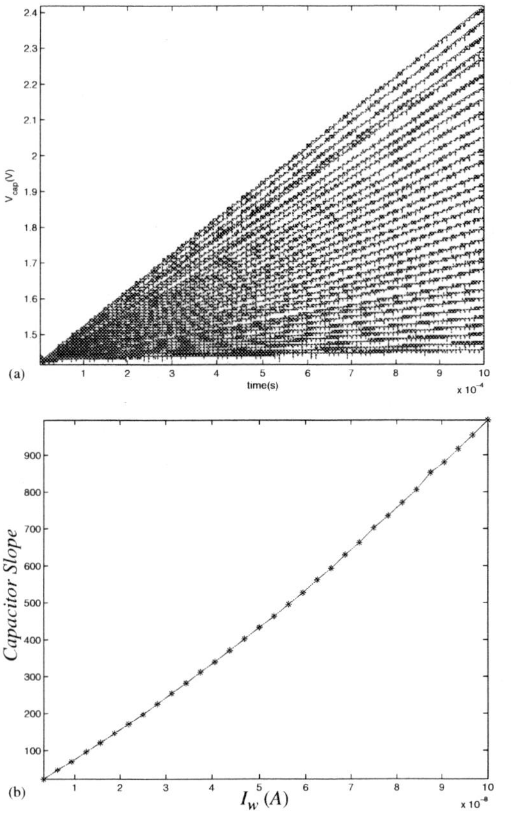

Fig. 6: (a) Experimental Measurements of Integration Capacitor voltage vs time, for different values of I_w (b) Capacitor Voltage Slope vs I_w

voltage when sign signal duty cycle is 20%, as well as the positive and negative output spikes. For this duty cycle only negative output spikes are produced at a fixed rate. Fig. 7(b) shows the same waveforms but when applying a sign signal with 50% duty cycle. In this case no output spikes are produced. Fig. 7(c) shows the results when applying a sign signal with 80% duty cycle. In this case only positive output spikes are produced.

4. Acknowledgements

This work was partially supported by spanish grants TIC1999-0446-C02-02, TIC2000-0406-P4-05 (Victor), FIT-07000/2002/921 (Arquimedes), TIC2002-10878-E, TIC-2003-08164-C03-01 (Samanta), and EU grant IST-2001-34124 (Caviar).

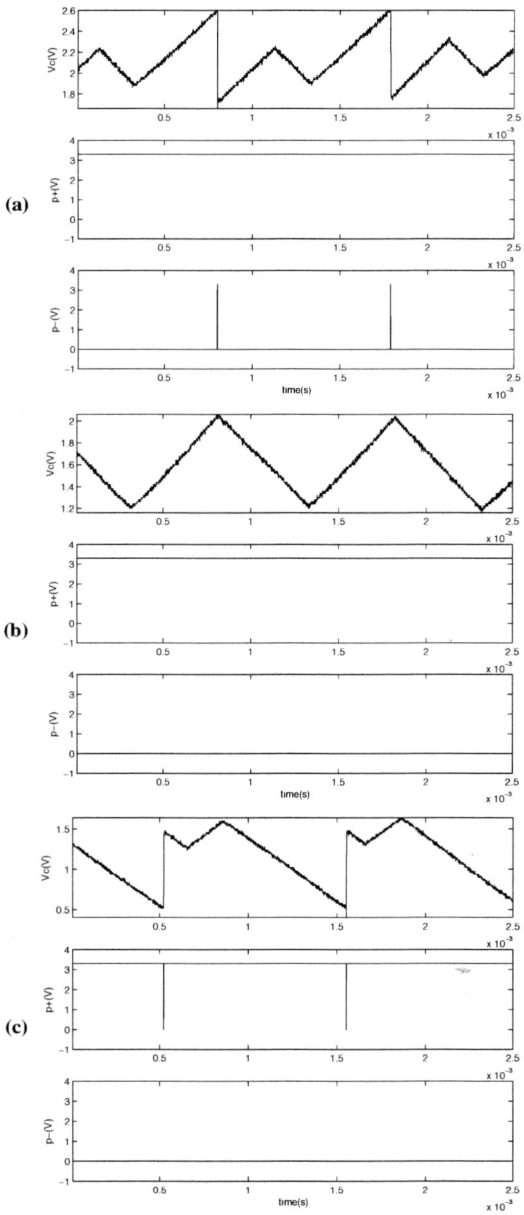

Fig. 7: Measured Capacitor Voltage Waveforms and Output Spikes. (a) Sign Signal with 20% Duty Cycle, (b) 50% Duty cycle, and (c) 80% Duty Cycle

5. References

[1] A. Cohen, R. Etienne-Cummings, T. Horiuchi, G. Indiveri, S. Shamma, R. Douglas, C. Koch, and T. Sejnowski, *Report on the 2003 Workshop on Neuromorphic Engineering*, 2003.

[2] M. Sivilotti, Wiring Considerations in analog VLSI Systems with Application to Field-Programmable Networks, *Ph.D. Thesis*, California Institute of Technology, Pasadena CA, 1991.

[3] J. P. Lazzaro and J. Wawrzynek, "A Multi-Sender Asynchronous Extension to the Address-Event Protocol," *16th Conference on Advanced Research in VLSI*, W. J. Dally, J. W. Poulton, and A. T. Ishii (Eds.), pp. 158-169, 1995.

[4] D. H. Goldberg, G. Cauwenberghs, and A. G. Andreou, "Analog VLSI spiking neural network with address domain probabilistic synapses," *Proc. of the IEEE 2001 Int. Symp. Circ. Syst.* (ISCAS 2001), vol. 2, pp. 241-244, May 2001.

[5] T. Serrano-Gotarredona, A. G. Andreou, and B. Linares-Barranco, "AER Image Filtering Architecture for Vision Processing Systems," *IEEE Trans. Circuits and Systems Part-II*, vol. 46, No. 9, pp. 1064-1071, September 1999.

[6] K. Boahen, "Point-to-Point Connectivity Between Neuromorphic Chips Using Address Events," *IEEE Trans. on Circuits and Systems Part-II*, vol. 47, No. 5, pp. 416-434, May 2000.

[7] T. Choi, B. Shi, K. Boahen, "An Orientation Selective 2D AER Transceiver," *Proc. of the 2003 IEEE Int. Symp. Circ. Syst.*, vol. IV, pp. 800-803, ISCAS'03, 2003.

[8] K. Boahen, "Retinomorphic vision systems," *Proc. Int. Conf. on Microelectronics for Neural Networks*, pp. 2-14, 1996.

[9] P. Häfliger, "A spike based learning rule and its implementation in analog hardware," *PhD Thesis*, ETH Zürich, Switzerland, 2000.

[10] G. Cauwenberghs and A. Yariv, "Fault-tolerant dynamic multi-level storage in analog VLSI," *IEEE Trans. Circ. Syst. Part-II*, vol. 41, No. 12, pp. 827-829, 1994.

[11] B. Linares-Barranco, T. Serrano-Gotarredona, and R. Serrano-Gotarredona, "Compact Low-Power Calibration Mini-DACs for Neural Massive Arrays with Programmable Weights," *IEEE Trans. on Neural Networks*, September 2003.

CAN SPIKE TIMING DEPENDENT PLASTICITY COMPENSATE FOR PROCESS MISMATCH IN NEUROMORPHIC ANALOGUE VLSI?

Katherine Cameron and Alan Murray

School of Engineering and Electronics
The University of Edinburgh
Edinburgh, EH9 3JL, UK
k.l.cameron@sms.ed.ac.uk, alan.murray@ee.ed.ac.uk

ABSTRACT

Analogue VLSI can be used to implement spike timing dependent neuromorphic training algorithms. This paper presents novel circuitry that uses spike timing to "adapt out" the effects of device mismatch in such circuits. Simulation results for the circuit implemented in a 0.35μm CMOS process are reported.

1. INTRODUCTION

As circuit geometries decrease, the problems caused by device mismatch increase [1]. When sub-micron devices are operated in sub-threshold mode the effect is exacerbated [2]. The effect of threshold (*Vth*) variation can clearly be minimised by the use of transistor dimensions greater than minimum size. However, other compensation mechanisms have been proposed, which allow the fabrication of devices such as current mirrors and dividers and comparators [3]-[5] with improved matching characteristics.

We present a new technique for device mismatch correction based upon a form of asymmetric Hebbian training, a neural algorithm implementing Spike Timing Dependent Plasticity (STDP). In STDP synaptic weight change is driven not by spike-rate correlations, but by *individual inter-spike timings*. This class of algorithms has been shown to recover different and potentially richer information from a spike train when compared to rate and population coding methods [6] and due to its relative simplicity has been successfully implemented in analogue VLSI [7]-[11].

We have chosen, as a context for this study, a spike-timing driven algorithm for visual scene analysis [12] which, if implemented in analogue focal-plane VLSI, will be degraded by unavoidable mismatches in the temporal characteristics of nominally identical circuits. The algorithm is thus an excellent demonstrator for any benefits that STDP adaptation may offer. The algorithm is based upon the comparison of two related inter-neural firing times, where the delay between them effectively measures the speed of motion of an object in the visual field. A simple analogue circuit has been designed that predicts a spike arrival time from the timings of two previous spikes. This circuit has been subjected to "process mismatch" simulation and an adaptation network has been added to attempt to compensate for the mismatch error.

This paper will present the circuit designed and Monte Carlo simulation results showing both the problems caused by mismatch and the benefits of the adaptation network. A 0.35μm CMOS chip is being fabricated to verify the simulation results.

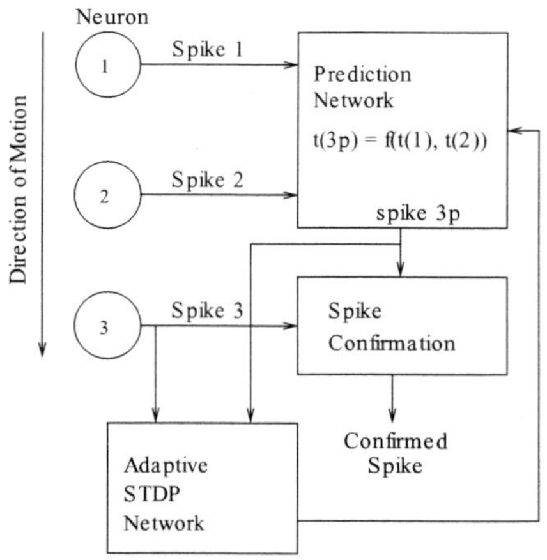

Fig. 1. Neuron firing time and prediction. Neurons 1 to 3 fire at predetermined intervals. Neuron 3p fires at a time predicted from the firing time of neurons 1 and 2. If the neurons fire at equal intervals $t(3p) = t(2) + \{t(2) - t(1)\}$ where t(1) is the firing time of spike 1, t(2) for spike 2 etc. Spike 3p is the prediction of spike 3.

2. PREDICTION CIRCUITRY

The system chosen to use spike timing dependent training to adapt out device mismatch is that of Wörgötter et al [12] in which the time at which an edge passes a pixel in a visual radial flow field is used to predict further such events. This can be used to eliminate noise by accepting only events that have been predicted using signals from previous edge-sensitive pixels. In this paper, neuron firing times will replace the edge times. The details of the vision-processing algorithm are not relevant as the presented circuit implements the noise reduction part of the work and not the actual depth-from-motion algorithm.

The system block diagram is shown in figure 1. The spike firing circuit is made up of three "integrate and fire" neurons that are set to fire at predetermined intervals. For example, the delay between neurons one and two firing may be the same as the delay between neurons two and three. In a complete system the intervals would be determined by pixel spacing. The spikes from neuron 1 and 2 are passed to the prediction network which generates another spike, spike 3p (i.e. "predicted", not actual), at a time determined

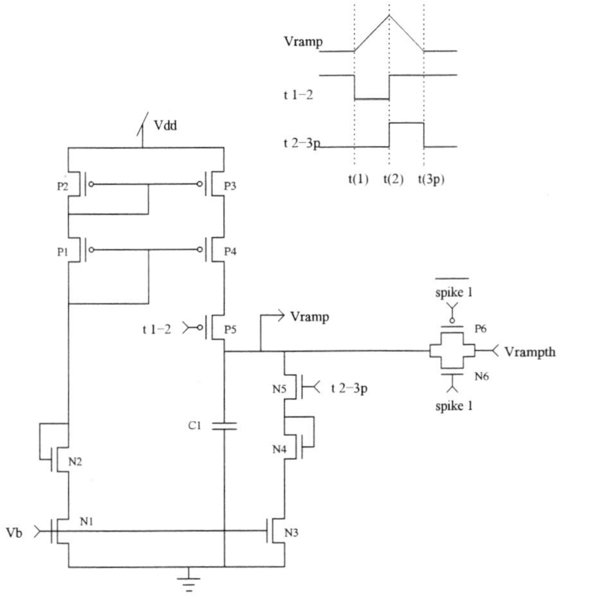

Fig. 2. Current sink/source circuit. C1 charges between $t(1)$ and $t(2)$ and discharges from $t(2)$ to $t(3p)$. The charging and discharging currents are matched to enable prediction of spike 3 at the time when C1 has returned to *Vrampth*.

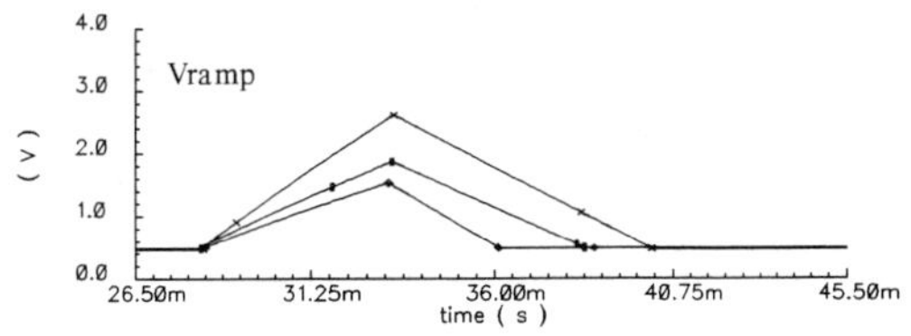

Fig. 3. Simulation outputs for *Vramp*. The results of three process mismatch Monte Carlo simulations show a variety of possible shapes for *Vramp*

by the firing times of the input spikes. The timing of this spike can then be compared to that of spike 3 by the spike confirmation block. Spike 3 will only be accepted as genuine if it arrives within a given time window of spike 3p. The adaptive STDP network is also shown. We will attempt to use it to adapt out the process-induced mismatch between the actual firing time $t(3)$ and its predicted time $t(3p)$. The circuit will be described in detail in section 3.

The circuit used to provide the prediction of $t(3)$ is shown in figure 2. Capacitor C1 is charged and discharged by a nominally matched current source and sink. When the voltage across capacitor C1 returns to its initial value a spike is fired, at time $t(3p)$, and if the process were perfect this spike would coincide with the actual spike at $t(3)$. The sequence of events is as follows:

t(1) Spike 1 is fired and P6 and N6 act as a switch which sets the initial value of *Vramp* to *Vrampth*.

t(1)-t(2) P5 is on and C1 is charged through P3 and P4 which supplies a mirrored version of the current set by *Vb*.

t(2) P5 turns off while N5 becomes active.

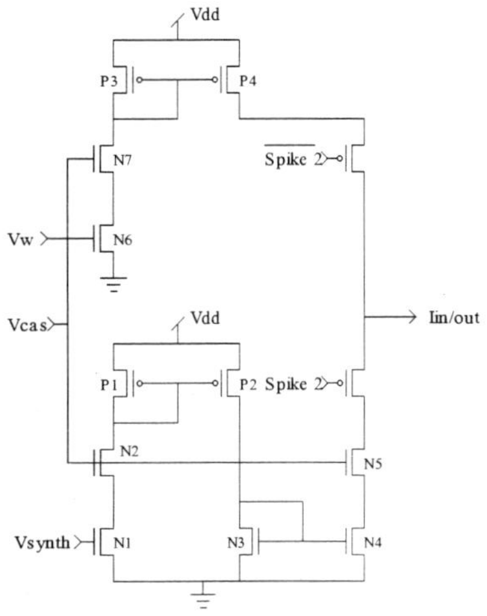

Fig. 4. Excitatory/inhibitory synapse circuit. The synapse is excitatory if *Vw* is greater than *Vsynth* and inhibitory if *Vsynth* is greater.

t(2)-t(3p) C1 is discharged through N3 and N4. If the current source and sink are matched the time taken to discharge C1 should match the charging time.

t(3p) *Vramp* is compared to *Vrampth* using a differential pair. When *Vramp* crosses it, neuron 3p is "fired" . This ends the discharge period.

The accuracy of the prediction depends on how well the current sink and source have been matched. A perfect match is not possible without post fabrication trimming which is expensive and time consuming. Figure 3 is the result of a process mismatch Monte Carlo simulation. The *Vramp* signal for three different simulation runs is shown. Clearly, process mismatch causes significant differences not only in the current value but also between pairs of charging and discharging currents leading to an inaccurate prediction. We will use the relationship between the actual occurrence of the spike at $t(3)$ and its corrupted prediction at $t(3p)$ as inputs to the STDP algorithm, aiming to "pull" the actual and predicted spikes back into coincidence. We will, effectively, adapt synaptic *delays*, as opposed to synaptic strengths, in response to the STDP algorithm. The ability to adapt synaptic delays can be found in other neuromorphic systems which use spike timing [13]-[15].

3. ADAPTIVE CIRCUITRY

To compensate for the mismatch in the current source and sink, at $t(2)$ a current path is opened and charge is injected/removed from the capacitor in figure 2. This effectively increases or decreases the discharge time by raising or lowering the peak signal voltage. The charge is injected/removed by the excitatory/inhibitory synapse designed in [16]. Figure 4 shows the circuit which is controlled by two inputs *Vsynth* and *Vw*. If they are equal the output current is zero. If *Vsynth* is greater than *Vw* the synapse is inhibitory. The synapse is excitatory when *Vw* rises above *Vsynth*. In this application *Vsynth* is set to 1.8V. *Vcas* is the bias for the cascoding transistors and is set at 1.37V.

The control of *Vw* is provided by the circuit in figure 5 which responds to the timing differences between $t(3)$ and $t(3p)$.

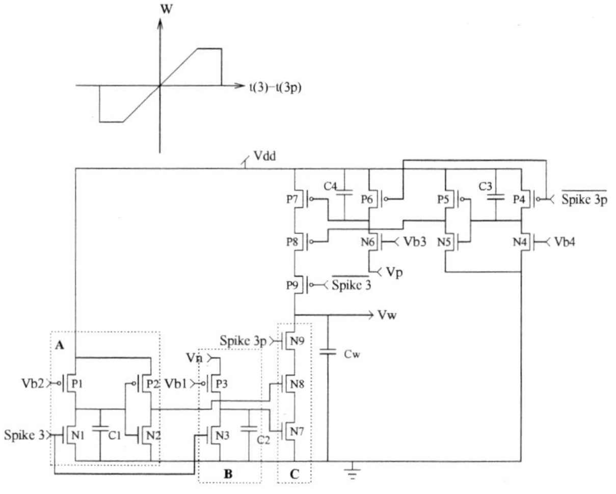

Fig. 5. **Weight adaptation circuit and algorithm.** *Vw* is increased if spike 3p occurs before spike 3 and decreased if neuron 3 fires first. Weight change only occurs if spikes 3 and 3p happen within a specified time window. The weight change (ΔW) is "capped" by *Vn* and *Vp* to prevent *Vw* moving from one supply rail to the other. Three circuit blocks are highlighted. **A:** The circuitry setting the time window. **B:** The circuitry setting the amount of weight change. **C:** Combining the two together to create the negative part of the ΔW graph.

The circuit in figure 5 has two distinct parts. MOSFETs N1-3 and P1-3 control reductions in *Vw* (depression) through N7, N8 and N9, while increases (potentiation) are achieved through transistors N4-6, P4-6, P7, P8 and P9. As the potentiation and depression circuits are mirror-images, with transistor polarities changed, we will present a detailed description of only the depression mechanism.

When spike 3 occurs C1 is discharged through N1 and then charges slowly to Vdd through P1. This results in a pulse at the gate of N8, whose width is determined by *Vb2*. This pulse defines the window within which weight change occurs. At $t(3)$ C2 is discharged through N3 and then charged slowly to *Vn* through P3. This voltage is N7's gate voltage and controls the size of the weight change. As the voltage across C2 increases with time, the amount of weight change will increase the longer the delay between $t(3)$ and $t(3p)$ until the voltage reaches *Vn*. This defines the maximum weight change. If spike 3p occurs within the time window, *Cw* is discharged through N7, N8 and N9 by an amount related to the voltage across C2. When the weight voltage is decreased, the peak of *Vramp* decreases. This reduces the discharge time and spike 3p will occur at a time $t(3p)$ that is closer to $t(3)$.

The weight change is "capped" by *Vn* and *Vp* to prevent *Vw* moving from one supply rail to the other. The window width is determined empirically from *a priori* knowledge as to when two spikes may be regarded as unrelated events.

When spike 3p is a good prediction of spike 3 no further weight change occurs. Therefore the adaptation network can be left active after calibration. It will then continue to compensate for any circuit drift caused by changes in environmental conditions or charge leakage from the capacitor. The latter is particularly important as many compensation techniques suffer from problems in the permanent storage of calibration values.

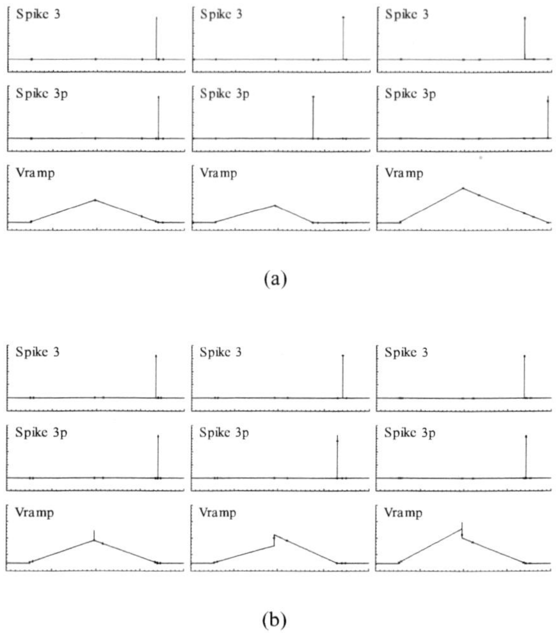

Fig. 6. **Pre and post adaptation results.** These results show the spike 3, spike 3p and *Vramp* signals during three process mismatch Monte Carlo simulations. (a) is pre-adaptation and the error in spike 3p is clear. (b) shows the signals post-adaptation where the error has been reduced. The alteration to *Vramp* can also be seen.

4. RESULTS

To test the circuit a 100-run Monte Carlo process mismatch simulation of the AMS 0.35μm CMOS CSI process was performed using Cadence Design Tools. During each run the transistor parameters are altered to simulate the process mismatch that occurs during manufacture. Usually this type of analysis is performed to check the robustness of a design but we use it to determine the circuit's ability to *compensate* for mismatch. During each run the difference between $t(3)$ and $t(3p)$ was measured before and after STDP adaptation.

Figure 6(a) shows example results for three runs before adaptation. The timing of spike 3p can be seen to vary widely. After adaptation, figure 6(b), the spread of spikes has been decreased and the peak ramp voltage can be seen to change, on charge injection, to alter the discharge time. The first example is a good match before adaptation and is not affected adversely by it.

While Figure 6(b) is encouraging, a more statistically significant result can be obtained by examining the results for the full 100 runs. Figure 7 shows the spread of the estimation error $t(3)-t(3p)$ before and after STDP adaptation. $t(2) - t(1)$ is 1ms and 5ms in (a) and (b) respectively.

It can be seen very clearly that the standard deviation of the prediction error is reduced dramatically, indicating that prediction after adaptation is far more reliable. The STDP adaptation network has reduced the effect of process mismatch by "pulling together"

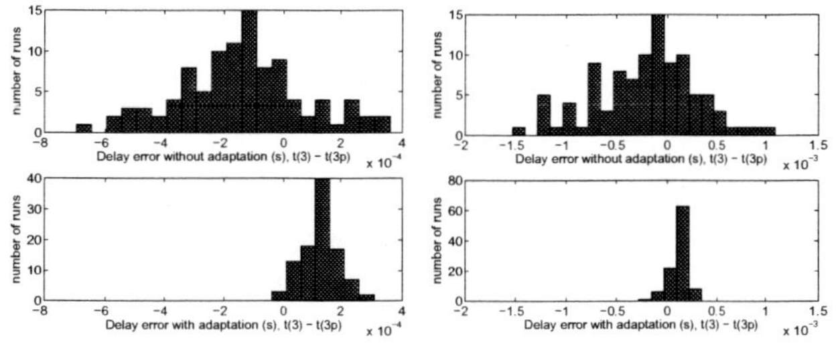

(a) (b)

Fig. 7. **Distribution of prediction errors.** The prediction error is defined as $t(3) - t(3p)$. Each graph has the results before adaptation at the top with after adaptation underneath. Results (a) were from a simulation when $t(2) - t(1) = 1ms$ and (b) $t(2) - t(1) = 5ms$

spikes which are effectively coincident in real time, but have been "spread" on silicon by process imperfections.

5. CONCLUSIONS

Simulation results show that spike timing adaptation can minimise the effects of process mismatch in the context of a spike-timing processing scheme for visual scenes.

Our circuit predicts the timing of a third spike from the timing of two previous spikes. This prediction circuit's accuracy depended on the degree of transistor parameter variation. To compensate for this an adaptation network was added to reduce the prediction error.

The circuit was tested using a process mismatch Monte Carlo simulation. Process mismatch degrades the accuracy of prediction significantly. However, STDP adaptation effects a significant and useful reduction in the error spread. The circuit is being fabricated using a CMOS 0.35μ process to verify the simulation results.

It should be noted that this technique differs from dynamic techniques in that it will correct for other imperfections in the circuit, for example variation in neuron firing thresholds or track delay, and not simply mismatch within the current mirror. This technique can also be used when the charging current is a fraction of the discharging current allowing the non-uniform placement of pixels.

Further work will include the integration of the network with a pixel array to demonstrate the potential of a complete system.

The implications of this work are wide-ranging. It is now clear that inter-spike timing can be used both to perform useful processing on silicon and that the same spike timings can drive circuits that adapt to compensate for the unavoidable and ever-increasing mismatches between elements in real silicon chip. This may well prove a major tool in the drive towards reliable computation in an increasingly unreliable medium.

6. REFERENCES

[1] Marcel J M Pelgrom, Aad C J Duinnaijer, and Anton P G Welbers, "Matching properties of MOS transistors," *IEEE Journal of Solid-State Circuits*, vol. 24, no. 5, pp. 1433–1440, October 1989.

[2] Eric Vittoz and Jean Fellrath, "CMOS analog integrated circuits based on weak inversion operation," *IEEE Journal of Solid-State Circuits*, vol. SC-12, no. 3, pp. 224–231, June 1977.

[3] G Wegmann and E A Vittoz, "Basic principles of accurate dynamic current mirrors," *IEE proceedings*, vol. 137, no. 2, pp. 95–100, April 1990.

[4] J Robert, P Deval, and G Wegmann, "Very accurate current divider," *Electronics Letters*, vol. 25, no. 14, pp. 912–913, 6th July 1989.

[5] Yen S Yee, Lewis M Terman, and Lawrence G Heller, "A 1 mV MOS comparator," *IEEE Journal of Solid-State Circuits*, vol. SC-13, no. 3, pp. 294–297, June 1978.

[6] Wulfram Gerstner, "What's different with spiking neurons?," in *Plausible Neural Networks for Biological Modelling*, pp. 23–48. Kluwer Academic Publishers, 2001.

[7] Philipp Häfliger, Misha Mahowald, and Lloyd Watts, "A spike based learning neuron in analog VLSI," in *Advances in Neural Information Processing Systems*. 1996, vol. 9, MIT Press.

[8] Christal Gordon and Paul Hasler, "Biological learning modeled in an adaptive floating-gate system," in *IEEE International Symposium on Circuits and Systems*, 2002, vol. 5, pp. 609–612.

[9] Adria Bofill, Alan F Murray, and Damon P Thomson, "Circuits for VLSI implementation of temporally-asymmetric Hebbian learning," in *Advances in Neural Information Processing Systems*. 2002, vol. 14, MIT Press.

[10] Adria Bofill-i-Petit and Alan F Murray, "Learning temporal correlations in biologically-inspired aVLSI," in *IEEE International Symposium on Circuits and Systems*, May 25-28 2003, vol. 5, pp. 817–820.

[11] Adria Bofill-i-Petit and Alan F Murray, "Synchrony detection by analogue VLSI neurons with bimodal STDP synapses," in *Advances in Neural Information Processing Systems*, December 2003, Accepted for publication.

[12] F Wörgötter, A Cozzi, and V Gerdes, "A parallel noise-robust algorithm to recover depth information from radial flow fields," *Neural Computation*, vol. 11, pp. 381–416, 1999.

[13] Wulfram Gerstner, Richard Kempter, J Leo van Hemmen, and Hermann Wagner, "Hebbian learning of pulse timing in the barn owl auditory system," in *Pulsed Neural Networks*, pp. 353–377. MIT Press, 1998.

[14] Harald Hüning, Helmut Glünder, and Günther Palm, "Synaptic delay learning in pulse-coupled neurons," *Neural Computation*, vol. 10, pp. 555–565, 1998.

[15] Christian W Eurich, Klaus Pawelzik, Udo Ernst, Jack D Cowan, and John G Milton, "Dynamics of self-organized delay adaptation," *Physical Review Letters*, vol. 82, no. 7, pp. 1594–1597, 15th February 1999.

[16] Sebastian Loeda, *Circuits to Support Antisymmetric Hebbian Training in Artificial Neural Networks*, BEng Hons Thesis, The University of Edinburgh, May 2002.

AN MDAC SYNAPSE FOR ANALOG NEURAL NETWORKS

Ryan J. Kier, Reid R. Harrison*, and Randall D. Beer***

*University of Utah, Department of Electrical and Computer Engineering
Salt Lake City, UT
**Case Western Reserve University, Department of Electrical Engineering and Computer Science
Cleveland, OH

ABSTRACT

Efficient weight storage and multiplication are important design challenges which must be addressed in analog neural network implementations. Many schemes which treat storage and multiplication separately have been previously reported for implementation of synapses. We present a novel synapse circuit that integrates the weight storage and multiplication into a single, compact multiplying digital-to-analog converter (MDAC) circuit. The circuit has a small layout area (5400 μm^2 in a 1.5-μm process) and exhibits good linearity over its entire input range. We have fabricated several synapses and characterized their responses. Average maximum INL and DNL values of 0.2 LSB and 0.4 LSB, respectively, have been measured. We also report on the performance of an analog recurrent neural network which uses these new synapses.

1. INTRODUCTION

Many implementations of analog neural networks have been developed over the past 15 years. Two critical issues in analog neural networks are weight storage and multiplication, often referred to collectively as a synapse. Synapse implementations vary widely among reported analog neural networks. Weight storage can be accomplished by either analog or digital circuit techniques while multiplication is usually performed by a Gilbert multiplier.

Analog weight storage is typically implemented by storing charge on a capacitor. This charge must be refreshed periodically and therefore requires additional programming circuitry that is constantly operating. However, dynamic capacitive memories have often been favored by designers interested incorporating on-line, on-chip learning [1,2]. These analog memories have the advantage of small layout area, but they achieve this layout savings at the cost of increased power dissipation in the required refresh circuitry.

Floating-gate synapses offer an alternative to capacitive analog memory [3,4]. Floating-gate storage also boasts small layout areas, but requires special high-voltage programming circuitry on chip. Furthermore, additional feedback circuitry is usually required for accurate programming.

Digital weight storage has been less popular, but it has the advantage of a simple programming interface. Digital weights can be stored in all of the familiar digital memory structures: DRAM, SRAM, or EEPROM. Since all computation is performed in the analog domain, digital weights must be converted to analog signals through the use of DACs [5].

The Gilbert multiplier circuit is the most popular technique used in analog neural networks. Variations of the Gilbert multiplier have been proposed for various neural networks [5,6,7]. Other transconductance multiplication techniques have also been reported. In [2], an extended-range transconductance amplifier is used to multiply three quantities by each other. All of these multiplication circuits behave as expected only for limited signal swings; for large inputs, all of these circuits exhibit saturating nonlinearity.

We propose a new synapse circuit to be used in analog neural networks. Our synapse employs digital weight storage and gives a response that is linear over all possible input levels using a current-mode multiplying digital-to-analog converter (MDAC). MDAC synapses have the advantage of being able to solve the weight storage and multiplication problems using a single piece of hardware.

In Section 2 we describe the design of this MDAC synapse. In Section 3 we present linearity measurements taken for a large number of our synapses. In Section 4 we demonstrate the use of the synapse in a small analog neural network chip and compare measured to simulated results. We conclude the report in Section 5.

2. MDAC SYNAPSES

2.1. Previously reported MDAC synapses

To our knowledge, MDAC synapses have been reported twice before in the literature. Boser et al. employed a current-mode MDAC which multiplied a digital input by an analog weight [8]. The weight was stored on a

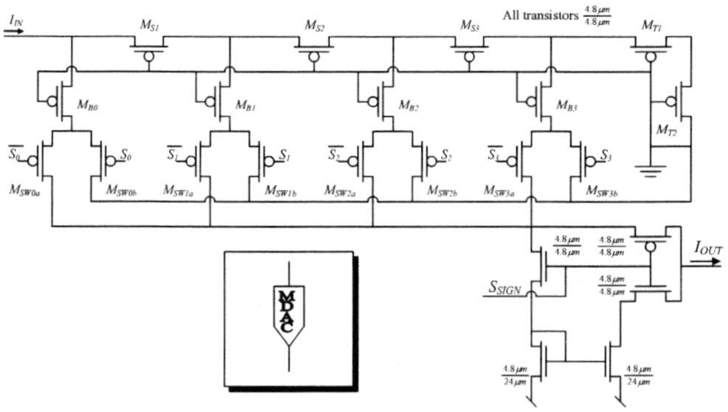

Figure 1 – Circuit diagram for a 5-bit *R*-2*R* pMOS MDAC and its symbol (inset).

capacitor which set the V_{GS} for a set of transistors with binary-weighted width-to-length ratios. The digital input was used to select which currents would sum to produce the output. More recently, a voltage-mode MDAC synapse that multiplies an analog differential voltage by a digital weight has been reported [9]. Unfortunately, this scheme occupies a prohibitively large amount of layout area (1.35-mm^2 in a 1.2-µm process).

2.2. Proposed MDAC synapse

Our proposed MDAC synapse is a compact current-mode circuit which multiplies an input current by a digital weight. The circuit diagram for our proposed synapse is shown in Fig. 1. The operation of the circuit is based on the familiar *R*-2*R* resistive current divider. Here, however, pMOS transistors are used in place of polysilicon resistors to save chip area.

In [10], Vittoz and Arreguit introduced the concept of a pseudo Ohm's law for MOSFETs. Simply stated, a network of MOSFETs sharing the same gate voltage is linear with respect to currents but not voltages. Further, the current through each transistor is determined only by its geometry. This allows one to borrow resistive current division networks and incorporate them directly into VLSI circuits without using large resistors.

Each transistor in Fig. 1 is drawn with identical width and length. The pseudo-resistance of a MOS transistor is determined only by its width-to-length ratio. We can denote the pseudo-resistance of each transistor as *R*. The specific value of *R* is not important. The 2*R* 'resistance' is provided by series combination of the switching transistor (M_{SWxa} *or* M_{SWxb}) and the branch transistor (M_{Bx}). Note that in each branching section, only one switching transistor is on at a time because the pair is driven with complementary signals. Therefore, each downward branch of the ladder provides a 'resistance' of 2*R* to ground. The current in each branch is switched to ground or onto the output node, which may not be at the same potential as the circuit ground. However, all that is required for MOS pseudo-resistor circuits is a large network voltage drop to keep the switch transistors M_{SWxy} in saturation. This concept is referred to as a 'pseudo-ground' [10]. This feature of MOS pseudo-resistive networks removes the need for an op-amp providing a virtual ground (which is required in resistive implementations).

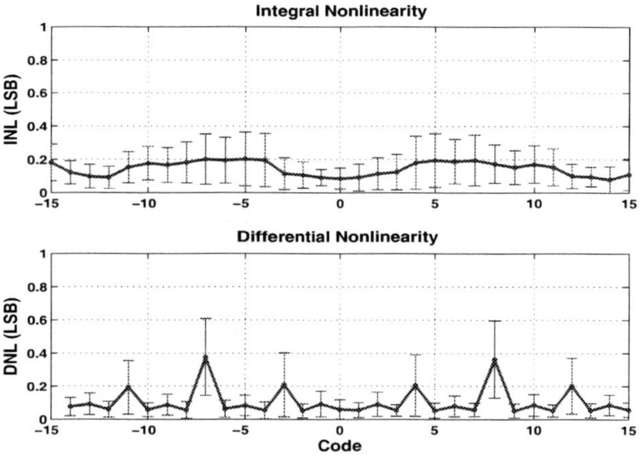

Figure 2 – Mean nonlinearity measures for 32 MDACs. Differential nonlinearity peaks at code changes from 7 to 8. This is caused by uneven current splitting at each stage of the *R*-2*R* structure.

Negative weights are realized by directing the output current of the MDAC into an nMOS current mirror, reversing the current flow. An additional set of switching transistors controlled by a digital input, S_{SIGN}, is used to direct the output of the MDAC either through an nMOS mirror or directly to the output node. The transfer characteristic for the 5-bit MDAC in Fig. 1 is given by

$$I_{OUT} = D \cdot I_{IN} \quad (1)$$

where

$$D = (-1)^{S_{SIGN}} \sum_{i=0}^{3} \frac{S_i}{2^i} \quad (2)$$

is the stored weight. It can be seen from (2) that weight magnitudes are always less than one. If larger weights are desired, additional current gain may be added before or after the MDAC circuit. This is most easily accomplished by increasing the size of the current mirror supplying I_{IN}.

The 5-bit MDAC described above occupies an area 90 µm by 60 µm (5400 µm^2 in a 1.5-µm process). The total synapse area depends on the choice of SRAM cells used to store the weight. Our initial implementation uses a SRAM cell based on a resettable D-latch. Consequently, the layout area required for weight storage is roughly five times the layout area for the MDAC. The area impact of weight storage circuitry will be reduced in future implementations by using a simpler SRAM cell.

This MDAC synapse has been incorporated into a four-neuron analog recurrent neural network chip. Each chip contains 20 MDAC synapses implemented in AMI's 1.5-µm two metal, two poly CMOS process through MOSIS. The design and operation of the neural network chip are described in Section 4.

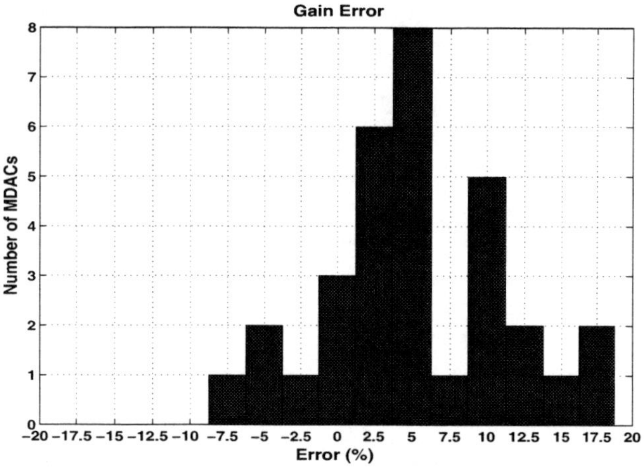

Figure 3 – Histogram of gain errors for 32 MDACs. The mean error is 5% and the maximum is 18.2%.

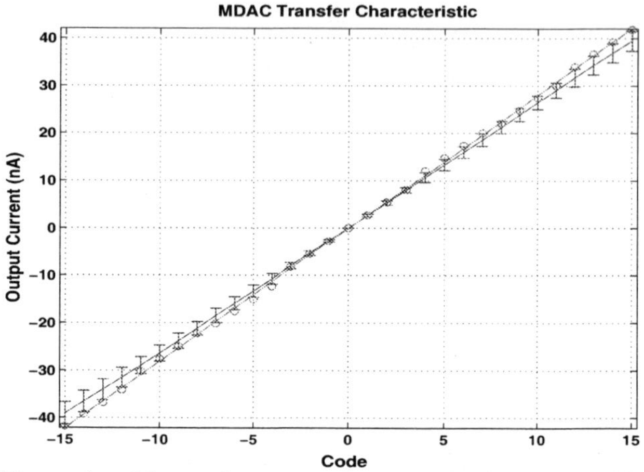

Figure 4 – Measured response for a randomly-chosen 5-bit MDAC. Also shown is the mean response of 32 MDACs.

3. CIRCUIT LINEARTIY

The response characteristics for 32 MDAC synapses were measured on two chips (16 synapses per chip). Standard linearity measures—integral nonlinearity (INL) and differential nonlinearity (DNL)—were computed for each MDAC. Fig. 2 shows plots of the average of the absolute value of the INL and DNL at each weight value. The error bars in Fig. 2 denote one standard deviation above and below the mean.

INL was computed as the difference between the least-squares linear fit and the measured data. Data was not normalized before computing INL; instead, a histogram of the gain errors is shown in Fig. 3.

The gain errors shown in Fig. 3 are not the result of the *R*-2*R* structure. Rather, the gain errors reflect device mismatch in three cascaded current mirrors through which input current must pass. Only the last current mirror, where gain is added, should be considered part of the MDAC circuit, but it was not possible to isolate every MDAC input during testing. Instead, a single master bias supplied the input current to all MDACs. As a result, large gain errors, which are not typical of the MDAC circuit, were measured.

The maximum INL was measured to be 0.7 LSB and the maximum DNL was 1.1 LSB. Only two MDACs had a maximum DNL value greater than 0.7 LSB. These linearity figures are not impressive when compared with those measured from general purpose DAC circuits. However, the layout area occupied by our DAC is only a small fraction of what more general purpose DACs require.

The mean response (with error bars denoting one standard deviation) of all 32 MDACs tested is shown in Fig. 4. Also shown in Fig. 4 is the response of a randomly-chosen MDAC along with its least-squares fit. The effects of the gain errors are clearly evident in this plot. The standard deviation grows large at the extremes of the input.

4. RECURRENT NEURAL NETWORK APPLICATION

We have fabricated a four-neuron chip which implements a continuous-time recurrent neural network (CTRNN). Each CTRNN neuron's behavior is described by the nonlinear differential equation

$$\tau_i \cdot \frac{dy_i}{dt} = -y_i + \sum_{j=1}^{N} w_{ij} \cdot \sigma\left(y_j + \theta_j\right) \qquad (3)$$

where

$$\sigma(x) = \frac{1}{1+e^{-x}} \qquad (4)$$

is the logistic sigmoid function [11].

The circuit diagram for a single CTRNN neuron is shown in Fig. 5. In this implementation, a subthreshold differential pair is used to realize the logistic sigmoid function. Its transfer function is described by

$$I_{out} = \frac{I_B}{1+e^{-\kappa V_{in}/U_T}} \qquad (5)$$

where $U_T = kT/q \approx 26$ mV is the thermal voltage, $\kappa \approx 0.7$ is the gate coupling coefficient, and I_B is the bias current [12]. Our 5-bit MDAC synapses provide programmable connections between neurons. External *RC* networks give the neurons the dynamic behavior described by (3). The time constant, τ_i, in (3) is set by the product of *R* and *C*. The value of the resistance *R*, is fixed and given by

$$R = 4 \cdot \frac{U_T}{\kappa I_B} \qquad (6)$$

This value corrects for the scaling introduced by (5) and provides an additional gain. The value of *C* is selected to give each neuron the desired time constant.

In this CTRNN implementation, we wish to use the network as a pattern *generator* as opposed to a pattern

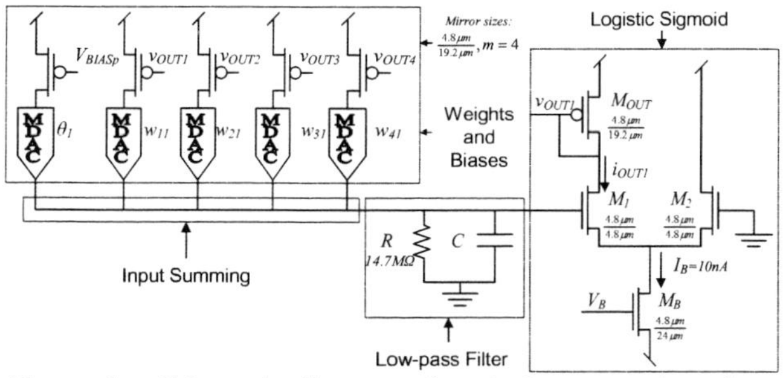

Figure 5 – Schematic diagram of a single neuron in a four-neuron CTRNN.

recognizer. Therefore, no external input connections to the neurons are available. The outputs of the neurons model the mean firing rate of motor neurons found most animal spinal cords. In animals, such temporal patterns are produced by central pattern generators [13], interconnected networks of nerve cells that autonomously generate rhythmic activity for behaviors such as walking and breathing.

4.1. CTRNN results

Fig. 6 shows a single cycle of the CTRNN chip output plotted on the same set of axes as the output of the simulated CTRNN. The network weights and biases were rounded to the nearest integer in the simulation to facilitate comparison. Time constants were selected using the nearest standard capacitance values.

The network shows good matching on neurons 1, 2, and 4. The output of neuron 3 does not match simulation as well, but the shape of the waveform is comparable to simulation. All of the on-chip neurons have amplitudes that differ from the simulation values, with neuron 3 showing the largest difference. This difference in amplitude is due to transistor mismatch on the output current mirrors.

5. CONCLUSIONS

We have presented a new synapse circuit design for use in analog neural networks. The synapse utilizes a current-mode multiplying DAC based on familiar *R-2R* circuits. The synapse is compact, requiring only 5400-μm^2 in a standard 1.5-µm CMOS process, and is linear over its entire input range. We have also shown that the synapse circuit is suitable for implementing continuous time recurrent neural networks and the behavior of such networks matches simulation well.

ACKNOWLEDGEMENTS

This work was supported by an NSF BITS award (EIA-0130773).

REFERENCES

[1] A.J. Montalvo; R.S. Gyurcsik; J.J. Paulos, "Toward a general-purpose analog VLSI neural network with on-chip learning," *IEEE Trans. Neural Networks*, **8:**413-423, Mar. 1997.

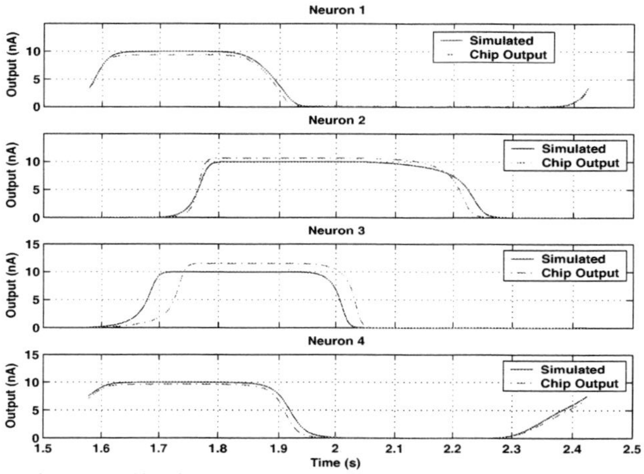

Figure 6 – Simulated vs. measured CTRNN patterns.

[2] G. Cauwenburghs, "An analog VLSI recurrent neural network learning a continuous-time trajectory," *IEEE Trans. Neural Networks,* **7:**346:361, Mar. 1997.

[3] B.W. Lee, B.J. Sheu, and H. Yang, "Analog floating-gate synapses for general-purpose VLSI neural computation," *IEEE Trans. on Circuits and Systems* **38:**654 -658, 1991.

[4] D. A. Durfee and F. S. Shoucair, "Comparison of floating-gate neural network memory cells in standard VLSI CMOS technology," *IEEE Trans. Neural Networks*, **3:**347-353, May. 1992.

[5] S. M. Gowda; B. J. Sheu; J. Choi; C. G. Hwang; and J. S. Cable, "Design and characterization analog VLSI neural network modules," *IEEE J. Solid-state Circuits*, **28:**301-313, Mar. 1993

[6] B. Linares-Barranco; E. Sanchez-Sinencio; A. Rodriguez-Vazquez; and J. L. Huertas, "A modular T-mode design approach for analog neural network hardware implementations," *IEEE J. Solid-state Circuits*, **27:**701-713, May 1992.

[7] D. Coue and G. Wilson, "A four-quadrant subthreshold mode multiplier for analog neural-network applications," *IEEE Trans. on Neural Networks*, **7:**1212-1217, Sept. 1996.

[8] B. E. Boser; E. Sackinger; J. Bromley; Y. L. Cun; and L. D. Jackel, "An analog neural network processor with programmable topology," *IEEE J. Solid-state Circuits*, **26:**2017-2025, Dec. 1991.

[9] M. Al-Nsour and H. Abdel-Aty-Zohdy, "ANN digitally programmable analog synapse," *42nd Midwest Symp. On Circuits and Systems*, **1:**489-492, Aug. 1999.

[10] E. A. Vittoz and X. Arreguit, "Linear networks based on transistors,"*Electron. Lett.*, **20:** 297–299, 1993.

[11] R. D. Beer, "On the dynamics of small continuous-time recurrent neural networks," *Adaptive Behavior*, **3:**471-511, 1995

[12] C. Mead, *Analog VLSI and Neural Systems*. Reading, MA: Addison-Wesley, 1989.

[13] F. Delcomyn, "Neural basis of rhythmic behavior in animals," *Science* **210:**492-498, 1980.

Supervised Learning in a Two-Input Analog Floating-Gate Node

Jeff Dugger and Paul Hasler
School of Electrical and Computer Engineering
Georgia Institute of Technology
Atlanta, GA 30332-0250

Abstract— **We pursue the realization of on-chip supervised learning networks for large-scale, real-time signal processing applications using arrays of analog floating-gate synapses. We present experimental data characterizing the performance of a two-input analog floating-gate pFET synapse network that implements a supervised learning algorithm similar to the Least-Mean-Square (LMS) learning rule; most other supervised learning algorithms possess a straight-forward relation to the LMS algorithm. Analog floating-gate synapses will enable larger-scale, on-chip learning networks than previously possible.**

Neurally-inspired analog VLSI learning systems promise tremendous computing capabilities while consuming four orders of magnitude less power than comparable digital approaches [1]. Floating-gate pFETs provide compact device level implementations of an adaptive analog gain, creating the possibility for large-scale, low-power analog VLSI adaptive signal processing systems, including neural networks with on-chip learning. Floating-gate synapses constructed from these devices provide simultaneous long-term storage of a weight value, multiplication of an input signal by the weight value, and adaptation of the weight value based on gate and drain signal voltages [2], [3]. These floating-gate synapses exhibit weight adaptation based upon correlations between input and error-voltage signals applied to these terminals [4].

Unfortunately, early versions of our synapse exhibited sizable weight decay that would cause large steady-state errors if used in an LMS learning system. Here we present a two-input LMS node based upon an improved continuously-adapting, analog floating-gate synapse circuit that minimizes weight decay, the source-follower floating-gate (SFFG) synapse, which addresses the five desirable properties of silicon synapses [2], [4]. Using this technology, we are building arrays on the order of 128×128 synapses in a 2mm×2mm area (0.35 μm process) that operate at under 1 mW of power; a custom digital processor or bank of DSP processors performing similar computations would consume 3-10 W. We present experimental data measured from ICs fabricated in a 0.5μm process available through MOSIS and compare these results with theoretical calculations.

I. Adaptive Computation

An adaptive node, such as that appearing in Figure 1, computes the inner product of an input signal vector and an adaptive weight as demonstrated by $y(t) = \mathbf{w}^T\mathbf{x}(t)$, where $\mathbf{x}$

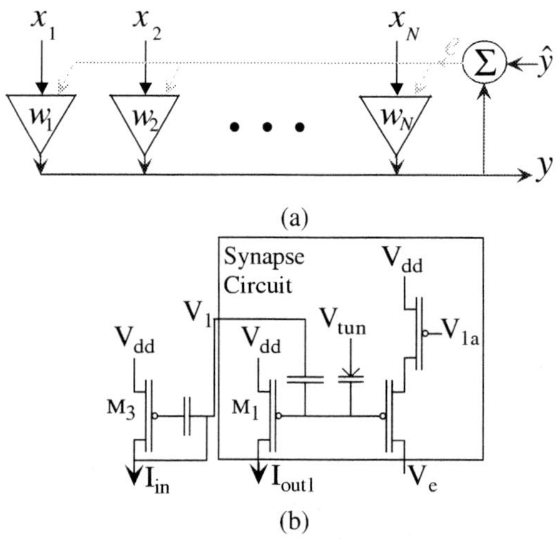

Fig. 1. (a) Block diagram of a single adaptive floating-gate node. Because the synapse elements far outnumber all of the other components, each synapse element must be as small as possible. Additional circuitry on the periphery allows us to appropriately pre- and post-process inputs and outputs and set the initial conditions and / or restart the weights in a known state when experimenting with various algorithms. (b) Our Floating-Gate LMS synapse. The source-follower circuit provides the learning rule; the extra transistor at the floating-gate computes the weight-signal multiplication and provides a current for summation via KCL. The pre-distort circuit increases the linear signal range to obtain a more ideal correlation learning rule. A similar post-distort circuit removes distortion terms in the weight update equation due to hot-electron injection nonlinearities.

denotes the input vector and $\mathbf{w}$ the weight vector. Because the weight, $\mathbf{w}$, is adaptive it is also a function of time, but the rate of adaptation is such that we can consider it a constant in computation.

Least-mean-square (LMS) learning rules result from minimization of a least-square-error objective function. Some LMS algorithms intentionally incorporate weight decay — a form of *forgetfulness* exhibited by the learning system — for better learning generalization and tracking of non-stationary signals [5], [6], [7]. Analytical modeling of learning rules yields the following weight dynamics for an LMS algorithm with weight decay

$$\tau\frac{d\mathbf{w}}{dt} = \mathbf{x}e - \epsilon\mathbf{w} \quad (1)$$

and steady-state solution [7]

$$\mathbf{w}_{ss} = (\mathbf{Q} - \mathbf{I}\epsilon)^{-1}\mathbf{r} \quad (2)$$

0-7803-8251-X/04/$17.00 ©2004 IEEE

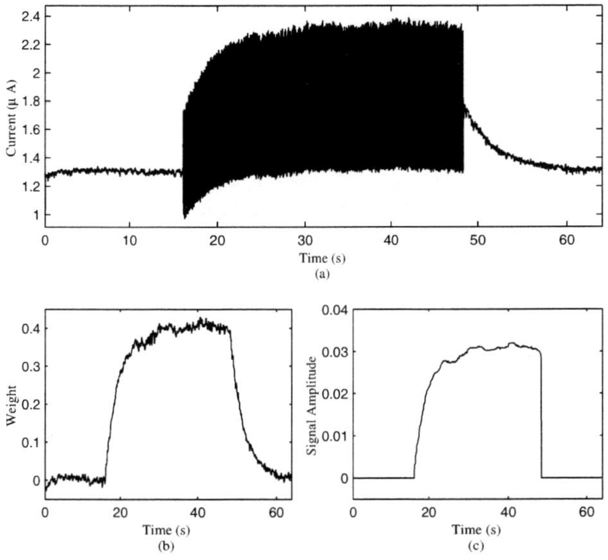

Fig. 2. Amplitude correlation results for sinusoids of same frequency and phase with input and error signal amplitudes of 0.3. (a) Synapse output current vs. time. Sinusoidal signals are turned on at 16 seconds and turned off at 48 seconds. (b) Extracted weight value vs. time, showing convergence to steady-state due to correlations affected by weight decay (c) Extracted signal amplitude vs. time, which follows a path similar to the weight, increasing from zero to a steady-state value. Without an input signal, the output is zero due to multiplicative effects.

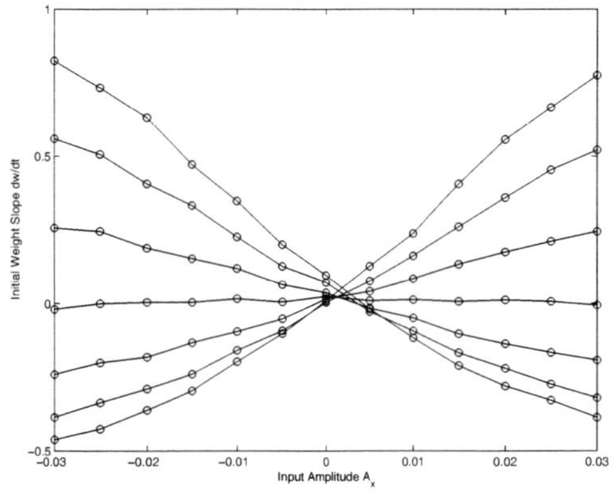

Fig. 3. Basic correlating floating-gate synapse measurements for an individual isolated synapse element. We measured the initial slope of the weight convergence (illustrated in Fig. 2)from this synapse for multiple input (sinewave) amplitudes near the zero-input steady-state condition. This initial slope shows the correlation function used by this LMS synapse; steady-state solution has less meaning for an LMS synapse with small weight decay.

The error signal, $e = \hat{y} - y$, results from the difference between the output and a target signal, $\hat{y}$. The steady-state solution, $\mathbf{w}_{ss}$, depends on the input auto-correlation matrix, $\mathbf{Q} = E\left[\mathbf{x}\mathbf{x}^T\right]$ and the input-output cross-correlation vector $\mathbf{r} = E\left[\mathbf{x}y\right]$. The expected value operation, $E[\cdot]$, represents ensemble averaging, however we will assume our signals are ergodic and use time-averaging for this operation. The strength of the weight decay, ϵ, should be small ($\epsilon << 1$) relative to the input and learning-signal amplitudes to minimize deviation from the ideal LMS solution.

II. The Source-Follower Floating-Gate Synapse

Figure 1 shows our improved LMS floating-gate synapse circuit. This core synapse circuit consists of three transistors: a floating-gate pFET in the middle computes the weight-signal product and the floating-gate and regular pFET pair on the right compose a source-follower circuit, enabling fairly ideal correlation behavior with minimal weight decay. The circuit is similar to our previous synapse element [4], except here we *linearize* and stabilize the synapse with a regular pFET current source rather than a DIBL pFET. Unlike the DIBL pFET, the regular pFET effectively increases the output resistance seen at the floating-node by the tunneling and injection mechanisms because this circuit behaves like a cascode configuration [8]. While the synapse presented here superficially resembles a previously proposed synapse [9], it operates in a fundamentally different way (continuous-time correlation) using analog instead of digital signals and is designed to yield minimal weight decay.

To realize continuous-time on-chip learning we take advantage of device physics, specifically, we take advantage of Fowler-Nordheim electron tunneling and hot-electron injection, which allow us to add charge to and remove charge from the floating gate, thereby providing us with a weight update mechanism. The currents resulting from each of these processes are approximately six-orders of magnitude smaller than signal currents, yielding a slowly adapting weight parameter which can be considered approximately constant during weight-signal computation. We assume the effect on tunneling current due to changes in the *weight* and *signal* voltages on the floating-gate is small in comparison to that due to hot-electron injection, and therefore treat it as a nearly-constant bias current for the adaptation operation. Nonlinearity inherent in the injection process provides a compact realization of the correlation learning rule at the heart of LMS, which is the key to implementing practical, large-scale on-chip learning systems. We extensively discuss the physics of floating-gate circuits elsewhere [2], [10]. Previous work illustrated how the weight update mechanism leads to a learning rule based on correlations between the gate and drain voltages of these devices [2], [4], [11]. We derived the correpsonding learning rule as

$$\tau \frac{dw}{dt} \approx -(\epsilon - \alpha E[xe])\, w + E[xe] \qquad (3)$$

where $\epsilon = \frac{U_T}{V_{inj} \| \kappa V_x}$, and $\alpha = 1 - \frac{U_T}{V_{inj}}$. Depending on the technology, we obtain ϵ in the range from 0.03 to 0.3.

We verify the correlation learning rule approximation, (3), through amplitude-correlation experiments. Experimental data illustrate this learning rule; to simplify our analysis, we approximate the weight decay solely as ϵ and point out deviations where necessary. Each experiment involves sinusoidal signals since they provide easily generated, commonly used test-signals that allow us to readily observe system nonlinearities and take advantage of Fourier theory to facilitate analysis of experimental results.

Figure 2 illustrates a typical synapse response for a simple amplitude correlation experiment, where we apply $x = A_x \sin(\omega t)$ and $e = A_e \sin(\omega t)$ as the input and error signals, respectively. The measured output current of the synapse

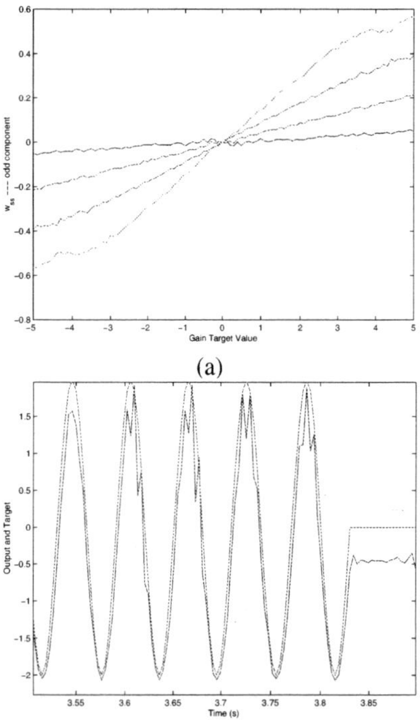

(a)

(b)

Fig. 4. Results for a one weight node used for LMS. (a) Steady-state results of a one weight node used for LMS versus target value; the target value is a linear function of the input signal, where we have plotted the results for several gain levels. (b) Transient solution for the one weight node used for LMS for a sine wave input.

appearing in Fig. 2a shows the fast-timescale output signal riding on a slow-timescale deviation in the current bias due to weight adaptation. Normalizing the current to its equilibrium value and low-pass filtering yields the extracted weight value shown in Fig. 2b; the time-constant, τ, was fit to an RC-model and is approximately 3 seconds. Removing all correlating signals, leaves only weight decay, which returns the synapse to its initial equalibrium. Finally, Fig. 2c exhibits an increase in output signal amplitude from zero to a steady-state value of $w_{ss}A_x$.

We observe that for $w \approx 0$, the initial slope of the weight dynamics, $\frac{dw}{dt}$, should yield correlations between the applied input and error signals scaled by the weight decay as $w_{ss} = E[xe]/\epsilon$. Fig. 3 demonstrates $\frac{dw}{dt}$ for $w \approx 0$ obtained from linearly fitting the initial slopes of each weight convergence. We see a compressive nonlinearity at high amplitudes due to higher-order terms in the Taylor series expansion of the learning rule which are not included in (3). Fig. 3 also shows an offset at $A_x = 0$ arising from variance of the error signal due to slight mismatch between the post-distort circuit and V_{inj} The size of this term tends to be similar to that of the weight decay for correlating synapses [11]. These results will be similar to the steady-state solutions when small amplitude signals are applied; for larger signals we will see nonlinearities due to the small amount of weight decay.

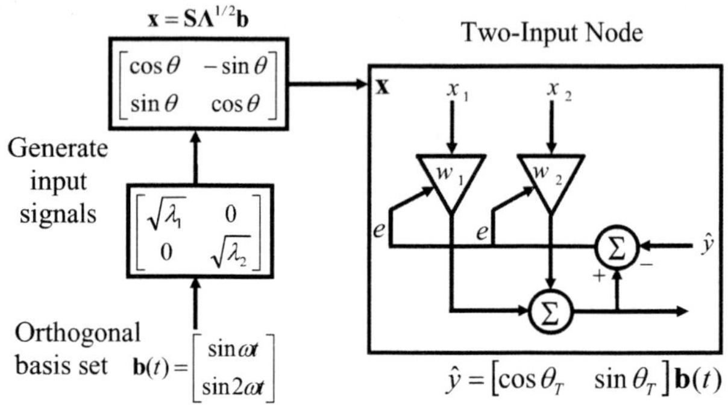

Fig. 5. Experimental setup for examining Least-Mean-Squares behavior in a two-input node. A scaling operation followed by application of a rotation matrix to an orthogonal signal-space basis of harmonically related sinusoids yields the system input signals; the target results from a linear combination of the signal-space basis controlled by the parameter θ_T. We performed two sets of experiments with this setup: in the first set, we chose $\lambda_1 = \lambda_2$, $\theta_T = 0$, and uniformly sampled θ from the unit circle; for the second set of data, we chose $\lambda_1 \neq \lambda_2$, $\theta = \pi/3$, and we uniformly chose samples of θ_T from a 0 to 2π.

III. Supervised Learning in SFFG Nodes

Our new synapse enables the first continuously-adapting floating-gate LMS algorithm. We previously demonstrated LMS learning for uncorrelated inputs [12]; we now demonstrate the effects of error-signal feedback results in a multiple-synapse circuit.

Unlike the pure correlation case, supervised learning requires feedback of an error signal defined as the difference between the computed output and a given target. However, before deriving the error from the single-node output current, we need to separate the output signal, $y = \mathbf{w}^T\mathbf{x}$, from the bias and slow-timescale weight terms $(2N + \sum w_i)$ by low-pass filtering, as illustrated in Fig. 2.

A. Single-Synapse Supervised Learning

The first LMS case concerns supervised adaptation of a single weight, which illustrates circuit dynamics for a manageable number of parameters. The output and target signals in this experiment are given by $y = wx$ and $\hat{y} = Ax$, which lead to the error signal $e = (A - w)x$. An ideal LMS response would force w to equal A. Instead, we find that with weight decay, as exhibited in (1) and (3), the LMS rule yields $w_{ss} = A/(1 - \epsilon/E[x^2])$. Including the correlation term of the weight decay in (3) adds terms on the order of $O(\epsilon^2)$, justifying our first-order approximation. Figure 4a shows the weight value tracking various target gains, and Fig. 4b shows the output sinusoid tracking the target signal for a single trial.

B. Two-Input Supervised Learning

Two-dimensional input data allow visualization of synapse weights as a set of points in the plane, making it easier to observe the learning behavior of the analog floating-gate node. Figure 5 illustrates the setup for this simplest possible multiple-input LMS experiment. First, we choose harmonically related sinusoids to obtain an orthonormal basis for the signal space. We then construct the input signals, $\mathbf{x}$, and target signal,

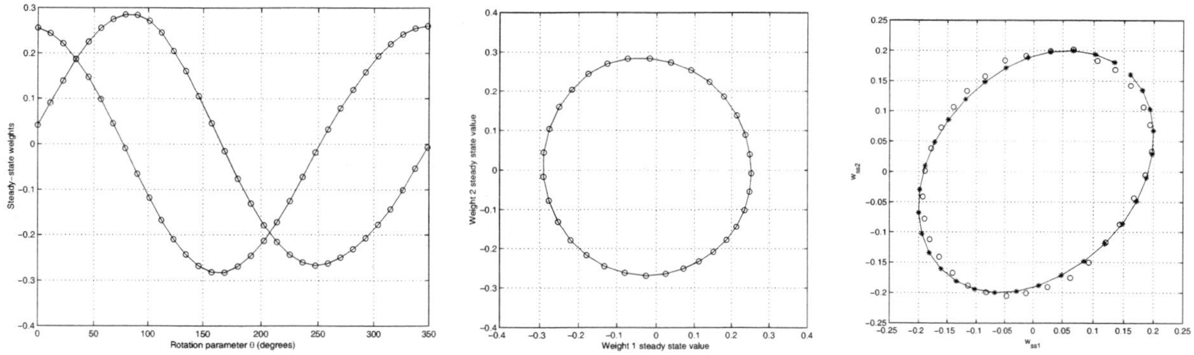

(a) (b) (c)

Fig. 6. Two-input Least-Mean-Squares experimental results for the source-follower floating-gate synapse circuit. (a) Measured data for the case $\theta_T = 0$ and $\lambda_1 = \lambda_2$ show steady-state weight dependence on the parameter, θ, of the two-dimensional input mixing-matrix. As expected, we get a cosine curve for the first weight, and a sine curve for the second weight. (b) Measured data for the same case is plotted as one steady-state weight versus the other. An ideal LMS rule would produce a circle similar to our circuit results. (c) Illustration of measured data compared with model results computed from equation (4) assuming non-zero constant weight decay for the case where $\lambda_1 = 1$, $\lambda_2 = 2$, $\theta = \pi/3$, and 32 values of θ_T uniformly sampled from 0 to 2π. The gain of the filter and the amplitude of the input signals determine the scale of the ellipse. The open circles show steady-state values measured from the actual circuit. The asterisks and the corresponding fit-line illustrate computed model results. We observe good agreement between the measured data and the computed values. As predicted in the text, the results form an ellipse.

$\hat{y}$, from a linear combination of the basis signals as illustrated in Fig. 5. Mathematical analysis of this experiment assuming constant weight-decay dynamics (the simplest case of (3) when the correlation term of the weight decay is ignored) and based on the results given in (1) yields the steady-state weight solution,

$$\mathbf{w_{ss}} = \mathbf{S}(\theta) \begin{bmatrix} \frac{\sqrt{\lambda_1}}{(\lambda_1+\epsilon)} \cos\theta_T \\ \\ \frac{\sqrt{\lambda_2}}{(\lambda_2+\epsilon)} \sin\theta_T \end{bmatrix}. \quad (4)$$

Measured results from the synapse circuit appear in Figure 6. We performed the first experiment with $\lambda_1 = \lambda_2$ and $\theta_T = 0$, uniformly sampling 32 trials of the rotation parameter, θ, from 0 to 2π. Figure 6a, shows the steady-state weight values for each of the two weights as a function of the rotation angle, θ. Steady-state values for the first and second weights approximate $\cos\theta$ and $\sin\theta$ respectively, as expected; a plot of the second vs. the first component of $\mathbf{w_{ss}}$ results in a circle as shown in Fig. 6b. In the second case where $\lambda_1 \neq \lambda_2$, and θ_T is swept from 0 to 2π with a fixed value of $\theta = \pi/3$, the result in (4) leads us to expect a weight-space plot of the first vs. the second node weight to yield an ellipse. Figure 6c exhibits the measured data compared with numerically computed results from (4) in a weight-space plot, which shows that the weights approximate an ellipse quite well. From the fit to the measured data, we estimate the weight decay, $\epsilon \approx 0.1$. The measured data exhibit a slight rotational difference from the computed results which is probably due to the higher-order terms in the weight decay which have been ignored in this analysis.

IV. Conclusion

We introduced a source-follower floating-gate synapse as an improvement over our earlier source-degenerated floating-gate synapse in terms of relative weight decay. It is impossible to build a physical system that does not have some form of decay inherent in the system. Weight decay adds some algorithmic benefits to the learning system, such as the ability to track non-stationary processes. We presented data from a two-input circuit and compared the results with a mathematical model of the system. We have developed a multiple-input, multi-node chip through MOSIS, in conjunction with an FPGA-based, real-time test set-up which can operate at audio frequencies. With this more advanced IC and test set-up, we are studying LMS for the case of more than two inputs (for example, Fourier approximation and adaptive equalization), as well as exploring multiple-node networks and unsupervised learning algorithms.

References

[1] C. Mead, "Neuromorphic electronic systems," in *Proceedings of the IEEE*, vol. 78, pp. 1629–1636, Oct 1990.
[2] P. Hasler, C. Diorio, B. A. Minch, and C. A. Mead, "Single transistor learning synapses," in *Advances in Neural Information Processing Systems 7* (G. Tesauro, D. S. Touretzky, and T. K. Leen, eds.), pp. 817–824, Cambridge, MA: MIT Press, 1995.
[3] P. Hasler, "Continuous-time feedback in floating-gate MOS circuits," *IEEE Transactions on Circuits and Systems II*, vol. 48, pp. 56–64, January 2001.
[4] P. Hasler and J. Dugger, "Correlation learning rule in floating-gate pFET synapses," *IEEE Transactions on Circuits and Systems II*, vol. 48, pp. 65–73, January 2001.
[5] P. M. Clarkson, *Optimal and Adaptive Signal Processing*. Boca Raton: CRC Press, 1993.
[6] J. Hertz, A. Krogh, and R. G. Palmer, *Introduction to the Theory of Neural Computation*. Addison-Wesley, 1991.
[7] S. Haykin, *Neural Networks: A Comprehensive Foundation*. Prentice-Hall, 1999.
[8] P. E. Allen and D. R. Holberg, *CMOS Analog Circuit Design*. Oxford University Press, 2 ed., 2002.
[9] A. P. Shon, D. Hsu, and C. Diorio, "Learning spike-based correlations and conditional probabilities in silicon," in *Advances in Neural Information Processing Systems 14* (T. G. Dietterich, S. Becker, and Z. Ghahramani, eds.), (Cambridge, MA), MIT Press, 2002.
[10] P. Hasler, B. Minch, J. Dugger, and C. Diorio, "Adaptive circuits and synapses using pFET floating-gate devices," in *Learning on Silicon* (G. Cauwenberghs and M. Bayoumi, eds.), pp. 33–65, Norwell MA: Kluwer Academic Press, 1999.
[11] J. Dugger and P. Hasler, "Improved correlation learning rule in continuously adapting floating-gate arrays using logarithmic pre-distortion of input and learning signals," in *2002 IEEE International Symposium on Circuits and Systems*, vol. 2, (Phoenix, AZ.), pp. II–536–II–539, 2002.
[12] J. Dugger and P. Hasler, "A floating-gate analog adaptive node," in *2000 IEEE Midwest Symposium on Circuits and Systems*, vol. 3, (East Lansing, MI.), pp. 1058–1061, August 8-11 2000.

MIXED-SIGNAL REAL-TIME ADAPTIVE BLIND SOURCE SEPARATION

Abdullah Celik, Milutin Stanacevic and Gert Cauwenberghs

Department of Electrical and Computer Engineering
Johns Hopkins University, Baltimore, MD 21218, USA
{*acelik,miki,gert*}*@jhu.edu*

ABSTRACT

A mixed-signal adaptive VLSI architecture for real-time blind separation of linear source mixtures is presented. The architecture is digitally reconfigurable and implements a general class of Independent Component Analysis (ICA) update rules in common outer-product form. In conjuction with *gradient flow*, a technique for converting time-delayed mixtures of traveling wave sources into equivalent linear instantaneous mixtures by observing spatial and temporal derivatives of the field over a miniature array, the ICA architecture allows to separate and localize multiple acoustic sources in the acoustic scene. Experimental results from VLSI implementation of the ICA architecture demonstrate 30 dB separation of two mixtures of two speech signals.

1. INTRODUCTION

The human auditory system performs remarkably well in segregating multiple streams of acoustic sources, even in significant presence of noise in the acoustic scene. To do so, it requires to resolve time delays and amplitude differences between sound waves entering both ears, and correlating and grouping these differences across the various source components [1]. Modern hearing aids utilize directional or multiple microphones to emulate some of the functionality of binaural sensing. However, they remain far from adequate in compensating the loss of functionality in impaired ears and their performance degrades significantly in the presence of multiple signal and noise sources in the acoustic scene. To be effective in resolving the signal of interest, both localization and separation of multiple acoustic sources are required.

In previous work we showed that the direction of wave propagation can be estimated obtained by differential spatial sensing of the field on sub-wavelength scale, using *gradient flow* [2, 3]. Mixed-signal VLSI implementation of the method [4] has demonstrated improved performance in terms of power dissipation and bearing resolution over conventional bearing estimation localizers.

This work was supported by ONR N00014-99-1-0612 and ONR/DARPA N00014-00-C-0315.

Besides its use in bearing estimation, gradient flow provides an efficient signal representation as a front-end for blind source separation. In the presence of multiple signal sources, gradient flow converts the problem of separating unknown delayed mixtures of independent signal sources, into a simpler problem of separating corresponding instantaneous mixtures of the time-differentiated signals [2]. This formulation is equivalent to the standard problem statement in Independent Component Analysis (ICA) [7]. Gradient flow and ICA combine to yield both separation and localization of multiple independent signal sources.

Various analog VLSI implementations of ICA exist in the literature, *e.g.,* [5, 6], and digital implementations using DSPs are common practice in the field. A general mixed-signal parallel architecture, that can be configured for implementation of various ICA update rules in conjunction with gradient flow, is presented. Experimental results from micropower VLSI implementation are included to demonstrate the approach.

2. GRADIENT FLOW

Gradient flow [2] is a signal conditioning technique for separating and localizing traveling wave sources by relating spatial and temporal derivatives of the field observed over a miniature sensor array. Observation of first order spatial gradients of the field $\xi_{10}(t)$ and $\xi_{01}(t)$ in perpendicular directions in the plane yield linearly mixed observations of the time-differentiated source signals $\dot{s}_\ell$ each scaled by propagation delays ${\tau_1}^\ell$ and ${\tau_2}^\ell$ along the gradient directions:

$$\begin{aligned} \xi_{10}(t) &\approx \sum_l \tau_1^l \dot{\xi}_{00}^l(t) \\ \xi_{01}(t) &\approx \sum_l \tau_2^l \dot{\xi}_{00}^l(t). \end{aligned} \quad (1)$$

In practice, the gradients $\xi_{10}(t)$ and $\xi_{01}(t)$ are estimated by finite differences of the field on the sensor grid [2]. Likewise, time differentiation of the observed spatial common mode of the field $\xi_{00}(t)$ yields further linearly mixed obser-

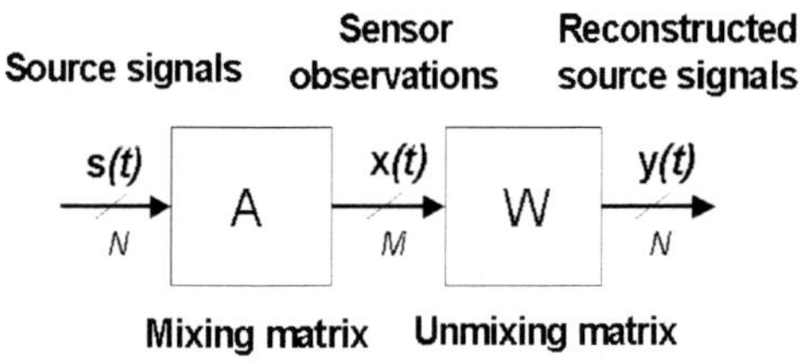

Fig. 1. Standard ICA problem definition. Linear instantaneous mixing, and unmixing, of independent sources.

vation of the time-differentiated source signals $\dot{s}_\ell$:

$$\dot{\xi}_{00}(t) \approx \sum_l \dot{s}^l(t). \tag{2}$$

For a single source $s(t)$, estimates of 3-D bearing angles are directly obtained by least squares adaptation of the coefficients τ_1 and τ_2 from (1) and (2) [3], efficiently implemented in a micropower mixed-signal VLSI architecture [3]. Multiple concurrent sources $s^\ell(t)$ ($\ell = 1, \ldots \mathcal{L}$) give rise to linear mixtures in the gradient observations ξ_{10}, ξ_{01} and $\dot{\xi}_{00}$ that can be blindly separated using independent component analysis (ICA) [7]. The identified source components from the gradient signals then yield the time-differentiated source signals $\dot{s}^\ell(t)$, and the unmixing coefficients yield the corresponding 3-D direction cosines in terms of inter-temporal differences ${\tau_1}^\ell$ and ${\tau_2}^\ell$ [2]. Therefore, a compact system composed by integration of gradient flow and ICA, can be used to achieve multiple source tracking.

3. INDEPENDENT COMPONENT ANALYSIS

The task of blind source separation (BSS) is to separate and recover independent sources from mixed sensor observations, where both sources and mixing matrix are unknown. Independent component analysis (ICA) offers a principled mathematical approach to solving the BSS problem under the assumption of signal independence across sources [7]. Figure 1 shows the standard problem setting of ICA, in which observations $\mathbf{x}(t)$ are unknown linear mixtures of the unknown independent sources $\mathbf{s}(t)$ and a linear transformation of them yields to a estimation of the independent sources $\mathbf{y}(t)$ (3).

$$\begin{aligned} \mathbf{x}(t) &= \mathbf{A}\mathbf{s}(t) \\ \mathbf{y}(t) &= \mathbf{W}\mathbf{x}(t) \end{aligned} \tag{3}$$

ICA minimizes higher-order statistical dependencies between reconstructed signals to estimate both unmixing matrix and independent signal sources.

3.1. Algorithms

Several approaches exist to solving the linear ICA problem, and the present treatment is limited to static (non-convolutive) ICA algorithms which assume linear instantaneous mixing. Four of such well known algorithms are presented here briefly, in light of the implementation architecture presented in Section 3.2.

Herault-Jutten (H-J) algorithm [8] : the first formulation of BSS as inspired by biomimetic principles, the H-J update algorithm is based on a feedback network topology $\mathbf{y} = -\mathbf{W}'\mathbf{y} + \mathbf{x}$ with zero diagonal terms ($w'_{ii} \equiv 0, \forall i$). An independence criterion based on nonlinear correlation yields the on-line learning rule for the off-diagonal terms

$$\Delta w'_{ij} = \mu\, f(y_i) g(y_j), \qquad i \neq j \tag{4}$$

where $f(.)$ and $g(.)$ are appropriately chosen, odd-symmetric functions.

Bell-Sejnowski (B-S) algorithm [9] : The derivation of the learning rule (5) is based on the information maximization (*InfoMax*) principle, and maximizes the output entropy of a neural network with nonlinear outputs:

$$\Delta \mathbf{W} = \mu\,([\mathbf{W}^T]^{-1} - f(\mathbf{y})\mathbf{x}^T) \tag{5}$$

Natural Gradient (NG) algorithm [10] : More robust and uniform convergence is obtained by using Amari's *natural gradient* of the *InfoMax* cost function, leading to the simple learning rule

$$\Delta \mathbf{W} = \mu\,[\mathbf{I} - f(\mathbf{y})\mathbf{y}^T]\mathbf{W} \tag{6}$$

The convergence of (6) implies $\mathrm{E}\{f_i(y_i)y_i\}{=}1$ as a constraint on reconstructed signals.

Cichocki-Unbehauen (C-U) algorithm [11] : To avoid numerical instability due to non-stationarity in the sources, the C-U algorithm introduces a non-holonomic constraint in the NG learning rule, fixing the diagonal of the unmixing matrix $\mathbf{W}$:

$$\Delta \mathbf{W} = \mu\,[\Lambda - f(\mathbf{y}) g(\mathbf{y}^T)]\mathbf{W} \tag{7}$$

where Λ is a diagonal scaling matrix. Convergence of the C-U algorithm implies $\Lambda_{ii} = E[f(y_i)g(y_i)]$.

3.2. General Outer-Product Formulation

Efficient implementation in parallel architecture requires a simple form of the update rule, that avoids excessive matrix multiplicatios and inversions. With simplification, the above ICA update algorithms can be cast in the common, unifying framework of the outer-product rule (4). To map

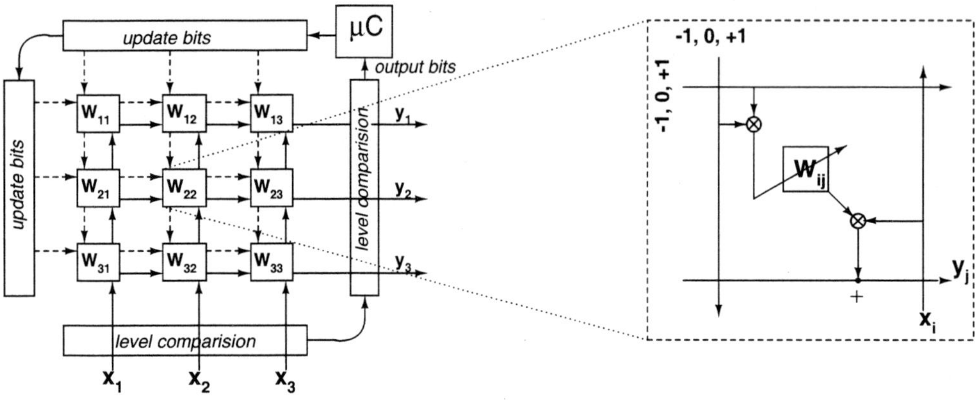

Fig. 2. Parallel implementation of ICA. General outer-product architecture.

the recurrent H-J network architecture onto a feedforward form, we apply the following approximation:

$$\mathbf{y} = (\mathbf{I} + \mathbf{W}')^{-1}\mathbf{x} \approx (\mathbf{I} - \mathbf{W}')\mathbf{x} \tag{8}$$

In other words, we choose to implement the H-J rule with linear feedworward networks of the type $\mathbf{y} = \mathbf{W}\mathbf{x}$ with fixed diagonal terms $w_{ii} \equiv 1$, and with off-diagonal terms adapting according to

$$\Delta w_{ij} = -\mu\, f(y_i)g(y_j), \qquad i \neq j \tag{9}$$

Equivalently, the implemented update rule can be seen as the gradient of *InfoMax* (5) multiplied by $\mathbf{W}^T$, rather than the natural gradient multiplication factor $\mathbf{W}^T\mathbf{W}$. To obtain the full natural gradient in outer-product form, it is necessary to include a back-propagation path in the network architecture, and thus additional silicon resources, to implement the vector contribution $\mathbf{y}\mathbf{W}$ [12].

4. IMPLEMENTATION

4.1. Architecture

Level comparison provides implementation of discrete approximations of any scalar function $f(y)$ and $g(y)$ appearing in different learning rules. Since speech signals are approximately Laplacian distributed, the nonlinear scalar function $f(y)$ is approximated by $\text{sign}(y)$ and implemented using single bit quantization. Conversely, a linear function $g(y) \equiv y$ in the learning rule is approximated by a 3-level staircase function $(-1, 0, +1)$ using 2-bit quantization. The quantization of the f and g terms in the update rule (9) simplifies the implementation to that of discrete counting operations.

The functional block diagram of a 3×3 outer-product incremental ICA architecture, supporting a quantized form of the general update rule (9), is shown in Figure 2. Unmixing coefficients are stored digitally in each cell of the architecture. The update is performed locally by once or repeatedly incrementing, decrementing or holding the current value of counter based on the learning rule served by the micro-controller. The 8 most significant bits of the 14-bit counter holding and updating the coefficients are presented to a multiplying D/A capacitor array [4] to linearly unmix the separated signal. The remaining 6 bits in the coefficient registers provide flexibility in programming the update rate to tailor convergence.

4.2. VLSI Implementation

A prototype 3×3 mixed-signal ICA processor was designed, fabricated, and tested. The differential analog input channels directly interface with gradient output signals from a previously developed gradient flow processor for acoustic localization [4], to extend its funcionality to joint separation and localization of up to three acoustic sources. The mixed-signal architecture is implemented using fully differential switched-capacitor sampled-data circuits. Correlated double sampling performs common mode offset rejection and 1/f noise reduction. An external micro-controller provides flexibility in the implementation of different learning rules. The ICA architecture is integrated on a single $3mm \times 3mm$ chip fabricated in 0.5 μm 3M2P CMOS technology. Details of circuit implementation and characterization of the chip will be presented elsewhere.

5. EXPERIMENTAL RESULTS

Initial experiments on the current prototype chip consisted of applying synthetic mixtures of speech signals as inputs. The original speech signals shown in Figure 3 (a) were mixed using the linear matrix:

$$\begin{bmatrix} 0.7 & 0.3 \\ 0.4 & 0.6 \end{bmatrix}$$

The synthetic mixture shown in Figure 3-(b) was applied to the ICA processor by using a 12 bit D/A converter. The on-line estimated signals, acquired from the output channels of the chip using 12 bit A/D conversion, are shown in Figure 3 (c). The estimates converge towards the original (unseen) sources in Figure 3 (a) over time, with 30 dB separation at convergence. Figure 3 (d) illustrates the dynamics of the recorded off-diagonal matrix elements w_{12} and w_{21} converging towards their ideal values (dashed lines).

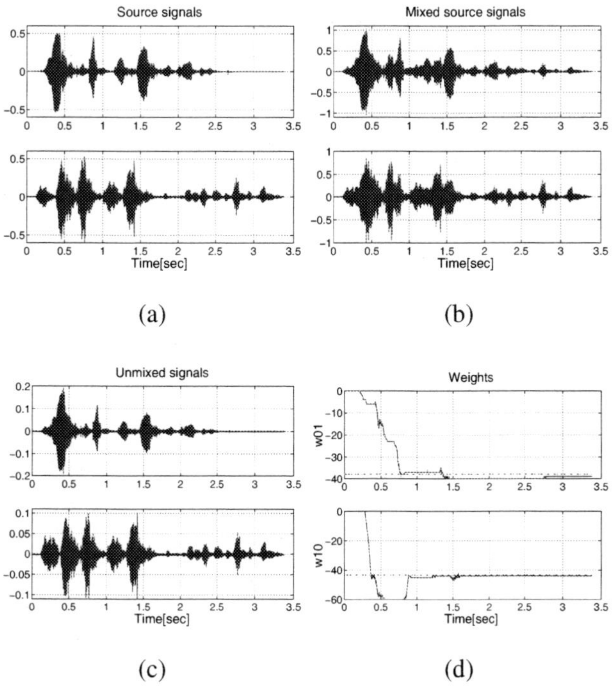

Fig. 3. Experimental blind separation of two speech sources by the VLSI chip. (a): Original speech signals; (b): Mixture of speech signals; (c): Recovered signals obtained from chip; (d): ICA weights recorded on-line from chip.

6. CONCLUSION

A mixed-signal parallel VLSI architecture for implementation of ICA in general outer-product form was presented. Initial experimental results from a prototype micropower chip demonstrated 30 dB real-time separation of mixed speech signals presented to the chip. Further experimentation is directed towards integrating the ICA VLSI architecture with previously developed circuits implementing gradient flow [4], for real-time adaptive acoustic source separation and localization, using miniature acoustic arrays. The ICA architecture is directly amenable to integration with sensor arrays for small, compact, battery-operated "smart" sensor applications in hearing aids, personal digital assistants, and surveillance networks.

7. REFERENCES

[1] A.S. Bregman, *Auditory Scene Analysis, The Perceptual Organization of Sound*, Cambridge MA: MIT Press, 1990.

[2] G. Cauwenberghs, M. Stanacevic, and G. Zweig, "Blind Broadband Source Localization and Separation in Miniature Sensor Arrays," *Proc.* IEEE *Int. Symp. Circuits and Systems (ISCAS'2001)*, Sydney, Australia, May 6-9, 2001.

[3] M. Stanacevic and G. Cauwenberghs, "Mixed-Signal Gradient Flow Bearing Estimation", *Proc.* IEEE *Int. Symp. Circuits and Systems (ISCAS'2003)*, Bangkok Thailand, May 25-28, 2003.

[4] M. Stanacevic and G. Cauwenberghs, "Micropower Mixed-Signal Acoustic Localizer," *Proc.* IEEE *Eur. Solid State Circuits Conf. (ESSCIRC 2003)*, Estoril Portugal, Sept. 16-18, 2003.

[5] Cohen, M.H., Andreou, A.G. "Analog CMOS Integration and Experimentation with an Autoadaptive Independent Component Analyzer," *IEEE Trans. Circuits and Systems II*, vol 42 (2), pp 65-77, Feb. 1995.

[6] Gharbi, A.B.A., Salam, F.M.A. "Implementation and Test Results of a Chip for the Separation of Mixed Signals," *Proc. Int. Symp. Circuits and Systems (ISCAS'95)*, May 1995.

[7] A. Cichocki and S. Amari, *Adaptive Blind Signal and Image Processing: Learning Algorithms and Applications*, New York: John Wiley, 2002.

[8] C. Jutten and J. Herault, "Blind Separation of Sources I. an Adaptive Algorithm Based on Neuromimetic Architecture," *Signal Proc*, vol. 24 (1), pp 1-10, 1991.

[9] A.J. Bell and T.J Sejnowski. "An Information-Maximization Approach to Blind Separation and Blind Deconvolution," *Neural Computation*, vol. 7, pp 1129-1159, 1995.

[10] S. Amari, A. Cichocki, and H. H. Yang. "A new learning algorithm for blind signal separation," *In NIPS 95*, vol. 8 pp 757-763. MIT Press, 1996.

[11] A. Cichocki, R. Unbehauen, "Robust neural networks with on-line learning for blind identification and blind separation of sources ," *IEEE Trans. Circuits and Systems I: Fundamental Theory and Applications* , vol. 43 (11), pp 894-906.

[12] M. Cohen and G. Cauwenberghs, "Blind Separation of Linear Convolutive Mixtures through Parallel Stochastic Optimization," *Proc. IEEE Int. Symp. Circuits and Systems (ISCAS'98)*, Monterey CA, vol. 3, pp. 17-20, 1998.

FAST NEURAL NETWORKS FOR SUB-MATRIX (OBJECT/FACE) DETECTION

Hazem M. El-Bakry, and Herbert Stoyan

Informatic Institute – Artificial Intelligence Dept. (Informatic 8)
Friedrich Alexander University of Erlangen-Nürnberg, Germany
helbakry1@hotmail.com

ABSTRACT- In recent years, fast neural networks for sub matrix (object/Face) detection have been introduced based on cross correlation in frequency domain between the input image and the weights of neural networks. In a previous paper [1], it has been proved that for those fast neural networks to give correct results as conventional neural networks, either the weights of neural networks or the input image must be symmetric. In case of converting the input image into a symmetric one, those fast neural networks become slower than conventional neural networks. In this paper, a new form of symmetry for the input image to fast the operation of neural nets is presented. Simulation results using Matlab confirm the theoretical computations.

Keywords: Fast Sub-Image (Object/Face) Detection, Neural Networks, Cross Correlation

I. Introduction

Recently, neural nets have shown good results for detecting an object or a face in a given image [2,3,5]. Some authors try to speed up the detection process of neural nets [4,6,7]. They have proposed a multilayer perceptron (MLP) algorithm for fast object/face detection. They claimed that, applying cross correlation in frequency domain between the input image and the neural weights is much faster than using conventional neural nets. They stated this without any conditions and introduced formulas for the number of computation steps needed by conventional and their proposed fast neural nets. Then, they deduced an equation for the speed up ratio. It has been proved in [1] that their equations contain many errors, which lead to invalid speed up ratio. Moreover, a symmetry condition is necessary and must be found either in the input image or in the neural weights so that those fast neural networks can give correct results as conventional neural network for detecting an object or a face in a given image. The main objective of this paper is to fast the operation of neural networks and speed up the detection process. This is done by converting the input image into a symmetric one and applying cross correlation in frequency domain between the new symmetric image and the neural weights. Mathematical prove and simulation results for fast testing of the new proposed symmetric image using Matlab are given. In section II, fast neural nets for object/face detection are described. A new symmetric form for the input image to speed up the detection process is presented in section III.

II. Theory of Fast Neural Nets Based on Cross Correlation in Frequency Domain For Object/Face Detection

Finding an object/face in the input image is a search problem. Each sub-image in the input image is tested for presence or absence of the required object/face. At each pixel position in the input image, each sub-image is multiplied by a window of weights which has the same size. The outputs of neurons in the hidden layer are multilpied by the weights of the output layer. When the final output is high this means that the sub-image under test contains the required object/face and viceversa. So, we may conclude that, such searching problem is cross correlation between the image under test and the weights of the hidden neurons.

The convolution theorem in mathematical analysis says that a convolution of f with h is identical to the result of the following steps: let **F** and **H** be the results of the Fourier Transformation of f and h in the frequency domain. Multiply **F** and **H** in the frequency domain point by point and then transform this product into spatial domain via the inverse Fourier Transform. As a result of this, these cross correlations can be represented by a product in frequency domain. So, by using cross correlation in frequency domain, speed up in an order of magnitude can be achieved during the detection process [1,2,3,5]. In the detection phase, a sub image **I** of size mxn (sliding window) is extracted from the tested image which has a size **PxT** and fed to the neural network. Let $\mathbf{X_i}$ be the vector of weights between the input sub image and the hidden layer. This vector has a size of mxn and can be represented as mxn matrix. The output of hidden neurons **h(i)** can be calculated as follows:

$$h_i = g\left(\sum_{j=1}^{m}\sum_{k=1}^{n} X_i(j,k)I(j,k) + b_i\right) \qquad (1)$$

where g is the activation function and **b(i)** is the bias of each hidden neuron **(i)**. Equation 1 represents the output of each hidden neuron for a particular sub-image **I**. It can be obtained to the whole image **Z** as follows:

$$h_i(u,v)= g\left(\sum_{j=-m/2}^{m/2}\sum_{k=-n/2}^{n/2} X_i(j,k)\, Z(u+j, v+k)+b_i\right) \qquad (2)$$

Equ.2 represents a cross correlation operation. Given any two functions **f** and **d**, their cross correlation can be obtained by:

$$f(x,y)\otimes d(x,y)= \left(\sum_{m=-\infty}^{\infty}\sum_{n=-\infty}^{\infty} f(m,n)d(x+m,y+n)\right) \qquad (3)$$

Therefore, equ. 2 may be written as follows:

$$h_i = g(X_i \otimes Z + b_i) \quad (4)$$

where h_i is the output of the hidden neuron **(i)** and $\mathbf{h_i}$ **(u,v)** is the activity of the hidden unit **(i)** when the sliding window is located at position **(u,v)** and **(u,v)** $\in$ **[P-m+1,T-n+1]**.

Now, the above given cross correlation can be expressed in terms of Fourier Transform:

$$Z \otimes X_i = F^{-1}(F(Z) \bullet F^*(X_i)) \quad (5)$$

Hence, by evaluating this cross correlation, a speed up ratio can be obtained compared to conventional neural networks. Also, the final output of the neural network can be evaluated as follows:

$$O(u,v) = g\left(\sum_{i=1}^{q} w_o(i)\, h_i(u,v) + b_o\right) \quad (6)$$

where **q** is the number of neurons in the hidden layer. **O(u,v)** is the output of the neural network when the sliding window located at the position **(u,v)** in the input image **Z**.

We can analayz the complexity of cross correlation in frequency domain as follows [1]:

1- For a tested image of NxN pixels, the 2D-FFT requires a number equal to $N^2\log_2N^2$ of complex computation steps. Also, the same number of complex computation steps is required for computing the 2D FFT of the weight matrix for each neuron in the hidden layer.

2- At each neuron in the hidden layer the inverse 2D FFT is computed, so, q backward and (1+q) forward transforms have to be computed. Therefore, for an image under test, the total number of the 2DFFT to compute is $(2q+1)N^2\log_2N^2$.

3- The input image and the weights should be multiplied in the frequency domain. Therefore, a number of complex computation steps equal to qN^2 should be added.

4- The number of computation steps required by fast neural networks is complex and must be converted into a real version. It is known that, the two dimensions Fast Fourier Transform requires $(N^2/2)\log_2N^2$ complex multiplications and $N^2\log_2N^2$ complex additions. Every complex multiplication is realized by six real floating point operations and every complex addition is implemented by two real floating point operations. So, the total number of computation steps required to obtain the 2D-FFT of an NxN image is [1]:

$$\rho=6((N^2/2)\log_2N^2) + 2(N^2\log_2N^2) \quad (7)$$

which may be simplified to:

$$\rho=5(N^2\log_2N^2) \quad (8)$$

Also, performing complex dot product in frequency requires $6qN^2$ real operations.

5- In order to perform cross correlation in frequency domain, the weight matrix must have the same size as the input image. So, a number of zeros = (N^2-n^2) must be added to the weight matrix. This requires a total real number of computation steps = $q(N^2-n^2)$ for all neurons. Moreover, after computing the FFT2 for the weight matrix, the conjugate of this matrix must be obtained. So, a real number of computation steps $=qN^2$ should be added in order to obtain the conjugate of the weight matrix for all neurons. Also, a number of real computation steps equal to N is required to create butterflies complex numbers $(e^{-jk(2\Pi n/N)})$, where 0<K<L. These (N/2) complex numbers are multiplied by the elements of the input image or by previous complex numbers during the computation of FFT2. To create a complex number, this requires two real floating point operations. So, the total number of computation steps required for fast neural networks becomes [1]:

$$\sigma=((2q+1)(5N^2\log_2N^2) + 6qN^2 + q(N^2-n^2) + qN^2 + N) \quad (9)$$

which can be reformulated as:

$$\sigma=((2q+1)(5N^2\log_2N^2) + q(8N^2-n^2) + N) \quad (10)$$

6- Using sliding window of size nxn, for the same image of NxN pixels, $(q(2n^2-1)(N-n+1)^2)$ computation steps are required when using traditional neural networks for object/face detection process. The theoretical speed up factor η can be evaluated as follows [1]:

$$\eta = \frac{q(2n^2-1)(N-n+1)^2}{(2q+1)(5N^2\log_2N^2) + q(8N^2-n^2) + N} \quad (11)$$

7- But as proved in [1], this cross correlation in frequency domain (Fast Neural Networks) gives the same results as conventional cross correlation (Conventional Neural Networks) only in two cases. Either the weights are symmetric or the input image is symmetric. It is very complex to allow the weights to be symmetric in the required form which is needed to be as follows [1]:

$$W = \begin{bmatrix} w & 0 \\ 0 & w_d \end{bmatrix} \quad (12)$$

Adding this constrain to the learning rules will cause many well known problems during the training process of the neural network. The probability that the network will not trained is very high. Another solution is to convert the input image into one of the required symmetric forms as shown in Fig. 1. As the input image has a dimension of (N), the new symmetric image will have a length of (2N). In this case, the number of computation steps required for fast neural networks can be calculated as follows [1]:

$$\sigma_{2N}=((2q+1)(5(2N)^2\log_2(2N)^2) + q(8(2N)^2-n^2) + 2N) \quad (13)$$

But, converting the non-symmetric input image into a symmetric one will slow down those fast neural networks more and more compared to conventional neural networks. In this case, for any size of the input image, dividing the number of operations required for conventional neural networks by those needed by fast neural networks (equ.11) gives a speed up ratio less than one as listed in table (1) [1].

III. A New Proposed Symmetric Form For the Input Image to Fast the Process of Neural Networks

In this section, another new symmetric form for the input image in order to reduce the number of computation steps required by neural networks for object/face detection is presented. As shown in Fig. 2, the input image will be converted into symmetric form by rotating it into down direction and test both up and down images as a one (symmetric) image consists of two images. In this case, the new symmetric image will have (2NxN) dimensions. Substituting in equ.9 for the new dimensions, the number of computation steps required for cross correlating this new image with the weights in frequency domain can be calculated as follows:-

$$\sigma=((2q+1)(5(2N^2\log_2N+2N^2\log_2 2N))+q6(2N^2)+ q(2N^2-n^2)+q(2N^2)+2N) \quad (14)$$

which can be simplified to:

$$\sigma=((2q+1)(10N^2(\log_2 2N+ \log_2N)) +q(16N^2-n^2)+2N) \quad (15)$$

The theoritical speed up ratio in this case with different sizes of the input image and different in size weight matrices is listed in table (2). Also, practical speed up ratio for manipulating images of different sizes and different in size weight matrices is listed in table (3) using 700 MHz processor and Matlab.

Moreover, this new configuration is useful for reducing the number of patterns that the neural network will learn. Because the image is rotated into down direction as shown in Fig.2, and then the up image and its rotated down version are tested together as one (symmetric) image. If an object/face is detected in rotated down image, this means that this object/face is found at the relative position in the up image. So, if conventional neural networks are trained for up and rotated down examples of the object/face, fast neural networks will train only to up examples in case of using the presented configuration for the input image. As number of trained examples is reduced, the number of neurons in the hidden layer will be reduced and the neural network will be more faster in the test phase compared with conventional neural networks.

IV. Conclusion

I have presented a new form of symmetry for the input image to speed up the detection process of an object/face in a given image. I have proved mathematically that, by converting the input image into the presented symmetric form, the speed of the detecion process becomes higher than conventional neural networks. Simulation results have confirmed this approval by using Matlab. Moreover, by using the presented new symmetric configuration for the input image, the number of neurons in the hidden layer hass been reduced and fast neural networks based on cross correlation in frequency domain have become more faster than conventional neural networks.

References

[1] Hazem M. El-Bakry, "Comments on Using MLP and FFT for Fast Object/Face Detection," Proc. of IEEE IJCNN'03, Portland, Oregon, pp. 1284-1288, July, 20-24, 2003.

[2] Hazem M. El-Bakry, "Human Iris Detection Using Fast Cooperative Modular Neural Nets and Image Decomposition," Machine Graphics & Vision Journal (MG&V), vol. 11, no. 4, 2002, pp. 498-512.

[3] Hazem M. El-Bakry, "Automatic Human Face Recognition Using Modular Neural Networks.", Machine Graphics & Vision Journal (MG&V), vol. 10, no. 1, 2001, pp. 47-73.

[4] S. Ben-Yacoub, B. Fasel, and J. Luettin, "Fast Face Detection using MLP and FFT," in Proc. Second International Conference on Audio and Video-based Biometric Person Authentication (AVBPA'99)", 1999.

[5] H. A. Rowley, and T. Kanade, " Neural Network - Based Face Detection, " IEEE Trans. on Pattern Analysis and Machine Intelligence, Vol. 20, No. 1, pp. 23-38, 1998.

[6] Beat Fasel, "Fast Multi-Scale Face Detection," IDIAP-Com 98-04, 1998.

[7] S. Ben-Yacoub, "Fast Object Detection using MLP and FFT," IDIAP-RR 11, IDIAP, 1997.

Table 1: A comparison between the number of computation steps (in millions) required for conventional and fast neural networks to manipulate images shown in Fig.1 with different sizes (n=20).

Image size	Conventional Neural Nets	Fast Neural Nets (2N)	Speed up ratio
100x100	157,267170	196,098091	.802
200x200	785,281170	822,028364	.890
300x300	189,2695170	2113,036584	.896
400x400	3479,509170	3918,549456	.888
500x500	5545,723170	6319,116413	.877
600x600	8091,337170	9330,582337	.867
700x700	11116,351170	12965,856005	.857
800x800	14620,765170	17235,856005	.848
900x900	18604,579170	21986,745146	.846
1000x1000	23067,793170	27716,501654	.832

Table 2: The theoretical speed up ratio in case of converting an image into symmetric one through rotation into down direction.

Image Size	Speed up ratio (n=20)	Speed up ratio (n=25)	Speed up ratio (n=30)
100x100	1.71	2.36	2.96
200x200	1.88	2.79	3.79
300x300	1.89	2.85	3.96
400x400	1.86	2.85	3.99
500x500	1.84	2.82	3.98
600x600	1.82	2.80	3.96
700x700	1.80	2.77	3.93
800x800	1.78	2.74	3.90
900x900	1.76	2.72	3.87
1000x1000	1.74	2.70	3.84

Table 3: Simulation results for Speed up ratio in case of converting an image into symmetric one through rotation into down direction.

Image Size	Speed up ratio (n=20)	Speed up ratio (n=25)	Speed up ratio (n=30)
100x100	5.29	6.74	11.16
200x200	4.46	6.24	9.94
300x300	4.17	5.08	8.66
400x400	3.59	4.78	7.45
500x500	3.40	4.34	6.87
600x600	3.30	4.42	6.16
700x700	3.12	4.20	5.74
800x800	2.60	3.58	4.76
900x900	2.97	4.10	5.38
1000x1000	2.57	3.47	4.63

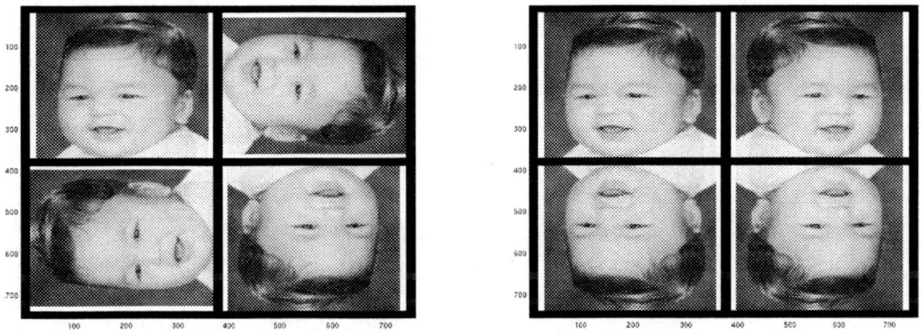

Fig. 1. Image conversion from non-symmetric to symmetric one.

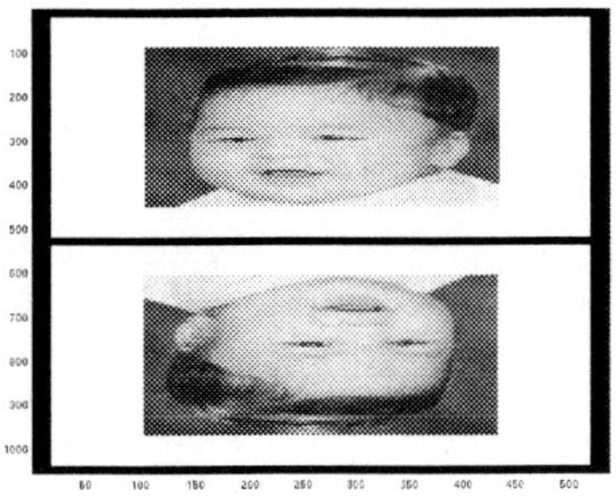

Fig. 2. Image conversion from non-symmetric to symmetric one through rotation into down direction.

NEURAL NETWORK SYSTEM FOR FACE RECOGNITION

Kussul E., Baidyk T., Kussul M.*

CCADET, UNAM
Cd. Universitaria, A.P. 70-186, C.P. 04510, Mexico, D.F.
ekussul@servidor.unam.mx tbaidyk@aleph.cinstrum.unam.mx
Tel.: (52) 5622-86-02 ext.1204, Fax: (52) 5550-06-54

*Beenet Computer Sytems Inc.
#115A-4471 No6 Road Richmond, British Columbia, Canada V6V 1P8
Tel.: (604) 270-4222 k_maximu@telus.net

ABSTRACT

The new image recognition method based on neural network system is proposed. This method uses the Permutative Coding technique for image preprocessing and neural classifier for image recognition. We have proposed the Permutative Coding technique to make recognition process invariant to small displacements of the object in the image. The system was tested on the ORL database. This database contains 400 face images of 40 persons. 200 images are used for training and 200 for recognition. The error rate of 0.1% for face recognition was obtained. This method was tested also with 40, 80, 120, and 160 images for system training and the rest images for recognition. The error rates 16.1%, 7.09%, 2.15%, 1.4% were obtained respectively.

1. INTRODUCTION

In this article we describe the new method of image recognition based on Permutative Coding Neural Classifier (PCNC). We apply this method to the face recognition problem.

The automatic face recognition could be used in different security systems (for building, banks, etc.), for passport and other document verification. Different approaches are investigated and proposed for this task solving [1]-[6]. They were tested on the base ORL (Olivetti Research Laboratory) [6],[8]-[10]. We have tested our classifier on this database and obtained one of the best results. The error rate was 0.1%.

2. PERMUTATIVE CODING NEURAL CLASSIFIER

The principles of the proposed method we worked out on the base of the associative-projective neural networks (APNN) [11]. We are developing the APNN since 80th. We use the APNN paradigm as generic structure for different applications such as random threshold classifier (RTC) [12] random subspace neural classifier, LIRA [13] and PCNC.

In this article we propose new algorithm, which outperforms the predecessors due to new elements (feature extractor and encoder) included into its structure.

2.1. PCNC structure

The PCNC structure is presented in Fig.1.

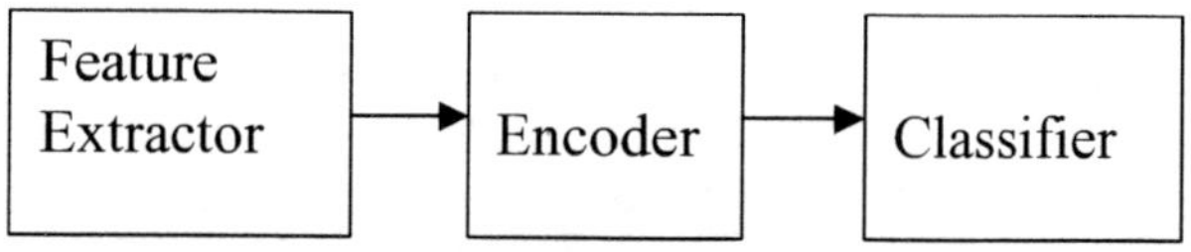

Fig. 1. Structure of Permutative Coding Neural Classifier

The image is input to the feature extractor. The extracted features are applied to the encoder input. The encoder produces the output binary vector of large dimension, which is presented to the one-layer neural classifier input. The output of the classifier gives the recognized class.

2.2. Feature extractor

Initial image (Fig.2) is to be input to the feature extractor. The feature extractor begins the work with the selection of the specific points in the image. In principle various methods of the specific points selection could be proposed. For example, the contour points could be selected as the specific points.

Fig. 2. Initial image

The rectangle of $h \cdot w$ size is formed around each specific point (Fig.3).

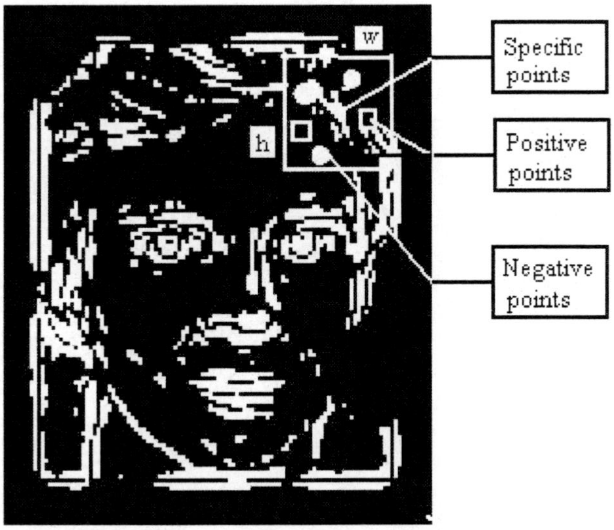

Fig. 3. Specific points selected by the feature extractor

Each feature is extracted from this rectangle. The p positive and the n negative points determine one feature. These points are randomly distributed in the rectangle. Each point P_{rs} has the threshold T_{rs} that is randomly selected from the range:

$$T_{min} \leq T_{rs} \leq T_{max}, \quad (1)$$

where s stands for feature number; r stands for point number.

The positive point is "active" if in initial image it has brightness:

$$b_{rs} \geq T_{rs}. \quad (2)$$

The negative point is "active" if it has brightness:

$$b_{rs} \leq T_{rs}. \quad (3)$$

The feature under investigation exists in the rectangle if all its positive and negative points are active. In other cases the feature is absent in the rectangle.

2.3. Encoder

The encoder transforms the extracted features to the binary vector:

$V = \{v_i\}$ $(i = 1, \ldots, N)$,

where $v_i = 0$ or 1. For each extracted feature F_s the encoder creates an auxiliary binary vector:

$U = \{u_i\}$ $(i = 1, \ldots, N)$,

where $u_i = 0$ or 1.

A special random procedure is used to obtain the positions of ones in the vector U_s for each feature F_s. This procedure generates the list of the positions of ones for each feature and saves all such lists in the memory. The vector U_s we term "mask" of the feature F_s.

In the next stage of encoding process it is necessary to transform the auxiliary vector U to the new vector U^* which corresponds to the feature location in the image. This transformation is made with permutations of the vector U components (Fig. 4).

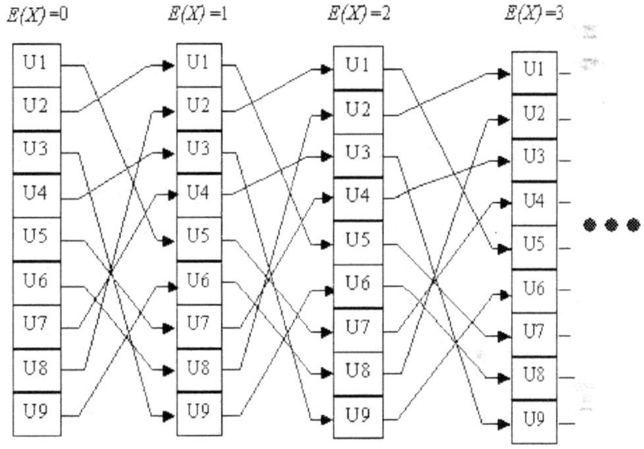

a

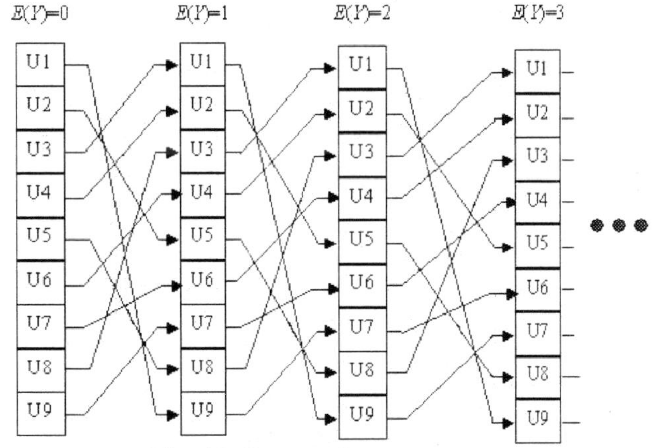

Fig. 4 b. Permutation pattern.

The number of permutations depends on the feature location on the image. The permutations in horizontal (X)

(Fig.4, *a*) and vertical (*Y*) (Fig.4,*b*) directions are different permutations. It is necessary to obtain such binary codes of the features that are strongly correlated if the distance between the feature locations is small and are weakly correlated if the distance is large. Such a property could be obtained with the following procedure.

To code the feature F_s location in the image it is necessary to select the correlation distance D_c and calculate the following values:

$$X = j / Dc,$$
$$E(X) = (int)X, \quad (4)$$
$$R(X) = j - E(X) \bullet Dc,$$
$$Y = i / D_c,$$
$$E(Y) = (int)Y, \quad (5)$$
$$R(Y) = i - E(Y) \bullet D_c,$$
$$P_x = \frac{R(X) \bullet N}{D_c}, \quad (6)$$
$$P_y = \frac{R(Y) \bullet N}{D_c}, \quad (7)$$

where *E(X)* is the integer part of *X*; *R(X)* is the fraction part of *X*; *i* – the vertical coordinate of the detected feature; *j* – the horizontal coordinate of the detected feature, *N* – the number of neurons.

The mask of the feature F_s is considered as a code of this feature located at the left top corner of the image. To shift the feature location in the horizontal direction it is necessary to make its permutations *E(X)* times and to make an additional permutation for P_x components of the vector. After that it is necessary to shift the code to the vertical direction making the permutations *E(Y)* times and an additional permutation for P_y components.

2.4. Neural Classifier

The structure of proposed recognition system is presented in Fig. 5.

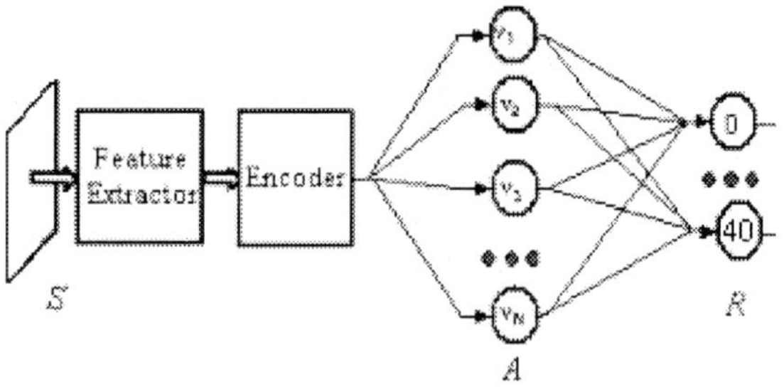

Fig. 5. Permutative Coding Neural Classifier.

The system contains sensor layer *S*, feature extractor, encoder, associative neural layer *A*, and reaction neural layer *R*. In the face recognition task each neuron of *R*-layer corresponds to one of the persons of data base. Sensor layer *S* corresponds to the initial image.

Associative neural layer contains "binary" neurons having the outputs 0 or 1. The values of its neuron outputs are produced as a result of the work of the encoder. The neurons of the associative layer *A* are connected to the reaction layer *R* with trainable connections having the weights w_{ji}. The excitations of *R*-layer neurons are calculated according the

$$E_i = \sum_{j=1}^{n} a_j * w_{ji} \quad (8)$$

where E_i – the excitation of the *i*-th neuron of the *R*-layer; a_j – the excitation of the *j*-th neuron of *A*-layer; w_{ji} – weight of the connection between *j*-th neuron of *A*-layer and *i*-th neuron of *R*-layer.

Neuron winner having maximal excitation is selected after the calculations of the excitations.

We use the following training procedure. Denote the neuron-winner number as i_w, and the number of neuron, which really corresponds to the input image, as i_c. If $i_w = i_c$ nothing to be done. If $i_w \neq i_c$

$$(\forall j)\left(w_{ji_c}(t+1) = w_{ji_c}(t) + a_j\right)$$
$$(\forall j)\left(w_{ji_w}(t+1) = w_{ji_w}(t) - a_j\right) \quad (9)$$
$$\text{if } (w_{ji_w}(t+1) < 0) \;\; w_{ji_w}(t+1) = 0.$$

where $w_{ji}(t)$ is the weight of connection between *j*-neuron of *A*-layer and *i*-neuron of *R*-layer before reinforcement, $w_{ji}(t+1)$ is the weight after reinforcement.

3. RESULTS OBTAINED ON THE ORL DATABASE

ORL data base contains 400 images of 40 persons (10 images of each person). Each image is full face (Fig.2).

The difference between 10 images of each person consists in shifts, head inclination, different face expression, presence or absence of glasses. As a rule the algorithms are tested on the ORL data base in the following manner: five images of one person are used for classifier training and other five are used for testing. There are two modes to make this partition.

In the first mode they use the first five images for training and last five images for testing. In the second mode they select randomly five images for training and the rest of images are used for testing. The first mode was good for comparison of the classifiers until classifiers with almost 100% recognition rate appeared. If error percentage is less than 2-3% it is difficult to compare the classifiers using this mode.

The second mode permits to make many different experiments with the same data base and to obtain statistically reliable results. In our experiments we use the second mode with 10 runs for each experiment. This method was used also in other works for estimation mean value of error rate [2]. They used 6 runs for each experiment. Their results are given in Table 1 (NOM – number of misclassifications or errors).

TABLE 1
Best performances on 6 simulations

Simulation	1	2	3	4	5	6
NOM	2	6	6	2	1	6

It is also interesting to investigate the cases where less then five images of each person are selected for training and the rest for testing. The data from such experiments are presented in [1].

The results of our experiments are presented in Table 2.

TABLE 2
Results of the experiments

Tr./ ex.	*NOM* Run 1	2	3	4	5	6	7	8	9	10	Total NOM	E_{avc} (%)
1/9	64	66	55	46	52	63	72	44	58	60	580	16.1
2/8	14	29	18	18	25	26	19	13	21	44	227	7.09
3/7	6	13	6	0	9	8	10	3	4	1	60	2.15
4/6	3	3	7	2	5	1	0	7	4	2	34	1.4
5/5	0	0	0	0	0	0	0	0	0	2	2	0.1

The note tr./ex. reflects how much images of each person were used for training (tr.) and how much for examine (ex.). The comparison of results show that our classifier in most of the cases gives the best recognition rate.

4. CONCLUSION

The new image recognition system was developed. The system is based on neural network with the permutation coding technique. This technique decreases the sensitivity of the system to the object displacements in the image. Tests of the system shows the error rate 0.1 % on ORL database. In the future the recognition methods, which are insensitive to the object rotation and scale change, would be developed.

5. ACKNOWLEDGMENT

This work was funded by Projects CONACYT 33944-U, NSF-CONACYT39395A, PAPIIT IN 112102.

6. REFERENCES

[1] Teewoon Tan, Hong Yan, Object recognition using fractal neighbor distance: eventual convergence and recognition rates. *Proceedings of the 15th IEEE International Conference on Pattern Recognition*, **V.2**, pp. 781-784, 2000.

[2] Meng Joo Er, Shiqian Wu, juwei Lu, Hock Lye Toh, Face recognition with radial basis function (RBF) neural networks, *IEEE Transactions on Neural Networks*, **V.13**, N 3, pp. 697-710, 2002.

[3] R Javad Haddadnia, Majid Ahmadi, Karim Faez, An efficient method for recognition of human faces using higher orders pseudo Zernike moment invariant, *Proceedings of the Fifth IEEE International Conference on Automatic Face and Gesture Recognition* (FGR'02), pp. 6, 2002.

[4] Victor-Emil Neagoe, Armand-Dragos Ropot, Concurrent self-organizing maps for pattern classification. *Proceedings of the First IEEE International Conference on Cognitive Informatics* (ICCI'02), pp. 9, 2002.

[5] Phiasai T., Arunrungrushi S., Chamnongthai K., Face recognition system with PCA and moment invariant method, *The 2001 IEEE International Symposium on Circuits and Systems*, pp. II-165 – II-168, 2001.

[6] Steve Lawrence, C.Lee Giles, Ah Chung Tsoi, Andrew D.Back, Face recognition: a convolutional neural-network approach. *IEEE Transactions on Neural Networks*, **V.8**, N 1, pp. 98-113, 1997.

[7] Guodong Guo, Stan Z. Li, Kapluk Chan, Face recognition by support vector machines. In: *proceedings of the fourth IEEE International Conference on Automatic Face and Gesture Recognition*, pp. 196-201, 2000.

[8] S.-H. Lin, S.-Y.Kung, and L.-J. Lin, Face recognition/detection by probabilistic decision-based neural network, *IEEE Trans. Neural Networks*, **V.8**, pp. 114-132, 1997.

[9] V. Brennan and J.Principe, Face classification using multiresolution principal component analysis, in *Proc. IEEE Workshop Neural Networks Signal Processing*, pp. 506-515, 1998.

[10] S.Eickeler, S.Mueller, and G. Rigoll, High quality face recognition in JPEG compressed images, in *Proc.IEEE Int. Conf. Image Processing*, pp. 672-676, 1999.

[11] E. M. Kussul, D. A. Rachkovskij, T. N. Baidyk, Associative-projective neural networks: architecture, implementation, applications. *Proc. of Fourth Intern. Conf. "Neural Networks & their Applications"*, Nimes, France (EC2 Publishing), pp. 463-476, 1991.

[12] E. Kussul., T. Baidyk, V. Lukovitch, D. Rachkovskij, Adaptive high performance classifier based on random threshold neurons, in R. Trappl (Ed.) *Cybernetics and Systems'94* (Singapore: World Scientific Publishing Co. Pte. Ltd., pp. 1687-1695, 1994.

[13] E. Kussul, T. Baidyk, Improved method of handwritten digit recognition tested on MNIST database, *Proceedings of the 15-th Intern Conf. on Vision Interface,* Calgary, Canada, pp. 192-197, 2002.

NO-REFERENCE QUALITY ASSESSMENT OF JPEG IMAGES BY USING CBP NEURAL NETWORKS

Paolo Gastaldo and Rodolfo Zunino

Department of Biophysical and Electronic Engineering, University of Genoa
Via Opera Pia 11a, 16145 Genova, Italy – e-mail: {gastaldo, zunino} @dibe.unige.it

ABSTRACT

Reliable methods for measuring perceived image quality are needed to evaluate visual artifacts brought about by digital compression algorithms such as JPEG. This paper presents an objective quality-assessment method based on a Circular Back-Propagation (CBP) neural structure: the network is trained to predict quality ratings, as scored by human assessors, from numerical features that characterize images. As such, the method aims at reproducing perceived image quality, rather than at defining a comprehensive model of the human visual system. The neural model allows one to decouple the task of feature selection from the mapping of these features into a quality score. Experimental results on a public database of test images confirm the effectiveness of the approach.

1. INTRODUCTION

In most applications, the effectiveness of compression methods for digital images (such as JPEG) depends on their impact on the visual fruition of pictures by consumers. Subjective testing [1] is the conventional approach for quality evaluation; these methods measure perceived quality by asking human assessors to score the overall quality of a set of test images. These tests yield accurate results; nonetheless, they are very difficult to model in a deterministic way.

Objective methods [2] instead, aim to estimate perceived quality, bypassing human assessors. These techniques measure image quality by processing numerical quantities ("objective features") extracted from images. To be effective, objective models must cohere with subjective opinions of quality perception. In most cases, the assessment system has to know the reference (uncompressed) image [3,4]. By contrast, no-reference (NR) methods assess perceived quality by receiving as input only the compressed image [5, 6].

This paper presents a method using neural networks for the objective assessment of JPEG compressed images. A Circular BackPropagation (CBP) feedforward network [7] processes objective features worked out from JPEG images, and returns the associated quality scores. As the neural-based set-up does not require any information about the original uncompressed image, the overall objective method follows a no-reference approach.

2. QUALITY ASSESSMENT OF JPEG IMAGES

Figure 1 shows a schematic representation of the proposed no-reference quality assessment system for JPEG compressed images: a CBP network directly yields the quality assessments associated with input vectors of features worked out form the image.

The design of the objective system takes into account that 1) several features characterizing images jointly affect subjective judgments, and 2) non-linear relationships may complicate the modelling process. The effectiveness of the neural-network approach lies in the ability to decouple the problem of feature-selection from the design of an explicit mathematical model; to this purpose, the CBP network provides a paradigm to deal with multidimensional data characterized by complex relationships. The function that maps feature vectors into quality ratings is learned from examples by use of an iterative training algorithm. Hence, the design of the objective metric is not involved in the set-up of the mapping function.

2.1. Feature-based description of the image

In the present research, the feature vector characterizing the digital image has been built up by following the approach that proved effectively in the quality assessment of digital images processed by an image-enhancement algorithm [8]. As a result, 1) objective features are extracted on a block-by-block basis from the picture (non-overlapping blocks of 32x32 pixel), and 2), the correspondent feature vector is built up by exploiting the statistical properties of the features over the blocks.

The objective metric set characterizing each block has been defined by applying a feature-selection criterion [8] to a quite large set of measures based on the luminance

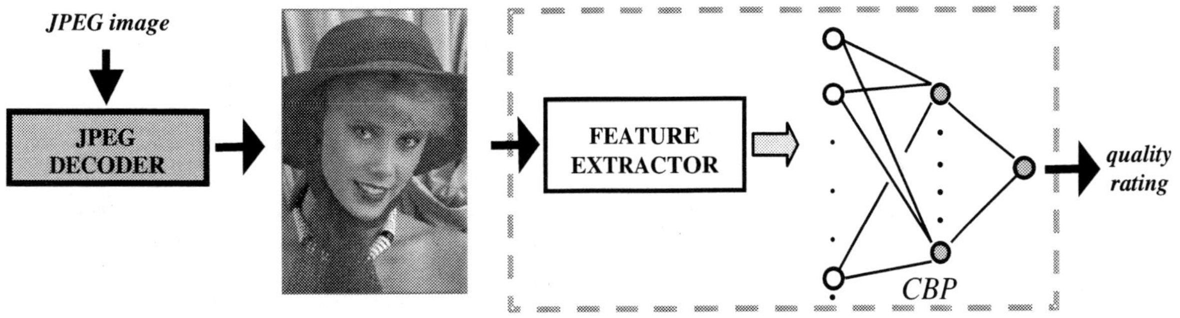

Fig. 1. The objective assessment system.

values of the images (see the Appendix). The selection algorithm is based on statistical analysis and uses skewness and kurtosis as a paradigm to characterize the statistical activity of the features. Given a library of test images $I = \{I_n; n = 1,..,n_p\}$, the feature-selection algorithm processes the sets Φ_h defined as:

$$\Phi_h = \{f_{h11},...,f_{h1q},...,f_{hn_p1},...,f_{hn_pq}\} \quad k = 1,..,n_f; \tag{1}$$

where f_{hnb} is the value of the feature f_k for the block b of the image n, n_f is the number of features, and q is the number of blocks extracted for each test image. As output, the algorithm yields the subset Z of n_s ($<=n_f$) features that are classified as statistically relevant.

2.2. Feature vector

Although images have to be split into non-overlapping blocks for the space-variant nature of the perceptual phenomenon, subjective ratings from human assessors necessarily end up with one overall quality score per image. Therefore, the block-based information is put together into one vector to be associated with the unique score. Such a vector is built up by global-level statistical descriptors of the features $f_h \in Z$, for a JPEG image I, as:

1) compute a statistical description of feature values:

$$f_h^M = median(f_{hI1},...,f_{hIn_b}); \tag{2}$$

$$f_h^S = stdev(f_{hI1},...,f_{hIn_b}); \tag{3}$$

2) generate the feature vector $\vec{x}_I$ for the image I as:

$$\vec{x}_I = \{f_h^M, f_h^S; h = 1,...,n_S\}. \tag{4}$$

3. CBP FOR IMAGE-QUALITY ESTIMATION

3.1. CBP Neural Network

In the proposed system the feed-forward neural network map feature-based image descriptions into the associated estimates of perceived quality, which, in the present formulation, are represented as scalar values.

Theory proves that feed-forward networks embedding a sigmoidal nonlinearity can support arbitrary mappings [9]. The MultiLayer Perceptron (MLP) model [9] belongs to this class of networks, and has been proved to perform effectively in those problems where the target-mapping function can be attained by a few computing units endowed with global scope. The "Circular Back Propagation" (CBP) network [7] extends the conventional MLP by adding one additional input, which sums the squared values of all the network inputs. By this formulation the properties of the MLP structure remain unaffected. At the same time, CBP theory shows that this additional unit enables the overall network either to adopt the standard, sigmoidal behaviour, or to drift smoothly to a bell-shaped radial function. More importantly, the selection between either model is entirely data-driven and stems from the empirical training process, hence model selection does not require any *a priori* assumption. Such a behavioural adaptiveness makes CBP networks quite interesting for application to perceptual problems, where even the domain structure is often obscure.

The CBP architecture can be formally described as follows. The input layer connects the n_i input values (features) to each neuron of the "hidden layer". The j-th "hidden" neuron computes a weighted combination of input values, with coefficients $w_{j,i}$ (j=1,…,n_h; i=1,…,n_i):

$$r_j = w_{j,0} + \sum_{i=1}^{n_i} w_{j,i}x_i + w_{j,n_i+1}\sum_{i=1}^{n_i} x_i^2 . \tag{5}$$

Each hidden neuron performs a non-linear transformation of the result:

$$a_j = \sigma(r_j); \qquad j = 1,..., n_h; \tag{6}$$

where $\sigma(x) = (1+e^{-x})^{-1}$, r_j is the *stimulus* and a_j is the *activation*. Likewise, the *output* layer provides the actual network responses, y_k, (k = 1,…, n_o):

$$r_k = w_{k,0} + \sum_{j=1}^{n_h} w_{k,j}a_j ; \qquad y_k = \sigma(r_k) . \tag{7}$$

The structural CBP enhancement still allows one to adopt conventional back-propagation algorithms [7] for weight adjustment. Hence an efficient tool is available for an effective training. A quadratic cost function measures

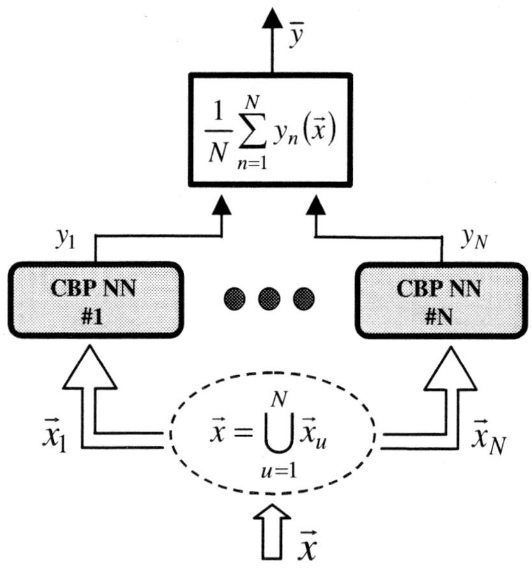

Fig. 2 Ensemble structure for the CBP-based estimator.

the distortion between the actual NN output and the expected reference score on a sample of training patterns. The cost is expressed as:

$$E = \frac{1}{n_o n_p} \sum_{l=1}^{n_p} \sum_{k=1}^{n_o} \left(t_k^{(l)} - y_k^{(l)} \right)^2 , \quad (8)$$

where n_p is the number of training patterns, and t_k are the desired training outputs. In the present application, k=1 and the expected output is given by the quality assessment (score) measured experimentally from a human panel.

3.2. Using Ensembles of CBP Networks

A neural network designed and trained as described in Section 3.1 operates as an estimator whose predictions are always subject to some error, due to statistical fluctuations in the empirical sample drawn from subjective tests. A typical approach to increasing the robustness of the system behavior and the reliability of the neural stage is to use an "ensemble" of different estimators [10] trained on the same problem instead of a single network.

To this purpose, it is known that the variance of an estimator can be reduced by simply taking its average over many independent realizations. The parallel use of several networks like that shown in Fig. 2 can help reduce the variance of the overall quality assessment; in principle, an ensemble of N statistically independent elements can ideally reduce the estimation variance to $\overline{\sigma}^2 = \sigma^2 / N$.

In the present application, the ensemble of estimators is obtained by dividing the data space into various subspaces, then training a specialized neural network for each space region. The overall estimate still results from averaging the outcomes of individual estimators. As a result, the feature vector, $\vec{x}$, associated with an input image is split into N lower-dimensionality subvectors, which form the training sets for each network in the ensemble (Fig. 2)). The rationale for this methodology is that the various data-space partitions contribute to the global estimation task in a coordinated but (ideally) independent, fashion.

4. EXPERIMENTAL RESULTS

The system for image quality assessment has been tested by using a database of images available at the LIVE Quality Assessment Database [11]. The test set included 175 test images generated by compressing (with JPEG) with different compression ratios 29 original color images.

Each image in the LIVE database is followed by the scores given to it by the observers that took part in a subjective quality experiment; their perception of quality is provided on a linear scale ranging from 1 ("bad") to 100 ("excellent"). In the present experiment, the subjective scores have been rescaled in the interval [-1,1]. Then, the scores of each image have been averaged in order to obtain a Mean Opinion Score (MOS) of the image.

4.1. Quality-Assessment Results

The feature-selection procedure described in Section 2 has been used to define the input vector for the neural-network system. It resulted in a set of four descriptive features: *abs_value*, *covariance*, *contrast*, *difference_entropy*. All these features were derived from the co-occurrence matrix, hence, they depend on two parameters, i.e. r and ω. A fixed value of r=2 was chosen, which means that the co-occurrence matrix was always computed within a neighborhood radius of two pixels. The angular orientation parameter ω instead, varied in the four principal directions, i.e. 0°, 45°, 90°, 135°. As a result, the design of the ensemble strategy led to N=4 different neural networks, mainly because this choice made it possible naturally to devote each ensemble element to a specific angular orientation of the co-occurrence matrix. As described in (4) the input-space dimension for each CBP network amounted to $d = n_y \times 2 = 8$.

The effectiveness of the neural quality-assessment system was measured by using a conventional cross-validation approach: the available sample was randomly divided into a training set and a test set, including 125 and 50 images, respectively. During the quality-assessment system set-up and training process only the training set was used, whereas the test set was applied only to predict the system generalization ability. For each ensemble element the number of neurons in the hidden layer was empirically set to N_h=7.

The scatter plot in Fig. 3 shows the results obtained for the test set, with the estimated objective quality as the x axis and the subjective scores as the y axis. The proposed system achieved an averaged prediction error $\hat{\mu}_{err} = 0.03$ ($\hat{\sigma}_{err}$ =0.14), an averaged value of the absolute prediction error $\hat{\mu}_{|err|} = 0.11$ ($\hat{\sigma}_{|err|}$ =0.09) and a Pearson's correlation coefficient ρ=0.95. The effectiveness of the proposed model is confirmed by the Root Mean Squared error, which took the value 0.14.

5. CONCLUSIONS

The paper presented a no-reference system for objective assessment of quality of JPEG compressed image. Objective features are extracted from the image and processed by an ensemble of CBP networks that yields as output the correspondent quality estimate. The objective metric is chosen through a feature-selection procedure based on statistical analysis, thus avoiding a priori hypotheses on the significance of the features.

The major result of the proposed model is the possibility of assessing human perception consistently by using a Circular Back-Propagation NN. Experimental evidence confirmed that the system provides a satisfactory approximation of the actual subjective scores.

APPENDIX

The set $\Phi = \{f_h ; h = 1,..,n_f\}$ of pixel-based objective features includes two different kinds of measures:

- **Features derived from the first-order histogram**: measures that are formulated as quantitative shape descriptors of the first-order histogram [8] derived from the luminance levels of a block *b*: *mean*, *standard deviation*, *skewness*, *kurtosis*, *energy* and *entropy*.
- **Features derived from the co-occurrence matrix**: texture is concerned with the spatial distribution and spatial dependence among gray tones in a local area. Co-occurrence matrices (also named "second-order histograms") define the joint probability distribution of pairs of pixel as a function of 1) the angular relationship between the neighboring cells (ω), and 2) the distance between the neighboring cells (r). This set of measures includes features [8] characterizing the co-occurrence matrix derived from the luminance levels of a block *b*: *autocorrelation*, *covariance*, *inertia*, *absolute value*, *inverse diff.*, *energy*, *contrast*, *diff. variance*, *diff. entropy*.

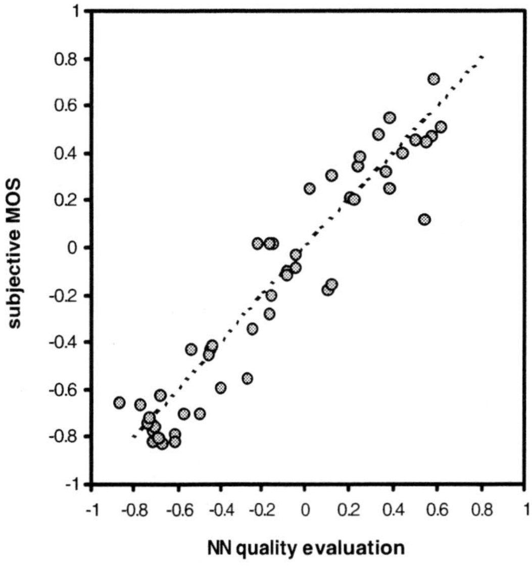

Fig. 3 Test results obtained for the ensemble estimator

REFERENCES

[1] "Methodology for the subjective assessment of the quality of television pictures," ITU, Geneva, Switzerland, ITU-R BT.500.

[2] T.N. Pappas and R.J. Safranek, "Perceptual Criteria for Image Quality Evaluation," in *Handbook of Image and Video Processing*, A. Bovik, ed. San Diego: Academic Press, 2000.

[3] M. Miyahara, K. Kotani, and V. R. Algazi, "Objective picture quality scale (PQS) for image coding," *IEEE Trans. on Communications*, vol. 46, no. 9, pp. 1215-25, Sept. 1998.

[4] Z. Wang and A. C. Bovik, "A universal image quality index," *IEEE Signal Processing Letters*, vol. 9, no. 3, pp. 81-4, March 2002.

[5] M. Jung, D. Léger, and M. Gazalet, "Univariant assessment of the quality of images," *Journal of Electronic Imaging*, vol. 11, no. 3, pp. 354-64, July 2002.

[6] Z. Wang, H. R. Sheikh, and A. C. Bovik, "No-reference perceptual quality assessment of JPEG compressed images," in *Proc. IEEE ICIP*, vol. 1, pp. 477-80, Sept. 2002.

[7] S. Ridella, S. Rovetta, and R. Zunino, "Circular back-propagation networks for classification," *IEEE Trans. on Neural Networks*, vol. 8, no. 1, pp. 84-97, Jan. 1997.

[8] P. Gastaldo, R. Zunino, E. Vicario, and I. Heynderickx, "CBP neural network for objective assessment of image quality," in *Proc. IJCNN*, pp. 194-99, July 2003.

[9] J. Hertz, A. Krogh, and R. Palmer, *Introduction to Neural Computation*. Englewood Cliffs, NJ: Prentice-Hall, 1992.

[10] M. Perrone, *Improving regression estimates: Averaging methods for variance reduction with extension to general convex measure optimization*, Ph.D. dissertation, Phys. Dep., Brown Univ., Providence, RI, 1993.

[11] H. R. Sheikh, Z. Wang, L. Cormack and A. C. Bovik, "LIVE Image Quality Assessment Database," at http://live.ece.utexas.edu/research/quality.

A NEURAL SYSTEM FOR RADIATION DISCRIMINATION IN NUCLEAR FUSION APPLICATIONS

Basilio Esposito[*], *Luigi Fortuna*[§], *IEEE Fellow*, *Alessandro Rizzo*[+], *IEEE Member*

[*]Associazione Euratom-ENEA sulla Fusione, C.R. Frascati
C.P. 65, I-00044 Frascati (Roma) Italy – email esposito@frascati.enea.it

Dipartimento di Ingegneria Elettrica, Elettronica e dei Sistemi – Università degli Studi di Catania
Viale A. Doria 6 – 95125 Catania (Italy) – email lfortuna@diees.unict.it

[+]DEE – Dipartimento di Elettrotecnica ed Elettronica – Politecnico di Bari
Via Re David 200 – 70125 Bari – email rizzo@deemail.poliba.it

ABSTRACT

This paper presents a novel approach to discriminate between neutrons and γ-rays in nuclear fusion applications, based on a neural network able to analyse the shape of light pulses produced by these ionizing particles in an organic liquid scintillator. Such approach is particularly promising especially for the possibility of classifying correctly (either as neutrons or as γ-rays) fast superimposed events (pile-ups). Satisfactory experimental results were obtained at the Frascati Tokamak Upgrade, ENEA-Frascati, Italy.

1. INTRODUCTION

An example of a nuclear fusion reaction is given by

$$D + T => {}^{4}He + n \tag{1}$$

in which two light hysotopes of hydrogen (Deuterium and Tritium) interact and release an alpha particle (Helium-4 nucleus) and a high energy neutron. The DT reaction (1) has the highest reaction rate at the plasma temperatures which are currently achievable in tokamak experiments; moreover, it has a very high energy release. These properties make it the most suitable reaction to use in a man-made fusion reactor. The alpha particle carries 3.5 MeV energy, whereas the neutron carries 14.1 MeV energy. The neutron escapes from the plasma (it has no charge and is not confined) and can be trapped in a surrounding "blanket" structure, where another reaction can be used to "convert" the neutrons back into tritium fuel. The trapped neutron energy can be exploited to generate heat and produce energy in a conventional way.

The measurement of the total number of neutrons emitted in a plasma discharge provides indication of the produced fusion power. Combined with the measure of the electron density, it is also possible to obtain the ion temperature [1]. Moreover, by measuring the neutron pulse height distribution, the energy distribution of the incoming neutrons can be determined.

One of the most used techniques for the detection of ionizing particles (both for counting and spectroscopy) is the measurement of the scintillation light which these particle produce by interaction with certain materials called *scintillators*[2]. In fusion applications, both γ-rays and neutrons are emitted in an experiment by the same source of radiation (plasma). As only neutrons contribute to the production of the fusion energy, it is very important to discriminate between these two kinds of particles in order to compute correctly the fusion power.

Neutrons and γ-rays interacting with organic scintillator detectors produce slightly different light pulses: such property is effectively used to discriminate between neutron and γ-ray events by means of a technique called pulse shape discrimination (PSD). Typical n/γ discrimination techniques are implemented in an analog fashion [3,4]; recently, digital pulse shape discrimination (DPSD) has also been developed [5,6].

In the last years, soft-computing techniques have been successfully applied to data analysis and fault diagnosis in nuclear fusion applications [7,8]. A novel approach based on DPSD and neural networks has been investigated and is presented in this paper for the analysis of pulse shapes in organic scintillators. Experimental results, obtained at the Frascati Tokamak Upgrade (FTU, ENEA-Frascati, Italy), are fully satisfactory and confirm the suitability of the approach.

2. SCINTILLATION DETECTOR PRINCIPLES

As stated above, ionizing radiations can be revealed by measuring the light emitted when they travel through certain types of materials. The light emitted by the material is captured and transformed in a voltage or current signal by a photomultiplier and some conditioning circuitry. The most common materials employed in scintillators are inorganic alkali halide crystals (sodium iodide), and organic-based liquids and plastics. An ideal scintillation material should be able of converting in a linear fashion the kinetic energy of particles into detectable light. A good scintillation material should also exhibit peculiar optical properties; in particular the decay time of the induced luminescence should be short enough so that fast events (particle impacts) can be revealed [2]. The typical response of a scintillator after interaction with a ionizing particle is illustrated in Fig. 1.

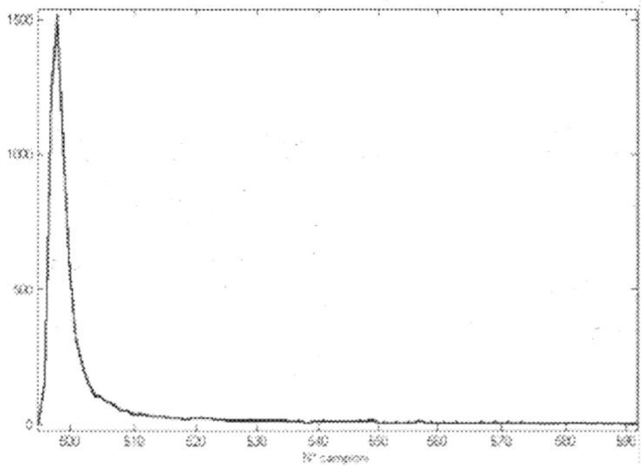

Fig. 1 – Typical response of a scintillator to a charged particle.

The response consists of a very fast peak, followed by an exponential decay. The amplitude of the peak is a function of the kinetic energy of the particle, and is caused by the *fluorescence* (a prompt emission of visible radiation emitted from an excited substance) phenomenon. The decay, on the other hand, is caused by the *phosphorescence* (light emission with a longer wavelength and a longer charachteristic time than fluorescence), and *delayed fluorescence* (same wavelength but much slower time constant than fluorescence) phenomena. It is clear that the slow exponential tail in the time response can prevent the detection of fast, subsequent events. When a second particle hits the scintillator liquid before the exponential decay corresponding to the first particle is completed, the two responses overlap. Such an event is called *pile-up*. An example of a pile-up is illustrated in Fig. 2.

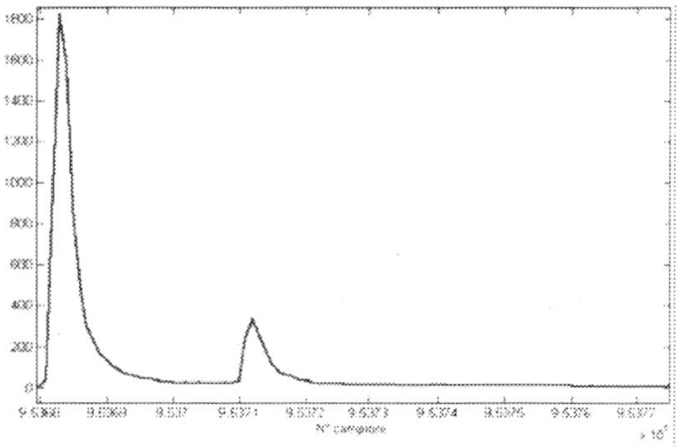

Fig. 2 – A pile-up; the typical response of the scintillator to two overlapping events.

In absence of pile-ups, the time response of a scintillator can be assumed as:

$$v(t) = V_0(e^{-t/\tau_s} - e^{-t/\tau_f}) \qquad (2)$$

where V_0 is proportional to the kinetic energy of the particle, and the two time constants τ_f and τ_s regulate the exponential decay due to the (fast) fluorescence and (slow) phosphorescence phenomena. Typical values of the time constants are $\tau_f \sim$ 3-4 ns and τ_s up to 270 ns. The slow component of the decay would be useless, except for a very important property: *the fraction of light appearing on the slow decay is representative of the nature of the particle.* It is therefore possible to discriminate between different particles of different nature, even if they have the same kinetic energy. This process is called pulse shape discrimination (PSD). In nuclear fusion applications, it is widely used to eliminate γ-ray induced events when scintillators are used as neutron detectors. PSD can be carried out with analog or, more recently, also with digital equipment. A DPSD system has been already developed at ENEA-Frascati Labs [5]. It is is based on the charge comparison method [4] applied to pulses acquired digitally: each pulse is integrated for two different time intervals (Δt_F and Δt_S), corresponding to the fast and slow scintillation components; the value of the ratio Q_S/Q_F (Q_S and Q_F are the charges corresponding to Δt_S and Δt_F) provides, independently on the amplitude of the pulse, the indication whether a neutron or a γ-ray event has taken place. Although this technique is rather accurate and fast, it does not allow so far the analysis of pile-up events. In the following, the neural network approach is presented as another method (alternative to charge comparison) for DPSD analysis which, in addition to neutron/γ-ray separation, also provides the classification of pile-up events.

3. NEURAL-NETWORKS-BASED DPSD

Artificial Neural Networks (ANN) are an effective tool for pattern recognition. In this section the design and implementation of a DPSD system based on ANN is described.

At first, experimental data have been pre-processed. The data file has been cut into separate events, events normalized (as we are interested in the nature of the particle, not in its energy), and the decay phase isolated. A pre-processed event is illustrated in Fig. 3.

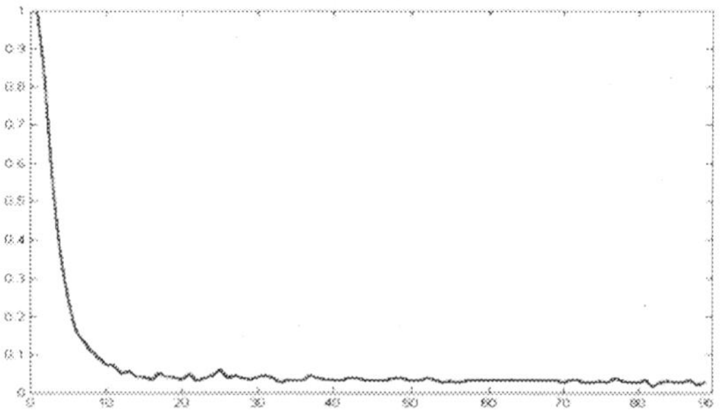

Fig. 3 – A preprocessed event.

The main difficulty of this work was to build learning and testing data set. While in fact it is possible to generate sets of γ–rays alone (for example, using a ^{60}Co source), neutron sources always emit also γ–rays. Consequently, a ^{60}Co source has been adopted as a γ–rays calibration source to reveal the main features of the response of the scintillator to these particles. The following parametric model has been adopted for the normalized decay phase:

$$v(t) = e^{-t/\tau_s} + Be^{-t/\tau_f} \tag{3}$$

This differs from model (2) for the presence of the weighting coefficient B. A wide number of events have been fitted with the MATLAB® Curve Fitting Toolbox to find an estimate for the parameters B, τ_f and τ_s for the γ–rays source. Subsequently, the same procedure has been applied to a source of γ–rays and neutrons. By comparing the experimental parameter distributions, it can be noticed that the most suitable parameter to discriminate between the two particles is the coefficient B rather than the slow time constant. Fig. 4 illustrates a comparison between the experimental distribution of the coefficient B for γ–rays only, and for a mixed source. It can be clearly observed that a separation between two classes is found in the case of a mixed source, corresponding to γ–rays and neutron events.

After the curve fitting phase, patterns for learning and testing have been selected. As a huge number of experiments was available, only events to which a class can be assigned without uncertainty have been selected (i.e. the events with a B value located around the peak of the distribution bell). Subsequently, pile-up events have been added to the pattern sets, as belonging to a third class to be detected.

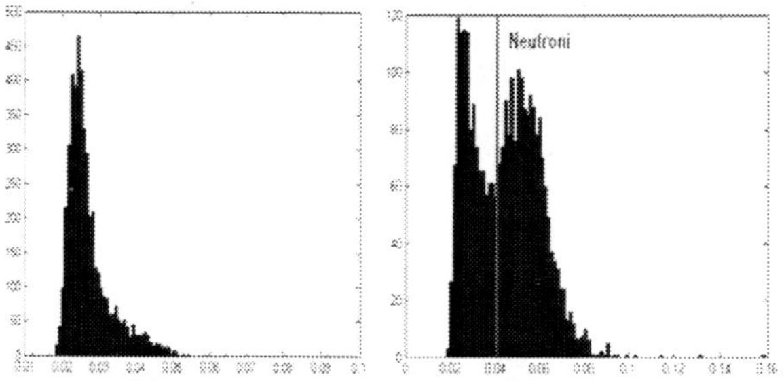

(a) (b)

Fig. 4 – Experimental distribution of parameter B, for a γ–ray (a) and a mixed neutron/γ–ray source (b), respectively.

A Multi Layer Perceptron (MLP) with 89 inputs and 1 output has been trained. After a trial and error phase, one hidden layer with five neurons has been selected. The network receives in input the time decay sequence (like that illustrated in Fig. 3), and returns a value which can be 1 for γ–rays, 2 for neutrons, 3 for pile-ups. Learning and testing performance is reported in the following table. The performance of the network is excellent: an on-line implementation has been carried out and it is described in the following section.

Total 2000	Classified as γ	Classified as n	Classified as P.U.
Emitted γ	1003 (100%)	0	0
Emitted n	0	873 (100%)	0
Emitted P.U.	0	0	124 (100%)

(a)

Total 470	Classified as γ	Classified as n	Classified as P.U.
Emitted γ	231 (99.1%)	2 (0.9%)	0
Emitted n	2 (0.9%)	212 (99.1%)	0
Emitted P.U.	0	0	23 (100%)

(b)

Table 1: Learning (a) and testing (b) results achieved by a 89-5-1 MLP for the discrimination of radiating particles.

4. EXPERIMENTAL SETUP AND RESULTS

The data analyzed here have been acquired using a DPSD system (described in detail in [5]) consisting of a NE213 organic scintillator detector coupled to a photomultiplier, whose output pulses are fed to 12-bit 200 MHz transient recorder card with 256 MSamples on-board memory. A mixture of neutron and γ-ray events has been obtained by placing the detector in front of an AmBe neutron source, while standard γ-ray calibration sources (i.e. ^{60}Co) have been used to produce γ-ray only events.

The trained MLP has been implemented in LabView®, and integrated in a DPSD software tool performing the discrimination via the charge comparison method.

Each event is preprocessed through the following phases: it is normalized and its amplitude stored (the amplitude is directly related to the kinetic energy of the particle), the decay phase is isolated and classified by the MLP. After the classification of the particles, two separate pulse height spectra are created. As the ANN-based tool is able to discriminate pile-up events, a third spectrum is also created. Although such pile-up spectrum has no physical meaning, these pile-ups can be subsequently processed for the discrimination of the constituent events (neutron and γ-ray): this will be the subject of future work.

Fig. 5 illustrates a comparison between the pulse height spectra performed via the charge comparison method and the ANN-based method. The spectra obtained using the two different methods are qualitatively equivalent, confirming the suitability of the ANN-based approach also in an on-line implementation. The third spectrum of the second row is the pile-up spectrum, performed only by the ANN-based tool. The performance of pile-up discrimination in the on-line version has been checked manually, and results are very satisfactory.

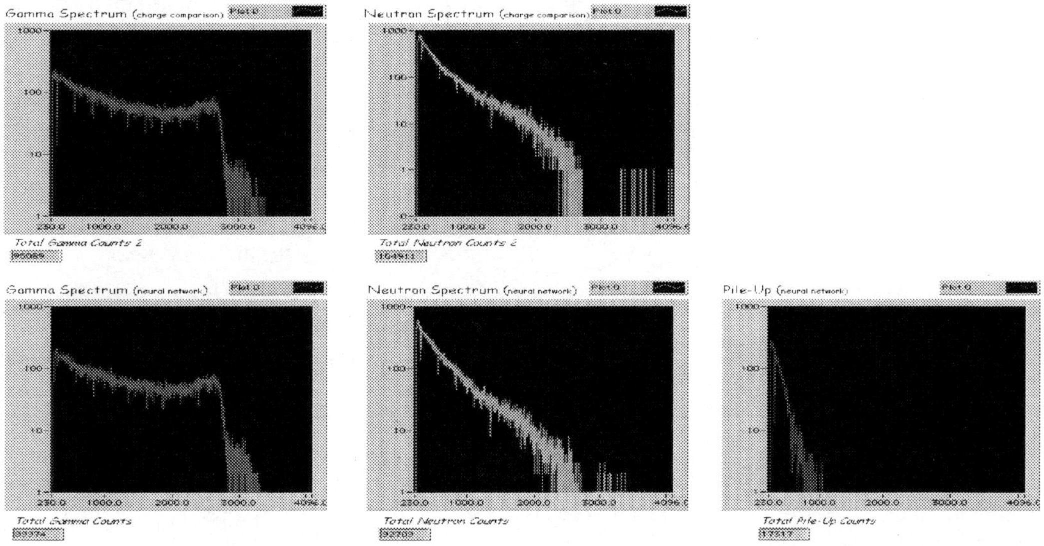

Fig. 5 – Comparison of spectra in the on-line implementation. First row refers to charge comparison method, second row to MLP-based method. Columns report γ-rays, neutron and pile-up spectra, from left to right.

5. CONCLUSIONS

This work presents a novel approach for the digital discrimination of ionizing particles in liquid scintillators. Neutron/γ–ray particle discrimination through the analysis of the pulse shape is extremely useful in nuclear fusion applications, in which neutron counting and spectroscopy are of fundamental importance. A ANN-based tool has been realized and implemented on-line in LabView® in a neutron/γ–ray digital pulse shape discrimination system (DPSD); on-line results are very satisfactory. At present pile-up events are also recognized and classified separately: further work will be devoted to the separation of each pile-up in its constituent events. This will allow to perform DPSD also at very high counting rates (> MHz).

6. REFERENCES

[1] J. Wesson, *Tokamaks*, Clarendon Press, 1987.

[2] G. F. Knoll, *Radiation Detection and Measurement,* Second Edition, JohnWiley and Sons, (1989).

[3] J. M. Adams and G. White, "A versatile pulse shape discriminator for charged particle separation and its application to fast neutron time-of-flight spectroscopy", *Nucl. Instr. and Methods*, **156** (1978) 459.

[4] J. Kasagi, T. Murakami, T. Inamura, "Use of charge integrating ADCs with pulse shape discriminators for neutron-gamma discrimination", *Nucl. Instr. and Method in Physics Research*, **A236** (1985) pp. 426-427.

[5] B. Esposito, Y. Kaschuck, A. Rizzo, L. Bertalot, A. Pensa, "Digital pulse shape discrimination in organic scintillators for fusion applications", *Nucl. Instr. and Method in Physics Research A,* Vol. 518 No.1-2, pp. 626-628.

[6] N.V. Kornilov et al. "Neutron Spectroscopy with Fast Waveform Digitizer", *Nucl. Instr. and Method in Physics Research,* **A497** (2003), pp. 467-478.

[7] A. Rizzo, M.G. Xibilia, "An Innovative Intelligent System for Sensor Validation in Tokamak Machines", *IEEE Trans. On Control Systems Technology*, Vol. 10, No. 3, May 2002.

[8] G. Buceti, L. Fortuna, A. Rizzo, M.G. Xibilia, "Automatic Validation of the 5-Channel DCN Interferometer in ENEA-FTU based on Soft Computing Techniques", *Fusion Engineering and Design, Elsevier Science*, Vol. 60, No. 3, pp.381-387, 2002.

MIXED ANALOG-DIGITAL IMAGE PROCESSING CIRCUIT BASED ON HAMMING ARTIFICIAL NEURAL NETWORK ARCHITECTURE

Stéphane Badel, Alexandre Schmid and Yusuf Leblebici

Swiss Federal Institute of Technology, Microelectronic Systems Laboratory
CH – 1015 Lausanne, Switzerland

ABSTRACT

A versatile integrated circuit based on the Hamming artificial neural network (ANN) architecture is presented. The circuit operation relies on capacitive processing of sum-of-products terms, complemented with digital post-processing allowing various complex functions to be processed on chip, with a minimal hardware overhead. VLSI realization and measurements are discussed.

1. INTRODUCTION

Over the past decades, artificial neural network (ANN) algorithms have proven to be efficient tools to solve some computationally intensive signal processing problems. The complexity of artificial neural network algorithms has opened the door to their hardware implementation with the main target of accelerating all computations involved.

The hardware Hamming ANN implementation proposed in this paper addresses the issues related to severe limitations in terms of circuit resolution, temporary memory size and their access schemes, realization of high interconnect density by the implementation of analog atomic elements, each processing some specific neural functions, and to be repeated into a regular structure forming a high-performance processing unit. Digital post processing is used in a hardware-friendly realization in order to enhance the functionality of the proposed device.

The Hamming ANN discussed in this paper is based on a compact implementation this approach, with the goal of constructing a building block for high-speed and low-power image processing applications.

2. A HAMMING ARTIFICIAL NEURAL NETWORK

A Hamming ANN is a two layer feed-forward neural network which has the ability of classifying input patterns, based on the criterion of the Hamming distance between previously stored patterns, and the actual input vectors [1]. The first layer – called the quantifier network - is composed of a number of neurons, which perform the Hamming distance computation. The second layer – called the discrimination layer – is traditionally composed of a feed-forward network, which performs the winner-take-all (WTA) operation, i.e. selects the first-layer neuron of smallest Hamming distance as the winner. Pattern classification applications based on the charge-based operation of the Hamming VLSI circuit have been previously demonstrated [2]-[3], where the quantifier networks is composed of capacitive-threshold logic (CTL) gates [4], and the discrimination network consists of an n-input version of a sense-amplifier, all processing being thus performed in the analog domain.

The quantification network consisting of CTL gates is depicted in Figure 1, where a number of capacitances make the bridge between a common node called row and the set of digital inputs and pre-charge circuitry. A perturbation column is added to the regular CTL gate as an analog input to modify the row voltage state [3]. The comparator and buffer stages build the decision network, which provides the outside world with a binary decision.

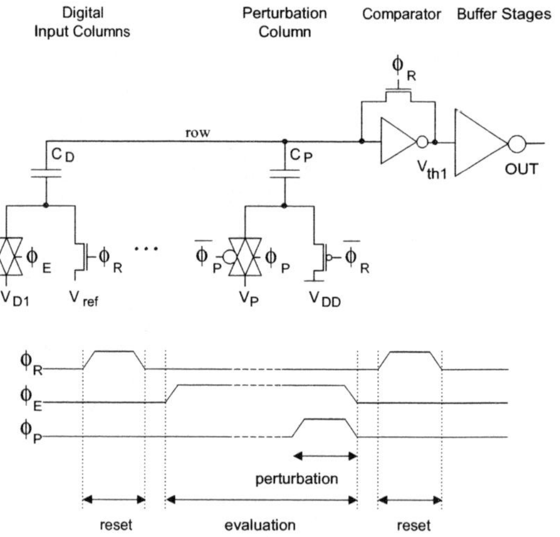

Fig. 1: The adapted CTL gate.

The circuit operation is based on a three cycles scheme, throughout which charge conservation applies to the row node. The quantification operation is realized with a very simple two phase non-overlapping clock scheme consisting of a pre-charge (Φ_R) and evaluation phase (Φ_E). All nodes are imposed a voltage during the pre-charge phase. The row voltage is set as the comparator circuit threshold voltage V_{th1} while the capacitances other node voltages are imposed as reference voltages set to V_{DD} or GND. The amount of charge transferred to the row is given in Equation 1. All nodes are set back into high impedance after completion of the pre-charge phase. The subsequent evaluation phase starts then, throughout which several vectors may be applied without performing any extra reset. The charge on the row node is considered as constant, the time constant of the leakage parasitic process being significantly larger than that of the system operation. However, the charge on the row capacitors nodes is affected as given in Equation 2. Assuming the equality of these two charges, the row voltage is forced to vary to ΔV_{row} as given in Equation 3. Eventually, the comparator circuit restores a binary voltage depending on the sign of the row voltage variation as given in Equation 4.

$$Q_{reset} = (V_{th1} - V_{ref})C_D^{TOT} + (V_{th1} - V_{DD})C_P \quad \text{with} \quad C_D^{TOT} = \sum_n C_{D_n} \tag{1}$$

$$Q_{eval} = \sum_n (V_{row} - V_{D_n})C_{D_n} - (V_{row} - V_P)C_P \tag{2}$$

$$\Delta V_{row} = \frac{\sum_n (V_{D_n} - V_{ref})C_D^{TOT} + (V_P - V_{DD})C_P}{C_D^{TOT} - C_P} \tag{3}$$

$$\begin{cases} \Delta V_{row} < 0 \Rightarrow V_{OUT} = V_{DD} \\ \Delta V_{row} > 0 \Rightarrow V_{OUT} = GND \\ \Delta V_{row} = 0 \Rightarrow \text{limit of the circuit precision} \end{cases} \tag{4}$$

A detailed overview of the perturbation process is given in [5] to which the reader is referred to for further details.

3. OPERATION MODES

The proposed circuit architecture is capable of implementing a number of distinct operation modes, which allows the synthesis of multiple functions. Each mode corresponds to a set of circuit operation characteristics that depend on the targeted application, and dictates a particular hardware configuration.

The circuit can be operated to detect relative or absolute Hamming distances. The neuron configurations for these two modes are explained in detail in [6].

In order to operate on relative distances, only the capacitances representing a stored "Logic 1" value (to be used for comparison) are integrated. Thus, the operation actually implemented is an AND function between the input and the pattern stored in the capacitances. Note that the operation of this kind of neuron results in a distance computation that is not absolute, i.e. it must be compared to the respective results of other neuron computations in order to produce meaningful information. Using a unit capacitance for each input, while processing a digital Hamming distance operation prior to the capacitive stage, allows the computation of the absolute Hamming distance. The logic operation to be performed is an XOR function between the input and stored data. The row voltage after evaluation of this kind of neuron is proportional to the Hamming distance, thus it is possible to use it either way, with or without any WTA unit.

The second proposed working mode is related to the kind of perturbation signal that is applied to the perturbation columns, and hence affect the row voltages. In this paper we consider the perturbation signal to be equal for each neuron, which is not restrictive and can be easily modified for further applications. Both a pulse and a ramp perturbation signal prove to be interesting candidates, targeting very different potential applications. Their respective effects on row voltage and/or output voltage can be seen on Figure 2 as SPICE simulations, and Figure 8-b as circuit measurements.

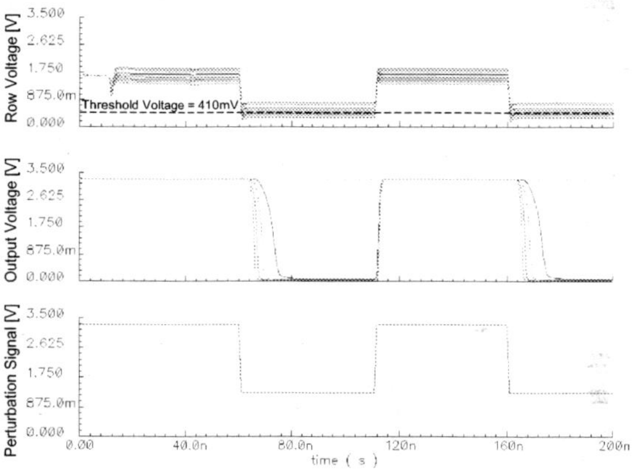

Fig. 2: Pulse perturbation of the neuron row voltage.

4. DIGITAL POST-PROCESSING AND APPLICATIONS

A number of functions can be addressed using the proposed circuit techniques. Post-processing can be added in order to enhance the possible functionality. In the following we restrict the range of applications to digital post-processing based systems.

Ranked order distance measurement based on the Hamming distance computation is depicted in Figure 3.

The distance computation is performed as described in the previous Section (3) by a Hamming ANN core. The output information triggers a pulse generator, which in turn allows a counter to switch into its next state. The rank is given as the value of the counter at the time the Hamming ANN output of a neuron switches. This information is stored into an array latches, shown in Figure 3.

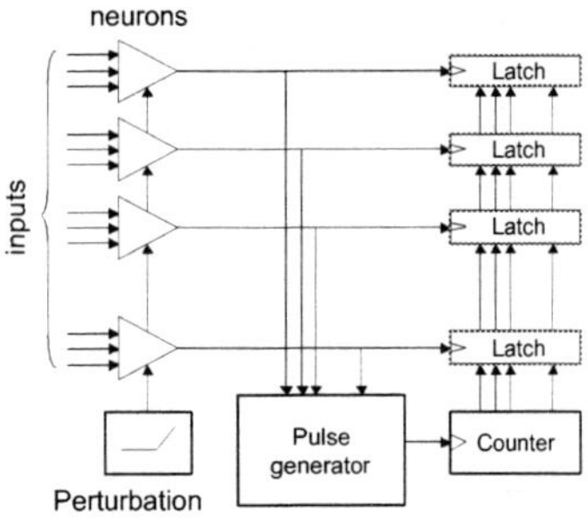

Fig. 3: Ranked order distance computation.

True Hamming distance can be extracted as the latched value of a counter sampled at the switching time of a neuron as depicted in Figure 4.

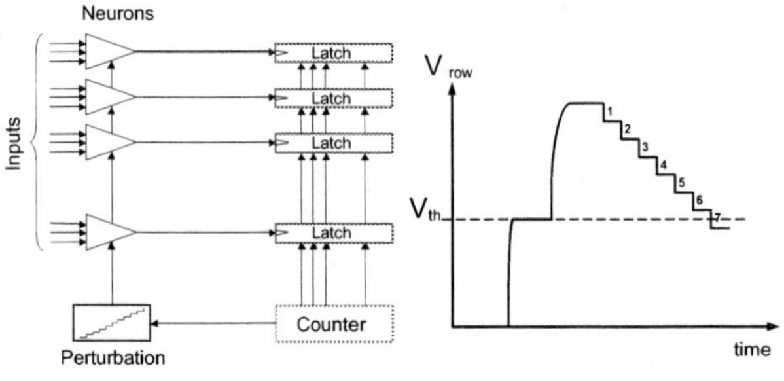

Fig. 4: True Hamming distance extraction. a) System description and b) example row voltage in the case of a Hamming distance equal to seven.

The perturbation signal is chosen as a pseudo-ramp signal, which is made of calibrated steps. Each of these steps is related to the unit voltage quantum required to discriminate to neurons with unit distance. The synchronization of the counter with the perturbation generation circuit guarantees the extraction of the correct value.

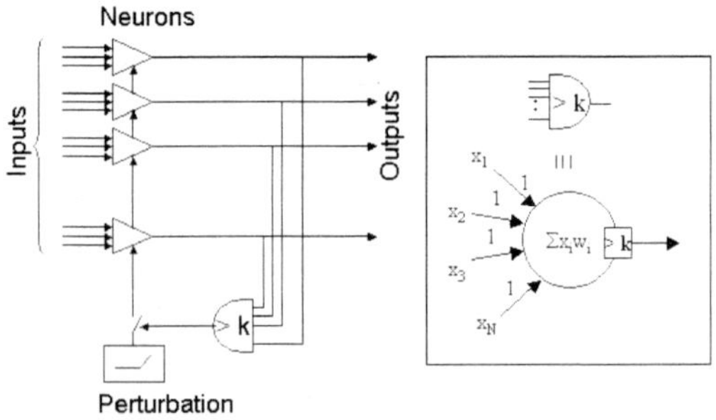

Fig. 5: a) Circuit-level k-winner-take-all description, and b) majority gate.

K-winner-take-all and k-loser-take-all circuits can be synthesized using a calibrated pulse system as described previously. A ramp perturbation signal can also be applied, in which case one additional majority gate has to be added as post-processing element. A CTL gate is proposed in Figure 5, where the output of the gate controls the end of the process. The winner neurons can be read out of the Hamming core output lines directly.

5. VLSI REALIZATION

An integrated circuit was realized in CMOS 0.35 μm double-polysilicon technology, including a complete 16x16 array Hamming ANN core. The circuit layout of one individual cell is depicted in Figure 6. Two CMOS memory bits have been included into each cell in order to increase the versatility of the circuit. Each of these memory points can be replaced by the application specific connection to the desired sensor input; as such the Hamming core can be modified to accommodate on-chip light sensors in order to construct an intelligent CMOS sensor with early processing of Hamming-based algorithms in the analog domain.

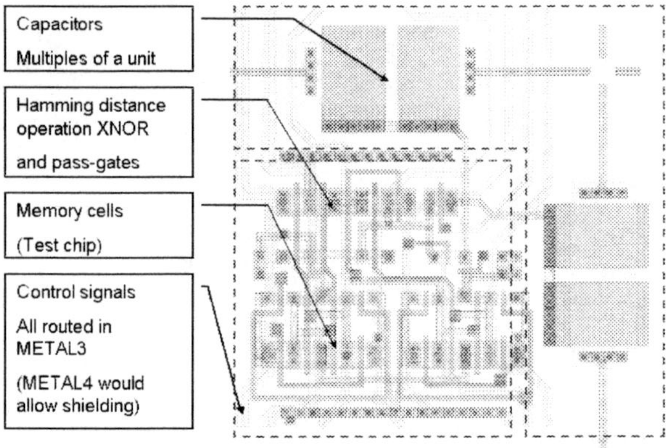

Fig. 6: Layout of one individual ANN cell.

The complete core consists of an array arrangement of the individual cells as depicted on Figure 7. The calibrated perturbation capacitors have been placed on two sides of the core, whereas the offset cells and the output driving circuitry has been placed on two opposed sides of the core, thus resulting in a (600x600) μm^2 circuit. The number of signals routed to I/O pads is 80, allowing easy direct and parallel access to most of the signals to be tested. The minimal number of pads allowing parallel access is only five more than the number of data I/Os.

The circuit was measured using a precision ramp generator and a high-speed oscilloscope. Figure 8-a shows the ramp perturbation test applied to sixteen neurons. Each neuron has a different Hamming distance to the stored pattern, ranging from zero to fifteen. Full operability in this mode is achieved, which is witnessed by the successive switching of all neurons.

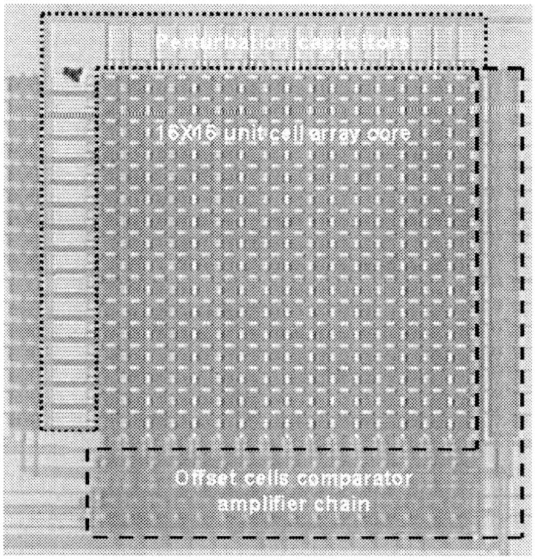

Fig. 7: Microphotograph of the complete Hamming ANN.

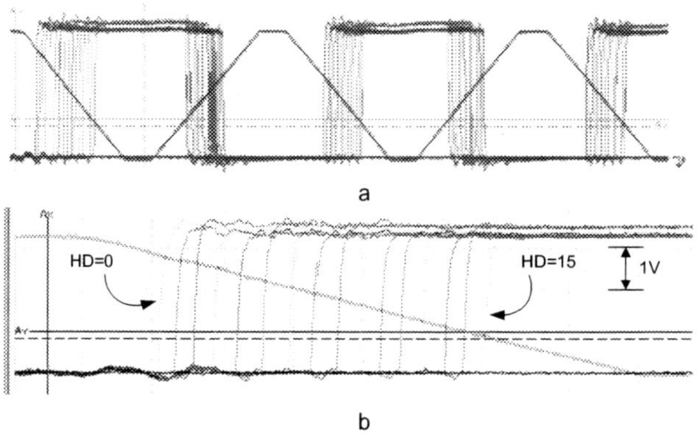

Fig. 8: Oscilloscope waveforms of the ramp test using persistence mode – three channels for each measurement. a) Several cycles, and b) detail of the measurement.

Unexpected coupling effects could be identified to cause a deviation of the absolute perturbation switching voltage from its expected value, as can be seen on Figure 9, where the dependence of the extracted Hamming distance on the stored pattern is evidenced to cause a distance error varying between zero and four, thus reducing the absolute circuit accuracy in pulse-perturbation mode. Two extreme cases are considered, where the reference image loaded into the network consists of "Logic 1" and "Logic 0."

Figure 10 shows the response linearity. Monotonic operation is confirmed by the lower limit of the acceptance interval being above 0V. In order to guarantee predictable step size of the perturbation, the upper limit of the acceptance interval should remain below a value equal to twice the lower limit. This is generally the case; however potential errors may result from extreme values, attributed to measurement artifacts.

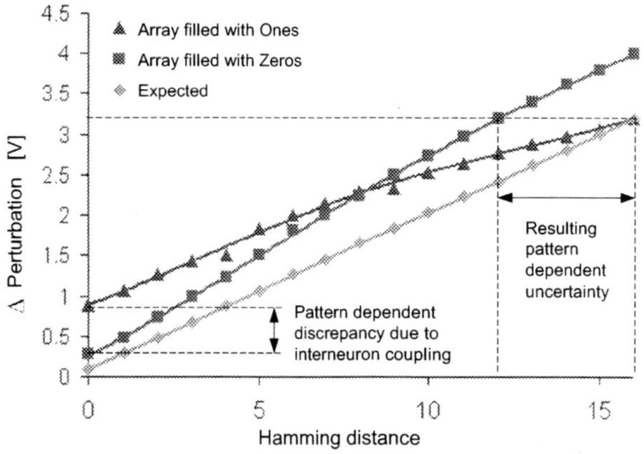

Fig. 9: Measured Hamming distance vs. perturbation voltage: expected, and two extreme patterns.

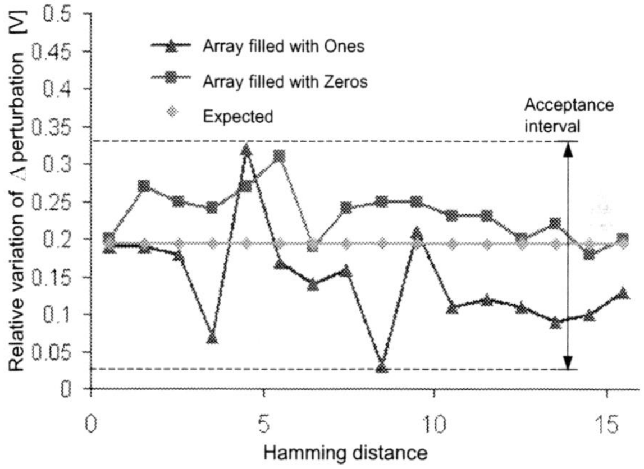

Fig. 10: Measured nonlinearity in the cases of Fig. 9.

The circuit fully qualifies for use with a ramp, or stepped ramp perturbation. A careful redesign is necessary for use in high-accuracy pulse-perturbation mode. Closed-loop operation and digital post-processing are available solutions to compensate for the observed excessive coupling.

REFERENCES

[1] Richard P. Lippmann, "An Introduction to Computing with Neural Nets," IEEE ASSP Magazine, April 1987.

[2] U. Cilingiroglu, "A Charge-Based Neural Hamming Classifier," IEEE Journal of Solid-State Circuits, Vol. 28, No. 1, January 1993.

[3] Alexandre Schmid, Yusuf Leblebici and Daniel Mlynek, "Hardware Realization of a Hamming Neural Network with On-Chip Learning," Proceedings of the 1998 IEEE International Symposium on Circuits and Systems ISCAS'98, Monterey, CA, 1998.

[4] H. Özdemir, A. Kepkep, B. Pamir, Y. Leblebici, U. Cilingiroglu, "A Capacitive Threshold-Logic Gate," IEEE Journal of Solid-State Circuits, Vol. 31, No. 8, August 1996.

[5] S. Badel, A. Schmid, Y. Leblebici, "VLSI Realization of a Two-Dimensional Hamming Distance Comparator ANN for Image Processing Applications," ESANN'03, Bruges, Belgium, April 2003.

[6] S. Badel, A. Schmid, Y. Leblebici, "A VLSI Artificial Neural Network with k-Winner-Take-All and k-Loser-Take-All Capability," IJCNN'03, Portland OR, July 2003.

On Synthetic AER Generation

Alejandro Linares-Barranco[1], Gabriel Jimenez-Moreno[1], Antón Civit-Ballcels[1], and Bernabé Linares-Barranco[2]

[1]Arquitectura y Tecnología de Computadores. ETSI Informática, Av. Reina Mercedes s/n, 41012 Sevilla, SPAIN. Phone: 95-455-6145, Fax: 95-455-6449, E-mail: alinares@atc.us.es

[2]Instituto de Microelectrónica de Sevilla, Sevilla, SPAIN.

Abstract

In this paper several software methods for generating synthetic AER streams from images stored in a computer's memory are proposed and evaluated. Evaluation criteria cover execution time, distribution error and how they perform with two receiver cell models. A hardware PCI to AER interface is presented.

1. Introduction

Address-Event-Representation (AER) was proposed in 1991 by Sivilotti [1] to transfer the state of an array of neurons from one chip to another. It uses mixed analog and digital principles and exploits pulse density modulation to code information. The state of the neurons is a continuous time varying analog signal.

Fig. 1 explains the principle behind the AER basics. The Emitter chip contains an array of cells (like, for example, a camera or artificial retina chip) where each pixel shows a continuously varying time dependent state that changes with a slow time constant (in the order of milliseconds). Each cell or pixel includes a local oscillator that generates digital pulses of minimum width (a few nanoseconds). The density of pulses is proportional to the state or intensity of the pixel. Each time a pixel generates a pulse (which is called "event"), it communicates with the array periphery and a digital word representing its code or address is placed on the external inter-chip digital bus (the AER bus). Additional handshaking lines (Acknowledge and Request) are also used to complete the asynchronous communication.

In the receiver chip, the pulses are directed to the pixels or cells whose code or address was on the bus. This way, pixels with the same code or address in the emitter and receiver chips will "see" the same pulse stream. The receiver cell integrates the pulses and reconstructs the original low frequency continuous-time waveform. Pixels that are more active are accessing the bus more frequently than those less active.

Transmitting the pixel addresses allows performing extra operations on the images while they travel from one chip to another. For example, inserting properly coded memories (ie. EEPROM) allows transformation (ie. shifting and rotation) of images. Also, the image transmitted by one chip can be received by many re- ceiver chips in parallel, by properly handling the asyn- chronous communication protocol. The peculiar nature of the AER protocol also allows for very efficient convolution operations within a receiver chip [2].

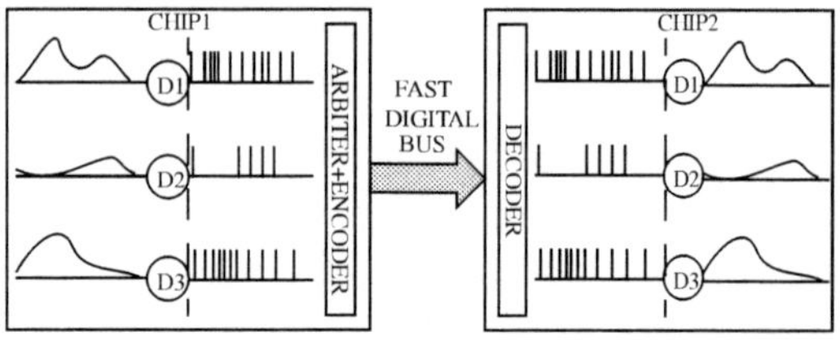

Fig. 1: AER inter-chip communication scheme

There is a growing community of AER protocol users for bio-inspired applications in vision and audition systems, as demonstrated by the success in the last years of the AER group at the Neuromorphic Engineering Workshop series [3]. The goal of this community is to build large multi-chip and multi-layer hierarchically structured systems capable of performing complicated array data processing in real time. The success of such systems will strongly depend on the availability of robust and efficient development and debugging AER-tools [4][5]. One such tool is a computer interface that allows not only reading an AER stream into a computer and displaying it on screen in real-time, but also the opposite: from images available in the computer's memory, generate a synthetic AER stream in a similar manner as would do a dedicated VLSI AER emitter chip [1][6][7].

In Section 2 we review some synthetic AER generation methods and present some improvements over earlier presented ones [4][5]. In Section 3 different methods are evaluated attending to three criteria: execution time, error of distribution and distance between ideal distribution in two kind of receptors, the Boahen integrator [8] and the Mortara integrator [9]. Finally, section 4 presents a hardware interface.

2. Synthetic AER Generation

One can think of many software algorithms to transform a bitmap image (stored in a computer's memory) into an AER stream of pixel addresses [4][5]. In all of them the frequency of appearance of the address of a given pixel must be proportional to the intensity of that pixel. If pixel signal time constant is much slower than inter-event timing, the precise location of the address pulses is not critical. The pulses can be slightly shifted from their nominal positions; the AER receivers will integrate them to recover the original pixel waveform.

Whatever algorithm is used, it will generate a vector of addresses that will be sent to an AER receiver chip via an AER bus. Let us call this vector the "*frame vector*". The *frame vector* has a fixed number of time slots to be filled with event addresses. The number of time slots depends on the time assigned to a frame (for example $T_{frame} = 40ms$) and the time required to transmit a single event (for example $T_{pulse} = 10ns$). If we have an image of NxM pixels and each pixel can have a grey level value from 0 to K, one possibility is to place each pixel address in the *frame vector* as many times as the value of its intensity, and distribute it with equidistant positions. In the worst case (all pixels with maximum value K), the *frame vector* would be filled with *NxMxK* addresses. Note that this number should be less than the total number of time slots in the *frame vector* T_{frame}/T_{pulse}. Depending on the total intensity of the image there will be more or less empty slots in the *frame vector*. Each algorithm would implement a particular way of distributing these address events, and will require a certain time.

A. The Scan method

In this method a frame is scanned many times. For each scan, every time a non-zero pixel is reached its address is put on the *frame vector* in the first available slot, and the pixel value is decremented by one. If a pixel value is zero, a blank slot is left in the *frame vector*. This method is very fast. However, the resulting event distribution is very different from the one an AER retina, for example, would produce. Particularly, the events of pixels with low intensity will appear only at the beginning of the *frame vector*.

B. The Uniform method

In this method, the objective is to distribute equidistantly the events of one pixel along the *frame vector*. The image is scanned pixel by pixel only once. For each pixel, the generated pulses must be distributed at equal distances. As the *frame vector* is getting filled, the algorithm may want to place addresses in slots that are already occupied. This situation is called a *'collision'*. In this case, we propose three solutions:

The *Back-Forward* (Uniform-BF method) solution will put the event in the nearest empty slot of the *frame vector*.

The *Forward* (Uniform-F method) solution will put the event in the following empty slot in the *frame vector*.

And the *Winner-Takes-All* (Uniform-WTA method) solution will put in the collision position of the vector the event that produces a lower error and will ignore the others. The winning event is the one of the pixel with the lowest intensity.

Uniform-BF, Uniform-F and Uniform-WTA methods, apparently, will make more mistakes at the end of the process than at the beginning. The execution time grows considerably because the collisions consume an important amount of time to be resolved.

C. The Random method

This method places the address events in the slots obtained by a pseudo-random number generator based on Linear Feedback Shift Registers (LFSR) [10]. Due to the properties of the LFSR used, each slot position is generated only once, except position zero, and no collisions appear. If a pixel in the image has intensity *p*, then the method will take *p* values from the pseudo-random number generator and places the pixel address in the corresponding *p* slots of the *frame vector*. They will not be equidistant but will appear along the complete address sequence randomly. This method is faster than any of the *Uniform* methods.

Note that by using an LFSR it would be possible to obtain two very close addresses in a few calls. This can be avoided using a *b*-bit counter for the most significant bits of the address. For each value of the LFSR, four addresses are generated by incrementing the counter. This ensures absence of collisions. Fig. 2 shows the LFSR structure with a 2-bit counter for a 128x128 frame with 256 grey levels.

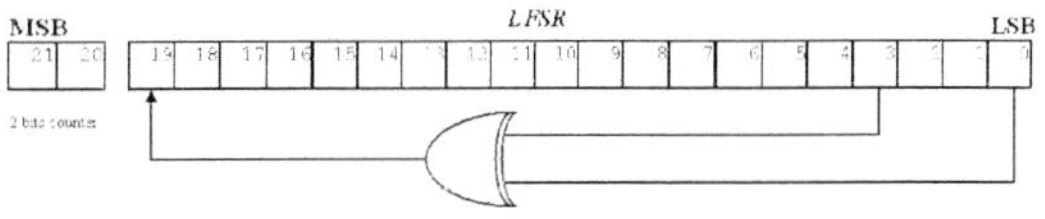

Fig. 2: Random method structure: LFSR with a 2-bit counter.

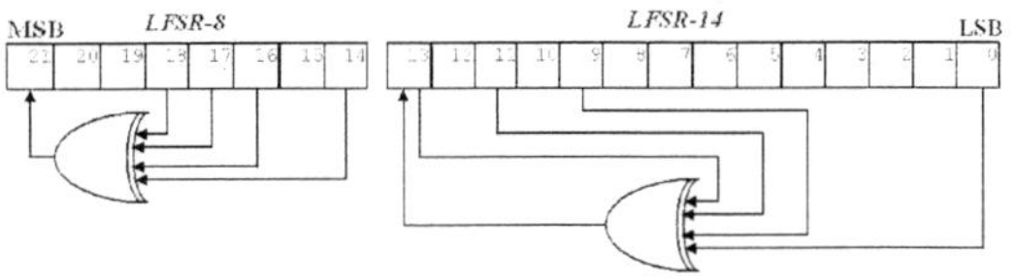

Fig. 3: Random-Square structure: LFSR-8 and LFSR-14 .

D. The Random-Square method

For the *Random* method with a fixed size counter from 1 to the maximum grey level, the event distribution for high activity pixels is acceptable, but poor for low level values. Substituting the counter by another LFSR, the distribution could be improved.

For a *128x128* frame with maximum grey level of *255*, an *8*-bit LFSR (LFSR-8) is used for selecting *255* slices of *128x128* positions, and another *14*-bit LFSR (LFSR-14) selects the position inside the slice. The image is scanned only once. For each pixel a *14*-bit number is generated by the LFSR-14, and the LFSR-8 is called as many times as the intensity level of the pixel would indicate. Fig. 3 shows the LFSRs used by this Random-Square method.

E. The Exhaustive method

This algorithm also divides the address event sequence into *K* slices of *NxM* positions for a frame of *NxM* pixels with a maximum grey level of *K*. For slice *k*, an event of pixel *(i,j)* is sent on time *t* if the following condition is asserted:

$$(k \times P_{i,j}) mod K + P_{i,j} \geq K \quad (1)$$

and

$$N \times M \times (k-1) + (i-1) \times M + j = t \quad (2)$$

where $P_{i,j}$ is the intensity value of the pixel *(i,j)* [4][5].

The Exhaustive method tries to improve the Random-Square one by distributing the events of each pixel into the *K* slices at equal distances. The algorithm scans the frame *K* times. In iteration *k*, if the previous condition is true, then the corresponding event is sent, otherwise the algorithm will wait for the following event (no event is sent at time *t*).

3. Evaluation Results

In this Section we compare the methods proposed above and estimate how the performance of the methods is affected by the traffic or load of events in the AER bus. To carry out this analysis a set of random images have been generated, which represent a population of images.

This set of images has been obtained considering two aspects: (a) its histogram must be close to a Gaussian distribution and (b) the number of events required to transmit them. This way, a 100% event load corresponds to an image with all pixels at maximum value. Consequently, an image with 10% of event load, represents an image that uses 10% of the possible events. Let us generate a 'Test Image Set' (TIS) composed of nine images with event load of 10%, 20%, 30%, ... and 90%. This set will be used to

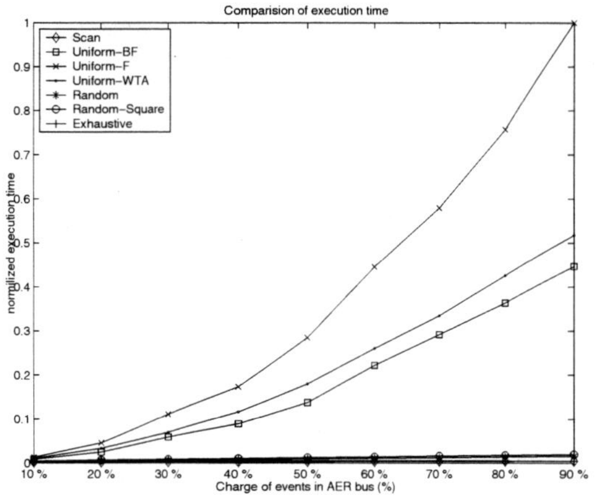

Fig. 4: Execution time comparision of sotware implementation

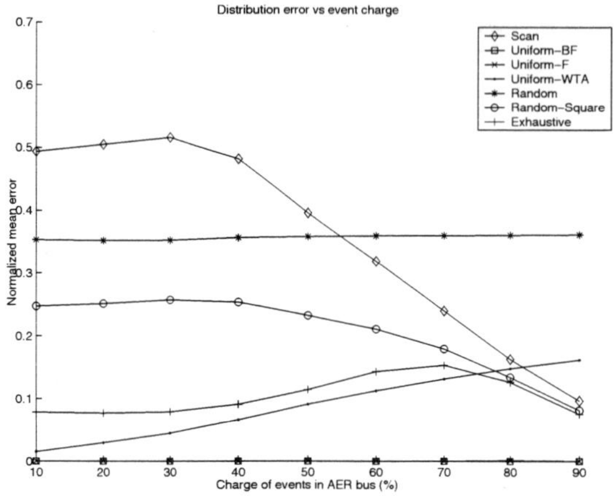

Fig. 5: Mean of NE matrix for methods along incremental charge of events in AER bus.

compare the algorithms according to the following criteria:

A. Execution Time

Fig. 4 shows the execution time versus the event load of the images. The *Scan* and *Exhaustive* methods follow an almost constant relation because the event load does not affect much the execution time for these algorithms.

B. Distribution Error

In an ideal AER distribution all events for one pixel are equidistant in time: constant frequency of events. In this section, the distribution of events obtained with each method is evaluated. Let us call '*Distribution Error*' how much the event distribution generated by a method deviates from the ideal distribution.

Let us suppose *Dij* is the ideal distance between events of pixel (*i,j*) of a *NxM* image with *K* grey level values. Then

$$D_{i,j} = (N \times M \times K)/P_{i,j} \tag{3}$$

where $P_{i,j}$ is the intensity value of pixel (*i,j*).

Let us suppose $d^k{}_{i,j}$ is the distance between the *k*-th event and the (*k*+1)-th one.

$$d^k{}_{i,j} = p^{k+1}{}_{i,j} - p^k{}_{i,j} \tag{4}$$

where p_k is the position of event *k* in the *frame vector*.Then we can measure the mean error for a pixel as the average of the differences between the ideal and real distance. The error expression is:

$$e_{i,j} = \frac{\sum_{k=1}^{P_{i,j}} \left|D_{i,j} - d^k{}_{i,j}\right|}{P_{i,j}} \tag{5}$$

It is easy to see that the worst case for this error measurement is when all the events are together in the address sequence. Therefore, in order to compare the error obtained for different methods and images, the error of each pixel must be normalized with respect to the maximum error associated to the pixel. The following expression is the maximum error for pixel (*i,j*):

$$me_{i,j} = 2 \cdot (D_{i,j} - 1) \cdot (1 - 1/P_{i,j}) \text{ with } P_{i,j} \neq 1 \tag{6}$$

For $P_{i,j} = 1$, the distribution error is zero, because only one event has to be sent.

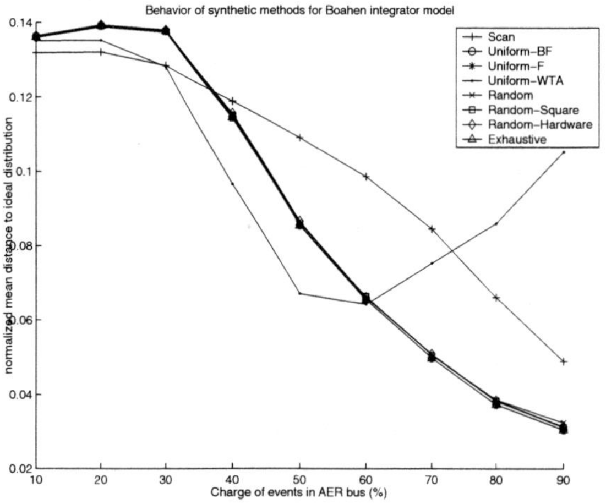

Fig. 6: Normalized mean distance between methods and ideal distribution for Boahen integrator.

Finally, we define a matrix (*NE*) with the same size of the test image, and where each element (*i,j*) represents the error normalized for pixel (*i,j*).

$$NE_{i,j} = e_{i,j}/(me_{i,j}) \tag{7}$$

Fig. 5 shows the measure of the *NE* matrix calculated for the nine test images using the methods proposed. The x-axis represents the image *event load* and the y-axis is the mean normalized error.

C. Integrator Cells

Consider the receptor cells proposed by Boahen [8] (diode-capacitor integrator) and by Mortara [9] (two capacitors working in two phases). We have modelled the ideal behavior of these cells in MATLAB. Then for each synthetic AER generation method, different *frame vectors* were obtained. These *frame vectors* were then used to feed an array of integrators of either the Boahen type or the Mortara type. Fig. 6 and Fig. 7 show the distance between the ideal distribution of events and the real distribution due to each method using our "Test Image Set" (TIS) and for each receptor model.

4. Hardware Interface

All simulations presented have been performed in software. However, the final goal is to build a dedicated

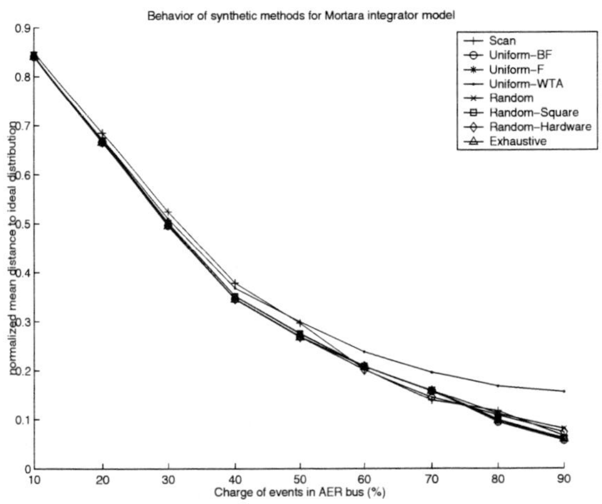

Fig. 7: Normalized mean distance between methods and ideal distribution for Mortara integrator.

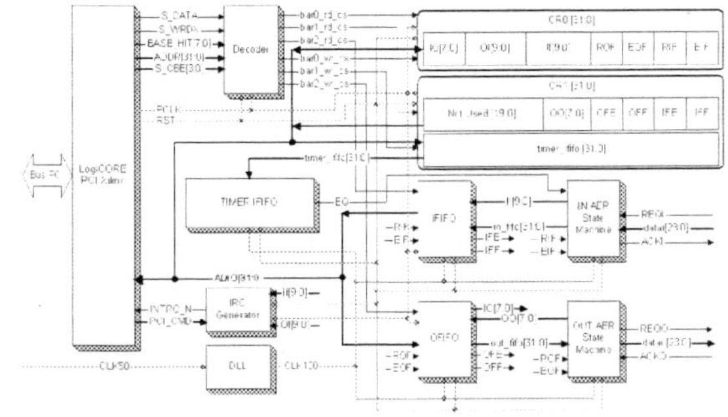

Fig. 8: Hardware Interface Architecture.

hardware that transforms a video frame sequence into an AER stream in real time. Such hardware is presently under development. At this moment a PC based system is available where the frame-AER transformations are performed in software but the resulting *frame vector* is dumped through the computer PCI bus on to an AER bus.

Fig. 8 shows the architecture of the present hardware interface. This is a PCI interface based on the LogiCORE PCI of Xilinx that uses I/O space for configuration and memory space for AER format reading and writing. It has two AER buses, one for incoming AER data and another for outgoing one. There are two FIFOs for both directions. It has a programmable timestamp assignment for incoming AER and programmable wait states for outgoing AER. There is an interrupt generation for avoiding overflows at the incoming FIFO.

The system has been implemented using VHDL and synthesized into a VirtexE 600 FPGA. It has been tested on a Nallatech Ballyinx prototyping board under Linux operating system. It can read or write an AER event every $T_{pulse} = 40ns$. If $T_{frame} = 40ms$, then this implies $N \times M \times K \le T_{frame}/T_{pulse} = 10^6$.

5. Conclusions

Algorithms for transforming synchronous frame based video streams in asynchronous address event streams are presented and evaluated. Three criteria (cxecution time, error distribution and distance to ideal behavior with two integrator models) have been evaluated for the seven software methods. A hardware interface between a computer (PCI) and a bioinspired system (AER) has also been presented.

The results presented in Section 3 show that: (a) Software based Unifom methods are not valid for real-time due to the overhead introduced by collision resolution. A hardware version is currently under development to solve these problems. (b) Uniform methods have lower distribution error than others for the test set (TIS). (c) Reconstruction of images, using two models of spike based integrators, show that any method could be valid with small differences among them. The Uniform-WTA has the worst results in this aspect due to the reduction of events by collisions.

6. Acknowledgements

This work was partially supported by spanish grants TIC1999-0446-C02-02, TIC2000-0406-P4-05 (Victor), FIT-07000/2002/921 (Arquimedes), TIC2002-10878-E, TIC-2003-08164-C03-01 (Samanta), and EU grant IST-2001-34124 (Caviar).

7. References

[1] M. Sivilotti, *Wiring Considerations in analog VLSI Systems with Application to Field-Programmable Networks*, Ph.D. Dissertation, Caltech, Pasadena CA, 1991.

[2] T. Serrano-Gotarredona, A. G. Andreou, B. Linares-Barranco, "AER Image Filtering Architecture for Vision-Processing Systems," *IEEE Trans. Circ. and Syst.* Part-I, vol. 46, No. 9, September 1999.

[3] A. Cohen, R. Etienne-Cummings, T. Horiuchi, G. Indiveri, S. Shamma, R. Douglas, C.Koch and T. Sejnowski, *Report on the 2003 Workshop on Neuromorphic Engineering*, Telluride, CO, June 29 to July 19, 2003. {www.ini.unizh.ch/telluride}

[4] A. Linares-Barranco, *Study and Evaluation of AER Interfaces for Neuromorphic Systems*, Ph.D. Dissertation, University of Seville, Spain, 2003. (In spanish)

[5] A. Linares-Barranco, R. Senhadji-Navarro, I. García-Vargas, F. Gómez-Rodríguez, G. Jimenez and A. Civit, "Synthetic Generation of Address-Event for Real-Time Image Processing," *Proc. ETFA 2003*, Lisbon, September, vol. 2, pp. 462-467.

[6] K. Boahen, "Communicating Neuronal Ensembles between Neuromorphic Chips," *Neuromorphic Systems*, Kluwer Academic Publishers, Boston 1998.

[7] M. Mahowald, *VLSI Analogs of Neuronal Visual Processing: A Synthesis of Form and Function*. Ph.D. Dissertation. Caltech, Pasadena, California 1992.

[8] K. Boahen, "Retinomorphic vision systems II: Communication channel design," *Proc. of the IEEE ISCAS, vol. supplement*, pp. 14-17. May 1996.

[9] A. Mortara, Eric A. Vittoz, Philippe Venier, "A communication Scheme for Analog VLSI Perceptive Systems," *IEEE Journal of Solid-State Circuits*, vol. 30, No. 6, pp. 660-669, June 1995.

[10] S.W. Golomb, *Shift Register Sequences*. Laguna Hills, CA: Aegean Park Press, 1982.

LOGIC COMPUTATION USING COUPLED NEURAL OSCILLATORS

Dongming Xu, Jose C. Principe and John G. Harris

Computational NeuroEngineering Laboratory
Department of Electrical and Computer Engineering
University of Florida, Gainesville, 32611 USA
{dmxu, principe, harris}@cnel.ufl.edu

ABSTRACT

This paper presents our recent work in using neural oscillators inspired by olfactory cortex models for boolean logic computation. Two such coupled oscillators are sufficient to implement all the boolean primitives (including AND, OR and XOR). Although the same computational model is being studied as a content addressable memory, these results indicate that additional computational power exists in this class of oscillatory neural networks. Detailed configuration of the proposed structure, experimental results and potential applications are presented.

1. INTRODUCTION

Oscillatory networks have been studied as information processors by many researchers because they can be constructed from realistic nonlinear dynamical systems and are biologically plausible. Examples such as relaxation networks [1, 2] that are generalized from natural physical systems have demonstrated their computational ability in image processing and as associative memories. Different structures of oscillatory networks such as Cellular Neural Networks (CNNs) [3, 4, 5], phase locked loop neural networks [6] and multi-resolution neural networks [7] have also been proposed for the purpose of pattern recognition and image segmentation. The dynamical behavior of the individual components in most of the networks are either single-frequency oscillations or limit cycles. In its most general structure, Freeman's computational model of the olfactory cortex [8] is a chaotic network. There is a significant difference between oscillatory networks and the classical Hopfield-type dynamical neural networks. Processed information is represented by either a limit cycle or nonconvergent dynamics [9], rather than by fixed-point solutions. As for the information processing, in most oscillatory networks the initial conditions are given as the inputs and the networks evolve trough time, where phase information are used to memorize and recall patterns. Freeman's network receives external input that is mixed with the states, and can work as either a dynamic autoassociative memory, feature extractor or a classifier [10, 11].

In this paper, we propose a different application of Freeman's network. Instead of using the network as a dynamical content addressable memory, we use the same coupled oscillatory components to perform logic computation. In [12], we studied the synchronization behavior of coupled neural oscillators in terms of coupling strength. With fixed coupling coefficients, a combination of external inputs will lead the coupled oscillators to two different states: synchronization and desynchronization. It turns out that with only two oscillators, by determining the proper values of the coupling coefficients, we can build functional blocks to implement two-input boolean logic gates including AND, OR, XOR, NAND, NOR, XNOR and NOT. The oscillator used here is named the reduced KII (RKII) set that is the fundamental building block in Freeman's model.

Freeman's model is a biologically realistic computational model of the olfactory cortex [8, 9] that has been used in various pattern-recognition applications [10, 11]. It is a locally stable but globally chaotic system with very high dimensionality. The complexity of the whole system is expressed by a hierarchical embedding of simpler structures. Four different levels are included as K0, KI, KII and KIII, where K stands for Katchalsky. The most basic processing element (PE) in the system is a K0 set which is a nonlinear second order dynamical system as shown in Fig. 1.

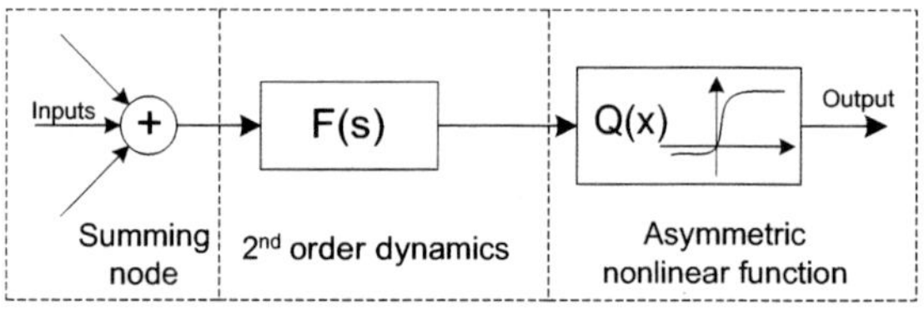

Fig. 1. An excitatory K0 set. An inhibitory K0 set has negative sign of nonlinear gain in the 3rd block.

The K0 has three major blocks: a summing node, second order dynamics and an asymmetric nonlinear block. Two kinds of K0 sets are distinguished by their signs of the nonlinear gain. The second level KI has interconnected K0 sets with only positive (excitatory) or negative (inhibitory) connections. A KIII network that consists of all the previous levels describes the behavior of the olfactory cortex [8]. A KII set is composed of four K0 sets. Two of them are excitatory while the other two are inhibitory. Basically, the KII set is an oscillator whose behavior is controlled by an external input. A KII network is a network of coupled oscillators (KII), in which different KII sets are interconnected through excitatory cells and inhibitory cells. The RKII set discussed in this paper is a simplified version but plays the same role as that of a KII set when the number of channels in the KII network increases [8, 13].

In Section 2, the basic dynamics of single and coupled RKII sets will be introduced. Based on the conclusions in Section 2, we propose in Section 3 the methodology to use coupled RKII sets as logic gates as well as basic structure in potential applications of pattern recognition. Section 4 presents the simulation results obtained through ODE solvers in MATLAB. Finally, in Section

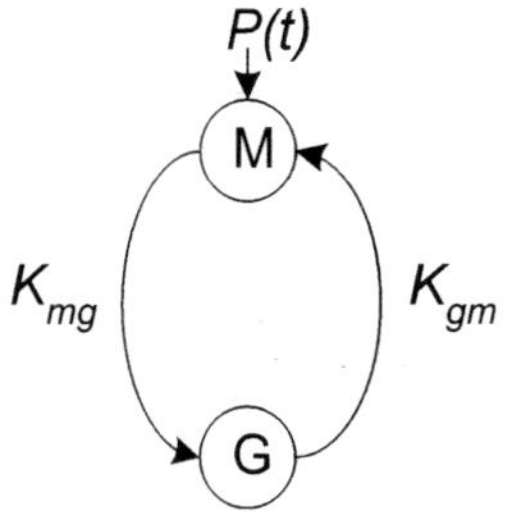

Fig. 2. A reduced KII set consists of one mitral cell and one granule cell that are coupled through $K_{mg}(>0)$ and $K_{gm}(<0)$.

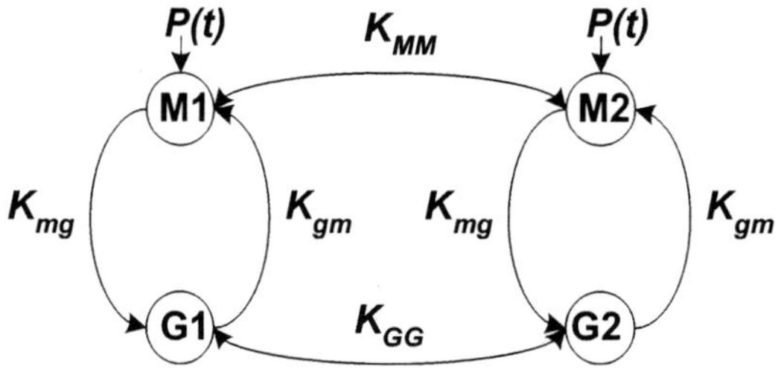

Fig. 3. Two coupled identical reduced KII sets. K_{mm} and K_{gg} are the interconnection weights among excitatory and inhibitory cells respectively.

5, we will give conclusions and discuss possible structures to use RKII networks with more computational power.

2. DYNAMICS OF SINGLE AND COUPLED REDUCED KII SETS

An RKII set is an input controlled oscillator [14, 15] that is described by the following ordinary differential equations (ODEs)

$$\begin{cases} \frac{1}{a\cdot b}\cdot\left(\frac{\mathrm{d}^2 m(t)}{\mathrm{d}t^2}+(a+b)\cdot\frac{\mathrm{d}m(t)}{\mathrm{d}t}+m(t)\right) \\ = K_{gm}Q(g)+P, \quad K_{gm}<0 \\ \frac{1}{a\cdot b}\cdot\left(\frac{\mathrm{d}^2 g(t)}{\mathrm{d}t^2}+(a+b)\cdot\frac{\mathrm{d}g(t)}{\mathrm{d}t}+g(t)\right) \\ = K_{mg}Q(m), \quad K_{mg}>0 \end{cases} \tag{1}$$

where $1/a$ and $1/b$ are time constants of the 2nd-order dynamics. They are given experimentally as $a = 220/s$ and $b = 720/s$ [8]. P is a time-invariant input. $Q(x)$ is the nonlinear function that models the spatio-temporal integration of spikes into mesoscopic waves measured in the cortex [8] and is defined by the following equation:

$$Q(x)=\begin{cases} Q_m\cdot(1-e^{-\frac{e^x-1}{Q_m}}) & x>x_0 \quad \text{(2a)} \\ -1 & \text{else} \quad \text{(2b)} \end{cases}$$

where $x_0 = \ln(1 - Q_m \cdot \ln(1 + 1/Q_m))$. Q_m is an adjustable parameter that controls the ratio between positive and negative saturation values of $Q(x)$.

As discussed in [15], when Eq. (3) is satisfied, the behavior of a RKII set is determined solely by the input.

$$\frac{1}{Q'(m^*)\cdot Q'(g^*)}\cdot\frac{(a+b)^2}{a\cdot b}<|K_{mg}\cdot K_{gm}|<\frac{(a+b)^2}{a\cdot b} \tag{3}$$

In Eq. (3), m^* and g^* are the fixed point solutions of Eq. (1), which are also determined by the external input $P(t)$. Under this condition, positive input will cause a RKII set to oscillate while with zero input, it will stay at equilibrium.

In order to help understand the more complicated dynamical behavior of Freeman's KII and KIII networks, in [12], a synchronization analysis is performed for two identical coupled RKII sets (Fig. 3). The configuration of coupling in terms of the first system is defined as

$$\begin{cases} \ddot{m}^{(1)}+(a+b)\cdot\dot{m}^{(1)}+m^{(1)} \\ \quad = ab\cdot(K_{gm}\cdot Q(g^{(1)})+P(t)+K_{mm}\cdot Q(m^{(2)})) \\ \ddot{g}^{(1)}+(a+b)\cdot\dot{g}^{(1)}+g^{(1)} \\ \quad = ab\cdot(K_{mg}\cdot Q(m^{(1)})+K_{gg}\cdot Q(g^{(2)})) \end{cases} . \tag{4}$$

where $K_{mm} > 0$ and $K_{gg} < 0$ are the interconnection weights. $m^{(2)}$ and $g^{(2)}$ are coupled from the second system. The term *identical* here means that the two RKII sets have exactly the same internal coupling weights (K_{mg} and K_{gm}) as well as external input $P(t)$. Analytical solutions are given in [12] on the interconnection weights K_{mm} and K_{gg} so that synchronized behavior could be achieved between the outputs of M_1 and M_2 of the coupled RKII sets. Given a specific set of external inputs and internal couplings, the synchronization boundary can be computed in the parameter space of K_{mm} and K_{gg}. Apparently, there are two typical behaviors in the state space of M_1 and M_2: identical synchronization (IS) and desynchonization (which may include 180 degree phase difference as well as quasi-periodic motion). Note that IS indicates a synchronization of both amplitude and phase.

3. COUPLED NEURAL OSCILLATORS AS LOGIC GATES

3.1. Design Methodology and Measure of Synchronization

From previous section, we know that there are two basic behaviors for two coupled RKII sets. Based on the analysis, if we denote synchronization as state *1* and desynchronization as state *0*, the coupled system is a logic function when defined as

$$f(\mathbf{I}) = S(m^{(1)}, m^{(2)}, \theta) \tag{5}$$

where $\mathbf{I} = \{0,1\} \times \{0,1\}$ is the input space and $S = \{0,1\}$ is any function that determines whether or not the outputs of the two systems $m^{(1)}$ and $m^{(2)}$ are synchronized. θ is the threshold used in S. Note that $\mathbf{I}$ is not limited to binary values of 0 and 1. Basically, any two values that represent higher and lower level inputs can form the input space. Function S is just a symbolic system that you can assign either 0 or 1 to any of the two states. This means that, if f can implement one set of the basic logic computations (for example, AND, OR, XOR), it should also be able to realize their complements with a different assignment of states in the actual measure of synchronization. The NOT function can also be implemented with many options. One example is to short the inputs of NAND gate or fix one of its inputs to 1. A two-input logic gate has a total of 16 boolean functions. In this paper, we only conisder the case that the coupled oscillators have a symmetric structure. So it cannot distinguish between the inputs (0,1) and (1,0). However, this could be solved by using weighted inputs. The weights then become the parameters in addition to K_{mm} and K_{gg} that need to be designed. By considering only the symmetric structure, without combination of different gates, the coupled oscillators have the ability to implement 8 out of the 16 logic functions that include all the boolean primitives. They cover most of the universal sets (for example, {AND, OR, NOT} and {NAND}) that can express any logic functions.

Synchronization can be measured using either correlations between the outputs of the two sets or direct detection of phase difference. Equation (6) calculates the correlations between the time averaging between the two state variables .

$$C(m_1, m_2) = \frac{< m_1 \cdot m_2 > - < m_1 >< m_2 >}{\sqrt{(< m_1^2 > - < m_1 >^2)(< m_2^2 > - < m_2 >^2)}} \quad (6)$$

Notice that when the two sets are synchronized, the phase plot is a straight line. When they are desynchronized, there may be either limit cycle or quasi-periodic motions in the phase plane. So another way is to plot the bifurcation diagram, so that in the case of desyncronization, multiple period could be detected.

3.2. Determining the Coupling Coefficients

The interconnection coefficients determine when the two oscillators will be synchronized. Given a desired logic function, how can we decide the values of K_{mm} and K_{gg}? There are two basic approaches depending on actual applications. If our purpose is to build each of the six gates given specified input values, we can simply search the parameter space to get the appropriate set of coupling strengths for every logic gate. After we have all the settings for the required functions, the couplings will never be changed.

A second alternative calls for the design of learning rules, as done in neural networks. The RKII can be thought as a learning machine implementing discriminant functions (DF) (such as the multiplayer perceptron), where the output space has only binary values and the coupled oscillators are actually dividing the input space in a particular way by means of DF. The purpose of setting the coupling coefficients is to move the DF around and to look for the best way of mapping the input space to the desired output space configuration. In such cases, an hypothetical learning algorithm should be able to train the interconnection weights based on different problem settings. From [12], the criterion to build synchronized oscillators is:

$$\Re[\Lambda_1] < 0, \quad (7)$$

where $\Re[\Lambda_1]$ is the largest real part of the eigenvalues of the Jacobian matrix computed from the dynamical system in terms of the differences between the two oscillators. It turns out that in our particular case, most of the synchronization boundaries are actually monotonically increasing functions with respect to K_{mm} and K_{gg}. So, for example, there exists a bifurcation point in the coupling space that once crossed for larger coupling produces synchronized outputs, while with a smaller coupling we will see desynchronized behavior. A rule of thumb is $sgn(\delta K) = D - S$, where D is the desired state of synchronization and S is the measured state. However, we have not yet devised an automated rule. It is also interesting to mention that a single layer of coupled RKIIs yields non convex regions that implement the XOR and XNOR, unlike topologies of static nonlinearities (the MLP requires a hidden layer to solve the XOR problem).

4. SIMULATIONS AND DISCUSSIONS

In the simulations, we fix $K_{mg} = |K_{gm}| = 3$ and use {0,1} as the input space. Different values of K_{mm} and K_{gg} are searched to see if all the six basic logic gates could be realized. Given the simple structure of just two coupled oscillators, the results are very promising. Using a synchronization criterion that assigns state 1 to synchronization and state 0 to desynchronization, the coupled RKII sets can implement AND, NOR and XNOR functions. That is, by assigning 1 to desynchronization and 0 to synchronization, we have the complementary functions, i.e. all the six logic functions. Figures 4 and 5 show the implemented XNOR function in the time domain and phase plane respectively. Transient behaviors are discarded to guarantee a steady-state of the outputs. In the case when inputs are [1 1] and [0 0], the two reduced KII sets are perfectly synchronized. When inputs are [1 0] and [0 1], the two oscillators are desynchronized and even present chaotic behaviors.

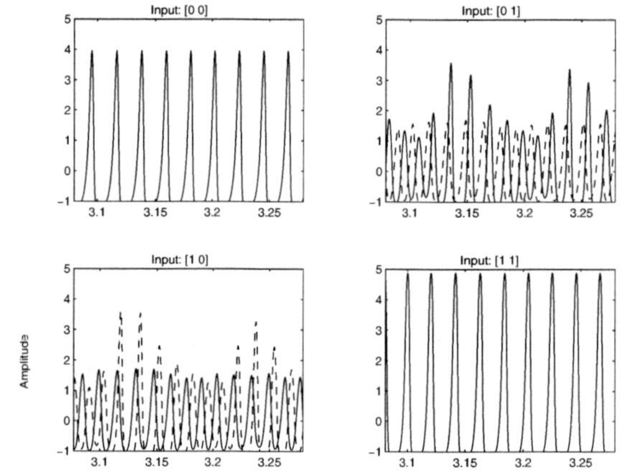

Fig. 4. Time domain response when implementing coupled RKII sets as XNOR gate. (Solid line shows the output from M_1 while dashed line shows the output from M_2.)

Table 1 gives the parameter values to build the logic gates. In the second column of the table, the label (0 or 1) defined for synchronized behaviors in S is indicated.

Table 1. Configurations for different logic functions

Function	Sync.	K_{mm}	K_{gg}	K_{mg}	K_{gm}
AND (NAND)	1(0)	1.75	-2.0	3	-3
NOR (OR)	1(0)	1.2	-0.4	3	-3
XNOR (XOR)	1(0)	1.5	-0.4	3	-3

One important question is how to extend this result to a network of oscillators. There are basically two structures. The first one is to implement the logic gates just as in digital design. That

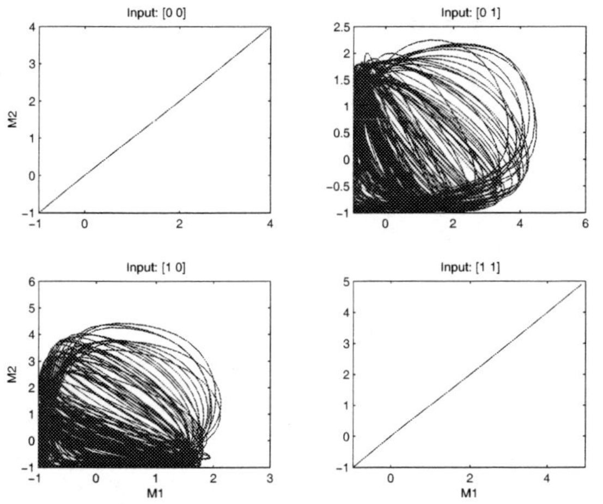

Fig. 5. Phase plot in the state space of M_1 and M_2 when implementing coupled RKII sets as XNOR gate.

is, outputs from one set of coupled oscillators will be the input of the next one. And combinations of different gates will implement more complicated logic functions. The other way is to fully connect N RKII sets. In this network, if we consider all the combinations of the synchronization between any subset of the N oscillators, the output space will be of the size 2^N. This shows the great potential of this network in terms of capacity.

We already successfully showed that the areas in the parameter space exist to get all desired logic functionalities. However learning is still an open question because we still need to show that the coupling coefficients are trainable regarding both computational efficiency and ability of convergence. Also, although the potential capacity of this network is very large, the controllability of all the 2^N attractors in the output space is not guaranteed. Theses are some of the questions that need to be further investigated.

5. CONCLUSIONS

Based on the synchronized behaviors of two coupled neural oscillators, we proposed a novel method to build basic logic gates from this structure. Experiments showed that the intrinsic structure of the coupled sets provide us with very flexible configurations regarding the coupling strengths so that all six logic gates can be implemented. The results and discussions demonstrate additional computational power that this particular oscillatory network has.

Acknowledgement

This work was partially supported by ONR N00014-1-1-0405.

6. REFERENCES

[1] D. Wang, "Relaxation oscillators and networks," in *Wiley Encyclopedia of Electrical and Electronics Engineering*, J. G. Webster, Ed. New York: Wiley Sons, 1999, pp. 396–405.

[2] K. Chen and D. Wang, "A dynamically coupled neural oscillator network for image segmentation," *Neural Networks*, vol. 15, no. 3, pp. 423–439, Apr. 2002.

[3] L. Chua, *CNN: A Paradigm for Complexity*. New York: World Scientific Pub Co, 1998.

[4] L. Chua and L. Yang, "Cellular neural networks: Theory and applications," *IEEE Trans. Circuits Syst.*, vol. 35, no. 10, pp. 1257–1290, Oct. 1988.

[5] L. Chua and T. Roska, "The cnn paradigm," *IEEE Trans. Circuits Syst.*, vol. 40, no. 3, pp. 147–156, 1993.

[6] F. Hoppenstadt and E. Izhikevich, "Pattern recognition via synchronization in phase-locked loop neural networks," *IEEE Trans. Neural Networks*, vol. 11, no. 3, pp. 734–738, May 2000.

[7] R. D. Henkel, "Segmentation with synchronizing neural oscillators," Center for Cognitive Science, Bremen, Tech. Rep. ZKW-Report, Apr. 1994.

[8] W. Freeman, *Mass Action in the Nervous System*. New York: Academic, 1975.

[9] J. Principe, V. Tavares, J. Harris, and W. Freeman, "Design and implementation of a biologically realistic olfactory cortex in analog VLSI," *Proc. IEEE*, vol. 89, no. 7, pp. 569–571, July 2001.

[10] R. Kozma and W. Freeman, "Chaotic resonance-methods and applications for robust classification of noisy and variable patterns," *Neural Networks*, vol. 4, no. 1, pp. 103–121, 2001.

[11] Y. Yao, W. Freeman, B. Burke, and Q. Yang, "Pattern recognition by a distributed neural network: An industrial application," *International Journal of Bifurcation and Chaos*, vol. 11, no. 6, pp. 1607–1629, 1991.

[12] D. Xu, J. Gao, and J. Principe, "Synchrnoization analsysis of neural oscillators in the olfactory system," to be submitted to Phys. Rev. E.

[13] V. Tavares, "Design and implementation of a biologically realistic olfactory cortex model," Ph.D. dissertation, Univ. of Florida, Gainesville, May 2001.

[14] D. Xu, L. Deng, J. Harris, and J. Principe, "Design of a reduced kii set and network in analog VLSI," in *Proc. IEEE International Symposium on Circuits and Systems (ISCAS'03)*, Bangkok, Tailand, May 2003, pp. 837 –840.

[15] D. Xu and J. Principe, "Dynamical analysis of neural oscillators in an olfactory cortex model," accepted by *IEEE Trans. Neural Networks*.

C^4 BAND-PASS DELAY FILTER FOR CONTINUOUS-TIME SUBBAND ADAPTIVE TAPPED-DELAY FILTER

Heejong Yoo, David Graham, David V. Anderson, and Paul Hasler

School of Electrical and Computer Engineering
Georgia Institute of Technology, Atlanta, GA, 30332

ABSTRACT

This paper describes the implementation of delay element using C^4 band-pass filter for subband analog tapped-delay adaptive filter, where implementation of larger group delay is required. Most analog delay elements have been implemented with low-pass or all-pass filters. While they can easily achieve constant group delay within pass band, maximum group delay is severely restricted by the corner frequency because group delay is inversely proportional to the corner frequency.

In this paper, we implemented a delay element with a Capacitively-Coupled Current Conveyer (C^4) band-pass filter to produce larger group delay required for analog subband adaptive filter. Experimental results from circuits fabricated in $0.5\mu m$ CMOS technology through MOSIS are also presented.

1. CONTINUOUS-TIME ADAPTIVE FILTER

Adaptive tapped-delay filter like LMS (Least Mean Square) has been widely implemented in discrete-time domain as well as in continuous-time domain for it's simple structure and stability [1]. Figure 1 shows the configuration of tapped-delay adaptive FIR filter.

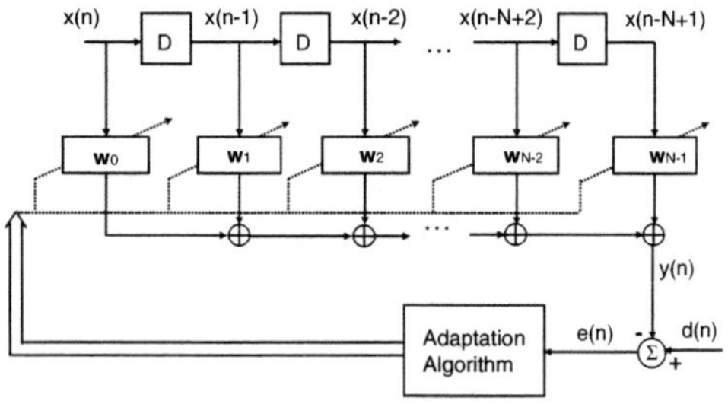

Fig. 1. Tapped-delay adaptive FIR filter structure. It consists of tapped-delay part, weight adaptation part, and multiplication/add part. Most analog delay elements in continuous-time mode have been implemented with first-order low-pass (Gamma filter), first-order all-pass filters, or the combination of first-order low-pass and all-pass filters (Laguerre filter).

This Research is sponsored by the Hewlett-Packard Corporation.

With all the development of adaptive algorithms for the last 30 years in discrete time domain, real-time implementation still remains as a hard problem because of the high computational complexity. The combination of subband processing with fast adaptive algorithms can be used for a real-time implementation. However, even with a high performance DSP hardware combining subband processing and fast adaptive algorithm, high power consumption is still unavoidable because of the DSP system's fast clock rate.

Many efforts have been made among analog circuit engineers to resolve real-time and low-power constraint by implementing adaptive LMS filters with analog circuitry. Sample mode delay elements using a CCD or a switched-capacitor were also investigated [2]. However, the necessity of clocking and the occurrence of aliasing effects still remain as drawbacks. Continuous-time LMS adaptive filter consists of tapped-delay part, weight adaptation part, and multiplication/add part. One of the difficulties we face in continuous-time implementation of tapped-delayed filter is that it requires ideal delay elements, which are impossible to implement with non-ideal circuitry.

2. LOW-PASS OR ALL-PASS DELAY ELEMENT

2.1. Constant Group Delay

Most analog delay elements in continuous-time mode have been implemented with first-order low-pass (Gamma filter) [3], first-order all-pass filters [4], or the combination of first-order low-pass and all-pass filters (Laguerre filter) [3]. While their group delay is generally constant within pass band, maximum group delay is severely restricted by the corner frequency because group delay is inversely proportional to the corner frequency.

The first-order low-pass filter transfer function and group delay become

$$H_l(s) = \frac{1}{\tau s + 1}, \quad D_l(w) = \frac{\tau}{\tau^2 w^2 + 1}, \tag{1}$$

and for the first-order all-pass filter,

$$H_a(s) = \frac{\tau s - 1}{\tau s + 1}, \quad D_a(w) = \frac{2\tau}{\tau^2 w^2 + 1}. \tag{2}$$

It is obvious from Eq. (1) and Eq. (2) that, when $w \ll \tau$, $D_l(w) \approx \tau$ and $D_a(w) \approx 2\tau$.

Constant group delay property of low-pass and all-pass filter is desirable in many applications. However, since their group delay is proportional to τ, when an input signal has a wide bandwidth ($\tau \ll 1$), maximum group delay that each delay element can produce is severely reduced as corner frequency increases. For example, when input signal is band limited by 8 KHz, τ should be less than 1/8000 and the maximum group delay of low-pass and all-pass delay filter is limited to 0.125 msec and 0.25 msec, respectively.

2.2. Delay of Subband Analog Adaptive Filter

Subband adaptive filter in discrete-time domain has gained much attentions these days because it enables faster convergence by reducing eigenvalue spread of input signal and reduces processing time when implemented with hardware [1]. From multi-rate signal processing theory, we know that signal can be decimated by either equal to M or smaller than M without having aliasing for M-ary filter bank. For example, when Fs = 8 KHz and the length of room impulse response is 200 msec, adaptive filter for AEC should have 1600 filter taps. In subband AEC, when number of subband, M, is 64 we only need 25 filter taps for each subband.

While decimation is advantageous in terms of decreasing computational complexity and number of cascade of delay element in discrete-time domain, if it could be directly implementable in continuous-time domain, M times bigger delay should also be implementable from one delay element. For above example, delay that each element generates should be 8 msec for each subband while it is 0.125 msec for full-band AEC. It is apparent from Eq. (1) and Eq. (2) that 8 msec group delay out of all-pass delay elements can be achieved at very lower subbands where $\tau > 1/250$.

3. BAND-PASS DELAY ELEMENT

3.1. Capacitively-Coupled Current Conveyer (C^4)

The C^4 was originally developed for separating corner frequencies, as its model system was the auto-zeroing floating-gate amplifier [5]. The frequency response of the C^4 is characterized by [5]

$$\frac{V_{out}}{V_{in}} = -\frac{C_1}{C_2}\frac{s\tau_l(1-s\tau_f)}{s^2\tau_h\tau_l + s(\tau_l + \tau_f(\frac{C_O}{\kappa C_2}-1)) + 1}, \quad (3)$$

where the time constants are given by

$$\tau_l = \frac{C_2 U_T}{\kappa I_{\tau_l}}, \ \tau_f = \frac{C_2 U_T}{\kappa I_{\tau_h}}, \ \tau_h = \frac{C_T C_O - C_2^2}{C_2}\frac{U_T}{\kappa I_{\tau_h}},$$

where the total capacitance, C_T, and the output capacitance, C_O, are defined as $C_T = C_1 + C_2 + C_W$ and $C_O = C_2 + C_L$. The currents I_{τ_l} and I_{τ_h} are the currents through M_2 and M_3, respectively in Fig. 2. Thus, the C^4 takes on the properties of a band-pass filter with first-order slopes and band-pass gain set by the ratio of the two coupling capacitors as $A_v = -C_1/C_2$. The overall time constant of the filter, which gives the center frequency, is

$$\tau = \sqrt{\tau_l \tau_h}. \quad (4)$$

The corner frequencies can be set independently of each other, and the bandwidth can therefore be tuned at will. If the corners are brought close together, a very tight bandwidth with a single Q peak can be developed and by setting the corners wide apart from each other, all-pass filter or low-pass filter can be implemented by setting lower corner frequency down to a few Hz and higher frequency up to a several hundred KHz.

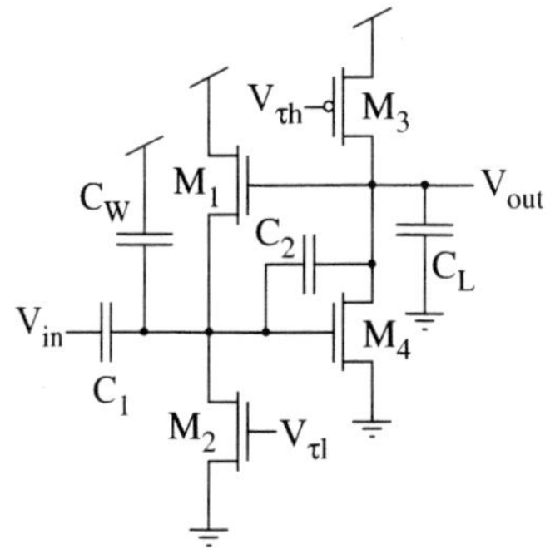

Fig. 2. Circuit diagram of a nFET-based Capacitively-Coupled Current Conveyer (C^4) circuit. C^4 allows both low-frequency and high-frequency cutoffs to be controlled electronically by changing the appropriate bias currents. The ratio of C_2 to C_1 sets the gain of C^4 and overall time constant is $\tau = \sqrt{\tau_l \tau_h}$.

The value of Q peak is

$$Q = \sqrt{\frac{\tau_h}{\tau_l}}\frac{1}{1 + \frac{I_{\tau_l}}{I_{\tau_h}}(\frac{C_O}{\kappa C_2} - 1)} \quad (5)$$

and the maximum value of the Q peak can be written as

$$Q_{max} = \frac{1}{2}\sqrt{\frac{\kappa(C_T C_O - C_2^2)}{C_2(C_O - \kappa C_2)}}. \quad (6)$$

3.2. Group Delay of Band-pass Filter

The general second-order band-pass filter transfer function is

$$H_b(s) = \frac{K(\frac{w_o}{Q})s}{s^2 + (\frac{w_o}{Q})s + w_o^2}, \quad (7)$$

where w_o is a center frequency, K is a gain, Q is a quality factor defined as w_o/BW, and BW is the bandwidth of the filter. Group delay can be written as

$$D_b(w) = \frac{(\frac{w_o}{Q})(w_o{}^2 + w^2)}{(w_o{}^2 - w^2)^2 + (\frac{w_o}{Q})^2 w^2}. \quad (8)$$

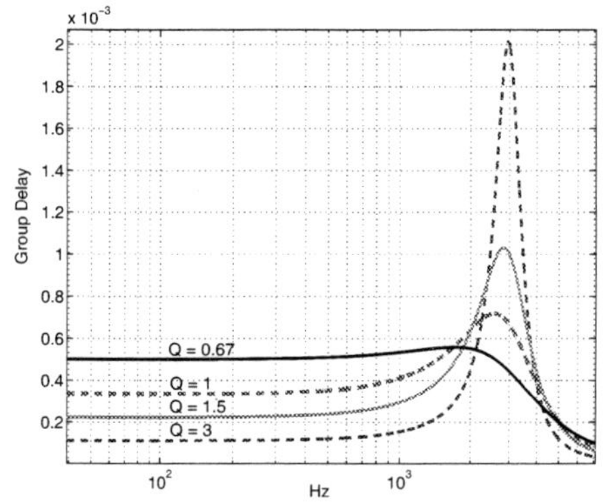

Fig. 3. Theoretical group delay of 2nd order band-pass filter for different Q values with $f_o = 3/2\pi$ KHz. When Q is low, group delay of band-pass filter has a constant group delay for the frequency lower than lower cut-off frequency. As Q increases, maximum group delay of the center frequency approaches to $2/BW$ and group delay at constant group delay region, where frequency is smaller than lower cut-off frequency, decreases.

Group delays of 2nd order band-pass filter for different Q values are shown at Fig. 3. When Q is low, group delay of band-pass filter is very similar to the group delay of 1st order low-pass filter having constant group delay for the frequency lower than w_o. When $w \approx w_o$, it is obvious from Eq. (8) that $D_b(w)$ is close to its maximum delay, $2/BW$, and is independent of center frequency. The property of band-pass delay filter that maximum delay is independent of the center frequency allows us to use band-pass delay element for the other subbands without increasing filter order to achieve same delay. As Q increases, maximum group delay at the center frequency approaches to $2/BW$ and group delay at constant group delay region, where frequency is smaller than lower cut-off frequency, decreases.

3.3. Group Delay of C^4 Band-pass Filter

High-Q and low-Q C^4 delay elements and C^4 SOS delay element have been implemented using MOSIS $0.5\mu m$ process technology. The C^4 is operated at the subthreshold region, where current and voltage has a exponential relationship, and consequently extremely low-power computation is achievable.

For the application to the audio signal processing, τ_f from Eq. (3) is so fast that the zero it produces lies far outside of the operating range and without loss of generality we can omit $s\tau_f$ from the numerator of Eq. (3). Then, the phase response of C^4 band-pass filter can be derived from this reduced transfer function as

$$\arg H(w) = -90° - \arctan\Big(\frac{\frac{\tau_l+\tau_f(\frac{C_O}{\kappa C_2}-1)}{\tau_l\tau_h} w}{\frac{1}{\tau_l\tau_h} - w^2}\Big), \quad (9)$$

and the constant group delay of C^4 is

$$D_b(w) = \frac{\alpha(w_o{}^2 + w^2)}{(w_o^2 - w^2)^2 + \alpha^2 w^2}, \quad (10)$$

where $\alpha = \frac{\tau_l+\tau_f(\frac{C_O}{\kappa C_2}-1)}{\tau_l\tau_h}$ and $w_o = 1/\tau_l\tau_h$. The maximum group delay at the center frequency, where $w \approx w_o$, is

$$D_b(w_o) \approx 2\frac{\tau_l\tau_h}{\tau_l + \tau_f(\frac{C_O}{\kappa C_2} - 1)}, \quad (11)$$

and group delay, where $w \ll w_o$, is

$$D_b(w) \approx \frac{\tau_l + \tau_f(\frac{C_O}{\kappa C_2} - 1)}{\tau_l\tau_h}\frac{1}{w_o^2}. \quad (12)$$

We designed the drawn capacitance ratio of C_O/C_2 to be 5 and extracted κ is close to 0.65 for the high-Q C^4 delay element. With these parameters Eq. (11) can be rewritten as

$$D_b(w_o) \approx \frac{2\tau_h}{1 + 6.7(\tau_f/\tau_l)}, \quad (13)$$

and Eq. (12) also as

$$D_b(w) \approx \frac{1 + 6.7(\tau_f/\tau_l)}{\tau_h}\frac{1}{w_o^2}. \quad (14)$$

Figure 4 shows group delays of measured data of low-Q C^4 band-pass filter of which magnitude and phase response are shown at Fig. 5. Group delay in Fig. 4 is similar to the group delay of typical low-pass or all-pass filter showing no peak around the center frequency and constant delay at $w \ll w_o$. The low-Q C^4 delay element is fabricated to have a unity gain for further cascade of delay element in delay networks and its Q_{max} defined in Eq. (6) is ≈ 0.828.

Group delays of the high-Q C^4 band-pass filters of Fig. 7 are shown at Fig. 6. As we have seen from Fig. 3, high peak delay around the center frequency is sharper as Q increases. High-Q C4 has a 20 dB magnitude gain and consequently the magnitude of input signal should be limited to a few tenth mV range for linearity. The input signal to the delay element should have a narrower bandwidth than the bandwidth of band-pass delay element in order not to have a spectral loss while exploring larger group delay. The peak of Fig. 6, where is pointed by A, is governed by Eq. (13) and the region B in the figure can be approximated by Eq. (14).

Group delays shown at Fig. 4 and Fig. 6 are reconstructed from measured data by taking numerical derivative of measured phase responses of C^4 and then adding negative sign. Since the measured phase responses, unlike ideal phase response, was not monotonically decreasing function of w, phase response was low-pass filtered for better approximation of the group delay.

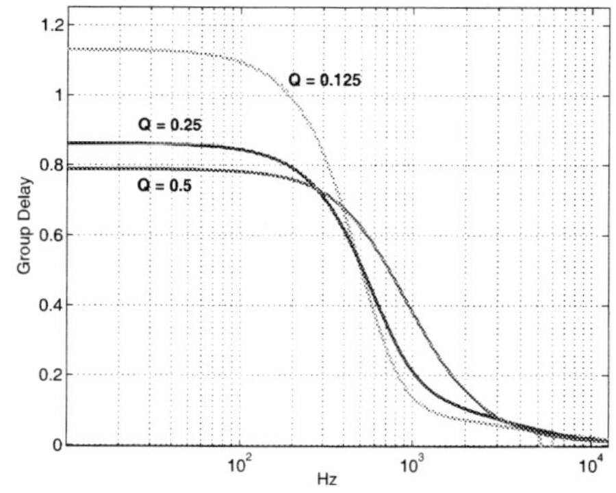

Fig. 4. Group delays of measured phase response of C^4 band-pass filter for low-Q values of Fig. 5. This figure suggests that group delays of C^4 band-pass delay element can be made similar to that of typical low-pass filter by having low Q.

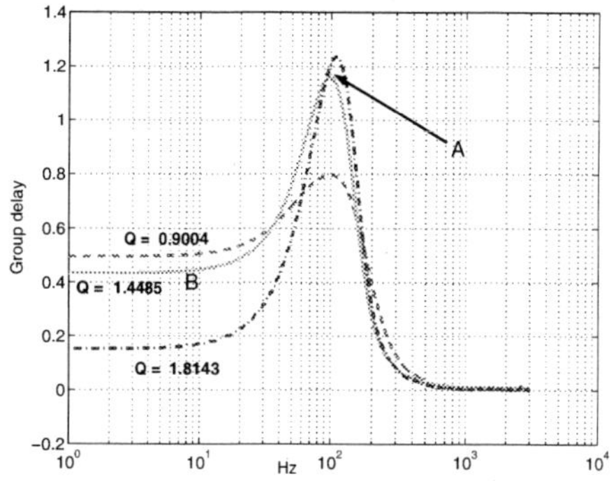

Fig. 6. Group delays of measured phase response of C^4 band-pass filter for high-Q values of Fig. 7. The peak pointed by A is governed by Eq. (13) and the region B can be approximated by Eq. (14). For Q = 1.4485, BW is 500 Hz and τ_l and τ_h are 0.32 msec and 0.16 msec, respectively.

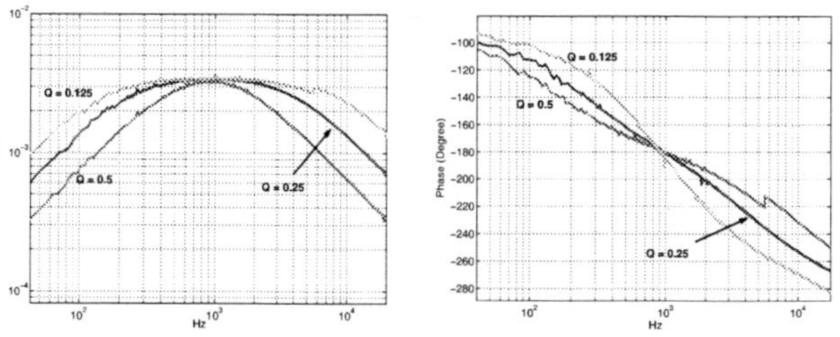

(a) (b)

Fig. 5. (a) Measured frequency responses for low-Q values. All the frequency response has a 1 KHz center frequency. The slope is ± 20 dB/dec. (b) Measured phase responses. The slope of phase response at the center frequency, which gives maximum group delay, increases as Q increases.

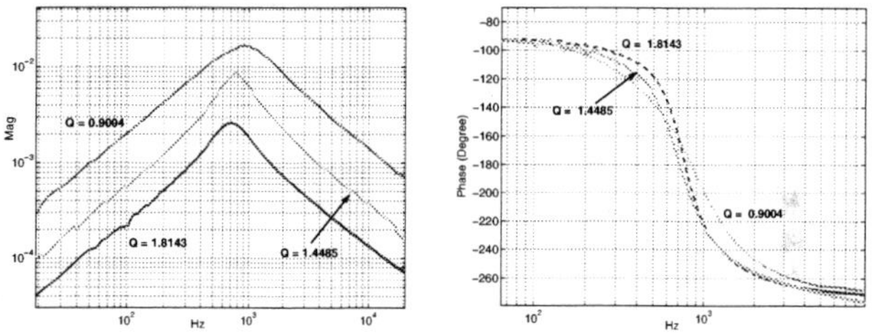

(a) (b)

Fig. 7. (a) Measured frequency responses. The frequency response of Q = 0.9004 has center frequency at 900 Hz and the center frequency of the rest two plots is close to 750 Hz. The slope is ± 20 dB/dec. (b) Measured phase responses.

4. CONCLUSIONS

We have described a C^4 second order band-pass delay element which can be used in continuous-time subband adaptive tapped-delay filter. C^4 band-pass filter has been investigated extensively, however, most of them were focused on the characteristics of magnitude response. In this paper, we consider C^4 as a delay element focusing on phase response and group delay. We showed that all-pass or low-pass delay element, in spite of their constant group delay property where $w \ll w_o$, are not suitable for the application where larger delay is required. Delay elements with high-Q and low-Q C^4 band-pass filter were fabricated and tested to meet the delay constraint of subband adaptive tapped-delay filter. Analysis of C^4 band-pass filter as a delay element was also presented.

5. REFERENCES

[1] S. L. Gay and J. Benesty, *Acoustic Signal Processing for Telecommunication*, Kluwer Academic Publishers, 2000.

[2] K. R. Laker, A. Ganesan, and P. E. Fleischer, "Design and implementation of cascaded switched-capacitor delay equalizers," *IEEE Transactions on Circuits and Systems*, vol. 32, no. 7, pp. 700–711, July 1985.

[3] J. Juan, J. G. Harris, and J. C. Principe, "Analog hardware implementation of adaptive filter structures," in *Proceedings of the International Joint Conference on Neural Networks*, June 1997, vol. 2, pp. 916–921.

[4] K. Bult and H. Wallinga, "A CMOS analog continuous-time delay line with adaptive delay-time control," *IEEE Journal of Solid-State Circuits*, vol. 23, no. 3, pp. 759–766, 1988.

[5] D. Graham and P. Hasler, "Capacitively-coupled current conveyer second-order section for continuous-time bandpass filtering and cochlea modeling," in *Proceedings of the IEEE International Symposium on Circuits and Systems*, Phoenix, AZ, May 2002, vol. 5, pp. 485–488.

NEURAL MODELING OF THE LARGE-SIGNAL DRAIN CURRENT OF THE DUAL-GATE MESFET WITH DC AND PULSED I-V MEASUREMENTS

M. Abdeen and M.C.E. Yagoub

SITE, 161 Louis Pasteur, University of Ottawa, Ottawa, ON, K1N 6N5, Canada
E-mail: mabdeen@site.uottawa.ca , myagoub@site.uottawa.ca

ABSTRACT

This paper presents a neural network model of the drain current for dual-gate MESFET. The model is a combination of two sub-models; a static model represented by DC IV characteristics and a dynamic model represented by pulsed IV characteristics. Pulsed measurements are performed at many bias points to capture the full dynamic behavior of the device. The final model has a total of 25 neurons and is generated in a few minutes. The measurement and model data are in very good agreement with model error of less than 1%.

1. INTRODUCTION

The increasing need for advanced communication technologies in the 21st century is leading to continuous development of new and more complex active devices and systems. Transistors such as BJTs, FETs, and HEMTs are fundamental components in today's personal, corporate and global communication systems.

The single gate Metal-Semiconductor FET (MESFET) is one of the most popular active devices with wide areas of RF/Microwave applications. The dual-gate MESFET (DGFET) is a variation of the single gate with a second gate between the first gate and the drain. Its advantages over the single gate of comparable size include higher gain, better input gate-to-drain isolation, and higher output impedance. That made it attractive for various nonlinear applications such as mixers, frequency multipliers, and power combiners and splitters. Despite the existence of some modeling work for the dual-gate MESFET, there is a need to develop more efficient, accurate, and faster nonlinear modeling methodologies.

Most existing MESFET large-signal models use DC I-V measurements. Device parameters such as the transconductance and the output conductance are then extracted form that model. However, the values obtained from DC I-V based models were not with a good agreement with their counterparts obtained from RF S-parameter measurements. This is due to low frequency dispersion, self-heating, and trap effects.

Many MESFET researches [1]-[3] have shown that pulsed I-V characterization renders better device parameters than the ones produced by DC I-V ones. Paggi *et. al.* [2], for example, showed that the value of the output resistance, R_o, extracted directly from pulsed I-V curves are much closer to the actual microwave values than the value taken from DC I-V curves. Pulsed measurements are, therefore, preferred if more accurate large-signal device characterization is desired.

The drain current is the most important non-linear component in the dual-gate FET model. Existing dual-gate FET drain current models, with the exception of [4] and [5], are based on considering the device as two single-gate FETs connected in cascode. This approximation did not result in a good agreement between the simulation and the measurement data.

In the model developed by Jenner [5] the nonlinear drain current is represented by a three-dimensional power series. However, obtaining good accuracy can require as many as 165 parameters. Moreover, the model is generated from DC measurements and does not take into account the frequency dispersion of the transconductance and the output conductance.

On the other hand, although the model developed by Ibrahim *et. al.* [4] required only 49 parameters, it still present a significant amount of model extraction efforts. Moreover, modeling the RF current component required the extraction of the transconductance from three-port S-parameter measurements. This presents a fairly complex and costly process.

Neural modeling of devices and circuits is one of the most recent trends in microwave CAD. Fast, accurate and reliable neural network models can be trained from measured or simulated data. Once developed, these neural models can be used in place of CPU-intensive physics/EM models of active/passive devices to speed up microwave design. Neural network (NN) techniques have been used to model a wide variety of microwave devices and circuits [6]-[8] with significant successes.

In this paper we present a large-signal drain current model for the dual-gate MESFET using neural networks and based on both DC and pulsed IV measurements.

2. MODEL DESCRIPTION

The dual-gate MESFET drain current model developed in this work has two components, a DC and an RF component. Static current component is modeled using the DC IV characterization. For the RF component, however, dynamic characterization is required. Most published work on the single and dual-gate FET approximates the dynamic behavior of the device with the static IV curves. This results in significant errors in calculating the trans- and output conductances of the device. Pulsed measurements, on the other hand, provide a better representation of the RF large-signal behavior of the transistor [9].

The total drain current is then given by [4]:

$$I_{ds} = I_{ds}^{DC} h(f) + I_{ds}^{RF} (1 - h(f)) \tag{1}$$

where I_{ds}^{DC} is the DC component of the drain current, I_{ds}^{RF} is the RF component, and $h(f)$ is a function that ensures a smooth transmission from DC to RF characteristics. For the model developed here, and for simplicity, $h(f)$ is chosen to be the unit-step function, i.e., the model represents either the DC or the RF current model.

$$h(f) = \begin{cases} 0 & f = 0 \\ 1 & f > 0 \end{cases} \tag{2}$$

where f is the frequency.

The DC current is a function of the static potential of the two gates and the drain, V_{gs1}, V_{gs2}, and V_{ds} respectively. On the other hand, pulsed IV characteristics depend on the bias point as well as the pulse amplitudes applied [10]. Therefore, the RF drain current is a function of the bias point (V_{gs1}, V_{gs2}, V_{ds}) as well as the pulsed voltages (v_{gs1}, v_{gs2}, v_{ds}) applied over the bias point. In other wards:

$$I_{ds}^{DC} = I_{ds}^{DC}(V_{gs1}, V_{gs2}, V_{ds}) \tag{3}$$

$$I_{ds}^{RF} = I_{ds}^{RF}(V_{gs1}, V_{gs2}, V_{ds}, v_{gs1}, v_{gs2}, v_{ds}) \tag{4}$$

The NN model generation process starts by assuming an initial model configuration. The "universal approximation theorem" states that a multilayer preceptron network (MLP) can approximate any arbitrary multidimensional function [11]. We started with a three layer neural network. The number of layers and neurons per layer are then changed until a satisfactory model training error is reached.

Measurement of data points are performed such that they cover the desired parameter range. The obtained data is divided into two sets, a training and a testing/verification set. The testing/verification data set is independent from the training set (never used in training).

3. RESULTS

We have modeled an 1x1.5x400 μm GaAs dual-gate MESFET. The DC and pulsed measurements are performed by a programmable DC/pulse measurement setup. Each measurement set is performed in a separate session. For the DC measurements, the bias range for the two gates is -1.0 ~ 0.25V (in steps of 0.25V) while that of the drain is 0~8V (steps of 0.25V). For the pulsed measurements, the DC bias point is scanned from (V_{gs1} = -1.0V, V_{gs2} = -1.0V, V_{ds} = 0.0V) to (V_{gs1} = 0.2V, V_{gs2} = 0.2V, V_{ds} = 5.0V) in steps of (ΔV_{gs1} = 0.3V, ΔV_{gs2} = 0.3V, ΔV_{ds} = 1.0V). The drain and the gates are pulsed simultaneously. The pulsed gate voltage levels are varied from -1.0 to 0.2V (in steps of 0.3V), while the drain pulsed voltage levels are varied from 0 to 8V (in steps of 0.4V). The drain current was averaged over many reading points under the same measurement conditions. This is all software programmed from a software user interface. The width of pulse is 1μs with pulse separation of 1 ms (a duty cycle of 0.1 %).

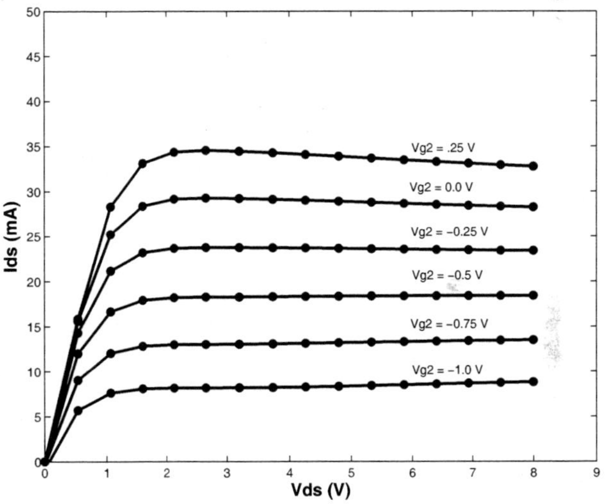

Figure 1. DC IV curves (• : measurement,– : modeled) for the dual-gate FET (V_{gs1} = 0.0 and V_{gs2} = -1.0 ~ 0.25 V)

3.1. The DC neural model

The DC neural model developed here is a three layer. We found no significant improvement of the model accuracy if more layers are used since the model is relatively simple. Ten hidden neurons were sufficient to obtain a very good accuracy of less than 1%. The input layer has 3 input parameters, the static voltages, (V_{gs1}, V_{gs2}, V_{ds}). The output layer has one output, the DC drain current, I_{ds}^{DC}. Figure 1 shows the measured and modeled data at V_{gs1} = 0.0 for different gate voltage (V_{gs2}).

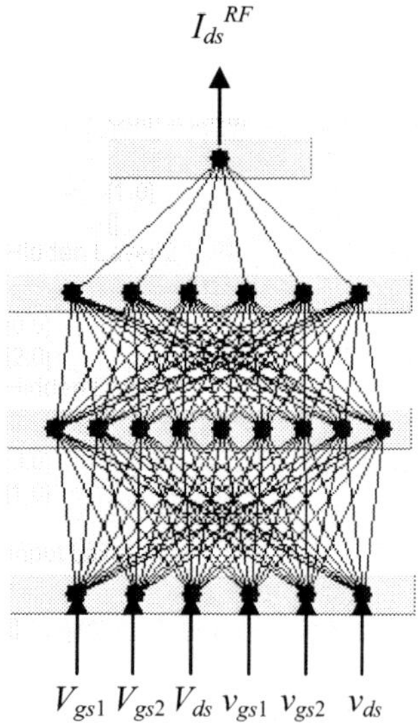

Figure 2. A Four-layer Neural Network model of RF current component of the Dual gate FET

3.2. The RF neural model

We have tested three different neural models for the RF current; a three-, a four-, and a five-layer models. We found a significant improvement of the four-layer model over the three-layered one. On the other hand, the model error improvement presented by the five-layer model did not justify the model complexity over the four-layered model. The model presented here is therefore of four layers; one input layer, one output layer, and two hidden layers. The input layer has 6 input parameters, the static voltages, (V_{gs1}, V_{gs2}, V_{ds}), and the dynamic voltages (v_{gs1}, v_{gs2}, v_{ds}). The output layer has one output, the RF drain current, I_{ds}^{RF}. The total number of hidden neurons is 15 (9, and 6 neurons for the first and second layer respectively). An illustration of the model is shown is Figure 2. Figures 3, 4, and 5 show the measured and modeled pulsed data for the dual gate FET for various DC and pulsed voltages. The model shows very good fitting of the data. The model error is less than 1%.

3.3. The large-signal drain current neural model

The large-signal drain current neural model is a combination of the two individual neural submodels developed in the previous two sections, i.e., the DC and RF neural models (Figure 6).

Before training, the measurement data is first randomized. It is then split into training and testing/verification parts. This step is important to ensure the model training and testing/verification data are well spread in the data range of interest. About 70% of the measurement data points are used to train the model. The remaining 30% data points are used in model testing/verification. Backpropagation with quasi-Newton algorithm is used for model training.

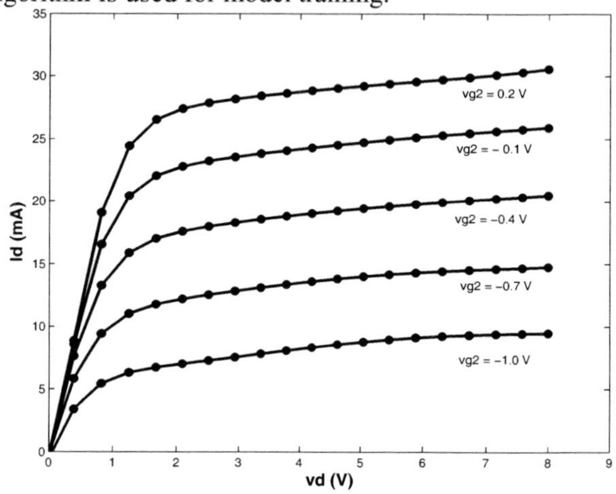

Figure 3 Pulsed IV curves (• : measured,– : modeled) for the dual gate FET at different gate pulse amplitudes (v_{g2})

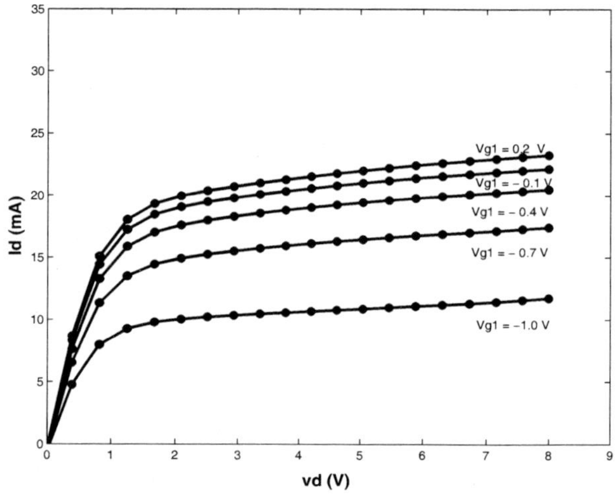

Figure 4. Pulsed IV curves (• : measured,– : modeled) for the dual gate FET at different gate bias voltages (V_{g1})

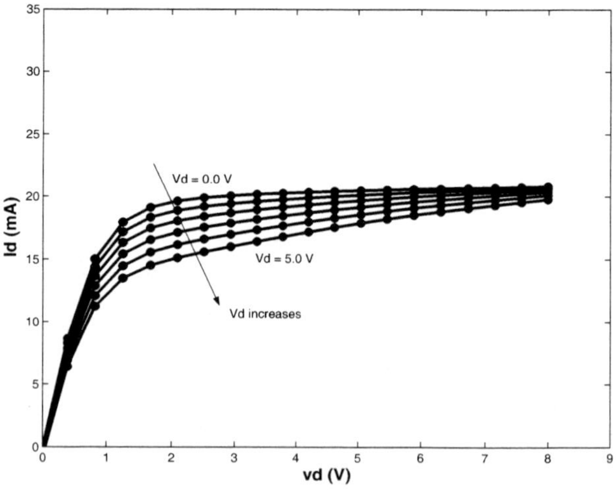

Figure 5 Pulsed IV curves (• : measured,– : modeled) for the dual gate FET at different V_d bias points

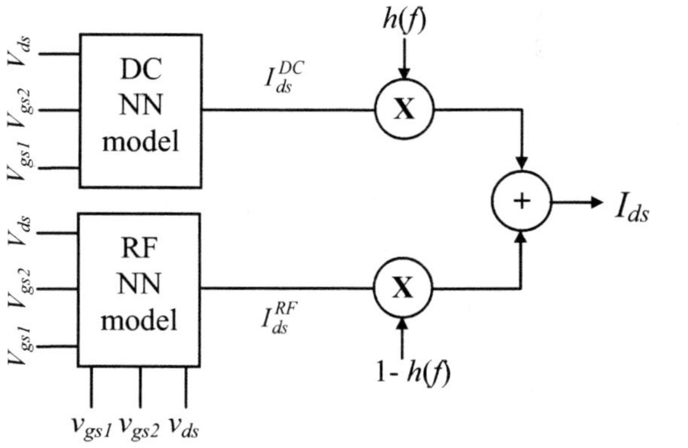

Figure 6 The overall large-signal drain current neural model for the dual-gate MESFET

The final model showed a very good agreement between the simulations and the measurement data. The average model error was less than 1%. The overall model generation takes only a few minutes on a typical computer.

It is to be noted that the pulsed IV characteristics do not show a negative output resistance effect. This effect is mainly due to the channel self-heating variation with the DC bias. The pulsed characteristics are iso-thermal (have a fixed DC bias current) and hence the channel temperature is constant for one given set of bias points. In addition, the pulsed IV characteristics change appreciably with the DC drain bias. As the DC drain bias increases, the drain saturation current decreases and the knee of the curve moves closer to the zero drain voltage. Similar effects have been noticed for the single gate FETs.

4. CONCLUSIONS

We have presented a large-signal drain current neural model for the dual-gate MESFET. The static behavior of the device is represented by the DC IV characteristics while the RF behavior is modeled using pulsed IV measurements at different bias points.

The model generation is relatively fast (a few minutes on a typical computer). The final model showed very good agreement between the measured and simulated data. The model average error is less than 1%. This model paves the way to including the device model in commercial circuit simulators (e.g. ADS [12]) with the help of user defined model capability. This is a topic of future research.

5. REFERENCES

[1] C. Fiegna, F. Filicori, G. Vannini, and F. Venturi, "Modeling the effects of the traps on the I-V characteristics of GaAs MESFETs", *Inter. Electron Devices Meeting*, Dec. 1995, pp. 773 -776.

[2] M. Paggi, P.H. Williams, and J.M. Borrego, "Nonlinear GaAs MESFET modeling using pulsed gate measurements" *IEEE Trans. Microwave Theory Tech.*, vol. 36, pp. 1593-1597, Dec. 1988.

[3] K. Koh, H.-M. Park, and S. Hong, "A spline large-signal FET model based on bias-dependent pulsed I-V measurement", *IEEE Trans. Microwave Theory Tech*, vol. 50, pp. 2598–2603, Nov. 2002.

[4] M. Ibrahim, B. Syrett, and J. Bennett, "Modeling the Drain Current of the Dual-Gate GaAs MESFET," *IEEE MTT-S Int. Symp. Digest*, Philadelphia, PA, June 2003, pp. 2113 -2116.

[5] M. B. Jenner, "A large-signal compatible dual-gate MESFET dc model," *Proc. IEEE European Conf. on Circuit Theory and Design (ECCTD)*, Stresa, Italy, Aug. 1999, pp. 888-891.

[6] J. Xu, M.C.E. Yagoub, R. Ding, and Q. J. Zhang, "Neural-based dynamic modeling of nonlinear microwave circuits," *IEEE Trans. Microwave Theory Tech.*, vol. 50, pp. 2769-2780, Dec. 2002.

[7] V. K. Devabhaktuni, M.C.E. Yagoub, and Q. J. Zhang, "A robust algorithm for automatic development of neural-network models for microwave applications," *IEEE Trans. Microwave Theory Tech.*, vol. 49, pp. 2282–2291, Dec. 2001.

[8] V. K. Devabhaktuni, B. Chattaraj, M.C.E. Yagoub, and Q. J. Zhang, "Advanced microwave modeling framework exploiting automatic model generation, knowledge neural networks, and space mapping," *IEEE MTT-S Int. Symp. Digest*, Seattle, WA, May 2002, pp. 1097–1100.

[9] T. Fernández *et al.*, "Extracting a bias-dependent large-signal MESFET model from I-V measurements," *IEEE Trans. Microwave Theory Tech.*, vol. 34, pp. 372–378, Mar. 1996

[10] M. Lazaro, I. Santamaria, and C. Pantaleon, "Neural Networks for Large and Small-Signal Modeling of MESFET/HEMT Transistors", *IEEE Transactions on Instrumentation and Measurement*, vol. 50, pp. 1587-1593, Dec. 2001.

[11] K. Hornik, M. Stinchcombe, and H. White, "Multilayer Feedforward networks are universal approximators," *Neural Networks*, vol. 2, pp. 359-366, 1989.

[12] *ADS* 2002, Agilent Technologies, Palo Alt.

HOPFIELD ASSOCIATIVE MEMORY ON MESH

R. A. Ayoubi
University of Balamand
Dept. of Computer Eng.
Tripoli, Lebanon
rafic@balamand.edu.lb

H. A. Ziade
Lebanese university
Faculty of Eng. I
Tripoli, Lebanon
hziade@ul.edu.lb

M. A. Bayoumi
University of Louisiana
Ctr. For Advanced Computer Studies
Lafayette, Louisiana 70504
mab@cacs.louisiana.edu

Abstract: The associative Hopfield memory is a very useful Artificial Neural Network (ANN) that can be utilized in numerous applications. Examples include, pattern recognition, noise removal, information retrieval, and combinatorial optimization problems. This paper provides an algorithm for implementing the Hopfield ANN on mesh parallel architectures. A Hopfield ANN model involves two major operations; broadcasting a value to a set of processors and summation of values in a set of processors. The main advantage of this algorithm is high performance and cost effectiveness. An iteration of an N-bit (neuron) Hopfield associative memory only requires $O(\log N)$ time, whereas other known algorithms in literature of similar topology require $O(N)$ time. Moreover, the proposed algorithm is cost effective because only higher dimension architectures were reported to achieve a complexity of $O(\log N)$ such as hypercubes.

1. INTRODUCTION

The basic associative memory problem can be defined as the storage of a set of patterns in such a way that if a new pattern X is presented, the response is a pattern among the stored patterns which most closely resembles X. This implies that it is possible to recall the complete pattern even if only a part of it is available. This is a powerful concept that can be utilized for many applications like, pattern recognition, image reconstruction from a partial image, noise removal, and information retrieval. The Hopfield discovery in the 1980s [1] produced a very interesting artificial neural network (ANN) model, which can be used as an associative memory and as a solver for combinatorial optimization problems.

ANNs share several common characteristics with massively parallel computers. The Hopfield network (and ANNs in general) consists of a (usually) large number of simple processing units (neurons) with small local memory per neuron. These neurons communicate with one another via links. This makes a highly parallel computing system with simple processing elements and point-to-point communication typical target architectures for efficiently implementing ANNs.

Several mapping schemes have been reported to implement neural network algorithms on parallel architectures. Examples can be found in [2-11]. In this paper, we present a technique to implement the Hopfield network on SIMD mesh massively parallel machines. A major advantage of this technique is high performance. A Hopfield ANN requires two major operations; broadcasting of a value to a set of processors and summation of values residing in a set of processors. These two operations are effectively performed by means of a special mesh communication feature called pipe communication. The proposed algorithm takes $O(\log N)$ time, where N is the size of the network. Another important feature of this method is cost effectiveness. The proposed algorithm achieves comparable performance to more complex architecture such as hypercubes without the added cost.

The rest of the paper is organized as follows. Section 2 provides a brief description of the Hopfield model as well as some terminologies to be used throughout the paper. Section 3 explains the implementation procedure. Section 4 compares our technique with other techniques proposed in the literature. Finally, Section 5 concludes the paper.

2. PRELIMINARIES AND TERMINOLOGY

In this section, we discuss the Hopfield model along with the introduction of some terminologies.

2.1 The Hopfield Associative Memory

A basic ANN model of computation consists of large number of neurons connected to each other by connection weights. Each neuron, say neuron i, has an activation value a_i. Associated with each connection from neuron j to neuron i, is a synaptic weight (or simply, a weight) w_{ij}.

A Hopfield ANN is a time iterative feedback network, which consists of a set of N neurons. The output of each neuron is connected to every other neuron. The connection weights are represented by a weight matrix $W = (w_{ij})$. A main feature of W is that $W = W^t$, that is, W is a symmetric matrix. Initially, a set of patterns $\{X^1, X^2, ..., X^r\}$ is stored in the neural network by setting the weights to appropriate values. The assigned weight values $w_{ij}, 1 \le i, j \le N$ are computed using the Hebbian rule,

$$w_{ij} = \sum_{p=0}^{r} x_i^p x_j^p \qquad i \neq j$$

$$w_{ij} = 0 \qquad i = j \qquad (1)$$

where, $X^P = (x_1^P, x_2^P, ..., x_N^P)$ and $x_i^p \in \{-1,+1\}$. The stored patterns form a set of attractors (local minima for an energy function) in the state space.

At the beginning, the neuron activation values are initialized to the values of an input pattern $I = \{I_1, I_2, ..., I_N\}$ in the state space. From this initial state, the system dynamics drive the ANN to a stable state (attractor). This allows the Hopfield ANN to work as an associative memory and to provide a basic mechanism for signal retrieval and error correction from partial information or from information distorted by noise. The system dynamics are described by:

$$a_i[t] = f(h_i[t]) = f(\sum_j w_{ij} a_j[t-1]) \qquad (2)$$

for $1 \le i \le N$. Here, t refers to time and the function f is usually a step function given by,

$$f(x) = 1 \qquad x \ge 0$$
$$f(x) = -1 \qquad x < 0 \qquad (3)$$

The application of (2) is repeated until all the activation values remain unchanged with further iterations. The activation values then represent the attractor, which best matches the input pattern.

3. MAPPING OF HOPFIELD ON MESH

In this section, we map the Hopfield Network on the SIMD mesh architecture. We start by describing briefly the mesh architecture, which is suitable for processing the Hopfield ANN. After that we show a procedure for mapping the network iterations on the mesh. Finally, we present a technique for stopping the iteration when the network output reaches equilibrium (i.e. does not significantly change with time).

3.1 The Mesh Architecture

The processor array is a simple 2-D mesh, where each processor has four neighbors: North, East, West, and South (NEWS). However, our algorithm does not require the use of the wrap-around connections among boundary processors, although it can execute on such a machine without any modifications. On the other hand, in addition to regular NEWS communications, the processor array can also perform a special kind of communication called pipe NEWS communications. This feature is available on some commercial machines (e.g. MasPar), which speeds up the communications among PEs. In pipe communications, any PE can communicate with another PE in any one of the four directions without the intervention of any intermediate PE along that direction, provided that all intermediate PEs are inactive. In other words, the communication between two processors looks like a direct link (i.e. pipe) regardless how far they are apart horizontally or vertically as long as there are no active PEs along the path. Thus, the communication time overhead becomes the initial setup time to send the packet plus the propagation delay. Since the setup time is the dominant factor, the propagation delay can be ignored and communication steps can be based on the communication setup steps. In contrast, regular NEWS communications between any two non-adjacent PEs require that data be passed to successive PEs along the horizontal or vertical path until its reaches destination. This means that each PE along the path has to check the packet and forward it to its neighbor PE until it arrives to the destination PE . Thus, the overhead of processing the packet at each node is accumulated, whereas in pipe communication there is only the initial overhead.

3.2 Hopfield on Mesh

Basic Iteration

To process an ANN of size N (N neurons) we use an $N \times N$ mesh. Initially, we assume that the weight matrix W , computed by the Hebbian rule, is pre-stored in the local memory of the weight processors such that W_{ij} is stored in processor PE_{ij} . Initializing the ANN state is done by setting the activation values $(a_i[0] \mid 1 \le i \le N)$ to the desired figures $(I_i \mid 1 \le i \le n)$. These activation values are entered into the PEs such that $a_i[0]$ is stored in the local memory of row PE_{1i} .

One network iteration corresponds to moving from row PE_{1i} to column PE_{i1} and vice-versa in the manner explained below. Eq. 2 describes the operations of one iteration at time t (iteration t). Iteration t computes the values $(a_i[t] \mid 1 \le i \le N)$ from the activation values $(a_i[t-1] \mid 1 \le i \le N)$ computed by the preceding iteration at time $t-1$. Equation 2 suggests that the steps involved in one iteration are as follows:

1. Distribute $a_j[t-1]$ to all the elements of column j in the weight matrix W . This is done for all $1 \le j \le N$. This step can be performed in $O(\log N)$ time units as explained below.
2. Multiply $a_j[t-1]$ and w_{ij} for all $1 \le i \le N$ and $1 \le j \le N$. All of these multiplications will be done simultaneously in the various PEs . Therefore, this step will require a single time unit.
3. Sum the results of multiplications of Step 2 along each row of W to compute the weighted sums $(h_i[t] \mid 1 \le i \le N)$. This summation can be computed in $O(\log N)$ time units by employing the method described below.
4. Apply the activation function $f(h_i[t])$ for all $1 \le i \le N$. The function f will be applied concurrently on all the h_i values in the different PEs .

Due to space limitations, the listing of the algorithm has been omitted. However, a description of the algorithm follows. Iterations are done in multiple of two. We start each two iterations at row PE_{1j} and finish the fist iteration at column PE_{j1} . The second iteration starts at column PE_{j1} and ends at row PE_{1j} . First, row PE_{1j} , $1 \le j \le N$, sends its activation value $a_j[t-1]$ column-wise, so that PE_{ij} $1 \le i \le N$, receives $a_j[t-1]$. Then, each PE_{ij} finds the product $w_{ij} a_j[t-1]$. After that, the sum of the product values is done row-wise, and then the result $h_i[t]$ is produced in column PE_{i1} . Finally, column PE_{i1} computes $(a_i[t] = f(h_i[t])$ where; $1 \le i \le N)$.
The next iteration is very similar to the first but travels from column PE_{j1} to row PE_{1j} . This is okay because of the important property of Hopfield networks that $W = W^t$. Therefore it does not matter if we move from rows to columns or the other way around. In both cases, a_i will meet the correct weight value (w_{ij} or w_{ji}). The iterations continue this way going back and forth from rows to columns and vice versa till the ANN reaches a stable state in which further iterations bring about no (or insignificant) change in the activation values.
Now let us take a look at Step 3. This could be best illustrated by an example. Assume that for iteration t , where t is odd, the number of PEs in each row is 16. Initially each even-numbered row sends its data to be summed to each adjacent odd-numbered

row. This could be represented as $PE_{i(2j-1)} \leftarrow PE_{i(2j)} + PE_{i(2j-1)}$ for all $1 \le i \le N$, $1 \le j \le N/2$, i.e. $PE_{i15} \leftarrow PE_{i15} + PE_{i16}$, $PE_{i13} \leftarrow PE_{i13} + PE_{i14}, \ldots, PE_{i1} \leftarrow PE_{i1} + PE_{i2}$. In the second step, for each row starting from the PE which is one before the farthest PE (in this case PE_{i15}) and taking PEs that are a distance of 4 from it, each of these PEs sends its data to be summed in its second east neighbor (i.e., $PE_{i13} \leftarrow PE_{i13} + PE_{i15}$, $PE_{i9} \leftarrow PE_{i9} + PE_{i11}, \ldots, PE_{i1} \leftarrow PE_{i1} + PE_{i3}$). This could be represented as $PE_{i(4j-3)} \leftarrow PE_{i(4j-3)} + PE_{i(4j-1)}$ for all $1 \le i \le N, 1 \le j \le N/4$. For step k, we have $PE_{i(2^k \times j-(2^k-1))} \leftarrow PE_{i(2^k \times j-(2^k-1))} + PE_{i(2^k \times j-(2^{k-1}-1))}$ for all $1 \le i \le N, 1 \le j \le N/2^k$. The final summation result is computed and stored in the first column of PEs $(PE_{i1}, 1 \le i \le N)$. The case for iteration t, for t is even, is similar to the previous case except that the summation is considered column-wise instead of row-wise. Since the weight matrix in Hopfield model is symmetrical (i.e., $W = W^t$), it does not matter if the summation is done row-wise or column-wise. In both cases, a_i will meet the correct weight value (w_{ij} or w_{ji}). Figs.1 and 2 show the weight matrix along with their corresponding activation values for an odd iteration and even iteration respectively on a 4x4 processor array. Fig.3 shows the summation method used for the case of eight data items. Notice that all PEs between the sending PEs and the PEs in a row or a column are inactive. Thus they act as pipes and the sending and receiving PEs act like two adjacent PEs. This leads to $O(\log N)$ in Step 3 for both summation and communication.

Distribution step (step 1) follows tree structure of Fig. 3 going backward. Following the same argument as before, the distribution step takes logarithmic time steps.

Row-Wise Summation ←

$w_{11}[1]*a_1[0]$	$w_{12}[1]*a_2[0]$	$w_{13}[1]*a_3[0]$	$w_{14}[1]*a_4[0]$
$w_{21}[1]*a_1[0]$	$w_{22}[1]*a_2[0]$	$w_{23}[1]*a_3[0]$	$w_{24}[1]*a_4[0]$
$w_{31}[1]*a_1[0]$	$w_{32}[1]*a_2[0]$	$w_{33}[1]*a_3[0]$	$w_{34}[1]*a_4[0]$
$w_{41}[1]*a_1[0]$	$w_{42}[1]*a_2[0]$	$w_{43}[1]*a_3[0]$	$w_{44}[1]*a_4[0]$
$w_{51}[1]*a_1[0]$	$w_{52}[1]*a_2[0]$	$w_{53}[1]*a_3[0]$	$w_{54}[1]*a_4[0]$

Fig. 1: Illustration of data mapping of the weight matrix and the activation values for iteration 1 (time step 1).

Stopping Iteration

In order to detect that there has been no change in any of the activation values, row PE_{1i} and column PE_{i1} keep the activation values from the previous row iterations. Row PE_{1i} compares the activation values from the previous two iterations to the current iteration. If there is a match, then this indicates that a solution is reached and iterations are terminated.

The advantage of our stopping method is that it avoids the communication step to send the activation values computed in column (row) PE_{i1} (PE_{1i}) to the activation values stored in row (column) PE_{1i} (PE_{i1}) from the previous iteration for every two iterations for comparison. This is achieved at the expense of executing an extra iteration.

Transposing a column vector to a row vector or vice versa is a costly operation in mesh architecture because data has to travel diagonally (i.e. horizontal and vertical movement).

$w_{11}[2]*a_1[1]$	$w_{21}[2]*a_1[1]$	$w_{31}[2]*a_1[1]$	$w_{41}[2*a_1[1]$
$w_{12}[2]*a_2[1]$	$w_{22}[2]*a_2 1]$	$w_{32}[2]*a_2[1]$	$w_{42}[2]*a_2[1]$
$w_{13}[2]*a_3[1]$	$w_{23}[2]*a_3[1]$	$w_{33}[2]*a_3[1]$	$w_{43}[2]*a_3[1]$
$w_{14}[2]*a_4[1]$	$w_{24}[2]*a_4[1]$	$w_{34}[2]*a_4[1]$	$w_{44}[2]*a_4[1]$
$w_{15}[2]*a_5[1]$	$w_{25}[2]*a_5[1]$	$w_{35}[2]*a_5[1]$	$w_{45}[2]*a_5[1]$

↑ **Column-Wise Summation**

Fig. 2: Illustration of data mapping of the weight matrix and the activation vector for the second iteration (time step 2).

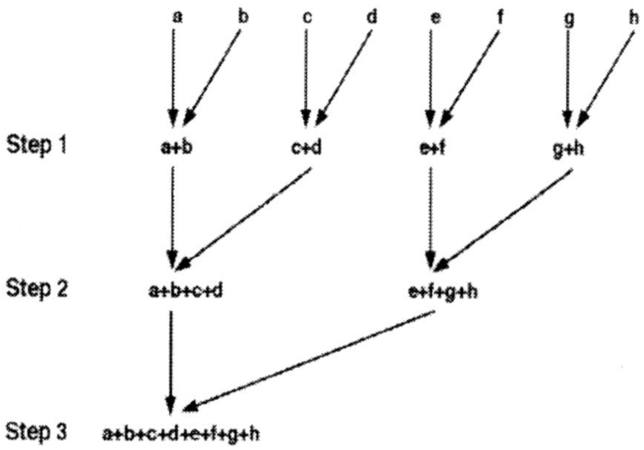

Fig. 3. Example of the summation method used to obtain the weighted sum.

4. PERFORMANCE EVALUATION AND COMPARISON WITH PREVIOUS WORK

Many techniques for mapping ANNs onto parallel architectures have been proposed in literature. In this section we compare our work with previous work in the same field. The first of these was introduced by [5, 6]. It maps ANNs onto systolic ring architecture. The second technique [10] is an extension of Hwang's method in order to make it possible to pipeline more than one pattern. The target machine for this mapping scheme is a two-dimensional SIMD processor array. The third technique [7] implements ANNs on fine grain, mesh-connected SIMD machine. The mapping is established based on a set of congestion-free routing procedures. The fourth technique [10] is another implementation by Shams and Gaudiot on mesh with an enhancement over the previous work. This architecture, called DREAM machine, which is a variation of mesh network with programmable communication links, has been specifically designed for the implementation of neural networks. The fifth technique [5] is based on mapping MLP onto memory-processor integrated architecture, called MPAA. Finally, a comparison with hypercube is provided. The comparison is summarized in Table 1.

Mapping scheme	Architecture	Communi-cations	Multipli-cation	Addition
[7]	Systolic ring	O(*N*)	O(*N*)	O(*N*)
[11]	SIMD 2D mesh	O(*N*)	O(*N*)	O(*N*)
[8]	SIMD 2D mesh	O(*N*)	O(1)	O(*N*)
[10]	SIMD 2D mesh	O(*N*)	O(*N*)	O(*N*)
[5]	SIMD MPAA	O(1)	O(*N*)	O(*N*)
[4, 9]	SIMD hypercube	O(*logN*)	O(1)	O(*logN*)
Our approach	SIMD 2D mesh	O(*logN*)	O(1)	O(*logN*)

Table 1: Comparison with other mapping schemes.

5. CONCLUSION

This paper presents a new efficient algorithm for mapping the operations of the Hopfield associative memory on mesh-connected SIMD machines. The time complexity is of order $O(\log N)$, which is superior to any known algorithm on planar architecture and achieve the same performance of higher degree architecture such as hypercubes without the added cost. This logarithmic complexity is achieved by taking advantage of pipe NEWS communication, which is available on some commercial machines (e.g. MasPar). Moreover, the new algorithm can execute efficiently on most of the known commercial and experimental mesh-connected SIMD machines even if they don't have pipe NEWS communication. However, the communication step would take $O(N)$, while the performance of other steps will not get affected. We should note here that even though our algorithm was shown on Hopfield network, it can be efficiently extended to accommodate other models such as feed forward networks, self-organizing map, and the Boltzmann machine.

REFERENCES

[1] J. J Hopfield, "Neurons with graded response have collective computational properties like those of two-state neurons," in *Proceedings of the National Academy of Sciences*, vol. 79, 1982.

[2] R. A. Ayoubi; M: A: Bayoumi, "An efficient implementation of multi-layer perceptron on mesh architecture," In *Proceedings of IEEE International Symposium on Circuits and Systems* ISCAS 2002., vol.: 2, pp.109-112, May 2002

[3] J. Hwang and S. Kung, "Parallel algorithms/architectures neural networks," *Journal of VLSI Signal Processing*, 1982.

[4] K. Kim and V. K .P .Kumar, "Efficient implementation of neural networks on hypercube SIMD arrays," in *Proceedings of International Joint Conference on Neutral Networks*, vol. 2, pp. 614-617, Washington, DC, 1989.

[5] Y. Kim, M. J. Noh, T. D. Han, and S. D. Kim, "Mapping of neutral networks onto the memory processor integrated architecture," *Neutral Networks*, no. 11, pp. 1083-1098,1988.

[6] S. Y. Kung, "Parallel architectures for artificial neural nets," in *Proceedings of International Conference on Systolic Arrays*, vol 1, pp. 163-174, San Diego, DC, CA, 1988.

[7] S. Y. Kung and J. N. Hwang, "A unified Systolic architecture for artificial neural networks," *Journal of Parallel and Distributed Computers*, no. 6, pp. 358-387, 1989.

[8] W. Lin, V. K. Prasanna, and K. W. Przytula, "Algorithmic mapping of neutral network models onto parallel SIMD machines," *IEEE Transactions on Computers*, vol. 40, no. 12, pp. 1390-1401, 1991.

[9] Q. M. Malluhi, M. A. Bayouni, and T. R. N. Rao, "Efficient mapping of ANNs on hypercube massively parallel machine," *IEEE Transactions on Computers*, vol. 44, no. 6, pp. 769-779, 1995.

[10] S. Shams and J. L. Gaudiot, "Implementing regularly structure neural networks on the DREAM machine," *IEEE Transactions on neural networks*, vol. 6, no. 2, pp. 407-421, 1995.

[11] S. Shams and W. Przytula, "Mapping of neutral network onto programmable parallel machines," in *Proceedings of IEEE International Symposium on Circuits and Systems*, pp. 2613-2617, New Orleans, LA, 1990.

FREQUENCY SENSITIVE SELF-ORGANIZING MAPS AND ITS APPLICATION IN COLOR QUANTIZATION

Chip-Hong Chang and Pengfei Xu

Centre for High Performance Embedded Systems, Nanyang Technological University
Nanyang Avenue, Singapore 639798

ABSTRACT

A new competitive learning algorithm named Frequency Sensitive Self-organizing Maps (FS-SOM) is proposed. It harmonically blends the neighborhood adaptation of the well-known Self-organizing Maps (SOM) with the neuron dependent frequency sensitive learning model. The net effect is an improvement in adaptation, a well-ordered codebook and the alleviation of underutilization problem. A global butterfly jumping sequence is used to permute the input data and the dead neurons are re-initialized in the early phase of the training process. Extensive simulations have been performed to analyze and compare the learning behavior and performance of FS-SOM against other vector quantization algorithms. The results show that FS-SOM is superior in reconstruction quality and topological ordering and its performance is highly robust against variations in network parameters.

1. INTRODUCTION

Neural-network based clustering and vector quantization methods generally fall in the competitive learning (CL) paradigms [4,5]. The computing mechanism of competitive learning scheme is inherently parallel, which promises significant speedup and real-time processing capability when it is realized in massively parallel hardware architecture. Another advantage of CL is its on-line updating property. Unlike the batch algorithms such as the LBG algorithm [9], CL schemes do not have to wait for all the training data to be stored in memory before processing can begin. The simplest and most widely used CL algorithms are based on the *winner-take-all* (WTA) [4,5] (or hard competitive learning) strategy, where adaptation is restricted to the *winner* that is the single neuron prototype best matching the input pattern. Unfortunately, the simple CL algorithm is susceptible to the *underutilization* problem [6]. When it occurs, some prototypes will never become a winner due to bad initialization, and has no contribution to the learning outcome.

To overcome this underutilization problem, Frequency Sensitive Competitive Learning (FSCL) [2,3,8] is proposed by incorporating a frequency sensitive component into the competitive learning rules. The idea is to enhance the winning capability of the less competitive neurons, i.e., potential dead neurons so that they too, have equal opportunity to adapt. Kohonen' Self-organizing Maps (SOM) [1,7,10] is another promising vector quantization technique which employs *winner-take-most* (WTM) strategy [4,5] at the early stage of training and gradually become a WTA when its neighborhood size shrinks to zero. With proper calibration, SOM is capable of achieving good reconstruction quality with low distortion. Besides, it also realizes a topology preserving mapping through the tapering neighborhood updating scheme [7]. The topology preserving property implies data that are close in the input space are mapped to nearby nodes in the output space. This is very useful in many applications, like coding, transmission and data visualization [7].

In this paper, a novel Frequency Sensitive Self-organizing Maps (FS-SOM) is proposed and it is applied to the quantization of color images. It inherits the neighborhood adaptation property of SOM so that greater efficiency will accrue to the coding and transmission of color images from the topology preserving codebook. In contrast to SOM, the learning rate of FS-SOM is adaptive to the winning frequency, or more precisely, the "update counter" of each neuron. Our proposed FS-SOM can be conceived as a generalization of FSCL with the notion of neighborhood adaptation, but it is much more intricate than a simple combination of FSCL and SOM. The adaptation rate and reconstruction performance of many SOM and FSCL algorithms are very sensitive to the setting of the initial state and the network parameters. Quality performance with acceptable training time is often accomplished through scrupulous tuning and customized tailoring of a set of network parameters. One unique feature of our FS-SOM is its robustness against key network parameter variations. In short, the training algorithm has been designed with a holistic consideration of convergent efficiency, computational complexity in parallel implementation and performance stability.

2. PROPOSED FREQUENCY SENTITIVE SOM

Fig. 1 shows the flow chart of the FS-SOM learning algorithm, which is in essence similar to SOM at block level [1,7]. The salient features and novelties that distinguish our proposed FS-SOM from SOM and FSCL are: first, an elegant formulation of a generalized global butterfly jumping sequence is introduced to permute the training data. Second, a new update rule, which combines the idea of neighborhood adaptation in SOM and frequency sensitive learning in FSCL is proposed to enhance its robustness. Third, the knowledge of update frequency is exploited to re-initialize the dead neurons. These three distinctive features will be elaborated in what follows.

2.1 Butterfly Permutation for Input Randomization

Competitive learning has the advantage of being online over the batch algorithms like LBG [9]. However, on-line learning has the order dependent problem, i.e., the final weight vectors are affected by the order of the input sequence. For example, in color quantization, the input data consists of pixel samples from a two dimensional coordinate system. If we input the pixels in a raster scanning order, the pixels towards the bottom-right of the trained image will not be well learned due to the diminishing learning rate. Here we propose a novel global butterfly jumping sequence to permute the images into quasi-repetitive patterns so that the online learning can also be fast and effective.

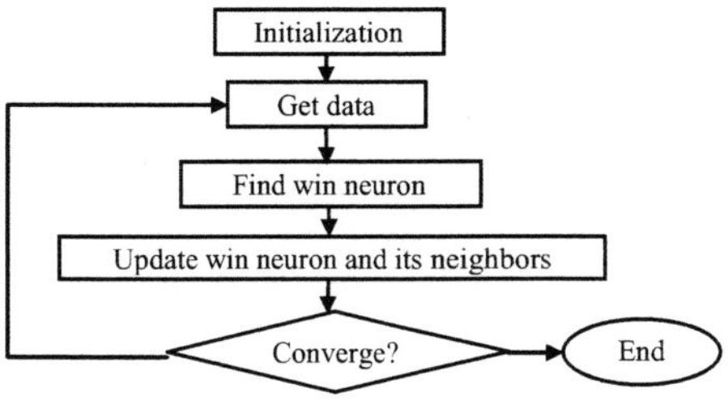

Figure 1. Flow chart of FS-SOM

The butterfly permutation is defined by a mapping, $\pi: Z \to Z^n$ of an input order number $r \in Z$ to a n-dimensional coordinate system, $(x_1, x_2, \ldots, x_n) \in Z^n$ where Z is a finite integer space. For $x_i \in [0, N)$, the butterfly permutation π is given by:

$$\pi(r)=\left\{(x_1,x_2,\ldots,x_n)\in Z^n \mid x_j=\sum_{i=0}^{\lceil \log_2 N\rceil-1} 2^{\lceil \log_2 N\rceil-1-i}\left(r_{ni}\oplus r_{ni+j}\right)\forall j\neq n, x_n=\sum_{i=0}^{\lceil \log_2 N\rceil-1} 2^{\lceil \log_2 N\rceil-1-i} r_{ni}\right\} \quad (1)$$

where r_i is the i-th bit of the binary representation of the decimal number r, and r_0 is its least significant bit.

This global butterfly jumping sequence is a generalization of the 2-D localized butterfly jumping sequence proposed in [10] to span the entire input space without any constraint on the dimensionality of the input vectors. The difference in the permutation effect of the sequences proposed in [10] and ours for a 2-D image is shown in Fig. 2. In Fig. 2, the permuted data are read back in the raster scan order to form a 2-D image to make the spatial characteristics visually perceivable. The results show that, the training data permuted by the global jumping sequence are more evenly distributed in the 1-D sequence. Thus, the main features of the image are learnt at the early possible encounters while the learning rate is still high and similar patterns are learnt repeatedly to enhance competitive learning.

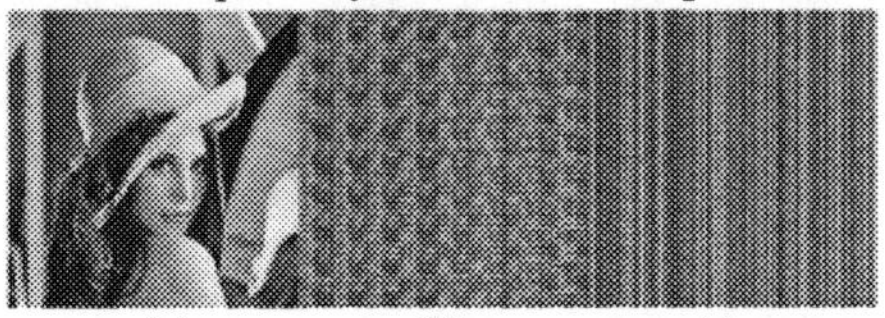

(a) (b) (c)

Figure 2. (a) the original image and the effects of (b) localized butterfly permutation (c) global butterfly permutation

2.2 Updating of Win Neuron and its Neighborhood

As each data point $x(t)$ is fed to the competitive layer, the neuron with the smallest distortion, measured by the L_2 norm, $\|x(t) - w(t)\|$ is deemed the winning neuron. The weight vectors, $w_c(t)$ and the update counters, u_c of the winning neuron, c and its neighborhood are adjusted according to the following equations:

$$w_i(n+1) = w_i(n) + G(u_c)H(i,c,m)[x(n) - w_i(n)] \quad (2)$$

$$u_i(n+1) = u_i(n) + H(i,c,m) \quad (3)$$

where $G(u_c)$ is the frequency sensitive learning rate function. It is a monotonic decreasing function of u_c. Its value is usually bounded in the range (0,1). $H(i, c, m)$ is a tapering neighborhood function which is a hallmark of the SOM algorithms. This neighborhood taper alters the update of each neuron, i within the neighborhood of c by a different amount at discrete time interval, m depending on the distance between the weight vectors w_i and w_c. It is this neighborhood taper that gives SOM its topological ordering property. The following neighborhood taper of [10] is used

$$H(i,c,m) = e^{-\frac{\|C_i - C_c\|^2}{\sigma(m)^2}} \quad (4)$$

where C_i denote the topological coordinate of the neuron i. $\sigma(m) = N\beta k^m$ and m represents the number of sweeps. A sweep is a period during which the updating parameters remain constant. In our simulations, we set the number of input data in one sweep to be 12 times the number of neurons, i.e.,

$$m = \left\lfloor \frac{n}{12N} \right\rfloor \quad (5)$$

$\sigma(m)$ is the neighborhood width at sweep number m. It defines the boundary of the lateral interaction between the winning neuron c and other neurons. β defines the extent of the initial neighborhood interaction. It is usually set to a value between 0 and 0.5, where $\beta = 0.5$ implies that the neighborhood interaction exists in the entire network initially, and $\beta = 0$ implies that there is no neighborhood interaction at all. The constant $k < 1$ and it is used to control the rate at which the boundary of the neighborhood shrinks.

As this mode of learning uniquely characterizes our FS-SOM from other SOM algorithms, it is important to establish a "fast adaptation and slow forgetting" learning function to ensure that the weights settle well during training. We experiment with the nonlinear reciprocal function, $G(u_c) = u_c^{-l}$ for different values of $l > 0$. It is found that $l = 0.5$ gives excellent results for different test images and it is set as the default value for our simulation. The comparison of the proposed learning rate function and some popular learning rate functions [2,10] used by other researchers is shown in Fig. 3.

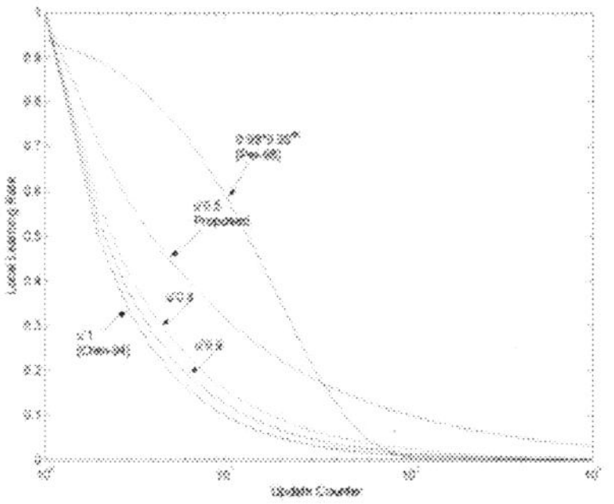

Figure 3. Comparison of different learning rate functions

The proposed learning rate function $G(u_c)$ is dependent on the update counter value of the winning neuron whereas the learning rate function for the SOM algorithms, such as that of [10], is time dependent. We expect that some minority neurons of FS-SOM are still able to make a big learning to the training data even at the later stage of training while the overall learning rate has largely reduced. This main difference in intrinsic learning mechanism will be visualized in the simulation results, and its influence on the results will also be shown.

2.3 Dead neuron reinitialization

To further increase the stability of the algorithm, we will look for dead neurons ($u_i = 0$) at the end of each predefined training interval. If dead neurons exist, they will be reinitialized. Otherwise, the learning process will be continued without further checking for dead neuron. By reinitialization, the weight of the most frequently updated neuron is copied to the dead neuron. The update counters of both neurons will be reset to half of that of the most frequently updated neuron. If the network parameters are well selected, this reinitialization operation will never be needed. However, if one or more of the parameters cause FS-SOM to adapt poorly, the existence of the dead neurons will quickly be detected and corrected by this reinitialization process. In [2], a similar concept but different approach is adopted. The major difference is that: in [2], after checking and reinitializing the dead and inactive neurons, the update counters of all neurons are refreshed to zero and a new learning cycle is started. In our proposed algorithm, we halve the update counter and continue with the learning instead of resetting them to zero and starting a new learning cycle. In this way the detection and reinitialization are more purposeful, faster and effective. Furthermore, we only detect and reinitialize the neurons with zero update rather than all neurons with low winning frequency. This is because the frequency sensitive competitive learning rate function has allowed the less competitive neurons to learn adequately. Only the dead neurons, which have no opportunity to update at all, need to be dealt with. Another reason is that, the minority color can be very important in some color quantization applications. For example, in the images of roadmaps or landscapes, it is useful to preserve the minority colors of prominent landmarks with less utilized but sufficiently trained neurons. It is unwise to sacrifice them completely to trade for a slightly lower global distortion.

3. SIMULATION RESULT AND DISCUSSION

We use color quantization of images to evaluate and demonstrate the performance and property of FS-SOM. First, we study the adaptation behavior of the weights at various discrete points by measuring the mean Euclidean distance, $\delta(m)$ between the weight vectors of FS-SOM calculated at successive sweeps during the training process, i.e.,

$$\delta(m) = \frac{\sum_{i=1}^{N} \left\| w_i(m) - w_i(m-1) \right\|}{N} \tag{6}$$

The mean change in Euclidean distance, $\delta(m)$ is also used as the convergent criterion. When $\delta(m) \leq 0.01$, the training is terminated. In Fig. 4, $\delta(m)$ is plotted against the sweep number for different number of neurons. The solid lines represent the $\delta(m)$ responses of FS-SOMs with $N = 16$ to 256 neurons, and the result of the 16-neuron SOM of [10] is plotted as dotted line for comparison. This figure demonstrates the difference between the learning philosophies of SOM and FS-SOM. In SOM, $\delta(m)$ reduces with time gradually and steadily under an optimally tuned learning rate function which is time-decreasing and neuron-independent. Consequently, the longer the elapse time, the smaller the aggregate change in neuron weights between successive sweeps. Depending on the descending rate of the learning function, such brute force convergence (learning rate diminishes to zero as time elapses) does not necessarily guarantee that all neurons are adequately trained upon convergence. On the other hand, the learning rate function of our FS-SOM is dependent on the winning neuron's update counter. The learning rate is large when a less frequently updated neuron wins at the time of observation and the learning rate is small when a more competitive neuron is being updated. The surges on the curves correspond to the updates of infrequently win neurons. The $\delta(m)$ responses undergo three different phases. During the initial fast adaptation phase, the competitive neurons corresponding to majority colors or denser clusters are rapidly updated and settled. During the equalization phase, the skimpily updated neurons correspond to minority colors or sparse clusters are given opportunity to order themselves, giving rises to a number of spikes. After which FS-SOM enters into a final settlement phase where the response of $\delta(m)$ dips steeply till convergence. This equalization phase and the occasional drastic fluctuations of $\delta(m)$ response clearly mark the unique neuron-dependent learning behavior of FS-SOM.

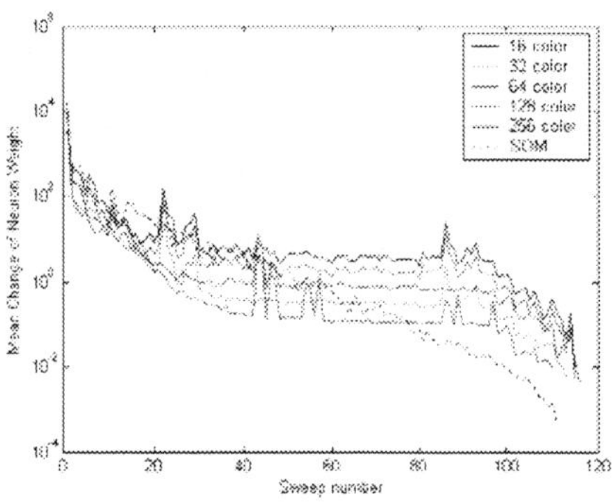

Figure 4. Mean change of the neuron weights during training.

The Peak Signal to Noise Ratio (PSNR) of the reconstructed image is used to measure the quality of the reconstruction during the training process. The PSNR is defined as follows:

$$PSNR = 10 \times \log\left(3 \times 255^2 / MSE\right)$$

$$MSE = \sum_{i=0}^{N_t - 1} \left(X_j - X'_j\right)^2 / N_t \tag{7}$$

where X_j and X_j' are the pixel values of the original and the reconstructed images, respectively, and N_t is the total number of pixels.

Table 1: Comparison of PSNR (in dB) and training time (in sec)

N	Image	CL		LBG [9]		SOM[10]		FS-SOM	
32	Lena	30.71	2.0	32.14	232.5	32.09	1.2	32.12	1.2
	Baboon	26.97	1.2	27.16	319.1	27.05	1.1	27.11	1.3
	Pepper	28.96	1.2	29.24	355.4	29.10	1.6	29.22	1.2
64	Lena	32.53	3.8	34.25	563.6	34.23	4.0	34.25	4.4
	Baboon	28.89	5.7	29.16	817.4	29.01	4.2	29.10	4.7
	Pepper	31.25	4.0	31.26	1484.3	31.40	4.3	31.59	4.3
128	Lena	33.94	13.8	36.08	1623.8	36.08	18.6	36.15	16.5
	Baboon	30.79	19.4	31.08	2009.4	30.91	18.7	31.04	16.7
	Pepper	32.79	16.5	33.49	2212.5	33.48	16.6	33.65	18.9
256	Lena	35.26	56.6	37.78	3783.6	37.74	61.2	37.97	58.8
	Baboon	32.61	57.8	33.00	4423.9	32.66	56.1	32.96	63.0
	Pepper	34.13	57.0	35.40	3806.5	35.14	77.0	35.44	61.0

The performance of FS-SOM in terms of PSNR (left) and training time (right) are compared against various competitive learning schemes in Table 1. The convergence criteria for different algorithms are kept the same and the training time is obtained from executing the algorithms on the Pentium Pro. 2.1 GHz PC with 512 Mbytes of system memory.

From Table 1, the basic competitive learning CL gives the worst performance due to severe underutilization problem. LBG delivers much better performance but consumes large memory and has long computation time. SOM and FS-SOM both have online learning capability and FS-SOM outperforms SOM slightly.

One distinct advantage of our FS-SOM algorithm is its performance robustness. As the performances of SOM and other quantization algorithms reach the margins of diminishing return, they become highly susceptible to the variation of parameters and input data distribution. Besides the parameters that control the learning rate function, two other most important parameters common to the FS-SOM and SOM [10] training algorithms are β and k. These parameters determine the neighborhood interaction and they need to be empirically tuned to achieve optimal quantizer performance. The performance variations over a reasonable range of these two parameters are studied for FS-SOM and SOM. The simulation results are compared in Fig. 5. The test image is Lena, and the size of the network is 16.

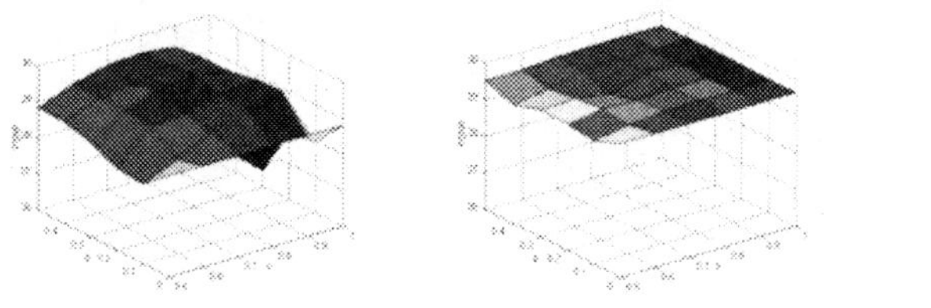

(a) SOM (b) FS-SOM

Figure 5. PSNRs for different combinations of β and k.

Fig. 5 shows that the performance of SOM fluctuates conspicuously with the training parameters. In the best case, it can achieve a PSNR value of 29.62 dB. Its worst PSNR can drop dramatically down by more than 3 dB to around 26.30 dB. On the contrary, our proposed FS-SOM algorithm has a very flat PSNR response at around 29.60 dB over all combinations of β and k. The difference between the peak and valley of Fig. 5(b) is less than 0.4 dB.

The topology mapping property of FS-SOM and the current art SOM in color quantization [10] is also compared with different sets of parameters. In the 1-D string neighborhood structure, the topological order can also be measured by an objective function J, which is defined as [7]:

$$J = \sum_{i=2}^{N} \|\mathbf{w}_i - \mathbf{w}_{i-1}\| - \|\mathbf{w}_N - \mathbf{w}_1\| \tag{8}$$

where N is the number of neurons and w_i is the weight vector of the i-th neuron. The lower the value of J, the better the codebook is ordered. The results are shown in Fig. 6.

From Fig. 6, the J value of SOM is very unstable, varying from 13 to 1173 in Fig. 6(a). On the contrary, the J value of our proposed FS-SOM in Fig. 6(b) is very stable, changing within a narrow range from 293 to 383. FS-SOM gives smaller J than SOM in 24 of the 30 combinations of β and k, and of the 6 combinations for which SOM has smaller J values, its PSNRs happen to be the worst as observed in Fig. 5(a).

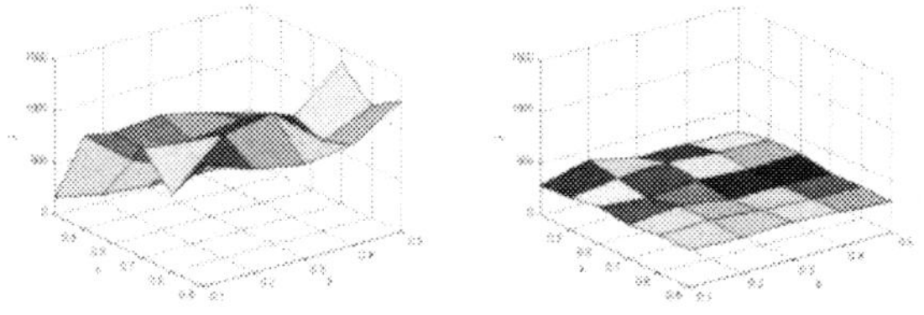

(a) SOM (b) FS-SOM

Figure 6. Topology quality J for different β and k.

4. CONCLUSION

This paper presents a new FS-SOM featuring a harmonious blend of global butterfly-jumping sequence for input data presentation, frequency sensitive learning scheme attuned to neighborhood adaptation and dead neuron reinitialization technique. This consortium of unique features contributes directly to the elevation of quantization performance and better topological order of the resultant codebook. The proposed FS-SOM is efficient and the performance on color quantization applications is proven to be more robust against variation in network parameters than the current art SOM.

REFERENCES

[1]C. H. Chang, R. Xiao and T. Srikanthan, "An adaptive initialization technique for color quantization by self organizing feature map," in *Proc. IEEE Int. Conf. on Acoustics, Speech and Signal Processing* (ICASSP-2003), Hong Kong, Vol. 3, pp. 477-480, April 2003.

[2] O. T.-C. Chen, B. J. Sheu, and W. C. Fang, "Image Compression Using Self-Organization Networks," *IEEE Trans. Circuits Syst. Video Techno.*, vol. 4, pp. 480-489, Oct. 1994.

[3] A. S. Galanopoulos, R. L. Moses, and S. C. Ahalt, "Diffusion Approximation of Frequency Sensitive Competitive Learning," *IEEE Trans. on Neural Networks*, vol. 8, no. 5, pp. 1026-1030, Sep. 1997.

[4] S. Haykin, *Neural Networks: A Comprehensive Foundation.* Prentice Hall, 1998

[5] J. Hertz, A. Krogh, and R. G. Palmer, *Introduction to the Theory of Neural Computation.* New York: Addison-Wesley, 1991.

[6] R. Hecht-Nielsen, "Counterpropagation networks," *Appl. Opt.*, vol. 26, pp. 4979-4984, 1987.

[7] T. Kohonen, *Self-organizing Maps.* Berlin, New York: Springer-Verlag, 2001.

[8] A. K. Krishnamurthy, S. C. Ahalt, D. E. Melton and P. Chen, "Neural Networks for VQ of Speech and Images," *IEEE Journal on Select. Areas in Commun.*, vol.38, no.1, pp. 25-29, Feb. 1992.

[9] Y. Linde, A. Buzo and R. M. Gray, "An algorithm for vector quantizer design," *IEEE Trans. Commun.*, vol. COM-28, pp. 84-95, 1980.

[10] S. C. Pei and Y.S. Lo, "Color image compression and limited display using self-organization Kohonen map," *IEEE Trans. Circuits Syst. Video Techno.*, vol. 8, no. 2, pp. 191-205, 1998.

LEARNING VECTOR QUANTIZATION: CLUSTER SIZE AND CLUSTER NUMBER

Christian Borgelt

School of Computer Science
University of Magdeburg, Universitätsplatz 2
D-39106 Magdeburg, Germany
e-mail: borgelt@iws.cs.uni-magdeburg.de

Daniela Girimonte, Giuseppe Acciani

Department of Electrotechnics and Electronics
Polytechnic of Bari, Via Re David 200
I-70125 Bari, Italy
e-mail: {girimonte,acciani}@deemail.poliba.it

ABSTRACT

We study learning vector quantization methods to adapt the size of (hyper-)spherical clusters to better fit a given data set, especially in the context of non-normalized activations. The basic idea of our approach is to compute a desired radius from the data points that are assigned to a cluster and then to adapt the current radius of the cluster in the direction of this desired radius. Since cluster size adaptation has a considerable impact on the number of clusters needed to cover a data set, we also examine how to select the number of clusters based on validity measures and, in the context of non-normalized activations, on the coverage of the data.

1. INTRODUCTION

Learning vector quantization (LVQ) [7] is a well-known prototype based clustering method, which describes a cluster by a center and possibly some size and shape parameters. It tries to adapt these parameters in order to fit the clusters to a given data set. Closely related approaches are k-means clustering [5, 4] and fuzzy clustering [2, 6].

In this paper we focus on LVQ variants that do not rely on a normalization of the activations. Such methods are especially desirable, because the assignments of data points to clusters that result from them are generally more intuitive. However, these methods suffer from the drawback that normalization is an important means to achieve a mutual dependence of the individual prototypes, without which several clusters may end up in the same position. To solve this problem, we employ a restricted hard clustering/winner takes all scheme, which leads to a (limited) dependence of the clusters without removing the possibility to have data points that belong to some degree to several clusters.

2. LEARNING VECTOR QUANTIZATION

Learning vector quantization (LVQ) [7] is a well-known method to form a quantized approximation of the distribution of an input data set $\mathbf{X} \subset \mathbf{R}^p$ using a finite number k of reference vectors $w_i \in \mathbf{R}^p$, $i = 1, 2, ..., k$. These vectors are stored in the connection weights of a neural network with two layers, which is trained with competitive learning: for each input $\vec{x}_j$ the closest reference vector $\vec{w}_c$ is determined (i.e. $c = \arg\min_i\{|\vec{x}_j - \vec{w}_i(t)|\}$). The corresponding neuron "wins" the competition and is updated according to

$$\vec{w}_c(t+1) = \vec{w}_c(t) + \alpha(t)(\vec{x}_j - \vec{w}_c(t)),$$

where α, $0 < \alpha < 1$, is a learning rate, which, starting from an initial value α_0, reduces monotonically to zero (for example, according to $\alpha(t) = \alpha_0 \cdot \eta^t$, $0 < \eta < 1$).

In the standard version of LVQ only the "winner neuron" is updated. However, the similarity of LVQ to k-means clustering [5, 4] and thus also to fuzzy clustering [2, 6] suggests a softened version, in which the weights are updated according to the activation of a neuron, which may be computed from a Cauchy ($f_{\text{Cauchy}}(x) = \frac{1}{d^2/\sigma^2+1}$) or Gaussian activation function ($f_{\text{Gauss}}(x) = \exp(-\frac{1}{2} \cdot d^2/\sigma^2)$). Both of these functions depend on the ratio between the Euclidean distance d and a (user-specified) reference radius σ. In this case several neurons may be updated for each data point.

3. NORMALIZATION AND DEPENDENCE

In softened LVQ (as well as in standard fuzzy clustering) the neuron activations are usually normalized to sum 1 over the neurons/clusters, so that each data point has the same weight. This normalization makes the clusters dependent on each other: whatever is gained in data point coverage by one cluster must be lost by another. The same holds, of course, also for hard LVQ with its "winner takes all" approach. Mutual cluster dependence is very important for successful clustering, because it drives the clusters apart and thus makes sure that all data points are covered. However, it also has its drawbacks. For example, situations can occur in which the (normalized) activation increases even though one moves away from a reference vector [6].

To obtain a more intuitive cluster description, different suggestions have been made. One of them is possibilistic

0-7803-8251-X/04/$17.00 ©2004 IEEE

fuzzy clustering [8], in which there is no normalization of membership degrees. Hence it can yield very intuitive cluster descriptions. Unfortunately, possibilistic fuzzy clustering suffers from the fact that its objective function is truly minimized only if all cluster centers are identical. In practice reasonable results are achieved only because the algorithm gets stuck in local optima. But even then, clusters tend to merge if they are not very well separated.

In this paper we draw on the idea of possibilistic fuzzy clustering and do not normalize the activations to sum 1. We try to overcome the resulting drawbacks by introducing a mutual dependence of the clusters through a restricted winner takes all scheme. If a data point has a distance from a reference vector that is less than a reference radius, it is assigned exclusively to the corresponding neuron (or, more generally, to the neuron yielding the highest non-normalized activation) and only this neuron is updated. For a data point outside the "winner takes all regions" non-normalized activations are computed and several neurons may be updated.

4. CLUSTER SIZE

To obtain a more flexible clustering scheme, one may make the reference radius σ in the activation functions neuron-dependent, updating it in each iteration, so that clusters of different size can be found. The general idea of the update is to compute a desired reference radius from the data points assigned to a cluster center/reference vector and then to

- set the reference radius to this desired radius or to
- change the current reference radius in the direction of the desired radius using a learning rate as for the update of the reference vectors.

The simplest choice for a desired radius is the average distance of the data points to a reference vector (or, alternatively, the square root of the average squared distance), with the data points weighted with the neuron activation. If online training is used for LVQ, a similar behavior can be achieved by updating the current radius according to

$$\sigma_i(t+1) = \sigma_i(t) + \alpha(t)(d(\vec{x}, \vec{w}_i(t)) - \sigma_i(t)),$$

where α may be the same learning rate as the one that is used for the neuron weights. In [1] a slightly more complex scheme is used, which distinguishes whether a data point is inside the (hyper-)sphere defined by the current radius (then only this radius is decreased) or outside the radius (hyper-)spheres of all clusters (then all radii are increased).

Other approaches are based on the relative weight of assigned data points, thus trying to find clusters that do not differ too much in the number of data points they cover. An example is frequency sensitive competitive learning [9], in which the distance to a reference vector is modified according to the number of data points that are assigned to this reference vector, i.e.

$$d_{\text{mod}}(\vec{x}_j, \beta_i) = \frac{n_i}{n} d(\vec{x}_j, \beta_i),$$

where n_i is the number of data points assigned to reference vector β_i in the previous epoch and n is the total number of data points. Obviously, this is equivalent to using a reference radius $r = \frac{n}{n_i}$ to modify the activation.

Drawing on this idea, we may also state explicitely that our goal is to assign (roughly) the same number of data points to each cluster. That is, we desire $\frac{n}{c}$ data points per cluster, where n is the total number of data points and c the number of clusters. If a given radius r leads to an assignment of n_i data points, the desired radius is computed as

$$r_{\text{desired}} = r \cdot \frac{n}{c \cdot n_i}.$$

The rationale is to decrease the radius if the desired number of data points is less than the current number and to increase it if it is greater, thus balancing the number of data points.

It should be noted that if we do not normalize the neuron activations, size adaptation can be slightly problematic, because in this case the sum of the activations over all neurons and all data points will, in general, differ from the total number of data points. Depending on the method to determine the desired radius, this can lead to collapsing clusters in some cases (e.g., if the average distance is computed from distances that are weighted with the activation). To cope with this problem, we introduce a parameter by which we multiply the computed desired radius before we use it to adapt the current reference radius.

5. CLUSTER NUMBER

One of the main problems in clustering is how to determine the optimal number of clusters. Usually the user has to specify how many clusters are to be found. Automatic approaches rely, for instance, on so-called *validity measures*, with which a given clustering result can be assessed, so that the best cluster number can be determined. Well-known validity measures, developed for classical and fuzzy clustering, are [6]: The fuzzy hypervolume (FH), the partition density (PD), and the average partition density (APD).

If we do not normalize the activations, another simple scheme suggests itself: With each new cluster some more data points should be covered. Hence we can compute the (absolute) coverage of the data as the sum of the activations over all data points and all clusters and stop adding clusters once the relative coverage (i.e. the absolute coverage divided by the number of data points) exceeds some user-defined threshold. Alternatively, we may stop adding another cluster if adding it increases the coverage only by a small amount, so that it is likely that the new clusters mainly steels data points from other clusters.

6. EXPERIMENTAL RESULTS

To illustrate the properties of different size adaptation methods in the context of non-normalized activations, we conducted experiments on simple artificial data sets with two point clouds of different size and with different numbers of points, each at two different distances from each other (see Figures 1, 2, and 3). To each of these data sets we applied five variants of generalized LVQ. The cluster are depicted as grey dots for the centers, dark grey circles at the reference radius and light grey circles at two times this radius.

Diagrams a_1 and a_2 show the result of standard LVQ (winner takes all) with adaptive cluster size. For diagrams b_1 and b_2 we used activation normalization to sum 1 and the same desired radius for all clusters (average of the individual desired radii). For diagrams c_1 and c_2 we also used activation normalization to sum 1, but individual sizes for each cluster. Diagrams d_1, d_2, e_1 and e_2 show results of clustering without activation normalization, but with a restricted winner takes all scheme (if a data point is inside the dark grey circle for the winner neuron, only this neuron is updated). In diagrams d_1 and d_2 the desired radius is twice the average distance of the data points (weighted with the activation) from the cluster center, in diagrams e_1 and e_2 we applied the scheme that balances the number of data points per cluster with reference radius factor of 0.8.

For the first two data sets (Figure 2, same cluster size, same number of points) all algorithms work very well, even though in the second data set the two clusters are very close together. Note that the restricted winner takes all approach can compete well with the other approaches.

For the second pair of data sets (Figure 2, different cluster size, same number of points) results are less good. A strict winner takes all almost fails since one cluster almost collapses (diagram a_2), i.e., the size adaptation interferes harmfully with the cluster detection. The approach that tries to capture the same number of points in each cluster (diagrams e_1 and e_2) yields the best result—not surprisingly, because it is tailored for such situations.

The last pair of data sets (Figure 3, different size, difference number of points, same density) leads to the biggest problems. Here only the strict winner takes all approach (diagrams a_1 and a_2) leads to acceptable results, while all other approaches have severe problems getting the cluster positions right. However, the results based on a restricted winner takes all scheme and no normalization can still clearly compete with the other approaches.

For our experiments on selecting the number of clusters by the coverage of the data points we chose the well-known wine data sets from the UCI machine learning repository [3]. We used the attributes 7, 10, and 13, which are most relevant. The results for clustering with restricted winner takes all and no normalization are shown in table 1. Clearly the coverage of the data points gives a very good indication of the correct number of clusters (there are three classes), because the coverage reduces if four clusters are used. For other data sets we tried, the coverage could also compete with the known validity measures.

wine	1	2	3	4	5
FH	9.35	10.6	6.65	6.63	6.77
PD	12.5	11.1	18.1	18.1	18.2
APD	12.5	20.8	3119	2575	4266
coverage	130	131	160	157	163

Table 1. Selecting the number of clusters.

7. CONCLUSIONS

In this paper we considered LVQ methods to adapt the size of (hyper-)spherical clusters, especially in the context of non-normalized activations. Our experimental results show that, although normalization clearly helps the clustering process and stabilizes it, usable results can often be achieved without normalization if the desired radius is multiplied by a user-specified parameter to achieve stability (where necessary). It also turns out that without normalization a consideration of the coverage of the data is a feasible method to determine a good number of clusters.

8. REFERENCES

[1] G. Acciani, E. Chiarantoni, G. Fornarelli, and S. Vergura. A Feature Extraction Unsupervised Neural Network for an Environmental Data Set. *Neural Networks* 16(3–4):427–436. Elsevier Science, Amsterdam, Netherlands 1999

[2] J.C. Bezdek. Pattern Recognition with Fuzzy Objective Function Algorithms. Plenum Press, New York, USA 1981

[3] C.L. Blake and C.J. Merz. UCI Repository of Machine Learning Databases. Department of Information and Computer Science, University of California, Irvine, CA, USA 1998. http://www.ics.uci.edu/~mlearn/MLRepository.html

[4] B.S. Everitt. *Cluster Analysis*. Heinemann, London, UK 1981

[5] J.A. Hartigan and M.A. Wong. A k-means Clustering Algorithm. *Applied Statistics* 28:100–108. Blackwell, Oxford, UK 1979

[6] F. Höppner, F. Klawonn, R. Kruse, and T. Runkler. *Fuzzy Cluster Analysis*. J. Wiley & Sons, Chichester, UK 1999

[7] T. Kohonen. *Self-Organizing Maps*. Springer-Verlag, Heidelberg, Germany 1995 (3rd ext. edition 2001)

[8] R. Krishnapuram and J. Keller. A Possibilistic Approach to Clustering. IEEE Transactions on Fuzzy Systems, 1:98–110. IEEE Press, Piscataway, NJ, USA 1993

[9] D. DeSieno. Adding a Conscience to Competitive Learning. *IEEE Int. Conf. on Neural Networks*, Vol. I, 117–124. IEEE Press, Piscataway, NJ, USA 1988

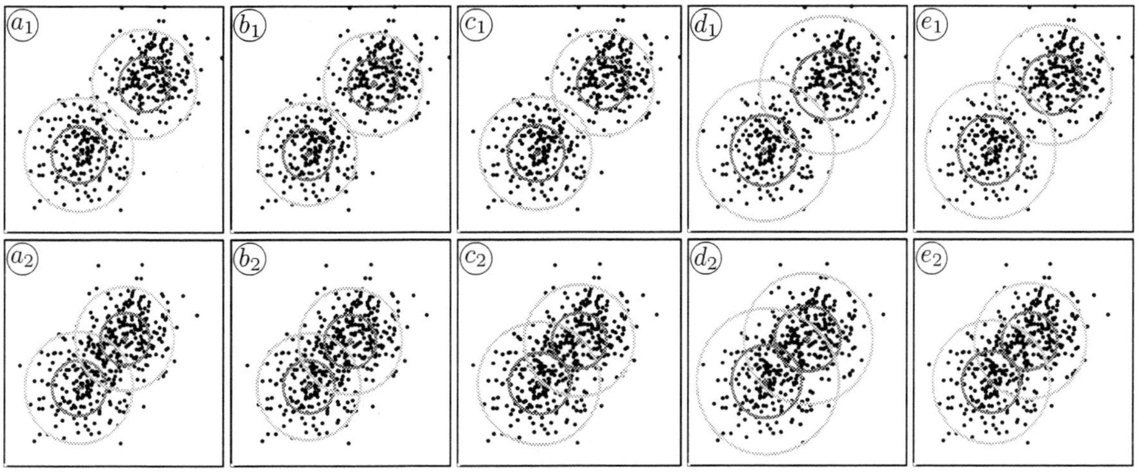

Fig. 1. Same cluster size, same number of points.

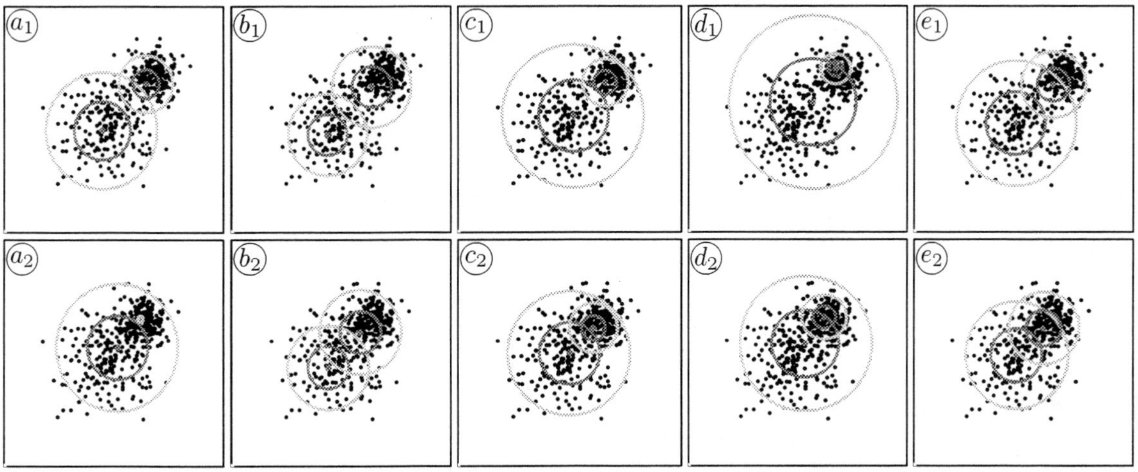

Fig. 2. Different cluster size, same number of points.

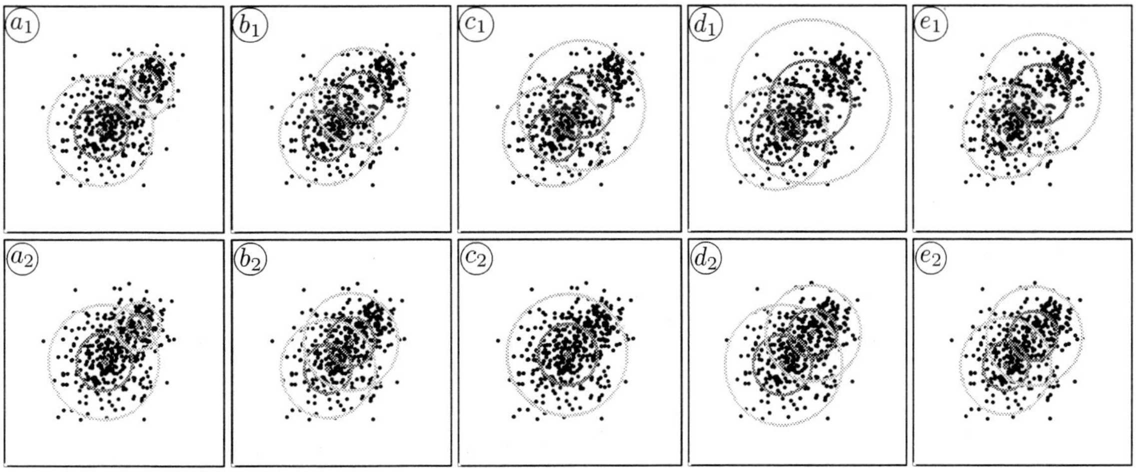

Fig. 3. Different cluster size, different number of points, same point density.

FAST LEARNING ALGORITHMS FOR NEW L2 SVM BASED ON ACTIVE SET ITERATION METHOD

Juan-juan Gu

Dept. of Computer & Information Engineering
Hefei Association University
Hefei, Anhui 230022, P. R. China
j_j_gu@hotmail.com

Liang Tao

Dept. of Electronic Engineering
Anhui University
Hefei, Anhui 230039, P. R. China
taoliang@ustc.edu

H. K. Kwan

Dept. of Electrical & Computer Engineering
University of Windsor
401 Sunset Avenue, Windsor, Ontario,
Canada N9B 3P4
kwan1@uwindsor.ca

ABSTRACT

A new L2 soft margin support vector machine (new L2 SVM) is introduced in this paper. What is unusual for the SVM is that the dual problem for the constrained optimization of the SVM is a convex quadratic problem with simple bound constraints. The active set iteration method for this optimization problem is applied as a fast learning algorithm for the SVM, and the selection of the initial active/inactive sets is discussed. For incremental learning and large-scale learning problems, a fast incremental learning algorithm for the SVM is presented. Computational experiments show the efficiency of the proposed algorithms.

1. INTRODUCTION

The support vector machine (SVM) developed by V. Vapnik and his team at AT&T Bell Labs can be seen as a new way to train polynomial, neural network, or Radial Basis Function classifiers, based on the idea of structural risk minimization rather than empirical risk minimization. Some classical problems such as multi-local minima, curse of dimensionality and overfitting in neural networks [1], seldom occur in support vector machines. In the past few years, support vector machines (SVM) have generated a great interest in the community of machine learning due to its excellent generalization performance in a wide variety of learning problems, such as handwritten digit recognition [2], face detection [3], and face recognition [4]. However, training support vector machines is still a bottleneck, especially for a large-scale learning problem [2]. Therefore, it is important to develop a fast training algorithm for SVMs to facilitate its applications to various engineering problems.

In this paper, a new L2 soft margin support vector machine (new L2 SVM) is introduced. What is unusual for the SVM is that the dual problem for the constrained optimization of the SVM is a convex quadratic problem with simple bound constraints, which enables the active set iteration method [5] to be applied as a fast learning algorithm for the SVM. Moreover, for incremental learning and large-scale learning problems, a fast incremental learning algorithm for the SVM is presented. Computational experiments are carried out to show the efficiency of the proposed algorithms

2. NEW L2 SVM AND ITS FAST LEARNING ALGORITHMS

2.1 New L2 SVM

Given that training samples $\{\boldsymbol{x}_i, y_i\}_{i\in N_{\mathrm{all}}}$ N_{all}:={1, 2, · · ·, N}, $\boldsymbol{x}_i \in \mathrm{R}^d$, $\boldsymbol{x}_i = [x_i^1 \ x_i^2 \ \cdots \ x_i^d]^{\mathrm{T}}$, $y_i \in \{+1, -1\}$, where y_i is the class label. In soft margin support vector machine, we consider the linear decision function

$$D(\boldsymbol{x}) = \sum_{j=1}^{M} w_j \varphi_j(\boldsymbol{x}) + b \ = \boldsymbol{w}^{\mathrm{T}} \varphi(\boldsymbol{x}) + b \qquad (1)$$

in the feature space, where $\boldsymbol{w} = [w_1 \ w_2 \ \cdots \ w_M]^{\mathrm{T}}$ is the weight vector, $\varphi(\boldsymbol{x})=[\varphi_1(\boldsymbol{x}) \ \ \varphi_2(\boldsymbol{x}) \ \cdots \ \varphi_M(\boldsymbol{x})]^{\mathrm{T}}$ is the mapping function that maps the input $\boldsymbol{x}$ into the feature space, b is a scalar. The L2 soft margin support vector machine (L2 SVM) presented in [6] requires the solution of the following optimization problem:

$$\begin{aligned} &\underset{\boldsymbol{w}}{\text{Min}} \ \frac{1}{2}(\boldsymbol{w}^{\mathrm{T}}\boldsymbol{w}) + \frac{C}{2}\left(\sum_{i=1}^{N}\xi_i^2\right) \\ &\text{s.t.} \ \ y_i(\boldsymbol{w}^{\mathrm{T}}\varphi(\boldsymbol{x}_i) + b) - 1 + \xi_i \geq 0 \qquad (2) \\ &\qquad \xi_i \geq 0, \ i = 1, 2, \cdots, N \end{aligned}$$

where $\xi = [\xi_1 \xi_2 \ \cdots \ \xi_N]^{\mathrm{T}}$, its elements are the positive slack variables, N is the number of training samples, C is the margin parameter.

A new L2 soft margin support vector machine (new L2 SVM) was presented in [4] and [7] via a little change of the cost function in (2).

Denote $W=[w_1\ w_2\ \cdots\ w_M\ b]^{\mathrm{T}}$, and $\Phi(x)=[\varphi_1(x)\ \varphi_2(x)\cdots\varphi_M(x)\ 1]^{\mathrm{T}}$. The linear decision function (1) can be rewritten as:

$$D(x)=W^{\mathrm{T}}\Phi(x) \tag{3}$$

We replace w in the cost function in (2) with W, and (2) becomes:

$$\begin{aligned}&\min_{w}\ \frac{1}{2}(w^{\mathrm{T}}w+b^2)+\frac{C}{2}\left(\sum_{i=1}^{N}\xi_i^2\right)\\ &\text{s.t.}\ \ y_i(w^{\mathrm{T}}\varphi(x_i)+b)-1+\xi_i\geq 0\\ &\qquad \xi_i\geq 0,\ i=1,2,\cdots,N\end{aligned} \tag{4}$$

The SVM described by (4) is called the new L2 soft margin support vector machine (new L2 SVM). The dual problem for the new L2 SVM can be written as

$$\begin{aligned}&\min_{\alpha}\ \frac{1}{2}\alpha^{\mathrm{T}}Q\alpha-\mathrm{e}^{\mathrm{T}}\alpha\\ &\text{s.t.}\ \ \alpha\geq 0\end{aligned} \tag{5}$$

where e is an $N\times 1$ vector with all the elements being 1; $\alpha=[\alpha_1\ \alpha_2\ \cdots\ \alpha_N]^{\mathrm{T}}$ is the Lagrange multiplier vector; $Q=[Q_{ij}]_{i,j\in N_{\text{all}}}$,

$$Q_{ij}=y_iy_j\big(K(x_i,x_j)+1\big)+\frac{\delta_{ij}}{C} \tag{6}$$

where $\delta_{ij}=1$ if and only if $i=j$, and $K(x_i,x_j)$ is the kernel function,

$$K(x_i,x_j)=\varphi(x_i)^{\mathrm{T}}\varphi(x_j)=\sum_{m=0}^{M}\varphi_m(x_i)\varphi_m(x_j)\quad \text{for}\ \ i,j=1,2,\cdots,N \tag{7}$$

Note that Q is a positive definite matrix, and the dual problem for the new L2 SVM described by (5) is a convex quadratic minimization problem with simple bound constraints.

Researches and experiments in [4] and [7] have shown that the generalization performance of the new L2 SVM is very close to that of traditional SVMs although they have many differences; therefore, in the remaining of this paper, we will focus on the research on the fast learning algorithm for the new L2 SVM.

2.2 Fast learning algorithm for new L2 SVM

For the convex quadratic minimization problems with simple bound constraints like (5), [5] presented a simple and fast algorithm based on the active set iteration method. We will apply it to the dual problem (5) of the new L2 SVM.

The following notations will be used throughout. For a subset $A\subseteq N_{\text{all}}:=\{1,2,\cdots,N\}$, we write α_A for the components of α indexed by A, i.e. $\alpha_A:=(\alpha_i)_{i\in A}$. The complement of A will be denoted by $\bar{A}$. If Q is a matrix and A and B are subsets of N_{all}, then $Q_{A,B}$ is the submatrix of Q, with rows indexed by A and columns indexed by B.

The Karush-Kuhn-Tucker (KKT) system for (5) is given by

$$Q\alpha-\mathrm{e}-z=0 \tag{8}$$

$$z\bullet\alpha=0 \tag{9}$$

$$z\geq 0 \tag{10}$$

$$\alpha\geq 0 \tag{11}$$

where $z=[z_1\ z_2\ \cdots\ z_N]^{\mathrm{T}}$ is the Lagrange multiplier vector, we write $z\bullet\alpha$ to denote the vector of element-wise products, i.e. $z\bullet\alpha:=(z_i\alpha_i)_{i\in N_{\text{all}}}$.

The crucial step in solving (5) is to identify those inequalities which are active, i.e. the active set $A\subseteq N_{\text{all}}$, where the solution to (5) satisfies $\alpha_A=0$. Then, with the inactive set $I:=\bar{A}=N_{\text{all}}\backslash A$, we must have $z_I=0$, and the elements of I are just the indexes of support vectors in $\{x_i\}_{i\in N_{\text{all}}}$.

To compute the remaining elements α_I and z_A of α and z, [5] proposed an active set iteration algorithm by using (8) and partitioning the equations and variables according to the active set A^k and the inactive set I^k at the k-th iteration:

$$\begin{bmatrix}Q_{A^k,A^k} & Q_{A^k,I^k}\\ Q_{I^k,A^k} & Q_{I^k,I^k}\end{bmatrix}\begin{bmatrix}0\\ \alpha_{I^k}\end{bmatrix}-\begin{bmatrix}\mathrm{e}_{A^k}\\ \mathrm{e}_{I^k}\end{bmatrix}-\begin{bmatrix}z_{A^k}\\ 0\end{bmatrix}=0 \tag{12}$$

The second set of equations can be solved for α_{I^k}, because Q_{I^k,I^k} is by assumption positive definite:

$$\alpha_{I^k}=Q_{I^k,I^k}^{-1}\mathrm{e}_{I^k} \tag{13}$$

and the first set of equations can be solved for z_{A^k},

$$z_{A^k}=Q_{A^k,I^k}\alpha_{I^k}-\mathrm{e}_{A^k} \tag{14}$$

If $\alpha_{I^k}\geq 0$ and $z_{A^k}\geq 0$, then stop the iteration; otherwise, let

$$A^{k+1}=\{j\,|\,\alpha_j<0\ \text{or}\ z_j>0\} \tag{15}$$

and the $(k+1)$-th iteration is continued.

[5] provided sufficient conditions for the iterations to converge in a finite number of steps with an optimal solution. Computational experiments in [5] indicated that this algorithm often requires only a few iterations to find the optimal solution.

2.3 Selection of the initial active/inactive sets A^0 and I^0

From (13), we note that the inverse matrix of Q_{I^k,I^k} has to be computed at each iteration, which needs huge computation when the size of I^k becomes very large. In fact, the size of I^k is usually small when the iteration algorithm applied to the SVM converges, because at the final iteration, the elements of I^k are just the indexes of support vectors in $\{\boldsymbol{x}_i\}_{i\in N_{\text{all}}}$ and the number of support vectors is usually much less than N. Therefore, a specific selection method of the initial active/inactive sets A^0 and I^0 is given as follows in order to control the size of I^k, especially, and to avoid the occurrence of the worse case I^k=N_{all} or I^k=∅ in the iteration process.

- Suppose that there are two classes of samples, C_a and C_b, with a total of N samples, indexed from 1 to N, and the sample number of C_a is less than or equal to that of C_b, i.e., the sample number of C_a is less than or equal to 0.5N.
- If the sample number of C_a is less than 0.25N, combine all the samples of C_a and some samples of C_b into a total of 0.5N samples, select the indexes of the 0.5N samples as an initial inactive set I^0, and the complement of I^0 in N_{all} as an initial active set A^0.
- If the sample number of C_a is more than or equal to 0.25N, combine 0.25N samples of C_a and 0.25N samples of C_b into a total of 0.5N samples, select the indexes of the 0.5N samples as an initial inactive set I^0, and the complement of I^0 in N_{all} as an initial active set A^0.

2.4 Fast incremental learning algorithm for new L2 SVM

In this section, a simple and fast incremental learning algorithm is presented based on the fast learning algorithm of the new L2 SVM. The incremental learning problem can be described in the following:

- Given an existing training sample set U_0, and a set of support vectors U_0^{sv} obtained by learning with the training set U_0; and an incremental training sample set H, and $U_0 \cap H=\varnothing$.
- The problem we want to solve is to quickly find a new set of support vectors U^{sv} corresponding to the new training sample set $U=U_0 \cup H$ based on the given support vector set U_0^{sv}.

Because the fast learning algorithm of the new L2 SVM is based on the simple linear algebra, a simple and fast incremental learning algorithm can be obtained as follows:

- In terms of the selection method of the initial active and inactive sets, combine half of the samples of H and all the samples of U_0^{sv} into a new sample set T;
- Select the indexes of T as an initial inactive set I^0, and the complement of I^0 as an initial active set A^0. Apply the fast learning algorithm of the new L2 SVM to the training sample set U, as a result, a new set of support vectors U^{sv} is solved.

In the incremental learning process, because the selected initial inactive set I^0 contains all the indexes of the given support vector set U_0^{sv}, I^0 is even closer to the indexes of U^{sv}, which leads to the fast convergence of the learning algorithm of the new L2 SVM.

For large-scale learning problems, we can separate the training sample set U into p subsets, i.e. $U = U_0 \cup H_1 \cup H_2 \cup \cdots \cup H_{p-1}$. First of all, we can apply the learning algorithm of the new L2 SVM to the initial training sample set U_0, then apply the incremental learning algorithm in series to the training sample set, $U_i = U_0 \cup H_1 \cup H_2 \cup \cdots \cup H_i$, until $i=p-1$. We call it batch-incremental learning algorithm.

3. COMPUTATIONAL EXPERIMENTS

We use a face database with 2240 face images (224 people, 10 frontal face images for each person) to evaluate the performance of the proposed algorithm. The training sample set U consists of 2240 facial feature vectors extracted from the 2240 face images. All the algorithms in the experiments are programmed by the code of Matlab 6.1 and work on a Pentium III/450M personal computer. The learning algorithms are designed to make a distinction between the 10 images of one specified person and others. Suppose $C=10^5$ in $\boldsymbol{Q}$, and the Gaussian function is used as the kernel function,

$$K(\boldsymbol{x}_i, \boldsymbol{x}_j) = \exp(-\|\boldsymbol{x}_i - \boldsymbol{x}_j\|_2^2 / \sigma^2) \tag{16}$$

where $\sigma^2 = 0.1$.

In the first experiment, the learning algorithm of the new L2 SVM is applied based on the common used optimization method by using the "fmincon" function in the optimization toolbox of Matlab 6.1. Unfortunately, it has not been able to reach the final result after one day running.

In the second experiment, the learning algorithm of the new-L2 SVM is applied based on the active set iteration method [5]. We test the algorithm under 3 different selections of the initial inactive sets:

1) I^0=∅;
2) I^0=N_{all};
3) Specific selection of I^0 presented in Section 2.3.

Table 1 shows the results on iteration time and number. We can see that the specific selection of I^k reached the best result.

In the third experiment, U is separated into 3 subsets: U_0 with 720 feature vectors, H_1 with 720 feature vectors, and H_2 with 800 feature vectors. The learning algorithm of the new-L2 SVM based on the active set iteration method is applied to the initial training subset U_0, then the incremental learning algorithm is applied in series to $U_1 = U_0 \cup H_1$ and $U = U_2 = U_1 \cup H_2$ respectively. Table 2 shows the results on iteration time and number. The total iteration time taken by the batch-incremental learning algorithm is 0.61s+1.48s+7.85s=9.94s, which is much less than the iteration time 64.92s taken by the learning algorithm in Table 1. Therefore, for large-scale learning problems, applying the incremental learning algorithm can save a lot of learning time.

Table 1. Iteration time and number under 3 different selections of initial inactive sets.

Iteration No. k	$I^0=\varnothing$	$I^0=N_{all}$	Specific Selection
	Size of I^k	Size of I^k	Size of I^k
0	0	2240	1120
1	2240	1201	1719
2	1201	702	974
3	702	426	582
4	426	277	368
5	277	193	238
6	193	154	177
7	154	139	146
8	139	135	139
9	135	134	134
10	134	133	133
11	133	133	133
12	133	--	--
Iteration time	130.33s	120.35s	64.92s

Table 2. Iteration time and number for incremental learning algorithm.

Iteration No. k	Learning U_0	Incr. Learning U_1	Incr. Learning U
	Size of I^k	Size of I^k	Size of I^k
0	360	402	504
1	371	568	801
2	217	343	463
3	133	226	301
4	83	168	219
5	54	133	166
6	45	113	142
7	42	108	133
8	42	104	133
9	--	104	133
Iteration Time	0. 61s	1.48s	7.85s

In the fourth experiment, to compare with the proposed algorithms, the well-known SVM learning algorithm --- SVM-Light [8] is applied to the classification problem using the Matlab code provided by Schwaighofer [9]. The running time is 17.50s, which is longer than the time (9.94s) taken by the proposed batch-incremental learning algorithm. Note that the KKT conditions are strictly satisfied in the proposed SVM algorithms while they are met with tolerance in the SVM-Light algorithm. In other words, the solution obtained by the SVM-Light algorithm is only approximately optimal.

4. CONCLUSIONS

We have introduced a new L2 soft margin support vector machine in this paper. Because the dual problem for the constrained optimization of the SVM is a convex quadratic problem with simple bound constraints, the active set iteration method can be applied as a fast learning algorithm for the SVM. To reduce the computation of the inverse matrix of Q_{I^k,I^k}, we presented a selection method for the initial active/inactive sets in order to control the size of I^k, especially, to avoid the occurrence of the worse case $I^k=N_{all}$ or $I^k=\varnothing$ in the learning iteration process. For incremental learning and large-scale learning problems, we proposed a simple and fast batch-incremental learning algorithm. Computational experiments showed the efficiency of the proposed algorithms.

REFERENCES

[1] C. M. Bishop. *Neural Networks for Pattern Recognition*, Clarendon Press, Oxford, 1995.

[2] D. DeCoste and B. Schölkopf, "Training Invariant Support Vector Machines," *Machine Learning*, vol. 46, No. 1-3, pp. 161–190, 2002.

[3] E. Osuna, R. Freund, and F. Girosi, "Training Support Vector Machines: An Application to Face Detection," *Proceedings of the 1997 conference on Computer Vision and Pattern Recognition* (CVPR'97), pp.130-136, 1997.

[4] L. Tao, "Researches on Face Recognition Algorithms for Human Identification," *Ph. D. Dissertation*, University of Science and Technology of China, 2003.

[5] K. Kunisch and F. Rendl, "An Infeasible Active Set Method for Quadratic Problems with Simple Bounds," *SIAM Journal on Optimization*, vol. 14, no. 1, pp. 35-52, 2003.

[6] S. Abe, "Analysis of Support Vector Machines," *Proceedings of the 2002 12th IEEE Workshop on Neural Networks for Signal Processing*, pp. 89-98, 2002.

[7] G. M. Wu, "Machine Learning Based on Kernels," *Ph. D. Dissertation*, University of Science and Technology of China, 2002.

[8] T. Joachims, SVM^{Light} Support Vector Machine, http://svmlight.joachims.org/

[9] Anton Schwaighofer, Software that might be of interest, http://www.cis.tugraz.at/igi/aschwaig/software.html

DESIGN AND SENSITIVITY ANALYSIS OF FEED-FORWARD NEURAL ADC'S

Hamid Movahedian, Mehrdad Sharif Bakhtiar

Electrical Engineering Department, Sharif Univ. of TechnologyTehran, IRAN
Email:*movahedian@sharif.edu*

ABSTRACT

In this paper, a one-layer and a two-layer Gray-coded feed-forward neural ADC with soft-limiting activation functions are introduced. The use of soft-limiting activation functions results in reduction of the total number of neurons, but at the same time it increases the sensitivity to various errors. The required number of bits for training the network derived from sensitivity analysis. A design example is also presented.

1-INTRODUCTION

After the first and probably the best known neural ADC was proposed by Hopfield [1], several schemes have been proposed to overcome the main shortcoming of the original work, namely the local minima and the limited speed due to multi-loop structure [2,3]. However, the increased complexity and even lower speeds caused both the original work and its extensions to become of less interest during the recent years.

Feed-forward neural ADC's which guarantees the global stability has been proposed in some references [4,5]. The reported feed-forward neural ADC's are generally based on neurons with hard-limiting or sigmoid activation functions. In the following sections, we show that the same function can be performed by neurons with soft-limiting activation functions. This will result in considerable reduction of the total number of the neurons of the hidden layer(s). The sensitivity analysis of neural ADC's, however, reveals that soft-limiting networks are more sensitive to errors. As a neural network, this shortcoming can be overcome by training. The required number of bits for training the network is also determined by the sensitivity analysis.

2- FEED-FORWARD NEURAL ADC'S

As the starting point, let's consider the simple 3-bit Gray-coded ADC of Fig.1-a [6]. In this one-hidden-layer perceptron, neurons have hard-limiting activation functions. For neurons with hard-limiting activation function, the output switches between two discrete values (i.e. 0 and 1) as the input passes the threshold limits. In general, for N-bit conversion, this network has 2^N-1 neurons at the hidden layer. This ADC is comparable to a Flash converter with 2^N-1 comparators.

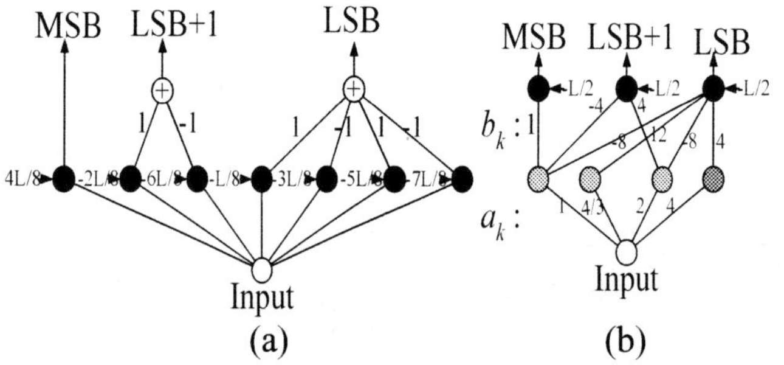

Fig.1. (a) 3-bit Neural ADC, (b) Alternative Structure

An alternative way to implement this ADC is shown in Fig.1-b, where neurons of the hidden layer (gray circles) have soft-limiting activation function defined as follows:

$$f(x)=\begin{cases} 0 & x<0 \\ x & 0>x>L \\ L & x>L \end{cases} \qquad (1)$$

With no loss of generality, we assume that the input range of the ADC is between 0 and L .The output neurons have hard-limiting activation functions. They act as comparators with thresholds of L/2.

In general, for N-bit conversion, in this structure the hidden layer has $m=2^{N-1}$ neurons. It can be shown that the weights of the synapses are calculated from following equations:

$$a_k = \frac{m}{k} \qquad k = 1:m \qquad (2\text{-}1)$$

$$b_k = \begin{cases} 2k.(-1)^{k+1} & k = 1:m-1 \\ k.(-1)^{k+1} & k = m \end{cases} \qquad (2\text{-}2)$$

The function of this network can be best explained by considering the waveforms at the input of the LSB neuron

(Fig.2). As the input varies between 0 and L, the input of the LSB neuron sweeps the same range N times and each time it passes L/2, the output of the neuron switches between 0 and 1.

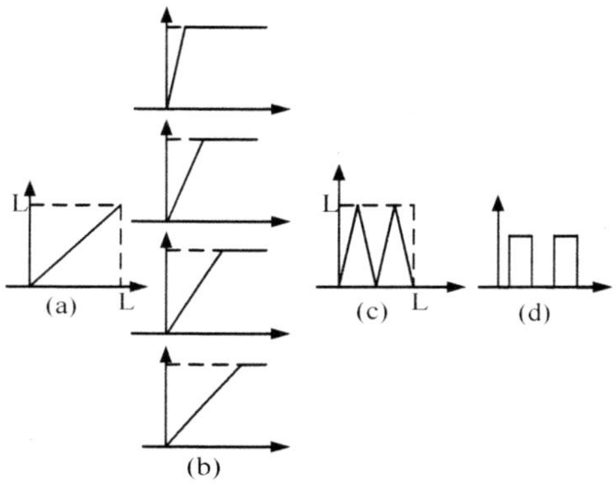

Fig.2. Waveforms of : (a) Input, (b) Outputs of Hidden-layer Neurons , (c) Input of the LSB Neuron and (d) LSB

Although the number of the neurons of the hidden layer of this network is almost half the previous case, it's still a large number for N=6 or more.

The number of neurons can be further reduced by the use of two hidden layers. Fig. 3-a shows such a network while the reduced form network , with exactly the same behavior, is shown in Fig.3-b. The higher order bits and also biases are not shown here.

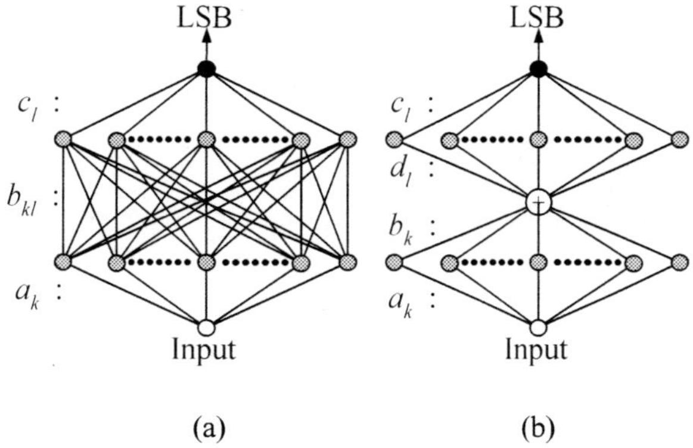

Fig.3. (a) Two Layer Network and (b) Reduced Form

For the network in Fig. 3-b with $m=2^M$ neurons at the first layer and $n=2^N$ neurons at the second layer, the weights a_k and b_k are the same as those given in Eq. (2) and c_l and d_l are as follows:

$$d_l = \frac{n}{l} \qquad l = 1:n \qquad (3\text{-}1)$$

$$c_l = \begin{cases} l.(-1)^{l+1} & l = 1:n-1 \\ l.(-1)^{l+1} & l = n \end{cases} \qquad (3\text{-}2)$$

In general, for $m=2^M$ and $n=2^N$, M+N+1 bit can be extracted from this network by adding M+N neurons to the output layer to realize higher order bits. For example, to implement a 6-bit converter with this network, the total number of neurons will be 12 which is far less than 63 and 32 neurons needed with previous networks. It must be noted that the output of this ADC is also Gray-coded.

3-SENSITIVITY TO ERRORS

To have a quantitative measure of the sensitivity, we first define the output error (returned to the input) as the difference between the actual value of the input which cause p^th^ output transition and its ideal value:

$$\Delta x_p = x_p - x_{p0} \qquad p=1,2,3,\ldots 2^{N\text{-}1} \qquad (4)$$

Where:

$x_{p0} = (2p-1)\Delta$ is the ideal input value corresponding to p^{th} output transition ,and:

$\Delta = \frac{L}{2^N}$ is the input equivalent to one LSB.

The normalized value of Δx_p is the Integral Non Linearity (INL) at the pth output transition:

$$INL_p = \frac{\Delta x_p}{\Delta} \qquad p=1,2,3,\ldots 2^{N\text{-}1} \qquad (5)$$

In general, the output error is a function of the errors in weights, limiting values and biases. Lets the relation between the actual value of any parameter (*P*) and its designed value be as follows:

$$P = P_0 + \Delta P = P_0(1+e_P) \qquad (6)$$

Considering only first order terms, the relation between the output error and errors in parameters can be shown as:

$$INL_p = SUM(S_{P_p} e_P) = \sum_{k=1}^{m} S_{a_{k_p}} e_{a_k} + \sum_{k=1}^{m} S_{b_{k_p}} e_{b_k} + \cdots\cdots \qquad (7)$$

Where S_{P_p} is the Sensitivity coefficient of the parameter *P* which is defined as the ratio of the normalized output error to the relative error in *P*, assuming all other parameters to be ideal. Note that S_{P_p} is a function of both the parameter *P* and the output code (indicated by *p*).

For the neural ADC of Fig.1-a, the output error is only due to error in biases of the neurons. Any deviation of the

p^{th} bias will cause an equal shift in p^{th} output transition and has no effect on other transitions:

$$\left|S_{bias_{kp}}\right| = \begin{cases} 0 & p \neq k \\ 1 & p = k \end{cases} \tag{8}$$

Note that both p and k vary between 1 and 2^{N-1} for LSB. For the general form of the neural ADC of Fig.1-b, output error is due to error in a_k's, b_k's, and L_k's:

$$INL_p = S_{a_{kp}} e_{a_k} + S_{b_{kp}} e_{b_k} + S_{L_{kp}} e_{L_k} \tag{9}$$

Between any two adjacent breaking points the input of the LSB neuron (y) can be written as a linear function of the input (x):

$$y = \sum_{k=p}^{m} a_k b_k x + \sum_{k=0}^{p} b_k L_k \qquad 2(p-1)\Delta \le x \le 2p\Delta \tag{10}$$

Sensitivity coefficients can be calculated using (2) to (7), (9) and (10):

$$\left|S_{a_{kp}}\right| = \begin{cases} 0 & p \ge k \\ 2(2p-1) & p < k < m \\ (2p-1) & k = m \end{cases} \tag{11-1}$$

$$\left|S_{b_{kp}}\right| = \begin{cases} 4k & p \ge k \\ 2(2p-1) & p < k < m \\ (2p-1) & k = m \end{cases} \tag{11-2}$$

$$\left|S_{L_{kp}}\right| = \begin{cases} 4k & p \ge k \\ 0 & p < k \end{cases} \tag{11-3}$$

The sensitivity coefficients of two-layer network can be calculated in a similar manner to the one-layer network. Between any two adjacent breaking points, the input of the LSB neuron (y) can be described as a linear function of the input:

For $2(p-1)\Delta \le x \le 2p\Delta$:

$$y = \sum_{l=j}^{n} \{c_l d_l (\sum_{k=i}^{m} a_k b_k x + \sum_{k=0}^{i-1} b_k L_k)\} + \sum_{l=0}^{j-1} d_l L_l \tag{12}$$

Where:

$$\Delta = \frac{L}{2m.n} \tag{13}$$

$$p.\Delta = (i.n - n + j)\Delta \qquad i = 1:m \qquad j = 1:n \tag{14}$$

Note that using above notations, p varies between 1 and m.n.

The relation between INL_p and various sources of errors is as follows:

$$INL_p = S_{a_{kp}} e_{a_k} + S_{b_{kp}} e_{b_k} + S_{c_{lp}} e_{c_l} + S_{d_{lp}} e_{d_l} + S_{L1_{kp}} e_{L1_k} + S_{L2_{kp}} e_{L2_k} \tag{15}$$

Using Eq.'s (12) to (15), sensitivity coefficients can be calculated as:

$$\left|S_{a_{kp}}\right| = \begin{cases} 0 & i > k \\ 2(2p-1) = 2[(2i-1)n - n + (2j-1)] & i \le k < m \\ (2p-1) = [(2i-1)n - n + (2j-1)] & k = m \end{cases} \tag{16-1}$$

$$\left|S_{b_{kp}}\right| = \begin{cases} 4nk & i > k \\ 2(2p-1) = 2[(2i-1)n - n + (2j-1)] & i \le k < m \\ (2p-1) = [(2i-1)n - n + (2j-1)] & k = m \end{cases} \tag{16-2}$$

$$\left|S_{c_{lp}}\right| = \begin{cases} 0 & j > l \\ 2(2j-1) & j \le l < n \\ (2j-1) & j = n \end{cases} \tag{16-3}$$

$$\left|S_{d_{lp}}\right| = \begin{cases} 4l & j > l \\ 2(2j-1) & j \le l < n \\ (2j-1) & j = n \end{cases} \tag{16-4}$$

$$\left|S_{L_{1kp}}\right| = \begin{cases} 4nk & i > k \\ 0 & i \le k \end{cases} \tag{16-5}$$

$$\left|S_{L_{2lp}}\right| = \begin{cases} 4l & j > l \\ 0 & j \le l \end{cases} \tag{16-6}$$

Comparing sensitivity coefficients of one and two layer networks indicates that for equal number of bits, sensitivity of two-layer network to the parameters of the first layer is of the same order as sensitivity of the one-layer network. Sensitivity coefficients of the second layer are in order of 1/n of those of the first layer.

Sensitivity coefficients can be used to determine the maximum allowable error of each parameter. As the most general case, we will show it for the two-layer network.

If parameter errors are considered as independent random variables with zero mean, the output INL would be also a random variable and its variance can be calculated as follows:

$$\sigma^2_{INL_p} = \sum_{k=1}^{m} S^2_{a_{kp}} \sigma^2_{e_{a_k}} + S^2_{b_{kp}} \sigma^2_{e_{b_k}} + S^2_{L1_{kp}} \sigma^2_{e_{L1k}} + \sum_{l=1}^{n} S^2_{c_{lp}} \sigma^2_{e_{cl}} + S^2_{d_{lp}} \sigma^2_{e_{dl}} + S^2_{L2_{lp}} \sigma^2_{e_{L2l}} \quad (17)$$

If we let all the various terms of (17) to have the same magnitude (i.e. each have equal contribution to the overall error), and noting that for any input value one of S_{akp} or S_{L1kp} (and one of S_{clp} or S_{L2p}) is zero, $\sigma_{e_{x_p}}$ can be computed for any input value (i.e. any p):

$$\sigma_{e_{x_p}} = \frac{\sigma_{INL_p}}{\sqrt{2(m+n)}} \frac{1}{S_{x_p}} \quad (S_{x_p} \neq 0) \quad (18)$$

In this way, the minimum value of $\sigma_{e_{x_p}}$ for any parameter can be found. The ratio of this value for any parameter to the intrinsic tolerance of that parameter will indicate the number of bits required to train that parameter:

$$N \geq \log_2 \frac{\sigma_{\text{int}_x}}{\min(\sigma_{e_{x_{ij}}})} \quad (19)$$

For example, for the case of a 6-bit converter with m=4 and n=8, if we consider : $\sigma_{INL} = 0.3LSB$ (which means in 99.5% of the cases INL is less than 1 LSB), and assuming that without any training all the parameters have a variance of 2%, the calculated values of $\min(\sigma_{e_{x_{ij}}})$ are shown in Table1 while minimum number of bits required for training each parameter is shown in Table 2.

After training, the overall INL is expected to be better than the presumed value because for many input values, most of the parameters are over adjusted.

Table1. Allowable error variances (%) for INL<1LSB

k,l	1	2	3	4	5	6	7	8
$\sigma_{e_{ak}}$	0.2	0.1	0.06	0.1				
$\sigma_{e_{bk}}$	0.19	0.10	0.06	0.1				
σ_{eL1_k}	0.19	0.10	0.06	0.19				
$\sigma_{e_{cl}}$	3.06	1.02	0.61	0.44	0.34	0.28	0.23	0.41
σ_{ed_l}	1.53	0.76	0.51	0.38	0.3	0.25	0.21	0.41
σ_{eL2_l}	1.53	0.76	0.51	0.38	0.3	0.25	0.21	1.53

Table2. Minimum No. of bits for training each parameter

k,l	1	2	3	4	5	6	7	8
N_{ak}	4	5	5	5				
N_{bk}	4	5	5	5				
N_{L1k}	4	5	5	4				
N_{cl}	-	1	2	3	3	3	4	3
N_{dl}	1	2	2	3	3	3	4	3
N_{L2l}	1	2	2	3	3	3	4	1

At last it must be noted that for one-layer network the maximum number of training bits is higher than two-layer network . That's because the total number of error sources is higher for one-layer network, so the maximum error of each source must be limited to lower value.

4- CONCLUSION

A one and a two-layer feed-forward neural network were introduced and studied from the sensitivity point of view. The results were used to evaluate the required No. of bits for training each parameter. The results shows that using two-layer networks will reduce both the complexity and sensitivity to errors.

5- REFERENCES

[1] Tank,D.W., and Hopfield,J.J.:"Simple neural optimization networks: An A/D converter, signal Decision circuit and a linear programming circuit" IEEEE Trans., May 1986, CAS-36

[2] Lee, B.W., and SHEU, B.J.: "Design of a neural-based A/D converter using modified Hopfield network" IEEE J. August 1989, SC-24,

[3] Dempsey, G.L., Pintoy, A.G. and Wood, J.A.:"A new design strategy for the Tank and Hopfield neural analog-to-digital converter" IEEE 1993

[4] Chigusa, Y., and Tanaka, M.:"A neural-like feed-forward ADC" Proc. IEEE Int. Symp. On circuits and Systems, May 1990

[5] Dapnte, P., Grimaldi, D. and Michaeli, L.:"A Full Neyral Gray Code Based ADC" IEEE Transactions on Instrumentation and Measurement, , Volume: 45 Issue: 2 , April 1996 Page(s): 634-639

[6] Martinelli, G. and Perfetti, R.:"Synthesis of feed-forward neural analog-digital converters" IEE Proccdings-G, vol. 138, No.5 October 1991

GLOBAL ASYMPTOTIC STABILITY OF A CLASS OF NEURAL NETWORKS WITH TIME VARYING DELAYS

Tolga Ensari[1], Sabri Arik[1] and Vedat Tavsanoglu[2]

[1] *Department of Computer Engineering,*

Istanbul University, 34320 Avcilar, Istanbul, TURKEY

email : {ensari,ariks}@istanbul.edu.tr

[2] *Department of Electronics, Yildiz Technical University, Besiktas, Istanbul,Turkey*

email : tavsanav@yildiz.edu.tr

ABSTRACT

This paper presents a new sufficient condition for the uniqueness and global asymptotic stability (GAS) of the equilibrium point for a larger class of neural networks with time varying delays. It is shown that the use of a more general type of Lyapunov-Krasovskii functional leads to establish global asymptotic stability of a larger class of delayed neural networks than the neural network model considered in some previous papers.

I. INTRODUCTION

In recent years, stability of different classes of neural networks with time delay, such as Hopfield neural networks, cellular neural networks, bidirectional associative neural networks, Lotka-Volterra neural networks, has been extensively studied and various stability conditions have been obtained for these models of neural networks,(see [1]-[28] and the references cited therein). In a recent paper [28], some elegant stability results for neural networks with time varying delays have been presented by using different types of Lyapunov-Krasovskii functionals and LMI (Linear Matrix Inequality) approach. It is also shown in [28] that the results of [28] generalises some previous stability results derived in the literature. In the present paper, by employing a more general Lyapunov functional, we will improve some of the results given in [28].

The delayed neural network model we consider is defined by the following state equations :

$$\frac{du(t)}{dt} = -Au(t) + W_o g(u(t)) + W_1 g(u(t-\tau(t))) + I \qquad (1)$$

where $u = [u_1, u_2, ..., u_n]^T$ is the neuron state vector, $A = diag(a_i)$ is a positive diagonal matrix, τ is the transmission delay, $W_o = (w^o_{ij})_{nxn}$ and $W_1 = (w^1_{ij})_{nxn}$ are the interconnection matrices representing the weight coefficients of the neurons, $I = [I_1, I_2, ..., I_n]^T$ is the constant external input vector, and the $g(u) = [g_1(u_1), g_2(u_2), ..., g_n(u_n)]^T$ denotes the neuron activations.

The assumptions on the activation functions are as follows :

$$(H) \qquad 0 \leq \frac{g_j(\xi_1) - g_j(\xi_2)}{\xi_1 - \xi_2} \leq \sigma_j, \quad j = 1, 2, ..., n$$

$$(H^*) \qquad 0 < \frac{g_j(\xi_1) - g_j(\xi_2)}{\xi_1 - \xi_2} \leq \sigma_j, \quad j = 1, 2, ..., n$$

for each $\xi_1, \xi_2 \in R$, $\xi_1 \neq \xi_2$, where σ_j are positive constants.

It should be noted that the class of functions satisfying the condition H is larger than the class of functions satisfying the condition H^*.

In the following, we will shift the equilibrium point $u^* = [u_1^*, u_2^*, ..., u_n^*]$ of system (1) to the origin. The transformation $x(\cdot) = u(\cdot) - u^*$ puts system (1) into the form

$$\frac{dx(t)}{dt} = -Ax(t) + W_o f(x(t)) + W_1 f(x(t-\tau(t))) \qquad (2)$$

where $x = [x_1, x_2, ..., x_n]^T$ is the state vector of the transformed system, and $f_j(x_j) = g_j(x_j + u_j^*) - g_j(u_j^*)$, with $f_j(0) = 0$, $\forall\, j$. Note that the functions $f_j(\cdot)$ satisfy the condition H, that is

$$0 \leq \frac{f_j(\xi_1) - f_j(\xi_2)}{\xi_1 - \xi_2} \leq \sigma_j, \quad j = 1, 2, ..., n$$

for each $\xi_1, \xi_2 \in R$, $\xi_1 \neq \xi_2$, where σ_j are positive constants.

Throughout this paper we will use the following notations : B^T, B^{-1}, $\lambda_m(B)$, $\lambda_M(B)$ denotes, respectively, the transpose of, the inverse of, the minimum eigenvalue of, the maximum eigenvalue of a square matrix B. The notation $B > 0$ ($B < 0$) means that B is symmetric and positive definite (negative definite).

0-7803-8251-X/04/$17.00 ©2004 IEEE

II. MAIN STABILITY RESULT

In this section, we present a sufficient condition for the uniqueness and GAS of the equilibrium point for the delayed neural system defined by (2). This result is given in the following :

Theorem 1 : Suppose that in system (2), τ satisfies $\tau'(t) \leq \eta < 1$. Let $\Sigma = diag(\sigma_i > 0)$. If the condition (H) is satisfied and there exist a positive definite matrix P, a positive diagonal matrix D and a positive constant β such that

$$\begin{aligned}\Omega &= -2DA\Sigma^{-1} + DW_o + W_o^T D + \beta W_1^T P W_1 \\ &\quad +\beta^{-1}(1-\tau'(t))^{-1} D P^{-1} D < 0\end{aligned}$$

then, the origin of system (2) is the unique equilibrium point and it is globally asymptotically stable.

Proof : We will first prove the uniqueness of the equilibrium point. Consider the equilibrium equation of system (2) :

$$-Ax^* + W_o f(x^*) + W_1 f(x^*) = 0 \qquad (3)$$

where x^* is the equilibrium point of (3). $f(x^*) = 0$ implies that $x^* = 0$. Let $f(x^*) \neq 0$. Multiplying both sides of (2) by $2f^T(x^*)D$, and then adding and subtracting the term $\beta^{-1}(1-\tau'(t))^{-1} f^T(x^*) D P^{-1} D f(x^*)$ results in :

$$\begin{aligned}&-2f^T(x^*)DAx^* + 2f^T(x^*)DW_o f(x^*) + f^T(x^*)DW_1 f(x^*) \\ &\quad -\beta^{-1}(1-\tau'(t))^{-1} f^T(x^*) D P^{-1} D f(x^*) \\ &\quad +\beta^{-1}(1-\tau'(t))^{-1} f^T(x^*) D P^{-1} D f(x^*) = 0 \qquad (4)\end{aligned}$$

where β is positive constant,D is a positive diagonal matrix and P is a positive definite matrix. Under condition H, we have

$$|f_i(x_i(t))| \leq \sigma_i |(x_i(t))|, \ \forall\, x_i(t) \in R \ , i = 1, 2, ..., n \quad (5)$$

Using (5), we can write

$$2f^T(x^*)DAx^* \geq 2f^T(x^*)DA\Sigma^{-1} f(x^*) \qquad (6)$$

We also have the following :

$$\begin{aligned}&-\beta^{-1}(1-\tau'(t))^{-1} f^T(x^*) D P^{-1} D f(x^*) \\ &\quad +2f^T(x^*(t)) D W_1 f(x^*) \\ &\quad \leq \beta f^T(x^*) W_1^T P W_1 f(x^*) \qquad (7)\end{aligned}$$

Using (6) and (7) in (4), we obtain

$$\begin{aligned}&-2f^T(x^*)DA\Sigma^{-1} f(x^*) + 2f^T(x^*)DW_o f(x^*) \\ &\quad +\beta f^T(x^*) W_1^T P W_1 f(x^*) \\ &\quad +\beta^{-1}(1-\tau'(t))^{-1} f^T(x^*) D P^{-1} D f(x^*) \geq 0\end{aligned}$$

or equivalently

$$f^T(x^*)\Omega f(x^*) \geq 0, \ \forall\, f(x^*) \neq 0 \qquad (8)$$

On the other hand, since Ω is a negative definite matrix, we have

$$f^T(x^*)\Omega f(x^*) < 0, \ \forall\, f(x^*) \neq 0 \qquad (9)$$

The contradiction between (8) and (9) implies that $f(x^*) \neq 0$ is not an equilibrium point of system (2), meaning that $x^* = 0$ is the only equilibrium point under the conditions given in Theorem 1.

In order to prove the GAS of the origin of (2), following Lyapunov-Krasovskii type functional will be used :

$$\begin{aligned}V(x(t)) &= x^T(t)Ax(t) + 2\alpha \sum_{i=1}^{n} d_i \int_0^{x_i(t)} f_i(s)ds \\ &+ \alpha\beta \int_{t-\tau(t)}^{t} f^T(x(\zeta)) W_1^T P W_1 f(x(\zeta)) d\zeta \\ &+ \frac{1}{1-\eta} \int_{t-\tau(t)}^{t} f^T(x(\zeta)) W_1^T W_1 f(x(\zeta)) d\zeta\end{aligned}$$

where $P = P^T > 0$, $d_i > 0$, $i = 1, 2, ..., n$, and α and β are positive constants.

The time derivative of the functional along the trajectories of system (2) is obtained as follows :

$$\begin{aligned}&\dot{V}(x(t)) = \\ &-x^T(t)A^2x(t) - x^T(t)A^2x(t) + 2x^T(t)AW_o f(x(t)) \\ &+2x^T(t)AW_1 f(x(t-\tau(t))) - 2\alpha f^T(x(t))DAx(t) \\ &+2\alpha f^T(x(t))DW_o f(x(t)) + 2\alpha f^T(x(t))DW_1 f(x(t-\tau(t))) \\ &+\alpha\beta f^T(x(t))W_1^T P W_1 f(x(t)) + \frac{1}{1-\eta} f^T(x(t)) W_1^T W_1 f(x(t)) \\ &-\alpha\beta(1-\tau'(t)) f^T(x(t-\tau(t))) W_1^T P W_1 f(x(t-\tau(t))) \\ &-\frac{1-\tau'(t)}{1-\eta} f^T(x(t-\tau(t))) W_1^T W_1 f(x(t-\tau(t))) \quad (10)\end{aligned}$$

Adding and subtracting the term $\alpha\beta^{-1}(1-\tau'(t))^{-1} f^T(x(t)) D P^{-1} D f(x(t))$ in the above equation results in

$$\begin{aligned}&\dot{V}(x(t)) \\ &-x^T(t)A^2x(t) - x^T(t)A^2x(t) + 2x^T(t)AW_o f(x(t)) \\ &+2x^T(t)AW_1 f(x(t-\tau(t))) - 2\alpha f^T(x(t))DAx(t)\end{aligned}$$

$$+2\alpha f^T(x(t))DW_o f(x(t)) + 2\alpha f^T(x(t))DW_1 f(x(t-\tau(t)))$$
$$+\alpha\beta f^T(x(t))W_1^T PW_1 f(x(t)) + \frac{1}{1-\eta} f^T(x(t))W_1^T W_1 f(x(t))$$
$$-\alpha\beta(1-\tau'(t))f^T(x(t-\tau(t)))W_1^T PW_1 f(x(t-\tau(t)))$$
$$-\frac{1-\tau'(t)}{1-\eta} f^T(x(t-\tau(t)))W_1^T W_1 f(x(t-\tau(t)))$$
$$+\alpha\beta^{-1}(1-\tau'(t))^{-1} f^T(x(t))DP^{-1}Df(x(t))$$
$$-\alpha\beta^{-1}(1-\tau'(t))^{-1} f^T(x(t))DP^{-1}Df(x(t))$$

We can write the following linear matrix inequalities :

$$-x^T(t)A^2x(t) + 2x^T(t)AW_o f(x(t))$$
$$\leq f^T(x(t)){W_o}^T W_o f(x(t))$$
$$-x^T(t)A^2x(t) + 2x^T(t)AW_1 f(x(t-\tau(t)))$$
$$\leq f^T(x(t-\tau(t))){W_1}^T W_1 f(x(t-\tau(t)))$$
$$-\alpha\beta^{-1}(1-\tau'(t))^{-1} f^T(x(t))DP^{-1}Df(x(t))$$
$$+2\alpha f^T(x(t))DW_1 f(x(t-\tau(t)))$$
$$\leq \alpha\beta(1-\tau'(t))f^T(x(t-\tau(t)))W_1^T PW_1 f(x(t-\tau(t)))$$

From (5), we obtain the following

$$-2\alpha f^T(x(t))DAx(t) \leq -2\alpha f^T(x(t))DA\Sigma^{-1}f(x(t))$$

Since

$$\frac{1-\tau'(t)}{1-\eta} \geq 1$$

we can write

$$-\frac{1-\tau'(t)}{1-\eta} f^T(x(t-\tau(t)))W_1^T W_1 f(x(t-\tau(t)))$$
$$+f^T(x(t-\tau(t))){W_1}^T W_1 f(x(t-\tau(t))) \leq 0$$

Using the above inequalities, we obtain

$$\dot{V}(x(t)) \leq$$
$$f^T(x(t)){W_o}^T W_o f(x(t)) + \frac{1}{1-\eta} f^T(x(t))W_1^T W_1 f(x(t))$$
$$-2\alpha f^T(x(t))DA\Sigma^{-1}f(x(t)) + 2\alpha f^T(x(t))DW_o f(x(t))$$
$$+\alpha\beta f^T(x(t))W_1^T PW_1 f(x(t))$$
$$+\alpha\beta^{-1}(1-\tau'(t))^{-1} f^T(x(t))DP^{-1}Df(x(t))$$

or equivalently

$$\dot{V}(x(t)) \leq$$
$$f^T(x(t))({W_o}^T W_o + \frac{1}{1-\eta} W_1^T W_1) f(x(t))$$
$$-\alpha f^T(x(t))(-\Omega) f(x(t))$$
$$\leq \lambda_M({W_o}^T W_o + \frac{1}{1-\eta} W_1^T W_1) f^T(x(t)) f(x(t))$$
$$-\alpha\lambda_m(-\Omega) f^T(x(t)) f(x(t))$$
$$= [\lambda_M({W_o}^T W_o + \frac{1}{1-\eta} W_1^T W_1) - \alpha\lambda_m(-\Omega)]||f(x(t))||_2^2$$

The choice

$$\alpha > \frac{\lambda_M({W_o}^T W_o + \frac{1}{1-\eta} W_1^T W_1)}{\lambda_m(-\Omega)} > 0$$

ensures that $\dot{V}(x(t)) < 0, \ \forall f(x(t)) \neq 0$. Note that $f(x(t)) \neq 0$ implies that $x(t) \neq 0$. Now let $f(x(t)) = 0$ and $x(t) \neq 0$. In this case, $\dot{V}(x(t))$ is in the following form :

$$\dot{V}(x(t)) = -x^T(t)A^2x(t) - x^T(t)A^2x(t)$$
$$+2x^T(t)AW_1 f(x(t-\tau(t)))$$
$$-\alpha\beta(1-\tau'(t))f^T(x(t-\tau(t)))W_1^T PW_1 f(x(t-\tau(t)))$$
$$-\frac{1-\tau'(t)}{1-\eta} f^T(x(t-\tau(t)))W_1^T W_1 f(x(t-\tau(t)))$$

Since $W_1^T PW_1 > 0$, $\beta > 0$ and $\alpha > 0$, $\dot{V}(x(t))$ satisfies $\dot{V}(x(t)) \leq -x^T(t)A^2x(t)$ Since A is a positive diagonal matrix, we can conclude that $\dot{V}(x(t)) \leq -x^T(t)A^2x(t) < 0, \ \forall x(t) \neq 0$. Now consider the case where $f(x(t)) = x(t) = 0$. In this case, it can be easily seen that $\dot{V}(x(t)) < 0, \ \forall f(x(t-\tau(t))) \neq 0$. Hence, we have proved that $\dot{V}(x(t)) = 0$ if and only if $f(x(t)) = x(t) = f(x(t-\tau(t))) = 0$, otherwise $\dot{V}(x(t)) < 0$. On the other hand, $V(x(t))$ is radially unbounded, that is, $V(x(t)) \rightarrow \infty$ as $||(x(t))|| \rightarrow \infty$. Thus, from standard Lyapunov Stability Theorems (see. e.g.,[29, Corollary 3.2, ch. 3]), it follows that the origin of system (2) is GAS.

We will now compare our result with the previously published results.

Theorem 2 [28] : Suppose that in system 2, τ satisfies $\tau'(t) \leq \eta < 1$. Let $\Sigma = diag(\sigma_i > 0)$. If the condition (H^*) is satisfied and there exist a positive definite matrix P, a positive diagonal matrix D and a positive constant β such that

$$\Omega^* = -2DA\Sigma^{-1} + DW_o + W_o^T D + \beta W_1^T PW_1$$
$$+\beta^{-1}(1-\tau'(t))^{-1}DP^{-1}D < 0$$

then, the origin of system (2) is asymptotically stable.

It should be noted that Theorem 1 and Theorem 2 impose the same constraint conditions on the network parameters of system (1). However, we have the following analysis to show the advantage of our result over the result given in [28]: Theorem 4 proves only asymptotic stability of the equilibrium with respect to the activation functions satisfying the condition (H^*). Theorem 1 proves global asymptotic stability of the equilibrium point with respect to the activation functions satisfying the condition (H). Since global asymptotic stability implies asymptotic stability (but the reverse implication may not always be true), and the functions satisfying the condition (H) form a larger class than those satisfying the condition (H^*), our condition given in Theorem 1 can be considered a generalization of the result of Theorem 2 obtained in [28].

III. CONCLUSIONS

We have studied the equilibrium and stability properties a class of neural networks with time varying delays. A new stability criteria has been derived by employing a more general type of Lyapunov-Krasovskii functional. It has been shown that the result obtained allows to employ a larger class of activation functions than the class of functions considered in some previous works.

REFERENCES

[1] T. Roska, C.W. Wu, and L.O. Chua , "Stability of cellular neural networks with dominant nonlinear and delay-type template," IEEE Trans. Circuits Syst. I, vol. 40, pp. 270–272, April 1993.

[2] T. Roska, C.W. Wu, M. Balsi and L.O. Chua , "Stability and dynamics of delay-type general and cellular neural networks," IEEE Trans. Circuits Syst. I, vol. 39, pp. 487–490, June 1992.

[3] H.Ye, A. N. Michel, and K. Wang, "Qualitative analysis of Cohen-Grossberg neural networks with multiple delays," Physical Review E, vol. 51, pp. 2611–2618, March 1995.

[4] M. Gilli and P.P. Civalleri and L. Pandolfi, "On stability of cellular neural networks with delay," IEEE Trans. Circuits and Syst. I, vol. 40, pp. 157–164, March 1993.

[5] S. Arik and V. Tavsanoglu , "Equilibrium analysis of delayed CNNs," IEEE Trans. Circuits and Syst.I, vol. 45, pp. 168–171, February 1998.

[6] J. D. Cao," Periodic-Solutions and Exponential Stability in Delayed Cellular Neural Networks", Physical Review E, vol.60, no.3, pp.3244-3248, 1999.

[7] S. Arik , "Stability analysis of delayed neural networks," IEEE Trans. Circuits and Syst.I, vol. 47, pp. 1089–1093, July 2000.

[8] S. Arik and V. Tavsanoglu,"On the global asymptotic stability of delayed cellular neural networks", IEEE Trans. Circuits and Syst.I, vol. 47, no. 5, pp. 571-574, April 2000.

[9] T-L. Liao and F-C. Wang, "Global Stability for Cellular Neural Networks with Time Delay", IEEE Trans. on Neural Networks, vol. 11 pp. 1481-1485, November 2000.

[10] N. Takahashi, "A New Sufficient Condition for Complete Stability of Cellular Neural Networks with Delay", IEEE Trans. Circuits and Syst.I, vol. 47, pp. 793-799, June 2000.

[11] X. F. Liao, K.W. Wong and J. Yu , "Novel Stability Conditions for Cellular Neural Networks with Time Delay", Int. Journal of Bifurcation and Chaos, vol.11, no. 7, pp.1853-1864, 2001.

[12] Z. Yi, P.A. Heng and K.S. Leung,"Convergence Analysis of Delayed Cellular Neural Networks with Unbounded Delay", IEEE Trans. on Circuits and Syst. I, vol. 48, pp. 680-687, June 2001.

[13] J.Y. Zhang and Y.R. Yang,"Global Stability Analysis of Bidirectional Associative Memory Neural Networks with Time Delay", Int. Journal of Circuit Theory and Applications vol. 29, no.2, pp. 185-196, March-April 2001.

[14] H. Lu," Stability Criteria for Delayed Neural Networks", Physical Review E, vol. 64, November 2001.

[15] J. Cao, "Global Stability Conditions for delayed CNNs", IEEE Trans. on Circuits and Syst. I, vol. 48, no.11, pp. 1330-1333, November 2001.

[16] H. Ye and A. N. Michel, "Robust stability of nonlinear time-delay systems with applications to neural networks", IEEE Trans. Circuits and Syst. I, vol. 43, pp. 532–543, July 1996.

[17] S. Arik , "An improved global stability result for delayed cellular neural networks," IEEE Trans. Circuits and Syst. I, vol. 49, pp. 1211–1214, August 2002.

[18] K. Gopalsamy and X. He , "Stability in asymmetric Hopfield nets with transmission delays," Phsica D, vol. 76, pp. 344–358, 1994.

[19] Y. J. Cao and Q. H. Wu, "A note on stability of analog neural networks with time delays," *IEEE Trans. Neural Networks,* vol. 7, pp. 1533–1535, 1996.

[20] Y. Zhang, "Global exponential stability and periodic solutions of delay Hopfield neural networks," *Int. J. Syst. Sci.,* vol. 27, pp. 227–231, 1996.

[21] J.D. Cao and L. Wang, "Exponential stability and periodic oscillatory solution in BAM networks with delays," IEEE Trans. Neural Networks, vol. 13, pp. 457–463, March 2002.

[22] Z. Yi, P. A. Heng and P. Vadakkepat, "Absolute periodicity and absolute stability of delayed neural networks ", IEEE Trans. Circuits and Syst. I, vol. 49, pp. 256–261, February 2002.

[23] J. Zhang,"Absolutely exponential stability in delayed cellular neural networks ", International Journal of Circuit Theory and Applications vol. 30, no.4, pp. 395-409, July-August 2002.

[24] H. Huang, J. Cao and J. Wang, "Global exponential stability and periodic solutions of recurrent neural networks with delays ", Physics Letters A, vol. 298, pp. 393-404, June 2002.

[25] C. Sun, K. Zhang, S. Fei and C. B. Feng, "On exponential stability of delayed neural networks with a general class of activation functions ", Physics Letters A, vol. 298, pp. 122-132, June 2002.

[26] X. F. Liao and J.B. Yu, "Robust stability for interval hopfield neural networks with time delay", *IEEE Trans. Neural Networks.,* vol. 9, pp. 1042–1045, 1998.

[27] X. F. Liao, K.W. Wong, Z. Wu and G. Chen "Novel robust stability for interval-delayed hopfield neural networks," IEEE Trans. Circuits and Syst.I, vol. 48, pp. 1355–1359, November 2001.

[28] X. Liao, G. Chen and E. N. Sanchez,"LMI-based approach for asymptotic stability analysis of delayed neural networks", IEEE Trans. Circuits and Syst.I,vol. 49, pp. 1033–1039, July 2002.

[29] H. K. Khalil, Nonlinear Systems, Mcmillan Publishing Company, New York, 1988

A 1.2-V BUCK CONVERTER WITH A NOVEL ON-CHIP LOW-VOLTAGE CURRENT-SENSING SCHEME

Chi Yat Leung, Philip K. T. Mok and Ka Nang Leung

Department of Electrical and Electronic Engineering
The Hong Kong University of Science and Technology
Clear Water Bay, Hong Kong SAR
Tel: (852) 2358-8517 Fax: (852) 2358-1485 Email: eemok@ee.ust.hk

ABSTRACT

A novel low-voltage on-chip current-sensing circuit for buck power converter is proposed. The minimum supply voltage of the proposed current-sensing circuit is 1.2V (where V_{th} = 0.85V at room temperature), and the sensing accuracy is higher than 94%. The proposed buck converter is able to operate at a 1.2-V supply with higher than 89% power-conversion efficiency and with a maximum output current of 120mA.

1. INTRODUCTION

In the past few years, industry has been paying more attention on the size, efficiency and reliability of the power converters in portable electronic equipment such as mobile phone, personal digital assistants (PDAs), internet audio players, pagers, portable medical diagnostic equipment, remote control and wireless headsets [1]. This motivates the need of low-voltage high-efficiency switch-mode power converters, which are suitable for single-cell battery applications to minimize the physical size and also maximize the operation time of the portable equipment.

In order to reduce the size and weight of the portable devices, miniaturization of the power modules is essential. One current trend is to lower the power supply voltage to reduce power consumption of the converters and hence the battery size. Moreover, according to the roadmap from Semiconductor Industry Association, the supply voltages of digital circuits have to be reduced to 0.9V-1.2V by the year 2005 [2]. Thus, the design of low-voltage buck converter is a challenging and important research.

In addition, it is well-known that current-mode buck converter takes the advantages of automatic over-current protection, better stability and faster dynamic responses compared with the voltage-mode control. Many different current-sensing schemes have been developed and implemented to sense the inductor current [3]-[5]. Other approaches include the use of a series resistor, the on-resistance of the power MOSFET and even an integrator. However, these schemes have their limitations such as high power dissipation, process-dependence, difficulty in control and also high complexity of implementation.

Therefore, a low-voltage simple on-chip current-sensing circuit for a buck converter implemented in a standard CMOS process is proposed. The proposed circuit provides solutions to many problems of the existing current-sensing schemes. This paper will first provide a brief review on an existing current-sensing scheme in Section 2, and then the proposed current-sensing circuit will be introduced in Section 3. Measurement results of the proposed current-sensing circuit and buck converter that using the proposed current-sensing circuit will be presented in Section 4. Finally, the conclusions of this paper will be given.

2. AN EXISTING CURRENT-SENSING CIRCUIT

An accurate on-chip current-sensing scheme has been proposed and implemented by Lee *et al.* as shown in Fig. 1 [5]. In this circuit, an error amplifier is used to ensure $V_A = V_B$. This circuit, providing a very high current-sensing accuracy, has already proven by experimental results [5]. However, this circuit cannot be used in low-voltage power converters due to the two V_{GS} drops of M_5 and M_R.

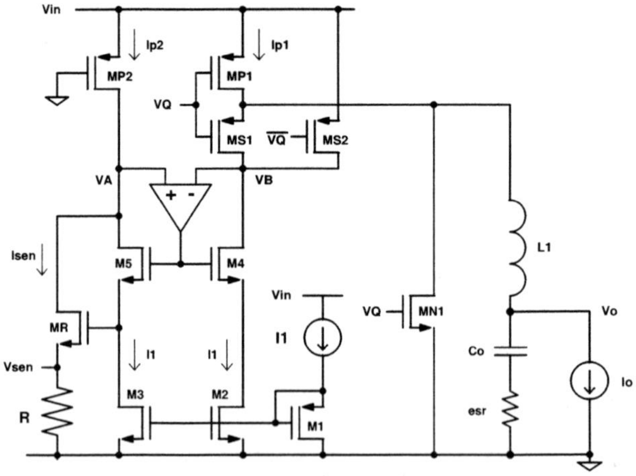

Fig. 1: An existing CMOS current sensing circuit [5].

This work was supported by the Research Grant Council of Hong Kong SAR Government, China, under Project no. HKUST6022/01E.

3. PROPOSED LOW-VOLTAGE ON-CHIP CURRENT-SENSING SCHEME

A low-voltage current-sensing circuitry for the PMOS power transistor in buck converter is therefore proposed and shown in Fig. 2. As the transistor M_R is now a PMOS transistor, the minimum supply voltage is less than the one shown in Fig. 1. The minimum supply voltage can be as low as 1.2V for a CMOS technology with $V_{th} = 0.85$V. A two-stage cascode NMOS input-stage amplifier shown in Fig. 3 is used as it has a high voltage gain and high output swing.

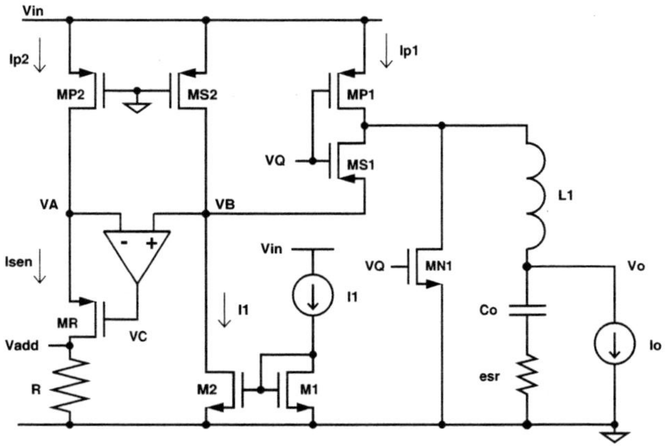

Fig. 2: Proposed low-voltage current-sensing circuit for buck converters.

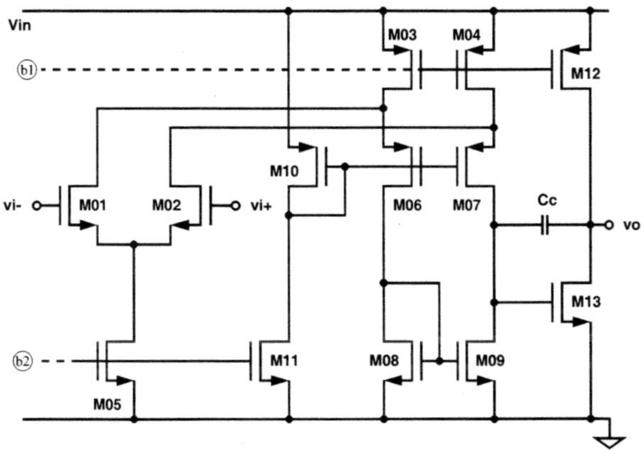

Fig. 3: A cascode amplifier for the proposed current-sensing circuit.

The operational principle is similar to the one proposed by Lee *et al.* The current flowing through the power MOSFET M_{P1} is sensed instead of the inductor current. When the power MOSFET M_{P1} is on ($V_Q = 0$V), as shown in Fig. 4, the error amplifier enforces the voltages at V_A and at V_B to be equal, and therefore the V_{DS} of M_{P1} and M_{P2} are approximately equal. In the design, the size ratio of M_{P1} to the size of M_{P2} is 2000:1, and therefore the current ratio of I_{P1}:I_{P2} is also 2000:1. Now, $I_{sen} = I_{p1}/K$ where $K = W_{MP1}/W_{MP2} = 2000$, and the sensing current can be used to add with ramp signal to form a compensation ramp for current-mode control [6].

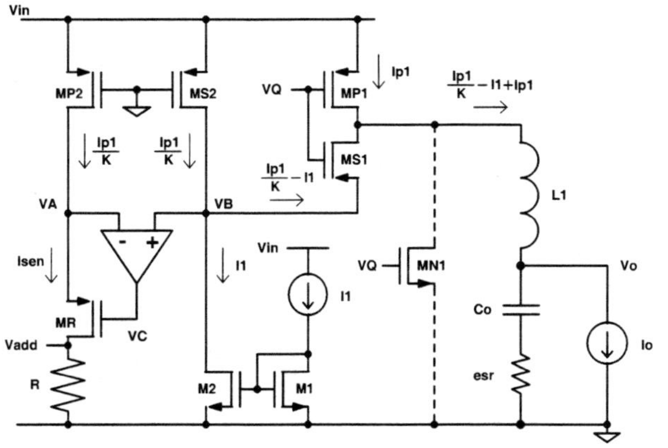

Fig. 4: Proposed current–sensing circuit for buck converter during the ON-period.

As a remark, when the switch M_{P1} is on, the drain current of M_{S2} is equal to that of M_{P2}, and M_2 is biased at a drain current of I_1, so there is a current of $I_{P1}/K - I_1$ flowing from node V_B through M_{S1} to the inductor of the power stage. However, this current has only a current level of $1/K$ of the inductor current. This current is so small that any effect due to this current on the operation of the converter can be ignored.

When the switch is off as shown in Fig. 5, the sensing circuit is disconnected from the power stage by switching off the transistor M_{S1}. The error amplifier and M_R form a loop such that the error amplifier provides V_C to control I_{sen} to achieve $V_A = V_B$. Therefore, I_{sen} is equal to I_1 in this case.

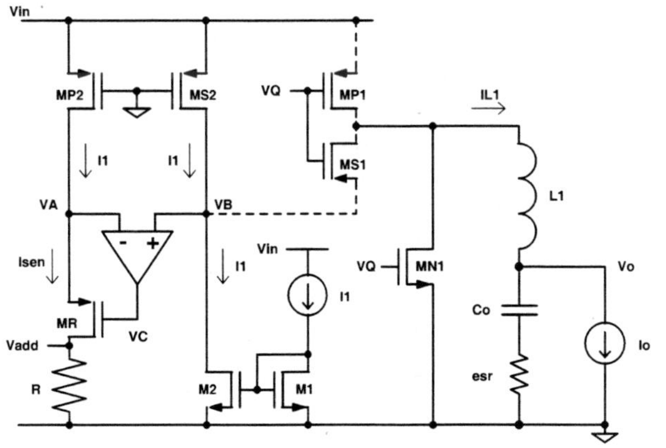

Fig. 5: Proposed current–sensing circuit for buck converter during the OFF-period.

Two poles are introduced at V_A and V_C. The impedance at V_A is dependent on the sensing inductor current which changes the impedance at V_A. Therefore, the pole at V_A varies with inductor current and hence the load current. Since the resistance at V_A is about $1/g_m$ of M_R and the corresponding pole is at a high frequency, the stability, which is compensated by the dominant pole at the output of the error amplifier, is not affected significantly at different load currents.

4. EXPERIMENTAL RESULTS

The schematic of the proposed buck converter using the proposed current-sensing circuit is shown in Fig. 6. This power converter has been implemented in AMS CMOS 0.6-μm technology, and the micrograph is shown in Fig. 7.

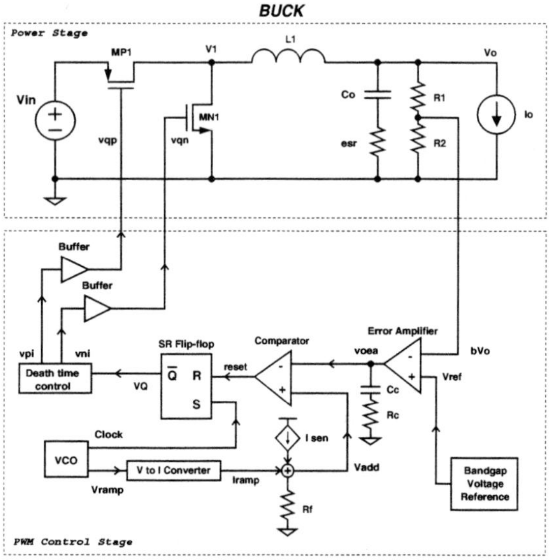

Fig. 6: Schematic of the proposed current-mode buck converter.

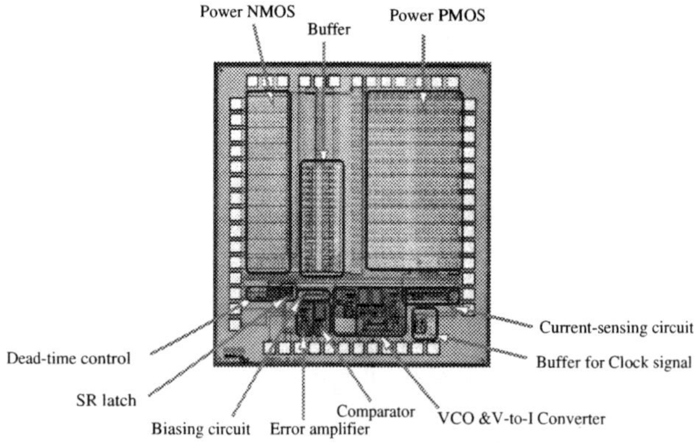

Fig. 7: Micrograph of the proposed current-mode buck converter.

The experimental results of the proposed inductor-current sensing scheme for the power PMOS transistor are shown in Figs. 8 and 9. The current-sensing circuit operates at 500kHz with the supply voltage ranging from 1.2V to 2V. The top graphs of each measurement show the inductor currents I_L, while the bottom graphs show the sensing voltage V_{add}.

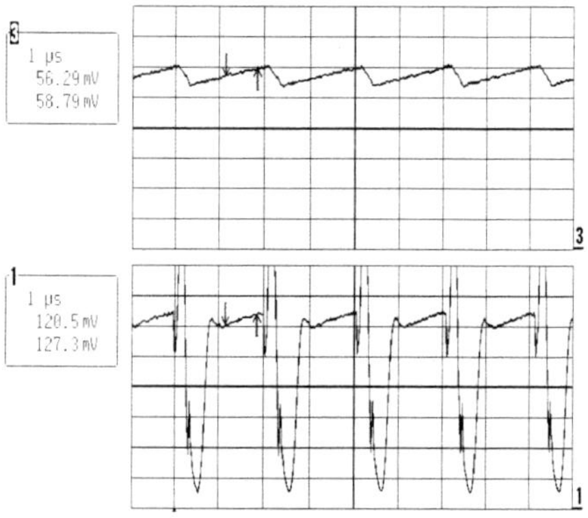

Fig. 8: Measurement results of the proposed current-sensing circuit at $V_{IN} = 1.2V$ $I_O = 120mA$ and $V_O = 1V$.

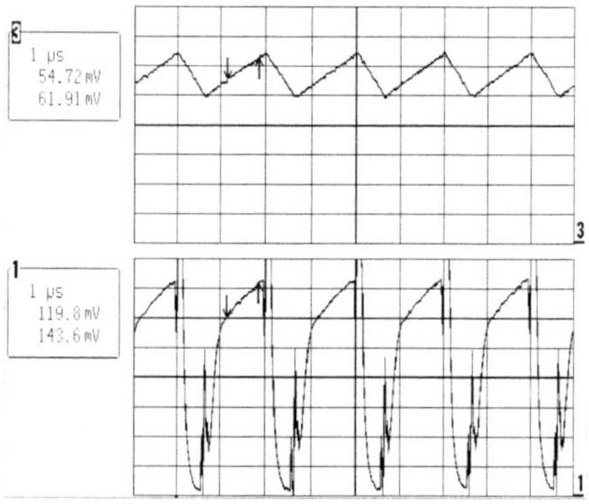

Fig. 9: Measurement results of the proposed current-sensing circuit at $V_{IN} = 1.6V$ $I_O = 120mA$ and $V_O = 1V$.

The accuracies of the proposed current-sensing circuitry at different supply voltages and two load currents are plotted in Fig. 10. From the measurement results, the accuracy of the proposed current-sensing circuit for the power PMOS transistor is higher than 94%, and it is sufficient for power converter design.

Based on the proposed current-sensing circuit, a current-mode buck converter is implemented. The minimum supply voltage is 1.2V and the converter is capable of delivering 120mA. The regulated output voltage is set to 1V in this measurement, and it can be changed to other values by different feedback factors. This converter provides line regulation of 29.1mV/V and load regulation of 0.31mV/mA. Figs. 11 and 12 show the measured signals of the inductor current, added sensing currents to ramp signal, and the voltage at V_1 in Fig. 6 with two different load currents.

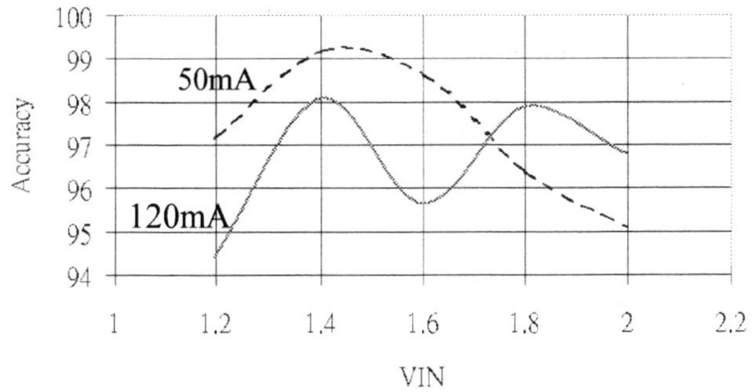

Fig. 10: Accuracy of the proposed current-sensing circuit for buck converter.

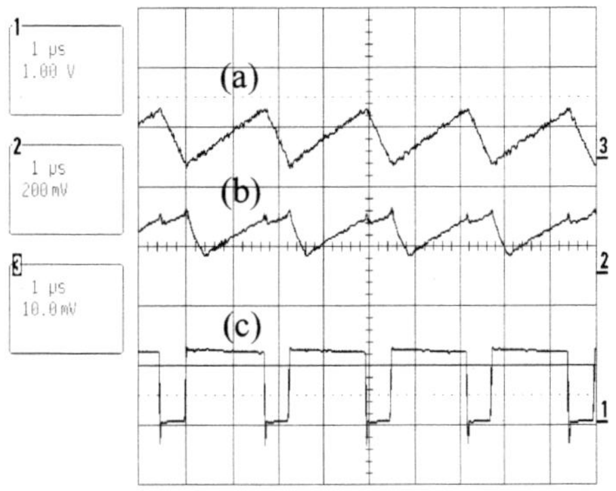

Fig. 11: V_{IN}=1.2V V_O=1V I_O=20mA (a) I_L (b) V_{add} (c) V_1.

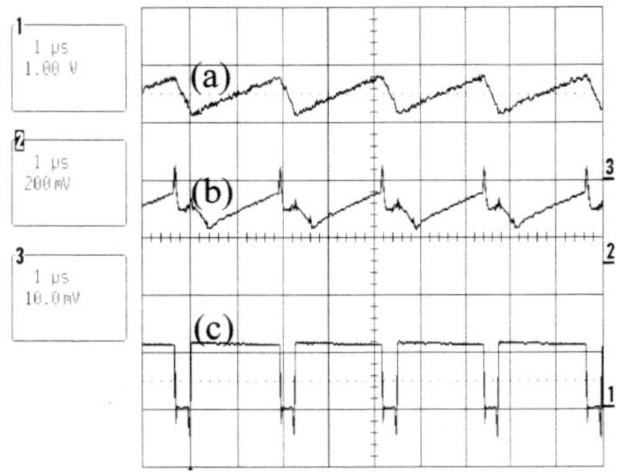

Fig. 12: V_{IN}=1.2V V_O=1V I_O=120mA (a) I_L (b) V_{add} (c) V_1.

Fig. 13 shows the measured power-conversion efficiency at V_{IN} = 1.2V and 1.6V with different load currents. From the results, the efficiency is higher than 89% even at a very low load-current level. This proves the effectiveness of the designs of the analog and digital building blocks, as well as the proposed current-sensing circuit.

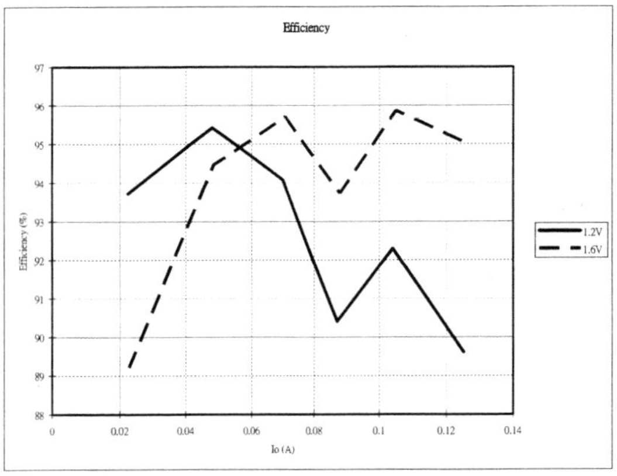

Fig. 13: Power-conversion efficiency.

5. CONCLUSION

A novel low-voltage current-sensing circuit for CMOS buck power converters has been introduced in this paper. The principle of operation has been presented. The sensing accuracy is proven by experimental results to be higher than 94%. Moreover, the proposed circuit has been applied in a buck converter. Experimental results verify the effectiveness of the current-sensing circuit for current-mode operation and stability of the power converter.

REFERENCES

[1] A. Stratakos, C. Sullivan, S. Sanders, R. Brodersen, "DC Power Supply Design in Portable Systems," *Technical Report ERL Memorandum Number M95/4*, University of California, Berkeley, 1995.

[2] *International Technology Roadmap for Semiconductors*, http://public.itrs.net.

[3] M. Corsi, "Current Sensing Schemes for use in BiCMOS Integrated Circuits," *IEEE Proceeding of the 1996 Bipolar/BiCMOS Circuits and Technology Meeting*, New York, USA, pp. 55-57, 1995.

[4] W. H. Ki, *Current Sensing Technique using MOS Transistors Scaling with Matched Bipolar Current Sources*, U.S. Patent 5,757,174, May 26, 1998.

[5] C. F. Lee and P. K. T. Mok, "On-Chip Current Sensing Technique for CMOS Monolithic Switch-Mode Power Converters," *IEEE International Symposium on Circuits and Systems*, Vol. V, pp. 265-268, May 2002.

[6] W. H. Ki, "Analysis of Subharmonic Oscillation of Fixed Frequency Current-Programming Switching Mode Power converters," *IEEE Transactions on Circuits and Systems I*, vol.45, pp.104-108, Jan., 1998.

LOOP GAIN ANALYSIS AND DEVELOPMENT OF HIGH–SPEED HIGH–ACCURACY CURRENT SENSORS FOR SWITCHING CONVERTERS

*Hylas Y. H. Lam, Wing-Hung Ki and Dongsheng Ma**

Dept. of Electrical & Electronic Engineering, Hong Kong University of Science & Technology, HKSAR, China
*Dept. of Electrical & Computer Engineering, Louisiana State University, Baton Rouge, LA 70803, USA
hylas@ee.ust.hk, eeki@ee.ust.hk, ma@ece.lsu.edu

ABSTRACT

Loop gain analysis for performance evaluation of current sensors for switching converters is presented. The MOS transistor scaling technique is reviewed and employed in developing high-speed and high-accuracy current sensors with offset-current cancellation. Using a standard 0.35μm CMOS process, an integrated full-range inductor current sensor for a boost converter is designed. It operated at a supply voltage of 1.5V with a dc loop gain of 38dB, and a unity gain frequency of 10MHz. The sensor worked properly at a converter switching frequency of 500kHz.

1. INTRODUCTION

Inductor current sensing is essential for high performance monolithic switch mode power converters. It is used for over-current protection and current-mode feedback control. It can also be used in fuel gauge applications for battery chargers. A rudimentary implementation is to insert a small resistor in series with the inductor and measure the voltage across it. Evidently, this sensing resistor consumes much power and degrades the efficiency of the converter [1]. Besides, the voltage generated by the small resistor is very small, and is difficult to be used as a control signal. A more energy efficient method is to monitor the switch and/or diode currents. For a boost converter (Figure 1), the inductor current ramps up when the NMOS M_{N1} turns on, and can be measured if the current passes through M_{N1} is known. Similarly, the inductor discharging current can be measured by monitoring the current passing through M_{P1}. The complete inductor current can be obtained by combining the currents in M_{N1} and M_{P1}.

An energy efficient sensor is to employ the MOS transistor scaling technique [2]. The PMOS transistors M_Y and M_X in Figure 2 have the same channel length but a channel width ratio of N : 1, with N >> 1, such that M_Y is a power transistor, and M_X is a sensing transistor. If v_x is set equal to v_y, then the current ratio $i_y : i_x$ is also $N : 1$. High-gain operational amplifiers (op amps) [3, 4] can be used to force v_x to be equal to v_y. The topology in [3] requires a V_{DD} of at least 3V. Ref.[4] rearranges the cascode stage such that a smaller V_{DD} can be used, by sacrificing the matching accuracy between M_X and M_Y.

This research is in part supported by the Hong Kong Research Grant Council under grant CERG HKUST 6209/01E and Innovative Technology Fund ITS/033/02.

Another way to make v_x and v_y equal is to employ matched current source technique using bipolar transistors [2] or MOS transistors [5] that is much simpler when compared with high-gain op amp designs. In this paper, loop gain analysis of various current sensors is presented. A current sensor which can sense switch and diode current at the same time with improved speed and accuracy is developed, analyzed and simulated.

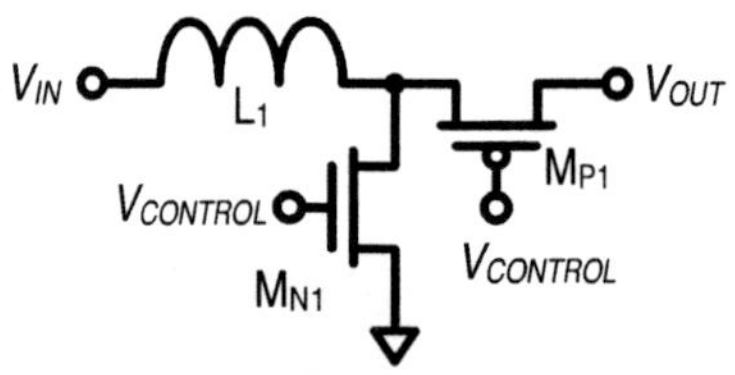

Figure 1. Boost Converter

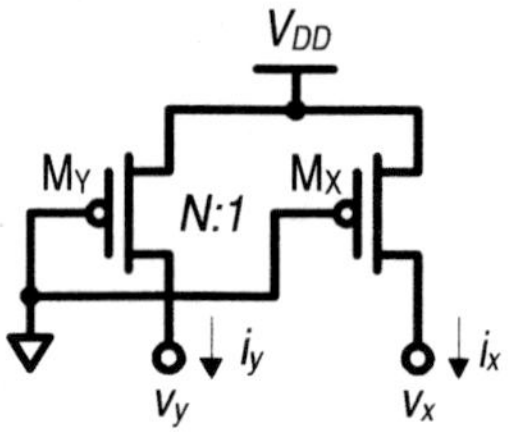

Figure 2. MOS transistor scaling technique

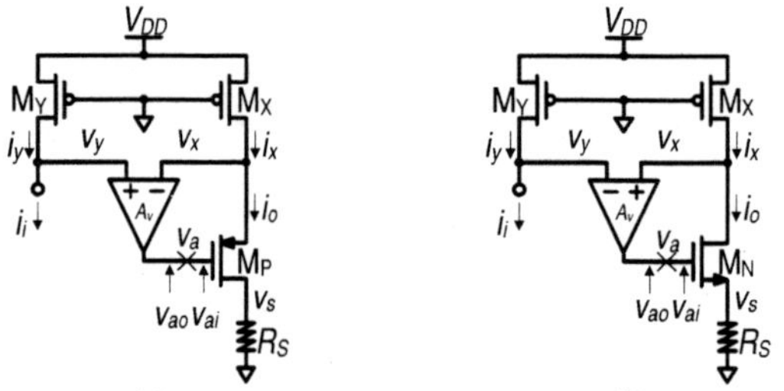

(a) (b)

Figure 3. An op amp based voltage mirror (a) with PMOS, and (b) with NMOS.

2. LOOP GAIN ANALYSIS OF CURRENT SENSORS

To evaluate various current sensors on the same platform, their loop gains have to be computed. We first consider a realization of current sensor with an op amp having a gain of A_v and the PMOS (Figure 3a) or NMOS (Figure 3b) transistor forming a voltage mirror to force v_x equal to v_y. A resistor R_S is used to convert the current signal to a voltage signal such that v_s is proportional to the sensing current. The loop gain can be computed by setting $v_y = 0$ and breaking the loop at the output of the op amp (v_a). The loop gains of the two cases are

$$T_{PMOS} = \frac{-v_{ao}}{v_{ai}} = \frac{g_{mp} R_X A_v}{1 + g_{mp} R_X} \tag{1a}$$

$$T_{NMOS} = \frac{-v_{ao}}{v_{ai}} = \frac{g_{mn} R_X A_v}{1 + g_{mn} R_S} \tag{1b}$$

where g_{mp} and g_{mn} are the transconductances of the PMOS and NMOS transistors, respectively, and R_X is the on-resistance of the sensing transistor M_X.

To derive the I/O relation between i_i and i_o for Figure 3a, we have

$$i_i = -v_y / R_Y \tag{2a}$$

$$i_o = -v_x / R_X \tag{2b}$$

$$v_a = A_v (v_y - v_x) \tag{2c}$$

$$i_o = g_{mp}(v_x - v_a) = g_{mp}[v_x(1 + A_v) - v_y A_v] \tag{2d}$$

$$\frac{i_o}{i_i} = \frac{R_Y}{R_X}\left(\frac{1}{1 + \frac{1 + g_{mp} R_X}{g_{mp} R_X A_v}}\right) = \frac{R_Y}{R_X}\left(\frac{1}{1 + \frac{1}{T_{PMOS}}}\right) \tag{2e}$$

Similarly, for Figure 3b:

$$\frac{i_o}{i_i} = \frac{R_Y}{R_X}\left(\frac{1}{1 + \frac{1 + g_{mn} R_X}{g_{mn} R_S A_v}}\right) = \frac{R_Y}{R_X}\left(\frac{1}{1 + \frac{1}{T_{NMOS}}}\right) \tag{3}$$

R_Y and R_X are the on-resistances of M_Y and M_X, respectively. Hence, $R_Y : R_X = 1 : N$. From equation (2e) and (3) we identify that if the loop gain of the current sensor is high (T_{PMOS} and $T_{NMOS} >> 1$), $i_o : i_i$ is equal to 1 : N. If parasitic capacitors are considered, a high impedance node is located at the output of the op amp, and depending on the design of the op amp, frequency compensation may be needed for stability.

If we compare equation (1a) and (1b), with $g_{mp} = g_{mn}$, T_{PMOS} is larger than T_{NMOS} if $R_X < R_S$. This is normally true since $|V_{DS}|$ of M_X ($= i_o \times R_X$) at the maximum current level should still be much smaller than V_{DD} for high efficiency, while v_s ($=i_o \times R_S$) should be large enough to serve as a control signal. Moreover, the minimum supply voltage using NMOS to drain the sensed current is higher as R_S is connected to the source of M_{S1}.

The op amp in the current sensors must be able to operate with a common mode input voltage close to V_{DD} since the $|V_{DS}|$ of the power transistor M_Y is very small (in the order of 10mV). At the same time, the output swing of the op amp has to be large such that the current passing through M_P or M_N can be controlled in the interested range. These requirements increase the difficulties in designing op amp for the current sensors.

Another way to realize the voltage follower is to use the matched current source technique [2, 5], as shown in Figure 4. The circuit is a modification of [2] by changing all BJT to MOS transistors, and can be implemented in a standard CMOS process. The two PMOS M_1 and M_2 are matched and are biased in the saturation region. Two small and equal biasing currents i_1 and i_2 ($=i_{bias}$) force M_1 and M_2 to have the same V_{SG} such that v_x and v_y are the same. Both i_x and i_y are required to be much larger than i_{bias} ($i_{bias} << i_x, i_y$) to give $i_x : i_y = 1 : N$. This simple implementation can fulfill the requirements for common mode input range and output voltage swing.

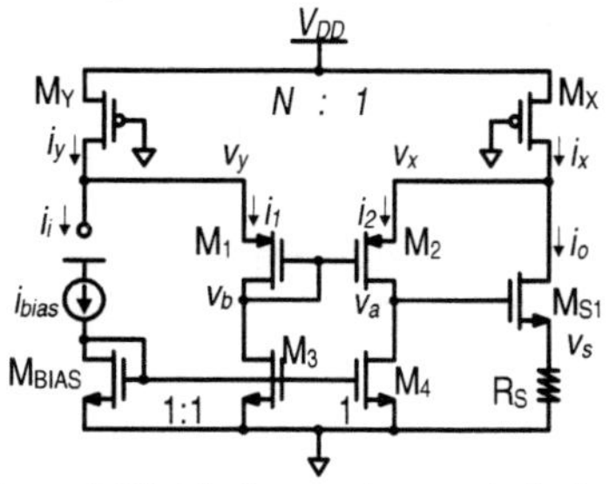

Figure 4. Matched current source technique

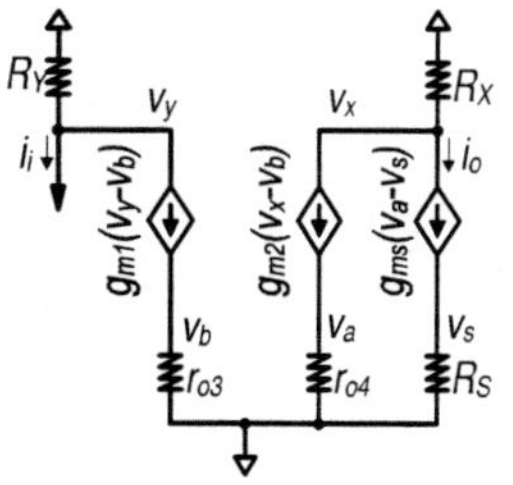

Figure 5. Small signal model of the sensing circuit

To investigate the dynamic performance of the sensing circuit, the small signal model is derived (Figure 5). To reduce complexity, the output resistors r_{o1}, r_{o2} and r_{os} are omitted. Their effects can roughly be reflected by changing r_{o3} to $r_{o1}//r_{o3}$, r_{o4} to $r_{o2}//r_{o4}$, and R_S to $R_S//r_{os}$. The KCL equations are:

$$@\ v_b: \quad g_{m1}(v_y - v_b) = \frac{v_b}{r_{o3}} \tag{4a}$$

$$\Rightarrow \quad v_y = v_b \frac{1 + g_{m1} r_{o3}}{g_{m1} r_{o3}} \approx v_b \qquad (\because g_{m1} r_{o3} >> 1) \tag{4b}$$

$$@\ v_y: \quad 0 = i_i + g_{m1}(v_y - v_b) - i_y \tag{4c}$$

$$\Rightarrow \quad 0 = i_i + \frac{v_b}{R_Y} \qquad (\because v_y \approx v_b) \tag{4d}$$

$$@\ v_s: \quad g_{ms}(v_a - v_s) = \frac{v_s}{R_S} \tag{4e}$$

$$\Rightarrow \quad v_a = i_o \frac{1 + g_{ms} R_S}{g_{ms}} \tag{4f}$$

$$@\ v_x: \quad 0 = \frac{v_x}{R_X} + g_{m2}(v_x - v_b) + i_o \tag{4g}$$

$$\Rightarrow \quad 0 = v_x \frac{(1 + g_{m2} R_X)}{R_X} + g_{m2} R_Y i_i + i_o \tag{4h}$$

$$@\ v_a: \quad g_{m2}(v_x - v_b) = \frac{v_a}{r_{o4}} \tag{4i}$$

$$\Rightarrow \quad v_x = \frac{i_o(1 + g_{ms} R_S)}{g_{ms} g_{m2} r_{o4}} - i_i R_Y \tag{4j}$$

By eliminating v_x in eq.(4h) and eq.(4j), i_o/i_i is given by:

$$\frac{i_o}{i_i} = \frac{R_Y}{R_X} \frac{1}{1 + \frac{(1 + g_{ms} R_S)(1 + g_{m2} R_X)}{g_{ms} g_{m2} r_{o4} R_X}} = \frac{R_Y}{R_X} \frac{1}{1 + \frac{1}{T}} \tag{5a}$$

$$T = \frac{g_{ms} g_{m2} r_{o4} R_X}{(1 + g_{ms} R_S)(1 + g_{m2} R_X)} \tag{5b}$$

Equations (5b) and (1b) have a similar form. Transistors M_1–M_4 form an amplifier with a dc gain of $A_v = g_{m2}r_{o4}/(1+g_{m2}R_X)$. In this current sensor, there is only one dominant high-impedance node at the gate of M_s in the loop. Hence, the circuit has excellent stability and no frequency compensation is needed. Since the current sensing circuit consists of only 4 transistors in the gain stage, on-chip implementation is easy. However, the use of NMOS in draining the sensed current has a smaller dc loop gain and a larger minimum supply voltage. Also, the speed of the gain stage is limited by the small biasing currents of M_1 and M_2.

3. LOOP GAIN ENHANCEMENT AND OFFSET-CURRENT CANCELLATION

Figure 6 shows a modification of the matched current source technique that improves both the dc loop-gain and speed. Dependency of R_S in the dc loop gain is eliminated by changing M_s to be a PMOS transistor. Negative feedback is maintained by connecting the gate of M_s to the M_1 and M_3 branch instead. The KCL equations for loop gain calculation are

$$@\, v_a : \quad g_{m2}(v_x - v_a) = \frac{v_a}{r_{o4}} \tag{6a}$$

$$@\, v_x : \quad 0 = \frac{v_x}{R_X} + g_{m2}(v_x - v_a) + i_o\,, \quad i_o = g_{ms}(v_x - v_b) \tag{6b}$$

$$@\, v_y : \quad 0 = \frac{v_y}{R_Y} + g_{m1}(v_y - v_a) + i_i \tag{6c}$$

$$@\, v_b : \quad g_{m1}(v_y - v_a) = \frac{v_b}{r_{o3}} \tag{6d}$$

By solving equations (6a–d),

$$\frac{i_o}{i_i} = \frac{R_Y}{R_X}\frac{1}{1+\dfrac{(1+g_{ms}R_X)(1+g_{m1}R_Y)}{g_{m1}g_{ms}r_{o3}R_X}} = \frac{R_Y}{R_X}\frac{1}{1+\dfrac{1}{T_{Psensor}}} \tag{7a}$$

and the loop gain is given by

$$T_{Psensor} = \frac{g_{ms}g_{m1}r_{o3}R_X}{(1+g_{m1}R_Y)(1+g_{ms}R_X)} \,. \tag{7b}$$

Equation (7b) and (1a) have a similar form, and the dc gain generated by (M_1–M_4) is equal to $g_{m1}r_{o3}/(1+g_{m1}R_Y)$, which is larger than the current sensor in Figure 4. If we consider the large signal behavior at v_a and v_b, $\Delta v_a/\Delta v_b$ is positive in this current sensor, and negative in the current sensor shown in Figure 4. Therefore, better matching of the V_{DS} of M_1 and M_2 is achieved, and sensor accuracy is enhanced.

To further enhance the matching accuracy, a PMOS transistor M_C matched to M_1 and M_2 is added to provide a compensation current i_{comp} (= i_{bias}) sourcing from the large switch M_Y. The current through R_S is now given by

$$i_s = i_x - i_{bias} + i_{comp} \tag{9}$$

with i_{comp} effectively canceling the offset-current of i_{bias}. Hence, the condition $i_x >> i_{bias}$ is relieved, and i_{bias} can be increased to speed up the current-sensing circuit.

To sense the switch current of the boost converter, an NMOS current-sensing circuit is needed (Figure 7). It has a similar structure as the PMOS counterpart, but the i_{bias} is generated with reference to v_x and v_y. This is a valid connection because v_x and v_y are equal, and both are very small. The output current i_o is mirrored by M_{S1} with R_S connected to ground. To better match M_{S1} and M_{S2}, a dummy resistor R_{dummy} is added.

The I/O relation between i_i to i_o is given by:

$$\frac{i_o}{i_i} = \frac{R_Y}{R_X}\frac{1}{1+\dfrac{(1+g_{m1}R_X)}{g_{ms}g_{m4}r_{o2}R_X}} = \frac{R_Y}{R_X}\frac{1}{1+\dfrac{1}{T_{Nsensor}}} \tag{8a}$$

The loop gain of the N current sensor is given by:

$$T_{Nsensor} = \frac{g_{ms}g_{m4}r_{o2}R_X}{(1+g_{m1}R_X)} \tag{8b}$$

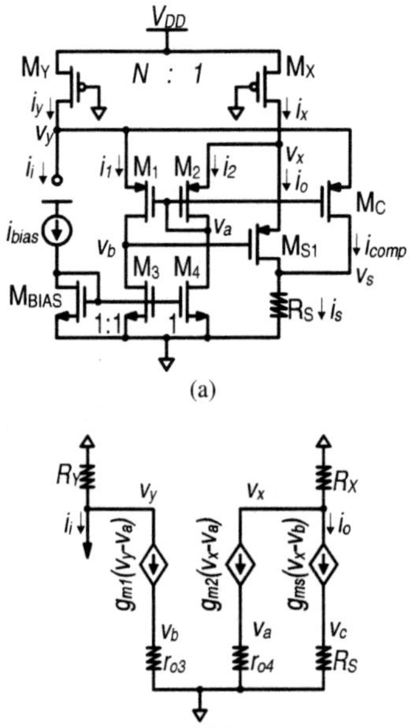

Figure 6. P current sensor with current-offset cancellation (a) circuit diagram and (b) small signal model

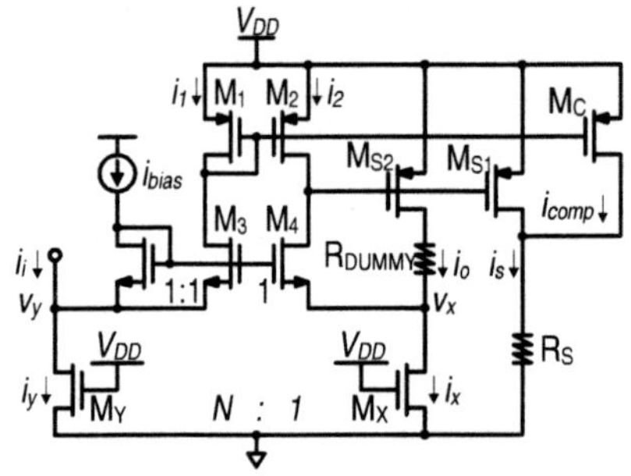

Figure 7. N current sensor with current-offset cancellation

4. FULL-RANGE INDUCTOR CURRENT SENSOR

For conventional converter designs, either the switch current or the diode current is required for current-mode control. But in some applications, the complete inductor current has to be known, such as fuel gauge application for battery chargers. Existing current sensors cannot switch over from sensing switch current to sensing diode current fast enough, and there could be a 'blackout' period in the order of μs. Our proposed P sensor and N sensor have the same core circuit. Hence, by combining the two, a high-accuracy high-speed full-range inductor current sensor can be achieved (Figure 8). M_{XN} measures the switch current and M_{XP} measures the diode current. The 6 switches (S_{D1}–S_{D2}, S_{P1}–S_{P4}) on the upper part of the circuit can be implemented by PMOS transistors, while the remaining switches (S_{N1}–S_{N4}) can be implemented by NMOS transistors. S_{D1} (matched to S_{P1} and S_{P2}) and S_{D2} (matched to S_{P3} and S_{P4}) are dummy transistors and they are closed all the time such that the matching between M_1 and M_2 and between M_{C1} and M_{C2} are better. During the charging phase of the inductor, power transistor M_{YN} turns on, and S_{N1}, S_{N2}, S_{P2} and S_{P4} are closed to form the N sensor of Figure 7. During the discharging phase, power transistor M_{YN} turns on, and S_{N3}, S_{N4}, S_{P1} and S_{P3} are closed to form the P sensor of Figure 6. In both cases, the sensed current passes through R_S, and the voltage across R_S (v_S) is a good replica of the full-range inductor current.

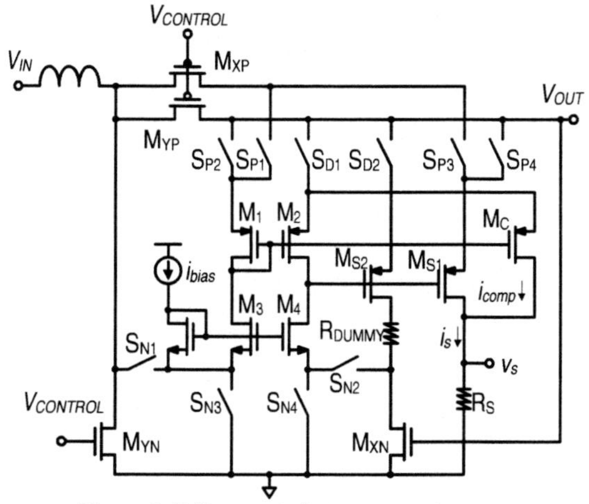

Figure 8. Full-range inductor current sensor

5. SIMULATION RESULTS

The proposed current-sensing circuits were designed using a standard 0.35μm CMOS process. The power transistors are designed to operate at a maximum current of 500mA, with a resistance of around 100mΩ. The sensing transistors have channel width 1/1000 of the power transistors, and the sensing resistors R_S are 1kΩ. The simulation results of the full-range inductor current sensor are discussed below. For an inductor current of 100mA, i_x = 100μA and V_S = 100mV is obtained. The biasing current i_{bias} is 50μA for high-speed current-sensing operation. Figure 9 shows the loop gain response of the sensor with different supply voltages (from 1.5V to 3V). The dc gain is 35dB at V_{DD} = 3V and 38dB at V_{DD}=1.5V, which corresponds to less than 2% error between v_x and v_y. The unity gain bandwidth is 10MHz with a phase margin of 90°, which should be fast enough for a switching converter operating at 500kHz. The dc gain increases with the decrease in the supply voltage, which is a result of a larger R_X and R_Y due to a lower gate drive voltage. Figure 10 shows the simulated transient response of the full-range inductor current sensing operation. The boost converter operated at 500kHz with a peak-to-peak inductor current of 260mA. The sensed voltage v_s tracked the inductor current closely with a small delay of less than 100ns when switching between the inductor charging and discharging phases.

6. CONCLUSIONS

High-accuracy and high-speed P and N current sensors are presented. A full-range inductor current sensor was designed, and simulated using a standard 0.35μm CMOS process. Simulation results show that smooth and accurate inductor current information can be provided by the proposed circuit. The simple circuit structure can easily be implemented fully on-chip and no compensation is needed while providing excellent stability.

REFERENCES

[1] H.P. Forghani-Zadeh and G.A. Rincon-Mora, "Current-sensing technique for DC-DC converters," *IEEE Midwest Symp. on Ckts. & Sys.*, Vol. II, pp.577-580, 2002.

[2] W.H. Ki, "Current sensing technique using MOS transistors scaling with matched bipolar current sources," *U.S. Patent 5,757,174*, May 26, 1998.

[3] C.F. Lee and P. Mok, "A monolithic current-mode CMOS DC-DC converter with on-chip current-sensing technique," *IEEE J. of Solid-State Ckts.*, Vol. 39, No. 1, pp. 3-14, Jan 2004.

[4] C.Y. Leung, P. Mok and K.N. Leung. "A 1.2-V buck converter with a novel on-chip low-voltage current-sensing scheme," *IEEE Int. Symp. on Ckt. & Sys.*, May 2004.

[5] D. Ma, W.H. Ki, C.Y. Tsui and P. Mok, "A 1.8V single-inductor dual-output switching converter for power reduction techniques," *IEEE VLSI Symp. on Ckts.*, June 2001, pp.137-140.

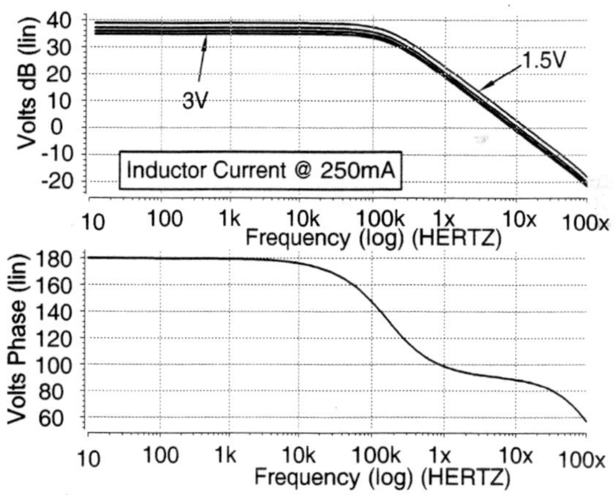

Figure 9. Frequency response of the current sensor

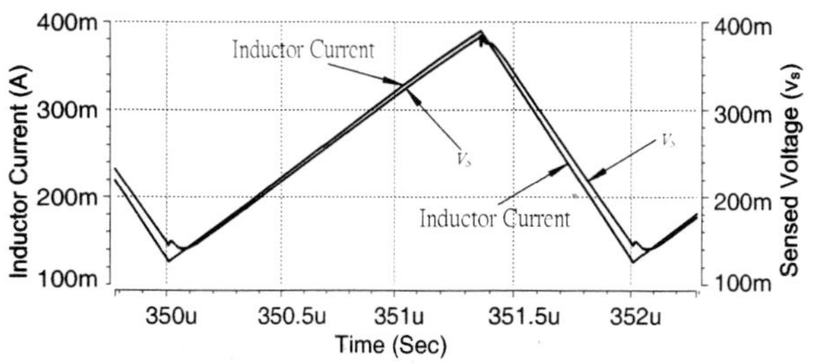

Figure 10. Transient response of the full-range inductor current sensor

5V-ONLY, STANDARD .5UM CMOS PROGRAMMABLE AND ADAPTIVE FLOATING-GATE CIRCUITS AND ARRAYS USING CMOS CHARGE PUMPS

Mark Hooper, Matt Kucic, Paul Hasler

Georgia Institute of Technology
School of Electrical and Computer Engineering
Atlanta, GA 30332

ABSTRACT

This paper presents the integration of high voltage charge pumps into floating-gate arrays in a standard 0.5um CMOS N-well double poly process. In this research two different Dickson charge pumps are presented for the control of tunneling and injection in floating gate circuits and arrays. A six stage design implimented with Schottky rectifiers is used to modulate tunneling and a three stage high voltage design is used to modulate injection. The Schottky configuration is able to boost the output voltage to approximately 18V while sourcing 1uA and the high voltage configuration is able to boost the output voltage to approximately 7V while sourcing 0.1uA. In both cases no voltage spiking is observed in the output voltage. Experimental results and analysis of each configuration are presented.

1. INTRODUCTION

Floating gate EEPROM type circuits have been steadily proliferating since their introduction in 1967 [1]. Since then, floating gate transistors have been widely used to store digital information for long periods of time in devices such as EEPROMS, EPROMs and flash memories. These digital nonvolatile memory technologies have been generally fabricated on specialized and expensive processes such as EEPROM and Flash processes. Fabricating floating gate circuits in standard CMOS processes is highly advantageous in terms of cost and mixed signal integration. Designing floating gate devices in a standard CMOS process uses several inherent mechanisms among which are electron tunneling and hot electron injection as described in section 3. Effecting these two processes requires boosting the input voltage at the gate and drain of the floating gate device since voltages of approximately 17V and 7V are needed respectively. To accomplish this in standard CMOS, a number of rectifying devices were designed, fabricated and characterized in [8]. Extrapolating from said work we arrived at two charge pump configurations for effecting electron tunneling and injection, i.e, six stage Schottky rectifier charge pump and a three stage high voltage charge pump respectively. It is emphasized that these charge pumps will be used to control floating-gate circuits in a commercially available standard CMOS process which would allow mixed signal integration.

2. PREVIOUS WORK

[2],[3],[4],[5],[6],[7] are papers which have specifically targeted the design of charge pumps for EEPROMS or flash memories. However, in two of these papers, non standard CMOS processes

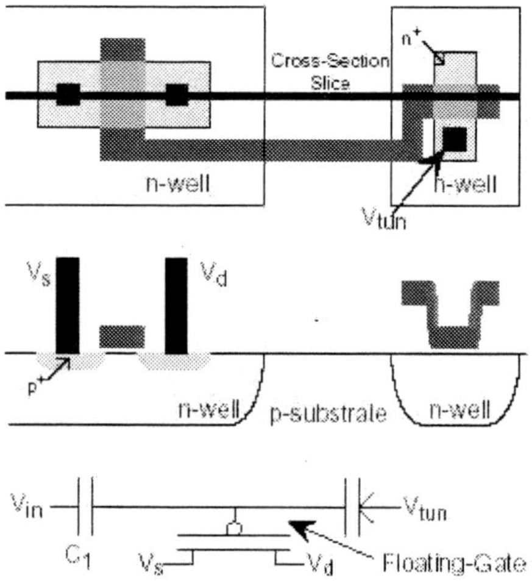

Fig. 1. Shown is the layout, cross section and circuit diagram for a floating gate pFET device in a standard double poly N-well MOSIS process. The cross section corresponds to the horizontal line slicing through the layout view. The pFET transistor is the standard one used in an N-well process.The gate input capacitively couples to the floating gate by a poly-poly capacitor. Floating gate charge is added by electron tunneling and removed by hot electron injection. Between V_{tun} and the floating gate is the symbol for a tunneling junction, a capacitor with an added arrow designating charge flow.

are used[3],[4] i.e., EEPROM process technology and Flash process technology. [2] thoroughly analyzes a Dickson charge pump but only the NMOS configuration and there is no discussion of integrating this configuration in a standard CMOS process to control floating gate circuits. There are essentially six points to [2]: (1) modeling the body effect with a formula which linearly –with a slope of somewhat less than one– relates the source voltage to the drain voltage; (2) using this model they analytically derive the DC transfer function and equivalent resistance; (3) state the limitations of the model: substrate bias of at least 3V; effects of leakage currents; range of output current; (4) measures $\text{Vout}|_{I_{out}=0}$ for different input voltages and compares results to model predictions; (5) measures equivalent resistance and compares to model predictions; (6) shows time domain degradation of current and discusses this phenomenon with respect to controlling nonvolatile memory circuits. Since the analysis is only valid for the NMOS rectifier

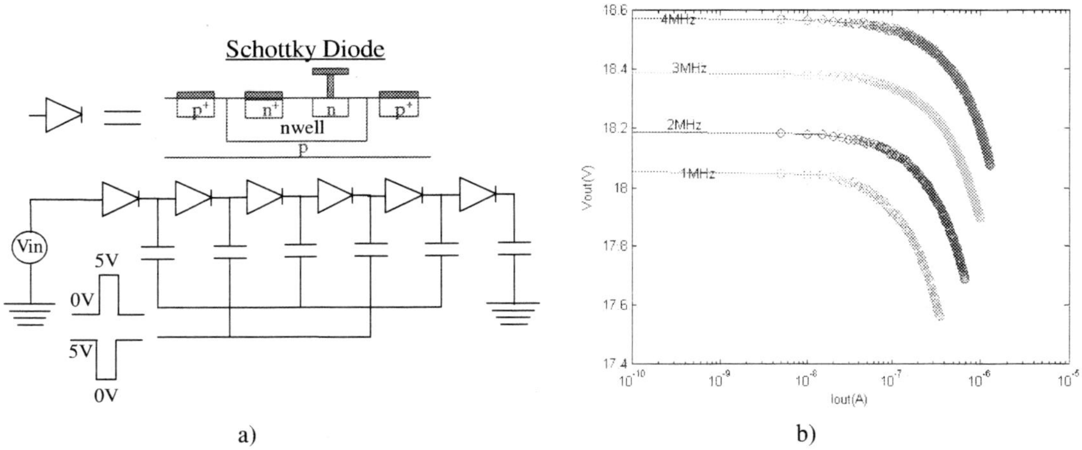

Fig. 2. Illustration of the Schottky charge pump used for tunneling and its characterization from fabrication results in a standard .5um CMOS process. (a) The Schottky charge pump configuration to effect tunneling. (b) Current loading characteristics for various frequencies. The output voltage of approximately 18V will be sufficient for tunneling in a .5um CMOS process. This configuration shows ample sourcing capability to drive an array of floating gate circuits sinking a total of 1nA where each floating gate transistor sinks 1fA.

configuration, parts of this analysis may not hold for other rectifying structures. For example they conclude that junction leakage currents are not a factor in the voltage gain, whereas in [8] we show that the junction leakage currents affect transfer voltage characteristics. [6] discusses charge pumps for 40V systems but in the application of TTL systems and there is no mention of the type of fabrication process used. [7] presents a new charge pump design for a CMOS submicron process, however it is not specifically designed to control floating gate circuits and it is exclusively designed for a low supply voltage(.9V to 2V).

3. FLOATING GATE OVERVIEW

Shown in Fig. 1 is the layout, cross section and circuit symbol for a floating gate pFET device. A floating gate is a polysilicon gate surrounded by SiO_2. Charge on the floating gate is stored permanently, providing long term memory because it is completely surrounded by a high quality insulator. From the layout it is seen that the floating gate is a polysilicon layer with no contacts to other layers; this floating gate can be the gate of a MOSFET and can be capacitively connected to other layers. The floating gate voltage, determined by the charge stored on the floating gate, can modulate a channel between a source and drain, and therefore, can be used in computation. Floating gate circuits provide IC design engineers with a practical, capacitor based technology; since capacitors, rather than resistors, are a natural result of a MOS process. There are two principle modes of operation for floating gate devices: electron tunneling and hot electron injection. Electron tunneling is used to increase the voltage on the floating gate by removing electrons from the floating gate and electron injection is used to add electrons to the gate thereby reducing the gate voltage. Therefore by using a combination of electron tunneling and injection one can change the charge on the floating gate to the desired value and with a varying gate voltage the floating gate device can be used in different applications. Floating gate circuits are used primarily in three areas: (1) analog memory elements and trimming techniques; (2) as part of capacitive based circuits and (3) as adaptive circuit elements.

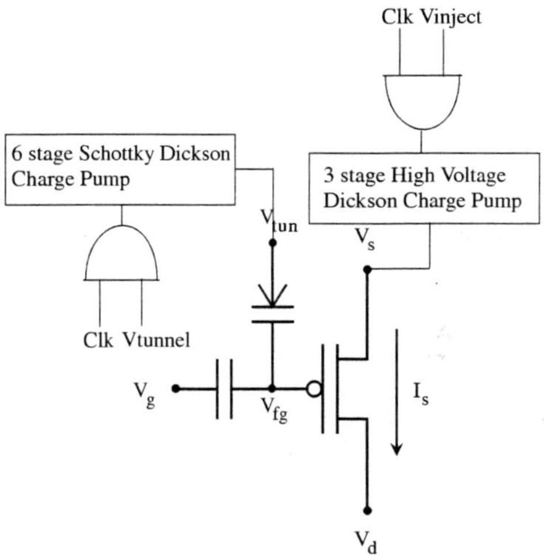

Fig. 3. Shown is the test set up for modulating electron tunneling and injection on a floating gate device in a standard CMOS process. The clock magnitude and input voltage would be 5V. The clock frequency would be 3-5MHz.

4. FLOATING GATE TUNNELING AND INJECTION

Shown in Fig. 3 is our test set up for implementing charge pumps for injection and tunneling to control floating gate devices in a standard CMOS process. Attached to the tunneling capacitor is a six stage Dickson charge pump composed of Schottky diodes. It is activated and de-activated by being 'AND' gated to a 5V clock. Similarly a 3 stage high voltage charge pump is activated and deactivated to control injection by being 'AND' gated with the clock

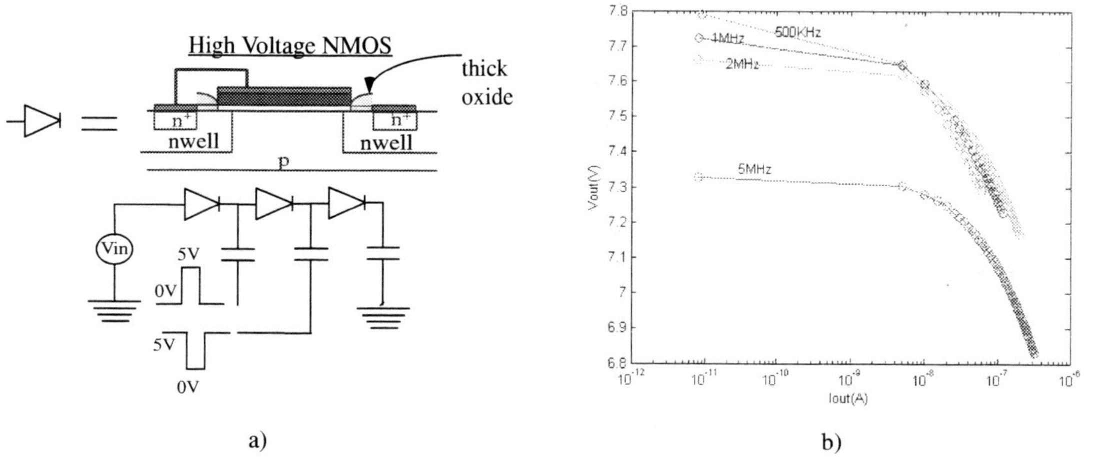

Fig. 4. Illustration of the high voltage charge pump used for injection and its characterization from fabrication results in a standard .5um CMOS process. (a) A 3 stage high voltage charge pump used to control injection in floating gate circuits. (b) The current loading characteristics for controlling injection on the drain of the floating gate device. The voltage interval of approximately 7V-8V will definitely accommodate the required interval 6-6.5V needed for injection.

to the drain of a PMOS transistor. The input voltage in both cases is 5V.

5. EXPERIMENTAL RESULTS

5.1. Schottky Charge Pump Current Loading Characteristics

It is seen in Fig. 2(b), that in general for higher frequencies, the output voltage is increased. This corresponds well with the output voltage v.s. frequency characteristics shown in Fig. 5(b). However at approximately 5-6 MHz it was found that the inverting clock signal significantly attenuated. The clock signal was passed through an inverter on the CD4007UB chip to generate the inverting clock signal needed for the charge-pump. The attenuation of the inverting clock signal most likely explains why the output voltage does not continue to increase with frequency. The clock signals were generated off chip to simplify testing of inital structures. Currently in fabrication is a design with the clock and inverted clock on-chip to prevent signal degradation. This charge pump will be used to control a floating gate array where each transistor draws approximately 1fA for a total of 1nA. Shown in Fig. 5(a) is the transient output of the Schottky charge pump boosting the output voltage from approximately 4V to 18V. This is a clean output in the sense that there are no voltage spikes. Voltage spikes need to be avoided as they can damage the oxide in the floating gate device. The diffusion dimensions of the Schottky diode were 2.4um x 3.9um. Metal1 to N-well contact was effected with non standard contacts to preclude an oxide interface to the N-well. All capacitors were 1pF.

The tunneling current from a 4-element floating-gate array was characterized and the results are shown in figure 6. The input voltage to the tunneling node was applied to a leak pad on which there is approximately a 1MΩ resistor to ground that provides protection while permitting input voltages higher than the chip's V_{dd}. The slope of the vout v.s. current line is approximately 1e6. This indicates the presence of the 1MΩ resistor in the path of the tunneling node. This protection resistor prohibited the direct connection of the charge-pump output directly to the tunneling node of another chip to demonstrate that the charge-pump could provide sufficient current for tunneling. However we know that changes in current due to tunneling would occur approximately between 12-15V. The smaller graph in 6 shows the derivative of the current v.s. voltage graph. It is observed that the change in current due to tunneling is never more than 10nA. Therefore the tunneling current never exceeds 10nA which means that the Schottky charge pump has sufficient current sourcing capability for tunneling.

5.2. High Voltage Charge Pump Current Loading Characteristics

Shown in Fig. 4(b) are the current loading characteristics for a 3 stage Dickson charge pump using high voltage transistors. Conforming with the results shown in Fig. 5(b) there is a tendency for the output voltage to decrease with frequency. However once again at higher frequencies, at approximately 5MHz, there is a sharp decrease in magnitude most likely due to the attenuation of the inverting clock signal magnitude. The charge pump used for injection would be used to source approximately a total of 1uA for an array of 100 elements, each sinking 10nA. Further research will be conducted to modify the design in order to increase current sourcing capability. The high voltage structure was an NMOS diode connected transistor except that the source and drain regions were surrounded by N-wells. Inputs to the charge pumps were attached to leak pads and outputs were taken from bare pads. The values of the capacitors were 1pF.

The current needed to provide injection is to some degree varied. The current that can be supplied will directly relate to how fast current can be injected onto the floating-gate in the device. The optimal current needed for injection is threshold current which is .8uA for the 0.5um process. Currents below this value can also be used for injection. Hence it appears that this charge pump will be able to supply the current needed for injection before the output

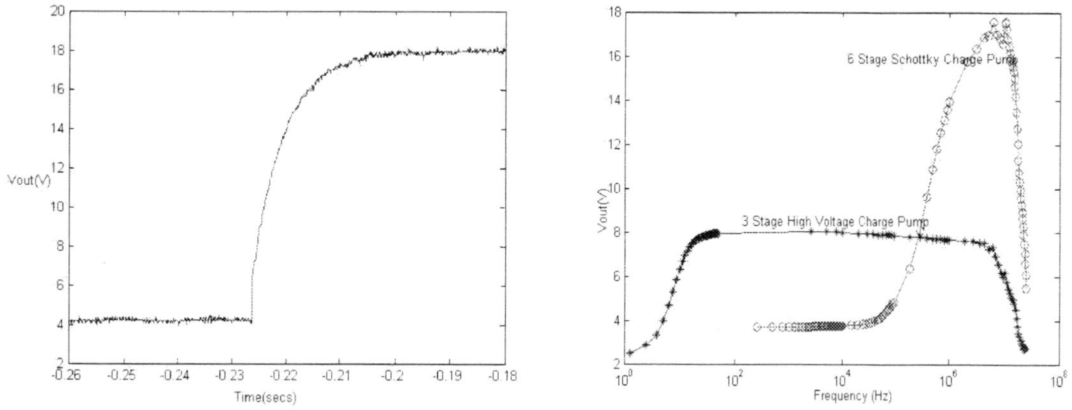

a) b)

Fig. 5. Illustration of transient and frequency output voltage characteristics from fabrication results in a standard 0.5um CMOS process. (a)Transient output of Schottky 6 stage charge pump boosting output voltage from approximately 4.3V to 18V. (b) Output voltage as a function of frequency for the high voltage and Schottky charge pumps. These tests were effected with a 5V input and a 5V clock operating at 5MHz for a no load case.

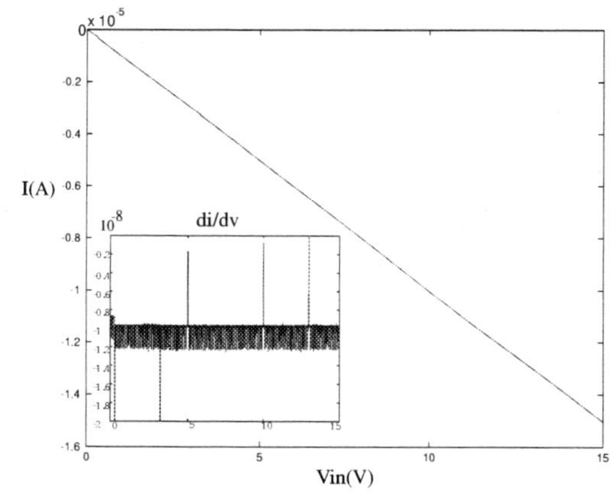

Fig. 6. Floating-gate tunneling current characterization from fabrication results in a standard 0.5um CMOS process.

voltage starts to drop. Again, as in the tunneling setup, a leak-pad prevented directly connecting the charge pump and floating gate circuit for testing purposes. A layout revision is in fabrication which has the charge-pump on the same chip as the floating-gate element to permit such testing. The approach was to first design charge pumps on a separate chip from the floating gate circuits to first make certain that they would meet the loading specifications. Once it was established that they would meet the loading specifications, they could then be integrated on the same chip as the floating gate circuit to control the operations of injection and tunneling.

6. CONCLUSION

Dickson charge pumps were designed, fabricated and characterized for controlling injection and tunneling for an array of floating gate circuits in a standard .5um CMOS process. These designs have been shown to be practical. The charge pumps were able to boost to the desired output voltages of 6-6.5V for injection and 15-17V for tunneling and no voltage spiking on the output was observed. Current sourcing of the charge pumps was also investigated as well as current characterization of the floating gate circuits for tunneling and injection. Designing the charge pumps on chip in a standard CMOS process to control floating gate circuits, reduces cost, allows mixed signal circuit integration, eliminates an external power supply, reduces pin count and power consumption.

7. REFERENCES

[1] D. Kahng and S.M. Sze, A Floating Gate and its Application to Memory Devices, *Bell Syst. Tech. J.*, vol. 46, no. 4, pp.1288-1295, 1967.

[2] J.S. Witters et Al., Analysis and Modeling of On-Chip High-Voltage Generator Circuits for Use in EEPROM Circuits, *IEEE JSSC*, Vol.24, No.5, 1989, pp.1372-1380.

[3] O. Khouri et Al., Improved Charge Pump for Flash Memory Applications in Triple Well CMOS Technology, *IEEE ISIE* 2002, Vol. 4 pp.8-11.

[4] O. J. Shor, Self Regulated Four-Phased Charge Pump with Boosted Wells, *IEEE ISCAS*, 2003, Vol.1, pp.241-244.

[5] K. Sawada et Al., An On-Chip high Voltage Generator Circuit for EEPROMs with a Power Supply Voltage below 2V, *1995 Symposium on VLSI Circuiuts Digest of Technical Papers*, 1995, pp.75-76.

[6] J.F. Dickson, On-Chip High-Voltage Generation in MNOS Integrated Circuits Using an Improved Voltage Multiplier Technique, *IEEE JSSC*, Vol. SC-11, No.3, 1976, pp. 374-378.

[7] K.H. Cheng et Al., A CMOS Charge Pump for Sub 2V Operation, *IEEE ISCAS*, 2003, Vol. 5, pp.89-92.

[8] M. Hooper, M. Kucic, P. Hasler, Characterization of Charge Pump Rectifiers for Standard Submicron CMOS Processes, *IEEE ISCAS* 2004.

ANALYSIS OF SWITCHED CAPACITOR DC-DC STEP DOWN CONVERTER

A.K.P. Viraj, Student Member, IEEE, G.A.J. Amaratunga, Member, IEEE,

Department of Engineering, Cambridge University, Cambridge, CB2 1PZ, UK

ABSTRACT

A 5V/1V Switched Capacitor (SC) dc-dc converter designed for a $0.18\mu m$ CMOS process is analysed in detail, in this paper. Analytical equations are derived for the voltages and currents through the main components of the SC converter. The model includes switches, capacitors, equivalent series resistances and the load. The switches in the converter are represented by MOSFETs in the UMC $0.18\mu m$ CMOS process. The impact of system parameters on output voltage ripple are studied using the analytical expressions.

1. INTRODUCTION

Many methods have been developed to analyse and understand the behavior of SC circuits. Among those the most common method is the State Space Averaging, SSA method. Other methods include models based on voltage waveforms, discrete-time analysis based on charge, energy balance and charge balance[1, 2]. Computer simulation is one method of analysing voltage and current waveforms. This is a useful tool for design validation, but it does not reveal analytical relations among the components in the circuit as a function of input and output. Therefore it is not possible to find an optimum solution for a particular design, except by using a trial and error approach.In this paper, analytical equations and SSA techniques are used to analyse SC power supplies.

2. STEP-DOWN SC DC-DC CONVERTER

An n-stage SC dc-dc step down converter topology is shown in Figure 1. The circuit consists of a control device S_c and n stages. The first stage consists of an output capacitor C_1, in parallel with the load resistor, R_L and all the other n-1 stages consist of one capacitor and three semiconductor devices. The semiconductor devices (M, N and S) are controlled by the complementary signals.

In the proposed SC dc-dc converter, capacitors C_1, and C_j are connected in series across the input voltage during the charging phase and are paralelly discharged to the load during the discharging phase. Since the voltage across the capacitors do not change (ideally) when the state changes, a simple form of voltage step down is realised. The number of capacitors (C_1,C_2,...,C_n) in series, n is selected according to the voltage step down ratio.

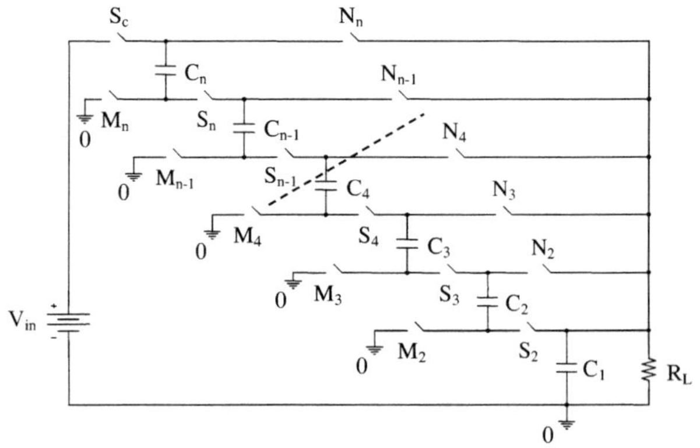

Fig. 1. An n-stage step down SC dc-dc converter.

State 1: Devices, S_j and S_c are ON while devices N_j and M_j are OFF. Capacitors C_1 and C_j are charged by the source. Where $j = 2, 3, ..., n$.
State 2: Blanking time; all devices are OFF. Load is supplied by the output capacitor, C_1.
State 3: Devices, S_j and S_c are OFF while devices N_j and M_j are ON. Capacitors C_j discharge to supply the load.
State 4: Blanking time, same as state 2.

The same SC dc-dc converter topology was presented by Ngo and Webster in [3]. In their analysis only two states (i.e., charging and discharging) were considered. The SC dc-dc converter discussed in this work is intended to be monolithically integrated and should be small in size. Moreover it operates at very high frequencies (i.e., very small switching periods). Therefore the added delay time, which is used to avoid shoot through between the two switching states is also significant compared to the switching period. As a result four switching states can be identified in a cycle. To simplify the analysis and design, S_c is controlled as the primary control device and all other devices are operated simply as ON/OFF switches. In the proposed topology, S_c is controlled as a current source. The other possible control mechanisms are resistance control and duty ratio control[3]. All semiconductor devices are implemented using n-channel MOSFETs.

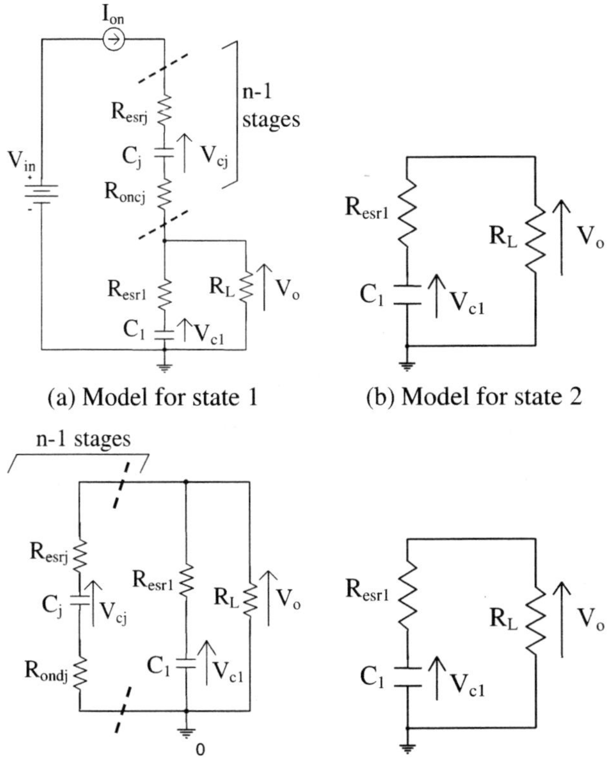

(a) Model for state 1 (b) Model for state 2

(a) Model for state 3 (b) Model for state 4

Fig. 2. Analysed model for States 1,2,3 and 4 respectively.

3. ANALYSIS

SSA is a useful method in the small signal analysis of charge pump and switching regulator dc-dc converters. Selection of the model for analysis is always a trade-off between accuracy and practicality of computation. All the capacitors are represented as a series connection of an equivalent ideal capacitor, C and an equivalent series resistance ESR during all the states. The equivalent resistance R_{oncj} (on state resistance during charging) characterizes resistive voltage drop through the semiconductor device S_j during the charging interval. And the equivalent resistance R_{ondj} (on state resistance during discharging) characterizes resistive voltage drop through the semiconductor devices N_j and M_j during the discharging interval. The control device S_c, is replaced by a constant current I_{on} in the proposed model as it is driven into the saturation region during the charging process. As shown later current control is simple to implement because the output voltage can be linearly adjusted by the choice of an appropriate constant charging current, I_{on}.

3.1. Small signal analysis of the power stage

If capacitor voltage ripple is linear, the SSA technique can be used for analysis[3]. An exponential current spike during the discharging interval results in a nonlinear voltage ripple. Therefore the Modified State-Space Averaging, MSSA technique is used for the analysis in[4]. A comparison of the two approaches in [5] confirms the steady state capacitor voltages obtained are very similar. Therefore, the simpler SSA technique is used to derive the steady state voltages. Figure 2(a), 2(b), 2(c) and 2(d), show the models used for the analysis during the states 1, 2, 3 and 4 respectively. For the switching states shown in Figure 2, the set of state space equations take the form,

$$\dot{x} = A_m x + B_m u \qquad y = C_m x + D_m u$$

where $m \in (1,2,3,4)$, x- state vector, $x^T = [V_{c1}\ V_{cj}]$ u- source vector, $u = [I_{on}]$, y- output vector, $y = [V_o]$ and A_m, B_m, C_m and D_m are state space matrices for the mth state. Note that input vector is taken as the constant charging current I_{on}. For the switching states shown in Figure 2 the state space equations can be derived by using classical KVL and KCL equations. Therefore, the average state space matrices A and B are obtained by averaging A_m and B_m with appropriate weighting factors. i.e.,

$$\dot{x} = Ax + Bu = (d_1A_1 + d_2A_2 + d_3A_3 + d_4A_4)x + (d_1B_1 + d_2B_2 + d_3B_3 + d_4B_4)u \quad (1)$$

Similarly, $$y = Cx + Du \quad (2)$$

Where d_1 is (charging period)/T_s, d_2 is (switch transition period)/T_s, d_3 is (discharging period)/T_s, d_4 is (switch transition period)/T_s and T_s is the switching period. Also

$$\frac{C_1}{C_j} = \frac{C_1}{C} = k = \frac{R_{esrj}}{R_{esr1}} = \frac{R_{esr}}{R_{esr1}}$$

3.2. DC analysis

Under steady state conditions, the Right Hand Side (RHS) of average state space equation (1) is zero. Then,

$$\overline{V}_{c1} = nd_1 I_{on} R_L = \overline{V}_o$$

Where $\overline{V}$ denotes a steady state value of the voltage.

3.3. Deriving analytical equations for the voltages and currents

Voltages and currents were derived as a function of time for the capacitor, C_1 during states 1,2,3 and 4. Assuming output current ripple, ΔI_o is negligible compared to the capacitor, C_1 current ripple. Where I_{c11} is given by current through capacitor C_1 during state 1 etc.
State 1: $0 \le t \le d_1T_s$

$$I_{c11}(t) = I_{on} - I_o \qquad V_{c11}(t) = \frac{I_{on}-I_o}{C_1}t + K_0$$

State 2: $d_1T_s \le t \le (d_1 + d_2)T_s$

$$I_{c12}(t) = -I_o \qquad V_{c12}(t) = \frac{-I_o}{C_1}t + \frac{I_{on}-I_o}{C_1}d_1T_s + K_0$$

State 3: $(d_1 + d_2)T_s \leq t \leq (d_1 + d_2 + d_3)T_s$
By using KVL for the discharging circuit and solving the first order linear equation,

$$I_{cj3}(t) = \frac{K_1 e^{\frac{-t}{R''C''}} + \frac{I_o C''}{C_1}}{n-1}$$

The above solution assumes the output current ripple, ΔI_o, is negligible compared to the discharge current ripple $(n-1)I_{cj3}(t)$. Under steady state conditions the average dc current through capacitors $C_{2,3,...,n}$ must be zero. It can therefore be shown that,

$$K_1 = \frac{\frac{I_o C''}{C_1} d_3 T_s - (n-1) I_{on} d_1 T_s}{R''C''\left(e^{\frac{-d_3 T_s}{R''C''}} - 1\right)} \quad \text{where,}$$

$$R' = \frac{R_{esr} + R_{ond}}{n-1} \qquad C' = (n-1)C$$
$$R'' = R' + R_{esr1} \qquad C'' = \frac{C' C_1}{C' + C_1}$$

Therefore,

$$I_{c13}(t) = (n-1) i_{cj3}(t) - I_o$$
$$V_{c13}(t) = \frac{-R''C'' K_1 e^{\frac{-t}{R''C''}} + (\frac{I_o C''}{C_1} - I_o)t}{C_1} + K_0$$
$$- \frac{I_o}{C_1} d_2 T_s + \frac{I_{on} - I_o}{C_1} d_1 T_s + \frac{K_1 R'' C''}{C_1}$$

State 4: $(d_1 + d_2 + d_3)T_s \leq t \leq (d_1 + d_2 + d_3 + d_4)T_s$

$$I_{c14}(t) = -I_o$$
$$V_{c14}(t) = \frac{-I_o}{C_1} t - \frac{I_o}{C_1} d_2 T_s + \frac{I_{on} - I_o}{C_1} d_1 T_s + K_0$$
$$+ \frac{-R''C'' K_1 e^{\frac{-d_3 T_s}{R''C''}} + (\frac{I_o C''}{C_1} - I_o) d_3 T_s}{C_1} + \frac{K_1 R'' C''}{C_1}$$

The sum of integrals of voltage of capacitor, C_1 over the respective states is equal to the area under the average voltage across capacitor, C_1 for one switching period. Therefore the only unknown parameter K_0 can be found. If K_0 is known then the voltage across capacitor C_1 can be obtained. Also, $V_o(t)$ can be written as,

$$V_o(t) = V_{c1}(t) + R_{esr1} I_{c1}(t)$$

4. SIMULATION RESULTS

Simulation in PSPICE was performed to verify the analysis discussed in section 3. The Spice model library for the devices used was for the UMC TECHNOLOGY $0.18\mu m$ mixed mode $1.8V$ CMOS process. The equations derived in section 3 were incorporated in to a MATLAB programme to determine the voltage and current waveforms with a constant load. The analytical and simulated waveforms with a constant load are shown in Figure 3. It can be seen that the simulated waveforms very closely match with the analytical solution.The analytically derived voltage and current waveforms were compared with the simulated ones at the same conditions as listed below.
$V_{in} = 5V,\ \overline{V}_o = 1V,\ n = 4,\ d_1 = 0.28,\ d_2 = 0.02,$
$d_3 = 0.68,\ d_4 = 0.02,\ T_s = 100ns,\ C = 30nF,$
$C_1 = 60nF,\ R_{esr} = 0.16\Omega,\ R_{esr1} = 0.08\Omega,$
$R_L = 10\Omega,\ R_{ond} = 4.06\Omega,$

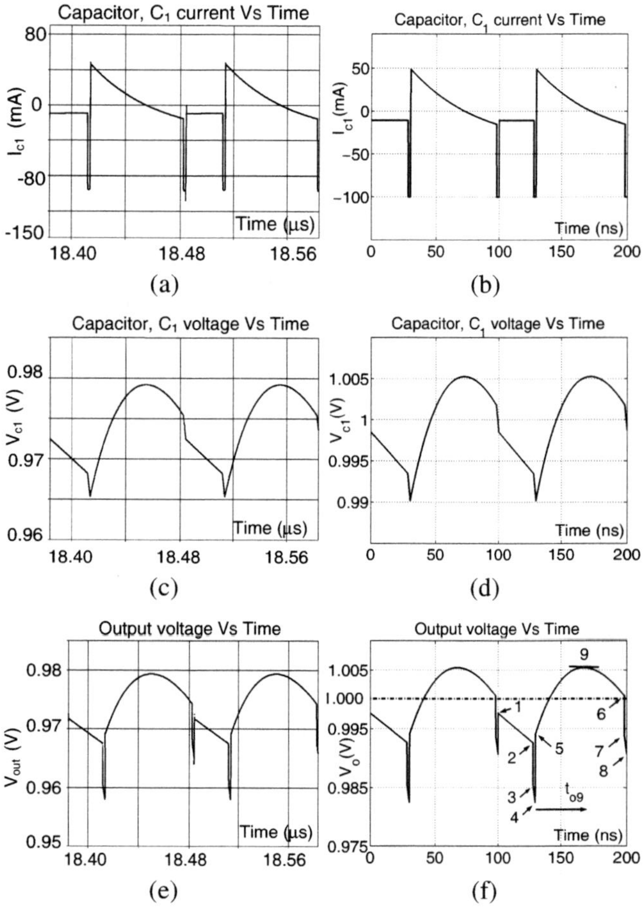

Fig. 3. Comparison of simulated (a),(c),(e) and analytical (b),(d),(f) waveforms

4.1. Output voltage ripple

The peak to peak output voltage ripple is defined as

$$\Delta V_o = max(V_{oi}) - min(V_{oi})$$

where $i \in (1, 2, ..., 9)$ as shown in Figure 3(f). Figure 4 illustrates the variation of output voltage ripple with duty ratio for different instances, i as shown in Figure 3(f). Accordingly an approximate equation for ΔV_o can be derived

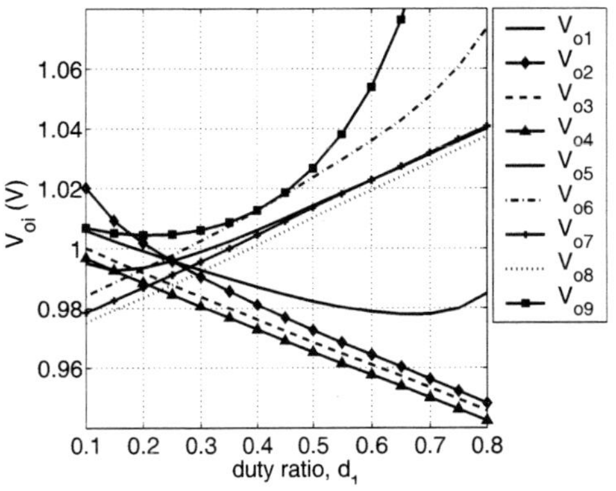

Fig. 4. Variation of output voltage ripple with duty ratio for different instances.

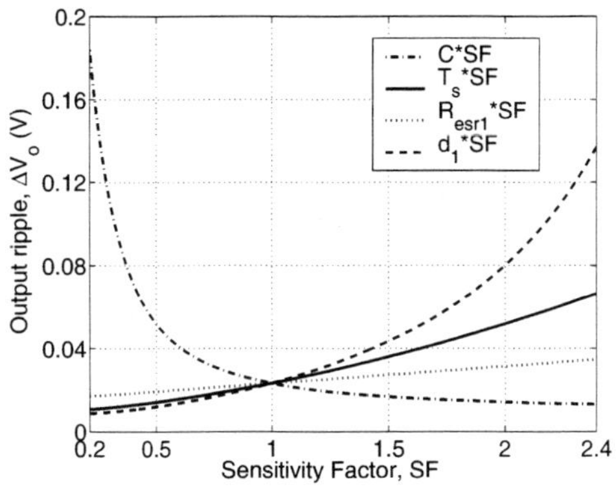

Fig. 5. Output voltage ripple variation with model parameters C, T_s, R_{esr1} and d_1. Where SF is given by $(C + \Delta C)/C$ etc.

as,

$$\begin{aligned}\Delta V_o &\approx V_{o9} - V_{o4} \\ &\approx \frac{-R''C''K_1 e^{\frac{-t_{o9}}{R''C''}} + (\frac{I_o C''}{C_1} - I_o)t_{o9}}{C_1} \\ &\quad + \left[K_1 e^{\frac{-t_{o9}}{R''C''}} + \frac{I_o C''}{C_1}\right] R_{esr1} + \frac{K_1 R''C''}{C_1}\end{aligned}$$

where t_{o9} as marked in the Figure 3(f) is,

$$t_{o9} = -R''C'' ln \frac{I_o - \frac{I_o C''}{C_1}}{K_1(1 - R''C''C_1 R_{esr1})}$$

This is true for all $d_1 > 0.23$ and the probability of having a $d_1 < 0.23$ is quite low. Further analysis was done to investigate how sensitive the output ripple is to model parameters. Figure 5 shows the sensitivity of ΔV_o to the variation of C, T_s, R_{esr1} and d_1. Accordingly parameter C and the duty ratio, d_1 should be selected with considerable care. Increase of d_1 will significantly increase the output voltage ripple. Also C is very sensitive to output ripple when it becomes less than the designed value. It can be seen that increase of C beyond the designed will not help to reduce the output ripple significantly. Variation of switching frequency and R_{esr1} have a lesser influence to the output voltage ripple compared to C and d_1.

5. CONCLUSION

The SC step down dc-dc converter with constant load is analysed. In the analysis, the behavior of the model during switching transition periods which has previously been neglected is considered . Analytical equations for the voltages and currents through components of the model are derived. The accuracy of the model and equations are confirmed by comparison with simulations in PSPICE.

An approximate equation for the output voltage ripple has been derived. It is shown that the output voltage ripple is very sensitive to the capacitance below the designed value and to the duty ratio. As seen in Figure 4 it is recommended to use duty ratio around the inverse of the number of stages to get lower ripple. Increasing switching frequency and decreasing parasitic resistances have lesser influence on reducing output voltage ripple. Integrating a converter in the proposed topology can be constrained mainly by the charge-up loss and parasitics. The models consider parasitics for a certain extent but further attention should be paid on any stray inductances present and any other significant parasitic elements on integration.

6. REFERENCES

[1] B Choi, W Lim and S Choi, "Control design and closed-loop analysis of a switched-capacitor dc-to-dc converter," *IEEE Transactions on Aerospace and Electronic Systems*, vol. 37, no. 3, pp. 1099–1107, 2001.

[2] W C Wu and R M Bass, "Analysis of charge pumps using charge balance," in *Power Electronics Specialists Conference*, 2000, vol. 3, pp. 1491–1496.

[3] K D T Ngo and R Webster, "Steady state analysis and design of a switched-capacitor dc-dc converter," *IEEE Transactions on Aerospace and Electronic Systems*, vol. 30, no. 1, pp. 422–426, 1994.

[4] W S Harris and K D T Ngo, "Power switched-capacitor dc-dc converter: Analysis and design," *IEEE Transactions on Aerospace and Electronic Systems*, vol. 33, no. 2, pp. 386–395, 1997.

[5] William Scott Harris, *Analysis and Design of a Power switched-capacitor dc-dc voltage converter:*, Ph.D. thesis, University of Florida, 1997.

OPTIMIZATION DESIGN OF THE DICKSON CHARGE PUMP CIRCUIT WITH A RESISTIVE LOAD

Ming ZHANG and Nicolas LLASER

IEF, AXIS
University of South Paris, 91405 Orsay, France
zhang@ief.u-psud.fr

ABSTRACT

In this paper, an optimization design of Dickson charge pump circuit with a resistive load in terms of circuit die area is proposed. The purpose is to minimise the circuit die area for a given performance. To do this, first a dynamic analysis of Dickson charge pump circuit with a resistive load is made, resulted from which is the equivalent circuit of Dickson charge pump with a resistive load. Then based on the equivalent circuit, an optimization design is developed. In the proposed design, both pump capacitors and the stage number are considered as design parameters. Compared with the minimum stage number design, the circuit die area can be greatly reduced, a 50% reduction of circuit die area is obtained in the design example given.

1. INTRODUCTION

More and more applications of charge pump circuits can be found in different areas: power IC, programming EEPROM, low voltage integrated circuits, control of micro-actuators in a micro-system etc. Although different structures have been proposed recently [1-3], Dickson charge pump circuit, which was the first integrated charge pump circuit and proposed by Dickson in 1976 [4], and the charge pump circuits derived from Dickson's structure [5-6] still remain dominant in the field mainly because of its simplicity in structure.

Many efforts have been made in analysis and design of Dickson charge pump circuit [7-9]. Cataldo proposed a dynamic analysis and an optimization design of Dickson charge pump circuit with a capacitive load and a current consuming load, but the stage number of charge pump circuits is limited no more than five. Tanzawa, however, has successfully made a dynamic analysis of an *N*-stage Dickson charge pump with a capacitive load. Based on this result, Tanzawa also proposed an optimization design to minimise the rise time of the output voltage.

However, in some applications where Dickson charge pump circuit works as a power supply, the Dickson charge pump circuit is equivalent to have a resistive load. Since the price of an integrated circuit is directly related with its die area, in order to reduce the circuit's price it is always preferable for the integrated circuit to have a minimum die area while keeping the same performance. For this purpose, an optimization design of an *N*-stage Dickson charge pump with a resistive load is proposed in this paper in order to minimise the circuit die area. Particularly in this proposed design, both the pump capacitors and the stage number are considered as design parameters.

2. DYNAMIC ANALYSIS

First, a brief recall of the dynamic analysis of Dickson charge pump with a resistive load is made [10]. An *N*-stage Dickson charge pump circuit is shown in Fig. 1. Here Clk_1 and Clk_2 represent two clock signals with opposite phase. To simplify the dynamic analysis of Dickson charge pump with a resistive load, the same assumptions as those given by Tanzawa in the case of only a capacitive load are taken, that is

1) Each diode and charge pump capacitor has constant values of Vt and C, respectively.
2) The parasitic capacitance is negligibly small compared with the charge pump capacitance.
3) The cycle time of the input clocks $Clk_{1,2}$ driving the charge pump is sufficiently large for all RC time constant to be negligible.

Besides, it is also supposed that the Dickson charge pump with a resistive load is in steady state. By steady state, we mean that the charge transferred from one capacitor to the next one at time j is the same as that transferred to the output at time j, $qout(j)$. In the following, the stage number N is supposed to be even but for an odd N, the same principle can be used as well.

0-7803-8251-X/04/$17.00

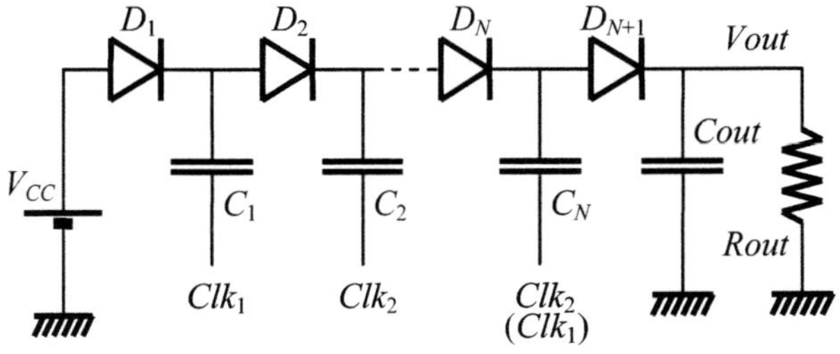

Fig. 1 Dickson charge pump circuit with a resistive load

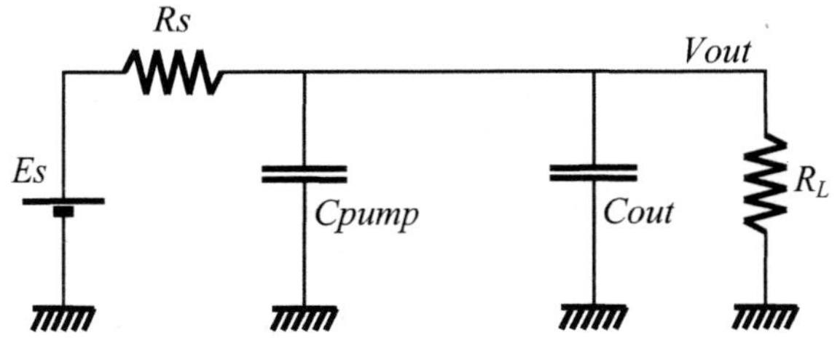

Fig. 2 Equivalent circuit for a capacitive and resistive load $Rs = N \cdot T/C$ and $Es = Vg \cdot (N+1) = V_M$

The approach used can be described as follows. In order to find the output recurrence formula, two methods are used to calculate the total stored charge on pump capacitors $C_k, (1 \le k \le N)$ and $Cout$: first, the charge pump with a capacitive load is considered. In this case, for a desired output voltage *Vout(j)* at time *j*, the total stored charge in capacitors is calculated. Then the charge pump with a capacitive and a resistive load in parallel is considered. In the second case, one part of charge is consumed by the resistive charge during each cycle time. As a result, to offer the same *Vout(j)* at time *j* mentioned above, the total stored charge, which equals to the difference of the total power supplied charge and the total consumed charge by the resistive load, must be equal to that found in the first case. By equalizing the two expressions of total stored charge and after simplification, the output recurrence equation given by equation (1) can be obtained.

$$Cload[Vout(j+1)-Vout(j)]+\frac{T}{R_L}Vout(j)=\frac{C}{N}[V_M-Vout(j+1)] \tag{1}$$

where

$$R_L=\frac{Rout}{\left[1-\frac{T}{8\cdot Rout}\left(\frac{1}{Cout}-\frac{1}{C+Cout}\right)\right]}$$

From (1), we can deduce the equivalent circuit of the Dickson charge pump circuit with a resistive load, shown in Fig. 2.

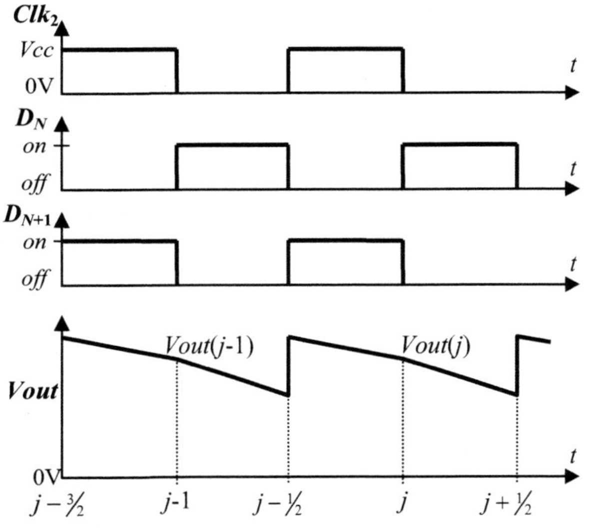

Fig. 3 Chronograms of Clock signal, states of diodes D_N and D_{N+1}, and *Vout*

3. OPTIMIZATION DESIGN

For a Dickson charge pump circuit, the output voltage required across a resistive load can be considered as the average output voltage because the output voltage varies around its average value once the charge pump circuit reaches its final value. Based on the equivalent circuit (Fig. 2) found for a resistive load, if the pump charge circuit is in its final regime, the capacitors *Cpump* and *Cout* can be considered open. As a result, from the voltage divider formed by *Rs*, R_L and *Es*, the average output voltage can be found and expressed by

$$\langle Vout\rangle=\frac{E_S}{1+\frac{T\cdot N}{C\cdot R_L}} \tag{2}$$

Replacing R_L into (2), we find the relation between the average output voltage and the pump charge circuit's parameters:

$$\frac{V_M}{\langle Vout\rangle}-1=\frac{N\cdot T}{Rout\cdot C}\left[1-\frac{T}{8\cdot Rout}\left(\frac{1}{Cout}-\frac{1}{C+Cout}\right)\right] \tag{3}$$

To simplify this expression, we define $\alpha=\frac{V_M}{\langle Vout\rangle}-1$ in the following.

This is the main equation based on which the optimization design will be made. In fact, in most applications such as power supplies for integrated circuits, the undulation $\Delta Vout$ of a power supply is considered as one of the main characteristics to be given. In this paper, we focus our design on this kind of application. In this case, the output capacitor *Cout* is mainly used to smooth the output voltage

Table I Parameters used in the design

Parameter	Value
Resistive load *Rout*	100 kΩ
Output voltage *Vout*	20V
Power supply *Vcc*	5V
Frequency *f*	10MHz
Vt	0.5V
ΔV	1V

Table II Parameters determined by design

Optimized		Without optimized	
N	7	N	4
C	9.2pF	C	32.5pF
C_T	64.6pF	C_T	130pF
$\Delta C_T = 65.4$pF (~50%)			
$Cout = 15.4$pF			

to satisfy the $\Delta Vout$ required. From Fig. 3, $\Delta Vout$ can be expressed as a function of *Cout* and given by [10]

$$\Delta Vout = \frac{Vout(j)\cdot T}{2Rout}\left(\frac{1}{C+Cout}+\frac{1}{Cout}\right) \quad (4)$$

In fact the approximation $\frac{\Delta Vout^2}{Vout(j)^2} << 1$ is usually valid. So after simplification, we obtain

$$C = \frac{N\cdot T}{2\alpha\cdot Rout}\left(1+\sqrt{1+\frac{2\alpha}{N}}\right) \approx \frac{N\cdot T}{\alpha\cdot Rout}\left(1+\frac{\alpha}{2N}\right) \quad (5)$$

and

$$Cout = \frac{\frac{2}{K} - C + \sqrt{C^2+\frac{4}{K^2}}}{2} \quad (6)$$

where $K = \frac{2Rout^2\cdot\alpha\cdot C}{N\cdot T^2}\frac{\Delta Vout}{\langle Vout\rangle}$.

As the charge pump circuit is mainly occupied by pump capacitors, the die area of charge pump can be considered proportional to the sum of pump capacitors C_T, i.e. $C_T = N\cdot C$. The minimum C_T can be found by making $dC_T/dN = 0$. This derivation has two solutions: one is negative, which is ignored and the other is given by

$$N = \frac{\frac{Vout}{Vg}-1}{1-\frac{1}{1+\sqrt{1+\frac{Vg}{2Vout}}}} \quad (7)$$

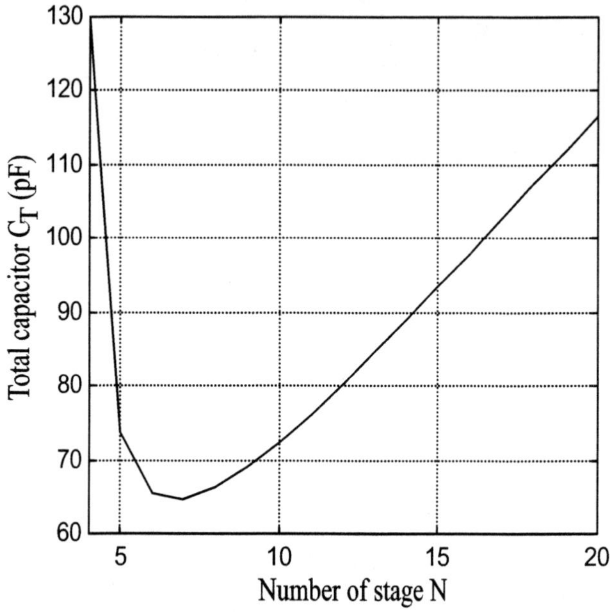

Fig. 4 Total Capacitor C_T versus the number of stage N

4. DESIGN EXAMPLE

In the following, a design example using the optimization design method proposed in this paper is given. In the example, a voltage multiplier with an output voltage of 20V across a resistive load is designed with a power supply of 5V. By using the design parameters summarized in table I, the sum C_T of pump capacitors is traced with the number of stage N and is shown in Fig. 4. From Fig. 4, which shows that for a N=7, a minimum C_T exists.

To see the die area gain by minimum die area design, two designs are made: minimum die area and minimum stage number. The parameters determined for the two designs are given in table II. We can see that a gain of half of the die area can be obtained by minimum die area design for this example.

To verify the minimum die area design proposed, SPICE simulation with the parameters given in table II is made. The simulation output voltage, shown in Fig. 5, reveals a relative error of about 3% compared to the required output voltage, plotted in Matlab by using equation (1). In fact, the relative error depends on the driving clock frequency and on the undulation of output voltage tolerated. The higher the driving clock frequency, the smaller the relative error will be. The smaller the undulation tolerated, the smaller the relative error.

5. CONCLUSIONS

In this paper, dynamic analysis of Dickson charge pump circuit with a resistive charge is made. The dynamic analysis results in the finding of the recurrence formula of the output voltage, from which the equivalent circuit of Dickson charge pump with a resistive load during boosting is deduced.

Then based on the equivalent circuit, an optimization design of Dickson charge pump in terms of die area is developed. The optimization study shows that for a given resistive load and an output voltage across it, an optimal stage number exists, which gives a charge pump circuit with a minimum die area. The example given shows a gain of 50% in die area compared to the minimum stage number design.

6. REFERENCES:

[1] P. Favrat, P. Deval and M. J. Declercq, "A high-efficiency CMOS voltage doubler", IEEE *J. Solid-state Circuits*, Vol. 33, No. 3, pp. 410-416, 1998.

[2] M. Zhang, N. Llaser and F. Devos, "Multi-value Voltage-to-voltage Converter Using a Multi-stage Symmetrical Charge Pump for on-chip EEPROM Programming", *Analog Integrated Circuits and Signal Processing*, Vol.27, pp. 85-95, April, 2001.

[3] M. Zhang, N. Llaser and F. Devos, "An optimized design of an improved voltage tripler", *IEEE Int. Conf. Electronics, Circuits and Systems*, Malta, pp. 353-356, Sept. 5-7, 2001.

[4] J. F. Dickson, "On-chip high-voltage generation in NMOS integrated circuits using an improved voltage multiplier techniques", *IEEE J. Solid-state Circuits*, Vol. 11, pp. 374-378, June, 1976.

[5] Jieh-Tsorng Wu and Kuen-Long Chang, "MOS charge pumps for low voltage operation", *IEEE J. Solid-state Circuits*, Vol. 33, No. 4, April, pp. 592-597, 1998.

[6] Jongshin Shin, In-Young, Young June Park, and Hong Shick Min, "A New Charge Pump Without Degradation in Threshold Voltage Due to Body Effect", *IEEE J. Solid-state Circuits*, Vol. 33, No. 4, April, pp. 592-597, 1998.

[7] T. Tanzawa and T. Tanaka, "A Dynamic Analysis of the Dickson Charge Pump Circuit", *IEEE J. Solid-state Circuits*, Vol. 32, No. 8, pp. 1231-1240, August, 1997.

[8] G. Di Cataldo and G. Palumbo, "Dynamic Analysis of 3 stage Dickson voltage multiplier for an optimized design", in *Proc. 7th Mediterranean Electrotechnical Conf.*, pp. 633-636, 1994.

[9] G. Di Cataldo and G. Palumbo, "Design of an *N*th Order Dickson Voltage Multiplier", *IEEE Transactions on Circuits and Systems*, Vol. 43, no.5 pp. 414-418, may, 1996.

[10] M. Zhang, and N. Llaser, "A dynamic analysis of Dickson charge pump circuits with a resistive load", *IEEE Int. Conf. Electronics, Circuits and Systems*, Sharjah, United Arab Emirates, Dec. 14-17, 2003.

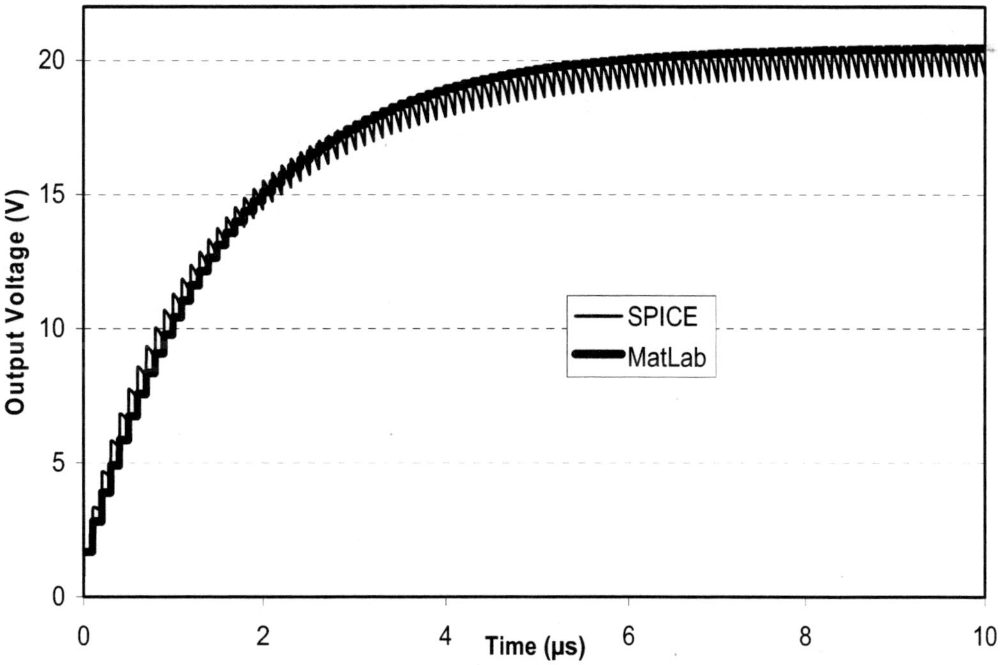

Fig. 5 Simulation SPICE and MatLab of output voltage

NOVEL D^2T CONTROL FOR SINGLE-SWITCH DUAL-OUTPUT SWITCHING POWER CONVERTERS

C. K. Tse, S. C. Wong and K. C. Tam

Department of Electronic and Information Engineering
Hong Kong Polytechnic University, Hunghom, Hong Kong

ABSTRACT

A method for programming current in dc/dc converters operating in discontinuous conduction mode is described in this paper. The control variable is the product of the square of the duty cyle and the switching period (D^2T), which is directly proportional to input and output currents of a discontinuous-mode converter. A method of controlling D^2T is applied to converters that utilize one switch (or one set of synchronous switches) for achieving two control functions. In particular a single-switch two-output boost converter is studied. In this system, current-mode control is used to regulate the output voltage of the continuous-mode converter and the proposed D^2T control is used to regulate the other discontinuous-mode converter. The result is a generic current-mode controlled dual-output converter.

1. INTRODUCTION

This paper describes a single-switch two-output regulator which comprises two boost converters, one operating in continuous conduction mode (CCM) and the other in discontinuous conduction mode (DCM) [1]. Previously reported control methods take advantage of the insensitivity of the CCM converter to switching frequency, and apply duty-cycle control to regulate the CCM converter while regulating the DCM converter by frequency modulation [1, 2]. In this paper, we consider application of generic current-mode control to both converters in order to achieve faster transient responses [3]. Specifically, we use a standard current programming for the CCM converter and a novel D^2T control for regulating the DCM converter. We will also discuss the local stability of the combined current-mode and D^2T control scheme, and present a practical circuit implementation of the controller.

2. OUTLINE OF THE PROPOSED CONTROL

Inspired by its averaged behaviour model, the DCM converter can be controlled by varying the control quantity D^2T which is proportional to the output current [4]. In principle, if the quantity D^2T is adjusted via a feedback mechanism, the output load voltage can be regulated. A special but practically important case is when either the on-time or off-time duration is already determined by another control law. We shall make reference to a single-switch two-output boost converter later in the paper.

Suppose the duty cycle is D, and the period begins with the switch turned off. Thus, after a duration of $(1-D)T$, the switch is turned on; and after another duration of DT, the switch is turned off again, completing one cycle.

Our objective is to derive a scheme, whereby the off-time duration $(1-D)T$ is pre-determined and the period T is adjusted to give the desired value of D^2T. The proposed controller consists of a periodic parabola generator $p(t)$ and a ramp generator $r(t)$:

$$p(t) = a(t-t_0)^2 \quad (1)$$

$$r(t) = b(t-t_0') \quad (2)$$

where t_0 and t_0' are arbitrary start instants for the two generators, and a and b can be considered as constant for the time being. Fig. 1 (a) shows these generator waveforms.

At the start of the switching cycle, say $t=0$, the ramp generator is triggered to start. At $t=(1-D)T$ (which is determined externally), the switch is turned on. At the same time, the parabola generator is triggered to start from zero. As soon as the outputs of the generators are equal, the switch is turned off whereby spawning a new cycle. Clearly, this condition forces

$$p(DT) = r(T) \Rightarrow a(DT)^2 = bT \Rightarrow D^2T = b/a. \quad (3)$$

Hence, the quantity D^2T can be made controllable by varying b/a. A conceptual implementation of this control scheme is shown in Fig. 1 (b). In this scheme, the off-time duration is externally determined, and the control circuit in turn produces a pulse to set the period. Thus, the proposed control can be viewed as a frequency modulator which, for each cycle, maintains a constant D^2T. If the current sources (i.e., a and b) are controllable, the quantity D^2T can be modulated via feedback.

This work is supported by Hong Kong Polytechnic University under Grant G-T662.

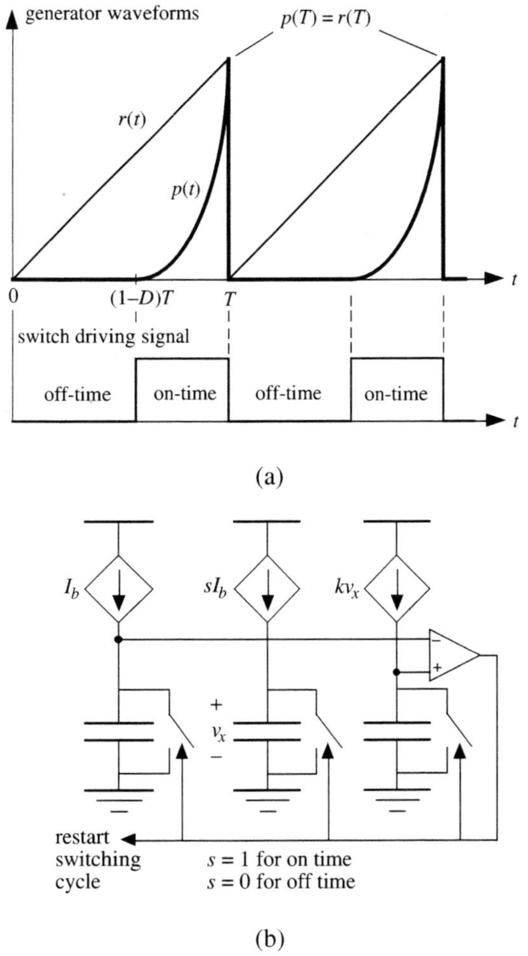

Fig. 1: (a) Waveforms of the parabolic and ramp generators in the conceptual D^2T control scheme; (b) conceptual implementation.

3. LOCAL STABILITY OF COMBINED CURRENT-MODE AND D^2T CONTROL

Our interest in this section is to study the local stability of the proposed control scheme when the off-time duration is pre-determined by a current-mode control scheme. Essentially, in a typical continuous-mode dc/dc converter under current-mode control, the turn-off instant is determined by comparing the inductor current with a reference level. Thus, effectively, this conventional scheme is controlling the on-time duration. In our scheme, however, we control the off-time duration instead. As we will see, this has an important implication on the stability. In brief, the switch is turned on at the instant the inductor current descends to a reference level. The turn-off instant, moreover, is determined by our proposed D^2T control. In other words, the repetition period is controlled by the D^2T controller. The stability issue therefore involves the consideration of the stability of the current-mode control with off-time duration being pre-determined and the period being controlled by the D^2T scheme.

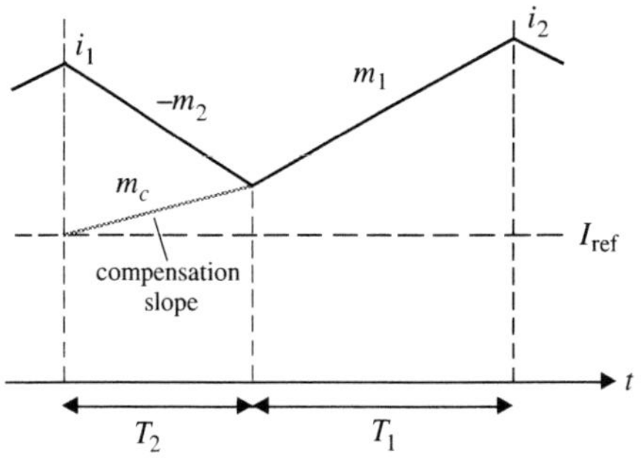

Fig. 2: Inductor current waveform for stability analysis.

First, referring to Fig. 2, we can write the inductor current values at the start and end of a period as

$$\begin{aligned} i_1 &= I_{\text{ref}} + (m_c + m_2)T_2 \\ i_2 &= I_{\text{ref}} + m_c T_2 + m_1 T_1 \end{aligned} \tag{4}$$

where T_1 is the on-time duration, T_2 is the off-time duration, m_1 is the on-time inductor current slope, m_2 is the off-time inductor current slope, and m_c is the compensation slope. Note that we consider here the general case where a compensation slope is included in the current-mode control. Upon differentiating (4), we get

$$\frac{\delta i_2}{\delta i_1} = \frac{m_1 \delta T_1 + m_c \delta T_2}{(m_c + m_2)\delta T_2} = \frac{m_c + m_1 \frac{\delta T_1}{\delta T_2}}{m_c + m_2}. \tag{5}$$

Let us now introduce the D^2T control to the abovementioned current-mode controlled converter. In the steady state, the aim is to fix the value of D^2T, i.e.,

$$\frac{T_1^2}{T_1 + T_2} = \text{constant}, \tag{6}$$

where the constant in the RHS is equal to the steady-state value of D^2T. Thus, differentiating (6), we get

$$\frac{\delta T_1}{\delta T_2} = \frac{D}{2 - D}. \tag{7}$$

Hence, from (5), and using $m_1 D = m_2(1 - D)$, we have

$$\frac{\delta i_2}{\delta i_1} = \frac{m_c + m_1 \frac{D}{2-D}}{m_c + m_2} = \frac{m_c + m_2 \frac{1-D}{2-D}}{m_c + m_2} \tag{8}$$

Local stability requires that the magnitude of the above expression be less than 1. Thus, we can see that stability is guaranteed for all D since

$$0 < \frac{1 - D}{2 - D} < \frac{1}{2} \quad \text{for all } D \in (0, 1), \tag{9}$$

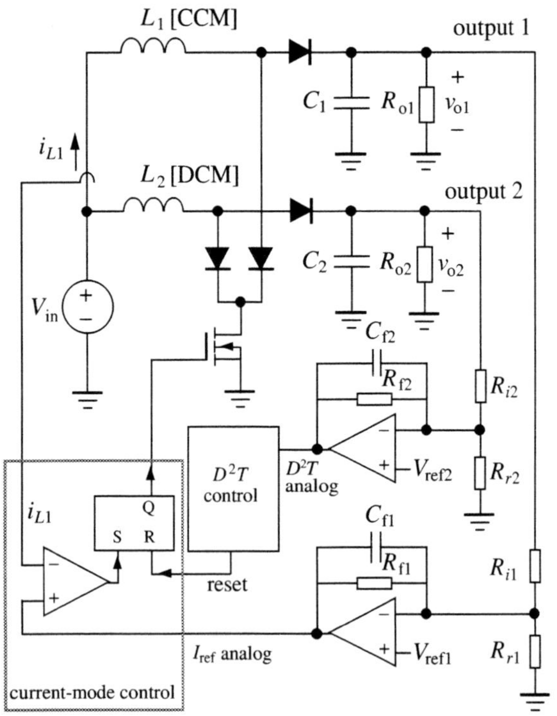

Fig. 3: Schematic of dual-output boost regulator under combined current-mode D^2T control.

which implies $|\delta i_2/\delta i_1| < 1$ for all $0 < D < 1$.

It is interesting to note that the stability of the system is unaffected even when the compensation ramp is zero. In other words, the D^2T control inherently stabilizes the current-mode control, eliminating the need for the use of ramp compensation.

4. APPLICATION TO DUAL-OUTPUT BOOST REGULATOR

In this section we present a single-switch two-output boost regulator, as outlined previously in the Introduction. Fig. 3 shows the schematic of the system. In brief, this regulator consists of a CCM boost converter and a DCM boost converter, sharing one common switch. The circuit design aspect has been studied extensively by Sebastián and Uceda [1] and Charanasomboon *et al.* [2]. Here, we focus on the application of the proposed combined current-mode and D^2T control for simultaneous regulation of the two outputs, and verify the control function by SPICE simulation. In the simulation, real devices are used for the boost converters, while the D^2T control block is constructed based on the conceptual circuit shown in Fig. 1 (b). Component and parameter values for simulation are summarized in Table 1.

It is worth noting that the inductor current of the CCM converter is referenced at the turn-on instant, and hence its peak is generally unlimited (actually determined by the D^2T control). It is thus necessary to impose a peak limiter to limit the maximum switching period. This arrangement has been incorporated in our simulation study.

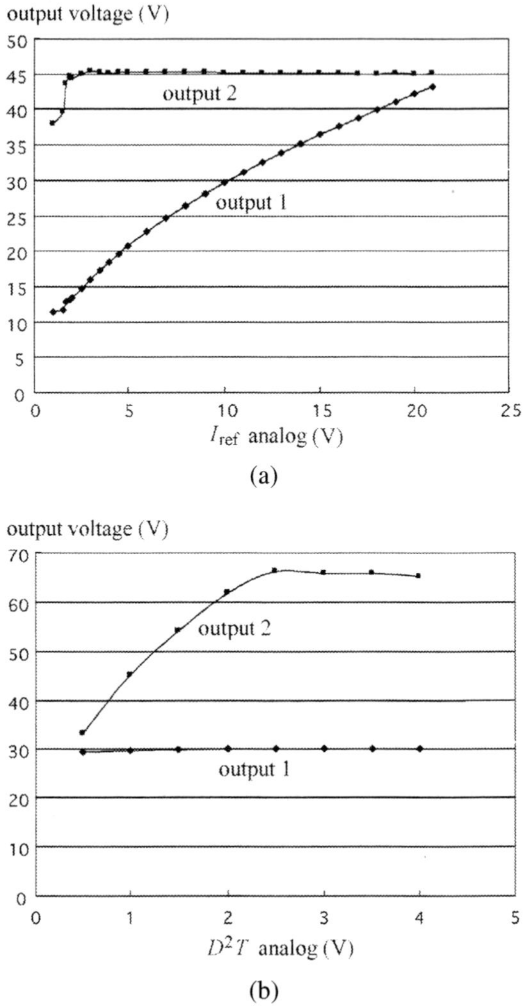

Fig. 4: Steady-state control-to-output characteristics. (a) Output voltages versus $I_{\rm ref}$ analog (control voltage of the CCM converter) with D^2T analog (control voltage of the DCM converter) kept at 1 V; (b) output voltages versus D^2T analog (control voltage of the DCM converter) with $I_{\rm ref}$ analog (control voltage of the CCM converter) kept at 10 V. Note that the abrupt drop or saturation of output 2 at either end of the control range is a result of the use of a maximum period limiter.

Several tests have been performed to verify the operation of the proposed control. First of all, the steady-state control-to-output transfer characteristics have been obtained by plotting the output voltages against the control voltage analogs. Results are shown in Fig. 4. Then, the transient performance has been evaluated, including input regulation, load regulation and cross regulation. Results are shown in Figs. 5 and 6. In Fig. 5, the transient responses are recorded when the input voltage is stepped, whereas in Fig. 6, the

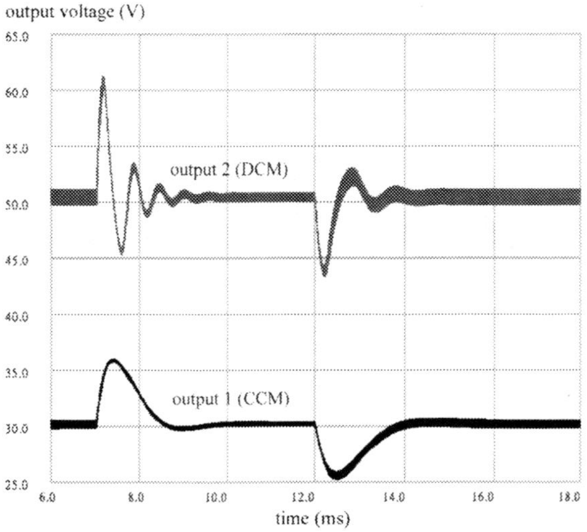

Fig. 5: Output transient responses for step changing input voltage. At $t = 7$ ms, V_{in} steps up from 8 V to 12 V, and at $t = 12$ ms, it steps down back to 8 V.

Table 1: Component and parameter values for simulation

Component/parameter	Value
Inductance 1 (CCM), L_1	100 μH
Inductance 2 (DCM), L_2	10 μH
Output capacitance 1, C_1	50 μF
Output capacitance 2, C_2	5 μF
Input voltage, V_{in}	10 V
Load resistance 1, R_{o1}	8–12 Ω
Load resistance 2, R_{o2}	80–120 Ω
On-resistance of MOSFET	0.005 Ω
Frequency	20–170 kHz

transient responses are recorded when the load resistances are stepped. Note that Fig. 6 reflects very satisfactory self load regulation performance as well as cross regulation performance, i.e., transient of v_{o1} (or v_{o2}) when R_{o2} (or R_{o1}) is stepped.

5. CONCLUSION

In a dual-output voltage regulator, where two dc/dc converters (one operating in CCM and the other in DCM) are sharing one switch, generic current-mode control can be achieved by applying conventional current-mode control to the CCM converter and a D^2T programming control to the DCM converter. In this paper we propose a strategy for controlling the D^2T quantity, whereby both outputs can be tightly regulated. Satisfactory cross regulation is possible by virtue of the CCM converter being insensitive to frequency changes and the DCM converter being directly current-programmed.

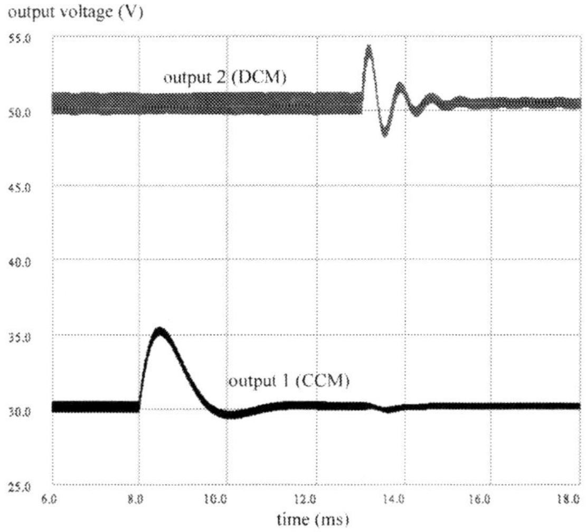

(a)

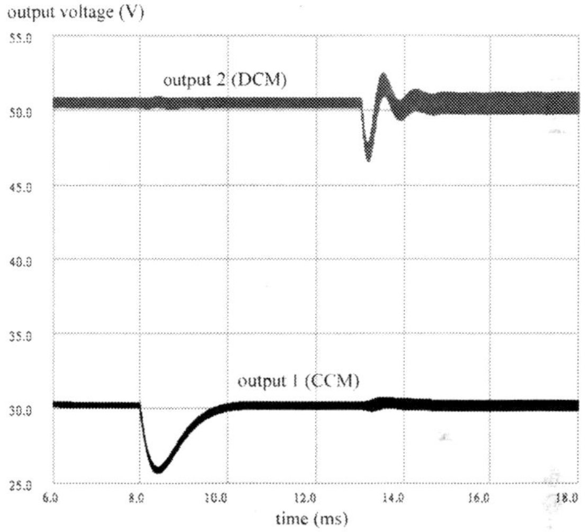

(b)

Fig. 6: Output transient responses for step changing loads. (a) At $t = 8$ ms, R_{o1} changes from 8 Ω to 12 Ω, and at $t = 13$ ms, R_{o2} changes from 80 Ω to 120 Ω. (b) At $t = 8$ ms, R_{o1} changes from 12 Ω to 8 Ω, and at $t = 13$ ms, R_{o2} changes from 120 Ω to 80 Ω.

6. REFERENCES

[1] J. Sebastián and J. Uceda, "Double converter: a fully regulated two output dc-to-dc converter," *IEEE PESC Rec.*, pp. 117–126, 1985

[2] T. Charanasomboon, M.J. Devaney and R.G. Hoft, "Single switch dual output dc/dc converter performance," *IEEE Trans. Power Electron.*, vol. 5, no. 2, pp. 241–245, 1990.

[3] R. Redl and N.O. Sokal, "Current-mode control, five different types, used with the three basic classes of power converters," *IEEE PESC Rec.*, pp. 771–775, 1985.

[4] P.R. Severns and E.J. Bloom, *Modern Switching DC-to-DC Power Converter Circuits*, New York: Van Nostrand, 1985.

TSK-FUZZY CONTROLLER DESIGN FOR A PWM BOOST DC-DC SWITCHING REGULATOR OPERATING AT DIFFERENT STEADY STATE OUTPUT VOLTAGES

S. Gomáriz, E. Alarcón*, F. Guinjoan*, E. Vidal-Idiarte**, L. Martínez –Salamero** and D.Biel**

*(gomariz@eel.upc.es) Dept de Ingeniería Electrónica. Universitat Politècnica de Catalunya (Spain)
** Dept de Ingeniería Electrónica, Eléctrica y Automática. Universitat Rovira i Virgili (Spain)

ABSTRACT

A Takagi-Sugeno-Kang (TSK) fuzzy controller design for a PWM boost DC-DC switching regulator operating at different steady-state voltages is presented. On one hand, the controller assigns different linear control laws according to the steady-state voltage value and limits their application to small-signal perturbations. On the other hand the controller ensures a proper start-up to reach the desired steady-state voltage by means of properly saturating the PWM according to the sliding-mode control principles. As a result, both an equalized small-signal behavior and a controlled start-up are achieved for different steady-state voltages, thus enhancing the regulator features. Simulation results for a current-controlled boost regulator operating at different steady-state voltages are presented to validate the approach.

1. INTRODUCTION

Traditionally, the design of control loops for switching dc-dc regulators has been tackled by means of linear control techniques for fixed-frequency PWM operation [1], [3] Although those techniques exhibit the advantages of using well-known linear control tools, their validity limits are located in a small neighborhood of the steady-state operating point. Therefore the linear control design based on the "small-signal" modeling of the power stage presents the following drawbacks concerning the switching regulator features:

- The regulator will exhibit the expected dynamical behavior in terms of frequency or time domain specifications (i.e, bandwidth, phase margin, maximum overshoot, damping factor..) only around the operating point for which the linear control was designed. Particularly, if the steady-state voltage is modified by means of the voltage reference, the design specifications will not be fulfilled and the regulator would even become unstable. For this reason, the regulator is designed to operate only at one steady-state voltage. It can be noted that the same reasoning can be applied in the case of operating at different steady-state loads.
- Large-signal perturbations such as in the case of the start-up transient process can also lead to an undesirable dynamical behavior since the linear model of the power stage from which the control is designed is out of its validity limits.

In order to extend the features of the regulator it would be therefore desirable to design a control strategy ensuring for different steady-state voltages:

- A similar behavior in front of small-signal perturbations, this being referred as "small-signal equalization".
- A controlled start-up.

Nonlinear control techniques such as global linearization of the closed loop system [7] can be applied to implement this strategy. However these techniques entail implementation pitfalls and are not as well-known as the corresponding linear ones among the designers.

In front of this scenario, this work proposes the design of a TSK-fuzzy controller which assigns different linear control laws in order to perform a small signal equalization response according to the desired steady-state voltage value, and limits their application to small-signal perturbations [8]. Additionally, a set of control rules to properly saturate the PWM according to the sliding-mode control principles, will be also embedded into the controller to ensure proper start-up whatever the desired output voltage is.

The work is organized as follows: section 2 recalls the basis of the linear current control technique applied to a boost converter, evidencing the steady-state dependence of the resulting control laws. Section III is devoted to the complete design of the fuzzy controller for the small signal equalization, whereas section 4 establishes the design of the rules ensuring a proper start-up. Finally, section 5 presents the simulation results to validate the proposed design and the conclusions of the work.

2. SMALL SIGNAL EQUALIZATION OF A CURRENT CONTROLLED BOOST REGULATOR

Fig.1 shows the power stage and the control loop of a PWM current-controlled boost switching regulator. The control loop includes the PWM modelled by means of a comparator and a switching ramp of fixed frequency $1/T_S$ with upper and lower saturation limits defined by V_M and V_m respectively.

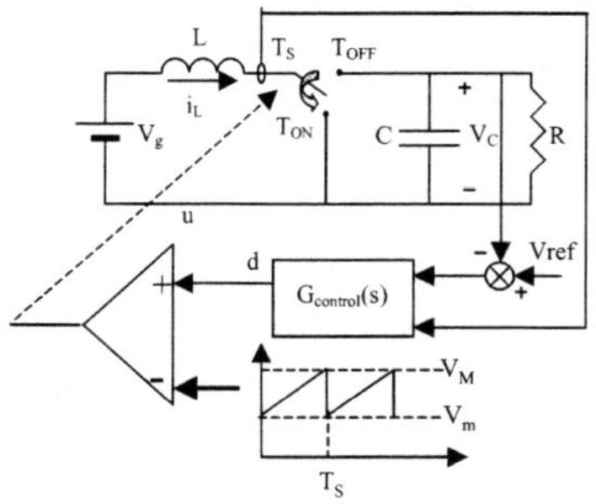

Fig. 1 Current-controlled boost switching regulator

In the linear approach, the control block $G_{control}(s)$ is designed from the linearized model of the boost stage according to the current-control technique, this including an inner current feedback loop of proportional type and an outer voltage feedback loop of PI-type [3]. The resulting block diagram of the current- controlled regulator is shown in Fig.2, where the linear model of the boost stage is defined by means of the transfer functions $G_v(s)$ and $G_i(s)$ and can be found in [1]:

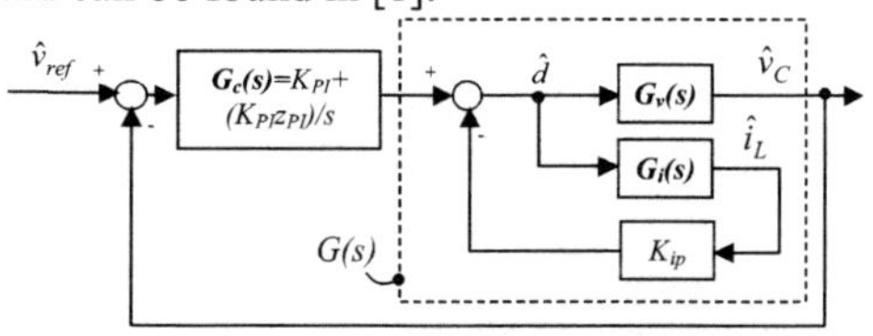

Fig.2 Block diagram of the current controlled regulator

The duty-cycle delivered to the modulator can be deduced from Fig.2 and can be written as:

$$\hat{d}(t) = K_{PI}\hat{e}(t) + K_{PI} z_{PI}\int \hat{e}(\tau)d\tau - K_{IP}\hat{i}_L(t) \qquad (1)$$

The equalization of the closed-loop small-signal response can be achieved by designing as many linear current-controllers as desired operating output voltages. In a previous work [4], two linear current controllers have been designed for a boost regulator with the following parameters V_g=10V, L=200µH, R=10Ω, C=200 µF and T=20 µs which has to operate at the output voltages V_{ref}=20V and V_{ref}=50V with the same damping factor fixed at $\xi_d=\sqrt{2}/2$ (i,e, damping factor equalization). These controllers referred as PWM_{20} (delivering the duty-cycle $d_{20}(t)$) and PWM_{50} (delivering the duty-cycle $d_{50}(t)$), exhibits the following parameters K_{PI}, z_{PI}, K_{IP} summarized in table I:

	α	K_{PI}	$z_{PI}(rad/s)$	K_{IP}
PWM_{20} ($d_{20}(t)$)	1,04	0,15	-1,6.10^3	0,14
PWM_{50} ($d_{50}(t)$)	3	0,0218	-1,3.10^3	0,0272

Table I: Parameter values of the current-controllers

Figs 3(a) and (b) show the regulator output voltage small-signal response for the two current-controllers PWM_{20} and PWM_{50} around V_{ref}=20V and V_{ref}=50V when small-signal perturbations are applied to the voltage reference. As it can be seen, the damping factor equalization can only be achieved by applying the controller corresponding to the steady-state output voltage for which it was designed.

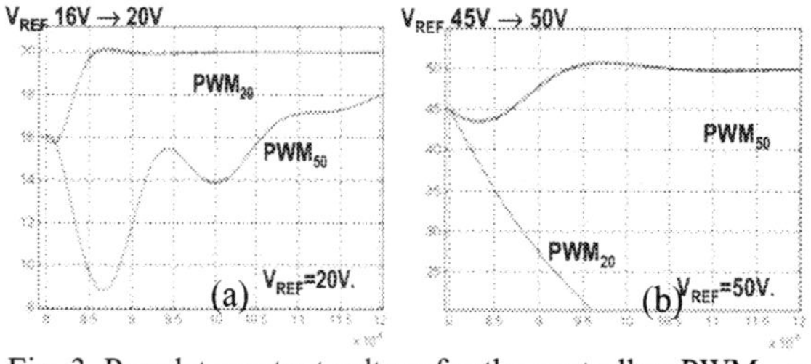

Fig. 3. Regulator output voltage for the controllers PWM_{20} and PWM_{50} (a) Around V_{ref}= 20V (b) Around V_{ref} =50V

3. TSK FUZZY CONTROLLER DESIGN

According to section 2, if small-signal damping factor equalization is desired, the control strategy has to:

- Assign the linear current-control laws to the corresponding desired steady-state output voltage and limit their application to small-signal perturbations.
- In the case of large signal perturbations such as in start-up operation, assign other control laws leading to the desired output voltage (see section 4).

These control actions can be summarized in Table II:

Perturbations → Output Voltage ↓	Small-signal ($\xi=\sqrt{2}/2$)	Large-signal
V_{ref}=20V	PWM_{20}	Other law
V_{ref}=50V	PWM_{50}	Other law

Table II: Control laws for damping factor equalization

This task is suited to the TSK fuzzy architecture [2], which is described by a set of if-then rules of the form:

$$IF\ \ x_1 \text{ is } \tilde{x}_{1i}\ \ AND \ldots AND\ x_n \text{ is } \tilde{x}_{ni}\ \ THEN\ \ u_i = \sum_{k=0}^{n} a_1 \cdot x_k + b_1 \qquad (2)$$

where $x_1 \ldots x_n$ are the physical inputs to the fuzzy controller and the rule consequents are defined as input-dependent linear singletons. Assuming the Mamdani inference, the fuzzy controller output is given by:

$$u = \sum_{\lambda=1}^{P} \omega_\lambda \cdot d_\lambda \Big/ \sum_{\lambda=1}^{P} \omega_\lambda \quad \text{being} \quad \omega_\lambda \cdot u_\lambda = \min_{j=1}^{n}\{\tilde{x}_{j\lambda}(x_j)\}\cdot d_\lambda \qquad (3)$$

where the rule antecedents and $\tilde{x}_{1i},\ldots\tilde{x}_{ni}$ are variables described by means of membership functions, which are in charge of locally and gradually map the corresponding variable into the [0,1] interval. It can be noted that if u designs the control variable, the linear control law:

$$u_i = \sum_{k=0}^{n} a_i \cdot x_k + b_i \qquad (4)$$

is assigned by the rule i in a region of the space $(x_1,\ldots x_n)$ of the input variables if:

$$\tilde{x}_{ji}(x_j) = 1 \quad for \quad j = 1,\ldots,n \qquad (5)$$

This property will be applied for achieving the control strategy of Table II. In this sense, after defining the controller input variables, the membership functions of the antecedents, the linear consequents and the set of control rules are defined to carry out the TSK-fuzzy controller design, as follows:

Input variables

On one hand, the voltage reference V_{ref} is chosen as an input variable in order to know the desired steady-state output voltage. On the other hand, the small or large signal operation can be detected by means of the normalized error of the converter state vector (e_{Vr}, e_{Ir}) defined as:

$$e_{Vr} = (V_{ref} - v_C)/V_{ref} \qquad (6)$$

$$e_{Ir} = (I_{ref} - i_L)/I_{ref} \quad \text{where } I_{ref} = V_{ref}^2/V_g \cdot R \qquad (7)$$

Rules consequents:

According to section 2, the consequents are given by:

$$u_{CD} = d(t) = K_{PI}e + K_{PI}z_{PI}\int e - K_{IP}i_L \qquad (8)$$

where the coefficients of (8) are given in Table I.

Accordingly, the block diagram of the fuzzy-controlled regulator is shown in Fig.4.

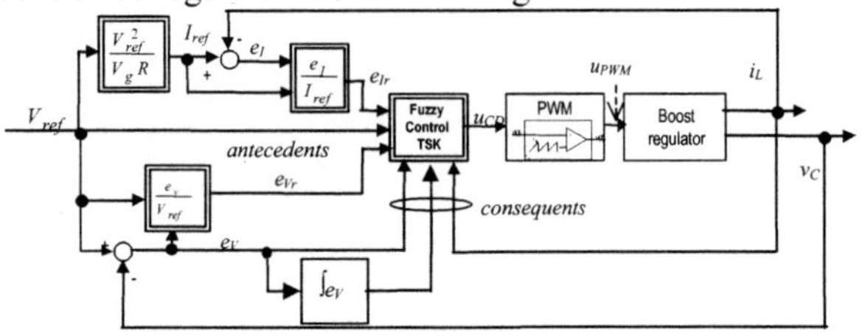

Fig.4. Block diagram of the fuzzy controlled boost regulator

Membership functions of the input variables (antecedents)

On one hand two trapezoidal memberships $\tilde{V}_{ref1}, \tilde{V}_{ref2}$ shown in Fig.5 are designed such that the maximum credibility (membership value=1) is guaranteed for the desired steady-state output voltages V_{ref}=20V and V_{ref}=50V, thus enabling the control to assign the laws of Table II when these reference voltages are applied.

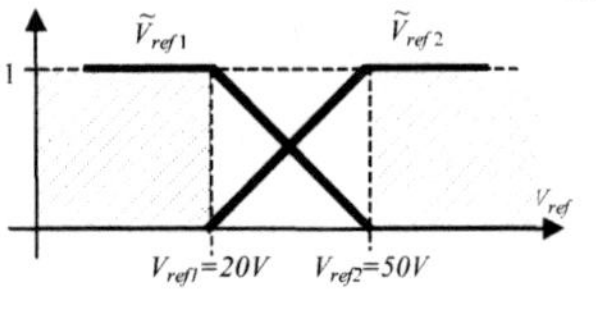

Fig. 5.Membership functions for V_{ref}

On the other hand, 3 trapezoidal membership functions, namely $\tilde{e}_{Vr1}, \tilde{e}_{Vr2}, \tilde{e}_{Vr3}$ and $\tilde{e}_{Ir1}, \tilde{e}_{Ir2}, \tilde{e}_{Ir3}$, are designed for each component of the normalized space vector error (e_{Vr}, e_{Ir}) as shown in Fig.6.

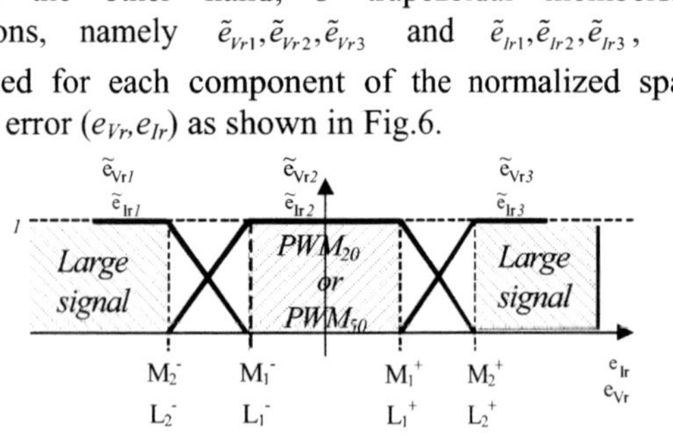

Fig. 6.Membership functions for e_{Ir}, e_{Vr}

These memberships define the following maximum credibility regions in which one control law can be assigned:

- $L_1^- < e_{Vr} < L_1^+$ and $M_1^- < e_{Ir} < M_1^+$: is the small-signal region where the control laws PWM_{20} or PWM_{50} have to be applied.
- $e_{Vr} > L_2^+$ or $e_{Vr} < L_2^-$ or $e_{Ir} > M_1^+$ or $e_{Ir} < M_1^-$: is the region corresponding to the large-signal behavior.

The design of the boundaries L_1^-, L_1^+, M_1^-, M_1^+ is performed by simulation and takes into account that perturbations considered as "small-signal" ones should not lead the normalized space vector error (e_{Vr}, e_{Ir}) out of these limits. Fig.7 shows the simulations of the transients of Fig.3 in the state plane (e_{Vr}, e_{Ir}):

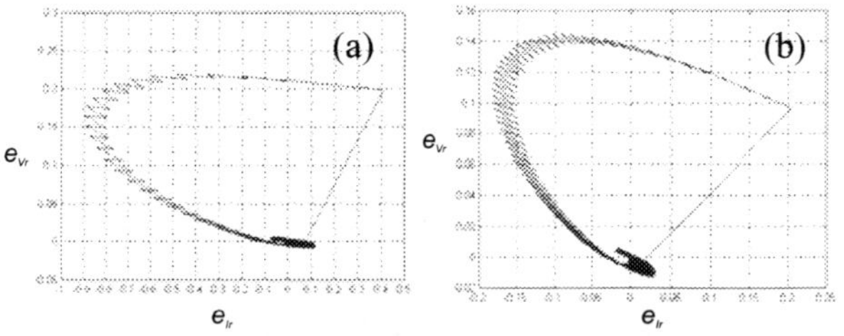

Fig. 7 Transient simulations in the state plane (e_{Vr}, e_{Ir})
(a) V_{ref} from16V to 20V. (b) V_{ref} from 45V to 50V

From these simulations, the following small-signal region boundaries have been selected:

$$M_1^+ = 0.4;\ M_1^- = 0.9;\ L_1^+ = 0.3;\ L_1^- = -0.3 \qquad (9)$$

It has to be noticed that these simulations correspond to instantaneous variables (not to the averaged ones), and therefore include the corresponding switching ripple.

4. START-UP CONTROL DESIGN

The previous TSK-fuzzy control laws only apply for the regulator small-signal operation. In order ensure a proper start-up leading to the desired steady-state output voltage from the start-up initial conditions (i.e, $(v_c, i_L = (0,0))$, the following strategy based on the sliding mode control technique [5] is proposed. Assuming that u stands for the boost converter control variable, and $u(t)$=1 or $u(t)$=0 correspond to the switch in T_{ON} or in T_{OFF} respectively (see Fig.1), the following switching control law:

$$\begin{cases} u(t) = 1 & if \quad \sigma(v_c, i_L) = i_L - K < 0 \\ u(t) = 0 & if \quad \sigma(v_c, i_L) = i_L - K > 0 \end{cases} \qquad (10)$$

leads to the steady-state $i_L = K$ whatever the initial conditions are [5], this meaning that any surface σ of constant inductor current combined with the previous switching action is a sliding surface. This control law can be applied for the start-up operation and embedded into the fuzzy controller by defining on one hand the "start-up" region as the large-signal region of Fig.6, in which the previous law will be applied: since the initial conditions corresponding to the start-up case are $(v_c, i_L = (0,0)$, the corresponding values of the normalized error state vector are given by:

$$e_{Vr}\big|_{(i_L,v_C)=(0,0)} = (V_{ref} - v_C)/V_{ref}\big|_{v_C=0} = 1 \quad (11)$$

$$e_{Ir}\big|_{(i_L,v_C)=(0,0)} = (I_{ref} - i_L)/I_{ref}\big|_{i_L=0} = 1 \quad (12)$$

Therefore the boundaries L_2^-, L_2^+, M_2^-, M_2^+ are fixed to ± 1. On the other hand, the sliding control law is carried out by properly saturate the PWM by fixing the modulator input u_{CD} defined in Fig.1 to V_M or V_m according to the mapping of Fig.8:

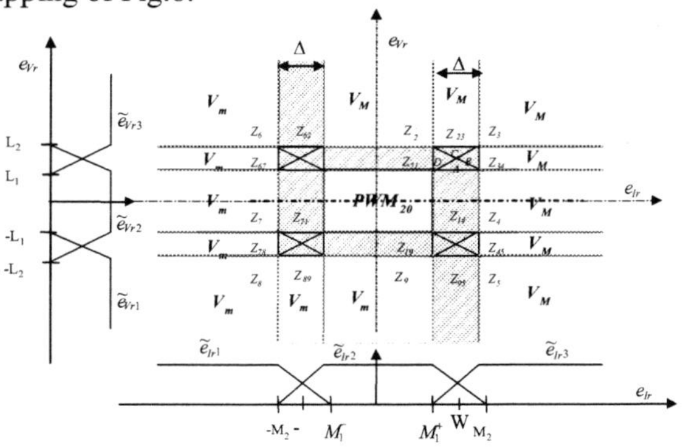

Fig.8 Control variable u_{CD} values in the plane (e_{Vr}, e_{Ir})

In can be pointed out that the small-signal control (PWM_{20} or PWM_{50}) is assigned only to the small-signal region, whereas sliding control prevails over the other ones. The control laws in the dashed areas result from a linear interpolation of the control laws assigned to the adjacent regions. Finally, the fuzzy controller exhibits the following set of control rules, shown in Table III:

e_{Vr}, e_{Ir}	$\tilde{e}_{Ir3}$	$\tilde{e}_{Ir2}$	$\tilde{e}_{Ir1}$
$\tilde{e}_{Vr3}$	V_M	V_M	V_m
$\tilde{e}_{Vr2}$	V_M	$d_{20}(t)/d_{50}(t)$	V_m
$\tilde{e}_{Vr1}$	V_M	V_m	V_m

Table III: Control rules (output value of the fuzzy controller)

5. SIMULATION RESULTS AND CONCLUSIONS

Fig.9 (a), (b), (c) show the simulation of the regulator output voltage with the TSK-fuzzy controller for a start-up and small-perturbations in the reference voltage. Figs 9(a) and 9(c) corresponding to the designed steady-state operating points V_{ref}=20V and V_{ref}=50V, evidences both the proper start-up and the small-signal equalization. Additionally, Fig.9(b) depicts the response for an intermediate reference voltage of V_{ref}=25V and shows also a proper start-up and a small-signal response around $\xi_d=\sqrt{2}/2$, this resulting from the interpolation properties of the fuzzy controller.

Finally, it can be concluded that the design presented in this paper has combined the properties of well-known analytical control techniques such as linear current-mode control and sliding-mode control with the ability of the TSK fuzzy controller in mapping the variables state space by means of the membership functions.

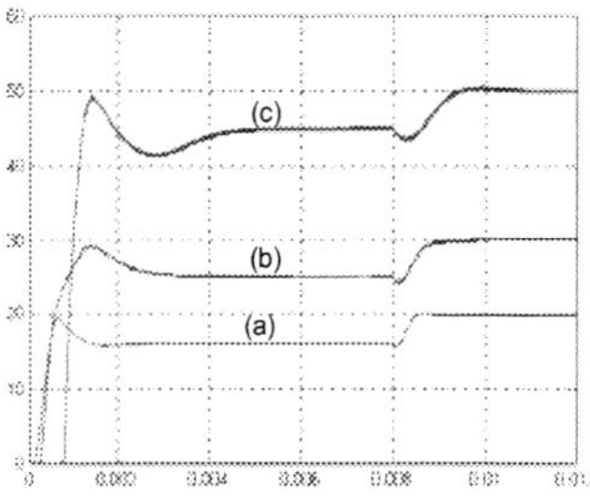

Fig.9 Regulator output voltage in front of reference voltage perturbations (a): Start-up at 16V and step to 20V at t=0.008s (b): Start-up at 25V and step to 30V at t=0. (c): Start-up at 45V and step to 50V at t=0.008s

As an example, a complete design of a Takagi-Sugeno-Kang (TSK) fuzzy controller for a PWM boost DC-DC switching regulator operating at two different steady-state voltages has been presented. Two sets of control laws, one for small-signal equalization based on the linear current control technique and the other for a proper start-up based on sliding-mode control principles, have been embedded into the TSK-fuzzy controller. Simulation results validating the proposed design confirm the extension of the regulator features. The current work is devoted to the TMDS-320006711 DSP-based implementation of the design presented in this work.

6. ACKNOWLEDGEMENTS

This work has been partially funded by the Spanish Ministry of Science and Technology and the European Union (FEDER), project nº: DPI2003-08887-CO3-01.

7. REFERENCES

[1] Middlebrook, R.D. and Cuk, S. "Advances in switched mode power conversion". Vol 1, *Teslaco*, pp 131-151.1981.

[2] Takagi, T. and Sugeno, M. "Fuzzy Identification of systems and its applications to modelling and control", *IEEE Trans. on SMC*, Vol 15, nº1., pp 116-132 January 1985.

[3] Capel, A. et al. "Current Control Modulators: general Theory on Specific Design" *IEEE Trans. on IECI*, Vol. IECI 28, nº 4 pp 293-307, Nov. 1981.

[4] S. Gomáriz, et al. "Analytical considerations in the design of a nonlinear state-dependent Takagi-Sugeno fuzzy controller for a boost switching power regulator" *9th Med. Conf. on Control and Automation (MED'01).* Dubrovnik, Croatia June 2001.

[5] Sira-Ramirez, H. "Sliding motions in bilinear switching networks". *IEEE Trans on CAS,* Vol.34, nº 8, pp 919-933, 1987.

[6] Sira-Ramirez, H. "Switched control of bilinear converters via pseudolinearization". *IEEE Trans. on CAS.* Vol. 36, nº6, pp. 858-865. June 89

[7] R A Viswanathan, K. Srinivasan, D. Oruganti,"Universal fuzzy controller for a non-linear power electronic converter". IEEE Int. Conf. on Fuzzy Systems, Vol 1, pp. 46 -51, 2002.

SPURIOUS MODULATION ON CURRENT-MODE CONTROLLED DC/DC CONVERTERS: AN EXPLANATION FOR INTERMITTENT CHAOTIC OPERATION

S. C. Wong, C. K. Tse and K. C. Tam

Department of Electronic and Information Engineering
Hong Kong Polytechnic University, Hunghom, Hong Kong
URL: http://chaos.eie.polyu.edu.hk Email: {scwong,cktse,enkctam}@eie.polyu.edu.hk

ABSTRACT

In this paper, we explain the mechanism that causes "intermittent" instability and chaos in a current-mode controlled switching converter. The circuit model used to study the phenomenon incorporates a coupling process through which a spurious signal is coupled to the current sensing and ramp compensation circuitry, resulting in a modulation of the compensation slope which causes the system to become unstable intermittently. We describe a way to find the parameter boundaries where intermittent chaotic operation emerges.

Keywords — Switching power converters, current-mode control, intermittent chaos.

1. INTRODUCTION

Power electronics engineers have frequently reported intermittent instability in switching power converters, especially when the converters are not properly protected against instrusion of spurious signals and noise [1]. The intrusion can take the form of coupling via conducted or radiated paths. Sometimes, the intruders (spurious signals) can live on the same circuit board or be present at a close proximity [2]. In this paper we show how the intermittent chaotic operation in a current-mode controlled switching converter can be properly modelled and explained in terms of intrusion of spurious signals. To facilitate design, we identify the critical parameters that affect intermittent chaotic operation and describe a way to calculate the parameter boundaries where intermittent chaotic operation emerges.

2. REVIEW OF CIRCUIT OPERATION

We first review the operation of the current-mode control boost converter [3, 4]. Refering to Fig. 1, when the switch turns on, the inductor current goes up linearly, and is compared with a reference level i_{ref}, which is given by

$$i_{\text{ref}} = I_{\text{ref}} - m_c \bmod(t, T) \tag{1}$$

where T is the switching period and m_c is the slope of the compensation ramp signal. When the inductor current

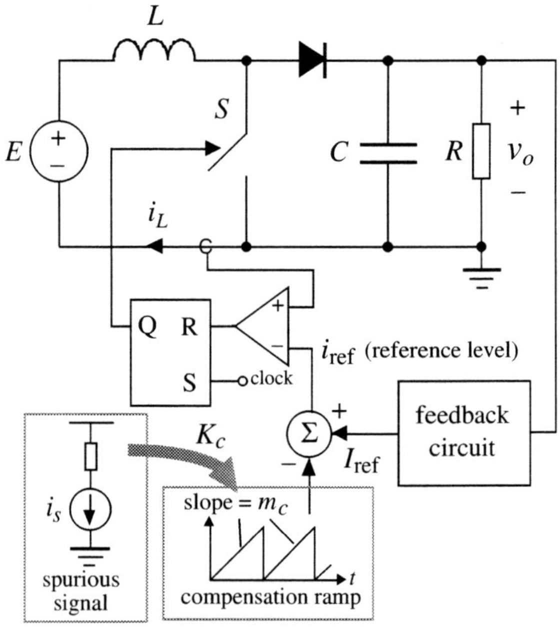

Fig. 1: Boost converter under current-mode control with ramp compensation.

reaches the reference level, the comparator resets the flip-flop, thereby turning off the switch. When the switch is off, the inductor current falls almost linearly. The cycle repeats when the flip-flop is set again by the clock. It should be noted that the inclusion of the compensation ramp is mandatory for maintaining stable operation [5], and stability may be affected when insufficient compensation is applied.

3. INTERMITTENT CHAOTIC OPERATION BY SIMULATION

We consider an additive process which injects the spurious signal directly to the compensation ramp signal of the power supply, as depicted in Fig. 1. Suppose the overall effect of intruding signals is lumped to one spurious source i_s, which

Table 1: Parameter values used in simulations.

Parameter	Value
Inductance L	200 μH
Capacitance C	100 μF
Load resistance R	47 Ω
Switching frequency f_o	25 kHz
Input voltage E	15 V
Reference current I_{ref}	8 A
Slope of compensation ramp m_c	75×10^3 A/s
Spurious signal frequency f_s	25.02 kHz

modulates the slope of the compensation ramp in such a way that the actual compensating slope varies in the range $m_c(1\pm\alpha)$, i.e.,

$$\alpha = \frac{\Delta m_c}{m_c} = \frac{K_c \hat{i}_s}{m_c} \quad (2)$$

with $\hat{i}_s$ being the amplitude of the effective intruding source and K_c being the coupling gain. Note that the overall effect is, at the end, reflected in the magnitude of Δm_c or α. We may therefore study the effect of the intruding source in terms of Δm_c or α.

We assume that the intruding source is sinusoidal of frequency f_s, i.e.,

$$m_c \mapsto m_c(1+\alpha\sin 2\pi f_s t). \quad (3)$$

It should be noted that the exact form of the intruding source is unimportant as long as the ramp signal is caused to change its slope, thereby destabilizing the inner current loop. Also, f_s is generally different from the switching frequency f_o. In our simulations, we take $f_s = 25020$ Hz and apply varying amplitudes of the spurious signal to the system. Other simulation parameters are shown in Table 1.

In order to reveal the periodicity of the operation and to facilitate investigation of the intermittent behavior, we examine the sampled waveform of the inductor current. Effectively, we sample the inductor current at the switching frequency. When we plot the sampled waveform as a function of time, we can observe how the operation changes from time to time. Such plots are called *time-bifurcation diagrams* [1].

From the simulated time-bifurcation diagrams shown in Fig. 2, the following observations can be made.

1. When the intruding signal strength is very weak (i.e., small α), the converter can still maintain its regular operation, though the steady-state operation point fluctuates. The effect is not significant.
2. As α increases, the converter experiences subharmonic operation intermittently with regular operation. For relatively small α, period-2 operation is observed intermittently with regular operation. Subharmonics of longer periods emerge as α increases further.
3. For a sufficiently large α, chaotic and subharmonic operations are observed between periods of regular operations.
4. The intermittent period is equal to $1/|f_s - f_o|$. Thus, if the intruding signal frequency is very close to the switching frequency of the power converter, the intermittency is long.

4. ANALYSIS

We let $i_{L,n}$ and $i_{L,n+1}$ be the inductor current at $t = nT$ and $t = (n+1)T$, respectively. Also, let the output voltage be v_o. Now, by inspecting the slopes of the inductor current and the compensation ramp (see Fig 3), we get

$$i_{L,n+1} = I_{\mathrm{ref}} - f(nT + d_nT) - m_2(1-d_n)T \quad (4)$$
$$i_{L,n} = I_{\mathrm{ref}} - f(nT + d_nT) - m_1 d_n T \quad (5)$$

where d_n is the duty cycle of the nth switching period, $f(t)$ is the displacement of the reference level due to the compensation ramp, and m_1 and m_2 are the rising slope and falling slope, respectively, of the inductor current, i.e.,

$$m_1 = \frac{E}{L} \quad \text{and} \quad m_2 = \frac{v_o - E}{L}. \quad (6)$$

Note that in the steady state, we have $v_o = E/(1-D)$, where D is the operating duty cycle, and hence,

$$\frac{m_2}{m_1} = \frac{D}{1-D}. \quad (7)$$

When analyzing the dynamics at the vicinity of the switching frequency, I_{ref}, m_1 and m_2 can be treated as constants. The variations of $i_{L,n}$ and $i_{L,n+1}$ can be calculated as

$$\delta i_{L,n+1} = -\frac{\partial f(nT+d_nT)}{\partial d_n}\delta d_n + m_2 T\delta d_n \quad (8)$$
$$\delta i_{L,n} = -\frac{\partial f(nT+d_nT)}{\partial d_n}\delta d_n - m_1 T\delta d_n. \quad (9)$$

Combining the above equations, we get the characteristic multiplier or eigenvalue, λ, as

$$\lambda = \frac{\delta i_{L,n+1}}{\delta i_{L,n}} = \frac{-\dfrac{\partial f(nT+d_nT)}{\partial d_n} + m_2T}{-\dfrac{\partial f(nT+d_nT)}{\partial d_n} - m_1T}. \quad (10)$$

For the sinusoidal intruding source under consideration, we have

$$\frac{\partial f(nT+d_nT)}{\partial d_n} = m_cT\,[1+\alpha\sin(\omega_s d_n T+\theta) \quad (11)$$
$$+\alpha\omega_s d_n T\cos(\omega_s d_n T+\theta)] \quad (12)$$

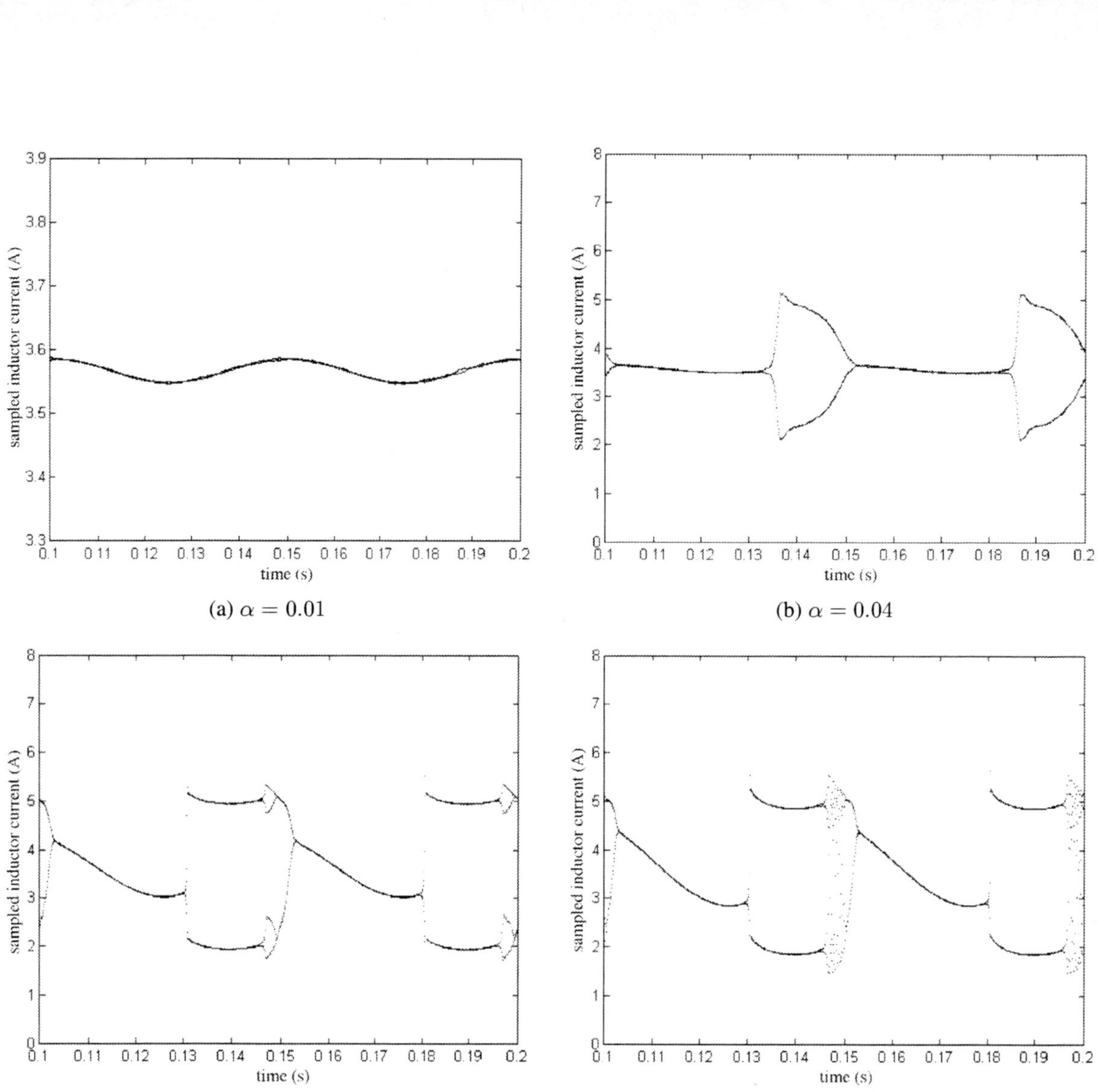

(a) $\alpha = 0.01$ (b) $\alpha = 0.04$

(c) $\alpha = 0.3$ (d) $\alpha = 0.4$

Fig. 2: Sampled inductor current waveforms (time-bifurcation diagrams) for the current-mode controlled boost converter with unintended coupling of sinusoidal source to the compensation ramp signal for different spurious signal strengths. (a) Regular operation with fluctuation of average value; (b)–(c) intermittent subharmonic operation; (d) intermittent chaotic operation.

where $\theta \in [0, 2\pi]$ can be regarded as a random variable for any particular switching period. Thus, we get

$$\lambda = -\frac{m_2 - g(\alpha)m_c}{m_1 + g(\alpha)m_c} = -\frac{\left(\dfrac{D}{1-D}\right) - g(\alpha)M_c}{1 + g(\alpha)M_c} \qquad (13)$$

where $M_c = \frac{m_c L}{E}$ and $g(\alpha) = 1 + \alpha \sin(\omega_s DT + \theta) + \alpha\omega_s DT \cos(\omega_s DT + \theta)$, with D being the operating duty cycle. Here, we note that M_c is the normalized slope of the compensation ramp and $g(\alpha)$ acts as an adjustment factor to the compensation slope due to the presence of the spurious signal. Thus, we may define an effective compensation slope M_{eff} as

$$M_{\text{eff}} = g(\alpha)M_c \qquad (14)$$

which is less than M_c if $g(\alpha) < 1$. In particular, $g(0) = 1$ corresponds to the case where the spurious signal is absent and $M_{\text{eff}} = M_c$.

Now, since subharmonics and chaos occur when $\lambda \leq -1$, we can find, from (13), the condition for maintaining regular operation as

$$M_{\text{eff}} > -\frac{1-2D}{2(1-D)}. \qquad (15)$$

Also, the extreme values of $g(\alpha)$ are given by

$$\sup_{\alpha} g(\alpha) = 1 \pm \alpha\sqrt{1 + \omega_s^2 D^2 T^2}. \qquad (16)$$

Here, we consider only the case where $g(\alpha) < 1$ since the effect of the spurious signal being considered is to reduce the effective compensation slope. Thus, the minus sign should be taken for the extreme value of $g(\alpha)$ given in

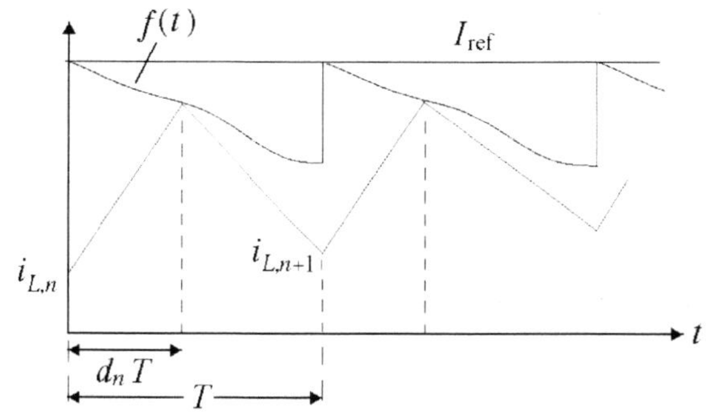

Fig. 3: Inductor current waveform and compensation ramp signal in the presence of intruding source.

(16). Furthermore, we limit ourselves to the practical situation where the amplitude of the spurious signal is relatively small such that the modulation effect does not completely cancel the slope compensation, i.e., $1-\alpha\sqrt{1+\omega_s^2 D^2 T^2} > 0$. Clearly, this condition is equivalent to

$$\alpha < \frac{1}{\sqrt{1+\omega_s^2 D^2 T^2}}, \tag{17}$$

which can also be translated into $D > 0.5$ for all $M_c > 0$. This is consistent with our usual understanding that the use of slope compensation is only needed for $D > 0.5$. We henceforth omit the discussion of the impractical case where (17) is not satisfied.

Now, from (15) and (16), we can find the critical spurious signal strength, α_{crit}, at which the first period-doubling occurs and regular operation fails intermittently, i.e.,

$$\alpha_{\text{crit}} = \frac{(1+M_c)(1-D)-0.5}{M_c(1-D)\sqrt{1+\omega_s^2 T^2 D^2}}. \tag{18}$$

Thus, from (18), we can compute the boundary of regular operation for any given set of steady-state operating parameters. Note that since (17) has to be satisfied, we restrict the plotting range within $\alpha > 0$ and $0.5 < D < 1$. For ease of visualization, we show in Fig. 4 a few specific boundary curves for some selected values of the compensating slope. We have also plotted the simulation data alongside the curves and found perfect agreement with the analysis.

5. CONCLUSION

It should be clear that intermittent operation occurs at a frequency which is simply given by the difference between the spurious signal and an integer multiple of the switching frequency, i.e., $|f_s - nf_o|$ for $n = 0, 1, 2, \ldots$. Also, intermittent operation can be observed only when the transient is sufficiently fast (or the frequency of intermittency sufficiently low) such that regular, subharmonic or chaotic operation can show up successively in time and be orchestrated as

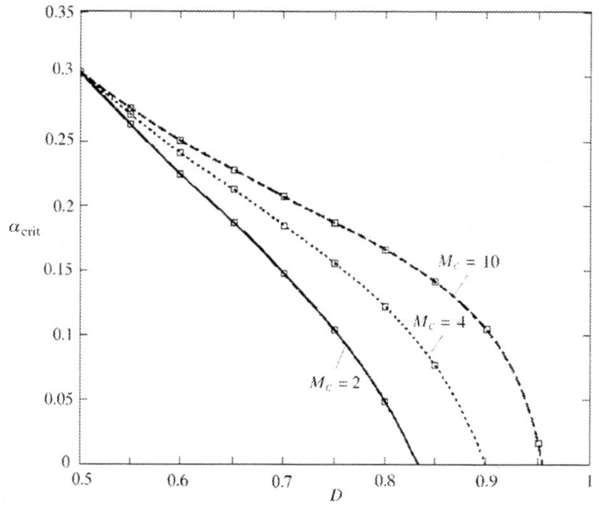

Fig. 4: Boundaries of regular operation. Analytical curves are plotted for $M_c = 2$ (solid curve), $M_c = 4$ (dotted curve) and $M_c = 10$ (dashed curve). Data obtained from cycle-by-cycle simulations are plotted as boxes. Region below the curve corresponds to stable regular operation, and region above the curve corresponds to intermittent subharmonic and/or chaotic operation.

an intermittent operation. In summary, when the spurious signal frequency is sufficiently close to an integer multiple of the switching frequency and the spurious signal is strong enough, intermittent operation occurs. The question of how close the two frequencies should be is therefore dependent upon the transient response of the converter.

Acknowledgments

This work was supported by Hong Kong Research Grants Council under Grant PolyU5241/03E.

6. REFERENCES

[1] C.K. Tse, Y. Zhou, F.C.M. Lau and S.S. Qiu, "Intermittent chaos in switching power supplies due to unintended coupling of spurious signals," *IEEE PESC Record*, pp. 642–645, 2003.

[2] J.A. Ferreira, P.R. Willcock, and S.R. Holm, "Sources, paths and traps of conducted EMI in switch mode circuits," *Proc., IEEE Ind. Appl. Conf.*, pp. 1584–1591, 1977.

[3] A. Capel, G. Ferrante, D.O. Sullivan and A. Weinberg, "Application of the injected-current control model for the dynamic analysis of switching regulators with a new concept of LC^3 modulator," *IEEE PESC Record*, pp. 135–147, 1978.

[4] R. Redl and N.O. Sokal, "Current-mode control, five different types, used with the three basic classes of power converters," *IEEE PESC Record*, pp. 771–775, 1985.

[5] P.T. Krein, *Elements of Power Electronics*, Oxford University Press, New York, 1998.

SMALL-SIGNAL DUTY CYCLE TO INDUCTOR CURRENT TRANSFER FUNCTION FOR BOOST PWM DC-DC CONVERTER IN CONTINUOUS CONDUCTION MODE

Brad Bryant and Marian K. Kazimierczuk, IEEE, Senior Member

Wright State University
Department of Electrical Engineering
Dayton, OH 45435
bbryant@cs.wright.edu, mkazim@cs.wright.edu

ABSTRACT

The small-signal transfer function from duty cycle to inductor current is derived for the boost PWM DC-DC converter operating in continuous conduction mode (CCM). The transfer function is found for two cases, the converter with a resistive load, and with a current sink load. The derivations are performed using an averaged circuit small-signal model of the boost converter for CCM. The transfer function frequency responses and transient responses are shown for a set of given component values. The transfer functions can be used to model the current loop of boost PWM converters with peak current-mode control (PCM).

1. INTRODUCTION

Development of small-signal transfer functions for non-linear pulse-width modulated (PWM) DC-DC converters can aid in ease of understanding circuit performance and control. Determining the open-loop transfer functions to properly describe a system is essential to obtain the correct plant for control design. In PWM DC-DC converters with current-mode control, the relationship between duty cycle and inductor current is an essential characteristic of the current loop. A transfer function between duty cycle and inductor current can be determined using an averaged linear time-invariant circuit model of the non-linear converter [1], [2]. The relationship between duty cycle and inductor current is dependent upon converter topology and mode of operation.

The objectives of this paper are to derive the small-signal duty cycle to inductor current transfer function for boost PWM DC-DCconverters operating in continuous conduction mode, to illustrate the transfer function frequency response, and transient response.

PWM converters can be modeled with a circuit averaging technique. The switching devices prevalent in PWM converters are transistors and diodes. The transistor can be modeled as a current-dependent current-source in CCM operation as [1], [2]

$$i_S = d_T i_L,$$

where the total duty cycle $d_T = D + d$, the switch current $i_S = I_S + i_s$, and inductor current $i_L = I_L + i_l$. The diode can be modeled as a voltage-dependent voltage-source in CCM operation as

$$v_D = d_T v_O,$$

where $v_D = V_D + v_d$, and $v_O = V_O + v_o$. The resulting transistor and diode models become

$$i_S = (D + d)(I_L + i_l) = DI_L + Di_l + dI_L + i_l d,$$

and

$$v_D = (D + d)(V_O + v_o) = DV_O + Dv_o + dV_O + v_o d.$$

The double small-signal terms $i_l d$ and $v_o d$ can be neglected when the following conditions are met, $i_l \ll I_L$, $v_o \ll V_O$, and $d \ll D$. This enables the derivation of small-signal transfer functions. Figure 1 shows a boost converter operating in CCM with the transistor and diode replaced with their respective models. In Fig. 1, r is given by [1], [2]

$$r = r_L + Dr_{DS} + (1 - D)R_F,$$

where r_L is the inductor ESR, r_{DS} is the transistor on-resistance, and R_F is the diode forward resistance.

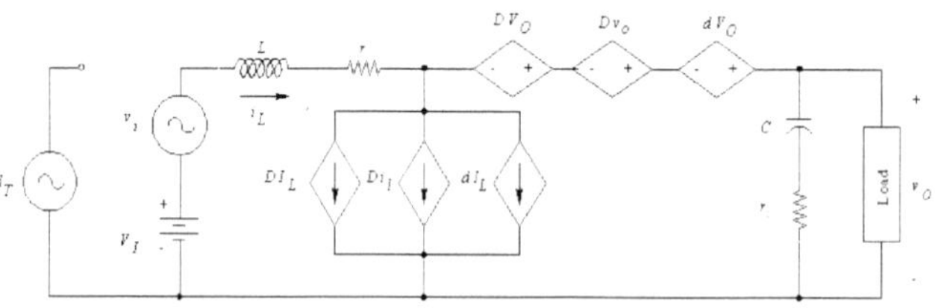

Fig. 1. Large-signal model of boost PWM converter for CCM operation.

2. DUTY CYCLE TO INDUCTOR CURRENT TRANSFER FUNCTION WITH A RESITIVE LOAD

The model of Fig. 1 can be separated into dc and ac models. Figure 2 shows the small-signal ac circuit model of a boost converter with resistive load operating in CCM. This small-

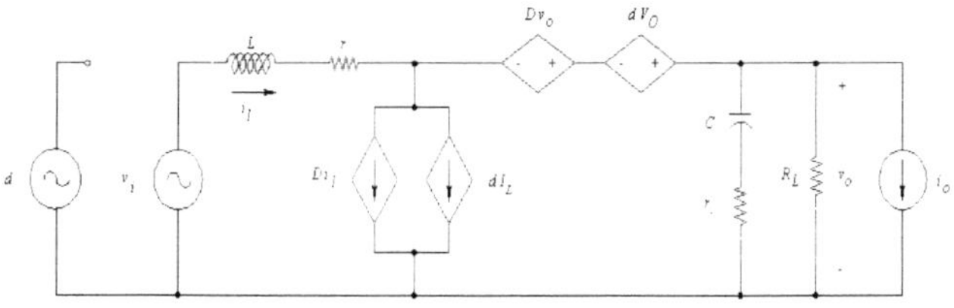

Fig. 2. Small-signal model of boost PWM converter with a resistive load operating in CCM.

signal model can be used to find the transfer functions from inputs d, v_i, and i_o to outputs v_o and i_l. The duty cycle to inductor current transfer function, T_{pi}, can be found by setting inputs v_i and i_o of Fig. 2 to zero, resulting in the circuit of Fig. 3, where Z_1 and Z_2 are

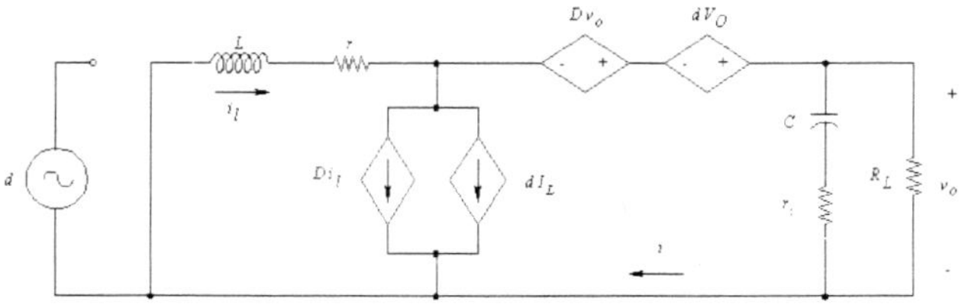

Fig. 3. Small-signal model to determine T_{pi} for boost convert with a resistive load operating in CCM .

$$Z_1 = sL + r,$$

$$Z_2 = \left(\frac{1}{sC} + r_C\right) \parallel R_L.$$

We use KCL to obtain

$$i_l = Di_l + dI_L + i,$$

$$i_l = Di_l + dI_L + \frac{v_o}{Z_2},$$

$$i_l(1-D) = d\frac{I_O}{(1-D)} + \frac{v_o}{Z_2},$$

$$v_o = i_l(1-D)Z_2 - d\frac{V_O Z_2}{R_L(1-D)}. \qquad (1)$$

From KVL, we get

$$-i_l Z_1 + Dv_o + dV_O - v_o = 0,$$

$$v_o = \frac{dV_O - i_l Z_1}{1-D}. \qquad (2)$$

Equating Eqs. (1) and (2) results in

$$i_l(1-D)Z_2 - d\frac{V_O Z_2}{R_L(1-D)} = \frac{dV_O - i_l Z_1}{1-D},$$

$$i_l\left[(1-D)Z_2 + \frac{Z_1}{1-D}\right] = d\left[\frac{V_O}{(1-D)} + \frac{Z_2 V_O}{R_L(1-D)}\right].$$

This gives T_{pi} as

$$T_{pi} = \frac{i_l}{d} = V_O\frac{\frac{1}{1-D} + \frac{Z_2}{R_L(1-D)}}{Z_2(1-D) + \frac{Z_1}{1-D}},$$

$$T_{pi} = V_O\frac{1 + \frac{Z_2}{R_L}}{Z_1 + Z_2(1-D)^2}.$$

Substituting for Z_1 and Z_2 gives

$$T_{pi}(s) = V_O\frac{1 + \frac{(\frac{1}{sC}+r_C)\parallel R_L}{R_L}}{sL + r + \left[\left(\frac{1}{sC} + r_C\right) \parallel R_L\right](1-D)^2},$$

$$T_{pi}(s) = \frac{V_O(R_L + 2r_C)}{L(R_L + r_C)}$$

$$\times \frac{s + \frac{1}{C(R_L/2 + r_C)}}{s^2 + \left[\frac{C[r(R_L+r_C)+R_L r_C(1-D)^2]+L}{LC(R_L+r_C)}\right]s + \frac{(1-D)^2 R_L + r}{LC(R_L+r_C)}}$$

$$= T_{pix}\frac{s + \omega_{zi}}{s^2 + 2\zeta\omega_n s + \omega_n^2},$$

where

$$T_{pix} = \frac{V_O(R_L + 2r_C)}{L(R_L + r_C)},$$

$$\omega_{zi} = \frac{1}{C(R_L/2 + r_C)},$$

$$\omega_n = \sqrt{\frac{(1-D)^2 R_L + r}{LC(R_L + r_C)}},$$

and

$$\zeta = \frac{C\left[r(R_L + r_C) + R_L r_C(1-D)^2\right] + L}{2\sqrt{LC(R_L + r_C)\left[r + (1-D)^2 R_L\right]}}.$$

For an example boost converter with $V_O = 20$ V, $I_O = 0.5$ A, $I_L = 1$ A, $D = 0.516$, $L = 156$ μH, $C = 68$ μF, $r_L = 0.19$ Ω, $r_{DS} = 0.18$ Ω, $R_F = 0.072$ Ω, and $r_C = 0.111$ Ω, we obtain $T_{pix} = 128559.9$, $f_{zi} = 116.38$ Hz, $f_n = 759.42$ Hz, and $\zeta = 0.2693$. Bode plots of T_{pi} for this example are shown in Figs. 4 and 5. Tpi is a second-order function with $\omega_{zi} \ll \omega_n$, resulting in significant peaking of the magnitude response. The phase response of T_{pi} creates the phase lead compensation effect present in the closed-current loop, compensating for the phase response of the non-minimal transfer function $T_p = v_o/d$. Figure 6 shows a transient response of i_L due to a step change in duty cycle $\Delta d = 0.1$.

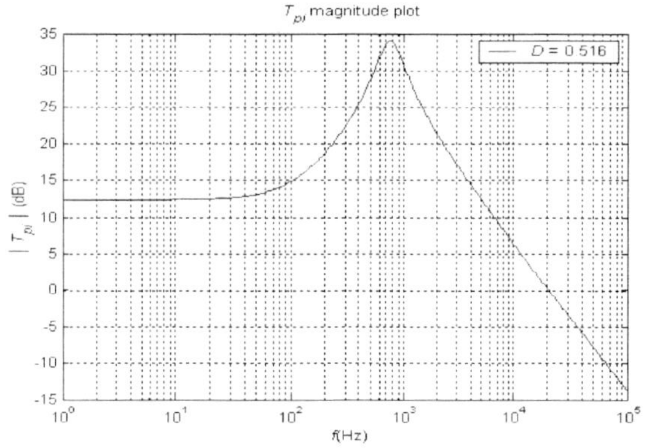

Fig. 4. T_{pi} magnitude response.

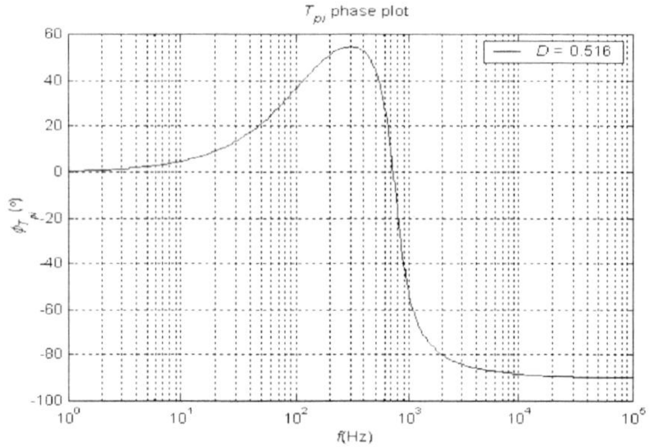

Fig. 5. T_{pi} phase response.

3. DUTY CYCLE TO INDUCTOR CURRENT TRANSFER FUNCTION WITH A CURRENT-SINK LOAD

The model of Fig. 1 can be separated into dc and ac models. The load for a DC-DC converter can behave as a current sink. Figure 7 shows the small-signal ac circuit model of a boost converter with a current sink load operating in CCM. The duty cycle to inductor current transfer function, T_{pi}, can be found by setting inputs v_i and i_o of Fig. 7 to zero, resulting in the circuit of Fig. 8, where Z_1 is as in Section 2, and Z_2 is

$$Z_2 = \frac{1}{sC} + r_C.$$

From KCL, we obtain

$$i_l = Di_l + dI_L + \frac{v_o}{Z_2} = Di_l + d\frac{I_O}{(1-D)} + \frac{v_o}{Z_2},$$

$$v_o = i_l(1-D)Z_2 - d\frac{Z_2 I_O}{(1-D)}. \tag{3}$$

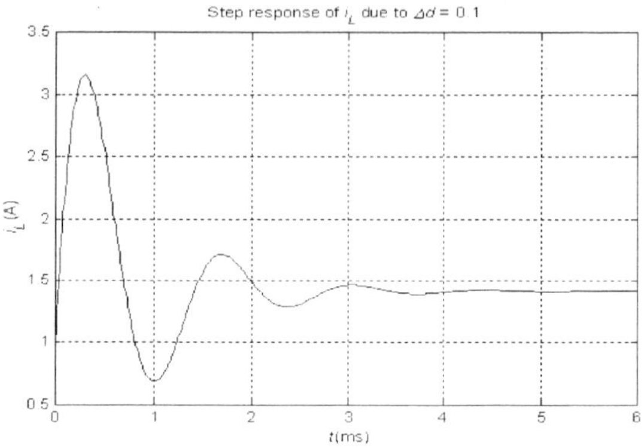

Fig. 6. Step response of i_L due to $\Delta d = 0.1$.

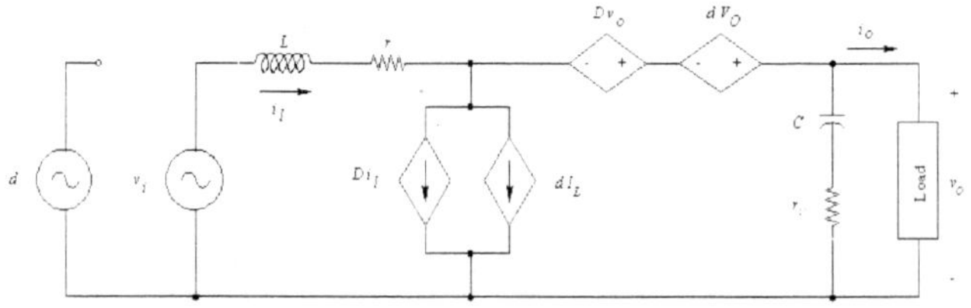

Fig. 7. Small-signal model of boost PWM converter with a current sink load operating in CCM.

Using KVL, we get

$$-i_l Z_1 + Dv_o + dV_O - v_o = 0,$$

$$v_o = \frac{dV_O - i_l Z_1}{1-D}. \tag{4}$$

Equating Eqs. (3) and (4) results in

$$i_l\left[(1-D)Z_2 + \frac{Z_1}{1-D}\right] = d\left[\frac{V_O}{(1-D)} + \frac{Z_2 I_O}{(1-D)}\right].$$

This gives T_{pi} as

$$T_{pi} = \frac{i_l}{d} = \frac{V_O + Z_2 I_O}{Z_1 + Z_2(1-D)^2}.$$

Substituting for Z_1 and Z_2 gives

$$T_{pi}(s) = \frac{V_O + \left(\frac{1}{sC} + r_C\right) I_O}{sL + r + \left(\frac{1}{sC} + r_C\right)(1-D)^2},$$

$$T_{pi}(s) = \frac{V_O + r_C I_O}{L} \frac{s + \frac{I_O}{C(V_O + r_C I_O)}}{s^2 + \left[\frac{r + r_C(1-D)^2}{L}\right] s + \frac{(1-D)^2}{LC}}$$

$$= T_{pix}\frac{s + \omega_{zi}}{s^2 + 2\zeta\omega_n s + \omega_n^2},$$

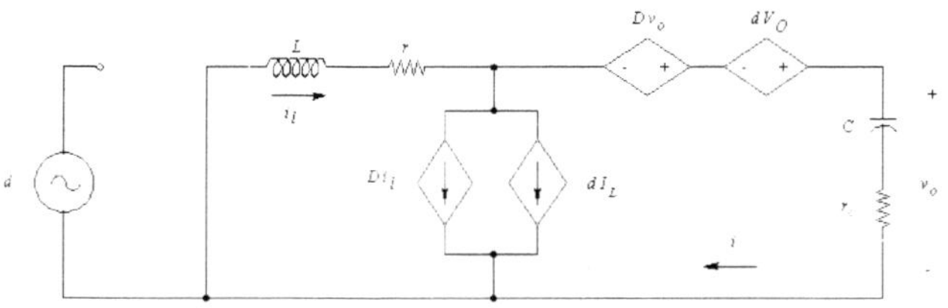

Fig. 8. Small-signal model to determine T_{pi} for a boost converter with a current sink load operating in CCM.

where

$$T_{pix} = \frac{V_O + r_C I_O}{L},$$

$$\omega_{zi} = \frac{I_O}{C(V_O + r_C I_O)},$$

$$\omega_n = \sqrt{\frac{(1-D)^2}{LC}},$$

and

$$\zeta = \frac{r + r_C(1-D)^2}{2(1-D)}\sqrt{\frac{C}{L}}.$$

For the boost converter example presented in Section 2, we obtain $T_{pix} = 128560.9$, $f_{zi} = 58.35$ Hz, $f_n = 747.92$ Hz, and $\zeta = 0.2344$. Bode plots of T_{pi} are shown in Figs. 9 and 10. Figure 11 shows a transient response of i_L due to a step change in duty cycle $\Delta d = 0.1$.

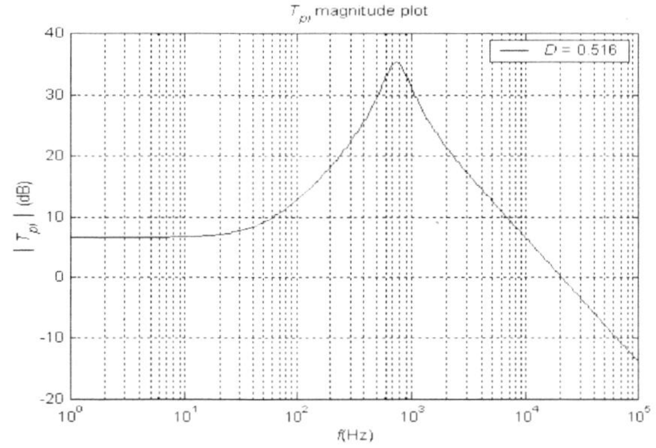

Fig. 9. T_{pi} magnitude response.

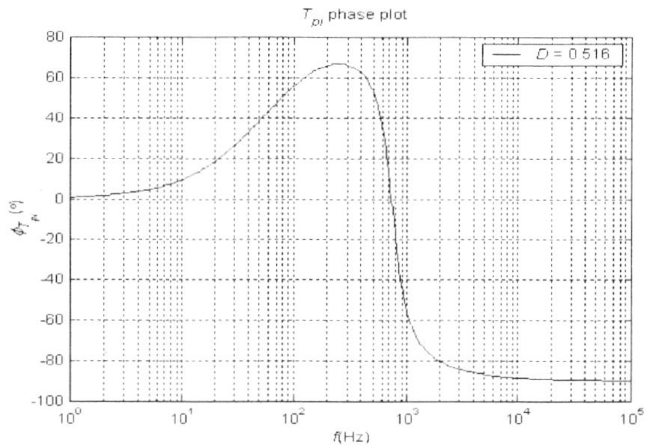

Fig. 10. T_{pi} phase response.

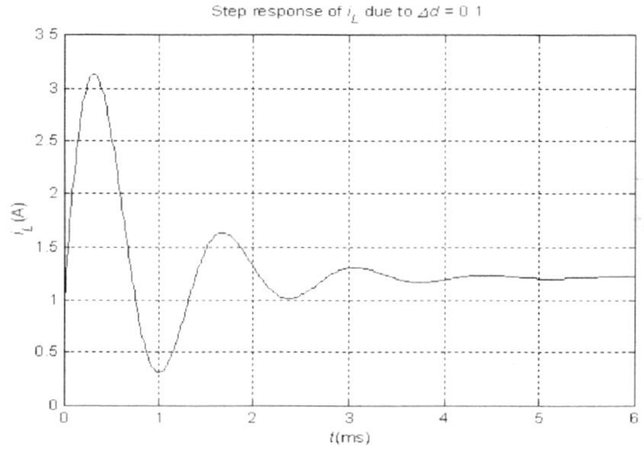

Fig. 11. Step response of i_L due to $\Delta d = 0.1$.

4. CONCLUSIONS

The small-signal transfer function of duty cycle to inductor current, T_{pi}, has been derived for boost PWM DC-DC converters operating in CCM. $T_{pi}(s)$ is a second-order function. When the load behaves as a current sink, f_{zi} is lowered by a factor of 2 compared to the case of a resitive load, and ζ decreases by 14% for the given examples. The frequency of the zero f_{zi} is much lower than the natural resonant frequency f_n in the given examples. Therefore, the magnitude response in both cases exhibits significant peaking at f_n, and the phase response increases until reaching approximately f_n, where it decreases to -90°. This transfer function is essential in characterizing the current loop and shows that the bandwidth of the current loop can be made wide, enabling a fast control response.

5. REFERENCES

[1] D. Czarkowski and M. K. Kazimierczuk, "Energy-conservation approach to modeling PWM dc-dc converters," *IEEE Trans. Aerospace Electron. Syst.*, vol. AES-29, pp. 1059-1063, July 1993.

[2] M. K. Kazimierczuk and D. Czarkowski, "Application of the principle of energy conservation to modelling the PWM converters," 2nd IEEE Conference on Control Applications, September 13-16, 1993, pp. 291-296.

SAMPLE AND HOLD EFFECT IN PWM DC-DC CONVERTERS WITH PEAK CURRENT-MODE CONTROL

Brad Bryant and Marian K. Kazimierczuk, IEEE, Senior Member

Wright State University
Department of Electrical Engineering
Dayton, OH 45435
bbryant@cs.wright.edu, mkazim@cs.wright.edu

ABSTRACT

The sample and hold effect, or quasi-digital (discrete) behavior, of constant-frequency peak current-mode control (PCM) in pulse-width modulated DC-DC converters is presented in a manner consistent with the physical behavior of the circuit. This approach developes a transfer function using control theory that captures the quasi-digital behavior encountered when the current control loop is closed in PCM. The new transfer function accurately captures the current loop behavior, leading the way for a complete linear time-invariant system circuit for PWM DC–DC converters. This work improves upon previous derivations by remaining true to the physical behavior of the circuit while resulting in a rational transfer function that can be used in a mutliple loop complete linear system circuit to accurately predict converter behavior due to changes in input voltage, output current, and reference voltage.

1. INTRODUCTION

A popular control technique in PWM DC-DC converters due to its fast response, accuracy, and overload protection is PCM [1]-[5]. The ability to account for the quasi-digital effect, or the process of natural sampling that occurs in constant-frequency current-mode control of PWM DC-DC converters, is essential to accurately capture high-frequency behavior. Past work accounting for this effect used methods that do not lead to a transfer function that can be placed in a multiple loop linear system circuit to predict instability and correctly capture converter behavior due to the input signals [2], [3]. This paper presents a transfer function for the natural sampling process that can be used to complete a multiple loop linear time-invariant system circuit that captures converter behavior due to all inputs to the DC-DC converter while using voltage and current feedback loops for the control of output voltage.

2. SAMPLE AND HOLD EFFECT

The procedure to determine the natural sampling process is to start with an examination of the relevant converter waveforms under current-mode control. Figure 1 shows waveform behavior due to a step change, a, in the control voltage, v_c. In Fig. 1, the solid lines are the initial steady-state

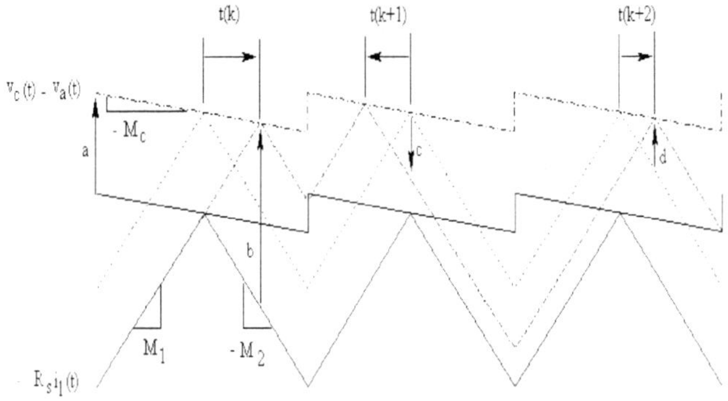

Fig. 1. Converter waveforms for closed-current loop.

waveforms, the dash-dot lines are the perturbed waveforms, the dotted line is where $R_s i_l$ will reach its new steady-state for a stable system, M_1 is the sensed inductor current on-time slope, M_2 is the sensed inductor current off-time slope, M_c is the compensation slope, $v_c(t) - v_a(t)$ is the continuous-time control voltage minus compensation voltage ramp, $R_s i_l(t)$ is the continuous-time sensed inductor current, a is the step change in v_c, b is the change in $R_s i_l$ due to a, c is the change in $R_s i_l$ from its eventual steady-state operating point, d is the subsequent change in $R_s i_l$ due to c, and $t(k)$, $t(k+1)$, and $t(k+2)$ are the cycle-by-cycle changes in the natural sampling instant. In PCM operation, $v_c(t)$, $R_s i_l(t)$, and $v_a(t)$ produce the small-signal error $v_{ei}(t)$, given as $v_{ei}(t) = v_c(t) - R_s i_l(t) - v_a(t)$. When $v_{ei}(t) = 0$, the process of natural sampling occurs and the period of transistor off-time is entered.

The waveforms in Fig. 1 are the result of closed-loop behavior and are used to find the closed-loop transfer function $H_{icl} = R_s i_l / v_c$. The derivation is performed by using the discrete-time relationships set by the natural sampling instants. The forced response is the change in $R_s i_l$ due to a change in v_c. From Fig. 1, the forced response is

$$\frac{R_s i_l(k)}{v_c(k)} = \frac{b}{a} = \frac{(M_1 + M_2)t(k)}{(M_1 + M_c)t(k)} = 1 + \alpha \quad (1)$$

resulting in $R_s i_l(k) = (1+\alpha)v_c(k)$, where α is defined as $\alpha = (M_2 - M_c)/(M_1 + M_c)$. The natural response, or subsequent cycle-by-cycle oscillatory behavior of the inductor current due to the step disturbance in v_c, is given by

$$\frac{R_s i_l(k+2)}{R_s i_l(k+1)} = \frac{d}{-c} = \frac{(M_2 - M_c)t(k+2)}{-(M_1 + M_c)t(k+2)} = -\alpha \quad (2)$$

resulting in $R_s i_l(k+2) = -\alpha R_s i_l(k+1)$. By combining the forced and natural responses to capture the complete waveform behavior, the discrete-time closed-loop transfer function H_{icl} is determined by

$$R_s i_l(k) = -\alpha R_s i_l(k-1) + (1+\alpha)v_c(k). \quad (3)$$

Taking the z-transform, we get

$$R_s i_l(z) + \alpha z^{-1} R_s i_l(z) = (1+\alpha)v_c(z). \quad (4)$$

This leads to $H_{icl}(z)$ as

$$H_{icl}(z) = \frac{R_s i_l(z)}{v_c(z)} = \frac{1+\alpha}{1+\alpha z^{-1}} = \frac{z(1+\alpha)}{z+\alpha}. \quad (5)$$

$H_{icl}(z)$ is as found in [2], [5]. We can see from (5) that there is a pole $z_p = -\alpha$. When $\alpha > 1$, the current loop is unstable as the pole moves outside of the unit circle. For the case of $M_c = 0$, the current loop is unstable for duty cycles $D > 0.5$. $H_{icl}(z)$ can be converted to the s-domain by using the definition of the z-transform $z = e^{sT}$ and multiplying by $(1 - e^{-sT})/(sT)$ [2], [5]. This results in $H_{icl}(s)$ as

$$H_{icl}(s) = \frac{R_s i_l(s)}{v_c(s)} = \frac{(1+\alpha)}{sT}\frac{e^{sT} - 1}{e^{sT} + \alpha} \quad (6)$$

where $T = 1/f_s$, and f_s is the clock switching frequency. It is clear that H_{icl} is independent of converter topology. The term e^{sT} in (6) can be replaced with a second-order Padé approximation,

$$e^{sT} \approx \frac{1 - \frac{sT}{2} + \frac{(sT)^2}{12}}{1 + \frac{sT}{2} + \frac{(sT)^2}{12}}. \quad (7)$$

This leads to

$$H_{icl}(s) = \frac{R_s i_l(s)}{v_c(s)} \approx \frac{12 f_s^2}{s^2 + 6\frac{1-\alpha}{1+\alpha} f_s s + 12 f_s^2}. \quad (8)$$

Bode plots of H_{icl} are shown in Figs. 2 and 3 for $f_s = 100$ kHz and $\alpha = 0, 0.5,$ and 1. H_{icl} is approximately 1 for low frequencies, as $R_s i_l$ follows v_c in PCM, which has been verified with non-linear PWM converter simulation in PSpice. As α approaches 1, H_{icl} exhibits significant peaking at approximately half the switching frequency.

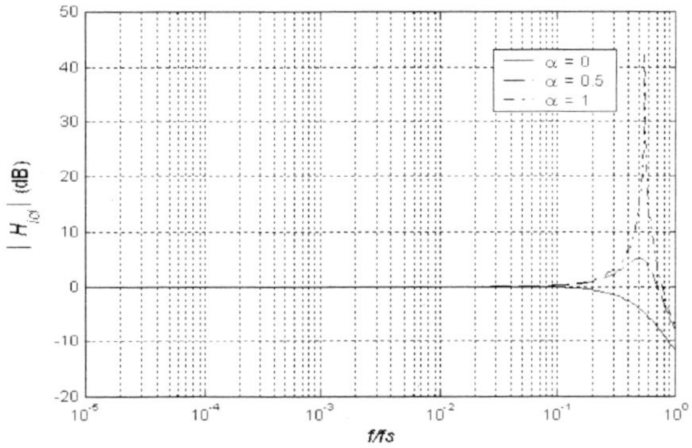

Fig. 2. H_{icl} magnitude response.

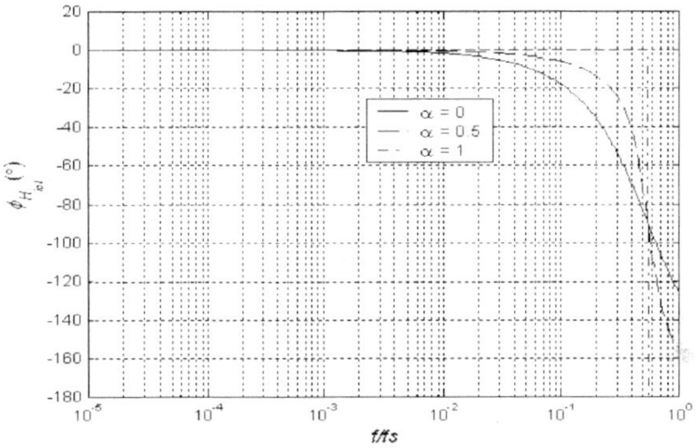

Fig. 3. H_{icl} phase response.

Figure 4 shows a block diagram for the current loop with $R_s i_l$ as the output. The closed-loop transfer function from

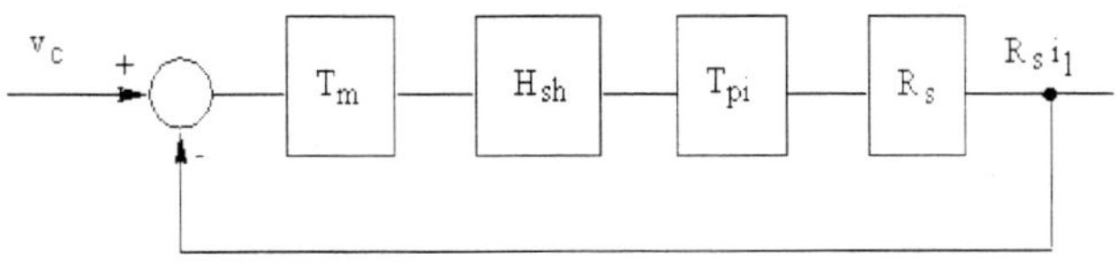

Fig. 4. H_{icl} block diagram.

Fig. 4 is defined as

$$H_{icl}(s) = \frac{R_s i_l(s)}{v_c(s)} = \frac{T_m H_{sh} T_{pi} R_s}{1 + T_m H_{sh} T_{pi} R_s} \quad (9)$$

where $T_m = d_o/v_{ei} = f_s/(M_1 + M_c)$ [1]. By equating (6) and (9),

$$\frac{(1+\alpha)}{sT}\frac{e^{sT}-1}{e^{sT}+\alpha} = \frac{T_m H_{sh} T_{pi} R_s}{1 + T_m H_{sh} T_{pi} R_s}. \tag{10}$$

The sample and hold function H_{sh} is determined as

$$H_{sh}(s) = \frac{d}{d_o} = \frac{1}{T_m T_{pi} R_s \left(\frac{sT}{1+\alpha}\frac{e^{sT}+\alpha}{e^{sT}-1} - 1\right)}. \tag{11}$$

By using the second-order Padé approximation in (7), H_{sh} is given by a rational function

$$H_{sh}(s) \approx \frac{12 f_s^2}{T_m T_{pi} R_s s(s + \omega_{sh})} \tag{12}$$

with $\omega_{sh} = 3\omega_s(1-\alpha)/\pi(1+\alpha)$. Therefore, H_{sh} is converter dependent, as it depends on the duty cycle to inductor current transfer function T_{pi}. From [6], T_{pi} is

$$T_{pi}(s) = \frac{i_l(s)}{d(s)} = T_{pix}\frac{s + \omega_{zi}}{s^2 + 2\zeta\omega_n s + \omega_n^2}. \tag{13}$$

Substituting (13) into (12) gives H_{sh} as

$$H_{sh}(s) = \frac{12 f_s^2 (s^2 + 2\zeta\omega_n s + \omega_n^2)}{T_m R_s T_{pix} s(s + \omega_{zi})(s + \omega_{sh})}. \tag{14}$$

For the boost converter example given in [6], the Bode plots of H_{sh} are presented in Figs. 5 and 6. H_{sh} behaves as an integrator for low frequencies.

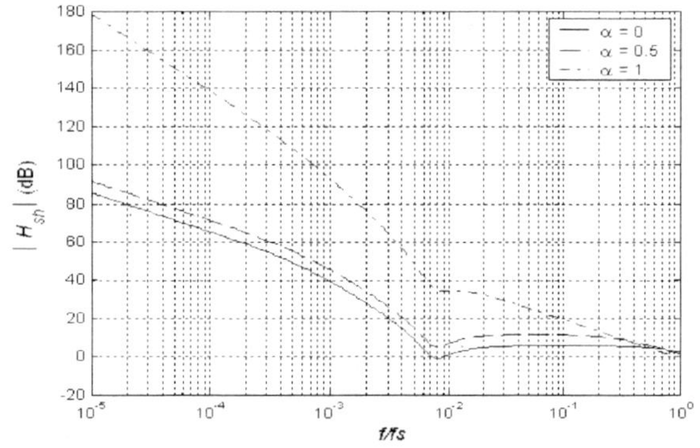

Fig. 5. H_{sh} magnitude response.

The current-loop transfer function of interest is $T_{icl} = d/v_c$, as shown by the block diagram in Fig. 7. The current-loop gain, T_i, is given by

$$T_i(s) = T_m H_{sh} T_{pi} R_s \approx \frac{12 f_s^2}{s(s + 6\frac{1-\alpha}{1+\alpha} f_s)} = \frac{12 f_s^2}{s(s + \omega_{sh})}. \tag{15}$$

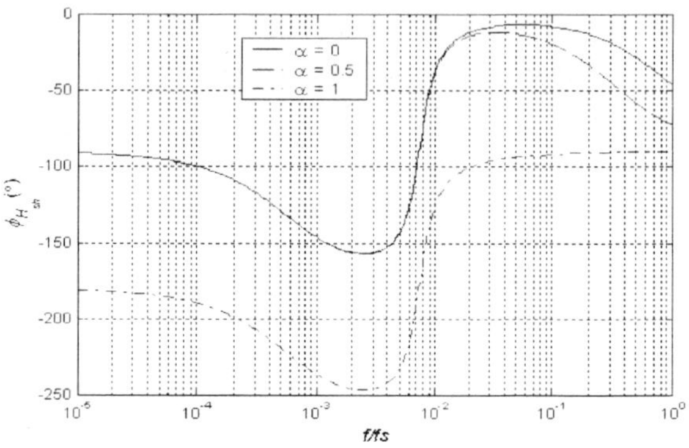

Fig. 6. H_{sh} phase response.

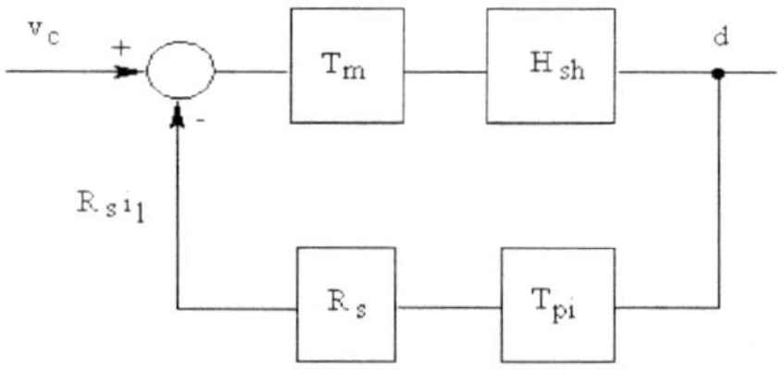

Fig. 7. T_{icl} block diagram.

Therefore, T_i is independent of converter topology, as it only depends on f_s and α. For $\alpha < 1$, the pole due to ω_{sh} is in the left-half s-plane. When $\alpha > 1$, the pole moves into the right-half s-plane creating an unstable current loop. Bode plots of $T_i(s)$ for the boost converter example in [6] are presented in Figs. 8 and 9. T_i behaves as an integrator, and as a double integrator for $f \geq f_{sh}$. T_{icl} is usefull in developing a complete multiple loop linear system circuit.

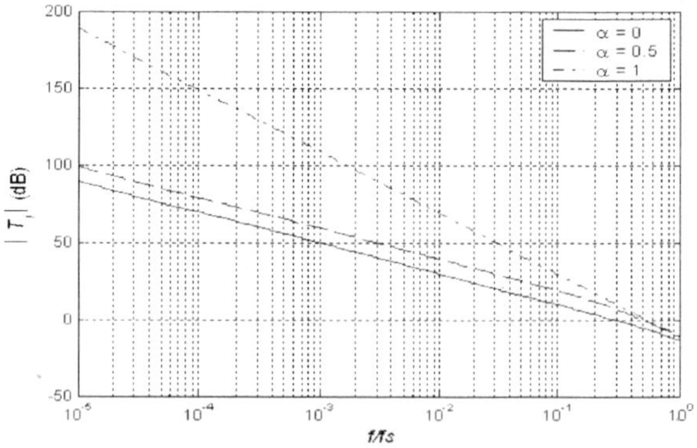

Fig. 8. T_i magnitude response.

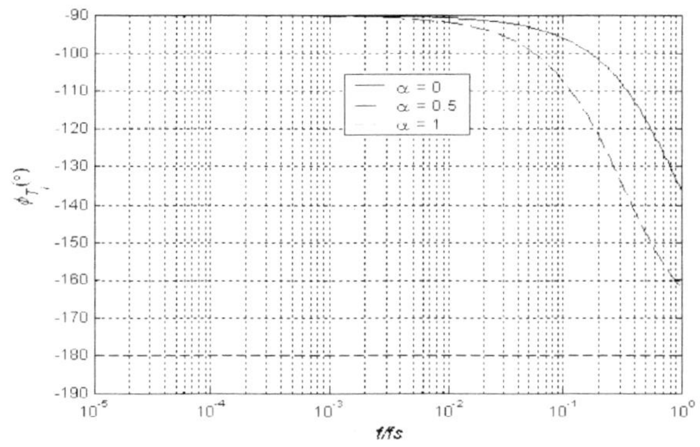

Fig. 9. T_i phase response.

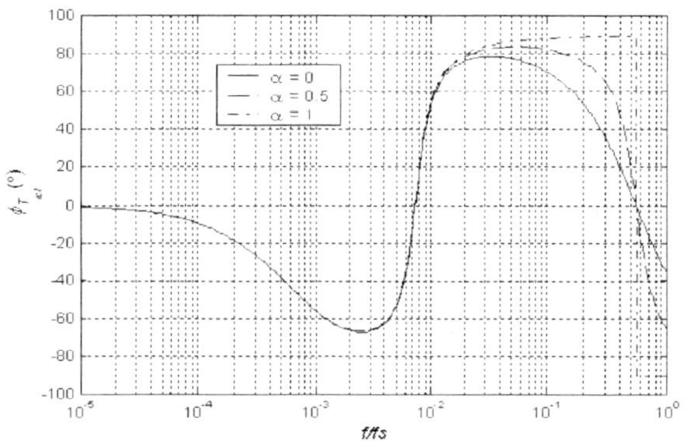

Fig. 11. T_{icl} phase response.

From Fig. 7, T_{icl} is defined as

$$T_{icl}(s) = \frac{d(s)}{v_c(s)} = \frac{T_m H_{sh}}{1 + T_m H_{sh} T_{pi} R_s} = \frac{T_m H_{sh}}{1 + T_i}. \quad (16)$$

We can reduce (16) to

$$T_{icl}(s) = \frac{H_{icl}(s)}{R_s T_{pi}(s)} = \frac{12 f_s^2}{R_s T_{pix}} \times \frac{(s^2 + 2\zeta\omega_n s + \omega_n^2)}{(s + \omega_{zi})(s^2 + 6\frac{1-\alpha}{1+\alpha} f_s s + 12 f_s^2)}. \quad (17)$$

For $\alpha > 1$, T_{icl} has two poles in the right-half s-plane. Therefore, both inner and outer loops are unstable. Bode plots of T_{icl} for the boost converter example given in [6] are presented in Figs. 10 and 11, and has been verified by non-linear PWM converter simulation in PSpice.

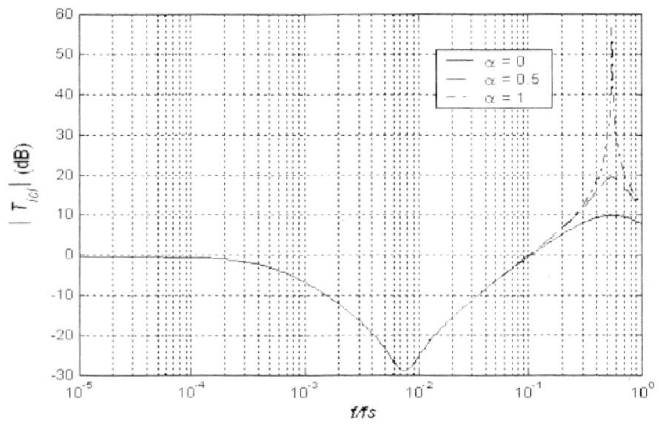

Fig. 10. T_{icl} magnitude response.

3. CONCLUSIONS

The sample and hold effect of constant-frequency PCM in PWM DC-DC converters has been presented in a manner consistent with the physical behavior of the circuit. Transfer function H_{sh} has been introduced to model this behavior, and is placed in the forward path of the current loop, as it must be to create an accurate multiple loop linear system circuit. The stability of the current loop is dependent only upon α in both z and s-domains, regardless of converter topology. The current loop is unstable for $\alpha \geq 1$. The current-loop gain T_i behaves as an integrator for all PWM converter topologies. The phase response of T_{icl} demonstrates the lead compensation given by the current loop.

4. REFERENCES

[1] M. K. Kazimierczuk, "Transfer function for current modulator in PWM converters with current-mode control," *IEEE Trans. Circuits Syst.*, Part 1, vol. 47, pp. 740-746, May 2000.

[2] R. B. Ridley, "A new, continuous-time model for current-mode control," *IEEE Trans. Power Electron.*, vol. 6, pp. 271-280, April 1991.

[3] R. Tymerski and D. Li, "State-space models for current programmed pulsewidth-modulated converters," *IEEE Trans. Power Electron.*, vol. 8, pp. 271-278, July 1993.

[4] G. Verghese, C. Bruzos, and K. Mahabir, "Averaged and sampled-data models for current mode control: A re-examination," *IEEE Power Electron. Specialists Conf.*, 1989, pp. 484-491.

[5] F. Dong Tan and R. D. Middlebrook,"A unified model for current-programmed converters," *IEEE Trans. Power Electron.*, vol. 10, no. 4, pp. 397-408, July 1995.

[6] B. Bryant and M. K. Kazimierczuk, "Small-signal duty cycle to inductor current transfer function for boost PWM DC-DC converter in continuous conduction mode," *IEEE ISCAS*, 2004.

The Implementation of a Transient DC-link Boost Based Digital Amplifier for Eliminating Pulse-Dropping Distortion

Y.C. Chiu*, B. Zhou**, Henry S.H. Chung*†, and W.H. Lau**
*Department of Electronic Engineering
**Department of Computer Engineering and Information Technology
City University of Hong Kong
83 Tat Chee Avenue, Kowloon Tong, Kowloon, Hong Kong
†Fax.: (852) 2788 7791
†Email: eeshc@cityu.edu.hk

Abstract – **A transient dc-link boost technique for avoiding pulse dropping in digital amplifiers is presented. The concept is based on increasing the dc link voltage and reducing the modulation index momentarily, when the modulating signal is large. Apart from effectively reducing harmonic distortion at the output, the proposed method can also widen the dynamic modulation range and thus reduce the required supply voltage. A 10W prototype with an H-bridge inverter has been built. Some pulses are artificially dropped in the gate signals to simulate the pulse-dropping effect. The inverter output harmonic spectra with and without the transient boost have been compared.**

I. Introduction

Nowadays, digital amplifier is widely used in consumer applications. The most common realization approach is to convert the pulse-code-modulated (PCM) digital audio data, as obtained from compact discs (CD), into its corresponding pulse-width-modulated (PWM) signal, which is then applied to an H-bridge inverter. The loudspeaker is connected to the inverter output via a low-pass filter [1]. In order to fulfill high-fidelity requirement, it requires performing a high-resolution PCM-to-PWM conversion. For the 16-bit audio data and 44.1 kHz sampling frequency used in CD, the resolution of the PWM signal is $(44.1 \times 10^3 \times 2^{16})^{-1}$ = 346 ps, which is extremely difficult to be achieved, even for low power applications. A viable solution is to apply oversampling and noise shaping to the PCM data [2]. For example, the one-bit digital-to-analog conversion (DAC) chip that uses an oversampling ratio of 256 to convert the 16-bit PCM data to a 1-bit output for audio application is commercially available and widely used in CD players. However, the switching frequency is too high for high power applications, due to the gate input capacitance, output capacitance, finite rise time and fall time of the semiconductor switches [3]. Those intrinsic parameters limit the minimum pulse width in the output waveform, and thus the maximum modulation index is confined. A high value of modulation index will make the pulse width of the gate signals narrow, which results in dropping pulses at the output and distorting the output spectra. Thus, the maximum modulation index used in many designs is usually to be between 0.1 and 0.2. The modulation range is not fully utilized and the required supply voltage level is higher than the one operating with a wider modulation range under the same output power.

This paper presents a transient dc-link boost technique to avoid pulse dropping in digital amplifiers. The concept is based on boosting the dc link voltage and reducing the modulation index momentarily, when the modulating signal is large. Apart from lowering harmonic distortion at the output, the method can also widen the dynamic modulation range and reduce the required supply voltage. A 10W prototype has been built. Some pulses are artificially dropped in the gate signals for simulating the pulse-dropping effect. The output harmonic spectra of the prototype with and without the transient boost have been compared.

II. Mathematical Analysis

A. Relationships among the supply voltage, modulation index, and minimum pulse width

Narrow pulses appear when the modulating signal v_m is at the maximum or minimum. Consider a switching cycle shown in Fig. 1, when v_m is at the maximum. The average voltage $\hat{v}_o$ of the pulse is equal to

$$\hat{v}_o = (2d-1)V_d \quad \Rightarrow \quad d = \frac{1}{2}(1+M) \tag{1}$$

where $d = t_1 / T$ is the ratio of the duration of the pulse at the positive value (i.e., t_1) to the period T, V_d is the dc link voltage, and $M = \hat{v}_o / V_d$ is the modulation index.

Thus, the minimum pulse duration t_{min} is equal to

$$t_{min} = [1 - \frac{1}{2}(1+M)]T = \frac{1}{2}(1-M)T \tag{2}$$

t_{min} decreases as M increases, giving the required minimum response time of the devices. If the switch cannot respond to t_{min}, pulse dropping occurs. For a given output power, a feasible way of reducing M is to increase V_d. Based on (2), the ratio of increment in t_{min} to the ratio of increment in V_d is

$$\phi = \frac{t_{min,2}}{t_{min,1}} = \frac{1 - M_1/\Psi}{1 - M_1} \tag{3}$$

where $\Psi = V_{d,2} / V_{d,1}$, M_1 is the modulation index when the supply voltage is $V_{d,1}$, $t_{\min,2}$ is the minimum pulse duration when the supply voltage is $V_{d,2}$.

Fig. 2 shows the relationships between ϕ and Ψ. Significant increment in ϕ appears, when Ψ is between 1 and 3.

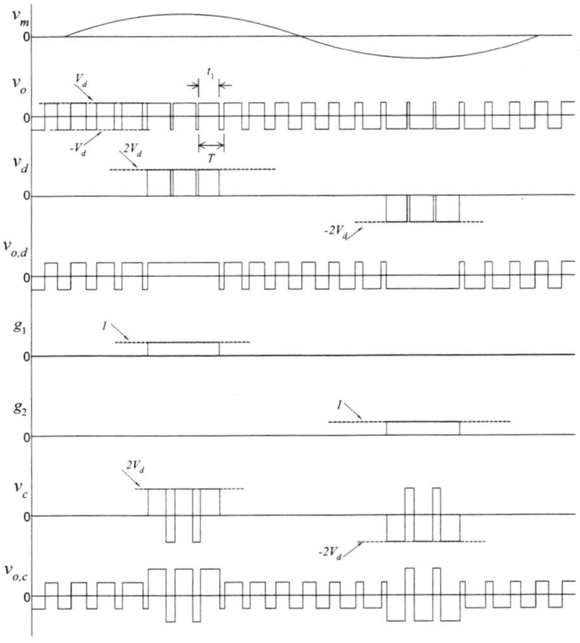

Fig. 1 Waveforms of v_m , v_o , v_d , $v_{o,d}$, g_1 , g_2 , v_c , and $v_{o,c}$.

B. Harmonic spectra of the output

Fig. 1 shows the ideal output waveform v_o and the one with dropped pulses $v_{o,d}$. The spectral function of v_o is

$$\begin{aligned} v_o(t) &= F(\omega_c t, \omega_v t) \\ &= V_d \; M \cos(\omega_v t) + A(t) \end{aligned} \tag{4}$$

where ω_c is the switching frequency and ω_v is the modulating signal frequency,

$$\begin{aligned} A(t) &= \sum_{m=1}^{\infty} \frac{4V_d}{m\pi} \sin\frac{m\pi}{2} J_0(\frac{m\pi M}{2}) \cos(m\omega_c t) \\ &+ \sum_{m=1}^{\infty} \sum_{n=\pm 1}^{\pm\infty} \frac{4V_d}{m\pi} \sin\frac{(m+n)\pi}{2} J_n(\frac{\pi m M}{2}) \cos(m\omega_c t + n\omega_v t) \end{aligned}$$

$v_{o,d}$ is derived by superimposing $v_o(t)$ and $v_d(t)$ in Fig. 2,

$$v_{o,d}(t) = v_o(t) - v_d(t) \tag{5}$$

where $v_d(t) = v_o(t)[g_1(t) + g_2(t)] - V_d [g_1(t) - g_2(t)]$.

The spectral function of $v_{o,d}(t)$can be obtained by (4) and (5), in which $g_1(t)$and $g_2(t)$can be expressed as

$$g_1(t) = \frac{\theta}{2\pi} + \sum_{k=1}^{\infty} \frac{2\sin(k\theta/2)}{k\pi} \cos(k\omega_v t) \tag{6}$$

$$g_2(t) = g_1(t + \frac{\pi}{\omega_v}) \tag{7}$$

where θ is the angular duration of the pulse dropping.

$g_1(t)$ and $g_2(t)$ generate harmonics of significant amplitude, which is due to the multiplication of V_d and $g_1(t)$ and $g_2(t)$. $t_{\min}$ can be effectively increased if V_d is increased. The proposed method momentarily double the supply voltage and reduce M by half, if the estimated pulse width is too narrow that pulse dropping may happen. The waveform becomes

$$\begin{aligned} v_{o,c}(t) &= v_o(t)[1 - g_1(t) - g_2(t)] + v_c(t)[g_1(t) + g_2(t)] \\ &= v_o(t) + [B(t) - A(t)][g_1(t) + g_2(t)] \end{aligned} \tag{8}$$

where $v_c(t)$is the waveform having a magnitude of $2V_d$ and modulation index of $M/2$. $v_c(t)$can be expressed as

$$v_c(t) = V_d \; M \cos(\omega_v t) + B(t) \tag{9}$$

where

$$\begin{aligned} B(t) &= \sum_{m=1}^{\infty} \frac{8V_d}{m\pi} \sin\frac{m\pi}{2} J_0(\frac{m\pi M}{4}) \cos(m\omega_c t) \\ &+ \sum_{m=1}^{\infty} \sum_{n=\pm 1}^{\pm\infty} \frac{8V_d}{m\pi} \sin\frac{(m+n)\pi}{2} J_n(\frac{\pi m M}{4}) \cos(m\omega_c t + n\omega_v t) \end{aligned}$$

Eq. (4) shows the spectral characteristics of ideal waveform. As shown in eq. (5), $v_d(t)$ will introduce considerable harmonic components at the output. Compared with (8), the product term $[B(t) - A(t)][g_1(t) + g_2(t)]$ significantly reduce the harmonic components due to inter-modulation. The frequencies with significant amplitude are just inside the high-order band (>300kHz), so it can be filter out by the low pass filter and the signal can retain high spectral behavior. A detailed spectral analysis of the proposed technique is given in [4].

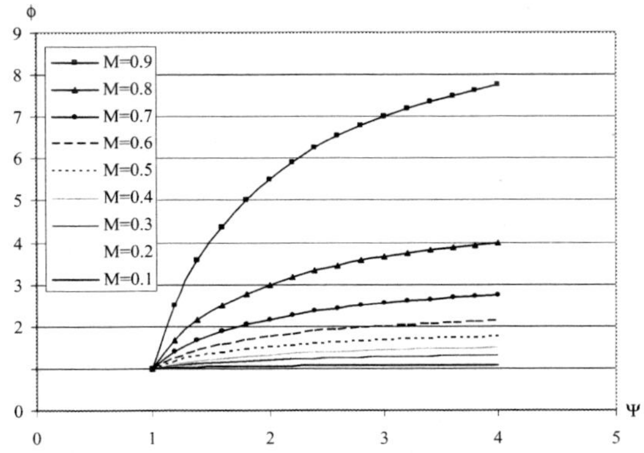

Fig. 2 Relationships between ϕ and Ψ.

III. Prototype

An experimental prototype has been built. Fig. 3 shows the schematic of the prototype. The component values are tabulated in Table I. For the sake of simplicity, two dc supply sources that provide 5V and 10V are used to achieve the transient boost function. The audio digital code is generated from a CD player. The audio data represents a 1kHz pure sinusoidal signal. A digital audio interface receiver CS 8412 converts the data to PCM code. The PCM code is supplied to a Field Programmable Gate Arrays (FPGA) MAX 9400 for voltage transformation and fed to digital audio PWM processor TAS 5000 to generate PWM signals for the switches. The FPGA will generate two sets of PCM coding and two corresponding sets of PWM signals will be generated by the two PWM processors in Fig. 3. The first set is a normal PWM signal with the original modulation index, while the second one is a modified PWM signal with the modulation index reduced by half. The FPGA will estimate the possible width of the pulse generated by the PWM processor. If the estimated pulse width is sufficiently wide, the boost function will not be activated. The first set of PWM signal will be used to drive the switches S_1 to S_4. Conversely, if the pulse width is too narrow (i.e., pulse dropping may occur), the FPGA will generate signals to the multiplexer to select the second set of PWM signal for the switches, as well as the switch S_d to double the dc-link voltage.

Table I – Component values of the prototype

S_1, S_2, S_3, S_4	IRL520N
L_1, L_2	10 μH
C_1, C_2	2.2 μF
S_d	IRF9630
D_d	1N4001
Impedance of the speaker	6Ω

Fig. 4(a) shows the output waveform without pulse dropping, Fig. 4(b) shows the waveform with pulse dropping, and Fig. 4(c) shows the waveform with the transient boost. As circled in Fig. 4(b), the waveform is distorted and suddenly increased at the peak and valley, because pulse droppings occur. With the proposed transient boost [Fig. 4(c)], the original waveform can be reconstructed, as compared with Fig. 4(a). A microscopic investigation into the waveforms is shown in Fig. 5. Pulse dropping causes more energy than required transferring to the load. This will thus increase the instantaneous value of the output [Fig. 5(a)]. With the transient boost, the dc link voltage is changed from 5V to 10V and M is halved [Fig. 5(b)]. As circled in Fig. 5(b), the neighborhood pulses are compared during the transient boost. The pulse width is increased. The total energy transferred to the load in the switching cycle is same as the one without pulse dropping.

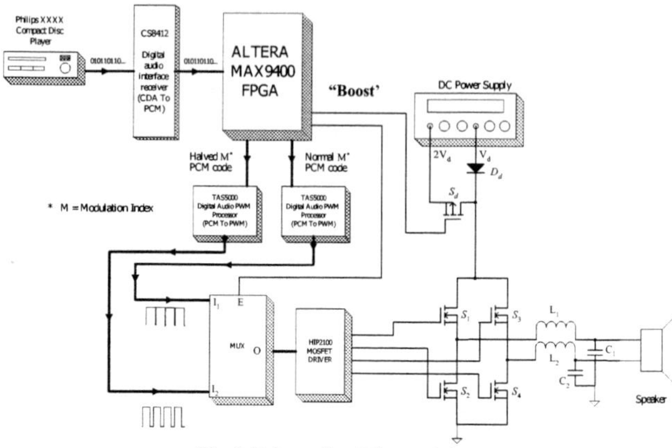

Fig. 3 Schematic of the prototype.

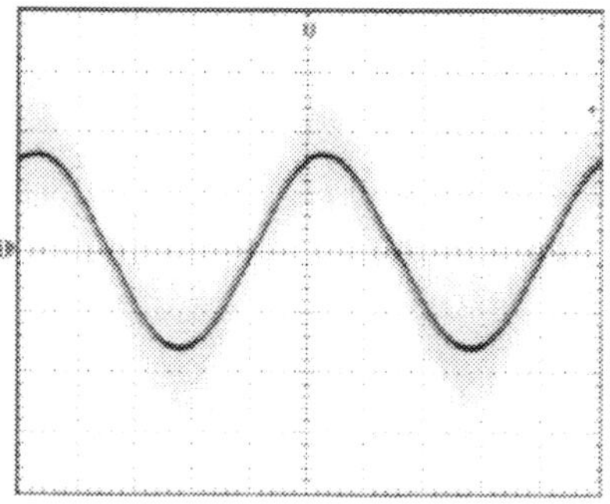

(a) Without pulse dropping.

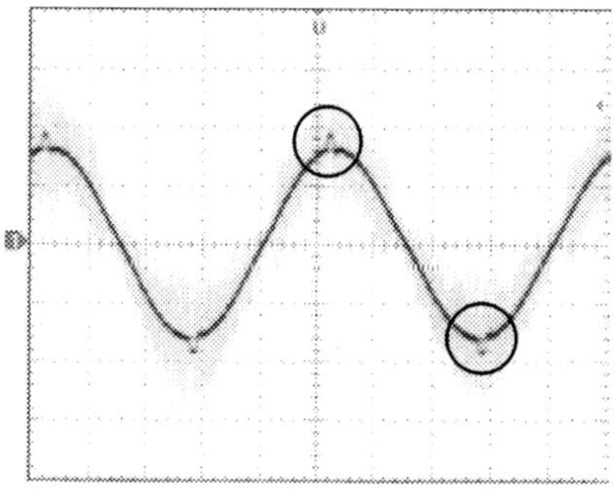

(b) With pulse dropping and without transient boost.

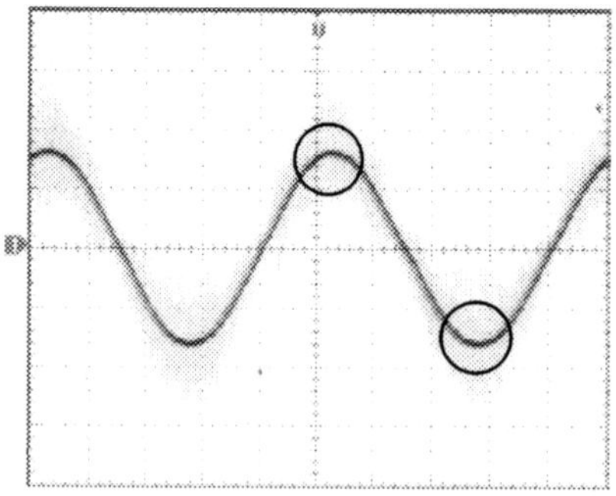

(c) With pulse dropping and transient boost.

Fig. 4 Comparisons of the output (Timebase: 200μs/div, 2V/div).

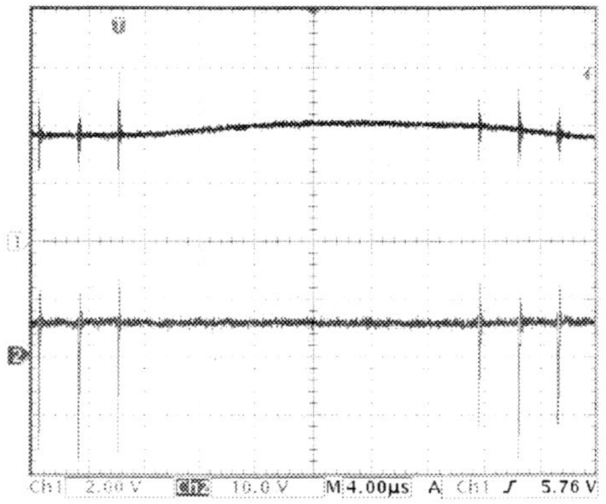

(a) With pulse dropping and without transient boost.

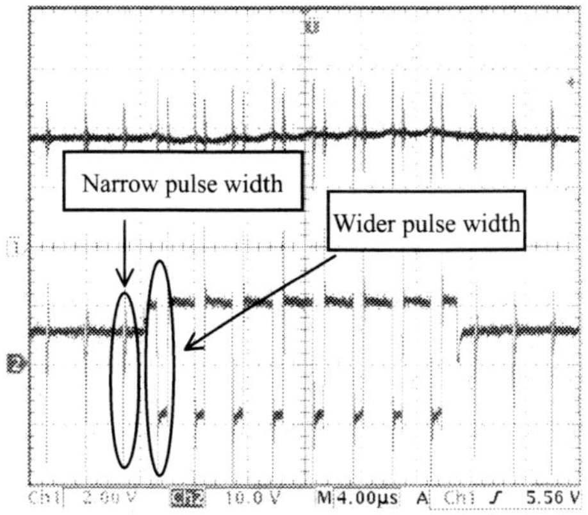

(b) With pulse dropping and with transient boost.

Fig. 5 Microscopic study of the two operations.

Fig. 6 shows the harmonic spectra of the original signal [Fig. 6(a)], with pulse dropping [Fig. 6(b)], and with the transient boost [Fig. 6(c)]. Compared Fig. 6(a) with Fig. 6(b), the noise level is increased by almost 30dB, due to the pulse dropping effect. With the transient boost [Fig. 6(c)], the harmonic spectrum can almost be recovered, demonstrating the effectiveness of the proposed method.

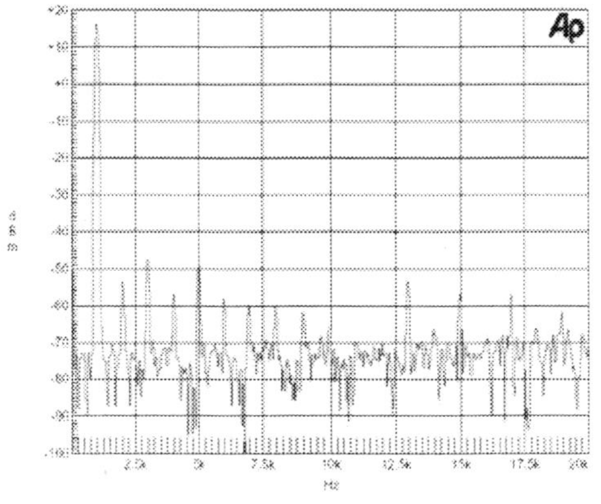

(a) Without pulse dropping.

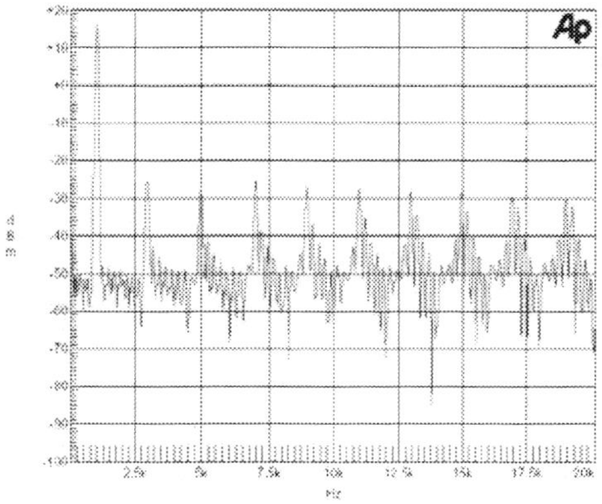

(b) With pulse dropping and without transient boost.

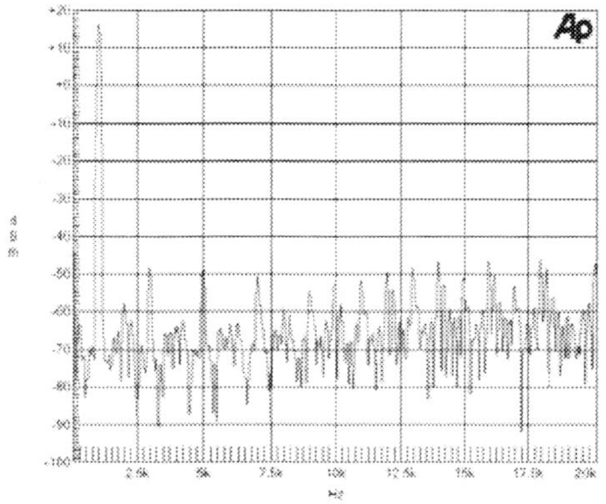

(c) With pulse dropping and with transient boost.

Fig. 6 Comparisons of the harmonic spectra.

IV. Conclusion

A new transient boost technique for digital amplifiers has been proposed. It can solve the problem of pulse dropping and widen the modulation range. Apart from effectively reducing harmonic distortion at the output, the proposed method can also widen the dynamic modulation range and thus reduce the required supply voltage. The effectiveness of the method has been supported by experimental measurements.

References

[1] C. Pascual, *et al*, "High-fidelity PWM inverter for digital audio amplification: spectral analysis, real-time DSP implementation," *IEEE Trans. Power Electron.*, vol. 18, no. 1, pp. 473-484, Jan. 2003.

[2] M. Pedersen and M. Shajaan, "All digital power amplifier based on pulse width modulation," in *The 96th AES Convention*, 1994, preprint 1809.

[3] B. Attwood, "Design parameters important for the optimization of very-high-fidelity PWM (class D) audio amplifiers," *Journal of Audio Engineers Society*, vol. 31, no. 11, pp. 842-853, Nov. 1983.

[4] B. Zhou, Y.C. Chiu, W.H. Lau, and Henry S.H. Chung "Spectral Analysis of A Novel Transient Dynamic Boost PWM Inverter Control For Power Amplifiers", submitted to ISCAS'2004.

ANALYSIS OF CLASS D INVERTER WITH IRREGULAR DRIVING PATTERNS

H. Koizumi, K. Kurokawa

Dept. of Electrical and Electronic Eng.
Tokyo Univ. of Agriculture and Technology
Koganei, Tokyo, 184-8588, Japan

S. Mori

Dept. of Elec. and Electronics Eng.
Nippon Institute of Technology
Minamisaitama, Saitama, 345-8501, Japan

ABSTRACT

This paper presents an analysis of Class D inverter when irregular driving patterns are given to the gate drive of the switch devices. The analysis has been carried out with focusing on the waveforms, harmonics, low frequency components, and equivalent dc resistance, which are numerically analyzed and discussed. Appropriate patterns are sorted based on the calculation results. The waveforms are compared to the circuit experiments. These results corresponded with the calculation. Analytical results show a possibility of a novel control method with irregular driving patterns. In spite of discontinuous control, the output power or voltage can be strictly changed using the selected driving patterns.

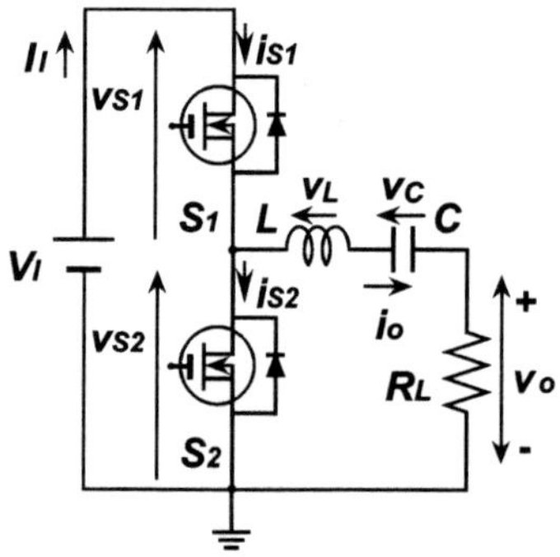

Fig. 1. Class D inverter.

1. INTRODUCTION

Class D Inverter is one of the high-frequency high-efficiency resonant power sources, which has been applied to dc/dc resonant converters, radio transmitters, and electronic ballasts for fluorescent lamps. Its high dc/ac power conversion efficiency is achieved by the low switching losses, which enables its operation at several hundred kHz. To control its output power or voltage, the operating frequency is varied (FM control). In this study, instead of a uniform driving pattern, irregular driving patterns are given to the gate driving of the switch devices (MOSFETs), which can be a novel method to control its output power or voltage. It has been pointed out that these patterns form various waveforms which include harmonics and low frequency components. In this paper, an analysis of Class D inverter with irregular driving patterns is carried out with numerical calculation method [1]. The waveforms are compared with circuit experiments. The various waveforms, THD (total harmonic distortion) and low frequency components are calculated and discussed. In addition, we focus on the equivalent dc resistance as an important parameter in case when an inverter is applied to nonideal dc power sources, for example photovoltaic systems. The variation of the equivalent dc resistance corresponding to the driving patterns is clarified. A similar control way was applied to Class DE inverter [2]. However, it couldn't prevent a small increase of power dissipation of switching loss, because its high efficient operation at a few mega hertz is on the assumption of zero-voltage and zero-current switching. On the contrary, Class D inverter doesn't require zero-voltage switching and its operating frequency is less than that of Class DE inverter. Therefore, it can operate with keeping high power conversion efficiency against various switching patterns. Analytical results show a possibility of a novel control method with irregular driving patterns.

2. CIRCUIT DESCRIPTION

A Class D voltage-switching series resonant inverter [3] [4], which is fed by a dc voltage source, is shown in Fig.1. It is composed of two switch devices S_1, S_2, a band pass filter (BPF) L-C, and a load resistance R_L. The pair of switch devices performs as one switch which alternately connects with the dc voltage source V_I and an earth with keeping 0.5 duty ratio; i.e. while one device is ON the other is OFF in a half period. In the regular operation, it keeps alternate operation by one period. This operation generates a uniform square waveform v_{S2} at the input terminal of the BPF at the operating frequency. An impedance of the BPF at the switching frequency is zero, therefore only the fundamental component of the square waveform flows to the load resistance in ideal. The output current i_o becomes sinusoidal and it flows through each switch by a half period. As shown in Fig. 2 while the current is flowing through one switch, the voltage across the switch device is zero, and the switching transition occurs at the zero current point, therefore the 100-percent power conversion efficiency can be achieved.

In proposed method, the circuit topology itself is the same to the conventional Class D inverter, however the uniform driving pattern is rearranged, i.e. ON or OFF state sometimes appears in series. Regarding the square waveform patterns as binary codes, in case of normal pattern, those can be written as [1 0 1 0....], which means [V_I 0 V_I 0 ....] of the bottom switch voltage v_{S2} in Class D inverter. In this study, instead of the regular pattern, irregular patterns are given, for example [1 1 0 0....] and [1 0 0 0....]. Each pulse keeps the half period of the operating cycle.

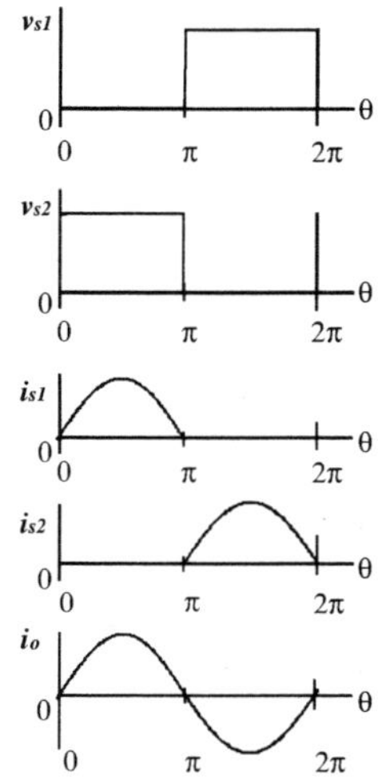

Fig. 2. Waveforms in Class D inverter.

3. DESIGN EQUATIONS

A Class D inverter is designed following the conventional design equations assuming 100 percent power conversion efficiency [4]. The following parameters are given; the switching frequency f=200 kHz, input voltage V_I=10.0 V, load resistance $R_L = 5.0\Omega$, and loaded quality factor of the series resonant circuit Q_L=10. The resonant inductance and capacitance are calculated by the given parameters; thus $L = Q_L R_L/\omega = 39.79\mu$H and $C = 1/\omega^2 L = 1/\omega Q_L R_L = 15.9$ nF where ω means operating angular frequency. These values are used in both the numerical analysis and the circuit experiments.

4. ANALYSIS

In regular operation, waveforms of Class D inverter are simply given based on the hypotheses of an infinite loaded quality factor Q_L and ideal sinusoidal output current. By the uniform driving operation, the input voltage of the BPF, i.e. v_{S2} is a square wave,

$$v_{S2} = \begin{cases} V_I & (0 \le t < \frac{\pi}{\omega}) \\ 0 & (\pi \le t < \frac{2\pi}{\omega}). \end{cases} \tag{1}$$

According to the assumptions the output voltage v_o becomes the fundamental component of the square wave, that is

$$v_o = V_o \sin \omega t, \tag{2}$$

where its amplitude V_o is obtained by expanding the square wave into Fourier series,

$$V_o = \frac{2V_I}{\pi} \simeq 0.637 V_I. \tag{3}$$

In case when the circuit is driven by irregular patterns, waveforms are calculated from the differential equations. In the circuit of Fig. 1, circuit equations are

$$C\frac{dv_C}{dt} = i_o, \tag{4}$$

$$L\frac{di_o}{dt} = -v_C - R_L i_o + v_{S2}, \tag{5}$$

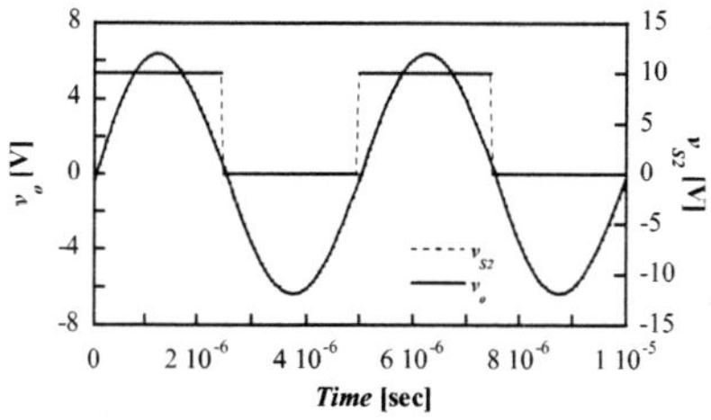

Fig. 3. Calculated waveforms in Class D inverter with the driving pattern [1 0 1 0]; switch voltage v_{S2}, and output voltage v_o.

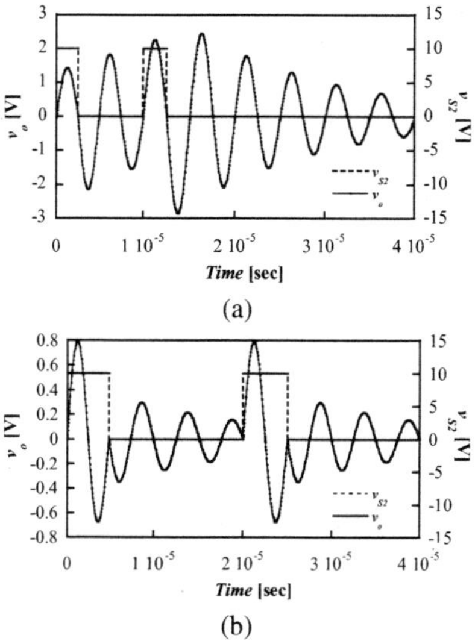

Fig. 4. Examples of calculated waveforms in Class D inverter, with the driving pattern (a) [1000100000000000] and (b) [1100000011000000].

where v_C and v_L are voltages across the capacitance C and the inductance L as depicted in Fig. 1. In (5), v_{S2} takes V_I and zero according to the drive pattern. Thus, there exist two states. that is 'State 1' for $v_{S2} = V_I$, and 'State 0' for $v_{S2} = 0$. These states are symbolized by binary codes '1' and '0' respectively for simplification. Each state is maintained within a half period that is $(0 < t \le \frac{\pi}{\omega})$. According to the continuity of inductor current and capacitor voltage, the initial values of i_o and v_C in a half period use the final values in the previous period. In a steady state, the last values should be equal to the initial values in one cycle. Using this condition, the circuit equations (4) and (5) can be numerically solved with Runge-Kutta formula, then voltage and current waves are obtained. Figure 3 shows the calculated waveforms with a uniform driving pattern [1 0 1 0]. The values of elements, the operating frequency, and the input voltage are based on the designed values in Sec. 3. These results correspond to the theoretical waveforms. Using the same algorithm, various waveforms have been calculated with driving patterns consisting of 16-figures binary codes. Figure 4 shows two examples of calculated waveforms. Obviously, these waveforms include the harmonics and low frequency components. To evaluate these waveforms, several parameters are calculated with considering the total period. Assuming that m-figures binary codes are given as driving patterns, the total period i.e. the sum of m half periods is $mT/2$, where $T = 1/f$ and m is an even number. During the total period, the averaged input current is

$$I_I = \frac{2}{mT}\int_0^{\frac{mT}{2}} i_{S1}(t)dt. \tag{6}$$

Using this, the input power is obtained as $P_I = V_I I_I$ then the equivalent dc resistance is given as $R_{DC} = V_I/I_I$. The output power P_o is

$$P_o = \frac{2}{mT}\int_0^{\frac{mT}{2}} i_o^2(t)R_L dt. \tag{7}$$

As shown in an example of Fig. 4 (b), some patterns make sharp turns. To quantify them, the harmonics and low frequency components are calculated. The fundamental component is on a basis of the operating frequency f, therefore the harmonics are obtained by the following equations with taking account of the total period $mT/2$,

$$a_n = \frac{2}{\frac{mT}{2}}\int_0^{\frac{mT}{2}} x(t)\cos(n\omega t)dt, \tag{8}$$

$$b_n = \frac{2}{\frac{mT}{2}}\int_0^{\frac{mT}{2}} x(t)\sin(n\omega t)dt, \tag{9}$$

where a_n and b_n are Fourier coefficients, and $x(t)$ can be v_o, v_C, v_{S2}, etc. Using (8) and (9), total harmonic distortion (THD) is

$$\mathrm{THD} = \sqrt{\frac{\sum_{n=2}^{\infty}(a_n^2 + b_n^2)}{a_1^2 + b_1^2}}, \tag{10}$$

where a_1 and b_1 are fundamental harmonics. Then, regarding the total period as a basic period, low frequency components are also calculated as

$$c_k = \frac{2}{\frac{mT}{2}}\int_0^{\frac{mT}{2}} x(t)\cos\left(\frac{k}{\frac{m}{2}}\omega t\right)dt, \tag{11}$$

$$d_k = \frac{2}{\frac{mT}{2}}\int_0^{\frac{mT}{2}} x(t)\sin\left(\frac{k}{\frac{m}{2}}\omega t\right)dt, \tag{12}$$

where k is a natural number from 1 to $m/2 - 1$. Referring to (10), a parameter of total low-frequency-components distortion (TLD) which is normalized by the fundamental component is assumed as,

$$\mathrm{TLD} = \sqrt{\frac{\sum_{k=1}^{\frac{m}{2}-1}(c_k^2 + d_k^2)}{a_1^2 + b_1^2}}. \tag{13}$$

In a similar way, total distortion (TD) is obtained with (10) and (13),

$$\mathrm{TD} = \sqrt{\frac{\sum_{n=2}^{\infty}(a_n^2 + b_n^2) + \sum_{k=1}^{\frac{m}{2}-1}(c_k^2 + d_k^2)}{a_1^2 + b_1^2}}. \tag{14}$$

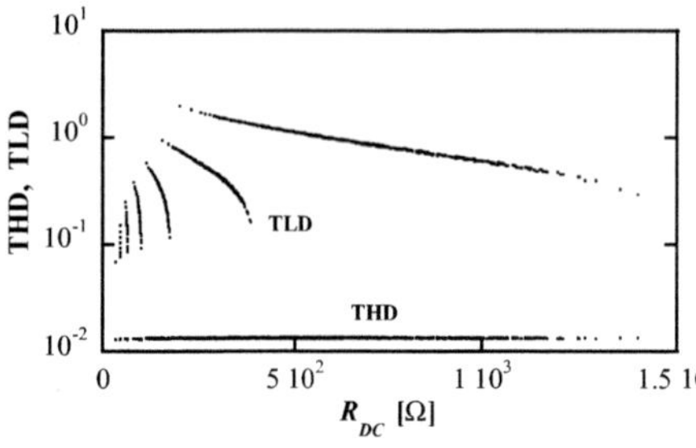

Fig. 5. The extracted THD and TLD data plots.

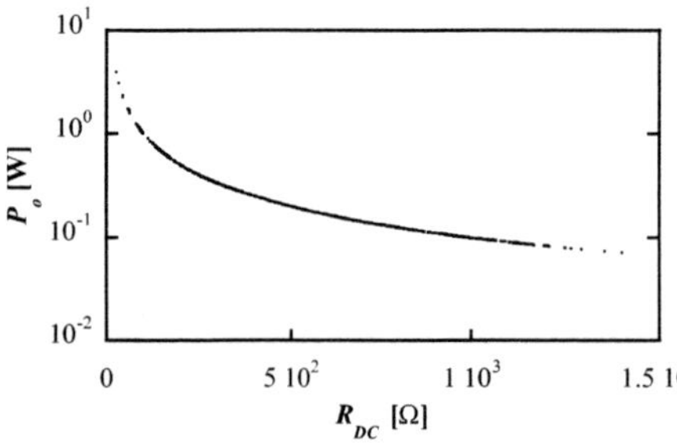

Fig. 6. Relationship between the output power P_o and R_{DC} with selected data plots.

The circuit operation has been analyzed using the above equations with $m = 16$. Each parameter has been calculated in 2^{16} patterns. Substituting the output voltage v_o for $x(t)$, THD, TLD and TD are obtained. The harmonics higher than 14th are ignored. According to the calculation results, these parameters are scattered in wide range, thus THD is from 1.31e^{-2} to 2.36e^3, TLD is from 6.92e^{-2} to 7.79e^9, TD is 7.05e^{-2} to 7.79e^9 and R_{DC} is 32.08 Ω to 9.58 kΩ, except the data of all '1', all '0', and regular '10' patterns. To select appropriate patterns, THD, TLD, and TD were mapped out with reference to a horizontal axis R_{DC}. TD-plots approximately overlap for TLD-plots. TLD is at least five times of THD in each pattern. High THD groups include sharp terns in the waveform as shown in Fig 4 (b), which also have high TLD. We have noticed the groups with lowest values about TLD and THD respectively. The data less than 0.014 for THD and less than 2.02 for TLD are extracted, which are shown in Fig. 5. Using the selected data, a relationship between the output power P_o and R_{DC} is shown in Fig. 6. As is widely alleged, around the regular operation point, that is small R_{DC} and large P_o range, distribution of the plots are obviously discontinuous. It seems impossible to control an inverter's operation smoothly. On the other hand, in larger R_{DC} and small P_o range, the plots exist in high density as if continuous. If an inverter was designed to supply a rated output power in this range, it could be controlled over one digit in case of Fig. 6. If the TLD and THD are within an allowance, above method can be applied to an inverter.

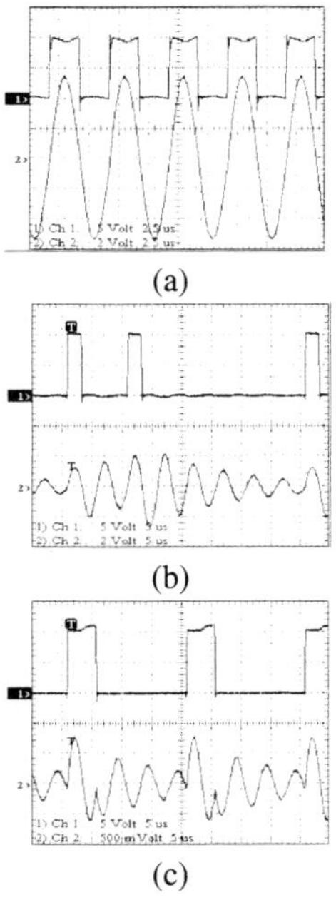

Fig. 7. Observed waveforms of v_{S2} (ch 1) and v_o (ch 2) with patterns of (a) [1010], (b) [1000100000000000], and (c) [11000000].

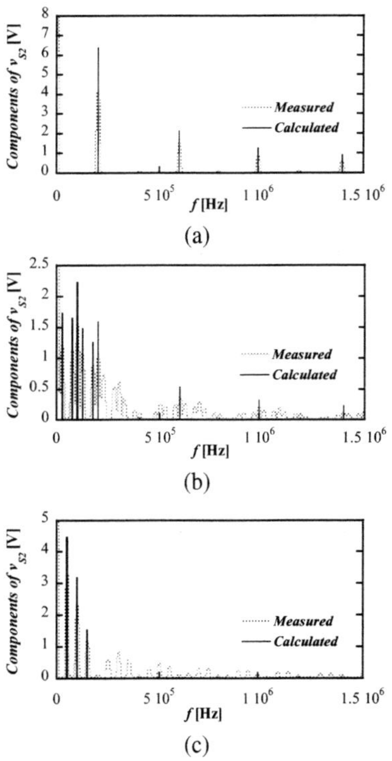

Fig. 8. Harmonics and low frequency components of v_{S2} with patterns of (a) [1010], (b) [1000100000000000], and (c) [11000000].

5. CIRCUIT EXPERIMENTS

An experimental circuit has been built and tested. Measured values of the elements were $L = 39.78$ μH, with 270 mΩ, $C = 16.02$ nF, with 200m Ω, and $R_L = 4.95$ Ω at 200 kHz. Two MOSFETs (IRF510) were used as switch devices. Various driving patterns were applied to the gate drive. Figure 7 shows the observed waveforms under the condition of $f = 200$ kHz, $V_I = 10.0$ V. The input port was connected to a dc-power supply with an input capacitor. The voltage source was kept at 10.0 V in each test. The upper switch S_1 was driven through a transformer. The gate driver was manually regulated to give an intended pattern. Observed waveforms under the regular operation with a pattern of [1 0 1 0] were shown in Fig. 7 (a), which are in agreement with Fig. 3. In this test, the output power 2.80 W was measured with power conversion efficiency 72.7 percent. Figure 7 (b) corresponds to Fig. 4 (a). These waveforms also show good agreement. In Fig. 7 (c), v_{S2} has distortions in the square wave which was caused by the distortions of gate drive waveform through a transformer. Therefore, the observed output voltage waveform does not fully accord with Fig. 4 (b). The output power and efficiency were 0.22 W with 71.4 percent and 0.020 W with 33.0 percent respectively. Except the patterns which cause sharp turns in the output current, the power conversion efficiency could be kept about 70 percent with adjusting the driving wave. In each case, harmonics and low frequency components were measured with FFT function in the oscilloscope. Figure 8 compares the measured data and the calculated ones. The difference of the amplitude is mainly caused by the accuracy of measurements and by using the flat top window. Main components are agreed each other in the outline.

6. CONCLUSIONS

An analysis of Class D inverter with irregular driving patterns has been carried out. The various waveforms, voltage and current parameters, THD, TLD, and equivalent dc resistance have been calculated and clarified. Observed waveforms in the circuit experiment showed good agreement with the calculated ones. The analytical results show that, using the selected low-THD and low-TLD patterns, the output power or voltage can be controlled in wide range as if continuous. The power conversion efficiency can be kept constant because the zero-current switching is maintained against these patterns. Those characteristics show that variation of driving pattern can be applied as a novel control method to Class D inverter.

7. REFERENCES

[1] M. Albulet, "An exact analysis of Class-DE amplifier at any output Q" *IEEE Trans. Circuits Syst. I*, vol. CAS-46, No. 10 pp. 1228-1239, Oct. 1999.

[2] H. Sekiya, M. Matsuo, H. Koizumi, S. Mori, I. Sasase, "New control scheme for class DE inverter by varying driving signals " *IEEE Trans. Industrial Elec.* , vol. 47, No. 6 pp. 1237-1248, Dec. 2000.

[3] P. J. Baxandall, "Transistor sine-wave LC oscillators, some general considerations and new developments, "*Proc. IEE*, vol. 106, pt. B, suppl. 16, pp. 748–758, May 1959.

[4] M. K. Kazimierczuk, D. Czarkowski, "Resonant Power Converters," Willey-Interscience, N.Y. 1995.

A NEW TOPOLOGY FOR A SIGMA-DELTA AUDIO POWER AMPLIFIER

Zorzano Martínez, Antonio

Dep. of Electrical Engineering, University of La Rioja, Logroño, Spain
antonio.zorzano@die.unirioja.es

Beltrán Blázquez, Fernando

Dep. of Electronic Engineering and Communications, University of Zaragoza, Zaragoza, Spain
beltran@posta.unizar.es,

Beltrán Blázquez, José Ramón

Dep. of Electronic Engineering and Communications, University of Zaragoza, Zaragoza, Spain
jrbelbla@posta.unizar.es

ABSTRACT

A new inverter topology based on high frequency power conversion and Sigma-Delta modulation is proposed. Sigma-Delta technique is particularly an attractive idea in audio power amplifying area. Analysis of this technique is presented. Power modulation is a way to obtain power amplifiers with better efficiency than conventional lineal power amplifiers. A quasi-resonant converter with Sigma-Delta modulation input, has been integrated to reduce switching losses, allowing high frequency operation. Audio power amplifier and isolated high power supply are integrated in one unit and analysed. Simulations are performed to backup the analysis.

1. INTRODUCTION

In the high fidelity audio power amplification field, class-D is currently adopted. Efficiency is the great advantage of this class. Amplification method is totally different than any other classes (class A, AB,...). In the figure Fig. 1, a general scheme of a Class D amplifier is showed.

Two are the most adopted methods to develop a class D amplifier. In the first one, the original analog line-level signal is voltage amplified and converted to a high frequency square wave, with the analog signal modulated in the duty cycle of the square wave (pulse width modulation, PWM). In the second one, a discrete sequence of values, with its time-density distribution represents the instantaneous amplitude of analog input signal (Sigma-Delta modulation, $\Sigma\Delta$) is voltage amplified.

Adoption of a high sampling frequency is desirable to improve base band performance. Power losses limit this performance. Soft switching schemes [1], and particularly quasi-resonant converter, enables high speed of switching and low losses ratings for the power devices. A single-cycle quasi-resonant converter (SC-CQR) is considered here. In the single cycle operating mode, free wheeling and the resonance (positive or negative) stage, appear in the power converter.

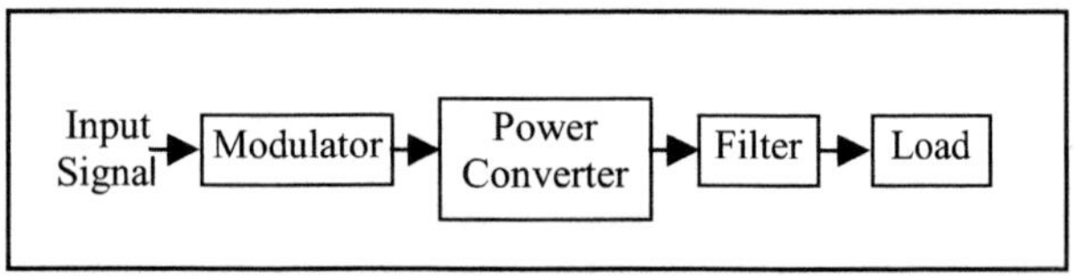

Fig. 1. General scheme of Class D amplifier

This paper proposes a new topology for a Sigma-Delta Audio Power Amplifier based on quasi-resonant converter with a high frequency operation. Section 2 is a background of knowledge of resonant converter and soft switching. Section 3. describes the new topology. Section 4. and section 5., show Sigma–Delta Modulator and Power Amplifier, respectively. Section 6 analyses results of the stage.

2. RESONANT CONVERTER

Resonant Power Converters are a class of converters, whose operation differs significantly from the rest of switching converters [2]. They contain a resonant L-C network, whose voltage and current waveforms vary sinusoidally during one or more intervals of each switching period. The network resonant frequency is tuned to the fundamental component of input network signal. Network response is negligible at the harmonics of fundamental component of input network signal.

Quasi-resonant converter (CQR) is obtained if resonant elements (L_r, C_r), are inserted into the switch network of a switching converter (PWM converter, $\Sigma\Delta$ converter, ...). Here, L_r and C_r are combined with the switch network. The resonant frequency of these elements is higher than the switching frequency. This causes voltage waveform $V(C_r)$ to become quasi-sinusoidal pulses.

The chief advantage of resonant and quasi-resonant converters, is their reduced switching loss, via Zero Voltage Switching (ZVS) and Zero Current Switching (ZCS). Main components of switching loss are the loss induced by the diode reverse recovery process and the loss of the energy stored in the MOSFET output capacitance. Soft switching of converters lowers switching losses allowing higher frequency operation and reduced electromagnetic interference [3]. For Single Cycle CQR [4], conversion ratio depends only of the ratio between switching frequency (f_s) and the resonance frequency (f_r). Positive and negative resonance stages will be performed according to the bit stream instantaneous modulation value. A power level reconstruction of the signal codified is obtained at the output of the stage.

For each switching time period a duty cycle δ (ratio between resonance and switching period), results [5]. So, if the input bit stream of the modulator is bipolar and its mean value is "m" ($-1 < m < 1$), converter gain will be

$$\delta = \frac{T_r}{T_s} \qquad d = \frac{\delta}{1-\delta} \qquad (1)$$

and output voltage will be

$$V_o = mdV_{dc} \qquad (2)$$

where V_{dc} is the DC voltage source.

3. NEW TOPOLOGY

Typical audio amplifier has two units: an isolated power supply and an audio power amplifier. It has a significant cost. Objective is to obtain a power supply isolated switch -mode audio power amplifier capable of driving any load with high bandwidth, low distortion, and high efficiency in one unit.

A new topology for Sigma-Delta Audio Power Amplifier has been designed (Fig. 2). Scheme comprising a modulator, logic controller, a pulse transmitting, a power switch, power transformer, a quasi-resonant converter and a low pass filter. Output section is electrically isolated from the power source.

Scheme satisfies ability of feeding a bipolar output voltage and bidirectional output current. Sigma-Delta Modulator produces pulses with a duration time of T_s (switching period). Pulses, with the same polarity of Sigma-Delta Modulator pulses, are sent to power switch (M1). Secondary section of transformer is split into two equal and opposite windings. Without a bridge type topology, load polarity can be selected.

The duration time of power switch pulses is the same of the resonance time period (T_r):

$$T_r = 2\pi\sqrt{L_r C_r} \qquad (3)$$

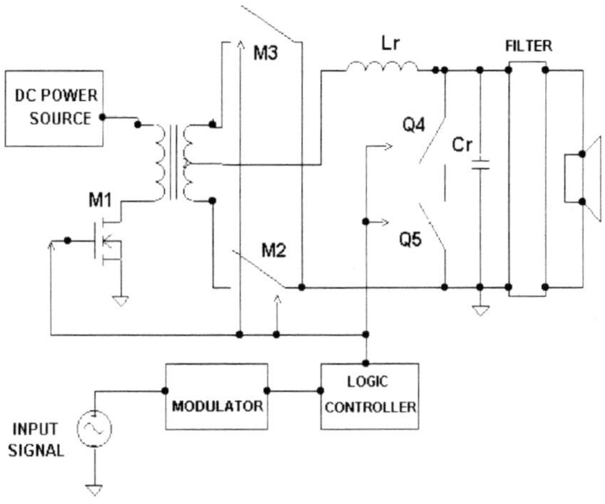

Fig. 2. New Topology for Sigma-Delta Audio Power Amplifier

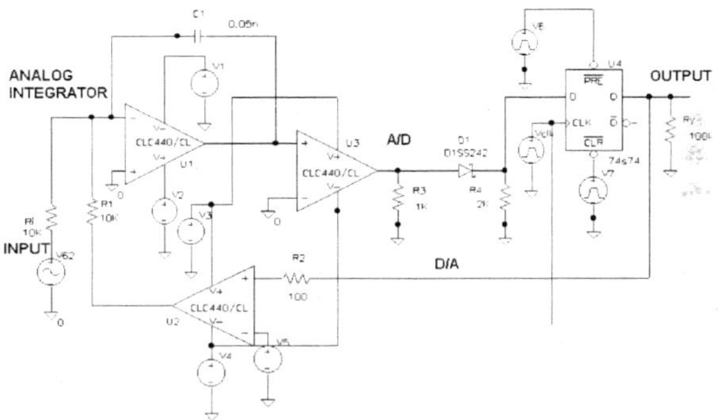

Fig. 3. First Order Sigma-Delta Modulator.

At the end of this time, power switch is switched off and a free wheeling period is performed.

4. SIGMA-DELTA MODULATOR

Scheme of a simple Sigma-Delta Modulator is simple [6]. The analog input is fed to an integrator, whose output is fed to a quantizer (A/D converter). The output of the quantizer is fed back and subtracted from the analog input. This feedback forces the average value of the quantized signal to track the average input. The output of the modulator is a digital representation of the signal, which can de demodulated by smoothing the impulses in a low pass filter. ΣΔ is an example of pulse-density modulation [7]. The number of pulses per unit time increases with the magnitude of the input.

A first order Sigma-Delta modulator circuit has been designed (Fig. 3). Modulator integrates an analog integrator, a comparator, a subcircuit for synchronized signal generator. Minimum period of output impulses is 500 ns. . Fig. 4 shows a sinusoidal input signal, and its Sigma-Delta modulated signal.

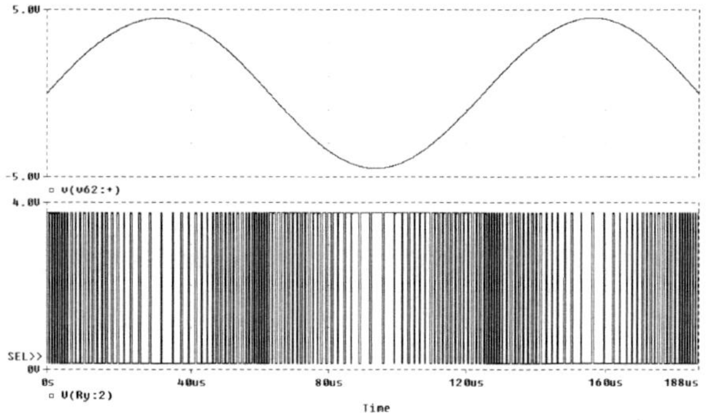

Fig. 4. Input Signal and Sigma-Delta Modulated Signal

5. POWER AMPLIFIER

A Sigma-Delta power amplifier based on the described topology is showed in Fig. 5. The amplifier operates at f_s = 2 MHz (T_s = 500 ns.). A series resonant circuit is created across L_r and C_r. The duration time of resonance cycle is 250 ns.. In the single cycle operation mode, free wheeling, positive resonance and negative resonance appear in the power amplifier. During each switching period, two subintervals then occur. In the first subinterval a free wheeling operation is performed (250 ns.). The second subinterval let generate an output power pulse which has the polarity required by the bit stream of Sigma-Delta modulator. At the end of the resonant time a free wheeling stage is performed again.

Free Wheeling: This subinterval begins when transistor M1 is switched off. Transistors M2, M3, Q4 and Q5, allow creation of zero voltage switching (ZVS) and zero current switching (ZCS) (Fig. 5 a).

Positive Resonance: When Sigma-Delta modulation is a positive value, a positive pulse is generated at the power amplifier output. A dc voltage is applied to the transformer primary winding. During this subinterval, transistor M1 conducts, while M3, Q4, Q5 are off. See Fig. 5. b. M2 conducts, hence current flowing into the positive polarity of load.

Negative Resonance: During this subinterval M1 conducts, while transistors M2, Q4 and Q5 are off. M3 conducts, hence current flowing in the negative polarity of load (Fig. 5. c).

A low pass filter must follow quasi-resonant converter to eliminate the high frequency harmonics and pass the audio signal to speaker. The filter chosen here is a 4 pole Butterworth low pass.

6. RESULTS

Fig. 6 presents detailed Power Amplifier Circuit. Decoupling capacitors C33 and C44 have been included in the design. High Switching MOSFETs were selected.

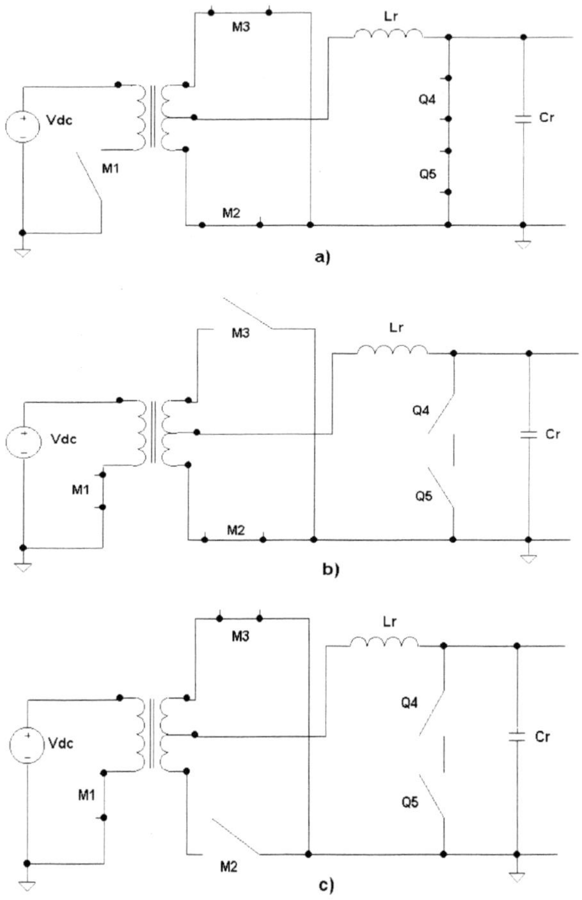

Fig. 5. a) Free Wheeling, b) Positive Resonance and c) Negative Resonance

Fig. 7 shows, for different values of resistive load, power in speaker (watts) and amplification ratio Vo/Vi (dB.). Either power or A_v, presents a similar characteristic performance for different resistive load. A sinusoidal signal had been used as input signal, with frequency 10KHz, and amplitude 4.5v. An unipolar dc power supply of 32 volts has been used in primary winding. Fig. 8 shows input signal and speaker current for a 20 watts output power case. A little delay can be observed between input signal and speaker current. Besides switching frequency (2 MHz) is higher enough input signal frequency (10KHz), spectrum of Sigma-Delta modulated signal contains harmonics in bandwidth between 200KHz and 2 MHz. Anyway, these harmonics have been eliminated by pass filter. Total Harmonic Distortion (THD) have reached similar levels to other switching audio power amplifiers (PWM or ΣΔ). For similar levels of dc power supply, this stage obtains a high level of power in speaker. All simulations have been made with PSPICE ®.

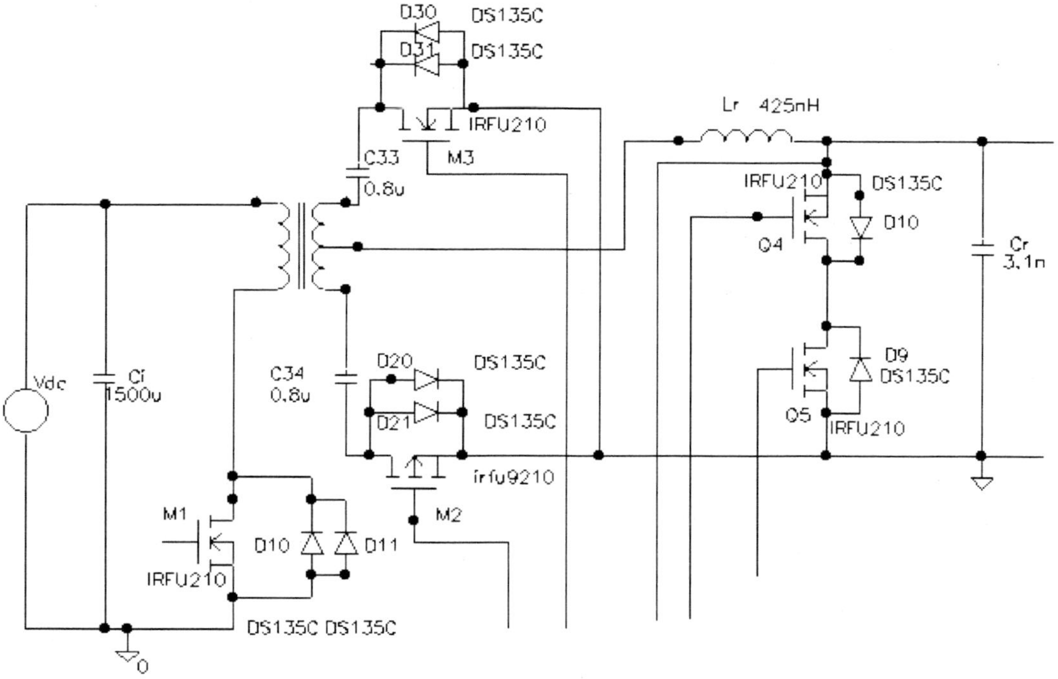

Fig. 6. Power Amplifier.

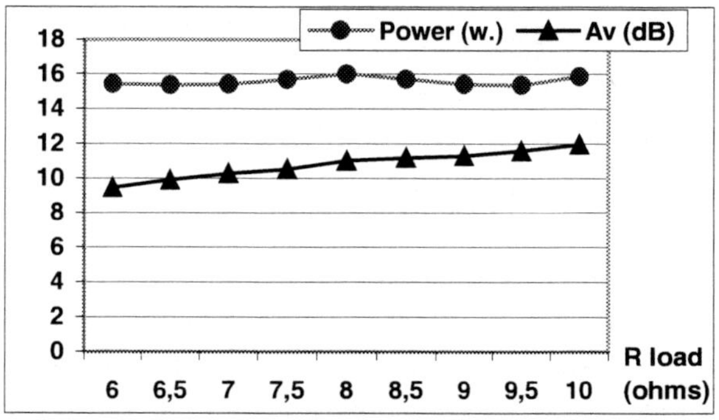

Fig. 7. Power and V_o/V_i (dB) in speaker for different values of resistive load.

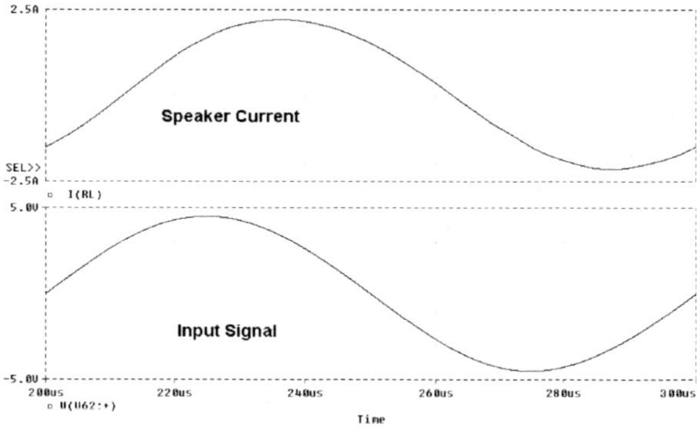

Fig. 8. Speaker Current and Input signal

7. REFERENCES

[1] K. Mark Smith and K M. Smeadly, "Properties and Synthesis of Passive, Lossless Soft Switching PWM Converters", EPMC '97, 1997.

[2] R. Erickson and D. Maksimovic, *Fundamentals of Power Electronics*, Editorial Kluwer Academic Publishers, ISBN 0-7923-7270-0. 2001.

[3] K. Mark Smith and K M. Smeadly, "Lossless, Passive Soft Switching Methods for Inverters and Amplifiers", *IEEE PESC Proceedings*, 1997.

[4] E. Dallago, R. Quaglino, and G. Sassone, "Single-Cycle Quasi-resonant Converter with Controlled Timing of the Power Switches", *IEEE Transactions on Power Electronics*, vol. 11, no. 2, March 1996.

[5] E. Dallago and G. Sassone, "Advances in High Frequency Power Conversion by Delta-Sigma Modulation", *IEEE Transactions on Circuits and Systems I, Fundamental Theory and Applications*, Vol. 44, no. 8, August 1997.

[6] J. Candy and O. J. Benjamin, "The Structure of Quantization Noise from Sigma-Delta Modulation" *IEEE Transactions on Communications*, vol. COM-29, no. 9, September 1991.

[7] J. Paramesh and A. Von Jouanne, "Use of Sigma-Delta Modulation to Control EMI from Switch–Mode Power Supplies", *IEEE Transactions on Industrial Electronics*, vol. 48 no. 1, February. 2001.

Spectral Analysis of A Novel Transient Dynamic Boost PWM Inverter Control For Power Amplifiers

B. Zhou*, Y.C. Chiu†, W.H. Lau*, and Henry S.H. Chung†
*Department of Computer Engineering and Information Technology**
Department of Electronic Engineering†
City University of Hong Kong
83 Tat Chee Ave., Kowloon, Hong Kong

Abstract - This paper presents a detail spectral analysis and the experimental results of a new topology to solve the distortion due to pulse dropping. Owing to the non-ideal switching characteristics of the components, pulse dropping occurs since the devices cannot produce a very narrow pulse at the output of a PWM inverter. This causes spectral distortion and is especially unacceptable for digital audio amplifier application. A new topology called Transient Dynamic Boost has been developed to solve this problem. By introducing a compensation signal when the amplitude of the modulating signal reaches a critical level, the output signal will remain undistorted with an appropriate DC link voltage is applied. Both the theoretical analysis and experimental results have verified the validity of the new technique.

1. Introduction

PWM inverter has found numerous applications in both industry and domestic market. In particular, PWM inverter becomes popular in digital audio power amplifier design owing to the rapid development in digital audio technology [1,2]. In order to reproduce the audio signal with high-fidelity, a 8 times oversampling will usually be applied to the PCM audio signal and the resulting sampling frequency becomes 352 kHz. The oversampled PCM data will then be converted PWM format to drive a high power H-bridge circuit. It should be noted that the PWM signal generated is uniformly sampled in nature and hence requires noise-shaping to suppress the in-band undesirable harmonics and quantization noise. In order to cater for the practical design problems, the highest resolution is usually restricted to 8-bit, representing a narrowest pulse width of 11 nsec.

One of the most important considerations in designing digital audio power amplifier is the switching characteristics of the gate drivers and power devices. It is well known that the higher the power rating of the device, the slower the switching response. If pulse is too narrow, it will just disappear at the inverter output and this phenomenon is called *pulse dropping*. One simple solution to this problem is to limit the modulation index, i.e., the maximum output voltage is restricted not to reach the DC link voltage and leaving a safe margin to avoid pulse dropping. Practically speaking, 11 nsec is an extremely narrow pulse for most of the power devices and some manufacturers actually restrict the signal amplitude and effectively reduce the modulation index.

The work described in this paper is fully supported by a research grant (CityU 1192/99E) from the RGC of the HKSAR, China.

In this paper, we will introduce a novel design, called *Transient Dynamic Boost*, to make full use of the DC link voltage and at the same time to avoid pulse drooping. The new design is for a 2-level PWM H-bridge inverter. A detail theoretical spectral analysis to examine the pulse dropping effects and the compensation due to the transient dynamic boost will be presented to justify the new topology. Various considerations for determining the control parameters are also discussed. Experimental results will be presented to justify the new design technique.

2. Spectral Analysis of Pulse Dropping

To examine the spectral characteristics of a PWM signal, lets consider that the modulating signal is represented by $v_s = \cos\omega_v t$ and ω_v is the angular frequency, and the angular frequency of the carrier is ω_c. The uniformly sampled PWM, converted from the CD PCM data, can be approximated by the naturally sampled PWM after oversampling and noise-shaping process. The ideal double-sided naturally sampled 2-level PWM [3] and is given in (1).

$$v_p(t) = F(\omega_c t, \omega_v t) = \frac{V_d M}{2}\cos(\omega_v t) + A(t) \tag{1}$$

with

$$\begin{aligned} A(t) = & \sum_{m=1}^{\infty} \frac{2V_d}{m\pi}\sin\frac{m\pi}{2} J_0\left(\frac{m\pi M}{2}\right)\cos(m\omega_c t) \\ & + \sum_{m=1}^{\infty}\sum_{n=\pm 1}^{\pm\infty} \frac{2V_d}{m\pi}\sin\frac{(m+n)\pi}{2} J_n\left(\frac{\pi m M}{2}\right)\cos(m\omega_c t + n\omega_v t) \end{aligned} \tag{2}$$

where M is the modulation index, V_d is the DC link voltage of the PWM inverter and $J_n(\bullet)$ is the n-th order Bessel function of the first kind.

Figure 1(a) shows the ideal PWM and Figure 1(b) shows the PWM with pulse dropping. Generally speaking, pulse dropping occurs when the magnitude of the modulating signal exceeds a critical value. For sinusoidal signal, when the signal is close to its peak level, the pulse width becomes too narrow to trigger the switching components.

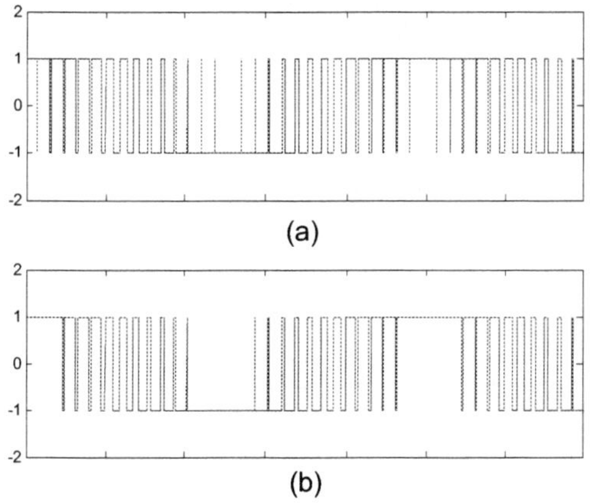

Figure 1. (a) The ideal PWM;
(b) PWM with pulse dropping

In order to obtain a theoretical description of the pulse dropping effects, two pulse trains $g_1(t)$ and $g_2(t)$ shown in Figure 2 will be used to mask the pulses. $g_1(t)$ is responsible for masking the +ve peak and $g_2(t)$ is for the -ve peak. Mathematically, this is equivalent to multiply $g_1(t)$ and $g_2(t)$ to the ideal PWM signal.

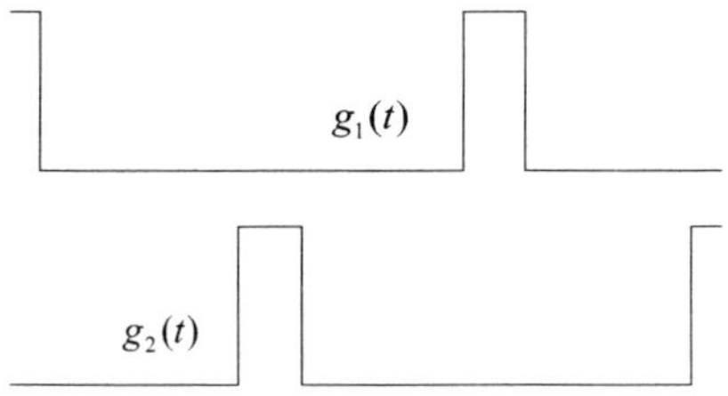

Figure 2. Pulse trains: (a) $g_1(t)$; (b) $g_2(t)$

$g_1(t)$ and $g_2(t)$ are given in (3) and (4).

$$g_1(t) = \gamma + \sum_{k=1}^{\infty} \frac{2\sin(k\pi\gamma)}{k\pi} \cos(k\omega_v t) \tag{3}$$

$$\begin{aligned} g_2(t) &= g_1\left(t + \frac{\pi}{\omega_v}\right) \\ &= \gamma + \sum_{k=1}^{\infty} \frac{2(-1)^k \sin(k\pi\gamma)}{k\pi} \cos(k\omega_v t) \end{aligned} \tag{4}$$

where γ represents the pulse dropping ratio defined as $\gamma = T_1/T_0$, T_1 is the pulse width of $g_1(t)$ and T_0 is the period of the modulating signal. For simplicity reason, the pulse width of $g_2(t)$ is made the same as $g_1(t)$. The pulse dropping effects can conveniently be examined by adjusting γ.

By multiplying $g_1(t)$ and $g_2(t)$ to (1), the pulse dropped PWM is given in (5):

$$\begin{aligned} v_{drop}(t) &= v_p(t)[1 - g_1(t) - g_2(t)] + \frac{V_d}{2}[g_1(t) - g_2(t)] \\ &= v_p(t) + v_{crre}(t) \end{aligned} \tag{5}$$

where,

$$v_{crre}(t) = -v_p(t)[g_1(t) + g_2(t)] + \frac{V_d}{2}[g_1(t) - g_2(t)] \tag{6}$$

Obviously, the second term in (6) is the major source of harmonic distortion. If we substitute (3) and (4) into (5) and expand the result, the pulse dropping effect to the spectral can be clearly seen:

$$\begin{aligned} V_{drop}(t) &= \frac{V_d M(1-2\gamma)}{2} \cos(\omega_v t) \\ &- \sum_{k=1}^{\infty} \frac{V_d M[(-1)^k + 1]\sin(k\pi\gamma)}{2k\pi} [\cos(k+1)\omega_v t + \cos(k-1)\omega_v t] \\ &+ \sum_{k=1}^{\infty} \frac{V_d[1-(-1)^k]\sin(k\pi\gamma)}{k\pi} \cos(k\omega_v t) + A(t)[1 + g_1(t) + g_2(t)] \end{aligned} \tag{7}$$

By comparing (1) with (7), it is observed that the output amplitude of the fundamental signal will be decreased by a factor of 2γ. A clear source of harmonic distortion is directly attributed to the second and third terms of in (7). In addition, the last term in (7) represents the inter-modulated harmonics which will also contribute to the harmonic distortion for certain k, m and n combinations. However, if the carrier switching frequency is very high compared with the modulating signal, the effects of the inter-modulated harmonics to the in-band harmonics will not be as significant as the harmonics generated by the second and third terms in (7). We have theoretically analyzed the distortion due to pulse dropping and the parameters used are $\omega_v = 1kHz$, $\omega_c = 300kHz$ and M=0.98. Figure 3 shows the distortion of the 3rd, 5th and 7th harmonics for different pulse dropping ratio γ. Significant distortions can clearly be seen and in particular the distortion increases substantially for γ rising from 0% to 3%. It is also interesting to observe that these harmonics actually suffer more or less the same distortion for γ up to 4%.

3. Spectral Analysis of the Transient Dynamic Boost

Let's define the pulse dropping ratio β as:

$$\beta = t_1/T = \frac{1}{2}(1+m) \tag{8}$$

where t_1 is the pulse duration in positive cycle, T is the carrier switching period, and $m = \hat{v}_o \Big/ \frac{V_d}{2}$ is the ratio of the average output voltage to the dc link voltage. Then the minimum pulse duration $t_{\min}$ is equal to:

$$t_{\min} = (1-\beta)\,T = \frac{1}{2}(1-M)\,T \tag{9}$$

If $t_1 = t_{\min}$, then $m=M$ which is the maximum modulation index. Equation (9) shows that $t_{\min}$ decreases as M increases and pulse dropping occurs if the switch cannot respond to a pulse with width $t_{\min}$. For a given output power, a feasible way to compensate for the distortion caused by pulse dropping is by increasing V_d and decreasing M simultaneously. This implies that we can replace the original PWM signal by $v_q(t)$ defined in (10) during the pulse dropping period as shown in Figure 4. This control strategy is called the *Transient Dynamic Boost* since the boosting of DC link voltage is only applied when the amplitude of the modulating signal reaches a critical value. The DC link voltage will fall back to the original level when the modulating signal amplitude reduces beyond the critical level.

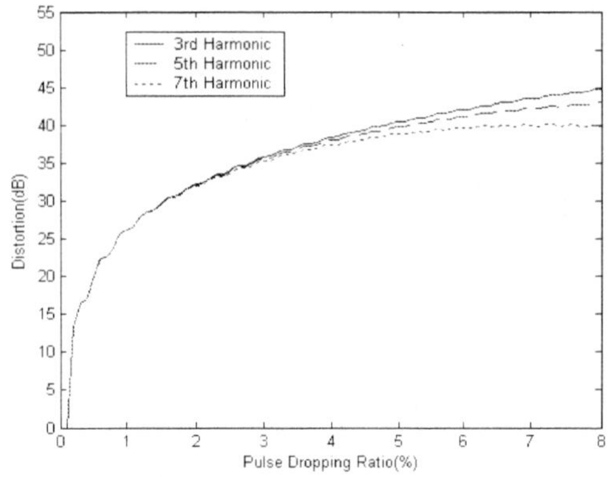

Figure 3. Spectral distortion due to pulse dropping

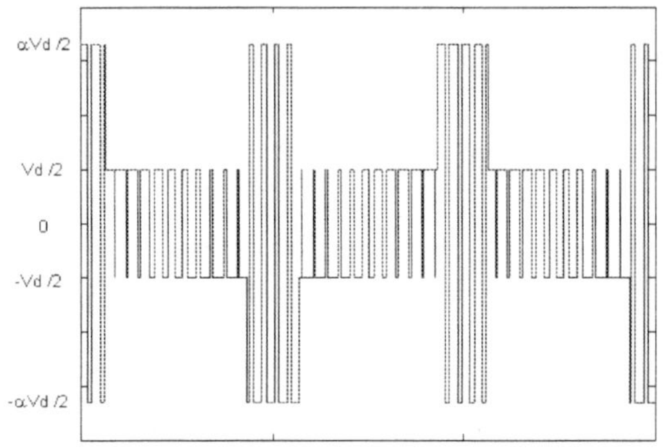

Figure 4. PWM with compensation

In order to maintain the output amplitude of the fundamental signal during the boosting, the DC link voltage will be set to αV_d if the modulation of $v_q(t)$ is set to M/α, where α is an arbitrary scalar. $v_q(t)$ can then be written as:

$$\begin{aligned} v_q(t) &= \frac{1}{2}\alpha V_d \frac{M}{\alpha}\cos(\omega_v t) + B(t) \\ &= \frac{V_d M}{2}\cos(\omega_v t) + B(t) \end{aligned} \tag{10}$$

where

$$\begin{aligned} B(t) = &\sum_{m=1}^{\infty} \frac{2\alpha V_d}{m\pi}\sin\frac{m\pi}{2} J_0\left(\frac{m\pi M}{2\alpha}\right)\cos(m\omega_c t) \\ &+ \sum_{m=1}^{\infty}\sum_{n=\pm 1}^{\pm\infty} \frac{2\alpha V_d}{m\pi}\sin\frac{(m+n)\pi}{2} J_n\left(\frac{\pi m M}{2\alpha}\right)\cos(m\omega_c t + n\omega_v t) \end{aligned} \tag{11}$$

Then the compensated output waveform in Figure 5 can be written as:

$$\begin{aligned} v_r(t) &= v_p(t)[1-g_1(t)-g_2(t)] + v_q(t)[g_1(t)+g_2(t)] \\ &= \frac{V_d M}{2}\cos(\omega_v t) + A(t) \\ &\quad + [B(t)-A(t)][g_1(t)+g_2(t)] \end{aligned} \tag{12}$$

By comparing (12) with the ideal PWM output given in (1), $[B(t)-A(t)][g_1(t)+g_2(t)]$ represents the distortion and they are actually the inter-modulated harmonics. The scalar α influences the magnitude of these harmonics in two ways. The first is related to the Bessel function $J_n(\bullet)$ and its value decreases monotonically with n and m and it becomes almost independent of α when n and m become large. The second is related to the linear component $2\alpha V_d / m\pi$. Though its value decreases as m increases, the magnitude tends to increase if we boost the scalar α. Theoretical analysis has been carried out to examine the effectiveness of the transient dynamic boost. The testing conditions are $\omega_v = 1kHz$, $\omega_c = 300kHz$, M=0.95 and γ = 2%. Figure 5 shows the pulse drop distortion effects to the spectrum and the compensated spectra for different α. It is clear that all the odd harmonic distortions caused by pulse dropping are suppressed. On the other hand, the distortion in even harmonics increases as α increases. This implies that the choice of α, in turn the DC link voltage, cannot be made arbitrary since it may cause significant even harmonic distortion.

Practically speaking, the choice of various parameters including α is really dependent upon the switching characteristics of the components. For a fixed $t_{\min}$, the corresponding maximum M can be calculated from (9). If one does not want to use the maximum M, α has to be adjusted accordingly. Figure 6 shows the relationship between $t_{\min}$ and α for various M. It should be noted that $t_{\min}$ is relatively to T which is normalized to 1 sec. Since $t_{\min}$ varies for different components, it will be very convenient for us to choose a proper α for a given M from Figure 6.

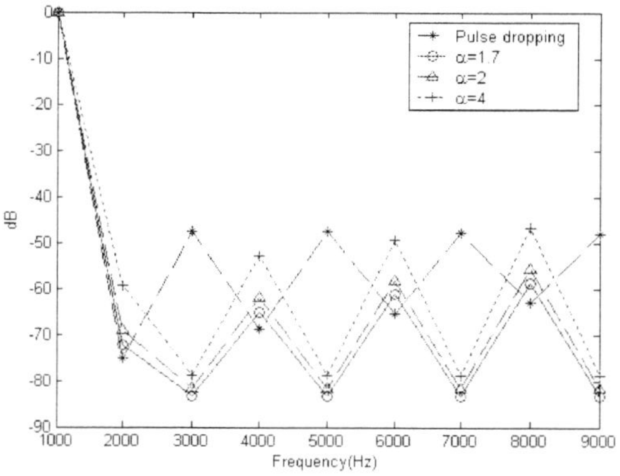

Figure 5. Spectra envelope due to different α ($\omega_v = 1kHz$, $\omega_c = 300kHz$, M=0.95, γ=2%)

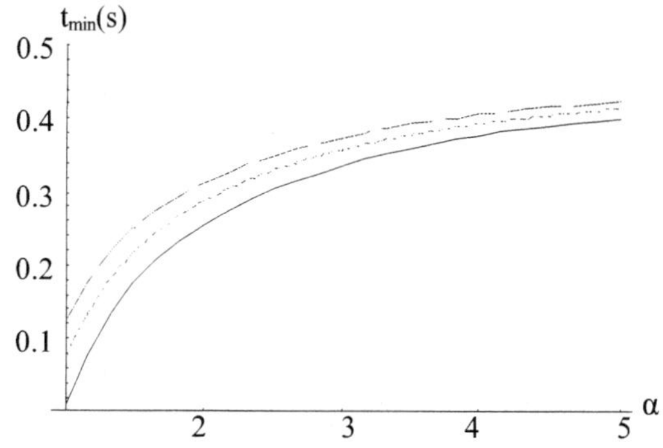

Figure 6 Relationships between t_{min} and α (—:M=0.98; ···:M=0.85; - -:M=0.75)

4. Experimental Results

An experimental 10W PWM inverter based prototype digital power amplifier has been implemented to demonstrate the effects of spectral compensation of the transient dynamic boost [4]. The parameters for the testing signal $v_p(t)$ is ω_v =1 kHz, ω_c =352 kHz and M_{Max} = 1. For easy implementation, α is chosen to be 2 since the DC link voltage can easily be doubled using a simple charge pump circuit. And $M/2$ is applied to $v_q(t)$ in order to obtain an undistorted output signal. A pulse dropping ratio of 2% is purposely introduced to evaluate the performance of the topology. Figure 7(a) shows the output spectrum with pulse dropping. The odd harmonic distortion can clearly be seen and an increase of noise floor is also observed. In fact the odd harmonic distortion is very close to the theoretical prediction as shown in Figure 5, despite the slight difference in *M*. The compensated output spectrum is illustrated in Figure 7(b) and it shows that the harmonic distortion has been suppressed and the noise has also been reduced.

5. Conclusion

This paper gives a comprehensive spectral analysis of PWM with pulse dropping and proposes a novel transient dynamic boost technique to compensate for the distortion caused by pulse dropping. The influence of various parameters has been studied in detail in order to facilitate for the correct choice. Both the theoretical analysis and experimental result demonstrate that the significant odd harmonic distortions can effectively be suppressed by using the proposed technique.

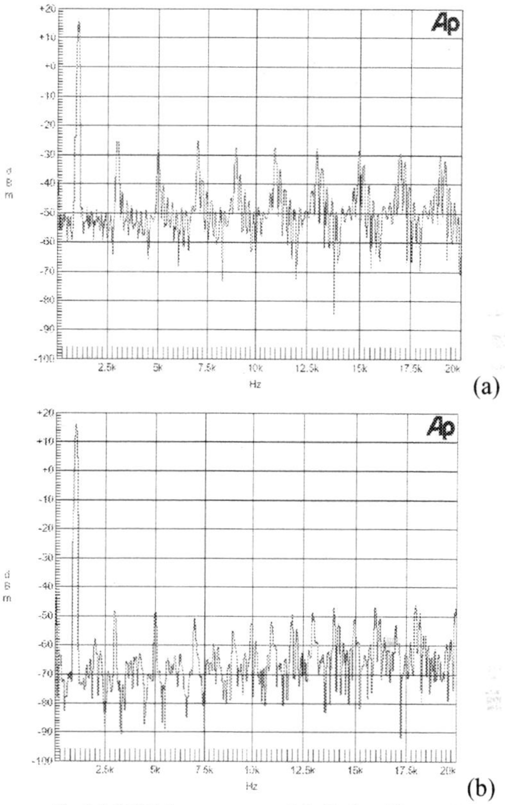

Figure 7 (a) PWM spectrum with Pulse Dropping (b) PWM spectrum with transient dynamic boost compensation

6. Reference

[1] R. E. Hlorns and M. B. Sandler "Power digital to analogue conversion using pulse width modulation and digital signal processing" *IEE Proceeding-G*, Vol. 140, No. 5, pp.329-338, 1993

[2] W.H. Lau, H. Chung, C.M. Wu and N.K. Poon, "Realization of Digital Audio Amplifier Using Zero-Voltage-Switched PWM Power Converter," *IEEE Trans. on Circuits and Systems, Part I*, Vol. 47, No. 3, pp. 303-311, Mar. 2000.

[3] H. S. Black, *Modulation Theory*. New York: Van Nostrand, 1953.

[4] Y.C. Chiu, B. Zhou, Henry S.H. Chung and W.H. Lau, "The implementation of a transient DC-link boost based digital amplifier for eliminating pulse-dropping distortion," to appear in ISCAS'2004.

CHARACTERISTICS AND MODELING OF PEM FUEL CELLS

S. Yuvarajan and Dachuan Yu

Electrical and Computer Engineering Department
North Dakota State University
Fargo, ND 58105 USA

ABSTRACT

PEM fuel cells are being used in a variety of applications including transportation and back-up power systems. The paper presents the complete experimental characteristics of a 1.2kW PEM fuel including its overload capability and dynamic performance. A novel circuit model for the fuel cell module that can be used to analyze and design fuel-cell power systems is also given. The model includes the mass transport effect and the pressure regulator effect. Simulated characteristics of the fuel cell are compared with the experimental results obtained on a commercial fuel cell.

1. INTRODUCTION

The Proton Exchange Membrane Fuel Cell (PEMFC) has received much attention during the last ten years because of its low working temperature, compactness, and easy and safe operational modes. There is a need to study the characteristics of the PEMFC and model it for developing power converters. Almost all of the models proposed for the PEMFC consist of mathematical equations and are not of much use in fuel cell system design and analysis [1]-[4]. Other models of PEMFC use Matlab-Simulink [5] and PSPICE [6], but they are still mathematical in nature.

This paper presents the complete V - I characteristics and a novel circuit model for a PEMFC using the nonlinearity of a diode and BJTs. In the model, a diode is used to model both the activation losses [7] and the ohmic losses in a PEMFC, while two BJTs are used for modeling the mass transport losses. The justification of this model is done through the experiments performed on a commercial PEMFC stack for both static and dynamic responses.

2. POLARIZATION CHARACTERISTICS OF A PEM FUEL CELL

Proton exchange membrane (PEM) fuel cells combine hydrogen and oxygen over a platinum catalyst to produce electrochemical energy with heat and water as the byproducts. Fig. 1 shows the V - I characteristic of a single cell operating at room temperature and normal air pressure [8]. The variation of individual cell voltage is found from the maximum cell voltage (or EMF) and the various voltage losses. Multiple factors contribute to the irreversible losses (voltage drop) in an actual fuel cell that causes the cell voltage to be less than its ideal potential [8]. The losses, which are also called polarization, irreversibility or overvoltage, originate primarily from three sources: a) activation polarization, b) ohmic polarization, and c) concentration (mass transport) polarization.

The ideal voltage is the maximum voltage that each cell in the stack can produce at a given temperature with the partial pressure of the reactants and products known. For the PEMFC (Commercial), where pure hydrogen and air are used, the ideal voltage can be calculated based on Gibbs free energy and it is equal to 1.2 V at 25°C [9].

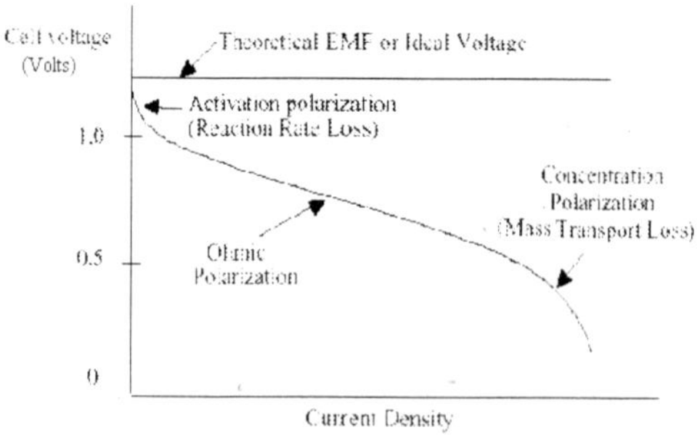

Fig. 1 V-I Characteristic of a single PEM Fuel Cell

Activation Polarization: The activation polarization loss (dominant at low current density) is present when the rate of the electrochemical reaction at an electrode surface is controlled by sluggish electrode kinetics [8]. Activation losses increase as current increases. The activation losses can be approximated by Tafel Equation [7]:

$$V = A.\ln(i/i_o) \qquad (1)$$

where A is a constant, V is the overvoltage, i is the current density, and i_o is the current density at which the voltage begins to drop.

Ohmic Polarization (Loss): The ohmic loss is due to the resistances of the polymer electrolyte membrane to the ions and of the imperfect electrodes. The loss in the fuel cell is approximately linear in this region.

Concentration Polarization (Mass Transportation Losses): The concentration polarization relates to the change in the concentration of the reactants at the surface of the electrodes as the fuel is used. The concentration of the fuel and oxidant is reduced at various points in the fuel-cell gas channels and is less than the concentration at the inlet value of the stack. This loss becomes significant at higher currents when the fuel and oxidant are used at higher rates and the concentration in the gas channel is a minimum.

In general, the mass transportation losses are given by the equation [7]:

$$V = -B.\ln(1 - i/i_1) \qquad (2)$$

where B is a constant that depends on the fuel cell and its operating state, i_l is the current density at which the fuel is used up at a rate equal to its maximum supply rate.

3. LOAD CHARACTERISTICS

In a commercial fuel cell module, several cells are connected in series giving a total voltage of the order of 40V [9]. The output current depends on the area of cross section. Fig. 2 shows the load characteristic which includes the variation of terminal voltage for currents well beyond the rated value. The characteristic was obtained by loading as well as by simulation. The unit has overcurrent protection built within the control section which has other functions like sensing hydrogen leakage, pressure, output voltage, and current. With the control section consuming a small power, the

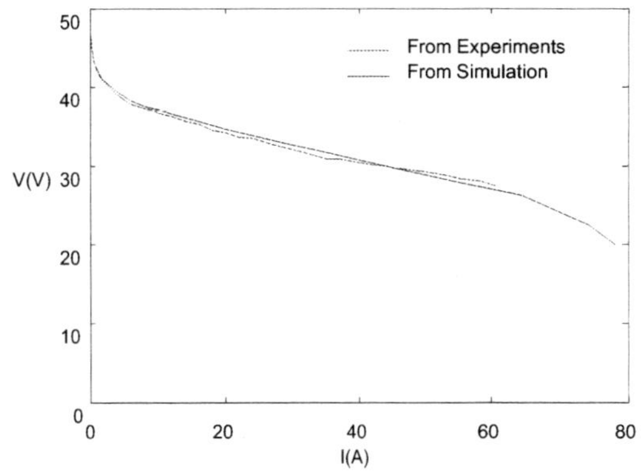

Fig. 2 Load characteristics of fuel-cell module

actual noload voltage observed is slightly below the theoretical no load voltage.

4. CIRCUIT MODEL OF FUEL CELL

Fig. 3 shows the proposed circuit model of a commercial PEM FC. The complete model is developed by modeling the different operating regions using elements from the simulation library. Both static and dynamic responses are briefly introduced.

Static Model

a. Activation Polarization

In the model proposed, a diode is used to model the activation polarization of the Fuel Cell. It is well known that, in a semiconductor diode, the depletion region provides a potential barrier to inhibit the migration of carriers across the junction, which is similar to the "sluggish electrode kinetics" in the fuel cell. Actually, the similarity can be seen by comparing equation (1) and the diode equation:

$$V_D = nV_T \ln(I_D / I_s), \quad V_T = kT/q \qquad (3)$$

where V_D is voltage across the diode and I_D is the current through the diode, n is an empirical constant between 1 and 2, I_S is the reverse saturation current, k is Boltzmann's constant, T is absolute temperature, and q is the electronic charge. In the simulation package PSPICE, the diode model can be adjusted to match the V-I characteristics of a fuel cell stack by choosing suitable values for the following parameters: I_S (Saturation current), R_S (Parasitic resistance), N (Emission Coefficient), I_{KF} (High-injection "knee" current), I_{SR} (Recombination current parameter), and I_{BV}

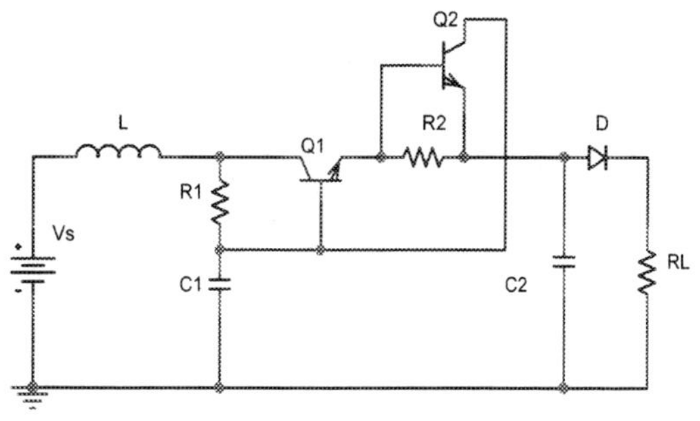

Fig. 3 Proposed circuit model of PEM Fuel Cell

(Reverse breakdown "knee" current) [10].

b. Ohmic Polarization

A resistor is used to model the ohmic polarization. Instead of using a separate resistor, the "Parasitic resistance (R_s) " in the diode itself is used in the model.

c. Concentration Polarization

To model the concentration or mass transport overvoltage, a "current limiting circuit" is used. In Fig. 3, the current limiting circuit is composed of two BJTs Q_1, Q_2 and the current sensing resistor R_2. When, the current through R_2 exceeds a set limit, Q_2 starts conducting and reduces the base voltage of Q_1. As a result, the emitter voltage of Q_1 (output voltage) will decrease at an exponential rate similar to equation (2). In the PSPICE model of the BJT, the area of the device can be changed to match the V-I characteristics of the fuel cell stack.

Dynamic Model

a. The Charge Double Layer

Due to diffusion effects and the reaction between the electrons (in the electrodes) and the ions (in the electrolyte), there is charge double-layer around the cathode in a fuel cell. The layer of charge on or near the electrode interface behaves much like an electrical capacitor. The result is that if the current suddenly changes, the operating voltage takes some time to arrive at its final equilibrium value [7]. Thus, it is quite reasonable to use a capacitor to model the capacitance effect resulting from the charge double layer.

b. Undershoot in stack voltage

The experimental step response of the commercial fuel-cell module to a step change in load shows a small undershoot. After a load step to full power, the air pump in the fuel cell takes about 0.5 second to reach a higher speed to provide a correspondingly higher airflow rate. In the mean time, the output current increases very quickly using up of the air around the electrodes. It is the temporary shortage of the air that causes the undershoot of the output voltage. This behaves like an inductor in an electric circuit. An inductor of suitable value is used to model the undershoot in stack output voltage.

5. RESULTS

To validate the proposed model, experiments were conducted on a commercial fuel-cell module [9]. The module is a fully integrated system that produces unregulated DC power from a supply of hydrogen (fixed pressure of 72psi) and air. It has a rated power of 1200W. The output voltage varies from 43V at system idle to about 26V at rated power (46A). The load was varied from 0.5Ω - 100Ω. Figure 2 shows the static characteristics obtained from simulation and experiment and the two agree. For the dynamic response, the load was changed suddenly from 108Ω to 0.55Ω. The simulation and experimental results are shown in Fig. 4.

5. CONCLUSION

The V – I characteristics of a PEM fuel cell is presented and the regions explained. A novel circuit model for the PEM fuel cell stack is also developed. The model is simple and uses a diode and a pair of BJTs for static conditions and a set of capacitors and an inductor for dynamic conditions. The model is validated by comparing the simulation and experimental results obtained on a commercial fuel-cell module.

6. REFERENCES

[1] S. Yerramalla, A. Davari, and A. Feliachi, "Dynamic modeling and analysis of polymer electrolyte fuel cell," Proc. of IEEE Power Engineering Society Summer Meeting, Vol. 1 , pp. 82 - 86 July 2002.

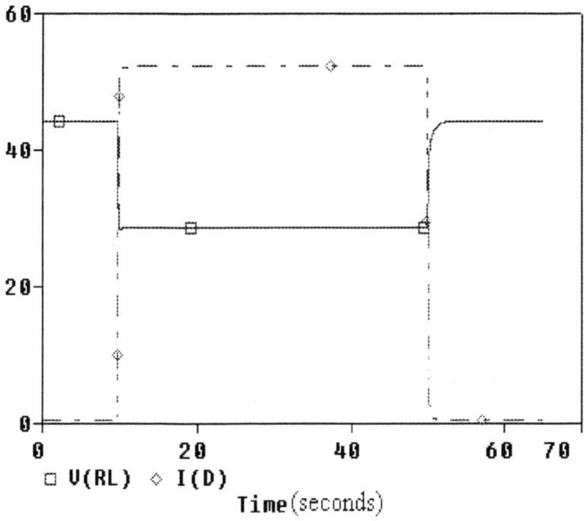

Fig. 4(a) Step response from simulation

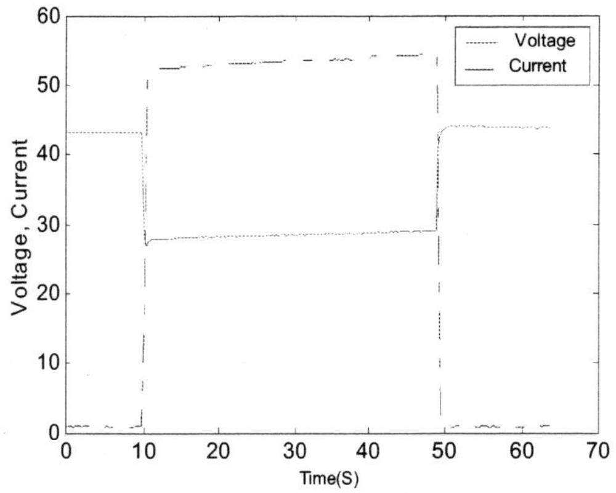

Fig. 4(b) Step response from experiment

[2] J.B. van der Merwe, C. Turpin, T. Meynard, and B. Lafage, "The installation, modeling and utilisation of a 200 W PEM fuel cell source for converter based applications, " Proc. of IEEE Power Electronics Specialists Conference, June 2002, pp. 333 -338.

[3] K. Dannenberg, P. Ekdunge, and G. Lindbergh, "Mathematical model of the PEMFC," Journal of Applied Electrochemistry, Vol. 30, pp.1377-1387, 2000.

[4] A.Kazim, H.T.Liu, and P. Forges, "Modeling of performance of PEM fuel cells with conventional and interdigitated flow fields," Journal of Applied Electrochemistry, Vol. 29, pp. 1409-1416, 1999.

[5] Turner, W.; Parten, M.; Vines, D.; Jones, J.; Maxwell, "Modeling a PEM fuel cell for use in a hybrid electric vehicle," Proc. of Vehicular Technology Conference, 1999 IEEE, pp. 16-20 May 1999.

[6] Randall Gemmen, Parviz Famouri, "PEM Fuel Cell Electric Circuit Model", Presented at Power Electronics for Fuel Cells Workshop, August 8-9, 2002

[7] James Larminie, Andrew Dicks, *"Fuel Cell Systems Explained",* John Wiley & Sons, 2000.

[8] National Energy Technology Laboratory, *"Fuel Cell Hand Book (6th Edition)",* Nov. 2002.

[9] Ballard Power System, "Nexa Power Module Integration Guide", 2003.

[10] Cadence Design Systems, "ORCAD PSPICE A/D Reference Manual", Version 9.2, October 2000.

A BOOST-SWITCHED CAPACITOR - INVERTER WITH A MULTILEVEL WAVEFORM

B. Axelrod, Y. Berkovich, *Member, IEEE* and A. Ioinovici, *Fellow, IEEE*

Holon Academic Institute of Technology,
Department Electrical and Electronics Engineering
52 Golomb St., Holon 58102, Israel
Tel: +972-3-5026632; Fax: +972-3-5026643
E-mail: Axelrod@hait.ac.il

***Abstract*-** Two structures, a switched-capacitor (SC)-based boost converter and a two-level inverter, are connected in cascade. The DC multilevel voltage of the first stage becomes the input voltage of the classical inverter, resulting in a staircase waveform for the inverter output voltage. Such a multilevel waveform is closed to a sinusoid; its harmonics content can be reduced by multiplying the stage number of the SC converter. The output low-pass filter, customary after a two-level inverter, becomes obsolete, resulting in a small size of the system, as the SC circuit can be miniaturized. Both stages are operated at a high switching frequency, resulting in a high-frequency inverter output, as required by some industrial applications. A Fourier analysis of the output waveform is performed. The design is optimized with reference to the nominal duty-cycle for obtaining the minimum total harmonic distortion (THD). Simulations and experiments on a prototype with a five-level output confirm the theoretical analysis.

1. MOTIVATION

The more and more stringent standards imposed today to the electrical energy, the requirement of clean non-polluted sinusoidal waveforms, with a minimal distortion factor, make the use of classical inverters obsolete. The output voltage of the two-level full-bridge inverter is a far cry from a sinusoid. A large filter is required to eliminate even a part of its harmonics.

In [1], a switched-capacitor (SC) circuit is used in order to get a multilevel DC voltage waveform which is applied at the input of a full-bridge inverter. As a result, a staircase output voltage is obtained at the load; the total harmonic distortion (THD) of the load voltage is extremely low. However, besides the complexity of the controller, this solution suffers from a discontinuous input current, what attracts EMI (electromagnetic interference) problems.

A simple solution is presented in [2]: a full-bridge of diodes, two capacitors and an active switch form a switching circuit which is inserted between the source and the inverter full-bridge of power switches. As a result, a five-level waveform is obtained at the inverter output. Unfortunately, the solution is limited to five levels, it can not be extended to a higher number of levels.

A still simpler solution is proposed in this paper: a multilevel DC voltage resulting as the output of a SC-circuit is applied to the input of a full-bridge inverter. For getting a five-level load voltage, the SC-circuit makes use of two diodes, two capacitors and an active switch, i.e. it requires two diodes less than the previous solution. The advantage of the solution proposed here is that by adding only a few elements to the SC-circuit, a load staircase voltage containing more levels is easily obtained.

Even if the power switches are submitted to the full bus voltage, the novel solution, due to its simplicity, is preferable to the multilevel inverters for low powers.

As, usually, the regulated inverter draws its power from the unregulated ac national grid, it is also of great importance **to get a high power factor in order to satisfy IEC 1000-3-2 requirements. The solution proposed here uses a boost-switched-capacitor converter for taking advantage of its property of drawing a nearly sinusoidal line current.**

Many industrial applications require ac power supplies providing a high-frequency voltage [3]. This is why, in the inverter proposed here, both the boost SC active switch and the inverter power switches are operated with a high switching frequency, resulting in a staircase load voltage of 25 kHz frequency.

II. NOVEL INVERTER WITH FIVE-LEVEL OUTPUT VOLTAGE

The proposed system is shown in Fig. 1. The input stage is a boost switched-capacitor-based converter. It is formed by L_{in} , S, an output capacitor split into C_{01} , C_{02} , the rectifier diode being replaced by two diodes D_{01}, D_{02} . The switch S together with C_{01}, C_{02} and D_{01}, D_{02} behave like a switched-capacitor circuit [4] – [6], operating with the duty-cycle D. The second stage is a classical inverter formed by a power switch bridge $S_1 – S_4$. The load is R_o, L_o, the output voltage being V_o. The timing diagram is given in Fig. 2. The boost stage is operated at the switching frequency f_s and the second stage at the switching frequency f_o, $f_o=f_s/2$. As a result, the output waveform will be a periodical voltage of frequency f_o.

2.1 Cyclical operation

The circuit goes through eight switching topologies. Let's consider that with switch S_3 on, S_1 is turned-on at t_0.

Mode 1, $[t_0, t_1]$: with S_1, S_3 on and S off, the circuit behaves like in Fig. 3(a): capacitors C_{01}, C_{02} are charged in parallel via diodes D_{01}, D_{02}, at the voltage V_C, i.e. in steady state:

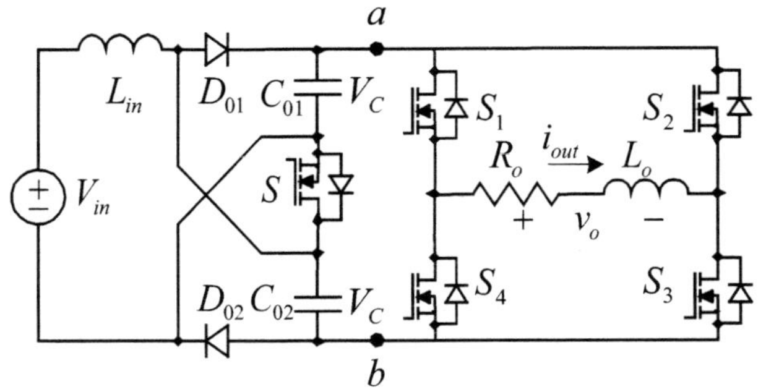

Fig. 1. Proposed inverter with five-level output voltage.

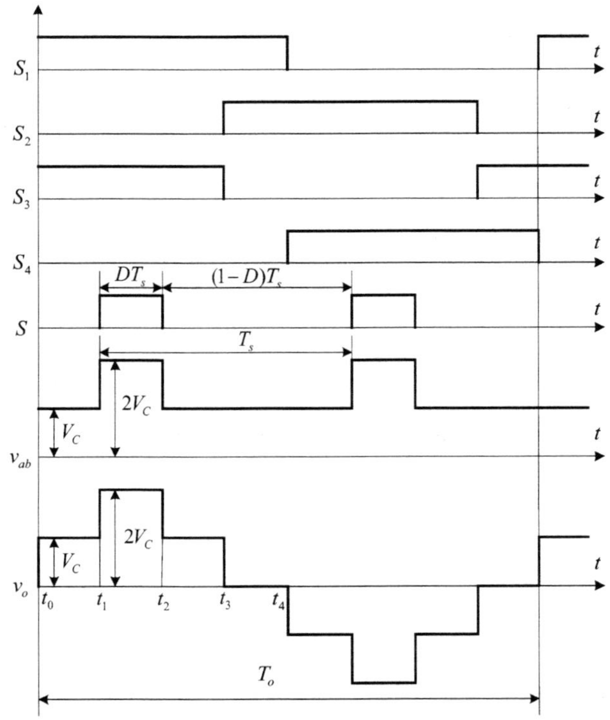

Fig. 2. Timing diagram and ideal waveforms of the novel inverter.

$$V_o = V_{ab} = V_C \,.$$

Mode 2, $[t_1, t_2]$: with S1, S_3 still on, S is turned-on (Fig. 3(b)). Capacitors C_{01}, C_{02}, connected in series, are discharged on the load. L_{in} is charged from the line (typical boost in the first topology). This is the controlled, "on-topology" of the switched-capacitor circuit. D_{01}, D_{02} are off due to the inverse voltage on the capacitors. In steady-state:

$$V_o = V_{ab} = 2V_C \,.$$

Mode 3, $[t_2, t_3]$: S_1, S_3 on, S off. This topology is identical to that in mode 1, i.e.

$$V_o = V_{ab} = V_C \,.$$

Mode 4, $[t_3, t_4]$: S_3 is turned-off and S_2 is turned-on, i.e. the second stage goes from the line-to-load transfer energy topology to the freewheeling topology. Even if V_{ab} is still equal to V_C, V_o=0

$$V_{ab} = V_C ,\ V_o = 0$$

The second half-cycle (Mode 5 – Mode 8) starts by turning-off S_1 and turning-on S_4, the switching topologies will be similar, but the current will circulate in an opposite direction through the load, giving the negative half-waveform of the load voltage.

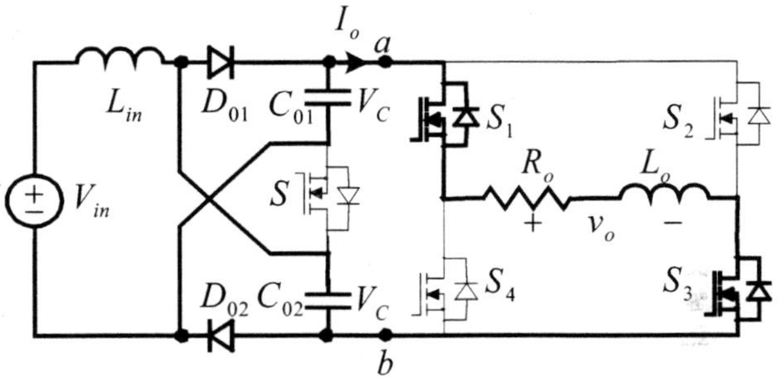

(a) Mode 1 and Mode 3.

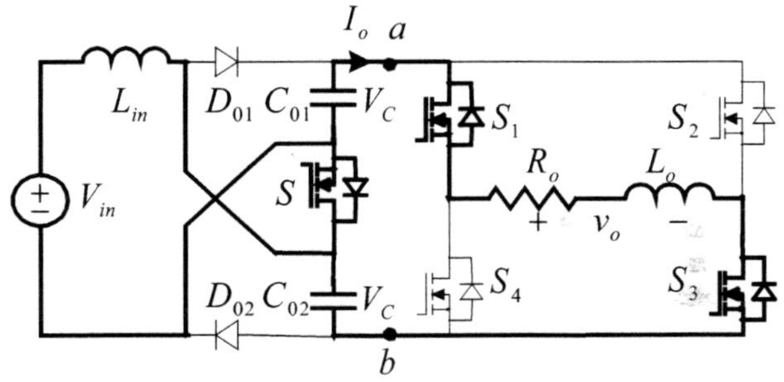

(b) Mode 2.

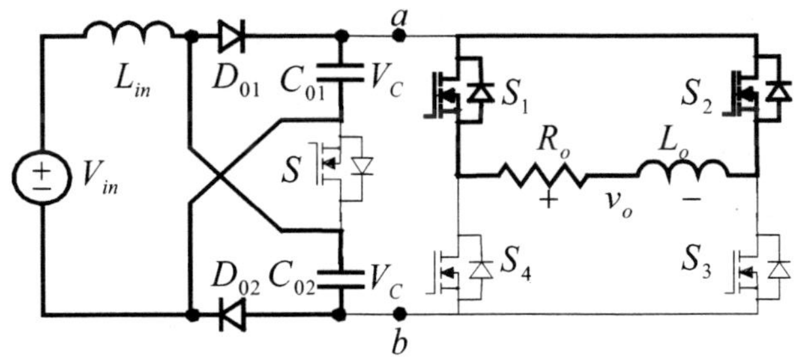

(c) Mode 4.

Fig. 3. Switching topologies.

From the energy balance on the input inductor, one gets

$$V_C = \frac{V_{in}}{(1-D)} \quad (1a)$$

and
$$V_{ab} = \frac{V_{in}}{1-D} \quad \text{in } \textit{Modes 1, 3, 4} \quad (1b)$$

$$V_{ab} = \frac{2V_{in}}{1-D} \quad \text{in } \textit{Mode 2} \quad (1c)$$

Therefore the output waveform of the boost SC stage is a controlled two-level voltage. The inverter stage is operated in a classical way. Thus, the multilevel load voltage is obtained due to the fact that a multi-level DC voltage is applied at the input of the inverter full-bridge.

The first advantage over [2] is that the same result is obtained here by using two diodes less.

2.2 Fourier analysis of the load voltage and optimized design

The staircase output voltage V_o is breaked down in two components in Fig. 4, $v_o^{'}$ and $v_o^{''}$. The duration of each step in V_o is determined by the nominal duty-cycle and the delay between turning-on the leading switch of the inverter full-bridge and turning-on *S*. With respect to a half-cycle, let's denote by *x* half of the duration of the inverter full-bridge freewheeling topology, by *y* the time between starting a new transfer energy topology and turn-on of *S*, and by *z* the interval in which *S* conducts (i.e. DT_s).

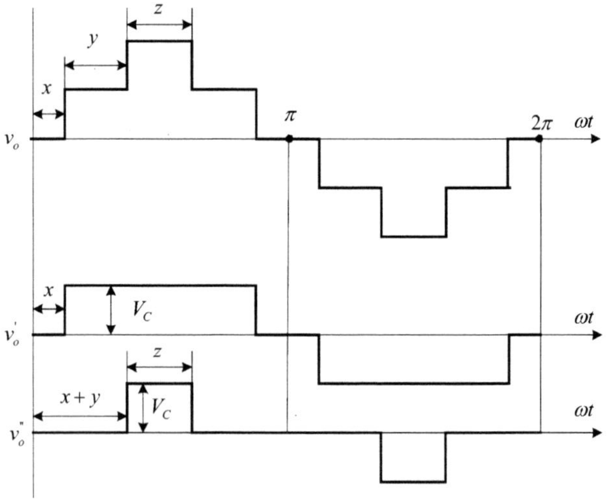

Fig. 4. Fourier-oriented decomposition of the five-level output voltage.

A Fourier analysis of the signal in Fig. 4 gives the magnitudes of the harmonics as functions of *x* and *y*:

$$V_n = V_C \frac{4}{n\pi}\left(\cos(nx) + \cos(n(x+y))\right); \quad (2)$$
$$n = 1, 3, 5...$$

The ideal is to get a "clean" sinusoidal output voltage, i.e. the content of high harmonics (*n=3, 5....*) to be zero.

The total harmonic distortion (THD) of V_o, according to (2), is represented in Fig. 5 as a function of the parameters *x* and *y*.

The computer simulation shows that the minimum THD (THD<10%) is obtained for $x \approx 0.23 rad; \; y \approx 0.48 rad.$

Accordingly, the optimized duty-cycle value of the boost SC stage is found to be

$$D_{opt} = \frac{z_{opt}}{\pi} = \frac{\pi - 2(x_{opt} + y_{opt})}{\pi} \approx 0.55 \quad (3)$$

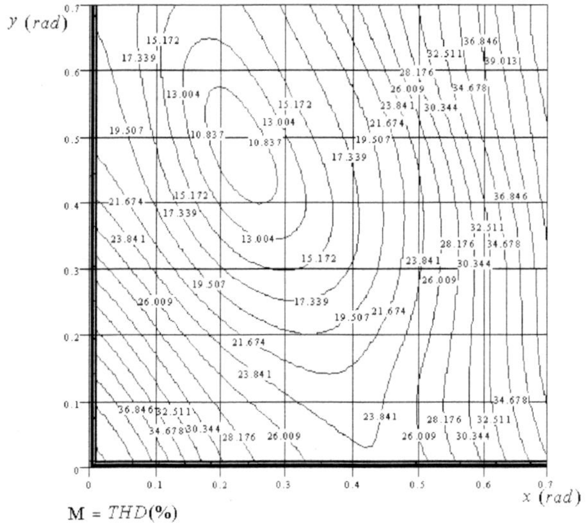

Fig. 5. THD of the output voltage as a function of the time ratios *x* and *y*.

This is the second advantage of the method proposed here: the design of the first stage being that of an SC-circuit, its parameters can be optimized.

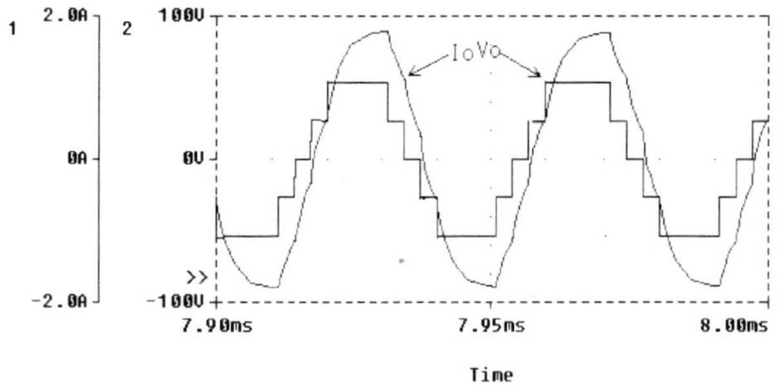

Fig. 6. PSPICE simulated waveforms of the output voltage and load current.

III. EXPERIMENTAL RESULTS

The proposed inverter whose data have been optimized in Section II has been implemented in the laboratory for $V_{in} = 12V$; $L_{in} = 600\mu H$; $C_{01} = C_{02} = 50\mu F$; $R_o = 30\Omega$; $L_o = 75\mu H$; f_o=25kHz, and output power $P_{out} = 50W$. The switches are MOSFET *IRF540* and the diodes D_{01} -D_{03} are diodes *MBR*.

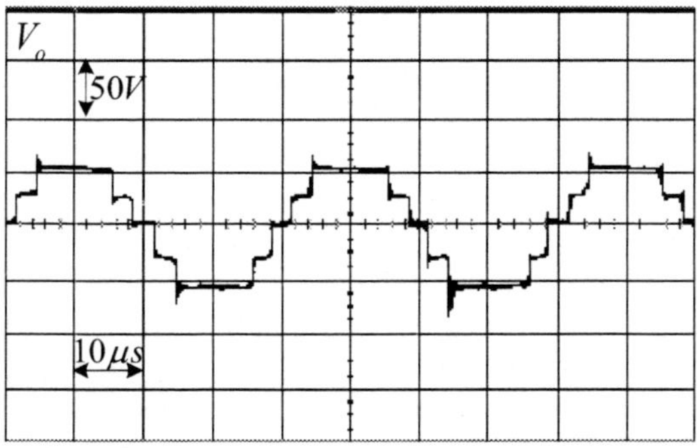

a.

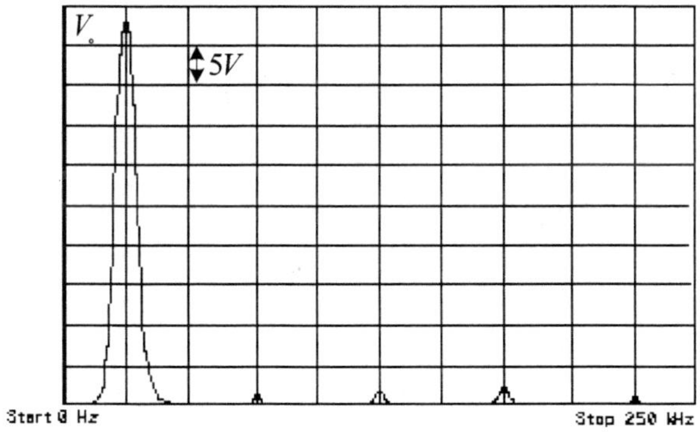

b.

Fig. 7. Experimental waveform of the output voltage (a); and its frequency spectrum (b) of the five-level inverter.

The oscillogram of the output voltage of the five-level inverter and its frequency spectrum are given in Fig. 7. The THD of this voltage is 11.3%.

CONCLUSIONS

A new idea is proposed in order to get inverters with a multilevel output waveform: a cascade between a boost-switched-capacitor converter and a classical full-bridge inverter.

The novel circuit contains less elements than the available solutions for the case of a five-level inverter. It can be easily generalized to a multilevel inverter, by simply adding more stages to the SC circuit. This systematic approach allowed for the development of a procedure of optimization of the nominal duty-cycle of the switches for getting an output staircase waveform with minimum harmonic distortion. The resulting load current and voltage are almost "clean" sinusoids.

In order to provide an output waveform with the same THD, the optimized inverters proposed here make use of less circuit elements than the available multilevel inverters.

For low power applications, where no constraints exist on the switch voltage stress, the new inverters have the edge.

The simulated and experimental results confirmed the theoretically expected high performance of the proposed circuit.

REFERENCES

[1] O.C. Mak and A. Ioinovici, "Switched-capacitor inverter with high power density and enhanced regulation capability, "*IEEE Trans. Circuits Syst.*, vol. 45, pp. 336-348, April 1998.

[2] S. J. Park, F. S. Kang, M. H. Lee, and C. U. Kim, "A new single-phase five-level PWM inverter employing a deadbeat control scheme," *IEEE Trans. Power Electron.*, vol. 18, pp. 831-843, May 2003.

[3] C. C. Liu, C. L. Chen, and K. M. Lee, "A novel energy-recovery sustaining driver for plasma display panel," *IEEE Trans. Ind. Electron.*, vol. 47, pp. 1271-1277, Dec. 2000.

[4] S. V. Cheong, H. Chung, and A. Ioinovici, "Development of power electronics based on switched-capacitor circuits," in *Proc. 1992 IEEE Int. Symp. Circuits and Systems,* San Diego, pp. 1907-1911, May 1992.

[5] G. Zhu and A. Ioinovici, " Steady-state characteristics of switched-capacitor electronic converters, "*J. Circuits, Systems, and Computers,* vol. 7, pp. 69-91, April 1997.

[6] B. Axelrod, Y. Berkovich, and A. Ioinovici, "Transformerless DC-DC converters with a very high DC line-to-load voltage ratio," in *Proc. 2003 Int. Symp. Circuits and Systems,* Bangkok, vol. 3, pp. 435-438, May 2003.

NOVEL MULTILEVEL CONVERTER FOR POWER FACTOR CORRECTION

Bor-Ren Lin, Senior Member, IEEE, Tsung-Yu Yang and Yung-Chuan Lee

Department of Electrical Engineering, National Yunlin University of Science and Technology
Touliu City, Yunlin 640, Taiwan, Grant: NSC 92-2213-E-224-051

ABSTRACT

An eight-switch three-phase neutral-point switch-clamped rectifier is presented to achieve the high input power factor and draw the sinusoidal currents from the AC mains. In the conventional three-phase neutral-point switch-clamped rectifier, the circuit configuration consists six power switches with voltage stress v_{dc} and three AC power switches with voltage stress $v_{dc}/2$. There are only four power switches with voltage stress of DC bus voltage and two AC power switches with voltage stress of half the DC bus voltage in the proposed rectifier. The main functions of the proposed control scheme are to achieve unity input power factor, to draw the sinusoidal line currents from the AC source, and to keep the DC link voltage constant. The analysis and mathematical model of the proposed rectifier is derived. Some simulations and experimental results are provided to verify the effectiveness of the proposed control scheme.

1. INTRODUCTION

Recently the multilevel converters or inverters are widely studied for high power and medium voltage applications. The advantages of multilevel techniques are low voltage stress of power semiconductors, low voltage harmonics and less electromagnetic interference. The topologies of multilevel converters or inverters can be classified into series H-bridge multilevel inverter [2, 6], flying capacitor multilevel inverter [2, 5], and neutral diode clamped (NPC) multilevel inverter [1-4]. Three-level neutral point clamped rectifier is proposed [3-4] to draw the sinusoidal line currents in phase with mains voltages. The input power factor is close to unity. This circuit topology has twelve power switches and six clamping diodes. However, the disadvantages of the multilevel converters are many power devices in the circuits. The control scheme of multilevel converters is also complicated. The low cost ac drives with four power switches are proposed [7-8] to achieve two-level PWM operation in the low power applications. This topology has the advantages of less power switches and small size of drive. Eight power switches are used in the [9] to perform rectifier/inverter system with power factor correction and to perform two-level PWM operation.

This paper proposes an eight-switch three-phase three-level rectifier to achieve power factor correction and current harmonics elimination. Only eight power switches and four clamping diodes are employed in the proposed rectifier. Three sinusoidal line currents with nearly unity power factor are drawn from the AC source. The proposed rectifier allows bidirectional power flow. The system analysis and mathematical model of the rectifier are presented. To verify the effectiveness of the proposed control strategy, simulation and experimental results are discussed and provided.

2. CIRCUIT CONFIGURATION

Fig. 1 shows the proposed three-phase switch-clamped rectifier. Three are only two rectifier legs in the proposed converter. Four power switches with voltage stress of DC bus voltage and two AC switches (four unidirectional power switches) with half the DC bus voltage are used in the proposed rectifier to achieve the three-level PWM operation, draw sinusoidal line currents, maintain the DC link voltage constant and achieve bidirectional power flow. The phase *c* is directly connected to the midpoint of the split DC capacitors. A proportional-integral based voltage controller and a carrier-based current controller are used in the outer and inner control loops respectively to achieve the DC link voltage regulation and to track the line current commands.

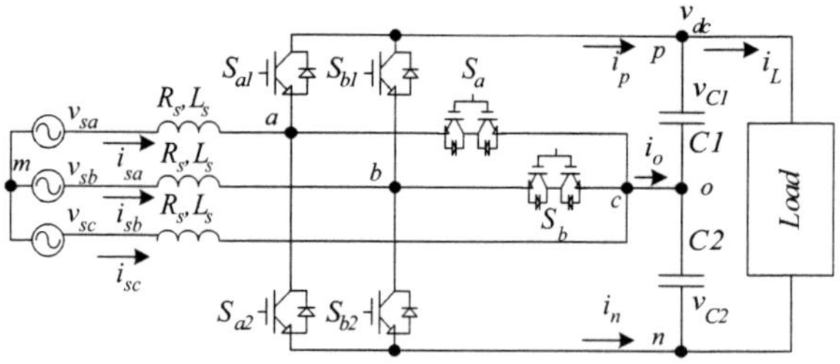

Fig. 1 Proposed three-phase neutral-point switch-clamped rectifier.

3. SYSTEM ANALYSIS AND CONTROL SCHEME

3.1 System Analysis

The main functions of the proposed three-phase rectifier are DC-link voltage regulation and the balanced and sinusoidal line current generation. There are two legs in the proposed rectifier (Fig. 1). The constraints of power switches are expressed as (1) to avoid the power switches conducting at the same time.

$$S_{a1}+S_{a2}+S_a=1, \quad S_{b1}+S_{b2}+S_b=1. \tag{1}$$

where S_{xy}=1 if switch S_{xy} is turned on, or S_{xy}=0 if switch S_{xy} is turned off, x=a,b and y=1, 2. Based on the valid switching signals in each rectifier leg, there are three voltage levels ($-v_{dc}/2$, 0 and $v_{dc}/2$) generated on the AC terminal to neutral point voltage and five voltage levels ($-v_{dc}$, $-v_{dc}/2$, 0, $v_{dc}/2$ and v_{dc}) generated on the AC terminal voltage v_{ab}. For example, if power switch S_{a1} is turned on, AC side voltage v_{ac} is equal to $v_{dc}/2$. In this switching state, line current i_{sa} decreases because half the DC link voltage is greater than peak line-to-line voltage. If power switch S_a is turned on, AC side voltage v_{ac} is equal to 0. In this state, line current i_{sa} increases or decreases if line voltage v_{sac} (=v_{sa}-v_{sc}) is positive or negative. The voltage v_{ac} is equal to $-v_{dc}/2$ if S_{a2} is turned on. In this switching state, line current i_{sa} increases because the boost inductor voltage is positive. Since there are three operating states in each rectifier leg, the total possible switching states of the proposed rectifier are nine. These nine valid switching signals and the corresponding switching states in the proposed rectifier are shown in Table 1.

Table 1: Possible switching states of the proposed converter.

	switching signals						AC side voltages		
state	S_{a1}	S_a	S_{a2}	S_{b1}	S_b	S_{b2}	v_{ac}	v_{bc}	v_{ab}
1	1	0	0	1	0	0	$v_{dc}/2$	$v_{dc}/2$	0
2	1	0	0	0	1	0	$v_{dc}/2$	0	$v_{dc}/2$
3	1	0	0	0	0	1	$v_{dc}/2$	$-v_{dc}/2$	v_{dc}
4	0	1	0	1	0	0	0	$v_{dc}/2$	$-v_{dc}/2$
5	0	1	0	0	1	0	0	0	0
6	0	1	0	0	0	1	0	$-v_{dc}/2$	$v_{dc}/2$
7	0	0	1	1	0	0	$-v_{dc}/2$	$v_{dc}/2$	$-v_{dc}$
8	0	0	1	0	1	0	$-v_{dc}/2$	0	$-v_{dc}/2$
9	0	0	1	0	0	1	$-v_{dc}/2$	$-v_{dc}/2$	0

Based on the Kirchhoff law applied on the AC and DC terminals, the system behavior of the proposed rectifier can be expressed by the following equations.

$$\frac{di_{sa}(t)}{dt}=[v_{sa}(t)-R_s i_{sa}(t)-v_{ac}(t)-v_{cm}(t)]/L_s$$
$$\frac{di_{sb}(t)}{dt}=[v_{sb}(t)-R_s i_{sb}(t)-v_{bc}(t)-v_{cm}(t)]/L_s$$
$$\frac{di_{sc}(t)}{dt}=[v_{sc}(t)-R_s i_{sc}(t)-v_{cm}(t)]/L_s \tag{2}$$

where v_{ac} and v_{bc} are AC side voltages, v_{cm} is the voltage between the neutral point c on the DC bus and the neutral point m on the AC source, i_{sa}, i_{sb} and i_{sc} are three-phase line currents, v_{sa}, v_{sb} and v_{sc} are three-phase mains voltages, R_s and L_s are series equivalent resistance and inductance of the boost inductor. If the three-phase supply voltages are balanced, the sum of instantaneous three-phase voltages and currents is zero.

$$i_{sa}(t)+i_{sb}(t)+i_{sc}(t)=0, \tag{3}$$
$$v_{sa}(t)+v_{sb}(t)+v_{sc}(t)=0, \tag{4}$$

where

$$v_{sa}(t)=V_s\sin(\omega t),\ v_{sb}(t)=V_s\sin(\omega t-2\pi/3),$$
$$v_{sc}(t)=V_s\sin(\omega t+2\pi/3) \tag{5}$$
$$i_{sa}(t)=I_s\sin(\omega t-\theta),\ i_{sb}(t)=I_s\sin(\omega t-2\pi/3-\theta),$$
$$i_{sc}(t)=I_s\sin(\omega t+2\pi/3-\theta) \tag{6}$$

where θ is the phase angle between the phase voltage and current. With (2), (3) and (4), one can obtain the voltage v_{cm} as follow:

$$v_{cm}(t)=-\frac{v_{ac}(t)+v_{bc}(t)}{3} \tag{7}$$

Based on (2) and (7), the ac side voltages of proposed rectifier can be rewritten as follows:

$$v_{ac}(t)=v_{sa}(t)-v_{sc}(t)-2L_s\frac{d}{dt}i_{sa}(t)-2R_s i_{sa}(t)$$
$$-L_s\frac{d}{dt}i_{sb}(t)-R_s i_{sb}(t)$$
$$v_{bc}(t)=v_{sb}(t)-v_{sc}(t)-L_s\frac{d}{dt}i_{sa}(t)-R_s i_{sa}(t)$$
$$-2L_s\frac{d}{dt}i_{sb}(t)-2R_s i_{sb}(t) \tag{8}$$

If one can neglect the series equivalent resistance on the boost inductor, (8) can be further expressed as follows:

$$v_{ac}(t)=v_{sac}(t)-2L_s\frac{d}{dt}i_{sa}(t)-L_s\frac{d}{dt}i_{sb}(t)$$
$$v_{bc}(t)=v_{sbc}(t)-L_s\frac{d}{dt}i_{sa}(t)-2L_s\frac{d}{dt}i_{sb}(t) \tag{9}$$

3.2 Required line current for the output power

The DC side power P_{out} required by the load can be expressed as the average of the measured DC bus voltage V_{dc} and the load current I_L.

$$P_{out}=V_{dc}\cdot I_L \tag{10}$$

One assumed that the system is power balance between the AC and DC sides. For unity power factor operation, the required line current command for the output power is given as:

$$I_s=\frac{2P_{out}}{3V_s}=\frac{2V_{dc}\cdot I_L}{3V_s} \tag{11}$$

Therefore, the required line currents for the output power are expressed as:

$$i_{sa}(t)=I_s\sin(\omega t)=\frac{2V_{dc}\cdot I_L}{3V_s}\sin(\omega t),$$
$$i_{sb}(t)=I_s\sin(\omega t-2\pi/3)=\frac{2V_{dc}\cdot I_L}{3V_s}\sin(\omega t-2\pi/3),$$
$$i_{sc}(t)=I_s\sin(\omega t+2\pi/3)=\frac{2V_{dc}\cdot I_L}{3V_s}\sin(\omega t+2\pi/3). \tag{12}$$

3.3 Required line current for the converter losses

There are some switching and conduction losses on the proposed rectifier because the semiconductors are not ideal. A proportional-integral (PI) based voltage compensator is used to obtain the required active current for converter losses compensation. The DC bus voltage is measured and compared with the DC link command. The DC bus voltage error is sent to the voltage controller to obtain the required line current for converter losses compensation.

$$I_{s,loss} = k_p \Delta v_{dc} + k_i \int \Delta v_{dc} dt \tag{13}$$

where k_p and k_i are the proportional and integral gain of voltage controller respectively, $\Delta v_{dc}=v^*_{dc}-v_{dc}$, v^*_{dc} and v_{dc} are reference and measured DC bus voltage. The required line currents for the compensation of converter losses are given as:

$$i_{sa,loss}(t) = I_{s,loss}\sin(\omega t),\ i_{sb,loss}(t) = I_{s,loss}\sin(\omega t - 2\pi/3),$$
$$i_{sc,loss}(t) = I_{s,loss}\sin(\omega t + 2\pi/3). \tag{14}$$

3.4 Compensated current for the neutral point voltage

To compensate the neutral point voltage balance, a neutral point voltage compensator is generated and added to the line currents i_{sa} and i_{sb}. A low pass filter with narrow bandwidth (10Hz) is used to obtain the average capacitor voltages V_{C1} and V_{C2}. The average voltage variation between these two capacitor voltages is used to obtain the compensated current to balance the neutral point voltage. The compensated current for neutral point voltage is expressed as:

$$I_{balance} = K[V_{C2} - V_{C1}] \tag{15}$$

where V_{C1} and V_{C2} are average voltage values of capacitors *C1* and *C2*.

3.5 Line current commands

The resultant line current commands of the proposed rectifier based on (12), (14) and (15) are expressed as:

$$i^*_{sa}(t) = i_{sa}(t) + i_{sa,loss}(t) + I_{balance},$$
$$i^*_{sb}(t) = i_{sb}(t) + i_{sb,loss}(t) + I_{balance} \tag{16}$$

3.6 Control Scheme

Fig. 2(a) gives the control block for the proposed rectifier to achieve power factor correction and make DC bus voltage constant. A line current computation unit is used to calculate the line current amplitude I_s based on the power balance between the AC source and DC side load. A proportional integral based voltage controller is used to achieve the compensated line current $I_{s,loss}$ for the converter losses. A neutral point voltage compensator is used to balance the neutral point voltage for the transient response. A phase locked loop circuit is used to generate the unity sinusoidal waves $\sin(\omega t)$ and $\sin(\omega t-2\pi/3)$. The AC terminal voltage commands $v_{ac}(t)$ and $v_{bc}(t)$ can be obtained based on the (9) from the measured line current errors and the line voltages. The PWM waveform generator based on fixed switching frequency scheme is used to generate the proper switching signals for the properly power switches. Fig. 2(b) gives the PWM signals for the rectifier leg-*a*. The duty cycle d_{ac} equals $v_{ac}/(v_{dc}/2)$. The duty cycles of power switches in each rectifier leg are expressed as:

$$d_{Sa1} = \begin{cases} \dfrac{v_{ac}(t)}{v_{dc}/2}, & if\ v_{ac}(t) > 0 \\ 0, & if\ v_{ac}(t) < 0 \end{cases},\quad d_{Sa2} = \begin{cases} 0, & if\ v_{ac}(t) > 0 \\ \dfrac{-v_{ac}(t)}{v_{dc}/2}, & if\ v_{ac}(t) < 0 \end{cases},$$

$$d_{Sa} = \begin{cases} 1 - \dfrac{v_{ac}(t)}{v_{dc}/2}, & if\ v_{ac}(t) > 0 \\ 1 + \dfrac{v_{ac}(t)}{v_{dc}/2}, & if\ v_{ac}(t) < 0 \end{cases},\quad d_{Sb1} = \begin{cases} \dfrac{v_{bc}(t)}{v_{dc}/2}, & if\ v_{bc}(t) > 0 \\ 0, & if\ v_{bc}(t) < 0 \end{cases},$$

$$d_{Sb2} = \begin{cases} 0, & if\ v_{bc}(t) > 0 \\ \dfrac{-v_{bc}(t)}{v_{dc}/2}, & if\ v_{bc}(t) < 0 \end{cases},\quad d_{Sb} = \begin{cases} 1 - \dfrac{v_{bc}(t)}{v_{dc}/2}, & if\ v_{bc}(t) > 0 \\ 1 + \dfrac{v_{bc}(t)}{v_{dc}/2}, & if\ v_{bc}(t) < 0 \end{cases}. \tag{17}$$

With the generated PWM signals for the proper power switches, there are three voltage levels on the AC terminal voltages v_{ac} and v_{bc} and five voltage levels on the AC side v_{ab}. The balanced three-phase line currents with nearly unity power factor will be drawn from the AC source and DC bus voltage is constant.

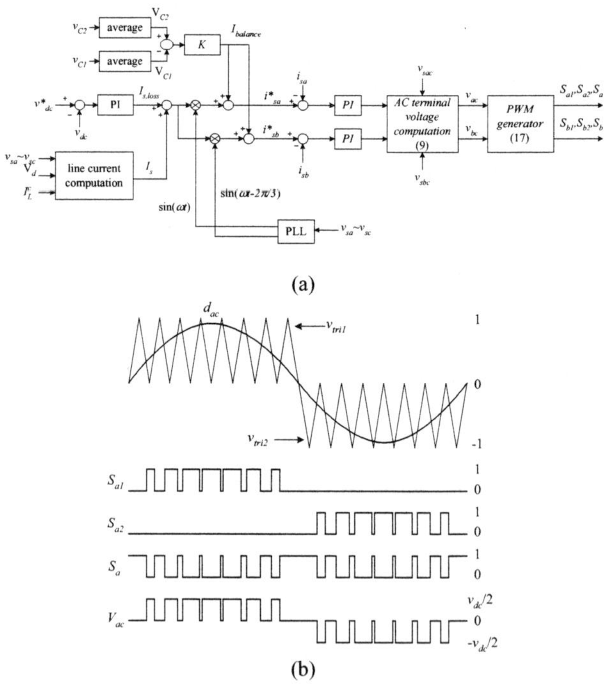

Fig. 2 (a) Control block of the proposed rectifier (b) PWM signals for the rectifier leg-*a*.

4. SIMULATION AND EXPERIMENTAL RESULTS

To confirm the validity of the proposed control scheme, some experimental results are provide based on a laboratory scale-down prototype. A digital signals processor (TMS320C32) is adopted as the kernel in the implementation of a digital controller. In the experimental tests, the boost inductors of the adopted converter are $5mH$, DC side capacitors are $C1=C2=2200\mu F$, and AC source line voltage is $110V_{rms}$/ 60Hz. The DC bus capacitor voltage is 400V. The carrier frequency is 10kHz. A phase-locked loop (PLL) circuit is employed to produce the reference unit sinusoidal waves $sin(\omega t)$ and $sin(\omega t\text{-}2\pi/3)$. Fig. 3(a) illustrates the simulated results of three-phase mains voltage and line current. The current and phase voltage are in phase each other. The simulated results of mains voltage v_{sa} and ac terminal voltages are given in Fig. 3(b). Fig. 4(a) gives the measured three-phase phase voltages and line currents. The balanced and sinusoidal line currents are drawn form the AC source. The measured input power factor of line current is about 0.992. The measured total harmonic distortion of line current is 4.2%. The measured efficiency of the proposed rectifier is about 87%. Fig. 4(b) gives the measured waveforms of three AC terminal voltages of the rectifier. Three voltage levels are shown on the AC terminal voltages v_{ac} and v_{bc}. Five voltage levels are given on the AC terminal voltage v_{ab}. Fig. 5(a) shows the measured results of mains voltage v_{sa}, line current i_{sa} and ac side voltage v_{ac}. Fig. 5(b) gives the measured capacitor ripple voltages and voltage variation. The ripple voltage is under 5V.

5. CONCLUSION

An eight-switch three-phase rectifier is proposed to achieve input power factor correction and to regulate the DC-link voltage. Two rectifier legs are used in the proposed circuit to reduce the number of the power switches compared to the conventional three-phase neutral-point switch-clamped rectifier. Three voltage levels are generated on the AC side voltages v_{ac} and v_{bc} and five voltage levels are shown on the voltage v_{ab}. Based on the proposed control scheme, three-phase sinusoidal line currents with nearly unity power factor are drawn from the AC source. The proposed circuit topology can be applied for the AC motor drives, AC voltage regulators and active power filters.

6. REFERENCES

[1] A. Nabae, I. Takahashi, and H. Akagi, "A new neutral-point clamped PWM inverter", *IEEE Trans. on Industry Applications*, vol. 17, no. 5, pp. 518-523, 1981.

[2] J. S. Lai, and F. Z. Peng, "Multilevel converters – a new breed of power converter", *IEEE Trans. on Industry Applications*, vol. 32, no. 3, pp. 509-517, 1996.

[3] G. Sinha, and T. A. Lipo, "A four level rectifier-inverter system for drive applications", in *Conf. Rec. IEEE-IAS Annual Meeting*, 1995, pp. 980-987.

[4] J. Rodriguez, J. S. Lai, and F. Z. Peng, "Multilevel inverters: a survey of topologies, controls and applications," *IEEE Transactions on Industrial Electronics*, vol. 49, no. 4, pp. 724-738, 2002.

[5] P. Carrere, J. Meynard, and P. Lavieville, "4000V-300A eight level IGBT inverter leg," in *Conf.-Rec., EPE'95*, 1995, pp. 106-111.

[6] F. Z. Peng, J. S. Lai, J. W. Mckeever, and J. Vancoevering, "A multilevel voltage-source inverter with separate dc sources for static VAR generation," *IEEE Trans. Ind. Applicat.*, vol. 32, no. 6, pp. 1130-1137, 1996.

[7] P. N. Enjeti, A. Rahman, and R. Jakkli, "Economic single-phase to three-phase converter tolopogies for fixed and variable frequency output", *IEEE Trans. on Power Electronics*, vol. 8, no. 3, pp. 329-335, 1993.

[8] G. A. Covic, G. L. Peters, and J. T. Boys, "An improved single phase to three phase converter for low cost ac motor drive", in *Proc. Power Electronics and Drive Systems*, 1995, pp. 549-554.

[9] G. T. Kim, and T. A. Lipo, "VSI-PWM Rectifier/Inverter System with a Reduced Switch Count", *IEEE Trans. on Industry Applications*, vol. 32, no. 6, pp. 1331-1337, 1996.

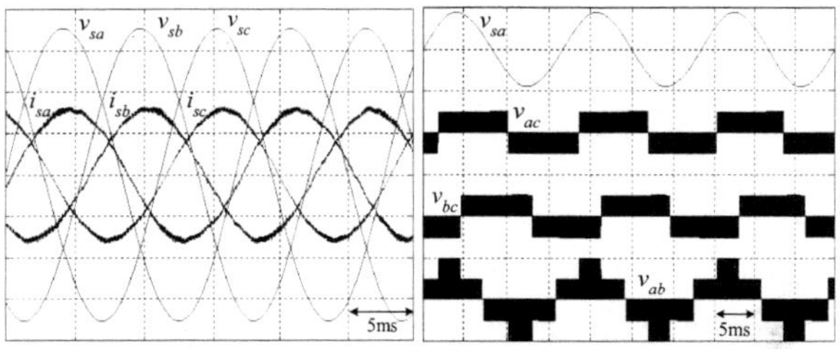

Fig. 3 Simulated results (a) three-phase source voltages and line currents [v_{sa}, v_{sb}, v_{sc}: 25V/div; i_{sa}, i_{sb}, i_{sc}:5A/div] (b) phase voltage and three AC terminal voltages [v_{sa}:100V/div; v_{ac}, v_{bc}, v_{ab}:400V/div].

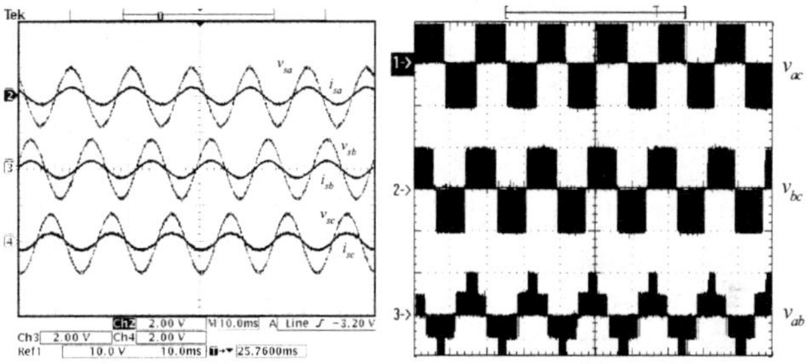

(a) (b)

Fig. 4 (a) Measured three-phase source voltages and currents of the proposed rectifier (b) measured three AC terminal voltage of the proposed rectifier [v_{sa}, v_{sb}, v_{sc}:100V/div; i_{sa}, i_{sb}, i_{sc}:20A/div; v_{ac}, v_{bc}:200V/div; v_{ab}:400V/div; time:10ms/div].

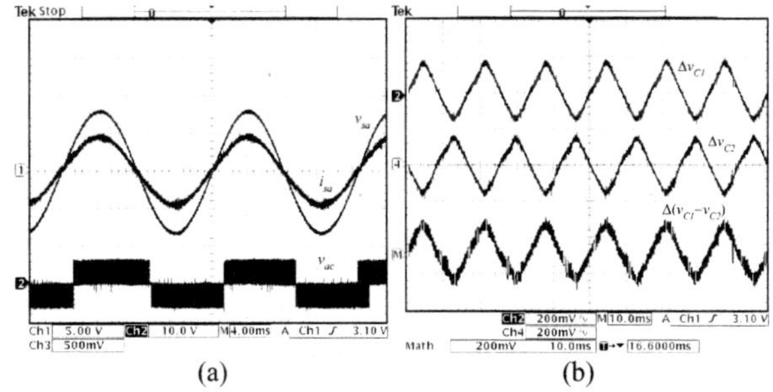

(a) (b)

Fig. 5 (a) Measured phase voltage, line current and ac side voltage in the rectifier leg *a* (b) measured capacitor ripple voltages and the voltage variation [v_s:50V/div; i_s:10A/div; v_{ac}:400V/div; Δv_{C1}, Δv_{C2}, $\Delta(v_{C1}\text{-}v_{C2})$:2V/div; time:10ms/div].

A Compact Generalized Solution to the Determination of Spectral Components for Multilevel Uniformly Sampled PWM

B. Zhou*, W.H. Lau* and Henry Chung[†]
*Department of Computer Engineering and Information Technology**
Department of Electronic Engineering[†]
City University of Hong Kong
Tat Chee Ave., Kowloon, Hong Kong

Abstract – The detailed derivation of a compact generalized solution to determine the spectral components for multilevel uniformly sampled PWM (MPWM) will be presented in this paper. By decomposing the multilevel PWM into a series of sub-PWM and making use of double Fourier series with variable translation, a compact solution without involving complex integral has been developed. Solutions for both singled-sided and double-sided sampled MUPWM are given in this paper. The compact solution has been verified with simulations.

1. Introduction

Owing to the limited power rating of switching devices, multilevel PWM (MPWM) inverters have been developed for high power applications. Typical topologies include neutral-point-clamped inverter [1], capacitor-clamed inverter [2] and cascaded inverter [3]. The spectral characteristics of the PWM output is one of the major concerns in choosing the control strategy. The classical multi-carrier based sinusoidal PWM (SPWM) [4] is one of the most common strategies to produce MPWM. Analytical solutions to determine the spectral components of multilevel naturally sampled PWM have been reported in [4-6]. The only solution for multilevel uniformly sampled PWM (MUPWM) is available in [6]. However, the solution is not easy to use since it involves complex integral in the final expressions.

In this paper, we present an analytical technique, based on our previous work in [5], to obtain a compact and generalized solution without involving complex integral to compute the spectral components for MUPWM. Simulations have been carried out and the results match very well with the proposed analytical solution.

2. The Generation of Multilevel PWM

Since the double-sided MUPWM can be formed from two single-sided ones, we will begin the discussion with the single-sided MUPWM. A typical example of a five-level PWM control strategy is shown in Figure 1

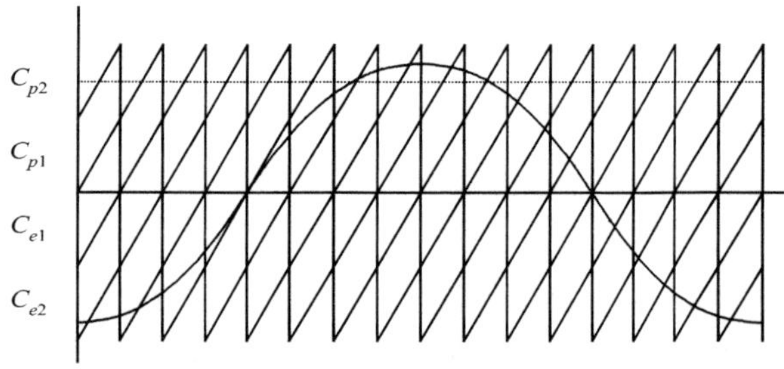

Figure 1. Generation of a five-level PWM signal

The four carriers are of the same angular frequency ω_c and with magnitude normalized to 1. The modulating signal is defined as $V_S = A_m \cos\omega_v t$ with angular frequency ω_v and magnitude A_m. The two carriers C_{p1} and C_{p2} placed in the positive band of V_S are called positive carriers. Similarly, the other two C_{e1} and C_{e2} are called negative carriers. When the instantaneous amplitude of V_s is larger than the carrier, the PWM output equals to 1, otherwise, the output is 0. By comparing the carriers and the modulating signal, a series of sub-PWM signals, v_{pi} and v_{ei}, will be generated. The final PWM output voltage v_o can then be obtained by summing the four sub-PWM signals as shown in Figure 2, i.e.,

$$v_o = v_{p1} + v_{p2} + v_{e1} + v_{e2} = \sum_{i=1}^{2}(v_{pi} + v_{ei}) \qquad (1)$$

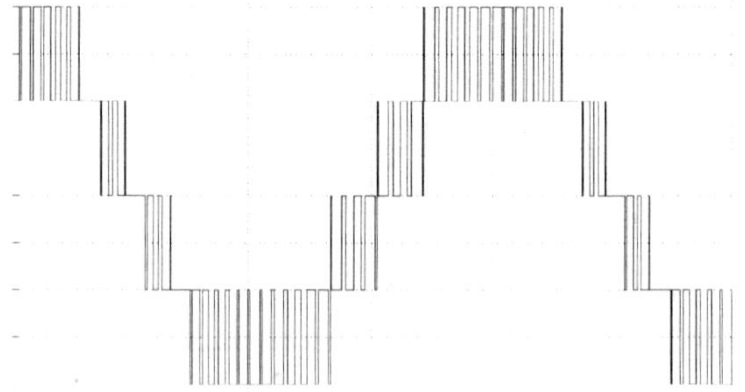

Figure 2. Output Waveform of the 5-level PWM

The work described in this paper is fully supported by a research grant (CityU 1192/99E) from the RGC of the HKSAR, China

In general, a $2N$+1 level PWM requires $2N$ carriers. The PWM output voltage will be the sum of all the $2N$ sub-PWMs.

2. The Generalized Solution for MUPWM

Since the analysis of uniform sampling involves discontinuous variable and the solution will be more complicated than that of the natural sampling. Here we extend the technique in [7] to derive the analysis of the single-sided MUPWM in detail.

A. Uniform sampling and 3-D PWM model

The interaction between the carrier and the modulating signal within one single switching cycle is shown in Figure 3. A sub-PWM signal is generated in the positive band by comparing C_{pi} with a modulating signal with a DC offset of 1-$2i$. The corresponding pulse width function $\Omega_p(\phi)$ can be defined as:

$$\frac{\Omega_p(\phi)}{k\pi} - 1 = A_m \cos\phi - 2i + 1 \qquad (2)$$

Similarly, the pulse width function for the negative band is given as follows:

$$\frac{\Omega_e(\phi)}{k\pi} - 1 = A_m \cos\phi + 2i - 1 \qquad (3)$$

Single-sided and double-sided PWM will be obtained if k=1 and k=1/2, respectively.

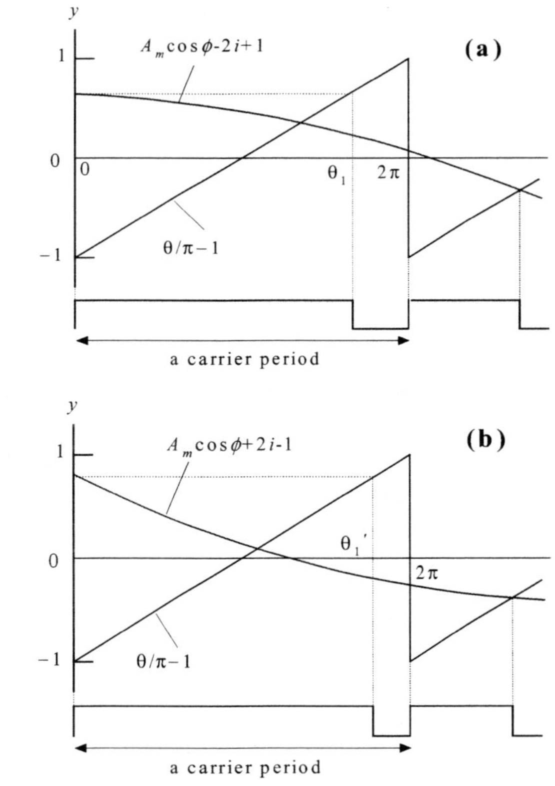

Figure 3. Pulse width for uniform PWM in one switching cycle: (a) generated by +ve carrier; (b) generated by -ve carrier

For natural sampling, the pulse width functions $\Omega_p(\phi)$ and $\Omega_e(\phi)$ are simply related to the continuous variable ϕ. However, they will become discontinuous for uniform sampling since the pulse width is defined at the beginning of each switching cycle as shown in Figure 3. The time axis represents two frequency variables: ϕ and θ are the angular frequencies of the modulating signal and switching carrier, respectively, and they are defined as:

$$\begin{aligned} \phi &= \alpha\,|\,\theta\,|_{2\pi} \\ \theta &= \omega_c t \end{aligned} \qquad (4)$$

where $\alpha = \omega_v / \omega_c$ and $|\,\theta\,|_{2\pi}$ denotes the nearest multiple of 2π less than or equal to θ.

Since ϕ is not a continuous variable, we need to make use of variable substitution technique to translate it to a continuous domain for carrying out the double Fourier series analysis [7]. A new variable Φ is defined as:

$$\Phi = \phi + \alpha(\theta - |\,\theta\,|_{2\pi}) \qquad (5)$$

Then equation (4) can then be rewritten as:

$$\begin{aligned} \Phi &= \omega_v t \\ \theta &= \omega_c t \end{aligned} \qquad (6)$$

Within 0 to 2π, Φ is continuous and can be divided into several sections according to the intersection points between the carrier envelope and modulating signal. The envelope of the carrier is given by two straight lines, y=-1 and y=1, and may or may not intersect with the modulating signal. Figure 4 shows the three possible cases for a positive carrier and the intersection points Φ_i is given in (7) with $\Phi_0 = \pi/2$.

$$\Phi_i = \begin{cases} \cos^{-1}\left(\dfrac{i}{NM}\right), & \dfrac{i}{NM} \le 1 \\ 0, & \dfrac{i}{NM} > 1 \end{cases} \quad (i=1,2,\ldots N) \qquad (7)$$

Similarly, $\pi - \Phi_i$ will be the projection of the intersection points between the modulating signal and a negative carrier signal.

By examining all the intersection points, a 3-D PWM function $F(\theta, \Phi)$ is defined in (8) within the region $-\pi \le \Phi \le \pi$ and $0 \le \theta \le 2\pi$. $F_p(\theta, \Phi)$ and $F_e(\theta, \Phi)$ represent the PWM output of the positive and negative bands, respectively.

$$F_p(\theta, \Phi) = \begin{cases} V & -\Phi_{i-1} \le \Phi \le -\Phi_i \text{ and } 0 \le \theta \le \Omega_p \\ V & -\Phi_i \le \Phi < \Phi_i \text{ and } 0 < \theta < 2\pi \\ V & \Phi_i \le \Phi \le \Phi_{i-1} \text{ and } 0 \le \theta \le \Omega_p \\ 0 & \text{otherwise} \end{cases} \qquad (8)$$

$$F_e(\theta,\Phi)=\begin{cases} V & -(\pi-\Phi_i)\le\Phi\le-(\pi-\Phi_{i-1})\ \&\ 0\le\theta\le\Omega_e \\ V & -(\pi-\Phi_{i-1})<\Phi<\pi-\Phi_{i-1}\ \&\ 0<\theta<2\pi \\ V & \pi-\Phi_{i-1}\le\Phi\le\pi-\Phi_i\ \&\ 0\le\theta\le\Omega_e \\ 0 & \text{otherwise} \end{cases} \tag{9}$$

where Ω_p and Ω_e are given in (2) and (3).

The final MUPWM waveform $v(t)$ can be described by:

$$\begin{aligned} v(t) &= F_p(\theta,\Phi)+F_e(\theta,\Phi) \\ &= F_p(\omega_c t,\omega_v t)+F_e(\omega_c t,\omega_v t) \end{aligned} \tag{10}$$

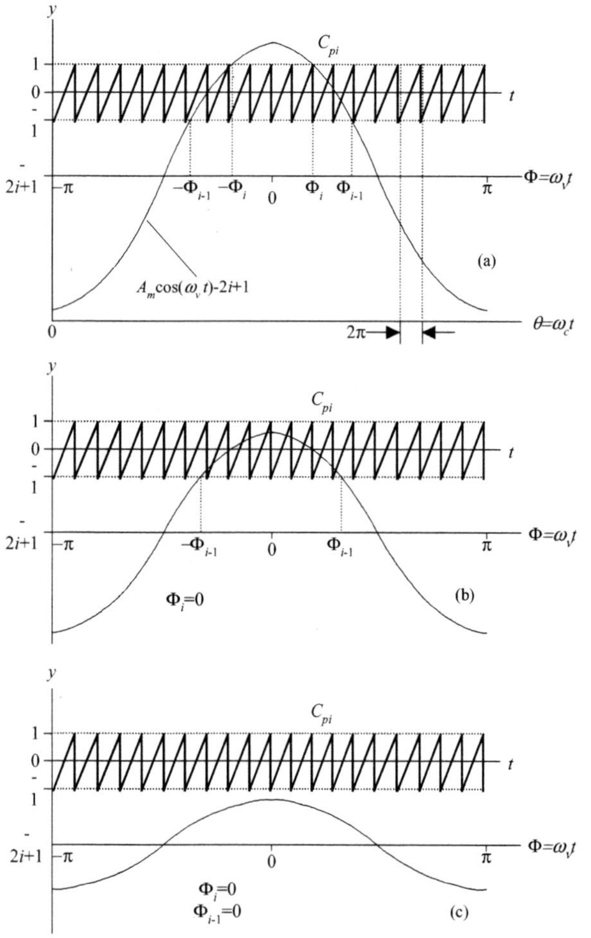

Figure 4. Different situations on the intersection between the modulating signal and a positive carrier (a) intersect with y=-1 and y=1; (b) intersect with y=-1; (c) no intersection

B. Spectrum of single-sided MUPWM

$F(\theta,\Phi)$ is a periodic function in both θ and Φ and can be represented by double Fourier series in (11).

$$\begin{aligned} F(\theta,\phi) &= \frac{1}{2}A_{00}+\sum_{n=1}^{\infty}\left[A_{0n}\cos n\phi+B_{0n}\sin n\phi\right] \\ &+\sum_{m=1}^{\infty}\sum_{n=\pm1}^{\pm\infty}\left[A_{mn}\cos(m\theta+n\phi)+B_{mn}\sin(m\theta+n\phi)\right] \end{aligned} \tag{11}$$

where

$$A_{mn}=\frac{1}{2\pi^2}\int_0^{2\pi}\int_0^{2\pi}F(\theta,\phi)\cos(m\theta+n\phi)\,d\theta d\phi$$

$$B_{mn}=\frac{1}{2\pi^2}\int_0^{2\pi}\int_0^{2\pi}F(\theta,\phi)\sin(m\theta+n\phi)\,d\theta d\phi$$

Substituting (6)-(10) into (11), we can obtain the output for single-sided MUPWM which is given in (12).

$$\begin{aligned} v_s(t) &= N+\sum_{m=1}^{\infty}\frac{2\sin(m\omega_c t)}{m\pi^2}\left[\int_{\Phi_0}^{\Phi_N}\cos(2\pi mNM\cos u)du-(\Phi_N-\Phi_0)\right] \\ &+\sum_{m=0}^{\infty}\sum_{n=\pm1}^{\pm\infty}\sum_{i=1}^{N}\frac{1}{2(\alpha n+m)\pi^2}\{P^-\int_{\Phi_{i-1}}^{\Phi_i}(\Psi_s^++\Psi_s^-)du-Q^+\int_{\Phi_{i-1}}^{\Phi_i}(\Psi_c^++\Psi_c^-)du \\ &+2\frac{\sin 2\pi\alpha n}{n(\alpha n+m)}[\sin n\Phi_i-(-1)^n\sin n\Phi_{i-1}]\}\cos(m\omega_c t+n\omega_v t) \\ &\sum_{m=0}^{\infty}\sum_{n=\pm1}^{\pm\infty}\frac{1}{2\pi^2}\{\frac{2}{n(\alpha n+m)}(\sin n\Phi_0-\sin n\Phi_N)[1+(-1)^n] \\ &+\sum_{i=1}^{N}\frac{1}{\alpha n+m}\{P^+\int_{\Phi_{i-1}}^{\Phi_i}(\Psi_c^++\Psi_c^-)du+Q^-\int_{\Phi_{i-1}}^{\Phi_i}(\Psi_s^++\Psi_s^-)du \\ &-\frac{2R}{n(\alpha n+m)}[\cos 2\pi(\alpha n+m)-1]\}\}\sin(m\omega_c t+n\omega_v t) \end{aligned} \tag{12}$$

where $P^{\pm}=(-1)^n\cos 2\pi\alpha ni\pm\cos 2\pi\alpha n(1-i)$,

$Q^{\pm}=(-1)^n\sin 2\pi\alpha ni\pm\sin 2\pi\alpha n(1-i)$,

$\Psi_s^{\pm}=\sin[2\pi NM(\alpha n+m)\cos u\pm nu]$,

$\Psi_c^{\pm}=\cos[2\pi NM(\alpha n+m)\cos u\pm nu]$ and

$R=\sin n\Phi_i-(-1)^n\sin n\Phi_{i-1}$.

C. Spectrum of double-sided MUPWM

For double-sided MUPWM, $F_d(\theta,\Phi)$ can be obtained by summing a trailing edge PWM $F_{dt}(\theta,\Phi)$ and a leading edge PWM $F_{dl}(\theta,\Phi)$, which is the reflection of $F_{dl}(\theta,\Phi)$.

$$\begin{aligned} F_d(\theta,\Phi) &= F_{dt}(\theta,\Phi)+F_{dl}(\theta,\Phi) \\ &= F_{dt}(\theta,\Phi)+F_{dt}(-\theta,-\Phi) \end{aligned} \tag{13}$$

By substituting k=1/2 in (2), (3) and (13), the double-sided MUPWM $v_d(t)$ is given as follows:

$$\begin{aligned} v_d(t) &= N+\sum_{m=1,3,5\ldots}^{\infty}\sum_{i=1}^{N}\frac{4(-1)^i}{m\pi^2}\int_{\Phi_{i-1}}^{\Phi_i}\sin(\pi mNM\cos u)du\cos(m\omega_c t) \\ &+\sum_{m=0}^{\infty}\sum_{n=\pm1}^{\pm\infty}\sum_{i=1}^{N}\frac{1}{\pi^2}\{\frac{\Gamma^+}{\alpha n+m}\int_{\Phi_{i-1}}^{\Phi_i}(\Psi_s^++\Psi_s^-)du \\ &-\frac{\Gamma^-}{\alpha n+m}\int_{\Phi_{i-1}}^{\Phi_i}(\Psi_c^++\Psi_c^-)du \\ &+2\frac{\sin\pi\alpha n}{n(\alpha n+m)}[\sin n\Phi_i-(-1)^n\sin n\Phi_{i-1}]\}\cos(m\omega_c t+n\omega_v t) \end{aligned} \tag{14}$$

where $\Gamma^+=(-1)^n\cos(\alpha n+m)i\pi-\cos(\alpha n+m)(1-i)\pi$ and $\Gamma^-=(-1)^n\sin(\alpha n+m)i\pi+\sin(\alpha n+m)(1-i)\pi$.

3. Simulation Results

Extensive simulations have been carried out to demonstrate the validity of our solution. Two examples used to illustrate the single-sided and double-sided MUPWM solutions given in (12) and (14) are shown in Figure 5. The simulation conditions are: f_v =50Hz, f_c=2100Hz, N=2 (5-level), and M=0.8. Figures 5 (a) and 5(b) are the simulation and analytical results for single-sided modulation, respectively. Figures 5 (c) and 5(d) are the simulation and analytical results for double-sided modulation, respectively. It can be seen that the simulation results and analytical results match very well.

4. Conclusion

This paper presents the detail derivation of a compact and generalized analytical solution to determine the spectral characteristics of a MUPWM. Our solution is much simple than other solution reported earlier since complex integral is not involved and it only requires simple numerical integration. The solutions for both single-sided and double-sided modulation are given and also verified with simulations. The proposed technique has already also been extended to obtain the solution for overmodulation.

5. References

[1] A. Nabae, I. Takahashi, H. Akagi, "A new neutral-point-clamped PWM inverter," *IEEE Trans. on Industry Application,* Vol. IA-17, No. 5, pp. 518-523, 1981

[2] Xiaoming Yuang, H. Stemmler and I. Barbi, "Self-balancing of the clamping-capacitor-voltages in the multilevel capacitor-clamping-inverter under sub-harmonic PWM modulation," *IEEE Trans. on Power Electronics*, Vol. 6, pp. 256-263, 2001

[3] F.Z. Peng and J.S. Lai, "Dynamic performance and control of a star var compensator using cascade multilevel inverters'" *Conf. Record of IEEE-IAS Annu. Meeting*, Vol. 2, pp. 1009-1015, San Diego, CA, Oct. 1996

[4] G. Carrara, S. Gardella, M. Marchesoni,R. Salutari, and G. Sciutto, "A new multilevel PWM method: a theoretical analysis," *IEEE Trans. on Power Electronics*, Vol. 7, No. 3, pp. 497-505, 1992

[5] W.H. Lau, Bin Zhou and H. Chung, "Compact analytical solutions for determining the spectral characteristics of multi-carrier based multilevel PWM," to appear in *IEEE Trans. on CAS-I.*

[6] B.P. McGrath and D.G. Holmes, "An analytical technique for the determination of spectral components of multilevel carrier-based PWM methods," *IEEE Trans. on Industrial Electronics*, Vol. 49, pp. 847-857, 2002

[7] Mellor, P.H.; Leigh, S.P.; Cheetham, B.M.G.; "Reduction of spectral distortion in class D amplifiers by an enhanced pulse width modulation sampling process", *IEE Proc: Circuits, Devices and Systems*, Vol. 138 , pp. 441 –448, 1991

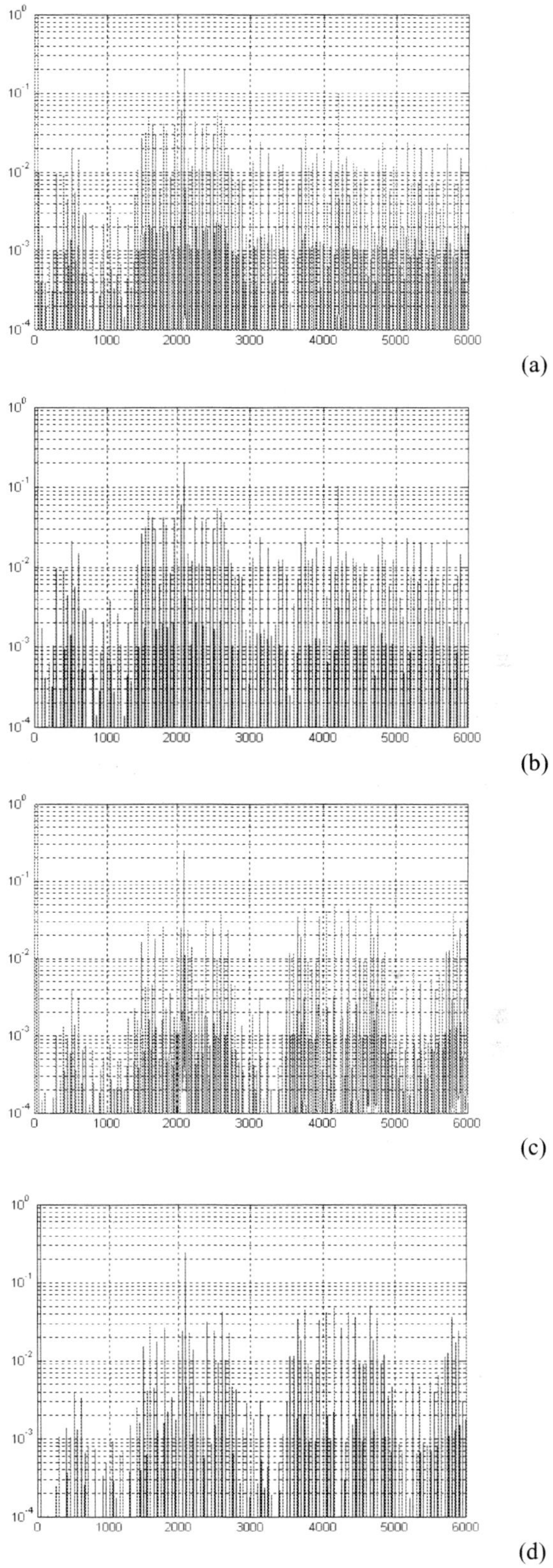

Figure 5 Simulation results for a 5-level MUPWM: (a) Single-sided modulation, (b)double-sided modulation; Analytical results: (c) Single-sided modulation, (d)double-sided modulation

CURRENT-MODE CONTROL TO ENHANCE CLOSED-LOOP PERFORMANCE OF ASYMMETRICAL HALF-BRIDGE DC-TO-DC CONVERTERS

Wonseok Lim, Byungcho Choi, and Jiemyung Ko

School of Electrical Engineering and Computer Science, Kyungpook National University
1370, Sankyuk-Dong, Buk-Gu, Taegu, 702-701, Korea
Fax:82-53-950-5505, Email:bchoi@ee.kyungpook.ac.kr

ABSTRACT

Asymmetrical half-bridge dc-to-dc converters exhibit the fourth-order power stage dynamics. When the conventional voltage-mode control is employed, the fourth-order dynamics impose certain constraints on the feedback compensation design and the result could be poor closed-loop performance. This paper demonstrates that the current-mode control could alleviate the detrimental effects of the power stage dynamics and offer an improvement in the closed-loop performance. The principles, performance, and experimental results of current-mode control are presented in comparison with those of voltage-mode control.

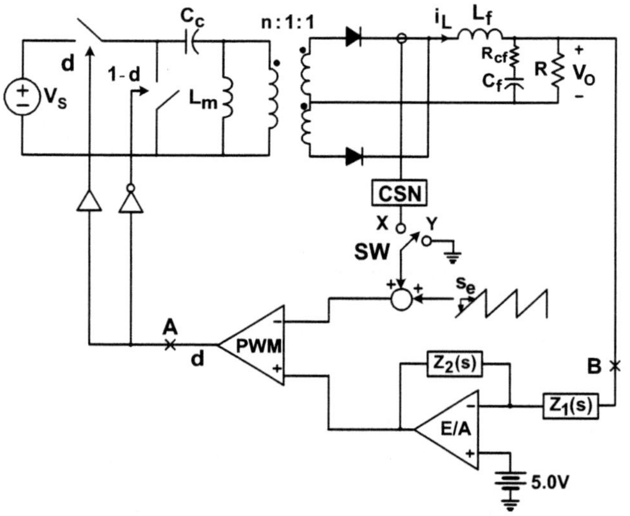

Fig. 1. Asymmetrical half-bridge dc-to-dc converter combined with PWM feedback controller: $V_s = 48\,V$, $V_o = 5\,V$, $C_c = 6\mu F$, $f_s = 200\,kHz$, $L_m = 18\mu H$, $n = 3$, $L_f = 4.6\mu H$, $C_f = 194\mu F$, $R_{cf} = 0.03\,\Omega$, $R = 0.5\,\Omega$, $CSN\ gain = 0.41$ $S_e = 0.36\ V/\mu s$ for voltage-mode control and $S_e = 0.1\ V/\mu s$ for current-mode control. The switch SW is at "Y" position for voltage-mode control and at "X" position for current-mode control.

1. INTRODUCTION

Recently, bridge-type pulse width-modulated (PWM) converters operating with asymmetrical duty ratios [1], [2] have received increasing research attention primarily due to their circuit characteristics offering zero-voltage switching conditions for active switches without any penalty of an increased conduction loss. As an example of such converters, Fig. 1 shows an asymmetrical half-bridge (ASHB) dc-to-dc converter combined with PWM feedback controller. In addition to the circuit components usually found in conventional half-bridge converters, the power stage of ASHB converter utilizes a clamp capacitor C_c and magnetizing inductor L_m to accommodate PWM operation with asymmetrical duty ratios.

Since the clamp capacitor and magnetizing inductance introduce an additional resonance [3], the ASHB converter exhibits the fourth-order power stage dynamics. When the conventional voltage-mode control is employed, the fourth-order dynamics impose certain constraints on the voltage feedback compensation design and the resulting design only offers limited closed-loop performance.

This paper demonstrates that the current-mode control could ease the detrimental effects of the power stage dynamics and enhance the closed-loop performance of ASHB converters. This paper presents theoretical and experimental details about principles, benefits, and performance of current-mode control. To highlight the benefits of current-mode control, the performance of the current-mode controlled ASHB converter is presented in parallel with that of the voltage-mode controlled converter.

This work was supported in part by EESR(00-Y-01), which is funded by MOCIE(Ministry of commerce, industry), and in part by grant No. R12-2002-055-02001-0 from the Basic Research Program of the Korea Science & Engineering Foundation.

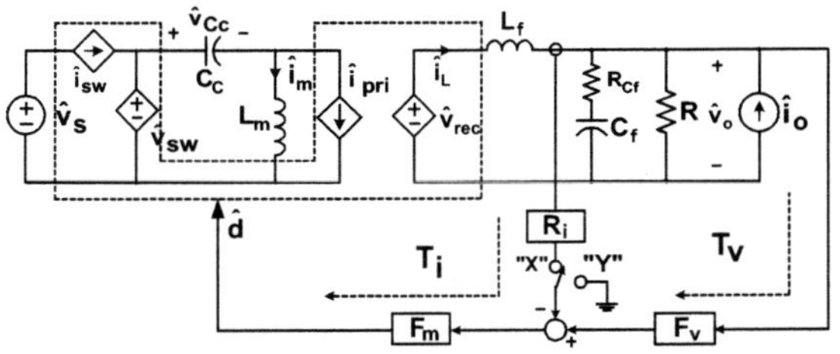

(a)

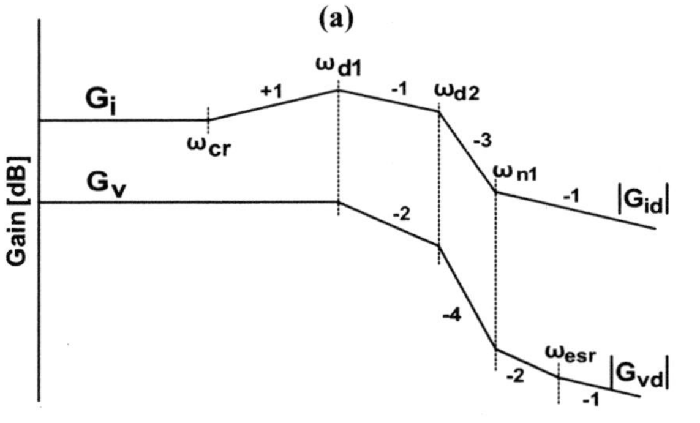

(b)

Fig. 2. Small-signal power stage model and transfer functions of ASHB converter. (a) Small-signal power stage model: $\hat{i}_{sw} = D\,\hat{i}_m + (I_M + I_L/n)\,\hat{d} + (D/n)\,\hat{i}_L$, $\hat{v}_{sw} = D\hat{v}_s + V_S\hat{d}$, $\hat{v}_{rec} = ((1-2D)/n)\hat{v}_{cc} + (D/n)\hat{v}_s + ((V_S - 2V_{Cc})/n)\hat{d}$, $\hat{i}_{pri} = ((2D-1)/n)\hat{i}_L + (2I_L/n)\hat{d}$ with $D = 0.21$, $I_M = 1.9A$, $I_L = 10A$, $V_s = 48V$, $V_{cc} = 10V$.
(b) Asymptotic plots for transfer functions: $G_i = (2(1-2D)V_s/(nR) = 36.6$, $\omega_{cr} = 1/(C_f R) = 9.7kr/s$, $\omega_{d1} \approx 1/\sqrt{L_f C_f} = 33.5kr/s$, $\omega_{d2} \approx 1/\sqrt{L_m C_c} = 95.4kr/s$, $\omega_{n1} \approx 1.4/\sqrt{L_m C_c} = 135kr/s$, $G_v = (2(1-2D)V_s)/n = 18.4$, $\omega_{esr} = 1/(C_f R_{cf}) = 172kr/s$.

2. SMALL-SIGNAL DYNAMICS AND LIMITATION OF VOLTAGE-MODE CONTROL

2.1. Small-Signal Power Stage Dynamics

Fig. 2(a) shows the small-signal model of the ASHB converter obtained by averaging and linearizing the power stage dynamics. The accuracy of the small-signal model for ASHB converters was already confirmed in [3]. The duty ratio-to-output voltage transfer function, $G_{vd}(s)$, can be derived from Fig. 2(a) and subsequently approximated into a factorized form

$$G_{vd}(s) \equiv \frac{\hat{v}_o}{\hat{d}} \approx G_v \frac{\left(1+\frac{s}{\omega_{esr}}\right)\left(1+\frac{s}{Q_{n1}\omega_{n1}}+\frac{s^2}{\omega_{n1}^2}\right)}{\Delta(s)} \qquad (1)$$

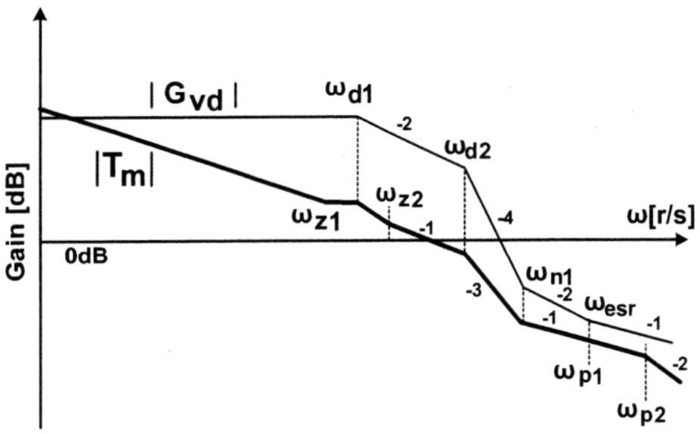

Fig. 3. Loop gain of voltage-mode control and duty ratio-to-output transfer function.

where

$$\Delta(s) = \left(1+\frac{s}{Q_{d1}\omega_{d1}}+\frac{s^2}{\omega_{d1}^2}\right)\left(1+\frac{s}{Q_{d2}\omega_{d2}}+\frac{s^2}{\omega_{d2}^2}\right) \qquad (2)$$

The first quadratic term in (2) is a result of the resonance between C_f and L_f, and the second quadratic term is due to the secondary resonance between C_c and L_m. The duty ratio-to-inductor current transfer function, $G_{id}(s)$, can also be derived from Fig. 2(a) as

$$G_{id}(s) \equiv \frac{\hat{i}_L}{\hat{d}} \approx G_i \frac{\left(1+\frac{s}{\omega_{cr}}\right)\left(1+\frac{s}{Q_{n1}\omega_{n1}}+\frac{s^2}{\omega_{n1}^2}\right)}{\Delta(s)} \qquad (3)$$

Fig. 2(b) shows the asymptotic plots for $|G_{vd}(s)|$ and $|G_{id}(s)|$ along with the expressions and values for dc gains, corner frequencies, and resonant frequencies of the respective transfer function. Derivation of those quantities have been already presented in [3].

2.2. Limitation of Voltage-Mode Control

When the switch "SW" in Fig. 1 is placed at "Y" position, the feedback controller becomes the conventional voltage-mode control. Fig. 3 shows the asymptotic plots for $|G_{vd}(s)|$ and loop gain, $|T_m| = |G_{vd}(s)F_mF_v|$, of the voltage-mode controlled converter. The asymptotic plot for the loop gain is constructed from the product of $G_{vd}(s)$, PWM gain $F_m = 1/(S_eT_s)$, and the voltage-feedback compensation of

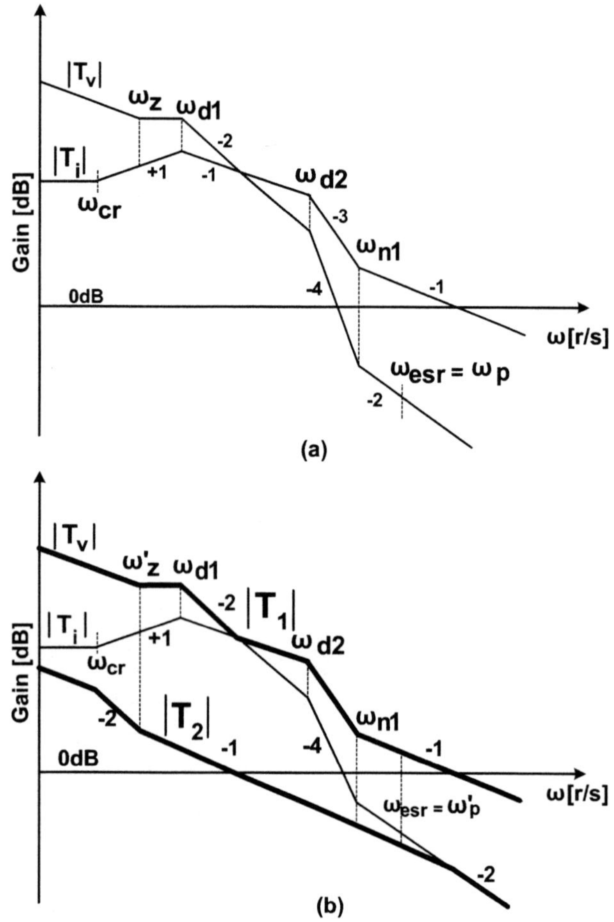

Fig. 4. Loop gains of current-mode control. (a) Individual loop gains. (b) Overall loop gain and outer loop gain.

$$\frac{Z_2(s)}{Z_1(s)} = \frac{K_v(1+s/\omega_{z1})(1+s/\omega_{z2})}{s(1+s/\omega_{p1})(1+s/\omega_{p2})} \tag{4}$$

whose compensation parameters are selected according to the guidelines given in [3]. The loop gain reveals the limitation of the voltage-mode control applied to ASHB converters. Around the secondary resonance at $\omega_{d2} \approx 1/\sqrt{L_m C_c}$, the loop gain undergoes an additional phase delay of 180° and the phase of the loop gain drops rapidly to -270°. To ensure stability with the presence of the secondary resonance, the voltage-mode control should place the 0-dB crossover frequency of the loop gain at the frequencies well before ω_{d2}. This constraint places the upper limit on the feedback gain and hinders further enhancement of the closed-loop performance.

3. CURRENT-MODE CONTROL

As shown in Fig. 1, the current-mode control is implemented by sensing the diode current, with the current sensing network (CSN) and feeding it to PWM block with "SW" at "X" position. Fig. 4 (a) illustrates the principles of current-mode control using individual feedback loops associated with respective feedback signals. The current loop, $T_i(s) = G_{id}(s)R_iF_m$ with R_i denoting the CSN gain, represents the feedback loop created by the diode current feedback. The voltage loop, $T_v(s) = G_{vd}F_m(Z_2(s)/Z_1(s))$, denotes the feedback loop associated with the output voltage feedback. The asymptotic plot of the $|T_v(s)|$ is constructed assuming

$$\frac{Z_2(s)}{Z_1(s)} = \frac{K_v'(1+s/\omega_z')}{s(1+s/\omega_p')} \tag{5}$$

for the voltage feedback compensation. The parameters for the voltage feedback compensation are selected as outlined below.

- Place ω_z' as high as possible, yet not exceeding ω_{d1}
- Place ω_p' at ω_{esr} to nullify the effects of ω_{esr}
- Adjust K_v' to trade-off stability margins and closed-loop performance.

Fig. 4(b) shows the loop gains of the converter. T_1 represents the overall loop gain [4] measured at 'A' point in Fig. 2(a). Being a vector sum of individual feedback loops, $T_1 = T_i + T_v$, this loop gain illustrates the design strategy for current-mode control. By designing $|T_v| \gg |T_i|$ at low frequencies and $|T_i| \gg |T_v|$ at high frequencies, the current-mode control could increase the feedback gain while locating the crossover frequency of T_1 at higher frequencies with good phase margin. T_2 is the outer loop gain [4] measured at 'B' point. The asymptotic plot of T_2 is constructed using the relationship of $T_2 = T_v/(1+T_i)$. As shown in Fig. 4(b), the effects of the secondary resonance do not show in T_2 and therefore the feedback compensation can be designed independently from the secondary power stage resonance.

4. PERFORMANCE ENHANCEMENT WITH CURRENT-MODE CONTROL

This section presents the performance of the current-mode control employed to the ASHB converter. To highlight the merits of current-mode control, the performance of the current-mode controlled converter is presented in parallel with that of the voltage-mode controlled converter.

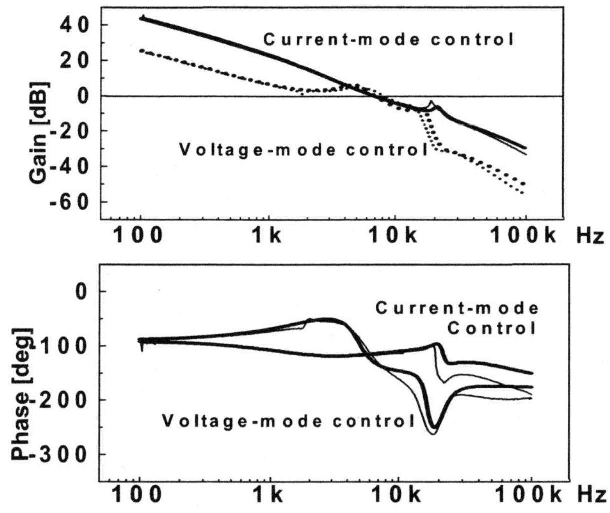

(a)

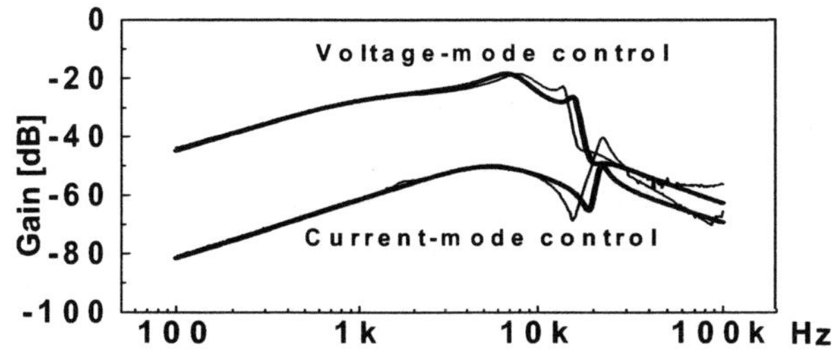

(b)

Fig. 5. Performance enhancement with current-mode control. (a) Loop gain. (b) Input-to-output transfer function. Thick lines are the analytical predictions and thin lines are experimental data.

Fig. 5 (a) compares the outer loop gain of the current-mode controlled converter with the loop gain of the voltage-mode controlled converter. The feedback compensation parameters are deliberately selected in such a way that both control schemes result in the same 0-dB crossover frequency at $16\pi\ kr/s$: $R_i = 0.41$, $F_m = 0.37$, $K_v' = 48.4$, $\omega_z' = 28.6\ kr/s$, $\omega_p' = 172\ kr/s$ for the current-mode control, and $F_m = 0.56$, $K_v = 17.7$, $\omega_{z1} = 24.8\ kr/s$, $\omega_{z2} = 37.2\ kr/s$, $\omega_{p1} = 172\ kr/s$, $\omega_{p2} = 296\ kr/s$ for the voltage-mode control. Both analytical predictions and experimental measurements are presented in parallel in Fig. 5. As shown in Fig. 5(a), the current-mode control offers the superior loop gain characteristics with enhanced phase margin and substantial gain boost for all frequencies. Fig. 5(b) compares the input-to-output transfer function of the converters. As the direct benefits of improved loop gain characteristics, the current-mode control provides an additional 40dB attenuation up to mid frequencies, compared with the voltage-mode control.

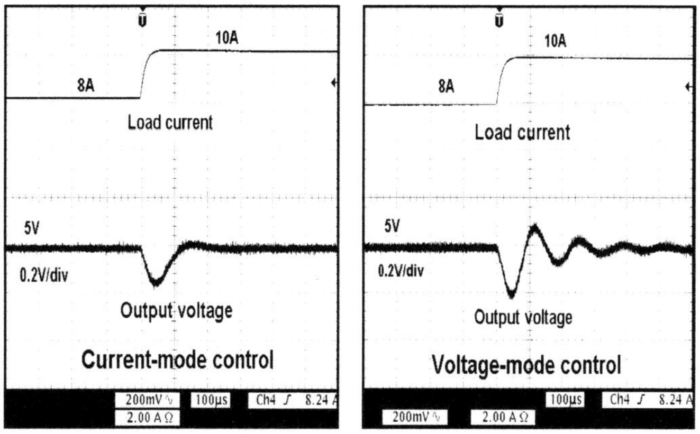

Fig. 6. Load transient responses.

Fig. 6 shows the measured load transient response of the experimental converters. The current-mode controlled converter exhibits smaller undershoot and well damped transient behavior while the voltage-mode controlled converter shows a larger undershoot and oscillatory behavior.

5. CONCLUSIONS

The voltage-mode control adapted to the ASHB converter is found only to offer limited closed-loop performance. To secure the stability with the presence of the secondary resonance in the power stage transfer functions, the voltage-mode control should sacrifice the closed-loop performance.

This paper demonstrated that the current-mode control could overcome the adverse effects of the secondary resonance and enhance the closed-loop performance of ASHB converters. By making the current feedback dominant at high-frequencies, the current-mode control could improve the loop gain, input-to-output transfer function, and load transient responses.

6. REFERENCES

[1] T. Ninomiya, N. Matsumoto, M. Nakahara, and K. Harada, "Static and dynamic analysis of ZVS half-bridge converter with PWM control," In proc IEEE Power Electronics Specialists' Conference, 1991, pp. 230-237.

[2] R. Oruganti, P. C. Heng, J. T. K. Guan, L. A. Choy, "Soft-switched DC/DC converter with PWM control," IEEE Trans. Power Electron., vol. 13, no. 1, pp. 102-114, Jan. 1998.

[3] S. Bang, W. Lim, B. Choi, T. Ahn, and S. Park, "Dynamic analysis and control design of asymmetrical half-bridge dc-to-dc converters," In Proc. IECEC'03, Aug. 2003, Portmouth.

[4] R. B. Ridley, B. H. Cho, and F. C. Lee," Analysis and interpretation of loop gains of multi-loop-controlled switching regulator," IEEE Trans. Power Electron., vol. 3, no. 4, pp. 489-498, Oct. 1998.

STEP-UP VERSUS STEP-DOWN DC/DC CONVERTERS FOR RF-POWERED SYSTEMS

Sean Nicolson and Khoman Phang
{nicolson, kphang}@eecg.utoronto.ca
The Edward S. Rogers Sr. Department of Electrical and Computer Engineering, University of Toronto
10 King's College Rd., Toronto, Ontario, M5S 3G4, Canada

ABSTRACT

We compare two designs for dual power supply RF-powered systems such as implantable stimulators and contactless smartcards. Specifically, we compare a traditional design using a step-up DC/DC converter versus a design that uses a step-down converter. We analyze which system can operate from a lower input source voltage, and show how non-idealities in DC/DC converters affect the relative merits of the step-up versus step-down designs. We demonstrate that the step-down architecture can increase maximum operable link distance and power available to the load in systems with high source resistances or inefficient DC/DC converters. Through simulation and experiment, we verify that source resistances greater than 250kΩ are possible in RF-powered systems, and that step-down converters can improve available load power by over 100%.

1. INTRODUCTION

RF-powered devices such as contactless smartcards and implantable neural prosthetics are increasingly pervasive and widely adopted. In some such applications, multiple power supply voltages are required, for example, to program EEPROM, or for electrical stimulation of muscle tissue. For systems with standard power supplies, the solution is clear: as shown in Figure 1 (a), a step-up DC/DC converter is used to generate the higher voltage from the lower system supply. In RF-powered systems however, the solution is less clear because the voltage available at the receiver coil is very sensitive to link distance and to loading conditions. Because DC/DC converters are in essence load converters [3], one method of optimizing the load on the receiver coil is to use a step-down converter in place of a step-up. In such a system, illustrated in Figure 1 (b), the higher supply voltage is generated directly from the receiver coil, and a step-down converter generates the low voltage system supply.

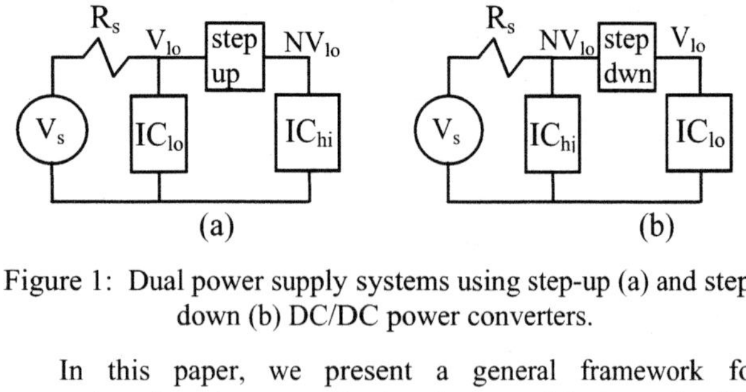

Figure 1: Dual power supply systems using step-up (a) and step-down (b) DC/DC power converters.

In this paper, we present a general framework for comparing the step-up and step-down architectures in RF-powered systems. In section II we describe the general characteristics of weakly coupled RF-powered systems. In section III, we begin our comparison of the systems in Figure 1 using an ideal DC transformer model of a power converter. In sections IV and V, we expand our model to account for non-idealities present in real systems. Through our analysis in sections IV-V, and our simulations in section VI, we show that in RF-powered systems, using a step-down converter in place of a step-up allows operation down to a lower receiver coil voltage. A lower receiver coil voltage implies that weaker coupling between the coils is required, and this translates into greater maximum operable link distance, or alternatively more available power for the same link distance.

2. WEAKLY COUPLED RF-POWERED SYSTEMS

Some RF-powered systems, particularly in biomedical applications, are very small (mm dimensions) and therefore also have small receiver coils. Typically, these systems are powered using much larger transmitter coils held at a range of several cm [1]. The physical arrangement is shown in Figure 2 (a).

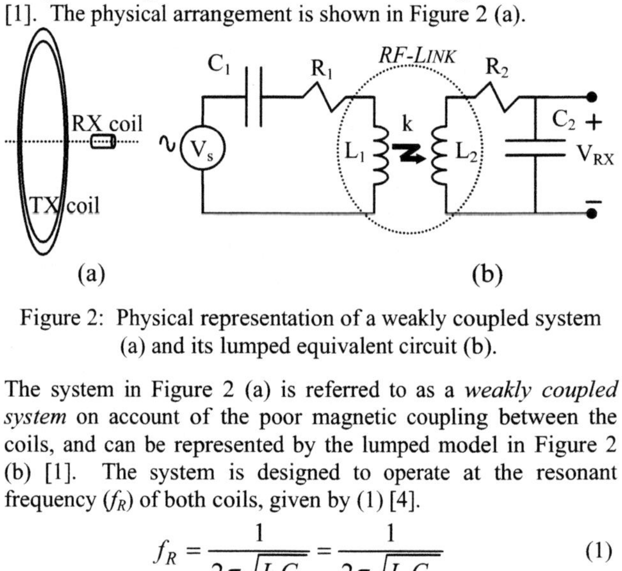

Figure 2: Physical representation of a weakly coupled system (a) and its lumped equivalent circuit (b).

The system in Figure 2 (a) is referred to as a *weakly coupled system* on account of the poor magnetic coupling between the coils, and can be represented by the lumped model in Figure 2 (b) [1]. The system is designed to operate at the resonant frequency (f_R) of both coils, given by (1) [4].

$$f_R = \frac{1}{2\pi\sqrt{L_1 C_1}} = \frac{1}{2\pi\sqrt{L_2 C_2}} \tag{1}$$

If the system in Figure 2 (b) is represented by the Thevinin equivalent (solid lines) in Figure 3, then R_s is given by (2) [1].

$$R_s = \frac{1}{(2\pi f_R C_2)^2 R_2} \tag{2}$$

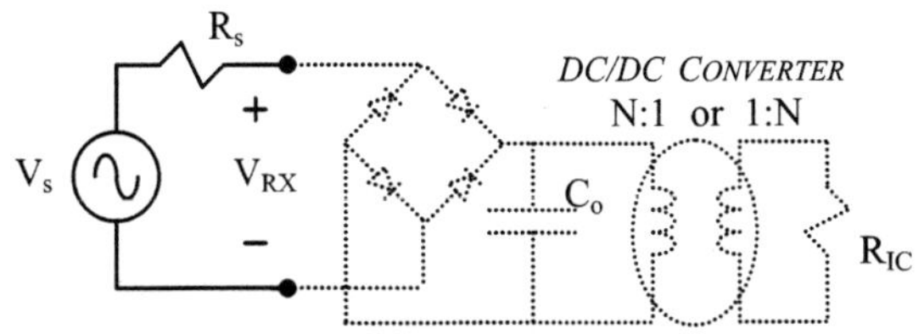

Figure 3: Thevinin equivalent of a weakly coupled RF-powered system, including a typical load.

Increasing power available to the load in weakly coupled systems allows either additional circuitry to be added to the remote portion of the system, or alternatively, greater operable link distance for the original level of power consumption. To maximize available load power, the mutual inductance between the coils should be as large as the physical constraints of the receiver will allow. To increase the mutual inductance, L_1 and L_2 should also be increased. However, as L_2 is increased, C_2 must be decreased to keep f_R at the desired frequency. Because R_s increases when C_2 decreases, *weakly coupled RF-powered systems have large source resistances by design*[1].

As shown in Figure 3, the transmitted RF signal is conditioned into a DC voltage using a full-wave rectifier and DC/DC converter. We have represented the DC/DC converter as an ideal DC transformer. Although DC/DC converters such as charge pumps and other switching converters are often presented as voltage converters, they are actually *power or load converters* (see Appendix I), [3]. Note that the small size of the receiver necessitates the use of power converters without inductors (such as charge-pumps). *Such converters often exhibit lower efficiency.*

3. STEP-UP VS STEP-DOWN: IDEAL CASE

Shown in Figure 4 is a model of the system in Figure 1 (a), where the system supply (V_{lo}) is generated directly from the source, and a step-up converter is used to generate the high voltage supply (V_{hi}). Seemingly, $V_{s(up)}$ is minimized by generating V_{lo} directly from the source.

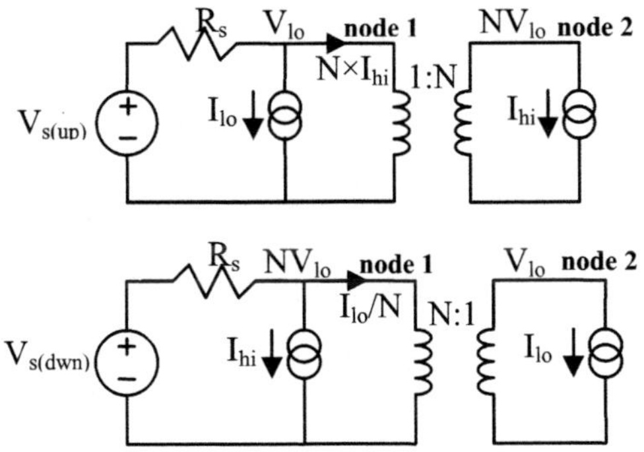

Figure 4: A dual power supply system can use either a step-up power converter (a) or a step down converter (b)

Shown in Figure 4 (b) is a model of the system in Figure 1 (b), where V_{hi} is generated directly from the source ($V_{s(dwn)}$), and a step-down power converter is used to generate V_{lo}.

[1] In [1], examples of weakly coupled systems have source resistances from 90Ω to 60kΩ. In the authors' experience, source resistances of over 250kΩ are possible.

At first glance, the step-down system of Figure 4 (b) appears to require a larger source voltage than the step-up system ($V_{s(dwn)} > V_{s(up)}$), and therefore is not suitable for low voltage applications. In reality, the values of N, V_{lo}, R_s, I_{lo}, and I_{hi} determine which system requires the lower source voltage; *the traditional step-up system is not always advantageous.*

To determine which system requires the lower source voltage, we define $V_{s(up)}$ and $V_{s(dwn)}$ using the voltages at nodes 1 and 2, and the current through R_s. For the step-up system, the current is given by (3), and the minimum $V_{s(up)}$ is given by (4).

$$I_{s(up)} = NI_{hi} + I_{lo} \tag{3}$$

$$V_{s(up)} = V_{lo} + R_s\left(NI_{hi} + I_{lo}\right) \tag{4}$$

Alternatively, for the step-down system, the current is given by (5) and the minimum source voltage is therefore (6).

$$I_{s(dwn)} = I_{hi} + I_{lo}/N \tag{5}$$

$$V_{s(dwn)} = NV_{lo} + R_s\left(I_{hi} + I_{lo}/N\right) \tag{6}$$

Examining (4) and (6), we note that although the voltage contribution is N times larger in (6), the load current contribution is N times smaller. Thus, for sufficiently large R_s, the step-down solution can function using a lower source voltage. To determine the range of R_s that results in superiority of the step-down solution, we solve the inequality given by (7), which yields (8).

$$V_{s(dwn)} \le V_{s(up)} \tag{7}$$

$$R_s \ge \frac{V_{lo}}{I_{hi} + I_{lo}/N} \tag{8}$$

Consider a system where I_{lo} = 200μA, I_{hi} = 50μA, and V_{lo} = 3V. In this case, for $R_s \ge$ 30kΩ, the step-down solution is advantageous. Although battery internal resistances are « 30kΩ, some RF power sources have internal resistances > 100kΩ. Now consider a system where I_{lo} = 200mA, I_{hi} = 50mA. In this case, the step-down system is advantageous for only $R_s \ge$ 30Ω.

4. A COMPLETE MODEL FOR THE DC/DC CONVERTER

Practical implementations of DC/DC converters have non-idealities which are not included in the DC transformer model presented in Section III [3,5,6]. These non-idealities affect which architecture (step-up or step-down) is advantageous. Here we summarize the main sources of power loss in CMOS DC/DC switching converters, and model their effects using 3 parameters: operating current losses (I_C), output resistance losses (R_o), and charge redistribution losses (η).

A. Operating Current Losses (I_c)

In some switching converters, integrated circuits biased with constant current are used to regulate the output voltage. In charge pumps, parasitic capacitances must be charged every cycle. A constant current source (I_c) in parallel with the charge pump input, as shown in Figure 5, can be used to model any of these losses.

B. Output Resistance Losses (R_o)

MOSFET switches, present in all CMOS DC/DC converters, are imperfect conductors and give rise to an output resistance. The switched capacitor nature of charge pump operation also introduces an equivalent output resistance [5]. The output resistance (R_o) appears in series with the charge pump output, as shown in Figure 5. Note that charge-pumps typically have higher output resistance because they contain more switches than converters relying on inductors (buck, boost, etc.). Also note that step-up converters typically have higher output resistance than step-down converters.

C. *Charge Redistribution Losses (η)*

If two capacitors with different voltages are connected in parallel, some energy is lost due to charge redistribution [5]. When a charge pump is loaded, the voltage on the output capacitor drops every cycle. When the pumping capacitors are connected and disconnected from the output capacitor, energy is lost due to charge redistribution. The larger the load current, the more energy is lost, and the less current efficient the charge pump becomes. We can represent charge redistribution losses by modifying the output current with a current efficiency factor η where $0 \leq \eta \leq 1$. Shown in Figure 5 is a complete models of a step-up DC/DC converter.

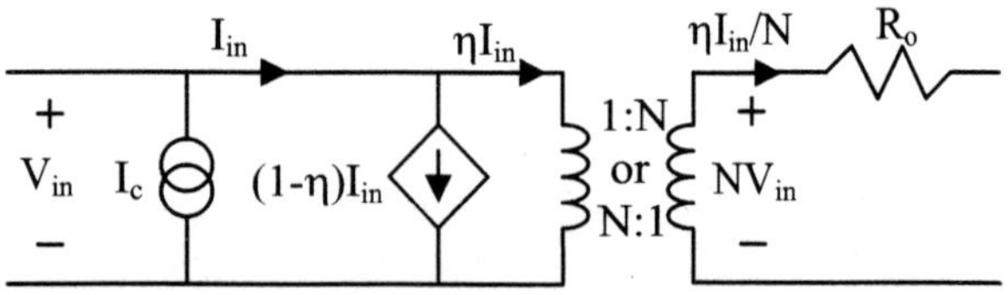

Figure 5: Step-up DC/DC converter model showing conduction (R_o), parasitic (I_c) and charge redistribution (η) losses.

Note that charge redistribution losses are significant only in converters that make parallel connections of capacitors with different voltages. Therefore, these losses may not be significant in lossless AC/DC capacitor-only converters [6], and switching converters which rely on inductors.

5. STEP-UP VS. STEP-DOWN: PRACTICAL CASE

A thorough analysis of the merits of step-down versus step-up power converters must use the model in Figure 5 in place of the ideal model in Section II. Therefore, we remove the ideal models from Figure 4 (a) and (b), and replace them with the model in Figure 5. The resulting systems are shown in Figure 6. Note that the $(1-\eta)I_{in}$ current sources in Figure 5 are represented implicitly by the ratios between the coil currents shown in Figure 6.

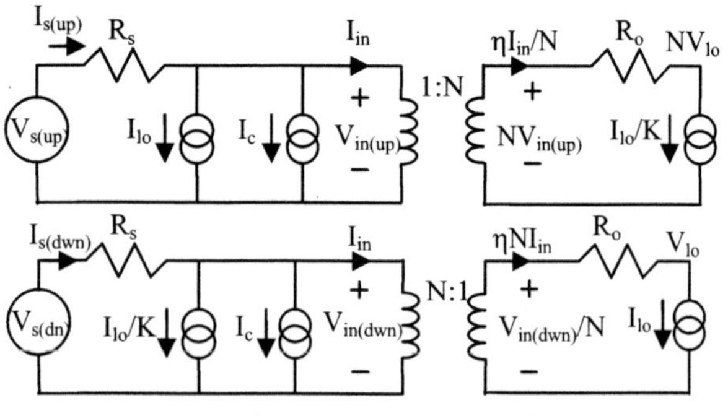

Figure 6: Step-up (a) and step-down (b) systems using the complete charge pump model of Figure 5.

When designing a system, the affects of R_o, I_c, and η must be considered simultaneously. However, to clearly demonstrate how each non-ideality affects the *relative* merits of the step-up and step-down systems, we have chosen to illustrate their effects separately.

A. *Parasitic Capacitance Losses (I_c)*

The current I_c is constant in the step-up and step-down systems, and flows through R_s. To determine how I_c affects each system, we split $I_{s(up)}$ and $I_{s(dwn)}$ into two contributions: I_c, and the remainder, which we label $I_{L(up)}$ and $I_{L(dwn)}$. Therefore, we write (9) and (10).

$$V_{s(up)} = V_{in(up)} + I_{L(up)}R_s + I_cR_s \qquad (9)$$

$$V_{s(dwn)} = V_{in(dwn)} + I_{L(dwn)}R_s + I_cR_s \qquad (10)$$

Once again, we solve the inequality in (7) to find the values of R_s that make the step-down system superior. Substituting (9) and (10) into (7) causes the terms containing I_c to cancel. Therefore, I_c does not affect the relative merits of the step-down versus the step-up system.

B. *Conduction Losses (R_o)*

Adding an output resistance to the charge pump produces an additional voltage drop. Note that more current flows through R_o (more voltage drop) in the step-down converter than the step-up converter. Therefore, the larger R_o, the more the step-up system will be favoured over the step-down system.

The minimum allowable voltages at the *outputs* of the charge pumps are NV_{lo} (step-up), and V_{lo} (step-down). Accordingly, the voltages at the secondary coils of the step-up and step-down systems are given by (12) and (13) respectively.

$$NV_{in(up)} = NV_{lo} + R_oI_{lo}/K \qquad (12)$$

$$NV_{in(dwn)} = V_{lo} + R_oI_{lo} \qquad (13)$$

To find voltages $V_{s(up)}$ and $V_{s(dwn)}$, we simply multiply equations (12) and (13) by the charge pump voltage gain (N or $1/N$) to obtain the voltage at the primary coils, and then add the voltage drop across R_s. The results are given by (14) and (15).

$$V_{s(up)} = V_{lo} + \frac{R_oI_{lo}}{NK} + R_sI_{lo}\left(1 + \frac{N}{K}\right) \qquad (14)$$

$$V_{s(dwn)} = NV_{lo} + NR_oI_{lo} + R_sI_{lo}(1/N + 1/K) \qquad (15)$$

Finally, we solve inequality (7) for R_s, by substituting (14) and (15). The result is given by (18).

$$R_s \geq \frac{V_{lo}}{I_{lo}}\left(\frac{NK}{N+K}\right) + R_o\left(\frac{N+1}{N+K}\right) \qquad (18)$$

As we predicted, R_o adds an additional term to the original expression (8) for R_s. Therefore, R_s must be even larger to make the step-down system superior to the step-up system. Note that in deriving (18) we have assumed that the output resistances of the step-up and step-down converters are equal. In reality, step-up converters have higher R_o.

C. *Charge Redistribution Losses (η)*

To investigate the effects of η, we ignore I_c (which does not influence the comparison), and we neglect R_o for simplicity. Once again, we write expressions (19) and (20) for $V_{s(up)}$ and $V_{s(dwn)}$ respectively, and solve inequality (7) for R_s. The result is given by (21).

$$V_{s(up)} = V_{lo} + R_sI_{lo}(1 + N/\eta K) \qquad (19)$$

$$V_{s(dwn)} = NV_{lo} + R_sI_{lo}(1/K + 1/\eta N) \qquad (20)$$

$$R_s \geq \left[(N-1)\frac{V_{lo}}{I_{lo}}\right] \div \left[\frac{1}{\eta}\left(\frac{N^2-K}{NK}\right)+1-\frac{1}{K}\right] \quad (21)$$

In Fig. 7 we have graphed (21) for the following parameters: $V_{lo} = 3V$, $I_{lo} = 400\mu A$, $N = 4$, and $K = 4$, 16, and ∞. Note that $K = \infty$ simply means that the high voltage supply (V_{hi}) is not used. The area above each line indicates combinations of R_s and η that make the step-down system superior for a given set of parameters.

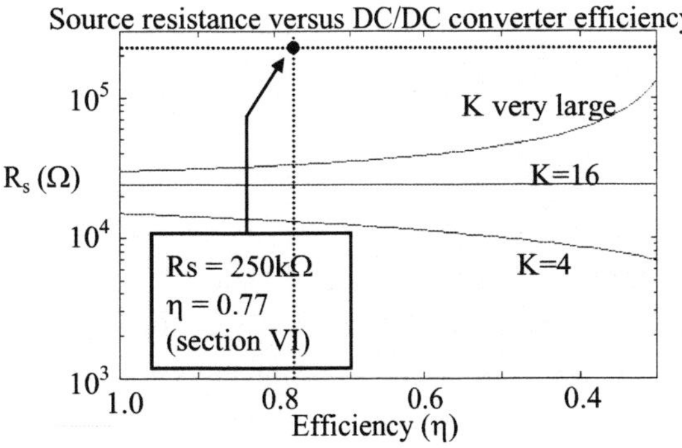

Fig. 7. R_s versus η for $N = 4$. For a given value of K, points above the line indicate combinations of R_s and η that make the step-down system superior.

D. Recommendations

Based on our analysis of the effects of charge pump losses on the relative merits of the step-up and step-down architectures, the step-down architecture is most useful in systems where a significant fraction of the total current is drawn from V_{hi}, the power source has a large resistance, and where the efficiency of the power converter is poor. These are precisely the characteristics the weakly coupled RF systems discussed in section II, and therefore step-down converters are ideally suited to this application.

6. SIMULATION AND EXPERIMENTAL RESULTS

Through simulation, we found that a step-down converter can improve total available load power by over 100%. To compare the step-up and step-down systems, we attached the loads in Figure 1 (a) and (b) to the RF frontend in Figure 2 (b). We designed a step-down charge pump in a 0.8μm process, and extracted η and R_o. We then modeled an equivalent step-up converter using the extracted values of η and R_o, and the equivalent circuit in Figure 5. For the component values and system parameters given in Table 1, $(P_{lo}+P_{hi})_{max}$=1.5mW for the load in Figure 4 (a) (step-up), whereas for Figure 4 (b) (step-down), $(P_{lo}+P_{hi})_{max}$= 0.7mW. Note that step-up charge pumps are more difficult to design than step-down, and therefore *our estimate of a 100% improvement is conservative.*

7. CONCLUSION

We have shown that step-down power converters in RF-powered systems can increase operable link distance and available load current while reducing minimum required source voltage. We recommend that step-down converters be used in systems that have high impedance sources, require a high voltage supply, and rely on converters with poor efficiency. Our analysis of power converters, and the resulting recommendations, can be generalized beyond RF-powered systems, leading to performance gains in other applications.

Table 1: Values used for comparative simulations. Parameters in **bold** were confirmed using experimental prototypes.

Parameter	Value	Parameter	Value
L_1	40μH	**k**	**0.0007**
C_1	2.5nF	**Rs**	**200kΩ**
R_1	2Ω	N	4
L_2	3mH	V_{LO}	3V
C_2	33pF	R_o	1.8kΩ
R_2	376Ω	η	0.77
f_R	500kHz	K	4

In Fig. 7, we have marked the location of the simulated system {η=0.77 and R_s=250kΩ} with a black dot. Clearly, the point is well above the line corresponding to $K = 4$, indicating that theory also suggests that the step-down system is superior.

8. ACKNOWLEDGEMENTS

The authors thank Ravi Ananth of the Alfred Mann Foundation for his ideas, suggestions, and experimental work.

9. REFERENCES

[1] W. Heetderks, "RF powering of millimeter and sub-millimeter sized neural prosthetic implants", *IEEE Trans. Biomed. Eng.*, vol. 35, no. 5.

[2] U. Gehrig, T. Roz, *RFID Made Easy*, EM Microelectronic-Marin SA, 2000.

[3] M. Makowski, D. Maksimovic, "Performance limits of switched-capacitor DC-DC converters," *IEEE PESC 1995.*

[4] Z. Tang, B. Smith, J. Schild, P. Peckham, "Data transmission from an implantable biotelemeter by load-shift keying using circuit configuration modulator", *IEEE Trans. Biomed. Eng.*, vol. 42, no. 5. May 1995.

[5] E. Bayer, H. Schmeller "A high efficiency single-cell cascaded charge pump topology". *IEEE PESC 2001*, vol. 1.

[6] C. Tse, S. Wong, M. Chow, "On lossless switched-capacitor converters," *IEEE Trans. On Power Elec.*, vol. 10, no. 3, May 1996.

APPENDIX

The IC load (R_{IC}) in Figure 3, referred through the rectifier and DC/DC converter, can be represented using an equivalent AC resistance [4]. Therefore, the system in Figure 3 can be reduced to the system in Figure 8 (a) if the DC/DC converter is step-up, or Figure 8 (b) if the DC/DC converter is step-down. Clearly, the load resistance ($R_{L(EQ)}$) is higher in Figure 8 (b) than in Figure 8 (a). In weakly coupled systems, the higher load resistance may be better matched to R_s, thereby increasing available load power, and/or operable link distance.

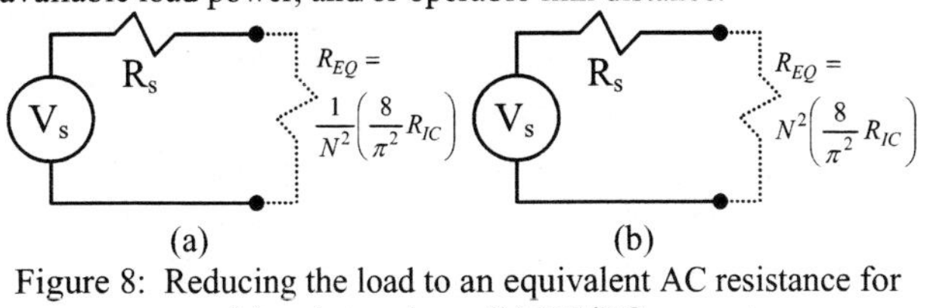

Figure 8: Reducing the load to an equivalent AC resistance for step-up (a) and step-down (b) DC/DC converters.

Hilbert Space Techniques for Reactive Power Compensation with Limited Current Bandwidth

Hanoch Lev-Ari

Northeastern University
ECE Department
Boston, USA
levari@ece.neu.edu

Aleksandar M. Stanković

Northeastern University
ECE Department
Boston, USA
astankov@ece.neu.edu

***Abstract*— The paper formulates and solves an optimization problem related to reactive power compensation with limited current bandwidth. The current bandwidth limitation is essential for all practical compensators that are built around switching power converters. We use the Hilbert space terminology of [1] to formulate and solve the corresponding constrained optimization problem. It turns out that our solution generalizes some results from the literature. Under operating conditions that are prevalent in practice, our solution leads to a decomposition of reactive power into three components - a component that can be compensated without energy storage, a component that can be compensated when energy storage is available, and a component that can not be compensated at all because of the limited bandwidth of the compensator.**

***Index Terms*— Reactive power compensation, Hilbert space optimization, band-limited compensator.**

I. Introduction

There exists a great practical interest to optimize the operation of power systems, in particular by minimizing the line losses and concomitantly reducing the reactive power flows. This issue is related to significant theoretical challenges, as there is still no consensus about the general polyphase case in the presence of harmonics. The main new enabling technology is provided by the so called active power filters (for a recent review see, for example, [2]). These compensators, connected typically in parallel with the load, can inject currents that reduce the overall line current drawn from the source. Active power filters are typically built around switching voltage sourced inverters, so the achievable current bandwidth is fundamentally limited, typically to approximately a decade below the switching frequency of the inverter. A practical problem that is especially illustrative of this aspect is the compensation of arc furnaces. This type of load has been proliferating in power systems worldwide, and it is a source of substantial power quality problems, in particularly light flicker [3], [4].

In this paper we extend the results of [1] to accommodate harmonic restriction on the compensator current. We again use the notion of a projection in a suitable Hilbert space to develop closed formulas for the optimal current, and to generalize some results from the literature. Our direct motivation stems from the fact that the so called instantaneous ("Depenbrock-Akagi-Nabae") compensator which matches all harmonics of the instantaneous power of the load may turn out to have infinitely many harmonics, even in case when the original voltages and currents are limited in bandwidth. While in [1] we derived expressions for the current $i_a^{(L)}$ which matches load power harmonics up to L-th, this current is not optimal under the added restriction on harmonic content of the compensator, as we show later.

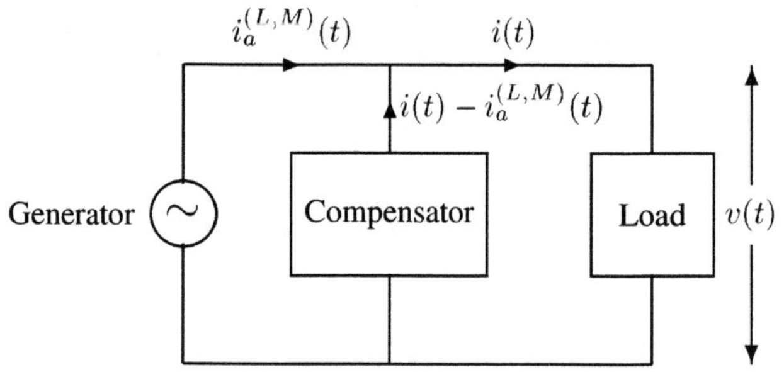

Fig. 1. Load compensation in a power delivery system.

Our objective is to determine the equivalent load current $i_a^{(L,M)}(t)$ that: (1) is as small as possible in the rms sense, (2) supplies the same power harmonics as the original load current $i(t)$, up to the L-th harmonic, and (3) results in a compensator current $i(t)-i_a^{(L,M)}(t)$ that has no more than M harmonics.

Some special cases are of particular practical interest: **1.** setting $M = \infty$ recovers the results of [1], which already include as a special case the Fryze compensator

0-7803-8251-X/04/$17.00 ©2004 IEEE

($i_f(t) \equiv i_a^{(0,\infty)}(t)$), and the instantaneous compensator ($i_a(t) \equiv i_a^{(\infty,\infty)}(t)$); **2.** setting $L = \infty$ and $M = 1$, assuming that both $v(t)$ and $i(t)$ are sinusoidal (and also that $v(t)$ is balanced), we recover the results of [5], as we demonstrate in Sec. III.

II. The Optimal Compensator

We use the Hilbert space terminology of [1] to formulate and solve our constrained optimization problem. Thus $v(t)$ and $i(t)$ are vectors representing load voltage and current, which we view as elements in a Hilbert space of n-phase T-periodic waveforms, with the inner product defined by

$$\langle\, x\, ,\, y\, \rangle \stackrel{\text{def}}{=} \frac{1}{T}\int_T x(t)y^\top(t)dt$$

Our choice of using row vectors (rather than column vectors) to describe polyphase waveforms turns out to be more convenient for direct evaluation of orthogonal projections, a calculation we carry out in Sec. III.

To facilitate a Hilbert space formulation of our optimization problem, we introduce two subspaces that describe the constraints on the optimal equivalent current, viz.,

$$\mathcal{H}_L \stackrel{\text{def}}{=} \textit{span}\left\{v(t)e^{j\ell\omega t}\,;\ -L \le \ell \le L\right\} \qquad (1)$$

and

$$\mathcal{B}_M \stackrel{\text{def}}{=} \textit{span}\left\{\text{polyphase T-periodic waveforms restricted to M harmonics}\right\} \qquad (2)$$

The optimization problem that we posed in Sec. I can now be restated as follows:

> Given a polyphase load current $i(t)$ and load voltage $v(t)$, find an equivalent load current $i_a^{(L,M)}(t)$ such that: (1) $||i_a^{(L,M)}||$ is minimized, (2) $i - i_a^{(L,M)} \perp \mathcal{H}_L$, and (3) $i - i_a^{(L,M)} \in \mathcal{B}_M$.

The optimal solution to this linearly constrained quadratic optimization problem consists, in general, of two mutually-orthogonal components: the *in-band* component (which belongs to $\mathcal{B}_M$), and the *out-of-band* component (which is orthogonal to $\mathcal{B}_M$). Since the compensator current $i(t) - i_a^{(L,M)}(t)$ can not have an out-of-band component, it follows that $i(t)$ and $i_a^{(L,M)}(t)$ must share the same out-of-band component.

Since the compensator is not allowed to modify the out-of-band component, our problem can be restricted to in-band components alone: minimize the norm $\left\|\left[i_a^{(L,M)}\right]_{\mathcal{B}}\right\|$, subject to the constraint $[i]_{\mathcal{B}} - \left[i_a^{(L,M)}\right]_{\mathcal{B}} \perp [\mathcal{H}_L]_{\mathcal{B}}$, where $[\cdot]_{\mathcal{B}}$ indicates the in-band component of a polyphase T-periodic waveform. In particular, we must restrict all waveforms in the subspace $\mathcal{H}_L$ to their in-band components (by removing all higher order harmonics). This results in the "band-limited" subspace

$$\mathcal{H}_{L,M} \stackrel{\text{def}}{=} \textit{span}\left\{\left[v(t)\,e^{j\ell\omega t}\right]_{\mathcal{B}}\,;\ -L \le \ell \le L\right\} \qquad (3)$$

Once all waveforms have been restricted to their in-band components, the resulting optimization problem has the same mathematical form as the one described and solved in [1]. As a result, the optimal equivalent load current is given by a sum of two mutually orthogonal components, viz.,

$$\begin{aligned} i_a^{(L,M)}(t) \;&=\; \text{orthogonal projection of } i(t) \text{ on } \mathcal{H}_{L,M} \\ &\oplus\ \text{out-of-band component of } i(t) \end{aligned} \qquad (4)$$

In the following section we use a simple example to illustrate the computational procedure for evaluating the optimal $i_a^{(L,M)}(t)$.

The subspaces $\mathcal{H}_{L,M}$ form a nested chain, in the sense that

$$\mathcal{H}_{0,M} \subset \mathcal{H}_{1,M} \subset\ \ldots\ \subset \mathcal{H}_{\infty,M} \subset \mathcal{B}_M$$

As a result, we obtain a decomposition of the original load current $i(t)$ into three mutually orthogonal components, viz.,

$$i(t) = i_f^{(M)}(t) \oplus i_x^{(M)}(t) \oplus i_y^{(M)}(t) \qquad (5)$$

where

$$\begin{aligned} i_f^{(M)}(t) &\stackrel{\text{def}}{=} i_a^{(0,M)}(t) \\ i_x^{(M)}(t) &\stackrel{\text{def}}{=} i(t) - i_a^{(\infty,M)}(t) \\ i_y^{(M)}(t) &\stackrel{\text{def}}{=} i_a^{(\infty,M)}(t) - i_a^{(0,M)}(t) \end{aligned}$$

This is a band-limited version of the decomposition $i(t) = i_f(t) \oplus i_x(t) \oplus i_y(t)$ that was discussed in [1]. A band-limited compensator can compensate the component $i_x^{(M)}(t)$ without using energy storage, because $i_x(t)v(t)^\top = 0$ for all t, so that there is no instantaneous power flow into the compensator. In contrast, energy storage is required for (band-limited) compensation of $i_y^{(M)}(t)$. Notice that we actually have four orthogonal components: according to (4), the current $i_f^{(M)}(t) \equiv i_a^{(0,M)}(t)$ consists itself of two mutually orthogonal components.

When the polyphase voltage $v(t)$ is band-limited, our orthogonal current decomposition (5) gives rise to a corresponding power decomposition, viz.,

$$\|v\|^2\|i\|^2 = P_0^2 + \left[Q_f^{(M)}\right]^2 + \left[Q_x^{(M)}\right]^2 + \left[Q_y^{(M)}\right]^2 \quad (6)$$

where $P_0 = \langle\, i\,,\, v\,\rangle$ is the real power delivered to the load, and

$$\begin{aligned} Q_f^{(M)} &\stackrel{\text{def}}{=} \|v\| \cdot \|\text{out-of-band component of } i(t)\| \\ Q_x^{(M)} &\stackrel{\text{def}}{=} \|v\| \cdot \|i_x^{(M)}(t)\| \\ Q_y^{(M)} &\stackrel{\text{def}}{=} \|v\| \cdot \|i_y^{(M)}(t)\| \end{aligned}$$

We can therefore interpret $Q_x^{(M)}$ as the reactive power component that can be compensated without energy storage, $Q_y^{(M)}$ as the reactive power component that can be compensated when energy storage is available, and $Q_f^{(M)}$ as the reactive power component that can not be compensated at all (because of the limited bandwidth of the compensator).

III. Example: Sinusoidal Waveforms

A case of particular interest is when both the load voltage $v(t)$ and the load current $i(t)$ are sinusoidal, viz.,

$$v(t) = \Re\{V\, e^{j\omega t}\} \quad , \quad i(t) = \Re\{I\, e^{j\omega t}\}$$

where $\Re\{\cdot\}$ denotes the real part of a complex number. In this case it is reasonable to require that the compensator current also be sinusoidal, by choosing $M = 1$ in our optimization problem. Since the given load current is sinusoidal, it has no out-of-band component, and the same is true for the optimal equivalent load current $i_a^{(L,M)}(t)$, for all L, M. Also, because both $v(t)$ and $i(t)$ are sinusoidal, we have

$$i_a^{(0,M)}(t) = i_a^{(0,\infty)}(t) = \frac{\Re\{IV^H\}}{\|V\|^2}\, v(t) \quad (7)$$

for all $M \geq 1$.

In order to evaluate $i_a^{(L,1)}(t)$ we need only to determine the projection of $i(t)$ on $\mathcal{H}_{L,1}$ (since there in no out-of-band component). The amount of effort involved in this evaluation depends on L. However, since $i(t)$ is sinusoidal, the instantaneous power $i^T(t)\, v(t)$ is bandlimited to two harmonics, so setting $L = 2$ has the same effect as choosing a larger value for L. In short, $i_a^{(2,1)}(t) \equiv i_a^{(L,1)}(t)$ for all $L \geq 2$, so that we only need to consider $L = 0, 1, 2$.

Because we are dealing with bandlimited waveforms, it is more convenient to carry out our calculations in the frequency domain, using the (complex) Fourier series representation, band-limited to the DC and first harmonic coefficients, viz.,

$$v(t) \Leftrightarrow [V^* \quad 0 \quad V] \quad , \quad i(t) \Leftrightarrow [I^* \quad 0 \quad I] \stackrel{\text{def}}{=} \mathcal{I}$$

where the asterisk $(*)$ denotes complex conjugation. Again, recall that all polyphase Fourier coefficients are $1 \times n$ vectors.

We start with $L = 2$, the highest L value of interest. The subspace $\mathcal{H}_{2,1}$ is spanned by the band-limited waveforms $\left\{\left[v(t)\, e^{j\ell\omega t}\right]_{\mathcal{B}} \,;\, -2 \leq \ell \leq 2\right\}$, which we represent by a matrix of (polyphase) Fourier coefficients, viz.,

$$\Phi = \begin{pmatrix} V & 0 & 0 \\ 0 & V & 0 \\ V^* & 0 & V \\ 0 & V^* & 0 \\ 0 & 0 & V^* \end{pmatrix}$$

Notice that Φ has $2M+1 = 3$ (block) columns and $2L+1 = 5$ rows. Increasing the value of L beyond $L = 2$ augments the matrix Φ with all-zero rows, so that

$$\mathcal{H}_{L,1} \equiv \mathcal{H}_{2,1} \Rightarrow i_a^{(L,1)}(t) \equiv i_a^{(2,1)}(t)$$

for all $L \geq 2$, as stated earlier.

Using the language and notation of [1], we can now write an explicit expression for the optimal equivalent load current. Its Fourier coefficient representation is $\langle\, \mathcal{I}\,,\, \Phi\,\rangle \langle\, \Phi\,,\, \Phi\,\rangle^{-1} \Phi = \mathcal{I}\, \Phi^H (\Phi\Phi^H)^{-1}\, \Phi$, and a direct calculation establishes that

$$\langle\, \mathcal{I}\,,\, \Phi\,\rangle = \Re\{IV^H\} \left[\xi^* \quad 0 \quad 2 \quad 0 \quad \xi\right]$$

$$\langle\, \Phi\,,\, \Phi\,\rangle = \|V\|^2 \begin{pmatrix} 1 & 0 & \nu & 0 & 0 \\ 0 & 1 & 0 & \nu & 0 \\ \nu^* & 0 & 2 & 0 & \nu \\ 0 & \nu^* & 0 & 1 & 0 \\ 0 & 0 & \nu^* & 0 & 1 \end{pmatrix}$$

where the superscript H denotes conjugate transposition, and

$$\nu \stackrel{\text{def}}{=} \frac{VV^\top}{\|V\|^2} \quad , \quad \xi \stackrel{\text{def}}{=} \frac{IV^\top}{\Re\{IV^H\}}$$

The final result is

$$i_a^{(2,1)}(t) = \frac{\Re\{IV^H\}}{\|V\|^2\,(1 - |\nu|^2)}\, \Re\left\{\left[\alpha\, V + \beta\, V^*\right] e^{j\omega t}\right\} \quad (8)$$

where $\alpha \stackrel{\text{def}}{=} 1 - \Re\{\nu\xi^*\}$, $\beta \stackrel{\text{def}}{=} \xi - \nu + j\nu\, \Im\{\nu\xi^*\}$, and $\Im\{\cdot\}$ denotes the imaginary part of a complex number.

When the voltage $v(t)$ is *balanced*, we get $\nu = 0$, so that $i_a^{(2,1)}(t) \equiv i_a^{(\infty,1)}(t)$ is represented, in the balanced case, by the phasor

$$\frac{\Re\{IV^H\}}{\|V\|^2} V + \frac{IV^\top}{\|V\|^2} V^*$$

This is precisely the result obtained in [5] (using $L = \infty$). In contrast, our result (8) holds for arbitrary (i.e., unbalanced) V and I.

Next, for $L = 1$ we have

$$\Phi = \begin{pmatrix} 0 & V & 0 \\ V^* & 0 & V \\ 0 & V^* & 0 \end{pmatrix}$$

so that

$$\langle \mathcal{I}, \Phi \rangle = \Re\{IV^H\} \begin{bmatrix} 0 & 2 & 0 \end{bmatrix}$$

$$\langle \Phi, \Phi \rangle = \|V\|^2 \begin{pmatrix} 1 & 0 & \nu \\ 0 & 2 & 0 \\ \nu^* & 0 & 1 \end{pmatrix}$$

$$\langle \mathcal{I}, \Phi \rangle \langle \Phi, \Phi \rangle^{-1} \Phi = \frac{\Re\{IV^H\}}{\|V\|^2} [V^* \quad 0 \quad V]$$

and $i_a^{(1,1)}(t) = \frac{\Re\{IV^H\}}{\|V\|^2} v(t)$. The same result is obtained for $L = 0$, namely $i_a^{(0,1)}(t) = i_a^{(1,1)}(t)$. Indeed, now

$$\Phi = [V^* \quad 0 \quad V] \Leftrightarrow v(t)$$

so that

$$\langle \mathcal{I}, \Phi \rangle = 2\Re\{IV^H\} \quad , \quad \langle \Phi, \Phi \rangle = 2\|V\|^2$$

and $\langle \mathcal{I}, \Phi \rangle \langle \Phi, \Phi \rangle^{-1} \Phi = \frac{\Re\{IV^H\}}{\|V\|^2} [V^* \quad 0 \quad V]$ as we have already obtained for $L = 1$.

Recall that in this example the Fryze component $i_f^{(M)}(t)$ is independent of M. As a result, we get from (5) the orthogonal decomposition

$$i(t) = i_f(t) \oplus i_x^{(M)}(t) \oplus i_y^{(M)}(t)$$

with $i_f(t)$ given by (7), for all $M \geq 1$, and with no out-of-band component. In particular, for $M = 1$ all three current components are sinusoidal. In the balanced case we also have the simple expression $i_y^{(1)}(t) = \Re\left\{\frac{IV^\top}{\|V\|^2} V^* e^{j\omega t}\right\}$.

Finally, we observe that the power decomposition expression (6) simplifies in this example to

$$\|v\|^2 \|i\|^2 = P_0^2 + \left[Q_x^{(M)}\right]^2 + \left[Q_y^{(M)}\right]^2$$

Here $Q_f^{(M)} = 0$ because the out-of-band component of $i(t)$ vanishes. Thus, in this example all the reactive power can be compensated (with energy storage).

IV. Concluding Remarks

The paper formulates and solves the following optimization problem related to reactive power compensation with limited current bandwidth – determine the equivalent load current that is: (1) is as small as possible in the rms sense, (2) supplies the same power harmonics as the original load current, up to the L-th harmonic, and (3) results in a compensator current that has no more than M harmonics. We used the Hilbert space terminology of [1] to formulate and solve the corresponding constrained optimization problem. Under operating conditions that are prevalent in practice, our solution leads to a decomposition of reactive power into three components – a component that can be compensated without energy storage, a component that can be compensated when energy storage is available, and a component that can not be compensated at all because of the limited bandwidth of the compensator.

V. References

[1] H. Lev-Ari and A.M. Stanković, "Hilbert Space Techniques for Modeling and Compensation of Reactive Power In Energy Processing Systems" *IEEE Trans. on Circuits and Systems-I: Fundamental Theory and Applications*, 50(4), Apr. 2003, pp. 540-556.

[2] B. Singh, K. Al-Haddad and A. Chandra, "A review of active filters for power quality improvement," *IEEE Trans. on Industrial Electronics*, 46(5), Oct. 1999, pp. 960-971.

[3] A. Garcia-Cerrada, P. Garcia-Gonzalez, R. Collantes, T. Gomez and J. Anzola, "Comparison of thyristor-controlled reactors and voltage-source inverters for compensation of flicker caused by arc furnaces," *IEEE Trans. on Power Delivery*, 15(4), Oct. 2000, pp. 1225-1231.

[4] A. Wolf and M. Thamodharan, "Reactive Power Reduction in Three-Phase Electric Arc Furnace" *IEEE Trans. on Industrial Electronics*, 47(4), Aug. 2000, pp. 729-733.

[5] J.L. Willems and D. Ayeles, "New Decomposition for 3-Phase Currents in Power Systems," *IEE Proceedings - C*, 140(4), July 1993, pp. 307-310.

A GEOMETRICAL STUDY ON VOLTAGE COLLAPSE MECHANISMS OF POWER SYSYTEMS

Yongqiang Liu
South China University of Technology
Guangzhou, P.R. China
epyqliu@scut.edu.cn

Zheng Yan *Yixin Ni* *Felix Wu*
The University of Hong Kong
Hong Kong SAR, P.R. China
zyan@eee.hku.hk

ABSTRACT

In this paper, the mechanism of voltage collapse is studied based on a nonlinear circuit model and transversality theory, and a unified explanation of the phenomena of voltage collapse is given. The research shows that the fundamental reason for voltage collapse is due to the non-transversal intersection between solution manifolds of the injection branch equations and the linear electrical network equations in the voltage-current space. It is also proved that this non-transversal intersection exactly corresponds to the static state bifurcation point. These conclusions are valid no matter differential algebraic equations or power flow equations being employed.

1. INTRODUCTION

The voltage collapse problem is an extremely concerned problem in power industry. In the literatures, there are many publications analyzing in details the phenomenon of voltage collapse[1~6]. References [1,2] summarized main results on voltage stability. Reference [3] pointed out that general voltage stability can be classified into three types, i.e., transient voltage stability, small-signal voltage stability and static voltage stability. References [4,5] studied voltage stability problems by employing characteristics of multiple power flow solutions. And reference [6] studied voltage instability by using static bifurcation theory. It was concluded that voltage instability is the consequence of the existence of static bifurcation, and at the bifurcation point the system will lose the local operation points. A general idea is that the loss of the local power flow solutions or equilibria means the occurrences of voltage collapse.

A real-world power system is a large-scale and special nonlinear circuit, in which there are two types of branches with completely different properties. One type of branches is transmission line, i.e., link branches, by which different buses are connected to form the intact system. Another type of branches is the injection branch, i.e., tree branches, including generator and load branches.

If the branch characteristic equations of all the injection branches in the system are known, then by combining the network equations, the operation point of the power system can be viewed as an intersection of the solution manifolds of the two typical equations. In this paper, by employing network-preserving power system model, intensive investigation of the power system voltage stability is conducted from the viewpoint of solution manifolds of system circuits, and the mechanism of power system voltage collapse is revealed. This paper shows that the voltage collapse is always associated with non-transversal intersection of solution manifolds of injection branch equations and network equations, and it also corresponds to the saddle-node bifurcation.

2. A CIRCUIT MODEL FOR POWER SYSTEMS

In this paper, we assume that all state variables, including load variables, are continuous variables. The equivalent circuit of a power system is a nonlinear circuit. The nonlinear characteristics of nonlinear elements such as generators and loads etc. form the nonlinear branch characteristics of injection branches or tree branches. While the linear voltage-current characteristics of linear elements such as transmission lines form the branch characteristic equations of link branches.

2.1. Characteristic Equations of Load Branches

Suppose that dynamics of loads may change continuously. Under the rectangular coordinates of synchronous rotation, load model may adopt motor convention and can be written as:

$$\dot{s}_i = f_{mi}(s_i, u_i) \tag{1}$$

$$i_i = i_i(s_i, u_i) \tag{2}$$

where , $i = 1, \cdots n_L$, n_L is number of load buses; $i_i = [i_{Di}, i_{Qi}]^T$, $u_i = [u_{Di}, u_{Qi}]^T$ are injection current and voltage vectors of the injection branch in i-th bus. s_i is slip of load model, or named as mechanical variable of load model. From the viewpoint of circuit theory, equation (2) represents branch characteristic of i-th branch which is controlled by the slip s_i . Thus load branch current vector i_L can be written as:

$$f_L(s, u_L, i_L) = i_L - i_L(s, u_L) = 0 \tag{3}$$

where u_L is bus voltage vector.

0-7803-8251-X/04/$17.00 ©2004 IEEE

For constant PQ load model, the following branch equations can be obtained:

$$\begin{pmatrix} i_{Di} \\ i_{Qi} \end{pmatrix} = \begin{pmatrix} (P_i/U_i^2)u_{Di} + (Q_i/U_i^2)u_{Qi} \\ (P_i/U_i^2)u_{Qi} - (Q_i/U_i^2)u_{Di} \end{pmatrix} \quad (4)$$

where $U_i = (u_{Di}^2 + u_{Qi}^2)^{1/2}$ is the magnitude of the bus voltage. Equation (4) is a special case of (3) by omitting induction motor slip.

2.2. Characteristic Equations of Generator Branches

A 3rd-order synchronous machine model without damping coils can be written as:

$$\dot{\delta}_i = \Delta\omega_i \quad (5)$$

$$\Delta\dot{\omega}_i = M_i^{-1}\{T_{mi} - D_i\Delta\omega_i + E'_{qi}(i_{xi}\cos\delta_i + i_{yi}\sin\delta_i) - \Delta x_{dqi}[(i_{xi}^2 - i_{yi}^2)\sin\delta_i\cos\delta_i - i_{xi}i_{yi}\cos 2\delta_i]\} \quad (6)$$

$$\dot{E}'_{qi} = T'^{-1}_{doi}[E_{fd} - E'_{qi} - \Delta x_{ddi}(i_{Di}\sin\delta_i - i_{Qi}\cos\delta_i)] \quad (7)$$

where $\Delta x_{dqi} = x_{qi} - x'_{di}$, $\Delta x_{ddi} = x'_{di} - x_{di}$. The generator branch equation with motor convention is:

$$\begin{pmatrix} i_{Di} \\ i_{Qi} \end{pmatrix} = \begin{pmatrix} G_{Di}(\delta_i) & B_{Di}(\delta_i) \\ B_{Qi}(\delta_i) & G_{Qi}(\delta_i) \end{pmatrix}\begin{pmatrix} u_{Di} \\ u_{Qi} \end{pmatrix} - \begin{pmatrix} C_{Di}(\delta_i) \\ C_{Qi}(\delta_i) \end{pmatrix} E'_{qi} \quad (8)$$

where, $i = n_L + 1, \cdots N$ and $N = n_L + n_G$ is number of system buses (here it is assumed that buses without injections, i.e. floating buses, have been eliminated from the system); and n_G is number of generators.

The generator branch equation can be written as:

$$f_G(\delta, E'_q, u_G, i_G) = i_G - i_G(\delta, E'_q, u_G) = 0 \quad (9)$$

where u_G and i_G are generator voltages and currents.

2.3. Network Equations

In a power network, transmission lines have linear characteristics. The injection branches can be viewed as vector of current sources and the bus voltage equations of the electrical network can be written as:

$$f_N(i,u) = Y_n u + Y_s u_s + i = 0 \quad (10)$$

where $u = [u_L^T, u_G^T]^T$ and $i = [i_L^T, i_G^T]^T$ are vectors of injected branch voltages and currents, u_s is the known voltage of the infinite bus. If we consider a multi-machine system without infinite bus, then the bus voltage equation of the electrical network can be written as:

$$f_N(i,u) = Yu + i = 0 \quad (11)$$

where $i, u \in R^{2N}$. This means that besides the slack bus, the system has N buses.

2.4. Nonlinear Model of Power Systems

It can be seen that the integrated model of a power system can be formed by two sets of equations. The first set is a set of differential equation formed by (1), (5)~(7); the second set is a set of algebraic equations, formed by (3), (9), (10). The dynamic behaviors of s, δ and E'_q are governed by differential equations, and the branch characteristics of these injection branches are maintly influenced by the dynamic behaviors of these variables. Since $i = [i_L^T \; i_G^T]^T$ hence (3) and (9) can be re-written as $f'(x,i,u) = 0$. Thus the power system model can be summarized as:

$$\dot{x} = f(x,u) \quad (12)$$

$$0 = f_N(i,u) \quad (13a)$$

$$0 = f'(x,i,u) \quad (13b)$$

where $x = [\delta^T, \Delta\omega^T, E'^T_q, s^T]^T$.

2.5. Nonlinear Circuit Model of Steady State

In steady state analysis of power systems, usually the load buses are treated as PQ constant buses whose branch characteristics are shown in (4). For deriving the branch characteristic corresponding to PV constant bus, we give the following assumption.

Assumption 1: There is no injection power in the system except generators associated with PV buses.

With this assumption we can eliminate generator buses the way shown in Figure 1.

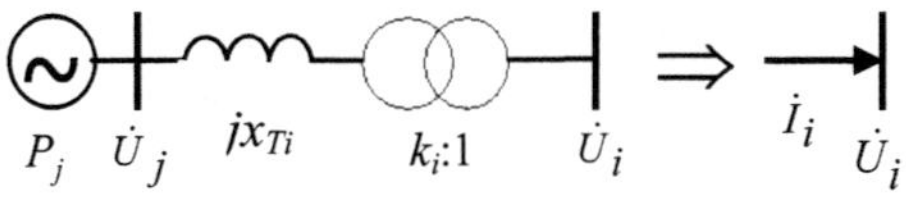

Fig.1 The elimination of generator bus

After eliminating the generator bus the characteristics equation of the generation branch can be written as

$$\begin{pmatrix} i_{xi} \\ i_{yi} \end{pmatrix} = -k_i^2 \begin{pmatrix} 0 & 1/x_{Ti} \\ -1/x_{Ti} & 0 \end{pmatrix}\begin{pmatrix} u_{xi} \\ u_{yi} \end{pmatrix} + k_i \begin{pmatrix} D_{xi}(\theta_j) \\ D_{yi}(\theta_j) \end{pmatrix} U_j \quad (14)$$

where θ_j is the phase angle of voltage $\dot{U}_j$, and

$D_{xi} = x_{Ti}^{-1}\sin\theta_j$, $D_{yi} = -x_{Ti}^{-1}\cos\theta_j$;

$$\sin\theta_j = \frac{u_{yi}\sqrt{U_j^2U_i^2 - (x_{Ti}P_j/k_i)^2}}{U_jU_i^2} + \frac{P_jx_{Ti}u_{xi}}{k_iU_jU_i^2};$$

$$\cos\theta_j = \frac{u_{xi}\sqrt{U_j^2U_i^2 - (x_{Ti}P_j/k_i)^2}}{U_jU_i^2} - \frac{P_jx_{Ti}u_{yi}}{k_iU_jU_i^2}.$$

For convenience the following assumption is given.

Assumption 2: $\partial i/\partial u$ of (14) is non-singular for given real power P_j and voltage magnitude U_j.

So after eliminating the floated buses, the power flow equation can be expressed as

$$f_N(i,u) = Y_n u + Y_s u_s + i = 0 \quad (15a)$$

$$F_{inj}(i,u) = \hat{f}_{inj}(u) - i = 0 \quad (15b)$$

3. GEOMETRICAL MEANING OF OPERATION POINT

Apparently the power system equilibrium is determined by equation (13) and (12) with $\dot{x} = 0$, the injection branch current vector can be written as:

$$f_{inj}(s(u_L), \delta(i_G), E'_q(i_G), u_L, u_G, i_L, i_G) = 0 \quad (16)$$

where $s(u_L)$, $\delta(i_G)$ and $E'_q(i_G)$ are solved from equations (1), (5)~(7) with $\dot{s} = 0$, $\dot{\delta} = 0$, $\Delta\dot{\omega} = 0$ and $\dot{E}'_q = 0$.

According to mathematical theory[7] if M is a m-manifold, $F: M \rightarrow R^n (n \leq m)$ is a smooth mapping, and F has maximum rank in $\overline{N} = \{x | F(x) = 0\}$, then $\overline{N}$ is a regular (m-n)-sub-manifold. And take note of both $\partial i_i / \partial u_i$ in (8) for any δ and $\partial f_L / \partial i_L$ being non-singular, then we have the following theorems

Theorem 1: The solution manifold of network equations (10) or (11) is regular in a $2N$-submanifold in the $4N$ injection current-voltage space.

Theorem 2: If the mapping f_{inj} is sufficiently smooth in the scope of discussion, then the solution manifold of $f_{inj} = 0$ is regular, which is a $2N$-submanifold in the $4N$ injection current-voltage space.

Let M_N and M_{inj} are solution manifolds of (10) and (16), then the equilibrium point $p = (i, u)$ of the system can be expressed as:

$$(i, u) = M_{inj} \cap M_N \tag{17}$$

The above equation shows that: if theorem 2's conditions are satisfied, the equilibrium point in the injection current-voltage space of nonlinear controlled circuits (controlled by s_i, δ_i, $\Delta\omega_i$ and E'_{qi}) is the intersection of the two regular sub-manifolds M_{inj} and M_N in the space.

Similar to above analysis the power flow solutions are also the intersection of the solution manifolds of (15a) and (15b). No matter differential algebraic model or power flow equation being employed the operation points of the power system can always be described by the intersection of the solution manifolds of injection branch equations and network equations.

4. VOLTAGE COLLAPSE AND TRANSVERSALITY

We consider a simple power system with a constant PQ load model and an infinite bus, shown in figure 2(a). The operation point is determined by

$$U_s^2 = (U \cos\varphi)^2 + (IX + U \sin\varphi)^2 \tag{18a}$$

$$0 = P^2 + Q^2 - U^2 I^2 \tag{18b}$$

where φ is power factor angle of the load. Setting $P=\alpha P_0$, $Q=\alpha Q_0$ then φ does not vary while load parameter α changes, hence (18a) only depends on line reactance X. And (18b) depends on load parameter α . It is easy to prove that (18a) is an elliptical curve and (18b) is a hyperbola in I-U space. The operation point of the system is the intersection point of the solution manifolds of (18a) and (18b), shown in figure 2(b). When $\alpha < \alpha_0$ the solution curve of (18b) is curve 1, the normal operation point is located at point a and other intersection point b corresponds to lower voltage solution. When increase α to α_0, then curve 1 move to curve 2, in the meantime the cross-point changes to a non-transversal intersection point (point c in Fig.2(b)). Once α is large than α_0 then the system will lose operation point, i.e., voltage collapse occurs.

Clearly, the loss of operation point is relevant to transversality from above example. For obtaining further results, we give the following definition and lemma.

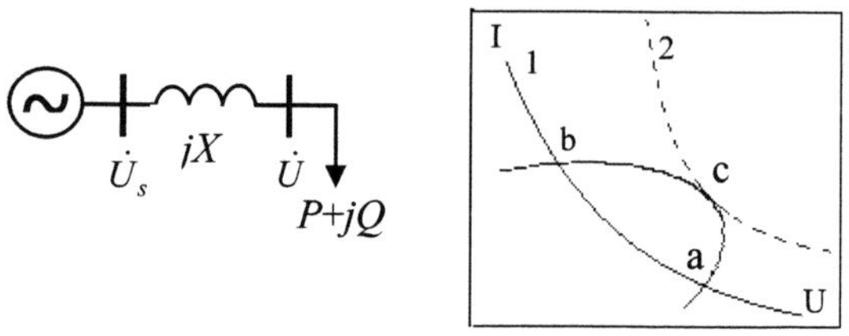

(a) (b)

Fig.2 A simple system and its operation point

Definition 1[8]**:** Let S and L are two sub-manifolds of N, if $p \in S \cap L$ and

$$T_p S + T_p L = T_p N \tag{19}$$

then S and L are transversal at point p; otherwise they are non-transversal, where $T_p S$, $T_p L$ and $T_p N$ represent tangent spaces of S, L and N, respectively.

Suppose $F(x,y) \in R^n$ and $G(x,y) \in R^m$ are two sufficiently smooth vector fields, here $x \in R^n$ and $y \in R^m$. By definition 1 we have the following lemma.

Lemma 1: Assume that both F and G have maximum rank. If (x_p, y_p) is the cross-point of F=0 and G=0, then (x_p, y_p) is a transversal point of the solution manifolds of F=0 and G=0 in R^{n+m}, if and only if:

$$rank \begin{pmatrix} \frac{\partial F}{\partial x} & \frac{\partial F}{\partial y} \\ \frac{\partial G}{\partial x} & \frac{\partial G}{\partial y} \end{pmatrix}_{(x_p, y_p)} = n + m \tag{20}$$

From lemma 1 we can obtain theorems 3, 4.

Theorem 3 Suppose (16) is smooth in the scope of interests. If the solution manifolds of (16) and (10) or (11) is non-transversal in the operating point (i_p, u$_p$)$\in M_{inj} \cap M_N$, then the matrix

$$\tilde{J} = \begin{pmatrix} I_{N \times N} & Y_n \\ \frac{\partial f_{inj}}{\partial i} + \frac{\partial f_{inj}}{\partial \delta}\frac{\partial \delta}{\partial i} + \frac{\partial f_{inj}}{\partial E'_q}\frac{\partial E'_q}{\partial i} & \frac{\partial f_{inj}}{\partial u} + \frac{\partial f_{inj}}{\partial s}\frac{\partial s}{\partial u} \end{pmatrix}_{(i_p, u_p)} \tag{21}$$

is singular at (i_p, u$_p$), and vice versa.

Theorem 4: Suppose $\hat{f}_{inj}$ (see (15)) is sufficiently smooth in the scope of interests. If the solution manifolds of (15a) and (15b) is non-transversal at the operation point (i_p, u$_p$), then the Jacobi of (15) is singular at (i_p, u$_p$), and vice versa.

Usually, we think that non-transversal intersection point means instability of geometrical structure, and once the injection branches parameters larger (less) than the critical parameters system will lose locally operation point, and voltage collapse occur. Theorem 3, 4 show that non-transversality of the solution manifolds M_{inj} and M_N results in voltage collapse no matter differential algebraic equations or power flow equations being employed to analyze voltage collapse. If the structure and parameters of transmission network do not change, then the non-linearity of injection branch determines the transversality of M_{inj}

and M_N, i.e. branch characteristics determines voltage stability. When turn ratios of transformers change, the solution manifold of transmission network equation will change, then the voltage collapse might occur under certain special situations.

5. TRANSVERSALITY AND BIFURCATION

It is well-known that the phenomenon of voltage collapse is relevant to static bifurcation[6]. In this section, we will study the relationship between static bifurcation and transversality of two solution manifolds M_{inj} and M_N.

Assumption 3 Singularity of the system Jacobian means static bifurcation.

Assumption 4 $\partial f / \partial x$ is non-singular in the scope of interests.

Theorem 5 If solution manifolds M_{inj} and M_N is non-transversal at $(i_p, u_p) \in M_{inj} \cap M_N$, then under the Assumptions 3, 4 the operating point (i_p, u_p) corresponds to static bifurcation point.

Proof If static bifurcation happens at the equilibrium point (i_p, u_p), the Jacobian matrices of (12) and (13)

$$J = \begin{pmatrix} \frac{\partial f}{\partial x} & \frac{\partial f}{\partial i} & \frac{\partial f}{\partial u} \\ 0 & I & Y_n \\ \frac{\partial f'}{\partial x} & \frac{\partial f'}{\partial i} & \frac{\partial f'}{\partial u} \end{pmatrix}_{(i_p, u_p)}$$

is singular. From Schur formula

$$|J| = \left|\left(\frac{\partial f}{\partial x}\right)\right| \left|\begin{pmatrix} I & Y_n \\ \frac{\partial f'}{\partial i} - \frac{\partial f'}{\partial x}\left(\frac{\partial f}{\partial x}\right)^{-1}\frac{\partial f}{\partial i} & \frac{\partial f'}{\partial u} - \frac{\partial f'}{\partial x}\left(\frac{\partial f}{\partial x}\right)^{-1}\frac{\partial f}{\partial u} \end{pmatrix}\right| \qquad (22)$$

by implicit function theorem we have

$$\frac{\partial f'}{\partial i} - \frac{\partial f'}{\partial x}\left(\frac{\partial f}{\partial x}\right)^{-1}\frac{\partial f}{\partial i} = \frac{\partial f_{inj}}{\partial i} + \frac{\partial f_{inj}}{\partial \delta}\frac{\partial \delta}{\partial i} + \frac{\partial f_{inj}}{\partial E'_q}\frac{\partial E'_q}{\partial i}$$

$$\frac{\partial f'}{\partial u} - \frac{\partial f'}{\partial x}\left(\frac{\partial f}{\partial x}\right)^{-1}\frac{\partial f}{\partial u} = \frac{\partial f_{inj}}{\partial u} + \frac{\partial f_{inj}}{\partial s}\frac{\partial s}{\partial u}$$

Comparing with (21), we can see that singularity of the Jacobi matrix J means $\tilde{J}$ is also singular. Contrarily, singularity of $\tilde{J}$ means singularity of J. Hence the non-transversal point of the solution manifolds M_{inj} and M_N corresponds to static bifurcation point.

Similar to the above analysis we can prove that the static state bifurcation point of the power flow equation exactly corresponds to non-transversal intersection point of the solution manifolds of injection branches equations and network equation.

6. CONCLUSIONS

The major conclusions of this paper are as follows:

(1) A power system model can be described in the semi-state-space formed by the system state variables (such as power angles, slips etc.) and circuit variables (injection currents and bus voltages). The solution manifolds of injection branch equations and network equations are all regular sub-manifolds in injection current and bus voltage space, and the operation point is the intersection of the two manifolds.

(2) That non-transversality of the solution manifolds of the injection branch equations and linear network equations at (i_p, u_p) means any small disturbance will cause non-equilibrium in the neighbor of (i_p, u_p). This means that the fundamental reason of voltage collapse is that the two solution manifolds are not intersected transversally, and the non-transversal intersection point exactly corresponds to the static bifurcation point of the system.

(3) The above conclusions are also valid for power flow equation (15), this means a unified explanation of voltage collapse phenomena is given in this paper.

REFERENCES

[1] T.Van Custem, C.D. Vournas, "Voltage stability of electric power system", Kluwer Academic Publisher, Boston, 1998.

[2] C. W. Taylor, Power System Voltage Stability, McGraw-Hill Inc., New York, 1994.

[3] T. Van Cutsem, "Voltage Instability: Phenomena, Counter-measures and Analysis Methods," Proceedings of IEEE, Vol. 88, No.2, 2000, pp.208-227.

[4] Y. Tamura, H. Mori and S. Iwamoto, "Relationship between Voltage Instability and Multiple Load Flow Solutions in Electric Power Systems," IEEE Trans. on Power Apparatus and Systems, Vol. PAS-102, May 1983, pp.1115-1125.

[5] Y. Tamura, K. Sakamoto and Y. Tayama, "Voltage instability proximity index (VIPI) based on multiple load flow solutions in ill-conditions power systems", *Proc. IEEE Conference on Decision and control*, Austin, Texas December1988, pp. 2114- 2119.

[6] H. G. Kwatny, A. K. Pasrija, L. Y. Bahar, "Static Bifurcations in Electric Power Networks: Loss of Steady State Stability and Voltage Collapse," IEEE Trans. on Circuits and Systems, Vol. 33, 1986, pp.981-991.

[7] Yozo Matsushima (translated by E. T. Kobayashi), "Differentiable Manifolds," Marcel Dekker Inc., New York, 1972, pp.65.

[8] Stephen Wiggins, "Introduction to Applied Nonlinear Dynamical Systems and Chaos," Springer-Verlag Inc., 1990, pp.101.

ACKNOWLEDGEMENTS

The authors are grateful to RGC of Hong Kong Government (Grant No. City/U 1/00C) and National Nature Science Fund Project of China (Grant No. 50337010) for the financial support to this research work.

PROBABILISTIC LOAD-DEPENDENT CASCADING FAILURE WITH LIMITED COMPONENT INTERACTIONS

Ian Dobson	*Benjamin A. Carreras*	*David E. Newman*
ECE Department	Oak Ridge National Lab	Physics Department
University of Wisconsin	Oak Ridge TN 37831	University of Alaska
Madison WI 53706	carrerasba@ornl.gov	Fairbanks AK 99775
dobson@engr.wisc.edu		ffden@uaf.edu

ABSTRACT

We generalize an analytically solvable probabilistic model of cascading failure in which failing components interact with other components by increasing their load and hence their chance of failure. In the generalized model, instead of a failing component increasing the load of all components, it increases the load of a random sample of the components. The size of the sample describes the extent of component interactions within the system. The generalized model is approximated by a saturating branching process and this leads to a criticality condition for cascading failure propagation that depends on the size of the sample. The criticality condition shows how the extent of component interactions controls the proximity to catastrophic cascading failure. Implications for the complexity of power transmission system design to avoid cascading blackouts are briefly discussed.

1. INTRODUCTION

Industrialized society depends heavily on complicated infrastructure systems with many interconnected components. These infrastructures can suffer widespread failures when stressed components fail successively, with each failure further stressing the system and making further failures more likely. For example, a long, intricate cascade of events caused the August 2003 blackout of a substantial portion of the electrical power system of Northeastern North America affecting fifty million people. The vital importance of the electrical power infrastructure to society motivates the study of models that capture salient features of cascading failure.

Previous work [3, 4, 5] introduced a probabilistic model of cascading failure with a large number of identical components called CASCADE. The components fail when their load exceeds a threshold, and become more loaded when any other component fails. The components initially have a random load and the cascade is started by an initial disturbance increasing the loading of all components. The number of components failed is a measure of the size of the cascade and it has an analytic probability distribution (a saturating form of the quasibinomial distribution). The CASCADE model can be approximated by a saturating Poisson branching process [6] and the relation of these models to cascading failure in simulated blackouts of power transmission systems is studied in [2, 3]. CASCADE is an abstract model of cascading failure and one of its purposes is helping to explain the results of power system models of cascading failure blackouts that represent the transmission network and generation redispatch [1, 3].

The CASCADE model (and its branching process approximation) show interesting behavior as the average initial component load is increased. In one scenario, as this loading is increased, the average number of failures sharply increases at a critical loading. Moreover, at this critical loading, the probability distribution of the number of failures has a power tail of exponent approximately −1.5. The critical loading marks a phase transition and an operational boundary with respect to cascading failure. That is, the risk of cascading failure becomes significant at or above the critical loading. Studying this criticality and finding ways to monitor and detect the corresponding criticality in more detailed simulation models or in real infrastructure systems is a promising new direction of research [6, 2].

One significant limitation of the CASCADE model is the assumption that all components interact. That is, when one component fails, the loading of *all* other components is increased. In applications such as blackouts, many thousands of components can interact by a variety of mechanisms and the interactions can sometimes span the entire system. However, it is more realistic to assume that when one component fails, it interacts with only a subset of the

I. Dobson and B.A. Carreras gratefully acknowledge coordination of this work by the Consortium for Electric Reliability Technology Solutions and funding in part by the Assistant Secretary for Energy Efficiency and Renewable Energy, Office of Power Technologies, Transmission Reliability Program of the U.S. Department of Energy under contract 9908935 and Interagency Agreement DE-A1099EE35075 with the National Science Foundation. I. Dobson, D.E. Newman, and B.A. Carreras gratefully acknowledge support in part from NSF grants ECS-0214369 and ECS-0216053. Part of this research has been carried out at Oak Ridge National Laboratory, managed by UT-Battelle, LLC, for the U.S. Department of Energy under contract number DE-AC05-00OR22725.

other components. This paper generalizes the CASCADE model to this limited interaction case and derives the new criticality condition from the branching process approximation to the generalized model. The result has implications for the interesting question of whether new system technologies that improve system performance by increased communication and coordination between system components introduce many unlikely failure modes that could increase the risk of catastrophic cascading failure [8].

2. CASCADE MODEL WITH k INTERACTIONS

This section summarizes the generalized CASCADE model. There are n identical components with random initial loads. For each component the minimum initial load is $L^{\min}$ and the maximum initial load is $L^{\max}$. Component j has initial load L_j that is a random variable uniformly distributed in $[L^{\min}, L^{\max}]$. $L_1, L_2, \cdots, L_n$ are independent.

Components fail when their load exceeds L^{fail}. When a component fails, a fixed amount of load P is transferred to k samples of the n components. The sampling is uniform so that the probability of choosing a particular component is $1/n$ and the components are sampled independently and with replacement. Moreover, the k samples are chosen independently for each failure.

To start the cascade, an initial disturbance loads k samples of the components by an additional amount D. Other components may then fail depending on their initial loads L_j and the failure of each of these components will distribute an additional load $P \geq 0$ that can cause further failures in a cascade.

It is useful to normalize the model so that $L^{\min}$ becomes zero and both $L^{\max}$ and L^{fail} become one [4, 5]. The normalized initial load $\ell_j = (L_j - L^{\min})/(L^{\max} - L^{\min})$ so that ℓ_j is a random variable uniformly distributed on $[0, 1]$. Let

$$p = \frac{P}{L^{\max} - L^{\min}}, \qquad d = \frac{D + L^{\max} - L^{\text{fail}}}{L^{\max} - L^{\min}} \tag{1}$$

Then p is the amount of load increase on any component when one other component fails expressed as a fraction of the load range $L^{\max} - L^{\min}$. d is the initial disturbance shifted by $L^{\max} - L^{\text{fail}}$ expressed as a fraction of the load range. (The shift ensures that the failure load is one [4, 5].)

The model produces failures in stages $i = 0, 1, 2, \ldots$ where M_i is the number of failures in stage i. It is convenient to state the normalized version of the algorithm. This can be obtained from [4] by adding the random sampling.

Algorithm for normalized CASCADE with k interactions

0. All n components are initially unfailed and have initial loads $\ell_1, \ell_2, \cdots, \ell_n$ determined as independent random variables uniformly distributed in $[0, 1]$.

1. Uniformly sample components k times independently with replacement and add the initial disturbance d to the load of a component each time it is sampled. Initialize the stage counter i to zero.

2. Test each unfailed component for failure: For $j = 1, \ldots, n$, if component j is unfailed and its load > 1 then component j fails. Suppose that M_i components fail in this step.

3. Independently for each of the M_i failures, uniformly sample components k times independently with replacement and add p to the load of a component each time it is sampled.

4. Increment the stage counter i and go to step 2.

3. BRANCHING PROCESS APPROXIMATION

In a Poisson branching process model of cascading failure, failures are produced in stages. Each failure at a given stage produces further next stage failures independently according to a Poisson distribution of rate λ. This section derives the Poisson branching process approximation of the generalized CASCADE model and shows that $\lambda = kp$. Thus $\lambda = kp$ governs the propagation of failures in the cascading process. The implications are discussed in section 4. Those readers interested in the details of the approximation in this section should read the simpler case in [6] first.

Consider the end of step 2 of stage $i \geq 1$ in the CASCADE algorithm. The failures that have occurred are $M_0 = m_0, M_1 = m_1, \ldots, M_i = m_i$, but component loads have not yet been incremented in the following step 3. Let T_{ji} be the number of times component j is sampled in the km_i samples of step 3 of stage i. Then the marginal distributions of $T_{ji}, j = 1, \ldots, n$ are binomial so that

$$P[T_{ji} = t \mid M_i = m_i] = \binom{km_i}{t} \left(\frac{1}{n}\right)^t \left(1 - \frac{1}{n}\right)^{km_i - t} \tag{2}$$

$$E[T_{ji} \mid M_i = m_i] = km_i/n \tag{3}$$

$$\text{Var}[T_{ji} \mid M_i = m_i] = (km_i/n)(1 - 1/n) \tag{4}$$

Write $\underline{T}_{ni} = (T_{1i}, T_{2i}, \ldots, T_{ni})$, $\underline{M}_i = (M_0, M_1, \ldots, M_i)$, $\underline{T}_i = (\underline{T}_{n0}, \underline{T}_{n1}, \ldots, \underline{T}_{ni})$, $S_i = M_0 + M_1 + \ldots + M_i$, $\Sigma_{ji} = T_{j1} + T_{j2} + \ldots + T_{ji}$, and use the corresponding lower case notation for the symbols $\underline{m}_i$, s_i, $\underline{t}_{ni}$, $\underline{t}_i$ and σ_{ji}. The complete history of the component sampling at step 3 of stage i is $\underline{T}_i = \underline{t}_i$.

Define α_{ji} and the saturation function ϕ as

$$\alpha_{ji} = \begin{cases} 0 & ;\text{component } j \text{ failed before stage } i \\ \dfrac{pt_{ji}}{1 - dt_{j0} - p\sigma_{j(i-1)}} & ;\text{component } j \text{ unfailed at beginning of stage } i \end{cases}$$

$$\phi(x) = \begin{cases} 0 \ ; x < 0 \\ x \ ; 0 \leq x \leq 1 \\ 1 \ ; x > 1 \end{cases}$$

Consider unfailed component j and suppose its total stage i, step 2 additional load $dt_{j0} + p\sigma_{j(i-1)} < 1$. Then, when conditioned on $\underline{T}_{i-1} = \underline{t}_{i-1}$, the load of component j is uniformly distributed in $[dt_{j0} + p\sigma_{j(i-1)}, 1]$. In the following step 3, the probability that the load increment of pt_{ji} causes component j to fail is $\phi(\alpha_{ji})$. Now suppose that $dt_{j0} + p\sigma_{j(i-1)} \geq 1$. Then the probability that component j fails is $\phi(\alpha_{ji}) = 1$.

When conditioned on $\underline{T}_i = \underline{t}_i$, the component failures in step 2 of stage $i+1$ are independent and hence M_{i+1} has generating function

$$Ee^{z[M_{i+1}|\underline{T}_i]} = \prod_{j=1}^{n}(1 + (z-1)\phi(\alpha_{ji})) \tag{5}$$

Since $P[M_{i+1} = m_{i+1}|\underline{M}_i]$

$$= \sum_{\underline{t}_i} P[M_{i+1} = m_{i+1}|\underline{M}_i, \underline{T}_i = \underline{t}_i]P[\underline{T}_i = \underline{t}_i|\underline{M}_i]$$

$$= \sum_{\underline{t}_i} P[M_{i+1} = m_{i+1}|\underline{T}_i = \underline{t}_i]P[\underline{T}_i = \underline{t}_i|\underline{M}_i] \ ,$$

$$\begin{aligned} Ee^{z[M_{i+1}|\underline{M}_i]} &= \sum_{\underline{t}_i} Ee^{z[M_{i+1}|\underline{T}_i]}P[\underline{T}_i = \underline{t}_i|\underline{M}_i] \\ &= \sum_{\underline{t}_{i-1}} A_i P[\underline{T}_{i-1} = \underline{t}_{i-1}|\underline{M}_i] \end{aligned} \tag{6}$$

$$\begin{aligned} \text{where } A_i &= \prod_{j=1}^{n}\sum_{\underline{t}_{ni}}(1 + (z-1)\phi(\alpha_{ji}))P[\underline{T}_{ni} = \underline{t}_{ni}|\underline{M}_i] \\ &= \prod_{j=1}^{n}\sum_{t_{ji}}(1 + (z-1)\phi(\alpha_{ji}))P[T_{ji} = t_{ji}|\underline{M}_i]. \end{aligned}$$

Define $X_{ji} = dt_{j0} + p\sigma_{ji}$. Then $X_{ji} \geq 1 \iff pt_{ji} \geq 1 - dt_{j0} - p\sigma_{j(i-1)} \iff \phi(\alpha_{ji}) = 1$. Using (3) and (4),

$$E[X_{ji} \mid \underline{M}_i] = \frac{kd + kps_i}{n} \tag{7}$$

$$\text{Var}[X_{ji} \mid \underline{M}_i] = \frac{kd^2 + kp^2 s_i}{n}\left(1 - \frac{1}{n}\right) \tag{8}$$

It is convenient to renumber the components so that components $1, 2, ..., S_i$ are the S_i components that have failed in previous stages. Then $\alpha_{ji} = 0$ for $j = 1, 2, ..., S_i$. Moreover $A_i = \prod_{j=S_i+1}^{n} B_{ji}$ where

$$\begin{aligned} B_{ji} &= \sum_{t_{ji}}\Big[(1 + (z-1)\alpha_{ji})P[T_{ji} = t_{ji}, X_{ji} < 1|\underline{M}_i] \\ &\quad + zP[T_{ji} = t_{ji}, X_{ji} \geq 1|\underline{M}_i]\Big] \\ &= \left(1 + (z-1)\frac{pE[T_{ji}|X_{ji} < 1, \underline{M}_i]}{1 - dt_{j0} - p\sigma_{j(i-1)}}\right) \\ &\quad P[X_{ji} < 1|\underline{M}_i] + zP[X_{ji} \geq 1|\underline{M}_i] \end{aligned} \tag{9}$$

Let $kp = \lambda$ and $kd = \theta$ and k/n be fixed and let $n, k \to \infty$ and $p, d \to 0$. If $E[X_{ji}] < 1$, using (7), (8) and (3),

$$\begin{aligned} P[X_{ji} \geq 1|\underline{M}_i] &\leq P[|X_{ji} - E[X_{ji}]| \geq |1 - E[X_{ji}]| \ |\underline{M}_i] \\ &\leq \frac{\text{Var}[X_{ji} \ |\underline{M}_i]}{(1 - E[X_{ji} \ |\underline{M}_i])^2} \leq \frac{(n/k)(\theta^2 + \lambda^2 s_i)}{(n - \theta - \lambda s_i)^2} \to 0 \end{aligned}$$

and $E[T_{ji}|X_{ji} < 1, \underline{M}_i] \to E[T_{ji} \mid \underline{M}_i] = km_i/n$. Similarly, if $E[X_{ji}] > 1$, $P[X_{ji} < 1 \mid \underline{M}_i] \to 0$. Thus $P[X_{ji} < 1|\underline{M}_i] \to I[E[X_{ji}] < 1]$ and

$$B_{ji} \sim \left(1 + \frac{z-1}{n}\lambda m_i\right)I[E[X_{ji}] < 1] + zI[E[X_{ji}] > 1].$$

Now $E[X_{ji}] < 1 \iff kd/n + kps_i/n < 1 \iff \theta + \lambda s_i < n$.

If $s_i < (n-\theta)/\lambda$, since $(1+\frac{z-1}{n}\lambda m_i)^{n-s_i} \to e^{\lambda m_i(z-1)}$, $A_i \to e^{\lambda m_i(z-1)}$. Moreover, since the limit of A_i is independent of t_{i-1}, (6) implies that $Ee^{z[M_{i+1}|\underline{M}_i]} \to e^{\lambda m_i(z-1)}$. If $s_i > (n-\theta)/\lambda$, $A_i \to z^{n-s_i}$. Therefore, similarly to [6], we can approximate

$$Ee^{z[M_{i+1}|\underline{M}_i = \underline{m}_i]} \approx \begin{cases} \left[e^{\lambda m_i(z-1)}\right]^{\dagger} + z^{n-s_i}\left(1 - \left[e^{m_i\lambda(z-1)}\right]^{\dagger}(1)\right); \\ \qquad\qquad s_i < (n-\theta)/\lambda, \\ z^{n-s_i}; \qquad s_i > (n-\theta)/\lambda. \end{cases} \tag{10}$$

where $[p(z)]^{\dagger}$ denotes terms of $p(z)$ of degree $\leq n - s_i - 1$.

Since $e^{m_i\lambda(s-1)} = \left(e^{\lambda(s-1)}\right)^{m_i}$, (10) is the distribution of the sum of m_i independent Poisson random variables with rate λ with saturation occurring when the total number of failures exceeds n [6]. Thus we can consider each failure as independently causing other failures in the next stage according to a saturating Poisson Galton-Watson branching process with rate $\lambda = kp$. (This result is the same for the original CASCADE model, except that in the original CASCADE model, $\lambda = np$ [6].)

The failures produced by the initial disturbance when $i = 0$ can also be approximated by a saturating Poisson distribution with rate θ.

4. CRITICALITY CONDITION & IMPLICATIONS

Galton-Watson branching processes proceed in stages to randomly generate an average of λ failures from each failure in the previous stage. It is well known [7] that the criticality condition for branching processes is $\lambda = 1$, and this conclusion also applies to saturating branching processes [6] and in particular to the saturating branching process derived in the previous section. λ governs the propagation of failures so that for $\lambda < 1$ the propagation of failures is likely to be limited, whereas for $\lambda > 1$ there is a high probability of propagation of failures to the entire system. Thus criticality

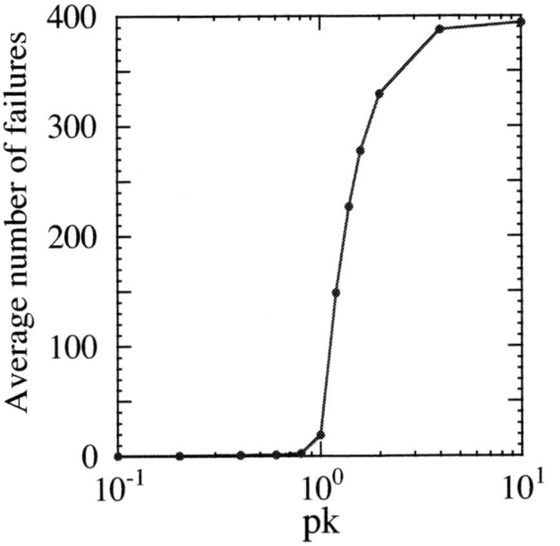

Fig. 1. Average number of failures versus pk as p is varied showing change in gradient at critical point pk=1. There are n=1000 components and the sample size k=100.

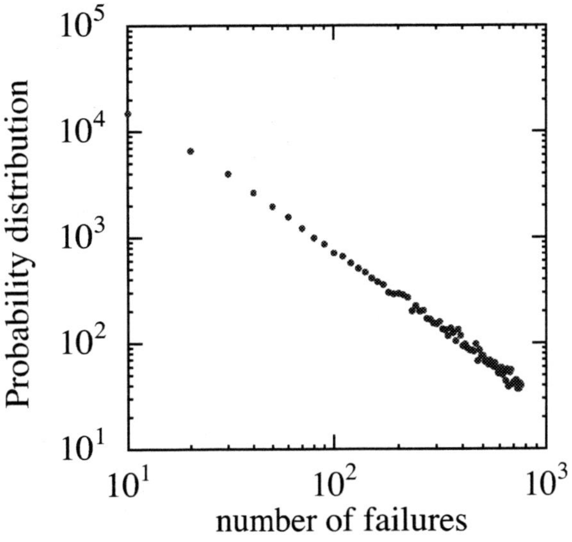

Fig. 2. Probability distribution of number of failures on log-log plot at criticality kp=1. There are n=1000 components.

in the generalized CASCADE model occurs approximately at

$$\lambda = kp = 1 \tag{11}$$

Simulations of the generalized CASCADE model confirm (11). Figure 1 shows the sharp change at $kp = 1$ in the rate of increase of average number of components failed as initial average load is increased. (According to (1), fixing $L^{\max}$ and increasing average initial load $(L^{\max} + L^{\min})/2$ by increasing $L^{\min}$ increases p.) Figure 2 shows the power tail at criticality at $kp = 1$.

As explained in [6], the risk of cascading failure in these models can be minimized by fixing a design limit $\lambda_{\max} < 1$ and requiring $\lambda = kp < \lambda_{\max}$. Then, even if p is very small, large k can cause cascading failure. This suggests that numerous rare interactions between many components can be equally influential in causing cascading failure as a smaller number of likely interactions. Indeed, one can deduce that a design change that introduces a very large number of unlikely failure interactions, thus greatly increasing k, could greatly increase the risk of cascading failure, despite the rarity of the failures (low p). It is conceivable that coupling infrastructures together such as controlling the power transmission system over an internet or certain types of global control schemes could make the system more vulnerable to cascading failure in this fashion. Note that many traditional power system controls are designed to reduce interactions by deliberate separation in distance, frequency, and time scale. Thus the reliability concerns for the effect on cascading failure risk of complicated interconnecting solutions raised by (11) may be consistent with traditional power engineering practice. Our analysis of cascading failure risk is indeed highly approximate and global in nature, but it starts to quantify trade-offs of complexity versus reliability in engineering large networked systems.

5. REFERENCES

[1] B.A. Carreras, V.E. Lynch, I. Dobson, D.E. Newman, Critical points and transitions in an electric power transmission model for cascading failure blackouts, Chaos, vol. 12, no. 4, December 2002, pp. 985-994.

[2] B.A. Carreras, V.E. Lynch, I. Dobson, D.E. Newman, Dynamical and probabilistic approaches to the study of blackout vulnerability of the power transmission grid, 37th Hawaii Intl. Conf. System Sciences, Hawaii, Jan. 2004.

[3] I. Dobson, J. Chen, J.S. Thorp, B.A. Carreras, D.E. Newman, Examining criticality of blackouts in power system models with cascading events, 35th Hawaii Intl. Conf. System Sciences, Hawaii, Jan. 2002

[4] I. Dobson, B.A. Carreras, D.E. Newman, A probabilistic loading-dependent model of cascading failure and possible implications for blackouts, 36th Hawaii Intl. Conf. System Sciences, Hawaii, Jan. 2003.

[5] I. Dobson, B.A. Carreras, D.E.Newman, A loading-dependent model of probabilistic cascading failure, preprint 2003. http://eceserv0.ece.wisc.edu/~dobson/PAPERS/publications.html

[6] I. Dobson, B.A. Carreras, D.E. Newman, A branching process approximation to cascading load-dependent system failure, 37th Hawaii Intl. Conf. System Sciences, Hawaii, Jan. 2004.

[7] T.E. Harris, Theory of branching processes, Dover NY 1989.

[8] C. Perrow, Normal accidents, updated edition, Princeton University Press, 1999.

METER PLACEMENT FOR LOAD ESTIMATION IN RADIAL POWER DISTRIBUTION SYTEMS

Jie Wan
jw33@drexel.edu
Areva T&D

Karen Nan Miu
karen@cbis.ece.drexel.edu
Drexel University, Philadelphia, PA 19104

ABSTRACT

This paper addresses the meter placement problem with respect to load estimation (LE) in radial power distribution systems. A problem formulation of meter placement for LE considering measurement costs, LE accuracy and reliability for a given network configuration is proposed. A heuristic incremental meter placement method is developed. The zonal method addressed in [1,2] is used for LE. The effectiveness of the proposed method is shown in a 20-bus radial distribution network.

1. INTRODUCTION

Up-to-date nodal load data is required to accurately represent customers for operation, control and planning of power distribution systems. Installation of digital multi-channel instruments such as remote terminal units (RTUs) and feeder terminal units (FTUs) makes possible new and more accurate approaches to obtain real-time load data. But economical and technical constraints may restrict widespread installation of measurement devices and polling of the devices from all nodal loads.

Therefore, in practice, nodal load data is estimated from limited available measurements and an immediate question is where measurement devices should be installed and/or where meter data should be polled for load estimation. Meanwhile, conflicting aspects such as

- measurement costs
- estimation accuracy
- reliability: maintain estimation accuracy under the loss of measurements

must be considered when answering this question.

Different from state estimation (SE), load estimation (LE) can provide the estimates of both loads and states. The zonal LE method can sustain measurement loss as long as there are measurements at the substation for radial networks [1,2]. But LE accuracy will change from different measurement schemes. Since there are normally direct measurements at substations, system observability is not a concern in meter placement for LE in this work.

Mathematically, a meter placement problem is a nonlinear combinatorial optimization problem. Global optimization techniques for combinatorial optimization such as simulated annealing and evolutionary algorithms are usually computationally intensive. As such, heuristic algorithms are normally adopted for local optimum or heuristic optimum.

Most studies in meter placement address state estimation. Meter placement methods considering system observability were addressed in [3,4,5]. Celi et al. [6] suggested a method which provided incremental enhancement of existing measurement by adding measurement to satisfying SE accuracy requirement. In [7], a meter placement method was presented to improve feeder voltage calculation with real-time measurements. A three-stage method was proposed by Baran et al [8] to incorporate cost, accuracy, reliability and bad data processing requirements sequentially.

As compared with meter placement for SE, a limited amount of work has been done for LE. Liu et al. [9] proposed a heuristic two-stage method for LE considering accuracy requirements. The method employs LE results for initial loads with uniform percent errors which may occur because of a common bias such as temperature. If the LE method, e.g., the zonal methods shown in [1], can estimate loads very well for initial loads with uniform percent errors, it may not perform well.

Therefore, this paper presents a meter placement method for radial systems with substation which performs consistently regardless of initial parameter estimates. The paper

- formulates the meter placement problem for LE as two sequential single objective optimization according to the priority order of the objectives. Other objectives are represented as constraints with satisfactory specified target levels.
- presents a heuristic incremental meter placement method for two sequential sub-problems: the basic meter placement problem and the advanced meter placement problem.

The zonal method presented in [1,2] is used for LE. The methodology can be generalized to other LE methods. The details are discussed next.

2. PROBLEM STATEMENT

In this work, the meter placement problem is formulated to find measurement buses, i.e., locations of

Mini-RTUs/FTUs so that the measurement cost is minimized with satisfactory LE accuracy and reliability. We formulate this problem under the following assumptions:

- there are measurement devices at substations. Hence, the substation is a default measurement location in the meter placement scheme
- real and reactive branch power flow and voltage magnitude measurements are taken at all existing phases of measurement buses.

Then, the problem is separated into two sequential sub-problems: (i) the basic meter placement problem aims to find the meter placement scheme which will provide the desired accuracy with the minimum cost; (ii) the advanced meter placement problem will enhance the meter placement scheme obtained from the basic meter placement problem so that LE accuracy is satisfactory under the loss of any measurements. The details of each sub-problem are now discussed.

2.1. The Basic Meter Placement Problem

The basic meter placement problem aims to decide locations where RTUs/FTUs need to be installed and/or meters are polled to satisfy specified LE accuracy level, ε, with the minimum cost. It can be formulated as follows:

$$\underset{x}{Min}\, C(x) = \sum_{i \in Loc} c_i x_i \qquad (1)$$

subject to

$$Acc(S_L^{est}) \le \varepsilon$$

where

Loc the measurement location candidate set;

c_i the cost of installing Mini-RTUs/FTUs and/or polling meters at measurement location candidate i;

x_i x_i=1 if the location candidate i is selected as a measurement point, otherwise x_i=0;

$S_L^{est} = [S_{L1}^{est} \;\; S_{L2}^{est} \;\; \cdots \;\; S_{Ln_n}^{est}]^T \in \mathbb{R}^{6n}$ load estimate vector;

$S_{Lk}^{est} = [P_{Lk}^{a,est} \;\; P_{Lk}^{b,est} \;\; P_{Lk}^{c,est} \;\; Q_{Lk}^{a,est} \;\; Q_{Lk}^{b,est} \;\; Q_{Lk}^{c,est}]^T \in \mathbb{R}^6$ the load estimate vector for bus k;

$Acc(S_L^{est})$ the positive scalar function indicating LE accuracy with a given meter placement scheme.

It is worth noting that the constraints in (1) also imply the observability of LE, i.e., there are enough measurements to perform LE. Since the zonal LE method can work as long as there are measurements at the substation for radial networks, LE observability holds true with the assumption that there are measurements at the substation. Also, ε has to be chosen carefully to make sure that feasible solutions to (1) exist.

The notation in (1) assumes that there are three-phase load at each load bus. For single-phase or two-phase load bus, the size of the variables introduced here can be adjusted accordingly.

2.2. The Advanced Meter Placement Problem

The basic meter placement scheme satisfies LE accuracy requirements. Here, advanced meter placement is concerned about the reliability of existing meter placement scheme. Specially, reliability of a meter placement scheme with respect to load estimation is defined as the ability of a meter placement scheme to keep the desired LE accuracy under the loss of measurements from any Mini-RTU/FTU. Next, a heuristic incremental method is developed to address the basic meter placement and the advanced meter placement problem.

3. SOLUTION ALGORITHM

In this section, a heuristic method is developed with the assumptions that all measurement location candidates have the same costs and all the measurements have the same accuracy. Thus, minimum measurement costs are equivalent to minimum number of measurement locations, which is achieved in a greedy way as follows: each time add one measurement location which can improve load estimation accuracy most. The zonal method presented in [1,2] can be employed to obtain the load estimation accuracy. The details are described next starting with LE accuracy merit function.

3.1. Load Estimation Accuracy

LE accuracy is represented by the error between load estimates and the *true* load which is the measured data. It can be quantified in many different ways. However, it is difficult to specify the required accuracy level by using absolute load error because of the large size difference in the loads. Thus, to avoid large individual relative load error, the maximum standard deviation of individual relative load estimation error (*IRLErr*) is adopted to quantify LE accuracy, that is,

$$Acc(S_L^{est}) = max(std(IRLErr(S_L^{est})))$$

where

$IRLErr(S_L^{est}) = (|\, S_L^{est} - S_L^{true} \,| \,./\, |\, S_L^{true} \,|) \times 100\% \in \mathbb{R}^{6n}$;

$std(x) \in \mathbb{R}^{6n}$ standard deviations of a random variable vector $x \in \mathbb{R}^{6n}$, an element-wise function;

$\max(x)$ a scalar, the maximum value over a vector x;

$./$ element-wise division.

3.2. The Basic Meter Placement Procedure

Starting with measurements at the substation, measurements are sequentially added to improve the load estimation accuracy (decrease the standard deviation of *IRLErr*). The procedure is summarized in Figure 1 where:

B_Mset the basic measurement locations;

Cset the remaining measurement location candidates.

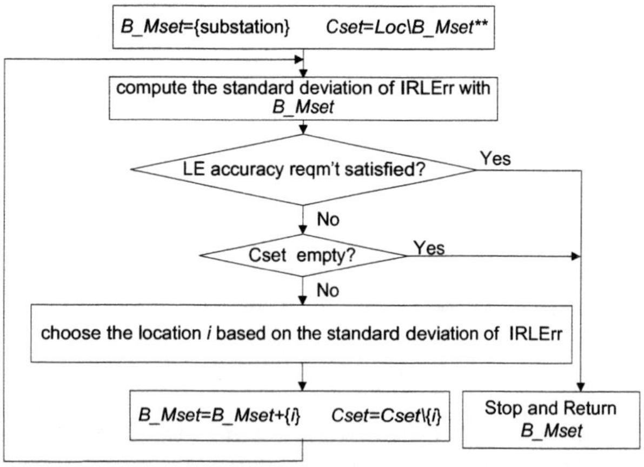

** A\B means the complement of set B in set A

Figure 1: The Basic Meter Placement Procedure

In the above procedure, the additional measurement location i is selected so that maximum improvement in load estimation accuracy is obtained with the addition of measurements at location i. The zonal method [1,2] is employed to provide load estimates. For a given meter placement scheme, the standard deviation of *IRLErr* of the load estimates obtained from the zonal method can be calculated through Monte Carlo simulation as follows:

S1. Generate initial load levels: perturb the *true* load by normally distributed random noise between $\pm b$.

S2. Obtain load estimates from the zonal method and S1.

S3. Go back to S1 until *N* runs have been performed.

S4. Compute the standard deviation of *IRLErr*

As discussed in [1], the zonal LE methods can sustain measurement loss as long as there are measurements at the substation for radial networks. But LE accuracy may suffer. Hence an advanced meter placement procedure is considered to enhance the basic meter placement scheme under the loss of any measurement location.

3.3. The Advanced Meter Placement procedure

Reliability requirements are checked assuming the loss of any measurement location from the basic meter placement scheme. If the LE accuracy requirement is not satisfied, additional measurement locations are selected using the same procedure stated in Section 3.2. The procedure stops when either the accuracy requirement is satisfied or no more candidate locations can be chosen. The procedure is summarized in Figure 2 where Adv_Mset represents the advanced measurement locations.

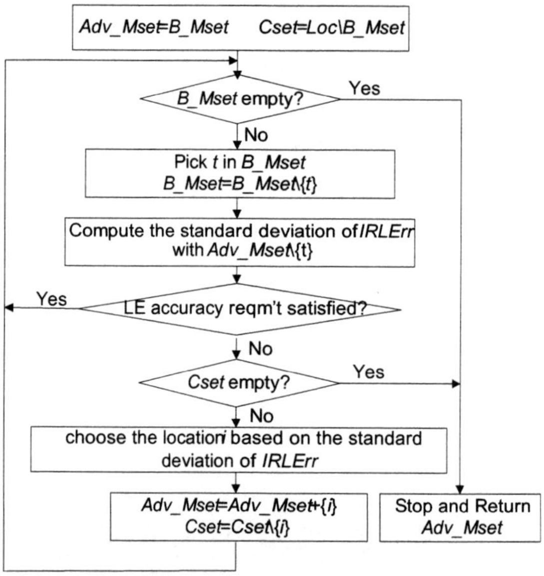

Figure 2: The Advanced Meter Placement Procedure

4. SIMULATION RESULTS

The proposed meter placement method is tested on the 20-bus radial distribution system shown in Figure 3. The basic and advanced meter placement procedures are performed starting with measurements at the substation. The details are discussed next starting with test setups.

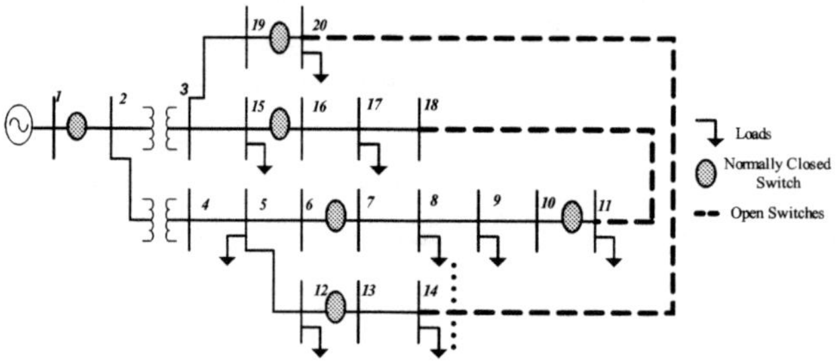

Figure 3: A One-line Diagram of the 20-bus Test System

4.1. Test Setups

Nominal loads are selected as the *true* loads in lieu of actual load measurements and the three-phase power flow solution using constant power load models is obtained. The subsequent voltages, power flows, etc. are treated as *accurate* measurements of the network. Initial load levels are obtained by perturbing the *true* loads with b=5%. For simplicity, the loads at each phase of a bus have the same percent perturbation. With these setups, case study is discussed in detail next.

4.2. Case Study

Meter placement starts with no measurements in the system except at the substation. The results of the basic meter placement procedure are illustrated in Table 1.

TABLE 1: THE BASIC METER PLACEMENT RESULTS ON THE 20-BUS SYSTEM WITH b =5%, N=1000 AND ε =1.7%

# of Iter.	Maximum Standard Deviation of *IRLErr* (%)	Selected Measurement Locations (Bus #)
1	1.77	substation
2	1.74	substation, 14
3	1.68	substation, 10, 14

Remark:

- Even starting with the maximum standard deviation as low as 1.77%, the proposed meter placement method can still decrease it to 1.68%.
- When a limited number of measurements are available, the zonal method still provides reasonable good results, which explains small numerical percent improvement in accuracy.

TABLE 2: THE ADVANCED METER PLACEMENT RESULTS ON THE 20-BUS SYSTEM WITH b =5%, N=1000 AND ε =1.7%

Lost Measurement Location (Bus #)	10	14
Maximum S.T.D. of *IRLErr* with the Loss of One Measurement Location (%)	1.74	1.76
Added Measurement Location (Bus #)	11	13
Maximum Standard Deviation of *IRLErr* with the Added Location (%)	1.68	1.68

Based on the basic meter placement scheme obtained before, the advanced meter placement procedure is performed. As shown in Table 2, two more measurement locations, bus 10 and bus 13, are added to maintain the load estimation accuracy. The advanced meter placement scheme is substation, bus 10, bus 11, bus 13 and bus 14.

4.3. Performance

The performance of the proposed method in terms of solution quality is evaluated with the results obtained from an exhaustive search, which checks all the possible subsets of the location candidates to solve (1). The results with the measurements at *k-subsets* of the location candidates excluding the substation (k is up to 3) and $\varepsilon = 1.7\%$ are checked. The minimum number of locations needed for the required load estimation accuracy is 3. Four possible solutions were found by the exhaustive search. The proposed method found a global non-inferior solution as shown in Table 1.

5. CONCLUSION

In this paper, meter placement for load estimation in radial power distribution systems is studied. A heuristic meter placement method considering measurement costs, LE accuracy and reliability was presented. The problem formulation was tailored for radial networks because the zonal LE methods can sustain measurement loss as long as there are measurements at the substation. Test results in a 20-bus radial distribution system show that the meter placement method is promising in its applications. Rigorous tests on large systems are currently under investigation. The maximum standard deviation of individual absolute load estimation error was chosen as the accuracy index. In this paper, we note that other error indices could also be used if the desired accuracy level can be specified.

6. ACKNOWLEDGEMENT

This work was supported in part by the National Science Foundation under Contract ECS-9984692 and ONR 0014-01-1-0760.

7. REFERENCES

[1] J. Wan and K. Miu, "A Zonal Load Estimation Method for Unbalanced, Radial Distribution Networks", *IEEE Trans. Power Delivery*, Oct. 2002, Vol. 17, No. 4, pp.1106-1112.

[2] J. Wan and K. Miu, "Load Estimation in Radial Electric Power Distribution Networks Using Limited Measurements," in *Proc. 2000 IEEE International Symposium on Circuits and Systems*, Geneva, Switzerland, 2000, Vol. 2, pp. 517-520

[3] K.A. Clements, G. R. Krumpholz and P.W. Davis, "Power System State Estimation with Measurement Deficiency: An Observability Measurement Placement Algorithm", *IEEE Trans. PAS*, Vol. PAS-102, pp. 2012-2020, July 1983.

[4] A. Monticelli, F.F. Wu, "Network Observability: Identification of Observable Islands and Measurement Placement", *IEEE Trans. PAS*, Vol. PAS-104, pp. 1035-1041, May 1985.

[5] F. H. Magnago and A. Abur, "A Unified Approach to Robust Meter Placement against loss of Measurements and Branch Outages", *IEEE Trans. on Power Syst.*, Vol. 15, No. 13, Aug. 2000, pp. 945-949.

[6] M. K. Celi and W.-H.E. Liu, "An Incremental Measurement Placement Algorithm for State Estimation", *IEEE Trans. Power Syst.*, Vol. 10, No.3, Aug. 1995, pp. 1698-1703.

[7] R.C. Leou and C. N. Lu, "Improve Feeder Voltage Calculation Results with Tele-metered Data", *IEEE Trans. Power Delivery*, Vol. 2, Oct. 1996 pp. 1914-1920.

[8] M.E. Baran, J. Zhu, H. Zhu and K.E. Garren, "A Meter Placement Method for State Estimation", *IEEE Trans. Power Syst.*, Vol. 10, No. 3, Aug. 1995, pp. 1704-1710.

[9] Haijun Liu, "Probabilistic Power Flow and Meter Placement in Distribution Systems", *Ph.D Dissertation*, University of Wisconsin-Milwaukee, 1999.

[10] Anssi Seppala, “"Load Research and Load Estimation in Electricity Distribution", *Ph. D Dissertation*, Helsinki University, Espoo, Finland, 1996.

A TOPOLOGICAL MEASUREMENTS AND RTUS DESIGN AGAINST A CONTINGENCY

Garng M. Huang, Jiansheng Lei

Department of Electrical Engineering, Texas A&M University, College Station, Texas 77843, USA

ABSTRACT

The measurements and remote terminal units (RTUs) placement design in an electric power system that minimizes the number of measurements and RTUs under the constraint that the system remains observable under a contingency is formulated. The contingency considered in this paper can be a pre-selected single branch outage or a single measurement loss. In addition, a single RTU loss is taken into account as a type of contingency. A two-stage topological heuristic approach is proposed to solve the problem based on the concept of measurement graph, and an algorithm to search for a valid measurement graph is developed in detail. Tests on the IEEE 30-Bus system demonstrate the validity and efficiency of the proposed approach.

1. INTRODUCTION

State estimation (SE) in power system is to calculate a reliable estimate of the power system state vector of bus voltage angles and magnitudes from tele-metered measurements on the system [1, 2]. Measurements can be power injection, power flow and bus voltage magnitude measurements. Note that except for bus voltage magnitude measurements, measurement devices always appear in pairs of active and reactive power measurements.

Regardless of the estimation algorithms, the configuration of measurements and remote terminal units (RTUs) are decisive factors for a successful state estimator. Therefore, measurements and RTUs placement design has been widely addressed [3-8], and the problem is formulated as an optimization problem, whose objective is to minimize the number of equipments or the installation cost while satisfying some performance constraints.

When there are enough measurements available for the state estimator to proceed, the network is said to be observable [1, 2]. As the power markets become more competitive, a measurement system that can maintain its observability under a contingency becomes more and more important. Recently a numerical algorithm is proposed to optimally upgrade the existing measurement system to be observable under any single branch outage or loss of single measurement [4].

A systematic two-stage measurements and RTUs placement algorithm is presented by the authors in [8] to minimize the number of equipments, where the constraints in [4] are all fully considered. In addition, different from the numerical approach in [4] which is based on the measurement Jacobian matrix and integer programming technique, a topological approach is proposed in [8]. Based on [8], we have the following new developments in this paper:

• The concept of measurement graph in [8] is further refined and an algorithm to search for a valid measurement graph is developed in detail.

• A single RTU loss is also included as a contingency in this paper. As pointed out in [3], all measurements at a substation can have their signals fed to a common RTU for transmission to the control center. Therefore, the failure of the RTU would make all the measurements that it serves unavailable, which is an important factor during the reliability evaluation for a measurement system.

This paper is organized as follows: We formulate the measurements and RTUs design problem in Section 2. In Section 3 the measurement graph concept is refined and an algorithm to find a measurement graph is proposed. The previous heuristic two-stage placement algorithm in [8] is modified accordingly in Section 4. Numerical tests on a standard IEEE 30-bus system are studied in Section 5. Then we conclude in the last section.

2. PROBLEM STATEMENT

2.1. Constraints: Maintaining Observability

The network is required to maintain its observability under a contingency, which can be either a measurement loss, a branch outage or an RTU loss.

It has been demonstrated in [8] that the minimum number of measurements is (2N+1) to maintain observability against a single measurement loss or a branch outage, where N is the total number of buses in the system. The issue of observability maintenance against an RTU loss is discussed as follows:

Meters measure power injections, power flows and bus voltage magnitudes, while RTUs' communication channels transfer the measured raw data from a substation to the control center for further processing. Recent technical advance makes abundant channels in one RTU, and thus all the measurement raw data from a substation can be transferred by a single RTU. However, to maintain observability under loss of an RTU, two RTUs are installed in one substation to transfer measurement data as shown in Fig.2. Then when one RTU is lost, not all measurement data of the substation are lost. For example, if RTU1 in Fig.2 is lost, then measurement data 1 and 2 are lost while measurement data 3 and 4 are still available to the control center. Most traditional RTU design algorithms [5, 6] are based on the architecture as Fig.2.

On the other hand, a new RTU architecture shown in Fig.3 is proposed in this paper with the following novelties:

1. As shown in Fig.3 all data is transferred into two RTUs simultaneously. Since all the devices are in the same substation, we can easily send the measurement data into more than one RTU at one time.

2. We take advantage of the recent technical advance to fully utilize the abundant channels. As discussed in [7], modern

0-7803-8251-X/04/$17.00 ©2004 IEEE

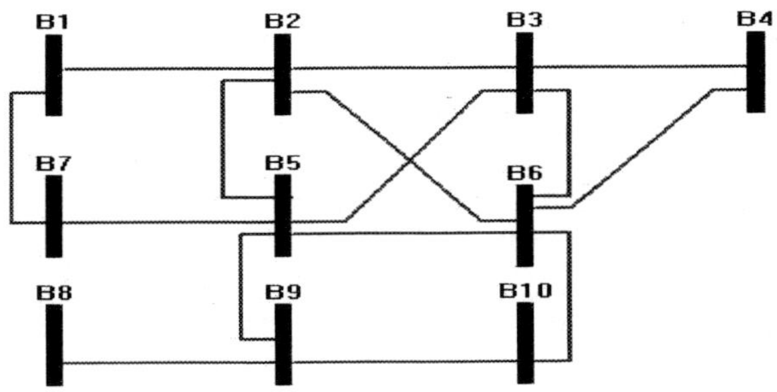

Fig.1 The sample system

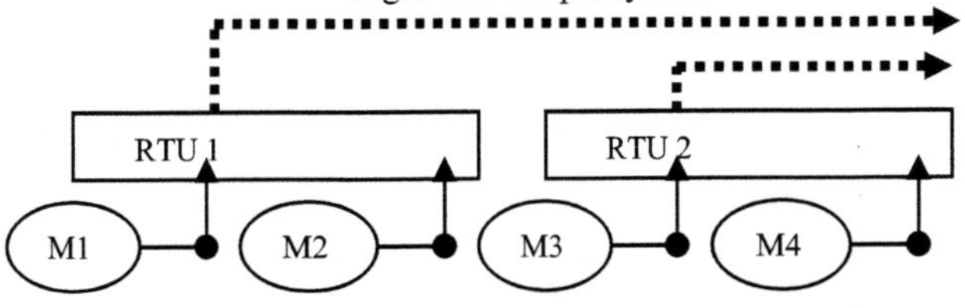

Fig.2 An RTU configuration scheme in a substation

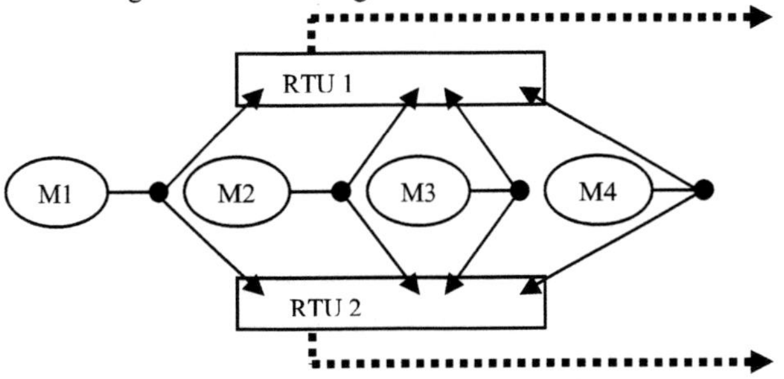

Fig. 3 A new RTU configuration scheme in a substation

analog-to-digital converter (A/D) in RTUs can deal with more than 100 analog measurement inputs per second.

The advantages of the novel scheme in Fig.3 are:

1. When one RTU is lost, all measurement data are still available to the control center. Therefore, the observability is maintained. By contrast, when one RTU is lost in Fig.2, some measurement data will be unavailable to the control center, which will jeopardize the system observability.

2. Note that the measurement data are transferred to the control center by channels of two independent RTUs. Therefore, we can detect RTU transmission errors from data inconsistency of these two RTUs. Schemes such as consistency polling can be devised to further identify the transmission errors.

Since the architectures in Fig.2 and Fig.3 both use two RTUs, the novel RTU architecture in Fig.3 is preferred and applied in this paper. In Section 5, the results are compared to those [5, 6] based on the old architecture in Fig.2.

2.2. Objective: Minimizing the Number of Equipments

The design objective in this paper is to minimize the numbers of both measurements and RTUs.

The heuristic optimization algorithm begins with the measurement graph concept as follows.

3. MEASUREMENT GRAPH

For a given network with a pre-selected contingency set, its measurement graph satisfies the following five conditions:

1. It consists of branch loops, radial branches and the corresponding buses located at the ends of these branches. Note that the in-loop/radial properties of the branches and buses are used in this paper to construct a measurement graph. We flag '*in-loop*' on the branches and their corresponding buses in branch loops; and we flag '*radial*' on the remaining branches and buses.

2. It is a connected graph containing all buses of the network.

3. Its radial branches do not contain any pre-selected single branch outage.

4. Its radial branches have at most one radial tier, i.e., it contains no branch whose two end buses are both *radial*.

5. The number of branches in the graph satisfying the above four conditions is minimal, i.e., removal of any branch in the graph will violate at least one of the four conditions.

A graph satisfying the above five conditions is called a valid measurement graph of the network. To describe an algorithm to find a measurement graph, the following terms are defined:

- The Kernel Graph of a measurement graph consists of all the branches and buses with *in-loop* property.
- The degree of a bus in a graph is the number of branches connected to the bus in the graph. Accordingly, there are two degrees for a bus: namely, α-degree is the degree in the measurement graph, and β-degree is the degree in the corresponding kernel graph. The β-degrees of radial buses are zero because these buses are excluded from the kernel graph, while the β-degrees of in-loop buses are at least two.

3.1. An Algorithm to find a Measurement Graph

The algorithm, described as follows, modifies the network graph by deleting branches until the above five conditions are met.

Step1: Begins with a candidate graph that includes all branches and buses of the network.

Step2: Find the degrees of buses in the graph.

Step3: Find the buses with α-degree=1 (i.e., only one branch is connected to the bus), flag these buses and its connected branch as '*radial*'.

Step4: Flag the rest of buses and branches as '*in-loop*'.

Step5: Start with a set of buses, say {Bus-Id}, with a rule that the smaller α-degree the higher priority; call function flagging-as-radial (Bus-Id) to tentatively remove some branches. Details of function flagging-as-radial (Bus-Id) are given at the end of this sub-section.

Step6: Update the degree information and re-flag all the relevant buses temporarily. Check if the conditions (1-4) of a valid measurement graph are satisfied in the current candidate graph. If yes, permanently retain the tentative operations and the updated degree information (accordingly the branches are permanently removed). Otherwise, null the tentative operations and the degree information.

Step7: Repeat Steps 5-6 until no more branches can be removed.

Step8: Check those branches, whose two end buses have β-degrees greater than 2, named branch-in-test; tentatively remove branch-in-test from the current candidate graph and check if the graph still satisfies conditions (1-4). If yes, permanently retain the removal; otherwise null the tentative removal of branch-in-test.

Step9: Repeat Step8 until no more branches can be removed. Then condition 5 is satisfied and a valid measurement graph is found accordingly.

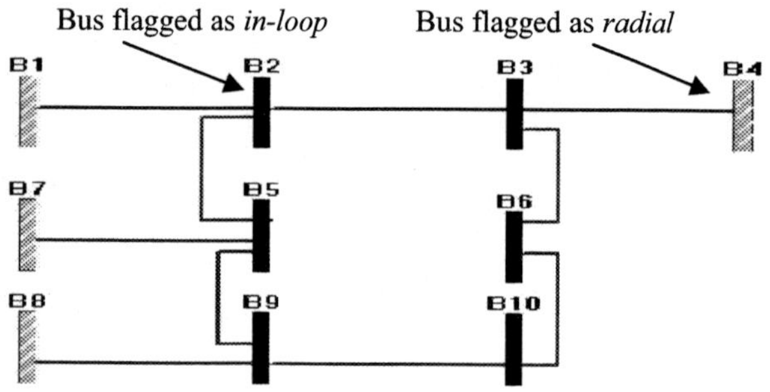

Fig.4 Measurement Graph for the sample system

Steps 1-4 simply initialize the measurement graph, while Steps 5-7 and Steps 8-9 are two procedures to remove the redundant branches at different situations. In Steps 5-7, we are trying to create as many radial branches and buses as possible by removing redundant branches. Therefore, the bus properties are changing from *in-loop* to *radial* in the final measurement graph. On the other hand, Steps 8-9 are to remove the redundant loop branches based on the fixed bus properties from Steps 5-7, and the bus properties will not change in Steps 8-9.

The function flagging-as-radial (Bus-Id) is described in the following C-style pseudo codes:

Function flagging-as-radial (Bus-Id)

{

Get all the buses linked to Bus-Id, say {Bus-1, ..., Bus-m};

If all the branches linked to Bus-Id are labeled as pre-selected outage, then exit;

Else, if a branch linked to Bus-Id, say branch-1, is flagged as radial, i.e. if a bus flagged as radial is connected to Bus-Id via radial branch branch-1, then exit;

Else, tentatively remove all branches linked with Bus-Id from the graph except Branch-1, where Branch-1 connects Bus-Id and Bus-1 satisfying the following two conditions:

1, Branch1 is not pre-selected outage;

2, Current β-degree of Bus-1 is greater than 2.

}

3.2. Illustrative Example

A 10-bus system, as shown in Fig.1 with pre-selected single branch outage from {B1B7, B2B3, B2B5, B3B6, B4B6, B5B9, B6B10, B9B10}, is used to illustrate the above algorithm.

First, a candidate graph is initialized as the network including all branches and buses. Obviously branch B8B9 and bus B8 are flagged as *radial* because current α-degree of B8 is one. Then the other buses and branches are flagged as *in-loop*.

In Step5, since the α-degree of B1 is as small as 2, we call function flagging-as-radial to first process B1. B1 is connected to B2 and B7, however only branch B1B2 (Here Bus-Id is B1, Bus-1 is B2, Branch-1 is B1B2) satisfies the two conditions, i.e.

1, B1B2 is not pre-selected outage;

2, Current β-degree of B2 is greater than 2.

Therefore, we tentatively remove all branches linked with B1 except branch B1B2, i.e. we remove B1B7.

In Step6, we update the degree information temporarily and re-flag *radial* on buses B1 and B7 and branches B1B2 and B7B5 because at this time the α-degrees of both B1 and B7 are one after the tentative removal of branch B1B7. We find the candidate graph satisfies conditions (1-4) after the above tentative operations; therefore we permanently keep all the tentative modifications.

The same process is repeated on other buses until no more modifications can be made. Branch B4B6 is removed during the process, and bus B4 and branch B3B4 are *radial* accordingly.

In Step8 and Step9, redundant branches such as B2B6, B5B3 and B5B6 are removed one by one. Note that β-degrees of the corresponding buses are all greater than two before the removal.

Finally, a valid measurement graph is found as Fig.4, which consists of a branch loop (B2B5, B5B9, B9B10, B10B6, B6B3, B3B2) and four radial branches {B2B1, B5B7, B9B8, B3B4}.

4. MEASUREMENTS PLACEMENT ALGORITHM

Based on the measurement graph concept, a two-stage measurement placement algorithm is proposed in [8]. In the first stage, based on the measurement graph, we place injection measurement pair on the buses located in the loops, place flow measurement pair on the radial branches and place a voltage magnitude measurement on a bus. In the second stage, in order to further reduce the number of RTUs, we replace the injection measurement pair on those buses whose α-degree in the measurement graph is 2 (i.e. only two branches are connected to the bus, thus these branches are in the same loop) by flow measurement pair on one of those two branches. The obtained measurement system has the following features [8]: First, the system is observable under any single measurement pair loss or a pre-selected single branch outage. Second, the number of measurements is (2N+1), which is the minimum number of measurements under the observability constraints.

Furthermore, compared to the original algorithm in [8], the following improvements have been made:

First, based on the novel architecture in Fig.3, one RTU is installed in substations with only one measurement pair and two RTUs are installed in substations with more than one measurement pair. Accordingly, loss of any single RTU has no impact on the system observability, since we place two RTUs when there is more than one measurement pair in the substation. If there is only one measurement pair and one RTU in the substation, losing one RTU leads to the loss of a single measurement pair, which will not jeopardize the system observability due to the first feature of the measurement system.

Second, the previous assumption that every bus stands for an independent substation is not always true. In fact, multiple terminal buses of a transformer are located in the same substation. For example, if branch B3B6 stands for a 2-winding transformer connecting B3 and B6 in Fig.1, then buses B3 and B6 are in the same substation. Accordingly, this substation is labeled as substation (B3, B6). Note that the measurements located on the same substations can always share one RTU, even though the measurements are located on different buses of the same substation. Accordingly, we need to assign RTU based on the substations instead of buses. For example, if branch B3B6 is a transformer, then B3 and B6 are in the same substation (B3, B6) and the measurements located on B3 and B6 can share the same RTUs in substation (B3, B6) accordingly.

5. NUMERICAL TESTS

The problem is to minimize the number of measurements and RTUs for the IEEE 30-Bus standard system shown in Fig.5, so that system remains observable under any single measurement loss, any single RTU loss or any pre-selected single branch outage from {B15B23, B2B5, B22B24, B10B20, B27B28, B16B17, B12B14, B12B15}. The pre-selected branches are labeled in Fig.5.

Here every substation consists of one bus except substations (B12, B4), (B6, B9, B10) and (B27, B28). Note: B6, B9 and B10 are the terminal buses of a three-winding transformer.

Based on the algorithm in Section3, a measurement graph in Table1 is found.

In stage1, the measurement placement, given in Fig.5, contains 30 measurement pairs and 1 voltage magnitude measurement, where the number of measurements is already minimized under the observability requirements.

In stage2, the original injection measurements located on buses {B22, B5, B18} are replaced by flow measurements located on buses {B24, B2, B15}. Then the RTUs located on buses {B22, B5, B18} are removed and the RTUs located on buses {B24, B2, B15} are more efficiently utilized. In addition, the voltage magnitude measurement on B7 is relocated to another substation with more than one measurement pair, say B24. Then only one RTU is needed in B7 because there is only one measurement pair on B7 now. At the same time, the number of RTU on B24 is still two as before.

Finally the number of RTUs is reduced to 17 as shown in Table 2. Note that the measurement number is unchanged and the observability requirements are still satisfied.

Compared with the result in [5-6], the number of RTUs is 17, same as 17 in [5] and smaller than 18 in [6] while all the constraints considered in [5-6] are satisfied. In addition, the number of measurements is minimized in this paper, while in [5-6] the number of measurements is much higher. Furthermore, the topological algorithm avoids the complicated numerical iteration process in [5-6].

Table1. Configuration of the Measurement Graph

Branch loops	Radial branches
B2B4, B4B6, B6B7, B7B5, B5B2	B2B1, B4B3, B6B8
B6B4, B4B12, B12B15, B15B18, B18B19,B19B20,B20B10,B10B6	B12B13, B15B14, B12B16, B10B17
B6B10, B10B9, B9B6	B9B11
B6B10, B10B22, B22B24, B24B25,B25B27,B27B28,B28B6	B10B21, B24B23, B25B26, B27B30,B27B29

Table2. Configuration of the RTUs

Substations with one RTU	Substations with two RTUs
B7, B19, B20	B2, (B4, B12), (B6, B9, B10), B15, B24, B25, (B27, B28)

6. CONCLUSIONS

A two-stage topological approach is proposed to minimize the number of both measurements and RTUs under the requirement that the system remains observable under any single contingency. The contingency can be a pre-selected single branch outage or a single measurement loss. In addition, a single RTU loss is taken into account as a type of contingency. The approach is based on the concept of measurement graph, and an algorithm to search for measurement graph is developed in detail. Numerical tests on the IEEE 30-Bus system demonstrate that the proposed algorithm is valid.

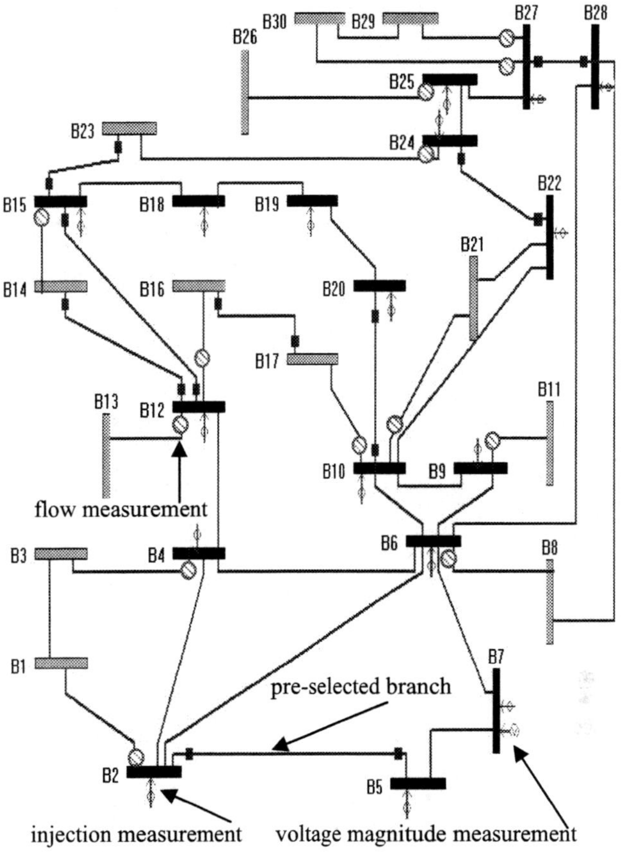

Fig.5 Measurement configuration after Stage1

7. REFERENCES

[1] A. Monticelli, "Electric Power System State Estimation", *Proceedings of IEEE*, Vol. 88, No.2, pp. 262-282, Feb. 2000.

[2] G. R. Krumpholz, K. A. Clements and P. W. Davis, "Power System observability: A Practical Algorithm Using Network Topology", *IEEE Trans. PAS*, Vol. 99, No. 4, pp. 1534-1542, Jul. 1980.

[3] K. A. Clements, G. R. Krumpholz and P. W. Davis, "State Estimation Measurement System Reliability Evaluation", *IEEE Trans. PAS*, Vol. 101, No. 4, pp. 997-1004, Apr. 1982.

[4] F. H. Magnago and A. Abur, "A unified approach to robust meter placement against loss of measurements and branch outages", *IEEE Trans. Power Systems*, Vol.15, No.3, pp.945-949, Aug. 2000.

[5] M. Yehia, R. Jabr, I. El-Bitar and R. Waked, "A PC Based State Estimator Interfaced with a Remote Terminal Unit Placement Algorithm", *IEEE Trans. Power System*, Vol. 16, pp. 210-215, May 2001.

[6] N. D. R. Sarma, V. V. Raju and K. S. P. Rao, "Design of Telemetering configuration for Energy Management Systems", *IEEE Trans. Power Systems*, Vol. 9, pp. 381-387, Feb. 1994.

[7] H. L. Smith and W. R. Block, "RTUs slave for supervisory systems", *IEEE Computer Applications in Power*, Vol. 6, No. 1, pp. 27-32, Jan. 1993.

[8] G. Huang, J. Lei and A. Abur, "A heuristic approach for power system measurement placement design", *IEEE ISCAS 2003*, Vol. 3, pp. 407-410, May 2003.

SINGLE-PHASE CAPACITOR CLAMPED INVERTER WITH SIMPLE STRUCTURE

Bor-Ren Lin, Senior Member, IEEE, and Chun-Hao Huang

Department of Electrical Engineering, National Yunlin University of Science and Technology
Touliu City, Yunlin 640, Taiwan
Grant: NSC 91-2213-E-224-045 & NSC 92-2213-E-224-008

ABSTRACT

A novel single-phase inverter with capacitor clamped topology is presented to generate a sinusoidal voltage to the critical load. There are two legs in the proposed inverter. One inverter leg is consisted of four active switches and one flying capacitor. This inverter leg is operated at high switching frequency to generate three voltage levels on the ac terminal. The other inverter leg is consisted of four active switches without flying capacitor. This inverter leg is operated at low switching frequency to generate two voltage levels on the ac terminal. Therefore, five voltage levels are generated on the ac terminal of the proposed inverter. The control scheme of proposed inverter is easy to implement compared with the conventional capacitor clamped inverter. The control goals of the proposed inverter are generating a sinusoidal output voltage to the critical load and to generate a multilevel PWM voltage waveform on the ac side to reduce the voltage harmonics. The system model and control algorithm are described and analyzed. Experimental results are presented to verify the effectiveness of the proposed control scheme.

1. INTRODUCTION

Recently multilevel inverters or converters have recently increased interest in the university and industry research [1-16]. The multilevel inverters improve the ac power quality by performing the power conversion in small voltage steps resulted in lower harmonics. The output voltage on the ac side can take several discrete levels of equal magnitude. The harmonic content of this output voltage waveform is greatly reduced, if compared with a two-level voltage waveform. Nowadays, there are three prominent multilevel topologies, diode-clamped multilevel inverter, flying-clamped inverter and cascaded multilevel inverter. The diode-clamped inverter and flying-clamped inverter are based on a serial connection of power semiconductors with low voltage stress. The cascaded multilevel inverter [14-16] is based on a serial connection of several two-level inverters. In this topology, each inverter bridge is supplied by a separate and isolated dc power supply, such as in automotive applications. Diode-clamped topology [1-5] is most commonly used in the inverter and converter systems. However, the voltage balance problem between each capacitor on the dc bus is much difficulty to control if the voltage level is beyond four. The flying-clamped inverter [6-13] is another topology to reduce the harmonics by increasing the inverter levels. The flying capacitors in this topology have smaller capacity than dc-link capacitors. There is a flying capacitor in every two-switch pair. Two different switching states can be selected to charge and discharge this flying capacitor such that the voltage balance between each flying capacitor is easy to achieve if the times for charging and discharging flying capacitor are equal. The self-balancing phase-shifted pulsewidth modulation (PWM) scheme was suggested in [8-12] to achieve the voltage balance of flying capacitors. Therefore it is easier to extend the flying clamped inverter to multilevel than the diode-clamped inverter if the voltage level is beyond four.

This paper presents a single-phase flying-clamped inverter with simple structure to generate a three-level voltage waveform on the ac terminal of the inverter and to generate a sinusoidal voltage waveform to the output load. There are four active switches and one flying capacitor in the inverter leg-*a*, and only four active switches in the inverter leg-*b*. The power switches in inverter leg-*a* are operated at high switching frequency PWM scheme and active switches in inverter leg-*b* are operated at low operating frequency. The self-balancing phase-shifted PWM is used in the proposed control scheme to achieve the voltage balance. Based on the proposed control strategy, the adopted flying-clamped inverter generates a sinusoidal voltage with high power quality to the critical load. The effectiveness of the proposed control scheme is verified by the experimental results.

2. SYSTEM ANALYSIS

For the conventional single-phase three-level capacitor clamped inverter, there are four power switches and one flying capacitor in each leg to generate three voltage levels on the ac side. The voltage stress of active switches is equal to half the dc bus voltage. The control scheme of this circuit configuration is complicated. The proposed single-phase three-level inverter is shown in Fig. 1 to reduce one flying capacitor and to make the control scheme easy. Four power switches and one flying capacitor are used in the leg-a to generate a three-level PWM waveform on the voltage v_{an}. The flying capacitor voltage v_2 is equal to half the dc bus voltage ($v_2=v_{dc}/2$). There are four active switches with voltage stress $v_{dc}/2$ in the leg-b to generate a two-level PWM waveform on the voltage v_{bn}. Power switches in the leg-b are operated at low switching frequency (60Hz). By the proper control, a five-level PWM voltage waveform is generated on the ac terminal voltage v_{ab}. There are eight power switches and one clamped capacitor in the adopted three-level inverter. Power switches have the following relationship to prevent the damage of active switches in the adopted inverter.

$$S_{xy}+S_{xy}'=1, \tag{1}$$

where S_{xy} (or S_{xy}')=1 if switch S_{xy} (or S_{xy}') is turned on, or S_{xy} (or S_{xy}')=0 if switch S_{xy} (or S_{xy}') is turned off, $x=a$ and b, y=1 and 2. In the inverter leg-b, the PWM signals for power switches are $S_{b1}=S_{b2}$ and $S_{b1}'=S_{b2}'$. Thus there are only three independent power switches S_{a1}, S_{a2} and S_{b1} in the proposed inverter. All elements are assumed ideal and the output voltage is constant during one switching period. The capacitor voltages v_1 and v_2 are assumed to be $v_2=v_1/2=v_{dc}/2$. According to the switching states of these three independent switches, there are eight valid switching states as shown in Table 1. Depending on the voltage level, there is one operating state to generate voltage $v_{ab}=v_{dc}$ and $-v_{dc}$, respectively, two operating states to achieve voltage $v_{ab}=v_{dc}/2$, 0 and $-v_{dc}/2$, respectively. According to the voltage level of ac terminal voltage v_{ab}, the operation of the adopted inverter can be classified into five operating modes. In the first operating mode, power switches S_{a1}, S_{a2}, S_{b1}' and S_{b2}' are turned on to achieve $v_{ab}=v_{dc}$. This operating state ($v_{ab}=v_{dc}$) can be used to increase the output voltage v_o because the voltage source inverter is worked in the buck operation. There are two switching states in the second operating mode to generate voltage $v_{ab}=v_{dc}/2$. For the positive output filter current i_{Lo}, the output current charges the clamped capacitor $C2$ as the power switches S_{a1}, S_{a2}', S_{b1}' and S_{b2}' are closed or discharges the capacitor $C2$ when the power switches S_{a1}', S_{a2}, S_{b1}' and S_{b2}' are closed. In the positive half cycle of output voltage, the output current i_{Lo} is linearly increasing if $v_{dc}/2>v_o$, or linearly decreasing if $v_{dc}/2<v_o$. These two switching states can be used to balance the capacitor voltage v_2. In the third operating mode, the ac terminal voltage v_{ab} equals zero. There are two valid states in this mode. The equivalent circuits cannot achieve voltage compensation, because output current i_{Lo} does not flow through any one capacitor. The output current i_{Lo} is linearly increasing (or decreasing), if the output voltage is negative (or positive). In the fourth operating mode, there are two valid states in the adopted inverter. The ac terminal voltage $v_{ab}=-v_{dc}/2$ in this mode. For the positive output current, power switches S_{a1}, S_{a2}', S_{b1} and S_{b2} are turned on to charge clamped capacitor $C2$. Power switches S_{a1}', S_{a2}, S_{b1} and S_{b2} are closed to discharge clamped capacitor $C2$. In the last operating mode, power switches S_{a1}', S_{a2}', S_{b1} and S_{b2} are turned on to generate voltage $v_{ab}=-v_{dc}$. It is clear that it is not possible to achieve capacitor voltage compensation, because output current does not flow through clamped capacitor. Modes 1, 2 and 3 ($S_{b1}'=S_{b2}'=1$) are used in the positive output voltage ($v_o>0$) to generate voltage $v_{ab}=v_{dc}$, $v_{dc}/2$ and 0, respectively. Modes 3, 4 and 5 ($S_{b1}=S_{b2}=1$) are selected in the negative output voltage ($v_o<0$) to achieve voltage v_{ab}=0, $-v_{dc}/2$ and $-v_{dc}$, respectively.

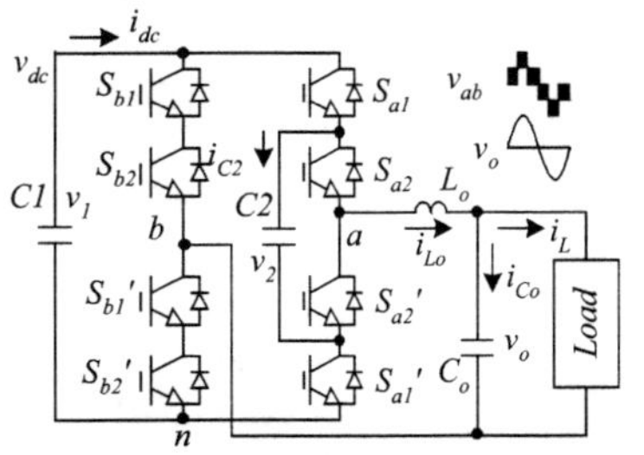

Fig. 1 Proposed single-phase voltage source inverter.

Table 1: Valid switching states of the proposed DC/AC inverter ($i_{Lo}>0$).

State	S_{a1}	S_{a2}	S_{b1}	S_{b2}	v_{an}	v_{bn}	v_{ab}	*Capacitor C2*
1	1	1	0	0	v_{dc}	0	v_{dc}	No change
2	1	0	0	0	$v_{dc}/2$	0	$v_{dc}/2$	Charge
3	0	1	0	0	$v_{dc}/2$	0	$v_{dc}/2$	Discharge
4	0	0	0	0	0	0	0	No change
5	1	1	1	1	v_{dc}	v_{dc}	0	No change
6	1	0	1	1	$v_{dc}/2$	v_{dc}	$-v_{dc}/2$	Charge
7	0	1	1	1	$v_{dc}/2$	v_{dc}	$-v_{dc}/2$	Discharge
8	0	0	1	1	0	v_{dc}	$-v_{dc}$	No change

Based on the eight valid switching states of the adopted inverter, the ac side voltages and dc side current of the adopted inverter are expressed as:

$$v_{an}=S_{a1}v_1+(S_{a2}-S_{a1})v_2 \tag{2}$$

$$v_{bn}=S_{b1}v_1 \tag{3}$$

$$v_{ab}=(S_{a1}-S_{b1})v_1+(S_{a2}-S_{a1})v_2 \tag{4}$$

$$i_{dc}=(S_{a1}-S_{b1})i_{Lo} \tag{5}$$

The differential equations of the adopted inverter can be expressed as follows:

$$L_o\frac{di_{Lo}}{dt}=v_{ab}-v_o-ri_{Lo} \\ =(S_{a1}-S_{b1})v_1+(S_{a2}-S_{a1})v_2-v_o-ri_{Lo} \tag{6}$$

$$C1\frac{dv_1}{dt}=-i_{dc}=(S_{b1}-S_{a1})i_{Lo} \tag{7}$$

$$C2\frac{dv_2}{dt}=(S_{a1}-S_{a2})i_{Lo} \tag{8}$$

$$C_o \frac{dv_o}{dt} = -i_L + i_{Lo} \quad (9)$$

where r is the equivalent resistor across filter inductor L_o. Based on equations (2)-(9), the proposed system can be simulated to provide a stable sinusoidal output voltage.

3. CONTROL SCHEME

The main functions of the proposed control scheme for the adopted inverter are to maintain the balanced capacitor voltages and generate a stable sinusoidal output voltage to the critical load. Power switches in the adopted inverter leg-b are operated at low switching frequency. The operating modes 1, 2 and 3 are selected to generate the ac terminal voltage $v_{ab}=v_{dc}$, $v_{dc}/2$ and 0, respectively for the positive output voltage v_o. On the other hand, operating modes 3, 4 and 5 are used to achieve ac terminal voltage $v_{ab}=0$, $-v_{dc}/2$ and $-v_{dc}$ respectively for the negative output voltage. Therefore, five voltage levels are generated on the ac terminal voltage of the adopted inverter. Fig. 2(a) gives the control block of the proposed control scheme. To maintain the constant root-mean-square (rms) output voltage, a proportional-integral (PI) based voltage controller is used in the outer control loop. The output root mean square voltage is measured. The error value between the reference output voltage $V^*_{o,rms}$ and the measured output voltage $V_{o,rms}$ is sent to the rms voltage controller to obtain the reference output voltage V*. The instantaneous output voltage command v^*_o is obtained from the multiplication of output voltage command V* and a sinusoidal waveform. The instantaneous output voltage variation is $\Delta v_o = v_o^* - v_o$. A proportional-integral (PI) instantaneous voltage controller is used to compensate this output voltage variation. The control signal for the PWM operation is expressed as:

$$d = [v_o^* + PI(\Delta v_o)] / v_{dc} \quad (10)$$

The duty cycle for power switches in each leg can be obtained by properly selecting the operating states in Table 1. Fig. 2(b) gives the PWM signals for power switches S_{a1}, S_{a2}, S_{b1} and S_{b2}. Two carriers with phase shift 180° are used to compare with the control signal d. The PWM signal for power switch S_{a1} is obtained by comparison the control signal d and carrier signal $v_{tri,1}$. The PWM signal for power switch S_{a2} is generated by comparison the control signal d and carrier signal $v_{tri,2}$. The PWM signal for the power switches S_{b1} and S_{b2} equals zero (or one) if output voltage command is positive (or negative). Based on the adopted PWM technique, two operating states in the mode 2 and mode 4 are generated in each half cycle of output voltage. Therefore, the capacitor voltage v_2 will be compensated in this PWM technique as shown in Fig. 2(c). For the positive output current i_{Lo} and ac side voltage $v_{ab}=v_{dc}/2$, power switches S_{a1}, S_{a2}', S_{b1}' and S_{b2}' are closed to charge capacitor $C2$ and power switches S_{a1}', S_{a2}, S_{b1}' and S_{b2}' are closed to discharge capacitor $C2$. Based on the proposed control scheme, the output voltage v_o is controlled to track the reference output voltage with constant rms value.

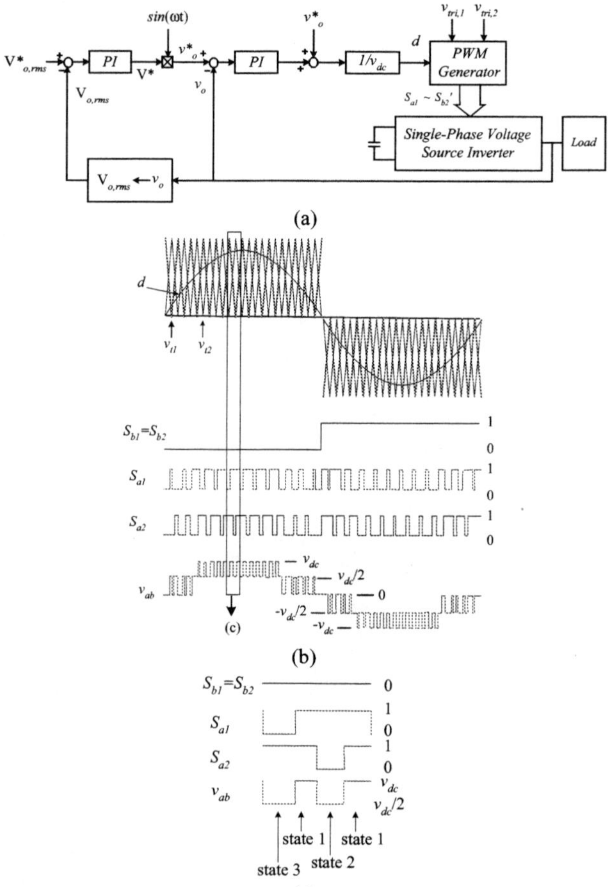

Fig. 2 Proposed inverter (a) control block diagram (b) PWM waveform in one output cycle (c) PWM waveform in positive output voltage.

4. EXPERIMENTAL RESULTS

The proposed capacitor clamped inverter with 1.5kW rating was built and tested in the laboratory to verify the operation and feasibility of the proposed control scheme. The inverter was implemented with IGBT (IRG4PC40W) and the main components were: $C1$=2200μF, $C2$=330μF, C_o=18uF, L_o=1mH and $v_{o,rms}$=110V/60Hz. The capacitor voltage v_2 equals 100V. The dc bus voltage is 200V. The triangular carrier frequency is 20kHz. The inverter is controlled by a digital signal processor (DSP) TMS320C32. Fig. 3 shows the experimental results of output voltage v_o and PWM signals S_{b1}', S_{b2}', S_{a1} and S_{a2}. Fig. 4 illustrates the measured results of output voltage v_o and ac terminal voltages v_{an}, v_{bn} and v_{ab}. Two, three and five voltage levels are generated on the voltages v_{bn}, v_{an} and v_{ab}, respectively. Fig. 5 gives the measured results of output voltage v_o and load current i_L under the linear and nonlinear loads. The sinusoidal and clean output voltage is provided by the proposed inverter. The total harmonic distortion of output voltage is about 1.5%. From the measured results the proposed inverter can provide a stable and sinusoidal output voltage to the output load and generate a three-level PWM voltage pattern on the ac terminal voltage v_{ab}.

5. CONCLUSION

A single-phase voltage source inverter based on capacitor clamped configuration is proposed to provide a sinusoidal voltage to the output load. Three-level PWM scheme is used in the proposed inverter to reduce voltage harmonics on the ac side and reduce the voltage stress of the power semiconductors. The voltage stress of power devices is clamped to half the dc-link voltage. Power switches in the inverter leg-b are operated at low switching frequency to generate two voltage levels on the voltage v_{bn}. Power switches in the inverter leg-a are operated at high switching frequency to generate three voltage levels on the ac side voltage v_{an}. Five voltage levels are achieved on the ac terminal voltage v_{ab}. According to the proposed control scheme and the tested results, a sinusoidal voltage with stable root-mean-square value is provided to the output load.

6. REFERENCES

[1] A. Nabae, I. Takahashi, and H. Akagi, "A new neutral-point-clamped PWM inverter," *IEEE Transactions on Industry Applications*, vol. 19, no. 5, 1981, pp. 518-523.

[2] J. S. Lai and F. Z. Penf, "Multilevel converters – a new breed of power converters," *IEEE Transactions on Industry Applications*, vol. 32, no. 3, 1996, pp.509-517.

[3] M. Marchesoni, M. Mazzucchelli ans S. Tenconi, "A non conventional power converter for plasma stabilization," in *Proc. IEEE Power Electronics Specialist Conference*, 1988, pp. 122-129.

[4] J. Steinke, "Switching frequency optimal PWM control of a three level inverter," *IEEE Transactions on Power Electronics*, vol. 7, no. 3, 1992, pp. 487-496.

[5] L. M. Tolbert, F. Z. Peng and T. G. Habetler, "Multilevel PWM methods at low modulation indices," in *Proc. IEEE Applied Power Electronic Conference*, 1999, pp. 1032-1038.

[6] Y. Chen and B. T. Ooi, "Regulating and equalizing dc capacitor voltages in multilevel converters," *IEEE Transactions on Power Delivery*, vol. 12, no. 2, 1997, pp. 901-907.

[7] T. A. Meynard and H. Foch, "Multilevel conversion: High voltage choppers and voltage source inverter," in *proc. IEEE Power Electronics Specialist Conference*, 1992, pp. 397-403.

[8] X. Yuan, H. Stemmler and I. Barbi, "Investigation on the clamping voltage self-balancing of the three-level capacitor clamping inverter," in *proc. IEEE Power Electronics Specialist Conference*, 1999, pp. 1059-1064.

[9] X. Yuang; H. Stemmler and I. Barbi, "Self-balancing of the clamping-capacitor-voltages in the multilevel capacitor-clamping-inverter under sub-harmonic PWM modulation," *IEEE Transactions on Power Electronics*, vol. 16, no. 2, 2001, pp. 256 –263.

[10] C. Turpin, L. Deprez, F. Forest, F. Richardeau, and T. A. Meynard, "A ZVS imbricated cell multilevel inverter with auxiliary resonant commutated poles," *IEEE Transactions on Power Electronics*, vol. 17, no. 6, 2002, pp. 874 –882.

[11] T. A. Meynard, H. Foch, P. Thomas, J. Courault, R. Jakob, and M. Nahrstaedt, "Multicell converters: basic concepts and industry applications," *IEEE Transactions on Industrial Electronics*, vol. 49, no. 5, 2002, pp. 955 –964.

[12] S.-G. Lee, D.-W. Kang, Y.-H. Lee and D.-S. Hyun, "The carrier-based PWM method for voltage balance of flying capacitor multilevel inverter," in *Proc. IEEE Power Electronics Specialists Conference*, 2001, pp. 126 –131.

[13] T. A. Meynard and H. Foch, "Imbricated Cells Multilevel VSI for High Voltage Applications", EPE Journal, vol. 3, no. 2, pp. 99-106, 1993.

[14] C. Hochgraf, R. Lasseter, D. Divan and T. A. Lipo, "Comparison of Multilevel Inverters for Static Var Compensation", IEEE Industrial Applications Society Annual Meeting, Conf.-Rec., pp. 921-928, 1994.

[15] F. Z. Peng, J. S. Lai, J. W. Mckeever and J. Vancoevering, "A multilevel voltage-source inverter with separate dc sources for static VAR generation," IEEE Transactions on Industry Application, vol. 32, no. 5, 1996, pp. 1130-1137.

[16] M. D. Manjrekar, P. K. Steimer and T. A. Lipo, "Hybrid multilevel power conversion system: a competitive solution for high-power applications," IEEE Transactions on Industry Application, vol. 36, no. 3, 2000, pp. 834-841.

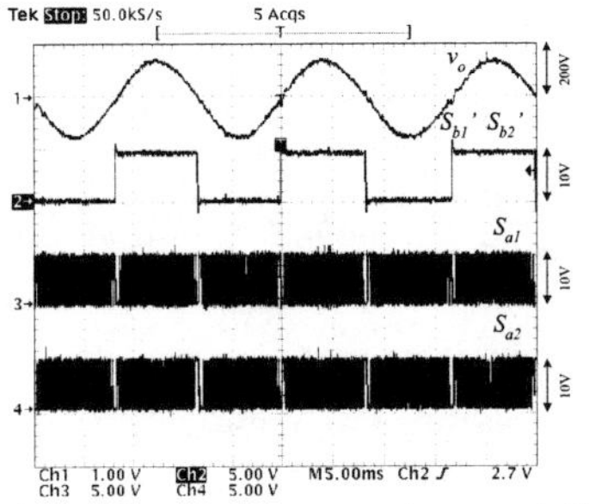

Fig. 3 Measured waveforms of output voltage and PWM signals of power switches S_{a1}, S_{a2}, S_{b1}' and S_{b2}'.

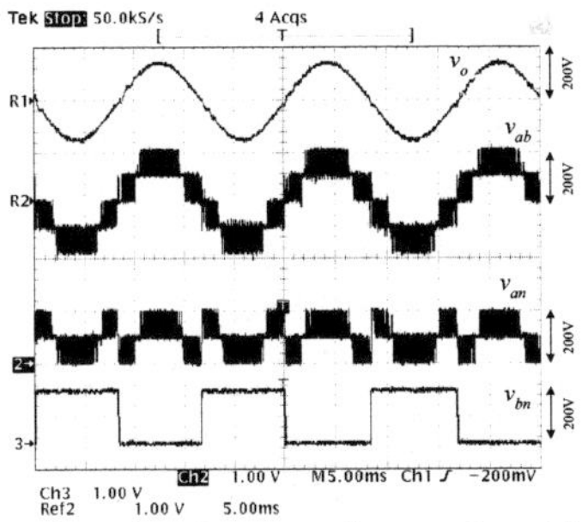

Fig. 4 Experimental results of output voltage and ac side voltages of the adopted inverter.

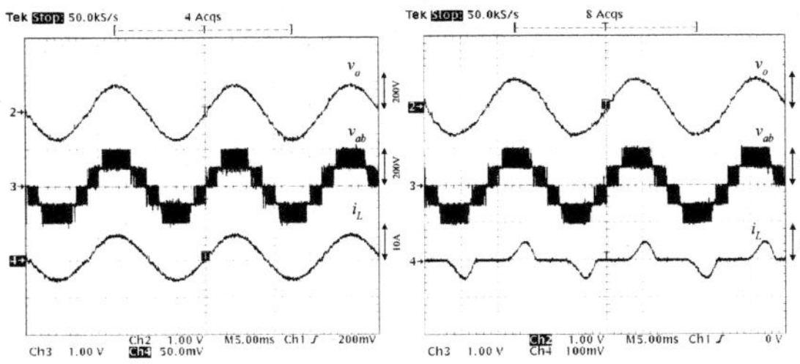

(a) (b)

Fig. 5 Experimental results of output voltage v_o, ac side voltage v_{ab} and output load current i_L (a) linear load (b) nonlinear load.

A GENERALIZED AVERAGING MODEL FOR FIXED FREQUENCY CONVERTERS

S.W. Ng and Y.S. Lee

The Hong Kong Polytechnic University

ABSTRACT

A 1st order approximated small signal response model which is suitable for all kinds of fixed frequency converter is introduced. Further simplification and extraction of characteristics, e.g. poles, are possible by using a symbolical package. By using this model, model development becomes much simpler and traceable. The accuracy of the model is also demonstrated on a ZCS and a ZVS converter.

1. INTRODUCTION

Modeling is an important step in the analysis and design of switching-mode power converters. However, the exact models of complex converters are in general unknown. Approximate models, developed by heuristic approaches, are used instead. The state-space averaging method [1] is the most successful and popular approach for deriving approximate models of simple converters like buck and boost. However, to model complex converters, e.g. the more popular ZCS [2] and ZVS [3] converter, by using the averaging method, are much complicated tasks. Their development and usage demand a higher level of expertise and can be very tedious and time consuming.

Recently, the exact and explicit expressions for the small-signal response of fixed frequency converters are derived [4]. This can be the basis for the derivation of simpler and symbolical approximate models for analysis and design. A symbolical model allows an engineer to understand the effect of every component on the performance of a converter. In [4], examples of deriving approximate models from the exact expressions are shown. However, the derivation steps in [4] are still tedious and heuristic. In this paper, the set of exact expressions in [4] are rearranged and a single 1st order approximated model is proposed. Since it is a 1st order approximated model, it is effectively a generalized state-space averaging model which is applicable to all kinds of fixed frequency converters. To keep this paper short, the mathematical proof will be omitted. Instead, this paper will concentrate on demonstrating the application of this model on two popular converters. Numerical results will show that the proposed model is accurate enough for practical use.

2. A GENERALIZED AVERAGING MODEL FOR FIXED FREQUENCY CONVERTERS

A switching converter is a simple switched circuit composed of linear components and ideal switches. At the steady state, the circuit will cycle through m different stages within a switching period (T). Let t_k be the time when the converter switches from the kth stage to the $k+1^{th}$ stage, the time domain behavior of the converter within the kth stage can be described by a linear state equation, $\dot{x} = A_k x + b_k u$. Its solution for a given initial condition, $x(t_{k-1})$, is

$$x(t - t_{k-1}) = e^{A_k(t-t_{k-1})} x(t_{k-1}) + A_k^{-1}(e^{A_k(t-t_{k-1})} - I) b_k u$$

for $t_{k-1} < t < t_k$. We will state without proof that (1) below models the 1st order approximated small signal line-to-output response h_u of all kinds of fixed frequency converters.

$$h_u \approx T \left\{ \sum_{k=p}^{m+p-1} d_k \prod_{j=p}^{k-1} M_j \right\} Z_p^q \{ Z_p^q - (1 - sT) H_{q+1}^{m+p} M_q \}^{-1} \times \left\{ \sum_{k=p}^{m+p-2} d_k H_{k+1}^{m+p-1} M_k b_k + d_{p-1} M_{p-1} b_{p-1} \right\} \quad (1)$$

where

$$H_j^i = M_i (I + A_i d_i T) M_{i-1} \ldots M_j (I + A_j d_j T)$$

$$Z_j^i = (I - A_j d_j T)(I - A_{j+1} d_{j+1} T) \ldots (I - A_i d_i T)$$

$$A_{k+m} = A_k \quad , \quad b_{k+m} = b_k \quad , \quad d_k = \frac{t_k - t_{k-1}}{T}$$

$$M_j = M_{j+m} = I - \frac{(\dot{x}_{t_j-\Delta} - \dot{x}_{t_j+\Delta})\beta}{\beta \dot{x}_{t_j-\Delta}}$$

$$= I - \frac{[A_j x(t_j) + b_j u - A_{j+1}x(t_j) - b_{j+1}u]\beta}{\beta[A_j x(t_j) + b_j u]}$$

The term M_j above is a modifier which models the effect of conditional (soft) switching [5] which is defined by β. The indexes p and q should be chosen to maximize the accuracy or the simplicity of the model. The engineer may do a trial-and-error to choose ones which suit his need. The control-to-output model may then be derived by

$$h_d = \frac{\partial h_u}{\partial d}$$

Since (1) is a 1st order approximation of the set of expressions in [4], it can be seen as a generalized state-space averaging model. Cases below will demonstrate its usage.

3. CASE I: A ZERO CURRENT SWITCHED CONVERTER

The circuit of the converter in [2] is shown in Fig.1. The switching frequency is 200kHz. This converter has 6 stages and their corresponding timing is shown in Fig.2. The derivations of their state matrixes are trivial and only their modifiers will be shown here. They are:

$$M_1 = M_3 = M_6 = I\,,\; M_2 = \begin{pmatrix} 1 & 0 & 0 & 0 \\ 0 & 0 & 0 & 0 \\ 0 & 0 & 1 & 0 \\ 0 & 0 & 0 & 1 \end{pmatrix},$$

$$M_4 = \begin{pmatrix} \frac{L_1}{L_1+L_2} & \frac{L_2}{L_1+L_2} & 0 & 0 \\ \frac{L_1}{L_1+L_2} & \frac{L_2}{L_1+L_2} & 0 & 0 \\ 0 & 0 & 1 & 0 \\ 0 & 0 & 0 & 1 \end{pmatrix} \approx \begin{pmatrix} 1 & 0 & 0 & 0 \\ 1 & 0 & 0 & 0 \\ 0 & 0 & 1 & 0 \\ 0 & 0 & 0 & 1 \end{pmatrix},$$

$$M_5 = \begin{pmatrix} 1 & 0 & 0 & 0 \\ 0 & 1 & 0 & 0 \\ 0 & 0 & \frac{C_1}{C_1+C_2} & \frac{C_2}{C_1+C_2} \\ 0 & 0 & \frac{C_1}{C_1+C_2} & \frac{C_2}{C_1+C_2} \end{pmatrix} \approx \begin{pmatrix} 1 & 0 & 0 & 0 \\ 0 & 1 & 0 & 0 \\ 0 & 0 & 0 & 1 \\ 0 & 0 & 0 & 1 \end{pmatrix} \text{ when}$$

$$C_2 >> C_1 \text{ and } L_1 >> L_2 \qquad (2)$$

In this case, we choose $q=1$ and $p=2$ because our experiments indicate the choice of p and q has no effect on the simplicity or the accuracy of the final model below. They also illustrate the case when $q<p$. When (2) is substituted into (1), we have

$$h_u \approx TP_Z\{I-(1-sT)H_2^7\}^{-1}P_b \qquad (3)$$

where

$$P_Z = d_1M_5M_4M_2 + d_6M_5M_4M_2 + d_5M_4M_2 + d_4M_2 + d_3M_2 + d_2I$$

$$P_b = d_1M_1b_1 + d_6h_1b_6 + d_5h_1h_6M_5b_5 + d_4h_1h_6h_5M_4b_4 + d_3h_1h_6h_5h_4b_3 + d_2h_1h_6h_5h_4h_3M_2b_2$$ and

$$h_i = M_i(I + A_i d_i T)$$

Obviously, (3) is still too complex to have any practical application. We should further simplify (3) by eliminating higher order terms which involve power of T. A possible simplification is to extract the zero order terms of P_b and the 1st order terms of H_2^7. This gives

$$P_b \approx d_1M_1b_1 + d_6b_6 + d_5M_5b_5 + d_4M_5M_4b_4 + d_3M_5M_4b_3 + d_2M_5M_4M_2b_2$$

$$H_2^7 \approx (I + A_1d_1T + A_6d_6T)M_5M_4M_2 + T\{M_5A_5d_5M_4M_2 + M_5M_4(A_4d_4 + A_3d_3)M_2 + M_5M_4M_2A_2d_2\}$$

While the steps of approximation and extraction above looks tedious and heuristic, they are carried out by the symbolical package Mathematica® and the engineers and researchers are spared from all the tedious manipulation of the equations. Because of the length of the Mathematica codes, they will not be shown here. The final simplified model equations are:

$$v_{C2} = \frac{N_C}{D_C} \qquad (4)$$

$$i_{L1} = \frac{N_I}{D_C} \qquad (5)$$

where

$$N_C = c_2(d_1 + d_6)(L_1 + [1 - d_5 - d_6]L_2)R$$

$$D_C = c_2(d_1 + d_6)(d_5 + d_6)L_1R + (c_2 + [1 - d_5 - d_6]c_1)L_1(L_1 + L_2)s + c_2(c_1 + c_2)L_1(L_1 + L_2)Rs^2$$

$$N_I = (L_1 + [1 - d_5 - d_6]L_2)$$
$$\times\{(c_2 + [1 - d_5 - d_6]c_1) + c_2(c_1 + c_2)Rs\}$$

Equations (4,5) are outputs directly from the Mathematica®. By assuming $c_1 << c_2$ and $L_2 << L_1$, we may further reduce (4,5) to

$$v_{C2} \approx \frac{c_2 d_6 L_1 R}{c_2 d_6^2 L_1 R + c_2 L_1^2 s + c_2^2 L_1^2 Rs^2} = \frac{d_6 R}{d_6^2 R + L_1 s + c_2 L_1 Rs^2} \quad (6)$$

$$i_{L1} \approx \frac{(1 + c_2 Rs)}{d_6^2 R + L_1 s + c_2 L_1 Rs^2} \quad (7)$$

Equations (6,7) above are simple and suitable for implementation into a PSPICE macro circuit model. Furthermore, one may derive the poles of the converter easily from (6,7). The small signal response predicted by (6) is plotted in Fig.3 along with the exact response calculated by the expression in [4]. The two overlap each other. Thus, while the equations (6,7) are simple, they accurately model the small signal response.

4. CASE II: A FULL BRIDGE ZERO VOLTAGE SWITCHED (FB-ZVS) CONVERTER

The circuit of the converter in [3] is shown in Fig.4. The switching frequency is 100kHz. This converter also has 6 stages in its continuous current mode. Their modifiers are:

$$M_1 = \begin{pmatrix} \frac{L_1}{L_1 + L_2} & \frac{L_2}{L_1 + L_2} & 0 \\ \frac{L_1}{L_1 + L_2} & \frac{L_2}{L_1 + L_2} & 0 \\ 0 & 0 & 1 \end{pmatrix},$$

$$M_2 = M_3 = M_5 = M_6 = I,$$

$$M_4 = \begin{pmatrix} \frac{L_1}{L_1 + L_2} & \frac{-L_2}{L_1 + L_2} & 0 \\ \frac{-L_1}{L_1 + L_2} & \frac{L_2}{L_1 + L_2} & 0 \\ 0 & 0 & 1 \end{pmatrix} \quad (8)$$

When (8) is substituted into (1), we have

$$h_u \approx TP_Z Z_2^4 \{Z_2^4 - (1 - sT)H_4^7 M_4\}^{-1} P_b \quad (9)$$

where $P_Z = (d_1 + d_5 + d_5)M_4 + (d_4 + d_3 + d_2)I$ and

$$P_b = d_1 M_1 b_1 + d_5 h_1 h_6 b_5 + d_4 h_1 h_6 h_5 M_4 b_4 + d_2 h_1 h_6 h_5 h_4 h_3 b_2$$

Similar to Case 1, Mathematica® is used to eliminate higher order terms in (9). The simplified small signal line-to-output model is

$$v_C = \frac{N_V}{D_V} \quad (10)$$

where

$$N_V = 2L_2[(d_1 + d_2)(L_1 + L_2) - 2d_2^2 L_1)RV_{in}$$
$$D_V = 4L_1 L_2 (L_1 + L_2)(1 + CRs)$$
$$+ [(L_2^2 + 2d_1 L_1^2 + 4d_1 L_1 L_2)R + (L_1 + L_2)(L_2^2 - 4L_1 L_2 + L_1^2)s$$
$$+ C(L_2 - L_1)^2 (L_1 + L_2)Rs^2]T$$

In comparison with the model in [3], the model here is a little bit more complex. However equation (10) is derived automatically using a software. Including the time spend on deriving (8), the whole model is derived in half a day. Thus, (1) is still a very useful analytical tool for average engineers and researchers.

5. CONCLUSION

A generalized averaging model for all kinds of fixed frequency converters is introduced. It merely demands trivial derivation of the state matrixes and modifiers of a converter circuit. The developments of simple analytical models for a ZCS and a ZVS converter are demonstrated. The outcome is accurate and the steps involved are simple and can be carried out automatically by software. The final software can even derive macro models for PSPICE simulation. Because of the limited space, this feature cannot be demonstrated here. Hence, the proposed model here will be a useful tool for engineers to derive analytical models for new converters which do not have known dynamic model.

6. REFERENCES

[1] R.D. Middlebrook and S. Cuk, "Modelling and analysis methods for DC-to-DC switching converters," IEEE International Semiconductor Power Converter Conference 1977 Rec., pp.90-111, 1977.

[2] L. Yang, D. Long, Y. Zhang, and C. Q. Lee, "A new

zero-current-switching constant switching frequency boost converter," Proceedings of IEEE International Symposium on Circuits and Systems, vol. 4, pp. 2335-2338, 1993.
[3] V. Vlatkovic, J.A. Sabate, R.B. Ridley, F.C. Lee, and B.H.Cho, "Small-signal analysis of the phase-shifted PWM converter," IEEE Transactions on Power Electronics, vol. PE-7, pp.128-135, 1992.
[4] S. W. Ng and Y. S. Lee, "Explicit expressions for the exact small-signal behavior of PWL switching converters," International Journal on Circuit Theory and Applications, vol. 32, pp. 1-10, 2004.
[5] S.W. Ng and Y.S. Lee, "Implementation of an Efficient Small Signal Simulation Algorithm", IEEE Transactions on Circuits and Systems I, vol. 50, No.19, pp1229- 1231, 2003.

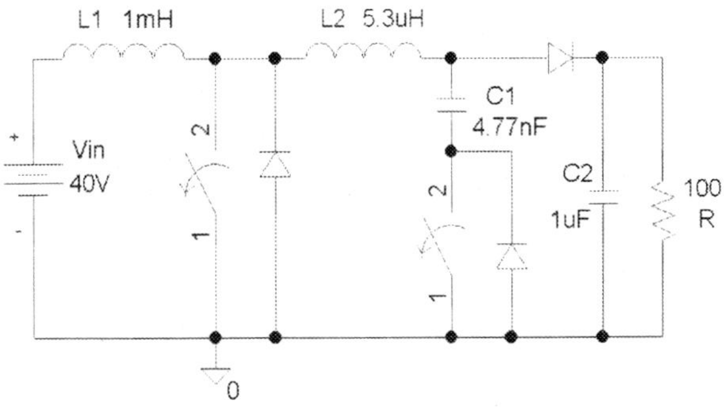

Fig. 1 A ZCS converter

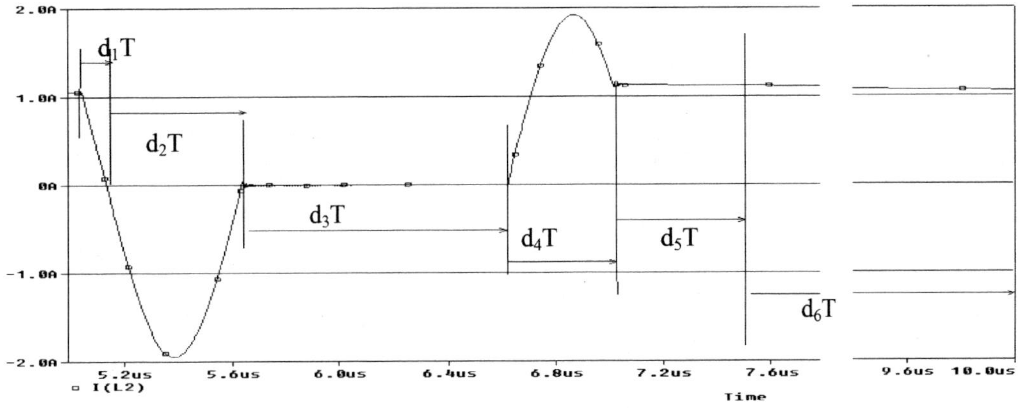

Fig. 2 The timing of the ZCS converter in Fig. 1

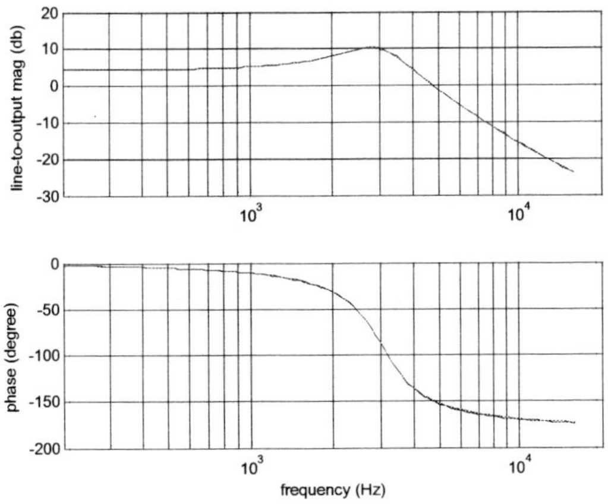

Fig. 3 The small signal line-to-output response of the ZCS Converter in Fig.1

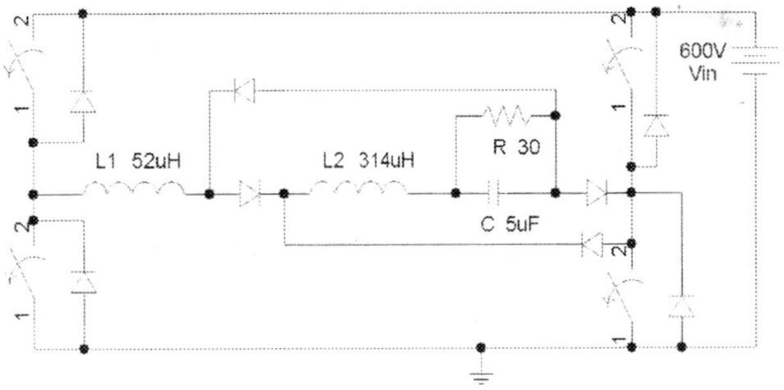

Fig.4 A FB-ZVS converter

CHARACTERIZING NONLINEAR LOAD HARMONICS USING FRACTAL ANALYSIS

K C Umeh, A Mohamed, R Mohamed, A Hussain
Department of Electrical, Electronics and System Engineering
Faculty of Engineering, Universiti Kebangsaan Malaysia
43600 Bangi, Selangor, Malaysia
e-mail: umehkingsley@ieee.org

Abstract
Power quality has become a major concern to both utilities and their consumers because equipments currently are more sensitive to power quality related disturbances. One of the power quality concerns that have received most attention is the problem of harmonics which are generated by widely dispersed nonlinear loads. In order to fully understand the problem of harmonic distortion, an effective means of identifying the harmonic patterns generated by different types of nonlinear loads is considered. This paper presents the application of fractal analysis for analyzing the various harmonic current waveforms generated by typical nonlinear loads such as personal computer, fluorescent lights, UPS, oscilloscope, monitor and laser jet printer. The fractal technique provides both time and spectral information of the nonlinear load harmonic patterns. The analysis results shows that the various harmonic current waveforms can be easily identified from the characteristics of the fractal features. This investigation proves that the fractal technique is a useful tool for identifying harmonic current waveforms and forms a basis towards the development of the harmonic load recognition system.

1. INTODUCTION

Power quality problems are increasing with the proliferation of non-linear devices, which draw non-sinusoidal current waveforms when supplied by a sinusoidal voltage source. When these devices are present in an electric power system, they cause harmonic distortion of voltages and currents. Individually, single phase non-linear load may not pose many serious harmonic problem but large concentrations of these loads have the potential to raise harmonic voltages and currents to unacceptable high levels which results in increased neutral currents in four wire system, over heating of distribution system components and mechanical oscillations in generators and motors. Other undesirable effects are capacitor and insulation failure due to harmonic resonance, unpredictable behavior of installed protection systems, rapid voltage fluctuations, over heating of transformer and cables, malfunction of power electronic equipments and telephone interference [1]. To solve the harmonic problems, one of the early task to be performed is to identify the various captured harmonic waveform in accordance to nonlinear load types. In an attempt to study the various waveforms generated by some of the typical nonlinear loads, laboratory test and measurements were performed on some of the loads.Among the nonlinear loads for which the waveforms were obtained are the personal computers, fluorescent lights of the electronic ballast and magnetic ballast types and uninterruptible power supply, oscilloscope and laser jet printer.
Nonlinear load current waveforms are non-stationary; any abrupt change is likely to spread over the entire frequency axis. In addition, majority of the modern electronic devices are switched mode powered causing them to exhibit similar distorted current waveforms.

The fast Fourier transform (FFT) method has been widely applied to extract the harmonic contents of voltage and current waveforms. The FFT method is effective in computing the harmonic distortion components of the current waveform but it is not very suitable for identifying the load harmonic patterns due to the presence of noise in terms of spike in real current waveforms and also its inefficiency in tracking the signal dynamics [2]. The wavelet transform emerged to overcome the problems of Fourier analysis method [3]. However, the efficiency and accuracy of the wavelet transform depends on the choice of the mother wavelet. The Daubechies'db4 wavelet is the most suitable Daubechies' wavelet for accurately detecting fast transient and short time information signals. However, like others in the Daubechies' wavelet family, its spectral information is not well localized in frequency rather it is provided in terms of frequency bands. The effectiveness of the wavelet transform is therefore highly dependent on factors such as the number of data points and sampling frequency. A poor selection of any of these may lead to inaccuracy in measurements.

In recent years, fractal techniques have attracted increased attention as a tool for image and signal processing. The method has been suggested in many applications as an alternative for analyzing time-varying signals where other techniques have not achieved the desired results [4]. Fractal theory is an extension of the classical geometry that can be used to make precise models of physical structures [5]. The applications of fractal geometry extend to biological modeling, geography, coastlines, computer graphics, images and so on. Different concept of fractal geometry has been applied for quantization of nonlinear systems in recent years [6]. Fractal techniques have been successfully applied in the analysis of chaotic systems [7], in biomedical signal analysis [8] in radar weak target detection [9] and image processing [10] In the field of power engineering, fractal based techniques has been used for the classification of impulse faults [11], analysis of chaotic properties of high impedance faults [12], pattern recognition of partial discharge [13] and power quality disturbances classification [14]. In this paper, an attempt has been made using the fractal technique in identifying and representing the nonlinear load harmonic patterns in terms of fractal features. The technique provides both time and spectral information which is useful in solving the frequency localization problem encountered using wavelet transform. Apart from increased speed, accuracy or efficiency, the features of the new method as compared to the other existing techniques are:

i. The approach is able to present both fractals and time information simultaneously.
ii. It is very effective in monitoring time-varying signal dynamics making it easier to locate the area of interest for proper investigation.
iii. It can be widely applied to localize various disturbance waveforms for better visualization of the signal characteristics

iv. It is a suitable candidate for pattern recognition of the various load harmonic waveforms.

2. HARMONIC MEASUREMENTS OF NONLINEAR LOADS

This section presents the experimental setup involved in conducting the harmonic measurement for the nonlinear loads. The purpose of the measurement is to characterize the behavior of harmonic generating loads and also to provide preliminary data for real-time harmonic signature recognition. The harmonic measurements being conducted is shown in Fig, 1 in terms of a schematic block diagram. A power recorder (RPM) compatible with a personal computer is placed between the single-phase AC mains and non-linear load to be measured. The readings on the recorder are stored with the load energized. These readings are then accessed and down loaded to the computer for report making, printouts and analysis of readings. A windows-compatible software package available with the power recorder enables the data to be studied and analyzed. The software program reads the recorded data files generated by the meter and performs the analysis and summary of the information stored. The harmonic information in terms of individual harmonic (up to sixty-third including DC and fundamental), total harmonic distortion, frequency, RMS amplitude, current crest factor and phase angle are recorded. Other power quality information available are VA, VARS., watts, power factor, voltage and current imbalance.

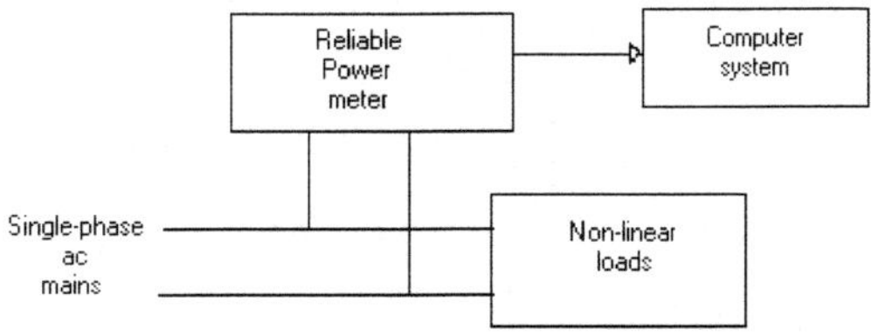

Fig .1 Experimental set up for nonlinear load harmonic measurement

Measurements were carried out by means of the RPM, which gave the spectrum and magnitude of the harmonics for each load. The single phase nonlinear loads that have been considered in the study are listed as follows:

- Personal Computer: AC 100-230V, 3.0/1.5A, 50/60 Hz
- Fluorescent Lamp (Electronic Ballast): AC 220/ 240, 0.1A, 50/60Hz
- Fluorescent Lamp (Magnetic Ballast): AC 220/240, 0.42A, 50Hz
- Uninterruptible Power Supply: AC 230V, 2.6A, 600VA, 50Hz
- Oscilloscope: AC 120-240V, 3.6A, 50/60Hz
- Printer: AC 220/240, 3.5A, 50/60Hz

The current absorption for the various loads was continuously monitored during typical operation phases.

3. FRACTAL DIMENSION

3.1 Theoretical Background

Mendelbrots observation on the existence of geometric nature led to the concept of fractal dimension [15]. Since then, it has evolved from mere intuitive ideas to a more elaborate algorithm in recent years. According to Barnsley, the fractal dimension of a set is that number which quantifies how densely the set occupies the metric space in which it lies [5]. Fractal dimensions provide an empirical means for comparing fractals between closely related objects through numbers associated with them. A fractal space is a complete metric space in which all functions defined in such space can be easily measured by fractal dimensions.

Consider a complete metric space (*X*, *d*). For each $\varepsilon > 0$, let $N(A, \varepsilon)$ denotes the smallest number of closed balls of radius ε needed to cover the set of A. If this assumption holds, then there exists the fractal dimension *D* given by:

$$D_B = \lim_{\varepsilon \to 0} \left\{ \frac{\log N(\varepsilon)}{\log(1/\varepsilon)} \right\} \tag{1}$$

The capacity fractal dimension, D_B provides a quantitative approach of describing a set. Several other definitions of fractal dimension as measures of chaotic system behavior exists. Some of the commonly used ones are the information and correlation fractal dimensions. The information dimension is a generalization of the capacity function having N_B replaced with the average information function $I(\varepsilon)$:

$$I(\varepsilon) = \sum_{i=1}^{N(\varepsilon)} - P(\varepsilon, i) \log(P(\varepsilon, i)) \tag{2}$$

where $P(\varepsilon, i)$ is the probability that a point of set S is in the ith box of size ε. Thus, the information dimension can be defined as:

$$D_I = \lim_{\varepsilon \to 0} \left\{ \frac{\log I(\varepsilon)}{\log(1/\varepsilon)} \right\} \tag{3}$$

The correlation dimension is given as:

$$D_C = \lim_{r \to 0} \left\{ \frac{\log C(r)}{\log r} \right\} \tag{4}$$

where $C(r)$ is the correlation integral defined as:

$$C(r) = \lim_{N \to \infty} \frac{1}{N^2} \sum_{i=1}^{N} \sum_{j=i+1}^{N} I\left(\left\| x_j - x_i \right\| \le r\right) \tag{5}$$

and *I* is an indicator function.

In general these three dimensions are related by the inequality [10].

$$D_c \le D_I \le D_B \tag{6}$$

3.2 Computation of Fractal Number

Different concept of fractal geometry has been employed in recent years for quantization of chaotic behavior of nonlinear systems. In recent years, the estimation of fractal dimensions has been widely applied in the studies involving time varying signals. The algorithm has been well tested on data sets ranging from a few hundreds to tens of thousands of data points [12]. In terms of computability, the correlation dimension is much easier to compute when compared to the capacity dimension. For a collection of *N* points in a trajectory either through simulations or measurements, the task of finding the correlation dimension simply reduces to finding the number, N_p of pairs of data points. Considing a given S set of N_p pairs of data points $\{x_J, x_j\}$ in a metric space, the correlation dimension can be calculated using the distances between each pair of the points in the set.

$$\left| x_i - x_j \right| = \left(\sum_{j=1}^{N} \left| x_i - x_j \right|^2 \right)^{\frac{1}{2}} \tag{7}$$

such that

$$\left\| x_i - x_j \right\| < \varepsilon \tag{8}$$

The equation for correlation dimension is reduced to:

$$D_C = \lim_{N \to \infty} \frac{1}{N^2} \left[N_p \right] \tag{9}$$

For our studies we have normalized and partitioned the current values into subsets with each data sets confined to a strict range [0, 1] having the Euclidian metric scaled down by the factor of 100. The process is quantified in terms of fractal numbers which is derived using:

$$F = \left[\frac{\sqrt{\sum_{j=1}^{N-l} \left\| x_{j+k} - x_j \right\|^2}}{\sqrt{\sum_{j=1}^{N-l} \left\| x_{j+l} - x_j \right\|^2}} \right] \times 100 \qquad (10)$$

where, F is the fractal number of m-th data set S_m containing N data points, k and l are small integer sampling steps size which specifies the time interval between close data points, such that l is greater than k. For the effectiveness of the equation, different norms can be selected, however for this particular work, the Euclidean norm is used due to its simplicity [14]. The fractal number computation turned out to be more efficient in this case if the ratio l to k is not an integer. The data points have to be properly classified such that data points belonging to k-step sampled subsets do not fall into the l-step subsets.

4. RESULTS AND DISCUSSION

Results of the Daubechies'db4 wavelet transform performed on the waveforms obtained from the single phase nonlinear loads are first presented followed by the fractal analysis.

4.1 Wavelet Analysis

Wavelet analysis on the harmonic current waveform is performed to derive a time-frequency description for the waveforms. Three level decompositions of the db4 wavelets have been applied to the captured harmonic current waveforms. Fig.1 shows the wavelet decomposition of the personal computer current waveforms. Subplot 1 shows the current waveforms; Subplot 2 shows a 1-level coefficient; Subplot 3 and 4 shows the 2-level and 3-level wavelet coefficient respectively. Fig.2 depicts the corresponding waveforms and wavelet coefficients of the florescent lamp with electronic ballast. Fig.3, Fig.4, Fig.5 and Fig.6 show the wavelet decompositions of the current waveforms belonging to the florescent lamp with magnetic ballast, UPS, oscilloscope and printer respectively.

4.2 Fractal Features of Nonlinear Loads Currents

A sampling rate of 6.4 *KHz* has been used in recording the current waveforms of the monitored nonlinear loads. A time interval step of 80*msec* has been used for the recordings. From the recorded waveform, the data are sampled such that the data sets consist of 512 data points, which are partitioned into 31 subsets containing 16 data points each. A fractal number computation is then performed for a set of collected data. The fractal analysis has been tested on harmonic current waveforms of various single phase nonlinear loads. From the analysis, the fractal features of the nonlinear loads in terms of fractal numbers can be obtained. The result shows a clearly unique characteristic fractal feature for each of the load

Fig .7 shows the current waveform and fractals of the PC load where for every half cycle, a repetitive fractal patterns with sharp peaks was observed. By applying the fractal analysis to waveforms obtained from other PC loads the method yielded patterns which are identical. The RMS value of the input AC current total harmonic distortion (THD) for the PC is 98.4%.

The triangular current waveform generated by a single fluorescent lamp with electronic ballast together with its fractal pattern is sshown in Fig 8. The current THD for this current is high, that is 102.8%.

Fig.9 shows the current waveform and fractals of the fluorescent lamp with magnetic ballast. The current yielded a repetitive

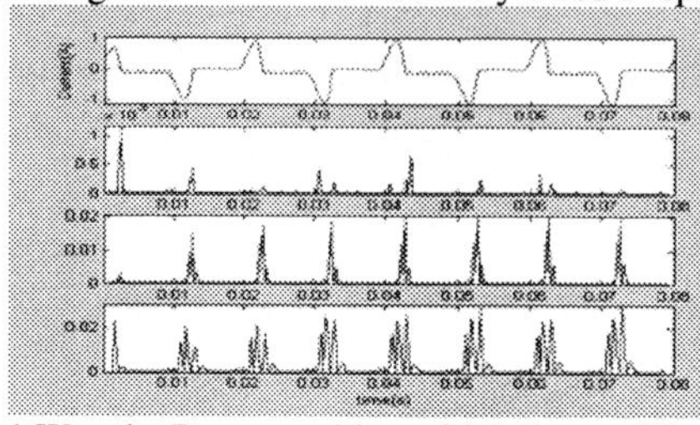

Fig.1 Wavelet Decomposition of PC Current Waveform

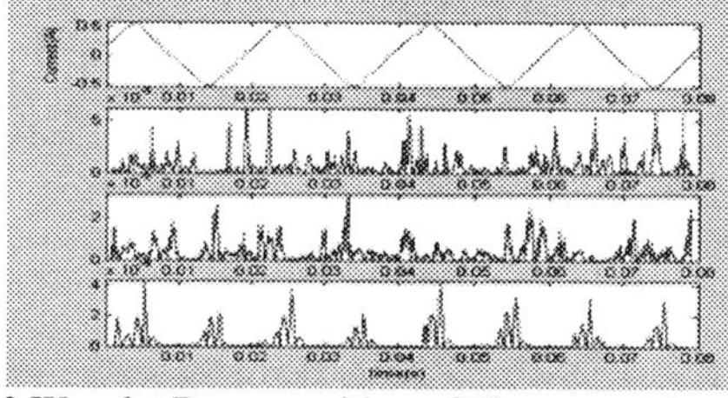

Fig.2 Wavelet Decomposition of Fluorescent Lamp with Electronics Ballast Current Waveform

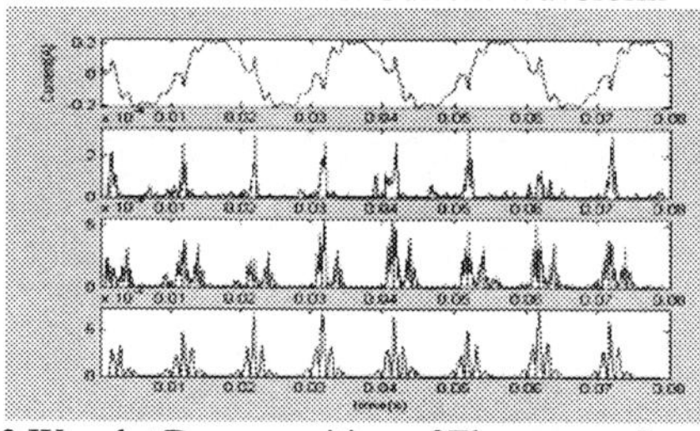

Fig.3 Wavelet Decomposition of Fluorescent Lamp with Magnetic Ballast Current Waveform

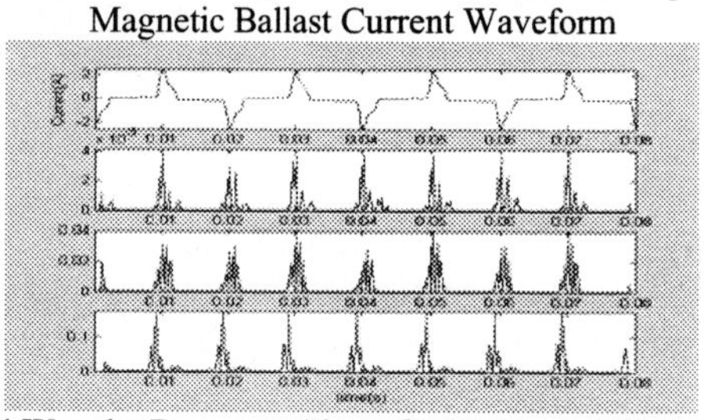

Fig.4 Wavelet Decomposition of UPS Current Waveform

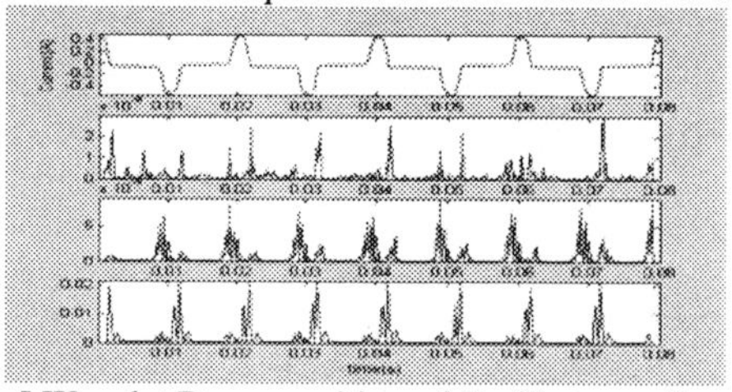

Fig.5 Wavelet Decomposition of Oscilloscope Current Waveform

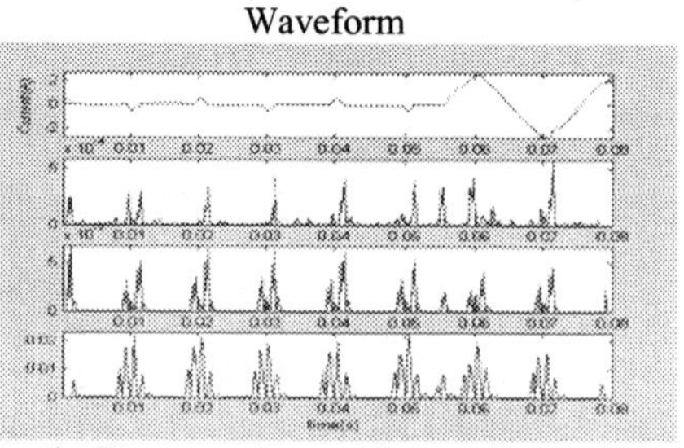

Fig.6 Wavelet Decomposition of Printer Current Waveform

fractal patterns with sharp peaks as shown in Subplot 2.The nonlinear nature of this current is as a result of the nonlinearity between the ballast and the lamp arc itself. The current THD value for this load is 22.26%.

The highly harmonically distorted current waveform of the UPS yielded a current THD of 101.6%. Due to the high distortion of the current waveform a chaotic but repetitive double–peaked fractal pattern is observed as shown in Fig.10

Fig.11 shows the current waveform and fractals of a single oscilloscope. The current waveform of the oscilloscope produced a repetitive fractal pattern having sharp peaks with slant base. The current THD value of 94.8% was recorded for the oscilloscope.

Fig.12 depicts the current waveform and fractals of a laser jet printer. The current drawn by the printer is high. The irregularities observed in both the current waveforms and fractal pattern are due to the reiterated movements of the printing head along with the paper sheet during the printing operation stages.

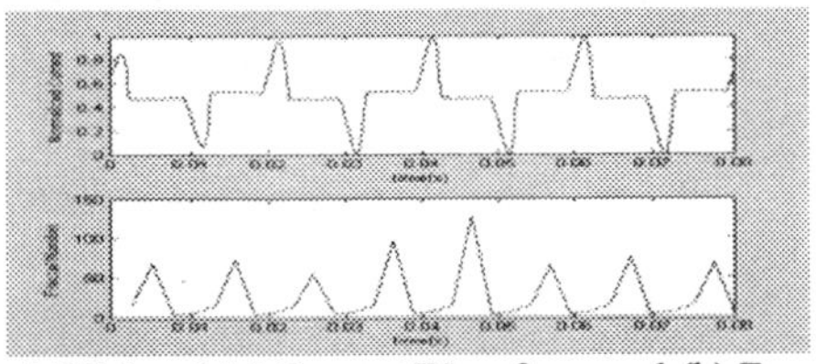

Fig. 7 PC Load (a) Current Waveform and (b) Fractals

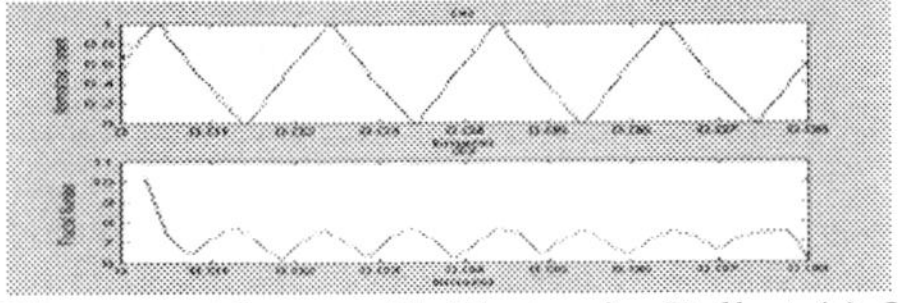

Fig. 8 Fluorescent Lamp with Electronics Ballast (a) Current Waveform (b) Fractals.

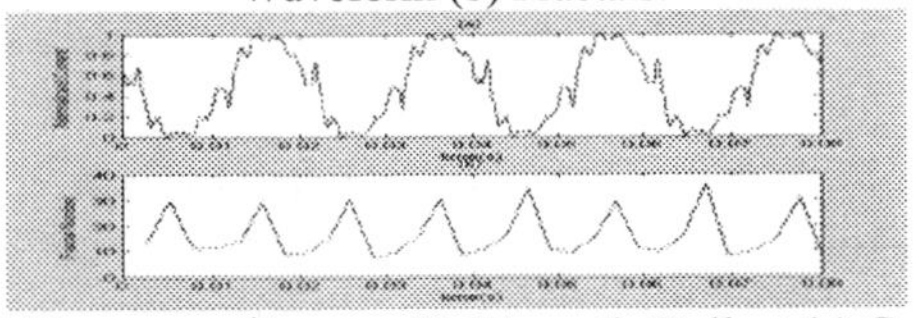

Fig. 9 Fluorescent Lamp with Magnetic Ballast (a) Current Waveform (b) Fractals

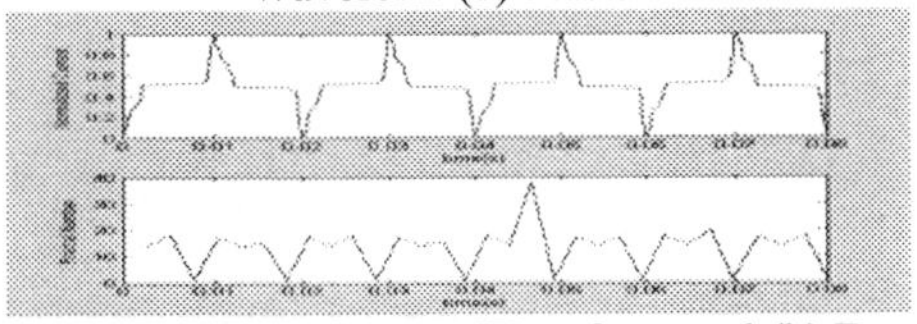

Fig.10 UPS load (a) Current Waveform and (b) Fractals

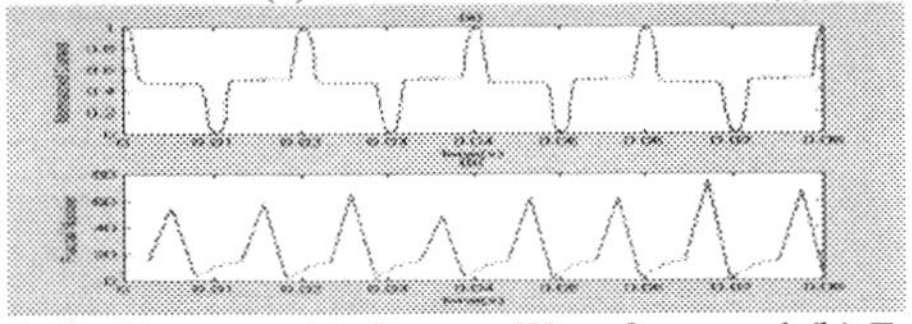

Fig. 11 Oscilloscope (a) Current Waveform and (b) Fractals

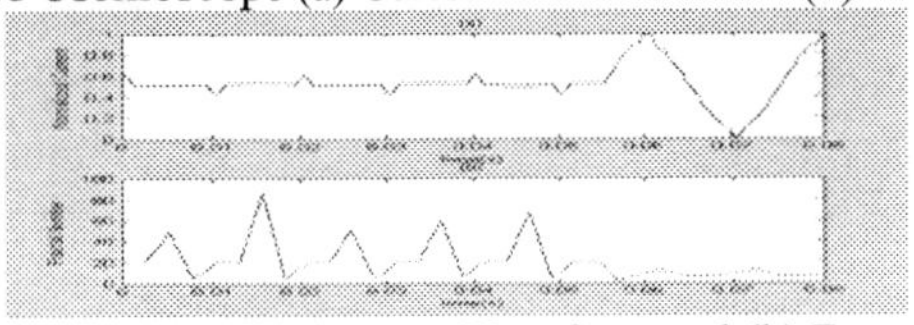

Fig. 12 Printer (a) Current Waveform and (b) Fractals

5. CONCLUSION

A fractal–based method for characterizing current waveforms from nonlinear harmonic loads has been proposed. From the experimental results, the harmonic current waveforms obtained from the various nonlinear loads are processed and represented by fractal numbers through fractal analysis. The result shows a clearly unique characteristic feature for each of the load. The application will be incorporated to an expert system to be developed for the automatic recognition of nonlinear harmonic load signature. This reveals the usefulness of fractal geometry in the area power quality diagnoses

6. ACKNOWLEDGEMENT

The authors would like to show appreciation to the Malaysian Ministry of Science, Technology and Environment for funding this work through research grant IRPA: 02-02-02-0014 EA 188.

7. REFERENCE

[1] IEEE Std. 519-1992, "IEEE Recommended Practices and Requirements for Harmonic Control in Electrical Power Systems,"1993.

[2] A.A Girgis, W. B Chang and E. B Makram, "Digital Recursive Measurement Scheme for On-line Tracking of Power System Harmonics. IEEE Transaction on Power Delivery, vol.6 NO.3, pp.1153-1160, 1991.

[3] W.L. Chan, A.T.P. So and L.L.Lai, "Harmonics Load Signature Recognition by Wavelets Transforms", Proceedings of the IEEE International Conference on Electric Utility Deregulation and Restructuring and Power Technologies, pp. 666 – 671, 2000.

[4] R.M Crownover, "Introduction to Fractals and Chaos", Boston, USA: Jones and Barlett Publishers, 1995.

[5] M.F Barnsley, "Fractals Everywhere", Second Edition, Academic Press, San Diego, 1993.

[6] M.J Feigenbaum, "The universal metric properties of nonlinear transformations", Journal of Statistical Physics, vol. 21, No.2, pp.669 – 706, 1979.

[7] E Ott, "Chaos in dynamical Systems", Cambridge, UK: Cambridge University Press, 1993.

[8] W.S. Tirsch, M.Keidel , S.Perz, H.Scherb, and G. Sommer, "Inverse Covariation of Spectral Density and Correlation Dimension in Cyclic EEG Dimension of Human Brain," Biological Cybernetics, vol. 82, pp. 1-14, 2000.

[9] Hu Weidong, Yu Wenxian and Luo Aimin, "Fractal Detection of Rader Weak Targets, "Proceedings of the IEEE 1995 National Aerospace and Electronics Conference, NECON 1995, vol.1, pp. 140-144, 1995.

[10] Q. Huang, J.R Lorch, and R.C Dubes, "Can the Fractal Dimension of Images Be Measured?" Pattern Recognition, vol.27, NO.3, pp. 339-349, 1994.

[11] P. Purkait, and S. Chakravorti, "Impulse Fault Classification in Transformers by Fractal Analysis," IEEE Transactions on Dielectrics and Electrical Insulation, vol.10 NO.1, pp. 109-116, 2003.

[12] A.V Mamishev, B.D Russell, C.L Benner, "Analysis of High Impedance Faults Using Fractal Techniques," IEEE Transactions on Power Systems, vol.11 NO.1, pp. 435-440, 1996.

[13] C. Sevcik, "A Procedure to Estimate the Fractal Dimension of Waveforms," Complexity International, J Vol.5, 1998.

[14] S.Huang, and C.Hsieh, "Feasibility of Fractal-based Methods for Visualization of Power System Disturbances," Electrical Power and Energy Systems vol.23, pp.31-36, 2001.

ANALYSIS OF NONLINEAR DYNAMICS IN DELTA MODULATORS FOR PWM CONTROL

Hiroshi Shimazu and Toshimichi Saito

EECE Dept., HOSEI University, Tokyo, 184-8584 Japan, *tsaito@k.hosei.ac.jp*

ABSTRACT

This paper studies dynamics of the delta modulators for PWM control of inverters. First, we introduce normal form circuit equation and 1-D return map. We then clarify that the system exhibits nonperiodic behavior in wide parameter region. Second, we apply compulsory switching to the circuit. This switching can stabilize the operation in some wide parameter region for controlling inverters and can cause chaotic phenomena in the other parameter region.

1. INTRODUCTION

The delta modulators (DMs [1]-[4]) are interesting systems from both practical and fundamental viewpoints. On one hand the DM can be used as the independent voltage-mode pulsewidth modulation (PWM) circuit for controlling inverters [5]. As compared with sine-triangle and space-vector modulators, the DM has several advantages and disadvantages: it is relatively easy to implement and provides low harmonics at the inverter output, however, it has asynchronous and unstable operations [2] [6] [7]. Although steady state of the DMs has been analyzed for improving the disadvantages [2]- [4], the nonlinear dynamics has not been analyzed sufficiently so far. On the other hand the DMs can be regarded as switched dynamical systems (SDSs) can provide rich nonlinear phenomena by interaction between the continuous and discrete states [8]-[11].

This paper studies the nonlinear dynamics of the DMs. First, we derive the normal form equation so as to extract essential parameters and construct 1-D return map [8] [12] from the normal form. We then clarify parameters regions for the following: 1) the return map is monotone and exhibits either periodic or quasiperiodic behavior, 2) the return map has extrema and exhibits either periodic or chaotic behavior. The nonperiodic behavior corresponds to the unstable operation of the DMs. Second, we apply a compulsory switching to the DM in order to suppress the unstable operation. In this case, we clarify that the system can exhibit periodic attractor in wide parameter region corresponding to practical PWM control: the compulsory switching can stabilize the PWM operation. We also clarify that the system can exhibit chaos and bifurcation phenomena in the other parameter region. Using a simple test circuit, typical phenomena can be verified in the laboratory. These results provide useful information for understanding dynamics of the DMs and for improving disadvantages of the DM. It should be noted that this paper is a developed version of [11].

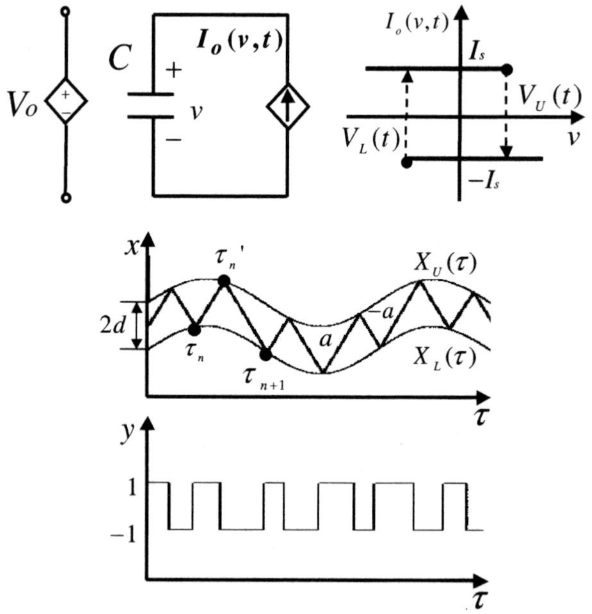

Fig. 1. An equivalent circuit of the delta modulator. v, I, V_U and V_L are proportional to x, h, X_U and X_L, respectively.

2. THE DELTA MODULATION CIRCUIT

First, we introduce an equivalent circuit of the delta modulator (ab. DMC). It is used as the independent modulator for voltage mode PWM control [1]-[3]. In Fig. 1 $I_o(v,t)$ is a voltage-controlled current source (VCCS) with time-variant hysteresis characteristics: the output current is switched from I_s to $-I_s$ (respectively, $-I_s$ to I_s) if the capacitor voltage v hits the upper periodic threshold $V_U(t)$ (respectively,lower threshold $V_L(t)$) holding the continuity property of v. V_o is a CCVS with characteristic resistance R_o that is used to control switches in the inverters [5]. An implementation example can be found in [11]. The system dynamics is de-

0-7803-8251-X/04/$17.00 ©2004 IEEE

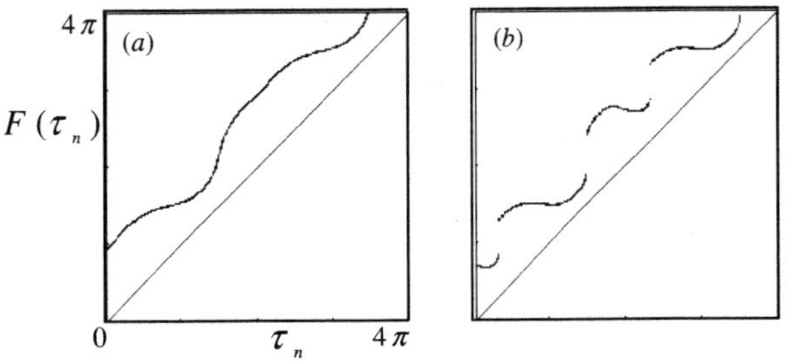

Fig. 2. Switching time maps. (a) Continuous map for (a, d)=(1.3, 0.5). (b) Discontinuous map for (a, d)=(0.9, 0.2).

scribed by Equation (1).

$$C\frac{d}{dt}v = I_o(v,t),\ \ V_o = R_o I_o(v,t), \tag{1}$$

$$I_o(v,t) = \begin{cases} I_s, & \text{for } v \leq V_U(t) \\ -I_s, & \text{for } v \geq V_L(t), \end{cases}$$

$$V_U(t) = V_m \sin \omega t + V_b,\ V_L(t) = V_m \sin \omega t - V_b,$$

Using dimensionless variables and parameters:

$$\tau = \omega t,\ x = \frac{v}{V_m},\ y = \frac{V_o}{R_o I_s},\ a = \frac{I_s}{\omega C V_m},\ d = \frac{V_b}{V_m},$$

Equation (1) is transformed into Equation (2).

$$\frac{d}{d\tau}x = ah(x,\tau) \tag{2}$$

$$y = h(x,\tau) = \begin{cases} 1, & \text{for } x \leq X_U(\tau), \\ -1, & \text{for } x \geq X_L(\tau), \end{cases}$$

$$X_U(\tau) = \sin\tau + d,\ X_L(\tau) = \sin\tau - d.$$

where $h(x,\tau) \equiv \frac{1}{I_s} I_o(V_m x, \frac{\tau}{\omega})$. Equation (2) has two parameters a and d which control slope of orbits and the hysteresis width, respectively. As shown in Fig. 1, x vibrate between upper threshold X_u and lower threshold X_L. x and $\sin\tau$ correspond to carrier and modulating signals, respectively. Let τ_n be the n-th switching time at X_L and let τ_n' be the successive switching time at X_U. An orbit started from $(x,\tau) = (X_L(\tau_n), \tau_n)$ increases with slope a and hits X_U at time τ_n'. Then the orbit decreases with slope $-a$, hits the lower threshold X_L at time τ_n, and the system repeats this manner. The switching time τ_n is governed by determined the 1-D map F from positive reals to itself:

$$\tau_{n+1} = F(\tau_n),\quad F(\tau_n + 2\pi) = F(\tau_n) + 2\pi. \tag{3}$$

Since the system is periodic with period 2π, we introduce an angular variable $\theta_n = \tau_n \bmod 2\pi$. Using this, we can define the return map f from $I_D \equiv [0, 2\pi)$ to itself:

$$\theta_{n+1} = f(\theta_n) \equiv F(\theta_n) \bmod 2\pi \tag{4}$$

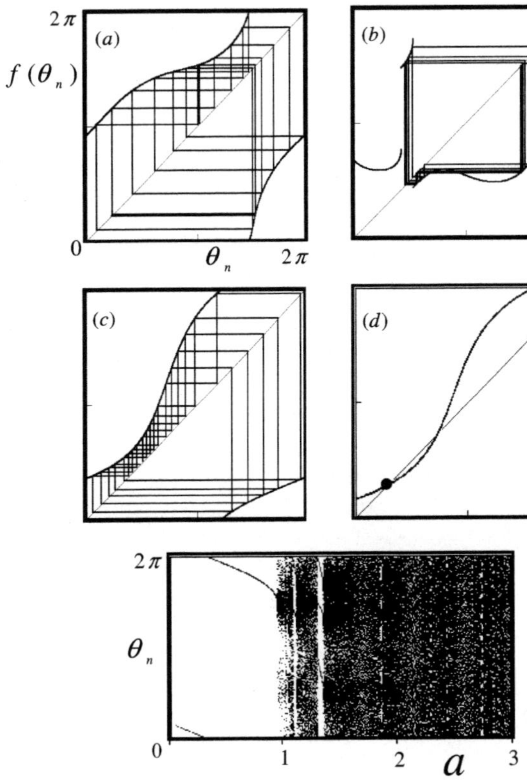

Fig. 3. Return maps and diagram of attractors. (a) Quasi-periodic attractor for (a, d)=(1.3, 0.5), $\lambda \doteq 0.0$. (b) Chaotic attractor for (a, d)=(0.75, 7.0), $\lambda \doteq 0.166$. (c) Quasi-periodic attractor for (a, d)=(3.4, 7.0), $\lambda \doteq 0.0$. (d) Periodic attractor for (a, d)=(4.0, 7.0), $\lambda \doteq -0.7$.(e) Bifurcation diagram for $d = 0.5$.

For the return map we define the Lyapunov exponent.

$$\lambda = \lim_{N\to\infty} \frac{1}{N} \sum_{n=0}^{\infty} |Df(\theta_n)| \tag{5}$$

This exponent can be calculated using exact piecewise solutions. Fig. 2 shows typical switching time maps. Fig. 3 shows typical return maps with Lyapunov exponents and diagram of attractor. Noting $|\frac{d}{d\tau}\sin\tau| \leq 1$, we can say that the switching time map F is monotone increasing for $1 < a$. In this case, the return map is equivalent to the circle map and exhibits either periodic attractor [1] characterized by $\lambda < 0$ or quasiperiosic attractor characterized by $\lambda = 0$. For $0 < a < 1$, F can have the discontinuity and the extremum points. In this case, the return map exhibits either periodic attractor or chaotic attractor characterized by $\lambda > 0$. The quasi-periodic and chaotic attractors correspond to unstable operation of the DMC. Fig. $3(e)$ shows one-parameter

[1] it corresponds to a stable periodic point. The periodic point is a point $\theta_* \in I_D$ such that $F^p(\theta_*) = \theta_*$ and $F^l(\theta_*) \neq \theta_*$ for $0 < l < p$, where F^p is the p-fold composition of F.

diagram of attractors: it suggests that the DMC operates unstably in wide parameter range.

3. SYSTEM WITH COMPULSORY SWITCHING

In order to suppress unstable operation, we apply compulsory switching to the DMC as illustrated in Fig. 4. In this case, the dynamics is described by adding the following condition to Equation (2).

The output of the hysteresis $h(x, \tau)$ is switched compulsory to -1 at period end $\tau = 2n\pi$ if $h(x, \tau) = 1$ and $X_L < x$, where n is a positive integer.

The switching time map F is given by implicit Equation (6). Using this F, Equation (4) gives the return map.

$$\tau_{n+1} = F(\tau_n)$$

$$X_L(\tau_{n+1}) - X_L(\tau_n) = a(4n\pi - \tau_{n+1} - \tau_n) \quad \text{for } \tau_c \le \tau_n \le 2n\pi$$

$$\begin{cases} X_U(\tau_n') - X_L(\tau_n) = a(\tau_n' - \tau_n) \\ X_U(\tau_n') - X_L(\tau_{n+1}) = a(\tau_{n+1} - \tau_n') \end{cases} \quad \text{otherwise} \tag{6}$$

where τ_c is the minimum time such that the orbit started from $(\tau_c, X_L(\tau_c))$, $\tau_c < 2n\pi$, is switched compulsory at time $2n\pi$ before hitting X_U (see Fig. 4). As the compulsory switching is applied, return maps in Fig. 3 (a) to (d) and the diagram in Fig. 3 (e) are changed into (a) to (e) in Fig. 5, respectively. Fig. 6 shows parameter regions of positive, zero and negative Lyapunov exponents for DMC with/without the compulsory switching. In the figures, we can see that the quasi-periodic behavior in Fig. 3 (a) and (e) is changed into periodic behavior in Fig. 5 (a) and (e): it corresponding to stabilization of the operation of DMC. Such stabilization is confirmed in the parameter region for

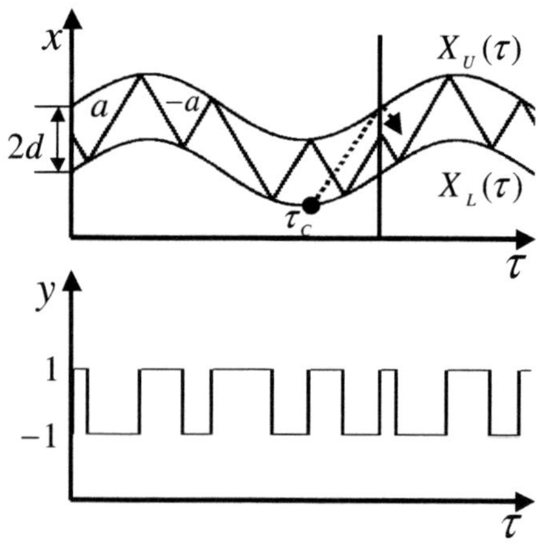

Fig. 4. Compulsory switching

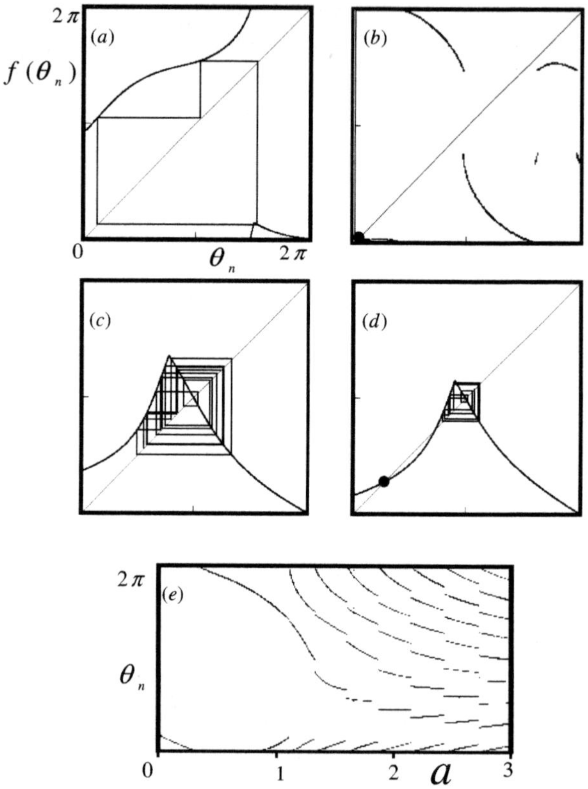

Fig. 5. Typical return maps and diagram of attractors. (a) Periodic attractor for (a, d)=(1.3, 0.5), $\lambda \doteq -0.472$. (b) Periodic attractor for (a, d)=(0.75, 7.0), $\lambda \doteq -1.94$. (c) Chaos attractor for (a, d)=(3.4, 7.0), $\lambda \doteq 0.48$. (d) Co-existence of periodic attractor and chaos attractor for (a, d)=(4.0, 7.0), $\lambda \doteq -0.7$ (periodic orbit), $\lambda \doteq 0.654$ (chaotic orbit). (e) Diagram of attractors for $d = 0.5$.

relatively small d which corresponds to practical voltage-mode control of the DMC. In Fig. 6 (a), as discussed in the preceding section, DMC exhibits either quasi periodic or periodic attractor for $a > 1$ and exhibits either periodic or chaotic attractor for $0 < a < 1$. Fig. 6 (a) and (b) show that the compulsory switching can cause the following four types of change of attractors at least:

(A) From quasi-periodic attractor to periodic attractor.
(B) From quasi-periodic attractor to chaotic attractor.
(C) From periodic attractor to chaotic attractor.
(D) From chaotic attractor to periodic attractor.

It should be noted that the compulsory switching can cause co-existence phenomena of chaotic and periodic attractors as shown in Fig. 5 (d), where the system exhibits either depending on the initial state. Such phenomena and related bifurcation are observed for relatively large d as shown in Fig. 6 (b).

Using a simple OTA-based test circuit in [11] we have confirmed typical phenomena in the laboratory. Fig. 7 shows

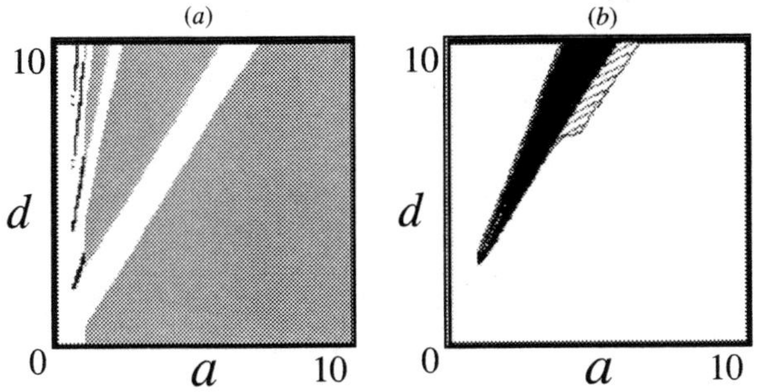

Fig. 6. Lyapunov exponent. The white, gray and black regions correspond to $\lambda < 0$, $\lambda = 0$ and $\lambda > 0$, respectively. In stripe region the map exhibits co-existence phenomena of chaotic and periodic attractors. (a) Without compulsory switching. (b) With compulsory switching.

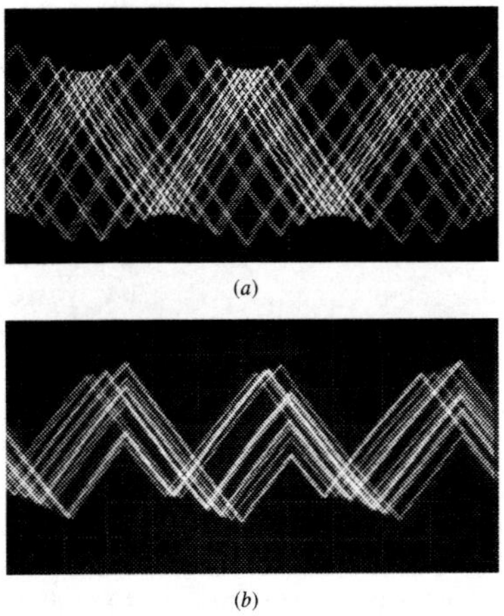

Fig. 7. Laboratory measurements. horizontal=t [1 ms/div.], vertical=v [2 V/div.], $R \doteq 60$ k Ω, $C \doteq 22$ nF, $V_m \doteq 0.4$ V, $RI_s \doteq 4.0$ V, $T \doteq 11$ ms, ($a \doteq 3.4$, $d \doteq 7.2$) (a) Quasiperiodic orbit without compulsory switching. (b) Chaotic orbit with compulsory switching.

some example of them: applying the compulsory switching, the quasi-periodic orbit (Fig. 7(a)) corresponding to Fig. 3(c) is changed into the chaotic orbit (Fig. 7(b)) corresponding to Fig. 5(c). Ref. [11] demonstrates laboratory data for changing from quasi-periodic orbit into the periodic orbit by the compulsory switching. It should be noted that Ref. [11] discusses neither Lyapunov exponent nor chaos and bifurcation phenomena.

4. CONCLUSIONS

We have studied basic dynamics of DMC with/without compulsory switching. Using return map and its Lyapunov exponent, the dynamics is analyzed. It is clarified that, depending on the parameters, the compulsory switching can suppress instability as practical voltage control and can cause interesting nonlinear phenomena as chaotic generator. Future problems include construction of practical circuits and detailed analysis of bifurcation phenomena.

5. REFERENCES

[1] P. D. Ziogas, Delta modulation technique in static PWM inverter, IEEE Trans. Ind. Applicat. pp. 199-204, 1981

[2] M. A. Rahman, J. E. Quaicoe and M. A. Choudhury, Performance analysis of delta modulated PWM inverter, IEEE Trans. Power Electron. pp. 227-233, 1987.

[3] R. Razali, V. Suddiah, M. A. Choudhury and R. Yusof, Performance analysis of online dual slope delta modulated PWM inverter, Proc. IEEE/ISCAS, pp. 845-848, 2002.

[4] K. M. Rahman, M. R. Khan and M. A. Choudhury, Implementation of programmed modulated carrier HCC based on analytical solution for uniform switching of voltage source inverters, IEEE Trans. Power Electron. 18, pp. 188-197, 2003.

[5] J. Vithayathil, Power electronics, principles and applications, McGraw-Hill, 1995

[6] F. Wang, Sine-triangle versus space-vector modulation for three-level PWM voltage-source inverters, IEEE Trans. Ind. Applicat., 38, pp. 199-204, 2002.

[7] S. K. Mondal, B. K. Bose, V. Oleschuk and J. O. P. Pinto, Space vector pulse width modulation of three-level inverter extending operation into overmodulation region, IEEE Trans. Power Electron., 18, pp. 604-611, 2003.

[8] T. Saito, H. Torikai and W. Schwarz, Switched dynamical systems with double periodic inputs: an analysis tool and its application to the buck-boost converter, IEEE Trans. Circuits Systs. I, 47, pp.1038-1046, 2000.

[9] C. K. Tse and M. di Bernardo, Complex behavior in switching power converters, Proc. IEEE, 90, pp. 768-781, 2002.

[10] H.H.C. Iu and B. Robert, Control of chaos in a PWM current-mode H-bridge inverter using time-delayed feedback, IEEE Trans. Circuits Systs. I, 50, pp. 1125-1129, 2003

[11] H. Shimazu, T. Saito & H. Torikai, Analysis of conventional/improved delta modulators as switched dynamical systems, Proc. IEEE/IECON, pp. 1814-1817, 2003

[12] A.Lasota and M.C.Mackey, Chaos, Fractals, and Noise - Second Edition, Springer-Verlag, 1994.

A CAD SIMULATOR BASED ON LOOP GAIN MEASUREMENT FOR SWITCHING CONVERTERS

Dongsheng Ma, Vincent H. S. Tam, Wing-Hung Ki** and Hylas Y. H. Lam***

Dept. of Electrical & Computer Engineering, Louisiana State University, Baton Rouge, LA 70803, USA
Artesyn Technologies Asia-Pacific Ltd., Hong Kong SAR*
Dept. of Electrical & Electronic Engineering, Hong Kong University of Science & Technology, Hong Kong SAR**
ma@ece.lsu.edu, vincent_tam@artesyn.com*, {eeki, hylas}@ee.ust.hk**

ABSTRACT

A CAD simulator for power converters on the MATLAB platform is presented. It employs an interpolation technique in determining switching instants in the time domain accurately. A small sinusoidal source is then injected to obtain the steady state time response and is followed by FFT to obtain the resultant amplitude and phase. By varying the injected frequency, loop gain frequency response is obtained. Issues on loop gain measurement are discussed. A cost-effective set-up is then presented. The methods of simulation and measurement can be applied in power converters of any topologies.

1. INTRODUCTION

A switch mode power converter, or switching converter for short, is a non-linear sampled-data system due to the presence of switches. Hence, analysis methods for linear systems are not applicable. With increasing popularity of switching converters due to their high efficiencies and flexible conversion manners, a unified methodology for analysis and design is in urgent need. Over the past thirty years, many researches have been conducted to tackle this problem. One appealing approach is to solve for the steady state solution of the converter, followed by perturbation analysis to obtain a linearized small signal model of the converter, from which the system loop gain can be determined for stability and transient response considerations [1] [2].

The common practice to verify that a model is valid is to perform measurement. If the measurement results cope well with analysis, the model is then concluded as valid. The problem is that every model is substantiated with measurement results, yet the models do not necessarily agree with each other. One problem lies in the lack of measurement details for fair comparison, another is due to variation in parameters, and parasitic elements that are unaccounted for. In loop gain analysis, every element assumes a constant value, e.g., ESR (equivalent series resistance) is 50mΩ, but their values are difficult to be measured accurately. In fact, they may vary according to operating condition and frequency. Despite the difficulty, design engineers of integrated switching converters still need the loop gain in advance for stability consideration. Hence, a good and efficient simulator is needed. Moreover, a good simulator may be used to evaluate the validity of a converter model, as all elements have ideal characteristics. For example, if the ESR is set at 50mΩ, then it stays constant throughout the simulation.

This research is in part supported by the Hong Kong Research Grant Council under grant CERG HKUST 6209/01E and the Innovative Technology Fund under grant ITS/033/02.

2. DEVELOPMENT OF THE SIMULATOR

2.1 Modeling of the Switching Converter

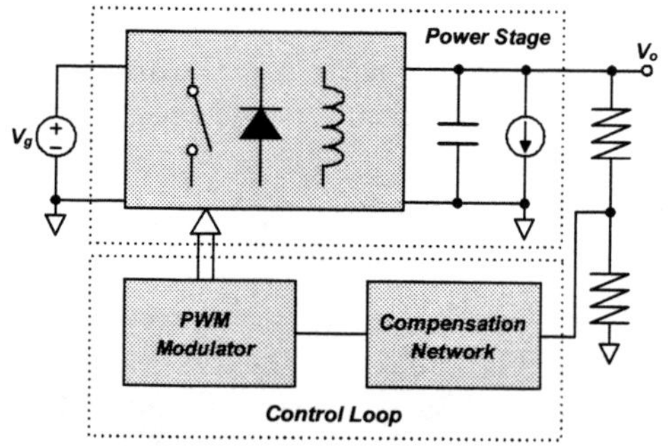

Fig. 1 Block diagram of a switching converter

Switching converters can be simulated using Hspice. However, if realistic models are used for transistors, the same problem of parameter variation arises as in measurement. For example, a transistor enters different modes of operation during both turn on and turn off, and the corresponding switch resistance cannot be a constant. If ideal elements are used, then the simulation may not converge. Hence, our simulator is developed on the MATLAB platform instead.

Fig. 1 shows a regulated switching converter that constitutes a loop composed of the power stage, the compensation network and the pulse width modulator. The power stage consists of inductor(s), capacitor(s) and switch(es). Due to the switching actions, it can be modeled by a set of piecewise linear differential equations for each state. Second order converters such as buck, boost or flyback converters have up to three states (two states for continuous conduction mode (CCM) operation, and three states for discontinuous conduction mode (DCM) operation). The state space equations for n=1, 2, 3 are

$$\dot{x} = A_n x + B_n u \tag{1}$$

$$y = C_n x + B_n u \tag{2}$$

where $x = [v_c\ i_l]^T$ is the state vector, with v_c being the voltage of the filtering capacitor and i_l the inductor current, and u and y are the input and output vector respectively. The solution to Eqn. (1) is

$$x(t+\Delta t) = A_n^{-1}Bu + e^{A_n \Delta t}(x(t) - A_n^{-1}Bu) \tag{3}$$

By approximating $e^{A_n \Delta t}$ as $I + A_n\Delta t$, the equation implemented by the simulator is

$$x(t+\Delta t) = x(t) + \dot{x}(t)\Delta t \tag{4}$$

The compensation network consists of linear elements such as op amp, resistors and capacitors in generating poles and zeros for loop compensation, which is also modeled as one set of linear differential equations for simulation. The pulse width modulator is to determine the switching instants (trip points) of the power stage. In the simulator, it determines the set of state space equations used at a particular time. Fig. 2 shows the state transition diagram of a 2nd order switching converter, where L_0~L_2 and T_0~T_2 represent leading- and trailing-edge modulation tripping conditions [3]. For example, for a voltage-programming converter with trailing-edge modulation, T_0 is the condition when the oscillator ramp hits the output of the error amplifier. The simulator follows the transitions in the state diagram strictly in simulating time domain dynamics.

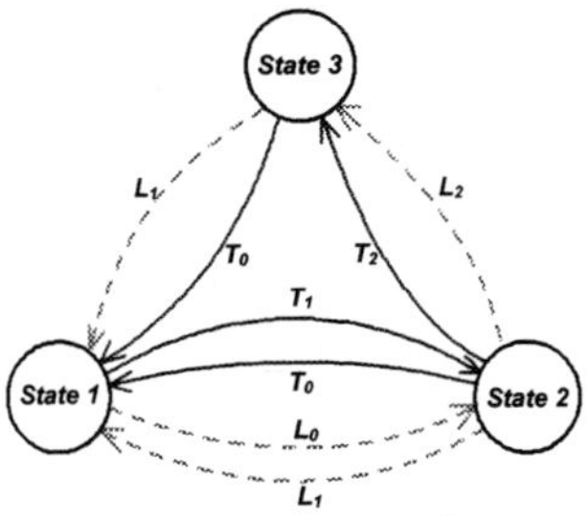

Fig. 2 State transition diagram of 2nd order converters

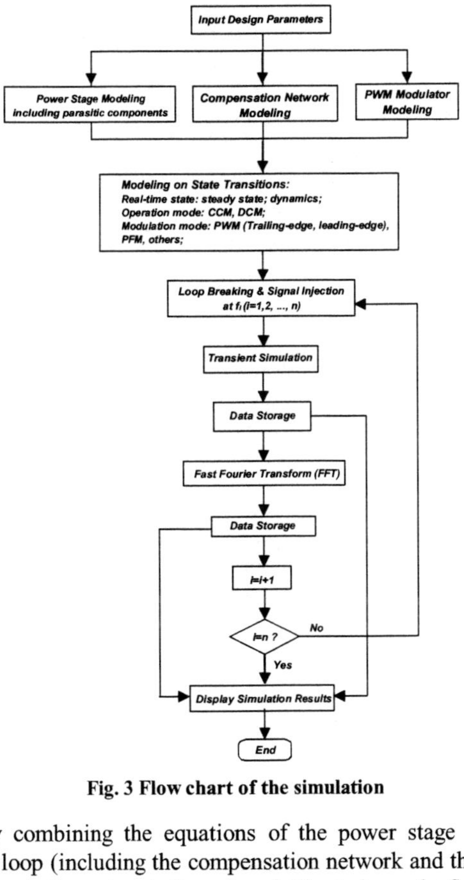

Fig. 3 Flow chart of the simulation

By combining the equations of the power stage and the control loop (including the compensation network and the PWM modulator), the system is then closed. Fig. 3 shows the flow chart of the simulation in finding the loop gain of the converter.

To find the loop gain of the converter, the loop is broken at a suitable location and a signal is injected, and the responses after (V_t) and before (V_r) the injection point are compared. For a linear circuit, the loop gain is given by $T(s) = -V_t(s)/V_r(s)$, but for a switching converter, the loop gain cannot be obtained directly in the frequency domain. Voltage injection is employed, and from the discussion in Sec. 3, it is required that $Z_r(s) >> Z_t(s)$ at the injection point. A suitable location is at the output of the error amplifier V_a that feeds into the PWM comparator. Here, V_a compares with a ramp to determine the trip point. An ideal signal $V_j = A_j \sin 2\pi f_j t$ is then injected across V_t and V_r. Time domain simulation is then performed until the converter has settled in the steady state. Fast Fourier Transform (FFT) is then performed on the settled time domain section of both $V_r(t)$ and $V_t(t)$. The frequency spectrum of $V_r(s)$ consists of the injected signal at f_j, the switching noise at the switching frequency f_s, mixed signals at $f_s - f_j$ and $f_s + f_j$, and high order harmonics [3]. The spectrum of $V_t(s)$ consists of the same components, but with different magnitudes. Yet, we only focus on the magnitude and phase at f_j to compute the loop gain at f_j. The simulation procedures will be repeated and we sweep f_j in the overall interested frequency range. The simulation algorithm mimics that of the loop gain measurement by a gain/phase analyzer. Hence, the simulation results can be used for validating both modeling and measurement results. Performance such as ripple voltages, line and load regulations can also be obtained.

2.2 Error Correction on Finite Duty Ratio

Suppose the total number of simulation steps within one switching cycle is n, then the resolution of duty ratio is 1/n. Let the actual duty ratio be D and $i/n < D < (i+1)/n$. In simulation, due to limited resolution, the system can only assign a duty ratio that toggles between i/n and $(i+1)/n$ to obtain an average duty ratio of D. For example, if a buck converter requires an actual duty ratio of D=0.5752, with a resolution of 1/200, the simulated duty ratio will toggle between 0.575 and 0.580, as shown in Fig. 4. Numerical oscillation of the duty ratio is reflected as fluctuation on v_C and i_L as shown in Fig. 5. When FFT is performed, inaccurate frequency information will be introduced.

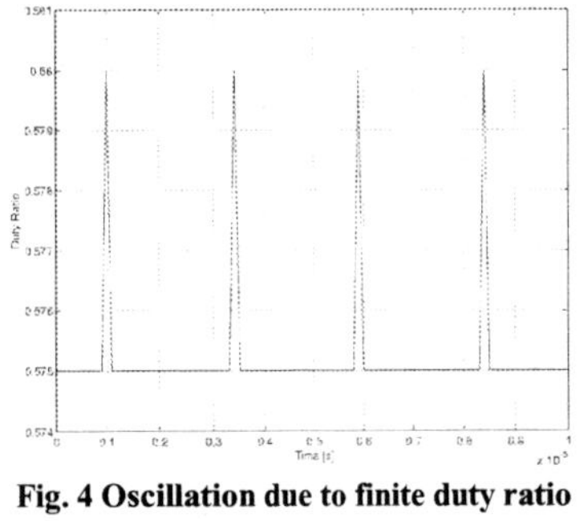

Fig. 4 Oscillation due to finite duty ratio

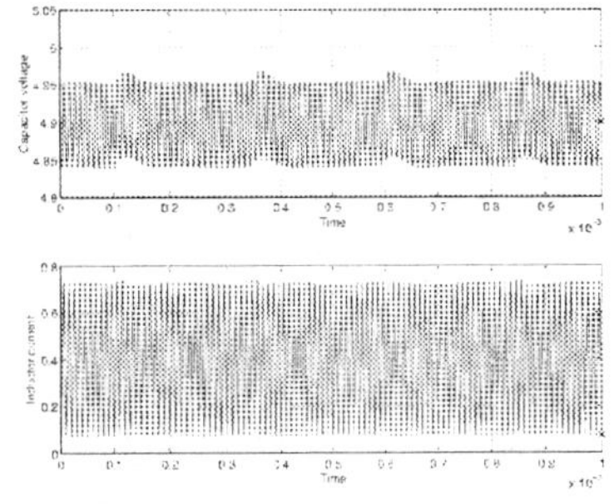

Fig. 5 Simulated effect of oscillation on v_C and i_L

To solve the problem of numerical oscillation while maintaining computation efficiency, an error correction technique is employed [3]. If a change in state is detected, the actual trip point is calculated by using the equations from the previous state to trace the system variables back to the trip point, and then using the equations of the current state to find the actual variables $x(t+\Delta t)$, as shown in Fig. 6. Hence, accuracy is much enhanced and numerical oscillation can thus be effectively eliminated.

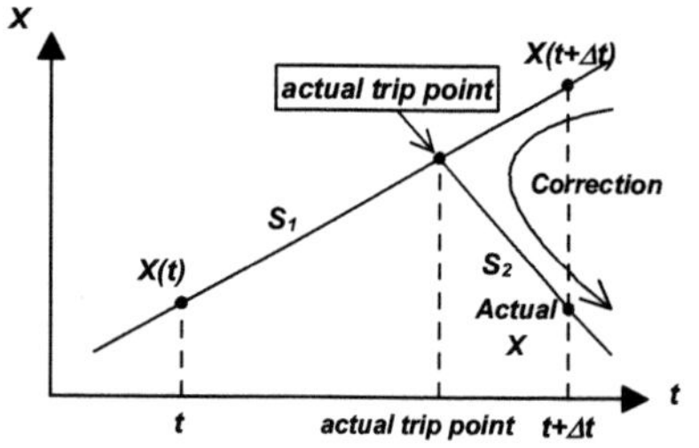

Fig. 6 Error correction on finite duty ratio

2.3 Signal Injection

The injection point of the simulator is chosen to be the output of the error amplifier, which is the same location as for loop gain measurement. The ratio of the reference node impedance Z_r and test node impedance Z_t (Sec. 3) is $Z_r/Z_t=\infty/0=\infty$, which indicates an ideal injection point. The injected signal should be chosen within a correct range of amplitude. With the injected signal, the trip point is determined by $V_{ramp} = V_a + V_j = V_a + A_j\sin(2\pi f_j t)$. A_j should be chosen such that $0 < V_a \pm V_j < V_m$ for proper comparison, where V_m is the amplitude of the ramp. A_j should not be too small, as numerical precision and approximation in modeling are considered as random noise, and A_j should be larger than the noise to minimize errors. Extensive simulation shows that $10\mu V < A_j < 100mV$ gives good simulation results.

3. LOOP GAIN MEASUREMENT AND THEORETICAL ANALYSIS

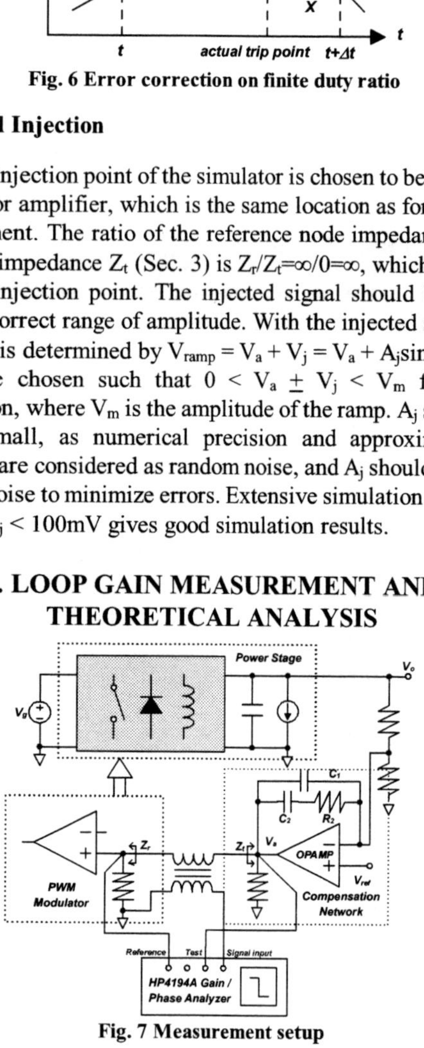

Fig. 7 Measurement setup

To verify the validity of the simulator, loop gain measurement is performed. In this paper, loop gain simulation and measurement serve as cross-checking mechanisms. However, loop gain is not as easy to measure as other quantities such as voltages and currents. The loop has to remain operative at DC such that operating points are maintained. Yet, it has to be broken at higher frequencies. In [4], a voltage source is inserted, but strict impedance constraints have to be satisfied. In [5], the loop is broken by an inductor. The drawback is that a very large inductor is needed and is not suitable for measuring loop gain at low frequencies. In [6], a digital modulation method is proposed. However, the loop gain thus measured may not be directly useful for designing compensation network for current programming switching converters [2].

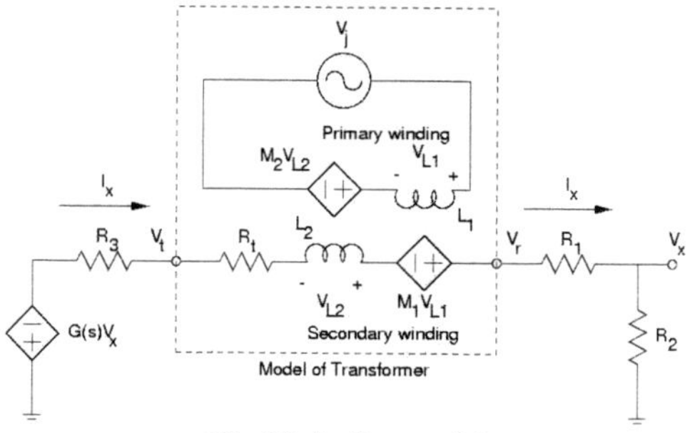

Fig. 8 Injection model

This research uses a measurement setup motivated by [5], as shown in Fig. 7. A feedback system with signal injection can be modeled in Fig. 8. Before inserting the transformer for signal injection, the original loop gain is

$$T(s) = G(s)\frac{R_2}{R_1 + R_2 + R_3} \tag{5}$$

The loop is then broken at the junction of R_1 and R_3, generating two nodes V_r and V_t. V_r is the reference node and V_t is the test node. Signal injection propagates from V_r, goes around the loop, and returns to V_t. The loop gain is given by

$$T'(s) = -V_t(s)/V_r(s) = -(-G(s)V_x(s) - I_x R_3)/V_r(s) = \frac{G(s)R_2 + R_3}{R_1 + R_2} \tag{6}$$

By substituting (5) into (6), we have

$$T'(s) = T(s)\left(1 + \frac{R_3}{R_1 + R_2}\right) + \frac{R_3}{R_1 + R_2}. \tag{7}$$

With $Z_r=R_1+R_2$ and $Z_t=R_3$, Eqn. (5) shows that a proper injection point is where $Z_t/Z_r(=R_3/(R_1+R_2))$ should be as small as possible. For this reason, the injection point in our measurement is located between the error amplifier and the comparator. An impedance/gain phase analyzer injects signal through a transformer to the converter, which propagates around the loop. The ratio of the signals at two terminals of the injection gives the loop gain.

It can be shown that the measured loop gain T' is the sum of the actual loop gain T plus noise divided by the injected signal V_j ($T'(f_j) = T(f_j) + \text{noise}(f_j)/V_j$). If V_j is too small, T' is dominated by noise, resulting in wrong simulation results. As an example, Fig. 9 shows a measured loop gain with unsuitable injection amplitude.

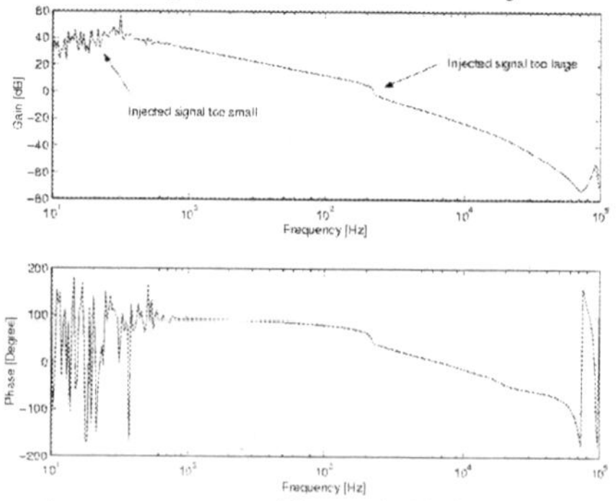

Fig. 9 Loop gain measurement with unsuitable injection amplitude

For better understanding the characteristics of switching converters, theoretical analysis has also been performed in this research. We primarily utilize state space averaging (SSA) and signal flow graph (SFG) techniques for analysis [2]. After SSA is performed on the sets of linear differential equations of the power stage and the feedback network, the averaged state equations are then perturbed, and the corresponding SFG is generated. The closed loop transfer functions can be obtained by solving the perturbed equations mathematically, or by applying Mason's Gain Formula to the SFG. This method is applicable to all kind of switching converters operating in either CCM or DCM.

4. EXPERIMENTAL VERIFICATION

Table 1. Configuration of the buck converter for verification

	CCM	DCM			CCM	DCM
V_g (V)	8	12		f_s (kHz)	97.3	97.3
C (μF)	10	10		R_1 (kΩ)	6.8	6.8
L (μH)	39	39		R_2 (kΩ)	2.2	2.2
R (Ω)	15	24.3		b	0.244	0.244
				V_{ref} (V)	1.224	1.224
R_C (Ω)	0.4	0.4		G_{EA} (dB)	100	100
R_L (Ω)	0.3	0.3		R_f (kΩ)	100	100
R_S (Ω)	1.5	1.5		C_f (nF)	1	1
R_d (Ω)	3.4	3.9		V_m (V)	3.61	3.61

The configuration of a buck converter for verification is listed in Table 1. Simulation and measurement results on the transient performance for both CCM and DCM are shown in Fig. 10, which cope with each other well. Loop gains by simulation, measurement and theoretical analysis are also performed and compared in Figs. 11 and 12. The derived loop gains in CCM and DCM based on the analysis in Sec. 3 are shown as Eqns. (8) and (9). As a whole, the simulation approximates the measurement well. The small phase deviation is mainly due to inaccurate values of the parasitics, especially R_C, as well as sampling problem at around the switching frequency. The results of converters with a different topology have also been obtained with satisfactory performance.

CCM:

$$T(s)=\frac{A(s)b}{V_m}\left(\frac{V_C}{d_1}\right)\frac{1}{1+\frac{1}{Q}\frac{s}{\omega_0}+\left(\frac{s}{\omega_0}\right)^2} \tag{8}$$

$$=\frac{676}{s}(8)\frac{1}{1+\frac{1}{7.6}\left(\frac{s}{2\pi 8.06k}\right)+\left(\frac{s}{2\pi 8.06k}\right)^2}$$

DCM:

$$T(s)=\frac{A(s)b}{V_m}\left(\frac{2V_C(1-M)^{3/2}}{(2-M)\sqrt{KM}}\right)\frac{1}{1+sRC\frac{1-M}{2-M}} \tag{9}$$

$$=\frac{676}{s}(12.1)\frac{1}{1+\frac{s}{2\pi 1.78k}}$$

5. CONCLUSIONS

In this paper, a unified simulation method for the loop gain of switching converters is introduced. Several techniques are proposed to improve simulation accuracy. Experimental measurement issues are also discussed. Theoretical analysis based on SSA and SFG is also addressed. The method is demonstrated to work satisfactorily for PWM switching converters. Experimental results fit well with the simulation and theoretical ones. The research contributes to a better understanding of switching converter loop gain and shortens the design time with a cost-effective simulation tool.

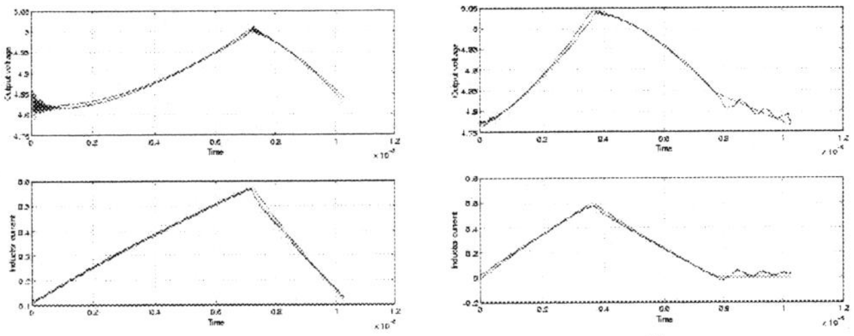

(a) (b)

Fig. 10 Transient simulations in (a) CCM and (b) DCM

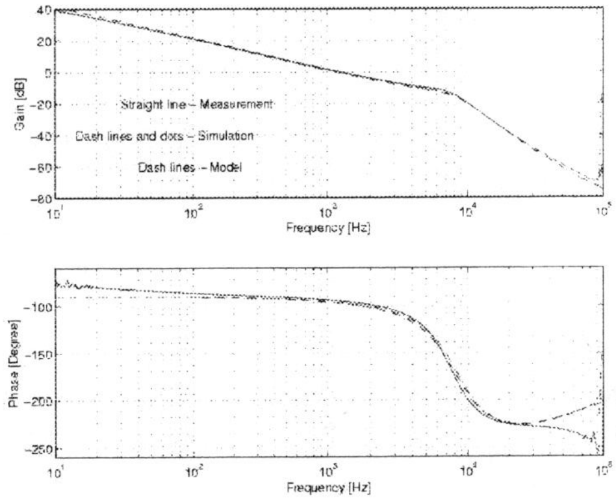

Fig. 11 Loop gain of the CCM buck converter

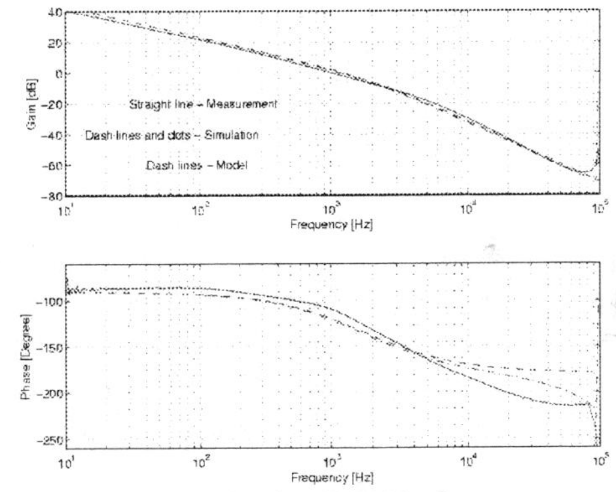

Fig. 12 Loop gain of the DCM buck converter

REFERENCES

[1] R. D. Middlebrook and S. Cuk, "A general unified approach to modeling switching-converter power stage," *IEEE Power Elec. Specialists Conf.*, pp.18-34, 1976.

[2] W-H Ki, "Signal flow graph in loop gain analysis of dc-dc PWM CCM switching converters," *IEEE Trans. on Ckts. & Sys. I*, no. 6, pp.644-654, June 1998.

[3] V. H.S. Tam, *Loop Gain Simulation and Measurement of PWM Switching Converters*, MPhil. Thesis, HKUST, 1999.

[4] R. D. Middlebrook, "Measurement of loop gain feedback systems," *Int. J. of Elec.*, vol. 57, no. 4, pp.485-512, 1975.

[5] S. Rosenstark, "Loop gain measurement in feedback amplifiers," *Int. J. of Elec.*, vol. 57, no. 3, pp.415-421, 1984.

[6] B. H. Cho and F. C. Lee, "Measurement of loop gain with the digital modulator," *IEEE Power Elec. Specialists Conf.*, pp.363-373, 1998.

DIGITALLY CONTROLLED BUCK CONVERTER

Dušan Gleich, Miro Milanovič, Suzana Uran, Franc Mihalič

University of Maribor
Faculty of Electrical Engineering and Computer Science
Smetanova 17, 2000 Maribor
Slovenia

ABSTRACT

Digitally controlled step down converter with different control law algorithms is studied in this paper. The state controller is designed to eliminate the start-up overshoot and reduce maximal dynamic error of load change response, but the state controller is not capable to eliminate the steady-state error under the load change condition. By using an additional discrete PID controller the steady state error could be eliminated. For the next consideration, the decomposed discrete fuzzy PID controller has been analyzed. All control algorithms are performed at a continuous current mode of operation. The experimental results are presented in the paper. The control algorithms were implemented on 16-bit microcomputer.

1. INTRODUCTION

DC-DC converters are nonlinear systems due to their inherent switching operation. To assure a constant output voltage, a classical linear design of a control is frequently used. The regulation is normally achieved by the pulse width modulation (PWM) at a fixed frequency. The switching device is a power MOSFET. The PWM linear control techniques are widely used [1]. Sliding-mode control based techniques [2] exhibit the property of robustness in front of large signal perturbations, leading to global stability. The sliding controller parameters leading to an optimal behaviour in large signal operation and do not yield a small-signal optimum behaviour. During the last decade, digital control in power electronics has been intensively used. The increasing performance and cost reduction of digital circuits has enable their application for power converters control. The control algorithms are implemented on digital signal processors (DSP), which are not very common in high switching frequency domain or low cost applications. Some control algorithms are implemented on a field of programmable gate array (FPGA) using hardware description language (VHDL) [4].

Three control structures are presented in this paper. The first one is based on ordinary state controller supported by static gain. The second one is based on state controller supported by discrete PID controller in order to eliminate the steady state error. The third one is based on state controller supported by decomposed fuzzy PID controller to improve the converter dynamics. The fuzzy set theory has evolved as a powerful modeling tool that can cope with the uncertainties and non-linearity of the control systems [3], [5].

2. THE STATE CONTROL OF BUCK CONVERTER

The buck converter is modeled using an averaged signal model [1]. From Fig. 1, a buck converter model is obtained using linear differential equations:

$$\begin{bmatrix} \frac{di_L}{dt} \\ \frac{du_0}{dt} \end{bmatrix} = \begin{bmatrix} -\frac{R_L}{L} & -\frac{1}{L} \\ \frac{1}{C} & -\frac{1}{RC} \end{bmatrix} \begin{bmatrix} i_L \\ u_0 \end{bmatrix} + \begin{bmatrix} \frac{1}{L} \\ 0 \end{bmatrix} u_d \delta \quad (1)$$

where i_L is the inductor current, u_0 is the output voltage, u_d is input voltage, δ is a duty cycle, L is the circuit inductance, R_L is the resistance of inductor, C is the output capacitance and R is the load resistance. In order to get the dynamic model the small signal perturbation must be introduced:

$$\begin{aligned} u_d &= U_d + \tilde{u}_d \\ u_C &= U_C + \tilde{u}_C \\ i_C &= I_C + \tilde{i}_C \\ \delta &= \Delta + \tilde{\delta} \end{aligned} \quad (2)$$

where capital letters have been used to describe the converter operating point. The corresponding transfer function

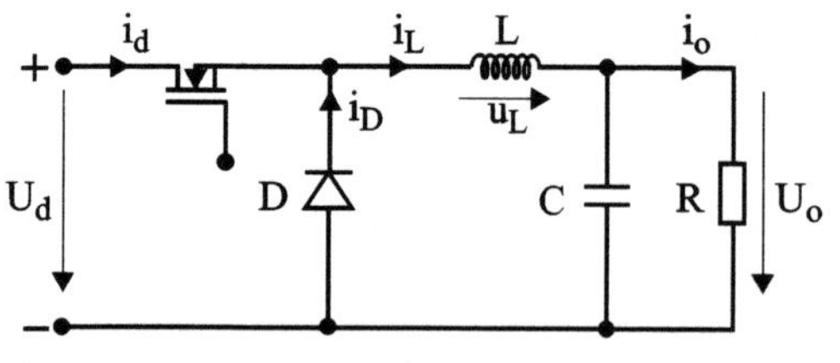

Fig. 1. Step down or Buck converter.

of the continuous system is given by:

$$H_1(s) = \frac{\tilde{u}_0(s)}{\tilde{\delta}(s)}\Big|_{\tilde{u}_d=0} = \frac{U_d}{as^2 + bs + c} \tag{3}$$

where $a = LC$, $b = \frac{L}{R} + R_L C$ and $c = 1 + \frac{R_L}{R}$. ($R = 6.8\Omega$, $L = 2.1mH$, $C = 100\mu F$, $R_L = 1.1\Omega$ and $U_d = 12V$). In order to control the state variables the output voltage u_0 and its derivative du_0/dt has been chosen. For this purposes from (3) the controllable canonical form is obtained:

$$\begin{bmatrix} \frac{d\tilde{x}_1}{dt} \\ \frac{d\tilde{x}_2}{dt} \end{bmatrix} = \begin{bmatrix} 0 & 1 \\ -\frac{c}{a} & -\frac{b}{a} \end{bmatrix} \begin{bmatrix} \tilde{x}_1 \\ \tilde{x}_2 \end{bmatrix} + \begin{bmatrix} 0 \\ 1 \end{bmatrix} \tilde{\delta} \tag{4}$$

$$\begin{bmatrix} \tilde{y}_1 \\ \tilde{y}_2 \end{bmatrix} = \begin{bmatrix} \tilde{u}_0 \\ d\tilde{u}_0/dt \end{bmatrix} = \begin{bmatrix} U_d/a & 1/a \end{bmatrix} \begin{bmatrix} \tilde{x}_1 \\ \tilde{x}_2 \end{bmatrix}$$

where $x_1 = au_0$ and $x_2 = a\frac{du_0}{dt}$. The above system could be written in a matrix form:

$$\begin{aligned} \frac{d\tilde{\mathbf{x}}(t)}{dt} &= \mathbf{A}\tilde{\mathbf{x}}(t) + \mathbf{B}\tilde{u}(t) \\ \tilde{\mathbf{y}}(t) &= \mathbf{C}\tilde{\mathbf{x}}(t) \end{aligned} \tag{5}$$

The continuous system is discretized by using a sample time $T_s = 50\mu$s and zero order hold element. The discretized system is:

$$\begin{aligned} \tilde{\mathbf{x}}(k+1) &= \mathbf{\Phi}\tilde{\mathbf{x}}(k) + \mathbf{\Gamma}\tilde{u}(k) \\ \tilde{\mathbf{y}}(k) &= \mathbf{C}\tilde{\mathbf{x}}(k) \end{aligned} \tag{6}$$

A discrete matrix $\mathbf{\Phi}$ and Γ can be evaluated by:

$$\begin{aligned} \mathbf{\Phi} &= \mathcal{L}^{-1}\left\{[sI - A]^{-1}\right\} = \begin{bmatrix} \phi_{11} & \phi_{12} \\ \phi_{21} & \phi_{22} \end{bmatrix} \\ \mathbf{\Gamma} &= \int_0^{T_s} \mathbf{\Phi}(\tau)\mathbf{B}d\tau = \begin{bmatrix} b_1 \\ b_2 \end{bmatrix} \end{aligned} \tag{7}$$

From (6) and (7) it follows:

$$\begin{aligned} \begin{bmatrix} \tilde{x}_1(k+1) \\ \tilde{x}_2(k+1) \end{bmatrix} &= \begin{bmatrix} 0.99 & 4.7e^{-5} \\ -261 & 0.89 \end{bmatrix} \begin{bmatrix} \tilde{x}_1(k) \\ \tilde{x}_2(k) \end{bmatrix} \\ &+ \begin{bmatrix} 1.208e^{-10} \\ 4.748e^{-5} \end{bmatrix} \tilde{\delta}(k) \end{aligned} \tag{8}$$

$$\begin{bmatrix} \tilde{y}_1(k) \\ \tilde{y}_2(k) \end{bmatrix} = \begin{bmatrix} U_d/a \\ 1/a \end{bmatrix} \begin{bmatrix} \tilde{x}_1(k) \\ \tilde{x}_2(k) \end{bmatrix} \tag{9}$$

The control scheme is shown in Fig. 2. The state controller was determined using a standard approach. The poles of the system were chosen in such a way, that damping factor $D = 1$ and frequency $\omega_0 = 8000$ of the characteristic polynom $s^2 + 2d\omega_0 + \omega_0^2$, in the continuous space. The parameters were transformed into discrete space and poles of closed loop discrete transfer function were evaluated as $z_1 = z_2 = e^{-d\omega_0 T} = 0.637$. According to this the discrete state controller gains have values:

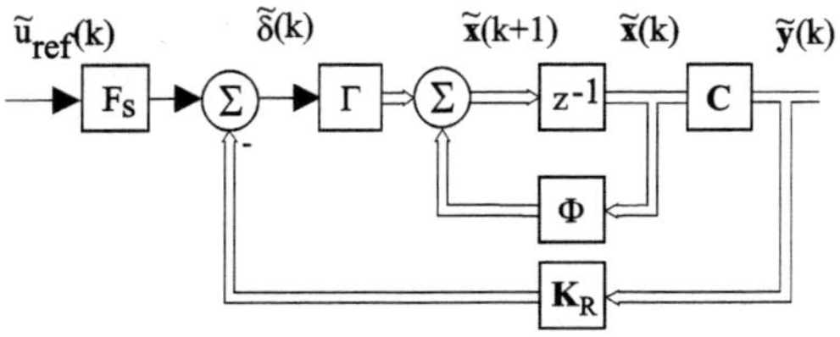

Fig. 2. The state controller scheme.

$$\mathbf{K}_R = \begin{bmatrix} K_{r1} & K_{r2} \end{bmatrix} = \begin{bmatrix} 4.99 & 0.0012 \end{bmatrix} 10^7. \tag{10}$$

According to the (8), (9) and (10) the closed loop discrete transfer function is:

$$H_2(z) = \frac{\tilde{u}_0}{\tilde{u}_{ref}} = \frac{0.6871z + 0.06646}{z^2 - 1.274z + 0.4058}, \tag{11}$$

In order to reduce the steady state error the static gain was introduced. The gain has been evaluated by:

$$F_S = \frac{1}{\lim_{z\to 1} H_2(z)} = 0.975. \tag{12}$$

The reference (output) voltage is $U_{ref} = U_0 = 3.3V$. The open loop start-up and the transient response when load resistance has been changed is shown in Fig. 3(a) and (b), respectively.

The start-up of the state-space controlled converter and the transient response when the load resistance has been changed are shown in Fig. 4(a) and (b), respectively. There is no overshoot at the start-up, but the steady state error is noticed, when the load resistance is changed. The over voltage at the load change is $500mV$ peak-to-peak (Vpp) and the steady state error is around $100mV$. The settling time of the step response was $1ms$ and the settling time at the load resistance change was $800\mu s$. To guarantee the fulfillment of the Nyquist criteria, the sampling rate was chosen as $F_s \geq 1/(2\pi\sqrt{LC})$. The switching frequency of PWM and sampling time in all presented experiments was $50\mu s$.

3. REDUCTION OF STEADY STATE ERROR USING A DISCRETE PID CONTROLLER

The state controller shown in Fig. 2 cause the static error under the dynamic response. To reduce or eliminate it, a discrete PID controller was used, as shown in Fig. 5. The closed loop control system can be written:

$$H_{CL}(z) = \frac{H_2(z)R_{PID}(z)}{1 + H_2(z)R_{PID}(z)} \tag{13}$$

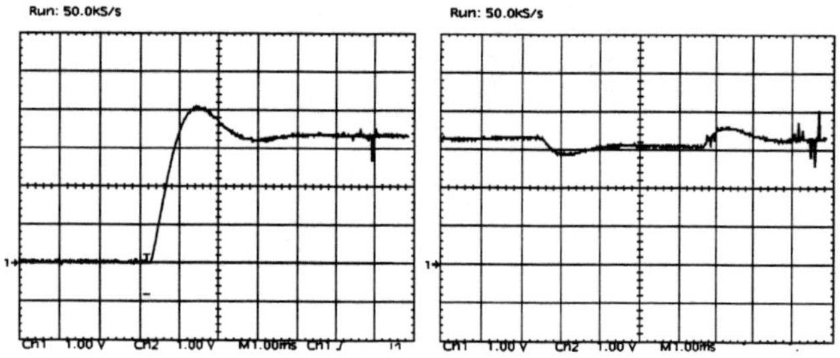

Fig. 3. The open loop response: (a) The start-up; (b) The load change response ($R = 6.8$ to $4.8\ \Omega$).

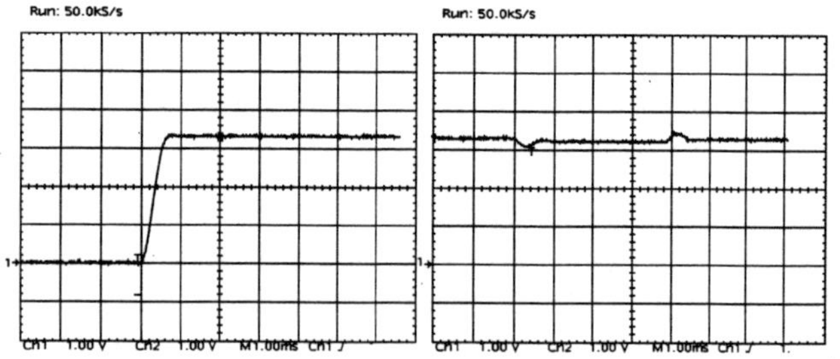

Fig. 4. The state space controller response: (a) The start-up; (b) The load change response ($R = 6.8$ to $4.8\ \Omega$).

where $H_2(z)$ is defined as follows from (11). According to Fig. 5 and equation (13), $H_2(z) = \tilde{u}_0/\tilde{u}$. The controller $R_{PID}(z)$ can be defined in two forms:

$$R_{PID}(z) = K_R \frac{(z - z_1)(z - z_2)}{z(z-1)} \tag{14}$$

or

$$R_{PID}(z) = Kp\Big(1 + \frac{T}{T_i}\frac{1}{z-1} + \frac{T_D}{T}\frac{z-1}{z}\Big). \tag{15}$$

The K_R is a gain, K_p is a proportional gain, $K_D = T_D/T$ differential gain and $K_I = T/T_I$ integral gain.
The Root-Locus method was used to define the controller zeros z_1, z_2 and gain K_R. Using a relative damping ζ, it follows:

$$z_1 = p_1 + (p_1 - 1)\frac{\zeta - 1}{2} \quad \text{and} \quad z_2 = p_2\frac{1+\zeta}{2} \tag{16}$$

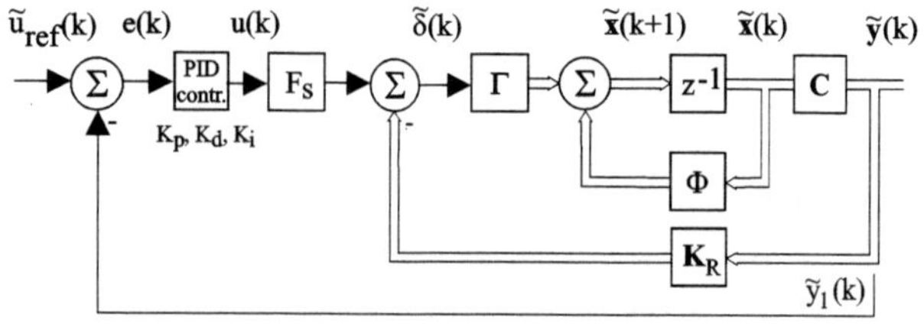

Fig. 5. The PID and state controller scheme.

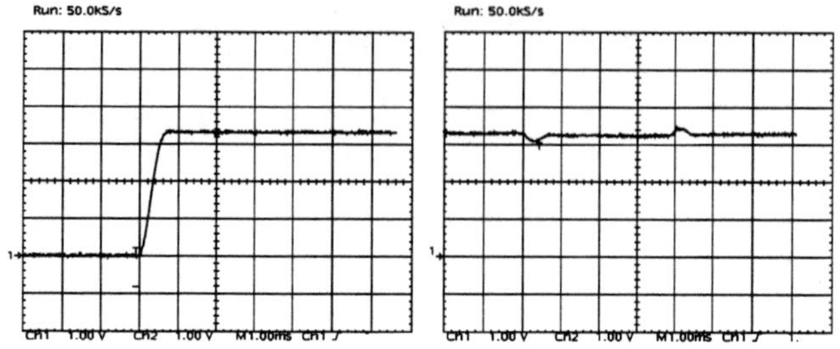

Fig. 6. The PID controller response: (a) The start-up; (b) The load change response ($R = 6.8$ to $4.8\ \Omega$).

where $p_{1,2}$ are the poles of $H_2(z)$ in (11). When the damping $\zeta = 0.9$ is chosen:

$$-\ln|H_{CL}(z_1)| + \angle H_{CL}(z_1)\frac{\zeta}{\sqrt{1+\zeta^2}} = 0, \tag{17}$$

from (13) and (17) the gain K_{PID} can be evaluated numerically. It follows:

$$R_{PID}(z) = 2.89\frac{(z - 0.6834)(z - 0.6427)}{z(z-1)}. \tag{18}$$

The (18) was transformed into another form denoted by (15). Using sampling time $T_s = 50\mu s$ the constants are $K_R = 1.592$, $K_I = 0.227e^{-4}$, $K_D = 0.138e^{-3}$. The start-up response is shown in Fig. 6(a). There is no overshoot and has settling time is $0.8ms$. The controller eliminates the steady state error as well. The overshoot is $400mVpp$, when the load resistance is changed. The settling time of the over and under-shoot caused by load resistance change is $0.85ms$.

4. DECOMPOSED FUZZY PID CONTROLLER

The dynamic response of the PID controller can be improved by adaptation of the constant K_p, K_i and K_d to the control plant parameter change. It is obvious that the parameters denoted by b and c in (3) will change the buck converter dynamics under the load change condition. Due to this the decomposed fuzzy PID controller can be used [5]. The output of the fuzzy PID controller is denoted by:

$$u(k) = N_p(e(k)) + N_D(de(k)) + N_I(ie(k)) \tag{19}$$

where N_P, N_D and N_I are the non-linear functions determined by a Fuzzy Rule-Based System (FRBS). A FRBS presents two main components: the Rule Base (RB), representing the knowledge about the controller being described in the form of the fuzzy IF-THEN rules, and the Fuzzy Inference System (FIS) needed to obtain an output from FRBS. The structure of the fuzzy PID controller is depicted

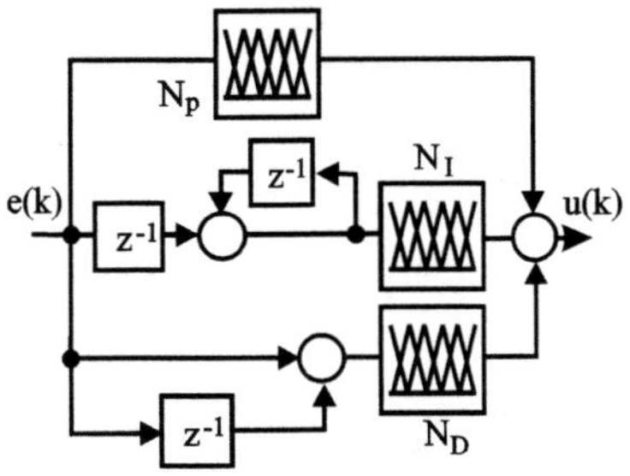

Fig. 7. A structure of a decomposed fuzzy controller

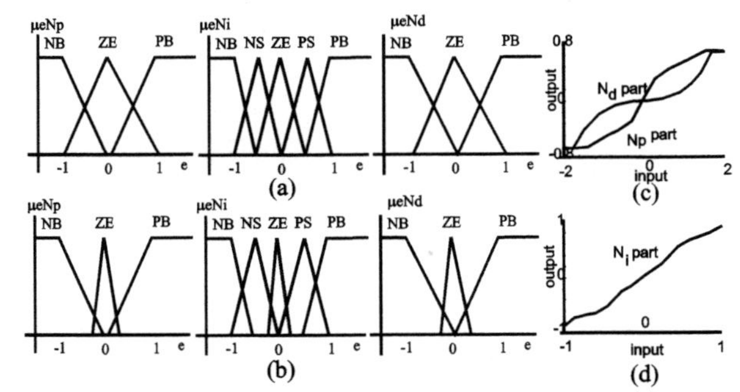

Fig. 8. Membership functions of a PID controller: (a) Input membership function, (b) Output membership function.

in Fig. 7. The linguistic description of the knowledge base is given by three RB's. The output signal is the sum of the defuzzified outputs of proportional FIS, differential FIS and integral FIS. The proportional part will produce linear effect and the integral and differential parts will produce the non-linear effect. The membership functions are shown in Fig. 8. The fuzzy rules were designed that for each input membership function the output membership function was assigned. For N_I part the fuzzy rules are:

$(1) if\ e(k) = NB\ then\ y = NB$
$(2) if\ e(k) = NS\ then\ y = NS$
$(3) if\ e(k) = ZE\ then\ y = ZE$
$(4) if\ e(k) = PS\ then\ y = PS$
$(5) if\ e(k) = PB\ then\ y = PB$

Similar fuzzy rules were assigned to the integral and differential part, where rules (1), (2) and (3) were applied. Fuzzy rules and membership functions form a nonlinear input-output function as show in Fig. 8. A look-up table was build from nonlinear input/output functions.

The start-up of the decomposed discrete fuzzy PID controller is shown in Fig. 9. There is no overshoot and the settling time is $0.8ms$. From Fig. 9 it is obvious that the controller eliminates the steady state error. The overshoot is $390mVpp$. The settling time is $0.8ms$.

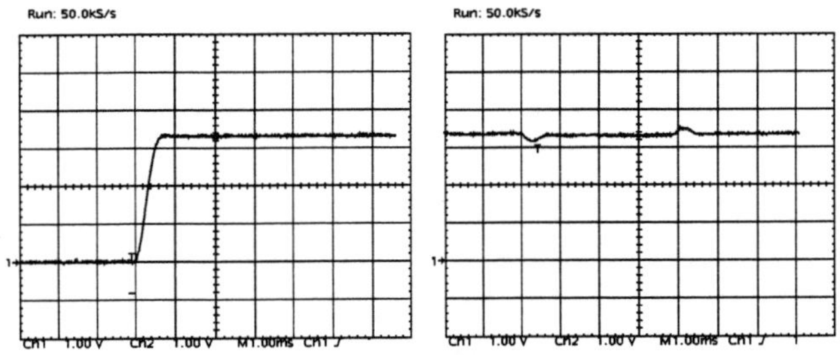

Fig. 9. (a) A step response of a decomposed fuzzy PID controller (b) Change of load from 6.8 to 4.8 Ω.

5. CONCLUSION

A digital linear and nonlinear algorithms using fuzzy logic were studied in order to control buck converter output voltage. The state controller, the discrete PID controller and decomposed fuzzy PID controller were presented. All algorithms were implemented on the 16-bit RISC microcomputer. Results shown in Figs. 4, 6, 9 look very similar, because differences in responses are very small. The results were measured using markers on digital oscilloscope. From experimental results the following conclusions can be made: the state controller eliminates the overshoot at start-up, but the steady state error appears. The steady state error is eliminated using a discrete PID controller. The discrete PID controller has faster response and smaller overshoots under the load change condition. A decomposed fuzzy discrete PID controller provide the better dynamic then the PID controller. All control algorithms executes less than $50\mu s$.

6. REFERENCES

[1] R. D. Middlebrook, S. Cuk, "A general unified approach to modeling switchnig-converter power stages", *PESC'76 Record*, pp. 19-34, 1976.

[2] V.I. Utkin, "Variable structure systems with sliding mode: a surey,", *IEEE Trans. Autimatic Control*, vol. 22, pp. 2121-222.

[3] W.-C. So, C. K. Tse, Y-S. Lee, "Development of a fuzzy logic controller for DC/DC converters: design, computer simulations, and experimental evaluation," *IEEE Trans. Power Electronics,* vol. 11, no. 1, Jan. 1996.

[4] A. de Castro, P. Zumel, O. Garcia, T. Riesgo, J. Uceda, "Concurrent and simple digital controller of an AC/DC converter with power correction based on FPGA," *IEEE Trans. Power Electronic*, vol. 18, no. 1, pp. 334-343, 2003

[5] M. Golob, B. Tovornik, Modeling and control of the magnetic suspension system, *ISA Trans.*, vol. 43, no. 1, pp. 89-100, 2003.

A 3-D PWM CONTROL, H-BRIDGE TRI-LEVEL INVERTER FOR POWER QUALITY COMPENSATION IN THREE-PHASE FOUR-WIRED SYSTEMS

Pui-In Mak †, *Man-Chung Wong* and *Seng-Pan U*

Faculty of Science and Technology, University of Macau, Avenida Padre Tomas Pereira, S.J., Taipa, Macao, China
E-mails –† - p.i.mak@ieee.org

ABSTRACT

A three-dimensional pulse width modulation (3-D PWM) control, H-bridge tri-level inverter, acts as a power quality compensator for three-phase four-wired systems, is proposed. Such compensator is an alternative of diode-clamped inverter based one, which generally suffers from the DC-link voltage unbalance problem and possesses low expandability in very high-voltage and high-power applications. This paper based on the concepts of switching function and state-variable formulation to deduce the mathematical model of H-bridge tri-level inverter, with also an enhanced switching table that balances the average switching frequency of the switches. The compensation is accomplished through a sign cubical hysteresis control strategy, which controls the total 27 space vectors in 3-D aspect to inject the nearest negative amplitude of harmonics into the load currents to compensate both the harmonic and in-phase currents into the systems to reduce the peak-load power supply from the generator. System-level simulation results demonstrate the feasibility and performance of such compensator.

1. INTRODUCTION

In power distribution networks, both three-phase three-wired and three-phase four-wired systems are particularly important. Due to the non-linear loading effects, the power quality of the neighbor systems will be degraded significantly. In order to minimize such disturbance, a power quality compensator (or an active filter), typically constructed by multilevel inverter, can be adopted. In the literature, different kinds and different levels inverter have been proposed for three-phase three-wired systems with two-dimensional pulse width modulation (2-D PWM) control [1]-[4]. There are seldom papers that focus on three-phase four-wired systems, in which the extra neutral line makes the traditional 2-D PWM not applicable. Recently, 3-D PWM has been proposed and demonstrated in a diode-clamped tri-level inverter [Fig. 1(a)] for power quality compensation, motor drivers or peak-load power supply [5]-[7]. The space vectors are considered in 3-D aspect to take consider both the three-phase power lines as well as the neutral one without the need of four arms-inverter. However, the diode-clamped inverter suffers from the natural problems of DC-link voltage unbalance, and the difficulty to alter the number of clamping diode as the voltage level changes, thus it is very inefficient in expansion to higher-level architecture based on low-level one.

In this paper, a 3-D PWM control, tri-level H-bridge inverter [8] as shown in Fig. 1(b) is proposed in order to attain the highest expandability for higher-voltage and higher-power applications, since the level of the inverter is proportional with the number of H-bridge cell that cascaded together. The disadvantage is obviously the increased number of DC sources, but it is still worth to make such choice in considering the global manufactory costs.

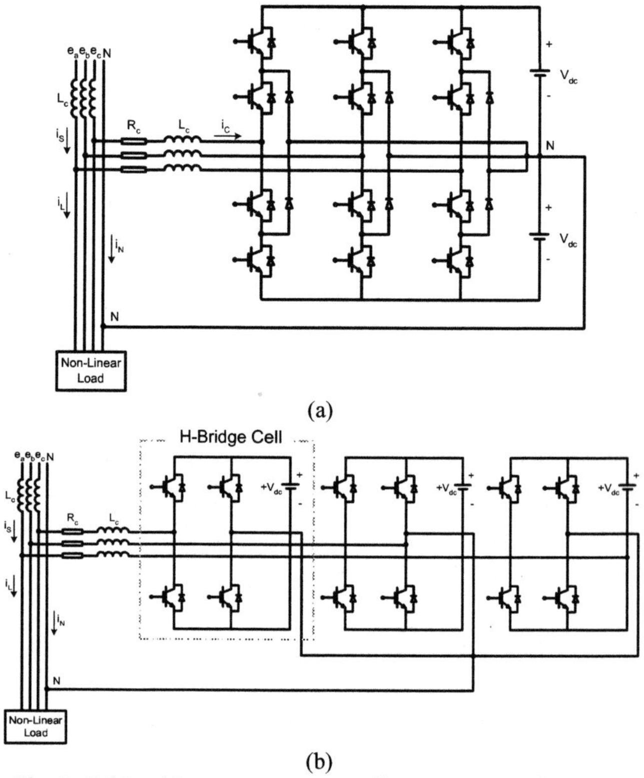

Fig. 1. Tri-level inverter as power quality compensator for three-phase four-wired systems. (a) Diode-clamped, and (b) H-bridge.

2. H-BRIDGE TRI-LEVEL INVERTER AND ITS MATHEMATICAL MODEL

This section analyzes the structure of an H-bridge tri-level inverter as a power quality compensator. The basic principle is to inject the same negative amplitude of harmonics into the load current in order to compensate the harmonic current and the in-phase current into the systems to reduce the peak-load power supply from the generator. The losses of the switching devices and snubber circuits, and process of commutation are ignored so that the equivalent switched-circuit can be obtained as shown in Fig. 2.

The model of H-bridge three-phase four-wired tri-level inverter is investigated in the *a-b-c* frame. Switching functions can be considered as the equivalent switched devices such as IGBT's, e.g., in phase **A**, S_a may be written as:

$$S_a = \begin{cases} 1, & when\ S_{a2}\ \&\ S_{a3}\ are\ closed \\ 0, & when\ S_{a1}\ \&\ S_{a2}\ or\ S_{a3}\ \&\ S_{a4}\ are\ closed \\ -1, & when\ S_{a1}\ \&\ S_{a4}\ are\ closed \end{cases} \quad (1)$$

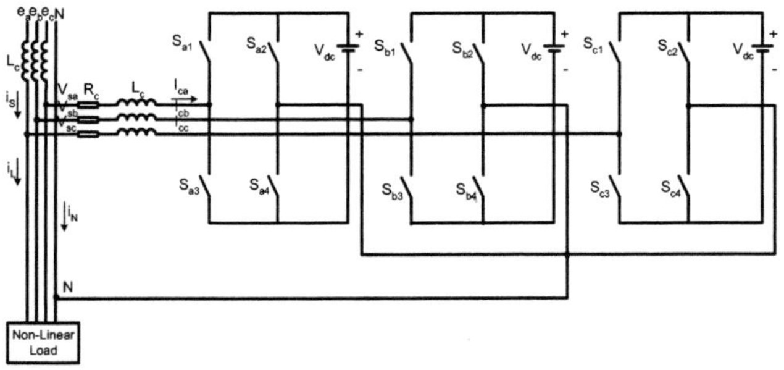

Fig. 2. Equivalent model of H-bridge tri-level inverter.

There are 3 cases in one arm of the tri-level converters such as positive, zero or negative switching function.

(i) if $S_a=1$, then $S_{a1}=0,\ S_{a2}=1,\ S_{a3}=1,\ S_{a4}=0$;
(ii) if $S_a=0$, then can be $S_{a1}=1,\ S_{a2}=1,\ S_{a3}=0,\ S_{a4}=0$ or $S_{a1}=0,\ S_{a2}=0,\ S_{a3}=1,\ Sa4=1$
(iii) if $S_a=-1$, then $S_{a1}=1,\ S_{a1}=0,\ S_{a1}=0,\ S_{a1}=1$;

Note that the possible values of S_a, S_b and S_c are *1*, *0* and *-1*. However, there are 2 possible situations when S_a changed from *1* to *0*. The possible results can be $S_{a1}=1,\ S_{a2}=1,\ S_{a3}=0,\ S_{a4}=0$ or $S_{a1}=0,\ S_{a2}=0,\ S_{a3}=1,\ S_{a4}=1$. In order to balance the switching activities in all the switches, a balanced switching table for phase **A** is obtained in Table 1, thus the average switching activities of the switches will be closer.

Table 1. Proposed switching table for phase **A**.

S_a	S_{a1}	S_{a2}	S_{a3}	S_{a4}	⇔	S_{a1}	S_{a2}	S_{a3}	S_{a4}
1⇔0	0	1	1	0	⇔	1	1	0	0
-1⇔0	1	0	0	1	⇔	0	0	1	1
1⇔-1	0	1	1	0	⇔	1	0	0	1

In Fig. 2, the boundary condition of S_{1a}, S_{2a} and S_{3a} is defined as:

$$\begin{cases} S_{a1}+S_{a2}+S_{a3}+S_{a4}=2 \\ S_{a1}=1 \ or \ 0, \ S_{a2}=1 \ or \ 0, \\ S_{a3}=1 \ or \ 0, \ S_{a4}=1 \ or \ 0. \end{cases} \tag{2}$$

It is worth to mention that when S_{a1} and S_{a2} are equal to 1, the values of S_{a3} and S_{a4} must be zero. The relationship among the ac-side compensating current and the terminal voltage of the inverter can be expressed in (3) according to Fig. 2.

$$\begin{cases} L_C\dfrac{di_{ca}}{dt}=-R_C\cdot i_{ca}-v_a+v_{sa} \\ L_C\dfrac{di_{cb}}{dt}=-R_C\cdot i_{cb}-v_b+v_{sb} \\ L_C\dfrac{di_{cc}}{dt}=-R_C\cdot i_{cc}-v_c+v_{sc} \end{cases} \tag{3}$$

By using the switching functions, the relation between the terminal voltage (V_a, V_b, V_c) and the dc-link voltage set $v_{dc1}=v_{dc2}=v_{dc}$ can be expressed in (4).

$$\begin{cases} V_a=S_{a2}\cdot S_{a3}\cdot v_{dc}-S_{a1}\cdot S_{a4}\cdot v_{dc} \\ V_b=S_{b2}\cdot S_{b3}\cdot v_{dc}-S_{b1}\cdot S_{b4}\cdot v_{dc} \\ V_c=S_{c2}\cdot S_{c3}\cdot v_{dc}-S_{c1}\cdot S_{c4}\cdot v_{dc} \end{cases} \tag{4}$$

A generalized mathematical model of the tri-level converter in three-phase four-wired systems can be established as follows:

$$Z\dot{X}=AX+BU \tag{5}$$

where

$$A=\begin{bmatrix} -R_C & 0 & 0 & -S_{a2}\cdot S_{a3} & S_{a1}\cdot S_{a4} \\ 0 & -R_C & 0 & -S_{b2}\cdot S_{b3} & S_{b1}\cdot S_{b4} \\ 0 & 0 & -R_C & -S_{c2}\cdot S_{c3} & S_{c1}\cdot S_{c4} \end{bmatrix}$$

$$X=[i_{ca}\ \ i_{cb}\ \ i_{cc}\ \ V_{dc1}\ \ V_{dc2}]^T \qquad B=diag[1\ \ 1\ \ 1]$$

$$U=[V_{sa}\ \ V_{sb}\ \ V_{sc}]^T \qquad Z=diag[L_C\ \ L_C\ \ L_C]$$

Due to three-phase symmetrical properties, the model equations are valid for both S_b and S_c in phase **B** and **C**, respectively.

3. 3-D PWM TECHNIQUE WITH SIGN CUBICAL HYSTERESIS CONTROL STRATEGY

In three-phase three-wired systems, there are many researches focusing on the PWM techniques such as the Sinusoidal PWM, Hysteresis Control PWM and Space Vector PWM etc. however, all of them are only investigated in 2-dimensional aspect. Only 2-D PWM cannot be utilized to solve the issues in three-Phase four-Wired Systems. The mathematical model of 3-D PWM is reviewed in this section.

The instantaneous voltage in *α-β-0* frame can be transferred from *a-b-c* frame, as expressed by

$$\begin{bmatrix} v_\alpha \\ v_\beta \\ v_0 \end{bmatrix}=\sqrt{\frac{2}{3}}\begin{bmatrix} 1 & -1/2 & -1/2 \\ 0 & \sqrt{3}/2 & -\sqrt{3}/2 \\ 1/\sqrt{2} & 1/\sqrt{2} & 1/\sqrt{2} \end{bmatrix}\cdot\begin{bmatrix} v_a \\ v_b \\ v_c \end{bmatrix} \tag{6}$$

The instantaneous voltage vector can be given by

$$V_s=\sqrt{\frac{2}{3}}(V_{SA}+\alpha\cdot V_{SA}+\alpha^2\cdot V_{SC}) \tag{7}$$

where

$$\alpha=e^{j(2\pi/3)}, \qquad \alpha^2=e^{-j(2\pi/3)}$$

According to the switching functions in (1), equation (7) can be expressed in *α-β-0* frame as given by:

$$V_s=V_{dc}\begin{bmatrix} i\sqrt{\dfrac{2}{3}}\left(S_a-\dfrac{1}{2}\cdot S_b-\dfrac{1}{2}\cdot S_c\right)+ \\ j\dfrac{1}{\sqrt{2}}(S_b-S_c)+k\left(\dfrac{1}{\sqrt{3}}(S_a+S_b+S_c)\right) \end{bmatrix} \tag{8}$$

Equation (8) can be re-expressed as:

$$V_s=V_{dc}\left[i\sqrt{\frac{2}{3}}S_\alpha+j\frac{1}{\sqrt{2}}S_\beta+k\frac{1}{\sqrt{3}}S_0\right] \tag{9}$$

where

$$S_\alpha=S_a-\frac{1}{2}S_b-\frac{1}{2}S_c$$
$$S_\beta=S_b-S_c$$
$$S_0=S_a+S_b+S_c$$

If the network has accessible neutral wire, a zero-sequence current component can exist. It is desired that the load current zero-sequence component can be compensated by the power quality compensator. For these cases, the zero-sequence converter current component, as well as the other components, must be controlled also. Fig. 3 shows the space vector allocation for tri-level inverter in *α-β-0* frame. According to (9), there are totally 27 possible vectors in tri-level inverter. They can be classified as four types: large-voltage, medium-voltage, small-voltage, and zero-voltage as listed in Table 2 [5]. Due to the special vectors problem [5], there are 6 operations that the compensation improves one direction will increase dedicated error in another direction since it is not a one-

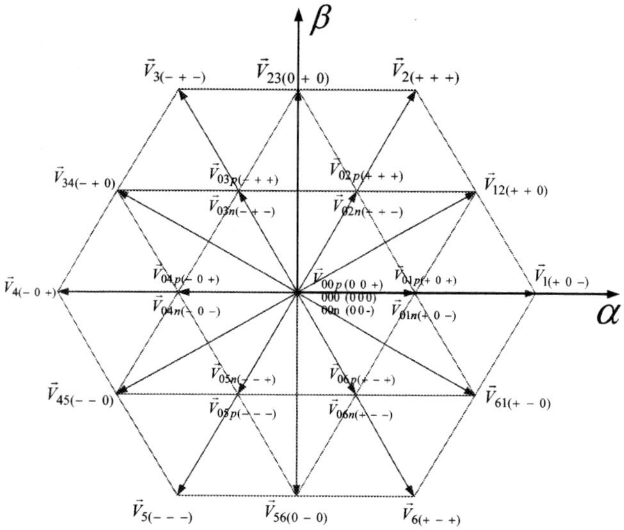

Fig. 3. Tri-level voltage vector allocation *α-β-0* frame.

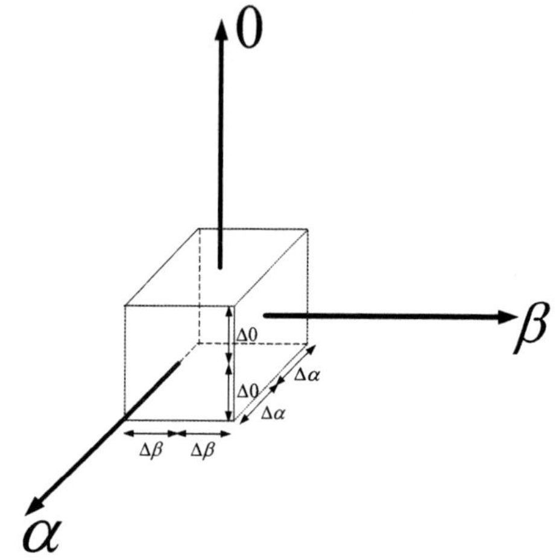

Fig. 4. Conception of sign cubical hysteresis control.

to-one mapping relationship. This error can be solved by using cylindrical coordinate control strategy [9].

The sign cubical hysteresis current controller is studied in this section to control a tri-level 3-D voltage inverter for three-phase four-wired systems. The conception of sign cubical hysteresis control technique is shown in Fig. 4. The hysteresis limits of $\Delta\alpha$, $\Delta\beta$ and $\Delta 0$ can be equal to each other ($\Delta\alpha=\Delta\beta=\Delta 0$) so as to have a cubical controlling environment. There are 3 voltage levels {-1, 0, 1} in a tri-level voltage inverter. When the difference between the reference signal and actual input signal is larger than the hysteresis limited value, it will trigger to either positive or negative. However, when the difference is tolerable in the hysteresis limit, there will be the zero level. Fig. 5 shows the entire control strategy. The difference between the current reference and the actual load current would be the signal to compare with the injected current signal,

$$i^*_{\alpha\beta 0} = i^{reference}_{\alpha\beta 0} - i^{load}_{\alpha\beta 0} \quad (10)$$

$i^{reference}_{\alpha\beta 0}$ is the reference signal that can be obtained by instantaneous reactive power compensation technique, $i^{load}_{\alpha\beta 0}$ is the actual load current that may be distorted by non-linear or unsymmetrical load, and the difference between the reference signal and load current signal will be the tracking current ($i^*_{\alpha\beta 0}$) that should be injected by the inverter,

$$\Delta i_{\alpha\beta 0} = i^*_{\alpha\beta 0} - i_{c\alpha\beta 0} \quad (11)$$

The difference between the tracking current ($i^*_{\alpha\beta 0}$) and the coupling current $i_{c\alpha\beta 0}$ between the inverter and the load terminal will be the control signal ($\Delta i_{\alpha\beta 0}$) to the controller to control the action of inverter.

Table 2. Space-vector control table.

	S_α	S_β	S_0	S_a	S_b	S_c		S_α	S_β	S_0	S_a	S_b	S_c	S_a	S_b	S_c
$\vec{V}_{02n}$	+	+	-	0	0	-1	$\vec{V}_{000}$	0	0	0	0	0	0			
$\vec{V}_{06n}$	+	-	-	0	-1	0	$\vec{V}_{1}$	+	0	0	1	-1	-1			
$\vec{V}_{03p}$	-	+	+	0	1	0	$\vec{V}_{4}$	-	0	0	-1	1	1			
$\vec{V}_{05p}$	-	-	+	0	0	1	$\vec{V}_{23}$	0	+	+	0	1	-1			
$\vec{V}_{12}$	+	+	0	1	0	-1	$\vec{V}_{23}$	0	+	-	0	1	-1			
$\vec{V}_{61}$	+	-	0	1	-1	0	$\vec{V}_{56}$	0	-	-	0	-1	1			
$\vec{V}_{34}$	-	+	0	-1	1	0	$\vec{V}_{56}$	0	-	+	0	-1	1			
$\vec{V}_{45}$	-	-	0	-1	0	1										
$\vec{V}_{23}$	0	+	0	0	1	-1	$\vec{V}_{2}/\vec{V}_{02p}$	+	+	+	1	1	-1	1	1	0
$\vec{V}_{56}$	0	-	0	0	-1	1	$\vec{V}_{6}/\vec{V}_{06p}$	+	-	+	1	-1	1	1	0	1
$\vec{V}_{01p}$	+	0	+	1	0	0	$\vec{V}_{3}/\vec{V}_{03n}$	-	+	-	-1	1	-1	-1	0	-1
$\vec{V}_{04n}$	-	0	-	-1	0	0	$\vec{V}_{5}/\vec{V}_{05n}$	-	-	-	-1	-1	1	-1	-1	0
$\vec{V}_{00p}$	0	0	+	1	1	1	$\vec{V}_{1}/\vec{V}_{01n}$	+	0	-	1	-1	-1	0	-1	-1
$\vec{V}_{00n}$	0	0	-	-1	-1	-1	$\vec{V}_{4}/\vec{V}_{04p}$	-	0	+	-1	1	1	0	1	1

4. SIMULATION RESULTS

In Fig. 5, there is a switch in the signal-path that the injected current from the three-level inverter into the network can be detected. In this case, that switch could determine the switching frequency of the switching device, such as: IGBT. The switching functions of S_a, S_b and S_c can be adjusted from one state to another, and the speed of state change depends on the sampling rate of this switch. However, the systems performance will be affected by the sampling rate, the inductance and resistance values of the coupling transformer, voltage difference between the terminals of the inverter, the coupling point with the network, and also the values of the hysteresis limit. The simulation is conducted in MATLAB/SIMULINK to verify the functionality and

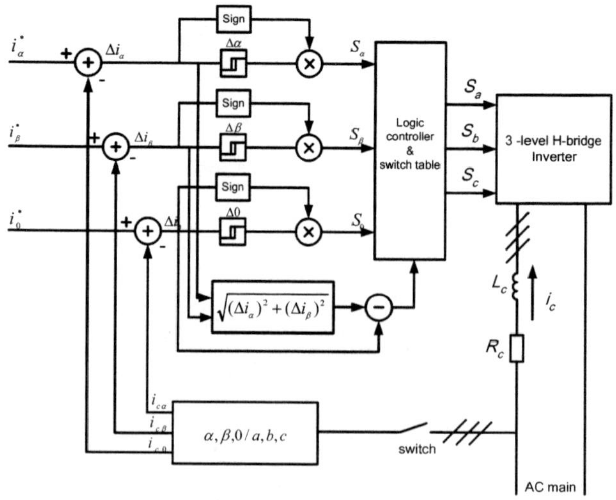

Fig. 5. Control strategy.

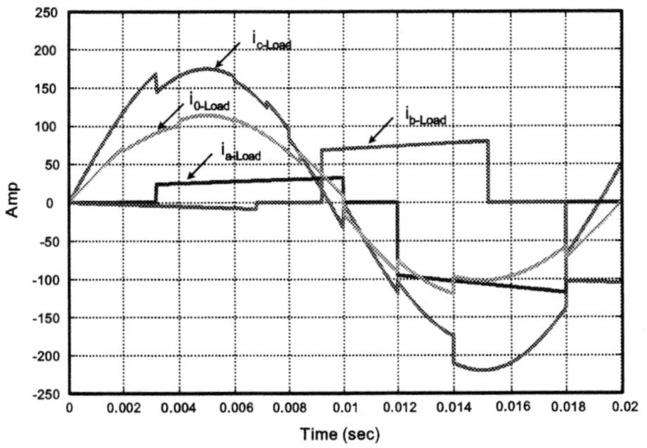

(a)

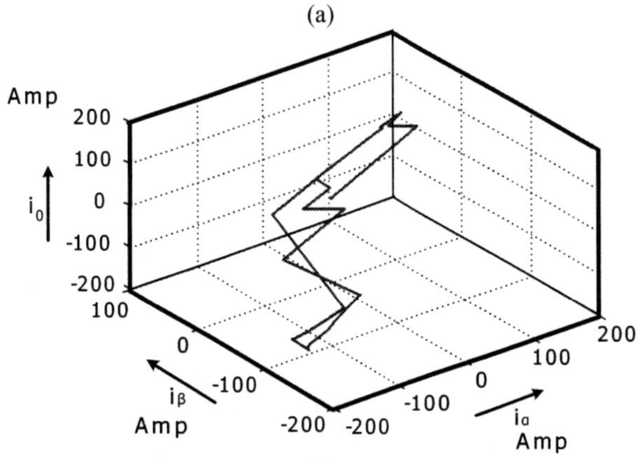

(b)

Fig. 6. Unbalance case of three-phase and neutral line load currents. (a) Time domain. (b) *α-β-0* domain.

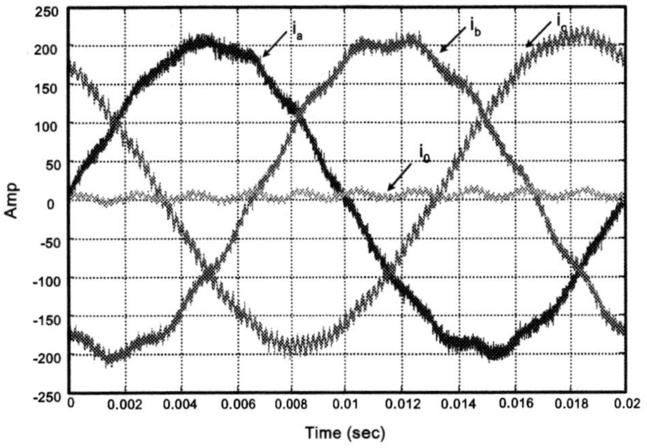

(a)

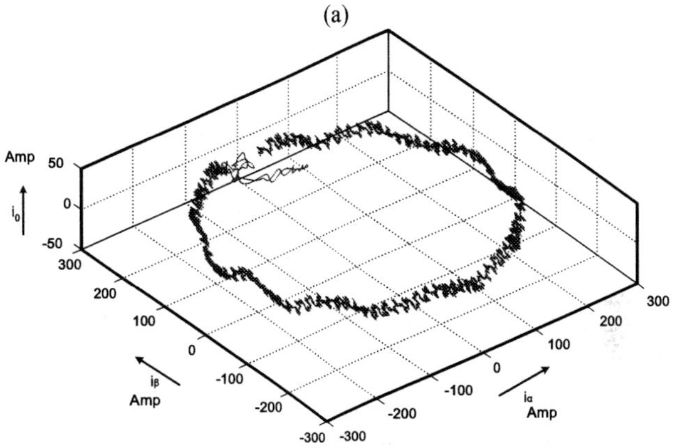

(b)

Fig. 7. Compensated three-phase and neutral line source currents. (a) Time domain. (b) *α-β-0* domain.

performance of the tri-level H-bridge inverters, which acts as a power quality compensator for three-phase four-wired systems. Suppose there are three-phase unbalance load currents with 50 Hz as shown in Fig. 6(a), then the corresponding expression in *α-β-0* domain will be no longer a pure horizontal circle as shown in Fig. 6(b). However, when a tri-level H-bridge inverter is applied to compensate the source currents by using 3-D PWM with sign cubical hysteresis control operating in 20 kHz. Both the three-phase and neutral lines are highly improved in terms of total harmonic distortion (THD) and ripple, as shown in Fig. 7(a) and 7(b) in time and *α-β-0* domain respectively. Thus, the result verifies both the proposed model and the controlling technique are capable to compensate the heavy distorted three-phase and neutral line source currents simultaneously for three-phase four-wired systems.

5. CONCLUSIONS

A novel power quality compensator, constructed by H-bridge tri-level inverter with 3-D PWM control, has been proposed for three-phase four-wired systems. The mathematical model of such compensator and an enhanced switching table that balances the switching frequency of the switches were presented. With the sign cubical hysteresis control strategy, the simulation results verified that the current imbalance between the three-phase power lines as well as the neutral line due to non-linear loadings was effectively reduced. Thus, for three-phase four-wired systems, this highly expandable H-bridge tri-inverter with 3-D PWM control can be an attractive alternative of the widely utilized diode-clamped one, which suffers from the nature problem of DC-source unbalancing.

REFERENCES

[1] M.P. Kazmierkowski and W. Sulkowski, "Novel space vector based current controllers for PWM-inverter", in *Proc. of PESC"89*, p.657-664, 1989.

[2] J. Holtz, "Pulsewidth modulation-a survey," *IEEE Transaction Ind. Electronics*, vol. 99, p.410-420, Dec. 1992.

[3] M.P. Kazmierkowski and M.A. Dzieniakowski, "Review of current regulation techniques for three-phase PWM inverters," in *Proc. Conference Rec. IEEE-IECON'94*, pp. 567-575, 1994.

[4] A. Trzynadlowski, "An overview of modern PWM techniques for three-phase, voltage-controlled, voltage-source inverters," in *Proc. Conference Rec. IEEE-ISIE'96*, pp. 25-39, 1996.

[5] Man-Chung Wong, *et al.*, "Three-dimensional pulse-width modulation technique in three-level power inverters for three-phase gour-wired systems," *IEEE Transactions on Power Electronics*, vol. 16, no.3, pp. 719-725, May 2001.

[6] Man-Chung Wong, Jing Tang, Ning-Yi Dai, Ying-Duo Han, "Fundamental study of 3 dimensional pulse width modulation," in *Proc. of International conference on Power Systems technology (PowerCon 2002),* vol. 1, pp. 560 –564, Oct. 2002.

[7] Changjiang Zhan, Arulampalam, A. Jenkins, N., "Four-Wire dynamic voltage restorer based on a three-dimensional voltage space vector PWM algorithm," *IEEE Transactions on Power Electronics*, vol. 18, no.4, pp. 1093-1102, July 2003.

[8] D.W. Kang, *et al.*, "A carrierwave-based SVPWM using phase-voltage redundancies for multilevel H-bridge inverter," in *Proc. of IEEE Annual Conference on Industrial Electronics IECON '99*, vol. 1, pp. 324 –329, 29 Nov.-3 Dec. 1999.

[9] Man-Chung Wong, *et al.*, "Cylindrical coordinate control of three-dimensional PWM technique in three-phase four-wired trilevel inverter," *IEEE Transactions on Power Electronics*, vol. 18, no.3, pp. 208-220, Jan. 2003.

AUTOMATED STATE-VARIABLE FORMULATION FOR POWER ELECTRONIC CIRCUITS AND SYSTEMS

Juri Jatskevich, *Member, IEEE* and Tarek Aboul-Seoud, *Student Member, IEEE.*

Electrical and Computer Engineering, University of British Columbia, Vancouver, Canada

ABSTRACT

State variable approach is often used for modeling power-electronic circuits at the system level with controls. In the approach considered herein, the minimal state-space representation of the overall system is generated from the circuit branch data and updated for each new the topology of the system. The network partitioning into optimal sets of tree and link branches is achieved using the spanning tree algorithms. For systems with variable parameters re-computing the entire state equation to reflect the parameter changes may be prohibitively expensive. In this paper, a computationally effective algorithm for updating the required terms based on topological information is presented.

Keywords *- power electronic system, switched circuits, switched systems, state-space methods, simulation.*

1. INTRODUCTION

Power-electronic circuits and components are often modeled using circuit simulators from the Spice family, wherein the detailed models of various switching devices can be used to predict the switching transients with a high level of accuracy [1]. Extensive component libraries of solid-state transistors, diodes, etc., have been developed for these languages. However, for the so called "system-level studies", wherein power electronic subsystems are modeled together with electromechanical subsystems and respective controllers, the details of individual switching transients may be neglected and the overall system dynamics or behavior can be represented using idealized "on/off" switch models [2], [3]. For the purposes of computing time-domain transient responses and system-level analysis, modeling of power-electronic systems and electromechanical energy conversion devices based on their state equation description is perhaps one of the most analytically meaningful and efficient techniques. Additionally, once the state-space model is established, well-established linear and nonlinear analysis techniques can be applied to the state-space model to investigate and characterize the system dynamics together with the corresponding controllers. Perhaps those are some of the reasons why simulation programs such as Matlab/Simulink are getting significant attention in modeling of power electronic systems, electrical machines, and drives [2]-[6].

When implementing switching devices, some simulation programs rely on fictitious resistances with on/off (very small/large) values and may require a non-zero inductance and/or capacitance augmented with each switch [5], [6]. The disadvantage of the later approach is that the resulted state equations have larger dimensions and may contain artificial high frequency modes that are undesirable from numerical considerations.

An alternative method for assembling governing differential and algebraic equations for power-electronics system based on their branch-node representation has been set forth [7] and since then significantly enhanced [8], [9]. This method produces the state and output equations for each encountered topology of the circuit considered. As the simulation proceeds, these equations are solved using an appropriate integration routine and the solutions for currents and voltages are concatenated across switching boundaries. However, when considering power electronic systems with variable parameters the corresponding state equations must be updated to reflect the parameter change. A brute-force re-assembling of state equations is computationally very expensive even for small circuits. In this paper, an efficient method for updating only the required entries of the matrices is proposed. The proposed method is based on utilizing the topological information about the respective circuit branch and updating only the impacted terms.

2. STATE MODEL FORMULATION

A large class of switched networks can be represented using a combination of the elementary branch models depicted in Fig. 1. The first two branches are topological duals. Simple resistors, inductors, or sources can also be represented by setting the appropriate parameters of the elementary branch types to zero. A network constructed from a finite number of elementary branches can be defined by a corresponding graph G, a set of branch parameters P, and a topological state vector $\mathbf{s} \in \Im^{k}$, where $\Im = \{0,1\}$ implying that each switch can be open or closed, and k is the number of switch branches [8].

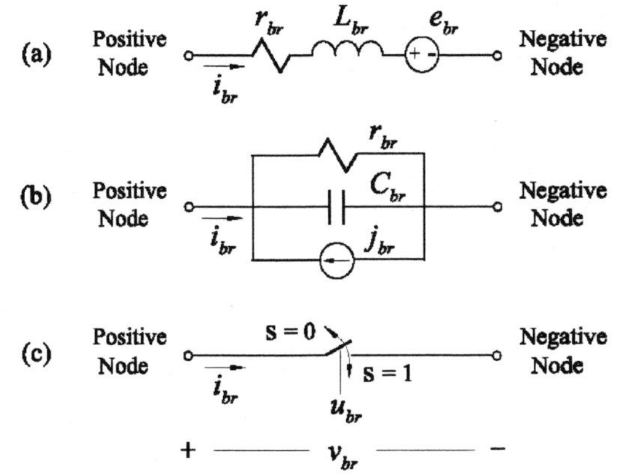

Fig. 1: Branches: (a) inductive, (b) capacitive, (c) switch.

Since each switch branch has only two states (open or closed), there exists 2^k potential instances (values) that $\mathbf{s}$ can take on. In general, however, not all of the topological states may be reachable. The state of a switch is controlled by the variable u_{br} that may be used to represent various power electronic devices. Symbolically, a network object is defined as the triplet

$$\mathbf{N} = \langle G, P, \mathbf{s} \rangle \quad (1)$$

The network graph G conveys the manner in which the elementary branches are connected in a circuit. The graph may be expressed as $G = \langle V, E \rangle$, where V is the set of vertices (nodes) and E is a set of edges (branches). The network graph may be represented using a node incidence matrix whose columns correspond to branches and rows to nodes [10]. The parameter set P can be conveniently expressed as a set of parameter matrices. For example, for a network with b branches, set is

$$P = (\mathbf{R}_{br}, \mathbf{L}_{br}, \mathbf{C}_{br}, \mathbf{E}_{br}, \mathbf{J}_{br}) \quad (2)$$

where, for example, the diagonal elements of the network inductance matrix $\mathbf{L}_{br} \in \Re^{b\times b}$ represent the branch self inductance, and the off-diagonal elements, if nonzero, represent mutual coupling between the branches. The nonzero elements of $\mathbf{E}_{br} \in \Re^{b}$ correspond to the independent sources e_{br} in Fig. 1 (a) for those inductive branches whose independent source is nonzero. The vector $\mathbf{J}_{br}$ and matrices $\mathbf{R}_{br}$ and $\mathbf{C}_{br}$ are defined similarly. In general, any of the parameters can vary with time; however, if the inductances or capacitances vary with time, the rates of change of the corresponding inductances or capacitances need to be defined in order to formulate a state model. Typically, the aforementioned parameters matrices/vectors are sparse and substantial computational savings can be achieved by employing sparse storage.

When formulating a state model for a given topological state $\mathbf{s}^i$, a composite electrical network can be viewed as a collection of interconnected subnetworks. This would result in equations of a smaller size. Partitioning also simplifies the development of a state model by considering each subnetwork separately and then accounting for their respective coupling. For example, one may assume partitioning into inductive, capacitive, and resistive subnetworks. Symbolically, this may be expressed as

$$\mathbf{N} = \mathbf{N}_L^i \cup \mathbf{N}_C^i \cup \mathbf{N}_R^i \quad (3)$$

where the superscript i is used to denote topological state $\mathbf{s} = \mathbf{s}^i$. The network partitioning may be performed very effectively by utilizing the so-called Spanning Tree algorithms and searching for a sub-tree with certain optimal properties [10]. In particular, for identifying the inductive subnetwork $\mathbf{N}_L$, it is convenient to define the following weight function

$$\omega_L = \begin{cases} 0, \mathbf{L}_{br}(j,j) = 0 \\ 1, \mathbf{L}_{br}(j,j) \neq 0 \end{cases}, \text{ for } j = 1,2,\ldots,m \quad (4)$$

Using weights ω_L and applying the Minimum Spanning Tree (MinST) algorithm the largest possible set of inductive link branches E_{link}^L is readily identified. This set, together with some tree branches in G, form the $\mathbf{N}_L$. Similarly, defining the weight function

$$\omega_C = \begin{cases} 0, \mathbf{C}_{br}(j,j) = 0 \\ 1, \mathbf{C}_{br}(j,j) \neq 0 \end{cases}, \text{ for } j = 1,2,\ldots,m \quad (5)$$

and applying the Maximum Spanning Tree (MaxST) algorithm the largest possible set of capacitive tree branches E_{tree}^C is identified. Set E_{tree}^C, together with some links in G, form the subnetwork $\mathbf{N}_C$. Although the above sets may not be unique, any such sets with the optimal property are equivalent. Such decomposition of the global system into three elementary networks enables the identification of minimal sets of state variables; inductive link currents of the branches in E_{link}^L and capacitive tree voltages of the branches in E_{tree}^C for the inductive and capacitive subsystems, respectively. The resistive subnetwork results in an algebraic equation. The state equations for the inductive and capacitive networks have the following general form

$$\mathbf{M}^i\left(\mathbf{x}^i,t\right)\frac{d\mathbf{x}^i}{dt}=\mathbf{F}^i\left(\mathbf{x}^i,t\right)+\mathbf{g}^i(\mathbf{u},t) \quad (6)$$

where $\mathbf{F}^i\left(\mathbf{x}^i,t\right)$ represents all terms that define the state self-dynamics. The forcing term $\mathbf{g}^i(\mathbf{u},t)$ accounts for the independent sources and coupling. Finally, $\mathbf{M}^i\left(\mathbf{x}^i,t\right)$ is a mass matrix. Since the resistive network does not affect the selection of the state variables and the capacitive network is a dual of the inductive network, only the inductive network is considered in further discussion. The results are readily extendable to the composite network (3). For $\mathbf{N}_L$, for example with time-varying parameters, the terms of the state equation may be expressed [8]

$$\mathbf{M}^i\left(\mathbf{x}^i,t\right)=\mathbf{B}_b^i\mathbf{L}_{br}\left(\mathbf{B}_b^i\right)^T \quad (7)$$

$$\mathbf{F}^i\left(\mathbf{x}^i,t\right)=-\mathbf{B}_b^i\left(\mathbf{R}_{br}+\frac{d\mathbf{L}_{br}}{dt}\right)\left(\mathbf{B}_b^i\right)^T\mathbf{i}_{link}^L \quad (8)$$

$$\mathbf{g}^i(\mathbf{u},t)=-\mathbf{B}_b^i\mathbf{E}_{br}-\mathbf{B}_C^i\mathbf{v}_{tree}^C-\mathbf{B}_R^i\mathbf{v}_{tree}^R \quad (9)$$

Here, $\mathbf{B}_b$ is a basic loop matrix that can be obtained by partitioning the graph $G=\langle V,E\rangle$ into a spanning tree and corresponding link branches by applying MinST and MaxST [10]. Similarly, matrices $\mathbf{B}_C^i$ and $\mathbf{B}_R^i$ account for the coupling with capacitive and resistive subnetworks, respectively. For a given topological state $\mathbf{s}^i$, these matrices are constant and sparse. However, the matrices $\mathbf{M}$ and $\mathbf{F}$ and are often not sparse. Finally, $\mathbf{i}_{link}^L$ and $\mathbf{v}_{tree}^C$ are the state vectors of independent inductive link currents and capacitive tree voltages, respectively.

When the circuit parameters change the corresponding terms in (6) must be updated accordingly. It can be noted that both $\mathbf{F}$ and $\mathbf{M}$ are defined in terms of a product of three matrices that will be expensive to evaluate according to the standard multiplication rules for the matrices with full storage. In particular, $\mathbf{M}$ is $l\times l$ matrix, where l is the number of links (loops), and its entries may be computed as

$$\mathbf{M}(i,j)=\sum_{m=1}^{b}\sum_{n=1}^{b}\mathbf{B}_b(i,m)\mathbf{B}_b(j,n)\mathbf{L}_{br}(m,n) \quad (10)$$

Here b denotes the number of branches. From (10) it is observed that the computational complexity grows as b^2l^2. If multiplication is carried out one matrix at a time the resulted complexity would be $bl(b+l)$, which is still very large. This fast grows of complexity with the circuit size may result in prohibitively long computing times for practical power electronic circuits. Moreover, because multiplication of sparse matrices may result in a non-sparse matrix, the sequential multiplication of these three matrices should be avoided. In order to further analyze the complexity, only (7) is considered; whereas the same technique applies for computing (8). An approach for reducing the computing time for the triple-matrix product presented in this paper is based on utilizing the topological information. In particular, the terms under summation in (10) are non-zero only when the internal indices m and n correspond to the inductive branches that are members of the i-th and j-th basic loops. Therefore, it is possible to define sets of respective *branch indices* and use only those when evaluating (10). Thus, defining loop set LS_i to be a set of all inductive branches in the i-th basic loop, the (i,j)-th element of $\mathbf{M}$ can be computed as

$$\mathbf{M}(i,j)=\sum_{(m\in LS_i)}\sum_{(n\in LS_j)}\mathbf{B}_b(i,m)\mathbf{B}_b(j,n)\mathbf{L}_{br}(m,n) \quad (11)$$

The computational complexity of (11) is significantly reduced compared to that in (10). However, there is still some inefficiency in (11) due to that fact that not always m-th and n-th inductive branches have mutual coupling. Additionally, in many practical applications only a specified number of parameters changes and the rest of the circuit stay the same. Therefore, it is desirable to design a scheme capable of updating only those elements of $\mathbf{M}$ that are affected by the changes of some specified parameters in $\mathbf{L}_{br}$. In order to accomplish this it is possible to define sets of *loop indices* for each inductive branch. In particular, defining loop participation set LPS_m to be a set of all loops (indices) in which the m-th inductive branch takes part, the contribution of each (m,n)-th parameter to the matrix $\mathbf{M}$ can be expressed as

$$\Delta\mathbf{M}(i,j)=\sum_{(i\in LPS_m)}\sum_{(j\in LPS_n)}\mathbf{B}_b(i,m)\mathbf{B}_b(j,n)\mathbf{L}_{br}(m,n) \quad (12)$$

It can be noted here that the indices (m,n) are the inputs, and $\Delta\mathbf{M}$ is the resulted contribution. Therefore, in order to perform a complete update, one should "undo" the contribution due to the old value and then re-compute the contribution due to the new value. Alternatively, it is advantageous to implement these two steps in the same loop, which is equivalent to updating the matrix $\mathbf{M}$ due to the change in the absolute value of the variable inductance.

In order to illustrate the difference among the three methods of evaluating the triple-product it is instructive to consider a simple circuit shown in Fig. 2. It is assumed that all branches are inductive of type Fig. 1(a) with no mutual coupling. The computational complexity of evaluating the matrix $\mathbf{M}$ is summarized in Table 1.

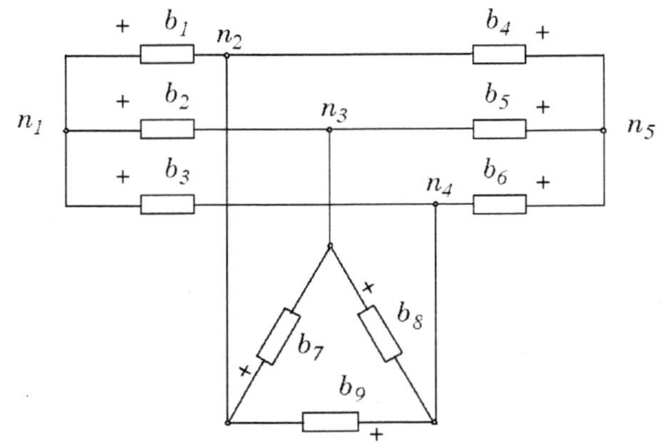

Fig. 2: Example circuit.

The complexity is expressed as the number of assignments made to $\mathbf{M}$ (which is the same as the total number of executing the lowest level *for* loop, if the multiplication is implemented for example in C). From these results it can be concluded that significant savings can be achieved by implementing the update according to (12). Similar technique based on topological sets can be used for implementing (9).

Table 1: Costs of computing matrix $\mathbf{M}$

Triple-matrix product, eq. (10)	Loop-sets, eq. (11)	Loop-participation -sets, eq. (12)
2025	289	43

If the circuit parameters are allowed to depend on time and/or on state variables, the matrix $\mathbf{M}$ will be positive definite and invertible. Thereafter, if desired, the state equation (6) may be converted into explicit form

$$\frac{d\mathbf{x}^i}{dt} = \mathbf{A}^i(\mathbf{x}^i, t) + \mathbf{B}^i(\mathbf{x}^i, t)\mathbf{u} \qquad (13)$$

The formulation (13) may be useful for circuits with constant parameters, or when the linearization is necessary for obtaining a local transfer-function, for example. However, for the numerical simulation of circuits with variable parameters it is advantageous to solve (6) and avoid the inversion of $\mathbf{M}$. There are special solvers designed to handle the variable mass matrix in an efficient manner [11]. If this is the case, the algorithm defined by (12) should be used to perform the update.

3. SUMMARY

In this paper, a method of assembling state equations for power electronic circuits based on branch information has been presented. The implicit state equation contain matrices that are often non sparse. For linear and time invariant circuits the resulted equations can be readily put into a standard state-space form. However, when the parameters (such as inductances and/or resistances) are variable it is more efficient to consider an implicit state equation and update the respective terms. Different update techniques have been discussed. The proposed method requires minimal computational costs and consists of utilizing the topological information in a form of branch/loop sets and performing the necessary updates only due to specific parameters that change. Based on the algorithm presented an alternative Matlab/Simulink toolbox for simulation and analysis of power electronic systems has been developed.

4. REFERENCES

[1] R.S. Ramshaw, D. Schuurman, "PSpice Simulation of Power Electronics Circuits," Chapman & Hall, 1996.

[2] D. Logue, P.T. Krein, "Simulation of electric machinery and power electronics interfacing using MATLAB SIMULINK, 7th Workshop on Computers in Power Electronics (COMPEL) 2000, p. 34 -39.

[3] H. Jin, "Behavior-mode simulation of power electronic circuits," IEEE Trans. on Power Electronics, Vol. 12, No. 3 , p. 443-452, May 1997.

[4] A. Monti, E. Santi, R.A. Dougal, M. Riva, "Rapid prototyping of digital controls for power electronics," IEEE Transactions on Power Electronics, Vol. 18, No. 3, p. 915-923, May 2003.

[5] L.A. Dessaint, K. Al-Haddad, H. Le-Huy, G.Sybille, P. Brunelle, "A power system simulation tool based on Simulink," IEEE Transactions on Industrial Electronics, Vol.46. No. 6, p. 1252-1254, Dec. 1999.

[6] G. Sybille, P. Brunelle, Le-Huy Hoang L.A. Dessaint, K. Al Haddad,"Theory and applications of power system blockset, a MATLAB/Simulink-based simulation tool for power systems," IEEE PES Winter Meeting, Vol. 1, p. 774-779, Jan. 2000.

[7] O. Wasynczuk, S. D. Sudhoff, "Automated State Model Generation Algorithm for Power Circuits and Systems", IEEE Transactions on Power Systems, Vol. 11, No. 4, November 1996, pp. 1951-1956.

[8] J. Jatskevich, S. D. Pekarek, T. Skvarenina, S. D. Sudhoff, E. A. Walters, O. Wasynczuk, Dynamic Simulation of High-Power Machinery Systems, SBIR Phase II Final Report to NSWC, May 1, 1999.

[9] O. Wasynczuk, J. Jatskevich, "Circuit simulation," United States Patent Application 10/043,981. Dec. 2002.

[10] K. Thulasiraman, M. N. S. Swamy, "Graphs: Theory and Algorithms", Wiley-Interscience, 1992.

[11] L. F. Shampine, M. W. Reichelt, "The Matlab ODE Suite", The MathWorks, Inc., 1995.

GLOBAL BEHAVIOR ANALYSIS OF A DC-DC BOOST POWER CONVERTER OPERATING WITH CONSTANT POWER LOAD

C. Rivetta, Geoffrey A. Williamson

Illinois Institute of Technology
Department of Electrical and Computer Engineering
Chicago Il 60616, U.S.A.

ABSTRACT

This paper addresses the closed-loop behavior of a boost power converter feeding a constant power load. This configuration is present in DC-DC converters feeding power converters tightly regulated in multi-converter power electronics systems. The dynamic properties of a boost converter operating with a constant power load differ from when the converter is loaded by resistors. Based on a large-signal model of the converter, the local behavior around the operation point and its basin of attraction are defined. This region of attraction, in general, excludes the low range of the output voltage affecting the transient stability and forcing proper coordination between the converter and loads during start-up.

1. INTRODUCTION

Distributed power systems, in general, consist of the cascade of DC-DC converters [1] [2]. The first stage provides a regulated DC voltage for an intermediate bus and the second stage converts this bus voltage to a tightly regulated low voltage for each load. In this topology, the first stage has power converters as the effective load. These DC-DC converters operating with tight closed loop regulation present a behavior, at the input terminals, similar to a constant power load (CPL) in a certain range of frequencies and input voltages. The dynamic behavior of converters loaded by constant power loads differs from the behavior of converters loaded by resistors or current generators. Under constant power loads, the converter is not globally stable for either switch position open or closed [1].

The large-signal analysis of power converters loaded by resistors or constant current generators has been addressed previously [3] [4]. DC-DC power converters operating in closed loop and loaded by CPL have been analyzed assuming the converter operates around the equilibrium point [5] [6] [7]. This paper presents the large-signal analysis of a boost power converter loaded with a constant power load and operating in closed loop using a state feedback controller. Based on an averaged model of the converter, both the closed-loop behavior of the converter loaded by a CPL and the determination of the basin of attraction of the operation point are described. Insights into converter behavior from this analysis are important for the design of converters loaded by CPL.

2. STATE FEEDBACK CONTROL OF BOOST CONVERTERS OPERATING WITH CONSTANT POWER LOAD

Fig. 1 shows the electrical circuit for a boost power converter loaded by a CPL. Assuming the converter operates

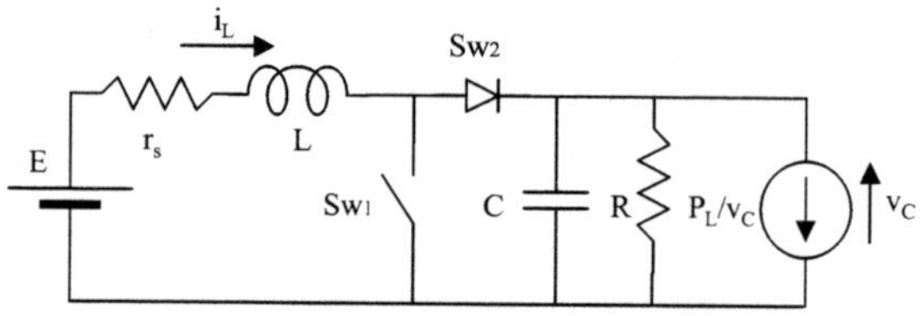

Fig. 1. Electric schematic of boost converter.

in continuous conduction mode (CCM), the switches can take only complementary positions, either S_{w1}:open and S_{w2}:closed or S_{w1}:closed and S_{w2}:open. For each status, the converter changes accordingly its circuit topology. The output v_C is regulated by switching between the topologies, with a fixed period T_s, and controlling the duty cycle. The behavior of the converter can be analyzed using averaging techniques if the switching frequency is high enough respect to the dynamics of the system. The averaged model is

$$C\frac{dx_C(t)}{dt} = (1-d(t))x_L(t) - \frac{x_C(t)}{R} - \frac{P_L}{x_C(t)}$$
$$L\frac{dx_L(t)}{dt} = E - r_s x_L(t) - (1-d(t))x_C(t); \quad (1)$$

where $x_C(t)$ and $x_L(t)$ are the averaged capacitor voltage and the averaged inductor current, respectively. C and L are the capacitance and inductance magnitude, E is the primary voltage, P_L is the constant power load, with $P_L \geq 0$, R an equivalent resistive load. The variable $d(t)$ is a continuous

0-7803-8251-X/04/$17.00 ©2004 IEEE

function with image in the real interval $[0, 1]$ that represents the averaged status of the switches.

We assume a simple state variable feedback law

$$d_c(x(t)) = u_{eq} + K_i(I_{Lss} - x_L(t)) + K_v(V_{Css} - x_C(t)); \quad (2)$$

where V_{Css} and I_{Lss} are the reference voltage and the nominal current, respectively, K_i and K_v are the controller gains and u_{eq} the nominal duty cycle. In this feedback law, it is assumed that $x_L(t) \simeq i_L(t)$ and $x_C(t) \simeq v_C(t)$ because the ripple in the state variables is negligible [8]. Since $d(t)$ is limited to $[0, 1]$, the complete feedback algorithm is defined by

$$d(x(t)) = \begin{cases} d_c(x(t)), & d_c \in (0,1) \\ 0, & d_c \leq 0 \\ 1, & d_c \geq 1 \end{cases} \quad (3)$$

We divide the first quadrant $X = \{x = (x_C, x_L)^T \in R^2; x_C > 0, x_L > 0\}$ into three regions or cells of operation as

$$\begin{aligned} X_0 &= \{x(t) \in X;\ d(x(t)) = 0\}; \\ X_{01} &= \{x(t) \in X;\ d(x(t)) \in (0,1)\}; \\ X_1 &= \{x(t) \in X;\ d(x(t)) = 1\}. \end{aligned}$$

These cells are limited by the boundary curves

$$\mathcal{L}_0 : x_L = \frac{1}{K_i}\Big(u_{eq} + K_i I_{Lss} + K_v(V_{Css} - x_C)\Big);$$
$$\mathcal{L}_1 : x_L = \frac{1}{K_i}\Big(u_{eq} - 1 + K_i I_{Lss} + K_v(V_{Css} - x_C)\Big).$$

The boundary curves are determined by the controller parameters K_i and K_v. The dynamics in X_{01} is defined by eqn. (1) after substituting the control law (2) and the behavior in X_1 and X_0 is defined by (1) after putting back $d(t) = 1$ and $d(t) = 0$, respectively. Before studying the complete behavior of the converter when $d(t) \in [0, 1]$, let us analyze the behavior of the converter assuming there is no limitation in the control variable $d(t)$. From this study, it is possible to characterize the performance of the converter in the region X_{01}. Then, we proceed to analyze the converter for $d(t) = 0$ and $d(t) = 1$, to finally assemble the individual behaviors in cells X_0, X_{01} and X_1 to obtain the global performance of the converter in the first quadrant and the region of converge of the operation point.

2.1. Behavior for $d(t)$ unconstrained

For any operation condition (open loop/closed loop) of the power converter loaded by a CPL, the equilibrium points of eqn.(1) are defined by the solution of the set of equations

$$\begin{aligned} (1 - d_o)x_L - \frac{x_C}{R} - \frac{P_L}{x_C} &= 0 \\ E - r_s x_L - (1 - d_o)x_C &= 0; \end{aligned} \quad (4)$$

For $x_C = 0$ there is no solution and for $x_C \neq 0$ the solutions to these equations lie on the ellipse, parameterized by the steady state duty cycle d_o, that define the energy-balance of the converter. The solution curve is

$$Ex_L - r_s x_L^2 - x_C^2/R - P_L = 0 \qquad \text{for all } x_C \neq 0 \quad (5)$$

Using as example a $100W$ boost converter with parameters $E = 12V$, $L = 100\mu Hy$, $C = 100\mu F$ and nominal values $P_L = 100W$, $R = 1K\Omega$ and $V_{Css} = 24V$, eqn.(4) has two equilibrium points in X for the particular steady state duty cycle $d_o = d(t) = 0$ and none for $d(t) = 1$. For any steady state duty cycle $d_o \in [0, d_{max})$, with $d_{max} < 1$, there exist two equilibrium points located on curve (5). The $100W$ boost converter will be used throughout the paper to carry out the analysis.

The equilibrium points for a particular feedback law as (2) can be calculated by replacing eqn.(2) into (1)

$$\begin{aligned} \Big(1 - u_{eq} - K_i(I_{Lss} - x_L) - K_v(V_{Css} - x_C)\Big)x_L - \tfrac{x_C}{R} - \tfrac{P_L}{x_C} &= 0 \\ E - r_s x_L - \Big(1 - u_{eq} - K_i(I_{Lss} - x_L) - K_v(V_{Css} - x_C)\Big)x_C &= 0; \end{aligned}$$

This set of equations have two equilibrium points on curve (5), one of them corresponds to the operation point, $x_1 = (x_{C1}, x_{L1}) = (V_{Css}, I_{Lss})$ and the other $x_2 = (x_{C2}, x_{L2})$, such that $x_{C2} \simeq 0$ and $x_{L2} \simeq E/r_s$. The solution x_1 defines the steady state duty cycle $d_o = u_{eq}$, while the solution x_2 gives a duty cycle d_o negative. The stability of these equilibria is defined by both the controller and converter parameters. The local performance of the converter operating in closed loop around the operation point is defined by the eigenvalues of x_1. These eigenvalues can be assigned linearizing eqn.(1) around x_1 and setting the controller gains K_i and K_v, such the converter operating in closed loop meets, locally, some performance or robustness criteria. The equilibrium point x_2 is normally a saddle point for regular controller parameters that define stability and good performance around the operation point x_1.

Choosing controller parameters such that x_1 is a stable node and x_2 is a saddle point, the phase portrait in the first quadrant for our example is shown in fig. 2.a. In this portrait, the unstable manifold of x_2, $W^u(x_2)$, joins the stable manifold of x_1, $W_s(x_1)$. Following the analysis in [9], the stable manifold of x_2, $W^s(x_2)$, is part of the boundary of the region of attraction of the equilibrium point x_1. All trajectories starting to the right of $W^s(x_2)$ converge asymptotically to the equilibrium point x_1, while all trajectories to the left of $W^s(x_2)$ diverge toward the axis $x_C = 0$.

A detail of the phase portrait around the equilibrium point x_1 is depicted in fig. 2.b. In this figure, the boundary curves $\mathcal{L}_0$ and $\mathcal{L}_1$ are included in the phase portrait to show the region X_{01} where this behavior is valid when the converter is operating with $d(t) \in [0, 1]$. Changing the controller parameters K_i and K_v not only impact locally the

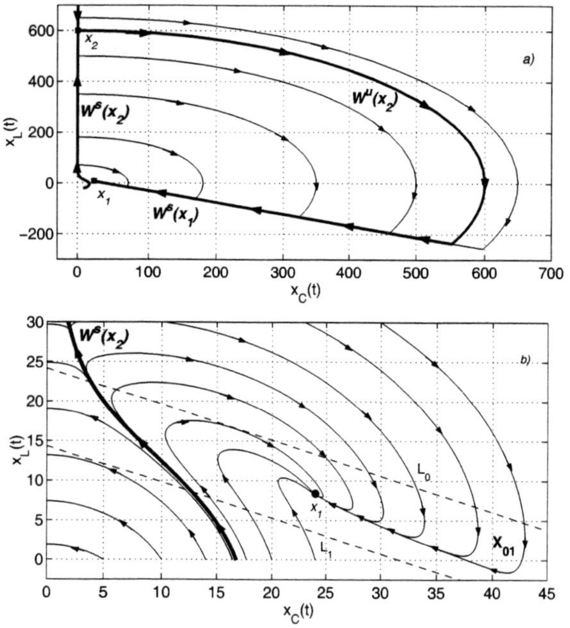

Fig. 2. a)- Phase portrait when the control variable $d(t)$ is not restricted to the set $[0, 1]$.- b)- Detail of fig. 2.a) around the equilibrium point x_1. The limit curves $\mathcal{L}_0$ and $\mathcal{L}_1$ define the cell X_{01} where these trajectories are valid.

dynamic behavior of the converter by altering the eigenvalues of x_1, but also change both the location of the boundary curves $\mathcal{L}_0$ and $\mathcal{L}_1$ and the position of the stable manifold $W^s(x_2)$ affecting the global behavior of the converter.

From Fig. 2.b, it is possible to observe that trajectories can reach the axis $x_L = 0$ and cross over to negative inductor currents. The converter does not operate in a region with negative currents and for low inductor currents it operates in discontinuous conduction mode (DCM). This mode has been neglected in previous analysis due to the assumption that the converter operates in CCM. When the converter operates in DCM, the system includes a new switch status S_{w1}:open, S_{w2}:open. Eqn. 1 no longer describes the behavior of the converter for DCM. Several reduced and full order models for DC-DC converters operating in DCM are analyzed in [10]. A full order model can be expressed by

$$\frac{dx_C(t)}{dt} = \frac{d(t)x_L(t)}{C} - \frac{x_C(t)}{CR} - \frac{P_L}{Cx_C(t)} \qquad (6)$$
$$\frac{dx_L(t)}{dt} = \frac{2x_L(t)}{d(t)T_s}\left(1 - \frac{x_C(t)}{E}\right) - \frac{r_s x_L(t)}{L} + \frac{d(t)x_C(t)}{L};$$

where T_s is the switching period. It is necessary to define a new cell X_{D01} inside the region X_{01}, limited by the boundaries $x_L = 0$ and $x_L = \frac{dT_sE}{2L}$ where the converter operates in DCM following eqn. (6).

2.2. Behavior for $d(t) = 0$ and $d(t) = 1$

When the controller saturates, the system operates in either cell X_0 or X_1.

2.2.1. *Behavior for* $d(t) = 0$

For this case, the converter is modeled by the dynamic equation (1) after substituting $d(t) = 0$. As it was described above, this differential equation has two equilibrium points. If we designate x_{0_1} to the equilibrium point $x_C \simeq E$, $x_L \simeq E/R + P_L/E$ and x_{0_2} to the equilibrium point $x_C \simeq 0$, $x_L \simeq E/rl$, the stability of x_{0_1} is defined by the relative magnitude of the constant power load P_L and the equivalent load resistance R. In general, the equilibrium point x_{0_1} can be an unstable focus or node or a stable focus or node, while the equilibrium point x_{0_2} is a saddle point. For our example, x_{0_1} is an unstable focus. Qualitatively, the phase portrait is similar to the one depicted in Fig. 2 for the case of the converter operating in closed loop with $d(t)$ unconstrained. There exists, in this case, the stable manifold of x_{0_2}, $W^s(x_{0_2})$ that divide the region X in two areas, where trajectories starting to the left of $W^s(x_{0_2})$ diverge toward $x_C = 0$ and trajectories starting to the right move to the right to encircle the equilibrium point x_{0_1}.

2.2.2. *Behavior for* $d(t) = 1$

When the switch S_{w1} is closed ($d(t) = 1$), the dynamic equation (1) has no equilibrium points in X. In this region, $dx_C(t)/dt$ is negative for any (x_C, x_L) and $dx_L(t)/dt$ is positive if $x_L \geq 0$ so that trajectories in this region move up and to the left.

2.3. Behavior for $d(t) \in [0, 1]$

The overall behavior of the boost converter is defined by combining in X the three regions previously described. As cell X_1 does not have equilibrium points in X and the equilibrium points of regions X_{01} and X_0 lie on the curve represented by (5), there exists in X two equilibrium points. One of them corresponds to the operation point x_1, which will be locally asymptotically stable for a properly designed system. This equilibrium point belongs to the cell X_{01}. The other equilibrium is the saddle point x_{0_2}, for $d(t) = 0$, that lies in region X_0 for which the current magnitude is impractically large.

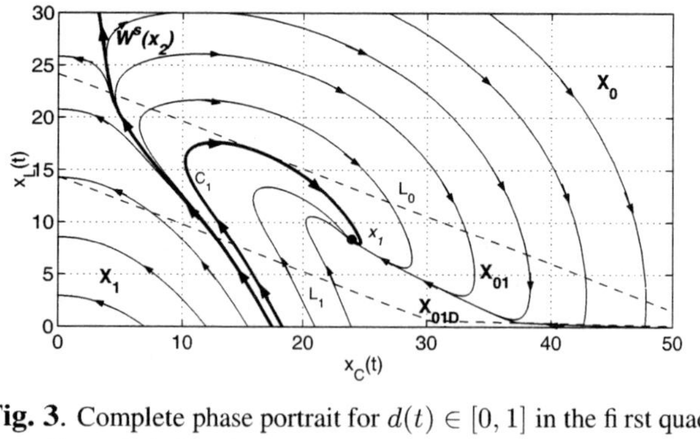

Fig. 3. Complete phase portrait for $d(t) \in [0, 1]$ in the first quadrant. Curve $\mathcal{B}$ define the boundary of the region of attraction of the equilibrium point x_1.

The phase portrait for the system operating with $d(t) \in [0, 1]$ is shown in fig. 3 by combining in region X the cells X_0, X_1 , X_{01} and X_{D01}. It is possible to observe that any trajectory starting to the right of the boundary curve $\mathcal{B}$ converges asymptotically to the operation point x_1. The curve $\mathcal{B}$ is mainly defined by the stable manifold $W^s(x_{0_2})$ and its extension, mapped back by the X_1 and X_{01} dynamics, respectively. The analysis and calculation of the region of attraction is addressed in next section.

3. REGION OF CONVERGENCE OF THE EQUILIBRIUM POINT

The region of convergence of the equilibrium point x_1 is defined by the union of sectors X_0, X_{01}, X_{D01} and X_1 containing points whose state trajectories tend to the operation point. This region of attraction depends on both the converter and controller parameters. Following an analysis similar to [11], the region of attraction can be qualitatively analyzed by dividing the region X_{01} in two sectors. The first sector $\mathcal{BA}_1$ contains points $x(t_o) \in X_{01}$ such that state trajectories $x(t)$ starting at $x(t_o)$ converge to x_1 with $x(t) \in X_{01}$ for all $t \geq 0$. This sector is limited in part by a trajectory (marked $\mathcal{C}_1$ in Fig. 3) that starts on the boundary curve $\mathcal{L}_1$ and is tangent to the boundary curve $\mathcal{L}_0$ before $x(t)$ reaches the equilibrium point. The tangent point divides the curve $\mathcal{L}_0$ in two parts where the vector field defined by (1) for $d(t) = 0$ is outward out of X_{01} above this tangent point and inward toward X_{01} for the segment below. On the bottom, this sector in X_{01} is limited by the axis x_C. Outside of $\mathcal{BA}_1$, trajectories to the left of curve $\mathcal{C}_1$ leave the region X_{01} toward the cell X_0. In this case, there exists two possibilities, trajectories can start in region X_0 to the right of the stable manifold $W^s(x_{0_2})$, in whose case converge back to region X_{01} to finally reach asymptotically the equilibrium point. Trajectories starting to the left of the separatrix $W^s(x_{0_2})$ will diverge toward $x_C = 0$. This analysis shows the boundary $\mathcal{B}$ of the basin of attraction of the operation point x_1 is mainly defined by the stable manifold $W^s(x_{0_2})$ and its extension, mapped back by the X_{01} and X_1 dynamics.

4. CONCLUSIONS

We have provided a large signal analysis of a boost converter operating with a state feedback control and a constant power load. We observe that boost converters loaded with CPL present two equilibrium points in the first quadrant. The boundary of the basin of attraction for the stable operating point is mainly defined by the stable manifold of the unstable equilibrium point. Based on an ideal CPL, this study shows that the region of convergence does not include the origin. DC-DC converters acting as loads do not behave as CPL at low voltages, behaving more as a combination of passive elements. This analysis shows that special consideration must be given to the start-up process due to any trajectory starting from $x = (0, 0)$ has to reach the basin of attraction of the operation point before the load behaves as a CPL. The size of the region of convergence also provides insight into the robustness of the converter's control. This type of analysis therefore provides much useful information for the analysis and design of control schemes used for boost converters operating with CPLs.

5. REFERENCES

[1] A. Emadi and M. Ehsani, "Multi-converter power electronic systems: definition and applications," in *IEEE Power Electronics Specialists Conference Record*, Vol. 2 pp. 1230-1236, 2001.

[2] B. Choi, B. Cho and S. Hong, "Dynamics and control of DC-to-DC converters driving other converters downstream," *IEEE Trans. Circ. Syst. I: Fund. Theory and Applic.*, vol. 46, pp. 1240-1247, Oct. 1999.

[3] R.M. Bass and P.T. Krein, "Large-signal design alternatives for switching power converter control," in *IEEE Power Electronics Specialists Conference Record*, pp. 882-887, 1991.

[4] M. Greuel, R. Muyshondt, and P.T. Krein, "Design approaches of boundary controllers," in *IEEE Power Electronics Specialists Conference Record*, vol. 1, pp. 672-678, 1997.

[5] A. Emadi and M. Ehsani, "Negative impedance stabilizing for PWM DC/DC converters using feedback linearization techniques," in *Proc. 35th Intersociety Conversion Engineering Conf.*, Las Vegas, NV, July 2000.

[6] J.G. Ciezki and R. Ashton, "The application of feedback linearization techniques to the stabilization of DC-to-DC converters with constant power loads," in *Proc. IEEE ISCAS 98*, vol. III, pp. 526-529, 1998.

[7] V. Grigore, J. Hatonen, J. Kyyra and T. Suntio, "Dynamics of a buck converter with a constant power load," in *IEEE Power Electronics Specialists Conference Record*, pp. 72-78, 1998.

[8] B. Lehman and R.M. Bass, "Extensions of averaging theory for power electronic systems," *IEEE Trans. on Power Electronics*, vol. 11, pp. 542-553, 1996.

[9] H.-D. Chiang, M.W. Hirsch, and F.F. Wu, "Stability regions of nonlinear autonomous dynamical systems," *IEEE Trans. on Automat. Contr.*, vol. 33, No. 1, pp. 16-27, 1988.

[10] J. Sun, D. Mitchell, M. Greuel, P.T. Krein and R.M. Bass, "Modeling of PWM converters in discontinuous conduction mode - a reexamination," in *IEEE Power Electronics Specialists Conference Record*, pp. 615-622, 1998.

[11] C. Rivetta and G.A. Williamson, "Large-signal analysis of a DC-DC buck power converter operating with constant power load," in *Proc. IEEE IECON 03*, pp. 732-737, Nov. 2003.

SINGLE-PHASE THREE-LEVEL CONVERTER FOR POWER FACTOR CORRECTION

Bor-Ren Lin, Senior Member, IEEE, and Tsung-Yu Yang

Department of Electrical Engineering, National Yunlin University of Science and Technology
Touliu City, Yunlin 640, Taiwan

ABSTRACT

A novel single-phase half-bridge switching mode rectifier is presented to draw a sinusoidal line current, to achieve power factor correction and to maintain the dc-link voltage constant. Four active switches with voltage stress of half dc bus voltage are used in the proposed rectifier to generate a unipolar PWM voltage waveform on the ac terminal voltage. There is no clamping diode in the proposed rectifier compared with the neutral point clamped converter to achieve three-level PWM operation. Two control loops are used in the proposed control scheme. In the outer control loop, a proportional integral voltage controller is used to regulate the dc-link voltage. A phase lock loop circuit is adopted to generate a sinusoidal waveform in phase with mains voltage to achieve power factor correction. In the inner control loop, a carrier-based current controller is used to track the line current command. The experimental results are presented to verify the effectiveness of the proposed control algorithm.

1. INTRODUCTION

The conventional power converters using the diode rectifiers or thyristor rectifiers followed by a bulk capacitor as a front-end stage draw the non-sinusoidal line current from the ac source. The undesired line current harmonics are generated in the distribution and transmission networks. Thus, international standards such as IEC 1000-3-2 or EN 61000-3-2 are defined to restrict the harmonic contents of the mains current. Passive filters are the easier way to reduce current harmonics. However, the bulk components, fixed compensated characteristics and series and parallel resonance with the system impedance are the main disadvantages of the passive filters. Power factor correctors [1-4] are the alternative approach to solve the harmonics elimination and reactive power compensation. Recently multilevel converters have been proposed for high power/voltage applications [5-10]. Single-phase and three-phase three-level rectifiers with hysteresis current technique were presented in [5-6]. These rectifiers have the features of lower blocking voltage of power semiconductors, less inductor size and power factor correction. The switching frequency is not fixed in this control scheme that will result in the difficulty of the design of the electromagnetic interference. Three-phase three-level rectifiers with fixed switching frequency have been proposed in [7-9] to achieve high power factor and generate a three-level PWM voltage waveform on the ac terminal of the rectifier.

A novel single-phase half-bridge switching mode rectifier with bi-directional power flow is proposed to achieve power factor correction and current harmonics reduction. There are four active switches in the proposed rectifier to perform three-level PWM operation. The circuit topology of the proposed rectifier has less power semiconductors compared with the conventional single-phase three-level half-bridge neutral-point-clamped (NPC) rectifier. A dc-link voltage controller is used in the outer control loop to maintain the dc bus voltage constant. The output signal of the dc-link voltage controller is multiplied by a unit sinusoidal wave in phase with mains voltage to obtain the reference line current. The current error between the reference line current and measured line current is sent to the current controller in the inner control loop to track the reference line current. The experimental results are provided to verify the validity of the proposed control strategy.

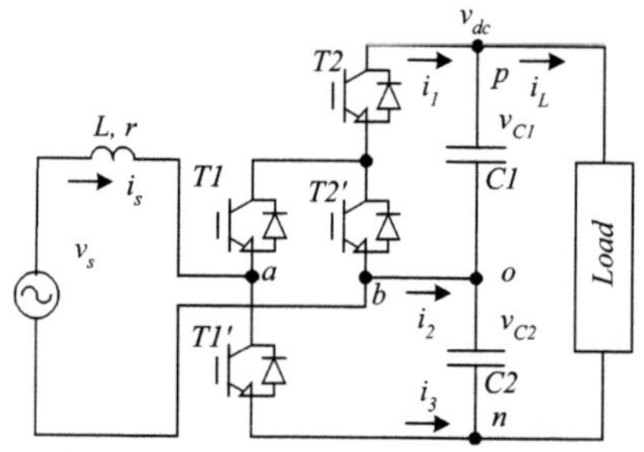

Fig. 1 Proposed single-phase half-bridge three-level PWM converter.

2. CIRCUIT DESCRIPTION

Conventional single-phase half-bridge neutral point diode clamped converter has four power switches and two clamping diodes in the circuit. All power semiconductors have voltage stress of half dc bus voltage. The single-phase half-bridge capacitor clamped converter has four power switches and one flying capacitor in the circuit. The flying capacitor voltage is equal to half dc-link voltage.

The voltage stress of four active switches is $v_{dc}/2$. Fig. 1 shows the proposed single-phase half-bridge converter based on only four active switches with voltage stress $v_{dc}/2$ to generate three voltage levels on the ac side of the converter. No clamping diode and flying capacitor is used in the proposed rectifier compared with neutral point clamped and capacitor clamped topologies. To avoid the damage of power switches in the proposed rectifier, power switches *T1* and *T1'*, *T2* and *T2'* are complement each other respectively. Therefore there are only two independent power switches *T1* and *T2* in the rectifier. Based on the on and off states of these two active power switches, four valid operating states are defined and shown in Table 1.

Table 1: Operating modes of the proposed rectifier.

states	power switches		v_{ab}
	T1	*T2*	
1	1	1	$v_{dc}/2$
2	1	0	0
3	0	0	$-v_{dc}/2$
4	0	1	$-v_{dc}/2$

In the first operating state power switches *T1* and *T2* are turned on to achieve ac terminal voltage $v_{ab}=v_{dc}/2$. Because the proposed rectifier is a topology of double boost operation, the dc bus voltage $v_{dc}>2v_{s,peak}$. Therefore, the line current is linearly decreasing if the operating state 1 is selected in the proposed rectifier. In the second operating state, power switches *T1* and *T2'* are turned on to achieve the ac terminal voltage $v_{ab}=0$. The line current is linearly increasing or decreasing if the line voltage is positive or negative. In the operating states 3 and 4, power switch *T1'* is turned on and *T1* is turned off. No matter what power switch *T2* is turned on or off, ac terminal voltage $v_{ab}=-v_{dc}/2$ in these two operating states. The line current can be controlled to increase if the line voltage is negative. Based on the above analysis of four operating states, a unipolar pulsewidth modulation (PWM) voltage waveform is generated on the voltage v_{ab}. The voltage stress of four active power switches in the proposed rectifier is $v_{dc}/2$.

3. SYSTEM ANALYSIS

According to the operating states of the proposed rectifier in Table 1, the ac side voltage and dc side currents of the proposed rectifier can be expressed as:

$$v_{ab} = (T1 \cdot T2)v_{C1} - (T1')v_{C2}\,. \tag{1}$$

$$i_1 = (T1 \cdot T2)i_s \tag{2}$$

$$i_2 = -(T1' + T1 \cdot T2)i_s \tag{3}$$

$$i_3 = (T1')i_s \tag{4}$$

where v_{C1} and v_{C2} are capacitor voltages on the capacitors *C1* and *C2* respectively, i_1, i_2 and i_3 are dc side currents flowing into the points *p*, *o* and *n* respectively. If two capacitor voltages v_{C1} and v_{C2} are equal, there are three voltage levels, $v_{dc}/2$, 0, and $-v_{dc}/2$, on the voltage v_{ab}. To make the control scheme easy, only three operating states (states 1, 2 and 3) are selected in the system analysis. In the positive half cycle of the mains voltage, the power switch *T1* is turned on. To generate voltage level $v_{ab}=v_{dc}/2$, power switch *T2* is turned on and the line current is decreasing. Power switch *T2'* is turned on to generate $v_{ab}=0$ and line current is linearly increasing. During the positive half cycle of mains voltage, two voltage levels, $v_{dc}/2$ and 0, are generated on the voltage v_{ab}. During the negative half cycle of mains voltage, the rectifier generates another two voltage levels, 0 and $-v_{dc}/2$, on the voltage v_{ab} by selecting the operating states 2 and 3 respectively. Power switch *T2* is turned off during the negative mains voltage. The Kirchhoff voltage and current laws applied on the ac and dc sides of rectifier are:

$$v_s = ri_s + L\frac{di_s}{dt} + v_{ab} \tag{5}$$

$$i_1 = (T1 \cdot T2)i_s = C1\frac{dv_{C1}}{dt} + \frac{v_{C1}+v_{C2}}{R} \tag{6}$$

$$i_3 = (T1')i_s = -C2\frac{dv_{C2}}{dt} - \frac{v_{C1}+v_{C2}}{R} \tag{7}$$

where *r* is the equivalent series resistance, *L* is the boost inductor, *C1* and *C2* are dc bus capacitors, *R* is the output resistor on the dc bus. The neutral point current i_2 can be expressed as (if $C1=C2=C$):

$$i_2 = -i_1 - i_3 = -(T1 \cdot T2 + T1')i_s = -C\frac{d(v_{C1}-v_{C2})}{dt} \tag{8}$$

Based on (8), the voltage variation between two capacitor voltages on the dc bus can be given as:

$$\Delta v = (v_{C1} - v_{C2}) = -\frac{1}{C}\int i_2 dt + \text{constant} \tag{9}$$

This means that a dc component in the neutral point current i_2 can be used to compensate the voltage unbalance between two capacitor voltages on the dc side. The state equations in the matrix form of the proposed rectifier based on (5)-(7) are given as:

$$\dot{\mathbf{x}} = \mathbf{f}(\mathbf{x}) + \mathbf{g}(\mathbf{x})\mathbf{u} \tag{10}$$

$$\mathbf{y} = \mathbf{h}(\mathbf{x}) \tag{11}$$

where

$$\mathbf{x} = \begin{bmatrix} x_1 \\ x_2 \\ x_3 \end{bmatrix} = \begin{bmatrix} i_s \\ v_{C1} \\ v_{C2} \end{bmatrix}, \mathbf{u} = \begin{bmatrix} u_1 \\ u_2 \end{bmatrix}, \quad u_1 = T1 \cdot T2, \quad u_2 = T1',$$

$$\mathbf{f}(\mathbf{x}) = \begin{bmatrix} \dfrac{v_s - ri_s}{L} \\ -\dfrac{v_{C1}+v_{C2}}{RC} \\ -\dfrac{v_{C1}+v_{C2}}{RC} \end{bmatrix}, \mathbf{g}(\mathbf{x}) = [\mathbf{g}_1 \quad \mathbf{g}_2] = \begin{bmatrix} -\dfrac{v_{C1}}{L} & \dfrac{v_{C2}}{L} \\ \dfrac{i_s}{C} & 0 \\ 0 & -\dfrac{i_s}{C} \end{bmatrix},$$

$$\mathbf{h(x)} = \begin{bmatrix} h_1 \\ h_2 \\ h_3 \end{bmatrix} = \begin{bmatrix} i_s \\ v_{C1} + v_{C2} \\ v_{C1} - v_{C2} \end{bmatrix}. \quad (12)$$

The controllability matrix **P(x)** [11] of the system (10) is given as

$$\mathbf{P(x)} = [\mathbf{g}_1, \ \mathbf{g}_2, \ [\mathbf{f(x)}, \mathbf{g}_1], \ [\mathbf{f(x)}, \mathbf{g}_2]]$$

$$= \begin{bmatrix} -\frac{v_{C1}}{L} & \frac{v_{C2}}{L} & \frac{v_{C1}L + v_{C2}L - rRCv_{C1}}{RL^2C} & \frac{-v_{C1}L - v_{C2}L + rRCv_{C2}}{RL^2C} \\ \frac{i_s}{C} & 0 & \frac{-rRCi_s + RCv_s + Li_s}{RLC^2} & \frac{-i_s}{RC^2} \\ 0 & -\frac{i_s}{C} & \frac{i_s}{RC^2} & \frac{rRCi_s - RCv_s - Li_s}{RLC^2} \end{bmatrix} \quad (13)$$

where

$$[\mathbf{f(x)}, \mathbf{g}_i] = \frac{\partial \mathbf{g}_i}{\partial \mathbf{x}}\mathbf{f(x)} - \frac{\partial \mathbf{f(x)}}{\partial \mathbf{x}}\mathbf{g}_i, \quad i=1, 2. \quad (14)$$

The rank of controllability matrix **P(x)** is three hence the system is controllable. To prove the observability of the system, the observability matrix **Q(x)** [12] of the system (11) is

$$\mathbf{Q(x)} = \begin{bmatrix} dh_1 \\ dh_2 \\ dh_3 \end{bmatrix} = \begin{bmatrix} 1 & 0 & 0 \\ 0 & 1 & 1 \\ 0 & 1 & -1 \end{bmatrix} \quad (15)$$

where

$$dh_i = \begin{bmatrix} \frac{\partial h_i}{\partial x_1} & \frac{\partial h_i}{\partial x_2} & \frac{\partial h_i}{\partial x_3} \end{bmatrix}, \quad i=1, 2, 3. \quad (16)$$

The rank of observability matrix **Q(x)** is three hence the system is observable. Based on the above analysis, the proposed rectifier is controllable and observable.

4. CONTROL SCHEME

The carrier-based PWM scheme is used in the inner control loop to track the line current command. If the switching frequency is higher than the line frequency of ac source, one can obtain the low frequency model of the rectifier by neglecting the high frequency switching terms. The ac side voltage equation of the proposed rectifier by neglecting the voltage drop across r can be expressed as:

$$v_s = L\frac{di_s}{dt} + v_{ab} = L\frac{di_s}{dt} + v_{con}k_{pwm} \quad (17)$$

where v_{con} is the control signal for pulsewidth modulation, $k_{pwm} = v_{dc}/(2\hat{v}_{tri})$ is the gain of PWM rectifier, and $\hat{v}_{tri}$ is the amplitude of the carrier signal. The bandwidth of the current loop control is related to the tracking performance of line current command. The current controller of the proposed rectifier based on (17) is shown in Fig. 2. The transfer function between actual line current and line current command is derived from Fig. 2 as

$$\frac{i_s}{i_s^*} = \frac{a}{s+a} \quad (18)$$

where $a = k_p / k_{pwm}$ is the bandwidth of the current loop control and can be adjusted by the gain k_p. The carrier-based sine-triangular PWM scheme is employed to generate the proper switching signals. Fig. 3(a) shows the switching signals and the corresponding PWM waveforms. In the positive value of control signal v_{con}, the power switch $T1$ is turned on. Line current is controlled to decrease or increase by turning on or off the power switch $T2$. If power switches $T1$ and $T2$ are turned on, the ac side voltage $v_{ab}=v_{dc}/2$ and line current is linearly decreasing. If the power switch $T2$ is turned off and $T2'$ is turned on, the voltage $v_{ab}=0$ and line current is linearly increasing. During the positive control signal ($v_{con}>0$), the ac side voltage can be expressed as $v_{ab} = (T1 \cdot T2)v_{C1}$. During the negative value of control signal v_{con}, active power switch $T2$ is turned off. By turning on or off the active switch $T1$, ac side voltage v_{ab} equals 0 or $-v_{dc}/2$ to track the line current command. During the negative value of control signal ($v_{con}<0$), the ac side voltage can be given as $v_{ab} = -(T1')v_{C2} = -(1-T1)v_{C2}$. If capacitor voltages v_{C1} and v_{C2} are equal, three voltage levels ($v_{dc}/2$, 0, and $-v_{dc}/2$) can be generated on the ac side voltage v_{ab}.

If one assumed that the line current tracking in the current controller is well, the line current dynamic can be neglected in the design of outer voltage control loop. Fig. 3(b) gives the control block for the dc-link voltage control and neutral point voltage compensation. The dc-link voltage controller is used to regulate the dc bus voltage and to supply power for the dc side load. Because the system input power factor is controlled to be unity, the regulated voltage signal obtained from the output of proportional-integral controller is multiplied by a sinusoidal wave in phase with mains voltage. The addition of the output of dc-link voltage controller and neutral point voltage compensator is the line current command for the inner current control loop.

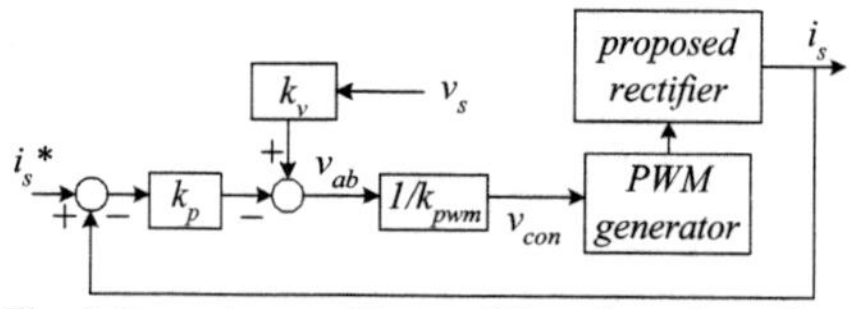

Fig. 2 Current control loop of the adopted rectifier.

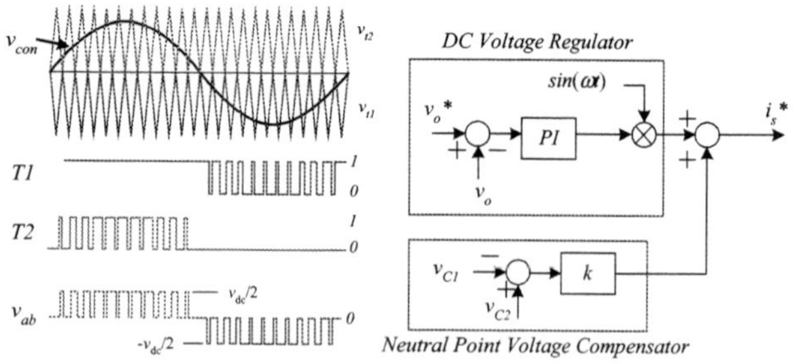

Fig. 3 (a) Carrier based PWM scheme and the corresponding switch gating signals (b) voltage control loop of the adopted rectifier.

5. EXPERIMENTAL RESULTS

An experimental prototype with 1kW rating was built to verify the effectiveness of the proposed control scheme. The boost inductor is 2mH and dc side capacitors *C1* and *C2* are 2200uF. The line voltage is 110V with 60Hz. The capacitor voltages v_{C1} and v_{C2} on the dc side are equal to 200V. The PWM frequency is set to 20kHz. A digital signal processor TMS320C32 was used as the controller to implement the proposed control algorithm. Fig. 4(a) gives the measured waveforms of line voltage, line current and ac terminal voltage v_{ab} under the rectification operation mode. The line current is a sinusoidal waveform in phase with line voltage. Based on the power meter measurement, the input power factor of rectifier is 0.994, and the total harmonic distortion is 3.5%. Two voltage levels 0 and $v_{dc}/2$ are generated on the ac terminal voltage v_{ab} during the positive half cycle of v_s. On the other hand, two voltage levels 0 and $-v_{dc}/2$ are shown on the voltage v_{ab} for the negative mains voltage. Fig. 4(b) gives the measured ac source voltage and line current under the inversion operation mode of the proposed rectifier. The measured line current is out of phase with the ac source voltage. Based on the simulations and experimental results, the sinusoidal line current with nearly unity power factor is drawn from the proposed rectifier. A three-level PWM voltage pattern is generated on the ac terminal of the adopted rectifier. Fig. 5(a) gives the measured results of four switching signals for power switches. Fig. 5(b) shows the measured capacitor ripple voltages and the voltage variation on the dc side. The voltage variation between two capacitors is below 10V.

6. CONCLUSION

A novel single-phase half-bridge rectifier with power factor correction is presented to generate a three-level PWM voltage waveform on the ac terminal of the rectifier, to draw a sinusoidal line current with low current harmonics from ac source, and to regulate the dc bus voltage. The proposed control scheme is based on the carrier-based PWM approach to track the line current command. The voltage stress of four power switches is clamped to half dc-link voltage. The proposed rectifier has less power semiconductors compared with the conventional NPC rectifier leg and flying capacitor rectifier leg. The performance of proposed control strategy is verified by the experimental results.

7. REFERENCES

[1] J. T. Boys, and A. W. Green, "Current-Forced Single-Phase Reversible Rectifier", IEE Proceedings-B, vol. 136, pp. 205-211, 1989.

[2] S. Manias, "Novel Full Bridge Semicontrolled Switch Mode Rectifier", IEE Proceedings-B, vol. 138, pp. 252-256, 1991.

[3] R. Martinez, and P. N. Enjeti, "A High-Performance Single-Phase Rectifier with Input Power Factor Correction", IEEE Trans. on Power Electronics, vol. 11, pp. 311-317, 1996.

[4] M. S. Dawande, V. R. Kanetkar, and G. K. Dubey, "Three-Phase Switch Mode Rectifier with Hysteresis Current Control", IEEE Trans. on Power Electronics, vol. 11, pp. 466-471, 1996.

[5] Bor-Ren Lin and Hsin-Hung Lu, "A new control scheme for single-phase PWM multilevel rectifier with power-factor correction", IEEE Transactions on Industrial Electronics, vol. 46, no. 4 , 1999, pp. 820 – 829.

[6] J. W. Kolar and F. C. Zach, "A novel three-phase utility interface minimizing line current harmonics of high-power telecommunications rectifier modules," IEEE Trans. Ind. Electron., vol. 44, pp. 456–467, Aug. 1997.

[7] Y. Zhao, Y. Li, and T. A. Lipo, "Force commutated three level boost type rectifier," IEEE Trans. Ind. Applicat., vol. 31, pp. 155–161, Jan./Feb. 1995.

[8] G. Sinha and T. A. Lipo, "A four-level rectifier-inverter system for drive applications," IEEE Industry Applications Magnazine, pp.66-74, 1998.

[9] Chongming Qiao and Keyue Ma Smedley, "Three-Phase Unity Power Factor Star Connected Switch (VIENNA) Rectifier With Unified Constant-Frequency Integration Control", IEEE Transactions on Power Electronics, vol. 18, no. 4, 2003, pp. 952-957.

[10] P. Ide, N. Froehleke, H. Grotstollen, W. Korb and B. Margaritis, "Operation of a three-phase/three-level-rectifier in wide range and single-phase applications", IEEE Industrial Electronics Conference, IECON '99, 1999, pp. 577 –582.

[11] A. Isidori, "Nonlinear Control Systems: an introduction," Springer-Verlag, 1991, second edition.

[12] S. H. Tsai and H. K. Chiang, "Transformations for a class of nonlinear timing-varying multivariable systems and their applications", Journal of Mathematical Control & Information, 1992, 9, pp. 61-87.

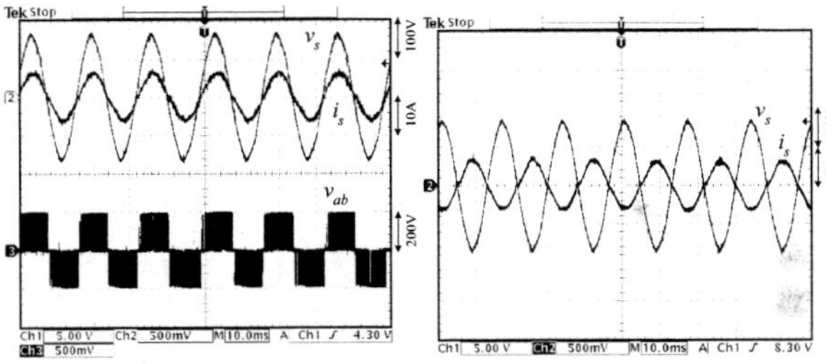

Fig. 4 Experimental results of the proposed rectifier (a) under rectification operation (b) under inversion operation.

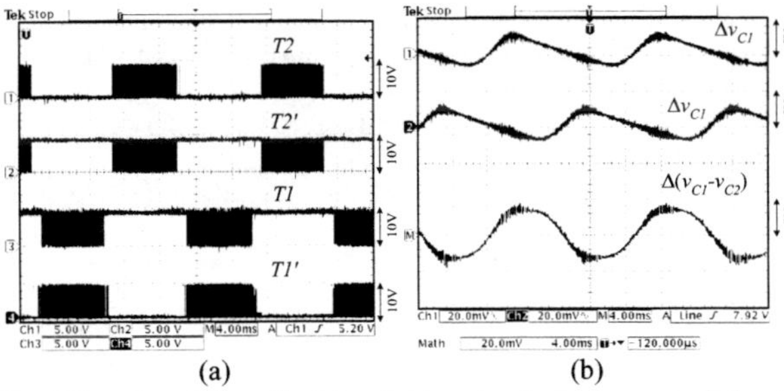

Fig. 5 Experimental results (a) switching signals for power switches (b) capacitor ripple voltages on the dc side.

CHARACTERIZATION OF CHARGE-PUMP RECTIFIERS FOR STANDARD SUBMICRON CMOS PROCESSES

Mark Hooper, Matt Kucic, Paul Hasler

Georgia Institute of Technology
School of Electrical and Computer Engineering
Atlanta, GA 30332

ABSTRACT

This work presents an innovative approach in designing CMOS charge pumps for a 0.6um CMOS process for the control of floating gate circuits. Central to this design are rectifier structures which would function best for both medium and high voltage applications. Shown in this paper are six different rectifier structures designed and fabricated in a standard .6um CMOS N-well double poly process along with experimental results. These rectifier structures are then implemented in two stage Dickson charge pumps in a standard .6um CMOS process to determine their effectiveness. Experimental results show by way of comparison charge pump performance with the experimental results of the rectifying structures.

1. INTRODUCTION

Two circuits that have been widely used to obtain a DC output voltage higher than the supply voltage or of opposite polarity, are the charge pump and the switching regulator. Both of such circuits provide DC conversion using a high switching frequency clock. The main difference between the two circuits is the element used to store energy during the switching cycles. Switching regulators use an inductor to store the energy whereas charge pumps use capacitors. In the case of the switching regulator, energy is stored by the change in current in the inductor whereas in the charge pump, energy is stored in the capacitors by the change in voltage. Capacitive energy storage circuits have the advantage of being able to be completely integrated on chip when the power requirements are relatively small thereby reducing cost, eliminating an external supply, reducing pin count and hence using less board space. Another advantage of on chip charge pumps is the feasibility for distributed power systems which can reduce overall power consumption. Consequently charge pumps are now being used in nonvolatile memory circuits, dynamic random access memory circuits, low voltage circuits, continuous time filters, RS-232C transceivers and they are now finding their way in floating gate circuits [2].

2. PREVIOUS WORK

There are no previous papers that have characterized rectifier structures in a standard CMOS process for use in charge pumps other than [2]. However [2], was focused on designing a rectifier structure for a negative output voltage. There have also been a couple of papers that have focused on designing and characterizing diodes in a standard CMOS process [1],[3], however for uses other than charge pumps: [1] addressed the design of Schottky diodes and its focus was on increasing the frequency range through a layout design solution;[3] was focused on the translation between stress and device characteristics as an explanation for leakage currents in a .12um CMOS process. There have also been a number of charge pump designs –[4],[5],[6],[7],[8],[9],[10],[11]– fabricated in standard submicron non triple well CMOS, however none of such designs provide details concerning the layout nor do they analyze the different rectifier structures for their effectiveness in fabrication.

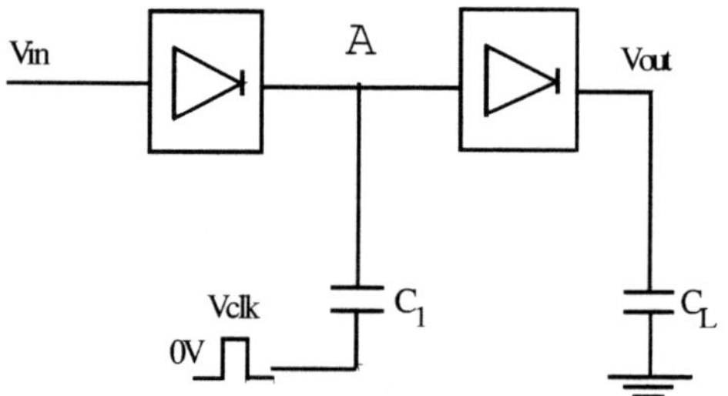

Fig. 1. This is a schematic representation of a charge pump. In our work we will be designing a Dickson charge pump. Our objective is to find the optimum rectifier structure which will yield the highest output voltage with the best efficiency. The value of the capacitor used in the designs presented in this paper equals 1pF.

3. CHARGE PUMP OVERVIEW

The schematic diagram of a CMOS charge pump is shown in Fig. 1. Most charge pumps are derived from the Dickson charge pump [12]. The operation of the Dickson charge pump can be thought of water being pumped down the capacitor chain where each successive capacitor represents a bucket of water of increasing volume. For example, assume in a two stage Dickson charge pump that the input is 3V and the clock magnitude is 3V. If there is no clock pulse and assuming ideal diodes then the voltage on each capacitor is 3V. The operation of the clock is responsible for the boosting operation. For example at point A, the voltage is boosted to 6V when the clock is pulsed high. This occurs because the voltage is set at 3V by the voltage source input and it is immediately elevated to 6V when the clock toggles high since the voltage across a capacitor cannot change instantaneously. Therefore the output capacitor is charged to 6V since this stage is pulled to ground by the clock which operates in antiphase. Thus each successive stage is increased by the magnitude of the clock, i.e., 3V. The output voltage is given as:

0-7803-8251-X/04/$17.00 ©2004 IEEE

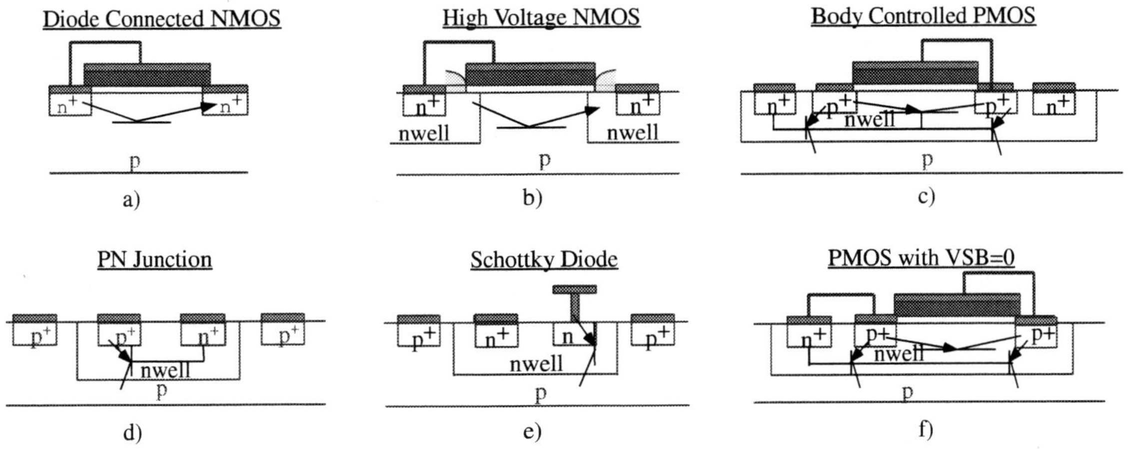

Fig. 2. Shown in this figure are the 6 possible rectifer structures designed in a standard .6um CMOS process.
a) Diode connected NMOS (DCN) transistor W=6um and L=.6um.
b) High voltage NMOS transistor (HV), W=6.75um. & L=.6um.
c) Body controlled PMOS transistor (BCP),W=6um and L=.6um.
d) PN Junction (PN), diffusion dimensions 5.4um x 2.4um.
e) The Schottky diode (SCH),diffusion dimensions 2.4um x 3.9um.
f) PMOS diode connected transistor with VSB=0 (VSB=0), W=6um & L=.6um.

$$V_{out} = V_{in} - V_d + n * [V_\phi^i - V_d - V_l] \quad (1)$$

where
V_{in} is the input voltage to the circuit;
V_ϕ^i is the voltage swing at each node i due to capacitive coupling of the clock;
V_d is the voltage drop across each rectifier;
V_l is the charging and discharging of the capacitors when the charge pump is sourcing current.

4. RECTIFIER DESIGNS

Six different rectifying structures were identified for the charge pump design: diode connected NMOS (DCN), high voltage NMOS (HV), body controlled diode connected PMOS transistor (BCP), PN junction (PN), Schottky diode (SCH) and PMOS diode connected transistor with source to body voltage equal to zero (VSB=0). All of these devices were fabricated in a standard .6um CMOS double poly process. In order to layout some of these rectifiers in standard CMOS some of the DRC violations were ignored. Simulations for most of these devices would not be possible since the modelling parameters for each device are not available. However, the experimental results of the rectifying structures presented in this paper, would help understand the simulation results of a charge pump utilizing these rectifiers.
Four of these were modified transistor configurations and two were modified diode configurations. Inputs were connected to leak pads and outputs were connected to bare pads. All diodes were surrounded by a P type guard ring. The DCN was simply a diode connected NMOS transistor effected by a metal1 poly contact with a channel length of .6um and width of 6um. The HV structure was an NMOS diode connected transistor except the source and drain regions were surrounded by N-wells. The N-wells are lower doped than conventional n diffusions thus allowing the possibility of a higher breakdown voltage. However the layout was atypical in that the active layer was added beneath the gate in order to achieve thin oxide in the gate to N-well overlap region. The channel width was 6.75um and the length was .6um as measured from N-well to N-well. The BCP structure was a diode connected PMOS transistor with an N-well contact thus allowing voltage change on the N-well. The N-well contact was placed as a ring around the N-well. The width of the channel was 6um and the length was .6um. The PN junction was created on an N-well using p+ and n+ diffusions. The diffusion dimensions were 5.4um by 2.4um. The SCH device was also fabricated on an N-well. Metal1 to N-well was effected with non standard contacts to preclude an oxide interface to the N-well. Finally the last configuration was a diode connected PMOS transistor with source to body voltage equal to zero. The diffusion dimensions were 5.4um by 2.4um.

Table 1. Parasitic β Measurements

DCN and HV	BCP	PN	SCH	VSB=0
N/A	Blat=1.3 BVert=.5	B=0.9	N/A	N/A

5. EXPERIMENTAL RESULTS

5.1. Bipolar Parasitics

Of particular interest were the parasitic bipolar transistors because these are in part responsible for the leakage currents. Investigating parasitics also provides a better idea of how to model the rectifiers as well as giving insight into determining which parasitics should be reduced. It was found in the DCN structure that the lateral BJT was essentially non parasitic. This could only be measured by grounding the left side –so that the channel remains off– but since the base is at ground, VBE would be equal to zero. The same situation occurred with the HV structure. The BCP structure exhibited a lateral β of approximately 1.3 for $2V < VCE < 10V$ and a vertical β of approximately .7 for $2V < VCE < 10V$

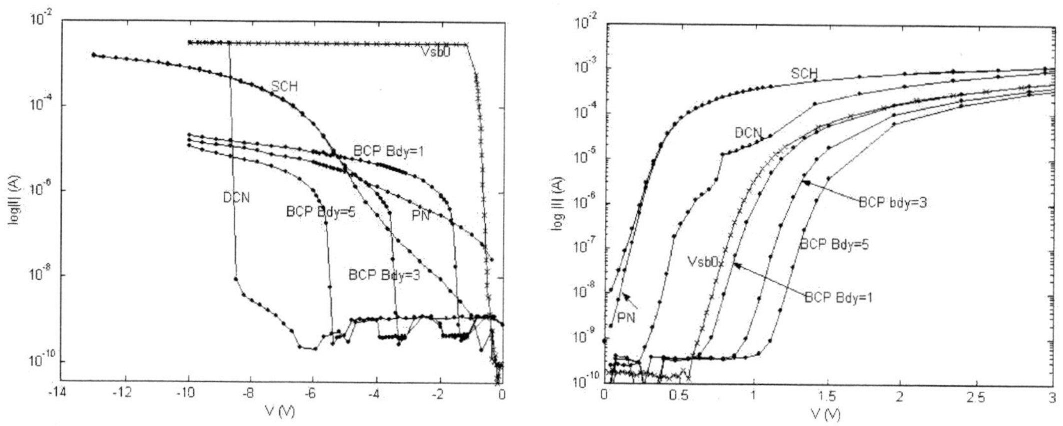

Fig. 3. Fabrication results from .6u standard CMOS process indicating the reverse and forward bias characterisitcs of the 6 different rectifiers. *Legend:* DCN ⇒ Diode Connected NMOS transistor; BCP Bdy=5 ⇒ Body controlled PMOS transistor with Vsb = 5V; BCP Bdy=3 ⇒ Body controlled PMOS transistor with Vsb = 3V; BCP Bdy=1 ⇒ Body controlled PMOS transistor with Vsb = 1V; PN ⇒ PN junction; SCH ⇒ Schottky diode; Vsb0 ⇒ PMOS diode connected transistor with VSB=0.

and for $0V < VBE < 1.5V$. β lateral was measured by applying a VEC sweep from the right side, i.e. increasing the voltage on the right side and grounding the left side, as in the DCN case to prevent the turn on of the channel. IC and IE were measured and IB was obtained from the equation IC=IB+IE. The vertical β was measured by applying a VEC voltage sweep on either the drain or source while maintaining the opposite side at the same voltage level to prevent current flow in the channel. IE and IB were measured. The β measured for the PN junction was approximately .9 for $0 < VCE < 10$ and for $0 < VBE < 1.5V$. IE and IB were measured. For the VSB=0 device, the left vertical BJT essentially draws no current since VBE=0. The right vertical BJT could not be measured since the left diffusion could not be maintained at the same voltage as the emitter on the right diffusion to prevent turn on of the channel. Finally the lateral BJT could not be measured when applying the VEC sweep from the right side because it becomes a resistor when the left side is grounded. For all measurements, $0 < VCE < 10V$ and $0 < VBE < 1.5V$. These results are summarized in Table 1.

5.2. IV Curves

Shown in Fig. 3 are the IV curves for both forward and reverse bias regions for five of the devices. The high voltage device was omitted because it was found to be inoperative, probably due to an inadequately long channel length which most likely resulted in punch through. The channel length was later doubled and as such was successfully used in subsequent multi-stage charge pump designs[13]. In the reverse bias regions it is seen that all the rectifying structures except for the DCN rectifier break down with a reverse bias of 1V or less. The IV curves for the BCP device were measured for body voltages of 1V(Bdy=1), 3V(Bdy=3) and 5V(Bdy=5) and its breakdown voltage also occurred when the reverse bias was approximately 1V or less. In the forward bias case, the Schottky diode structure is the most current prolific device and has the highest initial value of current. This explains in part its higher voltage gain as described in section 5.3. Shown in Fig. 4 is the test set up to measure the IV characteristics.

Test set up for measuring IV Curves:
Keithley 2400 SourceMeter

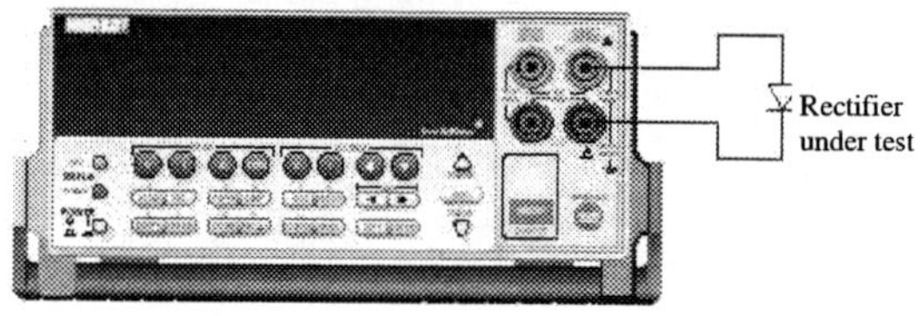

Fig. 4. Shown is the test set up for obtaining the IV curves.

Table 2. Leakage Current at Vin=-5V and Vin=-10V

I(A)@(-5V)		I(A)@(-10V)	
DCN	4.06e-10	BCP3	-1.14e-5
BCP3	-4.34e-7	BCP2	-1.56e-5
BCP2	-4.2e-6	BCP1	-1.96e-5
BCP1	-8.06e-6	PN	-7.55e-4
SCH	-1.89e-5	SCH	-7.774e-4
PN	-2e-5	VSB=0	-3e-3
VSB=0	-3e-3	DCN	-3e-3

5.3. Two Stage Charge Pumps

Shown in Fig. 5 is the transfer curve and Vout v.s. frequency curve for a no load case. The transfer curves were measured at the optimal frequencies for each device. The SCH and VSb=0 devices have the largest gains and the SCH device has the largest offset voltage. However, because of the initial large leakage currents in the VSB=0 and SCH devices, relatively higher frequencies were needed to pump these devices to their passband voltages. This is also seen in the Vout v.s. Frequency graph. We notice in this graph that that Vout begins to attenuate quickly at around 10 MHz; this is due to magnitude attentuation and phase shift of the inverting

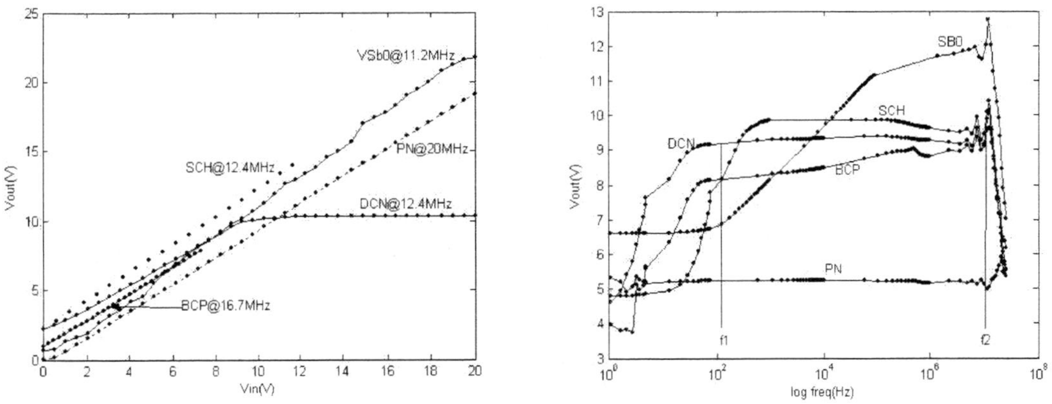

Fig. 5. Fabrication results from a standard .6u CMOS process: transfer curve (Vout v.s. Vin) and Vout v.s. frequency (f1→approx. lower 3dB freq., f2→approx. upper 3dB freq. These measurements were taken for no load on the output. *Legend:*
Vsb0 ⇒ PMOS diode connected transistor with VSB=0; PN ⇒ PN junction; SCH ⇒ Schottky diode; DCN ⇒ Diode Connected NMOS transistor; BCP⇒ Body controlled PMOS transistor.

Table 3. Frequency Charactersitics

Device	f1(KHz)	f2(MHz)	Vout Max.
DCN	0.075	13.6	5.07@12.4MHz
BCP	1.64	16.1	4.82@12.4MHz
SCH	0.756	13.6	5.2@12.4MHz

clock signal. Some frequency characteristics are shown in Table 3.

6. CONCLUSION

Six different rectifier structures were designed, fabricated and characterized in a standard CMOS process for use in CMOS charge pumps. The rectifying devices were subsequently implemented in two stage charge pumps which were also designed, fabricated and characterized in a standard CMOS process. The SCH and VSB=0 devices showed the most promise in terms of gain and maximum output voltage. The output voltage characterisitcs of the SCH and VSB=0 devices were among the highest as was their forward bias IV charactersitcs. In general it was found that the βs of the lateral and vertical parasitic bipolar transistors were not significant. The DCN structure proved to be the most efficient however the SCH and VSB=0 devices had the best gains.

7. REFERENCES

[1] B. Rivera, R. Jacob Baker et Al., Design and layout of Schottky Diodes in a Standard CMOS Process,*International Semiconductor Device Research Symposium*, 2001, pp.79-82.

[2] M. D. Ker, C. Y. Chang, H. C. Jiang, Design of a Negative Charge Pump Circuit with Polisilicon Diodes in a 0.25um CMOS Process, *Proceedings of the IEEE Asia Pacific Conference ASICS*, 2002, pp. 145-148.

[3] W.T. Toren, E. Perrin et Al., Characterization of the Diode leakage current in advanced .12um CMOS technology, *Extended Abstracts of the Second International Workshop on Junction Technology*, 2001, pp.15-18.

[4] J.F. Dickson, On-Chip High-Voltage Generation in MNOS Integrated Circuits Using an Improved Voltage Multiplier Technique, *IEEE JSSC*, Vol. SC-11, No.3, 1976, pp. 374-378.

[5] W. Chen et Al., Switched-Capacitor Power Converters with Integrated Low Dropout Regulators,*IEEE ISCAS*, 2001, p.293-296.

[6] G. Nicollini, A Nagari et Al., A -80 dB THD 4Vpp Switched Capacitor Filter for 1.5 V Battery-Operated Systems, *IEEE JSSC*, 1996, pp.1214-1219.

[7] S.C. Lee et Al., A Low Ripple Switched Capacitor DC-DC Up Converter for Low Voltage Applications, *Proceedings of the 2nd. IEEE Asian Pacific Conference on ASICs*, 2000, pp.13-16.

[8] T. Duisters and E. Dijkmans, A -90dB THD Rail-to-Rail Input Opamp Using a New Local Charge Pump in CMOS, *IEEE Journal of Solid State Circuits*, Vol. 33, No. 7, July 1998.

[9] C.H. Lee et Al., Supply Noise Insensitive Bandgap Regulator Using Capacitive Charge Pump DC-DC Converter, *IEEE ISCAS*, 1999, pp.89-92.

[10] J. Zhou et Al., Charge Pump Assisted Low-Power/Low-Voltage CMOS Opamp Design, *Proceedings of the 1997 ISLPED*, 1997, pp. 108-109.

[11] D. S. Hong, Low Operating Voltage and Short Settling Time CMOS Charge Pump for MEMS Applications, *IEEE ISCAS*,2003,Vol. 5, pp.281-284.

[12] S. H. Lai and J.S. Wang, A High Efficiency CMOS Charge Pump Circuit, *IEEE ISCAS*, 2001, pp.406-409.

[13] M. Hooper, M. Kucic and P. Hasler, 5V-Only Standard 0.5um CMOS Programmable and Adaptive Floating-Gate Circuits and Arrays Using CMOS Charge Pumps, *IEEE ISCAS*, 2004.

SUPPRESSION OF HARMONIC SPIKES IN ASYNCHRONOUS SIGMA DELTA MODULATION BY RANDOMIZING HYSTERESIS WINDOW

A. Aurasopon, P. Kumhom and K. Chamnongthai

Department of Electronic and Telecommunication Engineering, Faculty of Engineering,
King Mongkut's University of Technology Thonburi, Bangkok 10140, Thailand
E-mail: aurasopon@yahoo.com, ipinmhom@kmutt.ac.th and kosin.cha@kmutt.ac.th

ABSTRACT

This paper proposes to suppress the harmonic spikes in asynchronous sigma delta modulation by randomizing the hysteresis window for the distribution of the switching frequency and harmonic spectrum. The proposed scheme was examined in the dc/dc buck converter. The experimental results show that the proposed technique provides improved performance in terms of low peaks of harmonic while maintaining low output voltage ripple.

1. INTRODUCTION

Pulse width modulation (PWM) has been widely used in power electronic systems. It normally operates with a fixed switching frequency, which causes power spectrum to be concentrated at multiples of the switching frequency. This gives rise to harmonic spikes that can produce unwanted effects such as acoustic noise, torque ripple and electromagnetic interference (EMI) in the power converter. Many techniques were proposed to reduce the peaks of harmonic spectrum in PWM signal, frequency modulation (FM) [1] and random pulse width modulation (RPWM) [2][3]. In FM technique, the PWM switching signal is modulated with sinusoidal signal. As a result, the power is spreaded into some side band frequencies and the level of the emission spectrum is reduced. In RPWM techniques, a random noise signal is used to control switching frequency with a constant duty ratio, pulse position in each interval of switching time or duty ratio with a fixed switching frequency. The result is the continuous distribution of spectrum output voltage and the harmonic spikes are reduced. However, these spread spectrum techniques affect the efficiency of output regulation, current and voltage ripple, which are also important to consider in applications.

Sigma delta modulation (SDM) has been applied to generate PWM signal for switching power converters. The SDM has the advantage in keeping the regulation of EMI noise level with spread spectrum characteristics of second-order SDM [4]. However, the problem of noise is still exists in SDM. In the first-order SDM, there are some harmonic spikes in cases of a dc input and in cases of an ac input with a low modulation index. Therefore, the dither sigma delta modulation (DSDM) was used to avoid discrete spectrum [5][6]. In DSDM, a small dither signal is added to the input of the comparator, which produces a perturbation of the correlation between the quantization error, the input decreases, and the spectrum becomes continuous. Although the peaks of harmonic spectrum can be reduced by these techniques but it causes the distribution of noise on a pass band filter of a power converter and the duty cycle of switching period is not constant. As a result, low-frequency ripple appears in the converter output, which is not suitable for applications that require the tight output regulation, such as dc-dc converters.

This paper presents the suppression of harmonic spikes in asynchronous sigma delta modulation (ASDM) by using the random noise signal added to the hysteresis window of comparator. The advantages of this technique are the constant duty ratio in every switching cycle and easy to implement on analog devices without sampling clock. The proposed method was examined in the dc/dc buck converter. The results were shown that the reduction of harmonic spikes and can be achieved with low ripple of output voltage.

2. BACKGROUNDS AND BASIC CONCEPT

Fig. 1(a) shows the block diagram of ASDM, which consists of integrator and hysteresis comparator in the forward path, $x(t)$ as input signal, $y(t)$ as PWM signal. The ASDM is a unity feedback system. The output, and also the feedback, is a two-level digital signal $v_o = \pm V$. The switching points of the comparator are $\pm v_{ho}$ and the integrator has time constant of τ.

The input $x(t)$ has been assumed a constant dc voltage illustrated in Fig1 (b) assuming that the output of hysteresis comparator starts at $v_o = -V$. The error signal fed to the integrator is $v_{in} - (-V)$ and integrator output increases. The comparator output is turned "on," $v_o = +V$ when the input reaches $+v_{ho}$. The error signal applied to the integrator is $v_{in} - V$. The integrator output reduced until $-v_{ho}$ whereupon the comparator switches "off" and the cycle repeats. Equations (1)-(5) give the on time, off time and switching frequency f_s [7].

$$t_{on(m)} = \frac{-2\tau v_{ho}}{v_{in} - v_o} = t_l \frac{1}{1-m} s \qquad (1)$$

$$t_{off(m)} = \frac{2\tau v_{ho}}{v_{in}+v_o} = t_l \frac{1}{1+m} s \quad (2)$$

$$t_l = \frac{4\tau v_h}{v_o} = \frac{1}{f_l} \quad (3)$$

$$f_s(m) = \frac{1}{t_l}(1-m^2) = f_l(1-m^2) \quad Hz \quad (4)$$

$$D(m) = \frac{1}{2}(1+m), \quad (5)$$

where

f_s = Switching frequency
f_l = Central frequency
D = Duty ratio
m = Modulation index, v_{in}/V

The above equations show that ASDM's switching frequency depends on the modulation index m with central frequency f_l. When m is adjusted to control the output voltage, the f_s is reduced. As a result, the harmonic spectrum also varies with m. However, the peaks of harmonic are still high and concentrated at the f_s. To solve the problem, the random noise signal is added to the hysteresis window to distribute the harmonic spectrum, which causes the reduction of harmonic spikes.

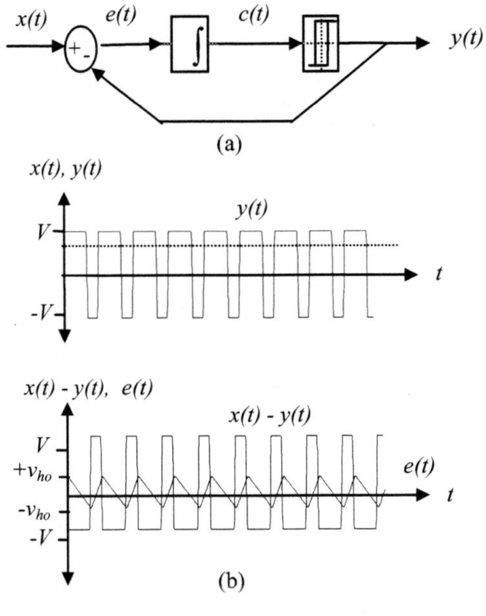

Figure 1 (a) The block diagram of ASDM
(b) The modulation signal of ASDM

3. PROPOSED SCHEME

Normally, the ASDM's hyteresis window is fixed with v_{ho} that causes the peaks of harmonic of output voltage. So, when the hysteresis is varied by controlling of the random noise signal, it results in the continuously distribution of harmonic spectrum and the peaks of harmonic are reduced. Fig. 2 shows modulated signals of random hysteresis asynchronous sigma delta modulation (RH-ASDM), when the random noise signal $v_n(t)$ is generated by 10-bit shift registers used for generating pseudo random binary sequence (PRBS) as shown in Fig. 2(a). Amplitude of noise signal $v_n(t)$ can be adjusted to design the range of switching frequency distribution. The $v_n(t)$ is added with the fixed hyteresis $+v_{ho}$. The result is that the hysteresis is distributed around $+v_{ho}$ as shown in Fig. 2(b). The hysteresis is varied with a positive side only; therefore, it produces the constant duty cycle in every switching cycle of output voltage when input is a dc voltage. So, it is suitable to control the dc/dc converter, which requires a stable output voltage [2].

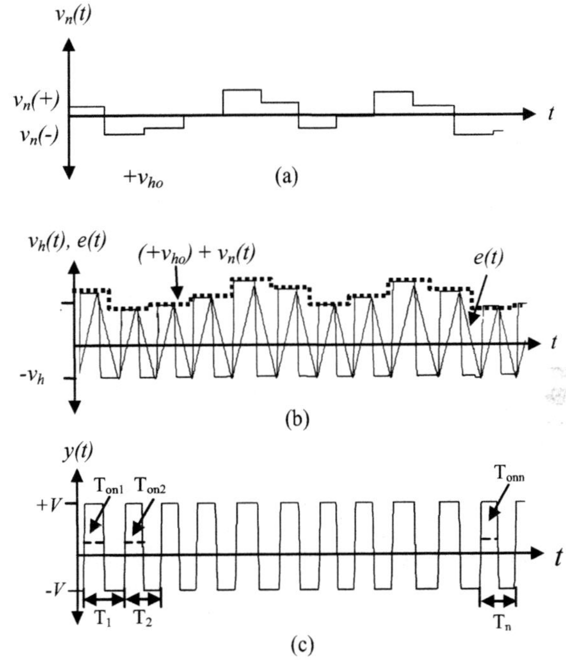

Figure 2 Modulated signals of RH-ASDM (a) PRBS signal (b) Integrator signal and hysteresis window
(c) The output voltage $y(t)$ with constant duty cycle $T_{on1}/T_1 = T_{on2}/T_2 = \ldots = T_{onn}/T_n$

The RH-ASDM circuit is shown in Fig. 3. The circuit was constructed with three op-amps and an analog switch. The summation and integrator circuits were combined, using 10 kΩ resistors and a 0.02 uF capacitor. The feedback-path used a comparator circuit, which was made using a TL082 op-amp. In the hysteresis circuit, the analog switch 74HC4316 operates when output voltage v_o is positive, $+V$. So, the hysteresis voltage $v_h(t)$ is generated by two voltage sources, output voltage $+V$ and noise voltage $v_n(t)$. Equations (6) and (7) give the negative and positive values of hysteresis voltage, respectively.

$$-v_h(t) = -v_{ho} = \frac{R_6}{R_5+R_6}.(-V) \quad (6)$$

$$+v_h(t) = \frac{R_6//R_7}{R_5+R_6//R_7}.(+V) + \frac{R_5//R_6}{R_7+R_5//R_6}.(v_n(t)+v_{set}) \quad (7)$$

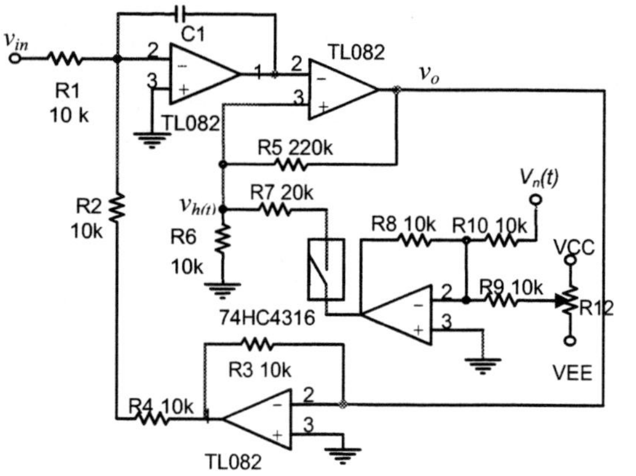

Figure 3 RH-ASDM circuit

4. EXPERIMENTAL RESULTS AND DISCUSSIONS

In this section we show the experimental results of standard PWM, DSDM and RH-ASDM. These schemes were practically examined in the same converter-load system as shown in Fig. 4. The RH-ASDM scheme's parameters values were determined from the results of PSPICE program simulation that indicates amplitude and speed of random noise $v_n(t)$ affect the distribution of harmonic. If we choose the too large value of $v_n(t)$, it will lead to harmonic components distributing in too low frequency, which produces the large output voltage ripple. If speed of the noise generator is too slow or too high, random property of the harmonic distribution of output voltage is not good. In many simulations it should be lower than the central frequency f_l.

In comparison of various PWM schemes, the standard PWM's switching frequency f_s is fixed at 20 kHz. In the RH-ASDM, $\pm v_{ho}$ = 600 mV, $v_n(t)$ = 300 mV so that the switching frequency distributed in the range of 15-24 kHz at D=0.5. The DSDM scheme is implemented on analog devices with sampling frequency 80 kHz, f_l 40 kHz and a small dither signal 200 mV so that the average of switching frequency is at 20 kHz. Magnitude spectrum of the diode voltage v_d is observed with a spectrum analyzer (HP 35665A).

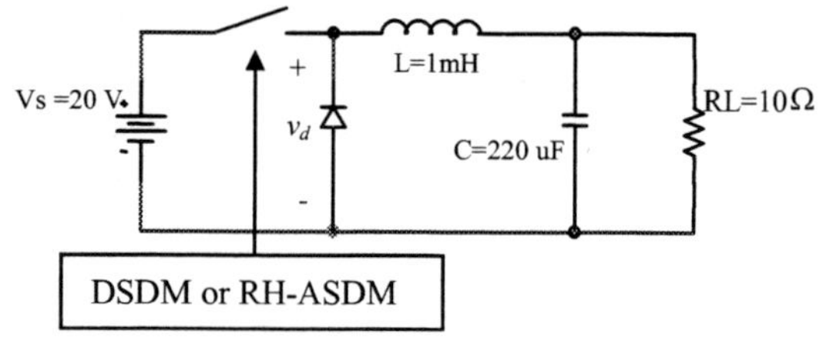

Figure 4 Experimental setup

Fig. 5 shows the modulated signals of RH-ASDM circuit; the integrator signal, $c(t)$ and hysteresis voltage, $v_h(t)$. Fig. 6-8 show voltage ripple, voltage across diodc v_d, its harmonic spectrum of the standard PWM, the DSDM, and the proposed, respectively. Fig. 6 shows that the peaks of harmonic are concentrated at switching frequency 20 kHz. Figs. 7 and 8 show the results of spread spectrum techniques, RH-ASDM with DSDM. The RH-ASDM technique shows that continuously distribution of harmonic in the range of switching frequency was designed and constant duty ratio in every cycle of switching time which produces low ripple of output voltage. In DSDM, there is low discrete harmonic peak amplitude. However, it shows that noise floor distributed widely range into low frequency so that the output voltage appears low frequency ripple.

Fig. 9 shows the comparison of measured of various PWM schemes in term of output voltage ripple and the maximum peak amplitude of the harmonic components at various duty cycles. As shown in Fig. 9(a), the standard PWM has low output voltage ripple, because its harmonics are concentrated at switching frequency. So it can be attenuated by low pass filter. In the case of DSDM, there is the large ripple because its harmonics are distributed into pass band filter and duty cycle is not constant. The RH-ASDM has voltage ripple lower than the DSDM, because its harmonic is distributed in the range of designed frequency. Fig. 9(b) shows that the DSDM has the lowest peaks of harmonic but it shows the worst output voltage ripple. So, the RH-ASDM is a choice for applications that require a spread spectrum while maintaining low ripple of output voltage, and the proposed scheme should be able to apply with the other dc/dc converters such as boost, and buck-boost converter.

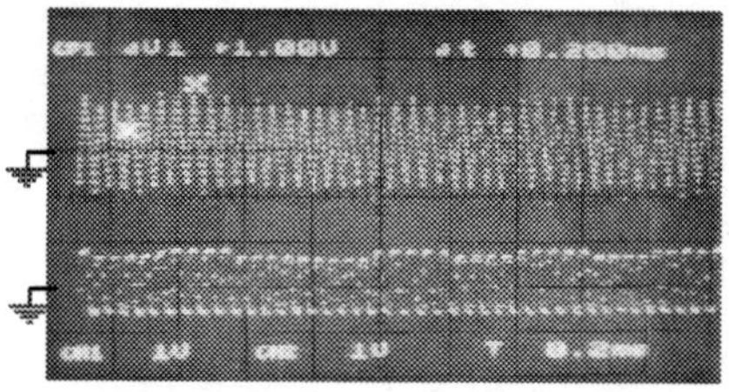

Figure 5 RH-ASDM signals: upper: Integrator voltage 1 V/div, lower: hysteresis voltage 1 V/div, time 0.2 mS/div.

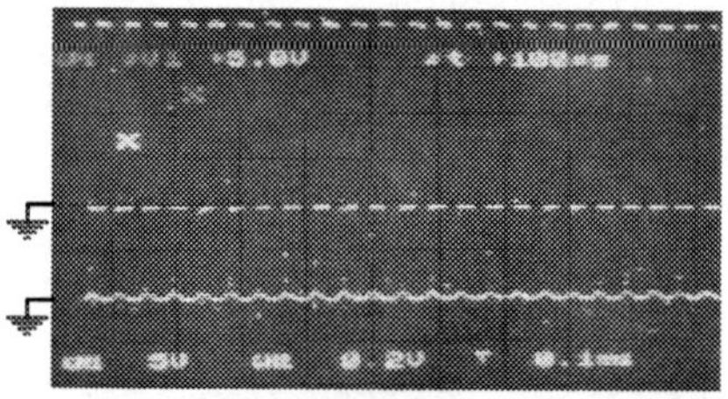

(a)

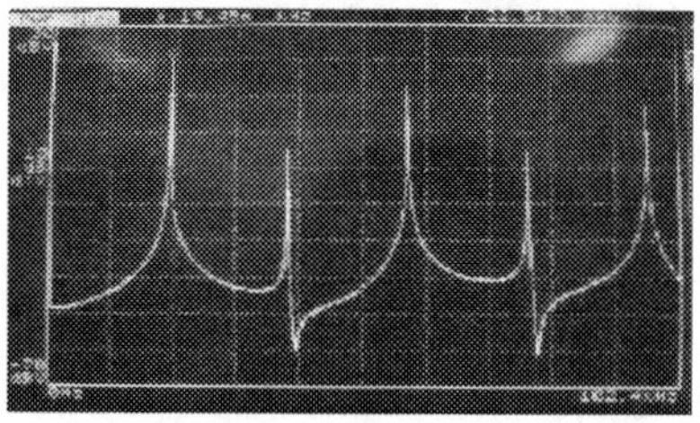

(b)

Figure 6 Experimental results of standard PWM: (a) The upper: voltage across diode v_d 5 V/div, lower: output voltage ripple 200 mV/div, time 0.1 mS. (b) Harmonic spectrum of v_d 10 dB/div, freq. 10.2 kHz/div.

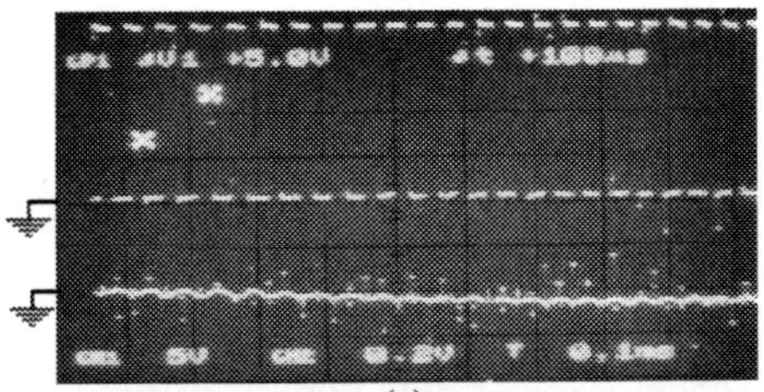

(a)

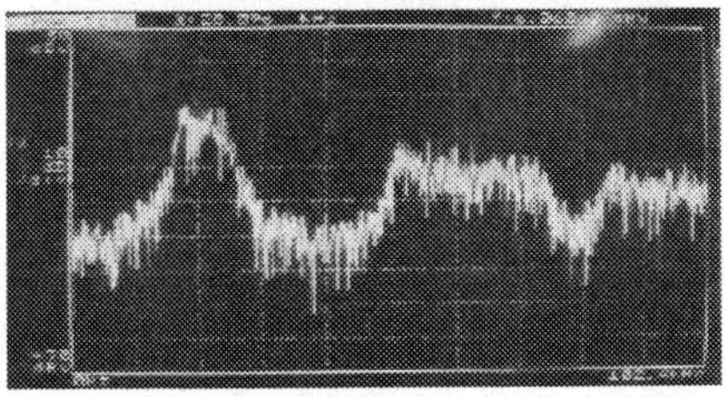

(b)

Figure 7 Experimental results of RH-ASDM (a) The upper: voltage across diode v_d 5 V/div, lower: output voltage ripple 200 mV/div, time 0.1 mS. (b) Harmonic spectrum of v_d 10 dB/div, freq. 10.2 kHz/div.

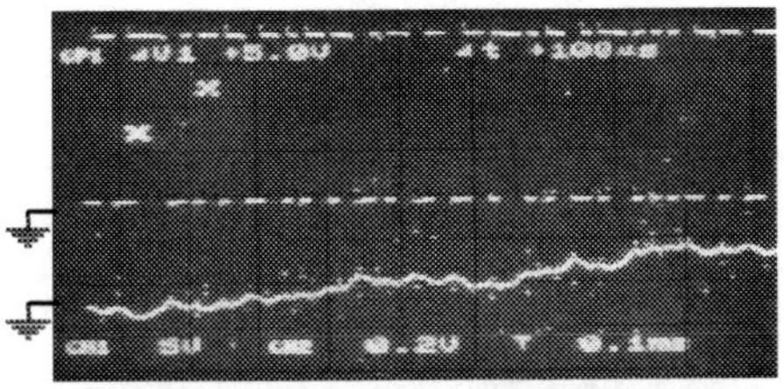

(a)

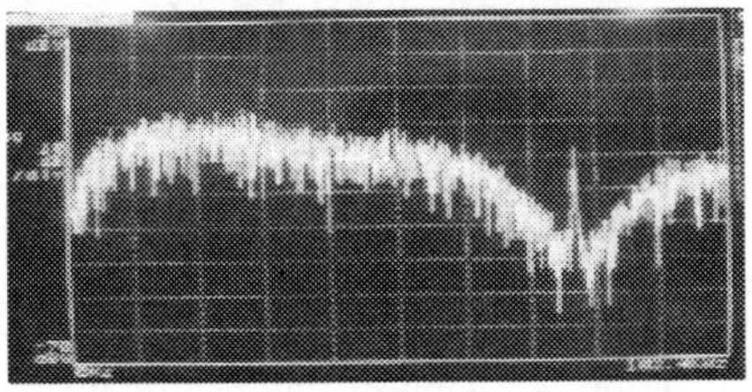

(b)

Figure 8 Experimental results of DSDM (a) Upper: voltage across diode v_d 5 V/div, lower: output voltage ripple 200 mV/div, time 0.1 mS. (b) Harmonic spectrum of v_d 10 dB/div, freq. 10.2 kHz/div.

5. SUMMARY

The problem of noises caused by the peaks of harmonic spectrum of output voltage in asynchronous sigma delta modulation is solved by the random hysteresis. The advantages of this method are that it is easy to implement by analog circuit without sampling clock and easier to design than a high-frequency sigma delta modulation. The experimental results show that the peaks of noise can be reduced while keeping the low ripple of output voltage. However, the switching frequency of asynchronous sigma delta modulation also varies with modulation index. So, the proposed technique is suitable for applications that require for fixed dc output voltage from a fixed dc supply.

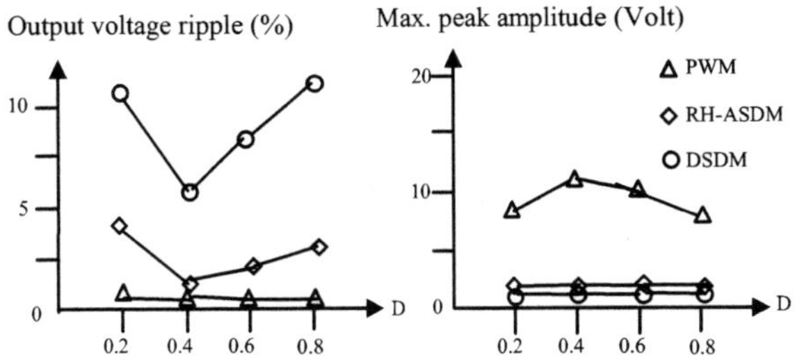

(a) (b)

Figure 9 Comparison of measured results of various PWM schemes (a) Output voltage ripple. (b) Maximum peak of amplitude of harmonic voltage.

6. ACKNOWLEDGMENTS

The authors would like to thank the Thailand Research Fund through the Royal Golden Jubilee, for the financial supporting this project under grant N0.PHD / 0080 / 2546.

7. REFERENCES

[1] F. Lin and D. Y. Chen, "Reduction of power supply EMI emission by switching frequency modulation," *IEEE Trans. Power Electron.*, vol. 9, pp. 132-137, Jan. 1994.

[2] K. K. Tse, H. Chung, S. Y. R. Hui, and H. C. So, "A Comparative Investigation on the Use of Random Modulation Schemes for DC/DC Converters," *IEEE Trans. Ind. Electron.*, vol. 47, pp. 253-263, April 2000.

[3] Y. Shrivastava, S. Y. R. Hui, S. Sathiakumar, H. Chung, and K. K. Tse, "Effects of conyinuous noise in randomized switching dc-dc converters," *Electron. Lett.,* vol. 33, pp. 919-921, 1997

[4] J. Paramesh and A. von Jouanne, "Use of sigma-delta modulation to control EMI from switch-mode power supplies," *IEEE Trans. Ind. Electron.*, vol. 48, pp. 111-117, Feb. 2001.

[5] C. H. Bae, J. H. Ryu, and K. W. Lee, "Suppression of Harmonic Spikes in Switching Converter Output Using Dithered Sigma-Delta Modulation," *IEEE Trans. Ind. Electron.*, vol. 38, pp. 159-166, Jan. 2002.

[6] A. Hirota, S. Nagai, and M. Nakaoka, "A Novel Delta-sigma Modulated DC-DC Power Converter utilizing Dither Signal," in *PESC.,* vol. 2, Galway, Ireland, pp. 831-836, 2000.

[7] T. C. Green and B. W. Williams, "Spectra of delta-sigma modulated inverters: An analytical treatment," *IEEE Trans. Power Electron.*, vol. 7, pp. 644-654, Oct. 1992.

A NEW RELIABLE SELF SUPPLIED GATE DRIVE CIRCUIT FOR SCRS WITH BREAKOVER DIODES FOR PROTECTION

Jun Zhang, Renjie Ding, Haitao Song

Department of Electrical Engineering, Tsinghua Univ. Beijing, P. R. China, 100084
Phone: +86 10 62773376-215; E-mail: zhangjuan02@mails.tsinghua.edu.cn

ABSTRACT

A new gate drive circuit for silicon controlled rectifiers (SCRs) is presented which provides the combined functions of reliable firing performance, overvoltage protection to the device with a breakover diode (BOD), and status report of both the firing circuit and the SCR to the controller for judgment. In order to save cost and space, the gate drive circuit is self supplied eliminating auxiliary power supply. The gate drive circuit is isolated from the controller and the gate pulse generator with optic fibers for safety. The design has already been successfully applied in a thyristor-controlled reactor (TCR), and some experimental results are presented.

1. INTRODUCTION

The resent increase in the application of static var compensators such as thyristor-controlled reactors has made it necessary to design a kind of reliable gate drive circuit with proper cost and size. A simple RC snubber circuit is normally used by the SCRs to limit the dv/dt within the maximum allowable rating [1]. Thus the idea about self supplied gate drive circuit for the SCRs is revived using the advantage of higher voltage across the SCR, with substantial savings in cost and space [2].

In the application of TCRs, overvoltage can in many cases arise from external sources (e.g., from the mains, lightning, etc.) or from within the electric equipment, e.g., due to the switching of breakers or as resulting from mishaps, therefore it is necessary to design the circuit in such a way that the rated device blocking voltage is never exceeded. This can be achieved using semiconductor devices such as BODs [3].

This paper introduces an improved self supplied gate drive circuit designed for TCRs. It receives a train of three pulses generated by the pulse generator through an optic fiber, and decodes the train to form a reliable firing pulse to the SCR. A status report pulse is generated by the circuit if the circuit works normally, and it is sent to the controller through another optic fiber to be monitored. The back to back SCRs are protected against overvoltage with a BOD. If one SCR is not switched on normally, the BOD fires it again. This action signal is fed back to the controller through the third optic fiber for judgment. In the following work, experimental results are also presented to confirm the theoretical analysis.

2. POWER SECTION OF THE GATE DRIVE CIRCUIT

2.1. System Configuration

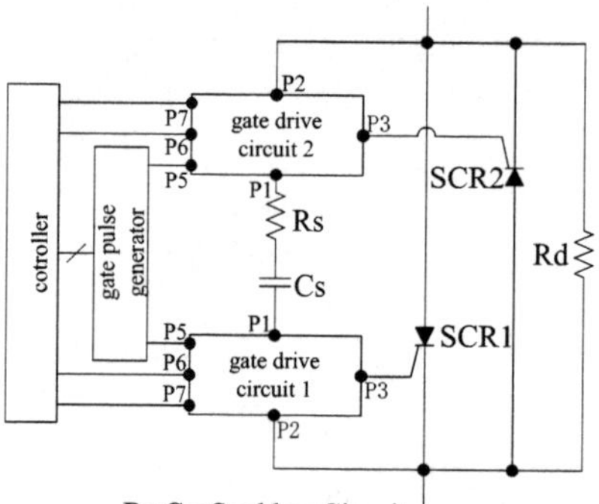

Rs,Cs; Snubber Circuit
Rd; Voltage Grading Circuit

Fig. 1. System configuration. For convenience, the BOD module is not drawn in this figure.

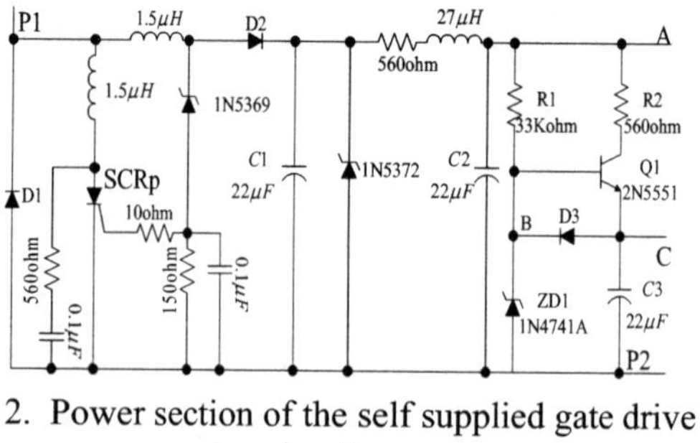

Fig. 2. Power section of the self supplied gate drive circuit.

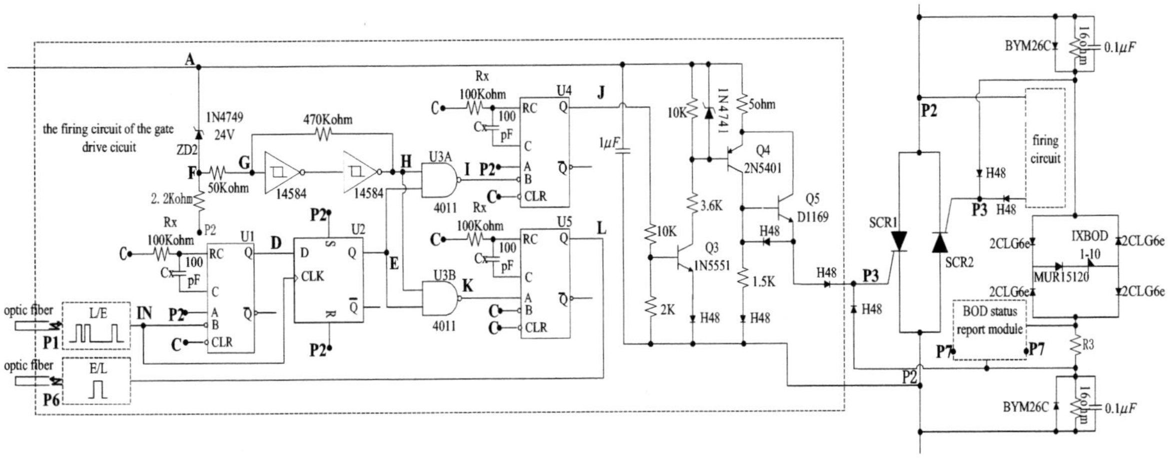

Fig. 4. Complete reliable firing circuit and the BOD module of the new gate drive circuit for SCRs.

Fig. 1 is the system configuration. It shows how the self supplied gate drive circuits are connected in the system.

2.2. Power Section

The power section is shown in Fig. 2. It derives power for decoding the train of pulses and gating the SCR from energy stored in the snubber capacitors at turn on and the voltage across the SCR at turn off [2]. C1 and C2 are storage capacitors. Thus the output voltage from point A to point P2 (say V_{AP2}) is set to a value from 31V to 49V. The clamping voltage at B provided by the Zener ZD1 is 11V while the output voltage from C to P2 is

$$V_{CP2} = V_{BP2} - V_{be}$$

where V_{be} is the voltage from base to emitter of the transistor Q1. If V_{CP2} is too higher, C3 discharges through D3. So V_{CP2} is set to approximately 10V.

3. RELIABLE FIRING CIRCUIT USING DECODING TECHNOLOGY

Reference pulses in synchronism with the fundamental component of system voltage are generated by the synchronizing system [4]. Then they are used by the gate pulse generator to time the train of three pulses sent to the gate drive circuits as shown in Fig. 3.

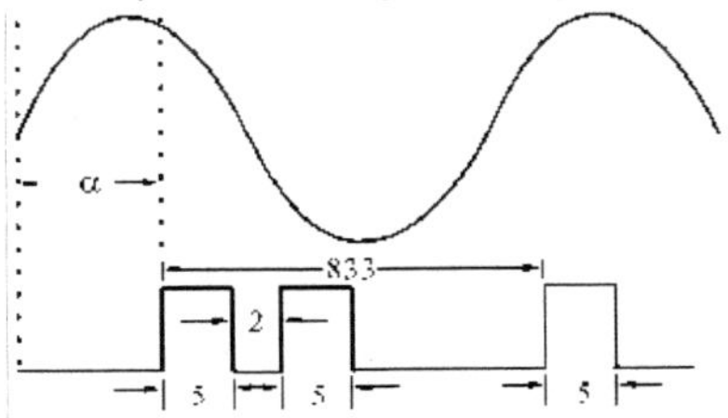

Fig. 3. The input train of pulses generated by the gate pulse generator (unit: μs). α is the firing angle. This figure is not drawn to scale.

The complete reliable firing circuit and the BOD module of the new gate drive circuit for SCRs are shown in Fig. 4. Some of the critical values of the final stage in the circuit shown, however, are selected for ABB 5STP 06D2200 phase control thyristors, which were used for the experiments.

At the negative-going edge of the first pulse of the input train, the Q output of the monostable multivibrator U1 at point D changes from low to high. Thus a positive pulse with a period T would be produced at D. The values of the timing capacitor Cx and the timing resistor Rx are chosen to make T about 14 μs (which is not required to

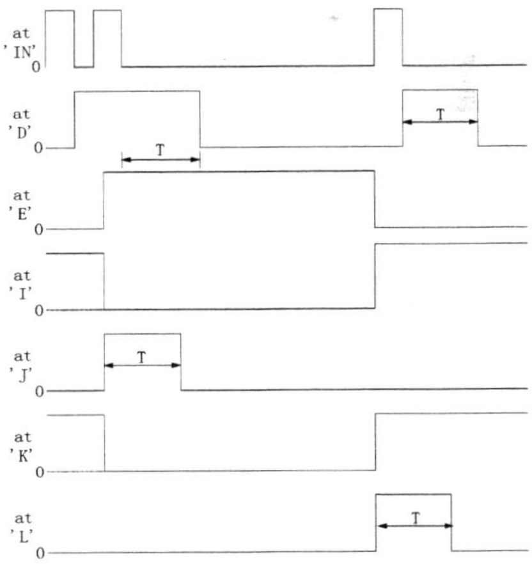

Fig. 5. Waveforms of various locations of drive circuit.

be accurate). At the positive-going edge of the second pulse which is input to the clock input of the D-type flip-flop U2, the logic high level present at the D input is transferred to the Q output at point E. Before the Q output

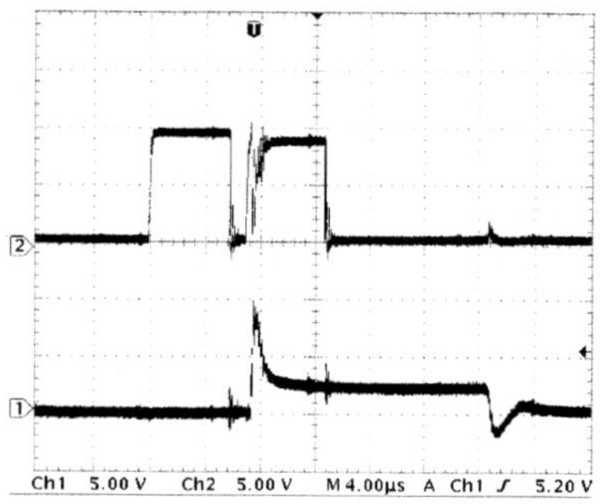

Fig. 6. Experimental result of decoding the first two pulses. Channel 2 shows the first two pulses of the input train and Channel 1 shows the gate-pulse.

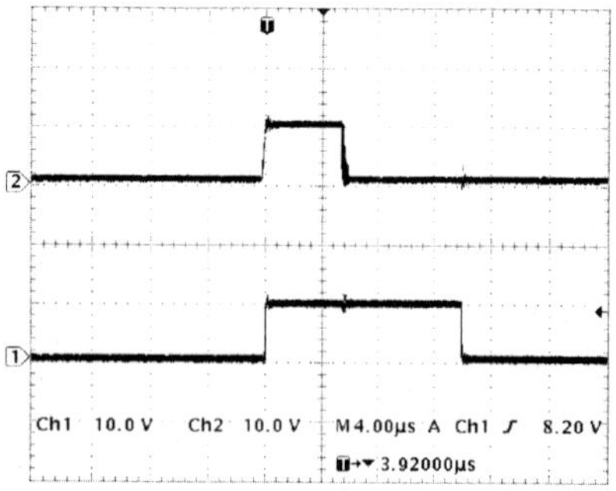

Fig. 7. Experimental result of producing the status report signal. Channel 2 shows the third pulse of the input train and Channel 1 shows the status report signal.

of U1 returns to the quiescent (zero) state, the negative-going edge of the second pulse causes an increase in output pulse width T at point D.

To ensure there is adequate energy supply to the pulse amplifier, the Zener ZD2 monitors the output voltage V_A of the power section. Under the normal condition, both the logic levels at point G and H remain high since the voltage at F is above 7V. Thus at the positive-going edge of the output signal at E, the output of the NAND gate U3A changes from high to low, providing a negative pulse input to the monostable multivibrator U4. At the negative-going edge of this input at point I, a positive pulse with the period T is input to the pulse amplifier at point J driving the transistor Q2 on. By the transistors Q3 and Q4, the pulse is amplified to fire the SCR. Accidental firing can be prevented reliably.

Before the third pulse of the train is input, the Q output of U1 has returned to the quiescent (zero) state. At the positive-going edge of the third pulse, the logic low level present at the D input is transferred to the Q output at point E making the output signal at I change from low to high which is not recognized by U4. At the same time the negative-going edge at E makes the output of the NAND gate U3B change from low to high, providing a positive pulse output at point L with the period T. Converted to light signal by an electric/light converter, this pulse is sent to the controller as a status report. If this signal is lost, the gate drive circuit could be considered aborted and certain measures would be adopted.

The waveforms at some important locations in the circuit are sketched in Fig. 5.

The experimental result of decoding the first two pulses is shown in Fig. 6. And the status report signal is shown in Fig. 7.As shown in Fig. 6, the gate-pulse lags behind the first pulse of the input train for 7 μs. Since in digital realization of the controller and the gate pulse generator of TCRs, the linearizing function is often realized through a lookup table [4], precorrection can be adopted to prevent errors from being produced here.

This implementation is more precise than those with pulse transformers that usually need modulated signals.

4. BOD MODULE FOR PROTECTION

The breakover diode is a small four-layer thyristor. When the applied voltage exceeds the breakdown voltage V_{BO} of the device, turn-on occurs in a controlled manner along the outer rim of the chip [3].

The circuit designed to protect the back to back SCRs (say BOD module) using one BOD with a rectifier is shown in Fig. 4. R3 is a current-limiting resistor and the series diode MUR15120 protects the asymmetric BOD against excessive reverse voltage.

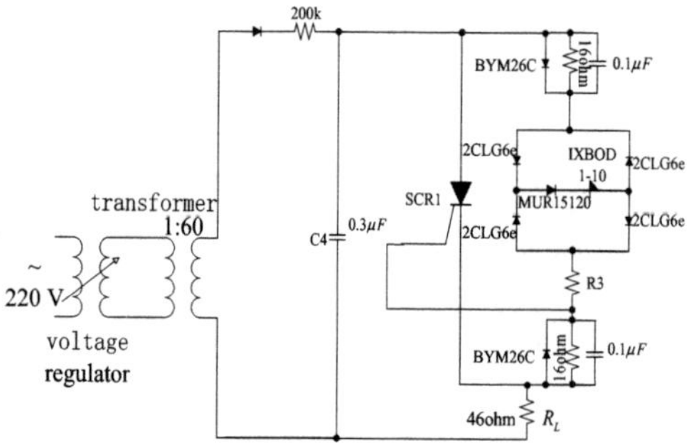

Fig. 8. Experimental circuit for testing of the BOD module.

The BOD module is tested with an IXBOD 1-10 and an ABB 5STP 06D2200 phase control thyristor in the circuit shown in Fig. 8. Adjusting the voltage regulator, capacitor C4 is charged by the output of the transformer. When the voltage across C4 exceeds the breakdown voltage V_{BO} (which is $1000 \pm 50V$), turn-on occurs. The peak value of the BOD current I_{TM} is determined by the resistors R_L and R3, namely

$$I_{TM} \approx V_{BO} / (R_L + R3)$$

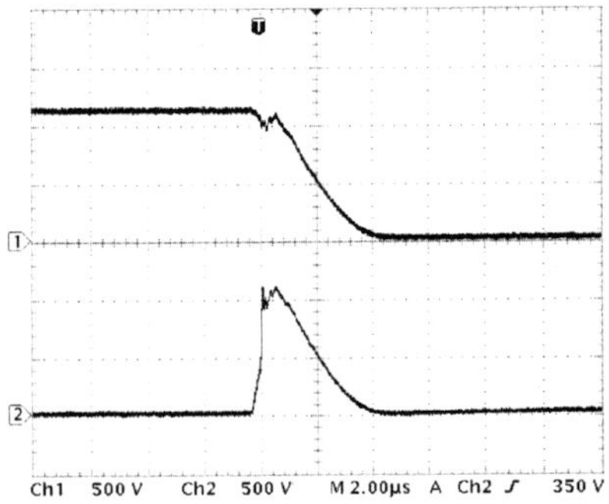

Fig. 9. Waveforms of the voltages across the SCR and R3. Channel 1 shows the voltage across the SCR. Channel 2 shows the voltage across R3 which is the same as the BOD current.

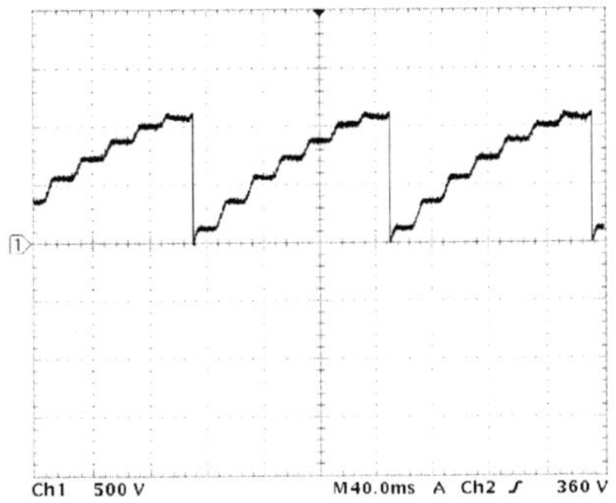

Fig. 10. Waveform of the voltage across the SCR for several periods.

The BOD current quickly goes below its holding current I_H because the current path through the BOD is shunted after firing the SCR. The waveforms of voltages across the SCR and R3 are shown in Fig. 9 and Fig. 10. With voltage dividers and electric/light converters, the turn-on signals of the BOD can be sent to the controller for judgment. If unduly frequent turn-ons occur to BODs, it is considered that something is wrong with SCRs or the gate drive circuits, and the shutdown inspection is necessary.

5. CONCLUSIONS

A reliable self supplied gate drive circuit for SCRs has been presented. The power section supplies adequate energy to the firing circuit, eliminating an isolated power supply board. Thus substantial cost and place are saved. The decoding technique generates gate-pulses significantly more reliable than ordinary firing circuits. With BODs, SCRs can be protected against overvoltage under most conditions. Also status reports of the firing circuits and the SCRs provide sufficient information to the controller for judgment.

This design is verified experimentally on a TCR.

6. REFERENCES

[1] Muhammad H. Rashid, *Power Electronics Circuits, Devices, and Applications*, 2nd ed. Englewood Cliffs, NJ: Prentice-Hall, Inc. A Simon & Schuster Company, 1993, pp. 597.

[2] Dusan M. Raonic, "SCR self-supplied gate driver for medium-voltage application with capacitor as storage element," *IEEE Trans. Industry Applications,* vol. 36, no. 1, pp. 212–216, Jan./Feb. 2000.

[3] Herbert M. Lawatsch and Janis Vitins, "Protection of thyristor against overvoltage with breakover diodes," *IEEE Trans. Industry Applications*, vol. 24, no. 3, pp. 444–448, May/June 1988.

[4] R. Mohan Mathur, Rajiv K. Varma, *Thyristor-based FACTS Controllers for Electrical Transmission System.* Piscataway: IEEE, New York: Wiley, 2002, pp. 124–127.

Hilbert Space Techniques for Evaluating Trade-Offs in Reactive Power Compensation

Hanoch Lev-Ari
Aleksandar M. Stanković
Kevin Xu
Milun M. Perišić

Northeastern University
Department of Electrical and Computer Engineering
Boston, USA
{levari,astankov,kxu,perisic}@ece.neu.edu

***Abstract*— The paper describes a set of Hilbert space-based analysis tools for reactive power compensation, and uses them to quantify the tradeoff between line losses and compensator cost. Our method is illustrated on two polyphase examples, one involving a resistive circuit, and the other involving a nonlinear circuit (diode bridge rectifier) which is often encountered in industrial practice. The paper also addresses the issue of an efficient (order-recursive) computational procedure for the considered class of compensators.**

***Index Terms*— Reactive power compensation, Hilbert space optimization, voltage sourced inverters, active filters, energy storage.**

I. INTRODUCTION

The compensation of reactive (inactive) power is of great interest in energy processing systems, as it directly affects efficiency. This issue is related to significant theoretical challenges, as there is still no consensus about the general polyphase case in the presence of harmonics. In industrial practice, the main new enabling technology is provided by so called active power filters (for a recent review see, for example, [1]). These compensators, connected typically in parallel with the load and typically built around switching voltage–sourced inverters, can inject currents that reduce the overall line current drawn from the source.

While active power filters can substantially reduce the source current (and consequently reduce the power losses in the line), they may require substantial energy storage for the task [2]. This storage is typically provided by capacitors whose cost may become prohibitive. In addition, if the storage is provided by a series bank of capacitors, control problems related to voltage balancing can become serious [3], [4].

A family of compensators was introduced in [5] that achieve power harmonic matching (PHM) on the line up to a selected (say L-th) harmonic. These "PHM-compensators" span the range between Fryze compensator (which matches only the DC-component and thus minimizes line losses while possibly requiring substantial energy storage) and the so-called instantaneous compensator that matches all harmonics (thus increasing power losses, but requiring no energy storage). PHM-compensators allow flexible control of the tradeoff between ohmic loss and hardware cost. As L increases, ohmic line losses increase as well, but hardware cost (which is roughly proportional to the required energy storage) typically decreases.

In this paper we use Hilbert space analysis tools to quantify the tradeoff between line losses and compensator cost, and illustrate the method on two examples. We demonstrate through examples that a modest value of L (i.e., $L = 1$, or $L = 2$) can be an attractive alternative to using either Fryze ($L = 0$), or instantaneous ($L = \infty$) compensators. We also address the issue of efficient computations for PHM-compensators which is paramount for actual implementations. We describe a lattice realization which is: (i) order-recursive, allowing to change the index L on-line, possibly in response to changing loads and onset of disturbances; (ii) characterized by lower computational cost and better numerical robustness, because we avoid explicit inversion of a Grammian matrix.

II. COST-PERFORMANCE TRADEOFF

We use the Hilbert space terminology of [5] to formulate our objectives and derive our results. Thus $v(t)$ and $i(t)$ are vectors representing load voltage and current, which we view as elements in a Hilbert space of n-phase T-periodic waveforms, with the inner product defined by

$$\langle\, x\,,\, y \,\rangle \overset{\text{def}}{=} \frac{1}{T}\int_T x(t)y^\top(t)dt$$

For instance, in this terminology the rms value of the polyphase voltage $v(t)$ is expressed as $\|v\| = \sqrt{\langle\, v\,,\, v\,\rangle}$. Our choice of using row vectors (rather than column vectors) to describe polyphase waveforms turns out to be more convenient for direct (algebraic and/or numerical) evaluation of orthogonal projections.

The *instantaneous active current* $i_a(t)$ is defined as:

$$i_a(t) \overset{\text{def}}{=} \frac{i(t)v^\top(t)}{v(t)v^\top(t)}v(t) = \frac{p(t)}{v(t)v^\top(t)}v(t) \tag{1}$$

The *instantaneous inactive current*

$$i_x(t) \overset{\text{def}}{=} i(t) - i_a(t) \tag{2}$$

satisfies the relation $i_a(t)i_x^\top(t) = 0$, so that $p(t) = i(t)v^\top(t) = i_a(t)v^\top(t)$.

0-7803-8251-X/04/$17.00 ©2004 IEEE

This definition of active current is distinct from a related definition due to Fryze

$$i_f(t) = \frac{\langle\, i\,,\, v\,\rangle}{\|v\|^2} v(t) \tag{3}$$

where we observe that $\langle\, i\,,\, v\,\rangle$ is simply the average value (= DC component) of the instantaneous power $p(t)$. In circuit terms, $i_a(t)$ is the smallest current (by rms) that supplies the same $p(t)$ as the load current $i(t)$, while $i_f(t)$ is the smallest current (by rms) that supplies the same DC component of $p(t)$ as the load current $i(t)$.

We quantify the performance of a compensator in terms of the (normalized) *wasted power coefficient*

$$\nu = \frac{\|i - i_{comp}\|^2 - \|i_f\|^2}{\|i\|^2 - \|i_f\|^2} \tag{4}$$

where $i_{comp}(t)$ is the compensator current, as shown in Fig. 1. With our definition of ν, a Fryze compensator achieves $\nu = 0$ because $i(t) - i_{comp}(t) = i_f(t)$. On the other hand, $\nu = 1$ when no compensator is used because $i_{comp}(t) = 0$ and $\|i - i_{comp}\|^2 - \|i_f\|^2 = \|i\|^2 - \|i_f\|^2$.

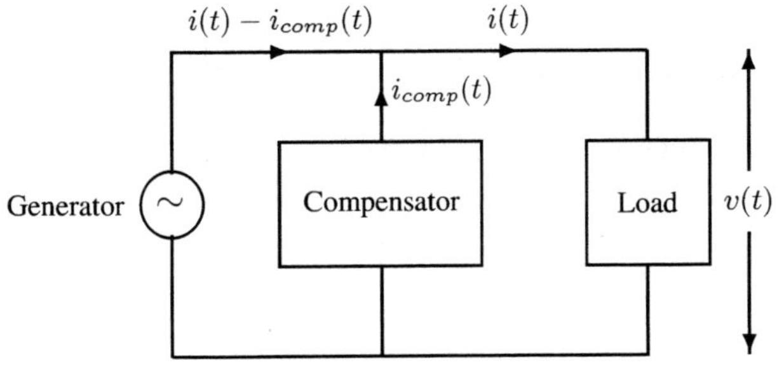

Fig. 1. Load compensation in a power delivery system.

The currently used technology for hardware implementation of compensators requires the use of a capacitor to store energy during each cycle. The instantaneous energy stored in the compensator is

$$E(t) = \int_0^t i_{comp}(\tau) v^\top(\tau) d\tau\,, \qquad 0 \le t \le T$$

and it fluctuates during a cycle, with a peak value

$$E_{max} = \max_{0\le t\le T} \left| \int_0^t i_{comp}(\tau) v^\top(\tau) d\tau \right| \tag{5}$$

The compensator uses a capacitor whose value is proportional to E_{max} for temporary storage of energy during the cycle. We use the "normalized" peak energy E_{max}/T as our measure of hardware (i.e., capacitor) cost. Thus, our basic problem is to obtain a desired tradeoff between wasted power and hardware cost. The two extreme choices are the Fryze compensator, which has the lowest wasted power coefficient ($\nu = 0$), and the instantaneous compensator, which has the lowest hardware cost ($E_{max} = 0$).

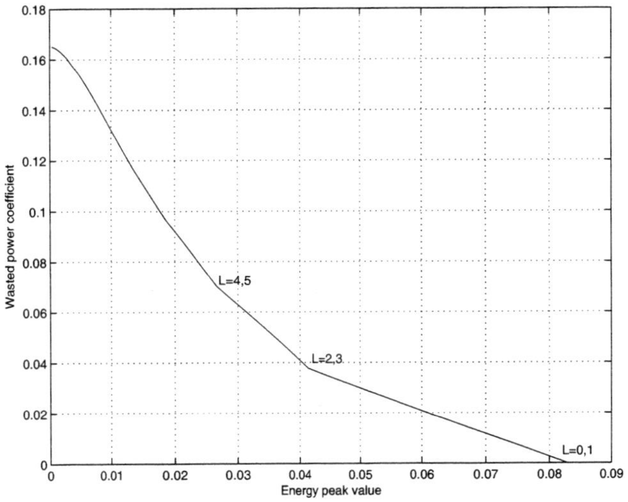

Fig. 2. Cost-performance tradeoff (Example 1).

The family of PHM-compensators, which was introduced in [5], spans the range between those two extremes. Its members are parametrized by an integer index L, such that $L = 0$ corresponds to the Fryze compensator, while the instantaneous compensator is obtained for $L = \infty$. Thus the cost-performance tradeoff can be controlled by varying the index L. We illustrate this capability via two examples.

EXAMPLE 1 - A RESISTIVE CIRCUIT

This is the same example that was analyzed in [5], namely

$$\begin{aligned} v(t) &= \left[\, V_1 \cos(\omega t) \qquad V_2 \sin(\omega t) \,\right] \\ i(t) &= \left[\, {}^{P_0}/_{V_1} \cos(\omega t) \qquad {}^{P_0}/_{V_2} \sin(\omega t) \,\right] \end{aligned}$$

The corresponding cost-performance tradeoff curve (with $V_1 = 1$, $V_2 = 0.1$, $P_0 = 1$) is shown in Fig. 2. Note that in this example $i_a(t)$ has all odd harmonics, despite the fact that $v(t)$ and $i(t)$ have only the fundamental. We observe that, for $L = 2$, hardware cost is reduced by 50% as compared with the Fryze compensator, while wasted power is reduced by 77% as compared with the instantaneous compensator.

EXAMPLE 2 - A NONLINEAR CIRCUIT

This example consists of an unbalanced three phase voltage source connected to a three-phase diode bridge supplying an RLC load, shown in Fig. 3. The circuit parameters are: L = $0.1H$, C = $40\mu F$, R = 80Ω, $V_a = 50\sin(\omega t)$, $V_b = 60\sin(\omega t + 90)$, $V_c = 40\sin(\omega t + 270)$. This is a simple model for loads that are common in industrial practice, and the corresponding cost-performance tradeoff curve is shown in Fig. 4.

Both examples illustrate and quantify the fact that PHM-compensators with fairly small values of L achieve substantial line loss reduction with only modest hardware costs.

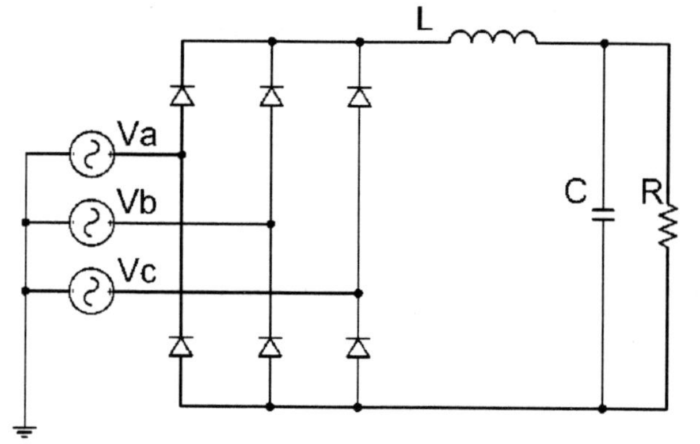

Fig. 3. Circuit schematic for Example 2.

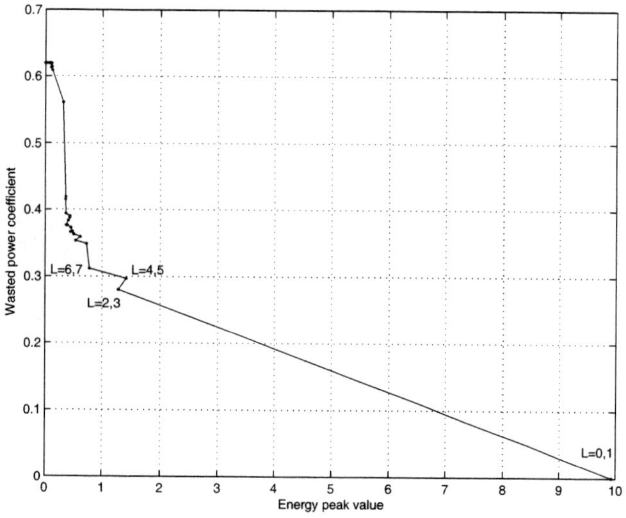

Fig. 4. Cost-performance tradeoff (Example 2).

These findings are in agreement with [2] which recommends (in qualitative terms) that "only power components above a certain frequency [should] be compensated by the filter".

III. Lattice Filter Implementation

As demonstrated in [5], optimal compensation is often synonymous with orthogonal projection on a suitable subspace. In particular, the compensated line current $i_a^{(L)}(t) = i(t) - i_{comp}(t)$ that minimizes ohmic line losses, while preserving the first L power harmonics, is the projection of the original load current $i(t)$ on the subspace

$$\mathcal{H}_L \stackrel{\text{def}}{=} span\ \{v(t)\, e^{j\ell\omega t}\ ;\ |\ell| \leq L\}$$

In particular, the Fryze compensator corresponds to $L = 0$, since it preserves only the average of the instantaneous power $p(t) = i(t)\, v^{\top}(t)$. On the other extreme, the instantaneous compensator, which preserves $p(t)$ itself, corresponds to $L = \infty$.

Direct evaluation of $i_a^{(L)}(t)$ involves the solution of a system of $(2L+1)$ equations, at a computational cost that is proportional to $(2L+1)^3$. A more efficient order-recursive construction of the compensated line current $i_a^{(L)}(t)$ was introduced in [6]. It uses a cascade configuration, akin to the well known lattice-filter [7], to determine the entire set $\{\, i_a^{(\ell)}(t)\, ;\ 0 \leq \ell \leq L\}$ recursively in ℓ (Fig. 5). Notice that the top part of the block diagram is self-contained, using only the polyphase voltage $v(t)$ as input. It serves to construct an orthogonal basis (i.e., the polyphase signals $b_0(t)$, $f_1(t)$, $b_2(t)$, $f_3(t)$, $b_4(t)$, etc.) for the subspace $\mathcal{H}_L$. This basis is used in the bottom part of the block diagram, together with $i(t)$, to determine the compensated line currents $i_a^{(0)}(t)$, $i_a^{(1)}(t)$, ..., $i_a^{(L)}(t)$, recursively in order. The detailed input-output relations are summarized in Table I.

Once the current $i_a^{(L)}$ has been evaluated via the lattice recursions, the required compensator current is given by $i_{comp}(t) = i(t) - i_a^{(L)}(t)$. The wasted power coefficient ν and the energy peak value E_{max} are then obtained via (4) and (5), respectively. Repeating this procedure for $L = 0, 1, 2, 3, \ldots$ produces the tradeoff curves in Figs. 2 and 4.

When the voltage $v(t)$ is band-limited, consisting of only a few harmonics, the recursive procedure of Table I can be efficiently implemented in the frequency domain, using as input the Fourier coefficients of $v(t)$ and $i(t)$ [6]. Further simplification is achieved by relying explicitly on the fact that $i(t)$ and $v(t)$ are real-valued, so that for $l = 1, 2, \ldots, L$,

$$b_{2l-1}(t) = f^*_{2l-1}(t) \quad , \qquad b_{2l}(t) = f^*_{2l}(t)$$

If, in addition, $v(t)$ has only odd harmonics (in particular, if it is sinusoidal) then more simplifications occur, viz.,

$$K^f_{2\ell-1} = 0 = K^b_{2\ell-1} \qquad \text{for } \ell = 1, 2,\ \ldots\ , L$$

These special relations reduce the overall computational cost by approximately a factor of four, making the lattice implementation especially attractive.

IV. Concluding Remarks

We have used the notion of cost-performance tradeoff to illustrate the capabilities offered by the new family of PHM-compensators. In particular, we have shown that choosing small L values often yields very attractive alternatives to both Fryze and instantaneous compensation, in the sense of significantly reducing both ohmic line losses and energy storage costs, as compared with their maximal values.

We have also introduced an efficient order-recursive lattice-type realization for PHM-compensators. In addition to reducing the computational burden to a level comparable with that of the instantaneous compensator, our lattice realization makes it possible to adjust the value of the compensator order L online (i.e., during the operation of the compensator) with very little additional computation. This capability could be used, for example, to adjust the performance to track load changes in utility applications.

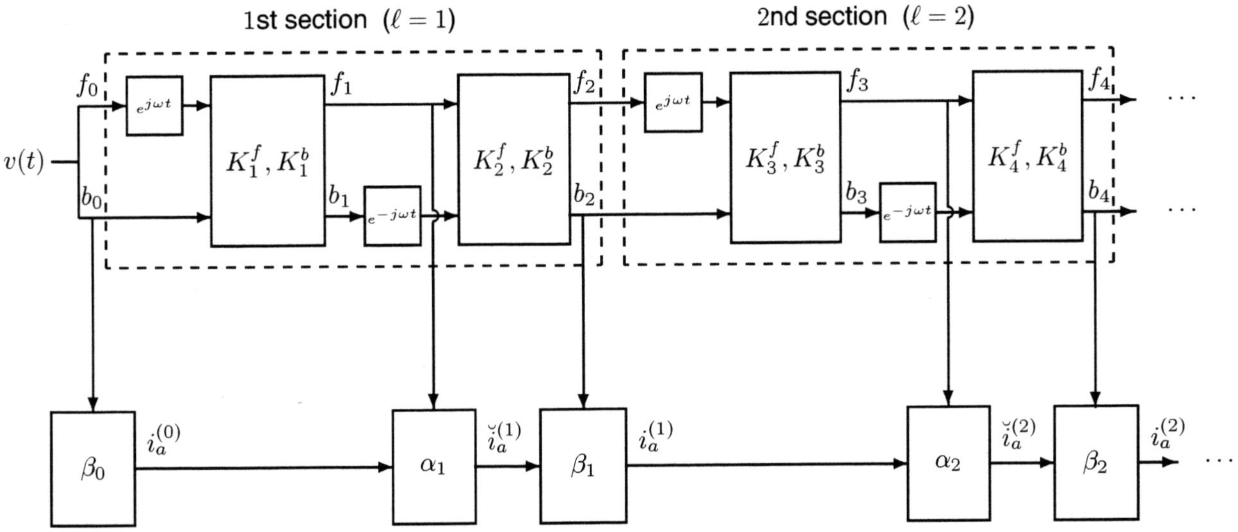

Fig. 5. Lattice filter realization of $i_a^{(L)}(t)$

1st cell:

$$\tilde{f}_{2\ell-2}(t) = e^{j\omega t} f_{2\ell-2}(t)$$

$$K^f_{2\ell-1} = \frac{\langle \tilde{f}_{2\ell-2}, b_{2\ell-2} \rangle}{\|b_{2\ell-2}\|^2} \quad , \quad K^b_{2\ell-1} = \frac{\langle b_{2\ell-2}, \tilde{f}_{2\ell-2} \rangle}{\|\tilde{f}_{2\ell-2}\|^2}$$

$$\begin{bmatrix} f_{2\ell-1}(t) \\ b_{2\ell-1}(t) \end{bmatrix} = \begin{bmatrix} 1 & -K^f_{2\ell-1} \\ -K^b_{2\ell-1} & 1 \end{bmatrix} \begin{bmatrix} \tilde{f}_{2\ell-2}(t) \\ b_{2\ell-2}(t) \end{bmatrix}$$

$$\alpha_\ell = \frac{\langle i, f_{2\ell-1} \rangle}{\|f_{2\ell-1}\|^2}$$

$$\check{i}_a^{(\ell)}(t) = i_a^{(\ell-1)}(t) + \alpha_\ell \, f_{2\ell-1}(t)$$

2nd cell:

$$\tilde{b}_{2\ell-1}(t) = e^{-j\omega t} b_{2\ell-1}(t)$$

$$K^f_{2\ell} = \frac{\langle f_{2\ell-1}, \tilde{b}_{2\ell-1} \rangle}{\|\tilde{b}_{2\ell-1}\|^2} \quad , \quad K^b_{2\ell} = \frac{\langle \tilde{b}_{2\ell-1}, f_{2\ell-1} \rangle}{\|f_{2\ell-1}\|^2}$$

$$\begin{bmatrix} f_{2\ell}(t) \\ b_{2\ell}(t) \end{bmatrix} = \begin{bmatrix} 1 & -K^f_{2\ell} \\ -K^b_{2\ell} & 1 \end{bmatrix} \begin{bmatrix} f_{2\ell-1}(t) \\ \tilde{b}_{2\ell-1}(t) \end{bmatrix}$$

$$\beta_\ell = \frac{\langle i, b_{2\ell} \rangle}{\|b_{2\ell}\|^2}$$

$$i_a^{(\ell)}(t) = \check{i}_a^{(\ell)}(t) + \beta_\ell \, b_{2\ell}(t)$$

TABLE I

INPUT-OUTPUT RELATIONSHIPS FOR THE ℓ-TH LATTICE SECTION

V. REFERENCES

[1] B. Singh, K. Al-Haddad and A. Chandra, "A review of active filters for power quality improvement," *IEEE Trans. on Industrial Electronics*, 46(5), Oct. 1999, pp. 960 -971.

[2] L. Malesani, L. Rossetto and P. Tenti, "Active Power Filter with Hybrid Energy Storage," *IEEE Trans. on Power Electronics*, 6(3), July 1991, pp. 392-397.

[3] H. Akagi, "Active Filters and Energy Storage Systems Operated under Non-Periodic Conditions," *IEEE Power Engr. Society Summer Meeting*, Seattle, WA, July 2000, pp. 965-970.

[4] H. Akagi, "Large Static Converters for Industry and Utility Applications," *Proceedings of the IEEE*, 89(6), June 2001, pp. 976-983.

[5] H. Lev-Ari and A.M. Stanković, "Hilbert Space Techniques for Modeling and Compensation of Reactive Power in Energy Processing Systems," *IEEE Trans. on Circuits & Systems- I: Fundamental Theory and Applications*, 50(4), Apr. 2003, pp. 540-556.

[6] K. Xu, "Lattice Filters for Efficient Reactive Power Compensation," M.S. Thesis, Dept. of Electrical and Computer Engineering, Northeastern University, Boston, MA, May 2003.

[7] C.W. Therrien, *Discrete Random Signals and Statistical Signal Processing,* Prentice-Hall, Englewood Cliffs, NJ, 1992.

PETRI NET BASED TRANSFORMER FAULT DIAGNOSIS

Pavlos S. Georgilakis, Member, IEEE, John Katsigiannis, Kimon P. Valavanis, Senior Member, IEEE

Department of Production Engineering and Management, Technical University of Crete, Greece

ABSTRACT

The reduction of the time needed for transformer fault diagnosis is an important task for transformer users. In this paper, Petri Nets are exploited, in order to simulate the transformer fault diagnosis process and to define the actions followed to repair the transformer.

1. INTRODUCTION

Power system reliability depends on components reliability. As the ultimate element in the electricity supply chain, the distribution transformer is one of the most widespread apparatus in electric power systems. During their operation, transformers are subjected to many external electrical stresses from both the upstream and downstream network. As a result, various types of faults (e.g. insulation failure, overloading, oil leakage, short-circuit, etc) can occur to the transformers of an electric utility. The consequences of transformer fault can be significant (damage, oil pollution, etc). Failure of these transformers is very costly to both the electrical utilities and their customers.

When a transformer fault occurs, it is important to identify the fault type and to minimize the time needed for transformer repair, especially in cases where the continuity of supply is crucial. Consequently, it should not come as a surprise that transformer fault diagnosis forms a subject of a permanent research effort.

Various transformer fault diagnosis techniques have been proposed in the literature, for different types of faults [1]. For thermal related faults, the most important diagnostic method is the gas-in-oil analysis, while other methods such as the degree of polymerization, the furanic compounds analysis and the thermography are also applicable. For dielectric related faults, it is necessary to localize and to characterize the partial discharge source, in order to give a correct diagnosis after receiving an alarm signal via sensors or via gas-in-oil sampling. For mechanical related faults, the frequency response analysis and the leakage inductance methods are the more frequently used transformer fault diagnosis techniques. Finally, for transformer general degradation, the dielectric response, the oil analysis and the furanic compounds analysis methods are applicable.

In spite of the wide range of the transformer fault diagnosis methods, the diagnostic criteria developed till today are not fully applicable to all faulty cases, and consequently, the experience of experts still play an important role in the diagnosis of the transformer faults. Dismantling the suspected transformers, performing internal examinations, and holding a group discussion are usually the procedure to conclude the diagnosis.

Expert systems and artificial intelligence techniques have already been proposed to understand the obvious and non-obvious relationships between transformer failures and the causes of failures (i.e. internal or external causes). Preliminary results, obtained from the application of these techniques, are encouraging, however some limitations exist [2]. Knowledge acquisition, knowledge representation and maintenance of a great number of rules in the expert systems require plenty of efforts.

In this paper, Petri Nets are proposed for modeling of transformer fault diagnosis process. Petri Nets are both a mathematical and graphical tool capable of capturing deterministic or stochastic system behavior and modeling phenomena such as sequentialism, parallelism, asynchronous behavior, conflicts, resource sharing and mutual exclusion. The proposed method offers significant advantages such as systematical determination of the sequence of fault diagnosis and repair actions, visual representation of the above actions, as well as estimation of the time needed for transformer repair.

The paper is organized as follows: Section 2 describes the Petri Nets methodology. The application of Petri Nets to transformer fault diagnosis and the obtained results are described in Section 3. Finally, Section 4 concludes the paper.

2. OVERVIEW OF PETRI NETS

Petri Nets (PNs) were introduced in Carl A. Petri's 1962 Ph.D. dissertation. Since that time, they have proved to be a valuable graphical and mathematical modeling tool applicable to many systems. As a graphical tool, PNs can be used as a visual communication aid similar to flow charts, block diagrams, and networks. As a mathematical tool, it is possible to set up state equations, algebraic

0-7803-8251-X/04/$17.00 ©2004 IEEE

equations, and other mathematical models governing the behavior of systems.

A PN is a particular kind of directed graph, together with an initial marking, M_0. The underlying graph of a PN is a directed, weighted, bipartite graph consisting of two kinds of nodes, called places and transitions, where arcs are either from a place to a transition or from a transition to a place. In graphical representation, places are drawn as circles, and transitions as either bars or boxes. If a marking (state) assigns to each place p a nonnegative integer k, it is called that p is marked with k tokens. Pictorially, k black dots (tokens) are placed in p.

Places are used to describe possible local system rates, named conditions or situations. Transitions are used to describe events that may modify the system state. Arcs specify the relation between local states and events in two ways: they indicate the local state in which the event can occur, and the local state transformations induced by the event.

The presence of a token in a place is interpreted as holding the truth of the condition associated with the place. The only execution rule in a PN is the rule for transition enabling and firing. A transition t is considered as enabled if each input place p of t is marked with at least $w(p,t)$ tokens, where $w(p,t)$ is the weight of the arc from p to t. An enabled transition may or may not fire. A firing of an enabled transition t removes $w(p,t)$ tokens from all its input places p, and adds $w(p,t)$ tokens to each of its output places, where $w(t,p)$ is the weight of the arc from t to p. The movement of tokens through the PN graph represents the flow of information or control in the system.

For problems that include the completion of an activity, it is necessary and useful to introduce time delays associated with transitions (activity executions) in their net models. Such a PN model is known as a deterministic timed net if the delays are deterministically given, or as a stochastic net, if the delays are probabilistically specified. In both cases, boxes of thick bars graphically represent transitions. The Stochastic Petri Net (SPN) model provides a more realistic representation of matter. In SPNs transitions are associated with random variables that express the delay from enabling to the firing of the transition. The type of distribution in random variables can be uniform, exponential, etc.

3. FAULT DIAGNOSIS USING PETRI NETS

This paper simulates the actions that are followed by the transformer maintenance personnel in order to diagnose the fault and repair the transformer. It is important to notice that the maintenance staff is not able to know the exact problem from the beginning of the diagnosis process; there is crucial information that is obtained during the whole tranformer fault diagnosis process.

To better model the transformer fault diagnosis process, stochastic PNs are used in this paper. These nets provide a structural tool, like flow charts, with the additional advantages of simulating dynamic and concurrent actions, and they provide the simulation results using stochastic times for a number of transitions.

Fig. 1 presents the proposed PN model for transformer fault diagnosis, Fig. 2 shows the "not on-site repair" subnet (i.e. in case that the transformer repair is implemented in the factory), and Table 1 describes all places and transitions that constitute the PN models of Fig. 1 and 2. Places in shadow boxes represent the crucial information that is obtained during the transformer fault diagnosis process; these places represent two opposite events, so tokens can be placed only in one of the places.

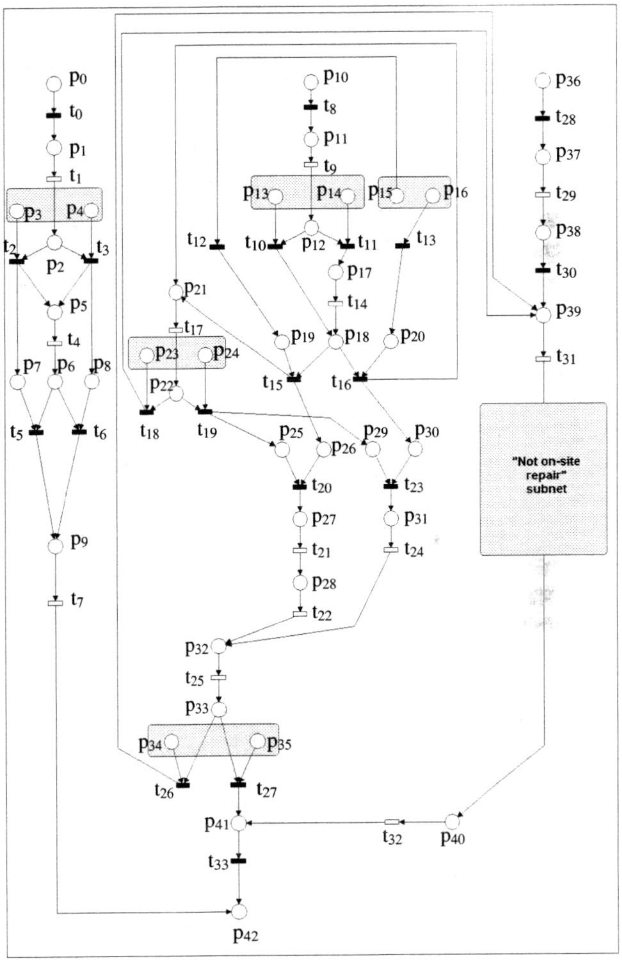

Fig. 1: PN model for transformer fault diagnosis.

The proposed PN models the following transformer faults: overloading, short-circuit, insulation failure and oil leakage. The protection equipment that is used in a typical distribution transformer for fault detection is the oil thermometer and the Buchholz relay. These protecting schemes may be alarmed or tripped with the appearance of a problem, and when this happens there is an immediate warning to the personnel. The possible initial warnings are a) alarm of the oil thermometer (oil thermometer cannot trip without earlier alarm), and b) alarm or trip of the Buchholz relay. In case of alarm, it can be a change to trip when the maintenance staff arrives to the transformer, depending on problem's seriousness and the time required arriving in transformer's area.

	Description of places and transitions	Duration (hours)
Main Petri net		
p_0	Oil thermometer alarms	
t_0	Alarm is activated	0
p_1	Personnel is notified	
t_1	Personnel is moving to transformer area	[0 2]
p_2	Existence of alarm or trip?	
p_3	Oil thermometer still alarms	
t_2	Alarm is still activated	0
p_4	Oil thermometer tripped	
t_3	Trip is activated	0
p_5	Need to check the loads	
t_4	Loads are checked	[1 5]
p_6	Does transformer need to restart?	
p_7	It doesn't need to restart	
t_5	No restart is needed	0
p_8	It needs to restart	
t_6	Transformer is restarting	0
p_9	Loads have to be reduced properly	
t_7	Loads are reduced properly	[1 3]
p_{10}	Buchholz relay alarms	
t_8	Alarm is activated	0
p_{11}	Personnel is notified	
t_9	Personnel is moving to transformer area	[0 2]
p_{12}	Existence of alarm or trip?	
p_{13}	Buchholz relay tripped	
t_{10}	Trip is activated	0
p_{14}	Buchholz relay still alarms	
t_{11}	Alarm is still activated	0
p_{15}	Low level of oil indicator	
t_{12}	Oil volume has reduced	0
p_{16}	Air bubbles in Buchholz relay's glass	
t_{13}	Air bubbles are observed	0
p_{17}	Transformer needs to stop	
t_{14}	Transformer is stopped	1
p_{18}	Existence of oil leakage or insulation failure?	
p_{19}	Oil leakage	
t_{15}	Existence of oil leakage	0
p_{20}	Insulation failure	
t_{16}	Existence of insulation failure	0
p_{21}	Check for the exact type of fault	
t_{17}	Transformer is checked	1
p_{22}	Is it possible repair fault on the spot?	
p_{23}	It is not possible to repair	
t_{18}	Fault cannot be repaired on the spot	0
p_{24}	It is possible to repair	
t_{19}	Fault can be repaired on the spot	0
p_{25}	Possibility for repairing oil leakage	
p_{26}	Problem of oil leakage	
t_{20}	Repair of oil leakage is possible	0
p_{27}	Personnel prepares to repair transformer	
t_{21}	Transformer is repaired	[2 5]
p_{28}	Lost oil needs to be replaced	
t_{22}	Lost oil is replaced	1
p_{29}	Possibility for repairing insulation failure	
p_{30}	Problem of insulation failure	
t_{23}	Repair of insulation failure is possible	0
p_{31}	Need to replace problematic external parts	
t_{24}	Parts are replaced	2
p_{32}	Check if everything works properly	
t_{25}	Transformer is checked	1
p_{33}	Is transformer working properly?	
p_{34}	It is not working properly	
t_{26}	Fault still exists	0
p_{35}	It is working properly	
t_{27}	Fault is repaired	0
p_{36}	Buchholz relay trips	
t_{28}	Trip is activated	0
p_{37}	Personnel is notified	
t_{29}	Personnel is moving to transformer area	[0 2]
p_{38}	Identification of transformer's fault	
t_{30}	Existence of a powerful short-circuit	0
p_{39}	Transformer needs to disconnect	
t_{31}	Transformer is disconnected	2
p_{40}	Transformer arrives in area of installation	
t_{32}	Transformer is reinstalled	2
p_{41}	Transformer is ready to work	
t_{33}	Transformer is restarted	0
p_{42}	Transformer reworks properly	
"Not on-site repair" subnet		
p_0	Transformer is sending to repairing area	
t_0	Transformer arrives to repairing area	[2 24]
p_1	Oil has to be removed	
t_1	Oil is removed	1
p_2	Inside search is needed	
t_2	Tank is opened	1
p_3	Check for the exact type of fault	
t_3	Check is done	1
p_4	Identification of fault	
t_4	Fault is repaired	[72 360]
p_5	Transformer has to be reassembled	
t_5	Transformer is reassembled	1
p_6	Oil has to be added	
t_6	Oil is added	4
p_7	Check for the proper operation	
t_7	Check is done	1
p_8	Is transformer working properly?	
p_9	It is not working properly	
t_8	Fault still exists	0
p_{10}	It is working properly	
t_9	Fault is repaired	0
p_{11}	Transformer is ready to be sent back in its area	
t_{10}	Transformer is transferred	[2 4]

Table 1: Description of PN places and transitions and duration of PN transitions.

When the oil thermometer alarms or trips, there is an overloading problem in the transformer. The maintenance staff has to check if the loads are over the transformer overloading limits, reduce the loads accordingly and restart the transformer (in case of trip).

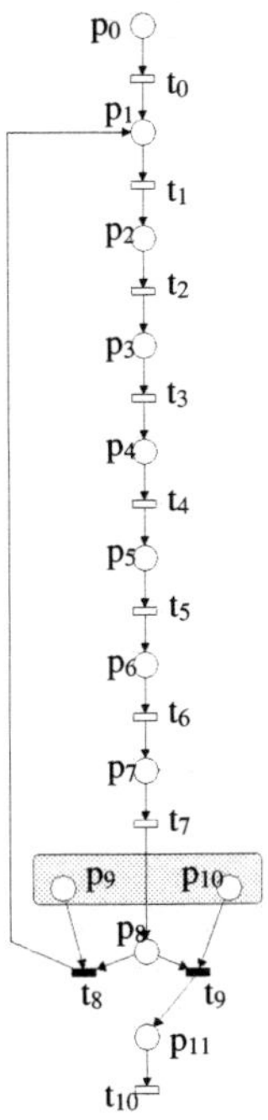

Fig. 2: PN model for the "not on-site repair" subnet

The handling of the maintenance staff is more complex, in case that the Buchholz relay is activated. The possible problems can be short-circuit, insulation failure or oil leakage. On the contrary to the activation of the oil thermometer, the initial warning of the Buchholz relay can be a trip. In this case, the problem is the appearance of a strong short-circuit. The repair of the damage can not be done in the transformer installation area; the transformer must be disconnected and transferred in a dedicated repairing area (e.g. in a transformer factory).

If the initial warning of the Buchholz relay is alarm, then the maintenance staff checks if the relay has been tripped, when they finally arrive in the transformer's area. They also check for the kind of damage. There are two possible contingencies: either the level of the oil indicator is low (p_{15}), or there are air bubbles behind the glass of the Buchholz relay (p_{16}). In the first case, the problem is oil leakage, otherwise there is insulation failure. The operation of transformer stops (in case of alarm) and it is checked if it is possible to repair the transformer on site. This depends on a) the type of problem: the repair can be done if the oil leakage is not wide (i.e. the size of hole in the tank is very small) or if the insulation failure is on a part outside the tank, and b) the existence of suitable tools. The capability of on site repair enables repairing possibilities for the two possible problems (p_{25} and p_{29}) and the specific type (p_{26} or p_{30}) enables the transition t_{20} or t_{23}. Then the staff works on the problem (in the case of oil leakage, the lost oil has also to be replaced). Finally, there is a check if everything works right. If there is still a problem, then the transformer must be sent to a dedicated repairing area (i.e. on site repair is not possible). The "not on-site repair" subnet of Fig. 2 is then models the transformer fault diagnosis and repair process.

Fault	Duration (hours)
Oil leakage (without trip)	8
Oil leakage (not on-site repair)	159
Overloading	4
Insulation failure (bushings, without trip)	6
Insulation failure (not on-site repair)	258
Short-circuit (not on-site repair)	275

Table 2: Simulation results

In the proposed PN modeling, immediate, deterministic and stochastic transitions are used, which take integer values that represent hours. For stochastic transitions, uniform distribution is considered. The duration of the transitions depends on the electric utility as well as the manufacturing plant that is in charge for the "not on-site repair" of the transformer. Table 1 presents indicative values for the duration of the transitions of the main Petri Net and the "not on-site repair" subnet. In Table 1, the duration of the immediate transitions is zero hours, the duration of the deterministic transitions is a positive integer number of hours and the duration of the stochastic transitions can take an integer value of hours from the interval that is specified by two integer numbers into square brackets. Table 2 presents the simulation results for fault diagnosis and repair using the main Petri Net and the "not on-site repair" subnet models as well as the duration of the transitions shown in Table 1. The heavy impact of the "not on-site repair" on the duration of the transformer repair is obvious, when analyzing the results of Table 2.

4. CONCLUSIONS

Transformer fault diagnosis and repair is a complex task that includes many possible types of faults and demands special trained personnel. This paper is concentrated on the investigation of the applicability of Stochastic Petri Nets in the modeling of transformer fault diagnosis and repair process. Simulation results for the most common types of transformer faults (overloading, oil leakage, short-circuit and insulation failure) are presented. The proposed methodology provides a systematical determination of the sequence of fault diagnosis and repair actions and aims at estimating the duration for transformer repair.

5. REFERENCES

[1] C. Bengtsson, "Status and trends in transformer monitoring," *IEEE Trans. Power Delivery*, vol. 11, pp. 1379-1384, July 1996.

[2] Z. Wang, Yilu Liu, P.J. Griffin, "A combined ANN and expert system tool for transformer fault diagnosis," *IEEE Trans. Power Delivery*, vol. 13, no. 4, pp. 1224-1229, October 1998.

EQUILIBRIUM ANALYSIS OF VOLTAGE-FED FIELD ORIENTED CONTROLLED INDUCTION MOTORS

*Rubén Salas-Cabrera**

Instituto Tecnológico de Cd. Madero
Div. de Estudios de Posgrado e Investigación
Cd. Madero, Tam., México

Claudio A. Cañizares

University of Waterloo
Dept. Electrical & Computer Eng.
Waterloo, ON, N2L-3G1, Canada

ABSTRACT

This paper presents an analysis of the equilibrium points of the induction motor when field oriented voltage control is used. Thus, a study of the steady state of the motor and controller dynamics is carried out for two cases. In the first one, all of the parameters are constant and known. In the second case, the rotor resistance is considered constant and unknown. In both cases, the resulting equations are nonlinear, however the equations for the second one are far more complicated since the field orientation is not achieved due to the parameter mismatch. Under the assumption of the knowledge of the state vector, this paper demonstrates that the equilibrium point is unique for the dynamical systems corresponding to each one of these cases.

1. INTRODUCTION

The field oriented control is the most popular technique in applications where induction motor control is used [1, 2]. For this reason, it is important to understand its behavior under different operating conditions. In particular, this paper presents a local bifurcation analysis of the steady state conditions of the system composed by the motor and controller dynamics. In this work, our main interest is to study the equilibria associated with the voltage-fed induction motor control operating with fixed control set points. Some parametric changes in the system are proposed, so that steady state stability and local bifurcation studies can be performed. Two cases are studied. In the first one, all of the parameters are constant and known. In the second case, the rotor resistance is constant and unknown to represent cases such as temperature variations that affect its nominal value during normal operating conditions. In these cases, measurement of the state vector of the induction motor is assumed. In papers [3] and [4], issues related to the equilibrium points of the current-fed field oriented control are discussed. In these papers, inputs are the stator currents instead of the stator voltages. Therefore, a fourth-order model is considered, without using the stator equations. The present paper discusses similar issues, but for a voltage-fed field oriented control, which requires the use of an eighth-order model.

*This work has been partially supported by the International Council of Canadian Studies, Promep México and a Research Grant from the National Engineering and Science Research Council (NSERC) of Canada.

2. FIELD ORIENTED CONTROL

Consider the following induction motor model expressed in the synchronously reference frame [5]:

$$\begin{aligned}
\frac{d\omega_r}{dt} &= \frac{P}{2J}(T_e - T_L) \\
\frac{di_{qs}}{dt} &= -\left(\frac{L_r^2 r_s + M^2 r_r}{L_r\sigma}\right) i_{qs} - \omega i_{ds} + \frac{M r_r}{L_r\sigma}\lambda_{qr} \\
&\quad - \frac{M}{\sigma}\omega_r\lambda_{dr} + \frac{L_r}{\sigma}v_{qs} \\
\frac{di_{ds}}{dt} &= -\left(\frac{L_r^2 r_s + M^2 r_r}{L_r\sigma}\right) i_{ds} + \omega i_{qs} + \frac{M r_r}{L_r\sigma}\lambda_{dr} \\
&\quad + \frac{M}{\sigma}\omega_r\lambda_{qr} + \frac{L_r}{\sigma}v_{ds} \\
\frac{d\lambda_{qr}}{dt} &= -(\omega - \omega_r)\lambda_{dr} - \frac{r_r}{L_r}\lambda_{qr} + \frac{r_r M}{L_r} i_{qs} \\
\frac{d\lambda_{dr}}{dt} &= (\omega - \omega_r)\lambda_{qr} - \frac{r_r}{L_r}\lambda_{dr} + \frac{r_r M}{L_r} i_{ds}
\end{aligned} \tag{1}$$

where $\tilde{x} = [\omega_r \ i_{qs} \ i_{ds} \ \lambda_{qr} \ \lambda_{dr}]^T$ is the state vector, v_{qs} and v_{ds} are the inputs, and $\sigma = L_s L_r - M^2$. Standard notation [5], for every variable and parameter, is used here. The outputs to be controlled are the electric rotor speed ω_r and the *d* component of the rotor flux λ_{dr}; therefore,

$$y = [\omega_r \ \lambda_{dr}]^T \tag{2}$$

is the output vector. The electric torque is given by

$$T_e = \frac{3PM}{4L_r}(i_{qs}\lambda_{dr} - i_{ds}\lambda_{qr}) \tag{3}$$

For the field oriented control, a particular model of (1) is selected, i.e. a model where the *d* axis of the rotating frame is aligned with the rotor flux vector. In practice, this orientation is maintained by employing the following expression for the slip frequency [6, 1]:

$$\omega - \omega_r = \frac{r_r M}{L_r}\frac{i_{qs}}{\lambda_{dr}} \tag{4}$$

where ω is the reference frame speed (stator frequency), and locking the phase of the reference system such that the *q* component of the rotor flux λ_{qr} is equal to zero [1]. Consider now the nonlinear feedback in [2],

$$\begin{aligned}
v_{qs} &= u_{qs} + \frac{\sigma}{L_r}\omega_r i_{ds} + \frac{\sigma M r_r}{L_r^2}\frac{i_{ds} i_{qs}}{\lambda_{dr}} \\
v_{ds} &= u_{ds} - \frac{\sigma}{L_r}\omega_r i_{qs} - \frac{\sigma M r_r}{L_r^2}\frac{i_{qs}^2}{\lambda_{dr}}
\end{aligned} \tag{5}$$

Since $\lambda_{qr} = 0$, and substituting (4) and (5) into (1), the state equations for the induction motor are

$$\begin{aligned}
\frac{d\omega_r}{dt} &= \frac{P}{2J}\left(\frac{3PM}{4L_r}i_{qs}\lambda_{dr} - T_L\right) \\
\frac{di_{qs}}{dt} &= -\left(\frac{L_r^2 r_s + M^2 r_r}{L_r\sigma}\right)i_{qs} - \frac{M}{\sigma}\lambda_{dr}\omega_r + \frac{L_r}{\sigma}u_{qs} \\
\frac{di_{ds}}{dt} &= -\left(\frac{L_r^2 r_s + M^2 r_r}{L_r\sigma}\right)i_{ds} + \frac{Mr_r}{L_r\sigma}\lambda_{dr} + \frac{L_r}{\sigma}u_{ds} \\
\frac{d\lambda_{dr}}{dt} &= -\frac{r_r}{L_r}\lambda_{dr} + \frac{r_r M}{L_r}i_{ds}
\end{aligned} \tag{6}$$

Observe that the electric torque (3) is now defined by

$$T_e = \frac{3PM}{4L_r}i_{qs}\lambda_{dr} \tag{7}$$

and u_{qs} and u_{ds} are the new inputs [6], which are defined as

$$\begin{aligned}
u_{qs} &= -k_{q1}(T_e - T_{ref}) - k_{q2}\int_0^t (T_e(\tau) - T_{ref}(\tau))\,d\tau \\
u_{ds} &= -k_{d1}(\lambda_{dr} - \lambda_{ref}) - k_{d2}\int_0^t (\lambda_{dr} - \lambda_{ref}(\tau))\,d\tau \\
T_{ref} &= -k_{q3}(\omega_r - \omega_{ref}) - k_{q4}\int_0^t (\omega_r(\tau) - \omega_{ref}(\tau))\,d\tau
\end{aligned} \tag{8}$$

These inputs depend on the controller gains (k_{di}, k_{qj}), the reference value for the rotor speed ω_{ref} and the reference value for the d component of the rotor flux λ_{ref}. Differentiating (8) and employing (6) and (7), the following state equations for the inputs are obtained:

$$\begin{aligned}
\frac{du_{qs}}{dt} =& \frac{3PMk_{q1}}{4L_r}\left[\frac{r_r}{L_r}i_{qs}\lambda_{dr} - \frac{r_r M}{L_r}i_{qs}i_{ds} + \frac{M}{\sigma}\lambda_{dr}^2\omega_r \right.\\
&\left. + \left(\frac{L_r^2 r_s + M^2 r_r}{L_r\sigma}\right)i_{qs}\lambda_{dr} - \frac{L_r}{\sigma}\lambda_{dr}u_{qs}\right] \\
&+ k_{q1}k_{q3}\left(\frac{PT_L}{2J} - \frac{3P^2M}{8JL_r}i_{qs}\lambda_{dr}\right) \\
&+ k_{q2}\left(T_{ref} - \frac{3PM}{4L_r}i_{qs}\lambda_{dr}\right) \\
&+ k_{q1}k_{q4}(\omega_{ref} - \omega_r)
\end{aligned} \tag{9}$$

$$\frac{du_{ds}}{dt} = k_{d1}\left(\frac{r_r}{L_r}\lambda_{dr} - \frac{Mr_r}{L_r}i_{ds}\right) + k_{d2}(\lambda_{ref} - \lambda_{dr}) \tag{10}$$

$$\frac{dT_{ref}}{dt} = k_{q3}\left(\frac{PT_L}{2J} - \frac{3P^2M}{8JL_r}i_{qs}\lambda_{dr}\right) + k_{q4}(\omega_{ref} - \omega_r) \tag{11}$$

Thus, if $\bar{x} = [u_{qs}\ u_{ds}\ T_{ref}]^T$ is defined as the state vector of the controller, $x = [\tilde{x}\ \bar{x}]^T$ can be defined as the state vector of the composed motor-controller system.

3. LOCAL BIFURCATION ANALYSIS

The equilibrium points of a dynamical system are determined by setting the derivative of the state vector equal to zero and solving the resulting nonlinear algebraic equations

$$f(x,p) = 0 \tag{12}$$

for the equilibrium point $x = x^e$, where superscript e is used to denote steady state quantities in this paper. Since $f(\cdot)$ is generally a nonlinear function, multiple solutions for x^e may be obtained. In (12), p represents a set of independent varying parameters used for bifurcation studies [7], which is used to represent the rotor resistance in this paper.

3.1. Constant and Known Rotor Resistance

The main goal in this case is to study the equilibria of the composed system when one consider that all of the parameters p are known and constant.

Proposition 1 *Consider the system (6) with the control defined by (9)-(11). Under the assumption of the exact knowledge of state vector of the induction motor $\tilde{x}$, with all of the parameters p known and constant, and a given set of reference values ω_{ref} and λ_{ref} ($\lambda_{ref} > 0$), the equilibrium point $x^e = [\tilde{x}^e\ \bar{x}^e]^T$ of the composed motor-controller system is unique.*

Under the above assumptions field orientation is achieved, therefore $\lambda_{qr}^e = 0$. Components of $\bar{x}^e$ and the remaining components of $\tilde{x}^e$ (i.e. $\omega_r^e, i_{qs}^e, i_{ds}^e, \lambda_{dr}^e$) are obtained by finding the real solution of $f(x,p) = 0$, where $f(\cdot)$ is the right side of the equations (6) and (9)-(11). A straightforward analysis yields the following expressions for the components of the state vector x^e:

$$\begin{aligned}
\omega_r^e &= \omega_{ref} & \lambda_{dr}^e &= \lambda_{ref} \\
i_{qs}^e &= \frac{4L_r T_L}{3PM\lambda_{ref}} & \lambda_{qr}^e &= 0 \\
i_{ds}^e &= \frac{\lambda_{ref}}{M} & T_{ref}^e &= T_L \\
u_{ds}^e &= \frac{r_s\lambda_{ref}}{M} & &
\end{aligned} \tag{13}$$

$$u_{qs}^e = \frac{4T_L(L_r^2 r_s + M^2 r_r)}{3PML_r\lambda_{ref}} + \frac{M\omega_{ref}\lambda_{ref}}{L_r}$$

Clearly, for each set of parameters and references, there is a unique equilibrium point. In order to illustrate the aforementioned result, a branching diagram is presented in Fig. 1, where the branching parameter is the rotor resistance r_r and the 2-norm of x^e is used as a scalar measure. Components of x^e are calculated by employing (13) with the parameters of the 3 HP induction motor given in [5] and considering $\omega_{ref} = 100$, $\lambda_{ref} = 0.3369$, $k_{d1} = 3230$, $k_{d2} = 90330$, $k_{q1} = 30$, $k_{q2} = 60$, $k_{q3} = 60$, $k_{q4} = 1765$ and $T_L = 1$. Local stability analysis can be carried out by linearizing (6) and (9)-(11), and calculating the eigenvalues of the resulting linear system at the equilibrium point for different values of r_r. In Fig. 2, the locus of the closest eigenvalues to the imaginary axis, with respect to the bifurcation parameter, are depicted; observe that each equilibrium point is asymptotically stable, as all of the eigenvalues lie in the left half complex plane. However, it is clear from Fig. 2 that the probability of a Hopf bifurcation occurring in the system can not be ruled out. That can be shown in the eigenvalue locus, obtained for the same motor but different controller gains ($k_{d1} = -1$ and $k_{d2} = 2300$ which are not realistic values but show the possibility of Hopf bifurcations). For the sake of space that result is not shown here. These types of bifurcations have also been observed for current-fed field oriented control [8, 9]. The results presented here suggest that the stability boundary of this system, as well as the one discussed below, is basically

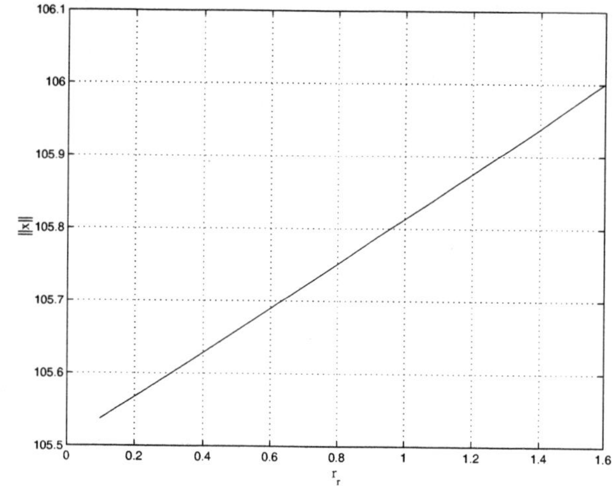

Fig. 1. Branching diagram for 3 HP motor example.

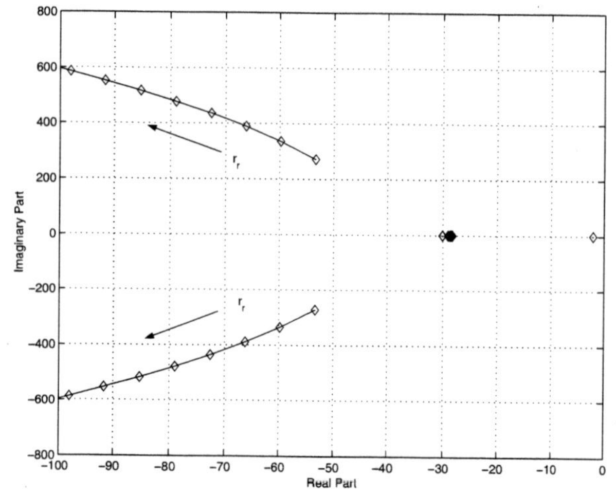

Fig. 2. Eigenvalue locus for 3 HP motor example.

defined by limit cycles [10], suggesting a complex stability behavior that cannot be analyzed using simple local approximations around equilibrium points.

3.2. Constant and Unknown Rotor Resistance

In this case, the knowledge of the state vector $\tilde{x}$ and constant and known parameters are assumed, with the exception of the rotor resistance, which is considered constant and unknown. Thus, let $\omega = \omega_r + r_{r_N} M i_{qs}/(L_r \lambda_{dr})$ be the reference frame speed expressed in terms of the nominal rotor resistance r_{r_N} (r_r denotes the actual rotor resistance). Feedback (5) can then be rewritten as

$$\begin{aligned} v_{qs} &= u_{qs} + \frac{\sigma}{L_r}\omega_r i_{ds} + \frac{\sigma M r_{r_N}}{L_r^2}\frac{i_{ds} i_{qs}}{\lambda_{dr}} \\ v_{ds} &= u_{ds} - \frac{\sigma}{L_r}\omega_r i_{qs} - \frac{\sigma M r_{r_N}}{L_r^2}\frac{i_{qs}^2}{\lambda_{dr}} \end{aligned} \tag{14}$$

Under the above assumptions, the field orientation is not achieved because of the incorrect estimation of the reference frame speed [11]. Consequently, the q component of the rotor flux vector is not equal to zero. Substituting the above expressions for ω, v_{qs} and v_{ds} into the induction motor model (1) yields

$$\begin{aligned} \frac{d\omega_r}{dt} &= \frac{P}{2J}\left[\frac{3PM}{4L_r}\left(i_{qs}\lambda_{dr} - i_{ds}\lambda_{qr}\right) - T_L\right] \\ \frac{di_{qs}}{dt} &= -\left(\frac{L_r^2 r_s + M^2 r_r}{L_r \sigma}\right) i_{qs} + \frac{M r_r}{L_r \sigma}\lambda_{qr} \\ &\quad - \frac{M}{\sigma}\lambda_{dr}\omega_r + \frac{L_r}{\sigma}u_{qs} \\ \frac{di_{ds}}{dt} &= -\left(\frac{L_r^2 r_s + M^2 r_r}{L_r \sigma}\right) i_{ds} + \frac{M r_r}{L_r \sigma}\lambda_{dr} \\ &\quad + \frac{M}{\sigma}\omega_r \lambda_{qr} + \frac{L_r}{\sigma}u_{ds} \\ \frac{d\lambda_{qr}}{dt} &= \frac{M}{L_r} i_{qs}\left(r_r - r_{r_N}\right) - \frac{r_r}{L_r}\lambda_{qr} \\ \frac{d\lambda_{dr}}{dt} &= \frac{M r_r}{L_r}\frac{i_{qs}\lambda_{qr}}{\lambda_{dr}} - \frac{r_r}{L_r}\lambda_{dr} \\ &\quad + \frac{M r_r}{L_r} i_{ds} - \frac{M}{L_r}\frac{i_{qs}\lambda_{qr}}{\lambda_{dr}}\left(r_r - r_{r_N}\right) \end{aligned} \tag{15}$$

From (3), (8) and (15), the state equations for the inputs are

$$\begin{aligned} \frac{du_{qs}}{dt} &= \frac{3PMk_{q1}}{4L_r}\left[\frac{r_r}{L_r} i_{qs}\lambda_{dr} - \frac{r_r M}{L_r} i_{qs} i_{ds} + \frac{M}{\sigma}\lambda_{dr}^2 \omega_r\right. \\ &\quad + \left(\frac{L_r^2 r_s + M^2 r_r}{L_r \sigma}\right) i_{qs}\lambda_{dr} - \frac{L_r}{\sigma}\lambda_{dr} u_{qs} \\ &\quad + \frac{M}{L_r}\frac{i_{qs}^2 \lambda_{qr}}{\lambda_{dr}}\left(r_r - r_{r_N}\right) - \frac{M r_r}{L_r}\frac{i_{qs}^2 \lambda_{qr}}{\lambda_{dr}} + \frac{M}{\sigma}\lambda_{qr}^2 \omega_r \\ &\quad + \frac{M}{L_r} i_{qs} i_{ds}\left(r_r - r_{r_N}\right) - \frac{r_r}{L_r} i_{ds}\lambda_{qr} \\ &\quad \left. + \frac{L_r}{\sigma}\lambda_{qr} u_{ds} - \left(\frac{L_r^2 r_s + M^2 r_r}{L_r \sigma}\right) i_{ds}\lambda_{qr}\right] \\ &\quad + k_{q1}k_{q3}\left[\frac{PT_L}{2J} - \frac{3P^2 M}{8JL_r}\left(i_{qs}\lambda_{dr} - i_{ds}\lambda_{qr}\right)\right] \\ &\quad + k_{q2}\left[T_{ref} - \frac{3PM}{4L_r}\left(i_{qs}\lambda_{dr} - i_{ds}\lambda_{qr}\right)\right] \\ &\quad + k_{q1}k_{q4}\left(\omega_{ref} - \omega_r\right) \end{aligned} \tag{16}$$

$$\begin{aligned} \frac{du_{ds}}{dt} &= k_{d1}\left[\frac{r_r}{L_r}\lambda_{dr} - \frac{M r_r}{L_r} i_{ds} - \frac{M r_r}{L_r}\frac{i_{qs}\lambda_{qr}}{\lambda_{dr}}\right. \\ &\quad \left. + \frac{M}{L_r}\frac{i_{qs}\lambda_{qr}}{\lambda_{dr}}\left(r_r - r_{r_N}\right)\right] + k_{d2}\left(\lambda_{ref} - \lambda_{dr}\right) \end{aligned} \tag{17}$$

$$\begin{aligned} \frac{dT_{ref}}{dt} &= k_{q3}\left[\frac{PT_L}{2J} - \frac{3P^2 M}{8JL_r}\left(i_{qs}\lambda_{dr} - i_{ds}\lambda_{qr}\right)\right] \\ &\quad + k_{q4}\left(\omega_{ref} - \omega_r\right) \end{aligned} \tag{18}$$

Proposition 2 *Consider the system (15) with the control defined by (16)-(18). Under the assumption of the exact knowledge of state vector of the induction motor $\tilde{x}$, with all of the parameters known and constant, except the rotor resistance ($r_r \neq r_{r_N}$), and for a given set of reference values ω_{ref} and λ_{ref} ($\lambda_{ref} > 0$), the equilibrium point $x^e = [\tilde{x}^e\ \bar{x}^e]^T$ of the composed motor-controller system is unique.*

From an analysis of the steady state of (16)-(18), the following expressions result:

$$\begin{aligned}\omega_r^e &= \omega_{ref} \qquad \lambda_{dr}^e = \lambda_{ref} \\ T_{ref}^e &= T_e^e = 3PM\left(i_{qs}^e\lambda_{ref} - i_{ds}^e\lambda_{qr}^e\right)/(4L_r)\end{aligned} \tag{19}$$

From the fourth equation in (15), it follows that

$$\lambda_{qr}^e = \frac{M}{r_r} i_{qs}^e \left(r_r - r_{r_N}\right) \tag{20}$$

Substituting (19)-(20) into the fifth equation in (15), one has that

$$i_{ds}^e = \frac{\lambda_{ref}}{M} - \frac{M r_{r_N}\left(r_r - r_{r_N}\right) {i_{qs}^e}^2}{r_r^2 \lambda_{ref}} \tag{21}$$

Using (19)-(21) in the first equation in (15), the following polynomial in i_{qs}^e can be obtained:

$$\mathcal{A} {i_{qs}^e}^3 + i_{qs}^e - \mathcal{B} = 0 \tag{22}$$

where $\mathcal{A} = \frac{M^2\left(r_r - r_{r_N}\right)^2}{r_r^2 \lambda_{ref}^2}$ and $\mathcal{B} = \frac{4L_r T_L r_r}{3PM\lambda_{ref} r_{r_N}}$. Polynomial (22) has three roots,

$$i_{qs(1,2,3)}^e = \begin{bmatrix} \frac{\mathcal{C}}{6\mathcal{A}} - \frac{2}{\mathcal{C}} \\ -\frac{\mathcal{C}}{12\mathcal{A}} + \frac{1}{\mathcal{C}} + j\frac{\sqrt{3}}{2}\left(\frac{\mathcal{C}}{6\mathcal{A}} + \frac{2}{\mathcal{C}}\right) \\ -\frac{\mathcal{C}}{12\mathcal{A}} + \frac{1}{\mathcal{C}} - j\frac{\sqrt{3}}{2}\left(\frac{\mathcal{C}}{6\mathcal{A}} + \frac{2}{\mathcal{C}}\right) \end{bmatrix} \tag{23}$$

where $\mathcal{C} = \sqrt[3]{108\mathcal{A}^2\mathcal{B} + 12\sqrt{3}\sqrt{4\mathcal{A}^3 + 27\mathcal{A}^4\mathcal{B}^2}}$. Only real solutions of (22) are important from the practical point of view. An analysis of the parameters involved in the roots (23) follows: since $r_r \neq r_{r_N}$, $T_L > 0$ and $\lambda_{ref} > 0$, then $\mathcal{A} > 0$ and $\mathcal{B} > 0$ for any set of practical parameters; hence, $4\mathcal{A}^3 + 27\mathcal{A}^4\mathcal{B}^2 > 0$, and thus $\sqrt{4\mathcal{A}^3 + 27\mathcal{A}^4\mathcal{B}^2}$ is always a real number. On the other hand,

$$\left(108\mathcal{A}^2\mathcal{B} + 12\sqrt{3}\sqrt{4\mathcal{A}^3 + 27\mathcal{A}^4\mathcal{B}^2}\right) > \left(108\mathcal{A}^2\mathcal{B} + 12\sqrt{3}\sqrt{27\mathcal{A}^4\mathcal{B}^2}\right) > 0$$

Therefore, $\mathcal{C}$ is always real; then, $j\frac{\sqrt{3}}{2}\left(\frac{\mathcal{C}}{6\mathcal{A}} + \frac{2}{\mathcal{C}}\right)$ is always imaginary, and hence the second and third roots in (23) are always complex. Thus, the only real solution of (22) is

$$i_{qs}^e = \frac{\mathcal{C}}{6\mathcal{A}} - \frac{2}{\mathcal{C}} \tag{24}$$

Uniqueness of i_{qs}^e (24) guarantees the uniqueness of λ_{qr}^e (20), i_{ds}^e (21) and T_{ref}^e (19). The rest of the components of x^e can be obtained from a straightforward analysis of the second and third equations in (15), i.e.

$$\begin{aligned} u_{qs}^e &= \frac{1}{L_r^2}\left[\left(L_r^2 r_s + M^2 r_r\right) i_{qs}^e - M r_r \lambda_{qr}^e + M L_r \lambda_{ref}\omega_{ref}\right] \\ u_{ds}^e &= \frac{1}{L_r^2}\left[\left(L_r^2 r_s + M^2 r_r\right) i_{ds}^e - M r_r \lambda_{ref} - M L_r \lambda_{qr}^e \omega_{ref}\right] \end{aligned} \tag{25}$$

Hence, it is clear that the equilibrium point x^e is unique.

For the 3 HP motor example [5], one can obtain rather similar branching and eigenvalue locus diagrams to those depicted in Fig. 1 and Fig. 2. It is important to point out that Proposition 2 is valid only when $r_r \neq r_{r_N}$, thus avoiding the singularity in (24). However, a connection between the ideal case in Proposition 1 and the case discussed here can be readily shown. It is interesting to note that the equilibrium point is unique for any set of practical parameters and references, including any value of the load torque; any value of the parameter mismatch; and even for a load torque model that depends on the rotor speed (e.g. $T_L = K\omega_r$).

4. CONCLUSIONS

In this paper, a local bifurcation analysis of an induction motor with field oriented voltage control is presented, demonstrating the uniqueness of the equilibria of the composed motor-controller system with respect to the rotor resistance variations. This particular characteristic plus the numerical analyses presented show that the stability features of this motor-controller system cannot be studied with the help of local linearizations around the equilibrium points, which in case of systems with multiple equilibria do yield enough information to give a basic idea of some of the basic stability characteristics of associated nonlinear system (e.g. [12]). Hence, additional nonlinear analysis tools should be used in this case to gain some insight on the stability behavior of this particular motor-controller system.

5. REFERENCES

[1] D. W. Novotny and T. A. Lipo, *Vector Control and Dynamics of AC Drives*, Oxford University Press, 1996.

[2] C. M. Ong, *Dynamic Simulation of Electric Machinery*, Prentice Hall, 1998.

[3] P. A. S. de Wit, R. Ortega, and I. Mareels, "Indirect field-oriented control of induction motors is robustly globally stable," *Automatica*, vol. 32, no. 10, pp. 1393–1402, 1996.

[4] A. S. Bazanella and R. Reginatto, "Robustness margins for indirect field-oriented control of induction motors," *IEEE Transactions on Automatic Control*, vol. 45, pp. 1226–1231, June 2000.

[5] P. C. Krause, *Analysis of Electric Machinery*, McGraw Hill, 1986.

[6] R. Marino, S. Peresada, and P. Valigi, "Adaptive input-output linearizing control of induction motors," *IEEE Transactions on Automatic Control*, vol. 38, pp. 208–221, Feb. 1993.

[7] R. Seydel, *Practical Bifurcation and Stability Analysis*, Springer-Verlag, 1994.

[8] A. S. Bazanella, R. Reginatto, and R. Valiati, "On hopf bifurcations in indirect field oriented control of induction motors: designing a robust pi controller," in *38th IEEE Conference on Decision and Control*. IEEE, Dec. 1999, pp. 689–694.

[9] A. S. Bazanella and R. Reginatto, "Robust tuning of the speed loop in indirect field oriented control of induction motors," *Automatica*, vol. 37, pp. 1811–1818, 2001.

[10] J. Guckenheimer and P. Holmes, *Nonlinear Oscillations, Dynamical Systems, and Bifurcations of Vector Fields*, Springer-Verlag, 1986.

[11] M. P. Kazmierkowski and H. Tunia, *Automatic Control of Converter-Fed Drives*, Elsevier Science Publishers, 1994.

[12] W. D. Rosehart and C. A. Canizares, "Bifurcation analysis of various power system models," *Electrical Power and Energy Systems*, vol. 21, pp. 171–182, 1999.

NEURAL NETWORK APPROACH FOR ESTIMATION OF LOAD COMPOSITION

J. Duan, D. Czarkowski, and Z. Zabar

Department of Electrical and Computer Engineering, Polytechnic University
Brooklyn, NY 11201 USA

ABSTRACT

A neural network methodology to solve the problem of estimation of modern electrical load distribution in typical commercial and residential areas is proposed in this paper. The inputs for the neural network are harmonic characteristics of each type of typical loads and possible combinations of these loads. The output is the estimation of load composition. The Multi-Layer Feed-Forward Back-Propagation neural network and Elman neural network are used in the paper to calculate the load distribution. A case study of a Manhattan area and two practical tests are presented to demonstrate the feasibility of this approach. The new method will be useful for electrical load monitoring and harmonic reliability assessment in the new utility environment.

1. INTRODUCTION

A new approach employing artificial neural networks to calculate load composition is developed in this paper. Neural computing has attractive features, such as ability to tackle new problems that are hard to define, robustness in dealing with incomplete or fuzzy data, processing speed, flexibility, and ease of maintenance [1]. Based on the load distribution of typical commercial and residential areas and harmonic characteristic of modern electrical loads, and the strong generalizing ability of the neural networks [2], the new approach can be used to quickly calculate load composition from a measured load current waveform.

Section 2 of the paper provides a brief review of the multi-layer feed-forward neural network and the Elm neural network used in this paper to calculate the load composition. In Section 3, the problem of load estimation is formulated in terms of harmonic content table and load distribution table. The proposed methodology is implemented in Section 4 and a case study is given in Section 5 to demonstrate the effectiveness of the presented method. Finally, conclusions are provided in Section 6.

2. NEURAL NETWORK

2.1. Multi-Layer Feed-Forward Neural Network

The multi-layer feed-forward neural network [3], also known as the multi-layer perceptron (MLP) network, was developed in the early 1970's and is the most popular topology in use today. A schematic diagram of the topology is given in Fig. 1. This network consists of a set of input neurons, output neurons and one or more hidden layers of intermediate neurons. Such networks are capable of approximating functions of almost any arbitrary complexity with a finite number of discontinuities.

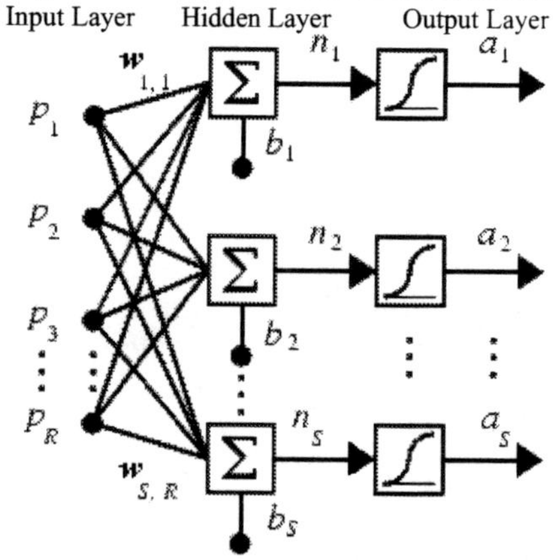

Figure 1: Topology of an MLP, where R is the number of elements in the input vector, S is the number of neurons in the layers, and W_{ij} are the weighting factors between different layers.

An MLP operates in two modes: training and testing. Training is a procedure used to minimize the difference between outputs of MLP and the desired values by adjusting the weights of the network. Sets of input vectors are presented to the network until training is completed. Then the network's weights are "frozen" in the trained state and the new input data are presented to the network to determine the appropriate output.

2.2. Elman Neural Network

The Elman neural networks are two-layer back-propagation networks, with the addition of a feedback connection from the output of the hidden layer to its input [4]. This feedback path allows Elman networks to learn to recognize temporal patterns as well as spatial patterns. Since Elman networks are an extension of the two-layer sigmoid/linear architecture, they inherit the ability to fit any input/output function with a finite number of discontinuities. A two-layer Elman network is shown in Fig. 2.

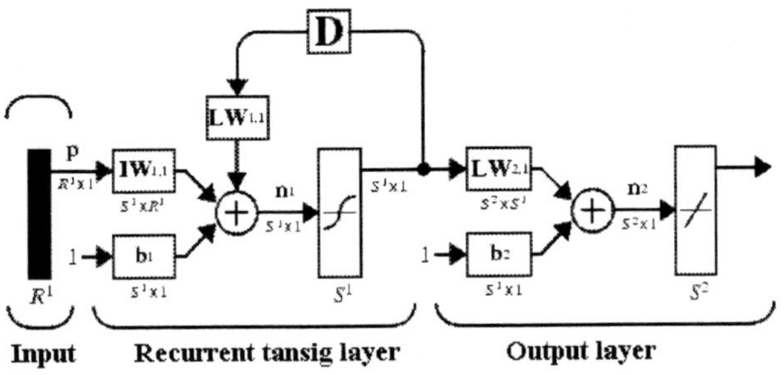

Figure 2: Topology of a two-layer Elman Network.

3. PROBLEM FORMULATION

The problem of the load distribution estimation is to determine the composition of a customer's load from the measured bus current waveforms [5]. The measured bus current waveform $I(t)$ can be approximated by a composite waveform which consists of the sum of a set of connected load current waveforms, weighted appropriately. The connected loads are characterized by the frequency spectrum of their current waveforms. The loads are assigned to groups based on their harmonic spectrum. For an individual load,

$$l_i(t) = \sum_{k=0}^{N}[c_{ki}\cos(k\omega t+\phi_{ki})]$$
$$=\sum_{k=0}^{N}[a_k\sin(k\omega t)+b_k\cos(k\omega t)] \quad (1)$$

where N is the highest harmonic order used for the load composition and $i=1, 2, \dots M$ is the number of the load. The total composite load will be:

$$L(t,\mathbf{w}) = \sum_{i=0}^{M} w_i l_i(t) \quad (2)$$

where w_i is the weighting factor of each load.

To match the shape of the composite waveform $L(t,w)$ to that of the actual bus waveform $I(t)$, the root mean square error can be used. This results in the minimization criterion for the following quantity:

$$J(t,\mathbf{w})=\sqrt{\frac{1}{T}\int_0^T[I(t)-L(t,\mathbf{w})]^2\,dt} \quad (3)$$

where $T = 2\pi/\omega$ is the period of the fundamental component of the power system and the physical constraint for w is that $w_i \geq 0$ for all i's.

It can be shown that expanding $I(t)$ into a Fourier series

$$I(t)= \sum_{k=0}^{\infty}[A_k\sin(k\omega t)+B_k\cos(k\omega t)]$$
$$= \sum_{k=0}^{N}[A_k\sin(k\omega t)+B_k\cos(k\omega t)]$$
$$+ \sum_{k=N+1}^{\infty}[A_k\sin(k\omega t)+B_k\cos(k\omega t)] \quad (4)$$

and using the orthogonality property, the requirement that $J(t,w)$ be a minimum can be transformed into the requirement that

$$J(\mathbf{w})=\left\{\frac{1}{2}\left[\sum_{k=0}^{N}(A_k-\sum_{i=1}^{M}w_i a_{ki})^2+\sum_{k=0}^{N}(B_k-\sum_{i=1}^{M}w_i b_{ki})^2 + \sum_{k=N+1}^{\infty}(A_k^2+B_k^2)\right]\right\}^{\frac{1}{2}} \quad (5)$$

be a minimum. Then, the calculation of load composition can be formulated as an minimization of $J(t,w)$. Minimization of $J(t,w)$ is a quadratic programming problem. Hence, it is theoretically possible to estimate percentage shares of various types of loads connected to this service point using the frequency spectrum of the current at this point. In practice, the goal function $J(t,w)$ is a complex nonlinear function of load harmonic characteristics, current waveform at the service point, and uncertain external factors like harmonic deviation, voltage disturbance from outside etc. With traditional methods such as linear or multiple regression methods and general exponential smoothing, the load models would be very complicated and cannot assure the accuracy of the estimation due to the local minima problem [6]. Thus, artificial neural network approach is considered to be a promising one because of its adaptability, generalization, and nonlinear character.

4. IMPLEMENTATION ALGORITHM

A neural network approach to solve the estimation of load composition problem is presented in this section.

4.1. Input Vector

The input vector is the harmonic spectrum at the service point and harmonic content table of the typical electrical loads. A number of field surveys were made at commercial and residential sites at Manhattan. From the results of the surveys, we constructed a list of 9 basic loads responsible for more than 90% of the power consumption of the total loads. The harmonic characteristic of an individual load was obtained through lab experiments. The list of 9 basic loads is given in Appendix Table A1. Table I is a harmonic spectrum example (Air Conditioner) of a basic load, the current waveform for 8 basic loads (except the resistive load) are shown in Figure 3.

4.2. Output Vector

Only one output quantity is used here: the set of weighting factors $\mathbf{w}$ for the composite load.

TABLE I: HARMONIC SPECTRUM OF AIR CONDITIONERS

Harmonic	Magnitude (%)	Phase (Degree)
1	100	-23
3	8	110
5	6.8	58
7	0.5	165
9	0.6	82
11	0.3	157
13	0.13	-79.6
15	0.04	16.7
17	0.2	87

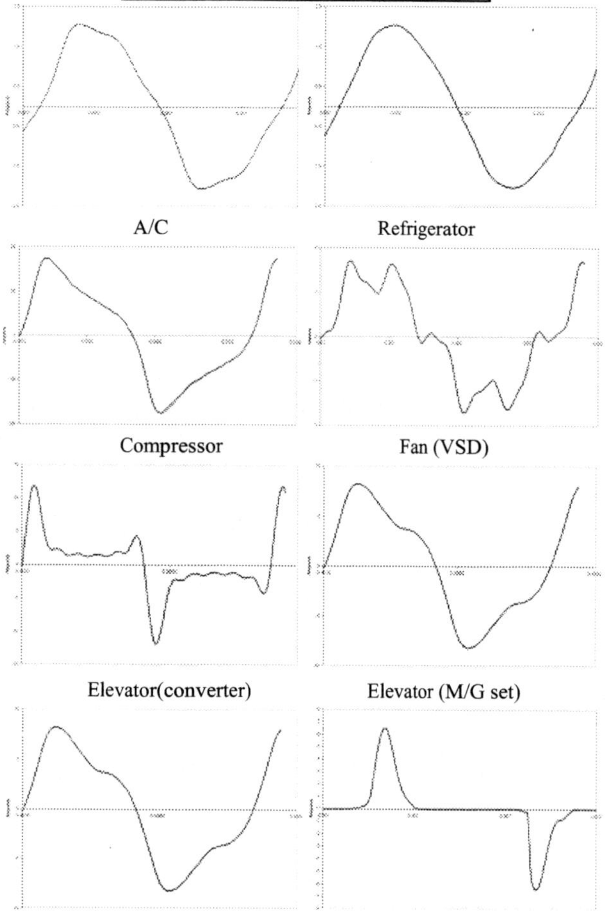

Figure 3: Current Waveform of 8 Basic Loads

4.3. Network Architecture and Training

The number of hidden layers and the number of neurons in each hidden layer characterize the complexity of a neural network. There is no general rule for selection of these parameters. The critical issue for developing a neural network is generalization. Like other nonlinear estimation methods such as kernel regression, neural networks can suffer from either underfitting or overfitting. A neural network with a small number of neurons may not be sufficiently powerful to model a complex function. For example, a network with no hidden layers actually models a simple linear function. On the other side, a neural network with too many neurons may lead to overfitting the training sets and lose its ability to generalize which is the main desired characteristic of a neural network. According to [4], five effective ways to avoid overfitting are: model selection, jittering, weight decay, early stopping and Bayesian estimation. Here, we will use "early stopping" in the training process as a means to avoid overfitting.

5. STUDIES

5.1. Test System

For the purpose of testing, 10% or 20% deviation at one harmonic parameter (amplitude or phase angle) was applied to the harmonic content table of typical commercial area. The evaluation of the calculation results is compared with Percentage Relative Error (PRE):

$$\varepsilon_r (\%) = [\sum_{i=1}^{N} (P_{ti} - P_i)^2]^{0.5}$$

where N is the number of devices. P_{ti} is the load i percentage calculated from ANN model and P_i is the actual percentage of load i.

5.2. Training Patterns of the Neural Network

Training sets provided to the neural network are representative of the whole state space of concern so that the trained neural network has the ability of generalization. The basic training data includes the 300 combinations of 9 basic loads according to result of field surveys. For each case the amplitudes and phase angles for each odd harmonics through the 29th and 2nd harmonic are taken into the training data to cover the possible deviation of the load composition. Thus, there are 17 training patterns in each case including the actual case itself and 5,100 training patterns in total. They cover the range of load composition for a typical commercial or residential area shown in the Appendix. Levenberg-Marguardt algorithm is used to train the neural network in this study.

5.3. Test Patterns of the Neural Network

The trained neural network was tested using 24 test cases, which are composed of different load composition. None of these test cases were used in the training of the neural network.

5.5. Network Topology

A neural network with one hidden layer and two hidden layers were tried and the neural network with two hidden layers was selected for further analysis because of more accurate results. Table II shows the average error at different numbers of iterations for six different neural network topologies. 300 valid data are used to apply "early stopping" method in the training process to avoid

TABLE II: COMPARISON OF PRE FOR SIX NEURAL NETWORK ARCHITECTURES

Name	Structure	Iterations No.	PRE
MLP17kA	9/20/5/1	24	13.10%
MLP17kB	9/50/5/1	525	10.06%
MLP17kC	9/50/6/1	472	10.48%
Elm17kA	9/20/5/1	53	22.17%
Elm17kB	9/50/5/1	1000	30.92%
Elm17kC	9/50/6/1	1000	26.72%

overfitting. Structure 9/20/5/1 means that there are 9 input neurons, 20 neurons in the first hidden layer, 5 neurons in the second hidden layer and one neuron for output. Sigmoid transfer functions are used for the hidden layer and a linear transfer function is used for the output layer.

5.6. Comparison of MLP and Elman neural network

The MLP has the smallest average error of 10.06% in comparison with 22.17% for Elman neural network. Hence, we have chosen MLP 9/50/5/1 as the network architecture for further tests.

5.7. Practical tests

Two practical tests with 4 and 5 electric loads were conducted in the lab and the combined current waveform was used to test the neural network. Table III shows the results for these two practical tests. It can be seen that the PRE is below 10% and particular loads are estimated with a few percent error.

TABLE III: TEST RESULTS OF PRACTICAL TEST CASE 1

	Equipment	Actual	Estimation
1	A/C	58.01%	51.63%
2	Refrigerator	0	1.51%
3	Compressor	0	2.05%
4	Fan (VSD)	0	0.54%
5	Elevator (M/G set)	0	0.94%
6	Elevator (converter)	0	1.31%
7	Fluorescent Lights	9.20%	14.51%
8	Linear load	17.36%	14.22%
9	Computers, TV etc	15.34%	13.24%
PRE	9.62%		

TABLE IV: TEST RESULTS OF PRACTICAL TEST CASE 2

	Equipment	Actual	Estimation
1	A/C	54.67%	51.17%
2	Refrigerator	20.22%	23.09%
3	Compressor	0	1.19%
4	Fan (VSD)	0	3.26%
5	Elevator (M/G set)	0	1.25%
6	Elevator (converter)	0	0.65%
7	Fluorescent Lights	10.58%	14.15%
8	Linear load	0.00%	0.71%
9	Computers, TV etc	14.54%	8.57%
PRE	9.13%		

6. CONCLUSIONS

A neural network based approach to estimate electrical load composition is proposed and verified. Using case studies for typical residential and commercial areas, it is shown that the method is capable of accurate estimation of load distribution. Although the chosen neural network architecture performs well in this study, it might be not optimal. Indeed, the process of choosing the number of hidden layers, the number of neurons for each layer, how to deal with overfitting and underfitting during the training procedure, remains an open research topic. Other training algorithms such as RPROP, SASS, and Conjugate Gradient may be chosen to train the neural network. It is difficult to predict in advance which training algorithm is more applicable. Also which physical variables are chosen as input variables and how many cases to use in the training patterns depend on experience and the availability of the data. We believe that the proposed method may have important applications in electrical loads monitoring calculation and harmonic reliability assessment in the new utility environment. The method would allow for obtaining detailed electric loads composition profiles of a certain area without time-consuming and expensive surveys.

7. REFERENCES

[1] T. S. Dillon and D. Niebur, *Neural Networks Applications in Power Systems*, London: CRL Publishing Ltd, 1996.

[2] R. Aggarwal, Y. Song, "Artificial neural networks in power systems. I. General introduction to neural computing," *Power Engineering Journal*, vol. 11, issue 3, pp. 129–134, Jun 1997.

[3] L. Fausett, *Fundamentals of Neural Networks*, Prentice-Hall Inc., 1994.

[4] J. L. Elman, "Finding structure in time," *Cognitive Science*, vol. 14, pp. 179-211, 1990.

[5] H. Yan, Z. Zabar, D. Czarkowski, L. Birenbaum, E. Levi and J. Hajagos, "Experimental test of a load model in the presence of harmonics", *Generation, Transmission and Distribution, IEE Proceedings*, vol. 146, issue 2, pp. 186 –192, Mar 1999

[6] S. B. Leeb, B. C. Lesieutre and S. R. Shaw, "Determination of Load Composition Using Spectral Envelope Estimates," *27th North American Power Symposium*, Bozeman, Montana, pp. 618-627, 1995.

8. APPENDIX

TABLE A1: LOAD DISTRIBUTION OF TYPICAL COMMERCIAL AND RESIDENTIAL AREAS

	Equipment	Comm. -Large	Comm. -Small	Resid.- Large	Resid.- Small
1	A/C	0	0	15.84	19.14
2	Refrigerator	6.27	12.54	10.56	0
3	Compressor	14.63	29.26	26.4	44.66
4	Fan (VSD)	8.1	0	0	0
5	Elevator (M/G set)	6.9	3.4	0	0
6	Elevator(converter)	2.2	0	8.8	0
7	Fluorescent Lights	19.4	8.2	2.5	0.3
8	Linear load	27.1	32.4	30.9	25.5
9	Computers, TV etc	5.2	1.8	1.3	2.7

MODIFYING EIGENVALUE INTERACTIONS NEAR WEAK RESONANCE

*Vincent Auvray** *Ian Dobson*

ECE Dept., University of Wisconsin
1415 Engineering Drive
Madison WI 53706 USA
dobson@engr.wisc.edu

Louis Wehenkel

EE&CS Dept., Université de Liège
Sart-Tilman B28
B-4000 Liège, Belgique
L.Wehenkel@ulg.ac.be

ABSTRACT

In electric power system instabilities such as subsynchronous resonance or interarea oscillations, two complex modes can approach each other in frequency and then interact by changing damping so that one of the modes becomes unstable. Selecting changes in parameters to minimize this interaction is difficult by trial and error. By analyzing the interaction as a perturbation of a weak resonance, we calculate sensitivities that indicate the parameters to be changed to minimize the interaction and stabilize the system. The method is illustrated with a simple example of two coupled linear oscillators. The use of sensitivity methods to change the type of the interaction is also demonstrated.

1. INTRODUCTION

A power system mode is an eigenvalue of the linearized system and its associated eigenvector. Since the modes determine the power system stability and small signal transient performance, understanding the behavior of the complex modes is fundamental to avoiding power system oscillations [9, 2, 8]. Indeed, a better understanding and suppression of low frequency interarea oscillations or subsynchronous resonance could relax some of the constraints on larger bulk power transfers and increase power system security.

An exact coincidence of complex eigenvalues in both frequency and damping is called a resonance. There are two types of resonance: strong and weak [3]. Strong resonance has a nondiagonalizable linearization and weak resonance has a diagonalizable linearization. As parameters vary, the power system modes vary and it is not uncommon for two complex eigenvalues to pass near resonance. For a comprehensive review up to 2000 see [3] and for more recent work see [6, 7, 4, 5]. Only a subset of oscillatory power system problems involve resonant eigenvalues and there are some established and useful mitigation methods for subsynchronous resonance and interarea oscillations.

Near a weak resonance, as a parameter t is changed, two eigenvalues initially approaching each other in frequency can interact and move in damping so that one of the eigenvalues becomes unstable as shown in Figure 1(a). Similar interactions cause subsynchronous resonance instabilities [4] and instabilities in models of interarea oscillations [5]. Let δ be the maximum separation between the eigenvalues in damping (maximum difference in their real parts) as the parameter t is changed. In this paper we compute the sensitivity of δ with respect to other parameters of the problem so that changes in these other parameters can be made to reduce δ, so as to minimize the interaction and stabilize the system as shown in Figure 1(b). This systematic procedure to minimize the interaction does not require insight into the relatively complicated eigenvalue movements near weak resonance and their relation to the parameters of the problem, nor does it require exhaustively recomputing the root locus for each parameter by trial and error.

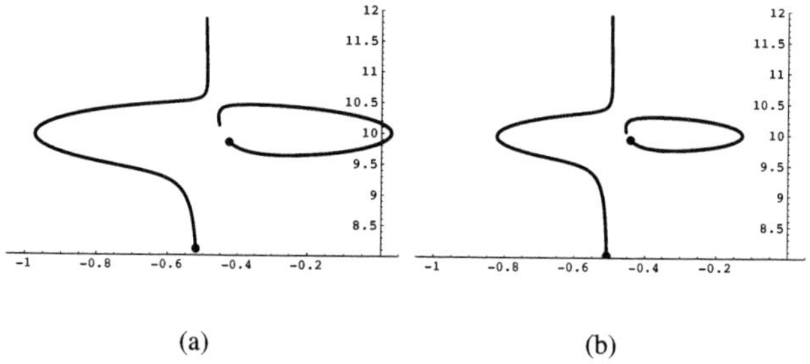

Fig. 1. Eigenvalue interactions as a parameter t is varied. The dots indicate the initial eigenvalue positions. (a) shows eigenvalues initially moving together in frequency interacting by changes in damping so that one eigenvalue becomes unstable. (b) shows a minimized eigenvalue interaction that retains stability.

*Vincent Auvray is a visiting scholar at the University of Wisconsin and a F.N.R.S. research fellow at the University of Liège, Belgium. Funding in part from the F.N.R.S., Belgium, the University of Liège, and the National Science Foundation, USA under grant ECS-9988574 is gratefully acknowledged.

2. INTERACTION NEAR A WEAK RESONANCE

The power system differential equations are assumed to depend smoothly on a real parameter t that parameterizes the eigenvalue movement in the complex plane. That is, t is the root locus parameter. The system Jacobian evaluated at the operating equilibrium is the real matrix $J(t)$ and has two complex eigenvalues of interest $\lambda_1(t)$ and $\lambda_2(t)$.

The eigenvalues do not vary smoothly at weak resonance. However, the following function can be shown to smoothly depend on the parameters [5, 3]:

$$\mu(t) = (\lambda_1(t) - \lambda_2(t))^2/4. \tag{1}$$

μ determines the separation of the eigenvalues in both magnitude and angle. That is, μ determines the relative position of the eigenvalues. The distance between the eigenvalues is $|\lambda_1 - \lambda_2| = 2\sqrt{|\mu|}$. In particular, the condition for coincident and resonant eigenvalues is $\mu = 0$. Moreover, the relative direction of the eigenvalues is $\angle(\lambda_1 - \lambda_2) = \frac{1}{2}\angle\mu + k\pi$, for some integer k. Dobson [5] shows that close to a weak resonance, the locus $t \mapsto \mu(t)$ for real t describes to second order a parabola in the complex plane passing near the origin. In this paper, we propose a method to modify the eigenvalue movement.

3. STABILIZING THE INTERACTION

We first introduce a measure of the interaction of the eigenvalues near weak resonance. Suppose that the two eigenvalues resonate weakly for some t, say $t = 0$. Dobson [5] shows that this implies $\mu_t(0) = \mu(0) = 0$, where the subscript denotes differentiation. A weak resonance is thus a root of μ of multiplicity at least two. We make the generic assumption that $A = \frac{1}{2}\mu_{tt}(0) \neq 0$; i.e., $t = 0$ is a double root. A Taylor expansion around $t = 0$ gives

$$\mu(t) = At^2 + o(|t|^2). \tag{2}$$

Equation (2) implies that the eigenvalues approach each other along a line of angle $\theta = \frac{1}{2}\angle A$ as $t \to 0$.

To model the proximity to a weak resonance, the Jacobian is supposed to depend smoothly on an additional parameter ϵ such that there is weak resonance at $(t, \epsilon) = (0, 0)$. For small fixed $\epsilon \neq 0$, the eigenvalues pass close to weak resonance as t passes through 0. Define d as the length of the projection of $\lambda_1 - \lambda_2$ on the direction perpendicular to θ. For $\theta = \frac{\pi}{2}$, d is the size of the difference in damping between the eigenvalues. We have

$$d = |\lambda_1 - \lambda_2||\sin(\angle(\lambda_1 - \lambda_2) - \theta)| \tag{3}$$

$$= 2\sqrt{|\mu|}|\sin(\frac{1}{2}\angle\mu - \theta)|. \tag{4}$$

Equation (4) shows that $d(t)$ can be defined as a function $d(\mu(t))$ of μ alone. One can see that $d(\mu) \leq K$ if, and only if, in the complex plane the point $\mu e^{-2i\theta}$ is contained inside the parabola parametrized by

$$\nu(s) = (s^2 - \frac{K^2}{4}) + iKs, \quad s \in \mathbb{R}. \tag{5}$$

We propose to measure the extent of the eigenvalue interaction by

$$\delta(\eta, \epsilon) = \max_{|t| \leq \eta} d(\mu(t, \epsilon)), \tag{6}$$

where $\eta > 0$.

Let us construct an approximation μ^a of μ that will allow us to modify the shape of the locus of μ and, in particular, influence the quantity (6). The system Jacobian evaluated at the operating equilibrium is supposed to depend on a vector of parameters $p = (p_1, p_2, ..., p_m)$ that can be varied to modify the eigenvalue interaction. The nominal parameter vector is $p = 0$ and the eigenvalues resonate weakly for $(t, \epsilon, p) = (0, 0, 0)$.

One key mathematical step is to consider the variables t, ϵ and p as complex. Recall that for $(\epsilon, p) = (0, 0)$, $t = 0$ is a real double root of μ. Let us show that for small (ϵ, p), this root generically perturbs to two complex single roots r_1 and r_2. We suppose that J, and thus μ, is analytic in t, ϵ and p. It follows from the Weierstrass preparation theorem [1] that there exist unique continuous functions $r_1(\epsilon, p)$ and $r_2(\epsilon, p)$ defined for (ϵ, p) in a sufficiently small neighborhood of $(0, 0)$ and such that $r_i(0, 0) = 0$, $\mu(r_i(\epsilon, p), \epsilon, p) = 0$ and r_i is analytic at the points (ϵ, p) where $r_1 \neq r_2$. We thus generically suppose that r_1 and r_2 are distinct at the points considered.

Define the following approximation to μ

$$\mu^a(t, \epsilon, p) = A(t - r_1(\epsilon, p))(t - r_2(\epsilon, p)). \tag{7}$$

For real t, $\mu^a(t, \epsilon, p)$ traces a parabola in the complex plane as t varies. The approximation (7) can be intuitively justified by the following reasoning. In a neighborhood of $(0, 0, 0)$, μ can be expanded as

$$\begin{aligned} \mu(t, \epsilon, p) &= At^2 + 2B(\epsilon, p)t + C(\epsilon, p) + o(|(t, \epsilon, p)|^2) \quad (8) \\ &= A(t - q_1(\epsilon, p))(t - q_2(\epsilon, p)) + o(|(t, \epsilon, p)|^2), \end{aligned}$$

where $q_1, q_2 = A^{-1}(-B \pm \sqrt{B^2 - AC})$ are the roots of the quadratic polynomial in t of (8). To obtain μ^a, we drop the high-order term $o(|(t, \epsilon, p)|^2)$ and replace q_1 and q_2 by r_1 and r_2.

Using approximation (7) we define

$$\delta^a(\eta, \epsilon, p) = \max_{\{t | t \in \mathbb{R}, |t| \leq \eta\}} d(\mu^a(t, \epsilon, p)) \tag{9}$$

and for large enough η, evaluate this quantity with geometric relation (5) as

$$\delta^a(\eta, \epsilon, p) = \sqrt{|A|}\sqrt{(\Re(r_1 - r_2))^2 + (\Im(r_1 + r_2))^2}. \tag{10}$$

We hence propose to use $\delta^a(\eta, \epsilon, p)$ and its vector of derivatives $\delta^a_p(\eta, \epsilon, p) = (\frac{\partial \delta^a}{\partial p_1}, \ldots, \frac{\partial \delta^a}{\partial p_m})$ as approximations of $\delta(\eta, \epsilon, p)$ and $\delta_p(\eta, \epsilon, p)$ for small enough p.

A, $r_1(\epsilon, p)$ and $r_2(\epsilon, p)$ can be approximated by fitting a parabola to the observed μ locus for fixed (ϵ, p). Section 6 discusses the computation of the root sensitivities $r_{ip_k} = \frac{\partial r_i}{\partial p_k}$. The sensitivity vector δ^a_p can be used to select and change parameters to stabilize the interaction, for example by a gradient descent incrementally minimizing δ^a.

4. EXAMPLE

This section gives a simple example to illustrate the use of the sensitivities. We incrementally reduce δ to stabilize a system, changing the interaction of Figure 1(a) into the one in Figure 1(b).

Following [4], consider the 2 coupled linear oscillators

$$\ddot{x} + \dot{x} + t^2 x = p_1 \dot{y}, \tag{11}$$

$$\ddot{y} + p_3 \dot{y} + 100y = p_2 \dot{x}. \tag{12}$$

The parameters p_1 and p_2 control their coupling, while p_3 controls the damping of the second oscillator. In state space form, (11) and (12) can be written as

$$\frac{d}{dt}\begin{pmatrix} x \\ \dot{x} \\ y \\ \dot{y} \end{pmatrix} = \begin{pmatrix} 0 & 1 & 0 & 0 \\ -t^2 & -1 & 0 & p_1 \\ 0 & 0 & 0 & 1 \\ 0 & p_2 & -100 & p_3 \end{pmatrix}\begin{pmatrix} x \\ \dot{x} \\ y \\ \dot{y} \end{pmatrix}. \tag{13}$$

Suppose that initially $(p_1, p_2, p_3) = (1, 1, 0.89)$ and t varies from 8 to 12. The locus of the eigenvalues of positive frequency is numerically computed and shown in Figure 1(a). Using equation (1), we can plot μ and observe that it describes a parabola around the origin. This suggests that the eigenvalues are passing close to a weak resonance. By numerically fitting a parabola to this observed μ, the roots r_1 and r_2 are found to be approximately $9.00448 - i0.104793$ and $11.0016 - i0.00997545$.

Let us iteratively minimize δ^a. The nth step of the optimization modifies the vector p of parameters as follows

$$p^{(n)} = p^{(n-1)} - c\delta^a_p(p^{(n-1)}) \tag{14}$$

where c is a constant step size. The eigenvalue loci for the first steps of this process are shown in Figure 2.

5. CHANGING INTERACTION TYPE

It is known that there are two types of perturbations from a weak resonance: type A and type B [5]. In type A the eigenvalues do not pass each other as the parameter is varied and in type B, the eigenvalues pass each other. Examples of type A and B interactions are shown in Figure 3. This

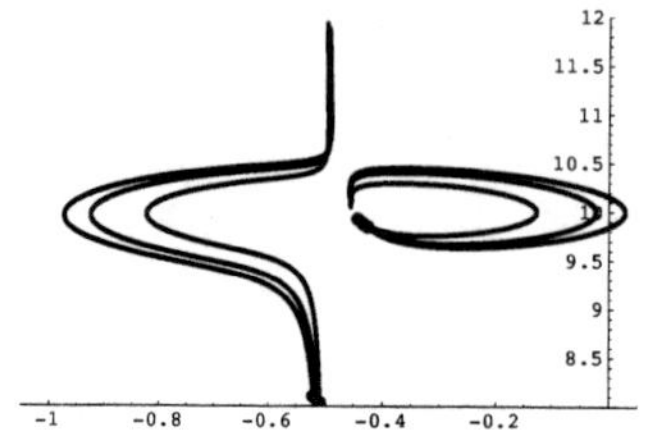

Fig. 2. First steps of gradient descent minimizing δ^a.

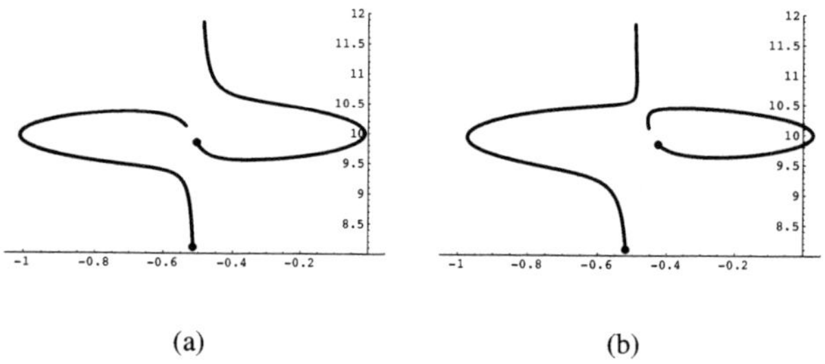

Fig. 3. Type A and type B eigenvalue interactions.

difference can be described in terms of the approximating parabola μ^a. The interaction is type A if the parabola does not pass around the origin and type B if the parabola does pass around the origin [5].

We show how the roots r_1 and r_2 of (7) determine the type of interaction. Following [5], consider the function $f : \mathbb{R} \to \mathbb{C};\ k \mapsto Ak$ that defines the line through the origin parallel to the principal axis of the parabola. The intersection of this line with the parabola generically occurs for $k = k^*$ and $t = t^*$. The eigenvalue interaction is type A if $k^* > 0$, while it is type B if $k^* < 0$. One can show that

$$k^* = -\Im(r_1)\Im(r_2)\left[1 + \left(\frac{\Re(r_1 - r_2)}{\Im(r_1 + r_2)}\right)^2\right]. \tag{15}$$

Hence, the two eigenvalues have a type A or B interaction if r_1 and r_2 are, respectively, on opposite sides or the same side of the real axis.

The sensitivities r_{ip_k} can be used to select parameters to move the roots r_1, r_2 from the same side of the real axis to opposite sides of the real axis, turning, for example, the type A interaction of Figure 3(a) to the type B interaction of Figure 3(b).

6. THE ROOT SENSITIVITIES r_{ip_k}

For notational convenience, we omit the dependence on ϵ. Taking the derivative of $\mu(r_i(p), p) = 0$ w.r.t. p_k gives

$$\mu_t(r_i(p), p) r_{ip_k}(p) + \mu_{p_k}(r_i(p), p) = 0. \tag{16}$$

If we make the generic assumption that $\mu_t(r_i(p), p) \neq 0$, we have

$$r_{ip_k}(p) = -\frac{\mu_{p_k}(r_i(p), p)}{\mu_t(r_i(p), p)}. \qquad (17)$$

Furthermore, the previous assumption that $r_1(p) \neq r_2(p)$ implies that $(r_i(p), p)$ are single roots and thus strong resonances.

Now we evaluate the sensitivity $\mu_s(s_0)$ where μ and J are analytic functions of a complex parameter s and where there is a strong resonance at s_0. Let $\lambda_1(s_0) = \lambda_2(s_0) = \lambda_0$ and let us make the generic assumption that the remaining eigenvalues $\lambda_i(s_0)$, $i = 3, \ldots, n$ are distinct. Let v_1 be a right eigenvector and v_{1g} a generalized right eigenvector of $J(s_0)$ related to λ_0 and let v_i, $i = 3, \ldots, n$ be right eigenvectors of $J(s_0)$ related to, respectively, $\lambda_i(s_0)$, $i = 3, \ldots, n$.

The theory of versal deformation presented in [10] states that there exists smooth matrices $C(s)$ and $V(s)$ defined in a neighborhood of s_0 such that

$$J(s) = C(s)V(s)(C(s))^{-1}. \qquad (18)$$

$C(s)$ and $V(s)$ are not explicitly specified, but we know that $C(s)$ is invertible, $C(s_0) = (v_1, v_{1g}, v_3, \ldots, v_n)$ and

$$V(s) = \begin{pmatrix} \alpha(s) & 1 & & & \\ \mu(s) & \alpha(s) & & & \\ & & \lambda_3(s) & & \\ & & & \ddots & \\ & & & & \lambda_n(s) \end{pmatrix}, \qquad (19)$$

where $\alpha(s_0) = \lambda_0$. One immediately notes that the first element of the second row is μ. Also, recall that $\mu(s_0) = 0$.

The left and right eigenvectors of $V(s_0)$ related to λ_0 are $l_{V(s_0)} = (0, 1, 0, \ldots, 0)$ and $r_{V(s_0)} = (1, 0, 0, \ldots, 0)^{\mathrm{T}}$. Hence, taking the derivative of (19),

$$\begin{aligned} \mu_s(s_0) &= l_{V(s_0)} V_s(s_0) r_{V(s_0)} \\ &= l_{V(s_0)} \left[C^{-1}J_sC + C^{-1}JC_s + (C^{-1})_sJC\right] r_{V(s_0)} \\ &= l_{J(s_0)} \left[J_s + JC_sC^{-1} + C(C^{-1})_sJ\right] r_{J(s_0)} \\ &= l_{J(s_0)} \left[J_s + \lambda_0 C_sC^{-1} + \lambda_0 C(C^{-1})_s\right] r_{J(s_0)} \\ &= l_{J(s_0)} J_s(s_0) r_{J(s_0)}, \end{aligned} \qquad (20)$$

where $l_{J(s_0)}$ and $r_{J(s_0)}$ are the left and right eigenvectors of $J(s_0)$ related to λ_0 obtained by transforming $l_{V(s_0)}$ and $r_{V(s_0)}$ according to

$$l_{J(s_0)} = (C(s_0))^{-1} l_{V(s_0)} \text{ and } r_{J(s_0)} = C(s_0) r_{V(s_0)}. \qquad (21)$$

Combining (17) and (20), we obtain

$$r_{ip_k}(p) = -\frac{l_i J_{p_k}(r_i(p), p) r_i}{l_i J_t(r_i(p), p) r_i}, \quad i = 1, 2, \qquad (22)$$

where l_i and r_i are the eigenvectors of $J(r_i(p), p)$ with their scaling specified by (21). However, the scaling of the eigenvectors cancels in the quotient of (22). Hence, it is not necessary to compute $C(s_0)$. The eigenvectors can be computed from $J(r_i(p), p)$ with any scaling.

7. CONCLUSION

We have shown that the interaction between two eigenvalues as they pass near weak resonance is governed by two complex roots. (The two roots r_1 and r_2 are complex values of the root locus parameter that zero the square of the difference between the two interacting eigenvalues and that determine strong resonances.) Using quadratic approximations, we compute the sensitivity of these roots to parameter variations so that parameter changes to move the roots can be selected to minimize the interaction and also to change the interaction type. The method is illustrated in a simple example. Application to more realistic examples, particularly to minimize loss of damping due to eigenvalue resonance, is the next step of research.

8. REFERENCES

[1] S-N. Chow, J.K. Hale, Methods of bifurcation theory, Springer NY 1982.

[2] Cigré Task Force 07 of Advisory Group 01 of Study Committee 38, Analysis and control of power system oscillations, Paris, December 1996.

[3] I. Dobson, J. Zhang, S. Greene, H. Engdahl, P.W. Sauer, Is strong modal resonance a precursor to power system oscillations?, IEEE Trans. Circuits and Systems, Part 1, vol. 48, no. 3, March 2001, pp. 340-349.

[4] I. Dobson, Strong resonance effects in normal form analysis and subsynchronous resonance, Bulk Power System Dynamics and Control V, August 2001, Onomichi, Japan.

[5] I. Dobson, E. Barocio, Perturbations of weakly resonant power system electromechanical modes, Bologna Power Tech, Italy, June 2003.

[6] L. Jones, G. Andersson, V. Knyazkin, On modal resonance and interarea oscillations in power systems, Bulk Power System Dynamics and Control V, Aug. 2001, Onomichi, Japan.

[7] B.M. Nomikos, C.D. Vournas, Modal interaction and PSS design, Porto Power Tech Conference, Portugal, Sept. 2001.

[8] K.R. Padiyar, Analysis of Subsynchronous Resonance in Power Systems, ISBN 0792383192, Kluwer, 1998.

[9] G. Rogers, Power system oscillations, Kluwer, 2000.

[10] S. Wiggins, Introduction to applied nonlinear dynamical systems and chaos, Springer-Verlag, 1990.

Author Index